METALLURGICAL TREATISES

METALLURGICAL TREATISES

Presentations at the USA-China Bilateral Conference Sponsored by The Metallurgical Society of AIME, American Society for Metals and The Chinese Society of Metals, to be held in Beijing, November 13-22 1981, will be in part based on the Treatises in this volume.

Edited by
JOHN K. TIEN
Columbia University
New York City, New York

JOHN F. ELLIOTT
Massachusetts Institute of Technology
Cambridge, Massachusetts

A Publication of The Metallurgical Society of AIME

A Publication of The Metallurgical Society of AIME
P.O. Box 430
420 Commonwealth Drive
Warrendale, Pa. 15086
(412) 776-9000

Printed in the United States of America.
Library of Congress Card Catalogue Number 81-83754
ISBN Number 0-89520-381-2

Foreword

The existence of this volume owes much to China. During the visit of the first U.S. Metallurgical Delegation to China in November 1978, highly specialized extractive, metal processing, physical, and mechanical metallurgists were forced by circumstance to listen to each other on the state of development in the many areas of metallurgy. We found the experience stimulating and valuable, and it clarified the need for specialists to broaden their knowledge in allied branches of metallurgy. We concluded that collection into a single volume of interpretive review treatises, spanning the major topical areas in metallurgy, would contribute to promoting a broadening of perspective among United States metallurgists. It should also serve as a convenient reference source for researchers and engineers alike. Finally, the treatises, taken in groups, could serve as up-to-date texts for graduate and advanced undergraduate courses.

The actual development of this volume also involved a U.S.A.-China exchange activity. In 1979, American Society for Metals, The Metallurgical Society of AIME, and the Chinese Society of Metals jointly agreed, in support of the scientific and scholarly exchange between the two countries, to hold a Bilateral Metallurgical Conference in 1981. The U.S. contributors would prepare scientific and interpretive reviews on major metallurgical topics, which encompasses extractive, metal processing metallurgy, physical, mechanical, and fracture metallurgy. The Chinese metallurgists would present original research papers in these same subjects. The Chinese Society had graciously offered to host the Conference in Beijing November 13-22.

In 1980, the Boards of Directors of the U.S. societies named a Joint Conference Organizing Committee to help select the topics and the leading authorities to write the U.S. papers. The Committee members were: John B. Ballance of The Metallurgical Society of AIME, Otto Buck of Rockwell International Corporation and of Iowa State University, John F. Elliott (Co-Editor) of the Massachusetts Institute of Technology, Allen G. Gray of American Society for Metals, William D. Manly of Cabot Corporation, Harris L. Marcus of the University of Texas, John W. Pridgeon of Special Metals Corporation, John K. Tien (Co-Editor and Chairman) of Columbia University, and Milton E. Wadsworth of the University of Utah.

Due to the constraint of delegation size, the spectrum of topics had to be covered in about 30 papers. It was decided that each paper would be a comprehensive treatise on the subject, in which the appropriate facts would be systematically presented, and the principles and conclusions discussed methodically. The breadth of coverage of each topic was left to the discretion of the authors. The results from the labors of many of this country's leading metallurgists are enclosed. Exigencies of time and distance made it impossible to include the Chinese papers in this volume or in a companion volume. Instead, the volume containing the Chinese papers (in English) will be published separately by the Chinese Society of Metals.

There are obvious gaps in the coverage of the field. With a limit on the number of papers, compromises had to be made with respect to selection, and priorities were given to active and forefront areas. In addition, a few papers were not submitted in time to be included. Nevertheless, we hope that this volume will be useful and that periodically it will be reedited, updated, and improved.

We are especially grateful to the Chinese Society of Metals, in particular, its Chairman of the Board, Mr. Ye Zhiqiang, and its Vice-Secretary General, Professor Fu Junzhao, for their efforts in promoting the Conference.

We are grateful to Mr. Allan Ray Putnam and Mr. Alexander R. Scott for their support of this project. We thank our staff assistants, Ms. Abby Peyton and Ms. Jane Benson, for performing the many thankless but critical tasks that were essential to the completion of this volume. Mr. John Ballance and Ms. Barbara Pontello and others on the publication staff of The Metallurgical Society of AIME are to be complimented for their diligence in organizing the efficient and speedy publication of the volume. Finally, we thank the authors for enthusiastically undertaking their tasks in spite of already full professional schedules.

John K. Tien
Columbia University
New York City, N.Y.

John F. Elliott
Massachusetts Institute of Technology
Cambridge, Massachusetts

June 1981

Table of Contents

ELECTROCHEMICAL REACTIONS IN HYDROMETALLURGY

Milton E. Wadsworth

State College of Mines
University of Utah
Salt Lake City, Utah 84112

Sulfide minerals and many heavy metal oxides are good electronic conductors. Consequently the kinetics of dissolution of these phases are controlled by electrochemical processes. The electrochemistry operative in the anodic dissolution of several sulfide, metal, and oxide systems is presented. Four general types of electrochemical rate processes are considered as special examples of importance. These are: 1) charge transfer rate control with two or more phases in contact with the aqueous electrolyte; 2) charge transfer rate control with a single solid phase in contact with electrolyte; 3) transport through boundary films separating the reacting mineral from the electrolyte, and 4) coupled diffusion. Recent studies have shown that the cathodic conversion of metal sulfides to other forms provide a means for pretreatment prior to anodic dissolution. Examples of cathodic phase alteration are presented.

INTRODUCTION

Several investigators have presented information on the electrochemical behavior of sulfide and oxide minerals. Metal sulfides have received the most attention because they are recognized as valuable sources of non-ferrous metal. Koch (1) has reviewed the electrochemistry of sulfide minerals and Shuey (2) and Vaughan and Craig (3) have reviewed the electronic properties of both oxide and sulfide minerals. Table I lists resistivities, electronic, and structural information for selected sulfide and oxide minerals. The covalent character of most sulfides provides non-localization of charge resulting in appreciable intrinsic electronic conductivity. The ability of many sulfides to form non-stoichiometric compounds results in increased conduction via electron holes or excess electrons. Oxides in general are more ionic in nature and usually have higher resistivities compared to sulfide minerals.

A fundamental property of semiconducting minerals is the characteristic "rest" potential. For interfacial electrode processes, the "rest" potential corresponds to the equilibrium (no net anodic or cathodic current) electrode potential. Rest potentials for several metal sulfides are shown in Table II. It is important to remember that a mineral electrode system will establish and maintain a certain equilibrium potential that depends not only on the solution composition but also on the composition of the solid phase.

Table I. Electronic and Structural Properties of Selected Sulfide and Oxide Minerals

Formula	Name	Resistivity OHM-M	Usual Conductor Type	Structure	Ionic Structure
Cu_5FeS_4	Bornite	10^{-3}-10^{-6}	p	Tetragonal	$(Cu^+)_5Fe^{3+}(S^{2-})_4$
Cu_2S	Chalcocite	$4x10^{-2}$-$8x10^{-5}$	p	Orthorhombic	$(Cu^+)_2S^{2-}$
$CuFeS_2$	Chalcopyrite	$2x10^{-4}$-$9x10^{-3}$	n	Tetragonal	$Cu^+Fe^{3+}(S^{2-})_2$
CuS	Covellite	$8x10^{-5}$-$7x10^{-7}$	metallic	Hexagonal	$(Cu^+)_2S_2^{2-}$
PbS	Galena	$1x10^{-5}$-$7x10^{-6}$	n&p	Cubic	$Pb^{2+}S^{2-}$
MoS_2	Molybdenite	7.5-$8x10^{-3}$	n&p	Hexagonal	$Mo^{4+}(S^{2-})$
FeS_2	Pyrite	$3x10^{-2}$-$1x10^{-3}$	n&p	Cubic	$Fe^{2+}S_2^{2-}$
ZnS	Sphalerite	$3x10^{-3}$-$1x10^{-4}$	-	Cubic	$Zn^{2+}S^{2-}$
SnO_2	Cassiterite	10^{+2}-10^{-2}	n	Tetragonal	$Sn^{4+}O_2^{4-}$
Cu_2O	Cuprite	10^{11}-10	p	Cubic	$(Cu^+)_2O^{2-}$
Fe_2O_3	Hematite	$2.5x10^{-1}$-$4x10^{-2}$	n&p	Trigonal	$(Fe^{3+})_2(O^{2-})_3$
Fe_3O_4	Magnetite	$2x10^{-4}$-$4x10^{-5}$	n&p	Cubic	$Fe^{3+}[Fe^{3+}Fe^{2+}](O^{2-})_4$
MnO	Pyrolusite	10^{-1}-10^{-3}	n	Tetragonal	$Mn^{4+}(O^{2-})_2$
TiO_2	Rutile	10^4-10	n&p	Tetragonal	$Ti^{4+}(O_2^{2-})_2$
UO_2	Uraninite	20-$4x10^{-1}$		Cubic	$(U^{+4})_{1-x}(O^{2-})_{2+x}$

Table II

Mineral	Rest Potential (V vs. SHE)	Conditions Solution	Temp	Ref.
FeS_2	0.63	1.0M H_2SO_4	25°C	4,5
$CuFeS_2$	0.52	1.0M H_2SO_4	20°C	6
Cu_2S	0.44	1.0M H_2SO_4	20°C	7
CuS	0.42	1.0M $HClO_4$	25°C	8
PbS	0.28	1.0M H_2SO_4	20°C	7
ZnS	-0.24	1.0M H_2SO_4	20°C	7
FeS	-0.28	1.0M H_2SO_4	20°C	7

The electronic properties of most metal sulfides and of certain oxides are such that corrosion and galvanic couples are readily established in aqueous systems. This is true for p-type, n-type or metallic like compounds for which $t_e \simeq 1$, where t_e is the electronic transport number. As a result of their electronic conductivity, certain minerals can participate in coupled charge transfer processes analogous to a metal corroding in an electrolyte, and the kinetics of leaching can be related to the potential of the solid in contact with the aqueous electrolyte. From a thermodynamic point of view most hydrometallurgical dissolution reactions are strongly favored such that back reaction kinetics normally are unimportant. Build up of ions in solution, however, may markedly affect the kinetics by influencing the potential at the mineral (electrode) solution interface.

Electrochemical reactions of importance to hydrometallurgists may be viewed in terms of the character of charge and mass transfer. In this presentation four types of coupled systems will be considered. These are:

1) Mixed potential corrosion couples
2) Mixed potential galvanic couples
3) Charge transfer through passivation layers
4) Voltage control by sequential mineral equilibria.

The concept of mixed potential is important in many anodic and cathodic processes of importance to hydrometallurgists since even a modest voltage may have an important influence on the observed kinetics where charge transfer is involved. Also slow discharge and coupled diffusion resulting in mixed chemical and diffusion overpotential are operative in many systems.

MIXED POTENTIALS IN ELECTROCHEMICAL PROCESSES

Electrochemical processes are unique in that the solid may assume a uniform potential throughout, providing ohmic resistance is negligible. Consequently the chemical reaction may be influenced over relatively long distances. The potential may directly affect the kinetics of the reaction, if slow discharge is involved, and may also serve to stabilize intermediate solid phases. If one or more electrode reactions occur simultaneously a mixed potential may result. Vetter (4) has discussed the concept of additive partial electrode processes important in mixed electrode systems. Concepts developed in the field of corrosion are translatable to the electron conducting solids in hydrometallurgical systems, interfacial charge-transfer reactions being the fundamental denominator of both. Apart from this important feature both behave as internally short circuited (no external applied voltage) systems, where the solid assumes a mixed potential determined by spatially separated anodic and cathodic processes. At the mixed potential, the sum of all anodic currents, ΣI_a is equal to the negative sum of all cathodic currents, ΣI_c, such that

$$\Sigma I_a = -\Sigma I_c$$

where $I_a = i_a A_a$ and $I_c = i_c A_c$. The terms A_a and A_c are the respective anodic and cathodic areas, and i_a and i_c are the corresponding current densities. As in corrosion processes, two types of mixed potential regimes may be operative (i.e. local cell and galvanic couples). The local cell type involves a single phase having both anodic and cathodic regions on its surface such that the total surface area of the solid is equal to the sum of anodic and cathodic areas. A galvanic couple is operative where two or more solid phases are in electrical contact. In this union each solid assumes either anodic or cathodic behavior and has its own separate surface area for reaction.

Individual electrode polarization curves are helpful in visualizing the concept of a mixed potential. Figure 1

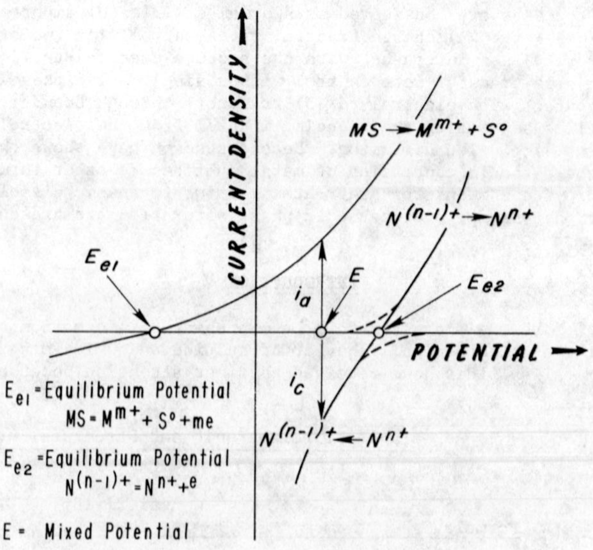

FIGURE 1. Current density potential diagram showing curves for metal sulfide with rest potential E_{e1} and redox ions with rest potential E_{e2} forming mixed potential E.

depicts hypothetical polarization curves for the dissolution of a metal sulfide in the presence of a cationic oxidant N^{n+}. If MS is placed in contact with a solution containing an oxidant N^{n+} (e.g. Fe^{3+} in acid solutions or $Cu(NH_3)_4^{2+}$ in alkaline solutions) having a more positive equilibrium potential, a corrosion cell will result in the anodic oxidation of MS and the cathodic reduction of N^{n+}. The half cell reactions are

$$MS = M^{m+} + S^o + me^- \tag{1}$$

and

$$N^{n+} + e = N^{(n-1)+} \tag{2}$$

The equilibrium potentials E_{e1} and E_{e2} for half-cell reactions (1) and (2) are illustrated in Figure 1. Equilibrium electrode potentials are independent of the properties of the electrode surface, however, electrode surface properties influence to a great extent the shape of the partial current-voltage curves. The value of the mixed potential is established by both the partial equilibrium potentials and the shape of the current-voltage curves. Hence, the mixed potential depends upon the properties of the electrode surface. Potentials more positive than the equilibrium potential drive the half-cell reaction in the net anodic direction and potentials more negative than the equilibrium potential drive the reactions in the net cathodic direction.

The net current density (i) for the system is the sum of the partial current densities i_a (anodic) and i_c (cathodic). The mixed potential (E) occurs where the net current density is zero; the values of the partial current densities therefore are equal and of the opposite sign ($i_a = -i_c$) as indicated in Figure 1. The anodic overvoltage η_a is defined as $(E-E_{e1})$ and is positive. The cathodic overpotential η_c is defined as $(E-E_{e2})$ and is negative.

The electrochemical bias for charge transfer in simple reversible reactions may be expressed quantitatively by the Butler-Volmer equation. This fundamental equation of electrodics relates the current density at a solid-solution interface to the established overpotential. Applied to the specific example of a metal sulfide undergoing anodic dissolution, Eq. (1), the Butler-Volmer equation, for a single electron transfer process, is

$$i = zF\overrightarrow{k}\exp\left[\frac{\beta FE}{RT}\right] - zF\overleftarrow{k}\,[M^{m+}]\exp\left[\frac{-(1-\beta)FE}{RT}\right] \tag{3}$$

The first term in the above equation is the partial current density for the anodic (forward) direction of Eq. (1) and the second term the partial current density for the cathodic (reverse) direction, F is the Faraday, $\overrightarrow{k}$ and $\overleftarrow{k}$ are the rates for the forward and reverse directions, respectively, in the absence of potential, and β is the transfer coefficient. The rate $\overleftarrow{k}$ may contain a concentration term for ad-ions on the mineral surface which may change if the stoichiometry of the mineral varies. At the equilibrium potential E_{e1}, the partial current densities of Eq. (1) are equal $|\overrightarrow{i}| = |\overleftarrow{i}|$ and are represented by the exchange current density, i_o. The exchange current density is given by the following relationships

$$i_o = zF\overrightarrow{k}\exp\left[\frac{\beta FE_{e1}}{RT}\right] = zF\overleftarrow{k}\,[M^{m+}]\exp\left[\frac{-(1-\beta)FE_{e1}}{RT}\right] \tag{4}$$

and it follows directly that

$$i = i_o\{\exp\left[\frac{\beta F\eta}{RT}\right] - \exp\left[\frac{-(1-\beta)F\eta}{RT}\right] \tag{5}$$

By definition the overvoltage η is equal to the difference between E and E_{e1}. Equation (5) is sometimes referred to as a simplified form of the Butler-Volmer relationship.

If the metal sulfide is allowed to contact an aqueous solution containing an oxidant, N^{n+}, a corrosion couple involving net anodic and net cathodic reactions will result and a steady-state mixed potential will be established. The Butler-Volmer expression for the net anodic, i_a, and net cathodic, i_c, current densities for reactions (1) and (2) are

$$i_a = z_aF\overrightarrow{k}_a\exp\left[\frac{\beta_a n_a F E}{RT}\right] - z_aF\overleftarrow{k}_a\,[M^{m+}]\exp\left[\frac{-(1-\beta_a)n_a E}{RT}\right] \tag{6}$$

and

$$i_c = z_cF\overrightarrow{k}_c\,[N^{(n-1)+}]\exp\left[\frac{\beta_c n_c FE}{RT}\right] - z_cF\overleftarrow{k}_c\,[N^{n+}]\exp\left[\frac{-(1-\beta_c)n_c FE}{RT}\right] \tag{7}$$

Coupling the anodic and cathodic half cell reactions provides a means for estimating the mixed potential, although there is no explicit solution without certain assumptions. For purposes of simplification the assumption that $\beta_a \cong \beta_c = \beta$ is usually a satisfactory approximation. Also rate controlling processes most commonly involve single charge transfer steps such that n_a and n_c may be taken as unity, an assumption made here only for purposes of simplification. Accordingly equations (6) and (7) may be combined giving

$$\exp\,[FE/RT] = \frac{A_a z_a \overleftarrow{k}_a\,[M^{m+}] + A_c z_c \overleftarrow{k}_c\,[N^{n+}]}{A_a z_a \overrightarrow{k}_a + A_c z_c \overrightarrow{k}_c\,[N^{(n-1)+}]} \tag{8}$$

where E is the mixed potential. Also

$$-\frac{dn_{MS}}{dt} = \frac{A_a i_a}{z_a F} \tag{9}$$

where n_{MS} refers to the moles of metal sulfide remaining at anytime t. Combining equations (6), (8), and (9) results in the general charge transfer rate expression

$$-\frac{dn_{MS}}{dt} = \frac{A_a A_c z_c \overrightarrow{k}_a \overleftarrow{k}_c [N^{n+}][1-\overrightarrow{k}_a\overleftarrow{k}_c[M^{m+}][N^{(n-1)+}]/\overleftarrow{k}_a\overrightarrow{k}_c[N^{n+}]]}{[A_a z_a \overrightarrow{k}_a + A_c z_c \overrightarrow{k}_c[N^{(n-1)+}]]^{\beta}[A_a z_a \overleftarrow{k}_a[M^{m+}] + A_c z_c \overleftarrow{k}_c[N^{n+}]]^{1-\beta}} \tag{10}$$

It may be shown that

$$\overrightarrow{k}_a\overleftarrow{k}_c\,[M^{m+}][N^{(n-1)+}]/\overleftarrow{k}_a\overrightarrow{k}_c\,[N^{n+}] = \exp\,(\Delta G/RT) \tag{11}$$

where ΔG is the free energy for the reaction producing a given final product. Equation (10) becomes

$$-\frac{dn_{MS}}{dt} = \frac{A_a A_c z_c \overrightarrow{k}_a \overleftarrow{k}_c [N^{n+}][1-\exp(\Delta G/RT)]}{[A_a z_a \overrightarrow{k}_a + A_c z_c \overrightarrow{k}_c[N^{(n-1)+}]]^{\beta}[A_a z_a \overleftarrow{k}_a[M^{m+}] + A_c z_c \overleftarrow{k}_c[N^{n+}]]^{1-\beta}} \tag{12}$$

It is apparent from equation (12) that for hydrometallurgical processes having a ΔG more negative than -4.0 kcal/mole ($\Delta G/RT<<1$) are essentially irreversible. Nevertheless products of reaction can influence the kinetics by their affect upon the associated half-cell reactions. Most sulfide and metal dissolution processes are irreversible. Equation (12) may have several forms depending upon the relative values of the terms in the denominator. Assuming irreversibility, special cases are:

1) Mixed potential within both the anodic and cathodic Tafel regions;

$$-\frac{dn_{MS}}{dt} = (A_a\vec{k}_a)^{1-\beta}\,(A_c z_c \overleftarrow{k}_c/z_a)^{\beta}\,[N^+]^{\beta} \tag{13}$$

2) Mixed potential within the cathodic Tafel region only;

$$-\frac{dn_{MS}}{dt} = \frac{(A_a\vec{k}_a)^{1-\beta}\,A_c z_c \overleftarrow{k}_c\,[N^{n+}]}{z_a^{\beta}\,[A_a z_a \overleftarrow{k}_a [M^{m+}] + A_c z_c \overleftarrow{k}_c [N^{n+}]]^{1-\beta}} \tag{14}$$

3) Mixed potential within the anodic Tafel region only;

$$-\frac{dn_{MS}}{dt} = \frac{A_a\vec{k}_a\,(A_c z_c \overleftarrow{k}_c)^{\beta}\,[N^{n+}]^{\beta}}{[A_a z_a \vec{k}_a + A_c z_c \vec{k}_c\,[N^{(n-1)+}]]^{\beta}} \tag{15}$$

The pseudo fractional orders observed in reactions of this type result from the transfer coefficient β and fall generally in the range 0.4 to 0.6, with values commonly near 0.5.

A special case is one in which one electrode is near equilibrium, i.e. has a large exchange current density. If, for example, the cathode is near equilibrium, such that E(mixed) ≅ E_{el}, and the anode is in its Tafel region, equation (12) becomes

$$-\frac{dn_{MS}}{dt} = A_a\vec{k}_a \left(\frac{\overleftarrow{k}_c}{\vec{k}_c}\right)^{\beta} \frac{[N^{n+}]^{\beta}}{[N^{(n-1)+}]^{\beta}} \tag{16}$$

If the cathodic reaction involves the Fe^{3+}/Fe^{2+} couple the rate of dissolution in this latter case is related to the solution E_h.

MIXED POTENTIAL CORROSION COUPLES

Oxidation of Uranium Dioxide in Acid and Alkaline Carbonate Solutions

Nicol and Needes (10,11) studied the leaching of UO_2 as an electrochemical reaction and developed models describing the leaching process based on a charge transfer mechanism. Detailed electrode kinetic measurements (12, 13) were obtained for the anodic oxidation of UO_2 and redox reactions at the UO_2 surface. The electrochemical mechanism correlated well with experimental data for both acid and alkaline systems. The dissolution mechanism was interpreted in terms of an anodic process involving two successive single electron-transfer reactions. The mechanism consistent with the electrochemical observations for acidic solutions is as follows

$$UO_2 + H_2O = UO_2OH + H^+ + e \tag{17}$$

$$UO_2OH = UO_3 + H^+ + e \tag{18}$$

$$UO_3 + 2H^+ = UO_2^{2+} + H_2O \tag{19}$$

For alkaline carbonate solutions the electrochemical data supported the following mechanism

$$UO_2 + HCO_3^- = UO_2HCO_3 + e \tag{20}$$

$$UO_2HCO_3 + OH^- = UO_2CO_3 + H_2O + e \tag{21}$$

$$UO_2CO_3 + 2CO_3^{2-} = UO_2(CO_3)_3^{4-} \tag{22}$$

Recently, Hiskey (14) presented chemical leaching results for the dissolution of UO_2 in ammonium carbonate. These results confirm an anodic dissolution sequence that includes both charge-transfer and chemical-reaction steps. Furthermore, these experiments supported a mechanism involving consecutive one-electron-transfer reactions. At low anodic potentials, the dissolution of surface product films (i.e. UO_3 or UO_2CO_3) is considered fast in comparison to charge transfer steps.

During leaching the anodic steps are coupled with the cathodic reduction of the oxidant. For alkaline carbonate systems oxygen is an important oxidant, the cathodic discharge of oxygen at the UO_2 surface occurs by the reaction

$$O_2 + 2H_2O + 4e = 4OH^- \tag{23}$$

With few exceptions, the leaching of UO_2 has been found to obey a half order dependence on oxygen partial pressure. Independent experiments have demonstrated that oxygen is reduced cathodically at a UO_2 surface with a first-order dependence on oxygen. This fact along with the irreversible nature of oxygen reduction at the UO_2 surface verifies the apparent one-half order with respect to oxygen partial pressure. Such a dependence was previously developed theoretically using the Butler-Volmer equation for the special case where i_a and i_c each fall within the corresponding Tafel region (see Eq. 13).

The leaching of a solid is considered electrochemically controlled when the rate of leaching is equal to the rate of anodic charge-transfer. For instance Figure 2

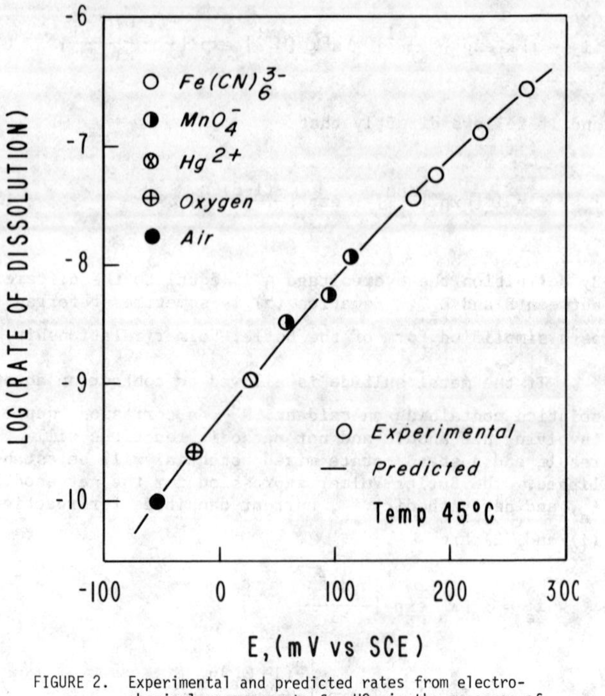

FIGURE 2. Experimental and predicted rates from electrochemical measurements for UO_2 in the presence of several oxidants.

Table III. Calculated Equilibrium Potentials for Several Cathodic Half-Cell Reactions, 25°C

Reaction	Conditions	E_c (V vs SCE)
$O_2 + H_2O + 4e = 4OH^-$	P_{O_2} = 1 atm, pH = 10	0.40
$Fe(CN)_6^{3-} + e = Fe(CN)_6^{4-}$	$[Fe(CN)_6^{3-}] = [Fe(CN)_6^{4-}]$ pH = 12	0.22
$Cu(NH_3)_4^{2+} + e = Cu(NH_3)_2^{+} + 2NH_3$	$[Cu(NH_3)_4^{2+}] = [Cu(NH_3)_2^{+}]$ $[NH_3] = 10^{-2}$	0.07
$H_2O_2 + 2e = 2OH^-$	$[H_2O_2] = 10^{-2}$, pH = 10	0.88

illustrates how closely the observed chemical leaching rates compare with the predicted electrochemical rates for the alkaline carbonate dissolution of UO_2 in the presence of various oxidants. Similarly, close agreement was found between chemical and electrochemical data for oxidants in acidic solutions. It is interesting to analyze the effectiveness of various oxidants in terms of their equilibrium potentials for several important cathodic half-cell reactions. Table III provides the equilibrium for several important cathodic half-cell reactions. If oxidants were selected solely by thermodynamic (equilibrium potential) criterion, then the predicted rates would reflect the following order

$$H_2O_2 > O_2 > Fe(CN)_6^{3-} > Cu(NH_3)_4^{2+}$$

Actually, these oxidants rank in the following order for the leaching of UO_2

$$Fe(CN)_6^{3-} > H_2O_2 > Cu(NH_3)_4^{2+} > O_2$$

Oxygen is a comparatively poor oxidant for the leaching of UO_2 even though it possesses a comparatively high equilibrium potential. This fact is clearly explainable in terms of the electrode kinetics of the half-cell reactions. The typical pattern of current-voltage curves is shown in Figure 3 for the anodic oxidation of UO_2 and cathodic reduction of O_2 and $Fe(CN)_6^{3-}$. It must be emphasized that the most effective oxidant develops the highest anodic mixed potential during leaching. To a degree a large equilibrium electrode potential helps by establishing a high overvoltage, however, the shape of the current-voltage curves (electrode kinetics) is probably more important since it establishes the ultimate mixed potential and the corresponding leaching current. From Figure 3, it is clearly evident that the slope of the cathodic reduction of oxygen is very low resulting in an extremely small leaching current corresponding to mixed potential position. The extremely reversible $Fe(CN)_6^{3-}/Fe(CN)_6^{4-}$ couple, unlike the irreversible discharge of oxygen, is characterized by a steep polarization curve which yields a mixed potential at E_M' with the magnitude of i_a' equivalently greater.

The dissolution of UO_2 in ammoniacal carbonate solution with hydrogen peroxide as an oxidant deserves special consideration. In comparison to oxygen, the H_2O_2 system exhibits a measurable increase in the rate of UO_2 leaching (11,15). This enhancement in the leaching kinetics can be explained by electrochemistry concepts which take into account both anodic and cathodic effects. Figure 4 compares the polarization curves for both oxygen and hydrogen peroxide systems. In general hydrogen peroxide has the ability to shift the mixed potential to more anodic values, thereby increasing the anodic current (i_a). The higher equilibrium potential for the H_2O_2 half-cell reaction (see

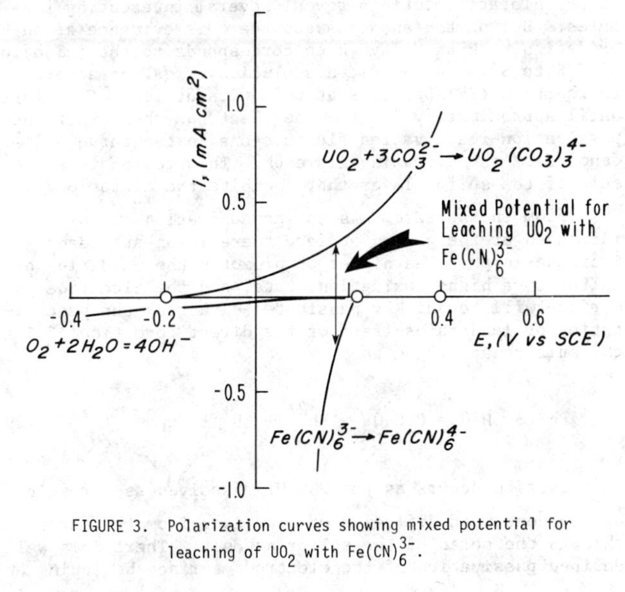

FIGURE 3. Polarization curves showing mixed potential for leaching of UO_2 with $Fe(CN)_6^{3-}$.

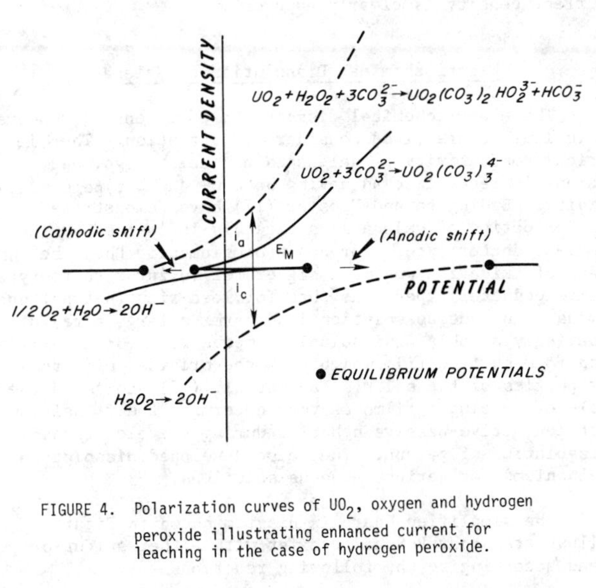

FIGURE 4. Polarization curves of UO_2, oxygen and hydrogen peroxide illustrating enhanced current for leaching in the case of hydrogen peroxide.

Table III) favors an anodic mixed potential shift. In addition, H_2O_2 can react with $UO_2(CO_3)_3^{4-}$ to form a $UO_2(CO_3)_2HO_2^{3-}$ species with an associated cathodic shift from

$$UO_2(CO_3)_3^{4-} + 2e = UO_2 + 3CO_3^{2-}$$

$$E = -0.22 - 0.03 \log \frac{[CO_3^{2-}]^3}{[UO_2(CO_3)_3^{4-}]} \tag{24}$$

to

$$UO_2(CO_3)_2HO_2^{3-} + HCO_3^- + 2e = UO_2 + 3CO_3^{2-} + H_2O_2$$

$$E = -0.38 - 0.03 \log \frac{[CO^{2-}]^3\,[H_2O_2]}{[UO_2(CO_3)_2HO_2^{3-}][HCO_2^-]} \tag{25}$$

The cathodic shift in potential is equal to

$$\Delta E_c = E_{25} - E_{24}$$

$$= -0.16 - 0.03 \log \frac{[UO_2(CO_3)_3^{4-}][H_2O_2]}{[UO_2(CO_3)_2HO_2^{3-}][HCO_3^-]}$$

where the logarithmic term is equivalent to the inverse equilibrium constant for the reaction

$$UO_2(CO_3)_3^{4-} + H_2O_2 = UO_2(CO_3)_2HO_2^{3-} + HCO_3^- \tag{26}$$

Komarov (16) reports an equilibrium constant of 1.6 x 10^2 for this reaction at infinite dilution. Therefore, the cathodic shift in the pH region where H_2O_2 is dominant is approximately -0.09 v. The dashed lines in Figure 4 are the respective polarization curves for H_2O_2 reduction and UO_2 oxidation in the H_2O_2 system. The mixed potential for these conditions is noted by E_M and the increase in current density is clearly apparent.

Electrochemical Dissolution of Galena

The electrochemical dissolution of galena in aqueous solutions has received considerable attention. The electrical conductivity of galena is unusually high, and galena is recognized as having both n- and p-type conductivity. Eadington and Prosser (17) have demonstrated that the oxidation of galena is noticeably influenced by the semi-conductor type. For non-stoichiometric PbS, the rate of oxidation was much higher for sulfur-rich (p-type semi-conductor) specimens than for lead-rich (n-type) specimens. In aqueous solution lead forms a large array of sparingly soluble or insoluble compounds; these materials can form surface film which mask the original electronic properties of the solid. Paul et al. (18) addressed the role of insoluble films as they determine the transition between active-passive behavior during the electrochemical dissolution of galena. They also developed dissolution mechanisms for various aqueous solutions.

The simplified Pourbaix diagram shown in Figure 5 illustrates that PbS can coexist with either sulfur or lead according to the following reactions

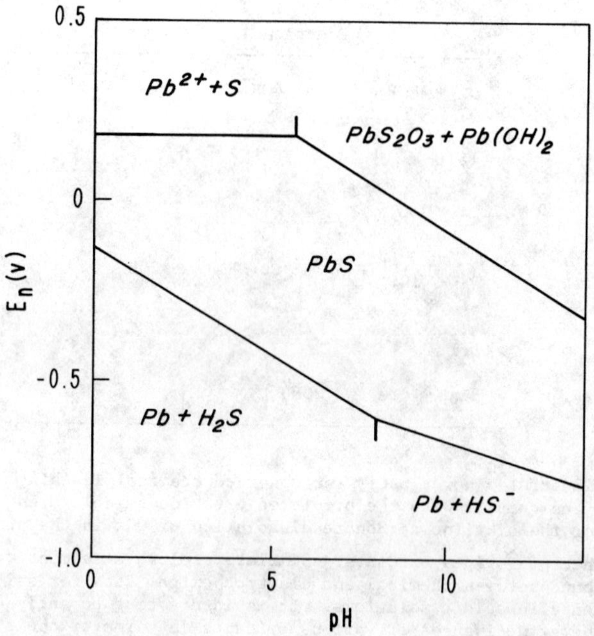

FIGURE 5. Simplified E_h-pH diagram for the Pb-S system.

$$PbS = Pb^{2+} + S° + 2e \tag{27}$$

or

$$PbS + 2H^+ + 2e = Pb + H_2S \tag{28}$$

The galena electrode is acknowledged as one of the most reversible among sulfide minerals. The voltammogram (18) appearing in Figure 6a illustrates the complex nature of the electrode kinetics. The anodic and cathodic sweeps in perchlorate solutions reveal several interesting features. During the anodic sweep there is evidence of an oxidation peak at 0.1 v which corresponds to the oxidation of H_2S to S°. The anodic dissolution of galena according to reaction (27) proceeds at pH 2 without loss of activity until approximately 0.5 v. The electrode then experiences passivation-reactivation fluctuations in the curve as denoted by the arrow in Figure 6b. This region is a result of the sulfur layer that inhibits the diffusion of Pb^{2+} from the surface. As the concentration of Pb^{2+} increases at the galena surface there is an attendant increase in pH. The higher pH promotes the oxidation of sulfur to a higher oxidation state, and the electrode surface will be quickly passivated either by the precipitation of lead thiosulfate or the direct formation of lead thiosulfate according to

$$PbS + 3H_2O = PbS_2O_3 + Pb^{2+} + 6H^+ + 8e \tag{29}$$

Reactivation occurs as the PbS_2O_3 dissolves as a result of the continued diffusion of lead ions from the surface through the pores of the sulfur deposit. There is a well defined passivation of the electrode surface beginning at

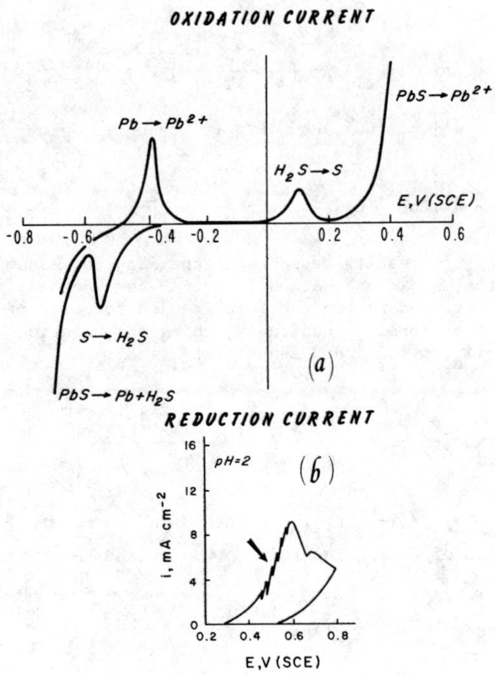

FIGURE 6. a) Polarization and reduction currents for the cyclic-voltammetric reactions of galena in a perchlorate solution; and b) anodic dissolution of galena showing sequential passivation and oxidation.

about 0.6 v caused by the precipitation of a basic lead sulfate ($PbO \cdot PbSO_4$ or $3PbO \cdot PbSO_4$ on the galena surface. On the cathodic side of the sweep, a peak which reflects the reduction of S° to H_2S appears near −0.55 v. The peak corresponding the the oxidation of Pb to Pb^{2+} can only occur following the reduction of galena to lead and hydrogen sulfide. It appears that the formation of lead on the galena surface does not reduce its cathodic activity. Similar cyclic-voltammetric behavior was observed for chloride and sulfide solutions. However, these systems exhibited a pronounced influence from both $PbCl_2$ and $PbSO_4$ formed from the respective solutions.

From the study described above Paul et al. (18) suggested a mechanism for the dissolution of galena in acidic perchlorate solutions. The overall anodic reaction is given in Eq. (27). Tafel slopes fail to support this simple two electron-transfer reaction. Therefore, the oxidation of galena is assumed to take place according to distinguishable one-electron-transfer steps,

$$PbS \rightarrow Pb^{2+} + S^- + e \tag{30}$$

$$S^- \rightarrow S° + e \tag{31}$$

The "S^-" sulfur intermediate has been proposed elsewhere (13). The Butler-Volmer treatment for steady state conditions ($d[S^-]/dt = 0$) yields the following equation for the net anodic current

$$i_a = \frac{2F\vec{k}\,k'\exp[(2-\beta_a-\beta'_a)FE/RT]}{\overleftarrow{k}\,[Pb^{2+}]_s\exp[-\beta_a FE/RT] + k'\exp[(1-\beta_a)FE/RT]} \tag{32}$$

where $[Pb^{2+}]_s$ is the concentration of lead ions at the electrode surface, $\vec{k}$ and $\overleftarrow{k}$ are the respective forward and reverse rate constants for reaction (30), k' is the specific rate of reaction (31), and the other terms are analogous to those previously defined. At sufficiently positive values of E or at low $[Pb^{2+}]$, since the first term of the denominator of Eq. (32) is small compared to the second term, the expression for i_2 reduces to

$$i_a = 2F\vec{k}\exp(1-\beta_a)FE/RT) \tag{33}$$

Conversely, when the $[Pb^{2+}]$ is high and E is low the second term in the denominator can be neglected and

$$i_a = \frac{2F\vec{k}\,k'}{\overleftarrow{k}\,[Pb^{2+}]}\exp[(2-\beta'_a)FE/RT] \tag{34}$$

Both cases cited above along with an additional condition when $[Pb^{2+}]_s$ approaches zero have been verified experimentally. Therefore, the proposed mechanism involving consecutive one-electron-transfer reactions is an accurate description of the anodic dissolution of galena.

Oxidation of Chalcopyrite in Ammoniacal Solutions With Oxygen

Beckstead and Miller (19) have explained the kinetics of the oxidation of chalcopyrite in oxygenated ammoniacal solution on the basis of an electrochemical model. The stoichiometry of the reactions involved were established by Forward and Mackiw (20) and more recently Kuhn et al. (21) have described the Arbiter process based upon the ammoniacal dissolution of copper sulfides using oxygen. The overall reaction for chalcopyrite is

$$CuFeS_2 + 4NH_3 + \frac{17}{4}O_2 + 2OH^- =$$

$$Cu(NH_4)^{2+} + \tfrac{1}{2}Fe_2O_3 + 2SO_4^= + H_2O \tag{35}$$

Thiosulfate is formed as an intermediate and is subsequently oxidized to the sulfate (20). Beckstead and Miller (19) found the kinetics to be approximately half order with respect to hydroxyl and cupric ions and oxygen at low pressures. At high pressures the rate became zero order with respect to oxygen. Sigmoidal reaction curves were explained on the basis of the autocatalytic effect of cupric ion. Two parallel reactions, one of which involves cupric ion, were postulated. Figure 7 illustrates typical rate data obtained showing the effect of cupric ion on the rate of reaction. The kinetic data were well correlated over broad ranges of temperature, pressure, and cupric ion concentration by the rate equations

Low Pressure (P_{O_2} < 1 atmosphere)

$$\frac{d\alpha}{dt} = \frac{k_\ell}{d_o}[OH^-]^{1/2}[O_2]^{1/2}(k_1 + k_2[Cu^{2+}])^{1/2}(1-\alpha)^{2/3} \tag{36}$$

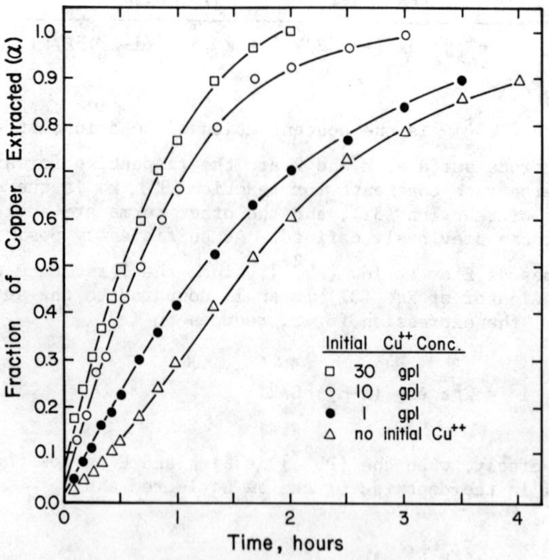

FIGURE 7. Fraction reacted versus time plots for different initial cupric ion concentrations at 90°C, 1% solids, 1.8 M ($NH_4^+ + NH_3$), pH = 10.2, 325 x 400 mesh particles, 2.18 atmospheres O_2 pressure.

High pressure ($P_{O_2} > 1$ atmosphere)

$$\frac{d\alpha}{dt} = \frac{k_h}{d_o}[OH^-]^{\frac{1}{2}}(k_1 + k_2[Cu^{2+}])^{\frac{1}{2}}(1-\alpha)^{2/3} \qquad (37)$$

where k_1, k_2, k_ℓ, and k_h are empirical rate constants, d_o is the initial particle diameter and α is the fraction of $CuFeS_2$ reacted. The approximate half-power orders are characteristic of electrochemical control in leaching. The half cell reactions proposed include the anodic dissolution of $CuFeS_2$

$$CuFeS_2 + 19OH^- =$$

$$Cu^{2+} + \tfrac{1}{2}Fe_2O_3 + 2SO_4^{2-} + \frac{19}{2}H_2O + 17e^- \qquad (38)$$

and two parallel cathodic reactions involving the discharge of oxygen without and with the cupric ion autocatalytic effect

$$O_2 + 2H_2O + 4e^- = 4OH^- \qquad (39)$$

$$O_2 + 2H_2O + Cu^{2+} + 4e^- = 4OH^- + Cu^{2+} \qquad (40)$$

The Butler-Volmer current density equations corresponding to the above half cell reactions were proposed to be

$$i_a = 17F[OH^-]\,k_a \exp\left[\frac{\beta_a FE}{RT}\right] - 17F[Cu^{2+}][SO_4^{2-}]\,k'_a \exp\left[\frac{-(1-\beta_a)FE}{RT}\right] \qquad (41)$$

$$i_{c1} = 4F[OH^-]\,k_{c1} \exp\left[\frac{\beta_{c1}FE}{RT}\right] - 4F[O_2]\,k'_{c1} \exp\left[\frac{-(1-\beta_{c1})FE}{RT}\right] \qquad (42)$$

$$i_{c2} = 4F[OH^-][Cu^{2+}]\,k_{c2} \exp\left[\frac{\beta_{c2}E}{RT}\right] - 4F[Cu^{2+}][O_2]\,k'_{c2} \exp\left[\frac{-(1-\beta_{c2})FE}{RT}\right] \qquad (43)$$

The kinetic results were consistent with the assumption that the anodic and cathodic partial current densities were large and within the Tafel region for each of the above reactions. Accordingly, using the relationship $i_a = -(i_{c1} + i_{c2})$, and assuming $\beta_{c1} = \beta_{c2}$

$$\exp\left[\frac{\beta_a FE}{RT}\right] = \left[\frac{4[O_2](k'_{c1} + [Cu^{2+}]k'_{c2})}{17[OH^-]k_a}\right]^m \qquad (44)$$

where $m = \beta_a/(1+\beta_a-\beta_c)$. Expressed in terms of α, the fraction of $CuFeS_2$ reacted

$$\frac{d\alpha}{dt} = -\frac{1}{n_o}\frac{dn}{dt} = \frac{i_a A}{17Fn} \qquad (45)$$

where n = moles $CuFeS_2$ remaining at time t

n_o = original number of moles of $CuFeS_2$ = w_o/M_c

M_c = molecular weight, $CuFeS_2$

A = area at time t, = $\frac{6w_o f}{\rho_c d_o}(1-\alpha)^{2/3}$

f = geometric shape factor

w_o = original weight, $CuFeS_2$

ρ_c = density, $CuFeS_2$

d_o = original particle diameter

The concentration of copper in solution is related to the solution volume, V, the fraction, α, reacted and the original concentration, $[Cu^{2+}]_o$ by

$$[Cu^{2+}] = [Cu^{2+}]_o + \frac{w_o}{M_c V}\alpha \qquad (46)$$

Experimental results indicate $m \approx 0.5$. If this value is used, the rate expression for the anodic dissolution of $CuFeS_2$ in oxygenated ammoniacal solution is given by

$$\frac{d\alpha}{dt} = K_o(k_1 + k_2[Cu^{2+}]_o + k'_2\alpha)(1-\alpha)^{\frac{1}{2}} \qquad (47)$$

where

$$K_o = \frac{127f}{d_o}[OH^-]^{\frac{1}{2}}[O_2]^{\frac{1}{2}} \qquad (48)$$

and $k_1 = k'_{c1}k_a^{\frac{1}{2}}$; $k_2 = k'_{c2}k_a^{\frac{1}{2}}$; and $k'_2 = k_2 w_o/VM_c$. The zero order kinetics at larger oxygen pressure was explained on the basis of Langmuir adsorption of oxygen where $[O_2]$ refers to the surface concentration and is given by

$$[O_2] = \frac{k_o K P_{O_2}}{1 + KP_{O_2}} \qquad (49)$$

where k_o is the total concentration of surface sites and

K is the equilibrium constant for adsorption. Eqs. (47), (48) and (49) explain the experimental results as depicted by Eqs. (36) and (37) for low and high pressures respectively. Experimentally K_o may be held constant. Figure 8 illustrates the theoretical correlation of Eq. (47) for three sizes of $CuFeS_2$ at 90°C, 1% solids, 1.8 M (NH_4^+ + NH_3), pH = 10.2, and P_{O_2} = 2.18 atm. The slope is equal to K_o and varies with d_o^{-1} as required by Eq. (48).

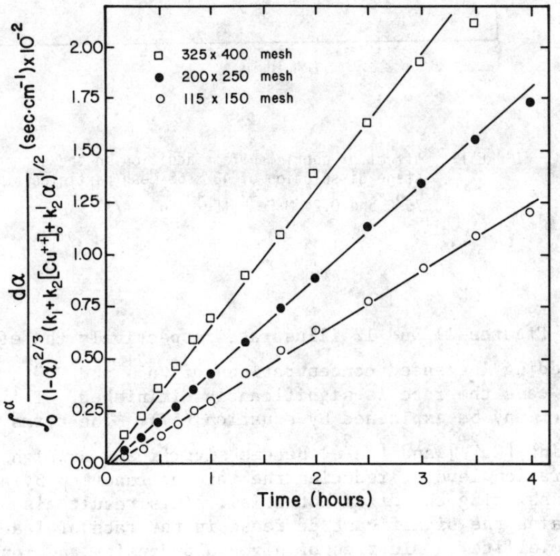

FIGURE 8. Plot of data for three sizes of $CuFeS_2$. Solid lines are theoretical, based upon the electrochemical model.

Second Stage Leaching of Chalcocite in Acid Ferric Sulfate Solutions

Sullivan (22) as early as 1930 identified two stages of leaching of Cu_2S. These were depicted as

$$Cu_2S + 2Fe^{3+} = Cu^{2+} + CuS + 2Fe^{2+} \quad (50)$$

and

$$CuS + 2Fe^{3+} = Cu^{2+} + S° + 2Fe^{2+} \quad (51)$$

Marcantonio (23) similarly identified two stages but provided greater detail regarding the kinetics of each stage. The following half-cell reactions are consistent with the observed results

Stage Ia:

$$5Cu_2S = 5Cu_{1.8}S + Cu^{2+} + 2e^- \quad (52)$$

Stage Ib:

$$\frac{5}{3} Cu_{1.8}S = \frac{5}{3} Cu_{1.2}S + Cu^{2+} + 2e^- \quad (53)$$

Stage II:

$$\frac{5}{6} Cu_{1.2}S = \frac{5}{6}S° + Cu^{2+} + 2e^- \quad (54)$$

Stage I proceeds through a series of compounds. The sequential formation of intermediates establishes the particle potential and controls the rate of ferric ion reduction. Stage I will be considered later under "voltage control by sequential mineral equilibria." Stage II leaching, at elevated temperatures, (Eq. (54)) occurs by a corrosion mechanism and the elemental sulfur produced does not passivate the surface. Stage I is rapidly completed at higher temperature resulting in the removal of approximately forty percent of the copper with the formation of blue remaining covellite, $Cu_{1.2}S$. The blue remaining covellite leaches at a much slower rate.

Figure 9 illustrates typical results obtained for

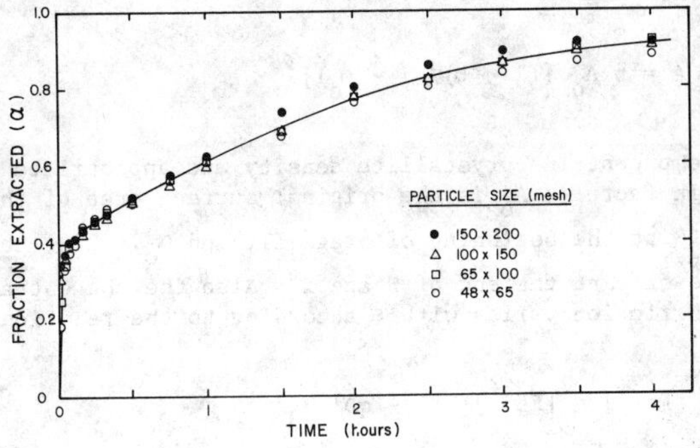

FIGURE 9. Effect of particle size on the second stage dissolution of Cu_2S at 75°C and 0.12 M Fe III (Fe III/Cu = 3/1).

Stage II leaching, at 75°C, for various initial Cu_2S particle sizes. The rate is independent of the initial Cu_2S particle size. The blue-remaining covellite formed in Stage I is porous, consisting of uniform columnar crystals. The rate is sufficiently slow that diffusion into the columnar array does not influence the kinetics. The rate was found to vary with [Fe III] with an apparent half-power order and the activation energy was 18.0 ± 1 kcal/mole. Consistent with the physical appearance of the crystals the kinetics were correlated by cylindrical crystalline geometry and a corrosion couple such that $A_a + A_c = A$, and $A_a/A_c = b$, where be remains constant, and A is the total area. The net anodic and cathodic processes may be expressed by

$$i_a = 2F\vec{k}_a \exp\left[\frac{\beta_a FE}{RT}\right] - [Cu^{2+}]\overleftarrow{k}_a \exp\left[\frac{-(1-\beta_a)FE}{RT}\right] \quad (55)$$

$$i_c = F[Fe^{2+}]\vec{k}_c \exp\left[\frac{\beta_c FE}{RT}\right] - F\overleftarrow{k}_c [Fe^{3+}] \exp\left[\frac{-(1-\beta_c)FE}{RT}\right] \quad (56)$$

Using the relationship $d\alpha/dt = i_a A_a/2\ n_o$, where α is the fraction of total copper reacted and n_o is the total moles of copper initially added to the system, equation (12) for the $Cu_{1.2}S/Fe^{3+}$ system becomes

$$\frac{d\alpha}{dt} = \frac{b\ A\ \vec{k}_a\ \overleftarrow{k}_c\ [Fe^{3+}]}{n_o(1+b)\ [2b\vec{k}_a + \vec{k}_o[Fe^{2+}]]^{\frac{1}{2}}\ [2b\overleftarrow{k}_a[Cu^{2+}] + \overleftarrow{k}_c[Fe^{3+}]]^{\frac{1}{2}}} \qquad (57)$$

Fow low concentrations of $[Cu^{2+}]$ and $[Fe^{2+}]$, equation (57) becomes

$$\frac{d\alpha}{dt} = A\ \left[\frac{b\ \vec{k}_a\ \overleftarrow{k}_c}{2}\right]^{\frac{1}{2}} \frac{[Fe^{3+}]^{\frac{1}{2}}}{n_o(1+b)} \qquad (58)$$

Using shrinking core cylindrical geometry, the total crystallite surface area is given by

$$A = p\ A_o\ [(1-\alpha)/(1-\alpha_o)]^{\frac{1}{2}} \qquad (59)$$

where p contains crystallite density and appropriate geometric factors, A_o is the original surface area of the $Cu_{1.2}S$ at the beginning of Stage II, and $\alpha_o = 0.4$, the value of α at the end of Stage I. Also the concentration of ferric ion varies with α according to the relationship

$$[Fe^{3+}] = [Fe^{3+}]_o\ [1 - q\alpha]\ \gamma_3 \qquad (60)$$

where $[Fe^{3+}]_o$ is the initial ferric ion concentration, q is the stoichiometry factor, and γ_3 is the activity coefficient for ferric ion. Using these relationships equation (58) becomes

$$\frac{d\alpha}{dt} = \frac{p\ A_o}{n_o(1+b)} \left[\frac{b\ \vec{k}_a \overleftarrow{k}_c\ [Fe^{3+}]_o\ \gamma_3}{2}\right]^{\frac{1}{2}} \left[\frac{1-\alpha}{1-\alpha_o}\right]^{\frac{1}{2}} [1 - q\alpha]^{\frac{1}{2}} \qquad (61)$$

The solid curves of Figure 10 were calculated by numerical integration of equation (61).

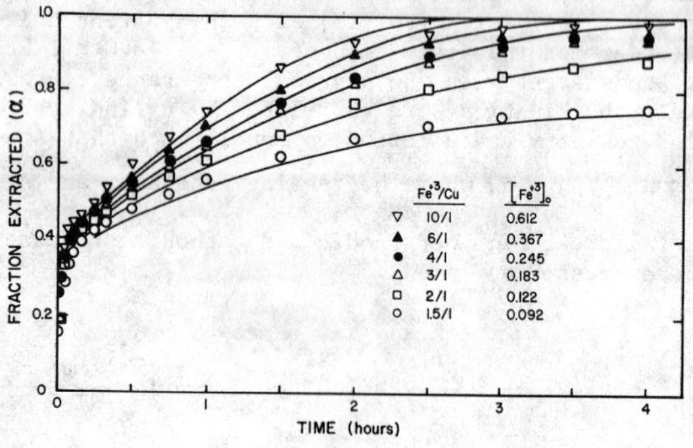

FIGURE 10. Effect of ferric concentration on the second stage dissolution of 48 x 65 mesh Cu_2S at 75^oC.

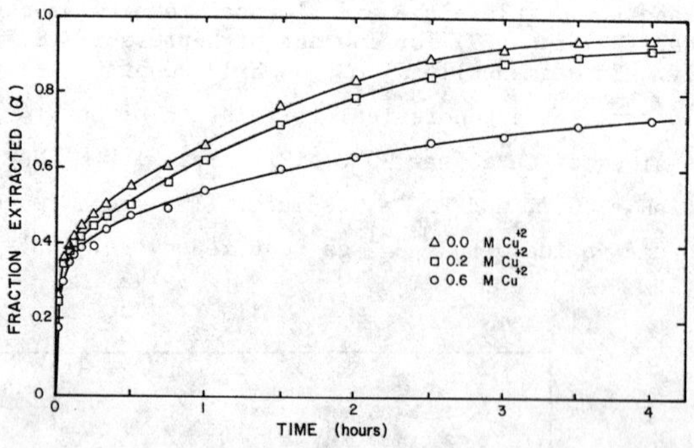

FIGURE 11. Effect of cupric sulfate addition on the second stage dissolution of 48 x 65 mesh particles at 75^oC and 0.24 M Fe^{+3} (Fe^{+3}/Cu = 4/1).

Figures 11 and 12 illustrate respectively the effect of adding increased concentrations of Cu^{2+} and Fe^{2+}. In each case the rate is significantly diminished. This effect may be explained by equation (57). The terms containing $[Cu^{2+}]$ and $[Fe^{2+}]$ become significant at high concentration levels, reducing the rate of reaction by adversely shifting the mixed potential. This result also explains the significant decrease in the rate of leaching of covellite in slurries of high pulp density and ionic concentration.

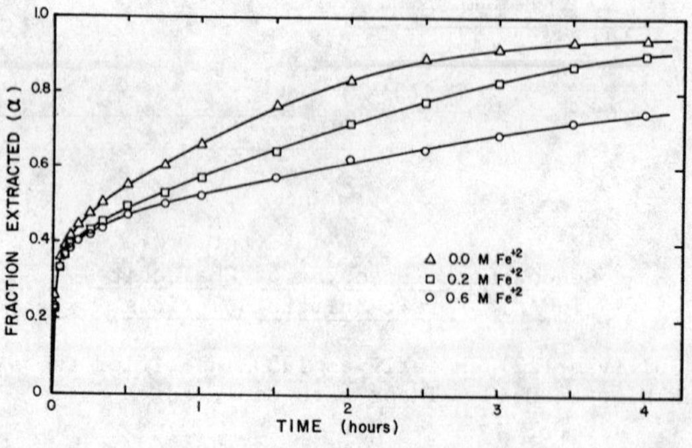

FIGURE 12. Effect of ferrous sulfate addition on the second stage dissolution of 48 x 65 mesh particles at 75^oC and 0.24 M Fe^{+3} (Fe^{+3}/Cu=4/1).

MIXED POTENTIAL GALVANIC COUPLES

When conducting or semi-conducting minerals in electrical contact are subject to leaching simultaneously in the same process, galvanic interactions may occur which can cause an enhanced corrosion of one or more members of the electrochemically coupled mineral regime, and cathodic protection of other mineral members. The mineral with the lower "rest" potential will react anodically, whereas, cathodic behavior (passivation) may be imposed on the mineral possessing the higher "rest" potential. As shown in Table II, pyrite possesses the highest "rest" potential of the mineral sulfides listed. If pyrite is brought into electrical contact with another sulfide mineral, the oxidation of the other sulfide mineral will be accelerated under leaching conditions.

Anodic Dissolution

Peters, et al. have demonstrated enhanced leaching of galena in the presence of pyrite. If pyrite is brought into contact with galena in the presence of oxygen the pyrite surface acts cathodically discharging oxygen according to the reaction

$$\tfrac{1}{2}O_2 + 2H^+ + 2e^- = H_2O \qquad (62)$$

and the galena dissolves anodically according to equation (27) releasing Pb^{2+} and forming elemental sulfur. Figure 13 illustrates the galvanic couple formed.

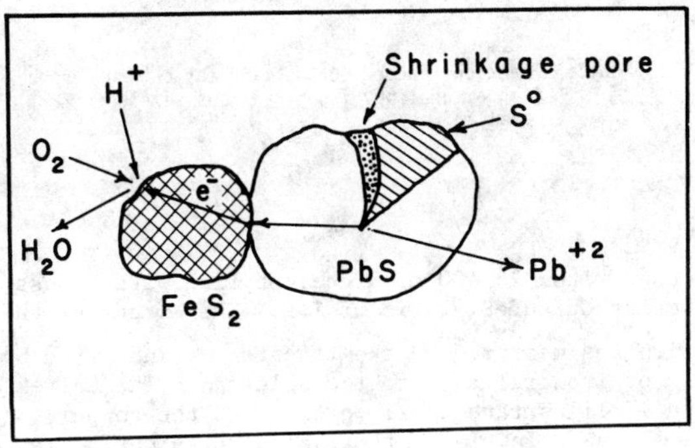

FIGURE 13. Galvanic leaching couple for galena in electrical contact with pyrite.

Another excellent example of this phenomena was recently observed in the leaching of a chalcopyrite bearing material (25). Figure 14 illustrates schematically the polarization curve for chalcopyrite superimposed on that for pyrite. In this example log |i| is plotted versus potential (SHE). Galvanic coupling between pyrite and chalcopyrite within the leaching environment will establish a mixed potential at the voltage represented by line a of Figure 14. At the mixed potential, pyrite is an inert electrode, and oxygen adsorbed on the pyrite surface is discharged cathodically according to reaction (62). Chalcopyrite reacts anodically according to the following reaction

$$CuFeS_2 = Cu^{2+} + Fe^{2+} + 2S^\circ + 4e \qquad (63)$$

The galvanic leaching mechanism is depicted in Figure 15 showing the flow of electrons from the anodic $CuFeS_2$ particle resulting in the discharge of oxygen at the particle surface.

Cathodic Conversion of Chalcopyrite

Shirts et al. (26) have demonstrated that chalcopyrite will react cathodically by bringing it into physical contact with a metal having a lower potential. Three metals, Cu, Fe, and Pb, were used. The reactions observed were

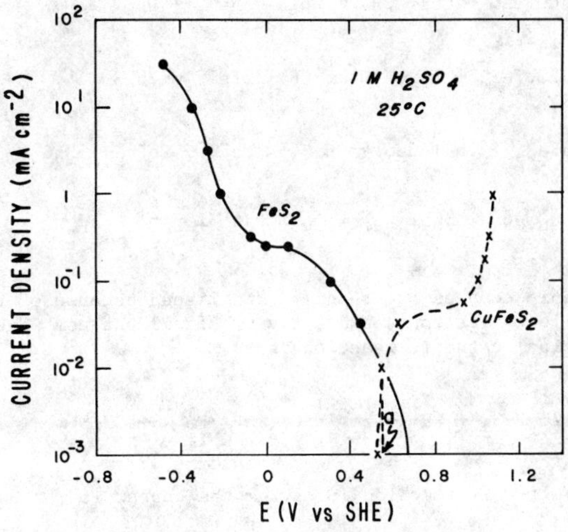

FIGURE 14. Polarization curves for pyrite (FeS_2, cathodic branch) and chalcopyrite ($CuFeS_2$, anodic branch) with mixed potential a.

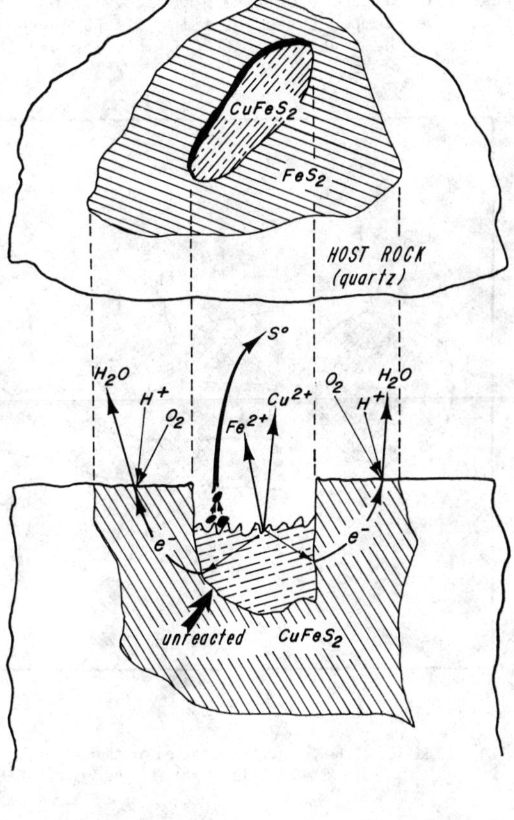

FIGURE 15. Galvanic coupling of pyrite (FeS_2) and chalcopyrite showing flow of electrons from anodic $CuFeS_2$ particles to cathodic FeS_2 particle.

$$CuFeS_2 + Cu + 2H^+ \rightarrow Cu_2S + Fe^{2+} + H_2S \tag{64}$$

$$2CuFeS_2 + Fe + 4H^+ \rightarrow Cu + 2Fe^{2+} + 2H_2S \tag{65}$$

$$2CuFeS_2 + Pb + 6H^+ \rightarrow Cu_2S + 2Fe^{2+} + Pb^{2+} + 3H_2S \tag{66}$$

$$CuFeS_2 + Pb + 4H^+ \rightarrow Cu + Fe^{2+} + Pb^{2+} + 2H_2S \tag{67}$$

Galvanic couples are readily established because of the excellent electronic conductivity of all phases. The principal cathodic reaction is

$$2CuFeS_2 + 6H^+ + 2e^- = Cu_2S + 2Fe^{2+} + 3H_2S \tag{68}$$

The half-cell reaction involving the copper is

$$Cu_2S + 2H^+ + 2e^- = 2Cu + H_2S \tag{69}$$

Reaction (69), having a more negative potential than reaction (68), will react anodically. Similarly metallic iron will reduce $CuFeS_2$ to metallic copper as will metallic Pb in acid chloride solutions.

Hiskey and Wadsworth (27) measured the kinetics of the galvanic conversion of $CuFeS_2$ using copper. Figure 16 illustrates the galvanic $CuFeS_2$-Cu couple. Both the cathodic and anodic reactions produce Cu_2S which welds and establishes good electrical contact. It was found that at lower temperatures, stirring disrupted the bond and retarded the reaction. At 90°C stirring had less effect on the overall kinetics. Figure 17 illustrates results obtained for an intimate mixture of copper powder and chalcopyrite concentrates. Several sizes of copper were used with 325/400 mesh chalcopyrite with excess copper, molar $Cu/CuFeS_2$ ratio of 1.5. The extent of the reaction was measured by the fraction α, indicated by the amount of iron released to the solution. The curves of Figure 17 illustrate the importance of the copper, anodic surface area. Initially the chalcopyrite surface area is larger than that of the copper and the rate is predominantly anodically controlled. As the chalcopyrite reacts, its area becomes smaller and the kinetics shift from predominantly anodic control to predominantly cathodic control. The levelling off below $\alpha = 1.0$ was attributed to lack of perfect mixing of the initial solid phases. Careful blending increased the extent of reaction. In a galvanic system it is necessary to normalize the data in terms of the anodic (A_a) and cathodic (A_c) surface areas by the relationships $I_a = i_a A_a$ and $I_c = i_c A_c$ where I_a and I_c refer to the total anodic and cathodic currents, respectively. Additional tests showed the rate of reaction to be essentially independent of [Fe II] in solution. The experimental activation energy was approximately 11.5 kcal/mole. Figure 18 depicts the initial rate versus initial hydrogen ion concentration. The rate order for hydrogen ion is in the range 0.48-0.52 or essentially one-half. Assuming single charge transfer in the activated state, the net cathodic and anodic electrode processes may be represented by the equations

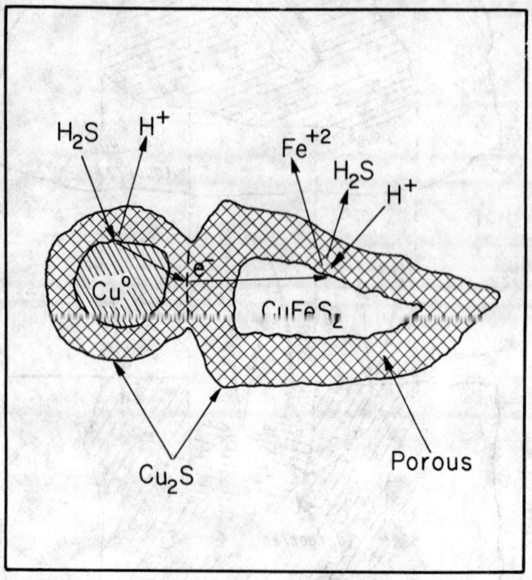

FIGURE 16. Schematic representation of the $CuFeS_2$-Cu galvanic couple in sulfuric acid.

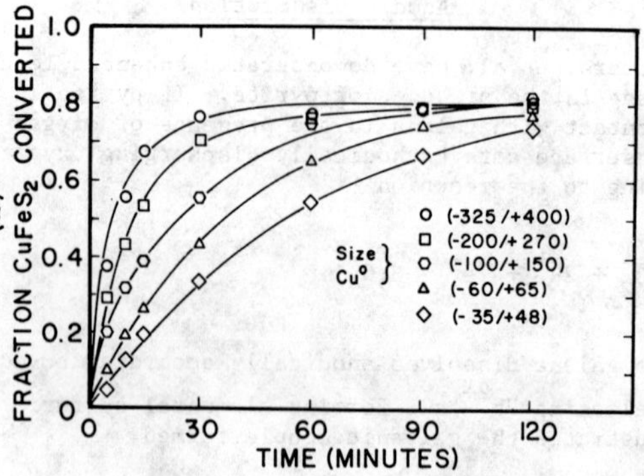

FIGURE 17. Fraction reacted versus time for cathodic reduction of chalcopyrite by various sizes of metallic copper.

$$I_c = A_c \vec{k}_c 2F \exp\left[\frac{\beta_c F E_c}{RT}\right] - A_c \overleftarrow{k}_c a_{H^+} 2F \exp\left[\frac{-(1-\beta_c) F E_c}{RT}\right],$$

$$I_a = A_a \vec{k}_a 2F \exp\left[\frac{\beta_a F E_a}{RT}\right] - A_a \overleftarrow{k}_a a_{H^+} 2F \exp\left[\frac{-(1-\beta_a) F E_a}{RT}\right] \tag{70}$$

The rates k_c and k_a may contain the activity of H_2S. In this case a_{H_2S} is constant since H_2S continually evolves at atmospheric pressure. Also k_c may conceivably contain [Fe II] although experimental results showed little dependence on [Fe II]. When steady-state currents are established E is the mixed potential such that $E_a = E_c = E$ and $I_c = -I_a$. An explicit solution for the mixed poten-

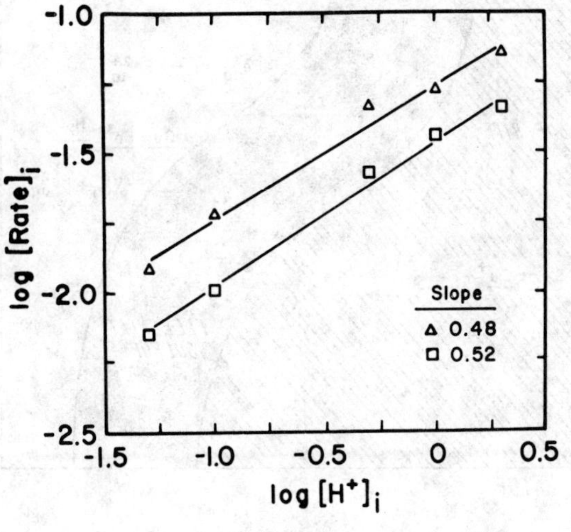

FIGURE 18. Plot of Log $[H^+]_i$ versus Log $[rate]_i$ to determine the hydrogen ion rate dependence. (Upper points: 270 x 325 Mesh Cu; Lower points: 150 x 200 Mesh Cu.)

tial is not possible unless the Tafel slopes are essentially equal, or $\beta_a = \beta_c = \beta$. Using this approximation as before, equation (12) becomes

$$\frac{d\alpha}{dt} = \frac{a_{H^+}^{1/2} A_c A_a \overleftarrow{k}_c \overrightarrow{k}_a}{(A_a\overrightarrow{k}_a + A_c\overrightarrow{k}_c)^{1/2}\,(A_a\overleftarrow{k}_a + A_c\overleftarrow{k}_c)^{1/2}} \qquad (71)$$

Equation (71) provides for the hydrogen ion dependence and anodic control for the initial condition $A_c\overrightarrow{k}_c \gg A_a\overrightarrow{k}_a$ and $A_c\overleftarrow{k}_c \gg A_a\overleftarrow{k}_a$ observed experimentally. Similarly cathodic control results when $A_a\overrightarrow{k}_a \gg A_c\overrightarrow{k}_c$ and $A_a\overleftarrow{k}_a \gg A_c\overleftarrow{k}_c$.

Nicol (28) has demonstrated the electrochemical character of the galvanic coupling of $CuFeS_2$ with metallic copper, iron, and lead.

CHARGE TRANSFER THROUGH PASSIVATION LAYERS

Often passivation layers form as intermediate products with diffusion of charged ions or electrons rate controlling. Three examples are presented here: 1) diffusion of ions through an intermediate defect phase during the anodic dissolution of chalcopyrite; 2) high temperature passivation of chalcopyrite with diffusion of electrons through elemental sulfur as rate controlling; and 3) cathodic reduction of chalcopyrite with ionic diffusion through a bornite intermediate rate controlling.

Anodic Dissolution of Chalcopyrite

The anodic dissolution of chalcopyrite in acidic ferric sulfate solutions occurs according to the half-cell reaction (63) forming elemental sulfur. Initially a defect chalcopyrite structure forms followed by passivation by a layer of elemental sulfur. Also, sulfate ion forms but in relatively smaller proportion than elemental sulfur. Initially ferrous ion is released much more rapidly than cupric ions (6)(29)(30) forming an intermediate structure according to the reaction

$$CuFeS_2 \rightarrow$$

$$Cu_{1-x}Fe_{1-y}S_{2-y-x} + xCu^{2+} + yFe^{2+} + (x+y)S° + 2(x+y)e^- \qquad (72)$$

where y > x initially. As the reaction proceeds x approaches y. Elemental sulfur forms according to the reaction

$$Cu_{1-x}Fe_{1-y}S_{2-x-y} \rightarrow$$

$$(1-x)Cu^{2+} + (1-y)Fe^{2+} + (2-x-y)S° + 2(2-x-y)e \qquad (73)$$

In the first step of the reaction the intermediate product serves as a passivation layer by limiting the reaction to ionic transport through the layer. Results obtained by Warren (6) are illustrated in Figure 19 where anodic polarization curves are shown for various ferrous ion levels. The plateau region, indicating diffusional control, moves up regularly with increasing $[Fe^{2+}]$. The electrode is not passivated by elemental sulfur at this point as noted by Jones and Peters (30) since sulfur, being an electronic insulator, would not respond to the ferrous ion present. The increased plateau current results from the oxidation of ferrous iron to ferric iron at the electrode surface, indicating good electronic conductivity through the passivating layers. Although Warren (6) found sulfur in this region it did not appear as a dense continuous layer but occurred as patches and around etched regions of the surface. Warren observed the kinetics to be typical of ionic transport in a high potential gradient in this region. These results are in contrast to those observed by Munoz et al. (31) who found massive layers of sulfur on fine chalcopyrite leached in acid ferric sulfate solutions at elevated temperatures. Under these conditions the rate of reaction was not

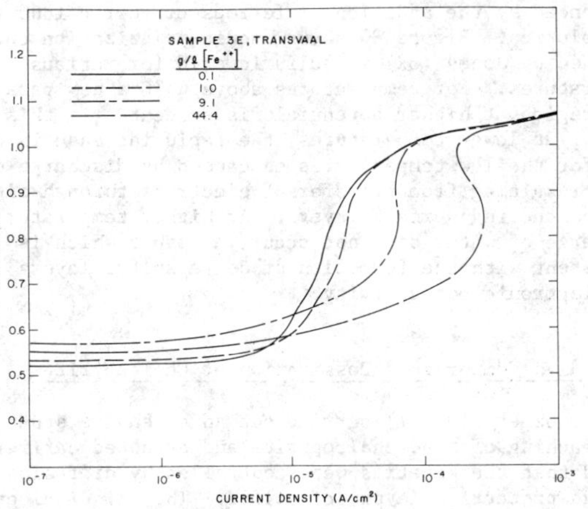

FIGURE 19. Effect of ferrous ion on the anodic polarization curve of Transvaal $CuFeS_2$ in 1 M H_2SO_4, 40 mv/min, 25°C. $[Cu^{++}]$ was held constant at 0.90 g/l.

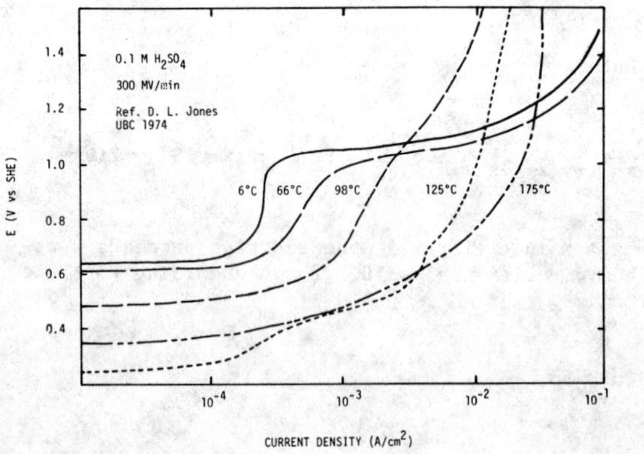

FIGURE 20. Polarization curves for Craigmont chalcopyrite at various temperatures showing passivation at higher temperatures.

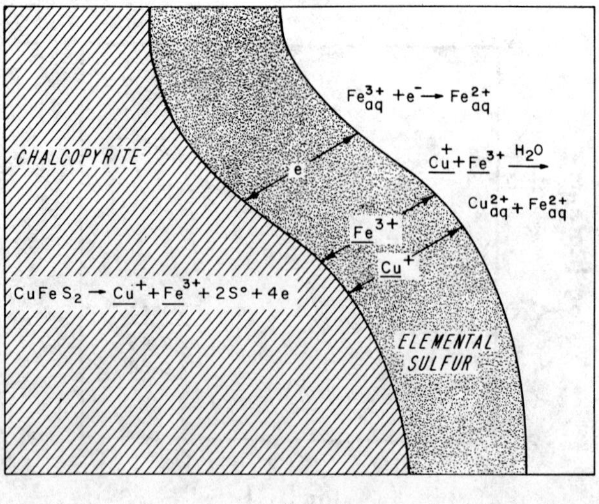

FIGURE 21. Schematic illustrating transport processes in acid ferric sulfate leaching of chalcopyrite.

influenced by the addition of ferrous or ferric ions to the solution. Figure 20 shows anodic polarization curves obtained by Jones (32) in sulfuric acid for various temperatures. For temperatures above 66°C a new passivation region at higher potentials is evident. In this region, at lower temperatures, the rapid increase in current for the low temperatures is caused by discharge of water resulting from transfer of electrons through electron conducting surface layers. At higher temperatures discharge of water does not occur, a result which is consistent with the formation of dense sulfur layers of low electronic conductivity.

High Temperature Passivation of Chalcopyrite

Munoz et al. (31) carried out an extensive study of the leaching of fine chalcopyrite and as noted earlier showed that the kinetics were controlled by diffusion through protective layers of sulfur. The principal overall reaction is

$$CuFeS_2 + 4Fe^{3+} \rightarrow Cu^{2+} + 5Fe^{2+} + 2S° \tag{74}$$

with the formation of some sulfate, particularly at high anodic over voltages. Munoz et al. (31) explained the kinetics of leaching of monosized fractions of chalcopyrite by a model in which the main rate limiting step is electronic transport, accompanied by the outward diffusion of coupled Cu^{2+} plus Fe^{2+} ions, through the sulfur layer. Equally plausible is the outward diffusion of coupled Fe^{3+} and Cu^+. Figure 21 illustrates the transport processes proposed. The activation energy was observed to be 83.7 kJ/mol (20 kcal/mole) which compares favorably with the activation energy (23 kcal/mol) for the electronic conductivity of elemental sulfur. Figure 22 presents the rate data for three monosized fractions of chalcopyrite. The data are well correlated by the diffusion equation for spheres, for diffusion of a reactant through the product layer

$$1-(2/3)\alpha - (1-\alpha)^{2/3} = \frac{k'_p}{d_o} t \tag{75}$$

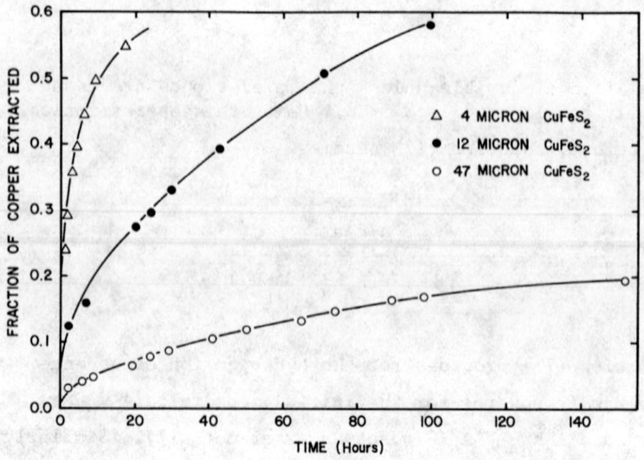

FIGURE 22. A plot of fraction of copper extracted from monosize chalcopyrite particles as a function of time for 1.0 M H_2SO_4, 0.25 M $Fe_2(SO_4)_3$, 90°C, 0.5 percent solids and 1200 rpm.

where α is the fraction reacted and d_o is the initial particle diameter. The rate constant k'_p is given by

$$k'_p = \frac{8\chi}{\rho} \left(\frac{D_i \varepsilon}{\tau}\right) a_i$$

where: $\frac{D_i \varepsilon}{\tau}$ = effective diffusivity, ε is the porosity and τ is the tortuosity

a_i = activity of reacting species (electrons) at the $CuFeS_2S°$ boundary

χ = stoichiometric factor, moles $CuFeS_2$ per mole of electrons

ρ = molar density of chalcopyrite

The experimental value of k_p' was determined to be 2.24×10^{-13} cm^2/s.

The rate of leaching was derived using Wagner's theory of oxidation which relates the intrinsic mass transport or electrical conductivity of products of reaction to rates of reaction. The flux of a charged diffusing anion, cation or electrons (mol cm^{-2} sec^{-1}) is given by

$$J_i = \frac{300}{96{,}500} \frac{t_i \sigma}{|z_i|^2 F} \left[\frac{d\mu_i}{dx} - z_i F \frac{d\phi}{dx}\right] \qquad (76)$$

for modest voltage gradients, where

$d\mu_i/dx$ = chemical potential gradient

$d\phi/dx$ = electric field strength

F = Faraday (23,060 cal/equiv. volt)

t_i, z_i = transport number and charge for ith species

σ = electrical conductivity

Electrical neutrality requires the condition

$$J_{\underline{Cu}^+} + 3J_{\underline{Fe}^{3+}} = J_e \qquad (77)$$

The coupling equation consistent with the observed stoichiometry of the reaction is

$$J_{\underline{Cu}^+} = J_{\underline{Fe}^{3+}} = \frac{1}{A}\frac{dn(CuFeS_2)}{dt} \qquad (78)$$

If transport of electrons through the sulfur layer is rate controlling then the transport number t_e for electrons is expected to be much smaller than the transport number $t_{\underline{Fe}^{3+}}$ for $\underline{Fe}^{3+}$. Combination of equations (76), (77), and (78) for the condition te << $t_{\underline{Fe}^{3+}}$, results in the general equation

$$\frac{dn}{dt}(CuFeS_2) = -A\frac{300}{96{,}500}\frac{t_e\sigma}{4F}\frac{d\mu_e}{dx} \qquad (79)$$

For spherical geometry equation (79) may be integrated, assuming quasi steady state, resulting in the equation

$$\frac{d\alpha}{dt} = \frac{3\times300}{96{,}500}\frac{M}{\rho}\frac{t_e\sigma}{Fd_o^2}\frac{\Delta G\,(1-\alpha)^{1/3}}{1-(1-\alpha)^{1/3}} \qquad (80)$$

where $\Delta G = \mu_e^\circ - \mu_e^i$,

μ_e, μ_e^i = the chemical potentials for the outer and inner boundary of the sulfur layer

ΔG = the calculated Gibbs molar free energy for the overall reaction

M, ρ = molecular weight and density of $CuFeS_2$

Integration of equation (80) assuming a constant average $\overline{\Delta G}$ results in the general diffusion equation for isometric particles, equation (73). Equation (80) may be integrated numerically, including corrected ΔG values with time.

From k_p' and calculated values for ΔG at temperature, the solid lines of Figure 22 were calculated using a best fit value of 7.6×10^{-13} ($ohm^{-1}cm^{-1}$) for $t_e\sigma$. This value compares to a measured value of ~10^{-13} for pure sulfur. The correlation was considered satisfactory since estimate values of the activity coefficients of Cu^{2+}(aq), Fe^{2+}(aq) and Fe^{3+}(aq) were used to calculated ΔG. Also, the sulfur produced contained impurities which have been noted to increase the electronic conductivity of sulfur. The proposed model also explains the modest response of the system to the solution concentration of ferric ion.

Cathodic Reduction of Chalcopyrite

Sohn (33,34) has investigated the reduction of $CuFeS_2$ in cupric sulfate solutions using SO_2 as the reductant. The important half-cell reactions involved are:

cathodic:

$$2CuFeS_2 + 3Cu^{2+} + 4e^- = Cu_5FeS_4 + Fe^{2+} \qquad (81)$$

$$Cu_5FeS_4 + 3Cu^{2+} + 4e^- = 4Cu_2S^* + Fe^{2+} \qquad (82)$$

anodic:

$$SO_2 + 2H_2O = HSO_4^- + 3H^+ + 2e^- \qquad (83)$$

In the sequence of reactions, initially a defect structure was observed with a main lattice parameter near that of bornite but slightly greater. As the reaction proceeded bornite formed followed by djurleite, $Cu_{1.96}S$. The defect structure $Cu_{1.96}S$ (djurleite) formed in solutions of modest acidity. Solutions containing one molar or greater H_2SO_4 formed digenite, $Cu_{1.8}S$. Figure 23 illustrates the fraction of contained iron released to solution as a function of time for three monosized samples of $CuFeS_2$. The rate of conversion increased with copper concentration but was only slightly affected by the partial pressure of SO_2 gas. Ferrous iron catalyzed the reaction with significant increases in reduction rate as Fe^{2+} was added to the solution. Figure 24 illustrates results for the 2.5 μm fraction with and without added Fe^{2+} to solution. When the iron was added after 45 minutes a sharp increase in rate was noted. Warren (6) noticed a similar effect when reducing $CuFeS_2$ cathodically in the absence of cupric ion. In this case H_2S gas was evolved. Sohn (33) carried out similar cathodic reduction experiments in the presence of Cu^{2+}. The catalytic effect has been attributed to the lowered activation energy for the surface charge transfer process involving both iron and copper according to the following reactions at the electrode surface:

$$Fe^{2+}(aq) + Cu^{2+}(aq) = Cu^+(aq) + Fe^{3+}(aq) \qquad (84)$$

$$Cu^+(aq) = Cu^+\ (lattice) \qquad (85)$$

$$Fe^{3+}(aq) + e^- = Fe^{2+}(aq)$$

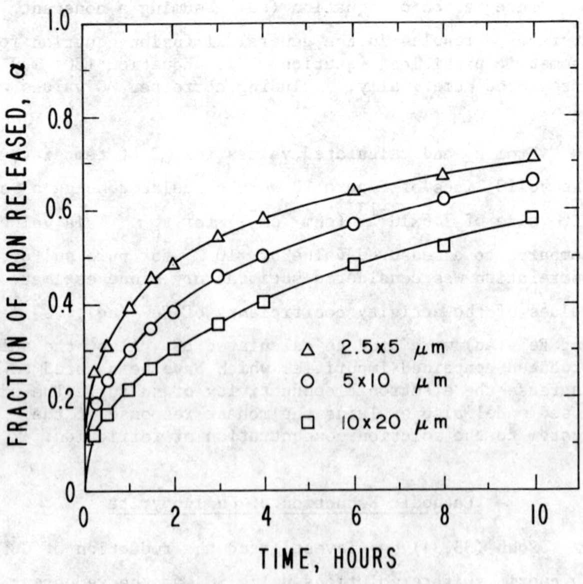

FIGURE 23. A plot of iron extracted from three monosize chalcopyrite fractions as a function of time for 90°C, 0.5% solids, molor ratio of $Cu^{++}/CuFeS_2$ = 7.5 and 1200 rpm.

Figure 25 illustrates the cathodic polarization curves obtained for various Fe^{2+} solution concentrations. The sharp passivation peak formed just above 0.4 volts was attributed to the formation of an initial defect structure. Appreciable enhancement of the current in this region resulted when the ferrous iron concentration was increased. At more negative potentials again the ferrous ion concentration increased the current although in this region

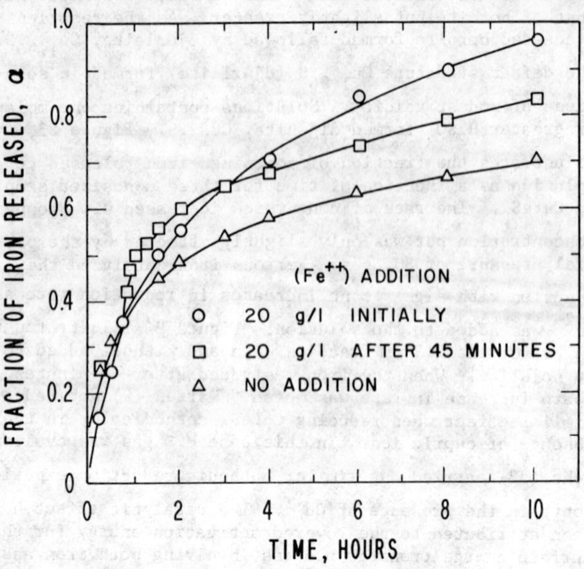

FIGURE 24. Fraction of 2.5 x 5 μm chalcopyrite reduced as a function of time, showing catalytic effect of Fe^{2+}.

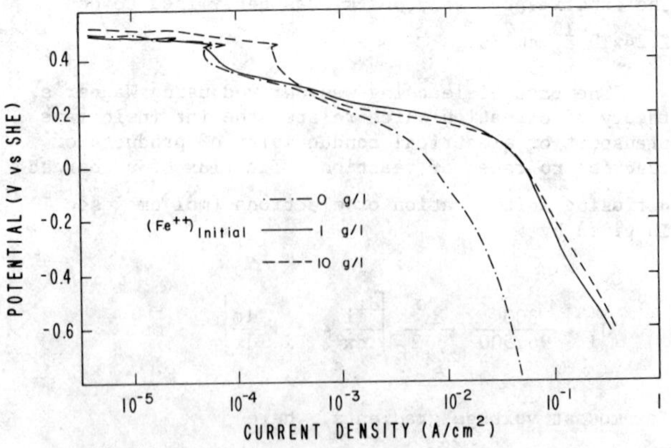

FIGURE 25. Cathodic polarization curves for Transvaal chalcopyrite for various iron concentrations for 10 g/l of Cu^{++}, pH = 1.8 and 25°C.

only a slight increase occurred above 1 gpl Fe^{2+}. In this region $Cu_{1.96}S$ forms.

The kinetics of the process follow logarithmic kinetics, usually observed for thin tarnish films in which high voltage gradients occur. In this system thick layers of $Cu_{1.96}S$ form with a visible band of bornite which moves through the particle. Microscopic examination of a partially reacted particle clearly showed a narrow band of bornite (less than 0.5 μm in thickness) adjacent to the unreacted core of $Cu_{1.96}S$. Figure 26 illustrates the data of Figure 23 plotted according to the

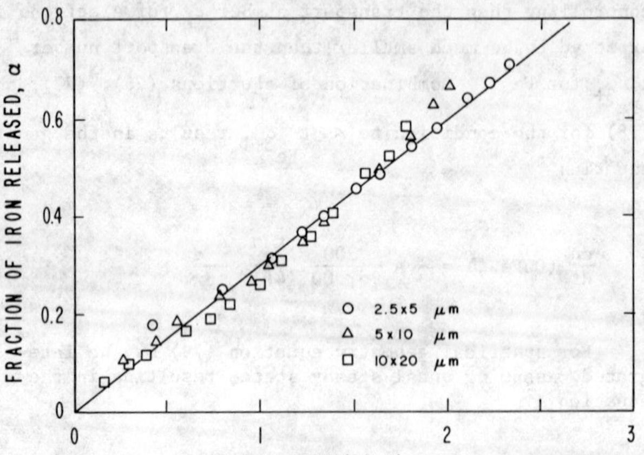

FIGURE 26. A plot of the logarithmic law for the data shown in Figure 23.

logarithmic rate law.

The electrochemical nature of the process was demonstrated (33) by using a diaphragm cell with graphite electrodes. The cathodic cell contained $CuFeS_2$ and graphite powder in a solution of acidified cupric sulfate. The anode compartment was fed SO_2 gas and contained a suspension of graphite powder. Each compartment had a solid graphite electrode. The effect of ferrous iron addition was examined with separate additions to each compartment. Ferrous iron added to the catholyte increased the current markedly while addition to the anolyte decreased the current slightly.

VOLTAGE CONTROL BY SEQUENTIAL MINERAL EQUILIBRIA

Two examples are presented illustrating the importance of sequential phase formation which establishes the reaction potential without surface passivation. This phenomenon is similar to the localized current flow in corrosion couples but differs in that two phases will establish a constant potential as long as the activity of each phase remains constant. The first example is that of chalcocite Stage I leaching (equations (52) and (53)) previously referred to. The second example given is for the anodic dissolution of chalcopyrite at high potentials where decomposition of water occurs with the formation of chemisorbed oxygen on the mineral surface forming an intermediate oxidized phase.

Anodic Dissolution of Chalcocite

Figure 27 illustrates results for the electrochemical oxidation of Cu_2S in an acid solution (35). The steps in the data have been explained on the basis of the sequential formation of a series of CuS compounds with corresponding rest potentials as follows:

$$Cu_2S \rightarrow Cu_{1.93}S \rightarrow Cu_{1.83} \rightarrow Cu_{1.67} \rightarrow Cu_{1.4}S \rightarrow Cu_{1.2}S$$

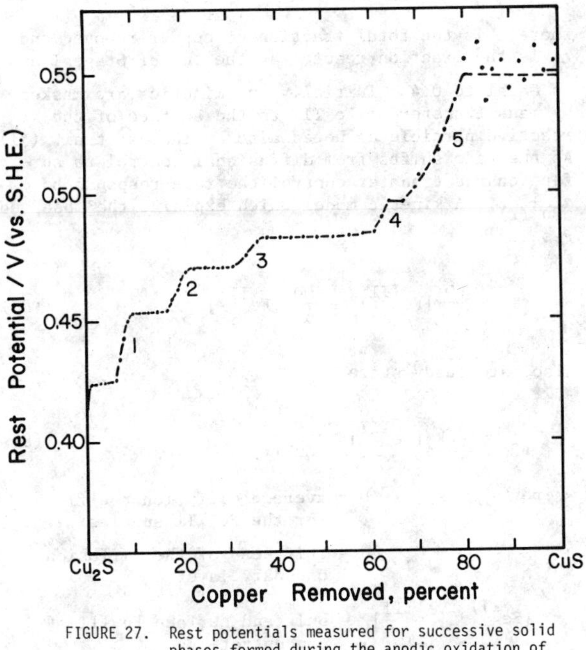

FIGURE 27. Rest potentials measured for successive solid phases formed during the anodic oxidation of Cu_2S to CuS in acid solution. The vertical steps correspond to: 1, $Cu_{1.93}S$; 3, $Cu_{1.67}S$; 4, $Cu_{1.4}S$; 5, CuS.

The flat (constant voltage) steps represent retions in which two mineral phases are in equilibrium and the rising voltages correspond to a change in composition of a single sulfide phase. Compounds generally designated $Cu_{2-x}S$ may be considered as copper deficient defect structures of chalcocite for x values, $1 \geq x > 0.6$. Blue remaining covellite may be viewed as a covellite lathed with excess copper, $Cu_{1+x}S$. Moh (36) reported the existence of blue remaining covellite, designated Type B, formed in the presence of oxidizing agents and having the composition $Cu_{1+x}S$ with variable values of x from approximately 0.09 to 0.43 at room temperature. Another form of blue remaining covellite having a different structure, designated Type A, formed in a pure Cu-S system, was reported to have a solid-solution region with x values from approximately 0.06-0.20. Frenzel (37) reported blue remaining covellite to be a defect structure with excess copper up to $x \simeq 0.4$. Potter (38) has identified two distinct types of blue-remaining covellite (blaubleiblender) formed using electrochemical cells. Instead of a solid solution the compounds blaubleiblender I, $Cu_{1.1}S$ and blaubleiblender II, $Cu_{1.4}S$ apparently form as distinct compounds. He established a series of Cu-S structures as follows:

$Cu_{1.97 \pm .003}S$	tetragonal djurleite
three metastable phases	proto-djurleite proto-hexagonal chalcocite proto-digenite
$Cu_{1.76}S$ to $Cu_{1.82}S$	low digenite
$Cu_{1.4}S$	blaubleiblender I
$Cu_{1.1}S$	blaubleiblender II

Cavallotti and Salvago (39) reported low digenite (face-centered cubic) to exist in the composition range $Cu_{1.76}S$ to $Cu_{1.79}S$ with a transformation to a high temperature form above 70°C. Brennet et al. (40) investigated the anodic dissolution of Cu_2S and observed the sequence: $Cu_2S \rightarrow Cu_{1.8}S \rightarrow Cu_{1.1}S \rightarrow CuS$. While discrepancies appear concerning the exact compositions it is clear that there are a series of identifiable Cu-S structures, some of which are metastable, with appreciable variation in copper content. Also the presence of copper and iron in the electrolyte influences the character and composition of the solid phases. Based on solid state properties those designated as copper deficient $Cu_{2-x}S$ extend from the high copper region down to approximately $Cu_{1.4}S$. Phases of lower copper (Cu/S < 1.4) content, extending to covellite are viewed as having excess copper, or as $Cu_{1+x}S$. Figure 28 illustrates results obtained by Warren (6) for the oxidation of a thin layer of Cu_2S formed by the cathodic reduction of chalcopyrite ($CuFeS_2$). The rest potential of the $CuFeS_2$ is +0.52 volts. As voltage was decreased the cathodic branch, indicated by the arrow on the rising curve, resulted and at the plateau value (-6×10^{-4} amp) Cu_2S formed in a thin layer. Toward the upper part of the loop at -0.20 volts the voltage was reversed and increased anodically. The sharp peaks indicate a series of Cu-S intermediate solid phases with rest potentials in the region of 0.40 to 0.55, essentially in the same region as those reported by Koch and McIntyre (35).

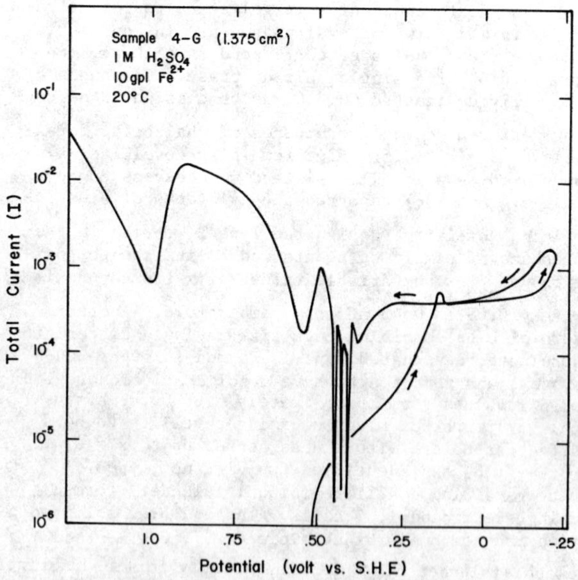

FIGURE 28. Potential-current diagram showing the formation of several intermediate Cu-S layers formed on the surface of chalcopyrite.

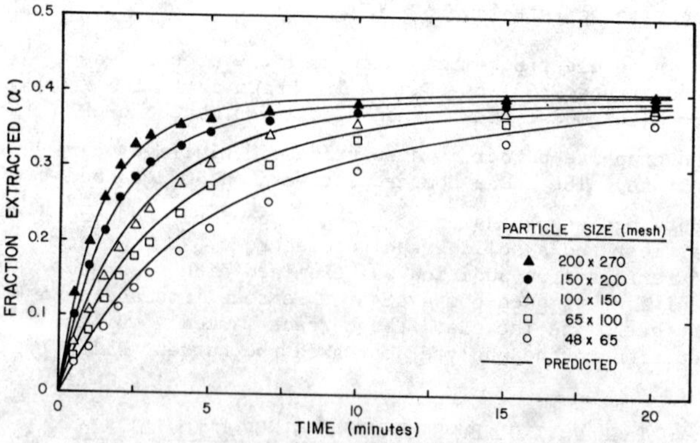

FIGURE 29. Effect of particle size on the first stage dissolution of Cu_2S at 30°C and 0.06 M Fe III (Fe III/Cu = 1/1).

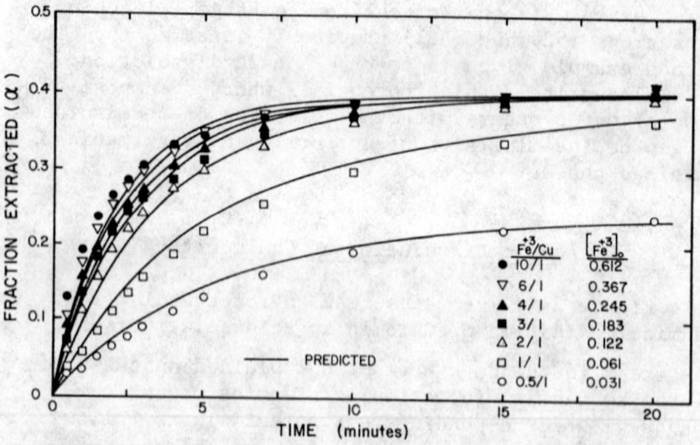

FIGURE 30. Effect of ferric concentration on the first stage dissolution of 48 x 65 mesh Cu_2S at 30°C.

As noted earlier, Marcantonio (23), in an analysis of the acid ferric sulfate leaching of natural Cu_2S, has identified a sequence of reactions in which the solid phases observed were $Cu_2S \rightarrow Cu_{1.8}S \rightarrow Cu_{1.2}S$. Digenite, $Cu_{1.8}S$, was found to form rapidly. At 30°C, 48x65 mesh Cu_2S particles were completely converted to $Cu_{1.8}S$ in 3.5 minutes in 0.03 M Fe III solutions, indicative of the rapid diffusion of Cu^+ through the sulfur lattice. The final product at 30°C, and low Fe III concentrations, was blue-remaining covellite or blaubleiblender of the composition $Cu_{1.2}S$. Normal covellite, CuS, was never observed as a product of reaction, in agreement with the findings of Thomas, et al. (41).

Figure 29 illustrates rate data for Stage I leaching at 30°C for various Cu_2S initial particle sizes under strong stirring conditions. Figure 30 illustrates the effect of Fe III concentration upon the kinetics of Stage I leaching. The kinetics were found to be first order with respect to [Fe III] up to approximately $\alpha = 0.27$ ($Cu_{1.46}S$), and mass transport controlled. For $\alpha < .27$, measured rates correspond to calculated rates based upon transport through a liquid limiting boundary film. Also initial mass transport control is consistent with the measured activation energy of 2.8 kcal/mole. For $\alpha > 0.27$ the rate diminished rapidly leveling off at $\alpha = 0.40$. For the range $.27 > \alpha > 0.40$ the kinetics were related to the Fe III solution concentration and the activity of $Cu^+(\ell)$ in the copper sulfide lattice.

The dependence of the rate on $a_{Cu^+(\ell)}$ is consistent with a model in which a series of compounds form sequentially with a systematic decrease in the activity of reactive monovalent copper in the lattice. If it is assumed that the activity of $Cu_\phi S$ ($Cu^+(\ell)$) goes from unity for $\phi = 2$ to zero for $\alpha = 0.4$ then the $a_{Cu^+(\ell)}$ is given by

$$a_{Cu^+(\ell)} = (1 - \frac{\alpha}{\alpha_o}) \tag{87}$$

where α is the total fraction of copper removed and α_o is the fraction reacted at the end of Stage I and is equal to 0.4. Initially the kinetics are masked by mass transfer of Fe III to the surface of the reactive particle as noted also by Thomas et al. (41). As the rate shifts from diffusional control to surface charge transfer control the rate responds to $a_{Cu^+(\ell)}$. A kinetic model which explains the observed results is

$$i_c = \frac{FD}{\delta}[Fe^{III}]_s - \frac{FD}{\delta}[Fe^{III}] \tag{88}$$

also, at steady state,

$$i_c = -Fk_s[Fe^{III}]_s\, a_{Cu^+(\ell)} \tag{89}$$

where

D = average diffusion coefficient for the Fe III species

δ = thickness of the diffusion boundary layer

$[Fe^{III}], [Fe^{III}]_s$ = bulk and surface Fe III concentrations

k_s = surface rate constant.

Consistent with a heterogeneous process the kinetics depend upon the initial surface area, A_o; and $A_o = 3w_o f/r_o \rho c$ where w_o is the original weight of Cu_2S, ρ_c is the density of Cu_2S, f is a geometric sphericity factor, and r_o is the particle radius. The ferric iron concentration is given by

$$[Fe^{III}] = [Fe^{III}]_o\ (1-b\alpha) \qquad (90)$$

where $[Fe^{III}]_o$ refers to the initial concentration, $b = 2/q$, q is the molar FeIII/Cu ratio and M_c is the molecular weight of Cu_2S.

The fraction reacted, α, is related to i_c by

$$\frac{d\alpha}{dt} = -\frac{1}{n_o}\frac{dn}{dt} = -\frac{i_c A_o}{2Fn_o} \qquad (91)$$

where n is the moles of Cu remaining in time t and n_o is the original number of moles of Cu present as Cu_2S. Under steady state conditions, equations (88) and (89) result in the rate equation.

$$\frac{d\alpha}{dt} = \frac{k'}{r_o}\frac{\gamma\pm[Fe^{III}]a_{Cu^+(\ell)}}{(a_{Cu^+(\ell)} + \frac{D}{\delta k_s})} \qquad (92)$$

where $k' = 3w_o fD/2\rho_c n_o \delta$. Equations (90) and (92) may be combined and integrated numerically resulting in the solid curves of Figure 29.

The apparent influence of $a_{Cu^+(\ell)}$ may reflect a variation in voltage due to the change in $[S_\ell^{2-}]$ which is related stoichiometrically. In CuS (2), cuprous ion is reportedly present co-valently bonded as $(Cu^+)_2S_2^{2-}$. Consistent with the solid state properties of $Cu_{1+x}S$ the reaction in Stage I may be viewed as

$$Cu^+(lattice) \rightarrow Cu^{2+}(aq) + e^- \qquad (93)$$

$$S^{2-}(lattice) \rightarrow \tfrac{1}{2}S_2^{2-}\ (lattice) + e^- \qquad (94)$$

$$Cu^+(lattice) + \tfrac{1}{2}S_2^{2-}\ (lattice) \rightarrow \tfrac{1}{2}(Cu^+)_2S_2^{2-} \qquad (95)$$

For each two moles of Fe III reduced, one mole of $Cu^+(\ell)$ appears in solution as Cu^{2+} and one mole of $Cu^+(\ell)$ becomes fixed due to the strong co-valent character of the $(Cu^+)_2S_2^{2-}$ bond.

Hepel and Hepel (42) have considered Cu_2S as an electrode of varying composition according to the reaction

$$(y-x)^{-1}\ Cu_{2-x}S = (y-x)^{-1}\ Cu_{2-y}S + Cu^{2+} + 2e \qquad (96)$$

where $Cu_{2-x}S$ and $Cu_{2-y}S$ are non-stoichiometric sulfides. Following suggestions of Hepel and Hepel and combining equations (93)(94) and (95) it is possible to express the equilibrium reaction, equation (96), in terms of the activities of $Cu^+(\ell)$, $S^{2-}(\ell)$ and $S_2^{2-}(\ell)$. The desulfide may alternately be considered as $Cu^{2+}(\ell)$ but more recent evidence (2) supports the concept that covellite is cuprous disulfide. For a sulfide designated as $Cu_\phi S$ the activity of S_2^{2-} would vary from zero for $\phi = 2$ to one for $\phi = 1.2$. If $a_{Cu^+(\ell)} = 1 - \alpha/\alpha_o$, then $a_{S_2^{2-}} = \alpha/\alpha_o$. The combined equations may be written

$$Cu^+(\ell) + S^{2-}(\ell) = Cu^{2+}(aq) + \tfrac{1}{2}S_2^{=}(\ell) + 2e \qquad (97)$$

or

$$2Cu^+(\ell) = Cu^{2+}(aq) + Cu^{2+}(\ell) + 2e^- \qquad (98)$$

where $a_{Cu^{2+}} = a_{\frac{1}{2}S_2}-2$. Accordingly the voltage would be given by

$$E = E° + .0295\ \log \frac{a_{S_2}{}^{-2}(\ell)}{a_{Cu^+}(\ell)a_{S}{}^{-2}(\ell)} + .0295\ \log\ [Cu^{2+}] \qquad (99)$$

Since $a_{Cu^+(\ell)} = a_{S^{2-}(\ell)}$ equation (99) becomes

$$E = E° + .0295\ \log \frac{a_{S_2}{}^{-2}(\ell)}{a^2_{Cu^+}(\ell)} + .0295\ \log\ [Cu^{2+}] \qquad (100)$$

Using the data of Koch (35) for the rising, single phase, segments of Figure 27, E should vary linearly with $\log a_{S_2}{-2}/a^2_{Cu^+(\ell)}$. Figure 31 illustrates the correlation obtained. The measured slope is 0.025 compared to the expected value of 0.0295. It is probable that the activity $a_{Cu^+(\ell)}$ in the rate equation (equation (92)) reflects a voltage contribution which influences the discharge of Fe III at the mineral surface.

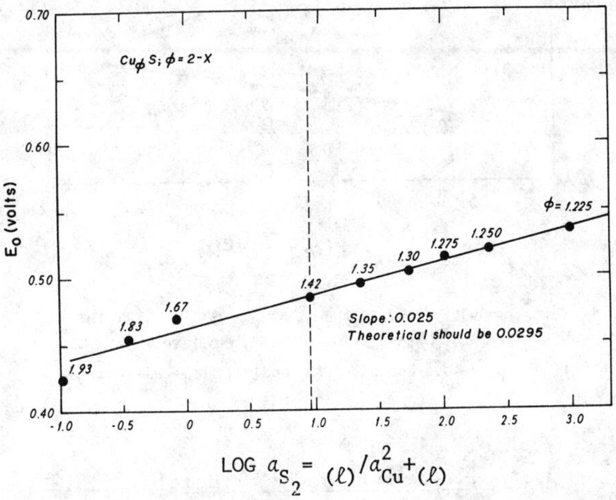

FIGURE 31. Standard rest potential versus Log $[a_{S_2^=(\ell)}/a^2_{Cu^+(\ell)}]$ for $Cu_\phi S$ compounds for various values of ϕ.

More recently Zhong (43) has examined the Cu_2S-Fe^{3+}/Fe^{2+} system using a two electrode system in a diaphragm cell. The anode consisted of Cu_2S and the cathode graphite exposed to a catholyte containing Fe III and Fe II in a sulfuric acid electrolyte. The Cu_2S was exposed to an acidic sulfate solution containing H_2SO_4. An external resistance in series was decreased in increments to determine the influence of the electrode reactions on the anodic and cathodic over-potentials. Figure 32 shows the results obtained for two catholytes; one 0.005 molar in Fe III and the other 1.0 molar. In each case the anolyte contained 1 M H_2SO_4, 0.5 M $MgSO_4$ and 0.1 M $CuSO_4$. The catholyte contained 0.1 M Fe^{2+} and 1 M H_2SO_4. It is apparent from these results that coupled mass transfer of Fe III is rate limiting and that the Cu_2S electrode is near its rest potential. These results generally are supportive of the model suggested by Marcantonio (23).

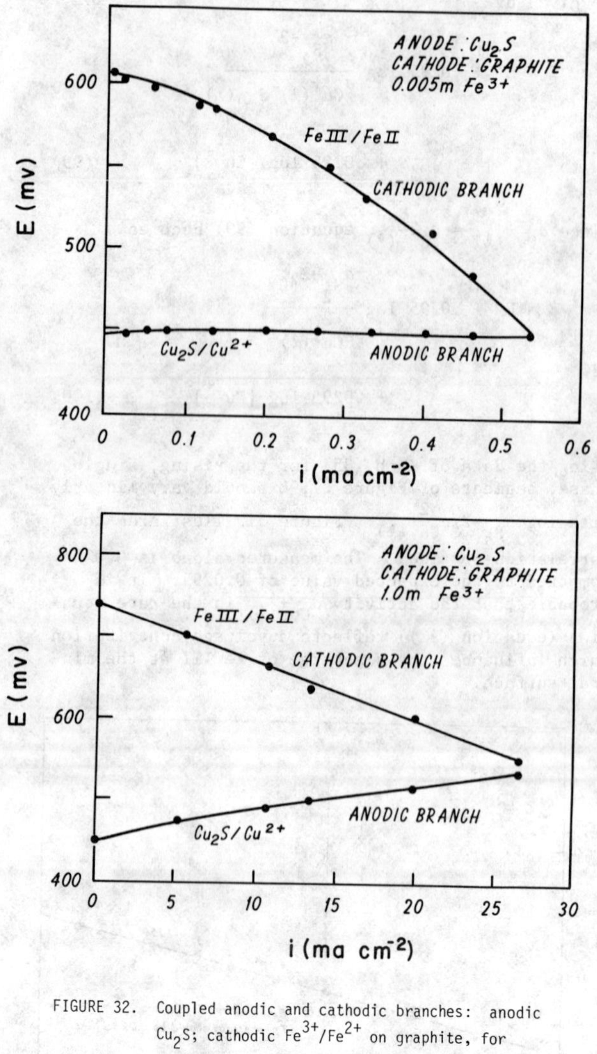

FIGURE 32. Coupled anodic and cathodic branches: anodic Cu_2S; cathodic Fe^{3+}/Fe^{2+} on graphite, for various concentration and external resistance.

Anodic Dissolution of Chalcopyrite

Warren (6) observed a rapid increase in potentials at about 1.09 volts (SHE)(Figure 19) for the anodic dissolution of $CuFeS_2$. As noted earlier the passivation region results in the formation of a defect structure deficient in iron. For simplicity the defect state will be referred to as CuS(n-s) where (n-s)(42) indicates the sulfide is non-stoichiometra. Warren (6) observed that the increase in current at 1.09 volts did not release oxygen but resulted in the release of copper in proportion to the current passed. Vetter (44) has noted that chemisorbed oxygen has been observed to form in this voltage regime.

Recently Jara (45) has carried out galvanostatic experiments for the anodic dissolution of $CuFeS_2$. Figure 33 illustrates the results obtained at four separate current densities ranging from 10 to 100 microamperes. The voltage increases due to the formation of the passive CuS(n-s). At 1.09 volts a constant voltage is maintained and is the same for the four current densities used. These results may be explained by the reactions

$$CuFeS_2 \rightarrow CuS(n\text{-}s) + Fe^{3+} + S° + 3e^- \qquad (101)$$

$$CuS(n\text{-}s) + H_2O = CuS\cdot O + 2H^+ + 2e^- \qquad (102)$$

$$CuS\cdot O = Cu^{2+} + \tfrac{1}{4}SO_4^{2-} + \tfrac{3}{4}S° + \tfrac{3}{2}e^- \qquad (103)$$

The plateau voltage was observed to vary with pH as suggested by equation (102) and the S° and SO_4^2 mass balances corresponded closely to those expected in the above equations.

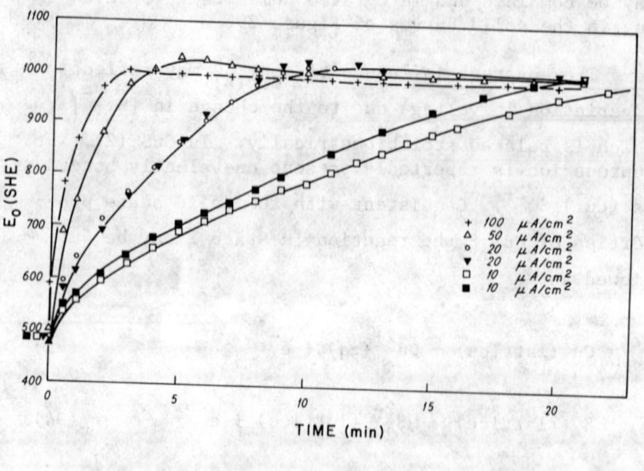

FIGURE 33. Galvanostatic curves for the anodic dissolution of $CuFeS_2$ in 1.0 molar H_2SO_4.

REFERENCES

1. G.H. Koch, Modern Aspects of Electrochemistry, No. 11, p. 211-237, Plenum Press, New York, 1975.

2. R.T. Shuey, Semiconducting Ore Minerals, Elsevier Publishing Company, New York, 1975.

3. D.J. Vaughan and J.R. Craig, Mineral Chemistry of Metal Sulfides, Cambridge University Press, New York, 1978.

4. T. Biegler and D.A. Swift, "Anodic Behavior of Pyrite in Acid Solutions", Electrochimica Acta, vol. 24 (1979), p. 415-420.

5. N. Masuko and Y. Hisamatsu, "On the Electrochemical Behavior of Pyrite", Denki-Kagaku, vol. 31 (1963), p. 907-12.

6. G.W. Warren, "The Electrochemical Oxidation of $CuFeS_2$", Ph.D. Dissertation, University of Utah, 1978.

7. D.M. Chizhikov and V.N. Kovylina, "Investigation of Potentials and Anodic Polarization of the Sulfides of Metals and Their Alloys", Trudy Chetvertogo Soveshchaniya po Electrokhimii (1956), p. 715-719.

8. H. Majima, "How Oxidation Effects Selective Flotation of Complex Sulphide Ores", Can. Met. Quart., vol. 8 (1969), p. 269-273.

9. K.L. Vetter, "Electrochemical Kinetics-Theoretical and Experimental Aspects, p. 732, Academic Press, New York, 1967.

10. M.J. Nicol, C.R.S. Needes, and N.P. Finkelstein, "Electrochemical Model for the Leaching of Uranium Dioxide: 1-Acid Media", p. 1-11, in Leaching and Reduction in Hydrometallurgy, A.R. Burkin, ed., IMM, London, 1975.

11. C.R.S. Needes, M.J. Nicol, and N.P. Finkelstein, "Electrochemical Model for the Leaching of Uranium Dioxide: 2-Alkaline Carbonate Media," p. 12-19, in Leach and Reduction in Hydrometallurgy, A.R. Burkin, ed., IMM, London, 1975.

12. M.J. Nicol, C.R.S. Needes, and N.P. Finkelstein, "The Mechanism of the Anodic Dissolution of UO_2 in Aqueous Solutions", Natn. Inst. Metall., Report PIN 171, Oct. 1973, 37 p.

13. C.R.S. Needes and M.J. Nicol, "The Study of Some Redox Reactions at a UO_2 Surface", Natn. Inst. Metall., Report PIN 165, Sept. 1973, 60 p.

14. J.B. Hiskey, "Kinetics of Uranium Dioxide Dissolution in Ammonium Carbonate", IMM Trans., Sect. C 1979, vol. 88, p. 145-152.

15. J.B. Hiskey, "Hydrogen Peroxide Leaching of Uranium in Carbonate Solutions", Submitted for publication.

16. E.V. Komarov, "Complex Formation in the $UO_2(NO_2)_2$-$KHCO_3$-H_2O_2-H_2O System", Russ J. Inorg. Chem., vol. 4, (1959), p. 591-596.

17. P. Eadington and A. Prosser, "Oxidation of Lead Sulphide in Aqueous Suspensions", IMM Trans., Sect. C 1969, vol. 78, p. 74-82.

18. R.L. Paul, M.J. Nicol, J.W. Diggle and A.P. Sanders", The Electrochemical Dissolution of Galena in Aqueous Solutions", Natn. Inst. Metall., Report No. 1878, May 1977.

19. L.W. Beckstead and J.D. Miller, "Ammonia, Oxidation Leaching of Chalcopyrite-Reaction Kinetics", Met. Trans. B, vol. 8B, (1977) p. 19-38.

20. F.A. Forward and V.N. Mackiw, "Chemistry of the Ammonia Pressure Process for Leaching Ni, Cu and Co from Sherritt-Gordon Sulphide Concentrates", Trans. AIME, 1955, vol. 203, p. 457-463.

21. M.C. Kuhn, N. Arbiter, and H. King, "Anaconda's Arbiter Process for Copper", Can. Min. Met. Bull. 1974, vol. 67, p. 62-73.

22. J.D. Sullivan, "Chemistry of Leaching Chalcocite", U.S. Bureau of Mines, Tech. Paper 473, 1930.

23. P. Marcantonio, "Chalcocite Dissolution in Acidic Ferric Sulfate Solutions", Ph.D. thesis, Department of Mining, Metallurgical and Fuels Engineering, University of Utah, 1976.

24. E. Peters, I.H. Warren, and H. Veltman, First Tutorial Symposium on Hydrometallurgy, The University of Denver, 1972.

25. V.K. Berry, L.E. Murr, and J.B. Hiskey, "Galvanic Interaction between Chalcopyrite and Pyrite during Bacterial Leaching of Low-Grade Waste", Hydrometallurgy, vol. 3, (1978), p. 209-326.

26. M.B. Shirts, J.K. Winter, P.A. Bloom, and G.M. Potter, "Aqueous Reduction of Chalcopyrite Concentrate with Metals", U.S. Dept. of Interior, Bureau of Mines, RI 7953, 1974.

27. J.B. Hiskey and M.E. Wadsworth, "Galvanic Conversion of Chalcopyrite", Met. Trans. vol. 6B, (1975), p. 183-190.

28. M.J. Nicol, "Mechanism of Aqueous Reduction of Chalcopyrite by Copper, Iron and Lead", IIM, 1975, vol. 84, p. C206-C209.

29. J.P. Baur, H.L. Gibbs, and M.E. Wadsworth, "Initial-Stage Sulfuric Acid Leaching of Chalcopyrite Using Radiochemical Techniques", U.S. Dept. of Interior, Bureau of Mines, RI 7823, 1974.

30. D.L. Jones and E. Peters, "The Leaching of Chalcopyrite with Ferric Sulfate and Ferric Chloride", Extractive Metallurgy of Copper, Vol. II, J.C. Yannopoulos and J.C. Agarwal, eds., American Institute of Mining, Metallurgical and Petroleum Engineers, 1976.

31. P.B. Munoz, J.D. Miller, and M.E. Wadsworth, "Reaction Mechanism for the Acid Ferric Sulfate Leaching of Chalcopyrite", Met. Trans. vol. 10B, (1979) p. 149-158.

32. D.L. Jones, "The Leaching of Chalcopyrite", Ph.D. thesis, University of British Columbia, 1974.

33. H.J. Sohn, "Chemical and Electrochemical Conversion of Chalcopyrite into Copper Sulfide", Ph.D. thesis, University of Utah, 1980.

34. H.J. Sohn and M.E. Wadsworth, "Reduction of Chalcopyrite with SO_2 in the Presence of Cupric Ions", Journal of Metals, vol. 32, (1980) p. 18-22.

35. D.F.A. Koch and R.J. McIntyre, "The Application of Reflectance Spectroscopy to a Study of the Anodic Oxidation of Cuprous Sulfide", J. Electroanal. Chem., vol. 17, (1976), p. 285-296.

36. G.H. Moh, "Mineral Soc. Japan", Spec. Paper 1, p. 226-232, 1971.

37. G. Frenzel, N. Jb. Miner, 1959, Abh 93, p. 87-132.

38. R.W. Potter II, "An Electrochemical Investigation of the System Copper-Sulfur", Econ. Geol., vol. 72, (1977), p. 1524-1542.

39. P. Cavallotti and G. Salvago, Electrochim Metal, vol. IV, no. 3, (1969), p. 181-210.

40. P. Brennet, S. Jafferalli, J.M. Vanseveren, J. Vereecken and R. Winand, Met. Trans., vol. 5, (1974), p. 127-134.

41. G. Thomas, T.R. Ingraham, and R.J.C. MacDonald, "Kinetics of Dissolution of Synthetic Digenite and Chalcocite in Aqueous Acidic Ferric Sulphate Solution", Can. Met. Quart., vol. 6, (1967), p. 281-292.

42. M. Hepel and T. Hepel, "The Anodic Dissolution of Chalcocite in Ammoniacal Environment", J. Electroanal. Chem., vol. 81, (1977), p. 161-170.

43. T.K. Zhong and M.E. Wadsworth, Coupled Electrode Process in the Anodic Dissolution of Chalcocite and Chalcopyrite, work in progress, University of Utah, 1981.

44. K.L. Vetter, Electrochemical Kinetics-Theoretical and Experimental Aspects, p. 615; Academic Press, New York, 1976.

45. J.O. Jara, "Galvanostatic Electrode Reactions of Chalcopyrite in Acid and Basic Solutions", M.S. thesis, University of Utah, Department of Metallurgy and Metallurgical Engineering, in preparation.

GAS-SOLID REACTIONS IN EXTRACTIVE METALLURGY

H.Y. Sohn

Department of Metallurgy and Metallurgical Engineering
University of Utah
Salt Lake City, Utah 84112

Gas-solid reations occupy an important position in extractive metallurgy. Examples include the gaseous reduction of metal oxides and sulfides, the roasting of sulfide minerals, the chlorination of minerals, the sulfation of sulfide and oxide minerals, and the calcination of limestone. Furthermore, certain reactions between two solid reactants such as the carbothermal reduction of metal oxides proceed through coupled gas-solid reactions.

This treatise summarizes recent advances in the analysis of such gas-solid reaction systems. The important interactions among geometrical factors, interfacial reactions, transport phenomena and thermodynamics in systematically describing these reaction systems are discussed. Theoretical development begins with the description of simple reactions in single solid particles, followed by the descriptions of more complex cases such as volume change reactions, nonisothermal systems, and simultaneous reactions involving more than one gas and/or one solid reactant. Finally, reference is made to the application of the results for single solids to the description and design of multiparticle systems.

1. INTRODUCTION

Most pyrometallurgical processes contain steps in which gas-solid reactions take place. Examples include the reduction of metal oxides and sulfides by gases, the roasting of sulfide minerals, the chlorination of minerals, the sulfation of sulfide and oxide minerals, and the calcination of limestone, to name but a few. Furthermore, many pyrometallurgical reactions between solid reactants proceed through gaseous intermediates and thus can be considered as coupled gas-solid reactions. The reduction of metal oxides by carbon is an example.

Gas-solid reactions usually take place at a phase boundary. Gaseous reactants must be transported to the reaction interface and gaseous products away from the reaction interface. For reactions involving substantial changes in enthalpy, heat must be supplied to or removed from it. Therefore, the transfer of mass and heat in addition to the chemical reaction constitutes an important aspect of gas-solid reactions. It follows that the analysis of gas-solid reactions must necessarily take into account the geometry and structure of the solid before, during, and after the reaction. The models of gas-solid reactions thus depends on the answers to the following questions:

1. Is the solid initially porous or nonporous?
2. Is a coherent solid product layer formed by the reaction?
3. If so, is the product layer porous or nonporous?
4. Does the solid product occupy more or less volume than the reactant solid?

Although reactions of other types do occur in extractive metallurgy, the most important group of gas-solid reactions in this regard is one in which a porous or a nonporous solid reacts to produce a coherent layer of porous solid product. Therefore, this paper will mainly be concerned with this type of reactions. Examples of such reactions are:

A. the reduction of metal oxides
B. the reduction of metal sulfides
C. the roasting of metal sulfides
D. the calcination of limestone
E. the volatilization of metals by chlorination
F. the decomposition of metal compounds

For the analysis of other types of gas-solid reactions, the reader is referred to other sources (1,2).

Many aspects of the treatment presented in this paper for gas-solid reactions are also applicable to liquid-solid reactions such as leaching and dissolution. Furthermore, many basic principles involved are applicable, directly or indirectly, to gas-liquid and liquid-liquid systems.

In the following the elements involved in the analysis of gas-solid reactions will be briefly discussed. Gas-solid reaction models will then be reviewed and examples of gas-solid reactions encountered in extractive metallurgy will be discussed.

2. ELEMENTS OF GAS-SOLID REACTIONS

The description of gas-solid reactions requires the examinatin of chemistry, thermodynamics and rate processes. In keeping with the above discussion, we shall emphasize the gas-solid reactions which can in general be expressed by the following chemistry:

$$A(g) + bB(s) = cC(g) + dD(s) \qquad (1)$$

Thermodynamics of most gas-solid systems of interest in extractive metallurgy is relatively simple. In most cases the reactant and product solids form separate, pure phases. The solids then are at unit activity. Furthermore, the reactions are carried out at relatively low pressures in which fugacities of gases can be approximated by partial pressures. Therefore, the equilibrium of reaction (1) can be expressed by

$$K = p_C^c/p_A \qquad (2)$$

Frequently, the same reactants may produce different gaseous and/or solid products depending on the partial pressures of gaseous species and temperature. The reduction of iron ore by gaseous reductants is an example. Hematite (Fe_2O_3) is reduced in steps to magnetite (Fe_3O_4), wustite (Fe_xO), and finally iron. Below 570°C wustite is unstable and magnetite is reduced directly to iron. The equilibrium diagram for the reduction of iron oxides by hydrogen and by carbon are given in Figures 1(a) and 1(b). As can be seen in these figures, the reduction of hematite to magnetite is essentially an irreversible reaction but the reductions of magnetite and wustite are limited by equilibrium.

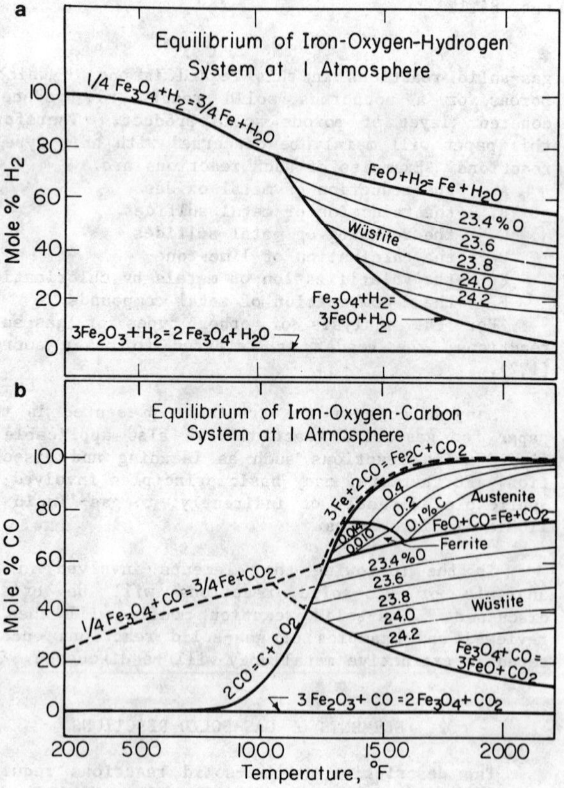

Figure 1. Equilibrium diagrams for the reduction of iron oxides by (a) hydrogen and (b) by carbon monoxide. [Adapted from Evans and Koo (3).]

The roasting of sulfides is another example. In this case the reactions are more complex. In order to examine the thermodynamics of such a system, one writes the various possible reactions in the form of Eq. (1) and their equilibria in the form of Eq. (2). Using the standard free energy of formation of various gaseous and solid species involved, one can determine which solid phase is most stable under certain conditions. The results of such calculations can most conveniently be presented in terms of predominance (or stability) diagrams (4-8). As an example, let us consider the roasting of NiS. Depending on temperature and the partial pressures of gaseous species, the solid product can be either NiO or $NiSO_4$. The following chemical reactions can be written for this system:

$$2\,NiS + 3O_2 = 2NiO + 2SO_2 \tag{3}$$

$$NiS + 2O_2 = NiSO_4 \tag{4}$$

$$2NiO + 2SO_2 + O_2 = 2NiSO_4 \tag{5}$$

By writing equilibrium relations similar to Eq. (3) for these reactions, relationships between p_{SO_2} and p_{O_2} for corresponding equilibria can be obtained. Such a predominance diagram at 1000°K is shown in Figure 2. The figure also shows a relatively small region of stability for Ni_3S_2 which was not considered in the reactions considered above. Thus, thermodynamic evaluation tells us which reaction is most likely to take place under given conditions.

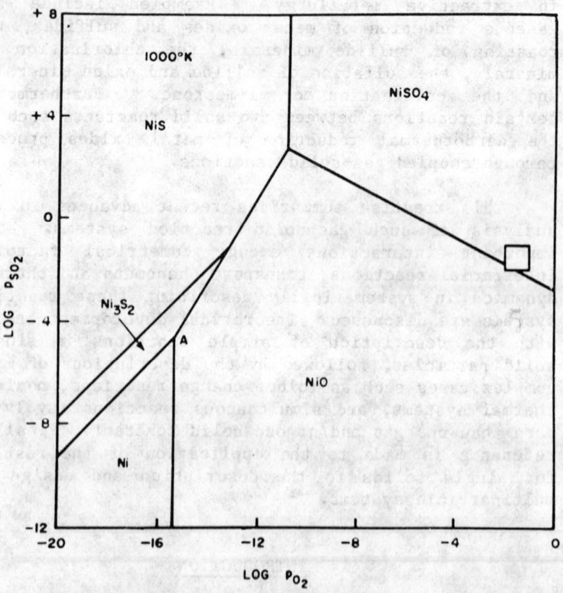

Figure 2. Predominance area diagram for the Ni-S-O System at 1000°K. [Adapted from Ingraham (5).]

Although thermodynamic data provide information on the stability of the phases, the state of equilibrium, and the heat of reaction, a complete analysis of metallurgical processes requires knowledge on the rates at which these processes proceed. Modern metallurgical processes are becoming increasingly complex and large in scale. Thus the tendency is toward the adoption of continuous operations and process control and optimization. The implementation of these modern technologies depends on the availability of reliable description of the rate processes. The following discussion will, therefore, emphasize recent advances in the analysis of the kinetics of gas-solid reactions encountered in extractive metallurgy.

A gas-solid reaction may involve all or a combination of the following individual steps, as illustrated in Figure 3:

1. mass transfer of gaseous reactants and products between the bulk gas phase and the external surface of the solid,
2. diffusion of gaseous reactants and products within the pores of the solid,
3. chemical reaction at the reaction interface between the gaseous and the solid reactants which may involve adsorption and desorption of species,

4. conduction of heat through the porous solid, and
5. heat transfer between the bulk gas phase and the external surface of the solid.

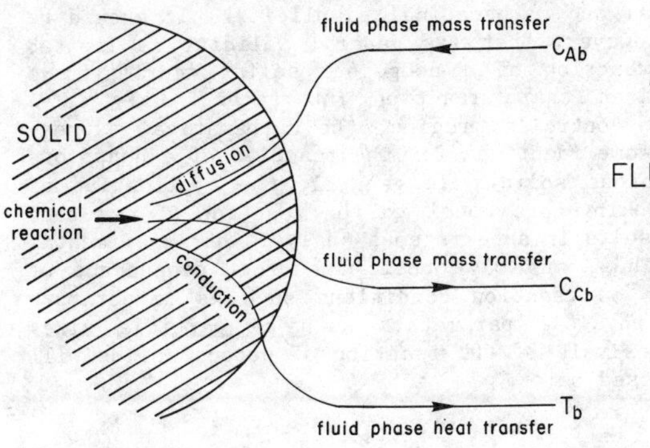

Figure 3. Schematic diagram of the overall reaction process.

These elementary steps and their roles in gas–solid reactions have previously been reviewed (1,2). Furthermore, as separate steps, they are not unique for gas–solid reactions and more comprehensive treatments on individual phenomena are available in the literature. The following references are provided here as representative examples, rather than an exhaustive list of literature sources for such treatments:

for gas–solid mass transfer, Refs. 1, 9–13;
for estimation methods of molecular transport properties, Refs. 9 and 14;
for diffusion through porous media, Refs. 1, 12, 15–17;
for chemical reaction kinetics on solid surfaces, Refs. 1, 18–20;
for conduction of heat in porous solids, Refs. 12, 21 and 22;
for gas–solid heat transfer, Refs. 1, 9 and 11.

3. REACTION OF A SINGLE NONPOROUS PARTICLE

If the solid reactant has no initial porosity, chemical reaction takes place at the surface of the unreacted solid. If the product is also nonporous, either the fluid species must diffuse into the solid by solid-state diffusion or a constituent species of the solid reactant must diffuse to the surface to react with the fluid reactant. If the solid product is porous, the fluid reactant can reach the reaction interface by diffusing through the pores of the product solid. In extractive metallurgy this type of reaction is more frequently encountered than the former.

Gas–solid reactions usually involve the adsorption of gaseous reactants at preferred sites on the solid surface and the formation of nuclei. For small particles the formation and growth of nuclei are important steps. The conversion-vs.-time relationship in this case is given by the nucleation-and-growth kinetics. A frequently used form of such a rate expression, which is attributed to Erofeev (23), is

$$\left[-\ln(1-x)\right]^{1/n} = mt \tag{6}$$

In the reaction of fine particles for which this type of rate equation is applicable, diffusion is usually unimportant or its effect is included in the kinetic parameter m.

As the solid size becomes larger, the time within which nuclei overlap and cover the entire surface forming a continuous reaction interface becomes a small portion of the total reaction time. Such reaction interface is usually parallel to the external surface, and advances into the interior of the solid as the reaction proceeds. Such a system, illustrated in Figure 4, is frequently termed the shrinking-core system.

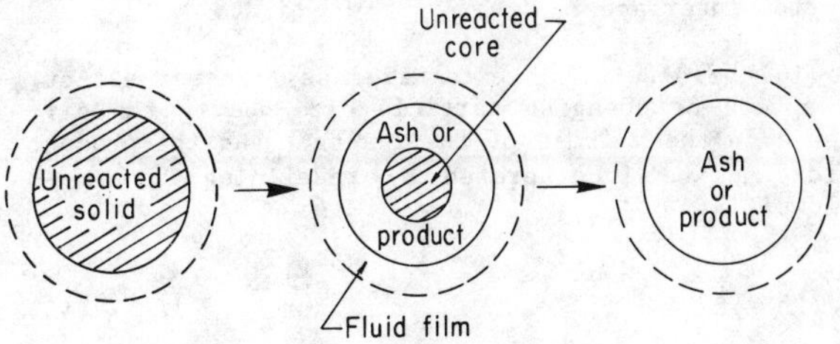

Figure 4. Schematic diagram of a shrinking-core system.

The shrinking core-system has been studied very extensively and applied to a wide range of reactions. These early investigations have been summarized and reviewed elsewhere (1,2,8). Szekely et al. (1) generalized the results of many previous investigators on reactions which are isothermal and of first order with respect to gaseous reactants. They formulated the following equation incorporating chemical kinetics, diffusion through porous product layer, and external mass transfer:

$$t^* = g_{F_p}(X) + \sigma_s^2 \left(p_{F_p}(X) + \frac{2X}{Sh^*}\right) \tag{7}$$

where F_p is the shape factor (= 1, 2, or 3 for a slab, a long cylinder, or a sphere) and

$$t^* \equiv \frac{bk}{\alpha_B \rho_B}\left(\frac{A_p}{F_p V_p}\right)(C_{Ab} - C_{Cb}/K)t \tag{8}$$

$$\sigma_s^2 \equiv \frac{k(1+1/K)}{2F_p D_e}\left(\frac{F_p V_p}{A_p}\right) \tag{9}$$

$$Sh^* \equiv \frac{k_m}{D_e}\left(\frac{F_p V_p}{A_p}\right) \tag{10}$$

$$g_{F_p}(X) \equiv 1 - (1-X)^{1/F_p} \tag{11}$$

and

$$p_{F_p}(X) \equiv X^2 \quad \text{for } F_p = 1 \tag{12}$$

$$\equiv X + (1-X)\ln(1-X) \quad \text{for } F_p = 2 \tag{13}$$

$$\equiv 1-3(1-X)^{2/3} + 2(1-X) \quad \text{for } F_p = 3 \tag{14}$$

It has been shown (1) that chemical reaction controls the overall rate when $\sigma_s^2 < 0.1$ or 0.01 depending on the range of tolerable error. On the other hand, pore diffusion and external mass transfer control the overall rate ($g_{F_p}(X)$ becomes negligible) when $\sigma_s^2 > 10$ or 100.

Individual forms of Eq. (7) were used by McKewan (24), Themelis and Gauvin (25), Kawasaki et al. (26), Lu (27), Lu and Bitsianes (28), Seth and Ross (29), and St. Clair (30), among others, for various extractive metallurgical reactions. Spitzer et al. (31) applied the shrinking-core concept to the reduction of a hematite pellet by considering three reaction interfaces.

In the absence of external mass transfer effect, the effect of changing particle size due to the difference in the volumes of the reactant and the product solids can be incorporated by redefining $P_{F_p}(X)$ as follows (1):

$$p_{F_p}(X) \equiv zX^2, \qquad \text{for } F_p = 1 \tag{15}$$

$$\equiv \frac{[z + (1-z)(1-X)]\,\ell n\,[z + (1-z)(1-X)]}{z-1} + (1-X)\ell n(1-X), \qquad \text{for } F_p = 2 \tag{16}$$

$$\equiv 3\left\{\frac{z - [z + (1-z)(1-X)]^{2/3}}{z-1} - (1-X)^{2/3}\right\}, \qquad \text{for } F_p = 3 \tag{17}$$

where z is volume of the product solid formed from unit volume of reactant solid. The rate of external mass transfer varies with particle size. If z is not too far from unity, as in most cases, an average Sh^* may be used even for particles of changing size. This will enable the use of Eq. (7) even in this case with $P_{F_p}(X)$ defined as in Eqs. (15)-(17).

Sohn and Szekely (32) analyzed the system in which the reaction order with respect to the gaseous reactant is different from unity. When there is a substantial change in the gas volume upon reaction, Sohn and Sohn (33) obtained the following approximate solution:

$$t^* \simeq g_{F_p}(X) + \sigma_s^2\left[\frac{\theta}{\ell n(1+\theta)}\,p_{F_p}(X) + \frac{2(1+\theta)X}{Sh^*}\right] \tag{18}$$

where

$$\theta \equiv (\nu-1)(C_{Ab}/C_T) \tag{19}$$

Natesan and Philbrook (34) and Haung and Bartlett (35) used special forms of equations analogous to Eq. (18) for the analysis of the diffusion-controlled regimes in the roasting of zinc sulfide and of mixtures of chalcopyrite and lime, respectively.

In many gas-solid reactions substantial enthalpy changes accompany the reactions. The analysis of nonisothermal systems can be substantially more complex than isothermal systems discussed above. Such a system may exhibit multiple steady states and thermal instability. These aspects have been reviewed elsewhere (1,2).

The shrinking-core model is attractive for its conceptual and mathematical simplicity. It should be noted, however, that its general validity is limited to the reaction of a nonporous solid occuring at a well-defined sharp reaction interface. Only in a diffusion-controlled regime, the mathematical expressions become identical for the reaction of a nonporous and a porous solid. In general, the application of the shrinking-core model to the reaction of a porous solid results in an erroneous analysis of experimental data. Thus, one may obtain incorrect dependence of the rate on reaction conditions such as temperature and on physical parameters such as particle size. Proper analysis of the reaction of porous solids will be discussed below.

4. REACTION OF A SINGLE POROUS PARTICLE

Many solid reactants have some initial porosity, allowing diffusion and chemical reaction to occur simultaneously in a diffuse zone. Most recent developments in the analysis of gas-solid reactions have been made in the reaction of a porous solid. The progress of reaction in a porous solid is illustrated in Figure 5.

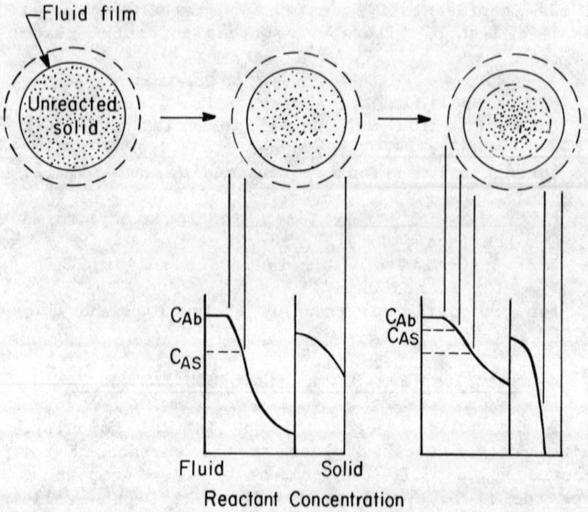

Figure 5. Schematic diagram of the reaction of a porous particle.

The reaction takes place throughout the solid. The concentration of gaseous reactant decreases as it diffuses into the interior of the solid while reacting on its path. In most cases, the external layer will react faster than the interior. The reaction given by Eq. (1) taking place in a porous solid can be described mathematically by formulating the conservation equations for the gaseous and the solid reactants and for energy if substantial heat effect is involved.

The conservation of gaseous species is given by

$$\underline{\nabla} \cdot \underline{N}_i = R_i; \quad i = A \text{ or } C \tag{20}$$

In Eq. (20) we used the pseudo-steady-state approximation which states that the accumulation term can be neglected for gas-solid reactions (36,37). The conservation of solid reactant is given by

$$\frac{\partial w}{\partial t} = k_a f(w) \tag{21}$$

Finally, the conservation of energy is described by

$$\rho c_p \frac{\partial t}{\partial t} = -\underline{\nabla} \cdot \underline{q} + (-\Delta H) \cdot (-R_A) \tag{22}$$

The fluxes of gaseous species, N_i, must in general by related to concentrations through the Stefan-Maxwell equations (9,38-40). However, for a binary gas mixture or by using an effective binary diffusivity (9), one may use the simpler relationship given below:

$$N_i = -CD_{ei} \nabla X_i + x_i \sum_{j=1}^{n} N_j \tag{23}$$

The rate of production of gaseous species, R_i, is related to the rate of reaction of the solid reactant through stoichiometry, as follows:

$$R_A = -\frac{\alpha_B \rho_B}{b} \frac{\partial w}{\partial t} = -\frac{\alpha_B \rho_B k_a}{b} f(w) \tag{24}$$

$$R_C = \frac{c\alpha_B \rho_B}{b} \frac{\partial w}{\partial t} = \frac{c\alpha_B \rho_B k_a}{b} f(w) \tag{25}$$

The apparent rate constant k_a contains the dependence of rate on gaseous species. For a gas-solid reaction, the following Langmuir-Hinshelwood kinetics generally apply:

$$k_a = \frac{k_p S_v (p_A^m - p_C^n/K)}{1 + K_A p_A^i + K_C p_C^j} \tag{26}$$

Many reactions have the simple kinetic expression with $m = n = 1$ and $K_A = K_C = 0$, which represents a first-order reversible reaction.

The solution of Eqs. (20) - (22) with appropriate boundary conditions yields the profile of gaseous species as a function of position inside the solid. Using these concentration profiles in Eq. (21) together with Eq. (26), the local conversion of solid, w, can be obtained as a function of time. The overall conversion is then calculated by integrating the local conversion over the entire solid volume.

Solutions to governing equations given by Eqs. (20) - (22) have been obtained for certain special cases. When the reaction involves a substantial heat effect, these equations must in general be solved using numerial techniques. Even in such a case, the conversion-vs.-time relationship for a diffusion-controlled irreversible reaction ($K \to \infty$) is given by Eq. (18) without the $g_{F_p}(X)$ term. Various models for a reaction between a porous solid and a gas are mainly concerned with the dependence of rate on solid conversion, f(w). We will disucss these models for an isothermal system.

4.1 Early Models

Ausman and Watson (41) proposed a model in which they assumed that the local rate of reaction was independent of solid conversion as long as some solid reactant remains. This is equivalent to assuming the following:

$$\begin{aligned} f(w) &= 1 && \text{when } w < 1 \\ &= 0 && \text{when } w = 0 \end{aligned} \tag{27}$$

They also assumed the equimolar counterdiffusion of gaseous species. Ishida and Wen (42) obtained an analytical solution for such a reaction taking place in a porous sphere. In their solution Ishida and Wen (42) incorporated the effect of structural changes in solid by assigning different diffusivities to the completely and the partially reacted zones. Szekely and Sohn (43) obtained similar solutions for reactions taking place in a slab and a long cylinder.

Although an analytical solution is helpful, the equations representing the solution are rather complex and cumbersome to use. Furthermore, not many gas-solid reactions follow the rate expressions given in Eq. (27).

4.2 The Grain Model

In this model the porous solid is pictured as an aggregate of small nonporous particles of a uniform size. This picture is physically reasonable especially for pellets made up of fine concentrate particles.

Sohn and Szekely (44) developed a generalized grain model which encompasses many other models to represent this system (45-47). These models were formulated for isothermal first-order reactions with equimolar counterdiffusion. They also assumed that the diffusion of the reactant gas through the product layer around the individual grain offers little resistance to overall rate. This assumption is valid for small grains. With these assumptions, the grain model includes the models of Ausman and Watson (41) and Ishida and Wen (42) as a special case when the grains can be considered as flat plates. Sohn and Szekely (1,44) considered porous solids in which the pellet and the grain may each have the shape of a slab, a long cylinder, or a sphere -- nine possible combinations. In their development, they used the following equation for the reaction of the grains in lieu of Eq. (21):

$$-\rho_B \frac{\partial r_c}{\partial t} = bk\,(C_A - C_C/K) \tag{28}$$

where r_c is the position of the reaction interface within the grain, ρ_B is the molar density of the solid reactant. It is noted that Eq. (28) is a special form of Eq. (21) for grains reacting according to the chemical-reaction-controlled shrinking-core scheme (with $K_A = K_C = 0$ in Eq. (26)). The term R_A now is given as

$$-R_A = (1-\varepsilon)k\,(A_g/V_g)\,(A_g r_c/F_g V_g)^{F_g - 1}\,(C_A - C_C/K) \tag{29}$$

The case of flat grains ($F_g = 1$) corresponds to models discussed in the previous section. When $F_g \neq 1$, Eqs. (20) and (21) must be solved numerically, even for an isothermal, first-order reaction. For systems in which the effective diffusivity remains unchanged upon reaction, Sohn and Szekely (44) obtained the following approximate solution for the conversion-vs.-time relationship:

$$t_g^* \simeq g_{F_g}(X) + \hat{\sigma}^2 \left[p_{F_p}(X) + \frac{2X}{Sh^*} \right] \qquad (30)$$

where

$$t_g^* \equiv \frac{bk}{\rho_B} \left(\frac{A_g}{F_g V_g}\right) (C_{Ab} - C_{Cb}/K)\, t \qquad (31)$$

$$\hat{\sigma} \equiv \frac{V_p}{A_p} \sqrt{\frac{(1-\varepsilon) F_p k}{2D_e} \left(\frac{A_g}{F_g V_g}\right) \left(1 + \frac{1}{K}\right)} \qquad (32)$$

and g(X) and p(X) have been defined in Eqs. (11) - (14). The results of these analyses were applied to the reduction of nickel oxide pellets with hydrogen by systematically designed experiments (48).

When intragrain diffusion is significant, Sohn and Szekely (4,9) have shown that the following equation is applicable:

$$t_g^* \simeq g_{F_g}(X) + \hat{\sigma}_g^2\, p_{F_g}(X) + \hat{\sigma}^2 \left(p_{F_p}(X) + \frac{2X}{Sh^*}\right) \qquad (33)$$

where

$$\hat{\sigma}_g^2 \equiv \frac{k}{2D_g} \left(\frac{V_g}{A_g}\right) \left(1 + \frac{1}{k}\right) \qquad (34)$$

For the reaction of a spherical pellet made up of spherical grains, Eq. (33) was found (49) to satisfactorily represent the exact numerical solution obtained by Calvelo and Smith (47). A closed-form solution such as Eq. (33) is much more useful than the numerical solution of different equations. The advantage increases greatly in the analysis of multiparticle systems.

The above analyses have been made for the case of a constant effective diffusivity. In many gas-solid reactions, the product solid layer has a substantially different effective diffusivity than the reactant. One of the simplified methods of incorporating this structural change is to assign different effective diffusivities to the completely and partially reacted zones. The case of flat grains was discussed in the previous section. Szekely and Sohn (43) reported the results of computation for pellets made up of grains of other geometries.

Other types of structural changes such as sintering, swelling and softening are very complex and not very well understood. If information is available, the effects of such changes on effective diffusivity and chemical reactivity could be incorporated into the governing equations.

When there is a substantial heat effect, the analysis becomes much more complex with the possible existence of multiple steady states and instability (1, 47). Many metallurgical systems involve solids with high thermal conductivities. In such cases temperature within the solid may be uniform and the resistance to heat transfer may be limited to the external heat transfer between the external surface and the bulk gas stream. For such systems the reaction rate in the form of dX/dt may still be obtained from Eq. (30) or (33) by writing it in a dimensional form and expressing the temperature-dependent parameters as a function of temperature. The validity of this approach has been verified (50).

4.3 The Law of Additive Reaction Times

Based on the conversion-vs.-time relationships for porous and nonporous solids discussed above, Sohn (50) proposed the "law of additive reaction times." This law is applicable for an isothermal reaction in which the effective diffusivity of the solid remains constant during the reaction. The law states that the time required to attain a certain conversion is the sum of the time required to attain the same conversion in the absence of resistance due to intrapellet diffusion of reactant gas ($\hat{\sigma} \to 0$) and the time for the same conversion under the control of intrapellet diffusion of reactant gas ($\hat{\sigma} \to \infty$) (50). Equations (7), (18), (30), and (33) are examples of the mathematical expressions of this law.

This law is exact for the reaction of a nonporous solid following the shrinking core scheme. It is approximately valid for the reaction of a porous solid with constant effective diffusivity. In the latter case, the exact solution must in general be obtained by numerical solution of the governing second-order differential equations. The law which results in a closed-form solution is even more useful in the analysis of multiparticle systems involving various particle sizes and changes in gas concentration and solid temperature.

It has been shown (50) that this law is valid regardless of the dependence of rate on solid conversion. The general applicability and usefulness of the law have been verified using a nucleation and growth kinetics as an example (50). By applying the law to a gas-solid reaction obeying the nucleation and growth kinetics, Sohn (50) obtained the following equation:

$$t_N^* \simeq g_N(X) + \hat{\sigma}_N^2 \left(p_{F_p}(X) + \frac{2X}{Sh^*}\right) \qquad (35)$$

where

$$t_N^* \equiv kS_v\,(C_{Ab} - C_{Cb}/K)\,t \qquad (36)$$

$$\hat{\sigma}_N \equiv \frac{V_p}{A_p} \sqrt{\frac{\alpha_B \rho_B F_p k S_v}{2bD_e} \left(1 + \frac{1}{K}\right)} \qquad (37)$$

$$g_N(x) \equiv \int_0^x (f(w))^{-1} dw = (-\ln(1-X))^{1/n} \qquad (38)$$

Here, Eq. (6) has been used as an example of the nucleation and growth kinetics. Other rate expressions for f(w) in Eq. (21) could be used giving rise to corresponding forms of $g_N(X)$ according to the first equality in Eq. (38).

Figure 6 illustrates how the law of additive reaction times can be applied in general if we have information on the chemical reaction rate in the absence of the resistance due to intrapellet diffusion and on the rate controlled by intrapellet diffusion.

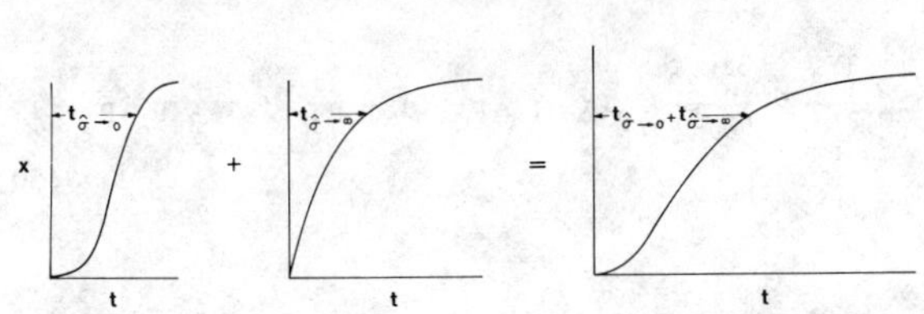

Conversion-vs-time when intrapellet diffusion is very fast (chemical reaction control)

Conversion-vs-time when chemical kinetics is very fast (pore diffusion)

Actual conversion-vs-time relationship

Figure 6. Graphical representation of the law of additive reaction times.

As discussed earlier, the treatment of nonisothermal reactions taking place in a porous solid is rather complex. A generalization would not be very meaningful. For nonisothermal systems in which temperature within the solid is uniform (large thermal conductivity) at any instance, Sohn (50) has shown that the differential forms of Eqs. (30), (33), and (35) remain valid. It has also been shown (50) that the variation of bulk gas concentration can be incorporated. When temperature and bulk gas concentration vary with time, we express the relationship as follows:

$$C_{Ab} = C_{Abi} \cdot Y(t^*_{Ni}) \tag{39}$$

and

$$k = k_i \cdot Z(t^*_{Ni}) \tag{40}$$

where C_{Abi} and k_i are the values of C_{Ab} and k at t = 0, and t^*_{Ni} is defined with C_{Abi} and k_i instead of C_{Ab} and k. Y and Z represent functional relations. Using Eq. (35) as an example, we can write it in a dimensional form, substitute Eqs. (39) and (40), differentiate with respect to time, and rearrange the resultant equation in the following form:

$$\frac{dX}{dt^*_{Ni}} \simeq \frac{Y(t^*_{Ni})}{g'_N(X)/Z(t^*_{Ni}) + \hat{\sigma}^2_{Ni}\,(p'_{F_p}(X) + 2/Sh^*)} \tag{41}$$

where $\hat{\sigma}^2_{Ni}$ is defined with k_i instead of k, and prime designates derivatives with respect to X. Here, we have neglected the variation of D_e and k_m with temperature. It is noted that the variations of concentration and time may be spatial rather than temporal in certain systems such as steady-state moving-bed reactors. The development discussed above will equally apply to such cases. It has been shown (50) that the conversion-vs.-time relationship obtained by integrating Eq. (41) gives an excellent approximation to the exact numerical solution of the governing mass balance equations. Although the solution of Eq. (41) requires numerical integration, it is much simpler than the numerical solution of a second-order differential equation at every time step plus the integration for conversion. The computational advantages of using Eq. (41) is much greater in the analysis of a multiparticle system with a wide size distribution.

5. COMPLEX GAS-SOLID REACTIONS

Above sections have described single one-step reactions. Certain gas-solid reactions encountered in extractive metallurgy are rather complex involving more than one gas and/or solid reactants. Examples of such reactions and mathematical models describing them will be reviewed in this section.

5.1 Staged Reaction of a Solid with a Gas

The reactions in this category include the reduction of metal oxides which proceeds through the formation of stable lower oxides. The gaseous reduction of hematite is the most important example. Thus, the following discussion will use this reaction as an example. Other reaction systems of similar nature can be analyzed in similar manner.

Earlier models of hamatite reduction assumed that the thicknesses of the intermediate oxides were very narrow. Thus, the multistep reactions were simplified as a one-step reaction. Spitzer et al. (31) proposed a variation of the simple shrinking-core model to allow for the existance of distinct layers of the intermediate oxides. All the layers of intermediate oxides and iron were considered to be porous but chemical reactions were assumed to be confined to the interfaces between two adjacent layers. This approach would be reasonable for the reduction of hematite having rather small initial porosity; in this case there would be large differences in porosity between layers and reactions would take place mostly near the interfaces. As the authors pointed out, in the case of a porous ore chemical reaction would take place in rather diffuse zones and the above approach would become less satisfactory.

In recent years pelletized concentrates of iron ore have become a substantial portion of the feed to blast furnaces and direct reduction plants. A considerable amount of high-porosity natural ore is also used in the direct reduction processes. Thus, much attention has recently been directed to the reduction rates of porous iron oxides. The reduction of hematite to iron may be represented by the following reactions:

$$A(g) + 3Fe_2O_3 = AO(g) + 2Fe_3O_4 \tag{42}$$

$$A(g) + \frac{x}{4x-3z}\,Fe_3O_4 = AO(g) + \frac{3z}{4x-3z}\,Fe_xO + \frac{3x(1-z)}{4x-3z}\,Fe \tag{43}$$

$$A(g) + Fe_xO = AO(g) + xFe \tag{44}$$

Here, A represents the gaseous reactant, either hydrogen or carbon monoxide, and AO the gaseous product, water vapor or carbon dioxide. In Eq. (43), z represents the fraction of iron in magnetite that forms wustite. Below 570°C wustite is unstable and magnetite is reduced directly to iron and hence z=0.

On the other hand, $z = 1$ when wustite is formed as an intermediate.

In what follows, a model for the reduction of a porous hematite pellet, incorporating the effects of intrinsic kinetics, pore diffusion and structural changes, will be described. The models of this nature are those of Trushenski et al. (51) and Sohn (52). These two models are somewhat different in details. We shall describe the model proposed by Sohn which allows the use of more general expressions for physical structure and kinetics. The model will be presented using the factors which can be determined by separate experiments.

Let us consider the reactions given by Eqs. (42) - (44) taking place in a porous pellet. The concentration of the gaseous reactant in the solid is governed by the following equation which can be derived from Eqs. (20) and (23):

$$\frac{1}{r^2} \cdot \frac{d}{dr}\left(D_e r^2 \frac{dp_A}{dr}\right) - RT \cdot v_A = 0 \qquad (45)$$

in which the first term represents the diffusion of the reactant gas and the second term the rate of consumption of the gaseous reactant per unit volume of the porous solid. We note that $v_A = -R_A$. Here, equimolar counterdiffusion of the reactant and product gases, which is applicable for iron oxide reduction, has been assumed. In order to obtain a solution, the following pieces of information are necessary:

1. the variation of the effective diffusivity with temperature, the local properties of the solid, and time, and
2. the relationship between v_A and the local concentrations of gaseous and solid reactants.

If the reaction is reversible, an equation similar to Eq. (45) should be considered for the product gas:

$$\frac{1}{r^2} \cdot \frac{d}{dr}\left(D_{eC} r^2 \frac{dp_C}{dr}\right) - RT \cdot v_C = 0 \qquad (46)$$

We note that, for the gaseous reduction of iron oxide,

$$v_C = -v_A \qquad (47)$$

Under certain conditions, e.g., when $D_{eA} = D_{eC}$, Eqs. (45) and (46) can be combined into one equation and only one of the gaseous species needs to be followed. The concentration profiles of gaseous reactant and product obtained from Eqs. (45) and (46) is used to calculate the local consumption rate of the solid reactant. In turn, the rate of consumption of the gaseous reactant, v_A, is related by stoichiometry to the consumption rate of the solid reactants.

The local reduction rate is related to the local concentration of gaseous species as follows:

$$n_{10} \frac{\partial Y_1}{\partial t} = -3n_{10} \cdot f_1(Y_1, p_A, p_C, T) \qquad (48)$$

$$n_{20} \frac{\partial Y_2}{\partial t} = 2n_{10} \cdot f_1(Y_1, p_A, p_C, T) - \frac{xn_{20}}{4x-3z} \cdot f_2(Y_2, P_A, P_C, T) \qquad (49)$$

$$n_{30} \frac{\partial Y_3}{\partial t} = \frac{3zn_{20}}{4x-3z} \cdot f_2(Y_2, P_A, P_C, T) - n_{30} \cdot f_3(Y_3, P_A, P_C, T) \qquad (50)$$

where f's represent the expressions for the intrinsic kinetics, which will be discussed subsequently, the indices 1,2, and 3 refer to the hematite, magnetite, and wustite, respectively, n_{10} is the number of moles of hematite initially present per unit volume of the porous solid, and n_{20} and n_{30} are the molar concentrations, respectively, of magnetite and wustite equivalent to n_{10} moles of hematite. Thus, from stoichiometry:

$$n_{20} = \frac{2}{3} n_{10} \qquad (51)$$

and

$$n_{30} = \frac{2z}{x} n_{10} \qquad (52)$$

The fractions Y_i are defined as

$$Y_i = \frac{n_i}{n_{io}} \qquad (53)$$

Equations (48)-(50) should be solved at various radial positions because the gas concentrations vary with r. Thus, Y_i's vary with both time and radial position. From stoichiometry, the rate of consumption of the gaseous reactant per unit volume of the porous solid in Eq. (45) is given by

$$v_A = n_{10} \cdot f_1(Y_1, p_A, p_C, T) + n_{20} \cdot f_2(Y_2, p_A, p_C, T) + n_{30} \cdot f_3(Y_3, p_A, p_C, T) \qquad (54)$$

The dependence of the intrinsic kinetics on the concentrations of the gaseous species has usually been considered to be of first order. At atmospheric or lower pressures this approximation is justified. The dependence on the amount of unreacted solid reactant has not been well-established. Turkdogan and Vinters (53) used a first-order dependence on the fraction of solid unreacted, but Tien and Turkdogan (54) used what is equivalent to a zeroth-order kinetics. Matyas and Bradshaw (55) employed an essentially curve-fitting technique to describe their intrinsic kinetics. No systematic experiments have been carried out to determine this dependence for each of the three reactions listed in Eqs. (42) - (44) using solids in the form of pellets. This requires further research. In view of the lack of accurate information Sohn (52) suggested the use of the nucleation and growth kinetics, which gives rise to S-shaped conversion curves and includes, as a special case, the first-order kinetics (50). Thus,

$$f_i(Y_i, p_A, p_C, T) = nkY_i\,(-\ln Y_i)^{1-1/n}\left(\frac{p_A - p_C/K_i}{RT}\right) \quad (55)$$

where n is a dimensionless number usually greater than unity and K_i is the equilibrium constant for the reduction of ith oxide. El-Rahaiby and RaO (56) used this type of kinetics to intrepret their data for the reduction of thin foils of wustite. Other forms of dependence on Y_i can of course be incorporated.

The boundary and initial conditions for Eqs. (45), (46), and (48) - (50) are:

$$\frac{dp_A}{dr} = \frac{dp_C}{dr} = 0, \quad \text{at } r = 0 \quad (56 \text{ and } 57)$$

$$\left.\begin{aligned} D_{eA}\frac{dp_A}{dr} &= k_{mA}(p_{Ab} - p_A) \quad (58) \\ D_{eC}\frac{dp_C}{dr} &= k_{mC}(p_{Cb} - p_C) \quad (59) \end{aligned}\right\} \text{at } r = r_p$$

$$\left.\begin{aligned} Y_1 &= 1 \quad (60) \\ Y_2 = Y_3 &= 0 \quad (61 \text{ and } 62) \end{aligned}\right\} \text{at } t = 0$$

Here, we assumed that the overall size of the solid remains constant during the reaction. The size change due to swelling is highly sensitive to the types and levels of the impurities in the iron oxide and thus not amenable to systematic predictions. Furthermore, it has been shown that volume increases up to 50% have only a small effect on the overall rate (1).

In view of the large change of porosity accompanying each reduction step, the effective diffusivity cannot realistically be assumed constant. Thus, the local effective diffusivity should be evaluated in terms of other variables. It has been observed that the structural changes during iron ore reduction is mainly a result of chemical reaction rather than a thermal effect (51, 54, 57). Therefore, the effective diffusivity may be assumed to be independent of reduction time for a given temperature. The effective diffusivity may then be expressed as

$$\frac{D_{ei}}{D_i} = f_d\,(\varepsilon_o, Y_1, Y_2, Y_3, T) \quad (63)$$

Turkdogan and coworkers (57, 58) obtained the ratio D_e/D at various temperatures for the following cases:

$$\varepsilon_o = 0.31, \quad Y_1 = Y_2 = Y_3 = 0$$

This represents the only known systematic measurement of the effective diffusivity. Much experimental work is needed to obtain the complete information.

The solution of Eqs. (45) - (50) with Eqs. (54), (55), and (63) together with the boundary and the initial conditions given by Eqs. (56) - (62) would yield the solution. The local value of the fractional oxygen removal is given by

$$F = 1 - \frac{3n_1 + 4n_2 + n_3}{3n_{10}}$$

$$= 1 - \left(Y_1 + \frac{8}{9}Y_2 + \frac{2z}{3x}Y_3\right) \quad (64)$$

The overall conversion is then obtained by the following integration:

$$X = \frac{3}{r_p^3}\int_o^{r_p} F\,r^2\,dr \quad (65)$$

In order to apply this model the following information needs to be determined experimentally, as noted earlier:

1. Rate expression for the individual reactions of hematite, magnetite, and wustite, as they exist in the pellet, that is, the functions f_i in Eqs. (48) - (50).

2. The effective diffusivity as a function of porosity, the composition of solid, and reduction temperature, as expressed by Eq. (63).

The rate expressions should be obtained by isolating each reduction step using gases with appropriate compositions and carrying out the reaction in the absence of diffusional effects. The latter may be achieved by using a thin porous sample or a thin layer of grains that make up the porous pellet. The effective diffusivity can be determined by applying the procedure of Turkdogan and coworkers (57, 58) to each of the solid reactants and the product iron. A suitably weighed average value should then be used for the effective diffusivity in the portion of the pellet containing a mixture of the solid species.

With these pieces of information, the model will yield the instantaneous rate of reaction not only for constant bulk conditions but also under the conditions of varying bulk gas concentration and temperature of the solid. The model will also predict the temperature at which rapid reaction starts when the solid experiences changing temperature. This information is important from the viewpoint of designing and analyzing a reactor.

5.2 Simultaneous Gas-Sold Reactions

Simultaneous gas-solid reactions can be between one solid and several gaseous reactants, between one gas and several solid reactants, or between several gases and solids.

5.2.1 Reactions Between One Solid and Two Gases. An important example of this type of reaction is the reduction of metal oxides by a mixture of hydrogen and carbon monoxide.

Wen and Wei (59,60) and Rehmat and Saxena (61) studied the problem of simultaneous gas-solid reactions in terms of the shrinking-core model. Szekely and coworkers (38, 39) have studied the reduction of hematite pellets by hydrogen-carbon monoxide mixture.

The reaction of a porous solid with two reactant gases can be described by writing Eq. (20) for the four gaseous species involved (two if both reactions are irreversible) and modifying Eq. (21) to add the contribution from the second reactant gas. This would involve a substantial amount of computation, especially when analyzing a multicomponent system. Sohn and Braun (62) obtained an approximate but much simpler solution by applying the law of additive reaction times described earlier. In order to present their result, we write the simultaneous reactions in the following form:

$$A_1(g) + b_1 \cdot B(s) = P_1(g) + d_1 D(s) \qquad (66a)$$

$$A_2(g) + b_2 \cdot B(s) = P_2(g) + d_2 \cdot D(s) \qquad (66b)$$

For an isothermal system with a constant effective diffusivity and first-order kinetics, Sohn and Braun (62) found that the application of the differential form of Eq. (35) to the two equations gave a result which satisfactorily represented the results of numerical integration of the governing differential equations. Thus, the overall rate of conversion is given by:

$$\frac{dX}{dt^*} \approx \frac{\psi_{1b}}{g_N'(X) + \hat{\sigma}_1^2 \left(p_{F_p}'(X) + 2/Sh^*\right)} + \frac{\psi_{2b}}{g_N'(X) + \hat{\sigma}_2^2 \left(p'_{F_p}(X) + 2/Sh^*\right)} \qquad (67)$$

where

$$t^* = \left(k_1(C_{A_1b} - C_{P_1b}/K_1) + k_2(C_{A_2b} - C_{P_2b}/K_2)\right) S_v t \qquad (68)$$

$$\psi_{1b} \equiv \frac{k_1(C_{A_1b} - C_{P_1b}/K_1)}{k_1(C_{A_1b} - C_{P_1b}/K_1) + k_2(C_{A_2b} - C_{P_2b}/K_2)} \qquad (69)$$

$$\psi_{2b} \equiv \frac{k_2(C_{A_2b} - C_{P_2b}/K_2)}{k_1(C_{A_1b} - C_{P_1b}/K_1) + k_2(C_{A_2b} - C_{P_2b}/K_2)} \qquad (70)$$

$$\hat{\sigma}_1^2 \equiv \frac{V_p}{A_p}\sqrt{\frac{\alpha_B \rho_B F_p k_1 S_v (1 + 1/K_1)}{2b_1 D_{eA_1}}} \qquad (71)$$

$$\hat{\sigma}_2^2 \equiv \frac{V_p}{A_p}\sqrt{\frac{\alpha_B \rho_B F_p k_2 S_v (1 + 1/k_2)}{2b_2 D_{eA_2}}} \qquad (72)$$

Furthermore, Sohn and Braun (62) showed that the rate of conversion of the solid due to each gaseous reductant was very adequately represented by each of the terms on the right hand side of Eq. (67). The overall conversion and the conversion by each gas can be obtained by numerically integrating the respective derivatives. They also described how to apply this approximate solution to a system in which concentrations and bulk temperature vary with time. Although $g_N(X)$ was used in the formulation Eq. (67) is expected to be valid with $g_{Fg}(X)$, with appropriate modifications in the dimensionless parameters.

The approximate solution given in Eq. (67) is exact when both reactions are controlled either by chemical reaction or by diffusion. Thus, when both $\hat{\sigma}_1$ and $\hat{\sigma}_2$ approach zero (chemical reaction control), we have:

$$g_N(X) = t^* \qquad (73)$$

When both $\hat{\sigma}_1$ and $\hat{\sigma}_2$ approach infinity (diffusion control), the asymptotic solution is:

$$p_{F_p}(X) + \frac{2X}{Sh^*} = \left(\frac{\psi_{1b}}{\hat{\sigma}_1^2} + \frac{\psi_{2b}}{\hat{\sigma}_2^2}\right) t^* \qquad (74)$$

5.2.2 Reactions Between One Gas and Two Solids. This problem can be treated by adding an additional term in Eq. (24) for the consumption of the gaseous reactant by the second solid. Since there is only one gaseous reactant, Eq. (20) with the modified R_i term remains valid. The local conversion rate of the second solid which will also depend on local concentrations of gaseous species can be obtained by writing another equation similar to Eq. (21). No general approximate solution for this system has been reported.

When both reactions are rapid the overall rate is controlled by the diffusion in the product layer. In this case, using Eq. (35), we have

$$t_N^* = \hat{\sigma}_N^2 \left(p_{F_p}(X) + \frac{2X}{Sh^*}\right) \qquad (75)$$

where $\hat{\sigma}_N^2$ is defined with $(\alpha_{B1}\rho_{B1}/b_1 + \alpha_{B2}\rho_{B2}/b_2)$ in lieu of $\alpha_B\rho_B/b$. Equation (30) could be used with a similar modification on $\hat{\sigma}^2$.

By extending the above arguments one could formulate a model for the reaction of several solid reactants simultaneously present in an ore by a mixture of gaseous reactants.

5.3 Successive Gas-Solid Reactions

In recent years much attention has been directed to successive gas-solid reactions taking place in a porous pellet made up of two solids. A major source of such reaction systems is the treatment of sulfide minerals. Until now large amounts of sulfur-containing pollutants have been emitted into the atmosphere. This cannot be tolerated any longer and, consequently, various new processes have recently been proposed as potential alternatives to the conventional processes for extracting metal values from sulfide minerals. Two notable examples, in which successive gas-solid reactions play a major role, utilize lime as a scavenger of sulfur-containing gases. The first is the hydrogen reduction of metal sulfides in the presence of lime (63-69) which can be represented by the following general scheme:

$$H_2 + Me_xS = H_2S + xMe \tag{76a}$$

$$H_2S + CaO = H_2O + CaS \tag{76b}$$

$$H_2 + Me_xS + CaO = H_2O + xMe + CaS. \tag{76c}$$

In this reaction scheme, lime serves two purposes: (1) It fixes sulfur in the solid, thereby reducing the emission of hydrogen sulfide into the atmosphere, and (2) the presence of lime improves the otherwise unfavorable thermodynamics of reaction (76a) by removing hydrogen sulfide from the gas phase. (Reaction (76b) is highly favorable thermodynamically.)

The second example is the oxidation roasting of sulfide minerals in the presence of lime (35, 70, 71), which may be represented by the following general scheme:

$$\frac{3}{2}O_2 + Me_xS = SO_2 + Me_xO \tag{77a}$$

$$SO_2 + \frac{1}{2}O_2 + CaO = CaSO_4 \tag{77b}$$

$$2O_2 + Me_xS + CaO = Me_xO + CaSO_4 \tag{77c}$$

In this case both component reactions are usually favorable thermodynamically, and thus, lime serves simply as a scavenger of sulfur dioxide to eliminate it from the effluent gas.

Sohn and Rajamani (67) presented a model for isothermal successive gas-solid reactions in a porous pellet. The model takes into consideration the effects of the relative amounts of the solids, grain sizes, the pellet size and porosity, and the diffusion of gaseous species. We will discuss this model below.

The successive gas-solid reactions exemplified by reactions (76) and (77) given above can be written in the following general form:

$$A(g) + bB(s) = cC(g) + fF(s) \tag{78a}$$

$$C(g) + dD(s) = eE(g) + gG(s) \tag{78b}$$

Let us now consider a porous pellet of volume V_p and of superficial surface area A_p, made up of a uniform mixture of grains of solids B and D, as shown in Figure 7.

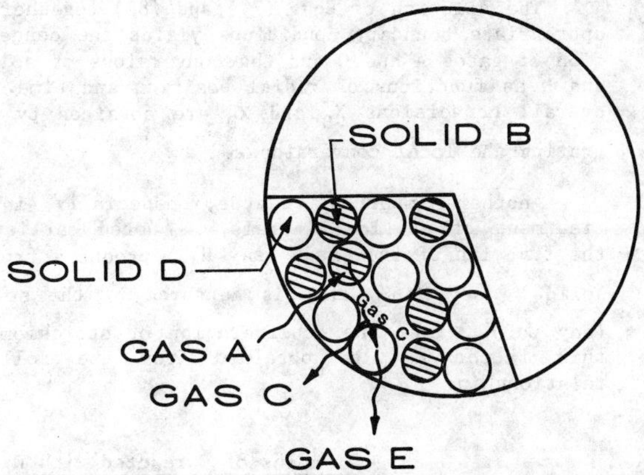

Figure 7. Schematic representation of a successive gas-solid reaction.

By applying Eq. (20) to the three gaseous species involved in the reaction and assuming equimolar counterdiffusion, the conservation of the gaseous species is given by the following:

$$D_{eA}\nabla^2 C_A - v_1 = 0 \tag{79a}$$

$$D_{eC}\nabla^2 C_C + cv_1 - v_2 = 0 \tag{79b}$$

$$D_{eE}\nabla^2 C_E + ev_2 = 0 \tag{79c}$$

where v_1 and v_2 are the local net forward rates per unit volume of the pellet of reactions (78a) and (78b), respectively.

As for the rates of the reaction for the solids, Sohn and Rajamani (67) used the rate expressions for a chemically-controlled shrinking-core system given by Eq. (28). Thus,

$$\frac{\partial r_B}{\partial t} = -\frac{bk_1}{\rho_B}(C_A - C_c/K_1) \tag{80a}$$

$$\frac{\partial r_D}{\partial t} = -\frac{dk_2}{\rho_D}(C_c - C_E/K_2) \tag{80b}$$

If the reaction of the solid follows other schemes, the more general relationship given by Eq. (21) could be used (68). Furthermore, a nonfirst-order kinetics can be incorporated using Eq. (26). Again noting that $v = -R$,

$$v_1 = \alpha_B \left(\frac{A_B}{V_B}\right) \left(\frac{A_B r_B}{F_B V_B}\right)^{F_B - 1} k_1(C_A - C_C/K_1) \quad (81a)$$

$$v_2 = \alpha_D \left(\frac{A_D}{V_D}\right) \left(\frac{A_D r_D}{F_D V_D}\right)^{F_D - 1} k_2(C_C - C_E/K_2) \quad (81b)$$

The solution of Eqs. (79) and (80) together with appropriate boundary conditions yields the concentration of gases A and C and the conversions of solids B and D as functions of radial position and time. The overall conversions X_B and X_D are obtained by integrating the local conversions.

Another important factor, especially in the treatment of sulfide minerals, as noted earlier, is the fraction of the Gas C (say H_2S) produced from the solid B (say Me_xS) that is captured by the solid D (say CaO). From the consideration of stoichiometry, this factor may be obtained from the following relationship:

$$F = \frac{\text{moles of C reacted with D}}{\text{moles of C produced by B reacted}}$$

$$= \frac{\alpha_D \rho_D X_D/d}{\alpha_B \rho_B X_B/b} \quad (82)$$

Conversion as a function of time depends in a rather complex way on the kinetics of the two reactions, effective diffusivities, and the relative amounts of the solid reactants. From the computed results Sohn and Rajamani (67) determined that the final fixation of gas C depends only on a single parameter $\hat{\sigma}_D$ defined as follows:

$$\hat{\sigma}_D \equiv \frac{V_p}{A_p} \sqrt{\frac{\alpha_D F_p k_2}{2D_e} \left(\frac{A_D}{F_D V_D}\right)} \quad (83)$$

This relationship is shown in Figure 8.

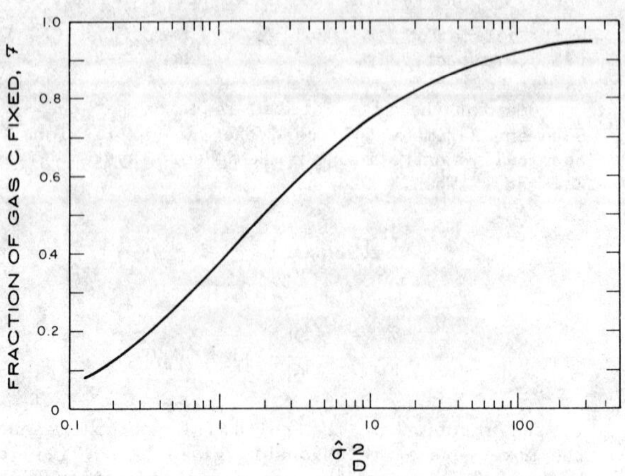

Figure 8. Fixation of gas C as a function only of $\hat{\sigma}_D^2$.

This result is very useful because the fixation of gas C can be predicted simply by computing the value of $\hat{\sigma}_D$ only. It would also be useful for designing pellets that would meet the required emission limits of the effluent gas containing gas C (sulfur-containing gas).

6. REACTIONS BETWEEN SOLIDS PROCEEDING THROUGH GASEOUS INTERMEDIATES

A number of reactions between solids are of considerable importance in pyrometallurgical processes. Some of these reactions are true solid-solid reactions which take place in the solid state between two species in contact with each other. Many other important reactions, however, occur through gaseous intermediates. These latter reactions may be considered as coupled gas-solid reactions, and can thus be analyzed in light of the mathematical analyses developed in the previous parts of this article.

6.1 Reactions With Net Production of Gases

Most important examples of such reactions are the carbothermal reduction of metal oxides. These reactions proceed through the intermediates of CO and CO_2, according to the following mechanism:

$$Me_xO_y(s) + CO(g) = Me_xO_{y-1}(s) + CO_2(g)$$

$$zCO_2(g) + zC(S) = 2z\ CO(g)$$

$$Me_xO_y(s) + zC(s) = Me_xO_{y-1}(s) + (2z-1)CO(g) + (1-z)\ CO_2(g) \quad (84)$$

Some examples of this type of reactions are: Reduction of iron oxides (72, 73);

$$Fe_xO_y + yCO = xFe + yCO_2$$

$$CO_2 + C = 2CO \quad (85)$$

Reaction between ilmenite and carbon (74);

$$FeTiO_3 + CO = Fe + TiO_2$$

$$CO_2 + C = 2CO \quad (86)$$

Reduction of stannic oxide (75);

$$SnO_2 + 2CO = Sn + 2CO_2$$

$$CO_2 + C = 2CO \quad (87)$$

Reaction between chromium oxide and chromium carbide (76);

$$\frac{1}{6} Cr_{23}C_6 + CO_2 = \frac{23}{6} Cr + 2CO$$

$$\frac{1}{3} Cr_2O_3 + CO = \frac{2}{3} Cr + CO_2 \quad (88)$$

Recent evidence (77) has shown that formation of silicon carbide proceeds according to the following reaction steps:

$$SiO_2 + CO = SiO + CO_2$$

$$CO_2 + C = 2CO \tag{89}$$

$$SiO + 2C = SiC + CO$$

In this system CO_2, CO and SiO are the gaseous intermediates.

In all these reactions, there is a net generation of gaseous species CO and CO_2 resulting in a bulk flow of the gas mixture from the reaction zone.

With the exception of the last system, these reactions can be expressed by the following general equations:

$$A(g) + bB(s) = cC(g) + eE(s) \tag{90a}$$

$$C(g) + dD(s) = aA(g) + fF(s) \tag{90b}$$

In order for this reaction to be self-sustaining without requiring external supply of gaseous reactants and with a net generation of gaseous species, the condition $ac > 1$ must be satisfied.

In spite of the practical importance of the reactions between solids, there have been few mathematical descriptions developed for these systems which would provide suitable bases for prediction and interpretation of their behaviors.

In early work the kinetics of solid-solid reactions were analyzed for specific conditions of one of the gas-solid reactions controlling the overall reaction. The studies on the carbothermic reduction of hematite by Rao (73), on the reaction between ilmenite and solid carbon by El-Guindy and Davenport (74), and on the reaction between chromium oxide and chromium carbide by Maru et al. (75) make use of the assumption of such a single controlling reaction.

Sohn and Szekely (78) developed a theoretical basis for analyzing a system in which both reactions affect the overall rate and for establishing a criterion for the controlling step. The analysis involves writing mass balance equations for the reactants involved.

Let us consider a porous pellet made up of uniformly mixed grains of B and D as sketched in Figure 9. Solids B and D may have different shapes and sizes. The particle size of either solid is assumed to be uniform. We shall consider isothermal system, which reacts under a uniform concentration of gaseous species.

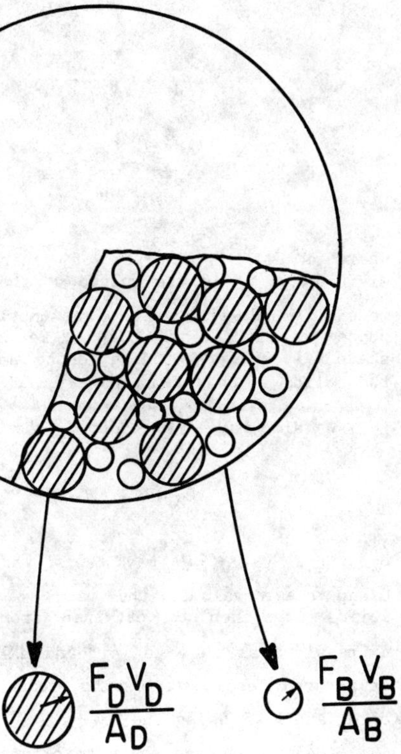

Figure 9. Schematic representation of a pellet made up of two solid reactants.

The rates at which reactions proceed, in terms of the generation of gaseous intermediates, are given as follows:

$$\frac{dn_A}{dt} = -v_1 + av_2 \tag{91a}$$

$$\frac{dn_c}{dt} = cv_1 - v_2 \tag{91b}$$

where v_1 and v_2 are the net forward rates of reactions (90a) and (90b), respectively, per unit volume of the pellet. These terms can be expressed, using Eq. (24), as follows:

$$v_1 = \left(\frac{\alpha_B \rho_B}{b}\right) \frac{dX_B}{dt} \tag{92a}$$

$$v_2 = \left(\frac{\alpha_D \rho_D}{d}\right) \frac{dX_D}{dt} \tag{92b}$$

The rate of reaction dX/dt term depends on the mechanism of reaction of each solid particle. It can be given by any of the solutions given earlier for the reaction of a single solid such as the differential form of Eq. (7), (21), (30), or (35). It must be noted that the bulk concentration terms in these equations are now the interstitial concentrations, which vary with time as the reaction progresses.

If the total pressure of the system is maintained constant, the following relationships hold:

$$\frac{dn_A}{dt} = \frac{C_A}{V_p} \frac{dV}{dt} \qquad (93a)$$

$$\frac{dn_C}{dt} = \frac{C_C}{V_p} \frac{dV}{dt} \qquad (93b)$$

where dV/dt is the rate of generation of the gas mixture and V_p is the volume of the pellet. We have made a psuedo-steady-state assumption that the gas-phase concentrations at any time are at the steady-state values corresponding to the amounts and sizes of the solids at that time, namely, $CdV/dt >> VdC/dt$.

We also have the condition that

$$C_A + C_C = C_T \qquad (94)$$

Equations (91)-(94) may be solved for C_A and C_C. Using these values, the rates of reaction of the solids may then be obtained from the appropriate expressions used for dX_B/dt and dX_D/dt. Integration of these rates over Δt will yield incremental values of X_B and X_D. Using the new values of X_B and X_D, Eqs. (91)-(94) are solved again to obtain C_A and C_C at the next time step. The procedure is repeated to give conversions as functions of time. In certain simple cases, analytical solutions are possible (78).

6.2 Reactions without Net Production or Consumption of Gases

This situation arises when ac = 1 in reactions represented by Eqs. (90a) and (90b). The preparation of metal carbides from metals and carbon is an example:

$$C(s) + 2H_2(g) = CH_4(g)$$

$$Me(s) + CH_4(g) = MeC(s) + 2H_2(g)$$

$$Me(s) + C(s) = MeC(s) \qquad (95)$$

Another example is the oxidation of certain metal sulfides with lime to produce the corresponding oxides, as recently proposed (79, 80). This reaction can in general be expressed by the following:

$$Me_xS + H_2O = Me_xO + H_2S$$

$$CaO + H_2S = CaS + H_2O \qquad (96)$$

$$Me_xS + CaO = Me_xO + CaO$$

Using this scheme, certain metal sulfides such as molybdenum disulfide and zinc sulfide can be transformed into the corresponding oxides without producing a sulfur containing gas. Chalcopyrite, treated by this process, is transformed into magnetite and bornite. Bornite is more readily leachable than chalcopyrite. The role of lime in this process is two-fold: (1) It improves the thermodynamics of the reaction between sulfide and steam, which has a very low equilibrium constant due to the positive Gibbs free energy change, by eliminating hydrogen sulfide, and (2) the reaction between hydrogen sulfide and lime fixes sulfur as calcium sulfide which can be further treated to recover sulfur.

Reactions (95) and (96) involve no net consumption or generation of the gaseous intermediates and the gaseous species do not appear in the overall stoichiometry. The gaseous species simply act as carriers of atoms. In this respect, the mechanism for this type of reactions might be termed "catalysis by gases" (2).

The oxidation of sulfide minerals without involving net generation or consumption of gases has an advantage over conventional roasting: The reaction can be carried out in a closed system. Consequently, the emission of sulfur containing pollutant can be eliminated.

The kinetics of this reaction can be described by Eq. (91). The difference between this case and the reactions with net generation of gases is that, since no gases are formed or consumed,

$$V_1 = cV_2 \qquad (97)$$

Equations (92) and (94) are valid also in this case. The solution of these equations gives the conversion as a function of time. Using this approach, Kim and Sohn (80) described the detailed kinetics of reactions of zinc sulfide and molybdenum disulfide with lime (80).

7. CONCLUDING REMARKS

In this treatise the analysis of gas-solid reactions most frequently encountered in extractive metallurgy has been discussed. The reactions in which a coherent solid product layer is formed have been emphasized because of their importance in extractive metallurgy. Early models for gas-solid reactions were based on the shrinking-core scheme. In the first half of the 1970s, the analysis of a single gas-solid reaction in a porous solid was developed. In the second half of the 1970s the techniques for analyzing the reactions of porous solids were applied to the analysis of more complex reactions involving simultaneous and successive gas solid reactions. Development of some of these complex gas-solid reactions in extractive metallurgy has been prompted due to the increasing concern for reducing environmental pollution associated with conventional processes.

In this article we have mainly treated the analysis of reactions taking place in a single particle or pellet of solid. The eventual objective for studying single particle systems is, of course, to apply the results of such a study to analyzing, predicting and designing multiparticle systems. Most practical fluid-solid systems of industrial importance consist of particulate assemblages. Examples of multiparticle fluid-solid contacting equipment include packed beds, moving beds, fluidized beds, and rotary kilns. The extension of single-particle studies to multiparticle systems will depend on the nature of the particulate assemblages, the mode of fluid-solid contacting, and the spatial variation of the fluid properties within the system. These problems are quite analogous to those involved in extending the catalytic reaction in a single pellet to an industrial catalytic reactor.

Since much work has been done on the general analysis of multiparticle reaction systems and space does not allow an in-depth discussion in this article, the reader is referrd to literature articles for previous work on gas-solid reactions in multiparticle systems (1,81-85).

NOMENCLATURE

A_g, A_p	external surface area of an individual grain and the pellet, respectively
b	number of moles of solid B reacted per mole of gaseous reactant A
c,d	stoichiometry coefficients
c_p	heat capacity
C_i	concentration of gaseous species i
C_T	total concentration of gas
D_e	effective diffusivity of gaseous species in the porous solid
D_g	effective diffusivity of gaseous species in the product layer around a grain
F	local value of the fractional oxygen removal in the reduction of hematite defined by Eq. (64), or degree of fixation of intermediate gas C in successive gas-solid reactions defined by Eq. (82)
F_g, F_p	shape factor for the grain and pellet, respectively (= 1, 2, or 3 for flat plates, long cylinders, or spheres, respectively)
$g_{F_g}(X)$ or $g_{F_p}(X)$, $g_N(X)$	conversion functions defined by Eqs. (11) and (38), respectively
$(-\Delta H)$	heat of reaction
k	reaction-rate constant
k_a	apparent rate constant defined by Eq. (26)
k_m	external-mass-transfer coefficient
K	equilibrium constant
n_i	number of moles of hematite, or those of magnetite and wustite equivalent to n_i moles of hematite
$\underline{N_i}$	molar flux vector for species i
p_i	partial pressure of species i
p(X)	conversion function defined by Eqs. (12) - (14) or Eqs. (15) - (17)
$\underline{q}$	heat flux vector
r	distance from the center of symmetry
r_c	position of reaction interface in a nonporous particle
r_p	radius of the pellet
R	universal gas constant
R_i	rate of production of species i per unit volume
S_v	surface area of solid per unit volume
Sh^*	modified Sherwood number defined by Eq. (10)
t	time
t^*, t_g^*, t_N^*	dimensionless times defined by Eqs. (8), (31), and (36), respectively
T	temperature
v	rate of consumption of gaseous species per unit volume
V_g, V_p	volume of a grain and the pellet, respectively
w	local fractional conversion of solid reactant
x_i	mole fraction of species i
X	overall fractional conversion of solid reactant
Y_i	fraction defined by Eq. (53)
z	volume of product solid formed from unit volume of reactant solid, or fraction of iron in magnetite that forms wustite

Greek Symbols

α_B, α_D	fractions of pellet volume initially occupied by solid reactants B and D, respectively
ε	porosity
θ	volume change parameter defined by Eq. (19)
ν	number of gaseous products formed from one mole of gaseous reactant
ρ	density
ρ_B, ρ_D	true molar densities of solid reactants B and D, respectively
$\hat{\sigma}, \hat{\sigma}_g, \hat{\sigma}_N, \sigma_s$	gas-solid rection moduli defined by Eqs. (32), (34), (37), and (9), respectively
ψ_i	dimensionless concentration of gaseous species i defined by Eqs. (69) and (70)

Subscripts

A,C	gaseous species A and C, respectively
b	bulk stream value
B,D	solid species B and D, respectively
i	species i, or based on initial conditions
o	original value

REFERENCES

1. J. Szekely, J. W. Evans, and H.Y. Sohn, Gas-Solid Reactions; Academic Press, New York, 1976.

2. H.Y. Sohn, "Fundamentals of the Kinetics of Heterogenous Reaction Systems in Extractive Metallurgy," p. 1-51 in Rate Processes of Extractive Metallurgy, H.Y. Sohn and M.E. Wadsworth, eds.; Plenum, New York, 1979.

3. J.W. Evans and C.-H. Koo, "The Reduction of Metal Oxides," p. 285-321 in Rate Processes of Extractive Metallurgy, op cit.

4. H.H. Kellog and S.K. Basu, Trans. Met. Soc. AIME, 218 (1960) p. 70-81.

5. T.R. Ingraham, "Sulfate Stabiity and Thermodynamic Phase Diagrams with Particular Reference to Roasting," p. 179-196 in Applications of Fundamental Thermodynamics to Metallurgical Processes, G. Fitterer, ed.; Gordon and Breach, New York, 1967.

6. C.B. Alcock, Principles of Pyrometallurgy; Academic Press, New York, 1976.

7. S.E. Khalafalla, "Roasting as a Unit Process," p. 245-285 in Rate Processes of Extractive Metallurgy, H.Y. Sohn and M.E. Wadsworth, eds.; Plenum, New York, 1979.

8. H.Y. Sohn and R.P. Goel, Minerals Sci. Eng., 11 (1979) p. 137-153.

9. R.B. Bird, W.E. Stewart, and E.N. Lightfoot, Transport Phenomena; Wiley, New York, 1960.

10. V.G. Levich, Physiochemical Hydrodynamics; Prentice-Hall, Englewood Cliffs, New Jersey, 1962.

11. D. Kunii and O. Levenspiel, Fluidization Engineering; Wiley, New York, 1969.

12. C.N. Satterfield, Mass Transfer in Heterogeneous Catalysis; MIT Press, Cambridge, Massachusetts, 1970.

13. J. Szekely and N.J. Themelis, Rate Phenomena in Process Metallurgy; Wiley-Interscience, New York, 1971.

14. R.C. Reid, J.M. Prausnitz, and T.K. Sherwood, The Properties of Gases and Liquids, 2nd ed.; McGraw-Hill, New York, 1977.

15. J.M. Smith, Chemical Engineering Kinetics, 2nd ed.; McGraw-Hill, 1970.

16. E.A. Mason and T.R. Marero, Adv. in Atomic and Molecular Physics, 6 (1970) p. 155-232.

17. T.R. Marero and E.A. Mason, J. Phys. Chem. Ref. Data, 1 (1972) p. 1-118.

18. J.M. Thomas and W.J. Thomas, Introduction to the Principles of Heterogeneous Catalysis; Academic Press, New York, 1967.

19. O.A. Hougen and K.M. Watson, Chemical Process Principles, Part 3 Kinetics and Catalysis; Wiley, New York, 1947.

20. K.J. Laidler, p. 75, 119, 195 in Catalysis, Vol. 1, P.H. Emmett, ed.; Reinhold, 1954.

21. J. Francl and W.D. Kingery, J. Am. Ceram. Soc., 37 (1954) p. 99.

22. A.V. Luikov, A.G. Sashkov, L.L. Vasillies, and Y.E. Fraimon, Int. J. Heat Mass Trans., 11 (1968) p. 117.

23. D.A. Young, in The International Encyclopedia of Physical Chemistry and Chemical Physics, F.C. Tompkins, ed., Topic 21, Vol. 1; Pergamon Press, 1966.

24. W.M. McKewan, Trans. Met. Soc. AIME, 212 (1958) p. 791; 218 (1960) p. 2.

25. N.J. Phemelis and W.H. Gauvin, Trans. Met. Soc. AIME, 227 (1963) p. 290.

26. E. Kawasaki, J. Sanscrainte, and T.J. Walsh, AIChE J., 8 (1963) p. 48.

27. W.-K. Lu, Trans. Met. Soc. AIME, 227 (1963) p. 203.

28. W.-K. Lu and G. Bitsianes, Can. Met. Quart., 7 (1968) p. 3.

29. B.B.L. Seth and H.U. Ross, Trans. Met. Soc. AIME, 233 (1965) p. 180; Can Met. Quart., 5 (1966) p. 315.

30. H.W. St. Clair, Trans. Met. Soc. AIME,. 233 (1965) p. 1145.

31. R.H. Spitzer, F.A. Manning, W.O, Philbrook, Trans. Met. Soc. AIME, 236 (1966) p. 726; p. 1715.

32. H.Y. Sohn and J. Szekely, Can. J. Chem. Eng., 50 (1972) p. 674-676.

33. H.Y. Sohn and H.-J. Sohn, Ind. Eng. Chem. Process Des. Dev., 19 (1980) p. 237-242.

34. K. Natesan and W.D. Philbrook, Trans. Met. Soc. AIME, 245 (1969) p. 2243.

35. H.H. Haung and R.W. Bartlett, Met. Trans. B, 7B (1976) p. 369.

36. K.B. Bischoff, Chem. Eng. Sci., 18 (1963) p. 711; 20 (1965) p. 783.

37. D. Luss, Can. J. Chem. Eng., 46 (1968) p. 154.

38. J. Szekely and Y. El-Tawil, Met. Trans. B, 7B (1976) p. 490.

39. J. Szekely, M. Choudary, and Y. El-Tawil, Met. Trans. B, 8B (1977) p. 639.

40. D.C. Lynch and J.F. Elliott, paper presented at the 107th AIME Meeting, Denver, Colorado, 1978.

41. J.M. Ausman and C.C. Watson, Chem. Eng. Sci., 17 (1962) p. 323.

42. M. Ishida and C.Y. Wen, AIChE J., 14 (1968) p. 311.

43. J. Szekely and H.Y. Sohn, Trans. Inst. Mining Met., 82 (1973) p. C92-C100.

44. H.Y. Sohn and J. Szekely, Chem. Eng. Sci., 27 (1972) p. 763-778.

45. J. Szekely and J.W. Evans, Met. Trans., 2 (1971) p. 1691-1698; Chem. Eng. Sci., 26 (1971) p. 1901-1913.

46. R.H. Tien and E.T. Turkdogan, Met. Trans. 3, (1972) p. 2039-2048.

47. A. Calvelo and J.M. Smith, Proc. Chemeca '70, no. 3, p. 1-24, Butterworth, 1972.

48. J. Szekely, C.I. Lin, and H.Y. Sohn, Chem. Eng. Sci., 28 (1973) p. 1975-1989.

49. H.Y. Sohn and J. Szekely, Chem. Eng. Sci., 29 (1974) p. 630-634.

50. H.Y. Sohn, Met. Trans. B, 9B (1978) p. 89-96.

51. S.P. Trushenski, K. Li, and W.D. Philbrook, Met. Trans., 5 (1974) p. 1149.

52. H.Y. Sohn, "Mathematical Models of Iron Ore Reduction," p. 3.8-3.11 in Physical Chemistry and Steelmaking, Proceedings of the International Conference on the Physical Chemistry of Steelmaking, Versailles, France, October 23-25, 1978, vol. 3, 1978.

53. E.T. Turkdogan and J.V. Vinters, Met. Trans., 3, (1972) p. 1561.

54. R.H. Tien and E.T. Turkdogan, Met. Trans., 3 (1972) p. 2039.

55. A.G. Matyas and A.V. Bradshaw, Ironmaking and Steelmaking, no. 3 (1974) p. 180.

56. S.K. El-Rahaiby and Y.K. Rao, Met. Trans. B, 10B (1979) p.257.

57. E.T. Turkdogan and J.V. Vinters, Met. Trans., 2 (1971) p. 3175.

58. E.T. Turkdogan, R.G. Olsson, and J.V. Vinters, Met. Trans., 2 (1971) p. 3189.

59. C.Y. Wen and L.Y. Wei, AIChE J., 16 (1970) p. 848.

60. C.Y. Wen and L.Y. Wei, AIChE J., 17 (1971) p. 272.

61. A. Rehmat and S.C. Saxena, Ind. Eng. Chem. Process Des. Dev., 16 (1977) p. 502.

62. H.Y. Sohn and R.L. Braun, Chem. Eng. Sci., 35 (1980) p. 1625-1635.

63. H. Kay, p. 33 in High Temperature Refractory Metals, W.A. Krivsky, ed.; Gordon and Breach, New York, 1968.

64. R.E. Chech and T.E. Tiemann, Trans. Met. Soc. AIME, 245 (1969) p. 1727.

65. F. Habashi and R. Dugdale, Met. Trans. 4 (1973) p. 1865.

66. F. Habashi and B.I. Yostos, J. Metals, 29 (7) (1977) p. 11-16.

67. H.Y. Sohn and K. Rajamani, Chem. Eng. Sci., 32 (1977) p. 1093-1101.

68. S. Won and H.Y. Sohn, paper presented at the 109th AIME Annual Meeting, Las Vegas, Nevada, February 24-28, 1980.

69. I.D. Shah and P.L. Ruzzi, Met. Trans. B, 9B (1978) p. 247-253.

70. F.P. Haver and M.M. Wong, Mining Eng., 24 (6) (1972) p. 52.

71. R.W. Bartlett and H.H. Haung, J. Metals, 25 (12) (1973) p. 28.

72. K. Otsuka and D. Kunii, J. Chem. Eng. Japan, 2 (1969) p. 46-50.

73. Y.K. Rao, Met. Trans., 2 (1971) p. 1439-1447.

74. M.I. El-Guindy and W.G. Davenport, Met. Trans., 1 (1970) p. 1729-1734.

75. R. Padilla and H.Y. Sohn, Met. Trans. B, 10B (1979) p. 109-115.

76. Y. Maru, Y. Kuramasu, Y. Awakura, and Y. Kondo, Met. Trans., 4 (1973) p. 2591-2598.

77. J.G. Lee, Carbide and Nitride Ceramics by Carbothermal Reduction of Silica, Ph.D. Dissertation, University of Utah, 1976.

78. H.Y. Sohn and J. Szekely, Chem. Eng. Sci., 28 (1973) p. 1789-1801.

79. H.Y. Sohn, A Process for Oxidizing Sulfide Ores Using Steam Without Emitting Sulfur-Containing Pollutants, patent pending, 1980.

80. D. Kim and H.Y. Sohn, "Treatment of Sulfide Ores with Steam to Produce Metal Oxides Without Emitting Sulfur-Containing Pollutants, p. 1-15, Technical Session E-2, Proc. Fourth MMIJ/AIME Joint Meeting, Tokyo, Japan, November 4-8, 1980.

81. M. Ishida and C.Y. Wen, Ind. Eng. Chem. Process Des. Dev., 10 (1971) p. 164.

82. J.W. Evans and S. Song, Ind. Eng. Chem. Process Des. Dev., 13 (1974) p. 146.

83. H.Y. Sohn, Hwahak Konghak (J. Korean Inst. Chem. Engrs), 14 (1976) p. 203-213.

84. J.A. Herbst, "Rate Processes in Multiparticle Metallurgical Systems," p. 53-112 in Rate Processes of Extractive Metallurgy, H.Y. Sohn and M.E. Wadsworth, eds.; Plenum, New York, 1979.

85. R.W. Bartlett, "Chemical Process Analysis and Design for Multiparticle Systems," p. 113 - 131 in Rate Processes of Extractive Metallurgy, H.Y. Sohn and M.E. Wadsworth, eds.; Plenum, New York, 1979.

PHYSICAL CHEMISTRY OF LIQUID METAL SOLUTIONS

John F. Elliott

Department of Materials Science and Engineering
Massachusetts Institute of Technology
Cambridge, Massachusetts 02139

There has been a great deal of interest and activity in the past 15 years regarding modeling and simulating thermodynamic properties of liquid metal solutions and of their components. A major motivation for this activity has been the desire to use computer methods for modeling phase diagrams of multicomponent systems. A consistent thread in the work on modeling has been concern about the nature of bonding in systems, especially those in which compounds appear in the solid state. A wide variety of methods have been employed to account for bonding. They include: (a) modification of the composition coordinate in formalisms, such as use of the atom ratio and lattice ratio rather than mole fraction; (b) modification of the quasi-chemical theory of solution behavior in various ways; (c) incorporation in a model of the concept of a cluster, or complex, consisting of a central atom and its cluster of nearest neighbor atoms, or its coordination cluster. The number of nearest neighbors may be different from the average number of nearest neighbor atoms of component atoms of the system, and may be the atoms surrounding an interstitial site in the liquid structure. (d) assignment of bond strengths differing from the average for certain of the atoms in the coordination cluster of the central atom. The difference may be related to spacial relationship of atoms in the cluster, to the type of atom occupying a given location, or to the population of atoms in the cluster; (e) assumption of the existence of molecular species (actual or hypothetical) in the solution. These varied approaches have led to a proliferation of models, most of which are somewhat empirical in character, but some of which are essentially totally empirical. Each type of model has its own adherents and particular use. It appears, however, that no model or formalism has been developed as yet which permits the prediction of the thermodynamic properties of a component of a solution, or of the solution itself, from first principles, or from a limited number of chemical and physical properties of the component elements of the system.

It is concluded that greater effort needs to be invested in developing and applying experimental methods which will provide information related to the nature of the bonding mechanism(s), the electron states, and properties of holes and vacancies, etc., in liquid metallic solutions. Because of the inherent uncertainties in the results of experimental methods for measuring thermodynamic properties of liquid alloys, it seems unlikely that it will be possible to continue to improve theoretical and empirical models of solution behavior of liquid alloys and to test one model against another through the use of thermodynamic properties alone. Thus it seems necessary that other properties be measured that are related to the structure, electronic states, and bond energies of alloys. Accordingly, there is a need for new experimental methods and for the application of nonthermodynamic methods employed at lower temperatures to the study of liquid metals.

INTRODUCTION

The liquid metal phase is found in one or more stages in producing, alloying, and primary forming (i.e., casting) most of our metals. There are many reasons for this, but the principal ones are that (a) it is relatively easy to separate gangue constituents in an ore from a liquid phase that carries the metal being extracted; (b) chemical reactions involved in smelting, refining, and alloying, and the attendant transport of heat and mass are highly favored when the metal is in liquid form. For example, it is relatively easy to disaggregate the liquid to obtain a very large ratio of surface area to mass for the phase, and it is also relatively easy to coalesce the liquid into a massive form once again; and (c) liquids may be poured, pumped, stirred, and mixed at very high rates if desired. Thus, there are many economic and technical advantages to converting a metal into the liquid state in the primary processing of metals.

As a consequence of the broad utilization of the liquid phase, there was a very early interest in applying the concepts of thermodynamics to help understand the chemical behavior of metals themselves, of alloying elements, and of refining agents in liquid metals. There are many papers in the metallurgical literature which report actual measurements and studies of that behavior. There are also many, and possibly even more, papers related to modeling, simulating, and interpreting the behavior of constituents of both dilute and concentrated solutions of liquid metals.

This paper is divided into two major parts: first is a discussion of the current state of interpretation, modeling, and simulation of the thermodynamic properties of these liquid systems; and second, a brief commentary on our current level of actual knowledge of the thermodynamic properties of liquid metals and alloys, their chemical constituents, and of the methods being employed to measure them.

COMPOSITION COORDINATES

It is first useful to discuss the use of composition coordinates that are employed when quantitatively describing the behavior of components of an alloy system. The mole and atom fraction

$$X_i \equiv n_i/\Sigma_j n_j; \quad j = 1 \text{ to } m \tag{1}$$

is a natural outgrowth of the chemist's treatment of chemical compounds. The n's are the number of moles or gram atoms of each species, i and j, in the solution which has m components. It has found extensive use in the metallurgical literature, especially for alloy systems which are substitutional in character in the crystalline form. Most metallurgists resort to this composition coordinate when thinking about the idealized behavior of an alloy system. The ideal solution is defined as the system in which all components follow Raoult's Law, i.e.,

$$a_i = X_i;\ 0 < X_i < 1 \tag{2}$$

where a_i is the thermodynamic activity of component i in the solution which Lewis (1) defined implicitly by the relationship

$$\mu_i = RT \ln a_i + \mu_i^\circ \tag{3}$$

where μ_i and μ_i° are, respectively, the chemical potentials of i in the solution and in the selected standard state for i. There are additional equations that are necessary to this definition that involve the reference and standard states; they are given below.

Intrinsic to the use of Raoult's Law are the reference state of the pure component at one atmosphere pressure for which

$$\lim_{X_i \to 1} a_i/X_i \to 1, \tag{4}$$

and the standard state for each pure component at one atmosphere pressure where

$$a_i = 1 \quad \text{when } X_i \to 1. \tag{5}$$

It is readily shown that a solution which obeys Raoult's Law will have values for all partial molar and the molar enthalpies of mixing (H_i^M and H^M, respectively) which are zero:

$$H_i^M \equiv \overline{H}_i - H_i^\circ = 0 \tag{6a}$$

and

$$H^M \equiv \Sigma X_i(\overline{H}_i - H_i^\circ) = \Sigma X_i H_i^M = 0 \tag{6b}$$

for conditions of $0 < X_i < 1$, where $\overline{H}_i$ is the partial molar enthalpy, and the superscript degree sign designates the value for the standard state of each component.

The partial molar and molar entropies of mixing (S_i^M and S^M, respectively) have the following definitions and values for Raoultian behavior as shown:

$$S_i^M \equiv \overline{S}_i - S_i^\circ = - R \ln X_i \tag{7a}$$

and

$$S^M \equiv \Sigma X_i(\overline{S}_i - S_i^\circ) = \Sigma X_i S_i^M = \Sigma - RX_i \ln X_i. \tag{7b}$$

Other interesting properties of an ideal solution are that there is no volume change, nor change in heat capacity, in the system from the unmixed to the mixed state (2). A point that is sometimes overlooked is that the equations for Raoultian behavior are based on the mixing of pure components in exactly the same state of aggregation as that of the solution that is formed. That is to say that the reaction involves all liquid components coming together to form a liquid solution, or all components in the body-centered cubic structure are mixed to form the solution which also is body-centered cubic in structure, etc.

In dilute solutions where the activity of a component in the solution is linear with composition,

$$a_i = b\, \xi_i\,, \tag{8a}$$

where ξ_i is generally a composition coordinate. It may be mole fraction, or weight percent, or any one of a number of others which will be discussed later. The term b is a constant. Because the arithmetic transformation of one compositional coordinate to another usually is not linear, a component which follows equation 8a for one composition variable may not follow it for another. The behavior described by equation 8a leads to the concept of Henry's Law, where

$$a_i/\xi_i = b\, f_i(\xi_i) \tag{8b}$$

and the activity coefficient $f_i(\xi_i)$ is defined

$$a_i/\xi_i \equiv f_i(\xi_i). \tag{9a}$$

It is convenient to carry the argument (ξ_i) with the activity coefficient to remind one of which composition coordinate is being used. The activity coefficient (γ_i) based on Raoultian behavior and the pure substance as the standard and reference states is defined:

$$a_i/X_i \equiv \gamma_i. \tag{9b}$$

A set of reference and standard states different from that described in equations 4 and 5 are often employed with dilute solutions, particularly when the solute in the pure form is a gas and the solution is a liquid or solid; or when the range of stability of the solution is quite limited. The reference state is the infinitely dilute solution of the solute constituent in the solvent

$$\lim_{\xi_i \to 0} a_i/\xi_i \to 1 \tag{10a}$$

and the standard state is

$$a_i = 1 \text{ when } \xi_i = 1. \tag{10b}$$

This standard state is hypothetical (3) in many cases because the behavior of component i may not follow equations 8a and 8b out to the concentration where $\xi_i = 1$. The use of the hypothetical standard state offers no difficulty because all partial molar and molar properties actually are related to the behavior of components in the infinitely dilute solution.

The compositional coordinate atom ratio, Y_i, is sometimes employed. Its definition and its relationship to atom fraction for a 1-2 binary solution is

$$Y_2 \equiv n_2/n_1 = X_2/(1-X_2) \; . \tag{11}$$

Component 1 is the solvent. This coordinate is frequently found in equations for modeling the behavior of interstitial solid solutions such as carbon dissolved in gamma-iron, where there is one interstitial (octahedral) site per lattice atom. It is the "natural" coordinate for such solutions as the equation for the ideal relative partial molar entropy of the interstitial component 2 is

$$S_i^M = -R \ln Y_2 \; . \tag{12}$$

Another coordinate that often finds use when treating interstitial solutions is the lattice ratio:

$$Z_2 \equiv n_2/(n_1 - n_2/d). \tag{13}$$

Z is proportional to the ratio of number of interstitial sites that are filled with atoms of component 2 to the number of unfilled interstitial sites, there being d interstitial sites of the type that atoms of component 2 may occupy. In the case that d equals 1

$$Z_2 = n_2/(n_1 - n_2). \tag{14a}$$

Also, it can be shown that for this condition

$$Z_2 = Y_2/(1-Y_2) = X_2/(1 - 2X_2). \tag{14b}$$

The use of Z as the composition coordinate with equations 8a, 8b, 9a, 10a, and 10b provides what can be considered to be Henry's Law for interstitial solutions. It is readily seen that it is practical to use the infinitely dilute solution as the reference state for these interstitial solutions, and the hypothetical standard state exists at $Z_2 = 1$.

There has been considerable interest in "molecular" or associated metal solutions recently. Examples of these solutions are metal-oxygen and metal-sulfur systems. For example, Belton and Tankins (4) and Jacob and Alcock (5) have shown that it is useful to consider liquid Fe-O and Fe-S systems in terms of the nonmetallic atom being bonded to one or more metallic atoms. The assumed associated species would have the formula MX_d. In relatively dilute solutions, the number of gram atoms of the uncombined metallic (and solvent) constituent would be $n_M - dn_X$ and the number of the associated molecules would be n_X/d. One then finds that the behavior of the solution could be correlated in terms of the general Z_2 composition coordinate as given in equation 13. Chipman (6) has found the composition coordinate Z_i to be a useful basis for correlating the behavior of solute elements in what is called "strongly-bonded" liquid solutions. Two examples of this are silicon in liquid iron and in liquid cobalt (7), and there are a number of others. It is interesting to note that these systems in most cases are not interstitial in character in the solid state. On the other hand, they all contain relatively stable solid compounds.

It is most convenient to employ the Y and Z coordinates at dilute concentrations. As noted earlier, however, there arises a problem of applying Henry's Law to the behavior of solute elements; if a solute element is Henrian with respect to one coordinate, it usually will not be with respect to another because of the nonlinearity that exists between the various possible pairs of coordinates. This is illustrated in Fig. 1 where the relationship between X_2 and Y_2, and X_2 and the various forms of Z_2 are shown graphically. The term r is 1/d in equation 13. If the uncertainty in experimental results is large, and/or if data at only dilute concentrations of a solution are available, it may not be possible to distinguish whether Henrian behavior is best described in terms of X, Y, or Z. There are some interesting ramifications to this matter that are discussed in reference 7.

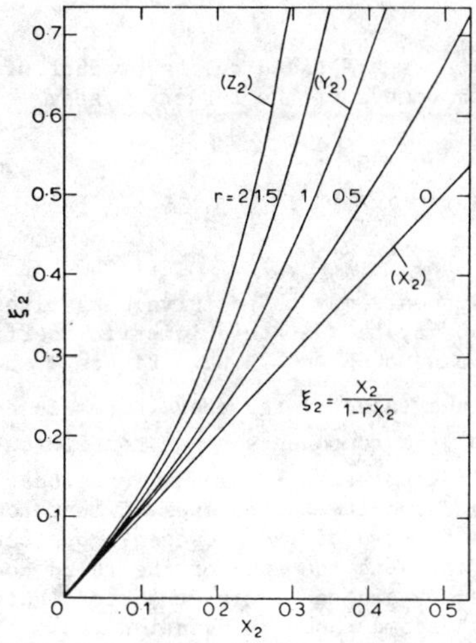

Fig. 1. Non-Linearity of Y and Z Coordinates Relative to X (7).

The atom or mole fraction is often employed for ternary and higher order, substitutional-type systems. Because

$$X_1 + X_2 + X_3 + \ldots + X_m = 1 \tag{15}$$

for an m component system, only m-1 variables are required to describe the composition of the system; for a ternary system, there are two. Examples of the sets that might be employed are:

$$X_1 \text{ and } X_2 \tag{15a}$$

$$X_1/X_3 \text{ and } X_2/X_3 \tag{15b}$$

$$X_1/(X_1 + X_3) \text{ and } X_2 \tag{15c}$$

$$X_1/(X_1 + X_3) \text{ and } X_1/(X_1 + X_3) \tag{15d}$$

and so forth. The set in equation 15b may be troublesome because each coordinate can take values from 0 to ∞. The others have the range from 0 to 1. This matter can be of importance when integrating the Gibbs-Duhem equation. An additional variable must be added to each of equations 15a through 15d for each additional component in the system. Recently Chou and Elliott (8) have proposed that more general coordinates can be devised for ternary and higher-order systems. For the ternary 1-2-3 system, there are three convenient sets:

$$k_1 = \frac{X_2}{\varepsilon X_3 - \nu X_2} \; ; \; \ell_1 = \frac{\varepsilon X_1 - \eta X_2}{\varepsilon X_3 - \nu X_2} \tag{16a}$$

$$k_2 = \frac{X_1}{\eta X_3 - \nu X_2} \; ; \; \ell_2 = \frac{\eta X_2 - \varepsilon X_1}{\eta X_3 - \nu X_1} \tag{16b}$$

$$k_3 = \frac{X_2}{\varepsilon X_1 - \eta X_2} \; ; \; \ell_3 = \frac{\varepsilon X_3 - \nu X_2}{\varepsilon X_1 - \eta X_2} \; . \tag{16c}$$

These coordinates are based on the product of activities, ϕ, of the components of the ternary system

$$\Phi = a_1^{\eta} \cdot a_2^{\varepsilon} \cdot a_3^{\nu} \tag{17}$$

where the exponents may be positive, negative, or zero; and they also may be the stoichiometric coefficients of a ternary compound such as $A_\eta B_\varepsilon C_\nu$. Figure 2 shows the range of values taken by the coordinates in equation 16a. If any one of the components has a coefficient equal to zero, the pivot point for ℓ_1 moves to a location on the binary system opposite to the apex of that component. If coefficients for two of the components are both zero, the pivot point moves to the apex of the third component. Thus, if η and ν are each zero, the coordinates in equation 16a become those in equation 15b. There are advantages to the use of these special sets of coordinates, particularly when dealing with a system containing a compound (8).

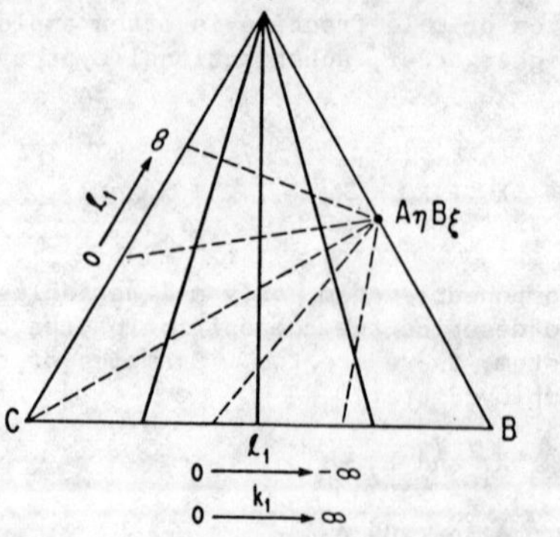

Fig. 2. Range of Values For Composition Coordinates k_1 and ℓ_1 See equation 16a, with $\nu = 0$.

MODELING

The simplest model of solution behavior is the ideal solution (equations 2 and 4) which has been described earlier. Few, if any, systems follow ideal behavior, but the model provides a convenient reference for judging the behavior of a component in a system. Often the comparison is quite general being "positive or negative, with respect to ideality." More specifically, the comparison is expressed in terms of excess properties:

$$\Psi_i^{EX} \equiv \bar{\Psi}_i - \bar{\Psi}_i^{ideal} - \Psi_i^{\circ} \tag{18}$$

and

$$\Psi^{EX} \equiv \Psi - \Sigma X_i \, (\bar{\Psi}_i^{ideal} - \Psi_i^{\circ}), \tag{19}$$

where Ψ is a state property of the system; $\bar{\Psi}_i$ is the partial molar and Ψ_i^{EX} is the excess partial molar property of species i; Ψ and Ψ^{EX} are molar and excess molar properties of the solution, and $\bar{\Psi}^{ideal}$ is the relative partial molar property of component i for the ideal solution. In the case of the Gibbs Free Energy

$$G_i^{EX} = RT \, \ell n \, \gamma_i \tag{20}$$

and

$$G^{EX} = RT \, \Sigma X_i \, \ell n \, \gamma_i \; . \tag{21}$$

A number of polynomial expressions have been proposed to describe empirically the relationship between the composition and activity coefficient of a component. An early one by Margules (9) has become a popular starting point for many modeling efforts in recent years. The relationship for the components 1 and 2 in the 1-2 binary system are

$$\ell n \, \gamma_1 = \alpha_2 X_2^2 + \alpha_3 X_2^3 + \alpha_4 X_2^4 + \cdots \tag{22a}$$

and

$$\ell n \gamma_2 = \beta_2 X_1^2 + \beta_3 X_1^3 + \beta_4 X_1^4 + \cdots \; , \tag{22b}$$

where

$$\alpha_0 = \beta_0 = \alpha_1 = \beta_1 = 0 \tag{22c}$$

$$\beta_2 = \alpha_2 + 3/2\alpha_3 + 2\alpha_4 \tag{22d}$$

$$\beta_3 = -(\alpha_3 + 8/3\alpha_4) \tag{22e}$$

$$\beta_4 = \alpha_4 + \cdots \; . \tag{22f}$$

It is to be noted that each of the coefficients of the zero and first order terms in both series is zero.

The regular solution model which was first defined by Hildebrand (10) assumes ideal mixing of the components (equations 7a and 7b) and that the partial molar heat of mixing of each component follows the relationship

$$H_i^M = (1 - X_i)^2 \, \Omega \tag{23a}$$

so that

$$G_i^{EX} = RT \, \ell n \, \gamma_i = (1 - X_i)^2 \, \Omega, \tag{23b}$$

and for a binary A-B system

$$G^{EX} = H^M = X_A X_B \Omega . \tag{23c}$$

The simple treatment of the regular solution by the pair-potential or quasi-chemical model of solution behavior gives

$$\Omega = Z N_o (W_{AB} - 1/2(W_{BB} + W_{AA})) . \tag{24}$$

In this model, the enthalpy change in the formation of the solution arises from breaking of A-A and B-B bonds in the pure substances and the formation of A-B bonds between dissimilar molecular entities. W_{AA}, W_{BB} and W_{AB} is the energy per bond, respectively, of each type of pair. There is bonding only between an atom and each of its nearest neighbors, where Z is the number of nearest neighbors. N_o is Avagadro's number. It is to be seen that important assumptions of the model are that the number of nearest neighbors is the same over the full range of composition of the system, that the only source of entropy is configurational, which is ideal, and that the bonding energy between two atoms is independent of the surroundings of the pair. This last assumption appears to be particularly limiting in the use of the model and there have been many efforts to lift this restriction by elaborating on the quasi-chemical model of solution behavior, an issue which will be discussed later. The cell model of the liquid state (11) that was first introduced by Eyring (12) provides a way of accounting for the vibrational contribution to the entropy of a solution and for variation of the pair-wise bond energies with composition of the solution and the population of the nearest-neighbor shell of an atom. The quasi-chemical model has been the basis for a number of efforts to produce more sophisticated models of the thermodynamic behavior of components of both substitutional and interstitial types of liquid metallic solutions.

There has been considerable interest in predicting the behavior of a component A in the A-B-C ternary system at dilute concentrations in the liquid phase along the B-C side. It is desired to predict the activity coefficient for A at the concentration X_B, $\gamma_{A(AB)}$, from the activity coefficients of A at infinite dilution in B and pure C, $\gamma_{A(B)}$ and $\gamma_{A(C)}$, and other parameters of the system. Alcock and Richardson (13) employed the quasi-chemical model to obtain a formulation that would be expected to apply for substitutional type systems where there is a close similarity in the sizes of the molecules of the three components. The final form of their equation is

$$\ell n\gamma_{A(BC)} = X_A \ \ell n\gamma_{A(B)} + X_B \ \ell n\gamma_{A(C)} - X_B \ \ell n\gamma_B - X_C \ \ell n\gamma_C . \tag{25}$$

The last two terms on the right are the properties in the B-C binary system at infinite dilution of A. Weinstein and Elliott (14) found that this equation fitted the behavior of manganese in the lead-bismuth system well up to approximately $X_{Mn} = 0.1$ (See Fig.3). Darken (15) has also obtained this relationship through manipulation of the α-function $[G_i^{EX}/(1-X_i)^2]$ without using any model.

Later, Alcock and Richardson (16) modified their earlier model to account for the behavior of the solute atom A where the energy of interaction of it with the solvent atoms B and C is relatively high. Specific examples are sulfur and oxygen dissolved in binary alloy systems. They assumed that there was a preferential bonding of B and C atoms with A atoms rather than with each other. The new relationship is

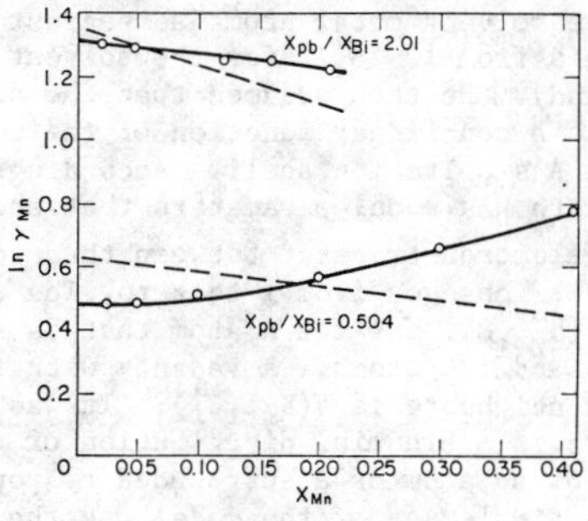

Fig. 3 Predicted Value of γ_{Mn} in the Liquid Phase on the Pb-Bi Side of the Mn-Pb-Bi System (14); See equation 25.

$$\left[1/\gamma_{A(BC)}\right]^{1/Z} = X_B \ (\gamma_B/\gamma_{A(B)})^{1/Z} + X_C \ (\gamma_C/\gamma_{A(C)})^{1/Z} \tag{26}$$

where Z is the number of nearest neighbors of component A; it is assumed to be the same for atoms of B and C. For the special case where A is only slightly soluble in the B-C system, which he assumed to be ideal, Wagner (17) obtained the related solution for the saturation of A in the BC system

$$(X_{A(BC)})^{1/Z} = X_B \ (X_{A(B)})^{1/Z} + X_C \ (X_{A(C)})^{1/Z} . \tag{27}$$

An example would be the solubility of graphite in a binary transition metal system.

In discussing the applicability of their equation, Alcock and Richardson (16) noted the systematic way in which the behavior of real systems deviated from the predictions of the equation. They hypothesized that the A-B and A-C interaction energies may vary with the proportions of B and C atoms near these pairs and that the population of the nearest neighbors to the pairs may be different from that of the average for the solution. The assumption of strong pairwise interactions of A with B and C and the reduction of the number of pairwise interactions to be less than the coordination number has led a number of authors to postulate the presence of AB and AC molecules in the solution. Jacob and Jeffes (18) developed a model based on A_2B and A_2C molecules and Jacob and Alcock (5) have proposed a more general treatment. Wagner (19) protested the assumption of molecular entities in metallic solutions and its use in models of metallic solution behavior when there is as yet no evidence for the presence of such entities. Nevertheless, this practice has become quite common both in modeling of the type being discussed here and in what might be called empirical modeling in which the computer is used extensively, which will be discussed later. It is reasonable to suppose that Wagner was concerned with the use of modeling to enhance our understanding of the nature of solutions, whereas many authors are seeking mathematical means for describing the behavior of solutions empirically.

Wagner (19) modified the use of the quasi-chemical theory for modeling the behavior of a non-metallic element such as oxygen (A) dissolved in a binary system of transition elements (B and C); Examples of the latter are Ag, Cu, Fe, Ni etc. From the behavior of oxygen in the crystalline metals, he assumed that the non-metallic atoms in the liquid state are present as quasi-interstitials. Thus in liquid structures, the interstitial is accommodated in octahedral sites for which Z equals 6 of the solvent metal atoms as nearest neighbors. (Presumably Z is from 10 to 12 for the solvent metals among their kind). He then assumed that the solvation energy for A is a non-linear function of the number of B and C atoms in A's solvation shell. Accordingly he incorporated into his model parameters that account for the change in electron transfer between the atom pairs A–B and A–C as X_B changes from 1 to zero. The symbols employed are $A(B_{Z-i}C_i)$ for the A atom that is surrounded by Z-i B atoms and i C atoms. A vacancy with the same set of nearest neighbors is $V(B_{Z-i}C_i)$. It was assumed also that there is a binomial distribution of complexes which consist of an atom of A surrounded by populations of B and C. A simple use of the model was the solution of a diatomic gas A_2 to occupy a vacancy (19):

$$1/2A_2(g) + V(B_{Z-i}\,C_i) = A(B_{Z-i}C_i)\ . \tag{28}$$

The enthalpy change for this reaction is ΔH_i. The result of the treatment gives the solubility of A, X_A, in the B-C solution to be

$$X_{A(BC)} = K\sqrt{P_{A_2}}\sum_{i=0}^{Z}\binom{Z}{i}(1-X_C)^{Z-1}X_C^{i}\exp(-\Delta H_i/RT)\ . \tag{29}$$

K is the solubility constant.

The reaction for the transfer of an A atom from a quasi-interstitial site with a set of $B_{Z-i}C_i$ nearest neighbors to a vacant site with the set of $B_{Z-i-1}\,C_{i+1}$ nearest neighbors is

$$A(B_{Z-i}C_i) + V(B_{Z-i-1}C_{i+1}) = V(B_{Z-i}C_i) + A\,(B_{Z-i-1}\,C_{i+1}). \tag{30}$$

The enthalpy change for the reaction as written is designated $\Delta H(i\rightarrow i+1)$. It was then assumed that the difference in the change in enthalpy for reactions where i = 0, i = 1,···etc. is constant having the value of h. Wagner's equation for the activity coefficient of A at infinite dilution in the B-C system for the condition of h being a constant is

$$\frac{\gamma_{A(BC)}}{[\gamma_{A(B)}\cdot\gamma_{A(C)}]^{1/2}} = \{\sum_{i=0}^{i=Z}\binom{Z}{i}\left[\frac{(1-X_C)}{\phi^{1/2Z}}\right]^{Z-i}\cdot[X_C\phi^{1/2Z}]^{i}x$$

$$\exp\frac{(Z-i)ih}{2RT}\}^{-1}\ . \tag{31}$$

The term ϕ is equal to $\gamma_{A(B)}/\gamma_{A(C)}$. For the condition of h = 0, i.e. where ΔH_i is linear with i,

$$1/(\gamma_{A(BC)})^{-Z} = (1-X_C)/(\gamma_{A(B)})^{1/Z} + X_C/(\gamma_{A(C)})^{1/Z}. \tag{32}$$

In applying Wagner's relationship (equation 31), Chiang and Chang (20) found that it would fit the behavior of oxygen in the O-B-C systems well by the assumption that h is a linear function of i. That is

$$(\Delta H_{i+2} - \Delta H_{i+1}) - (\Delta H_{i+1} - \Delta H_i) = h_1 + h_2 i \tag{33}$$

The resulting equation is

$$\frac{\gamma_{A(BC)}}{(\gamma_{A(B)}\cdot\gamma_{A(C)})^{1/2}} = \left\{\sum_{i=0}^{Z}\binom{Z}{i}\left[\frac{(1-X_C)}{\phi^{1/2Z}}\right]^{Z-i}\right.$$

$$\left.x[X_C\phi^{1/2Z}]^{i}\exp\left[\frac{(Z-i)i}{2RT}(h_1-h_2)+\frac{(Z^2-i^2)i}{6RT}h_2\right]\right\}^{-1}. \tag{34}$$

A comparison of the applicability of Wagner's relationship (equation (31) with the modification of it by Chiang and Chang (equation (34) for oxygen in liquid copper-cobalt alloys is shown in Fig. 4.

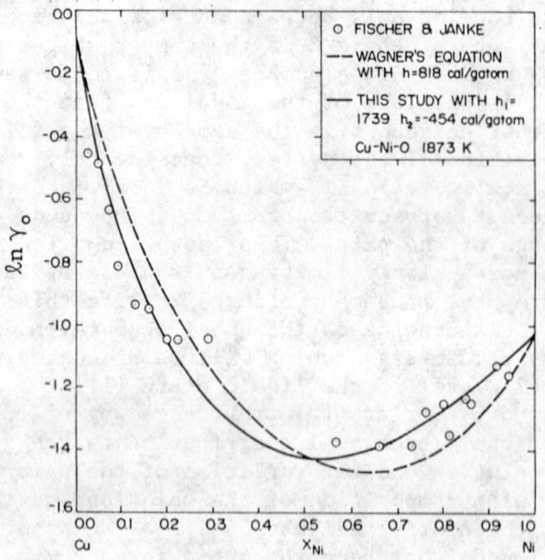

Fig. 4 Activity of Oxygen in Liquid Copper-Cobalt Alloys at 1600 C. Data of Fischer and Janke (21). Curve - - - by equation 31, curve —— by equation 34. From Chiang and Chang (20).

It is interesting to note that these developments of a model which involves the quasi-chemical treatment of a complex of an interstitial atom A surrounded by B and C atoms are similar in many respects to the "Central Atoms Model" of solution behavior of Lupis and co-workers (22). This model will be discussed later.

Blander and co-workers (23) have applied the coordination cluster theory to the thermodynamic behavior of a dilute component in a liquid binary metallic solution. The treatment is statistical mechanical in nature and applies both to interstitial and substitutional alloys. The theory includes the energetic effects accompanying the solution of solute A in the binary B-C solvent. There are three types which are due to (1) interactions of the solute atom with the solvent atoms, (2) interactions among the solvent atoms that are neighbors of the solute atom, and (3) changes in the interactions between these nearest neighbor solvent atoms and those in the bulk solution. Two regions are assumed to exist in the solution; one region labelled b which consists of all atoms which are not neighbors of A atoms (other C atoms are included) and the other region labelled c consists of all B and C atoms which are nearest neighbors of A atoms. Again, each A atom has Z-i B atoms and i C atoms as nearest neighbors. The exchange reactions considered are

$$A(B_{Z-i+1}C_{i-1}) + C(b) = A(B_{Z-i}C_i) + B(b) \tag{35}$$

where $0 < i < Z$. The reaction involves the exchange of a solvent atom B from the nearest neighbor (solvation) shell of the solute atom with a C atom from the "b" region of the solution.

In their model, Blander and co-workers calculate the dependence of the activity coefficient of the solute element A on the composition of the B-C solvent from knowledge of the activity coefficients of the two solvent components at the composition of interest, the activity coefficients of the solute at infinite dilution in each of the solvent components, a coordination number, a geometric factor that depends on the type of solution, and a term that represents the nonadditivity of pair-bond interactions within the complex of the solute atom and its neighboring solvent atoms. The formulations obtained are too elaborate to be discussed here, but it is of interest that the equations developed for the dependence of the activity coefficient of the solute element A on the solvent composition with successively greater use of simplifying assumptions as to solution behavior will be reduced to those of Chiang and Chang, Wagner, and Alcock and Richardson, equations 34, 31, and 26, respectively, and also to that by Jacob and Alcock (5).

As can be seen, in recent years there has been a rapid development of increasingly complex models to describe the behavior of a solute element in a liquid binary solution. In all cases, the behavior of the solute element at infinite dilution in each of the solvent components must be known. It is also necessary that a choice be made as to the coordination number for the solute-solvent complex. Additional terms must be established by the fit of the calculated results with experimental measurements on a given system to obtain a parameter, or parameters, that relate to the enthalpy change (or free energy change (20)) for reactions involving atoms in the coordination cluster (nearest neighbor shell) of atoms around the solute atom. It would appear that we are as yet a very long way from predicting the behavior of the solute in this type of system from first principles, or even from outside thermodynamic properties of the components of the system of interest.

Multi-Component Dilute Solutions

The behavior of solute components at dilute concentrations in liquid metals is of particular interest in process pyrometallurgy. Reactions among these components and with constituents in other phases are involved in refining, degasing, etc. Thus the development of mathematical formalisms and models to describe the behavior of solute species has been of considerable interest in the field for many years.

Figure 5 shows the effects of alloying elements on the activity coefficient of sulfur in liquid iron. The results are obtained from the Fe-S and Fe-S-j systems, where j is the dilute alloying element. This figure is an example of the kind of information that is now available from many sources on the thermodynamic properties of solute constituents in a number of liquid metals (25, 26, 27).

Wagner (17) first proposed the Taylor series formalism for describing the functional relationship between the excess partial molar free energy, or logarithm of the activity coefficient and composition of a dilute constituent, i, in an alloy system consisting of 1, ···, j, ···, m components, where 1 is the solvent element, and 2 through m are solute species. He wrote down the formal statement of the series and included the zero and first order coefficients in terms of mole fraction as follows:

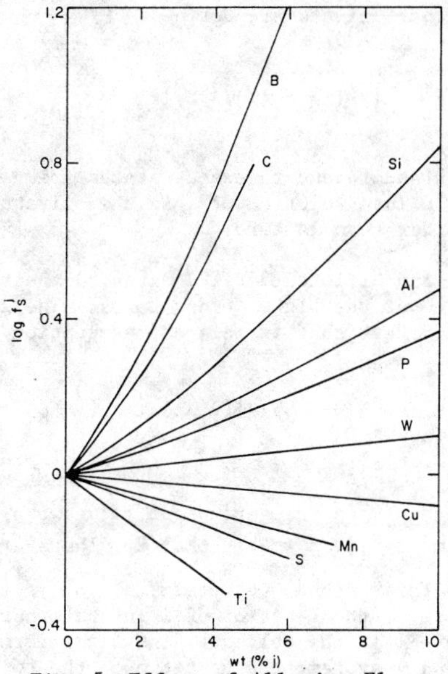

Fig. 5 Effect of Alloying Elements on the Activity Coefficient of Sulfur in Liquid Iron at 1600 C. (See ref. 25 for sources.)

$$\ell n\gamma_i = \ell n\gamma_i^{\circ} + \sum_{j=2}^{m} \varepsilon_i^j \cdot X_j \ . \qquad (36)$$

The expansion is about the pure component, i.e. in the limit $X_1 \to 1$, so that the first term on the right refers to the activity coefficient of species i at infinite dilution in 1, and the epsilons are also evaluated at the same point:

$$\varepsilon_i^j \equiv (\partial \ \ell n \ \gamma_i / \partial \ X_j)_{X_1} \to 1 \cdot \qquad (37)$$

The self-interaction coefficient, ε_i^i, pertains to the behavior of i in the 1-i binary system whereas the cross-interaction coefficient, ε_i^j, relates to the behavior of i in the 1-i-j system at infinite dilution in pure 1. If the reference and standard states for i are based on the infinitely dilute solution of i in 1, equation (36) becomes

$$\ell n \ f_i(X) = \sum_{i=2}^{m} \varepsilon_i^j . X_j . \qquad (38)$$

The zero order term, $\ell n \ f_i(X=0)$, is zero. In technical work, it is usually the practice to use weight percent as the composition coordinate so that the Taylor expansion for $\log f_i(\%)$ becomes

$$\log f_i(\%) = \sum_{j=1}^{m} e_i^j \cdot \%_j , \qquad (39)$$

where the coefficients are defined

$$e_i^j \equiv [\partial \log f_i(\%)/\partial \%_j]_{\%1 \to 100} \cdot \qquad (40)$$

Again the standard and reference states are based on the infinitely dilute solution of i in the solvent 1. Also, the zero order term is zero.

Lupis and Elliott (28, 29) extended the treatment of Wagner to cover the higher order terms. The general second order term that is related to equation (36) is

$$\rho_i^{j,k} = \frac{1}{n_j \cdot n_k} \{\partial^2 \ln \gamma_i/[(\partial X_j)^{n_j} \cdot (\partial X_k)^{n_k}]\} \qquad (41)$$

where n_j and n_k are integers whose values run from 0 to 2 and the sum of $n_j + n_k$ must total 2. There are four classes of this second order term, ρ_i^i, ρ_i^j, $\rho_i^{i,j}$, $\rho_i^{j,k}$, which relate to the behavior of i in quaternary 1-i-j-k system where 1 is the solvent. The first term applies to the 1-i binary system, the next two to the two ternary systems (1-i-j and 1-i-k) and the fourth to the quaternary system. The equation for $\ln\gamma_i$ with all the first and second order terms is then

$$\ln\gamma_i = \ln\gamma_i^\circ + \varepsilon_i^i X_i + \rho_i^i \cdot X_i^2 + \sum_{j=2}^{m}[\varepsilon_i^j \cdot X_j + \rho_i^j \cdot X_j^2 + \rho_i^{i,j} \cdot X_i \cdot X_j] + \sum_{j=2}^{m}\sum_{k=2}^{m} \rho_i^{j,k} \cdot X_j \cdot X_k \ ; \ (i \neq j \neq k). \qquad (42)$$

There are also second order terms for weight percent as the composition coordinate. The general definition is

$$r_i^{j,k} \equiv \frac{1}{n_j \cdot n_k} \{\partial^2 \log f_i(\%)/[(\partial\% j)^{n_j} \cdot (\partial\% k)^{n_k}]\} \ , \qquad (43)$$

and the types are r_i^i, r_i^j, $r_i^{i,j}$ and $r_i^{j,k}$. The terms in the series expansion for $\log f_i(\%)$ are the same as those shown in equation (42) except that again the zero order term is omitted.

With relatively dilute solutions, the behavior of a solute element usually can be expressed satisfactorily by the use of the first order terms ε_i^i, ε_i^j, e_i^i and e_i^j, only. It is to be seen however that there may be considerable curvature of the plot of the logarithm of the activity coefficient vs. composition. This is evident for the curves for carbon, boron and silicon in Fig. 5. Thus second order coefficients ρ_i^i and ρ_i^j in particular are often required. Although the interaction coefficients are defined at infinite dilution, it is necessary to evaluate the second order terms from data for higher concentrations. Also as a practical matter, the first order term ε_i^i (or e_i^i) is determined first, followed by ρ_i^i, ε_i^j and so on. As a consequence, uncertainties in the values of these coefficients tend to increase with the second order terms. If the uncertainty in the data for $\ln \gamma_i$ or $\log f_i$ is large, evaluation of the second order terms may not be warranted, particularly $\rho_i^{i,j}$ and $\rho_i^{j,k}$.

There are few if any systems where use of third order terms is warranted. Many first order terms have been determined for elements dissolved in the common metals as solvents (25, 26, 27, 29, 30, 31), as well as a few of the second order terms. The Taylor expansion has also been applied to the excess partial molar enthalpy and enthalpies of solute elements in the same fashion that has been employed for the free energy (i.e. $\ln \gamma_i$ and $\log f_i$) (22). From the relationship between the Gibbs' free energy, enthalpy and entropy, the equation may be written

$$RT\ \varepsilon_i^j = \eta_i^j - T\sigma_i^j \qquad (44)$$

where η_i^j and σ_i^j are the first order enthalpy and excess entropy interaction coefficients, respectively. The definitions are equivalent to that for ε_i^j (see equation 37). Very few of these coefficients have been evaluated. Because of the uncertainties in experimentally determined values of the partial molar enthalpies and excess partial molar entropies of solute elements, no second order interaction coefficients have been evaluated. An interesting observation of Chipman and Corrigan (32) was that there is a linear relationship between ε_i^j and η_i^j for several non-metallic solutes such as oxygen, nitrogen or sulfur in solutions of iron (See Fig. 6).

Jacob and Jeffes (18) observed the same relationship for oxygen dissolved in liquid copper alloys. The equation of the line is different for different solutes (i) and solvents. No adequate explanation of this linear relationship is yet available.

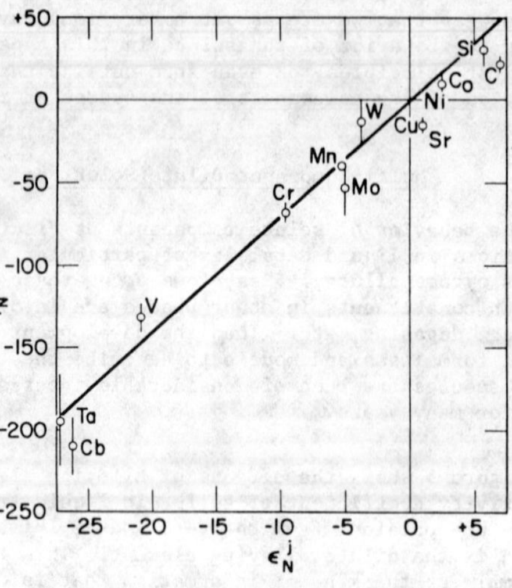

Fig. 6 Comparison of Enthalpy and Free Energy Interaction Coefficients of Nitrogen in Liquid Iron at 1600 C (32).

The formalism of the Taylor series has been applied to the Gibbs free energy function in a number of ways, such as for a solution saturated with the solute element i (33). An example is the effect of solute alloying elements on the concentration of carbon in liquid iron at saturation with graphite. There are a number of others (34), most of which have remained thermodynamic curiosities. Also, it has been noted that the reference solution for the Taylor series expansion need not be the infinitely dilute solution of the solute in the solvent (35). Again, there has been little or no use of these treatments as yet.

Chipman (6) has utilized the Taylor series formalism for describing the relationship between the activity coefficient of an "interstitial type" solute expressed as a_i/Z_i as a function of the lattice ratio, Z_j. The advantage of this procedure is that log a_i/Z_i is a linear function of Z_i and the series only includes the first order terms. The value of this use of composition in terms of Z rather than X or Y is shown in Fig. 7 where the data of Schwerdtfeger and Engell (36) are plotted. It is to be seen that the use of Z gives a linear relationship for a larger composition range than do the other two composition coordinates. The avoidance of the higher order terms inherent when X is used (equation 42) is a significant improvement. However, there is the difficulty that most of the data in the literature has been tabulated in the ε-ρ and e-r formalism. It is also somewhat confusing to mix the X and Z coefficients in dealing with an alloy system. However, relationships for the transformation of data expressed in terms of one composition coordinate to that expressed in another have been developed (7).

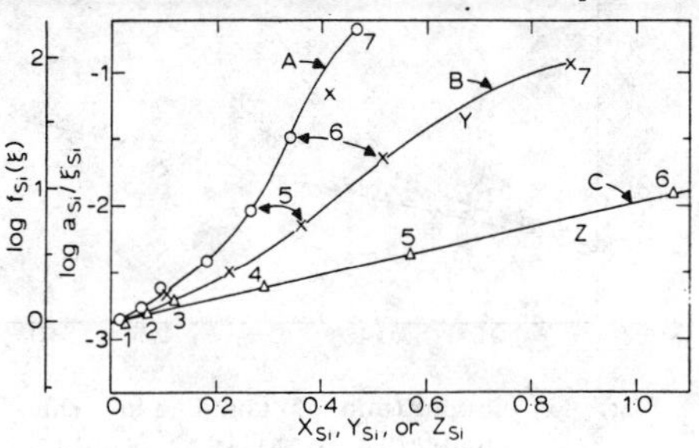

Fig. 7 Correlation of Log (a_{Si}/ξ_{Si}) For the Co-Si System at 1600 C (7). Data from Schwerdtfeger and Engell (36). Numbers show equivalent data points.

The Quadratic Formalism. Drawing on earlier composition models of Marqules (equations 22a and b) and others, Darken (37) proposed the use of a quadratic formalism to express the dependence of the activity coefficient on composition. The form for solute species B in solvent A is

$$\log (\gamma_B/\gamma_B^\circ) = \alpha_{AB} (X_A^2 - 1) = \alpha_{AB}[1 - (1 - X_B)^2] \quad (45)$$

where α_{AB} is a coefficient for the system and the range of composition is $0 < X_B < X_B'$. From equation 45, the behavior of species A in this range that is obtained from the Gibbs-Duhem equation is

$$\log \gamma_A = \alpha_{AB} (1 - X_A)^2. \quad (46)$$

On the other side of the binary system for the composition range of $X_B'' < X_B < 1$, there are the symmetrical relationships

$$\log (\gamma_A/\gamma_A^\circ) = \alpha_{BA} (X_B^2 - 1) \quad (47)$$

and

$$\log \gamma_B = \alpha_{BA} (1 - X_B)^2 . \quad (48)$$

Turkdogan and Darken (24, 37, 38, 39) have applied this formalism to a number of systems with reasonably good success. An example is the iron-silicon system shown in Fig. 8 for which $\alpha_{FeSi} = -3.10$ at 1600 C. They also found that $\alpha_{SiFe} = -3.20$ for this system. Darken also defined the properties stability and excess stability; the latter is most conveniently defined as follows:

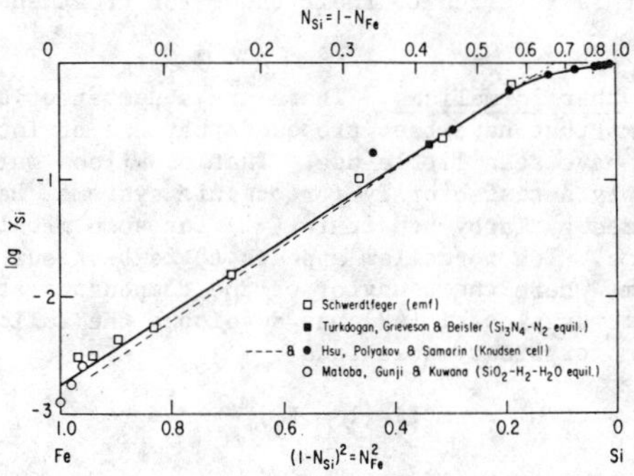

Fig. 8 Quadratic Plot of Activity Coefficient of Silicon in the Liquid Fe-Si System at 1600 C. From reference 37.

$$\text{excess stability} = - 4.605 \text{ RT d} \log \gamma_B/d [1-X_B)^2]. \quad (49)$$

Stability is the same except that the activity of B replaces the activity coefficient. Plots of excess stabilities for the iron-silicon and iron-nickel systems at 1600 C are shown in Fig. 9. It is to be seen that by this treatment, a binary system can be considered to have three regions, the two terminal regions $0 < X_B < X_B'$ and $X_B'' < X_B < 1$, where equations 45, 46, 47, and 48 apply and a central region where there is a sharp change in the excess stability of the system (Fig. 9).

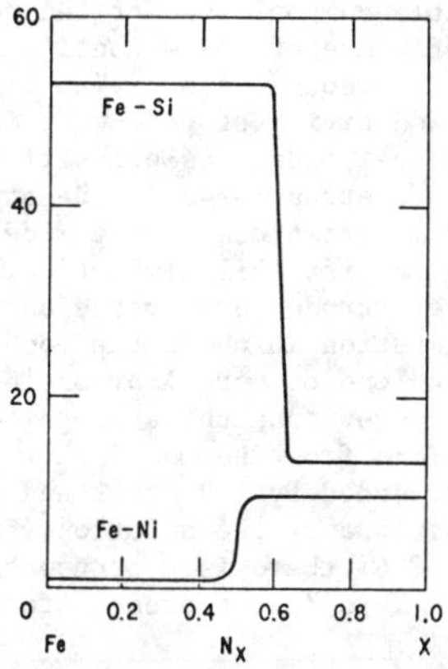

Fig. 9 Excess Stability of Fe-Si and Fe-Ni Alloys at 1600 C (37). N is Mole Fraction. X is Silicon or Nickel.

The quadratic formalism may be extended to ternary and higher order systems. The formulation for the behavior of B in the ABC system with A the solvent is (37)

$$\log (\gamma_B/\gamma_B^\circ) = - 2\alpha_{AB} X_B + (\alpha_{BC} - \alpha_{AB} - \alpha_{AC})X_C$$
$$+ \alpha_{AB}X_B^2 + \alpha_{AC}X_C^2 + (\alpha_{AB} + \alpha_{AC} - \alpha_{BC})X_B X_C . \quad (50)$$

This relationship is very similar in form to equation 42, but it is not satisfactory to equate corresponding coefficients in the two equations. This is because the coefficients in the Taylor series formalism are all evaluated at infinite dilution of solute components in the solvent. Turkdogan has tabulated some values of α for several iron-X systems (40). Turkdogan and Darken (38) have extended the use of the quadratic formalism to the relative partial molar enthalpy and entropy. The reader is referred to their paper for treatment of the topic.

Other Formalisms. There are a number of other formalisms that have been proposed that are of interest, but which have seen little use. That of Wilson which appears to apply satisfactorily for organic systems, has been utilized by Tarby and Stein (41) for some metallic systems. The formalism appears to be best suited to systems where the behavior of the components are nearly ideal. Krupkowski (42) has developed the following formalism for binary solutions

$$\ln \gamma_1 = \omega(T)\ (1 - X_1)^m \tag{51}$$

and by the Gibbs-Duhem equation

$$\ln \gamma_2 = \omega(T)\ [(1 - X_1)^m - \frac{m}{m-1}(1 - X_1)^{m-1} + \frac{1}{m-1}]. \tag{52}$$

The values of ω and m are evaluated from a plot of $\ln(\ln\gamma_1)$ versus $\ln(1-X_1)$. The coefficient ω is assumed to be dependent on temperature and m may be. Recently Moser and Smith (43,44) have applied this formalism to a number of binary systems. The formulations are most satisfactory for systems showing positive departures from ideality where the value of m ranges from 1 to 2. A number of other empirical correlations based on power series have been developed for describing thermodynamic properties of components of solutions. These correlations have been used in computer modeling of metallic solutions and will be discussed in that context later.

The "central atoms" theory (22) and the related "surrounded atom" treatment (45) of solution behavior move beyond the additivity of pairwise bonding of the simple quasi-chemical model (46). These newer treatments describe the partition function in terms of probabilities of various possible compositions of the nearest neighbor shell and the influence of these configurations on the field acting on the central atom. The linear case of these models of additivity of pair-wise interactions, as in the quasi-chemical model, is not satisfactory for solutions showing a strong negative departure from ideal behavior. Several variations of the model have been proposed (22,45) which lift this limitation. Lupis and coworkers (47) have introduced into the analysis a preferred configuration in which the bonding between the central B atom and one or more A atoms in the surrounding shell is much stronger than the average for all A-B bonds. Calculations from the model show the probability of an A atom surrounded by i B atoms and Z-i A atoms at different compositions of the solution (Fig. 10). At the composition $X_B = 0.67$ there is a pronounced maximum at $i \rightarrow 8$ (Z = 12). This is not present for the linear case.

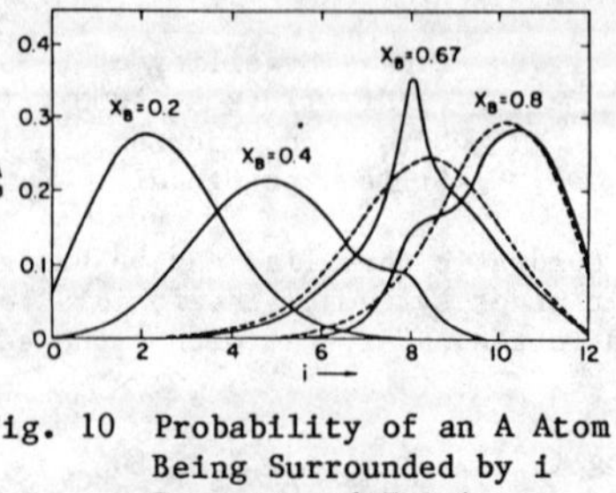

Fig. 10 Probability of an A Atom Being Surrounded by i B atoms and Z-i A Atoms at Various Concentrations X_B For Central Atoms Model (47). - - - Lines for quasi-chemical and linear model, —— Lines for preferred configuration AB_2.

Use of Formalisms. It seems clear that no one formalism that has been proposed is suitable for use with all types of liquid metallic systems. All of the formulations become more or less cumbersome when treating ternary and higher-order systems. There are of course proponents of each method, but it is possible to determine which formalism is most suitable for use with a given system by statistical tests of how well each formalism represents the original data. Unfortunately few such tests have been made. It is likely that in many cases the uncertainties inherent in the data may preclude making a clear choice among the various formalisms. One basis for choice is the width of the composition range over which each formalism will represent the experimental data by a straight line. Figure 7 provides a test of the use of X, Y and Z as composition coordinates. It is to be seen that a plot of log a_{Si}/Z_{Si} versus Z_{Si} is superior to the others for this system. A similar comparison is made in Fig. 11 between the quadratic formalism and the relationship log a_{Si}/Z_{Si} versus Z_{Si}. Both are quite

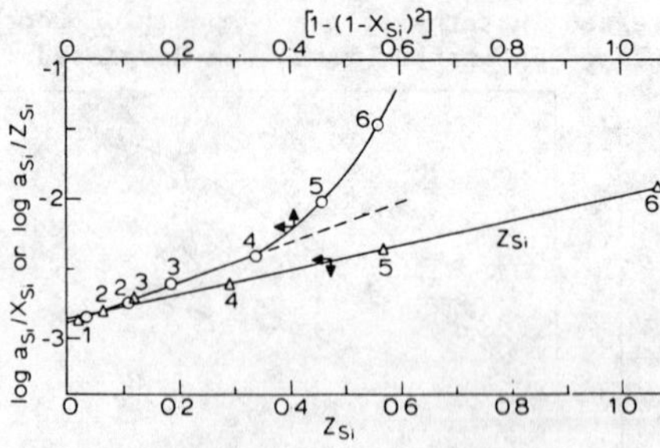

Fig. 11 Comparison of the Use of the Quadratic Formalism (Equations 45 and 46) in X With Log a_{Si}/Z_{Si} for the Co-Si System at 1600 C (7). Data from Schwerdtfeger and Engell (36).

satisfactory at low concentrations, but the latter provides a longer linear range for the data. It would appear that there is some advantage to the use of this formalism over the quadratic one for a system that shows a very large departure from ideality (log a_i/X_i vs X_i). This is verified by transforming the correlations from Fig. 7 back to the plot of log a_{Si}/X_{Si} versus X_{Si} for the silicon-cobalt system. The results are shown in Fig. 12.

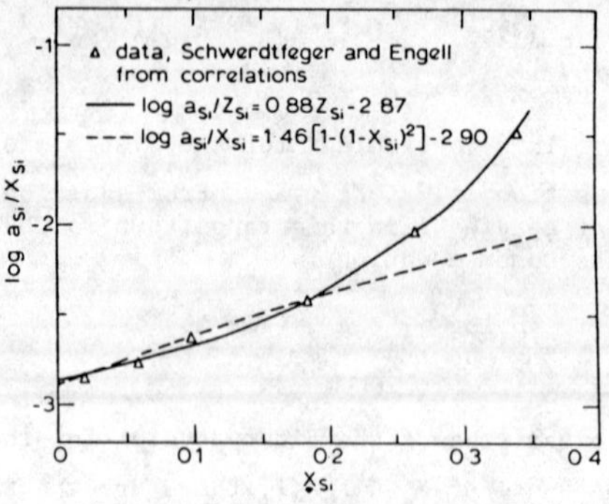

Fig. 12 Transformation of Linear Plots From Fig. 7 to the Plot of Log a_{Si}/X_{Si} Versus X_{Si} (7).

Of course, the final choice of which formalism is for general use depends on a number of issues that are not thermodynamic in character; such matters as which method is more generally used and which is the least complex for use by people who are not immersed in the thermodynamics of solutions. At present, the choice seems to be the Taylor series in X and weight percent. Unfortunately, this means that the tendency for the practical world will be to use only first-order terms, a practice which produces satisfactory results only for low concentrations of solute elements. For the researcher in the field, the most satisfactory procedure is perhaps to find statistically the best linear correlation for a set of data and then transform the data by suitable means (7) to the Taylor coefficients in X and weight percent for general use.

Many authors have attempted to predict values of the Taylor interaction coefficients from parameters in their models. Except for solutions which show only modest departures fron ideality, these predictions have not been especially successful when compared with experimentally derived values. It has been possible to show that there is a regularity by position in the periodic table of the element j for values of the Taylor series first order interaction coefficients, ε_i^j, and others, for a given pair of solvent and solute element i. An example is Fig. 13. The interaction parameter ω_C^j is defined

$$\lim_{X_j \to 1} (\partial \ln X_C / \partial X_j)_{a_C=1} = \omega_C^j . \qquad (53)$$

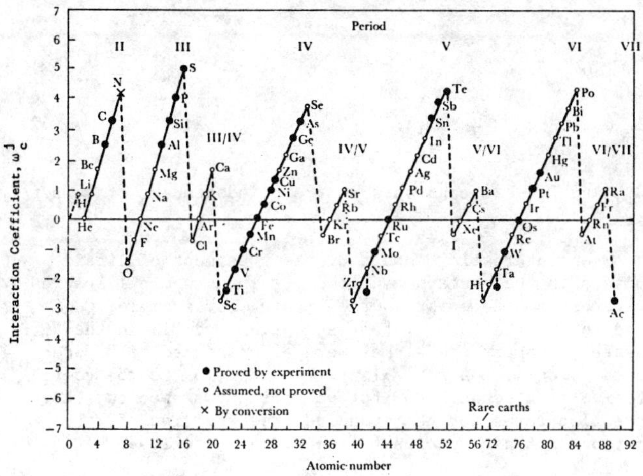

Fig. 13 Relationship Between the Values of Taylor Series Interaction Coefficients for Carbon in Iron at Graphite Saturation and Position of j Elements in the Periodic Table of Elements (48). Interaction coefficient is defined in equation 53.

Another example is the correlation of ε_H^j for iron versus the number of free electrons in the j element as shown in Fig. 14.

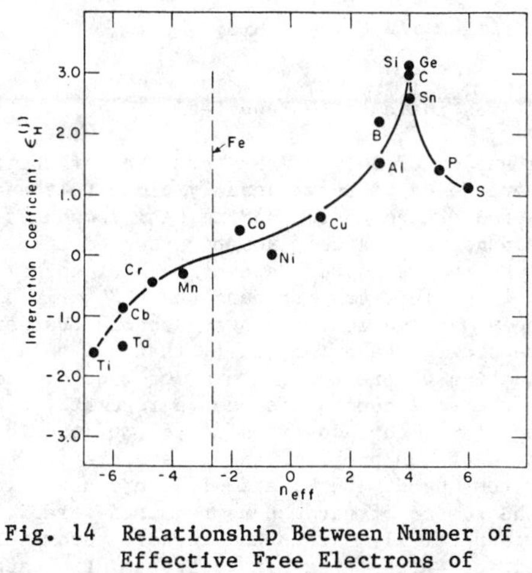

Fig. 14 Relationship Between Number of Effective Free Electrons of Element j and Value of Interaction Coefficient, ε_H^j, in Liquid Iron (14).

Most experimentalists have studied the thermodynamic properties of components of solutions at constant pressure. However, most theoretical models are derived at constant volume. Wagner (49) has pointed out that this difference may be of considerable importance. Lupis (50) has continued treatment of the matter. He concludes that there may be a marked difference in composition dependence of the chemical potential and fugacity at constant pressure as compared to conditions of constant volume. He shows that the difference of activity and activity coefficient for the two conditions may be slight to negligible. The issues involved are subtle and may need to be considered carefully for some investigations.

Computers In Modeling

High capacity computers and sophisticated software are being utilized extensively in describing the thermodynamic properties of many types of inorganic solutions and compounds and in calculating phase diagrams of multicomponent systems of all types. The methods employed have been reviewed recently by Ansara (51). Analysts have made use of all of the various models that have been discussed earlier for describing the thermodynamic properties of components in the liquid phase (as well as others) and of the phase itself. Also, many empirical power series have been employed. Issues as to ease of programming and simplicity in obtaining a solution of compositions of the phases at equilibrium are often deciding issues as to the formalisms to be used. Two to five term equations are employed to represent the excess thermodynamic properties of phases and components of those phases with high precision relative to the uncertainties in the data on which the correlations are based. Several advantages accrue from the thermodynamic point of view to the use of such equations. First, the number of terms can be increased almost without limit to obtain a precise representation as long as the number of coefficients does not exceed the number of data points that are available. Second, a properly constructed set of equations avoids the need for integrating the Gibbs-Duhem equation in the analysis. This is because the equations for the components satisfy that relationship as was pointed out for the Margules equations (equations 22a through f). (There is confusion on this point in the literature as some authors say that the use of these equations makes the Gibbs-Duhem relationship superfluous; this, of course, is not the case.)

The relationship for the regular solution

$$G^{EX} = \sum_{i=1}^{m-1} \sum_{j=i+1}^{m} \Omega_{ij} X_i X_j \qquad (54)$$

has often been employed, particularly by Kaufman and coworkers (52). The parameters, Ω_{ij}, are often expressed as a linear function of temperature

$$\Omega_{ij} = C_{ij} + D_{ij} T. \qquad (55)$$

Also, higher order terms of the form $\Omega_{ijk} \cdot X_i X_j X_k$, etc., are often included in equation 54 for a "quasi-regular" solution model. These added terms may also be a linear function of temperature. In our work on the phase relationships in the quinary Fe-Cr-Ni-Mn-C system (53), we have found that the equations for the excess molar free energy of the liquid phase of the Fe-Mn-C, Fe-Cr-C, and Fe-Ni-C systems require third order terms, but they do not for the Fe-Cr-Mn and Fe-Mn-Ni systems. However, the fourth order coefficient, Ω_{ijkl}, is required for the liquid phase of the Fe-Cr-Ni-C system of this quinary system, but not for the other quaternary system.

Hillert and co-workers (54,55,56) have employed a "subregular" model to describe the thermodynamic properties in the liquid field of the Fe-S, Mn-S, and Fe-Mn-S systems. In the treatment, it is assumed that sulfur is an interstitial species and that interstitial vacancies constitute a component of a system in addition to the metals and sulfur. The composition coordinates are the lattice ratios for the interstitial species for which

$$Y_j \equiv X_j/(X_{Fe} + X_{Mn}) \quad . \tag{56}$$

j is either sulfur or the vacancy. Apparently an additional assumption is that there is one vacancy per "lattice" atom (iron and manganese). Thus by simple stoichiometry

$$X_V = (X_{Fe} + X_{Mn}) - X_S \quad . \tag{57}$$

The composition of a lattice component, i, is

$$Y_i = X_i/(X_{Fe} + X_{Mn}) \quad . \tag{58}$$

The excess molar free energy of the liquid phase may be written

$$G^{EX} = Y_1 \cdot Y_2 \sum_{i=1}^{2} \sum_{j=3}^{4} Y_i Y_j L_i^j + Y_3 \cdot Y_4 \sum_{j=3}^{4} \sum_{i=1}^{2} Y_i Y_j L_j^i \quad . \tag{59}$$

Species 1, 2 and i are the component metals, and 3, 4 and j are the interstitials. The L-parameters were evaluated empirically from available experimental data. As with other models, each parameter may be expressed as a linear function of temperature (equation 55). Six of the 8 parameters were evaluated from the constituent binary systems, Fe-Mn, Fe-S, Mn-S; and the other two, L_{Fe}^{S} and L_{Mn}^{S}, from the FeS-MnS system. Because the assumption of ideality for this system seemed reasonable, these two parameters were set to zero. Only one of the 8 parameters, L_S^{Fe}, was found to be a function of temperature. Considering the uncertainties in the available data on the behavior of the system, this model and the parameters obtained provide a reasonably good description of the behavior of the liquid phase of the Fe-Mn-S system.

The variety of polynomial expressions that has been employed in the empirical description of the thermodynamic properties of phases and components of phases is large (51,57-60). The coefficients in these expressions are usually determined by statistically fitting the equation to the data so as to minimize the overall imprecision of the line of the equation relative to the experimental data. Kellogg and co-workers (60, 61) have employed the "three-suffix" Margules equation in their treatment of several systems: Cu-S, Ni-S, Ni-Cu-S, Fe-O, and Fe-O-SiO_2. The general relationship for the activity coefficient of species i in the system consisting of 1 to m species is

$$\ell n\gamma_i = 1/2 \sum_{i=1}^{m} (k_{ij} + k_{ji})X_j - 1/2 \sum_{j=1}^{m} \sum_{p=1}^{m} k_{jp}X_jX_p + \sum_{j=1}^{m} (k_{ij} - k_{ji})X_j(X_j/2 - X_i) + \sum_{j=1}^{m} \sum_{p=1}^{m} (k_{jp} - k_{pj})X_j^2X_p \ . \ . \ . \ . \tag{60}$$

where

$$k_{ii} = k_{jj} = \cdot \cdot \cdot \cdot = 0 \tag{61}$$

and

$$k_{ij} = C_{ij}/T + D_{ij} \quad . \tag{62}$$

C_{ij} and D_{ij} are constants that are independent of temperature.

An interesting aspect of the treatment of liquid phases in the systems mentioned is that chemical species were assumed to exist in the liquid; for example, Cu_2S and CuS in the Cu-S system, and Ni_3S_2 and NiS in the Ni-S system. Copper and nickel were also assumed to be species as well, but not sulfur. In addition to the coefficients in equation 60 for each of these species, there are also terms for the standard free energies of formation of these species by reactions as follows:

$$Cu(\ell) = \underline{Cu} \tag{63}$$

$$2Cu(\ell) + 1/2\ S_2(g) = \underline{Cu_2S} \tag{64}$$

$$Cu(\ell) + 1/2\ S_2(g) = \underline{CuS} \tag{65}$$

$$Ni(\ell) = \underline{Ni} \tag{66}$$

$$3Ni(\ell) + S_2(g) = \underline{Ni_3S_2} \tag{67}$$

$$Ni(\ell) + 1/2\ S_2(g) = \underline{NiS} \quad . \tag{68}$$

The underlines indicate the species.

The values of ΔG° for the formation of the compound species were found to be reasonably close to those for the formation of the corresponding real compounds. On the other hand, they should not be expected to be the same because the standard state of each of these species is the pure hypothetical compound where $X_i = 1$, a state that is only hypothetical. It was assumed that each species is regular in behavior, i.e., that all values of D_{ij} in equation 62 are zero. This, of course, does not mean that the real components behave regularly because of the entropy terms for equations 64, 65, 67 and 68. In the solution of the problem, the parameters for the binary systems Cu-Ni, Cu-S, and Ni-S were determined first. The number of parameters required were 2, 8 and 10 each, respectively. Then the ternary parameters were determined, of which there are thirteen: two each for the pairs Cu-Ni_3S_2, Cu-NiS, Ni-Cu_2S, Cu_2S-Ni_3S_2, and Cu_2S-NiS; and one each for the pairs CuS-Ni, CuS-Ni_3S_2, and CuS-NiS. The number of parameters employed are shown in Fig. 15, the total being 33.

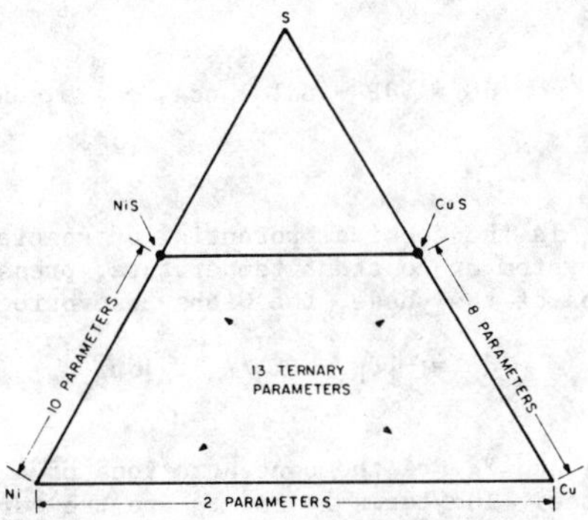

Fig. 15 Number of Parameters for Model of Behavior of the Liquid Phase in the Cu-Fe-S System. The total is 33. Kellogg (60).

By use of the model, the mole fractions of the hypothetical species may be determined for any real composition of the liquid. Figure 16 shows the results in the region of Cu_2S in the Cu-S system. It is to be noted

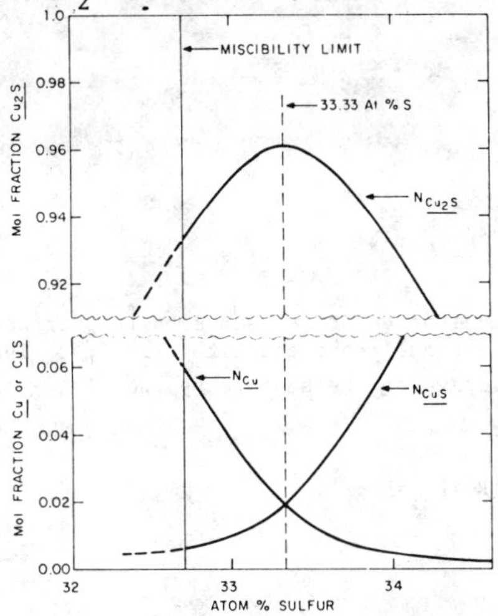

Fig. 16 Mole Fractions of Assumed Species in the Cu-S System in Vicinity of The Compound Cu_2S, 1200 C. Kellogg (60).

that the model gives values of the activity coefficients and concentrations of the hypothetical species. From these values, the activities of the hypothetical species can be calculated. Because of equations 63 and 66, these values for Cu and Ni also apply for the real constituents Cu(ℓ) and Ni(ℓ). The activity coefficients of these constituents must then be obtained from the real concentration of these elements in the solution. An example of how the equilibrium sulfur pressure is to be obtained is to combine equations 63 and 64 into the reaction

$$\underline{2Cu} + 1/2\ S_2(g) = \underline{Cu_2S} \tag{69}$$

and insert the calculated activity coefficients and mole fractions of the species Cu and Cu_2S. Other equations in the set of equations 63 through 68 could be combined in the same way.

Recently Goel, Kellogg and Larrain (61) have modeled the behavior of the liquid phases in the Fe-O-SiO_2 system. The basis for the model was equation 60, and 16 parameters were required; two additional ones should be included if the symmetrical pairs are counted.

MOLAR VOLUME

The molar volume of most pure liquid metals near their melting points have been determined, and Lucas (62) and Allen (63) have compiled the available information relatively recently. Values for a number of metals at the melting point of each taken from those compilations are shown in Table I. It also has been observed that the temperature coefficient of expansion for the liquid, $[1/V \cdot (\partial V/\partial T)_p]$, is nearly inversely proportional to the temperature of melting of the metal.

It appears that the molar volume of a liquid alloy can be estimated and modeled with reasonable accuracy from the molar volumes of the component species by assuming idea mixing:

$$V_{molar} \sim \sum X_j V_j \tag{70}$$

TABLE I Molar Volumes of Selected Metals at Their Melting Points (62, 63)

Metal	Melting Point, K	V, cm^3/g·atom	Coef of Exp., $K^{-1} \times 10^6$
Ag	1234	11.54	97
Al	933	11.33	114
Au	1336	11.35	70
Bi	544	20.75	125
Ca	1112	28.86	162
Cd	594	13.94	134
Co	1768	7.37	160
Cr	2130	8.11	---
Cu	1356	7.91	100
Fe	1811	7.96	119
Hf	2500	14.87	---
Hg	234	14.64	174
K	336	47.18	268
Mg	922	15.29	167
Mn	1517	9.41	127
Mo	2890	10.26	---
Na	371	24.81	237
Ni	1726	7.56	142
Pb	601	19.41	123
Pt	2042	10.32	153
Si	1685	11.12	145
Sn	505	16.96	88
Ti	1943	11.65	170
U	1405	13.30	58
W	2680	10.51	---
Zn	693	9.94	177
Zr	2125	16.58	---

SURFACE TENSION AND CHEMISORPTION

Processes involving the transport of matter, heat and momentum across surfaces and interfaces are of great importance in pyrometallurgy. Interest in surface and interfacial phenomena has always been high in the field. Unfortunately, experimental measurements of the surface and interfacial tensions of liquid metals are difficult so that progress in our understanding of the role of surfaces in the field has been slow.

The interface between two phases has a tension (or surface energy), σ, which is defined in terms of the Gibbs Free Energy, G, and area of the surface, A_s:

$$\sigma \equiv (\partial G/\partial A_s)_{T,P,n_i} \tag{71}$$

There is available information on the surface tensions of many metals (63,64), some of which are shown in Table II.

Table II Surface Tensions of Selected Pure Metals at Their Melting Points (63, 64)

Metal	σ J/m^2	$-d\sigma/dT$ $(J/m^2 \cdot K) \times 10^3$
Ag	0.903	0.16
Al	0.914	0.35
Au	0.140	0.52
Bi	0.378	---
Ca	0.838	0.10
Cd	0.570	0.26
Co	1.873	0.49
Cr	1.700	0.32
Cu	1.360	0.21
Fe	1.872	0.49
Hf	1.630	---
Hg	0.498	0.20
K	0.115	0.08
Mg	0.559	0.35
Mn	1.090	0.20
Mo	2.250	---
Na	0.191	0.10
Ni	1.778	0.38
Pb	0.468	0.13
Pt	1.800	---
Si	0.865	---
Sn	0.544	0.07
Ti	1.650	---
U	1.550	0.14
W	2.500	---
Zn	0.782	0.17
Zr	1.480	---

From empirical considerations, Allen (63) has proposed the equation for the surface tension of pure liquid metals:

$$\sigma = 0.0036\ T_m \cdot V^{-2/3} (J/m^2) \tag{72}$$

where T_m is the absolute melting temperature of the metal, and V is the molar volume in cc/gm mole. Lennard-Jones and Corner (65,66) have developed a theory of surface tension based on the Lennard-Jones 6-12 intermolecular energy function. Coupled with a rigid-sphere model for the structure of liquid metals, this theory leads to the same relationship for the surface tension of a liquid metal, but the coefficient is 10 percent lower. These relationships are reasonably good approximations for many metals, but they predict values for metals such as uranium, aluminum, tin, gallium, indium, mercury, etc., that are considerably lower than the measured ones.

An equation by Grosse (64), based on the bonds broken in transferring an atom from the surface into the vapor is

$$\sigma = 6.6 \times 10^{-5} \Delta H_v \cdot V^{-2/3} \tag{73}$$

where ΔH_v is the heat of vaporization of the metal in calories/gm mole. This equation predicts satisfactorily the surface tension near the melting point of liquid metals that solidify to form cubic or tetragonal structures.

The differential change in free energy of a multicomponent system may be written:

$$dG = VdP - SdT + \sigma dA_s + \sum_{j=1}^{m} \mu_j\, dn_j \tag{74}$$

where μ_j is the chemical potential of species j. For a binary system at constant temperature, pressure and surface area of the phase, the Gibbs adsorption isotherm is:

$$d\sigma = -[\Gamma_B - (X_B/X_A)\Gamma_A]d\mu_A \tag{75}$$

where X_A and X_B are the concentrations of A and B in the bulk phase. The terms Γ_A and Γ_B are the surface excess for each of the two species; that is the concentration of a species in the surface "layer" in excess of that in the bulk phase. If the position of the interface is located where Γ_A (the solvent) is zero, equation 75 becomes

$$d\sigma = -\Gamma_B d\mu_B\ . \tag{76}$$

From the definition of activity of species B

$$d\sigma = -\Gamma_B\ RT\ d\ \ell n\ a_B \tag{77}$$

and

$$d\sigma/d\ \ell n\ a_B = RT\ \Gamma_B\ . \tag{78}$$

The total number of active sites on the surface is Γ°_B and the number at full coverage with B is Γ_i. Thus the fractional coverage of the surface by the single solute B, Θ_B, is

$$\Gamma_B/\Gamma^\circ = \Theta_B \tag{79}$$

and

$$\Theta_B + \Theta_S = 1 \tag{80}$$

where Θ_S is the fraction of active sites not covered.

The reaction for ideal adsorption of species B from the bulk phase onto an unoccupied surface site, S, to form the adsorbed species, B(ads), is

$$\underline{B} + S = B(ads);\quad K = \frac{\Theta_B}{\Theta_S \cdot a_S} \tag{81}$$

and by manipulation

$$Ka_S = \Theta_B/(1 - \Theta_B)\ . \tag{82}$$

Combination of equations 78, 79, and 82 and integrating between σ°, $a_B = 0$ and σ and a_B gives an expression that relates the activity of species B in the bulk solution with the surface tension of the liquid;

$$\sigma^\circ - \sigma = RT\ \Gamma^\circ_B\ \ell n\ (1 + Ka_B)\ . \tag{83}$$

Belton (67) has shown that the surface tension of liquid Fe-C-S melts can be correlated well by equation 83, where a_S is the activity of sulfur in the melt (See Fig. 17).

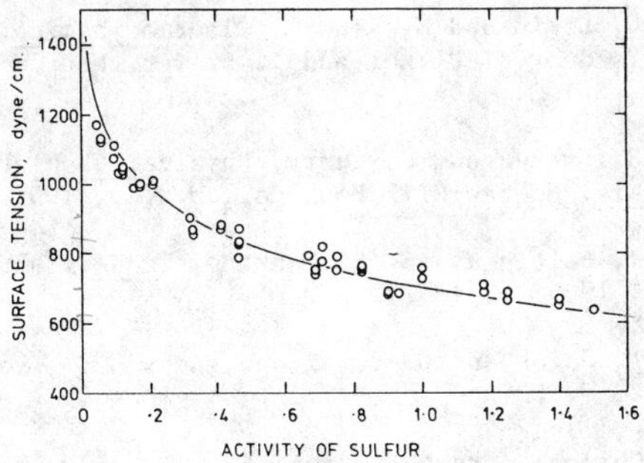

Fig. 17 Surface Tensions of Liquid Iron Alloys Containing Carbon and Sulfur at 1450 C. Belton (67). Data are from Kozakevitch, Chatel, Urbain and Sage (68).

Sulfur is surface active in this system whereas carbon is not. Belton also showed that the surface tensions of a number of liquid alloys may be expressed by equation 83: Cu-S, Ag-O, Fe-O, and Fe-S. Turkdogan (24) has summarized data on the effects of gas species and alloying elements on the surface tensions of liquid iron, copper, and aluminum. The data for iron are shown in Fig. 18.

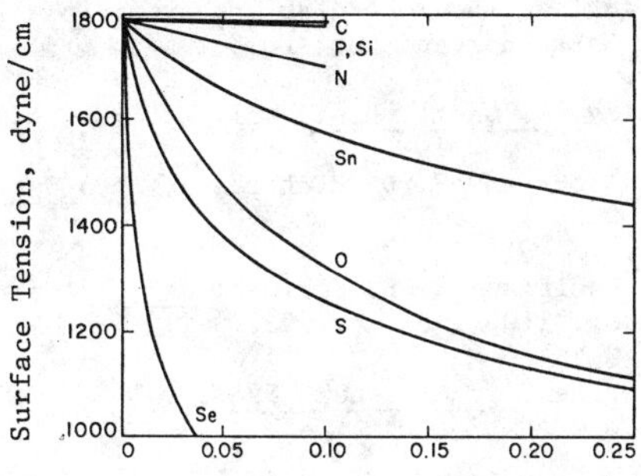

Fig. 18 Effects of Solute Species on the Surface Tension of Liquid Iron at 1550 C. Turkdogan (24).

When treating systems where there are two surface-active species, it is necessary to consider the competition for surface sites by the two, B and C:

$$-d\sigma = \Gamma_B d\mu_B + \Gamma_C d\mu_C \ . \qquad (84)$$

Walen and co-workers (69) have shown that equation 84 may be transformed into two separate equations of the form

$$\Gamma_i = (\alpha_i + \beta_{ij}\alpha_j)\ (1 - \beta_{ij}\beta_{ji}) \qquad (85a)$$

where

$$\alpha_i = -(\partial\sigma/\partial\mu_i)_{X_A}/X_j \qquad (85b)$$

and

$$\beta_{ij} = (\partial X_i/\partial X_j)_{\mu_i, X_A} \ . \qquad (85c)$$

Here i and j are B and C, or C and B, and A is the solvent. It is difficult to obtain measurements of good accuracy of surface tensions in multicomponent systems and over a broad enough range of compositions so equation 85a has not been utilized very much. Belton (70) has derived the following relationship for the A-B-C system:

$$-[(\partial\ \sigma/\partial\ \ln X_B)_{X_C}]/RT = \Gamma_B(1 + \varepsilon_B^B X_B) + \Gamma_C\ \varepsilon_C^B\ X_C, \qquad (86)$$

where ε_B^B and ε_C^B are Taylor series interaction coefficients for the system. He was successful in treating the surface tensions of Fe-Cr-C alloys from Walen and co-workers (69) with available data on the interaction coefficients.

EXPERIMENTAL METHODS

Methods for measuring the thermodynamic properties of liquid metal solutions and their components have been treated in considerable detail in the volumes edited by Bunshaw and Rapp (71). Accordingly, they will not be discussed here. There are a great many data available in the literature on the free energy properties of components of solutions; however, there are several serious deficiencies in these data. The first is that often they are not of high enough accuracy, nor do they cover a wide enough range of composition. So it is not possible to test the applicability and reliability of the various models that are being employed for describing the various properties that are related to the Gibbs free energy of these components. The second is that we lack measurements of enthalpy and entropy properties of many solutions for which we have free energy data. Often it would be very useful to have such information on systems because of the need in models for parameters that relate to bond energies and mixing characteristics of the solutions. All too often we adopt simple models or we simplify models to forms based on regular solution behavior because of the limitations in our knowledge of the properties of solutions.

A third deficiency is that very little data are available, nor do we have facts that shed light on the nature of the bonding and electronic states of components in liquid alloys. In the past three decades, much has been learned about the behavior of solid compounds and solid solutions from methods of solid-state physics and chemistry. Those methods have not been particularly successful when applied to liquid solutions. There are fundamental reasons why this is so. On the other hand, it would appear that there are some properties of solutions that we need to know to be able to advance the use of models for predicting the properties of solutions and their components. Information on the behavior of vacancies, electron holes, valence electrons, etc., should be of great help in this quest. We need to improve our basis for determining the nature of the bonding between nonmetallic components, such as oxygen and sulfur, and the metallic components in liquid alloys, rather than estimating the nature of the bond from the nature of the model of the assumed interstitial behavior which gives the best fit to the experimental data. To an important degree, Wagner (19) is right in his protest against the use of assumed "molecular entities" in solutions when there is no fundamental justification for such species being present. It would appear that such empiricism tends to deflect many investigators from searching for experimental methods which will reveal the fundamental behavior of liquid solutions that illustrate why such assumptions provide us with empirical models that are reasonably successful. There is a need for generic models of solution behavior which permit us to predict the behavior of components of liquid solutions from first principles.

One has to be realistic about the accuracy of currently available experimental methods that are being employed. All measurements of free energies, activities, activity coefficients, etc., are relative to a standard state of some sort. If the experimentalist is clever, skillful, or just lucky, he or she may be able to make measurements relative to gaseous species where the standard free energies of formation are known with uncertainties of from 800 J/mole to as low as 400 J/mole. On the other hand, if the standard state involves the liquid state or some solid state, the uncertainties may be 5 to 30 times greater than those for the gases. It is these facts that result in a serious barrier to our being able to improve the quality of data that are used to test the applicability of theoretical models. It is also these facts that make it essential that new and different experimental methods be applied to obtain different kinds of information on the behavior of liquid metallic solutions and their components.

DISCUSSION

Modeling of the behavior of metallic solutions currently is receiving a great deal of attention. Much of this attention has been directed to the liquid phase of binary and higher-order systems. An important trend in these efforts is the incorporation in some manner of the assumption of interstitial and compound complexes in models. Perhaps the simplest technique has been to modify the composition coordinate by use of the atom ratio or lattice ratio to reflect the assumed interstitial behavior or the possibility of compound formation in the liquid (6). There are a number of ways that compound-like complexes have been incorporated into models. One is the more advanced modifications of the quasi-chemical approximation of solution behavior that includes some form of complexes in the structure of the solution. Examples are a preferred arrangement of atoms around the central atom for the "central atom" model (47), and assumptions ranging from strong pair-wise interactions between metal ions and nonmetallic species in liquid alloys, to the reduction in the number of nearest neighbors around an atom and to the presence of simple AB and A_2B complexes by Alcock and Richardson (16), Jacob and Jeffes (18), and Jacob and Alcock (5). Wagner (19) has proposed a modification which accounts for a difference in solvation energy depending on the population of the solvation shell of an atom. Blander and co-workers (23) have approached this same issue from a very different point of view.

On the other side, Kellogg and co-workers (60, 61) have arbitrarily assumed that certain compound species exist in a solution, and have utilized a subregular solution model that includes these species to describe the behavior of the solution. The species are chosen to provide the best fit of the results of the model with available data on the system.

It would appear that co-valent as well as metallic bonding needs to be considered in modeling the behavior of liquid alloy systems. Thus experimental information on the states of the bonding electrons is needed before it will be possible to move on from what might be called adaptive modeling to predictive modeling so that it will be possible to predict the thermodynamic properties of a liquid alloy.

With this perspective, it is somewhat anomalous that the only "fundamental" relationship that has been developed thus far for the Taylor series interaction coefficients, ε_i^j, has been to rank them by the atomic number of the interacting species j, or the effective free electrons of j (See Fig. 13 and 14).

Clearly, there is much yet to be done by both the experimentalists and the theoreticians before we have a complete understanding of the thermodynamic behavior of liquid alloy systems.

List of References

1. G. N. Lewis, J. Am. Chem. Soc., vol. 35, 1913, p. 17.

2. G. N. Lewis and M. Randall, Thermodynamics, 2nd ed., Revised by K. Pitzer and L. Brewer, McGraw-Hill Book Co., N.Y.C., 1961.

3. L. S. Darken and R. Gurry, Physical Chemistry of Metals, McGraw-Hill Book Co., N.Y.C., 1953.

4. G. R. Belton and E. S. Tankins, Trans., AIME, vol. 233, 1965, p. 1892.

5. K. T. Jacob and C. B. Alcock, Acta Met., vol. 20, 1972, p. 221.

6. J. Chipman, Trans., TMS-AIME, vol. 239, 1967, p. 1332.

7. J. F. Elliott and J. Chipman, p. 121 in Proc., of Int. Symposium on Met. Chem., Sheffield, Iron and Steel Inst., London, 1971.

8. K-C Chou and J. F. Elliott, Proc., Carl Wagner Symposium, Annual AIME Meeting, Chicago, Ill., 1981.

9. S. B. Margules, Sitzber. Akad Wiss. Wien, vol. 104, 1895, p. 1243.

10. J. H. Hildebrand, J. Am. Chem. Soc., vol. 51, 1929, p. 66; Proc., Nat. Acad. Sci., vol. 13, 1927, p. 267.

11. I. Prigogine, The Molecular Theory of Solutions, p. 115ff, North Holland Pub. Co., 1957.

12. H. Eyring, J. Chem. Phys., vol. 4, 1936, p. 283.

13. C. B. Alcock and F. D. Richardson, Acta Met., vol. 6, 1958, p. 385.

14. M. Weinstein and J. F. Elliott, Jnl., Electrochem. Soc., vol. 110, 1963, p. 792.

15. L. S. Darken, J. Am. Chem. Soc., vol. 72, 1950, p. 2909.

16. C. B. Alcock and F. D. Richardson, Acta Met., vol. 8, 1960, p. 882.

17. C. Wagner, Thermodynamics of Alloys, Addison-Wesley Press, No. Reading, Mass., 1952.

18. K. T. Jacob and J. H. E. Jeffes, Trans. Inst. Min. Metall., Sec. C, vol. C80, 1971, p. 32; p. 181.

19. C. Wagner, Acta Met., vol. 21, 1973, p. 1297.

20. T. Chiang and Y. A. Chiang, Met. Trans., vol. 7B, 1976, p. 453.

21. W. A. Fischer and P. Janke, Arch. Eisenhuttenw., vol. 39, 1968, p. 89.

22. C. H. P. Lupis and J. F. Elliott, Acta Met., vol. 15, 1967, p. 265.

23. M. Blander, M-L. Saboungi, P. Cerisier, Met. Trans., vol. 10B, p. 613.

24. E. T. Turkdogan, Phy. Chem. of High Temperature Technology, Academic Press, New York, 1980.

25. G. K. Sigworth and J. F. Elliott, Met. Sci., vol. 8, 1974, p. 298.

26. G. K. Sigworth and J. F. Elliott, Can. Met. Q., vol. 13, 1974, p. 455; vol. 15, 1976, p. 123; (and G. H. Geiger) vol. 16, 1977, p. 104.

27. E. T. Jacob and J. H. E. Jeffes, Trans. Inst. Min. and Met., Sec. C. vol. C80, 1971, p. 181.

28. C. H. P. Lupis and J. F. Elliott, Acta Met., vol. 14, 1966, p. 529.

29. C. H. P. Lupis and J. F. Elliott, Ibid, p. 1019.

30. J. M. Dealy and R. M. Pehlke, Trans. Met. Soc., AIME, vol. 227, 1963, p. 88.

31. C. Bodsworth, Physical Chemistry of Iron and Steel Manufacture, Longmans, London, 1963.

32. J. Chipman and D. A. Corrigan, Applications of Fundamental Thermodynamics to Metallurgical Processes, G. R. Fitterer, ed., p. 23, Gordon and Breach, N.Y.C., 1971.

33. C. H. P. Lupis, Acta Met., vol. 16, 1968, p. 1365.

34. H. Schenck and E. Steinmetz, Supplemental Edition of Stahleisen- Sonderberichte, vol. 7, 1968, p. 1.

35. T. Mori and A. Moro-oka, Trans., J.I.S.I., vol. 6, 1966, p. 9.

36. K. Schwerdtfeger and H-J. Engell, Arch. Eisenhutten, vol. 35, 1964, p. 533.

37. L. S. Darken, Trans. Met. Soc., AIME, vol. 239, 1967, p. 80.

38. L. S. Darken, Ibid, p. 90.

39. E. T. Turkdogan and L. S. Darken, Ibid, vol. 242, 1968, p. 1997.

40. E. T. Turkdogan, BOF Steelmaking, Volume II, Chapter 4, J. M. Gaines, Ed., Iron and Steel Soc., AIME, 1975.

41. S. K. Tarby and F. P. Stein, Met. Trans., vol. 1, 1970, p. 2354.

42. A. Krupkowski, Bull. Int. Acad. Pol. Sci. Lett. Cl., Sci., Math. Nat., Ser. A1, 1950, p. 15.

43. Z. Moser and J. F. Smith, Met. Trans., vol. 6B, 1975, p. 457.

44. Z. Moser, Ibid, p. 657.

45. P. Hicter, J. C. Mathieu, F. Durand and E. Bonnier, Adv. in Phy., vol. 16, 1967, p. 523.

46. E. A. Guggenheim, Mixtures, Clarendon Press, Oxford, 1952.

47. C. H. P. Lupis, H. Gaye and G. Bernard, Scripta Met., vol. 4, 1970, p. 497.

48. H. Schenck and M. G. Frohberg, Steelmaking, The Chipman Conf., p. 95, M.I.T. Press, 1965.

49. C. Wagner, Acta Met., vol. 19, 1971, p. 843.

50. C. H. P. Lupis, Ibid, vol. 25, 1977, p. 751.

51. I. Ansara, International Metals Reviews, No. 1, 1979, Review No. 238, p. 20.

52. L. Kaufman and H. Bernstein, Computer Calculation of Phase Diagrams, Academic Press, N.Y.C., 1970.

53. D. Kundrat, Sc.D. Thesis, M.I.T., 1980.

54. M. Hillert and L-I. Steffanson, Met. Trans., vol. 7B, 1976, p. 203.

55. M. Hillert and L-I. Steffanson, Ibid, vol. 6B, 1975, p. 37.

56. L-I Steffanson, Ibid, vol. 7B, 1976, p. 131.

57. J. Thomiska, p. 41 in CALPHAD III, Proceedings of Conference, Stockholm, May 21 - 25, 1979, Royal Inst. Tech., Stockholm, B. Uhrenius and M. Hillert, ed.

58. T. Chart, F. Putland, and A. Dinsdale, Ibid, p. 183.

59. A. D. Pelton and C. W. Bale, CALPHAD, vol. 1, no. 3, 1977, p. 253.

60. H. H. Kellogg, Physical Chemistry in Metallurgy, Proceedings of The Darken Conf., U.S. Steel Corp., Pittsburgh, PA, 1976, p. 49.

61. R. P. Goel, H. H. Kellogg, and J. Larrain, Met. Trans., vol. 11B, 1980, p. 107.

62. L. D. Lucas, Chapter 7B in Physicochemical Measurements in Metals Research, R. A. Rapp, ed., vol. IV, part 2, from Techniques of Metals Research, R. Bunshaw, ed., John Wiley, N.Y.C., 1970.

63. B. C. Allen, Liquid Metals - Chemistry and Physics, S. Z. Beer, ed., M. Dekker, N.Y.C., 1972.

64. A. V. Grosse, Jnl. Inorg. Chem., vol. 26, 1964, p. 1349.

65. J. Lennard-Jones and J. Corner, Trans. Faraday Soc., vol. 36, 1940, p. 1156.

66. J. Corner, Ibid, vol. 44, 1948, p. 1036.

67. G. R. Belton, Met. Trans., vol. 7B, 1976, p. 35.

68. P. Kozakevitch, S. Chatel, G. Urbain, and M. Sage, Rev. Met., vol. 52, 1955, p. 139.

69. T. J. Walen, S. M. Kaufman, and M. Humenik, Trans. ASM vol. 55, 1962, p. 778.

70. G. R. Belton, Met. Trans., vol. 3, 1972, p. 1465.

71. Physicochemical Measurements in Metals Research, vol. IV, parts 1 and 2, R. A. Rapp, ed. of Series Techniques of Metals Research, R. F. Bunshaw, ed., John Wiley, N.Y.C., 1970.

THE THERMODYNAMIC PROPERTIES AND STRUCTURES OF SLAGS

D. R. Gaskell
University of Pennsylvania
Philadelphia, Pennsylvania 19174

INTRODUCTION

The properties of metallurgical slags are determined by the nature and forms of the ionic entities occurring and by the interactions among these entities. Traditionally, the component oxides of slags are classified in two groups; acidic or "network-forming" oxides and basic of "network-modifying" oxides. Oxides which do not fit obviously into either group are termed amphoteric. The fundamental property which determines the relative acidity or basicity of a metal oxide is the difference between the electronegativities of the metal and oxygen. Large differences cause significant transfer of negative charge from the metal to the oxygen and result in metal-oxygen bonds which are primarily ionic in character, whereas small differences result in primarily covalent bonds which are usually directional in nature. A difference in electronegativity of about 2.1 produces a bond which is 50% ionic and 50% covalent in character (1).

The acidic oxides of metallurgical interest are those of elements which are neither particularly electropositive nor particularly electronegative, and this gives rise to a peculiar complexity in their chemistry. Silica, which is the most important of the acidic oxides, has several polymorphic forms, in all but one of which the silicons are tetrahedrally-coordinated by oxygens and each oxygen is joined to two silicons. The various crystal structures can thus be regarded as infinite three-dimensional networks of SiO_4 tetrahedra joined at their corners. The oxygens are not collinear with their neighboring silicons, which, being impossible in a truly ionic structure, indicates that the Si-O bonds have a significant covalent character. The degree of covalency calculated from the difference between the electronegativities of Si and O is 63% (1). Other acidic oxides have similar network structures; for example, at normal pressures GeO_2 has the silica structure, P_2O_5 has a crystal form in which PO_4 tetrahedra share three of their four corners to form a coherent three-dimensional network, and B_2O_3 glass is a network of BO_3 triangles joined at their corners. The degrees of covalency in the P-O, Ge-O and B-O bonds, calculated from the electronegativities of the elements are, respectively, 60%, 63% and 68% which, taken at face value, indicates that the order of increasing acidity is P_2O_5, $SiO_2 = GeO_2$, B_2O_3. A major characteristic of acidic oxides is the ease with which they can occur in a vitreous state.

Basic oxides are those of the more electropositive elements and the greater extent of charge transfer from the metal to the oxygen results in simple ionic crystal structures in which positively-charged cations are symmetrically-coordinated by O^{2-} anions and vice versa. The actual crystal structure of any basic oxide is that which, for the particular cation size, maximizes the number of M-O contacts and hence minimizes the coulombic energy of the structure. The bonds in the basic oxides MgO, CaO and Na_2O have calculated ionic characters of 55%, 62%, and 65%, respectively. Amphoteric oxides can be classified generally as those which cannot exist in the vitreous state and yet do not have the simple crystal structures of the distinctly-basic oxides. Examples of metallurgical importance are TiO_2, Fe_2O_3 and Al_2O_3 which, respectively, have their own peculiar rutile, hematite and corundum crystal structures.

The majority of metallurgically-important slags form as the result of mutual solution of acidic and basic oxides and an understanding of the properties of slag melts requires knowledge of both the structural rearrangements that occur as a result of the solution process and the nature of the interactions which occur among the structural entities produced. Mutual solution of a basic and an acidic oxide causes depolymerization of the latter, which results in a breakdown of its network structure. In the case of silicate formation, the depolymerization reaction can be expressed as

$$-\overset{|}{\underset{|}{Si}}-O-\overset{|}{\underset{|}{Si}}- \;+\; M^{2+}O^{2-} \;\rightarrow\; -\overset{|}{\underset{|}{Si}}-O^{-}\;M^{2+}\;{}^{-}O-\overset{|}{\underset{|}{Si}}-$$

or, in shorthand notation

$$O^{o} + O^{2-} \rightarrow 2O^{-} \qquad |1|$$

where O^o denotes oxygen bonded to two silicons (the so-called bridging oxygen shared by two SiO_4 tetrahedra), O^- denotes oxygen bonded to a single silicon (so-called non-bridging oxygen formed by the rupture of a common corner shared by two SiO_4 tetrahedra) and O^{2-} denotes free oxygen ions. The extent to which the depolymerization reaction occurs depends primarily on the difference between the acidities (or basicities) of the component oxides. As a general rule, the greater this difference, which is referred to as the strength of the acid-base pair, the greater the extent of depolymerization, the greater the decrease in free energy due to the solution process and the greater the mutual decrease in the activities of the component oxides.

THE THERMODYNAMIC PROPERTIES AND STRUCTURES OF BINARY OXIDE MELTS

The molar free energies of formation of various binary acid-base melts, from the component liquid oxides, expressed as $\Delta G^M/RT$, are shown in Fig. 1 (2-12). In comparing the alkali and alkaline earth silicates, the magnitude of the free energy of solution increases in the order MgO, CaO, Na_2O, K_2O, which is the order of increasing basicity of the basic oxide. The curve drawn for the system K_2O-SiO_2 is for formation of the melts from solid SiO_2 and solid K_2O. Among the transition metal silicates, the magnitudes of the free energies of solution increase in the order FeO-SiO_2, CoO-SiO_2, MnO-SiO_2, which is in agreement with the relative basicities of the three transition metal oxides, and the free energies of formation of melts in the system PbO-SiO_2 are greater than in SnO-SiO_2 melts, which suggests that PbO is more basic than SnO. Comparison among the free energies of formation of melts in the systems PbO-Bi_2O_3, PbO-SiO_2, PbO-B_2O_3 and PbO-GeO_2 suggest that the acidity of the acidic oxide increases in the order Bi_2O_3,SiO_2,B_2O_3,GeO_2. However, as Si, Ge, Pb and Sn have the same value of electronegativity, this comparison illustrates the danger of placing too much emphasis on the calculated electronegativities as indicators of relative basicity. The activities of the component oxides, relative to pure liquid, in the

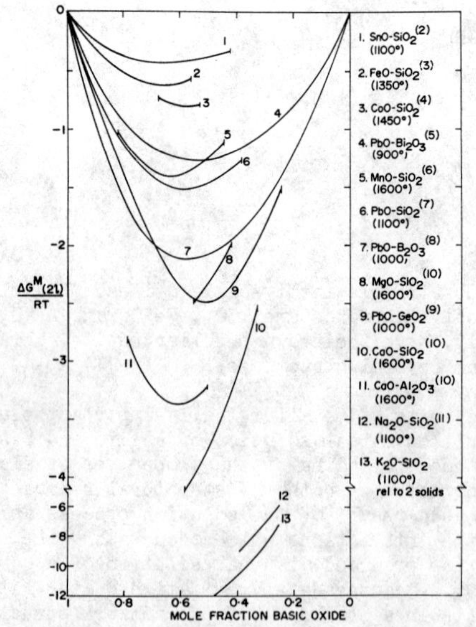

Figure 1. The molar free energies of binary oxide melts relative to the unmixed liquid oxides.

systems $PbO-B_2O_3$, $PbO-P_2O_5$, $PbO-GeO_2$, $PbO-SiO_2$ and $PbO-Bi_2O_3$ at 1000°C are shown in Fig. 2. This behavior suggests that the acidity of the acid oxides decreases in the order given.

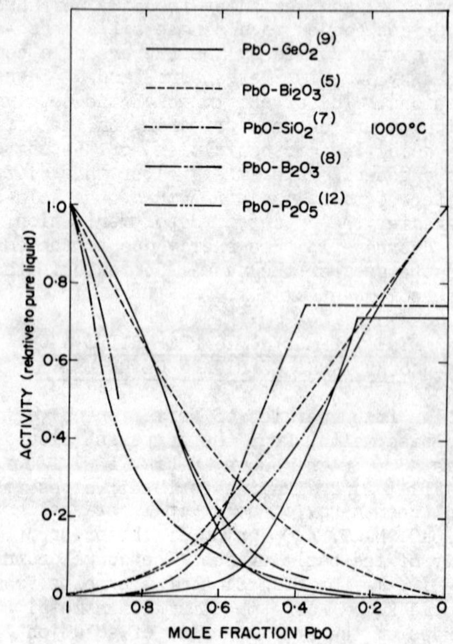

Figure 2. The activities of the component oxides in binary oxide melts containing PbO at 1000°C.

Considerable speculation has been made as to the nature of the structures formed on depolymerization of an acidic oxide by a basic oxide. The currently-accepted general theory was developed from interpretation of experimental measurements of structure-sensitive physical properties such as viscosity (13) and molar volume (14, 15). The variations of these properties with composition and temperature in alkali silicate melts indicate that, on adding M_2O to silica, when the mole fraction reaches 0.12 the melt changes abruptly from a randomly-damaged three-dimensional network to a system of discrete silicate anions in which the mean anion size decreases with increasing content of basic oxide. Simple stoichiometric and steric considerations suggest that SiO_4^{4-} is the predominant silicate anion in orthosilicate melts ($2MO.SiO_2$), the trimeric ring $Si_3O_9^{6-}$ or tetrameric ring $Si_4O_{12}^{8-}$ is the predominant ion in metasilicate melts ($MO.SiO_2$) and that pencil ions of formula $Si_{3x}O_{3(2x+1)}^{6-}$ or $Si_{4x}O_{4x(2x+1)}^{8-}$, formed, respectively, by face-to-face polymerization of x $Si_3O_9^{6-}$ or x $Si_4O_{12}^{8-}$ metasilicate ring ions, predominate at the compositions $MO.xSiO_2$. The orthosilicate composition, $2MO.SiO_2$, is thus of fundamental significance in that, in melts of $N_{SiO_2} < 1/3$, SiO_4^{4-} and O^{2-} could be the sole ionic species whereas, in melts of $N_{SiO_2} > 1/3$ silicate ions of greater complexity must occur. These more complex anions are formed by polymerization of SiO_4^{4-} ions and the actual forms and distribution of these complex ions occurring in the range $0<N_{SiO_2}<0.5$ are of particular metallurgical interest.

Thermodynamic models of binary silicate melts, which consider that the depolymerization reaction represented by eq. |1| occurs to the extent required to minimize the free energy of the system, occur in two groups; structure-based models which consider that depolymerization produces specific discrete silicate anions, and structureless models which consider mixing processes in quasi-lattices. A feature common to all the proposed thermodynamic models is the assumption that the configurational entropy arises from random mixing processes. The most successful of the structure-based theories are the polymerization models of Masson (16,17) and Whiteway, Smith and Masson (18,19). In the earlier model, Masson considered silicate melts to be systems of metal cations, free oxygen ions and linear chain silicate ions (16) and in the later model allowance was made for the formation of both linear and branching chain silicate ions (18). In both models the most probable distribution of silicate anion sizes is calculated from application of the principles of polymer theory. In a binary silicate melt of mole fraction of silica N, when N<1/3, complete depolymerization of the silica results in the elimination of O^o groups and the formation of N SiO_4^{4-} ions and (1-3N) O^{2-} ions. In the polymerization models the SiO_4^{4-} ion is considered to be the monomeric unit which can polymerize to form chain silicate anions according to

$$SiO_4^{4-} + SiO_4^{4-} = Si_2O_7^{6-} + O^{2-} \qquad |2|$$

$$SiO_4^{4-} + Si_2O_7^{6-} = Si_3O_{10}^{8-} + O^{2-} \qquad |3|$$

and $$SiO_4^{4-} + Si_xO_{3x+1}^{2(x+1)-} \quad Si_{x+1}O_{3x+3}^{2(x+2)-} + O^{2-} \qquad |4|$$

The principal parameter is the degree of polymerization, α, which is defined as the fraction of potentially-reactive O^- groups present in the completely-depolymerized state which has reacted according to the reverse of Eq. |1| to produce O^o groups in the $Si_xO_{3x+1}^{2(x+1)-}$ anions and additional O^{2-} ions. The major difference between the linear chain and branching chain models is that, in the former, it is considered that only two of the four O^- groups on any SiO_4^{4-} monomer may react according to the reverse of Eq. |1| whereas, in the latter case, all four of the O^- groups on a monomer may undergo the polymerization reaction. In the linear chain model the most probable distribution of silicate anion sizes is one in which the anion fractions of the silicate x-mers occur in a geometric array, i.e.

$$N_x = N_{Si_xO_{3x+1}^{2(x+1)-}} = \frac{N(1-\alpha)^2\alpha^{x-1}}{1-2N} \qquad |5|$$

and the anion fraction of O^{2-}, $N_{O^{2-}}$, is

$$N_{O^{2-}} = \frac{(1-3N + \alpha N)}{1-2N} \qquad |6|$$

The degree of polymerization, α, has that value which minimizes the free energy of the system. If it is considered that the silicate and free oxygen ions mix randomly, the equilibrium constant for reaction |4|, k, can be written in terms of the anion fractions of the species involved in the equilibrium as

$$k = \frac{N_{O^{2-}}.N_{x+1}}{N_x.N_1} = \frac{(1-3N + \alpha N)\alpha}{N(1-\alpha)^2} \qquad |7|$$

The independence of k on x in Eq. |7| indicates that the standard free energy change for reaction |4| is independent of x, and, furthermore, as reaction |4| is identical with the reverse of reaction |1|, the standard free energy change for the latter, ΔG^o_1, is also independent of x. Eq. |1| can be written as

$$\tfrac{1}{2}SiO_{2(1)} + MO_{(1)} = \tfrac{1}{2}M_2SiO_{4(1,\text{completely depolymerized})} \qquad |8|$$

As the completely-depolymerized liquid orthosilicate is a non-physical state, ΔG^o_1 is approximated by the standard free energy change for the corresponding all-solid reaction. Thus

$$k = \exp(\Delta G^o_1/RT) \qquad |9|$$

and, from Eq. |7|, the value of α which minimizes the free energy of the melt is obtained from

$$\alpha^2 N(1-k) + \alpha(1-3N + 2kN) - kN = 0 \qquad |10|$$

As random mixing of the anions is assumed, the activity of MO in a hypothetical linear chain melt, relative to pure liquid, is

$$a_{MO} = N_{O^{2-}} \qquad |11|$$

where, from Eqs. |6| and |10|, $N_{O^{2-}}$ is obtained from

$$N^2_{O^{2-}}(1-k)(1-2N) + N_{O^{2-}}[3N-1 + 2k(1-2N)]-k(1-2N) = 0 \qquad |12|$$

The activity of SiO_2 in the hypothetical melts, relative to liquid, is obtained by writing eq. |1| as

$$SiO_{2(1)} + 2O^{2-} = SiO_4^{4-} \qquad |13|$$

for which

$$\Delta G^o_{13} = 2\Delta G^o_1$$

Thus

$$K_{13} = 1/k^2 = \frac{N_{SiO_4^{4-}}}{a_{SiO_2}.N^2_{O^{2-}}}$$

and substitution from Eqs. |5|, |6| and |7| gives

$$a_{SiO_2} = \frac{\alpha^2(1-2N)}{N(1-\alpha)^2} \qquad |14|$$

The integral molar free energy of mixing liquid MO and liquid SiO_2 to form hypothetical linear chain silicate melts, given as

$$\Delta G^M = RT[N\ln a_{SiO_2} + (1-N)\ln a_{MO}]$$

is thus

$$\Delta G^M/RT = N\ln\frac{\alpha^2(1-2N)}{N(1-\alpha)^2} + (1-N)\ln\frac{(1-3N + \alpha N)}{(1-2N)} \qquad |15|$$

Eq. |15| is the sum of a chemical contribution to the free energy change due to the breaking of double oxygen bonds, $\Delta G_{chem}/RT$, given by

$$\Delta G_{chem}/RT = \frac{n_{O^-}}{2}.\frac{\Delta G^o_1}{RT} = N(2-\alpha)\Delta G^o_1/RT \qquad |16|$$

and a configurational contribution arising from random mixing of the anions, $\Delta G_{conf}/RT$, given by

$$\Delta G_{conf}/RT = n_{O^{2-}}\ln N_{O^{2-}} + \sum_1^\infty n_x \ln N_x$$

$$= (1-3N+\alpha N)\ln\frac{(1-3N+\alpha N)}{1-2N}$$

$$+ N(1-\alpha)\ln\frac{N(1-\alpha)^2}{1-2N} + N\alpha\ln\alpha \qquad |17|$$

The variations of a_{MO} and a_{SiO_2} with k and N, given by Eqs. |11| and |14| are shown in Fig. 3 and the corresponding molar free energies of mixing, given by Eq. |15| are shown as the full lines in Fig. 4.

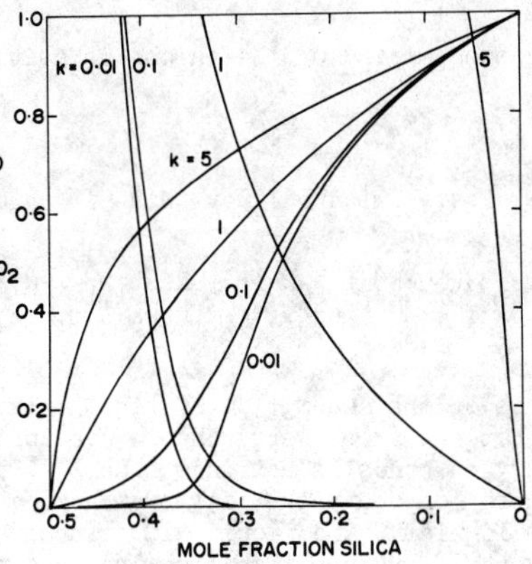

Figure 3. The activities of MO (Eq. 11) and SiO_2 (Eq.14) in Masson's hypothetical linear chain silicate melts as functions of k and composition.

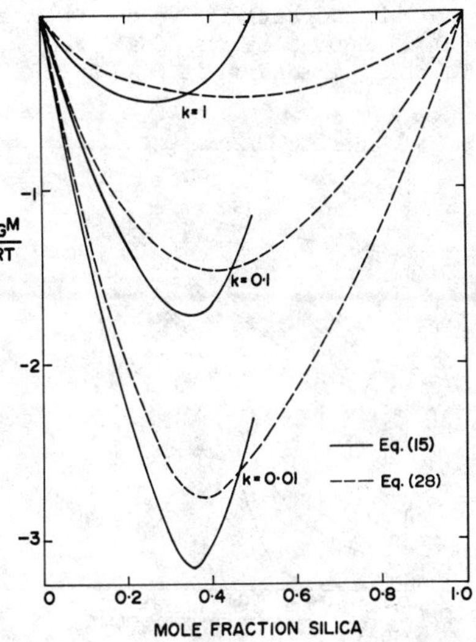

Figure 4. The molar free energies of formation of Masson's hypothetical linear chain silicate melts, given by Eq. (15), and Yokokawa and Niwa's hypothetical quasi-lattice melts, given by Eq. (28) as functions of composition and k (= exp $(\Delta G^o_1/RT)$).

In the branching chain model, in which all of the O^- groups on any SiO_4^{4-} monomer can react, the most probable size distribution of silicate anions is that in which

$$N_x = \frac{N(1-2\alpha)}{1-2N}\cdot\frac{(3x)!}{(2x+1)!x!}\left(\frac{2}{3}\alpha\right)^{x-1}\left(1-\frac{2}{3}\alpha\right)^{2x+1} \qquad |18|$$

and

$$N_{O^{2-}} = \frac{(1-3N+2\alpha N)}{1-2N} \qquad |19|$$

In this case the ion fractions of the silicate x-mers do not occur in a geometric array and hence, if random mixing of the anions is assumed, the equilibrium constant for reaction |4| is not independent of x, and consequently, ΔG_1^o is not independent of x. In this case α is obtained from consideration of the equilibrium constant for reaction |2|, $k_{1,1}$, as

$$k_{1,1} = \frac{N_{O^{2-}}\cdot N_2}{N_1^2} = \frac{6\alpha(1-3N+2\alpha N)}{N(1-2\alpha)(3-2\alpha)} \qquad |20|$$

Again, as random mixing of the anions is assumed,

$$a_{MO} = N_{O^{2-}}$$

where, from Eqs. |19| and |20|, $N_{O^{2-}}$ is obtained as

$$N^2_{O^{2-}}(k_{1,1}-3)(1-2N) + N_{O^{2-}}[3-9N-2k_{1,1}(1-N)] + k_{1,1} = 0 \qquad |21|$$

A consequence of the assumption of random mixing of the anions is that the free energy change due to the breaking of double oxygen bonds is not independent of x and hence the free energy of a hypothetical branching chain melt, relative to liquid MO and liquid SiO_2 cannot be calculated. However, if, rather than assuming random mixing of the discrete anions, it is assumed that the O^{2-}, SiO_4^{4-} and SiO_3 segments of x-mers mix randomly, the configurational entropy can be calculated from Guggenheim's equation (20), in which case it can also be assumed that ΔG_1^o is independent of x, and hence $\Delta G^M/RT$ can be calculated as the sum of the chemical and configurational contributions. This procedure leads to values of $\Delta G^M/RT$ which are not significantly different from those given by Eq. |15| (21). The variations of a_{MO} with $k_{1,1}$ and N, given by Eq. |21|, are shown in Fig. 5.

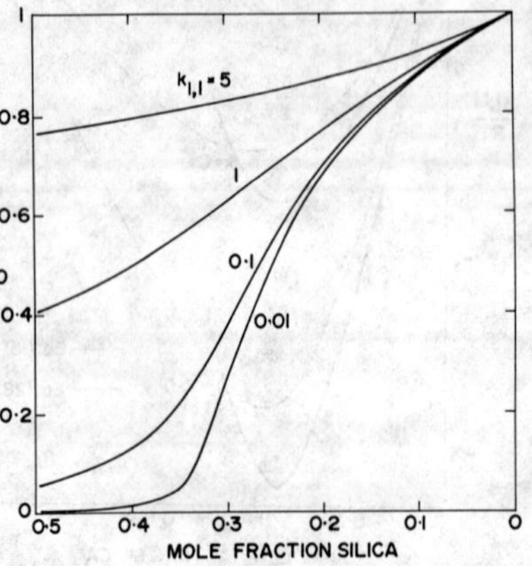

Figure 5. The activities of MO (Eq. 21) in Whiteway, Smith and Masson's hypothetical branching chain silicate melts as functions of $k_{1,1}$ and composition.

The polymerization models of Masson and Whiteway, Smith and Masson, can be adapted to consider polymerization in systems of the type $MO-A_2O_3$ in which the polymerizing unit is the ion AO_3^{3-}. For linear chain formation in such systems and the assumption of random mixing of the anions.

$$a_{MO} = N_{O^{2-}}$$

and

$$a_{A_2O_3} = \frac{k^3(1-N_{O^{2-}})^2}{(N_{O^{2-}}+k-kN_{O^{2-}})^2} \qquad |22|$$

where

$$N^2_{O^{2-}}(1-k)(1-2N) + N_{O^{2-}}[2k-1+4N(1-k)] - k(1-2N) = 0 \qquad |23|$$

N is the mole fraction of A_2O_3 and k is the equilibrium constant for the reaction

$$AO_3^{3-} + A_xO_{2x+1}^{(x+2)-} = A_{x+1}O_{2x+3}^{(x+3)-} + O^{2-}$$

Thus, in such systems

$$\Delta G^M/RT = N\ln\frac{k^3(1-N_{O^{2-}})^2}{(N_{O^{2-}}+k-kN_{O^{2-}})^2} + (1-N)\ln N_{O^{2-}} \qquad |24|$$

For branching chain polymerization

$$N^2_{O^{2-}}(2-k_{1,1})(1-2N) + N_{O^{2-}}[2k_{1,1}-2-2N(k_{1,1}-4)] - k_{1,1} = 0 \qquad |25|$$

where $k_{1,1}$ is the equilibrium constant for the reaction

$$AO_3^{3-} + AO_3^{3-} = A_2O_5^{4-} + O^{2-}$$

As the polymerization models only consider the formation of chain polyanions, they are only applicable to melts in which the mole fraction of the acidic oxide, SiO_2 or A_2O_3, is less than 0.5. Curve-fitting of Eq. |15| with experimental data for binary silicates and PbO-GeO_2 is shown in Fig. 6, in which it is seen that the agreement is good up to some limiting value of N_{SiO_2},

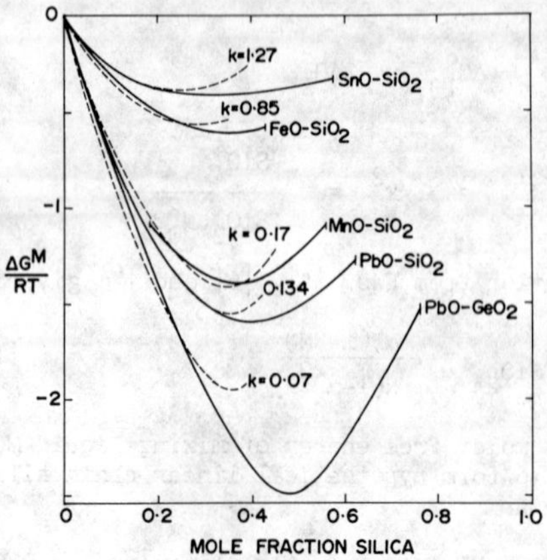

Figure 6. Curve-fitting of Eq. 15, (the free energy of formation of Masson's hypothetical linear chain melts) with experimental data.

beyond which the free energies of the actual melts become significantly more negative than those calculated for the hypothetical linear chain melts. The mole fraction of the acidic oxide at which this deviation begins can be taken as being that at which the formation of ring ions, which is specifically neglected in the polymerization models, becomes significant in the melts. It is to be expected that the formation of ring ions would increase the configurational entropy and hence increase the free energy of mixing. In comparing the free energy curves for the various binary silicate systems, it is seen that, in agreement with theoretical prediction, the position of the minimum in the free energy curve shifts to higher values of N as the basicity of the basic oxide increases, and hence as the degree of depolymerization increases. The excellent fits of Eq. |21| with the activities of the basic oxide in the systems $MnO-SiO_2$, $PbO-SiO_2$, and $PbO-GeO_2$ and Eq. |25| with a_{PbO} in $PbO-Bi_2O_3$ are shown in Fig. 7. Curve-fitting of Eq. |24| with experimental data for the

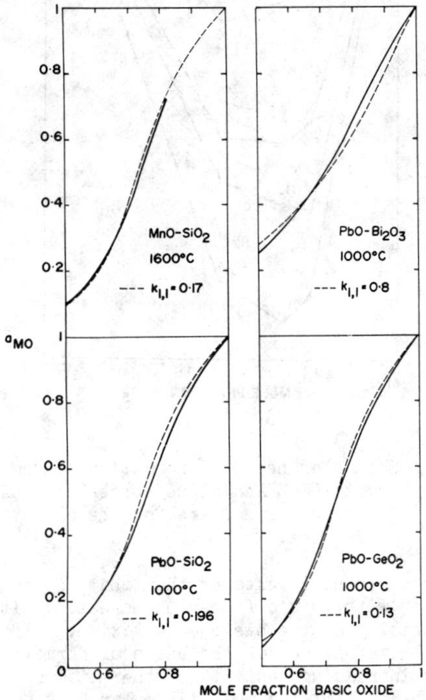

Figure 7. Curve-fitting of Eq. 21 (the activity of MO in Whiteway, Smith and Masson's hypothetical branching chain melts) with experimental data.

systems $PbO-Bi_2O_3$ and $PbO-B_2O_3$ is shown in Fig. 8 in which it is again seen that the agreement is only good within the range $0<N_{A_2O_3}<0.2$. The activities of PbO measured in the systems $PbO-P_2O_5$ and $PbO-B_2O_3$ and shown in Fig. 9 are of particular interest as they are lower than the theoretically-allowed minimum values obtained with k = 0, i.e. assuming a value of minus infinity for the free energy change for reaction |1|. This behavior could be explained thermodynamically in terms of the existence of an attractive interaction between the O^{2-} ions and the oxyanions PO_4^{3-} and BO_3^{3-}, or could be accounted for structurally by assuming that anions such as PO_5^{5-} and BO_4^{5-} are formed. A characteristic process of an acid-base reaction is the transfer of oxygen from one state of polarization to another (22) and the resultant decrease in free energy is related to the extent to which the state of polarization of the oxygens is changed. P_2O_5 and B_2O_3 are more acidic that SiO_2 and hence the oxygens which coordinate P and B in phosphate and borate

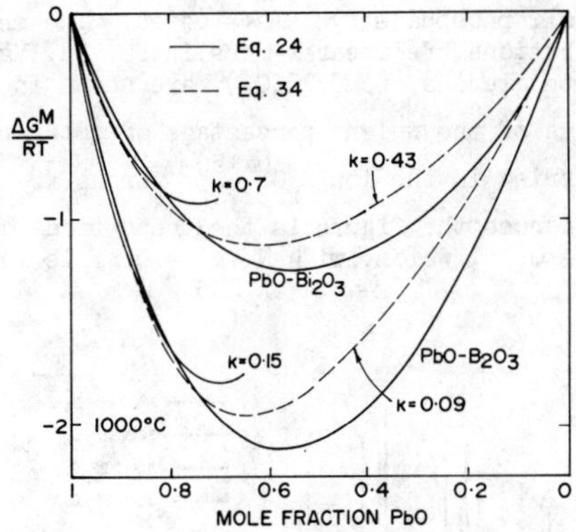

Figure 8. Curve-fitting Eqs. 24 and 34 with experimental free energies of formation of melts in the systems $PbO-Bi_2O_3$ and $PbO-B_2O_3$.

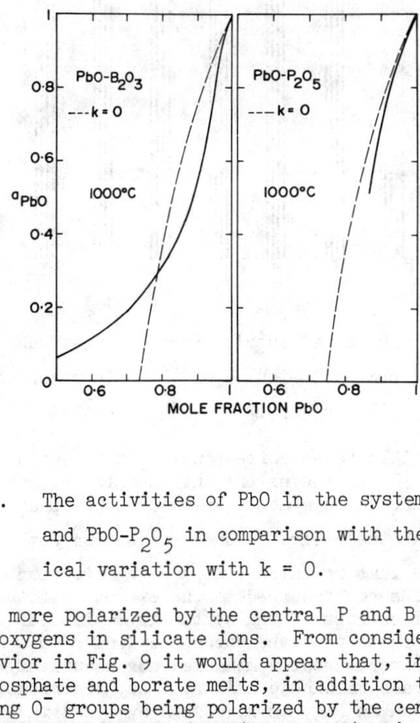

Figure 9. The activities of PbO in the systems $PbO-B_2O_3$ and $PbO-P_2O_5$ in comparison with the theoretical variation with k = 0.

ions are more polarized by the central P and B atoms than are the oxygens in silicate ions. From consideration of the behavior in Fig. 9 it would appear that, in highly basic phosphate and borate melts, in addition to the coordinating O^- groups being polarized by the central atom, the O^{2-} ions outwith the immediate coordination shell also come under the polarizing influence of the P and B atoms. Structurally, this attractive interaction is equivalent to an increase in the coordination number of the P and B atoms.

Major evidence supporting the basic assumptions of the polymerization models is provided by chromatographic studies of quenched acid-base glasses. As a result of the relative ease with which phosphate glasses can be dissolved in water and other room temperature solvents and the slow rate of hydrolysis of the dissolved phosphates, many paper chromatography studies of the ionic distributions occurring in quenched phosphate glasses have been conducted (23-25). These studies have shown that glasses between the compositions $2MO.P_2O_5$ and $MO.P_2O_5$ contain mainly linear chain polyphosphate ions, although appreciable concentrations of trimeric and tetrameric ring ions occur in systems of $N_{P_2O_5} \geq 0.4$ and branching chain ions begin to appear as $N_{P_2O_5}$ approaches 0.5. Linear chain ions containing up to 11 phosphorus atoms have been detected at the composition $7MO.5P_2O_5$. Of particular note is the observation that the concentrations of orthophosphate ion PO_4^{3-} are negligibly small in all the alkali phosphate systems studied. The measured size distributions of linear x-mers in Pb, Cd, Mg and Zn pyrophosphate glasses, ($3MO.2P_2O_5$), are shown in Fig. 10 as histograms of the weight percentage of total phosphorus occurring in the ion $P_xO_{3x+1}^{(x+2)-}$ versus x. The line drawn across the figure is the theoretical variation for $\alpha = 0.75$, which, with $N_{P_2O_5} = 0.4$, is the

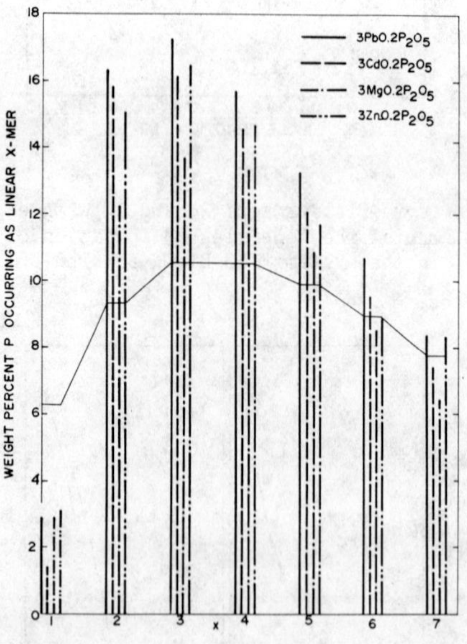

Figure 10. The size distributions of linear chain phosphate ions $P_xO_{3x+1}^{(x+2)-}$ in pyrophosphate glasses (25).

minimum possible value corresponding to the occurrence of reaction |1| to the extent that O^{2-} is eliminated. The ion occurring in a monodisperse pyrophosphate is $P_2O_7^{4-}$ and the theoretical size distribution with α = 0.75 is thus that in which random reorganization of the phosphate ions has occurred to the extent that the configurational entropy of the system is maximized. In view of the facts that the concentrations of O^{2-} can neither be experimentally measured nor accurately estimated from mass balance considerations, distributions of the type shown in Fig. 10 have been analyzed in terms of the equilibrium

$$2P_xO_{3x+1}^{(x+2)-} = P_{x-1}O_{3x-2}^{(x+1)-} + P_{x+1}O_{3x+4}^{(x+3)-} \qquad |26|$$

for which $$K_x = \frac{N_{P_{x-1}} \cdot N_{P_{x+1}}}{N_{P_x}^2}$$

If the system conforms with the predictions of Masson's model of linear chain polymers, in that the ion fractions of the x-mers occur in a geometric series and ΔG_1^o is independent of x, the free energy of reaction |26| is zero and hence $K_x = 1$. From the data in Fig. 10, as x increases from 2 to 6, K_x approaches unity and, for x>2, the values of K_x are not significantly dependent on $N_{P_2O_5}$ in any system. However, the value of K_2 in each system increases with increasing $N_{P_2O_5}$ which, in view of the absence of such an effect on K_3 and K_4, indicates an non-ideal behavior of the monomer PO_4^{3-}. The influence of this non-ideal behavior has been discussed in terms of the activity of PbO in highly basic $PbO-P_2O_5$ melts.

Further evidence in support of the general validity of the thermodynamic treatment of hypothetical polyionic melts is provided by the results of calorimetric studies of the heats of formation of crystalline and glassy phosphates. Fig. 11 shows the molar heats of formation of several crystalline and glassy phosphates (26). In each system the heat of formation of the

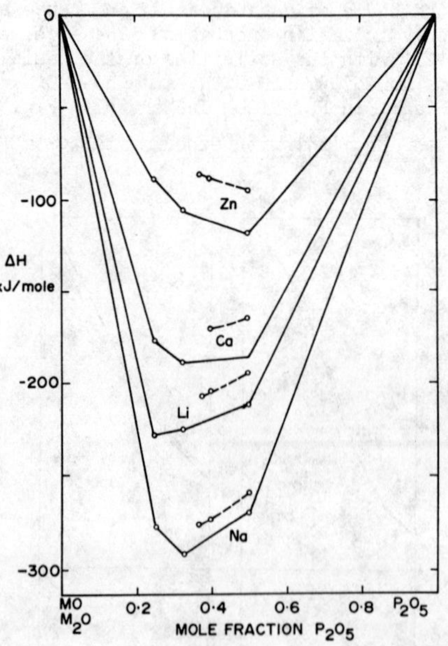

Figure 11. The molar heats of formation of some crystalline and glassy phosphates (26). The broken lines are for the glassy phosphates.

crystalline compound is greater than that of the corresponding glass with this difference decreasing with increasing basicity of the basic metal oxide. Also, as expected, the magnitudes of the heats of formation increase with increasing basicity of the metal oxide. By analogy with Eq. |16| the integral molar heat of formation of a hypothetical linear chain phosphate melt is

$$\Delta H^M = n_{O^-} \cdot \Delta H_1^o/2 = N(3-2\alpha)\Delta H_1^o \qquad |27|$$

where N is the mole fraction of P_2O_5 and ΔH_1^o is the standard heat for the reaction

$$MO + 1/3\, P_2O_5 = M(PO_4)_{2/3}$$

The crystalline phosphate compounds are monodisperse systems in which n_{O^-} is maximized, and the corresponding glasses are less-than-completely-depolymerized states in which the extent of the depolymerization of P_2O_5 by the basic metal oxide increases with increasing basicity of the oxide. Thus $|\Delta H^M_{crystalline}| > |\Delta H^M_{glass}|$ with this difference decreasing with increasing $|\Delta H_1^o|$, which is in agreement with Fig. 11.

The distribution of silicate ions extracted, as their silyl derivatives, from $PbO-SiO_2$ glasses is shown in Fig. 12 in comparison with the size distribution occurring in hypothetical branching chain melts with $k_{1,1} = 0.196$ (27). As a result of undesirable side-reactions which occur during the silylation of silicate glasses (28), the measured distribution is not completely quantitative but an obvious similarity exists between the measured and the theoretical distributions. Of particular note is the existence of the tetrameric ring

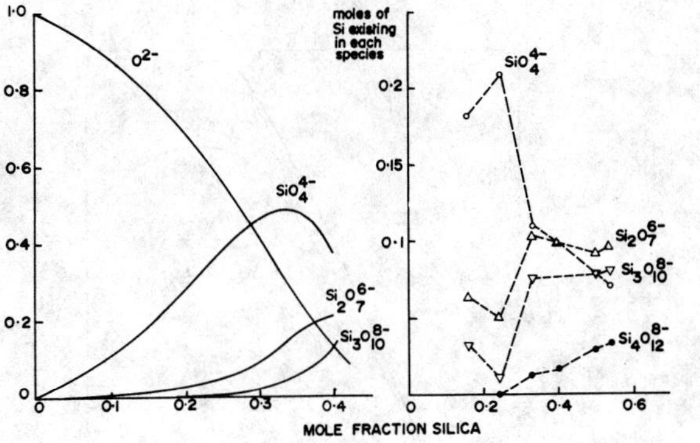

Figure 12. The measured size distribution of silicate anions in $PbO-SiO_2$ glasses (27) in comparison with the theoretical distribution of linear chain silicate ions in a system with $k_{1,1} = 0.196$.

ion $Si_4O_{12}^{8-}$ in relatively basic melts. The ions extracted from glassy samples of blast furnace and open hearth slags (29) are listed in Table I and the concentrations of the major ions found are listed in Table II.

Table I. Ions detected in Basic Open Hearth Steelmaking Slags and Iron Blast Furnace Slags (29)

monomer	SiO_4^{4-}	PO_4^{3-}	
dimer	$Si_2O_7^{6-}$	$SiPO_7^{5-}$	
linear trimer	$Si_3O_{10}^{8-}$	$Si_2PO_{10}^{7-}$	
linear tetramer	$Si_4O_{13}^{10-}$	$Si_3PO_{13}^{9-}$	*
trimeric ring	$Si_3O_9^{6-}$		
tetrameric ring	$Si_4O_{12}^{8-}$	$Si_3PO_{12}^{7-}$	
tetrameric ring with 2 internal bridges	$Si_6O_{16}^{8-}$		
pentameric ring	$Si_5O_{15}^{10-}$	$Si_4PO_{15}^{9-}$	*
pentameric ring with 1 internal bridge	$Si_6O_{17}^{10-}$	$Si_5PO_{17}^{9-}$	*
pentameric ring with 2 internal bridges	$Si_7O_{19}^{10-}$	$Si_6PO_{19}^{9-}$	*
pentameric ring with 3 internal bridges	$Si_8O_{21}^{10-}$	$Si_7PO_{21}^{9-}$	*
pentameric ring with 4 internal bridges	$Si_9O_{23}^{10-}$	$Si_8PO_{23}^{9-}$	*

* not detected in blast furnace slag

In Table II it is seen that, in both types of slag, the linear chain ions are the major species with their concentrations decreasing with increasing chain length. It is, however, of interest to note that quite complex ions were detected in the open hearth slag even at the high basic oxide to acid oxide ratio of 4.7. Also, only 36% of the phosphorus occurs as the orthophosphate ion PO_4^{3-}, with the remainder occurring as silicophosphates.

Table II. The concentrations of ions detected in Open Hearth and Blast Furnace Slags (29)

		O.H. Slag	B.F. Slag
monomer	SiO_4^{4-}	56.6%	43.6%
	PO_4^{3-}	3.2%	
dimer	$Si_2O_7^{6-}$	19.3%	26.6%
	$SiPO_7^{5-}$	2.9%	
trimer	$Si_3O_{10}^{8-}$	6.7%	15.2%
	$Si_2PO_{10}^{7-}$	1.8%	
linear chains (total)		90.5%	85.4%

At present, melts of composition outwith the range of applicability of the polymerization models must be dealt with by quasi-lattice models. The first of these models is that of Yokokawa and Niwa (30) who considered a melt containing n "molecules" of basic metal oxide MO and m "molecules" of SiO_2 in which the 2m+n oxygens are arranged such that they tetrahedrally-coordinate m+1/2n sites available for the m silicons, of which m are occupied by silicons and 1/2n are vacant. The occurrence of a single vacant site corresponds to the four coordinating oxygens being O^- and the occurrence of a pair of adjacent vacant sites corresponds to the oxygen between the sites being an O^{2-} and the other six coordinating oxygens being O^-. The distribution of the silicons on the available sites is considered in terms of the occupancies of pairs of sites. From the stoichiometry of Eq. |1|, the occurrence of r O^- groups, i.e. r vacancy-silicon pairs, requires a decrease in the number of O^{2-} from n to n-r/2, giving thus n-r/2 vacancy-vacancy pairs, and a decrease in the number of O^o from 2m to 2m-r/2, giving thus, 2m-r/2 silicon-silicon pairs. Taking the free energy change for reaction |1| as ΔG_1^o and calculating the configurational entropy from consideration of both the distribution of the silicons on the available sites and the distribution of the three types of occupancy of pairs of sites, Yokokawa and Niwa calculated the state of minimum free energy and derived the molar free energy of mixing N moles of liquid SiO_2 and (1-N) moles of MO as

$$\Delta G^M/RT = N\ln\left[\frac{1+N}{2N}\left(1-\frac{r'}{4N}\right)^2\right] + (1-N)\ln\left[\left(\frac{1+N}{1-N}\right)^{\frac{1}{2}}\left(1-\frac{r'}{2(1-N)}\right)\right] \qquad |28|$$

where r', the number of moles of O^- per mole of binary silicate is obtained from

$$[2(1-N)-r'][4N-r'] = r'^2\exp(\Delta G_1^o/RT) \qquad |29|$$

As it is assumed that ΔG_1^o is independent of configuration, a direct comparison can be made between the free energy of mixing in a system of linear chain polysilicate anions given by Eq. |15| and the free energy of mixing given by Eq. |28|. This comparison is shown in Fig. 4 and the variations of a_{SiO_2} and a_{MO} in Yokokawa and Niwa's hypothetical melts, given by the terms in the logarithms in Eq. |28| are shown in Fig. 13.

The free energy of mixing in Yokokawa and Niwa's hypothetical melts is the sum of a chemical contribution due to the breaking of O^o bonds, $\Delta G_{chem}/RT$, given by

$$\Delta G_{chem}/RT \; = \; \frac{r'}{2}\,\frac{\Delta G_1^o}{RT} \; = r' \ln \frac{\left[(2-2n-r')(4N-r')\right]^{\frac{1}{2}}}{r'} \qquad |30|$$

and a configurational contribution, $\Delta G_{conf}/RT$, given by

$$\Delta G_{conf}/RT = -\frac{3}{2}(1-N)\ln(1-N) - 3N\ln 2N + \frac{(1+N)}{2}\ln(1+N)$$
$$+ (1-N-\tfrac{1}{2}r')\ln(1-N-\tfrac{1}{2}r') + (2N-\tfrac{1}{2}r')\ln(2N-\frac{r'}{2})$$
$$+ r'\ln\frac{r'}{2} \qquad |31|$$

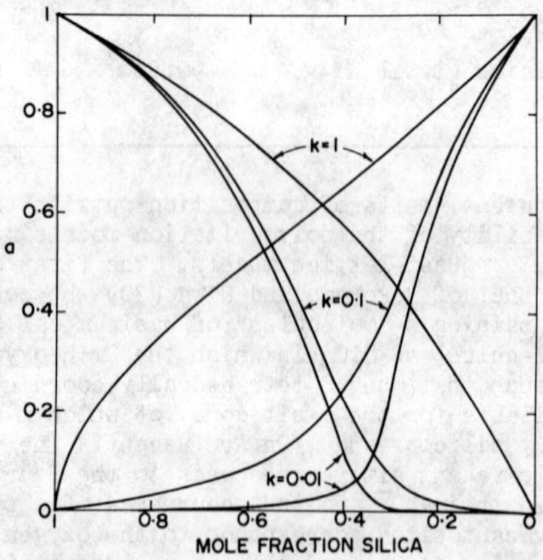

Figure 13. The activities of MO and SiO_2 in Yokokawa and Niwa's hypothetical quasi-lattice melts as functions of composition and $k = \exp(\Delta G_1^o/RT)$.

A comparison between the respective chemical and configurational contributions to the free energies of mixing in Masson's hypothetical linear chain melts and Yokokawa and Niwa's hypothetical quasi-lattice melts is shown in Fig. 14 for a system in which $k = 0.1$ ($\Delta G_1^o/RT = -2.303$). The states of minimum free energy in the chain systems are less polymerized than those in the corresponding quasi-lattice systems and hence the chemical contribution to the free energy of mixing in the former is greater than in the latter. Also (for $k = 0.1$), in hypothetical melts of $N<0.42$, the configurational contribution to the free energy of mixing in chain systems is greater than in quasi-lattice systems. However, with $N>0.42$, the configurational contributions in the quasi-lattice system increasingly exceeds that in the corresponding chain systems and, with $k = 0.1$, the free energies of formation of both types of hypothetical melt are equal to one another at $N = 0.46$, as is seen in Fig. 4. Thus generally, the influence of the configurational contribution is such that, in systems of low N, the free energy of formation of a hypothetical chain melt is greater than that of a quasi-lattice melt, and at higher values of N the reverse is the case.

Curve-fitting of Eq. |28| with experimental data for six binary silicate systems is shown in Fig. 15 and curve-fittings with activities in the systems $CaO-SiO_2$ and $PbO-SiO_2$ are shown in Figs. 16 and 17 respectively. The excellence of the fit with the experimental data for the system $CaO-SiO_2$ is of particular interest as the derived value of 0.0009 for k is considerably less than that at which Eq. |31| begins to give negative, and hence impossible, entropies of mixing in the hypothetical melts. With decreasing k the entropy of mixing in Yokokawa and Niwa's model first becomes zero at N=1/3 with k = 0.0153.

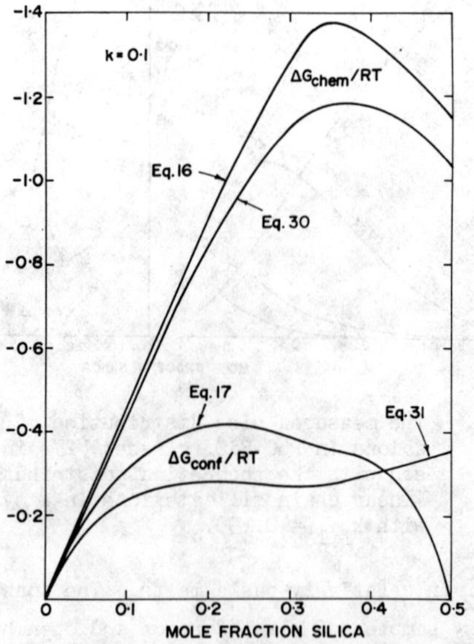

Figure 14. Comparison between the chemical and configurational contributions to the free energy of formation of Masson's hypothetical linear chain melts and Yokokawa and Niwa's hypothetical quasi-lattice melts in a system for which $k = 0.1$ ($\Delta G_1^o/RT = -2.3$).

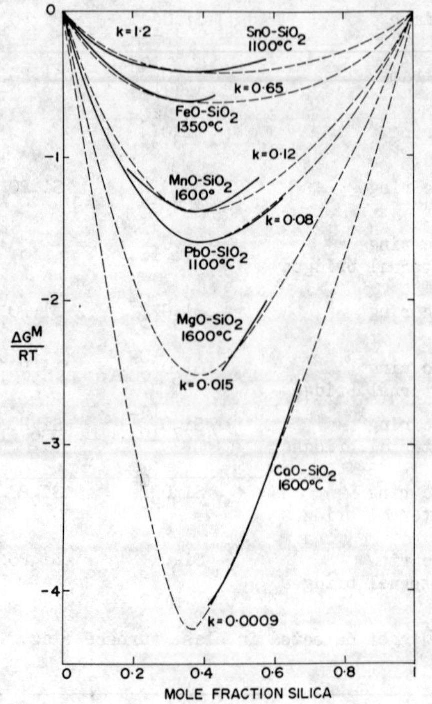

Figure 15. Curve-fitting of Eq. 28 (the free energy of formation of Yokokawa and Niwa's hypothetical quasi-lattice melts) with experimental data.

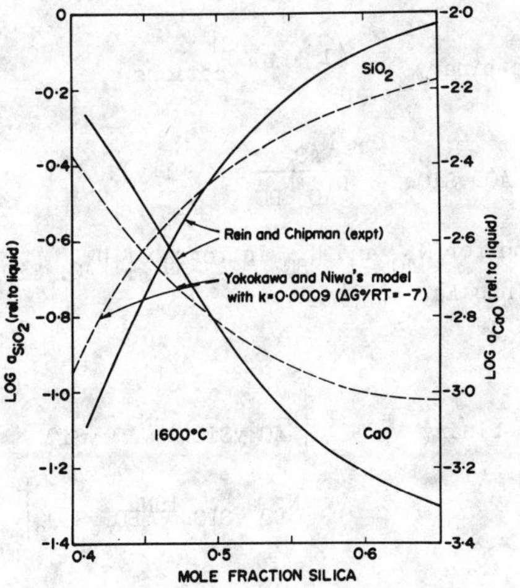

Figure 16. Comparison between measured activities in the system $CaO-SiO_2$ (10) and activities in Yokokawa and Niwa's hypothetical quasi-lattice melts calculated using the value of k obtained for $CaO-SiO_2$ in Fig. 15.

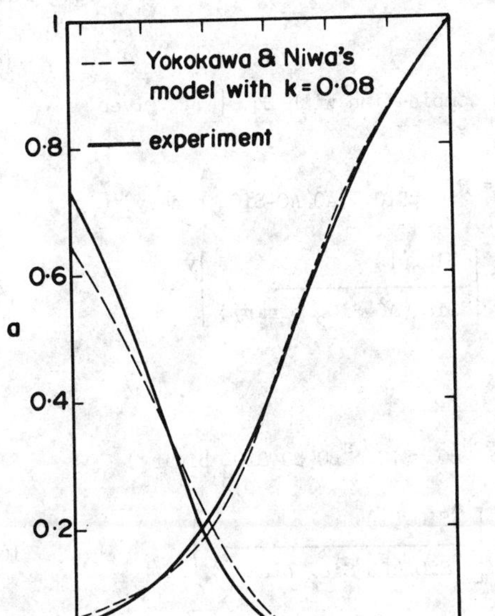

Figure 17. Comparison between measured activities in the system $PbO-SiO_2$ (7) and activities in Yokokawa and Niwa's hypothetical quasi-lattice melts calculated using the value of k obtained for $PbO-SiO_2$ in Fig. 15.

Adapting Yokokawa and Niwa's model to hypothetical melts in the system $MO-A_2O_3$ in which A is coordinated by three oxygens gives, for a system of N moles of A_2O_3 and 1-N moles of MO,

$$a_{MO} = \frac{(1+2N)^{1/3}(1-N-\frac{r'}{2})}{(1-N)^{4/3}} \qquad |32|$$

and

$$a_{A_2O_3} = \frac{(1+2N)(3N-\frac{r'}{2})^3}{(3N)^4} \qquad |33|$$

where r′ is obtained from

$$(2-2N-r')(6N-r') = r'^2\exp(\Delta G_1^o/RT)$$

and ΔG_1^o is the standard free energy change for the reaction

$$MO + 1/3\,A_2O_3 = 1/3\,M_3(AO_3)_2$$

The free energies of mixing in such melts, given by

$$\Delta G^M/RT = N\ln a_{A_2O_3} + (1-N)\ln a_{MO} \qquad |34|$$

are fitted to experimental free energies in the systems $PbO-B_2O_3$ and $PbO-Bi_2O_3$ in Fig. 8. As is seen, attempts to fit exactly at the ortho composition (N = 1/4) result in reasonable fits in melts of N<1/4 and poor fits in melts of N>1/4. In the case of the borate melts this poor fit in the more acidic melts is not unexpected and can be qualitatively explained in terms of the boron anomaly, whereby, at higher values of N, the coordination number of boron increases from 3 to 4 (31).

In contrast with acid-base systems such as binary silicates and binary phosphate melts, the thermodynamic properties of which can be described in terms of depolymerization of the acid by the base and mixing of the discrete polymerized oxyanions, it is found that the thermodynamics of mixing in the system $MnO-TiO_2$ can be closely represented by the regular solution formalism (32, 33). Applied to such melts, regular solution theory assumes, (i) a molten salt structure in which O^{2-} is the only anionic species, (ii) that the entropy of mixing arises solely from random mixing of the Mn^{2+} and Ti^{4+} cations on the cation sublattice, and (iii) that pairwise interaction between cations gives rise to a parabolic variation, with composition, of the molar heat of mixing. Consequently, in terms of the regular solution theory, the excess molar free energy of mixing, G^{xs}, is

$$G^{xs} = RT\alpha\, N_{MnO}\cdot N_{TiO_2}$$

where α is the regular solution parameter for interaction between the Mn^{2+} and Ti^{4+} cations. The variations, with composition, of the excess molar free energies of mixing liquid MnO and liquid TiO_2 at 1500° and 1550°C are closely represented by the equation (33).

$$G^{xs} = -55280N_{MnO}\cdot N_{TiO_2} \text{ joules}$$

This molten salt-like behavior is caused by the small difference between the electronegativities of Mn and Ti. The original application, by Lumsden (34), of the regular solution formalism to iron silicate melts required postulation of both an arbitrary disproportionation of Fe^{2+} ions at low silica contents and a hypothetical form of liquid silica in which Si^{4+} ions are randomly embedded in a matrix of O^{2-} ions. The free energy of this "ionic silica" is 21 kJ/mole higher than that of actual silica. Attempts to describe mixing in iron-saturated $FeO-TiO_2$ melts using the regular solution formalism were made difficult by the unknown concentrations of Fe^{3+} ions in these melts (35).

Few other binary systems containing a distinctly acidic or a distinctly basic oxide and an amphoteric oxide have been studied. The free energy of formation of $CaO-Al_2O_3$ melts at 1600°C (10), shown in Fig. 1, is much greater than would be expected for formation of a simple binary molten salt, and mutual solution of CaO and Al_2O_3 to form liquids which, on cooling, can easily be retained in a glassy state, has not yet been satisfactorily explained in structural terms.

THE THERMODYNAMIC PROPERTIES AND STRUCTURES OF TERNARY OXIDE MELTS

The best-known thermodynamic model of mixing in ternary silicate melts is Richardson's model of ideal silicate mixing (36). The single assumption in this model is that when two binary silicate melts, containing the same mole fraction of SiO_2, are mixed, the free energy decrease is due solely to random mixing of the cations. Thus, for convenience, any melt in the ternary system $AO-BO-SiO_2$ is considered to occur in the pseudo-binary system $AO.ySiO_2-BO.ySiO_2$ and the free energy of formation of the ternary melt is the sum of the free energies of formation of the pseudo-binary components and the free energy change due to ideal mixing of the cations when the pseudobinary components are mixed. The former free energy change, termed the excess free energy G^{xs}, is

$$G^{xs} = \frac{N_{AO}}{N_{AO}+N_{BO}} \Delta G^M_{AO-SiO_2\ binary} + \frac{N_{BO}}{N_{AO}+N_{BO}} \Delta G^M_{BO-SiO_2\ binary} \qquad |35|$$

where $\Delta G^M_{AO-SiO_2\ binary} = RT\left[\frac{1}{1+y}\ln a_{AO} + \frac{y}{1+y}\ln a_{SiO_2}\right]$

and $\Delta G^M_{BO-SiO_2\ binary} = RT\left[\frac{1}{1+y}\ln a_{BO} + \frac{y}{1+y}\ln a_{SiO_2}\right]$

and the latter free energy change, termed $\Delta G^M_{cations}$, is

$$\Delta G^M_{cations} = RT\left\{N_{AO}\ln\frac{N_{AO}}{N_{AO}+N_{BO}} + N_{BO}\ln\frac{N_{BO}}{N_{AO}+N_{BO}}\right\} \qquad |36|$$

The free energy of formation of the ternary melt, per mole of oxide, is thus

$$\Delta G^M = \Delta G^M_{cations} + G^{xs} \qquad |37|$$

Differentiating Eq. |37| with respect to N_{SiO_2} at constant N_{AO}/N_{BO} and substituting, with Eq. |37| into

$$\Delta \bar{G}^M_{SiO_2} = RT\ln a_{SiO_2(t)} = \Delta G^M + (1-N_{SiO_2})\left.\frac{\partial \Delta G^M}{\partial N_{SiO_2}}\right|_{N_{AO}/N_{BO}}$$

gives
$$\ln a_{SiO_2(t)} = \frac{N_{AO}}{N_{AO}+N_{BO}}\ln a_{SiO_2}(AO\text{-}SiO_2\ binary) + \frac{N_{BO}}{N_{AO}+N_{BO}}\ln a_{SiO_2(BO-SiO_2\ binary)} \qquad |38|$$

The free energy of mixing, per mole of silicate, in the pseudo-binary system, is

$$\Delta G^M_{pseudo-binary} = (1+y)\Delta G^M_{cations} \qquad |39|$$

and, as $N_{AO.ySiO_2} = \frac{N_{AO}}{N_{AO}+N_{BO}} = (1+y)N_{AO}$

substitution for N_{AO} and N_{BO} in terms of $N_{AO.ySiO_2}$ and $N_{BO.ySiO_2}$ into Eq. |39| gives

$$\Delta G^M_{pseudo-binary} = RT\left[N_{AO.ySiO_2}\ln N_{AO.ySiO_2} + N_{BO.ySiO_2}\ln N_{BO.ySiO_2}\right] \qquad |40|$$

from which, $a_{AO.ySiO_2} = N_{AO.ySiO_2}$ |41|

and $a_{BO.ySiO_2} = N_{BO.ySiO_2}$ |42|

In the pseudo-binary,

$$a_{AO.ySiO_2} = \frac{a_{AO(t)}\cdot a^y_{SiO_2(t)}}{a_{AO(AO-SiO_2\ binary)}\cdot a^y_{SiO_2(AO-SiO_2\ binary)}} \qquad |43|$$

which, in combination with Eq. |41|, gives

$$a_{AO(t)} = N_{AO.ySiO_2}\cdot a_{AO(AO-SiO_2\ binary)} \times \left[\frac{a_{SiO_2(t)}}{a_{SiO_2(AO-SiO_2\ binary)}}\right]^y \qquad |44|$$

Similarly,

$$a_{BO(t)} = N_{BO.ySiO_2}\cdot a_{BO(BO-SiO_2\ binary)} \times \left[\frac{a_{SiO_2(t)}}{a_{SiO_2(BO-SiO_2\ binary)}}\right]^y \qquad |45|$$

Thus, from knowledge of the free energies of mixing in the constituent binary silicate systems, $AO-SiO_2$ and $BO-SiO_2$, and the assumption of random mixing of the cations in ternary melts, the activities of SiO_2, AO and BO in any ternary melt are obtained, respectively, from Eqs. |38|, |44| and |45|.

The variation of a_{FeO} in $FeO-CaO-SiO_2$ melts and a_{MnO} in $MnO-CaO-SiO_2$ melts, calculated assuming ideal silicate mixing at 1600°C, are shown in Fig. 18(a) and 19(a), and the corresponding experimentally-measured activities are shown in Figs. 18(b) (37) and 19(b) (38) respectively. The origin of the liquid immiscibility, predicted by the ideal silicate mixing model, is illustrated in Fig. 20, which shows the calculated variations of G^{xs} and $\Delta G^M_{cations}$ with composition in the system $2/3CaO.1/3SiO_2-FeO$. As is

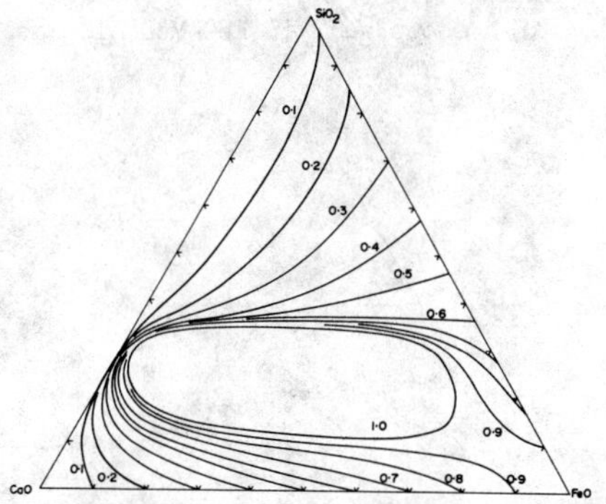

Figure 18(a). Iso-a_{FeO} contours calculated assuming ideal silicate mixing in the system FeO-CaO-SiO_2 at 1600°C.

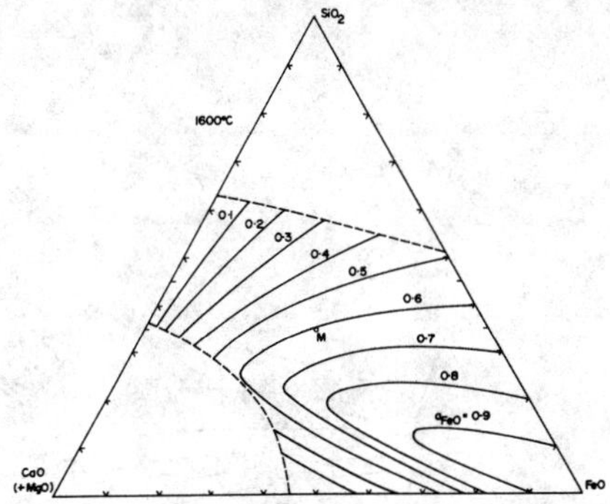

Figure 18(b). Experimentally determined iso-a_{FeO} contours in FeO-CaO(+ MgO)-SiO_2 melts at 1600°C (37).

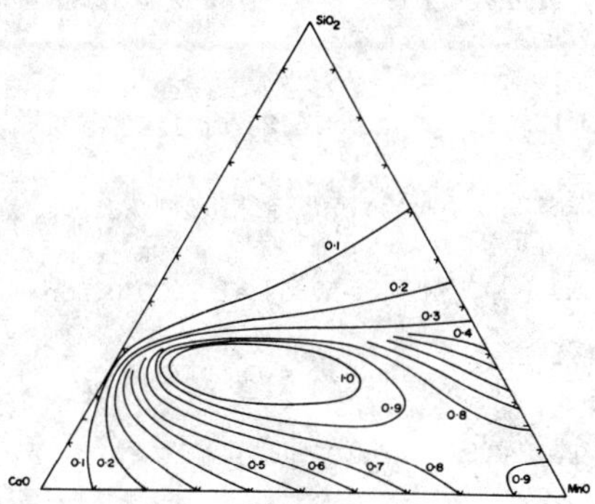

Figure 19(a). Iso-a_{MnO} contours calculated assuming ideal silicate mixing in the system MnO-CaO-SiO_2 at 1600°C.

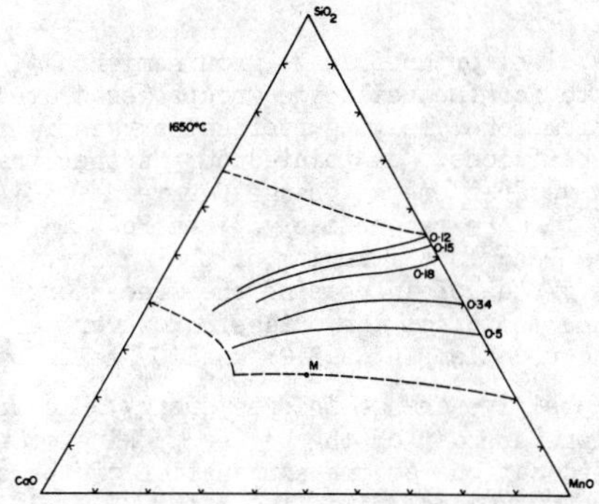

Figure 19(b). Experimentally-determined iso-a_{MnO} contours (relative to solid MnO as the standard state) in MnO-CaO-SiO_2 melts at 1650°C.

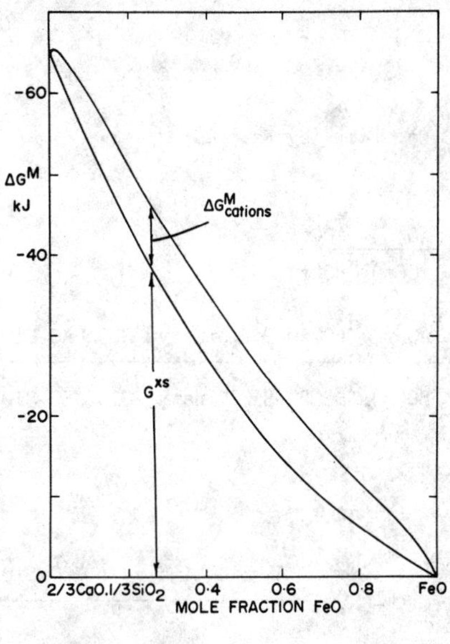

Figure 20. The variations of G^{xs} and $\Delta G^{M}_{cations}$ with composition in the system

FeO-$\frac{2}{3}$CaO$\frac{1}{3}SiO_2$ at 1600°C.

seen, the concave nature of the G^{xs} curve is not fully compensated for by $\Delta G^{M}_{cations}$ and hence, in the absence of consideration of the stability of any solid phase, liquid immiscibility is predicted. Richardson pointed out that although large deviations from ideal silicate mixing occur in orthosilicate mixtures, the activities of FeO and MnO are closely reproduced by the model near the metasilicate composition, i.e. in the composition range $0.42 \leq_{SiO_2} \leq 0.58$. He suggested two possible reasons for the deviations occurring in the orthosilicate mixtures; the anion matrix may maintain its same general structure from one limit of composition to the other but permit better packing of cations of differing size than of uniformly-sized cations, or a marked change may occur in the anionic structure across the composition range. With respect to the latter possibility, Richardson suggested that, in orthosilicate mixtures, reactions such as

$$Ca_2SiO_4 + Fe_2SiO_4 = 2CaSiO_3 + 2FeO \qquad |46|$$

may occur, i.e. orthosilicate groups might join together in rings to form metasilicate groups associated mainly with calcium ions, leaving residual oxygen ions associated with the Fe^{2+} ions. He pointed out further that although the negative free energy change for reaction |46| would give rise to more recognizably "free FeO" in the melt, it could also lead to a lowering of the activity coefficient of FeO by virtue of increasing the free energy of formation of the melt from the oxides. However, as the corresponding reaction in Ca_2SiO_4-Mn_2SiO_4 melts has a slightly positive change in free energy, Richardson stated a preference for the better packing of differently-sized cations as the explanation of the observed deviations from ideal silicate mixing.

In considering mixing in a pseudo-binary system, a more appropriate virtual equilibrium for consideration is that for the exchange reaction

$$AO + BO.ySiO_2 = BO + AO.ySiO_2 \qquad |47|$$

for which

$$\Delta G^o_{47} = (1+y)\left[\Delta G^M_{AO\text{-}SiO_2\ binary} - \Delta G^M_{BO\text{-}SiO_2\ binary}\right] \qquad |48|$$

Division of Eq. |44| by Eq. |45| gives

$$\frac{a_{AO(t)}}{a_{BO(t)}} = \frac{\gamma_{AO(t)}\cdot N_{AO(t)}}{\gamma_{BO(t)}\cdot N_{BO(t)}} = \frac{N_{AO(t)}\cdot a_{AO(AO\text{-}SiO_2binary)}\, a^y_{SiO_2(AO\text{-}SiO_2binary)}}{N_{BO(t)}\cdot a_{BO(BO\text{-}SiO_2binary)}\, a^y_{SiO_2(BO\text{-}SiO_2binary)}} \qquad |49|$$

from which

$$\ln\frac{\gamma_{AO(t)}}{\gamma_{BO(t)}} = (1+y)\left[\frac{\Delta G^M_{AO\text{-}SiO_2binary} - \Delta G^M_{BO\text{-}SiO_2binary}}{RT}\right] \qquad |50|$$

Combination of Eqs. |48| and |50| gives, for ideal silicate mixing,

$$\Delta G^o_{47} = RT\ln\frac{\gamma_{AO(t)}}{\gamma_{BO(t)}} \qquad |51|$$

From consideration of the exchange reaction given by Eq. |47|, the expectation of random mixing of the A and B cations is only reasonable with small values of $\Delta G^o_{47}/RT$, and the limit of $\Delta G^o_{47} = 0$ requires,

$$\Delta G^M_{AO\text{-}SiO_2\ binary} = \Delta G^M_{BO\text{-}SiO_2\ binary} \qquad |52|$$

Thus, the greater the difference between $\Delta G^M_{AO\text{-}SiO_2\ binary}$ and $\Delta G^M_{BO\text{-}SiO_2\ binary}$, the greater the expectation of deviation from ideal silicate mixing. This is illustrated by comparing Fig. 22, which shows a_{PbO} in PbO-Na_2O-SiO_2 melts at 1000°C (39), with Fig. 18(b), Fig. 19(b) and Fig. 21, which shows a_{FeO} in FeO-MnO-SiO_2 melts at

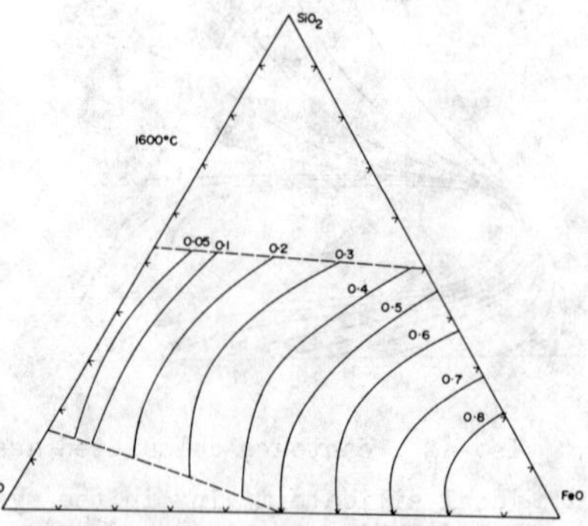

Figure 21. Experimentally-determined iso-a_{FeO} contours in the system MnO-FeO-SiO_2 at 1600°C (14).

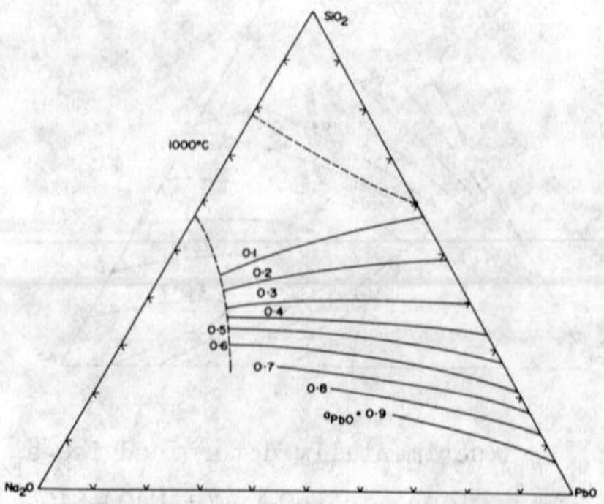

Figure 22. Experimentally-determined iso-a_{PbO} contours in the system Na_2O-PbO-SiO_2 at 1000°C (39).

1600°C (40). From the data presented in Fig. 1, the free energy changes for the appropriate exchange reactions are

$\Delta G^o_{47} = -185$ kJ ($\Delta G^o_{47}/RT = -16.2$) at 1100°C for

$$PbO.1/2SiO_{2(l)} + Na_2O_{(l)} = Na_2O.1/2SiO_{2(l)} + PbO_{(l)}$$

$\Delta G^o_{47} = -88$ kJ ($\Delta G^o_{47}/RT = -5.65$) at 1600°C for

$$FeO.1/2SiO_{2(l)} + CaO_{(l)} = CaO.1/2SiO_{2(l)} + FeO_{(l)}$$

$\Delta G^o_{47} = -69$ kJ ($\Delta G^o_{47} = -4.43$) at 1600°C for

$$MnO.1/2SiO_{2(l)} + CaO_{(l)} = CaO.1/2SiO_{2(l)} + MnO_{(l)}$$

$\Delta G^o_{47} = -19$ kJ ($\Delta G^o_{47}/RT = -1.22$) at 1600°C for

$$FeO.1/2SiO_{2(l)} + MnO_{(l)} = MnO.1/2SiO_{2(l)} + FeO_{(l)}$$

Bell (40) has shown that a_{FeO} in FeO-MnO-SiO_2 melts at 1600°C conforms exactly with ideal silicate mixing, which indicates that a value of $\Delta G^o_{47}/RT = -1.22$ for the iron silicate-manganese oxide exchange reaction is small enough to permit essentially-random mixing of the cations.

An origin of non-random mixing of the cations in ternary silicate melts is suggested by the polymerization models of binary silicate melts. If the oxide AO is more basic than BO, $AO.ySiO_2$ is more depolymerized than $BO.ySiO_2$, and hence, in the former, the ratio $n_{O^-}/n_{O^{2-}}$ is greater than in the latter. This arises because the difference in energy between the bond $-O^-M^{2+}\,^-O-$ and the bond $M^{2+}O^{2-}$ is greater for M ≡ A than for M ≡ B. Thus, on mixing $AO.ySiO_2$ and $BO.ySiO_2$, the A^{2+} cations are preferentially-associated with the O^- groups on the silicate anions and the B^{2+} cations are preferentially-associated with the O^{2-} ions. Consequently, random mixing of the cations does not occur, the differences in bond energy cause ΔG^o_{47} to have a negative value and ideal silicate mixing does not occur. Figs 18(b), 19(b) & 22 show that the addition of a basic oxide such as Na_2O or CaO to the silicate of a less basic oxide, actually increases the activity of the less basic oxide. With sufficiently large differences in bond energy this ionic microsegregation leads to macrosegregation (liquid immiscibility), as occurs in ternary systems containing P_2O_5, FeO and an alkali or alkaline earth oxide. The liquid immiscibility field in the system Na_2O-FeO-P_2O_5 is shown in Fig. 23, in which the conjugate liquids are "sodium phosphate-rich" and iron oxide-rich" (41).

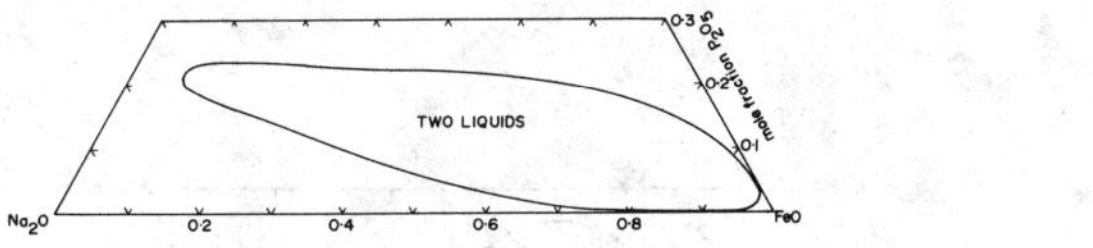

Figure 23. The liquid immiscibility field in the system Na_2O-FeO-P_2O_5.

The effect of the exchange reaction given by Eq. |47|, which is always to decrease the free energy of the system, would be expected to be greatest in the melt which contains equal concentrations of A^{2+} and B^{2+} and equivalent concentrations of O^{2-} and O^-. The branching chain polymerization model predicts that, at 1600°C, equivalent concentrations of O^{2-} and O^- occur at N_{SiO_2} = 0.45, 0.23 and 0.2 in the systems FeO-SiO_2, MnO-SiO_2 and CaO-SiO_2 respectively and thus the points labelled "M" in Figs. 18(b) and 19(b) are the estimated compositions at which the greatest influence of microsegregation on the thermodynamic behavior would be expected. With increasing N_{SiO_2}, $n_{O^{2-}}/n_{O^-}$ decreases, and hence the influence of microsegregation decreases.

Eq. |51| has been used to examine mixing in the systems $2MnO.SiO_2$-$2FeO.SiO_2$ (42), $2CoO.SiO_2$-$2FeO.SiO_2$ (43) and pseudo-binary sections in the system K_2O-Na_2O-SiO_2 (44-45). In the system $2MnO.SiO_2$-$2FeO.SiO_2$, $\gamma_{FeO(t)}/\gamma_{MnO(t)}$ is independent of composition at 1450°C, which, by virtue of Eq. |50|, indicates that this system exhibits ideal silicate mixing. However, the measured variation of $\gamma_{FeO(t)}/\gamma_{CoO(t)}$ with composition in the system $2CoO.SiO_2$-$2FeO.SiO_2$ at 1450°C, corresponds to the excess free energy of mixing of the pseudo-binary components shown in Fig. 24 and the activities shown in Fig. 25. This behavior indicates that, as-yet unexplained, complex interactions are occurring in the melts.

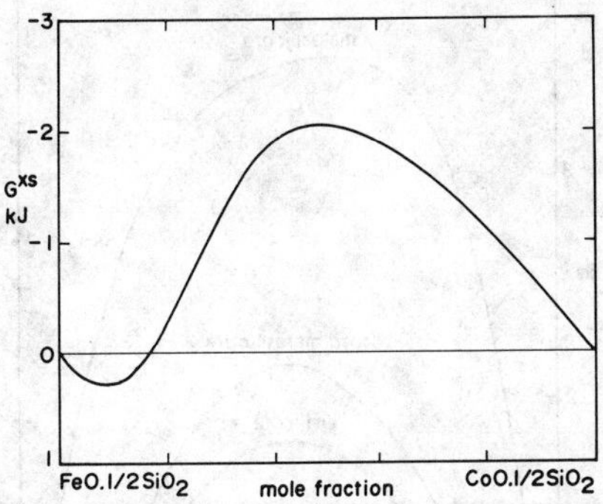

Figure 24. The excess free energy of mixing per mole of $MO.\frac{1}{2}SiO_2$ in the system $FeO.\frac{1}{2}SiO_2$-$CoO.\frac{1}{2}SiO_2$ at 1450°C (43).

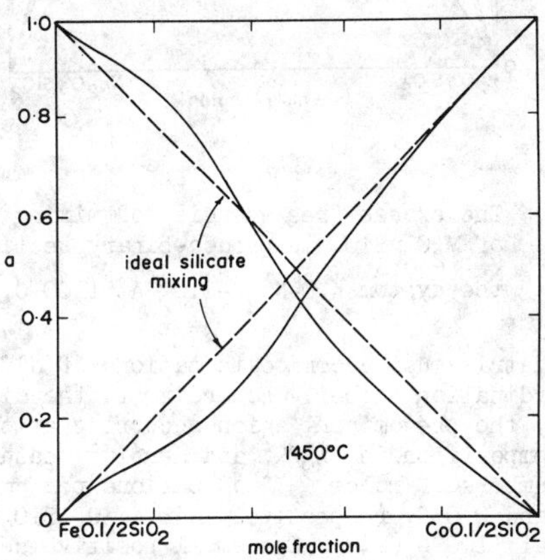

Figure 25. The activities of $FeO.\frac{1}{2}SiO_2$ and $CoO.\frac{1}{2}SiO_2$ in melts in the system $FeO.\frac{1}{2}SiO_2$-$CoO.\frac{1}{2}SiO_2$ at 1450°C (43).

The excess free energies of mixing per mole of $M_2O.ySiO_2$ in pseudo-binary sections in the system K_2O-Na_2O-SiO_2 of N_{SiO_2} = 0.75, 0.667 and 0.5 at 1100°C are shown in Fig. 26. As is seen, the excess free energies are positive quantities and the magnitude of the deviation from ideal silicate mixing increases with decreasing N_{SiO_2}. In the systems $K_2O.4SiO_2$-$Na_2O.4SiO_2$ and $K_2O.2SiO_2$-$Na_2O.2SiO_2$ the excess free energies are parabolic functions of composition, which is characteristic of regular solution behavior. The structural interpretation of this behavior is as follows. In alkali silicate melts the bonding between the cations and the O^- groups on the silicate ions is predominantly ionic and hence complex interactions of the type inferred in the system CoO-FeO-SiO_2 are not to be expected. In view of the predominantly ionic bonding, it is reasonable to assume that the difference between the structures of sodium and potassium silicate melts of the same silica content arises solely from the difference between the sizes of Na^+ and K^+. The size of the cation determines the oxygen coordination number of the cation and the energy of the M^+-O^- bond. X-ray diffraction studies of alkali silicate melts in the temperature 1100° - 1700°C indicate that the oxygen coordination numbers of Na^+ and K^+ are, respectively, 6 and 8 (46) and, thus in mixtures of $Na_2O.ySiO_2$ and $K_2O.ySiO_2$,

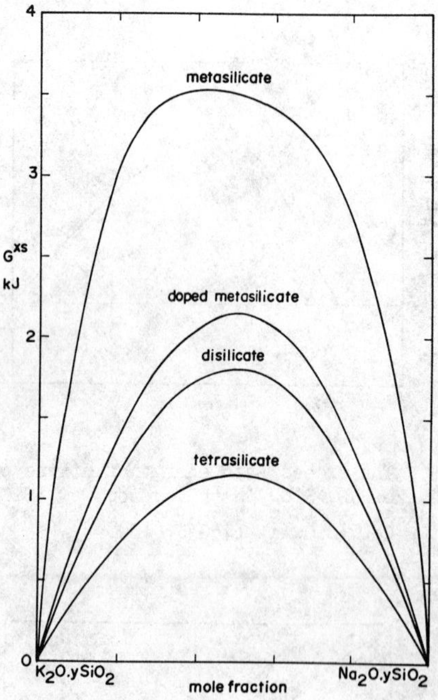

Figure 26. The excess free energies of mixing per mole of $M_2O.ySiO_2$ in pseudo-binary sections of the system K_2O-Na_2O-SiO_2 at 1100°C.

the anion matrix must accommodate cations of differing oxygen coordination number. According to the discrete ion theory, the predominant anion occurring at the tetrasilicate composition is $Si_9O_{21}^{6-}$ and density measurements give the numbers of moles of cations per cm^3 of melt as 0.0149 and 0.013, respectively, in $Na_2O.4SiO_2$ and $K_2O.4SiO_2$ at 1100°C (14). The small positive deviations from ideal silicate mixing, observed in the tetrasilicate pseudo-binary, suggest that the local distortion of the silicate anions, required for accommodation of cations of differing oxygen coordination numbers, produces the equivalent of a weak cation-cation interaction; although the parabolic form of the consequent deviation from ideal silicate mixing indicates that this interaction is not of sufficient magnitude to cause non-random mixing of the cations.

In moving from the tetrasilicate to the disilicate composition, the postulated mean silicate ion decreases in size to $Si_6O_{15}^{6-}$ and the concentrations of cations in the pure sodium silicate and pure potassium silicate melts increase to 0.0250 and 0.0209 moles/cm^3 respectively. Thus, in the pseudo-binary disilicate mixtures, the mean distance between the cations is smaller and hence their accommodation by the anion matrix is more difficult. The greater cation-cation interaction gives rise to greater positive deviation from ideal silicate mixing, although, again, the parabolic form of the excess free energy indicates that random mixing of the cations still occurs. In moving from the disilicate to the metasilicate composition, the postulated mean anion size is halved, to $Si_3O_9^{6-}$, and the concentrations of the cations in the pure sodium silicate and pure potassium silicate melts increase to 0.0374 and 0.0286 moles/cm^3, respectively. The observed behavior in the pseudo-binary metasilicate mixtures indicates that the cation-cation interaction is sufficiently strong, and accommodation of the cations sufficiently difficult, that random mixing of the cations does not occur. It is possible that the mean distance between the cations is small enough that each cation type influences the oxygen coordination number of the other cation type. The observed thermodynamic behavior suggests that the pseudo-binary metasilicate system can be divided into two terminal regions, each of which extend some 25% of the way into the system, and a central region. This behavior suggests two possibilities when K^+ is substituted for Na^+ in $Na_2O.SiO_2$ and Na^+ is substituted for K^+ in $K_2O.SiO_2$; (i) a limit is reached in accommodation when approximately 25% replacement has occurred and the central region is thus a near-ideal mixture of the two limiting groupings and (ii) the average oxygen coordination numbers reach the same value at the two limiting compositions, and thereafter, the melt has a reasonably constant structure. The positive deviations from ideal silicate mixing observed in the pseudo-binary disilicate and metasilicate mixtures are not inconsistent with the ideal volume and expansivity behavior observed in the pseudo-binary $2Na_2O.3SiO_2$-$2K_2O.3SiO_2$ at 1400°C (14).

The activities of Na_2O, K_2O and SiO_2 in the system $Na_2O.2SiO_2$-$K_2O.2SiO_2$, calculated from combination of the measured activities in Na_2O-SiO_2 melts (11) and the measured free energies of mixing in the system Na_2O-K_2O-SiO_2 are shown in Figs. 27 and 28. The behavior corresponding to ideal silicate mixing is shown by the broken lines in these Figs. and, as is seen, Na_2O and K_2O show positive deviations and SiO_2 shows a negative deviation. Figure 29 shows the excess free energy of mixing, G^{xs}, in the system $Na_2O.2SiO_2$-$K_2O.2SiO_2$, taken from Fig. 26, together with the contributions from the individual oxides. The contributions due to Na_2O and K_2O are essentially equal to one another and are also essentially equal, although opposite in sign, to the contribution due to SiO_2.

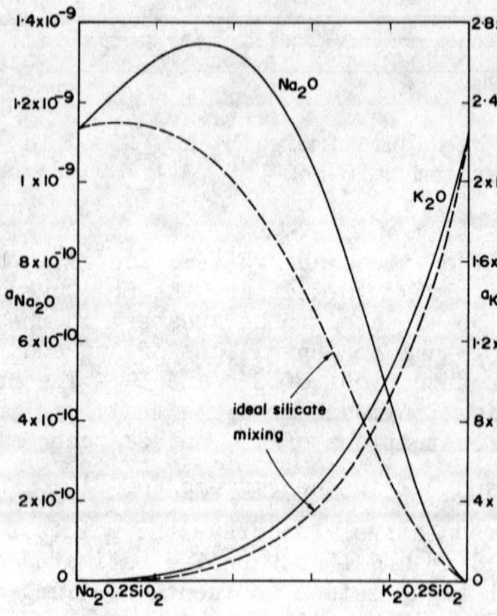

Figure 27. The activities of Na_2O (relative to liquid Na_2O) and K_2O (relative to solid K_2O) in melts in the system $Na_2O.2SiO_2$-$K_2O.2SiO_2$ at 1100°C in comparison with the variations calculated assuming ideal silicate mixing.

The validity of the structural interpretation in terms of the "cation accommodation effect" has been examined (45) by replacing a fraction of the cations in the pseudo-binary metasilicate mixture with Ca^{2+} which, being intermediate in size between Na^+ and K^+, should have an intermediate oxygen coordination number. By causing an increase in the average oxygen coordination number, it would be expected that a small replacement of Na^+ in $Na_2O.SiO_2$ by Ca^{2+} would cause the average structure of

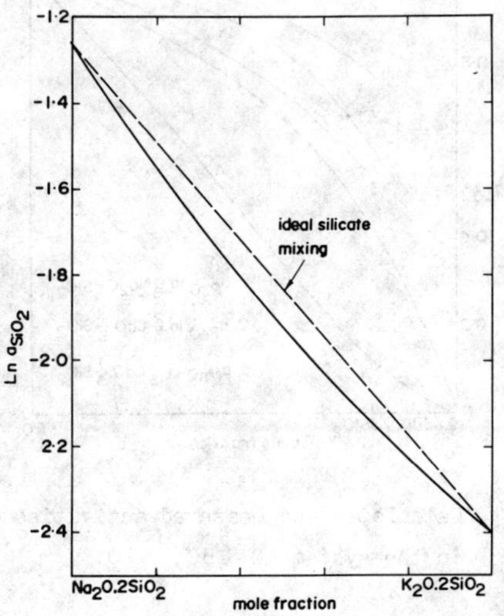

Figure 28. The activity of SiO_2 (relative to solid SiO_2) in melts in the system $Na_2O.2SiO_2$-$K_2O.2SiO_2$ at 1100°C in comparison with the variation calculated assuming ideal silicate mixing.

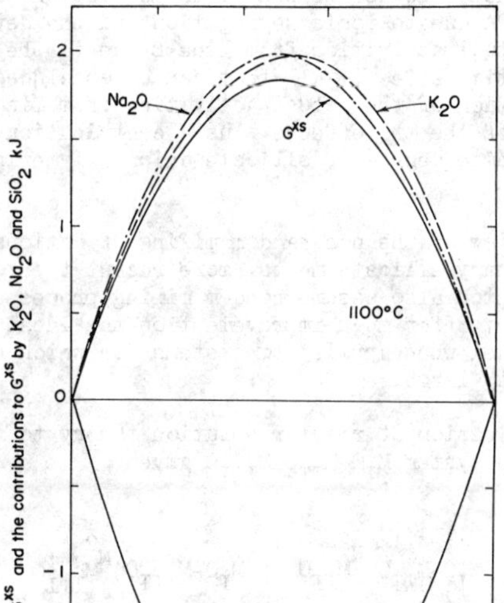

Figure 29. Contributions to the excess free energy of mixing in the system $Na_2O.2SiO_2$-$K_2O.2SiO_2$ at 1100°C from K_2O, Na_2O and SiO_2.

the melt to shift towards that typical of $K_2O.2SiO_2$, and, similarly, it would be expected that a small replacement of K^+ in $K_2O.2SiO_2$ by Ca^{2+} would cause the average structure to shift towards that typical of $Na_2O.2SiO_2$. As a result, it would be expected that the Ca^{2+}-doped metasilicate mixtures would exhibit smaller deviations, per mole of alkali oxide, from ideal silicate mixing than the undoped system. Measurements made on the system $(0.15CaO.0.85Na_2O).SiO_2$-$(0.15CaO.0.85K_2O).SiO_2$ verified this prediction. The excess free energy of mixing, per mole of alkali oxide, in the doped system, shown in Fig. 26, is almost half that in the undoped system and parabolic behavior is restored.

As a first approximation, deviation from ideal silicate mixing in ternary silicate melts can be considered in terms of the "preferred ionic association effect", which causes negative deviations, and the "cation accommodation effect", which causes positive deviations, as follows. Consider the composition path SiO_2 - a in the hypothetical system AO-BO-SiO_2, shown in Fig. 30(a).

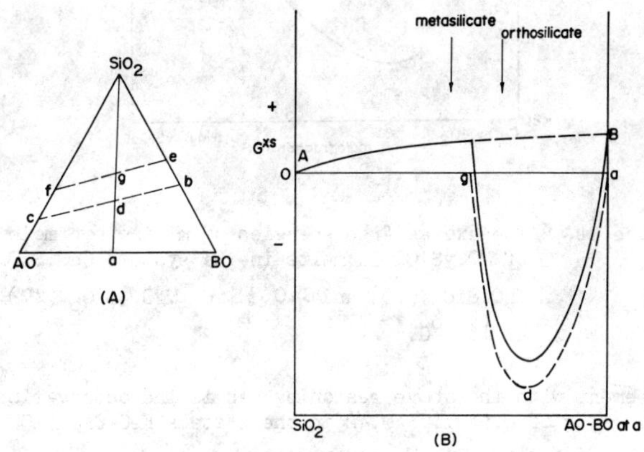

Figure 30. Illustration of the preferred ionic association and cation accommodation effects on the excess free energies of mixing in ternary silicate melts.

Let it be that AO is significantly more basic than BO, such that b and c are the compositions, in the binary systems BO-SiO_2 and AO-SiO_2 respectively, at which $n_{O}2-$ = $2n_{O}-$, and e and f are the values of N_{SiO_2} in the respective binary silicates above which $n_{O}2-$ is negligibly small. Thus, in the system SiO_2 - a, the points d and g represent the approximate compositions at which, respectively, it can be expected that $n_{O}2-$ = $2n_{O}-$ and $n_{O}2-$ → 0. The variation of G^{xs} per mole of oxide in the system SiO_2 - a, shown in Fig. 30(b) can thus be considered as follows. The broken line AB represents the contribution to G^{xs} due to the "cation accommodation effect". This effect increases with decreasing N_{SiO_2} due to the decreasing size of the silicate anions and the consequent increasing difficulty of cation accommodation. In the binary system AO-BO, the size difference between A^{2+} and B^{2+} gives rise to a positive deviation from Raoultian mixing typical of that of a fused salt mixture containing a common anion. The broken line adg in Fig. 30(b) represents the contribution to the deviation from ideal silicate mixing due to the "preferred ionic association effect" which, on adding SiO_2 to the binary melt a, increases to a maximum at d and, thereafter decreases and vanishes at g where $n_{O}2-$ becomes negligibly small. The full line drawn in Fig. 30(b) is the simple sum of the two effects. Thus in the vicinity of N_{SiO_2} = 0.5, the cation accommodation effect predominates, such that positive deviations occur, and, in the vicinity of $N_{SiO_2} = 1/3$ the preferred ionic association effect predominates, such that negative deviations occur. Between these compositions the effects compensate one another and apparent ideal silicate mixing occurs. The excess free energies, at 1600°C, measured in the metasilicate system $CaO.SiO_2$-$MgO.SiO_2$ (10) and calculated from the phase diagram for the orthosilicate system $2CaO.SiO_2$-$2MgO.SiO_2$ (47), shown in Fig. 31, are in qualitative

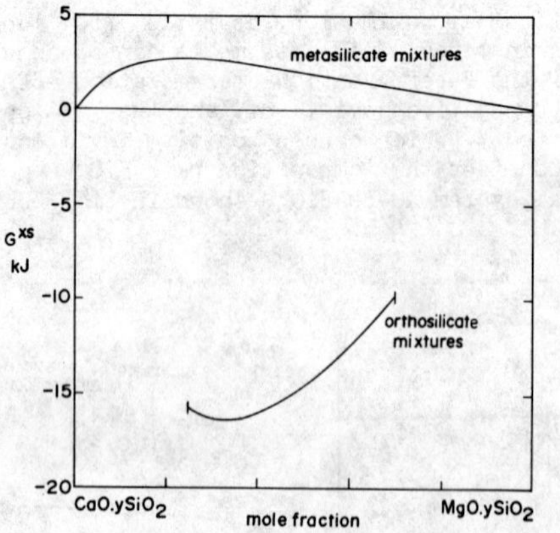

Figure 31. The excess free energies of mixing per mole of $MO.ySiO_2$ in melts in the systems $CaO.SiO_2$-$MgO.SiO_2$ (10) and $CaO.\frac{1}{2}SiO_2$-$MgO.\frac{1}{2}SiO_2$ (47) at 1600°C.

agreement with the above reasoning, as is the observation that melts of $0.5 > N_{SiO_2} > 0.42$ in the systems FeO-CaO-SiO_2 and MnO-CaO-SiO_2 exhibit ideal silicate mixing.

The occurrence of the cation accommodation effect and the preferred ionic association effect in basic ternary silicate melts can be demonstrated as follows. If the pseudo-binary AO-$2/3BO.1/3SiO_2$ is considered to be a Temkin mixture of the ions A^{2+}, B^{2+}, O^{2-} and SiO_4^{4-}, in that the free energy of mixing is determined solely by random mixing of A^{2+} and B^{2+} on the cation sublattice and random mixing of the O^{2-} and SiO_4^{4-} on the anion sublattice, the activity of AO is obtained as

$$a_{AO} = N_{A^{2+}} . N_{O^{2-}} = \frac{9N^2}{(2+N)(1+2N)} \qquad |49|$$

where $N_{A^{2+}}$ and $N_{O^{2-}}$ are, respectively, the cation fraction of A^{2+} and the anion fraction of O^{2-}, and N is the mole fraction of AO in the system AO-$2/3BO.1/3SiO_2$, Eq. (49) is drawn as line A in Fig. 32. If, however, as suggested by Kapoor and Frohberg (35), it is considered that the extent of the preferred ionic association of A^{2+} with O^{2-} and B^{2+} with the O^- on the SiO_4^{4-} is maximized, the free energy of mixing is that obtained by assuming random mixing of AO groups and B_2SiO_4 groups, in which case,

$$a_{AO} = \frac{n_{AO}}{n_{AO} + n_{B_2SiO_4}} \qquad |50|$$

$$= \frac{N}{N + 1/3(1-N)}$$

Eq. |50|, which is drawn as line B in Fig. 32, illustrates the extent to which a_{AO} can be increased by the occurrence of preferred ionic association. Also shown in

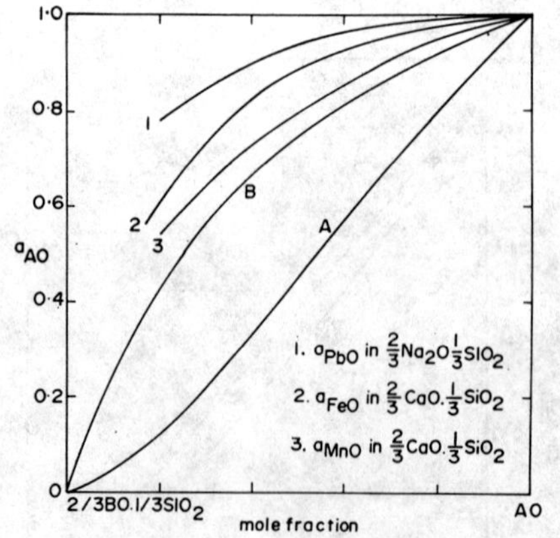

Figure 32. Calculated and measured activities of AO in the system AO-$\frac{2}{3}BO.\frac{1}{3}SiO_2$

Fig. 32 are the variations of a_{PbO} in the system PbO-$2/3Na_2O.1/3SiO_2$, a_{FeO} in the system FeO-$2/3CaO.1/3SiO_2$ and a_{MnO} in the system MnO-$2/3CaO.1/3SiO_2$. Kapoor and Frohberg suggested that the positive deviation of the actual activities from line B is due to the cation accommodation effect arising from interaction among the cations. The magnitudes of the deviations can be correlated with the difference between the sizes of the A^{2+} and B^{2+} cations, (the ionic radii of Pb^{2+}, Ca^{2+}, Na^+, Mn^{2+} and Fe^{2+} are 1.20, 0.99, 0.95, 0.80 and 0.76 Å.) The deviation of the actual activities from line A cannot be considered as the sum of the deviation of line B from line A due to ionic segregation and the deviation of the actual activities from line B due to the cation accommodation effect. All that can be concluded is that the deviation of the actual activities from line A is the sum of the two effects plus a contribution arising from a finite degree of silicate anion polymerization in the melts.

In view of the non-random mixing of cations in most basic ternary silicate melts, more recent thermodynamic models, which also assume random mixing processes (48, 49), do not offer much improvement on the ideal silicate mixing model when applied to systems for which ΔG^o_{47} is appreciably large.

Application of regular solution theory to the ternary oxide system M_IO-$M_{II}O$-$M_{III}O$ gives

$$\ln \gamma_{M_IO} = \alpha_{M_IO-M_{II}O} . N^2_{M_{II}O} + \alpha_{M_IO-M_{III}O} . N^2_{M_{III}O}$$

$$+ (\alpha_{M_IO-M_{II}O} + \alpha_{M_IO-M_{III}O} - \alpha_{M_{II}O-M_{III}O}) N_{M_{II}O} . N_{M_{III}O} \qquad |51|$$

where α_{i-j} is the regular solution parameter for the binary system i-j. Bell and coworkers have shown that activities in melts in the systems FeO-MnO-TiO_2 (50), FeO-MnO-Al_2O_3 (51) and CaO-MnO-TiO_2 (52) can be represented by the regular solution formalism with the values of α_{i-j} listed in Table III.

The observation, that the experimentally-measured activity of an oxide in a ternary oxide melt can be represented by the regular solution formalism, does not necessarily validate the structural assumptions inherent

Table III

system	α	temperature, °C
$FeO-TiO_2$	-1.25	1500
$MnO-TiO_2$	-1.95	1500
FeO-MnO	-0.1	1500
$CaO-TiO_2$	(-6.0)	1500
$FeO-AlO_{1.5}$	-0.05	1550
$MnO-AlO_{1.5}$	-0.6	1550

in regular solution theory. This is made evident in the work of Fujita and Maruhashi, who have shown that the activity of FeO in $FeO-MnO-SiO_2$ melts at 1560°C can be represented by the regular solution formalism with

$$\alpha_{FeO-MnO} = 0,\ \alpha_{FeO-SiO_2} = -1.84 \text{ and } \alpha_{MnO-SiO_2} = -4.28 \qquad (53).$$

In fact, if the interaction parameter for the binary system containing the two basic oxides is taken as being zero, the expressions given by the regular solution formalism are identical in form with those obtained assuming ideal silicate mixing, e.g. from the regular solution formalism, with $\alpha_{AO-BO} = 0$ in the system $AO.ySiO_2$-$BO.ySiO_2$

$$\ln. \frac{\gamma_{AO(t)} \cdot \gamma^{y}_{SiO_2(t)}}{\gamma_{AO(AO-SiO_2\ binary)} \cdot \gamma^{y}_{SiO_2\ (AO-SiO_2\ binary)}} = 0$$

which, on substitution into Eq. |43|, generates Eq. |41|. Also, with $\alpha_{AO-BO} = 0$, the regular solution formalism gives Eq. |38|, which was also derived assuming ideal silicate mixing. Thus, if the system exhibits ideal silicate mixing, some of its thermodynamic properties can be represented by the regular solution formalism if it is assumed that α_{AO-BO} is zero. However, the regular solution formalism does not reproduce the activities of FeO and MnO in the constituent binary silicate systems, and also the value of $\ln(\gamma_{FeO}/\gamma_{MnO})$ in any pseudo-binary section of constant N_{SiO_2}, given by the regular solution formalism as

$$\ln(\gamma_{FeO}/\gamma_{MnO}) = 2.44 N^2_{SiO_2}$$

is significantly different from the accurate value given by Eq. |50|. Thus the observation that a_{FeO} can be represented by the regular solution formalism does not require that the thermodynamic properties be determined by random mixing of Mn^{2+}, Fe^{2+} and Si^{4+} cations.

Structural interpretation of viscosities measured in $CaO-Al_2O_3-SiO_2$ melts indicates that, in this system, Al_2O_3 exhibits classical amphoteric behavior, i.e. behaves as an acidic or "network-forming" oxide in melts of high $N_{CaO}/N_{Al_2O_3}$ and behaves as a basic or "network-modifying" oxide in melts of low $N_{CaO}/N_{Al_2O_3}$. The iso-viscosity contours in the system at 2000°C, shown in Fig. 33, exhibit a ridge along the line $N_{CaO}/N_{Al_2O_3} = 1$ and, in any pseudo-binary system of constant N_{SiO_2}, the activation energy for viscous flow exhibits a maximum near the composition $N_{CaO}/N_{Al_2O_3} = 1$ (54). If it is assumed that the viscosity of a melt is determined primarily by the average anion size, then along lines of

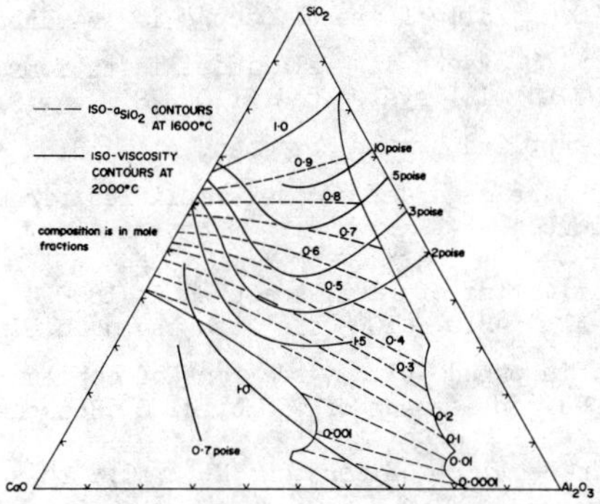

Figure 33. Iso-viscosity contours at 2000°C (54) and iso-a_{SiO_2} contours at 1600°C (10) in melts in the system $CaO-Al_2O_3-SiO_2$.

constant N_{SiO_2}, the average anion size increases as $N_{Al_2O_3}/N_{CaO}$ increases from zero to unity and the average anion size decreases as the ratio $N_{CaO}/N_{Al_2O_3}$ decreases from unity to zero. The composition $CaO.Al_2O_3.2SiO_2$, occurring at $N_{CaO}/N_{Al_2O_3} = 1$, is that of anorthite, which, in the solid state, has a 3-dimensional network structure of alternating SiO_4 and AlO_4 tetrahedra sharing common corners. Of every 32 oxygens, 20 are linked to one Ca, 4 are linked to two Ca and 8 have no links with Ca; and each Ca is located in an irregular cavity bounded by 10 oxygens (55). The structure is thus similar to that of silica in which half of the silicons are isomorphously replaced by aluminums and electroneutrality is maintained by Ca; and in the liquid state the structure can be regarded as a 3-dimensional network which is extensively damaged by the Ca - O bonds.

The variation, with composition, of viscosity in the system $CaO.SiO_2$-$Al_2O_3.SiO_2$, which contains the anorthite composition, can be considered as follows. According to the discrete ion theory the most probable ion in liquid $CaO.SiO_2$ is $Si_3O_9^{6-}$ or $Si_4O_{12}^{8-}$. As the CaO in $CaO.SiO_2$ is replaced by Al_2O_3, it would appear that aluminum enters the anion structure and the degree of polymerization, and hence the size of the aluminosilicate anions formed, are increased. This increase, which causes an increase in the viscosity of the melt, proceeds smoothly until the anorthite composition $CaAl_2Si_2O_8$ is reached. Thereafter, as CaO is further replaced by Al_2O_3, the added Al_2O_3 behaves as a basic oxide, providing Al^{3+} cations and O^{2-} ions which depolymerize the aluminosilicate anions in accordance with Eq. |1|. Thus, with further replacement, the average anion size and hence, the viscosity of the melt, are decreased. The degree of covalency of the bond between a metal atom and an oxygen is determined by the extent of charge transfer between them and hence it is likely that, in strong acid-base oxide systems, the amphoteric nature of aluminum allows it to exist in a range of states which vary continuously between "covalently-bonded" Al in an aluminosilicate anion and "cationic" Al^{3+}. The distribution of Al over the available states will be determined by the relative energies of the states, which, in turn, should be determined primarily by the ratio N_{CaO}/N_{SiO_2} and the value of $N_{Al_2O_3}$ in the melt, e.g. in melts of high N_{CaO}/N_{SiO_2}, Al-O bonds involving a lesser extent of charge transfer should be of lower energy than those involving a greater extent of charge transfer, and in melts of low N_{CaO}/N_{SiO_2} the reverse should be the case. In view of this, the presumptions that, in the partial system $CaO.SiO_2$-$CaO.Al_2O_3.2SiO_2$, all of the Al occurs in tetrahedral coordination with oxygen in aluminosilicate ions, and that, in the partial system $CaO.Al_2O_3.2SiO_2$-$Al_2O_3.SiO_2$, all the Al supplied by the component $Al_2O_3.SiO_2$ occurs as Al^{3+}, must be regarded as oversimplified first approximations.

The activities of SiO_2, measured at 1600°C in the system CaO-Al_2O_3-SiO_2 (10) are also shown in Fig. 33. As is seen, in pseudo-binary sections of constant N_{SiO_2}>0.6 the replacement of CaO by Al_2O_3 decreases a_{SiO_2} and in sections of constant N_{SiO_2}<0.6 this replacement increases a_{SiO_2}. In their present state of development, theories relating the thermodynamic properties of oxide melts with ionic structure and interaction are not capable of providing a quantitative explanation of the influence of the amphoteric nature of Al_2O_3 on the thermodynamic behavior of blast furnace slags.

ACKNOWLEDGEMENT

This work was supported by the National Science Foundation under Grant Number DMR77-03426.

REFERENCES

1. L. Pauling, The Nature of the Chemical Bond, 3rd ed., Chap. 3, Cornell Univ. Press, 1960.
2. A. E. Grau and S. N. Flengas, "Activities of SnO in the SnO-SiO_2 System", J. Electrochem. Soc., 123 (1976), p852.
Z. Kozuka, O. P. Siahaan and J. Moriyama, "Thermodynamic Study of Molten SnO-SiO_2 System", Japan Inst. Metals, 200 (1969), p9.
3. R. Schuhmann, Jr. and P. J. Ensio, "Thermodynamics of Iron Silicate Slags; Slags Saturated with Gamma Iron", J. Metals, 3 (1951), p401.
C. Bodsworth, "The Activity of Ferrous Oxide in Silicate Melts", JISI 193 (1959), p13.
P. A. Distin, S. G. Whiteway and C. R. Masson, "Thermodynamics and Constitution of FeO-SiO_2 Melts", Can. Met. Quart., 10 (1971), p73.
4. I. B. Smith and C. R. Masson, "Activities and Ionic Distributions in Cobalt Silicate Melts", Can. J. Chem., 49 (1971), p683.
5. G. M. Mehrotra, M. G. Frohberg and M. L. Kapoor, "Thermodynamic Properties of the System PbO-Bi_2O_3", Can. Met. Quart., 15 (1976), p215.
6. B.K.D.P. Rao and D. R. Gaskell, "The Thermodynamic Properties of Melts in the System MnO-SiO_2", Met. Trans. B, in press.
K. P. Abraham, M. W. Davies and F. D. Richardson, "Activities of MnO in Silicate Melts", JISI, 196 (1960), p82.
7. F. D. Richardson and L. E. Webb, "Oxygen in Molten Lead and the Thermodynamics of PbO-SiO_2 Melts", Bull. Inst. Min. Met., 64 (1955), p529.
R. Sridhar and J.H.E. Jeffes, "Thermodynamics of PbO and PbO-SiO_2 Melts", Trans. Inst. Min. Met., 76 (1967), pC44.
Z. Kozaka and C. S. Samis, "Thermodynamic Properties of Molten PbO-SiO_2 Systems", Met. Trans., 1 (1970), p871.
G. G. Charette and S. N. Flengas, "Thermodynamic Properties of PbO-SiO_2 Slags by EMF Measurements", Can. Met. Quart, 7 (1969), p191.
E. Kato and Y. Ohuchi, "Mass-Spectrometric Determination of Activities in the Liquid PbO-SiO_2 System", Chemical Metallurgy of Iron and Steel, p26, Iron and Steel Inst. London, 1973.
8. M. L. Kapoor and M. G. Frohberg, "Thermodynamic Properties of the System PbO-B_2O_3", Can. Met. Quart, 12 (1973), p137.
9. A. Leung and W. T. Thompson, "A Thermodynamic Study of PbO-GeO_2 Melts", Can. Met. Quart., 15 (1976), p227.
10. R. H. Rein and J. Chipman, "Activities in the Liquid Solution SiO_2-CaO-MgO-Al_2O_3 at 1600°C, Trans. Met. Soc. AIME, 233 (1965), p415.
11. D. A. Neudorf and J. F. Elliott, "Thermodynamic Properties of Na_2O-CaO-SiO_2 Melts at 1000° to 1100°C", Met. Trans. B, 11B (1980), p607.
12. W. F. Caley and C. R. Masson, "Thermodynamics of Lead Phosphate Slags by an Electrochemical Method", Metal-Slag-Gas Reactions and Processes, p140, Electrochem. Soc. Princeton, N.J., 1975.
R. Sridhar and J.H.E. Jeffes, "Thermodynamics of PbO-B_2O_3 and PbO-P_2O_5 Melts", Trans. Inst. Min. Met., 78 (1969), pC14.
13. J. O'M. Bockris, J. D. Mackenzie and J. A. Kitchener, Viscous Flow in Silica and Binary Liquid Silicates", Trans. Faraday Soc., 51 (1955), p1734.
14. J. O'M. Bockris, J. W. Tomlinson and J. L. White, "The Structures of Liquid Silicates; Partial Molar Volumes and Expansivities", Trans. Faraday Soc., 52 (1956), p299.
15. J. W. Tomlinson, M.S.R. Heynes and J. O'M. Bockris, "The Structure of Liquid Silicates. Part 2; Molar Volumes and Expansivities", Trans. Faraday Soc., 54 (1958), p1822.
16. C. R. Masson, "An Approach to the Problem of Ionic Distributions in Liquid Silicates", Proc. Roy. Soc., A287 (1965), p201.
17. C. R. Masson, "Ionic Equilibria in Liquid Silicates", J. Amer. Ceram. Soc., 51 (1968), p134.
18. S. G. Whiteway, I. B. Smith and C. R. Masson, "Theory of Molecular Size Distribution in Multichain Polymers", Can. J. Chem., 48 (1970), p33.
19. C. R. Masson, I. B. Smith and S. G. Whiteway, "Activities and Ionic Distributions in Liquid Silicates; Application of Polymer Theory", Can. J. Chem., 48 (1970), p1456.
20. E. A. Guggenheim, Mixtures, Chap. 10, Oxford Press, 1952.
21. D. R. Gaskell, "Thermodynamic Models of Liquid Silicates", Can. Met. Quart., in press.
22. H. Flood and T. Forland, "The Acidic and Basic Properties of Oxides", Acta Chem. Scandanavica, 1 (1947), p592.
23. A.E.R. Westman and P. A. Gartaganis, "Constitution of Sodium, Potassium and Lithium Phosphate Glasses", J. Amer. Ceram. Soc., 40 (1957), p293.
24. T. R. Meadowcroft and F. D. Richardson, "Structure and Thermodynamic Aspects of Phosphate Glasses", Trans. Faraday Soc., 61 (1965), p54.
25. C. J. Cripps-Clark, R. Sridhar, J.H.E. Jeffes and F. D. Richardson, "Chain Distributions and Transition Temperatures for Phosphate Glasses", Phys. Chem. of Process Metallurgy; the Richardson Conference, p233, Inst. Min. Met. London, 1974.
26. T. R. Meadowcroft and F. D. Richardson, "Heats of Formation of Some Crystalline and Glassy Phosphates", Trans. Faraday Soc., 59 (1963), p1564.
27. O. Ogawa and D. R. Gaskell, unpublished research, Univ. of Pennsylvania, 1975.
28. J. Gotz and C. R. Masson, "Trimethylsilyl Derivatives for the Study of Silicate Structure. Part I. A Direct Method of Trimethyl Silylation", J. Chem. Soc. Section A (1970), p2683.
29. C. R. Masson, W. D. Jamieson and F. G. Mason, "Ionic Constitution of Metallurgical Slags", Phys. Chem. of Process Metallurgy; the Richardson Conference, p223, Inst. Min. Metallurgy, London, 1974.
30. T. Yokokawa and K. Niwa, "Free Energy of Solution in Binary Silicate Melts", Trans. Japan Inst. Metals, 10 (1969), p1.
31. R. Kruh and S. H. Stern, "The Effects of Solutes on the Properties and Structure of Liquid B_2O_3", JACS, 78 (1956), p278.
32. B.K.D.P. Rao and D. R. Gaskell, "The Activity of MnO in Melts containing SiO_2, B_2O_3 and TiO_2", Met. Trans. B, in press.

33. E. Martin and H. B. Bell, "Activities in the System $MnO-TiO_2$ at 1500°C", Trans. Inst. Min. Met., 83 (1974), pC193.
34. J. Lumsden, "The Thermodynamics of Liquid Iron Silicates", Phys. Chem.Process Met. Part 1, p165, Interscience N.Y., 1961.
35. I. C. Smith and H. B. Bell, "Thermodynamics of $FeO-TiO_2$ Melts Saturated with Iron at 1475°C", Trans. Inst. Min. Met., 79 (1970), pC253.
36. F. D. Richardson, "Activities in Ternary Melts", Trans. Faraday Soc. 52 (1956), p1312.
37. J. F. Elliott, "Activities in the Iron Oxide-Silica-Lime System", Trans. AIME, 203 (1955), p485.
38. K. P. Abraham, M. W. Davies and F. D. Richardson, "Activity of MnO in Silicate Melts", JISI, 196 (1960), p82.
39. M. L. Kapoor and M. G. Frohberg, "The Activity of Lead Oxide in the System Sodium Oxide-Lead Oxide-Silica", Met. Trans. B, 8B (1977), p15.
40. H. B. Bell, "Equilibrium Between $FeO-MnO-MgO-SiO_2$ Slags and Molten Iron at 1550°C", JISI, 199 (1963), p116.
41. A. Muan and E. F. Osborne, Phase Equilibrium Among Oxides in Steelmaking, Chap. 8, Addison-Wesley Pub. Co., Reading, Mass, 1965.
42. K. S. Song and D. R. Gaskell, "The Free Energies of Mixing in Melts of the Systems $2FeO.SiO_2-2MnO.SiO_2$ and $2.33FeO.TiO_2-2.33MnO.TiO_2$", Met. Trans. B, 10B (1979), p15.
43. G. R. Belton, H. Suito and D. R. Gaskell, "Free Energies of Mixing in Liquid Iron-Cobalt Orthosilicates at 1450°C", Met. Trans., 4 (1973), p2541.
44. G. R. Belton, U. V. Choudary and D. R. Gaskell, "Thermodynamics of Mixing in Molten Sodium-Potassium Silicates", Phys. Chem. Process Metallurgy; the Richardson Conference, p247, Inst. Mining and Metallurgy, London, 1974.
45. U. V. Choudary, D. R. Gaskell and G. R. Belton, "Thermodynamics of Mixing in Molten Sodium-Potassium Silicates at 1100°C; The Effect of a Calcium Oxide Addition", Met. Trans. B, 8B (1977), p67.
46. Y. Waseda and H. Suito, "The Structure of Molten Alkali Silicates", Trans. ISIJ, 17 (1977), p82.
47. K. P. Abraham and F. D. Richardson, "The Mixing of Silicates in Molten Slags", Phys. Chem. Process Met. Part 1, p263, Interscience, N.Y., 1961.
48. T. Yokokawa and K. Niwa, "Free Energy and Basicity of Molten Silicate Solutions", Trans. Japan Inst. Metals, 10 (1969), p81.
49. M. L. Kapoor and M. G. Frohberg, "Theoretical Treatment of Activities in Silicate Melts", Chem. Met. Iron and Steel, p17, Iron and Steel inst. London, 1973.
50. I. C. Smith and H. B. Bell, "Thermodynamics of $FeO-MnO-TiO_2$ Melts Saturated with Iron at 1475°C", Trans. I.M.M., 80 (1971), pC55.
51. I. D. Sommerville, I. Ivanchev and H. B. Bell, "Activity in Melts Containing FeO,MnO,Al_2O_3 and SiO_2", Chem. Met. Iron and Steel p23, Iron and Steel Inst. London, 1973.
52. E. Martin, O.I.H. Abdelkarim, I. D. Sommerville and H. B. Bell, "The Thermodynamics of Melts Containing MnO,FeO,CaO,TiO_2 and SiO_2", Metal-Slag-Gas Reactions and Processes, p1, Electrochem. Soc., Princeton, N.J., 1975.
53. H. Fujita and S. Maruhashi, "Equilibrium between $FeO-MnO-SiO_2$ Slags and Molten Iron", Tetsu-To-Hagane, 56 (1970), p830.
54. P. Kozakevitch, "Viscosity of Lime-Alumina-Silica Melts between 1600° and 2100°C", Phys. Chem. Process Met. Part 1, p97, Interscience, N.Y., 1961.
55. L. Bragg and G. F. Claringbull, Crystal Structures of Minerals, p325, Cornell Univ. Press, 1965.

SLAG-METAL REACTION EQUILIBRIA

E. T. Turkdogan
United States Steel Corporation,
Research Laboratory,
Monroeville, Pa. 15146

The present state of knowledge of slag-metal reaction equilibria pertinent to ironmaking and steelmaking processes is critically reviewed. The equilibrium data are presented for silicon, manganese, and sulfur reactions under conditions appropriate for the blast furnace. Comments are made on departures from slag-metal equilibrium in the blast furnace. The equilibrium data are presented for steelmaking reactions in a form readily applicable to industrial slags, which are saturated with dicalcium (magnesium) silicate and magnesio-wustite. In most steelmaking processes there is close approach to slag-metal equilibrium with respect to silicon, manganese, sulfur, and phosphorus reactions at the time of furnace tapping. However, the iron oxide content of the slag is two to six times higher than that for the $\underline{C}$ - (FeO) equilibrium at 1 bar CO. Attention is drawn to special features of reactions in steelmaking with oxygen bottom blowing, as in Q-BOP and AOD. The equilibrium data are presented also for slag-metal reactions in the ladle deoxidation and desulfurization of steel.

INTRODUCTION

The present state of knowledge of slag-metal reaction equilibria discussed in this presentation is confined to systems and reactions that are relevant to ironmaking and steelmaking processes. The large volume of papers published on this subject will not be reviewed in any chronological order. The selected papers and text books to be cited in the course of the discussion may be consulted for references to numerous other papers on slag-metal reactions. With the objective of developing a better understanding of the workings of the pyrometallurgical processes, the fundamental aspects of slag-metal reactions will be discussed, with examples of key reactions encountered in iron and steelmaking. The discussion is divided into three topic areas in accord with the processing sequence of smelting and refining: (1) blast-furnace reactions, (2) steelmaking reactions, and (3) reactions in ladle treatment of steel. The state of reactions in industrial iron and steelmaking processes will be evaluated by comparing the plant data for compositions of metal and slag samples with the equilibrium data for slag-metal reactions.

A general comment should be made here on the chemical formulation of reactions involving slags. Although molten slags are ionic in nature, the chemical reactions are formulated in terms of the constituent oxides of the slag, because the thermodynamic activity of an individual ion cannot be determined. The thermodynamic quantities can be measured only for uncharged and electroneutral systems; therefore, only the activity of a neutral entity made up of at least two ionic species can be measured. The activity of an oxide in a polymeric melt is proportional to the product of the activities of the constituent ions

$$a_{MO} \propto (a_{M^{2+}})(a_{O^{2-}}) \qquad (1)$$

and is readily measured relative to the pure solid or liquid oxide MO.

However, numerous attempts have been made in the past to interpret the thermodynamic properties of slags in terms of the concentrations of ionic species derived from assumed ionic models of the polymer structures. The constitutional and structural models have been proposed by many to represent the equilibrium relations for slag-metal exchange reactions in terms of the concentrations of oxides or ionic species in slags. For a comprehensive review of this subject, reference may be made to a book by Richardson [1]. Although the structural models are helpful in improving our understanding of the structure of polymers, their application to slag-metal reaction equilibria is not particularly rewarding when compared with the much simpler conventional methods of representing the slag-metal equilibria in terms of the measured thermodynamic quantities. Unless stated otherwise, the equilibrium constants for the reactions considered in this paper are computed from the free energy and activity data compiled in a book by the author [2].

PARTIAL EQUILIBRIA AND COUPLED REACTIONS

The pyrometallurgical processes are dynamic processes in which three or more multicomponent phases interact. For the process to work, some reactions are made to go fast while other reactions are deliberately impeded. For example, the oxygen blowing in steel refining maintains a large driving force for rapid oxidation of impurities in the metal. On the other hand, the choice of the refractory materials and construction of the furnace lining are such that there is minimum refractory erosion due to reaction with liquid metal and slag. Although a state of nonequilibrium is maintained in the system as a whole, during ore smelting and metal refining some exchange reactions between slag and metal may approach equilibrium. Because of the sluggish approach to complete equilibrium in pyrometallurgical reactions involving several multicomponent phases, such as gas, slag and metal, the state of equilibrium for a particular heterogeneous reaction cannot be evaluated accurately from the experimental data without giving due consideration to interrelations among different reactions occurring simultaneously.

In conventional laboratory experiments with, for instance, blast furnace-type systems at unit activities of carbon and carbon monoxide, many hours of reaction time are needed to reach the equilibrium distribution of silicon and manganese between slag and metal to satisfy the three-phase equilibria in the reactions

$$(SiO_2) + \underline{2C} = \underline{Si} + 2CO \qquad (2)$$

$$(MnO) + \underline{C} = \underline{Mn} + CO \qquad (3)$$

In the above and subsequent equations, the underscore and squared brackets indicate elements dissolved in the metal and round brackets indicate oxides and sulfides dissolved in the slag. We see from the experimental data of Filer and Darken [3] on remelting of blast furnace metal and slag samples in a graphite crucible at 1 bar CO, Fig. 1, that it took about 18 to 20 hours at 1500°C for the manganese reaction (3) to approach equilibrium; for the silicon reaction (2), the time of equilibration was about 28 hours.

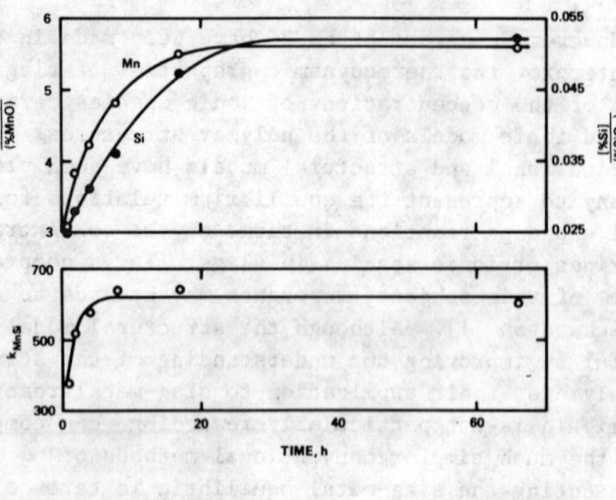

Figure 1. Approach to equilibrium for silicon and manganese reactions and change in the value of k_{MnSi} for manganese-silicon reaction with time of reaction of blast furnace slag and metal samples at 1500°C and unit activities of carbon and carbon monoxide; data are from Filer and Darken [3].

There is also the manganese-silicon exchange reaction between metal and slag

$$2(MnO) + \underline{Si} = (SiO_2) + \underline{2Mn} \qquad (4)$$

for which the equilibrium relation may be represented by

$$k_{MnSi} = \left(\frac{[\%Mn]}{(\%MnO)}\right)^2 \frac{(\%SiO_2)}{[\%Si]}, \qquad (5)$$

for a given temperature and slag basicity. As is seen from the data in Fig. 1, the value of k_{MnSi} remains essentially constant after about 5 hours reaction time, indicating partial equilibrium in the system relative to reaction (4). Although the manganese and silicon distribution ratios between metal and slag continue to increase with the reaction time, a state of equilibrium is maintained for the coupled reaction (4).

Another example of a coupled reaction and partial equilibrium is that observed by Turkdogan and Bills [4] in the simultaneous sulfurization and oxidation of iron silicate melts with gases containing CO, CO_2, and SO_2. Two reactions occur simultaneously:

$$\frac{1}{2} S_2 + (O^{2-}) = (S^{2-}) + \frac{1}{2} O_2 \qquad (6)$$

$$\frac{1}{2} O_2 + 2(Fe^{2+}) = (O^{2-}) + 2(Fe^{3+}). \qquad (7)$$

As suggested by the results in Fig. 2, reaction (6) is faster than reaction (7) with the result that the sulfur content of the melt reaches a maximum at some stage of the reaction when the following equilibrium is approached

$$\frac{1}{2} S_2 + 2(Fe^{2+}) = (S^{2-}) + 2(Fe^{3+}). \qquad (8)$$

However, the system as a whole is not at equilibrium.

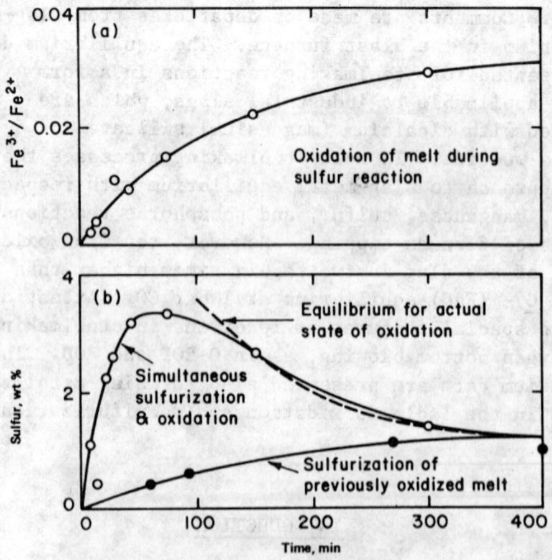

Figure 2. Oxidation and sulfurization of an iron silicate melt (40% SiO_2) in a CO-CO_2-SO_2 gas mixture at 1550°C with ingoing gas composition p_{CO}/p_{CO_2} = 3.0 and p_{SO_2} = 0.028 bar. After Turkdogan and Bills [4].

As the oxidation of divalent iron in the melt continues, the concentration ratio Fe^{3+}/Fe^{2+} increases, and because the sulfur potential in the gas is maintained constant, the sulfur content of the melt must decrease as dictated by reaction (8). As the whole system is approaching equilibrium, reaction (8) remains at a state of pseudo-equilibrium; that is, the system is at a state of partial equilibrium. When the ferrous silicate melt is first equilibrated with the prevailing oxygen partial pressure of the gas, the sulfurization proceeds in a normal manner, as shown by the lower curve in Fig. 2.

BLAST FURNACE REACTIONS

Relevant to the conditions in the bosh and hearth zones of the blast furnace, we may consider two types of overall reactions: primary reactions and coupled reactions. In the reaction zones considered, the partial pressure of CO (∿ 1 bar) and the activity of carbon (∿ 1 relative to graphite) remain essentially constant. The reactions involving two invariant components in the system, in this case C and CO, are called the primary reactions of which reactions (2), (3) and

$$\underline{S} + \underline{C} + (CaO) = (CaS) + CO \quad (9)$$

are the most important. The reactions that do not involve the invariant components of the system are called the coupled reactions, as in reaction (4) and

$$\underline{S} + \frac{1}{2}\underline{Si} + (CaO) = (CaS) + \frac{1}{2}(SiO_2) \quad (10)$$

$$\underline{S} + \underline{Mn} + (CaO) = (CaS) + (MnO). \quad (11)$$

These overall coupled reactions are the sum of transient electrochemical reactions. For example, the transfer of sulfur from metal to slag, requiring electrons, is accompanied by anodic oxidation of elements dissolved in the metal, leading to transient electrochemical reactions such as

$$\underline{Fe} + \underline{S} \rightarrow Fe^{2+} + S^{2-} \quad (12)$$

$$\underline{Mn} + \underline{S} \rightarrow Mn^{2+} + S^{2-} \quad (13)$$

$$2\underline{Si} + \underline{S} + 2(\equiv Si{-}O^-) \rightarrow 2(\equiv Si{-}O{-}Si\equiv) + S^{2-}. \quad (14)$$

As the slag and metal move towards equilibrium with respect to the primary sulfur reaction, for example, the distribution of silicon, manganese, and iron between slag and metal phases initially moves away from equilibrium. This phenomenon is well demonstrated experimentally by King and Ramachandran [5], and others.

Equilibria for Primary Reactions

The equilibrium constant for reaction (3) is represented by the equation

$$K_{Mn} = \frac{[\%Mn]\,\gamma'_{Mn}}{a_{MnO}}\,\frac{p_{CO}}{a_C} \quad (15)$$

where γ'_{Mn} is the activity coefficient of manganese as affected by carbon, defined such that $\gamma'_{Mn} \rightarrow 1$ when $\%C \rightarrow 0$; for graphite-saturated iron $\gamma'_{Mn} = 0.8$. In most blast furnace slags, the total number of moles of the constituent oxides per 100 g of slag is about 1.65. Therefore, the activity of manganese oxide, a_{MnO}, in the slag may be represented in terms of the weight percent MnO and its activity coefficient γ_{MnO}:

$$a_{MnO} = \frac{(\%MnO)\,\gamma_{MnO}}{70.94 \times 1.65}. \quad (16)$$

For unit activity of carbon in graphite-saturated melts ($a_C = 1$), equations (15) and (16) give

$$\frac{[\%Mn]}{(\%MnO)}\,p_{CO} = 1.07 \times 10^{-2}\,K_{Mn}\,\gamma_{MnO} \quad (17)$$

where p_{CO} is in bar. The free energy data give for the temperature dependence of the equilibrium constant

$$\log K_{Mn} = \frac{15{,}183}{T} + 10.874. \quad (18)$$

Abraham et al. [6] determined the activity coefficient of MnO in $CaO{-}MnO{-}Al_2O_3{-}SiO_2$ slags at 1500 and 1650°C. The values of γ_{MnO} taken from their data are plotted in Fig. 3 as a function of the weight percent ratio $\%CaO/\%SiO_2$ for 0, 10, and 20% Al_2O_3 in the slag. Because MgO behaves much like CaO in its effect on γ_{MnO}, the relation in Fig. 3 may be used for the blast furnace slags for which the basicity B is defined by the weight ratio

$$B = \frac{\%CaO + \%MgO}{\%SiO_2}. \quad (19)$$

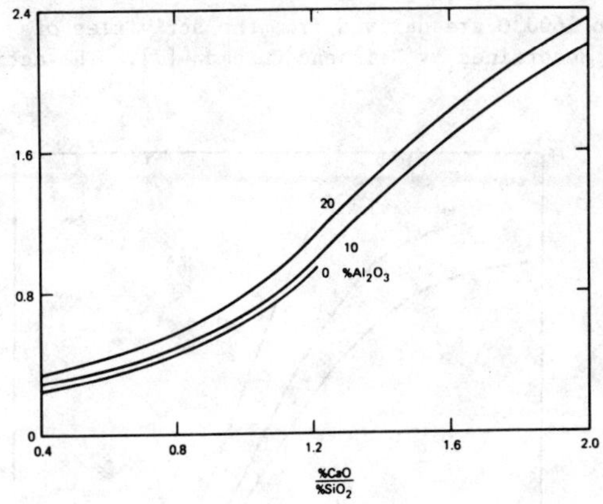

Figure 3. Activity coefficient of MnO, relative to pure solid oxide, in $CaO{-}Al_2O_3{-}SiO_2{-}MnO$ slags, with MnO<8%, at 1500 to 1650°C, derived from the data of Abraham et al. [6].

From equations (17) and (18) and Fig. 3 we can now compute the equilibrium distribution of manganese between graphite-saturated liquid iron and slag for any given partial pressure of CO and the basicity B. This equilibrium relation is shown in Fig. 4 for blast furnace-type slags, which usually contain 10 to 20 percent Al_2O_3. These calculated data agree well with values determined experimentally over limited ranges of slag compositions.

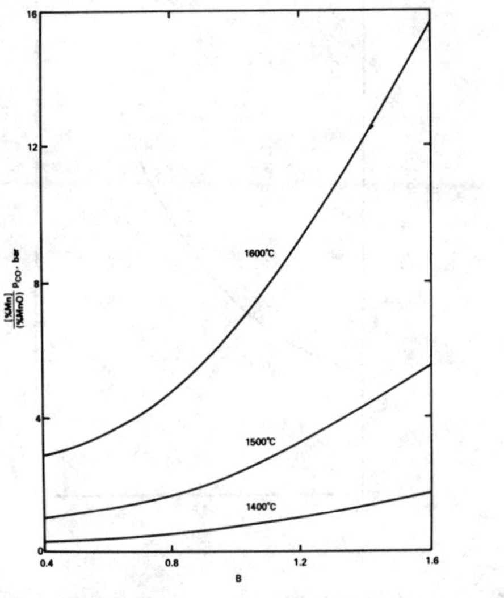

Figure 4. Calculated equilibrium metal/slag manganese distribution for graphite-saturated blast furnace-type slags containing 10% to 20% Al_2O_3.

Similar calculations will give the equilibrium distribution of silicon and sulfur between graphite-saturated liquid iron and slag for the primary reactions (2) and (9). The activity coefficients of the oxides and sulfides needed for such calculations are readily obtained from the available activity data. As indicated by the results of Rein and Chipman [7], the activity of SiO_2 is not affected much when CaO is substituted by MgO, at least up to about 20% MgO. The values of γ_{SiO_2}, relative to β-cristobalite, in Fig. 5 for the CaO-MgO-Al_2O_3-SiO_2 melts at temperatures of 1500 to 1600°C are derived from the activities of silica determined by Rein and Chipman [7]. The activity-coefficient ratio $\gamma_{CaO}/\gamma_{CaS}$, relative to solid oxide and sulfide, derived from the gas-slag equilibrium data of Sharma and Richardson [8], is shown in Fig. 6.

The equilibrium silicon and sulfur distribution ratios for graphite-saturated melts, shown in Figs. 7 and 8, were obtained by combining the equilibrium constants of various coupled reactions determined experimentally [9]. These equilibrium relations are considered more reliable than those calculated from the free energy and activity data.

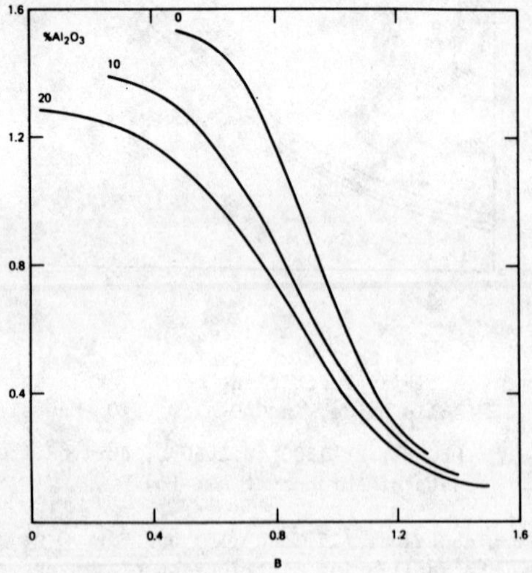

Figure 5. Activity coefficient of SiO_2, relative to β-cristobalite, in CaO-MgO-Al_2O_3-SiO_2 melts at 1500-1600°C, derived from the activity data of Rein and Chipman [7].

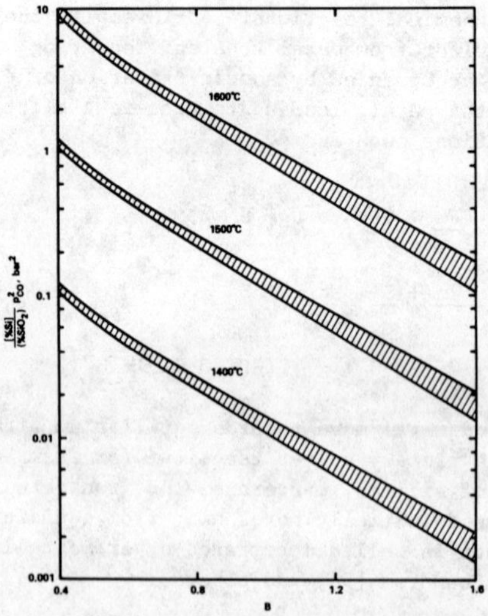

Figure 7. Equilibrium metal/slag silicon distribution for graphite-saturated blast furnace-type slags. After Turkdogan, et al. [9].

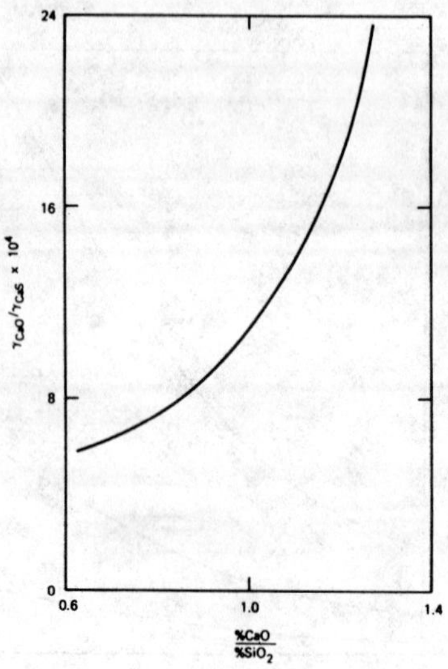

Figure 6. Lime/sulfide activity-coefficient ratio at 1500 to 1550°C in calcium silicate melts saturated with CaS, derived from the data of Sharma and Richardson [8].

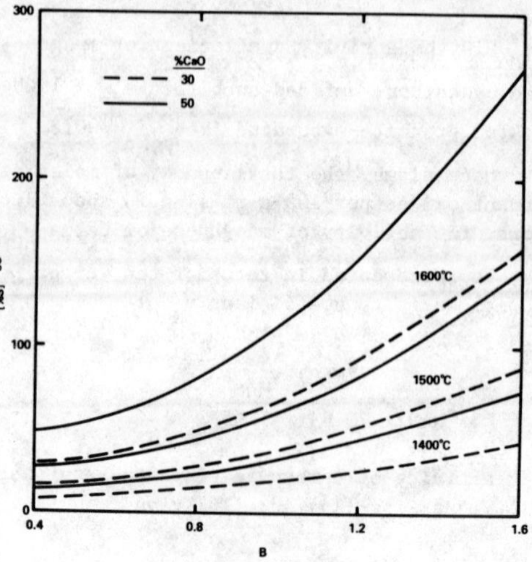

Figure 8. Equilibrium slag/metal sulfur distribution for graphite-saturated blast furnace-type slags. After Turkdogan, et al. [9].

Equilibria for Coupled Reactions

For small variations in the concentrations of Mn and Si in graphite-saturated iron, their activity coefficients, as affected by carbon, will remain essentially constant. Therefore, for a given temperature and slag basicity the equilibrium relation k_{MnSi} may be represented by

$$k_{MnSi} = \left(\frac{[\%Mn]}{(\%MnO)}\right)^2 \frac{(\%SiO_2)}{[\%Si]} . \tag{20}$$

Because the activity coefficients of MnO and SiO_2 are simple functions of the slag basicity B, the equilibrium relation k_{MnSi} should also depend on the slag basicity. The results of two different sets of experiments [3,9] are shown in Fig. 9 for liquid iron equilibrated with blast furnace-type slags in a graphite crucible at 1 bar pressure of CO. Because the

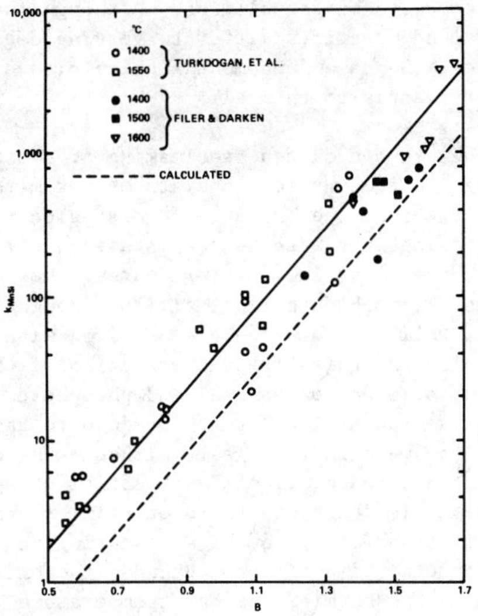

Figure 9. Variation of the equilibrium relation k_{MnSi} for the manganese-silicon coupled reaction (4) with slag basicity B for unit activity of carbon. After Turkdogan, et al. [9].

temperature coefficients of the equilibrium constants for reactions (2) and (3) (for 2 moles of MnO) are similar, a small temperature effect on k_{MnSi} is masked by the scatter in the experimental data. The average line indicating the equilibrium relation for the manganese-silicon reaction at unit activity of carbon is represented by the equation

$$\log k_{MnSi} = \log\left(\frac{[\%Mn]}{(\%MnO)}\right)^2 \frac{(\%SiO_2)}{[\%Si]} = 2.8B - 1.16. \tag{21}$$

For melts not saturated with graphite, equation (21) should be adjusted for the decrease in γ'_{Si} with decreasing concentration of carbon. The dotted line in Fig. 9 is calculated from the free energy and activity data. The difference of the calculated line from the average experimental values by a factor of 2.7 is attributed to the accumulated uncertainties in the free energy data and also to the approximations made in the calculations. An error in the calculated data may arise from the use of the free energy equations for the solution of silicon in pure liquid iron for reactions with graphite-saturated liquid iron.

The equilibrium constant for the silicon-sulfur coupled reaction (10).

$$K_{SiS} = \frac{(a_S)}{[a_S]} \left[\frac{(a_{SiO_2})}{[a_{Si}]}\right]^{1/2} \frac{1}{(a_{CaO})} \tag{22}$$

may be converted to a more practical form, thus

$$K'_{SiS} = \frac{(\%S)}{[\%S]} \left[\frac{(\%SiO_2)}{[\%Si]}\right]^{1/2} \frac{1}{(\%CaO)} \left(\frac{\gamma_{CaS}}{\gamma_{CaO}}\right) (\gamma_{SiO_2})^{1/2}. \tag{23}$$

Because of small variations in the concentrations of S and Si in the metal, their activity coefficients will remain essentially constant and therefore need not be included in the above equation. Because γ_{SiO_2} and the ratio $\gamma_{CaO}/\gamma_{CaS}$ vary primarily with the slag basicity, the equilibrium relation for a given temperature

$$k_{SiS} = \frac{(\%S)}{[\%S]} \sqrt{\frac{(\%SiO_2)}{[\%Si]}} \frac{1}{(\%CaO)} \tag{24}$$

will depend essentially on slag basicity only. For the manganese-sulfur coupled reaction (11) a similar expression may be written

$$k_{MnS} = \frac{(\%S)}{[\%S]} \frac{(\%MnO)}{[\%Mn]} \frac{1}{(\%CaO)} . \tag{25}$$

In a recent experimental study of blast furnace reactions, together with other experimental data, Turkdogan et al. [9] derived the following relations for the temperature range 1400 to 1600°C and for basicity B in the range 0.5 to 1.7,

$$\log k_{SiS} = \frac{6327}{T} - 4.43 + 1.4B$$
$$\log k_{MnS} = \frac{6327}{T} - 3.85. \tag{26}$$

Observations from Plant Data

From a study of blast furnace plant data, Turkdogan [10] noted that the manganese and silicon distribution ratios between metal and slag are two to four times lower than the equilibrium values for the primary manganese and silicon reactions for an average temperature of 1500°C and p_{CO} about 1 bar. As is seen from the plot of plant data in Fig. 10, the equilibrium is not reached even for the manganese-silicon coupled reaction. For all furnaces, the data points are below the equilibrium line, indicating that as the metal droplets pass through the slag layer, the direction of the reaction is

$$2(MnO) + \underline{Si} \rightarrow \underline{2Mn} + (SiO_2). \tag{27}$$

In most blast furnaces the sulfur distribution between slag and metal samples, taken at tap, is in the range 20 to 80, which is about one-third or half the equilibrium values for the primary sulfur reaction (9). However, the plant data indicate that there is closer approach to equilibrium for the manganese-sulfur coupled reaction.

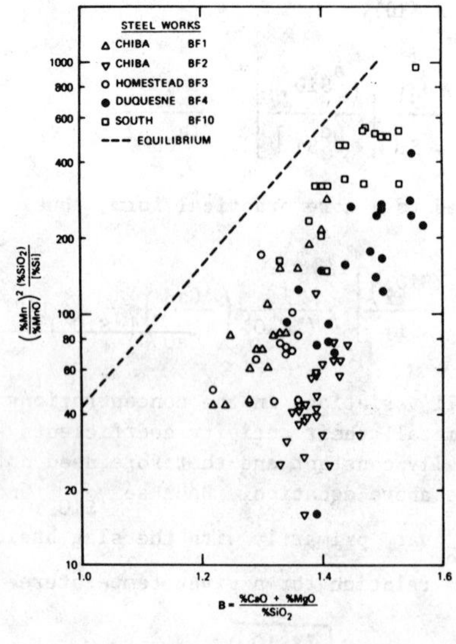

Figure 10. Product of manganese and silicon distribution ratios as a function of slag basicity for daily average blast furnace data compared with the equilibrium relation for silicon-manganese reaction. After Turkdogan [10].

Reactions in the Slag Layer

The laboratory experiments simulating the operating conditions in the combustion zone, the bosh, and the hearth zone of the blast furnace have shown that the silicon and sulfur in the coke ash are transferred to the metal and slag droplets in the bosh region via the volatile species SiO and SiS [9, 11]. Most of the slag-metal reactions in the blast furnace occur as the metal droplets pass through the slag layer. In experiments [9] to simulate the reactions that occur during the passage of liquid-iron droplets in the slag layer of the blast furnace, the rates of silicon and sulfur transfer from metal to slag and manganese transfer from slag to metal were found to be several orders of magnitude faster than ever observed in conventional experiments in graphite crucibles. It has been postulated [9] that a vaporization-absorption process facilitated by a thin gas film around the metal droplets in the slag could account for the observed extremely fast slag-metal reactions. Coupled with such a reaction mechanism, the diffusion in the metal droplets is believed to be enhanced by solute-induced interfacial turbulence.

Vaporization and Condensation of Alkalies

The alkali silicates in the ore and coke ash decompose under reducing conditions at elevated temperatures in the lower part of the bosh and the combustion zone of the blast furnace. Some of the alkali vapors carried away with the ascending gas react with the slag and are removed from the furnace. Some are converted to carbonates and cyanides and are deposited on the burden in the stack and the refractory lining; and some are carried away as dust with the furnace off gas. Deposited alkali carbonates and cyanides carried down with the burden decompose again in the upper part of the bosh. Because of this cycle of vaporization and condensation, the alkalies accumulate in the stack and cause scaffolding, gas channeling, and refractory wear. The alkali buildup in the stack is curtailed by making periodic additions of sand or olivine to the charge so that the concentration of alkali vapors may be lowered by the formation of alkali silicates, which subsequently dissolve in the slag and are removed from the furnace. Because the activity of alkali oxides in the slag decreases with increasing silica content, the lower the slag basicity, the higher will be the amount of alkalies carried away with the slag. For further reading on alkali problems in the blast furnace, reference may be made to a review paper by Abraham and Staffansson [12].

STEELMAKING REACTIONS

Primarily, four types of processes are currently employed in the production of low-alloy steels: (1) oxygen top blowing, known as LD, BOF, or BOP; (2) oxygen and lime bottom blowing, known as OBM or Q-BOP; (3) open-hearth furnace; and (4) electric-arc furnace. The AOD (argon-oxygen decarburization) is a bottom-blowing process used primarily for stainless steelmaking. Even in open-hearth and electric-furnace steelmaking much of the refining is done with oxygen blowing. Although the steelmaking methods and practices differ to varying degrees in these processes, the same fundamental principles of slag-metal equilibria apply to them all.

In open-hearth and oxygen-steelmaking processes the metallic charge to the furnace consists of hot metal from the blast furnace and steel scrap. In most electric-furnace steelmaking, the steel scrap is used entirely as the metallic charge material. In all cases, the impurities in the metal, such as carbon, silicon, manganese, and phosphorus, as well as some iron are oxidized with oxygen blowing. A molten slag is formed by reaction of the oxides of iron, silicon, manganese, and phosphorus with the calcium oxide (burnt lime) that is added to the furnace together with the metallic charges. Although the oxidizing conditions in the furnace do not favor efficient desulfurization of the steel, about one-third or half of the sulfur in the metal is transferred to the slag during the course of oxidation of other impurities. Because the oxidation of impurities is exothermic, the bath temperature increases with the progress of metal refining and brings about the melting of steel scrap in the charge and the formation of molten slag. Another important feature of the self sufficiency of the process is the evolution of carbon monoxide during decarburization of the metal, resulting in vigorous mixing of slag and metal phases, hence increasing the rate of refining and melting of scrap.

To facilitate a quantitative presentation of the compiled data on slag-metal equilibria, each reaction will be considered separately, although during steelmaking most of these reactions occur together. First, we shall discuss some specific features of the composition of steelmaking slags, as derived from laboratory equilibrium measurements.

Composition of Basic Slags

In low-phosphorus practices, the steelmaking slag consists primarily of CaO, MgO, SiO_2, and 'FeO', where '' indicate total iron, as oxides in the slag converted to the stoichiometric composition FeO. The total of these oxides in the slag is about 90 percent, the remainder being essentially MnO, P_2O_5, Al_2O_3, and S. Basic slags used in most steelmaking processes are saturated with CaO and often also with MgO, so that there would be minimum erosion of the furnace refractory lining, which is made of either calcined dolomite (CaO·MgO) or magnesia (MgO).

The effect of MgO on the liquidus isotherms of calcium silicates and calcium oxide at 1600°C is shown in Fig. 11 from the work of Tromel et al. [13].

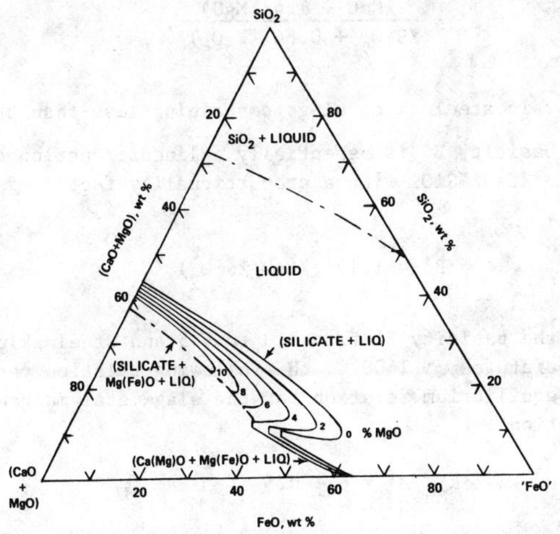

Figure 11. Projection of liquidus isotherms (1600°C) of dicalcium silicate in CaO-MgO-SiO_2-'FeO' quaternary system on the composition diagram of pseudoternary system (CaO+MgO)-SiO_2-'FeO'. After Tromel et al. [13].

In a slag containing, for example, 20% SiO_2, the dicalcium silicate saturation is reached at 35% CaO in the absence of MgO; with the addition of 8% MgO, the solubility of lime increases to 40% CaO. The dot-dash curve delineates the region of double saturation of the slag with solid calcium (magnesium) silicates and solid magnesio-wustite. The solubility of MgO at magnesio-wustite saturation is shown in Fig. 12 as a function of slag basicity %CaO/%SiO_2 for various concentrations of iron oxide in the range 10 to 60%. The dotted curves delineate double saturation of the slag with two solid phases. The addition of CaF_2 up to about 10% has no perceptible effect on the solubility of MgO in basic slags.

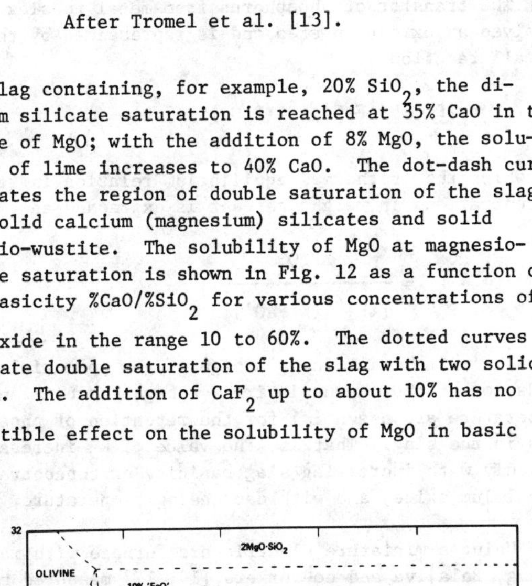

Figure 12. Solubility of MgO, as magnesio-wustite, in CaO-MgO-SiO_2-'FeO' system at 1600°C as a function of slag basicity and FeO concentration, derived from the data in Fig. 11.

The phase boundaries in the composition diagrams in Figs. 11 and 12 are for slags in equilibrium with liquid iron at 1600°C. At higher oxygen activities, i.e., higher Fe^{3+}/Fe^{2+} ratios, the solubility isotherms shift to higher concentrations of CaO. Perhaps it is for this reason that a faster lime dissolution in the slag (rapid slag formation) is achieved by oxidizing the slag with a high oxygen-lance practice in BOF steelmaking. Faster rise in slag temperature in the high oxygen-lance practice further aids the rate of dissolution of lime.

Oxidation of Iron

The equilibrium concentration of iron oxide in the slag depends on temperature, oxygen content of the metal and composition of the slag. From the results of the earlier [14] and later [15] studies, the solubility of oxygen in liquid iron in equilibrium with liquid iron oxide at temperatures of 1530 to 1960°C is represented by the equation

$$\log [\%O] = -\frac{6329}{T} + 2.734. \tag{28}$$

The iso-activity curves for 'FeO' are shown in Fig. 13 for the CaO(MgO)-'FeO'-SiO_2 melts in equilibrium with liquid iron at 1600°C. The concentration and activity of 'FeO' for slags saturated with CaO and MgO, derived from the data in Figs. 11 and 13, are shown in Fig. 14 as a function of the slag basicity CaO/SiO_2.

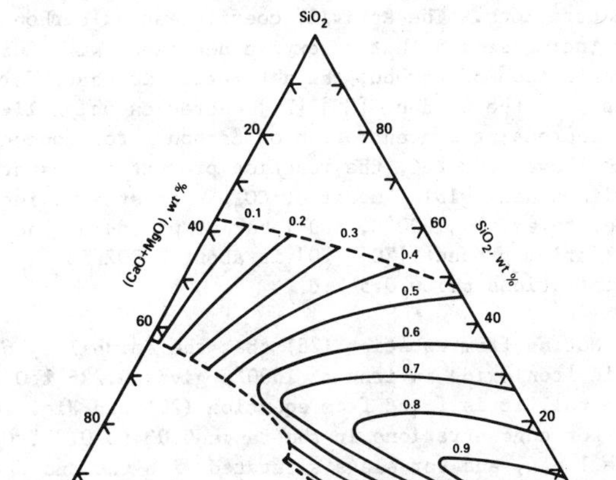

Figure 13. Iso-activity curves for 'FeO' in CaO(MgO)-'FeO'-SiO_2 melts in equilibrium with liquid iron at 1600°C. After Taylor and Chipman [14].

Oxidation of Carbon

The equilibrium constant for the carbon-oxygen reaction in liquid iron is represented by

$$CO(g) = \underline{C} + \underline{O}$$

$$K = \frac{[a_C][a_O]}{p_{CO}}, \quad \log K = -\frac{1168}{T} - 2.07 \tag{29}$$

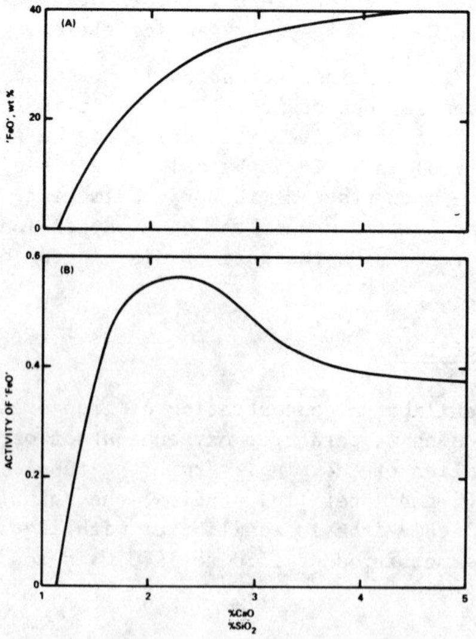

Figure 14. MgO and $2Ca(Mg)O \cdot SiO_2$-saturated CaO-MgO-'FeO'-SiO_2 melts in equilibrium with liquid iron at 1600°C: (a) concentration of 'FeO' and (b) activity of 'FeO'.

where p_{CO} is in bar and the activities a_C and a_O are taken equivalent to [%C] and [%O], respectively, at low solute concentrations. The activity coefficient of carbon in iron increases and that of oxygen decreases with increasing concentration of carbon; the net result is that, for a given p_{CO}, the product [%C] [%O] decreases only slightly with increasing concentration of carbon. For concentrations above 0.03 % C, the reaction product is essentially CO with a negligible amount of CO_2. For steelmaking temperatures, ∿ 1600°C, and 1 bar CO pressure, the equilibrium product [%C] [%O] is about 0.002 for concentrations below 0.5 % C.

Noting from equation (28) that the solubility of liquid iron oxide in iron at 1600°C gives 0.226 % O in the metal, it is found from equation (29) and Fig. 14 that for concentrations in the range 0.03 to 0.3 % C, p_{CO} = 1 bar, and for slags saturated with CaO and MgO, the slag-metal equilibrium with respect to the oxidation of carbon at 1600°C is simplified to

$$(\%'FeO')[\%C] = 0.33 \pm 0.02 \tag{30}$$

Oxidation of Manganese

The oxidation of manganese in iron by the slag is an iron-manganese exchange reaction between slag and metal,

$$\underline{Mn}\ [1\%] + (FeO)(\ell) = (MnO)(s) + Fe(\ell). \tag{31}$$

For the indicated standard states, the equilibrium constant and its temperature dependence are

$$K = \frac{(x_{MnO})}{(x_{FeO})} \frac{1}{[\%Mn]} \frac{\gamma_{MnO}}{\gamma_{FeO}}, \quad \log K = \frac{7784}{T} - 3.590 \tag{32}$$

The definition of basicity of complex slags is somewhat arbitrary. If we assume that in steelmaking-type slags the concentrations of CaO and MgO are equivalent on a molar basis, and similarly on a molar basis $\frac{1}{2} P_2P_5$ is equivalent to SiO_2, in terms of weight percentages of oxides, the slag basicity B' maybe defined by the ratio

$$B' = \frac{\%CaO + 1.4(\%MgO)}{\%SiO_2 + 0.84(\%P_2O_5)} . \tag{33}$$

In basic steelmaking slags containing less than 3% P_2O_5, the basicity B' is essentially a linear function of the ratio $\%CaO/\%SiO_2$ with a proportionality factor of about 1.17;

$$B' = 1.17\ (\%CaO/\%SiO_2) \tag{34}$$

For the basicity B' from 1.0 to 4.5 and steelmaking temperatures, ∿ 1600°C, the following relation represents the equilibrium relation for the slag-metal manganese reaction

$$k_{Mn}\ B' = 6 \pm 0.5 \tag{35}$$

where k_{Mn} = (%MnO)/(%FeO)[%Mn].

Oxidation of Phosphorus

The transfer of phosphorus from metal to slag involves an oxidation step and is represented by the overall reaction

$$2\underline{P} + 5('FeO') = (P_2O_5) + 5Fe, \tag{36}$$

for which the isothermal equilibrium relation in terms of concentrations in weight percent is expressed as

$$k_p = \frac{\%P_2O_5}{[\%P]^2\ (\%'FeO')^5} .$$

In addition to maintaining a high oxygen potential for the oxidation of phosphorus in iron, a high basicity and low temperature are essential for the retention of phosphate ions in the slag. That is, the value of k_p increases markedly with increasing slag basicity or concentration of calcium oxide, and with decreasing temperature.

Using a miniature electric-arc furnace with magnesia lining, Balajiva and co-workers [16, 17] measured the equilibrium distribution of phosphorus between liquid iron and steelmaking-type slags saturated with MgO at temperatures of 1550 to 1635°C. Of the various methods of representing the state of slag-metal equilibrium tested, with respect to the phosphorus reaction, they found that the experimental data were best represented by the following empirical relation

$$\log k_p = 10.78 \log (\%CaO) - 0.00894t - 6.245, \tag{38}$$

where t is temperature in °C. The lower the temperature the greater the slag/metal phosphorus distribution. The results of subsequent equilibrium measurements obtained by Kor [18] with steelmaking-type slags agree well with those calculated from equation (38). The increase in the value of k_p with increasing concentration of CaO is intuitively attributed to the decrease in the activity coefficient of P_2O_5 with increasing concentration of CaO.

Before pursuing this thought further we should take a closer look at the composition of slags used in these laboratory experiments.

The plots of %'FeO' vs %CaO in Fig. 15 are for the slags used by Balajiva et al. [16] in their experiments at 1585°C. The decrease in the iron oxide content with increasing concentration of lime follows a path close to that for slags saturated with dicalcium silicate and magnesio-wustite, as shown in Fig. 11.

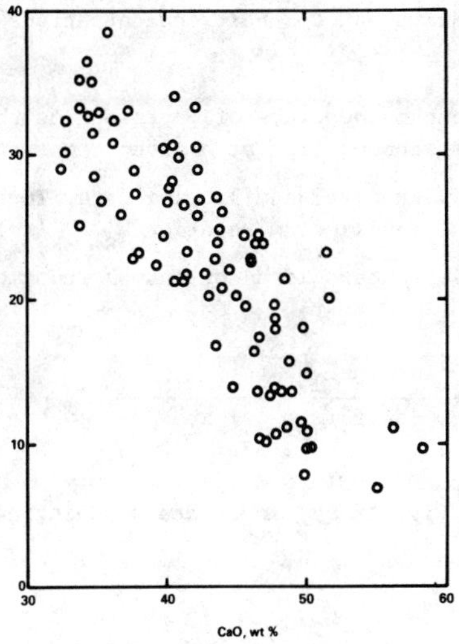

Figure 15. Variation of %'FeO' with %CaO in the slags used by Balajiva et al. [16] in their slag-metal equilibrium measurements at 1585°C.

Along the isotherm for dicalcium silicate saturation, the activity of CaO decreases and the activity of SiO_2 increases with increasing concentrations of CaO and SiO_2, and the corresponding decreasing concentration of 'FeO', accompanied by an increase in the value of k_p. Because the activity coefficient of iron oxide changes little with variation in the composition of slags saturated with dicalcium silicate, in similar slags an increase in the value of k_p is a direct indication of a decrease in the value of $\gamma_{P_2O_5}$ with increasing concentration of CaO and with the corresponding increase in the concentration and activity of silica.

We may surmise from these observations that the phosphate capacity of the slag increases, i.e., $\gamma_{P_2O_5}$ decreases, when the phosphate anions are incorporated in the silicate anion network in molten slags. Such a view would be consistent with the existence of a series of solid solutions in the mixtures of dicalcium silicate and tricalcium phosphate as, for example, silico-carnotite ($5CaO \cdot P_2O_5 \cdot SiO_2$) and nagelschmidtite ($7CaO \cdot P_2O_5 \cdot 2SiO_2$), both of which have been observed in crystallized basic slags.

As indicated by the experimental work of Kor [18], the phosphate capacity of basic slags increases with the addition of fluorspar to the slag. The fluoride ions evidently stabilize the phosphate anions in the slag, as would be anticipated from the existence of the mineral fluorapatite ($3Ca_3P_2O_8 \cdot CaF_2$).

Oxidation of Silicon

In the presence of a basic slag and under oxidizing conditions of steelmaking, the silicon in the metal is readily oxidized to levels below 0.005 percent. The oxidation of silicon is more important in the deoxidation of steel; this is discussed later in the paper.

Oxidation of Chromium

The reaction of particular importance in stainless steelmaking is

$$Cr_2O_3(s) + 3\underline{C} = 2\underline{Cr} + 3CO. \tag{39}$$

The solubility of Cr_2O_3 in oxidizing slags is relatively small, about 5%. For melts saturated with solid Cr_2O_3, the equilibrium constant for reaction (39) is given by the equation

$$K = \frac{[a_{Cr}]^2 \, p_{CO}^3}{[a_C]^3} \tag{40}$$

where a's are activities. The equilibrium relation between the concentrations of carbon and chromium in liquid iron saturated with Cr_2O_3, determined by Richardson and Dennis [19], is shown in Fig. 16 for p_{CO} = 1 bar and temperatures of 1550 to 1850°C.

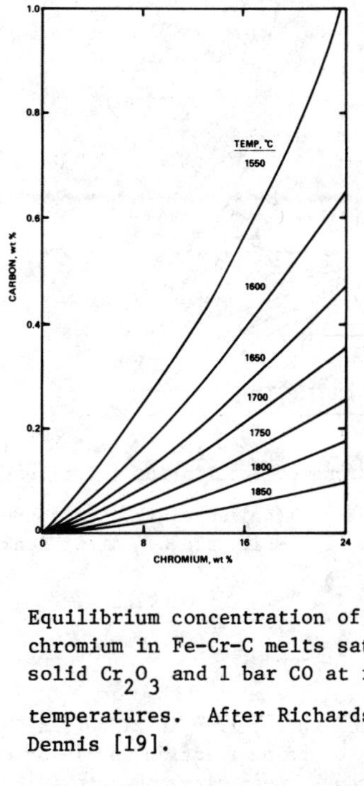

Figure 16. Equilibrium concentration of carbon and chromium in Fe-Cr-C melts saturated with solid Cr_2O_3 and 1 bar CO at indicated temperatures. After Richardson and Dennis [19].

The chromium lost to the slag during the refining of stainless steel under an oxidizing slag is finally recovered by reducing the slag with the addition of silicon. In the reduced slag the chromium is primarily in the divalent form, and the overall reactions may be represented by

$$(CrO_x) + x\underline{Fe} = \underline{Cr} + x(FeO) \quad (41)$$

$$2(CrO_x) + x\underline{Si} = 2\underline{Cr} + x(SiO_2) \quad (42)$$

In an experimental study of reaction of Fe-Cr-Si melts with $CaO\text{-}Al_2O_3\text{-}SiO_2\text{-}CrO_x$ slags at 1600°C, Rankin and Biswas [20] estimated x to be about 1.07; because the activity coefficients of all the reactants are not known, the state of the equilibrium was represented by the following relations:

$$k_1 = \frac{[\%Cr]}{(\%Cr)} \left[\frac{(\%'FeO')}{[\%Fe]}\right]^{1.07} \quad (43)$$

$$k_2 = \frac{[\%Cr]^2}{(\%Cr)} [\%Si]^{-1.07} \quad (44)$$

The activity of SiO_2 was not included in the expression for k_2, because k_2 was evaluated for slag compositions of constant silica activity. Variations of k_1 and k_2 with the slag compositions projected on the ternary $CaO\text{-}Al_2O_3\text{-}SiO_2$ system, as derived by Rankin and Biswas from their experimental data and those of others, are shown in Fig. 17 for temperatures of 1600 to 1640°C.

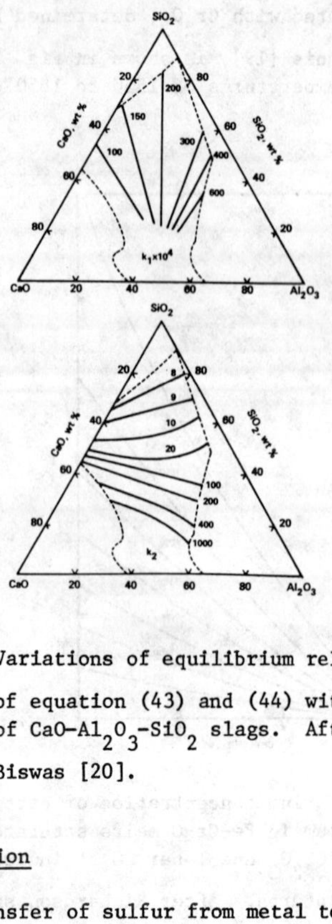

Figure 17. Variations of equilibrium relations k_1 and k_2 of equation (43) and (44) with the composition of $CaO\text{-}Al_2O_3\text{-}SiO_2$ slags. After Rankin and Biswas [20].

Sulfur Reaction

The transfer of sulfur from metal to slag during steelmaking is a reduction reaction involving exhange of sulfur and oxygen between slag and metal, thus

$$\underline{S} + (O^{2-}) = (S^{2-}) + \underline{O}. \quad (45)$$

Because the activity of oxygen ion in the slag cannot be determined, the equilibrium relation for reaction (45) is often expressed in the form

$$k_S = \frac{(\%S)}{[\%S]} [a_O] \quad (46)$$

where a_O is the activity of oxygen (≡ wt %O in the absence of alloying elements). Because of the low concentrations of alloying elements in the metal during steelmaking, the activities of sulfur and oxygen in the metal are essentially equivalent to their concentrations in weight percent.

There are numerous ways of representing the variation of k_S with slag composition at a given temperature. Using the available slag-metal equilibrium data, Turkdogan [21] showed that the equilibrium relation $k'_S = (x_S)[\%O]/[\%S]$ for a given temperature is a continuous function of a basicity index λ

$$\lambda = \frac{1}{x_{SiO_2} + 1.5\left(x_{P_2O_5} + x_{Al_2O_3}\right)}. \quad (47)$$

As is seen in Fig. 18, k'_S increases with increasing λ and temperature.

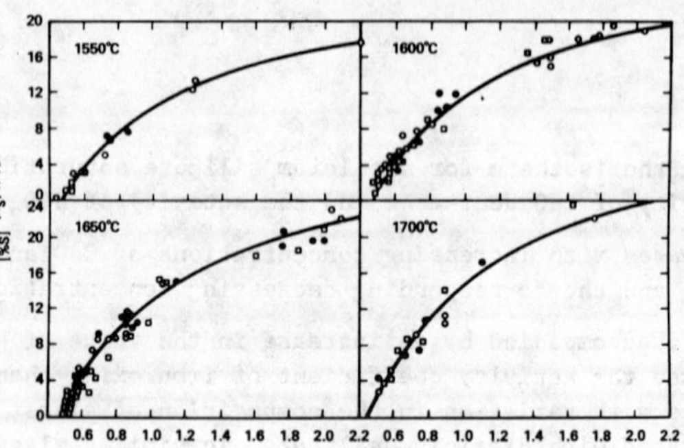

Figure 18. Slag-metal sulfur equilibrium relation as functions of temperature and slag basicity index λ. After Turkdogan [21].

The same equilibrium data for steelmaking temperatures may be represented by the relation shown in Fig. 19a; the values of k_S from various basic open-hearth data [2] also lie within the shaded area delineating the equilibrium relation. The simplest form of the equilibrium relations at 1600 ± 20°C is as shown in Fig. 19b, where the experimental data of Kor [18] are included. It is seen that the addition of small percentages of fluxing agents CaF_2, B_2O_3, and TiO_2 do not affect this equilibrium relation when k_S is plotted against the sum of the concentrations of acidic oxides, $\%SiO_2 + \%P_2O_5 + \%B_2O_3 +$

$\%TiO_2$. Because of the interrelation between the basicity and the concentration of iron oxide in saturated slags, the sulfur distribution ratio (%S)/[%S] also increases with the basicity CaO/SiO_2. In such a plot, however, the data points are scattered over a wider range, particularly at high basicities.

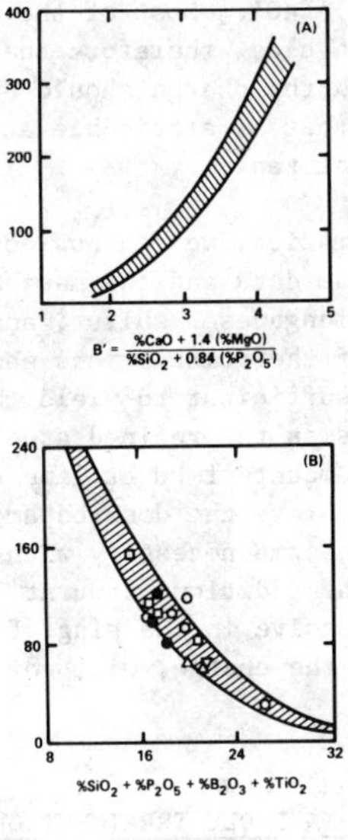

Figure 19. Variation of k_S (for 1600 ± 20°C) with (a) slag basicity B' and (b) sum of acidic oxides. Flux: (O) none, (●) CaF_2, (Δ) B_2O_3, (∇) 'Sorel', (□) MnO_2.

<u>Observations from Plant Data</u>

The amount of lime charged to the furnace in steelmaking is often in excess of that which is soluble in the slag, even at the end of refining. Although the slag is heterogeneous because of the presence of undissolved lime, particularly in BOF and Q-BOP processes, in the usual method of taking slag samples by immersing a steel rod into the bath, only the liquid portion of the slag is sampled. However, the tests have shown that even these slag samples contain some metallic iron shots and finely dispersed undissolved lime particles for which due corrections must be made in reporting the slag analysis. That is, the total iron oxide reported as 'FeO' is that for iron in solution as oxides in the slag. Also, the slag is analyzed for free lime which is deducted from the total lime to give percent CaO in solution. In some practices fluorspar is added to facilitate the rate of lime dissolution; the concentration is usually below 2% CaF_2. All the calcium in solution in the slag is reported as CaO, including that part associated with CaF_2.

The compositions of most steelmaking slags are in the ranges: 50 to 60% CaO, 3 to 8% MgO, 3 to 8% MnO, 6 to 26% 'FeO', 15 to 25% SiO_2, 1 to 5% P_2O_5, <1% Al_2O_3, and <0.2% S. In some European steelmaking practices with a high-phosphorus hot metal charge, the steelmaking slag at tap may contain 20% P_2O_5 or more, but much less SiO_2.

The variation of the total iron oxide content of the slag with basicity shown in Fig. 20 for BOP and Q-BOP slags is similar to that for open-hearth and electric-arc furnace slags. Although the industrial steelmaking slags are saturated with dicalcium (magnesium) silicate and magnesio-wustite phases, for a given iron oxide concentration, the basicity of industrial slags is noticeably higher than that of the saturated CaO-MgO-'FeO'-SiO_2 slag, because the presence of MnO and P_2O_5 and higher concentrations of Fe_2O_3 in industrial slags shifts the saturation isotherm to higher concentrations of CaO, that is, to higher basicities.

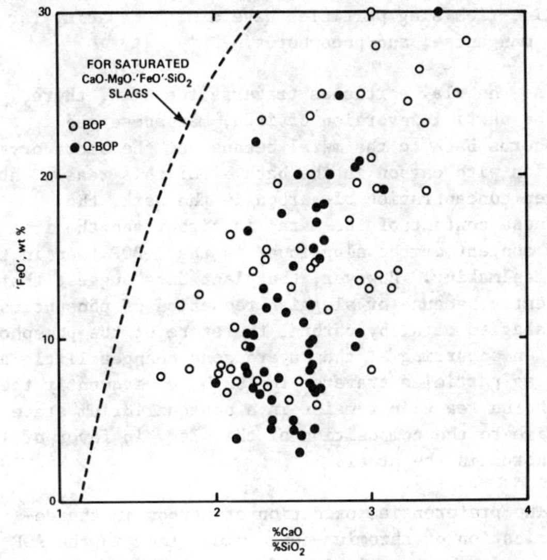

Figure 20. Variation of the iron oxide content of BOP and Q-BOP slags with basicity compared with that for CaO and MgO saturated CaO-MgO-'FeO'-SiO_2 slags at 1600°C.

The ratio Fe^{3+}/Fe^{2+} in slag samples from steelmaking furnaces is within the range 0.4 to 0.8, which is two to three times higher than for slags in equilibrium with liquid iron. Also, the total iron oxide contents of industrial slags are two to six times higher than that given by equation (30) for <u>C</u> - (FeO) equilibrium at 1 bar pressure of CO, the departure from slag-metal equilibrium being greater at higher concentrations of carbon. During refining the oxygen content of the steel increases with decreasing concentration of carbon, this increase being more noticeable below 0.2% C. At all carbon levels the oxygen content of the steel is 1.5 to 2 times higher than the equilibrium values for 1 bar pressure of CO in most steelmaking processes. In Q-BOP steelmaking, however, the oxygen content of the steel at tap is close to or below the equilibrium value for 1 bar CO, particularly at carbon levels below 0.08%. This is due to the carbon deoxidation of the steel when nitrogen or argon is blown during rotation of the vessel to the tapping position.

As indicated by reaction (29), lowering the partial pressure of carbon monoxide in the gas stream by dilution of the bottom-blown oxygen with argon or nitrogen will lower the residual oxygen in the steel and extent of oxidation of the iron, and hence will increase the yield when blowing steel to exceptionally low levels of carbon. In fact in some steelmaking practices now being developed, the refining is done with oxygen top blowing and argon bottom blowing to increase the production yield and decrease the residual concentrations of oxygen, nitrogen, and hydrogen in the steel.

From numerous previous studies of plant data, reviewed elsewhere [2], it has been concluded that in most steelmaking practices the slag/metal distributions of manganese, sulfur, and phosphorus at the time of tapping are close to equilibrium values, with a few exceptions for certain operating conditions. For example, in the Q-BOP process with oxygen plus lime bottom blowing, the steel is dephosphorized to levels below those predicted for slag-metal equilibrium from the slag composition. With oxygen plus lime bottom blowing, the reactions occur in the tuyere zone where the lime-rich slag particles have a high capacity for iron, manganese, and phosphorus.

As the slag particles traverse the bath, there will be partial reversion of iron, manganese and phosphorus back to the metal because of the secondary reaction with carbon in the bath. For this reason, at a given concentration of carbon in the bath, the manganese content of the metal is higher and the iron oxide content of the slag lower in the Q-BOP than in the BOF steelmaking. However, the plant data suggest that apparently because of sluggish reduction of phosphorus from slag to metal by carbon, the state of the phosphorus reaction occurring at the tuyere zone changes little as the slag particles traverse the bath, consequently the phosphorus reaction remains in a nonequilibrium state relative to the composition of the slag, in favor of low phosphorus in the steel.

The preferential oxidation of carbon in the decarburization of chromium-containing steel in the AOD process with bottom blowing of argon-oxygen mixtures is another example of the secondary reaction with carbon in the bath. In the tuyere region of the submerged gas stream, the chromium and carbon are oxidized independently of each other. As the chromic oxide particles float out of the melt, they are reduced by carbon in the bath, resulting in extensive recovery of chromium back into the metal. Similar to the events in the Q-BOP steelmaking, two primary reaction zones can be considered for the AOD:

(1) Near oxygen entrance

$$2\underline{Cr} + \frac{3}{2}O_2 = Cr_2O_3 \text{ and } \underline{C} + \frac{1}{2}O_2 = CO. \qquad (48)$$

(2) In metal bath

$$Cr_2O_3 + 3\underline{C} = 2\underline{Cr} + 3CO \qquad (49)$$

With argon + oxygen bottom blowing, the partial pressure of CO is lowered and consequently more Cr_2O_3 is reduced as indicated by the equilibrium relation in equation (40). The longer the residence time of chromic oxide particles in the melt, i.e., the deeper the metal bath, the greater the recovery of chromium on the reduction cycle, and hence the lower the chromium losses to the slag during decarburization of the bath. Reference may be made to papers by Fruehan [22] and Roy and Robertson [23] for mathematical modeling of the AOD process based on reaction (49) and the rate of mass transfer to CO + Ar bubbles traversing the bath.

Greater use is now being made of the slag-metal equilibrium data in controlling slag formation, hence steelmaking. The methods of static and dynamic charge control formulated for various steelmaking practices will not be discussed here. Only a brief reference will be made to the basic concept of static charge control for BOF steelmaking. Because of the interrelation between the carbon content of the bath at tap and the basicity and iron oxide content of the slag, the prevailing slag basicity is determined by the carbon content of the steel to be made. For example, the basicity CaO/SiO_2 will be about 2 for steel containing more than 0.2% C at tap; for 0.03 to 0.05% C in the steel at tap, the basicity will be in the range 3.2 to 3.5, the latter being the highest basicity attainable in lime and magnesia-saturated slags containing 25 to 30% 'FeO'. Most of the silicon in the charge ends up in the slag, therefore the amount of burnt lime added with the charge should be just enough to obtain the basicity that is attainable at the level of carbon in the steel at tap.

With this information, we can now compute from the slag-metal equilibrium data and the mass of the slag the residual amounts of manganese, sulfur, and phosphorus in the steel at tap. If the calculations show that the amount of slag is insufficient to yield the desired low sulfur and phosphorus in the refined steel, some sand plus an appropriate amount of burnt lime should be added with the charge to achieve the desired amount of slag. Such a practice may become necessary with a low-silicon hot metal charge. The addition of burnt lime in excess of that which can dissolve in the slag, for a given total amount of silicon in the charge, will not yield the desired end result.

REACTIONS IN LADLE TREATMENT OF STEEL

During the past decade we have seen major strides being made in the development of ladle steelmaking practices to improve steel cleanliness and control the shape and size distribution of inclusions necessary to achieve the desired mechanical properties in the high-strength, low-alloy (HSLA) steels. At a recent symposium on "Clean Steel," the author [24] discussed the present state of knowledge on the fundamental aspects of ladle deoxidation, desulfurization, and inclusions in steel, to which reference may be made for the compilation of data on slag-metal equilibria pertinent to reactions in the ladle-treatment of steel. The presentation here is confined to selected slag-metal equilibria for primary reactions in the ladle deoxidation and desulfurization of steel.

Ladle Deoxidation

The silicon/manganese deoxidation is used in the production of rimmed and capped steels. The equilibrium data for simultaneous deoxidation of steel with silicon and manganese at 1600°C are given in Fig. 21 [25]. The critical silicon and manganese contents of iron in equilibrium with silica-saturated manganese silicate melts are given in Fig. 21 for several temperatures. For alloy compositions above a given isotherm, the manganese does not participate in the reaction; instead silica is formed as the deoxidation product. In the region below the isotherm, the reaction product is molten manganese silicate (with little iron oxide), the composition of which is determined by the ratio $[\%Si]/[\%Mn]^2$ in the metal as deduced from the equilibrium constant for the reaction

$$\underline{Si} + 2(MnO) = 2\underline{Mn} + (SiO_2),$$

$$K = \frac{[\%Mn]^2}{[\%Si]}\frac{(a_{SiO_2})}{(a_{MnO})^2} = 118 \text{ at } 1600°C. \qquad (50)$$

and the activities of oxides in the $MnO-SiO_2$ melts.

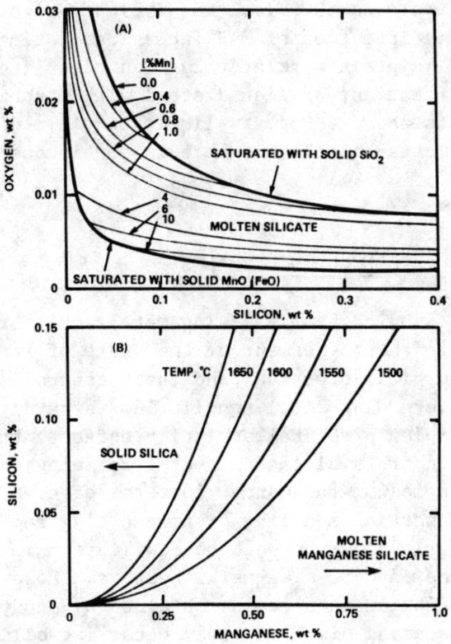

Figure 21. Equilibrium relations for deoxidation of steel with silicon and manganese at 1600°C. After Turkdogan [45].

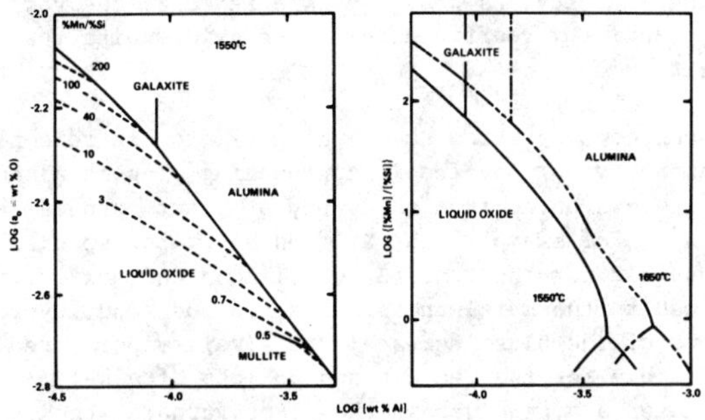

Figure 22. Equilibrium relations for deoxidation of liquid steel with Al-Si-Mn for [%Si] + [%Mn] = 1. After Fujisawa and Sakao [26].

The equilibrium relations for simultaneous deoxidation of steel (semikilled) with silicon, manganese, and aluminum are shown in Fig. 22 [26]. For a semikilled steel to contain, for example, 1% Mn, 0.025% Si, and 40 ppm O in solution, and a liquid aluminosilicate deoxidation product, the residual aluminum in solution would be negligibly small, below 2 ppm.

Complete deoxidation of steel with aluminum to produce killed steel is much facilitated by ladle addition of burnt lime, because of the formation of molten calcium aluminate as the reaction product,

$$(CaO) + 2\underline{Al} + 3\underline{O} = (CaO{\cdot}Al_2O_3). \qquad (51)$$

The calculated equilibrium relations between the oxygen activity in the deoxidized steel and the composition of the molten calcium aluminate are shown in Fig. 23 [24] for 1600°C and concentrations of dissolved Al in the range 0.001 to 0.05%. The deoxidation with a mixture of aluminum and burnt lime is particularly helpful in the production of semikilled or killed steel low in residual (dissolved) silicon and aluminum. For example, steel deoxidized with aluminum alone to a level of 4 ppm oxygen activity will contain 0.02% Al as residual aluminum in solution. On the other hand, with aluminum and lime deoxidation, forming lime-saturated aluminate liquid reaction product, the same level of deoxidation will be achieved with only 0.001% Al left in solution. For 100% usage of the additives in deoxidation, the recommended mixture is 3 parts CaO + 1 part Al by weight to give lime-saturated liquid calcium aluminate as the reaction product.

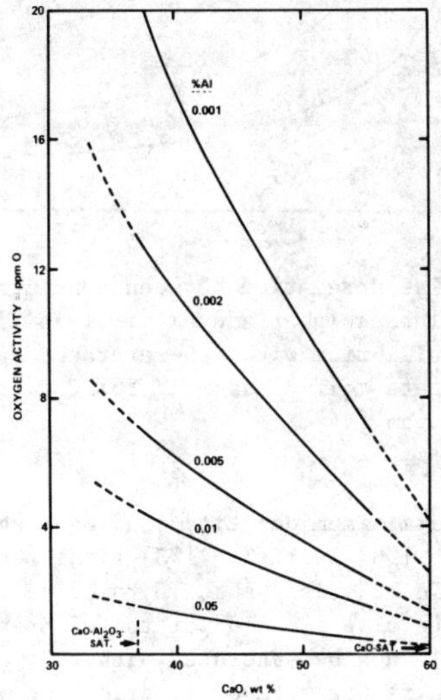

Figure 23. Deoxidation equilibrium with aluminum and calcium aluminate at 1600°C. After Turkdogan [24].

Ladle Desulfurization

The ladle desulfurization of hot metal, usually from 0.04 to about 0.02% S, is relatively easy because of the prevailing reducing conditions in the hot metal and a high sulfur activity in the presence of carbon and silicon in the iron. Magnesium, calcium carbide, and soda ash are most effective in removing sulfur from the hot metal.

More stringent conditions are required for ladle desulfurization of steel from concentrations of 0.015 to 0.010% S in the steel at tap to less than 0.002% S in the product. Lime and aluminum are the two essential ingredients of ladle additions for effective desulfurization of steel accompanied by extensive deoxidation. The overall reaction is

$$(CaO) + \frac{2}{3}\,\underline{Al} + \underline{S} = (CaS) + \frac{1}{3}\,(Al_2O_3) \qquad (52)$$

From the known free energy data, the sulfide capacities and activities of oxides in the calcium aluminate melts, the equilibrium relations between dissolved sulfur and aluminum have been calculated [24]. The equilibrium relations shown in Fig. 24 for 1550, 1600, and 1650°C are for CaS-saturated calcium aluminate melts containing 56% CaO. At steel refining temperatures of about 1600°C, the solubility of CaS is about 2% S in lime-saturated molten calcium aluminate. The minimum quantity of slag required to achieve the desired amount of sulfur removal is that for CaS saturation of the slag. For this

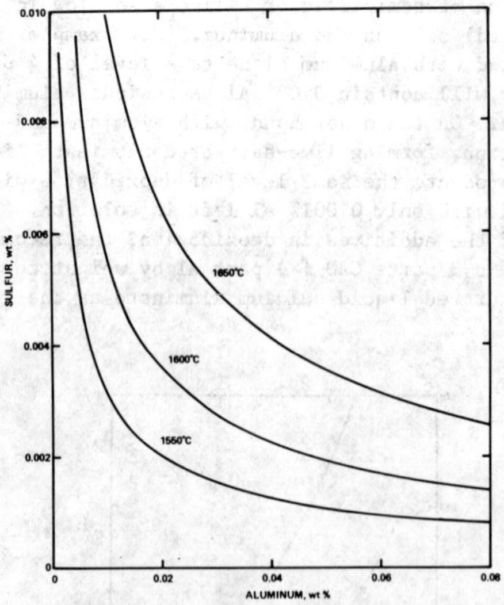

Figure 24. Calculated relation between concentration of dissolved sulphur and aluminum in liquid iron in equilibrium with CaS-saturated calcium aluminate melt containing 56% CaO. After Turkdogan [24].

limiting case, the minimum quantity is 5 kg slag per ton of steel for $\Delta\%S$ = [%S] initial - [%S] final (equilibrium) = 0.01%. In practice, about 8 to 10 kg lime-saturated aluminate slag is used per ton of steel, so that the slag would not be saturated with CaS, which in turn would allow a faster rate of reaction and more extensive sulfur removal.

Efficient ladle deoxidation and desulfurization is brought about by intimate mixing of molten calcium aluminate based slag with liquid steel by argon stirring. The injection of magnesium or calcium compounds, though not prerequisite for the process, does assist desulfurization by increasing the intensity of stirring and counteracting air oxidation of the melt. Various process developments have been made to curtail air oxidation of the bath during the ladle treatment of steel. For example, the ladle top is contained completely under a cover that is flushed with argon. In another single-ladle practice, the ladle is placed in a tank that is maintained at a low pressure during argon stirring of the bath for desulfurization with the top slag. Then there is the double-ladle practice where the desulfurization is incorporated with the ladle-to-ladle degassing. The added advantage of desulfurization under a reduced pressure is the removal of some hydrogen and nitrogen from the steel. Also, under a reduced top pressure, vigorous slag-metal mixing is achieved even with a slow rate of argon injection through the porous plug.

With lime-saturated top slag, a good practice of argon stirring, and minimum air oxidation of the bath, the steel can be desulfurized to levels below 10 ppm S. The amount of calcium aluminate inclusions, equivalent to 40 to 60 ppm Ca, dispersed in the melt at the time of casting is sufficient to absorb the residual sulfur during cooling of the steel so that, at sulfur levels below 20 ppm, there is no precipitation of MnS during solidification.

When the residual sulfur in the steel after the ladle desulfurization is in the range 20 to 60 ppm, mischmetal should be added to the steel in the ladle just before casting, the recommended amount being about 6 times the total concentrations of residual oxygen and sulfur in the melt. Such a practice will ensure even distribution of spheroidal rare-earth oxysulfide inclusions thus prevent the precipitation of MnS inclusions during solidification, yet maintain a relatively high cleanliness rating. The treatment of liquid steel with mischmetal at sulfur contents above 60 ppm is likely to yield dirty steel and may cause nozzle blockage in continuous casting.

CONCLUDING REMARKS

Accurate knowledge of gas-slag-metal equilibria is prerequisite to the assessment of the state of reactions in pyrometallurgical processes and their control. Although complete slag-metal equilibrium is hardly ever reached in complex pyrometallurgical processes, the state of reactions in ore smelting and metal refining can be described in a meaningful manner in terms of coupled reactions and partial equilibria between slag and metal phases. There is slow approach to equilibrium in three-phase reactions in C-CO-slag-metal systems. However, accurate equilibrium data can be obtained by studying interrelations among simultaneously occurring silicon-manganese, silicon-sulfur, and manganese-sulfur reactions in blast furnace-type slag-metal systems. Within the range of practical interest, the metal and slag compositions do not change much; therefore, the isothermal equilibrium relations, represented in terms of weight percentages of reactants, turn out to be simple functions of slag basicity, as would be expected from the variation of the activity coefficients of the oxides with the slag basicity.

Due consideration should be given to the role played by gas and vapor species in approaching slag-metal equilibria, as evidenced from the study of blast furnace reactions. For example, the SiO and SiS vapor species are responsible for the transfer of silicon and sulfur from coke ash to the metal and slag in the bosh and tuyere regions of the blast furnace; the alkali vapors are responsible for the recycle and buildup of alkali carbonates and cyanides in the blast furnace stack.

The slags used in basic steelmaking processes are essentially saturated with dicalcium (magnesium) silicate and magnesio-wustite solid phases. For this reason the basicity and iron oxide content of the slag, and hence the carbon content of the steel at tap, are interrelated; the concentration of iron oxide increases with increasing slag basicity, which can be represented simply by the ratio $\%CaO/\%SiO_2$. Therefore, a slag of low basicity (1.8 to 2.0) is attained in the production of high-carbon steels (0.2 to 0.5% C), whereas a slag of high basicity (3.2 to 3.5) is attained in making low-carbon steels (<0.05% C). An oxidizing slag of high basicity is essential for extensive dephosphorization of steel, a condition that is readily attainable in the production of low-carbon steels. When refining steel with slags of low basicity, as in the production of high-carbon steels, the slag volume may have to be increased, either by using a high-Si hot metal charge or by the additions of sand together with an appropriate amount of burnt lime, to bring about sufficient phosphorus and sulfur removal from the steel. Although the oxygen activity of the bath is low at high carbon levels, there can still be effective dephosphorization of steel because of attainment of low bath temperatures in the production of high-carbon steels.

In all steelmaking processes, the iron oxide content of the slag is much higher than that predicted by the $\underline{C}$ - (FeO) equilibrium at the prevailing CO pressure in the melt ($\sim$ 1 bar). However, in most processes there is partial slag-metal equilibrium with respect to the distributions of manganese, sulfur, and phosphorus between slag and metal at the end of refining.

With oxygen + lime or oxygen + argon bottom blowing, as in the Q-BOP and AOD processes, respectively, the slag and oxide particles forming at the tuyere zone subsequently react with carbon as they traverse the metal bath. This secondary reaction is responsible for the oxidation of iron and manganese in the Q-BOP being less than that in oxygen top blowing processes. Similarly, in the AOD process the Cr_2O_3 forming in the tuyere zone is subsequently reduced by carbon in the bath, and furthermore, the chromium losses to the slag are kept to a minimum by argon dilution of CO in the melt.

The slag-metal reactions are also important in the ladle treatment of steel, as in deoxidation and desulfurization. The steel can be desulfurized effectively by argon stirring and mixing of the melt with lime-saturated calcium aluminate slag. Low levels of sulfur ($\sim$ 10 ppm S) achieved in practice by ladle treatment of aluminum-killed steel are in close agreement with predictions made from data on slag-metal equilibria.

REFERENCES

1. F. D. Richardson, Physical Chemistry of Melts in Metallurgy; Academic Press, London, 1974.

2. E. T. Turkdogan, Physical Chemistry of High Temperature Technology; Academic Press, New York, 1980.

3. E. W. Filer and L. S. Darken, "Equilibrium Between Blast Furnace Metal and Slag as Determined by Remelting", Trans. AIME, 194 (1952), p. 253-257.

4. E. T. Turkdogan and P. M. Bills, cited in E. T. Turkdogan and M. L. Pearce, "Kinetics of Sulfur Reaction in Oxide Melt-Gas Systems", Trans. Met. Soc. AIME, 227 (1963), p. 940-949.

5. T. B. King and S. Ramachandran, "Electrochemical Nature of Sulfur Transfer in the System Carbon-Saturated Iron-Slag" in Physical Chemistry of Steelmaking (J. F. Elliott, ed.), p.125-133; Wiley, New York, 1958.

6. K. P. Abraham, M. W. Davies, and F. D. Richardson, "Sulphide Capacities of Silicate Melts", J. Iron Steel Inst. 196 (1960), p.309-317.

7. R. H. Rein and J. Chipman, "Activities in the Liquid Solution SiO_2-CaO-MgO-Al_2O_3 at 1600°C", Trans. Met. Soc. AIME, 233 (1965), p.415-425.

8. R. A. Sharma and F. D. Richardson, "The Solubility of Calcium Sulphide and Activities in Lime-Silica Melts", J. Iron Steel Inst., 200 (1962), p.373-379.

9. E. T. Turkdogan, G. J. W. Kor, and R. J. Fruehan, "Blast Furnace Reactions", Ironmaking & Steelmaking, 7 (1980), p. 268-280.

10. E. T. Turkdogan, "Blast Furnace Reactions", Met. Trans., 9B (1978), p. 163-179.

11. N. Tsuchiya, M. Tokuda, and M. Ohtani, "The Transfer of Silicon from the Gas Phase to Molten Iron in the Blast Furnace", Met. Trans., 7B (1976), p. 315-320.

12. K. P. Abraham and L. J. Staffansson, "The Alkali Problem in the Blast Furnace", Scan. J. Met., 4 (1975), p. 193-204.

13. G. Tromel, K. Koch, W. Fix, and N. Grobkurth, "The Effect of Magnesium Oxide on the Equilibria in the System Fe-CaO-FeO-SiO_2 and on the Sulphur Distribution at 1600°C", Arch. Eisenhuettenwes., 40 (1969), p. 969-984.

14. C. R. Taylor and J. Chipman, "Equilibria of Liquid Iron and Simple Basic and Acid Slags in a Rotating Induction Furnace", Trans. AIME, 154 (1943), p.228-245.

15. P. A. Distin, S. G. Whiteway, and C. R. Masson, "Solubility of Oxygen in Liquid Iron from 1785 to 1960°C; A New Technique for the Study of Slag-Metal Equilibria", Can. Met. Quarterly, 10 (1971), p. 13-18.

16. K. Balajiva, A. G. Quarrell, and P. Vajragupta, "A Laboratory Investigation of the Phosphorus Reaction in the Basic Steelmaking Process", J. Iron Steel Inst., 153 (1946), p. 115-150.

17. K. Balajiva and P. Vajragupta, "The Effect of Temperature on the Phosphorus Reaction in the Basic Steelmaking Process", J. Iron Steel Inst. 155 (1947), p. 563-567.

18. G. J. W. Kor, "Effect of Fluorspar and Other Fluxes on Slag-Metal Equilibria Involving Phosphorus and Sulfur", Met. Trans., 8B (1977), p. 107-113.

19. F. D. Richardson and W. E. Dennis, "Effect of Chromium on the Thermodynamic Activity of Carbon in Liquid Iron", J. Iron Steel Inst. 175 (1953), p. 257-266.

20. W. J. Rankin and A. K. Biswas, "Oxidation States of Chromium in Slag and Chromium Distribution in Slag-Metal Systems at 1600°C", Inst. Min. Metall. Trans. Section C, 87 (1978), p. 60-70.

21. E. T. Turkdogan, "Sulphur Partition Between Gas, Slag and Metal Phases", J. Iron Steel Inst., 179 (1955), p. 147-154.

22. R. J. Fruehan, "Reaction Model for the AOD Process", Ironmaking & Steelmaking, 3 (1976), p.153-158.

23. T. D. Roy and D. G. C. Robertson, "Mathematical Model for Stainless Steelmaking", Ironmaking & Steelmaking, 5 (1978), p. 198-206.

24. E. T. Turkdogan, "Ladle Deoxidation, Desulphurization and Inclusions in Steel - Fundamentals and Observations in Practice" to be published in Proc. Clean Steel Conference; Metals Society, London.

25. E. T. Turkdogan, "Causes and Effects of Deoxidation Occurring During Cooling and Solidification of Steel", Trans. Met. Soc. AIME, 233 (1965), p. 2100-2112.

26. T. Fujisawa and H. Sakao, "Equilibrium Relations Between the Liquid Iron Alloys and the Deoxidation Products Resulting from Mn-Si-Al Complex Deoxidation", Tetsu-to-Hagané, 63 (1977), p. 1494-1503.

SOLUTION CONCENTRATION AND PURIFICATION

J. D. Miller

University of Utah
Salt Lake City, Utah 84112

INTRODUCTION

An important part of any hydrometallurgical process is the treatment of the leach liquor by solution concentration and/or purification techniques prior to product recovery. The strategy used varies considerably depending on the composition of the solution and other factors, not the least of which is the subsequent product recovery process. Concentration and purification processes include crystallization, precipitation, cementation, ion exchange, and solvent extraction. These reactions are, for the most part, heterogeneous reactions involving particulate systems which vary from solubility phenomena in the case of crystallization to complex electrochemical phenomena operative in cementation systems. The attendant reaction kinetics are varied and the rate of reaction can be limited by diffusion through the mass transfer boundary layer (MTBL), pore diffusion, and/or surface reaction.

As in other heterogeneous reaction systems, study of agitated particulate or droplet assemblages is important from a pragmatic standpoint for it can provide empirical correlations for scale-up procedures. However, the determination of intrinsic kinetic parameters by such experimentation is, at best, difficult and the results may not be of great accuracy. Consequently, many experimental techniques have been devised to provide more accurate information regarding intrinsic reaction rate constants and to provide a convenient way of analyzing the relative importance of mass transfer in these systems.

Because of space restrictions imposed on this manuscript, review has been limited to advances in the understanding of one important solution concentration and purification process - cementation. In the case of cementation reaction kinetics, particular attention has been focused on the electrochemical features of the process and the nature of the surface deposit. This review is an extension and update of a contribution originally prepared in 1974 (1).

RATE PROCESSES IN CEMENTATION

Cementation is used both for concentration and purification. In the case of concentration, an intermediate product is generally obtained and the cement product requires further refining. For example, cementation is used for the recovery of copper from dump leach liquors with scrap iron, the recovery of gold from cyanidation liquors with zinc dust and the recovery of cadmium from flue dust leach liquors with zinc. On the other hand, cementation is used for electrolyte purification in the electrolytic production of zinc and nickel. In such cases a number of metallic impurities, particularly Cu and Cd, are removed either by zinc dust for electrolytic zinc production or by metallic nickel for electrolytic nickel production. Reactor design can be classified according to whether the reactive metal is stationary (launders, cementation cones, fixed plates in STR) or present as suspended particles (STR or tube reactor).

Undoubtedly copper cementation from dump leach liquors represents the greatest production of any industrial cementation system. As can be seen from the production data presented in Table I, the total annual production in 1979 amounts to almost 200,000 metric tons of copper. In the case of U.S. copper production the cementation operation accounts for about 20 percent of the total copper produced. A discussion of copper cementation practice has been given in which the influence of solution chemistry on plant performance is considered (2). The industrial performance of both cementation cones and launders is discussed and compared to results from continuous flow laboratory experiments in a column reactor. At the Beijing Mining and Metallurgical Research Institute, Peoples Republic of China, studies have been done on reactor design and process strategy for the cementation of copper from certain leach liquors. These studies included the evaluation of a column reactor with steam injection and the evaluation of two sizes of laboratory scale cementation cones.

Electrochemical Reactions

An electrochemical reaction is distinct from a chemical redox reaction in that charge transfer does not occur at the same physical position but rather the half-cell reactions are separated by some finite distance; consequently the solid phase must be an electrical conductor. The kinetics of electrochemical reactions, such as cementation, not only involve the activation energy associated with the chemical process but also an activation energy effect associated with the charge-transfer process. This latter effect arises because a charge carrier (metal ion or electron) is transported through an interfacial electrical potential gradient developed by the system.

In its most elementary form, the overall cementation reaction, a net cathodic reaction with respect to the noble metal M_2,

$$\frac{Z_1}{Z_2} M_2^{+Z_2} + M_1^0 \rightarrow \frac{Z_1}{Z_2} M_2^0 + M_1^{+Z_1} \qquad (1)$$

can be represented by the two half-cell reactions:

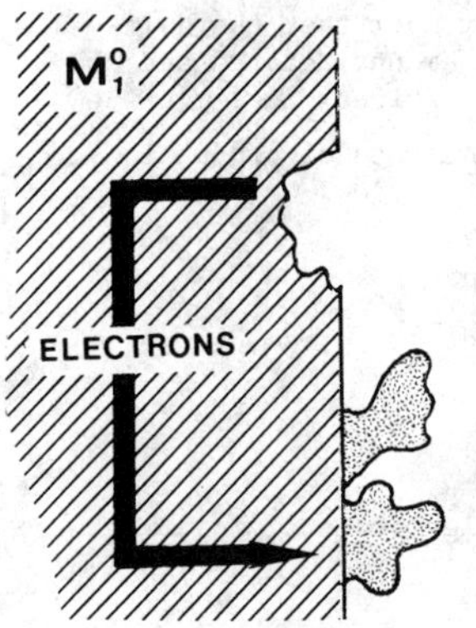

$$M_1^0 \rightleftarrows M_1^{+Z_1} + Z_1 e \qquad (2)$$

$$\frac{Z_1}{Z_2} M_2^0 \rightleftarrows \frac{Z_1}{Z_2} M_2^{+Z_2} + Z_1 e \qquad (3)$$

Table I. Cement Copper Production

Country	Company	Plant/Location	Ore	1979 Annual Prdt. (metric tons)
USA	Anaconda	Berkeley	dumps	7,700
	Asarco	Silver Bell	"	14,600
	Duval	Battle Mt.	"	3,500*
		Mineral Park	"	1,500
		Esperanza	"	2,700
		Sierrita	"	
	Inspiration	Inspiration	"	8,000*
	Phelps Dodge	Bisbee	"	3,400
		Morenci	dumps & tailings	33,900
		Tyrone	dumps	4,400
	Ranchers	Old Reliable	"	900
	Kennecott	Ray	"	9,000*
		Utah	"	25,000
		Chino	"	16,000
	Cities Service	Miami	"	1,900*
		Pinto Valley	"	450*
Chile	Sagasca	Sagasca	oxide ores	8,500
	Mantos Blancos	Mantos Blancos	" "	5,000
	Michilla	Michilla	" "	7,000
Peru	Centromin	Cerro de Pasco	ores & in situ leaching	4,500*
Mexico	Cananea	Cananea	dumps & in situ leaching	12,300*
Israel	Timna Copper	Timna	oxide ores	
Philippines	Marcopper	Taipan	" "	1,550
PNG	Bougainville	Bougainville	oxide & sulphide ores	
Spain	RTM	Cerro Colorado, Alfredo, Costa Atalaya	oxide ores	3,400
	Other small cement producers in Huelva region		" "	9,000
			TOTAL	184,200

*Now replaced by SX-EW plants.

Now each of these half-cell reactions can be considered as separate reactions involving both a forward and reverse direction. However, remember that the half-cells are short-circuited so that the amount reacted and the rate of consumption of M_1 for this ideal case must be related to the amount reacted and rate of production of M_2 by the stoichiometry of the reaction.

$$\frac{dn_{M_1}}{dt} = -\frac{Z_1}{Z_2}\frac{dn_{M_2}}{dt} \qquad (4)$$

where

n_{M_1}, n_{M_2} = number of moles of metal ion,

t = time, and

Z_1, Z_2 = valence.

Detailed analysis of electrochemical half-cells has been given by Bockris and Reddy (3). Application of these concepts to electrochemical cementation reactions has been presented by Wadsworth (4). More recently Power and Ritchie (5-7) have extended this analysis by the use of Evans' diagrams.

For *either one* of the separate half-cells the reaction kinetics can be considered in terms of absolute reaction-rate theory. Absolute reaction-rate theory considers the rate process, in which the reactant species is transformed to the product species, to involve an intermediate high-energy state. This high-energy state is referred to as the activation state and any chemical species which acquires this energy of activation is referred to as the activated complex. Only a small number of reactant or product species acquire this energy at any instance and the activated complex cannot be isolated or observed directly by experimentation.

The reactants are assumed to be in equilibrium with the activated complex and the rate at which the reaction proceeds in a given direction is determined by the concentration of the activated complex, which in turn is determined by the magnitude of the energy barrier.

$$\text{reactants} \underset{\text{equilibrium}}{\rightleftarrows} \text{activated complex}^{\ddagger} \xrightarrow[\text{limiting}]{\text{rate}} \text{products}$$

Butler-Volmer Equation. The Butler-Volmer equation applicable to electrolytic as well as electrochemical reactions, can be derived from these basic postulates of absolute reaction-rate theory. The rate process for a half-cell written as an oxidation reaction can be formulated in terms of the cathodic (reverse) and anodic (forward) direction of the half-cell reaction. It can be shown for a unit area that

$$\left(\frac{dn}{dt}\right)_{\text{cathodic direction}} = \frac{i_-}{ZF} = \overleftarrow{C}^{\ddagger}\lambda\frac{kT}{h}\kappa \quad (5)$$

$$\left(\frac{dn}{dt}\right)_{\text{anodic direction}} = \frac{i_+}{ZF} = \overrightarrow{C}^{\ddagger}\lambda\frac{kT}{h}\kappa \quad (6)$$

where

$\overleftarrow{C}^{\ddagger}$, $\overrightarrow{C}^{\ddagger}$ = the concentrations of the activated complex for the cathodic and anodic directions, respectively,

k = the Boltzmann constant, 1.38×10^{-16} erg/°K,

T = absolute temperature,

h = the Planck constant, 6.6×10^{-27} erg/sec, and

κ = the transmission coefficient, usually unity.

The net current density, i, for this half-cell reaction

$$M \underset{i_-}{\overset{i_+}{\rightleftarrows}} M^+ + e^- \quad (7)$$

is the difference between the partial current density in the anodic direction and the partial current density in the cathodic direction,

$$i = i_+ - i_- \quad (8)$$

provided that the anodic and the cathodic areas are equivalent. After substitution for the partial current density expressions, equations (5) and (6), the net current density for the half-cell can be written as,

$$i = (\overrightarrow{C}^{\ddagger} - \overleftarrow{C}^{\ddagger})zF\lambda(kT/h) \quad (9)$$

Next, substitution of the expressions for the concentrations of the activated complexes, from absolute reaction rate theory results in one form of the Butler-Volmer equation,

$$i = \overrightarrow{k}C_a \exp\left(-\frac{\phi zFE}{RT}\right) - \overleftarrow{k}C_c \exp\left[\frac{(1-\phi)}{RT}zFE\right] \quad (10)$$

where

$$\overrightarrow{k} = zF\lambda\overrightarrow{k}' \quad (11)$$

$$\overleftarrow{k} = zF\lambda\overleftarrow{k}' \quad (12)$$

E = potential

ϕ = transfer coefficient

In this expression C_c is the concentration of the metal ion and C_a is a constant characteristic of the solid metal and representing, in essence, the surface concentration of discharge sites.

At some half-cell potential, E_o, the partial current densities are equal in magnitude and opposite in direction. The value of either partial current density is designated the equilibrium exchange-current density, i_o,

$$i_o = i_- = \overleftarrow{k}C_c \exp\left[\frac{(1-\phi)zFE_o}{RT}\right] = i_+ = \overrightarrow{k}C_a \exp\left(-\frac{\phi zFE_o}{RT}\right) \quad (13)$$

The Butler-Volmer equation can then be written in terms of the exchange-current density, i_o, and the overpotential, $\eta = E_o - E$. The net current for the half-cell reaction is

$$i = i_o\left\{\exp\left(-\frac{\phi zF\eta}{RT}\right) - \exp\left[\frac{(1-\phi)zF\eta}{RT}\right]\right\} \quad (14)$$

Now in a system such as cementation the hypothetical half-cell which we have been discussing is short-circuited to another more reactive half-cell which undergoes a net anodic, or oxidation reaction [equation (3)]. The theoretical discussion given for equation (2) applies equally well to the more reactive half-cell, equation (3). Of course, it is apparent under these conditions that the net current for the cathodic half-cell must be equal to the net current for the anodic half-cell. One of the most important aspects of this type of electrochemical reaction is the fact that, unlike electrolytic reactions in which the potential can be fixed externally, the short-circuited cementation reaction and other electrochemical reactions may acquire a mixed potential, similar to what is known as a corrosion potential. This is not an equilibrium potential such as is observed at the exchange current density for a single half-cell reaction, but rather a steady-state mixed potential the magnitude of which arises from the relative activation energy barriers for each of the half-cells. The steady-state mixed potential is a complex function of the forward and reverse rate constants for each half-cell, as well as reactant and product concentrations and the cathodic/anodic area ratio. In the absence of significant internal and ohmic resistance, each half-cell sees the same mixed potential.

Cementation reaction kinetics can be analyzed in terms of electrochemical theory, using the concept of a steady-state mixed potential, E_m, which the system develops. The half-cells of the hypothetical cementation reaction under consideration, equation (1), are presented in the current density-potential diagram in Figure 1. The position of these curves may be dependent on the composition of the electrolyte, the concentrations of $M_1^{+Z_1}$ and $M_2^{+Z_2}$. Thus, the current-potential diagram for the short-circuited system, shown by the dashed line, also varies as the sum of the respective half-cell current potential curves. Under these circumstances the mixed potential can vary between the two reversible potentials, E_{o,M_2} and E_{o,M_1}. Its exact position is determined by the fact that $I_{M_1} = -I_{M_2}$ for a system at steady state. Furthermore, as the reaction proceeds, the half-cell current-potential plots change and the mixed potential drifts.

The mixed potential can be treated quantitatively by using the Butler-Volmer equation for each half-cell

$$I_{M_1} = A_c\overrightarrow{k}_c \exp\left(-\frac{\phi zFE_m}{RT}\right) - A_c\overleftarrow{k}_c(M_1^{+z_1})\exp\left[\frac{(1-\phi)\ FE_m}{RT}\right] \quad (15)$$

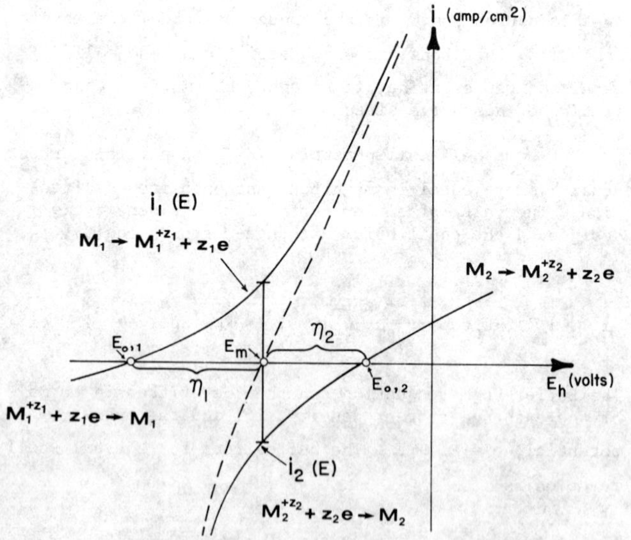

Figure 1. Current density-potential diagram (3) for a short-circuited system which achieves a mixed potential, E_m.

$$I_{M_2} = A_a\vec{k}_a \exp\left(-\frac{zFE_m}{RT}\right) - A_a\overleftarrow{k}_a(M_2^{+Z_2})\exp\left[\frac{(1-\phi)zFE_m}{RT}\right] \quad (16)$$

For a symmetrical energy barrier, the transfer coefficient equals 0.5. Setting, $I_{M_1} = I_{M_2}$, equations (15) and (16) can be solved to give

$$\exp\left(\frac{FE_m}{RT}\right) = \frac{A_c\vec{k}_c + A_a\vec{k}_a}{A_a\overleftarrow{k}_a(M_2^{+Z_2}) + A_c\overleftarrow{k}_c(M_1^{+Z_1})} \quad (17)$$

and the complex nature of the mixed potential is revealed. The expression can be greatly simplified if the back reaction is considered negligible, that is large overvoltages exist. In this case the expression becomes

$$\exp\left(\frac{FE_m}{RT}\right) = \frac{A_a\vec{k}_a}{A_c\overleftarrow{k}_c(M_1^{+Z_1})} \quad (18)$$

In an actual system it is conceivable that not only the concentrations of the metallic ions could vary, but also the anodic and cathodic areas.

Finally it is worthwhile to consider what the rate expression might be for a cementation reaction controlled by the cathodic discharge of the noble metal ion $M_1^{+Z_1}$. In terms of the rate of change in the number of moles of M_1, the rate expression is

$$\left(\frac{dn}{dt}\right)_{M_1} = \frac{I_{M_1}}{Z_1F} = A_c\left[\vec{k}_c \exp\left(\frac{FE_m}{2RT}\right) - \overleftarrow{k}_c(M_1^{+Z_1})\exp\left(\frac{FE_m}{2RT}\right)\right] \quad (19)$$

Evans Diagrams. A convenient way to think about cementation reaction kinetics is to plot the current-potential curves for the respective half-cells involved in the cementation reaction, as shown by Power and Ritchie (5-7). These diagrams are referred to as *Evans diagrams* and have been used in the analysis of corrosion reactions. In essence, the diagrams are simply the super-position of two current-potential (polarization) curves. A typical polarization curve for an arbitrary metal is shown in Figure 2. At very low

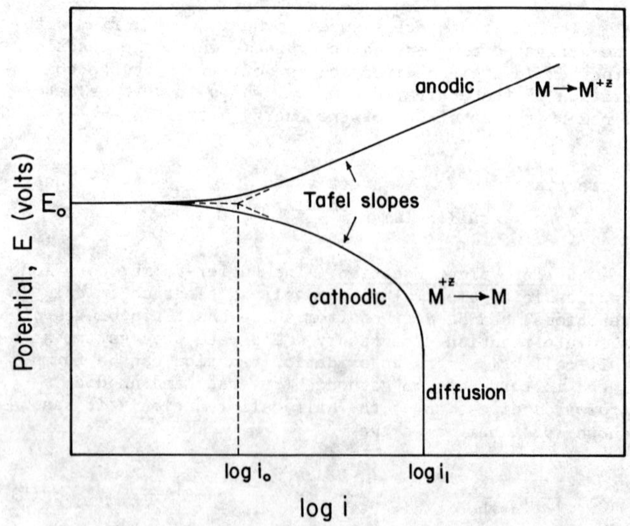

Figure 2. Polarization curves for an arbitrary metal showing cathodic and anodic reaction (7).

currents the potential, E_o, is equivalent to the reversible electrode potential. As E is made more negative (cathodic) than E_o, the metal ions begin to be reduced. At some point the potential begins to vary linearly with log i. This region of the polarization curve is known as the *Tafel region* and the slope of the linear portion of the polarization curve is known as the *Tafel slope*. For large negative values of potential, the current density reaches a maximum value, that is, the *limiting current density*. Under these conditions the current is determined by the rate of transport of metal ions to the surface. Similar regions can be identified for the anodic behavior of the system when the potential is made more positive than E_o. In this case, however, the limiting current density is reached only when the metal surface becomes saturated with respect to a salt of the metal. From the linear region of polarization curves, the Tafel equation is defined,

$$\eta = a \pm b \log i \quad (20)$$

where a and b are constants, and η is the overpotential (the difference between the potential at some current i and the reversible electrode potential, $E_o - E$). Alternatively, the Tafel equation can be written from the Butler-Volmer equation (14) in terms of the exchange-current density,

$$\eta = b \log (i/i_o) \quad (21)$$

The polarization curves can be determined experimentally or calculated, provided the exchange-current density and transfer coefficients are known.

The use of Evans diagrams in cementation systems is best illustrated by some examples taken from Power and Ritchie (5-7). In these systems it is assumed that the polarization curves are independent, the cathodic and anodic areas are equal, and there is no ohmic resistance between cathode and anode. Furthermore, the diagrams are constructed for mass transfer rates calculated for a rotating disk at 100 rad/sec. Details of the rotating disk system are discussed in the section on mass transfer. The most important feature of these diagrams is that they demonstrate quite clearly whether the reaction will be controlled by an electrochemical surface reaction.

For the first example, consider the (Cu^{2+}/Fe) Evans diagram presented in Figure 3. The point of intersection of the polarization curves determines the mixed potential

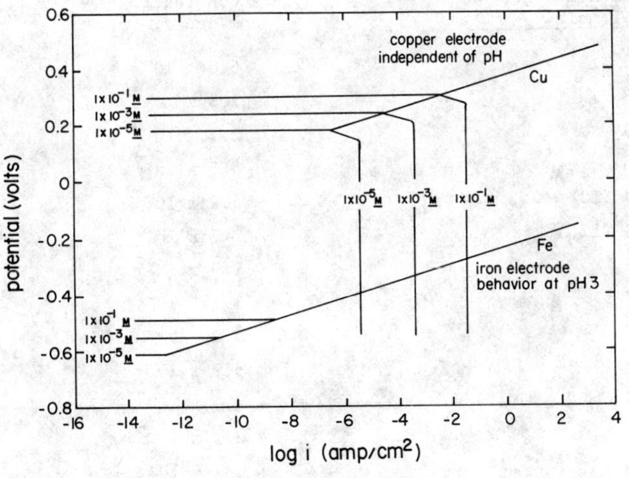

Figure 3. Evans diagram for the Cu^{2+}/Fe system, mass transfer controlled. Polarization curves constructed with data from Hurlen (8).

for the reaction and the rate-controlling step. Because the polarization curves intersect the cathodic copper curve in the diffusion-limited region, the reaction would be expected to be controlled by mass transfer of cupric ion to the reaction surface. Many investigators have found this to be true from experimental cementation studies.

As the other example, consider the (Fe^{2+}/Zn) system shown in Figure 4, for which a different conclusion would be reached. In this case the point of intersection may occur in the Tafel region of the polarization curve and under these conditions the reaction would be electrochemically controlled.

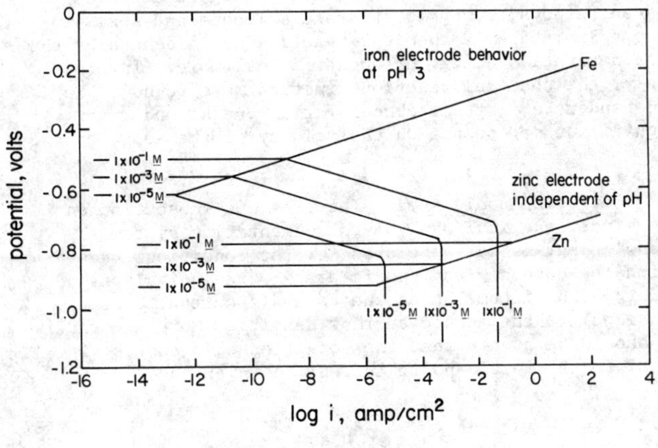

Figure 4. Evans diagram for the Fe^{2+}/Zn system, electrochemical reaction controlled. Polarization curves constructed with data from Hurlen (8).

The criterion for chemical control is intersection of the anodic and the cathodic curves in the Tafel region. An idealized Evans diagram can be constructed as shown in Figure 5 and the point at which rate control switches from mass transfer to chemical reaction can be evaluated. Solution of the respective Tafel equations for cathodic and anodic half-cells at the transition mixed potential results in

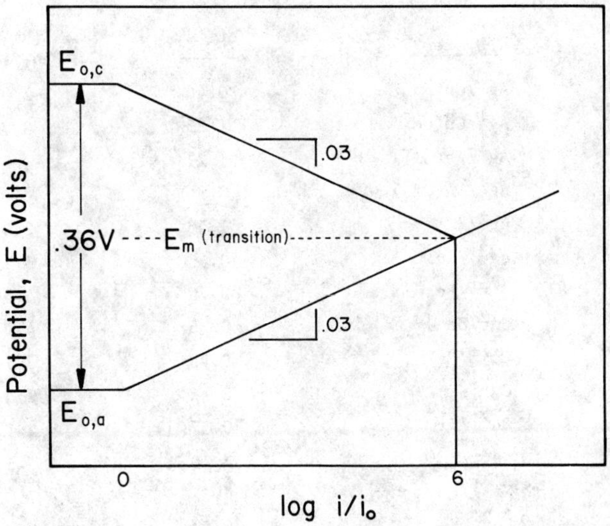

Figure 5. Idealized Evans diagram which represents the transition from mass transfer control to electrochemical reaction control (7).

$$\Delta E_o = b[\log (\frac{i_\ell}{i_o})_c + \log (\frac{i_\ell}{i_o})_a] \tag{22}$$

The variables in equation (22) take on the following values:

$b \simeq 0.03$ V/decade current density for many metals (Cu, Fe, Zn)

$i_\ell \simeq 10^2$ amp/m² (0.1 M solution)

$i_o = 10^1$ amp/m² (Cu) $\rightarrow 10^{-4}$ amp/m²(Fe)

Consequently, for cementation reactions a good rule-of-thumb would be:

$\Delta E_o < 0.06$ V for electrochemical control

$\Delta E_o > 0.36$ V for mass transfer control

Between these values, the exchange-current densities must be known in order to predict the rate-limiting step. As a further approximation, the standard electrode potentials ΔE_o^o of the respective half-cells can be used to approximate ΔE_o. When these values are calculated it is found that most cementation systems have a difference in standard-electrode potential of greater than 0.36 V. Furthermore, experimental investigations indicate that most cementation reactions are controlled by mass transfer processes as indicated by the results presented in Table II.

Generally, then, cementation reactions obey a first order-rate law of the following form, as would be expected for mass transfer control:

$$dn_1/dt = -kAC_1 \tag{23}$$

The disappearance of the noble metal ion can be studied in a variety of experimental geometries, an important consideration due to the importance of mass transfer in these systems. The experimental geometries will be discussed in the section on mass transfer. The integrated rate expression then assumes the following common form in terms of the noble metal ion concentration,

Table II. Data for Selected Cementation Systems at 25°C (1)

System	ΔE_o^o, V	Activation Energy, kcal/mole	Rate Constant, cm/sec
Ag^+/Cu	0.46	2.0 - 5.0 (12)	2.5 - 6.0 x 10^{-2}
$Ag^+/Cu(CN^-)$	1.83	3.7 - 5.8	1.5 x 10^{-2}
$Ag^+/Fe(Cl^-)$	1.29	3.0	2.2 x 10^{-2}
$Ag^+/Zn(CN^-)$	0.95	5.5	5.5 x 10^{-2}
Ag^+/Zn	1.56	2.0 - 6.0 (12)	2.6 - 5.2 x 10^{-2}
Ag^{3+}/Cd	0.65	3.1 ± 1.3 (12.3)	7 x 10^{-3}
Bi^{3+}/Fe	0.76	4.5 - 7.6	2.9 x 10^{-2}
Cd^2/Zn	0.36	4.0 - 4.7	0.54 - 1.1 x 10^{-2}
Cu^{2+}/Fe	0.75	3.1 - 5.1	0.6 - 0.9 x 10^{-2}
Cu^{2+}/In	0.83	2.3	5.9 x 10^{-2}
Cu^{2+}/Ni	0.57	2.7 - 3.7 (14.2 - 19.0)	0.25 - 1.0 x 10^{-2}
Cu^{2+}/Zn	1.10	3.1	1.6 - 2.1 x 10^{-2}
Ni^{2+}/Fe	0.21	7.0	1 x 10^{-4}
Pb^{2+}/Fe	0.31	12.0	---------
Pb^{2+}/Zn	0.64	----	0.64 x 10^{-2}
Pd^{2+}/Cu	0.49	9.5 - 7.4	0.36 - 2.3 x 10^{-2}

$$\log\left[\frac{C_1(t)}{C_1(0)}\right] = \frac{-kA}{2.303\ V}\, t \qquad (24)$$

where

$C_1(0)$ and $C_1(t)$ = initial concentration and concentration at time t of noble metal ion
k = rate constant (cm/sec)
A = reaction surface area (cm^2)
V = solution volume (cm^3)
t = reaction time (sec)

From the analysis of cementation reactions in terms of Evans diagrams it appears that the reaction kinetics of only a few systems should be controlled by an electrochemical reaction mechanism. Furthermore, the general irreversibility of these electrochemical reactions was previously established by Wadsworth (4), in which it was shown that if $Z_1\Delta E_o^o$ is greater than 0.3 V, the back reaction kinetics should not be important. Experimental results confirm these conclusions, although, as will be seen later, surface deposit effects can lead to inaccurate interpretation of kinetic data. Notice in Table II that both systems (Ni^{2+}/Fe and Pb^{2+}/Fe), which exhibit a difference in standard-electrode potentials of less than 0.36 V, have activation energies which support a surface reaction control mechanism. Further evidence in the case of the Ni^{2+}/Fe system includes a rather small rate constant of less than 1 x 10^{-4} cm/sec. All other systems have a difference in standard-electrode potentials, ΔE_o^o, in excess of 0.36 V and the reaction kinetics appear, at least under most circumstances, to be limited by aqueous-phase mass transfer, as supported by measured activation energies and rate constants.

However, there are circumstances, even though mass transfer control would be predicted, $\Delta E_o^o > 0.36$ V, which can result in a switch from mass transfer control to chemical control. General situations for which unexpected chemical control can result are presented in subsequent sections together with examples from specific systems. These situations include; Cathodic/Anodic Area, Ohmic Resistance, and Codeposition. Just by way of example, one system which exhibits this unexpected chemical control is the As^{3+}/Cd system(54). The difference in standard electrode potentials is 0.65 V (see Table II) and at low initial arsenic concentrations, 0.1 gpl, the rate is found to be controlled by mass transport as evidenced by the activation energy of about 3 kcal/mole and reaction velocity constant on the order of 10^{-2} cm/sec at high stirring speeds. However, at higher arsenic concentrations, 1.0 gpl, the rate switches and seems to become limited by a chemical reaction as evidenced by the high activation energy of about 12 kcal/mole. Interestingly, the chemical reaction kinetics were found to be independent of arsenic concentration but showed a very significant dependence on the acidity of the solution.

Cathodic/Anodic Area. The previous subsection on the electrochemical analysis of cementation systems by Evans' diagrams was developed for the conditions under which the cathodic and anodic areas were equivalent and under which circumstances the cathodic current density was equal to the anodic current density. More accurately it should be stated that the total cathodic current is equal to the total anodic current in which case,

$$A_c i_c = A_a i_a \qquad (25)$$

and the respective current densities could differ from one another and not necessarily be determined by the point of intersection fo the respective cathodic and anodic polarization curves on the Evan's diagram. Not only may the cathodic and anodic areas differ from one another at the time of reaction initiation but in addition the fraction of the surface acting cathodically or anodically may change during the reaction progress. Such conditions are difficult to quantify and complicate the electrochemical analysis considerably.

The example given by Power and Ritchie (5) illustrates how the Evans' diagram analysis must be modified. Consider the case where the cathodic and anodic reactions are independent and the cathodic area is larger than the anodic area, $A_c > A_a$. Further, assume as is frequently the case that the rate is limited by transport, i.e. the limiting current density for the cathodic reaction. Such a system is represented by the Evans' diagram in Figure 6. Now if the total cathodic current is fixed, as it must be under diffusion control, the result of having a significantly smaller anodic area is to necessitate a higher anodic current density and an increased mixed potential. In the extreme situation, the anodic area could become so small that the anodic current density and the mixed potential could increase to the point where the reaction comes under chemical control. Such a decrease in anodic area may occur if a tight and coherent cathodic deposit forms over anodic sites.

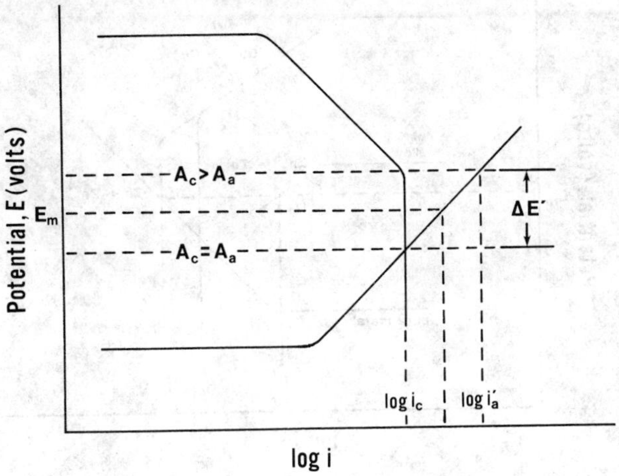

Figure 6. Polarization diagram showing the shift in the mixed potential E_m for the case when $A_c > A_a$. $\Delta E'$ is the potential through which E_m is raised in order for chemical control to supplant diffusion control (5).

Chemical control results when the mixed potential is raised by an amount $\Delta E'$ as shown in Figure 6. The magnitude of this potential rise required for transition to chemical reaction control would be,

$$\Delta E' = b\ (\log i_a' - \log i_c) \qquad (26)$$

where b is the Tafel slope for the anodic polarization curve. Thus the magnitude for the potential rise is dependent on the ratio of the anodic/cathodic current densities, i_a'/i_c, and generally it can be expected that $\Delta E'$ would not exceed 0.15 volt.

Another possibility is that as the anodic area decreases, the limiting anodic current density is reached before reaching the point of chemical reaction control. Under these circumstances as indicated in Figure 7 the anodic metal salt crystallizes at the surface and the anodic current density will be fixed by the concentration gradient between the saturated solution surrounding the salt at the surface and the bulk solution. Further decrease in the anodic area and the increase in mixed potential will result in a transition to chemical reaction rate control.

For cementation systems which show marked decrease in rate from initial first order reaction kinetics, a tight, coherent surface deposit may reduce the anodic area and the kinetics explained in terms of the foregoing electrochemical analysis. The results of Von Hahn and Ingraham

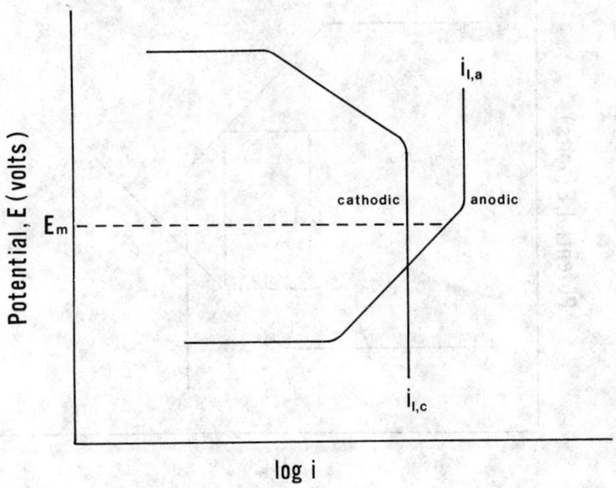

Figure 7. Polarization diagram showing the effect of a diffusion-limited anodic reaction on E_m (5).

(9) in the study of the Ag^+/Cu (cyanide) system might be interpreted in this fashion. The opposite situation in which the cathodic area is decreased is difficult to imagine but would suggest that a transition from chemical reaction control to diffusion control is possible. It is evident that the cathodic/anodic area and how it might change during the course of the reaction is an important factor in the electrochemical analysis of cementation kinetics. In this regard, measurement of the mixed potential in cementation reactions can be useful in determination of the intrinsic reaction kinetics.

Ohmic Resistance. Another phenomenon which may be important in cementation systems is the possible existence of an ohmic resistance between cathodic and anodic sites. Such a resistance and the attendant potential drop may arise from the presence or formation of oxide films during the reaction progress. Consider the case where $A_c = A_a$ and the reaction is limited by transport of the noble metal ion to the surface as shown in Figure 8. If the insulating character of the oxide film provides sufficient resistance the reaction system will passivate, the potential at the cathodic sites will not equal the potential at the anodic sites, and the rate will become limited by chemical reaction control. The situation is depicted in Figure 8. Examples of such behavior are found for the Cu^{2+}/Ni system (10) and perhaps the Cd^{2+}/Zn system (11). A general analysis of this phenomenon has been published by Hope, *et al.* (12). During the course of the cementation reaction, an oxide film may form such that the kinetics initially limited by transport, become chemically controlled. High anodic current densities developed due to the formation of a coherent deposit and a reduction in the anodic area (see previous subsection) may account for the system's eventual passivation.

The rates of deposition of cadmium on to zinc under inert and oxygen containing atmospheres have been investigated (11). In the absence of oxygen, the reaction was found to be mass-transfer controlled. In the presence of oxygen, however, the rate controlling mechanism was found to change from mass transfer to mixed or chemical control as the oxygen content in solution increased. The change in reaction kinetics has been attributed, in part, to the formation of $Cd(OH)_2$ and the ohmic resistances arising from the surface deposit.

In another system, Co^{2+}/Zn, the importance of ohmic resistance significantly influences the cementation reaction kinetics (13). The difficulties encountered during the cementation of cobalt on zinc in concentrated

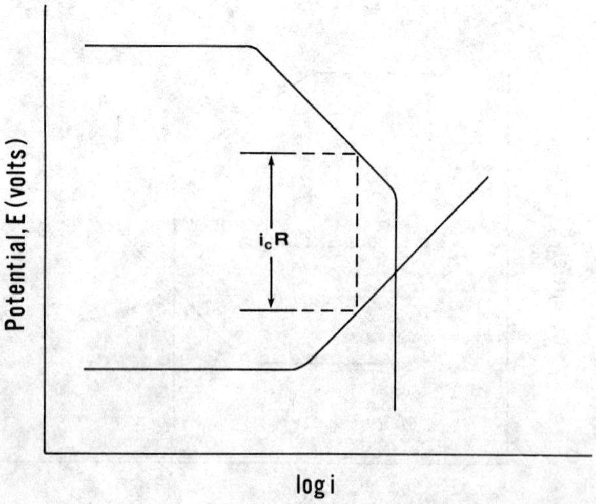

Figure 8. Polarization diagram representing the case in which a resistive oxide film has brought a diffusion-controlled reaction under chemical control. The potential drop across the oxide of resistance R is i_cR (5).

$ZnSO_4$ solutions (160 g/l of Zn^{2+}) are due to the Zn^{2+} ions and not to the overpotential of cobalt deposition on a metallic zinc substrate being particularly high, as numerous authors have assumed (14-18).

This inhibition of cobalt deposition by zinc ions was identified by Yunus *et al.* (19) who suggested that in a concentrated Zn^{2+} solution the local increase in pH due hydrogen evolution on the reaction surface could lead to the formation of a layer of zinc hydroxide which would retard the discharge of the cobalt ions. This type of inhibition has been clearly shown by the results of de Blander and Winand (13). In their investigation the presence of basic zinc sulfate was found in the deposit at higher pH values. Under these conditions the deposit contained consistently less cobalt than the deposits obtained in more acidic solution.

Passivation effects arising from the formation of oxide or hydroxide films have been found in other systems in which the contribution of ohmic resistance may explain the chemical rate control observed. For example, in the Ni^{2+}/Fe system (20) the formation of iron and nickel oxides was found to be a function of the local surface mixed potential and the local oxygen potential. Similarly, in the Bi^{3+}/Fe system, the oxy chloride has been found in the deposit (21) presumably as a result of local excursions of pH in certain surface regions of the deposit. Consideration of these findings, together with other results (22) which show an activation energy from 4.5 - 7.6 kcal/mole and a reaction velocity constant of $\sim 10^{-2}$ cm/sec, suggest that the rate is under mixed control and that the oxychloride film contributes significantly to the observed kinetics. The significance of aluminum oxide films in the Cu^{2+}/Al system (23-26) has been studied extensively. In order for this reaction to proceed the compact protective oxide layer must be broken. Chloride ion at relatively low concentrations (15 ppm) destroys the oxide film and allows the reaction to proceed.

Codeposition. For systems containing more than one noble metal ion, alloy formation may be possible and in such cases codeposition occurs (5-7). Further, it is possible that the more active metal, even in a binary system, could codeposit with the noble metal either as the metal itself or as an alloy. If codeposition of the active metal itself would occur the mixed potential would have to be close to the equilibrium potential. Such an event would be favored for small noble metal ion concentrations, large active metal ion concentrations, and low stirring speeds.

An important system which exhibits the case of alloy formation is the Cu^{2+}/Zn system. If the activity of the zinc in the alloy is less than that of pure zinc, the cathodic polarization curve for the zinc in the alloy will be at a higher potential more readily reduced as shown in Figure 9. The active metal, the pure zinc, will then tend to dissolve and redeposit as the alloy phase. Under these circumstances the anodic current for the pure zinc half cell must be balanced by the cathodic half cell currents for copper and zinc deposition in the alloy phase.

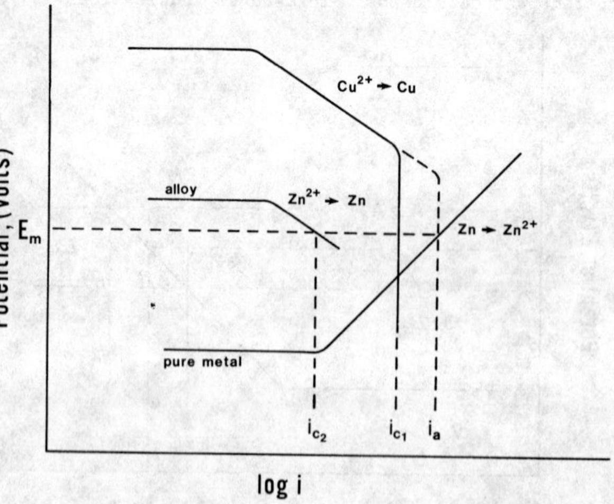

Figure 9. Polarization diagram showing the case in which zinc is cocemented with copper as an alloy during the copper (II)/zinc reaction (5).

For the mixed potential indicated in Figure 9, the current would be balanced in the following manner:

$$A_a i_a = A_{c1} i_{c1} + A_{c2} i_{c2} \tag{27}$$

Of course the position of all polarization curves will be a function of the aqueous phase composition. Furthermore, the composition of the alloy deposit will be a function of the aqueous phase composition. The numerous partial current expressions that might exist in a real cementation system makes such an analysis exceedingly complex. Nevertheless Evans' diagrams are instructive for a basic understanding of cementation reactions.

More recently, the codeposition study of copper and cadmium on to zinc under various oxygen potentials was carried out (27). Unlike the case of single deposition, the reaction controlling mechanism changed from the initial rate controlled by mass transfer to a mixed control brought about by the formation of an alloy. In the enhanced rate period, the rate of copper cementation was found to be higher than when it was cemented by itself. However, the rate of cadmium deposition in the presence of copper was reduced due partly to the formation of an alloy and partly to the resolution of cadmium from loose cadmium-bearing deposits.

The electrochemical behavior of cobalt, antimony and copper as they relate to electrolyte purification for zinc production has been studied extensively (13). Polarization curves and potentiostatic measurements were used to characterize this cementation system which typically contains 160 gpl zinc as zinc sulfate. The complex composition of this multicomponent system is intentionally prepared in order to achieve the desired rate of removal of cobalt impurity which may be present initially at about 15 mg/l. The increased rate and extent of cobalt removal was found by de Blander and Winand (13) to be due to the formation of intermetallic phases. This accelerating effect on the rate of cobalt cementation was attributed principally to the copper addition, whereas it seemed the antimony addition served to stabilize the deposit and protect it from redissolution.

The classical cobalt cementation processes use rather high concentrations of Cu^{2+} (250 mg/l) and Sb_2O_3 (10 mg/l) or As_2O_3 (100 mg/l). However, the obtained deposits show a very marked tendency for the redissolution of cobalt. Figure 10 shows the cobalt concentration as a function of cementation time during cobalt cementation at the Akita Electrolytic Zinc Plant, Mitsubishi, Japan (28). As can be seen the redissolution of cobalt occurs together with that of copper and arsenic. This behavior is typical of the cement having a high copper content, the redissolution being due either to the presence of oxygen dissolved in the solutions or to the corrosion of the zinc particles under hydrogen evolution (14, 15).

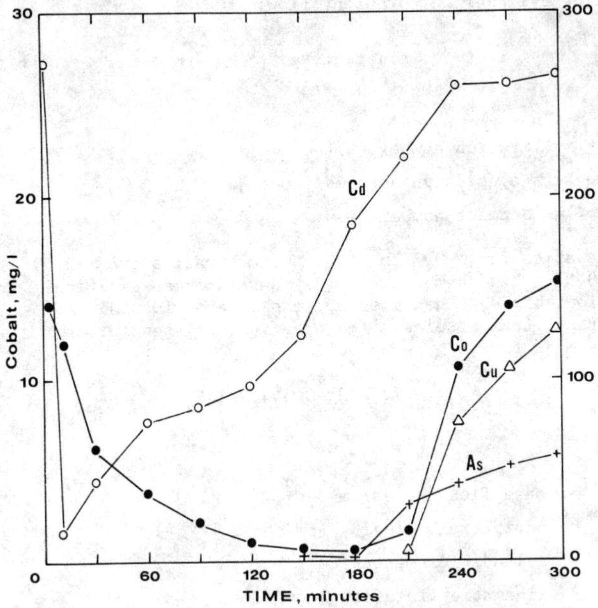

Figure 10. Time dependence of the concentration of cobalt, arsenic, copper and cadmium during cementation on zinc powder (1 g/l) in a $ZnSO_4$ solution (110 g/l of Zn^{2+} initially containing 15 mg/l of Co^{2+}; 100 mg/l of As_2O_3; 210 mg/l of Cu^{2+} and 276 mg/l of Cd^{2+} - Akita Electrolytic Zinc Plant, Mitsubishi, Japan (28).

In the cementation process used at the Balen (Belgium) plant of Vieille Montagne (29) the low residual copper concentration (10 mg/l) of the solution entering the cobalt removal step means that only a small amount of antimony has to be added to stabilize the cement: the redissolution of cobalt is no longer observed, even after the cemented zinc has been in contact with the solution for a long time. This plant observation has been confirmed in the laboratory. Even sparging of the solution with oxygen did not lead to redissolution.

If the conditions of cementation are such that the deposits obtained are sufficiently stable - this is the case if the deposits are made from a solution which has a relatively low copper content - one can accelerate the cementation of cobalt by working with slightly acid solutions, but concommitantly must accept an increase in consumption of zinc. It may be that the reaction could take place in two reactors connected in series: in the first reactor the pH would be kept low by adding depleted solution coming from the electrolysis and in the second one, the pH would be increased to the buffer value of 5.4 in order to protect the cement from redissolution by the formation of basic zinc sulphate. An economic evaluation would determine whether the time gained outweighs the additional consumption of zinc powder and of Sb_2O_3 (13).

Other studies which have identified the importance of codeposition or activation by foreign metal ions include a Russian study of the Cd^{2+}/Zn system (30) in which it is reported that the presence of Ni^{2+} and Co^{2+} reduces the rate of cadmium cementation due to the formation of a hard (tight?) layer on the surface of the zinc. Further, the enhanced resolution of any cadmium sponge contributes to the negative effect of these ions. In the $Au(CN)_2^-/Zn$ system (31) the addition of as little as 4 ppm lead nitrate results in a significant shift in the aurocyanide polarization curve, facilitating reduction. Similar depolarizing effects on the deposition of gold were observed in the presence of thallium and mercury ions. A possible explanation for the effect of these additives has been suggested to be that they prevent passivation due to adsorption of the intermediate AuCN at the zinc surface (32). In addition, lead serves to remove any poisonous sulfide ions from solution. However, excessive lead addition can inhibit cementation, perhaps due to the difficulty of gold nucleation at a lead surface.

Mass Transfer

For reactions limited by diffusion through the mass transfer boundary layer, the hydrodynamic flow regime of the reacting surface is of considerable importance in understanding the reaction kinetics. Mass transfer in an aqueous phase has been analyzed in detail by Newman (33) and Bockris and Reddy (3). Basically in these systems mass is transferred by,

1. diffusion, i.e., motion of both charged and uncharged species due to concentration gradients,
2. migration, i.e., motion of charged species in an electric field, and
3. convection, i.e., transfer of mass due to the motion of elements of the liquid phase.

Because the kinetics of most cementation reactions appear to be controlled by a mass transfer process, several experimental geometries and methods which allow for the evaluation of the respective mass transfer coefficients will be considered. The principles discussed, of course, are not limited to cementation reactions but could apply to any heterogeneous reaction in which mass transfer in the aqueous phase limits the reaction rate.

Rotating Disk. The rotating disk system has been used with good success in the quantitative study of electrode kinetics and more recently in the study of cementation kinetics. The rotating disk system is unique for kinetic studies, laminar flow being maintained even to Reynolds numbers ($r^2\omega/\nu$) of 2×10^5, where

r = the radial position on the disk (cm),

ω = angular velocty (rad/sec), and

ν = kinematic viscosity (cm^2/sec).

Furthermore, the hydrodynamic flow in the system is well defined and can be calculated from first principles as done by Levich in 1942 (34). As a result of this well-defined hydrodynamic flow, a theoretical expression can be developed from which the rate of mass transfer to the disk surface can be calculated. This important fact provides additional insight into distinguishing the rate-limiting step for heterogeneous reactions such as cementation.

A theoretical treatment of the rotating-disk system is given by Riddiford (35) and is reviewed with respect to cementation by Miller (1). From the mass transfer expression and the necessary velocity vectors determined by the Navier-Stokes equation, the mass flux at the disc interface can be shown to be,

$$J_{y=0} = -D\left(\frac{dC}{dy}\right)_{y=0} = 0.62D^{2/3}\nu^{-1/6}\omega^{1/2}C \qquad (28)$$

A number of cementation-rate studies have been reported in which the reaction kinetics were studied with a rotating disk. The results of these studies invariably support the contention presention in the section Electrochemical Reactions: for most cementation systems the rate is limited by mass transfer. According to equation (28) the mass transfer coefficient should be,

$$k = 0.62\ D^{2/3}\nu^{-1/6}\omega^{1/2} \qquad (29)$$

A plot of k vs. $\omega^{1/2}$ is presented in Figure 11 for a number of cementation systems which have different noble metal ion diffusivities (36). The lines represent a plot of equation (29) with the following values for diffusion coefficients at 25°C selected from the literature:

	Cation		
	Cu^{2+}, Cd^{2+}	Pb^{2+}	Ag^{+}
$D, cm^2/sec$	7.3×10^{-6}	9.5×10^{-6}	16.5×10^{-6}

The symbols represent the experimental data. Excellent agreement between theory and experiment is observed and was expected from electrochemical theory since for each of the systems reported the difference in the standard-electrode potential of the respective half-cells exceeds 0.36 V. (See Table II, p. 6).

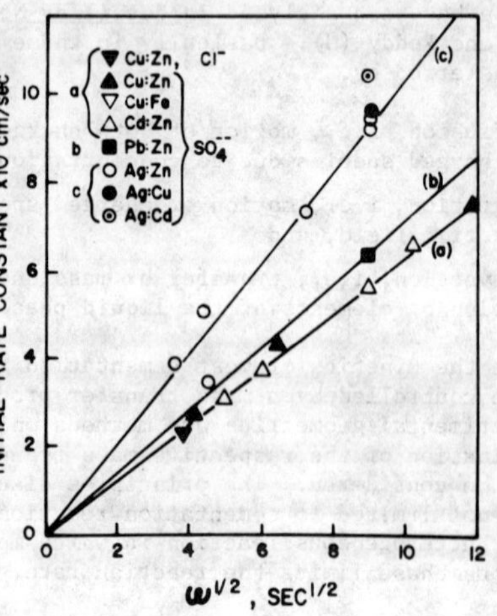

Figure 11. Demonstration of the half-order dependence of the angular velocity on the mass transfer coefficient as predicted by the Levich equation for rotating disks for a number of cementation systems (36).

Generally, the systems are more complicated than has been indicated and some of these complications such as cathodic/anodic area, oxide films and codeposition were mentioned previously. Further complications include competing reactions, such as equilibrium complexation, hydrolysis, and chemical redox reactions which make the analysis of kinetic data difficult. Finally, the formation of a surface deposit can cause an increase in the apparent reaction rate constant. The nature of this deposit can be particularly sensitive to the temperature and electrolyte composition. Details of this effect are considered in the section Surface Deposit Effects on p 13.

Rotating Cylinder. The major difficulty encountered for this particular geometry, unlike the rotating disk system, is that the flow regime becomes turbulent at very low Reynolds numbers,

$$Re = Vd/\nu$$

where V is the peripheral velocity of the cylinder, d is the cylinder diameter, and ν is the kinematic viscosity. On this basis the transition to turbulent flow for a rotating cylinder begins to occur at $Re \simeq 10^1$, whereas for the rotating disk the transition occurs at $Re \simeq 10^5$. For a rotating cylinder, turbulent flow would certainly be established at Re = 20, which, for a cylinder of 2.0 cm diameter in an aqueous solution at room temperature ($\nu = 10^{-2}\ cm^2/sec$), would correspond to a peripheral speed of only 0.1 cm/sec, or approximately 1.0 rpm.

Empirical correlations of mass transfer data have established the following dimensionless relationship for a rotating cylinder in turbulent flow:

$$St = \text{constant}\ Re^a Sc^b \qquad (31)$$

where

St = Stanton number, k/V,

Re = Reynolds number, Vd/ν, and

Sc = Schmidt number, D/ν,

and a varies from -0.3 to -0.4, while b varies from -0.59 to - 0.67. Eisenberg *et al.* (37) developed an empirical equation which has had good success in experimental electrochemical studies. The equation, written in terms of the mass flux, J, is as follows

$$J = 0.079V^{0.7}d^{-0.3}\nu^{-0.344}D^{0.644}C \qquad (32)$$

where

J = mass flux ($moles/cm^2 sec$),

V = peripheral velocity (cm/sec),

d = cylinder diameter (cm),

ν = kinematic viscosity (cm^2/sec),

D = diffusion coefficient (cm^2/sec), and

C = concentration ($moles/cm^3$).

Cornet and Kappesser (38) have independently verified this empirical equation in other electrochemical studies and it should be adequate to evaluate mass transfer coefficients for cementation reactions.

Gabe and Robinson (39) have derived a mass transfer equation quite similar to that which has been determined empirically. The derivation is based on the Prandtl-von Karman model which divides the fluid regime into three zones: I. a laminar flow layer next to the surface-Nernst diffusion layer; II. a viscous damping transition layer which damps out eddy current effects; and III. completely developed turbulence far from the surface-Prandtl's hydrodynamic layer. The theory for turbulent mass transfer at a rotating cylinder basically involves the determination of a turbulent diffusion coefficient, D_{turb}, analysis of Fick's law for the concentration gradient in each zone, and demonstration that the greatest resistance to mass transfer occurs in the laminar flow layer (Zone I) in which discrete molecular interactions prevail. The equation derived from this analysis is,

$$J = b'V^{0.67}\, d^{-0.33}\, \nu^{-0.33}\, D^{0.67}\, (C_b - C_s) \quad (33)$$

which agrees with the empirical equation (32) presented previously.

A number of cementation rate studies have been made using rotating cylinders but few investigators have analyzed their results in terms of predicted mass transfer rates for turbulent flow at a rotating cylinder. Power (7) was one of the first investigators to examine experimental cementation rate data in this fashion. However, there are two major complications which must be considered. The first factor which may lead to erroneous interpretation of cementation rate data is the surface deposit and its morphology. As mentioned previously, this effect is common to all experimental techniques and will be discussed in the next section. The second factor which may cause difficulty in the analysis of cementation data for a rotating cylinder is the end effect. This becomes important only when the length-to-diameter ratio, ℓ/d, reaches a critical value. Power (7) examined this effect for the Cu^{2+}/Zn system and found that for 25°C and a peripheral speed, V = 138 cm/sec, the critical ℓ/d ratio is approximately unity, as shown in Figure 12. Notice that for ℓ/d ratios of less than unity the experimental mass transfer coefficient increases significantly, almost linearly with decreasing ℓ/d.

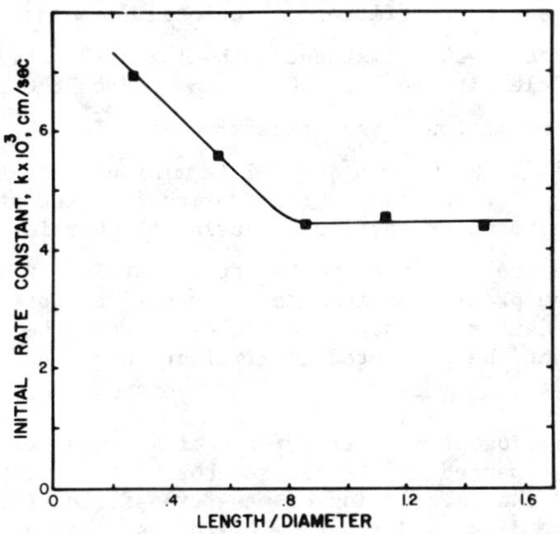

Figure 12. The effect of length/diameter ratio on the experimental mass transfer coefficient for the Cu^{2+}/Zn rotating cylinder system (7).

According to equation (32), the mass transfer coefficient for rotating cylinders, k, should be,

$$k = 0.079\, V^{0.7}\, d^{-0.3}\, \nu^{-0.344}\, D^{0.644} \quad (34)$$

provided the aforementioned complications are avoided. The dependence on the peripheral velocity to the 0.7 power has been demonstrated for the Cu^{2+}/Zn system, as shown in Figure 13. A review of the literature indicates that for studies in which surface deposit effects have been eliminated the calculated mass transfer coefficient agrees well with the experimental values as shown in Table III.

To further illustrate the critical nature of the ℓ/d ratio, it can be seen that for the last two systems listed in Table III the ℓ/d ratio is significantly less than the critical value of unity determined by Power (7) and, as might be expected, the experimental mass transfer coefficient is greater by a factor of two than the calculated value. From the data in Figure 12 an end-effect correction factor of 0.58 can be calculated for an ℓ/d ratio of 0.13. Applying this empirical correction factor to the measured mass transfer coefficients, corrected values of 7.2×10^{-3} cm/sec for Cd^{2+}/Zn and 1.5×10^{-2} cm/sec for Ag^{+}/Zn are obtained which agree with the mass transfer coefficients calculated from equation (34).

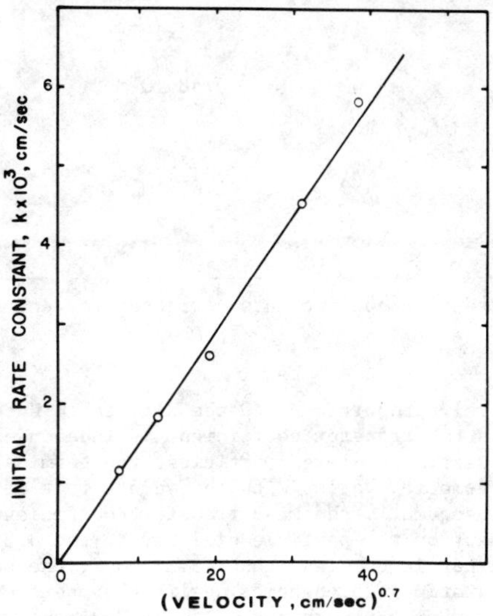

Figure 13. Demonstration of the 0.7 power dependence of the peripheral velocity on the mass transfer coefficient for the Cu^{2+}/Zn rotating cylinder system (7).

Suspended Particles. Cementation on suspended particles is of considerably greater interest than the rotating disk or the rotating cylinder systems in that frequently cementation reactions on an industrial scale are accomplished in stirred reactors with particulate metal powder, e.g., Cd^{2+}/Zn, Cu^{2+}/Fe, Ag^{+}/Zn, and Cu^{2+}/Ni. However, the theoretical treatment of mass transfer to particles suspended in a stirred reactor has not been fully developed. Nevertheless, the minimum mass transfer coefficient, k^*, for which the slip velocity would be equivalent to that for free-falling particles, can be calculated on a semitheoretical basis for mono-size spheres from the physical properties of the aqueous and solid phase and the diffusion coefficient of the reactant. The free-fall mass transfer coefficient can then be empirically corrected for design and operating variables of the reactor.

Basically, the semitheoretical correlation (43) takes the following form for free-falling spherical particles:

$$Sh = 2 + 0.6\, Re^{0.5} Sc^{0.33} \quad (35)$$

where

Sh = Sherwood number, $k^* d/D$,

Re = Reynolds number, Vd/ν, and

Sc = Schmidt number, D/ν.

Figure 14 is a plot of the above correlation in terms of the mass transfer coefficient vs. particle size. Notice that in the range of 100 to 1000 microns the mass transfer coefficient is nearly constant because the terminal settling velocity in this particle-size region is

Table III. Comparison of Experimental vs. Calculated Mass Transfer Coefficients for Selected Cementation Systems Using Rotating Cylinder Geometry

System	Reference	D cm^2/sec	ν cm/sec	d cm	ℓ/d	Mass Transfer Coefficient, cm/sec Experimental	Calculated
Cu^{2+}/Zn	7	$0.67x10^{-5}$	19.5	3.5	1.14	$1.1x10^{-3}$	$1.0x10^{-3}$
Cu^{2+}/Fe	40	$0.67x10^{-5}$	99.7	0.64	40	$4.2x10^{-3}$	$5.1x10^{-3}$
Ag^{+}/Zn	41	$1.65x10^{-5}$	166	1.75	1.47	$8.3x10^{-3}$	$9.7x10^{-3}$
Cd^{2+}/Zn	42	$0.48x10^{-5}$	497	7.6	0.13	$1.23x10^{-2}$ $(7.2x10^{-3})^*$	$6.0x10^{-3}$
Ag^{+}/Cu	9	$1.65x10^{-5}$	500	7.4	0.13	$2.6x10^{-2}$ $(1.5x10^{-2})^*$	$1.4x10^{-2}$

*Mass transfer coefficients corrected for end effects.

approximately proportional to the diameter. Under these conditions the transfer coefficient is independent of particle size. For large particles, the terminal settling velocity varies with the square root of particle size. Consequently the mass transfer coefficient should be dependent on the particle diameter to the 0.25 power. For small particles, less than 10 microns, the second term of equation (35) becomes negligible and the Sherwood number approaches a value of 2.0, the values for diffusion in an infinite stagnant medium. Under these conditions the mass transfer coefficient varies inversely with particle size to the first power.

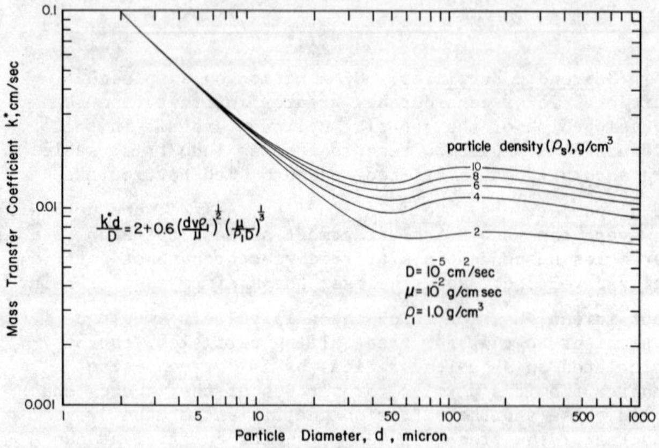

Figure 14. Plot of mass transfer coefficient vs. particle diameter for free-falling spheres; minimum mass transfer coefficient, k^* (1).

The diagram shown in Figure 14 is constructed using the terminal settling velocity to calculate the Reynold's number and should give the minimum transfer coefficient for a suspended particle. The actual coefficient is dependent on reactor design and stirrer speed, or power input, whereas the effect of the physical properties can be predicted from equation (35) or Figure 14. Harriot (44) described a method for prediction of mass transfer coefficients in stirred reactors which has been reviewed (1).

It appears from an examination of the literature that cementation reaction kinetics have not been extensively studied using monosize particles and consequently the correlation for suspended particles remains untested in cementation systems. Almost all cementation studies involving particulate metal in a stirred reactor have used an unspecified particle size distribution. However, in one study of the Cu^{2+}/Fe system, Nadkarni *et al.* (45) did report cementation kinetics with monosize particles of granulated cast iron. The mass transfer coefficient calculated from the reported data at 26°C is found to be 1.2×10^{-2} cm/sec for a particle diameter of 200 microns. From the plot in Figure 14 the minimum mass transfer coefficient k^* is found to be 1.4×10^{-2} cm/sec. This minimum mass transfer coefficient can be transformed into the predicted mass transfer coefficient using the Harriot correlation and was found to be approximately 2×10^{-2} cm/sec. The prediction is made for spherical particles and for irregularity shaped particles it would be expected that the predicted coefficients would be somewhat smaller.

Calculation of mass transfer coefficients for particles suspended in stirred reactors is not nearly as accurate as the calculation of mass transfer coefficients for rotating disks and rotating cylinders. Again, for cementation systems, as always, complications associated with surface deposit formation must be carefully considered in the analysis of rate data.

Apparent Activation Energy - Effect of Hydrodynamics

The temperature coefficient for heterogeneous reactions is frequently referred to as an "apparent" activation energy or the critical increment of energy, because of the complications that may arise in the analysis of the temperature dependence of the rate process. In the next section it will be seen how surface deposit effects cloud the significance of the activation energy. At this point, however, in view of the fact that the reaction kinetics for many cementation systems are limited by transport in the aqueous phase other considerations should be given to the activation energy associated with transport controlled reactions. The questions that should be examined are inter-related; what is the expected magnitude of the activation energy for rate control mass transfer? and is the activation energy related to the hydrodynamic flow regime of the system?

An analysis of such effects was made by Riddiford and Bircumshaw in the early 1950's (46-49) and explains certain results that have been observed in heterogeneous reactions limited by transport in the aqueous phase. It appears that this analysis may explain, in part, the range of apparent activation energies (3-6 kcal/mole) observed in cementation systems. As we have seen for transport controlled reactions, the mass transfer coefficient, and hence the rate of reaction, is dependent on the diffusion coefficient of the reacting solute, D, and the kinematic viscosity of the solution, ν.

It is generally accepted for liquid systems that,

$$D = D_o e^{-E_D/RT} \tag{36}$$

and that,

$$\nu = \nu_o e^{E_\nu/RT} \tag{37}$$

The temperature dependency of these parameters is satisfactory from an intuitive standpoint inasmuch as the diffusion coefficient is known to increase with an increase in temperature, whereas, the kinematic viscosity decreases with an increase in temperature. Both E_D and E_ν when determined independently are found to be on the order of 4 kcal/mole.

Rotating Disc. Now recalling the Arrhenius expression for the transport process in the liquid phase,

$$k_T = Ae^{-E_T/RT} \tag{38}$$

as well as the expression for transport to a rotating disc in a laminar flow regime,

$$k_T^{disc} = 0.62D^{2/3}\nu^{-1/6}\omega^{1/2} \tag{39}$$

The apparent activation energy can be represented in terms of its component energy increments by taking the logarithm of equations (36, 37 and 38) and after differentiation making appropriate substitutions into equation (39). The results show that for *laminar flow* the apparent activation energy is described by

$$E_{T(laminar)}^{disc} = (4E_D + E_\nu)/6 \tag{40}$$

For *turbulent flow*, the dependence of k_T on ν is more significant (34), such that in the limit k_T is proportional to $\nu^{-2/3}$, in which case,

$$D_{T(turbulent)}^{disc} = 2(E_D + E_\nu)/3 \tag{41}$$

The difference in component energy increments, (E_D - E_ν) is typically a small value on the order of 0.5 kcal/mole for aqueous systems. Therefore as a first approximation, if it is assumed that $E_D = E_\nu$, the apparent activation energy for transport in laminar flow would be expected to be,

$$E_{T(laminar)}^{disc} = \frac{5}{6} E_D \tag{42}$$

and the apparent activation energy in turbulent flow would be expected to be,

$$E_{T(turbulent)}^{disc} = \frac{4}{3} E_D \tag{43}$$

so that,

$$\frac{E_{T(laminar)}}{E_{T(turbulent)}} = \frac{5}{8} \tag{44}$$

In the case of rotating disc systems, the activation energy for transport control in the aqueous phase would be expected to vary from 3 kcal/mole in a laminar flow regime to 6 kcal/mole in a turbulent flow regime.

Rotating Cylinder. For the case of a rotating cylinder, the flow regime is invariably turbulent and, in addition, the mass transfer coefficient is described by the following relationship,

$$k_T^{cylinder} = 0.079V^{0.7}d^{-0.3}\nu^{-0.344}D^{0.644} \tag{45}$$

in which it is seen that k_T varies approximately with ν to the -1/3 power. Under these circumstances, the apparent activation energy for transport can be expressed in terms of its component energy increments as

$$E_T^{cylinder} = (2E_D + E_\nu)/3 \tag{46}$$

Notice in the case of rotating cylinder systems that if E_D is approximately equal to E_ν then the apparent activation energy for transport, $E_T^{cylinder}$, is equal to the activation energy for diffusion, E_D, and should be on the order of 4 kcal/mole.

Cementation Systems. Apparent Activation Energies for various cementation reactions limited by mass transfer in the aqueous phase are reported in Table IV. It should be noted that the apparent activation energies for rotating disc studies are usually lower (3-4 kcal/mole) than the apparent activation energies for rotating cylinder studies (5-6 kcal/mole) (50). The data in Table IV support, in semi-quantitative fashion, the theoretical analysis discussed previously. That is, under most circumstances rotating disc studies are done under laminar flow conditions and as such apparent activation energies would be expected to be less than the activation energy for diffusion. On the other hand, studies involving the rotating cylinder geometry are in a turbulent flow regime and apparent activation energies would be expected to be at least equal to the activation energy for diffusion if not greater. This last statement is qualified because it is not clear if the limiting value of -2/3 for the kinematic viscosity suggested by Levich for highly developed turbulence is restricted only to rotating discs.

Surface Deposit Effects

During the past decade the importance of the surface deposit in the interpretation of cementation rate data has been clearly demonstrated in a quantitative manner. Unlike many other heterogeneous reaction systems, cementation reactions are unique in that the reaction product usually does not impede the reaction progress but rather frequently enhances the reaction kinetics. This phenomenon can be attributed to the electrochemical nature of the reaction and the structure of the surface deposit.

Similar to electrolytic deposition processes the cementation surface deposit may be influenced by hydrodynamics, concentration, temperature, and ligand and organic colloid addition. In electrolysis, of course, the current density may be an independent variable and it in itself can determine the nature of the deposit.

Table IV. Variation of Activation Energies in Cementation Systems Limited by Aqueous Phase Mass Transfer

System		Geometry	Activation Energy; kcal/mole	Reference
Ag^{+}/Fe	(Cl^{-})	Disc	3.0	52
Ag^{+}/Fe	$(SO_4^{=})$	Disc	3.1	53
Ag^{+}/Zn	(CN^{-})	Disc	3.1	51
As^{+3}/Cd	$(SO_4^{=})$	Disc (N_{Re} ≃ 39,000)	3.1 ± 1.3	54
Au^{+}/Zn	(CN^{-})	Disc	3.1	51
Cd^{++}/Zn	$(SO_4^{=})$	Disc	3.8	11
Cu^{++}/Al	$(SO_4^{=},Cl^{-})$	Disc (N_{Re} = 16,000)	2 - 3 (in absence of protective oxide film)	23
Cu^{++}/Fe	(Cl^{-})	Disc	3.1	52
Cu^{++}/Fe	$(SO_4^{=})$	Disc	2.6 - 3.7	
Cu^{++}/Fe	$(SO_4^{=})$	Disc (N_{Re} ≃ 32,000)	3.2	53
Cu^{++}/Fe	$(SO_4^{=})$	Disc (N_{Re} ≃ 200,000)	4.9 ± 0.5	55
Cu^{++}/Zn	$(SO_4^{=})$	Disc	3.0	56
Ag^{+}/Cu	(ClO_4^{-},CN^{-})	Cylinder	5.0 ± 0.05	9
Ag^{+}/Zn	(NO_3^{-})	Cylinder	2 (for non-protective deposit)	57
Ag^{+}/Zn	(ClO_3^{-},CN^{-})	Cylinder	6.0	58
Bi^{+3}/Fe	(Cl^{-})	Cylinder	4.5 - 7.6	22
Cd^{++}/Zn	$(SO_4^{=})$	Cylinder	4.7 ± 0.8	42
Cu^{++}/Al	$(SO_4^{=})$	Cylinder	5.8	59
Cu^{++}/Fe	$(SO_4^{=})$	Cylinder	4.7 - 4.9	40
Pd^{+}/Cu	(ClO_4^{-})	Cylinder	7.4 - 9.5	60
Ag^{+}/Fe	(Cl^{-})	Powder (Laminar)	3.4	61
Cu^{++}/Fe	$(SO_4^{=})$	Suspended Plate (Turbulent)	5.0 ± 0.7	62

Surface deposits formed in cementation systems are quite similar to electrolytic deposits because of the electrochemical nature of the reaction. In cementation systems, however, the current density is not an independent variable. In many instances the rate of cementation is increased due to the rough nature of the surface deposit. On the other hand, conditions which lead to a tight, smooth, coherent deposit may in fact inhibit the reaction as discussed previously in the subsection on Cathodic/Anodic Area. Generally, however, the surface deposit is either dendritic or botryoidal and the reaction kinetics are enhanced significantly due to the presence of the deposit. The significance of the surface deposit and this rate enhancement, which has been observed by many investigators, will be discussed in terms of hydrodynamics, concentration, temperature, and particle-particle interaction effects.

Hydrodynamics. Solution flow is an important variable which controls the type of deposit obtained in electrolytic cells. It has been shown (63) in the case of zinc deposition from cyanide solution that, as the flow rate of solution over the cathode increases at constant current density, the electrolytic deposit changes from a dendritic, to a botryoidal, to a smooth deposit. Similar effects are observed in cementation systems and a surface deposit phase diagram can be imagined such as shown in Figure 15. This surface deposit phase diagram characterizes the nature of the surface deposit with respect to cementation rate and Reynolds number.

The effect of fluid flow on the nature of the surface deposit in cementation systems has not been studied extensively. In the Cu^{2+}/Fe rotating-disk system the surface deposit has been shown to vary with respect to radial position on the disk's surface (55). For laminar flow, $(r^2\omega/\nu)<10^5$, there was a gradual change from a coarse deposit at the disk center (small Reynold's number approaching dendrite formation) to a finer textured deposit at the edge of the disk (large Reynold's number).

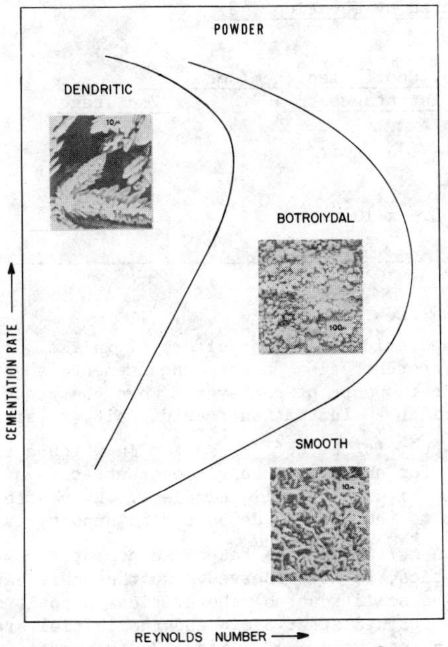

Figure 15. Surface deposit phase diagram for cementation reactions.

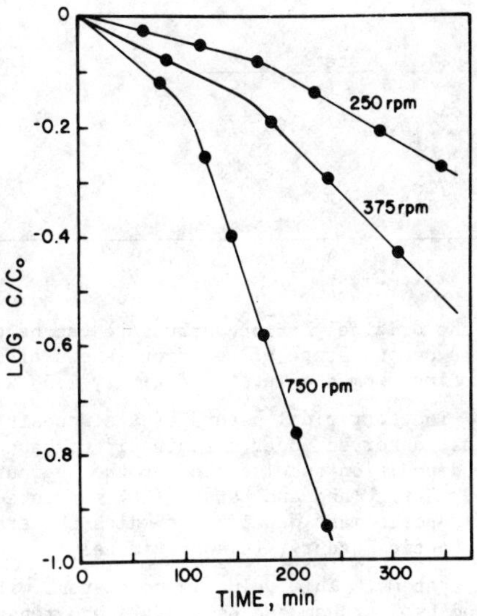

Figure 16. First-order reaction plot for rotating disk; Cd^{2+}/Zn system (36).

For the case in which the flow changes from laminar to turbulent at some radial position of the disk, there is an abrupt rather than a gradual change in surface-deposit character, and the influence of the transition to the turbulent flow regime on the surface deposit is most pronounced. Under certain circumstances, the coarse deposit at the center of the disk does not shear off but rather grows continually to form a cone which, in one instance, is reported to reach a height of 0.5 cm (55). The results from this study indicate in a qualitative fashion that fluid flow not only determines the extent of surface deposit removal by shear forces on a macroscopic basis, but also determines at a microscopic level the basic character and structure of the surface deposit, namely nucleation and growth phenomena.

Concentration. In many systems a characteristic effect of the noble metal ion concentration on cementation reaction kinetics has been observed. These systems include Cd^{2+}/Zn, Pb^{2+}/Zn, Ag^{+}/Zn, Cu^{2+}/Zn, and Cu^{2+}/Fe. Without exception, for small initial concentrations of the noble metal ion first-order plots of a type similar to that obtained for the Cd^{2+}/Zn system (rotating disk), shown in Figure 16, are obtained (36). Two stages of reaction are apparent: the initial stage of reaction corresponds to deposition on a relatively smooth surface. During this stage of reaction the deposit structure and thickness has not significantly affected the reaction kinetics and the reaction kinetics follow Levich's equation for a rotating disk system:

$$J = 0.62\, D^{2/3}\nu^{-1/6}\omega^{1/2}C \tag{47}$$

For example, theoretically calculated mass transfer coefficients,

$$k = 0.62D^{2/3}\nu^{-1/6}\omega^{1/2} \tag{48}$$

agree quite well with experimental values from initial-stage kinetics for a number of reactions studied with the rotating disk as shown by the values presented in Table V.

At some critical point of reaction progress, which corresponds to a critical deposit mass per unit area, a second stage of enhanced first-order kinetics is observed as shown in Figure 16. The increased rate of reaction is due to the development of a structured surface deposit which acts as though the surface area available for reaction has been increased. Under these circumstances theoretical mass transfer coefficients, calculated on the basis of the geometric area of a flat disk are smaller than those determined experimentally, as shown in Table V. The change in surface character is vividly illustrated by the photographs (400x) for the Cu^{2+}/Zn system presented in Figure 17. The photograph of the surface during the initial stages of reaction (shown after 5 min reaction from a solution initially 1.56×10^{-4} $\underline{M}$ Cu^{2+}) reveals the original structure of the zinc surface with scratches present from the polishing preparation before reaction.

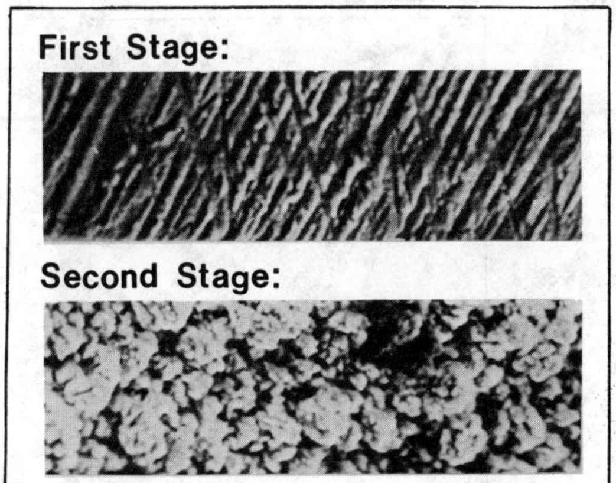

Figure 17. Comparison of the nature of the surface deposits for 1st stage (top) and 2nd stage (bottom) reaction kinetics; rotating disk Cu^{2+}/Zn system (7).

Table V. Comparison of Mass Transfer Coefficients at 750 rpm Determined by Rotating Disk Data (36) with Theoretical Values Predicted by Equation (48)

System	Mass Transfer Coefficient, cm/sec		
	Theoretical	1st stage	2nd stage
$Cu^{2+}/Zn, Cu^{2+}/Fe, Cd^{2+}/Zn$	4.5×10^{-3}	5.6×10^{-3}	2.15×10^{-2}
$Ag^{+}/Zn, Ag^{+}/Cd, Ag^{+}/Cu$	7.2×10^{-3}	8.9×10^{-3}	

No trace of the original surface structure can be observed during the second stage of reaction (shown after 130 min reaction time from a solution initially 1.56×10^{-4} M Cu^{2+}). The botryoidal nature of the deposit is similar to that observed under certain circumstances in electrolytic deposition as discussed at the beginning of this section. Strickland and Lawson (36) suggested that the critical deposit mass density, at which the transition to enhanced kinetics occurs, is approximately 0.3 mg/cm^2 for most systems. This value is consistent with the value obtained by Ingraham and Kergy (42) who report that the transition to enhanced, second-stage, first-order kinetics occurs at 0.25 mg/cm^2. These results should be contrasted with those for the Cu^{2+}/Al system in which case the critical specific mass was found to be 2.6 mg/cm^2 for 100 ppm copper and 12.9 mg/cm^2 for 500 ppm copper (23). The wide range in critical specific mass appears to be related to the initial concentration of noble metal ion and any other variable that influences the structural nature of the deposit.

The rate enhancement continues to increase with deposit mass density up to 5 mg/cm^2 in the case of Cu^{2+}/Zn, above which a steady-state surface profile may be reached under certain circumstances and the second-stage kinetics no longer increases with the extent of reaction. This phenomenon is relected in Figure 18 by the fact that as the initial cupric concentration increases, the second-stage reaction rate increases and is reached in a much shorter time interval. At initial concentrations above 100 ppm, rate enhancement is no longer found to be a function of initial concentration. Consequently, for high initial concentrations of noble metal ions, the first-stage reaction kinetics may not be seen, the reason being, of course, that the critical deposit mass density is reached almost immediately.

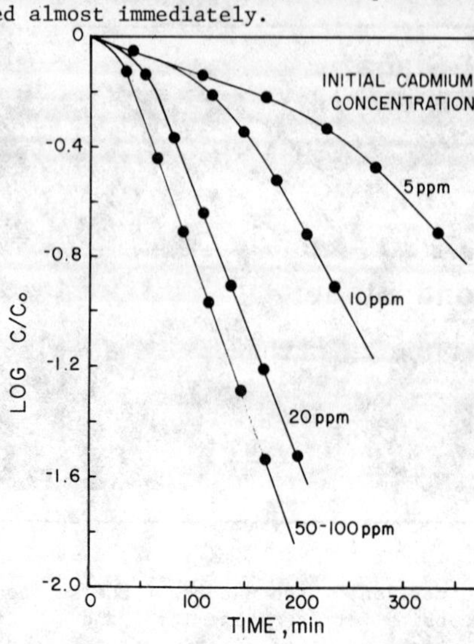

Figure 18. The effect of initial concentration on 1st stage reaction kinetics; rotating disk Cd^{2+}/Zn system (36).

Further evidence of the effect of initial noble metal ion concentration on rate enhancement is given by the kinetic response of the system when the reductant metal is given an initial surface deposit, a "strike". The case study is the Cu^{2+}/Fe system in which a first-order plot for an initial cupric concentration of 10 ppm is shown in Figure 19 (solid circles). Even after 60 min reaction time the surface deposit mass density is approximately 0.6 mg/cm^2 and the rate enhancement, or second-stage kinetics, is not observed. On the other hand, for 200 ppm (see solid squares) the critical deposit mass density is reached after 3 min and the initial stage of reaction in the absence of surface deposit effects is not observed. Notice, importantly, that when the iron surface is given an initial "strike" several times greater than the critical deposit density, and the reaction then followed, the observed reaction kinetics for an initial cupric concentration of 10 ppm are almost equivalent to the observed reaction kinetics for an initial cupric concentration of 200 ppm.

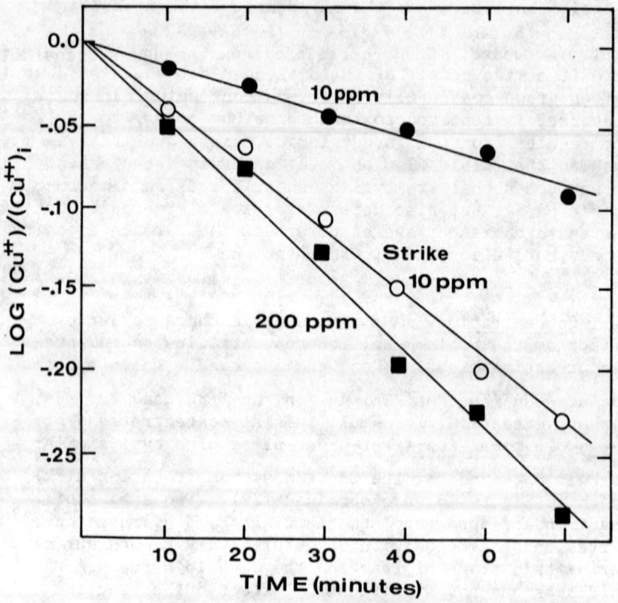

Figure 19. Influence of a surface deposit (strike), produced at 200 ppm Cu^{2+}, on the cementation rate at 10 ppm Cu^{2+}, a temperature of 45°C, and a pH of 1.90. Key: , 1 hr at $(Cu^{2+})_i$ = 10 ppm; o, 10-min "strike at 200 ppm (Cu^{2+}), then 1 hr at $(Cu^{2+})_i$ = 10 ppm; 1 hr at $(Cu^{2+})_i$ = 200 ppm (40).

The effect of precipitant surface roughness on cementation kinetics has continued to be analyzed by the rotating disk technique (64). The systems chosen for this more recent study included Cd^{2+}/Zn, Cu^{2+}/Zn and Cu^{2+}/Cd. As observed in previous studies, (11,50) each of these three different systems showed the distinctive two regions of cementation rate, namely the initial slow deposition rate and the final enhancement rate. Empirical correlation equations were developed and are as follows:

For the initial rate period for all three systems,

$$Sh = 0.62\ (Re)^{0.5}\ (Sc)^{0.333} \tag{49}$$

For the enhanced rate period,

$$Sh = A(Re)^{0.74}\ (Sc)^{0.29} \tag{50}$$

where

$Sh = rk/D$ $\qquad A = 0.37$ for Cd^{2+}/Zn

$Re = r^2\omega/\nu$ $\qquad 0.15$ for Cu^{2+}/Zn

$Sc = \nu/D$ $\qquad 0.11$ for Cu^{2+}/Cd

The copper-aluminum cementation reaction has been studied by Annaimalai, Hiskey, and Murr (23-26). These investigators examined various aspects of the Cu^{2+}/Al cementation system including: the fundamental kinetics of the cementation reaction; the structural characteristics of the copper deposit; the use of aluminum canning alloys; and the role of nucleation and growth of the deposit. The overall Cu^{2+}/Al cementation reaction is represented by,

$$Cu^{2+} + 2/3\ Al \rightarrow 2/3\ Al^{3+} + Cu \tag{51}$$

In most cases the Cu^{2+}/Al cementation reaction is characterized by two distinct kinetic regions, an initial slow period followed by a final period where the rate is observably faster. Enhancement is thought to be induced by the deposit and associated surface roughening. The region of enhanced rate was analyzed as a function of chloride ion concentration, agitation, hydrogen ion concentration, copper ion concentration, and temperature. Copper cementation on pure aluminum was examined to determine the influence of substrate conditions (i.e., grain size, grain boundaries, and dislocations) on the nucleation and growh of the deposit. Conceivably, a small grain size might result in a large number of Cu precipitate nuclei per unit area of Al surface. No observable effect was found for variations in grain size and "nucleation" was not visibly controlled by grain boundaries. In addition, for single-crystals containing dislocations there was no significant effect of the dislocations on the copper precipitation (i.e., the emergence and number of individual copper nuclei was not related to the presence of dislocations). The copper deposit seems to nucleate and grow at isolated locations on the substrate surface, with dendritic deposit structure prevailing for optimum kinetic conditions.

The concentration of noble metal ion, and probably the ionic strength in general, can influence cementation reaction kinetics not only as they relate to the structure and mass density of the surface deposit, but also as these parameters influence the diffusion coefficient and activity of diffusing species. Ideally, it would be expected that the kinetics of diffusion through the MTBL would be independent of initial concentration of noble metal ion in the absence of a time-dependent surface deposit effect. However, it appears that such is not necessarily the case. For example, consider the Cu^{2+}/Fe system shown in Figure 20. Clearly, these data are for the second stage of reaction in which the surface deposit has been fully developed to some steady-state profile. Consequently, the minor effect of the initial cupric concentration must be accounted for in some manner such as activity or diffusion coefficient corrections. In the case of activity coefficients, according to absolute reaction-rate theory, the reactants are in equilibrium with the activated state and the apparent rate constant for a nonideal system can be evaluated by the following expression:

$$k_{apparent} = \frac{kT}{h} K^{\ddagger} \frac{\pi_i \gamma_i^{J_i}}{\gamma^{\ddagger}}$$

where (kT/h) is the ideal rate constant, $K^{\ddagger}$ is the equilibrium constant between reactants and activated species, T is the absolute temperature (°K), h is Planck's constant, k is Boltzmann's constant, $\pi_i \gamma_i^{J_i}$ is the product of the activity coefficients of the reactants, $\gamma^{\ddagger}$ is the activity coefficient of the activated species, and $k_{apparent}$ is the apparent reaction-rate constant.

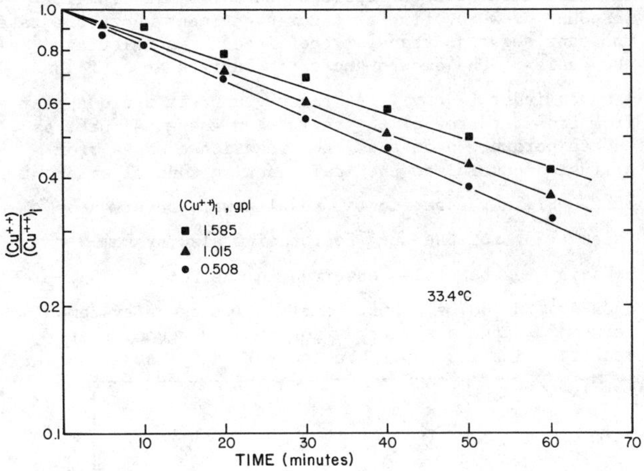

Figure 20. Initial concentration effect on 2nd-stage reaction kinetics; Cu^{2+}/Fe system (45).

For a diffusion controlled reaction such as the Cu^{2+}/Fe cementation system the only reactant is the diffusing species, i.e., the cupric ion. Furthermore, because of the normally low concentrations of the activated species it is assumed that it behaves ideally, and the activity coefficient for the activated complex, $\gamma^{\ddagger}$, is unity. The above expression can then be reduced to,

$$k_{apparent} = k_{ideal}\ \gamma_{Cu^{2+}} \tag{53}$$

where k_{ideal} is the rate constant predicted by ideal behavior. This expression can be used to correct for non-ideal conditions and accounts successfully for the observed decrease in rate constant at higher initial copper concentrations, as shown in Table VI.

Alternatively, the concentration dependence of the mass transfer coefficient can be rationalized in terms of the concentration dependence of the self-diffusion coefficient,

The decrease in the rate constant at higher concentrations can also be explained in terms of a changing deposit morphology. At higher concentration of noble metal ion (higher cementation rate) and/or higher Reynold's numbers, surface deposits can change from a rough or rigid dendritic structure to a smooth or powdery deposit as discussed with respect to the surface deposit phase diagram shown in Figure 15.

Table VI. Second-Stage Rate Constants for the Cu^{2+}/Fe System Taken from Figure 20 and Corrected for Nonidal Behavior

Initial Copper Concentration, gpl	$k_{apparent}$, cm/sec	$\gamma_{Cu^{2+}}$	k_{ideal}, cm/sec
0.508	2.3×10^{-2}	0.48	4.85×10^{-2}
0.015	1.66×10^{-2}	0.44	4.25×10^{-2}
1.585	1.67×10^{-2}	0.39	4.28×10^{-2}

Temperature. Surface deposit effects on cementation reaction kinetics have been shown to have a significant temperature dependence. Consider the Arrhenius plots for the Cu^{2+}/Fe rotating cylinder system presented in Figure 21. The open squares represent the standard kinetic response of a rotating cylinder experiment where the rate constant, or mass transfer coefficient, is calculated on the basis of the smooth geometric surface area of the iron cylinder (7 cm^2). At first glance it would appear from consideration of the activation energies that, as the temperature decreases, the reaction changes from MTBL diffusion control to chemical reaction control at about 35°C. Similar experimental results were observed by Calara (55) for the Cu^{2+}/Fe rotating disk system. This behavior has been also observed in the Cu^{2+}/Ni (65) and Pd^{2+}/Cu (60) systems. Analysis of these systems in terms of standard electrode potentials suggest that reaction kinetics should be controlled by mass transfer.

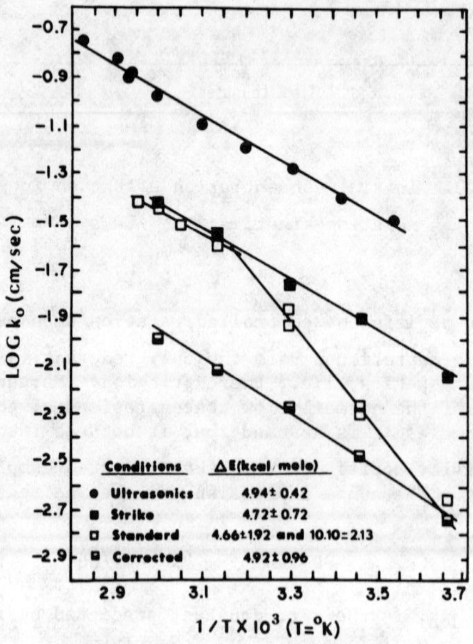

Figure 21. Arrhenius plots for an initial copper concentration of 2.00 g/liter and a pH of 1.90. The *strike* data are for cementation tests using iron rods that had been given an initial strike for 6.5 min at 2.00 g/liter, pH of 1.90, and 45°C. The corrected data are for standard tests whose rate constants have been corrected by a surface roughness factor - see Table VII (40).

Calculation of the theoretical mass transfer coefficients are much lower than those observed experimentally. For example, in the case of the Cu^{2+}/Fe system at 45°C the calculated mass transfer coefficient for rotating cylinders is 7.7×10^{-3} cm/sec [see equation (34)] compared to the 2.5×10^{-2} cm/sec observed experimentally, which suggests that the reaction is three times as fast as theoretically possible. The reason for this apparent anomalous behavior is the effect of the surface deposit. The particular experiments were performed at a cupric concentration for which enhanced, second-stage reaction kinetics is observed. The temperature dependence of the surface deposit is substantiated by both direct and indirect evidence.

As for indirect evidence, it is seen from Figure 21 that when the reaction is studied in an ultrasonic field the mass transfer coefficient is increased significantly because of a reduction in the effective MTBL thickness from the microstreaming which accompanies the ultrasonic field. In addition, the dissipation of ultrasonic energy at the cylinder surface continually removes the surface deposit and the apparent region of chemical reaction control is not longer observed; an activation energy of approximately 5 kcal/mole is obtained over the entire temperature range studied as shown by the solid circles in Figure 21. Furthermore, the effect of temperature on the surface deposit and how it affects reaction kinetics is illustrated by experiments in whicn the iron cylinder is given a "strike", an initial surface deposit, of copper at 45°C. The established surface maintained its form at all other temperatures for which the reaction was studied, even at 0°C, and an activation energy of 5 kcal/mole is observed over the entire temperature range (solid squares).

Direct evidence of the surface deposit's temperature dependence is revealed by microscopic examination of the surface for the standard cementation experiment. A distinct change in the nature of the deposit at a critical temperature of 35°C is observed in the set of microphotographs shown in Figure 22 (approximately equivalent deposit mass densities, 10 mg/cm^2, in considerable excess of the critical deposit mass density). Notice that below 35°C for the cross sectional and surface photographs the botryoidal deposit becomes more dense and consolidated as the temperature is lowered, whereas above 35°C the deposit maintains a constant uniform porosity and surface area at all temperatures. Basically, these observations support the concept that the area term, A, in the rate expression is a function of temperature, depending, under certain circumstances, on the relative magnitudes of the growth and nucleation rates. From the cross sectional photographs, the actual surface area can be measured and the rate constants corrected with respect to the true surface area. The results of these calculations are presented in Table VII.

When the standard rate constants are corrected in this fashion and plotted according to the Arrhenius equation, an activation energy of 4.9 kcal/mole is obtained as shown by half-filled squares in Figure 21; and thus, the activation energy for the standard cementation data actually supports boundary layer diffusion as the rate limiting step in the cementation reaction. Furthermore, when the rate data are corrected for the area increase because of the surface deposit the

Table VII. Rate Constants Corrected for a Surface Roughness Factor (40)

Temperature °C	$k_{apparent}$ cm/sec	Roughness Factor	$k_{corrected}$ cm/sec
0	0.20×10^{-2}	1.0	0.20×10^{-2}
15	0.52×10^{-2}	1.5	0.34×10^{-2}
30	1.32×10^{-2}	2.4	0.54×10^{-2}
45	2.70×10^{-2}	3.6	0.75×10^{-2}
60	3.10×10^{-2}	3.4	0.92×10^{-2}

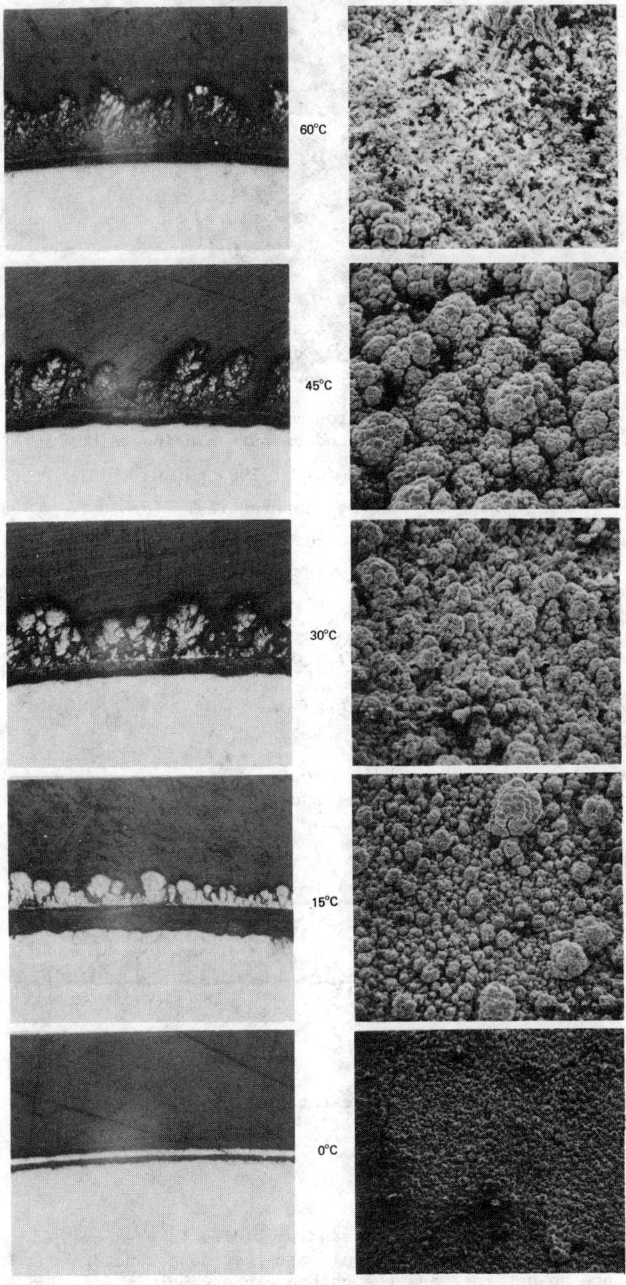

Figure 22. Photomicrographs taken using an optical microscope (left column) and a scanning electron microscope (right column) of copper deposits produced at a Cu^{2+} concentration of 2.00 g/liter and a pH of 1.90 at various temperatures, 200X (40).

experimental rate constants, or mass transfer coefficients, agree quite well with the mass transfer coefficients that would be predicted from semitheoretical considerations. For example, at 45°C the corrected rate constant is 7.5×10^{-3} cm/sec (Table VII) which compares well with the calculated theoretical value of 7.7×10^{-3} cm/sec (see p. 18). These experimental results illustrate vividly how surface deposit effects can influence the interpretation of rate data for heterogeneous reactions.

On the basis of the foregoing, it would be expected that an Arrhenius plot at low cupric concentrations, where the first-stage reaction response is observed, should be linear over the entire temperature region. The expected simultaneous interaction of temperature and concentration is revealed by Arrhenius plots in Figure 23. Again, the effect of the surface deposit is influenced by temperature and concentration is markedly demonstrated.

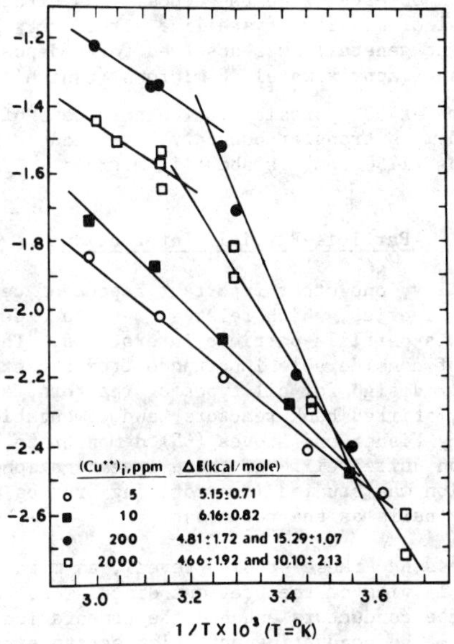

Figure 23. Arrhenius plots for various initial Cu^{2+} concentrations (40).

Similar findings have been reported by Majima *et al.* who studied the Cu^{2+}/Fe system in a suspended plate reactor and found an activation energy of 4.6 kcal/mole for both the initial and enhanced rate regions. The rate enhancement phenomenon was qualitatively explained with the "Penetration Model" of Bockris and basically the authors considered that the flux varies exponentially with time for certain regions of the surface in the presence of a surface deposit. In essence, this means that the effective mass transfer boundary layer thickness is reduced and hence the rate enhanced.

The activation energy for the Cu^{2+}/Fe rotating disk system has been shown in the NIM laboratories to be influenced by the specific density of the copper deposit, the concentration of the Cu^{2+}, the stirring rate and the addition of thiourea, i.e., by all the variables having an influence on the deposit morphology (67). Activation energies determined under conditions where deposit effects are absent ranged from 2.6 to 3.7 kcal/mole whereas with high specific deposit densities (3.63 mg/cm^2) values of up to 8.5 kcal/mole were found. Under certain conditions of high temperature and concentration the rigid deposit structure can pass into a powder deposit phase region (see Figure 15). It is suggested that the rate decrease under these circumstances reflects the fact that the soft powder deposit is unable to penetrate or alter the mass transfer boundary layer and hence eliminates the rate enhancement normally found for rigid deposits.

In the case of the Cu^{2+}/Al system (24) nature and morphology of the copper deposit were investigated as a function of various kinetic parameters. Kinetic data correlate well with the structure of the copper deposit. Deposit morphology was greatly modified by variations in chloride ion concentration and temperature. In general, the deposits varied from botryoidal (or massive crystalline) to dendritic to fine dendritic (approaching powder). Structural changes in deposit morphology corresponding to different temperatures at 75 ppm chloride are shown in the scanning electron micrographs in Figure 24 (a-c). Similar variations in deposit character were observed for increases in chloride ion concentration at constant temperature. Maximum cementation rates occurred normally when the copper deposit was a rigid fine dendritic texture, and lower rates were observed when the deposit was either a botryoidal or massive crystalline structure. Rate enhancement generally was observed for a deposit layer thickness of approximately 6 microns (equivalent to 2.6 mg/cm^2 at 50% porosity). The thickness of the idealized mass transfer boundary layer under these conditions (1250 ppm) is about 12 microns.

Figure 24. Scanning electron micrographs of copper deposits cemented on pure aluminum discs at pH 2.5, 75 ppm (Cl^-), 100 ppm Cu^{2+}, 131 rad/sec, and at temperatures: (a) 10°C, (b) 27°C, (c) 60°C. Magnification 1000X.

Particle-Particle Interactions

Finally, one other important aspect of cementation reaction kinetics, which relates to surface deposit effects, is particle-particle interaction. This phenomenon is of considerable importance from the standpoint of reactor design for all types of reactors, such as launders, stirred tank reactors, and cementation cones. Studies by Fisher and Groves (68) demonstrate this phenomenon quite well. In these investigations copper cementation was studied in a rotating drum using iron finishing nails as the reductant.

In essence these studies reveal, as might be expected in view of the previous discussion, that particulate conductors enhance the cementation reaction kinetics. The conducting particles act to extend the cathodic surface area and promote copper deposition even though the cupric ion may never reach the iron surface. Rate enhancement in the presence of particulate conductors occurs because of a "fluidized cathode mechanism", i.e., an effective increase in cathodic area, provided good particle-particle contact is maintained for electron conduction. See the schematic presented in Figure 25.

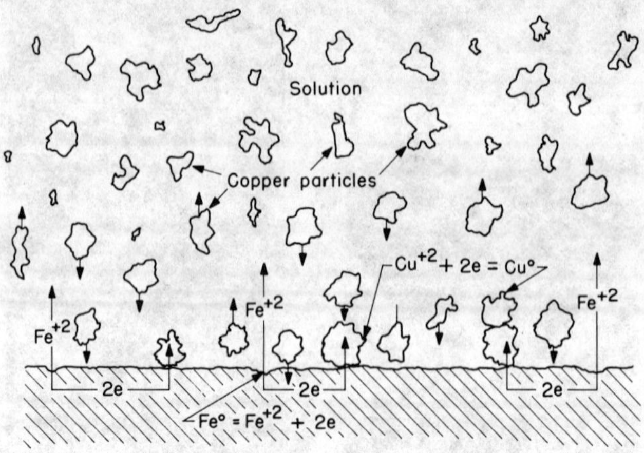

Figure 25. Schematic representation of *fluidized cathode mechanism* operative in cementation systems (68).

Typical reaction kinetics for the drum reactor in the absence of conducting particles are shown by the solid circles in Figure 26. Addition of particulate conductors, such as cement copper and gold powder, significantly increase the reaction kinetics, whereas the addition of nonconductors causes a decrease in the reaction kinetics. The decrease in rate observed in the presence of nonconducting sand particles is presumably due to the fact that not only are they nonconductors and hence do not increase the cathodic area, but in addition the sand particles abrade the iron surface removing the surface deposit and tend to eliminate the second-stage rate enhancement attributed to surface deposit effects. Furthermore, in the presence of nonconducting particulates, short-circuiting of the fluid bed is prevented and the "fluidized cathode mechanism" is not operative.

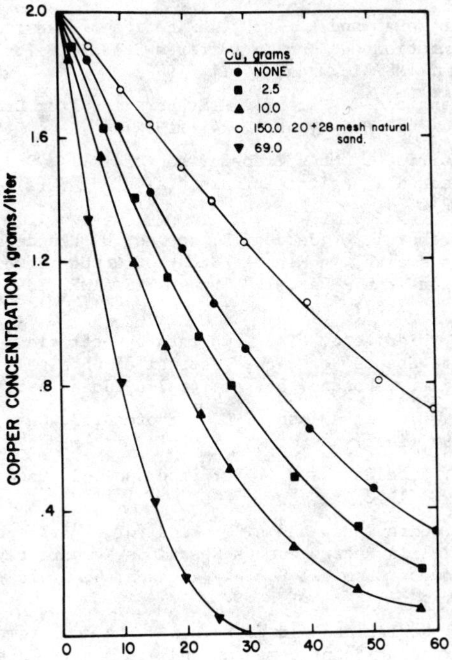

Figure 26. The effect of conducting and nonconducting particles on the reaction kinetics of the Cu^{2+}/Fe system (68).

As would be expected, the rate enhancement is directly related to the weight of particulate conductors up to some saturation point. More accurately, the rate enhancement should be proportional to the number of conductive particles, the collision rate, and time of contact. The result of these studies indicate that packed-bed reactors such as cementation cones or any other system which would provide for good particle-particle, electrochemical interaction are ideal. Further consideration should be given to designing with inert electronic conductors to catalyze the electrochemical reaction in general.

In summary, surface deposit effects demand careful consideration in the interpretation of cementation rate data. These principles have been found to be important in the analysis of industrial operations as reported by Schlitt and Richards (2). Several explanations for the rate enhancement due to surface deposit effects have been tendered:

1. decrease in diffusion path length, MTBL,
2. microturbulence in the interstices of the surface deposit, and
3. increase in the effective area (distortion of the MTBL profile) for the mass flux.

Of these, it appears that the first can be discounted because the shape of the effective MTBL is distorted when the scale of roughness is comparable to the thickness of the MTBL, which is the case for second-stage cementation kinetics. The second explanation may be reasonable provided the interstices are not too small to be able to accommodate eddy currents of the size normally encountered in turbulent flow systems. Finally, rate enhancement may be due to an increase in the effective area (distortion of the MTBL profile), as demonstrated by the behavior of the Cu^{2+}/Fe system (40).

ACKNOWLEDGEMENTS

Review of certain systems and data on operating cementation plants were provided by K. Han, J. B. Hiskey, B. R. Palmer and J. W. Schlitt. Their contribution is gratefully acknowledged as well as the assistance of Rong-Yu Wan.

REFERENCES

1. M. E. Wadsworth and J. D. Miller, "Hydrometallurgical Processes", Rate Processes of Extractive Metallurgy, Plenum Press, 133 (1979).

2. W. J. Schlitt and K. J. Richards, "Chemical Aspects of Copper Cementation," Solution Mining Symposium, SME/AIME, 401-421 (1974).

3. J. Bockris and A. Reddy, Modern Electrochemistry, 1 and 2, Plenum Press, New York (1973).

4. M. Wadsworth, "Reduction of Metals in Solution," Extractive Metallurgy Lecture, Trans. TMS/AIME 245, p. 1381 (1969).

5. G. P. Power and I. M. Ritchie, "A Contribution to the Theory of Cementation (Metal Displacement) Reactions," Aust. J. Chem. 29, p. 699-409 (1976).

6. G. P. Power and I. M. Ritchie, "Metal Displacement Reactions," Modern Aspects of Electrochemistry, II, p. 199 (1975).

7. G. Power, "Cementation Reactions," Ph.D. Thesis, University of Western Australia, Nedlands, Western Australia (1975).

8. T. Hurlen, "Electrochemical Behavior of Iron," Acta Chem. Scand. 14, p. 1533 (1960); "Study on the Kinetics of the Cu/Cu^{++} Electrode," 15, p. 630 (1961); and "Study on the Kinetics of the Zn/Zn^{++} Aq. Electrode," 16, p. 1337 (1962).

9. E. von Hahn and T. Ingraham, Trans. TMS/AIME, 239, 1895 (1967).

10. D. J. Mackinnon, T. R. Ingraham and R. Kerby, Can. Metall. Quart., 110, p. 165 (1971).

11. E. C. Lee, F. Lawson and K. N. Han, "Precipitation of Cadmium with Zinc from Dilute Aqueous Solutions: 1 - Under Inert Atmosphere," Trans. I.M.M. Vol. 84, 1975, C87-91; E. C. Lee, F. Lawson and K. N. Han, "Precipitation of Cadmium with Zinc from Dilute Aqueous Solutions: 2 - Under Oxygen-Containing Atmospheres, Trans I.M.M. 84 (1975) C149-53.

12. G. A. Hope, I. M. Ritchie and J. E. Wajon, "The Effect of Oxide Films on Cementation Reactions," Trends in Electrochemistry, Plenum Press, 341 (1976).

13. F. de Blander and R. Winand, "Influence of Antimony and Copper on the Cementation of Cobalt by Zinc," Electrochimica Acta 20, p. 839-852 (1975).

14. A. Fontana and R. Winand, 1st part Rev. A.T.B. Metall. 11, p. 162-179 (1971).

15. A. Fontana et al., 2nd part Rev. A.T.B. Metall. 11, 162-79 (1971).

16. A. Grevel, Wiss. Veröff. Siemens-Konzern 13, 61-71 (1934).

17. A. B. Pomosov and A. I. Pochinok, Tsvet. Metally. Mosk. 9 (6) 48-52 (1966).

18. C. Fischer-Bartelk et al., Electrochim. Acta 14, (9) p. 829-844 (1969).

19. M. Yunus et al., Electrochim. Acta 10, p. 885-900 (1965).

20. N. A. Sareyed-Dim and F. Lawson, "The Cementation of Nickel onto Iron," SME Preprint No. 75-B-63, presented at the Annual SME/AIME meeting (1975).

21. L. I. Korzhenevskii, "Cementation of Bismuth with Iron," The Bulletin of the Bismuth Institute, Brussels, Belgium, 29, Third Quarter 1980, 4-6.

22. D. J. MacKinnon, "Recovery of Bismuth from HCl Leach Solutions by Cementation on Iron," Department of Energy, Mines and Resources, Ottawa, Research Report R252, September 1972, 18 p.

23. V. Annamalai and J. B. Hiskey, "A Kinetic Study of Copper Cementation on Pure Aluminum," Min. Eng., 30, No. 6, June 1978, p. 650-659.

24. V. Annamalai, J. B. Hiskey and L. E. Murr, "The Effects of Kinetic Variables on the Structure of Copper Deposits on Pure Aluminum Discs: A Scanning Electron Microscopic Study," Hydrometallurgy, 3, p. 163-180 (1978).

25. V. Annamalai, J. B. Hiskey and L. E. Murr, "Copper Cementation onto Beverage Can Aluminum Alloys," Trans. SME-AIME, 266, p. 1963-1967 (1980).

26. L. E. Murr and V. Annamalai, "An Electron Microscopic Study of Nucleation and Growth in Electrochemical Displacement Reactions: A Comparison of the Cu/Fe and Cu/Al Cementation Systems," Met. Trans. B., 9B, p. 515-525 (1978).

27. E. C. Lee, F. Lawson and K. N. Han, "Codeposition of Copper and Cadmium onto Zinc." Trans I.M.M., 87, 1978, C170-179.

28. AIME World Symposium on Mining and Metallurgy of Lead and Zinc, St. Louis, Mo., AIME (1970).

29. Belgian Patent, "Purification Process for the Zinc Sulphate Solutions Produced by the Leaching of Zinc Minerals", No. 783,549 of 15 June, 1972 (Vieille Montagne).

30. L. M. Kabanova and O. A. Khan, "Cementation of Cadmium from Concentrated Sulfate Solutions," Trudy Altaiskogo Gorn Metallurgisheskoyo Nauchno Issledovatelskogo Insttitita Akademiya Nauk Kazakh SSR, 11, p. 34-47 (1961).

31. M. J. Nicol, E. Schalach, P. Balestra, H. Hegedres, "A Modern Study of the Kinetics and Mechanism of the Cementation of Gold," Report 9354, NIM, Colloquium on Recent Advances in the Extractive Metallurgy of Gold, September 1978; South African Institute of Mining and Metallurgy, Johannesburg, South Africa, vol. 79, p. 191, 1979.

32. E. T. Eisenmann, "Kinetics of the Electrochemical Reduction of Dicyanoaurate," J. Electrochem. Soc. 125, p. 717 (1978).

33. J. Newman, Electrochemical Systems, Prentice-Hall, Englewood Cliffs, N.J. (1973).

34. V. Levich, Physiocochemical Hydrodynamics, Prentice-Hall, Englewood Cliffs, N. J., 1962.

35. A. Riddiford, "The Rotating Disc System," Ad. Electrochem. Electrochem. Eng., 4, 47 (1966).

36. P. Strickland and F. Lawson, "The Cementation of Metals from Dilute Aqueous Solution," Proc. Aust. Inst. Mining Met. 237, p. 71 (1971).

37. M. Eisenberg, C. Tobias, and C. Wilke, "Ionic Mass Transfer and Concentration Polarization at Rotating Electrodes," J. Electrochem. Soc. 101, 306 (1964).

38. I. Cornet and R. Kappeser, "Cathodic Protection of a Rotating Cylinder," Trans. Inst. Chem. Eng. 47, p. 194 (1969).

39. D. Gabe and D. Robinson, "Mass Transfer in a Rotating Cylinder Cell—1. Laminar Flow," Electrochemica Acta, 17, p. 1121 (1972).

40. J. D. Miller and L. W. Beckstead, "Surface Deposit Effects in the Kinetics of Copper Cementation by Iron," Trans. TMS/AIME 4, p. 1967 (1973).

41. R. Glickman, H. Mouguin, and C. King, "Rate of Displacement of Silver from Aqueous Silver Nitrate by Zinc and Copper, J. Electrochem. Soc. 100, p. 580 (1953).

42. T. Ingraham and R. Kerby, "Kinetics of Cadmium Cementation on Zinc in Buffered Sulfate Solutions," Trans. TMS/AIME 245, p. 17 (1969).

43. W. Ranz and W. Marshall, "Evaporation from Drops," Chem. Eng. Prog., 48 p. 141 (1952).

44. P. Harriott, "Mass Transfer to Particles, Part 1. Suspended in Agitated Tanks," AIChE J., 8 (1), p. 93 (1962).

45. R. Nadkarni, C. Jelden, K. Bowles, H. Flanders and M. Wadsworth, "A Kinetic Study of Copper Precipitation on Iron --Part 1," Trans. TMS/AIME, 239, p. 581 (1967).

46. A. C. Riddiford, "The Temperature Coefficient of Heterogeneous Reaction," Journal of Physical Chemistry, London, 56, p. 745-748, June (1952).

47. L. L. Bircumshaw and A. C. Riddoford, J. Chem. Soc., p. 598 (1951).

48. A. C. Riddiford and L. L. Bircumshaw, J. Chem. Soc., p. 698 (1952).

49. L. L. Bircumshaw and A. C. Riddiford, "Transport Control in Heterogeneous Reactions," Quarterly Review of Chemical Society, London, 6, p. 157-185 (1952).

50. P. H. Strickland and F. Lawson, "The Measurement and Interpretation of Cementation Rate Data," International Symposium on Hydrometallurgy, SME, TMS/AIME, p. 293-330 (1973).

51. I. A. Kakovskii and O. K. Scherbakov, "Kinetics of Noble Metal Cementation from Cyanide Solution," Izv. Akad. Nayk. S.S.S.R. Metall., No. 1, p. 76-85, (1967).

52. M. L. Episkoposyan, and I. A. Kakovskii, "An Examination of Copper and Silver Cementation Kinetics with Metallic Iron from Chloride Solutions," Soviet, J. Non Ferrous Metals, 38 (10) p. 16-21 (1965).

53. M. L. Episkoposyan and I. A. Kakovskii, "An Examination of Copper and Silver Cementation Kinetics with Metallic Iron from Sulphate Solutions," IZV, UYSCH UCHEBN ZaVED. Tsvet. Metall; 9 (1) p. 34-40.

54. B. R. Palmer, G. Gutierrez and M. C. Fuerstenau, "Electrochemical Reduction of Arsenic (III) with Cadmium Metal," Met. Trans. B. 6B, p. 557-563 (1975).

55. J. V. Calara, "Kinetics of Copper Cementation on a Rotating Iron Disc," Master Thesis, University of the Phillipines, Department of Metallurgy, Manila, Phillipines (1970).

56. P. H. Strickland and F. Lawson, "Cementation of Copper with Zinc from Dilute Aqueous Solutions," Aust. Inst. Min. Met., No. 236, Dec. 1970.

57. R. Glickman, H. Magquin and C. V. J. King, Electrochemical Soc., 100, p. 580-585 (1953).

58. E. A. von Hahn and T. R. Ingraham, "Kinetics of Silver Cementation on Zinc in Alkaline Cyanide and Perchloride Acid Solutions," Can. Met. Quart., 7, p. 15-26 (1968).

59. D. J. MacKinnon and T. R. Ingraham, "Copper Cementation on Aluminum Canning Sheet," Canadian Metallurgical Quarterly 10, No. 3, p. 197-201 (1971).

60. E. A. Von Hahn and T. R. Ingraham, "Kinetics of Pd (II) Cementation on Copper Sheet in Perchlorate Solutions," Trans TMS/AIME, 236, p. 1098-1103 (1966).

61. C. J. Hamdorf, "Precipitation of Lead and Silver from Brine Solutions by Metallic Iron," Proc. Aust. Int. Min. and Met., 199, p. 19 (1961).

62. R. M. Nadkarni and M. E. Wadsworth, "Kinetic Study of Copper with Iron," Advances in Extractive Metallurgy, IMM, 1967, p. 918-941.

63. R. Naybour, "The Effect of Electrolyte Flow on the Morphology of Zinc Electrodeposited from Aqueous Alkaline Solution Containing Zincate Ions," J. Electrochem. Soc. Electrochem. Technol. 116, p. 520 (1969).

64. E. C. Lee, F. Lawson, and K. N. Han, "Effect of Precipitant Surface Roughness on Cementation Kinetics," Hydromet. 3, p. 7-21 (1978).

65. R. Miller, Ph.D. Thesis, "A Kinetic Study of the Cementation of Copper with Nickel," Ph.D. Thesis, University of Utah, Salt Lake City, Utah (1968).

66. H. Majima, M. Mamiya, H. Tanaka and K. Natsuda, "Physical Chemistry of Copper Cementation," World Mining and Metals Technology, AIME, p. 567-584 (1976).

67. E. Schalch, M. J. Nicol, "An Electrochemical Investigation of Copper Cementation on Iron," Trans. Institute of Mining and Metallurgy, vol. 87, p. C16, 1978.

68. W. Fisher and R. Groves, "Copper Cementation in a Revolving Drum Reactor, A Kinetic Study," U.S.B.M. RI. 8098 (1975).

TRENDS IN NON-FERROUS PYROMETALLURGY

Herbert H. Kellogg
Henry Krumb School of Mines
Columbia University
New York, N.Y. 10027

Pyrometallurgy accounts for more than half of non-ferrous metal production, and is likely to retain this position because of its many inherent advantages. Large volumes of waste gas, and fugitive gas released to the work-place are now recognized as serious disadvantages of most conventional smelting processes. Modern process designs, however, can significantly reduce the volume of waste gas that must be cleaned and processed, reduce the escape of fugitives, and, at the same time, further exploit the inherent ability of pyrometallurgy to yield energy-efficient processes of high specific capacity.

A number of processes used, or proposed, for smelting copper, lead and zinc are reviewed and compared, in order to identify shortcomings of current technology and trends in modern process design. Significant advantages accrue to those processes that:

§ Make maximum use of commercial purity oxygen to replace air.

§ Employ a reactor capable of high specific-capacity.

§ Make full use of the fuel value in sulfide concentrates.

§ Replace batch by continuous operation.

INTRODUCTION

Pyrometallurgical processes (i.e., chemical processes for metal recovery from ore that are carried out at elevated temperature, in the absence of an aqueous phase) have served mankind since the dawn of recorded history. In the modern world of instant communications, space exploration, and atomic energy, pyrometallurgy still accounts for more than 95% of the tonnage of all metal produced. Although iron and steel production, which is entirely pyrometallurgical, accounts for the vast majority of the total metal made each year, there are some twenty important non-ferrous metals, with a combined world production of nearly 50-million tons/year. More than half of this non-ferrous metal production comes from pyrometallurgical processing.

Clearly, pyrometallurgical processes have been predominant in the past and are still predominant today. This paper will attempt to explore the reasons for this, and the forces that may alter the role of pyrometallurgy in the future--with special emphasis on the major non-ferrous metals. The intrinsic properties of high-temperature processes, which lead to both advantages and disadvantages for pyrometallurgy, will be examined, as well as trends in modern process design that seek to maximize advantages and minimize disadvantages.

The alternatives to pyrometallurgical processes are hydrometallurgy (chemical reactions in aqueous systems) and electrometallurgy (electrochemical reactions in either aqueous or fused-salt systems). Both of these are late 19th-century developments, for the most part, although some very simple hydrometallurgy (cementation of copper, for example) is of much earlier date. Hydrometallurgy and electrometallurgy have their own advantages that supplement those of pyrometallurgy, and the availability today of all three processing types greatly enhances the prospects for improved metal production processes in the future.

Development of hydro-electrometallurgy during the 20th century has given us superior new processes that broaden the capabilities of metal extraction and refining. Electrolytic reduction is ideally suited to the reactive metals (aluminum, sodium, magnesium) which are difficult to produce by pyrometallurgy. Selective leaching can be used to recover metals from low-grade ores (gold, uranium) that are unsuitable for pyrometallurgical treatment and which cannot be concentrated by mineral processing. More than 75% of the world's zinc and cadmium is now made by the roast-leach-electrowinning process, which has been rapidly replacing outmoded horizontal-retort smelting. Difficult metal separations, such as cobalt/nickel, zirconium/hafnium, and columbium/tantalum, have yielded to new hydrometallurgical methods.

Many have been tempted to generalize from these successes of hydro-electrometallurgy that the demise of pyrometallurgy is only a matter of time. Pyrometallurgy is likened to medieval alchemy or 'blacksmith' metallurgy--an ancient and somewhat mysterious art that will ultimately yield to the enlightment of modern 'chemical' processing. Others, the writer among them, believe that pyrometallurgical processing will retain a monopoly on iron and steel production and an important position in non-ferrous production far into the future.

It is critical for industry, government, and academia to decide which of these views has greater validity; if hydro-electrometallurgy is the wave of the future for non-ferrous processing, then our funding for research and development should be directed that way; if pyrometallurgy is here to stay, with future development of modern processes that combine pyro, hydro, and electrometallurgy in the most efficient manner, then both our thinking and our funding of R&D must recognize the tripartite nature of non-ferrous production metallurgy. In the section that follows I give my reasons for believing that most of pyrometallurgical processing is here to stay.

CHARACTERISTICS OF PYROMETALLURGY

Intrinsic Properties

Table I lists nine properties of the high-temperature systems of pyrometallurgy that I believe help to explain both the advantages and disadvantages of this means of metal processing. These are *intrinsic* properties--they remain true regardless of development of new technology. Most are self evident, but I discuss each briefly in the following paragraphs.

High Reaction Rate. Typical pyrometallurgical processes operate at temperatures between 800 and 1600°C, and at these temperatures most chemical reaction rates are so high that overall process kinetics are controlled by mass-transfer rates (diffusion and convection) to the reaction site. This must be contrasted with leaching reactions of hydrometallurgy near ambient temperature (25-150°C), where slow chemical-kinetics frequently control the process rate, and lead to the need for long residence time in the reactor.

Table I. Some Intrinsic Properties of High-Temperature Processes

a. High chemical reaction rates at high temperature.

b. Temperature variation can be used to alter reaction equilibrium.

c. Metal sulfides are also 'fuels'.

d. Process streams have high concentration of metal.

e. Many molten metals are immiscible with molten oxides (slags).

f. Precious metals are soluble in molten lead, copper, or nickel.

g. Vapor pressures are often large at high temperature.

h. Processes, almost invariably, produce a 'waste-gas' stream.

i. Metallurgical slags are relatively stable in the natural environment.

Reaction Equilibrium. The freedom, in high-temperature processing, of selecting the process temperature makes possible the adjustment of the equilibrium state of a given reaction to favor a desired result. Thus conversion of white metal to blister copper by blowing with air depends upon the equilibrium state of the reaction

$$Cu_2S + 2Cu_2O \underset{1200^\circ C}{\overset{800^\circ C}{\rightleftarrows}} 6Cu + SO_2$$

This reaction equilibrium is favorable for copper production at 1200°C, and unfavorable at 800°C. Similar examples are legion in pyrometallurgy. The severely restricted range of temperature in hydrometallurgy precludes the use of temperature variation to significantly alter reaction equilibrium except in a few cases.

Sulfide Fuel. That raw material sources of many non-ferrous metals are sulfides ($CuFeS_2$, Zn(Fe)S, PbS, Fe(Ni)S, etc.) is a fact of nature. That oxidation of the sulfur and iron in these minerals releases considerable heat is both a fact of nature and a big advantage to pyrometallurgical processes that can use this heat to replace that from fossil fuels. Some processes (for copper, nickel and lead) also use the sulfur as a 'no-cost' reducing agent. Some new hydrometallurgical processes for direct oxidation of sulfide concentrates may also find it possible to use part of the heat released, but matching of heat release to heat requirement is less satisfactory than in pyrometallurgy.

Metal Concentration. The process stream in pyrometallurgy--be it the concentrate and flux mixture, or metal and slag mixture--is highly concentrated with respect to metal; typical values range from 500-2,000 grams of metal per liter. This contrasts with hydrometallurgical systems, where metal concentrations usually range from 10-100 grams of metal per liter. The absence of the large volume of diluent water in pyrometallurgical practice accounts for this difference.

Phase Equilibria. The immiscibility that exists between molten metal and oxide slag, and between sulfide matte and slag, for many non-ferrous systems is a fact of nature that lends itself to simple, inexpensive phase separations in pyrometallurgy. Likewise, the preferential solubility of the precious metals in molten base-metal (Pb, Cu, or Ni) compared to their solubility in slag or matte is an inherent property of high-temperature systems that has few counterparts in hydrometallurgical processing.

Vapor Pressure. Pyrometallurgical processes often operate at temperatures where the vapor pressure of metals and metal compounds is appreciable (.001 to 1.0 atm). This can lead to a desired result, such as selective vaporization of magnesium in the silico-thermic reduction process, or vaporization of the zinc in the Imperial Smelting Process. It can, and frequently does, lead to unwanted vaporization with the production of fume and dust that must be collected and returned to the process (e.g., vaporization of PbS, PbO, and Pb in lead smelting).

Waste Gases. The ubiquity of waste-gas streams in pyrometallurgy resembles the ever-present aqueous phase in hydrometallurgy. High-temperatures are achieved and maintained by combustion of fuel--either intentionally added or present in the raw material treated--and combustion gases are the result. Only when expensive electric heating is employed are combustion gases absent, and even in such cases a significant volume of waste gas is usually generated by the smelting reactions themselves. By the very nature of things, pyrometallurgy will always be faced with waste gases containing dust, fume, and noxious elements; the volume and composition of the waste gas produced, however, will vary significantly with the specific design of the process.

Stable Slags. Slags, the solid waste from many pyrometallurgical processes, are close relatives of natural rocks, and are relatively resistant to leaching and weathering in the natural environment.

Advantages of Pyrometallurgy

These intrinsic properties of high-temperature systems help to explain six practical advantages and two disadvantages of pyrometallurgical processes, as listed in Table II and briefly discussed in the following paragraphs.

Table II. Advantages and Disadvantages that Result From Intrinsic Properties of High-Temperature Processes*

Advantages:

1. Energy requirements are low (a, b, c, d).

2. Inexpensive reducing agents can be used (a, b).

3. High specific-capacity leads to low capital and labor cost (a, d).

4. Separation of metal and residue is simple and inexpensive (e).

5. Traces of precious metals are inexpensively recovered as byproducts (f).

6. Solid waste disposal is usually simplified (i).

Disadvantages:

1. Treatment of dusty, noxious waste gases is costly (a, g, h).

2. Collection and treatment of fugitive gases are costly (a, g).

*Letters following each item refer to intrinsic properties in Table I.

Energy Requirements. It comes as a surprise to some of us that most pyrometallurgical processes use less energy per unit of metal output than competing hydro-electrometallurgical processes. Table III documents this conclusion for three metals where comparison can be made. As indicated in Table II, the reasons for low energy consumption in pyro-processes relate to the intrinsic properties of high reaction rate (low residence time in the reactor), ability to adjust equilibrium to the most favorable condition by temperature selection, the inherent fuel-value of sulfide concentrates, and the high concentration of metal in the process stream. A simple illustration of the great importance of metal concentration on energy requirements is the following fact: the enthalpy requirement to heat a ton of copper from 25 to 75°C, in the form of an aqueous copper sulfate solution at 50 gm Cu/1000 gm H_2O, is about five times larger than the enthalpy to heat a ton of copper in the form of pure Cu_2S from 25 to 1150°C. The high concentration of copper in pure Cu_2S, compared to the aqueous solution, explains this unexpected behavior.

Table III. Energy Requirements for Metal Production*

Metal, Nature of Process	Ref.	Energy Used GJ/1000 kg M°
Copper: sulfide conc. to cathode copper		
Outokumpu Flash Smelt (Pyro)	1.	22.0
Mitsubishi Cont. Process (Pyro)	1.	22.9
Roast-Leach-Electrowin (Pyro-Hydro-Electro)	1.	35.3
Arbiter Ammonia Leach (Hydro-Electro)	1.	72.0
Zinc: sulfide conc. to SHG slab		
Imperial Smelting Process (Pyro)	2.	44.8
Roast-Leach-Electrowin (Pyro-Hydro-Electro)	2.	50.1
Magnesium: raw material to ingot		
Magnetherm Process (Pyro)	**	321.
Dow Sea-Water Process (Hydro-Electro)	4.	416.

* Energy used includes fuel, the fuel equivalent of electrical energy (at the U.S. average rate of 11.07 x 10^6 joules/kwh), the fuel equivalent of supplies (reagents, oxygen, chlorine, etc.) and credit for the fuel equivalent of byproducts.

**Calculations of the author, based on process description in ref. 3.

Inexpensive Reducing Agents. Carbon, CO, and hydrogen, derived from fossil fuels, are far less expensive reducing agents than the electricity most commonly used to reduce metal compounds to metal in aqueous processing. These cheap reducing agents are effective and simple to use at the high temperatures characteristic of pyrometallurgy, but find little use near ambient temperature in hydrometallurgy (hydrogen, under pressure, at 150°C or higher, can be used to reduce some metals from aqueous solution). Part of the natural sulfur in copper, nickel, and lead concentrates can be used to effect reduction to crude metal in pyrometallurgy (e.g., converting of molten copper-sulfide or nickel-sulfide matte to metal) at no cost for the reducing agent. Few comparable reactions are possible in aqueous processing.

High Specific-Capacity. Most pyrometallurgical processes possess specific capacities (throughput per day per unit of reactor volume) that are an order of magnitude larger than for hydro-electrometallurgical processes (5). This advantage of high-temperature processing stems from the high concentration of metal (absence of diluent water) and inherent high reaction-rates. High specific capacity translates to smaller equipment volume and a more compact plant per unit of metal output. Construction of some high-temperature reactors (furnaces) is more costly per unit volume than for common aqueous processing equipment, but the balance between reactor volume and construction cost per unit volume usually results in lower capital cost for pyrometallurgical processing.

Simple Separation of Metal and Residue. The immiscibility of molten metal and slag (or matte and slag), coupled with the large density difference between these phases, makes possible a very simple and effective separation of product from residue in many pyrometallurgical systems. A short period of settling, followed by separate tapping of the two liquid layers suffices in most cases. No special equipment (or, at most, a small forehearth) is required to effect the separation. The liquid phases in pyrometallurgy can also be cleanly separated from the gas phase by no more than the induced draft of the gas-handling system. In contrast to these simple separations, the common liquid/solid separations of hydrometallurgy require separate thickeners, filters and driers, and not infrequently the thickening and filtration characteristics of the pulp are so poor as to entail considerable expense for this purpose.

Collection of Precious Metals. The sulfide ores of lead, copper and nickel commonly contain valuable trace amounts of gold, silver, and platinum-group metals. Pyrometallurgical processing of the base metals leads to a natural concentration of the trace precious metals in the molten base-metal. In all three cases the precious metals can be separated from the crude base-metal by well established and inexpensive technology. In contrast, most proposed hydrometallurgical processes for treating sulfide concentrates of copper, lead, and nickel recover little or none of the precious metals, unless relatively major and costly processing steps are added to the base-metal treatment scheme.

Waste Disposal. When slag is the solid waste from pyrometallurgy, the small volume and environmental stability of slag, compared to hydrometallurgical wastes (e.g., 'red-mud' or jarosite), usually results in lower costs and improved environmental protection.

Disadvantages of Pyrometallurgy

In thinking of the disadvantages of pyrometallurgy one might be tempted to list the difficulty of maintenance of refractories for high-temperature service. This feature of pyro-processes, although it represents a significant cost, does not differ in magnitude from similar problems of equipment corrosion in aqueous processing. The real disadvantages of pyrometallurgy, listed in Table II and discussed below, represent behavior that places pyro-processes at a significant disadvantage relative to aqueous processing.

Waste Gas from Smelting. Except in special cases, pyrometallurgical processes will always produce a waste-gas stream. The gas will carry with it dust (particularly when the process feed is a fine concentrate), and vapors and fumes arising from the various substances that have appreciable vapor pressure at the process temperature. At least partial dedusting of waste gas, to recover metal values from coarser particles, has been common practice in smelting for centuries, and such gas treatment was economically justified by the value of metal recovered. Fine particulates, fume and noxious gases (SO_2 NO_x, CO, etc.) were released to the atmosphere, via a stack, in times past when the atmosphere could be used as a convenient sewer.

Burgeoning population and industrialization throughout the world are now placing so heavy a stress on the natural environment that, one by one, governments are finding it necessary to regulate effluents from industrial and even some consumer sources. It doesn't require a seer to predict an inevitable tightening of such regulations, and their spread to other countries, as population and industrial capacity continue to expand. This is the reality that faces pyrometallurgy, and all industry, for the foreseeable future--what was acceptable for the environment in 1900, or even in 1950, may be quite unacceptable in 1981, or 1990, or 2000.

Without question, 'clean air' regulations pose a serious new problem for pyrometallurgical operations. The gravity of the problem (i.e., its cost) will depend on the nature of particulates and noxious gases in the waste stream, but even more importantly on the volume of that stream. Waste gas may be ubiquitous in high-temperature processing, but the particular process design can significantly alter the volume of gas produced and its composition. A dramatic example of this may be seen by comparison of two methods for smelting copper concentrate to matte: the conventional green-charge reverberatory furnace produces about 18 tons of waste gas, containing about 1.5% SO_2, for each ton of copper processed; the INCO Oxygen Flash smelting process produces about 1.3 tons of gas, containing 80% SO_2, for the same amount of copper processed (6).

The cost of dedusting and fume collection from gases is roughly proportional to the volume of gas to be treated, and, on this basis, gas cleaning for the reverberatory will be 14 times more costly than for flash smelting. Sulfur fixation, now required in many localities, provides an even greater advantage for the flash-furnace gas: the 80% SO_2 gas can be converted either to liquid SO_2 or sulfuric acid at modest cost; removal of sulfur from the dilute reverberatory gas requires fixation with lime, or concentration of the SO_2 by an absorption process, at a cost considered to be unacceptably high in most countries.

The need for cleaning and de-toxification of waste gas from smelting is here to stay, and has rendered obsolete many existing pyro-processes--such as the conventional reverberatory furnace--that were designed for the requirements of a bygone era. Novel process design, however, can sharply mitigate the burden of waste-gas treatment, and oxygen flash-smelting of copper concentrate provides an excellent example of the direction that new pyrometallurgical processes must follow.

Fugitive Gases. In addition to the principal waste-gas streams, discussed above, smelting operations emit fugitive gases into the work-place atmosphere from a variety of sources--from tapping and pouring of furnaces, during ladle-transfer, from launders, from incompletely-sealed furnaces, etc. Fugitive gases may endanger worker health, and eventually they find their way outside the plant and contribute to general atmospheric pollution. Those of us who grew up with the fume-laden, eye-watering, work-place of smelters, earlier in this century, may pride ourselves on our stamina, and ridicule the necessity of improving these conditions. We would be shortsighted to do so, for the evidence for long-delayed damage to health from such exposure accumulates as each year passes. When modern pyrometallurgical plants can be said to provide as favorable working environments as a modern zinc-electrowinning plant, then pyrometallurgy will have rid itself of this disadvantage that it now suffers.

Not aesthetics, but hard facts on micrograms of various pollutants per cubic meter of the work-place atmosphere provide the measure of a favorable environment. The world's work-force deserves no less than a safe and healthy work-place, and if it can be achieved only by the hydro-electrometallurgy route then this writer will switch his allegiance to that technology.

But, as was the case for waste-gas effluents, there is ample reason to believe that pyrometallurgy's reputation for unhealthy working conditions also results from obsolete process design, rather than an inescapable shortcoming of this technology. Thirty years ago we did not know of the delayed danger to health from particulates and low levels of many air-borne pollutants. We did not understand the need to avoid fugitive emissions, and so our plants failed to incorporate design features that would have minimized their occurance. Recognition of this problem is of quite recent date, and the basic design of pyrometallurgical processes has only begun to adapt to this new design criterion. Rather the problem has been temporarily ameliorated by provision of costly and cumbersome hooding and venting of major fugitive sources from our obsolete current processes. A major task, involving innovative re-design to minimize fugitive emissions, faces the pyrometallurgical industries.

COPPER SMELTING

The remaining sections of this paper present an analysis of the state-of-the-art in pyrometallurgical production of three major non-ferrous metals--copper, lead, and zinc. Particular attention will be directed at weaknesses of current technology that call for imaginative new process design.

Copper smelting provides an appropriate starting place because it has produced more innovative technology in recent years than other metal industries. Thirty years ago the standard practice of the industry--the method that would be adopted for new plant construction, almost without question--was reverberatory smelting to low-grade (30-40% Cu) matte followed by Peirce-Smith converters. Today, although many reverberatory plants still operate, the prevalent choice for new plants is flash smelting with oxygen-enriched air to high-grade (55-65% Cu) matte followed by Peirce-Smith converters. But the choice of process is not as clear-cut as it was thirty years ago. If we consider only those processes that have been proved on an industrial scale, besides the Outokumpu-design flash-smelter (7) with about 30 years of experience, we should look with equal care at INCO Oxygen Flash smelting (8), also with 30 years of experience, the Noranda Process (9), with 8 years of experience, and the Mitsubishi Continuous Process (10), with 7 years of experience.

Table IV compares these four new processes, and the conventional reverberatory, with respect to several features that put the new processes in a class by themselves. The matte-smelting stage of the new processes oxidizes much more of the iron and sulfur in the concentrate, thereby producing a much higher grade of matte, and sharply reducing the requirements for smelting fuel. Furthermore, the liberal use of oxygen to replace part or all of the air needed for oxidation greatly reduces the volume of smelter gas produced. The INCO Oxygen Flash smelting process carries this trend to the extreme by use of undiluted oxygen (95% O_2) for smelting, eliminating the need for any fuel (except possibly for drying concentrate), and reducing smelter gas to a very small volume of highly concentrated SO_2.

The high-grade matte from the new processes greatly reduces the blowing time in converters, and the volume of converter gas produced (alone among the four new processes, the Mitsubishi Process has abandoned batch converting in Peirce-Smith vessels, and uses a unique design of continuous converter, producing a continuous stream of SO_2-rich converter gas). Total gas volumes from smelting and converting are only 20 to 40% of those from reverberatory smelting, and all of this gas contains sufficient SO_2 for inexpensive acid manufacture.

Any of the four new copper-smelting processes may be said to have successfully minimized the problem of waste-gas treatment in copper smelting through redesign of the basic smelting reactor. At the same time, they have made copper smelting more energy efficient than the reverberatory, as evidenced by the sharply reduced values of Process Fuel Equivalent shown in Table IV. Process Fuel Equivalent (11) sums the energy of the fuel used, plus the fuel equivalent of electric energy and supplies (oxygen, flux, etc.), and credits the surplus steam produced by the process. Thus, the new copper smelting processes also conserve overall energy supplies, compared to reverberatory smelting, and provide an even more elusive target for rival hydro-electrometallurgical process concepts.

Table IV. Comparative Data for Copper Smelting (Note a)

Item	A Reverb.	B Flash I	C Flash II	D Noranda	E Mitsub.
	Smelting Process (Note a)				
For Matte Smelting Only					
Fuel, GJ/1000 kg Cu	19.2	3.5	0.0	3.4	4.8
Matte grade, % Cu	35.	61.4	54.3	75.	65.5
kg 98% O_2/1000 kg Cu	0.	480.	790.	760.	390.
Process gas, % SO_2	1.8	18.	77.	10.	14.
Nm^3/1000 kg Cu	11,600.	2,700.	570.	5,500.	2,900.
Specific Capacity: tons conc./d/m^3	0.81	3.5	2.3	6.4	4.3
Convert. gas (note b), % SO_2	7.6	8.6	9.3	9.4	14.
Nm^3/1000 kg Cu	6,100.	2,600.	3,200.	1,800.	2,200.
Total gas, smelt + conv.					
Nm^3/1000 kg Cu	17,700.	5,300.	3,770.	7,300.	5,100.
Overall PFE (note c)					
GJ/1000 kg Cu anodes	21.5	14.2	11.6	14.3	16.2

Notes: a) Based on data from reference 6. A is green-charge reverberatory furnace; B is Outokumpu flash-smelter with oxygen enrichment; C is INCO Oxygen flash-smelter; D is Noranda process with oxygen enrichment, producing matte; E is Mitsubishi Continuous process. All are compared for smelting the same concentrate.

b) Assumes the use of tight-fitting converter hoods and waste-heat boilers; common U.S. practice would yield 4-5% SO_2 and larger gas volumes.

c) PFE = Process Fuel Equivalent (see ref. (11)) includes all fuel, and fuel equivalent of electric energy, for drying, smelting, converting, anode refining, oxygen production, and acid-production, less fuel credit for byproduct steam. Energy for fugitive gas control not included.

Deep oxidation of iron and sulfur in the concentrate, and effective use of strong oxygen-enrichment are only possible if accompanied by greatly enhanced mixing of oxidant and concentrate, so as to achieve high rates of heat and mass transfer. The new processes achieve intimate mixing either by flashing (dispersion of fine concentrate in the oxidizing gas stream) or by injection of concentrate and oxidant into the vigorously agitated bulk of molten slag and matte. A corollary advantage follows from the intensive smelting reactions: The specific capacity (tons conc./day/m^3 furnace vol.) of the new processes is 3-8 times greater than the reverberatory furnace (see Table IV), which results in smaller furnace size for a given smelting rate, lower capital investment for the furnace, and lower heat-loss per ton of concentrate processed (6).

Copper Converting

The new matte-smelting processes discussed above show very significant improvements over conventional reverberatory smelting, but only the Mitsubishi Process has provided a new design for conversion of matte to blister copper. The other processes continue to employ the conventional Peirce-Smith (P-S) converter with all of its inherent disadvantages: a batch process, requiring filling and emptying with crane-transported ladles; intermittent production of SO_2-bearing waste-gas, thus greatly complicating both recovery of heat in waste-heat boilers and production of acid from the gas; serious problems with fugitive gases during charging and emptying the converter, from the intermittent tapping of the matte furnace, and from the ladles during transport; large capital investment and labor requirement for the elaborate auxiliaries required by the batch system.

The P-S converter has the advantages of proven performance and an unusually high specific-capacity for oxidation of iron and sulfur, and these have been of sufficient importance to encourage attempts to control fugitive emissions so as to meet environmental regulations. Details of the elaborate enclosures, hooding and ducting systems that have been employed for this purpose (12) are beyond the scope of this paper. Suffice it to say that these contribute significantly to the cost of the installation and its operation, and are an inconvenience to the operators.

Copper smelting has no greater need than for a reliable, continuous process for conversion of matte to blister. Although such a design presents some difficult problems, they can be surmounted, and the Mitsubishi Continuous Process offers the first design of this kind to be operated on a commercial scale. The Mitsubishi converter approximates to a well-mixed continuous reactor, fed with a continuous stream of high-grade matte and blown with oxygen-enriched air from top lances, with continuous tapping of slag and blister copper, and a continuous stream of SO_2-rich gas exiting to a waste-heat boiler, gas cleaning and acid production.

Table V gives some comparative data for the Mitsubishi and P-S converters that shows what the Mitsubishi

Table V. Comparison of Copper Converters

Item	Peirce-Smith (a) 13'x30'	Mitsubishi Continuous (b)
Internal volume (inside brickwork), m^3	67.	99.
Blowing rate, Nm^3/min	570.	138.
Specific blowing rate, $Nm^3/min/m^3$ vol.	8.50	1.39
% O_2 in blast	21.	26.
Oxygen-utilization efficiency, OUE, %	90.	87.
Stack time, %	60. (c)	100.
Specific oxygen-capacity (note d): Tons O_2 burned/day/m^3 vol.	1.98	0.65

Notes: a) Based on data from reference 13.

b) Based on data from reference 14.

c) May vary from 35-80%, but 60% is a reasonable average.

d) Specific oxygen-capacity (SOC) is the number of tons (1000 kg) of pure oxygen actually used to burn sulfur and iron, per day, per cubic meter of furnace volume. It is calculated from: S.O.C. = 2.058 x specific blowing rate x % O_2/100 x % OUE/100 x % stack time/100.

converter has sacrificed to achieve the advantages of a continuous process, and that suggests ways to improve its performance. The Mitsubishi converter is 50% larger in internal volume than a standard 13'x30' P-S converter, and is blown at one-fourth the air rate. These two factors combine to yield a specific blowing-rate ($Nm^3/min/m^3$) for the Mitsubishi converter that is only one-sixth that of the Peirce-Smith. Oxygen-utilization efficiency--the percentage of oxygen blown that actually reacts with iron and sulfur in the matte--is about the same in the two processes. In contrast, the Mitsubishi converter has the distinct advantage of operating continuously (100% stack time), whereas the P-S converter is only blowing for a fraction of the time (60% stack-time, for typical good operation).

The combined influence of the factors cited above is best illustrated by the item called specific oxygen-capacity (S.O.C.) in Table V. This parameter (rather than specific capacity expressed as tons of matte (or tons of copper) per day per cubic meter) is particularly suitable for comparison of converting performance because it is independent of matte grade; it simply measures how much oxygen is actually used to burn iron and sulfur, per day, per unit volume of the converter. Except for the units employed, S.O.C. is the same measure that Johnson, Themelis, and Eltringham (13) called the % Converter Production Efficiency, and it combines the effects of specific blowing rate, % oxygen in the blast, oxygen-utilization efficiency and % stack time.

According to this measure, the Mitsubishi converter burns oxygen only one-third as rapidly, per unit of furnace volume, as the P-S converter. This lower specific capacity means that the Mitsubishi unit must be three-times larger (and correspondingly more costly) than the P-S unit to accomplish the same converter work, and this is the sacrifice that has been made to achieve a continuous operation. What S.O.C. does not measure are the very important cost reductions for the Mitsubishi converter, resulting from continuous operation, the absence of a converter aisle, cranes, ladles, tuyere punchers, etc., the smaller volume of converter gas requiring cleaning and sulfur fixation, and the absence of fugitive gases that must be controlled. A recent study (1) estimates that the Mitsubishi process saves about 3 GJ of energy per ton of copper that processes employing batch converting must must use for fugitive gas control. This is one quantitative measure of the value of continuous operation, not reflected in the energy figures given in Table IV, which makes the Mitsubishi Process more attractive relative to the other new processes.

One wonders whether the Mitsubishi converter can be blown at a higher specific rate, and whether oxygen-enrichment can be carried beyond 26% O_2. Increase in either of these factors would increase the specific oxygen-capacity. Perhaps the present operation of the furnace simply reflects the caution of designers in development of a radical new technology. Top-blown lances for oxygen steelmaking operate with 99.5% oxygen at much higher furnace temperatures. This writer expects to see improvements in the Mitsubishi converter, particularly in increased oxygen-enrichment, that will result in at least a doubling of the present specific oxygen-capacity, although it may require a new, water-cooled lance design to realize this goal.

The Mitsubishi continuous converter is the essential element that makes possible continuous copper production by the Mitsubishi Process. Because of this unit, the matte-smelting and slag-cleaning furnaces can be continuously tapped, thus avoiding the pyrotechnic display (and fugitive gases) associated with intermittent furnace tapping. Because of this unit, the converter aisle with all of its drama (and fugitive gases) is absent. Instead, the Mitsubishi Process has three well-sealed furnaces, connected by launders, through which flow small, continuous streams of molten matte, slag, and blister. This process may not prove to be the ultimate design for the copper smelter of the future, but it certainly represents major progress toward the goal of a continuous, low-pollution, low-energy means for extraction of copper.

Other Converter Processes. Besides the Mitsubishi continuous converter and the conventional P-S converter, two designs--the Hoboken converter and the Top-Blown Rotary Converter (TBRC)--are in use today. The Hoboken (siphon) converter (15) is best considered a variation on the P-S design, with the modifications directed toward producing smaller gas volume (and higher concentration of SO_2) by reduction in the amount of infiltrated air. It retains the other basic disadvantages of the P-S batch process, and probably sacrifices somewhat in productivity and campaign life to achieve its advantage in gas volume and strength.

The TBRC (16) is also inherently a batch process, and, for this reason, is not the ultimate answer to copper converting. It possesses, however, a versatility that probably assures it a useful role in smelting technology for many years to come. It combines the high productivity (high specific-capacity) of the P-S converter, with the flexibility characteristic of top-blowing through a lance. Top-blowing permits the starting or stopping of the blow, or change from oxidizing to neutral or reducing conditions, or change in temperature of the bath, at any point in the batch cycle. It becomes possible, therefore, to combine smelting of concentrate, converting, and even fire-refining into a single batch-cycle. The lance can accomodate any degree of oxygen-enrichment, including use of undiluted oxygen, provided the charge in the TBRC is capable of dissipating the heat generated. Rotation of the vessel provides mixing, and distributes the heat uniformly to the charge and the refractories, and this permits operation at higher temperatures than can be successfully employed in P-S converters.

The 'Ideal' Copper Smelter

Although an 'ideal' copper-smelting process may never be achieved, the concept of such a design can serve as a yardstick against which to evaluate present and potential processes. The ideal goals that this author seeks for copper smelting include the following:

a) A fully continuous process from concentrate to blister copper, producing a finished slag (<0.5% Cu) and a low sulfur (<0.5% S) blister copper. A continuous process would obviate the difficulties and expense now encountered from intermittent tapping of matte, ladle transfer, batch converting, and cleaning and sulfur fixation of gas streams that are variable in volume and sulfur dioxide content.

b) The process gas should contain the maximum practical percentage of sulfur dioxide, so as to minimize the gas volume that must be cleaned and subjected to sulfur fixation. Gas of very high sulfur-dioxide content (>70% SO_2) has the added advantage of providing greater flexibility in the means of sulfur fixation (liquid SO_2, reduction to elemental sulfur, as well as acid production).

c) The process should be tightly sealed, so as to avoid infiltration of air that would dilute the process gas, and so as to avoid escape of fugitive gases that must be captured and treated.

d) The process should require no fuel, other than the inherent fuel value of sulfur and iron in the concentrate. Typical copper concentrate contains more than enough heating value to accomplish all major smelting reactions, provided it is used wisely.

e) The process should have the maximum possible specific capacity (tons of copper produced per day, per cubic meter of furnace volume). This would minimize equipment size, capital cost, labor requirements and heat loss per ton of copper produced.

These goals, and a variety of secondary goals, will be examined more closely by comparing them to the four new processes of Table IV. As these processes are operated today, only the Mitsubishi Process satisfies the first goal of continuous copper production. Any one of the matte-smelting processes could, however, be operated continuously, if coupled with a continuous converting process (like the Mitsubishi converter). Thus, all four matte-smelting processes remain as candidates for a part of the continuous copper smelter of the future. Each will be examined, therefore, with respect to its ability to meet the remaining four goals of the ideal smelter.

INCO Oxygen Flash Smelting. This process ranks first in ability to produce a small volume of very rich, sulfur-dioxide gas from a tightly sealed furnace. It is the only process among the four that requires no fuel for smelting. Its specific capacity, although improved over that for the reverberatory, is lower than that of the other new processes. As operated today, it is not capable of producing high-grade matte (65-75% Cu) and this limitation would need to be overcome if it were to be coupled to a single-stage continuous converter, rather than batch converting.

These characteristics all relate to the INCO operating mode, which is the flashing of dried concentrate with undiluted oxygen (95% O_2) in a tightly sealed furnace. In the absence of nitrogen and water vapor to absorb heat, the oxidation of copper concentrate with oxygen is capable of producing more heat than is required to maintain optimum furnace temperature, if sufficient iron and sulfur is oxidized to produce high-grade matte. Not only does the process require no fuel, but it must limit the degree of oxidation of the concentrate (limit the matte grade) in order to avoid overheating the furnace. Control of the oxygen/concentrate ratio in the flash burner must be such as to produce intermediate matte-grades (46-54% Cu) in order that a proper balance between heat generation and heat losses is maintained.

Several practical means are available to free the INCO process of this limitation on matte grade. a) The oxygen for flashing can be diluted with cooled and recycled process gas (80% SO_2); the increased volume of furnace gas would abstract more heat from the process, and thereby permit use of more oxygen and higher matte grade without sacrifice of the advantages of the process; in fact, the larger volume of hot furnace gas would result in more heat recovery from a waste-heat boiler (17). b) A portion of the concentrate could be roasted in a fluid-bed, prior to smelting, so as to remove part of its heating value. Proper proportioning of roasted and green concentrate in the flash-furnace feed would permit high-grade matte production (18). The only disadvantage would be the need for added roasting capacity. c) The flash furnace side-walls and roof could be constructed of welded water-tubes (similar to a radiation boiler), in which refractories are replaced by frozen slag on the water-tube shell. Such a furnace design has about ten times the heat-loss rate per unit area of the typical refractory-lined furnace; the extra heat transfered to the water tubes would permit increased use of oxygen and higher matte grade, and the heat abstracted could be recovered as steam. This last alternative would seem to hold the most promise for improved furnace performance. A water-tube shell design has been successfully tested for the reaction shaft of an Outokumpu flash furnace (19); it holds the promise of reduced maintenance, and might make possible the use of a smaller furnace volume, with increased specific capacity, because of the freedom from refractory failure.

Outokumpu Flash Smelting. This process began with use of preheated air, supplemented by fuel, for flash smelting. In recent years there has been a trend toward more oxygen-enrichment, less air preheat, less fuel use, higher grade matte production, and increased specific capacity (19). Unlike the INCO process, high matte grades can be produced, since the process uses considerable nitrogen (from air) that dissipates a large part of the heat of oxidation, and a heat deficit must usually be made up by burning a small amount of fuel.

The Outokumpu process, as currently operated, does not approach the ideal goals of no-fuel and low-volume, high-strength sulfur dioxide as closely as does the INCO process. The question arises whether the Outokumpu process could be operated with much higher oxygen-enrichment, or even with undiluted oxygen, and thus approach the ideal more closely. To the author's knowledge, undiluted oxygen has not been used in the Outokumpu process, except on an experimental basis (19), and then only in combination with a reaction shaft constructed of welded water-tubes, without a refractory lining (see above, under the INCO process).

This experience suggests that the localization of exothermic heat release in the vertical reaction-shaft of the Outokumpu furnace is not as well adapted to use of undiluted oxygen as is the INCO design, wherein the flashing flame is directed horizontally above the bath of the slag and matte. In the latter case, the flashing flame can radiate heat to the entire bath surface, thereby avoiding local overheating.

It is the author's opinion that the experimental water-cooled reaction shaft (without refractory lining) will prove practical, and will permit high oxygen-enrichment or undiluted oxygen to be used in the Outokumpu process. Besides resulting in smaller gas volume and higher strength of sulfur dioxide, this should also make possible operation without fuel, along with high-grade matte production, because the reduction in heat lost to nitrogen will be balanced by the increased heat lost to the water-cooled shell.

Noranda Process. The Noranda Process started operation as an air/fuel fired process, producing blister copper and a copper-rich slag (~12% Cu) that was externally cleaned and recycled. Today, both plants that use the process operate with considerable oxygen enrichment, and produce very high-grade matte (70-75% Cu) and slag with about 5% Cu, which is cleaned and recycled. As shown in Table IV, this current version of the process possesses the highest specific capacity of any of the new smelting processes. This is achieved by the excellent mixing that results from feeding concentrate pellets into the highly turbulent molten-bath created by the action of submerged, side-blown tuyeres. In this way, most of the heat of oxidation is generated and dispersed directly in the molten bath, where it is needed, thus avoiding the requirement for radiative and convective heat transfer to the bath in flash smelting.

Despite use of almost as much commercial oxygen per unit of copper as the INCO Process, the Noranda Process produces the largest volume of gas, with the lowest percentage of sulfur dioxide of any of the new processes, and also requires a small amount of fuel for smelting (see Table IV). These departures from the ideal smelter goals result from the following features: a) The submerged tuyeres cannot operate with more than modest oxygen-enrichment (~30% O_2) without excessive attack of refractories at the tuyere line. b) The submerged-tuyere design requires a tilting vessel, and this results in a less than gas-tight seal between the reactor mouth and the hood for process gas; significant air-infiltration occurs at the mouth-hood juncture and through the concentrate feed-port. c) Concentrate is fed to the furnace as damp pellets (8-10% H_2O) so as to minimize dusting; the smelting of water from the damp feed adds to the fuel requirements of the process and further dilutes the process gas.

Modifications that would bring the Noranda Process closer to ideal may be suggested, as follows: a) Use of gas-shielded tuyeres, similar to those employed in steelmaking (20) or the QSL process (21); this would permit use of undiluted oxygen, and the shielding gas could be recycled process gas (high-strength SO_2). b) A feeding system for dry concentrate; this might be injection through submerged tuyeres with a small amount of recycled process gas as a carrier, or use of top lances as in the Mitsubishi process. c) A tightly-sealed gas-exit, perhaps located in the end-wall of the reactor, rather than in the cylindrical shell.

The Mitsubishi Continuous Process. Referring here only to the matte-smelting unit of the Mitsubishi Continuous Process, so as to make comparisons with the other matte-smelting processes discussed above, Table IV shows a process with a specific capacity second only to Noranda, using less oxygen and more fuel than the other new processes, but producing a smaller volume of richer sulfur dioxide than the Noranda Process. The latter advantage, relative to the Noranda Process, is achieved by the absence of moisture in the feed and reduction of infiltrated air to a negligible amount. The high specific capacity follows from the intense mixing of concentrate with slag and matte, by action of the high-pressure, top-blown, lance-feeding system--a mixing action closely analogous to that in the Noranda Process.

It would seem possible for the Mitsubishi matte smelter to more closely approach the ideal goals of copper smelting simply by use of more oxygen--perhaps, even undiluted oxygen. This should result in elimination of the need for fuel, considerable reduction in gas volume (and increased sulfur-dioxide content), and probably a further increase in specific capacity. Limitation on oxygen enrichment may result from excessive burn-off of the unprotected steel lances now employed. A water-cooled lance design might be necessary for use of high strength oxygen, but the advantages to be gained therefrom would more than justify the necessary development effort.

Continuous Copper Smelting

Returning now to the primary goal of the ideal copper smelting process--a fully continuous process--it is well understood from the work of Yazawa (22) and others that production of finished slag (<0.5% Cu) requires an oxygen potential of about 10^{-8} atm, and that production of low-sulfur blister copper requires an oxygen potential of about 10^{-6} atm. A single, well-mixed reactor cannot, therefore, satisfy both requirements. If vigorous agitation is employed in the furnace design, with the laudable aim of increasing heat and mass transport and specific capacity, the furnace is likely to approximate to a well-mixed reactor. Considerations such as this probably led to design of the Mitsubishi Continuous Process, in which three separate furnaces (smelting, slag-cleaning, and converting), each approximating a well-mixed reactor of uniform oxygen-potential, are linked by continuous flow through launders.

Alternatively, a continuous process could consist of a single reactor, with counter-current flow of slag and matte (and copper), in which an oxygen-potential gradient (also, a corresponding gradient in slag and matte composition) is maintained from one end of the reactor to the other. In principle, this could yield a superior process, in that: it requires only one reactor; it could incorporate any desired degree of slag cleaning or sulfur elimination from the blister; it eliminates transfer of material by launders, and the fugitive gases produced therefrom; cleaning of converter slag is accomplished internally and automatically, whereas the Mitsubishi, 3-stage process requires granulation, drying, and external recycle of this slag to the smelting unit.

The three-furnace Mitsubishi process has been operating successfully for seven years, whereas the single reactor with oxygen-potential gradient remains a concept only. The reason for this probably lies in the greater difficulty of achieving the design goals of the single-reactor. If separate well-mixed reactors are employed, the required oxygen potentials are easily achieved by proper proportioning of oxidant and concentrate in smelting, or oxidant and matte in converting; mixing should be as intense as practical so as to approach the well-mixed concept as nearly as possible; quiescent conditions are only required for phase separation (matte from slag, in the slag-cleaning furnace). In contrast, if the single-reactor concept is employed, lateral mixing must be avoided in order to achieve the end-to-end concentration gradient on which the concept depends, and this requirement tends to conflict with the need of thorough mixing to achieve high specific-capacity; the ideal conditions would be thorough mixing of concentrate, matte, slag and oxidant in any vertical plane through the reactor, so as to achieve rapid approach to chemical equilibrium, with little or no mixing in the horizontal plane, so that the counter-current flow of slag and matte can establish the necessary concentration gradients; such a design has yet to be demonstrated on a commercial scale.

The Worcra process (23) and the earliest concept of the Noranda process (24) were both attempts to design a single-reactor continuous copper smelter that produced both finished slag and blister copper. The Worcra process, extensively piloted from 1963-1972, and now apparently abandoned, employed a long reactor of U-shaped plan*, with vigorous agitation from converting lances in one limb of the U, and relatively quiescent smelting and slag cleaning zones in the remainder of the reactor. Important lessons were learned from the work on this process, but performance was erratic and difficult to control--fuel consumption was higher, and specific capacity lower, than expected.

* The U shape was adopted because of space restriction, rather than for basic design reasons.

In the original concept of the Noranda process it was proposed to produce a finished, low-copper slag by blowing reducing gases through submerged tuyeres in the slag-discharge end of the reactor, at the same time that the converting zone was operated at higher oxygen potential to produce blister. Attempts to counteract the vigorous mixing action of submerged tuyeres (which tended to destroy the required lateral composition gradients), by introduction of internal dams to separate the zones of different composition were unsuccessful. In the eventual design of the Noranda process, production of finished slag was abandoned in favor of what is in essence a single, well-mixed reactor producing either high-grade matte or blister, and a slag of high copper-content that must be cleaned externally.

Q-S Process. Despite these less-than-auspicious beginnings, the potential advantages of a single-reactor, continuous copper smelter continues to attract the attention of process designers. A very promising concept for such a process is that proposed in 1974 by P.E. Queneau and R. Schuhmann, Jr.--the Q-S continuous process (25). This concept has yet to be tested for copper smelting, but a quite similar reactor for lead smelting began operation in March 1981, following several years of pilot work by Lurgi (21). Successful lead smelting by the QSL process, wherein an oxygen-potential gradient is also a necessity, would lend strong support to the validity of the copper smelting design.

The most recent description of the Q-S copper smelting concept (26) envisions a long cylindrical reactor (20' dia. x 150' long), sloping downward (1% slope) toward the blister-discharge end. Dried concentrate would be flash smelted with oxygen (or oxygen-enriched air) through multiple burners in the top of the shell. Converting of matte would be accomplished with oxygen or oxygen-enriched air injected through gas-shielded, bottom-tuyeres (Savard-Lee injectors) shielded by low-purity nitrogen. Slag cleaning would be accomplished at the slag-discharge end of the reactor by the joint action of flash-melted iron sulfide from a top-mounted burner, and the reducing action of powdered coal, SO_2 and O_2 injected through bottom tuyeres. It is claimed that spacing of the bottom tuyeres sufficiently far apart creates moderately turbulent zones in the melt, separated by relatively stagnant zones, and that this design permits the establishment of the necessary oxygen-potential gradient which earlier process designs failed to achieve.

Assuming that the Q-S copper smelter operates sucessfully at its design capacity* it would have a specific capacity of about 1.3 tons of concentrate per day per cubic meter of internal volume. This is only 57% of the specific capacity of the Mitsubishi Continuous process (2.3 tons conc/d/m^3, if the volume of all three furnaces is included). Lower specific capacity may be the price that must be paid to achieve the other goals of the single-reactor concept; the wide spacing between bottom tuyeres--necessary to minimize lateral mixing--adds to the furnace volume, but detracts from specific capacity.

* Dr. Queneau has estimated that the reactor capacity should be at least 1200 tons of concentrate per day, and that it conceivably might be as high as 2000 tons per day (private communication).

Design goals of the Q-S copper smelter are continuous production of blister copper with less than 1% sulfur, slag with less than 0.25% copper and gas with more than 25% SO_2, with no direct use of fuel, except for small amounts of coal used for slag cleaning. If these can be achieved in a single, tightly-sealed furnace, with no external treatment and recycle of slags, the relatively low specific capacity may be a modest price to pay. This writer awaits disclosure of the performance of the QSL lead-smelting process with anticipation that it may signal a long-awaited breakthrough in design of continuous smelting.

'Oxide' Routes to Copper

The lengthy discussion of copper smelting presented above all relates to new design concepts applied to the traditional sulfide (matte-smelting) chemistry for copper production. Non-sulfide chemistry also offers new and attractive possibilities for pyrometallurgical copper production. Several process concepts start with dead-roasting of copper concentrate in a fluid bed to yield a fully oxidized calcine, containing CuO and $CuFe_2O_4$, and a continuous stream of gas (~12% SO_2) that contains virtually all of the sulfur from the concentrate. Copper in the oxidized calcine is then reduced by one of several schemes to blister copper. Potential advantages and disadvantages of these oxide processes for copper production will be briefly considered.

Dead-Roasting. Dead-roasting of zinc concentrate, in preparation for leaching and electrowinning of zinc, is a highly developed and successful pyrometallurgical operation (27). The train of processes, consisting of roaster, waste-heat boiler, cyclones, electrostatic precipitator, gas cooling and cleaning and contact sulfuric acid plant, are all continuous operations and tightly-sealed against escape of fugitive gas. A large part of the heat of oxidation of the sulfide is recovered as by-product steam, that can either be used elsewhere in the process or converted to electric energy. The energy recovered as steam exceeds the total energy requirement for operation of the roaster, gas train and sulfuric acid manufacture, so that the total process operates as a net energy-producer (2).

Experience with dead-roasting of chalcopyrite concentrates is meagre. In particular, we need more knowledge of the optimum roasting temperature (probably 850-900°C), the amount of excess oxygen needed to achieve low sulfide-sulfur in the calcine, the amount of sulfation of the dust that will occur on cooling in the boiler, whether it is feasible to employ hot cyclones before the boiler in order to reduce sulfation of dust, and many similar practical process parameters.

Assuming that those practical problems can be resolved, it should be possible to dead-roast a typical chalcopyrite concentrate in a fluid-bed of Vieille-Montagne-Lurgi design, produce sulfuric acid by the double-contact process, raise about one ton of steam/ton of concentrate in the boiler and convert the steam to electric energy, all with a net credit of about 40 kwh/ton concentrate roasted (135 kwh/ton copper processed) of surplus electric energy. Thus the roasting step of oxide copper-smelting concepts is attractive from both energy and environmental viewpoints.

Segregation Roast: The most novel (and to this author, the most attractive) concept for recovery of blister copper from oxide calcine is application of segregation roasting. The calcine (recovered hot from roasting) is mixed with 1-2% sodium chloride (or other chloride) and 8-10% ground coal or coke breeze, and the mixture allowed to 'cook' at 700-750°C for 30-60 min. During this period a complex series of cyclic reactions occurs, involving chlorination of copper oxides, transport of cuprous chloride through the vapor phase, and reduction of copper metal onto the surface of the carbon particles. The principal overall reactions occurring are

$$3CuFe_2O_4 + 2C \rightarrow 2Fe_3O_4 + 3Cu + 2CO_2$$

$$2CuO + C \rightarrow 2Cu + CO_2$$

but these reactions fail to indicate the essence of the process, which is 'segregation' of metallic copper on to carbon particles. A copper-carbon concentrate is readily produced from the cooled reaction-mass by proper combination of gravity separation and flotation, for the coarse and fine particles of copper-coated carbon, respectively.

In laboratory-scale equipment, under carefully controlled conditions, concentrate grades of 82-88% Cu (<0.5% Fe, remainder carbon and ash) with copper recoveries exceeding 98% can readily be produced (28). The fuel content (carbon) of the concentrate is more than adequate to melt the concentrate and produce blister copper and a small amount of slag. An excellent review of segregation chemistry, and its potential application to roasted copper-concentrate, was prepared by Marcuson (29).

Many unresolved problems remain before this concept can be designed into a process: how to utilize the maximum amount of sensible heat in the hot roaster-calcine for the segregation reactor; how to thoroughly mix the coal and salt with calcine; the best design of reactor to provide adequate residence time at temperature; the unknown behavior of many impurities. Search for the answers to these questions would seem well worth the effort because the apparent advantages of the process are impressive. It should recover as much copper as conventional smelting for less total energy expenditure than the most energy-efficient of today's sulfide-smelting processes (INCO Oxygen Flash smelting); moreover, the energy used is mostly inexpensive coal. It requires no fluxes or oxygen, produces no slag, and is free of problems with refractory maintenance. Almost all (~97%) of the sulfur in the concentrate should be totally controlled in the roaster-acid-plant circuit without fugitive problems. The required capital investment will remain a question until a firm design has been established, but indications are that this should be less than for sulfide smelting.

Other Means for Oxide Reduction. Almost ten years ago the author urged consideration of blast-furnace reduction of copper-oxide calcine (30). Since then, AMAX Base Metals R&D, Inc. has described a low-shaft blast furnace technique for smelting of briquetted or pelletized oxide-calcine (31) with low-quality coke that appears to have many attractive features. Carbon reduction of oxide calcine in an electric furnace is practiced at Brixlegg, Austria (32), but the high cost of electric energy renders this process unacceptable except for localities with inexpensive hydropower. Many other schemes for carrying out selective reduction of copper from oxide calcine containing copper ferrite, by use of solid carbon, CO, H_2 or mixtures thereof, are certain to occur to clever process designers.

LEAD SMELTING

Three intrinsic properties of lead and its compounds --two of a physical-chemical nature, and the third physiological--are responsible for difficulties that must be surmounted in design of a modern, low-pollution lead-smelter. These are: a) the high volatility of lead, lead oxide, and particularly, lead sulfide; b) the high stability of the lead sulfates, particularly the basic sulfates; c) the toxicity of lead and its compounds.

Lead and lead oxide have effective vapor pressures that reach 10^{-3} atm. at temperatures near to 950°C, and lead sulfide is about 10-times more volatile (33). At temperatures where common iron-silicate slags are fluid (1150-1200°C), the effective vapor pressure of lead and lead oxide approaches 0.025 atm. (19 mm. Hg) and that of PbS is about 0.2 atm. These high vapor pressures mean that significant volatilization of lead will occur in lead smelting, with the problem increasing in severity with increase in temperature, or when lead sulfide (rather than the metal or oxide) is present. Volatilized lead compounds will enter the process gas, and be converted to fume (fine particles of oxide or sulfate) as the gas is cooled. Fume production creates both a metallurgical problem (i.e., the need to collect and recycle the fume to achieve acceptable lead recovery) and an environmental problem (i.e., lead and its compounds are toxic, and this requires extraordinary measures to assure that traces of fume are not allowed to escape and contaminate the environment during smelting, fume collection, and subsequent handling).

The great stability of the lead sulfates, compared to the sulfates of iron, copper, or zinc, makes it unusually difficult to separate lead from its associated sulfur in PbS by oxidation. ZnS can be roasted at 900°C to produce an almost sulfur-free oxide calcine, because the sulfates of zinc are unstable at this temperature. Much higher temperature would be necessary to decompose lead sulfates, and formation of a molten oxy-sulfate phase, at temperatures between 890 and 1000°C, effectively prevents the use of such high temperatures in a fluid-bed roaster, or similar equipment (34).

The stability of the lead sulfates imparts a thermodynamic behavior to the Pb-S-O system such that, in direct smelting of PbS concentrate, one is faced with the choice of producing either a low-lead slag with a high-sulfur lead bullion, or a lead-rich slag with a low-sulfur lead bullion (35). This means that in lead smelting, as in copper smelting, a single-stage, well-mixed reactor of fixed oxygen-potential cannot hope to simultaneously produce a finished, low-lead slag and a low-sulfur bullion suitable for refining. Two or more separate stages, or a single reactor with counter-current flow and an oxygen potential gradient, will be required for smelting.

Conventional Smelting

The conventional sinter-roast, blast-furnace flowsheet is a continuous process with two separate stages--the sinter machine and the blast furnace. The sintering process operates at high temperature (over 1200°C in the narrow combustion zone) and high oxygen-potential so as to reject most (80-85%) of the sulfur from PbS as dilute (4-7%) sulfur dioxide. The sinter-cake contains the remainder of the sulfur as very stable sulfates and unburned lead sulfide. The highly-oxidized sinter cake is then reduced with carbon (coke) in the lead blast-furnace to lead bullion and slag, at temperatures again exceeding 1200°C in the hottest part of the furnace.

Both the sintering and blast-furnace processes produce large volumes of fume-laden gases that must be captured and cleaned in order to meet environmental regulations. The sintering process, with its associated equipment for mixing and pelletizing feed materials, and crushing, sizing, and transportation of sinter, presents numerous opportunities for escape of dust and fume as fugitives to the workplace. Problems of control of toxic lead particulates in the workplace atmosphere are so severe that most operators doubt that proposed U.S. standards of less than 50 micro-grams of lead per cubic meter of air can ever be achieved in a conventional lead smelter. For this reason alone the sinter-roast, blast-furnace flowsheet for lead production may be forced into obsolescence.

Metallurgical features, as well as environmental, suggest the need for basic changes in lead smelting technology. The sinter-roast stage of conventional smelting dissipates into the waste gases virtually all of the fuel value inherent in lead concentrates. The blast furnace cannot handle sinter containing much more than 50% lead, so that typical concentrates, containing 60-75% Pb, must be diluted with recycled slag, thus increasing the coke requirement for smelting, and the slag and gas volumes. Lead blast-furnace slags contain 2-4% lead, and this represents a significant loss of lead unless the slag is subjected to further treatment (e.g., slag-fuming, which recovers both zinc and lead).

Conventional lead smelting finds itself in a position today very much like that of the copper reverberatory-furnace ten years ago--for a variety of reasons (metallurgical as well as environmental) it appears that new plant construction should seek alternative processes. Existing smelters will undoubtedly continue to operate for many years, with extraordinary (and expensive) efforts directed to environmental control. The future, however, belongs to novel processes, and both pyrometallurgy and hydrometallurgy vie to fill this new role.

New Lead Smelting Processes

Two new pyrometallurgical developments appear, to this writer, to offer the most promise at this time--the Kivcet process and the QSL process. Both feature a single, continuous, well-sealed reactor with zones of different oxygen potential--a design feature of particular importance in lead smelting, where toxic fugitives must be reduced to a minimum. Both employ commercial purity oxygen to burn the sulfur from lead concentrate, and thereby usefully employ this inherent fuel-value, and drastically reduce the volume of waste gas produced. Both dispense with the sintering operation--the most serious source of fugitives in conventional smelting--by direct smelting of fine lead-sulfide concentrate.

Kivcet Process. This process, developed on a pilot scale in the USSR, will be employed on a small commercial scale (about 85 tons of lead bullion per day) for the first time in the plant at Potosi, Bolivia, which should commence operations some time during 1981 (36). It is also rumored that at least two other large lead smelting companies are actively studying the feasibility of Kivcet smelters for their operations. The process is designed for recovery of both lead and zinc, so that it is particularly suitable for lead concentrates that are rich in zinc.

The Kivcet design features a vertical shaft in which dry lead-sulfide concentrate is flash smelted with sufficient oxygen (95%) to yield a low-sulfur lead bullion and lead-rich slag (~50% Pb) which are collected in the furnace hearth. The flash smelting is very nearly autogeneous (a small amount of fuel may be required for concentrates of low sulfur content), thus the fuel value of the concentrate is fully utilized for smelting. By operating at sufficiently high temperature (at Potosi the gas exits at 1350°C), and high oxygen-potential, most of the sulfur can be oxidized to a sulfur-rich gas (40-50% SO_2). An indirect, air-cooled heat exchanger cools the gas to about 550°C, after which it is dedusted in a specially designed electrostatic precipitator. Dust and fume are recycled to the flash smelter.

The zone of high oxygen-potential (the flash-smelting chamber) is separated from the reduction zone by a water-cooled barrier-wall immersed in the slag bath. Reduction of lead and zinc from the slag is accomplished with carbon (coke or charcoal) added to the slag bath; heat for the reduction zone is supplied by three electrodes immersed in the slag. The claim is made that zinc can be recovered either as metallic zinc condensed from the gases that leave the reduction zone, or as zinc-oxide fume collected subsequent to combustion of the gas in an after-burner. The plant at Potosi will employ the latter method of zinc recovery. A final slag containing 1.5-1.9% Pb and 3-4% Zn is expected, with recovery of 92% of the lead in bullion, and 6.3% of the lead and 85% of the zinc in the oxide fume.

QSL Process. This process was piloted by Lurgi during the period 1978-79 (21). A small commercial-scale demonstration plant (150 tons of lead bullion/day) began operation at Duisberg, West Germany in late March 1981, but details of its operation are not yet available. This installation will recovery lead only; the small amount of zinc in the feed materials will be retained in the slag, as far as possible.

The reactor at Duisberg is a horizontal, cylindrical vessel (inside dimensions, 1.8 m dia. by 21.3 m long), which can be tilted about its horizontal axis to lift the tuyeres out of the bath. The feed (damp pellets of lead concentrate, flux and recycled dusts) is introduced through ports in the top of the reactor at one end, and is rapidly oxidized to low-sulfur bullion and lead-rich slag (40-50% Pb) by oxygen (95% O_2) injected through gas-shielded tuyeres (Savard-Lee injectors) in the bottom of the reactor. At this end of the reactor, the high content of lead oxide in the slag lowers the melting point of the slag sufficiently to permit operation at 900-950°C. The low temperature, and the rapid oxidation of PbS by envelopment of feed pellets into the highly-oxidized slag, help to minimize volatilization of lead sulfide. As in the Noranda reactor, oxidation of sulfides, and release of heat, occur mostly within the agitated slag and lead bullion, thus providing ideal conditions for distribution of heat to the bath.

Low-sulfur lead bullion continuously overflows from a lead-well at the feed end of the reactor, and the lead-rich slag flows counter-current to the reduction zone at the opposite end. There, reduction of lead from the slag results from injection of powdered coal or reducing gases through tuyeres in the furnace bottom. The temperature at the reduction end is raised sufficiently (~1150°C) to permit production of a low-lead slag (~1.% Pb), by a fuel burner located in the end wall of the reactor. As described for the Q-S copper smelter (see above), the claim is made that spacing of the reducing tuyeres sufficiently far apart creates alternate stagnant and well-mixed zones that permit establishment of gradients of oxygen potential and slag composition. Gas from both the reduction and oxidation zones mixes, and exits through the feed-end of the reactor to a waste-heat boiler and electrostatic precipitator.

Comparison of Lead Smelting Processes

Table VI attempts to compare a number of performance parameters for conventional lead smelting and the two new processes. This material is offered with some misgivings because neither the QSL nor Kivcet process has yet been operated on a commercial scale, and when they commence operations the results may be quite different than anticipated. In the case of the QSL process, few details have been published regarding even the anticipated performance, and this necessitated many calculations (of doubtful accuracy) by the author in order to fill what otherwise would show as blank spaces in Table VI. Despite these uncertainties, such marked differences exist between conventional smelting and the new processes that the general conclusions to be drawn from Table VI should be valid.

Yet another difficulty in comparison of these processes stems from the fact that the Kivcet plant at Potosi will treat an unusually low-grade lead concentrate (47.4% Pb, 7.5% Zn) and produce both lead and a zinc-oxide fume. This task is significantly more difficult than recovery of lead alone from high-grade concentrate (74.% Pb, 2.3% Zn) by the conventional or QSL processes, and this will require special consideration in comparison of Kivcet to the other processes. The design of the Kivcet process makes it particularly suitable for high-zinc lead concentrate, just as the Imperial Smelting Process (37) is particularly suitable for simultaneous recovery of zinc and lead from high-lead zinc concentrate.

Table VI. Comparison of Lead Smelting Processes (note a)

Item	Sinter Blast Furnace	QSL (Duisberg)	Kivcet (Potosí)
Overall Process			
Concentrate: % Pb, % Zn, % S	74., 2.3, 15.7	74., 2.3, 15.7	47.4, 7.5, 18.8
Capacity: tons bullion/day	300.-500.	150.	85.
Recovery of metal: % Pb, % Zn	96., 0.	99., 0.	98.5, 85.
Slag, % Pb	2.-4.	1.	1.5-1.9
Specific Capacity: t bullion/day-m^3	(1.1)	(2.8)	(0.82)
Energy and Reagents			
Fuel, GJ/ton bullion	7.7	(2.3)	2.9
95% oxygen, kg/ton bullion	0.	(350.)	540.
Electric energy, kwh/ton bullion (b)	185.	(275.)	1616. (c)
Electrode carbon, kg/ton bullion	0.	0.	8.6
Process Fuel Equivalent, GJ/ton bullion	10.0	(5.5)	(21.7)
Oxidation Stage			
Reactor type	sinter	inject-smelt	flash smelt
Specific capacity, t bullion/day-m^3	(1.7)	(5.6)	(3.0)
Waste-gas: Nm^3/t bullion	2400.	(330.) (d)	545.
% SO_2	5.5	(45.) (d)	40.-48.
Reduction Stage			
Reactor type	blast fur.	pneum. smelt	electric fur.
Specific capacity, t bullion/day-m^3	(3.0)	(5.6)	(1.1)
Waste-gas: Nm^3/t bullion	2000.-3000.	(170.) (d)	220.
% SO_2	0.1-0.2	~0. (d)	~0.

Notes: a) Values enclosed in parentheses ( ) are the author's estimates, and may be uncertain to ±30%. Values for the sinter-blast-furnace are for good modern practice (recirculation of weak gas in sintering is assumed). Kivcet performance is based on data from reference 36. QSL performance based on incomplete data in reference 21, supplemented by calculations of the author.

b) Includes electric energy for smelting, gas cleaning, air compression, oxygen production, and acid manufacture.

c) 89 kwh/ton bullion was added to the reported value of 1527 kwh/t bullion (ref. 36) to allow for manufacture of acid, which was not included in the Potosi flowsheet.

d) The waste gases from the oxidation and reduction zones of the QSL reactor join together to give a total gas stream estimated to be 500 Nm^3/t bullion, at 30% SO_2.

Oxidation Stage. Each of the processes begins with oxidation of a substantial part of the sulfur in the concentrate. The conventional process employs air and updraft sintering for this purpose, whereas the new processes employ commercial oxygen with either injection smelting or flash smelting. Most of the heat of oxidation is lost in sintering, but both new processes use this heat for smelting and are able to operate autogenously, or with very minor fuel addition, to yield bullion and lead-rich slag. This is only possible by use of oxygen to replace air, but use of oxygen also results in reduction of gas volume by a factor of 4.4 (Kivcet) to 7.3 (QSL), compared to the conventional process. The greatly reduced gas volumes, and higher-strength sulfur dioxide translate to significant savings in gas-cleaning equipment and acid manufacture.

The QSL oxidation stage operates at much lower temperature (900-950°C) than Kivcet (gas temperature stated to be 1350°C at Potosi), and, since this is the stage where volatile lead sulfide is present, this should result in less fume production for the QSL process. The lower temperature in QSL oxidation should also make possible autogenous operation with less sulfur in the concentrate. At Potosi, the Kivcet operation requires 18% S in the total feed (i.e., concentrate plus recycled fume, etc.) to achieve autogenous smelting. According to the author's calculations the QSL operation is autogenous with 15.7% S in the concentrate (about 13.2% S in the total feed).

The oxidation stage in both the QSL and Kivcet process operates at considerably higher specific-capacity than the sinter-roast operation in conventional smelting. Intensive reactors, that employ intimate mixing of concentrate and oxidant throughout the reactor volume, are responsible for the high specific-capacity of the new processes.

Either flash smelting (Kivcet) or injection smelting (QSL) with oxygen seems to offer tremendous advantages over sinter-roasting for the oxidation stage of lead smelting: tightly-sealed reactors of high specific-capacity, with excellent control of fugitives; greatly reduced gas volume and much higher strength of sulfur dioxide; utilization of the inherent fuel value of the lead concentrate to produce molten phases for the subsequent reduction stage.

Reduction Stage. In the conventional process, the sinter from the oxidation stage is reduced with coke in the blast furnace. The QSL and Kivcet processes produce a lead-rich (40-60% Pb) slag that must be reduced to complete the recovery of lead. Kivcet employs an electric furnace (separated from the oxidation stage by a water-cooled barrier wall), in which the slag is reduced by coke or charcoal. As shown in Table VI, this furnace operates at low specific-capacity, because it lacks any means for intimate mixing of slag and reductant, and consumes a large amount of electric energy. In the author's view the electric furnace represents the weak link in the Kivcet process, and may only be justified in the special case where it is used to produce a zinc-rich vapor that is condensed to metallic zinc.

At Potosi the electric-furnace vapor will be burned in an after-burner to make zinc-oxide fume; for this purpose, the far less costly and less energy-intensive conventional slag-fuming process (38) would seem a better alternative. Slag-fuming injects powdered coal and air (or oxygen) into the slag through tuyeres, instead of using electric energy and coke, and is a highly intensive reactor well suited for reduction of both lead and zinc. Conventional slag-fuming is a batch process, but a single-stage, continuous slag-fuming unit that reduces the slag to 1-2% lead, without significant zinc reduction, would appear to be a feasible alternative to the Kivcet electric furnace. Such a design would also enable the Kivcet process to compete for treatment of high-grade, low-zinc lead concentrate.

Reduction of lead-rich slag in the QSL process operates not unlike the slag-fuming mechanism recommended above. Powdered coal or reducing gas is injected into the slag through tuyeres in the bottom of the reactor, in amounts sufficient to reduce lead, but not zinc. Table VI shows that this is accomplished at five-times the specific capacity of the Kivcet electric-furnace, because of intensive mixing of reductant and slag, and without the excessive electric energy used by the Kivcet process. There can be little doubt that injection of reductant into the slag is the easy way to reduce lead slags.

QSL employs no barrier wall between the oxidation and reduction zones of the reactor, as does Kivcet. Absence of the wall obviates one source of possible maintenance trouble, but it introduces the possiblity that the necessary gradients in slag composition and oxygen potential, between the oxidation and reduction zones, will be disturbed by lateral mixing. Wide spacing of tuyeres and low volume of gas flow through the tuyeres are claimed to minimize mixing in the QSL reactor. Furthermore, the oxidation and reduction zones in the reactor may be separated by a central, quiescent zone that aids establishment of concentration gradients. Proof of the concept should be available from operation of the smelter at Duisberg by the time this manuscript reaches the reader.

A common advantage of the reduction stage for both new processes is reduction in gas volume by a factor of ten or more compared to conventional blast-furnace reduction (see Table VI). The blast furnace produces a large volume of gas because it must heat a cold charge to smelting temperature, employing air-combustion of coke for this purpose. In contrast, the new reduction processes receive an already molten charge; only a small amount of CO_2 and H_2O result from the reduction of PbO by carbon or hydrocarbons, plus a small amount of fuel combustion in the QSL to maintain temperature.

Overall Energy Requirement. Table VI shows estimated values for Process Fuel Equivalent (PFE) that are considerably less for the QSL, and considerably greater for Kivcet, than for conventional smelting. PFE sums the energy content of fuels and electricity used by the process (including production of oxygen, gas cleaning and acid manufacture) and the energy content of fluxes, electrode carbon, etc. (11).

PFE for conventional smelting (10 GJ/ton bullion) is reasonably well established, although the most efficient modern smelter may operate at 9 GJ/ton, whereas many less efficient smelters consume 11 or 12 GJ/ton (4). The major energy item for conventional smelting is coke for the blast furnace (7.3 GJ/ton), but electricity for compression of air, handling and cleaning the huge volumes of waste gas, and acid manufacture contributes 2 GJ/ton to the total.

The QSL process uses 70% less fuel for smelting, compared to the conventional process, because of efficient use of the fuel value in the concentrate. It uses 50% more electric energy than conventional smelting because of the need to produce oxygen. The very low PFE value for QSL in Table VI may be revised somewhat when results of the commercial operation are known, but this author believes that the value will remain significantly lower than for conventional smelting.

The high PFE value for the Kivcet process results from excessive use of electric energy for the electric furnace. Of the total electric energy (1616 kwh/t bullion shown in Table VI) it is estimated that 1250 kwh are used for the electric furnace. Two considerations that help to justify the large PFE are the simultaneous recovery of zinc fume and the unusually low grade of the lead concentrate. Any process (QSL, conventional, or other) would show higher PFE for treating concentrate with 47.4% Pb than for one with 74% Pb, perhaps approximately in the ratio 74/47.4 = 1.56. Moreover, the definition of PFE requires that byproducts produced (such as the oxide fume in the Kivcet process) should show as an energy credit for the process (11), and this has not been done for the values in Table VI. At Potosi, it is expected to produce 0.21 tons of oxide fume (24% Pb, 54% Zn) per ton of bullion from the electric furnace. If this fume is assigned an energy value similar to that of roasted zinc-concentrate (about 12 GJ/ton), a credit of 2.5 GJ/ton of bullion should be subracted from the PFE value in Table VI. If both corrections are applied, a revised PFE for the Kivcet process may be calculated as follows:

$$PFE = \frac{21.7}{1.56} - 2.5 = 11.4 \text{ GJ/ton bullion}$$

Even with these generous allowances, the Kivcet process consumes more energy than conventional smelting and much more than the QSL. The inherent inefficiency in generation of electric energy and the low specific capacity of the Kivcet electric-furnace are responsible for the poor energy-efficiency of this process.

The next few years should prove an exciting time for lead metallurgists, as the details of commercial operation of QSL and Kivcet become known. Neither may prove to be the ultimate design for new lead smelters, but both have features that are major steps in the right direction, and both will provide invaluable lessons for future design engineers.

ZINC SMELTING

Unlike copper and lead smelting, pyrometallurgy stands today as the underdog in zinc extraction. More than 75% of the world's zinc is produced by the roast-leach-electrowinning process and this process is the preferred choice for new plant construction. The initial roasting operation is pyrometallurgical, and highly successful, as discussed above under 'Oxide' Routes to Copper, but the remainder of the process employs hydro-electrometallurgy in a very effective manner to produce high-purity zinc.

The only serious environmental problem facing the roast-leach-electrowinning process is disposal of large volumes of solid wastes (jarosite, goethite, etc.) that are unstable in the natural environment. Further processing of these residues to recover some metal values, and produce an environmentally stable material for discard, may be the ultimate solution, but it will add to the cost of an already costly process. The electrolytic process consumes considerable energy--about 50 GJ/ton zinc (2)--and most of this is electricity which has increased in price dramatically in recent years.

Imperial Smelting Process

A few pyrometallurgical retort processes for zinc remain in production, but most have been abandoned in favor of the electrolytic process over the past two decades. Of the pyrometallurgical processes, only the Imperial Smelting Process (ISP) seems to be holding its own in competition with the electrolytic process. This blast-furnace process recovers lead as well as zinc, and consequently is most advantageous for treatment of high-lead zinc concentrates and a wide variety of lead and zinc-bearing residues from other sources--including residues from the electrolytic zinc process (37). It directly produces the lower grades of zinc (prime western) with about 25% less energy consumption than the electrolytic process; if this zinc must be distilled to produce high-purity zinc (special high grade) the energy consumption increases to about 45 GJ/ton zinc--still 10% less than for the electrolytic process (2).

The major weaknesses of the ISP, that, in this author's view, will prevent it from offering greater competition to the electrolytic process, are environmental in nature. The sinter-roast operation, that prepares the blast-furnace feed, is almost identical to that employed in conventional lead smelting, and suffers from the same difficulties of control of fugitive dust and fume, and handling and cleaning of excessive volumes of process gas. The blast furnace delivers its huge volume of gas (containing about 7% Zn) to the lead splash-condensers which are, at the same time, the essential feature that makes possible a zinc blast-furnace process, and a source of particularly bad working conditions during regularly scheduled maintenance for clean-out of accumulated dross and accretions. The cooling launders that cool the recirculated lead, and from which crude zinc is recovered, are open to the atmosphere and are a source of fugitive lead-fume.

Control of lead fugitives in the ISP can probably be achieved to a level acceptable in most countries by extensive hooding and venting of the major sources, but the cost of accomplishing this will detract from the advantages of the process. It is reasonable to expect that the process will continue for many years to serve the useful function of co-production of zinc and lead from high-lead zinc concentrates and a wide variety of residues and intermediates. The ISP is not likely, however, to displace the electrolytic process for the major role of treating high-grade zinc concentrates.

New Zinc Smelting Processes

The writer has long believed that development of a modern high-intensity smelting process for recovery of zinc is both feasible and overdue. Such a process should be a real competitor for the electrolytic process in production of zinc from high-grade concentrate. For this purpose, it must incorporate the advantages of lower capital investment, lower energy-intensity, and use of less expensive energy forms, compared to the electrolytic process. Like any modern pyrometallurgical process it must minimize waste-gas volumes and control fugitive emissions by virtue of the basic design, rather than by hooding and venting.

Direct Smelting. Direct smelting of zinc-sulfide concentrate, in analogy to direct smelting of copper or lead concentrates, offers the most elegant concept for such a process--a one-step process from concentrate to a gas containing zinc, SO_2, CO, CO_2, and N_2. Yazawa (39) and Davey and Turnbull (40) have shown the thermodynamic conditions necessary to achieve this result. The process parameters discussed below are those calculated by Davey and Turnbull.

The zinc concentrate would be flash smelted at 1327°C with powdered coal and air, with careful control of the O_2/Zn and C/Zn ratios, to yield a large volume of gas (8.5 tons of gas per ton of zinc volatilized) containing more than 95% of the zinc in the feed, along with CO, CO_2, SO_2, H_2, H_2O and a large amount of nitrogen. Gangue oxides, flux and a small amount of zinc oxide would form a discard slag. There seems little doubt that this vaporization stage of the potential process can be achieved in practice. For treatment of high-grade zinc concentrate about 0.38 ton of coal (83.5% carbon on an ash-free basis) would be used per ton of zinc vaporized, and this fuel requirement is far below that of the ISP or any other zinc smelting method.

Condensation of zinc from a lean gas, expected to analyze 6.1% Zn, 6.5% SO_2, 4% H_2O, 1.8% CO and 9.2% CO_2, constitutes the unsolved part of this process concept. Davey and Turnbull (40) have proposed an ISP lead splash-condenser for this purpose, and point out that the gas has about the same zinc content as the ISP blast-furnace gas. The gas from the flash furnace concept, however, is 300° hotter than the ISP gas, considerably more oxidized (i.e., $P_{CO_2}/P_{CO} = 5.1$ for the flash furnace, compared to $P_{CO_2}/P_{CO} \cong 0.5$ for the ISP, means that the oxygen partial-pressure is about 100 times higher in the flash-furnace gas), and contains SO_2 which may or may not result in rapid oxidation or sulfidization of the zinc vapor on cooling.

Further investigations to determine the feasibility of condensation of zinc by shock-cooling a gas of this composition are certainly warrented; the outcome, however, will depend on complex and unpredictable kinetic behavior. Even if the condensation can be achieved with reasonable zinc recovery (85-90%), the heat load on the condenser will be higher than for the ISP because of the higher temperature of the gas. This will require recirculation of more lead coolant than than the already huge amount (~400 tons of lead per ton of zinc condensed) used by the ISP. The condensation problems for direct smelting will not be easy to solve.

On first glance, it would appear possible to reduce the flash-furnace gas volume, and increase the zinc content of the gas, by use of oxygen-enrichment of the air for flash smelting. However, closer examination of the chemistry, and application of equilibrium calculations, show that the diluent effect of N_2 in air is advantageous in this case. Much higher temperatures would be required to vaporize 95+% of the zinc from the concentrate if less diluent nitrogen were present, and this unexpected result follows from the pressure-sensitive equilibria that control zinc vaporization:

$$ZnS(c) + O_2(g) = Zn(g) + SO_2(g)$$

$$ZnO(c) + CO(g) = Zn(g) + CO_2(g)$$

Reduction of total pressure, or dilution with inert gas (N_2), drives both equilibria toward the right--toward greater production of Zn(g). However, this analysis suggests an alternative to that of N_2 dilution--namely, operation at reduced total pressure. Rough calculations of the author indicate that operation near to 0.25 atm. total pressure, and the same temperature (1327°C), should make possible the use of commercial oxygen for flash smelting, with much lower gas volumes (measured at standard temperature and pressure), and much higher percentage of zinc in the vapor. Speculations of this sort, and hard facts on condensation of zinc from gases containing SO_2, are needed for further development of this process concept.

Roast-Smelt Process. The author has favored a modern zinc-smelting process based on intensive smelting of fine zinc-oxide calcine (2). This would require a roasting step prior to smelting, but, as already discussed (see 'Oxide' Routes to Copper) , roasting technology for zinc concentrate is well developed, environmentally practical and a net producer of useful energy. The reduction chemistry would be similar to that in the zinc blast-furnace (ISP), in that burning carbon with oxygen would serve both to supply heat and reducing agent (CO gas).

Unlike the blast-furnace that reacts lump coke and sinter with air, a modern, intensive process (flash smelting or injection smelting) would react finely-divided calcine and carbon with oxygen (95% O_2). In this way five important advantages are obtained, relative to the ISP: the sintering process with its environmental difficulties and poor thermal efficiency is replaced by efficient fluid-bed roasting; the need for expensive lump coke is replaced by a requirement for less-expensive finely-divided carbon (coke breeze, petroleum coke or anthracite); gas volume is reduced and zinc content increased due to the absence of nitrogen, and this facilitates condensation of zinc; the process is freed from the need to co-produce lead (i.e., the ISP requires lead in the feed in order to produce high-quality sinter); the condenser off-gas would have 3 to 4 times the heating value per unit volume of the ISP condenser gas, because of the absence of N_2, and all of it would represent a valuable surplus byproduct (about 60% of the ISP condenser gas--a fuel of low heating-value--must be used for preheating blast air and coke for the blast furnace).

The author has made numerous calculations regarding the potential behavior of the process, a few of which have been published (2). These serve to show the general validity of the concept, but cannot be relied upon to establish the details of fuel and oxygen consumption, best temperature of operation, recovery of zinc, specific capacity, etc. Answers to such questions can only come from mini-plant test programs. Flash smelting, injection smelting into a slag bath, and cyclone smelting* all deserve to be tested to find the best design for carrying out the process. The critical parameters that need to be determined are the temperature and residence time of ZnO particles in the flame (or slag bath) required to achieve satisfactory (>95%) volatilization of zinc, and the carbon and oxygen requirements necessary to achieve this.

* Cyclone smelting (41) is another form of modern, high-intensity smelting process that is under development by several companies.

Condensation of zinc from the smaller volume of zinc-rich gas should present fewer difficulties than condensation of ISP blast-furnace gas. A lead splash-condenser, or possibly even a zinc splash-condenser (N.J. Zinc Co. design (42)) might be employed. Prime western zinc, the direct product of such a process, could be refined to the high grades by distillation.

Preliminary estimates of the process fuel equivalent for this concept indicate that it should be possible to produce high-grade zinc (SHG) for about 35 GJ/ton, compared to about 45 GJ/ton for the ISP and 50 GJ/ton for the electrolytic process. Environmental control in the roasting step is known to be good; it should be equally good in a well-sealed intensive flash or injection-smelting unit; if a lead splash-condenser must be used, we can expect some environmental problems similar to that for the ISP condenser, but reduced in magnitude because of the smaller gas volume to be condensed. A possible zinc splash-condenser would avoid recirculating large tonnages of lead, but might be found to have its own environmental and maintenance problems.

Rumors surface, from time to time, that a process of this nature is being developed, but no announcement has yet appeared. Delay in a development that must have been conceptualized by many metallurgical designers over the past decade may result from lack of incentive, due to the wide acceptance and satisfactory performance of the electrolytic process, and the generally poor market for zinc.

CONCLUSION

Many interesting and important features of both new and old pyrometallurgical processes have been omitted from this review in order that attention could be directed to the major theme--the influence of environmental control, energy consumption and specific capacity on process design. New developments in smelting tin, nickel, titanium, magnesium, and in metal refining processes have been omitted for the same reason.

I hope to have made a case for the urgent need of innovative design in pyrometallurgy. The inherent characteristics of high-temperature processing yield important advantages for pyrometallurgy, but also entail production of process gas streams and fugitive gas and fume. These latter features of pyrometallurgy have only recently been recognized as serious environmental disadvantages for conventional practice, designed for the requirements of an earlier era.

Innovative design can reduce process gas volumes and fugitive emissions to manageable proportions. Happily, the very design features that accomplish environmental control tend also to increase specific capacity and reduce energy consumption relative to conventional pyrometallurgy. Modern pyrometallurgy should be less costly and less energy-intensive than before, at the same time that environmental control is greatly improved.

The design features that lead to these improved performance characteristics are:

§ Maximum use of commercial purity oxygen to replace air for oxidation of fuel, or iron and sulfur in the concentrate.

§ Use of intensive reactor design--flash smelting, injection smelting, cyclone smelting.

§ Full utilization of the fuel value of sulfide concentrates to supply energy for the process.

§ Design of continuous processes to replace batch processes.

One final comment regarding the future: Development of new pyrometallurgical processes entails costs, experienced personnel and support from management beyond the capabilities of most non-ferrous metal companies. Unfortunately, this seems to be particularly true of U.S. companies, who have followed, rather than led, in process development over the past several decades. To establish this point, one need only recapitulate the source of the developments cited in the text: INCO oxygen flash smelting (Canada), Outokumpu flash smelting (Finland), Noranda process (Canada), Mitsubishi process (Japan), Kivcet process (USSR), Q-S and QSL processes (conceived in the U.S., but brought to realization in West Germany), Imperial Smelting Process (England).

U.S. companies, in particular, must recognize the urgent need to constantly seek improved processes, and develop a mechanism to finance long-range process research. If this is not feasible for individual companies then legal methods for cooperative research should be sought. U.S. industry has been dormant in process research for much too long.

REFERENCES

1. C.H. Pitt and M.E. Wadsworth, "An Assessment of Energy Requirements in Proven and New Copper Processes," final report, Dec. 31, 1980, to U.S. Dept. Energy, Contract No. EM-78-S-07-1743, Univ. of Utah, Salt Lake City, Utah.

2. H.H. Kellogg, "Energy Use in Zinc Extraction," p. 28-47 in Lead-Zinc-Tin '80, J.M. Cigan, T.S. Mackey, T.J. O'Keefe, ed.; TMS (AIME), Warrendale, Pa., 1979.

3. M.P. Lugagne, "The Magnetherm Process for the Production of Magnesium," Erzmetall, 31 (7/8) (1978) p. 310-313.

4. "Energy Use Patterns in Metallurgical and Nonmetallic Mineral Processing," reports to U.S. Bur. Mines by Batelle-Columbus Laboratories, Columbus, Ohio, "High Priority Commodities," NTIS PB 245 759/AS, 1975, and "Intermediate Priority,Commodities," NTIS PB 246 357/AS, 1975.

5. H.H. Kellogg, "Conservation and Metallurgical Process Design," Trans. Inst. Min. Met. (London), 86 (1977) p. C47-C57.

6. H.H. Kellogg and J.M. Henderson, "Energy Use in Sulfide Smelting of Copper," p. 373-415 in Extractive Metallurgy of Copper, J.C. Yannopoulos and J.C. Agarwal, ed.; TMS (AIME), New York, N.Y., 1976.

7. S. Harkki, O. Aaltonen, and T. Tuominen, "High Grade Matte Production with Oxygen Enrichment by the Outokumpu Flash Smelting Method," p. 488-507 in ref. 6.

8. S. Merla, C.E. Young, and J.W. Matousek, "Recent Developments in the Inco Oxygen Flash Smelting Process," p. 19-35 in Environmental Control, C. Rampacek, ed.; TMS (AIME), New York, N.Y., 1972.

9. L.A. Mills, G.D. Hallett, and C.J. Newman, "Design and Operation of the Noranda Process Continuous Smelter," p. 458-487 in ref. 6.

10. T. Nagano and T. Suzuki, "Commercial Operation of Mitsubishi Continuous Copper Smelting and Converting Process," p. 439-457 in ref. 6.

11. H.H. Kellogg, "Energy Efficiency in the Age of Scarcity," Jour. of Metals, 26 (6) (1974) p. 25-29.

12. S.N. Sharma, K.M. Ogilvie, R.A. Jimenez, and A.H. Hanssen, "Control of Secondary Emissions from Copper Converters," p. 312-335 in Copper and Nickel Converters, R.E. Johnson, ed.; TMS (AIME), New York, N.Y., 1979.

13. R.E. Johnson, N.J. Themelis, and G.A. Eltringham, "A Survey of Worldwide Copper Converter Practices," p. 1-32 in ref. 12.

14. M. Goto and K. Kanamori, "Converting Furnace Operation of Mitsubishi Process," p. 210-224 in ref. 12.

15. J.D. Gomez, "Paipote Smelter: Seven Years Operating Hoboken Converters," p. 291-311 in ref. 12.

16. W.J. Thoburn and P.M. Tyroler, "Optimization of TBRC Operation and Control at INCO's Copper Cliff Nickel Refinery," p. 274-290 in ref. 12.

17. A. Yazawa and K. Tozawa, "Oxygen Smelting of Copper Concentrate with Exhaust SO_2 Gas Recycling," paper presented at AIME Annual Meeting, Chicago, Ill., Feb. 1981. TMS Paper Selection A81-26.

18. G.S. Victorovich, C. Diaz, and J. Raskauskas, "Impurity Distribution Dusting and Control of Matte Grade in INCO Oxygen Flash Smelting," paper presented at 15th Anniversary Meeting of Chilean Inst. of Min. Eng., Santiago, Chile, Nov. 1980.

19. B. Andersson, Y. Anjala, and T. Mantymaki, "Development Trends in Outokumpu Flash Smelting Technology with Particular Reference to the Requirements on Furnace Reaction Shaft Cooling Techniques and Refractories," paper presented at AIME Annual Meeting, Las Vegas, Nev., Feb. 1980. TMS Paper Selection A80-26.

20. R.D. Pehlke, "Steelmaking--The Jet Age," Metall Trans, 11B(4) (1980) p. 539-562.

21. W.H. Schwartz, P. Fischer, P.E. Queneau, and R. Schuhmann, Jr., "QSL--A Continuous Process for Environmentally Clean Lead Production," p. 394-406 in ref. 2.

22. A Yazawa, "Thermodynamic Consideration of Copper Smelting," Can. Met. Quart. 13 (1968) p. 443-453.

23. H.K. Worner and B.S. Andrews, "Continuous Copper Smelting-Converting by the WORCRA Process," p. 133-146 in The Future of Copper Pyrometallurgy, C. Diaz, ed.; Chilean Inst. Min. Eng., Santiago, Chile, 1974.

24. N.J. Themelis, G.C. McKerrow, P. Tarassoff and G.D. Hallett, "The Noranda Process," Jour. of Metals 24(4) (1972) p. 25-32.

25. P.E. Queneau and R. Schuhmann, Jr., "The Q-S Oxygen Process," Jour. of Metals, 26(8) (1974) p. 14-17.

26. P.E. Queneau, "Coppermaking in the Eighties--Productivity in Metal Extraction from Sulfide Concentrates," Jour. of Metals, 33(2) (1981) p. 38-46.

27. A. Berg and F. Pape, "Recent Practice in the Fluid Bed Roasting of Zinc Concentrates," Erzmetall, 31(5) (1978) p. 224-226.

28. S.W. Marcuson and H.H. Kellogg, "Studies on the Segregation of Roasted Copper Concentrates," Trans. Inst. Min. Metall. (London), 86 (1977) p. C195-C201. See also: A. del Campo, "Influence of the Reducing Agent in the Segregation of Roasted Copper Concentrates," Eng.Sci.D. dissertation, Columbia Univ. N.Y., N.Y., 1980.

29. S.W. Marcuson, "Application of the Segregation Process to Roasted Copper Concentrates," Minerals Sci. Engng., 12(1) (1980) p. 21-36.

30. H.H. Kellogg, "Prospects for the Pyrometallurgy of Copper," p. 13-29 in ref. 23.

31. W.R. Opie, H.P. Rajcevic, and E.R. Querijero, "Dead Roasting and Blast-Furnace Smelting of Chalcopyrite Concentrate," Jour. of Metals, 31(7) (1979) p. 17-22.

32. P. Kettner, C.A. Maelzer, and W.H. Schwartz, "The Brixlegg Electro-Smelting Process Applied to Copper Concentrates," paper presented at AIME Annual Meeting, San Francisco, Calif., Feb. 1972. TMS Paper Selection A72-48.

33. H.H. Kellogg, "Vaporization Chemistry in Extractive Metallurgy," Trans. Met. Soc. AIME, 236(5) (1966) p. 602-615.

34. H.H. Kellogg and S.K. Basu, "Thermodynamic Properties of the System Pb-S-O to 1100°K," Trans. Met. Soc. AIME, 218(2) (1960) p. 70-81.

35. T.R.A. Davey, "Developments in Direct Smelting of Base Metals--A Critical Review," paper presented at Int. Symposium on Chemical Metallurgy, Bombay, India, Jan. 1979. See also: G.M. Willis, "The Physical Chemistry of Lead Extraction," p. 457-476 in ref. 2.

36. K.B. Chaudhuri, M. Koch, and J.L. Patino, "The Technical-Scale Realization of the Kivcet Process for Lead," Can. Inst. Met. Bull. 73(817) (1980) p. 146-150.

37. C.F. Harris, A.W. Richards, and A.W. Robson, "The Zinc-Lead Blast Furnace Into the 1980s," p. 147-260 in ref. 2. See also: P. Mead, "Treatment of a Diverse Range of Low Grade Feed Materials in the ISF," p. 278-289 in ref. 2.

38. G.A. Yurko, "Slag Fuming Process at the Cominco Smelter," p. 330-347 in Lead and Zinc, Vol. II, C.H. Cotterill and J.M. Cigan, ed.; AIME, New York, N.Y., 1970.

39. A. Yazawa, "Thermodynamic Evaluations of Extractive Metallurgical Processes," Metall. Trans. 10B (1979) p. 307-321.

40. T.R.A. Davey and A.G. Turnbull, "The Direct Smelting of Zinc Sulfide Concentrate," p. 23-29 in Australia Japan Metallurgy Symposium 1980, Australasian Inst. Min. Met., Melbourne, 1980.

41. H. Hilbrans and F. Sauert, "The Humboldt Cyclone Smelting and Top Blowing Process," paper presented at Annual Meeting, AIME, Chicago, Ill., Feb. 1981.

42. E.H. Bunce and W.M. Peirce, "New Jersey Zinc Develops a New Condenser," Eng. Min. Jour., 150(3) (1949), p. 56-62.

ADVANCES IN THE SMELTING OF ALUMINUM

Noel Jarrett
W. B. Frank
Rudolf Keller

Aluminum Company of America
Alcoa Laboratories, Alcoa Center, PA 15069

A better understanding of the fundamentals of the electrochemical reduction of alumina, applications of computer technology, and advances in materials science will insure that the Bayer-Hall-Héroult Process remains the major aluminum production method during the 1980's. The technology of the Bayer Process to extract pure alumina from bauxite, the manufacture of prebaked carbon anodes, and theoretical aspects of electrowinning aluminum are summarized. Physicochemical properties of specific electrolytes influence the efficiency of the Hall-Héroult Process. The design of modern electrolysis cells places emphasis on energy efficiency and productivity. Metal recycling and purification of metal before ingot casting are discussed.

Further improvements in conventional aluminum smelting can be expected in anode manufacture, automated cell control, lining life, insulation, heat recovery, and current efficiency. The advantages that can be achieved from the development of dimensionally stable electrodes are discussed. Alternate processes for aluminum production are considered.

A profitable future is predicted for the aluminum industry. Industrialized nations will expand to meet home markets and third world needs will be met by expansion where both bauxite and electrical energy resources are found.

Introduction

Aluminum (Al) comprises 8% of the earth's crust and is its most abundant structural element. It is present largely as aluminum oxide or other combined oxides in a wide variety of minerals. Bauxite consists of several hydrous aluminum oxide phases containing 40-60% alumina (Al_2O_3) and is the most important aluminous ore. In high quality bauxite deposits, most abundant in Australia, Brazil, Guinea, and Jamaica, tropical leaching produced high Al_2O_3 concentration. The aluminous phases of gibbsite ($Al[OH]_3$), boehmite (AlOOH), and diaspore ($Al_2O_3 \cdot H_2O$) are associated with small amounts of silicon (Si), iron (Fe), and titanium (Ti) compounds along with other trace impurities.

In 1888 a German patent was issued to Karl Joseph Bayer for an improved process for making pure alumina from bauxite, and since that time bauxite has been the preferred ore for aluminum production. Adequate supplies of affordable bauxite assure continued economic preference for this material. Since 1966, proven reserves of bauxite containing 40-60% Al_2O_3 have tripled from $6x10^9$ to $17x10^9$ tons. During that period 20 times more bauxite was found than used, resulting in a 300-yr supply at 1978 consumption rates, and a 1000-yr supply if one simply drops to the next level of alumina bearing materials having 35-40% Al_2O_3 (1).

The first recorded success in producing aluminum was by Sir Humphry Davy in 1808, when he made a material he called "alumium", which was probably an Fe-Al alloy. From this point, in three successive stages, the present commercial process was developed. In 1827, Wöhler produced aluminum by reducing aluminum chloride ($AlCl_3$) with potassium (K). Sainte-Claire Deville, in 1854, showed that cheaper sodium (Na) could also be used, and the first commercial plant producing small quantities of aluminum was begun in 1855. Since potassium and sodium were produced electrolytically, the process was quite expensive. In 1886, following the development of large scale equipment for generating electric power, Heroult in France, and Hall in the United States, independently developed a process for the direct electrolytic decomposition of Al_2O_3. They discovered that when an electric current is passed through molten cryolite (Na_3AlF_6) containing dissolved Al_2O_3 at 980-1000°C, molten aluminum is deposited at the cathode and carbon dioxide (CO_2) is liberated at the carbon anode. This discovery, coupled with the Bayer Process resulted in the Bayer-Hall-Héroult Process for the prodeduction of aluminum. With minor modifications this process is still used worldwide for the production of aluminum.

The two types of smelting cells in use today, Soderbergs and "prebakes," differ only in their anode design. Soderbergs use a single continuous anode baked in place by heat from the cell; "prebaked cells" use multiple prebaked carbon anode blocks. Figure 1 depicts the Bayer-Hall-Héroult Process using prebaked anodes.

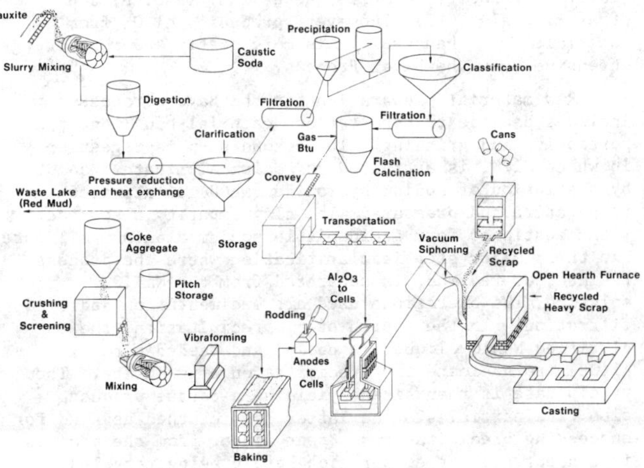

Figure 1. Flowsheet of Bayer-Hall-Héroult Process

As shown in the diagram, there are four steps in the production of aluminum ingot: (a) extraction of pure Al_2O_3 from bauxite, (b) production of carbon anodes from petroleum coke and coal tar pitch, (c) use of the Al_2O_3 and carbon anodes in electrolytic cells to produce molten aluminum, and (d) mixing of recycle scrap with cell metal, purification, alloying, and casting. There are several steps for caustic digestion of crushed bauxite at high temperature and pressure followed by separation to produce pure anhydrous Al_2O_3. Carbon anodes are made by formulating and mixing sized petroleum coke and coal tar pitch, forming green anodes, baking, and making poured cast iron electrical connections to the anode socket. The third step is carried out in smelting cells where Al_2O_3 is added to Na_3AlF_6, modified by the addition of aluminum fluoride (AlF_3), calcium fluoride (CaF_2), and other additives. Electrolysis at 950-970°C yields molten aluminum and carbon oxides (CO_2, CO). The carbon oxides, diluted with ~100 parts of air and containing fluoride values, are passed through wet or dry scrubbing systems where the fluorides are collected and returned to the cell. Carbon oxides are discharged to the atmosphere. Molten aluminum is tapped from the cell and transported to open hearth holding furnaces where recycled aluminum scrap is melted and alloyed. Purification systems remove oxides, unwanted trace elements, and hydrogen prior to ingot casting for fabrication.

Bayer Process

The Bayer Process was first developed industrially in France in the early 1890's and remains the dominant Al_2O_3 production method, with bauxites virtually the only commercial ores, Figure 2 (2-5).

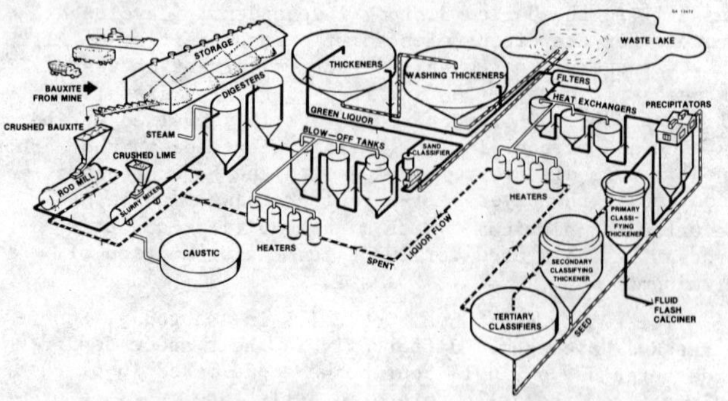

Figure 2. Bayer Process Flowsheet

Technology has been developed to produce Al_2O_3 from clays, anorthosite, alunite, nepheline, leucite, and other minerals (6-7). However, producing Al_2O_3 from materials other than bauxite is more costly and more energy-intensive than the Bayer Process.

Raw material preparation for the Bayer Process includes bauxite transportation, material handling, proportioning and grinding. The second step is digestion in which Al_2O_3 is extracted as sodium aluminate ($NaAlO_2$) by a solution of sodium hydroxide (NaOH) at elevated temperature and pressure. The clay impurity also reacts with caustic to form an insoluble sodium aluminum silicate. The third major step is clarification where the process residue, or red mud, is separated from the $NaAlO_2$ solution. The operations used are sedimentation and filtration. In the fourth step, precipitation, the clarified $NaAlO_2$ liquor is cooled and seeded with $Al(OH)_3$ to nucleate the precipitation of gibbsite. The precipitate is then sized, yielding a coarse product plus fine particles. The latter are returned as seed for succeeding precipitations. Spent liquor from the process is concentrated by evaporation before being recycled. After washing, the hydrate is calcined to essentially anhydrous alumina.

Product Properties and Specifications. Every operation of the Bayer Process is designed and controlled to produce a consistent sandy Al_2O_3 that meets stringent composition and physical specifications. Chemical purity specifications for metallurgical Al_2O_3 on material heated to 300°C are:

SiO_2	0.03% max
Fe_2O_3	0.03% max
TiO_2	0.005% max
MnO	0.001% max
Na_2O	0.35-0.55%
CaO	0.03% max
P_2O_5	0.003% max
Al_2O_3	98.35% min

Low levels of silica and iron oxides are essential to produce metal of the required purity. However, even very low levels of phosphorus compounds in the electrolyte lower smelting cell current efficiency. A controlled content of soda in the feed Al_2O_3 assists in control of the NaF/AlF_3 ratio of the electrolyte. Unavoidable calcium oxide (CaO) impurity in Al_2O_3 is converted to CaF_2 in the electrolyte.

A number of physical characteristics of Bayer Al_2O_3 must be controlled to assure proper handling and cell feeding, satisfactory performance in dry fluoride scrubbing systems, and an adequate rate of dissolution in the electrolytic bath (8). A median particle size of 60-100 microns with a relatively narrow size distribution is desirable. For ease of solution in cell electrolyte the proportion of α-Al_2O_3 should be less than 20%. Other specifications include:

Loss on Ignition (from 300-1200°C)	0.5-0.9%
Sorption at 44% Relative Humidity	2.75-3.75%
+100 Mesh (148 microns)	10% max
+325 Mesh (44 microns)	92% min
Surface area	$\geq$45 m^2/g min

Ore Preparation. Bauxite received from the mines is crushed in two stages to less than 25 mm pieces. The product from the secondary crusher (usually a hammermill) is stored using a blending bridge to obtain uniform feed to the process.

From storage, bauxite is metered into wet grinding rod mills where spent liquor from final product classification is used as a wet grinding medium. Lime is generally added to the process at this stage to assist in extraction of Al_2O_3, scavenge impurities, and enhance clarifiation. At a controlled percent solids, the bauxite slurry flows from the rod mills to agitated storage tanks. Storage for 20-30 hours permits preliminary desilication. From the storage tanks the slurry is metered directly to the digesters.

Digestion. The goal of digestion is to dissolve as much Al_2O_3 and as little impurity as possible. The concentration is determined by the amount that can be held in solution throughout the balance of processing until precipitation. The caustic in the spent liquor reacts with the hydrous Al_2O_3 in the bauxite to produce sodium aluminate.

$$AlO_nH_{(2n-3)} + NaOH \rightarrow NaAlO_2 + (n-1)H_2O \qquad [1]$$

The solubility of $Al(OH)_3$ is such that the desired concentration can be achieved easily at temperatures of about 145°C using a coarse bauxite grind ($\leq$1600 microns). To dissolve AlOOH to the desired concentration requires a digestion temperature of 230°C or higher. If AlOOH in the bauxite is not completely dissolved, it serves as seed for back precipitation in subsequent process steps. Thus, Al_2O_3 losses result not only from AlOOH insolubility, but also from back precipitation on the AlOOH during subsequent processing. In some Mediterranean and Chinese bauxites, Al_2O_3 may be present as diaspore ($Al_2O_3.H_2O$). Diaspore has the same chemical composition as boehmite, but it is denser and requires even more severe conditions of digestion. Among these conditions are finer grinding, higher caustic concentrations, longer digest periods, and a digest temperature of 260°C or higher.

In the traditional American Bayer digestion system, spent liquor is heated through a series of tubular heat exchangers and then brought into contact with the bauxite slurry and live steam in a contact heater. Slurry from the contact heater at 143°C passes through the series of digestion vessels to attain the required ~30 minutes holding time, needed to accomplish acceptable desilication of the process liquor. Desilication is necessary for Al_2O_3 with low levels of silica.

Most of the reactive silica in bauxite is present as kaolinite clay, which reacts with caustic to form sodium silicate and sodium aluminate. These two compounds react at digestion conditions to produce a desilication product having the approximate composition $3Na_2O \cdot 3Al_2O_3 \cdot 5SiO_2 \cdot 2H_2O$. Thus, each unit of combined silica in bauxite reduces the yield of Al_2O_3 as well as consuming a significant amount of soda.

At the end of digestion, the liquor stream is too hot for subsequent processing, so it is flash evaporated through a number of steps, driving off steam and reducing the slurry temperature. The steam is recycled for liquor heating to conserve energy. This is the continuous regenerative heat recovery system. The liquor is then heated to 150°C in tubular heaters using live steam from the power house. The same principles are applied in high-temperature digestion. At a digest temperature of 230°C, heat to be recovered is more than doubled, requiring more steps.

Bauxite slurry from the preliminary desilication tanks is pumped to digesters where it mixes with the hot liquor. Final heating to digestion temperature takes place in a contact heater where the slurry flows downward countercurrently through the hot steam. The contact is intimate so the heat is transferred in a relatively small volume. The internal heater configuration and large clearances make it relatively insensitive to scale accumulation.

Problems have been encountered in designing a digestion system for diasporic bauxites. At the higher required temperatures 260°C, pressures are increased considerably. The use of high temperatures with higher concentrations of caustic can cause embrittlement of the vessels. Furthermore, the amount of heat to be recovered has been increased so the number of heat exchange stages must be increased.

The Al_2O_3 concentration in solution is limited by the lowest temperature at which all the Al_2O_3 can be kept in solution for an appreciable time. This occurs in clarification where the temperature drops to ~105°C. Any Al_2O_3 that precipitates during sedimentation of the red mud is lost with the residue. If aluminum trihydroxide starts to precipitate during filtration it is removed with the red mud, overloading (blinding) the filters and stopping the flow. In spite of this, the liquors normally used in Bayer plants are saturated with $Al(OH)_3$ at ~130°C. The inherent stability of supersaturated $NaAlO_2$ solutions is exploited in clarification by using a degree of supersaturation that has been empirically established.

If the caustic concentration of the solution is low, the yield of Al_2O_3 per unit volume is decreased. This results in increased cost for equipment, steam, power, labor, and other variables. At high caustic concentrations the Al_2O_3 is generally fine, giving rise to problems in handling and further processing. Actual plant conditions are set by interplay of these factors.

Clarification. After the solution of dissolved Al_2O_3 has been flashed to atmospheric pressure, clarification can proceed. Generally, a relatively small proportion of the residue in the blow-off slurry from digestion is coarse enough to be called sand (≥149 μm), and this is removed in a sand trap--a slender, conical bottomed tank into which the liquor and residue are circulated. The sand is allowed to settle to the bottom and is withdrawn continuously as a 20-35% solids slurry. The bulk of the residue requires additional clarification and remains with the liquor that overflows the sand trap.

The sand slurry is pumped into classifiers where it is separated from entrained mud and then washed to remove residual liquor before the mud is pumped to the lake. The reason for removing this coarse sand is to lighten the load on the main clarification equipment and to prevent rapid settling of coarse material in locations where the velocity is lower.

Blow-off slurry is usually clarified in three steps. First, the coarse sand fraction is removed by sand traps. Second, the bulk of the red mud residue is settled out of the liquor in large diameter, low-profile mud thickener tanks. Starch solution or other flocculants are added to the sand trap overflow to flocculate the fine red mud particles. These particles are usually quite small; for some bauxites the median particle diameter is about two microns. Without flocculation, mud separation in 30-45 m diameter thickeners could not be achieved. In this operation the mud content of the liquor is usually decreased to less than 0.3 g/l. The mud that settles to the bottom of the tank becomes compressed into a thick slurry, from 20-30% solids, and is raked to the center of the tank, from which it is discharged. Third, the thickener overflow liquor is pumped to a filter to remove final traces of mud.

At many plants, polish filtration is accomplished by Kelly filter presses. In these filters the pressure in the vessel forces the liquor through filter cloths, which retain the solids. These are semicontinuous units in which filtration is carried out for a set period after which the filter is drained and the mud is washed from the cloth.

The fine mud from the bauxites tends to pack tightly, reducing the filtration rate, but the rate is often improved by precoating the filter cloths with materials that are effective in preventing the tight packing of these fine particles. Filtration is aided by continuously adding small quantities of materials such as calcium aluminate (formed by the reaction of hydrated lime with alumina in plant liquor) to the filter.

However, some bauxites produce mud particles of submicron size, which often penetrate the fibers of the filter cloth, effectively closing the filtration area. Once a filter cloth has been blocked it must be removed and replaced. To avoid this condition, which greatly increases the cost of clarification, the sand filters were adapted to clarify red mud slurries. The slurry is filtered through a bed of closely sized sand that entraps and retains the fine mud particles. The bulk of the mud removed from the liquor stream by the thickeners and discharged as a 20-30% solids slurry must be washed to remove $NaAlO_2$ liquor. This is accomplished by passing the mud through a series of washing tanks countercurrent to the flow of wash water. Most of the $NaAlO_2$ liquor associated with the mud is recovered in this countercurrent decantation system by washing with water from a mud lake. The overflow from the first stage of the mud washing, a partially diluted $NaAlO_2$ solution, is added to the blow-off slurry to recover valuable constituents.

Underflow from the final stage of mud washing is slurried in raw water and miscellaneous wash water from the plant condensate, and then pumped to a mud lake. The mud lake acts as an additional washing stage. The mud in the slurry pumped to the lake settles to ~60-70% solids, leaving a clear supernatant liquor containing from 10-20 g/L soda. Much of the soda and Al_2O_3 in the supernatant liquor is recovered as the solution is returned to the plant.

Precipitation. Precipitation can be represented by the equation

$$NaAlO_2 + 2H_2O \xrightarrow{\text{Seed}} NaOH + \underset{\text{gibbsite}}{Al(OH)_3} \qquad [2]$$

The clarified, supersaturated $NaAlO_2$ solution from filtration, known as green (or pregnant) liquor, is cooled to the precipitation temperature and fed to the precipitation system.

The basic requirements for precipitation are: aluminum trihydroxide with the proper particle size and strength; and the maximum yield of Al_2O_3 from the liquor stream while maintaining an economically viable seed balance. Changes in a number of operating parameters in precipitation are used to achieve these goals. However, the primary control tools are changes in the liquor temperature and in the particle size and surface area of the seed charged with the green liquor.

Cascade precipitation permits automation. Green liquor with seed slurry is continuously pumped through a series of precipitators. Slurry from the first precipitation tank overflows through a trough, cascading into the second tank and continues through the series. Control of the seed content of each precipitator is maintained by transporting a portion of the slurry through a transfer pipe into the trough leading to the next precipitator. Since the slurry contains a high concentration of solids, control of them throughout the system can be maintained by selecting the amount to be transferred from one unit to another. This has reduced labor cost, increased yield from precipitation, and at the same time produced a stronger, more nearly spherical particle. After the liquor and seed slurry have passed through the precipitators for a holding time of 30-50 hours, the whole slurry of spent liquor and precipitated gibbsite is pumped to a classification system.

Classification systems consist of three stages. First, primary thickeners provide hydraulic elutriation so that only the coarsest particles settle. The remaining hydrate, with the spent liquor, overflows the primary thickeners into secondary thickeners where a large area is provided so that the medium particle size fraction of seed settles. The very fine material overflows with the liquor to tray thickeners where final clarification takes place and the very fine seed settles. The clear liquor then overflows, ready for recycling to the digesters. The seed, collected in secondary and tray thickeners, is recycled to precipitation where fine particles eventually grow to product size.

The product underflows the primary thickener into a countercurrent washing system. The stages in this system are: wash tank, high tank, and hydrate filter. The hydrate is moved countercurrently to the wash water introduced on the filters. Overflow from the first wash tank joins the rest of the spent liquor going to evaporation.

Effective washing is essential to control the soluble soda content of the product to the desired level.

Evaporation. The washing of sand and mud residues recovers $NaAlO_2$ values and involves the addition of substantial amounts of water to the system. Water is also introduced by way of gland water for pump packings, instrument purge water, etc. While some of the water is removed by flashing in digesters and flash cooling of pregnant liquor, the rest of the water must be removed by an evaporator system to obtain the target soda concentration required for digestion and to maintain liquor volume balance. The location and type of evaporators vary from plant to plant. In most locations the spent liquor is concentrated with flash-type evaporators, utilizing recycled heat.

Calcination. Following washing, the aluminum trihydroxide is discharged at a free moisture content of ~8-10%. This moisture, as well as the chemically bound water, must be driven off to provide anhydrous Al_2O_3 for smelting. This is accomplished by calcination at temperatures of 950-1100°C At these temperatures the hydroxide is converted to a mixture of α and γ-Al_2O_3.

$$2Al(OH)_3 \rightarrow Al_2O_3 + 3H_2O \qquad [3]$$

The fluid flash calcination process (8) is represented schematically in Figure 3.

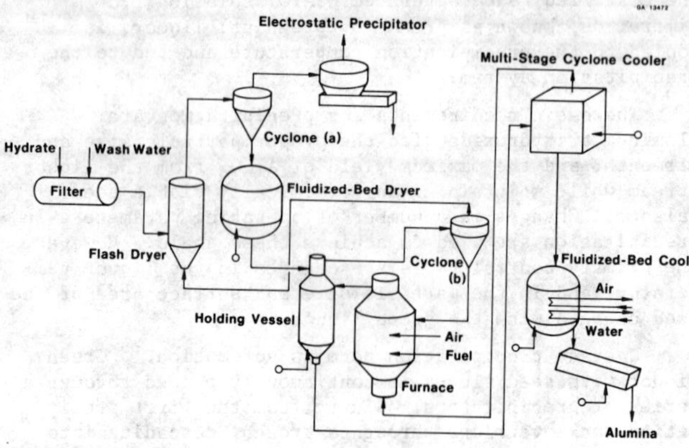

Figure 3. Alcoa Fluid Flash Calciner

Hydrate $Al(OH)_3$ is passed countercurrent to the flow of a hot gas stream. For example, the moist cake from the hydrate Filter is dried as it is conveyed to the Fluidized Bed Dryer through Cyclone a by the still warm gases exiting the Flash Dryer to which it was originally added. The dried hydrate is partially calcined by hot gases from the Holding Vessel as it is transported in a fluid stream from the Fluidized Bed Dryer to Cyclone b through the disengaging section of the Holding Vessel where the solids are separated from the gases that are returned to the Flash Dryer. The partially calcined solids drop from Cyclone b into the Furnace where calcination is completed. The calcined Al_2O_3 from this unit is blown over to the Holding Vessel where it is held long enough at the temperature necessary to achieve the desired degree of calcination. Calcined Al_2O_3 from the Holding Vessel passes through a multistage Cyclone Cooler and a Fluidized Bed Cooler where it heats the incoming air used for Furnace fluidization and combustion. Moist gas leaves the system through the Electrostatic Precipitator, removing entrained dried hydrate fines, after giving up its remaining heat in the Flash Dryer and Cyclone a.

This system of heat recuperation and regeneration through successive countercurrent contacting of the gas and solids streams at the temperature level useful at each processing step has resulted in highly efficient use of the gas energy entering the Furnace. The heat required to calcine Al_2O_3 using fluid flash calcination is only 3 million Btu per short ton.

The average energy required to produce Al_2O_3 by the Bayer Process in the United States is about 17 million Btu per short ton (10). With all internal energy recovery systems properly designed and installed, energy consumption of 11-12 million Btu/ton of Al_2O_3 is being achieved with gibbsitic bauxites. The Bayer Process, modified and improved through the years, remains the best method to produce metallurgical Al_2O_3 from a variety of bauxites.

For diasporic bauxites having high reactive silica contents, a combination Bayer-sinter process would probably be more economical. However, this would drive up both installation and operating costs (particularly energy) for Al_2O_3 production.

Carbon Anode Manufacture

Two types of commercial reduction cells are used--"prebake" and Soderberg. They differ in the manufacture and connection of the carbon anode.

Prebaked anodes are formed by blending sized petroleum coke aggregate and coal tar pitch, molding this into blocks (with preformed electrical connection sockets) by pressing or vibration, and firing to ~1100°C in oil or gas-fired ring furnace pits. A typical block is 70 cm wide x 125 cm long x 50 cm high. Petroleum coke and coal tar pitch are used because of their high purity. The more noble impurities (Fe, Si) deposit on the aluminum while less noble impurities (Ca, Mg) accumulate as flurides in the bath. Electrical contact is made through aluminum or copper rods welded or bolted to steel stubs fixed in the anode socket by poured cast iron.

The electrical resistivity of prebaked anodes ranges from 0.005-0.006 ohm-cm. Anode current density ranges from 0.6-1.3 A/cm^2.

Soderberg anodes are baked by waste heat from the cell. A steel casing or mold 6-8 M long x 2 M wide x 1 M high containing the aggregate-pitch mixture at varying stages of baking, is mounted over the cell. The carbon mass is made to slip through the casing at the rate of oxidation in the electrolyte. In the vertical spike version electrical contact is made by steel-aluminum spikes, each pulled (by special tool) and set to a higher level as its tip approaches the electrolysis.

Soderberg anode resistivity is about 30% higher than that of prebaked anodes. Soderberg cells save capital, labor, and the energy required to manufacture the prebaked anodes, but they are being phased out because of lower power efficiency and the greater difficulty in collecting and disposing of baking fumes.

Anode quality affects both the energy efficiency and productivity of smelting cells. About 10% of cell power is used to overcome the electrical resistance of the anodes. Anode carbon that enters the cell electrolyte as dust is unavailable for electrolysis and adversely affects the current efficiency of the cell. Dusting is caused by selective oxidation of the binder coke at the anode-bath interface, which releases aggregate particles into the electrolyte. Anode electrical conductivity and dusting are a function of coke characteristics, aggregate sizing, pitch content, mixing technique, forming, and baking. Other aspects of anode quality that affect productivity are resistance to cracking and burnoffs. Of the total energy consumed in aluminum smelting, 12% is spent in ring furnace baking of the anodes. Useful anode life is about three weeks, at which time the butt is removed and crushed, sized, and recycled into the new anode aggregate. Coke and pitch specifications are:

Table I

Specification for Coke Oven Pitch*

Softening Point	108-113°C
Xylene Insoluble	25-35%
Quinoline Insoluble	10-15%
Coking Value (Conradson Carbon)	54% Minimum
Distillation 0-355°C	4% Maximum
Ash	0.4% Maximum
Specific Gravity	1.30 minimum

Table II

Specifications for Calcined Petroleum Coke*

Sizing, cm	2.54	maximum
Bulk Density, g/cm^3	0.80	minimum**
Real Density, g/cm^3	2.03	minimum
Moisture, wt%	0.5	maximum
Volatile Matter, wt%	0.5	maximum
Ash, wt%	0.5	maximum
Silicon, wt%	0.02	maximum
Iron, wt%	0.03	maximum
Vanadium, wt%	0.045	maximum
Nickel, wt%	0.04	maximum
Sulfur, wt%	3.0	maximum

*Values indicated are those determined from representative Alcoa samples taken from individual shipping increments (i.e., individual railroad car, barge, truck, etc.).

**As determined on -28 +48 Tyler Mesh fraction of a representative sample by Alcoa Standard Method of Analysis--Determination of Bulk Density of calcined Coke -424-B (1979/01/29)

Formulation. One of the most important coke properties is porosity, which manifests itself as quantifiable bulk density. Bench scale tests have been developed that correlate well with plant experience and are amenable to scale-up and automation. Figure 4 is a plot of bulk density versus particle size for four different cokes and illustrates their wide differences (70).

Figure 4.
Bulk Density vs Particle Size

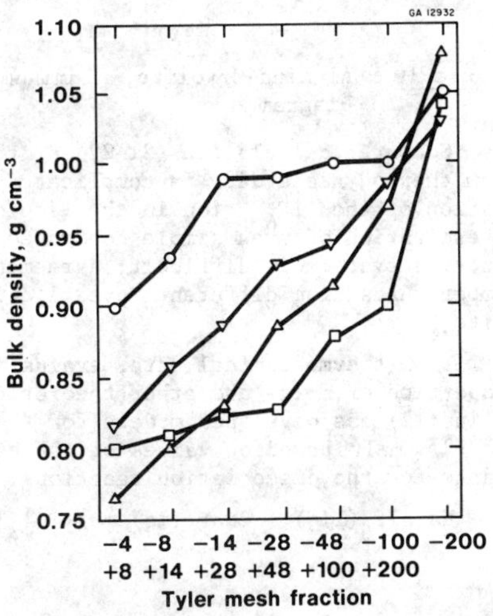

The average density for each coke is a good indicator of expected anode performance. In petroleum coke, bulk density increases as particle size decreases; some pores are large relative to the particle size, and grinding increases coke density. Figure 5 depicts how baked apparent density and flexural strength increase and the electrical resistivity and carbon consumption decrease with coke bulk density.

Figure 5.
Physical Prop. vs Bulk Density

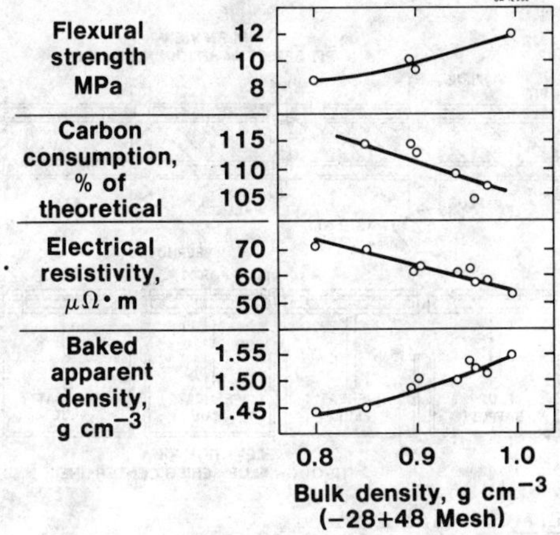

Each of these values is at optimum formulation for that coke. By calculating the effects of these and other anode differences, cokes may be compared on a cost effective basis, that is, operational penalties or gains versus purchase price and shipping cost.

Other major aspects of anode formulation are coke particle sizing and binder pitch level. Figure 6 indicates the effect of these variables.

Figure 6.
Physical Prop. vs % Pitch

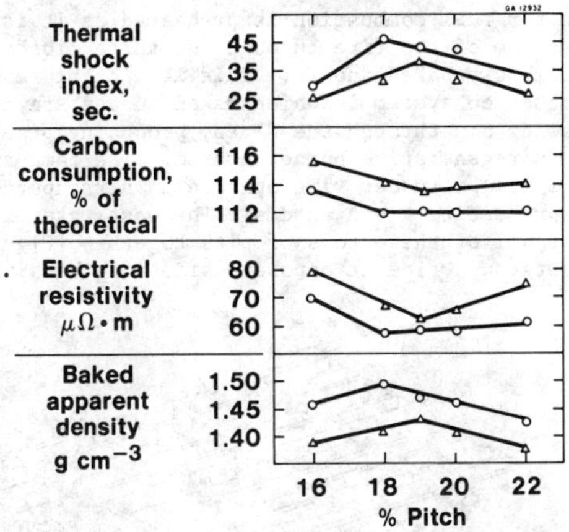

Both curves represent the same coke. The line connecting the circles shows the particle sizing that has the highest bulk density for the coke used; the line connecting the triangles represents lower density sizing. Note that all properties peak at a given pitch level for each sizing. It is apparent that the higher density coke has better properties and requires less pitch. From many curves of this type an empirical formula has been developed to calculate optimum pitch level for any coke, sizing, and pitch. Terms include coke bulk density and geometric surface area calculated from particle size distribution. The plant can vary the proportions of coarse, intermediate and fine coke, and crushed butt fractions to optimize density. The region of maximum density sizing differs from coke to coke and is a compromise between two factors, using as high a fraction of coarse particles as possible to reduce interstices and the highest fraction of fines possible as they are densest. Determining these for a given coke requires experimentation to develop triangular plots of the three sizes, establishing the area where maximum density can be obtained.

Ring Furnace Baking. The production of a good green anode does not insure a good finished anode; it must then be baked under the proper conditions (71).

The basic unit of a ring furnace is a ceramic lined pit 305 cm deep x 305 cm long and ~91 cm wide. Anodes are vertically stacked in these pits in rows separated by a thin layer of petroleum coke with a heavier layer of coke on the top row to reduce heat loss. Pits are placed end-to-end in long rows, arranged side-by-side to fill the ring furnace building. Figure 7 illustrates how these furnaces operate.

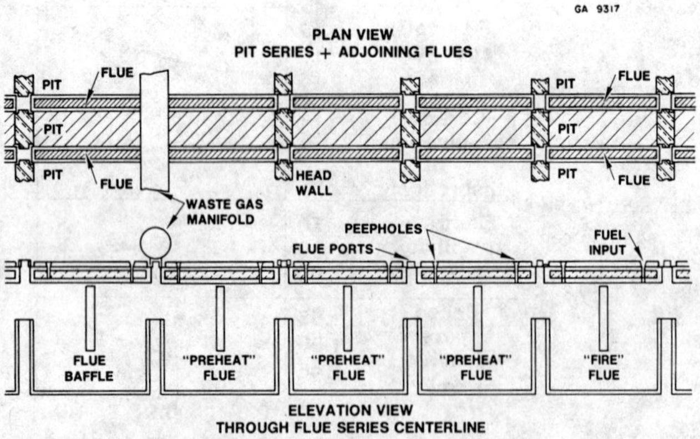

Figure 7. Ring Furnace

In the plan view the pits are flanked on either side by combustion flues. Although the pits are separated by ceramic wall headers, the flues are interconnected, as shown in the elevation view. Before passing to the next flue, combustion products are forced to the bottom of the flue by ceramic baffles. An oil or gas air mixture is injected at the fuel input of the pit under fire and the combustion products pass through the next three pit flues before exiting by way of the waste gas manifold. Air for fuel combustion is preheated as it is drawn upstream of the fire through the three flues, containing previously baked anodes. This is a highly efficient combustion system in which baked anodes are cooled by drawing air through the flues, producing preheated air for a regenerative burner system. The combustion products from the pit under fire operate in a recuperative manner by preheating green anodes. The waste gas manifold ties all of these rows of pits together (Figure 8) by an underground flue at opposite sides of the pit assembly.

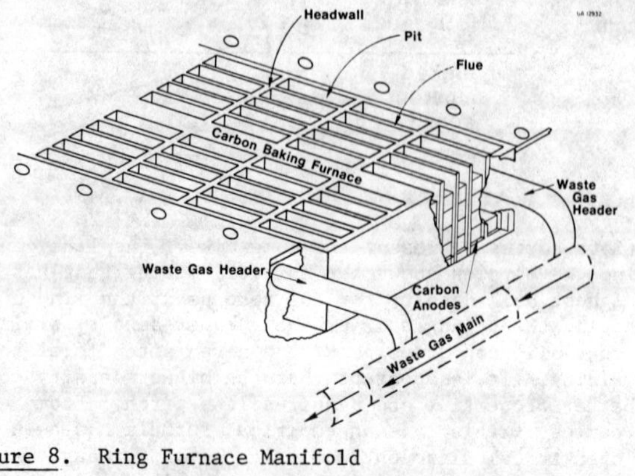

Figure 8. Ring Furnace Manifold

When the anodes in a given pit reach the finishing temperature, the fire and manifold are moved to the next pit and the process is repeated. The result of heat balance studies on these furnaces is shown in Figure 9.

Maintaining uniform temperature distribution so that all anodes in a pit attain the desired temperature is an important requirement if anode dusting into the electrolyte is to be avoided. Even though this is an efficient system having both regenerative and recuperative elements, only about 30% of the energy entering the furnace is utilized in baking anodes. Forty-six percent of energy input to the system comes from volatiles in the pitch, and theoretically these volatiles should provide more than enough to bake anodes. The remaining 54% of the energy input is injected as gas or oil.

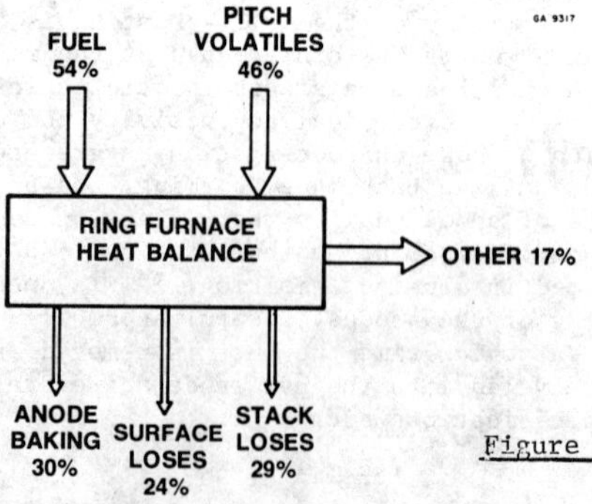

Figure 9.

Electrolytic Smelting

Thermodynamics and Phase Relations. The thermochemistry of electrolyte constituents and the electrolyte-aluminum system is treated comprehensively in a book by Grjotheim et al (11), the treatises of Rolin (12-13), and a series of publications from Dewing (14-18). The application of thermodynamics to the system and to the electrochemistry of the process has been useful in developing the theory underlying the Hall-Heroult Process.

Thermodynamic data for the important compounds in the Hall-Héroult Process are reviewed in Grjotheim et al (11) and Rolin (12). The JANAF project provides critically evaluated thermodynamic data for some of the important species in the JANAF "Tables" and "Supplements" (19).

The phase diagram of the $NaF-AlF_3-Al_2O_3$ system (Figure 10) indicates condensed phase relationships for the basic constituents of the electrolyte system. Craig et al (20) and Grjotheim et al (11) have compiled and reviewed other binary, ternary, and higher order phase diagrams involving the constituents Na_3AlF_6, AlF_3, Al_2O_3, CaF_2, and LiF.

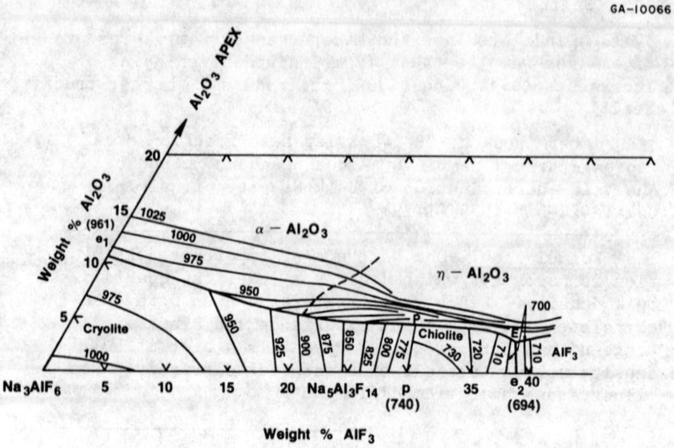

Figure 10. Alumina-Cryolite, Aluminum-Fluoride Phase Diagram

Computer calculation (21-22) of these phase diagrams from thermodynamic data is complicated by partial dissociation of the AlF_6^{3-} ion in the liquid state. Precise determination of even simple condensed phase diagrams of fluoride systems is difficult; agreement in liquidus temperatures from different investigations is sometimes quite poor.

From thermochemical data, evaluations of the physical properties of melts and other theoretical considerations, Rolin (12) has developed models for the constitution of $NaF-AlF_3$ melts based on values of the equilibrium constants for the dissociation reactions:

$$Na_3AlF_6(liq) \rightleftarrows 2NaF(liq) + NaAlF_4(liq) \qquad [4]$$

and

$$Na_3AlF_6(liq) \rightleftarrows 3NaF(liq) + 3AlF_3(liq) \text{ or } (s) \quad [5]$$

These models express the activities of the species NaF, $NaAlF_4$, Na_3AlF_6, and AlF_3 as functions of temperature and melt composition. The models are useful for understanding and predicting vaporization phenomena and various reactions of the electrolyte system in the environment of an operating cell.

From potentiometric measurements of Na_3AlF_6-Al_2O_3 melts, the following expression for the activity of Al_2O_3 has been developed (12):

$$a_{Al_2O_3(liq)} = N^3_{Al_2O_3} \quad [6]$$

where:

$a_{Al_2O_3(liq)}$ is the activity of supercooled liquid alumina

$N_{Al_2O_3}$ is the mol fraction alumina

The activity of Al_2O_3 in Na_3AlF_6-Al_2O_3 melts at 1300 K with respect to solid α-Al_2O_3 as reference state can be expressed by the relation (12)

$$\log a_{Al_2O_3} = 2.77 \log C_{Al_2O_3} - 3.042 \quad [7]$$

where:

$a_{Al_2O_3}$ is the activity of alumina (reference state α-Al_2O_3)

$C_{Al_2O_3}$ is concentration of alumina in wt%

Equation [7] has been modified (23) to express the activity of Al_2O_3 as a function of Al_2O_3 content of a melt and the saturation solubility for Al_2O_3, assuming that Equation [7] is valid at other temperatures and for $NaAlF_6$-base melts with additives

$$a_{Al_2O_3} = (C_{Al_2O_3}/C^*_{Al_2O_3}, \text{sat.})^{2.77} \quad [8]$$

where:

$^*C_{Al_2O_3}$, sat. is the saturation solubility of Al_2O_3 in wt% for the melt at a given temperature and melt composition.

The equilibrium partial pressures of the gaseous species Na and AlF from the reaction of aluminum with NaF-AlF_3 melts have been calculated by Rolin (12). The results of these calculations agree with Kvande's determinations (24) of the partial pressures of aluminum monofluoride and sodium in the system NaF-AlF_3-Al. Measurements of the sodium content of aluminum in equilibrium with NaF-AlF_3 melts (25) along with values for the activity coefficients of sodium in liquid aluminum (14) lead to similar values for the partial pressure of sodium in the system. For melts near the composition and temperature of the industrial electrolyte held with aluminum, the equilibrium partial pressure of sodium is much greater than that of aluminum monofluoride.

With completely nonreactive anodes and cathodes, the net of electrode reactions in electrolyzing Al_2O_3 from a melt saturated with α-Al_2O_3 can be represented by the reaction

$$Al_2O_3(\alpha) = 2Al(liq) + 3/2\ O_2(g) \quad [9]$$

$$\Delta G^\circ_{1233} = 306.93 \text{ kcal} \quad [10]$$

The reversible cell potential (decomposition voltage) for the above reaction can be calculated from the relation

$$\Delta G^\circ = -nFE^\circ \quad [11]$$

where: ΔG° is the standard free energy change for the reaction,

n = the number of mols of electrons involved in the cell reaction,

F = the Faraday constant, 23,061 cal/v equivalent,

E° = the standard electrode potential for the reaction.

Substituting in Equation [11]

$$E^\circ_{1233} = -2.218\text{V} \quad [12]$$

From the standpoint of thermodynamic equilibrium, the probable overall reaction for the electrolysis of dissolved Al_2O_3 with carbon electrodes is

$$Al_2O_3(\alpha) + 9/2C(s) = 1/2\ Al_4C_3 + 3CO(g) \quad [13]$$

$$\Delta G^\circ_{1233} = 131.06 \text{ kcal}$$

$$E^\circ = -0.95\text{V}$$

For industrial aluminum production, the cathodic product is predominantly molten aluminum and the product discharged on the anode is largely CO_2. Thus, the overall cell reaction may be represented

$$Al_2O_3(\alpha) + 3/2\ C(s) = 2Al(liq) + 3/2\ CO_2(g) \quad [14]$$

$$\Delta G^\circ_{1233} = 164.90 \text{ kcal} \qquad E^\circ = -1.192\text{V}$$

$$\Delta H^\circ_{1233} = 262.57 \text{ kcal}$$

For the discharge of CO on the anode

$$Al_2O_3(\alpha) + 3C(s) = 2Al(liq) + 3CO(g) \quad [15]$$

$$\Delta G^\circ_{1233} = 148.73 \text{ kcal} \qquad E^\circ = -1.075\text{V}$$

$$\Delta H^\circ_{1233} = 217.70 \text{ kcal}$$

Electrolyte Chemistry

Composition. Electrolyte is essentially a solution of aluminum oxide dissolved in molten Na_3AlF_6. The Al_2O_3 content normally ranges from 2-6%. Aluminum fluoride is usually added to the electrolyte in excess of the Na_3AlF_6 composition. The relative amounts of NaF and AlF_3 in the electrolyte are expressed as the cryolite ratio (CR), the molar ratio of NaF to AlF_3, or as the bath ratio, the mass ratio of NaF to AlF_3. Because the molecular weight of AlF_3 is almost twice that of NaF, CR is double the value of bath ratio. Addition of AlF_3 improves current efficiency. Industrial cells operate at CR between 2 and 3. Operating temperature is usually in the range 940-980°C.

Calcium fluoride (CaF_2) is always present in the electrolyte, arising from low levels of CaO impurity in the Al_2O_3. It usually attains a steady state concentration of 4-8%, at which level calcium is removed in the aluminum product and emitted in the off-gases at a rate essentially equal to its introduction. Calcium fluoride has a beneficial effect in that it reduces the freezing point of the electrolyte.

Lithium fluoride added as the carbonate is often used to increase electrolyte conductivity and to further reduce the freezing point.

Addition of virtually any compound to molten Na_3AlF_6 reduces the equilibrium solubility of Al_2O_3 and its rate of solution. Because rapid dissolution and high activity of Al_2O_3 in the electrolyte are desirable, the additive is usually controlled to less than ten mass percent (as fluoride).

The primary ionization of molten Na_3AlF_6 on melting can be represented as

$$Na_3AlF_6 \rightleftarrows 3Na^+ + AlF_6^{3-} \quad [16]$$

There is general agreement that the hexafluoroaluminate ion undergoes partial dissociation in the melt

$$AlF_6^{3-} \rightarrow AlF_4^- + 2F^- \quad [17]$$

$$K = (a_{AlF_4^-})(a^2_{F^-})/a_{AlF_6^{3-}} = 4\alpha^3/[(1-\alpha)(1+2\alpha)^2] \quad [18]$$

The mechanism and the degree of dissociation, α, of the hexafluoroaluminate anion have been developed from crycoposcopy, density and viscosity studies, and Raman spectroscopy. Values for the dissociation constant, K, and its temperature dependence developed from these studies vary considerably. The models for melt constitution developed by Paucirova et al (26) and Sato et al (27) probably best represent the ionic structure of NaF-AlF_3 melts near the composition $3NaF.AlF_3$. The degree of dissociation, α, of hexafluoroaluminate anion at the melting point of Na_3AlF_6 is about 0.3. The major ions present in fused Na_3AlF_6 are Na^+, AlF_6^{3-}, F^-, and AlF_4^-. Below a CR of 3, lowering the ratio decreases AlF_6^{3-} and F^- and increases AlF_4^- concentrations.

When Al_2O_3 dissolves in molten Na_3AlF_6, there is a strong chemical interaction between the solute and the solvent. There is evidence that more than one oxygen-containing anion is present in the electrolyte. Complex rather bulky oxyfluoroaluminates $Al_xO_yF_z^{(3x-2y-z)}$, are the likely structures rather than simpler aluminates such as AlO_2^-, AlO^+, $Al_2O_4^{2-}$, and AlO_3^{3-}. There is probably a single oxygen atom per complex ion, particularly at low concentration of Al_2O_3. At infinite dilution, three foreign particles are created for each Al_2O_3 dissolved. Gilbert et al (28) report that the presence of species such as $AlOF_2^-$, and $AlOF_3^{2-}$ and $AlOF_5^{4-}$ is not consistent with their results by Raman spectroscopy; only those species with bridging Al-O-Al bonds should be considered. Reasonable dissolution mechanisms for low Al_2O_3 contents are:

$$4AlF_6^{3-} + Al_2O_3 \rightarrow 3Al_2OF_6^{2-} + 6F^- \qquad [19]$$

and for more concentrated solutions,

$$AlF_6^{3-} + Al_2O_3 \rightarrow 3AlOF_2^- \qquad [20]$$

Physical Properties. The specific electrical conductivity of the electrolyte is important for the energy efficiency of the process. It is typically 2 to 2.2 ohm^{-1} cm^{-1} in industrial cells. Based on a correlation derived by Choudhuri (29), most measurements found in the literature fit the relationship:

$$K = Exp[2.0156 - 0.0207(\%Al_2O_3) - 0.005(\%CaF_2) - 0.0166(\%MgF_2) + 0.0178(\%LiF) + 0.0063(\%NaCl) + 0.217(CR) - 2068.4/T] \qquad [21]$$

where: K is in ohm^{-1} cm^{-1} and T is in °K.

In cryolite-rich melts, sodium ions provide most of the current transport.

The density of liquid aluminum at 950-1000°C is 2.3 g/ml. Experience has shown the density of the molten electrolyte should be less than 2.1 g/ml to maintain good separation between metal and fused salt phases. Additions that decrease the density of Na_3AlF_6-Al_2O_3 melts are beneficial.

Other electrolyte physical properties of importance in cell operation are vapor pressure, viscosity, and surface tension. The vapor pressure of the electrolyte and its reactivity with water vapor are important with respect to fluoride emissions. Grjotheim et al (11) and Rolin (30) provide comprehensive reviews of the physiochemical properties of the electrolyte and their implications.

Alumina Solubility and Dissolution. The solubility of Al_2O_3 depends on electrolyte composition and temperature. Na_3AlF_6 and Al_2O_3 are eutectic at 960°C and 11 wt% Al_2O_3 (31), the solubility limit of Al_2O_3 at 1000°C is 13 wt% Al_2O_3.

The solubility limit of Al_2O_3 in typical industrial electrolytes is lower, ~8%.

The rate of dissolution of Al_2O_3 is primarily governed by mass transfer limitations (32) rather than by differences in Al_2O_3 modification. Nevertheless, different aluminas may show different dissolution behavior as agglomeration occurs when Al_2O_3 is brought in contact with the molten electrolyte. Alumina added to a bath surface is wet and penetrated by electrolyte, partial freezing occurs, and the Al_2O_3 is not readily dispersed. A transformation of active Al_2O_3 modification into α-Al_2O_3 can occur and may form an interlocked structure of recrystallized α-Al_2O_3 (33). Pieces structured in this manner may enter the bath and dissolve with the mass transfer limitations valid for the aggregate. They may also form part of the frozen crust above the molten electrolyte.

Alumina is added in increments, by automation, every few minutes. With manual feeding, intervals extend to several hours. In such cases crust pieces enter the bath and may dissolve very slowly; they and bottom sludge (34) may supply Al_2O_3 to the electrolyte on a somewhat continuous basis.

Excessive bottom sludge, a viscous mass consisting of Al_2O_3 and electrolyte, is formed by improper feeding. It increases the electrical resistance between cathode carbon and metal and it may form solidified muck that is hard to remove.

Metal Solubility. Aluminum interacts with the electrolyte to form reduced species in solution. This occurs at the cathodic metal surface. There is still some disagreement over whether the reduced species is dissolved metallic sodium, metallic aluminum, monovalent aluminum, subvalent sodium, or a colloidal suspension of aluminum metal.

Recent investigations show the existence of two different dissolved species that can be visually observed and detected electrochemically (35). The issue may have been clouded in the past because reduced impurities might have been taken as reduced aluminum or sodium, or because the metal fog observed may really form from dissolved metal as a secondary product (36).

Metal solubility depends on electrolyte composition and temperature. Yoshida and Dewing (37) describe the relationship with the metal solubility expressed as aluminum in solution:

$$\text{Wt\% Al in sol.} = + 0.0003(°C) - 0.2877 + 0.0134(CR) - 0.0019(\%CaF_2) - 0.0017(\%Li_3AlF_6) - 0.0029(\%NaCl) \qquad [22]$$

These values are somewhat lower than those obtained by other investigators. One may assume that the solubility for a typical industrial electrolyte is about 0.1 wt%. Lower temperature and lower cryolite ratio (CR) favor lower metal solubility, as do various bath additives. The metal solubility affects the current efficiency, as discussed in another section.

Electrode Reactions and Polarization

Cathode. The cathode reaction is sometimes represented as:

$$Al^{3+} + 3e \rightarrow Al \qquad [23]$$

There are no Al^{3+} ions present in the electrolyte, and complexed ions such as AlF_6^{3-} and AlF_4^- are involved in the discharge. The following, therefore, may be more representative:

$$AlF_6^{3-} + 3e \rightarrow Al + 6F^- \qquad [24]$$

Bowman (38), using cyclic voltammetry, stationary electrode polarography, differential pulse polarography, and chronopotentiometry at tungsten and glassy carbon electrodes, found evidence only for a reversible 3-electron transfer process. There was no evidence for a chemical reaction following the electron transfer process, ruling out discharge of Na^+ at low activity followed by chemical reaction to form aluminum. Bowman also saw no evidence for a chemical reaction preceding electron transfer, which would be present if Al^{3+} discharge were preceded by dissociation:

$$AlF_6^{3-} \rightleftarrows Al^{3+} + 6F^- \text{ and/or } AlF_4^- \rightleftarrows Al^{3+} + 4F^- \qquad [25]$$

Such a dissociation step, however, is proposed by Rolin (39) to occur in a zone of the diffusion layer.

Laboratory measurements by various investigators indicate that the cathode polarization in an industrial cell amounts to about 100 mV or somewhat less, and is primarily concentration polarization. For practical purposes, the following empirical relationship has been used (23).

$$\eta = [RT(1.375 - 0.125CR)/(1.5F)] \ln (i/0.257) \qquad [26]$$

where: CR = cryolite ratio (mol ratio NaF/AlF_3)
i = cathodic current density (A/cm^2)

The activation overpotential at the cathode is low. Thonstad and Rolseth (40) measured the charge transfer process by a double pulse technique and found the exchange current density to be 12 A/cm^2 for η = 3 (36 A/cm^2 for η = 1).

There is some concentration polarization. Since sodium ions carry practically all the current, they accumulate near the cathode, and aluminum ions reach the metal-electrolyte interface only by diffusion or convection. Thonstad (41) calculated changes in electrolyte composition at the metal-electrolyte interface caused by: (a) inward migration of Na^+ ions, (b) inward diffusion of $3Na^+.AlF_6^{3-}$ and $Na^+.AlF_4^-$, (c) discharge of AlF_6^{3-} and AlF_4^- to F^- ions and aluminum at the interface, and (d) the resultant outward diffusion of Na^+F^-.

As may be expected from thermodynamic consideration, sodium is co-deposited with aluminum. The difference between the decomposition potentials for producing aluminum and sodium (at activity one) has been calculated from thermodynamic data as ~0.24v for the industrial range of electrolyte compositions and temperatures. This will be higher for an electrolyte of lower CR and less for an electrolyte with a higher sodium fluoride activity and at higher temperature.

It may be expected that sodium will be found in the aluminum at a concentration or activity corresponding to the thermodynamic equilibrium condition at the metal-electrolyte interface. The sodium content is lowered as the metal is processed because sodium is readily and preferentially oxidized by air. Sampling of the metal directly from an operating cell, however, yields sodium concentrations in the order of 100 ppm, indicating coformation of sodium.

Anode. The primary anode reaction may be represented by:

$$C + 2O^{2-} - 4e \rightarrow CO_2 \qquad [27]$$

As oxygen is complexed in the electrolyte, the anode reaction presumably involves complex ions, e.g., according to:

$$2Al_2OF_6^{2-} + 2AlF_6^{3-} + C \rightarrow 6AlF_4^{-} + CO_2 + 4e^{-} \qquad 28]$$

Thermodynamically, oxygen depositing onto carbon at the cell operating temperature should form essentially all CO with very little CO_2. However, based on net carbon consumption, the primary anode product is essentially all CO_2.

This has been verified by Ginsberg and Wrigge (42), who obtained essentially pure CO_2 except at very low current densities (<0.06 A/cm^2). They used graphite anodes and isolated the anode compartment by means of an oxygen-ion-conducting diaphragm.

Reaction kinetics are the cause for this surprising displacement of the primary anode gas composition from thermodynamic equilibrium. The desorption of CO from the carbon surface must be kinetically hindered. One may assume, with Thonstad and Hove (43), that the initial oxygen-containing ions are transported through the double layer and discharged with comparatively little overvoltage. This oxygen is chemisorbed on the carbon surface as proposed by Blyholder and Eyring (44) for ordinary combustion Figure 11.

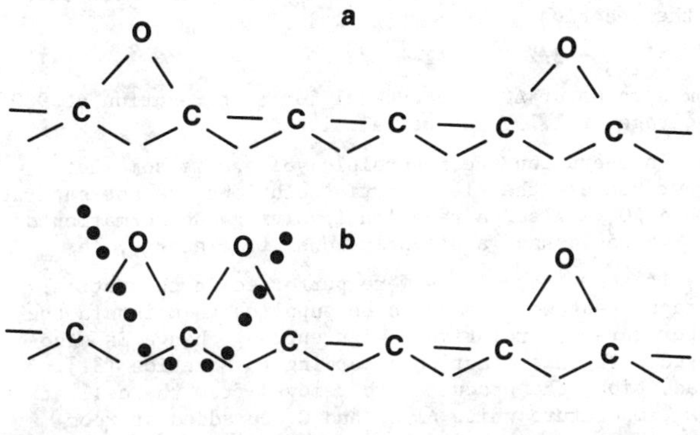

Figure 11. Formation of CO_2 on Carbon Anode Surface

A carbon site occupied by oxygen is not available for further oxygen discharge; in effect the site is electrically insulated. As sites become occupied, it takes additional energy (overvoltage) to deposit oxygen at increasingly less favorable sites. The surface compounds are quite stable and break down to CO only at a very slow rate. Thus, the surface is increasingly covered with oxygen and CO_2 eventually desorbs.

In commercial cells, at current densities of 0.7 to 1 A/cm^2, overpotentials are typically 0.5 to 0.6V. Many authors have studied anodic overvoltage on carbon and graphite in cryolite melts as summarized by Grjotheim et al (45). Virtually all of them found a Tafel behavior within a certain range. Thonstad (46) presents coefficients a = 0.4V and b = 0.22-0.28V for the Tafel equation $\eta = a + b \log i$; for prebaked carbon in a range of 0.03-1.0 A/cm^2 at 1010°C, the exchange current density is 0.024 A/cm^2.

At least four components contribute to the anode overpotential: (a) a reaction overvoltage, (b) an increase of local current density due to gas bubbles on the electrode surface, (c) an ohmic component resulting from an effective increase in resistivity of the electrolyte caused by gas bubbles, and (d) a concentration overvoltage.

Anode Effect. An anode effect is a phenomenon common to many fused salt electrolytic processes. Its physical manifestation is the growth of larger and larger bubbles on the anode as the contact angle of the bubble diminishes. Finally, a few large bubbles coalesce, forming a single large bubble that covers most if not the entire surface of the anode. With a constant potential source, the current falls to a low value. With a constant current source, as used industrially, the applied potential rises to about 30V and the current penetrates the gas film by a multitude of small electric arcs.

In aluminum electrolysis, the anode gas composition changes on anode effect from CO_2 to CO, with significant quantities (3-25%) of CF_4 and minor amounts of C_2F_6. Many investigators find that fluorine reaction precedes anode effect, producing a carbon-fluorine surface compound that causes dewetting of the anode. The critical current density leading to anode effect is not necessarily the diffusion limiting current density for the oxygen-containing ions.

While there is no general agreement among various investigators, the following mechanism is proposed as plausible (47). As Al_2O_3 is depleted, overvoltage increases. The increase in surface tension of the bath with decreasing Al_2O_3 concentration coupled with the electrocapillary effect, resulting from the higher anode overvoltage, reduces anode wetting and the bubble contact angle decreases. At ~1.2V anode overvoltage, sufficient fluorine activity is produced to result in fluorine bonding to the carbon. Fluorocarbon compounds have low surface energy and promote dewetting and growth of large bubbles. When a few very large bubbles are present it is easy for them to coalesce, producing anode effect. Once the cell is on anode effect the production of significant quantities of CF_4 further promotes dewetting; and even if the Al_2O_3 concentration is restored, the cell will remain on anode effect until the current is interrupted to allow the bubble to collapse.

Fluidynamics

Magnetic Forces. Magnetic fields are associated with electrical currents. At the high currents of industrial cells, magnetic forces are considerable; the field can reach 150 gauss in modern cells. Magnetic fields exert Biot-Savart forces on current conductors and may induce fluidynamic movements in liquid conductors such as the molten metal or, to a lesser extent, the electrolyte.

In aluminum cells, three types of magnetic effects on the metal have been observed: (a) a vertical, static displacement leading to a greater depth of the metal pad by as much as 5 cm in the center of the pot; (b) a wave-like change in metal depth, circling the pot with a frequency, of e.g., 1 cycle/30 sec; and (c) metal flow with flow rates of 10-20 cm/sec being common.

Figure 12 illustrates the forces, F, and the flow

pattern caused by the interaction of the magnetic field, B, resulting from an overall longitudinal cell current, I, and horizontal, perpendicular current components, i, in the metal.

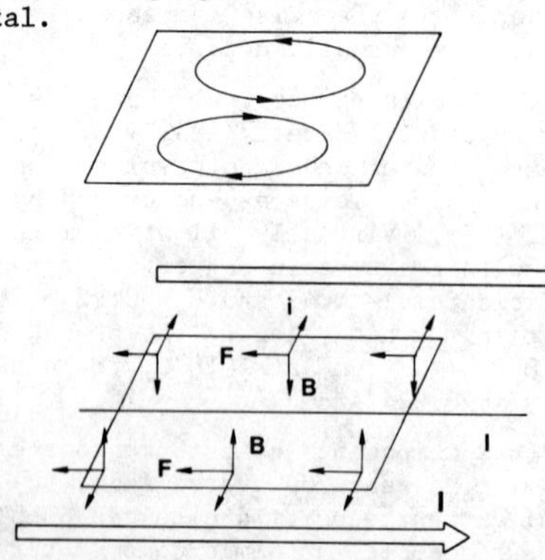

Figure 12. Magnetic Fields - Current Interaction

Such horizontal currents may result from the extension of the side ledge to underneath the metal pad. Other basic flow patterns result from other orientations of overall cell current.

Current flow patterns, magnetic shielding by steel components, and magnetic interactions with other pots are complicated, and intricate computations are necessary to describe adequately the fluidynamic consequences. Several approaches have been refined to the point where effects of deviating from optimum cell operation, such as caused by anode changes and muck formation, can be predicted (48,49,50,51,52).

Gas Bubble Convection. Natural convection due to temperature or composition differences in the electrolyte are not significant compared to the movement induced by magnetic forces. More important is bath movement resulting from the formation and escape of gas bubbles. Dernedde and Cambridge (52) observed bath movement under the anode and gas-induced sidewall circulation by simulating the metal with a mixture of silicone oil with perchloroethylene and the electrolyte by water.

Current Efficiency.

Reoxidation. Metal can be oxidized by anodically produced CO_2, resulting in a loss in current efficiency. CO_2 is thereby reduced to CO:

$$2Al + 3CO_2 \rightarrow Al_2O_3 + 3CO \qquad [29]$$

This is the predominant reaction that causes current efficiency losses in conventional aluminum electrolysis, although some reduction all the way to carbon also occurs (53) and there are other current efficiency losses from impurities and electrical shorting.

The ratio of CO to CO_2 in the off-gasses has been used to monitor "instantaneous" current efficiency. Other methods require more time and produce results integrated over a longer time. In the dilution method, a metal tracer is added and the metal inventory determined from its resulting concentration in the metal. Further tracer dilution by metal can then be monitored with time. Limitations of the analytical methods require a week of normal production to determine a current efficiency with an accuracy of about 1%. Determination from weighing the metal allows for a valid current efficiency determination only over 1-month periods.

Metal dissolution and reoxidation occur in several steps, but it is not entirely clear to what extent the individual steps contribute to the reoxidation rate. Figure 13 shows these steps and a concentration gradient for the dissolved metal. Dissolved metal moves through the diffusion layer at the metal-electrolyte interface, I. The rate is determined by the thickness of the diffusion layer, the concentration gradient, the diffusion coefficient, and the surface area. Metal is transported through the bulk of the electrolyte, II, mainly by bath motion. In a gas-filled bath layer below the anode, III, metal reacts quantitatively with CO_2; no metal seems to reach the anode surface. According to this model, the overall rate is essentially determined by the diffusion rate at the interface and by convection through the bulk. Depending on which of the two rates is decisive, different concentration profiles are to be expected as indicated in Figure 13. Which best represents normal electrolysis conditions is yet to be established.

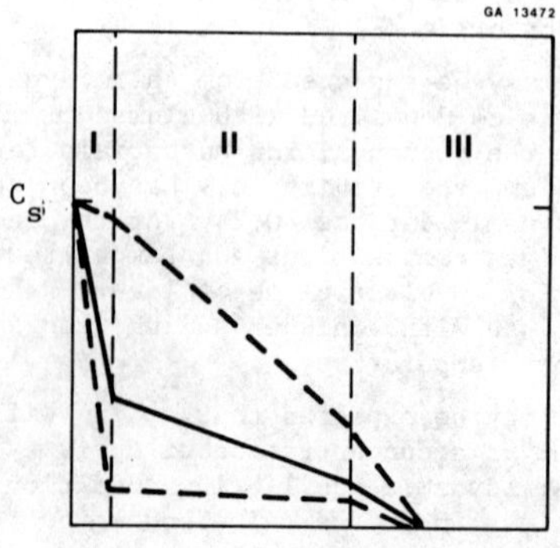

Figure 13.

Impurities. Impurities such as Fe, Si, Ti, and P are reduced by Al at the cathode or by dissolved metal. Metallic impurities are, to some extent, reduced to noncoherent form rather than to metal alloying directly with the aluminum (36). These noncoherent deposits can be detached from the cathode and subsequently reoxidized by anodically produced CO_2 or by atmospheric oxygen when they are carried to the bath surface. Reoxidized impurities that stay in the bath can be reduced again and in this manner cause a current efficiency loss many times above the one calculated on a stoichiometric basis.

Shorting. Metal movement and agitation may lead to localized shorting between the cathode metal and anodes. This effect is more pronounced when an effort is made to decrease the A-C distance in order to achieve lower ohmic losses.

Achieved Current Efficiency. In the 1970's current efficiencies of 85-90% were typical for cell operation. Recently, values above 90% have become fairly common and values above 94% have been achieved with prototype cells (54). Improvements have been made mainly by changes in bulk composition and by decreasing magnetic disturbances.

Power Efficiency.

Theoretical. The electric energy component in a kilogram of aluminum depends on cell voltage and current efficiency. It can be expressed as

$$P(\text{in kWh/kg}) = 2.981 \times E_{cell}/CE \qquad [30]$$

the current efficiency, CE, expressed as a fraction. Power efficiency can be obtained by comparing this figure with the theoretical minimum amount of energy required. Minimum decomposition voltage is given by the free energy of the reaction

$$Al_2O_3 + 3/2\ C \rightarrow 2Al + 3/2\ CO_2 \qquad [31]$$

From a value of $\Delta G = 164.9$ kcal for this reaction at 960°C, a voltage of 1.19V can be calculated.

In operation the reversible voltage is somewhat higher because the electrolytic solutions are not saturated with Al_2O_3. Also, a reaction involving the formation of some Co would have a slightly lower free energy.

If the electrolysis were performed at theoretical voltage, heat would have to be supplied to maintain the temperature. A relatively high entropy change is associated with this reaction according to Equation [31]. In addition, the products are removed from the cell at high temperature while Al_2O_3 and C are added at room temperature.

Considering the enthalpy of the reactants and products, Al_2O_3 and C at 298°K and aluminum and CO_2 at 1233°K,

$$Al_2O_3(s,\gamma) + 3/2\ C(s) \rightarrow 2Al(liq) + 3/2\ CO_2(g),$$

$$\Delta H = 289.9 \text{ kcal} \qquad [32]$$

A cell voltage of 2.095V would be necessary to supply this reaction enthalpy at a current efficiency of 100%. Figure 14 shows the correlation between the voltages corresponding to the free energy and the enthalpy, respectively.

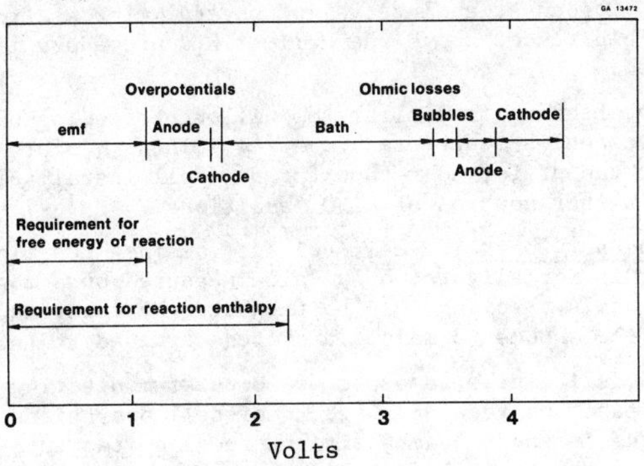

Figure 14. Voltage Components

The voltage necessary to produce aluminum and oxygen at 960°C would be 2.18V. The carbon anode has a depolarizing effect and acts, in essence, as a fuel cell electrode with carbon as the fuel.

Cell Voltage. As shown in Figure 14, the cell voltage is composed of the electrode potentials (reaction voltage + anode and cathode overpotentials, discussed above) and ohmic losses.

The hypothetical, bubble-free potential drop through the electrolyte is given by:

$$E_{bath} = IL/KA \qquad [33]$$

where: I is current, L the A-C spacing (3-5 cm), K the electrical conductivity of the melt (2.0-2.4 ohm^{-1} cm^{-1}), and A the effective bath area.

Typically, E_{bath} is about 1.6V. The actual voltage drop across the inter-electrode gap is increased 0.15 to 0.2V because of gas bubbles. The potential drop through the anode, its connecting stubs, and the contact between them ranges from 0.25-0.30V with prebaked anodes and 0.45-0.55V in Soderberg cells. The cathode voltage, which includes the voltage loss in the lining, collector bars, and the contact between the two, typically ranges from 0.45-0.55V. An additional voltage loss in external electrical conductors, not included in Figure 14, ranges from 0.1-0.2V. The collector bars are selected such that there is an optimum balance between investment and power costs due to ohmic losses.

Achieved Power Efficiency. Power efficiency can be calculated by dividing theoretical (minimum) energy consumption by actual energy consumption. Table III lists the power efficiencies for cell voltages of 4.0, 4.5 and 5.0V and current efficiencies of 87, 90 and 93%.

Table III

Power Efficiency at Various Cell Voltages and Current Efficiencies

Cell Voltage	% Current Efficiency	% Power Efficiency	kWh/kg
4.0	87	45.6	13.7
4.0	90	47.2	13.2
4.5	87	40.5	15.4
4.5	90	41.9	14.9
4.5	93	43.3	14.4
5.0	87	36.5	17.1
5.0	90	37.7	16.5
5.0	93	39.0	16.0
~2.1	100	100.0	6.2

The highest power efficiency publicly reported for an industrial plant (55) was achieved for a low current density operation at a cell voltage of 3.87V and a current efficiency of 89.5%; the energy consumption was 12.9 kWh/kg and the power efficiency 48%.

Design of Hall-Héroult Cells

Description. Figure 15 is a schematic of a typical reduction cell. It consists of an insulated rectangular steel shell. Carbon and molten aluminum act as the cathode. Anodes are prebaked carbon blocks. Frozen electrolyte forms ledges on the carbon sides of the cell and is a component of the top crust. Current carrying steel bars are set in the cathode, and anode rods supply the current and support the anodes. The anodes are mounted on a beam that can be adjusted vertically. Alumina feeders are placed in the center aisle between two rows of anodes. The cell is covered to permit collection of off-gases for processing in a gas-purifying system where volatile fluorides and other impurities are adsorbed on Al_2O_3.

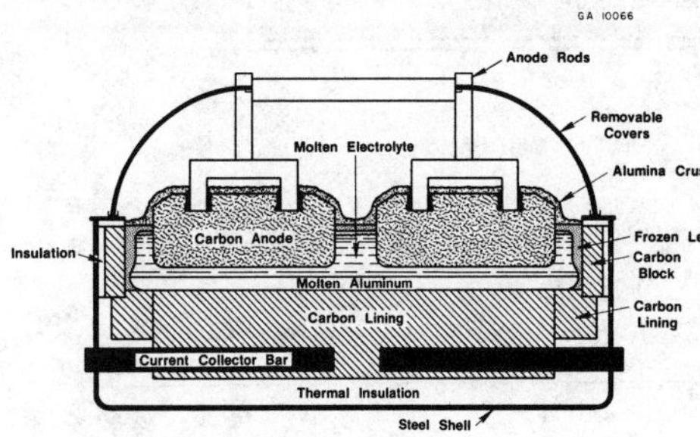

Figure 15. Hall-Héroult Cell with Prebaked anodes

The current capacity of today's cells ranges from 50-250 kA. The steel shell dimensions of large cells are 9-12 m long x 3-4 m wide x 1.12 m high.

There are alternatives to this cell design, with self-baking Soderberg anodes most noteworthy. Figure 16 represents a Soderberg cell.

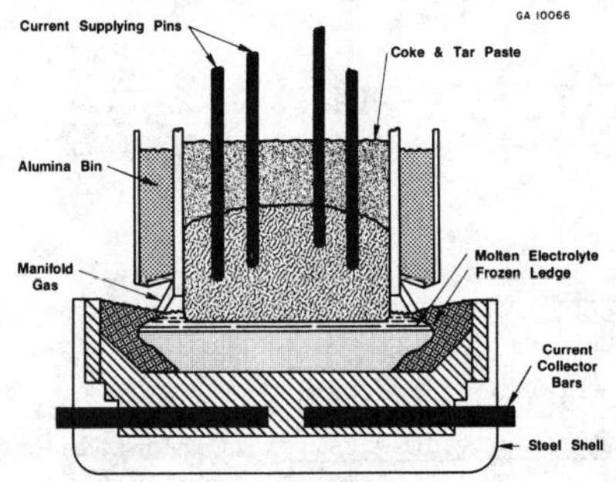

Figure 16. Soderberg Hall-Héroult Cell

Heat Balance. Mathematical models exist today that permit calculation of heat balance and heat losses for a variety of cell designs (48,57). Measurement methods of heat fluxes in operating cells are feasible (56). Figure 17 shows heat loss distribution for a typical center-fed cell.

In a side-break cell with prebaked anodes the sides are more heavily insulated, and >70% of the heat loss may be through anodes and top crust. Heat balance considerations are of prime importance in designing and operating a cell if an optimum temperature is to be maintained. At excessive temperature the ledges melt and the sidewalls are not protected. Current efficiency drops because of increased metal reoxidation. At low temperatures, ledge formation is excessive and Al_2O_3 does not dissolve as readily. Figure 18 shows the freeze profile of a side ledge and the prediction based on model calculations (57). The temperature profile in the cell bottom determines the penetration of liquid bath into the cathode.

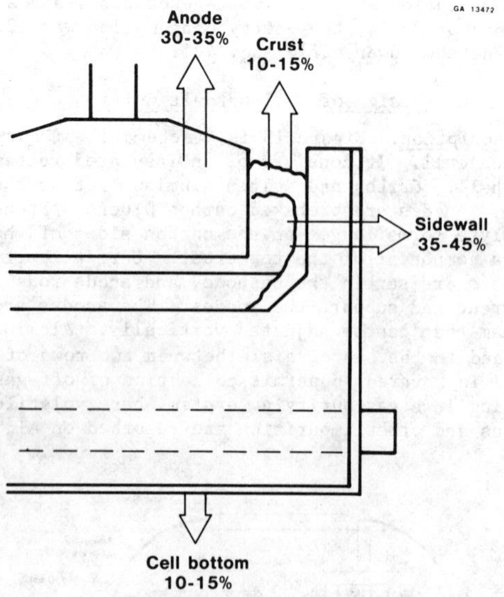

Figure 17. Cell Heat Balance

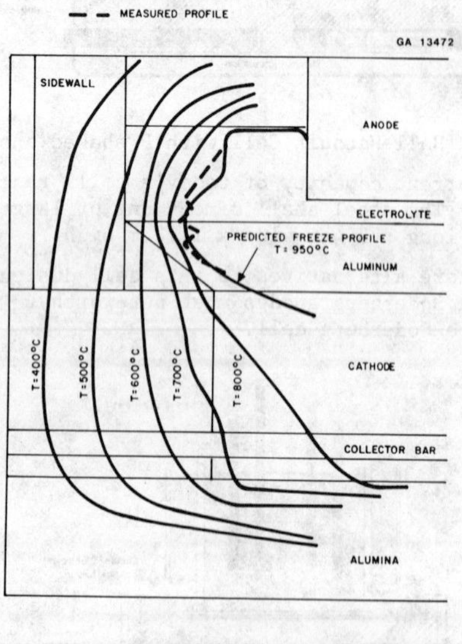

Figure 18. Cell Lining Isotherms (57)

Selection of Current Density. Current density conventionally refers to normal horizontal anode surface; it ranges between 0.7 A/cm^2 for a typical low current density cell to about 1 A/cm^2. The selection is largely governed by economics. Lower current density causes lower energy cost per unit of metal because of lower ohmic losses in the bath. It also requires higher investment related costs because of the lower product rate per electrode area. These costs can be optimized according to various criteria such as minimum product cost, maximum return on investment, or limited supply of energy (58). Today's trend is toward lower current density cells because energy cost is rising more rapidly than other costs.

Alumina Feed. The control of the Al_2O_3 concentration in the electrolyte is difficult because no continuous determintion of the oxide content is available. The occurrence of an anode effect is still the major reliable indication for Al_2O_3 concentration, although some indication is available from changes in voltage.

Essentially three methods of Al_2O_3 addition are used.

(a) For crust feeding the crust is broken after an interval of several hours. Pieces of crust containing a high percentage of Al_2O_3 enter the bath along with Al_2O_3 resting on top. Not all the Al_2O_3 immediately dissolves in the electrolyte but rather forms a deposit from which the electrolyte is fed continuously over an extended period.

(b) Guillotine feeders are arranged in the center of the cell. An opening along the center line is cleared and Al_2O_3 added at an interval of several minutes from a storage bin placed above the cell. This procedure is automated.

(c) Point feeding is similar to guillotine feeding but replenishment is restricted to a few points. Point feeding, the preferred method for new cell installations, permits better control of Al_2O_3 additions.

Process Control. Cell voltage, besides the line current, is normally the only value measured on a continuous basis. Voltage can be regulated by lowering or raising the anodes to maintain a predetermined value.

The cell can be assessed by computer monitoring cell voltage and keeping a record of cell operations such as changes in anode beam position, feeding, tapping, anode changes, etc. Information on electrode separation and changes in Al_2O_3 concentration in the bath can be obtained. Anode effects can be predicted and the feeding changed to prevent them. Electronic shorting can be discovered and pot operation corrected.

Environmental Considerations. The emission of fluorine-containing compounds is the predominant environmental concern associated with the Hall-Héroult Process.

Dry scrubbing is the most efficient way to remove fluorides. The cell off-gases are collected and sent through a fluidized bed of Al_2O_3, which is used to feed the cells, and valuable fluoride is returned. However, undesirable impurities are also fed back to the cell.

The present installations are very efficient and environmental requirements can be met.

Metal Purification and Recycling

Metal Purification for Casting In many ingot casting facilities primary metal is remelted with mill or obsolete scrap, and alloying materials are added to produce an alloy of the desired composition. Oil and gas fired reverberatory furnaces are commonly used to melt ingot for casting.

To produce sound ingot, molten aluminum must be treated to remove nonmetallic inclusions, dissolved hydrogen, sodium, calcium, and other impurities. This purification is accomplished by a variety of in-line systems where molten metal is filtered and/or fluxed during transfer between the holding furnace and the casting station. The Alcoa 622 Process is an advanced system of this type, purifying molten metal efficiently and in an environmentally acceptable manner (59,60).

The heart of the Alcoa 622 Process is a rotating, vaned gas disperser attached to a graphite shaft and powered by an air motor. Fluxing gas (Ar-Cl_2 mixture) is introduced through the shaft and exits beneath the disperser. The rotating action of the disperser forms tiny bubbles of fluxing gas that move upward through the molten metal, creating an efficient cleansing action. Sodium and calcium levels are readily controlled in the 1-2 ppm range at metal flow rates in excess of 454 kg per minute. Dissolved hydrogen is removed as a result of the activity gradient across the interface. Hydrogen reduction from 0.4-0.5 ml/100 g down to 0.1-0.15 ml/100 g are standard for most alloys at common metal flow rates.

One or more dispersers may be used, depending on the metal flow rate and quality requirements. In a single disperser unit the metal flow is countercurrent to the flow of the fluxing gas. In a multistage unit the metal flow will be either countercurrent or cocurrent to the gas flow.

The operating unit is a refractory lined steel shell divided into two or more zones with refractory baffles. It is heated with gas fired immersion burners in an attached side well. The basic box serves the dual purpose of insulating the unit against heat loss and providing a means of disperser assembly support. Doors installed along one side of the cover provide access to the treatment zone for skimming.

The fluxing gas is premixed before being introduced to the disperser system. The mixture consists of argon and chlorine in a ratio determined by the alloy and volume of metal being processed. The chlorine content of the mixture is the stoichiometrically required amount for reaction with sodium, calcium, lithium, and hydrogen.

Two basic sizes are used. The one rotor reactor is ~170 cm long x 147 cm wide and has a maximum production of 454 kg per min. The 3-rotor system is ~340 cm long x 170 cm wide and processes up to 760 kg per minute. Two rotors would be adequate for this metal volume but the third serves as insurance that the casting operation does not have to be interrupted should one rotor fail. Where it is not necessary to remove sodium and calcium, the argon flow rate would be typically 5.7 std m^3/hr, with the addition of chlorine at 0.06-0.28 std m^3/hr. With this amount of chlorine it is usually satisfactory to discharge the exhaust gases into the casting room atmosphere. For requirements where the chlorine flow must be increased, an exhaust system is necessary for pollution control.

The units are operated by the casting crew with no additional labor. This includes startup, changing rotors, and skimming dross. Preparation of the boxes and burner changes are the responsibility of the casting mold preparation crew. Alloy changes require flushing, although with appropriate arrangements the boxes could be drained if refilled promptly. Total operating costs for the metal purification process are about $0.0007/kg of metal treated.

The Alcoa 622 Process is used to treat molten metal for a wide range of aluminum alloy products. Refining processes may be employed or the equipment may be modified to provide still higher quality metal.

In addition, with minor engineering alterations the Alcoa 622 Process can be transferred into a highly efficient magnesium removal unit. Again the appropriate chlorine volume is determined stoichiometrically by the desired magnesium removal. As chlorine is added to the system past the point required for sodium and calcium removal, magnesium is reacted to form $MgCl_2$. This is at operating temperatures and $MgCl_2$ readily separates from the metal stream and floats to the surface where it is contained and removed.

The strength of the Alcoa 622 unit lies in its ability to address a wide spectrum of metal quality needs in one efficient, relatively inexpensive, rugged and reliable unit.

Scrap Recycling. Aluminum scrap recovery has become an important low energy source of metal that requires only 5% as much energy to recycle as it requires to produce it from bauxite. In 1978, aluminum recovered from scrap represented 22% of the supply available to the U.S. industry. A major recycling effort is aimed at beverage containers where scrap can be melted and recycled into cans. In 1980 the aluminum industry recycled 38% of manufactured cans and is aiming for 40-45% in 1981. New or prompt scrap, generated by plants making wrought and cast products, and old scrap recovered from obsolete products discarded by consumers, is returned to plants that cast ingot to produce mill products. The growth of secondary recovery in recent years is given in Table IV. Recovered aluminum values from old and new scrap represent ~26% of the domestic ingot supply. This does not include runaround or home scrap recycled internally by producers of castings and mill products. Cycles of scrap generation and use within the aluminum industry are represented in Figure 19. All 12 primary aluminum producers in the United States participate in either the collection or use of new scrap. They frequently purchase it from their industrial customers on a contract basis. There are two approaches to recovering aluminum from consumer solid wastes. One is the collection and sale by individuals and/or organizations of a single product, such as beverage cans. The second approach involves recovery of old aluminum scrap as a byproduct from such processed wastes as municipal refuse. The quantities of metal recovered from both systems are increasing.

Table IV

Aluminum Supply and Scrap Recovery in the U.S. (Million Short Ton) (66)

	Total Supply All Sources	Secondary Recovery	New Scrap	Old Scrap	Used Cans
1975	5.614	1.236	0.899	0.337	0.090
1976	6.393	1.471	1.062	0.409	0.106
1977	6.897	1.605	1.074	0.531	0.140
1978	7.470	1.673	1.098	0.575	0.170
1979	7.584	1.777	1.163	0.614	0.180
1980*	7.59	1.79	1.06	0.73	0.29

*Estimated

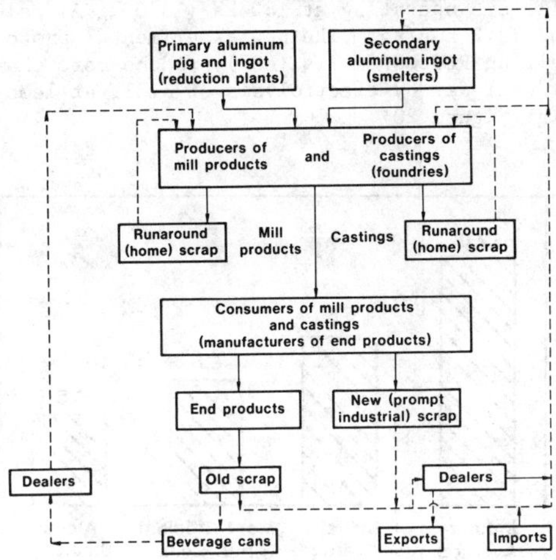

Figure 19. Scrap Recyling Paths

Secondary aluminum producers use more than half of the aluminum scrap generated. Secondary alloys are produced to specified compositions. Because the scrap that reaches the secondary producer is often a mixture of alloys, it cannot be indiscriminately remelted. Incoming scrap must be sampled, characterized, and chemically analyzed. The scrap is then processed as required to produce secondary ingot.

The consumers of aluminum scrap must convert it into ingot or mill product of well defined composition and prescribed physical and mechanical properties. The value of aluminum scrap is directly related to conversion efficiency of the process employed to convert to a commercial product (61). Factors that determine the value of scrap are relative freedom from contamination, proper segregation by alloy, packing, and potential melt loss, which is increased by coatings and other contaminants.

Before and during shipment scrap should be kept segregated, clean, and dry. In most remelt furnaces scrap is charged into a molten aluminum bath. Water charged into molten aluminum along with scrap can result in a violent explosion.

Only 5060-13420 Btu/kg are required for secondary processing (10). This is less than 6% of the energy required to produce primary aluminum from bauxite. Production cost for recycled aluminum is usually less than for primary metal. Effective recycling of old scrap also lessens consumption of bauxite, petroleum coke, and aluminum fluoride.

State of the Art in Hall-Héroult Cells

High capital intensity and low power efficiency have been problems in the electrochemical industry since its beginning. Because the cell current determines rate of aluminum production as well as power efficiency, smelting cell design has always been a compromise between capital expenditure and energy costs. Amperage higher than the early 10 ka cells was desirable in any case, but the selection of current density was dictated by the cost of power relative to the cost of capital and labor. Cells were designed to the economics of the region in which they were to be operated.

In the period 1935-1973 the relatively low cost of power in the United States caused cell development to be concentrated on the productivity and low capital investment, at the expense of power efficiency, by using high current densities (0.95 A/cm^2). In many other parts of the world, high energy costs caused increased capital and labor costs to be traded off for higher power efficiency. In all cases, however, as cell size increased, significant gains were made in improving power efficiency.

By the end of World War II the aluminum industry had reduced smelting electrical requirements from the 1940 level by one-third, and today high capacity energy efficient cells are operating at 180 ka (1370 kg Al/cell day) and expend 13.2 kWh/kg. The new experimental prototype Alcoa Smelting Process cells (ASP) produce more than 13,500 kg/cell day by electrolysis of $AlCl_3$ at less than 10 kWh/kg (Figure 20).

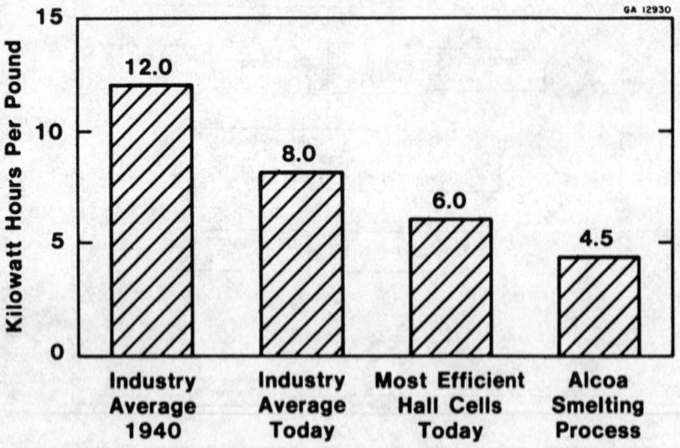

Figure 20. Power Reduction in Aluminum Smelting

The improvement in Hall-Héroult technology is the cumulative result of advancements in anode size and composition; control of cell magnetic turbulence; the introduction of solid state rectifiers; and computer control of cell operation. By 1945 prebaked Hall cells had been developed from the early 10-ka cells to greater than 75 ka. To this point scale-up had been relatively simple and developed on a trial and error basis. As cell amperages approached 100 ka, it became apparent that magnetic effects and thermal balance were factors of such significance that further advantages of scale could not be taken without more fundamental knowledge. Studies on 1/16 scale models and full scale prototypes, and field measurements on existing cells were made at Alcoa. These investigations, coupled with published data, resulted in several computer models used in cell design and operating analysis. They include a one-dimensional equilibrium model and a three-dimensional finite element heat and current flow model as well as a one-dimensional, dynamic model that can predict relative operating and economic results for varying power inputs and cell operating conditions. Using this sophisticated computer modeling, it is now possible to design cells with greater trade-off capabilities between capital and energy costs without extensive laboratory or pilot scale experimentation. Table V illustrates the Alcoa 697 cell and the 10% variation in production and 5% variation in power consumption possible by current adjustment alone. Cathode bus sizing, lining insulation, and magnetic patterns were predetermined, and the predicted stability of the metal pool at depths of only 10-12 cm verified by measurements made in the cells. Flow vectors are below 0.1 m/sec and the pool is flat within nominally ± 6 mm. Changes in insulation allow even greater flexibility in the trade-off between capital investment and power cost.

The highest productivity design known to be in operation today is the Alcoa P-225 cell (Table VI).

Table V (54)

Alcoa 697 Design and Operating Data*

	Design	170 kA	180 kA	190 kA	Average Total Data
Amperes	180,000	170,590	180,190	189,670	181,210
Volts/Pot	4.2	4.13	4.17	4.20	4.18
C.D. A/cm^2	0.72	0.68	0.72	0.76	0.73
Kg Al/day	1350	1302	1369	1402	1366
kWh/kg Al	13.5	13.0	13.2	13.6	13.3
CE, %	93	94.6	94.1	91.6	93.5
Days Operated	--	122	296	218	1322
Gross Carbon	0.60	NA	NA	NA	NA
Net Carbon	0.45	NA	NA	NA	NA

*2 test cells at Badin (North Carolina) Works, 77/05-79/04.

Table VI (54)

Alcoa P-225 Design and Operating Data

	Design	Actual Performance*
Amperes	225,000	231,200
Volts/Cell	4.5	4.6
Anode Current Density (A/cm^2)	0.87	0.89
Prod.(kg/Cell Day)	1620	1698
Power Consumption (kWh/kg)	14.9	15.1
Current Eff. (%)	89.4	91.0
Gross Carbon (kg C/kg Al)	0.595	0.581
Net Carbon (kg C/kg Al)	0.450	0.453

*Massena (New York) 1978 year

This cell was designed to operate at 225 ka for high productivity and reasonable power consumption using a high degree of automation to maintain optimum operating conditions and achieve low labor requirements. Other features include a tight hooding system with dual volume draft, automatic alumina feeders, and an air conveying system for alumina delivery to the cells. At these high current loadings, severe magnetic disturbances required intricate bus arrangements to insure smooth operation. Better operation was achieved by computer control of 16 pairs of anodes moving independently. This required an enormous amount of data, accomplished by using three Mod Comp II computers on line at all times (backups were also maintained). This enables these cells to be operated with the same degree of magnetic control as the 180 kA cells. A lower current density version of even higher power and current efficiency is under development.

Future Developments in Hall-Héroult Cells

Opportunities exist for improved efficiency throughout the aluminum manufacturing process from mine to product, ranging from increased Bayer Process production to energy and productivity improvements in ingot casting, rolling, and other fabrication processes. Because 75% of the total process energy is used in smelting, it is in cell design and operation that we look for major improvements in power efficiency and capital intensity; and the scope of this section is confined to smelting process development.

Electrolytic processes are inherently slow and rate of production is determined by the ampere load. Relatively small unit production of no more than 1727 kg/day for the larger aluminum cells (234 kA) accounts to a large extent for the inherently high capital intensity in smelting. The need to replace the consumable carbon anodes is another important factor affecting capital intensity as it requires the cell to be easily accessible from above. The metal must be made at a single horizontal surface, and the advantage of the third dimension, height, is denied.

The high cost and possible shortage of electrical energy have forced the aluminum industry from small high current density cells aimed at the lowest possible capital intensity to large low current density designs of higher capital intensity but lower energy consumption. A major challenge to the industry is to learn how to increase power efficiency while decreasing capital intensity; and much effort is being expended in this area.

Other factors affecting the direction of smelting R&D are the projected 1985 shortage of low sulfur petroleum coke and coal tar pitch and regulations setting the permissible evolution of fluorides and sulfur compounds.

The driving forces directing aluminum smelting development then are: (a) energy cost and availability; (b) capital intensity; (c) low sulfur coke, pitch cost and availability; (d) environmental constraints.

Current aluminum industry R&D programs are addressing these problems. Capital intensity and high power costs are being approached through improved cell and magnetic field control; closer control of anode formulation and baking; the development of inert anodes and cathodes; and waste heat recovery. Solutions for the shortage of low sulfur coke and the attendant environmental issues may be found in the blending of petroleum coke with low sulfur coke and the development of an inert anode, leading to a totally enclosed cell.

Anode Cathode Separation. As shown in Figure 21, 38% of cell power (voltage) is lost due to the electrical resistance of the electrolyte.

GA 11994

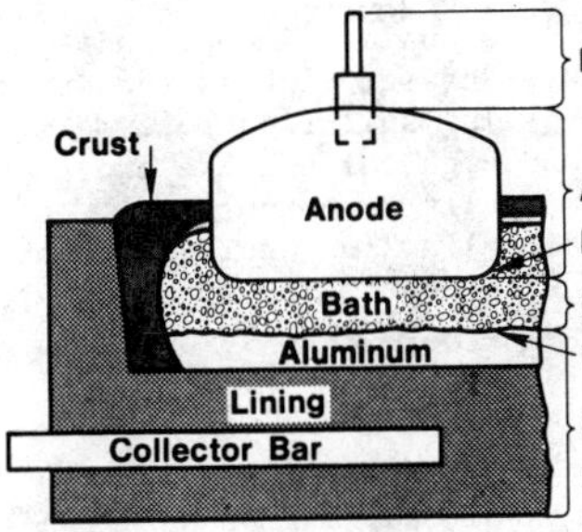

Voltage Component	Volts	% Of Total
External	.16	3.4
Anode	.32	6.9
Polarizations	.60	12.9
Bath	1.76	38.0
Decomposition	1.20	25.9
Cathode	.47	10.1
Other	.13	2.8
Total	4.64	100.0

Figure 21. Cell Voltage Drops

Reducing the A-C separation is one way to decrease this loss. There are limits, however. To prevent shorting due to magnetic turbulence, the A-C separation must always be slightly greater than the peak height of the molten aluminum waves. This minimum separation (4.0-4.5 cm) results in waste heat, generated by the electrolyte resistance, that must be discarded and is largely responsible for the low power efficiency.

Solving this problem would yield the greatest improvement of power efficiency of any of the development programs discussed. If the interelectrode spacing could be reduced from 4.5 cm to 2.1 cm without loss in current efficiency (and reduction in cell life), there would be a 20% savings of the energy input to the cell, and it would theoretically result in a reduction of smelting power consumption from 13.2 kWh/kg to 10.6 kWh/kg of aluminum in the most modern Hall-Héroult cells. The effect of further reduction in the interelectrode spacing is shown below.

Table VII

Effect of Interelectrode Spacing on Power Consumption (Current Eff., No Change Assumed)

Interelectrode Spacing, cm	% Power Savings	kWh/kg Industry Average	kWh/kg Modern Cells
4.5	Present practice	17.6	13.2
2.13	20	14.1	10.6
1.27	27	12.9	9.6
0.96	30	12.3	9.2

It is not possible to fully realize these savings for several reasons. The influence of interelectrode spacing on current efficiency has yet to be resolved, but loss can occur if careful attention is not given in cell design to minimizing shorting and reoxidation of reduced species at the anode. A certain rate of heat transfer through the cell walls is also needed to maintain ledge.

During the last 20 years, control of magnetics through bus and collector bar design has allowed cell current capacity to reach 234 kA at heretofore acceptable power efficiencies, but a breakthrough in design for metal motion control or cathode configuration is needed for the larger cells necessary for profitable operation at future projected capital and energy costs.

Figure 22 represents a cell, illustrating the interaction of the current path and magnetic fields.

GA 12932

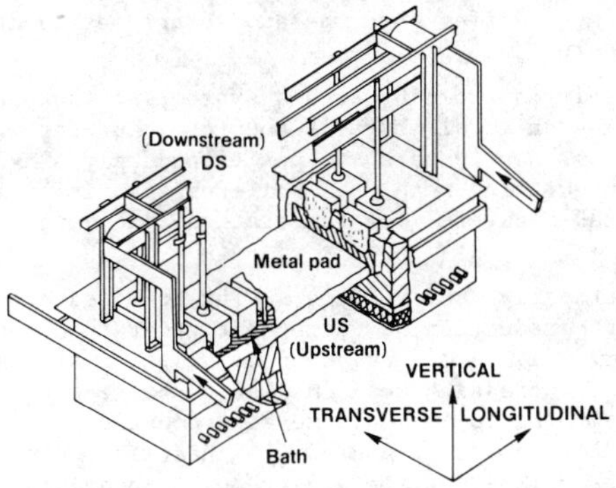

Figure 22. Cell Magnetic Circuit

The current enters through the end risers (see arrows), crosses longitudinally to the anodes, passes down through the electrolyte to the metal, through the lining to the collector bars, and out through the bus to the next cell. This complex current pattern is the source of a complcated pattern in magnetic fields

In recent years, measurements on many cell designs have been correlated into mathematical models that permit the generation of diagrams of these fields (72) as shown in Figure 23.

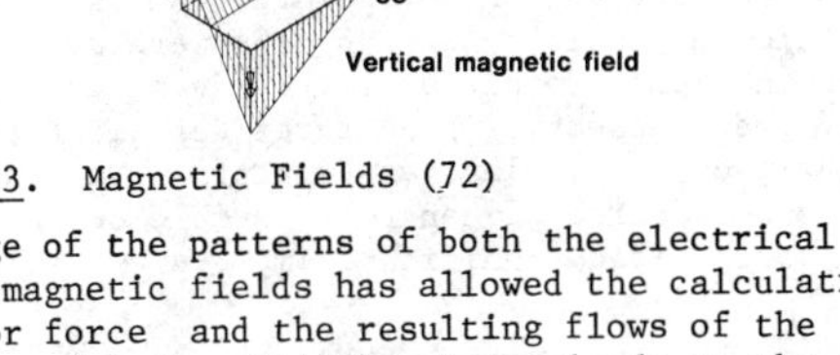

Figure 23. Magnetic Fields (72)

Knowledge of the patterns of both the electrical currents and the magnetic fields has allowed the calculation of the motor force and the resulting flows of the aluminum and bath. This knowledge leads to the development of the next generation of conventional Hall-Heroult cells that, with a relatively high chance of success, could be designed to operate at 300+ kA and >92% current efficiency at a calculated A-C spacing of <2.5 cm. If successful, this design could yield a reduction in power consumption and a significant decrease in capital intensity.

In addition to the development of new cells, this technology is particularly applicable to older cell designs operating at high current density for maximum productivity. Current loadings are frequently as much as 50% over the original design value. For existing cells, the goal is to reduce interelectrode spacing without loss in current efficiency with minimum additional capital investment. It may be possible to increase current efficiency as much as 2% and reduce energy consumption by ~1 kWh/kg while obtaining a productivity increase of as much as 40%.

Anode-Cathode Separation Control. To derive the benefit of reduced A-C spacing, the need for precise control cannot be overemphasized. The consumable anode is a deterent to precise control of A-C distance. In a cell with 24 anodes, each is at a different stage of consumption and therefore has unique electrical properties. While oxidation of the anodes according to the overall reaction $Al_2O_3 + 3/2\ C \rightarrow 2Al + 3/2\ CO_2$ is largely offset by metal accumulation, individual anodes must still be adjusted so that each operates at the same amperage. Since new and partially consumed anodes have different internal electrical resistances and different areas, however, the A-C separation must differ when the anodes are set to draw the same current.

Precisely controlling such a system, so that each anode carries an equal share of the total current with the anode most nearly consumed positioned only slightly above the undulating metal pad while substantially avoiding shorting, takes careful instrumentation and computer control.

Theoretically, the voltage saving from reducing A-C spacing (the reduction of waste heat generation) can be used either: (a) to reduce power consumption per kg of aluminum, increasing cell insulation to reject less heat, or (b) to increase metal production by increasing cell amperage and holding heat loss constant. In practice, neither of these extremes is possible. In any particular cell design a calculable rate of heat transfer (40% of the total heat rejected from the cell) is necessary to have a layer of frozen electrolyte on the sidewalls for economical cell life and maintenance of a proper heat balance for temperature control. Therefore, for each cell design with its bus bar and rod and clamp current capacity, one must accept a compromise. The voltage saving must be taken as a combination of increased production and power savings and results in both the reduction of capital intensity, through greater metal production, and improved power efficiency.

Selection and Control of Bath Composition. Modification of electrolyte composition offers another opportunity to improve power efficiency and productivity. Current efficiency and electrolyte resistance are influenced by AlF_3, LiF, and Al_2O_3 concentration. The effect of the variables on current efficiency and bath conductivity (voltage loss) were discussed under Physical Properties and Current Efficiency. This information has been incorporated into the cell design and operations computer models (State of the Art), which are used to predict the optimum operating conditions for a particular cell.

Although average smelting plant current efficiencies in excess of 90% are attained regularly, still higher averages would be possible if each cell were at optimum operating conditions. To discover the advantages to be obtained if the necessary expenditures were made to secure tighter control, plant experiments were conducted to determine the degree of control possible on the variables that would yield significant improvement. Results are shown below.

Table VIII

Controlled Experiment

Variables	Mean Value %, Target	Range %, Target
NaF/AlF_3 ratio	99.2	99.4-105
% Al_2O_3	92.5	42.0-207
% CaF_2	98.0	83.0-117
Bath temp.	99.9	98.8-106

Control in maintaining the mean value percent of target was good, but the range indicated that for a significant fraction of time cells were not operating under optimum conditions. This indicates that just maintaining the desired mean value of the variables at all times would have yielded a 1% current efficiency gain, and holding the optimum combination could yield as much as a 4% increase.

The ratio of NaF to AlF_3 and the concentration of such additives as LiF and CaF_2 are important, but the greatest effect is that of Al_2O_3 concentration. At present, anode effect is used to control Al_2O_3 concentration and avoid exceeding the saturation point of the bath and sludging the cell. After anode effect, Al_2O_3 is added at a rate slightly under that required to maintain concentration until anode effect again occurs. If anode effect occurs every 16 hours, however, 1% of aluminum production is lost. Controlling anode effects by computers coupled to automated Al_2O_3 feeders reduces production loss to <0.3%, and there is an excellent possibility anode effect could be eliminated during well controlled normal cell operation. while maintaining optimum Al_2O_3 concentration.

Continuous on-line analysis of electrolyte composition, accurate, dependable economic Al_2O_3 weigh feeders, and continuous temperature and heat loss measurements are needed if ultimate use is to be made of computer control and the control loop is to be closed. Research is underway to develop the sensors that could result in consistent commercial cell operation at current efficiencies in excess of 95% with significant increase in productivity.

Prebaked Anodes. Greater than 75% of the world's petroleum coke is made in the United States; it is a by-product of upgrading heavy oil residuals. Recent indications are that with the greater use of high sulfur crudes there may be a world shortage of low sulfur petroleum coke by 1985. Presently low and high sulfur coke are blended to achieve an environmentally acceptable cell effluent (~3.0% sulfur in the coke). Since sulfur is not removed by the present dry fume scrubbers it may be necessary to develop techniques for removing the sulfur from the coke prior to anode manufacture or for modifying or adding to the present fume scrubbers to avoid discharge of sulfur compounds to the atmosphere.

The perfect anode is low in Si, Fe, and S; has high electrical conductivity, high density for extended life; is crack resistant under thermal shock; and is resistant to dusting during electrolysis. At total conversion to CO_2 and 100% current efficiency, 0.33 kg of carbon is consumed for each kilogram of aluminum produced. Average American carbon consumption, however, is 0.45 kg C/kg Al, making carbon use about 72% efficient, but since this is influenced by current efficiency, only about 15% improvement is possible. These losses are due equally to air burning and carbon dust that enters the electrolyte. Dusting not only causes inefficient carbon utilization but also reduces current efficiency, limiting amperage and increasing unwanted heat generation. Although only about 10% of the unwanted waste heat generated in the Hall-Héroult cells is due to anode resistance (Figure 21), it is significant and good electrical conductivity is important.

Bench scale studies have correlated well with plant practice, and optimum anode manufacturing technique is known. To implement this information, anode formulation and manufacturing practice must be modified as raw material quality varies. For different cokes the maximum density occurs at different aggregate sizing and pitch concentration, and a time-consuming trial and error procedure is required to determine optimum composition. Because the need to change formulation is more rapid than the control laboratory determination, the optimum green anode is rarely achieved. Eliminating the quality monitoring barriers largely depends on the successful development of continuous sampling of coke particle size, control of mixing and forming temperatures and pitch content, and the development of a procedure to rapidly predict the maximum density composition region from selected densities of individual coke fractions. This sophisticated closed loop feed-back system, when linked to automated mixing equipment, yields the best green anode composition for any specific raw material.

To reduce dusting and loss of cell current efficiency, anodes should be baked to ≥1100°C. Although recent advances in the design and insulation of ring furnaces in which green anodes are baked have resulted in significant improvements in fuel efficiency and productivity, some anodes in corner zones are still baked as low as 900°C, and overall fuel efficiency is <40%. The volatiles alone leaving the binder pitch in each anode during baking contain enough energy to bake the anode and account for >45% of the energy entering the baking furnace. Water modeling of gas patterns in ring furnaces has shown that further improvements in productivity (tons C baked/furnace yr), fuel efficiency (tons C baked/1 x 10^6 Btu), and heat distribution are possible through furnace design changes, insulation, and sealing to reduce air in-leakage. Sufficient air must be drawn into the system, however, to avoid fuming at the fluid coke cover on the baking pits, and hydrocarbon fumes leaving the exhaust fans must be treated as in the Alcoa 446 unit.

These programs should result in significant energy savings and cost reduction in manufacture, as well as better utilization of raw materials through reduced dusting. It is expected that significantly improved current efficiency in cells, as a result of this program, should increase production and reduce power costs. The quantified extent of these gains must await quantity production of the optimum industrial anode.

Nonconsumable Anodes. From the very inception of Hall-Héroult smelting, the aluminum industry has searched for an anode material with all of the attributes shown in the following table.

Table IX

Requirements for Inert Anode Material

Low reactivity with oxygen and bath.
Adequate mechanical strength and resistance to heat shock.
Does not degrade aluminum product
Stable electrical connections to bus.
Ability to fabricate into large shapes.
Relatively low cost, readily available.
High electronic conductance.

The only known materials with any chance of success are metallic oxides, but a suitable oxide or combination of oxides remains elusive. The advantages of inert anodes are so great, however, that substantial research effort continues. Some of these benefits are reduced electrode separation; closed-loop recycling of fluoride emissions; low labor intensity; accurate and equidistant A-C separation for large, multielectrode cells; elimination of baking ovens; freedom from petroleum coke; and the production of a moderately valuable byproduct, oxygen.

Attaining uniform individual anode current loading with consumable anodes is difficult, and only the most modern, well instrumented cells operate continuously near optimum conditions. Permanent anodes would retain their shape and present a uniform surface to the electrolyte. Such an assembly, with each of the anodes carrying the same current, would be more amenable to adjustment for minimum voltage drop.

When CO_2 is formed at the carbon anode, the anode can be considered to act as a fuel cell electrode, reducing the decomposition voltage for Al_2O_3 from .2V to 1.2V at 960°C. When oxygen reacts with a carbon anode, however, the overvoltage is ~0.5V higher than for the simpler discharge of oxygen on an inert anode. Thus, the penalty in substituting an inert anode for a carbon anode is ~0.5 volt. This translates to ~1.7 kWH/kg increase in electrical energy consumed per kg of aluminum results in ~3 kg of coal burned at the power plant in place of 1 kg of anode carbon burned in the cell.

Table X shows a comparison of the cell voltages in retrofitted cells under these considerations and also at reduced A-C distance.

Table X

Voltages for Retrofitted Smelting Cells

Present Hall Cell[a] A-C = 4.45 cm	Inert Anode[b] A-C = 4.45 cm	Inert Anode[c] A-C = 1.9 cm
4.64	5.19	4.18

There is an overall cell voltage saving of 10% if the A-C separation can be reduced by the expected 57%.

A total energy balance, including expended gas or oil and electrical power in anode carbon manufacture and the energy content of the carbon, indicates an overall decrease in energy consumption of ~5% if interelectrode spacing remains unchanged. If a 57% reduction in the interelectrode spacing can be achieved, the decrease in total energy consumption is ~22%, including an electrical energy reduction of ~1.5 kWh/kg below present practice in the best cells (Table XI).

Table XI

Energy Savings Using an Inert Anode

	% Decrease	Energy Savings x 10^{15} Btu/yr Industry	Alcoa
Standard A-C Distance (4.45 cm)	5	0.04	0.010
Reduced A-C Distance (1.9 cm)	22	.170	0.05

Research and development of nonconsumable anodes has been reviewed (73).

Cathode Development. In a modern Hall-Héroult cell the bottom lining consists of prebaked carbon blocks joined by a carbonaceous seam mix placed on ceramic insulation. Electrical connection is made through steel collector bars, carbon cemented or metal cast in grooves in the bottom of the lining blocks and protruding through the sides of the steel shell. Cells must be relined when the electrolyte penetrates the cathode block and insulation below, and causes expansion of the lining, seam cracking, and aluminum penetration through chemical reaction.

Some of the factors affecting lining life are quality of the carbon block, care in insulation and seam mix, formulation and tamping, method of baking, method of startup, sodium intercalation, and erratic cell operation. Design changes resulting from careful autopsies of failed cells have resulted in a 25% increase in cell life over the past 10 years. Increased life also reduces energy costs through maintaining the integrity of collector bar contact. With this rate of improvement it may be possible to design the next generation of conventional Hall-Héroult cell, previously discussed, with an expected life of ≥2000 days.

Stabilized Cathode. So long as interelectrode spacing is dictated by the magnetic turbulence in the aluminum pad, wide A-C spacing will impair power efficiency. Although magnetic turbulence and flow controls may be successful in producing a more quiescent metal pad to permit some reduction in A-C spacing, stabilization of the metal pad through such control will never be totally successful.

An absolutely stable cathode can only be realized if the cell is operated without a molten metal pad on the cathode. Unfortunately, bare cathode operation is prevented by dissolution of the carbon cathode when alternately exposed to metal and electrolyte. The probable mechanism is the formation of an aluminum carbide layer that subsequently dissolves. Concentrated research is being directed toward the development of a thin-metal-layer cathode. The desirable properties of the new cathode material are shown in Table XII.

Table XII

Desirable Cathode Properties

A. Inert to the environment at 960-1000°C Electrolyte (Na_3AlF_6-AlF_3-CaF_2-Al_2O_3) and molten aluminum
B. Wet by aluminum
C. Good electrical conductor resistivities less than 150 μΩ-in.
D. Withstand severe differential thermal gradients imposed during cell startup and operation.

In 1957, C.E. Ransley discovered a class of materials that seemed to meet all of these properties except resistance to thermal shock. They are the borides and carbides of Ti and Zr refractory hard metals (RHM). Such materials are costly and must be fabricated using powder metallurgy technology with all its inherent disadvantages. Potential RHM parts suppliers are working with individual aluminum producers to develop inert cathodic surfaces with sufficient life and low enough cost to be economically viable. There are many different designs, but all are directed toward using the minimum amount of costly materials in the simplest configuration while gaining maximum advantage of interelectrode space reduction (74).

Alcoa has received a patent on hollow shapes filled with molten aluminum that rise above the metal pad (Figure 24) (75).

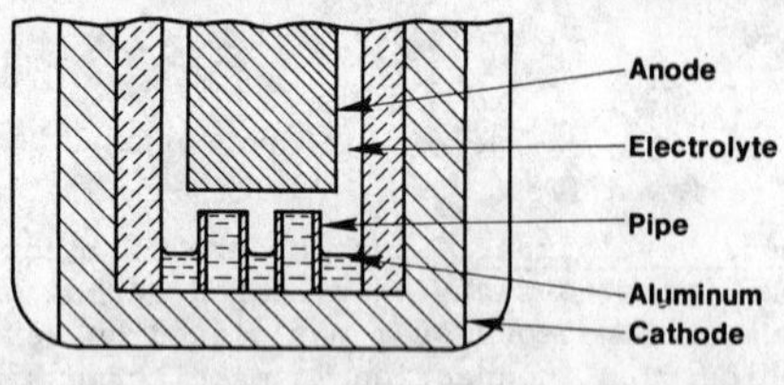

Figure 24. Low Energy Hall-Héroult Cathode Design

Should this or another design prove successful, A-C separation could be more than halved and energy savings would approach 23.%. If both the inert anode and the stable cathode were successful, we might expect to see power consumption in smelting cells reduced as much as 25% and the industry average approach 11 kWh/kg.

Heat Recovery and Sidewalls. Some method of recovering excess heat from the smelting cells must be developed to further improve power efficiency. All steps possible to avoid generating unwanted heat must be taken, but no matter how successful such projects are, heat rejection will always be necessary. The goal is to recover a large fraction of this heat at the highest possible temperature, directly from the cell. If heat exchangers that use a gas or low melting metal could be developed for the cell sidewalls to extract the required amount of heat directly from the electrolyte, temperature control independent of the electrical system could be achieved. This would allow operation at optimum energy efficiency and provide a cell that could take a more variable ampere load than presently possible. Even though up to 50% of the excess heat leaves by way of the anodes and the top of the cell, sufficient variability remains to operate a well insulated cell at a minimum base load with just enough heat being removed by the sidewall heat exchangers to maintain a frozen, thin protective bath layer. If the sidewall heat exchanger could be fabricated from inert material, this bath layer would be unnecessary. When a change in ampere load was desired, more or less heat would be removed by increasing or decreasing the velocity of the heat transfer fluid or the surface area of the heat exchanger sides by opening and closing channels. The cell could then be operated at minimum electrode spacing while maintaining temperature control, regardless of amperage within the design range, The heat recovered at the base load could be used for cogeneration, reducing cell power requirements. Excess heat at the higher load could be used in some other process that operated intermittently.

Two major barriers to such heat recovery are the large number of cells used for smelting and a suitable sidewall material. Costly and elaborate collection systems would be required for this. A truly variable heat balance cell may have to await the very large cells used in the Alcoa chloride electrolysis process (ASP).

Alternate Processes

Aluminum Chloride Electrolysis. Aluminum was prepared by electrolysis for the first time in 1854 when Bunsen and Deville independently obtained solid aluminum from fused $NaAlCl_4$. Since 1976, Alcoa has operated a smelting plant in Anderson County, Texas, based on electrolysis of $AlCl_3$.

Special Bayer alumina is reacted with chlorine and carbon at temperatures above 600°C to produce $AlCl_3$ and CO_2.

$$\gamma Al_2O_3(s) + 3/2\ C(s) + 3Cl_2(g) \rightarrow 2AlCl_3(g) + 3/2\ CO_2(g) \qquad [34]$$

Reaction [34] is exothermic (ΔH_{900K} = -27.3 kcal) and the equilibrium is quite favorable (ΔG_{900K} = -70.5 kcal).

The aluminum chloride is dissolved in molten chlorides and is electrolyzed at ~700°C to molten aluminum and chlorine, with the chlorine being recycled to the chemical reactor. The cell reaction can be represented

$$AlCl_3(liq) \rightarrow Al(liq) + 3/2\ Cl_2(g) \qquad [35]$$

$$\Delta H_{1000K} = 156.8\ \text{kcal},\ \Delta G_{1000K} = 121.2\ \text{kcal}$$

The reversible cell potential at 1000K, calculated from the relation $\Delta G = -nFE°$ becomes $E°_{1000K}$ = -1.75V.

Since $AlCl_3$ dissolves in the electrolyte as complex anions, its activity is less than unity and the observed voltage for electrolysis is somewhat greater.

The electrolytic cell is represented schematically in Figure 25.

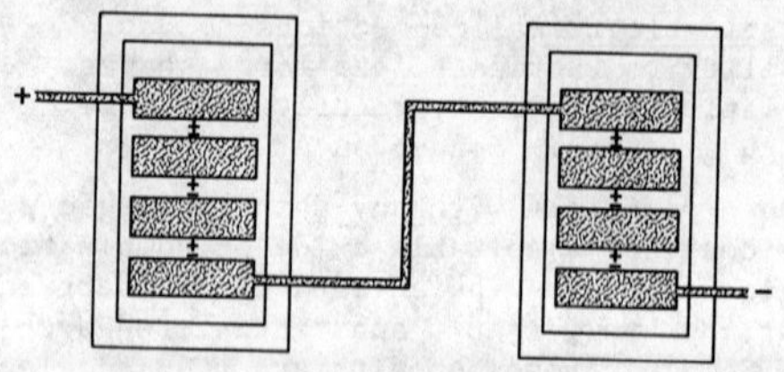

Figure 25 Bipolar Electrode Cells

The carbon electrodes are of bipolar configuration; on each electrode, chlorine is discharged on the bottom and aluminum on the top surface, except for the terminal anode and cathode. The small A-C separation, high electrical conductivity of the electrolyte, and low voltage drop in the electrode system make the process energy efficient. More than 13,500 kg of Al/cell day have been produced at electrical power consumption of less than 9.5 kWh/kg. The nonconsumable electrode makes a completely closed system possible.

Cell operation has been effective with respect to power efficiency, productivity, and cost. The productivity of the chemical train must be increased to reduce capital intensity and make $AlCl_3$ production more cost effective.

Carbothermic Reduction. Aluminum or aluminum alloys may be produced by direct, carbothermic reduction of Bayer Al_2O_3 or aluminous ores and SiO_2 at 1950-2200°C (62). Yield and composition depend on the silica-to-alumina ratio (63), the concentration of other metal oxides, and contact efficiency of the carbon to the ore.

Carbothermic reduction provides several advantages over electrolysis. Capital and operating costs are lower since direct reduction reactors typically have high space time yields in contrast to electrolytic cells. Fluorides are not required. Bauxite can be reduced directly and domestic ores can be used. The carbon may be derived from coal. Total electrical energy is less than the Bayer-Hall-Héroult Process if credit is taken for byproducts. Carbon monoxide may be used to generate power or produce gasoline or other hydrocarbons. Direct reduction requires efficient, cost effective refining steps to recover product and byproduct values.

To date, direct reduction has not been commercially successful due to low yields, difficulty of containing reactants at high temperatures, inefficient refining techniques, and long term commitment of money and manpower necessary to develop a feasible process. The only known commercial aluminum carbothermic operations are located in the USSR, where aluminosilicate ores are smelted in electric arc furnaces to produce 60% Al-Si alloys (64-65).

In cooperation with the U.S. Department of Energy, Alcoa initiated a program to determine the technical feasibility of a 3-step direct reduction process for aluminosilicate ores, including reduction, alloy refining, and aluminum purification. Three different reduction concepts were evaluated: a combustion heated shaft reactor in which oxygen was injected to burn excess coke; a submerged arc electric furnace; and a combination blast arc reactor utilizing coke combustion to supply heat for prereduction of silica to silicon carbide, followed by metal production in an arc furnace.

The aluminum blast furnace was not feasible at atmospheric pressure due to hold-up of volatiles by the high counter-current flow of combustion gases. The best empirical result was 18% aluminum recovery, with 23% combustion heat. Calculations predicted that a pressure of 12 atm would produce 48% aluminum recovery at 2300°C, utilizing all combustion heat. An electrically heated shaft reactor operated at atmospheric pressure and 2100°C in which complete reflux of volatiles was attained resulted in 100% aluminum recovery.

Production of Al-Si alloy was demonstrated in a 100-kW submerged arc furnace on pellets containing kaolin clay, Al_2O_3, and metallurgical coke. The alloy product contained 1-2% carbon as compared to 5-10% in the electrically heated shaft reactor product. Simulation of the use of the shaft furnace for prereduction of silica was also demonstrated, resulting in 70-90% conversion to silicon carbide at 1600°C. Comparison of energy requirements for these processes is shown in Table XIII. The alloy is 85% Al-15% Si with energy charges for purification and credits for CO combustion value and energy content of byproduct silicon and ferrosilicon. For cell grade aluminum, additional energy charges were made for second stage purification. All electrical energy was charged to the main products 85% Al-15% Si alloy and cell grade Al(99.85) and assumed to have been generated by fossil fuel at 10,500 Btu/kWh.

Table XIII

Energy Requirements for Carbothermic Reduction In Ground Ore to Ingot

	85% Al-15% Si Alloy			Cell Grade Al(99.85)		
	Arc	Blast Arc	Hall + Si Arc	Arc	Blast Arc	Bayer Hall
Total Energy 10^6 Btu/st	111	122	220	128	149	227
Electrical portion only kWh/kg	13.6	11.4	18.0	18.5	15.8	18.7

An Al-Si arc furnace is most efficient in total energy but the blast-arc is most efficient in electrical energy. If the CO byproduct is used to generate recycle electrical energy the requirement for 99.85% Al is reduced to 7.1 kWh/kg.

The Aluminum Industry

Extimated 1979 world production of primary aluminum totaled ~14.5 x 10^6 tonnes (Table XIV)(66). Over the 1969-1979 period, world aluminum production increased at an annual growth rate of 4.9%. The corresponding annual growth rate for the preceding 1955-1969 period was 6.2%.

North America accounted for 37.6% of the world 1979 production and Europe, 40.2%. The remainder came from Asia (12.4%), South America (4.1%), Oceania (2.9%), and Africa (2.7%). The countries with the largest production were, in descending order: United States, USSR, Japan, Canada, West Germany, and Norway. Together, these six nations produced about 9.6 x 10^6 tonnes of aluminum in 1979, over three-fifths of the total. There were 41 countries throughout the world producing primary aluminum in 1979, up from 33 a decade ago.

The world's primary aluminum smelting capacity is expected to be 17.8 x 10^6 tonnes/yr by 1985, reflecting an annual growth rate of 3.4% for 1979-1985 period (67).

Table XIV

World Aluminum Production (Thousands of Metric Tons)(66)

Region/Country	1979	1974	1969	1955
North America - Total	5,460	5,521	4,462	1,834
Canada	850	1,023	981	509
United States	4,566	4,457	3,448	1,325
South America - Total	602	212	127	1
Brazil	240	126	56	1
Asia - Total	1,807	1,524	840	65
Japan	1,013	1,120	570	53
Africa - Total	395	280	161	--
Oceania - Total	425	331	126	--
Australia	270	220	126	--
Europe - Total	5,836	5,235	3,270	920
West Germany	744	690	264	129
Norway	674	664	509	61
USSR	1,727	1,436	1,054	360
World - Total	14,525	13,103	8,986	2,820

Major primary aluminum smelting expansions in the 1980's will occur in Australia and Brazil where electrical energy and bauxite are relatively inexpensive and abundant. In the 1979-1985 period, Australia's annual smelting capacity is expected to increase 31%/yr from 0.27 x 10^6 to 1.36 x 10^6 tonnes. Over the same time period, Brazil's annual capacity is expected to increase 16%/yr from 0.24 x 10^6 to 0.58 x 10^6 tonnes (67,68,69).

A bright future is predicted for the aluminum industry, with the industrialized nations expanding capacity to meet their home markets and the rapidly increasing demand by the undeveloped nations being met by large expansions in those countries where both raw material and hydropower are found.

Increased recycling will provide a low energy source of an increased supply of metal for the future needs of the automotive industry, affordable bauxite will be plentiful, and environmental problems will be overcome by development programs now underway.

REFERENCES

1. D. Altenpohl, Materials in World Perspective, Springer-Verlag; Berlin, Heidelberg, New York, 1980.
2. T.G. Pearson, The Chemical Background of the Aluminium Industry, The Royal Institute of Chemistry, London, 1955.
3. Extractive Metallurgy of Aluminum, Vol. I, Alumina, Edited by Gary Gerard, Interscience Publishers, 1963.
4. A.N. Adamson, "Alumina Production: Principles and Practice," The Chemical Engineer, 239 (171) 1973, p. 156-164.
5. Light Metals 1981, Proceedings of Technical Sessions Sponsored by TMS Light Metals Committee, 110th AIME Annual Meeting, 1981 Feb. 22-26.
6. D.R. Swinburne, "Alternative Techniques for Alumina Production," Technologist (Minden Malays) 4, (1980) p. 38-49.

7. C. Abbruzzese, "New Processes for Aluminium Production," Ind. Min. (Rome) 1 (3)(1980) p. 49-58.

8. E. Barrillon, "The Merits and Demerits of Various Types of Bayer Aluminas," Erzmetall 3 (1978) p 519-522.

9. E.W. Lussky, "Experience with Operation of the Alcoa Fluid Flash Calciner," Light Metals (1980) p. 69-79.

10. S.W. Boercker, "Energy Use in the Production of Primary Aluminum," Materials and Society 3 (1979), p. 153-161.

11. K. Grjotheim, C. Krohn, M. Malinovsky, K. Matiasovsky, and J. Thonstad, Aluminium Electrolysis - The Chemistry of the Hall-Héroult Process, Aluminium Verlag GmbH, Dusseldorf, 1977.

12. M. Rolin, Les Equilibres Chimiques dans les Bains d'Electrolyse de l'Aluminium, Institut National des Sciences Appliquées de Lyon, 1973.

13. M. Rolin, Le Procédé Héroult, Etude de l'Electrolyse, Institut National des Sciences Appliquees de Lyon, 1977; English translation by N. Philips, The Electrochemistry of the Hall-Héroult Process, American Society for Metals, 1980.

14. E.W. Dewing, "Thermodynamics of the System NaF-AlF_3. Part I: The Equilibrium 6NaF(s) + Al = Na_3AlF_6(s) + 3Na," Met. Trans. 1 (1970), P. 1691-94.

15. E.W. Dewing, "Thermodynamics of the System NaF-AlF_3. Part II: The Free Energies of Formation of Cryolite (Na_3AlF_6) and Chiolite ($Na_5Al_3F_{14}$)," Met. Trans. 1, (1970), p. 2211-15.

16. E.W. Dewing, "Thermodynamics of the System NaF-AlF_3. Part III: Activities in Liquid Mixtures," Met. Trans. 3 (1972), p. 495-501.

17. E.W. Dewing, "Thermodynamics of the System NaF-AlF_3. Part IV: The Cryolite Liquidus Curve," Met. Trans. 3 (1972), p. 2699-2701.

18. E.W. Dewing, "Thermodynamics of the System NaF-AlF_3. Part V: Solid Solution in Cryolite, Met. Trans 9B (1978), p. 687-90.

19. JANAF Thermochemical Tables, 2nd Edition, National Standards Reference Data Service, National Bureau of Standards, June 1971; and JANAF Thermochemical Tables, 1978 Supplement; J. Phys. Chem. Ref. Data 7 (3) 793 (1978).

20. D.F. Craig, R.T. Cassidy, and J.J. Brown, Jr., "A Review of Phase Equilibria in the Na_3AlF_6-LiF-CaF_2-AlF_3-Al_2O_3 System," National Bureau of Standards SP-496, Applications of Phase Diagrams in Metallurgy and Ceramics 1, Edited by G.C. Carter, U.S. Department of Commerce, 1978.

21. P. Fellner, M. Chrenkova-Paucirova, and K. Matiasovsky, "Calculation of Phase Diagrams of Multicomponent Systems Containing Cryolite," Chem. Zvesti 30 (6), (1976).

22. M.L. Saboungi, P.L. Lin, P. Cerisier, A.D. Pelton, "Computer Analysis of Phase Diagrams and Thermodynamic Properties of Cryolite Based Systems, Part I: The AlF_3-LiF-NaF System," Met. Trans. B 11B (Sept. 1980) p. 493-501.

23. W.E. Haupin and W.B. Frank, Chapter 5, "Electrometallurgy of Aluminum," Comprehensive Treatise of Electrochemistry, Vol. 2, Electrochemical Processing, Plenum Publishing Corp. 1981, p. 301-325.

24. H. Kvande, "Thermodynamics of the System NaF-AlF_3-Al_2O_3-Al Studied by Vapor Pressure Measurements", The University of Trondheim, Norway, 1977.

25. W.H. Tingle, J. Petit, and W.B. Frank, "Sodium Content of Aluminum in Equilibrium with NaF-AlF_3 Melts", Aluminium 57 (4) (1981) 286-288.

26. M. Paucirova, M. Malinovsky, and K. Matiasovsky, Rev. Roum. de Chim 17 (1972), p. 807-817.

27. V. Sato, A. Kojima, and T. Ejima, J. Japan Inst. Metals 41 (12), (1977), p. 1249.

28. B. Gilbert, G. Mamantov, and G.M. Begun, Inorg. Nucl. Chem. Letters 12 (1976), p. 415.

29. G. Choudhuri, J. Electrochem. Soc. 120 (1973), p. 381.

30. M. Rolin, "La Physicochemie des Bains d'Electrolyse de l'Aluminium," Institut National Des Sciences Appliqueies de Lyon, Villeurbanne, France, 1975.

31. W.E. Haupin, J. Metals 23 (10), (1971). p. 46. P.A. Foster, Jr., J. Am. Cer.Soc 58 (7-8) (1975) p 288.

32. J. Gerlach, U. Hennig, and H. D. Poetsch, Erzmetall 31 (1978), p. 496.

33. L.N. Less, Light Metals 1973, Proceeding of the 105th AIME Annual Meeting, February 1976, p. 315.

34. J. Thonstad, P. Johansen, E.W. Kristensen, Light Metals 1980, Proceedings of the 109th AIME Annual Meeting, February 1980, p. 227.

35. J. Thonstad, R. Oblakowski, Electrochim. Acta 25, (1980) p. 223.

36. R. Keller, O. Frei, Extended Abstracts of the 3rd Aluminium Symposium in Czechoslovakia, Banska Bystrica, Sept. 1976, p. 46.

37. K. Yoshida and E.W. Dewing, Met. Trans. 3 (1972), p. 1817.

38. K.A. Bowman, PhD Dissertation, Univ. of Tennessee, Knoxville, March, 1977.

39. J.J. Camp, J.P. Millett, and M. Rolin, Electrochim. Acta 26 (1981).

40. J. Thonstad and S. Rolseth, Proceedings of 3rd ICSOBA Conference, Nice, Sept. 1973, p. 657.

41. J. Thonstad, "Model of Diffusion Layer of Cathodic Deposition of Aluminum," Molten Salt Electrolysis in Metal Production, Inst. of Mining and Metallurgy, London (1977) p. 1-6.

42. H. Ginsberg and H.C. Wrigge, Metal 126 (1972), p. 997.

43. J. Thonstad and E. Hove, Canadian J. Chem. 42 (1964), p. 1542.

44. G. Blyholder and H. Eyring, J. Phys. Chem. 61 (1957), p. 682.

45. K. Grjotheim, C. Krohn, M. Malinovsky, K. Matiasovsky, and J. Thonstad, Aluminium Electrolysis, The Chemistry of the Hall-Héroult Process, Aluminium Verlag GmbH, Dusseldorf (1977), p. 211.

46. J. Thonstad, Electrochim. Acta 15, (1970) p. 1569.

47. G.A. Solinas, Light Metals, 1980, Proceedings of 109th AIME Annual Meeting, Feb. 1980, p. 243.

48. T. Hasimoto, H. Ikeuchi, Light Metals, 1980, Proceedings of 109th AIME Annual Meeting, Feb. 1980, p. 273.

49. J.W. Evans, Y. Zundelevich, E.D. Tarapore, D. Sharma, Extended Abstracts, 30th Meeting of ISE, Aug. 1979, p. 8.

50. J.M. Blanc, P. Entner, Light Metals 1980, Proceedings of 109th AIME Annual Meeting, Feb. 1980, p. 285.

51. E.D. Tarapore, Light Metals 1981, Proceedings of 110th AIME Annual Meeting, Feb. 1981, p. 341.

52. E. Dernedde, E.L. Cambridge, Light Metals 1975, Vol. 1, AIME (1975) p. 111.

53. J. Thonstad, J. Electrochem. Soc. 111 (1964), p. 955.

54. G.T. Holmes, D.C. Fisher, J.F. Clark, W.D. Ludwig, Light Metals 1980, Proceedings of 109th AIME Annual Meeting, p. 401.

55. J.P. Dugois, Light Metals 1975, Proceedings of 104th Annual AIME Meeting, p. 3.

56. W.E. Haupin, J.W. Luffy, Light Metals 1977, Vol. 1, Proceedings AIME Annual Meeting, p. 125.

57. J.Q. Peacey, G.W. Medlin, Light Metals 1979, Vol. 1, Proceedings AIME Annual Meeting, p. 475.

58. R. Keller, Proceedings 89th AIChE National Meeting, Aug. 1980.

59. U.S. Patent 3,839,019 (1974-10-01).

60. R.E. Miller, L.C. Blayden, M.J. Bruno and C.E. Brooks, "In-Line Fumeless Metal Treatment," Light Metals 2, (1978) p 491-505.

61. Aluminum Recycling Casebook, The Aluminum Association, 1979 March.

62. U.S. Patent 4,053,303, 1977 Oct. 11.

63. U.S. Patent 4,046,558, 1977 Sept. 06.

64. S.I. Khitrik, B.I. Emlin et al, "Smelting Silicon Aluminium Alloys from Nonenriched Kaolin," Soviet Journal of Nonferrous Metals (Tsvetnye Metally) 7, (6) 1966, June, p. 66-70.

65. Canadian Patent 977,157, 1975 Nov. 04.

66. Aluminum Statistical Review 1979, The Aluminum Association, p 54-55.

67. R.G. Adams, The TMS Light Metals Session, AIME Annual Meeting, Chicago, Illinois, 1981 Feb. 24.

68. T.G. Langton, J. Metals, Aug. 1980, p 12-16.

69. J.L. Mazel, 33 Metal Producing, Jan. 1980, p 56-61.

70. D.L. Belitskus, Process Chemistry and Physics Div., Alcoa Laboratories, Aluminum Company of America, private communication.

71. B.J. Racunas, Light Metals, 1980, Proceedings of 109th AIME Annual Meeting, Feb. 1980, p 525.

72. R.F. Robl, Light Metals, 1976, Proceedings of 105th AIME Annual Meeting, Vol. 1, p 97.

73. K. Billehang, H.A. Øye, Aluminium 57, (1981), p. 146-228.

74. K. Billehang, H.A. Øye, Aluminium 56, (1980), p. 642-713.

75. U.S. Patent 4,071,420 (1978).

Acknowledgements

The authors wish to acknowledge the valuable advice and critique of members of Alcoa Laboratories in the following areas: T.J. Sivakumar, Bayer Refining; Dr. R.E. Miller, Metal Purification; M.J. Bruno, Direct Reduction; and Dr. S.K. Das, Industry Projections.

ADVANCES IN THE SMELTING OF MAGNESIUM

Noel Jarrett
Aluminum Company of America
Alcoa Laboratories
Alcoa Center, PA 15069

Magnesium production is approached by illustration of major smelting methods with emphasis on electrolysis of magnesium chloride and metallothermic reduction of dolomite. Means for improving efficiency and cost effectiveness of competing processes now in use are discussed. Cells using pure anhydrous magnesium chloride have the greatest possibilities for electrolysis. Shortened cycle time, continuous operation, and use of a less energy intensive reductant show the greatest opportunity for the Magnetherm Process. Comparison of processes shows the greatest economic opportunity for: Magnetherm in capital saving; and electrolysis of pure anhydrous magnesium chloride in energy efficiency. A promising new market is automotive castings.

Introduction

Magnesium (Mg) is the eighth most abundant element and the third most abundant structural metal in the earth's crust. It is also found in seawater at a concentration of ~0.13% and in brackish wells and lakes. The concentration in the Great Salt Lake in Utah, U.S.A., is ~0.45%. Since Mg is widely distribured and is easily extracted commercially from all sources, supply is virtually unlimited, making it a strong candidate for substitution for less abundant elements. Substitution is desirable where density is important, and in many cases (particularly in transportation applications) plastics, aluminum, and ferrous alloys can be replaced with magnesium.

The first recorded preparation of magnesium was by Sir Humphry Davy in 1808 (1,2). He electrolytically reduced a moist paste of magnesium and mercuric oxides to form an amalgam, and then distilled away most of the mercury. In 1828 the first pure magnesium metal was produced by A. Bussy (3), who reduced fused magnesium chloride with metallic potassium. In 1833, Michael Faraday developed an electrolytic process for producing magnesium from fused anhydrous magnesium chloride (1). These early experiments laid the ground work for the two processes in use today, thermal reduction by metals and the electrolysis of magnesium chloride.

Currently there are four magnesium manufacturers in the United States, accounting for >50% of the world's production. In 1980, the American industry operated at ~85% of rated capacity, producing 163,000 Mt of magnesium, which along with the USSR's estimated 80,000 Mt and Western Europe's 64,000 Mt accounted for 97% of the world's production of ~300,000 Mt that year. (Figure 1) (4,5,6,7).

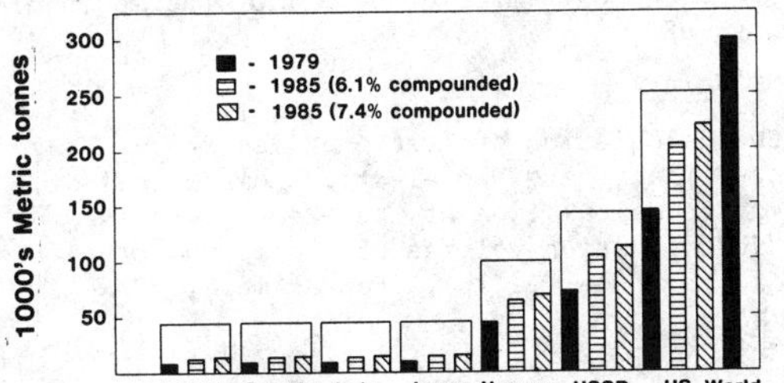

Figure 1. World Magnesium Production

Approximately 50% of the magnesium produced in the United States is used for Al-Mg alloys. Major uses are beverage containers, die castings, and light structural applications. Magnesium itself is used in flash photography, flares and munitions as well as chemical manufacturing, desulfurization of nodular cast iron, metal oxide reduction, and cathodic protection.

The greatest opportunitues for increased magnesium applications are in automotive die castings and light structural uses.

Overall growth rate of world production is estimated at 6.1% from 1977-86 or 7.4% from 1980-1985 (8). By that time, die castings for the automotive industry are projected to be 24% of the total, and if each of the present producers maintains his share of the market, production will be distributed as shown in Figure 1. With such a growth rate, magnesium has a bright future that could be accelerated by reduction process improvements (9,10).

Chemistry

Magnesium has an electrical equivalent of 0.455 Kg/1000 ampere hours, and at 174 Kg/m^3 is the lightest structural metal.

Reactions important to commercial magnesium extractive metallurgy are:

Electrolytic Processes

$$Mg + Cl_2 \rightarrow MgCl_2 \quad [1.0]$$

$$MgCl_2 + Ca(OH)_2 \rightarrow Mg(OH)_2 + CaCl_2 \quad [1.1]$$

$$MgSO_4 + Ca(OH)_2 \rightarrow Mg(OH)_2 + CaSO_4 \quad [1.2]$$

$$1/2\ C + MgO + Cl_2 \rightarrow MgCl_2 + 1/2\ CO_2 \quad [1.3]$$

$$MgO + Cl_2 + C \rightarrow MgCl_2 + CO \quad [1.4]$$

Thermal Processes

$$2MgO + Si \rightarrow 2Mg + SiO_2 \quad [1.5]$$

$$3MgO + 2Al \rightarrow 3Mg + Al_2O_3 \quad [1.6]$$

Some reactions for processes not in use but technically feasible are:

$$Mg(OH)_2 + Ca(OH)_2 + CaCl_2 + 2CO_2 \rightarrow MgCl_2 + 2CaCO_3 + 2H_2O \quad [1.7]$$

$$MgO + C \rightarrow Mg + CO \quad [1.8]$$

$$MgO + CaC_2 \rightarrow Mg + CaO + 2C \quad [1.9]$$

Eighty percent of the world's production is made by two methods for the electrolysis of $MgCl_2$. The first, the Dow Process, uses a semihydrous marine $MgCl_2$. The second, the I.G. Farben process, uses an anhydrous $MgCl_2$ made from magnesium bearing ores and seawater or brine from wells. A third process being developed by the Amax Specialty Metals Corp. is aimed at producing an anhydrous $MgCl_2$ from Great Salt Lake brine.

Reaction [1.0] is the basic equation that must be driven backwards electrolytically to form magnesium and chlorine from $MgCl_2$. This is the basis for both electrolytic processes. Equations [1.1] and [1.2] show the mechanism in the Dow process for producing $Mg(OH)_2$ from seawater and uses sea shells or dolomite as the source of $Ca(OH)_2$. The $Mg(OH)_2$ is neutralized with hydrochloric acid and dried to produce the semihydrated $MgCl_2$, which is then fed to the Dow cells.

Equations [1.3] and [1.4] are the mechanism by which the anhydrous $MgCl_2$ is produced for the I.G. Farben electrolytic process. Equation [1.5] is for the Pidgeon process and its modern counterpart, the larger scale, more energy efficient Magnetherm process that accounts for 20% of the world's production. In this process silicon reduces MgO in a vacuum furnace, from which magnesium vapor is removed and condensed. Ferrosilicon is the most economical source of silicon, even though at equilibrium all of it is not used. In Equation [1.6] aluminum is substituted in the form of low grade scrap.

Other extraction methods have been shown to be technically feasible and some have seen commercial application, but they are no longer in operation because they were not economically competitive. Equation [1.7] is a means of producing semihydrated $MgCl_2$ for electrolysis as a byproduct of the Solvay process and was developed by the Diamond Alkali Company.

Equation [1.8] is the basis for the Hansgirg (2,4,11) or carbothermic reduction process, which achieved technical success in the United States in a wartime economy but was unable to compete economically with the electrolytic and other thermal reduction processes after World War II.

Thermodynamics

The free energy change, ΔG, for a given isothermal, isobaric process represents the maximum work in addition to expansion work that can be obtained. In practice, the Gibbs free energy change is used as a general criterion of thermodynamic equilibrium, as a measure of the driving force of a chemical reaction, and as a tool in predicting phase equilibrium. No attempt is made to give broad treatment to chemical equilibrium, or to vigorously define every term and equation. Excellent texts are available for this purpose and may be consulted for more details (12,13,14). This section is aimed at illustrating the application of some relevant thermodynamic principles important to magnesium smelting technology.

The relationship of the Gibbs free energy, ΔG, to the thermodynamic activities of reactants and products for a chemical reaction is given by the equation

$$\Delta G = \Delta G° + RT\ln Q \qquad [2.1]$$

where ΔG° is the standard free energy change, term RTlnQ is the free energy change needed to transform reactants and products from their standard state conditions to a given set of states (activities), and R is the ideal gas constant, T is temperature °K and Q is the activity quotient. Both terms on the right side in Equation [2.1] are dependent on temperature, the chemical reaction and the standard states chosen for reactants and products; the second term is also dependent on pressure and composition. For an isothermal, isobaric process when the reactants and products are in their defined standard states, they are at unit activity and the second term in Equation [2.1] becomes zero. Under these conditions ΔG = ΔG°.

When, by the proper selection of standard states, these special well defined conditions exist,

$$\Delta G = \Sigma\Delta G \text{ products} - \Sigma\Delta G \text{ reactants} \qquad [2.2]$$

and at 700°C, 1 atm pressure, the ΔG of Equation [1.4a] is

$$MgO_{(s)} + Cl_{2(g)} + C_{(s)} \rightarrow MgCl_{2(l)} + CO_{(g)} \qquad [1.4a]$$

$$\Delta G = 116.07 + 47.29 - 118.52 - 44.84 \text{ kcal/mol}$$

of $MgCl_2$

For an isothermal, isobaric process in the equilibrium state ΔG = 0, the activity quotient Q has a unique value dependent on the same variables that fix ΔG°, is called the equilibrium constant K, and

$$\Delta G° = -RT\ln K \qquad [2.3]$$

$$\log_{10}K = \frac{-\Delta G°}{4.576T} \qquad [2.4]$$

with ΔG° measured in cal/mol and T (absolute temperature) in °K. Substituting ΔG° from Equation [1.4a]

$$\log_{10}K = \frac{-(-44840)}{(4.576)(973)} \approx 10$$

In the case of gaseous species, the conventional reference state chosen is the gas at one atmosphere pressure, in which case its partial pressure in atmospheres is equivalent to its mol fraction for equilibrium constant calculations. For Equation [1.4a] MgO in saturated solution may be considered to be reacting with Cl_2 in a solution of molten $MgCl_2$ at 700°C in the presence of excess carbon in suspension at one atmosphere.

Since carbon and MgO are assumed to be in excess as separate solids, their activities (equivalent to mol fractions for ideal solutions) are unity. We then have

$$K \approx \frac{(P_{CO})(X_{MgCl_2})}{(1)(1)(P_{Cl_2})}$$

MgO is only slightly soluble in $MgCl_2$, therefore, it may be assumed that the mol fraction of $MgCl_2$ is close to unity. Further, $P_{CO} + P_{Cl_2} = 1$ for atmospheric operation and $P_{Cl_2} = 1 - P_{CO}$ resulting in

$$K \approx \frac{(P_{CO})}{(1 - P_{CO})} = 1 \times 10^{+10}$$

$P_{CO} \approx 1$ atm (reaction essentially complete)

Differentiation of the free energy equation describes the effect of temperature on chemical equilibrium if applied to a change between two states

$$d(\Delta G/T)/dT = -\Delta H/T^2 \text{ (Ref. 12, p. 103-151)} \qquad [2.5]$$

where ΔH is the heat energy required to carry out, or derived from, a reaction as determined by calorimetric measurements of heat of reaction at constant temperature and pressure (same units as ΔG). Where ΔG refers to products and reactants at standard states

$$d(\Delta G°/T)dT = \Delta H°/T^2$$

substituting ΔG° = -RTlnK (12),

$$d\ln K/dT = \Delta H°/RT^2 \qquad [2.6]$$

These methods may be used to calculate the minimum energy required and the maximum conversion to be expected in the thermal processes.

For the electrolytic processes, free energy G may be defined as

$$\Delta G° = -nFE° \text{ (13) at standard states} \qquad [2.7]$$

where: n = number of g equivalents involved in the cell reaction, E = volts, F = Faraday's constant (23,061 cal/V equivalent), and ΔG° (standard free energy change for the reaction) in calories.

If the cell is reversible (E_r) and at equilibrium, $\Delta G° = nFE°_r$ and ΔG may be replaced in most thermodynamic equations by $-nFe_r$

$$E_r = -\Delta G_T/nF \qquad [2.8]$$

allowing the calculation of free energies from voltage measurements of a reversible electrochemical cell or, conversely, the calculation of the electromotive force of a cell or decomposition voltage of a compound from free energies of formation (12,13). For example,

$$MgCl_2 \rightarrow Mg + Cl_2$$

$$\Delta G°_{700°C} = -116.07 \text{ kcal/mol} \quad (14)$$

Substituting in Equation [2.8]

$$E_r = 2.52V$$

In the Gibbs-Helmholtz equation [2.5], $-nFE_r$ may be substituted for ΔG_T when calculating the effect of temperature and pressure on electrochemical equilibria at standard states (13).

$$\Delta H^\circ = nF\left(T\frac{dE^\circ_r}{dT} - E^\circ_r\right)$$

$$nFE^\circ_r = RT\ln Ka \qquad [2.9]$$

Anhydrous Magnesium Chloride Electrolytic Process.
(I.G. Farbenindustries' Cell)

In 1853, Bunsen discovered that if the $MgCl_2$ melt was not anhydrous, electrolysis would not proceed smoothly. Metal coalescence was poor, and there were losses due to basic chloride formation. Anhydrous $MgCl_2$, above its melting point of 713°C, slowly hydrolizes in moist air to form hydrochloric acid and MgO. Some decomposition takes place even in dry air, accounting for the poor metal coalescence caused by the resulting oxide films. Precipitation of the decomposition product MgO was responsible for the sludging observed by Bunsen and remains one of the main problems in commercial cells.

The composition of the electrolyte (bath) is dictated primarily by the source of $MgCl_2$; if the $MgCl_2$ is manufactured from seawater it will have a high level of $CaCl_2$. In this case a typical bath would be 40% $CaCl_2$, 30-40% NaCl, 5-20% KCl, and 10-15% $MgCl_2$. If the $MgCl_2$ has been made from magnesite or $Mg(OH)_2$ derived from brine wells, however, a bath composition of 3% $CaCl_2$ 50% KCl, 35% NaCl, and 15% $MgCl_2$ might be used. A wide variation of bath compositions is possible as long as $MgCl_2$ is held above 10%, and components may be selected on an economic basis to the point where unavoidable impurities dictate the composition.

Cell Design. Figure 2 represents the basic I.G. Farbenindustries cell, used by all manufacturers operating the anhydrous $MgCl_2$ processes. The cell is a steel tank that is lined with refractory insulation and filled with electrolyte. Monopolar electrode assemblies are set in a sealed lid and dipped side-by-side into the bath. These assemblies (Figure 2) consist of a graphite anode plate with cast steel cathode plates on either side. Each cathode contains slots for bath circulation and magnesium collection (Figure 3).

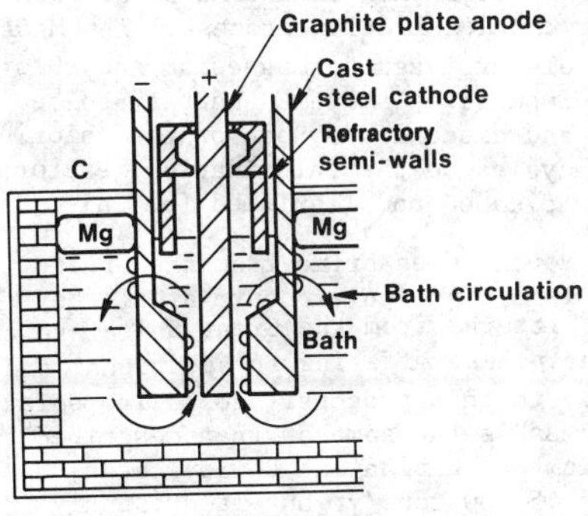

Figure 2 I.G. Farben Cell

Figure 3

Between the anode and the cathode are refractory semiwall partitions that dip into the electrolyte (see diagram) to permit separation of the chlorine and magnesium and prevent rechlorination and current efficiency loss. Chlorine bubbles are formed on, and rise in close proximity to, the anode. They are captured in the space provided between the anode and the semiwalls, and in the channel above. Magnesium formed on the cathode is carried by the bath circulation (resulting from the pumping action of the chlorine bubbles) through the slots to a zone outside the cathode assemblies. To effect proper separation and attain high current efficiency, the relationship between the depth of immersion of the semiwalls and the interelectrode spacing is closely controlled. Anode-cathode distance is ~140 mm. This wide separation is necessary to avoid recombination and is one of the reasons for the relatively low power efficiency of the cells.

Operation. Molten anhydrous $MgCl_2$ is fed to the cell three times a day; metal is either pumped or ladled from the cell once a day. The molten metal is readily wetted at the surface by the electrolyte, which prevents oxidation when the cell is opened to remove metal. All anhydrous $MgCl_2$ manufacturing methods result in some $CaCl_2$, NaCl, and KCl impurities. Since all of these accumulate in the bath, the cell gradually fills, and some bath must be removed to avoid overflow.

Once a week sludge must be removed from the cell bottom. Typical sludge components are: $MgCl_2$ (5-15%), $CaCl_2$ (30-40%), NaCl (15-30%), KCl (5-20%), MgO (5-15%), SiO_2 (1-5%), other oxides (1-5%), carbon (~0.10-1%), Mg (~0.50-2%). The magnesium is a valuable constituent and can be economically recovered.

Modern cells operate at 100,000 ampere capacity and produce ~908 kg/cell day. These cells are typically operated at 740°C at 90% current efficiency and use 12 kWh/kg magnesium. Graphite anode consumption is low, 0.015 kg/kg Mg, indicating a very nearly anhydrous $MgCl_2$ feed. With present current density, the cells are markedly free of polarization, and when the cells are switched off, back EMF shows only a few tenths of a volt higher than the $MgCl_2$ reversible decomposition potential of 2.52 volts. The overall cell voltage, however, is about 7 volts, reflecting the very wide AC distance and the higher than necessary electrolyte resistance resulting in roughly 34% power efficiency. There is much room for improvement.

Special Problems: If magnesium chloride concentration is allowed to fall too far, sodium or calcium is decomposed and adversely affects metal coalescence by covering the small droplets with metallic films. If sodium or potassium is made it burns readily at the surface and results in severe loss of current efficiency. The presence of boron and sulfate also result in poor metal coalescence by forming emulsions. Excess carbon dust from anode attack can cause electronic shorting and loss of current efficiency as well as a phenomenon called "boiling." High maintenance costs are due to erosion of the chlorine collecting semiwalls around the anode and corrosion of the anode in the cool region near the packing gland (2, 15).

Cell Feed Preparation (4,11,15). All producers using anhydrous magnesium chloride operate variations of the above cell. Differences are the source and preparation of the cell feed and production of byproduct chlorine. Magnesium chloride hexahydrate ($MgCl_2 \cdot 6H_2O$) cannot be dried directly without the formation of some MgO, so the critical step is dehydration or the formation of anhydrous $MgCl_2$ through chlorination of the oxide. Natural carnallite, $KCl.MgCl_2.6H_2O$, can be dried and electrolyzed to produce magnesium and chlorine, a process commercially operated in 1866 in Germany. There were many drawbacks, however. The cell feed contained large quantities of potassium chloride that remained in the bath, requiring constant tapping of the electrolyte to prevent the concentration of $MgCl_2$ dropping below 10% where the formation of potassium caused a loss in current efficiency. Other drawbacks were; small density variations resulting from composition and temperature changes that caused the magnesium to alternately sink or float; water remained and caused low current efficiency; poor metal coalescence; and anode wear.

$MgCl_2$ may be successfully dried in the presence of NH_4Cl or HCl. Although several methods have been attempted commercially, until recently none have been economical. Magnesium chloride hexahydrate may be dehydrated by electrolysis, and this method was commercially practiced in France using an electrically heated cell of molten artificial carnallite ($KCl \cdot MgCl_2 \cdot 6H_2O$). Although technically successful, it was not economical. Difficulties in dehydration of $MgCl_2 \cdot 6H_2O$ led to the development of the direct chlorination process by Equations [1.3, 1.4]. Chlorination is carried out in refractory lined vertical shaft furnaces heated by 3-phase alternating current passing through carbon resistor blocks. Magnesium oxide is pellitized with coal and peat, using the $MgCl_2$ solution as a binder, coked and fed to the shaft furnace. Chlorine is introduced and molten $MgCl_2$ is tapped from the bottom. Peat, originally used to create a porous pellet, eventually became unnecessary, but the highly reactive coal may produce environmentally unacceptable polychlorinated hydrocarbons due to the high content of volatile matter. Petroleum and metallurgical cokes are too low in activity for this process, and one low volatile carbon source suitable for reduction is calcined lignite. This lignite, or brown coal, comes from fields at Morewell, Australia. The material has highly developed microporosity and high reactivity, low sulfur content (<0.2%), and a low ash content (2.5%). Since lignite, unlike coal, does not pass through a fluid stage during the calcination process, microporosity is preserved and the carbon keeps its amorphous orientation through calcination. The low ash is believed to have improved metal purity, the high reactivity to have increased productivity in the chlorination units, and low sulfur to have reduced chlorine losses.

A flowsheet of the processes presently practiced at Norsk-Hydro is shown in Figure 4.

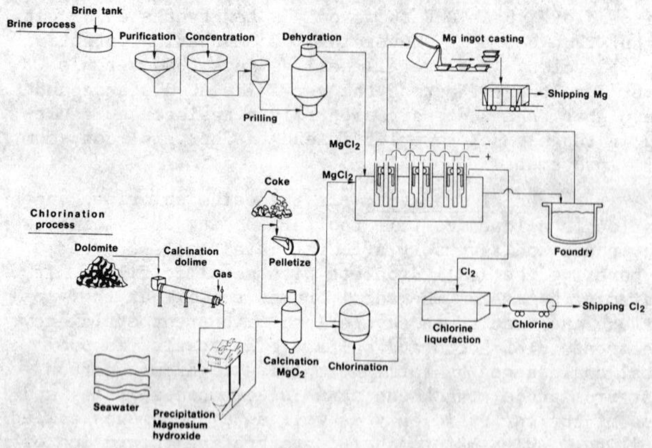

Figure 4. Norsk-Hydro Magnesium Processes

In the chlorination process seawater is mixed with calcined dolomite to precipitate $Mg(OH)_2$, which is calcined to MgO and pelletized with lignite coke using $MgCl_2$ brine as a binder. It is then chlorinated, producing molten $MgCl_2$, which is fed to modern I.G. Farbenindustries cells. Norsk-Hydro cells, operating at 700°C and 80,000 amperes, have been expanded to four anode-cathode assemblies. Chlorine is recycled to the chlorinators and magnesium is automatically swept into launders, or wells, from which it is removed each day. The chlorinators are operated at 1100-1200°C in the reaction zone and molten $MgCl_2$ is tapped from the bottom at 800-850°C. Seawater is roughly 0.13% $MgCl_2$ and 0.5 kg Cl_2/kg magnesium is required to make up for that lost in the process.

The new brine process developed by Norsk-Hydro has passed through the pilot stage and a 100,000 tons per year plant had been announced for the early 1980's but was recently delayed. This process, also shown in Figure 4, is based entirely on the evaporation of $MgCl_2$ bitterns. These bitterns are a byproduct of the potash industry and are imported from Germany. They contain sulfates and small amounts of boron and bromine. The bitterns are concentrated by evaporation to 30-36% $MgCl_2$ and desulfided by the addition of $CaCl_2$. Since both bromine and boron interfere with the electrolysis, they are removed by a proprietary process prior to evaporation of the saturated solution. The saturated $MgCl_2$ brine is then partly evaporated and converted to solid semihydrated $MgCl_2$ in large vertical shaft spray dryers followed by prilling towers. The remaining residual moisture is removed by drying in the presence of HCl gas. The HCl gas prevents MgO formation and is dried and recirculated. Anhydrous $MgCl_2$ is hygroscopic and must be stored under carefully controlled conditions. Feeding cold, solid anhydrous $MgCl_2$ allows increased production in the cells because of the cooling effect of solution. Much lower energy requirements are expected of this new process because it avoids the chlorination stage and produces chlorine as a byproduct. Further advantages from the elimination of the conventional chlorination stage are the absence of polychlorinated biphenyls and the ability to totally enclose the system. A 15,000 ton/yr plant has been operating on this new process at Porsgrunn, Norway, since 1977.

American Magnesium at Snyder, Texas, uses subsurface brine with a concentration of 11% $MgCl_2$ (Figure 5).

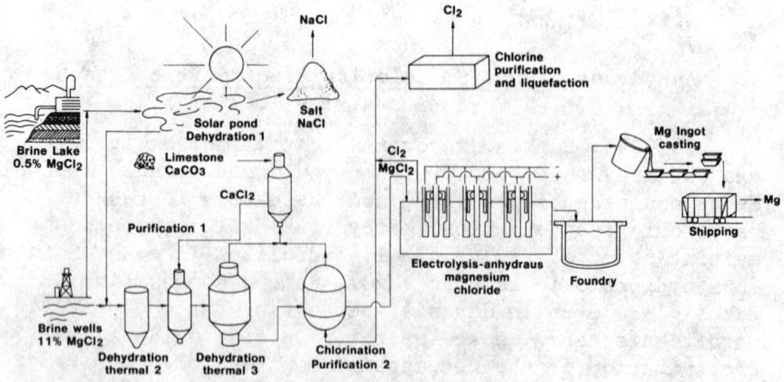

Figure 5. American Magnesium/AMAX Magnesium Brine Processes

It is apparent that a brine of this concentration presents a significant advantage over seawater where the concentration is only 0.13% $MgCl_2$. The brine is first concentrated to about 30-35% $MgCl_2$, using fossil fuels. Primary purification consists of desulfating by the addition of $CaCl_2$ followed by a series of ion exchange vessels to remove traces of bromine and boron. This step can be omitted and the trace impurifies removed in the last stage of purification, if desired. Two stages of crystallization to remove NaCl and KCl, complete primary purification. The saturated $MgCl_2$ solution (36%), leaves the second crystallizer stage, is heated to about 90°C, and fed to spray dryers (thermal 3) producing a solid product containing about 5% water, 5% MgO, and 5% KCl and NaCl. This material is fed to the second stage of purification where it is dissolved in molten $MgCl_2$ and chlorinated in the presence of carbon, usually in the form of a slurry. The molten anhydrous $MgCl_2$ is tapped from these reacting vessels through a filter, or decanted and sent directly to the electrolytic cells. The off-gasses from the spray dryer and the chlorinating reactors contain HCl gas and are reacted with limestone to produce the $CaCl_2$ needed for the desulfating process. The $MgCl_2$ fed to the cells is electrolyzed to magnesium and chlorine. The 5% KCl and NaCl impurities stay in the bath to be periodically tapped and discarded. Some of the chlorine from the cells is recycled to the chlorinating reactors and the remainder is cleaned and liquified for sale.

AMAX Specialty Metals uses brine from the Great Salt Lake. Located in a high sun exposure area, they can concentrate the bitterns from the Great Salt Lake to 30-35% $MgCl_2$ in solar ponds, offering an additional and significant advantage to this process. Following solar evaporation the process is the same as that described for American Magnesium. The plant at Rowley, Utah, has a stated capacity of 45,000 tons/yr but is presently operating at 24,000 tons/yr. Although difficulties with the last stage of purification were experienced during plant start-up these problems have been overcome, with increased production possible when the market warrants (15).

Semihydrated Magnesium Chloride (Dow Process) (16).

This electrolytic process accounts for 35% of the world's and 80% of United States' production of magnesium. The Dow cell uses a 73-75% $MgCl_2$ feed, with the remainder water. It combines thermal and electrolytic removal of water with the electrolysis of $MgCl_2$ to produce magnesium metal. Such a cell avoids the high temperature, capital intensive, high energy consuming chlorination process, and preparation of cell feed remains low temperature and totally aqueous.

The key development of this process was a cell able to overcome the problems discovered by Bunsen when electrolysis is carried out in a system that is not water free. Poor coalescence was discovered to be due to metal oxides (largely boron oxide) deposited on the metal droplets. This not only prevents coalescence but supplies sufficient weight to sink the magnesium droplets. A means was developed to remove these trace impurities prior to feeding the cell. Coalescence was further improved by designing the cell to allow the droplets to rise slowly along a surface, coalescing as they progressed. A feeding method was developed that permitted between 75-90% of the water to flash off as the $MgCl_2$ was added to the cell, thereby using cell heat to thermally remove much of the water. The remaining water was removed by electrolysis at the same time the magnesium was produced. The cell was also designed for easy removal of MgO sludge formed from back reaction of moisture with $MgCl_2$.

As in the design of the anhydrous $MgCl_2$ electrolytic cell, it is desirable to operate at minimum power consumption through close interelectrode spacing. Recombination of the chlorine gas anode product and magnesium, both of which rise to the surface of the cell, must be avoided. The need to design for easy sludge removal required wider interelectrode spacing and lowered power efficiency. Using salts with a decomposition potential higher than $MgCl_2$, an electrolyte, with a lower melting point and density than magnesium and high electrical and thermal conductivity, was developed to operate at low $MgCl_2$ concentrations.

Cell Design (4,11,16,17). There are many proprietary aspects of the Dow cell, and details are unknown, particularly the internal configuration of the cathodic surfaces. Some data have been published, however, and other details may be speculated upon. The main body of the cell is a rectangular steel tank 1.5 m wide x 3.9 m long x 1.8 m deep. It holds ~908 kg of fused electrolyte, is set in a brick enclosure, and is externally heated by gas or oil. Although early designs may have had as few as 12 cylindrical anodes in two rows of six, most modern cells are believed to have 30 graphite anode cylinders, 52 cm diameter and 2.47 m long, in two rows of 15, suspended through the arched refractory lid.

Present cells operate at 90-115 kA and are arranged in 600-650 volt lines of 108 cells per line. Eight cells are considered to be spares and off the line at any given time, giving 93% utilization of the capital equipment. Cell rooms are ventilated at a rate of about 1700 m^3/min/cell. Cells are held under about 0.63 cm of water vacuum while 10-12 m^3 of air is passed through each cell to capture the ~1 m^3/min of chlorine produced during the electrolysis. This dilutes the cell gas to about 4% Cl_2, 4% HCl, 1% H_2 and 1% CO_2. The diluted cell gas passes through brick ducts and water quenched graphite pipes into Haveg acid resistant collecting ducts running the length of the building. One of the most important and least known aspects of this cell is the avoidance of recombination with anode chlorine while coalescing and collecting molten magnesium in front wells. Metal may be dipped from these wells and cast directly into ingot, as no further purification is needed. It is known that the steel shell and metal collecting configuration around each anode are all cathodic. Figure 6 represents what is believed to be the internal structure of this cell.

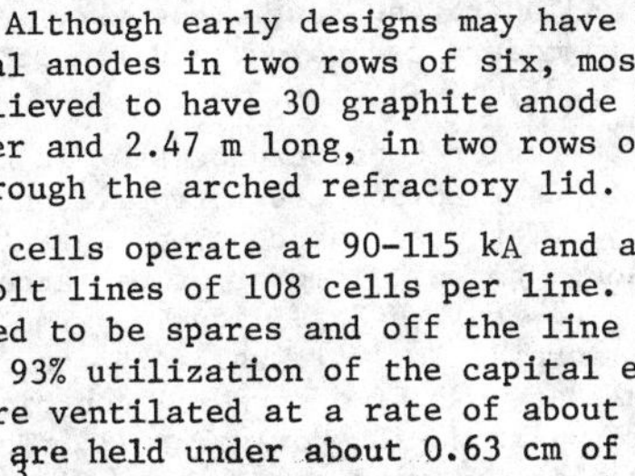

Figure 6. Conventional Dow Cell

Around each anode is a louvered steel spiral cathode connected to the steel shell. Magnesium plates on the steel louvers, forms droplets that grow until the gravitational force overcomes the interfacial tension and they begin spiraling to the surface in the inverted troughs provided for them. Initially the particles or droplets are coated with oxide, but as they coalesce they are freed from much of it and spiral to the surface. Some of the magnesium becomes so laden with oxide that it sinks and rests on the top side of the louvers; but as it moves downward it further coalesces, displacing oxide until its density is such that it rises and joins with the other magnesium spiraling upwards. Sludge settles to the bottom of the cell. Chlorine, rising along the anode, 38 mm from the cathode, does not back react with the magnesium droplets in the bath because the interfacial tension of the salt keeps chlorine bubbles and magnesium droplets apart. Back reaction is controlled at the surface by air dilution, which keeps the concentration of chlorine and hydrochloric acid below 10%. Cathodic contact to the system is made by bus welded to the outer shell.

Operation. By external heating, the temperature is controlled between 700-720°C. The temperature control, being independent of the electrical system, allows considerable flexibility in ampere load adjustment and cells may be operated at the closest anode-cathode distance commensurate with avoiding recombination of magnesium and chlorine. This allows operation on interruptible power and these cells may be used as peak leveling devices. Even a complete electrical shutdown can be tolerated for as long as a week by adjusting external heating. Sludge is removed by dipping, and metal is ladled into ingot from the front wells two to three times daily. Electrolytic drying of $MgCl_2$, electrolysis of soluble MgO, and removal of trace sulfates results in the consumption of considerable anodic graphite. This necessitates daily adjustment of the anodes to maintain close A-C distance for good power efficiency. Slow, uniform $MgCl_2$ feeding is necessary to insure maximum water removal while avoiding excess anode consumption and spitting and splashing of the salt. Good adjustment of high quality graphite anodes is necessary to avoid dusting, which causes violent agitation of the bath. Since desludging removes bath as well as MgO and other constituents, $CaCl_2$ and NaCl must be added periodically to maintain the proper composition of 60% NaCl, 25% $MgCl_2$, and 15% $CaCl_2$.

These cells are operated only by the Dow Chemical Co. Modern cells are thought to operate at about 6.3 v at 80+% current efficiency, producing 99.8% magnesium metal at 18.0 kWh/kg for about 33% total power efficiency. The HCl acid furnaces, for treatment of cell off-gas, and the evaporators are thought to use about two-thirds and one-third as much energy as the cell, respectively. The cell produces ~1000 kg of magnesium and 180-320 kg of sludge (18% MgO, 2-3% Mg, 1% Fe, balance electrolyte) per day. Each kilogram of magnesium produced requires 1.27-1.32 kg of feed material containing ~1.0 kg of $MgCl_2$, 0.52 m^3 of methane, 0.05 kg of graphite electrode, 0.23 kg of makeup chlorine for HCl production and ~0.013 man hours. Cell life is about 1000 days in two 500-day campaigns. Cells are taken out of operation and refurbished at the end of 500 days of operation (1,4,11,16).

Cell Feed Preparation. In modern practice the source of $MgCl_2$ feed for this process is seawater and dolomite (MgO.CaO). Although both the Norsk-Hydro and Dow $MgCl_2$ cells use seawater as the basic raw material, the processes are quite different. In the former, $MgCl_2$ is made by the direct combination of MgO, derived from calcining $Mg(OH)_2$ precipitate, and cell chlorine. In the latter case, $MgCl_2$ is made in aqueous media by reacting HCl with $Mg(OH)_2$ precipitate to produce $MgCl_2$ hexahydrate containing 47% $MgCl_2$. Further evaporation is necessary to produce 57% $MgCl_2$ as $MgCl_2.4H_2O$, which must be carefully dried to produce a cell feed of 73-75% $MgCl_2$ essentially free of MgO. Although the most important and difficult goal in this process is to avoid oxide formation in the final drying step, the production of the tetrahydrate must also be carefully controlled to remove certain impurities that are deleterious to cell operation. Figure 7 is a diagram of the process for producing semi-hydrated $MgCl_2$ feed for the Dow cells and may be considered to consist of four basic steps: a) precipitation and filtration to produce $Mg(OH)_2$; b) neutralization with hydrochloric acid to produce $MgCl_2$ hexahydrate; c) evaporation to $MgCl_2$ tetrahydrate (35% $MgCl_2$); and, d) final drying for the production of cell feed (73% $MgCl_2$).

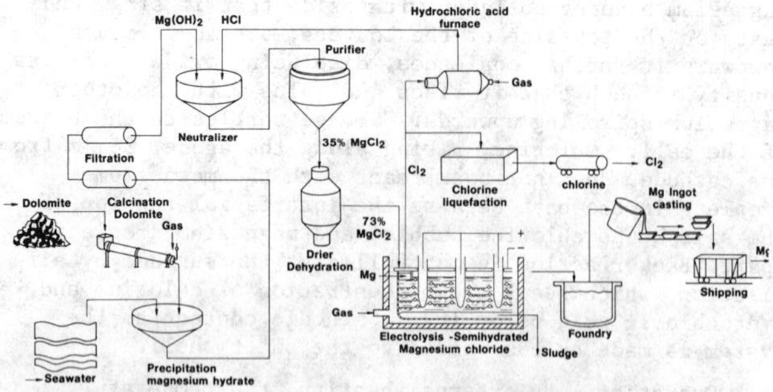

Figure 7. Dow Semihydrated Magnesium Chloride Process

Further steps are electrolysis and treatment of the cell Cl_2 for recycling as HCl. Two necessary peripheral steps are the calcination of dolomite to make the dolime reactant for precipitation of $Mg(OH)_2$, and electrolysis of NaCl to produce makeup chlorine for that lost in the process. Milk of dolime for treatment of seawater is made by calcining crushed dolomite at 1000-1100°C in rotary kilns to produce dolime, which is fed to rotary slakers for hydration to $Mg(OH)_2$ and $Ca(OH)_2$ by seawater. A small portion of the dolime is slaked in fresh water and used to remove $NaHCO_3$ from incoming seawater. Precipitation is carried out in concrete flocculators by vigorous mixing with dolime. The presence of excess lime, pH control (NaOH from Cl_2 production), and proper agitation are necessary to provide a $Mg(OH)_2$ precipitate that can be filtered while remaining free of boron hydroxide, which if introduced to the cell creates severe operating difficulties. The $Mg((OH)_2$ suspension is thickened to a 25% $Mg(OH)_2$ slurry while excess seawater is decanted and returned to the bay. The remaining slurry is filtered in Moore leaf filters. The filter cake is washed from the filters, diluted to a 50% slurry by recycled $MgCl_2$ solution, and fed to the neutralizers. In the neutralization step, $Mg(OH)_2$ is reacted with HCl acid (recycled from the cell off-gas) to produce $MgCl_2$ solution. A small amount of sulfuric acid is added to remove any $CaCO_3$ remaining after filtration. The solution is evaporated to 34-36% $MgCl_2$ by spraying into a stream of hot HCl gas. After cooling and crystallization to remove unwanted impurities ($CaSO_4$ and NaCl), the solution is filtered again and evaporated to 55% $MgCl_2$ solution.

A number of choices are available for the drying operation, and selection is economically determined. Although it is possible to spray dry 45% $MgCl_2$ solution to 72-73%, shelf drying is preferred to avoid briquetting of the fine particulate this method produces. In shelf driers circular shelves are stacked vertically on a central shaft. The solution is evaporated on the top shelf to layers of solids by an updraft of hot dry air and combustion products. The evaporated solid is raked to lower drying shelves as flakes. An evaporation step may be avoided by refluxing the major portion of the feed reaching the 70% $MgCl_2$ level to the top shelf after mixing with incoming 40% $MgCl_2$ solution to produce a 66% $MgCl_2$ stream. That which is not drawn off passes through the remaining shelves and is taken from the bottom of the unit at 73-75% $MgCl_2$ in a granular form suitable for cell feed. Modern practice suggests spraying 35% $MgCl_2$ solution directly into a bed of 75% $MgCl_2$ solid particles fluidized and heated by combustion products. All of these processes produce an acceptable cell feed.

The electrolysis step and collection of metal were described in the sections, Cell Design and Operation. In one technique some of the chlorine leaving the cells is extracted from the dilute gas and liquefied for sale. The remainder is burned with moist air in a HCl furnace where the HCl is extracted as a solution to be used in the neutralizers.

Thermal Reduction Processes

Magnesium oxide can be reduced at high temperatures by a number of common reductants, including ferrosilicon, aluminum, calcium carbide, aluminum-silicon alloys, and carbon. Although carbon would seem to be the economically preferred reductant, this is not the case because reduction with carbon presents a special problem. Carbon reduction proceeds according to Equation [1.8], and at temperatures approaching 1900°C MgO is reduced to Mg and CO. At these temperatures both the Mg and CO are gaseous, and as the temperature is lowered to condense and collect magnesium they back react to form MgO and carbon. If, after initial reaction, the temperature can be reduced rapidly enough, little back reaction takes place and a high yield is possible. Such a process was operated during World War II with technical success, but cost efficiency was not high enough to be commercially competitive with other processes after the war.

In all of the other processes the products are separated by magnesium vaporization, leaving a solid residue behind. Since these all involve the replacement of magnesium by another metal they are called "metallothermic reduction." According to Equation [1.9], metallothermic reduction with calcium carbide takes place at a much lower temperature than with carbon. Since the calcium combines with all of the oxygen there is no gaseous carbon product to back react with the vaporized magnesium. The reaction occurs readily at temperatures of 1120-1140°C in steel retorts under vacuum, and a 10% addition of CaF to the MgO improves the yield. This is the Murex Process and was commercial in the United Kingdom during World War II, producing 5000 tons per year. Capital and operating costs, however, were not competitive with modern technology and the process was abandoned (2,19,20).

The only thermal reduction processes in operation today are metallothermic and use ferrosilicon as the reductant [1.5]. The basic process was invented by Dr. L.M. Pidgeon, who was director of research of Dominion Magnesium Ltd., Haley, Ontario. The Pidgeon process was commercial in 1941, using 75% ferrosilicon to avoid the higher cost of a more silicon rich reductant. Six plants were built during World War II. The process is highly flexible, allowing small production units to be operated at points convenient to fuel and dolomite with low overall power requirements. However, low unit capacity retorts and the high cost of maintaining them resulted in production costs that were not competitive with other technology.

Pidgeon Process (4,11,21,22,23). Basically this is a 5-step process: a) arc furnace production of the ferrosilicon reductant; b) preparation of the reacting charge (calcined dolomite mixed with 75% ferrosilicon at a 5:1 ratio--ground, heated, and briquetted); c) reaction of the charge to produce magnesium; d) removal of the product from the retort and disposal of waste residue; and e) fluxing and casting magnesium into ingot. Briquettes are charged to tubular retorts, which are externally heated to 1150-1175°C for 8-9 hours under 0.1-0.2 mm mercury vacuum. Magnesium oxide is reduced by the silicon and magnesium vapor is evolved, leaving behind a residue of calcium silicate. Magnesium is collected as a solid of 99.98% purity in a water-cooled zone at the cool end of the reactor.

Magnetherm Process (24,25.26). All of the electrolytic processes, regardless of the raw material source--seawater, concentrated lake brines, or dolomite--require extensive, complicated and capital intensive pretreatment to provide a suitable cell feed material. There are significant potential advantages, therefore, in reducing calcined dolomite metallothermically in a system in which the product is easily separated from the waste residue by evaporation and condensation, even though stored electrical energy in the form of ferrosilicon must be used. Considerable attention was given, therefore, to determining the reasons for the economic disadvantage of the Pidgeon Process. Capital investment and low production rates resulting from small plant size scale-up limitations of retorts, and long turn-around times were determined to be the major road blocks to the development of a competitive process. These problems could be solved by extending reactor life, increasing the rate of heating of the reactants, operating continuously at higher temperatures to improve silicon efficiency and avoiding remelting and fluxing cost by condensing directly to liquid magnesium. With these goals, research and development was begun in France in 1948 by the Societe Generale du Magnesium to develop what is called the Magnetherm Process. Magnesium is obtained by reducing calcined dolomite with ferrosilicon, as in the Pidgeon Process, according to reaction [1.5]. This is an endothermic reaction, and in the case of the Pidgeon Process, heat must be transferred through a stainless steel wall to a self-insulating static charge. In the Magnetherm Process reaction heat is introduced in an electric furnace by resistance heating a mass of molten slag to which the reactants are added. Heat transfer to the finely divided reactants is very rapid, and the elimination of the stainless steel wall and the mobile nature of the charge allows for near unlimited scale-up. These characteristics represent the fundamental distinction between the Magnetherm and Pidgeon Processes and improves the economics of silicothermic reduction by bringing the reactants into direct contact with the energy source in compact equipment. The Magnetherm reactor attains high energy efficiency and allows easy removal of the product by in situ vacuum distillation. Electric current is passed through the molten slag from a water-cooled, submerged electrode to the carbon lining in which other electrodes are imbedded.

Many difficulties were overcome before the process became commercial. The reaction goes readily at ~1550°C, but the melting point of the slag resulting from the reaction was 2130°C, putting a severe burden on construction materials. It was discovered that the addition of ~15% Al_2O_3 gave an apparent melting point of ~1500°C. At operating temperature, some particles remain in suspension, however.

Pilot plant development was begun in the French Pyrenees in 1955 with a 400 kVA single phase furnace producing 0.23 Mt/reactor day at 30 kWh/kg. Commercial production was begun at Marignac, France, in 1964 with 2300 kVA reactors producing 3.4 Mt of Mg/day at 10.5 kWh/kg. By 1975 development of a 4500 kVA reactor with a design capacity of 7.5 Mt reactor day at 9.6 kWh/kg was underway. Alcoa licensed the Magnetherm Process from SOFREM and built a plant at Addy, Washington, using furnaces of 4400 kVA producing 6.0-7.5 Mt/day at a projected <9.3 kWh/kg.

Process Description. As shown in Figure 8 (Alcoa practice), the Magnetherm Process consists of four steps: a) ferrosilicon production, b) raw material calcination, c) magnesium production, and d) melting and ingot casting.

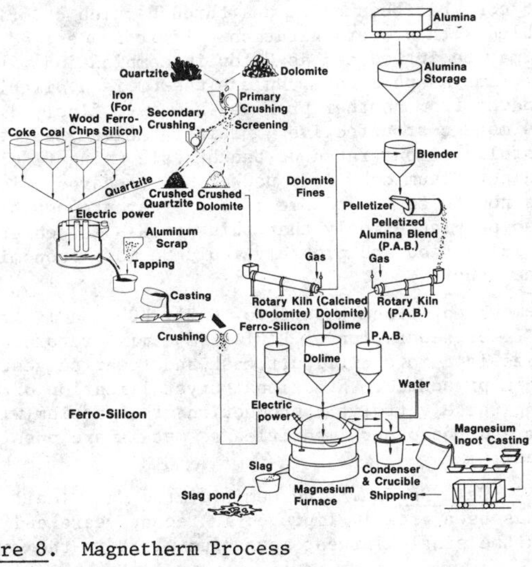

Figure 8. Magnetherm Process

Ferrosilicon is produced in a conventional 3-phase arc furnace from a mixture of coke, coal, wood chips, crushed quartzite, and scrap iron or taconite to produce ferrosilicon (75% Si), which is cast and crushed with a gyrotory cone and jaw crusher and stored for transport to the magnetherm furnace day tanks.

After primary and secondary crushing, dolomite, a mineral containing calcium carbonate and ~45% magnesium carbonate, is fed to a gas fired rotary kiln ~6-1/2 m in diameter and 8.5 m in length where it is heated to 1360°C to decompose the carbonates to calcium oxide and magnesium oxide, with 47% of the original dolomite weight leaving as CO_2. In the French process, all of the dolime is fed hot to the Magnetherm reduction furnace day tanks and Al_2O_3, to reduce the melting point of the slag, is added in the form of bauxite. Bauxite is crushed, sized, and calcined in the same manner as dolomite, and fed hot, as needed, to reduction furnace day tanks. In the Alcoa process finely divided Bayer metal grade Al_2O_3 is blended with lime (CaO) and pelletized before calcination in a rotary kiln. These pellets are also fed hot to the reduction furnace day tanks. The removal of water from the bauxite and Al_2O_3 and CO_2 from the dolomite is necessary to avoid residual amounts of CO and water adversely affecting the reduction furnace pressure.

Magnesium is produced in a furnace having three major parts: a hemispherical dome, a cylindrical tank, and a condensing system. The water-cooled refractory lined dome is made of corrosion resistant high strength steel. It has three apertures: a centrally located entry for the heating electrode, a small feed port entry, and a large ceramic lined magnesium vapor duct. The main furnace tank to which this hemispherical dome is attached is a carbon lined tank of a steel similar to that of the dome. Electrodes are sealed in the carbon bottom hearth into which tapholes are installed to remove the slag and 20% silicon ferrosilicon byproduct. The condensing system is a crucible connected to the reduction furnace vapor duct by way of a condensing section. Magnesium vapor is condensed and collected in the crucible as a solid. The top of the condenser is connected to the vacuum system by a cooling trap that condenses any remaining magnesium vapors to avoid plugging the vacuum pumps. Melting and ingot casting are accomplished by transferring the crucible from the condensing system to a remelting furnace. The metal is melted and purified under inert atmosphere or with water-free fluxes in the same crucible in which it was collected. After fluxing and degassing the metal is pumped to ingot casting machines where it is poured under protective atmosphere.

Operation. This is a semi-batch process in which all of the feed containers, the furnace, and the condensing system are sealed and evacuated at the start of a run. Ferrosilicon, dolime, and bauxite or pellitized Al_2O_3 containers are placed and sealed on their weigh-feeders. The slag is drawn down to the lowest level, permitting electrical contact to the anode before a campaign is begun. All cooling systems maintaining the temperature of the hemispherical head, the condenser, and the condensing crucible are activated and vacuum pumping is begun. When the desired pressure (~35 torr) has been attained, continuous feeding of raw materials is begun. Previous analysis of the raw materials and furnace slag permit accurate feeding and temperature is maintained through control of the electrical system. Magnesium vapors leaving the reaction zone are cooled in the condenser and magnesium is collected as a solid in the externally water-cooled crucible. The proportion of MgO to slag constituents is such that it is necessary to tap slag twice to obtain a full magnesium crucible. When the magnesium crucible is half full the slag has risen to the maximum level. The vacuum is broken with inert gas and low grade ferrosilicon and slag may be decanted into separate crucibles. When the slag has reached the minimum level the taphole is plugged, vacuum is restored, and raw material feeding begins again and the sequence is repeated. The total operating cycle requires about 20 hours, using roughly nine hours for each reduction, including one hour for intermediate slag tapping and one hour for crucible removal and final tapping. In the crucible removal step the entire condenser, crucible, and trap are removed in one piece and transported to a special stand for disassembly. This reduces turnaround time and allows a complete condensing assembly to be clamped in place for immediate resumption of production. The previously used condenser and trap are sent for cleaning and the crucible is taken to the foundry for remelting, purification, and casting of the magnesium.

Economical operation of this process requires maximum productivity since capital intensity is high. Automation is necessary and computer control is essential. Poor control during any part of the process can cause reduced productivity. For example, poor control of calcination can cause serious pressure unbalance in the furnace, resulting in plugging of the vacuum and feeder pipelines. Slugging of feed materials instead of smooth continuous feed causes the same problem by excess magnesium vapor production. X-ray analysis of raw material and slag allows computer regulation of raw material feeding and of the power supply, resulting in controlled composition. A mass spectrometer leak detector shortens turnaround time by quick detection for correction of assembly defects. Preassembly, preheating, and leak testing for the condenser crucible trap system prior to installation, as well as immersion cooling of the crucible, have reduced turnaround time and improved collection efficiency of the magnesium.

Northwest Alloys Plant.

Equipment Specifications and Capacity:

Inner diameter	317 cm
Height	424 cm
Slag weight - maximum	45,000 kg
- minimum	20,000 kg
Slag height - maximum	226 cm
- minimum	110 cm
Production capacity	6000-7500 kg Mg/day
Mg crucible - diameter	206 cm
- height	261 cm
Cycle time	20 hrs
Plant design capacity	21.8×10^6 kg/yr
Mg reduction furnace,	4.4 MW
Fe-Si furnace	25 MW

Materials Balance, Mt/Mt Mg Ingot

Feed:	FeSi (75% Si)	1.0-1.1
	Blended Al_2O_3	1.0-1.2
	Calcined dolomite	6.0-6.7
	Electrical Energy	10,000-12,000 kWh
Products:	Slag	5.5-710
	Residual FeSi (20% Si)	0.23
	Flux	0.25
	Sludge	0.5-0.6

Fluxing and Casting of Metal:

Ingots are cast in 50-lb ingots under gas cover consisting of SF_6 and air.

Flux:	Anhydrous carnalite: $MgCl_2$-KCl	0.17 t/t Mg
	NaCl	<u>0.08</u> t/t Mg
Total flux		0.25 t/t Mg

Slag Disposal: Presently, slag and residual FeSi alloy are poured back into dolomite quarry. Two potential uses currently being investigated for calcium-aluminate slag are: soil nutrient and cement component (requires granulation).

Environmental Control: Essentially, the plant operates as a zero discharge facility, especially for on site water.

Fabric filter units (baghouses)	28
Plant emission particulate	.023 g/m^3
	850 kg/day
Total volume	29,875 m^3/min

Future Developments

Magnesium has high potential as a competitor for aluminum where light die castings are desired *if* production cost can be reduced. There are opportunities for improving the operation efficiency of both the electrolytic and metallothermic processes.

Electrolytic Processes. Nearly 50% of the cost of producing magnesium is charged to cell feed preparation; this is primarily due to the high energy cost of dehydrating $MgCl_2$, even to the semihydrated level used in the Dow cell. It is apparent that if the electrolytic processes are to remain competitive with the improving metallothermic processes and the possibilities presented by an improved carbothermic process, much better commercial dehydrating techniques must be developed. Nearly all cell improvements depend on a water-free $MgCl_2$ feed. Water causes corrosion problems leading to high maintenance cost, sludge formation and anode consumption. Even higher operating efficiencies would be possible if the feed were not only anhydrous but also relatively free of impurities such as NaCl, KCl, and $CaCl_2$. If the total of these impurities were below 0.2%, LiCl cells in which magnesium settled could be economically operated. This low impurity level is necessary to avoid a change in bath density, which would allow the magnesium to float. If this is to be avoided, bath must be removed and LiCl added periodically. Because of the high cost of LiCl, $MgCl_2$ feed must be pure enough for the quantity of bath removed to be economically discarded or sold as a flux. Conversely, a method could be developed for low cost recovery of LiCl from the bath. Dow Chemical Company patented a low density LiCl cell many years ago, and it is believed it has not been commercialized due to the feed purity problem (27). Such a cell, operating sludge free, with permanent anodes (since there would be no oxide decomposition to consume them) and no Cl_2-Mg separation problem could be designed with close A-C distance for high energy efficiency and low maintenance and labor cost.

Feed Preparation. Many techniques for the development of a commercial process to make totally anhydrous $MgCl_2$ have been proposed. They are: a) the production of artificial carnalite or other alkali metal and alkaline earth chloride mixtures, b) dehydration in the presence of ammonium chloride and hydrochloric acid, c) electrolysis, d) filtration and settling after treatment with molten aluminum or $AlCl_3$, and e) fluidized bed chlorination of partially dried $MgCl_2$. Norsk-Hydro's successful new process involves dehydration in the presence of anhydrous HCl. Some believe the most cost effective technique is a slurry reactor. Such a reactor lends itself readily to the chlorination of the oxide remaining when $MgCl_2$ has been dehydrated to 1.7 mols water by conventional techniques. In this system semihydrated $MgCl_2$ is fed continuously to a resistively heated reactor containing molten $MgCl_2$ into which finely divided coke has been added and through which chlorine is bubbled with proper agitation. If it is desired, the carbon may be introduced as CO or in combination with the chlorine as phosgene. Unless the CO is available as a byproduct from another process, however, finely divided coke is more cost effective. This technique has been successfully demonstrated on bench scale using finely divided petroleum coke or a packed bed of carbon pieces. This is not to say that some of the other proposals may not also be viable, only that this process is nearer to commercialization and probably would be more economical than the others.

Removal of trace amounts of NaCl, KCl, $CaCl_2$ from the $MgCl_2$ or separation of LiCl from a mixture of such chlorides is a more difficult task and fewer suggestions have been presented. Fractional crystallization or coupling the electrolytic production of magnesium with lithium and the other impurities as metals are possibilities.

Cell Design. The semihydrated $MgCl_2$ cell (Dow cell) has been greatly improved in recent years and is used in the single largest magnesium plant in the world.

It operates at somewhat lower current efficiency than the anhydrous $MgCl_2$ cells but makes up for this by the lower cost of dehydration. In spite of these advantages it is believed that the potential for improvement in the design of an anhydrous $MgCl_2$ cell, coupled with the limited ability of the Dow cell to make byproduct chlorine seems to make the anhydrous cell the more likely economic winner if sufficient research and development is applied. Many improvements are possible. Increased energy efficiency and higher productivity could be achieved through closer A-C spacing if semiwall corrosion were prevented. This could be accomplished by the elimination of moisture and the selection of improved construction materials. Maintenance could be reduced by changing anode contact design to eliminate the critical temperature in the packing glands that cause graphite deterioration. The availability of anhydrous $MgCl_2$ would go a long way toward improving the existing cell design and is necessary if electrolytically produced magnesium is to become a competitive structural material.

The electrolytic industries have always been plagued by high capital investment and low unit productivity because of the low rate of metal production per square foot of electrode surface. Designers must leave the primitive monopolar electrolytic cell, where low space time yield insures inferiority to other types of chemical reactors, and enter into a new innovative design that is competitive with such reacting systems. A bipolar system of closely spaced electrodes converts the electrolytic cell to an electrochemical reactor where products are made in narrow slots with low waste heat generation and in which the gaseous and liquid products are swept from the slots for separation and collection. The Alcoa Smelting Process (ASP) cell is such a system and the production of a Al-Mg alloy up to 65% Mg has been demonstrated using anhydrous $MgCl_2$ feed. This cell could be adapted to 100% Mg production using impure <u>anhydrous</u> $MgCl_2$ through the development of a system of separation of magnesium and chlorine at the top of the cell. The most advantageous application of the bipolar system, when both anhydrous and pure $MgCl_2$ is available, however, would be a LiCl bipolar ASP cell making 100% magnesium that settled to the bottom and was operated exactly as for aluminum production. Such a system would produce magnesium under 8.0 kWh/kg at space time yields exceeding five times that of the best anhydrous monopolar cell that could be developed.

<u>Pidgeon Process</u>. At the end of World War II this process became non-competitive with the electrolytic processes because of the small size and high maintenance cost of the retorts. Reactor size was limited by the difficulty of heating a solid self-insulating static charge through the retort walls. Low production rates were also due to the lengthy interval required for recycling a batch process. Efforts were directed, therefore, to selecting the retort materials that would allow higher temperature operation (shorter reaction time and improved silicon efficiency), internal heating to permit larger retorts, and easier assembly for the reduction of cycle time. Designs were suggested using internal resistance heating that could process seven tons of briquettes per cycle, using scrap aluminum as a more active reductant. Such developments might reduce capital and operating costs sufficiently to be competitive with the present electrolytic processes since the complexities of $MgCl_2$ production and dehydration are avoided and dolomite can be used directly. The Pidgeon Process survives at a plant in Bolzano, Italy, where many of the above improvements are reputed to have been made. Dominion Magnesium Ltd. still operates the original plant at Haley, Ontario. It is doubtful, however, that the Pidgeon Process can survive with the rapid improvements being made in the Magnetherm Process.

<u>Magnetherm Process</u>. The Magnetherm Process resulted from efforts to solve the shortcomings of the Pidgeon Process. Internal resistive heating of the slag allows higher temperature operation, longer reactor life, reduced cycle time, and near unlimited scale-up possibilities for productivity increases. Condensation of the magnesium product as a liquid and continuous operation through tapping of slag and magnesium through barometric legs will no doubt be accomplished in the future.

The process is still being rapidly improved. Since 1975 energy consumption has been reduced 15% and labor 50% while progressing to a 4500 kW reactor producing 2500-3200 Mt/reactor year (26,28).

Short term improvements may be classed as operational control, and selection and preparation of raw materials.

<u>Operation</u>. Approximately 70% of the magnesium fed to the reactor in the form of dolime is recovered as magnesium ingot. As much as 60-70% of the magnesium lost is due to reaction of magnesium vapor with air and water leaks while the reactor is under vacuum. The resultant MgO, as well as some fine raw material feed and slag spray are carried to the condenser by the high velocity magnesium vapor. This results in 30% as much sludge production as magnesium metal. The sludge, separated by salt fluxing in the purification and casting steps, contains 14% magnesium. Recovering this magnesium represents a costly additional step in which only 87% of magnesium remaining in the sludge is actually recovered. Better, faster sealing techniques are being developed to reduce air and water leaks as well as downtime during sealing and leak testing. These improvements could result in a 24% reduction in the time required for sealing, leak testing, and repair. Research is being conducted on more efficient and cost effective techniques for metal recovery from sludge.

<u>Control</u>. Through thermodynamic and kinetic measurements, a mathematical model is being developed that describes the relationships between temperature, pressure, and feed composition as they relate to productivity. The viscosity, surface tension, and melting point covering a range of slag compositions are being investigated because electrode life and slag carry-over are affected by these properties. Properly proportioned uniform feed of dolime, ferrosilicon, and alumina pellets is necessary to avoid excessive slag carry-over. Since the reaction is nearly instantaneous, excessive or nonuniform feeding results in high velocity magnesium vapor that carries noncondensible material into the condenser. Using this information in a computer control model is expected to significantly improve magnesium collection efficiency.

<u>Raw Material Selection and Preparation</u>. Proper sizing of the three solid streams, fed continuously to the reactor, is necessary to avoid raw material carry-over into the condenser at normal magnesium vapor velocity. Pelletizing ferrosilicon fines removed after crushing and grinding is expected to increase ferrosilicon use efficiency and magnesium collection.

Substituting aluminum for a portion of the ferrosilicon reductant could yield some operating and economic advantages. Aluminum is more active, allowing a lower operating temperature. The denser slag resulting from aluminum should allow a reduction of power and result in a 2-5% increase in productivity. If a low value source of aluminum could be used there may also be significant capital investment and power cost savings in ferrosilicon production. The 0.3 kg Al/kg Mg required for reduction results in 10.3 kWh/kg magnesium for the reductant if the aluminum is made at 13.2 kWh/kg. At 10,500 Btu/kWh, 108 million Btu's/Mt of magnesium is used for aluminum reductant instead of 117 million Btu's/Mt of magnesium when ferrosilicon is used. This results in an 8% energy savings. Selection of the most cost effective reductant is very important as at present it is ~35% of production cost and consumes 48% of the energy in the final product. If low grade unuseable aluminum alloy, such as the final metallic residue from municipal refuse, were used, it would take 75% less energy in the reductant preparation stage, assuming the energy in the aluminum would not be recovered otherwise. This would result in a 36% reduction in the energy content of the magnesium product. If ferrosilicon were entirely replaced by low grade aluminum alloy with careful attention to purity levels a saleable calcium aluminum silicate cement byproduct might be produced.

Long term improvements are progressing toward continuous operation, design of more efficient condensing systems, and an increase in furnace size.

Continuous Operation. Present operation requires two slag accumulation cycles from the minimum, where the center electrode still barely maintains contact, to the maximum, when the vessel is full enough to have filled the magnesium crucible once. The first time the vessel fills with slag it is tapped and ferrosilicon and slag are decanted. First the vacuum must be broken and the tap hole opened and replugged. The vessel is then resealed, retested for leaks, and operation begun again. If the slag could be removed without breaking the vacuum and allow continuous operation until the magnesium crucible were full, a 14% increase in productivity might be possible while reducing capital intensity and labor. Power consumption could be lowered by as much as 30%. Further advantages might be increased magnesium collection efficiency due to less air leakage, and greater recovery efficiency in sludge treatment.

Condensing System. The magnesium collecting crucible and connecting pipes to the main reactor are all water cooled, resulting in solid magnesium contaminated to some extent with MgO and slag carry-over. If the crucible could be operated at the melting point of magnesium with slag carry-over reduced by operating at higher inert carrier gas pressure to sweep the magnesium to the crucible, the metal could be removed through a barometric leg, purified, and cast in continuous operation. This would yield an even more significant gain in economics and power saving, although it will require much more development. Experiments should be conducted on the condensing morphology of magnesium vapor to determine whether coalescence is possible on a solid condensing surface or whether it would be necessary to create additional surface for condensation and coalescence by forming a dispersed phase of molten magnesium.

Reactor Design. It should be relatively easy to scale a Magnetherm reactor up to the electrical capacity of the largest arc furnaces. A logical next step would be 10,000-kW 3-phase reactors yielding all the scale-up advantages of doubling production per unit.

Although there are predictions that atmospheric pressure operation has the potential of increasing productivity while decreasing power by as much as 25%, it would be difficult to operate at the higher required reaction temperature. Experimental evidence shows that at present operating temperatures production rate falls drastically if a ferrosilicon reductant is used at atmospheric pressure. While an aluminum reductant offers some temperature advantage, magnesium production at atmospheric pressure is not a promising approach to increased efficiency and reduced cost.

Carbothermic Reduction. While carbothermic reduction is considered to be a thermal process, more than 50% of the energy consumed has been electrical. The critical step is shock cooling of the gaseous products to avoid back reaction of the magnesium vapor and CO by rapidly condensing the magnesium vapor to a powder. The Permanente Company in California, used natural gas as the cooling media and produced a fine powder. This powder was then converted to ingots through briquetting, a process characterized by its explosion hazard. Studies have shown that while this process may be less capital intensive and more energy efficient than the Magnetherm Process, the advantage was less than 10%. Several researchers have expressed interest in the development of this technique, and in view of the continued development of plasma arc systems and the demonstrated successful commercial quenching, its viability is entirely possible. As in most new processes the actual economics must await reduction to practice.

Comparison of Magnesium Fxtraction Processes

In selecting a magnesium manufacturing process, the decision to use either electrolysis or thermal reduction will depend primarily on two variables, the choice of raw materials and the cost and availability of electricity. Table I compares the energy components of five candidate processes.

Table I

Comparison of Magnesium Producing Processes (4,11)

	Btu x 10^6 Mt	kWh/kg	% Elec.
Dow	374	18	50
Norsk-Hydro* Anhydrous	292	12	43
Pidgeon	358	17	50
Magnetherm	358	27	80
Carbothermic	358	18	53

*Anhydrous

All but carbothermic reduction are in commercial operation and apparently competitive--each in its own economic environment. It is clear, however, that the Norsk-Hydro anhydrous $MgCl_2$ process, using bitterns, dolomite, and a newly developed drying process, has a clear advantage over the others. It shows a 22% advantage in energy use over the Dow semihydrated $MgCl_2$ process and an 18% reduction over the contemporary Magnetherm Process. That only 43% of the energy is electrical is a further advantage. The cells operate at 12 kWh/kg, the lowest reported electrical energy consumption. Electrolytic processes require a greater capital investment per ton, however, and the Magnetherm Process enjoys a 25% reduction in capital intensity over the Norsk-Hydro Process.

When looking into the future, cost reductions for monopolar electrolytic cells, even with perfectly anhydrous feed, are limited by the inherent high capital intensity and low power efficiency of electrolytic processes. They are slow and must operate at low current densities. In space-time-yield the Magnetherm Process can far outstrip monopolar electrolytic systems as reactor size increases from 4500 to perhaps 15,000 kW furnaces. This triples present production per unit. To compete electrolytically with this scale-up, a bipolar system such as the Alcoa Smelting Process might be adapted to magnesium production. A single-unit production already equals the equivalent of more than five of the largest Hall cells and magnesium production would have the same advantage. This system has the potential of equalling the production of more than five of the 200,000amp cells presently being operated by Norsk-Hydro in one reactor. This could result in >60% reduction in capital investment over the monopolar systems while reducing electrical consumption >30% to ~8 kWh/kg. To accomplish this, an absolutely anhydrous feed with <0.2% other chloride impurities will be necessary. The Magnetherm reactor could equal such a cell in capital investment but would require 20 kWh/kg if ferrosilicon is the reductant. The Magnetherm reactor itself uses <10 kWh/kg of magnesium produced. Therefore, if a cheaper reductant, such as aluminum scrap from municipal waste were available, and the energy content of the reductant was not chargeable to the magnesium, since it would be unavailable if not used in this manner, then a large Magnetherm reactor 10,000-15,000 kW becomes highly competitive with the best conceived electrolytic process.

The 10% cost reduction in magnesium production required to win the transportation castings market is clearly within grasp and only awaits the decision to expend the necessary R&D effort and capital investment to bring this to pass.

References

1. E. Chin, "Magnesium," Mineral Facts and Problems, Bureau of Mines Bulletin 667, 1975, p. 637-651.
2. E.F. Emley, "Principles of Magnesium Technology," Pergamon Press, New York.
3. Kirk-Othmer Encyclopedia of Chemical Technology, 2nd Ed., John-Wiley & Sons, Inc., 1967, p. 661-708.
4. "A Survey of Electrochemical Metal Winning Processes," Battelle Columbus Laboratories, Contract No. 31-109-38-42k2, ANL/OEPM-79-3.
5, George B. Kenney, "An Analysis of the Energy Efficiency and Economic Viability of Expanded Magnesium Utilization," Massachusetts Institute of Technology, Feb. 1979.
6. B. Petkof, "Mineral Commodity Summaries - 1980, Bureau of MInes, p. 92-95.
7. "Magnesium Profile," Metal Statistics, 1980, American Metal Market, 1980.
8. "Chementator," Chemical Engineering, June 16, 1980, p. 69.
9. "Non-Ferrous Metal Data - 1979," American Bureau Metal Statistics.
10. Dow Chemical, American Metal Market, Sept. 18, 1979, p. 1, 48.
11. "The Economics of Magnesium Metal," 2nd Ed. 1978, Roskill Information Services Ltd., London.
12. B.F. Dodge, "Chemical Engineering Thermodynamics," 1st Ed., 1944, McGraw-Hill Book Co., Inc., New York.
13. J.H. Perry, "Chemical Engineers' Handbook," 4th Ed., McGraw-Hill Book Col, New York.
14. JANAF Thermochemical Tables, 2nd Ed., National Standards Reference Data Service, National Bureau of Standards, June 1971; and JANAF Thermochemical Tables, 1978 Supplement; J. Phys. Chem. Ref. Data 7 (3).
15. N. Høy-Petersen, "Some Aspects of the Electrolytic Production of Magnesium in I.G. Cells," Journal of Metals, 1969.
16. R.M. Hunter, "The Electrochemistry of the Dow Magnesium Process," Transactions of Electrochem Society, 1944, p. 86-21.
17. D.A. Elkin, K.C. Dean, and J.B. Rosenbaum, "Economic Aspects of Magnesium Production," Proceedings of Extractive Metallurgy Division Symposium on Electrometallurgy, AIME, Cleveland, Ohio, Dec. 1968, p. 173-184.
18. T.R. Beck, "Improvements in Energy Efficiency of Industrial Electrochemical Processes," Electrochemical Technology Corp., Argonne National Laboratory, ARgonne, Ill., Report ANL/OEPM-77-2.
19. G.L. Miller, Vacuum, 1952, 2, 19.
20. "Production of Magnesium and Magnesia," 2nd Nat. Expend. Rep., HMSO, 1943-44.
21. "Trends in Usage of Magnesium," Publication NMAB-322, National Academy of Sciences, Washington, D.C., 1975.
22. K.C. Deans, D.A. Elkins, and S.J. Hussey, "An Economic and Technical Evaluation of Magnesium Production Methods," Part 1, Metallothermic, Bureau of Mines, RI6656, 1965.
23. "Electrometallurgy," Proceedings of Extractive Metallurgy Division Symposium on Electrometallurgy, Cleveland, Ohio, Dec. 1968, American Institute of Mining, Metallurgical, and Petroleum Engineers, Inc., New York.
24. C. Faure and J. Marchal, "Magnesium by the Magnetherm Process," J. Metals, Sept., 1964.
25. M. P. Lugagne, "The Magnetherm Process for the Production of Magnesium," Erzmetall, Bd. 31(1978)H, 7/8.
26. Jean-Marie Logerot and André Michel Mena, "Extractive Metallurgy, Latest Developments of the Magnetherm Process," Société Francaise d'Electrométallurgie, France.
27. Unites States Patents 2,888,839; 2,398,891.
28. Pechiney Ugine Kuhlmann, Annual Report, 1979, p. 17.

Acknowledgements

The author wishes to acknowledge the valuable assistance provided by members of Alcoa Laboratories in the following areas: W.E. Haupin, cell design; Dr. W. E. Wahnsiedler, Thermodynamics; Dr. J.H. Blanks, Magnetherm.

THE PRODUCTION OF TITANIUM, ZIRCONIUM AND HAFNIUM

Ward W. Minkler
Transition Metals Associates, Inc.
Sewickley, Pennsylvania 15143

Edmund F. Baroch
Consultant, Albany, Oregon 97321

HISTORY

While titanium metal is one of the newest additions to the family of structural materials, it was first identified as a constituent of the earth's crust at the end of the eighteenth century. Two pioneer investigators, William Gregor, a clergyman in England, and M. H. Klaproth, a chemist in Germany, simultaneously reported the existence of a new element based upon examinations of mineral samples. Gregor called the new element menaccanite, found in black magnetic ilmenite sands of Cornwall. Klaproth, after examining a sample of rutile, concluded that it was an oxide of a new element he named titanium.

In 1910, Matthew Hunter and his associates at the General Electric Company, searching for a new lamp filament material, produced moderately pure titanium by combining titanium chloride and sodium in an evacuated steel bomb. However, the low melting point of the product, 1800°C, discouraged further work for this use. Eventually, the elements of this reaction were to form the basis for a commercial reduction process.

In 1925, Van Arkel and de Boer, in using an iodide dissociation method combined with the sodium reduction, prepared highly pure titanium. Samples produced by this method provided the material for the properites of titanium in Holland.

In the 1930's, Dr. W. S. Kroll, while conducting his investigations of vacuum metallurgy, produced titanium metal by both the sodium and magnesium process. However, his development of the pressureless reduction of the titanium tetrachloride with magnesium led to the development of the first practical production method and is still the most widely used process for the manufacture of titanium metal today.

In 1941, the United States Bureau of Mines initiated a program to develop a commercial process to manufacture titanium metal. By 1944 the Bureau of Mines was producing 130-pound batches at its Boulder City, Nevada, station using the Kroll Process.

Stimulated by the unique structural material requirements of the rapidly developing aircraft gas turbine engine, the United States government initiated a program to encourage a U.S. manufacturing capability for the production of titanium. In response, in 1948 the DuPont Company started a small production unit, producing 100 pounds of sponge per day. Shortly thereafter the National Lead Company started a similar activity in its Sayreville, New Jersey, titanium pigment plant.

By 1952, both companies had semi-production operations which combined, produced 225 mt of titanium sponge. Because of increasing demand from aircraft manufacturers, the U.S. government placed contracts with three other producers, Union Carbide, Dow Company, and Crane Company. These contracts, in combination with the DuPont and National Lead operations were supposed to provide 22,500 tons of titanium sponge using both the sodium and magnesium reduction processes. Simultaneously reduction plants using both the magnesium and sodium reduction processes were being constructed in Japan, the U.K., and the U.S.S.R.

In 1957 the switch in emphasis from military airplanes to missiles significantly reduced the demand for titanium metal. Only two sponge U.S. manufacturing plants survived, Reactive Metals, Ashtabula sodium reduction plant, and Titanium Metals Corporation, Herderson, Nevada, magnesium operation. These represented a U.S. production capacity of 11,000 annual tons, less than half the 1955 government goal. During much of the 1960's, even this was excess to demand; both Reactive Metals and Titanium Metals Corporation operated at a reduced rate until 1966.

Based upon the extensive use of titanium in wide-body commercial transports, demand for titanium metal accelerated rapidly after 1966, and new incremental capacity was added to both RMI and TIMET's sponge plants. In 1966, a third U.S. sponge plant was announced - Oregon Metallurgical Corporation in Albany, Oregon. Unlike the U.S., U.S.S.R., and Japan extraction plants which used either acid leaching or vacuum distillation to remove the excess magnesium chloride and unreacted magnesium from the sponge, argon gas was employed in the Oregon Metallurgical process to remove the magnesium chloride and magnesium from the heated sponge.

By 1970, U.S. sponge capacity had grown to 17,500 metric tons. Titanium mill products were available in a diversity of shapes, including extrusions, strip, seamless and welded tubing, casting plate, bar, and billet. A manufacturing base and the technology was accumulated in the melting and shaping of titanium and titaniumbase alloys. This supported the growth that occurred during the decade of the 70's. U.S. mill product shipments grew from 13,167 mt in 1970 to 24,730 mt in 1980, a compounded annual growth rate of 6.5%. Based upon the extensive new capacity under construction in 1981, it would appear that titanium producers anticipate a similar growth in the 1980's.

TITANIUM ORES

Occurrence

Titanium is the ninth most abundant element in the earth's crust and the fourth most abundant metallic element. It is widely distributed and economically exploitable ores are scattered in many places throughout the world. The primary mineral sources are the titaniferrous ore ilmenite, $FeTiO_3$ containing about 50% TiO_2. In sand deposits ilmenite has been upgraded by weathering and may contain 65% to 70% TiO_2; leucoxene fractions may contain up to 87% TiO_2.

Natural rutile, 96% TiO_2 is the most desirable ore for the production of titanium metal; however, gradual depletion of rutile resources and the increasing price have led to the development of methods for manufacturing synthetic rutile from ilmenite, either by smelting or chemical methods. Table I is a recent estimate (1) of world ilmenite and rutile resources.

Table I. World Ilmenite and Rutile Resources, Million Metric Tons Contained TiO_2

Country	Ilmenite	Rutile
India	197	26
Canada	130	
Norway	68	
South Africa	227	5
Australia	41	11
United States	145	13
USSR	30	4
Finland	6	
Sri Lanka	3	
Brazil	5	151
New Zeland	11	
Molaysia	2	
Sierra Leone		27
Italy		11
Other	62	7
Total	930	256

Australia leads the world in the production of both ilmenite and rutile. In 1979 Australia produced 645 thousand metric tons contained TiO_2 ilmenite and 263 thousand metric tons contained TiO_2 rutile. Only a small portion, less than 10%, of the titanium ore was used for metal production, the bulk being used to manufacture TiO_2 pigment.

Concentration

Rutile. Rutile deposits are the simplest to concentrate in a form adaptable for metal production. As late as 1950 deposits containing 30% rutile were common; now the average grade being worked is less than 1% requiring the development of special concentrating methods and equipment.

Rutile is mined in coastal deposits by a dredge operating in a dredge pond. After a primary concentration on the dredge, the monozite, zircon, rutile and ilmenite are separated by wet gravity concentrators as well as electrostatic and magnetic separation. The resulting rutile concentrate contains 95% or more TiO_2. A typical specification for rutile concentrate to be used in metal production is shown in Table II.

Table II. National Stockpile Purchase Specification

Compound	Percent by Weight
TiO_2	95.00 (1)
Fe_2O_3	1.00
SiO_2	1.25
P_2O_5	0.05
S	0.03
ZrO_2	1.00
Al_2O_3	0.75
V_2O_5	0.75
Cr_2O_3	0.75
MgO + CaO	0.25
Sn	0.10
Mn	0.75

(1) minimum - all others maximum

Control of MgO, CaO, Mn and Cr_2O_3 are most critical.

Ilmenite. The product of processes to reduce the iron oxide and concentrate TiO_2 content of ilmenite are becoming a more common feed stock for the production of titanium metal.

The oldest concentration process has been electric furnace smelting of ilmenite, in electric arc furnaces to produce pig iron and a high titanium slag. Table III illustrates the analysis of typical slag concentrate.

Table III. Typical Analysis of Richards Bay Slag

Compound	Weight %
TiO_2	85.5
Ti_2O_3	25.0
FeO	7.5
Fe (metallic)	0.2
SiO_2	1.5
Al_2O_3	2.0
MgO	0.9
CaO	0.14
MnO	1.4
Cr_2O_3	0.22
V_2O_5	0.40
C	0.07
S	0.06

(1) 30% of the titanium content is Ti_2O_5

Slag concentrates have been extensively used in Japan and the U.S.S.R. for metal production. However, in North America, the slags are produced from hard rock ilmenite containing significant quantities calcium and magnesium oxides which make them difficult to chlorinate. Rather, these slags are used for the production of titanium pigment using the sulfate process.

Synthetic Rutile. Efforts to develop beneficiation processes as an alternate to slag manufacture were accelerated during the 1973 - 1974 period when rutile prices tripled. Three types of process have reached varying degrees of commercial status.

1. Iron reduced to metal and separated from the TiO_2 chemically and physically.
The Western Titanium NL process involves a preoxidation step followed by the reduction of almost all the iron content to metallic iron by coal in a rotary kiln. The iron is then removed by agitating the metallized ore in aerated water in the presence of a catalyst, NH_4CL, so that the iron oxide is separated from the leached ilmenite in washing cyclones. The TiO_2 is acid leached and dryed. The iron oxide is sent to tailing ponds.

2. Iron reduced to the ferrous state and chemically leached.
The Benelite process is a commercial process used to beneficiate ilmenite. Ilmenite is partially reduced with carbonaceous material in a rotary kiln to convert the ferric oxide to ferrous oxide. The reduced ilmenite is cooled and charged into the digester for leaching with 18-20% H01. After leaching the beneficiated ilmenite (synthetic rutile) is separated. The leach liquor is sent to the HCl regeneration plant where the iron chloride is converted back to HCl and iron oxide (Fe_2O_3) produced. The byproduct iron oxide can be sold for steel manufacture or other applications.

3. Chlorine beneficiation
The Dunn process is based upon the fact that iron in ilmenite chlorinates more readily than titanium. After a pretreatment the ore is chlorinated in a fluidized bed with an excess of ilmenite so that any titanium that chlorinates will react with the iron oxide and remain in the concentrate as TiO_2. The iron chloride gas leaving the fluid bed ore reacted with oxygen to form iron oxide and chlorine gas for recycle.

The present commerical installations are shown in Table IV.

Table IV. Production Installation-Synthetic Rutile

Process Type	Company	Location	Capcity-mt
Iron reduced to metal then separated chemically and physically	AMC/Western Titanium	Capel, Australia	52,000
Iron reduced ferrous state, chemically leached	Ishihara-Sangyo Karsha	Yorrich, Japan	48,000
	Malaysian Titanium	Lakat, Malaysia	
	Kerr McGee Chemical	Mobile, Alabama	90,700
	Taiwan Alkali	Taiwan	30,000
	Dhrangadhra Chemicals	Sahupurasaa, India	20,000
			240,700

Source: U.S. Bureau of Mines

No commercial sized plant based upon selective chlorination has been constructed, despite extensive plant scale pilot size projects.

A typical analysis of the AMC/Western Mining, Capel plant is shown in Table V.

Table V. Typical Analysis of Synthetic Rutile - Western

Compound	Weight%
TiO_2 total	92.8
Ti_2O_3	10.0
Fe Total	2.1
MnO	1.0
Al_2O_3	1.1
SiO_2	1.2
ZrO_2	0.15
V_2O_5	0.02
Cr_2O_3	0.03
NbO_3	0.20
MgO	0.30
CaO	0.03
P	0.02
S	0.30
C	0.30

Unfortunately, the operation of even these commercial plants has been sensitive to the price of natural rutile. When in 1977, the demand for TiO_2 pigment fell and additional rutile supplies became available, these plants were closed or operated at reduced capacity. Now that natural rutile prices have again climbed, synthetic rutile is once more a viable raw material for the production of titanium tetrachloride. An increasing amount of synthetic rutile is being imported by the U.S. pigment and metal producers. The Kerr McGee plant in Mobile, Alabama, USA, resumed operations in 1980.

In addition, during the 1970's, other developments have occurred which will reduce the natural rutile dependence of the titanium sponge producer using a fluidized bed for production of titanium tetrachloride.

Foremost is the ability of chloride producers to use feed stocks containing mixtures of ilmenite, leucoxene and natural or synthetic rutile. While the DuPont Company has been a leader in these activities, other chloride producers have the ability to use feed stocks containing 80-90% TiO_2.

Summary. While natural rutile and high titanium content slags will continue to be important feed stocks for titanium metal, the product of other titanium ore forms, synthetic rutile, leucoxene and ilmenite will be used in increasing quantities to manufacture titanium tetrachloride. This will have the advantage of expanding the supply of titanium ore with the correlary of making titanium less likely to be subject to rapid price escalations due to ore supply/demand fluctuation or politically based forces which influence the cost of other metallic feed stocks. Secondly, it will make many countries independent of imports for their titanium feed stock.

TITANIUM TETRACHLORIDE

The first critical step in the manufacture of titanium metal is the production of titanium tetrachloride of adequate purity to produce acceptable metal. The basic reaction is a follows:

(1) $TiO_2 + 2\ Cl_2 + 2C \rightarrow TiCl_4 + 2\ CO$
$\Delta F = -10.9$ k cal at 1000°C

During commercial processes the following reaction also occurs:

(2) $TiO_2 + 2\ Cl_2 + C \rightarrow TiCl_4 + CO_2$

Three processes have been developed and are in commercial use:

1. The combination of coke, chlorine and titanium ore (rutile, ilmenite, synthetic rutile and leucoxene) in a fluidized bed reactor.

2. Pellitized charges of ore plus coke in a shaft furnace.

3. A ground charge of ore feedstock mixed with powdered coke introduced into a molten salt bath with chlorine bubbled through the melt.

In both the U.S.S.R. and Japan briquettes of coke and titanium are chlorinated in a shaft furnace, particularly in the production of $TiCl_4$ from slag produced by electric furnace smelting of ilmenite ores.

In the U.S.S.R. the use of molten salt chlorination has been reported. To chlorinate feedstocks containing an increased amount of alkali earth elements, CaO and MgO, it is possible to chlorinate the TiO_2 in a liquid bath of molten chlorides of alkali and alkali-earth elements.

The selection of the process is influenced by the type of ore being chlorinated and the impurities present. For example, it is difficult to chlorinate ores containing a significant quantity, above 1%, of CaO + MgO in fluidized bed reactors. The chlorides of calcium and magnesium have boiling points above 1400°C which plug the chlorinator beds and off gas piping. In addition, calcium oxide reduces the chlorination of TiO_2 and SiO_2 by forming $CaOTiO_2$.

The other metal oxides in the ore will be readily chlorinated with the titanium and exit from the chlorinator. Tetrachloride purity is of great significance in the production of titanium metal exhibiting acceptable commercial quality. Therefore, these impurities must be reduced to exceptionally low levels.

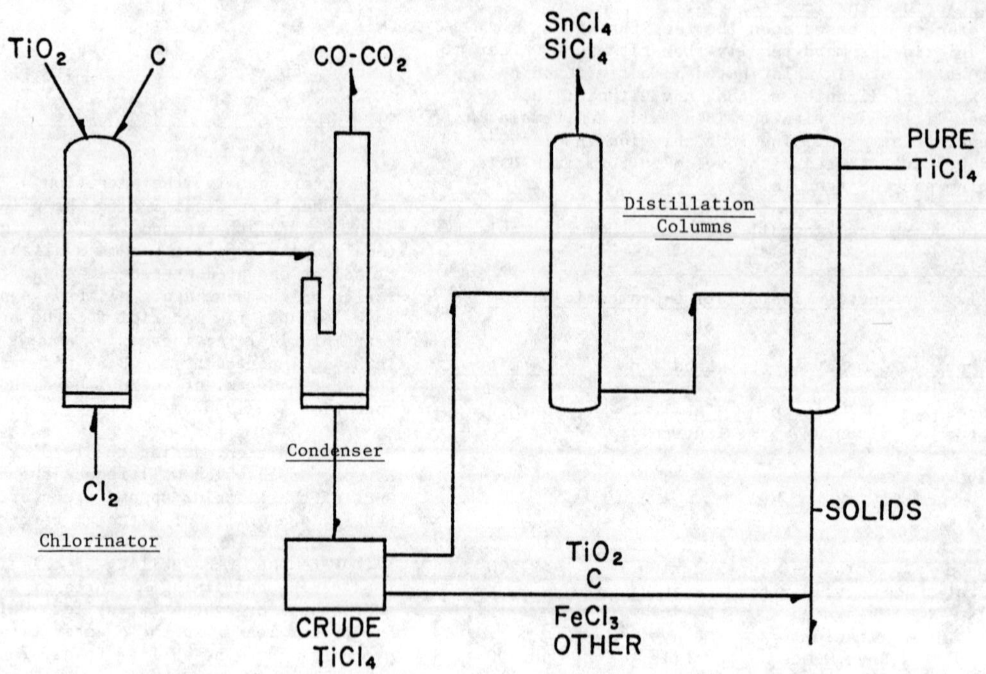

Figure 1. Flow diagram of tetrachloride production using the fluidized bed method with subsequent purification of the $TiCl_4$.

In the United States the only method used commercially to produce titanium tetrachloride is the fluidized bed reactor. In one commercial process a chlorinator approximately 8 meters high and 3.7 meters diameter is used. Chlorine gas is introduced in the bottom of the reactor. Calcined petroleum coke, containing 98.5% carbon and the ore feed stock are fed to the chlorinator. Initially oxygen, air, is introduced with the chlorine to start the reaction. Once started the rate of chlorination is controlled by monitoring the chlorine feed rate. These chlorinators operate at approximately 950°C with a charge of approximately 50% TiO_2 and coke. Sufficient free board is maintained so that most of the unreacted rutile and coke are not carried out with the $TiCl_4$. Gaseous reaction products and entrained dust exit at the top of the chlorinator.

After exiting from the chlorinator the gas stream is ducted to a cyclone where the solids, rutile and coke are removed. The gas is then cooled to approximately 30°C, condensing the $TiCl_4$ and high boiling point impurities such as $FeCl_3$, $AlCl_3$, and $MnCl_2$. Ore and coke dust not collected by the cyclone form a slurry with the condensed $TiCl_4$. The residual ore and coke dust, together with the high boiling point impurities are concentrated by distilling off product $TiCl_4$ or centrifugation. The concentrate is then dried by various means to produce a waste product relatively free of $TiCl_4$.

One of the major problems is the separation of the

vandium oxychloride ($VOCl_3$) from the $TiCl_4$, since they have similar boiling points. Several methods have been used commercially.

White mineral oil is added to the crude $TiCl_4$. The oil complexes with, or reacts with, the vandadium so that it can be removed in a single distillation. Hydrogen sulfide has also been used. Vanadium oxytrichloride in a $TiCl_4$ solution is reduced by hydrogen sulfide to $VOCl_2$.

(3) $VOCl_3 + 1/2\ H_2S + VOCl_2 + HCl + 1/2\ S$

The VOC_2 precipitates. The primary problem is the contamination of the $TiCl_4$ with elemental sulfur which can be detrimental for metal production unless carefully controlled.

In the U.S.S.R. minor additions of copper, 0.25-0.5 percent, of the amount of purified $TiCl_4$ are used. An added advantage is that copper removes not only the vanadium, but aluminum chloride and sulfur having impurities.

After removal of the vanadium, the $TiCl_4$ is reheated and fed to a distillation column where chlorides having a lower boiling point than $TiCl_4$ are removed, primarily gases, $SiCl_4$ and $SnCl_4$. The stream is then fed to a second distillation column where traces of the elements having a higher boiling point than $TiCl_4$ are eliminated. This distillation tower, and subsequent storage tanks are all maintained under a positive helium pressure to avoid contaminating the pure tetrachloride with air.

The residuals in a typical $TiCl_4$ used for the production of titanium metal is shown in Table VI.

Table VI. Titanium Tetrachloride Used To Produce Titanium Metal

Element	ppm
Sn	16
Al	6
Fe	5
V	7
Si	5
C	2.5
Cr	3
Residue	90
Color	Clear

REDUCTION OF TITANIUM TETRACHLORIDE

Titanium tetrachloride is a stable compound with the following decomposition reaction:

(4) $TiCl_4 \rightarrow Ti + 2\ Cl_2$

$\Delta F° = 175{,}600 - 29.11\ T\ (°K)$

Since this is possible only at very high temperatures, it is yet commercially impractical. Early titanium investigators explored the use of hydrogen in an electric arc as a method for reducing the $TiCl_4$ to titanium metal; however, because of equipment and product purity problems, the process was abandoned.

In 1980 two industrial processes and their variations are used to produce the known world production of titanium metal. As shown in Table VII, in excess of 80% of the identified world's titanium plants employ the magnesium reduction of titanium tetrachloride. As yet, no large-scale electrolytic extractive titanium manufacturing plant has been announced. One company, the D-H Titanium Company in the United States, has stated an intention to construct an electrowinning plant capable of producing 1,350 to 2,270 tons annually of electrowon product.

Table VII. Estimated World Sponge Production Capacities

	Process	Capacity - mt 1970	1980
U.S.			
TIMET	Mg-leach	10,000	12,705
RMI, Inc.	Na-leach	4,550	7,940
OreMet	Mg-Argon distillation	2,275	2,275
		16,825	22,920
U.K.			
ICI	Na-leach	3,000	3,000
Japan			
Osaka	Mg-distillation	6,370	14,500(1)
Toho	Mg-distillation	5,824	12,000
New Metals	Na-leach	1,820	180
		14,014	26,680
U.S.S.R.	Mg-distillation Electrowinning	30,000	Unknown
PRC	Mg-distillation Na-leaching		2,000

(1) Includes New Metal plant operated of Osaka Titanium

Magnesium Reduction

The fundamental reaction that occurs in the reduction of $TiCl_4$ with magnesium is:

(5) $TiCl_4\ (gas) + 2\ Mg\ (liquid) \rightarrow$

(6) $Ti\ (solid) + 2\ MgCl_2\ (liquid)$

The reaction is highly exothermic, -206 Kcal/mole at 900° C/with a free energy of -76.2 Kcal/mole at 900° C.

In commercial practice control of the rate at which the $TiCl_4$ is fed to the reactor along with the control of the external heat applied to the reactor are of primary importance, since they influence the quality of the titanium sponge formed during the reaction and the efficiency of the reaction.

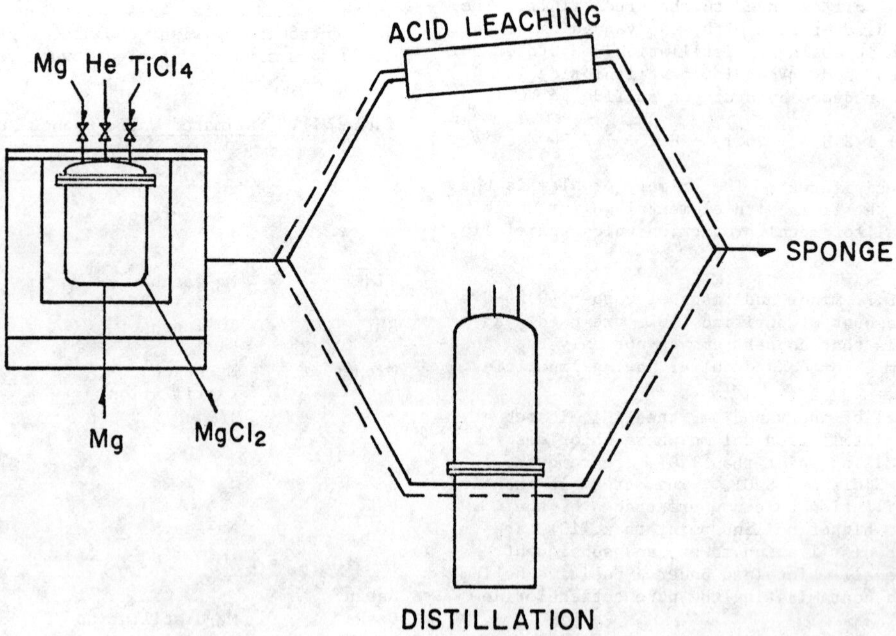

Figure 2. Flow sheet of Kroll process and sponge purification. Either leaching or distillation is used to remove excess magnesium and magnesium chloride. Either argon or helium, or mixtures are used in reaction pot.

As shown in Figure 2, magnesium is placed in the reactor either in the molten or solid state. Reactors vary in size such that individual sponge lots are from less than one metric ton to over 5 tons. Reactors are constructed from both carbon and high-strength stainless steels. After charging the magnesium the reactor top is securely fastened by welding or bolting. The reactor is purged with a helium or argon gas; then placed in a furnace to melt the magnesium prior to introduction of the $TiCl_4$. Both electric and gas heating of the reactor are used.

Titanium tetrachloride is fed into the reactor at a rate such that the reactor temperature does not exceed 900°C. As the reaction proceeds, molten liquid magnesium chloride is periodically removed by bottom taping or top decanting. In most plants excess magnesium is added, above the amount required stochiometrically to form the titanium sponge based upon the quantity of $TiCl_4$ fed to the reactor. This reduces the amount of lower chlorides, $TiCl_2$ and $TiCl_3$, which if present in the final sponge product degrade sponge quality.

After completion of the reaction, the sponge mass must be removed from the reactor and cleansed of any residual magnesium chloride and unreacted magnesium. There are significant differences between both of these operations in the various sponge manufacturing plants. Their efficiency will have a significant effect on both the sponge quality and economics of the manufacturing operation.

TIMET, the largest producer of titanium in the United States, uses a leaching process. After completion of the reaction, the vertical reactors are removed from the furnace and cooled. The welded top is removed and the sponge extracted by boring the sponge from the reactor walls in the form of chips. These chips are washed in a continuous counter-flow leacher using an acid solution, primarily nitric and hydrochloric. The sponge is washed, dried and packaged for melting.

At the Oregon Metallurgical plant, the reaction occurs in a large horizontal reactor, 6.7m in length by 1.8m in width. After completion of the reaction and draining the magnesium chloride, the sponge mass is swept by argon at 1,000°C. The vaporized magnesium chloride and magnesium are separated from the argon in a condenser external to the reaction unit. The sponge cake is removed from the reaction tray by exerting a slight mechanical pressure on the bottom.

In titanium plants in the U.S.S.R. and Japan the sponge mass is cleansed of the residual magnesium chloride and magnesium by vacuum distillation. This process is based upon the significant differences that exist between the equilibrium vapor pressure of magnesium, magnesium chloride and titanium in the range of 960-1020°C. Control of the porosity of the sponge mass during the reduction is extremely important as is the magnesium chloride withdrawal schedule. Unless these are carefully controlled abnormally long times are required to remove the last 5 to 10% of the magnesium chloride and magnesium. At a temperature of 950-1000°C and a vacuum of 10^{-1} to 10^{-3} mm Hg satisfactory separation is reported in 60 hours. Properly conducted, the vacuum distilled product displays the lowest residual gas content of any sponge commercially produced.

Two production distillation methods have been used. In the U.S.S.R., after completion of the reaction, an empty second reactor is placed on top of the reactor containing the hot sponge mass. A vacuum is induced in the system pulling the magnesium and magnesium chloride into the top reactor where it condenses on the walls. Upon completion of the distillation cycle, the top reactor is removed and is used for a subsequent reduction.

In an alternate method, the cooled reactor is placed in a furnace in an inverted position with the condensing unit below it and external to the furnace. A vacuum is applied to the bottom vessel; the top reactor containing the sponge mass is heated to 1000°C and the distillation proceeds to completion.

In both processes electrical heating is used and stainless steel reactors are required to withstand the vacuum pressure at 1000°C and temperature control is critical.

The cooled sponge mass is removed from the reactors by extruding the sponge in a horizontal press. The block of titanium sponge is crushed by pressing and cutting into a chip size adaptable for the melting operation.

Magnesium chloride formed during the reaction is recycled using conventional magnesium reduction methods including I. G. Farben, Alcan and U.S.S.R. VAMI cells. Modern titanium sponge plants use cells of 80,000 amperes or greater. Magnesium purity influences sponge quality; therefore, steps are taken to refine the magnesium after separation. Cold magnesium chloride can be recycled; however, it does pose operating problems. For this reason, coordination of the operating rate of the reduction and magnesium-chlorine recovery plants is important.

Sodium Reduction

The Hunter process is used in plants in Japan, the United States and the United Kingdom. The basic reaction is:

(7) $TiCl_4$ (gas) + 4 Na (liquid) $\rightarrow$
(8) Ti (solid) + 4 NaCl (liquid)

The reaction is highly exothermic (-270 kcal/mole) at 900°C. The free energy of the reaction is -143 kcal/mole at 900°C and at high temperatures back reactions do not occur.

Like the magnesium process, there are several variations of processes utilizing the sodium reduction reaction. The first large plant was constructed by ICI at Deeside, United Kingdom. The process was single-step process reacting 4 moles of sodium with one mole of $TiCl_4$. In the ICI plant the sodium chloride was not recycled, but discharged into the ocean.

In 1959, based upon work by the U.S. Bureau of Mines, a two-step process was developed. The first is the reduction of $TiCl_4$ to dichloride ($TiCl_2$) and trichloride ($TiCl_3$). During the second stage the lower chlorides are reduced to titanium metal in a salt bath (NaCl) to which free sodium is added.

The RMI, Inc., plant built in 1961 in the United States utilized this two-step process as shown in Figure 3. Titanium tetrachloride is fed into a continuous reactor operating at 232°C under an argon atmosphere. The $TiCl_4$ reacts with molten sodium fed simultaneously to form dichloride:

(9) $TiCl_4$ (gas) + Na (liquid) $\rightarrow$
$TiCl_2$ (solid) + 2 NaCl (solid)

The $TiCl_2$ + NaCl mixture is transferred to a reactor, called a sinter pot by means of a screw-type conveyer. The pot initially contains molten sodium under an argon atmosphere. The sealed vessel containing the sodium, sodium chloride and titanium dichloride are placed in a furnace and heated to 1000°C. The following reaction occurs:

(10) $TiCl_2$ + 2 Na $\rightarrow$ Ti + 2 NaCl

As the reaction proceeds, the titanium forms as fine particles which agglomerate into coarser sponge at the center of the sinter pot. After cooling, the sponge mass is surrounded by the sodium chloride.

The salt and titanium mixture are chipped from the pot by remote controlled pneumatic hammers. This mixture, termed spalt, is leached in a dilute acid after crushing to a 0.95 cm maximum particle size. After washing and vacuum drying, the product is ready for melting. Like the leached magnesium-reduced sponge, this product contains higher gas content than the vacuum distilled product. By proper control, the purity of the sponge product can exceed that of vacuum-distilled magnesium-reduced sponge.

Electrowinning

Investigations of the electrowinning of titanium metal from titanium tetrachloride using fused salt electrolysis have been conducted since the middle 1950's. Electrowinning and its companion electrorefining of titanium wastes have been the subject of numerous published papers. Despite the theoretical advantages in both product quality and economics, no production-sized plant equivalent to the Kroll or Hunter process plants has been reported. The highly cyclical market conditions and the adequacy of sponge plants constructed in the 1960's to fulfill demand has made an investment in large scale electrowinning financially impractical.

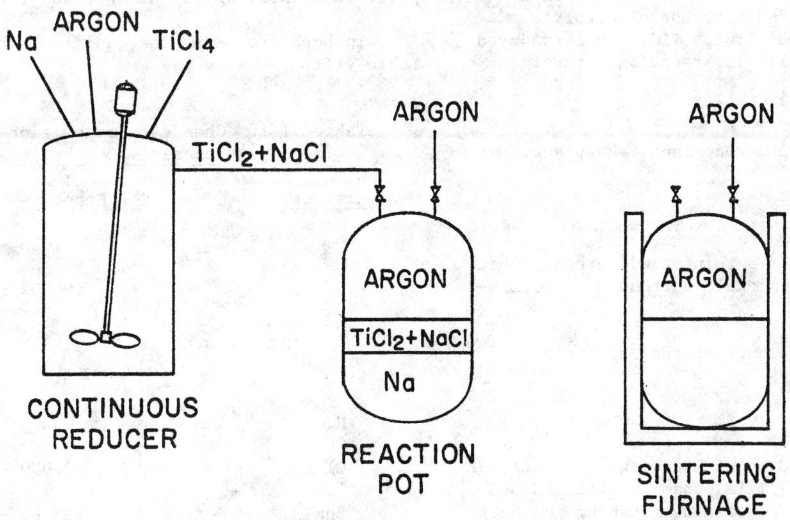

Figure 3. Two step sodium reduction process; the reactor (sinter) pot containing the lower chlorides and salt are transferred to a sintering furnace. The titanium sponge and salt mixture (spalt) are removed and water leached to cleanse the salt.

In 1967 the Titanium Metals Corporation of America operated a semi-works electrowinning plant producing at an an annual rate in excess of 150 metric tons. Industry economic conditions forced the pilot plant's closing in 1969. In 1980 the D-H Titanium Company announced its plan to build a production-sized plant in Freeport, Texas, based upon the process developed by the U. S. Bureau of Mines.

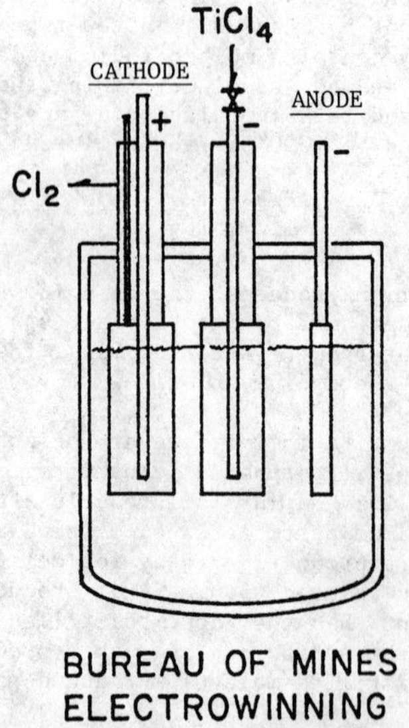

Figure 4. The electrolyter reduction cell utilized by D-H titanium based upon a concept developed by the U. S. Bureau of Mines.

The TIMET and D-H processes utilize both TiCl as the feed material and molten salt electrolysis. Titanium is deposited on the cathode and the chlorine liberated at the anode is recycled for the manufacture of tetrachloride.

The primary differences in these processes are in the electrolyte, diaphragm, internal cell design and cathode removal system.

The D-H cell uses a KCl-LiCl eutectic electrolyte operating at 520°C. After adjusting the electrolyte composition by the addition of $TiCl_2$, $TiCl_4$ is introduced at the feed cathode. The following reaction occurs:

(11) $Ti^{+4} + 2\ C \rightarrow Ti^{+2}$

At the anode the following reactions take place:

(12) $Cl^{-} \rightarrow 1/2\ Cl_2 + C$
(13) $Ti^{+3} \rightarrow Ti^{+4} + C$

A diaphragm consisting of a reinforced screen, electroless plated with cobalt or nickel, surrounds the cylindrical graphite anode.

The following reaction occurs at the deposition cathode:

(14) $Ti^{+2} + 2e \rightarrow Ti$

The deposits are removed by withdrawing the steel bar stock cathodes into an argon-filled chamber after 20,000 ampere-hours. The deposits, containing titanium and NaCl are stripped from the cathode. The mixture is removed from the chamber, crushed, leached in a dilute HCl solution, washed and air dried.

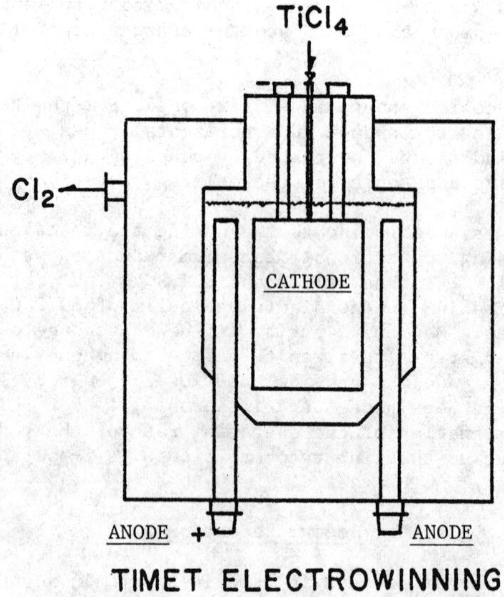

Figure 5. TIMET electrolyter cell where the cathode basket acts as a diaphragm.

In TIMET's process purified $TiCl_4$ is fed into a cell containing molten NaCl operating at 900°C. The anodes are graphite rods at the sides of the cell. The cathode consists of steel rods surrounded by a wire screen basket. The $TiCl_4$ is fed continuously into the center of the basket.

The initial deposit occurs as a result of sodium, formed from the NaCl on the cathode basket. This reacts with $TiCl_4$ to form a porous titanium deposit on the basket which acts as the diaphragm separating the catholyte and anolyte. The major part of the deposit now occurs on the steel rod cathodes.

Upon completion of the deposition an argon-filled chamber is moved over the cell. The cathode assembly is pulled into the chamber and a fresh cathode assembly lowered into the cell. The hot cathode is placed in an argon-filled cooling chamber. After cooling to room temperature, the cathode basket is disassembled and the deposit is washed, dryed and is ready for melting.

In both processes the quality is high as shown in Table VIII.

Table VIII. Chemical Composition of Electrowon Deposit - D-H Process

Element	D-H Deposit- Weight Percent
O_2	0.035 - 0.065
N_2	0.002 - 0.003
C	0.003 - 0.011
Fe	0.010 - 0.020
H_2	0.003 - 0.005
Cl	0.010 - 0.035
Bhn	60 - 90

Alternate Processes

There have been suggestions and reports of alternate processes to those discribed above. One that has moved into modest production is the manufacture of alloyed between sponge by the co-reduction of other metallic chlorides simultaneously with titanium tetrachloride with magnesium. This method shows commercial promise where a highly alloyed wrought product is desired. With the continuing interest in alloys containing large quantities of molybdenum, up to 30%, producing alloyed sponge can be technique for producing ingot free of segregation.

A process which had been proposed is the reduction of titanium droxide in a hydrogen plasma. The extremely high temperatures required make design of equipment difficult and expensive. The U. S. Bureau of Mines, as well as others, have investigated the possibility of reducing TiO_2 with carbon followed by electrorefinning step to obtain acceptable commercial purity. No commercial plants employing these processes have been reported.

MELTING

The combination of titanium's high melting point, 1704°C, and the reactivity of molten titanium with refractory mold materials initially posed a major problem for the production of commercial sized ingots. The first ingots were produced by side feeding the sponge and alloy additions into a tungsten electrode arc. The crucible was a water-cooled copper mold and the melting was carried out in an argon atmosphere. This method had disadvantages, including limited ingot size, slow melting and the tendency of the ingot to contain undissolved particles of tungsten.

In 1951, the Titanium Metals Corporation adapted a method previously used to melt refractory metals previously used to melt refractory metals such as tungsten and molybdenum. The technique, termed consumable-electrode melting, consists of the manufacture of a sponge electrode from pressed compacts containing a mixture of sponge and alloy additions. Individual compacts are welded together to form a single electrode. The electrode is placed in a consumable electrode furnace under a vacuum or inert gas atmosphere. The sponge electrode is lowered into an externally water-cooled cylindrical copper crucible and an arc struck on a starter charge, or plate placed in the bottom of the crucible. The sponge electrode is then consumed as the metal, in the form of droplets, is transferred to the water-cooled mold. The transferred metal solidifies and the ingot is gradually built up in the crucible.

The initial ingot, called a metal electrode, is remelted either once, or twice, to achieve greater size and chemical homogenity. While furnace equipment, arc controls, ingot size and quality today bear little resemblence to the consumable electrode ingots manufactured in 1952, the consumable electrode arc melting process is the method used to produce all but a small quantity of the titanium ingots produced in 1981. Ingots weighing up to 10 mt, meeting exacting specifications, are routinely melted in production melt shops.

Electrode Manufacture

After cleansing the sponge of all but minor amounts of chlorides and magnesium, it is segregated into lots by chemical content (purity) and particle size. Depending upon the desired ingot composition, these lots are blended and mixed with alloy additions. Alloys are added as elements such as copper, aluminum, or iron or master alloys, whose melting point is similar to that of titanium, i.e., 60% aluminum - 40% vanadium used in the manufacture of the titanium alloy containing 6% aluminum and 4% vanadium. If fine scrap is to be used, such as turnings, sheet clippings or hydrogenated crushed scrap (1) it is blended with the sponge and alloy additions. The mixture is loaded into a mold and pressed into shapes called electrode compacts. The compacts vary in size and geometry, octagonal, round or retangular.

A series of individual compacts are welded together to form a single sponge electrode, one or more of which is used to produce the first melt metal electrode. Compacts are welded using a variety of methods; plasma arc, or metal inert gas using local gas shielding, are commonly used. In some instances, where superior ingot quality is required, the welding operation is carried out in a vacuum chamber to avoid any possibility of weld contamination. Compact strength is determined by controlling the blending and pressing operations. This is critical in producing sound metal electrodes.

In the U.S.S.R., sponge electrodes are formed by extruding the sponge and alloy additions in a press. Electrodes, up to 800 mm diameter, are formed continuously. Because of limitations, including limited scrap additions, high labor intensity, round cross section ingot only, the extrusion method is being replaced by a new process utilizing melting into a skull of titanium, and casting square and slab, as well as round, shaped metal electrodes.

Revert scrap is an important component of ingot produced in the United States representing 40% of the ingot charge. In addition to scrap blended into the compact, massive pieces, called bulk weldables, are welded together to form a first melt electrode. Careful control of the surface condition and cleanliness of such revert scrap is mandatory and assurance of the chemical content of each piece requires individual piece-sampling to avoid ingot contamination and failure to meet chemical content requirements. Frequently, pressed sponge compacts are combined with compacts of welded scrap to form the first electrode.

Melting

During the first melt, volatiles remaining in the sponge - hydrogen, chlorine, magnesium or magnesium chloride - boil off, particularly when melting Kroll-leached sponge. This forms a rough ingot surface and requires sufficient pumping and filtering equipment to maintain a proper vacuum. In such instances, it has been found necessary to clean the metal electrode surface by acid dipping and/or high pressure water spray to remove the magnesium chloride which has condensed on the metal electrode surface after the first melt.

If more than one first melt metal electrode is to be used in the second melt, the individual metal electrodes are connected by welding. The assembly is loaded into a second consumable-electrode furnace. Depending upon the composition, the metal electrode is melted under vacuum, usually less than 60 microns, or in some instances, in a partial argon environment to form the second melt. In most instances, this is the final melt. However, for reasons of ingot size or quality, a third melt may be conducted, called triple melting.

In the last melt, the current is carefully reduced at the end of the melt cycle, to bury the shrinkage cavity below the solidified top of the ingot. If properly done, the cavity will seal during the first forging operation, allowing for a significant yield improvement when hot, worked to form a small product.

Final ingot sizes vary in diameter and weight depending on the individual melt shop. The most dominant commercial size ingots vary from 28 " diameter - 8,000 lbs. to 36" diameter - 15,000 lbs.

(1) Titanium scrap solids are heated in a hydrogen atmosphere. The resultant brittle titanium-hydrogen pieces are crushed into particles similar in size to sponge, less than 1.25 cm inch.

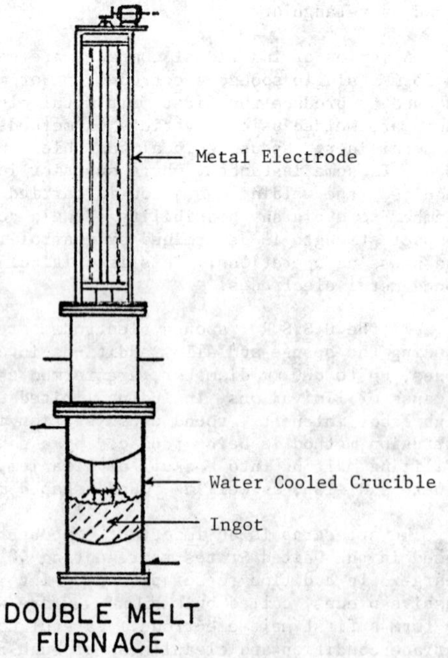

Figure 6. Consumable-electrode furnace used to melt titanium ingots.

In most production melt shops, water flooding of the external copper mold surface prevents the molten titanium from digesting the mold. In some instances, a combination of molten sodium and potassium, called NaK, is used to cool the copper mold. This removes the explosion hazard that plagued ingot producers in the 1950's and 1960's. There is some indication this also improves heat removal.

Quality Considerations

As titanium and titanium-base alloys have been increasingly used in sophisticated aerospace structural applications, freedom from any possibility of defects has been required. Titanium structures have displayed an unusually good record of reliability with respect to metal related failures; however, in those few instances where such has occurred, the failure has frequently been traceable to ingot manufacture. For this reason, melting methods and process control techniques have been developed to meet rigorous quality demands.

Melting related quality problems are of four basic types:

Ingot Surface - This affects yield of metal through excessive losses during the manufacture of mill products. Sponge quality, pool depth, amount of pool super heat, clearance between ingot mold and metal electrode are among the factors influencing ingot surface.

Inclusions - Inclusions have been observed in titanium ingots. These consist of refractory titanium compounds, (LDI), and particles of metals having a significantly higher melting point than titanium (HDI).

The HDI type of inclusion consists of discreet, unmelted particles of metals having a melting point significantly higher than titanium, such as molybdenum, niobium, tantalum, tungsten or alloys of these. A typical example is shown in Figure 7. In most instances, the source of such defects is scrap; either through cross contamination or from improper preparation and inspection. Chips produced by machining with carbide tools or parts having wear resistant coatings are the most frequent source of HDI's. Control of scrap recycling preparation is necessary to avoid this type of defect.

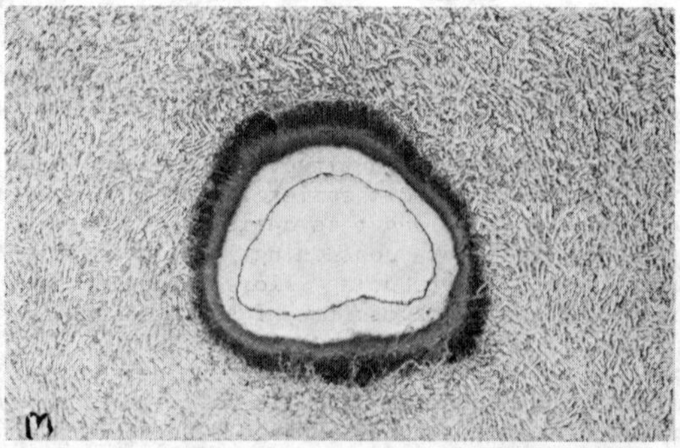

75X

Figure 7. High density inclusion in a Ti-6Al-4V alloy matrix. Center section consists of pure niobium surrounded by titanium-niobium diffusion zone. Source of this inclusion was recycled scrap.

The LDI type of defect has been of most concern since this defect has been related to the premature failure of highly stressed parts, such as compressor discs in aircraft gas turbine engines. Two types have been identified.

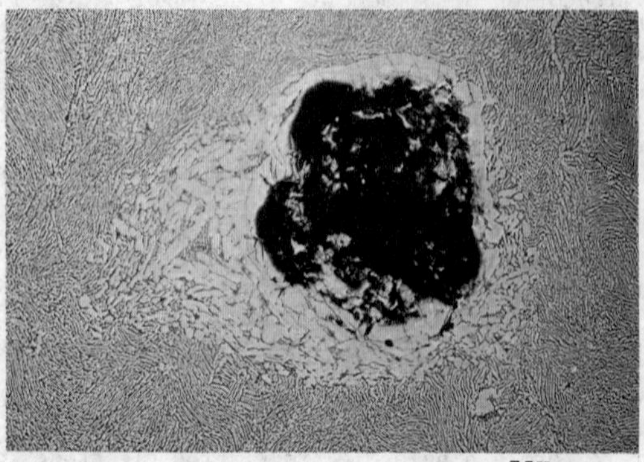

75X

Figure 8. Type I low density inclusion in a Ti-CAl-2Sn-4Zr-2Mo matrix. The major comtaminant is generally nitrogen. The surrounding stabilized alpha is harder than the matrix and exhibits an appreciably lower alloy content.

Type I, shown in Figure 8 generally consists of a void or crack region surrounded by nitrogen stabilized alpha. Several sources of such inclusion have been identified: nitrogen contaminated sponge particles; air contamination of welds during sponge electrode manufacture; air leaks during melting; improperly prepared scrap; poor furnace cleanliness. Elimination or control of raw materials and the conditions causing this type of defect is the most effective way for eliminating Type I LDI's. However, a third melt has been found to be effective for digesting LDI's and is specified where the metal parts are to be highly stressed.

Type II LDI's consist of streaks or bands of alpha stabilized areas containing slightly higher aluminum, up to 1.5% These have been associated with the shrinkage cavity which has healed during ingot forging. Proper hot topping will eliminate Type II LDI's.

Macrosegregation

Macrosegregation can come from many sources: mixed scrap, misidentified or incorrectly formulated compacts; improper stirring; improper pool depth. Macrosegregation can occurr along the length of the ingot; but most frequently appears as unacceptable center-to-outside variation of those elements that tend to segregate. These include oxygen, aluminum or iron.

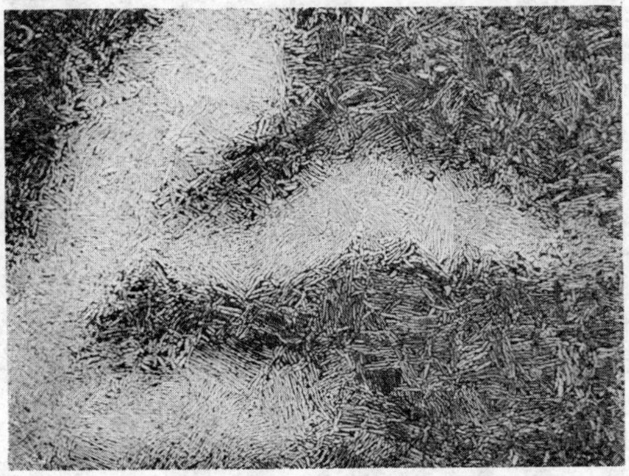

50X

Figure 9. Type II low density inclusion in a Ti-6Al-4V metrix. The inclusion consists of a band or streak of alpha stabilized by a somewhat higher aluminum content. Hardness is only slightly high than the matrix.

Microsegregation

This condition has received a great deal of attention of late, primarily because of their similarity in appearance to the "freckled" condition observed in nickel-base super alloys. In most instances, no effect on mechanical properties has been reported. The condition can be minimized by control of melting current, stirring and hot topping procedures.

Other Melting Practices

Several other commercial methods for manufacturing titanium ingot have been developed and are used on a limited scale. The non-consumable method employing a water cooled or tungsten electrode is used to melt a metal electrode. The trade names for this process are Durarc or Rototrode. Solid metal, generally light scrap, is melted using the nonconsumable electrode in a tilting water cooled copper or skull mold. After melting, the tun dish is tipped pouring the molten metal into a water cooled cylindrical mold. Another charge is melted and added to the top of the first. The ingot is lowered an equivalent length into a withdrawal chamber until the final ingot length has been completed. The advantages of this melting method are its ability to consume light weight scrap; the elimination of a press required to form the sponge compacts, and the possibility of higher pool temperatures. The disadvantages are high power consumption, and high maintenance of the non-consumable electrode.

The electron beam melting technique also is used by one U. S. producer to melt light grade scrap such as turnings to form segments that can be welded to form a metal electrode. After melting the molten metal cascades down a hearth where high density inclusions are separated by settling from the molten titanium. The metal flows into a mold which forms an electrode segment for second melting. The disadvantage is loss of volatile alloy additions such as aluminum which requires compensation in the construction of the metal electrode for the second melt.

A hollow electrode plasma electron beam furnace is being used in Japan to produce both metal electrodes and slab ingot castings for producing sheet and strip. It is claimed the single melt product can be used more economically than a slab freezed from a conventional cylindrical double melt ingot. Very low volatile containing sponge is necessary in this process which requires an additional vacuum degassing cycle, even when using vacuum distilled sponge.

In the U.S.S.R., the use of electroslag remelting, in combination with a skull melted metal electrode is finding increasing use. The melting takes place in an argon atmosphere using pure calcium flouride as the slag and an alternating current. A significant surface improvement is reported which produces a yield improvement, up to 15%, compared to vacuum-arc remelted ingots.

PRIMARY FABRICATION

Titanium ingots are shaped into mill products on equipment designed for the processing of stainless and carbon steel. However, heating and surface removal methods have evolved specifically for processing titanium.

The development of processes capable of manufacturing mill products meeting industry technical requirements yet consistant with an acceptable economic framework, must be based upon the basic principles applicable to hot and cold working titanium.

1. Titanium, when heated to a temperature sufficient to forge or roll, performs as a reactive metal absorbing hydrogen and oxygen from the hot atmosphere. To minimize this, furnace temperatures, heating times and atmospheres must be carefully controlled.

2. The mechanical properties of titanium and titanium alloy mill products are influenced by the thermal-mechanical working history of the metal. Temperature and working schedules must be designed to provide acceptable properties, yet be within the mechanical constraints of available equipment.

3. Titanium, and titanium alloys, display two crystallographic structures, hexagonal close packed alpha and body centered cubic beta. Each structure displays its own deformation characteristics and resultant textures. Such textures, particularly when formed as a result of working in the alpha phase, exert a strong influence on both mechanical properties and subsequent formability of the alloy. This requires process designs for hot and cold deformations, as well as recrystallization cycles and annealing so that anisotropic characteristics are controlled.

4. Titanium and titanium alloys tend to gall when in intimate contact with other metals such as those encountered in drawing. Steps must be taken to minimize this in such operations.

Hot Working

The first step in forming a mill product from a cast ingot is forging. Furnaces for heating the ingot can be electrical, gas or oil fired. To prevent hydrogen contamination, reducing atmospheres are maintained. If steel is forged in the same furnaces, steps must be taken to prevent contact between any residual iron oxide and the titanium. Heating times are held as short as possible to avoid excessive oxidation of the titanium. Close temperature control must be maintained to avoid excessive cracking, if too cold; and undesirable microstructures and surface contamination, if metal temperature constraints are exceeded.

Table VIV. Typical Temperatures Used to Forge Ti-6Al-4V

Beta transus	993°C
Ingot breakdown	1095 - 1150°C
Intermediate	980 - 1040°C
Finish forging	900 - 925°C

Ingots are upset to provide the desired width when making slabs or to provide metallurgical structure refinement. If surface rupturing occurs, intermediate grinding, prior to further forging, is used.

Both hammers and presses are used to forge titanium ingot. In general, presses 2500 to 3000 mt are most widely utilized to forge titanium alloy ingots weighing up to 100 mt. For most titanium alloys, higher forging pressures are employed to obtain the same reduction compared to carbon or stainless steel. For this reason, only the most rigid equipment, capable of withstanding these forces, is practical.

Bar and smaller cross section billets can be rolled on conventional mills used for stainless or tool steels. Minor changes in individual pass designs are required to obtain optimum structures and reduce cracking.

Titanium plate and strip are hot rolled on mills producing stainless steels. Adjustments in metal working schedules are required, particularly when rolling alloy grades, to compensate for the stiffness of the titanium alloys at the temperatures necessary to achieve acceptable structures.

Table IX. Rolling Temperature Ranges For Titanium Alloys.

Alloy	Bar °C	Plate °C	Sheet °C
Commercially-pure	760-815	760-790	1300-1400
Ti-5Al-2.5 Sn	1010-1065	980-1040	980-1010
Ti-5Al-4V	955-1010	925-980	900-925
Ti-13V-11Cr-3Al	955-1065	980-1040	730-900

Cold Working

Cold strip rolling is most widely used to produce commercially pure strip; although modest quantities of alpha stabilized Ti-5Al-2.5Sn, beta lean Ti-3Al-2.5 Sn, and metastable beta alloys, such as Ti-13V-11Cr-3Al are also strip rolled cold. Slabs are rolled on continuous hot strip mills, either tandem or reversing. The hot band, 3.8 to 5 mm thick in widths up to 1220 mm are rolled from starting temperatures above the beta transus. The cooled hot band is descaled, then given an anneal and light flattening pass, if necessary. In some instances, the coil is ground to remove surface imperfections.

The majority of cold rolling is performed on cluster type mills, such as Sendzimer, although commercially-pure grades are successfully produced on four-high Schloemann mills. The amount of reduction between anneals is determined by the grade of titanium being rolled and the type of mill. When rolling the softest commercially pure grades, ASTM Grade 1, severe reductions are possible; for example, reductions approaching 85% are routinely employed. However, much smaller reductions used when cold rolling the harder commercially-pure grades ASTM Grades 3 and 4, or Ti-6Al-4V. In such compositions reductions from 10 to 25% are required. Even if rolling forces were sufficient to deform the metal, excessive edge cracking leading to excessive coil breakage limits rolling reductions. Another condition governing the amount of cold reduction is internal shear cracks which form when the metal is over rolled. These degrade mechanical properties.

In the alpha-beta alloys, such as Ti-6Al-4V, directionability, the difference between longitudinal and transverse mechanical properties, and the tendency to shear crack, are a function of strating microstructural condition which must be controlled.

Continously cold rolled strip down to foil thickness, are commercially available in the softer grades.

In-process and final annealing are done using three different methods:

a. Continuous vacuum annealing

b. Batch vacuum annealing

c. Continuous air or partial inert gas atmosphere annealing, followed by chemical scale removal and acid pickle to eliminate any embrittling diffused oxygen rich layer.

Vacuum annealing eliminates metal loss by eliminating chemical surface removal. This has a major effect on yield, hence costs, particularly in the thinner thicknesses.

Summary. Titanium and titanium alloy mill products are produced on equipment intermixed with other metals, particularly stainless and alloy steels. In addition to those described, such products include extrusions, tubing, both welded and seamless, and wire. The primary manufacturing differences are in process design and equipment; modifications to compensate for the reactivity and metallurgical structural dependence of titanium and titanium alloys.

ZIRCONIUM & HAFNIUM

Introduction

While zirconium was known to exist in 1789, having been discovered by Klaproth, confusion existed over its atomic weight until it was discovered that hafnium coexisted with zirconium in all ores and was present in all metal produced until that time.

Van Arkel, DeBoer and Fast were the first to produce ductile metal using a hot filament reduction method followed by an iodide decomposition process which alternately bears their names. It is also commonly known as the iodide or crystal-bar process and is still employed for the purification of hafnium metal.

Except for relatively small quantities used for incindiary purposes, and made by a calcium reduction in a 'bomb' reactor, no commercial development occurred until Admiral Rickover, in 1949, chose zirconium--actually an alloy of zirconium--to use as the cladding material for uranium in the nuclear submarine program which he commanded. This choice was based upon zirconium's low neutron absorption, its resistance to corrosion in hot water and steam, its relatively good strength at moderately high temperatures (300-400°), its high ductility, and its reasonably good fabricability.

There ensued an intensive development program involving the U. S. Bureau of Mines, Oak Ridge National Laboratories, Battelle Memorial Institute, Westinghouse Electric Co., General Electric Co., and a large number of other organizations. A parallel program was undertaken in the United Kingdom.

Ultimately, a separation process conceived by W. Fischer in West Germany, and perfected at Oak Ridge National Laboratory, was chosen for the method of separating zirconium from hafnium. The reduction method developed by W. J. Kroll for titanium, working with the U. S. Bureau of Mines, was chosen as the reduction method for both zirconium and hafnium. Initially, it was envisioned that the iodide purification step would be required for both zirconium and hafnium, but improvements in the process rendered this step unnecessary for zirconium. However, further purification of the primary hafnium metal is still necessary. The iodide process remains the main method of achieving this, although electron-beam refining is a widely-used alternate.

Consolidation of zirconium and hafnium, as in the case of titanium, must be done in a protective atmosphere in order to prevent contamination from oxygen and nitrogen in the air, as well as from crucible materials. The multiple consumable arc method, then already under development for titanium at the U. S. Bureau of Mines in Albany, was utilized for zirconium and hafnium.

The first production facility for zirconium, at the U. S. Bureau of Mines in Albany, Oregon became operational in 1950. It attained a capacity of 135 mt per year by 1953. Additional production facilities were next developed through contracts with the U. S. Atomic Energy Commission. A total of four facilities were established: National Distillers, National Research Corporation, Carborundum Company, and Wah Chang Corporation.

At the conclusion of these contracts, two of the facilities were shut down, leaving only the Carborundum, by then acquired by AMAX, and Wah Chang facilities in operation by 1965. Since that time, the AMAX facility has ceased operation.

Uses

Concurrent with the development of zirconium for nuclear reactor use, investigation of other uses began. The excellent resistance to corrosion in many organic and inorganic acids was soon recognized, although cost and lack of understanding of the fabrication techniques limited applications until the early 1970's. Zirconium replaced magnesium and aluminum for photoflash applications and has continued to be the primary material for this application; hafnium has also received limited application in this area.

Originally, zirconium was felt to be an unnecessary expense for commercial water-cooled nuclear reactors, and slightly-modified 304 and 316 stainless steels were utilized. But at high irradiation levels, these steels fail through stress corrosion cracking. In 1964, soon after the failure mechanism was analyzed, a rapid shift to the use of zirconium alloys occurred, ensuring the major use of zirconium to the present time.

The use of zirconium in the chemical processing industry (CPI) is currently being expanded at a rapid rate, due to increased relative prices for alternate materials, especially nickel alloys, and also due to the need for producers to expand their markets.

The moderate neutron absorption of hafnium, coupled with its excellent corrosion resistance, formed the basis of hafnium's primary use, as a control rod material for nuclear reactors. Just as in the case of zirconium, initial use was limited to the military reactors where reliability considerations outweighed the cost disadvantage. In recent years, hafnium is being considered for the same use in commercial reactors. Hafnium is also used, as a carbide, as a substitute for tantalum carbide, and often in conjunction with niobium (columbium) carbide. It is also used as an alloy constituent in superalloys where it improves high-temperature fatigue and creep resistance by stabilizing the phase composition and grain structure.

At the present time, three major facilities exist in the western world: Teledyne Wah Chang Albany and Western Zirconium, Inc. in the U. S., and Pecheney Ugine Kuhlmann in France. Additional facilities exist in Japan, at least one in Russia, and one each in the Peoples Republic of China, Argentina and India.

The total capacity of these facilities--about 6,000 mt, is considerably larger than current demand, which is about 4,000 mt. Consequently, any further capacity increases in the near future will be related more to meeting national objectives rather than lack of capacity on a world-wise basis.

Occurrence

The primary ore of zirconium and hafnium is zircon--$(Zr\text{-}Hf)SiO_4$. This is usually found in the same ancient beach sands which are mined for their rutile and/or ilmenite content. Such deposits are found extensively in Australia, the Southeastern United States, Nigeria, and India, although many low-grade deposits are located throughout the world.

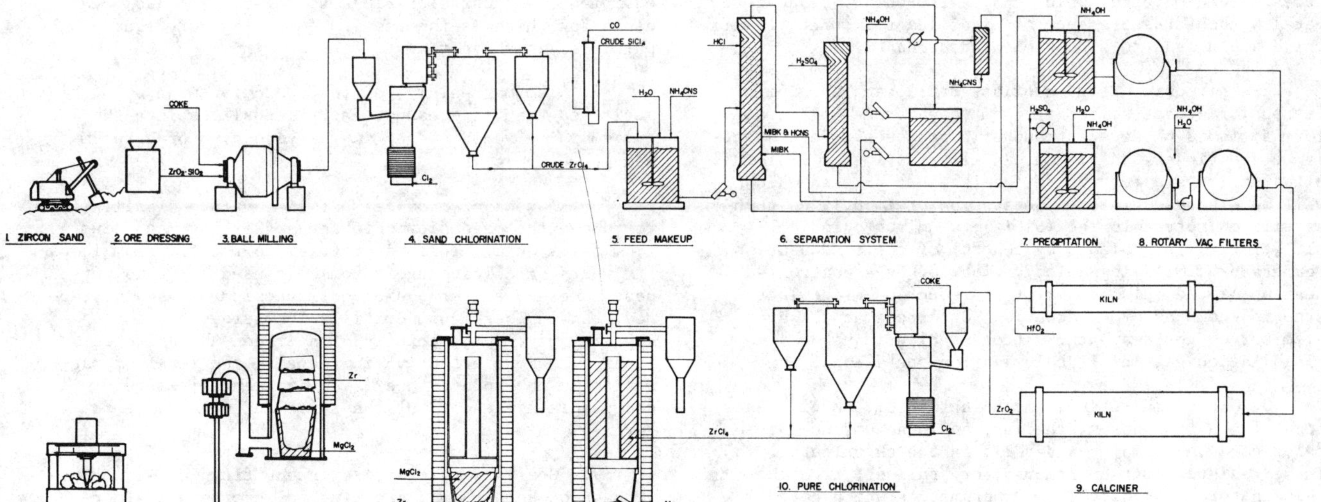

Figure 10. Schematic of the basic zirconium extractive cycle.

The only deposits which have been extensively utilized over a long period of time are those located in Australia. Most of the zircon derived from these sources is used for non-metallic purposes such as foundry sands.

While the availability for metallic reduction is usually not a problem, the price varies widely due to the prevailing market condition for all of the co-products of a given source.

Beneficiation of the deposit to produce a high-grade feed for metal production involves gravity, magnetic, flotation, and/or electrostatic methods, usually on the unmilled sand. The quality of the resultant concentrate depends largely upon the structure of the individual sand particles. The presence of iron, niobium, phosphorus, thorium, and uranium complicates the purification processes, so a deposit consisting of particles of 100% zircon are the most desirable. Even then, some phosphorus and thorium substitute in an isomorphous manner with the components of the zircon, and removal of these materials must be accounted for in the design of the facility.

There are no commercial deposits in which hafnium is a major constituent. It is usually present in the zircon, substitutionally with the zirconium, in a ratio of 2-4 parts of hafnium for each 100 parts of zirconium. Higher ratio ores, such as those in Nigeria in which the ratio is about 7 parts of hafnium for each 100 parts of zirconium, are usually complex ore bodies in which the other impurities cannot be readily removed.

EXTRACTIVE METALLURGY

Primary Process. A schematic diagram of the basic extraction cycle used by all current manufacturers in the western world, is shown in Figure 10, the basic operational processes have been well described in the literature. There are several notable features in this process.

1. There are two separate chlorination steps (three counting the one for the separated hafnium). The first, usually termed Crude Chlorination, cracks the ore and yields a product which is water soluble, a condition necessary in order to employ the MIBK-thiocyanate process for separating the hafnium from the zirconium. The second chlorination steps, termed Pure Chlorination, done separately using zirconium oxide and hafnium oxide, produces a feed material which can be thoroughly reduced to metal without the presence of significant impurity levels which would lead to a contaminated product.

2. Zirconium tetrachloride, $ZrCl_4$, and hafnium tetrachloride are solids at room temperature and do not form a liquid phase at atmospheric pressure. Consequently, conventional fractional distillation is not employed, although it is a feature of the Ishizuka process which will be discussed later. Separation of impurities such as silicon tetrachloride ($SiCl_4$), aluminum chloride, $AlCl_3$, etc., is achieved by control of the condensation temperature following chlorination, or by a controlled heating of the chloride prior to reduction. The latter step is known as pre-evacuation (or 'pre-evac'), which also serves to remove the nitrogen which is used as a fluidization gas in all chlorination operations.

3. Some impurities, including sodium, calcium and some of the phosphorus, thorium, and uranium, do not chlorinate and remain as a solid in the chlorinator. These residues must be removed, or "pulled", from the chlorinators periodically. Other metals which have chloride vapor pressure curves slightly lower than that of pure zirconium tetrachloride are retained in the chloride container following the reduction step and are discarded when the reduction assembly is taken apart.

In spite of all of these purification steps, it is common to resublime the 'pure' zirconium tetrachloride, allowing one more opportunity to leave the impurities behind. The major production facilities have such resublimers, although only a portion of the process stream flows through this step.

Phosphorus has shown a surprising ability to remain with the zirconium through all of these process stages through formation of a series of compounds, not all of which have been positively identified, which allow such mobility. Once in the metal, the phosphorus cannot be removed but forms a compound, probably Zr_3P, which is extremely insoluble in the metal but which leads to a loss in ductility at all fabrication and application temperatures. It also causes a poor corrosion appearance in hot water and steam, although no long-term detrimental effect has been documented. Surprisingly, few common-use or customer specifications limit the phosphorus content, although the major manufacturers are well aware of the problem and limit the content to about 20 parts per million (ppm) for nuclear uses and 40 ppm (0.004%) for other uses.

4. The chemical separation process will remove reasonable levels of iron, aluminum, uranium, niobium, etc., although some tailoring of the process is required to accommodate various starting materials. In the separation step, a large portion of the impurities present flow with the hafnium into the MIBK (methyl isobutyl ketone) phase. These are subsequently extracted by contacting the hafnium-laden MIBK stream with an aqueous phase which has been adjusted chemically to maximize extraction of the impurities while retaining most of the hafnium in the MIBK. The hafnium is stripped back into the aqueous phase; zirconium and hafnium are separately precipitated as either hydroxides or sulfates and burned to their oxides.

5. The large volume of silicon tetrachloride generated in crude chlorination is a by-product useful as a starting material for the silicon industry. It must first be purified to a high level; this is done using fractional distillation techniques. Other impurities such as $VOCl_3$ and $TiCl_4$ can be recovered for use in their respective industries.

6. Most of the chlorides are hygroscopic and must be protected from contact with the atmosphere to prevent formation of oxides and oxichlorides which would lead to oxygen contamination of the metal. Closed containers are used throughout and only limited exposure to the atmosphere is allowed. Hydrogen chloride is usually the product of reactions; this is objectionable environmentally and also leads to accellerated corrosion of the equipment. Exposure of $SiCl_4$ to moist air leads to a dense fog which is the result of forming very fine SiO_2 particles in the air.

7. The magnesium used for reduction must be of high purity so that it does not carry impurities into the system which would end up in the zirconium or hafnium sponge.

8. The major producers utilize vacuum distillation to remove the magnesium chloride and excess magnesium from the sponge. This is contrary to the titanium industry where acid leaching is commonly used for this purpose, whether either sodium or magnesium is used as the reductant. The vacuum distillation serves to reduce impurity pick-up especially of hydrogen and nitrogen. It also is more effective in reducing the level of magnesium in the sponge, and is environmentally more acceptable since more of the salts and metals are recovered.

9. Contrary to practice in the titanium industry, it has not been common practice to incorporate electrolytic recovery of magnesium and chlorine on the site of the zirconium/hafnium sponge facility. Instead, the excess magnesium is recovered for reuse, but the magnesium chloride is shipped offsite to be used for electrolytic recovery or for other uses. But there is no technical reason a recovery facility could not be eatablished on site.

10. As mentioned earlier, hafnium sponge requires further purification before it can be fabricated into final product shapes. This is done either using the

iodide decomposition process, or electron beam melting. The iodide process is effective in removing essentially all impurities in the sponge, other than zirconium and titanium. However, it is a slow and expensive process, requiring a large quantity of electrical energy, expensive capital equipment, close control of the process, and a large in-process inventory. The crystal-bar product of this step is generally too pure, and controlled amounts of oxygen, for strength, and iron, for corrosion resistance, must be added during arc melting in order to meet the final product properties. This alloying is sometimes achieved through combining crystal bars with controlled additions of sponge and scrap on the electrode prior to double arc melting.

During electron beam melting all impurities other than carbon, nitrogen, and the refractory metals can be removed. Through careful control of this step, only iron need be added after melting, since it is volatilized under the high temperature, high vacuum conditions which exist in the electron beam furnace.

Commercially both iodide and electron beam processes are employed.

Alternate Production Methods. The complexity, cost and environmental factors associated with the MIBK-thiocyanate system have stimulated a good deal of research aimed at alternate processes. Several of these alternates have been employed commercially. But, at this time, none have proven to be more commercially viable. It is interesting to note that Western Zirconium, the newest as well as the second largest producer in the world, chose the MIBK-thiocyanate process for their plant, which began production in 1980.

Nevertheless, several processes bear examination:

Sodium reduction. Just as in the case of titanium, it is possible to reduce zirconium tetrachloride with sodium rather than magnesium. The reduction can take place in much the same fashion as in the case of magnesium reduction. Furnace configuration and operating temperatures can be identical. But the sodium chloride has a relatively low vapor pressure at reasonable temperatures, and cannot be removed by vacuum distillation. It must either be removed by leaching, as is done in the production of titanium, or separated by a method such as "drip melting". Leaching is unsatisfactory since it leads to a high level of oxygen. Even then, a large amount of sodium and chlorine remain.

In the drip melting operation, the raw sponge, still containing residual sodium chloride, is formed into a long compact and passed through an induction field while under an inert gas. The bottom of the compact melts off, separating the zirconium and sodium chloride. The drops impinge on a rotating water-cooled copper plate where they solidify. The resulting "chunklets" can be compacted into electrodes using sponge consolidation method.

TBP - Nitric Acid Separation. The only other separation process which has been used extensively is the TBP - nitric acid process which was employed by Columbia National (formerly National Research Corp.) and also by Eldorado Nuclear in Canada. This is also a solvent (or liquid-liquid) extraction process; in this case, the metal-bearing starting material can be either a chloride or zircon which has undergone caustic fusion. Both zirconium and hafnium are first extracted into the tributyl phosphate/n heptane organic. Hafnium is then back-extracted and later, separately, the zirconium. The hafnium-containing phase contains too much zirconium to be used for most purposes. The zirconium is first converted to a hydroxide and then calcined to an oxide.

The Eldorado facility was different in that the zirconium compound was converted to a fluoride rather than an oxide. This fluoride was batch-reduced with calcium in a closed vessel or "bomb".

Other Fractionation Processes. Many other processes have been studied in which either the relative vapor pressure differences, or the relative solubilities of zirconium and hafnium form the basis for a separation of the two. These involve both molten salt and aqueous systems, with and without the addition of other salts, and with and without the presence of reducing agencies such as hydrogen or magnesium. But most are complicated by high temperatures, corrosive environments, and low separation coefficients; the latter restraint requires numerous stages of separation to attain the impurity level required. For this reason, none of these processes have been applied commercially on a large scale.

Separation by Distillation. It has long been known that there is a difference in the triple points between $ZrCl_4$ and $Hfcl_4$. They are:

	Melting Point	Vapor Pressure
$ZrCl_4$	437°C	14,200 mm Hg
$HfCl_4$	434°C	24,200 mm Hg

The ration of these pressures give a separation coefficient of 1.70 for a single-stage sublimation at 437°C. On this basis, a multi-unit facility should be able to produce separate high purity chlorides, bypassing the MIBK-thiocyanide separation step, as well as the second chlorination operation. However, the high temperature, coupled with the high pressure and extremely corrosive environment has, to date, prevented exploitation of this fact. Recent work at Ishizuka Research Institute in Japan has enabled fractional separation of the two tetrachlorides to yield material capable of being further processed to reactor-grade zirconium and hafnium, respecively. The operation is performed in a batch mode.

The key to the success of this operation is based upon design of a fractional distillation column in which the proper conditions of temperature and pressure are achieved; this is done in a double-shelled reactor. The chamber between the inner and outer vessels, made of stainless and mild steel, respecively, is pressurized with nitrogen gas to a level such that the stresses on the inner stainless steel chamber are very low, avoiding the possibility of stress corrosion cracking.

Primary Fabrication

The fabrication of titanium and titanium alloys has been described earlier. There are many similarities for the fabrication of zirconium and hafnium, including the double-arc melting step in which alloying if required, is achieved, and fabrication using conventional metalworking equipment and processes like those used in steel working. This discussion will be limited to differences between these metals and titanium. These differences include:

- Zirconium and hafnium ingots, following double arc melting, are machine-scalped to remove solidification shrinkage porosity and residual magnesium and magnesium chloride to reduce metal losses during hot forging. The turnings produced by this scalping are of high quality and can be combined into subsequent consumable-arc melt electrodes.

- The strength of most zirconium alloys, at all temperatures, is lower than the common titanium alloys. Thus lower unit stresses are required for all working operations.

- While all metals which have a hexagonal-close-packed crystal structure develop antisotropic (differing properties in different working directions), utilization and control of this anisotropy has been developed to a greater extent for zirconium than for any of the other metals. The fabrication methods which develop the optimum textures are employed, especially for tubing, in spite of significant economic disadvantages. For example, tube reducing (also known as 'rocking' or 'pilgering') is used exclusively for production of fuel cladding for nuclear reactors, rather than drawing or strip-welding methods. These criteria do not dominate the production as extensively when the tubing is intended for chemical processing equipment.

• Like titanium, the oxide which forms when zirconium or hafnium are heated in air is very tenacious, hard, and chemically inert. Sandblasting and grinding are the only reasonable methods for this removal, which is necessary to prevent crack initiation during subsequent working. Following blasting or grinding, etching in H_2O-HNO_3-HF is employed to ensure removal of any remaining oxygen-contaminated layers. Salt baths which would effectively remove the oxide layer without either excessive metal loss, or hydrogen embrittlement, have not been developed.

• Zircaloy 4, the most commonly used zirconium alloy, while apparently simple in formulation, as shown in Appendix D, is actually a complex alloy system in which all three alloy types--solid solution, interstitial, and second-phase formation--operate to control the material properties. For example, insufficient carbon content (or possibly other second-phase formers) will lead to uncontrolled grain growth and loss of ductility.

• Hafnium ingots, due to their high purity and high melting point, exhibit very large grains. The initial hot working operation must break up these large grains to achieve subsequent ductility. This is achieved through minor amounts of work during the initial forging, coordinated with frequent reheats.

• While all forms which are produceable in titanium can be accomplished with zirconium and hafnium and their alloys, the product mix used varies greatly. Tubing, in the size range of 12mm diameter with a 0.7mm wall, and strip in the range of 2 to 3 mm thick represent the largest volume of use for zirconium. For hafnium, plate about 6mm thick and rod in the 10mm diameter range are the primary forms.

<u>Zirconium for Chemical Process Equipment</u>

The purity requirements for chemical process equipment (CPI) are not greatly different from those of nuclear requirements, with one major exception: hafnium has virtually no effect on the corrosion resistance, strength, ductility, or fabricability of zirconium at the quantities encountered in commercial areas. Consequently, the expensive and complex hafnium separation step is not necessary when metal is intended for such uses. To date, virtually all zirconium has been made using a separation process since the nuclear market has been dominant. But as non-nuclear applications grow, this will not be the case.

However, zirconium metal made by the Kroll process and simply bypassing the separation step is not pure enough to be fabricated to the required shapes. Several impurities, notably iron, silicon, aluminum and phosphorus, are carried through the single chlorination and reduction steps to the metal sponge.

There are several choices for attaining the additional purification required; they include:

--Multiple resublimation of the chloride, utilizing careful temperature controls on the sublimer and condenser, as well as thorough filtration of the gaseous chloride to remove entrained solids.

--Crystallization of the oxychloride in an aqueous solution, followed by conversion to an oxide and rechlorination; and

--Fused salt purification of the gaseous chloride and using sodium, potassium and other chloride salts, either continuously in a counter-current column or else batch-wise.

While each of these processes involving purification of the chloride prior to Kroll reduction are capable of producing chemical grade zirconium, alternate processes appear to be even more promising. The major processes which have received attention include:

--Fused salt reduction into a metal bath:

Two variants of this process have been patented--the Parlee-Anderson (P-A) reduction using carbon as a reducing agent (8) and the Teledyne Wah Chang Albany (TWCA) process developed by Megy (9) which uses aluminum as the reductant. The TWCA process will be described in a more detail here; the P-A process parallels it in concept, although the specifics vary to a large extent:

Zirconium and hafnium can be reduced by aluminum in an oxygen-free environment. However, for the reaction product to be readily removed from the other reaction products, it is best to depress the melting point of the metal. In the TWCA process, molten zinc is used as a bath to receive the zirconium. A composition of about 10% zirconium, by weight, is chosen primarily due to melting point considerations; at this composition, the activity level of zirconium is reduced so that impurities such as nitrogen and oxygen which leak into the containment, as well as corbon from the vessel, do not react as readily with the bath. The double salts, Na_2ZrF_6 or K_2ZrF_6 are satisfactory starting materials for this reduction since they can be readily formed by fusion of zircon sand; they also melt at a reasonable temperature.

Starting materials for this process may be zircon which is converted to the double salt in a fusion process, or the double salts can be generated from zirconium tetrachloride.

Reduction takes place in a graphite crucible which is enclosed in an inert gas chamber at a temperature of about 900°C. The aluminum metal which is added reduces the zirconium, which forms an alloy with the zinc, which is in the bottom of the chamber. The aluminum combines with the salt to form a pseudo cryolite, $(NaF)_{1.5}AlF_3$, assuming the sodium double salt is used at the start of this reaction. Hafnium is not present in the metal at a high level unless all of the zirconium is first reduced.

The zinc-zirconium alloy is next transferred to a two-zone vacuum still and the zinc vaporized from a heated zone to a cooled chamber. A very porous solid mass, similar to the sponge generated in the Kroll process, remains.

The TWCA process has the advantage of removing the hafnium, contained in the ore to extremely low levels. A separation factor of 300 or more per stage has been reported. Consequently, a two-stage reduction process, following formation of the double salt and pre-melting of the components, is capable of producing zirconium meeting the nuclear requirements with respect to zirconium. No recovery of hafnium is achieved. This process is not currently being used.

<u>Electrolytic reduction</u>. Zirconium and hafnium can be directly reduced electrolytically (10, 11, 12) in a manner similar to that described previously for titanium. Many of the problems encountered in the titanium process are encountered here as well. To date, no production facility has utilized such a method, but if the current efforts on titanium are successful, one could anticipate renewed efforts in the direct electrolytic reduction of zirconium and/or hafnium.

One can view methods such as the Kroll reduction of zirconium, using either magnesium or sodium, as a two-stage electrolytic reduction, since the electron transfer takes place, alternately, in the Mg, or Na, electrolytic cell, and then in the Kroll reduction furnace.

Aluminothermic reduction. Aluminum will partically reduce zirconium oxide to metal; the amount of oxygen remaining depends upon the excess aluminum used for reduction. It is possible to remove the residual oxygen by heating the product to a very high temperature, and under a very high vacuum. At, for example, 2500°C, oxygen will be removed provided there is sufficient Al, Ti, Si, or other metal present to form a sub-oxide which is volatile at that temperature (3). Fletcher (13) reports attaining an oxygen of 238 ppm (0.0238 wt. %) and a nitrogen level of 95 ppm (0.0095%). But the metal recovery from the combined process is not stated. At such temperatures, one would expect a high loss of zirconium due to its volatilization. There would also be a very high recycle of aluminum through the system.

Conclusions

The basic Kroll process, or some relatively minor variant of it, will probably be the major process used for production of nuclear grade zirconium and hafnium for many years to come. The cost and time required to verify the quality of material produced by an alternate method is, of itself, a major impediment to change. There are many minor improvements which could be made in existing Kroll facilities, especially integrated recycle of magnesium and chlorine. The fact that both metals are currently in oversupply will discourage research and development, and also investment, in new processes; the oversupply could well last through much of the decade of the '80's.

The situation is less clear with respect to zirconium for non-nuclear uses, where the elasticity of demand serves as a major incentive to produce an acceptable product at the lowest possible price. The economics of scale favor large operations, but the capital investments required to achieve such a scale are very large. A facility using the Kroll process and integrated through mill product costs well in excess of $10 per lb. per year of capacity.

SELECTED REFERENCES - TITANIUM

- McQuillan and McQuillan; Titanium, Academic Press, Inc., April 1955.
- Lippert, TW, "Titanium in the USA"; The Technology and Application of Titanium; Pergamin Press, Ltd., 1968.
- W. W. Minkler; "Titanium Extractive Metallurgy Status"; Titanium '80, Proceedings of the Fourth International Conference of Titanium; The Metallurgical Society of AIME, 1980.
- H. Y. Lee and D. Pozzi, "Mine, Mill and Smelting Complex at Richards Bay, South Africa"; Annual Volume Featuring, "Hydrogen in Metals and Titanium"; The Canadian Institute of Mining and Metallurgy, November, 1978.
- Langtrey E. Lynd, "Titanium Raw Materials and Products - Supply and Demand"; SME-AIME Fall Meeting, Minneapolis, Minn. October 22-24, 1980.
- V. V. Rodyak, V. E. Geyer and V. M. Skrypnyuk; "Magnesiotheromic Production of Sponge Titanium", Metallurgy Publishing House, Moscow, April, 1971.
- Toshio Noda; "Titanium from Slag in Japan"; Annual Meeting AIME, New York, New York, February, 1964.
- W. W. Minkler, "Critical Review - Titanium Extractive Metallurgy Status"; Vol. 1, Proceedings of the Fourth International Conference Titanium, The Metallurgical Society of AIME, February 1981.
- S. V. Ogurtsov, "State and Prospects of the Development of Industrial Methods for the Production of Titanium", Third International Conference on Titanium, Moscow, May 1976.
- G. Cobel, J. Fisher, L. E. Snyder; "Electrowinning of Titanium from Titanium Tetrachloride: Pilot Plant Experience and Production Plant Projections"; Vol.3, Proceedings of the Fourth International Conference on Titanium; The Metallurgical Society of AIME, February 1981.
- V. I. Dobatkin and N. F. Anoshkin; "Choice of Vacuum-Arc Melting Parameters to Insure Ingot Chemical Homogeniaty"; The Science, Technology and Application of Titanium; Pergaman Press, 1970.
- Y. Kotani, K. Murase, F. Shimizu, T. Suzuki, T. Yamoto; "Production of Titanium Slab Ingots in a Plasma Electron Beam Furnace"; Titanium '80, Science and Technology, The Metallugical Society of AIME, February 1981.
- A. F. Belov; "Principle Trends of Investigation in the Field of Titanium Production"; Titanium '80, The Metallurgical Society of AIME, February 1981.
- ASM Committe on Titanium and Titanium Alloys; "Relation of Properties to Processing for Wrought Titanium Alloys", Metals Handbook, Ninth Edition; American Society of Metals, Metals Park, Ohio; December 1980.

SELECTED REFERENCES - ZIRCONIUM AND HAFNIUM

- Van Arkel, A. E., Preparation of Pure Zirconium, Titanium, Hafnium and Thorium Metal; Zeitschrift fur anorganische Chemie, v. 148, 1925, p. 345.
- Lustman, B. and Kerze, F., The Metallurgy of Zirconium, McGraw Hill, N. Y. 1955.
- Thomas, D. E. and Hayes, E. T. (Ed.), The Metallurgy of Hafnium, Navel Reactors, Division of Reactor Development, USAEC, 1960.
- Spink, D. R., "Extractive Metallurgy of Zirconium and Hafnium", Canadian Mining and Metallurgical Bulletin, November, 1977, p. 145.
- Stephens, W. W., "Extractive Metallurgy of Zirconium - 1945 to the Present", ASTM Symposium, Boston, August 1980. (to be published)
- Rolsten, R. F., J. Electrochem Soc., 106, p. 975, November, 1959; 215, p. 472, June 1959, and 215, p. 478, June, 1958. (All three of these papers deal with various aspects of iodide purification of reactive and refractory metals.)
- Tenckhoff, E., "The Development of the Deformation Texture in Zirconium During Rolling in Sequential Passes", Metallurgical Transactions A, Vol. 9A, October, 1978, p. 1401.
- Anderson, R. N., et al, U. S. Pat. No. 3,794,482, "Carbothermic Reduction Method for Converting Metal Oxides to Metal Form", February 26, 1974.
- Megy, J., U. S. Pat. No. 4,072,506. "Method of Separating Hafnium from Zirconium, February 7, 1978.
- Mellors, G. W. and Senderoff, S., Canadian Pat. No. 688,546, "Electrodeposition of Refractory Metals", June 9, 1964.
- Martinex, G. M., et al, "Investigation of Cell Design for Electrowinning Zirconium Metal from Zirconium Tetrachloride"; USBM RI 8125, 1976, Washington, D. C.
- Baker, D. H., et al, "Electro-refining Zirconium", U. S. Bureau of Mines RI 5758 (1961), Washington, D. C.

Appendix A. Commonly Used U. S. Commercial Titanium Alloys

Common Designation	Major Alloy Content(1) Percent	Beta Transus °C	Tensile Strength MPa - Minimum	Typical Applications
COMMERCIALLY PURE				
ASTM Grade 1	0.18 max O	890	241	Dimensionally stable anode-plate and frame heat exchangers
ASTM Grade 2	0.25 " "	912	345	Condenser tubing - chemical process equipment
ASTM Grade 3	0.35 " "	921	448	Aircraft ducting and firewalls
ASTM Grade 4	0.40 " "	950	552	Aircraft skins
ALPHA AND NEAR ALPHA				
ASTM Grade 7	0.2 Bd	912	348	Heat exchanger tanks in reducing acids
Ti Code 12	0.3Mo-0.8Ni	887	483	Salt evaporators - resistance to crevice corrosion
Ti-5Al-2.5Sn		1038	758	Welded structures aircraft engines and airframes
Ti-8Al-1Mo-1V		1038	896	Compressor discs and blades
Ti-6Al-2Sn-4Zr-2Mo		993	896	Compressor discs and blades
ALPHA BETA				
Ti-6Al-4V		999	896	Compressor discs and blade, airframe structures
Ti-6Al-6V-2Sn	0.7Fe-0.7Cu	946	1069	Airframe structures and forgings
Ti-6Al-2Sn-4Zr-6Mo		932	1016	Compressor discs and blades
Ti-3Al-2.5Sn		908	690	Hydraulic tubing
Ti-10V-2Fe-3Al		800	1207/1380	Airframe forgings (experimental)
BETA ALLOYS				
Ti-13V-11Cr-3Al		705	1240	Airframe structures
Ti-15-3	15V-3Cr-3Al-35N	732	1240	Airframe structures (experimental)

(1) Elements not shown in designation.

Appendix B. Properties of Zirconium and Hafnium

Property	Zirconium	Hafnium
Chemical symbol	Zr	Hf
Atomic number	40	72
Atomic weight	91.22	178.49
Density, g/cc	6.494	13.09
Melting point, °C	1865	2231
Crystal structure:		
Low temperature	Hexagonal close packed	Hexagonal close packed
High temperature	Body centered cubic	Body centered cubic
Transformation temperature , °C	872	1742
Thermal neutron cross-section, barns	0.18	105
Modulus of elasticity, psi	19.8×10^6	13.8×10^6
Coefficient of linear thermal expansion	5.8×10^{-6}/°C	5.8×10^{-6}/°C
Specific heat, cal/g/°C	0.67	0.035

Appendix C. Effect of Temperature on the Strength of Zircoloy -4
(Transverse tests on moderatly high oxygen material)

Temperature, °C	Tensile Strength MPa
20	609
150	435
260	356
315	338
360	313
400	309
425	286

Appendix D. Primary Alloys of Zirconium
Alloy composition, % by weight

Element	Pure Zr	Zircoloy 2	Zircoloy 4	Zr-2.5Nb	Zr - 1%Nb
Sn	-	1.20-1.70	1.20-1.70	-	-
Fe	-	0.07-0.20	0.18-0.24	-	-
Cr	-	0.05-0.15	0.07-0.13	-	-
Ni	-	0.03-0.08	-	-	-
O(1)	-	0.09-0.14	0.09-0.14	0.08-0.12	0.08-0.12
Nb	-	-	-	2.40-2.80	0.90-1.10

(1) Varies depending upon customer and application, not specified as an alloy constituent; a maximum limit as an impurity usually applies.

IRON BLAST FURNACE

Gordon H. Geiger
Inland Steel Company
East Chicago, Indiana 46312

INTRODUCTION

The modern blast furnace is the result of a long history of development, dating well back into the Middle Ages. While there is little physical resemblance between the modern furnace with its supporting auxiliaries, and ancient smelting furnaces, the essence of the process remains the same: use a carbonaceous material to provide heat and to reduce natural iron oxide to metal.

In the treatise T'ien-Kung K'ai-Wu, written in 1637, the Chinese author Sung Ying-Hsing described the iron industry of 17th Century China. Both magnetite (Fe_3O_4) and hematite (Fe_2O_3) ores were surface mined and smelted in blast furnaces such as that illustrated in Figure 1.

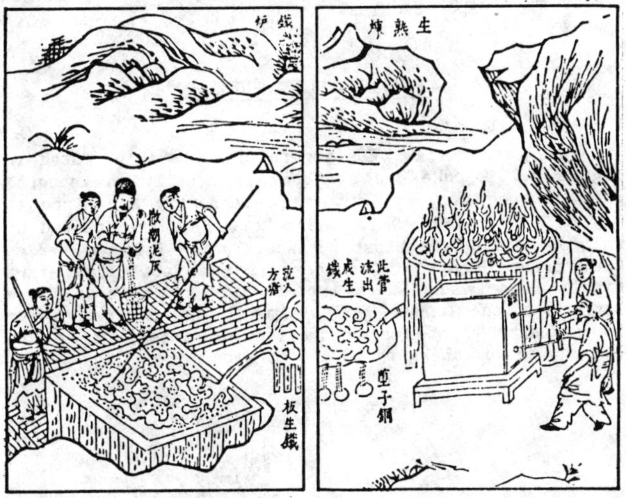

Figure 1. Drawing of 17th Century Chinese blast furnace (right) and wrought iron puddling pit (left). (1).

"The furnace for smelting iron is built of a salt and mud mixture and is usually contructed against a hillside, or surrounded by a large wooden frame. Great care and much time must be spent on fashioning the salt mud (furnace wall), as the slightest crack in it will bring the whole work to failure. Each furnace has the capacity for some 2,000 catties of ore together with hard wood, coal, or charcoal, depending on the local condition in south or north China. Usually four or six men are required to work the bellows. When the ore melts and becomes iron, it flows out from a hole (i.e. iron notch) on the side of the furnace which has been sealed with mud until casting time. A complete day and night is divided into six equal periods, and one casting takes place in each period. The discharge hole in the furnace is sealed up again after each casting, and the smelting is resumed.

If the pig iron is intended for casting purposes, the molten iron is discharged directly into pig beds molded in the form of long bars and round pieces. If it is intended for the making of wrought iron, the molten iron is run into a square ditch built a few inches below the furnace and a few feet away from it, surrounded by a low wall, where a few persons holding willow sticks and a certain amount of dry and finely screened earth stand ready. When the molten iron flows into the square ditch, one man quickly throws the earthen material into the molten mass, while the others vigorously stir it with the willow sticks, and soon wrought iron is obtained. Two or three inches of each willow stick will be burnt off at the end of an operation, so that new ones will have to be used at the next turn. When the wrought iron is somewhat cooled (and becomes pasty) it is either cut into square blocks in the ditch, or lifted and hammered into round pieces before it is sent to market.(1)"

By the latter part of the 19th Century, the lines of the blast furnace had become much more recognizable, but clearly the principles had not changed! In the United States, the carbonaceous material was now coke, usually made in non-recovery beehive ovens, and rich, natural ores were used. The Cowpers regenerative hot-blast stove, invented in 1857, was in use. At the beginning of the 20th Century, the by-product, slot-type, coke oven was introduced into operation with greater frequency. In order to achieve higher temperatures at the tuyere, and thus to drive the furnace harder, the endothermic affect of the presence of moisture in the air blast was recognized and moisture was removed by cooling the blast in ice houses prior to heating it in stoves. The Dwight-Lloyd traveling grate machine was developed and began to be used to sinter fine ore and waste material into agglomerates usable in blast furnaces, along with lumps of natural ore.

This state of affairs, charging of by-product coke, limestone, natural ore, and some sinter into the furnace, and blowing air preheated to temperatures on the order of $600^{\circ}C$, was common for several decades, until shortly after World War II. In the United States, at that time, it became clear that the reserves of natural ore were rapidly diminishing and their grade was approaching the point where the slag volumes generated required the expenditure of ever increasing quantities of fuel (Figure 2). This situation resulted in the development of new, efficient, processes for concentrating low-grade iron ores and pelletizing the concentrates into a uniformly sized, high grade product for blast furnace feed. This, along with other improvements in the charge burden and the blast, such as higher blast temperatures and injection of oil, gas and/or oxygen through the tuyeres, resulted in considerable improvement in the productivity of a given furnace, so that by 1970 a typical, modernized, furnace could produce more than twice as much iron in a day as it had in 1950.

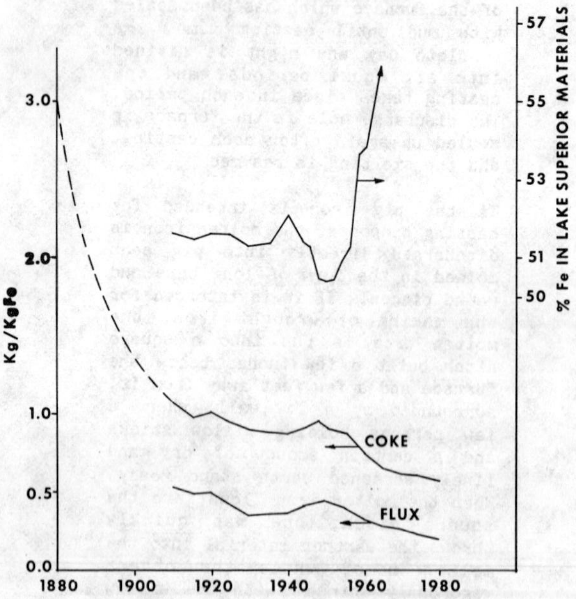

Figure 2. Historical trend of coke and limestone consumption in North American blast furnaces and of Mesabi Range ore iron contents.

This brings us to the 1970's, which was the decade of the giants. The Japanese steel industry, using all of the concepts developed over several previous decades, built many large, modern, blast furnaces that set one production record after another. This was accomplished by meticulous attention to operating details, and very sohisticated control of raw materials. Figure 3 illustrates the economy of scale achieved by the increased size of these furnaces. It also illustrates, however, that when a smaller furnace is properly modernized, and fed a well-prepared burden, it too can be highly productive.

It is the purpose of this paper to describe a large modern blast furnace, using Inland No. 7 as the example, and to discuss our understanding of how it functions.

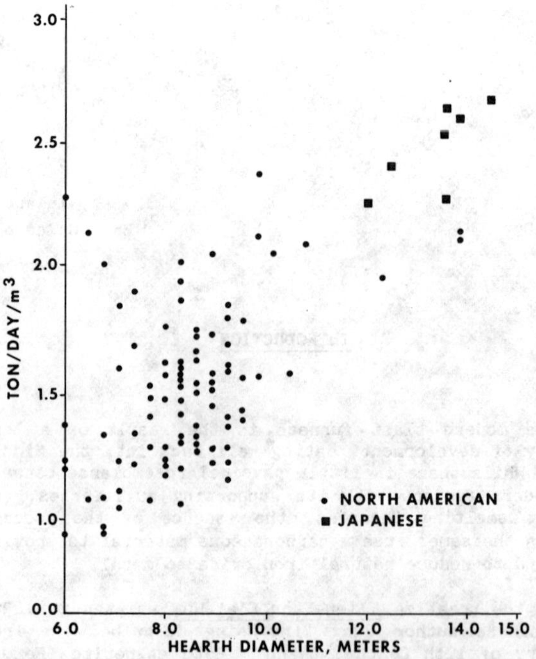

Figure 3. Volumetric productivity of all North American blast furnaces operating in fourth quarter 1979, and of representative large Japanese blast furnaces.

A MODERN BLAST FURNACE

The general view of a modern plant is illustrated in Figure 4. A coke oven battery is usually an integral part of such a plant, but not necessarily so, since it may be located elsewhere and coke may be transported by rail or ship to the blast furnace plant. Thus, as far as the furnace proper is concerned, the facilities start with the ore yard, where iron-bearing raw materials are stored, often blended, and ultimately transferred either to a sinter plant for production of sinter, or to the stockhouse for charging to the furnace.

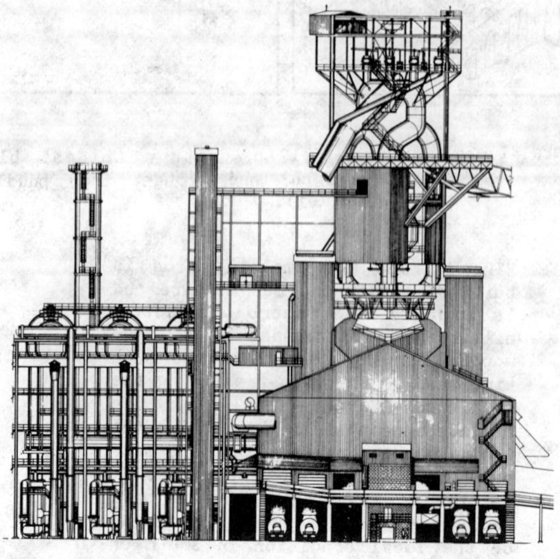

Figure 4. Side view of Inland No. 7 blast furnace.

Blending of ores is extremely desirable when natural ores are being used, and is most often practiced when preparing feed for sinter machines. There are several modes of operation of blending piles, "windrow" and "chevron". By laying the incoming ore either across the bed (windrow) or longitudinal along the centerline (chevron) and then reclaiming across the face or across the bed, significant blending can be achieved. Both stacking out and reclaiming can be done with the stacker/reclaimer illustrated in Figures 5 and 6.

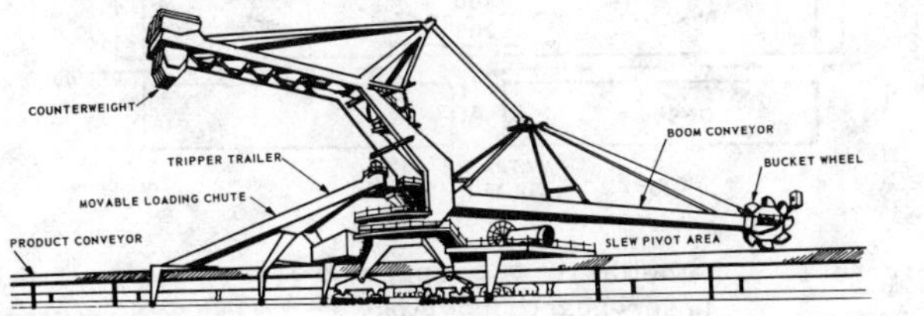

Figure 5. General arrangements of stacker/reclaimer.

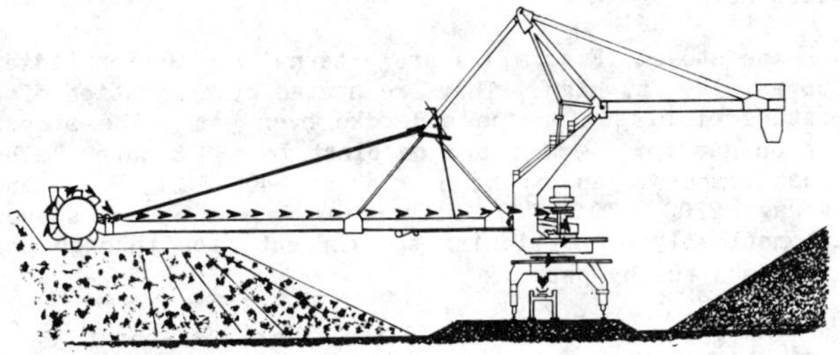

Figure 6. Stacker/reclaimer reclaiming from stockpile.

From either the ore piles, or the sinter plant, iron-bearing materials are sent to the stockhouse where about an 8-hour supply of feed materials is stored. Limestone, if used, is also stored in the stockhouse. Coke is brought to the stockhouse from either the coke plant, directly, or from a coke storage pile, via a coke screening station where undersize coke is screened out and returned to the sinter plant as fuel. The charge to the furnace is drawn from the stockhouse storage bins, and is screened on its way to the charging point, in order to eliminate fines, usually at 1/4-inch.

Burden material is raised to the top of the furnace via either a skip hoist or a belt conveyor. Modern, large furnaces, requiring a flow on the order of 5 tonnes/min. to the furnace top, almost all use belt conveyors. In the case of the Inland No. 7 furnace, this conveyor is 1130 ft. long, 5 ft. wide and rises at an angle of 16 degrees to the top of the furnace, 312 feet above the ground.

At the top of the furnace, the charge materials, which have been placed on the belt sequentially, are discharged into the receiving hopper, which feeds into one of two lock hoppers, (Figure 7). The lock hoppers each have a volume of 51 m^3. This type of top is called a Paul Wurth bell-less top, and differs considerably from the traditional two-bell top (Figure 8) in that the solids are distributed within the furnace by a rotating chute, whose rotational speed and angle can be controlled. In this way, material can be layered in any pattern desired within the furnace, once the solid trajectories have been correlated with the angle and speed. Charging is essentially continuous, with different materials entering alternately from one lock hopper and then the other, allowing furnace pressure to be maintained on the order of 30 psi.

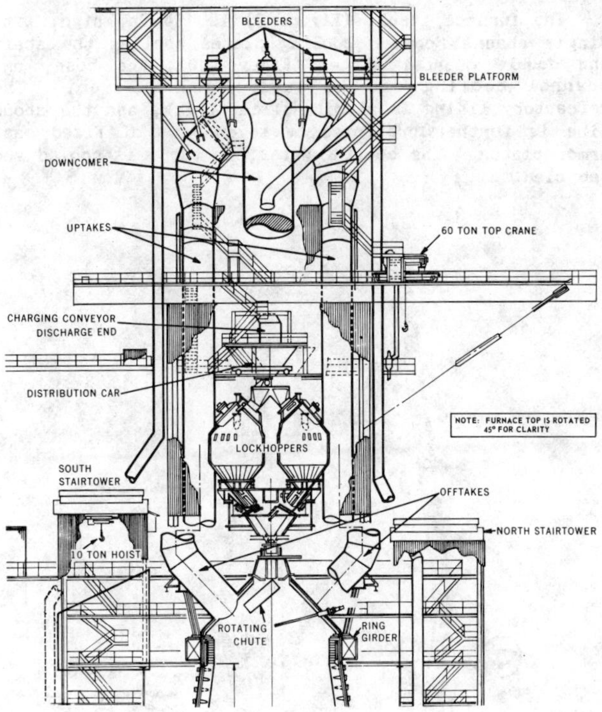

Figure 7. Detail of top of No. 7 blast furnace. Note that chute rotates about furnace axis and changes angle relative to vertical.

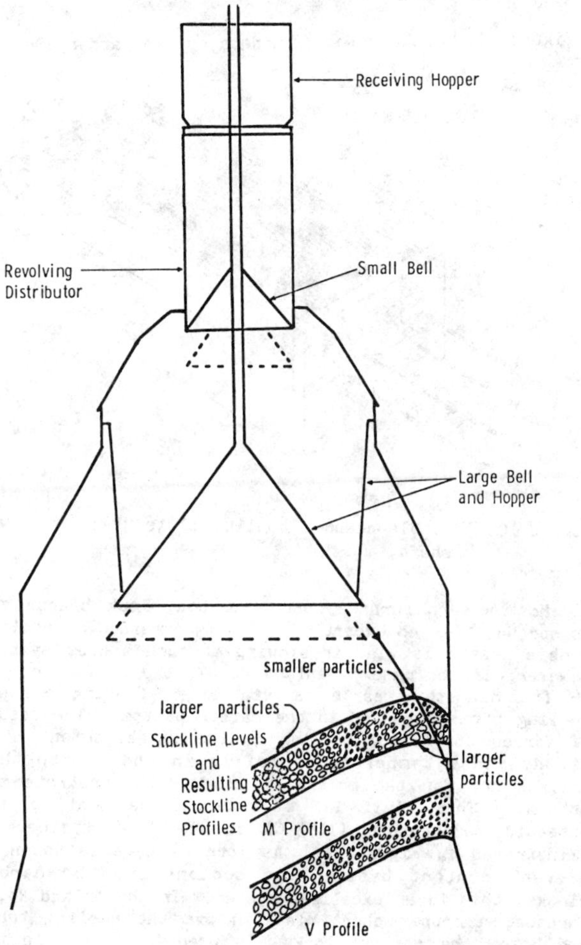

Figure 8. Two-bell top, with burden trajectory and two different burden distributions.

The furnace itself (Figure 9) is 154 ft. high, with single channel copper cooling plates cooling the shaft and double channel plates (Figure 10), the bosh, and channel cooling of the tuyere jacket area. The refractory lining is alumina-silica brick, and the stock line is further protected by seven rows of fixed cast armor plates. The cooling water, in turn, is cooled and recycled.

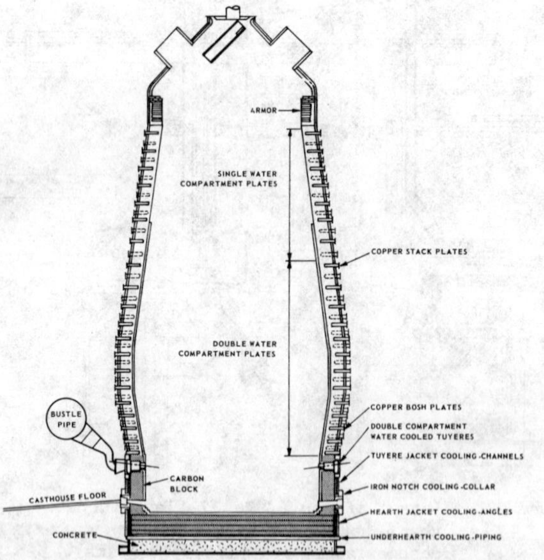

Figure 9. Inland No. 7 furnace cooling arrangement.

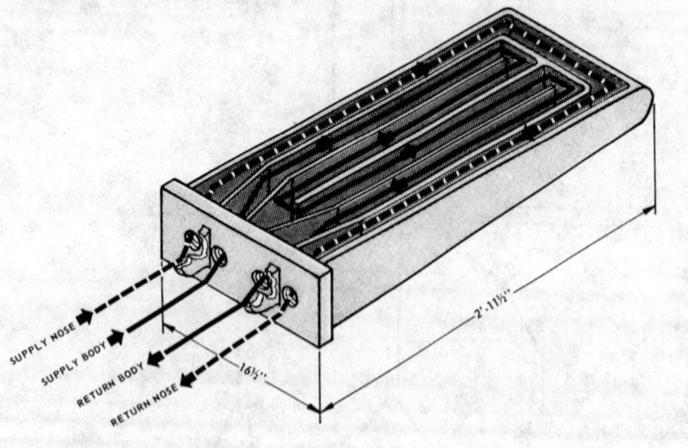

Figure 10. Double-channel cooling plate used in lower shaft.

Most modern furnaces use a carbon block hearth and Inland No. 7 is no exception. A common method of cooling such a hearth is via air blowing through a duct system. However, in the case of Inland No. 7, the cooling of the 45 ft. diameter hearth is via water flowing through cooling pipes embedded in the carbon bottom. The result of forced convection cooling is the establishment of a steady-state temperature profile in the hearth that defines how far the molten material will penetrate the bottom. This limit is, of course, the loci of the eutectic temperature (1153°C) in the Fe-C system, as illustrated in Figure 11. As long as care is taken to prevent contact between the coolant and the carbon blocks, this is an excellent system. In the Inland No. 7 furnace, a copper sheet placed above the cooling tubes and below the carbon blocks, is used to assist in the sealing process.

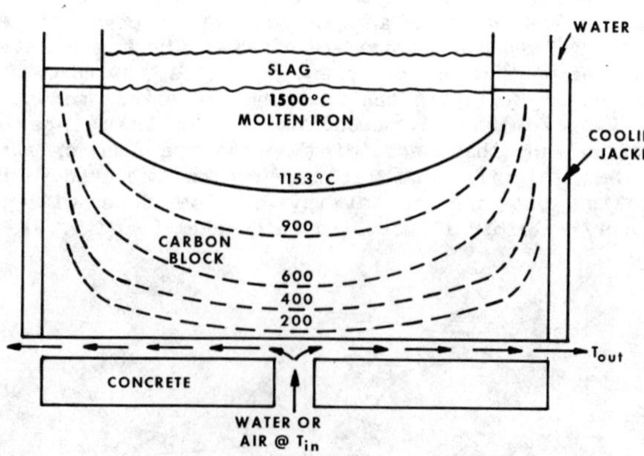

Figure 11. Schematic temperature profile with under-hearth cooling.

The blast for the furnace is furnished by two axial-flow 150,000 scfm compressors, which provide blast air at 65 psi. The heat evolved in the compression stage is retained in the "cold" blast by insulation of the cold blast main.

The stoves (Figure 12) are external combustion Didier stoves, 140 ft. high. They are heated by combustion of a mixture of blast furnace and coke oven gas. The stoves are on gas for 52 min. and on blast for 30 minutes. The blast temperatures can be as high as 2400°F (1315°C), and average 2200°F (1204°C), with mixer valves at the stoves automatically proportioning the correct flow through the stove and the by-pass.

Top gas leaves the top of the furnace at about 350°F (176°C) and at a maximum pressure of 37.5 psi. This high pressure gas is used to drive an expansion turbine coupled to an electrical generator, capable of producing 15 MW of electrical power at peak output. The gases then proceed into a dustcatcher and a Bischoff variable orifice scrubber where 99% of the dust is removed. The scrubber water is clarified and recycled. The blast furnace gas is blended with coke oven gas and used to underfire the coke oven batteries and to fire the stoves.

Casting of metal and slag takes place from four tapholes. Figure 13 illustrates one of the two cast houses; the other is a mirror image on the other side of the furnace. Since the production rate from such a large furnace is on the order of 5 ton/min, if the flow of hot metal in the runner is to be kept under control, it must be almost continuous. A given runner, or pair of opposing runners is used until the clay lining is eroded to the point where tapping must be started from the adjacent hole. This changeover typically takes place every three or four days.

Slag is skimmed from hot metal in the runners, and diverted to slag pits, where it is water-cooled. Recently, techniques have been developed to either atomize the slag or to recover the latent heat in the slag by running water through pipes buried in the slag, creating steam.

The hot metal is collected in 240-ton capacity Pugh ladles for transportation to the steelmaking shops. Flow is diverted from one ladle to the next using a tilting device, illustrated in Figure 14.

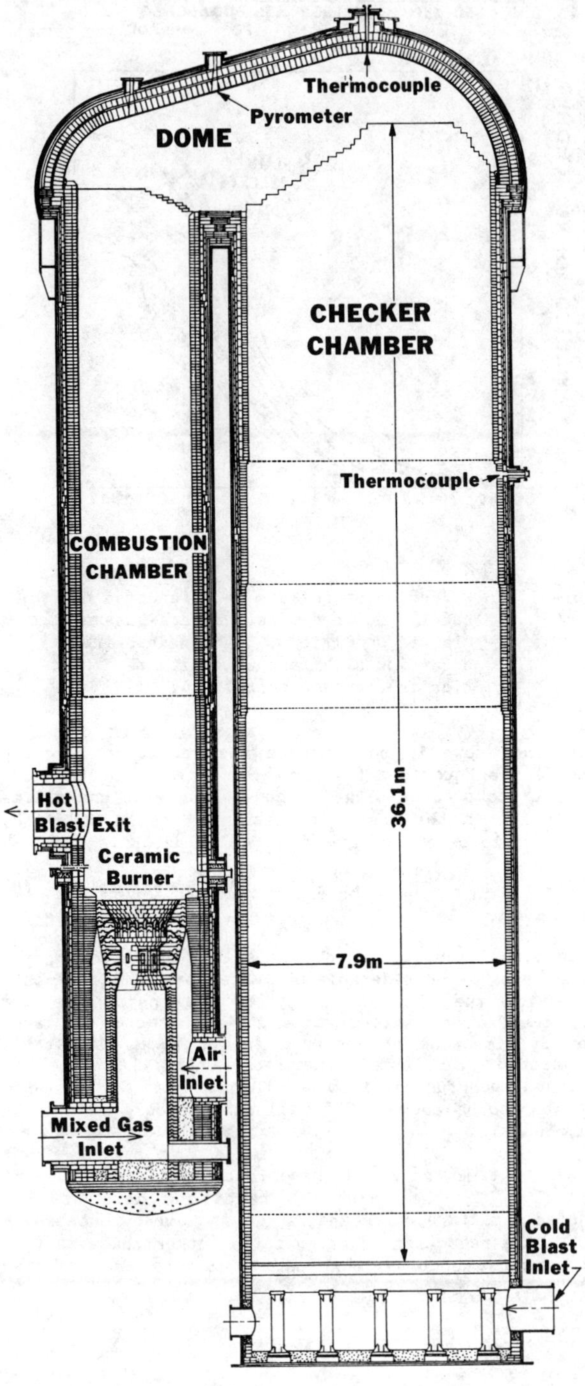

Figure 12. Didier external combustion stove.

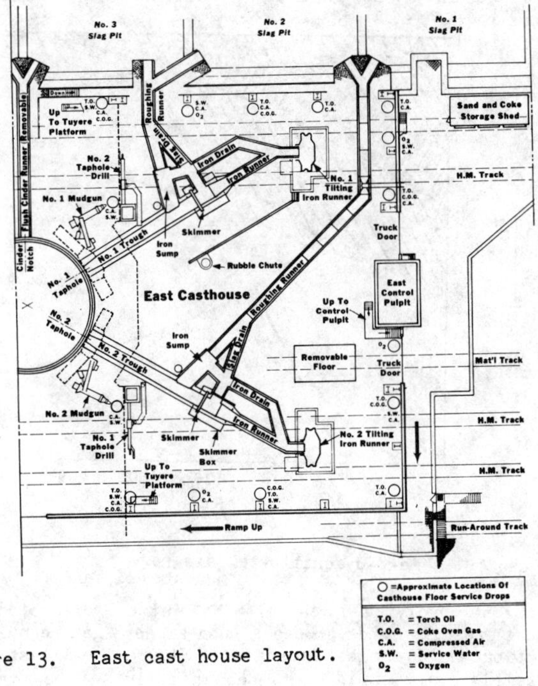

Figure 13. East cast house layout.

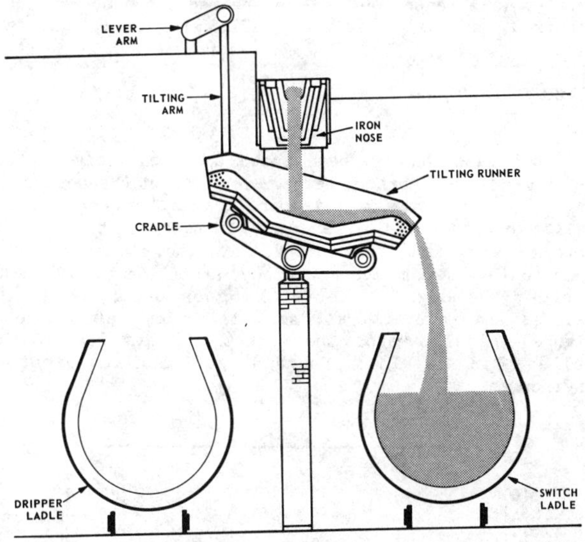

Figure 14. Distribution mechanism for diverting continuous tapping stream from one ladle to another.

CHEMISTRY OF THE PROCESS

The heart of the process is its ability to reduce iron oxide to metallic iron. In order to do this, conditions must exist that drive the reduction reactions to the right:

	ΔH^{o}_{298}	
$3Fe_2O_3 + CO \rightarrow 2Fe_3O_4 + CO_2$	-12,830 cal/mol	(1)
$Fe_3O_4 + CO \rightarrow 3Fe_xO + CO_2$	+72,570 cal/mol	(2)
$Fe_xO + CO \rightarrow Fe + CO_2$	- 3,240 cal/mol	(3)
$Fe_xO + C \rightarrow Fe + CO$	+38,010 cal/mol	(4)

Figure 15 illustrates the equilibrium conditions involved in these reactions. Of course, analogous

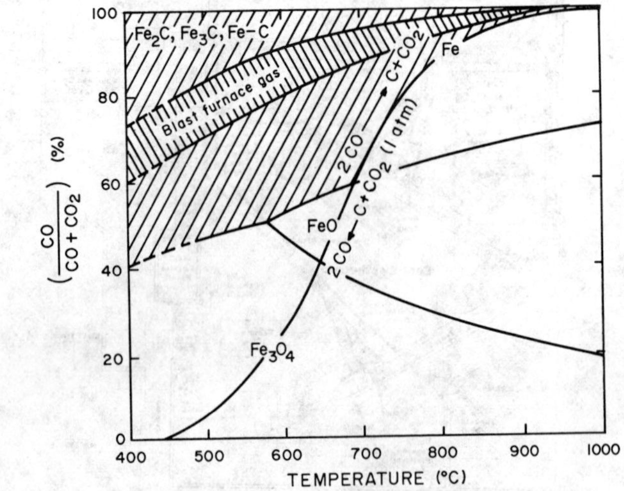

Figure 15. Fe–C–O equilibrium diagram.

reactions involving H_2 can also be written. Reaction (1) occurs readily and proceeds under any gas condition existing in any blast furnace. Figure 16 illustrates that this reaction is completed in pellets in a typical blast furnace gas and temperature environment in about 30 minutes (11% reduction). Once the reduction to Fe_3O_4 has occurred, another reaction (Bells' reaction) can take place:

$$2CO \rightarrow CO_2 + C \qquad \Delta H^o_{298} = -41{,}250 \text{ cal/mol.} \qquad (5)$$

This reaction has been shown to precipitate as much as 11% of the carbon needed for direct reduction of Fe_2O_3, in the time and temperature history of an acid pellet descending into a blast furnace. Its reaction rate is a maximum near 550°C. On the other hand, the reaction does not occur to any significant extent on sinter, although it certainly does occur on fluxed pellets and natural ores, as well as on acid pellets. Figure 17 illustrates the extent of this reaction on pellets in a blast furnace gas and temperature environment.

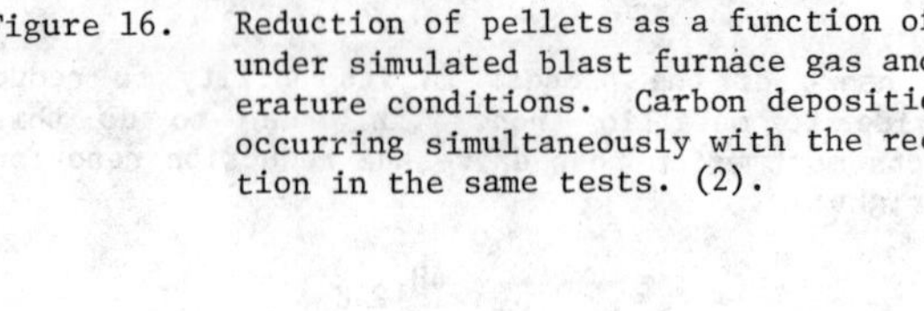

Figure 16. Reduction of pellets as a function of time under simulated blast furnace gas and temperature conditions. Carbon deposition occurring simultaneously with the reduction in the same tests. (2).

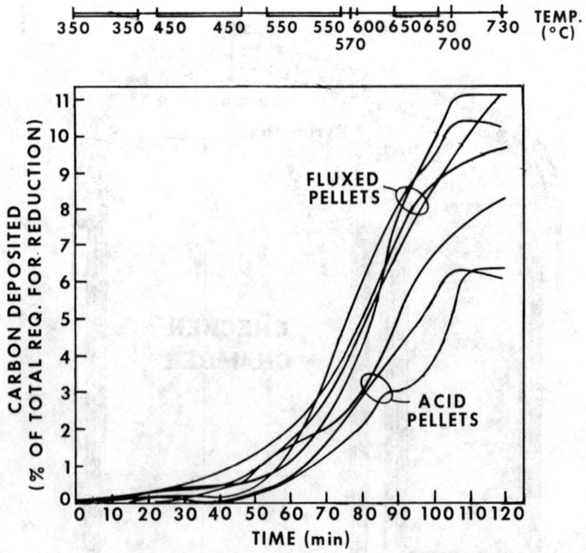

Figure 17. Reduction of pellets as a function of time under simulated blast furnace gas and temperature conditions. Carbon deposition occurring simultaneously with the reduction in the same tests. (2).

Further down in the furnace the reduction reaction proceeds via Equations (2) and (3). Any limestone added as a flux to produce slag in conjunction with the silica and alumina in the ore or pellets and in the coke ash, will start to decompose at about 900°C, according to

$$CaCO_3 \rightarrow CaO + CO_2 \qquad \Delta H^o_{298} = +288{,}450 \text{ cal/mol.} \qquad (6)$$

This reaction, of course, is highly endothermic, and is the cause of considerable ultimate consumption of coke to provide the thermal energy to compensate for the cooling effect of this reaction. Furthermore, if large lumps of limestone are charged (e.g., 7 cm. diameter), the reaction will not be completed until the stone has descended much further into the furnace, and CO_2 released at temperatures above 1000° will consume coke carbon via Reaction (7).

By the time the solids have reached a temperature of about 1000°C, the reduction reactions have usually produced a mixture of Fe and Fe_xO. At temperatures above 1000°C, the reduction reaction is no longer Reaction (3), but becomes Reaction (4), which is the sum of two reactions, from a mechanistic viewpoint:

	ΔH^o_{298}	
$Fe_xO + CO \rightarrow Fe + CO_2$	−3,240	(3)
$CO_2 + C \rightarrow 2CO$	+41,250	(7)
$Fe_xO + C \rightarrow Fe + CO$	+38,010	(4)

Reaction (7) is known as Boudouard's reaction or the Endothermic Gasification of Carbon (EGC) reaction, for obvious reasons, or as "solution loss", even though carbon is not being dissolved and is not lost, although it certainly does not reach the tuyeres. Reaction (4) is referred to as "direct" reduction, and is strongly endothermic.

Physical changes occur as a result of reduction and temperature changes, which ultimately affect gas flow and productivity. As the reduction reactions proceed, either topochemically in the case of ore, sinter, or dense pellets, or homogeneously in the case of porous pellets, the various phase transformations involve changes in specific volume. The result is a certain degree of cracking of the iron oxide charge, particularly when extensive reduction takes place at low temperatures.

At temperatures above 1000^oC, the iron-bearing materials begin to soften, and of course, eventually melt. The melting point of iron depends on its carbon content, ranging from 1535^oC for pure Fe to 1153^oC for Fe containing 4.3% C. Fe_xO melts at 1377^oC. Other minimum eutectic melting points are 1206^oC for the Fe_2O_3-CaO system, 1436^oC for the CaO-SiO_2 system, and 1175^oC for the FeO-SiO_2 system. Softening temperatures, defined as the temperature at which the material has contracted 10% under a standard test load, have been determined for various burden materials and some are illustrated in Table I.

TABLE I

Burden Material	Softening Temp., ^oC
Acid Pellets	1130 - 1150
Lime-fluxed Pellets	1155
Dolomite-fluxed Pellets	1230
Basic Sinter	1095
Acid Sinter	1070

Slag is formed from silica and alumina found in both coal ash (incorporated into the coke) and gangue in the ore or pellets, and from lime and magnesia added as either stone to the furnace or via the sinter or pellets as fluxed sinter or pellets. The overall composition of the slag must be such that it falls into the region of the CaO-SiO_2-Al_2O_3 phase diagram where it is fully liquid at the temperature of the slag phase in the furnace, Figure 18. That is, it must exist above the liquidus. When the temperature/composition combination is such that the slag is outside of the "eutectic valley", then its effectiveness is much reduced, because its effective liquid, or reactive, volume is decreased by the presence of solids, and its viscosity is greatly increased, making it more difficult to bring into contact with a second phase. In a large diameter furnace, it is entirely possible to not match the acids and bases properly across a diameter, with the result that the slag varies in composition across the diameter as it collects. This tendency diminishes as the hearth diameter decreases.

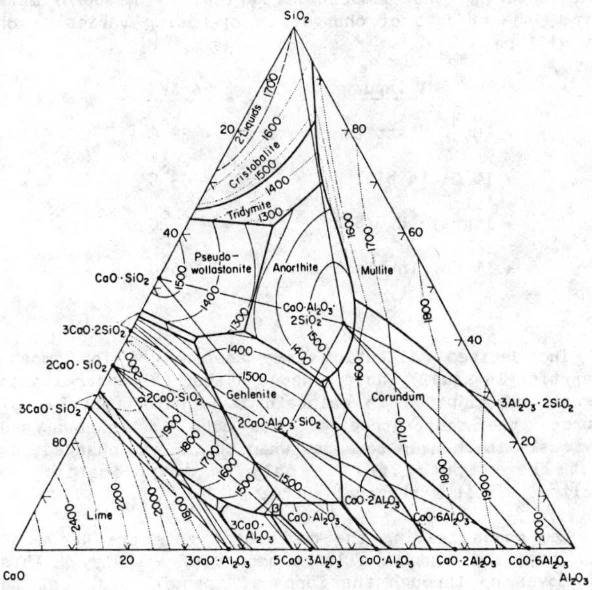

Figure 18. Liquidus temperature for the CaO-SiO_2-Al_2O_3 system. (3).

The typical blast furnace slag thus consists of SiO_2 (37%), CaO (45%), Al_2O_3 (10%), MgO (4%), S (1.5%) MnO (0.5%), and FeO (0.2-0.5%). It should be noted that the typical FeO content is not in equilibrium with solid carbon and this has an important effect on the extent of slag-metal equilibria reached.

The other primary chemical feature of the blast furnace process is the combustion of coke carbon (or injected fuel) at the tuyere by blast oxygen to produce high temperatures and a stream of reducing gas. This reaction,

$$1/2O_2 + C \rightarrow CO \qquad \Delta H^o_{2000K} = -28{,}450 \text{ cal/mol.} \qquad (8)$$

is completed within a very short distance from the tuyeres (1-3m) and largely within a short distance (0.5m) beyond the "raceway" immediately in front of the tuyeres. The size of the raceway is not influenced by the size of the furnace, but varies linearly only with the blast velocity, V_o, and kinetic energy of the blast, E_k, with the result that the proportion of hearth area that is occupied by the raceways decreases as furnace diameter increases.

In addition to Reaction (8), moisture in the blast will react according to

$$H_2O + C \rightarrow CO + H_2 \qquad \Delta H^o_{2000K} = +31{,}800 \text{ cal/mol.} \qquad (9)$$

absorbing heat. Similarly, injected hydrocarbons react and absorb heat: eg.

$$CH_4 + 1/2O_2 \rightarrow CO + 2H_2 \qquad \Delta H^o_{2000K} = +7{,}840 \text{ cal/mol} \qquad (10)$$

The temperature of the reaction zone in front of the tuyere is important, because it is the hottest point in the furnace and influences the temperature profile in the bed descending toward it from above. This temperature is not directly measured, but instead is computed as an adiabatic flame temperature (AFT), which assumes that all combustion reaction heat plus the blast enthalpy and coke enthalpy are absorbed by the reaction products raised to the AFT. Many versions of this computation have been made, differing only by the assumed temperature of the coke entering the combustion region. In approximate terms, the effects of changes in operating variables on AFT will be:

Δ input	Δ AFT
+ 100°C blast temperature	+ 82°C
+ 1% O_2 in blast	+ 53°C
+ 1 grain H_2O/SCF	- 13°C
+ 1% CH_4 in blast	- 56°C

The desired adiabatic flame temperature for smooth operation is usually determined and then the controllable parameters adjusted to maintain it. Unfortunately, of course, the temperature of the coke is unknown and obviously it changes somewhat when the AFT is changed, so that the computed AFT is only an approximation to reality, albeit a reasonable one.

The gases leaving the combustion zone are known as bosh gas, and contain a large amount of enthalpy. This gas moves up through the furnace, transferring heat to the descending solid coke, liquid iron, and slag-forming constituents. During combustion of the coke, the coke ash, which is a primary source of acid constituents of the slag, is melted and superheated. Several reactions take place. Pyrrhotite in the coke ash is burned to form possible species, SiS, or CS, or in the gas phase. Some silica reacts to form SiO in the gas phase.

$$SiO_2 + C \rightarrow SiO(g) + CO(g) \qquad (11)$$

These gaseous constituents rise into the descending column of solids; most react with the solids and liquids and some exit the furnace. Figure 19 illustrates the partial pressure of various sulfur-bearing gas species and SiO that exist at equilibrium with hot carbon and iron as a function of temperature, assuming that the initial concentration of SiO_2 and FeS are typical of that in a good coke.

The ascending gases give up their thermal energy as they rise until they have reached the same temperature as the descending solids. This occurs when the temperature of both gases and solids is about 1000°C. SiO(g) becomes less stable as the temperature decreases. It is easily reduced by carbon in iron to put silicon in solution in the metal:

$$SiO(g) + \underline{C} \rightarrow \underline{Si} + CO(g) \qquad (12)$$

This is the probable mechanism by which silicon enters the iron, since it has been shown (4) that the slag-metal reaction,

$$(SiO_2)_{slag} + 2\underline{C} \rightarrow \underline{Si} + 2CO(g) \qquad (13)$$

is much too slow to account for the amount of silicon present in most hot metal. Note, however, that from a thermochemical standpoint the net reaction heat is equivalent to that for Reaction (13).

Observations of composition changes in the iron-bearing material as it descends toward the hearth have been made on material collected from quenched blast furnaces. These observations show (Figure 20) that the Si content does indeed increase prior to reaching the slag level and decreases passing through the slag.

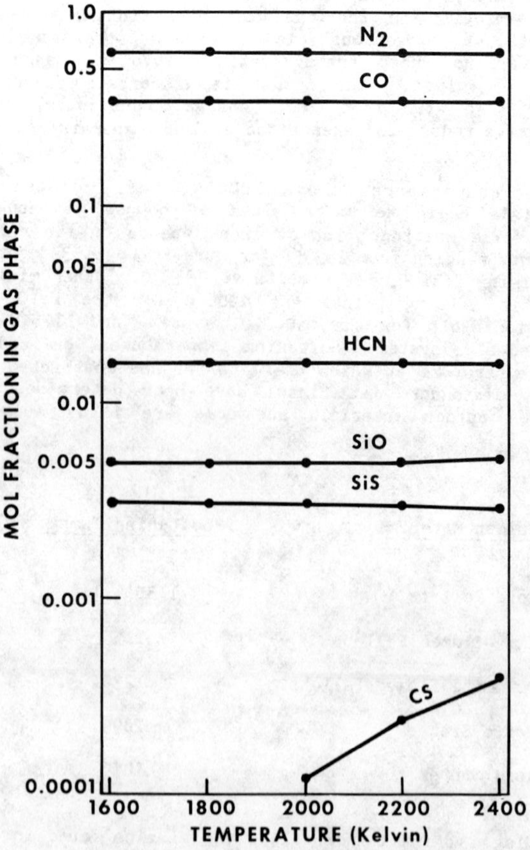

Figure 19. Gas phase concentration of various gas species formed by reaction of an initial mixture of 0.01 moles SiO_2, 0.004 moles FeS, 0.01 moles of H_2O and 1mol of air in the presence of excess C and Fe. Partial pressures are the total blast pressure times these mol fractions.

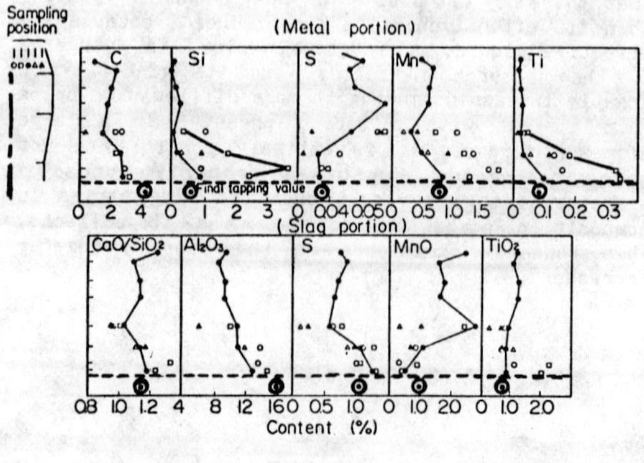

Figure 20. Changes in composition of metal and slag droplets as they descend to the hearth, as a function of both height and radial position. (5).

ENERGY BALANCE

Figure 21 is a schematic diagram of a blast furnace with the principle reactions and typical charge and product materials shown. Based on analyses weights, and temperature of input and output materials, of which the numbers in Figure 21 are typical, overall material and energy balances may be prepared. On an overall basis the heats of reaction can be written so that the product stream analyses reflect the proportions of each reaction that occur relative to 1 mol Fe and are shown in Table II.

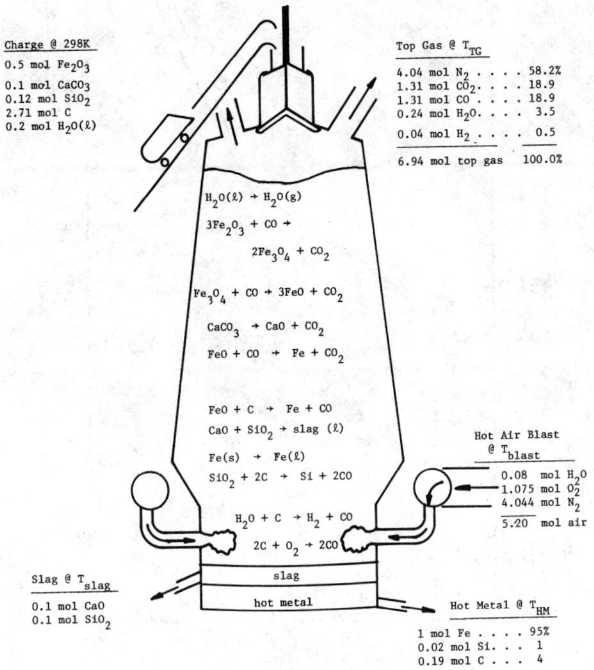

Figure 21. Schematic diagram of example blast furnace.

TABLE II

Overall Reactions in Example Blast Furnace

mol reactant / mol iron	Reaction	ΔH^o_{R298}/mol reactant	ΔH_{R298}
1.31	$C + 1/2\, O_2 \rightarrow CO$	- 26,400	- 34,584
1.21	$C + O_2 \rightarrow CO_2$	- 94,050	-113,800
0.04	$C + H_2O \rightarrow CO + H_2$	+ 32,150	+ 1,285
0.2	$H_2O(\ell) \rightarrow H_2O(g)$	+ 10,510	+ 1,260
0.10	$CaCO_3 \rightarrow CaO + CO_2$	+288,450	+ 28,845
0.50	$Fe_2O_3 \rightarrow Fe + 3/2O_2$	+196,800	+ 98,400
0.02	$SiO_2 \rightarrow Si + O_2$	+209,500	+ 4,190
		$\Sigma\Delta H_{R298}$	- 14,405cal.

The sensible heats contained in the air blast, slag, hot metal, and top gas are estimated to be:

$$\Delta H'_{blast} = -n_{air}(T_{blast} - 298)C_{p,air} = 40.56(T_B-298),\text{cal.}$$

$$\Delta H'_{slag} = n_{slag}(T_{slag}-298)C_{p,slag} = 3.3\ (T_{slag}-298),\ \text{cal.}$$

$$\Delta H'_{hot\ metal} = n_{HM}(T_{HM} - 298)C_{p,HM} = 19.36(T_{HM}-298),\ \text{cal.}$$

$$\Delta H'_{top\ gas} = n_{TG}(T_{TG} - 298)C_{p,gas} = (48.58)(T_{top}-298)$$

If T_B = 1000K, T_{slag} = 1620K, T_{HM} = 1620K, and T_{TG}=473K;

then, the overall heat balance is

$$\Delta H'_{HM} + \Delta H'_{slag} + \Delta H'_{top\ gas} + \Delta H'_{blast} + \Sigma\Delta H_{R,298} = -\Delta H'_{loss}$$

or

$$25{,}206 + 4{,}296 + 8{,}500 - (28{,}473) + (-14{,}405) = -\Delta H'_{loss}$$

or $\Delta H'_{loss}$ = 4,876 cal/mol Fe

The heat loss is 2.75% of the energy into the process, which in this case would be 176,857 cal, the sum of all exothermic reactions plus the sensible heat in the air blast.

Nowhere in the preceding calculation was there any mention of having to provide any temperature above that at which hot metal and slag emerge from the furnace. But, if some phase (gas, in this case) was not present above that temperature to transfer heat to the iron and slag, the iron and slag would never reach the necessary temperatures. Therefore, thermal energy must be provided with the proper quality, i.e, temperature. The overall balance does not even hint at this!

To see these needs more clearly, staged heat balances have been developed by many engineers and researchers in this field. Each differs slightly from the others based on the assumption of temperatures at which various reactions are assumed to occur, releasing or absorbing heat.

Based on a wide variety of measurements and observations made on operating furnaces, and laboratory simulations of the furnace shaft, the general picture that has emerged is that three major temperature regions within the furnace should be considered, with certain thermal needs and certain reactions occurring within each region.

In construction of staged heat balances, what is done is to calculate the net heat requirements of the solids within each zone, including the heat losses from the furnace in that zone and see what energy is therefore required to be given up by the hotter gases as they move countercurrently through the zone. In effect, we make a heat balance on each zone.

Figure 22, called a Reichardt diagram, shows the enthalpy needs of the condensed phases graphed as a function of temperature. In temperature regions where there are large heat requirements, such as in the intermediate region, the slope decreases, indicating that considerable energy is needed in those temperature regions.

TABLE III

Heat Requirements of Burden and Furnace by Stages for Example Blast Furnace

Temperature Region	Moles	Reaction	ΔH(cal/mol Fe)	
298	0.2	$H_2O(l) \rightarrow H_2O(g)$	+ 1,260	
	0.5	$Fe_2O_3 + CO \rightarrow FeO + CO_2$	+ 925	
	0.71	$FeO + CO \rightarrow Fe + CO_2$	- 2,300	
	0.05	$CaCO_3 \rightarrow CaO + CO_2$	+14,425	
		+Sensible heat absorbed by solids	+25,940	
		+Heat Losses (20%)	+ 975	
			+41,225	+41,225
1273K (1832°F)				
	0.27	$FeO + C \rightarrow Fe + CO$		
	0.02	$FeO + H_2 \rightarrow Fe + H_2O$	+11,470	
	0.05	$CaCO_3 \rightarrow CaO + CO_2$	+14,425	
	1.0	$Fe(S) \rightarrow Fe(l)$	+ 3,700	
	0.1	$CaO + SiO_2 \rightarrow slag(l)$	- 2,000	
	0.19	$C \rightarrow \underline{C}$	+ 970	
	0.02	$SiO(g) + \underline{C} \rightarrow \underline{Si} + CO(g)$	- 760	
		+ Sensible heat in condensed phases	+ 5,890	
		+ Heat Losses (40%)	+ 1,950	
			+35,645	+76,870
1500K (2240°F)				
	0.02	$SiO_2 + C \rightarrow SiO + CO$	+ 3,200	
		+ Superheating hot metal	+ 2,320	
		+ Superheating slag	+ 450	
		+ Heat Losses (40%)	+ 1,950	
1620K (2456°F)				
			+ 7,920	+84,790

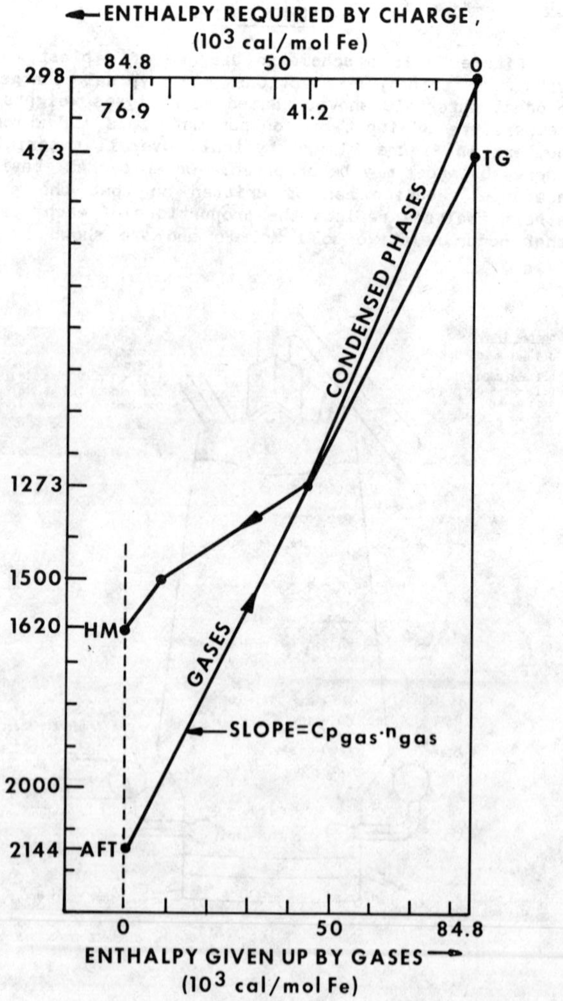

Figure 22. Reichardt diagram of example blast furnace.

The total needed is 84,790 cal/mol Fe, if the metal and slag only reach 1620K and 0.02 mol SiO_2 is reduced. If the metal and slag are to be hotter and more SiO_2 is to be reduced, more total energy will be needed and more high temperature energy will be needed.

On an overall basis, enough energy has to be supplied by the blast, through combustion and sensible heat in the preheated air, to equal this total required. In addition, enough energy has to be available to bring the condensed phases up to the necessary intermediate temperatures. This is tested by plotting, also on Figure 24, the change in sensible heat of the gases as they decrease from the AFT. Since the heat capacity and volume of the gases doesn't change significantly as the gas cools from the AFT, the result is essentially a straight line with slope equal to $C_p.n_g$. Since the gases are the heat source for the condensed phases, <u>this line can never be at a lower temperature than the condensed phase line</u>. Because of the shape of the condensed phase enthalpy plot, the gas line will touch the solid line only at a point, known as the <u>pinch point</u>. This occurs at about 1000°C (1273K) and is found in vertical temperature probes of operating furnaces as a region of the furnace where both gas and solid are at essentially the same temperature. This zone is known as the <u>thermal reserve zone</u> as indicated in Figure 23.

The interpretation of Figure 22, is that the gases must initially be at an AFT high enough to give up 43,565 cal/mol Fe between the AFT and 1273K, (this is the energy required to raise the condensed phases from 1273K to

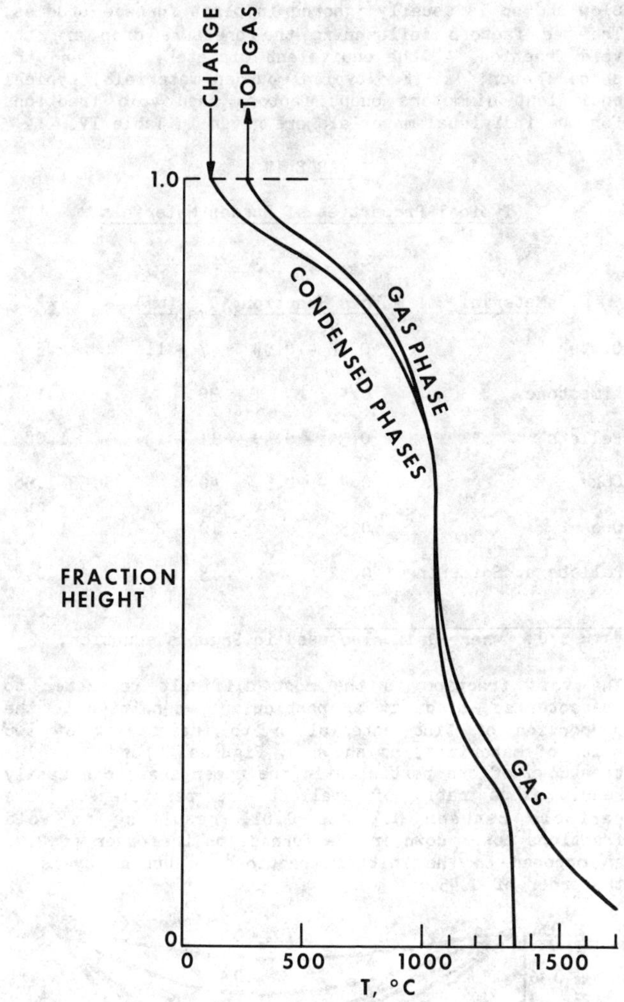

Figure 23. Schematic diagram of typical measured vertical temperature profiles of blast furnaces showing "thermal reserve zone" at 1000°C.

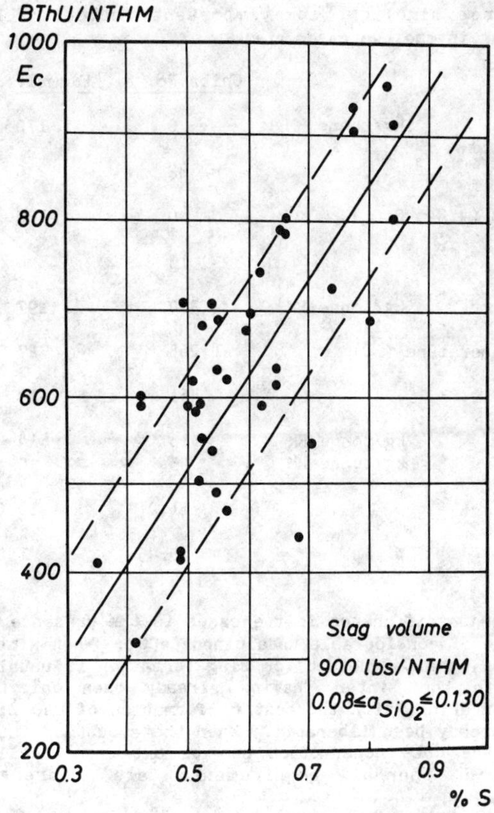

Figure 24. High temperature energy availability, E_c, versus %Si in hot metal. (6).

1620K) in order to end up with a product in usable form (liquid) and composition. If not enough energy is available, for example, because of low blast preheat, the AFT will be decreased. The slope of the gas line, however, will be the same and so will the solid line. All that will happen will be that the metal and slag will not reach as high a temperature, since less energy will be available, above 1273K, from the gases.

Thus, by making three stages of heat balance, the presence of a pinch point, and the need for sufficient high quality energy has been demonstrated. Computer control of blast furnaces has relied, in some cases, on continuous computation of energy available from gases at temperatures above 1425K, where the reduction of SiO_2 to SiO takes place. This energy, has been correlated with the silicon content of the hot metal, as shown in Figure 24.

These example results can be compared to energy balances reported on an operating furnace, Chiba No. 6 at Chiba Works, Kawasaki Steel Co.(7)

Thermal Input:(Kcal/tonne HM)	Chiba No. 6 Oct., 1978	Example
Fuel combustion with blast	536	938
Sensible heat of blast	479	484
Formation of slag	0	34
Indirect reduction	47	23.4
Sensible heat of burden	12	0
Total	1,074	1,479
Thermal Requirement		
Direct reduction of FeO	337	195
Sensible heat of HM and slag	441	502
Reduction of SiO_2 + MnO	38	42
Sensible heat of top gas	89	145
Evaporation of H_2O	11	21
Decomposition of $CaCO_3$	0	491
Heat Loss	158	83
	1,074	1,479

The two cases are not exactly comparable, and the differences highlight some important factors. The conditions in the two cases are:

		Chiba No. 6	Example
Limestone	kg/tonne HM	0	170
% Sinter		82	0
% Pellets (4.84% SiO_2)		0	100
% Ore		18	0
Slag Volume	kg/tonne HM	300	197
Blast temperature (°C)		1,268	727
Fuel rate		461	614
coke	kg/tonne HM	416	614
oil	kg/tonne HM	45	0
Coke Ash	%	11.2	10.0
Si in HM	%	0.45	1.00

The major difference in the cases is the presence in the charge of considerable uncalcined $CaCO_3$. When sinter is charged, the CaO needed for slag formation is usually present in the sinter, having already been calcined during sintering, with the heat of formation of $CaO \cdot SiO_2$ having already been liberated. When the amount of $CaCO_3$ is cut, the coke consumption can be cut dramatically, since the thermal requirements are decreased substantially.

Another difference lies in the proportion of indirect and direct reduction, and on the utilization of the CO for reduction. This is largely influenced by the gas distribution and the thermal profile in the furnace. Furthermore, the heat losses are also functions of the furnace profile. It should be noted that at various other times, the Chiba No. 6 furnace had lower heat losses, ranging from 43 to 65 Kcal/tonne HM so that the example heat loss is a reasonable one.

Finally, it should be added that, recognizing the high cost of coke consumption by H_2O in the blast, the blast for Chiba No. 6 is now dehumidified to 6 g/Nm^3, following the original practice in the U.S. at the turn of the century.(7) Other Japanese furnaces are also dehumidified to this level or below. Using these low moisture, high temperature (1350°C) blasts and close to 100% sinter burdens, the projected lower limits of coke rates in Japanese blast furnaces are in the neighborhood of 420 kg/kgHM.(11)

GAS DISTRIBUTION

The production of hot metal is proportional to the amount of reducing gas supplied, as well as the thermal energy carried with it. The ability to increase productivity requires that gas flow be as easy as possible, so that a given blower can put the maximum blast through the system at or below the maximum blast pressure that the blower can operate against. This, in turn, requires that $\Delta P/L$ for the furnace as a whole be the minimum possible.

To a first approximation, the flow of gases through a packed bed is governed by Ergun's equation, which relates the pressure drop per unit of bed height, $\Delta P/L$, to the characteristics of the bed and of the gas:

$$\frac{\Delta P}{L} = \frac{150 \eta V_o \lambda^2 (1-\varepsilon)}{\bar{D}_{vs}^{\,2} \; \varepsilon^3} + \frac{1.75 \rho_g V_{og}^{\,2} \lambda (1-\varepsilon)^2}{\bar{D}_{vs} \; \varepsilon^3} \qquad (14)$$

The first term is only important if the flow of gas is slow and so is usually ignored in blast furnace studies. The bed factors influencing the pressure drop are the void fraction, ε, the equivalent diameter, $\bar{D}_{vs}$, and the shape factor, λ. For typical burden materials, typical equivalent diameters shape factors, and void fractions for the individual materials are given in Table IV.

TABLE IV

Typical Properties of Burden Materials

Material	Void Fraction	$\bar{D}_{vs}$, (mm)	λ*
Sinter	0.48 - 0.54	7 - 11	2.3
Limestone	0.4	30	1.15
Pellets	0.34 - 0.40	11 - 13	1.08
Coke	0.4 - 0.5	40	1.38-1.56
Ore	0.35	10 - 20	1.1
Pellets at Softening	0.32	13	~ 1.0

*$\lambda = 1/\theta_j$, where θ is also used in Ergun's equation.

The void fraction is the most difficult parameter to characterize, since it is particularly sensitive to the proportion of fine material or to intermixing of two sizes of materials, as shown by Figures 25 and 26. The breakdown of ore particles in the upper shaft can easily result in a ratio of smallest ore particle to coke particle between 0.1 and 0.01, resulting in void fractions lower down in the furnace on the order of 0.2, as opposed to the initial fractions in upper layers on the order of 0.45.

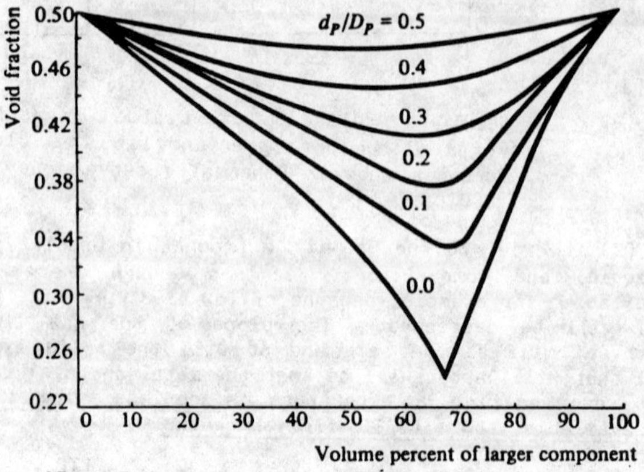

Figure 25. Experimental voidage of two component particle mixtures, both having initial void fractions of 0.5, one with diameter d_p and the other D_p. (8).

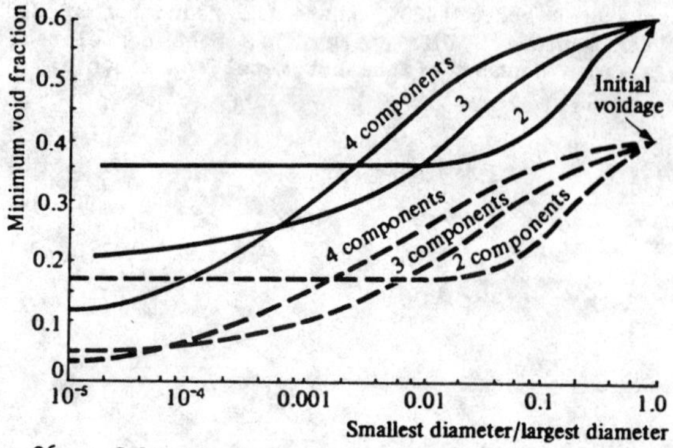

Figure 26. Calculated minimum voidage in two-, three-, and four-component particle mixtures. (8).

Measurements of pressure distribution both radially and vertically have shown that the pressure in any horizontal plane through the furnace is essentially uniform: i.e., the pressure drop per unit of height is essentially constant radially. However, when the burden is charged, the solids, with the divergent properties shown in Table IV do not distribute themselves uniformly in a radial direction. To make matters even more difficult, the combustion of coke takes place within an annular space between the tuyeres and an inner radius about 1-2m in front of the tuyeres.

The result of combustion in the raceway annulus is that the coke descends much faster from above it than it does above the center of the furnace, where coke is primarily consumed only by the Boudouard reaction resulting from direct reduction of FeO, MnO, and SiO_2. There is a conical-shaped portion of coke, known as the "deadman", that remains unreacted for a long time in the center of the furnace in front of the tuyeres.

The total pressure drop is made up of the sum of pressure drops through the various zones and layers within the zones in the furnace. The effect of the inhomogeneity of the solids is to cause changes in the relative amounts of gas that pass through two adjacent regions. For example, if two regions lying parallel to the overall direction of gas flow, have different particle sizes, void fractions, and/or shape factors, the ratio of volumetric flows through the two cross-sectioned areas can be shown to be

$$\frac{Q_1}{Q_2} = \left[\frac{\lambda_2(1-\varepsilon_2)}{\lambda_1(1-\varepsilon_1)} \cdot \frac{(\varepsilon_1)^3}{(\varepsilon_2)^3} \cdot \frac{(D_1)}{(D_2)} \cdot \frac{(A_1)^2}{(A_2)^2}\right]^{1/2} \tag{15}$$

provided that the pressure at the top and bottom of both layers is the same. A few examples of the magnitude of the ratio, which ideally is 1.0, are given in Table V.

TABLE V

Flow Ratios Between Adjacent Packings With Pressure Drops the Same

Situation	A	$\bar{D}$	λ	ε	$\frac{Q_1}{Q_2}$
Cohesive Zone:					
1 - Coke	1	40	1.5	0.4	2.60
2 - Soft ore	1	12	1.1	0.3	
Upper Shaft:					
1 - Coke-rich central core out to half-radius	1	35	1.4	0.35	2.25
2 - Pellet-rich annulus	3	18	1.2	0.25	

Resistance to flow increases when the gas density is high, or when the particles are small, or when the void fraction is low. Both of the latter conditions increase as the burden descends into the bosh and hearth. In particular, when the softening temperature is reached, the already less porous ore burden material becomes still less porous and the gases have an increased incentive to pass instead through the neighboring coke layers. However, once the descending burden has reached the melting temperatures or above, the gas is essentially flowing countercurrent to the downward flowing liquid through a bed of solid coke.

The result is a pattern of gas flow indicated in

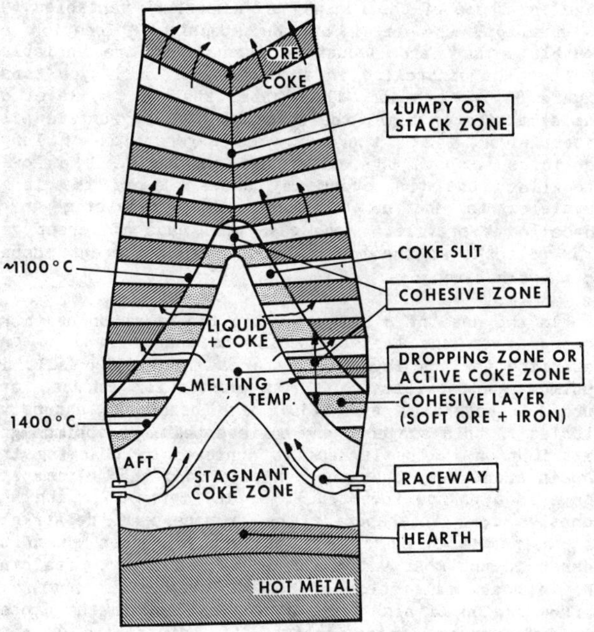

Figure 27. Schematic cross-section showing various zones and basic gas flow pattern in furnace.

Figure 27. The gases leave the tuyeres and immediately start to rise towards the top. Momentum, however, carries them towards the center, as well. They transfer heat and rise through the active coke zone. As the gas temperature decreases, they eventually encounter the region of the bed where melting is just beginning to occur, the lower boundary of the cohesive zone, defined as the zone where the temperatures are between the softening and melting temperatures. At this point, the gases encounter alternating layers of solids (coke and softened ore) which have greatly different resistances to flow. The gases channel through the coke layers, known as "coke slits", and then rise through the remainder of the burden. They also may channel as they rise in the upper shaft or granular zone. This channeling is what effects the efficiency with which the reducing power of the gas is utilized.

Overall pressure drops are often measured as functions of blast volume, Q, and the rest of Ergun's equation lumped together as a coefficient of blast resistance, K. It has been found that

$$\frac{\Delta P}{L} = KQ^{1.64} \tag{16}$$

is typical, which is not surprising, since Ergun's equation has two terms, one with V(=Q/A) raised to the first power and the other with V^2.

If the charge materials segregate at the top, as illustrated in Figure 8, where the large particles (i.e., coke) have collected in the center and the smaller ore particles are at the outside, the gas will tend to channel up the middle and not bring CO into contact with iron oxides, resulting in poor utilization of the CO. On the other hand, if charging is done in such a way as to allow large particles, with high void fraction, to collect and descend near the walls, then the hot gases will channel along the wall, increasing the heat losses to the stave coolers, and wasting heat that might otherwise go to heat the burden.

Thus, the practical problems facing the furnace operator include deciding how to charge the furnace at the top and blow it at the bottom so that the gases are used effectively for both chemical and thermal transport. With respect to blowing, the effect of blast temperature, wind rate, moisture content and injected fuels on the size and shape of the raceway are important variables.(9) At the top, the level of the stockline, location of movable armor, and sequence of charging are variables that can be controlled in furnaces with bell-type tops. Figure 8, for example, illustrates that if the level of the stockline is high, the shape of the top surface will resemble an M and there may be segregation of large particles to both the center and outside. At a lower stockline, the trajectory may cause all of the large particles to end up in the center, resulting in a so-called V profile. Each case will have different gas flow patterns at the top and result in different shapes to the cohesive zone.

In the case of a bell-less top the operator has more control over the trajectory of the charge, and so may determine various layering patterns that will result in optimum gas utilization for a particular furnace and burden. Wayakama, et al(10) have conducted extensive studies on this subject, and believe that the optimum gas distribution in the furnace is achieved by charging the top in such a way that the cohesive zone that ultimately forms is a narrow inverted V, as in Figure 27, with the cohesive zone intersecting the furnace wall relatively far down the stack. This pattern has been shown to result in the most uniform gas-flow, radially, obtaining the highest utilization of the CO, yet still having a narrow region of high-temperature gas flow in the center of the furnace, which assists in maintaining a good overall permeability and smooth burden descent. Figure 28 shows the results of tests using various charging

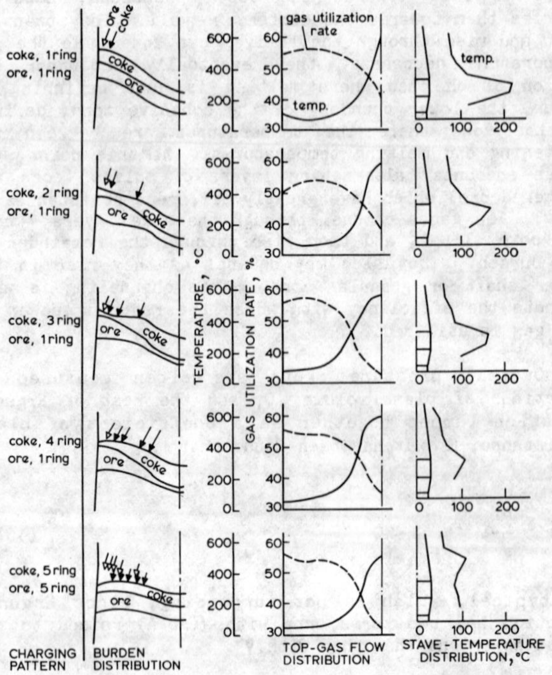

Figure 28. Correlation between charging pattern and top gas temperature and CO utilization and heat loss distribution. (10).

patterns, with the resulting gas temperature and utilization rate profiles, as well as the temperature distributions to the cooling system. The recommended charging pattern, for the particular furnace being studied, was, in this case, the bottom pattern: a 5-ring spiral of ore followed by a 5-ring spiral of coke. Figure 29 shows that the lower the 1100°C isotherm intersects the wall, the lower the overall heat loss. Figure 30 illustrates that the larger the zone of solid material, the better the gas utilization.

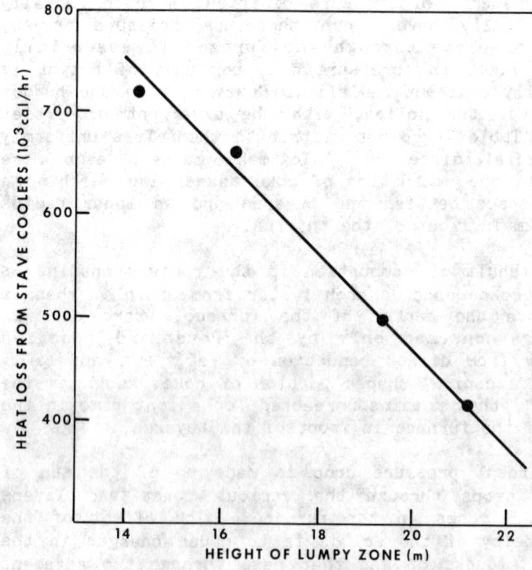

Figure 29. Relation between height of cohesive zone and total heat loss. (11).

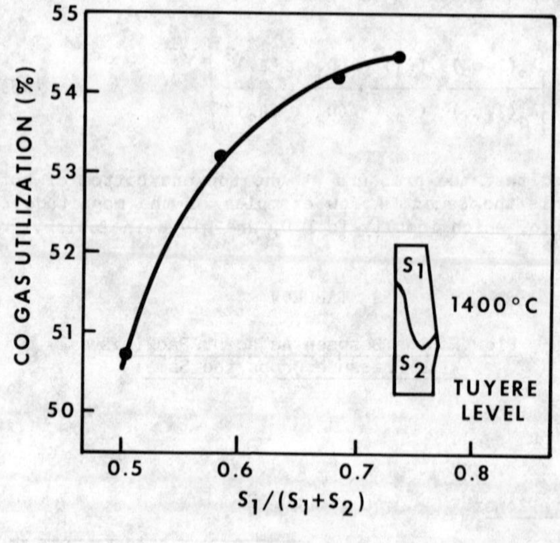

Figure 30. Relation between height of cohesive zone and gas utilization. (11).

It should be emphasized that the distribution needed for one furnace is not necessarily transferrable to another, since the properties of the raw materials can vary, and these play an important role in determining the effectiveness of a given charging pattern.

INTERNAL RECYCLING

It has become more apparent in recent years that some materials recycle internally within the furnace. Among these are sulfur, silicon, alkalies, and carbon.

In the case of sulfur, virtually all the sulfur enters the furnace as pyrrhotite, FeS, in the coke. Since it is finely divided and imbedded in the coal particles, it does not react with anything until the coal combusts. At that point it reacts with C, SiO_{21} and CO to form CS, SiS, and COS in the gas stream. These rising gases cool, becoming less thermodynamically stable, and as they encounter CaO or Fe, react to form CaS (in the slag forming phases) and S in solution in the iron. Thus, the descending iron and slag phases reabsorb the sulfur and bring it back down to the hearth level, in a thermodynamically much more stable form, Figure 20. Figure 31 illustrates the internal recycling of sulfur.

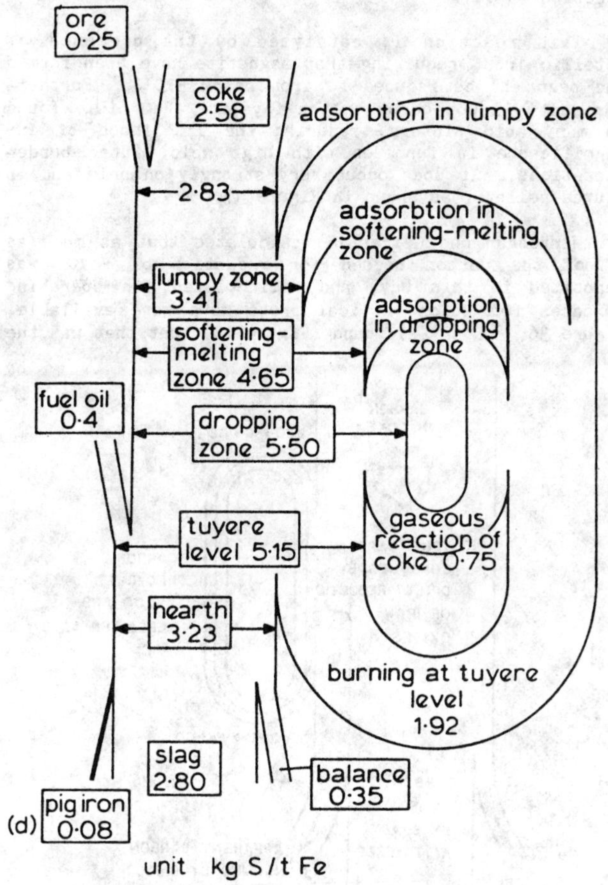

Figure 31. Recycling of sulfur inside Hirohata No. 1 blast furnace. (12).

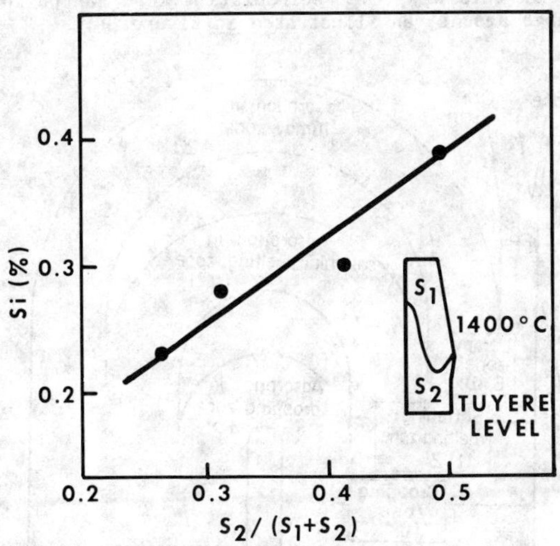

Figure 32. Ratio of dropping zone as a fraction of total furnace area to hot metal silicon content. (11).

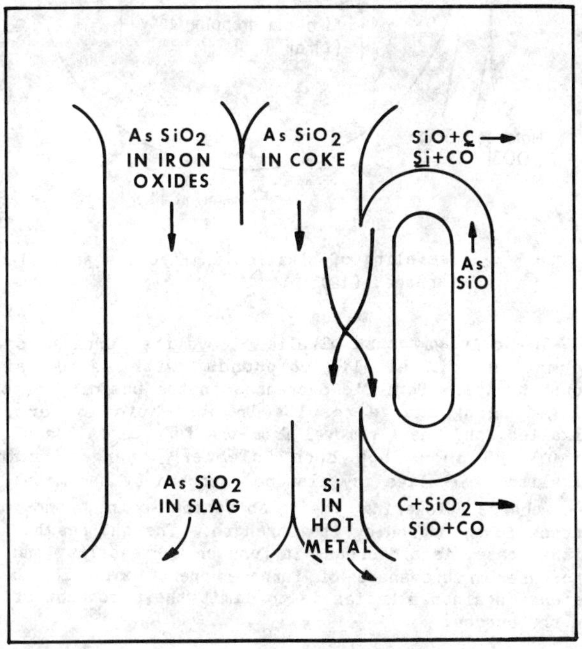

Figure 33. Schematic diagram of internal recycling of silicon.

As already mentioned, the majority of silicon probably enters the metal phase via the gas phase according to Reaction 12. The pressure of SiO in the gas phase depends on the ratio of SiO_2 in the coke to carbon in the coke. While SiO_2 in slag may also vaporize to some degree, decomposition in this fashion is not as likely nor as rapid as the vaporization that takes place in contact with carbon just above the tuyeres. Since the reduction of SiO to put Si into the metal requires that carbon be dissolved in the metal first, this precludes the reaction from occurring at temperatures much below 1400°C. Thus, the critical factor that determines how Si enters the metal is the residence time of the iron between the tuyere level and the 1400°C isotherm in the furnace. If the furnace has an excess of thermal energy, the isotherm will be higher in the furnace, the residence time will be longer, and the Si content increased. Okamura, et al,(11) have demonstrated this by plotting the ratio of the cross-sectional area of the dropping zone (temperatures greater than 1400°C) relative to the total working cross-sectional area of a furnace versus the averge hot metal Si content, Figure 32. The final result is a schematic picture of silicon recycling as shown in Figure 33.

Alkalies (Na and K) are troublesome elements that enter in coke and ores. While they usually enter as complex silicates, the only stable gas species that form are the cyanides, KCN and NaCN, and elemental Na and K, (13). The pressure of these species increases with temperature and as the minerals descend in the furnace they react, vaporize, and the alkalies are swept up in the furnace. As they cool, KCN and NaCN recondense and are again carried down. Some $NaCO_3$ may form in the upper shaft. In this way, the recirculating load can build up to a large amount, as illustrated in Figure 34.

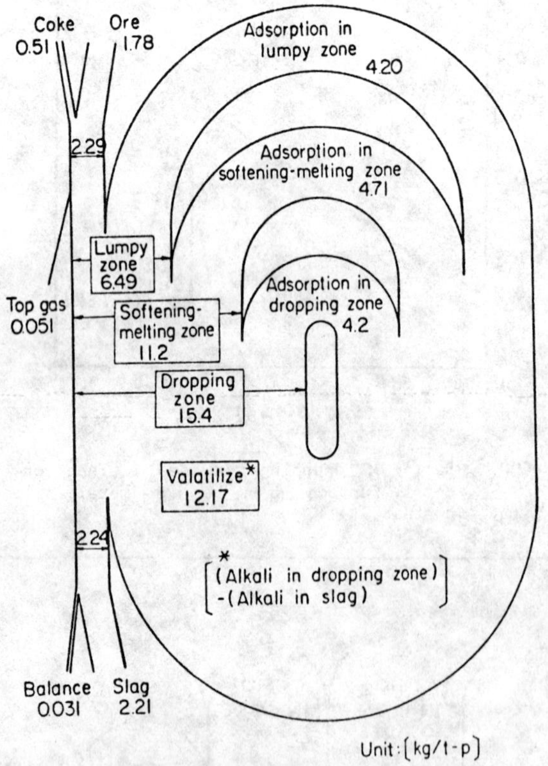

Figure 34. Recycling of alkalies in Hirohata No.1 blast furnace. (12).

The only way that alkalies leave the furnace is by formation of alkali compounds with acid slag constituents. Periodic decreases in the basicity of the slag-forming charge results in absorbtion of excess alkalies, and their removal from the furnace in the slag. Figure 35 shows how such "cleaner" charges remove alkalies. Alkalies may also be absorbed by the metallic ore charge, lowering their softening point, and may attack coke, degrading its strength. The net result, in either case, is a decrease in furnace permeability and an increase in uneveness of furnace operation. The best defense against alkalies is to limit their concentration in the burden.

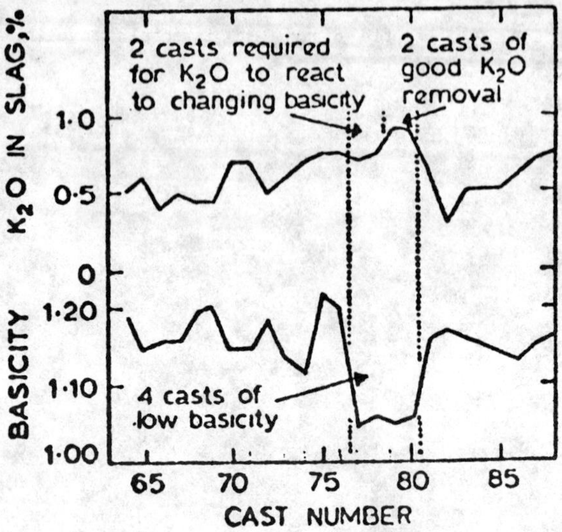

Figure 35. Removal of alkalies by addition of acid constituents (olivine) to charge. (14)

Finally, there is evidence that in furnaces charged with ores, or pellets, there is significant internal recycling of carbon. This is caused by the deposition, in the temperature range 450°-650°C, of carbon via Reaction (5).

This reaction is catalyzed by the presence of metallic iron, requiring that magnetite have been formed and begun to be reduced to iron below 570°C. Furthermore, it wll not occur on the fayalite (FeO SiO_2) found in many acid sinters, reducing the likelihood of its significance in furnaces with high acid sinter burden operations. It does occur very strongly on acid and on fluxed pellets, as shown in Figure 17.

Mertdogan and Geiger(15) calculated that as much as 5% of the carbon needed for reduction of Fe_2O_3 was deposited in this way, and recirculated, in operating furnaces for which vertical probe data were available, Figure 36. Lowry(2) found (Figure 17) that that was the

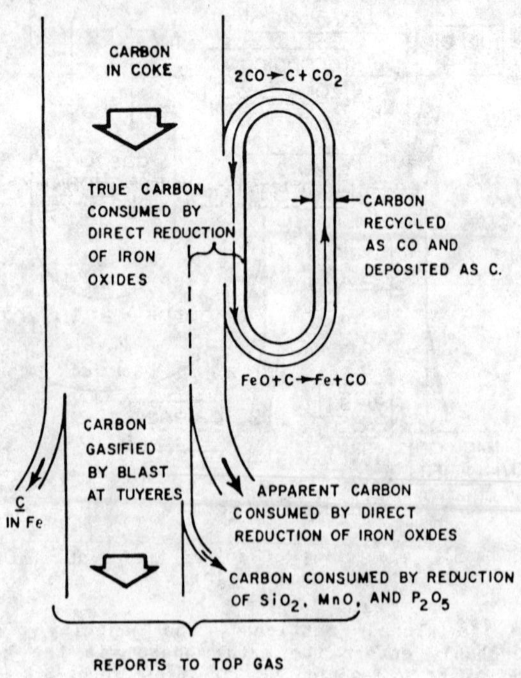

Figure 36. Calculated recycle of carbon based on data from furnaces operating on ore burdens. (15).

approximate amount of carbon that was deposited on single pellets during simulated blast furnace atmosphere - temperature cycles. The net result of such recycling is a change in the ore/coke ratio entering the lower portion of the furnace from that which is based on the charge ratio. Ultimately this leads to a redistribution of thermal effects in the furnace. Furthermore, it was shown that there could be significant changes in the proportion of direct reduction occurring, that could not be detected from changes in top gas analysis because that method explicitly depends on the assumption that reaction 5 does not occur. This would also render the method of Rist and co-workers(16,17), used to define the "chemical reserve zone", invalid, as a measure of chemical efficiency, since it depends on a heat balance on the lower part of the furnace to establish the location of the "operating line", and the heat balance depends largely on the assumed amount of direct reduction occurring.

Consider the evidence for cyclic behaviour in a blast furnace.(18,19) Nakatani et al., after analyzing computed values of carbon consumption, etc. from an operating blast furnace, Figure 37 suggested the presence of a "coke reserve zone", fluctuating in amount and alternately supplying an excess or deficiency of coke in the bosh of the furnace, Figure 38. They suggested no reason for the fluctuation of this zone, other than an "external disturbance".

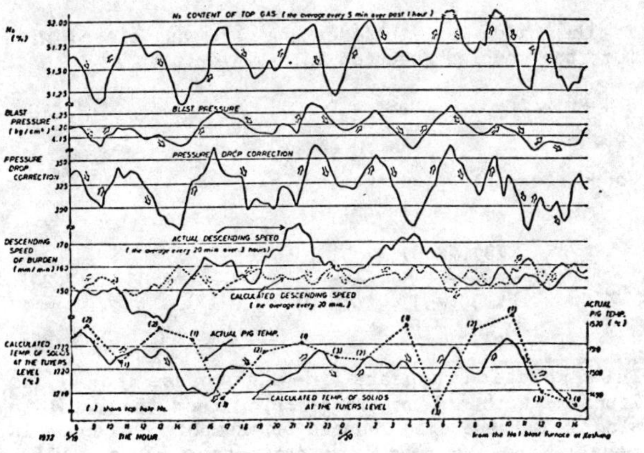

Figure 37. Cyclic behavior of furnace parameters during operation of Sumitomo blast furnace. (18).

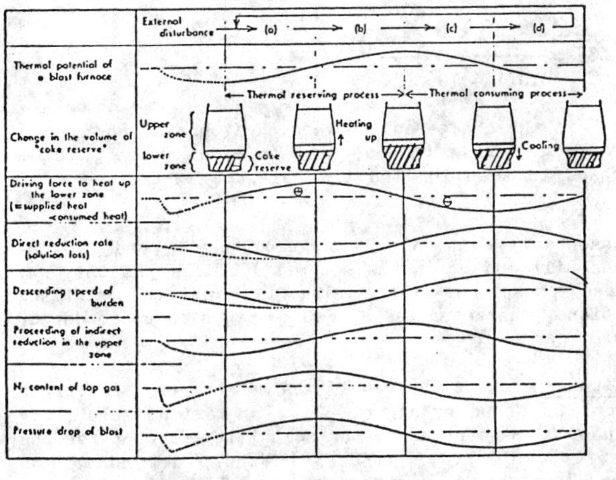

Figure 38. Hypothetical changes occurring during a cycle in the volume of "coke reserve". (18).

The sequence of a disturbance which affects the formation of the carbon regeneration zone can be explained. As the carbon regeneration zone is developed, possibly by a change in catalytic properties of the ore, carbon would be deposited by the decomposition of carbon monoxide and the temperature of the gases would be increased. The carbon monoxide content of the top gas would be decreased and N_2 increased, indication of a better gas "utilization".

Figure 38 shows that the pressure drop reaches a maximum at the same time that the N_2 content of the top gas is also at a maximum. This would be possible if carbon was being deposited out, since a maximum N_2 corresponds to a minimum carbon content of the top gas, which corresponds to carbon deposition and a resulting decrease in permeability.

During the time that the burden is descending about one-third of the way through the furnace, the change in top gas analysis and temperature will be continuously changing.

After the next one-third of the total retention time, the solids, with an increased carbon-to-ore ratio, will descend to the bosh where the temperature will start to increase and direct reduction begins.

If indirect reduction has not changed, then there is an excess of carbon available in this portion of the burden as it reaches the tuyere level and an excess of CO and heat is produced relative to the iron present, resulting in higher bosh gas temperatures and CO content which passes up through the burden, increasing the potential for indirect reduction, but also extending the temperature region in which carbon deposition will occur. Hot metal temperature and silicon will probably increase during this initial transition.

Later, if indirect reduction has decreased due to CO being lost by carbon deposition, then more heat will be required to carry out a larger proportion of direct reduction and the temperature of the metal and silicon content will decrease. Now, the coke-to-iron ratio will be lower than before and the bosh will cool down and the sensible heat in the bosh gases will decrease. Eventually this will translate into a shrinkage of the temperature zone between 925°K and 725°K where carbon deposition can occur, and the whole process will repeat itself. The result will be a constant oscillation of the thermal condition in the bosh and stack and therefore of hot metal temperature and silicon content.

HOT METAL CHEMISTRY

The principal constituents of hot metal are the carbon, silicon, manganese, sulfur, phosphorus and titanium dissolved in it. Hot metal is essentially saturated with carbon. Its concentration is affected by temperature and the presence of other constituents, and may be calculated from the following equation:

$$\%C = 1.34 + 2.54\times10^{-3}T(^{\circ}C) - 0.35[\%P] + 0.17[\%Ti]$$
$$-0.40[\%S] + 0.04[\%Mn] - 0.30[\%Si]$$

Manganese is charged to the furnace as MnO in the ore and is reduced by carbon in a manner similar to iron oxide. The degree to which Mn is distributed between the slag phase, where it presumably exists as Mn^{+2} ions, and the hot metal, as $\underline{Mn}$, is determined by the effective oxygen partial pressure in the slag-metal system according to the reaction

$$(MnO) \rightarrow \underline{Mn} + 1/2O_2(g) \quad (17)$$

The reaction

$$\underline{C} + 1/2O_2(g) \rightarrow CO \quad (18)$$

can be combined with Reaction (17) to yield

$$(MnO) + \underline{C} \rightarrow \underline{Mn} + CO \quad (19)$$

The hot metal manganese content is thus dependent on the temperature, the slag composition, the pressure of CO, and the activity of carbon.

$$\log[\%\underline{Mn}] = -\log P_{CO} + \log a_C + \log a_{MnO} - \log f_{Mn}$$
$$- 15226/T + 10.876 \quad (20)$$

where

$$a_{MnO} = 10^{-3}\left(\frac{1.6 + 5.9(\%CaO) + 1.4(\%MgO)/\%SiO_2}{(\%\ MnO)}\right)$$

and $\log f_{Mn} = -0.009[\%C] - 0.0043[\%S] - 0.0022[\%P]$

Assuming $a_C = 1.0$, the manganese contents observed are on the order of 50% less than what would be predicted if P_{CO} is taken to be the partial pressure in the bosh. However, since carbon saturation is not necessarily reached in metal drops falling through the slag, (Figure 20) the activity of carbon may be less than 1.0, accounting for the difference.

Manganese is often transferred from the slag to the metal as the metal descends through the slag to the hearth. On the other hand, both silicon and sulfur, having been absorbed into the metal from the gas phase as it traversed the active coke zone, are transferred from the metal to the slag during passage through the slag.

Silicon transfer between metal and slag is affected by a similar reaction to that for manganese

$$(SiO_2) + 2\underline{C} = \underline{Si} + 2CO \quad (13)$$

In this case,

$$\log[\%Si] = -2 \log P_{CO} + \log a_{SiO_2} - \log f_{Si} + \log a_C - 31220/T + 20.70$$

Plant data result in actual values of hot metal silicon being less than one-half of what would be expected. Equilibrium with respect to primary silicon reactions is clearly not achieved. Furthermore, analysis of blast furnace data shows that the reaction

$$2\,MnO + \underline{Si} = SiO_2 + 2\,\underline{Mn} \quad (21)$$

is not in equilibrium either.

Sulfur transfer between slag and metal may be considered to occur via the reactions.

$$1/2\,S_2(g) + (O^{2-}) = (S^{2-}) + 1/2\,O_2\,(g) \quad (22)$$

and

$$\underline{S} = 1/2S_2(g) \quad (23)$$

The thermodynamic behavior of sulfur in blast-furnace slags is best characterized by use of the sulfide capacity, C_s, defined by

$$C_s = (\%S)P_{O_2}^{1/2}/P_{S_2}^{1/2} \quad (24)$$

Sulfide capacities have been determined for many slag systems and may be calculated from:

$$\log C_s = 1.35 \cdot \frac{1.79(\%CaO)+1.24(\%MgO)}{1.66(\%SiO_2)+0.33(\%Al_2O_3)} - \frac{6911}{T} - 1.649 \quad (25)$$

Using the equilibrium constant for Reaction (23),

$$K = \frac{P_{S_2}^{1/2}}{f_s[\%S]}$$

and the definition of C_s, the distribution ratio of sulfur between slag and metal,

$$R = \frac{(\%S)}{[\%S]} = K\,C_s f_s/P_{O_2}^{1/2} \quad (26)$$

If the proper values of f_s and P_{O_2} are used, R can thus be calculated. For f_s, it is suggested that

$$\log f_s = 0.15\,[\%C] + 0.063[\%Si] - 0.026[\%Mn]$$

is satisfactory. Several studies of blast furnace data(20,21) have determined that the manganese equilibrium represented by Equation (17) is a good measure of the oxygen potential involved with slag-metal distributions in the hearth. When this reaction is combined with Reactions (22) and (23), the net result is

$$\underline{S} + \underline{Mn} + (O^{2-}) = (S^{2-}) + (MnO) \quad (27)$$

If the oxide and sulfide ions are associated with calcium, the coupled reaction is often written as

$$\underline{S} + \underline{Mn} + (CaO) = (CaS) + (MnO) \quad (28)$$

For carbon-saturated iron,

$$\log k_{MnS} = \log \frac{(\%S)(\%MnO)}{[\%S][\%Mn]} \cdot \frac{1}{(\%CaO)} = \frac{6327}{T} - 3.85 \quad (29)$$

This reaction appears in many cases to be in equilibrium, with sulfur transferring from the metal to the slag as manganese is reduced (presumably by carbon in the metal) from the slag into the metal. Thus, sulfur distribution is controlled by local oxygen potential in the slag, which is much higher than that associated with C-CO equilibria, as illustrated by the previously mentioned slag FeO contents of 0.2-0.5%, as opposed to equilibrium FeO values on the order of 0.01%.

The actual analysis of sulfur may be related to R by

$$[\%S] = \frac{L}{10(1 + 10^{-3}VR)} \quad (30)$$

where L is the charge sulfur load, kg/tonne of hot metal, and V is the slag volume, in the same units. It is important to keep in mind the relative importance of V and R.

Essentially all of the phosphorus charged to the furnace will end up in the hot metal, since the only way to keep phosphorus in a slag is to have oxidizing conditions, characterized by FeO contents on the order of 10 to 20 percent.

Titanium enters the metal in a manner very similar to silicon, as shown by Figure 20. The maximum solubility of titanium is governed by the solubility product for the formation of TiC, $[\%\underline{Ti}][\%\underline{C}]$ and, depending on temperature, is on the order of 0.25% Ti for carbon-saturated iron. Titanium levels above this usually imply the presence of TiC or Ti(C,N) in suspension in the iron.

CONCLUSIONS

The blast furnace is a very complicated countercurrent packed bed reactor. While it is very efficient from a thermal and chemical standpoint, it is inherently unstable and control of the composition of the hot metal product is difficult because the reduction of silica to silicon is not an equilibrium reaction. Furthermore, sulfur control is affected by local oxidation states and local basicities of the slag, which can be affected by driving rates as well as charge composition. The role of internal recycling is not entirely understood, particularly as it affects the thermal stability of the furnace, and therefore the chemistry of the product. This remains an area for further research.

REFERENCES

1. T'ien-Kung K'ai-Wu (Chinese Technology in the Seventeenth Century), Sung Ying-Hsing; translated by E-Tu Zen Sun and Shiou-Chuan Sun, publ. by Penn. State Univ. Press, University Park, PA, 1966.

2. M. Lowry and G. H. Geiger, Proc. Ironmaking Conference., ISS-AIME, 40, (Toronto, 1981).

3. E. F. Osborn, R. C. DeVries, K. H. Gee, and H. M. Kraner, Trans. AIME, 200, (1954), p. 33.

4. N. Tsychiya, M. Tokuda, and M. Ohtani, Met. Trans. B, 7B, (1976), p. 315.

5. M. Sasaki, K.Ono, A. Suzuki, Y. Okuno, and K. Yoshizawa, Trans. ISIJ, 17, (1977), p. 391.

6. J. M. Vanlangen, A. Poos and R. Vidal; J. Metals, (December, 1965), p. 9.

7. J. Kurihara, S. Hashizume, H. Takahashi, H. Marushima, K. Okumura, and H. Takashi, Kawasaki Steel Technical Report No. 1, Sept., 1980, p.14-26.

8. C. C. Furnas, Ind. & Engr. Chem., 23, (1931), p. 1052.

9. M. Kondo, T. Inatani, and K. Okabe, Trans. ISIJ, 14, (1974), p. 269.

10. S. Wayakama, Y. Kanayama, and Y. Okuno, Ironmaking and Steelmaking, 6, (1979), p. 261.

11. K. Okumura, T. Kawai, H. Marushima, H. Takahashi, and J. Kurihara; Trans. ISIJ, 66, (1980), p. 176-185.

12. Y. Shimoura, K. Nishikawa, S. Arino, T. Katayama, Y. Hida and T. Isoyama; Trans. ISIJ, 17, (1977), p. 381.

13. K.-P. Mey and W-K. Lu, Proc. Ironmaking Conf., ISS-AIME, 37, (Chicgo, 1978), p. 582.

14. J. D. Ashton, C.U. Gladysz, G. H. Walker, and J. Holditch, Ironmaking and Steelmaking, 2, (1974).

15. A. Mertdogan and G.H. Geiger, Proc. Ironmaking Conf., ISS-AIME, 37, (Chicago, 1978), p. 612.

16. A. Rist and G. Bonnivard, Revue de Metallurgie, 60, (1963), p. 23; 63, (1966), p. 197, 63, (1966), p. 296.

17. A. Rist, and N. Meysson, Proc. Ironmaking Conf., AIME, 25, (1966), p. 88.

18. S. Nakatani, Y. Matoba, and S.Tokura, Sumitomo Search, No. 9, May, 1973.

19. R. Vidal, R. Colin, N. Poughis, G. More, and R. Delacharlerie, CRM No. 36, Sept., 1973.

20. E. T. Turkdogan, G. J. W. Kor, and R. J. Fruehan, Ironmaking and Steelmaking, 7, (1980), p. 257.

21. D. J. Young and C. J. Cripps Clark, Ironmaking and Steelmaking, 7, (1980), p. 209.

DIRECT-REDUCTION TECHNOLOGY

Jerome Feinman
Occidental Oil Shale, Inc.
Grand Junction, Colorado 81501

The technology is reviewed of the commercially successful processes for producing direct-reduced iron including those based on gaseous reductants and carried out in fluidized beds, moving-bed shafts, and fixed-bed retorts, and those based on solid reductants and conducted in rotary kilns. The chemistry of direct reduction is discussed including equilibrium and energy considerations and kinetic factors as they apply to the various process concepts. The use of direct-reduced iron in steelmaking and ironmaking is reviewed. Proposed extensions of the technology to the production of liquid iron by applying plasma energy and by incorporating generation of the needed electric energy for smelting within the process are also described. The recent growth of direct reduction as an established technology is reviewed and prospects for the future are discussed.

Direct-reduction (DR) processes produce metallic iron ores by removing the associated oxygen at temperatures below the melting temperature of any of the materials in the process. Historically, DR processes are represented by the early methods for making iron before the blast furnace was developed. These early processes could not be carried out on a large scale and were eventually supplanted by higher capacity indirect processes based on the blast furnace. Through the years since the advent of the blast furnace, the increase in basic knowledge about the thermodynamics and kinetics of blast-furnace reactions, combined with the development and application of process-engineering technology relating to the dynamics of heat, mass, and momentum transport, has contributed to remarkable improvements in blast-furnace efficiency and productivity. Because the same principles underlie the DR processes, this increase in basic knowledge and technology accounts, in part, for their successful concurrent development during the past three decades as an alternative or supplement to blast-furnace ironmaking. Products from the DR processes, now generally referred to as direct-reduced iron (DRI), may be refined to steel in any conventional steelmaking process or used to increase the productivity of blast-furnace or other smelting processes with decreased coke consumption. In the most recent adaptations, still under development, DR technology has been extended to the production of liquid iron by applying plasma energy and incorporating generation of the needed electrical energy for smelting. These adaptations offer considerable promise for simplifying operations and easing quality requirements for iron oxide feeds, fuels, and reductants.

Attempts to develop large-scale DR processes have embraced practically every known type of apparatus suitable for contacting the reactants - pot furnaces, reverberatory furnaces, regenerative furnaces, shaft furnaces, shaft retorts, rotary-hearth furnaces, rotary and stationary kilns, traveling grates, fluidized beds, and various combinations (1). Out of the many projects and the intensive activity, the moving-bed shafts, fixed-bed retorts, fluidized beds, and rotary kilns have emerged as the only successful commercial operations - each with its own set of special characteristics, requirements, and advantages for a particular situation. Many different kinds of reducing agents have also been tried including coal, coke, char, fuel oil, tar, natural gas, producer gas, coal gas, water gas, carbon monoxide, and hydrogen. Ultimately, the actual reduction reactions in all DR processes involve carbon monoxide and hydrogen, and the reductant and energy requirements for all the successful commercial operations are based on natural gas, coal, or petroleum derivates.

CHEMICAL, PHYSICAL, AND PROCESS CONCEPTS

The reduction of iron ore in any DR process is accomplished by the same reactions that occur in the blast-furnace stack, including reduction by CO and H_2 through successive oxidation states to metallic iron.

$$Fe_2O_3 \rightarrow Fe_3O_4 \rightarrow FeO \rightarrow Fe^0$$

Where the reduction reactions take place below about 1000C, the reducing agents are restricted to CO and H_2 and the DRI produced will be porous and have essentially the same size and shape as the original iron-ore particle or agglomerate. Above 1000C, the reduced material begins to sinter, an important consideration in defining operating limits because of attendant difficulties in material movement. At about 1200C, a pasty porous mass forms. This latter temperature represents an upper limit for the DR processes because metallic iron formed above 1200C will absorb any carbon present, with resultant fusing or melting of the solid even though the melting temperature of pure iron is 1530C. A lower temperature limit for final reduction in any DR process may be established by considering that when iron oxide is reduced below about 650C, it is usually pyrophoric when cooled to ambient temperature, which presents difficult and costly handling and storage problems.

At this point it will be useful to consider in a preliminary way the essential features of the four commercial-process concepts - fluidized beds, (2, 3, 4); moving-bed shafts (5,6,7); fixed-bed reactors (8); and rotary kilns (9, 10, 11, 12). Figure 1 presents schematic flowsheets of these process concepts. Effective performance of these processes depends on sound process design and engineering to ensure that good contact is obtained between reacting gases and solids. Ultimately, their performance depends on how closely they can approach the theoretical energy requirements for the reaction system at throughputs high enough to provide economical plant productivity.

The fluidized beds, in which fine solids are main-

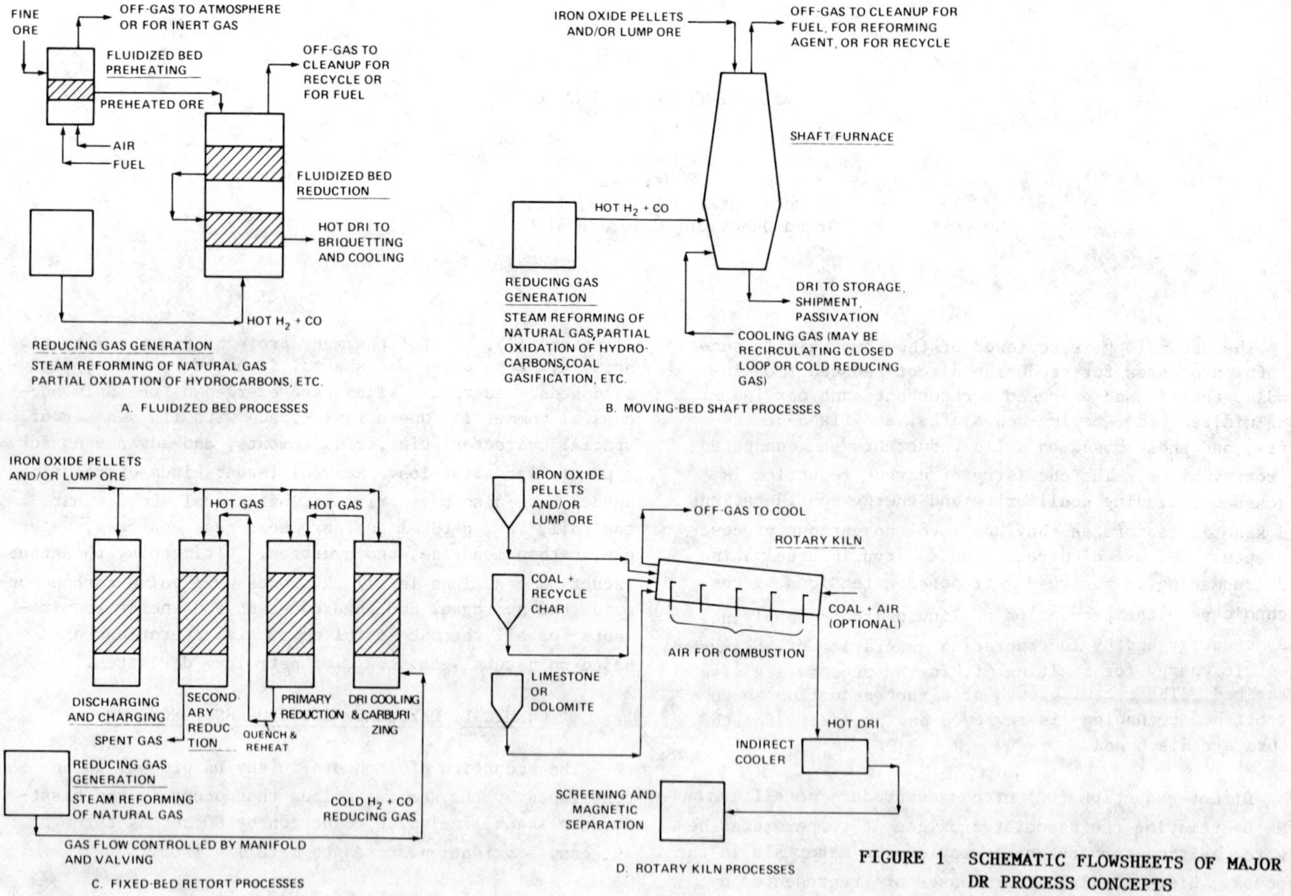

FIGURE 1 SCHEMATIC FLOWSHEETS OF MAJOR DR PROCESS CONCEPTS

tained in a state of "active" suspension in an uprising gas stream, provide very high heat-transfer rates and excellent gas/solid contact and offer a superior way to reduce some naturally fine iron ores. The energy requirements for fluidized-bed processes may be estimated very closely from material and energy balances based on the operating conditions, and some limited knowledge of process kinetics and equilibrium requirements, covered in detail later.

The advent of pelletizing stimulated subsequent development work on moving-bed shafts. They have the potential to maximize energy efficiency because they are truly countercurrent and can also be designed to recycle unused reductants in the off-gas.

In the processes based on fixed-bed reactors, the efficient countercurrent contact afforded by the blast furnace stack is approached ingeniously by arranging several reactors in series so that preheating, reduction, and cooling occur concurrently with continuous flow of the same process gas stream. In this way sensible energy in the hot reduced product is transferred to the reducing gas before it reaches the reduction stage, and sensible energy in the hot spent reducing gas is transferred to incoming ore before the spent gas is exhausted.

In the rotary kilns, the ore, coal, recycle char, and limestone and/or dolomite for sulfur control are charged together at the feed end and heated to reaction temperature by burning volatiles and CO generated during reduction with controlled amounts of air introduced at appropriate locations throughout the kiln length. Kilns are attractive because they represent a simple and effective way to contact ores and agglomerates with solid reductants at temperatures high enough to provide relatively rapid reduction of FeO with concomitant gasification of fixed carbon. Because the heat-transfer rates and reactant-contact effectiveness are not as good as for the other processes, energy consumption per unit of DRI produced is generally higher.

With this very brief description of the four major commercial process concepts in mind, a more meaningful discussion of their relevant equilibrium, energy, and process-kinetic factors is possible.

Equilibrium Considerations

The equilibrium for the reactions of CO and H_2 with the oxides of iron is well established (13) and is shown in Figures 2 and 3 for the systems Fe-O-C and Fe-O-H, respectively, over the temperature range of interest in DR processes, 500 to 1200C. Hematite, Fe_2O_3, is not shown because equilibrium mole fractions of CO_2 and H_2O on an inerts-free basis for the reactions

$$3Fe_2O_3 + \left\{ \begin{matrix} H_2 \\ CO \end{matrix} \right\} = 2Fe_3O_4 + \left\{ \begin{matrix} H_2O \\ CO_2 \end{matrix} \right\}$$

are very close to unity (nearly complete oxidation of CO and H_2). In other words, all DR processes will provide sufficient CO and H_2 in contact with Fe_2O_3 to effect complete reduction to Fe_3O_4; the reduction of Fe_2O_3 to Fe_3O_4 is not a limiting factor in establishing the total reductant required. Below 570C, Fe_3O_4 is reduced directly to Fe^0 by CO and H_2 so that the single-pass conversion for a process operating within this range would be limited by the equilibrium ratio in contact with Fe^0, a relatively low conversion because there is essentially no staging.

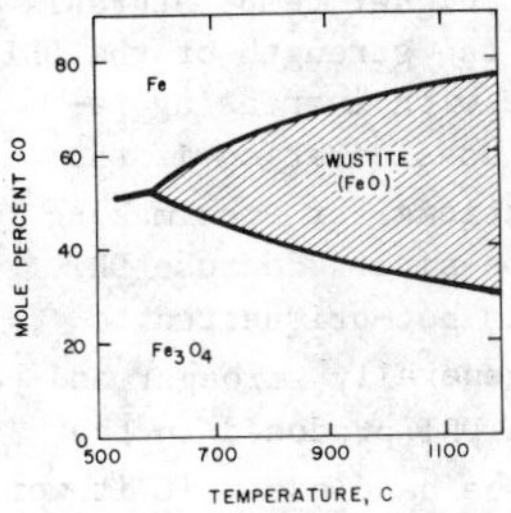

FIGURE 2 EQUILIBRIUM FOR IRON - OXYGEN - CARBON SYSTEM

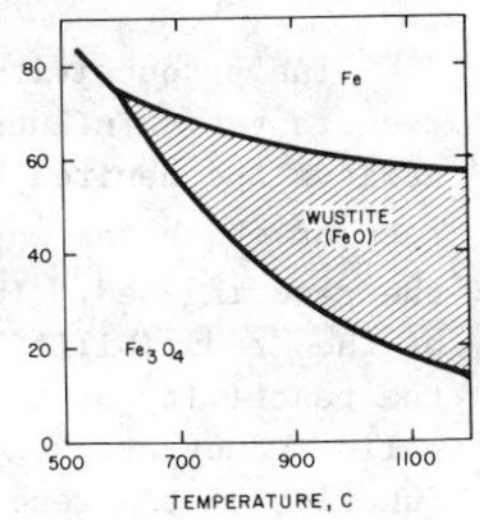

FIGURE 3 EQUILIBRIUM FOR IRON - OXYGEN - HYDROGEN SYSTEM

Above 570C, Fe_3O_4 is first reduced to FeO, which is then reduced to Fe^O so that the single-pass conversion for a process operating within this range would not be limited by the equilibrium ratio in contact with Fe^O, and much higher per-pass conversions could be obtained. For reduction of Fe_3O_4 to FeO, the equilibrium conversions of CO and H_2 increase with increasing reaction temperature; however, for the reduction of FeO to Fe^O, the equilibrium conversion for H_2 increases with increasing reaction temperature, whereas that for CO decreases. This decrease of equilibrium CO conversion with increasing temperature for reduction of FeO to Fe^O is not a limitation on the overall conversion because most DR processes operate with countercurrent flow of solids and reducing gases. Because the spent reducing gas leaves in contact with entering solids in their highest oxidation state, equilibrium for the reduction of Fe_3O_4 to FeO would represent the sole limitation on conversion or final-gas composition. The exit-gas composition for DR processes based on reduction with mixtures of CO and H_2 will usually satisfy the equilibrium for the water-gas reaction at the exit temperature, that is,

$$CO_2 + H_2 = CO + H_2O$$

$$K_{EQUILIBRIUM} = \frac{[CO]}{[CO_2]} \frac{[H_2O]}{[H_2]}$$

Energy and Process Considerations

In the processes based on direct use of solid reductants such as coal, usually in rotary kilns, the reduction is accomplished first by volatiles and reducing gases released as the coal is heated and then by CO formed during reduction by gasification of char with CO_2 produced during reduction by CO at the higher temperatures. Some CO may also be formed by direct contact of residual oxide with char. The endothermic heat of reduction and sensible energy required to heat the reactants in these processes are provided by combustion of volatiles and CO leaving the bed with air introduced in the free space above the bed. In these processes the highest utilization of the energy contained in the solid reductant would be obtained when all combustible gases leaving the reduction bed are burned to CO_2 and H_2O in the free space. Thus, equilibrium considerations do not govern the performance of the processes based on the direct use of solid reductants. Rather, the extent of combustion in the free space and the effectiveness of the transfer to the kiln bed of the energy released influence, if not control, the energy requirements and the productivity per unit of reactor volume. For the DR processes based on gaseous reductants in fluidized beds, shafts, and fixed-bed reactors, this heat transfer is very efficient as long as the gas-flow distribution and gas/solid contact are maintained uniform by good process engineering design and operation.

The energy requirements for DR processes are best discussed by referring to the heats of reactions for the reduction reactions over the temperature range of practical interest for DR processes. Table I presents a summary of heats of reaction for reduction of Fe_2O_3 and FeO by CO, H_2, and carbon. (14)

The reduction of Fe_2O_3 to Fe_3O_4 with H_2 and CO is moderately exothermic, and the reduction of FeO to Fe^O with CO is also moderately exothermic. All the other reactions are moderately endothermic, except the reduction of Fe_3O_4 to FeO and of FeO to Fe by carbon, both highly endothermic. These data show that the heat of reduction is not a major factor in establishing the total energy requirements for the gas-based DR processes. Instead, the energy requirements for these processes are comprised mainly of the energy used to generate the reducing gas (in most cases, catalytic steam reforming of natural gas), the energy required to heat the reactants to the reaction temperature, and the chemical energy in the reductants consumed in the process. For the DR processes that are conducted in moving-bed shafts and fixed-bed reactors, total energy requirements are minimized by recovering most of the sensible heat in the DRI and in the spent reducing gases. Such recovery is not practical in the fluidized-bed processes and accounts for the lower total energy requirements possible with the moving-bed shafts and fixed-bed reactors - 12.0 million kJ per metric ton compared with about 18.6 million kJ per metric ton for DR in fluidized beds.

TABLE I

HEATS OF REACTION OF REDUCTION REACTIONS IN DR PROCESSES

kJ/kg mole Fe*, ***

Temperature			$Fe_2O_3 \rightarrow Fe_3O_4$**			$Fe_3O_4 \rightarrow FeO$**			$FeO \rightarrow Fe$**		
°K	°C	°F	H_2	CO	C	H_2	CO	C	H_2	CO	C
900	627	1161	-733	-6722	21,888	17,573	5606	62,814	16,292	-19,597	152,027
1000	727	1341	-2105	-7920	20,539	17,154	5513	62,430	16,282	-18,631	152,109
1100	827	1521	-2628	-8269	20,027	16,736	5443	62,023	17,119	-16,747	152,981
1200	927	1701	-2791	-8269	19,841	16,608	5652	61,883	17,166	-15,101	152,981
1300	1027	1881	-2803	-8129	19,818	16,375	5722	61,604	16,038	-15,910	151,725

*Multiply by 0.4299 to convert to Btu/lb. mole Fe.

**All heats of reaction are based on stoichiometric equation for one pound atom Fe in reactant oxide.

***Negative values are exothermic; positive values are endothermic.

For the DR processes based on the direct use of solid reductants, the energy for the high-temperature reduction of wüstite by the reactions

$$FeO + \left\{ \begin{matrix} H_2 \\ CO \end{matrix} \right\} = Fe + \begin{matrix} H_2O \\ CO_2 \end{matrix}$$

$$\left\{ \begin{matrix} H_2O \\ CO_2 \end{matrix} \right\} + C = \left\{ \begin{matrix} H_2 + CO \\ 2CO \end{matrix} \right\}$$

is provided by the combustion reactions in the free space above the reacting bed. There is no recovery of sensible energy in the DRI or in the kiln off-gas, and the total energy requirement for the processes based on the direct use of solid reductants is also about 18.6 million kJ per metric ton.

The energy and material balances for the DR processes are established as functions of the chemical equilibrium, stoichiometry, and thermodynamic energy requirements for the reactions involved. The productivity of the DR processes depends on process kinetics, which integrate these energy and material balances with heat, mass, and momentum transport factors that combine to establish the overall rate and extent of reduction of the ore charged. Process kinetics comprise essentially two elements - preheating the reactants to the required reaction temperatures, $\Sigma_i M_i \overline{Cp}_i \Delta t$, and maintaining the reactants in effective contact so the reduction reactions can proceed to the desired extent. The rate and extent of reduction are complex functions of the temperature, pressure, gas composition, feed rate, feed properties, gas rate, and factors that influence the contact between the solid and gaseous reactants.

In fluidized-bed processes, the solids are preheated in a separate fluidized bed by in-situ combustion of a suitable fuel with air. The reducing gases are preheated in an indirect-fired heat exchanger. In this way, the endothermic heat of reduction is introduced as sensible heat in the entering solids and gases. The rate of the reduction reactions will then be a function of the temperature and pressure in the reduction beds, the porosity and size distribution of the ore, the composition of the reducing gas, the ore/gas ratio, the depth of the reduction beds, and the effectiveness of gas/solid contact. The reduction rate generally increases with increasing temperature and pressure up to about 500 kPa. When the reduction is done below about 650C, the DRI will be pyrophoric. Above about 800C, the beds tends to defluidize because of sticking or bogging of the reduced material. The fluidized-bed processes are therefore operated between these temperature limits.

In the moving-bed shafts and fixed-bed reactors, the solids are preheated by countercurrent contact with partially spent reducing gas, and the reducing gas is preheated either during generation (when reforming natural gas with stoichiometric steam) or by an indirect-fired heat exchanger (when reforming natural gas with excess steam). In shafts and reactors designed to provide uniform descent of the solids (in the case of the shafts) and uniform distribution of the reducing gases (in both cases), the rate and extent of reduction will generally increase with increasing reducing-gas temperature. The upper limit on temperature is determined by the resistance of the solids to sticking and clustering, which would interfere with effective solids movement and/or discharge. This upper temperature limit is about 1000C for the iron oxide feeds most resistant to sticking. Some materials cannot be reduced above about 800C because they tend to stick at higher temperatures. The resistance to reoxidation and the strength of the DRI from these processes also increase with increasing reduction temperature, an important consideration if the DRI is to be shipped to other locations for steelmaking or stored in the open before use on-site. Because DRI from the fluidized-bed processes is hot-briquetted to over 5.0 specific gravity, it is generally stronger and more resistant to reoxidation than DRI produced in the other gas-based processes and may be handled as if it were an inert material.

The productivity of the DR processes based on solid reductants is influenced by the temperature required to achieve the desired high degree of reduction and by the rate at which the energy required can be transferred to the reacting bed. The reduction temperature is a function of the "reducibility" of the iron ore or agglomerate and the reactivity of the char formed from the coal or other solid reductant. The rate of heat transfer will be a function of the reacting-bed temperature and the ash-softening temperature of the solid reductant. This is so because the ash-softening temperature will govern the maximum flame temperature (kiln-lining temperature) that can be used without formation of rings in the kiln that would impair the operability of the process. Thus, effective operation of these solid-based DR processes depends on careful selection of the iron oxides and coals used, as well as on good temperature control in the reduction zone of the kiln. Suitable coals will generally have ash-softening temperatures at least 100C higher than the reduction temperatures required for good productivity with high metallization, be relatively noncoking, and contain less than 1.5 percent sulfur to enable production of DRI with sulfur concentrations low enough to be acceptable for use in steelmaking. Coals that are relatively low in volatile matter may also be preferred in countercurrent kilns because energy losses due to incomplete combustion of volatiles released close to the feed end of of the kiln will be minimized. The processes based on the direct use of solid reductants usually operate with reduction temperatures greater than 1000C so that the DRI produced is also relatively strong and resistant to reoxidation.

The foregoing general discussion provides a proper foundation for more detailed descriptions of the specific commercial processes that represent the most highly developed examples of their type.

Fluidized-Bed DR Processes

The fluidized-bed DR processes were first implemented commercially in 1971 when U.S. Steel Corporation's High-Iron-Briquette (HIB) plant was commissioned in Puerto Ordaz, Venezuela. Figure 4 shows a flowsheet of this plant, which was designed to produce 1 million metric tons per year of 75 percent reduced briquettes*, primarily for blast-furnace feed.

The HIB plant comprises four distinct processing sections - ore preparation, ore preheating and reduction, briquetting, and reducing-gas production.

In the ore-preparation section, raw ore is dried in a gas-fired rotary drier and then processed to minus 10-mesh ore feed in a closed-circuit crushing and screening plant. To provide good fluidization without excessive

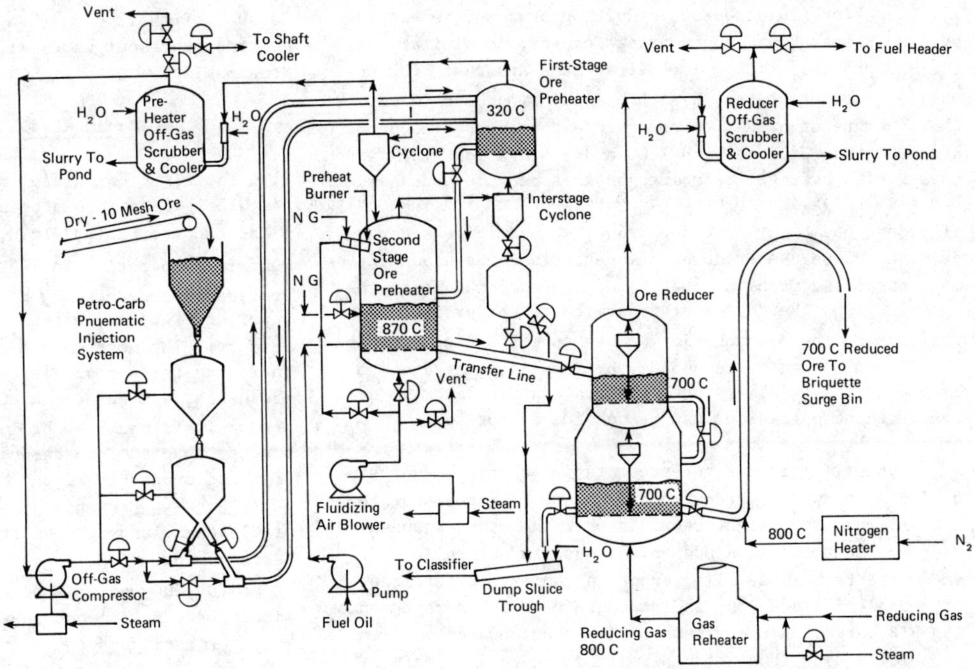

FIGURE 4 FLOWSHEET OF HIB PLANT

dust losses, the dry ore feed should contain a maximum of about 20 percent minus 200-mesh material.

$$\text{*Percent Reduction} = \frac{(O/Fe_T \text{ in ore} - O/Fe_T \text{ in DRI})}{O/Fe_T \text{ in ore}} \times 100$$

and

$$O/Fe_T \text{ in ore} = 48/111.7 = 0.4297$$

The dry minus 10-mesh ore is fed pneumatically into the first bed of a two-stage fluidized-bed preheater where it is heated continuously to about 320C by sensible heat exchange with hot gas from the second-stage preheater. The ore leaving the first-stage preheater is heated to about 870C in the second-stage preheater by direct combustion of natural gas with air in this bed.

Preheated ore is fed continuously to the first-stage of a two-stage fluidized-bed ore reducer where Fe_2O_3 is reduced to FeO (about 30 percent reduction) at about 700C by partially spent gas leaving the final reduction stage. The reducers have an inside diameter of about 7 meters. The partially reduced ore from this reduction stage is further reduced to about 75 percent in the final reduction stage at about 700C by contact with hot reducing gas from the gas-generation section. Gas leaving the first reduction stage is scrubbed and enters a fuel-gas manifold. Hot reduced product is conveyed pneumatically to a surge bin directly above the briquetting presses by inert gas produced by cooling, cleaning, and compressing the off-gas leaving the first-stage ore preheater.

The hot reduced ore discharges from the surge bin into screw feeders above the briquetting-press rolls. The screw feeders convey, precompact, and force-feed the hot material between briquetting rolls, which exert compressive forces transmitted by a hydraulic system and form the material into strips that discharge into a strip breaker located below the rolls. In the most recent modifications to the briquetting section of the HIB plant (15), the briquettes and leakage fines pass into a trommel screen. The briquettes are separated and conveyed by a vibrating conveyor directly to the shaft cooler; fines and undersize material discharge into a bucket elevator and are recycled to the surge bin. Briquettes flowing down through the shaft cooler are cooled by inert gas from the ore preheaters and are discharged onto conveyor belts that carry them to storage. Off-gas from the shaft cooler passes through a dry-dust collector, which returns dust to the surge bin, and the gas then goes to a wet scrubber. Part of the cooled and scrubbed gas is vented and the remainder is recompressed and recycled.

In the gas-generation section, natural gas is mixed with preheated steam at 400C; the mixture is further preheated to about 540C in heat-exchange coils in the refomer convection section and then passes through alloy tubes filled with catalyst and direct-fired so that the outlet temperature is maintained at about 850C. The gases leaving the reformer pass through a series of coolers, a knockout drum, and a gas chiller to remove water. The cool, dried gas containing about 72 percent hydrogen is preheated in a heat exchanger to 455C, and then enters the gas reheater where it is further preheated in direct-fired alloy tubes to about 800C before introduction to the final reduction stage.

The HIB plant at Puerto Ordaz includes three identical trains comprising ore preheating and reduction, reducing gas generation, and product-briquetting systems. A single ore-drying and closed-circuit-crushing system prepares the ore feed for the three reduction trains. The plant was designed for a nominal production capacity of 1 million metric tons per year - about 1000 metric tons per day for each train with a total operating time of 330 days per year expected for each train. The actual results since start-up in late 1971, (2, 15, 16) have fallen far short of the design goal, primarily because of mechanical and process problems that decreased the availability and productivity of the plant. The experience and knowledge obtained through the years have provided the basis for equipment modifications and operating and maintenance procedure modifications that are expected to result in substantial improvements (15, 16). By the end of 1980, campaign durations exceeding about eight months had been demonstrated and additional modifications to internal cyclones in the final stage reducer were expected to provide campaigns lasting at least a year. When these modified internal cyclones are installed in all the trains, the HIB plant is expected to sustain production of a least 600,000 metric tons per year.

The Fior fluidized-bed reduction process, developed by Esso Research and Engineering Company, is similar to the HIB process. The major differences are that Fior uses only one ore preheater, has three reduction beds in series that operate at higher temperature and pressure (about 800C and 1000 kPa compared with 700C and 300 kPa for HIB), and uses a high-hydrogen reducing gas in a recycle arrangement. The Fior plant, which started operations in 1976, was originally designed to produce about 400,000 metric tons per year of 90 to 92-percent-reduced briquettes and has experienced many operating problems that have limited both plant productivity and the degree of reduction of the product. Experience at the HIB and Fior plants suggests that although the fluidized-bed DR technology has been demonstrated to be technically viable, further refinements will be needed before these operations can be considered optimum.

Table II presents a representative chemical analysis of fresh briquettes from the HIB plant. In open storage HIB will undergo slight reoxidation with a corresponding decrease in total iron and an increase in oxygen associated with Fe. Table III presents a summary of pertinent size-analysis, density, and strength data for a representative HIB cargo. A comparison of HIB chemical and physical properties with those of other direct-reduced materials shows that HIB (and Fior briquettes) are a strong, dense, and stable form of iron units. These characteristics along with the availability of nearly two-thirds of the contained iron in the metallic form make the HIB an excellent blast-furnace burden material and a suitable source of iron units for other iron and steelmaking processes.

TABLE II

REPRESENTATIVE CHEMICAL ANALYSIS OF HIB

	Percent by Weight
Fe^{0} (Metallic Iron)	53.4
Fe Total	85.5
Fe^{++}	32.1
O (Associated with Fe)	9.2
SiO_2	1.8
Al_2O_3	2.1
CaO	0.022
MgO	0.017
P_2O_5	0.34
S	0.020
C	0.19
Percent Metallization	∿ 62.5
Percent Reduction	∿ 75.0

TABLE III

PHYSICAL PROPERTIES OF REPRESENTATIVE HIB CARGO

Size Analyses, Cumulative Percent Retained

Screen Size	
1-1/2 Inch	0
1 Inch	86.3
1/2 Inch	94.0
1/4 Inch	96.8
20 Inch	98.9
Average Density, grams/cc	5.1
ASTM Tumble Index % + 3 mesh after 200 revolutions	97.0
Bulk Density, kg/m^3	∿ 3000.0

Because of their higher degree of reduction, Fior briquettes represent an excellent source of iron units for steelmaking.

Moving-Bed Shaft DR Processes

The first commercial moving-bed shaft DR process was the Wiberg-Söderfors process in Sweden (17). The largest plant had a capacity of 20,000 metric tons per year and went into operation in 1952. The reducing gas for this process is generated by carburetion of coke with water vapor and recycled off-gas in an electrically heated carburetor. The most modern moving-bed shaft DR processes, including the Armco, Midrex, Purofer, and HyL III (18) processes, are based on reformed hydrocarbons and have productive capacities at least an order of magnitude greater than the Wiberg-Söderfors. The most widely used of these processes is the Midrex process developed by the Surface Combustion Division of Midland Ross Corporation of Cleveland and subsequently sold to Korf Industries of West Germany. The largest Midrex plant has a capacity of about 700,000 metric tons per year. The detailed discussion of the moving-bed shaft reduction processes will be based primarily on the Midrex process.

The first commercial Midrex plant was built adjacent to the Oregon Steel Mills plant in Portland, Oregon in 1969 and had a 3.7-meter I.D. shaft. The largest Midrex plant has a 5.5-meter I.D. shaft and is located at Sidbec-Dosco's Contrecour Works in Quebec, Canada. Figure 5 shows a schematic flowsheet of the Midrex process.

The moving-bed shaft DR processes comprise essentially two distinct processing sections - reducing-gas generation and ore reduction. The iron-bearing raw materials can be fired pellets and/or lump ore, screened just prior to introduction to the top of the shaft reducer to minimize the fines and attendant problems with gas-flow distribution and solids movement. Excessive fines in the shaft, either in the feed or from decrepitation during reduction, will usually result in poor gas-flow distribution and erratic solids descent in the shaft reducer. The shaft furnace in the Midrex process comprises an upper reduction zone and a lower cooling zone. The hot reducing gases enter the shaft near the bottom of the reduction zone, flow through the descending solids, and leave at the top of the shaft. The solids are charged to the shaft by means of feed pipes at the top that distribute them uniformly across the top. The top of the furnace is sealed by keeping the feed pipes full and bleeding inert gas to prevent leakage of off-gas and infiltration of air. The solids leaving the reduction zone are cooled by contact with cooling gas that enters near the bottom of the conical section and leaves at the top of the cooling section. This gas is cleaned and cooled in high-efficiency scrubbers and packed cooling towers before it is recycled by means of a blower.

The spent reducing gas leaving the shaft is cleaned and cooled in high-efficiency scrubbers and packed cooling towers, and then used as a fuel and reforming gas in the natural-gas reformer. The Midrex reformer is a catalytic system wherein the natural gas is reformed with a stoichiometric quantity of the recycled off-gas (the mole ratio of the CO_2 and H_2O in the off-gas to the carbon in the hydrocarbon portion of the natural gas is close to unity). The reformer is equipped with a convection section for preheating the feed-gas mixture. A portion

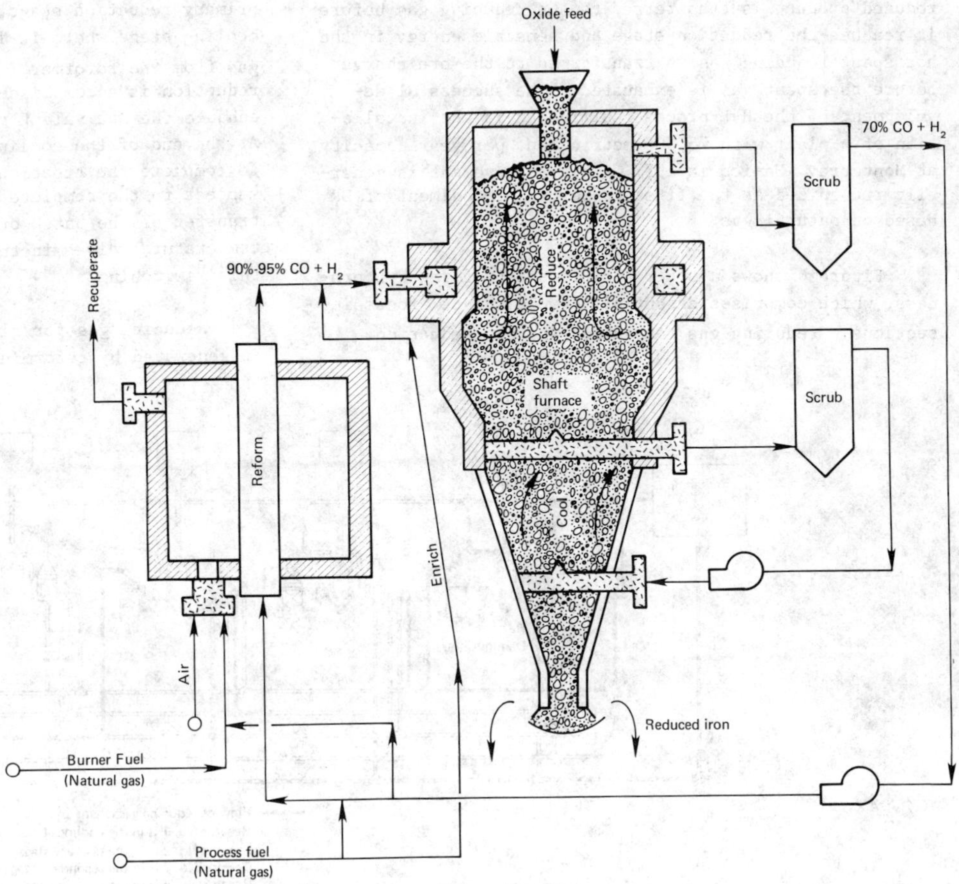

FIGURE 5 SCHEMATIC FLOWSHEET OF THE MIDREX PROCESS

of the reformed gas may be cooled to adjust the temperature of the reducing gas entering the shaft.

In the Armco moving-bed shaft DR process, the reduced product is cooled by cleaned and cooled off-gas from the shaft that mixes with the hot reducing gas entering the reduction zone. In this way the Armco process achieves maximum utilization of the reducing components of the reformed natural gas for reduction. The reducing gas for the Armco process is generated by reforming natural gas with close to the stoichiometric mole ratio of steam to carbon in the natural gas. The reducing gas leaving the reformer is therefore used directly. The temperature of the reducing gas entering the reduction zone is moderated by mixing a portion of the cooled off-gas with the reformer effluent.

In the Purofer moving-bed shaft DR process, the reduced product is discharged from the shaft hot and either charged directly into electric-arc steelmaking furnaces or cooled batchwise in sealed containers. Reducing gas for the Purofer process is also generated by stoichiometric reforming of natural gas; however, in this case the reforming is done at a higher temperature in regenerative reformers that are noncatalytic (thermal reforming).

Although the Armco and Purofer moving-bed shaft DR processes are in commercial operation, they have not achieved the general acceptance that the Midrex process has. Armco has one commercial plant in Houston; Purofer has a plant in Brazil based on partial oxidation of heavy fuel oil, and one in Iran based on reformed natural gas. In contrast, Midrex had installed 21 plants by the end of 1980 with a total capacity of about 8 million metric tons per year.

All the moving-bed shaft DR processes achieve optimum performance from the standpoint of productivity and energy consumption when the charge material comprises special fired pellets and possibly some lump ore that are resistant to degradation and sticking or clustering during reduction. When such feeds are used, higher inlet reducing-gas temperatures can be used (up to about 950C), with an attendant decrease in total energy consumption per unit of product and increase in throughput per unit reactor area. Another advantage of the higher temperature operation is that the DRI tends to be more resistant to reoxidation and may be shipped and stored in covered systems without special treatments (passivation) to make them resistant to reoxidation.

Fixed-Bed Reactor DR Processes

The fixed-bed reactor processes are best exemplified by the HyL process developed by Hojalata y Lamina SA (HyLSA) in Mexico in the early 1950's. The other fixed-bed process, the Höganas process, developed in Sweden in the early 1900's, is based on solid reductants and is a relatively low-capacity process used primarily to produce iron powder for powder-metallurgy applications.

In the HyL process, the efficient countercurrent contact afforded by the blast-furnace stack is approached by arranging four fixed-bed reactors in series so that preheating, reduction, and cooling take place concurrently with continuous flow of the same process-gas stream through the reactors. In this way, sensible energy in the hot reduced product is transferred to the reducing gas before it reaches the reduction stage and sensible energy in the hot spent reducing gas is transferred to the ore charge before the spent gas is exhausted. The successful development of the HyL process culminated in the installation of a plant with 75,000 metric tons per year capacity at Monterrey, Mexico in 1957. This achievement is generally recognized as the first commercial embodiment of DR based on natural gas.

Figure 6 shows a schematic flowsheet of the HyL process, which comprises essentially two distinct processing sections - reducing-gas generation and ore reduction.

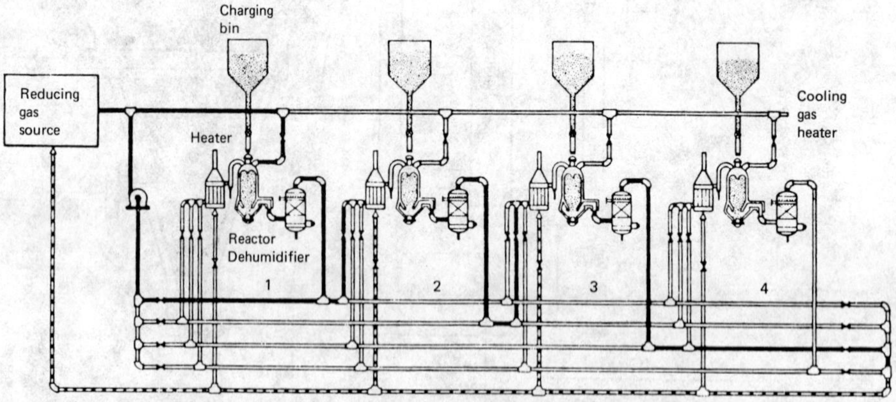

FIGURE 6 SCHEMATIC FLOWSHEET OF THE HyL PROCESS

As in the moving-bed shaft processes, the iron-bearing raw materials can be fired pellets and/or lump ore screened just prior to introduction to the first reactor to minimize the fines and attendant problems with gas-flow distribution and discharge of the charge after reduction and cooling. Fresh feed is charged to a reactor that has just been emptied. This discharge/charging stage, during which the reactor is not connected to the gas-distribution manifold, comprises about three hours of the total treatment cycle. During this time the other reactors are undergoing preheating and prereduction, primary reduction, and final cooling respectively. Each reactor passes through these stages successively so that the total cycle for each charge is about 12 hours. At the end of a three-hour period the reactors are switched to the next stage of the cycle by means of gas-flow control valves in the gas-distribution manifold. The fresh-feed charge is then contacted by hot spent reducing gas from the reactor currently undergoing primary or final reduction. The gas leaving the primary reduction stage is first cooled to condense water vapor formed during reduction, then heated to about 850C in a fired heater, and finally heated to about 1050C by partial combustion with air in a chamber at the reactor inlet. At the completion of this so-called secondary reduction stage, this reactor is switched to the primary reduction stage where it is contacted with strong reducing gas from the reactor undergoing cooling. This gas is also cooled to condense water vapor formed by reduction of residual oxide during the final cooling and then heated by indirect exchange and partial combustion in the same way that the gas leaving this stage is treated prior to introduction to the secondary reduction stage. At the completion of this primary reduction stage, the reactor is switched to the cooling stage where it is contacted directly by quenched gas from the reformer. During the cooling, additional reduction is accomplished and iron carbide is formed to enhance the equivalent reduction of the product (19).* At the end of the cooling stage, this reactor is discharged to complete the treatment for this batch. Each charge is subject to the complete cycle in the same reactor without transfer of the batch or moving of the reactor; the high-temperature valves in the reaction gas manifold accomplish this "switching".

Reducing gas for all the operating HyL-process plants is generated by reforming natural gas with excess steam in a conventional catalytic reformer. Other hydrocarbon feedstocks can also be reformed to generate a suitable reducing gas. Coke-oven gas and gas produced by partial oxidation of coal or fuel oil are also suitable reductants (20).

In the most recent improvements to the HyL process, the reactors are arranged in a compact cluster in a common tower with a single feed hopper, the gas heaters are constructed with special alloy tubes that permit indirect preheating to 950C and minimize the partial combustion needed for providing the final temperature, and the preheaters are modified so that only two are provided for the four reactors (20).

Acceptance of the HyL process has been comparable with that of the Midrex process for gas-based DR. By the end of 1980 15 HyL plants had been installed with a total capacity of about 10 million metric tons per year.

Rotary-Kiln DR Processes

The development of rotary-kiln DR processes based on the direct use of noncoking coals started in the 1920's. In the rotary kilns the ore, coal, and any recycle char are charged together at the feed end and heated to reaction

---

*Equivalent reduction is the reduction that would be obtained if the carbon in the DRI reacted with oxygen associated with iron in the DRI to minimize Fe^{+++} and Fe^{++} and maximize Fe^{0}, according to the general reaction

$$[C] \text{ in DRI} + Fe_xO_y \rightarrow CO + Fe_xO_{y-1} + Fe^0$$

temperature by burning volatiles and CO generated during reduction with controlled amounts of air introduced at appropriate locations along the kiln. Energy may also be introduced by burning pulverized coal or other fuels at the discharge end of the kiln. Kilns are attractive because they represent a simple and effective way to contact ores with solid reductants at temperatures high enough to provide rapid reduction of FeO with concomitant gasification of fixed carbon. Successful commercial operations have been limited to date to processing lower grade ferrous materials or to low-volatile coals in special situations. Examples are the operations on vanadiferous ores at Highveld Steel and Vanadium Company in South Africa, started in 1968 (21), and on titaniferous iron sands at New Zealand Steel Company, commissioned in 1970 (22). The Highveld kiln operates with cocurrent flow of solids and gases. The New Zealand Steel operation is an adaptation of the SL/RN process, which was developed cooperatively in the United States and Canada (22). Also, a plant at Dunswart Iron and Steel Works in South Africa is based on the Krupp process and uses a high-grade native ore and nonvolatile anthracite coal (23); this plant started operations in 1973.

The Krupp CODIR (for coal ore direct iron reduction) process, shown schematically in Figure 7, is representative

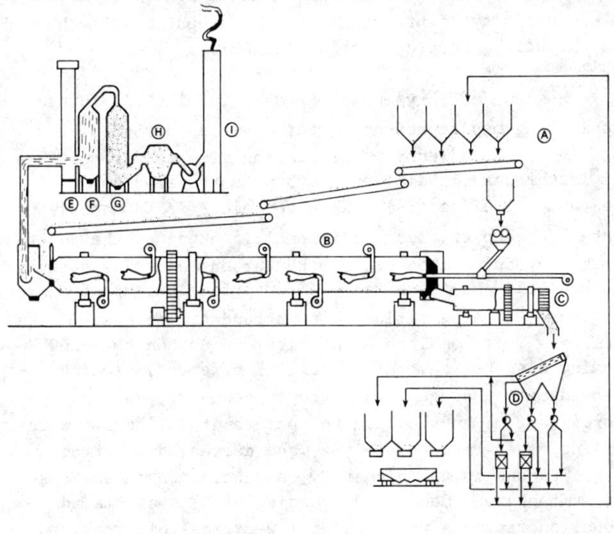

1. Lump ore/pellets	A. Day bins (24 hr cap.)
2. Recycled char	B. Rotary kiln
3. Desulphurizer	C. Cooler
4. Cool	D. Discharge separation
5. Pulverized cool	E. Emergency stack
6. Air	F. Afterburning chamber
7. DRI	G. Conditioning tower
8. DRI	H. Electrostatic precipitator
9. Waste material	I. Clean gas stack

FIGURE 7 SCHEMATIC FLOWSHEET OF THE KRUPP CODIR PROCESS

of the kiln processes and is the only one of them based on high-grade ores reported to have achieved sustained successful operation (23). In the Krupp plant in South Africa the kiln is 73.5 meters long and 4.6 meters I.D. Six fans mounted on the shell blow air into the kiln through alloy tubes that introduce the air along the center line of the kiln counter to the solids flow. The ore and a desulfurizing agent such as dolomite are fed into the kiln with fresh coal and recycle char (optional). The ore is 25x5 mm and the coal and desulfurizing agent are minus 10 mm and minus 3 mm respectively to facilitate subsequent separation operations. The kiln is heated by combustion of volatiles and CO during reduction with air introduced through the shell blowers and by a pulverized coal burner at the discharge end. The ratio of burner coal to feed coal depends on the volatile contents of the respective fuels. In general, a low-volatile coal is preferred for the feed coal because volatiles are released near the feed end of the kiln and are only partially combusted. The energy contained is not recovered because the capital cost of waste-heat burners for energy recovery has not been justified. The distribution of air introduced through the shell blowers is adjusted to maintain the last 50 to 60 percent of the kiln length in the range 950 to 1050C; most of the reduction takes place in this zone.

The product is discharged at a temperature close to the high temperature in the kiln and is cooled to about 150C by direct and indirect water sprays in a rotary cooler. The cooled product is screened into four fractions - a plus 5-mm fraction that is essentially all DRI, a minus 5 mm plus 3-mm fraction that is separated magnetically into a magnetic portion that is DRI and a nonmagnetic portion that is mostly char and may be recycled, and a minus 3-mm plus 1-mm fraction that is also separated magnetically into a magnetic portion that is DRI and a nonmagnetic portion that is a mixture of char, coal ash, and desulfurizing agent. The latter is usually rejected with the minus 1-mm material that is also primarily ash.

The kiln off-gas leaves at about 800C and, after being treated in an afterburner to burn any residual combustibles, is passed through a cooling/conditioning tower, a bag filter, and electrostatic precipitator before exhausting to the atmosphere through a stack.

In the Krupp plant in South Africa, the high reduction temperature and the formation of a thin surface oxide layer during quenching result in DRI that is resistant to atmospheric oxidation and has been shipped in ocean-going vessels without problems.

The rotary-kiln processes based on high-grade feeds and high-volatile noncoking coals have still not been demonstrated successfully in commercial operation. The SL/RN process has experienced inordinately high energy consumption and excessive accretion formation. The ACCAR process, developed by Allis-Chalmers Company, uses a kiln modified for supplementary under-bed injection of hydrocarbon gases or liquids (24). ACCAR claims to achieve efficient and reliable operation with relatively high coal-to-hydrocarbon ratios; however, the concept has not been demonstrated on a commercial scale. In the DRC process (Direct Reduction Corporation), the rotary-kiln reduction technology developed by Western Titanium Pty., Ltd. of Australia for producing synthetic rutile from ilmenite has been adapted to iron oxide pellets and ore (25). In the DRC process, which has been demonstrated in a large pilot plant at Rockwood, Tennessee, a portion of the reduction coal is injected into the discharge end of the kiln to release volatiles in the discharge half of the kiln and promote both fuel efficiency and relatively accretion-free operation. DRC operations at Rockwood have reportedly been very promising and the process appears to have attracted considerable worldwide interest.

General Process Discussion

DR processes were originally intended to supplement conventional iron and steelmaking operations in locations with established plants. Later they were intended to provide original facilities for locations lacking coking coals, but with abundant natural gas or noncoking coal and a desire to provide competitive upgraded supplementary iron units for the established plants. Concurrent technological improvements that resulted in a 3 to 4-fold increase in blast-furnace productivity and a 40 to 50-percent decrease in fuel consumption account for the lack of implementation of DR at existing integrated steel mills. The DR processes, with single-line capacities limited to about 700,000 metric tons per year in the case of the gas-based processes and about 250,000 metric tons per year in the case of the kiln processes, are not competitive with modern blast furnaces with capacities of at least 2,000,000 metric tons per year. The ready availability of cheap natural gas in some industrial countries in the early 1950's has also given way to relative scarcity and high cost. These factors account for the fact that almost all recent DR capacity is being installed where natural gas is abundant and cheap - Venezuela, the Middle East, Mexico, and Indonesia. These factors also account for the recent interest in basing shaft DR on coal gasification and in improving the efficiency and operating reliability of the kiln processes that use noncoking coals directly. Up to now, coal gasification requires too much capital investment to be economical for DR, and the low operating reliability and high energy consumption of the kiln processes operating on high-volatile noncoking coals have been a barrier to their acceptance.

The successful development and demonstration of technology for removing sulfur from coal-gasification gas while it is still hot would improve the economics of coal-based DR plants using moving-bed shafts (26, 27). Improvements in the energy efficiency and operating reliability of the kiln processes based on high-volatile noncoking coals will require engineering ingenuity in developing cost-effective waste-heat recovery equipment and control systems as well as operating practices to minimize the accretion-formation problems that have plagued most of these operations. New approaches to applying noncoking coals to DR for ironmaking are being pioneered in Sweden and will be discussed later.

USES OF DRI

The major uses of DRI are as metallic iron units for steelmaking in electric-arc furnaces, open hearths, basic-oxygen furnaces, and induction furnaces. Minor uses are as melting stock in foundry cupolas, as a super agglomerate for increasing productivity and decreasing energy consumption in blast furnaces and in submerged arc smelting furnaces, as a chemical reagent for copper cementation, and for the manufacture of iron powder for powder-metallurgy applications. Recently, the solid-based DR processes have been adapted to the reduction of agglomerates produced from waste iron-bearing materials generated in steel plants; the reduced agglomerates are used primarily for recycle to the blast furnace.

Electric-Furnace Steelmaking With DRI

DRI has been demonstrated to be an excellent electric-furnace charge because it is free of undesirable metallic impurities, has an uniform and precisely known chemical composition, and is easy to transport and handle (28). These properties enable dilution of the charge with low-cost scrap without sacrificing quality, and permit continuous charging with attendant increased productivity, decreased power consumption, and lower electrode and oxygen consumption. These advantages are partially offset by the presence of gangue in the DRI, which imposes an additional power requirement. DR processes retain essentially all the original gangue in the iron oxide charged. The presence of gangue (generally less than 5 percent), of unreduced oxides of iron (generally 5 to 15 percent of the original contained oxygen associated with iron), and of known amounts of carbon distinguishes DRI from scrap as electric-furnace charge.

Gangue in DRI used in electric-furnace steelmaking influences slag weight and composition and usually causes an increase in energy consumption and a concomitant loss in productivity in proportion to the amount of DRI charged and the acid gangue content of the DRI.

The residual oxygen in DRI used in electric-furnace steelmaking may represent either a yield loss or, more usually, an added energy and melting-time requirement. In the latter case, additional carbon is required to reduce the residual oxide (29). When the electric furnace is operated to recover these additional iron units, the reaction of the unreduced iron oxide with carbon maintains a strong "carbon boil" at the slag-metal interface throughout any continuous feeding period, with attendant increase in slag foaming. This combination of vigorous carbon boil and foaming slag has several beneficial effects including decreased bath hydrogen and nitrogen contents, enhanced energy capture and ability to operate at full power without increased refractory loss, and decreased electrode oxidation due to the higher CO concentration in the furnace atmosphere. Continuous feeding of DRI has the additional operating advantages of lower line-voltage disturbances and lower acoustic-noise generation.

The use of DRI in the electric-arc furnace also permits the manufacture of steel grades that would be difficult to produce from available scrap. These grades include high-quality flat products with superior formability and aging properties, and very low sulfur steels needed for high-strength plate and pipe intended for severe service conditions. Because the general historical trend is toward an increase in scrap residual-element content, the use of DRI in electric-furnace steelmaking for traditional grades is also expected to increase. The use of DRI in producing the traditional grades also results in improved fabrication yields, improved mill throughputs, and lower processing costs, primarily as a consequence of the unique and predictable chemical composition of DRI (30).

Use of DRI in Basic-Oxygen Furnaces and Open-Hearth Furnaces

DRI has been demonstrated to be a very effective

metallic charge in basic-oxygen and open-hearth steelmaking (31). In the basic-oxygen furnaces, up to 70 percent of the scrap used as a metallic coolant was replaced by DRI with no operating difficulties that could not be compensated by adjustments in charging practice. The predicted effect of DRI gangue content on product yield and lime consumption was obtained and, in all cases, the concentration of residual tramp elements in the finished steel decreased with an increase in the percent of DRI used. In the open hearth, DRI serves primarily as a replacement for scrap and provides advantages because of its lower sulfur and tramp element content. With the continuing decline in open-hearth steel production worldwide, significant amounts of DRI will not be consumed in these operations.

Use of DRI in Foundries

The application of DRI in ferrous-foundry operations started in the early 1960's and has proved to be very beneficial in special applications where it is especially important to eliminate undesirable contaminants from the charge. The major difference between DRI and conventional foundry charge materials, such as pig iron and clean scrap, is the relatively high gangue content and unreduced oxide content of DRI. For foundry applications, it is preferable to use DRI with a low gangue content and high degree of reduction to minimize energy requirements.

DRI has been used to replace up to 20 percent of the scrap portion of the charge in foundry cupolas without creating excessive windbox pressures (32). With DRI in the charge, the tramp-element content of the product is decreased; however, additional silicon and manganese are generally required. DRI can also be used as part of the charge in electric-arc furnace melting and electric-induction furnace melting associated with foundry operations. In all these foundry applications, the importance of using DRI with low gangue content and high degree of reduction cannot be overemphasized.

Use of DRI in Ironmaking

The first reported tests of DRI as a replacement for part of the normal blast-furnace burden were conducted in 1964 by the U.S. Bureau of Mines in an experimental blast furnace with a 1.2-meter hearth diameter at Bruceton, Pennsylvania (33). The results showed that as the percentage of DRI pellets (90-percent reduced) increased to 85 percent of the burden, there was a nearly linear increase of 75 percent in the production rate. The coke ratio decreased 47 percent over the same range; however, the rate of decrease of coke ratio decreased as the proportion of DRI increased. Similar increases in production rate and decreases in coke ratio were obtained during subsequent experimental blast-furnace tests conducted by U.S. Steel Corporation using burdens of various proportions of 70-percent reduced briquettes as well as burdens containing scrap to simulate very highly reduced DRI (34). These results have also been substantiated during full-scale blast-furnace tests using DRI pellets conducted by The Steel Company of Canada (35) and by Altos Hornos de Mexico (36). Analyses of these results of tests using DRI as part of the blast-furnace burden shows that for every 10 percent of metallic iron on the burden, that is, for every 100 kg of metallic iron per metric ton of hot metal, the coke ratio decreases about 35 kg per metric ton and the hot-metal production rate increases about 9 percent. The results of these blast-furnace trials with DRI in the burden are summarized in Figure 8, which shows the percent change in production rate and coke ratio as a function of the fraction of metallic iron in the charge (Fe^0/Fe_T).

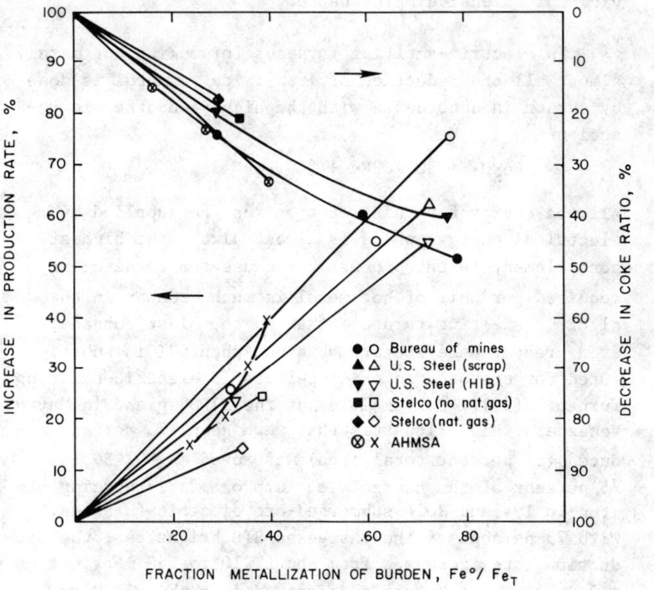

FIGURE 8 EFFECT OF FRACTION METALLIC IRON IN BLAST FURNACE BURDEN ON PRODUCTION RATE AND COKE RATIO

In a blast furnace operating with an iron oxide burden, about half the iron charged will be reduced to metal by CO at temperatures in accordance with the reactions.

$$Fe_2O_3 + CO = 2FeO + CO_2,$$

$$Fe_3O_4 + CO = 3FeO + CO_2,$$

and,

$$FeO + CO + Fe + CO_2$$

which are moderately exothermic, as was discussed earlier. The remaining iron will be reduced to metal at temperatures higher than 900C where the CO_2 formed will gasify carbon endothermically in accordance with the reactions

$$FeO + CO = Fe + CO_2,$$

$$CO_2 + C = 2CO, \quad \text{or}$$

$$FeO + C = Fe + CO.$$

The reduction of iron with endothermic gasification of carbon accounts for a large part of the energy required for making hot metal. When DRI is included in the burden, the amount of endothermic reduction is decreased because part of the burden is already completely reduced, and there is an equivalent coke savings. This reduction in coke ratio also translates to an equivalent increase in production rate because less carbon is burned at the tuyeres for each unit of iron produced. If the percentage of DRI in the burden is increased beyond the amount required to minimize the endothermic reduction, the fuel savings per unit of DRI charged are not as great. This explains why the rate of decrease of coke ratio decreased as the proportion of DRI increased in the tests run by the Bureau of Mines. A more complete discussion of the mechanism and thermodynamics of using DRI in the blast furnace is presented in the literature (37, 38).

In accordance with these results, it has been recommended that DRI be used to increase blast-furnace productivity during periods of peak demand, when metallurgical-coke availability is limited, and when hot-metal capacity is reduced because blast furnaces are out of service for extensive maintenance.

In electric-smelting furnaces for making hot metal, almost all the reduction of iron oxide to metal is done by carbon in accordance with the highly endothermic reaction

$$Fe_2O_3 + 3C = 2Fe + 3CO$$

All the energy for this reaction must be supplied by electrical energy, which is almost always the highest cost element in the process. Because the total energy required per unit of hot metal is much greater in the electric-smelting furnace than in the blast furnace, even greater benefits are obtained when DRI is substituted for ore or iron oxide pellets in electric-smelting furnaces (39). For example, at the SIDOR plant in Matanzas, Venezuela, HIB briquettes that were about 73 percent reduced (84 percent total iron) were used as 25, 50, and 75 percent of the charge with lump ore for producing pig iron in Tysland-Hole submerged-arc electric furnaces. With 75 percent of the charge as HIB briquettes, the production rate increased from about 210 to 350 metric tons per day. The coke ratio decreased from about 390 to 165 kg per metric ton of hot metal and the electrical energy consumption decreased from about 2600 to 1750 kwh per metric ton of hot metal. The results of these tests are presented in Figure 9, which shows the percent change in production rate, coke requirement, and electrical-energy requirement as a function of the fraction of metallic iron in the charge.

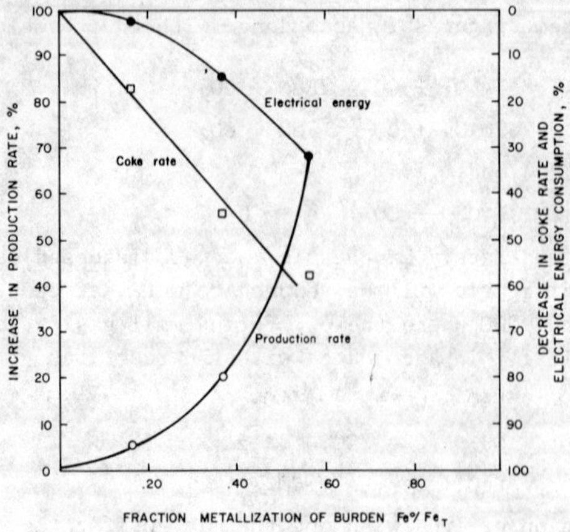

FIGURE 9 EFFECT OF FRACTION METALLIC IRON IN SMELT FURNACE CHARGE ON PRODUCTION RATE AND COKE RATIO

These results show that for the same increase in the fraction of metallic iron in the charge, there was a greater increase in production rate for the electric-smelting furnace than for the blast furnace. It was also observed that with the higher fraction of metallic iron in the charge, the silicon content of the hot metal decreased from about 1.4 to 0.7 percent. This result is to be expected because the amount of silica reduced to silicon should stay nearly constant with time, while the amount of hot metal produced increases with time as the fraction of metallic iron in the charge increases.

APPLICATIONS OF DR TO THE PRODUCTION OF LIQUID IRON WITH ELECTRICAL ENERGY (OR OXYGEN)

In the late 1950's and early 1960's the concept of producing liquid iron or semi-steel directly from ore fines or concentrates and an inexpensive and plentiful fuel and reductant such as pulverized coal was the subject of considerable pilot-plant development activity. In the United States both Inland Steel Company (40) and United States Steel Corporation (41) undertook large-scale projects based on the use of liquid-iron baths as reactors in which the ore and coal could be brought into intimate contact. Both projects were abandoned because of bath-containment problems and other very difficult process and mechanical problems. In the United Kingdom, the British Iron and Steel Research Association (BISRA) conducted an extensive pilot-plant program to develop a similar process, first called Cyclosteel and then Flame Smelting, in which partially reduced iron-ore fines would be smelted to liquid iron by injection into a reducing flame based on partial combustion of a carbonaceous or hydrocarbon fuel with oxygen (42). The partial reduction of the iron-ore fines or concentrates would be accomplished in a two-stage fluidized-bed reduction system fluidized by the off-gas from the smelting furnace. BISRA attempted to couple the two steps of the process without removing the dust entrained with the smelter off-gas, and experienced severe plugging problems in the gas-distribution system for the final reduction bed of the fluidized-bed reducer. This problem and several other technical problems encountered in the pilot plant, coupled with continuing improvements in blast-furnace efficiency during the course of the development program, led to termination of the project in 1963.

Other similar projects that involved partial reduction in rotary kilns or traveling grates followed by direct hot charging into electric-arc furnaces for final reduction and smelting included the Strategic Udy process (43) and the DLM process (44). Commercial operations based on these processes were implemented in Venezuela and Mobile, Alabama, respectively. However, serious process and mechanical problems associated with coupling the direct-reduction and smelting steps prevented successful operation.

In recent years interest has been renewed in developing such concepts to commercial viability, stimulated in part by prospects of long-term shortages of high-quality metallurgical coals, by increasing environmental-pollution problems associated with coke ovens, agglomeration plants, and other operations ancillary to blast-furnace ironmaking, and by strong incentives to decrease the high capital requirements associated with blast-furnace ironmaking. The most prominent projects representing this renewed interest (and those that appear to have the best prospects for succeeding) are the ELRED process being piloted by Stora Kopparbergs Bergslags AB, Falun, Sweden and ASEA AB, Västerås, Sweden (45, 46, 47) and the PLASMA-SMELT reduction process being pioneered by SKF Steel Engineering AB, Hofors, Sweden (48). It is interesting to note that both these promising new approaches to the elusive goal of producing liquid iron directly from ore, low-grade coal, and electrical energy are being pioneered to Sweden, which has suitable iron ores and concentrates and plentiful, relatively inexpensive hydroelectric energy, but must import almost all its fossil energy.

The ELRED Process

The ELRED process includes pre-reduction of fine-grained iron ore or concentrates with pulverized coal in a single stage, fast fluidized bed, followed by final reduction and smelting to liquid iron of the pre-reduced material with associated char in the plasma beneath the electrode in a direct-current (DC) arc furnace. The combustible flue gases from both reduction stages are cleaned and used to generate enough electricity in a combined-cycle power plant to provide the power requirements of the process and a small surplus for other use.

To provide a basis for evaluating the ELRED process, a typical iron-ore concentrate and high-volatile steam coal available in the United States were selected, and material and energy balances were estimated for a plant designed to produce 700,000 metric tons per year of liquid iron. Table IV presents typical chemical and size analyses for the ore concentrates used in this study; ELRED recommends that the ore feed have an average grain size below 1 mm and that the iron content be at least 65 percent.

TABLE IV

CHEMICAL AND SIZE ANALYSIS FOR TYPICAL IRON-ORE CONCENTRATE FOR ELRED

COMPONENT	WEIGHT PERCENT
Fe Total (Hematite)	66.5
SiO_2	5.0
Other Gangue	0.2
P	0.01
S	0.01

SIZE, MESH (TYLER)	CUMULATIVE WEIGHT PERCENT
+ 20	10
35	27
65	63
100	79
150	90
-150	10

TABLE V

CHEMICAL ANALYSIS AND PHYSICAL DATA FOR TYPICAL STEAM COAL FOR ELRED

PROXIMATE ANALYSIS	WEIGHT PERCENT (DRY)
Volatile Matter	40.0
Fixed Carbon	54.4
Ash	5.6
ULTIMATE ANALYSIS	**WEIGHT PRECENT (DRY)**
C	76.6
H	4.7
N	1.6
O	11.0
S	0.4
Ash	5.7
CAL/GRAM (DRY)	7309
ASH FUSION TEMPERATURE, C (REDUCING)	
Initial Deformation	1438
Softening	1480+
Hemispherical	1480+
Fluid	1480+
ASH COMPOSITION	**WEIGHT PERCENT**
SiO_2	58
Al_2O_3	27
CaO + MgO	5
Fe_2O_3	5
Other	5

Table V presents a typical chemical analysis and physical properties of the representative steam coal used in this study; ELRED prefers coal with less than 1 percent sulfur, a maximum of about 15 percent ash, and an ash-softening temperature under reducing conditions of at least 1100C. The coal should be ground to an average grain size of 0.2 to 1 mm. Major process assumptions, based on information presented by ELRED include

- Pre-reduction to 65 percent.
- Carbon for smelting includes stoichiometric for removal of oxygen as CO plus 3.5 percent in liquid ELRED iron.
- 95 percent iron recovery; losses split between pre-reduction and smelting stages.
- 90 percent overall plant availability.
- Sulfur content of ELRED iron 0.11 percent.

Pre-reduction Stage of ELRED

Figure 10 is a schematic flowsheet showing flows on an hourly basis for the single-stage, fast fluidized bed pre-reduction stage. The iron ore is fed pneumatically into a cyclone-type heat exchanger where it is preheated to about 790C by contact with the pre-reducer off-gas. Any moisture in the ore feed would be driven off in this heat exchanger, although for this study the ore is assumed to enter dry. The transport gas for the feed is assumed to be a portion of cleaned off-gas compressed to a higher pressure than the recycle gas. This transport gas therefore represents a closed recirculating loop that does not significantly affect the overall material and energy balances or the sizing of the downstream gas-treatment equipment. For example, a conservative loading of about 500 kg per actual cubic meter of carrier gas would require about 130 kg moles per hour of carrier gas, less than one percent of the total off-gas flow.

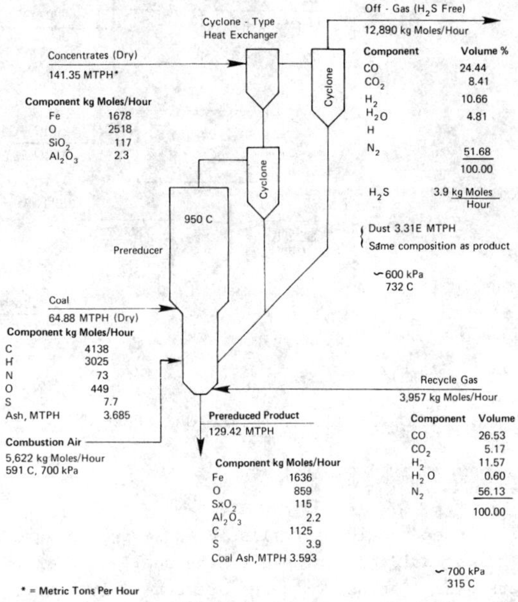

FIGURE 10 ELRED PREREDUCTION – SCHEMATIC FLOWSHEET & HOURLY BALANCES

To estimate the size of the fast fluidized-bed reactor, the superficial velocity in the lower section was assumed to be about 1 meter per second and in the upper section, 1.3 meters per second; the respective ID's are 5 meters and 8 meters. ELRED stated that the total height of the reactor is about 25 meters.

Because clean recycle gas is used to fluidize the lower section of the reducer, there are no plugging problems associated with the gas-distribution system in this part of the system, which was one of the major problems with the Flame Smelting process. Also, the high concentration of char in the bed of almost 11 percent is claimed to prevent defluidization or bogging of the bed and the formation of material buildups. The assumed dust losses are 23 grams per normal cubic meter, which is believed to be achievable in a high-efficiency hot-cyclone system, unless the high recirculation rates result in the generation of excessive submicron particles.

Gas Preheating and Gas Cleaning

The pre-reducer off-gas is used to preheat the combustion air, the recycle gas for the pre-reducer, and the boiler feed water for the steam-turbine plant before it is combusted and expanded in the gas-turbine plant. Figure 11 is a schematic flowsheet showing flows on an hourly basis for one possible conception of this section of the process. For this preliminary study, accurate pressure drops were not calculated because a detailed engineering design of the heat exchangers and scrubbers would be required. electrical power. Figure 12 is a schematic flowsheet showing flows on an hourly basis for the gas turbine. The off-gas is burned with sufficient excess air in the combustor to provide gas to the turbine at about 980C, a practical upper temperature limit with conventional turbine rotor materials. The turbine was assumed to be 92 percent efficient. The air compressor driven by the tur-

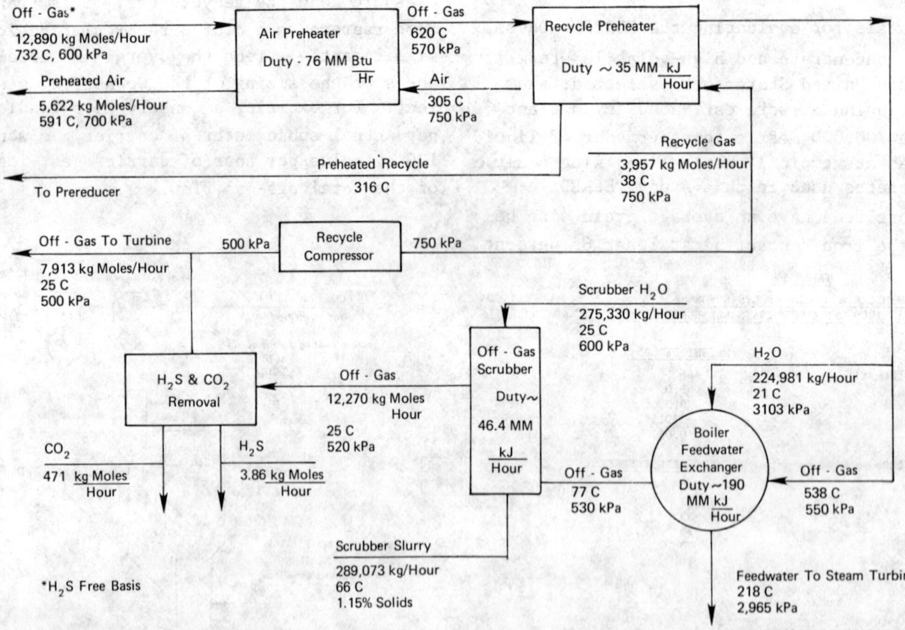

FIGURE 11 ELRED GAS PREHEATING & GAS CLEANING - SCHEMATIC FLOWSHEET & HOURLY BALANCES

The drops used should be realistic, however, because the dirty gases would be confined to the shell side of the exchangers, where pressure losses are generally lower.

Gas Turbine

The clean off-gas from the gas preheating and gas cleaning section is split, with about one-third recycling to the pre-reducer after recompression to about 750 kPa and the rest going to a gas-turbine plant for recovery of

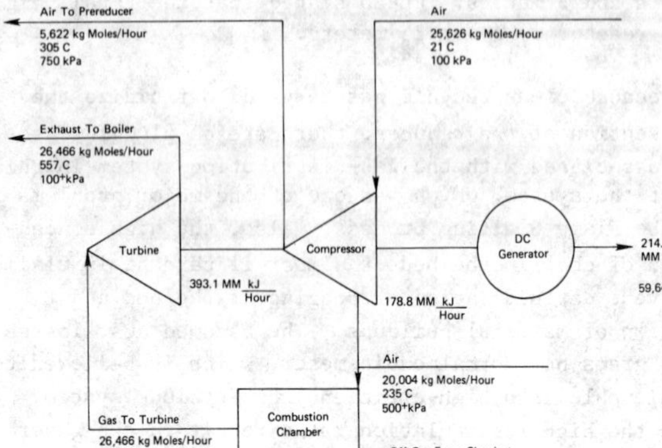

FIGURE 12 ELRED GAS TURBINE - SCHEMATIC FLOWSHEET & HOURLY BALANCES

bine provides the air for the combustor and the air for the pre-reducer at 500 and 750 kPa, respectively. The compressors were assumed to be 83 percent efficient. A single machine could probably be designed to satisfy these requirements; however, using separate machines for each air stream might provide more flexibility, particularly if this section of the plant is designed with turbines in parallel. Another alternative would be to compress all the air to 500 kPa in one machine and then boost the air for the pre-reducer to 750 kPa in a separate machine on the same shaft. About 55 percent of the total turbine output is available as shaft work to the DC generator.

DC Arc Smelting

The pre-reducer concentrate and associated char for smelting and carbon solution are assumed to be introduced into the DC arc furnace through a hollow electrode at 700C, along with the limestone required for a slag basicity of 1.2. Other assumptions made to define this operation include:

- Bath temperature ∿ 1450C
- Off-gas CO_2/CO ∿ 0
- Si in liquid ELRED iron ∿ 0.05 percent
- S in liquid ELRED iron ∿ 0.11 percent
- Energy losses ∿ 10 percent of total input

Figure 13 is a schematic flowsheet showing flows on an hourly basis for the DC arc smelter. The assumption of 90 percent utilization of total energy input is higher than the utilization factor of about 80 percent usually associated with electric-furnace steelmaking operations.

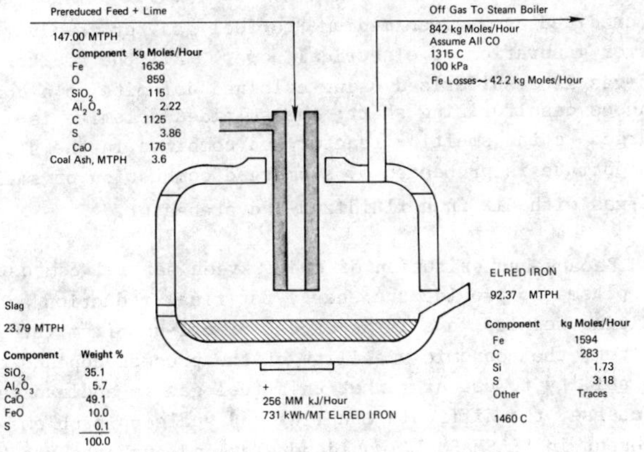

FIGURE 13 ELRED DC ARC SMELTER - SCHEMATIC FLOWSHEET & HOURLY BALANCES

This assumption is justified by ASEA and Stora Kopparberg on the basis that the ELRED DC arc furnace operates with a lower bath temperature, benefits from an insulating or energy-capturing effect of the "soufflé slag" formed throughout the operation, and operates essentially continuously so there are no hold periods with attendant uncompensated heat losses.

The off-gas from the arc furnace is assumed to be essentially pure CO. Although it is recognized that air infiltration will burn some of the CO and dilute the stream with N_2. In essence, air infiltration is transferred to the steam boiler without affecting the energy balance for any downstream operations.

Steam-Turbine Plant

The off-gas from the DC arc smelter and the exhaust gases from the gas turbine provide the energy for raising the steam for driving the steam turbine and generate the electric power for the smelter and for other uses. Figure 14 is a schematic flowsheet showing flows on an hourly basis for the steam-turbine plant. The steam boiler efficiency is about 0.69 (based on the total energy to the boiler above 25C), the turbine engine efficiency is assumed to be 0.75, and the generator efficiency is assumed to be 0.95. The total electric power generated in the steam-turbine plant is 48,300 kw and a preliminary balance for the plant is:

	kw
Total electric power, steam turbine	48,300
Total electric power, gas turbine	59,600
Grand Total Generated	107,900
Total Consumed in DC arc smelter	71,200
Gross Power For Other Uses	36,700

$$\text{Equivalent to } \frac{36{,}700 \text{ kwh per hour}}{92.37 \text{ metric tons of ELRED iron per hour}}$$

$$\sim 400 \frac{\text{kwh}}{\text{metric tons of ELRED iron}} \text{ for other uses}$$

When it is recognized that a small portion of this surplus electric power will be used to drive the boiler exhaust blower, the boiler feedwater pumps, and the boiler combustion air blower, the gross surplus power estimated in this analysis represents a good check on the figure of 300 to 400 kwh per metric ton of liquid iron reported in the lastest ELRED paper (47).

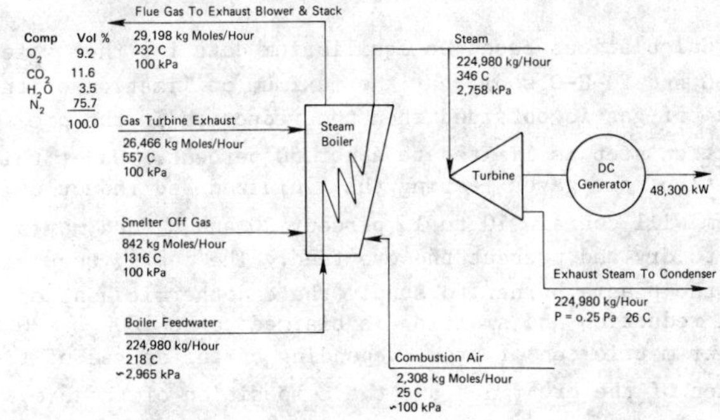

FIGURE 14 ELRED STEAM TURBINE PLANT - SCHEMATIC FLOW-SHEET & HOURLY BALANCES

The PLASMASMELT Process

In SKF Steel's PLASMASMELT process, iron-ore fines or concentrates are reduced to 40 to 50 percent reduction in a two-stage fluidized-bed system operated at about 800C, followed by final reduction and smelting in a low-shaft reactor filled with coke. Figure 15 is a schematic flow-sheet that shows the major features of a possible process design of PLASMASMELT as described by SKF Steel(48).

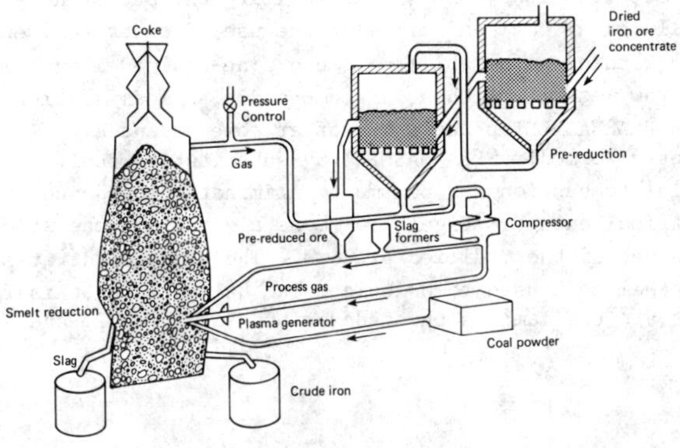

FIGURE 15 SCHEMATIC ILLUSTRATION OF THE PLASMAMELT PROCESS

To provide the energy and reductant for the final reduction and smelting, plasma generated from a portion of the shaft off-gas is injected into the lower part of the shaft reactor along with pulverized coal and/or liquid hydrocarbon. The partially reduced iron ore is introduced downstream of the plasma burner, and shaft off-gas, comprising essentially CO plus H_2 at about 1200C, is used directly in the fluidized-bed reduction system after cleaning in hot cyclones and cooling to about 800C.

The heart of the process is the smelt-reduction zone. This consists of a hollow or raceway in the column of coke into which the pre-reduced iron ore, pulverized coal and/or hydrocarbons, and limestone or dolomite are injected together with the hot plasma. Coke consumption in the shaft is minimized by injecting sufficient coal, oil, or slurry to complete the reduction. The coke serves essentially to form a permeable reaction chamber large enough to retain all solid material injected, to reduce all CO_2 and H_2O generated to CO and H_2, and to allow settling and collection of the liquid hot metal and slag in a hearth similar to the blast furnace.

Calculations based on equilibrium data for the systems Fe-C-O and Fe-H-O show that the maximum utilization of the shaft off-gas is obtained when the reduction in the pre-reduction step is limited to about 50 percent. Under this condition, the gases leaving the fluidized-bed reduction system will contain 10 to 15 percent CO and H_2 and may be used to dry and preheat the ore feed. The energy required from the plasma burner to supply the endothermic heat of final reduction and smelting is claimed to be 1000 to 2000 kwh per metric ton of iron, depending on the degree of reduction of the ore fines and the composition of the ore and coal used.

SKF Steel demonstrated the operation of the smelt reducer in a pilot plant based on one plasma burner and using ground partially reduced ore produced in a Wiberg direct-reduction plant. A pre-reduction pilot plant based on the HIB fluidized-bed reduction technology is now reportedly being designed and constructed for operation with a pilot smelt-reduction shaft. Effective coupling and scale-up of these disparate operations represents a difficult and challenging engineering and operating assignment, particularly in view of the lack of success the Flame-Smelting, Strategic-Udy, and DLM projects.

Smelt Reduction Based on Coal/Oxygen Partial Combustion

For locations that should maximize the use of nonmetallurgical coal and minimize the use of electrical energy while retaining the ability to use fine-grained iron oxide as a raw material without agglomeration, a modification of SKF's PLASMASMELT process may offer some potential. It is proposed to modify PLASMASMELT by substituting coal/oxygen partial combustion for plasma to eliminate dependence on electrical energy. Figure 16 shows a detailed schematic flowsheet of the modified process. The ore concentrates and steam coal used to calculate the balances for this flowsheet are the same as those described in Tables IV and V.

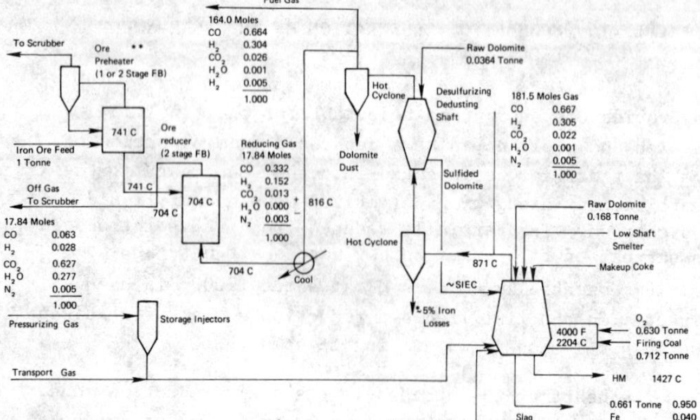

FIGURE 16 SCHEMATIC FLOWSHEET OF SMELT REDUCTION BASED ON COAL/OXYGEN PARTIAL COMBUSTION

Partial reduction is accomplished in a two-stage fluidized-bed reducer using about 10 percent of the off-gas from the smelting shaft. The remainder of the smelter off-gas would be credited as a clean medium-Btu fuel gas for in-plant use or generation of electrical power. All the smelter off-gas is desulfurized using calcined dolomite in a continuous desulfurizing shaft; the sulfided dolomite is charged to the smelting reactor and combined in the slag. Concentrate is preheated by submerged combustion of smelter off-gas with air in a fluidized-bed preheater.

Because substitution of coal/oxygen partial combustion for plasma to provide the energy for final reduction and smelting results in a large excess of off-gas from the smelter, the economic viability of the process depends on the ability to use or market this fuel gas as a clean fuel. Of course, the difficult coupling and scale-up problems inherent in PLASMASMELT would persist and perhaps even be compounded in this modification, primarily because of the larger volume of total gas flow being handled per unit of hot metal produced and because of the interposition of the desulfurizing/dedusting shaft (26) between the smelting reactor and the fluidized-bed reducer. This gas-cleanup step is considered essential to provide reducing gas sufficiently low in dust for relatively good performance in the reducer gas distribution system and to provide fuel gas low enough in H_2S and COS for use in subsequent operations without requiring additional treatment to satisfy environmental requirements.

It is important to reemphasize that although the separate operations comprising PLASMASMELT and the coal/oxygen partial-combustion modification have been demonstrated successfully in commercial operation, their coupling still represents a difficult engineering and operating challenge; the total process can only be as effective as the weakest link.

RECENT GROWTH OF DR AND PROSPECTS FOR THE FUTURE

The steady growth of DR processes and DRI in steelmaking is best illustrated by tracing the increase in total installed DR capacity and displaying the growth of the two major accepted processes. Table VI is a summary of installed commercial DRI capacity through 1975 with projected estimates for 1980 and 1985 by region and by type of fuel and reductant(48). Through 1980, the capacity data are based on projects that were under contract for completion by this date. The 1985 projections include proposed projects that are considered to have a good probability of eventually being implemented, although recent conditions of overcapacity throughout the world and a general slowdown of industrial activity may retard the schedule. Most of the capacity to be installed in the future will be in Latin America, Mexico, and the Middle East and will be based on natural gas. In 1975, the installed plants were operated at about 60 percent of rated capacity, an indication that many of the plants had serious operating and maintenance problems to solve before they would achieve the availability and reliability associated with fully established technologies. This is understandable when it is recognized that many of the plants are located in developing countries where adequately trained operating and maintenance personnel are in short supply. More recently, plants located in Canada and Mexico have demonstrated the long campaign durations and high productivities that represent a standard which the other operations can strive to attain (50, 51).

TABLE VI

SUMMARY OF INSTALLED AND PROJECTED COMMERCIAL CAPACITY FOR DRI (MILLION ANNUAL METRIC TONS OF DRI CAPACITY)*

BY REGION

YEAR	NORTH AMERICA	LATIN AMERICA	WESTERN EUROPE	ASIA	MIDDLE EAST	OCEANIA	AFRICA
1960	0	0.4	0.1	0	0	0	0
1965	0	0.4	0.1	0	0	0	0
1970	0.4	0.9	0.1	0.1	0	0.1	0
1975	2.0	2.1	0.1	0.9	0	0.1	0.2
1980	3.0	9.0	1.6	1.7	3.8	0.1	0.5
1985	4.0	16.0	3.1	4.2	5.4	0.2	3.7

BY FUEL REDUCTANT

YEAR	GOAL	GAS
1960	0.1	0.4
1965	0.1	0.4
1970	0.3	1.3
1975	1.9	4.1
1980	2.4	17.3
1985	4.0	32.6

*Based on Skillings Mining Review, 3/17/79.

The growth of the HyL fixed-bed retort process and the Midrex moving-bed shaft process is summarized in Table VII. The rapid growth in installed capacity for both processes after 1975 is clear evidence that they have established the reliability and economy of operation that will make it difficult for other processes to compete successfully for future new capacity. It is also evident that processes based on the direct use of coal are not expected to be a major contributor to new capacity after 1980.

TABLE VII

SUMMARY OF GROWTH OF INSTALLED CAPACITY FOR HYL AND MIDREX DR PROCESSES

	HyL		MIDREX	
YEAR	CAPACITY, MMMT*/Yr	NO. OF LINES	CAPACITY MMMT*/Yr	NO. OF LINES
1960	0.4	2	-	-
1965	0.4	2	-	-
1970	0.9	4	0.3	1
1975	1.6	6	1.5	4
1980	9.9	15	7.3	19

*EQUALS: Million metric tons per year.

In this context it is reasonable to predict that the smelt reduction processes based on nonmetallurgical coals and electrical energy or oxygen will require considerable additional development work to demonstrate their commercial viability. The economic and environmental incentives are there, but the coupling and scale-up problems are difficult and will require considerable sustained effort. These concepts are therefore not expected to be ready for full commercial implementation for at least another ten years.

In the perspective of the projected total world steel production of about 1 billion tons by 1985, the rapid growth in DR capacity projected during this intervening period, while spectacular in its own right, still leaves DR as a relatively minor factor. From the viewpoint of the developing countries where much of this capacity is being installed, however, DR represents the only way that local steel industries can be established with economic justification. Thus, DR will be making a major contribution to improved living standards and effective utilization of mineral and human resources in these developing countries. The very real need for increased amounts of higher purity iron units for future steelmaking operations can best be satisfied by increased DR capacity. The growth of this important new technology may be slow after the next decade, but it will probably be steady.

REFERENCES

1. E. P. Barrett, "Sponge Iron and Direct Reduction Processes," Bulletin No. 519, United States Bureau of Mines, Washington, D.C., 1954.

2. W. L. Davis, Jr., J. Feinman, and J. H. Gross, "Briquetacion del Mineral de Hierro de Alto Tenor por el Processo de Reduccion Directa en Lechos Fluizados," Instituto Latinoamericano del Fierro y el Acero (ILAFA), Memoria Tecnia del Seminario Latinoamericano, La Reduccion Directa de los Minerales de Fierro y Su Applicacion en America Latino, Ciudad de Mexico, Mexico, November 1973.

3. D. C. Violetta, "FIOR - The Process For Producing Highly Metallized Feed For Steelmaking," ibid, Reference 2.

4. A. M. Squires and C. B. Johnson, "The H-Iron Process," AIME Blast Furnace, Coke Oven and Raw Material Conference Proceedings, 16, 1957, p. 32-42.

5. C. L. Cruse, A. P. Kerschbaum, W. E. Marshall, and R. E. Moon, "Development of ARMCO's Direct Reduction Process," Journal of Metals, 21 (1969), p. 55-59.

6. E. R. Morgan, "Reduccion Directa Continua de los Oxidos de Ferro por el Process Midrex," ibid, Reference 2.

7. H. D. Pantke and G. H. Lange, "The Purofer Process," U. N. Economic Commission for Europe, Steel Committee, Seminar on Direct Reduction of Iron Ore: Technical and Economic Aspects, Bucharest (Romania), September 1972.

8. R. Lawrence, Jr., "The HyL Direct-Reduction Process-Past, Present, and Future," ibid, Reference 7.

9. G. Meyer and U. Bongers, "Krupps Direct-Reduction Route to Sponge Iron Goes Commercial," Engineering and Mining Journal, Vol. 174, No. 6, 1973, p. 159-164.

10. H. Serbent and W. Janke, "SL/RN Process - Production of Sponge Iron Using Solid Reductants," ibid, Reference 7.

11. D. A. Bold and N. T. Evans, "Direct Reduction Down Under: The New Zealand Story," Iron and Steel International, Vol. 50, 1977, p. 145-152.

12. D. W. Rierson and A. A. Albert, Jr., "Development of the ACCAR Process at Allis Chalmers, "AIME Ironmaking Conference Proceedings, Vol. 36, 1977, p. 455-467.

13. J. F. Elliot and M. Gleiser, "Thermochemistry for Steelmaking," Vol. 1, Addison-Wesley Publishing Co., Reading, PA 1960

14. J. P. Coughlin, Bulletin No. 542, United States Bureau of Mines, Washington, D.C. 1954.

15. S. E. Luchsinger, W. H. Harrell, and J. Feinman, "A Review of HIB Performance With The 1977-1979 Plant Modifications," ILAFA Meeting on Direct Reduction, Buenos Aires, Argentina, July 1980.

16. S. E. Luchsinger, "A Review of Experience in the Production and Use of HIB," IISI Committee on Technology Meeting, Rio de Janiero, Brazil, April 1979 (Written by J. Feinman).

17. J. Stalhed, "Die Herstellung von Eisenschwann Nach dem Wiberg-Söderfors-Verfahren," Stahl and Eisen, 72, 1952, p. 456-466.

18. R. Quintero, "First Year's Operational Results of the 2M HyL III Continuous Reactor," Presented at the 40th Ironmaking Conference of ISS/AIME, Toronto, Canada, March 30, 1981.

19. R. Quintero, "Equivalent Metallization of Sponge Iron," XVII Congreso Latinoamerciano del Fierro y el Acero, ILAFA, Caracas, Venezuela, 1976.

20. R. Quintero, "New HyL Technology," HyLSA, Monterey N.L., Mexico, August 1979.

21. I. F. Cairns and G. Langton, "The Integrated Steel Plant of the Highveld Steel and Vanadium Corporation," Steel Times, June 1970, p. 378-403.

22. M. J. Fraser and C. R. Gregg, "The SL/RN Process - Its Place in the Iron and Steel Industry," Eng. (Engineering Institute of Canada), Vol. 48, No. 6 1965, p. 29-34.

23. K. Schermer and A. Dalrymple, "Significance of Dunswart's Rotary Kiln Direct Reduction Experience to the Steelmaker in U.S.A.," Radex Rundschun, 1, 1979, p. 893-924.

24. J. A. Lepinski, "The ACCAR System and Its Application to Direct Reduction of Iron Ores," Association of Iron and Steel Engineers, Annual Convention, September, 1979.

25. V. P. Keran, A. C. Baker, A. J. Ridley, and G. N. Boulter, "The Direct Reduction Corporation's Process Technology," ISS/AIME Ironmaking Proceedings, Vol. 39, 1980.

26. J. R. Guseman and L.G. Hendrickson, "Fuels for the Direct Reduction of Iron Ore," 1st Process Technology Conference, AIME, Washington, D.C., March 24, 1980.

27. J. Feinman and J. E. McGreal, "Hot Desulfurization of Coal Gasification Gas by a Continuous Moving Bed of Calcined Dolomite and Application to Direct Reduction," 88th National Meeting, American Institute of Chemical Engineers, Philadelphia, PA, June 12, 1980.

28. R. M. Smailer, H. B. Jensen, and W. W. Scott, Jr., "Utilization of DRI in Electric Steelmaking," AIME Electric Furnace Procedures, 32, 1974, p. 29-36.

29. E. A. Elsner and H. Voss, "New Results Obtained in the Operation of UHP Electric Arc Furnaces Charging Scrap and Sponge Iron," Procedures of the International Iron and Steel Congress, Dusseldorf, May 1974.

30. R. L. Reddy, "Electric Arc Furnace Steelmaking with Sponge Iron," Canadian Metallurgical Quarterly, 18(2), 1979, p. 245-250.

31. J. M. Pena and D. Radke, "HyL Sponge Iron as a Coolant in BOF Steelmaking," Journal of Metals, Vol. 23, No. 8, 1971, p. 27-32.

32. R. H. Harner and S. C. Clow, "Melting Pre-Reduced Iron Ore Pellets in a Cupola," Modern Castings, 53, 1968, p. 73-80.

33. P. L. Woolf, "Blast Furnace Operations with Prereduced Burdens," TMS-AIME Ironmaking Proceedings, 24, 1965, p. 73-77.

34. D. G. White, P. B. Stubbs, R. J. Pygiel, and M. M. Harris, "Improving Blast Furnace Performance with HIB-Experimental Blast Furnace Results," TMS-AIME Ironmaking Proceedings, 32, 1973, p. 228-233.

35. J. A. Peart and F. J. Pearce, "The Operation of a Commercial Blast Furnace with a Pre-reduced Burden," TMS-AIME Ironmaking Proceedings, 24, 1965, p. 67-71.

36. B. H. Hector, A. D. Ramiro, and A. C. Rudolfo, "Utilization Experimental de Prereducidos en el Alto Horno No. 2 de Altos Hornos de Mexico SA (AHMSA)" Sierugico Latinoamericano/No. 222, October 1978.

37. J. C. Agarwal and H. R. Pratt, "The Thermodynamic Aspects of Using Partially Reduced Burdens," TMS-AIME Ironmaking Proceedings, 24, 1965, p. 33-36.

38. J. Michard, "Use of Pre-reduced Products in the Blast Furnace," Congres International sur la Production et L'Utilization des Minerals Reduits, Evian, France, 1967, p. 207-226.

39. O. Dam, "Steelmaking in Venezuela: Trials with HIB Briquettes at Siderurgica del Orinoco (SIDOR)," Technical Reports of the 13th Latin American Congress on Ferrous Metallurgy, 1973, p. 16 (Met. A., 7502-720022).

40. J. W. Halley, United States Patent 2, 919,983, Iron Ore Reduction Process, January 5, 1960.

41. J. C. Agarwal and W. L. Davis, Jr., United States Patent 3, 264,096, Method of Smelting Iron Ore, August 2, 1966.

42. T. W. Johnson and J. Davidson, "Reduction of Iron Ore by the Flame-Smelting Process," Journal of the Iron and Steel Institute, May 1964, p. 406-419.

43. M. J. Udy and R. A. Blackburn, "The Strategic-Udy Direct Iron Reduction Process, "Yearly Proceedings, Association of Iron and Steel Engineers, 1959, p. 921-928.

44. T. E. Ban and D. C. Violetta, "DLM, New Commercial Ironmaking Process," Iron and Steel Engineer, 45, September 1968, p. 101 et seq.

45. P. H. Collin and B. Widell, United States Patent 4,094,665, Method For Simultaneous Combined Production of Electrical Energy and Crude Iron, June 13, 1978.

46. P. Collin and H. Stickler, "ELRED - A New Process For The Cheaper Production of Liquid Iron," Association of Iron and Steel Engineers 1979 Annual Convention, Cleveland, Ohio.

47. E. Bengtsson and B. Widell, "The Chemistry of the ELRED Process," Presented at the 40th Ironmaking Conference of ISS/AIME, Toronto, Canada, March 30, 1981.

48. S. Santen, "What Can Be Done to Improve Iron and Steelmaking in the Future?" 1978, SKF Steel Engineering AB, Hofors, Sweden.

49. R. L. Stephenson, Editor, "Direct-Reduced Iron - Technology and Economics of Production and Use," The Iron and Steel Society of AIME, Warrendale, PA, 1980.

50. A. Claveau, A. Faucher, and M. A. Lafreniere, "Performance Review of Sidbec-Dosco's Reduction Plant After Four Years of Operation," Canadian Metallurgical Quarterly, Vol. 18, 1979, p. 85-95.

51. R. G. Quintero, "Up Date on HyL," Ironmaking Proceedings," The Iron and Steel Society of AIME, Vol. 37, 1978, Chicago, IL.

STEELMAKING PROCESSES

Robert D. Pehlke
Materials & Metallurgical Engineering
The University of Michigan
Ann Arbor, Michigan 48109

Modern steelmaking processes are described. Recent evolution and development of pneumatic, open hearth and electric furnace steelmaking processes are briefly reviewed, and their current status is outlined in terms of production of various types of steel, flexibility in raw materials consumed, variations in operating practices, energy conservation and process control. Recent developments and innovations in steel production are described including bottom injection of gases and gas-solid mixtures. Improvements in control of pneumatic steelmaking are considered. Electric furnace steelmaking improvements via increased power input and productivity are described, along with recently developed water cooling systems for shell, walls and roofs.

The use of ladle metallurgy or secondary steelmaking is reviewed. This technique has been rapidly adopted as a means for producing low sulfur-low oxygen steels as well as controlling steel analysis and ladle temperatures. Argon stirring and injection of deoxidation and desulfurization agents are described in terms of operating results. Furthermore, the ladle metallurgy for treatment of liquid pig iron prior to the steelmaking process is considered relative to final steelmaking chemistry objectives and the extent of steel ladle processing required.

The current state of continuous casting is detailed and advanced practices in this technology are outlined. Prevention of reoxidation and separation of inclusions is described. The objectives of higher yield and energy conservation provide strong incentives for greater adoption of this process. Horizontal continuous casting as a future process is described.

INTRODUCTION

The continuing evolution and development of steelmaking technology is a very active field. Beginning particularly with the availability of large scale supplies of oxygen, many advances in steelmaking have followed.

OXYGEN IN THE OPEN HEARTH AND ELECTRIC FURNACE PROCESSES

Until the period following World War II when tonnage supplies of oxygen became available, the use of oxygen in the open hearth or electric furnace had been extremely limited. With the availability of oxygen, injection using steel pipes as lances was initiated in both types of furnaces either to increase refining rates or to create localized pools of hot iron oxide-rich slag to flux refractory accretions. In 1947, some early trials with oxygen supplied as a jet in conjunction with the end burner fuel stream to produce a hotter flame in the open hearth furnace were conducted at the Steel Company of Canada. Jets of oxygen were then widely tested in the open hearth as gas injected into the bath to speed up refining, as a burner additive and as directed jets of oxygen to melt scrap. The availability of tonnage oxygen allowed development of high purity oxygen refining practices which greatly reduced heat times and improved thermal efficiencies. Oxygen burners were used to supplement conventional end-wall burners. Roof burners of the oxy-fuel type were also developed. In 1952, movable, water-cooled, roof-mounted oxygen lances were used for top blowing open hearth heats. Multiple orifice lances were used to reduce splash which improved roof life and also reduced the chance of bottom damage.

The use of high purity oxygen in the open hearth produced large quantities of iron oxide fume which required more extensive gas-cleaning equipment. Average oxygen consumption in open-hearth steelmaking has continually increased. This increase has been documented in Chapter 2 of the *BOF Steelmaking Monograph* (1) and shows that intensive use of oxygen can even exceed that required for BOF steelmaking because of the additional combustion oxygen used in end-burner or oxy-fuel roof burner practices.

In the electric furnace, oxygen is used principally for decarburization and is most commonly accomplished with a bare or refractory insulated steel pipe which is submerged in the bath for refining. A few arc furnaces have been equipped with movable, water-cooled oxygen injection lances through the roof.

Electric furnace process consumption of oxygen varies widely depending upon the steelmaking practice. Substantial amounts of oxygen (up to or over 1000 std cubic feet per ton) can be employed in stainless production and (over 300 std cubic feet per ton) in carbon steel practice, although with molten pig iron in the charge, these quantities can vary widely.

GROWTH OF BOF STEELMAKING

The first commercial steelmaking facility utilizing top blown oxygen converters was established at Linz, Austria, in 1952, and was followed by a sister plant at nearby Donawitz one year later. Within five years, nine additional plants in seven other countries had been erected. In 1960, thirteen million tons of capacity had been constructed and this increased to 75 million tons in 1965 and 240 million in 1970. The rapid increase was due not only to additional plants being constructed, but also to the fact that vessel size dramatically increased. In the decade from 1960 to 1970, total steel production increased by 285 million metric tons, of which about 80 pct was accounted for by the increased production of the BOF (LD or BOP, and so forth). During this period open hearth steel production held its own but began to show a decline on a global basis as indicated in Fig. 1.

The major growth in BOF steelmaking took place in the United States, Japan and, to some extent, Europe. These developments were based on larger vessels. Relative vessel size and shape for several American BOF plants are illustrated in Fig. 2. In particular, in 1961, Jones and Laughlin started up the Cleveland Plant with 225 ton vessels, and the following year in Australia, West Germany, and Ecorse, Michigan, plants were started at 235, 290, and 300 tons capacity, respectively. Concurrent with the move to larger vessels was the development in the United States of the multi-orifice lance (2). In Europe, during this period of time the requirement for refining a phosphorus pig iron was accomplished by the development of the LD-AC

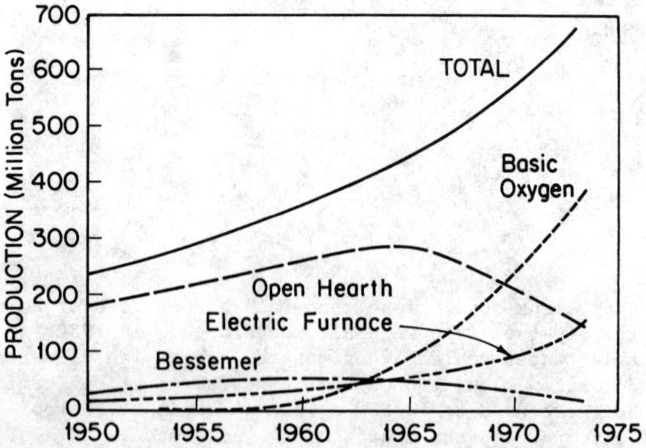

Figure 1. Steel production by process.

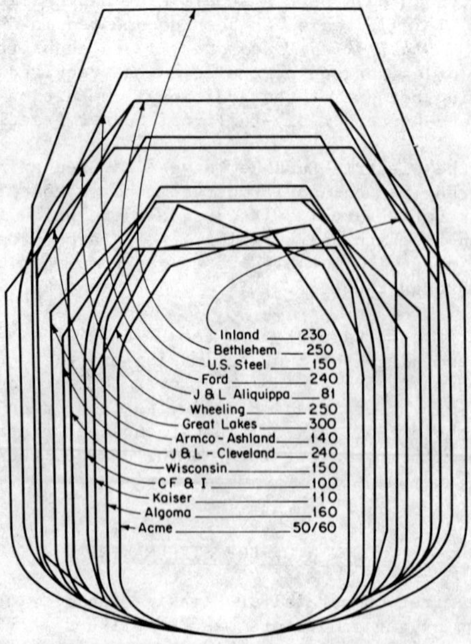

Figure 2. Relative size and shape of various oxygen converters (33 Magazine).

and OLP processes, in which lime powder was directly injected into the vessel along with use of a two slag practice. Subsequent to this period, expansion of the top blown oxygen process has involved practice improvements, including the capability of handling various raw materials, improvement in design and plant layout, the utilization of various waste gas cleaning systems and, subsequently, waste gas confinement with the objective of utilizing not only process heat, but also the combustion capability of the approximately 90 pct CO off-gas. Furthermore, the process reached across a broad product line, including the making of alloy grades, high carbon grades and tool steels such that by the early 1960's all open hearth grades could be made, as well as some arc furnace products. Several attempts were made to develop top blown processes for the production of stainless grades.

This dramatic growth of the top blown oxygen steelmaking process has made it the dominant force in steelmaking in the world today. Several challenges have been presented to the BOF including expansion of electric steelmaking, development of continuous steelmaking which has not been commercially successful but continues to have great technological interest, and submerged oxygen injection, the latter being a continued evolution of development of the bottom blown pneumatic processes.

SUBMERGED INJECTION PROCESSES

The early development stages of submerged injection processes were directed toward several European plants who converted their bottom blown Thomas (basic Bessemer) converters to use oxygen and hydrogen bearing gases such as hydrocarbons, steam or natural gas.

In 1972, a patent was issued to Maxhütte (Eisenwerk-Gesellschaft Maximilianshütte GmbH of Sulzback-Rosenberg) which describes the "OBM" process (3). This process was implemented in western Europe to provide bottom blowing capability with oxygen for Thomas converters and then subsequently in late 1971 was adopted by United States Steel Corporation under an agreement with Maxhütte concerning certain rights under the "OBM" process and the American version "Q-BOP" process. A schematic arrangement is shown in Figure 3.

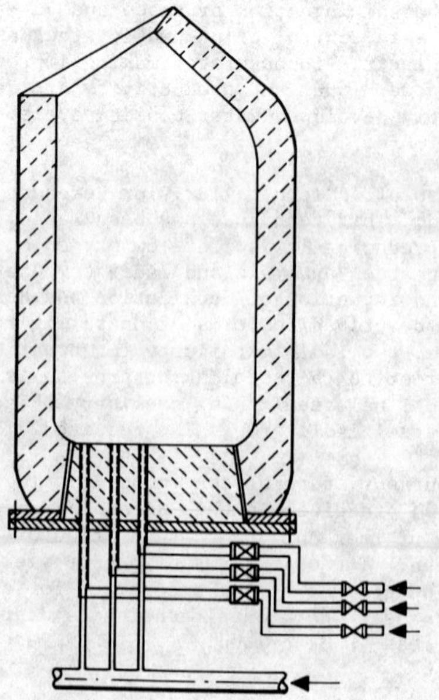

Figure 3. Longitudinal section through a bottom blown Q-BOP converter.

In a comparable time frame, Creusot-Loire filed patent applications in France and in the U.S. on the "LWS" process (4,5). The LWS process involves injection of a hydrocarbon liquid in the outer tuyere as shown in Figure 4.

Following the development work at Maxhütte, several conversions of Thomas converter plants were made to the bottom blown oxygen process using natural gas in the outer surrounding tuyere. The steelmaking units in these early installations in Europe ranged in size from 5 to 50 tons. Then, in 1973, U.S. Steel Corporation installed 200 ton converters using this process in the partially completed oxygen top blown steelmaking plant at Gary Works (6,7). This scaleup step for the bottom blown oxygen process paralleled similar developments in the United States a decade earlier for the top blown oxygen process.

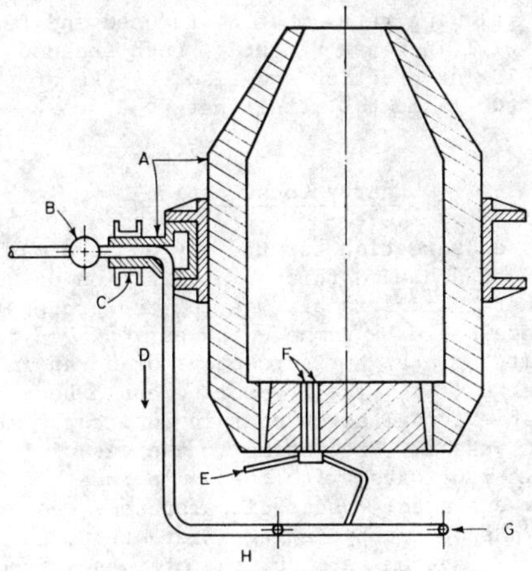

Figure 4. Longitudinal section through a bottom-blown LWS converter.

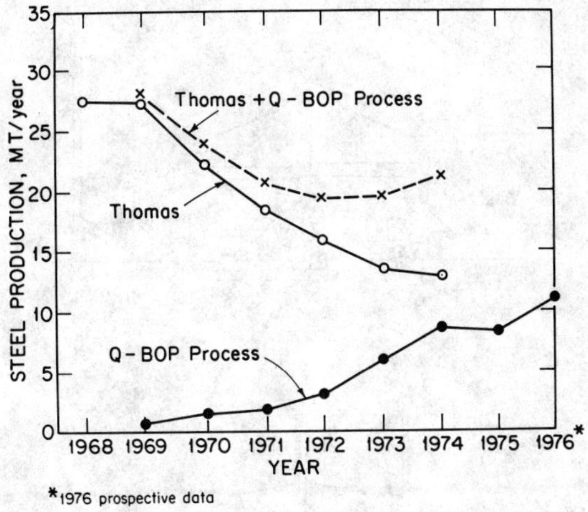

Figure 5. Growth of Thomas and Q-BOP processes.

The bottom blown oxygen steelmaking process uses a converter with a replaceable bottom containing tuyeres through which oxygen is introduced. The number of tuyeres depends upon the size of the furnace, the number increasing with the size. Tuyere placement is critical to take advantage of the stirring capability of bottom injection, particularly under conditions where oxygen flow rates are decreased. Replacement of the bottom refractory plug for the vessel which contains the tuyeres was required 4 to 5 times as often as replacement of the remaining vessel lining in the early stages of the process. However, with improved operating and refractory practices, this ratio has been decreased to 1 with recent operating practice indicating up to 1000 heats for the bottom plug and working lining of the vessel. The absence of the lance in the bottom blowing process has offered opportunities to decrease total plant height and has also allowed relatively easy conversion of older open hearth plants such as at the U.S. Steel Fairfield Works (8) and Republic Steel Chicago Works (9).

With the flexibility in plant design and lower conversion costs on a brown field site, and particularly considering the potential metallurgical advantages, the bottom blown process has shown rapid adoption as indicated in Figure 5 (10).

The rapid growth of the submerged tuyere processes is also reflected in the parallel growth of post refining processes. One specific example is the argon oxygen decarburization process (AOD) which uses an inert gas in the outer tuyere and an oxidizing gas in the inner tuyere. This process has rapidly achieved adoption for duplex refining of stainless steel and more recently for alloy grades. In addition, the process has gained rapid adoption throughout the foundry industry as a means for completing refining of small heats of a wide range of steels and alloys.

Comparing the top and bottom blown processes from an operational viewpoint brings out certain advantages for each (11). Because of lower plant height required for the bottom blown process, the minimum cost is less either for a greenfield site or the conversion of an open hearth or Bessemer shop. Several bottom blown installations representing various sizes and general shop configurations are shown in Figure 6. Figure 6(a) represents a conversion of an open hearth shop; Figure 6(b) is a small vessel simple plant; Figure 6(c) involves placement of gas cleaning equipment within the converter building and includes overhead addition facilities; and Figure 6(d) is a large and highly advanced arrangement, including a sublance control system.

However, the scrap melting capability of the bottom blown process is less, compared to the top blown process, because of the lower iron oxide content of the slag, a factor which can be looked upon as representing an increase in process yield.

Combined top and bottom blowing are currently being explored to combine the advantages of each type of oxygen steelmaking. These early tests have shown higher yield, improved manganese recovery, better control of oxygen content and lower oxygen concentrations, as well as improved dephosphorization.

CURRENT DEVELOPMENT ACTIVITY

Several specific areas of process development are being actively explored and a few examples are briefly described below.

Flexibility of Operation

As an autogenous process, the mass-energy balance for oxygen steelmaking is a relatively fixed quantity. However, for operating flexibility, changes in hot metal to scrap ratio, as well as heat size and converter operation would be highly desirable. The optimum hot metal ratio depends on hot metal and scrap prices, material availability and product demand. The optimum conditions can be more closely approached with the flexibility of utilizing higher or lower hot metal ratios.

Operation at lower hot metal ratios can be achieved through the use of auxiliary fuels. This subject has been explored in detail by Groen, Holmes, Onuschek and Pehlke (12). These operating practices include preheating in various forms and the use of hot scrap from the rolling mills, solid fuels including calcium carbide, silicon carbide, ferrosilicon, coke and cold iron, as well as addition of liquid or gaseous fuels during process blowing. More recently, operating experience within Nippon Steel Corporation has shown the viability of several operating strategies to permit oxygen steelmaking at low hot metal to scrap ratios (13). The use of cold pig iron as a heat

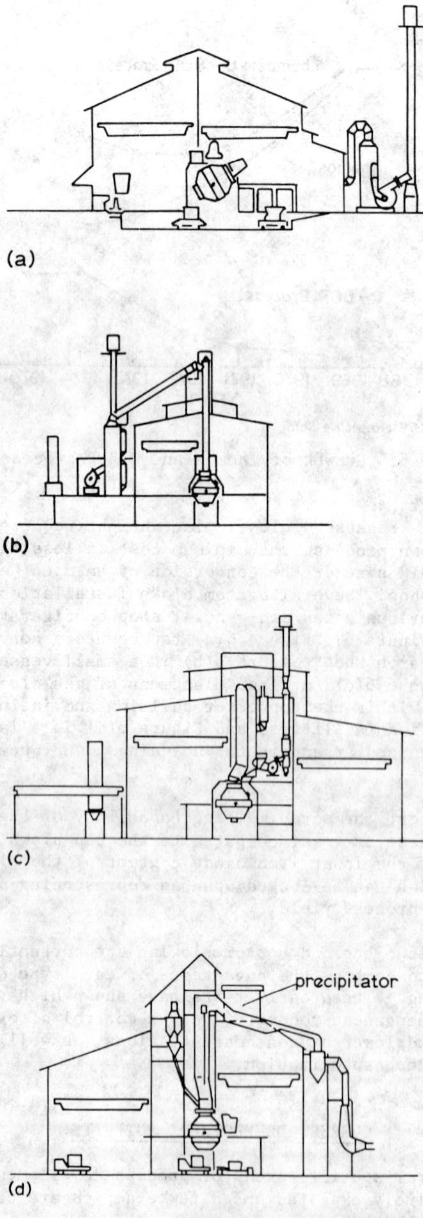

Figure 6. Some modern OBM/Q-BOP plants.

source has shown that 1 pct cold pig iron in the furnace charge provides a decrease of 0.3 to 0.4 pct in the hot metal to scrap ratio.

Preheating of scrap and cold pig iron either in or out of the furnace can reduce the hot metal to scrap ratio by as much as 10 pct. Certain operating difficulties are encountered from scrap oxidation which limits preheating temperatures and can cause increased lining wear in the case of in-furnace preheating.

Currently interest is increasing in using the bottom tuyeres of the Q-BOP as oxy/fuel burners during a brief scrap preheating stage, and also in employing side or top (in the cone section of the vessel) tuyere blowing during the regular oxygen bottom blow to induce afterburning of carbon monoxide within the vessel. The combined top and bottom blowing technique, referred to as the OBM-S or KMS process, offers the potential for decreasing the hot metal to scrap ratio (14,15).

Operating at high hot metal to scrap ratios has been achieved through the use of iron oxide coolants. This mode of operation, including an increase in the lime charge, leads to problems with slopping which can be mitigated by continuous charging of iron oxide coolant during the blowing operation. Vessels with greater height to diameter ratios are more amenable to this type of operation. Blowing rates must be reduced and furnace cycle time will increase slightly, both because of the change in blowing practice and also because of the requirement for a two ladle hot metal charge.

Energy Conservation

Energy conservation for the oxygen steelmaking process can be achieved through utilization of both the chemical energy and sensible heat of the process off-gases. The off-gases could be burned with pure oxygen to preheat scrap permitting a higher percentage of scrap in the furnace charge (16). The preheating could be carried out either in an external scrap charging unit or within the steelmaking vessel. Preheating in the vessel is more effective in many ways, with the exception that heat time is normally extended. Under circumstances where two vessels in a shop are operating alternately, and where the gas collection system permits, the off-gases from one steelmaking heat could be utilized to preheat a subsequent charge in the second vessel. This process would be reversed during a subsequent heat. The direct combustion of hot off-gases has the advantage of utilizing the sensible heat of the gases, as well as the heat of reaction available from combustion of CO to CO_2. The amount of sensible heat available depends upon the temperature at which the off-gases leave the steelmaking furnace. Assuming that the effluent gas is 90 pct CO, the energy available as sensible heat in the effluent gas as compared to the total energy available in the gas (reaction heat plus sensible heat) would be 14.8 pct if the gases left the vessel at 2240°F and 18.4 pct if the gases left the vessel at 2960°F.

Process Control

The dynamic control of oxygen converter steelmaking will continue to develop and should be adopted by most operations in the near future with the most vaible systems appearing to involve the use of a sensor or sublance.

Since the early commercialization of the top blown oxygen process, process control has been an important aspect of the steelmaking operation and has received considerable research and development attention. Steelmaking converter control technology has gone through a number of stages starting with early static models and including various sensor technologies relating to bath temperature and off-gas composition. More recently, attention has focused on successful utilization of a sublance which is capable of measuring carbon content and temperature of the bath without interrupting the blow. One system in current use is shown schematically in Figure 7. The response to these measurements can be rapid through utilization of computers and allows the achievement of dynamic control to attain the proper bath temperature and turndown chemistry which is particularly important for higher carbon grades.

ELECTRIC FURNACE STEELMAKING

The rapid development of electric arc furnace (EAF) steelmaking technology, not only for alloy and stainless steels, but especially for carbon steel production, has increased the fraction of production capacity of the EAF to 20 pct of the steel industry in the United States. This increase in capacity reflects the ascension of the "mini-mill" as a major steel supplier. The size of the "mini-mill" also has dramatically increased to the point where this label is not descriptive of this type of steel plant, several such mills being located within large integrated plants and having two or more 200 ton electric arc furnaces.

The modern "mini-mill" will include one or more

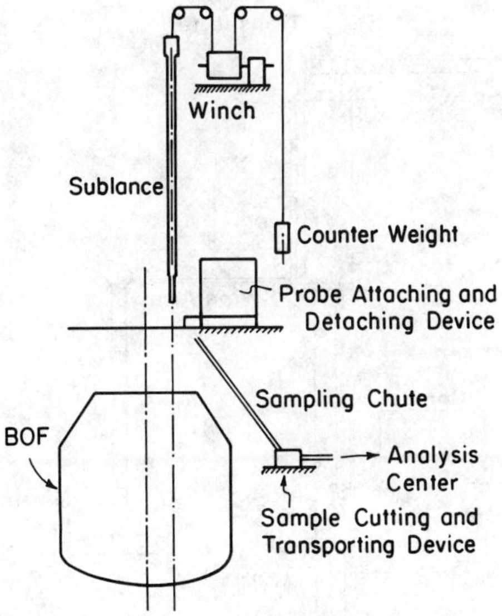

Figure 7. Sublance system.

electric furnaces with high power energy sources; a 100 ton electric furnace may have a 75 megawatt transformer. These furnaces are water-cooled, at least in the upper wall where cooled panels substitute for refractory above the slag line. Water-cooled roofs are being tested and water-cooled electrode holders are under development to provide reduced electrode consumption.

The product line of the "mini-mill" now includes special quality bar and rod, as well as reinforcing bar and small structural shapes. Some mills are producing plate from electric furnace remelting operations. This technology which may depend on direct reduced pellets in the future is now well advanced and has assumed an important position in the world of steel production.

LADLE METALLURGY

A major aspect of steel development during the past decade has been the effort directed at property improvement through steel cleanliness and reduced anisotropy which are largely dependent upon the amount and form of oxygen and sulfur in the steel. This interest has been brought about by the quality requirements for thin sheet materials, heavy plate steels, and HSLA steels for line pipe. The requirements for thin sheet steels has been good formability; for heavy plate, guaranteed through-thickness properties; and for line pipe steels, high transverse impact energy values.

During the course of deoxidation of molten steel, removal of reaction products is most critical. Nonmetallic inclusions in the form of oxides of silicon, manganese, and aluminum may adversely influence final properties.

Sulfur in steel has presented a problem in steelmaking from the very beginning. Inferior surface quality can be achieved by lowering sulfur levels to .015 pct or below. The achievement of low sulfur contents in steelmaking practice, for example with two slag practices in electric furnaces, has presented operating difficulties and production delays.

Steel was produced three decades ago primarily by the open hearth furnace where sulfur levels were normally in the range from 0.03 to 0.04 pct. Lower sulfur contents of 0.02 to 0.03 pct could be achieved with lower sulfur fuels and oxygen roof lances, permitting shorter heat times and improved mixing of metal and slag. The basic oxygen steelmaking process brought about an even further improvement within the range of 0.015 to 0.025 pct sulfur, and with the bottom blown Q-BOP, sulfur contents in the 0.01 to 0.02 pct range can be achieved. However, even lower sulfur contents are required.

In killed steels with low oxygen content, such as when aluminum is used for deoxidation and grain size control, sulfur combines with manganese as highly deformable manganese sulfides. These manganese sulfides have a low melting point and, as the last liquid to solidify in the steel, collect as films at grain boundaries. During hot rolling the manganese sulfides are plastically deformed into elongated stringers extending parallel to the rolling direction. This shape and distribution of sulfides can have a marked effect on the directional properties of steel products. Two methods can be utilized to minimize the directionality problem. One is to reduce the sulfur content in the steel product to a minimum, lowering the total volume of manganese sulfide inclusions. This can be accomplished by decreasing the total sulfur input, treating the hot metal to remove sulfur and controlling the scrap charge; by adjusting melting practices, including the use of additional quantities of flux; and by direct treatment of the liquid steel with desulfurizing agents, such as calcium or magnesium. A second approach involves modifying the shape or morphology of sulfide inclusion by producing a relatively nondeformable complex sulfide and/or oxysulfide by addition of calcium or rare earth metals. These effects are illustrated in Figure 8 (17).

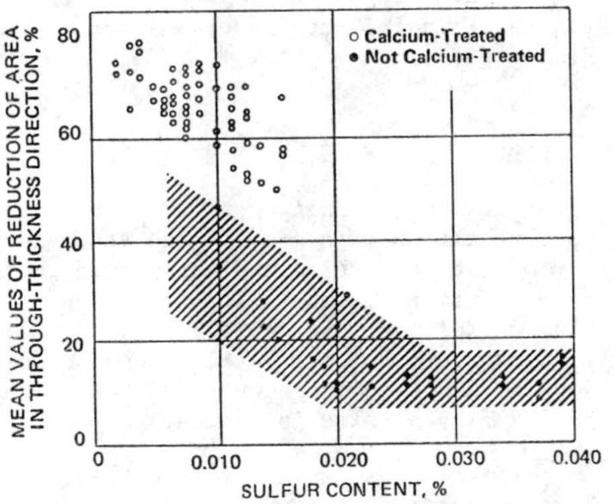

Figure 8. Relationship between the sulfur content and the mean values of reduction of area of tensile test specimens in the through-thickness direction of steel grade FG 36.

Ladle metallurgy was used first to produce high quality steels, but is now being utilized for many grades of steel because of the economic advantages of higher productivity. The purpose of these ladle treatments is to:

1. Produce "clean" steel.
2. Introduce reactive additions, such as calcium or rare earths.
3. Add alloying additions, as for microalloyed steels, with high recovery.
4. Increase furnace utilization, allowing higher productivity smelting operations of the blast furnace, and melting and refining operations in steelmaking, followed by final ladle treatment of the steel to achieve controlled low levels of contained inclusions.

Ladle treatments of steel can be described generally in five categories:

1. Synthetic slag systems.
2. Gas stirring or purging.
3. Direct immersion of reactants, such as rare earths.
4. Lance injection of reactants.
5. Wire feeding of reactants.

These are often used in combination to produce synergistic effects, e.g., synthetic slag and gas stirring for desulfurization followed by direct immersion, injection or wire feeding for inclusion shape control.

Synthetic Slag Systems

Addition of a prepared flux layer to the steel ladle during or immediately following tapping offers an opportunity to accomplish specific refining reactions. To prevent reversion of undesirable elements, such as phosphorous, sulfur and oxygen from the steelmaking slag, as little slag as possible should be allowed to escape from the furnace into the ladle. However, prevention of heat loss requires that some thermal insulation cover the steel in the ladle. This can be achieved with considerable advantage using a fluid, sulfur-free, highly reactive synthetic slag. The basicity can be controlled and desulfurization achieved. Materials currently in use are usually based on the calcia-alumina system with additions of fluorspar and perhaps the presence of a small amount of a reducing agent, such as carbon, aluminum or calcium silicide. The promotion of desulfurization can be achieved by good mixing and stirring of the system by a gas stream or by induction stirring. These materials must be dry to minimize hydrogen pick-up.

Gas Stirring and Purging

Bubbling of gas such as nitrogen, or in appropriate circumstances more expensive argon, in the steel ladle will promote flotation of inclusions and provide homogeneity of temperature and composition following the addition of ladle additives and deoxidizers. Gas or induction stirring also creates local metal currents flowing vertically downward, aiding in recovery of bouyant additions made by plunging or wire injection.

Immersion

Reactants can be added to liquid steel by direct immersion, for example, by plunging a metal cannister containing the reagent.

Lance Injection of Solids

Several processes are currently in use for injection of solid particulate metallic reactants such as magnesium, calcium, calcium-silicon, ferrosilicon, aluminum, or elements for microalloying; non-metallic fluxes, such as CaO or CaO-Al_2O_3; or mixtures of these additions. One is illustrated in Figure 9 (17). These systems generally use a refractory coated lance immersed deep in the ladle, and are based on dense phase transport. They utilize only 30 pct gas by volume which causes relatively little disturbance of the bath for injection of a substantial amount of reactant. The capital and operating costs for lance injection are generally higher than for alternative systems, and may show lower efficiencies because the lighter reactants tend to float out in the gas bubble stream.

Wire Feeding

Solid reactants in the form of wire or as a core in a wire can be directly fed into the steel ladle, continuous casting tundish, or under some circumstances directly into the mold; all with high recovery of reactant. The operating cost for these systems is relatively low, depending primarily upon the economics of reactant preparation.

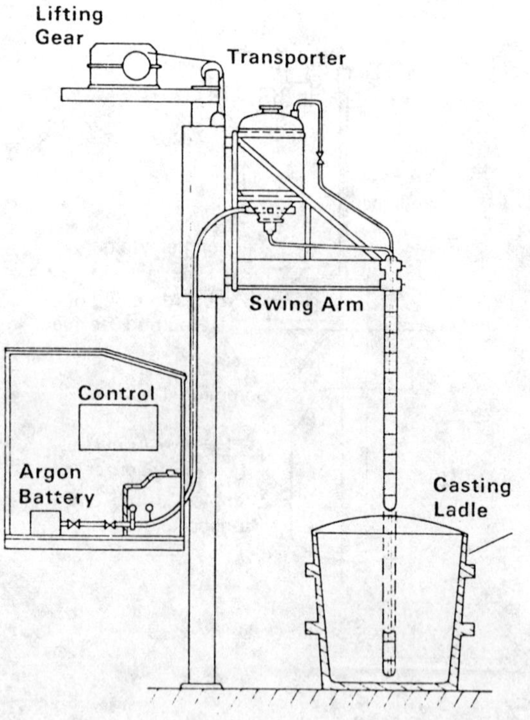

Figure 9. Schematic section of desulfurizing station.

A basic or neutral (high alumina) ladle lining is required for desulfurization to low levels, since these reactants will usually attack acidic linings, such as fireclay. High alumina refractories, 70 to 90 pct alumina, are typically used. A suggested installation is a 70 pct alumina lining with fine-grained magnesite refractory at the slag line.

The adoption of steel refining by ladle metallurgy permits the steelmaking processes to operate in a lower cost and higher productivity mode while simultaneously ensuring production of high quality steels.

REOXIDATION AND INCLUSION MINIMIZATION

On the basis of the strong affinity between oxygen and liquid steel, and noting the necessity to restrict pickup of oxygen during teeming operations, several practical steps can be taken to control reoxidation. The first of these is the protection of the steel pouring stream, whether from ladle to ingot mold or tundish, or from tundish to continuous casting mold. Furthermore, recognizing that reoxidation may occur from a number of sources, the separation of reoxidation products, or deoxidation products formed during cooling of the steel in the teeming operation, can be promoted. In particular, in continuous casting the appropriate design of tundish in terms of residence time of the steel in the tundish and the appropriate fluid flow patterns can enhance separation of oxides from the liquid steel. Similarly, appropriate flow patterns in the mold of the continuous casting machine can be developed under conditions where the liquid metal stream into the mold is diverted through a pouring tube with suitably directed exiting streams.

Protection of Pouring Streams

Recognizing that oxygen from the ambient atmosphere will readily react with liquid steel, the steelmaker can avoid oxygen pickup by shrouding the liquid steel pouring stream, and in point of fact can minimize reoxidation by maintaining a smooth compact pouring stream, in contrast to one which is rippled, ragged or spraying (18) as indicated in Figure 10. A mechanical shroud, such as the pouring tube used between tundish and mold in the continuous casting of large steel sections, can be applied as illustrated in Figure 11 (19). A similar shroud of even larger diameter has been utilized between ladle and ingot mold or tundish (Figure 12). One of the difficulties of this type of shroud is in the mechanical manipulation during the teeming operation, recognizing that incomplete seals can provide an opportunity for ingestion of even greater amounts of oxygen than the stream would be exposed to under normal conditions.

Figure 10. Water model study of smooth and rough streams showing the effects of gas entrainment on conditions in the mold pool.

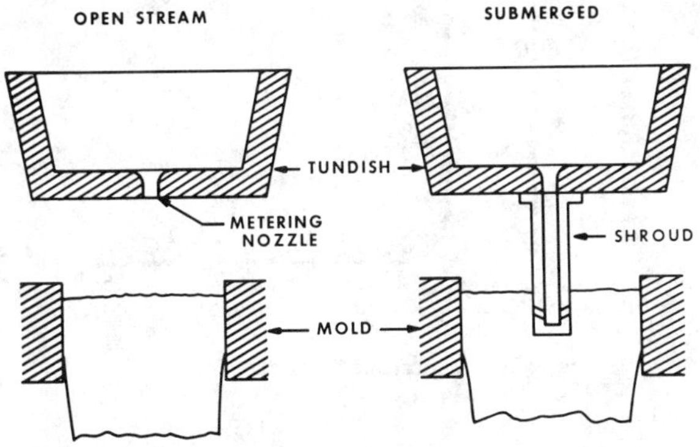

Figure 11. Casting techniques.

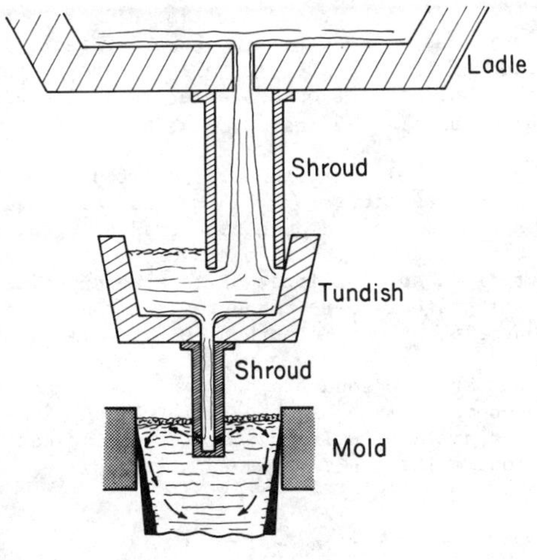

Figure 12. Pouring shrouds--ladle to tundish and tundish to mold.

Gas shrouds have been applied, particularly between tundish and continuous casting mold, utilizing inert gases such as argon or nitrogen or the products of combustion of a reducing flame which supply some heat at a critical area of the pouring system. Flooding the mold with inert gas can limit the available oxygen for reoxidation of the steel as shown in Figures 13 and 14 (20). Also, liquid nitrogen has been used to generate a gas blanket around the molten steel pouring stream to protect the stream and the metal surface.

Figure 13. Combustion of mold lubricant within the mold in the absence of a protective gas.

Figure 14. With protective gas injected through the manifold, the flames within the mold are extinguished.

One of the difficulties in continuous casting of small sections has been the protection of the pouring stream from mold to tundish because of the inapplicability of a pouring tube and the mechanical difficulty caused by the oscillation of the mold relative to the fixed tundish. In one instance, this has been overcome by use of a flexible bellows, as illustrated in Figure 15, and successfully applied to the continuous casting of special product quality steel bars.

Protection of the surface of the molten steel pool in the continuous casting tundish has been achieved through the use of suitable synthetic slags. Metal level in the tundish, as well as fluid flow pattern, is important in avoiding the ingestion of the slag layer into the metal stream flowing downward into the mold.

The use of protective slag or mold powder layers on top of the metal in the mold can limit reoxidation. However, with open stream pouring these protective layers

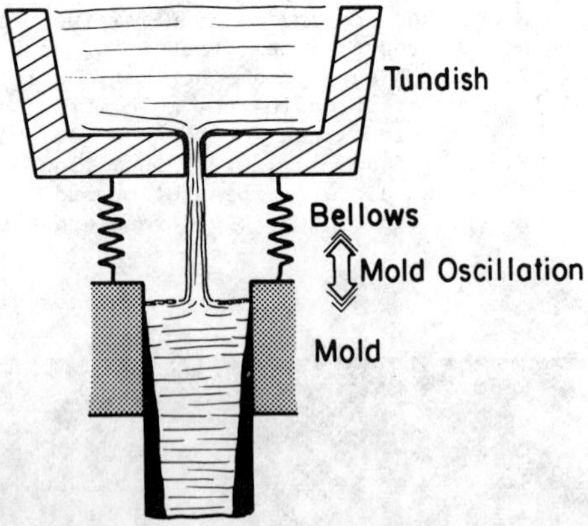

Figure 15. Bellows between tundish and mold for casting billets.

would be impacted into the body of the molten pool and are normally not used in open stream pouring.

Separation of Inclusions

The mechanisms and mechanics of separation of deoxidation and reoxidation products from liquid steel is a very extensive subject. With practical aspects in mind, the present discussion will be limited to a brief description of fluid flow in continuous casting tundishes as related to separation of reoxidation and deoxidation products. Also, separation of inclusions in large section molds has been studied in detail with the result that suitably directed streams using a bifurcated nozzle, rather than a straight through bore, are acknowledged to improve inclusion separation. The utilization of tundish reservoir capacity, as well as fluid flow patterns developed through the use of dams and weirs, and control of liquid steel level can promote separation of inclusions during passage of the steel through the tundish system (21), as illustrated in Figures 16-18.

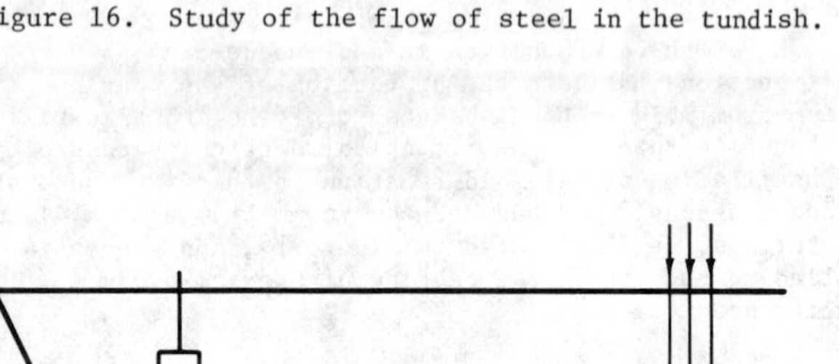

Figure 16. Study of the flow of steel in the tundish.

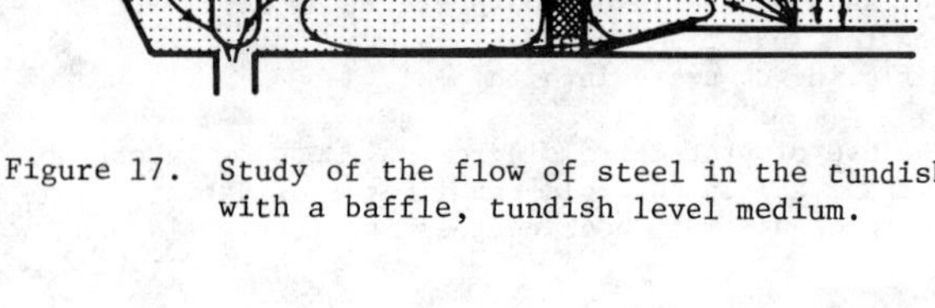

Figure 17. Study of the flow of steel in the tundish, with a baffle, tundish level medium.

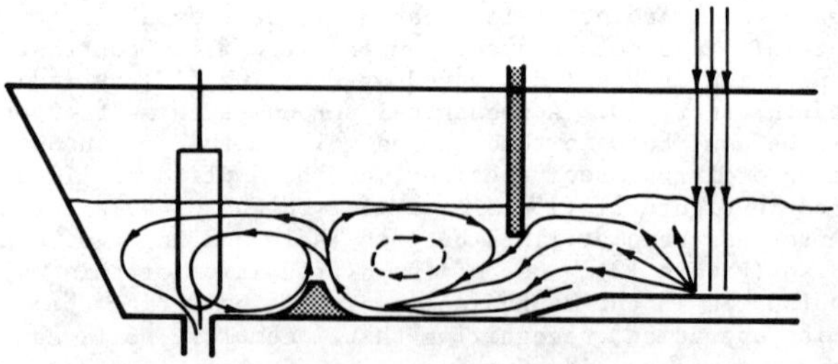

Figure 18. Study of the flow of the steel in the tundish, with a baffle and a deflector, tundish level medium.

CONTINUOUS CASTING

Several arrangements are now in commercial use for the continuous casting of steel. The principal types of continuous casting machines are vertical, vertical with bending, curved or S-strand with either straight or curved mold, and curved strand with continuous bending. Examples of these types of continuous casting machines for slabs are presented in Figure 19 (19).

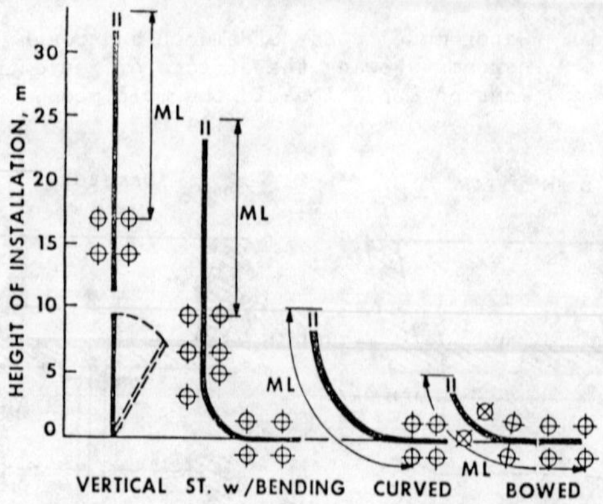

Figure 19. Four basic caster designs.

Most of the original continuous casting machines for steel were vertical machines. The vertical machines with bending, and also curved strand machines, although more complicated in their construction, were developed to minimize the height of the machine and allow installation in existing plants without modification of crane height. The four basic caster designs are shown in Figure 19 with an indication of the required installation height and the corresponding solidification distance or metallurgical length (ML) on a relative basis for slabs.

Nearly all steels continuously cast today are killed with typical compositions ranging from low carbon grades for sheet and strip to medium carbon grades in the range of 0.2 to 0.6 pct carbon, with manganese contents over 0.6 pct for plate applications. Extensive experiments for casting rimming steels were conducted by Atlas Steel at Welland, Ontario, and by McLouth Steel as well as the Steel Company of Wales. However, the cost and the precise control necessary to produce a sound rim zone made this process non-competitive. In the Soviet Union, some semi-killed steels are cast using a practice of low casting rates and long molds, up to 5 feet in length.

Mold Design

Special attention has been given to problems

associated with the design and material requirements for molds. A number of different designs have been used including thin-walled tube-type molds, solid molds and molds made from plate. The use of plate molds was found to provide excellent mold life and to avoid the necessity for manufacture of molds from solid copper blocks with marked reduction in costs. Steel and brass, as well as copper, have been used for molds, but the outstanding material is nearly pure copper with small additions of elements that promote precipitation hardening or raise the recrystallization temperature, since both effects apparently provide longer mold life.

The most suitable length for a continuous casting mold has been found to be between 24 and 30 inches, a range which seems to hold true regardless of section size. This surprising result can be explained by the fact that with smaller sections and higher casting rates, higher heat removal rates are achieved. Also, a thinner skin on exiting from the mold can be permitted for smaller sections than for larger sections. At higher casting rates the use of tapered molds is necessary to increase heat removal rates. Recent development work has provided movable mold sections which allow section size changes during casting.

Oscillating molds, as used by Junghans, have been adopted almost universally, although fixed molds can be employed successfully with efficient lubricating systems. The oscillation is usually sinusoidal, a motion which can be achieved easily with a simple mechanical arrangement. A fairly short stroke and a high frequency are used to provide "negative strip" in which the mean velocity of the mold movement is greater than the speed of withdrawal of the casting strand.

Secondary Cooling

The amount of heat removed by water spray depends on the volume of the water, its temperature, and in particular the method of delivery, including spray pressure. This latter variable of pressure is important in that the spray should be sufficiently intense to penetrate the blanket of steam on the surface of the solidifying strand. The thermal conductivity of steel is relatively low, and consequently as the surface temperature decreases and the shell thickness increases, cooling water has a limited influence on the solidification characteristics. The interrelationship between support rolls, particularly for the casting of wide slabs, and the spray water and its delivery characteristics is particularly important.

Quality Considerations

The operation of a continuous casting strand to produce good quality steel on a routine basis, uniformly and reliably, can be the most valuable asset of the process. Development work continues in an effort to improve the control and production reliability, particularly with regard to internal and surface cracking and steel cleanliness.

Internal quality is less important for those products which have large reduction ratios. Radial cracks, center looseness or centerline cracking, minor amounts of gas evolution, and other internal defects are not as deleterious in rolled products. However, in the manufacture of heavy plate, these conditions can represent serious product defects.

For aluminum killed steel, subsurface inclusions are usually in the form of aluminum oxide. In some cases surface scarfing can be effective in removing these inclusions which could provide a surface defect in a final heavily rolled product. Unfortunately, this surface conditioning can result in substantial yield losses.

Cracking, depending upon steel grade, can be the result of intensive cooling or deformation in the casting strand. Surface cracking can be minimized by proper control within the mold and the upper spray zone.

HORIZONTAL CONTINUOUS CASTING

Horizontal casting systems have been explored for many years, but only in the recent past have they been successfully applied to steels. Oldsmobile Division of General Motors Corporation has been casting 95 mm diameter steel bar in a horizontal mold since 1969 (22,23). Casts of up to 24 hours without interruption were made which is believed to be the longest sustained continuous casting operation for steel at that time.

The Nippon Kokan (NKK) Steel Company of Japan has recently produced horizontally cast steel sections up to 210 mm square (24). Both systems depend on (a) a refractory nozzle or "break ring" at the entrance end of a stationary water cooled metallic mold, and (b) an intermittent motion for translating the bar, each forward stroke being followed by a discrete rest or dwell.

A cross section of the tundish and mold arrangement of the NKK machine is shown in Figure 20 (24).

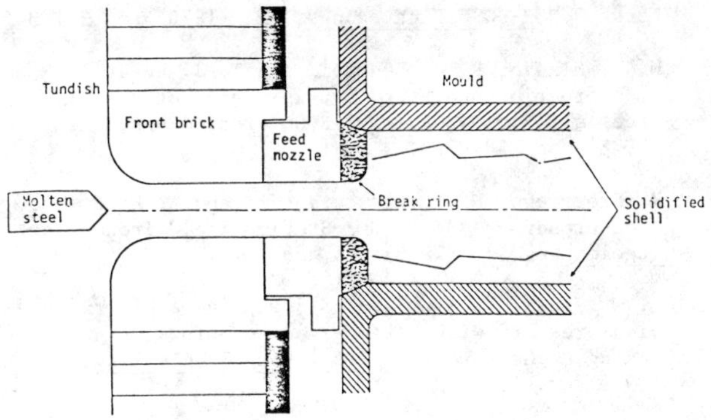

Figure 20. Tundish and mold/arrangement.

Process Advantages

The primary advantages of continuous casting include a substantially higher yield, a uniform product of excellent quality, and a higher overall productivity for capital investment and labor. These advantages coupled with the compatability and ease of integration with a steelmaking operation offer considerable potential for the process. Several large plants in Japan have "moth-balled" ingot facilities and are operating at 80 to 90 pct continuous casting for a wide variety of products. A number of plants, including most "mini-mills" have been built with continuous casting machines as the only solidification facility.

SUMMARY

This brief overview of the current status of technology in steelmaking offers a basis for detailed discussion of specific operating practices. Steelmaking processes have been outlined and areas where new inovations are being adopted have been described. Ladle metallurgy as a subsequent refining step for liquid steel has been presented in its various forms. Protection of steel from reoxidation and the promotion of separation of inclusions have been described. Finally, a brief review of the current status of continuous casting of steel has been presented, including horizontal continuous casting, a new and highly promising casting method. (See also refs. 25 and 26.)

REFERENCES

1. Editorial Board: R. D. Pehlke, Chairman; W. F. Porter, R. F. Urban, BOF Steelmaking, 5 volume Monograph Series, J. M. Gaines, ed., ISS-AIME, 1977.

2. R. S. Miltenberger, R. B. Morgan. R. Duffett, Jr. and W. E. Shepherd, U.S. Patent 3,323,906, June 6, 1967.

3. H. Knuppel, K. Brotzmann, H. G. Fossbinder, G. Savard and R. Lee, U.S. Patent 3,706,549, December 19, 1972.

4. Pierre J. Leroy, U.S. Patent Application, Serial No. 280,203, filed August 14, 1972.

5. Pierre J. Leroy, "Oxygen Bottom Blowing by the LWS Process," AISE, Pittsburgh, PA, Sept. 1972.

6. R. L. Deily, "Q-BOP: Year II," Journal of Metals, 25, (1973) pp. 33-41.

7. A. R. Parish and A. E. S. Pengelly, "BOS is a new technology emerging," Ironmaking Steelmaking, 6, (1979) pp. 114-122.

8. J. Hill and J. M. Edge, "The Q-BOP at Fairfield Works," NOH-BOSC Proceedings, 59, (1976) pp. 341-346.

9. W. F. Hladio, R. A. Bechtold and T. J. Herd, "Construction and Start Up Experience of Republic Steel's First Q-BOP," NOH-BOSC Proceedings, 61, (1978), pp. 384-393.

10. K. Brotzmann, W. T. Lankford, Jr. and A. H. Brisse, "Recent Advances in Q-BOP Steelmaking," Ironmaking Steelmaking, 3, (1976) pp. 259-267.

11. L. W. Frech, "Comparison of operating and metallurgical results with bottom- and top-blown oxygen steelmaking at Gary Works," Ironmaking Steelmaking, 6, (1979) pp. 110-113.

12. R. G. Groen, R. L. W. Holmes, J. W. Onuscheck and R. D. Pehlke, "Auxiliary Fuels," Chapter 16 in BOF Monograph Series, Vol. 5 (ibid., Ref. 1).

13. K. Wakabayashi, "The flexibility of LD operation," Ironmaking Steelmaking, 3, (1976) pp. 252-258.

14. E. G. Schempp, "Increased Scrap Melting Capacity for Oxygen Steelmaking--The New OBM-S Process," Iron Steel Eng., 56, (Dec. 1979) pp. 43-47.

15. 33 Metal Producing, 17, (Nov. 1979) p. 63.

16. W. M. Mahan, "Basic Oxygen Furnace Off Gas As A Source for Preheating of Scrap," AIME Steelmaking Proc., 62, (1979) pp. 119-125.

17. H. Pircher and W. Klapdar, "Controlling Inclusions in Steel by Injecting Calcium into the Ladle," Microalloying 75 Proc., Union Carbide Corporation, 1977.

18. W. Maddever, A. McLean, J. S. Luckett, G. Forward, Can. Met. Quart., No. 1, Vol. 12, (1973) pp. 79-88.

19. R. W. Joseph, N. T. Mills, "A Look Inside Strand Cast Slabs," AISI, New York, May 1975.

20. W. J. Maddever, L. Heaslip, A. McLean, J. Beaton, V. G. Remeika, L. C. Hutchinson, AIME NOH-BOS Steelmaking Proceedings, 59, (1976).

21. C. R. Jackson, "Design, Start-Up and Operation of Continuous Casters," University of Michigan Summer Conference on Continuous Casting of Steel, Ann Arbor, 1979.

22. F. J. Webbere, R. G. Williams and R. McNitt, "Steel Scrap Reclamation Using Horizontal Strand Casting," GM Research Publication GMR-11, October 20, 1971.

23. W. G. Patton, "GM Casts In-Plant Scrap Into In-Plant Steel," Iron Age, (December 9, 1971) pp. 53-55.

24. Private communication, Y. Miyashita, NKK Nippon Kokan, Tokyo, Japan, June 1981.

25. R. D. Pehlke, "Steelmaking--The Jet Age," Met. Trans. B, 11B, (Dec. 1980) pp. 539-562.

26. R. D. Pehlke, "Reoxidation of Liquid Steel," Radex Rundschau, Heft 1/2, (Jan. 1981) pp. 349-367.

THE ARGON OXYGEN DECARBURIZATION AND ELECTROSLAG REMELTING PROCESSES

H. J. Klein, B. H. Rosof,
C. L. Jeanfils, and J. H. Chen
Cabot Corporation
Kokomo, IN 46901

I. GENERAL INTRODUCTION

The secondary refining processes of argon oxygen decarburization and electroslag remelting are reviewed from both the technical and operational standpoints. Both processes have, in recent years, been extensively utilized in the production of high quality materials in the specialty metals industry. The AOD process which was developed primarily to enable one to produce low-carbon materials, initially stainless steels, without suffering the drastic loss in chromium from the alloy has been found capable of producing much higher quality products than originally conceived. In particular, due to the good mixing between the molten metal bath and the slag, excellent desulfurization is obtained. Also, the removal of unwanted impurities such as lead and bismuth as well as interstitials such as hydrogen is facilitated by the AOD process. In order to effectively utilize the AOD process, an understanding of the behavior of the various elements present in the bath, particularly carbon and chromium, is required. This is covered in the first section of this paper, which is a review of the AOD process. Also included is an AOD melt practice, the features of which can be understood in terms of not only the behavior of the various elements, but also the behavior of refractories and the need to control the process.

The electroslag remelting process has been shown to produce specialty steels and alloys which exhibit properties which are superior to those of single-melted materials and, in many cases, superior to those obtained when other remelting techniques are employed. The second section of this paper will review the important characteristics and parameters of electroslag remelting. The various process models proposed to predict pool profiles and liquid flow and, thus, segregation for the ESR process will be reviewed; also, the various slag-metal reactions leading to the removal of sulfides, oxides, and other unwanted inclusions as well as the retention of the desired reactive elements will be discussed. Special emphasis will be placed on the mode of heat generation and slag selection for optimum remelting.

In addition, the utilization of electroslag remelting for the production of novel shapes and components, some of which can be utilized either in the as-cast or wrought form, will be reviewed. Lastly, the properties obtained as a result of electroslag remelting of these specialty alloys will be reviewed and compared with properties generated from material produced by other techniques.

THE AOD PROCESS

II. INTRODUCTON TO THE AOD PROCESS

The argon-oxygen-decarburization (AOD) process had become in less than ten years after its commercial introduction the principal means in the world for refining stainless steel and other high chromium alloys. It had its origin in the observation that in the competition of chromium and carbon in a molten bath of stainless steel for oxygen, the equilibrium is shifted towards carbon oxidation if the carbon monoxide partial pressure is lowered, for example, by diluting it with argon. Over ten years of effort was expended before this simple observation found its commercial expression in the form of the AOD process.

The process was invented by W. Krivsky and developed by him and co-workers at the Union Carbide Corporation and by J. Saccamano and co-workers at Joslyn Stainless Steel (1,2,3). It was at the Joslyn plant that the final development of the practical process took place during the latter half of the 1960s. The process and various improvements are patented and use is by license from Union Carbide (1,4,5,6). The license generally includes the transfer of necessary technical details for practicing the process.

The AOD process was originally developed for the production of stainless steel. It rapidly found use in the production of many non-stainless grades. Almost immediately it was applied to the production of high chromium, nickel-base superalloys (7). Since then it has been used for the production of almost all classes of steels including those low in chromium (8), cobalt-base alloys (7), chromium-free nickel grades, and alloys used for their expansion, resistive, and magnetic properties. The incentive for use of this process for the production of low-chromium steels is its good degassing characteristics, permitting in some cases the pouring of grades sensitive to hydrogen cracking without the need for more expensive vacuum degassing, and its excellent sulfur and good inclusion removal characteristics, giving mechanical properties superior to those obtained when only arc furnace melting is used (8,9).

The excellent carbon removal characteristics of this process, even in the presence of high chromium, has permitted the economical production of a whole class of low-carbon alloys. Originally this included extra-low carbon versions of conventional stainless steels and nickel-base, corrosion-resistant superalloys. More recently, so-called super stainless steels have been developed. These alloys are characterized by higher chromium levels than in the traditional stainless grades, in some cases over 26%, as well as very low carbon contents, usually less than 0.02%. The low level of carbon in the extra-low carbon stainless steels and nickel-base alloys eliminates the need for post-weld heat treatment of fabricated structures, which is required in conventional grades in order to resolution the carbides precipitated in the heat affected zone.

A significant change in worldwide ferrochromium production has occurred due to the large-scale application of the AOD process to stainless steel production. Large quantities of expensive low-carbon ferrochromium are needed when producing stainless steel exclusively in the arc furnace, while virtually none is needed in AOD-processed heats. Inexpensive charge chromium, a low-grade, high-carbon ferrochromium, and scrap are the sources of chromium used in the charge of AOD-processed heats.

The original application of this process was in mill operations. Recently, foundries have started to install AOD vessels, which in cases have been as small as five tons (10).

In what follows, the emphasis will be placed on the production of stainless steels and other high-chromium alloys.

III. DESCRIPTION OF EQUIPMENT

Hot metal, produced in the electric arc furnace, is processed in what roughly can be described as a pear-shaped, refractory-lined vessel. The metal and slag fill the bottom third to half of the vessel, the large freeboard being needed in order to contain the metal and slag during the rapid injection of argon-oxygen mixtures into the bath. The vessel is mounted on a trunion ring, and the entire assembly can be tilted for tapping or sampling. Attachment of the vessel to the ring is such that a vessel with worn refractories can be quickly removed and a relined vessel rapidly put in its place. The use of multiple shells permits refractory work to take place without interrupting production.

Gas is injected into the bath through two or more tuyeres located in the side and near the bottom of the vessel. They are located along the back side of the vessel so that when the top of the vessel is tilted forward, the tuyeres are uncovered. This permits sampling and temperature measurement without blowing metal and slag through the mouth of the vessel onto operating personnel.

Each tuyere consists of two concentric tubes. Argon-oxygen mixtures or pure argon is blown into the bath through the inner tube, and pure argon is blown into the bath through the annulus formed by the two tubes. Only a relatively small amount of argon is blown into the vessel through the annulus; its function is to prevent the tuyeres from "burning up."

Most AOD installations are equipped with gas-handling systems capable of delivering, when desired, nitrogen in place of argon, permitting substantial savings in the cost of gas. Further savings are obtained by blowing oil-free, dry air through the tuyere when there is no metal in the vessel or when the vessel is tilted for sampling.

A pit large enough to hold the teeming ladle is situated in front of the vessel. A movable platform can be placed across the pit to permit sampling and temperature measurement through the mouth of the vessel.

IV. PHYSICAL CHEMISTRY

Little has been published on the physical chemistry of the AOD process. Most of what has been published is a description of the observed behavior of various elements in the AOD. Below, the behavior of various elements is considered in turn, as well as related topics.

A. Reactions with Oxygen

Reactions with oxygen can be divided into two groups: the reactions between carbon, chromium, and oxygen, which is at the heart of the process, and the reactions between other metallic elements and oxygen. We consider the latter first.

Most elements can conveniently be divided into two groups, those which are less and those which are more oxidizable than chromium. The less oxidizable elements include iron, nickel, cobalt, molybdenum, tungsten, and copper and are usually protected from oxidation by the large amount of chromium in the bath. (In iron-base alloys some of the iron oxidizes.) When these elements oxidize, they are readily recovered from the slag when the silicon or aluminum is added at the end of the heat for chromium reduction. Recoveries of 98% to 100% are reported (9). The elements which are more oxidizable than chromium include aluminum, silicon, and titanium. These elements must be oxidized before chromium and carbon can be oxidized. Finally manganese, at the levels usually found in most stainless steel grades, partially oxidizes when chromium oxidizes. It is also recovered at the end of the heat when the reductant is added.

The basic AOD reaction, the competition of carbon and chromium for oxygen, can be written as:

$$3\underline{C} + Cr_2O_3 = 3CO + 2Cr. \tag{1}$$

The equilibrium constant K is given by:

$$K = \frac{a_{Cr}^2 \, P_{CO}^3}{a_{Cr_2O_3} \, a_C^3},$$

where a_i and P_i are the activity and partial pressure of i, respectively. K markedly increases with temperature, and, all other factors being equal, it is generally easier to remove carbon and attain lower carbon levels at higher temperatures. Many authors take Cr_3O_4 as the appropriate chromium oxide. The actual oxide formed probably varies with alloy and slag system. In any case, the stoichiometry of the oxide formed does not affect the substance of what is presented below.

Reaction 1 is operative in both the arc furnace and the AOD vessel. The thermodynamic analysis of this reaction that is presented below is based on the work of Krivsky (1) and Rosof (7). In the arc furnace the minimum partial pressure of carbon monoxide needed for rapid carbon removal is one atmosphere, and the maximum chromium oxide activity is unity, which occurs when the slag is saturated with chromium oxide. One sees on examining Equation 1 that the minimum carbon content attainable in the arc furnace is determined by the chromium content. The only way to achieve a lower carbon content is to reduce the chromium content and increase the temperature. Thus, for example, at constant temperature and ignoring the effect of composition on activity coefficients, the chromium content has to be lowered by roughly $0.5^{1.5}$ in order to cut the carbon content in half during an oxygen blow once chromium oxide saturation is reached. Practices used to produce stainless steels in the arc furnace make use of this fact. The chromium content of the charge is less than the final chromium content desired; extended oxygen blows are used which oxidize some of the chromium in the bath, both lowering the chromium activity and raising the temperature, and a reducing mix is added after the oxygen blow which not only reduces the oxidized chromium from the slag, but also contains a significant amount of low carbon chromium units to raise the chromium content to the desired level.

It is usually stated that the AOD process works by diluting the carbon monoxide with an inert gas, shifting the equilibrium in Reaction 1 to the right, and thus preventing the oxidation of chromium. We note that blows using two parts inert gas to one oxygen, which would yield a gas containing one part inert gas to one part carbon monoxide if no chromium was oxidized, that is a carbon monoxide partial pressure of 0.5 atm., are capable in practice of lowering the carbon content to very roughly a tenth of the level attainable in the arc furnace. Yet on applying the same reasoning as used for carbon removal in the arc furnace, lowering the CO partial pressure by half should permit only a carbon level half of that attained in the arc furnace. As will be seen below, a more correct statement of how the process works is that it oxidizes chromium while oxidizing carbon, but that it limits the rate of chromium oxidation. This can be seen by using a simple equilibrium sparging model.

The rate at which carbon is removed from the bath, $\dot{n}_C$, is equal to the rate at which carbon monoxide leaves the vessel, $\dot{n}_{CO}$. The rate at which argon leaves the vessel, which is to all paractical purposes the same as the rate at which argon is injected into the vessel, $\dot{n}_{Ar}$, is related to $\dot{n}_{CO}$ by

$$\frac{\dot{n}_{CO}}{\dot{n}_{Ar}} = \frac{P_{CO}}{P_{Ar}} , \tag{2}$$

where P_i is the partial pressure of i in the effluent gas. Using Equation 2 and $P_{Ar} + P_{CO} = 1$ atm., the rate at which carbon is oxidized is given by

$$\dot{n}_C = \dot{n}_{Ar} \cdot \frac{P_{CO}}{1 - P_{CO}} . \tag{3}$$

The rate at which chromium is oxidized, $\dot{n}_{Cr}$, is given on using Equation 3 by

$$\dot{n}_{Cr} = \frac{2}{3}\left[2\dot{n}_{O_2} - \dot{n}_C\right] = \left[\frac{2}{3}\,2\dot{n}_{O_2} - \dot{n}_{Ar} \cdot \frac{P_{CO}}{1 - P_{CO}}\right], \tag{4}$$

where $\dot{n}_{O_2}$ is the rate at which oxygen is blown into the vessel. It is interesting to note that the rate of carbon oxidation given in Equation 3 does not depend on the rate at which oxygen is blown into the vessel as long as there is sufficient oxygen available, i.e. $\dot{n}_C < 2\,\dot{n}_{O_2}$. If sufficient oxygen is not available, the carbon oxidation rate is then limited by the rate at which oxygen is introduced into the vessel.

Equations 3 and 4 can be cast into the forms

$$\text{CRE} = \frac{\dot{n}_C}{2\dot{n}_{O_2}} = \frac{R}{2}\left[\frac{P_{CO}}{1 - P_{CO}}\right], \tag{5}$$

and

$$\frac{\dot{n}_{Cr}}{\dot{n}_{O_2}} = \frac{2}{3}\left[2 - R\,\frac{P_{CO}}{1 - P_{CO}}\right], \tag{6}$$

where R is the ratio of argon to oxygen injected into the vessel, and CRE is the carbon removal efficiency, defined as the percentage of the injected oxygen which combines with carbon. Because P_{CO} is determined by the carbon content, temperature, and other factors affecting the activity coeficients, as can be seen on examining Equation 1, the CRE is also a function of these factors. Several values of the CRE and the rate of chromium oxidation to rate of oxygen injection are given in Table 1. CREs of over 100% are not achievable due to the lack of available oxygen.

Table 1. Rates of Oxidation, Equilibrium Sparging Model

R	P_{CO}	CRE (%)	$\dot{n}_{Cr}/\dot{n}_{O_2}$
1/3	0.857	100	0
2	0.666	200*	- 1.333*
2	0.5	100	0
2	0.333	50	0.666
2	0.25	33.3	0.888
2	0.10	11.1	1.185
3	0.25	50	0.666
3	0.10	16.6	1.111

*Carbon removal efficiencies of over 100% are not attainable.

The carbon content is related to the carbon monoxide partial pressure by Equation 1. Ignoring the minor change in carbon activity with composition and taking the chromium oxide activity as unity, P_{CO} is proportional to the carbon content at fixed chromium content and temperature. In addition, P_{CO} is equal to the ratio of the carbon content, $\%_C$, and the carbon content, $\%^\circ_C$ corresponding to a carbon monoxide partial pressure of one. For example, when the carbon monoxide partial pressure is 0.1 atm, the carbon content is equal to $0.1\%^\circ_C$. $\%^\circ_C$ is a strong function of temperature since K is a strong function of temperature.

On examining Table 1 one can see that as the carbon content is lowered, chromium does oxidize. However, not much is oxidized. For example, when producing an alloy for which $\%^\circ_C = 0.24\%$ and for which the desired carbon content is 0.02%, and when using a gas ratio R of 2, the CRE will be 100% above 0.12% carbon and gradually fall to 11% by the time the carbon content reaches 0.02%. If the above equations are solved, it can be shown that the amount of chromium oxidized will be less than 1.25%. This is in marked contrast to the arc furnace where massive amounts of chromium have to be oxidized in order to lower the carbon content to this very low level.

In practice the temperature in the AOD rises with time during an oxygen blow due to chromium oxidation resulting in a decrease in $\%^\circ_C$. For this reason, analytical solutions to the above equations are of no interest. If the above model is correct, it should be possible to continuously adjust the gas ratio so that no chromium oxidizes. This cannot be done in practice. However, some oxidation of chromium is desirable in order to maintain, or even increase, the bath temperature.

B. Kinetics of Carbon Removal

Early developmental work on the AOD process consisted of injecting argon-oxygen mixtures into arc furnaces (1). No appreciable improvement in carbon removal was found in these shallow bath trials. It was only after argon-oxygen mixtures were injected near the bottom of a relatively deep bath that the desired effects were observed. The need for a deep bath was understood in terms of the underlying physical chemistry only after the process was commercialized. The kinetic model presented below was proposed by Freuhan (11).

When the argon-oxygen mixture is injected into the bath, the gas breaks up into small bubbles which rise to the surface. Because the rate at which mass transport of carbon and other constituents of the bath to and from the bubble is finite, the bubble can at first only react with a very thin layer of the surrounding bath. The oxygen not only oxidizes all the carbon in the layer but also chromium and other even less oxidizable elements such as iron. As the bubble rises, there is sufficient time for it to react with a thicker layer of the surrounding bath which supplies additional carbon to reduce at least part of the chromium oxide as well as the less oxidizable elements. The need for a deep bath in order to obtain the benefit of argon reducing the carbon monoxide partial pressure is now apparent.

The slag on top of the bath has no role in this model other than as a repository for unreduced chromium oxide. As soon as silicon and other highly oxidizable elements are oxidized, the process should behave as though the chromium oxide activity was unity. This is what seems to occur at lower carbon concentrations. On the other hand, if the slag, metal, and bubbles were in equilibrium, one would expect a gradual rise in carbon monoxide partial pressure as the chromium oxide activity in the slag increased to its saturation limit.

C. Chromium Reduction

The hallmark of the AOD process is the excellent chromium recovery attained. Recoveries in the range of 98 to 100 percent are common. The excellent recovery is due not only to the relatively small amount of chromium oxidized during carbon removal, but also to the very efficient reduction of chromium from the slag at the end of a heat with silicon or aluminum.

A reducing mix consisting of a reductant and lime is added to the vessel at the end of the heat to give the basic reducing conditions needed for good chromium recovery. The contents of the vessel are vigorously stirred by blowing argon into the vessel, giving excellent slag-to-metal contact and a high rate of solution of the added lime into the slag, conditions ideal for good chromium recovery.

D. Sulfur

The basic, reducing conditions and the vigorous stirring used at the end of a heat for chromium recovery are also ideal for sulfur removal. Appreciably lower sulfur levels than those obtained in the arc furnace are routinely obtained by AOD processing (9, 12).

E. Tramp Elements

Lead, bismuth, and possibly other low melting point, high vapor pressure elements, which have a deleterious effect on mechanical properties, are removed by AOD processing (9,12). Tin is not removed, however. Removal of these elements is particularly important when producing superalloys, since these alloys are used at high temperatures, and their high temperature mechanical properties are adversely affected by even small concentrations of these elements.

F. Gases

The constant flushing of the bath with gases in the AOD process leads to some particularly useful processing characterisitcs, especially for the control of dissolved gases.

Hydrogen pickup by the bath in the electric furnace is always a problem. Hydrogen contents of 2 to 4 ppm have been reported for AOD-processed material (9, 10). In at least some cases this is low enough to permit processing of hydrogen crack sensitive ferritic steels without vacuum degassing.

Flushing the bath with argon and argon-oxygen mixtures also removes nitrogen, enabling the production of alloys with far lower nitrogen levels than attainable in the electric furnace (5). Columbium-bearing alloys show far lower levels of nitrides in the final product when processed through the AOD vessel, for example. The low nitrogen levels attainable in the AOD vessel are beneficial in the case of the extra low carbon grades used for corrosion applications. The purpose of producing extra low carbon grades is to prevent the formation of carbides, which have a deleterious effect on corrosion properties. To some extent nitrogen can substitute for carbon in the carbides increasing the amount of carbides that can precipitate. AOD processing minimizes this problem.

Nitrogen can also be added to an alloy by substituting nitrogen for argon for part of the time the heat is in the vessel (13, 13a). Although nitrogen can also be added by using high nitrogen ferrochromium, adding nitrogen by flushing with nitrogen is less expensive. In addition, carbon pickup is not a problem in low-carbon alloys when nitrogen is added as a gas.

G. Models

Models of the AOD process which are based on the underlying physical chemistry have been published by Freuhan (11), Asai and Szekely (14, 15), Deb Roy, Robertson, and Leach (16, 17), and Bauer, et al. (18). The Freuhan and Deb Roy, et al. models are based in part on the concepts presented in Section B above. The Bauer, et al. model is an equilbrium model combined with a heat balance; it can be used for process control. Further discussion of these models is beyond the scope of this paper.

V. SOME THERMAL CONSIDERATIONS

Good control of temperature is important in any AOD operation. Factors which determine the desired operating temperature include the following: The temperature should be as high as possible in order to promote the oxidation of carbon relative to chromium (K for Reaction 1 increases with increasing temperature). The temperature should be kept low in order to obtain better refractory life. The bath must be sufficiently hot after the addition of the reducing mix at the end of the heat to permit the inevitable temperature loss which occurs while waiting for the results of chemical analysis of the pre-tap sample and the temperature loss due to final additions. The temperature within the ladle should be within the desired teeming temperature range.

Changes in the bath temperature occur due to oxidation of elements such as silicon, aluminum, and chromium which heat the bath; the addition of scrap, nickel oxide, lime and dolomitic lime, which cool the bath; the cooling effect of flushing the bath with inert gases; and general heat loss from the vessel to the surroundings. The amount of heat added due to the oxidation of carbon is relatively small.

There are two factors which make the simple calculation or experimental determination of the effect of any of the above variables on temperature somewhat less than completely straightforward. The first is refractories, and the second is the change in the CRE during a blow, or equivalently the varying rate of chromium oxidation during a blow.

In the case of refractories, several different effects are operative. Because of their finite thermal conductivity, only a fraction of the refractories dynamically change temperature with the bath. (Even this is an approximation.) Due to refractory wear, the total thermal conductivity of the refractories increases during a campaign. Finally, the starting refractory temperature varies from heat to heat due, among other factors, to the differing amounts of time between heats.

VI. REFRACTORIES

Short refractory life is the most important operating problem encountered by AOD operators. Improvements in refractories and operating practices have extended average refractory life from around 20 heats per campaign when the process was first introduced to over 100 heats per campaign in many shops. A relatively large literature exists on AOD refractories. Most of the material presented in this section can be found in References 19, 20, and 21.

The main body of the vessel is lined with bricks. The rate of brick wear is not uniform, being highest in the zone around and above the tuyeres, up to the trunion. It is believed that the high wear is due to mechanical erosion caused by the fluid flow pattern found in the vessel. In order to achieve balanced refractory wear, that is, the wearing out of all the bricks at the same time, the vessel is lined in the different wear zones with bricks of different thicknesses and grades, the thickest and highest grade brick being used in the zone around the tuyeres.

Two essentially different types of basic bricks are used, chrome-magnesite and dolomite. Basic bricks are used in order for the lining to be compatible with the basic slag used at the end of the heat. The basic slag is needed in order to obtain good chromium recovery. Both types of brick are in common use today, though chromium-magnesite bricks were introduced first. Most grades can be produced in vessels lined with either type of brick, but dolomite linings are preferred when producing chromium-free grades. The main disadvantage in using dolomite bricks is that the bricks hydrate when exposed to the atmoshphere when cold. This requires extra care in storing bricks and bricked vessels before use.

The key to good refractory life is control of slag chemistry and brick temperature.

Unless good control of slag chemistry is attained, the slag will rapidly attack the lining. The slag should be basic and preferably contain an appreciable amount of MgO, which is charged to the vessel in the form of dolomitic lime. It is not unusal for operators to aim for a V ratio of about 1.4 when using chrome-magnesite linings, and somewhat higher when using dolomite linings. Excessively high basicities have to be avoided when using chrome-magnesite bricks, or the slag will attack the acid chromite phase that is present. Since silicon, and, if used, aluminum, are oxidized during the first-stage blow, and additional silicon or aluminum is oxidized when added for chromium reduction towards the end of the heat, lime and dolomite must be added to the vessel both before the metal is charged to the vessel, after the first-stage blow, and with the reducing mix in order to maintain the desired slag chemistry.

The maximum and minimum brick temperatures are set by phase transformations which occur within the brick. Temperatures much above 1700°C are to be avoided with chrome-magnesite bricks, for example, due to liquidation of the silicate phase which acts as a binder, while repeated cycling through 900°C is to be avoided due to a volumetric phase transformation which leads to thermal spalling. Gas flames or insulating covers are normally used to keep the refractory face hot between heats.

VII. A SIMPLE AOD PRACTICE

A simple practice suitable for the production of one of the standard grades of stainless steel is given below. It is expected that many actual practices will differ, in some cases markedly, from the practice given here. The present intention is to capture the spirit of how the AOD is used. The large variability in AOD practices is due to different local conditions, such as type and availability of scrap and raw materials, as well as different solutions to the same operating problems. Comments have been added to the practice where appropriate. Process control, however, is discussed separately in the next section.

Additional descriptions of operating practices can be found in References 22, 23, 24, 25.

A. Arc Furnace

1. Charge the arc furnace with a combination of scrap and raw materials so that when it is melted, it will yield the desired final composition with the exception of the following major elements:

   a. Carbon - This element is high due to the inexpensive high-carbon ferrochromium used in the charge.
   b. Silicon - This element also comes in with the inexpensive ferrochromium. At least some is needed, however, to provide heat in the AOD vessel.
   c. Sulfur -This element may be high due to raw materials and is readily removed in the AOD process.
   d. Highly oxidizable elements which are added after AOD processing.
   e. Nitrogen - If called for in the alloy, it can be added in the AOD vessel by blowing with nitrogen.

   Also included with the charge is lime and fluorspar.

2. Melt down and back charge as necessary.

3. The remaining operations which occur in the arc furnace are highly dependent on the alloy and shop. These operations are directed at adjusting the composition of the bath. In many cases, however, the heat is transferred to the AOD vessel without any chemistry control in the arc furnace.

B. Transfer

4. Add lime, dolomitic lime, and fluorspar to the vessel. Additions to adjust the metal composition may also be put into the vessel at this time. The bath should contain at least some silicon and, at times, aluminum at this point. They oxidize before chromium and carbon and are needed for their ability to provide heat and raise the bath temperature. If the bath temperature is not sufficiently high after the silicon and aluminum are oxidized, carbon removal will be retarded, and chromium will be oxidized until the needed temperature for effective carbon removal is reached. Excessive chromium oxidation early in the heat will give hard-to-work slags and require more reductant in the reducing mix.

5. Tap the heat from the arc furnace into a top-pour ladle.

6. Remove the slag from on top of the metal in the ladle.

7. Teem the heat into a slightly tilted vessel.

8. Take a temperature reading and in some cases a chemistry sample. This information is used for process control.

C. First-Stage Blow

9. Inject a one argon to three oxygen mixture into the vessel. At first this oxidizes aluminum and

silicon. When these elements have been oxidized, oxidation of carbon and chromium takes place.

10. Continue the first-stage blow until a temperature of 1700°C is reached. (This may require the temperature to be measured several times before the desired temperature is reached.)

11. Sample for carbon. The result is used for process control.

12. Switch to argon on the tuyeres and add lime, dolomitic lime, and fluorspar. These additions are made to adjust the slag basicity and to cool the bath. The amount of the additions depends on the amount of silicon and aluminum oxidized. They are also added here because all of the slag-forming materials cannot be added before the start of the first-stage blow.

D. Second-Stage Blow

13. Inject a two argon to one oxygen mixture until the desired carbon content is reached.

14. Take a sample and determine its carbon content. If it is high, reblow the heat.

E. Reduction

15. Add the reducing mix. This consists of silicon and additional lime. Stir in by injecting argon at a moderate rate for several minutes. The lime is added to obtain the desired basicity for good chromium recovery and good refractory life.

16. Sample for composition. While waiting for the chemistry results, the vessel may be held in the sampling position to minimize heat losses due to flushing gas through the bath. A chemistry adjustment may be made if needed by additions to either the vessel or the ladle.

F. Tapping and Teeming

17. Check the temperature and, if need be, cool by argon stirring.

18. Tap the heat. Called for additions can be made to the vessel just prior to tapping or to the ladle while tapping.

19. Teem the heat.

Several common variations to the above practice include the following:

Use of nitrogen in place of argon until it is time to add the reducing mix. The argon used at the end of the heat flushes excess nitrogen from the bath. (The argon does not completely remove the nitrogen. Even heats made without nitrogen substitution contain some nitrogen.)

A three-stage instead of a two-stage blow. This is particularly useful when making the extra low carbon grades, but is also used in some shops for the production of the conventional grades. A one argon to one oxygen mixture would typically be used for the second-stage blow, and a three argon to one oxygen mixture, for the third-stage blow. Third-stage blows leaner in oxygen are sometimes used, but below three or four to one oxygen there is inadequate chromium oxidation to maintain the temperature.

A one-stage instead of a two-stage blow for low-chromium alloys.

VIII. PROCESS CONTROL

Virtually nothing has been published about process control. Private discussions with AOD operators reveal a wide variety of methods in use, running the gamut from statistical models and sophisticated models based on the physical chemistry (see e.g. Reference 18) of the process on one hand, to schemes based on experience and simple rules of thumb. A simple control strategy for a two-stage blow practice is presented below:

1. First-Stage Blow

The general strategy is to "blow for temperature." The aim temperature, which depends on the alloys and shop, usually is in the range of 1675°C to 1725°C. The amount of oxygen needed to achieve the aim temperature will depend on the bath composition, i.e. how much carbon, silicon, aluminum, and chromium are present; the transfer temperature; the temperature of the refractories (e.g. time lapse between heats); and even the refractory wear (rate of heat loss). An estimate of oxygen needed to reach temperature can be made on the basis of experience. The amount of oxygen blown can then be set somewhat below the estimated oxygen needed, a temperature reading taken, and the blow continued. The amount of additional oxygen needed can then be calculated on the basis of a rule of thumb such as an x°C temperature rise per 10 m^3 oxygen blown.

A control strategy such as this is particularly easy to use if the conditions at the start of the blow vary little from heat to heat. One can correct the amount of oxygen that needs to be blown for large deviations from standard conditions on the basis of the enthalpy of reaction or experience or both.

2. Second-Stage Blow

A carbon sample is usually taken at the end of the first-stage blow, and the carbon content used in the calculation of the length of the second-stage blow. Again experience is a useful guide for calculating the length of the blow. A useful tool for adjusting the length of the blow is the carbon removal efficiency, CRE, which, as covered above, is the percentage of oxygen blown into the vessel that is used to oxidize carbon. The CRE is a function of carbon content and temperature as well as other factors, but if these factors do not differ greatly from some "standard," the CRE is a useful control tool.

Because the carbon content, temperature, and other factors determine the CRE, one might expect the CRE to change during the blow. In practice the effect of increasing temperature tends to compensate for the decreasing carbon content, and the CRE tends to be nearly constant over a large part of the blow. In many cases, the blow can be divided into two parts, early, with a high carbon content and a large CRE, and late, with a low carbon content and a low CRE.

A carbon check is run after the second-stage blow. If the carbon is within acceptable limits, the reducing mix is added to reduce the chromium from the slag. If the carbon is high, the heat must be reblown. If reblown, the CRE is a useful tool for calculating the length of the blow.

3. Reduction

The amount of reductant that should be put in the reducing mix is a function of the amount of oxygen used and the composition of the bath at the start of the blow. The "excess oxygen calculation" is a useful tool. The excess oxygen can be defined as the amount of oxygen used to oxidize chromium, manganese, and, in iron base alloys, iron, or equivalently, the amount of oxygen blown into the bath which oxidizes elements other than carbon, silicon, and aluminum. Since the starting and pre-reduction carbon, silicon, and aluminum contents are known, the stochiometric amount of oxygen needed to oxidize these elements can readily be calculated. This amount of oxygen is then subtracted from the total oxygen blown into the bath to obtain the excess oxygen. The calculation, as presented here, is based on the assumption that 100% of the oxygen injected into the bath is used for oxidation, generally a good assumption. Enough reductant is then put into the reducing mix to reduce the oxidized chromium, manganese, and iron, that is to "combine" with the excess oxygen, plus an additional amount sufficient to give the desired residual of the reductant in the alloy. As before, corrections may have to be made to this calculation based on experience.

4. Final Check

After the reducing mix is added and stirred in, a final sample is taken and adjustments are made to the vessel and/or ladle as needed. If the residual amount of reductant is high, the excess can be oxidized with a short oxygen blow.

IX. OTHER PROCESSES

Other processes have been developed for the production of stainless steels (26, 27, 28). They also are based on shifting the equilibrium in the competition of chromium and carbon for oxygen towards carbon oxidation by lowering the carbon monoxide partial pressure.

In most of these processes a vacuum is used to reduce the carbon monoxide partial pressure (26, 27). Generally oxygen is blown onto the surface of the bath under vacuum in order to effect decarburization. An auxiliary means of heating the bath is provided in some variants. Bubbling argon introduced via a porous plug through the bath during and after the oxygen blow under vacuum is still another variant.

The CLU process (28) is a pneumatic process similar to the AOD, but in which steam is injected into the bath to both oxidize carbon and to give hydrogen, which acts as the dilutant instead of argon.

The current trend is clearly towards using the AOD process for the production of most chromium-bearing grades. The vacuum-type processes may, however, offer advantages when particularly low hydrogen levels or unusually low carbon plus nitrogen contents are needed.

THE ESR PROCESS

X. INTRODUCTION TO THE ELECTROSLAG REMELTING PROCESS

The development of the electroslag remelting (ESR) process essentially started with Hopkins's(29) early work in 1935. In the late forties, the Soviets, apparently independently, initiated work on ESR principally as a result of their electroslag welding developments. It is not the intent of this paper to review the early development of electroslag remelting which has adequately been covered in a number of papers (30, 31, 32,), but to discuss the current understanding, operation, and areas of utilization of this process.

Before proceeding into detailed discussions, a brief introduction to electroslag remelting will be given. A schematic of a basic electroslag remelting furnace is shown in Figure 1. The material (electrode) to be refined is remelted by passing a current through it into a molten slag which is resistivly heated and which, in turn, melts the electrode. Molten metal droplets form on the end of the electrode and fall through the slag forming an ingot in the water-cooled copper crucible. The process continues until the electrode is consumed forming the ingot. ESR is employed in order to improve the material quality by: (1) removal of oxide and sulfide inclusions, (2) control of the solidification rate, thus eliminating shrinkage porosity and minimizing or eliminating segregation, and (3) improving surface quality. In general, the ESR process has been shown to be capable of accomplishing these objectives, and materials produced by employing the ESR technique normally exhibit better property levels and, more significantly, less scatter in these properties, i.e., a lower standard deviation. In addition, the hot workability of the materials is significantly improved and, in some cases, superalloys which were not previously forgeable can be produced after ESR.

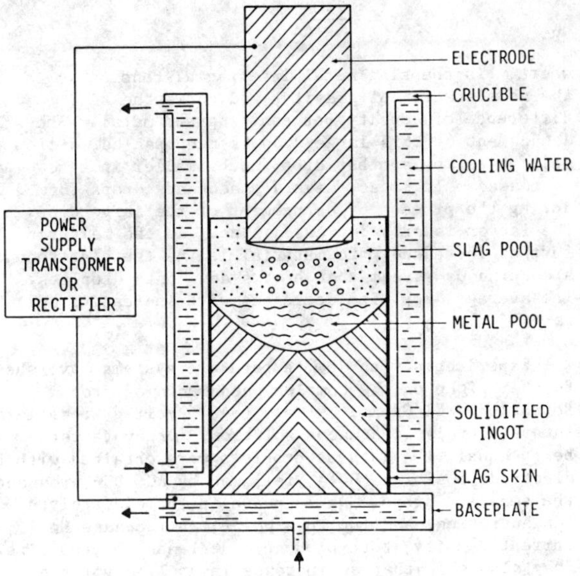

Figure 1. Schematic of the ESR process.

Due to the independent development of ESR in the U.S. and USSR, two different classes of alloys were initially remelted. In the U.S., high performance specialty alloys for the aerospace industry were processed by ESR instead of vacuum arc remelting because of the enhancement in product properties and improved forgeability. In the USSR low-alloy steels were remelted initially because of the association with the electroslag welding technology. Today, throughout the world a wide variety of materials including copper-base alloys, steels, superalloys, and refractory metals have been produced successfully using the ESR process. In the succeeding sections, the paper will concentrate on the metallurgical and operational characteristics of the ESR process which must be considered when processing specialty materials such as high performance steels and superalloys. However, the discussion will, of course, be applicable in most cases to electroslag remelting of all materials. The latter sections of the paper will discuss, in some detail, the possible ESR variants which can be utilized to produce hollow and shaped products and, finally, the resultant properties of electroslag remelted materials.

XI. TRANSFER OF METAL FROM ELECTRODE TO INGOT

One of the keys to the ESR process is the method of molten metal transfer from the electrode to the molten metal pool. Several stages can be considered in the transfer step. They include:

1. Heating the electrode material to its incipient melting point.

2. Remelting the alloy over its melting range. This is accompanied by the redissolution of many of the endogenous non-metallic inclusions.
3. Liquid flow along the electrode tip to a site where a drop is growing.
4. Detachment and fall of the drops.
5. Incorporation of the droplet in the liquid metal pool.

The droplet size is a function of a number of parameters. Campbell (33) showed that, above a certain diameter, the electrode size has no effect on the drop volume. Working with 50 Hertz AC, he showed that three regimes of drop formation could be distinguished as the melting rate was increased. At low melt rate, the drop radius r is given by:

$$r^2 = \frac{2.04\ \gamma}{g\ \Delta\rho} \tag{7}$$

where γ is the slag-metal interfacial tension, g is the acceleration of gravity, and $\Delta\rho$ is the difference of density between slag and metal. The detachment of this large drop is accompanied, at low melt rates, by another drop, much smaller in size. At increased melt rates, several secondary drops form during the process of detachment of the large drop. The latter is of a decreased size. As the melt rate is further increased, the material leaves the electrode as a continuous stream that breaks down into droplets. The average drop size decreases with increasing melt rate.

Experiments with low temperature systems have shed further light on the complex phenomenon of drop formation. El Gammal, et al. (34) remelted Woods metal into electrolytic aqueous solution. Other factors being equal, the largest droplets were obtained with DC electrode positive (DC+), followed by AC; DC- produced the smallest drops. An increase in the electrolyte concentration, and, up to a point, an increase in current density, both produce a decrease in drop size. They also show that an increase in applied voltage leads to a decrease in drop size. As the AC frequency is increased from 10 to 10^5 Hertz, the drop size first increases to reach a local maximum in the 200 Hertz region; it then reaches a local minimum in the 1000 Hertz region and finally increases to a steady value above 10^4-10^5 Hertz. The authors note that, at low frequencies, a pulsation of the forming drop can be detected with the eye.

The rankings of the current modes according to their effect on drop size may differ with the experimental systems. Whittaker, et al. (35) and Kojima, et al. (36) both report larger drop sizes with DC- than with DC+. These studies indicate that the factors that determine the drop size are many and complex. Medovar, et al. (37) discuss the relation between drops formed in small scale studies and larger operations. In large electrodes that are remelted flat, there are several centers of drop formation. In small-scale units, the current densities used, the configuration, and the slag motion often favor the formation of small drops at the tip of the electrode.

XII. SLAGS

Although the slag cap plays a number of roles in ESR, one of its main characteristics must be a high ohmic resistance in the electrical circuit. A slag of suitable resistivity and the slag volume are selected so that almost all of the electrical power is dissipated within the slag cap. The latter also constitutes:

1. A protection of the molten metal against exposure to the air.
2. A heat reservoir during operation, during a change of electrode, and during hot topping.
3. A supply of matter from which the slag skin is formed.
4. In slag-metal-gas reactions, the slag can provide reactants, incorporate reaction products, or serve as a transfer medium between the metal and the gas phase.

For the slag to operate properly in these various roles, the following slag characteristics are considered necessary:

1. An electrical resistivity such that the power supply can provide efficiently and in a stable manner enough Joule heat to reach the desired melt rate.
2. A low rate of volatilization.
3. A high degree of chemical stability. Neither should it contaminate the metal with undesirable elements (hydrogen, lead, often carbon, sometimes aluminum, etc.), nor should it oxidize the more reactive alloying elements: aluminum, titanium, silicon, ...
4. A low toxicity.
5. Availability and reasonable cost.

It is also considered desirable that the slag have:

1. The ability to incorporate inclusions and to remove sulfur and other impurities.
2. Properties which are not too sensitive to the changes of slag composition that take place in the course of the heat.
3. A liquidus temperature slightly below that of the alloy.
4. A low concentration in multi-valency cations.
5. The ability to form a skin that has some mechanical strength.

For most alloy-slag combinations, the requirement that the slag density be less than that of the alloy is not a problem. Considerations of density differences between alloy and slag, of interfacial tension, and of slag properties such as viscosity and heat capacity intervene also as secondary criteria in the slag selection.

To meet the requirements listed above, mixtures of CaF_2, CaO, and Al_2O_3 are frequently selected as process slags. These slags with additions of other oxides or fluorides, and pure or mixed fluorides are also used in some cases. In recent years, due to environmental considerations and problems in obtaining fluorspar, a significant level of interest has been given to low-fluoride slags, and to slags containing no fluorides.

Duckworth and Hoyle (31) provide an extensive review of the physical and chemical properties of slags used in ESR. Nafziger and co-workers review the phase diagrams and physical properties of fluoride (38) and of oxide-fluoride (39) slag systems. Their ternary phase diagram of CaF_2-CaO-Al_2O_3 has been reviewed and re-examined recently by Ries and Schwerdtfeger (40). Ogino and Hara have made recent determinations on the density, surface tension, and electrical conductivity of these slags (41). Hajduk and El Gammal give a compilation of electrical conductivity determinations in CaF_2-containing slags (41). These reviews present most of the currently available data on the physical properties of fluoride-based slags used in ESR. They also provide extensive literature surveys.

A. Volatilization of Fluorides

One drawback of fluoride-based slags is their relatively high vapor pressures at high temperatures. Schwerdtfeger and Klein (43) examined the volatilization rate of CaF_2, SiF_4 and HF in CaF_2 - CaO and CaF_2 - CaO - SiO_2 slags at 1600°C. The binary slag CaF_2 - CaO loses fluorspar by volatilization when exposed to a dry gas. In the presence of a moist gas, the rate of weight loss is significantly increased due to the formation of HF. For example, at a water vapor pressure of 15 mm Hg, the weight loss due to the formation of HF is three times that due to the vaporization of fluorspar. CaF_2 - CaO - SiO_2 slags behave similarly if the silica content is low. At high contents of silica, the weight loss is due primarily to the evaporation of SiF_4 if the gas is dry, and to that of HF if the gas is moist. The vapor pressure of AlF_3 is several orders of magnitude that of fluorspar (31). Povolotskii et al. (41) have measured the weight losses of slags of the CaF_2 - CaO - Al_2O_3 systems in dry argon at 1600°C. The highest weight losses for times less than thirty minutes occurred for slags of original composition 55% CaF_2, 45% Al_2O_3. Low weight losses correspond to slags rich in lime. After thirty minutes of exposure, a slag of original composition 70% CaF_2, 30% Al_2O_3 has lost twelve times more weight than the slag of initial composition 70% CaF_2, 30% CaO.

B. Low Fluoride Slag Systems

The evolution of environmental standards towards lower emission levels has fostered interest in ESR slags containing little or no fluorides (45-47). El Gammal and Hajduk (45) discuss the changes that have to be made to the operating parameters when a 50% CaO - 50% Al_2O_3 slag is substituted for a conventional one. The viscosity of the binary slag is more than ten times that of traditional high fluoride fluxes. The lesser contribution of convective transport in the slag leads to higher operating temperatures. The electrical resistivity of the binary slag is high: 1.5 ohm cm at 1600°C, 0.31 ohm cm at 1900°C. To operate in the usual voltage range, the slag cap height and the current intensity are lowered. Substantial reductions in specific energy consumption were achieved in a pilot scale 100 mm diameter mold. The degree of desulfurization was 60% provided that the iron oxide content of the slag was kept down to about 0.5% by deoxidizing the slag.

Pateisky (46) presents results of industrial trials with fluoride free slags. A good surface quality and a low specific energy consumption (1.1 kwh/kg) are achieved. Very high temperatures (1900-1950°C) are generated in the slag. The metallurgical quality is similar to that achieved with traditional slags, except for an increase in the amount of globular oxide inclusions. Rawson et al. (47) also found high levels of globular calcium aluminate inclusions in steels remelted with a 50% CaO/50%Al_2O_3 slag. Additions of 13% MgO to this slag were found effective in promoting the formation of submicron size inclusions and in preventing the formation of large (> 12 μm) globular oxide inclusions. Additions of magnesia were found necessary to maintain low oxygen levels in remelted titanium stabilized stainless steels. Desulfurization was also greatly improved with the MgO addition.

Rawson et al. (47) provide interesting data on the evolution of the slag composition with time. The slag composition changes because of three factors:

1. Partial oxidation of some alloying elements.
2. Evaporation of some fluorides.
3. Difference in composition between the slag skin and the liquid slag.

The slag cannot therefore be recycled entirely in the ESR process. Yakovlev et al. (48) indicate that 30% of recycled slag could be used without quality problems in plant trials with three different steel grades. Aside from the economic benefit, the use of recycled slag can be advantageous when low hydrogen contents are desired.

C. Hydrogen

Duckworth and Hoyle (31) examine the several ways by which hydrogen is introduced into the ESR process. The formation of hydroxides by reaction of atmospheric moisture with the basic oxides is a prevalent source of hydrogen. Lime, after exposure to the air, may contain 2 to 4 percent water. Prefused slags, recycled slags, and the other slag constituents are far less hygroscopic. Moisture has to be largely eliminated from the lime prior to its use in the ESR furnace. This is accomplished by heating the lime above 580°C, the decomposition temperature of the hydroxide. This operation is generally combined with that of prefusing the slag. Hydrogen can be further removed from the slag through the following reactions:

$$(OH^-) = \frac{1}{2}\{H_2O\} + \frac{1}{2}(O^=) \qquad (8)$$

$$(OH^-) + (F^-) = \{HF\} + (O^=) \qquad (9)$$

In an open system from which the gaseous reactants can escape, the reactions (8) and (9) lead to the complete elimination of the hydrogen. After sufficient time, the hydroxyl ion concentration becomes low enough for Reaction (8) to proceed to the left if the surrounding air is moist. At some point, the reaction rates of Reactions (8) and (9) will be equal, so that there is no further net elimination of hydrogen.

$$\{H_2O\} + 2\,(F^-) \rightarrow 2\,\{HF\} + (O^=) \qquad (10)$$

which is equivalent to:

$$\{H_2O\} + (CaF_2) \rightarrow 2\,\{HF\} + (CaO) \qquad (10a)$$

Masui, et al. (49) present extensive experimental data on ingot hydrogen contents. Except for special cases, the ingot hydrogen level decreases with time and reaches a steady value after times on the order of 30 to 90 minutes after the start of the remelt. With very low initial hydrogen levels in the slag, e.g. with recycled slags, the steady value is reached much faster. These authors show that, other factors being equal, the hydrogen in the ingot is higher the higher the lime content of the slag. This is true at the bottom of the ingot and when the steady value is reached. They develop a time-dependent mathematical model that predicts well the ingot hydrogen level in small and in large ESR ingots. The steady hydrogen level is the sum of a term proportional to the electrode hydrogen content and of a term proportional to the square root of the water vapor pressure. The physical reason for this is that, at very large melt rates, the ingot hydrogen level should be equal to the level in the electrode. For very low melt rates, the ingot hydrogen should reach a level which corresponds to chemical equilibrium with the hydroxyl ion concentration in the slag. The latter is proportional to the square root of the water vapor pressure.

Schwerdtfeger and Schubert (50) determined the solubility of hydrogen in $CaO-Al_2O_3$ slags. They show that the equilibrium hydrogen content is proportional to the square root of the water vapor pressure and that the constant of proportionality increases with the ratio % $CaO/\%Al_2O_3$. They suggest that, in $CaF_2-CaO-Al_2O_3$ slags, the solubility of hydrogen will be close to a constant at a given water vapor pressure and CaO/Al_2O_3 ratio. For a relative humidity of 80% in air at 20°C, the slag hydrogen contents would be 200 and 300 ppm at 1600°C and CaO/Al_2O_3 weight ratios of 0.68 and 1.21 respectively.

Rawson et al. (47) show that the equimolar replacement of CaO by MgO leads to a small decrease in the ingot hydrogen level (e.g. from 3.5 to 2.7 ppm). Practical measures to achieve extra low ingot hydrogen levels require a very dry atmosphere and a protective hood. The hydrogen levels in the starting electrode and slag should be low. These methods allow the producers of very large size ingots to achieve hydrogen levels of 1 or 2 ppm with lime-containing slags.(51,52).

D. Slag Skin Formation

Hoyle (53) reviews the factors affecting the surface quality of ingots. Under normal conditions in a static mold, the slag level progresses upwards steadily. The slag skin starts forming at this level. Its subsequent thickness is a function of opposing factors: while in contact with the wall, the skin has a tendency to grow, but once it is thick enough, the skin retracts and, because of the decreased heat transfer to the wall, it can redissolve partially into the slag cap. The decreased thickness of the solid skin and the increased hydrostatic pressure promote the closure of the air gap between skin and mold. This process can be repeated. Cracks in the slag skin are self-healing if they occur above the liquid metal pool. The slag composition is usually selected so that its liquidus is slightly below that of the metal. In such a case, and for not too low a melt rate, the slag skin in contact with the liquid metal will remelt or redissolve partially. The ingot solidifies against a molten film of slag and thereby a smooth surface is obtained. For this procedure to take place without pronounced ripple marks, it is necessary to have, below the metal meniscus, a "liquid head" or "cylindrical head," where the metal in contact with the slag skin is above the slag liquidus temperature. Ono and Takagi (54) showed that the height of this head increases with the melt rate and is somewhat affected by the slag composition.

Mitchell and co-workers have shown some interactions between electrochemical effects and slag skin formation (55). They also showed (56) that part of the remelting current does travel across the slag skin even in a nominally insulated mold. Hoyle (57) discusses the topic of current paths in ESR. This aspect of ESR appears to be far from being fully elucidated.

XIII. HEAT TRANSFER AND FLUID FLOW

Heat transfer and fluid flow in ESR have been studied with the help of analog and numerical models intended to simulate certain parts of the process. These models require assumptions and the knowledge of physical data which are rather scarce. Inferences from plant results and measurements performed on pilot-scale ESR units are rather straightforward. The high temperature and current intensity, the corrosiveness of the slag and metal to measurement probes, the opacity of the system, and the localized nature of some phenomena limit the amount of information that can be acquired by direct experimentation. So modeling and experimentation complement each other.

Kawakami and Goto (58) present their own experimental results and review the literature on temperature and voltage distribution measurements taken inside the mold in the slag and solidifying ingot. Tungsten-tungsten rhenium thermocouples and boron nitride sheaths are generally used. The authors make a distinction between the slag temperature distribution in the electrode negative mode (DC-) and that in AC or DC+. In these two modes, the slag temperature field is very uniform. There are steep gradients near the mold wall and in the vicinity of the gas and of the electrode. The uniform zone of temperature extends down to the pool interface. A small increase in temperature is sometimes observed at the interface. In the DC- mode, the temperature field is less uniform, and the gradients are not so steep. The hottest zone again appears to be the slag-pool interface. The temperatures observed in the hot zone are in the range of 1700°C to 1750°C for ferrous alloys melted in fluoride-based slags.

The voltage distribution in the slag is also affected by the electrical mode. In DC+, the voltage gradient near the electrode is high. This leads to a large density of heat generation in this region and is beneficial to the specific energy consumption. The heat generation pattern is more uniform in DC-. In AC, the voltage distribution is intermediate between that of DC- and that of DC+. Near the mold wall, the voltage profiles indicate that, even in the case of nominally insulated molds, there is a flow of current to the wall.

The flow pattern of the liquid slag has been studied experimentally in transparent molds by Campbell (33) and by Rawson et al. (59). In circular molds, the slag has a toroidal flow pattern. The flow is towards the electrode at the slag-gas interface. It is expected that the convective motion of the slag plays a major role in heat transfer in the slag cap. Numerical heat transfer models in the slag therefore combine the calculation of the electromagnetic field, of the temperature, and of the velocity. (60-63) The electromagnetic field provides the Joule heat and the electromagnetic body force. The velocity field is calculated through the solution of the turbulent Navier-Stokes equation. The calculation of the temperature field yields a prediction of the melt rate by integration of the heat flow at the electrode tip. A difficulty in solving this combined model is the proper formulation of the boundary conditions at the slag-pool interface since the ingot temperature field is not known a priori. Choudhary and Szekely (63) therefore combine an ingot heat transfer model with the slag model.

In some cases, the thermocouple that was used to determine the slag temperature could be left in place to determine the temperature of the liquid pool and of the solidifying ingot as solidification proceeded (58). These measurements imply a steep temperature gradient upwards in the liquid pool just below the slag. Between this zone and the liquidus, the temperature gradient is shallow. Campbell (33) observed a toroidal motion in the liquid pool with motion upwards against the transparent mold wall. Using an idealized pool shape, Dilawari and Szekely (60) calculate the velocity in this region and predict the possibility of two toroidal patterns in the liquid pool. Inouye (64) presents results of calculations showing one toroidal pattern with motion downwards at the wall. He also reports on results by Masui, et al. who showed that the observed range of angles between the columnar grains and the liquidus line goes from 75 to 115 degrees. Since the deviation of the columnar orientation from the vertical to the liquidus is an indication of the flow direction (65), both orientations of the toroidal motion take place under different conditions.

A method which is frequently used to determine a temperature isotherm inside the ingot is the marking of the pool profile by solid additions, such as tungsten powder. Solidification characteristics such as the dendrite arm spacing or the carbide primary precipitate size can be measured, and the local solidification time is calculated from them.

The mold wall temperature and the heat flux through the wall vary considerably with location. Pocklington and Patrick (66) relate the shape of the mold wall temperature versus time profile to the slag skin thickness. Mellberg (67) reports a peak temperature in the mold wall along the length of the liquid metal head. Typical peak heat fluxes are in the range of 700 to 2000 kw/m^2 (66, 68).

The modeling of heat transfer inside the ingot has been reviewed by Ballantyne and Mitchell (69). The heat conduction equation contains a term that accounts for the release of latent heat. It is solved by numerical methods. The heat flux boundary condition at the slag-pool interface contains a term that accounts for the sensible heat of the falling drops. The heat transfer coefficient at the mold wall is taken to be a function of position to reflect the formation of the air gap. The purpose of the heat transfer calculation in the solidifying ingot is to calculate the shape of the pool. Deep pools, or pools for which the slope changes abruptly at about mid-radius are considered undesirable.

XIV. SOLIDIFICATION

One of the major advantages of electroslag remelting is the soundness of the solidifed structure. The hot topping cycle and the insulating quality of the slag lead to the complete elimination of a center shrinkage pipe. Shrinkage microporosity is very small. For example, Choudhury et al. (70) report an as-cast density of 7.84 g/cm^3 at the center of a 2300 mm diameter ingot for which the maximum density is 7.845 g/cm^3. This corresponds to an as-cast microporosity of 0.06%. A forging ratio of 1.5 was sufficient to reach the maximum density.

Under steady conditions, columnar grains generally extend up to the axis of the ingot. Depending on the manner in which the hop topping is conducted, an equiaxed or a fully columnar structure can be obtained in the hot top. A small value for the angle between the columnar grains and the axis is considered desirable for good hot forgeability. Klein et al. (71) showed that this angle could be kept small in ESR and that the use of a large fill ratio (electrode to ingot area ratio) was, through its effect on pool shape, beneficial in keeping the cast structure oriented axially.

The secondary dendrite arm spacing and, if a primary phase forms during solidification, the size of this precipitate is determined by the local solidification time. For ingots of comparable sizes, the dendrite arm spacing is in an ESR ingot two to three times smaller than in a static cast ingot (68). The difference becomes smaller for very large size ingots. The range of variation of the dendrite arm spacing from ingot center to surface is not too large in an ESR ingot (68). Under steady conditions, the local solidification time is the ratio of the vertical distance separating liquidus and solidus to the ingot growth velocity. Mitchell et al. (72) compare two ingots of differing diameters produced at the same melt rate. The growth rate is faster for the smaller ingot, and, although a longer vertical distance separates solidus and liquidus in the small ingot, the solidification time is shorter in the small ingot. If it is required to keep the primary precipitate size or the dendrite arm spacing below a certain value, there is an upper limit to the diameter of ingot at which the alloy can be produced. For a given ingot size, there is a range of melt rate for which the local solidification time is not very sensitive to the level of growth rate.

In remelting, the achievement of a constant composition from ingot bottom to top is done by reaching a steady chemical regime as soon as possible. The liquid pool acts as a mixer where small variations in electrode composition are smoothed out. For normal size ingots, the center to surface differences in composition are usually below the limits of accuracy of conventional analytical chemistry. Even for a 2300 mm diameter ingot of 0.26% carbon low alloy steel, the difference in composition was about 0.01% carbon between center and surface (70).

A form of segregation sometimes encountered in consumable electrode remelting has been termed "freckle" segregation (73). Such a defect is characterized by a channel which is rich in alloying elements that segregate positively to the interdendritic regions. Some alloys are prone to this defect. The remelting of such an alloy is done under a narrowly defined set of operating parameters that have proven satisfactory. To better define the acceptable range of remelting parameters, mathematical models have been developed to predict stable and unstable interdentritic fluid flow conditions (74-77). Instabilities in this flow have been shown to result in channel segregation (78). While such models can predict instabilities when gross macrosegregation is present (74), they sometimes predict stable interdendritic fluid flow conditions in cases where the modeled ingot shows channel segregates (77).

XV. CHEMICAL REACTIONS

Chemical reactions take place during the remelting. It is desirable to maintain constant the composition of the solidifying metal. The goal is therefore to provide conditions for which a steady regime of production of refined metal is reached quickly, and is maintained. The slag will change in composition and in volume. Reactions of oxygen, of sulfur, and of some oxidizable elements will be discussed.

A. Oxygen

Oxygen is one of the major elements to be controlled because it affects cleanliness, sulfur removal, and losses of reactive elements. Miska and Wahlster (79) have reviewed the behavior of oxygen in the remelting of steels. They show the beneficial effect of lime and alumina, both of which decrease the activity of FeO in fluorspar-based slags. In aluminum-killed steels, the oxygen and aluminum contents correspond to equilibrium with alumina in the temperature range of 1750 to 1850°C. Miska et al note that, if the slag is saturated in alumina, a large number of small (< 14 μm) alumina inclusions remain trapped in the steel. Holzgruber and Plöckinger (80) show that the reaction product for silica formation in silicon deoxidized steels corresponds to equilibrium with the slag for a temperature range of 1700°C to 1800°C. The value of the product is strongly affected by the slag basicity.

The mechanisms of transfer of oxygen to the metal has been investigated. Miska and Wahlster (79) suggest that the oxidation of multi-valency cations by oxygen at the slag-air interface is one major way in which oxygen is introduced. Kojima et al. (36) compared the slag FeO levels for coated and uncoated electrodes and concluded that, in small ESR units, the oxidation of the electrode above the slag bath caused a several-fold increase in the slag FEO content. These authors investigated the effect of the electrode polarity on the oxygen behavior in DC ESR. They developed a mathematical model which is consistent with a rate of oxidation that increases as the current density increases (82). Since in practice, small molds are operated at high current densities, there is a difference in specific oxidation losses between units of different sizes.

Electrochemical reactions are known to take place at both the anode and the cathode. Because of the differences of deoxidation behavior in the DC- and DC+ modes, these reactions affect the oxygen content of the slag and of the metal. Workers who have studied the electrode processes in ESR (55, 83, 84) point out that many of these phenomena are only partially understood.

An example where electrochemical reactions affected the oxygen content and cleanliness of a nickel-base alloy was given by Klein et al. (71). Electrodes containing 20 ppm of oxygen, 0.15% aluminum, and 0.45% silicium were remelted in the AC and DC- modes. Ingots remelted in the AC mode showed a decrease in aluminum content but virtually no increase in oxygen content. Inclusions were of micron size and contained alumina and some magnesia. The ingots remelted in the DC- mode showed decreases in silicon, and major losses in aluminum and gains in oxygen. Aside from the micron size inclusions, larger (6 to 9 μm) inclusions were found. They contained aluminum silicates and calcium aluminates. It is thought that the small, micron and sub-micron size inclusions form mostly after solidification has started. The agglomeration and flotation behavior of oxide inclusions in electroslag remelted steels was recently studied by Elliott et al. (89).

B. Sulfur

The conditions in ESR usually favor an excellent desulfurization. The slag is normally very low in FeO and contains lime. Temperatures are also very high. Holzgruber et al. (80) observe that the ingot sulfur content increases from bottom to top if the slag is not exposed to gaseous oxygen. This indicates that the gas phase plays a role in the removal of sulfur. It is also observed that, in air, the quantity of slag has only a secondary effect on the degree of desulfurization achieved. Miska et al. (81) perform mass balances for sulfur and observe that, at high sulfur levels, the gas phase removes up to 80% of the sulfur introduced in he process. For slags containing silica, a basicity ratio of at least 1.5 is necessary to achieve maximum desulfurization. The latter is then in the range of 70 to 85% for high (0.045%) sulfur contents in the electrode.

Roshchin et al. (85) show that sulfide inclusions remelt and dissolve gradually while still in the electrode. Mellberg (86) shows that in the AC mode, the electroslag interface is the active site for transfer of sulfur from the metal to the slag. Inouye (64) and Schwerdtfeger (84) present mathematical models to calculate the evolution of the sulfur content during ingot growth.

C. Reactive Elements

Provided oxygen is present in the gas phse, and the slag is basic and low in FeO, there will be a natural tendency for the sulfur level of the growing ingot to stay constant. Hydrogen exhibits a similar tendency. The reason behind it is that both hydrogen and sulfur participate readily in reactions with a gas phase of constant composition. Once it has been oxidized, an atom of reactive element will normally migrate to the slag phase. To prevent the formation of certain inclusions, it is undesirable for the slag to be saturated in its oxides. It is recalled that, for thermal efficiency reasons, the slag weight is limited to perhaps five percent of the ingot weight. For these reasons, small oxidation losses can cause, with time, significant changes in the chemical activites of the oxides present in the slag.

If the rate of oxidation of a reactive element is decreased when the activity of the oxide produced increases, then it is probably advantageous to include some of that oxide in the initial slag mix. If on the other hand, the oxidation of the reactive element takes place away from the slag, then it is likely that a cleaner alloy will be obtained if the oxide produced is readily soluble in the slag. One would then avoid in the starting slag a composition too close to saturation in this oxide. So an exact knowledge of the reaction mechanisms that take place in ESR would be a valuable tool when an initial slag composition is to be specified. Mathematical models that use equations of the correct form with parameters adjustable to fit experimental data are also useful (84) especially since it often occurs that more than one element is competing for oxygen.

XVI. ESR PROCESS CONTROL

It is vitally important to control closely the melting of an ESR operation so that metallurgically sound and reproducible ingots can be produced. Effect of melting rate on heat transfer phenomena in a remelting ingot has been clearly illustratesd in early part of this paper. It is demonstrated, based on both experience and theory that defects such as segregation, freckles, etc., can be directly related to improper control of the melt rate. Once an optimum melt rate has been selected, it is critical that the optimum melt rate be approached in a controlled manner and, once the target rate has been reached, followed by a controlled melting and hot top cycle to ensure the soundness of the product.

Traditionally, under a given alloy and slag system, the control of melt rate is effected by the control of voltage and current by manipulating the set points of these two variables relying on the operator's judgement and experience. Lately, small computer systems, based either on microcomputer or minicomputer hardware, have been applied to assist the operator in such a critical and complicated task. The difficulties in obtaining good control without automation become ever more pronounced in situations where multiple furnaces are involved.

A. Control Philosophy

To optimize the energy efficiency during ESR it is desirable to operate the electrode at near the surface of the slag pool. Increasing the electrode penetration depth will displace the slag to the space in between the crucible and the electrode resulting in a greater heat loss through the crucible wall due to an increased area of exposure of slag to the cooling water. Under given conditions (ingot diameter, the electrode diameter, the resistance of the slag and its depth), the depth of penetration is dependent on the current density and the voltage. Increasing the current or decreasing the voltage will increase the depth of penetration and vice versa.

To minimize the penetration depth, ideally the voltage should be selected as high as possible, but this has to be balanced by the possibility of arcing between electrode and the surface of the slag pool.

The arcing will cause the vaporization of the slag leading to undesirable changes of the slag system, at the same time causing the oxidation of the electrode tip. This, therefore, requires maintaining a balance between the two contradicting factors. This means a strict requirement in the controlling of both voltage and current for refining to maintain the least penetration without causing arcing phenomenon.

This requirement and the melt rate requirement constitutes the basic control philosophy adopted by ESR operators. The control philosophy can be implemented in many ways, but in general the following approach has been found to be satisfactory.

B. Control of Melt Rate (Current Loop)

Both the current and voltage of a furnace must be mainipulated in order to control the melt rate to a predetermined value. Therefore, two control loops are used per furnace, one for current and the other for voltage. Due to the coupled nature of these two variables, adjusting one necessitates adjusting the other if stable melting conditions are required.

Traditionally analog controllers are used to control both current and voltage. These controllers perform the P & I logic and actuation of the control elements, which for the current loop is the saturable core reactor (SCR) and the hydraulic servo or screw motor for the voltage loop.

The block diagram of the current control loop is shown in Figure 2, where it can be seen that either the load cell or electrode position readings are used as the feedback of this loop.

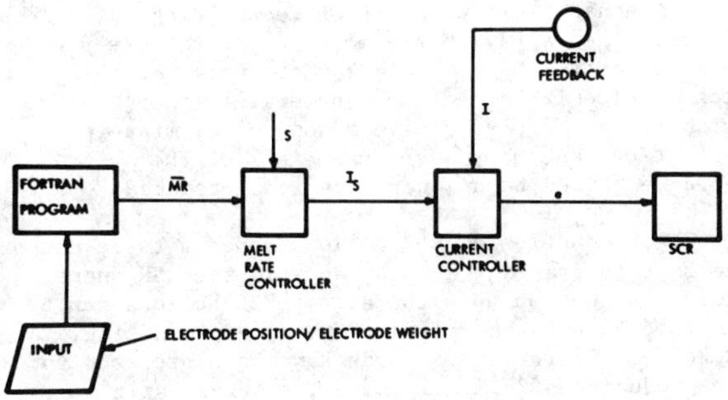

Figure 2. Current control block diagram

The current is manipulated according to the error between the measured melt rate and the melt rate set point by either continuous operation or by operation at a prefixed interval.

C. Control of Electrode Penetration (Voltage Loop)

While the current is being adjusted, the electrode position in the slag will change according to Ohm's Law, that is, the system voltage drop is equal to the current multiplied by resistance. This implies that an increase in current due to melting demends will necessitate the moving of the electrode toward the top of the ingot if the voltage is to remain constant. A decrease in current requirement on the other hand will move the electrode up if the voltage is to be held constant.

A change in the electrode immersion in the slag pool due to this current change is undesirable since it usually results in a poor ingot surface. The up-and-down motion of the electrode is replaced by a control scheme, where a change in current will result in a change in voltage but no change in the relative position of the electrode in the slag pool.

The instability of the current and voltage observed when the electrode is operated near the top of the slag has been postulated to be related to: (1) non-linearity in the resistance of the slag near the top of the slag pool, (2) arcing of the electrode to the slag pool, and (3) the formation of metal droplets at the electrode surface.

It has been reported by Thomas (87) that the resistance of the molten slag during ESR is a non-linear function with respect to position in the slag. According to Thomas, resistance, when plotted against distance from the top of the ingot, exhibits a curve as shown in Figure 3. This observed non-linearity which is responsible for the instability of the electrical parameters near the top of the slag pool is not well understood but is thought to be due to the non-linear relationship between resistivity and temperature as reported (88).

Alternately, when the electrode has a shallow penetration, it is thought that the unstable electrical condition can be caused by the occasional arcing between the electrode and slag, However, since the instability usually exhibits a regular cyclical pattern, this explanation seems unlikely unless the arcing is sustained as reported by Cameron, et al. (56).

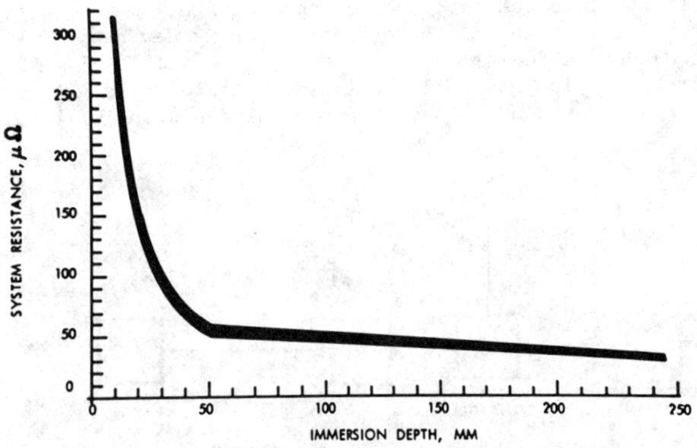

Figure 3. System resistance vs. immersions depth

A study by Kojima, et al. (90) suggests that the instability is caused by the formation and the subsequent detachment of the melted droplets during ESR. The formation of metal droplets decreases the resistance, but the detachment of the droplets, on the other hand, increases the resistance. This process repeats itself and causes the current instability.

Some experiments were done by Chen (91) both in experimental ESR and production ESR furnaces to correlate voltage and current instability to electrode submersion. Figure 4 shows the result of a submersion measurement test.

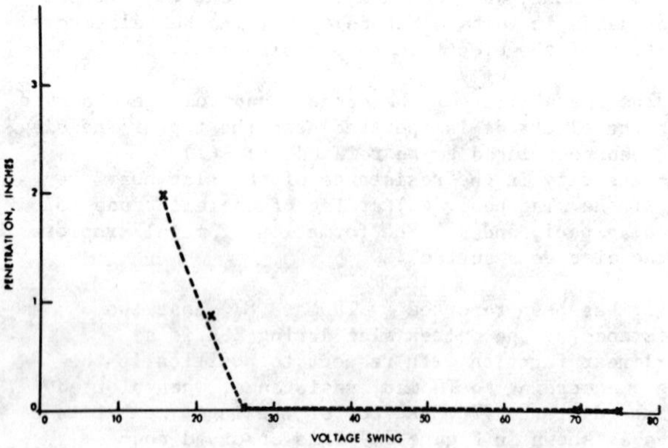

Figure 4. Penetration of electrode vs. voltage swing

This was done by interrupting the melting by withdrawing the electrode from the pool and measuring the slag meniscus on the perimeter of the electrode. A voltage instability in terms of amplitude versus electrode penetration could thus be established and used for the basis of devising a voltage control strategy.

Figure 5 then is a typical diagram representing the voltage control loop. It can be seen that the furnace voltage is used as an input signal to a capacitor amplitude amplifier. The amplifier basically acts as an integrator with a fixed time constant. A signal proportional to the amplitude of the voltage instability (swing) is the output from this device. This output is fed as a measurement to a P & I controller. The controller calculates the error signal as a result of summing the measured voltage swing with

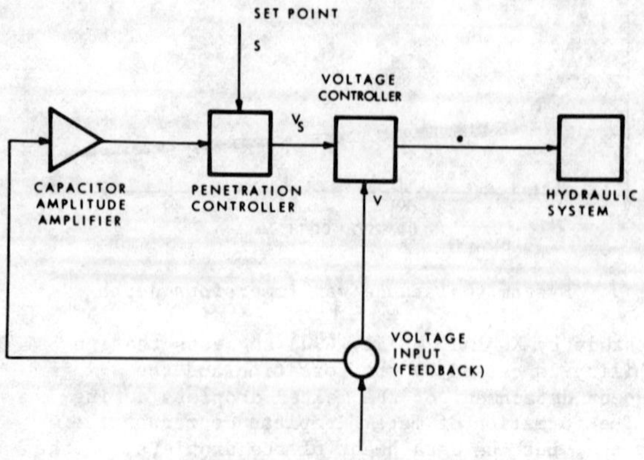

Figure 5. Voltage control block diagram

the voltage swing set point representing a certain electrode submersion. A new voltage set point is then calculated using the P & I algorithm and sent ot the downstream analog controller which in turn performs the actuation of the hydraulic servo or motor drive of the furnaces to adjust the position of the electrode.

In summary, the sequence of events is such that when a current change is made in order to satisfy the melt rate requirement, the voltage will need to be changed correspondingly in order to maintain a constant penetration of the electrode at all times.

D. Computer Control of ESR

The computer can be used to replace some of the control functions of the above scheme. The computer can control the melt rate by more frequently calculating the melt rate and comparing it with the melt rate set point or profiles. The usual means of achieving this is by current controller set point manipulation by the computer followed by an adjustment of voltage by monitoring the voltage swing as described above. Several computer systems have been successfully implemented and reported (92,93). It was reported that the computer improved the ESR operation by tightening the controllability of the system leading to better yield and energy efficiency.

XVII. ESR VARIANTS

During the development of the ESR process it was recognized that a number of variants were possible. Among the most popular were the production of shapes, slabs, hollows, and even more complex shapes by ESR methods. The production of slabs by electroslag remelting is a relatively straightforward procedure. The basic consideration is the fabrication of a suitable mold which will withstand the ESR process conditions described previously without distortion. In general, three approaches have been utilized in the design of these molds. The first is the use of machined, high velocity water passages which increases the heat transfer rate keeping the mold at a temperature which minimizes thermal stress, thus minimizing distortion. The utilization of this design allows for adequate mold strengths; however, fabrication is expensive. A second alternative would be to utilize what is known as a bookcase mold where the mold is free to expand or contract as required during the process without undergoing damaging distortion. In this design, the mold is constructed of four separate pieces with the ends being positioned inside the large, flat faces of the mold the desired distance to produce a slab of the desired width. The last alternative is to position restraining bolts in a predetermined array on the mold to prevent creep and mold distortion by mechanical restraint. At the present time it has not been determined completely which is the most effective method of eliminating distortion, and many slabs using all of the above-described techniques have been produced.

The production of ESR hollows appears to be one of the most interesting developments in the ESR shape field. Depending upon the alloy, ESR hollows can be utilized both in the as-cast or worked condition. A number of alternative methods have been proposed for the producion of ESR hollows. The three basic production processes in use today are the multiple electrode methods, Ujiie et al. (94) and Paton et al. (95), and the "single" electrode method, Klein et al. (96). The basic difference in the two methods can be seen in the schematic of the two processes as shown in Figure 6. The multiple electrode technique requires the preparation of special small electrodes and the use of a moving mold system which results in increased costs and poorer ingot surfaces. The method developed by Klein et al., allows the use of a standard ESR electrode and does not require the use of a moving mold, thus reducing the ESR hollow's cost. It should be pointed out that the latter method can also utilize multiple electrodes or a moving mold if the heat transfer conditions make their use advantageous. The increased solidification rate due to the reduced cross section of the ESR hollow results in a finer, more homogenous, ingot. This is illustrated by the significantly reduced secondary dendrite arm spacings of 41 and 56 microns for 16" O.D. X 10" I.D. and 20" O.D. X 10" I.D. HAYNES alloy No. 718 ESR hollow ingots as compared to a secondary dendrite arm spacing of 97 microns for a 20" diameter solid ingot. As previously mentioned, depending upon the application and alloy, ESR hollows can be utilized either in the as-cast or worked condition. Figure 7 shows the results obtained by Venal et al. (97) for 4330 modified ESR hollows in both the as-cast condition are comparable to those which are fully wrought. In addition, due to the absence of inclusions, the properties do not decrease as the forging ratio is increased beyond the 3:1 optimum for conventional ingots. The extremely good workability of the ESR hollows is illustrated in Figure 8 by the good surface quality of the HAYNES alloy No. 718 casings produced from a 0.16" O.D. x 7-1/2" I.D. ESR hollow (98). Mechanical property results have shown that more than adequate property levels can be generated by hot working these superalloys greater than 50%. The production of more complex shapes has also been reported in the literature by Paton (99), and this area is expected to develop in the future. It will be governed by the need for exceptional properties as well as the process economics of shape production.

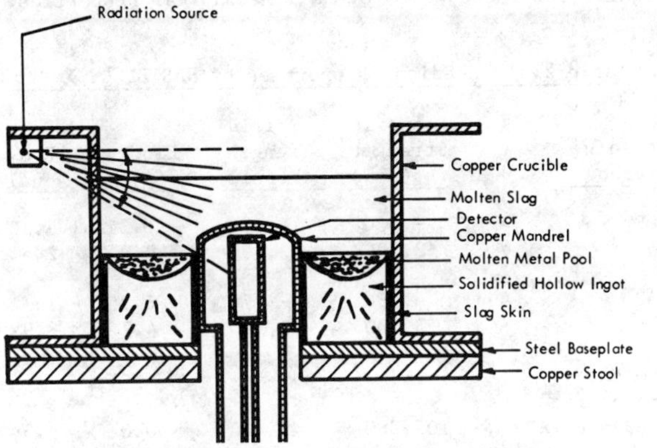

Figure 6a: Schematic diagram of single electrode ESR hollow unit.

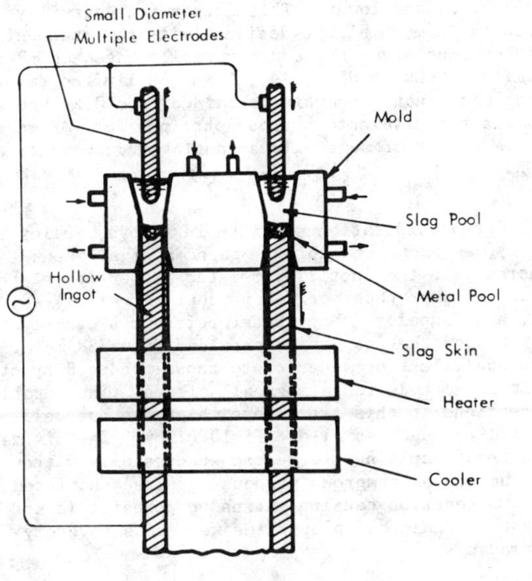

Figure 6b: Schematic diagram of multiple electrode ESR hollow unit.

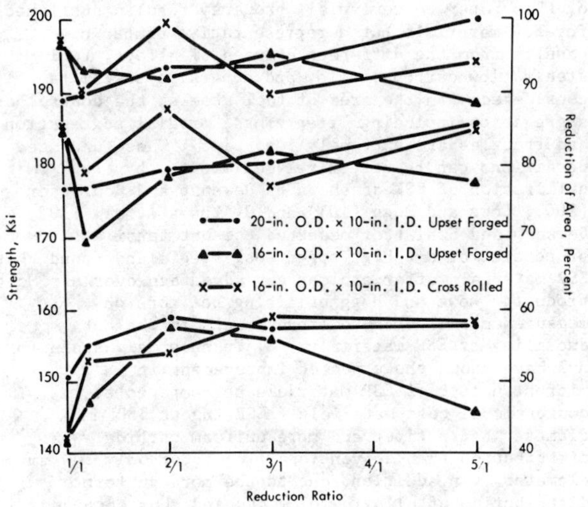

Figure 7: Effect of reduction ratio on room temprature tensile properties of AISI 4330 Mod ESR hollows.

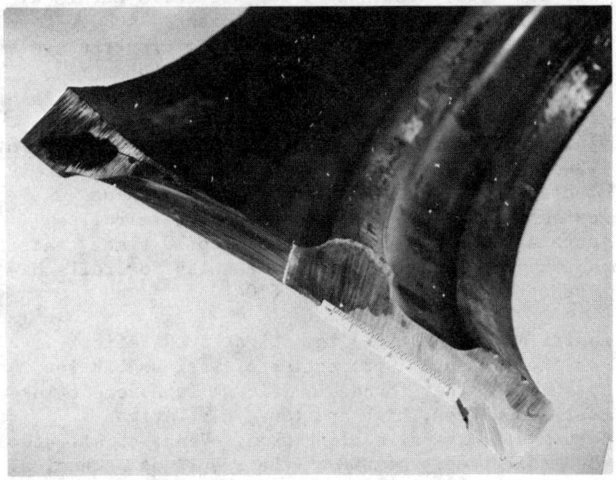

Figure 8. HAYNES alloy No. 718 casing produced from ESR hollow starting stock.

The last area which should be reviewed is that of a molten metal feed to ESR. This has been the subject of much speculation and work in recent years. From the energy standpoint, this development would be of great significance. However, it must be mentioned that one of the basic advantages of ESR is its controlled uniform solidification rate, and if one changes this rate in order to keep pace with a primary melting furnace, this main benefit would be lost. Thus, we must consider greatly reduced cross sections in order to accommodate the increased throughput. Secondly, the basic refining mechanisms have been shown to occur mainly at the electrode tip due to the increased temperature, surface area, and contact time in this zone. Therefore, any molten metal to ESR process must also take this into account.

XVIII. PROPERTIES OF ESR MATERIALS

As discussed in previous sections, the utilization of electroslag remelting is justified by the improvement in ingot quality which is reflected in better material yields and more consistent and often better mechanical properties. The nature and magnitude of these quality improvements obviously depends upon the chosen reference point and the primary melting procedures employed, i.e. open hearth, electric furnace, vacuum degassed, AOD, etc. to produce the original electrodes. In most cases the improvement in overall workability can be related to the reduction in inclusion content, less chemical segregation, and better as cast surface quality. It is beyond the scope of this paper to review all property results published for ESR materials but a representative number of results from the different classes of alloys, tool steels, low carbon steels, and superalloys will be considered. In the area of tool steels, the control of segregation including "tree rings," radial segregation and freckles are extremely important. These unwanted conditions can be effectively eliminated by the utilization of ESR as shown by Dewsnap and Schlatter (100), Lowe and Hogg (101) and G. Yuasa et al. (102). Dewsnap and Schlatter measured the hot tensile properties of M-2 high speed tool steels and found that ESR materials offer a clear cut advantage over Air Melt product. Lowe and Hogg utilizing hot torsion measurements also found improvements in the hot ductility of ESR materials. G. Yuasu et al. again for M-2 have shown the expected improvement in microstructure of ESR materials as compared to conventional cost materials. ESR ingots 350 mm diameter had a finer and more uniform carbide distribution than conventional ingots of only 200 mm diameter. In addition, due to the more uniform distribution of alloying elements and thus carbides, the maximum tempering hardness of the ESR-produced material was 15% higher.

For low alloy steels used for workrolls Muchizuki and Ohgu (103) have shown that ESR materials have a significantly more uniform macro- and microstructure with much less chemical segregation than in conventional steels. The secondary dendrite arm spacing was less than one half of that of a conventional ingot. The sulfur and silicon contents of the ESR refined steels have been reduced approximately 1/3 indicating a reduction in sulfide and oxide inclusions. Operating experience for these rolls indicate an increase in life from 3 to 10 times that of conventional materials. Similar results for rolls have been reported by Latash et al. (104).

Swift and Gulya (105) investigated the effect of ESR on the mechanical properties of various ESR low alloy steels. They found significant improvement in fracture toughness, high and low cycle fatigue properties, and tensile elongation. Venal et al. (106) studied the effects of ESR on the resultant mechanical properties of AISI 4330 Mod steel. They found the properties of the as cast ESR ingot to be comparable to those obtained on fully wrought materials. In addition, due to the reduction in sulfide and oxide inclusions, they found that excellent mechanical properties, including Charpy impact energies, were obtained over the entire forging reduction range of up to 12:1. In contrast for conventionally cast materials a forging ratio of 3:1 is considered optimum due to the elongation of inclusions during working which lowers transverse impact properties. They did, however, find an anomaly in properties at a ratio of 1.2:1 associated with a duplex microstructure at this low forging ratio. These results are however consistent with the findings of Sonomoto and Okuda (107). Wagner and Bor Avi (108) have reported using as cast ESR hollows for gun tube applications. In proof tests up to 143% of normal pressures no cracks or flaws were found. For large diameter ingots, those greater than 30" in diameter, both Takada (109) and Kriani (110) have found improvements in the wrought properties over as cast properties of ESR steels. This is presumably due to somewhat more segregation in these larger sizes. However, again the properties remained consistently good even at higher forging ratios, indicating that inclusion size and distribution was still excellent.

Property comparisons for superalloys has been the subject of much research and discussion over the last 15 years. In particular, most of the work has been aimed at comparing ESR and VAR results in order to justify the use of ESR in many applications previously specified as vacuum arc remelted material only. In the highly alloyed superalloys the criticality of the elimination of segregation and inclusions and their influence on resultant hot workability and final mechanical properties are of great significance. In early work on superalloys, Kelley (111), using the Gleeble test, found a five-fold increase in the hot working range of a U-700 ESR ingot as compared to VAR. Elliott (112) also found similar results for René 95.

In general, properties of ESR superalloys have been shown to be equivalent to or better than those of VAR materials. For HASTELLOY alloy X, Kelley et al. (113), as shown in Table 2, found that there was a small but significant increase in overall mechanical properties.

Table 2. Mechanical Properties of HASTELLOY X Bar

Refining Method		Tensile Strength (psi)	Yield Strength (psi)	Elong. (%)	R.A. (%)
ESR	Avg.	+ 530	+ 270	+ 0.31	+ 1.10
	σ	2,200	2,000	2.9	2.4
VAR	Avg.	- 580	- 270	- 0.25	- 1.14
	σ	3,400	2,300	4.7	3.9
Average Properties (ESR and VAR)		107,980	48,400	50.0	54.3

However, even more importantly, they determined that the standard deviation associated with these properties was lower. This latter fact actually results in much higher design limits for these critical applcations since these are normally based on -3 sigma limits. Cook et al. (114), based on limited data, concluded that properties obtained from ESR ingots were at least equivalent to those obtained for VAR material for a number of superalloys ranging from L605 to alloy 600.

For precipitation strength in alloys, Kelley et al. (115), as shown in Table 3, found a significant increase in the ductility related properties of ESR material over those of VAR product for alloy 718, René 41, and Waspaloy. Many other references can be sited which show that poperties obtained from ESR material are equivalent or superior to those produced by other melting methods for almost all alloys. The notable exceptions to this are alloys which are not well deoxidized such as alloy AF1410 (116). In this case, the ingot could not be melted successfully without the production of numerous gaseous defects associated with the CO reaction causing extensive porosity in the ingot. Mechanical properties were thus severely degraded.

Table 3. Comparison of AMS 5596C[a] Mechanical Property Test Results for ESR and VAR Flat Products of Solution Treated and Annealed HAYNES alloy No. 718

Mechanical Property	Overall Average	$\bar{X}_{ESR}$	$\bar{X}_{VAR}$	σ_{ESR}	σ_{VAR}	F	t	DF
R.T. UTS (ksi)	201.01	+ 0.38	- 0.38	3.57	4.95	1.918[b]	1.209	173
R.T. 0.2% YS (ksi)	168.70	- 0.32	+ 0.32	4.27	5.18	1.472[c]	- 0.914	186
R.T. El (%)	17.71	+ 0.38	- 0.38	0.89	0.89	1.013	5.812[b]	186
1200°F UTS (ksi)	163.86	+ 0.12	- 0.12	5.41	5.06	1.142	0.318	186
1200°F 0.2% YS (ksi)	140.56	- 0.40	+ 0.40	4.68	4.74	1.027	- 1.168	186
1200°F El (%)	13.48	+ 0.65	- 0.65	2.74	3.19	1.354[d]	2.987[c]	186
1200°F S/R Life (hrs)	4.67	+ 0.04	- 0.04	0.39	0.40	1.036	1.476[b]	186
1200°F S/R El (%)	1.93	+ 0.14	- 0.14	0.40	0.38	1.104	5.094[b]	186

There were 93 ESR tests, with 51 heats and 93 ESR ingots; and 95 VAR tests, with 40 heats and 80 VAR ingots.

[a]1775°F Solution Treatment.

[b]Rej. H_o: $\Delta\bar{X} = 0$ at 90% C.L.

[c]Rej. H_o: $\Delta\bar{X} = 0$ at 93% C.L.

[d]Rej. H_o: $\Delta\bar{X} = 0$ at 99% C.L.

REFERENCES

1. W. A. Krivsky, "Preparation of Metals and Alloys," U.S. Patent 3,252,790, May 24, 1966.

2. W. A. Krivsky, "The Linde Argon-Oxygen Process for Stainless Steel; A Case Study of Major Innovation in a Basic Industry," Metallurgical Transactions, 4 (1973), p. 1439-1447.

3. J. D. Ellis and J. M. Saccomano, "Recent Developments in Argon-Oxygen Reactor Stainless Making at Joslyn," p. 76-78 in Electric Furnace Proceedings, Vol. 27 (1969), TMS-AIME, New York, NY.

4. E. C. Nelson and N. R. Griffing, "Decarburization Process for High-Chromium Steel," U.S. Patent 3,046,107, July 24, 1962.

5. E. C. Nelson, "Decarburization, Deoxidation, and Alloy Addition," U.S. Patent 3,169,058, Feb. 9, 1965.

6. J. M. Saccomano, and J. D. Ellis, Jr., "Reactor," U.S. Patent 3,703,279, Nov. 21, 1972.

7. B. H. Rosof, "The Argon-Oxygen-Decarburization Process--A Special Melting Technique," p. 595-607 in Proceedings of the Fifth International Symposium on Electroslag and Other Special Melting Technologies, G. K. Bhat and A. Simkovich, eds., Carnegie-Mellon Institute of Research, Pittsburgh, PA, 1974.

8. D. A. Heckel, "AOD Operations at Allegheny Ludlum Steel Corporation, p. 127-128 in "Electric Furnace Proceedings, Vol. 32 (1979), ISS-AIME, New York, NY.

9. R. R. Netoskie, D. A. Heckel, and J. P. Wiencek, "Metallurgical Aspects of the AOD," p. 289-293 in Electric Furnace Proceedings, Vol. 35 (1977), ISS-AIME, Warrendale, PA,.

10. P. A. Tichauer and L. J. Venne, "AOD, a New Process for Steel Foundries, Promises Better Properties, Less Repair," Foundry Operations Planbook, April 1977, McGraw Hill, p. 35-38.

11. R. J. Fruehan, "Reaction Model of the AOD Process," Ironmaking and Steelmaking, 3 (1976), p. 153-158.

12. J. C. Hartman, "Metalloid Removal in the AOD," p. 286-288 in Electric Furnace Proceedings, Vol. 35 (1977), ISS-AIME, Warrendale, PA.

13. R. B. Moffett and J. M. Saccomano, "Use of Gaseous Nitrogen in the Argon-Oxygen Process," p. 59-60 in Electric Furnace Proceedings, Vol. 30 (1972), TMS-AIME, New York, NY.

13a. R. N. Dokken, Discussion of "Use of Gaseous Nitrogen in the Argon-Oxygen Process," p. 60-63, ibid.

14. S. Asai and J. Szekely, "Decarburization of Stainless Steel: Part I. A Mathematical Model for Laboratory-Scale Results," Metallurgical Transactions, 5 (1974), p. 651-657.

15. J. Szekely and S. Asai, "Decarburization of Stainless Steel: Part II. A Mathematical Model and a Process Optimization for Industrial-Scale Systems," Metallurgical Transactions, 5 (1974), p. 1573-1580.

16. T. Deb Roy and D. G. C. Robertson, "Mathematical Model for Stainless Steelmaking: Part 1 Argon-Oxygen and Argon-Oxygen-Steam Mixtures," Ironmaking and Steelmaking, 5 (1978), p. 198-206.

17. T. Deb Roy, D. G. C. Robertson, and J. C. C. Leach, "Mathematical Model for Stainless Steelmaking: Part 2 Application to AOD Heats," Ironmaking and Steelmaking 5 (1978), p. 207-210.

18. E. F. Bauer, R. N. Dokken, R. A. Hard, and U. K. Malhotra, "Process for the Decarburization of Molten Metal," U.S. Patent 3,816,720, June 11, 1974.

19. Y. Kinoshita, S. Oguma, and K. Harada, "Effect of Slag Composition and Temperature on the Wear of Refractories in AOD Vessels," TAIKABUTSU (Japan Refractories), 28 (1976), p. 12-18. [Henry Brutcher Translation No. 98401]

20. J. W. Kaufman and C. E. Aguirre, "Refractory Wear by Transfer, Decarburization, and Reduction Slags in the AOD Process," p. 74-79 in Electric Furnace Proceedings, Vol. 35 (1977), ISS-AIME, Warrendale, PA.

21. K. B. Yeo and R. A. Borowski, "Optimization of Slag Composition and Temperatures for Attaining Improved AOD Lining Life in the Production of Type 409 Stainless Steel," p. 286-292 in Electric Furnace Proceedings, Vol. 32 (1974), ISS-AIME, New York, NY.

22. W. C. Tobiason, "Experience with the Argon-Oxygen Process at Armco, Baltimore--Progress and Problems," p. 122-126 in Electric Furnace Proceedings, Vol. 32, (1974), ISS-AIME, New York, NY.

23. T. G. John and C. E. Aguirre, "Argon-Oxygen Practices at Jones and Laughlin Steel Corporation," p. 130-132 in Electric Furnace Proceedings, Vol. 32, (1974), ISS-AIME, New York, NY.

24. A. L. Hodge, "AOD Process and Its Eighty-Six Months of Growth," Ironmaking and Steelmaking, 4 (1977), p. 81-87.

25. J. C. C. Leach, A. Rodgers, and G. Sheehan, "Operation of the AOD Process in BSC, "Ironmaking and Steelmaking, 5 (1978), p. 107-120.

26. J. C. C. Leach, "Secondary Refining for Electric Steelmaking," Ironmaking and Steelmaking, 4 (1977), p. 58-65.

27. Y. Suzuki and T. Kuwabara, "Secondary Steelmaking: Review of the Present Situation in Japan," Ironmaking and Steelmaking, 5 (1978), p. 80-88.

28. T. O. Norrman and O. Johnsson, "The CLU-Method for Stainless Steel Refining," p. 135-136 in Electric Furnace Proceedings, Vol. 32 (1974), ISS-AIME, New York, NY.

29. R. K. Hopkins, Manufacture of Metal Articles, U. S. Patent No. 2,191,475,M.W., Kellogg Company (1940).

30. R. K. Hopkins, "The Electric Ingot Process," in Proc. Electric Furnace Steel Conference, pp. 91-105, New York: American Institute of Mining, Metallurgical and Petroleum Engineers, 1948.

31. W. E. Duckworth and G. Hoyle, Electroslag Refining Technology British Iron and Steel Research Association, 1969.

32. B. I. Medovar, V. Yu. Latash, B. I. Muksimovich, and L. M. Stupak, Electroslag Remelting, Translation No. JPRS 22217, Arlington, Virginia, Joint Publications Research Service, 1963.

33. J. Campbell, "Fluid Flow and Droplet Formation in the Electroslag Remelting Process," Jl. of Metals, (7) (1970), p. 23-35.

34. T. El Gammal, I. vonHagen, and R. Müllenberg, "The Role Played by the Electric Current on Metal Droplet Formation in the ESR Process," p. 45-54 in Proceedings of the Fourth International Symposium on Electroslag Remelting Processes, ISIJ, Tokyo, Japan, June 1973.

35. D. A. Whittaker, R. G. Ward, and G. R. Purdy, "Effect of Electrode Polarity and Power Input on Melt rate, Pool Depth, Fall Space, and Electrode Droplet Size," in Part III of The Collection of Technical Papers Presented at the Second International Symposium on Electroslag Remelting Technology, Carnegie-Mellon, Sept. 1969.

36. Y. Kojima, M. Kato, T. Toyoda, and M. Inouye, "The Formation of Liquid Droplet and the Behavior of Oxygen in Direct Current Electroslag Remelting Process," Trans. ISIJ, 15 (1975), p. 397-406.

37. B. I. Medovar, G. A. Boiko, and V. P. Serdyukova, "The Physical Modeling of the Process of the Electroslag Remelting of Consumable Electrodes," Problems of Special Electrometallurgy, Naukova Dunka, Kiev, No. 9, 1978, p. 34-48. [V. E. Riecansky Translation No. 1425/79]

38. R. H. Nafziger, R. L. Lincoln, and N. Riazance, "Fluoride Flux Systems for Electroslag Melting," p. 38-54 in The Electroslag Melting Process, R. H. Nafziger, ed., U.S. Bureau of Mines Bulletin No. 699, 1976.

39. R. H. Nafziger, "Oxide-Fluoride Flux Systems for Electroslag Melting," p. 55-75 in The Electroslag Melting Process, R. H. Nafziger, ed., U.S. Bureau of Mines Bulletin No. 669, 1976.

40. R. Ries and K. Schwerdtfeger, "Contribution to the Phase Diagram CaF_2-CaO-Al_2O_3," Arch. Eisenhüttenwes, 51 (4) (1980), p. 123-129.

41. K. Ogino and S. Hara, "Density, Surface Tension, and Electrical Resistivity of Calcium Fluoride-Based Fluxes for Electroslag Remelting," Tetsu-to-Hagane, 63 (13) (1977), p. 2141-2151. [BISI Translation No. 17049]

42. M. Hajduk and T. El Gammal, "Compilation of Conductivity Measurements in CaF_2-Containing Slags,"Stahl u. Eisen, 99 (3) (1979), p. 113-116. (In German)

43. K. Schwerdtfeger and K. Klein, "Rate of Fluorine Volatilization from Fluoride-Containing Liquid Slags, p. 81-90 in Proceedings of the Fourth International Symposium on Electroslag Remelting Processes, ISIJ, Tokyo, Japan, June 1973.

44. D. Ya. Povolotskii, et al., "Volatility of Melts of the CaO-Al_2O_3-CaF_2 System," Izv. VUZ Chern. Met., (2) (1977), p. 40-42. [BISI Translation No. 15833]

45. T. El Gammal and M. Hajduk, The Use of CaF_2-free Slags in the Electroslag Remelting Process," p. 141-144 in Proceedings of the Fifth International Conference on Vacuum Metallurgy and Electroslag Remelting Processes, Leybold-Heraeus GmbH, Munich, W. Germany, Oct. 1976.

46. G. Pateisky, "The Application of a Fluoride-Free Slag in a Production ESR Plant," ibid., p. 145-147.

47. J. D. W. Rawson, G. Jeszensky, and A. W. Bryant, "Novel Slags for ESR," p. 848-863 in Proceedings of Sixth International Vacuum Metallurgy Conference on Special Melting. G. K. Bhat and R. Schlatter, eds., San Diego, CA, April 1973.

48. N. F. Yakovlev, et al., "Effect of Used Flux on Quality of Electroslag Metal," Steel in the USSR, (6) (1978), p. 326-328.

49. A. Masui, Y. Sasacima, and M. Yamamura, "Mathematical Model for Hydrogen Behavior in Electroslag Remelting," p. 284-305 in Proceedings of the Fifth International Symposium on Electroslag and Other Special Melting Technologies, G. K. Bhat and A. Simkovich, eds., Carnegie-Mellon, Pittsburgh, PA, Oct. 1974.

50. K. Schwerdtfeger and H. G. Schubert, "Solubility of Water in CaO-Al_2O_3 Melts at 1600°C," Metallurgical Transactions, 9B (1978), p. 143-144.

51. Y. Hirose, K. Okohira, T. Shimuzu, N. Sata, M. Hirai, and N. Nishiwaki, "Refining Efficiency and Quality in 40 Tonne ESR Slabs and Plates," Tetsu-to-Hagane, 63 (13) (1977), p. 2208-2223. [BISI Translation No. 18732]

52. R. Jauch, A. Choudhury, F. Tince, and W. R. Lemor, "Present State of Electroslag Remelting Process in Röchling-Burbarch for Heavy Forging Ingots with 2300mm Diameters," p. 718-726 in Proceedings of the Third International Iron and Steel Congress, ASM-AIME, Chicago, IL, April 1978.

53. G. Hoyle, "Surface Phenomena in ESR," p. 624-640 in Proceedings of the Sixth International Vacuum Metallurgy Congerence on Special Melting, G. K. Bhat and R. Schlatter, eds., San Diego, CA, April 1979.

54. K. Ono and M. Takagi, "Factors Influencing the Depth of Molten Pool in ESR Ingot," presented at the 98th ISIJ Meeting, Nagoya, Japan, Oct. 1979. [Trans. ISIJ, 20 (1980), p. B311]

55. A. Mitchell and G. Beynon, "Electrode Polarization in the DC Electroslag Melting of Pure Iron," Metallurgical Transactions, 2 (12) (1971), p. 3333-3345.

56. J. Cameron, M. Etienne, and A. Mitchell, "Some Electrical Characterisitcs of a DC Electroslag Unit," Metallurgical Transactions, 1 (7) (1970), p. 1839-1844.

57. G. Hoyle, "Some Observations on ESR Practice," p. 272-278 in Electric Furnace Proceedings, Vol. 35, ISS-AIME, Chicago, Dec. 1977.

58. M. Kawakami and K. S. Goto, "A Survey of Profiles of Temperature and Electrical Potential in the Flux and Metal Pool During Electroslag Remelting," p. 500-513 in Proceedings of the Sixth International Vacuum Metallurgy Conference on Special Melting, G. K. Bhat and R. Schlatter, eds., San Diego, CA, April 1979.

59. I. D. W. Rawson, D. I. Dawson, and N. Kirkham, "Slag and Metal Flow in Electroslag Remelting," p. 55-68 in Proceedings of the Fourth International Symposium on Electroslag Remelting Processes, ISIJ, Tokyo, Japan, June 1973.

60. A. H. Dilawari and J. Szekely, "A Mathematical Model of Slag and Metal Flow in the ESR Process," Metallurgical Transactions, 8B (1977), p. 227-236.

61. J. Kreyenberg and K. Schwerdtfeger, "Stirring Velocities and Temperature Field in the Slag During Electroslag Remelting," Arch. Eisenhüttenwes, 50 (1) (1979), p. 1-6.

62. O. W. Stenzel and F. W. Thomas, "Factors Influencing Slag and Metal Flow Patterns in ESR Furnaces," p. 522-534 in Proceedings of the Sixth International Vacuum Metallurgy Conference on Special Melting, G. K. Bhat and R. Schlatter, eds., San Diego, CA, April 1979.

63. M. Choudhary and J. Szekely, "The Modeling of Pool Profiles, Temperature Profiles, and Velocity Fields in ESR Systems," Metallurgical Transactions, 11B (1980), p. 439-453.

64. M. Inouye, "Recent Fundamental Studies of the ESR Process in Japan," p. 51-59 in Proceedings of the Sixth International Conference on Special Melting, G. K. Bhat and R. Schlatter, eds., San Diego, CA, April 1979.

65. M. C. Flemings, C. M. Adams, E. E. Hucke, and H. F. Taylor, "Metal Solidification in a Flowing Stream," Trans. Am. Foundrymen's Soc., 64 (1956), p. 636-639.

66. D. N. Pocklington and B. Patrick, "Heat Flow in 3-Phase ESR Production," Metallurgical Transactions, 10B (1979), p. 359-366.

67. P. D. Mellberg, "Thermal Conditions in the ESR Process," p. 535-542 in Proceedings of the Sixth International Vacuum Metallurgy Conference on Special Melting, G. K. Bhat and R. Schlatter, eds., Am. Vac. Soc., San Diego, CA, April 1979.

68. W. Holzgruber, "Possibilities and Limitations in Controlling Solidification Structure of Alloy Steels During Electroslag Remelting, Radex Rundschau, (1975) No. 3, p. 409-421. [H. Brutcher Translation No. 9763]

69. A. S. Ballantyne and A. Mitchell, "Modeling of Ingot Thermal Fields in Consumable Electrode Remelting Processes," Ironmaking and Steelmaking, (4) (1977), p. 222-239.

70. A. Choudhury, R. Jauch, and H. Löwenkamp, "Primary Structure and Internal Properties of Conventional and Electroslag Remelted Ingots with Diameters of 2000 mm and 2300 mm resp.," p. 233-238 in Proceedings of the Fifth International Conference on Vacuum Metallurgy and Electroslag Remelting Processes, Leybold-Heraeus, GmbH, Munich, W. Germany, Oct. 1976.

71. H. J. Klein and J. W. Pridgeon, "Effective Electroslag Remelting of Superalloys," p. B1-B26 in Superalloys--Processing, Proceedings of the Second International Conference, TMS-AIME, Champion, PA, Sept. 1972.

72. A Mitchell and R. M. Smailer, "Practical Aspects of Electroslag Remelting Technology," International Metals Reviews, 24 (1979), p. 231-264.

73. J. Preston, "Segregation in Vacuum-Arc-Melted Alloy Steel Ingots," p. 569-588 in Transactions of the International Vacuum Metallurgy Conference, E. L. Foster, ed., Am. Vacuum Society, New York, N.Y., 1967.

74. S. Kou, D. R. Poirer, and M. C. Flemings, "Macrosegregation in Rotated Remelted Ingots," Metallurgical Transactions, 9B (1978), p. 711-719.

75. S. D. Ridder, F. C. Reyes, S. Chakravorty, R. Mehrabian, J. D. Nauman, J. H. Chen, and H. J. Klein, "Steady State Segregation and Heat Flow in ESR," Metallurgical Transactions, 9B (1978), p. 415-425.

76. C. L. Jeanfils, J. H. Chen, and H. J. Klein, "Modeling of Macrosegregation in Electroslag Remelting of Superalloys," p. 119-130 in Proceedings of the Fourth International Symposium on Superalloys, J. K. Tien, et al., eds., ASM-AIME, Champion, PA, Sept. 1980.

77. C. L. Jeanfils, J. H. Chen, and H. J. Klein, "Modeling Solidification in an Electroslag Remelted Ingot," Engineering Foundation Conference on Modeling of Casting and Welding Processes, D. Apelian and H. D. Brody, eds., Rindge, NH, Aug. 1980 (in press).

78. R. Mehrabian, M. Keane, and M. C. Flemings, "Interdendritic Fluid Flow and Macrosegregation; Influence of Gravity," Metallurgical Transactions, 1 (1970), p. 1209-1220.

79. H. Miska, and M. Wahlster, "Behaviour of Oxygen in Electroslag Remelting," Archiv Eisenhüttenwesen, 44 (1) (1973), p. 19-25. [H. Brutcher Translation No. 8993)

80. W. Holzgruber and E. Plöckinger, Metallurgical and Technological Fundamentals of the Electroslag Refining of Steel," Stahl u. Eisen, 88 (12) (1968), p. 638-648. [BISI Translation 6617]

81. H. Miska, and M. Wahlster, "Behavior of Sulfur in the Electroslag Remelting Process," Archiv Eisenhüttenwesen, 44 (2) (1973), p. 81-85. [Henry Brutcher Translation No 9003]

82. M. Inouye, "Recent Fundamental Researches on the ESR Reactions in Japan," p. 327-336 in Proc. of Third International Iron and Steel Congress, ASM-AIME, Chicago, Illinois, April, 1978.

83. M. E. Peover, "Electroslag Remelting: A Review of Electrical and Electrochemical Aspects," Jl of Institute of Metals, 100 (1972), p. 97-106.

84. K. Schwerdtfeger, "Some Aspects on the Kinetics of Reactions Occurring in the ESR Process," p. 133-142 in Proc. of Fifth International Conference on Vacuum Metallurgy and Electroslag Remelting Processes, Leybold-Heraeus, GmbH, Munich, W. Germany, Oct., 1976.

85. V. E. Roshchin, N. V. Mel'kov, D. Ya. Povolotskii, Sh. N. Gimadeev, and P. P. Biryukov, "Transformation of Non-Metallic Inclusions in Metal of Consumable Electrodes During Remelting," Steel in the USSR, 8 (12) (1978), p. 679-681.

86. P. O. Mellberg and H. Sandberg, "On the Mechanism of Desulfurization During AC and DC Electroslag Remelting, Scandinavian Jl. of Metallurgy, 2 (1973), p. 121-124.

87. F. W. Thomas, "Apparatus for Regulating the Immersion Depth of Electrodes in Electrode-Melting Furnaces," U.S. Patent No. 4,075,414, Feb. 21, 1978.

88. W. E. Duckworth and G. Hoyle, Electroslag Refining, p. 35, Chapman & Hall, Ltd., 1969.

89. J. F. Elliott, Y. Iguchi, and T-H Chiang, "The Behavior of Oxides in Electroslag Remelting of Steels," p. 208-220 in Electric Furnace Proceedings, Vol. 35, ISS-AIME, Chicago, Illinois, Dec., 1977.

90. Y. Kojima, M. Kato, T. Toyoda, and M. Inouye, "A Model Study on the Melting Phenomena of Consumable Electrodes in the ESR Process," p. 35-44 in Proceedings of the 4th International Symposium on ESR Processes, ISIJ, Tokyo, 1973.

91. Private communication, J. H. Chen, Cabot Corporation, 1978.

92. J. H. Chen, D. R. Engel, and R. C. Myers, "Computer Control of the Electroslag Remelted Process," p. 831-847 in Proceedings of the Sixth International Vacuum Metallurgy Conference on Special Melting, G. K. Bhat and R. Schlatter, eds., San Diego, CA, 1979.

93. F. W. Thomas and H. Birck, "Digital Control of Electroslag Furnaces," ibid., p. 820-830.

94. A. Ujiie and J. Nagata, U. S. Patent No. 3,683,997 (1972).

95. B. E. Paton et al., U. S. Patent No. 3,721,286 (1973).

96. H. J. Klein, W. V. Venal, and K. L. Love, "Production of Electroslag Remelted Hollow Ingots," p. 55-66 in Superalloys: Metallurgy and Manufacture, B. H. Kear, et al., eds., TMS-AIME, Seven Springs, PA, Sept. 1976.

97. W. V. Venal, R. T. Gross, H. J. Klein, and V. J. Colangelo, "Electroslag Remelting of Hollow Ingots at Cabot Corporation in Pampa, Texas," p. 264-271 in Electric Furnace Proceedings,ISS-AIME, Chicago, IL, 1977.

98. M. F. Rothman, unpublished research, Cabot Corporation, 1981.

99. B. E. Paton, B. I. Medovar, G. A. Bojko, I. I. Koomish, V. F. Karpov, "New Potentialities of Electroslag Shaped Casting," p. 169-173 in Special Electrometallurgy, Reports of the International Symposium, Part 1, Kiev, 1972.

100. P. Dewsnap and R. Schlatter, "Process and Product Characteristics of DC Electroslag Remelting of Alloy Steel," p. 91-114 in Proc. Fifth International Symposium on Electroslag and Other Special Melting Technology, G. K. Bhat and A. Simkovich, eds., Carnegie-Mellon, Pittsburgh, PA, Oct. 1974.

101. E. M. Lowe and A. Hogg, "Application of ESR to Alloy-Steel Forgings," p. 68-79, in Electroslag Refining, The Iron and Steel Institute, London, 1973.

102. G. Yuasa, T. Yajima, and K. Shiina, "Quality Improvements of High Alloy Steels by ESR," p. 229-239 in Proc. of the Fourth International Symposium on Electroslag Remelting Processes, ISIJ, Tokyo, Japan, June, 1973.

103. T. Mochizuki and E. Ohga, "On the Quality of Forged Hardened Rolls Made by Electroslag Remelting," p. 253-265 in Proc. of the Fourth International Symposium on Electroslag Remelting Processes, ISIJ, Tokyo, Japan, June, 1973.

104. V. Yu. Latash and B. I. Medovar, "Electroslag Melting," in Metallurgy, Moscow, 1970.

105. R. A. Swift and J. A. Gulya, "Property Evaluation of Electroslag Remelted A533B Plate," Welding Journal, 52 (1973): p. 537-542s.

106. W. V. Venal, H. J. Klein, E. E. McDowell, and W. D. Forsha, "The Effect of Electroslag Remelting on the Properties of AISI 4330 Mod," p. 173-176 in Proceedings of Fifth International Conference on Vacuum Metallurgy and Electroslag Remelting Processes, Leybold-Heraeus GmbH, Munich, W. Germany, Oct. 1976.

107. T. Sonomoto and K. Okada, Tetsu to Hagane, Vol. 55, 1969, No. 3 p. 269.

108. H. J. Wagner and K. Bar Avi, Proceedings of the International Conference on Applied Metallurgy, Sponsored by ASM and Israeli Metallurgical Society, Tel-Aviv, April, 1976.

109. M. Miura, H. Takada, and Y. Fukuhara, "An Evaluation of the Electroslag Remelted Ingot," in Part III of The Collected Papers of the Second International Symposium on Electroslag Remelting Technology, G. K. Bhat, ed., Mellon Institute, Pittsburgh, PA, Sept. 1969.

110. I. R. Krianin, et al., "The Electroslag Remelting of Steels for the Forgings of the Large Discs and Rotors," p. 73-84 in Special Electrometallurgy, Reports of the International Symposium, Part 1, Kiev, 1972.

111. E. W. Kelley, Manufacturing Process for Improved High-Strength Superalloy Sheet, Report No. AFML-TR-69-114, Wright-Patterson Air Force Base, Ohio, Air Force Materials Laboratory, 1969.

112. C. F. Elliott, J. W. Vorberger, M. G. Staton, and G. L. Mills, "Superalloy Melting Using an Ingot Bottom Withdrawal AC ESR Furnace," p. 298-310 in Proc. Fourth International Symposium on Electroslag Remelting Processes, ISIJ, Tokyo, Japan, June 1978.

113. T. N. Kelley, D. A. Junker, and R. C. H. Chen, "The Quality of Electroslag Melted HASTELLOY alloy X," p. 125-141 in Proc. Third International Symposium on Electroslag and Other Special Melting Technology, G. K. Bhat and A. Simkovich, eds., Mellon Institute, Pittsburgh, PA, June 1971.

114. R. L. Cook, G. W. Reese, Jr., and P. M. Gadsby, "Electroslag Remelting of Nickel Base Alloys," ibid., p. 142-175.

115. T. N. Kelley, H. J. Klein, R. C. Sun, and W. V. Venal, "Electroslag Remelting of Precipitation Strengthened Alloys," p. 150-179 in Proc. Fifth International Symposium on Electroslag and Other Special Melting Technology, G. K. Bhat and A. Simkovich, eds. Carnegie Mellon, Pittsburgh, PA, Oct. 1974.

116. R. R. Shiring and J. L. Milavec, "Effect of Production Melting Practices on the Strength and Toughness Properties of AF1410--An Aircraft Structural Steel," p. 291-314 in Proceedings of the 1977 Vacuum Metallurgy Conference, R. C. Krutenat, ed., Science Press, Princeton, 1977.

PRINCIPLES AND PRACTICES OF VACUUM INDUCTION MELTING AND VACUUM ARC REMELTING

J. W. Pridgeon, F. N. Darmara, J. S. Huntington, W. H. Sutton
Special Metals Corporation
New Hartford, New York 13413

The Vacuum Induction Melting (VIM) and Vacuum Arc Remelting (VAR) processes are reviewed. The VIM process is utilized to minimize the reaction of reactive alloying elements with the atmosphere, and to produce heats of exceptional compositional control and purity. A brief description of the VIM process, equipment, and controls is presented along with theoretical and experimental data. The importance of various melt-refractory reactions and their effect on alloy cleanliness are detailed. Post processing of VIM products (ingots and electrodes) is discussed also.

The Vacuum Arc Remelting Process is utilized primarily to produce macrosegregation-free superalloys, titanium and its alloys, the refractory metals and their alloys, and high performance steel products. The process is described for a number of alloys. Particular emphasis is placed on equipment and on the control of the process through the analysis of drip shorts, arc gap, and other burn-off phenomena. New control concepts are also discussed along with a new arc melting process (VADER).

VACUUM INDUCTION MELTING

Vacuum induction melting (VIM) is the most flexible of the vacuum melting processes, since it allows for the independent control of temperature, pressure, and mass transport by means of stirring. As a consequence, VIM provides the greatest degree of control over alloy composition of all the melting processes, which is probably its most important metallurgical justification. The term "composition" includes not only the desired alloy additions, but impurities (beneficial and harmful) as well. An example of the excellent control over chemistry is shown in Table I for the superalloy, IN-718. A similar degree of control is cited in the literature for waspaloy (2) and U-700 (3).

TABLE I. Chemical Composition Control of IN-718 Based on 100 Heats (1).

ELEMENT	SPEC. RANGE (%)	FINAL RANGE (%)	ANALY. FREQ. (%)	ANALYTICAL ACCURACY (%) *	**
C	0.02-0.08	0.04-0.05	95	±0.006	±.003
Ti	0.80-1.15	0.90-1.10	97	±0.05	±.03
Cb	4.75-5.50	5.05-5.40	99	±0.12	±.08
Al	0.30-0.70	0.50-0.60	95	±0.04	±.02

IN-718: C-0.04, Cr-19.0, Fe-18.0, Mo-3.0, Cb-5.2, Ti-0.8, Al-0.6, Ni-53.8 w/o
* - 1971 data (Ref. 1); ** - typical 1980 data

Many of the current superalloy specifications contain restrictions on trace elements. For example, AMS specification 2280 (4) limits Te and Bi to 0.5 ppm max. each, Se to 3 ppm max., and Pb and Tl to 5 ppm max. Fifteen additional elements are also limited to a maximum of 50 ppm each and to a combined maximum of 400 ppm. Not all trace or minor elements are harmful, and elements such as B, Zr, and Mg are purposely added during the VIM melting of superalloys (2,5-7). Thus the advent of VIM has made possible a class of high-purity alloys having a range of carefully controlled compositions that otherwise could not be produced on a large scale.

The benefits and limitations of the VIM process have been reviewed in the literature (1,2,8-11), and these are summarized in Table II. The types of alloys which are produced by VIM include: superalloys, corrosion-resistant alloys, special steels, maraging steels, nuclear alloys, tool and bearing alloys, special magnetic and electrical alloys, and other non-ferrous alloys for specialized applications.

TABLE II. BENEFITS AND LIMITATIONS OF THE VIM PROCESS

Condition	Results
Benefits	
1. Isolation of the Melt	. Prevents contact and reactions with air (oxygen and nitrogen). . Produces clean melts; minimizes inclusions. . Provides control over pressure and gases within the system; may use argon or other gases. . Products have superior properties over air melted products.
2. Accelerates Refining Reactions	. Removes dissolved gases and volatile constituents. . Provides high degree of purity. . Favors dissociation of compounds (oxides, nitrides, etc.). . Drives certain reactions to completion, (i.e., deoxidation, an advantage over inert atmosphere). . Makes practical scrap remelting. . Alloying elements become more effective (not combined in unwanted compounds).
3. Induction Stirring	. Homogenizes melt composition. . Brings reactants to melt surfaces (also accelerates reactions). . Provides superior composition control. . Provides excellent reproducibility of composition and properties.
Limitations	
1. Casting of Electrodes (or ingots)	. Segregation of solutes on a macro and micro scale during solidification. . Solidification produces large and nonuniform grain structure, and large shrinkage pipe.
2. Melt/Refractory Reactions	. Contaminates melt due to reoxidation of alloying elements. . Introduces inclusions (but at much lower level than air melts).

Brief History of Vacuum Induction Melting

It is hard to say how far in the past the concept of melting in a vacuum goes. A patent was taken in the U.S. in 1890 for the Colby furnace which is a form of coreless vacuum induction furnace. The first commercial application seems to have been done by W. Rohn during the 1920's. The lack of pumps of adequate capacity precluded the attainment of vacuums below 2-5 Torr.* The products from this furnace were high chrome steels and magnetic materials. Even though many well tailored laboratories did have small vacuum furnaces devoted to metallurgical research, as an industrial process it was merely a gadget. At that time there was no economically sound reason for achieving a lower pressure in a production furnace.

Two developments occurring during the second war provided both the means and the need for vacuum melting. The need was the precipitation hardening of alloys which was necessary to provide the stress resistance at high temperatures so essential to the success of the jet engine. The means were provided by the high capacity pumps developed for the nuclear programs.

In 1945, two of the authors (Darmara and Huntington) needing to do research on enhancing the properties of the Nimonic type of alloys then coming into use, built a small vacuum induction furnace at the Lewis Laboratories of NACA (now NASA). About the same time J. W. Moore at National Research Co. had a small furnace in operation for the production of pure iron and copper.

By 1950-1951 the needs of the newer, higher-thrust jet engines were straining the possibilities of conventional air melting methods to the limit, and eventually went past those limits. Some of the newer alloys just could not be produced with any acceptable consistency by air melting. Vacuum melting was tried with spectacular success and the process has been growing both in economic importance and technical complexity ever since (2,8,11,12). By 1958 1370 Kg (3,000 lb.) furnace was in operation (2). However demand was rising at such a rate that in 1961 a 5455 Kg furnace was put in operation. By today's standards these would be considered small, but at the time it took great courage to undertake the construction of these furnaces. Today it can be said that there are no technical inhibitions on the size of furnaces. Size is dictated by reasons other than technical feasibility.

VIM Equipment

Production vacuum induction melting furnaces range in size from less than one ton to sixty tons in melt capacity, and yet their basic components are essentially the same. Refer to Figure 1 for the basic arrangement.

Induction Furnace and Power Supply. The basic charge is melted in an induction furnace with a rammed or bricked lining. Most furnaces with melting capacities greater than 4 tons (3636 Kg) are brick-lined. The frequency and power requirements are similar to air melting counterparts, but every attempt is made to keep coil voltage as low as practical to avoid corona and arcing in certain ranges of reduced pressure. Special insulating techniques are also used for this purpose.

The input power for melting the charges varies from a few kilowatts for a small furnace, e.g., 30 Kw for 22.7 Kg (50 lb.) melt to over 6600 Kw for a 27,300 Kg (60,000 lb.) melt. For the larger furnaces, (viz., $\geq$ 2300 Kg) the available melting power ranges between 0.11 to 0.29 Kw/Kg (or 0.05 to 0.13 Kw/lb.).

Because of the necessity to expose all increments of the melt to the vacuum environment above the

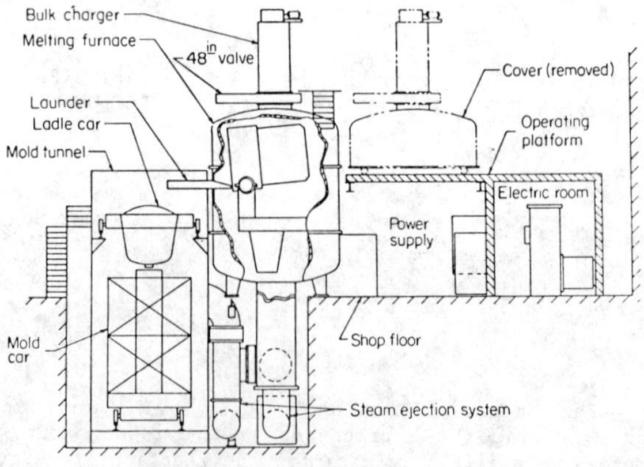

Figure 1. Schematic Drawing of a 15-Ton VIM Furnace.

melt surface, stirring is a necessity in furnaces larger than 455 Kg. This is usually obtained by electromagnetic movement of the metal and involves either dual frequency or the use of coil-capacitor arrangements to obtain "two-loop" instead of the usual "four-loop" motion within the melt.

Furnace Envelope. Obviously a water cooled metal tank is required to enclose the melting furnace. The configuration varies widely from furnace-to-furnace depending on the production requirements. Most large scale furnaces today employ a two-chamber arrangement; one to enclose the melting furnace and the other to house the molds and their handling mechanism (as shown in Figure 1). These chambers are usually connected by means of large scale vacuum valves, and each chamber has its own access doors and vacuum pumps. In this way melting can proceed while the molds are being set up or removed, so that "turn around time" can be greatly reduced.

Vacuum System. In the early days of production vacuum melting, the pumping systems consisted of mechanical pumps to rough down the chambers from 760 to 1 Torr. Ejector type diffusion pumps then reduced the pressure to the operating range of approximately 10^{-2} Torr. The ultimate pressure was an equilibrium between the quantity of gases evolving and the through-put characteristics of the pumps.

As furnaces increased in size, steam ejectors were used to handle the larger gas loads, and installations of twin six-stage systems with inlet sizes of 1200 to 1500 mm were not uncommon. Because of the very low thermal efficiency of the steam ejector, the operating costs have risen sharply during the last 8 years, as energy costs have increased, and there is a trend back to the mechanical booster combination using mechanical blowers with speeds to more than 5000 l/sec in the 10^{-2} to 20 Torr range (13).

Mold Arrangements. The mold configuration, arrangement, and means for introduction and removal depend on the types of ingots required. It is not uncommon to see two furnaces side-by-side with the product from one being scores of small ingots 75 mm round by 750 mm long and the other being two ingots 600 mm round by 3000 mm long.

Cars are frequently used in large furnaces to transport the molds into the furnace, to shift them under the tundish during pouring, and to remove them after solidification.

Bulk Charger. Since the packing factor of most charges prohibits preloading the full melt capacity into the crucible before melting, a means is required to add major fractions of the charge while melting is

*1 Torr = 1mm Hg = 1.32 x 10^{-3} atm pressure. (1 micron = 10^{-3} Torr).

proceeding under vacuum. For small furnaces, this is not a major problem, but for large furnaces with a wide range of charge geometries, the bulk charger becomes a critical item. Furnace productivity can be needlessly impaired if the bulk charger cannot keep up with the melt rate of the induction melting system.

Auxiliary Apparatus. Besides the above major components, a variety of other equally important equipment is necessary, depending on the basic purpose of the melting process. These include:

- Sample takers to remove molten metal through a vacuum lock for in-process control of chemistry.
- Temperature measuring equipment of optical and thermocouple devices to monitor the melt temperature.
- A variety of pressure gages to indicate and/or record pressures, not only over the melt but in critical parts of the vacuum system.
- Means of introducing controlled flow of selected gases for augmented stirring or certain chemical reactions.

Description of the Process

In theory the process is quite simple. The charge to be melted is put into a crucible surrounded by water cooled copper coils. Sufficient pumping capacity has to be provided to reduce the working pressure during melting to the range of 1-10 microns. The power supply must be of sufficient capacity to melt the charge within a reasonable time, usually of the order of one hour, and the frequency must be such that adequate stirring is achieved within the melt. A device for tilting the crucible is employed so that the melt can be poured into a mold of desired shape and weight (see Figure 1).

The progress in the melting and pouring of a typical VIM heat can be followed by observing the total furnace pressure as it changes with time, which is shown in the upper portion of Figure 2. At the start, the furnace is completely evacuated and a leak rate is taken. This is the rate at which the pressure in an evacuated furnace rises without the pumps operating, and is usually expressed as microns per minute. Tolerable leak rates are determined empirically and are part of the operating procedures.

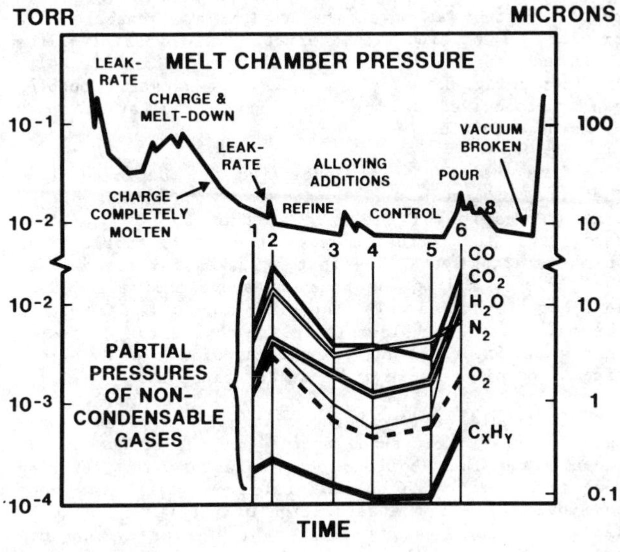

Figure 2. The Upper Curve Shows how Furnace Pressure Changes with Time During a Typical VIM Heat. The Lower Curves Show the Partial Pressures of Several Gases Recorded at Specific Time Intervals (Points 1-6). (from Johnson & Sutton, Ref. 14).

After establishing a satisfactory leak rate, the power is turned on and the heating commences. The charge can be either scrap to be remelted or virgin metals. The latter case will be considered here since it is more descriptive of the intricacies of the process. Let it be assumed that the final alloy will consist of 2% Al, 2% Ti, 5% Mo, 15% Cr, .06% C, .04% Zr, .015% B, and the remainder Ni. The initial or base charge will contain the Ni, Cr, Mo and sufficient excess carbon to finish the deoxidation with a carbon within the specification range of 0.02-0.08%.

As the melting begins, the volume of the charge will be reduced, so that more of the base charge can be added. As the charge is introduced into the melt, outgassing occurs, causing pressure spikes in the furnace (see Figure 2). As melting progresses, a boil will occur caused by the evolution of CO due to the following reactions*:

$$[O] + [C] = (CO) \quad . \; . \; . \; . \; . \; . \; . \; . \; . \; . \; . \quad (1)$$

$$\langle M_x O_y \rangle + y[C] = x[M] + y(CO) . \; . \; . \; . \; . \quad (2)$$

The CO gas results from the reaction of the dissolved carbon with the dissolved oxygen (equation 1) or with oxide films on the charge materials (equation 2).

For the reactions in equations 1 and 2, there is an equilibrium at each pressure, and if the pressure is held constant, eventually the bubbling will stop long before the lowest possible oxygen level is reached. Furthermore, it is universally agreed that bubbles are heterogeneously nucleated and that there is strong evidence to suggest that bubbles help to nucleate other bubbles. Therefore bubbling must not be allowed to terminate at too high a pressure as it may not start again as the pressure is lowered. Hence, the usual practice is to begin to lower the furnace pressure as soon as the boil is under control, and then achieve a balance between violence and vigor.

Eventually the chamber pressure will be at the lowest attainable level for the equipment and the boiling will stop. However the final and optimum equilibrium is not reached as yet. During this quiescent period the CO is being removed by surface desorption of the melt surface and refining continues.

At this stage the reaction of the melt with the crucible begins to assume importance. As the carbon content drops, at some point, determined by the effect of the carbide formers present in the melt on the activity of carbon, there will be a pick-up of oxygen from the crucible. Hence the melt has a rather short time in which the oxygen is at its lowest. To establish this end point, by inference, a leak rate can be taken (see point 2, Figure 2) until a steady state is reached, or if a probe O_2 analyzer is operative, this can be used to determine that the final level of deoxidation is attained (15).

The Al, Ti, Zr, and B are added at this stage (points 3-4, Figure 2) and after thorough mixing and final analytical check, the heat is brought to the proper pouring temperature and is poured into the mold (point 6, Figure 2).

By analyzing the composition (fractional pressures) of the gases in the melt chamber, the progress of the various stages during the VIM process can be monitored. The lower portion of Figure 2 shows the fractional pressures, as measured by a residual gas analyzer, of

*< > = solid phase; [ ]= liquid or solution; ( ) = gas.

several gases in the furnace of a 136 Kg (300 lb.) melt during 6 intervals (or points) of processing (14), and are described as follows:

Point 1 - Meltdown: The melt has stabilized after initial melt-down, outgassing, and boil. Gases such as N_2, O_2, and H_2O are present in appreciable quantities, and CO and CO_2 are present indicating refining and melt-refractory reactions, as well as the oxidation of hydrocarbons (C_xH_y) which are present in the system (pump oil residual, etc.).

Point 2 - Leak Rate Test: The pressure rise of all gases increases proportionately over the pressure shown at Point 1, indicating that these gases are being generated at approximately the same rate as they were prior to the test. There does not appear to be a significant external air leak, otherwise the O_2 and N_2 peaks would be disproportionately higher.

Point 3 - Pumps Restored: The pressures are reduced to those shown for Point 1.

Point 4 - After Reactive Alloy Adds (Al, Ti, Zr, B): Note the sharp drop of H_2O, CO_2 and O_2 pressures. The presence of hydrocarbons is negligible.

Point 5 - Late in the Heat: The CO pressure has begun to diminish. The other gases, N_2, O_2, H_2O, CO_2 are at pressure levels which are similar to those before the reactive adds. This indicates that the origin of these gases primarily results from a continuous outgassing of the total furnace system, since the refining reactions within the melt are essentially complete.

Point 6 - Pouring of the Heat: All gases monitored, except N_2, show a sharp increase in pressure, due mainly to the outgassing of the pour system refractories and of the steel mold, which are located within the furnace chamber of this 136 Kg heat.

Thermodynamic and Kinetic Considerations

During the past two decades, a great deal of theoretical work has been done to lay the thermodynamic foundations for the VIM practice (10,16-23). In fact, so much work has been done on deoxidation, decarburization, degasification, volatilization, desulfurization, and melt-crucible reactions, that only a brief discussion of the subject is possible here.

In vacuum melting, the kinetics of the reactions and reaction mechanisms are of crucial importance, so that the thermodynamic calculations which are based on equilibrium states, are insufficient to describe the termination point of the reactions. For example, thermodynamics would predict a final (equilibrium) oxygen content approximately 10^{-8} weight percent (0.0008 ppm) at 0.004% C at 1500°C in a nickel alloy melted under a pressure of 1 micron (24,25). In practice, this value is many orders of magnitude greater, viz., 5-20 ppm. Reasons for this discrepancy will be discussed later. How close to equilibrium the reaction approaches during a reasonable time is a matter of kinetics. Furthermore, during the melting cycle, most of the conditions deal with steady-state conditions and not true equilibrium.

Deoxidation Reactions. In VIM, the major means for deoxidation via the gas phase is by a reduction with carbon, as shown in equation 1. The fact that the product (CO) is a gaseous oxide allows for its removal in the vacuum environment. The final oxygen content for a given carbon content (or more rigorously, the C and O activities) will be proportional to the partial pressure of the CO, where K is the equilibrium constant:

$$\%O \approx \frac{P_{co}}{K[a_c]} \approx \frac{PCO}{K[\%C]} \quad \text{. . . . . (3)}$$

As was mentioned previously, the theoretical values of equilibrium oxygen in nickel (or iron melts) containing carbon at temperatures of 1500°C to 1600°C, and at furnace pressures of about 10^{-3} torr, are well below the 0.0001% (1ppm) level. In practice, the final (steady-state) oxygen levels are typically in the 2-20 ppm range. These higher levels of oxygen result from external (to the melt) sources, such as: impurities in the raw materials (26-28), from impurities in the crucible (10,19,23,29-31), from the outgassing of crucible materials (10,27,28,30,32,33), and from leaks within a furnace lining (10,27,28).

The CO deoxidation reactions proceeds in two stages:

(1) The Boiling Stage - During the first stage of melting bubbles of CO are heterogenously nucleated at a melt/gas interface. Initially this interface is largely provided by bubbles nucleated in crevices in the crucible, and when the bubbles become detached, they rise through the melt causing the "boil."

(2) The Desorption Stage - When the CO pressure (in equation 1) is insufficient to nucleate bubbles, CO will form only at the melt surface, where it desorbs into the vacuum.

Since the major quantities of gases are eliminated during the boil, the mechanism by which bubbles are formed was one of the first concern of various authors during the early days of vacuum induction melting (16-20). One of the latest papers to deal theoretically with the subject is that of Szekely and Fang (34). The first aspect of the problem of bubble formation is that of the nucleation mechanism involved. Machlin was the first to point out his view, which is almost universally accepted today, that bubbles are nucleated heterogeneously (17,35,36). The most commonly held belief is that the sites at which the bubbles are generated are crevices in the crucible material. Other possible sites could be refractory particles stirred up in the melt. An intriguing potential means for generating bubbles can be conceived as the creation of cavitation zones by means of violent stirring.

Once the bubbles are nucleated they begin to grow. The rate of growth will depend: on the extent of the supersaturation of the melt by the gas (CO); on the decrease of the ferrostatic pressure as the bubble rises in the melt; on the time that it takes for it to reach the surface; and on the pressure above the melt. Even in a supersaturated melt, the diffusion of the CO within the melt to the bubble-melt interface and the diffusion across the interface has to be considered. An excellent discussion of these is given in Reference 34.

It should be remembered that below a given depth, the ferrostatic pressure cancels out the effect of the vacuum, and that boiling occurs because the stirring constantly brings fresh material from below this zone to above it. As the saturation of the CO in the melt decreases to a critical level, depending on the external pressure, boiling will cease, and the surface desorption stage begins. Many have studied its mechanism (37-39). The important factor to be considered is the mass transport coefficient from the melt to the gaseous phase, which determines the speed with which desorption takes place. The mean mass transfer coefficients which are given in the above references, were obtained using laboratory-size furnaces, and can serve as approximate guides for application to full scale production heats.

A brief mention should be made of one of the most original approaches to deoxidation. This is the theoretical work of Meysson and Rist (39) and the experimental results of Moore (40) dealing with the deoxidation of iron and nickel alloys with methane. This technique is specially suitable to vacuum induction melting as there is no danger of the contamination of the melt with hydrogen. At a pressure of 10 microns the dissolved value of this element would be of the order of 0.05-1 ppm. Moore observed that a temperature of 1600°C was required for the reaction:

$$(CH_4) = [C] + 2(H_2) . . . . . . (4)$$

and the rate of carbon deposition on the melt increased rapidly above this temperature. If the gas was not shut off at the precise end point, the carbon content of the melt rose rapidly. This end point was determined by a change of color of the gas burning at the exit end of the vacuum pumps.

By using this end point, iron heats containing 10-20 ppm of oxygen with 10-20 ppm of carbon, and nickel heats with 10 ppm oxygen and 10-20 ppm carbon were obtained. The pressure of the gas in the chamber is not mentioned, but since reference is made to the gas being pumped one can surmise that the pressure was below one atmosphere.

There are great advantages to this method, for one there is a complete absence of boiling, which in practice would be of great help. Some of deoxidation would be due to the presence of hydrogen. Since the deoxidation is being carried out at the extreme surface layer mass transfer between liquid-gas interface would be less critical. The extreme surface layer being the first to come in contact with the methane, it should have a high carbon potential thus insuring a more efficient deoxidation. Furthermore if as is surmised, evaporation reactions are at times slowed by surface active substances (41), such as an ionic double layer $Fe^{+2}O^{2-}$, then the high carbon potential at the very surface would tend to reduce this layer and thus restore the full activity to the surface.

Removal of Nitrogen. Numerous investigators have studied the removal of nitrogen through vacuum melting (23,35,42,43). Providing that the melt does not contain strong nitride formers, reducing nitrogen to low levels, of the order of 20 ppm, does not present any particular problem. On the other hand, in alloys such as iron-chrome and nickel-iron-chrome there is an order of magnitude difference between the calculated nitrogen equilibrium values and values actually obtained. The removal of nitrogen is at first quite fast, but slowly declines and becomes asymptotical with increasing time. The presence of alloying elements such as Cr, V, Al and Ti appreciably reduce the activity (increase the solubility) of nitrogen in the molten alloy. Since the formation of brittle nitrides and carbonitrides in alloys containing strong nitride formers, such as Ti, the nitrogen content in these alloys must be low (1). Simkovich (44) found that alloys containing up to 15% Cr contained about the same residual nitrogen level as the chrome-free alloys. However, in the alloys containing Cr plus Al, Cb, and Ti, the nitrogen levels were much higher after vacuum refining. The combination of Al+Ti was particularly effective in retaining nitrogen.

Elimination of Hydrogen. The solubility of hydrogen in all alloys varies with the square root of the partial pressure:

$$[\% H] = K\sqrt{P_{H_2}} . . . . (5)$$

At about 1600°C and at atmospheric pressure, the hydrogen solubility in pure nickel is about 30 ppm, and in pure iron the solubility is slightly higher (1). However, in vacuum induction melting, very low levels, (>1 ppm) are routinely achieved. Some nitrogen and most of the hydrogen is removed during the CO boil.

Evaporation of Minor Elements. In the production of high temperature alloys for critical applications (e.g., aircraft jet engines), the presence or absence of minute amounts of certain elements can have profound effects on the properties of these alloys. The effect of trace elements on the properties of nickel-base alloys has been reviewed by Holt and Wallace (5).

With the advent of very precise analytical instruments, it has been possible to ascertain accurately and quantitatively these effects, and the removal of these elements has assumed a major importance. For example, Figure 3 shows how trace quantities of Ag, Se, Pb, and Bi can adversely affect the stress rupture life of IN-718 (45). Fortunately most of the harmful trace elements have relatively high vapor pressure and are distilled under vacuum. It is safe to say that as the quality of raw materials deteriorates this problem will become more and more acute.

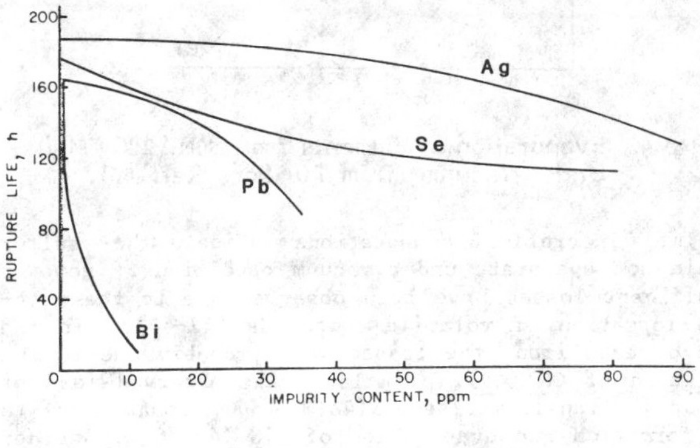

Figure 3. Effect of Various Trace Elements on the 649°C/ 690 MPa Stress-Rupture of IN-718 (from Kent, Ref. 45).

An excellent theoretical discussion of this process of vacuum distillation is given by Olette (38); calculations of the distillation coefficient and comparison between these and experimental values obtained by various authors is given. There is disagreement between the experimental and computed values, the experimental values being lower. He speculates that this may be due to the surface temperature being lower than the bulk temperature, to the possible existence of a condensation coefficient, to the presence of surface active substances, and to the surface concentration being less than the bulk concentration because of inefficient mass transfer in the liquid. In reality the last two factors may be the dominant ones. Some empirical observations in our plant tend to support the depletion of the surface argument. In several instances in one of the furnaces equipped with a most efficient stirring device, when the stirring is turned on to full power, the rate of evaporation seems to increase by at least two orders of magnitude.

Because of the practical importance of the deleterious effect of some minor elements on the properties of superalloys, many experimental evaporation results have been obtained. One set of results is shown in Figure 4. Fortunately two of the most nefarious elements, Pb and Bi evaporate readily. It is wise to highlight again that although experimental curves of this nature are extremely valuable, they should be used cautiously in practice. Olette's speculations mentioned above regarding the disparity between calculated and measured values applies for these experimental curves too. In the sense that the evaporation curves obtained for one furnace most likely will not match the performance of another equipment.

Sulfur Removal. Darmara (11) and Schlatter (1) have discussed the removal of sulfur during the vacuum induction refining of high temperature alloys and

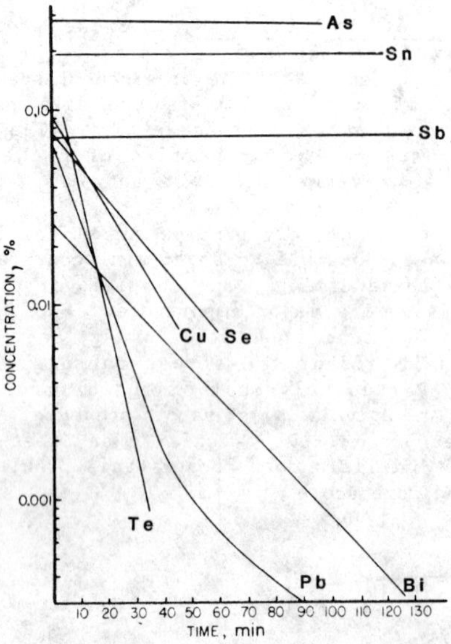

Figure 4. Evaporation of Elements from 80Ni-20Cr Melts Under Vacuum (from Turillon, Ref. 46).

steels. Theoretical computations indicate that sulfur should not evaporate under vacuum conditions. However significant losses have been observed due to the probable formation of volatile compounds (47-50). In the case of cast iron, the losses were probably due to the formation of CS (49). Similarly the observed loss of sulfur in transformer steel (4% Si) was probably due to the formation and evaporation of SiS (49,50). Another approach was studied by Machlin (17), where the gaseous oxides of sulfur were produced by raising the oxygen in a nickel base alloy melt to as high as 0.48%. The sulfur was reduced from 0.03% to 0.005% in 60 minutes. Theoretically an additional hour under vacuum should have dropped the sulfur to below 0.001%, but the experiments were not carried out long enough to substantiate this. In general, the removal through volatile compounds (CS, SiS, COS) are slow and tedious. A more practical approach is by the formation and removal of stable insoluble sulfides such as MgS (51), CaS (49,50), and CeS_2/LaS_2 (50,52-55).

The use of lime (CaO) has been studied by several authors (1,11,49,50), and it has been used extensively in practice. It is quite effective in reducing sulfur to the .002%-.004% levels, when the charge materials do not generally exceed 0.010 sulfur. Unfortunately in induction furnaces, the lime soon forms an unsightly slag, and can pose problems in maintaining melt cleanliness and in reducing refractory life. The desulfurization process is the most effective when the oxygen levels in the melt are low.

It would seem that calcium carbide additions could be effective in simultaneous deoxidation and desulfurization but there does not appear to be any evidence that there is practical application of this technique.

The additions of rare earth elements have also proven to be very effective in removing sulfur, however, the levels of residual RE elements in the melt must be carefully controlled. Bailey and co-workers (53) demonstrated that La was very effective in achieving low levels of both oxygen and sulfur in 901. Similar results were achieved when Ce was added (52). Rothman (54) demonstrated that misch metal (54% Ce, 24% La, 13% Nd, 5% Pr) is also an effective desulfurizing addition. In a 454 Kg (1000 lb.) heat of IN-718, the addition of 0.2% misch metal reduced the sulfur level from 220 ppm to <40 ppm in 10 minutes.

In a recent study by Li and co-workers (55), the addition of 0.07% Ce to an iron-nickel base superalloy, 901, after refining, reduced the oxygen level from 12-13 ppm to 4-8 ppm, and the sulfur level from 11-22 ppm to 4-8 ppm. The nitrogen level was not affected by the Ce additions.

Selenium is similar chemically to sulfur, and Burton (56) showed that the selenium level of a nickel-base superalloy (U-700) could be reduced from about 120 ppm to 10 ppm by volatilization (see Figure 4). The Se level was further reduced to less than 1 ppm by adding lime or CaNi to the melt. Such a low level would meet the current specifications.

VIM Melt Cleanliness

Melt cleanliness, as it relates to the inclusion content of the final product, is probably one of the most important quality considerations today. This is especially true for the newer nickel-base superalloys, which are subjected to much higher stresses and temperatures than their predecessors. As the stress, temperature, corrosion-resistance, and use-time requirements increase, the tolerance level for defects in the alloys decrease, which is indicated in Figure 5. This means that more severe restrictions on the allowable size of defects are being continually imposed (57,58). The defects may manifest themselves as microporosity, intermetallic segregates, nitride and carbide stringers, and other non-metallic inclusions (57-60). The first two types of defects can be controlled by improved processing methods (i.e., better solidification control). The latter types can be better controlled by composition modifications (i.e., lower carbon contents), and by improved melt and refractory practices.

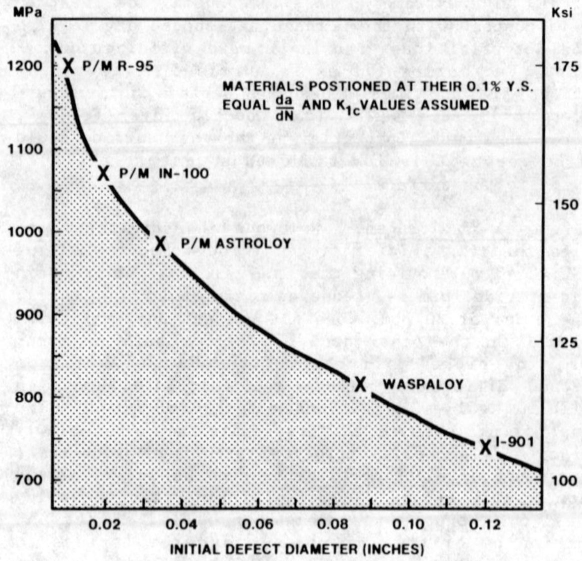

Figure 5. Stress Required to Propagate Defects in Superalloy Disc Materials to Critical Crack Size in 10^4 Cycles (from Meetham, Ref. 57).

In characterizing the defects in directionally solidified (DS) superalloy castings, Marder and Kortovich (60) indicated that the dross (oxide) inclusions were found to be the single most frequent cause for casting rejections. Studies on P/M nickel-base superalloy materials have shown that the minimum low cycle fatigue life (LCF) is set by the presence of small, non-metallic oxide inclusions in the material (61,62). Oxide inclusions having cross-sectional areas as small as 8 mils^2, or less, have caused LCF failures in test specimens of P/M alloys (63).

Several papers have discussed the influence of melt-cleanliness on casting quality (26,60,64-65). Obvious improvements can be made in the selection of raw materials, in minimizing the exposure of hot charges or melts to oxygen or nitrogen (such as preventing furnace leaks or premature furnace openings), and in good housekeeping practices (i.e., use of clean melt-pour systems and molds). However, the introduction of oxygen (or oxides) into the melt via melt/refractory reactions is an intrinsic problem that is often more difficult to overcome.

As was discussed previously, many investigators have noted or predicted a significant pick-up of oxygen in VIM melts of iron-(1,8,10,11,19,23,29,36,67-71) and nickel-(11,20,24,27,30,72) base alloys as a result of dissociating (decomposing) refractories. This oxygen is then recombined with the reactive elements in the melts to form primary oxide inclusions. The decomposition of the refractories also leads to the undercutting of crucible grains, which are swept into the melt and form larger, exogenous inclusions (65) during the stirring or pouring stages of the process.

In addition to oxygen (and oxides), being released into the melt, other products of the refractory decomposition process are also introduced, including Zr (27,66,73,74), B (74), Al (24,65,67,68), and Mg (24,27), depending upon the composition of the refractories used. Figure 6 shows how two crucibles of different materials can influence the composition of trace impurity content of the melt. The left hand portion represents a MgO crucible, which is typically used in the primary melting of superalloys. As Figure 6 indicates, Mg is introduced into the melt at the crucible/melt boundary, and because of its high vapor pressure, much of the Mg is vaporized and later condenses on cooler surfaces in the furnace system.

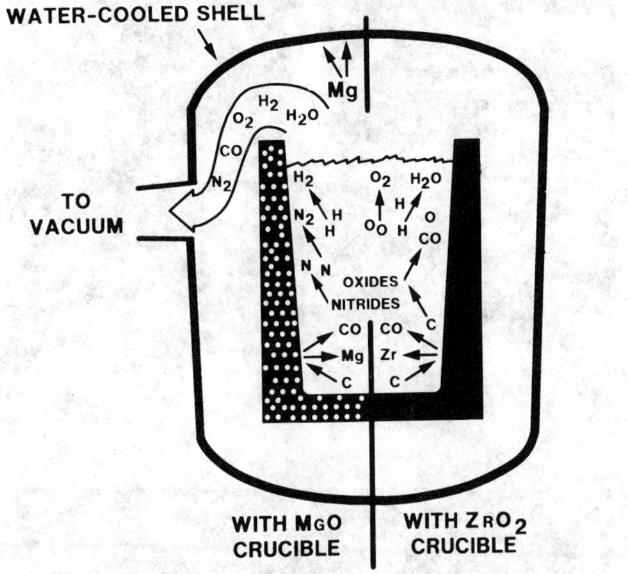

Figure 6. Schematic Diagram of Reactions Taking Place During Vacuum Induction Melting (after Moore, Ref. 75).

The right hand portion of the crucible in Figure 6 represents a ZrO_2 crucible, which is commonly used in the remelting process for investment castings. Some Zr is introduced into the melt as a result of melt-crucible reactions. The non-condensable gases shown in Figure 6 (N_2, O_2 H_2O, H_2, CO) are removed through the pumping system (also refer to Figure 2).

Since the dissociation and reduction of the oxides in the crucible with carbon involve the formation of CO, several investigators have suggested that the stability of the refractories could be measured by the steady-state CO pressure over the melt (10,76,77). Table III lists some predicted CO pressures over VIM melts contained in various crucibles.

TABLE III. Reduction Pressures of Various Crucible Materials with Iron Melts (.1% C) at 1600°C (after Krone, ref. 77)

Crucible Materials	Pco (torr)
CaO	0.3
ZrO_2	1
MgO	4
Al_2O_3	4
SiO_2	610

As the table indicates, the CaO crucible should be the most stable, and the SiO_2 crucible by far, the least stable. Crucibles of MgO and Al_2O_3 would be expected to be less stable than ZrO_2.

The changes in the fractional pressures of non-condensable gases in a small laboratory furnace during the melting of a nickel-base superalloy sample in contact with stabilized ZrO_2 and pure (single crystal) MgO were studied by Johnson and Sutton (14). They found marked differences in the composition of the gases during heat-up, melting, and near-steady state conditions, as shown in Figure 7. The final CO pressure was nearly double in the case of the MgO sample (Fig. 7A). Also, the removal of adsorbed moisture and chemical water (OH) from the MgO was measurable throughout the test; whereas, the initial moisture adsorbed on the ZrO_2 sample was removed during the initial heating period, and no H_2O was detectable after the alloy had melted (Fig. 7B). Thus, the evolution of other gases over the melt in addition to CO are also influenced by crucible composition.

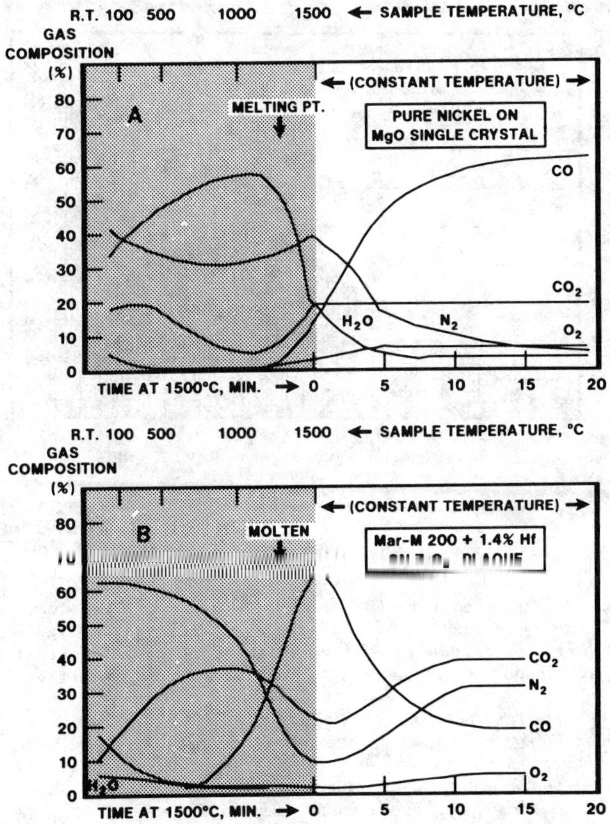

Figure 7. Changes in Furnace Atmosphere Composition During the Heating to 1500°C (gray portion) and During Exposure at Constant Temperature of a Nickel-Base Superalloy Sample in Contact with Single Crystal MgO (Fig. 7A) and with ZrO_2 (Fig. 7B). The Total Furnace Pressure During the Experiment was 10^{-5} Torr (from Johnson & Sutton, Ref. 14).

Measurement of Inclusion Content. One of the most significant advances in the quantitative determination of the inclusion content in superalloys was developed at Pratt and Whitney Aircraft (62). The technique involves the following steps:

1. Systematic sampling of a heat to provide a 1.4 Kg (3 lb.) sample for evaluation.

2. Electron beam skull remelting (no contact with refractories) to float and concentrate the oxides on the melt surface.

3. Electrochemical extraction and filtration of these oxides.

4. Gravimetric measurement of the oxide content.

This procedure and facility were used to measure the inclusion content of small 136 Kg heats of Merl 76, an advanced Hf-bearing superalloy, which were melted in crucibles of different compositions. Figure 8 shows the results. Note that the heats from prefired ZrO_2 and rammed Al_2O_3 crucibles contained the highest level of inclusions (>8 ppm oxides). The VIM melting conditions and the starting raw materials were identical for all the heats, so that the differences in the inclusion levels resulted from differences due to crucible materials (viz., difference in purity, composition, structure, and fabrication process). Detailed structural examinations of the melt-refractory interfaces (solidified skull) and the used crucibles indicated that numerous chemical reactions had occurred, and that the major oxides that formed in the melt at the refractory

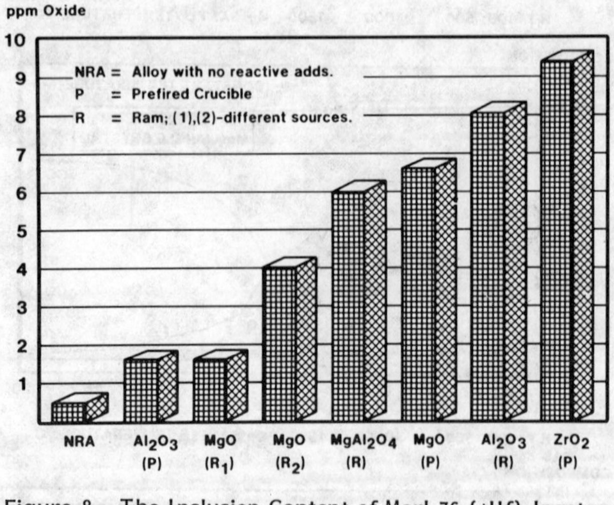

Figure 8. The Inclusion Content of Merl 76 (+Hf) Ingots Cast from Heats Melted in Crucibles of Different Compositions (from Sutton & Maurer, Ref. 80).

surface were mostly HfO_2 and Al_2O_3 (62,66). These were the same oxides found in the EB-inclusion test. Figure 9 summarizes some of the major chemical reactions that were found at or near the melt-refractory interface.

The sessile drop technique (73,78,79) has been used to study the melt/ refractory interfacial reactions in greater detail as well as to measure the tendency of a molten alloy to wet the refractories. Figure 10 shows the melting of a small cube sample of Merl 76 (Figure 10A and 10B) and the formation of a molten drop (Figure 10B & C, 10E & F) on a pure Al_2O_3 substrate (66,80) in a resistance heated vacuum furnace. Note that the alloy sample without the reactive adds (no Ti, Al, Hf, Zr) showed no evidence of oxides on the surface. This composition, after being melted as a 136 Kg (300 lb.) heat also exhibited the lowest inclusion level (viz., <0.3 ppm), of the heats shown in Figure 8. In contrast

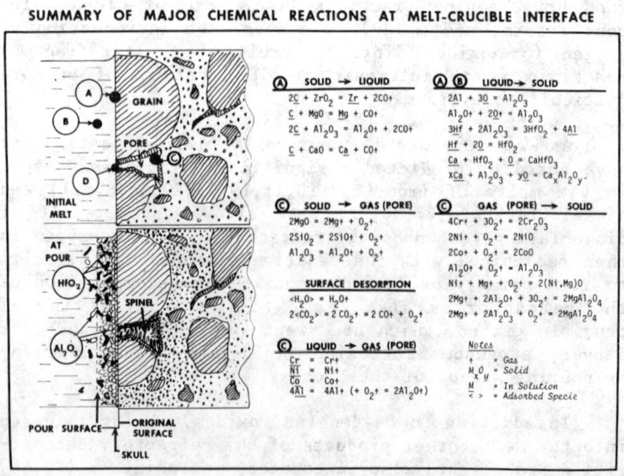

Figure 9. Sequence of Chemical Reactions Occurring at the Melt/Crucible Interface During Vacuum Induction Melting and Refining (from Sutton & Maurer, Ref. 66).

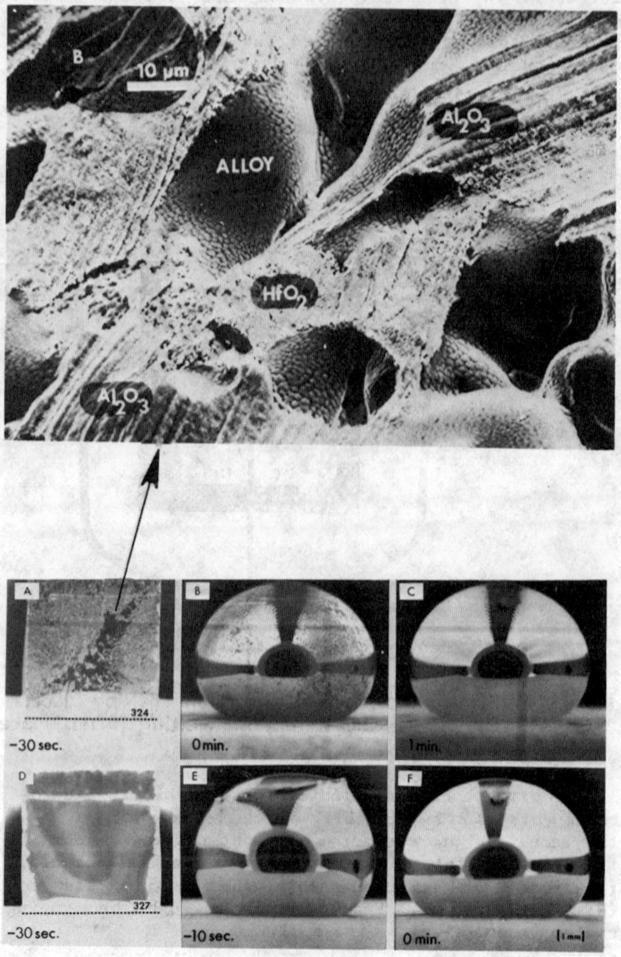

Figure 10. Formation of Al_2O_3 and HfO_2 Dross During the Melting of Merl 76 + Hf at 1500°C and 10^{-4} Torr. The Upper Photo Shows a Magnified View of the HfO_2 and Al_2O_3 Layers (Dross) on the Melted Alloy Surface. Note that there is No Dross on the Lower Sample (D, E, F) in Which the Al, Ti, Hf, and Zr were Absent (from Sutton & Maurer, Ref. 66,80).

to Figure 10A, Figure 10D shows oxides on the melted surface of the alloy containing the reactive adds, indicating that some portion of reactive alloying elements were oxidized prior and during melting. A magnified SEM view of the solidified surface shown in the upper photo of Figure 10 reveals the breakdown of a continuous network of oxides consisting of Al_2O_3 and HfO_2. These same oxides were found at the melt-refractory interface.

As was mentioned earlier, ZrO_2 refractories should be more stable than MgO refractories (because of the lower steady state CO pressure over the melts), and thus produce cleaner melts. However, this is not always true. The heats melted in the MgO crucibles were cleaner (Figure 8). This apparent anomaly is due to many other variables affecting the inclusion content in a melt or casting, including: the purity, the structure, and the open porosity (and density) of the crucible material; the wettability of the refractory; and the tendency of the oxides in the melt to cluster (81) and become attached to the refractory surface. If the oxides which form at the melt-refractory surface become easily detached and are pulled into the melt during VIM stirring or during pouring, the inclusion content is likely to be high in the cast material.

It should be emphasized that hafnium, like carbon, is a powerful deoxidizer, and when in solution in the melt, it will reduce Al_2O_3, MgO, and many other oxides. Consequently Hf-bearing alloys often pose problems in castings because of the presence of HfO_2 inclusions. The reactions between Hf-bearing superalloys and Al_2O_3 (66,80,82-83), ZrO_2 (66,84), MgO (66,80,82-84), and $MgAl_2O_4$ (66,83) refractories has been discussed in detail elsewhere.

Producing Cleaner Melts. Several methods have been used (or suggested) for producing cleaner heats, and some of these are summarized as follows:

1) Electric furnace or AOD melting of charge materials or scrap prior to VIM (1,85).

2) Minimizing the time that the melt is in the crucible (1,27,86).

3) Melting at higher VIM pressures to reduce the dissociation of refractories (19, 24).

4) Selecting more stable refractories (76,87, 88).

5) VIM/ESR remelting to remove inclusions (62,89).

6) VIM/EB remelting to remove inclusions (90).

7) Filtration of melts to remove inclusions (91).

VIM PRODUCTS

Figure 11 shows several processing routes for VIM ingots and electrodes depending on the alloy and application. These routes are applicable to superalloys and other alloys as well. The cast VIM ingots or electrodes are produced in a wide range of sizes (usually cylindrical in shape), and may weigh as little as 13.6-45.5 Kg (30-100 lbs.), and as much as 6,800-13,600 Kg (15,000-30,000 lbs.). The smaller ingots are used primarily by casting industry, and are remelted and cast into investment molds (jet engine blades, vanes, etc.). The conventional castings are cooled in air while still in the molds, and generally have an equiaxed grain structure. Some of the highly alloyed castings are prone to microshrinkage, and so may be hot isostatically pressed (HIP) in order to close the micro pores. Many of the advanced superalloy castings are

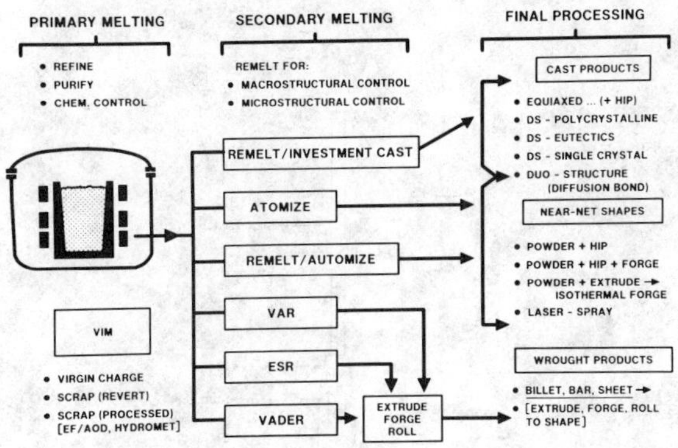

Figure 11. Processing Routes for Products Cast from VIM Ingots or Electrodes.

directionally solidified (DS) (as indicated in Figure 11) to produce blades having all grains oriented parallel to the direction of cooling (viz., along the blade length), or having eutectic structures where the fibrous or lamellar phases run parallel to direction of cooling, or having only one grain (single crystal) oriented along the blade length.

As Figure 11 also shows, the VIM heat may be directly atomized to make powders, or VIM ingots may be remelted at another location and atomized to make powders. The heats for producing powders generally weigh between 136-909 Kg (300 and 2,000 lbs.). The powder material is subsequently consolidated by HIP'ing or extrusion, and then may be further wrought by normal or by isothermal forging. Powders may also be melted by a laser beam and "sprayed" onto a mandrel or form to produce a special shape (such as a turbine disc). Another innovative process has been to combine a casting and a powder part into one integral article. An example of this duo-structure is a set of cast blades, in the form of a wagon wheel, which is diffusion bonded (by HIP'ing) onto a P/M center hub or disc, thereby producing an integral wheel with unique characteristics.

The large VIM ingots (used as electrodes) are remelted through VAR or ESR (electroslag remelting) to produce ingots having the desired soundness and macro- and micro-structures. The structure of these ingots is further refined by extrusion, pressing, rolling, etc., to form billets or bars which are then forged to the final shape or piece. The VAR and VADER processes are further described in the next section.

VACUUM ARC REMELTING

Consumable vacuum arc remelting is a casting process carried out in vacuum with the objective of remelting the consumable electrode to produce an ingot having improved chemical and physical homogeneity. The energy which promotes the remelting of the electrode is supplied by an arc which occurs between the bottom surface of the electrode (usually the cathode) and the ingot top surface/crucible wall (usually the anode) as shown in Figure 12. Noesen (92) presented a detailed historical review of vacuum arc remelting and, therefore, very little historical detail will be given in this paper. Originally, vacuum arc remelting was used to melt reactive metals such as zirconium, titanium and other high-melting metals such as tungsten, tantalum and molybdenum. In the case of the reactive metals (viz., Ti, Zr) there are no suitable ceramic crucible materials that will contain the molten metal without considerable contaminating reactions. Similarly, there are no suitable ceramic materials for containing the refractory metals, such as tungsten because of the extremely high melting temperatures, therefore the VAR process is used to melt these alloys in water-cooled (non-ceramics) metallic molds. The electrodes of the reactive and refractory metals are prepared by consolidating powders

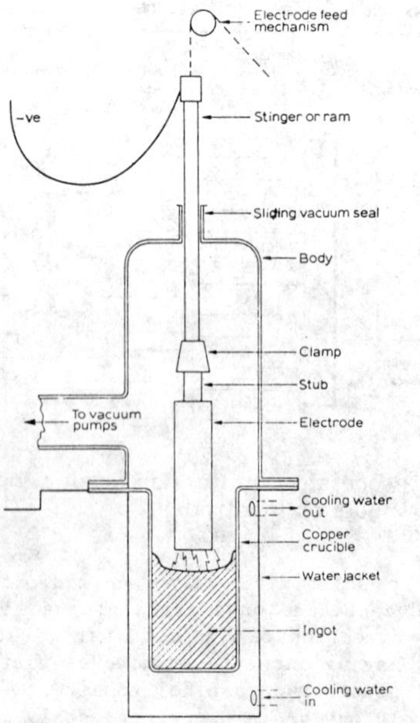

Figure 12. Schematic Illustration of the Principal Features of the Vacuum Consumable Arc Furnace (from Winkler & Bakish, Ref. 10).

or metallic chunks into solid forms of the desired geometry. The initial consumable vacuum arc melting is followed by a second VAR, and sometimes even a third VAR remelting step, to assure the required chemical and structural homogeneity.

In the case of the iron and nickel-base superalloys, the electrodes are cast either from air electric, air induction, or AOD furnaces if the alloying elements are non-reactive. In the case of superalloys which contained considerable amounts of reactive elements such as titanium, aluminum and columbium, the electrodes are cast from vacuum induction melts, which was discussed previously.

VAR Equipment

The basic features of a production vacuum arc furnace are shown in Figure 12 and include:

Crucible. A copper crucible with an inside dimension equal to the final ingot is surrounded by a water cooling system. This system usually includes a water guide to restrict the annulus around the crucible through which the cooling water flows, and thereby increase its velocity. This is done to promote maximum heat transfer from the outside of the crucible to the cooling water. The crucible is attached to the positive leads from the power supply. The input power, which is usually a dc source, can vary from 100 Kw for a small laboratory VAR furnace to over 1000 Kw for the large furnaces, with electrodes weighing up to 14,000 Kg or more.

Body or Upper Chamber. A cylindrical body is located above the crucible and attaches to it with a flange and seal arrangement. This body encloses the electrode to be melted and has provisions for a sliding seal, vacuum ports and sight ports.

Electrode - Ram Arrangement. The electrode (cathode) is usually held by a stub of a similar alloy which is welded to the electrode either manually or by using the arc power of the furnace. A stinger or ram extends outside the body through a sliding seal and at its uppermost end is attached to the feed mechanism. At this end is also attached the negative leads from the power supply and the cooling water system for the ram. At the lower end of the ram, a clamping arrangement inside the body makes electrical contact with the stub to transmit the melting current.

Vacuum System. Depending on the desired vacuum level within the furnace, various combinations of mechanical and diffusion ejector pumps are employed. These are attached to the vacuum port.

VAR Controls. Originally, vacuum arc furnaces were operated via voltage controls, where the electrode feed was controlled by a servomechanism that drove the electrode in relation to the feedback signal of the arc voltage. As the arc gap grew too large, the voltage went up and the electrode was driven downward. However, when low melting currents were required, the slope of the voltage line with respect to arc gap was found to be too small and most operators went to controlling arc length through a form of hash control. Hash is a general term for positive voltage spikes superimposed on the arc voltage. Their duration of only a few milliseconds requires the use of oscillography for examination. A relationship exists in the form of an S-shaped curve between frequency of hash pulses and arc gap as shown in Figure 13. Normally the arc gap is small enough so that it is controlled by the left hand slope of the S-curve. Overriding voltage limiters are required to insure that the control slope does not gradually shift to the right.

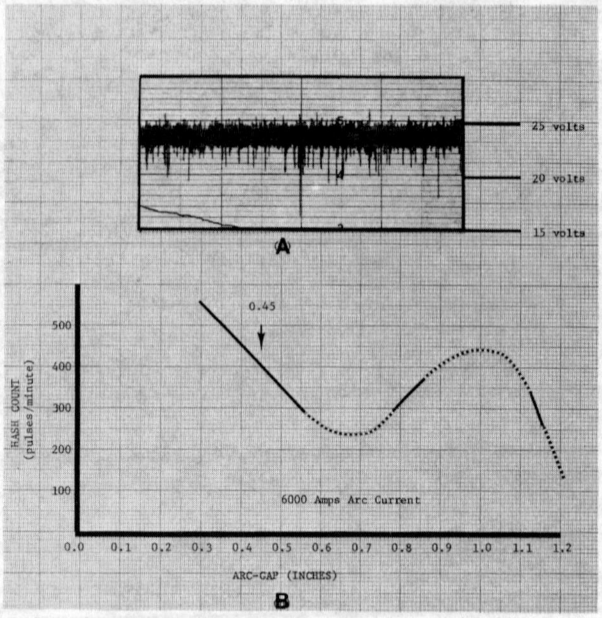

Figure 13. Voltage Trace (A) Produced at an Arc Gap of 0.45 in. (11 mm), and the Hash Count Curve (B) vs. Arc Gap Length. Note the Presence of Many Downward Voltage Spikes (Drip Shorts) in the Voltage Trace (from Rothman, Ref. 93).

The Vacuum Arc Remelting Process

The heat generated in the vacuum arc remelting process results from the electric arc between the two metallic conductors. An arc is an intensely hot body of ionized elemental species through which a relatively high current is passing. This high current arc (often 10,000 amps and above) is characterized by a total voltage drop ranging between 20 and 45 volts. Cobine (94), Somerville (95), and Wood (96) have discussed the arc phenomena in detail. Recently, however, Zanner has done interesting work in the area of the vacuum arc remelting phenomena as far as the arc and the metal transfer are concerned. Some of his work will be reviewed in this paper.

In Zanner's (97,98) work, visual comparison was made simultaneously with electrical measurements of the arc to determine the phenomena that occurred during vacuum arc remelting. Maraging steel (grade 250) was arc-melted into a 152 millimeter diameter mold using a nominal current of 3 kiloamperes. Zanner (97,98) described the wave forms associated with the phenomena of drop shorts. A drop short occurs when a column of molten metal bridges the gap between the anode and cathode causing the arc to short out. Zanner described three classes (Class A, B, and C) electrical wave forms associated with these drop shorts. The differentiation of each class is determined by the duration of the short and the presence of a voltage spike at arc reignition. Class A shorts are of "long" duration (0.010 sec) and have a voltage spike at arc reignition. The current increases during the period (0.010 sec) that the voltage has dropped to a low value. Upon reignition of the arc, the current returns to its value prior to the short. Class B shorts are similar to Class A shorts, except that the response time is shorter and there is usually little change in current. The Class C shorts differ from the other two by not having the large rise in voltage after arc reignition.

A typical Class B drop short is illustrated in Figure 14 where the voltage suddenly drops at 0.003 sec. and remains low until after 0.006 sec. The voltage suddenly peaks at about 0.0073 sec. and then returns to a steady level at about 30 volts. Several high speed motion picture frames are positioned at appropriate locations in Figure 14, and show the molten metal (bright region) attached to the cathode and the presence of liquid protuberances, two of which have made contact with the molten ingot surface (Frames 1-4).

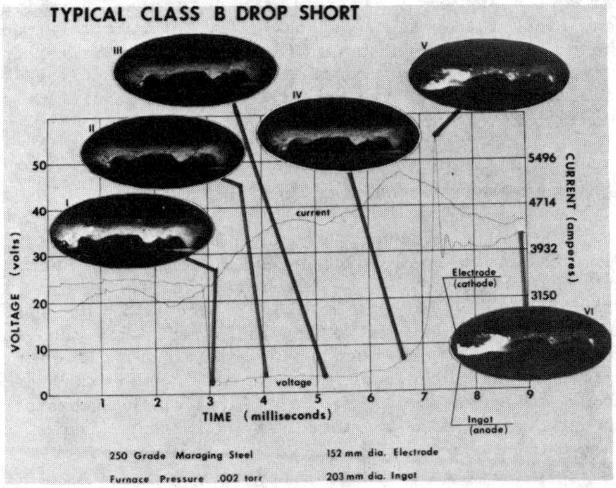

Figure 14. Montage of a Typical Class B Drop Short (from Zanner, Ref. 97).

Zanner (97) concluded from his overall observation of high speed motion pictures used in conjunction with electronic instrumentation, the following (using his words):

1. The vast majority of time during vacuum arc remelting is spent heating the cathode and anodes. This observation is supported by evidence wherein it was determined that the total time spent in the drop short mode is less than 3% of the total melt time. In other words, only about 3% of the total time is metal being tansferred from the cathode to anode.

2. The major mode of metal transfer from cathode to anode occurs by molten column rupture in relatively large agglomerations of transfer at each time.

3. The arc observed under the experimental conditions described previously behaves in the same manner as the classical manner as the low current mode vacuum arc. Kimblin (99) describes this mode as a multiplicity of cathode spots, a diffuse arc column, and a diffuse current collection at the anode.

4. A significant particulate ejection occurs only when a column is in the rupture states. This result was obtained from an inspection of over 15,000 motion picture frames.

5. Particulate ejection appears to be almost horizontal, and a major amount of this particulate sticks to both the anode and cathode surfaces in equal proportions. The remainder escapes the electrode surfaces and strikes the furnace wall where it is subsequently remelted.

6. A typical distribution of ejected particulate size shows particulate sizes ranging from 0.3 millimeters to 2 millimeters.

7. Significant vaporization of non-metallic inclusions does not occur during vacuum consumable arc remelting of fully killed steels. This result was also determined from careful inspection of over 15,000 frames. The electrode surface appeared to be very smooth, and there was no evidence of gas bubbles on the surface. Also, since the major transfer mode is by a bulk process, it follows that entrapped inclusions would be transferred from the electrode to the molten pool via the liquid metal drops.

8. The molten metal at the cathode tip appears to have the same uniform color, and by subjective judgement the temperature appears to be close to the liquidus temperature of the alloy.

9. Cathode spots travel in clusters and these clusters migrate (radially) to the outside surface of the cathode and are subsequently extinguished.

10. Events in the arc region appear to be controlled by surface properties. Protuberances reach to within very close proximity of the anode surface with little or no influence on the shape and magnitude of the electrical wave form as determined by visual inspection.

11. After the interior melting chamber surfaces were baked out, the chamber pressure during the melting was of the same order of magnitude as when measured with the arc off (viz., 0.001 Torr vs. 0.004 Torr).

Segregation and Structure

As was stated earlier, vacuum arc remelting of nickel- and iron-base alloys is performed for structure control. The progressive directional solidification that occurs in VAR allows for the formation of a nearly shrinkage-free ingot. In addition, homogeneous structures can be formed that possess a minimum amount of macrosegregation when compared to directly poured ingots. However, producing a segregation-free structure is not an easy task in alloys like IN-718, even in vacuum arc remelting. Figure 15 shows freckles in a billet which had been forged from a freckled ingot. In a macroetched surface, the freckles appear as dark-etched spots (Figure 15A). Figure 15B is a scanning electron microscope image of a "freckle" defect in an IN-718 billet. The typical structure for the same billet outside the "freckle" area is shown in Figure 15C. The defect area contains accicular delta phase (orthorhombic Ni_3Cb), numerous primary carbides (Cb-rich MC) and carbides that have precipitated on prior grain boundaries. Also, some globular Laves phase (A_2B containing Ni, Cb, Fe, Cr, Mo, Ti and Si) is present.

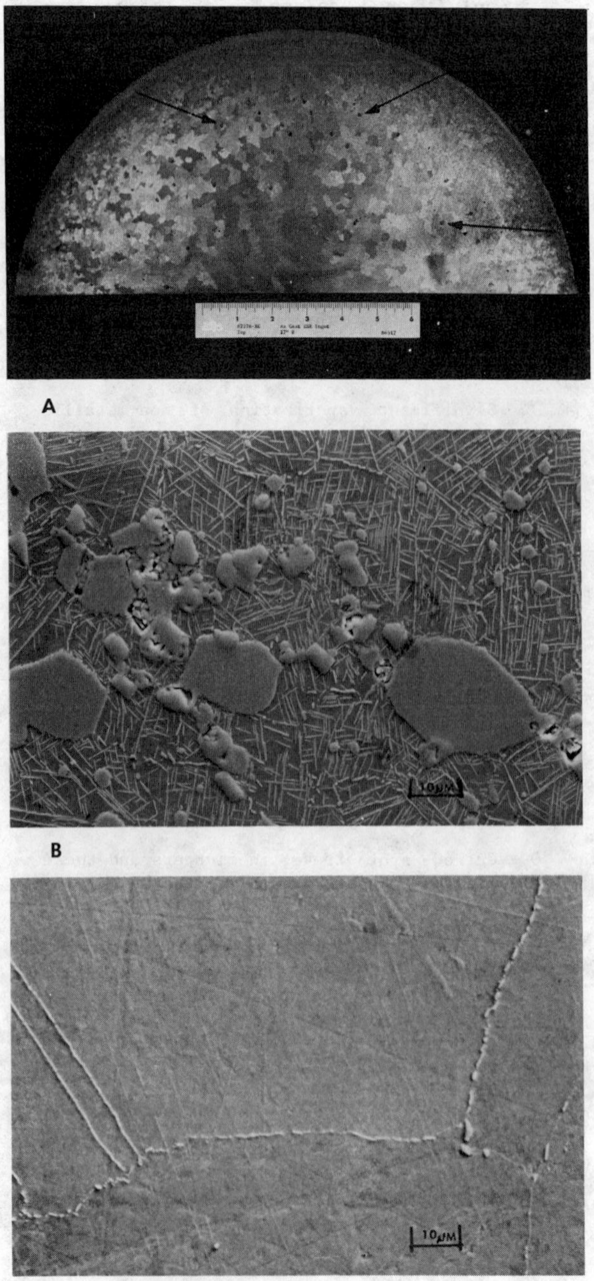

Figure 15. Photograph Showing Freckles (Dark-Etched) Spots; See Arrows in (A) in the Half Cross-Section of a Billet, and SEM Micrographs of Freckled (B) and Freckle-Free (C) Regions (from Jackman, Ref. 100).

It is postulated that the formation of macrosegregates in VAR melting is caused by interdendritic fluid flow and by fluid flow in the pool caused by electromagnetic stirring and by convective movement of the liquid. If the liquid had remained perfectly still, only interdendritic segregation would occur and the size of this interdendritic segregation would be controlled by the freezing rate at the solid-liquid interface; i.e., the faster the freezing rate, the smaller the secondary interdendritic space.

It is normal practice to homogenize VAR ingots prior to forging in order to eliminate the microsegregation. However, this technique will not remove macrosegregation by high temperature heat treatment (homogenization) since the required diffusion distances are so great on an interatomic scale. It is also possible to control ingot grain size through the melting practice. For example, stirring the material as Troutman (101) described can lead to equiaxed grains; Pridgeon (102) also showed similar results by stirring U-700. However, stirring in one direction for too long a period, also may lead to macrosegregation. Therefore common practice is to use reverse stirring techniques which refine the grains without causing macrosegregation.

In alloys of uranium and columbium, where the columbium level is about 6 wt.%, macrosegregation occurs as periodic rings at the solid-liquid interface. The columbium segregation in the ring can be as much as 2 wt.% difference from the average base chemistry. Again, it is believed that this difference is caused by adverse fluid flow and solid transport within the melt pool and the adjacent interdendritic region. Zanner and Bertram (103) have attempted to model this alloy through heat transfer, fluid flow and solidification theory.

Their mathematical model simulates the thermodynamics, magnetohydrodynamics, and fluid flow conditions ahead and within the solid-liquid region of the solidifying ingots (103). While a part of an on going program, the computer modeling code (ZAP) developed thus far and based on experimentally determined boundary conditions is showing good qualitative agreement between the simulated and experimental results. Further experiments and refinements of the model are being pursued in order to more sharply detail the internal conditions of the solidification process.

A New VAR Melting Technique (VADER)

The VADER (Vacuum Arc Double Electrode Remelting) Process has been developed and patented by Special Metals Corporation (104). A schematic of the process is shown in Figure 16. In the VADER process, two consumable electrodes are arc'ed against each other, and the molten metal is allowed to drip into a stationary, rotating, or withdrawal mold located under the electrode. The metal drips from the faces of the two arcing electrodes as soon as there is sufficient liquid for them to form and fall. This is a highly controlled temperature process because it is controlled by nature;

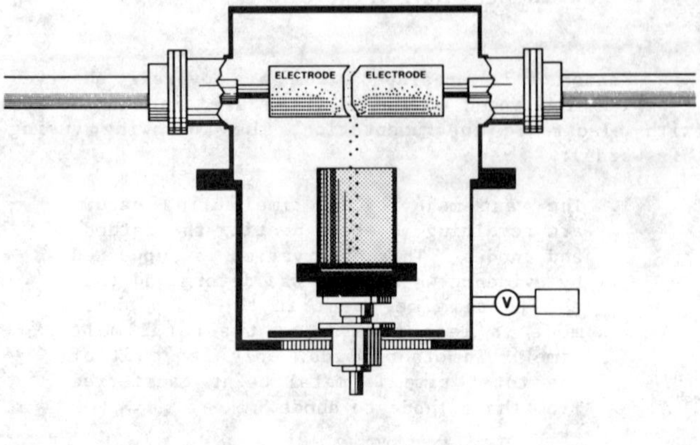

Figure 16. Schematic of the VADER (Vacuum Arc Double-Electrode Remelting) Casting Process (from Soykan & Huntington, Ref. 104).

i.e., the composition of the material and the force of gravity cause the molten metal to run off the electrode always at the same temperature. Measurements on the temperature of the drops coming off the electrode indicate that it is slightly under the liquidus temperature of the alloy. These measurements were made on IN-718 and were found to be about 20°F below the liquidus as measured by the differential thermal analysis techniques. The process, although not scaled up completely through production, has the following benefits compared to normal vacuum arc remelting:

1. The liquid is very viscous because it contains considerable amounts of solid nuclei which hinder fluid flow. In addition, since the arc is located above the melted material in the crucible, the electromagnetic effects of stirring are minimized.

2. Unlike normal VAR, there are no vapor deposits on the mold wall that can fall into the molten pool and cause surface or internal defects in the ingot.

3. The process requires only about 50% of the energy utilized in normal vacuum arc remelting.

4. The process produces grains of 4-8 ASTM, which are equiaxed in nature. The process has been found to be useful in melting highly alloyed materials that are normally processed as powder to be hot isostatically pressed or extruded.

5. The ingots are amenable to ultrasonic inspection.

SUMMARY AND CONCLUSIONS

The vacuum induction process is the most flexible of the vacuum melting processes, since it allows for the independent control of temperature, pressure, and mass transport by means of stirring. Metallurgically, it provides the greatest degree of control over alloy composition and purity (extremely low trace elements and gas contents). It is used for the primary melting of superalloys, special high-performance and corrosion resistant steels, special electrical and magnetic alloys, high purity metals, nuclear and other aerospace alloys. Vacuum induction melting is a relatively new industrial process; before 1955 the total U.S. VIM capacity was less than 10,000 lbs. (4550 Kg)/year, which is equivalent to one heat of a medium-sized, modern VIM furnace. The present U.S. VIM capacity is well over 100 million lbs. (45,450 Kg)/year.

The vacuum arc remelting process has proven to be a viable process to produce segregation-free ingots of superalloys, titanium, molybdenum, tungsten and tantalum. In the case of the high melting and reactive elements, it is the only process that is economically feasible to consolidate the materials by melting. The process produces dense, homogeneous, high purity materials that are readily wroughtable through forging extrusion and rolling. The structure that is built into the ingot helps to produce the excellent properties found in the materials after wroughting. The new process, VADER, may provide even better structural and grain size control in vacuum arc remelted materials.

REFERENCES

1. R. Schlatter, "Vacuum Induction Melting", J. Metals 24, 1972.

2. F. N. Darmara, J. S. Huntington, & E. S. Machlin, "Vacuum Induction Melting", J. Iron & Steel Inst., 191, pp. 266-275, Mar., 1979.

3. W. J. Boesch & J. F. Radavich, "Production and Properties of the Superalloy Udimet-700", Trans. 1960 Vac. Met. Conf., Inter. Sci. Pub., N.Y., pp. 119-130, 1961.

4. Aeorspace Material Specification, AMS-2280, "Trace Element Control", Soc. Auto. Eng., Inc., June 15, 1975.

5. R. T. Holt and W. Wallace, "Impurities and Trace Elements in Nickel-Base Superalloys", Int. Metals Rev., Review 203, The Metals Soc., London, pp. 1-24, Mar., 1976.

6. C. G. Bieber & R. F. Decker, "The Melting of Malleable Nickel and Nickel Alloys", Trans. AIME, 221, pp. 629-636, June, 1961.

7. J. F. Radavich, "The Effect of Minor Elements B, Zr, C and Hf on Superalloys", Proc. 3rd Ann. Purdue Univ. Student-Ind. High Temp. Mat'l Seminar, Purdue Univ., Dec. 8, 1975.

8. J. H. Moore, "Some Aspects of Vacuum-Melted Metal", Met. Progress, 64 (10), pp. 103-105, 1953.

9. R. Schlatter, "Progress in Vacuum Melting of Aerospace Alloys", Trans. 1968 Am. Vac. Soc. Met. Conf., Amer. Vac. Soc, pp. 333-354, 1969.

10. O. Winkler and R. Bakish, Vacuum Metallurgy, Elsevier Scientific Publishing Co., Amsterdam, Netherlands, 898 pp., 1971.

11. F. N. Darmara, "Vacuum Induction Melting: the Revolutionary Influence in Steelmaking", J. Metals, 19, pp. 42-48, 1967.

12. Britton, W. E. et al, "Vacuum Melting", Harvard Business School Report, April, 1957.

13. J. S. Huntington & F. H. Soykan, "Energy Savings with a Dual Pumping System for Vacuum Induction Melting", Proc. 6th Int. Vac. Met. Conference, San Diego, CA, pp. 450-457, April 23-27, 1979.

14. W. E. Johnson & W. H. Sutton, "Residual Gas Analysis Applied to Vacuum Induction Melting", Proc. 35th Electric Furnace Conf., Chicago, IL, Iron & Steel Soc. AIME, pp. 201-207, 1977.

15. C. E. Fitterer, "Determining Duplicate Oxygen Readings of Liquid Alloys with a Single Probe in the Vacuum Induction Process", 5th Int. Symp. on Electroslag and Other Special Melting Tech., Carnegie-Mellon Ins. of Res., Pittsburg, PA, 2, 679-688, 1975.

16. T. B. King, "Thermodynamics and Kinetics in Vacuum Metallurgy", Vacuum Metallurgy, Reinhold Pub. Corp., New York, pp. 35-58, 1958.

17. E. S. Machlin, "Kinetics of Vacuum Induction Theory", Trans. AIME, 218, pp. 314-326, April, 1961.

18. A. M. Aksoy, "Thermodynamics and Kinetics in Vacuum Induction Melting", Vacuum Metallurgy, Reinhold Pub. Corp., N.Y., pp. 59-78, 1958.

19. A. M. Samarin, "Deoxidation of Steel in Vacuum", Vacuum Metallurgy, Reinhold Pub. Corp., N.Y., pp. 255-265, 1958.

20. O. Winkler, "The Theory and Practice of Vacuum Melting", Metallurgical Reviews, 5 (17), pp. 1-117, 1960.

21. R. G. Ward, "Aspects of Modern Ferrous Metallurgy", University of Toronto Press, (1964).

22. J. F. Elliott, Le Vide, No. 135, p. 123, 1968.

23. W. A. Fischer & A. Hoffman, "Properties of Iron and Steel Melts in High Vacuum", Archiv fur das Eisenhuttenwesen, 29, pp. 339-359, June, 1958.

24. G. A. Vaughn and G. H. Geiger, "Melt-Crucible Interactions and Inclusion Formation in Vacuum Melted Ni-Al-Ti Alloys", Proc. 6th Int. Vacuum Met. Conf., San Diego, CA, pp. 357-380, April 23-27, 1979.

25. G. K. Sigworth, J. F. Elliot, G. A. Vaughn, and G. H. Geiger, "The Thermodynamics of Dilute Nickel Alloys", Trans. Met. Soc. Canadian Inst. Met., p. 104, 1977.

26. J. D. Barber and R. C. Hambleton, "Investigation Into Cleanliness of Nickel-Rich Vacuum-Casting Alloys", Foundry Trade J., 131, pp. 107-112, 1970.

27. V. I. Kashin and A. M. Samarin, "The Vacuum Melting of Nickel-Base Heat-Resisting Alloys", The Uses of Vacuum in Metallurgy, trans. by E. Bishop, London, pp. 3-12, 1964.

28. D. A. Ford, "Inclusion-Free Vacuum Melting of Superalloys", Foundry Trade J., 134, pp. 71-75, 1973.

29. P. P. Turillon & E. S. Machlin, "Deoxidation of Pure Iron by Vacuum Induction Melting", Trans. 1959 Vacuum Met. Conf., Amer. Vac. Soc., N.Y. Univ. Press, pp. 81-98, 1960.

30. A. Simkovich, "Refractory-Metal Studies in Vacuum Induction Melting", Trans. Vac. Met. Conf., Amer. Vac. Soc., pp. 361-375, 1967.

31. T. J. Bosworth, "Vacuum Melting: Influence of Hydrogen Pretreatment of Crucibles on Oxygen Content", Trans. AIME, 212, pp. 489-493, 1958.

32. K. M. Bonar, J. L. Cunningham & F. H. Walther, "Stability of Commercial Refractories in Vacuum", J. Amer. Ceramic Soc., 46 (7), pp. 683-687, 1967.

33. B. E. Hopkins, G. G. H. Jenkins & H. E. N. Stone, "Production of High Purity Iron and Iron Alloys on a 25-lb. Scale", J. Iron and Steel Inst., 168, p. 371, 1951.

34. J. Szekely & S. D. Fang, "The Dynamics of Gas Bubble Growth at Reduced Pressures", Proc. 4th Int. Conf. of Vac. Met., Japan. Proceedings Section 1, pp. 52-56, 1974.

35. B. V. Linchevskii, "Behavior of the Components of Stainless Steel during Vacuum Melting", Izvest, VUZ-Chernaya Met., pp. 70-76, March, 1963.

36. N. A. Parlee, S. R. Seagle & R. Schuhmann, Jr., "Rate of Carbon-Oxygen Reactions in Liquid Iron", Trans. AIME, 212 (2), pp. 132-138, 1958.

37. T. B. King, R. A. Karasec & P. Dastur, Heterogeneous Kinetics at Elevated Temperatures, Plenum Pub. Co., New York, 1970.

38. M. Olette, "Some Fundamental Aspects of Metallurgical Processes in Vacuum", Proc. 4th Int. Conf. Vac. Met., Japan, Section 1, pp. 29-34, 1974.

39. N. Meysson & A. Rist, "Theoretical Calculations of the Deoxidation of Liquid Steel by Bubbling of Methane", Rev. Met., 62 (2), pp. 121-126, Feb., 1965.

40. W. F. Moore, "Deoxidation Techniques for Vacuum Induction Melting", J. of Metals, 15 (12), pp. 918-921, Dec., 1968.

41. P. Kozakevich, "Surface Phenomena of Metals", Soc. of Chem. Ind., Monograph, No. 28, London, p. 223, 1967.

42. W. A. Fischer, "Effect of High Vacuum on the Metallurgy of Iron", Archiv fur das Eisenhuttenwesen, Vol. 31, pp. 1-9, Jan., 1960.

43. G. A. Garnyk & A. M. Samarin, "Vacuum Metallurgy (of Steel)", Izvest. Akademii Nauk SSSR, Otdelenie Tekhnicheskih Nauk, (May, 1957).

44. A. Simkovich, "Variables Affecting Nitrogen Removal in the Vacuum-Induction Melting of Iron-Base and Nickel-Base Alloys", J. Metals, 253 (4), pp. 504-512, 1966.

45. W. B. Kent, "Trace Element Effects in Vacuum-Melted Alloys", J. Vac. Sci. Technol., 11 (6), pp. 1038-1046, Nov./Dec., 1974.

46. P. P. Turillon, "Evaporation of Elements from 80/20 Nickel-Chromium During Vacuum Induction Melting", Trans. 6th Int. Vac. Met. Conf., p. 88, 1963.

47. R. Ohno, "Thermodynamic Considerations on Desulphurization of Cast Iron in Vacuum Melting", Tohoku University, Science Reports of the Res. Inst., Series A, Vol. 12, pp. 353, Aug., 1960.

48. W. Janiche & H. Beck, "Desulfurization and Deoxidation of High Purity Iron in the Vacuum Melting Process", Archiv fur das Eisenhuttenwesen, Vol. 29, pp. 631-642, Oct., 1958.

49. R. G. Ward & R. Hall, "Desulfurization of Molten Steel by Solid Lime During Vacuum Melting", J. Iron and Steel Inst., 195, pp. 75-78, May, 1960.

50. S. E. Volkov, et al, "Desulfurization of Steel in Vacuum Induction Melting Furnaces", Stal (in English), pp. 115-118, Feb., 1965.

51. D. J. Wilson, J. J. deBabarbadillo & E. Snape, "Desulfurization, Deoxidation and Sulfide Shape Control with Nickel-Magnesium", Proc. 33rd Electric Furnace Conf., Iron & Steel Soc., AIME, pp. 196-209, 1975.

52. J. G. Cannon, R. S. Cremisio, & C. F. Elliott, "Sulfur-Its Effects Removal or Modification in Vacuum", Proc. 3rd Int. Symp. ESR & Other Spec. Melt Tech., Part III, pp. 1-30, 1971.

53. R. E. Bailey, R. R. Shring, & R. J. Anderson, "Effects of a Rare Earth Addition on Unitemp 901", Superalloys, Metallurgy and Manufacture, Proc. 3rd Int. Symp. on Superalloys, Seven Springs, PA, Claitor's Pub. Div., Baton Rouge, LA, pp. 109-118, 1976.

54. M. F. Rothman, "Desulfurization of a Nickel-Base Superalloy During Vacuum Induction Melting by Means of Rare Earth Additions", Trans. 1977 Vac. Met. Conf., Amer. Vac. Soc., pp. 49-76., 1977.

55. D. Li, G. E. Maurer, F. Cosandy, R. Foote, & J. K. Tien, "Role of Cerium in the Refining Process of a Nickel-Chromium Alloy and an Iron-Nickel Base Superalloy", to be pub.

56. C. J. Burton, "Selenium Control in Vacuum Induction Melting", Trans. 1977 Vac. Met. Conf., Amer. Vac. Soc., pp. 76-89, 1977.

57. G. W. Meetham, "Superalloy Processing and Its Contributions to the Development of the Gas Turbine Engine", High Temperature Alloys for Gas Turbines, Applied Science Pub., London, pp. 837-859, 1978.

58. S. R. Houldsworth, "The Significance of Defects in Nickel-Base Superalloys", Superalloys '80, Proc. 4th Int. Symp. of Superalloys, ASM, pp. 375-383, 1980.

59. E. Bachelet, "Quality Castings of Superalloys", High Temperature Alloys for Gas Turbines, Applied Sci. Pub., London, pp. 665-699, 1978.

60. J. M. Narder & C. S. Kortovich, "Characterization of Casting Defects in Typical Castings of a Directionally Solidified Superalloy", Final Rept. AFML-TR-79-4060, TRW Materials Tech. Lab., Cleveland, OH, Contract F33615-76-5373, June, 1979.

61. C. E. Shamblin, R. F. Halten, & W. R. Pfouts, "Manufacturing Methods for Improved Superalloy Powder Production", AFML F33615-78-e-5225 Second Interim Report.

62. E. E. Brown, J. E. Stulga, L. Jennings, & R. W. Salkeld, "The Influence of VIM Crucible Composition, Vacuum Arc Remelting, and Electroslag Remelting on the Non-Metallic Inclusion Content of Merl 76", Superalloys '80, Proc. 4th Int. Symp. on Superalloys, ASM, pp. 159-162, 1980.

63. J. D. Buzzanell & L. W. Lherbier, "Processing Effects on the Properties of P/M Rene 95 Near Net Shapes", Superalloys '80, Proc. 4th Int. Symp. on Superalloys, ASM, pp. 149-158, 1980.

64. J. M. Drapier, "Progress in Advanced Directionally Solidified and Eutectic High Temperature Alloys", High Temperature Alloys for Gas Turbines, Applied Sci. Pub., London, pp. 701-736, 1978.

65. W. H. Sutton, "Influence of Metal-Ceramic Reactions on Casting Quality", 25th Ann. Mtg. of the Investment Casting Institute, Denver, CO, Oct. 5-8, pp. 10:1 to 10:29, 1977.

66. W. H. Sutton & G. E. Maurer, "Influence of VIM Crucible Composition on Non-Metallic Content of an Advanced Hafnium-Bearing Nickel-Base Superalloy", Proc. 6th Int. Vacuum Metallurgy Conference on Special Melting, San Diego, CA, pp. 340-356, April 23-27, 1979.

67. K. R. Olen, "Vacuum-Carbon Deoxidation of Iron Melts in a 70% Alumina Brick Crucible", J. Vac. Sci. & Tech., 7 (6), pp. S149-S153, 1970.

68. C. W. Hunter & S. K. Tarby, "Carbon Deoxidation of Steel Melts in Al_2O_3 Crucibles Under Reduced Pressures", Trans. Vac. Met. Conf., Amer. Vac. Soc., pp. 355-388, 1968.

69. J. F. Elliott, "Metal-Refractory Reactions in Vacuum Processing of Steel and Superalloys", 29th Electric Furnace Proceedings, AIME, pp. 12-18, 1971.

70. C. B. Griffith & M. P. Fedock, "Gas Evolution from Refractories During Ladle Degassing", Trans. Vac. Met. Conf., Amer. Vac. Soc., pp. 463-478, 1968.

71. G. H. Bennett, H. T. Protheroe & R. G. Ward, "The Role of Vacuum-Melted Steel", J. Iron & Steel Inst., 195, Part 2, pp. 174-180, 1960.

72. Syan-Khua Shao, Chzhi Shui, Chzhao Zhen' - Chuan' and Tsui Pei-Syum, "Oxidation of Nickel with Oxygen from the Crucible Material During Melting in a Vacuum Induction Furnace", Acta. Met. Sinica, 8 (3), pp. 292-294, 1965.

73. E. Snape and P. R. Beeley, "Refractory-Melt Reactions in Vacuum Induction Melting of Nickel-Base Alloys," J. Amer. Ceramic Soc., 50 (7), pp. 349-358, 1967.

74. R. F. Decker, J. P. Rowe & J. W. Freeman, "Relations of High-Temperature Properties of a Ti+Al Hardened Nickel-Base Alloy to Contamination by Crucibles", Trans. AIME, 212, pp. 686-694, 1958.

75. W. F. Moore, "Vacuum Melting-Induction", General Electric Co., Rpt. No. 66-C-386, Nov., 1966.

76. J. Zotos, "Thermodynamic Reactivity of Carbon in Steel with Past, Present, and Future Steel Processing Refractories", Amer. Foundry Soc. Trans., 71, pp. 689-692, 1963.

77. K. Krone, J. Krüger and H. Winterhager, "Beitrag zum Schmelzen von NiCr-Basislegierungen in Hockvakum, Forschungsber des. Landes Nordrhein-Westfalen", No. 1825, Westdeutscher Verlag, Köln und Opladen, 1967.

78. W. H. Sutton, "Investigation of Vacuum Melt-Refractory Reactions Using Sessile Drop Techniques", Proc. 5th Int. Symp. on ESR & Other Spec. Melt. Tech., Part II, Pittsburgh, PA, pp. 648-678, Oct. 16-18, 1974.

79. W. H. Sutton, "Investigation of Vacuum Melt Refractory Reactions Using Sessile Drop Techniques", Proc. 1975 Vac. Met. Conf., Vac. Met. Div. of AVS, Columbus, OH, Microfiche, pp. 160-182, June 23-25, 1975.

80. W. H. Sutton & G. E. Maurer, "Formation of Inclusions in Hf-Bearing Superalloys", 28th Ann. Mtg. Investment Casting Inst., San Francisco, CA, pp. 18:01-18:19, Oct. 2-4, 1980.

81. T. B. Braun, J. F. Elliott & M. C. Flemings, "The Clustering of Alumina Inclusions", Met. Trans., AIME, 10B (2), pp. 171-184, 1979.

82. W. H. Sutton & D. R. Green, "Influence of Hafnium on the Wetting Characteristics of MAR-M 200," Superalloys: Metallurgy and Manufacture, Proc. 3rd Int. Symp., Claitor's Pub. Div., Baton Rouge, LA, pp. 171-179, 1976.

83. W. H. Sutton & W. A. Johnson, "Reactions Between Al_2O_3/MgO Crucible Materials and a Vacuum Melted Hf-Bearing Alloy", Superalloys 1980, Proc. 4th Int. Symp. on Superalloys, ASM, pp. 311-320, 1980.

84. W. H. Sutton & Dr. R. Green, "Interfacial Reactions Between a Hafnium Containing Nickel-Base Superalloy and Zirconia Ceramics", Ceramic Microstructure '76. Ed. by R. M. Fulrath and J. A. Pask, Westview Press, Boulder, CO, pp. 520-531, 1977.

85. M. J. Woulds, "Recycling of Engine Serviced Superalloys", Superalloys 1980, Proc. 4th Int. Symp. on Superalloys, ASM, pp. 31-41, 1980.

86. Anon., "Vacuum Melting: Now Look What They're Doing", Steel, pp. 39-40, July 21, 1969.

87. W. H. Sutton, "Selection of Refractories for Vacuum Cast Alloys: Importance of Refractory Stability and Chemical Inertness", Proc. 26th Ann. Mtg. Invest. Casting Inst., Oct. 5-7, Scottsdale, AZ, pp. 12-1 to 12-33, 1978.

88. K. Harris & R. E. Schwer, "The Influence of Nitrogen and Oxygen on the Castability of IN-100", Proc. 1977 Vac. Met. Conf., Amer. Vac. Soc., pp. 389-422, 1977.

89. K. Harris & L. L. Gill, "Development of a VIM + ESR Process Route for Direct Forging of Components from Ingot in Nickel-Base Precipitation Hardened Superalloys", Union Carbide UK Ltd., Superalloys Div., ESR Conf., Sheffield, Eng., Trans. 151, pp. 95-96, Jan. 10-11, 1973.

90. G. P. Holoman, "Advanced Methods Improve Turbine Output", Precision Metal, pp. 47-50, March, 1980.

91. D. Apelian & R. Muthanson, "Filtration: A Melt Refining Method", J. Metals, 32 (9), pp. 14-18, 1980.

92. S. J. Noesen, "Vacuum Arc Melting from Robert Hare 1839 to the Present", Trans. 1967 Vac. Met. Conf., Amer. Vac. Soc., N.Y., pp. 503-550, 1968.

93. M. F. Rothman, Report H8-T-T2-T1, Special Metals Co., May 23, 1971.

94. J. D. Cobine, Gaseous Conductors, Dover Publications Inc., New York, 1953.

95. J. M. Somerville, The Electric Arc, Methuen & Co., Ltd., London, 1959.

96. F. W. Wood & R. A. Beale, "Studies of High-Current Metallic Arcs", Bulletin 625, Bur. Mines, U.S. Dept. Interior, 1965.

97. F. J. Zanner, "Observations of the Vacuum Arc and Metal Transfer During Vacuum Consumable Arc Remelting", Proc. 6th Int. Vac. Met. Conf., Amer. Vac. Soc., San Diego, CA, pp. 417-427, April 23-27, 1979.

98. F. J. Zanner, "Metal Transfer During Consumable Arc Remelting", Met. Trans. AIME, 10B, pp. 133-142, June, 1979.

99. C. W. Kimblin, J. Applied Physics, 40, p. 1744, 1969.

100. L. A. Jackman, unpublished data.

101. J. W. Troutman, "Ingot Structure Control of Vacuum-Arc-Melted High-Temperature Alloys", Trans. 1967 Vac. Met. Conf., Amer. Vac. Soc., p. 599-613, 1968.

102. J. W. Pridgeon, "Manufacturing Process for Improved High-Strength Superalloy Sheet", Union Carbide Corp., Technology Dept., Materials Syst. Div., Kokomo, IN, U.S. Air Force Interim Report, AFML IR-9-101 (No. 1), 1966.

103. F. J. Zanner & L. A. Bertram, "Computational and Experimental Analysis of a U-6w/o Nb Vacuum Consumable Arc Remelted Ingot - A Progress Report for the Sandia Macrosegregation Study", Sandia Nat'l. Lab., Rept. SAND 80-1156, April, 1981.

104. F. H. Soykan & J. S. Huntington, U.S. Pat. 4,261,412, April 14, 1981.

THE MATHEMATICAL AND PHYSICAL MODELLING OF
METALS PROCESSING OPERATIONS

Julian Szekely

Department of Materials Science
and Engineering, Massachusetts
Institute of Technology,

Cambridge, Massachusetts 02139

1. INTRODUCTION

The purpose of these notes is to introduce the reader to the mathematical and physical modelling techniques that have been developed for representing primary metals processing operations. These techniques have been proven to be very effective both for diagnostic purposes, such as trouble shooting, process improvement process optimization and in the development of new processes. The application of mathematical and physical models for the solution of specific problems in metals processing operations is quite well documented; for this reason in the present treatment emphasis will be placed on the principles and on the discussion of the strategies that may be employed in a given situation. Detailed illustrations of actual applications are available in the references and in recommended reading given in the Appendix.

In the discussion that follows we shall examine the use of

mathematical models

physical models and

pilot plants

in process evaluation, process optimization and new process development, and will then explore possible strategies that may be employed.

2. MATHEMATICAL MODELS

A mathematical model consists of a set of equations, algebraic or differential equations which quantitatively represents a process, or some aspect of a process. Such a mathematical model may be a relationship that defines the time required to empty a ladle holding molten steel, or the time required for an ingot to solidify. Yet other examples would include the calculation of the amount of oxygen that needs to be blown to complete a BOF heat, the cooling rates that are permissible in a continuous casting operation to prevent cracking velocity field in an ESR unit and the like.

Classification of Mathematical Models

Mathematical models may be classified into the following three main groupings:

(i) Fundamentally based theoretical models

(ii) Semi-empirical models

(iii) Input - Output type or "black box" models.

(i) Fundamentally Based Theoretical Models are derived from basic physical laws, with a minimum "fudging" or empirical adjustment. In certain physical situations it is possible to construct such models, or more properly termed a quantitative mathematical representation of the process. Many heat transfer problems, electromagnetic force field calculations, and laminar fluid flow problems fall in this category. In the majority of cases such a representation will appear in the form of partial differential equations, which together with the appropriate boundary conditions, have to be solved numerically.

(ii) Semi-Empirical Models are in essence based on physical laws, but at the same time a certain amount of empiricism is incorporated into the representation of these systems, either because some of the data are lacking or due to the complexity of the modelling equations which would be difficult to solve. The vast majority of the mathematical models used in practice fall in this category.

(iii) Input - Output Type "Black Box" Models are not based on physical laws, but rather represent a totally empirical relationship between the key process variables. The relatively simple "black box" type models are helpful in process control.

The difference between these three types of models is illustrated in the following example.

(a) Let us consider a spherical metal particle, of radius R, which is initially at a temperature T_i; from time = 0 this particle loses heat by radiation to the environment which is at a temperature T_e (both on the absolute scale).

The time dependent temperature of this spherical particle is given by the following differential equation: (1,2,3)

$$\frac{\partial T}{\partial t} = \alpha\left[\frac{2}{r}\frac{\partial T}{\partial r} + \frac{\partial^2 T}{\partial r^2}\right] \qquad (1)$$

with

$$T = T_i \text{ at } t = 0 \qquad (2)$$

$$\frac{\partial T}{\partial r} = 0; \quad r = 0 \qquad (3)$$

$$k\frac{\partial T}{\partial r} = \sigma\varepsilon(T^4 - T_e^4) \qquad (4)$$

at r = R

where T is the (absolute) temperature

t is time

$\alpha = k/\rho C_p$ is the thermal diffusivity

k is the thermal conductivity

σ is Boltzmann's constant

ε is the emissivity of the surface

This system of equations may be solved numerically. [3]

(b) Let us now consider the case, sketched in Fig. 1, where spherical particles, contained in a fluidized bed, undergo a radiative heat exchange with a wall.

For argument's sake, let us assign a temperature T_e to the wall and an initial temperature T_i to the particles.

Let us assume, in the first instance, that the principal mechanism for heat transfer is radiation, between the wall and the layer of particles, which are immediately adjacent to the wall, as sketched in Fig. 1(b)

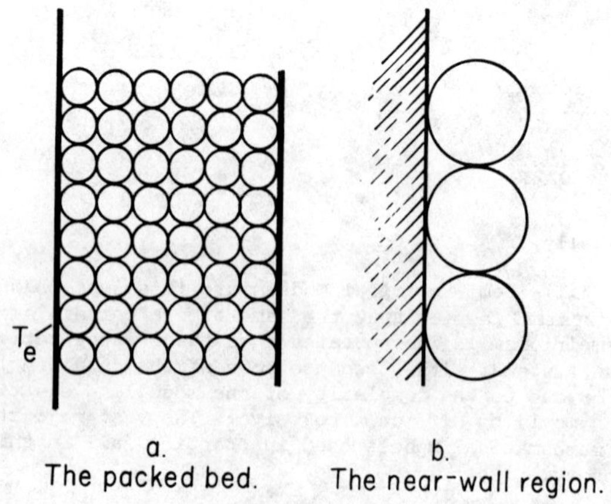

Fig. 1 Radiative Heat Transfer in a Fluidized Bed.

Under these conditions the rigorous formulation of the problem would require the statement of the truely three dimensional equation of conduction for the spherical particle, together with the boundary conditions, which would have to reflect the fact that there is a complex, spatially non-uniform radiative interchange between the particle, the wall and its neighbors. The spatial non-uniformity should be stressed here, that is, that the heat flux falling on the surface of the particle, will be strongly position dependent. The resultant differential equation:

$$\frac{\partial T}{\partial t} = \alpha \nabla^2 T \tag{5}$$

with $q = f(x,y,z,t)$ at the surface would be quite cumbersome to solve. Furthermore, there is some question, as to how accurately one would be able to define the relationship that has been expressed by Eq. (5). For this reason one may wish to simplify the statement of the problem by writing:

$$\frac{\partial T}{\partial t} = \alpha\left[\frac{2}{r}\frac{\partial T}{\partial r} + \frac{\partial^2 T}{\partial r^2}\right] \tag{6}$$

with

$$T = T_i \text{ at } t = 0 \tag{7}$$

$$\frac{\partial T}{\partial r} = 0 \text{ at } r = 0 \tag{8}$$

and

$$k\frac{\partial T}{\partial r} = F\,\sigma\,F(T^4 - T_e^4) \tag{9}$$

$$\text{at } r = R \tag{10}$$

Here F is some sort of a "view factor" which incorporates the radiative interaction between the sphere and its surroundings.

The modelling of the system through Eqs. (6-10) is not quite rigorous because we glossed over the fact that the radiative interchange between the particle and its surroundings is spatially non-uniform. In other words the isotherms would not be symmetrical about the center.

Under certain conditions yet another simplification may be possible, if we postulate that the temperature field within the spherical particle would be uniform. This "lumped parameter" approach would yield the following, much simpler representation of the heat flow problem:

$$4R^3\pi\,\rho\,C_p\frac{dT}{dt} = 4\pi R^2 \varepsilon\sigma F(T^4 - T_e^4) \tag{11}$$

$$T = T_i \text{ at } T = 0 \tag{12}$$

In essence the two latter sets of equations may be regarded as semi-empirical representations of the problem. The key physical features of the system are incorporated into the model (i.e. the radiative heat exchange) but certain simplifications have been introduced (i.e. by neglecting the spatial non-uniformity of the flux.)

A more full discussion of this specific problem is available in the literature. (4) We should remark here that most of the mathematical models used in metallurgical practice tend to fall in the semi-empirical category.

The degree of simplification that is desirable must necessarily depend on the nature of the problem, on the time scale available for generating a solution and on the computational skills of the investigator.

(c) The modelling of the system, depicted in Fig. 1 through the use of the "black box" approach, would be to postulate that:

$$T = f\,(T_e,\ T_i,\ \{k,\ C_p,\ \rho, \varepsilon\},\ t\,) \tag{13}$$

and then to find some empirical way to express the relationship; perhaps initially postulating a linear dependence between temperature and time (which would be certainly wrong) and then adjusting this relationship, e.g. by the use of series expansion. The use of satistical data, time series techniques, etc. could be useful in the development of models of this type.*

In some instances the models discussed under (i) and (ii) i.e. the fundamentally based and the semi-empirical models are termed "Mechanistic Models," to reflect the fact that a certain transport or reaction mechanism has been postulated in the development of the model.

The development of these mechanistic models usually relies on the establishment of differential or integral balances over control volumes such as sketched in Fig. 2. Here (a) corresponds to an infinitesimal volume element, (b) represents a differential slice, while (c) involves a macroscopic balance.

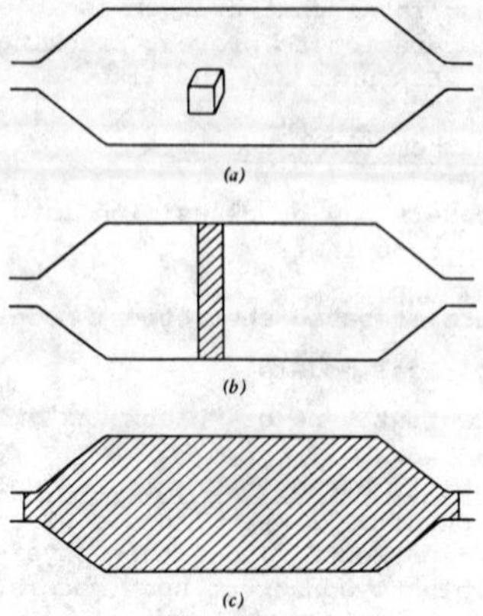

Fig. 2 The establishment of a balance; (a) over an infinitesimal element, (b) over a slice element, (c) over a macroscopic control volume.

*see e.g. W.H. Ray and J. Szekely, Process Optimization, Chapter 1, John Wiley, 1973. G.E.P. Box and R.E. Draper, Evolutionary Operation, John Wiley, 1969.

In a mathematical sense balances of the type shown in (a) will result in complex differential equations, (b) will give simpler differential equations, while the macroscopic balance should result in algebraic equations or simple differential relationships.

The selection of the approach taken will have to depend on the nature of problem, on the resources that may be available, and last but not least on the skills of the investigator.

In the following we shall indicate some groups of mathematical modelling problems which have received attention from the metallurgical community.

Fluid Flow Problems

The usual starting point in a modelling approach to fluid flow problems tends to involve:

(i) A mass balance, that is either an overall mass balance, or a differential mass balance.

The overall mass balance may be put in the following form:

$$\oint_A \rho \underset{\sim}{U} dA = \frac{d}{dt}\iiint \rho dV \equiv \frac{dM_T}{dt} \qquad (14)$$

which expresses the fact that if we consider a control volume V, then the net rate of outflow of material over the area covering this control volume must equal the rate at which the mass changes within the control volume.

The differential mass balance may be written as:

$$\underset{\sim}{\nabla} \cdot \rho \underset{\sim}{U} = \frac{\partial \rho}{\partial t} \qquad (15)$$

which is usually termed the equation of continuity, here K is the velocity vector.

(ii) The second component in the statement of fluid flow problems is the overall or the differential momentum balance.

As a practical matter, regarding overall balances, it is more convenient to work in terms of the energy, per unit mass of the fluid. Thus the overall mechanical energy balance will take the followign form:

$$\frac{d}{dt}(K_{ET}+Z_t) = w[(\frac{1}{2}U_m^2 + Z_T)_{A_1} - (\frac{1}{2}U_m^2 + Z_T)_{A_2}]$$

rate of accumulation of kinetic energy + potential energy — net input of kinetic + potential energy

$$+ w[\Pi|A_1 - \Pi|A_2]$$

net input of pressure energy

$$-W \qquad\qquad -E_f$$

rate of work done on the surroundings — rate of work done against fractional forces

Here $K_{E,T}$ is the kinetic energy

Z_T is the potential energy

This expression is often termed the "Engineering Bernoulli Equation".

The differential form of the momentum balance equation, or the Navier Stokes equation may be written as follows:

$$\rho \frac{DU}{Dt} = -\underset{\sim}{\nabla} P + \nabla\mu_e \underset{\sim}{\nabla U} + F_b \qquad (17)$$

accumulation + convective transport of momentum — pressure gradient — viscous transfer of momentum — body force

Here μ_e is the effective viscosity, which incorporates both the molecular and the turbulent terms.

Considerable success has been achieved in the use of Eqs. (14-18) for the modelling of fluid flow phenomena in metals processing operations. (5,6)

The overall balance equations have been useful to calculate:

the time required to empty ladles

the power requirements for pumping

the behavior of supersonic nozzles

flow in ducts and in open channels and the like

The (turbulent) Navier-Stokes equations have been used quite extensively in recent years for the modelling of argon stirred ladles, (7,8,9) fluid flow phenomena in the filling of ladles. (12)

Yet other interesting applications involve the modelling of electromagentically driven flows in

aluminum refining (13)

inductively stirred ladles (14,15,16)

fluid flow in electroslag refining and electro-slag welding (17,18,19)

As an illustration Fig. 3 shows a comparison between the experimentally measured and the theoretically predicted velocity fields in a water model of an argon stirred ladle, while Fig. 4 shows a comparison between the theoretically predicted and the experimentally measured velocity fields in an inductively stirred low alloy melt.

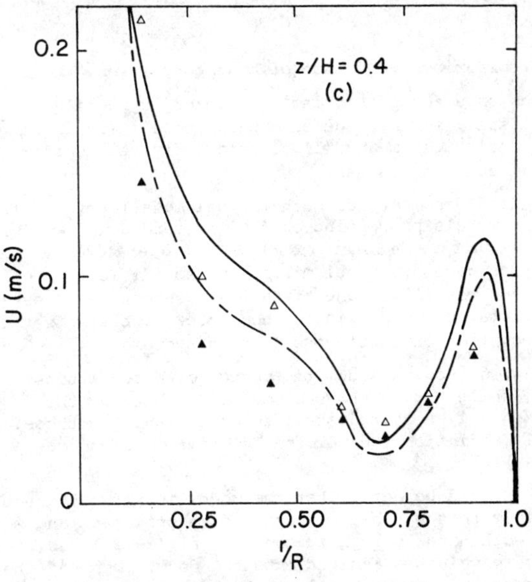

Fig. 3 A comparison of experimentally measured velocities (discrete data points) with the theoretical predictions (curves).

PROBLEMS IN HEAT TRANSFER

The usual starting point in the modelling of heat transfer phenomena is the statement of the differential thermal energy balance equation, which takes the following form:

$$\frac{DT}{Dt} = \nabla\alpha\nabla T + q'''/\rho C_P \qquad (18)$$

where $\alpha = k/\rho C_P$ is the thermal diffusivity

k is the thermal conductivity

C_P is the specific heat

is the rate of heat generation

In the absence of convective motion Eq. (18) will simplify to:

$$\frac{\partial T}{\partial t} = \nabla\alpha\nabla T \qquad (19)$$

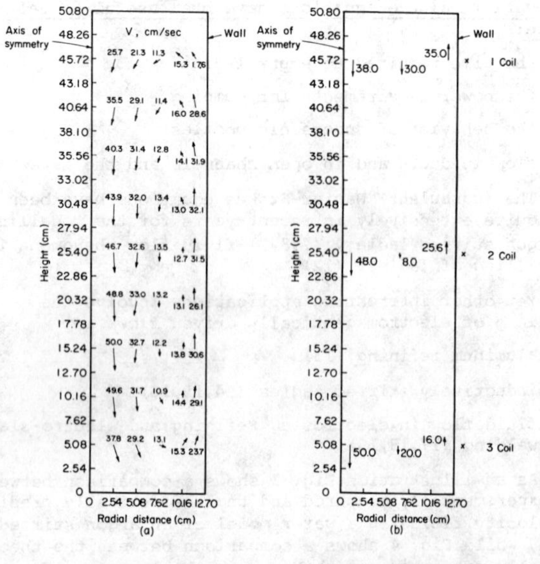

Fig. 4 Comparison of the experimentally measured velocity fields in an inductively stirred low alloy system.

which is Fourier's equation of heat conduction.

The solution of Fourier's equation is readily accomplished, for a range of boundary conditions, using either analytical means (1) or relatively straightforward computation schemes.

Indeed, the earlier mathematical modelling efforts aimed at metals processing operations tended to consider mainly heat flow problems, such as the operation of reheating furnaces in steel mills, the solidification of ingots and the like. One of the real success stories in the mathematical modelling of heat flow problems is the representation of continuous casting systems.

Fig. 5 shows a schematic sketch of a continuous caster, while Fig. 6 shows a comparison between the experimentally measured shell thickness and the theoretical predictions, developed by Brimacombe and coworkers. (21)

It should be noted that the modelling of continuous casting systems has become highly sophisticated, and more recent work notably by Brimacombe et al (22,23) has provided very valuable insight into the development of thermal stresses in continuously cast slabs and billets.

MASS TRANSFER AND REACTING SYSTEMS

The general representation of convective mass transfer in a turbulent fluid flow field takes the following form:

$$\frac{DC_A}{Dt} = \nabla D_e \nabla C_A + r_A''' \qquad (20)$$

where C_A is the concentration of the transferred species

D_e is the effective diffusivity and

r''' is the rate at which the reacting species is being consumed by a chemical reaction

D_e is the effective diffusivity, which is usually dominated by the turbulent (eddy) component.

A great deal of work has been done using Eq. (20) of some modification thereof to represent the dispersion of tracers, and hence to characterize the state of mixing within the system.

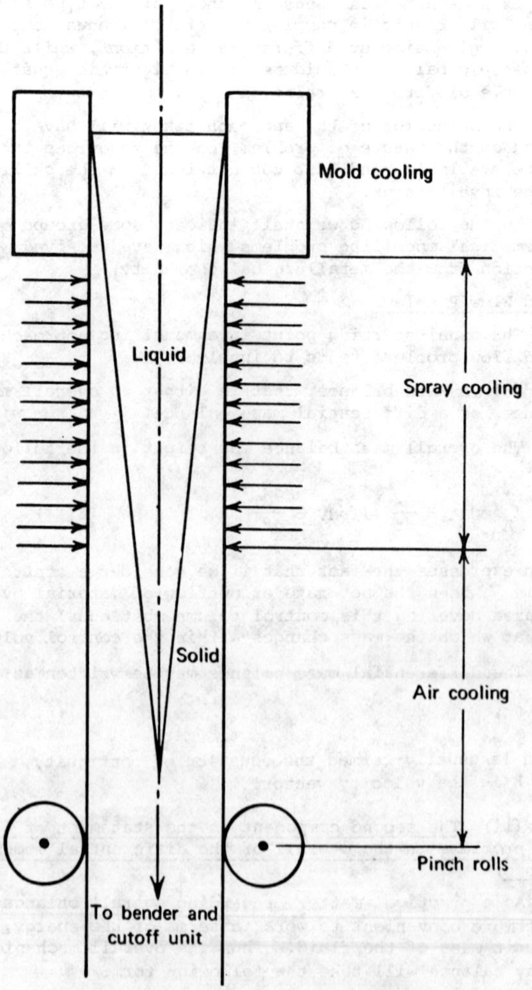

Fig. 5 Schematic diagram of a continuous casting system.

As an example Fig. 7 shows a comparison between the experimentally determined dispersion of a radioactive tracer and the theoretical predictions for the operation of an ASEA - SKF furnace. Yet other metallurgical examples of the use of Eq. (20) include deoxidation kinetics (24) and desulfurization kinetics. (25)

A rather special case of diffusion and mass transfer is posed by the reaction between gases and solids, exemplified by iron oxide reduction, the gasification of coke, the roasting of sulfides and the like.

In these systems the key diffusion process takes place not in the bulk of the fluid phase, but rather in the pores of the solid.

It is of interest to note that under these conditions the effective diffusivity D_e will be smaller than the molecular value. Gas solid reactions will be discussed in detail by another contribution to this volume.(26)

To summarize this brief overview of mathematical models, a mathematical model represents a physical process in terms of a set of algebraic or differential equations. The main function of a mathematical model is to provide a working relationship between key process parameters, such as the time required for an ingot to solidify, the time required empty a ladle, the blast pressure required to maintain a given hot blast flow rate, and the like.

The principal usefulness of mathematical models lies in the following capabilities:

For existing processes the model

(i) can improve our general understanding of the

behavior of a process and the interrelationships between key process variables.

(ii) Can perform paper or computer studies exploring the effect of changing process variables and thus ultimately provide a means for optimizing the process.

(iii) Can aid in the evaluation and the planning of plant trials.

For processes under development

(i) The model can help in the assessment of the feasibility of the process.

(ii) The model may be helpful in the planning of laboratory scale experiments and

(iii) The model may be very helpful in the evaluation of pilot scale trials and scale up.

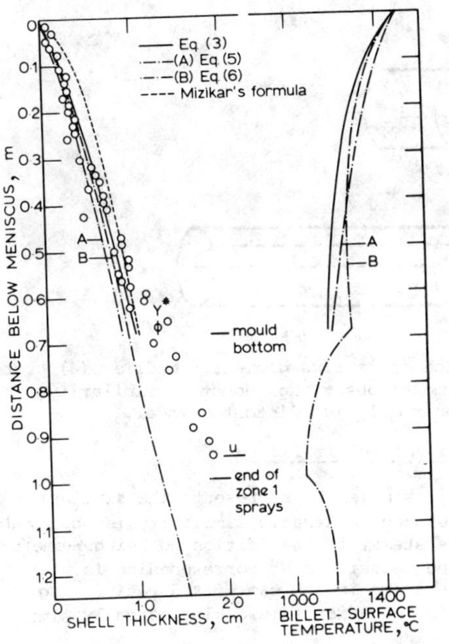

Fig. 6 A comparison of the theoretically predicted and the experimentally measured liquid pool profiles in a continuous casting system, after Brimacombe et al.

THE PHYSICAL MODELLING OF METALS PROCESSING OPERATIONS

In the previous section the reader was introduced to the concepts of mathematical modelling and to the techniques that are available for the development of mathematical models.

It must have become apparent from the previously presented discussion that in order to construct a realistic mathematical model, which is based on the fundamentals of transport phenomena (and kinetics) a necessary pre-condition is that we must have a good physical picture of the system. (e.g. we may be concerned with the flow of a gas through a packed bed, a fluidized bed operation, interaction between gas bubbles and melts or natural convection driven flows in melts, etc.) Then the mathematical model may be constructed by combining existing known relationships in order to represent the system.

In many cases these relationships exist, however, in many other instances one does not have an adequate physical basis for the construction of a rigorous or even approximate mathematical model; furthermore, the very nature of the real industrial system may preclude direct observations from which such a physical picture may be deduced. (e.g. it is difficult to observe the hearth region of the iron blast furnace, or the interior of an argon stirred ladle or the metal flow inside an aluminum cell, etc.)

In yet other cases, while one could represent the system rigorously, in terms of the appropriate differential equations, the system of equations that would ultimately result would be too complex or too expensive to solve.

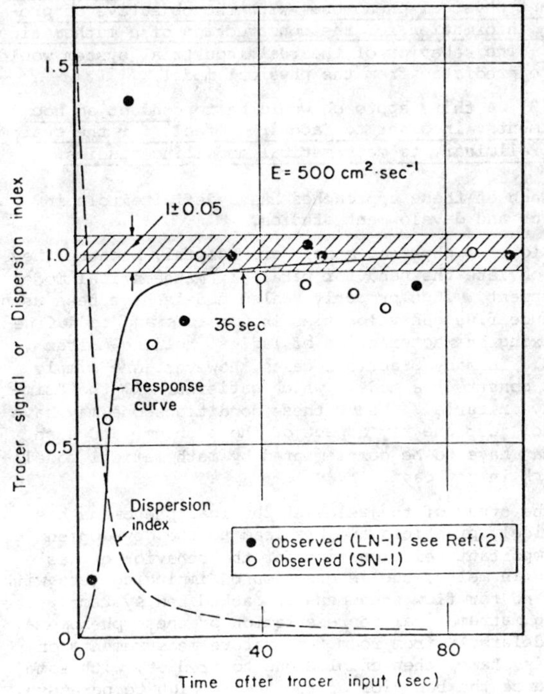

Fig. 7 A comparison of the experimentally measured (discrete data points) and the theoretically predicted tracer dispersion in an ASEA-SKF furnace, after Szekely and Nakanishi.

Under these conditions we have another very useful tool at our disposal, namely - the construction of a physical model.

DEFINITION AND CLASSIFICATION OF PHYSICAL MODELS

In constructing physical model of a process or of some particular aspect of a process, we seek to represent or model the system by changing the materials that are handled and also, frequently the scale of the operations.

The main objective of modelling is to achieve a realistic representation of the system using materials and equipment such that measurements may be made more conveniently and in a cost effective manner.

The key phrases in the above statement are realistic representation and cost effective manner. The actual criteria for realistic representation, more properly stated, the similarity criteria will be discussed and quantified subsequently.

We just note here that typical examples of physical modelling include the modelling of argon stirred ladles holding molten steel, using the air - water system, the use of water in plexiglass containers to represent the behavior of ladle - turndish systems in continuous casting, the modelling of the flow of hot combustion gases, using water, and the like.

In these cases mentioned molten steel or the hot combustion gases were modelled, using water, at room temperature, so that the velocity measurements or the flow visualization studies could be readily made. Furthermore, the reduction in linear scale, allowed these measurements to be made in a more cost effective way.

Physical models may be classified into the following

three main groups:

(1) Rigorous physical models are constructed by following certain strict rules or similarity criteria. Under these conditions the quantitative measurements taken in course of the model experiments (e.g. velocity fields, temperature profiles, reaction rates, etc.) may be translated directly, through appropriate scaling to describe the behavior of the real systems.

(2) An alternative approach would be to study the relevant physical phenomena, with the objective of providing information for the construction of a mathematical model. The behavior of the real industrial system would then be predicted from the physical model.

(3) A third approach would be to conduct ad hoc experiments, in order to "acquire a feel" for the system, as a preliminary to mathematical modelling studies.

Each of these approaches has a definite role in research and development studies.

Rigorous physical models are helpful, because they often obviate the need for parallel, mathematical model development. Thus properly scaled models have been used to deduce flue gas velocities in furnaces and to define the mixing characteristics of ladles, holding molten metals. In many practical cases, however, one simply cannot construct a model, which satisfies all the "similarity criteria." Under these conditions one may have to model just one key aspect of the problem, and such work may have to be complemented by mathematical modelling work in any case.

The study of the relevant physical phenomena has been widely practiced and has lead to the development of very important relationships for the behavior of gas bubbles in melts, the interaction of impinging jets with melts and for flow phenomena in packed bed systems. The ensuing mathematical representation of these phenomena, deduced largely from room temperature measurements on "model systems" then enabled one to predict, with some confidence the behavior of the "real" high temperature systems.

Finally, ad hoc experiments, aimed at flow visualization, the preliminary assessment of kinetics, or the assessment, whether a system would fluidize satisfactorily can be very useful preliminaries to a more serious research program.

THE DEVELOPMENT OF RIGOROUS PHYSICAL MODELS

The key requirement for the model to represent the real system, or the prototype is to observe the quantitative criteria for similarity.

There exist many states of similarity; however, here we shall confine our attention to the following:

geometric similarity

kinematic similarity

dynamic similarity

thermal similarity

chemical similarity.

(i) Geometric Similarity

Geometric similarity is the similarity of shape. Systems are geometrically similar when the ratio of any length in one system to the corresponding length in the other system is everywhere the same. This ratio is usually termed the scale factor. While geometric similarity is one of the most obvious requirements in modelling, often it may not be possible to attain perfect geometric similarity. Under such conditions certain critical dimensions of the model are scaled to the prototype while other dimensions are not. This procedure is called the use of a distorted model. Examples where distorted models have to be used include studies where the surface roughness may not be appropriately represented or when the proper scaling of rivers, lakes or processing equipment would result in too shallow a liquid level.

A special case of a distorted model is the use of a section model e.g., in representing a kiln or a converter the diameter and liquid (or solid) depth in the model may be appropriately scaled, but the model would have only a fraction of the (scaled) length of the prototype.

Fig. 8 shows various types of geometric models representing the behavior of a liquid in a horizontal, cylindrical reactor.

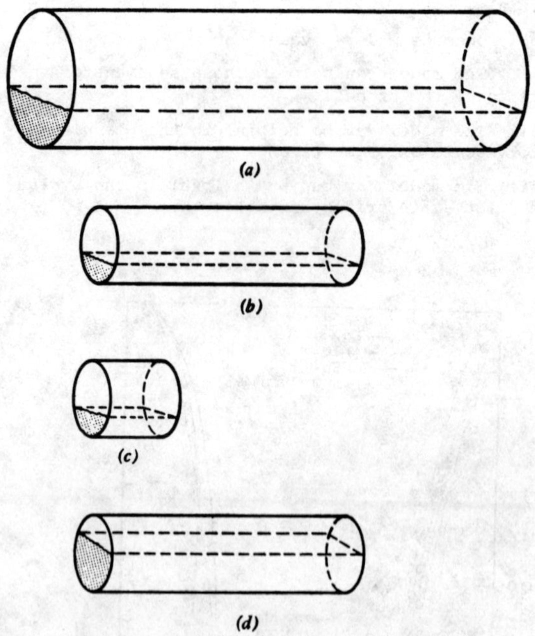

Fig. 8 Sketch of various geometric models: (a) prototype, (b) model observing geometric similarity, (c) section model, (d) distorted model.

(ii) Kinematic Similarity

Kinematic similarity represents the similarity of motion. In essence kinematic similarity is observed between two systems, if in addition to being geometrically similar the velocities at corresponding locations in the two systems are in the same fixed ratio. Fig. 9 illustrates the behavior of two kinematically similar systems.

(iii) Dynamic Similarity

Dynamic similarity represents the similarity of forces. Dynamic similarity is observed between two systems when the magnitude of forces at corresponding locations in each system are in a fixed ratio.

In the modelling of fluid flow problems it is essential that both geometric and dynamic similarity be observed if quantitative deductions are to be made of the results.

The principal forces to be considered in obtaining dynamic similarity are listed in the following:

inertia forces	gravity
pressure forces	surface tension
viscous forces	elastic forces
	electromagnetic forces

(iv) Thermal similarity

In non-isothermal systems, we must also satisfy the criteria for thermal similarity, in other words the rate of heat transfer by the various mechanisms, such as conduction, convection and readiation must be of a fixed ratio for the model and the prototype. Under these conditions, when the systems are geometrically and kinematically similar, then the temperature profiles in the model and in the prototype would be in a fixed ratio throughout.

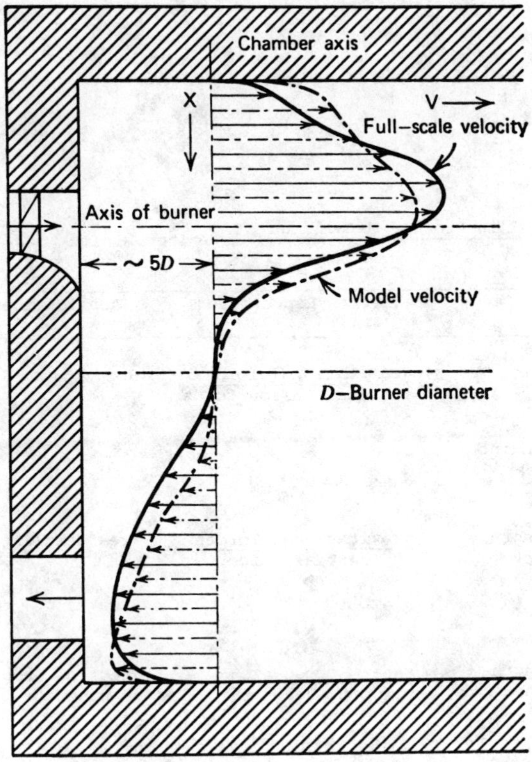

Fig. 9 Velocity profiles in a Heat Treatment Furnace and in its 1/10 scale flow model

(v) Chemical Similarity

Chemical similarity is concerned with the establishment of the necessary conditions in the model so that the rate of chemical reaction at any location will be proportional to the rate of the same reaction at the corresponding time and location in the prototype. It is therefore necessary that all conditions that contribute to the overall rate of the chemical reaction be simulated in the model.

In practice the chemical simulation of a processs presents a number of problems because it requires the proportionality of time, temperature, and concentration between model and prototype. Usually, the rate of the chemical reaction is very sensitive to temperature, and this sensitivity limits the range of operating temperatures in the model to that of the prototype.

In the following we shall discuss the means available for ensuring that the similarity criteria are indeed satisfied, thus the prototype and the model correspond to each other. It should, however, be apparent that in many cases it is simply impossible to satisfy all the similarity criteria simultaneously. Under these conditions certain compromises will have to be made, in the modelling just one key aspect of the process and then using some other means, such as mathematical modelling or complementary physical modelling techniques, to obtain a clear definition of the solution to the problem.

As an example we may wish to study macrosegration in the continuous casting of steel. Such macrosegration may depend on the fluid flow field in the model region, on the fluid flow field in the mushy zone, on the structure of the mushy zone, on the solidification characteristics of the system and on the chemical nature of the impurities. It would be clearly impossible to model all these aspects of the problem at once. However, one may make a useful start by physically modelling the fluid flow field in the in mold region, and using this information to construct a thermal mathematical model of the system, which may then be extended to regions below the mold level.

Let us now proceed by discussing the techniques available for defining the scaling factors and for the precise definition of the experimental conditions within the model to ensure its similarity to the prototype.

TECHNIQUES AVAILABLE FOR ENSURING SIMILARITY

(1) Geometric Similarity

Geometric similarity of the model and the prototype is the first necessary requirement, which has to be satisfied. This is readily done, for many systems, by simply keeping the ratio of any length in a system to a corresponding length in the other system, the same. In other words we use the same scale factor for each linear dimension.

It follows that if the prototype is a cylindrical vessel, of height H and of diameter D, and is filled with melt to a height H_o, then the corresponding dimensions for the prototype have to be:

αH, αD, αH_o where α is the scale factor.

It should be noted that in most cases the quantity α cannot be selected arbitrarily, but this magnitude will be defined as a result of the other constraints imposed on the system.

In some cases one cannot scale every dimension exactly; for instance, we may use a porous plug to introduce a gas into a molten metal. It may be not possible to scale the size of the pores in the porous plug in the prototype and in the model.

(2) Dynamic Similarity

Since most of the modelling work, involving metallurgical systems concerns fluids, the need to observe dynamic similarity is perhaps one of the most important criterian that has to be met.

Dynamic similarity represents the similarity of forces. The principal forces to be considered in obtaining dynamic similarity are listed in the following:

inertia forces	surface tension
pressure forces	elastic forces
viscous forces	electromagnetic forces
gravity	

Upon forming the ratios of some of these forces there result certain key dimensionless groups or dimensionless numbers. By ensuring that the values of these are the same in the model and in the prototype, one can ascertain that dynamic similarity is observed.

Let us now proceed by defining these forces in terms of the principal variables.

The inertial force acting on a fluid element is the mass of the element, multiplied by its acceleration. Thus on designating the characteristic length as L, the characteristic velocity as U, and the density as ρ, we may write:

$$\text{inertial force } \alpha(\rho L^3) \times U^2/L = \alpha(\rho L^2 U^2)$$

$$(\text{mass}) \times (\text{acceleration})$$

The viscous force acting on the fluid may be characterized as (shear stress) x (area) i.e., $\mu(du_x/dy)L^2$. On noting that $u_x\ \alpha U$, $y\ \alpha\ L$, we have viscous force $\alpha(\mu UL)$.

By similar arguments we have that

Gravity force $\alpha\ \rho L^3 g$

Surface tension force $\alpha\ \sigma\ L$

Electromagnetic (Ponderomotive) force $\alpha\ \tilde{\sigma}^*\ U\ B^2$

From the combination of the ratios of these (and other) forces there results the following key dimension-

*here $\tilde{\sigma}$ is the electric conductivity
B is the characteristic magnetic flu density

From the combination of the ratios of these (and other) forces there result the following key dimensionless groups, listed in Table I.

It is noted that the dimensionless groups listed in Table I could also be obtained from the differential equations describing these various processes. Upon making these differential equations dimensionless, through the introduction of the characteristic variable, these dimensionless groups would be readily formed.

Table I Dimensionless Groups Useful in Process Metallurgy

Group	Formula	Nomenclature		Quantities Represented	Application
Bond number, N_{Bo}	$g(\rho-\rho_f)d_p$	d_p	drop diameter	$\frac{\text{gravitational force}}{\text{surface tension force}}$	Spraying
		ρ	drop density		
		ρ	density of surrounding fluid		
		σ	surface tension		
Drag coefficient, C_d	$\frac{g(\rho-\rho_f)L}{\rho\mu^2}$	L	characteristic length dimension of object	$\frac{\text{gravitational force}}{\text{inertial force}}$	Free settling
		ρ	density of object		
		μ	relative velocity		
Friction factor, f	$\frac{nL(\Delta Pf\|\rho}{2\mu\ell}$	L	characteristic length dimension of conduit cross section	$\frac{\text{shear stress}}{\text{velocity head}}$	Friction drop in conduits
		$\Delta Pf\|\rho$	friction head		
		ℓ	length of pipe		
Froude number, N_{Fr}	u^2/gL	L	characteristic length dimension of system	$\frac{\text{inertial force}}{\text{gravitational force}}$	Wave and surface behavior pouring streams
Froude number, modified, N'_{Fr}	$\frac{\rho_g\mu^2}{(\rho l-\rho g)gL}$	ρ_g	gas density	$\frac{\text{inertial force}}{\text{gravitational force}}$	Fluid behavior of gas-liquid systems
		ρl	liquid density		
Galileo number, N_{Ga}	$\frac{gP^2L^3}{\mu^2}$	μ	viscosity	$\frac{\text{(inertial)x(gravitational}}{\text{(viscous force)}^2}$	Flow in baths of viscous liquids
Mach number, N_{Ma}	u/u_s	u	fluid velocity		
		u_s	velocity of sound in fluid		High-speed flow
Morton number, N_{Mo}	$\frac{g\mu_L^4}{\rho L(\sigma)^3}$	μ_L	liquid viscosity	$\frac{\text{(gravitational) x(viscous) forces}}{\text{surface tension force}}$	Velocity of bubbles in liquids
Power number, N_p	$\frac{P'}{\rho n^3L^5}$	P'	power input to agitator	$\frac{\text{drag force on paddle}}{\text{inertial force}}$	Power consumption in agitated vessels
		L	characteristic dimension of agitator paddle		
		n	angular speed of rotation		
Reynolds number, N_{Re}	$Lu\rho/\mu$	L	characteristic length dimension of system	$\frac{\text{inertial force}}{\text{viscous force}}$	Fluid flow
		ρ	fluid density		
		μ	fluid viscosity		
Weber number, N_{we}	$\rho Lu^2\|\sigma$	σ	surface tension	$\frac{\text{inertial force}}{\text{surface tension force}}$	Bubble formation, atomization of liquid jet
Grashof number, N_{Gr}	$\frac{g\rho^2L^3\beta\Delta T}{\mu^2}$	L	characteristic length dimension	$(N_{Re})(\frac{\text{buoyancy force}}{\text{viscous force}})$	Free convection
		ρ	density of fluid		
		β	coefficient of thermal volume expansion $-(1/\rho)(\partial\rho\partial T)_p$	$=(N_{Ga})\beta\Delta T$	
		ΔT	temperature difference		
Peclet Number, N_{Pe}	$\frac{Lu\rho C_p}{k}=\frac{Lu}{\alpha}$	u	fluid velocity	$\frac{\text{heat transfer by bulk motion}}{\text{conductive heat transfer}}$ $= N_{Re}N_{Pr}$	Forced convection
		C_p	specific heat		
		α	thermal diffusivity		

Group	Formula	Nomenclature	Quantities Represented	Application
Prandtl number, N_{Pr}	$\frac{C_p \mu}{k}$	C_p specific heat of fluid μ viscosity k thermal conductivity	$\frac{\text{momentum diffusivity}}{\text{thermal diffusivity}}$	Forced and free convection
Rayleigh number, R^1	$\frac{\rho^2 L^3 C_p \beta \Delta T}{k}$	β coefficient of thermal volume expansion ΔT temperature difference across film L Characteristic length dimension	$\frac{\text{heat transfer by convection}}{\text{heat transfer by conductivity}}$	Natural convection
Hartmann Number, N_{Ha}	$LB(\sigma_e/\mu)^{1/2}$	L characteristic length B characteristic magnetic flux density σ_e electric conductivity	$\frac{\text{(electromagnetic force)}}{\text{viscous force}}$	MHD
Grashof number, N'_{Gr} (mass transfer)	$\frac{g\rho^2 L^3 \beta' \Delta X}{\mu^2}$	β' coefficient of density change with concentration $-(1/\rho)(\partial\rho/\partial X)_T$ ΔX concentration of driving force	$(N_{Re})\left(\frac{\text{buoyancy force}}{\text{viscous force}}\right)$ $=N_{Ga}\beta'\Delta T$	Natural convection
Lewis number, N_{Le}	$\frac{\rho C_p D_{A-B}}{k}$	ρ density of fluid C_p specific D_{A-B} molecular diffusivity k thermal conductivity	$\frac{\text{molecular diffusivity}}{\text{thermal diffusivity}}$	combined mass and heat transfer
Schmidt number, N_{Sc}	$\frac{\mu}{\rho D_{A-B}}$	D_{A-B} molecular diffusivity	$\frac{\text{momentum diffusivity}}{\text{diffusivity}}$	Mass transfer
Sherwood number, N_{Sh}	$\frac{k_d L}{D_{A-B}}$	L characteristic dimension D_{A-B} molecular diffusivity k_d mass transfer coefficient	$\frac{\text{total mass transfer}}{\text{molecular diffusivity}}$	forced and free convection

Because of the many forces that may occur in a given metals processing operation, there are no hard and fast rules as to which of the criteria have to be satisfied. Some useful metallurgical examples are discussed in the recommended reading appended to this chapter.

However, the following general comments should be appropriate:

(1) In modelling a system consisting of just one continuous phase, e.g., a gas jet in a gas or a liquid jet in a liquid that results from forced flow, geometric similarity must be observed and the nozzle Reynolds numbers must be the same.

(2) In modelling a system consisting of one continuous phase, resulting from gravity driven flow, e.g., a teemed stream, using a submerged nozzle in continuous casting, then in addition to geometric similarity, the Froude numbers must be the same.

(3) In thermal natural convection driven flows geometric similarity should be observed, in addition to the (Grashof) (Prandtl) products, and the Froude numbers should be the same.

(4) While in principle it should be possible to model dispersed systems, such as atomization processes, the breakup of gas jets in liquids (melts) the behavior of bubble swarms or slag - metal emulsions through the simultaneous adjustment of the Froude and the Weber numbers, there are many practical difficulties, posed by the very high surface tension of liquid metals.

The application of these ideas are illustrated in the following simple examples:

Example 1:

Molten steel is teemed into a billet caster 0.15x0.15 m square through a 0.05 m diameter submerged nozzle at a mean linear velocity of 0.5 m/s.

If we wish to construct a water model for this process twice the size of the prototype, define the conditions in the model.

Solution:

Since the flow is gravity driven the only criteria to be considered are that:

(i) The model and the prototype be geometrically similar

(ii) The Froude number must be the same for the model and the prototype

(iii) Conditions must be such that the flow is turbulent in the nozzle for both the model and the prototype.

Geometric similarity is readily observed. We note here that there are many occasions when the model is larger than the prototype, especially when the prototype itself is not very large and when one wishes to make precise measurements.

Conditions (ii) may be written as:

$$\frac{u_m^2}{gL_m} = \frac{u_p^2}{gL_p}$$

thus,

$$\frac{u_m}{u_p} = \frac{L_m}{L_p}$$

Thus the linear velocity exiting the nozzle has to be $u_m = \sqrt{2} \times u_p \sim 0.71$ m/s.

In the prototype

$$N_{Re} = \frac{0.5 \times 7.1 \times 10^3 \times 5 \times 10^{-2}}{5 \times 10^{-3}} = 3.55 \times 10^4$$

i.e., turbulent

In the model

$$N_{Re} \frac{0.71 \times 10^3 \; 0.1}{10^{-3}} \simeq 7.1 \times 10^4$$

i.e. turbulent

Example 2:

In the design of a leaching plant for the purification of molybdenite concentrate, the vortex configuration in the agitated autoclaves must be estimated (Figure 10). In order to do so, the flow conditions in the industrial autoclave are simulated in a small flow model. What should be the various scaling factors in the model so as to produce a vortex similar to that in the industrial vessel?

Let us select the impeller diameter and its peripheral velocity as the characteristic length and velocity in the system. The geometric similarity is observed as:

$$\left(\frac{d_t}{d_i}\right)_m = \left(\frac{d_t}{d_i}\right)_p$$

and

$$\left(\frac{H}{d_i}\right)_m = \left(\frac{H}{d_i}\right)_p$$

where d_t = tank diameter

d_i = impeller diameter

H = depth of liquid in tank at rest

m,p = subscripts denoting the model and the prototype, respectively.

The forces acting on the liquid are the inertia forces due to the rotation of the impeller, the shear stress due to viscosity, and the gravity force which tends to restore the configuration that the liquid has at rest. The ratios of these forces can be expressed by the following dimensionless groups (Table 1).

$$\text{Reynolds number} = \frac{\text{inertial force}}{\text{viscous force}} = \left(\frac{d_i^{\,2} N \rho}{\mu}\right)_m = \left(\frac{d_i^{\,2} N \rho}{\mu}\right)_p \qquad (21)$$

$$\text{Froude number} = \frac{\text{inertial force}}{\text{gravity force}} \quad \left(\frac{d_i N^2}{g}\right)_m = \left(\frac{d_i N^2}{d}\right)_p \qquad (22)$$

where N = speed of impeller rotation, rev/sec (thus Nd corresponds to a velocity term)

ρ = liquid density

g = gravitational acceleration

We can start the design of the model by selecting a convenient ratio for length scale of model/prototype ($d_{i,m}/d_{i,p}$). We may then use equation (22) to calculate the proper impeller speed rationto satisfy the Froude number criterion:

$$\frac{N_m}{N_p} = \left(\frac{d_{i,p}}{d_{i,m}}\right)^{1/2} \qquad (23)$$

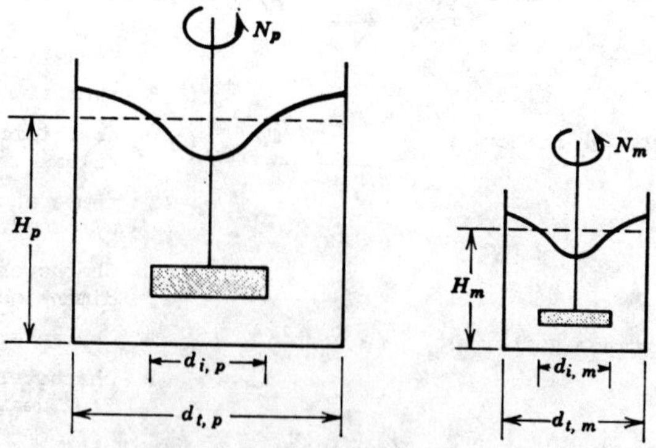

Fig. 10 Illustration of the prototype and the model in Example 2.

Finally, by substituting for N_m/N_p, from equation 23 in (1) we obtain

$$\left(\frac{\mu}{\rho}\right)_m = \left(\frac{\mu}{\rho}\right)_p \left(\frac{d_{i,m}}{d_{i,p}}\right)^{3/4} \qquad (24)$$

Equation (24) shows that the only way of obtaining similar vortices in the model and the prototype is to use a model fluid of lower kinematic viscosity (μ/ρ) than in the prototype.

The consideration of force balances is just one of the techniques that one may use in establishing the criteria for dynamic similarity. The other techniques available include dimensional analysis, and the use of differential equations, in order to identify the key dimensionless groups. (3)

SEMI-RIGOROUS MODELS, STUDY OF PHYSICAL PHENOMENA

The development of rigorous models is often not possible, because one just cannot satisfy all the similarity criteria. As an example, if we were interested in modelling the behavior of a gas bubble swarm in molten metal pool, we would need to satisfy simultaneously the Froude and the Weber numbers, which would not be possible. In a similar way, it may not be practicable to model the atomization of molten metals, using room temperature liquids, because the high surface tension could not be reproduced.

It may be also very difficult to represent chemical similarity, or to model electromagnetically driven flows using room temperature fluids.

In this latter case another complication may arise, because in the majority of electromagnetically driven flows thermal energy is also dissipated in the system, giving rise to buoyancy driven flows as well. In order to reproduce the relative importance of the Lorentz forces and the buoyancy forces, the electric conductivities of the model and the prototype fluids would have to be appropriately scaled.

The difficulties inherent in satisfying all the similarity criteria have stimulated alternative approaches, which in essence involve the identification of key phenomena within the system and then the development of an experimental - analytical study of these phenomena, with the objective of obtaining a quantitative relationship between the key process variables. Such a relationship may be extrapolated under certain circumstances, to the real system, (i.e. the prototype) alternatively the experimental data thus obtained may be used for verifying a mathematical model, which then can be used with some confidence for representing the industrial system.

Some simple examples of this approach are the following:

The pressure drop through packed beds of solids is given by the Ergun equation.

$$\frac{\Delta P}{Z} = 150 \left(\frac{1-\varepsilon_v}{\varepsilon_v}\right)^2 \frac{\mu U_o}{(\phi_s d_p)^2} + 1.75 \frac{1-\varepsilon_v}{\varepsilon_v^3} \frac{\rho U_o}{\phi d_p} \quad (25)$$

where

Δp is the pressure drop N/m^2

Z is the bed height m

ε_v is the void fraction in the bed

μ is the viscosity of the fluid kg/m-s

U_o is the fluid velocity through the empty column, m/s

ϕ is the shape factor of the particles

d_p is the particle diameter m

ρ is the fluid density kg/m^3

While this equation has been developed from measurements obtained with glass beads and air at room temperature, it has been shown to apply to high temperature systems.

Yet another example is that the rising velocity of large (d_p >2cm) spherical cap shape gas bubbles in liquids is given by the relationship:

$$U_b = 1.02\ (g d_B/2)^{1/2} \quad (26)$$

where

U_b is the rising velocity, m/s

g is the acceleration due to gravity m/s^2

d_B is the bubble diameter m

It has been shown that this relationship applies equally well to molten metals. In fact one may represent the map of gas bubbles, disregarding their shape and general behavior, on a plot of the Reynolds number and the bubble Eotvos number, (Fig. 11), where the Eotvos number is defined as:

$$N_{Eo} = g d_e (\rho_L - \rho_G/\sigma$$

A final example, of systems, where physical measurements, using room temperature liquids have been helpful in elucidating the behavior of molten metals system in the field of gas jets impinging onto liquid surfaces.

Bradshaw and Wakelin carried out extensive experiments, studying the depth of the cavity generated by a gas jet, impinging onto a liquid surface, as sketched in Fig. 12 where H_o is seen to be the distance from the jet orifice and the liquid level, while H_c is the depth of the cavity.

The experimental results obtained (using the air water and the mercury CO_2 systems) are shown in Fig. 13 where the quantity-

$P_j = \pi d_o \rho_G U_o^{2/4}$ is the jet momentum

It has been found subsequently, that this relationship holds remarkedly well for the BOF (LD) steelmaking system.

The important conclusion that one must draw from this section is that in many instances one cannot construct a complete physical model of the system to be studied. Under these conditions one must use engineering judgement to define the key components of the system (e.g. flow through packed beds, agitation of liquids, using gas bubbles, impinging gas jets, rising gas bubbles) and then proceed to characterize these through suitable experiments.

The experimental conditions to be used need not satisfy all the similarity criteria, but it is helpful if the range of operating parameters, or at least those of the key dimensionless groups does show an overlap with that expected for the prototype.

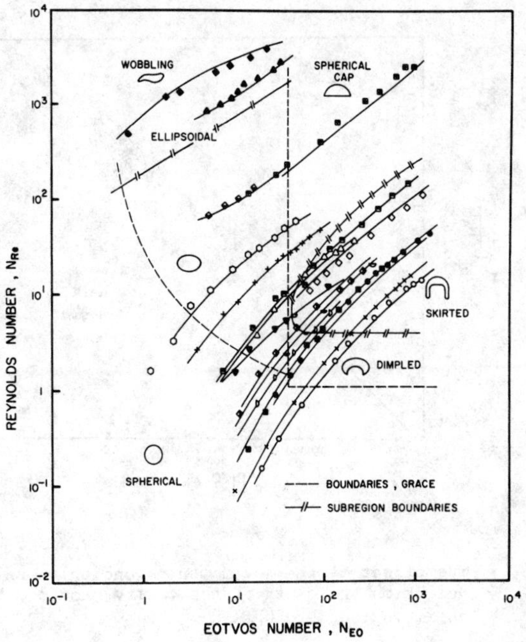

Fig. 11 Regions of bubble behavior on a plot of the Reynolds number against the Eotvos number.

Fig. 12 Schematic representation of a gas jet impinging onto a liquid surface. (27)

PRELIMINARY AD HOC MEASUREMENTS

The development of semi-rigorous models or the detailed study of key physical phenomena might require quite an extensive commitment of time and resources. In some cases it may be helpful to conduct a preliminary investigation, by performing some simple, ad hoc measurements.

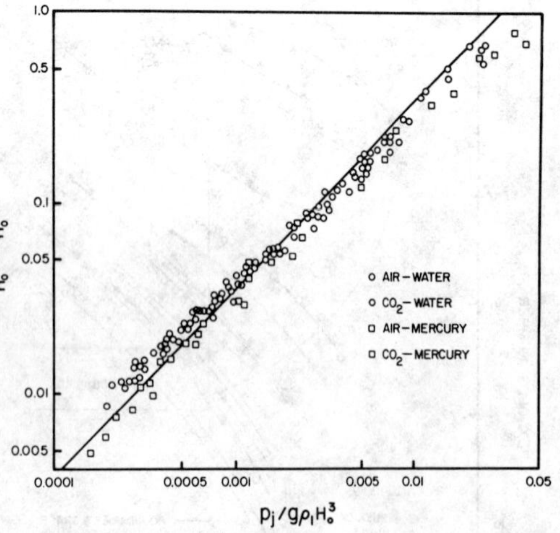

Fig. 13 The effect of the jet momentum and the nozzle height on the depth of the cavity produced by an impinging jet. (28)

Such ad hoc measurements may be particularly helpful, as a preliminary to mathematical model development, when the investigator does not have a good physical feel for the system, so that the governing equations would have to be set up quite arbitrarily.

It is often tempting to construct a simple physical model of a system just to develop a qualitative understanding, how the key process parameters may be related. However, unless care is taken in the design of the experiments the information thus obtained may be of little value.

In the design of ad hoc experiments we have to start with a clearcut definition of the objectives; e.g. we wish to obtain a qualitative feel for the factors that govern the erosion patterns in AOD systems.

In the definition of these objectives it may be helpful to go through the exercise of formulating a mathematical model, thus identifying the "missing relationships" between the important process variables.

Then one may design an apparatus (taking advantage of all possible shortcuts, e.g. the availability of existing equipment etc.) bearing in mind the similarity criteria. It should be helpful if a geometric similarity is attained, or at least approached, and also there must be an awareness as to how close one is to meeting the criteria for dynamic similarity.

There are no clear quidelines regarding the planning of preliminary, ad hoc measurements, except for the fact that these will be less rigorous than the experimental programs discussed in the preceeding two sections. Nonetheless even in this case careful planning, the clear definition of objectives, and a good level of awareness of the limitations are thought to be essential components.

The choice of the approach taken will necessarily have to depend on the nature of the system, on the nature of the problem and on the time scale available. One should note that many real life metallurgical systems are just too complex to allow their rigorous modelling.

PILOT PLANTS

Pilot plants constitute the third category in the classification of physical and mathematical modelling of metallurgical processes.

In essence a pilot plant is an exact replica of a full scale installation except that it operates on a smaller scale, usually 1/20 - 1/5 of the full scale unit. Pilot plants are primarily used in new process development and usually represent the final stage, before the construction of a demonstration scale, or commercial scale unit. The objective of pilot plants may be classified into the following categories:

(i) In new process development, laboratory scale measurements, physical and mathematical modelling studies usually do not enable one to design a full scale plant with confidence. Thus the construction of a processing unit, having all the features, of a full scale installation, except size, is a necessary stage in process development. In essence a pilot plant in this context should be used to provide the data for the design of the full scale installation or a larger pilot unit. Frequently the pilot plant data will influence the decision, whether to proceed with the project.

(ii) Pilot plants may also be used in the study of existing processes. Often it is impractical to carry out large scale plant trials on fully operating commercial processes, because these would be costly and may interrupt production schedules. Pilot plants may be useful, therefore for carrying out extensive tests, with the view of optimizing the process.

(iii) Finally, pilot plant operations may provide product material, suitable for a market survey; an example of this would be the use of FMC's pilot plant to provide form coke for blast furnace tests.

CONCLUDING REMARKS

THE RELATIONSHIP BETWEEN MATHEMATICAL MODELS, PHYSICAL MODELS AND PILOT PLANTS

The use of mathematical models, physical models and pilot plants has been an indispensable component of virtually all metallurgical process development during the past decades. Indeed, the progressively increasing use of mathematical modelling techniques and physical models has greatly aided the traditional route of proceeding from pilot plants to full scale operations.

The development and the subsequent improvements in oxygen steelmaking, the Q - BOP operation, continuous casting, ladle metallurgy, the new ALCOA smelting process for aluminum the Noranda process for continuous copper smelting have all benefitted immesurable from the use of both mathematical and physical modelling techniques. While process development and process improvements have been effected in the past without recourse to mathematical or physical modelling, the more widespread application of these techniques has made R&D efforts much more cost effective, by minimizing the hit or miss and trial and error aspects of these programs.

At this stage it may be worthwhile to comment briefly on the proper place of mathematical and physical modelling techniques and pilot plant operations within the general hierarchy of process development and process improvement.

It is perhaps a truism to say that most engineering problems have a multitude of solutions. Thus, in this regard one should consider mathematical and physical modelling techniques which, in conjunction with pilot operation and full scale plant trials represent a set of tools that one might select, depending on the specific nature of the problem. In many instances one may wish to combine mathematical modelling with physical modelling work, before a pilot plant is constructed; then use a more extensive mathematical modelling effort for the interpretation of the pilot plant results and their ultimate scale up.

In some instances the pilot plant work will point to some additional problems, which may have to be resolved through additional physical modelling work, aimed at some specific aspect of the process.

A note should be made of another group of experimental measurements that may be needed, in order to make the mathematical modelling work meaningful, which is the development of physical and chemical information on the system, such as the basic thermodynamic data, data on reaction kinetics, and information on the property values, such as the thermal conductivity and the like.

A properly constructed mathematical model could be invaluable at the very beginning of the investigation to put these matters into proper perspective and to draw attention to the need for specific information that has to be developed.

RECOMMENDED READING

(A) Modelling Techniques

Mathematical Process Models in Iron and Steelmaking, Proceedings of a Symposium held in Amsterdam in 1973, The Metals Society. - excellent review papers dealing with both mathematical and physical modelling.

Pilot Plants in the Iron and Steel Industry, The Iron and Steel Institute, London, Special Report No. 96.

The Mathematical and Physical Modelling of Metals Processing Operations, J. Szekely, J.W. Evans and K. Brimacombe, Lecture Notes, MIT Short Course, 1981.

(A) Transport Processes

J. Szekely and N.J. Themelis, Rate Phenomena in Process Metallurgy, Wiley, 1971.

J. Szekely, Fluid Flow Phenomena in Metals Processing, Academic Press, 1979.

R.B. Bird, W.E. Stewart and E.N. Lightfoot, Transport Phenomena, Wiley, 1960.

G.H. Geiger and D.R. Poirier, Transport Phenomena in Metallurgy, Addison Wesley, 1973.

R. Clift, J.R. Grace and M.E. Weber, Bubbles, Drops and Particles, Academic Press. 1978.

REFERENCES

(1) H.S. Carslaw and J.C. Jaeger, Conduction of Heat in Solids, Oxford, 1959.

(2) E.R.G. Eckert, and R.M. Drake, Heat and Mass Transfer, McGraw Hill, 1959.

(3) J. Szekely and N.J. Themelis, Rate Phenomena in Process Metallurgy, Wiley, 1971.

(4) J. Szekely and R.J. Fisher, Chem. Eng. Sci., 24, 833 (1969).

(5) J. Szekely, Fluid Flow Aspects of Metals Processing, Academic Press, 1979.

(6) J. Szekely, J.W. Evans and K. Brimacombe, Mathematical and Physical Modelling of Metals Processing Operations, Lecture Notes, Short Course, MIT, 1981.

(7) J. Szekely, J. Wang and K.M. Kiser, Met. Trans. 7B, 287 (1976).

(8) J. Szekely, Proc. Int. Conf. on Injection Metallurgy, Lulea, Sweden, 1977.

(9) J. Szekely, J.H. Grevet and N. El-Kaddah, Proc. Scaniject, Lulea, Sweden, 1980.

(10) J. Szekely and R.T. Yadoya, Proc. Int. Conf. on Math. Methods in Iron and Steelmaking, 165, Amsterdam, 1973.

(11) J. Szekely and R.T. Yadoya, Computer Solution of the Turbulent Flow Equations, Met. Trans. 4, 1379, (1973).

(12) M. Salcudean and R.I.L. Guthrie, Met. Trans. 10B 423 (1979.

(13) J. Szekely, J.W. Evans and K. Brimacombe, cited under #6 above.

(14) J.W.Evans and A. Tarapore, Met. Trans. 7B, 343 (1977).

(15) K. Nakanishi, J. Szekely, T. Fujii and Y. Mihara, Met. Trans. 6B, 111 (1975).

(16) J. Szekely, K. Nakanishi and S. Iwaoka, Proc. Electric Steelmaking Conf. Pittsburgh, pp 247, (1974).

(17) A.H. Dilawari and J. Szekely, Met. Trans . 8B, pp 123-138, (1977).

(18) A.H. Dilawari J. Szekely and T.W. Eagar, Met. Trans. 9B, p 339, (1978).

(19) M. Choudhary and J. Szekely, Ironmaking and Steelmaking, 8, (1981) - in press.

(20) Mathematical Models in Iron and Steelmaking, Proceedings of a Symposium held in London, 1969.

(21) J.K. Brimacombe, J.E. Lait and F. Weinberg, Mathematical Process Models in Iron and Steelmaking, The Metals Society, London, 1975.

(22) I.V. Samarasekara and J.K. Brimacombe, Canadian Metallurgical Quarterly, 18, 251, (1979).

(23) J.E. Lait, J.K. Brimacombe and F. Weinberg, Ironmaking and Steelmaking, 1, 35, 90, (1974).

(24) K. Nakanishi and J. Szekely, Trans. ISIJ, 15, 522, (1975).

(25) N. El-Kaddah and J. Szekely, Ironmaking and Steelmaking, 1981 - in press.

(26) H.Y. Sohn, Gas-Solid Reactions, these proceedings.

(27) A.V. Bradshaw and D. Wakelin , A.W.D. Hilld ed. Heat and Mass Transfer in Process Metallurgy, The Metals Society, 1966.

(28) S.K. Sharma, J.W. Hlinka and D.W. Kern, Ironmaker and Steelmaker, 4, No. 7, 1977.

SEGREGATION AND STRUCTURE IN RAPIDLY SOLIDIFIED CAST METALS

M. C. Flemings
Materials Processing Center
Massachusetts Institute of Technology
Cambridge, Massachusetts 02139

SUMMARY

This paper discusses dendrite arm spacing and segregation in metal alloy castings and ingots over a range of solidification rates up to that of very rapidly solidified powders, foils, or layers. Al-4.5% Cu alloy is used as example. In rapid solidification of commercial alloys such as this, we are rarely aiming for nonequilibrium phenomena, but rather for a closer approach to homogeneous, equilibrium, solidification structures. Effects of increasing solidification rate which produce this greater homogeneity include reduced dendrite arm spacing, reduced dendrite tip temperature, reduced eutectic growth temperature, elimination of dendrites, and undercooling before nucleation of primary and/or secondary phases.

INTRODUCTION

Solidification processes involving dendritic solidification are carried out commercially over a range of cooling rates of almost 15 orders of magnitude - from 10^{-6}K/s for large ingots to nearly 10^{9}K/s for surface treatments. Dendrite tip velocities, or isotherm velocities, range from as little as 10^{-5}cm s^{-1} to as much as 10^{4}cm s^{-1}. Table I, from H. Jones(1) lists examples of the various solidification processes and their respective regimes of cooling rates.

Table I. Ranges of Cooling Rate in Rapid Solidification Processing and Their Characteristics

Range of Cooling Rate Limits (K/s)	Designation	Characteristic Production Processes	Characteristic Limiting Thickness*	Characteristic Dendrite Spacing** (μ)
10^{-6} to 10^{-3}	Very Low	Large Sand Castings and Ingots	>6m	5 to 0.5
10^{-3} to 10^{0}	Low	Standard Castings, Ingots, and Strands	6 to 0.2m	500 to 50
10^{0} to 10^{3}	Medium Rapid	Thin Strip and Die Casting, Normal Atomization	200 to 6 mm	50 to 5
10^{3} to 10^{6}	Rapid	Fine Powder Atomization, Melt Extrusion/Extraction	6 to 0.2 mm	5 to 0.5
10^{6} to 10^{9} and above	Ultra-rapid	Spray Deposition, Melt Spinning, Electron Beam or Laser Glazing	200 to 6μm	0.5 to 0.05

* For metal slab cooled on both sides by cold mold.

** e.g., for Al-4.5wt% Cu alloy.

At the lower ranges of cooling rates, we believe we understand the formation of dendritic structure and segregation fairly well in usual metal alloys. This understanding extends up to about 10^{2} to 10^{3}K/s^{-1} cooling rate, or up to a dendrite tip velocity of about one cm s^{-1}. It is based on two assumptions that are certainly valid in nonfaceting alloys at least up to these rates: that interface kinetics are very rapid, and that liquid composition is essentially uniform within interdendritic regions, except in the vicinity of the dendrite tips (2).

As technical and commercial interest increases in products produced in solidification processes at rates above 10^{3}Ks^{-1}, and there is much impetus to extend our understanding of solidification mechanisms to the regime of these more rapid rates.

In this paper, a summary is first given of some aspects of structure, segregation, and properties of castings and ingots solidified at ordinary cooling rates (below about 10^{2}Ks^{-1}). Al-4.5% Cu alloy is used as example. The paper then extrapolates these ideas to rapid rates and considers other phenomena that become important at these rates. Our experimental and conceptual understanding of solidification at rapid rates has improved significantly in the last few years but there is much left to be done before we understand solidification at these rates to the degree that we do at the much slower rates of usual castings and ingots.

SEGREGATION AT 10^{-6} TO 10^{0}K/s

At the relatively slow rates of solidification of castings and ingots, nucleation generally occurs with little or no undercooling. It has also been long known experimentally that dendrite tips grow into the melt with little undercooling so that:

$$T_t \simeq T_L \tag{1}$$

where T_t is temperature of the dendrite tips and T_L is liquidus temperature. With rapid infinite interface kinetics, uniform interdendritic liquid, and constant partition ratio, k, the microsegregation which results from the dendritic growth is then described by the widely used "local solute distribution" equation, often referred to as the Scheil equation (2):

$$C_s^* = kC_o (1 - f_s)^{k-1} \tag{2}$$

where C_s^* is the composition of the isoconcentrates surrounding f_s, fraction solid, both during and after solidification, and C_o is the initial fraction liquid. Figure 1 illustrates the solidification model schematically for an Al-4.5% Cu alloy, and Figure 2 shows the composition variation to be expected from microprobe traces across dendrite arms. Data used for this and

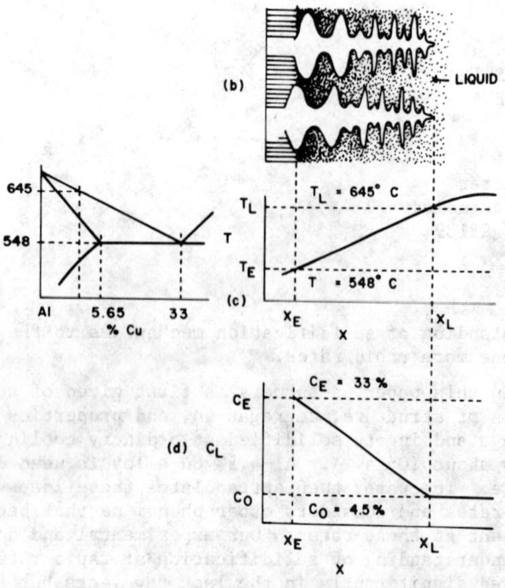

Figure 1. The basic model for dendritic growth in usual castings and ingots, illustrated for Al-4.5% Cu alloy. (a) Phase diagram, (b) schematic illustration of growth, (c) temperature distribution, (d) liquid composition.

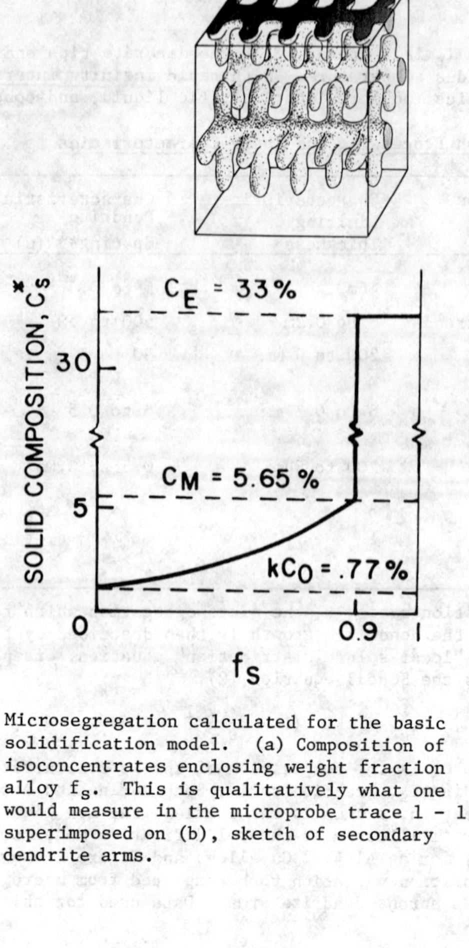

Figure 2. Microsegregation calculated for the basic solidification model. (a) Composition of isoconcentrates enclosing weight fraction alloy f_s. This is qualitatively what one would measure in the microprobe trace 1 - 1 superimposed on (b), sketch of secondary dendrite arms.

subsequent calculations are given in Appendix A. Quantitative metallographic techniques permit quantitative comparison of experiment with Equation 2.

The Scheil equation almost always predicts somewhat more microsegregation than is observed experimentally and this is due to three factors. One is diffusion in the solid; a second, perhaps more important, is "ripening" of secondary dendrite arms - i.e., the dissolution of small dendrite arms during solidification and the consequent more rapid growth of larger arms. This is shown schematically in Figure 1 by the increasing coarseness of the dendrite arms as the eutectic isotherm is approached. It is also shown schematically in the "volume elements" of Figure 3. A third factor reducing extent of microsegregation is "temperature gradient migration" of arms (the dissolution and reprecipitation of solid as a result of improved temperature gradient(4).

If dendrites are grown slowly but with a steep temperature gradient, as in a crystal growing furnace, dendrite tip temperature no longer remains near the liquidus temperature but is depressed as a result of solute diffusing down the concentration gradient from behind the dendrite tips. The dendrite tip temperature is now closely given by the simple relation (2):

$$\Delta T = \frac{DG}{R} \qquad (3)$$

where ΔT is tip undercooling, G is thermal gradient, and R is growth rate (dendrite tip velocity, isotherm velocity). A more exact and general equation is used to derive Figure 12, but Equation 3 gives the same result for sufficiently low rates and high gradients. Equation 3 is valid only up to a tip undercooling ΔT equal to the temperature range between the liquidus and solidus. At this undercooling, it becomes the constitutional supercooling criterion for plane front growth.

The solute diffusion down the thermal gradient, as shown schematically in Figure 4, modifies the Scheil equation and the new "local solute redistribution equation" now becomes (2):

$$C_o^* = kC_o \left[\frac{a}{k-1} + \left(1 - \frac{ak}{k-1}\right)(1-f_s)^{k-1}\right] \qquad (4)$$

where $a = -\frac{DG}{mRC_o}$, and m = liquidus slope. Figure 4 also shows the resulting microsegregation for several different values of a, including the critical value where the constitutional supercooling criterion is met, solidification is plane front, and no microsegregation forms.

In addition to the microsegregation, there is macrosegregation which must be contended with at low rates, and this results from one or more of three causes: (1) liquid flow caused by solidification shrinkage, (2) liquid flow caused by gravity driven convection, and (3) solid movement caused by thermal or mechanical stress. The first two of these mechanisms have been discussed in detail. The dimensionless parameter determining the extent of this macrosegregation is $\nu_x G/\dot{\varepsilon}$, where ν_x is component of flow velocity perpendicular to isotherms, G is thermal gradient, and $\dot{\varepsilon}$ is cooling rate. Solid movement has been less well studied but such movement has been amply demonstrated to be an important source of macrosegregation (2). Figure 5 is a schematic diagram illustrating the model for formulation of a mathematical description of effects of the liquid flow on solidification.

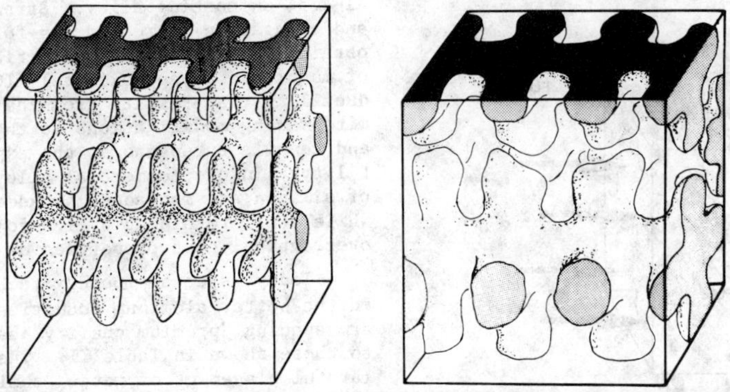

Figure 3. A "volume element" from a liquid-solid zone showing material at about 20% solid on the left, and about 50% solid on the right. Note coarsening of the structure that occurs as solidification progresses.

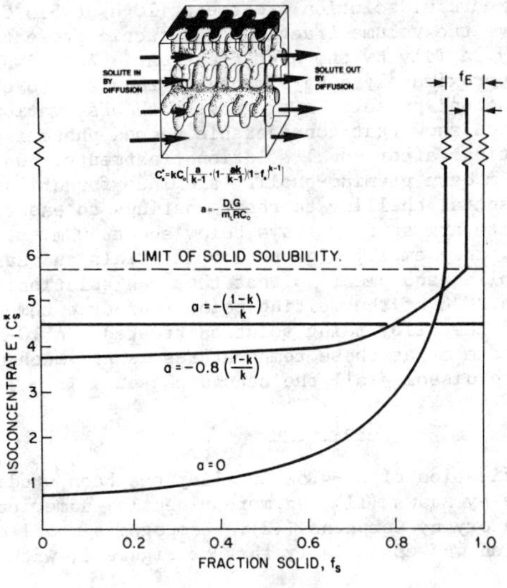

Figure 4. Microsegregation in solidification at low growth rate and high gradient. Top, volume element used for mathematical model. The graph plots microsegregation according to Equation 4.

DENDRITE ARM SPACING

It has been apparent now for over twenty years that increasing the rate of solidification reduces dendrite arm spacing. Figure 6 presents results for a series of aluminum alloys, including Al-4.5% Cu alloys. The relationship plotted here is similar to that found for dendrites of all nonfaceting metallic alloys, i.e.:

$$d_2 = a\dot{\varepsilon}^n \qquad (5)$$

where d_2 is secondary dendrite arm spacing and a and n are constants; n is usually about 1/3 as would be expected from the "ripening" mechanism that controls the secondary dendrite arm spacing as discussed earlier(2,5).

When solidification is by columnar grains, a "primary" dendrite arm spacing can also be measured and this is found experimentally in usual castings and ingots to follow a relationship similar to Equation 5 but with a slightly higher exponent - generally n approaches 1/2. Only in recent years have we developed a reasonable semi-quantitative model for formation of this spacing. The method employed to date is to meld the tip curvature required at the dendrite tip by surface tension and diffusional requirements to that required back of the tip so as to satisfy solute conservation according to Equation 2 or 4. Hunt (6) has done this using the spherical cap approximation and Wasson (7) carried out the analysis using the Trivedi parabolic tip. Both workers employed the minimum undercooling principle to choose tip curvature; more exact treatments can be expected. The equation developed by Wasson is:

$$d_1 = 4.76\left(\frac{T_M\gamma}{L_v}\right)^{.26} G^{-.50} \left[\frac{m(1-k)DC_o'}{R}\right]^{.24} \qquad (6)$$

where d_1 = primary dendrite arm spacing, L_v is volumetric heat of fusion, γ is surface energy and:

$$C_o' = C_o + \frac{k}{1-k}\,\frac{DG}{mR} \qquad (6a)$$

Approximate agreement is obtained between Equation 6 and experiment, as illustrated in Figure 7 for aluminum-copper alloys. In castings or ingots where no external heat is added during solidification, average thermal gradient in the liquid-solid zone depends on growth rate R. For high values of the metal-mold heat transfer coefficient, it is readily shown, as in a later section of this paper, that the two are linearly related. In these cases, a simple substitution of $\dot{\varepsilon} = G \cdot R$ in Equation 6 yields a relation of the form

$$d_1 \simeq a\dot{\varepsilon}^n \qquad (7)$$

where $n \simeq 3/8$. Figure 8 shows the curves representing available data on primary and secondary dendrite arm spacings of Al-4.5% Cu alloy (Figures 6 and 7). The solid lines represent the range of available data and the dashed portions represent extrapolations. Plots such as this are often used to estimate cooling rates in rapidly solidified alloys although we do not yet have a way of being sure they are as linear as they are drawn here. This plot is drawn for castings and ingots solidifying with no added heat input during solidification. It shows dendrite arm spacings versus both cooling rate and dendrite tip velocity, and further comments regarding its construction are made in a later section of this paper on heat flow.

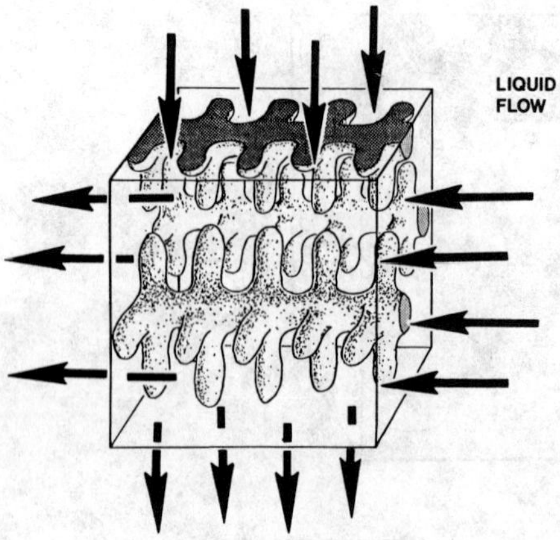

$$\bar{v} = -\frac{K}{\mu g_L}\left(\nabla P + \rho_L \bar{g}_r\right)$$

Figure 5. Fluid flow through a solidifying volume element. The component of flow perpendicular the isotherm results in local composition change.

MEDIUM-RAPID SOLIDIFICATION (10^0 - 10^3 K/s)

Up to about 10^3 K/s solidification rate, as will be seen later, tip undercooling in our model will remain small (less than one-tenth the total solidification range). The Scheil equation remains applicable and in fact appears to more accurately predict microsegregation at the more rapid rates, perhaps as a result of reduced solid diffusion as shown by data of Michaels and Bever (9). The spacing of this segregate (the dendrite arm spacing), however, is reduced dramatically as illustrated in Figure 8.

The understanding of the critical importance of cooling rate to obtaining a fine cast structure has led to development in both ingot casting and in shape casting of processes to increase solidification rate to obtain a finer, more homogeneous structure. Historically, this increase has been from the low range of solidification rates to the lower end of the medium-rapid range (i.e., to about 10^0 to 10^2). In many materials, the property advantages to be gained by this increase in solidification rate are very large indeed. Table II, for example, shows the increase obtainable in Al-4.5% Cu casting alloy. Strength is increased by 50% and elongation increased ten-fold, and this increase is obtained by increasing solidification only about 2 orders of magnitude to approximately 10 K/s. The greatly improved ductility that results from the rapid solidification permits making modifications to the alloy to raise its yield and tensile strengths at the expense of the excess ductility. In one commercial alloy, in which a small amount of silicon and manganese is added, strength is doubled, while still retaining a five-fold increase in ductility over typical cast properties of the Al-4.5% Cu alloy.

Today, special foundry techniques are widely used in the better aluminum foundries to produce fine dendrite arm spacing "premium quality" castings with properties up to those shown in Table II. These techniques involve the careful placement of metal "chills" in the mold to obtain a fine dendritic structure while also achieving soundness through directional solidification. The improved properties obtained from the finer structure result from a number of factors, including finer distribution of impurity phases and finer microporosity, but one important factor is certainly the ability to obtain greater homogeneity in the fine dendrite arm structure through heat treatment. This is illustrated in Figure 9, calculated from work of Singh et al (10). A moderate size sand casting with a dendrite arm spacing of 300μm and given a standard commercial solution treatment (10 h at 515°C), will not have the volume fraction of eutectic present reduced very rapidly by the heat treatment. According to Figure 9, approximately 4 days would be required to eliminate the second phase. This agrees with observations in practice which show that considerable second phase is left in sand castings after usual solution treatments. On the other hand, modern premium-quality aluminum foundries apply substantial chilling to their castings to ensure their dendrite arm spacing stays below about 50μm. Furthermore, they employ high-purity materials and carefully controlled equipment so that they can solutionize within 10 to 20°C of the melting point (eutectic temperature) of the alloy being solution treated. A 10-h solution treatment at these temperatures is now much more than ample to dissolve all the second phase.

HEAT FLOW

Solidification of Al-4.5% Cu alloy has been studied analytically by Adams (11) and, more recently, numerically in my laboratory by Campagna (12). Campagna's model of solidification was essentially that of Figure 1, with

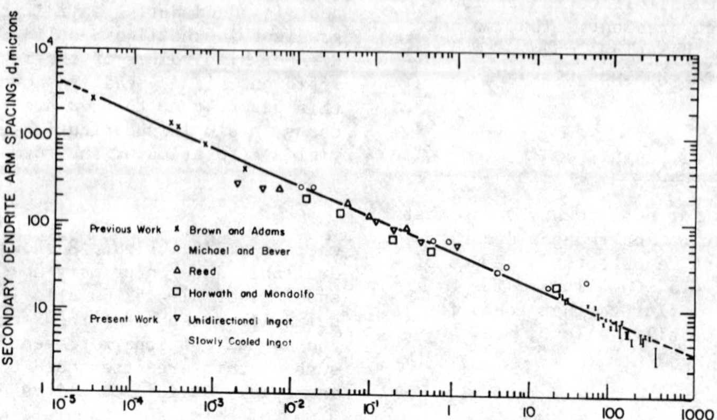

Figure 6. Experimental data on Al-4.5% Cu alloy from various investigators compiled by Bower et al (22) and on commercial aluminum alloys from Spear and Gardner (23).

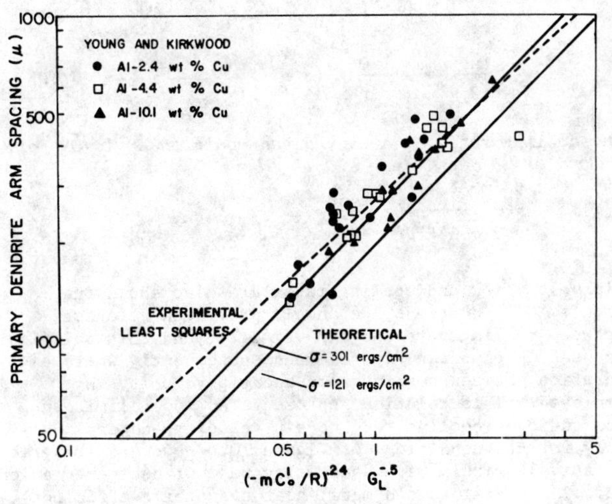

Figure 7. Correlation of primary dendrite arm spacing measurement of Young and Kirkwood (8) with Equation 5, from Wasson (7).

fraction solid in the liquid-solid zone varying according to the Scheil equation. Two pertinent results of his work will be considered here.

The first result places an absolute upper limit on the tip velocities and thermal gradients that can be obtained in directional solidification of nonsupercooled, nonheated, metal. This is shown in Figure 10, and it is for the ideal case of the alloy solidified against a metal mold of infinite thermal conductivity, and with no interface resistance between metal and mold. The tip and eutectic isotherms, X_t and X_E, respectively, move linearly with the square root of time until the liquidus isotherm reaches the end (or center line):

$$X_t = m\sqrt{t} = 1.85\sqrt{t} \quad (8a)$$

$$X_E = \hat{m}\sqrt{t} = 1.18\sqrt{t} \quad (8b)$$

Average thermal gradient behind the dendrite tip, $\bar{G}$, is readily calculated from Equations 8a and 8b to be:

$$\bar{G} = \frac{m\Delta T_s}{(m-n)X_L} = \frac{268}{X_L} \quad (9)$$

where ΔT_s is difference between the liquidus and eutectic temperature velocities of the tip and eutectic interface, R_t and R_E are, respectively:

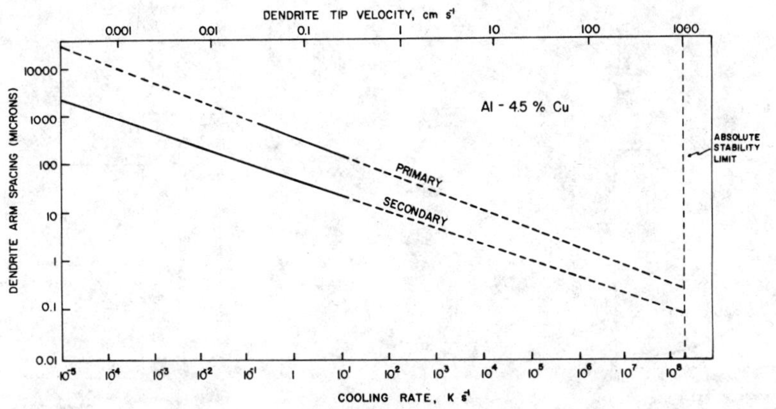

Figure 8. Primary and secondary dendrite arm spacings of Al-4.5% Cu alloy versus cooling rate, and versus dendrite tip velocity for unidirectional solidification with h = ∞ and no heat input during solidification. (Dashed portions of curve are extrapolations of experimental data.)

Table II. Mechanical Properties of Cast Al-4.5% Cu Alloy

	Yield Strength (psi)	Tensile Strength (psi)	Elongation (%)
Typical properties from commercial sand castings	22,000	26,000	2
Rapidly solidified	35,000	55,000	20
Rapidly solidified modified composition*	55,000	65,000	10
Guaranteed minimum properties in premium quality castings, modified composition*	50,000	60,000	5

* Composition modified by adding small amounts of Si and Mn as strengthener

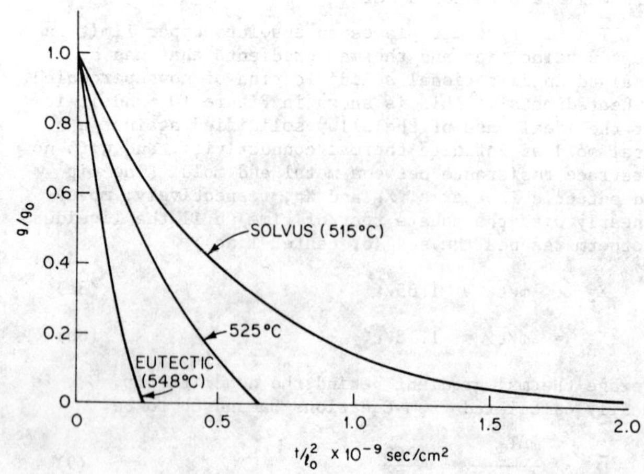

Figure 9. Rate of homogenization of Al-4.5% Cu alloy. g/g_o is residual second phase after homogenizing a cast structure of dendrite arm spacing $2\ell_o$ for time t.

$$R_t = \frac{m^2}{2X_L} = \frac{1.71}{X_L} \tag{10a}$$

$$R_E = \frac{n^2}{2X_L} = \frac{.70}{X_L} \tag{10b}$$

and cooling rate, ε, at any given position X_t is $\bar{G}R_t$ or:

$$\varepsilon = \frac{m^3 \Delta T_s}{2(m-n)X_L^2} = \frac{458}{X_L^2} \tag{11}$$

Tip velocities and cooling rates are also shown in Figure 10. Equations 10a and 11 provide a relation between cooling rate and tip velocity, and this relation is used in plotting Figure 8 and subsequently where a relation between cooling rate and tip velocity is employed. This relation represents an upper limit on the ratio of cooling rate to tip velocity and applies only for the conditions of Figure 10. However, it varies relatively slowly with decreasing rate of heat extraction (e.g., decreasing mold-metal heat transfer coefficient). Also to be noted for later use in discussing dendrite tip temperatures is that $\bar{G}/R_t$ for this type of solidification is a constant small number. From Equations 9 and 10a, it is:

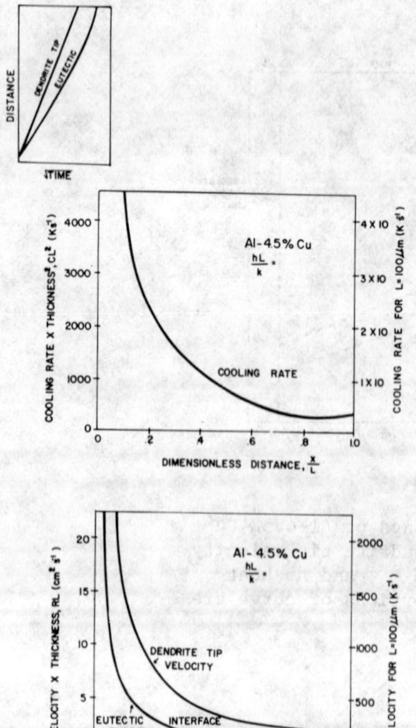

Figure 10. Calculated cooling rate, dendrite tip velocity, and eutectic interface velocity versus distance in unidirectional solidification, no superheat, h = ∞. Inset at upper left shows movement of liquidus and eutectic isotherms.

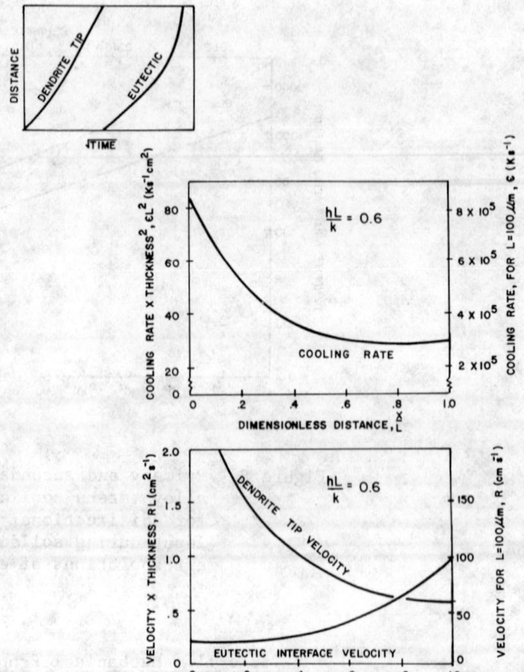

Figure 11. Calculated cooling rate, dendrite tip velocity, and eutectic interface velocity for Al-4.5% Cu alloy versus distance in unidirectional solidification, zero superheat. Inset at upper left shows movement of liquidus and eutectic isotherms.

$$\frac{\bar{G}}{R_t} = \frac{2\Delta T_s}{m(m-n)} = 157 \qquad (12)$$

Figure 11 presents results of calculations similar to those of Figure 10 except for the case of a metal-mold heat transfer coefficient h such that the Biot Number, hL/k equals 0.6 (L is casting thickness if solidification is from one side only, or half thickness if for both sides; k is thermal conductivity, taken in these calculations as average thermal conductivity of the liquid-solid zone.)

We may use Figure 11 to gain some insight into the solidification of the chilled "premium quality" castings discussed above. Consider a 10 cm casting, chilled on both sides: L = 5 cm, h = .04 cal/cm^{2}sK, k = 0.34 cal/cm sK. The Biot Number is 0.6 so Figure 11 applies. Note from the sketch at the upper left that the liquid-solid "mushy" zone extends throughout much of the casting for much of the period of solidification. At the centerline, cooling rate is approximately 1.0 K/s and at the surface, it is 3 K/s. From Figure 6, the resulting dendrite arm spacing is 50 microns at the center and 30 microns at the surface, a suitable range for good mechanical properties.

MACROSEGREGATION AT INTERMEDIATE AND RAPID SOLIDIFICATION

In the range of cooling rates of 10^0 to 10^3 (and above), there is no guarantee that macrosegregation will be absent. Gravity driven convection is no longer important at these rapid rates, but liquid flow caused by solidification shrinkage and solid movement can introduce serious segregation. At intermediate rates of solidification, these macrosegregation phenomena go under names such as "inverse segregation" and "filled hot tears" and there is ample driving force for similar phenomena to occur at very rapid rates as in splat cooling. For example, experiments in my laboratory demonstrate that at high pressures in "hammer and anvil" type splat cooling, solute-rich liquid can be exuded to the outer periphery of the sample.

RAPID SOLIDIFICATION (10^3 - 10^9 K/s)

In dendritic growth at rapid rates, we can no longer necessarily assume the dendrite tip is growing at or close to the liquidus temperature. Mathematical treatments of this problem have been carried out by Burden and Hunt (13) and by Jin and Purdy (14,15) who determined tip radius by assuming growth of dendrites at minimum undercooling, and by Trivedi[18] who defined the stable dendrite tip radius using the Mullins and Sekerka stability analysis (16) following Langer and Muller-Krumbhaar (17).

Calculations of tip temperature of Al-4.5% Cu alloy using the Trivedi analysis are given in Figure 12. Note from this that in castings and ingots where $G/R \lesssim 1000$ Ks/cm^2, and solidification rate is low, dendrite tip temperature is very near the liquidus. Tip depression reaches 1/10 the solidification range only above a growth velocity, R, of about 1 cm/s, or at a cooling rate about where the "rapid solidification" range begins. For slow growth rates where steep temperature gradient is maintained by heat input, large tip undercoolings are obtained. Curves for this case are given for G = 10^2 and 10^4 K/s; these are nearly exactly as would also be calculated by the approximate Equation 3.

Looking towards the regime of interest of higher tip velocities and cooling rates, tip temperature drops precipitously as tip velocity increases. It drops again to the equilibrium solidus at sufficiently high velocity (R = 1200 cm/s, $\dot{\varepsilon}$ = 10^8-10^9 K/s). This is exactly the "absolute stability limit" of Mullins and Sekerka. Figure 13 (top) shows how composition of the growing dendrite tip changes with solidification rate at some constant thermal gradient. At sufficiently low velocity,

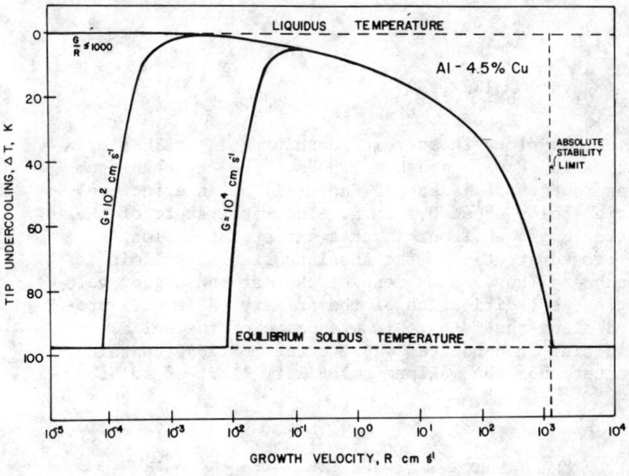

Figure 12. Calculated tip undercooling versus growth rate for Al-4.5% Cu alloy.

solidification becomes plane front and at the transition tip composition is C_o = 4.5% Cu. As velocity increases, tip composition drops to nearly kC_o = 0.77% Cu and it then increases again to C_o as velocity reaches that of the absolute stability limit. Figure 13 (bottom) from Coriell and Sekerka (19) shows these two stability limits and also indicates how the two stability limits approach each other as composition decreases.

Before turning to how this tip depression alters solute redistribution and microsegregation, we need to consider growth of the eutectic front. From the Jackson -Hunt analysis (20) and from ample experimental measurements at modest cooling rates, we expect the following relationships between lamellar spacing, λ, tip undercooling ΔT_t and growth velocity, R:

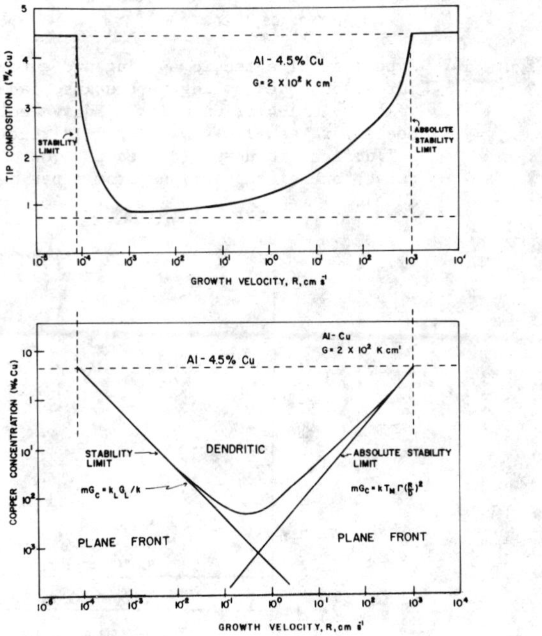

Figure 13. Top is calculated dendritic tip composition versus solidification rate. Bottom is interface stability plot, Al-Cu alloys, from Coriell and Sekerka (19). Note that at the gradient shown, dendrites (or cells) grow at velocities between about 10^{-4} and 10^3 cm s^{-1}.

$$\lambda^2 R = a$$
$$\Delta T = bR^{1/2} \qquad (13)$$

Figure 14 plots these relationships using values of $a = 1.25 \times 10^{-10} cm^3/s$ and $b = 9.5\ Ks^{1/2}/cm^{1/2}$ taken from measurements of Tassa and Hunt (21) at interface velocities below about 0.1 cm/s. The temperature of the eutectic growth front in this extrapolated plot is depressed greatly as the absolute stability limit is reached. Thus, we may expect that at the higher velocities, solidification of the primary phase will proceed along the metastable extension of the solid, resulting in dendrites whose outer portions contain greater than the maximum solubility limit of solute.

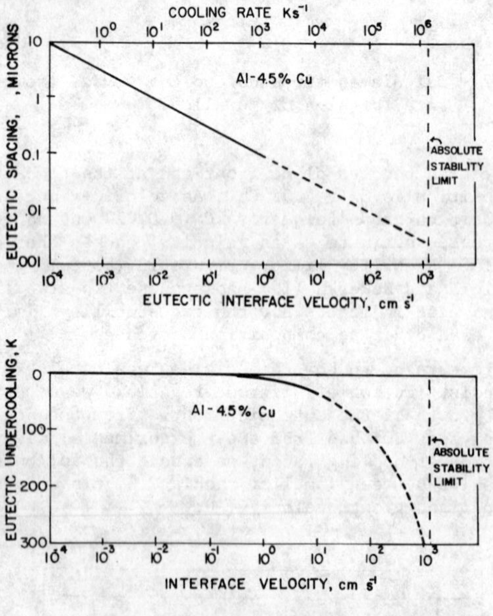

Figure 14. Experimental eutectic spacing and eutectic interface undercooling versus interface velocity, Al-4.5% Cu alloy. Equivalent cooling rates also shown for unidirectional solidification $h = \infty$ (dashed portion of curves are extrapolations from experimental data).

The effect of dendrite tip and eutectic temperature depression on local solute distribution is shown in Figure 15. For interface velocities at or below about 1 cm/s, the solute redistribution is that given by the Scheil equation with a fraction eutectic $f_e = .091$ expected. For solidification at very rapid rates ($R > 1200$ cm/s), growth is with a plane front; the entire structure is of uniform composition at $C_o = 4.5\%$ Cu.

Consider now, as example of an intermediate case, rapid solidification of a 100μm thick splat of Al-4.5% Cu. If very good contact with the substrate is maintained, h could be as high as 20 cal/cm^{2}sK giving a Biot Number of 0.6. At 40μm for the chill surface, $R_t = 100$ cm/s and cooling rate is 4×10^5K/s. For dendrites here, tip composition is increased above kC_o to 2.4% Cu (Figure 11). Assuming undercooling is dissipated just back of the tip, the equilibrium lever rule then gives the amount of solid that forms of this composition (approximately 0.8). Subsequent solidification is by the Scheil equation, following along the metastable solidus extension to 8.2% copper which is reached because of depression of eutectic formation at the eutectic front velocity, $R_E = 28$ cm/s. A very small amount of residual eutectic then forms. We will expect, from Figure 8, that this sample should have a primary dendrite arm spacing of about 1μm and a secondary spacing (if any exist) of 0.6μm. Of course, the heat flow calculations of Figure 11 can apply only approximately here since the Scheil equation was assumed in its derivation.

NUCLEATION AND UNDERCOOLING BEFORE NUCLEATION

Discussion in the preceding section implicitly assumed columnar growth; i.e., growth without nucleation ahead of advancing dendrite tips. In many instances, such nucleation does occur, and it is then possible to have very rapid isotherm movement, and very rapid cooling rates, while still having very low tip velocities. In such cases, because of random grain orientation, it is also usually impossible to define a primary as distinct from a secondary arm. If sufficient nucleation occurs near the liquidus, we may expect that solute redistribution, even at rapid cooling rates, will approach that of the Scheil equation except perhaps with some supersaturation at the outer parts of the dendrite due to lagging of the eutectic front.

In other instances of rapid solidification, nucleation occurs only much below the liquidus and the liquid may undercool significantly before any solid at all forms. Rapidly solidified droplets or "splats" can undercool significantly as indeed can bulk metals if one is sufficiently careful to avoid nucleating agents. Below the

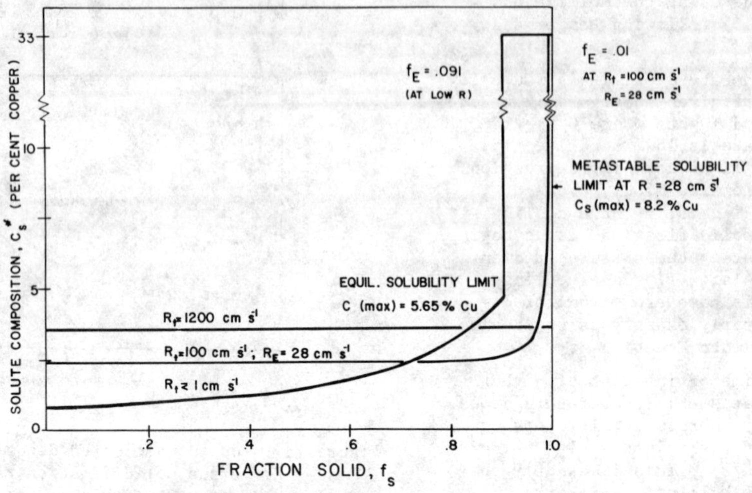

Figure 15. Solute redistribution in Al-4.5% copper dendrites solidified at three different rates, R.

solidus temperature, the solid forming will be of composition C_o, and from a simple heat balance, the fraction solid of composition C_o that will be obtained in this way is:

$$f_s^\circ = \frac{C_p(T_s - T_N)}{\Delta H} \quad (14)$$

The homogeneous nucleation temperature is sufficiently high in most metal alloys that f_s° is not likely to equal unity if recalescence is adiabatic, and so to get a fully homogeneous specimen, we must count on withdrawing heat sufficiently rapidly to prevent the metal from recalescing to its solidus temperature.

A model to calculate solute redistribution when recalescence is sufficiently rapid to be adiabatic has been developed by Kattamis and Flemings (22) and by Kattamis (23) who assumed (1) all solid that forms below the equilibrium solidus, T_s, is of composition C_o, (2) recalescence above T_s and subsequent cooling is sufficiently rapid that no remelting occurs, and (3) solidification after the maximum recalescence temperature occurs according to the basic Scheil solidification model. Then, the fraction solid, f_s°, which forms of composition C_o is given by Equation 14. As recalescence proceeds above T_s, the solid forming follows the solidus line, its composition being given by:

$$C_s^* = C_o + \frac{(f_s - f_s^\circ)H}{C_p m} \quad (15)$$

and the maximum recalescence temperature is reached when the liquid composition (calculated from a mass balance) reaches the liquidus curve and is therefore no longer undercooled. Subsequent solidification is by the Scheil equation. Figure 16 shows the calculated result for Al-4.5% Cu alloy. Note that centers of dendrites are predicted to be richer in solute than their periphery. Such solute-rich cores have been found in undercooled and splat-cooled metals (24,25).

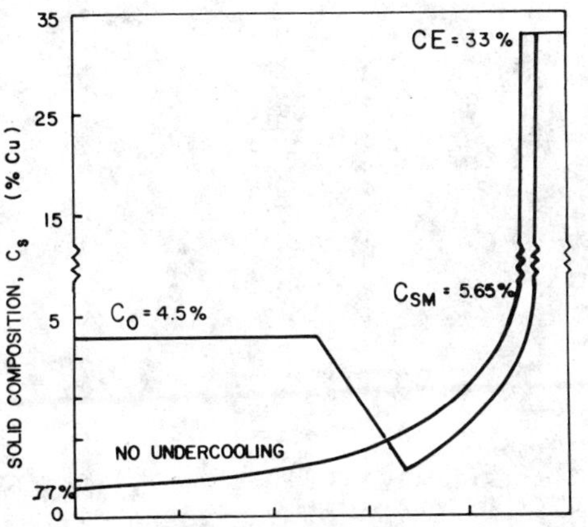

Figure 16. Calculated solute distribution in undercooled At-4.5% Cu alloy.

With the assumption of very rapid interface kinetics, undercooling before nucleation can only fully eliminate the second phase in an alloy such as this by sufficiently rapid heat extraction so that the homogeneous nucleation temperature is lowered as discussed by Hirth (26), or if rate of heat removal is sufficiently rapid after nucleation to prevent re-reaching the solidus temperature.

ACKNOWLEDGEMENTS

The author is indebted to Lawrence J. Masur for assistance and advice. This work was supported by the National Science Foundation, Grant. No. DMR-78-10905.

References

1. H. Jones: Rapid Solidification Processing, p. 28 R. Mehrabian, B. Kear, M. Cohen, Eds. Claitor's Publishing Div., Baton Rouge, La., 1978.

2. M. C. Flemings: Solidification Processing, McGraw-Hill, New York, 1974.

3. C. W. White, S. R. Wilson, B. R. Appleton, J. Narayan: Laser and Electron Beam Processing, p. 124, C. W. White and P. S. Peercy, Eds., Academic Press, New York, 1980.

4. D. J. Allen and J. D. Hunt: Solidification and Casting of Metals, p. 39, The Metals Society, Book 192, London, 1979.

5. T. Z. Kattamis, J. Coughlin, M. C. Flemings: Trans. Met. Soc. AIME, 1967, vol. 239, p. 1504.

6. J. D. Hunt: Solidification and Casting of Metals, p. 3, The Metals Society, Book 192, London, 1979.

7. R. A. Wasson: "Primary Dendrite Arm Spacing and the Cell-Dendrite Transition", Ph.D. Thesis, Department of Materials Science and Engineering, Massachusetts Institute of Technology, Cambridge, Massachusetts, June, 1978.

8. K. P. Young and D. H. Kirkwood: Met. Trans., 1975, vol. 6A, p. 197.

9. A. B. Michael and M. B. Bever: "Solidification of Al-Rich Al-Cu alloys", J. Metals, January, 1954.

10. S. N. Singh, B. P. Bardes, and M. C. Flemings: Met. Trans., 1970, vol. 1, p. 1383.

11. C. M. Adams: Liquid Metals and Solidification, p. 187, ASM, Cleveland, Ohio, 1958.

12. A. J. Campagna: Heat Flow in Solidification of Alloys, Ph.D. Thesis, Department of Materials Science and Engineering, Massachusetts Institute of Technology, 1970.

13. M. H. Burden and J. Hunt: J. Crystal Growth, 1974, vol. 23, p. 109.

14. I. Jin and G. R. Purdy: J. Crystal Growth, 1974, vol. 23, p. 29.

15. I. Jin and G. R. Purdy: J. Crystal Growth, 1974, vol. 23, p. 37.

16. R. F. Sekerka: Crystal Growth, Ed. H. S. Peiser, Pergamon, Oxford, 1967.

17. J. S. Langer and H. Muller-Krumbhaar: Acta Met., 1978, vol. 26, p. 1681.

18. R. Trivedi: Jl. Crystal Growth, 1980, vol. 49, p. 219.

19. S. R. Coriell and R. F. Sekerka in Rapid Solidification Processing, R. Mehrabian, B. H. Kear, and M. Cohen, Eds., Claitor's Publishing, Baton Route, Louisiana, 1980.

20. K. A. Jackson and J. D. Hunt: Trans. AIME, 1966, vol. 236, p. 1129

21. M. Tassa and J. D. Hunt, Jl. Crystal Growth, 1976, vol. 34, p. 38.

22. T. F. Bower, H. D. Brody, M. C. Flemings: "Measurements of Solute Redistribution in Dendritic Solidification", Trans. Met. Soc., AIME, vol. 236, 1966, p. 624.

23. R. E. Spear and G. R. Gardner: Trans. AFS, 1963, vol. 71, p. 209.

24. T. Z. Kattamis and M. C. Flemings: Met. Trans., 1966, vol. 236, p. 1523.

25. T. Z. Kattamis: Zeitschrift fur Metallkunde, 1970, vol. 61, p. 856.

26. J. P. Hirth: Met. Trans., 1978, vol. 9A, p. 401.

APPENDIX A

Values Used for Constants in Calculations (for Al-4.5% Cu)

T_m = 660.5 K

k = .17, partition ratio

m = 3.4°C/o/o liquidus slope

C_E = 33% eutectic composition

C_M = 5.65% maximum equilibrium solid solubility of Cu in α aluminum

σ = 121 erg/cm^2

m = 3.4°C/wt%

T_L = 645°C = 918°K, liquidus temperature

ρ_s = 2.7 gms/cm^3, solid density

H = 93 cal/gm, heat of fusion

D = 3 x 10^{-5}cm^2/sec, diffusion coefficient of Cu in liquid Al

C = .23 cal/gm°C, specific heat of liquid and solid aluminum

L = 10

METALLIC GLASSES: FORMATION, STRUCTURE AND PROPERTIES

L. A. Davis and R. Hasegawa
Materials Laboratory
Corporate Research and Development
Allied Corporation
Morristown, NJ 07960

INTRODUCTION

Amorphous metallic solids are materials characterized by metallic bonding, and hence high electrical conductivity and optical reflectivity, and the absence of three dimensional long range structural periodicity. Such solids can be produced by a variety of techniques, including molecular deposition, such as by physical vapor deposition, sputtering and electrolytic or electroless deposition from solution, by irradiation and by quenching from the liquid. One cannot necessarily produce an amorphous solid of the same composition by all these techniques, nor are the structures and properties of the resulting materials likely to be precisely the same when more than one fabrication technique is applicable. For example, samples produced by vapor deposition on a cold substrate will contain more structural disorder than those produced by deposition from solution or by quenching from the melt; foils or ribbons produced from the melt are ductile, whereas those produced by deposition from solution are highly brittle. The compendium of metallic compositions open to amorphous solid formation is enormous, if no limitation is imposed as to the method of synthesis. If one is limited to melt quenching processes, the list is markedly reduced. However, it is the latter alloys which are of the most practical interest and which will be considered, for the most part, below. (Some of the structural characterization studies discussed were performed on solution deposited samples.) The synthesis of an amorphous metallic solid was achieved by electroless deposition in 1947 (1). However, it was the work of Duwez and colleagues (2,3) who synthesized an amorphous $Au_{75}Si_{25}$ alloy (subscripts in atomic percentages) from the melt, that catalyzed the present large and rapidly growing interest in amorphous metals. Standard terminology holds that a noncrystalline solid produced by continuous cooling from the melt is a glass. Accordingly our comments below pertain primarily to metallic glasses. We discuss their formation, structure and mechanical and magnetic properties.

FORMATION

Kinetics

Metallic melts exhibit a considerable resistance to homogeneous nucleation of crystallization; undercoolings ΔT_r prior to crystallization of the order of 1/5 to 1/3 are observed, where $\Delta T_r = 1-T_r = 1-(T/T_m)$, T = temperature and T_m = the liquidus temperature. It is considered that this nucleation resistance arises, even though the coordination numbers and atom distances are nearly the same in the melt and crystal, because of a large negative entropy of interface formation as the disordered atoms of the liquid rearrange themselves into the ordered structure of the crystal at the crystal-melt boundary; this change contributes to an increased crystal-melt surface tension σ_{1s} {see Spaepen and Turnbull for discussion (4)}. According to simple homogeneous nucleation theory, the rate of nucleation I (in $cm^{-3}\ sec^{-1}$) is given by

$$I \simeq (10^{30}/\eta)\ \exp\ [-b\alpha^3\beta/T_r(\Delta T_r)^2] \qquad (1)$$

where η = viscosity, b = a geometrical factor = $16\pi/3$ for a spherical nucleus, $\alpha = (N_oV)^{1/3}\sigma_{1s}/\Delta H_f$, $\beta = \Delta S_f/R$ and N_o = Avogadro's number, V = the molar volume in the crystal, ΔH_f = the molar heat of fusion, ΔS_f = the molar entropy of fusion and R = the gas constant. It is apparent that a low nucleation rate, and hence, a high probability of bypassing crystallization to form a glass by rapid quenching, is favored by large α (i.e. σ_{1s}), β and η. For metals α is 1/2, while for non-metals it is $\sim$1/3 {see Chen (5)}. For metals, SiO_2, GeO_2, etc., $\beta \simeq 1$, while $\beta \simeq 4$-10 for organic and inorganic compounds and, indeed, the latter materials are easily prepared in glassy form. Network liquids, such as SiO_2, are ready glass formers, primarily because of a high viscosity at T_m, e.g. $\sim 10^7$ poise (P) for SiO_2, as compared to 10^{-2} poise for a typical metal.

One can estimate the cooling rates required for formation of a metallic glass beginning with Eq. 1. The variation of α with temperature may be introduced as $\alpha = \alpha_m(T/T_m)$, so that the nucleation rate for spherical nuclei becomes

$$I \simeq (10^{30}/\eta)\ \exp\ [(-16\pi\alpha_m^3\beta/3)(T_r/\Delta T_r)^2] \qquad (2)$$

The temperature dependence of the viscosity may also be included as

$$\eta = A\ \exp\ [a/(T_r-T_{rg})] \qquad (3)$$

where T_{rg} = the reduced glass transition temperature = T_g/T_m and T_g = the glass transition temperature. Choosing reasonable values for the various parameters α_m, β, A and a, one can calculate I vs. T_r for various values of the reduced glass temperature, as shown in Fig. 1 (4). The indications of this figure are as follows: If T_{rg} is low, <<1/2, nucleation rates of the order 10^{20}-10^{30}/cm^3sec will be attained over an appreciable temperature range and no practically achievable cooling rate would permit glass formation. If $T_{rg} \simeq 2/3$, the peak nucleation rate is of the order $10^{-30}/cm^3$ sec and any reasonable laboratory scale cooling rate should yield a glassy solid. [Near T_m the thermodynamic driving force for crystallization is low; at lower temperatures, where the driving force increases, the high viscosity, or low diffusivity of the melt, inhibits nucleation.] If $T_{rg} \sim 1/2$, a peak nucleation rate of the order of $10^6/cm^3$ sec is possible; assuming that T_m is $\sim$1500 K and recognizing that the peak rate obtains only over a small range of T_r, a cooling rate of $\sim 10^6$K/sec should allow glass formation. These results are only approximate, of course. For example, when $T_{rg} \sim 2/3$, cooling rates of the order of 10^2-10^3 K/sec are actually required for glass formation.

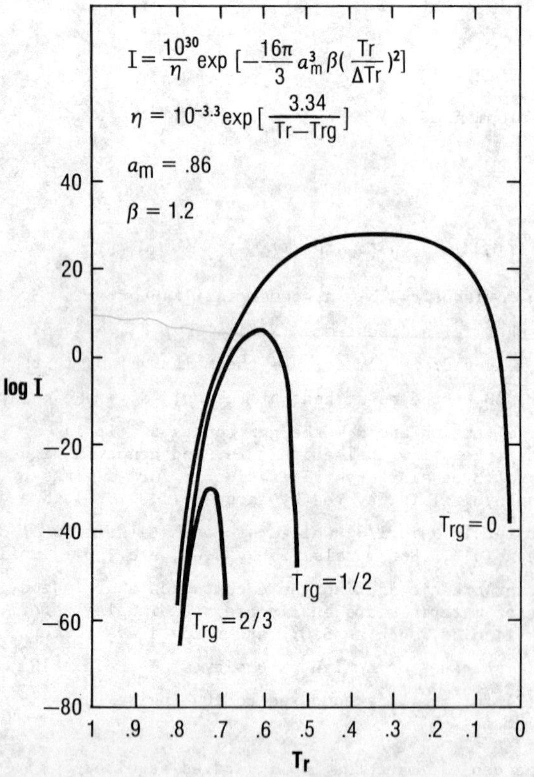

Figure 1. The variation of the nucleation frequency (I, in $cm.^{-3}$ $sec.^{-1}$) with reduced temperature (T_r), as calculated according to the formulae and parameters shown in the figure. (4)

However, a cooling rate of $\sim 10^6$ K/sec can be achieved when a thin layer of melt is brought into intimate contact with a Cu substrate at room temperature (6) and this approach has been used to produce a multitude of glassy alloys. In essence, the treatment of Eq. 2 indicates that metallic glass formation is fundamentally consistent with nucleation theory.

One may also consider crystal growth rates when estimating the critical cooling rates required for glass formation; a less rigid criterion is then established in that one allows some minimal volume fraction, say 10^{-6}, of crystalline material {see Davies for discussion (7)}. The fraction of crystalline material generated in time t is given by

$$X \simeq (\pi/3)\ I\ u_c^3\ t^4 \qquad (4)$$

where I, again, is the homogeneous nucleation rate and u_c is the crystal growth velocity; $u_c = (fD_g/a_o)[1-\exp(-\Delta T_r \Delta H_f/RT)]$ (7), where f = the fraction of sites at the crystal-melt interface where atom transfer can occur, D_g is the average diffusion coefficient across the interface and a_o is the mean atomic diameter. Using $f = 0.2\ \Delta T_r$, the Stokes-Einstein equation ($D = kT/3\pi a_o \eta$) to convert D_g to η, and allowing for a temperature dependence of η (in I) similar to that above, Davies solved Eq. 3 for t vs. T_r to produce time-temperature-transformation curves. At high and low T_r transformation times are long; at intermediate T_r the curve has a "short" time nose. The cooling rate required to just avoid this nose is a measure of the critical cooling rate, R_c, for glass formation. Table 1 shows the rates for $X = 10^{-6}$ for some selected alloys, as calculated by Davies (7). As the various materials parameters in Eqs. 2 to 4, save for T_g and T_m, were taken, essentially, as independent of composition, it is not surprising that R_c depends almost monotonically on T_{rg}. In other words, the smaller the reduced temperature range [$\Delta T_{rg} = (T_m - T_g)/T_m$] one must quench through, the lower the cooling rate required and the more likely will be glass formation.

Table 1

Critical cooling rates, R_c, for glass formation ($X = 10^{-6}$) (7)

Material	T_m(K)	T_g or T_c(K)	T_g/T_m	R_c(K sec^{-1})
$Fe_{83}B_{17}$	1448	760	0.52	1 x 10^6
$Ni_{75}Si_8B_{17}$	1340	782	0.58	1.1 x 10^5
$Pd_{82}Si_{18}$	1071	657	0.61	1.8 x 10^3
$Pd_{77.5}Cu_6Si_{16.5}$	1015	653	0.64	3.2 x 10^2

The rates indicated by Table 1 are not verifiable in detail; rapid cooling rates are not easily measured and calculations are inexact because the heat transfer coefficients between melts and quenching media are unknown. Nonetheless, the rates shown are generally considered to be approximately correct; they are certainly crudely consistent with the relative ease with which one can make glassy samples of Fe,Ni and Pd based alloys. An indication of this ease is given by the thickness of cast section that one can produce in a substantially glassy state; Fe-B based alloys can be cast on a metal substrate to a thickness of about 50 to 75 μm, some Ni-B type alloys can be cast in the same fashion to about 300 μm thickness, while Pd-Cu-Si rods up to about 2000 μm in diameter have been produced by plunging a quartz tube containing the melt into water. The same effect can be seen in more detail for alloy compositions adjacent to a eutectic composition (liquid ⇄ multiphase solid transition). The glass transition temperature (or crystallization temperature) often varies moderately with composition, while the liquidus temperature rises steeply on either side of

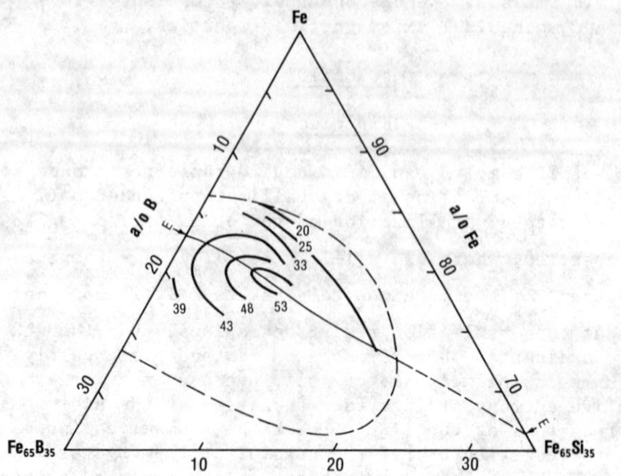

Figure 2. Contours of maximum ribbon thickness consistent with high bend ductility, i.e. ribbons can be bent 180° without failure. Compositions close to the eutectic trough (stretching between the locations marked E) require lower critical cooling rates and, hence, can be made in thicker sections. (8)

the eutectic; hence T_{rg} is a maximum at the eutectic composition and glass formation is easiest there. Fig. 2 shows the thickness variations observed for the Fe-B-Si system (8); the greater ease of glass formation for compositions adjacent to the eutectic trough is evident. The limits shown are somewhat more restrictive in that they refer to the maximum ribbon thickness consistent with high ductility, i.e. the ribbon can be bent back on itself without snapping. Brittleness can occur before the onset of crystallization.

Alloys

The above treatments of nucleation and nucleation and growth rationalize the ease of glass formation based on measured T_g and T_m. They have no capability with respect to predicting T_g and, thus, alloy systems where glass formation is easy and where the glass formed will be stable at reasonable temperatures. {We consider, now, melts which can be quenched to a glass at $\lesssim 10^6$ K/sec and, hence, exclude slightly impure elemental metals, which require high rates and are retained in the amorphous state only at low temperatures e.g. ∿40 K for Co (9)}. Identifying such systems is largely a matter of trial and error, although, as discussed by Chen (5) they have some features in common; they exhibit negative heats of mixing, indicative of strong interactions between unlike atoms and unusually deep eutectics, i.e. deeper than those which would be expected based on ideal solution theory (10). A typical example is shown in Fig. 3 for Fe-B (11). Readily producible glassy alloys have compositions at and around a eutectic composition. It has also been noted that glass formation appears favored when the constituent atoms differ in size by greater than about 12% (5,6,12). This size difference is best estimated from the near neighbor distances in the glass or from an interstitial compound of the atoms in question.

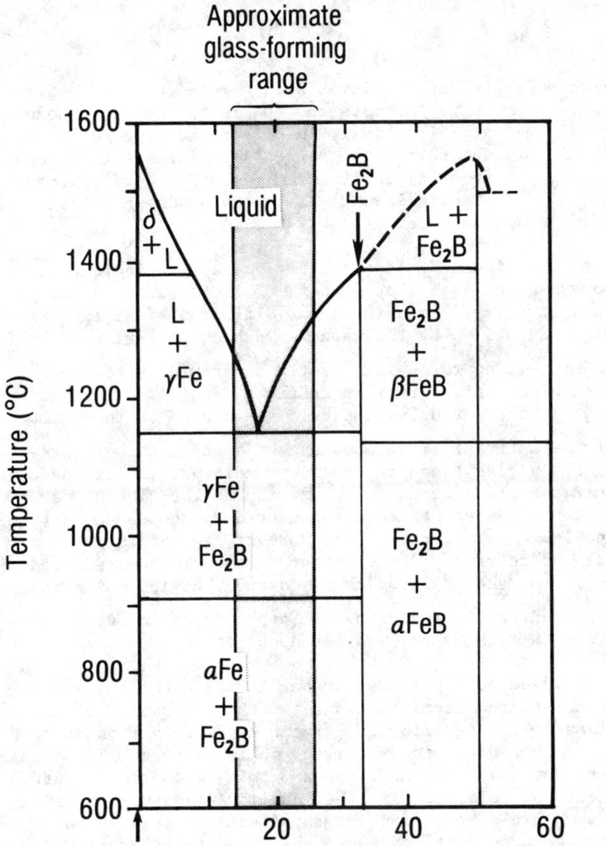

Figure 3. Portion of the iron-boron phase diagram. The glass-forming region spanning the deep eutectic composition at 17 atomic percent boron is indicated. (11)

Alloy glasses have been formed now from a wide variety of elements (5,7). The majority of them can be grouped in three categories: metal-metalloid alloys containing e.g. Au, Fe, Ni, Co, Pd and/or Pt and 10-30 at.% metalloids, e.g. B, C, Si, P and/or Al; early transition metal-late transition metal alloys, containing e.g. Zr, Nb and/or Ta and ∿30-60% Fe, Co, Ni, Cu and/or Pd {see for example Ray et al. (13)}; and alloys containing group IIA metals, Mg, Ca and Be, e.g. Mg-Zn (14a), Ca-Al (14b), Be-Ti, Be-Zr and Be-Hf (15). The first group of alloys has been studied most extensively and those based on the ferromagnetic elements Fe, Ni and Co are, by far, of the most practical interest. Accordingly, the sections below on structure and properties are restricted to consideration of metal-metalloid alloys.

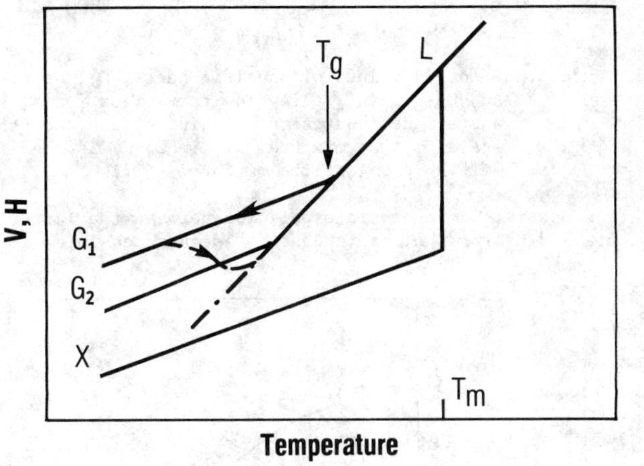

Figure 4. Schematic drawing of specific volume, V, and heat content, H, vs temperature, for a liquid (L), glass (G) and crystal (X). T_m is the melting temperature; T_g is the glass temperature. (5)

Glass Transition

A liquid at its melting point is in internal equilibrium. As it is undercooled in the absence of crystallization it remains in internal equilibrium as long as the structural relaxation (or rearrangement) time is short compared to $(\dot{T})^{-1}$, where $\dot{T}$ is the cooling rate. When this is no longer true the material remains in a nonequilibrium state as the temperature falls. From the point at which it leaves internal equilibrium it is said to be configurationally frozen, i.e., it becomes a glass (16). The variation of volume or enthalpy is shown schematically in Fig. 4 (5). At higher cooling rates the material will leave equilibrium (G_1) at higher temperatures. Hence, glasses with different internal energies and specific volumes can be produced and T_g is essentially a kinetic parameter. The specific heat also changes sharply at T_g. Fig. 5 shows C_p vs. T for an $Au_{77}Si_{14}Ge_9$ glass (17). This glass is sufficiently resistant to crystallization that C_p may be measured into the supercooled liquid state; the more common situation is that crystallization will intervene a few tens of degrees below T_g. As is evident from the figure, C_p in the crystalline state is close to the classical value of 6 cal/gm atom K; C_p of the glass is slightly higher. On heating through T_g (≃297 K) an increase of C_p ∿5 cal/gm atom K occurs, due to the translational degrees of freedom available in the liquid state; C_p of the supercooled liquid can be measured ∿3 K above T_g before crystalliza-

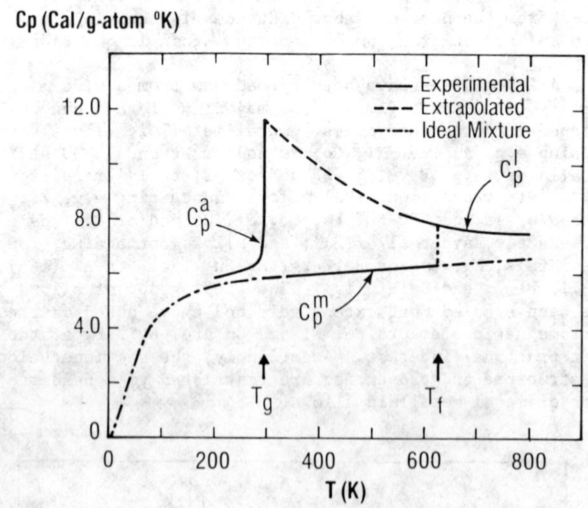

Figure 5. Smoothed values of specific heats, C_p, of a $Au_{77}Ge_{13.6}Si_{9.4}$ alloy in its various forms, where l designates the liquid; a, the amorphous state; m, the mixture. T_f (=625 K) is the melting point of the mixture. (17)

tion occurs. A smooth interpolation between the latter data and C_p for the liquid near T_m appears possible.

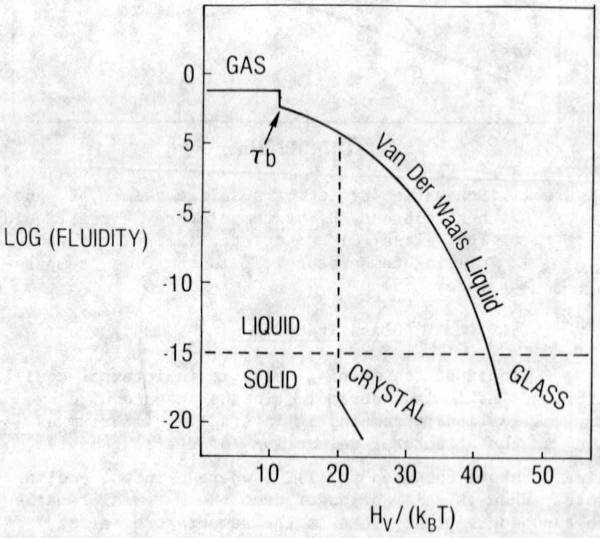

Figure 6. Fluidity ($\phi \equiv 1/\eta$) -reduced temperature (τ) relation for simple molecular substances, where H_V is the molecular heat of vaporization. The liquid to glass transition is arbitrarily shown here as at $\eta_{g} = 10^{15}$P. (16)

In the vicinity of T_g the viscosity of the liquid shows a continuous rapid increase (shown schematically in Fig. 6). For the case of the $Au_{77}Si_{14}Ge_9$ alloy, η is $\sim 10^{10}$P at T_g and rises to 1.4×10^{13}P at 285°C [as compared to $\sim 10^{-1}$P near T_m (18)]. This rapid rise of η is generally thought to occur due to increasingly strongly correlated atomic movements in the material. The correlations become so strong that it is no longer possible to measure η directly at $T \ll T_g$. That is, just above T_g the flow rate of the material responds instantaneously to a step-wise change of T or stress. Below T_g a delay occurs, while the structure relaxes to its new equilibrium state, before a new steady state flow rate is established. The delay time is approximated by $\tau \simeq 10^{-10}$ η sec (17) and obviously becomes extremely long as η increases (T decreases). Hence at $T \ll T_g$ creep measurements sample so-called isoconfigurational flow of the unrelaxed glass structure, with apparent viscosities orders of magnitude below the equilibrium η.

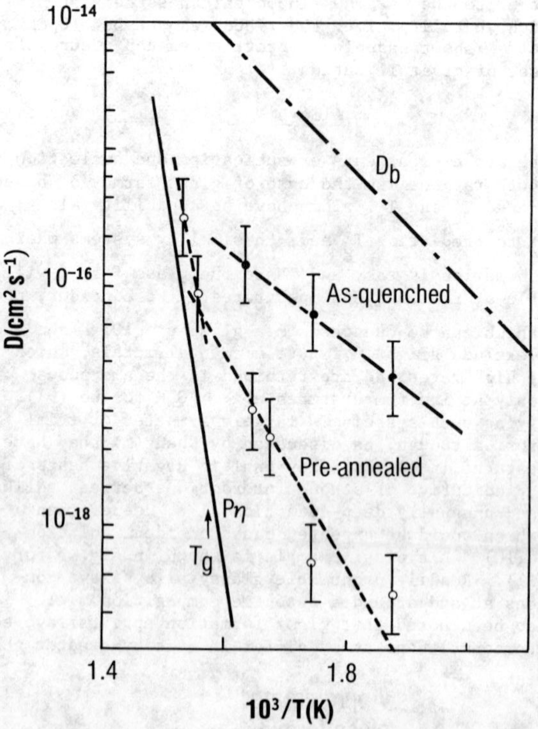

Figure 7. Au diffusivities D in a $Pd_{77.5}Cu_6Si_{16.5}$ glass in various states. D_b is the Au grain boundary diffusivity in polycrystalline Pd films. The solid line represents diffusivities deduced from the equilibrium viscosity data (extrapolated below T_g) using the Stokes-Einstein equation. (5)

The liquid $\rightleftarrows$ glass transition is also indicated, somewhat obliquely, by measurements of atomic mobility. Chen et al. (19) monitored the diffusion of Au in $Pd_{77.5}Cu_6Si_{16.5}$ by the Rutherford backscattering of He ions; this approach is expected to provide a good measure of Pd self-diffusion in the glass. As evidenced by Fig. 7, the diffusivities below T_g depend on the state of the material, i.e. as-quenched specimens show larger diffusivities than preannealed specimens, indicating that annealing leads to relaxation of the glass to a somewhat denser state, as suggested by the dashed curve in Fig. 4. In either case, however, the diffusivities are orders of magnitude larger than would be predicted (D_η) from the equilibrium viscosity behavior using the Stokes-Einstein equation. In other words, above T_g, D scales with η; below T_g it does not.

These observations further indicate the necessity of correlated atomic motions for the relaxation of the glass to its equilibrium structure. Observation of measurable diffusion requires, on the average, 100 jumps per atom (5). If independent atomic motions were sufficient for relaxation to the equilibrium structure, 100 jumps should be sufficient to effect the relaxation. The measured D values would then have to be orders of magnitude lower than those observed.

To summarize: Metallic melts exhibit a natural aversion to homogeneous nucleation. If the glass transition temperature is one-half to two-thirds of the melting temperature, crystallization may be bypassed and a glass formed by rapid quenching at a rate $\lesssim 10^6$ K/sec. Metallic melts susceptible to glass formation at such cooling rates are invariably alloys and commonly have near-eutectic compositions. The melt leaves internal equilibrium on cooling through the glass temperature and does so at higher temperature, the higher the cooling rate. On reheating, the glass to supercooled liquid transition is marked by a sharp increase in the specific heat of the material. The diffusivities in the glass are orders of magnitude larger than would be predicted based on the extrapolated equilibrium viscosity behavior.

STRUCTURE

Experimental Characterization

Metallic glasses possess a disordered structure which is experimentally impossible to characterize in precise detail; this follows inevitably from the absence of three dimensional long range periodicity. The electron, neutron or x-ray diffraction signature of a crystalline material, as recorded on a sheet of appropriately located film, is a series of spots (single crystal), or sharp rings (powder), which are indicative of the precise locations of the atoms in the material. The corresponding signature of a metallic glass is a series of diffuse rings. From these data one may calculate the radial distribution function of the glass (RDF), which is equal to $4\pi r^2 \rho(r)$, where $\rho(r)$ is the average number density of atoms at a distance r from any atom chosen as the origin (20). The RDF provides an average representation of the structure and indicates the approximate distance r (peak positions) and numbers (peak areas) of first, second, etc. nearest neighbors of a central atom. It does not, however, indicate the angular positions of the neighbors, wherein lies the lack of detail. Accordingly, the key results provided by RDF's (first peak) of a variety of TM-M glasses are numbers of nearest neighbors (of the order of 11.5 to 13) and near neighbor distance (∿2% greater than the respective 12-coordinated Goldschmidt diameter of the metal specie) (20). From these results it is apparent that the glasses are close packed materials; measured densities are, in fact, only 1 to 2% less than for the corresponding crystalline phase(s) (21,22).

An alternative representation of the structure is provided by the reduced radial distribution function $G(r) = 4\pi r[\rho(r)-\rho_o]$, where ρ_o is the average number density of atoms. Figure 8 shows this function for a variety of TM-M glasses, plotted vs. r/r_1, where r_1 is the position of the first peak maximum. From the figure it is evident that the similarities in the structures of the TM-M glasses extend beyond those noted above. The relative positions and heights of higher order peaks are approximately the same and the second peak is always split, with the higher spacing peak diminished relative to the first. The latter split is a key feature distinguishing the RDF's of TM-M glasses and liquids (split absent) of the same composition. It is also important to note that no peaks appear at $\sqrt{2}r_1$, a characteristic distance in face centered cubic and hexagonal structures, corresponding to the spacing between two close packed planes (23). This is key evidence that the structures of metallic glasses cannot be characterized by microcrystalline models, wherein diffuse rings are imagined to arise due to line broadening resulting from diffraction from crystals less than ∿20 Å in diameter (23). Cargill has also noted that such a model would require a large amount of boundary volume and an accompanying density deficit larger than the 1-2% observed (24).

The comments of the preceding two paragraphs largely dismiss the obviously important fact that we are attempting to characterize structures containing two (or more) elements. The scattered radiation intensity from a single wavelength (λ) experiment actually derives from the weighted sum of coherent scattering from

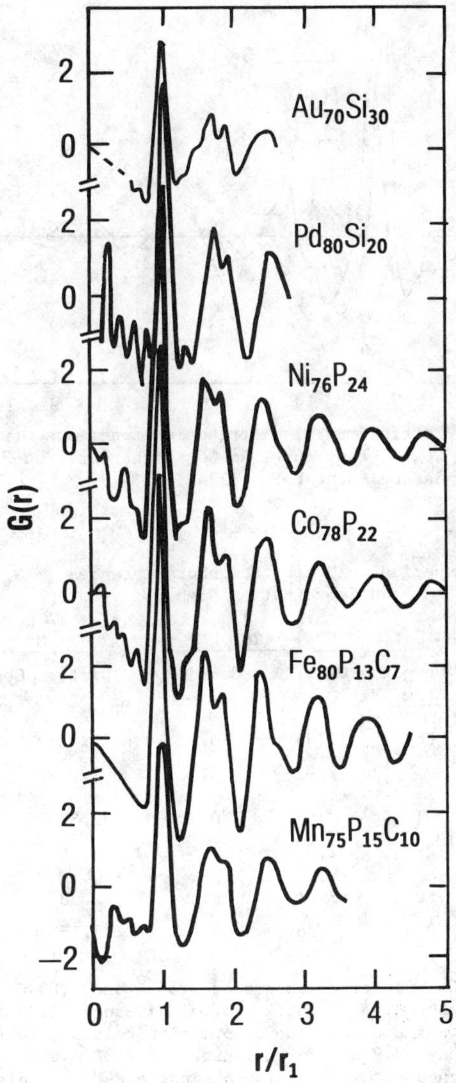

Figure 8. G(r) vs. r/r_1 for several metal-metalloid glasses. (20)

[n(n+1)/2] pair-wise interactions, where n = the number of components in the glass. For x-irradiation the weighting factors for TM-TM scattering are 3 to 5 times those for TM-M and ∿50 times those for M-M scattering (20). Consequently the curves of Figure 8 are dominated by the TM-TM contribution and indicate, essentially, the average number density of near neighbors of a TM atom, but not the relative proportion of TM and M neighbors; the curves provide no information concerning the environment of a metalloid atom.

In order to more fully characterize the average environment of each atom in an n component system, it is necessary to determine n(n+1)/2 partial pair correlation functions {see Boudreaux for discussion (25)}. To do so one must perform separate scattering experiments utilizing n(n+1)/2 different wavelength radiations. The three experiments required for a two component system have been performed for only a few glasses; e.g. Sadoc and Dixmier examined $Co_{80}P_{20}$ (23), using x-rays and polarized and unpolarized neutrons and Waseda examined $Ni_{75}P_{25}$ (26) and $Fe_{75}P_{25}$ (27), using anomolous x-ray scattering. The three partial radial distribution functions $[G(r)/4\pi\rho_o]$ for $Co_{80}P_{20}$ are shown in Figure 9. The results derived from the figure are summarized in Table 2; it is evident from the table that the first peak of the composite x-ray RDF includes contributions

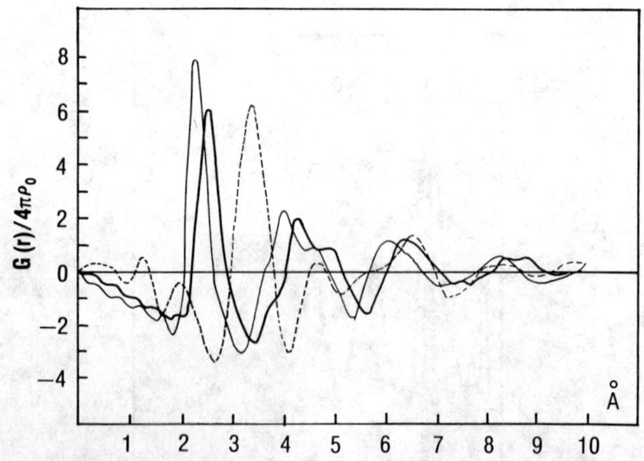

Figure 9. Partial radial distribution functions $G(r)/4\pi\rho_0$: heavy curve Co-Co; light curve Co-P; dashed curve P-P. (23a)

Table 2

Experimental First Neighbor Distances and Coordination Numbers

	Origin Atom	TM Dist. Å	TM Coord.#	M Dist.	M Coord.#
$Co_{80}P_{20}$	Co	2.55	10.1	2.32	2.09
(23)	P	2.32	8.9	2.09	0
$Ni_{75}P_{25}$	Ni	2.55	10.7	2.35	2.2
(26)	P	2.35	8.5	2.15	0
$Fe_{75}P_{25}$	Fe	2.61	10.4	2.38	2.6
(27)	P	2.38	8.1	2.15	0

from 10.1 Co-Co near neighbors at 2.55 Å and 2.09 Co-P near neighbors at a slightly shorter distance, 2.32 Å. Of more significance is the information relative to P; each atom has ∿8.9 Co near neighbors at 2.32 Å and no P near neighbors. The first peak in the P-P pair distribution function occurs at 3.34 Å, which corresponds to P atoms separated by a Co atom. Similar results are observed for $Fe_{75}P_{25}$ and $Ni_{75}P_{25}$. It should be noted that the TM-P distances are considerably less than the sum of the Goldschmidt radii [∿1.25 (TM) + 1.28 (P) = 2.53Å]; the diameters deduced for P are close to the tetrahedral covalent diameter {2.2Å; see e.g. Spaepen (28)}.

These results, then, provide additional key information: While the metal atoms lie in close packed configurations, the metalloid atoms do not; the metalloid atoms are preferentially coordinated with metal atoms at a foreshortened spacing. The latter details are indicative of a strong chemical interaction between the M and TM atoms, as is also evident from the negative heats of mixing of the elements. Our knowledge of the structure is still incomplete, however; the precise spatial arrangements of neighboring atoms remain inaccessible. For this reason there has been strong interest in developing structural models from which the local atomic placements can be postulated.

Models

Modeling studies relevant to the structures of metallic glasses began, coincidentally, in 1959 with the work of Bernal (29), which was directed towards a description of the structures of monatomic liquids. An outgrowth of this work was the physical construction of an 8000 "atom" (ball bearing) dense random packed hard sphere (DRPHS) model which was analyzed by Finney (30). The balls of the model were collected in a rubber bladder and immobilized with wax; their coordinates were recorded as they were removed one by one. A histogram for G(r) for the structure could then be calculated.

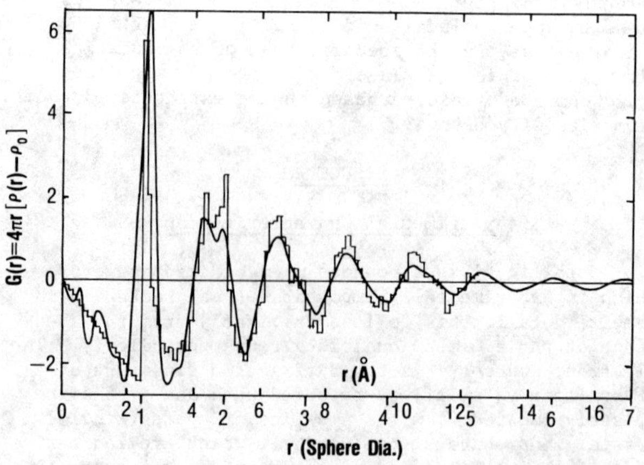

Figure 10. Comparison of reduced radial distribution functions G(r) for Finney's DRPHS structure and for amorphous $Ni_{76}P_{24}$. (20)

In his pioneering work on the experimental determination of the structures of metallic glasses, Cargill (31) first noted that G(r) for a TM-M glass (specifically $Ni_{76}P_{24}$) is rather similar to that for Finney's DRPHS model (Fig. 10); this comparison assumed that the TM and M atoms could be placed interchangeably on sites in the DRPHS structure. Significant differences exist, however, in that the peak intensities of the split second peak are reversed and the hard sphere diameter required for the best fit to the Ni-P data makes the Ni-P packing density too large. It was to resolve this latter discrepancy that Polk (32) first made his enlightened suggestion, in advance of the above experimental findings, that the metalloid atoms may reside within the canonical holes contained in a DRPHS structure made up of metal atoms; rather than on the DRPHS sites themselves. These holes, with idealized shape, are shown in Figure 11. Polk (33) suggested that the "larger" holes (a, b, c) in the struc-

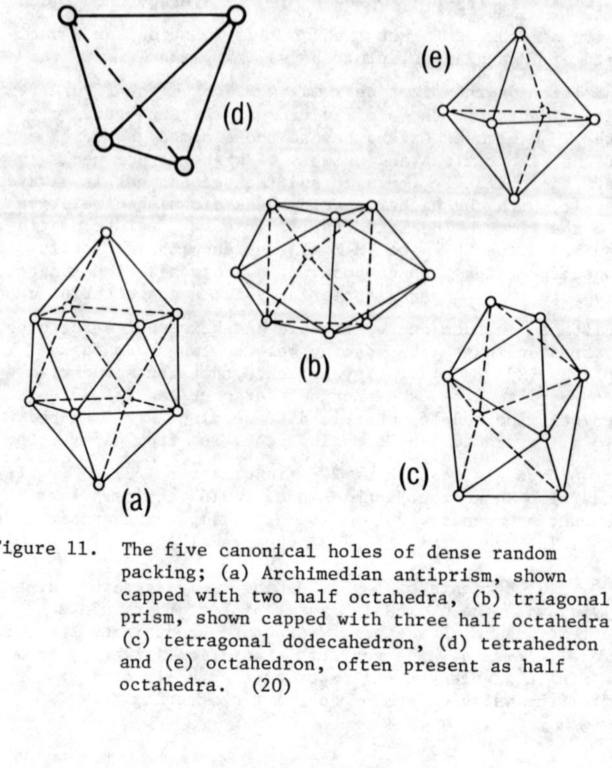

Figure 11. The five canonical holes of dense random packing; (a) Archimedian antiprism, shown capped with two half octahedra, (b) triagonal prism, shown capped with three half octahedra, (c) tetragonal dodecahedron, (d) tetrahedron and (e) octahedron, often present as half octahedra. (20)

ture would contain the metalloid atoms and noted that they occur at the rate of ∿21 per 79 atoms, matching rather well the typical atom ratio characteristic of TM-M glasses. It was subsequently recognized, however, that these holes are somewhat too small to contain the typical metalloid atom, such as P, without relaxation of the model (20,34) and that this might require gross rearrangement of the DRPHS skeleton.

The notion of packing a DRPHS metal structure with metalloid atoms provided, nonetheless, a clear insight into the structures of metallic glasses. It is consistent with the lack of metalloid-metalloid near neighbors and it also provides for a metalloid environment similar to that which exists in interstitial TM-phosphides, borides, etc. (35). In some of these compounds (e.g. metastable Fe_3B) the M atom resides in a slightly distorted trigonal prism of TM atoms; in others (e.g. Co_3P, Fe_3P) the M atom resides in a tetrakaidecahedron, a distorted version of a trigonal prism capped with three half octahedra (c of Figure 11). Hence, depending upon the relative placement of the capping atoms, polyhedron b can allow for 6 to 9 nearest metal neighbors of a metalloid atom. One can build a glassy structure based on distorted trigonal prisms, as suggested by Gaskell (36), and indicated in Figure 12.

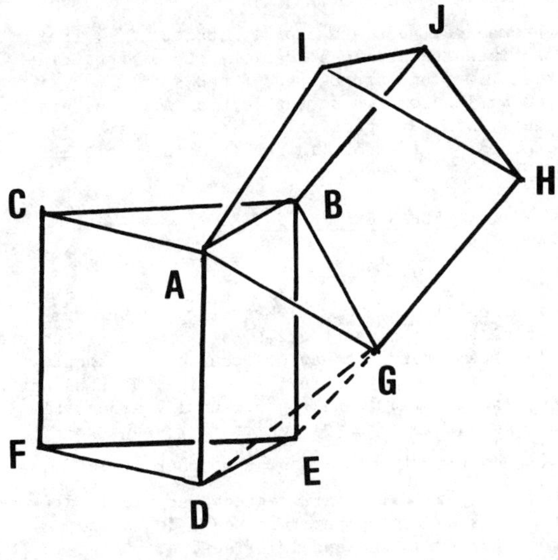

Figure 12. Structure building by interconnection of adjacent trigonal prisms. (36)

The notion of "quasi" crystalline chemical and topological short range order is plausible and instructive but one must look to computer modeling studies to provide an indication of whether it might exist in detail. An initial attempt at constructing a DRPHS structure atom by atom in a computer was that of Bennett (37). This was a single constituent structure and, hence, not directly applicable to TM-M glasses. Sadoc, Dixmier and Guinier (38) constructed a two constituent model, but it suffered from being too small to provide reliable statistics and it was constructed by an algorithm which is non-space filling. Hence it must have a basic deficiency in common with the Bennett model, which was progressively less dense from center to outside (39), a decidedly unphysical arrangement. Nonetheless, G(r) for this two atom model matched G(r) obtained for $Ni_{85}P_{15}$ reasonably well. In effect, such a match is not a particularly stringent test of a computer built model. Boudreaux (25) has noted that the proper test is to calculate the reduced interference function from G(r) of the model i.e. $F(k) = \int G(r) \sin kr\, dr$ and compare this with F(k) determined directly from the experimentally measured scattering data.

Boudreaux and Gregor (39) undertook the difficult manipulations required to build a binary computer model which would simultaneously maintain uniform and correct density, uniform composition and prohibit M-M near neighbors. Through appropriate choice of the TM-M size ratio they simulated a model of $Pd_{80}Si_{20}$ in order to compare the calculated F(k) with the data of Suzuki et al. (40), which are the most definitive scattering data presently available. X-ray scattering experiments are commonly carried out to a scattering vector $k(=4\pi\sin\theta/\lambda)$ of ∿7 to 15 Å^{-1} (26): scattering at larger k values is experimentally limited by the radiation wavelength conveniently available [0.71 Å for a Mo tube] and fundamentally limited by low scattering intensities at large k. The results of Suzuki et al. derive from neutron diffraction, where the scattered intensity is independent of k, and extend to 30 Å^{-1}. The experimental $F(k)[\equiv S(Q)]$ thereby deduced is compared with the theoretical binary hard sphere reduced interference function in Figure 13b (39); it is evident that the fit is not particularly good beyond the third peak. As the important near neighbor information in G(r) is heavily weighted by the Fourier transform of the large k regime of reciprocal space, it is clear that the model is deficient. To correct for this the atoms were made to interact through Lennard-Jones potentials, the M-TM interaction being taken twice as strong as the TM-TM interaction, and the structure was allowed to relax (39). The experimental and theoretical F(k) then calculated compare extremely well, as indicated in Figure 13a, and one can then conclude that the model is viable.

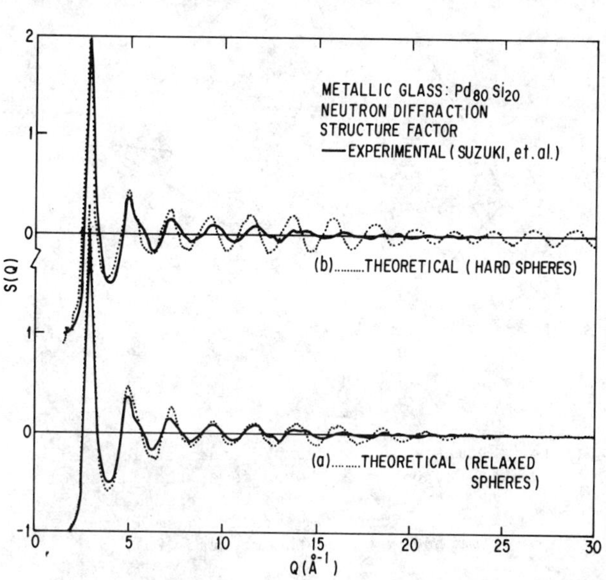

Figure 13. Model calculations for $Pd_{80}Si_{20}$ as compared to high resolution neutron diffraction measurements. (39)

Results for average near neighbor coordination numbers for Fe-P (70-90%Fe) and Fe-B (76-85%Fe) models are shown in Table 3 (25). Models for Fe-B were constructed using an appropriately reduced M/TM size ratio; the size ratio for P/Fe is the same as that for Si/Pd and, hence, the same models apply nominally to both systems. The most interesting contrast concerns the coordination

Table 3

Theoretical Near Neighbor (Average) Coordination Numbers {Boudreaux (25)}

Fe-P	$\langle N_{Fe-Fe}\rangle = 10.49$	$\langle N_{P-Fe}\rangle = 8.33$
Fe-B	$\langle N_{Fe-Fe}\rangle = 10.99$	$\langle N_{B-Fe}\rangle = 6.62$

numbers of the metalloid atoms with metal atoms; B is ∿6.9 fold coordinated with Fe and P is ∿8.2 fold coordinated, consistent with the 6 fold and 9 fold coordinations observed in crystalline Fe_3B and Fe_3P, and with the coordination experimentally observed for $Fe_{75}P_{25}$ glass (Table 2). For PdSi the predicted 8.2 fold coordination is larger than that observed experimentally for the glass (≈6.7) (41) and crystalline Pd_3Si (≈6). The latter discrepancy reflects, to some degree, uncertainty in designating distances of near neighbors in the model. However the model still predicts the same behavior for Fe-P and Pd-Si, suggesting, perhaps, that the strengths of TM-M interactions and/or TM/M size ratios should be modeled somewhat differently for the two cases.

In order to more precisely characterize the Fe-B models, Boudreaux and Frost (42) made a detailed analysis employing computer graphics, wherein the surroundings of individual atoms were displayed on a cathode ray screen and inspected. The neighbors of an Fe atom were commonly in an icosahedral arrangement. Most of the B atoms were observed to reside in mildly distorted trigonal prisms of Fe atoms as indicated in Figure 14. Hence it appears reasonable to expect that

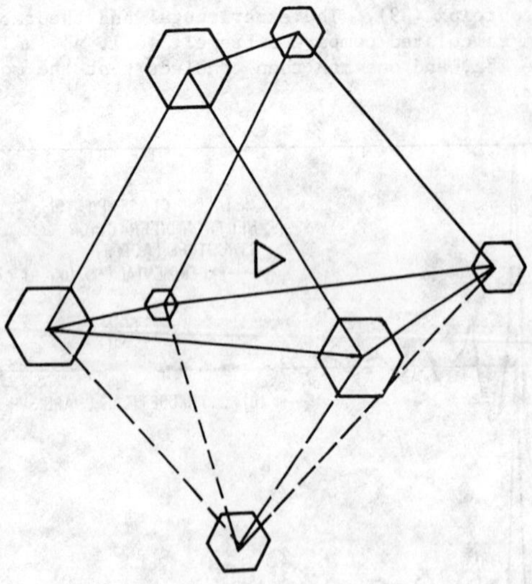

Figure 14. Geometry of short-range order around B sites in $Fe_{75}B_{25}$. The boron is at the center of the structure shown by the triangle. The surrounding iron atoms are drawn smaller if further away from the viewer. (25)

the local topology of metalloid near neighbors is, indeed, very similar to that in the crystalline interstitial phases. Examination of the change of B environment with composition indicates some tendency of the B atoms to reside in octahedra as the Fe/B ratio is increased. Boudreaux (25) notes that an octahedron provides more faces (eight vs. five for the trigonal prism) with which 2nd near neighbor Fe atoms can coordinate as the Fe concentration increases. This structural change presumably is associated with the composition-density (ρ) relation observed for FeB alloys; as indicated by Figure 15 there is little change in ρ for B$\lesssim$19 at.% but a sharp linear decrease at higher B content (22). Similar "breaks" are not observed for other TM-M glasses and one presumes that local structural variations for these glasses are less pronounced as composition varies.

To summarize: transition metal-metalloid glasses are characterized by disordered structures which are very similar for a wide variety of compositions; they are close packed, with TM radii only slightly less than the 12 fold coordinated Goldschmidt radii; chemical short range order exists, in that M atoms have only TM

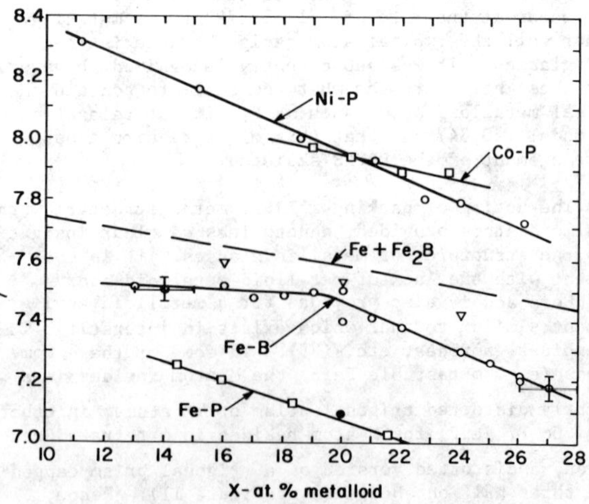

Figure 15. Densities of metallic glasses as a function of composition. The error bars represent estimated absolute uncertainties of ±1 at.% for composition and ±0.5% for density. The dashed line is for an equilibrium crystalline mixture of Fe and Fe_2B. (22)

near neighbors and these are closely spaced, i.e. their separation distances are less than the sum of the Goldschmidt radii; M atoms are between 6 and 9 fold coordinated by TM atoms whose topological arrangements are apparently similar in detail (trigonal prism; tetrakaidecahedron) to those in the TM_3M interstitial compounds.

MECHANICAL PROPERTIES

Plasticity

The mechanical behaviors of metallic glasses are indexed relative to the glass transition temperature; accordingly, "high" temperature behavior occurs above about ∿0.5 to 0.8 T_g. Deformation in this regime, at constant stress, occurs by creep. The experimentally accessible stress (σ) and strain rate ($\dot{\varepsilon}$) ranges correspond to apparent viscosities [$\eta = \sigma/3\dot{\varepsilon}$] of the order of 10^{15} to 10^{17} Ns/m^2(P); these values are centered at widely different positions on the temperature scale depending on the thermal history of the specimen (see, e.g. Fig. 16

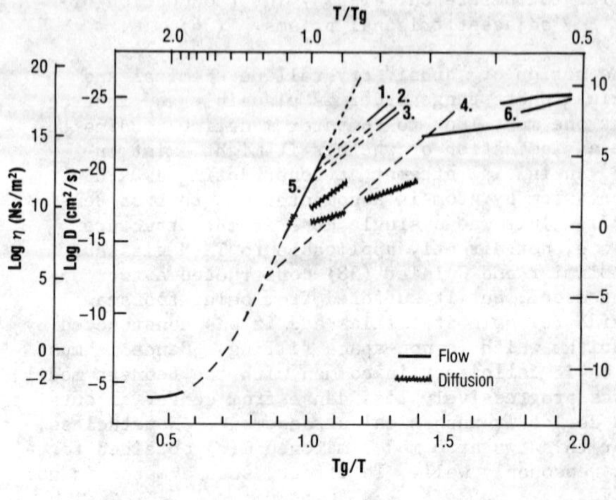

Figure 16. Creep flow & diffusion data for $Pd_{80}Si_{20}$(43). T_g = 642 K by calorimetry. Annealing conditions are, 1) external 279±3°C 242 hours, 2) external 240±5°C 432 hours, 3) in situ, 227°C 115 hours, 4) in situ 150°C 70 hours. Curve 5 is equilibrium flow of $Pd_{77.5}Cu_6Si_{16.5}$; curves 6 are for as-quenched samples.

for $Pd_{80}Si_{20}$ (43)}. For the as-quenched glass these values are located around 0.55 T_g; the temperature dependence of η is characterized by an activation energy ΔG* [$\eta = \eta_0 \exp(-\Delta G^*/kT)$] of the order of 0.5 eV (5). If the sample is preannealed at a temperature T<T , but beyond the subsequent creep testing range, these values of η shift to higher T, the higher the preannealing temperature, but ΔG* is approximately the same {∿2 eV for $Pd_{80}Si_{20}$ (43)} for each preannealing condition. These variations in viscosity are indicative of relaxation of the structure of the glass; the higher the annealing temperature, the more the relaxation and the higher the viscosity. If relaxation occurs during the creep test, as for the as-quenched glass, the apparent activation energy is reduced; if testing is conducted below the preannealing temperature, the structure remains "constant" during the test, i.e. one measures isoconfigurational viscosities. The approximate constancy of ΔG* for all isoconfigurational (preannealing) conditions indicates that the flow mechanism is the same in all cases. As noted above and shown by Fig. 16 the isoconfigurational viscosities are orders of magnitude lower than the (extrapolated) equilibrium viscosities; this suggests that the atomic correlations are so strong that flow must occur by the motion of quasi-linear defects. In addition, we note that the activation energies for diffusion for $Pd_{77.5}Cu_6Si_{16.5}$ (5) are close to those for creep for $Pd_{80}Si_{20}$, nominally the same alloy, and exhibit the same thermal history dependence. This further suggests that creep occurs by a stress assisted, diffusion controlled process analogous to dislocation climb in a crystalline material.

When glassy ribbons are tested under constant extension rate conditions ($\dot{\varepsilon} \sim 10^{-3}$/sec), "high" temperature behavior initiates at ∿0.8 Tg (44). Above this temperature samples "neck", i.e. exhibit a considerable local reduction in area, prior to failure. This behavior has often been incorrectly labeled as "homogeneous" deformation (28); the standard mechanics definition holds that any localized deformation mode is inhomogeneous. The temperature regime where necking occurs is marked by a rapidly decreasing yield strength, which tends toward zero at T_g, as indicated in Fig. 17.

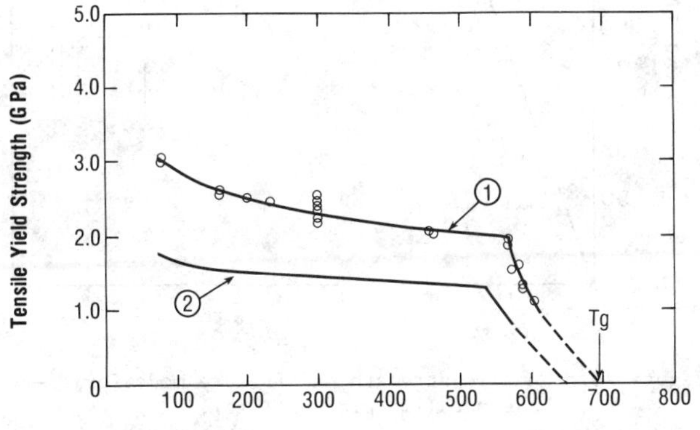

Figure 17. Tensile yield strength versus temperature: curve 1 for $Ni_{49}Fe_{29}P_{14}B_6Si_2$ glass; curve 2 for $Pd_{77.5}Cu_6Si_{16.5}$ glass from Pampillo and Chen (48). The solid portion of each curve covers the range of the measured experimental data (44).

Deformation at low temperature, $T \ll T_g$, occurs in the form of highly inhomogeneous shear bands. Such bands are readily observed on the surface of a bent ribbon (Fig. 18), on specimens deformed by cold rolling (45), on drawn wires (46), adjacent to hardness indentations (47) and on specimens deformed in uniaxial compression (48).

Figure 18. Shear offsets on the tension side of a bent metallic glass strip. Scanning electron micrograph (SEM). (55)

According to the transmission electron microscopy (TEM) observations of Masumoto and Maddin (49), bands ∿20 nm wide can sustain shear displacements of ∿200 nm, for a true shear of 10, or more. Such enormous strains are indicative of the intrinsic ductilities of metallic glasses and mark a sharp contrast with the behaviors of oxide glasses, such as SiO_2, which are brittle at $T \ll T_g$. Glassy metals are ductile because metallic bonds are nonstereospecific, allowing the angular locations of atoms to readily shift, whereas, in covalently bonded glasses, the bond angles are fixed, and bond breakage, leading to fracture, occurs preferentially.

As discussed by Cottrell (50) the occurrence of highly localized shear deformation is indicative of the absence of work hardening. If even a trace of hardening is possible the bands will rapidly broaden because deformation is easier in nonhardened material. The lack of work hardening in metallic glasses was further demonstrated by Pampillo and Chen (48). A rectangular cross section specimen of $Pd_{77.5}Cu_6Si_{16.5}$ was deformed in compression and the surface shear bands were photographed. (As noted above this alloy can be prepared in glassy form as rods 1-2 mm in diameter suitable for compression experiments.) The surfaces of the specimen were polished smooth to eliminate any stress concentrations associated with the shear offsets. On subsequent reloading the same shear bands were active, indicating the absence of hardening in the bands.

Energy is stored in the bands in the form of structural disorder, as evidenced by heat capacity measurements (51) and changes in the structure factor of the glass (52); especially, a broadened and diminished first peak, indicative of reduced long range structural correlations. It has been suggested (28) that the concentration of strain in bands, the attendant disorder and subsequent rupture through the bands (see below) imply that the bands are softened. This need not be the case. Rupture is a function of strain induced shear crack formation and would occur in an intense band, and commonly does in crystalline metals, even if it were hardened. Disorder is also a function of strain, not stress, and true stress-true strain measurements in compression (48) or tension (53) give no indication of the implied flow stress decrease due to strain softening.

Deformation Mechanism

Gilman (54) first proposed that plastic flow of glasses occurs via the motion of dislocations; due to the disordered structure the slip vectors of such dislocations would vary in magnitude and direction along the dislocation line, about a mean value indexed to the near neighbor distance in the glass. The absence of a periodic structure precludes the imaging of such dislocations by TEM diffraction contrast and in the absence of direct observation their existence remains controversial. It is becoming increasingly apparent, however, that the macroscopic deformation characteristics of metallic glasses are most sensibly rationalized as manifestations of the collective behavior of microscopic dislocations (55). We note, for example, that deformation occurs by the propagation of shear bands; the shear front can be formally described as a pile-up of microscopic dislocations. It was the observation of such shear bands that first indicated the presence of dislocations in crystalline materials (56). Also, metallic glasses are exceptionally strong, but the observed strengths (see below) are far less than the theoretical strengths predicted for homogeneous deformation (57). This discrepancy also exists, of course, for crystalline metals and is reconciled by the existence and propagation of dislocations. Morris has modeled the glass structure as a disclination-dislocation network (58); deformation at low temperatures would occur due to the glide of dislocations pinned at disclinations (rotational defects), while, at high temperature, nonconservative motion (climb) of the disclinations would also occur. Based on this model one can readily understand, for example, the lack of work hardening in glassy metals as being due to the nucleation and/or propagation of dislocations in a dislocation saturated structure.

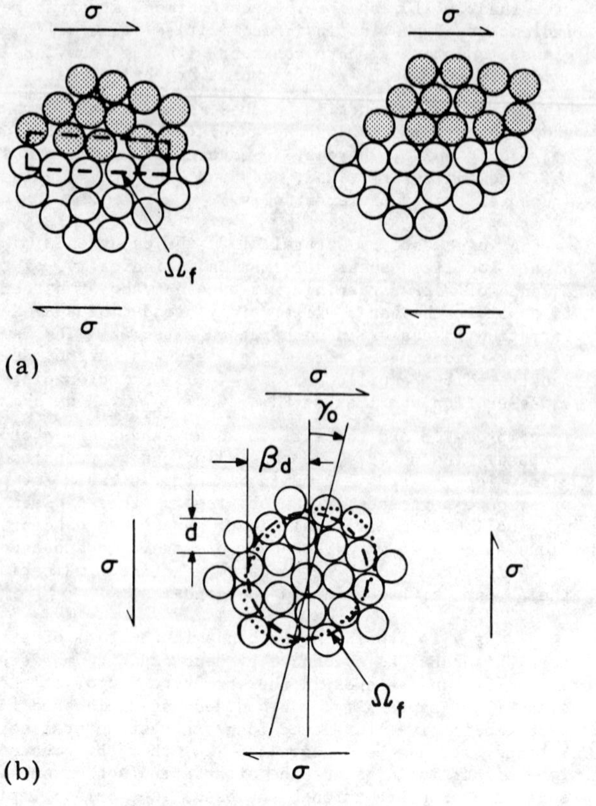

Figure 19. Idealization of the two types of shear transformations observed in sheared bubble rafts: (a) dislocation pair formation; (b) diffuse shear transformation. (60)

Argon and Kuo (59,60) have examined the deformation of a "glassy" bubble raft model, constructed by mixing soap bubbles from two sources with a size and concentration ratio of 0.75/1. A frame placed in the bubble raft imposed a controlled uniform shear strain. Macroscopic shear bands were, thereby, precluded, but overall deformation was effected through a collection of two kinds of local shear events, which could be roughly characterized as diffuse and sharp. The sharp shear event (Fig. 19a) is equivalent to the nucleation of a dislocation pair separately by ∿5 atom diameters; the analogous event in three dimensions would be the nucleation of a dislocation loop. The diffuse shear event (Fig. 19b) occurs where the local packing is somewhat looser. The shape of the active region is roughly circular with a diameter covering, again, ∿5 atoms; a spherical region would be expected in three dimensions. One can readily imagine that shear bands would form at low temperature due to a cascade of sharp shear transitions. At higher temperature the more diffuse transitions would be favored, rendering the deformation somewhat more homogeneous on the microscale. One is especially struck by the occurrence of microscopic dislocation pairs. These are, obviously, of limited extent in comparison to crystalline metals, but this is clearly not an important qualitative difference, lending further support to the belief that microscopic dislocations play a role in the deformation of metallic glasses.

Strength and Hardness

The requirement of high cooling rates to effect glass formation generally dictates that metallic glasses will be prepared as thin foils or ribbons. Strength tests are, therefore, limited to tensile tests, except in the case of Pd based glasses. Due to the absence of work hardening, the maximum tensile stress which can be achieved is the yield stress. However, most investigators have simply conducted tensile tests on uniform ribbons, which invariably fail by tearing at a stress less than the yield stress (Fig. 20). The tearing stress

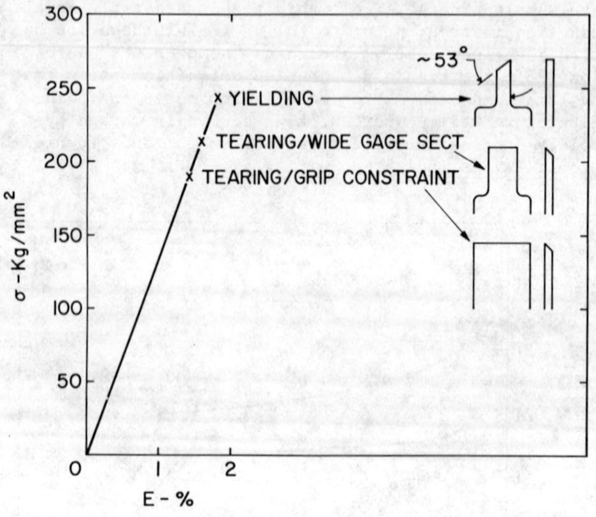

Figure 20. Tensile strength and failure mode for Ni-Fe base metallic glasses. (67)

can vary erratically depending on testing conditions, e.g. grip alignment, gripping technique, etc. If testing is conducted on a specimen with a relatively narrow reduced width gage section, premature failure can be avoided and failure will occur coincident with yielding. This latter event is signaled by the mode of failure, which occurs by shear instability through an incipient, but imperceptible (at room temperature), neck, running obliquely across the face of the specimen (Fig. 20) at an angle of ∿53° (44,61). Yield strengths so determined (at ∿295 K) are collected in Table 4. A brief inspection of this table suggests the following trends: strength increases as one moves up and to the left in the periodic table, i.e. Pd→Ni→Fe→Cr; the metalloid component plays a dominant role, Fe-B glasses being far stronger than Fe-P glasses.

Table 4

Mechanical Properties of Metallic Glasses (55)

Alloy	H GPa	σ_y GPa	H/σ_y	E GPa
$Ni_{36}Fe_{32}Cr_{14}P_{12}B_6$ (METGLAS® 2826A)	8.63	2.73 tension	3.16	141
$Ni_{49}Fe_{29}P_{14}B_6Si_2$	7.77	2.38 tension	3.26	129
$Fe_{80}P_{16}C_3B_1$	8.19	2.44 tension	3.35	
$Fe_{80}B_{20}$ (METGLAS® 2605)	10.79	3.63 tension	2.97	166
$Pd_{77.5}Cu_6Si_{16.5}$	4.88	1.54 comp.	3.17	88.0
		147 tension	3.39	
$Pd_{64}Ni_{16}P_{20}$	4.43	1.44 comp.	2.86	91.9

H and σ_y have uncertainties of the order of ±5%. Metglas® is a registered trademark of the Allied Corporation.

An interesting sidelight is the information provided by consideration of metallic glasses as model materials. Oblique necking is observed in rolled sheets of crystalline metals and its occurrence has been attributed to mechanical constraint associated with the sheet geometry (62). In axisymmetric bars necking occurs in the "normal" fashion, i.e. perpendicular to the tensile axis. In metallic glasses oblique necking can occur regardless of the sample shape (44,53); the controlling factor is that shear instabilities (failures) occur along directions of zero extension, i.e. in directions of pure shear deformation (not directions of maximum shear stress). In view of this, it becomes clear that normal necking is more the "anomaly" than oblique necking and its occurrence must be attributed to the ubiquitous existence of work hardening in crystalline materials; work hardening suppresses shear instabilities by dispersing slip, i.e. by making it more homogeneous. Even in crystalline sheets, work hardening causes the oblique neck to be slightly closer to "perpendicular" than expected {included angle of 60° rather 53°; see Fig. 20}(63,64).

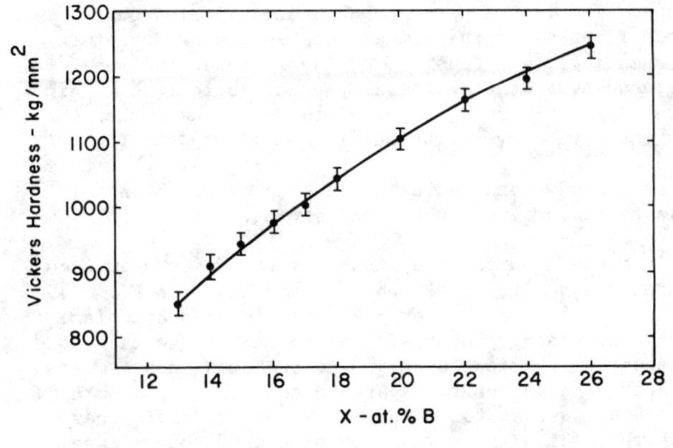

Figure 21. Vicker's diamond pyramid microhardness as a function of B content for Fe-B metallic glasses. The error bars (∿±2.5%) approximate the random variation for a given sample; absolute uncertainties are of the order of ±5%. (22)

Strength can be more conveniently sampled by hardness tests, wherein yielding occurs under triaxial constraint. As indicated by Table 4, the ratio of Vickers hardness to yield strength is ∿3.2 (65), approximately the same ratio which is observed for crystalline metals. The importance of the metal-metalloid interaction is further demonstrated (Fig. 21) by the strong increase of hardness as the B/Fe ratio is increased in Fe-B glasses (22).

Stiffness

The calculations of Weaire et al. (66) indicate that the bulk stiffnesses of amorphous (elemental) metals should be close to those in the crystalline state, whereas their shear stiffnesses should be of the order of 25%, or more, less, because the atoms are not constrained to move in a self-similar fashion in the glass, i.e. they can shift in somewhat random directions under the action of a shear stress. It is not clear how relevant these calculations are to alloys; it is especially unclear how one should choose the reference crystalline state. Nonetheless, we note, as a matter of practical interest, that the Young's modulus of $Fe_{80}B_{20}$ glass is ∿20% less than that for crystalline Fe and steel.

One may conveniently determine Young's moduli of ribbon shaped specimens using ultrasonic pulse-echo or vibrating reed techniques. A collection of such values for metallic glasses is given in (67). A complete elasticity characterization requires two (isotropic material) elastic constants, which are most commonly obtained by bulk acoustic wave technqiues. Such experiments are precluded for metallic glasses, except for alloys (Pd, Pt based) readily prepared in rod form (68). For alloys of more practical interest, e.g. Fe, Ni or Co based, one can determine the desired shear and bulk stiffnesses by calculation from Young's moduli, determined acoustically, and Poisson's ratios, determined by strain gage techniques (69,70), although this has been done for only a few glasses. Pertinent data are summarized in Fig. 22. The shear moduli (μ) of $(FeCo)_{80}B_{20}$ and $(FeNi)_{80}B_{20}$ glasses are relatively insensitive to composition (Fig. 22a), whereas the bulk stiffnesses (K) are strongly composition dependent (Fig. 22b). The shear moduli are, obviously, relatively insensitive to the near neighbor interactions which dominate K. The peak bulk moduli occur at about the same electron/atom ratio, indicating an important influence of the electronic structure on K.

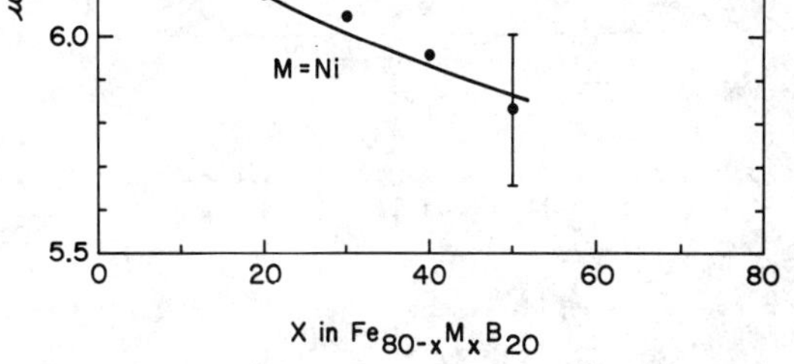

Figure 22a. Shear stiffnesses for $(FeNi)_{80}B_{20}$ and $(FeCo)_{80}B_{20}$ glasses vs Ni or Co content in atomic percent. Solid symbols are for Ni alloys, open symbols are for Co glasses. (70)

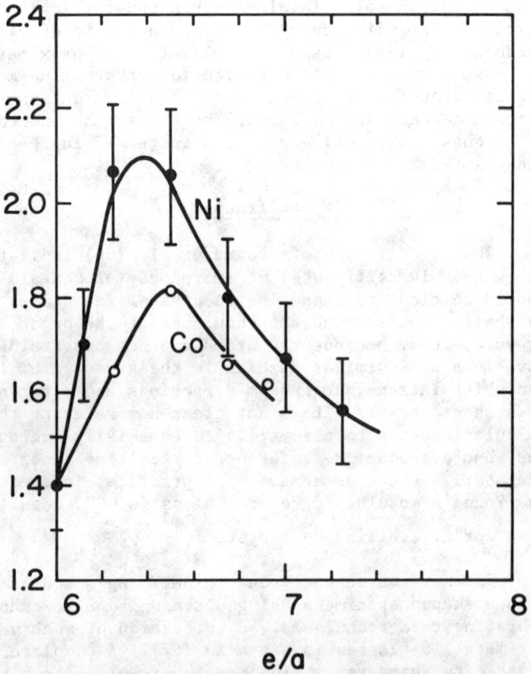

Figure 22b. Bulk stiffnesses for $(FeNi)_{80}B_{20}$ and $(FeCo)_{80}B_{20}$ glasses versus d-electron-to atom ratio. Solid symbols are for Ni glasses; open symbols are for Co glasses. (70)

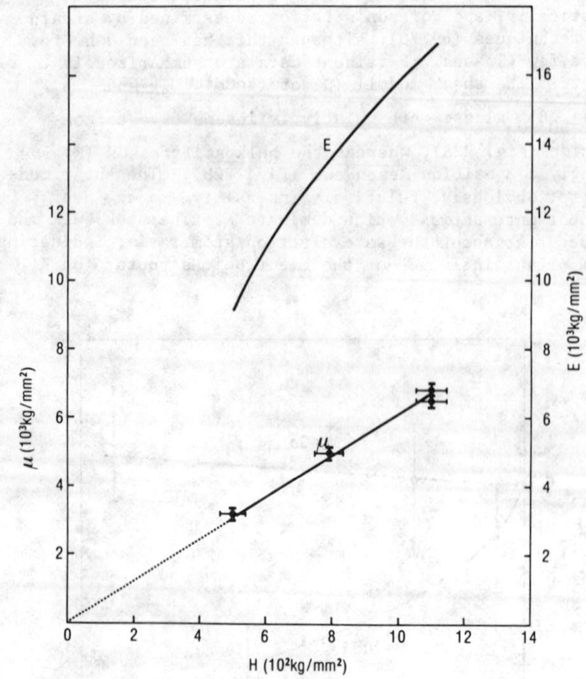

Figure 23. Linear correlation of the Vicker's hardnesses (H) and shear moduli (μ) of transition metal-metalloid glasses. (69)

For these same alloys and others for which both H and μ have been determined, the variation of microhardness tracks the variation of μ; as evidenced by Fig. 23, μ≃6.05 H (69). This indicates that the shear yield strains of TM-M glasses are all about the same, which, presumably, is a reflection of their structural similarities. This relation also is important with respect to the nature of the deformation mechanism. It has been noted that dilatation could play a role in the plastic flow process (59,71). The proportionality between H and μ indicates that this is not the case.

Based on the above relation and the relation between the yield strength and H, we find that $\sigma_y \simeq \mu/19.4$. It is worth noting that Morris' disclination-dislocation model predicts $\sigma_y = \mu/19.5$, when the disclination spacing (hence, the glide dislocation length) is taken as 5 metal atom diameters. This latter distance is observed for dislocation loop formation in the bubble raft model of Argon and Kuo and it is the experimentally observed limit of pair correlations in the RDF's of TM-M glasses.

Fracture Mode and Fracture Toughness

When metallic glasses fail coincident with yielding they do so by shear rupture through the prior shear instability zone. At $T \ll T_g$, e.g. room temperature, this zone is very narrow and the fracture surface is rather flat, though marked by features called veins (72), as shown in Fig. 24. A shear disk crack nucleates inside

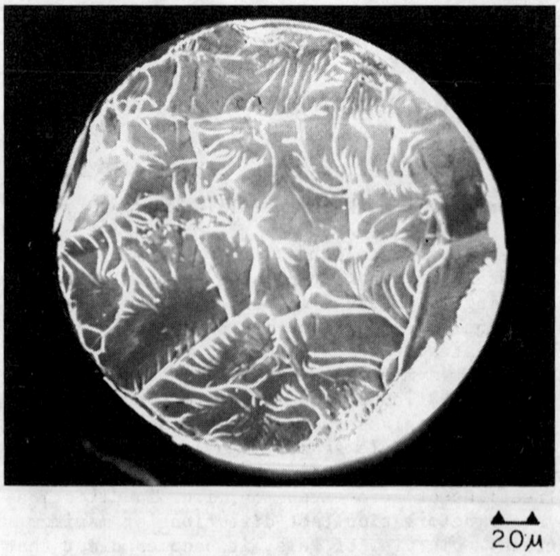

Figure 24. Fracture surface of a glassy $Pd_{77.5}Cu_6Si_{16.5}$ wire (SEM) with the direction of view parallel to the specimen axis. The smooth shear offset region is an arc to the right of the micrograph. The fracture surface is inclined at ∿45°, sloping down from left to right. (72)

each area enclosed by veins; adjacent cracks propagate and final separation occurs (locally) by necking down of the material to form the veins where the cracks impinge. The frequency of crack nucleation increases, and the fineness of the veins decreases, as temperature decreases (44). If tensile testing is conducted under superposed hydrostatic pressure, crack nucleation within the shear zone is suppressed. Rather, cracks nucleate where the periphery of the shear zone intersects the surface of the specimen (73). It should be noted that this suppression of crack nucleation increases the plastic strain to failure, not the yield stress.

If a specimen fails by tearing, the failure surface features are qualitatively the same as shown in Fig. 24. Samples may also fail under plane strain or quasi-plane strain conditions. In this case the failure surface is macroscopically oriented at 90° to the tensile axis ("square") and a "chevron" pattern appears on the surface (74). Just as for crystalline sheets, the apices of the chevrons point toward the origin of the failure. These latter failure modes are brittle, i.e. they occur prior to general yielding, and are best characterized by measurements of fracture toughness.

When the load is increased on a fatigue precracked specimen, the stress field adjacent to the crack will sample the state of the material. If inhomogeneities exist, such that voids can be nucleated under the action of triaxial stress ahead of the main crack, a square (plane strain) fracture will occur. If such inhomogeneities do not exist, the stress will increase further until failure occurs by tearing. Hence, for the same crack length the tearing mode is the higher fracture toughness mode. (In the simplest case toughness, $K_Q = \sigma_f \sqrt{\pi a}$, where σ_f is the nominal fracture stress calculated from the failure load and the uncracked area of the specimen and a is the half length of the preexisting crack.) Toughness values observed for metallic glasses range from ∿10 MN/m$^{3/2}$ for plane strain failure to ∿50 MN/m$^{3/2}$ for tearing failures (75). Samples of the same alloy can exhibit values anywhere in this range (mixed mode failures) depending on sample condition. In principle a metallic glass is an elastic continuum and should always fail by tearing from an existing crack, regardless of ribbon thickness. The fact that square failures do occur indicates the presence of structural inhomogoeneities, which can be gross, such as preexisting voids or isolated crystallites or subtle, such as, perhaps, fine scale density fluctuations resulting from thermal annealing (55).

The fracture toughnesses observed are about two orders of magnitude larger than those for silica glass (76). Accordingly, for the same flaw size the failure stresses for metallic glasses will be 10^2 times larger; or to achieve the same failure stress the flaws in the silica glass must be 10^4 times smaller, a clear indication of its intrinsic brittleness. The plane strain toughness values for two metallic glasses are compared with those for steels in Fig. 25. On an absolute scale the values are low; as suggested by the figure, however, they are, at least, consistent with the exceptionally high yield strengths of the glasses. Tearing toughnesses, though up to five times larger than plane strain toughnesses, are still relatively low compared to the toughnesses of steels, especially soft steels. This may seem surprising, in view of the "enormous" shear ductilities cited above; low toughnesses derive from the fact that these large shear strains are so intensely localized.

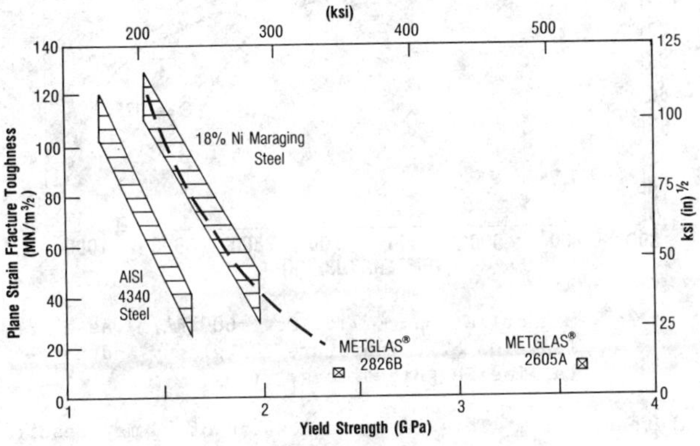

Figure 25. Plane-strain fracture toughness vs yield stress for ferrous materials. (55)

Fatigue

Figure 26 shows a collection of tension-tension fatigue data for Ni-Fe base metallic glasses (77). The ordinate is the stress ratio $S = \sigma_a/(\sigma_y - \sigma_m)$, where the stress amplitude $\sigma_a = (\sigma_{max} - \sigma_{min})/2$ and σ_{max} and σ_{min} are the maximum and minimum cyclic stresses, σ_y = the true tensile strength, which equals the yield strength and σ_m = the mean stress = $(\sigma_{max} + \sigma_{min})/2$. The upper dotted line, emanating from S = 1.0, is comparison data for steels. For a detailed discussion of the data shown, the reader is referred to references (55) and (77). We simply note here that, depending on sample condition, the stress ratio S_L for infinite lifetime ($N_f \gtrsim 10^6$ cycles) can be as low as 0.1. Higher values of S_L can be achieved, somewhat artificially, when the mean stress is set high and the stress amplitude is kept low. A more normal situation, however, would be zero-tension testing; in this case for $S_L \simeq 0.1$ the maximum permissible tensile stress will be ∿0.18 σ_y. The corresponding peak value for steel would be ∿0.6 times the tensile strength. This considerable departure is once again thought to reflect different work hardening characteristics. Fatigue lifetimes in steel are commonly about 90% consumed by the crack inititation step (78); crack growth to the critical flaw size takes the remaining time. Crack nucleation is relatively difficult because work hardening disperses localized slip. In the absence of work hardening, as in metallic glasses, cracks can nucleate rather easily by localized slip at small discontinuities and the lifetime will be short {crack propogation rates are comparable to steels (77)}, unless the stress amplitude is kept rather low. Such small discontinuities can be avoided, in principle, but appear unavoidable in practice.

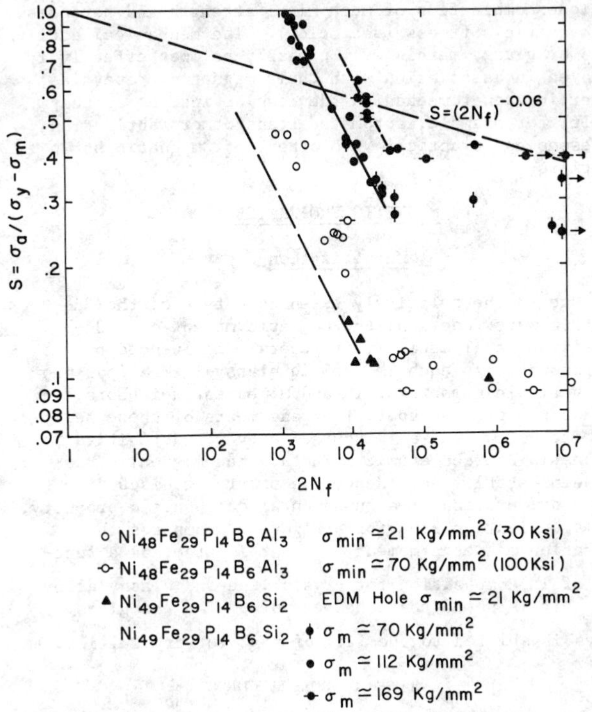

Figure 26. Stress ratio vs reversals to failure for Ni-Fe-P-B metallic glasses (77). The dotted line given by $S = (2N_f)^{-0.06}$ is the upper limit of fatigue lifetimes observed for high-strength steels.

Endurance limits can be improved (S_L from 0.1 →0.3) if gross external stress concentrations, such as holes, notches or grip constraints, are avoided. For S_L = 0.3, the endurance ratio = $\sigma_{max}/\sigma_y \simeq 0.46$ (σ_{min} = 0). It was hoped that these conditions would apply for metallic glass ribbon reinforced epoxy laminates. Unfortunately, endurance ratios ($N_f \gtrsim 10^6$) of only ∿0.15 are observed, i.e. the peak cyclic stress is ∿0.15 of the static strength of the laminate (79). This compares rather unfavorably with a ratio of ∿0.7 for graphite/epoxy laminates (80).

To summarize: "High temperature" mechanical behavior occurs above 0.5-0.8 T_g. Apparent viscosities determined from creep tests are orders of magnitude smaller than the expected equilibrium viscosities. The activation energy for isoconfigurational viscous flow (∿2 eV for $Pd_{80}Si_{20}$) is approximately the same as that for diffusion. Above 0.8 T_g, specimens deformed at constant extension rate exhibit diffuse necking at the yield stress; at lower temperatures tensile yielding (incipient necking) occurs coincident with shear instability. At $T \ll T_g$ deformation occurs in the form of extremely intense shear bands, indicative of an absence of work hardening. The deformation characteristics of metallic glasses, in general, appear to be readily understandable in terms of the existence of atomic scale dislocations. The yield stresses of metallic glasses are large (up to 3.63 GN/m^2, or more); they scale as ∿H/3.2 and ∿μ/19.4. Preyield fracture is characterized by fracture toughnesses of the order of 10 to 50 $MN/m^{3/2}$. These values are much higher than those for oxide glasses, as a result of the capability of metallic glasses to sustain plastic flow, but relatively low compared to those for crystalline metals such as steel, because the shear flow which occurs does so in such intensely localized fashion. Fatigue lifetimes at a given stress and the endurance limit (stress) for infinite cyclic lifetime tend to be low, because the absence of workhardening favors easy crack initiation. Metallic glasses possess a unique combination of mechanical strength and hardness and magnetic softness (see below). The mechanical attributes of glassy metals will probably be most effectively employed in association with their magnetic properties; either in magnetomechanical devices or insofar as they permit, for example, automated transformer fabrication processes not practical with other soft magnetic strip materials.

MAGNETIC PROPERTIES

Magnetization

Due to the relatively extended nature of the 3d-electron wave function, the magnetic moment of a 3d-transition metal atom is, in general, influenced by its environment. As each TM atom in glassy TM-M alloys may have variable numbers of TM and M nearest neighbors, and, as the precise spatial arrangements of these neighbors are not known, as discussed above, a definitive relationship between atomic structure and magnetism has not been established. Hence, we start our discussion with consideration of a fundamental macroscopic property, the magnetization behavior. Figure 27 compares the saturation magnetic moment, μ_s, per TM atom, as a funcrion of TM composition for crystalline and glassy alloys (81). It is noted that the peak of μ_s for the glassy alloys is shifted to the left of that for crystalline

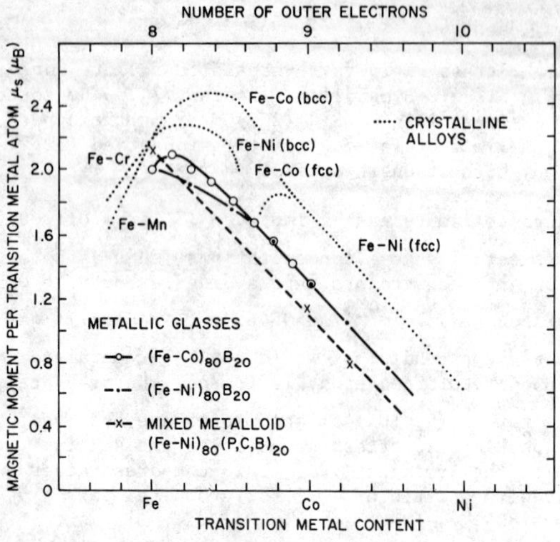

Figure 27. Saturation magnetic moment per TM atom for glassy and crystalline alloys. (81)

alloys. As the data are plotted as a function of TM composition, or number of outer (4s+3d) electrons per TM, the observed shift implies electron charge transfer from M atoms to TM atoms in glassy alloys. These data are consistent with the trends of Mössbauer isomer shifts (I.S.) and nuclear Knight shifts (K). For example, I.S. increases from 0.07 mm/sec for $Fe_{80}B_{20}$ to 0.14 mm/sec for $Fe_{80}P_{16}C_3B_1$ (81). A larger I.S. implies a larger TM d-electron concentration and, hence, these results further imply that M atoms with more outer sp electrons (e.g. P) donate more electrons to the d-band of TM atoms. The positive shift of K for ^{195}Pt in glassy Ni-Pd-P, to -0.2% from -3.5% for ^{195}Pt in pure Pt, also suggests that the d-bands of Pt atoms tend to be filled by the outer electrons of P atoms (82). Hence a simple charge transfer (rigid band) mechanism appears to qualitatively explain the observed trends of the atomic magnetic moments. However, the precise details of the mechanism are uncertain at present.

One of the more interesting properties of Fe-base alloys is the reduction of the ferromagnetic Curie temperature θ_f in the glassy state. An extreme example is shown in Fig. 28; θ_f = 878 K for crystalline (single phase) $Fe_{88}B_{12}$, as compared to 509 K for its noncrystalline

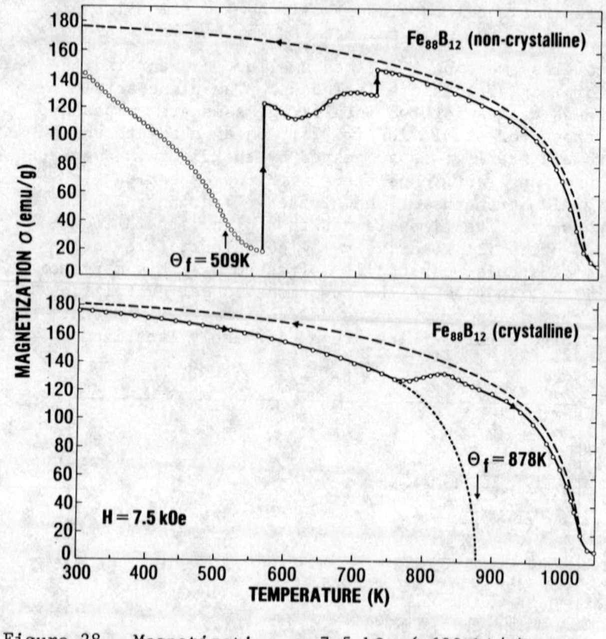

Figure 28. Magnetization at 7.5 kOe (=600 kA/m) as a function of temperature for glassy and crystalline $Fe_{88}B_{12}$ alloys. (83)

counterpart (83). This large depression of θ_f may result from a change in local structure from dense-random-packing to fcc-like (84) or bcc-like arrangements (83). The former interpretation assumes a drastic change in the exchange interactions and the latter a change in the number of magnetic nearest neighbors without modification of the interatomic exchange interactions. For an unambiguous interpretation, more precise structural determinations would be required. Iron base alloys of practical interest (see below) exhibit Curie temperatures of the order of 750 K. In addition to the reduction of θ_f, it has been found that the rate of decrease of the magnetization, ΔM(T)/M(0), with increasing T from T = 0 K obeys Bloch's $T^{3/2}$ - law and is faster by a factor of 5-10 in Fe-base glasses than in the corresponding crystalline alloys (85). Examples are shown in Fig. 29 (86). The applicability of Bloch's law indicates that the magnetization in glassy ferromagnets towards T = 0 K is governed by spin wave excitations as in crystalline ferromagnets. Furthermore, as indicated in Fig. 29, the $\Delta M(T)/M(0) = -b\ T^{3/2}$ -- dependence persists up to $T/\theta_f = 0.3$ in glassy alloys in

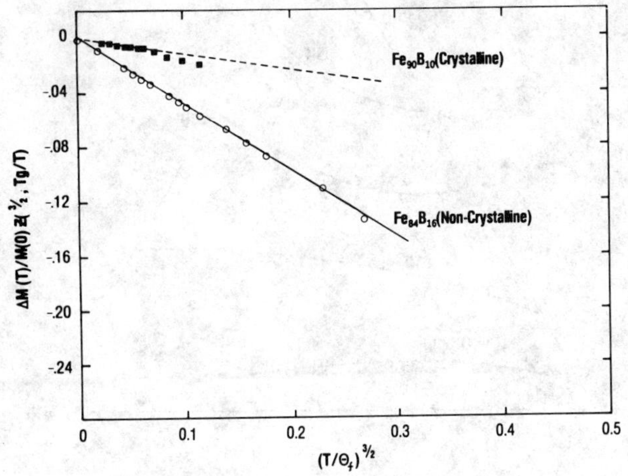

Figure 29. Change of the magnetization as a function of $(T/\theta_f)^{3/2}$ for crystalline $Fe_{90}B_{10}$ and glassy $Fe_{84}B_{16}$ alloys. (86) Z is defined in Ref. 86.

contrast to $T/\theta_f = 0.15$ for crystalline ferromagnets. The quantity b is related to the spin wave stiffness constant D through $b = 2.612\ [g\mu_B/M(0)]\cdot[k_B/4\pi D]^{3/2}$ where g is the g factor ($\sim$2), μ_B, the Bohr magneton and k_B, Boltzmann's constant. We thus find $D/\theta_f = 0.27$ and 0.12 meVA^{2}K^{-1} for crystalline and glassy Fe-B alloys, respectively (86). Smaller D/θ_f ratios imply shorter ranged exchange interactions in the glasses. While this comparison seems to be generally true for iron-base alloys, it does not hold for Co-base alloys. Using reported values of D (87,88) and θ_f, we obtain $D/\theta_f = 0.25$ and 0.29 meVA^{2}K^{-1} for crystalline cobalt and glassy $Co_{74}B_{26}$, respectively. These results are consistent with the findings that atomic magnetic moments, Curie temperatures and hyperfine field distributions are similar for crystalline and glassy Co-base alloys (89,90).

Extrapolation of the μ_s vs. TM composition curve for glassy alloys (Fig. 27) toward TM=Ni gives $\mu_s \sim 0$ for

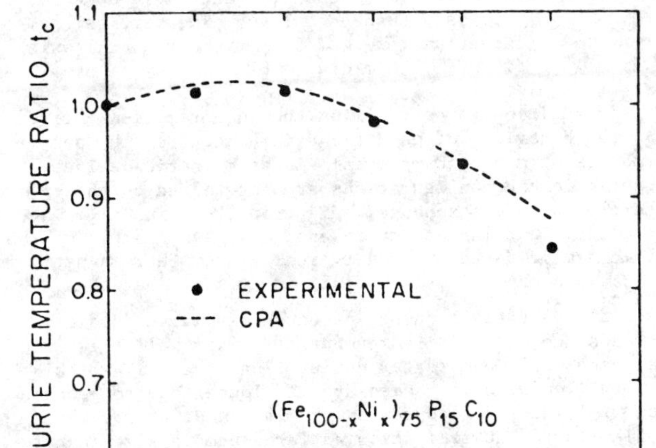

Figure 30. Reduced Curie temperature $t_c = \theta_f(x)/\theta_f(0)$ for glassy $(Fe_{100-x}Ni_x)_{75}P_{15}C_{10}$ alloys. The value for $\theta_f(0)$ is taken as 585 K. The abscissa denotes values of x. (91)

$Ni_{80}B_{20}$. It is also found that the experimental Curie temperature versus x behavior for $(Fe_{100-x}Ni_x)_{75}P_{15}C_{10}$ alloys can be fitted very well by results obtained by a coherent potential approximation (CPA) assuming reduced exchange constants of $J_{Fe-Fe} = 1$, $J_{Fe-Ni} = 1.2$ and $J_{Ni-Ni} \sim 0$, as shown in Fig. 30 (91). Hence, these data suggest that Ni is nonmagnetic in some TM-M base glassy alloys, as assumed by some investigators (92). However, this is not certain; recent studies combining NMR and magnetic susceptibility data on glassy Ni-P, Ni-Pd-P and Ni-Pt-P indicate that Ni tends to be paramagnetic with a high density of states at the Fermi level (93).

Magnetic Anisotropy

If magnetic atoms are randomly distributed, one would expect no bulk magnetic anisotropy in a non-crystalline ferromagnet, in the absence of shape anisotropy. This situation can be experimentally realized in a ferromagnetic glassy foil, made by quenching a liquid droplet between a piston and anvil (94); such a foil exhibits an in-plane anisotropy energy less than $\sim$50 J/m^3 (95). Since one can observe magnetic domains in the foil (96), the local magnetic anisotropy energy must be substantially larger.

Continuous chill-casting techniques, as applied to form glassy alloy ribbons, appear to introduce uniaxial magnetic anisotropy slightly oblique to the ribbon axis (97). The anisotropy energy, $K_u \sim 10^3$ J/m^3, is smaller than that ($\sim 10^4$ J/m^3) induced in sputtered amorphous Gd-Co-Mo films (98), but is substantial, nonetheless. It is also found that magnetic field-induced uniaxial anisotropy can reach the order of 10^3 J/m^3 in melt quenched glassy alloys. Uniaxial anisotropy is induced in order to generate a "square" B-H loop (see Fig. 34, below) and hence to minimize core losses and exciting powers (see below). Figure 31 represents the maximum induced K_u values for the Fe-B system (99). It has been speculated that the induced anisotropy originates from atomic pair ordering.

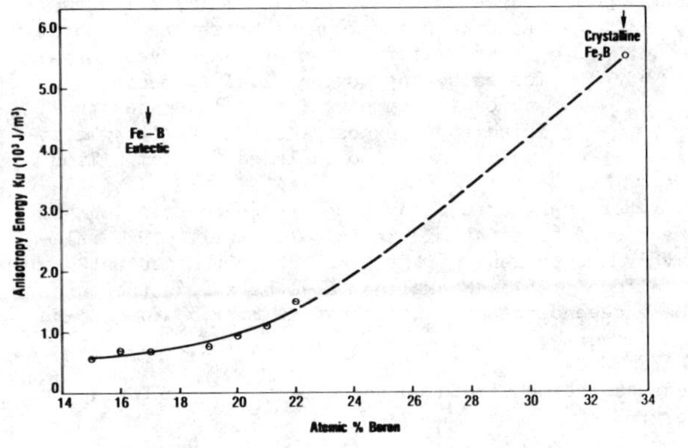

Figure 31. Variation of anisotropy energy vs boron content in Fe-B glassy alloys. (99)

Magnetostriction

A phenomenon closely related to magnetic anisotropy is magnetostriction, a magneto-mechanical effect. In an effort to obtain information about the nature of the mechanism of magnetostriction, experimental results have been discussed (100) in terms of the theory developed by Callen and Callen (101). Their theory describes the temperature dependence of the magnetostrictive strains in terms of thermal averages of single-ion and two-ion spin correlations functions. As shown in Fig. 32, data for three representative glassy alloys follow the curve based on single-ion anisotropy of uniaxial symmetry, $I_{5/2}$

(X), over a wide range of magnetization, m(T). The behaviors of alloys dominated by Co, however, can be described in terms of the two-ion anisotropic exchange of uniaxial symmetry $[m(T)]^2$. Figure 32 suggests that the linear saturation magnetostriction value, λ_s, should scale with the value of the saturation magnetization. Such a correlation has been found for a number of Fe-base glassy alloys, as reported in Ref. 102. While some deviations from this general trend are observed in glassy Fe-Be-B (103) and low-boron Fe-Ni-B (104) alloys, toward lower values of λ_s, no Fe-rich alloy has been identified which exhibits $\lambda_s = 0$. Soft magnetic properties are improved (lower core loss, higher permeability) when magneto-elastic effects are eliminated ($\lambda_s = 0$).

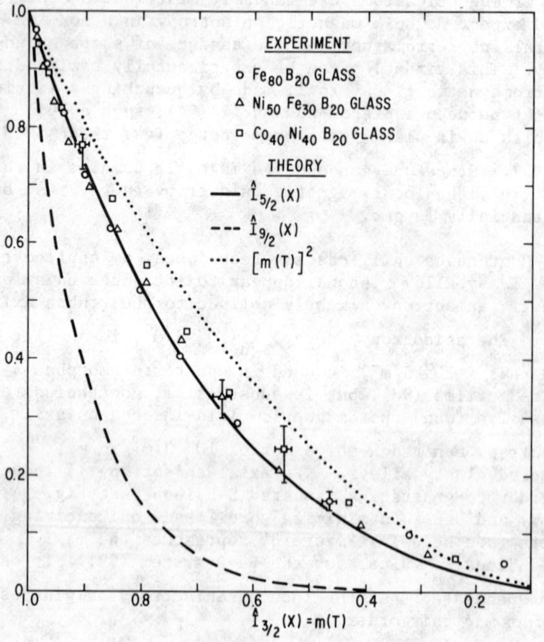

Figure 32. Reduced magnetostriction $\lambda_s(T)/\lambda_s(0)$ vs reduced magnetization for three representative glassy alloys. Theoretical curves for single ion anisotropy of uniaxial symmetry $\hat{I}_{5/2}$ (x) and cubic symmetry $\hat{I}_{9/2}$ (x) and for two-ion anisotropic exchange of uniaxial symmetry $[m(T)]^2$ are also included for comparison. (100)

For Co-Fe base alloys, zero magnetostriction has been found at an atomic ratio Co/(Co+Fe) ~0.94, relatively independent of the type and/or the content of the metalloid (105). Nickel may also be added; the ternary phase dependence of λ_s is shown in Fig. 33 for $(Co\text{-}Fe\text{-}Ni)_{78}Si_8B_{14}$ alloys (106). Again the type and/or the content of the metalloid do not affect the diagram appreciably.

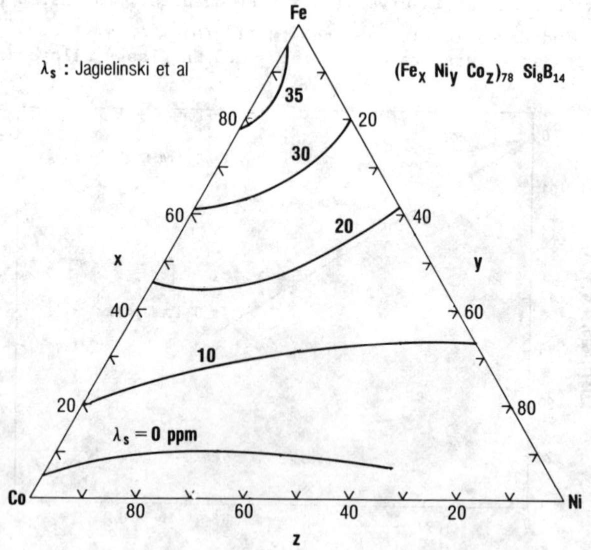

Figure 33. Saturation magnetostriction in glassy $(Co\text{-}Fe\text{-}Ni)_{78}B_{14}Si_8$ alloys. (106)

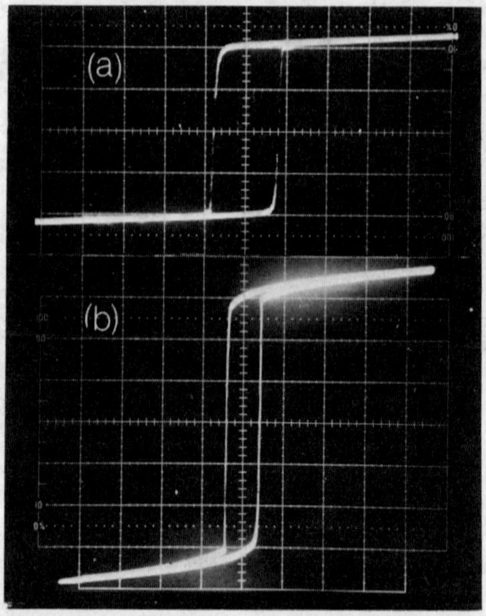

Figure 34. B-H loops for as-cast (a) and field-annealed (b) glassy $Fe_{80}B_{20}$ alloys. The vertical axis (induction): 0.4 T/div; the horizontal axis (field): 8 A/m/div. (81)

Magnetization Processes and Domains

A technically important magnetic property of glassy ferromagnets is magnetic softness. To be especially noted are their extremely low coercivities; $H_c \lesssim 8$ A/m can be achieved in as-cast materials. This follows due to the absence of obstacles such as grain boundaries, which act as domain wall pinning centers. The value of H_c can be further reduced to ~2 A/m by annealing. An example is shown in Fig. 34, in which B-H loops of as-cast and field annealed $Fe_{80}B_{20}$ are compared. The reduction of H_c (loop half-width) is significant because it leads to reduced hysteresis losses when the material is used for cores in magnetic devices.

In addition to controlling the magnetization (Fig. 34), the behaviors of magnetic domains control the performances of magnetic materials in ac magnetic devices. Magnetic domain configurations are determined by the interplay between exchange, anisotropy, domain wall, magnetoelastic and magnetostatic energies. An example of the static domain pattern for a typical glassy ferromagnet, $Fe_{80}B_{20}$, is shown in Fig. 35; it is evident that the material has a small number of unusually large domains. The presence of a limited number of domain walls requires that each wall move faster during flux reversal, which, in turn, leads to increased dynamic losses due to generation of larger eddy currents. While not an appreciable problem for 60 Hz excitation, a large domain size is not favorable for minimum core losses at high frequency ($\gtrsim$ 1 kHz) excitation.

For a given domain structure the eddy current losses in metallic glasses are reduced as compared to crystalline metallic magnetic materials, because the glasses have higher resistivities (ρ~150 μΩ-cm, as compared to ~50 μΩ-cm for crystalline metals).

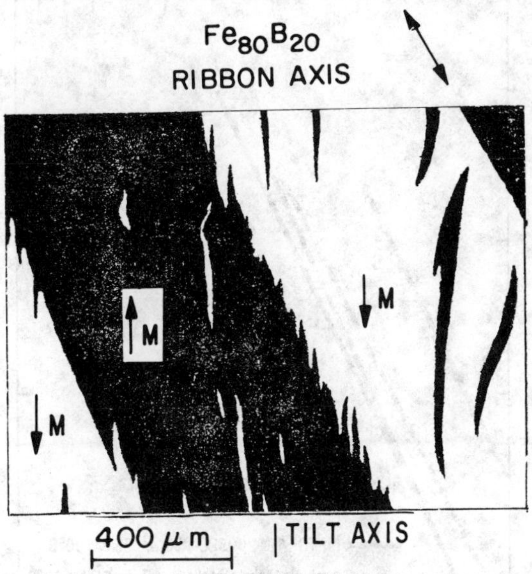

Figure 35. Domain configuration of as-cast $Fe_{80}B_{20}$ ribbon observed by scanning electron-microscopy. (81)

Low Frequency Properties

Hysteresis loss and saturation magnetic induction, B_s, are two major properties to be considered in selecting core materials for devices. Hysteresis loss is the major source of loss when a magnetic device is operated at low frequency. Higher operating flux values mean physically smaller devices, resulting in lower copper losses (lower joule heating). The unusually low values of H_c attainable in metallic glasses make these materials very attractive for 50-60 Hz distribution transformer cores. A disadvantage of glassy alloys for such applications is their lower room temperature saturation inductions, B_s(RT), as compared to competing materials, such as silicon steel. Efforts to increase B_s(RT) have been extensive (107) and some relative success has been achieved. For example, one can attain $B_s(RT) \simeq 1.8$ T in $Fe_{67}Co_{18}B_{15}$ (108); this compares with $B_s(RT) \sim 2$ T for grain-oriented silicon steel. For obvious economic reasons, however, Co-containing alloys cannot be considered for wide usage. Major efforts, therefore, have been directed toward development of high induction Fe base alloys. These efforts have focused on the Fe-B and Fe-B-Si systems. An alloy identified early on was $Fe_{82}B_{12}Si_6$ (Metglas® 2605S), which showed an ac core loss at B = 1.6 T of about 0.24 W/kg (108). This value is about 1/5 that exhibited by Si-steel. Unfortunately this alloy proved difficult to fabricate reliably. Ribbon samples commonly had trace crystallinities on the surface adjacent to the heat-extracting substrate (109). Auger electron spectroscopy studies indicated that this crystallization occurs in boron-deficient regions just below the ribbon surface; such regions have lower crystallization temperatures (83). Heat-treatment response (to induce uniaxial anisotropy) is very poor when the as-quenched material contains crystallites.

The above noted difficulties were overcome by increasing the boron content slightly and adding carbon at a ratio of about Si/C ∿2 (109,110). An optimum composition is $Fe_{81}B_{13.5}Si_{3.5}C_2$ (Metglas® 2605SC). The increased B/Fe ratio is desirable because it increases the crystallization temperature; addition of carbon may inhibit loss of B by oxidation at the surface (109); furthermore, this composition is a near-eutectic composition. Minimum H_c values are observed for eutectic alloys in the ternary Fe-B-Si system (111). The ac core losses, L, and exciting powers, P_e, for 2605SC are compared in Fig. 36 with those for Si-steel. The quantity P_e (f,B) is the power required to attain a given flux density B at a given frequency f, and is therefore an important factor in designing transformers. The glassy alloy exhibits L and P_e so much lower than those for Si-steel that it will probably replace the latter alloy in many applications, in spite of its lower B_s (∿1.6 T).

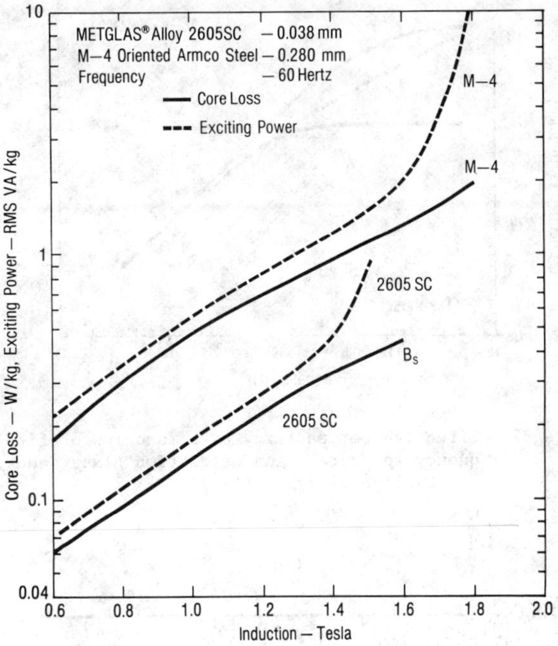

Figure 36. AC core losses & exciting powers for a glassy alloy (Metglas® 2605SC) and crystalline M-4 oriented Armco steel. (109)

High Frequency Properties

Although high induction level is desirable, high permeability μ(f,B), low ac core loss L(f,B) and low exciting power P_e(f,B) are more critical for many high frequency magnetic core applications. For these purposes, metallic permalloys and supermalloys [nominally 80%Ni-20%Fe, with zero magnetostriction] (112) and ferrites, which are insulators, are widely used. Ferrites are commonly used at f $\gtrsim$ 100 kHz where eddy current losses in metallic alloys are excessive; the metallic alloys are preferred at 1 kHz $\leq$ f $\leq$ 100 kHz, where their core losses are lower and permeabilities higher than those for ferrites.

Attempts to obtain glassy alloys with properties similar to those of 4-79 Mo permalloy have met with some success. For example, field-annealed $Fe_{40}Ni_{38}Mo_4B_{18}$ exhibits B_s = 0.88 T, L(10 kHz, 0.1 T) = 1.9 W/kg and μ(500 kHz, 0.01 T) = 700, which compare with L = 1.3 W/kg and μ∿100 for 4-79 Mo permalloy under the same conditions (113). The glassy alloy has a higher value of magnetostriction ($\lambda_s \sim 9 \times 10^{-6}$); it is expected that zero-magnetostriction glassy alloys (see Fig. 33) should exhibit properties comparable to those of the supermalloys. This is indeed the case, as shown in Fig. 37, in which effective permeabilities are compared as a function of frequency for zero-magnetostriction glassy and crystalline alloys (114).

During the course of optimizing the high frequency properties of Metglas® 2605SC, it was found that cores annealed at temperatures near 400°C exhibited better high frequency properties than those annealed at lower temperatures (∿350°C) (115). This has been related to the appearance of small crystallites which act as nucleation sites for domain walls and thereby increase the number of magnetic domains, resulting in a reduction of eddy current losses (109). Some of the high frequency properties obtained through this effort are shown in Fig. 38. It appears that Fe-rich glassy alloy containing small amounts of α-Fe crystallites performs better than any existing crystalline material at f ∿ 50 kHz and B ≳ 0.2 T.

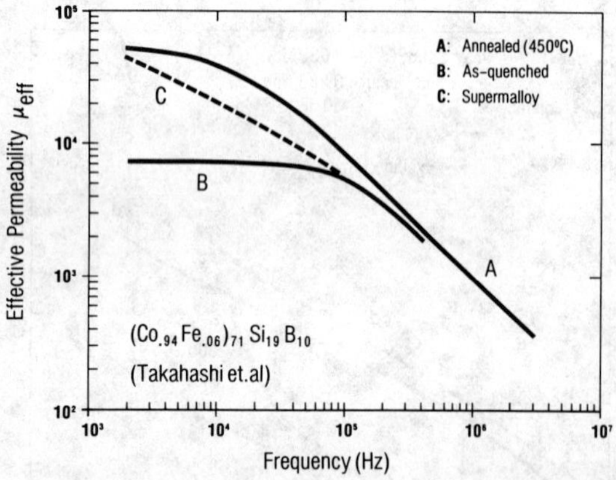

Figure 37. Effective permeability as a function of frequency for zero magnetostriction glassy and crystalline alloys. (114)

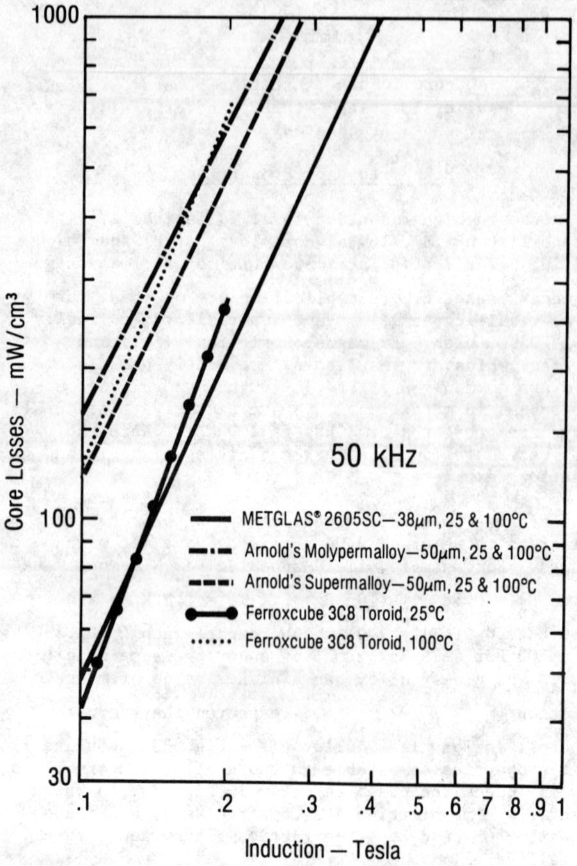

Figure 38a. AC core losses for Metglas® 2605SC & crystalline soft magnetic materials.

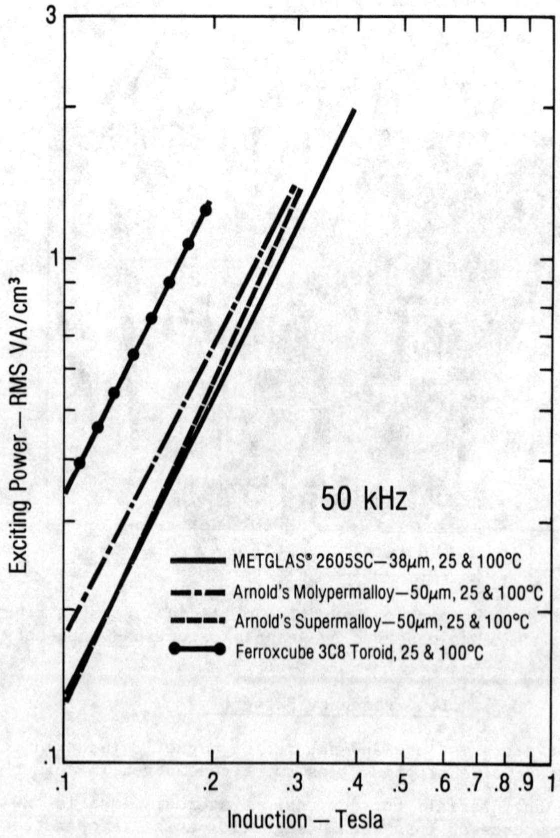

Figure 38b. AC exciting powers for Metglas® 2605SC and crystalline soft magnetic materials. (115)

To summarize: Metallic glasses exhibit saturation magnetic moments at 0 K (per transition metal atom) systematically lower than those for the corresponding crystalline alloys, due, apparently, to charge transfer of sp electrons from the metalloids into the unfilled TM d-electrons bands. Curie temperatures are also lower in Fe-rich metallic glasses, although they are sufficient (of the order of 750 K) for practical applications. The spin wave stiffnesses are less in glassy Fe-base alloys than in crystalline alloys; hence, for glasses, the decrease of the magnetization towards zero at the Curie temperature tends to be somewhat more rapid at intermediate temperatures. A combination of these three factors places the saturation inductions of metallic glasses somewhat lower than those of similar crystalline materials (e.g. 1.5-1.6 T for Fe base glasses vs. 1.97 T for crystalline 3 wt.% Si-Fe) in the temperature region of technological interest (RT to ∿100°C). Though nominally isotropic materials, metallic glasses can sustain field induced uniaxial magnetic anisotropy, apparently due to magnetic pair ordering. This anisotropy favors easy magnetization in a particular direction and is required for optimum soft magnetic properties (low core losses, high permeabilities). Zero magnetostriction glasses, rich in Co, have been identified; elimination of magnetoelastic effects promote easy magnetization. Metallic glasses are characterized by extremely low coercivities; these follow in the absence of grain boundaries, second phase particles, etc., which act as domain wall pinning sites. As a consequence, very low core losses can be obtained for low frequency (e.g. 50-60 Hz) excitations. Losses for Fe base alloys are 1/3 to 1/5 those of Si-Fe and, hence, in spite of their lower saturation inductions, glasses will replace this material in power transformer applications. Metallic glasses exhibit resistivities ∿3 times those of crystalline metallic ferromagnetic alloys. This favors lower eddy current losses associated with moving domain walls. High frequency properties (at <100 kHz) superior to those for crystalline materials may be achieved, provided that domain configurations are sufficiently refined that individual domain walls are not required to move at excessive speeds. Domain refinement is controlled by partial crystallization of the glass, the crystallites acting as domain wall nucleation sites.

CONCLUSION

Metallic glasses have proven to be and continue to be fascinating objects of scientific investigation. Extensive studies have been conducted on subjects not considered above; for example, on corrosion behavior (116), low temperature specific heat (117), resistivity (118), superconductivity (119), electronic structure (119), and so on. We chose to consider mechanical and magnetic properties because they are of the most technological interest; it is, of course, of general importance to understand the principles of formation and structure of these materials so that, for any property, processing-property-structure relationships can be established. To some degree, scientific curiosity fuels the expanding interest in metallic glasses. The key driving force, however, is the expectation that these materials will be used in huge volumes to replace Si-Fe and other soft magnetic materials in power and electronic applications. Towards that end, major development programs are going forward worldwide. The technological obstacles to the large scale processing of metallic glasses are formidable ones. Nonetheless, there are enormous driving forces to overcome them, provided by opportunities for improved performances; especially, energy savings.

REFERENCES

1) A. Brenner and G. Riddell, J. Res. Natl. Bur. Stds. 39, 385 (1947)
2) W. Klement, R. H. Willens and P. Duwez, Nature 187, 869 (1960)
3) P. Duwez and R. H. Willens Trans AIME 227, 362 (1963)
4) F. Spaepen and D. Turnbull, in "Rapidly Quenched Metals", Sec I (N.J. Grant and B.C. Giessen, eds) MIT Press, Cambridge (1976), p. 205
5) H. S. Chen, Rep. Prog. Phys. 43, 353 (1980)
6) D. E. Polk and B. C. Giessen, in "Metallic Glasses" (J.J. Gilman and H.J. Leamy eds) ASM, Metals Park, OH (1976), p. 1
7) H.A. Davies, in Rapidly Quenched Metals III, Vol. 1 (B. Cantor ed.) The Metals Soc., London (1978), p. 1
8) N. DeCristofaro, A. Freilich, G.E. Fish, submitted to J. Mat. Sci.
9) M.R. Bennett and J.G. Wright, Phys. Stat. Sol. (a) 13, 135 (1972)
10) D. Turnbull, J. de Phys., Colloque 4 p. 1 (1974)
11) J.J. Gilman, Science 208, 856 (1980)
12) B.C. Giessen and S. Whang, J. de Phys. Colloque C8 41 C8-95 (1980)
13) R. Ray, B.C. Giessen and N.J. Grant, Scripta Met. 2, 357 (1968)
14a) A. Calka, H. Matyja, D.E. Polk, B.C. Giessen, J. van der Sande, and M. Madhava, Scripta Met. 11 65 (1977)
14b) B.C. Giessen J. Hong, L. Kabacoff, D.E. Polk, R. Raman and R. St. Amand, Rapidly Quenched Metals III (B. Cantor, ed.) Metals Soc. London (1978), p. 249
15) L.E. Tanner and R. Ray, Scripta Met. 11, 783 (1977)
16) D. Turnbull, Trans. AIME 221, 422 (1961)
17) H.S. Chen and D. Turnbull, J. Chem. Phys. 48 2560 (1968)
18) D.E. Polk and D. Turnbull, Acta Met. 20, 493 (1972)
19) H.S. Chen, L.C. Kemmerling, J.M. Poate and W.L. Brown, Appl. Phys. Lett. 32, 461 (1978)
20) G.S. Cargill, III, "Solid State Physics", 30, 227 (1975)
21) H.S. Chen and B.K. Park, Acta Met. 21, 393 (1973)
22) R. Ray, R. Hasegawa, C.P. Chou and L.A. Davis, Scripta Met. 11 973 (1977)
23a) J.F. Sadoc and J. Dixmier, Mat. Sci. Eng. 23, 187 (1976)
b) J. Dixmier and J.F. Sadoc in Metallic Glasses (J.J. Gilman and H.J. Leamy, eds.), ASM Metals Park, OH (1978), p. 97
24) G.S. Cargill, III, J. Appl. Phys. 41, 12 (1970)
25) D.S. Boudreaux, in "The Magnetic, Chemical and Structural Properties of Glassy Metallic Alloys", (R. Hasegawa, ed.), CRC Press, Boca Rato, FL (1981) in press
26) Y. Waseda, H. Okasaki, M. Naka and T. Masumoto, Sci. Rep. Res. Inst. Tohoku Univ. 26A, 12 (1976)
27) Y. Waseda, H. Okasaki and T. Masumoto, Sci. Rep. Inst. Tohoku Univ. 26A, 202 (1977)
28) F. Spaepen, in "Rapidly Quenched Metals III", (B. Cantor, ed.) The Metal Soc., London (1978), V. 2, p. 253
29) J.D. Bernal, Nature 183, 141 (1959); 185 68 (1960)
30) J.L. Finney, Proc. Roy. Soc. A319, 479 (1970); 495 (1970)
31) G.S. Cargill, III, J. Appl. Phys. 41, 2249 (1970)
32) D.E. Polk, Scripta Met. 4, 117 (1970)
33) J.D. Bernal, Proc. Roy. Soc. A280, 299 (1964)
34) D.E. Polk, Acta Met. 20 485 (1972)
35) S. Rundqvist, Arkiv Kemi 20, 67 (1963)
36) P.H. Gaskell, in "Rapidly Quenched Metals III", (B. Cantor, ed.) The Metals Soc. London (1978) V. 2, p. 277
37) C.H. Bennett, J. Appl. Phys. 43 2727 (1972)
38) J.F. Sadoc, J. Dixmier and A. Guinier, J. Non-Cryst. Solids 12, 46 (1973)
39) D.S. Boudreaux and J. M. Gregor, J. Appl. Physics 48, 152 (1977)
40) K. Suzuki, K. Fukunaga, T. Masumoto, Sci. Rep. Res. Inst. Tohoku Univ. A26, 1 (1976)
41) T. Fukunaga, M. Misawa, K. Fukamichi, T. Masumoto and K. Suzuki, in "Rapidly Quenched Metals III", (B. Cantor, ed.) The Metals Soc. London (1978) V. 2, p. 325.
42) D.S. Boudreaux and H.J. Frost, Phys. Rev. B, 23, 1506 (1981)
43) A.I. Taub and F. Spaepen, Scripta Met. 13, 195 (1979)
44) L.A. Davis and Y.T. Yeow, J. Mat. Sci. 15, 230 (1980)
45) S. Takayama and R. Maddin, Acta Met. 23, 943 (1975)
46) S. Takayama, Mat. Sci. Eng. 38, 41 (1979)
47) T. Masumoto, Sci. Rep. Res. Inst. Tohoku U. 26A, 246 (1977)
48) C.A. Pampillo and H.S. Chen, Mat. Sci. Eng. 13, 181 (1974)
49) T. Masumoto and R. Maddin, Acta Met. 19, 725 (1971)
50) A.H. Cottrell, "The Mechanical Properties of Matter", John Wiley, NY (1964) p. 322
51) H.S. Chen, Appl. Phys. Lett. 29, 328 (1976)
52) Y. Waseda, in "Rapidly Quenched Metals III", V. 2 (B. Cantor, ed.) The Metals Soc., London (1978) p. 352
53) T. Murata, T. Masumoto and M. Sakai, ibid, p. 401
54) J.J. Gilman, J. Appl. Phys. 44, 675 (1973)
55) L.A. Davis in "Metallic Glasses" (J.J. Gilman and H.J. Leamy, eds.) ASM Metals Park, OH (1976), p. 190
56) J.P. Hirth and J. Lothe, "Theory of Dislocations", McGraw-Hill, NY (1968) p. 1
57) J.C.M. Li, in "Frontiers in Materials Science", (L.E. Murr and C. Stein, eds.) Marcel Dekker, NY (1976) p. 527
58) R.C. Morris, J. Appl. Phys. 50, 3250 (1979)
59) A.S. Argon and H.Y. Kuo, in "Rapidly Quenched Metals III", V. 2 (B. Cantor, ed.) The Metals Soc., London (1978) p. 269
60) Idem, Mat. Sci. Eng. 39, 101 (1979)
61) L.A. Davis, Scripta Met. 9, 939 (1975)
62) R. Hill, "The Mathematical Theory of Plasticity", Oxford Univ. Press, London (1967) p. 300
63) A.K. Chakrabarti and J.W. Spretnak, Met. Trans. A, 6A, 733 (1975)
64) Idem, ibid. 6A 737 (1975)
65) L.A. Davis, Scripta Met. 9, 431 (1975)
66) D. Weaire, M.F. Ashby, J. Logan and M.J. Weins, Acta Met. 19, 779 (1971)
67) L.A. Davis, in "Rapidly Quenched Metals", Sec. 1 (N.J. Grant and B.C. Giessen, eds.) MIT Press, Cambridge (1976), p. 369
68) H.S. Chen, J.T. Krause and E. Coleman, J. Non-Cryst. Solids 18, 157 (1975)

69) C.P. Chou, L.A. Davis and M.C. Narasimhan, Scripta Met. 11, 417 (1977)
70) C.P. Chou, L.A. Davis and R. Hasegawa, J. Appl. Phys. 50, 3334 (1979)
71) J.J. Gilman, Physics Today 28, 46 (1975)
72) H.J. Leamy, H.S. Chen and T.T. Wang, Met. Trans. 3 699 (1972)
73) L.A. Davis and S. Kavesh, J. Mater. Sci. 10, 453 (1975)
74) L.A. Davis, J. Mat. Sci. 10, 1557 (1975)
75) L.A. Davis, Met. Trans. A, 10A, 235 (1979)
76) G.T. Hahn, M.F. Kanninen and A.R. Rosenfield, Ann. Rev. Mat. Sci. 2, 381 (1972)
77) L.A. Davis, J. Mat. Sci. 11, 711 (1976)
78) R.K. Ham, in "Thermal and High-Strain Fatigue", The Institute of Metals, London (1967) p. 55
79) J.R. Strife and K.M. Prewo, Rep. #AFWAL-TR-80-4060, Air Force Wright Aeronautical Labs, Wright Patterson Air Force Base, OH (1980)
80) B.G. Agarwal and L.J. Broutman, "Analysis and Performance of Fiber Composites", John Wiley, NY (1980) p. 241
81) R. Hasegawa, R.C. O'Handley and L.I. Mendelsohn, AIP Conf. Proc. No. 34, 298 (1977)
82) R. Hasegawa, W.A. Hines, L.T. Kabacoff and P. Duwez, Solid State Commun. 20, 1035 (1976)
83) R. Hasegawa and R. Ray, J. Appl. Phys. 49, 4174 (1978)
84) J. Durand and M. Yung, in "Amorphous Magnetism II", (R.A. Levy and R. Hasegawa, eds.) Plenum Press, NY (1977) p. 275
85) C.L. Chien and R. Hasegawa, Phys. Rev. B 16, 2115 (1977)
86) R. Hasegawa and R. Ray, Phys. Rev. B 20, 211 (1979)
87) V. Jaccarino, Bull. Am. Phys. Soc. 4, 462 (1959)
88) H. Watanabe, H. Morita and H. Yamauchi, IEEE Trans. Mag. MAG-14, 944 (1978)
89) R. Hasegawa and R. Ray, J. Appl. Phys. 50, 1586 (1979)
90) J. Durand, B. Lémius, R. Hasegawa, D. Aliaga-Guerra and P. Panissod, J. Mag. Mag. Mat. 15-18, 1373 (1980)
91) R. Hasegawa and J.A. Dermon, Phys. Lett. 42A, 407 (1973)
92) R. Malmhäll, G. Bäckström, K.V. Rao and T. Egami, J. Appl. Phys. 50, 7656 (1979)
93) W.A. Hines, C.U. Modzelewski, R.N. Paolino and R. Hasegawa, submitted to Solid State Commun.
94) P. Duwez, in "Techniques of Metal Research", Vol. 1, (R.F. Bunshah, ed.) Interscience Pub., New York (1968) p. 347.
95) R. Hasegawa, unpublished data (1970).
96) D.I. Paul, J. Marti and L. Valadez, AIP Conf. Proc. No. 18, 1377 (1974)
97) R. Hasegawa, AIP Conf. Proc. No. 29, 216 (1976)
98) R. Hasegawa, J. Appl. Phys. 46, 5263 (1975)
99) E.A. Nesbitt, R. Hasegawa, R.C. O'Handley and M.O. Sullivan, J. Appl. Phys. 51, 3328 (1980)
100) R.C. O'Handley, Phys. Rev. B 18, 930 (1978)
101) E.R. Callen and H.B. Callen, Phys. Rev. 129, 578 (1963); Phys. Rev. A139, 455 (1965); E. Callen, J. Appl. Phys. 39, 519 (1968)
102) S. Ito, K. Aso, Y. Makino and S. Uedaira, Appl. Phys. Lett. 37, 665 (1980)
103) R. Hasegawa, to be published in J. Appl. Phys. 52 (March, 1981)
104) R. Hasegawa, J. de Phys. 41, C8-701 (1980)
105) R.C. O'Handley, L.I. Mendelsohn and E.A. Nesbitt, IEEE Trans. Mag. MAG-12, 942 (1976)
106) T. Jagielinski, K.I. Arai, N. Tsuya, S. Ohnuma and T. Masumoto, IEEE Trans. Mag. MAG-13, 1533 (1977)
107) R. Hasegawa and R.C. O'Handley, J. Appl. Phys. 50, 1551 (1979)
108) R.C. O'Handley, C.P. Chou and N. DeCristofaro, J. Appl. Phys. 50, 3603 (1979)
109) L.A. Davis, N. DeCristofaro and C.H. Smith, in "Metallic Glasses: Science and Technology", Budapest (1980) in press
110) M. Mitera, T. Masumoto and N. Kazama, J. Appl. Phys. 50, 7609 (1979)
111) F.E. Luborsky, J.J. Becker, J.L. Walter and H.H. Liebermann, IEEE Trans. Mag. MAG-15, 1146 (1979)
112) See for example, Arnold Engineering Company Catalog TC-101B (1972)
113) R. Hasegawa, M.C. Narasimhan and N. DeCristofaro, J. Appl. Phys. 49, 1712 (1978)
114) T. Takahashi, H. Fujimori and T. Masumoto, presented at the 80th Annual Meeting of the Japan Institute of Metals (1974)
115) D. Nathasingh and C.H. Smith, Proceedings of POWERCON 7, B2-1 (1980)
116) T. Masumoto, K. Hashimoto and M. Naka, in "Rapidly Quenched Metals III", Vol. 2 (B. Cantor, ed.) The Metals Soc., London (1978) p. 435
117) D. Onn, Proc. 26th Annual Conf. Magnetism Mag. Materials (to be published in J. Appl. Phys. March, 1981)
118) A. Fert and R. Asomoza, J. Appl. Phys. 50, 1886 (1979)
119) W.L. Johnson, J. Appl. Phys. 50 (1557) (1979)

SURFACE TREATMENTS USING LASER, ELECTRON AND ION BEAM PROCESSING METHODS

B. H. Kear
Exxon Laboratories
Linden, NJ 07036

J. W. Mayer
Cornell University
Ithaca, NY 14853

J. M. Poate
Bell Laboratories
Murray Hill, NJ 07976

P. R. Strutt
University of Connecticut
Storrs, CT 06268

I. INTRODUCTION

The use of directed energy and ion beam processing methods for the purpose of modifying the surface properties of metallic materials is currently attracting widespread interest. Rapidly solidified surface structures have been produced by laser (1-10) and electron beam (11-16) surface melting techniques. In addition, surface alloying has been accomplished by making controlled additions of pre-alloyed powder, wire, or ribbon to the interaction, or melt zone (17-20). High concentrations of one or more alloying elements have also been introduced into the surface layers by ion implantation and by ion beam mixing methods (21-24). Such treatments may involve subsequent laser annealing in order to restore crystalline perfection, while preserving the benefits of alloy enrichment. This article will review these developments from both the fundamental and applied perspectives, with the emphasis on fundamental aspects.

Amongst the specific areas of fundamental interest that will be considered are: (i) influence of cooling rate, temperature gradient and solidification rate on the microstructures developed in thin rapidly quenched surface layers; (ii) comparison of laser and electron beam surface melting techniques; (iii) segregation at high interface velocities in rapidly solidified surface layers; (iv) formation of extended solid solutions, refined structures, metastable phases and amorphous solids by rapid solidification and ion beam processing; and (v) ion beam mixing phenomena, including cascade effects.

Although all these considerations are generally relevant to any crystalline material, the discussion will have as its focus materials that are primarily of metallurgical interest. This will apply also to the brief descriptions given of applications for surface treatments by directed energy, ion implantation and ion beam processing methods.

II. LASER GLAZING

The interaction between a laser beam and a material surface is controlled by a number of variables, including the wavelength of the radiation, the incident power density, and the available interaction time. For the case of a high power CO_2 gas laser with a wavelength of 10.6 μm the laser-material interaction as a function of power density and interaction time is as shown in Figure 1. The rapid solidification laser processing region, here defined as 'laser glazing', requires high incident power density of the order of that utilized for welding, but with substantially shorter interaction times. With this combination, energy inputs may range from 1 to 100 J/cm^2, and the heating effect is concentrated in a very thin region at the material surface. Rapid surface melting occurs in a time period during which little thermal energy penetrates into the base material. This leads to the establishment of extremely high thermal gradients, which promote rapid solidification of the melt. As indicated below, the quench rate is ultimately dependent on melt layer thickness, with cooling rates of 10^4-10^8°C/sec readily attainable in appropriately thin sections. As a consequence of the extremely rapid solidification, a variety of interesting metallurgical microstructures can be produced, some of which are unique to the laser glaze process.

A continuous multikilowatt CO_2 laser is normally employed for laser glaze processing. A convenient experimental arrangement is depicted schematically in Figure 2. A nominal 7.5 cm diameter laser beam is directed towards and focussed upon the workpiece by reflective optics. In a typical test, a 46 cm focal length mirror is used to provide an effective minimum spot diameter of 0.05 cm at the workpiece surface. At 3.0 kW, these optics provide a maximum incident power density of approximately 1.5×10^6 W/cm^2. Such a high power density is essential for localizing the energy input at the material surface, and promotes effective coupling of the laser energy with the material, despite the initial high reflectivity of metallic surfaces to the 10.6 μm wavelength of carbon dioxide laser radiation. The 3 kW level is a convenient one, in that it promotes effective beam coupling, but does not create significant plasma generation problems. As indicated in Figure 2, plasma suppression is accomplished by means of an inert gas shield, which further prevents atmospheric contamination of the melt. Cooling due to the inert gas is negligible, in comparison to the quenching effect of the unheated substrate material.

A range of melt depths may be achieved by varying the translational speed of the workpiece under the focussed beam. Linear speeds of from 1.5 to greater than 60 m/min are attained by using a variable speed rotating disc with specimens located at a fixed radius on the disc surface. Such a speed range yields energies per unit length of melt ranging from 1200 down to 30 J/cm. Higher energy inputs are employed as initial homogenizing passes, and the lower energy inputs for subsequent rapid solidification processing.

Rotating disc equipment of the type described above facilitates the irradiation of multiple samples in a single pass. Other types of equipment can be used, including linear traverse equipment, and motion of the beam instead of the workpiece; however, due to the high traverse speeds involved, rotating equipment has proved to be the most convenient for laboratory work. Samples are arranged adjacent to each other with their top surfaces located at the focal plane of the beam focussing optics. A smooth machined surface is presented to the focussed laser beam. In order to facilitate rapid self-quenching, the sample thickness should be at least four times the melt depth.

A considerable amount of work has been done to determine the utility of the laser glaze process as a means to improve the corrosion, erosion and wear properties of metallic surfaces. Some good examples of such improvement have been obtained by different research groups, and will be described later. Much less work has been reported on laser glaze processing in conjunction with surface alloying. Two distinct approaches are being investigated: (i) pre-placement of alloying material on the workpiece surface prior to laser melting;

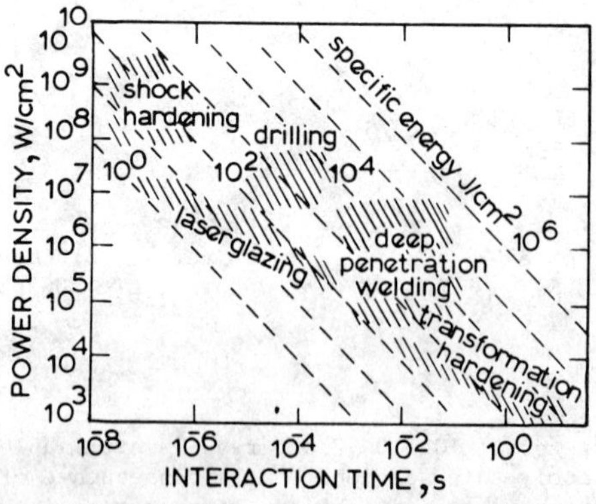

Figure 1. Operational Regimes for Laser Materials Processing Techniques.

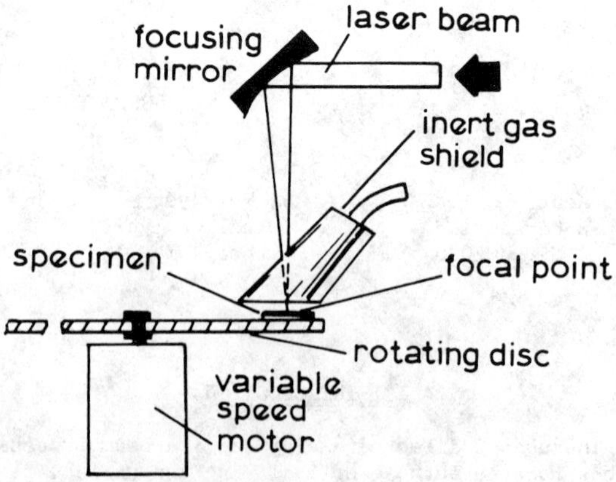

Figure 2. Schematic of Laser Glazing Apparatus

and (ii) continuous delivery of alloying material to the laser-material interaction, or melt zone. The pre-placement method has been exploited as a hard-facing treatment e.g. to harden critical regions of wear in a part (17). The continuous feed method is being developed for the purpose of bonding thin layers of alloying material to the workpiece surface, and also for fabricating bulk rapidly solidified structures in near-net shapes (20). A schematic of the incremental solidification process, so-called 'layer glaze process', for making bulk rapidly solidified structures is shown in Figure 3. Feedstock can be in the form of wire, ribbon or powder. Wire and ribbon are convenient to feed but often difficult to obtain in the required alloy compositions. Powder is relatively easy to procure, and can be conveniently fed into the interaction zone. The laser beam melts both the feedstock material and a thin layer of the substrate, which produces good interlay bonding and expitaxial growth from layer to layer. There is a critical relationship between the location of the feedstock impingement point and the laser heated zone. Since the substrate is rotating, feedstock impingement must occur slightly ahead of the laser beam for stable, steady-state material deposition. The feedstock is then carried into the laser beam by the rotation of the substrate. The part being fabricated may be internally water-cooled during material deposition in order to prevent excessive heat buildup, and thereby maintain high cooling rates.

Heat Transfer Analysis

In order to determine the effects of different process parameters on the melting and rapid-cooling processes resulting from high power density laser glazing, a one dimensional, finite-element heat transfer analysis has been applied to the transient cooling processes (25). Results of an analog computation(26) were used for the surface melting process occurring on heating. Using this method, if the laser power density and interaction time are specified, the melt depth can be calculated. Temperature profiles in the solid and molten regions were also obtained, and used as initial conditions in the finite-element program in order to compute the transient cooling of the material for the specified power and interaction time. The average cooling rate was calculated over the time from the cessation of power input to the time at which 800°C was reached.

Transient surface melting characteristics for nickel are shown in Figure 4. The solid lines represent melt depth as a function of interaction time for a variety of power densities. Interaction time is defined as the time required for the laser beam to traverse one spot diameter, and is computed by dividing the spot diameter by the surface traverse velocity (D/V). The period initiates from a uniform initial temperature of 21°C and thus includes the time that is required to heat the material in the solid state to its melting temperature. After the onset of melting, the advance of the melt interface was found to be linear with time. The dashed lines in Figure 4 represent constant specific energy inputs and are computed as the product of the absorbed specific power and the interaction time. It is obvious from Figure 4 that as the power density increases the specific energy necessary to obtain a given melt depth decreases, and the melting is also accomplished in a shorter time. This is the result of the fact that higher melting efficiencies are obtained at higher power densities, since less time is available for conduction into the substrate material. In addition to promoting higher melting efficiencies, the reduction in conduction to the substrate also leads to steeper temperature gradients in the alloys following melting and thereby increases cooling rates in the glazed samples.

In Figure 5, the effect of melt depth on cooling rate is shown for selected power densities. The maximum cooling rate which can be attained at each power density occurs for the limiting case in which the melt depth approaches zero. This maximum cooling rate increases with the square of the power density. For a given melt depth, the increase in cooling rate with absorbed specific power is a consequence of the increased temperature gradients; doubling the power density approximately doubles the cooling rate. A maximum melt depth occurs for a given power density, after which surface vaporization begins, and the lowest cooling rates are realized under these conditions. A theoretical maximum cooling rate for nickel is also included in Figure 5, as well as typical results for splat cooling of iron on a copper substrate (27). The theoretical maximum cooling rate is obtained by considering that all of the energy is initially concentrated in the melt and that a temperature discontinuity equal to the melting temperature exists at the solid/liquid interface. It is noteworthy that the splat cooling rate, the average cooling at the threshold of surface vaporization, and the theoretical maximum cooling rate are all inversely proportional to the square of the melt depth. The possibility of contributing effects from convection and radiation at the surface on cooling rates has also been investigated. Even assuming a radiative emissivity of 1.0 and a high convective heat transfer film coefficient (0.57 W/cm^2 °C), the average cooling rate is not changed by more than a few per cent for absorbed power densities in excess of 10^5 W/cm^2. It may be concluded, therefore, that at the high power densities employed for laser glazing, rapid cooling occurs primarily through conductive heat transfer to the substrate.

Figures 4 and 5 represent convenient 'working diagrams', from which an estimate can be made of the beam power density and interaction time required to achieve a specified melt depth and average cooling rate. From a strictly metallurgical viewpoint, such information is of limited utility. More pertinent to an analysis of the

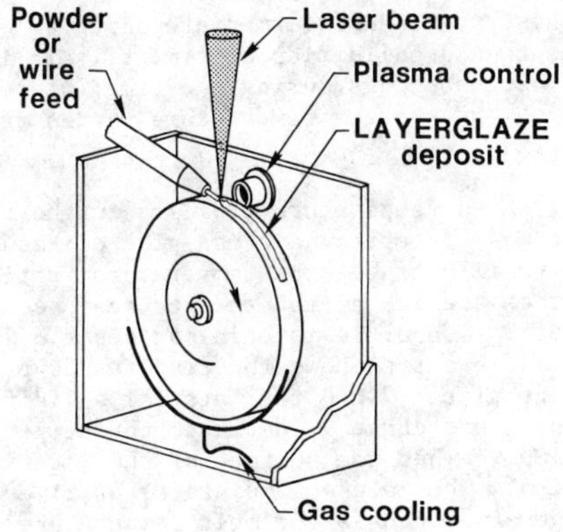

Figure 3. Schematic of the Layer Glaze Process

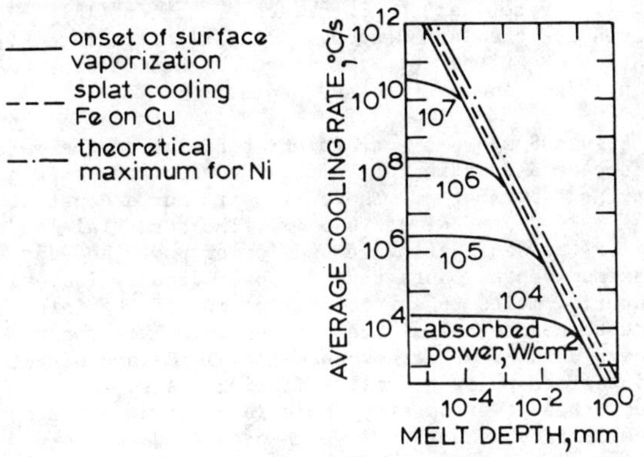

Figure 5. Effect of Melt Depth and Power Density on Average Cooling Rate.

metallurgical structures produced by laser glaze processing are the thermal properties at the solidification interface. The critical solidification parameters are the temperatures gradient (G), the solidification rate (R), and the cooling rate ($\dot{T}$) at the advancing solid/liquid interface. Calculations of the transient behavior of these parameters, including the gradient-rate ratio (G/R), have been reported.(28-30) An entirely analytical treatment, due to Greenwald et al(29) will be presented here. The analysis is approximate in that it neglects latent heat, but it does possess the virtue of avoiding the complexities of numerical solutions, and provides results rapidly and economically. For the purpose of the analysis the material is assumed to be pure nickel, with the properties given in Table I.

Table I

Properties of Pure Nickel

Melting Temperature	1455°C
Vaporization Temperature	2732°C
Conductivity	0.915 Watts/cm,°C
Density	8.89 gm/cm^3
Specific Heat	0.105 cal/gm,°C

The one dimensional conduction heat transfer equation applying to this problem is given in Equation (1), and the boundary and initial conditions are defined by Equations (2) and (3), respectively.

$$\frac{\partial^2 T(x,t)}{\partial x^2} + \frac{g(x,t)}{K} = \frac{1}{\alpha}\frac{\partial T(x,t)}{\partial t} \quad \text{for } 0 \leq x < \infty,\ t < 0 \qquad (1)$$

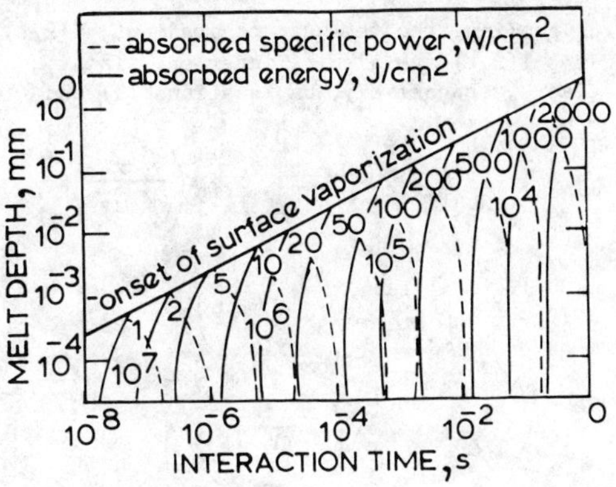

Figure 4. Transient Surface Melting Characteristics of Nickel.

$$\frac{\partial T(x,t)}{\partial x} = 0 \text{ at } x = 0,\ t > 0 \qquad (2)$$

$$T(x,0) = T_0 = \text{constant for } 0 \leq x < \infty,\ t = 0 \qquad (3)$$

The heat input to the material is defined as an internal heat generation term g(x,t) in Equation (1) where

$$g(x,t) = q_0\ \delta(x-0)\eta(\tau - t) \qquad (4)$$

The parameters δ and η are the delta and Heaviside functions, respectively. Equation (4) defines a surface heat input of magnitude q_0 for a time τ and of magnitude zero thereafter.

An integral transform technique (31) is used to solve Equations (1)-(4), yielding the following closed-form solution for the transient temperature field during heating and cooling.

During heating:

$$T(x,t) = \frac{q_o}{k}\left[\sqrt{\frac{4\alpha t}{\pi}}\ e^{-\left(\frac{x}{\sqrt{4\alpha t}}\right)^2} - x\ \mathrm{erfc}\left(\frac{x}{\sqrt{4\alpha t}}\right)\right] + T_o \qquad (5)$$

During Cooling:

$$T(x,t) = \frac{q_o}{k}\left[\sqrt{\frac{4\alpha t}{\pi}}\ e^{-\left(\frac{x}{\sqrt{4\alpha t}}\right)^2} - \sqrt{\frac{4\alpha\gamma}{\pi}}\ e^{-\left(\frac{x}{\sqrt{4\alpha\gamma}}\right)^2} - x\left(\mathrm{erfc}\left(\frac{x}{\sqrt{4\alpha t}}\right) - \mathrm{erfc}\left(\frac{x}{\sqrt{4\alpha\gamma}}\right)\right)\right] + T_o \qquad (6)$$

At the surface of the material (x=o) Equation (5) reduces to

$$T(t) = \frac{q_o}{k}\left(\sqrt{\frac{4\alpha t}{\pi}}\right) + T_o \qquad (7)$$

and Equation (6) becomes

$$T(t) = \frac{q_o}{k}\left(\sqrt{\frac{4\alpha t}{\pi}} - \sqrt{\frac{4\alpha\gamma}{\pi}}\right) + T_o \qquad (8)$$

During cooling, the temperature gradient, G(x,t), and cooling rate, T(x,t) are also of interest. These quantities are given, respectively, by Equations (9) and (10).

$$G(x,t) = \partial T/\partial x = -\frac{q_o}{k}\left[\operatorname{erfc}\left(\frac{x}{\sqrt{4\alpha t}}\right) - \operatorname{erfc}\left(\frac{x}{\sqrt{4\alpha \gamma}}\right)\right] \quad (9)$$

$$\dot{T}(x,t) = \partial T/\partial t = \frac{q_o}{k}\left[\sqrt{\frac{\alpha}{\pi t}}\; e^{-\left(\frac{x}{\sqrt{4\alpha t}}\right)^2} - \sqrt{\frac{\alpha}{\pi \gamma}}\; e^{-\left(\frac{x}{\sqrt{4\alpha\gamma}}\right)^2}\right] \quad (10)$$

Finally, the solidification rate, R, is related to G and $\dot{T}$ by the relation

$$\dot{T} = GR \quad (11)$$

where G and R are evaluated <u>at</u> the melt interface.

To better understand the head transfer during rapid solidification, the cooling process is examined for a constant depth of melt with absorbed power density as the primary variable. The approximate analysis does not include an actual change of phase; the material is assumed to be "melted" at a given depth if the temperature is above the melting point. The algorithm used to calculate the transient values of the solidification parameters at the melt interface during solidification is as follows:

For a given depth of melt, d, an absorbed power density, q_o, is chosen. Since the condition of melting implies a melt temperature T_{melt}, at depth d, Equation (5) can be solved iteratively for the time required to produce this condition. Let this time be τ. Thus, the material is surface-heated from time-zero to time-τ with a power density q_o and melted to depth d, at which depth the temperature is T_{melt}. The surface temperature, T_{max}, at this time (τ) is found from Equation (7). This is the highest temperature that occurs anywhere in the material at any time. With the material melted to depth d at this time τ, the energy input ceases and the material begins to cool. As solidification proceeds, the depth at which the temperature equals T_{melt} moves toward the surface. The surface temperature drops and when it reaches T_{melt} the solidification is complete. The time at which this occurs, τ', can be determined from Equation (8). The temperature field in the material during the cooling process (between times τ and τ') is defined by Equation (6).

To determine the temperature gradient and cooling rate at the melt interface as solidification proceeds towards the surface of the material, it is first necessary to establish the position of the interface as a function of time. This is accomplished by selecting, in the time interval τ'-τ (during which solidification occurs), a number of discrete points, t_i, and at each point solving Equation (6) iteratively for the depth, s_i, at which the temperature equals T_{melt}. This procedure is illustrated schematically in Figure 6.

Figure 6 shows the temperature history at different depths in the material. The upper curve shows how the temperature at the surface starts at the initial temperature T_o, begins to melt at time t_a, and reaches a temperature of T_{max} at time τ when energy input ceases. The surface temperature then falls until it reaches T_{melt} at time τ' at which time solidification is complete. The lower curve of Figure 6 shows the temperature history at depth d. The time τ at which the temperature at this depth reaches T_{melt} defines the required heating time. The middle curve shows the temperature history at an intermediate depth s_i which melts at time t_b and solidifies at time t_i. It is seen that solidification occurs during the time interval τ'-τ over which the surface temperature varies from T_{max} to T_{melt}.

Figure 7 shows temperature profiles in the material at selected times, in effect a cross-plot of Figure 6. The upper curve illustrates the temperature profile at time τ when the material at the surface reaches a temperature of T_{max} and the material at depth d just melts. The middle curve shows the temperature profile at a subsequent time t_i when the material solidifies to depth s_i. The lower curve illustrates the temperature profile at time τ' when the surface of the material just solidifies. The locus of points s_i at times t_i when the temperature equals the melt temperature, defines the position of the melt interface as a function of time during solidification. Knowing the depth of the melt interface as a function of time Equations (9) and (10) are evaluated to yield the temperature gradient G and cooling rate $\dot{T}$ at the interface; Equation (11) is solved for the instantaneous solidification rate R. The ratio G/R at the melt interface is thus determined both as a function of time and depth.

Figure 8 shows the transient behavior of the melt interface as a function of time, with absorbed power density as a parameter. The curves are for a constant initial melt depth, d, of .025 mm. The word "initial" refers to the depth at the cessation of power and not the maximum depth. Note that as power density increases the material melts to a greater fraction of the initial melt depth before solidification begins. Even for very low power densities there is an initial advance of the melt interface prior to solidification. Also, for each absorbed power density there is a characteristic maximum surface temperature which occurs at the cessation of power. An absorbed power density of 550,000 W/cm^2 is assumed to be the maximum practical value for surface-melting a depth of 0.025 mm, since it results in a surface temperature equal to the vaporization temperature of nickel.

The variation of the temperature gradient, G, at the melt interface with melt depth is shown in Figure 9 for various absorbed power densities. The gradient is a strong function of the absorbed power density, and hence, so is the cooling rate. The curves of Figure 9 are obtained by evaluating Equation 9 at the appropriate values of depth and time. The temperature gradient at the melt interface scales with initial depth such that if the depth d is increased by a factor M and the power q_o is reduced by the factor M, the resulting gradients are the same shape as those for Figure 9, except that they will be lower by a factor of M if plotted vs fractional melth depth, s/d. For example, using Figure 9, if it is desired to obtain the temperature gradient at a value of s/d = 0.6 for a .25 mm melt depth (M=10) at a power density of 5000 W/cm^2, one would read from Figure 9 the value 20,000°C/cm at a power density of 50,000 W/cm^2 and divide by 10 to obtain 2,000°C/cm.

The variation of the cooling rate, $\dot{T}$, at the melt interface is shown in Figure 10 with absorbed power density as a parameter. The cooling rate, evaluated using Equation 10, is a strong function of absorbed power q_o, as would be expected. The cooling rate must start at zero as solidification begins since melting continues after the cessation of power. Hence $\dot{T}$ at the interface, initially positive, must pass through zero to become negative during cooling. As solidification approaches the surface of the material, the cooling rates approach a constant value. The cooling rate at the melt interface scales with initial melt depth, d, such that if d is increased by a factor M and the power is reduced by the factor M the resulting cooling rates will have the same shape as those of Figure 10, except that they will be lower by a factor of M^2 if plotted as a function of fractional melt depth.

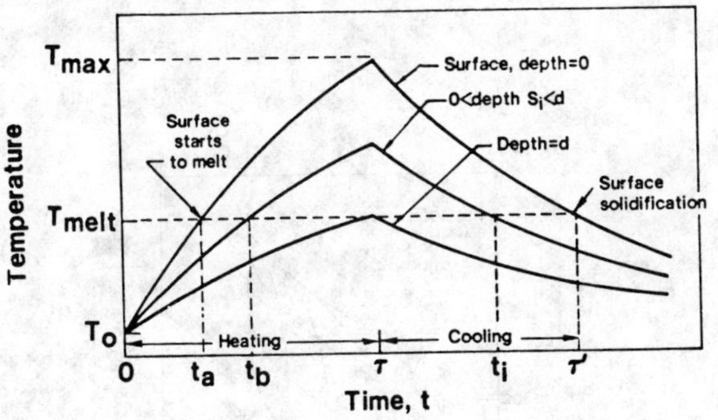

Figure 6. Variation of Temperature with Time at Selected Melt Depths.

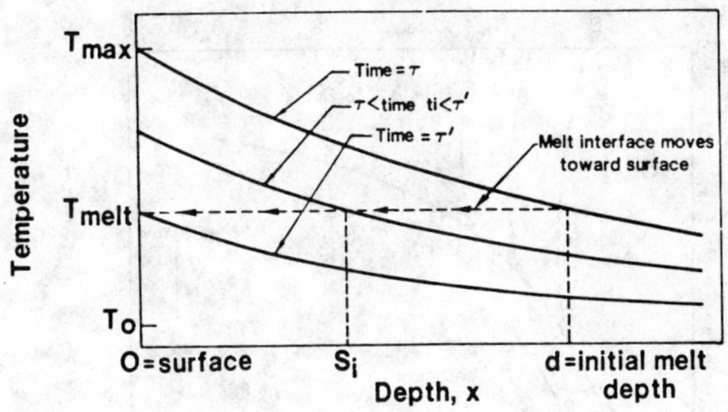

Figure 7. Temperature Profiles at Selected Times.

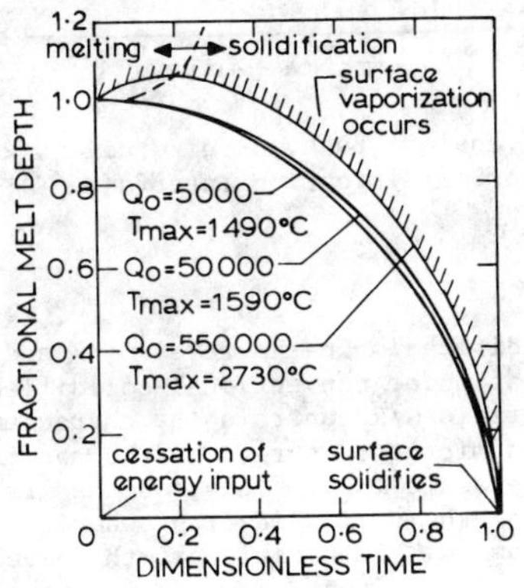

Figure 8. Transient Behavior of Solidification Interface.

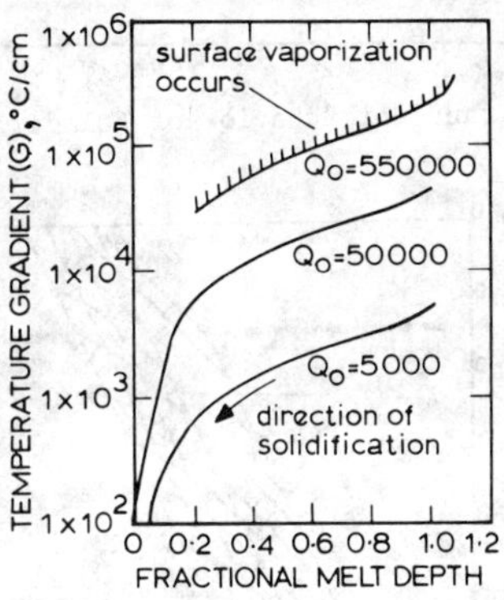

Figure 9. Transient Behavior of Temperature Gradient, G.

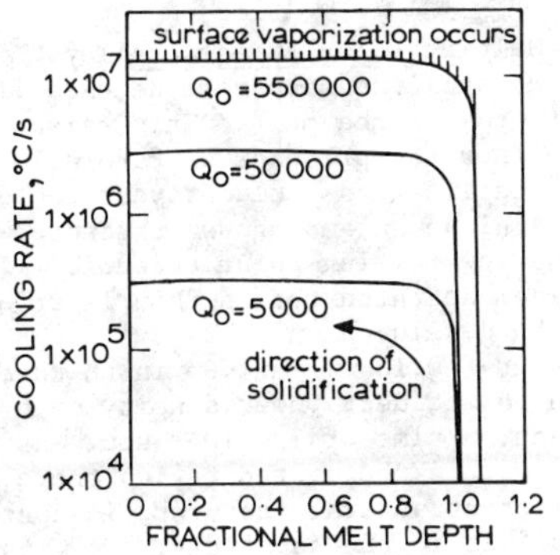

Figure 10. Transient Behavior of Cooling Rate, $\dot{T}$.

The slope of the melt-interface vs time curve is proportional to the solidification rate, R. This quantity, evaluated by means of Equation 11, is plotted in Figure 11 as a function of melt depth, with absorbed power density as a parameter and for a constant initial melt depth of .025 mm. The freezing rate tends to increase as the solidification front approaches the surface, and, after an initial transient (which begins at different depths) is insensitive to power density. There is a definite scaling law which relates solidification rate to initial depth: if the depth, d, is increased by a factor, M, and the power, q_o, is reduced by the factor M the resulting freezing rates will have the same shape as those in Figure 11, except that the freezing rates will be lower by a factor of M if plotted as a function of fractional melt depth.

The parametric dependence of G/R at the melt interface is shown in Figure 12 with absorbed power as a parameter for an initial melt depth of .025 mm. Characteristically, the curves begin at infinity at the onset of solidification and fall off rapidly as solidification approaches the surface. This behavior can be inferred from the curves of the gradient G (Figure 9) and the solidification rate R (Figure 11) for a constant absorbed power. Power density exerts a strong influence on G/R for a given melt depth as well as on the maximum surface temperature, T_{max}. Interestingly, the time required to solidify a given depth of melt is nearly independent of absorbed power as shown by the dashed lines of constant time, measured from the cessation of energy input. The variation of G/R with dimensionless melt depth, s/d, scales in an interesting way with melt depth. If the melt depth is increased by a factor M and the power density is decreased by the factor M the ratio G/R does not vary. Thus, the shape and magnitude of the G/R profiles is preserved as a function of dimensionless depth.

Referring to Figure 10, it can be seen that for a given absorbed power there is a characteristic, almost constant cooling rate which is attained as the final eighty percent of melt solidifies. Figure 13 shows the effect on the final (or characteristic) cooling rate of absorbed power for different values of melt depth. The line of melt depth equal to 0.025 mm corresponds to the case illustrated in Figure 10. For comparison, the three dark circles on the 0.025 mm melt depth line of Figure 13 correspond to the final cooling rates shown on Figure 10 for values of absorbed power of 5,000, 50,000 and 550,000 W/cm^2. The cooling rates for these cases are seen to be about $3x10^5$, $3x10^6$, and $1.3x10^7$°C/sec.

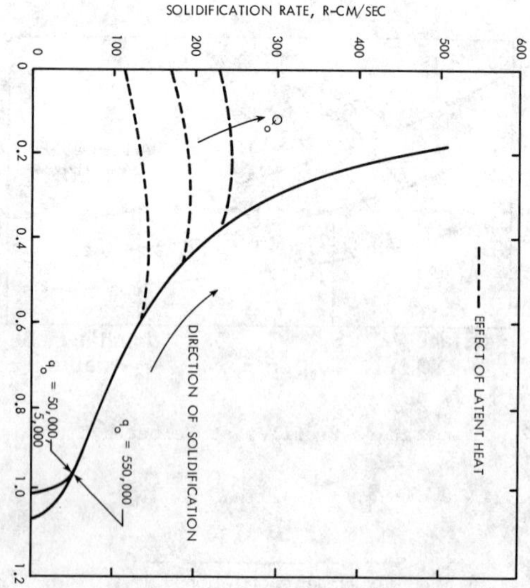

Figure 11. Transient Behavior of Solidification Rate, R.

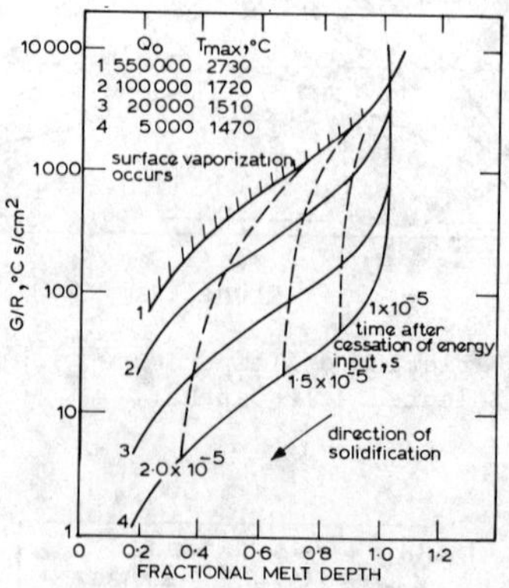

Figure 12. Transient Behavior of Gradient/Rate Ratio, G/R, at Solidification Interface.

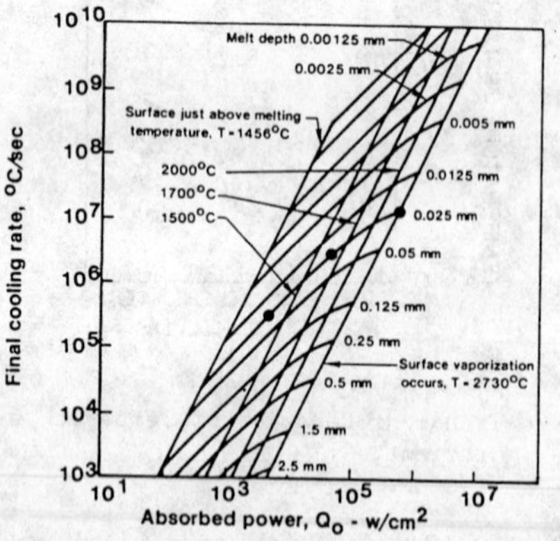

Figure 13. Effect of Absorbed Power and Melt Depth on Final Cooling Rate.

More definitive calculations by Mehrabian et al (28), which take into account the latent heat evolved at the solidification interface, show that the above conclusions are qualitatively correct, except for the solidification rate. As indicated in Figure 11, the solidification rate undergoes a radical change in behavior due to the latent heat effect, and tends to a finite (even decreasing) value as the solidification interface approaches the surface. The solidification rate increases with absorbed power density, presumably due to the higher temperature gradients in the superheated melt, despite the 'dam like' effect of latent heat in preventing energy from being conducted into the solid. On the other hand, the temperature gradient still tends to zero as the interface approaches the surface, so that the G/R parameter falls off sharply near the surface, even though the solidification rate remains finite.

Microstructural Aspects

The microstructural consequences of rapid solidification laser glaze processing have been examined in a wide variety of materials. The range of microstructures observed include refined dendritic structures, ultrafine eutectics, extended solid solution phases, metastable phases and amorphous solids. Systematic variations in microstructure within the melt zone have also been reported. As will be shown, these effects can be explained in terms of the calculated variations in the critical solidification parameters, G, R and $\dot{T}$. It is well known that under conventional solidification conditions, the ratio G/R controls the character of the solidification microstructure, whereas the product G.R determines its scale (32). Thus, as the gradient-rate ratio increases the solidification behavior changes progressively from fully dendritic growth to cellular dendritic and planar front growth. On the other hand, increasing the gradient-rate product, or cooling rate, since $\dot{T}$ = G.R, gives rise to shorter diffusion paths and finer structures. Such considerations apply also to rapid solidification processing, except under extremely high solidification rates where planar front growth can be achieved more or less independent of the imposed thermal gradient (33). Some typical results will be considered.

1. Low Melting Point Eutectic Alloys. In laser glazing, the crystalline substrate necessarily is in intimate contact with the melt. This raises the question as to whether the presence of favorable nucleation sites for crystallization would prevent the undercooling necessary to achieve an amorphous structure. Test experiments with the low melting point eutectic alloy (Pd-4.2 Cu-5.1 Si) have demonstrated that a thin layer of amorphous material can be developed on a crystalline substrate. The melted layer (~0.18 mm maximum in thickness) was obtained with an incident power density of 6 x 10^6 W/cm^2 and an interaction time of 3 x 10^{-5} sec. As shown in Figure 14(a), within the melted region there are no discernible microstructural features, in contrast to the eutectic structure in the substrate. One manifestation of the non-crystalline amorphous nature of the glazed material is the symmetrical pattern of curved shear bands formed around the hardness indentations, Figure 14(c). Another manifestation is the vein-like character of the fracture surface, Figure 14(b). Confirmatory evidence was obtained by TEM; the thin foils had a speckled appearance and the electron diffraction pattern showed diffuse rings, Figure 14(d). Similar tests have been performed on other low melting point eutectic alloys, and with comparable results. A good example is provided by the work on the technically interesting alloy $Fe_{40}Ni_{40}P_{14}B_6$, which exhibits exceptional mechanical and corrosion resistance properties (34). Because of the much higher critical cooling rates required (~10^6 °C/sec compared with 10^2 °C/sec for Pd-Cu-Si) the amorphous layer was very thin and could be detected only by TEM.

Recent experiments on rapidly solidified Pd-Cu-Si alloys have shown that the microstructure is controlled essentially by the interface velocity, or solidification

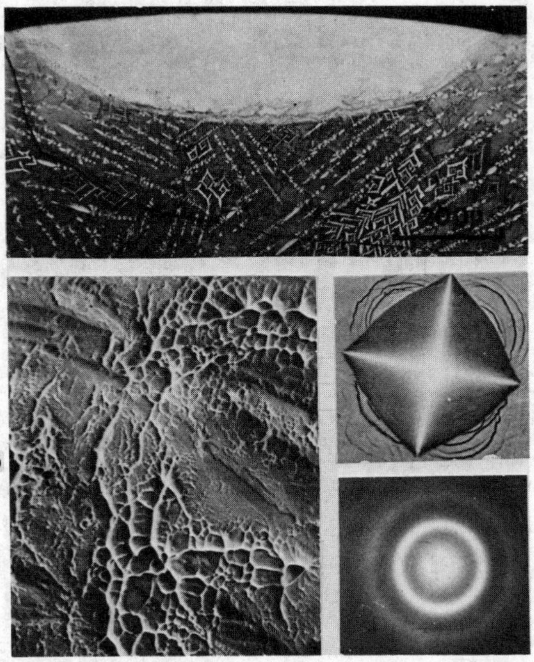

(a)

(b)

(c)

(d)

Figure 14. Amorphous Structure in Glazed Pd-4.2 Cu-5.1 Si Alloy; (a) Cross-Section of Glazed Region, (b) Fracture Surface, (c) Hardness Indentation, and (d) Electron Diffraction Pattern.

rate (35). For example, in off-eutectic alloys, a structure composed of dendrites and interdendritic eutectic at low interface velocity (~0.25 mm/sec) becomes a fine eutectic-like structure at velocities of ~1 mm/sec. At even higher velocities approaching 2.5 mm/sec, an abrupt transition occurs to the amorphous state. Because of the need for diffusional sorting of the components and the creation of solid/solid interfaces, the kinetics for eutectic solidification are relatively slow. Thus, the transition to the amorphous state at a critical interface velocity can be attributed to the diffculty of maintaining coupled growth, i.e. eutectic solidification, at high interface velocities. In this alloy system, crystallization into a metastable single phase solid with the same composition as the liquid is not possible, because of the special nature of the phase diagram.

In other low melting point eutectic alloys, notably the commercial Ni and Fe-base brazing alloys, the typical result of laser glazing is to form a featureless crystalline layer, at least when viewed under the optical microscope. More careful examination of such layers by TEM techniques show that they are composed of ultrafine eutectics, Figure 15. A noteworthy feature is the relatively large thickness (~250μm) that has been processed in the fine eutectic-like condition, again illustrative of the rather sluggish kinetics for eutectic crystallization. Under higher cooling rates, corresponding with much thinner glazed layers, the resulting structures are extended solid solutions, or amorphous solids, depending primarily on composition. Eutectic-like structures reflecting coupled growth can occur at high interface velocities even for radical departures in composition from the exact eutectic composition (36), i.e. for off-eutectic compositions that under conventional solidification conditions yield primary dendrites and interdendritic eutectic.

2. Nickel-Base Superalloys. Laser glazing studies of the superalloys have been restricted to two main classes of materials; namely the γ' precipitation hardened superalloys, and the new class of filamentary reinforced eutectic superalloys, such as CoTaC and γ/γ'-δ. In such studies, a relatively deep penetration welding (homogenizing) pass is usually applied to the surface prior to one or more superimposed rapid solidification laser glazing passes. Without such an homogenizing treatment the glazed layers exhibit incomplete dissolution of MC carbide particles and other refractory constituents in the initial microstructure.

Figure 16 shows the typical response to laser glazing of a superalloy crystal with the <100> orientation parallel to the direction of the incident beam. As indicated, laser glazing produces expitaxial growth in the resolidified layers, accompanied by a marked refinement in the scale of the microstructure. Within a given re-cast layer, the scale of the dendritic structure remains reasonably constant, although it undergoes obvious discontinuous changes with varying melt depth. This is to be expected, since Figure 10 shows that the cooling rate, which controls the scale of the microstructure, is relatively constant, after an initial transient stage. Close examination of Figure 16 shows a very thin layer of plane-front solidified material in contact with the melt/substrate interface. With increasing distance from this interface, the structure becomes cellular (one-dimensional dendritic), cellular dendritic, and finally fully dendritic. Such changes are clearly in accord with the calculated high initial G/R ratio and its rapid fall-off as solidification proceeds towards the free surface, Figure 12. The second pass in Figure 16 shows that cellular growth extends to the free surface, which is indicative of a persistently high G/R ratio in this high cooling rate regime of solidification, Figure 17.

Figure 16 shows that the region encompassing the second pass, including a heat-affected zone (HAZ) in the underlying first pass, does not etch-up as well as the rest of the material. The reason for this is that the cooling rate is fast enough to prevent precipitation of γ' particles only in the second pass and its associated HAZ. Apparently, in the HAZ, the elimination of the γ' particles is a consequence of much faster kinetics for the solution of the γ' phase than for its precipitation. The region including the HAZ and the second pass exhibits a striking difference in size, distribution and morphology of MC carbides. In the second pass, the carbides occur as small platelets, unidirectionally aligned within the cellular boundaries. On the other hand, in the first pass, the more massive carbide platelets are distributed in an irregular manner, following the tortuous paths defined by the interstices of a fully developed dendritic structure. The step change in the scale of the carbides in going from the first pass to the second pass is to be expected, since their formation is governed by the degree of segregation, and scale of the interdendritic interstices, both of which diminish with increasing cooling rate.

In laser glazed CoTaC-type eutectic superalloys, essentially two types of microstructures are observed. Under low-moderate cooling rates, the microstructure is distinctly dendritic, and is epitaxially related to the substrate, Figure 18(a). As the cooling rate increases, the microstructure becomes less obviously dendritic, at least under the optical microscope, and eventually all semblance of a dendritic structure is lost. In the dendritic regime of solidification, the dendrites are composed of a γ-solid solution phase, whereas the interdendritic regions are made up of a fine filamentary eutectic of γ + TaC, Figure 18(b). In the 'dendrite-free' regime of solidification (high cooling rate condition), the TaC phase assumes a more discrete particulate morphology, rather than a filamentary morphology. Dendritic growth is also characteristic of glazed γ/γ' - δ eutectic superalloys, at least under moderate cooling rates. All types of dendritic structure exhibit high hardness, e.g. 600-700 DPH for CoTaC-3 and 750-850 DPH for γ/γ' - δ

3. High Speed Steels. The most detailed investigation has been performed on M2 steel (5,8). In the

as-glazed condition, M2 consists of a two-phase mixture of δ-ferrite in an austenitic matrix, interspersed with a low concentration of fine carbide particles. A typical micrograph of a cross-section of a laser melted zone is shown in Figure 19. The maximum melt depth corresponds to the lower interface of the relatively featureless region (c). The heat-affected zone is below this interface, where the grain boundaries and undissolved carbides are clearly evident; less obvious is the highly refined martensitic structure of appreciable hardness. The martensitic structure of the parent material is visible at the bottom of the micrograph. Figures 19(a)-(c) are higher magnification scanning electron micrographs of regions close to (a), (b) and (c) in the adjacent figure. The dark etching phase in Figure 19, with increasing predominance towards the surface, is δ-ferrite, and the light-etching phase is γ-austenite. The morphology and distribution of these two phases are indicative of a peritectic reaction. Thus, in the lower melt region (c), the peritectic reaction δ + liquid $\rightarrow \gamma$ + carbide goes to completion. With increasing distance from the maximum melt depth, the reaction becomes more effectively arrested, such that at the surface the structure is composed primarily of δ-ferrite. Corresponding with these microstructural changes, hardness values range from 650 DPH in the upper δ-ferrite region to 950 DPH in the region near the maximum melt depth; the hardness of the parent material is 850 DPH. Aging at the normal tempering temperature of 560°C for 7×10^3 sec results in appreciable hardening of the maximum melt depth region, which attains a value of 1050-1100 DPH. Precipitation of a fine carbide dispersion accounts for the high hardness. A rather different type of micro-structure is formed by austenitizing the material at 1230°C for 300 secs, and then quenching into liquid nitrogen. A rather well-defined sub-structure is formed within larger size grains, which is decorated with extremely fine carbide particles. This fine scale carbide dispersion accounts for the high hardness values (1030-1100 DPH), which show only a small variation throughout the melt zone.

Considering the distribution of phases throughout the melt zone, it seems clear that this is determined by the varying solidification conditions. At the maximum melt depth, fine scale δ-dendrites form initially, due to the high G/R. These undergo complete transformation to γ because (i) the solidification rate (R) is low, and (ii) the fine scale microstructure provides short diffusion paths for redistributing alloying elements. As the liquid/solid interface moves towards the external surface the solidification rate increases, and a larger size equiaxed cellular structure develops. The effect is to reduce the time for the δ-ferrite/liquid reaction and to increase diffusion distances for redistributing alloying elements. As a consequence, the peritectic reaction is arrested , and the type of microstructure seen in Figure 19 is formed. The proportion of residual δ-ferrite increases as R increases, such that near the surface the δ-ferrite is nearly completely retained. The long term stability of the δ-ferrite at room temperature suggests enrichment in ferrite-stabilizing elements. Similar microstructural effects have been found in splat-quenched alloy steels (37,38), in that the microstructures are composed of a fine carbide dispersion in a matrix with varying proportions of an fcc phase and a bcc phase with no detectable tetragonality.

In complete contrast to M2, alloy 440C when subjected to laser glazing develops a mixed δ-ferrite/Cr_7C_3-carbide dendritic microstructure, somewhat similar in appearance to Figure 18. The volume fraction of carbide assuming all the carbon partitions to the carbide phase is 20 pct. The measured hardness of the melt (dendritic structure) and heat-affected zones are respectively 470 and 900 DPH; the carbide strengthened martensitic substrate is 825 DPH. Thus, the effect of glazing in this alloy is to refine the carbide structure and to soften the material. Subsequent heat treatment is necessary to recover the initial strength of the material.

4. Fluid Flow. The microstructural consequences of fluid flow during laser glazing have not yet been studied in detail, but some interesting preliminary findings have been reported.

Figure 20 shows cross-sectional views of a laser glazed region in an Al-4.5 wt.% Cu alloy. The unidirectionally solidified microstructure of the substrate, consisting of dendrites decorated with interdendritic eutectic, is clearly evident in these micrographs. As shown in Figure 20(a), the interdendritic eutectic extends into the laser melted region, albeit in a somewhat distorted pattern. This appears to be clear evidence for the influence of fluid flow on segregation behavior in the original melt pool. Under higher magnification, Figure 20(c), it can be seen that there is a much higher concentration of eutectic in the light-etching regions than in the regions immediately surrounding them. An electron microprobe trace, Figure 20(d), shows that the light-etching regions contain ~20 wt.% Cu, which is substantially less than 33 wt.% Cu in the α + Cu Al_2 eutectic. Clearly, some lateral diffusion of Cu away from the original eutectic must have occurred, but not sufficient to produce a homogeneous melt. This is understandable, since the calculated characteristic diffusion distance in this instance is much smaller than the measured primary dendrite arm spacing in the substrate. Depth profiling of the segregation pattern indicates that the most vigorous convection occurs in the middle of the melt pool.

At least two mechanisms have been proposed to account for the observed fluid flow during laser glazing. Under conditions where the melt surface experiences a substantial depression, the melt flows freely from front to back of the pool, moving around and beneath the actual depression, or trough in the pool (39). This phenomenon can give rise to various topographical features including shallow troughs with uniform cross-sectional profiles, and deeper troughs with central crests. When deep surface depressions are avoided, as under optimum laser glazing conditions, surface tension gradients due to non-uniform heating under the laser beam provide the driving force for convective flow (40). This flow process is not nearly as vigorous as the depression-induced flow, but nevertheless it is significant. Further discussion on this subject is presented in Section III.

Potential Applications

Much work has now been done to determine the utility of the laser glaze process as a means to improve surface properties of materials, such as corrosion, erosion and wear properties. Some promising examples of such improvements have been obtained by different research groups. A good example is the use of laser glazing to combat stress corrosion cracking in 'sensitized' 304 stainless steel (16). This is achieved by adjusting the cooling rate to avoid the formation of a harmful carbide phase at the grain boundaries. Another examples is surface hardening of M2 high speed steel to improve its cutting characteristics, e.g. in applications such as saw-blades, drill bits and end mills (8). An optimum hardening treatment involves laser glazing to achieve cooling rates ~10^4°C/sec, followed by heat treatment to develop a fine distribution of hard carbide particles in an austenitic/martensitic matrix. Laser glazing has also been applied to plasma-sprayed coatings to improve corrosion and erosion properties. The usefulness of the process in this case derives from its ability completely to eliminate porosity in the coating, and at the same time to provide a sound metallurgical bond between coating and substrate. Specific applications in this case have not yet been reported in the literature, because the work is of a proprietary nature.

Very little work has been reported on laser glaze processing in conjunction with surface compositional modifications, although this constitutes an area of obvious high potential. Nor for that matter has much

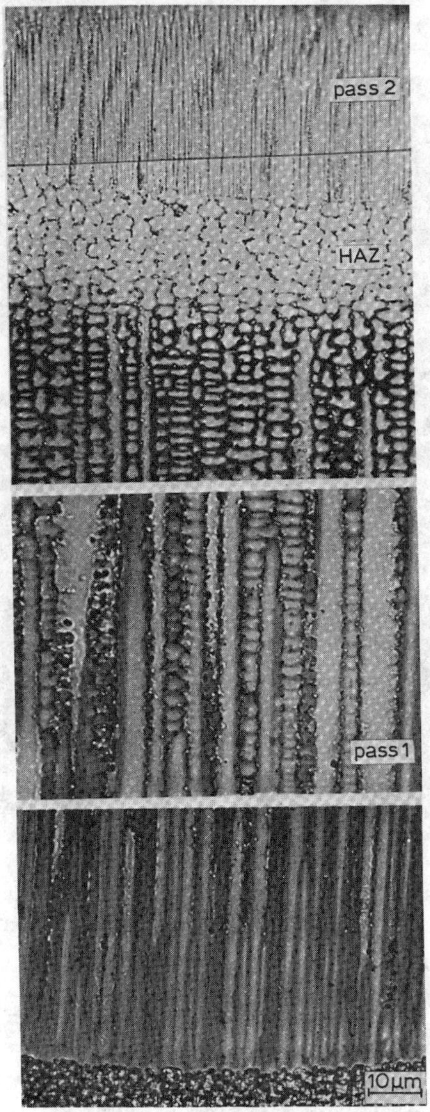

Figure 16. Refined Dendritic Structure in Glazed B1900 Superalloy Crystal (Double Pass); the (100) Surface is Normal to the Direction of the Incident Beam. Note Evidence for Epitaxial Growth Between Substrate and Pass 1, and Between Superimposed Passes 1 and 2.

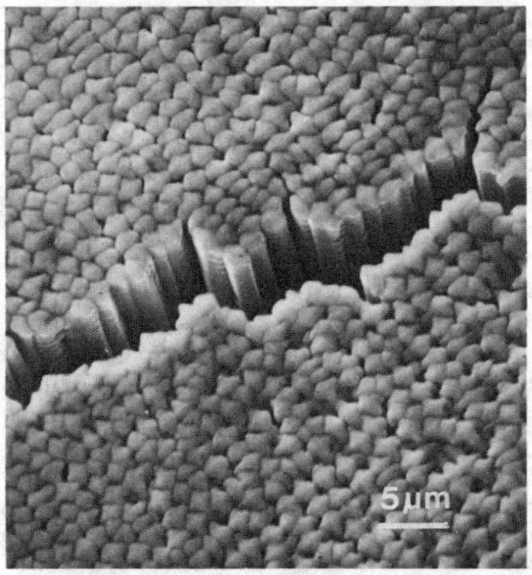

Figure 17. External Surface of Pass 2, Figure 16, Showing Cellular or Cellular-Dendritic Structure, Due to Inadequate Melt Feeding During the Last Stage of Solidification.

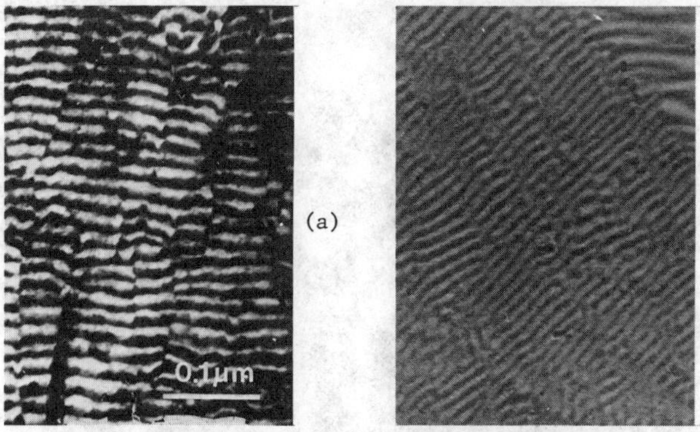

Figure 15. Ultra-Fine Lamella Eutectic in Glazed Nickel-Base Brazing Alloy (Ni-15 Co-15 Cr-5 Mo-4 B); (a) Thin Foil, (b) Replica.

(a)

(b)

(c)

Figure 18. Dendritic Structures in Glazed Eutectic Superalloys; (a) CoTaC-3, (b) γ/γ'-δ, (c) SEM Photomicrograph of Glazed Region in (a), After Partial Dissolution of Matrix. Note Network of Interdendritic Filamentary Carbide Phase.

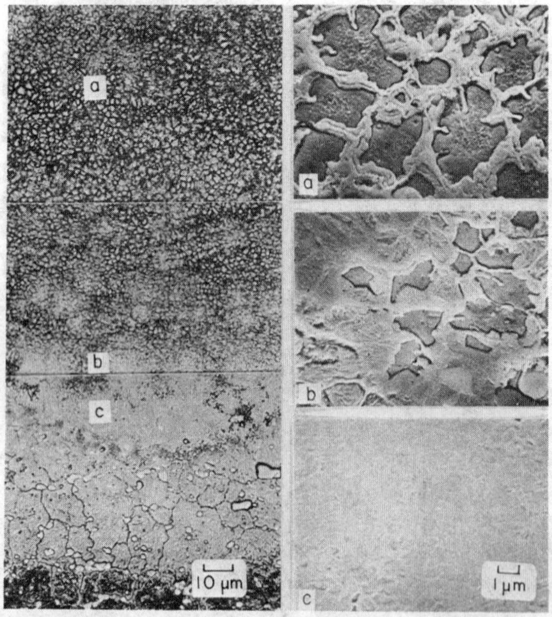

Figure 19. Microstructural Changes Within the Laser Glazed Region of Alloy M2. (a), (b), and (c) are Higher Magnification SEM Micrographs of the Regions Indicated in the Adjacent Optical Micrograph.

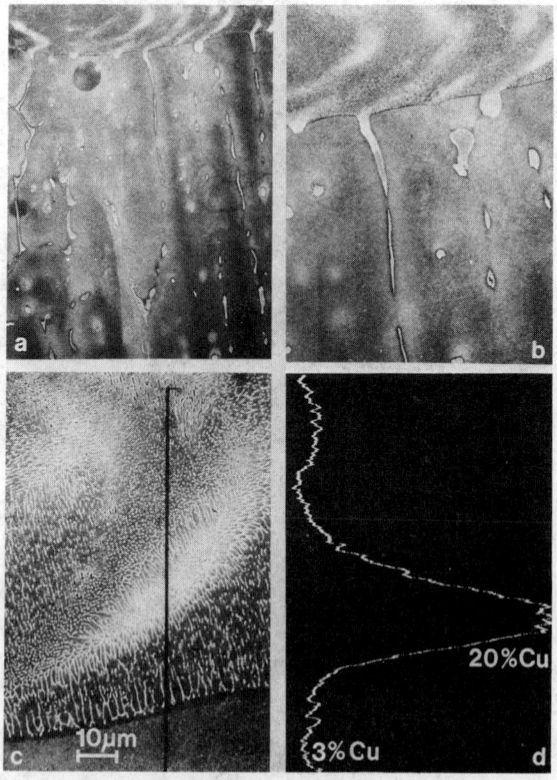

Figure 20. Microstructures of Laser Glazed Al-4.5 pct. Cu Alloy, Showing Evidence for Fluid Flow in the Melted Region; (d) is an Electron Microprobe Trace of a Copper-Rich Region Along the Path Indicated in (c).

been done on post-glaze heat treatment, with or without the benefit of mechanical deformation, so as to further modify the structure and properties of the material surface. In some limited experiments, pre-placement of alloying material on the workpiece surface prior to laser melting has been used, primarily to harden specific regions of wear in a part. More interesting, and potentially more useful, is the process that incorporates continuous delivery of alloying material to the workpiece surface. A good example is provided by Figure 21, where hard particles have been injected into the alloy substrate during laser glazing to produce a wear resistant surface (19).

By delivering material to the interaction zone on a continuous basis, as in the layer glaze process, surfaces may be coated with any desired composition, or bulk rapidly solidified material may be fabricated to 'near-net' shape. Surface alloying, using the layer glaze process with continuous pre-alloyed wire, or powder feed, is being developed for a wide range of applications, including hardfacing of valve seats, turbine blade tips, bearing surfaces and gas-path seals. Experimental work is also being conducted on layer glaze fabrication of bulk rapidly solidified structures. Simple axisymmetric shapes have already been produced by the layer glaze process, Figure 22(a). Typically, the deposited material exhibits a pronounced columnar-grained dendritic structure, Figure 22(b), with grains extending through many successive layers of material. The inherently strong tendency for epitaxial growth between layers ensures good mechanical strength at the interfaces between layers, even when the composition is deliberately changed, e.g. by changing the composition of the powder feed. Applications for this process are currently limited by the requirement that the deposited material must possess good weld-cracking resistance, and by the need to improve the shape-defining capabilities of the process. As indicated in Figure 23, the fabrication of more complex shapes requires the use of a numerically controlled work station, which is capable of simultaneous motion about at least three axes. The simple case of translation back and forth on a straight edge has obvious implications for hardfacing of cutting tools.

III. ELECTRON BEAM GLAZING

Since electron beam processing technology is well established, the requisite equipment for controlled surface melting, or 'electron beam glazing', is readily available. Furthermore, the electron beam is a particularly versatile directed energy source for such surface treatment, since a variety of melting patterns can be produced by programmed electromagnetic beam deflection.

A typical experimental arrangement is shown in Figure 24. It consists of two sets of deflection coils which permit controlled scanning of the focussed beam in two dimensions. The type of surface scanning pattern indicated is obtained by applying two nearly equal triangular waveform signals to the deflection coils. The rectangular Lissajous figure results from the progressive amount of phase lag of the y-axis signal behind the x-axis signal. In practice, various pairs of closely related frequencies f and Δf are selected from sets of frequencies derived from quartz crystal oscillators. With several possible choices of Δf at any particular frequency, f, the beam velocity and the line spacing can be varied independently.

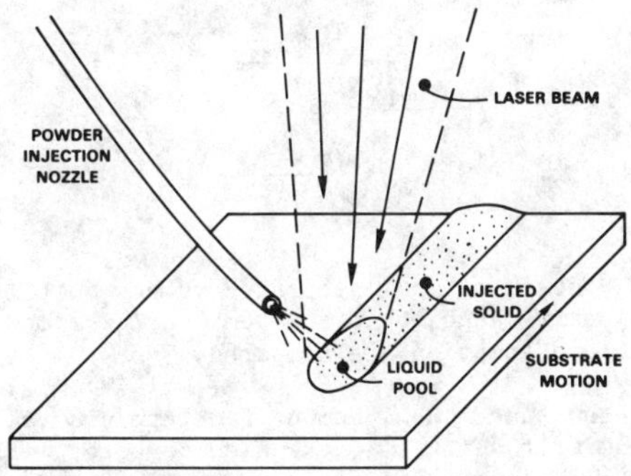

(a)

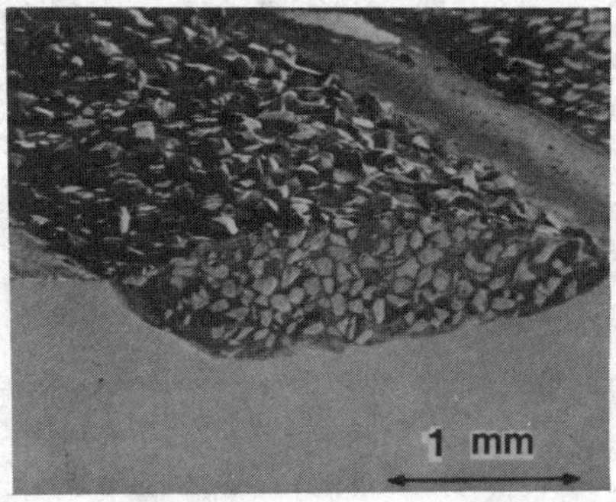

(b)

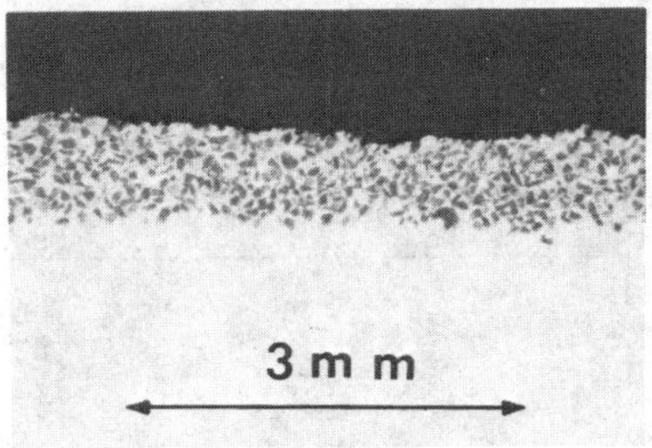

(c)

Figure 21. Surface Melting and Particle Injection; (a) Schematic of Apparatus, (b) Ti-6Al-4V Injected with TiC Particles, (c) 5052 Aluminum Injected With TiC Particles.

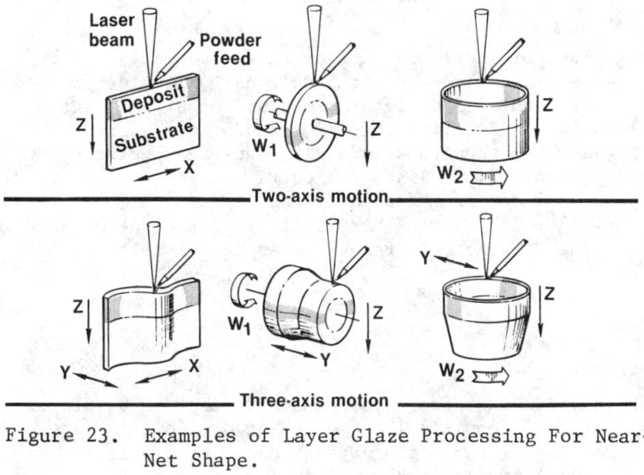

Figure 23. Examples of Layer Glaze Processing For Near-Net Shape.

(a)

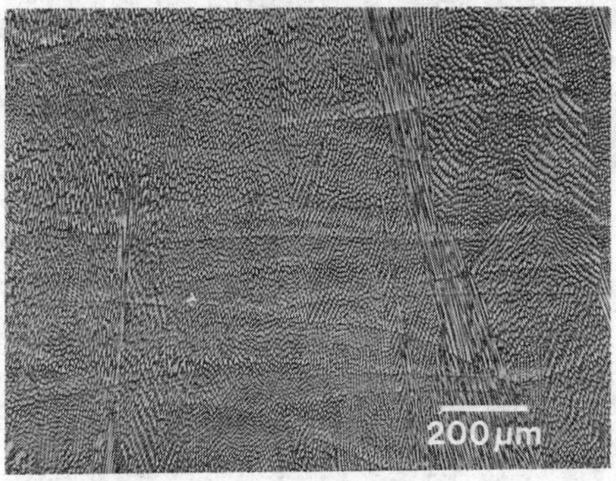

(b)

Figure 22. (a) Layer Glaze Test Sample, Produced From Pre-Alloyed Powder Feed, (b) Microstructure of Test Sample, Showing Epitaxial Growth Between Successive Deposited Layers.

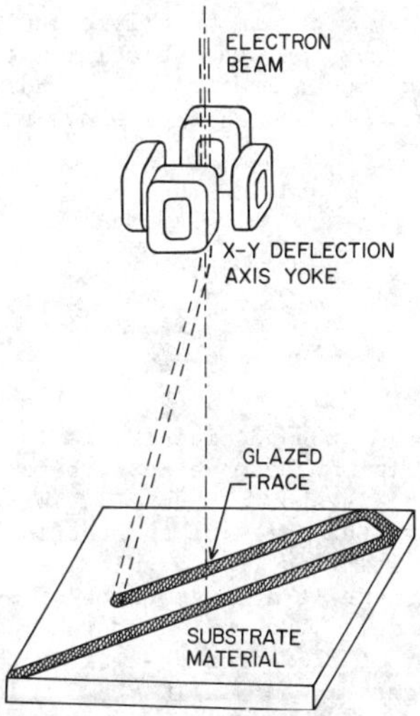

Figure 24. Schematic of Electron Beam Glazing, Utilizing Electromagnetic Beam Deflection.

(a)

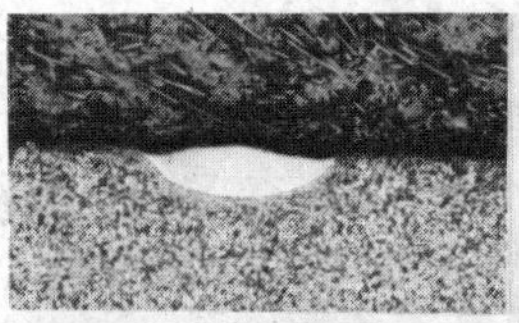

(b)

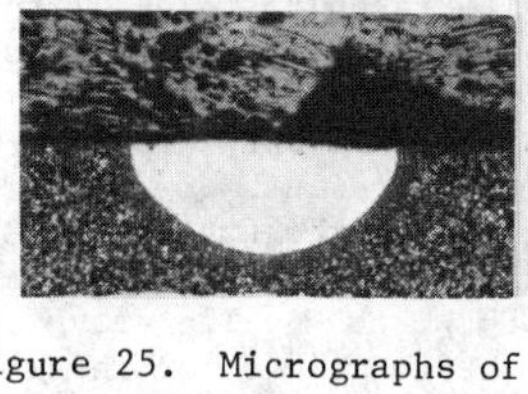

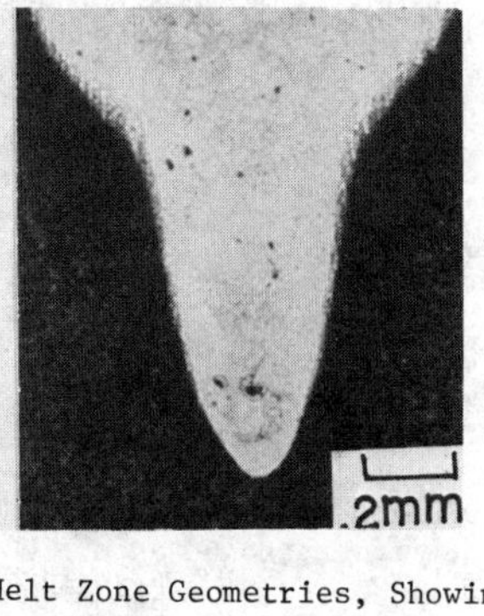

(c)

Figure 25. Micrographs of Melt Zone Geometries, Showing Influence of Beam Interaction Time; (a) 1.3×10^{-3} s, (b) 6×10^{-3} s, and (c) 3.1×10^{-2} s.

Melt Zone Geometry

The cross-sectional views in Figure 25 show the strong dependence of the melt zone geometry on the beam-material interaction time, defined as the ratio of the beam diameter to the velocity. In Figure 26(a) the maximum melt depth is shown as a function of the interaction time both for electron beam (Curve A) and laser (Curve B) surface melting of M2 steel. The full curves through the experimental points are computed from the moving point source model of Cline and Anthony (40), and the broken curves at small interaction times from the semi-infinite plate model of Greenwald et al. (29). The incident power required for laser melting is significantly greater than that required for electron beam melting, as discussed later. At present, however, it is interesting to note that in Figure 26(a), curves A and B correspond to an incident power of 375 W and 5.1 kW, respectively.

In considering the three distinct regions of curve A, Figure 26(a), the melt depth at short interation times ($\tau < 1.3 \times 10^{-3}$ s; region 1) is a fraction of the width, Figure 26(b). At higher values of τ (region 2) the melt depth increases slowly with τ, although the depth becomes about one-third of the width when τ is 6×10^{-3} s, i.e. near the termination of region 2. With this melt zone geometry, Figure 25(b), the heat flow is now essentially radially outwards. When τ increases further (greater than 1.2×10^{-2} s), surface vaporization becomes appreciable. In Figure 25(c), $\tau = 3.1 \times 10^{-2}$ s, the beam has clearly penetrated, as in the 'deep penetration mode', to produce a maximum melt depth of about 1200 µm.

On the basis of a heat flow model, such as that in Reference (29), the range of interaction times in which surface melting can occur is readily calculated. Defining V_v and V_m as the beam velocities corresponding to the maximum surface temperatures T_v and T_m respectively, the analysis in reference (29) shows that the ratio V_v/V_m is given by

$$\frac{V_v}{V_m} = \frac{(T_m - T_o)^2}{(T_v - T_o)^2} \qquad (12)$$

For alloy M2 $V_v/V_m \approx 3$ and since V_m is in excess of 100 cm s^{-1}, the minimum beam velocity before the surface begins to vaporize is about 30 cm s^{-1}. With a beam diameter of 0.5 mm the range of interaction times for actual surface melting is from $5 \times 10^{-4} - 1.7 \times 10^{-3}$ s. This restricted range is a consequence of using the semi-infinite plate model. The moving source model, which involves radial heat flow from the melt zone, gives a considerably larger range, $5 \times 10^{-4} - 2 \times 10^{-2}$ s.

As is evident in Figure 26(a), the simple model fits the experimental points particularly well for short interaction times. To derive some further useful results from the expressions in Reference (29), it is convenient to introduce the parameters a and b:

$$a = (4\alpha/\pi)^{1/2} \qquad b = q_o/k \qquad (13)$$

where α is the thermal diffusivity taken to be equal to 0.22 cm^2 s^{-1} and k is the thermal conductivity taken to be equal to 0.76 J cm^{-1} s^1 K^{-1}; the difference $T_m - T_o$ between the melting point T_m and the ambient temperature T_o is taken to be 1723 K. From measurements of the maximum melt depth and the length $\bar{L}$ of the trailing liquid zone the absorbed power for electron beam melting is determined to be about 100,000 W. cm^{-2}. The value of $\bar{L}$ is $V(\tau' - c)$, where c and τ' are the times for initial melting and final solidification at a point on the surface from the moment the power is applied. It may be shown that

$$c^{1/2} = \frac{T_m - T_o}{ab} \qquad (14a)$$

$$\tau' = \frac{(\tau + c)^2}{4c} \qquad (14b)$$

In Table I, observed and calculated values of $\bar{L}$ in the higher velocity range where the semi-infinite model is a reasonable assumption are compared.

Using the same model, computer-determined values of the maximum melt depth z_{max} are also compared with experimental values obtained from cross-sectional micrographs in Table I. On the basis of this comparison between prediction and observation, the energy coupling efficiency between the beam and the material is deduced to be about 55% for the electron beam and about 5.5% for the laser. Because of the uncertainty in the values of thermal parameters under actual melting conditions the absolute values of efficiency are, at best, approximate. This, however, does not detract from the fact that the surface-melting efficiency with an electron beam is about a factor of 10 greater than that with a laser. A similar result is obtained using the Cline and Anthony analysis, which assumes a moving source; in this case the electron beam is about 7 times more efficient than the laser.

Table I

Values of molten zone length $\bar{L}$, cooling rate $\dot{T}$, and maximum melt depth z_{max} for various beam velocities.

V(cm s^{-1})	$\bar{L}$ (µm)		$\dot{T}$(°C s^{-1})x10^6		z_{max} (µm)	
	Obs.	Calc.	Emp.	Calc.	Obs.	Calc.
38.0	10.0	9.9	1.6	0.5	108	108
57.5	8.5	6.9	2.7	1.1	67	60
76.0	5.0	5.3	3.7	2.1	23	34
97.5	3.0	4.1	5.0	3.8	18	17

From the model of Greenwald et al. (29), it may be shown that the solidification rate $\dot{T}$ is given by

$$\dot{T} = \frac{2abc^{3/2}}{(c^2 - \tau^2)} \qquad (15)$$

The values of $\dot{T}$ calculated from this equation are shown in Table I and compared with empirical values determined from the mean cell diameter λ measurements. The actual values of λ, given in Table II, are for six glazed samples, each scanned with a different beam velocity V. The adjacent column lists the corresponding values of $\dot{T}$ as calculated from the empirical relation $\dot{T} = (4.57 \times 10^{-8})\lambda^{-n}$ of Matyja et al. (41), where $1/n \approx 0.32$. From these values an empirical relation

$$\dot{T} = (2.07 \times 10^4)V^{1.2} \qquad (16)$$

is obtained with a correlation coefficient of 0.991. The comparison of empirical and calculated vlues of $\dot{T}$ in Table I is for the region where the model of Greenwald et al. (29) is reasonably valid, i.e. region 1 in Figure 26(a). The agreement is fairly good since on

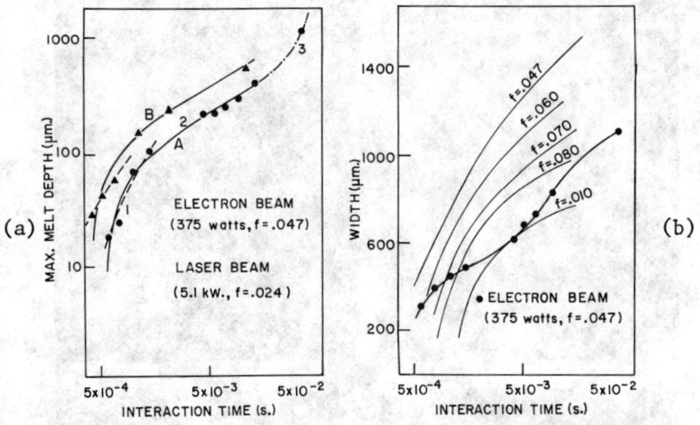

Figure 26. (a) Maximum Melt Depth, and (b) Maximum Melt Width as a Function of Interaction Time in Glazed M2 Steel; the Parameter f is Defined in Reference (40).

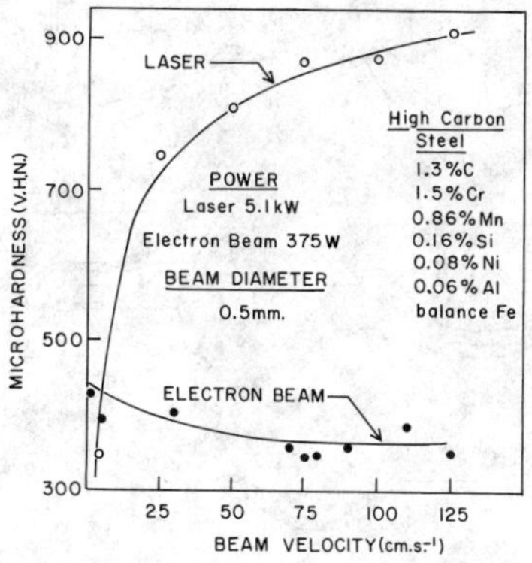

Figure 27. Comparison of Laser and Electron Beam Glazing of a High Carbon Steel, Showing Difference in Hardening Response.

average the empirical and calculated values of $\dot{T}$ differ by only a factor of 2.

Table II

The measured cellular diameter or secondary dendrite arm spacing for various beam velocities.

$V(cm\ s^{-1})$	$\lambda(\mu m)$	$\dot{T}(°C\ s^{-1})$
1.70	1.4	3.9×10^4
4.25	1.2	1.2×10^5
8.50	0.8	2.7×10^5
12.75	0.7	4.4×10^5
38.00	0.5	1.6×10^6
97.50	0.3	5.0×10^6

In addition to the intrinsic difference in beam coupling efficiency between laser and electron beam glazing treatments, a surprising finding has been the observation of a significant difference in the as-glazed hardness of a high carbon steel (42), Figure 27. The effect is evidently related to differences in the susceptibility to the formation of martensite during rapid quenching, but the mechanism is far from understood. It has been proposed that differences in environmental interactions in the two cases are responsible for this effect. In considering this aspect, it may be noted that the protective He atmosphere used in laser glazing, while greatly reducing exposure to oxygen, does not entirely eliminate it. On the basis of models for martensite nucleation (43,44), small oxide particles formed during the melting process may serve as martensite nucleation sites during cooling in the solid state. Since oxidation is much reduced during melting in the electron beam unit, martensite nucleation of oxide particles will be insignificant, and hence the microhardness is low. If other types of inclusions and small precipitates are capable of acting as martensite nucleation sites in the case of electron beam melting, then electromagnetic stirring may be effective in further retarding the formation of a martensitic structure.

Fluid Flow Effects

As mentioned briefly in Section II, amongst the many factors controlling the surface topography and microstructure of rapidly solidified layers is fluid flow in the melt zone. Of considerable interest are the surface ripple markings characteristic of glazed layers, and their sensitivity to solidification, or processing parameters.

Figure 28 shows the progressive change in the appearance of ripple markings on e.b. glazed M2 steel. Two distinct ripple periodicities are evident. The mean spacing of the large amplitude ripples is from 100 to 300 μm, whereas that between the fine ripples (clearly evident in Figure 28(b)) is about 10 μm. Recently Anthony and Cline (40) and Copley et al. (45) have studied this phenomenon, and it has been proposed (45) that the coarse ripples are created by a periodic overflow of liquid at the trailing edge of the molten pool. The fine ripples presumably result from the freezing in of surface waves, and their absence at higher velocities may well arise from attenuation in the shallow molten pool. An unusual feature noted at beam velocities of 4.0 and 8.5 cm s^{-1} is the highly serrated nature of the fine solidification ripples, with the segments aligned along specific directions, Figure 28(b). This intriguing effect suggests that the solidification surface is stabilized by becoming faceted (on a macroscopic scale) on specific crystallographic planes.

Some fluid mechanics considerations that give rise to ripples of high amplitude and low frequency, and also the central ridge/side trough topography, are summarized in Figure 29. This diagram combines and extends concepts and observations previously discussed. A critical feature is the surface depression formed beneath the incident beam; contributory causes are (i) the progressive increase in the surface tension between the center and edge of the molten pool (40), and (ii) the 'keyhole' due to the actual beam penetration into the material. Motion of the material, relative to the beam, results in formation of a liquid bulge ahead of the beam, Figure 29(a), which causes liquid flow beneath the beam. This results in liquid overflow at the trailing edge (45), Figure 29(b). The formation rate of liquid in the bulge ahead of the beam is determined by the velocity and beam power density. For instance, with a low power incident beam it takes a short time before a sufficient liquid bulge forms for liquid flow to the trailing edge to occur. Overflow at the trailing edge is thus periodic and results in ripple formation. However, with a high power incident beam the excess liquid ahead of the beam forms at a

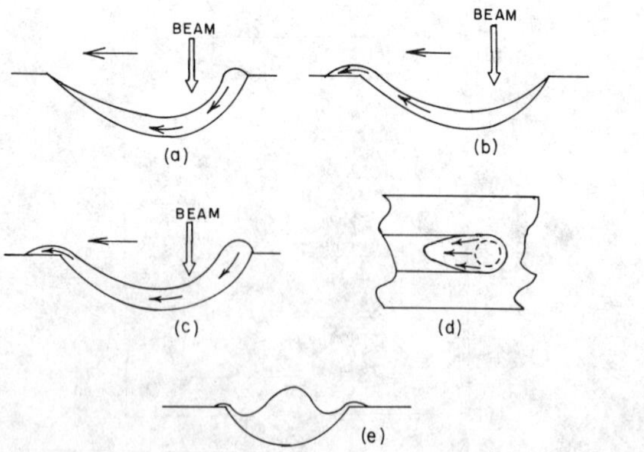

Figure 29. Schematic Diagrams Showing the Formation of the Depression Beneath the Incident Beam and Liquid Flow to Form (a) and (b) Surface Ripples, (c)-(e) Central Ridge and Side Trough Morphology.

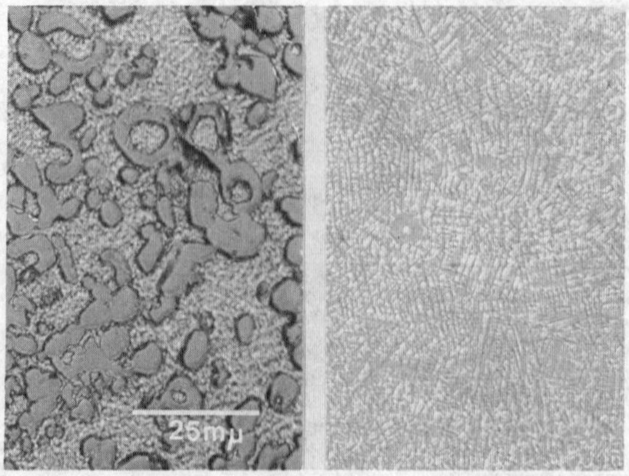

Figure 30. Microstructures of Ferro-TiC Alloy (a) As-Sintered, (b) After e.b. Glazing, Showing Refined Dendritic Carbides.

sufficient rate that there is continuous liquid flow beneath the beam, Figure 29(c). This results in the formation of a central ridge with accompanying side troughs (39), Figure 29(e). In the plan view, Figure 29(d), the arrows show the major flow directions of molten material within and around the beam depression.

In practical applications, surface smoothness may be of overriding importance, in which case the type of profile in Figure 29(e) would be highly undesirable. Fortunately, as shown in Figure 28(d), the effect can be reduced until it is almost negligible by appropriate selection of processing parameters. In other applications, however, where surface smoothness is not crucial, then the extensive fluid flow and relatively high superheat characteristic of the depression-mode of glazing can be exploited to facilitate the dissolution of refractory phases. This is shown in Figure 30 for a Ferro-TiC alloy, which normally contains large TiC particles embedded in a steel matrix. Following electron beam glazing, the carbide phase is redistributed in the form of a highly refined dendritic structure.

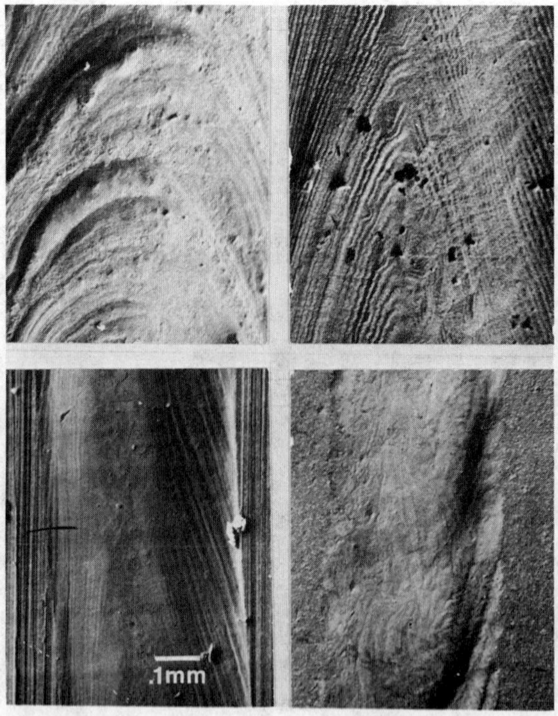

Figure 28. Scanning Electron Micrographs of Surface Ripple Structures in M2 Steel for Beam Velocities (a) 1.7 cm s^{-1}, (b) 4.25 cm s^{-1}, (c) 57.5 cm s^{-1}, and (d) 97.5 cm s^{-1}.

Beam Scanning Effects

Typical micrographs showing sequential changes in surface topography in the beam scanning mode of e.b. glazing are presented in Figure 31. Figure 31(a) shows part of a glazed surface for $\Delta f/f$ equal to 1/150. With this value the number of cycles per frame is 75. The surface finish obtained after a complete frame is shown in Figure 31(b). The effect of decreasing the ratio $(\Delta f/f)$ is to reduce the spacing distance d, and as a result produce overlapping of melted zones when $w > d$, where w is the beam diameter. An interesting finding is the remarkable increase in surface uniformity as $(\Delta f/f)$ is progressively decreased. This can be seen by comparing Figure 32(a), (b), and (c) where the corresponding values of $(\Delta f/f)$ are 1/150, 1/450, and 1/650.

The overlapping of glazed zones is shown in cross-ectional view in Figure 33. It seems likely that the increase in surface uniformity achieved by repetitive melting and solidification arises from the dispersion or evaporation of surface contaminants. This cleansing promotes progressively better wetting of the solid surface by molten liquid during each successive cycle. Thus, it is possible to eliminate the ripple structure and obtain a surface with minimal periodic topographical variations.

IV. ION BEAM METHODS

Ion beams are now used extensively to modify the surface layers of solids. Their most important application is presently in the implantation doping of semiconductors, where the constraints, such as precise spatial control, are well matched by the capabilities of ion beam technology. The range and doses of implanted atoms can be controlled with considerable accuracy. In fact, ion beam technology gives us the opportunity to tailor or modify solid state environments on the atomic

Figure 31. As-Glazed Surfaces of M2 Steel Showing (a) 14 Overlapping Glazed Traces, and (b)-(d) Entire Glazed Surfaces, With $\Delta f/f$ Respectively Equal to 1/150, 1/450 and 1/650: in Each Pass f is 60 Hertz.

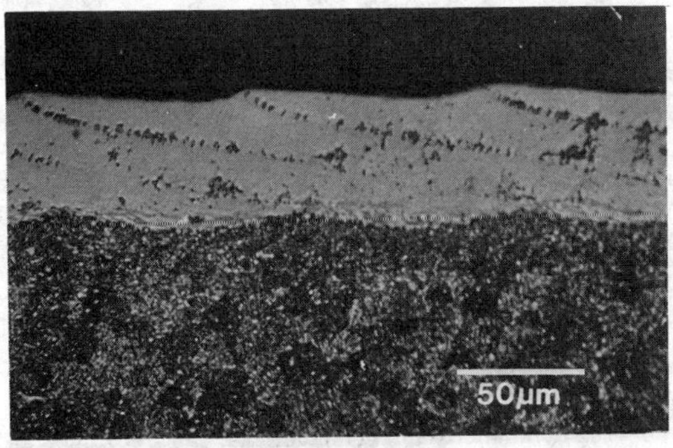

Figure 33. Cross-Sectional View of Glazed M2 Steel Showing Overlapping Passes.

scale. This has obvious implications for surface treatments of metallic materials.

In recent years unique metastable phases have been produced in metals by ion implantation. Because of sputtering limitations, these implanted metastable alloys are usually fairly dilute. Recent developments of ion beam mixing techniques, however, have permitted the study of phases over the complete concentration range of the phase diagram. The concept underlying ion beam mixing is to deposit thin films of the materials of interest, and then to mix or react them at the interface by the collision cascade generated, for example, by an inert gas atom.

Figure 32. Higher Magnification Views of (b), (c) and (d), Figure 31, Showing the Progressive Increase in Surface Smoothness Obtained by Decreasing $\Delta f/f$.

Ion Implantation

Although ion implantation is a well developed technique for modifying surface layers, it is only recently that the concepts of metastable phase formation during the implantation process have been recognized. This is perhaps surprising as ion implantation is a grossly non-equilibrium process. It can be understood, however, in light of the development of the subject. Implantation into semiconductors is by far the most widely studied and utilized branch of the subject. Typical implants into Si will result in the production of an amorphous surface layer. As the implanted dopant concentrations are usually much less than 1 at.%, the amorphous implanted layer cannot be usefully regarded as a metastable alloy but rather as doped amorphous Si. Crystallinity can be restored by annealing and the dopant atoms may then reside on regular lattice sites depending on equilibrium solid solubilities. Indeed equilibrium solid solubilities can be exceeded in this case but the systems are such exceedingly dilute solutions that they have not attracted much attention as metastable solid solutions.

In the past few years, implantation into metals at low and high concentrations has produced alloys of unique metallurgical interest (46). Such alloys are produced by implantation at room temperature without any heat treatment of the host. In metals, each host atom can be displaced many times and the lattice can still retain its basic structural integrity after the implantation. This resistance to damage is due to the lack of directionality of the metallic bond and the fact that damage anneals out easily. The situation is different for covalent semiconductors, such as Si, which can be made amorphous relatively easily by ion bombardment. The tetrahedral bonding is still essentially maintained but the atoms are assembled in a metastable amorphous network.

Dilute Alloys. Early measurements of Au implanted into single crystal Ni, Cu, Pd and Ag at concentrations less than 1 at.% established that gold is completely soluble in these fcc metals at these concentrations at room temperature. These results represent the first measurements of complete solubility formed by implantation, and are to be expected if equilibrium phase arguments are applicable to the implantation process.

On the other hand, when Ag, Sb, W and Pt are implanted into single crystal Cu the results are not interpretable in terms of equilibrium concepts, as all of the implanted species reside on substitutional sites. The Cu-Pt system forms a complete series of solid solutions, but Ag, Sb and W exhibit limited solubilities in Cu at room temperature. In fact W and Cu are immiscible in the melt. These are examples, therefore, of metastable solid solutions formed by ion implantation. How can these alloys formed by implantation be systematized? The most widely used rules of metallic alloying are those of Hume-Rothery (47), which state, in their most simplified form, that if two elements have the same crystal structure, and small differences in atomic size and electronegativity, they should form a complete series of solid solutions. One way of displaying these conditions is by a Darken-Gurry plot. Figure 34 shows such a plot for all the room-temperature low dose (<1 at.%) implantations in Cu (48). If the Hume-Rothery rules are obeyed then those elements which form substitutional alloys with the host should lie within an ellipse or circle whose major axis or radius is ≈15% of the host radius (and electronegativity difference of ±0.4). Clearly, for the case of Cu, many elements lie outside this circle but are substitutional - all the elements within the dashed box. Similar behavior is also observed for implantation of Ni and Fe. One explanation for these zones of substitutional metastability is that the final site is mainly determined by the probability of a replacement collision. For most of the atom-target combinations the probability will be quite high. Indeed channeling measurements on implantations into Al, where such wide ranges of replacement collisions are not possible, do not show extended zones of metastability. However, the replacement collision argument cannot be pursued too far, since a dense collision cascade can probably erase most memories of the replacement collisions and subsequent defect diffusion will surely influence the final lattice site location.

There is one metallic host, Be, which proves to be a nearly ideal model system. Its light mass guarantees that replacement collisions are impossible for most projectiles. The light mass is also helpful in insuring a minimum defect density in the collision cascade. In particular, heavy projectiles will create no vacancies at the end of the range. Moreover, the defects are not thought to be very mobile at room temperature. The final impurity lattice site should therefore represent the metallurgical or chemical propensities of the species involved.

Kaufmann and his colleagues, in a now classic series of experiments (49,50), implanted 25 metallic elements into Be at concentrations of ~0.1 at.%. The implanted atoms were found to occupy one of three sites in the host lattice: regular substitutional, or one of the regular octahedral or tetrahedral interstitial sites. These sites are metastable by any usual alloying definition. The first attempt to systematize the Be data involved a Darken-Gurry plot. A rough separation was obtained between the substitutional and interstitial cases, with the larger atoms being forced to remain in the interstices of the lattice. However, no separation between the octahedral and tetrahedral sites was discerned. There has been a recent advance in the theory of alloying due to Miedema and co-workers (51,52). They have developed thermochemical parameters for the elements with which signs of the heats of formation of over 500 alloy systems have been successfully predicted. Miedema's scheme is based on two parameters, ϕ^* and n_{ws}, related, respectively, to the electronic work functions of the binary alloy and the electron density in the metal at the boundary of the Wigner-Seitz cell. Both these parameters can be related to the heats of formation for binary alloys. A difference in ϕ^* between the constituents of the binary alloy, in the cellular model, results in a charge transfer across boundaries between dissimilar cells, thus providing a negative contribution to the heat of formation. A difference in n_{ws} implies the need to provide energy to smooth a discontinuity in electron density at that boundary, thus giving rise to a positive contribution to the heat of formation. The parameter n_{ws} can be related, in a more macroscopic sense, to the surface tension of the elemental liquid metal. Miedema has determined the variables ϕ^* and n_{ws} for more than 50 metallic elements by fitting the signs of ΔH_f for more than 500 binary metallic alloys with a formula of the form

$$\Delta H_f = -P(\Delta\phi^*)^2 + Q(\Delta n_{ws}^{1/3})^2 \qquad (17)$$

with the constants P, Q > 0.

Figure 35 shows a plot of the Miedema variables relative to the Be host. It is immediately obvious that the substitutional, octahedral and tetrahedral sites lie in separate domains. This is quite convincing evidence that this approach is applicable to these metastable implanted configurations. Kaufmann, et al. have used the Miedema coordinates in a modified Landau-Ginsberg expansion of the site energy differences for each element to determine the boundaries of the site regions. Thus the substitutional region is an ellipse and the octahedral and tetrahedral regions are separated by a hyperbolic contour which passes through the substitutional region. The one exception to this scheme is Li (53) which resides on a substitutional site and not the predicted tetrahedral site. This behavior is not surprising as Li implanted in Be is a case where ballistic considerations can play a significant role; the replacement collision probability is high and vacancies can be produced at the end of the range.

These experiments and theoretical developments represent significant developments in the field of implantation and alloying. It would appear that an ideal model system has been chosen, where the implanted atoms finally reside on sites determined by metallurgical constraints and not by the dynamics of the implantation process. These sites are metastable by any usual alloying definition. It is remarkable, therefore, how well the Miedema model, based on heats of formation obtained from equilibrium alloys, fits the data.

Concentrated Alloys. There is much current interest concerning high dose implantation into metals, for several reasons. The surface composition and microstructure of metals can control such important properties as corrosion, hardness and oxidation, and there is a strong technological driving force to investigate the obvious applications of implantation in metals. This subject is reviewed in the book by Hirvonen (23). Another reason for the interest is that the concentrated, metastable alloys formed by implantation can be compared with the alloys prepared by the more conventional rapid quenching techniques (46).

In 1959, Duwez, Willens and Klement (54) developed the first rapid quenching technique (splat cooling) and were able to produce a complete series of Ag-Cu solid solutions. The technique consisted of shooting molten alloys onto cooled Cu strips and, in this manner, quenching rates of 10^6 K/sec were achieved. The equilibrium alloys of Ag-Cu represent something of an anomaly as the maximum concentration of Ag in Cu is 4.9 at.% although these elements are prime candidates for obeying the Hume-Rothery rules. Duwez, et al. reasoned that if the Ag-Cu melt can be quenched rapidly enough, solid solubility may be extended across the composition range of the phase diagram.

In a series of experiments (55,56), attempts were made to produce similar metastable solid solutions by implantation. The maximum concentration of Ag that could be implanted was 16.6 at.%. At this concentration the Ag is completely substitutional even though the Cu lattice is considerably damaged. The implanted Ag is clearly in a metastable configuration. This metastability is demonstrated by the TEM diffraction pattern, Figure 36(a), of the 8 at.% implanted Ag in Cu sample. The as-implanted Cu specimen exhibits high quality single crystal diffraction spots. The diffraction pattern also

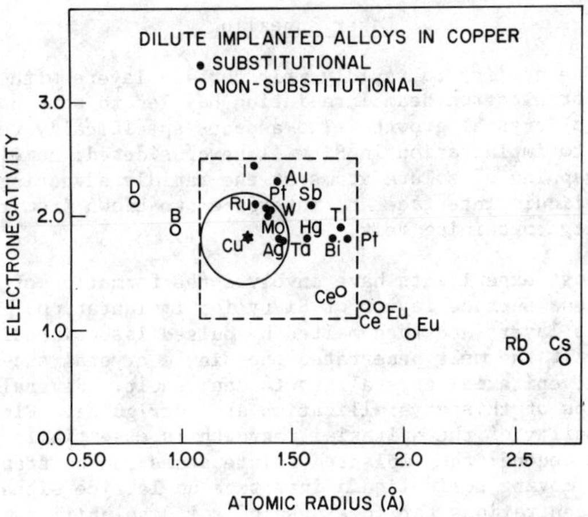

Figure 34. Darken-Gurry Plot for Dilute Implanted Alloys in Cu. Cerium and Eu are Plotted Twice According to Two Valence Values. Circle is Hume-Rothery Zone, Dashed Rectangle is Modified Hume-Rothery Zone.

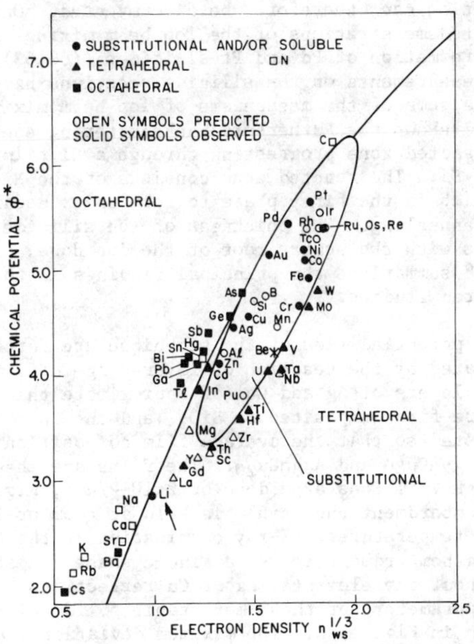

Figure 35. Miedema Plot. The Boundaries for the Site Regions Were Obtained From a Landau-Ginsberg Expansion of the Site Energy Differences.

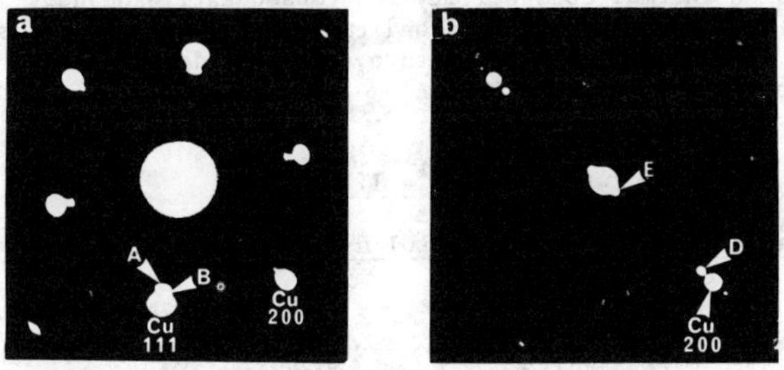

Figure 36. TEM Diffraction Patterns; (a) Of As-Implanted Ag in Cu at Concentrations of 8 at.%. Note Well Defined Cu Pole Pattern With Weak Subsidiary Spots Arising from Small Amounts of Crystalline Cu_2O (A) and Ag (B); (b) Sample Following Annealing at 400°C for 15 mins. Note Strong Metallic Ag Spots (D) and Double Diffraction (E).

shows weak subsidiary spots (A) due to epitaxial Cu_2O and very small amounts of oriented metallic Ag (B) present within the Cu matrix, although the lattice parameter of the Ag is reduced by about 1% from its equilibrium value. When the sample is annealed at 400°C for 15 min. the Cu_2O diffraction spots disappear. Moreover the diffraction pattern, Figure 36(b), exhibits extremely strong Ag spots (D), and when the corresponding Cu matrix spots are excited, pronounced double diffraction effects (E) are evident. The Ag precipitates had dimensions of the order 200 Å with their lattices parallel to that of the Cu matrix. These TEM results convincingly demonstrate that the as-implanted Ag is in a metastable solid solution, which returns to the equilibrium two-phase structure of Ag precipitates in Cu upon heating at 400°C. Many other concentrated, equilibrium and metastable alloys have been produced by implantation and these are reviewed elsewhere (46).

It might reasonably be argued that a rapid quench from the liquid should freeze in the liquid-like structure and produce an amorphous solid. The first metallic system to be made amorphous by splat-cooling from the melt was Au-Si (57). Many other metallic systems have been made amorphous by rapid quenching but they usually involve one non-metallic element (58). These structures are termed glassy metals as their electrical conductivity is metallic. One of the first metallic systems to be made amorphous by implantation appears to be W in Cu (59). Tungsten and Cu are immiscible in the melt, however, at low implantation concentrations metastable solid solutions are formed. At concentrations $\geq$ 10 at.%, amorphous surface W-Cu alloys are formed. There now exists quite a large amount of data on the formation of amorphous surface alloys by the implantation of metalloids (B or P) into Fe, Co and Ni polycrystalline foils (60). There are clear similarities between the metallic glasses formed by implantation and those formed by splat cooling.

The formation of quenched metastable alloys can be visualized by using the concept of rapidly collapsing spikes or cascades. The collapsing spikes need not revert to a crystalline structure but can retain a high degree of disorder if assisted by the presence of a large concentration of implanted 'impurity' atoms. Moreover spikes will spatially overlap at high implant concentrations. At low implant concentrations the spike can collapse to a crystalline structure and the implanted ion will find itself in a stable or metastable configuration. The phase change from metastable solid solutions to amorphous alloys can therefore be explained within the context of collapsing spikes. The role of sputtering can also be seen in the high concentration regime. The surface is continually receding, permitting the spikes to collapse in regions of high impurity concentration.

Ion Beam Mixing

Another development in ion-beam and thin-film science in recent years has been the reaction and interdiffusion of thin film structures by energetic ion beams (61). The concept behind this technique is to deposit the materials of interest in the form of thin films and then to react the films using the collision cascades generated by heavy ions. It turns out in many cases that this process is remarkably efficient and the number of mixed atoms can greatly exceed the number of bombarding atoms. In this technique concentrations are not limited by sputtering and one of the fundamental limitations of implantation can, therefore, be circumvented. One of the first demonstrations of the ion beam mixing phenomenon was the formation of Pd and Pt silicides (62,63). More recent measurements on the silicide reactions have helped elucidate some of the mechanisms of ion beam mixing (61). Figure 37 shows the Rutherford backscattering spectra of the reacted zone progressing through a Ni film evaporated on Si. The reacted zone consists of the Ni_2Si phase which is the first phase to form under usual thermal annealing. The thickness of the silicide phase increases with the square root of the ion dose. Figure 38 summarizes the principal findings in these and related studies.

The potentialities of the technique are further demonstrated by the reaction of layered Ag-Cu films (64). Multiple layers of Ag and Cu (of approximate thickness 200 Å) are first deposited on SiO_2, and the thicknesses are adjusted so that the average film composition varies between $Ag_{20}Cu_{80}$ and $Ag_{80}Cu_{20}$. The films are then mixed with 300 keV Xe ions at a dose of $2x10^{15}/cm^2$, Figure 39. During bombardment the sample is held at room or liquid nitrogen temperatures. X-ray examination of the films following bombardment showed a single phase fcc structure without any elemental Ag or Cu reflections. The lattice parameters of these metastable solid solutions are shown in Figure 40. The maximum deviation from linearity (Vegard's law for ideal solid solutions) is about 1%. These measurements of the lattice parameters of the Ag-Cu solutions produced by mixing agree with the splat cooling results of Duwez et al. (54) . It is remarkable that this single-phase metastable solid solution is produced by bombardment at liquid nitrogen. Moreover, a concentrated 1000 Å thick structure is formed by a bombardment dose of only $2x10^{15}$ Xe/cm^2. This result should be compared with the implantation of Ag into Cu, discussed previously, where the maximum attainable Ag concentration is 16.6 at.% over a depth of some 500 Å. To achieve this concentration, Ag doses in excess of $10^{17}/cm^2$ must be used.

The amount of mixing or reaction as a function of temperature for Cr films on Si is shown in Figure 41. The mixing is fairly insensitive to temperature below room temperature. But above room temperature the reaction increases very rapidly with temperature. This division into athermal and temperature activated regimes can be explained in terms of the different driving forces. In the athermal regime the mixing is dominated by the dynamic collision processes such as recoil mixing. The dynamic processes depend mainly on the efficiency of the nuclear stopping (i.e. on ion energy and mass) and should be fairly insensitive to temperature. In the high temperature regime, chemical driving forces and enhanced diffusion become dominant. Discrete phases can thus be formed. The interaction between the cascade processes and temperature of the local solid-state environment are clearly synergistic.

Novel metastable phases have been observed in the Au-Si, Au-Ge and Pt-Si systems. At present it does not appear possible to predict from the equilibrium phase diagrams which systems will show metastable phase formation. The large concentration gradients impose quite different thermodynamic driving forces than those present in implantation.

Ion beam mixing is a powerful new tool for the formation of metastable and equilibrium thin film structures. Compositions can be controlled accurately by depositing films of different thickness. The mechanisms of the mixing process are not fully understood, but the fact that mixing essentially occurs in the solid phase means that the process is very controllable. In some ways the technique is more versatile than either direct high dose implantation or the melt-quench technique.

Laser Annealing

The ability to rapidly melt surface layers with laser or electron beam irradiation has led to some novel areas of crystal growth. Two aspects specifically related to implantation in Si will be considered; namely the trapping of solute atoms at the rapidly advancing solid-liquid interface and interface breakdown from the impurity containing melt.

Most experiments have involved the formation of amorphous surface layers of Si by ion implantation. The surface layers are then melted by pulsed laser irradiation. If the melt penetrates the single crystal substrate, epitaxial crystal growth can result. Several features of this crystallization are intriguing. First, the quality of the epitaxial regrowth is essentially perfect. Second, the implanted solute atoms can be trapped by the moving solid-liquid interface on lattice sites at concentrations far in excess of solid solubility. Third, if the solute concentration in the liquid is sufficiently high, the interface can become unstable.

To quantify the processes of pulsed heating, the details of the heat flow must be understood. There now appears to be quite good agreement between calculated melt depths and velocities of regrowth and experiment. Table III shows calculations (65) of the velocity of regrowth of Si as a function of laser parameters and Si substrate temperatures. Changing substrate temperature changes thermal conductivity and consequently the heat flow. At the irradiation conditions of 1.7 J/cm^2, 15 nsec and RT Si background temperature, the melt depth penetrates approximately 3,500 Å.

Table III

Irradiation parameters and solid-liquid interface velocities.

$E(J/cm^2)$	Ruby Pulse τ(nsec)	Si Temperature	v(m/s)
1.0	15	RT	5.2
1.45	25	77K	4.2
1.5	50	RT	3.1
1.7	15	RT	3.0
2.5	50	RT	2.0
1.45	25	600K	1.8

Solute Trapping. A sensitive parameter of the crystallization process is the redistribution and trapping of impurities by the advancing solid-liquid interface. Figure 42 shows Rutherford backscattering measurements of the depth profiles of Bi implanted in (111) Si before and after laser annealing (65). The dashed line in the lower part of the figure shows the implanted distribution and the data points show the Bi distribution after regrowth at a velocity of 0.8 m/s. The smooth line is a fit to the data assuming a unique interfacial segregation coefficient k'. The profile is fitted on the basis of the calculated resolidification velocities and known liquid phase diffusivities. The fit is made under the tacit assumption that the dopant atoms are frozen in the solid and do not diffuse in the solid after the solid-liquid interface has passed; this assumption is perfectly reasonable considering the known quench rates and maximum solid state diffusivities. This value of the interface segregation coefficient of $8x10^{-2}$ is considerably higher than the equilibrium segregation coefficient of $7x10^{-4}$, as witnessed by the fact that a considerable fraction of the atoms are trapped in the lattice at concentrations far in excess of solid solubility. The fact that equilibrium solid solubilities can be exceeded by such large amounts can be explained (66-68) in terms of competition between the velocity of the moving solid-liquid interface and diffusive motion of the impurities in the liquid away from the interface. Impurities will be

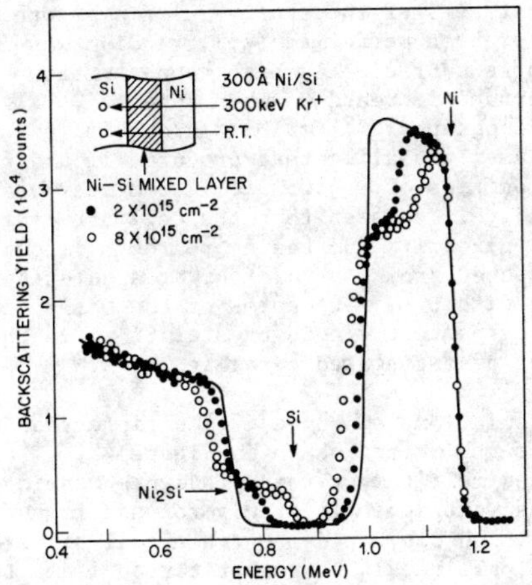

Figure 37. Formation of Ni_2Si by Ion Beam Mixing.

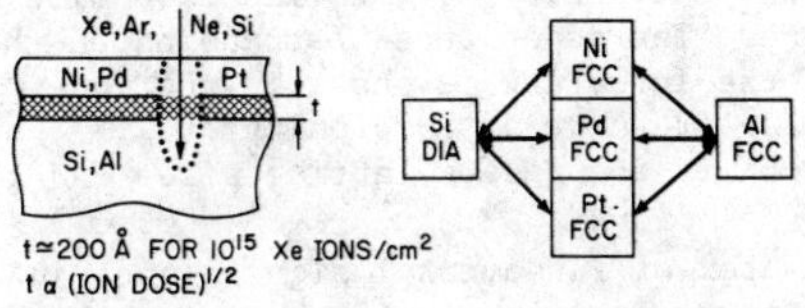

SYSTEM	THERMAL	ION BEAM (Xe AT R.T.)
Ni/Si	$Ni_2Si \rightarrow NiSi$	$Ni_2Si \rightarrow$ AMORPHOUS Ni–Si
Pd/Si	Pd_2Si (STABLE)	$Pd_2Si \rightarrow$ Pd Si
Pt/Si	$Pt_2Si \rightarrow PtSi$	$Pt_2Si \rightarrow$ AMORPHOUS Pt–Si
Ni/Al	Al_3Ni	Al_3Ni
Pd/Al	Al_3Pd	Al_3Pd_2
Pt/Al	Al_4Pt	Al_2Pt

Figure 38. Ion-Induced Compound Formation in Thin Film Structures of Ni, Pd, or Pt in Contact With Si or Al.

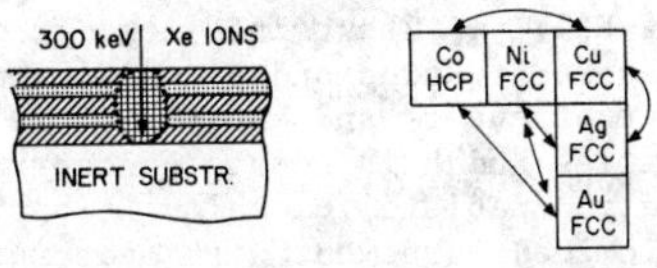

SYSTEM	THERMAL EQU.	ION BEAM MIX AT R.T.	T_T*
Ag – Cu Au – Co	EUTECTIC	(SOLID SOLUTION, COMPLETE SERIES)	200 300
Au – Ni Co – Cu	EUTECTIC	(SOLID SOLUTION, 50 – 50 COMPOS.)	350 >500
Ag – Ni Ag – Co	≈IMMISCIBLE, LIQ. & SOLID	(Ag – RICH, Ni – RICH SOLUTION (Ag, Co PHASE SEPARATION	400

*T_T ≈ TRANSITION TEMP. (°C), METASTABLE → EQUILIBRIUM

Figure 39. Ion-Induced Metastable Phase Formation for Ag-Cu, Au-Co, Au-Ni, Ag-Ni and Ag-Co Multilayered Structures on Most Substrates.

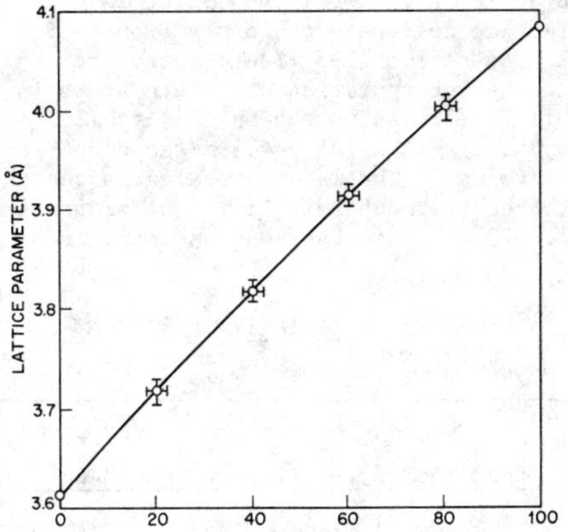

Figure 40. Lattice Parameters as Function of Composition for Ag-Cu Solid Solution Formed by Ion Beam Mixing.

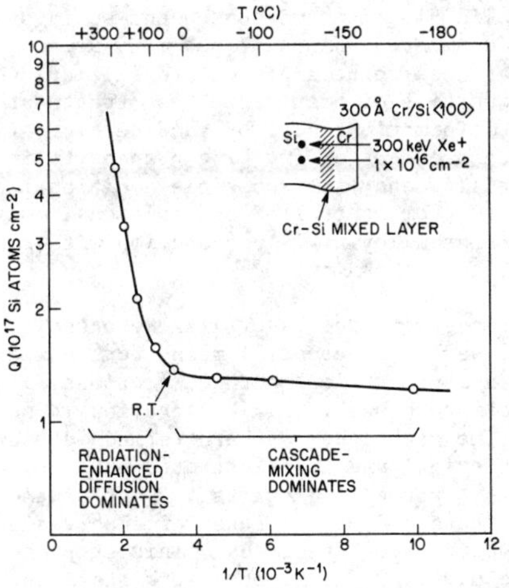

Figure 41. The Number of Si Atoms, Q, Contained in the Cr-Si Mixed Layers Versus Reciprocal Implantation Temperature.

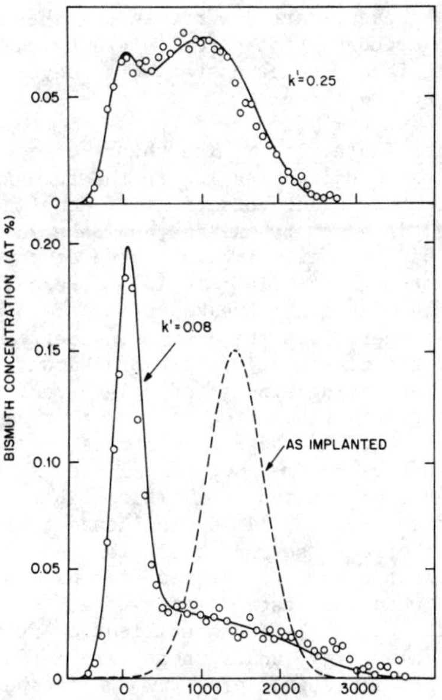

Figure 42. Bi Depth Profiles for Irradiation of (111) Si. Velocity of Regrowth in Upper Part is 2.3 m/s and 0.8 m/s in Lower. Smooth Curves are k' Fits.

trapped if they reside at the interface for greater times than the intrinsic recrystallization time of the interface. The dependence of trapping on the velocity of the interface is shown in the upper part of Figure 42 where at a velocity of 2.3 m/s most of the Bi is trapped on lattice sites and k'=0.25.

A parameter of fundamental significance in characterizing crystal growth is the orientation of the growth interface. Figure 43 shows the fitted k' values for both the orientation and velocity dependence of Bi in Si. The critical dependence of trapping in the velocity range 1-5 m/sec can be intuitively understood in terms of the near equivalence of interface velocities and diffusive velocities. The increased trapping at (111) planes over the (100) planes can also be understood intuitively in terms of greater undercooling at the (111) interface. Recently Gilmer (69) has been able to simulate the results of Figure 43 using Monte Carlo calculations of the kinetic Ising model in the presence of a large driving force.

What are the limits of trapping in a crystalline lattice? At the highest quench rate or regrowth velocities approaching 100 m/sec, there is evidence (70,71) that the lattice collapses to form amorphous Si. However, in the velocity regime (1-5 m/sec) there are other constraints to the incorporation of solute atoms in the lattice. Table IV (72) shows the solubility limits of several impurities in Si at v=4.5 m/sec compared to equilibrium values. The laser annealed As concentration is close to the thermodynamic limit, as proposed by Cahn et al. (73). The observed concentrations

Table IV

Solubility Limits

Dopant	Equilibrium (cm^{-3})	Laser Annealing (cm^{-3})
As	$1.5x10^{21}$	$6x10^{21}$
Sb	$7x10^{19}$	$1.3x10^{21}$
Ga	$4.5x10^{19}$	$4.5x10^{20}$
In	$8x10^{17}$	$1x10^{20}$
Bi	$8x10^{17}$	$4x10^{20}$

for the other dopants are lower than the thermodynamic limits. The concentrations are limited by the fact that the interface becomes unstable and develops a cellular structure.

Interface Instability. The theory of the conditions for interface breakdown, leading to the formation of a cellular structure under conditions of constitutional supercooling, is given by Mullins and Sekerka (74). This theory has been extensively tested at conventional rates of crystal growth (10^{-5} to 10^{-2} cm/sec). Under these conditions cellular breakdown occurs on a scale of order D/v, where D is the diffusion coefficient of the solute and v the growth velocity. Breakdown occurs when the destabilizing effects of solute redistribution outweigh the stabilizing effect of the thermal gradients in the system. At very high growth rate, this wavelength scale (D/v) becomes very small and the deformed interface has a high curvature. This high curvature provides a strongly stabilizing influence through the Gibbs-Thomson effect, so that breakdown now occurs only under conditions of a large degree of constitutional supercooling when the destabilizing effect of the solute redistribution outweighs the stabilization provided by the interfacial energy; under these circumstances the temperature gradients will play only a secondary role. Hitherto this prediction of the theory has not been amenable to test. However, the advent of pulsed laser annealing has made accessible recrystallization velocities of several meters per second and has opened up a regime where interfacial energy effects can dominate.

The first test of the theory in this growth velocity regime has now been performed (75). Indium was implanted in (100)Si at a sufficiently high concentration to produce interfacial breakdown when annealed. Figure 44 shows TEM micrographs of 3×10^{15} In/cm^2 implantations after Ruby laser annealing that produced liquid phase regrowth velocities of v=4.0, 3.0, 2.0 m/sec, respectively. It can immediately be seen that the cell structure decreases in size with increasing regrowth velocity, as would be expected from the first order arguments of cell size scaling with D/v. The cores of the cells consist of essentially perfect single crystal Si while the walls consist of segregated metallic In.

The calculations of Hurle (76) on interface stability for this regime are shown in Figure 45. The growth velocities and melt temperature gradients were computed. The k' values were determined only for specimens implanted at low dose which did not demonstrate cell structure. In this case, all the In was substitutional and the depth distributions could be fitted with a single k' value; such a fitting could not be made for specimens with a cellular structure. The plot gives the critical interface parameter mC_s/DG as a function of k', where m is the slope of the liquidus line, C_s is the solute concentration in the crystal and G is the weighted temperature gradient. The curves, for 3 velocities, represent the boundaries of stability. The points come from the TEM and Rutherford backscattering and channeling measurements. The agreement of theory and experiment is remarkable. At low velocities and high solute concentrations the interface becomes unstable. The calculations of cell size are also in good agreement with measurement.

V. FUTURE PERSPECTIVE

The ion beam processing methods, because of their great precision, will continue to play an important role in the further development of our understanding of surface modifications and alloying. The techniques are widely used in the semiconductor industry and the prospects for further extension to metallic systems look very promising. For example, there is much current interest in modifying such surface properties as hardness, wear and corrosion resistance, and other chemical and physical properties.

Although laser and electron beam glazing treatments have been shown to be capable of producing novel surface structures and properties in a variety of materials, much remains to be done to optimize these processes for specific applications. Future work is likely to be concerned primarily with this problem, now that the major effects of rapid solidification and subsequent solid state cooling have been documented. Amongst the many aspects of the problem that need to be explored are (i) influence on surface properties of post-glaze heat treatment and/or mechanical deformation, e.g. by shot peening, (ii) role of substrate pre-heat on cracking propensity in susceptible materials, and (iii) possible additional benefits to be derived from controlled compositional modifications, e.g. by surface alloying in conjunction with rapid solidification processing.

Experience with the layer glaze process has also pointed the way to a potential means for controlling structures and properties in thick sections, and even bulk parts, more closely than has previously been possible. The technique has been termed 'directed energy processing' and is illustrated schematically in Figure 46. Processing starts at work station #1 with material being added by the laser or electron beam melting and deposition of feedstock. This step essentially creates the part. At additional work stations, other materials processing operations can be sequentially accomplished on each incremental layer. These include inspection, mechanical deformation and/or heat treatment. Like deposition, which derives advantages from rapid solidification of a thin section, each of these subsequent operations also derives advantages from being applied to a small volume element. Incremental inspec-

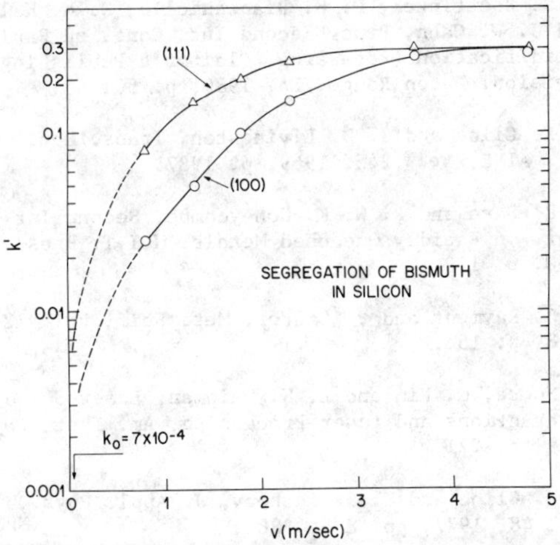

Figure 43. Bi Interface Segregation Coefficients in Si as a Function of Velocity and (100) and (111) Orientation.

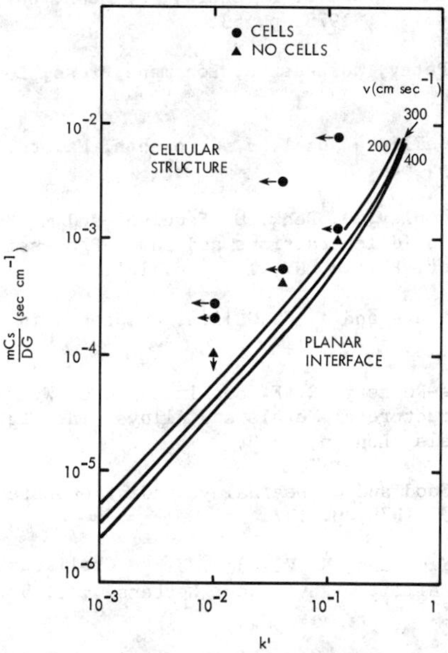

Figure 45. Plot of the Critical Parameter Specifying Interface Breakdown vs. Segregation Coefficient. The Points are Experimental Values (▲=Planar Interface; ● = Cellular Structure).

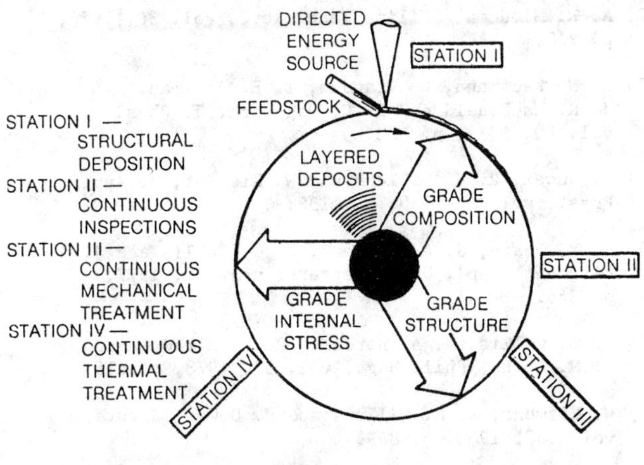

Figure 46. Schematic Representation of Directed Energy Processing.

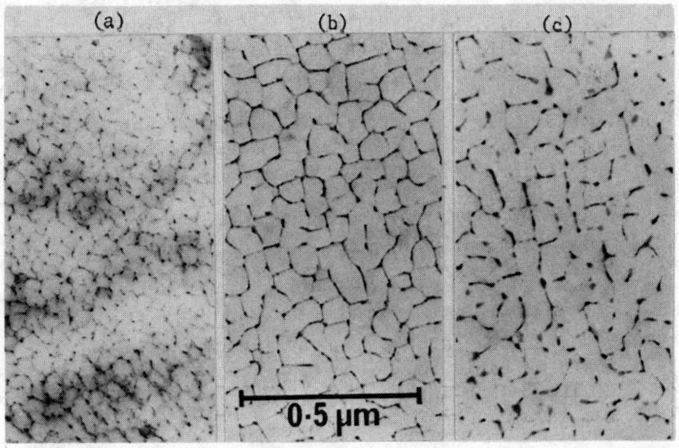

Figure 44. TEM Images of Si Layers Implanted with $3 \times 10^{15} cm^{-2}$ In and Ruby Laser Annealed at 1.5 J/cm^2. Resolidification velocities of a) 4m/sec, b) 3m/sec and c) 2m/sec.

tion can be performed without destructively testing the part. Mechanical processing can be applied to small increments without resulting in significant distortion, and thermal treatments, or even remelting can be carried out by interrupting the material feed.

REFERENCES

(1) W. A. Elliott, F. P. Gagliano and G. Krauss, Met. Trans., vol. 4, 1973, p. 2031.

(2) E. M. Breinan, B. H. Kear, C. M. Banas and L. E. Greenwald, Proc. Third Int. Symp. on Superalloys, Claitor's Publishing Division, Baton Rouge, LA, 1976, p. 435.

(3) B. H. Kear, E. M. Breinan and L. E. Greenwald, Metals Technology, April, 1979, p. 121.

(4) L. S. Weinman and C. Kim, Proc. First Int. Conf. on Rapid Solidification Processing, Claitor's Publishing Division, Baton Rouge, LA, 1977, p. 117.

(5) P. R. Strutt, M. Tuli, H. Nowotny and B. H. Kear, Mat. Sci. Eng., vol. 36, 1978, p. 217.

(6) T. R. Anthony and H. E. Cline, J. Appl. Phys., vol. 49, 1978, p. 1248.

(7) S. C. Hsu, S. Chakravorty and R. Mehrabian, Met. Trans., vol. 9B, 1978, p. 221.

(8) Y. W. Kim. P. R. Strutt and H. Nowotny, Met. Trans. vol. 10A, 1979, p. 881.

(9) S. L. Narasimhan, S. M. Copley, E. W. Van Stryland, and M. Bass, Met. Trans., vol. 10A, 1979, p. 654.

(10) K. Lipscombe, W. M. Steen and D. R. F. West, Proc. Second Int. Conf. on Rapid Solidification Processing, Claitor's Publishing Division, Baton Rouge, LA, 1980, p. 189.

(11) W. Gruhl, B. Grzemba, G. Ibe and W. Hiller, Aluminum, vol. 53, 1977, p. 177.

(12) B. G. Lewis, D. A. Gilbert and P. R. Strutt, Proc. Second Int. Conf. on Rapid Solidification Processing, Claitor's Publishing Division, Baton Rouge, LA, 1980, pp 221, 225.

(13) R. J. Schaefer, T. R. Tucker, and J. D. Ayers, Laser and Electron Beam Processing of Materials, Academic Press, NY, 1980, p. 749.

(14) P. R. Strutt, Mat. Sci. Eng., vol. 44, 1980, p. 239.

(15) B. G. Lewis, D. A. Gilbert and P. R. Strutt, Laser and Electron Beam Processing of Materials, Academic Press, NY, 1980, p. 747.

(16) S. Kadalbal, J. Montoyo-Cruz and T. Z. Kattamis, Proc. Second Int. Conf. on Rapid Solidification Processing, Claitor's Publishing Division, Baton Rouge, LA, 1980, p. 195.

(17) D. S. Gnanamuthu, Applications of Lasers in Materials Processing, American Society for Metals, Ohio, 1979, p. 177.

(18) E. M. Breinan and B. H. Kear, Proc. First. Int. Conf. on Rapid Solidification Processing, Claitor's Publishing Division, Baton Rouge, LA, 1978, p. 87.

(19) J. D. Ayers, T. R. Tucker and R. J. Schaefer, Proc. Second Int. Conf. on Rapid Solidification Processing, Claitor's Publishing Division, Baton Rouge, LA, 1980, p. 212.

(20) E. M. Breinan, D. B. Snow, C. O. Brown and B. H. Kear, Proc. Second Int. Conf. on Rapid Solidification Processing, Claitor's Publishing Division, Baton Rouge, LA, 1980, p. 440.

(21) G. Dearnaley, J. A. Freeman, R. S. Nelson and J. Stephen, "Ion Implantation", North Holland, Amsterdam, 1973.

(22) G. Carter, J. J. Colligon and W. A. Grant (eds.), Applications of Ion Beams to Materials, Institute of Physics, London, Conf. Series #28, 1975.

(23) J. K. Hirvonen (ed.), "Ion Implantation", Mat. Sci. and Tech., vol. 18, 1980.

(24) C. M. Preece and J. K. Hirvonen (eds.) Ion Implantation Metallurgy, AIME, 1980.

(25) L. E. Greenwald, United Technologies Research Center Report, R75-111321-1, Oct. 1975.

(26) M. I. Cohen, J. Franklin Inst., vol. 283, 1967, p. 271.

(27) R. C. Ruhl, Mat. Sci. and Eng., vol. 1, 1967, p. 313.

(28) R. Mehrabian, Proc. First Int. Conf. on Rapid Solidification Processing, Claitor's Publishing Division, Baton Rouge, LA, 1977, p. 9.

(29) L. E. Greenwald, E. M. Breinan and B. H. Kear, Laser-Solid Interactions and Laser Processing, Am. Inst. of Physics, NY, 1978, p. 189.

(30) R. Mehrabian, S. Kou, S. C. Hsu and A. Munitz, Laser-Solid Interactions and Laser Processing, Am. Inst. of Physics, NY, 1978, p. 129.

(31) M. N. Ozisik, Boundary Value Problems of Heat Conduction, Inter. Textbook Co., 1968.

(32) M. C. Flemings, Solidification Processing, McGraw-Hill, NY, 1974.

(33) S. R. Coriell and R. F. Sekerka, Proc. Second Int. Conf. on Rapid Solidification Processing, Claitor's Publishing Division, Baton Rouge, LA, 1980, p. 35.

(34) R. Becker, G. Sepold and P. L. Ryder, (private communication).

(35) W. J. Boettinger, F. S. Biancaniello, G. M. Kalonji and J. W. Cahn, Proc. Second Int. Conf. on Rapid Solidification Processing, Claitor's Publishing Division, Baton Rouge, LA, 1980, p. 50.

(36) H. E. Cline and J. D. Livingston, Trans. Met. Soc. AIME, vol. 245, 1969, p. 1987.

(37) I. R. Sare and R. W. K. Honeycombe, Second Int. Conf. on Rapidly Quenched Metals, M.I.T. Press, 1960, p. 179.

(38) J. J. Rayment and B. Cantor, Met. Sci., vol. 12, 1978, p. 156.

(39) P. Moore, C. Kim and L. W. Weinman, Laser-Solid Interactions and Laser Processing, Am. Inst. Phys., 1979, p. 221.

(40) H. E. Cline and T. R. Anthony, J. Appl. Phys., vol. 48, 1977, pp. 895, 3888.

(41) H. Matyja, B. C. Giessen and N. J. Grant, J. Inst. Met., vol. 9, 1968, p. 30.

(42) J. Wadsworth and O. D. Sherby, J. Mech. Working Tech. vol. 2, 1978, p. 53.

(43) E. R. Petty, Martensite, Longmans Press, London, 1970.

(44) J. M. Galligan and T. J. Garosshen, Nature, vol. 274, 1978, p. 664.

(45) S. M. Copley, D. Beck, O. Esquivel and M. Bass, Laser-Solid Interactions and Laser Processing, Am. Inst. Physic, NY, 1979, p. 161.

(46) J. M. Poate and A. G. Gullis, Chapter 4 in ref. 23, 1980.

(47) W. Hume-Rothery, R. E. Smallman and C. W. Haworth, The Structure of Metals and Alloys, Institute of Metals, London, 1969.

(48) D. K. Sood and G. Dearnaley, Radiation Effects, vol. 39, 1978, p. 157.

(49) E. N. Kaufmann, R. Vianden, J. R. Chelikowsky and J. C. Phillips, Phys. Rev. Letters, vol. 39, 1977, p. 1671.

(50) R. Vianden, E. N. Kaufmann and J. W. Rodgers, Phys. Rev., vol. B22, 1980, p. 63.

(51) A. R. Miedema, F. R. DeBoer and P. F. DeChatel, J. Phys. F, Metal Phys., vol. 3, 1973, 1558.

(52) A. R. Miedema, Philips Tech. Rev., vol. 36, 1976, p. 217.

(53) E. N. Kaufmann, R. Vianden, T. E. Jackman, J. R. MacDonald and L. C. Haggmark, J. Phys., vol. F9, 1979, p. L23.

(54) P. Duwez, R. H. Willens and W. Klement, J. Appl. Phys., vol 32, 1960, p. 1136.

(55) J. M. Poate, J. A. Borders, A. G. Cullis and J. K. Hirvonen, Appl. Phys. Letters, vol. 30, 1977, p. 365.

(56) A. G. Cullis, J. A. Borders, J. K. Hirvonen and J. M. Poate, Phil. Mag., vol. 37, 1978, p. 615.

(57) W. Klement, R. H. Willens and P. Duwez, Nature, vol. 187, 1960, p. 869.

(58) P. Duwez, Annual Review of Material Science, vol. 6, 1976, p. 83.

(59) A. G. Cullis, J. M. Poate and J. A. Borders, Appl. Phys. Letters, vol. 28, 1976, p. 316.

(60) P. J. Grundy, A. Ali and W. A. Grant, Thin Film Phenomena, The Electrochemical Society, vol. 78-2, 1978, p. 244.

(61) J. W. Mayer, B. Y. Tsaur, S. S. Lau and L. S. Hung, Nucl. Instr. Methods (in press), 1981.

(62) W. F. Van Der Weg, D. G. Sigurd and J. W. Mayer, Applications of Ion Beams to Metals, Plenum, NY, 1974, p. 209.

(63) J. M. Poate and T. C. Tisone, Appl. Phys. Letters, vol. 24, 1974, p. 391.

(64) B. Y. Tsaur, Ph.D. Thesis, California Institute of Technology, 1980.

(65) P. Baeri, J. M. Poate, S. U. Campisano, G. Foti, F. Rimini and A. G. Cullis, Appl. Phys. Letters, vol. 37, 1980, p. 912.

(66) D. Turnbull, J. DePhysique, vol. 14, 1980, p. C4-109.

(67) K. A. Jackson, G. H. Gilmer and H. J. Leamy, Laser and Electron Beam Processing of Materials, Academic Press, NY, 1980, p. 104.

(68) R. F. Wood, Appl. Phys. Letters, vol. 37, 1980, p. 302.

(69) G. H. Gilmer (private communication), 1981.

(70) R. Tsu, R. T. Hodgson, T. Y. Tan and J. E. Baglin, Phys. Rev. Letters, vol. 42, 1979, p. 1356.

(71) P. L. Lui, R. Yen, N. Bloembergen and R. T. Hodgson, Laser and Electron Beam Processing of Materials, Academic Press, NY, 1980, p. 156.

(72) C. W. White, S. R. Wilson, B. R. Appleton and F. W. Young, J. Appl. Phys., vol. 50, 1980, p. 3261.

(73) J. W. Cahn, S. R. Coriell and W. J. Boettinger, Laser and Electron Beam Processing of Materials, Academic Press, NY, 1980, p. 89.

(74) W. W. Mullins and R. F. Sekerka, J. Appl. Phys. vol. 35, 1964, p. 44.

(75) A. G. Cullis, D. T. J. Hurle, H. C. Webber, N. G. Chew, J. M. Poate, P. Baeri and G. Foti (in press), 1981.

(76) D. T. J. Hurle (to be published), 1981.

POWDER METALLURGY

F. V. LENEL
G. S. ANSELL
Rensselaer Polytechnic Institute
Troy, NY 12181

In powder metallurgy metallic products are produced by the basic steps of

producing metal powders
pressing the powders and
sintering the compacts

rather than by melting and casting or melting, casting and subsequent working. Technological applications of powder metallurgy are:

Products which cannot be or are not readily produced by fusion metallurgy including refractory metals, cemented carbides, porous products, combinations of metals insoluble in each other in the liquid state and combinations of metals and non-metals, i.e. "cermets."

Products which are fabricated more economically by powder metallurgy than by fusion metallurgy processes. These include structural parts and magnetic parts produced to close dimensional tolerances and products from high cost alloys which can be produced to "near net shape" by powder metallurgy.

Products whose properties are superior when produced by powder metallurgy because of limitations of the fusion metallurgy processes.

As in other metal working technologies, the basic understanding of the processing steps involved in powder metallurgy is lagging behind the development of new practical applications in the field. To discuss only the advances which have been made in the science of powder metallurgy without a brief description of some of what is new in the art of powder metallurgy would give an unbalanced picture of the progress made in recent years. For this reason only the first portion of this article will be concerned with recent work on the scientific aspects of powder metallurgy, i.e. sintering of single phase powders and liquid phase sintering.

In the second part of the paper, the production of powder metallurgy structural parts by hot forging preforms will be discussed. In this area considerable progress has been made not only in the basic technology, but also in the understanding of the factors which are important in the design of forging preforms.

The subject of part three is the area of powder metallurgy technology which is currently advancing most rapidly, the production of components by hot consolidation of powders. Methods of producing alloy powders suitable for hot consolidation will be discussed. The processing steps in hot consolidation, hot isostatic pressing, hot pressing in rigid dies and hot extrusion of metal and alloy powders, will be treated. The manufacture and the properties of products fabricated by these techniques will be described. They include beryllium, aluminum alloys, dispersion strengthened copper alloys, superalloys, tool steels and titanium alloys.

SINTERING THEORY

Single Phase Powders

The mechanism, by which particles of a single phase metal powder sinter together when they are brought into contact as loose powder particles or when they are pressed into a compact, has been under investigation for more than thirty years. The principal approach taken has been by the use of model experiments, in which spherical particles are sintered together, a spherical particle is sintered to a flat plate, two or more wires are sintered together or an array of wires is wound on a core and sintered to it. The phenomena observed are characteristic of the early stage of sintering, i.e. the increase in contact area between particles or wires and the decrease in distance between centers of spheres or wires.

These model experiments serve, first of all, to define the principal driving force in sintering which is the decrease in surface free energy, often called surface tension. A gradient in surface tension stress, which is equivalent to a chemical potential gradient, exists between highly curved surfaces at the necks between particles and adjacent surfaces with lower curvature. This gradient will cause a gradient in vacancy concentration. Kuczynski(1) developed rate equations for material transport by volume diffusion and by surface diffusion under the influence of this vacancy concentration gradient. There is another vacancy concentration gradient between the highly curved neck surface under a tensile stress and the grainboundary formed at the contact area, which is generally under a compressive stress. This gradient causes a material flux from the grainboundary by either grainboundary or volume diffusion(2). An effort has been made to distinguish between the contributions of the various fluxes to neck growth and shrinkage on the basis of scaling laws(3). Another approach has been to develop analytical techniques by which the contributions of the various fluxes may be added up(4). A principal problem in this development has been to correctly identify the geometrical changes taking place when a sphere is sintered to a sphere or a wire to a wire(5).

In addition to diffusional flux, plastic flow has been suggested as a material transport mechanism in sintering(6). Several experiments, such as the influence of second phase particles upon the rate of neck growth(7) and the difference in the rates with which wires with different crystallographic orientations with respect to each other sinter together(8), seem to point toward a contribution by plastic flow to material transport. Another evidence of plastic flow are the results of model experiments in which single crystal copper spheres are sintered to single crystal copper plates(9). The dislocation concentrations in the contact area after sintering are found to be much higher than those in the original well annealed material.

Both the model experiments and theoretical calculations of the stresses necessary to generate and move dislocations, seem to indicate that dislocation motion may be important in the early stages of sintering, but that diffusional transport dominates over most of the sintering process. There are still problems involved in developing analytical expressions for the rates of sintering of individual spherical particles or wires, but a solution does not seem too far off[10].

However, when a large number of particles and the contacts between them are involved, as in the sintering of actual compacts, the constraints which the sintering of one neck region exerts upon that of other neck regions make it very difficult to treat analytically the flow of material in the contacts between particles[11]. Constraints are important, even when only three instead of two spherical particles are sintered together. It is found that not only the distances between particles, but also the angles between the lines connecting the centers of the particles change, as is seen in Figure 1[12].

Figure 1. Sintering effect on an array of three spherical copper particles sintered at 1020°C, 300 x (a) 1 minute, (b) 2 hours, (c) 50 hours. The angle between the particles, initially larger than 90°, increases with increasing sintering time.

How the constraints affect the sintering of planar arrays of spherical particles is shown in Figure 2.

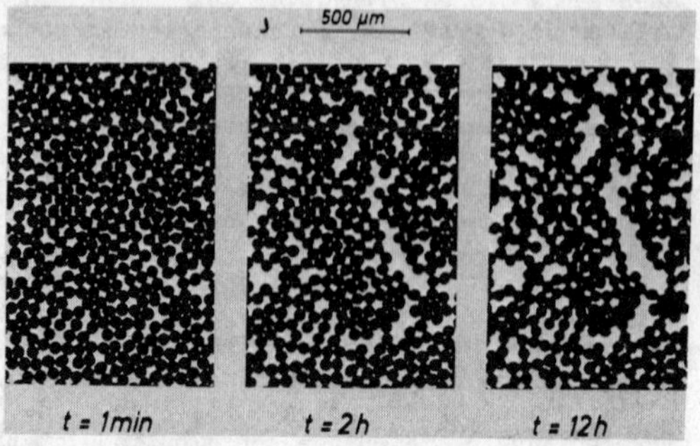

Figure 2. Sintering of planar arrays of spherical copper particles at 1020° in hydrogen, 65 x (a) original arrangement, (b) sintered 2 hours, (c) sintered 12 hours.

Instead of the uniform shrinkage which would be expected on the basis of the two particle model, it is found that the area of small pores between particles decreases - most of the triangular pores disappear - but the area of larger pores grows steadily. Centers of densification are formed, where rapid shrinkage takes place, but this shrinkage is partially compensated by the opening of large pores[12].

The problem of translating the results of calculations for the sintering of arrays of two spherical particles or two wires to the sintering of large assemblies of particles is somewhat analogous to that of extending the treatment of plastic deformation in single crystals to that applicable for polycrystalline aggregates. The constraints and the compatibility requirements which must be taken into account in treating sintering of large particle aggregates loosely correspond to those in the plastic deformation of polycrystalline bodies.

Model experiments are particularly valuable to analyze the initial stage of sintering, in which the necks between particles grow, the pores between particles become rounded and the powder aggregate begins to densify, but where the particles in the original aggregate are still distinguishable. In the later stages of sintering, the intermediate stage, in which pore channels between particles close, and the final stage, in which the pores become isolated, the same driving forces, i.e. surface tension forces, and the same material transport mechanisms, primarily diffusional flow, control the sintering process. These stages cannot be readily studied by model experiments. Instead, theoretical geometrical models have been constructed to calculate shrinkage during these stages[13]. In order to make the calculations manageable, geometrically simple models have been used. As in the case of initial stage sintering, further studies of these simple models have shown that they are not realistic. At the present time, the principal effort to understand the intermediate and final stages of sintering are based on a statistical theory of sintering, in which a statistical distribution of the diameters of the pore channels in the second stage and of the pore diameters in the final stage is assumed[14].

Liquid Phase Sintering

The best understood type of liquid phase sintering is the one in which a limited amount of liquid phase is present during the entire time while the compacts are at the sintering temperature. It has been called the "heavy alloy mechanism," because it was first investigated in alloys of tungsten with up to ten percent of lower melting alloying ingredients, such as nickel and copper or nickel and iron. The alloys are technologically important because of their high density. They are produced by compacting a mixture of the elemental powders of tungsten and of the alloying ingredients and sintering the compacts. Phenomena observed during the sintering of heavy alloy are the formation of a characteristic microstructure of relatively large single crystal grains of tungsten in a matrix of the alloy phase (see Figure 3) and rapid and complete densification at temperatures

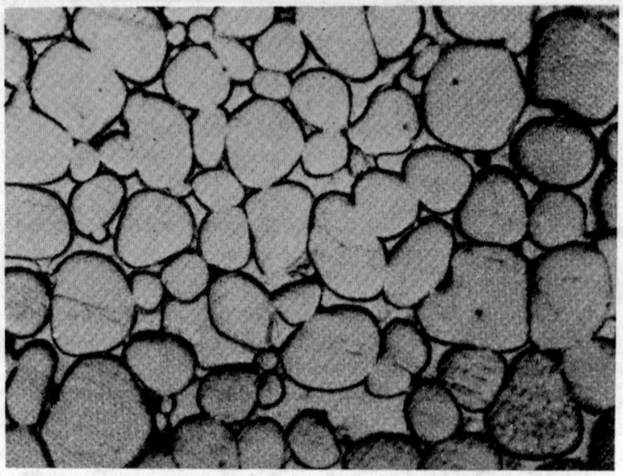

Figure 3. Microstructure of heavy alloy with 90% W, 7.5% Ni, 2.5% Cu, sintered 1 hour at 1450°C, x 500.

above the liquidus of the matrix phase. The smaller the tungsten particles in the original powder mixture, the more rapid the densification process. These phenomena were already described by Price and his coworkers[15] who developed the alloys. They ascribed the rapid growth of the tungsten particles to solution and reprecipitation of tungsten in the liquid matrix, a mechanism now generally called "Ostwald ripening," as first suggested by Wagner[16].

Densification and growth of the solid particles in the heavy alloy mechanism has been analyzed by dividing it into three stages[17].

The liquid flow or rearrangement stage in which densification is brought about by collapse of melt bridges between the particles and by rearrangement, in which the solid particles slide over each other.

The solution and reprecipitation or accommodation stage in which further densification and growth of the particles takes place by solution and reprecipitation or by coalescence.

If the compacts do not completely densify during the first two sintering stages, a third stage, the solid state sintering stage is postulated.

Attempts have been made to analyze the kinetics of densification in both the liquid flow and accommodation stages and to determine to what extent Ostwald ripening, on the one hand, and coalescence of the particles, on the other hand, contribute to the formation of the characteristic microstructure. In the sintering of the actual W - Cu - Ni and W - Ni - Fe alloys there is no sharp dividing line between liquid flow and accommodation. It is possible, however, to limit the study of the kinetics of densification to the liquid flow stage by investigating the sintering of alloys of tungsten and copper[18]. Since tungsten is practically insoluable in liquid copper, solution and reprecipitation does not play any role in the sintering of these alloys. It was found that, when arrays of spherical tungsten particles coated with uniform layers of copper are sintered, the length of time during which the arrays shrink, after liquid is formed, is extremely short and takes only seconds. The amount of shrinkage depends upon the packing density of the particles. The closer the packing and the more ideal the distribution of tungsten and copper, the greater the shrinkage. The total shrinkage is, however, always less than that which the amount of liquid phase would permit, if the liquid flowed into all the pores. The results indicate that the kinetics of densification in this stage depend primarily upon geometrical factors. It cannot be described, as was suggested by Kingery[19], by a process in which the driving force for densification, i.e. surface tension, is balanced by an intrinsic frictional force, i.e. the viscosity of the melt.

Densification during the second stage of the heavy alloy mechanism is a process of rearrangement of particles and is more pronounced when the conditions are conducive to grain growth. When the quantity of liquid phase is small, complete densification is impossible, if the solid phase particles maintain a spherical shape. Instead, as was already shown by Price and coworkers (see Figure 3)[15], the grains form polyhedra with rounded corners which accommodate themselves to each other; hence the term accommodation stage for the second stage of sintering. In a model designed to explain the kinetics of densification in this stage, Kingery[19] suggested that the surface tension forces, arising from the presence of pores in the sintering compact, produce high stresses across the thin liquid films between adjacent solid tungsten particles. This stress would increase the solubility of tungsten in the matrix in the area of near contact and develop a concentration gradient by which the tungsten would be transported in solution in the matrix from the areas near contact with their high solubility to regions with lower solubility. One would therefore expect a flattening of the tungsten particles in the areas of near contact. As seen in Figure 4, no such flattening is observed in experiments in which compacts made from a mixture of large spherical tungsten particles, fine tungsten powder and nickel powder were sintered[20]. Instead the growth of the large tungsten particles occurs preferentially away from the areas where the grains are in close contact. The shape accommodation occurring during particle growth appears to make complete densification possible.

With regard to the mechanism by which the tungsten particles grow during sintering, one theory postulates that growth takes place exclusively by the solution and reprecipitation process, i.e. Ostwald ripening, as Price and coworkers had suggested. That would mean that the dihedral angle in the solid - liquid system, i.e. the angle which the liquid makes at the boundary between two grains of the solid phase is zero or near zero. In this case the solid - liquid interfaces could be maintained at a lower energy than a grainboundary between two solid particles. In such a case coalescence between solid particles would not be possible.

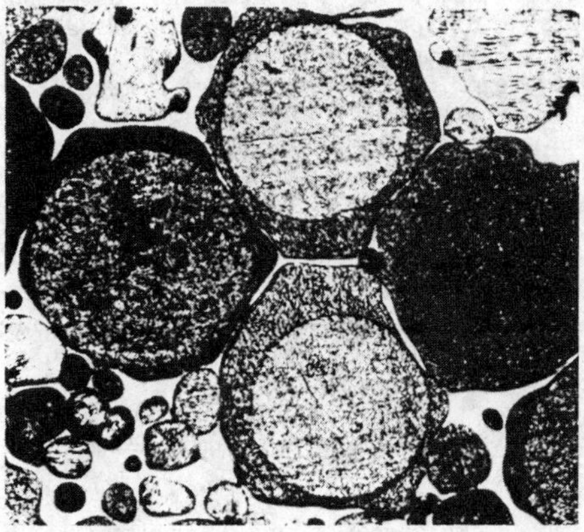

Figure 4. Microstructure of a mixture of 48 weight % large W spheres, 48 % of fine W powder and 4 weight % Ni, sintered at 1670°C for 120 minutes, 92 x.

On the other hand, a coalescence mechanism of particle growth has also been postulated. An indication that such coalescence does take place under certain conditions was given in experiments, in which large single crystal tungsten particles were sintered[21].

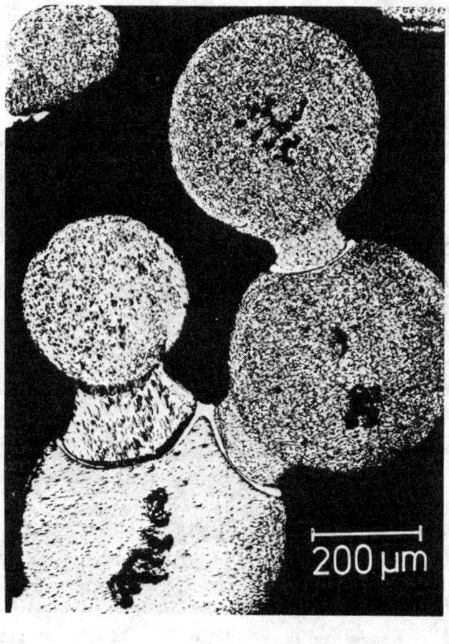

Figure 5. Microstructure of tungsten spheres of 200 - 250 µm diameter sintered with 2 weight % nickel for 60 minutes at 1650°C, etched in Murakami's solution.

As shown in Figure 5, the particles grow into their neighbors during sintering pushing a layer of liquid ahead of them. The migration of the layer of liquid takes place by solution and reprecipitation across the boundary. The composition of the particles on either side of the advancing liquid layer differs, with pure tungsten on one side and a solution of 0.15% of nickel in tungsten on the other side. The advance of the boundary and the eventual coalescence of the particles is, therefore, driven by a chemical potential gradient, rather than by capillary forces.

Quantitative data of the contributions which solution and reprecipitation, on the one hand, and coalescence, on the other hand, make to the process by which tungsten particles in heavy alloy grow are not yet available.

HOT FORGING OF PREFORMS

"P/M structural parts," a term used for components which are produced from metal powders to closely controlled dimensions, are one of the most important technological applications of powder metallurgy. They are produced from metal powders, because the technique is more economical than competing techniques, such as casting, machining or cold forming from bar stock. The basic method for producing "P/M structural parts" involves cold pressing metal powders in rigid dies and sintering in continuous furnaces. To improve the strength of the material used in the parts, alloy compositions and heat treatments after sintering have been developed. For many applications of structural parts it is, however, not just the strength which is important, but such other mechanical properties, as ductility, impact energy and fatigue resistance. The principal method to improve these properties is to raise the relative density of the parts so that it approaches as nearly as possible the density of wrought material of the same composition. Higher relative densities may be obtained by repressing and resintering the parts or by infiltrating which applies particularly to sintered steel parts whose porosity is filled with a liquid copper alloy. However, even with these techniques full density and therefore properties equivalent to those of wrought materials are generally not achieved. To obtain these properties the porosity must be completely or almost completely eliminated. A technique developed for this purpose is to first produce preforms from metal powder and then hot forge these preforms.

Three approaches to hot forging preforms have been investigated. The first approach developed from hammer forging, one of the types of conventional forging[22]. In hammer forging the stock to be forged is heated to the forging temperature and manipulated through a die with a series of three or four impressions with multiple hammer blows in each impression. The success of this type of forging depends on the skill of the hammerman. By replacing the stock used in hammer forging by a preform, compacted and sintered from metal powder, the forging stock has a shape nearer to that of the final product. As a result, the forging temperature may be lowered, the several impressions of the forging die may be replaced by a single impression and the multiblows by a single blow. Even though the dimensions of the forging produced from the preform may not be as closely controlled as those of conventional P/M structural parts, the simplification of the forging operation may make the process economically viable.

The second approach, developed from the operation of cold repressing a sintered compact in conventional powder metallurgy, involves hot repressing, sometimes called hot densification. The shape of the preform is close to that of the final component except for its length in the forging direction. In this process friction between die and preform during hot forging is high and therefore the pressure necessary to get complete densification would also be high, which would involve rapid wear of the forging tools. Hot repressing is therefore generally used in applications where densities on the order of 98% of theoretical are satisfactory. An example of a hot pressed component, both the preform and the final forging, shown in Figure 6, is a sprocket for a chain saw[23].

The third approach is the one most widely used industrially. It is a precision forging process without a flash in which the shape of the preform is simpler than that of the final part, so that the desired final shape is produced to closely controlled dimensions in the hot forging step. As an example, Figure 7 shows the preform and the forged shape of the differential side pinion gear whose pilot plant production was developed in the late 1960's[24]. The figure illustrates that this is not a hot repressing operation but that the preform is upset (and extruded) in hot forging. The advantages of hot forging by upsetting (and extrusion) over hot forging by repressing may be seen on the basis of Figures 8 and 9[25].

Figure 6. Hot repressed sprocket, preform and forging (Courtesy Gould, Inc.). (a) sintered preform, (b) final forging.

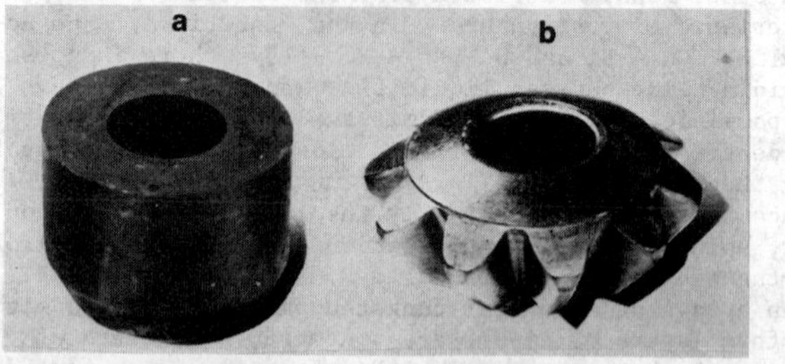

Figure 7. Differential pinion gear, preform and forging. (a) sintered preform, (b) finished part.

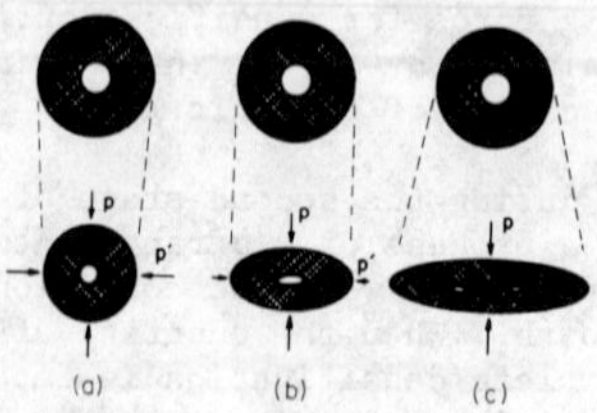

Figure 8. Schematic of void deformation under (a) isostatic pressure, p' = p; (b) repressing p' < p (no lateral deformation); (c) true forging p' = 0 (lateral flow).

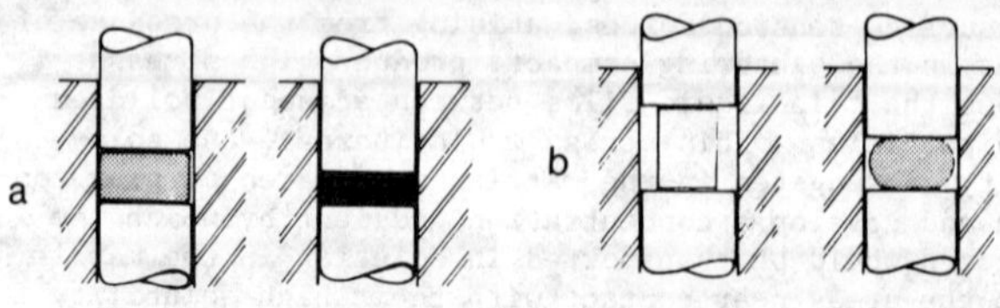

Figure 9. Schematic of overall deformation in (a) repressing, (b) forging (upsetting in trap die).

Figure 8 shows schematically the deformation of a porous solid represented by a material element with a void in it under:

a. condition of hydrostatic pressure,
b. repressing condition with little lateral flow and
c. true forging condition involving considerable lateral flow.

In Figure 10 the overall deformation of a preform during

a. repressing and
b. hot forging by upsetting

are illustrated. The latter involves much more lateral flow especially in the beginning of deformation. This leads to more rapid initial densification and also involves more shear stress at pore surfaces, producing relative motion between opposite sides of the collapsed pore. Mechanical rupturing of any oxide film present at the pore surface exposes the metal and insures a sound metallurgical bond across collapsed pore surfaces. Upset forging also produces fibering of inclusions in the lateral direction. Toward the end of the forging stroke by upsetting when the preform has reached the diewall, the mode of deformation becomes the same as in repressing.

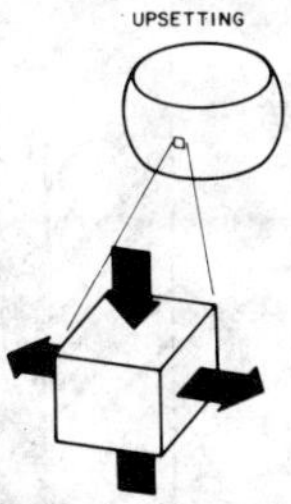

Figure 10. Circumferential tensile stress and axial compressive stress at the bulge surface of cylinders compressed axially with high contact friction.

A principal problem in this type of powder forging is the design of the preform. Its deformation during forging should lead to rapid densification and at the same time avoid strains which cause fracture. Kuhn and his associates have developed a theory of deformation processing porous materials which has been of considerable help in the design of preforms and has made it possible to avoid some of the pitfalls involved in design merely by trial and error[25][26].

Two important aspects of deformation processing of porous materials can be illustrated on the basis of an axial compression (upset) test: the Poisson ratio and a fracture criterion.

When a cylinder is compressed axially, it is strained in compression in the axial direction and strained in tension in the verticle direction. The ratio of the transverse strain to the axial strain is Poisson's ratio for plastic deformation. For a fully dense material this ratio is 0.5, which can readily be shown as a result of the fact that the volume of the cylinder remains constant. During compressive deformation of a porous cylinder some material flows into the pores, the volume decreases and the density increases. The ratio of expansion of the cylinder in the radial direction to its reduction in height, Poisson's ratio for plastic deformation for porous bodies, will be less than the value of 0.5 for solid bodies. Densification and lateral flow occur simultaneously rather than in sequence. From experiments on frictionless plastic deformation of porous bodies both at room and at elevated temperature, Kuhn[25] derived a relationship between Poisson's ratio ν for plastic deformation and ρ, the relative density of the porous body:

$$\nu = 0.5\ \rho^a$$

The exponent a was found to be equal to 1.91 for cold deformation of sintered bodies from iron powder and 2.0 for deformation at 700°F for sintered bodies from the aluminum alloy 601 AB.

Axial compression of a cylinder, the upset test, may also be used to develop a fracture criterion for porous compacts. During this test, friction at the top and bottom surfaces of the cylinder where the die contacts the cylinder retards radial outflow and leads to barreling, as illustrated in Figure 10. The greater the friction because of absence of lubrication and the greater the aspect ratio, i.e. the ratio of height to diameter, the more will the circumferential free surface bulge. The bulging causes a secondary tensile stress in the circumferential direction, as seen in Figure 10. The more the surface bulges, the greater will be the circumferential tensile stress. One type of fracture observed in upset forging, free surface fracture, is a direct consequence of these tensile circumferential stresses. Figure 11 gives the height strain at fracture as a function of the aspect ratio of sintered aluminum alloy powder 601 AB compacts during hot upsetting. The figure indicates that the height strain at fracture is considerably larger for upsetting in a lubricated than in an unlubricated die and that it increases with increasing aspect ratio. Surprisingly the height strain at fracture appears to be little affected by original porosity of the compacts. This may

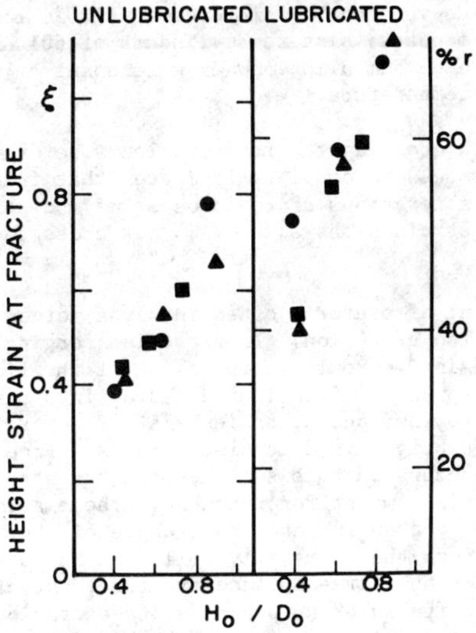

Figure 11. Overall height strain at fracture for hot upsetting of 601 AB aluminum alloy powder compacts, sintered at 370°C, ■ 76%, ● 88%, ▲ 93% relative density.

be explained by two opposing effects which compensate each other. As initial pore volume increases, the ability of the material to withstand circumferential tensile stress decreases, but, at the same time, the Poisson's ratio for plastic deformation decreases with increasing porosity and therefore the amount of lateral spread also decreases.

The technique for establishing a fracture criterion consists of measuring the axial and circumferential strain on small grid marks at the equator of the bulge surface of the upset cylinder, as shown in Figure 10. The ratio of these strains during compression is affected by the friction conditions, the aspect ratio of the cylinder and also the temperature of the test. When the compressive strain is plotted vs the tensile strain, a straight line with a slope of one half is obtained. This is illustrated in Figure 12 in which the locus of surface strains at fracture during upsetting at 980°C of 4620 low alloy steel powder compacts is plotted. In this experiment the method of sintering the cylinders prior to upsetting does not appear to affect the locus of fracture strains. It so happens that the locus of fracture strains during upsetting at room temperature of aluminum alloy 601 AB powder compacts fits on the same straight line.

The fracture strain loci can be considered as fracture criteria for evaluating the deformation to fracture in more complex deformation processes. Kuhn[25] suggests that "the progressing deformation strain paths in potential fracture regions of the process under consideration are first determined through plasticity analysis or measurements on a model material. These strain paths are then compared with the fracture locus of the material.

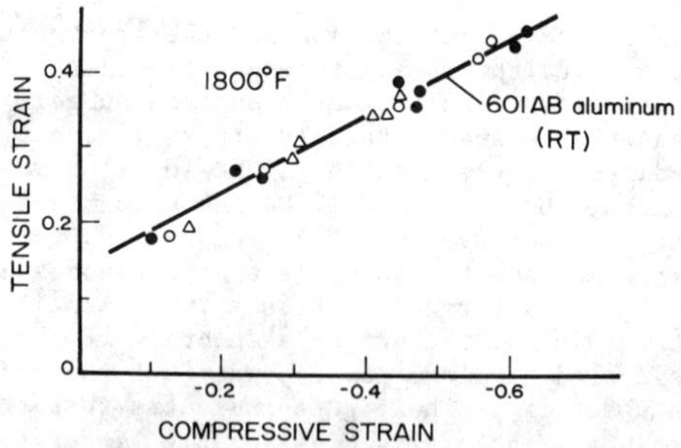

Figure 12. Locus of surface strain at fracture during upsetting at 980°C of sintered 4620 alloy steel powder cylinders.
● Induction sintered at 1290°C for 3 minutes
o Induction sintered at 1120°C for 3 minutes
Δ Conventionally sintered at 1120°C for 30 minutes; sintered cylinders of 601 AB aluminum alloy powder were upset at room temperature.

Any strain path crossing the fracture locus before the deformation process is complete indicates that fracture is likely and alterations of the process may be utilized to change the strain paths so they do not cross the fracture locus."

Hot forging of preforms often involves not only upsetting, but also extrusion. In extrusion forging a hub, as illustrated in Figure 13, compression of the flange section involves radial flow inward and up into the hub section. The top surface of the hub is a free surface which undergoes bulging and tensile strains. Free surface fracture occurs on this surface when the strains reach the critical amount for fracture. These strains may be altered by changing the draft angle of the hub or by using a preform that partially fills the hub section of the die. Not only free surface fracture, but also contact surface fracture may occur during extrusion forging of a hub on a cylinder. Again Kuhn and associates[27] analyzed the conditions under which such cracks will be found and how they may be eliminated by design changes. A third type of crack formation during hot forging, which

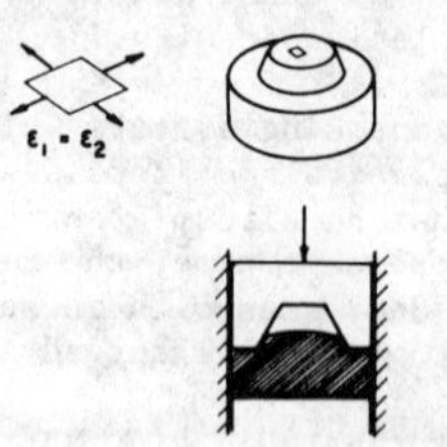

Figure 13. Schematic of hub extrusion, indicating tensile strain at top free surface.

fortunately is much less frequent than the other types of fracture is internal fracture which cannot be detected by surface observation; it has also been investigated by Kuhn and associates[27].

Kuhn[25] has applied his concept of workability analysis to the design of preforms for actual parts to be forged. Only one example will be mentioned here, which involves the forging of differential side pinion gears, the prototype of hot forging of preforms illustrated in Figure 7. The preform shown in this figure is flat-topped with a beveled bottom surface. When this preform is forged, cracks are developed in the partially formed gear during tooth formation, as shown in Figure 14. These cracks are closed up in the final stages of the forging stroke as the material is pressed against the die surface. However, oxidation of the crack surface and trapping of lubricant in the cracks may lead to structural weakness of the tooth surfaces. Kuhn showed that a modification of the preform contour shown in Figure 15 involving a slightly tapered top surface and a rounded instead of a beveled bottom surface will eliminate crack formation. Such a preform may be produced by isostatic compression.

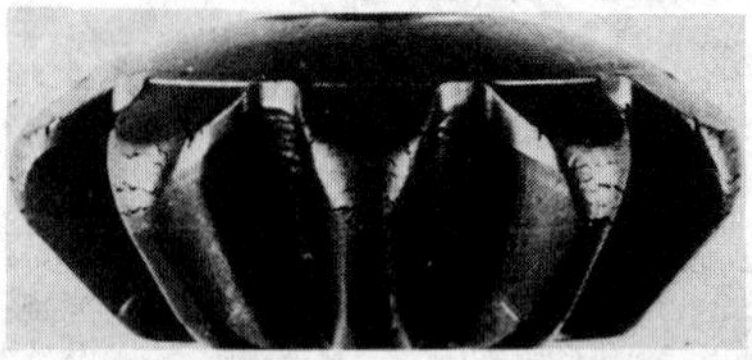

Figure 14. Partially formed pinion gear showing cracks forming on the teeth.

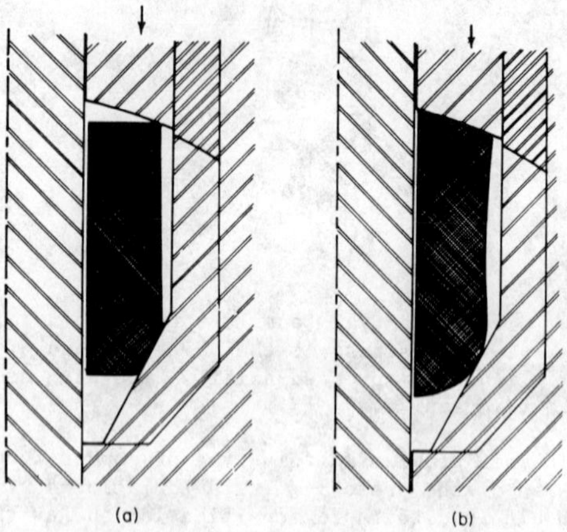

Figure 15. Longitudinal section view of preforms and tooling for forging pinion gear: (a) original preform which led to crack formation at teeth (Figure 14), (b) modified preform which led to defect-free gears.

The powder forging technique is used for structural parts with mechanical properties superior to those attainable with conventional P/M techniques. For this reason, mixtures of prealloyed atomized steel powders and graphite have been widely chosen as the raw material for preforms. With these powders, homogeneous solid solutions of carbon and the alloying elements in austenite at the forging temperature are obtained even without long time sintering at high temperature. Prealloyed atomized steel powders are not as widely used for conventional P/M structural parts, because their compressibility, i.e. the density of pressed compacts relative to the density of solid material, is not as high as for a mixture of elemental powders. For compacts, however, which obtain their final density by the hot forging step, the lower compressibility of the prealloyed powders is not important. The most common composition of these powders contain nickel and molybdenum as alloying elements, such as compositions with 0.4% nickel and 0.6% molybdenum and those with 2% nickel and 0.5% molybdenum. The advantage of these compositions is that they contain low contents of alloying elements, which have a tendency to oxidize, e.g. only 0.2 - 0.3% Mn and less than 0.1% Cr.

The same methods used for conventional compacting and sintering of P/M structural parts may be used in the fabrication of powder forgings. After sintering the preforms would be cooled in the cooling zone of the sintering furnace and later reheated to the forging temperature. Powder forgings should have low oxygen contents preferably in the range of a few hundred ppm, in order to give them good hardenability and good impact properties. Control of sintering temperature and time is necessary to obtain these low oxygen contents. For applications where obtaining the highest mechanical properties is less critical, omitting a separate reheating step has been suggested[28]. The compacts are transferred from the end of the hot zone of the sintering furnace directly to the forging die.

Instead of conventional pressing in rigid dies and sintering in continuous furnaces, the preforms may alternatively be pressed by dry bag isostatic pressing and sintered in a high frequency induction coil.

Forging machines for powder forging should have force-displacement characteristics matching the deformation characteristics of the preforms[29]. The time during which the workpiece and the tools are in contact with each other should be as short as possible. The machines should be stiff and the ram should have good guidance to obtain good tolerances in the powder forgings. Mechanisms for the ejection of the forged parts must be provided. For these reasons mechanical presses, in particular crank presses with short fast strokes, are generally used for forging powder preforms. Relatively simple tool design is satisfactory for forging powder preforms, an example of which is shown in Figure 16[29].

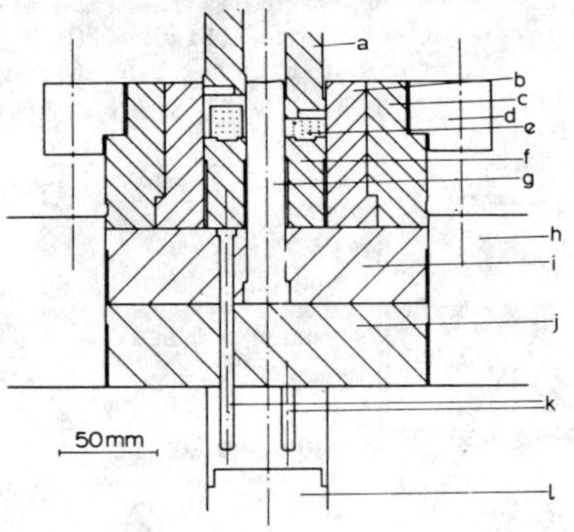

a hollow punch; *b* die; *c* prestressing ring; *d* clamp ring; *e* workpiece; *f* lower punch; *g* core rod; *h* die housing; *i* die pressure pad; *j* support pad; *k* ejector pins; *l* ejector of press

Figure 16. Example of powder forging tool construction.

Before a preform is forged it must be heated to the forging temperature. For preforms having low alloy steel compositions the forging temperature is in the same range as for conventional forgings, i.e. from 800 to 1200°C. With increasing temperature the flow stress decreases and the formability increases, but higher forging temperature generally means decreased tool life in spite of the lower force. Preforms may be heated in a continuous furnace or by high frequency induction heating, in both cases with protective atmosphere. In order to produce forgings to close dimensional tolerances, the forging temperature must be carefully controlled. This requires

1. automatic temperature controls of the equipment in which the preforms are heated to the forging temperature,
2. automatic control of the temperature of the forging tools which are generally operated at an elevated temperature,
3. automated operation of the transfer unit by which the preheated preforms are transported from the heating equipment to the preheated die.

Another important requirement in preform forging is lubrication. Systems in which the preform is lubricated by dipping it into or spraying on a colloidal suspension of graphite in a water carrier have been used successfully. They provide not only lubrication, but also protection against oxidation. On the other hand, spray lubrication of the forging tools has the advantage of keeping the lubricant from penetrating into the structure of the porous preform.

One problem peculiar to preform forging is the avoidance of local porosity near the surface of the forging. Surfaces which are in contact with the diewalls may be chilled and may not be completely densified exhibiting surface porosity. Development of surface porosity may be overcome by rapid deformation rate, by minimum contact time between preform and tooling, by using relatively high forging temperatures and by providing adequate lubrication.

Powder forgings are used for applications where they compete with wrought materials. Therefore, the design engineer will compare the mechanical properties of wrought and powder forged materials. One important difference in their properties is that those of wrought materials tend to be anisotropic, while those of powder forged materials are more isotropic[30]. The anisotropy in wrought materials is due to the working operations by which ingots are transformed into wrought billets or bars. The working produces banding of interdendritically microsegregated alloying elements and elongation and microstringering of non-metallic inclusions. For this reason the mechanical properties and, in particular, the fatigue strength of wrought steels are considerably better in the longitudinal than in the transverse direction. The anisotropy is less pronounced in powder forgings.

Porosity affects the impact energy values[31] and fatigue strength[32] of forged powder preforms more strongly than their tensile and yield strength[33]. However, porosity is not the only factor which determines mechanical properties. The amount of lateral flow during forging has a direct effect upon endurance limit and impact energy values. In order to get equivalent impact energy values as for wrought materials height strains of 60 to 80% during hot upsetting are necessary[34]. Residual oxides[35] and other inclusions[31] in the microstructure are also found to strongly influence the impact values of powder forgings.

Examples of powder forgings are shown in Figures 17, 18 and 19.

Figure 17. Preform and final forging of the lower roller clutch cam for General Motors automatic transmission (Courtesy: Federal Mogul Corporation).

Figure 18. Powder Forgings; inner and outer race of ball bearings at right. (Courtesy: Federal Mogul Corporation).

Figure 17 shows both the preform and the final forging of a lower roller clutch cam for an automobile automatic transmission produced from 4600 prealloyed powder. Figure 18 represents, on the left, various transmission parts and, on the right, an inner and an outer race for a ball bearing. Figure 19 is a connecting rod for the

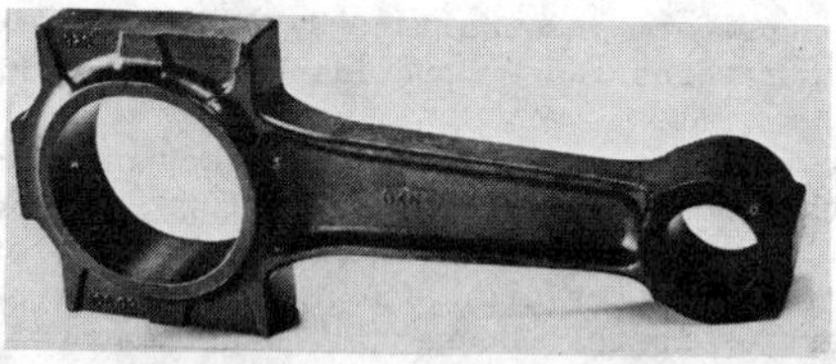

Figure 19. Connecting rod for Porsche Model 928 automobile forged from P/M preform.

Porsche Model 928 automobile produced by powder forging.

HOT CONSOLIDATION OF METAL POWDERS

Production of Spherical Metal Powders

The conventional sequence of production steps in powder metallurgy consists of

1) metal powder production
2) consolidation of the powder at room temperature
3) sintering.

This procedure may be contrasted with one in which powder production is followed by hot consolidation of the powder, combining steps 2) and 3), consolidation at room temperature and sintering. Laboratory investigations of hot consolidation go back to the earliest days of powder metallurgy. A few production applications of the technique, e.g., hot pressing of cemented carbides, vacuum hot pressing of beryllium powder in graphite dies and hot pressing of metallic friction material with simultaneous bonding to a steel backing, were also developed many years ago, but the commercial utilization of hot consolidation remained limited. In recent years it has come very much to the fore. Devising and perfecting new hot consolidation techniques for metal powders and using them for a large number of new applications is probably the most important aspect of powder metallurgy technology in the 1970's and 1980's.

Metal powders, which are to be hot consolidated, frequently have particle size distributions and, particularly, particle shapes different from those used in conventional powder metallurgy. Here, the compacts produced from powder by cold pressing must have adequate strength, so that the green compacts can be transferred from the press to the sintering furnace without crumbling. Except for soft metals, such as aluminum, tin or lead, this means that spherical powder particles are undesirable, because compacts from such powders have poor green strength. Powders with spherical particle shape were only used for such special applications as loose powder sintered bronze filters. With the advent of hot consolidation techniques, particularly hot isostatic pressing and hot extrusion of powders, the production of powders with spherical particle shape became of great commercial interest and new techniques for such powders were developed. Spherical powders have high apparent and tap densities; when they are filled loose into a container and particularly when they are tapped, the powder mass reaches densities in the range from 60 to over 65% of the density of the solid metal, while apparent and tap densities of powders with irregular particle shapes are much lower.

Production of Spherical Metal Powders

A widely used technique for metal powder production is atomizing[36]. A stream of molten metal issuing from an orifice in the bottom of a tundish is broken up into droplets by an impinging stream of either a liquid or a gas. The droplets freeze into powder particles. Many base metal powders, such as iron, copper and brass are produced by atomizing. For conventional powder metallurgy production, i.e. cold compacting followed by sintering, the variables of the process are controlled so as to produce powders with the desired particle size distribution and irregular, rather than spherical, particle shape. The most important variable used to produce irregular particle shape is the use of a liquid atomizing medium, generally water.

Among the commercial processes for producing metal powders with spherical particle shape for hot consolidation techniques is an atomizing process, in which a molten metal stream is broken up by a steam of inert gas, generally argon, under conditions where the molten droplets have a chance to take on spherical shape and freeze undisturbed. The alloy is melted, generally from a prealloyed vacuum cast ingot, tapped into a tundish on top of an atomizing chamber, high enough so that the molten droplets freeze by the time they reach the bottom. The atomizing chamber is equipped with one or more nozzles through which high velocity gas jets atomize the molten metal stream. Particle size and size distribution are controlled by the atomizing conditions. Figure 20 is a schematic drawing of the apparatus and Figure 21 a photomicrograph of the cross section of powder particle, so produced.

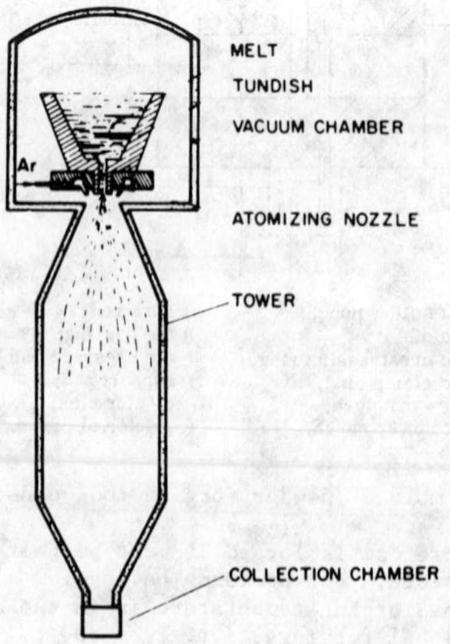

Figure 20. Schematic of argon atomizer.

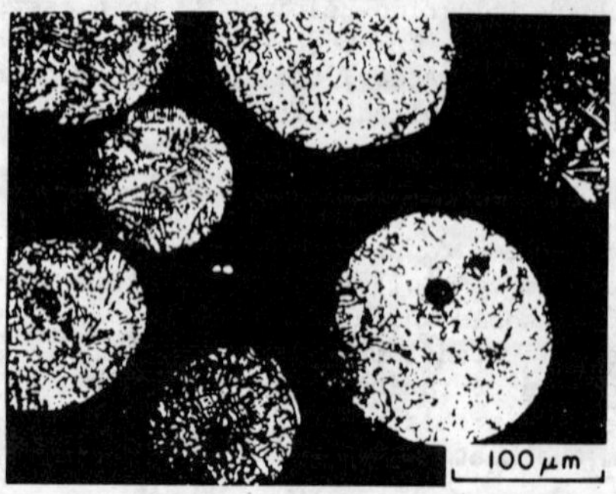

Figure 21. Cross section of IN-100 alloy powder particles produced by argon atomization.

A variation of inert gas atomization is a vacuum atomizing process developed by the Homogeneous Metals Corporation, primarily for nickel superalloy powders. The alloy is melted under vacuum. The charge is superheated before the melting chamber is pressurized with hydrogen which dissolves in the molten alloy. The melt is released upwards through a ceramic tube into the atomizing chamber held under vacuum. The stream of molten metal rising through the tube explodes into fine spherical powder particles when it hits the vacuum of the atomizing chamber. Figure 22 shows the vacuum atomizing apparatus and Figure 23 the shape of the powder particles produced by the technique.

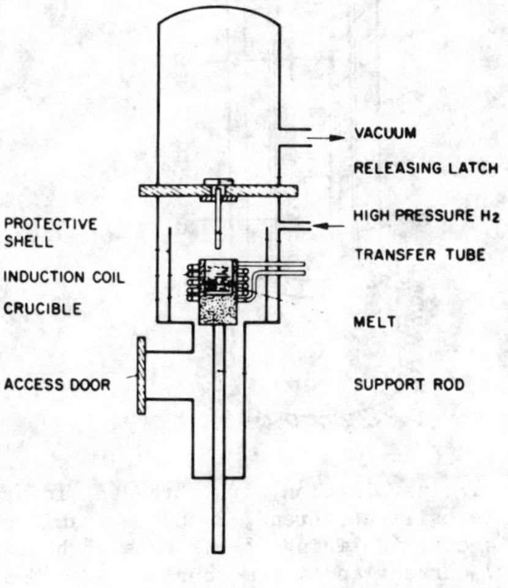

Figure 22. Schematic of vacuum atomization technique using hydrogen.

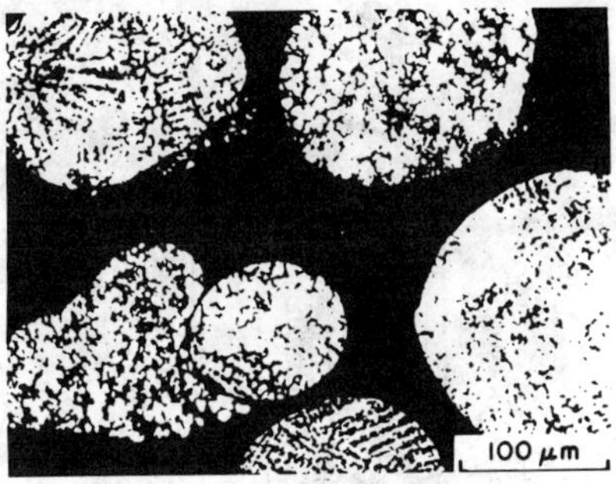

Figure 23. Cross section of IN-100 alloy powder particles produced by vacuum atomization.

A third commercial method of producing metal powder with spherical particle shape is the rotating electrode process(37). A bar of the metal or alloy to be produced as powder is rotated at high speed about its longitudinal axis. The end of the bar is melted by an electric arc and the molten material is ejected centrifugally as droplets which rapidly solidify as powder particles. The chamber in which the bar rotates is filled with helium. A schematic of the apparatus is shown in Figure 24 and the powder produced with it in Figure 25.

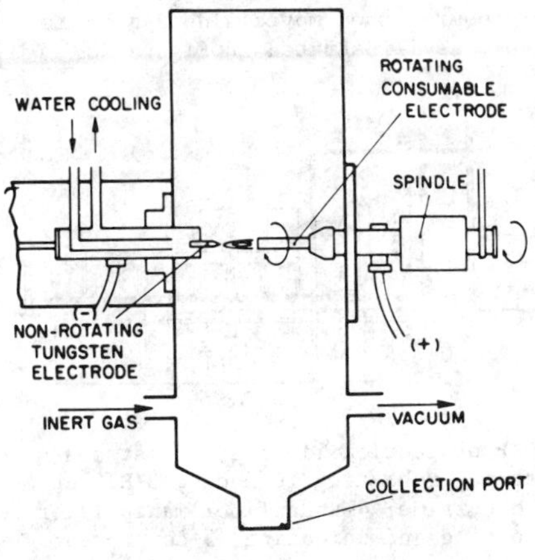

Figure 24. Schematic of rotating electrode process.

In a variation of this process which was developed to avoid the occasional contamination of the powder by particles of tungsten from the electrode, from which the

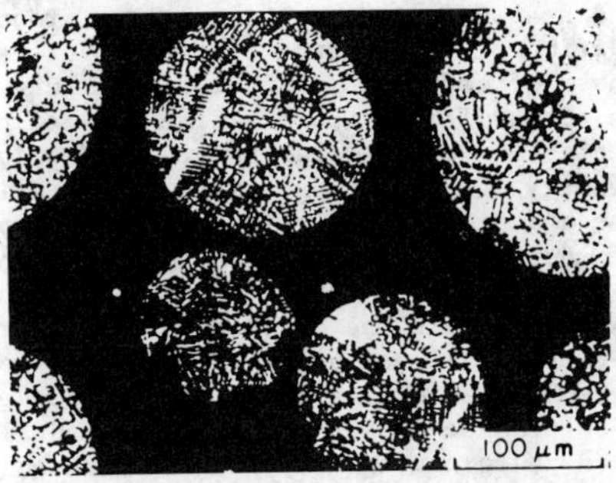

Figure 25. Cross section of IN-100 alloy powder produced by the rotating electrode process.

electric arc melting the alloy issues, a plasma arc is used as a heat source. The rotating electrode process is particularly valuable for titanium alloys, because they cannot be readily melted in crucibles and as a result not be produced by the inert gas and the hydrogen atomizing processes. The French PSV (pulverisation sous vide) process(38) is similar to the rotating electrode process. An ingot of the alloy to be converted into powder is placed into a mold rotating vertically at high speed. The top of the ingot is bombarded by an electron beam from an electron gun. The alloy melted at the top of the ingot is ejected and solidifies into spherical particles.

In addition to these three commercial processes for producing spherical alloy powders, several others are under development. They include the electron beam rotating disk process(39), which combines a vertical drip melting process and atomizing with a rotating disk. In the Colt Titanium process(40), specially developed for titanium powder, a billet of hydrogenated titanium or titanium alloy is slowly rotated in a vacuum chamber, a small volume is melted by an electron beam gun and the molten pool generated is disintegrated by the escaping hydrogen gas.

The spherical powders produced by the processes described are relatively coarse with particle sizes in the range from 50 up to 400 μm in diameter. A process which produces finer powder with particle sizes less than 50 μm is of particular interest for nickel base superalloys. It is the RSR (Rapid Solidification Rate) process(36), in which a stream of molten metal falls on a disk coated with a high temperature ceramic, rotating at a speed of about 400 revolutions per second. The fine molten particles are ejected from the periphery of the disk and quenched by a blast of helium gas which solidifies the particles at rates up to 10^6°/second. The process is used for powders having compositions different from those of the usual cast superalloys, in particular with higher contents of refractory metals. Because of the rapid rate of solidification the powders also have metastable metallographic structures. They are of particular interest in the development of nickel-base superalloys with superior mechanical properties at elevated temperatures.

Hot Consolidation Techniques

Hot consolidation techniques have been developed primarily for two reasons. One reason is to produce components from relatively expensive alloys, in particular superalloys and titanium alloys, to "near net shape." In the conventional production of these components the ratio of the weight of the cast ingot to that of the finished component is quite high. In view of the high cost of the alloys used, powder metallurgy hot consolidation techniques, in which this ratio is lowered materially, may result in an overall saving of the cost of production, even though the technqiues in themselves are more expensive than the conventional techniques they replace. The properties and particularly the mechanical properties of the alloys produced by hot consolidation from powder must, of course, be at least as good as those of conventionally produced alloys.

A second reason is that certain alloys produced by hot consolidation of metal powders have properties superior to those of conventional fusion metallurgy alloys. This applies to beryllium, where a fine-grained product having appreciable ductility can be produced only by powder metallurgy. It also applies to highly alloyed aluminum base and nickel base alloys and to alloy steels, in which the macrosegregation common in cast ingots can be avoided and in which the dispersion of carbides in the microstructure of steels and of second phase particles in that of non-ferrous alloys can be more readily controlled. Finally the techniques of hot consolidating powders make it possible to produce dispersion strengthened alloys, in which a second phase having very little or no solubility in the matrix at the solidus temperatures of the alloys is finely dispersed in the matrix.

The important characteristic of dispersion strengthened alloys is high strength at elevated temperature. The strengthening mechanism is related to that in precipitation strengthened alloys, in which a submicroscopically fine dispersed second phase, whose crystal structure is generally coherent with that of the matrix, is dispersed in the matrix and produces the strengthening effect. However, the strength of precipitation strengthened alloys is limited with regard to the temperature range in which it applies. The dispersed alloy phase is produced by quenching from a temperature where the alloy is a single phase solid solution and then heating to an intermediate temperature to cause precipitation. When the alloy is further heated to a temperature above that at which precipitation took place, the precipitated phase will coarsen ("overaging of the alloy") and eventually redissolve. On the other hand, in dispersion strengthened alloys, the dispersed phase should be stable up to the solidus temperature of the alloys. The mechanisms by which fine dispersions alter the yield strength, the work hardening, the creep and the fracture behavior of an alloy have been studied in a large number of investigations[41]. High temperature strength properties, in particular creep rate, are affected by parameters related to dispersion geometry, such as particle spacing, and by those related to grain size and to grain shape (ratio of grain length to grain width, called grain aspect ratio). The results indicate that creep both by dislocation climb over second phase particles and by grainboundary sliding are important in explaining the properties of the alloys.

Two methods of hot consolidating metal powders aimed at components produced to "near net shape" have been hot isostatic pressing and hot pressing in rigid dies. Hot isostatic pressing of titanium powder has been described by Kelto et al.[42], but the description also fits the "HIP"ing of other alloy powders: "A container with the general (oversize) shape of the final desired article is filled with prealloyed powder and evacuated using a combined cold and warm pumping cycle. The container is then hermetically sealed and placed into a pressure vessel (autoclave). Pressure is transmitted to the powder through the container either directly or indirectly through a secondary pressing medium. At the same time, heat is applied (by a furnace surrounding the container). The combination of high temperature and pressure cause the container to contract and exert isostatic pressure onto the powder. The powder is squeezed into a fully dense article with a smaller size than the original container, but with the same general shape." The shrinkage depends upon the density to which the powder is packed in the container. A high density of the powder increases its heat conductance and facilitates hot isostatic pressing. For this reason some powders, particularly tool steel powder and beryllium powder, are densified by cold isostatic pressing before being subjected to hot isostatic pressing.

The cold wall autoclaves used for hot isostatic pressing are quite expensive. Principal problems in their design are providing means for opening the vessel quickly for loading and unloading and insuring uniform temperature distribution within the hot zone. Two methods of opening and closing the pressure vessel are shown schematically in Figures 26 and 27[43]. One has a threaded closure,

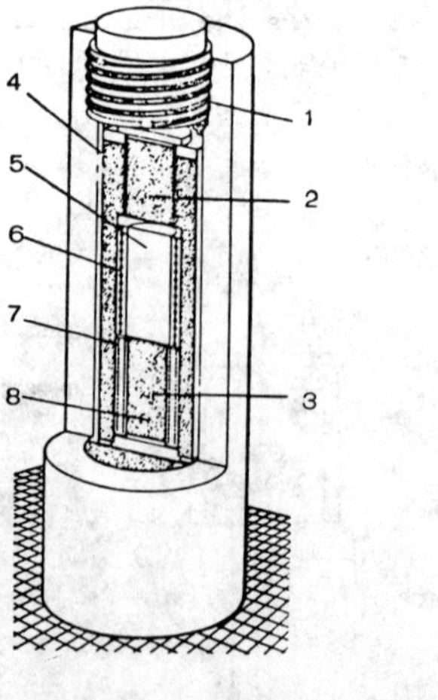

Figure 26. HIP installation with threaded closure; 1-resilient thread, 2 and 3-ceramic plugs, 4-ceramic lining, 5-HIP zone, 6-heater, 7-current and thermocouple leads, 8-heater support.

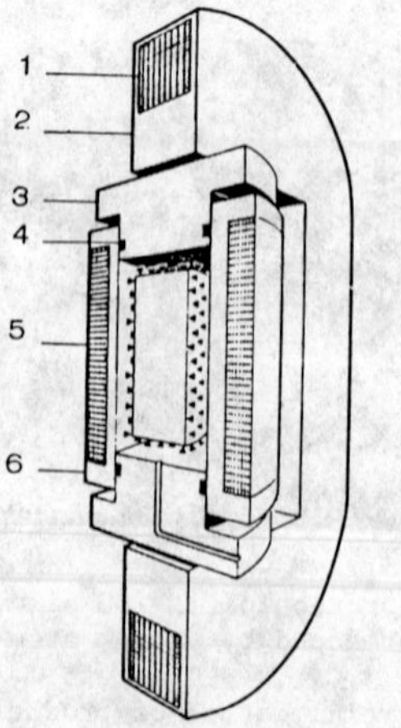

Figure 27. HIP installation with yoke closure; 1-wire wound furnace, 2-yoke segment, 3-end closure, 4-seal, 5-wire winding on HIP chamber, 5-steel core of pressure vessel.

into which the load is introduced from the top; the other has yoke closures. With yoke closures the load may be introduced into the pressure vessel from the bottom while it is separated from the top and the bottom yoke closures and the yoke closures then moved sideways to close the pressure vessel, as illustrated in Figure 28. With yoke

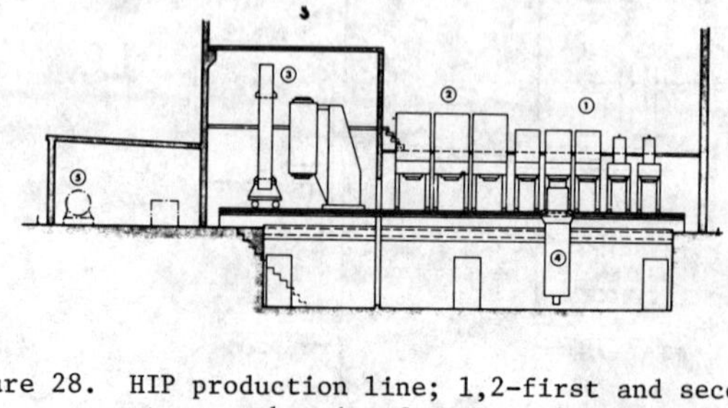

Figure 28. HIP production line; 1,2-first and second stage preheating furnaces, 3-HIP chamber and yoke frame, 4-underfloor manipulator for loading and unloading, 5-compressor.

closures the pressure vessel as well as the closures may be provided with a wire wound frame to take up the high pressures exerted by the gas. The difficulties with nonuniform temperature distribution is due to rapid heat convection in the dense pressurized gas between the cold vessel wall and the furnace. It can be overcome by compartmentalizing the space near the heater and by providing suitable insulation between furnace and wall, and furnace and work.

Several methods have been developed for encapsulating the powder for hot isostatic pressing. One method is a metal can, which can be readily designed to give the component to be hot pressed its "near net shape." This method has been used quite successfully for hot isostatic pressing jet engine compressor disks to "near net shape" from nickel base superalloy powders. A schematic of the shape of the can is shown in Figure 29[44].

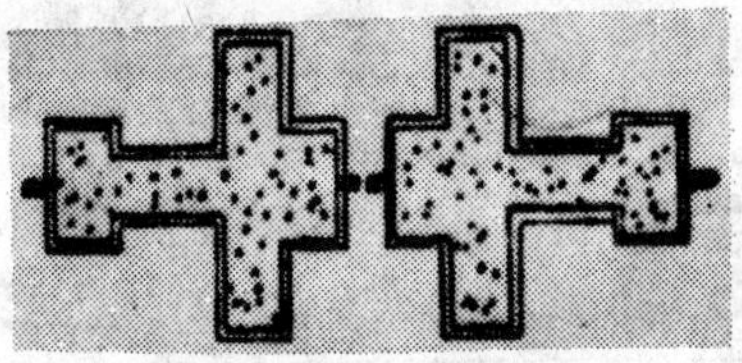

Figure 29. Schematic of shape of mild steel can for hot isostatic pressing of jet engine compressor disk.

An alternative method is encapsulation in a glass container. The glass is chosen so that it deforms uniformly by viscous flow at the pressing temperature. For superalloy powder, Vycor has the necessary viscosity. Containers are produced by slip casting glass powder and firing the casting. After filling with powder the glass container is evacuated and sealed. Glass containers strip themselves automatically from the hot pressed compacts by shattering when they cool.

In a variation of hot isostatic pressing in a glass container, spherical tool steel powders are consolidated without applying external pressure[45]. The difference in pressure between the atmosphere at the outside and the vacuum inside the thoroughly degassed and evacuated glass container, when it is introduced into a furnace, is sufficient to consolidate the powder.

Two other methods developed particularly for hot isostatic pressing titanium alloy powder into intricate "near net shape" components are the ceramic mold method[42] and the fluid die method[42] illustrated in Figures 30 and 31.

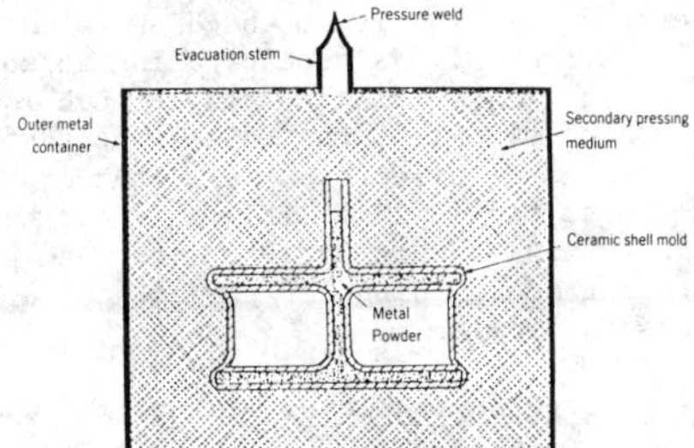

Figure 30. Schematic of powder-filled ceramic mold within outer metal container, ready for hipping.

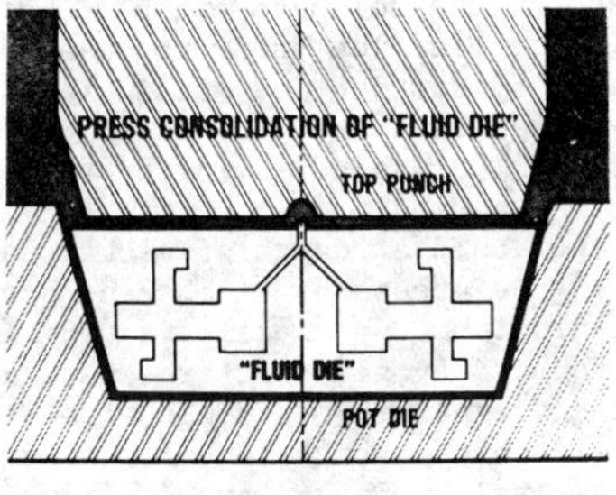

Figure 31. Schematic of fluid die process.

In the ceramic mold method a ceramic mold of precise configuration is produced by the lost wax process. A wax pattern is embedded in a thin-walled ceramic shell mold. After melting the wax the shell mold is fired, filled with the powder to be isostatically pressed and supported in a secondary pressing medium inside a welded evacuated steel can. At the HIP temperature the pressure applied by the gas to the outer metal container is transmitted to the ceramic shell mold which is sufficiently ductile to allow transfer of the pressure to the powder while maintaining the required shape. After "HIP"ing the part is easily extracted from the pressed can assembly. The sheet metal can is cut away and the collapsed shell broken off the part surface.

In the Fluid Die process the thick-walled container having an interior cavity with the enlarged profile of the part to be pressed is machined or cast. The cavity is filled with powder. During hot isostatic pressing the relatively large mass of the die becomes fluid under the combined effect of heat and pressure. Removing the hot pressed compact from the die, which is not reusable, requires rough machining of the steel followed by chemical milling.

Vacuum hot pressing in rigid molds made of graphite is a technique for hot consolidating beryllium powder[46]. This technique has recently been suggested for consolidating titanium alloy powders to "near net shape."[42] The molds are made of nickel base superalloy or of a molybdenum alloy. Titanium alloy powders are hot pressed at temperatures near 950°C and pressures of approximately 700 MPa. In order to avoid reaction of the titanium alloy powder with the mold material a parting compound must be applied to the mold wall.

The technique used for hot consolidating most of the dispersion strengthened alloys is extrusion[47]. It has the advantage that not only hydrostatic compressive forces are exerted upon the component in the can, but that it is made to flow through the extrusion die where frictional forces produce a shear component which greatly aids in the consolidation of the powder.

In the case of aluminum alloys, the extrusion of powder metallurgy alloys is similar to that of cast ingots, since the compacts produced from powder are hot pressed in tool steel dies prior to extrusion. For the alloys of the higher melting metals, copper, iron and nickel, the powders are encapsulated in an evacuated and hermetically sealed can before being extruded. This encapsulation closely resembles that in hot isostatic pressing of powders. To prevent turbulent flow during extrusion, the end of the can is made conical and fits into an extrusion die with a conical opening as illustrated in Figure 32.

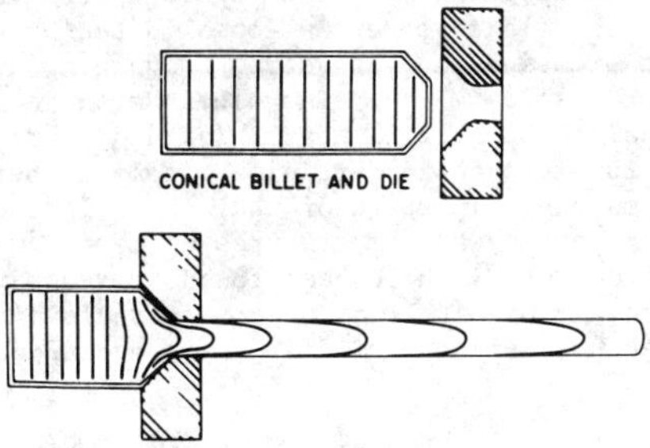

Figure 32. Extrusion with conical billet and die.

The material of the can must have as close as possible the same stiffness as the powder to be extruded at the extrusion temperature. After extrusion the can is removed by machining and/or chemical milling.

Products by Hot Consolidation of Powders

Beryllium. Beryllium has been hot consolidated by vacuum hot pressing in graphite molds for many years[46].

Some of the largest compacts produced by powder metallurgy, 1900 mm in diameter, 750 mm high, weighing 4000 kg for use in nuclear reactor cores have been fabricated by this method. In order to get optimum mechanical properties in beryllium[48] the grain size of the consolidated products should be as small as possible and its purity as high as possible. These requirements are difficult to combine, because disintegration into powder, size classification and the tendency of the powder to react with the atmosphere are apt to introduce impurities. For the best properties beryllium flakes deposited from a molten electrolyte are impact attrited in an inert gas stream against a beryllium target and classified in cyclone separators. The powder is first cold isostatically pressed in evacuated elastomeric bags and then hot isostatically pressed in evacuated and sealed steel cans. It has a yield strength up to 460 MPa, tensile strength up to 600 MPa, elongation up to 6-1/2%.

Aluminum Alloys. Two types of aluminum alloys are being considered for hot consolidation of powders into aircraft and aerospace components. One type has compositions close to those of conventional aluminum alloys. By producing the alloys by powder metallurgy properties superior to those of conventional alloys may be obtained. These properties are related to the finer microstructure in the alloys produced from powder, but the microstructural differences have not yet been investigated in detail. The other type of aluminum alloys are dispersion strengthened alloys. Those in which the second phase is aluminum oxide are called SAP (sintered aluminum powder). They were studied intensively in the 1950's. Those in which the second phase is an insoluble intermetallic compound were also developed 25 years ago, but are now again considered for aircraft and aerospace applications.

An example of the first type of material is the alloy MA 67, developed by the Aluminum Company of America[49]. The alloy with a composition of 8.0% Zn, 2.5% Mg, 1.0% Cu and 1.5% Co, balance Al is atomized into powder. The powder is cold compacted, encapsulated, preheated in vacuum and hot pressed at a pressure of 300 MPa. The hot pressed compact is extruded at 325°C and subsequently solution heat treated, quenched and aged. The alloy produced from powder can be aged to a yield strength of 523 MPa without losing its stress corrosion resistance, while the wrought alloy 7075 becomes prone to stress corrosion, when it is aged to values of yield strength above 450 MPa. The alloy MA 67 also has better fatigue strength than the corresponding wrought alloy.

The oxide dispersion strengthened alloys SAP[50] are produced by first milling atomized aluminum powder in ball mills. The flakes in the milled powder have thicknesses in the range from 1 μm down to 0.1 μm, depending upon the milling process. Each flake is covered on both sides by surface layers of aluminum oxide approximately 10 nm thick. The flake powder is consolidated by cold pressing, hot pressing and extrusion, techniques similar to those for the alloy MA 67. The aluminum oxide layers on the original flake particles are broken up during working into submicroscopic particles finely dispersed in the aluminum matrix. In spite of their excellent creep and rupture properties at temperatures up to 500°C not many applications of SAP have been found, because the material is more expensive than other materials with similar strength - weight ratios at the temperatures in question.

Dispersion strengthened aluminum alloys with an intermetallic compound as the dispersed phase are alloys of aluminum with iron, nickel and cobalt[51]. $FeAl_3$ and $FeNiAl_9$ are examples of these phases, which are soluble in the liquid metal, but practically insoluble in solid aluminum, even near its melting point. The alloys are hot consolidated from powders produced by atomizing the liquid aluminum alloys containing the alloying additions which form the intermetallic compounds.

Dispersion Strengthened Copper Alloys. A suitable dispersoid for dispersion strengthened copper alloys is aluminum oxide[52]. The alloys are used for applications in which the high electrical conductivity of copper is combined with elevated temperature strength and with resistance to softening at elevated temperatures of cold worked alloys. The material produced by atomizing a copper-aluminum alloy, mixing it with cuprous oxide and aluminum oxide and internally oxidizing the copper-aluminum alloy producing a fine dispersion of aluminum oxide in copper. The cuprous oxide acts as a source of oxygen; any excess is reduced with hydrogen. The powder is enclosed and sealed in copper containers and extruded at 925°C (1700°F) into the desired shape.

In Figure 33 the tensile properties and the electrical conductivity of copper dispersion strengthened by Al_2O_3 are plotted as a function of the aluminum oxide content. Relatively small amounts of oxide, less than 1 volume %, produce large increases in hardness and tensile strength, while the decrease in electrical conductivity is quite modest. The strength and hardness of dispersion strengthened copper may be further enhanced by cold reduction. The cold worked alloys show excellent resistance to softening after exposure to temperatures of up to 930°C. The drop in tensile strength upon annealing lessens with increasing oxide content. The yield strength of a cold worked alloy with 0.45 vol. % Al_2O_3 decreases 30% from 470 MPa to 330 MPa upon annealing one hour at 930°C, while that of an alloy with 2.68 vol. % oxide decreases only 17% from 580 MPa to 480 MPa with the same annealing treatment. This resistance to softening at elevated temperature distinguishes the dispersion strengthened copper alloys from alloys precipitation strengthened with chromium or zirconium, which soften at much lower annealing temperatures. Also, the tensile strength at elevated temperatures of dispersion strengthened copper is higher and its stress rupture life is longer than those of precipitation strengthened alloys. The principal uses are for resistance welding electrodes and for lead wire in incandescent lamps.

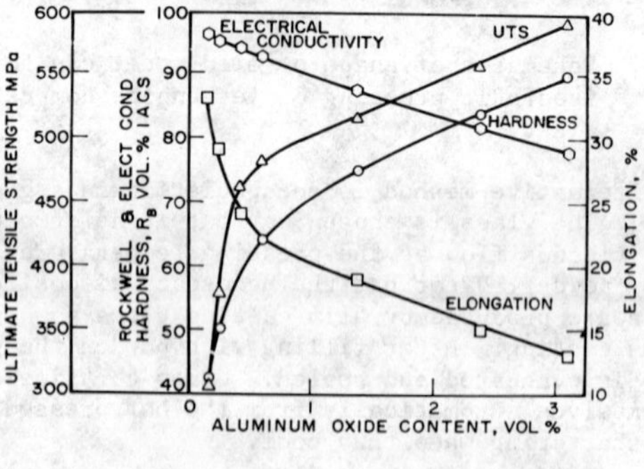

Figure 33. Properties of as-extruded copper dispersion strengthened with Al_2O_3 ("Glidcop") as a function of aluminum oxide content in volume percent.

Powder Metallurgy Superalloys. The term "superalloys" is applied to alloys of iron, nickel and cobalt which have high strength at temperatures of 600°C and above. They are of primary interest for components in jet aircraft engines and for aerospace applications. A large amount of work has gone into the development of superalloys by powder metallurgy[53]. The development work is concerned, on the one hand, with nickel-base superalloys having compositions near those of alloys produced by casting or by casting and working and, on the other hand, with dispersion strengthened alloys. Both groups of alloys will be discussed in the following.

The high temperature strength of nickel base superalloys is based on the presence of coherent precipitates, which are nickel-aluminum and nickel titanium intermetallic compounds, produced by solution treatment and aging. In addition to high temperature strength the alloys must have corrosion resistance, which requires sufficient chromium in the analysis. The advanced alloys developed to obtain balanced stress rupture and corrosion resistance exhibit in the cast condition gross segregation and structural inhomogeneity, which is a principal reason why these alloys are being produced by powder metallurgy methods. Segregation and inhomogeneity are then limited to the dimensions of individual powder particles. A principal application of powder metallurgy superalloys has been for turbine disks in jet engines. In this application stresses up to 480 MPa (70,000 psi) and temperatures up to 760°C (1400°F) are encountered. Up to this temperature a relatively fine grained material has better strength than coarse grained material.

The methods of producing superalloy powders by inert gas atomization, vacuum atomization or the rotating electrode process and the consolidation of the powders to "near net shape" by hot isostatic pressing in mild steel cans have been described.

Producing superalloys by powder metallurgy involves certain problems. The powders have to be very low in both oxygen and nitrogen in order to avoid films on the powder particles interfering with bonding. Another problem is the formation of carbides with the formula MC on prior powder particle boundaries in the consolidated material, which have a deleterious effect on the forgeability and the mechanical properties of the alloys. The precipitation of these carbides may be minimized by controlling the carbon content of the alloys and by hot isostatic pressing at low temperature.

When RSR (Rapid Solidification Rate) powder containing higher percentages of the refractory alloying components molybdenum and tungsten is hot consolidated, the resulting material has higher creep strength than material hot consolidated from coarser powders with conventional analysis.

Oxide dispersion strengthened superalloys have been intensively studied in the last 20 years. The aim is to produce alloys which retain their strength at higher temperatures than the nickel-base superalloys strengthened by precipitation strengthening with the gamma prime phase. The oxides in the fine dispersion in nickel, iron and cobalt are generally thorium or yttrium oxide which have been found to be stable while aluminum oxide is not stable.

The early approaches to oxide dispersion strengthened nickel-base alloys used wet methods of producing a powder consisting of a fine dispersion of the oxide in the metal matrix. For TD-nickel[54] (Thoria dispersed nickel), the best known of the alloys produced by wet methods, the powder was produced by precipitating nickel hydroxide from a nickel salt solution upon a colloidal suspension of thorium oxide in the solution. The precipitate was dried and reduced with hydrogen to a fine nickel powder, in which thorium oxide was dispersed. The powder is isostatically pressed into compacts, which are then sintered, canned and extruded. The extrusions may be rolled into sheet.

At the present time, the principal interest in dispersion strengthened superalloys is based on alloys fabricated by a new method, "mechanical alloying," rather than on alloys produced by the wet methods[55].

The metal powder and oxide to be dispersed in the metal matrix are introduced into an attritor grinding mill which is a high energy driven ball mill with the powder and balls held in a stationary tank and agitated by rotating impellers. During milling the ingredients of the powder mixtures are reduced in size and brought into intimate contact by flattening and crushing the particles, welding them together and repeating the process again and again.

Mechanical alloying is being used to produce a type of nickel alloy which can be both precipitation and dispersion hardened. A mixture of metal and alloy powders, typically carbonyl nickel powder, chromium powder, refractory metal powders, nickel-aluminum-titanium, nickel-zirconium and nickel-boron master alloy powders and yttrium oxide powder are processed into composite powder particles. The powders are consolidated by packing the powder in cans and extruding the canned powder. The extrusions may also be rolled into sheet.

The alloys contain aluminum and titanium in the matrix and can therefore be strengthened by precipitation of the gamma prime phase. They also contain very finely divided yttrium oxide which imparts dispersion strengthening at temperatures where conventional superalloys overage.

The latest of these alloys, MA 6000 E[56] developed by the International Nickel Company contains 15% Cr, 1.1% Y_2O_3, 4.5% Al, 2.5% Ti, 2.0% Ta, 2.0% Mo, 4.0% W, 0.1% Zr, 0.01% B, balance Ni. It is fabricated by extrusion and rolling and may be subsequently zone annealed to provide a grain structure with a grain aspect ratio greater than 10. The stress for 1000 hour stress rupture life of this alloy is compared in Figure 34 with that of TD-nickel, of the superalloy IN 792 and of a directionally solidified alloy DS-MAR M200+Hf. Its outstanding stress rupture properties over the temperature range from 800 to 1150°C are evident. Above 1000°C it is much superior to the common nickel base superalloys. It has good oxidation, sulfidation and carburization resistance and is intended for rotating, centrifugally loaded parts, such as jet engine blades.

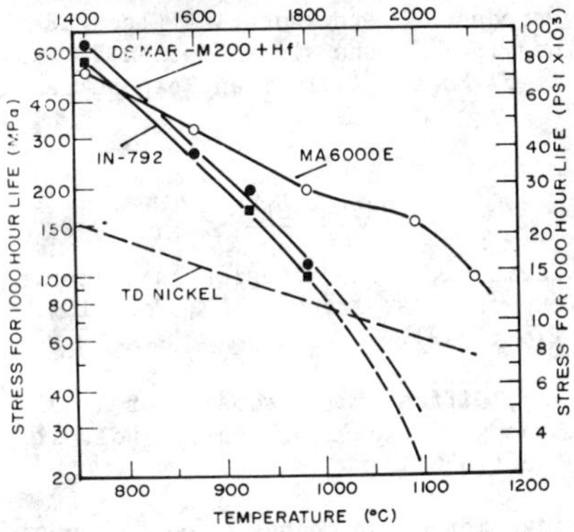

Figure 34. Stress for 1000 hours life to rupture as a function of temperature for alloys MA6000E, DS MAR M 200 (a directionally solidified alloy) and TD nickel.

Tool Steels. Tool steels are produced by hot consolidating spherical tool steel powder particles in place of the conventional technique of casting an ingot and forging it[57]. This technique avoids carbide segregation and results in a material with a uniform macro-and microstructure having fine, evenly sized and distributed carbide particles and a fine and uniform grain size. The size and distribution of the carbide particles is illustrated in Figure 35.

Titanium Alloys. In the case of titanium alloys the technique of hot consolidating powder is used primarily as a means of lowering the high cost of titanium alloy aircraft parts when they are produced by conventional manufacturing methods. At present, the principal emphasis is on hot isostatic pressing by the ceramic mold process described previously. Active trials are under way to produce large aircraft engine and airframe parts as well as missile engine parts by this technique[58]. The savings possible are illustrated by the relative weights in the case of one large airframe part, an arrester hook support. The finished part weighing 12.9 kg is conventionally produced from a 116 kg forging, while by hot isostating pressing, using the ceramic mold process, a part weighing only 25 kg is produced, which is machined into the final product.

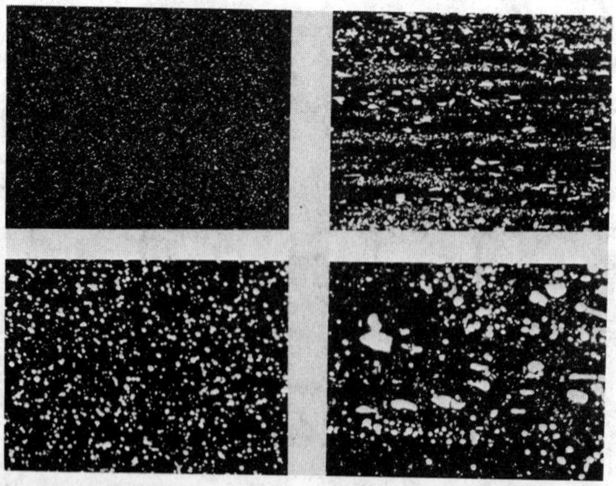

Figure 35. Longitudinal microstructures of T-15 tool steel; on the left: by hot isostatic pressing of powder; on the right: by conventional processing. Magnification: upper row 95x lower row 380x. (Courtesy: Crucible Inc., Specialty Metals Division).

The alloy most commonly produced in powder form by the Rotating Electrode Process contains 6% Al, 4% V, balance Ti, the same analysis as that of the widely used wrought alloy. However, other compositions have also successfully produced by the Rotating Electrode Process(59). They include the alloy "Corona 5," which contains 4.5% Al, 5% Mo and 1.5% Cr and can be readily welded.

REFERENCES

(1) G. C. Kuczynski, Self-diffusion in sintering of metallic particles, Trans. AIME, Vol. 185, p. 169-178 (1949).

(2) C. Herring, Diffusional viscosity of a polycrystalline solid, J. Appl. Phys., Vol. 21, p. 437-445 (1950).

(3) C. Herring, Effect of change of scale on sintering phenomena, J. Appl. Physics, Vol. 21, p. 301-303 (1950).

(4) D. L. Johnson, New method of obtaining volume, grainboundary and surface diffusion coefficients from sintering data, J. Appl. Phys., Vol. 40, p. 192-200 (1969).

(5) H. E. Exner, Principles of single phase sintering, Reviews on Powder Metallurgy and Physical Ceramics, Vol. 1, No. 1-4, p. 32-39 and 205-210 (1979).

(6) F. V. Lenel, G. S. Ansell and R. C. Morris, Theoretical considerations and experimental evidence for material transport by plastic flow during sintering, in "Modern Developments in Powder Metallurgy," Vol. 4, ed. by H. H. Hausner, Plenum Press, NY, p. 291-302 (1971).

(7) F. V. Lenel, G. S. Ansell and D. D. Barron, Marker Transport during the early stages of sintering, Metallurgical Transactions, Vol. 1, p. 1772-1773 (1970).

(8) J. J. Nunes, F. V. Lenel and G. S. Ansell, The influence of crystalline anisotropy on neck growth during the sintering of zinc, Acta Met., Vol. 19, p. 107 (1971).

(9) E. Friedrich and W. Schatt, Sintering of one-component model systems: nucleation and movement of dislocations in necks, Powder Metallurgy, Vol. 23, p. 193-197 (1980).
W. Schatt and E. Friedrich, Self-activation of sintering process in one-component systems, Powder Metallurgy International, Vol. 13, p. 15-20 (1981).

(10) H. E. Exner, Principles of Single Phase Sintering, Reviews on Powder Metallurgy and Physical Ceramics, Vol. 1, No. 1-4 (1979).

(11) P. C. Eloff and F. V. Lenel, The effect of mechanical constraints upon the early stages of sintering, Mod. Dev. in Powder Metallurgy, Vol. 4, ed. by H. H. Hausner, NY, p. 291-302 (1971).

(12) H. E. Exner, G. Petzow and P. Wellner, Problems in the extension of sintering theories to real systems, in "Sintering and Related Phenomena," ed. by G. C. Kuczynski, NY, p. 351-362 (1973).

(13) R. I. Coble and J. E. Burke, Sintering in crystalline solids, in "Reactivity of Solids," ed. by J. H. DeBoer, Amsterdam, p. 38-51 (1961).

(14) G. C. Kuczynski, Statistical approach to the theory of sintering, in "Sintering and Catalysis," ed. by G. C. Kuczynski, NY, p. 325-337 (1975).

(15) G. H. S. Price, C. J. Smithells and S. W. Williams, Sintered alloys Part 1: Copper-nickel-tungsten alloys sintered with a liquid phase present, J. of the Inst. of Metals, Vol. 62, p. 239-264 (1938).

(16) Carl Wagner, Theory of the aging of precipitates by solution and reprecipitation, Z. of Elektrochemie, Vol. 65, p. 581-591 (1961).

(17) H. S. Cannon and F. V. Lenel, Some observation on the mechanism of liquid phase sintering, Proc. 1st Plansee Seminar, ed. by F. Benesovsky, Reutte, Austria, p. 106-121 (1953).

(18) W. J. Huppmann and H. Riegger, Modelling of rearrangement processes in liquid phase sintering, Acta Met., Vol. 23, p. 965-971 (1975).

(19) W. D. Kingery, Densification during sintering in the presence of a liquid phase, I Theory, J. Appl. Phys., Vol. 30, P. 301-306 (1959).

(20) D. N. Yoon and W. J. Huppmann, Grain growth and densification during liquid phase sintering W-Ni, Acta Met., Vol. 27, p. 693-698 (1979).

(21) D. N. Yoon and W. J. Huppmann, Chemically driven growth of tungsten grains during sintering in liquid phase, Acta Met., Vol. 27, p. 973-977 (1979).

(22) R. A. Huseby and M. A. Scheil, Forgings from P/M preforms, Mod. Dev. in Powder Metallurgy, Vol. 4, ed. by H. H. Hausner, NY, p. 395-417 (1971).
R. H. Hoefs, Manufacturing and economic aspects of hammer forging of P/M preforms, Mod. Dev. in Powder Metallurgy, Vol. 7, ed. by H. H. Hausner and W. E. Smith, Princeton, p. 457-484 (1974).

(23) S. Altemeyer, Density Economy relationships of P/M parts, Machine Design, June 29, 1972.

(24) R. F. Halter, Pilot production systems for hot forging P/M preforms, Mod. Dev. in Powder Metallurgy, Vol. 4, ed. by H. H. Hausner, NY, p. 385-394 (1971).

(25) H. A. Kuhn, Deformation processing of sintered powder materials, ch. 4, p. 99-138 in "Powder Metallurgy Processing," ed. by H. A. Kuhn and A. Lawley, Academic Press, NY (1978).

(26) H. A. Kuhn and C. L. Downey, Deformation characteristics and plasticity theory of sintered powder metal materials, Int. J. of Powder Metallurgy, Vol. 7, No. 1, p. 15-25 (1971).

(27) S. K. Suh and H. A. Kuhn, Three fracture modes and their prevention in forming P/M preforms, Mod. Dev. in Powder Metallurgy, Vol. 9, ed. by H. H. Hausner and P. W. Taubenblat, Princeton, p. 407-426 (1977).

(28) G. Bockstiegel, Some technical and economic aspects of P/M hot forging, Mod. Dev. in Powder Metallurgy, Vol. 7, ed. by H. H. Hausner and W. E. Smith, Princeton, p. 91-127 (1974).

(29) W. J. Huppmann and M. Hirschvogel, Powder forging, Int. Metals Reviews, No. 233, Vol. 23, No. 5, p. 209-239 (1978).

(30) G. T. Brown and T. B. Smith, The relevance of traditional materials specifications to powder metal products, Mod. Dev. in Powder Metallurgy, Vol. 7, ed. by H. H. Hausner and W. E. Smith, Princeton, p. 9-31 (1974).

(31) G. Bockstiegel and C. A. Blande, The influence of slag inclusions and pores on impact strength of powder forged iron and steel, Powder Metallurgy Int., Vol. 8, No. 4, p. 155-160 (1976).

(32) T. Krantz, J. C. Farge, and P. Chollet, Hardenability and mechanical properties of hot forged Mn-Mo steels made from prealloyed powders, Mod. Dev. in Powder Metallurgy, Vol. 10, ed. by H. H. Hausner and P. W. Taubenblat, Princeton, p. 15-41 (1977).
C. Razim, Fatigue behavior of sintered ferrous materials with particular reference to their use in automobiles, ZwF, Vol. 73, p. 451-456 (1978).

(33) S. M. Kaufman and S. Mocarski, The effects of small amounts of residual porosity on the mechanical properties of P/M forgings, Int. J. of Powder Metallurgy, Vol. 7, No. 3, p. 19-30 (1971).

(34) H. A. Kuhn and C. L. Downey, How flow and fracture affect designs of preforms for powder forgings, Int. J. of Powder Metallurgy, Vol. 10, No. 1, p. 59-66 (1974).

(35) P. Lindskog and S. E. Grek, Reduction of oxide inclusions in powder preforms prior to hot forging, Mod. Dev. in Powder Metallurgy, Vol. 7, ed. by H. H. Hausner and W. E. Smith, Princeton, p. 285-301 (1974).

(36) A. Lawley, An overview of atomization processes and fundamentals, Int. J. of Powder Metallurgy, Vol. 13, p. 169-188 (1977).

(37) P. R. Roberts, P. Loewenstein, Titanium alloy powders made by the rotating electrode process in "Powder Metallurgy of Titanium Alloys," ed. by F. H. Froes and J. E. Smugeresky, Metallurgical Society of AIME, Warrendale, PA, p. 21 (1980).

(38) J. Devillard and J. P. Herteman, Evaluation of Ti-Al-4V powder compacts fabricated by the PSV process, in "Powder Metallurgy of Titanium Alloys, ed. by F. H. Froes and J. E. Smugeresky, Metallurgical Society of AIME, Warrendale, PA, p. 59 (1980).

(39) H. Stephan and J. K. Fischhof, Production of high purity metal powders by EBRD-electron beam rotating disk process, "Mod. Dev. in Powder Metallurgy," Vol. 9, ed. by H. H. Hausner and P. W. Taubenblat, p. 183-190 (1977).

(40) R. E. Peebles and C. A. Kelto, Investigation of method for the production of high quality, low cost titanium alloy powders in "Powder Metallurgy of Titanium Alloys," ed. by F. H. Froes and J. E. Smugeresky, Metallurgical Society of AIME, Warrendale, PA, p. 47 (1980).

(41) G. S. Ansell, The mechanism of dispersion strengthening: a review, in "Oxide Dispersion Strengthening," ed. by G. S. Ansell, T. D. Cooper and F. V. Lenel, NY, p. 61-141 (1968).
See also the contributions of M. F. Ashby, L. L. J. Chin and N. J. Grant, R. L. Jones and A. Kelly, I. G. Palmer and G. C. Smith, M. C. Inman and P. J. Smith, B. A. Wilcox and A. H. Clauer, R. Grierson and L. J. Bonis, R. W. Fraser and D. J. I. Evans, P. Guyot in Part II, Mechanism of dispersion strengthening, ibid., p. 143-427.
B. A. Wilcox and A. H. Clauer, Grain-size in dispersion hardened nickel alloys, Acta Met., Vol 20, p. 743-757 (1972).

(42) C. A. Kelto, B. A. Kosmal, D. Eylon and F. H. Froes, Titanium powder metallurgy - a prospective, in "Powder Metallurgy of Titanium Alloys," ed. by F. H. Froes and J. E. Smugeresky, Metallurgical Society of AIME, Warrendale, PA, p. 1 (1980).

(43) H. Fischmeister, Isotatic hot compaction - a review, Powder Metallurgy Int., Vol. 10, p. 119-123 (1978).

(44) S. H. Reichman, Low cost P/M superalloy applications in turbines, Int. J. of Powder Metallurgy, Vol. 11, No. 4, p. 277-284 (1975).

(45) U. S. Patent 3,704,508, V. N. diGiambattista, Process for compacting metallic powders, Dec. (1972).

(46) W. W. Beaver and F. H. Larson, The powder metallurgy of beryllium in "Powder Metallurgy," ed. by W. Leszynski, New York, p. 743-773 (1961).

(47) P. Loewenstein, L. R. Aronin and A. L. Geary, Hot extrusion of metal powders in "Powder Metallurgy," ed. by W. Leszynski, New York, p. 563-585 (1961).

(48) G. L. London, G. H. Keith and N. P. Pinto, Grain size and oxide content affect beryllium properties, Metals Eng. Quart., Vol. 16, No. 4, p. 45-57 (1976).

(49) J. P.Lyle, Jr., and W. S. Cebulak, Properties of high-strength aluminum products, Metals Engineering Quarterly, Vol. 14, No. 1, p. 52-63 (1974).
W. S. Cebulak, E. W. Johnson and W. Markus, High-strength aluminum P/M mill products, Int. J. of Powder Metallurgy, Vol. 12, No. 4, p. 299-310 (1976).

(50) E. A. Bloch, Dispersion strengthened aluminum alloys, Metallurgical Reviews, Vol. 6, No. 22, p. 193-239 (1961).

(51) R. J. Towner, APM alloys, Metals Engin. Quart., Vol. 1, No. 2, p. 24-40 (1961).

(52) A. V. Nadkarni, E. Klar and W. M. Shafer, A new dispersion-strengthened copper, Metals Eng. Quart., Vol. 16, No. 3, p. 10-15 (1976).

(53) G. H. Gessinger and M. J. Bomford, Powder metallurgy of superalloys, Internat. Metall. Reviews, Review 181, Vol. 19, p. 51-76 (1974).
J. S. Benjamin and J. M. Larson, Powder metallurgy techniques applied to superalloys, J. of Aircraft, Vol. 14, p. 613-623 (1977).

(54) F. J. Anders, G. B. Alexander and W. S. Wartel, A dispersion strengthened nickel alloy, Metal Progress, Vol. 82, No. 6, p. 88-91 and 122 (1962).

(55) J. S. Benjamin, Dispersion strengthened superalloys by mechanical alloying, Metall. Trans., Vol. 1, p. 2943-2951 (1970).

(56) Anonymous, Mechanically alloyed high temperature materials, J. Metals, Vol. 29, No. 12, p. 9-11 (1977).

(57) E. J. Dulis, High-speed tool steels by powder metallurgy in "Powder Metallurgy for High-Performance Applications," ed. by J. J. Burke and V. Weiss, Syracuse, p. 317-329 (1972).
P. Hellman, H. Larker, J. B. Pfeffer and I. Stromblad, the ASEA/Stora process, a new process for the manufacture of tool steel and other alloy steels from powder, Mod. Dev. in Powder Metallurgy, Vol. 4, ed. by H. H. Hausner, NY, p. 573-582 (1971).

(58) V. C. Petersen, V. K. Chandhok, C. A. Kelto, Hot isostatic pressing of large titanium shapes, in "Powder Metallurgy of Titanium Alloys," ed. by F. H. Froes and J. E. Smugeresky, Metallurgical Society of AIME, Warrendale, PA, p. 247 (1980).

(59) D. W. Becker, W. A. Baeslack III and F. H. Froes, Welding of Corona 5 PM products, in "Powder Metallurgy of Titanium Alloys," ed. by F. H. Froes and J. E. Smugeresky, Metallurgical Society of AIME, Warrendale, PA, p. 217 (1980).

METALWORKING AND THERMOMECHANICAL PROCESSING

N.E. Paton and A.K. Ghosh
Rockwell International Science Center
Thousand Oaks, CA 91360

Microstructure and properties of meallic components are intimately tied to the thermal and deformation histories imparted to the workpiece during shaping operations. Physical metallurgists have advanced this technology to a state where a wide variety of materials and processes are now available for optimizing cost, part geometry, and service properties. In this chapter, a review is presented on the mechanics of metal flow in the deformation, and microstructural changes resulting from deformation. In considering the various deformation processes, a brief description of the conventional metalworking technique of rolling, is followed by more detailed descriptions of specialized processes such as forging, sheet metal forming, superplastic forming, and high energy rate forming. The interrelationships between material properties such as strain hardening, strain ate sensitivity, anisotropy, and the processing conditions that lead to strain localization and fracture are emphasized.

Among the various thermomechanical processing techniques, certain key ones that lead to toughening, as well as improving workability, are addressed. Processing of titanium alloys for toughness and fine grain processing of aluminum alloys for improved superplasticity are discussed.

INTRODUCTION

Mechanical and thermal processing of metals is carried out with two principal objectives: (1) to impart a desired shape and (2) to alter properties. These operations are carried out simultaneously or consecutively in many situations, and the present paper is devoted to reviewing some of these operations. Shaping is normally carried out by passing a metal workpiece through a shaped die or a punch in the shape of desired product. The shape change is accommodated by internal shears as well as surface deformation due to frictional constraints, and the plastic work leads to property changes through work hardening, dynamic recovery and recrystallization. The microstructures produced by the various deformation operations are effectively utilized during thermomechanical processing steps to impart improvements in properties and performance. An adequate review of this vast field of metallurgy and materials science demands the length of this entire book. In the short span of this chapter, however, only a few selected topics are covered. For a more detailed review, the reader is referred to Refs. 1-5. We will first cover the field of deformation processing, and subsequently give a synopsis on the area of thermomechanical processing.

DEFORMATION PROCESSING

The field of deformation processing of metals may be broadly classified into two categories: (1) primary processing, i.e., those processes that are achieved at the metal producers end, and (2) secondary processing, i.e., those that are practiced by the metal user or manufacturing industry for making goods and articles. The primary processing operations are generally bulk metal shaping operations, such as rolling of blooms and billets, and secondary operations include all others, e.g., rolling of shaped sections such as I-beams, rails, etc., or extrusion of various section shapes, rolling of sheet and plate products and further fabrication of products using these materials.

An understanding of material response due to shape change is important in deformation processing. The stress states for deformation processes are best illustrated by means of a yield locus taken from an article by Backofen (3) (Fig. 1). The process of rolling, sheet bending, drawing of the cup wall are all plane strain processes for which the intermediate principal strain is zero.* The stress state, however, differs somewhat depending on the presence of absence of anisotropy. The highest yield resistance is offered on the loading path for plane strain. Rolling, even though a plane strain compression process, can be represented in the tension quadrant, since hydrostatic pressure causes negligible effect on yielding. Plate or sheet extrusion follow similar conditions while rod extrusion and drawing are axisymmetric operations along the compression or tension axis. The flange region of the cup deforms under stress states from almost all of the tension-compression quadrant where it is located; in an average sense, however, it preserves its thickness on a loading path near the middle of the quadrant. The shape of yield locus in the neighborhood of the applied stress state controls the tendency for strain localization, and its magnitude dictates the load requirements in shaping operations.

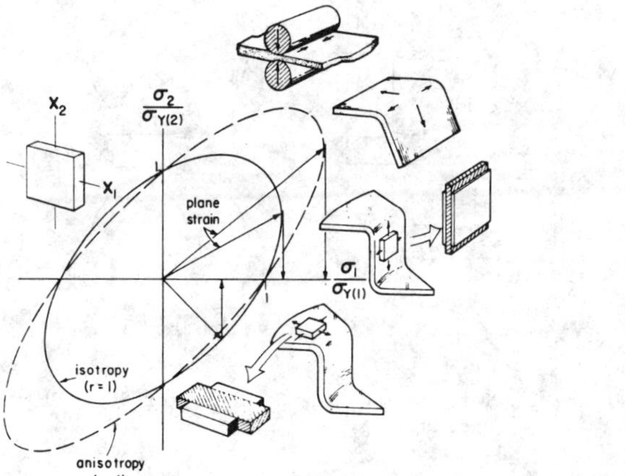

Figure 1. Some Processing Operations Related to Yield Loci for Isotropic and Anisotropic Materials.

*Strain increments are given by vectors normal to the yield locus; the principal directions for stress and strain are coincident.

Deformation Zone Geometry

Material emerging from the deformation zone is changed in shape, surface condition, structure and properties. Its reaction to that change appeared as load and elastic distortion along the die walls. Both the change and its reaction are sensitive to a host of processing variables, but chief among them is the geometry of the deformation zone. In detail, the zone can be geometrically complex, especially for non-steady-state operations with irregular channel walls, but it is useful to characterize the zone with a quantity Δ = the ratio of its average thickness, h, to its length L.

Two geometrically simple but practical steady-state reduction processes are given in Fig. 2. Drawing through a die of semiangle α is represented in Fig. 2a. In this example h is taken to be the circular arc midway through the deformation zone, normal to each die face, and L is the interface length. For an axisymmetric reduction,

$$\Delta = \frac{\bar{h}}{L} = \frac{\alpha}{r_1}\,[1 + \sqrt{1 - r_1}]^2 \qquad (1)$$

in which $\bar{h}$ is essentially a mean diameter, α is a radians and r_1 is the reduction in the pass. For a plane-strain reduction, e.g., strip drawing,

$$\Delta = \frac{\alpha}{r_1}\,(2 - r_1) \qquad (2)$$

For very small r_1, the axisymmetric Δ tends towards $4\alpha/r_1$ and the plane-strain toward $2\alpha/r_1$. In either case, Δ is proportional to α/r_1 which emphasizes its basic dependence on the slope of the channel wall. For strip rolling, as in Fig. 2b,

$$\Delta \simeq \sqrt{\frac{h_o}{4Rr_1}}\,(2 - r_1) \qquad (3)$$

in which R is roll radius. With the approximation that $L \simeq \sqrt{Rr_1h_o}$ the equivalent die angle in a rolling reduction is $\alpha_e \simeq \sqrt{r_1h_o/4R}$.

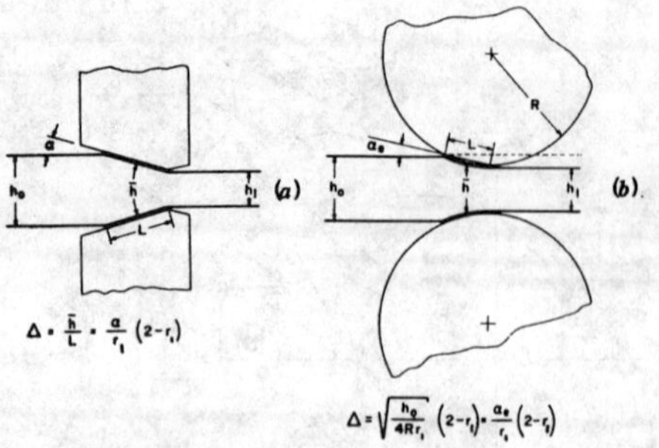

Figure 2. Deformation Zone Geometry for Strip Drawing and Rolling.

The Δ influence on the inhomogeneity of straining is great. As Δ becomes larger than 1, the gradient of strain over the reduced cross section increases, since this minimizes external work. Strain is generally more concentrated near the surface, and consequently lead to a gradient in grain size, texture and hardness. For smaller Δ, the gradient is limited and often undetectable, although shear-band formation may contribute to some volumetric nonuniformity; practical problems under these conditions of geometry are more commonly associated with the friction hill over the tool-metal interface (discussed later).

The Plastic Flow Properties of Materials

The distribution of plastic flow in metals during a deformation process is controlled primarily by the hardening processes accompanying their deformation. The processes of stain hardening and strain-rate hardening can reduce the severity of deformation through hardening of the deforming region. Another property that is also important in certain sheet metal forming processes is the anisotropy of plastic flow, which determines the resistance to thinning. Mechanical anisotropy causes anisotropy in fracture behavior (discussed later). All of these properties are controlled through the chemical composition and thermomechanical processing, and microstructure of the alloys, their interrelationship being known only in a semiquantitative manner.

Strain Hardening and Strain Rate Sensitivity

Shaping of metals require crystallographic slip which occurs by dislocation motion. This leads to strain hardening as dislocations are impeded by intersections with other dislocations as they glide through a crystal. This resistance also depends on the type of crystal structure and originates from obstacles, such as crystal boundaries, foreign atoms, faults in crystal stacking, and second phase particles. In brass and bronze, for example, the extended stacking faults can produce very strong obstacles and resist cross-glide of dislocations from one slip plane to another. As a result, the dislocations start piling up in planar arrays and their continued packing leads to rapid hardening. At the other extreme of the hardening behavior, aluminum and iron exhibit ease of cross glide creating a more three-dimensional cellular array of dislocations and produce a lower hardening rate.

Strain rate hardening (or sensitivity) depends mostly on the short range disturbance caused by the internal obstacles. The short-range disturbance can be overcome by combined thermal fluctuations and mechanical stress. Such obstacles include the inherent resistance of the lattice to dislocation motion, individual dislocation interactions (such as cross slip and climb) and solute atoms or clusters.

The determination of the plastic flow properties is normally carried out in an uniaxial tensile or compression test. A basic assumption utilized here (for isotropic materials) is that the effective stress-strain-strain rate relationship under multiaxial loading is the same as the one in uniaxial test. While there is need for a precise description of the stress-strain-strain rate relation for modeling and prediction purposes, at room temperature the stress plastic strain curve for most metals is roughly described by Ludwick's law: $\sigma = K\varepsilon^n$, where σ = true stress, ε = true strain, K = strength coefficient and n = strain hardening exponent. Low carbon steels exhibit a single value of n in the large strain region, which is not observed in most materials over a wide strain range. Most materials exhibit a three-stage n behavior, a high value of n bounded between lower values at both low and high strain regions. The low n for small strain is important for computing springback, however, for fracture or necking problems, it does not have a significant effect. The terminal value of n that is utilized in these calculations.

Many different constitutive laws combining strain hardening and strain rate hardening have been used for quantitative prediction of plastic flow in metals. Some of these are:

$$\sigma = K\varepsilon^n\dot{\varepsilon}^m \qquad (4)$$

$$\sigma = K(\varepsilon_o + \varepsilon)^n\dot{\varepsilon}^m \qquad (5)$$

and $$\sigma = [\sigma_\infty - (\sigma_\infty - \sigma_o)e^{-\varepsilon/k}]\dot{\varepsilon}^m \qquad (6)$$

where m = strain rate sensitivity index. All of these equations give a product-type relationship between the strain hardening and strain rate hardening terms. The other equations are of the type

$$\sigma = K\varepsilon^n(1 + \alpha\dot{\varepsilon}^m)^p \qquad (7)$$

$$\sigma = K[\varepsilon^n + \alpha(\dot{\varepsilon}/\dot{\varepsilon}_o)^m] \qquad (8)$$

and $$\sigma = K[\varepsilon^n + m\ell n(\dot{\varepsilon}/\dot{\varepsilon}_o)] \qquad (9)$$

using an additive stress model. These latter equations (7-9) seem to apply well to deformation near ambient temperatures, while the product-type relations seem to describe elevated behavior better.

ROLLING

Bulk metal rolling is conducted at a temperature greater than the recrystallization temperature (~Tmelt/2) with two primary aims: 1) to reduce its size to a more easily workable one, and 2) to break down the cast, columnar grain structure present in ingot materials. This step is normally preceded by long soaking cycles to aid homogenization, i.e., reduction in segregation of constituent materials. After hot rolling, the resulting bloom or billet does exhibit somewhat finer grain structure. The influence of simultaneous precipitation is also utilized to further refine grain structure and enhance properties (6).

Bulk metal rolling mills demand high temperature and load capability of the rolls. Even though rolls are not heated (actually copious water cooling), the load shock and temperature conditions can be quite severe, and both elastic and plastic deformation of rolls can occur if proper care is not exercised. The load requirements are somewhat eased in continuous casting operations since the ingots are not large. Also in a four-high mill, in which small rolls in contact with the workpiece can impart large deformation per pass on the plate, larger backup rolls can prevent bending of the smaller work rolls.

Cold rolling consists of passing unheated, previously hot-rolled bars, sheets or strip (cleaned of scale) through a set of rolls, often many times, until the final size is obtained. Cold working is employed to obtain the following effects: improved mechanical properties, better machinability, special size accuracy, bright surface, and the production of thinner gages than hot work can accomplish economically.

Mechanical Aspects of Rolling

The process of shaping metal by rolling consists essentially of passing the material between two rolls revolving at the same peripheral speed and in clockwise and counterclockwise directions, and spaced so that the distance between them is somewhat less than the height of the section entering them (Fig. 3). Under these conditions, the rolls grip the piece of metal and deliver it, reduced in cross-sectional area and increased in length, with only small amount of lateral spread. The turning of the rolls in contact with the work introduces a frictional force which acts along the contact arcs, and is proportional to the pressure between the rolls and the piece. This pulls the work into the opening between the rolls, against the wedging action of the tapered section entering the rolls. The piece enters the roll at a speed lower than the roll-surface speed, and leaves at a speed higher than the roll-surface speed. Evidently, there must be some point between A and B where the speed of the bar is equal to the roll-surface speed. This point, indicated by N, is called the neutral point, which coincides with the point of maximum pressure.

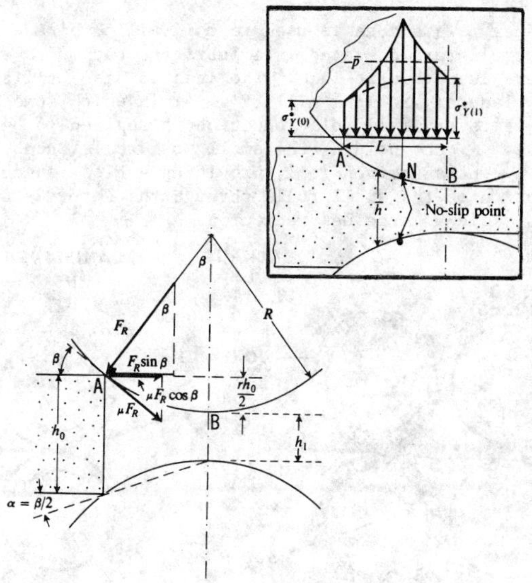

Figure 3 Forces Associated with Entering the Roll Gap. A Schematic View of the Roll Pressure Distribution Over the Contact Arc.

When the contact angle, β, is the maximum at which the piece will enter (without pushing), it is called the angle of bite. The radially directed reaction (shown in Fig. 3) to billet pressing on the rolls is F_R at angle β; the component pushing back is $F_R \sin\beta$. For unaided entry, i.e., without much external pushing or pulling,

$$\mu F_R \cos\beta > F_R \sin\beta$$

$$\tan\beta < \mu \quad .$$

since $\tan\beta \simeq \beta$ for μ up to at least 0.4 (when $\beta \simeq 22°$), the entry condition is

$$\text{Angle of bite} = \beta_{max} = \mu \quad , \qquad (10)$$

which shows the importance of friction control for successful rolling operation. (For further details see Ref. 7).

In cold rolling, material strain hardens significantly while internal stresses are reduced by recovery and recrystallization during hot working. For strain-hardening material, the yield strength of the deforming material rises through the roll gap which in turn raises the base level for the friction hill shown in inset of Fig. 3. The peak of the friction hill lies at the no-slip point, where roll and strip share the same speed. Because the contact angle in cold rolling is usually no more than 3-5°, the radially directed roll pressure and its vertical component are essentially the same. Under a back-tension stress, the base level for the friction hill at the entry is lowered by that amount and the no-slip point is moved forward; a front tension similarly lowers the exit level and moves the no-slip point backward.

Rolls are generally made of high grade alloy steels or chilled cast iron, and produced with a slight camber to prevent product from developing a crown. Small diameter rolls require less force than rolls of larger diameter to produce a given reduction. Smaller separating force is possible with small-diameter rolls for two reasons: first, the area of contact is less so that, the total force required is less; and second, the required average pressure is less because the smaller area of contact reduces the total frictional forces. Thus reduction in μR causes a drop in the friction hill, reduces roll flattening and raises deformation efficiency. This idea is used in foil mills where cluster rolls are operated in a lubricant bath with roll diameters less than an inch. The idea is also utilized in hot planetary rolling mill (Fig. 4) in which some of the benefits of small roll reduction is applied to heavy sections. A cage holding the small work rolls turns around a central driven roll, providing a close successive motion of the small rolls through the deformation zone to yield a large net reduction.

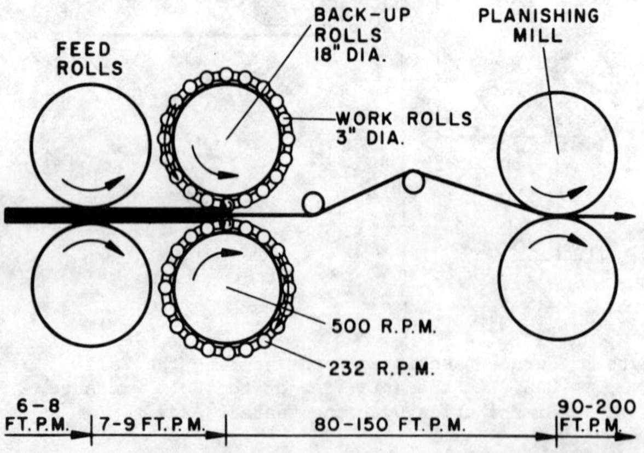

Figure 4. Schematic Representation of a Planetary Hot-Rolling Mill. (H.W. Ward, "Iron and Steel," January 1958).

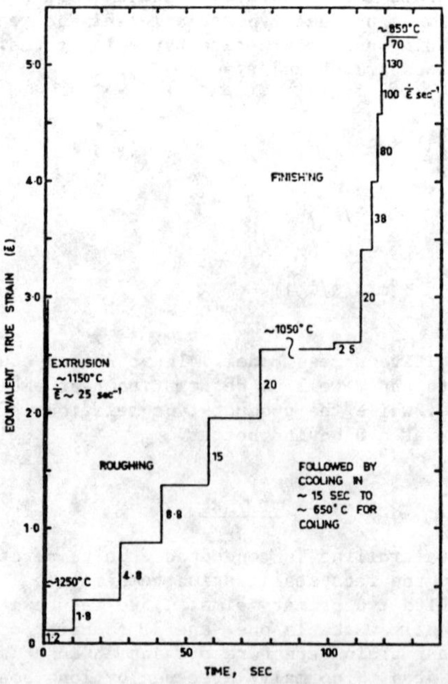

Figure 5. Typical Deformation/Time Schedules for Extrusion of Steel and for Rolling of 230 mm (9 in.) Steel Slab to Strip.

Strip tension effectively softens material in the roll gap. It does this with little influence on efficiency, since the total work within the system is not much changed. Tension is especially important in thickness- or gage-control practice as a means for maintaining essentially constant roll pressure (and roll force) in a given mill stand. The tension is regulated in response to a continuous measurement of roll-force, with a goal to stabilize the total elastic mill distortion, and to preserve more nearly uniform delivered gage. With steel-strip delivery speeds approaching a mile a minute, there is much economic pressure for the rapid control allowed by tension adjustment, to avoid potentially enormous material losses in distributing a product purchased to a thickness specification.

Table I

Possible Softening Processes Associated with Hot Working

Group	Example	Dynamic	Static
A	Al, α-Fe ferritic	Recovery (all strains)	Recovery followed by recrystallization
B	Cu, Ni, γ-Fe austenitic alloys	Recovery (small strains) Recrystallization (large strains)	Very limited recovery followed by recrystallization

Metallurgical Considerations in Hot Working

Hot workability relates to the ability of a metal or alloy to be deformed under conditions of high temperature >0.6 T_m (where T_m is the melting temperature in degrees Kelvin) and relatively high strain rates (10^{-1} to 10^3 s^{-1}). The two characteristics that govern hot workability are strength and ductility at the working temperature (2). Steady-state conditions are achieved only in a limited number of hot-working operations, e.g., extrusion or planetary rolling, where roughly isothermal deformation conditions exist. The major quantities of metals and alloys are hot worked under interrupted, nonisothermal deformation conditions, e.g., forging or rolling, where the incremental strains are smaller and may not reach those required to achieve steady-state conditions. The distinction between these two cases is shown in Fig. 5, where extrusion is compared with a typical hot-strip mill schedule. There is a complicated interaction between dynamic and static softening processes and this interaction becomes particularly important under interrupted, nonisothermal deformation conditions. The softening processes, are summarized as suggested by Sellars and McG. Tegart (2), as shown in Table I.

These differing processes lead to differing structures, strengths, and ductilities. Torsion of solid cylindrical specimens is a popular method of hot-workability testing since measurement of flow stress at high strains can be attained at constant surface rate, and quantitative measures of both flow stress and ductility can be obtained (8). Also, programmed deformation to simulate rolling or forging schedules can be applied relatively simply (9). Other methods include compression tests using cam plastometer for strain rate control (10). The stress/strain data is expressed in terms of effective stress and effective strain, utilizing von Mises relations, and equations are fitted through a family of such curves to obtain generalized description.

All rolling operations in practice require multiple passes in a series of rolling mills. The interpass microstructure is a function of the earlier deformation schedule. and the interpass times determine the extent of static recrystallization occurring between passes and the load requirements in subsequent rolling operations. The rolling schedules, and alloying additions which influence them are intimately tied to the metallurgy of the product. For example, in aluminum alloys, much recrystallization does not take place between passes unless the highest rolling temperature is used, when, of course, dynamic recrystallization can occur. In hot-working operations successive passes usually take place at continually decreasing temperatures (9), though in continuous rolling of rods and shapes the temperatures may rise during the latter passes. The effect of decreasing temperature during interrupted deformation on the stress/strain behavior is illustrated for superpurity aluminum in Fig. 6 (11). At 500° C (and above) the stress/strain curve obtained by programmed deformation is similar to that for initially annealed material, as complete recrystallization has occurred in the ~ 35 s delay period between successive deformation passes. At lower temperatures recrystallization is incomplete, so that successive cycles of deformation take place in material that retains the worked structure introduced by the prior deformation at higher temperature. As indicated by the curves at ~ 400° C, this results in much lower flow stresses than expected from initially annealed material.

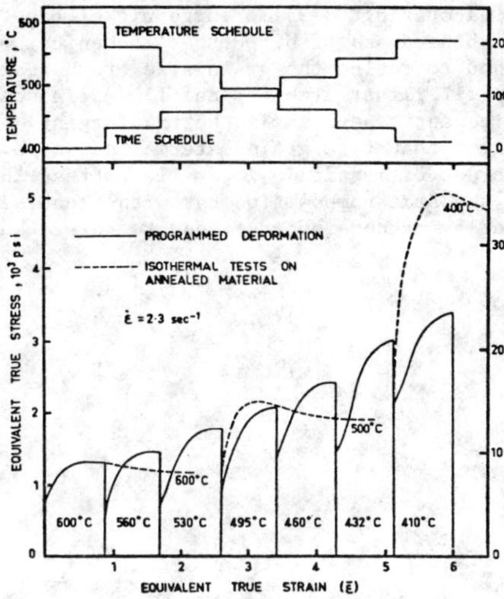

Figure 6. Deformation Temperature/Time Schedule and Resulting Stress/Strain Behavior of Superpurity Aluminum Deformed in Torsion at an Equivalent Strain Rate of 2.3 s^{-1}. (After Farag et al, (11)).

Low finish rolling temperature and the presence of alloying elements retard the recrystallization kinetics between passes. This effect is utilized in the manufacture of ultrafine grain microalloyed high strength steels. Small amounts of Nb reduces the recrystallization rate of the austenite such that substantial amount of rolling deformation could be imparted to develop elongated austenite grain structures prior to transformation into ferrite (12). The resulting microstructure after such controlled rolling contains ultrafine grain ferrite, which is partly responsible for the high strength of these alloys.

It is of interest to characterize hot workability of metals before undertaking the design and development of rolling, forging or extrusion procedures. Hot workability depends on a variety of parameters and properties. Microstructure, diffusivity and crystal structure influence hot workability to a significant degree (2). Crystal structure effects are generally interrelated with diffusivity effects. In titanium alloys, processing is performed in the β-range where formability is greater than in α or (α + β) range. This effect might primarily be that due to increased diffusivity in the b.c.c. β phase than in the h.c.p. α phase. In low carbon steels such an effect is graphically demonstrated in Fig. 7. A transformation from b.c.c. α to f.c.c. γ is found to reduce hot workability and therefore, it is wise to roll or forge below or around A_{c1} temperature than above it. To obtain equivalent workability in the γ-range, temperature needs to be significantly high which results in increased oxidation and energy expenditure. Duplex microstructure also generally reduces the hot ductility.

The deleterious effect of certain inclusions on hot ductility is well known. However, the effects are dependent upon the inclusion composition and distribution. If a material containing inclusions is plastically deformed, it is possible that: (a) the inclusion will fracture, or (b) the inclusion will deform

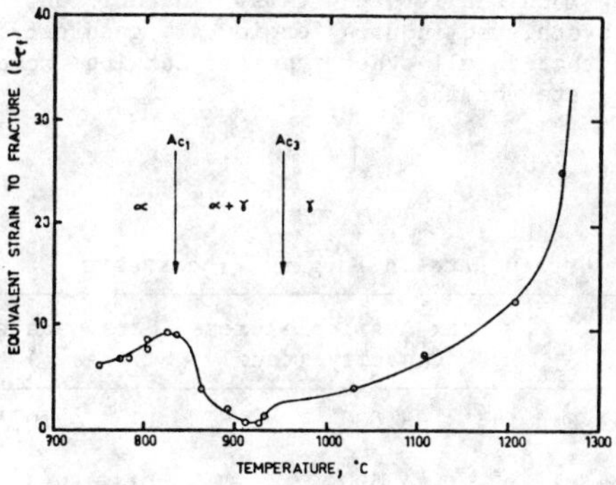

Figure 7. Ductility vs Temperature for a High-Oxygen Swedish Iron Tested in Torsion at an Equivalent Strain Rate of 0.5 s^{-1}. (After Reynolds and Tegart).

plastically with the matrix, and (c) a separation between the matrix and the particle will occur producing a cavity. Each of these cases can lead to void growth, and the formation of edge cracks during hot rolling. Sulphur content in steel is a problem with its hot ductility, however, the major reduction in ductility with increased volume fraction is attributable to oxides (13). Increased sulphur contents also give rise to reduced ductility in low carbon steels (14). At lower sulphur contents this is presumably due to the increase in volume fraction of sulphides but at high sulphur contents the formation of low-melting-point iron sulphide films at grain boundaries can lead to "hot-shortness" (15). The addition of manganese improves ductility (16), presumably by altering the nature of the sulphide inclusions. A Mn:S ratio of > 4:1 is needed to obtain adequate ductility in the critical temperature range of 950-1150° C. The beneficial effect of manganese is however reduced somewhat by an increase in the oxygen content of steels.

FORGING

Forging is one of the oldest of the metal working processes but because of the complexity of deformation generally occurring during forging it is one of the least well understood analytically. In addition, the wide range of forging processes commonly in use contributes to this complexity, making analytical approaches all the more limited in scope and application. Nevertheless considerable progress has been made in recent years in the treatment of forging processes and several excellent reviews have been published (17-18). Since the various characteristics of different kinds of forging equipment have been covered adequately elsewhere (see for example Ref. 17). We will attempt to cover in this section, the effects of metal flow characteristics on the forging process together with the effect that microstructural chracteristics have on forging operations.

Table II shows the strain rate ranges which are achieved by the various common forging processes. It can be seen that these range from a low of 10^{-4} per sec for isothermal forging to a high of 10^{4} per sec in high energy rate forming. The strain rate at which a forging process occurs, has a great deal of influence on metal flow characteristics and also on materials properties after the forging operation is completed. It can be seen from Table II that a wide range of forging press types is avalialble to the industry and that a choice must therefore depend on the particular application. Generally speaking the largest presses available both in the U.S. and in Europe are in the order of 50 thousand tons and are the hydraulic type. Strain rates achieved in hydraulic press forgings are generally low (order of 10^{-1} to 10/s). At the other end of the spectrum is the drop or hammer forging press the largest of which can generate equivalent loads in the order of 1,500 tons, but most are much smaller than this. In addition, strain rates achieved in drop forging are generally much higher than that in all other types of forging except high energy rate forming.

Table II

Strain Rate in Forging Processes

	Max Available Press Capacity (tons)	Strain Rate (sec^{-1})
Isothermal Forging	8,000	10^{-4} - 10^{-1}
Drop (Hammer) Forging	1,500 (equivalent)	10 - 10^{2}
Mechanical Press Forging	12,000	10 - 20
Screw Press Forging	8,000	10^{-1} - 10
Hydraulic Press	50,000	10^{-1} - 10
High Energy Rate Forging	750 (equivalent)	10^{4}

The importance of the strain rate achieved in a forging process is two fold. Firstly, high strain rates generate large amount of adiabatic heating which raises the component temperature thereby lowering the flow stress and making a part easier to be forged on a given size press. On the other hand, a very low strain rate can be used to advantage if a material has a high strain rate sensitivity. The importance of this aspect of the strain rate achieved in a forging process will be covered more fully in a later part of this chapter. At this point it is sufficient to say that the isothermal forging process can be used to form very complex components and that the low strain rates enable a lower forging capacity to be used than would be required if the part were formed at conventional strain rates.

Forging methods are generally described as being of the open die or closed die type. In open die forging nominally flat dies are generally used and components are upset forged by moving the component after each stroke to achieve the final desired shape. Open die forging requires considerable skill on the part of the operator and is generally used only for large components and small runs. Closed die forging describes a process in which a cavity in the die is used to shape the part geometry. The die cavity may have the final net shape dimensions of the desired component in which case the forging is referred to as a precision forging, or alternatively considerable machining may be required in which case a blocker forging results. In either case, a preform of some sorts is required for closed die forgings. Preformed shape may have major influence on the press load required to form a given part and is discussed in some detail later in this section.

Isothermal Forging

Isothermal forging is a process whereby the dies and the part are heated in a controlled manner to a constant temperature and the part formed at a somewhat slower strain rate (several minutes) than is normally used in conventional forging. The objective of the heated dies and the slow strain rates is to lower the forging load required to make a part of given complexity. Many components which are required for lightweight structures such as those used in the ground transportation and aircraft industries require complex components. These features normally require high press loads, or they have to be machined from a blocker forging which can result in as much as 80% of the forging part weight being removed by machining at considerable cost. Consequently, isothermal forging has become an important method of making complex components. Isothermal forming requires greater control over temperature, pressure, press and die conditions and cycle time than normally used in forging processing.

In addition to controlling press conditions in order to lower forging loads, some recent progress is also being made in controlling microstructure of the component of material to be forged in order to lower its flow stress. This has generally been achieved by refining the grain size of material either by a powder method as used in "Gatorizing" (19) to forge nickel base superalloy gas turbine disks or by further mechanical processing method to refine the grain size of the material (20). Table III illustrates typical flow stresses of materials frequently used in isothermal forging. The table indicates that fine grain material can have as much as an order of magnitude lower flow stress than a material of the same composition but with a coarse grain size produced by methods not designed to lower the flow stress.

Table III

Flow Stress of Materials for Isothermal Forging

Alloy	Temperature	Strain Rate	Fine Grain			Coarse Grain		
			Flow Stress ksi	Flow Stress MPa	Grain Size	Flow Stress ksi	Flow Stress MPa	Grain Size
Ti-6Al-4V	927°C	10^{-4} sec^{-1}	0.5	3.5	6	3.3	23	35
IN 100	1038°C	10^{-4} sec^{-1}	0.86	6	2.1	28.6	200	25
Aluminum 7075	450°C	10^{-4} sec^{-1}	0.9	6.3	10	3.5	24.5	40

Figure 8 shows the flow stress vs strain rate plot for two microstructures of Ti-6Al-4V illustating that the flow stresses differ widely at low strain rates but by a far lesser amount at the higher strain rates (21). This illustrates that the microstructure control is especially important in isothermal forging where the strain rates are low, but not nearly as important in conventional forging processes where the strain rates tend toward the higher end of the spectrum shown in Fig. 8. The microstructures of the two materials tested in Fig. 8 are illustrated in Fig. 9. The coarse microstructure in the beta processed material in Fig. 9b gives rise to the high flow stresses observed.

A method of making structures similar to forged ones which has been employed with great success on titanium alloys is diffusion bonding. Diffusion bonding or solid state joining as it is sometimes called can produce highly complex structures usually from sheet and plate starting materials in a wide variety of different materials (22). Properly utilized, the method can produce joints with parent metal strength but control of the process is critical in order to avoid defects in the bond line. The diffusion bonding process requires first that two cleaned interfaces be brought into intimate contact. This process requires first interfacial creep

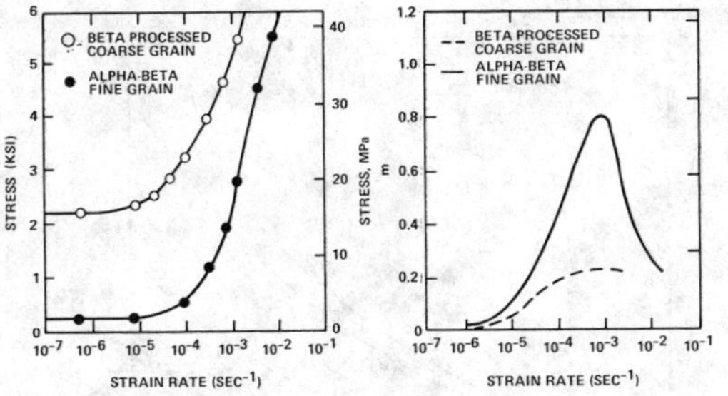

Figure 8. Strain Rate vs Flow Stress for Two Grain Sizes of Ti-6Al-4V.

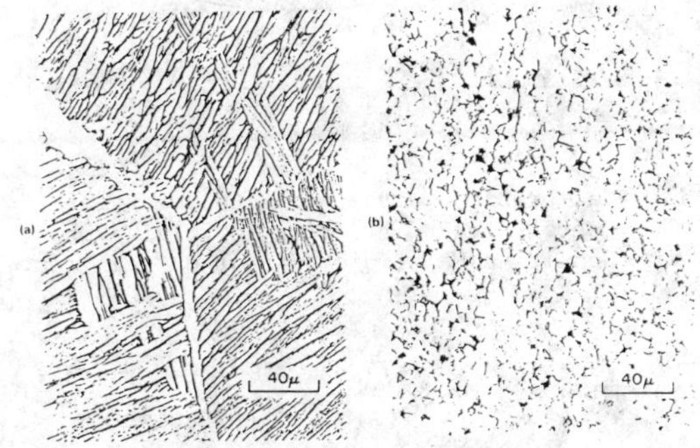

Figure 9. Micrographs of Ti-6Al-4V (a) Beta Processed, (b) Alpha-Beta Processed.

to smooth out the asparities in the interface and finally avoid shrinkage to eliminate microvoids which still remain as a result of fine scale surface discontinuities. Models for this process have been proposed by Hamilton (23) and by Garmong et al. (24) and both have been found to predict the time and pressure required to achieve interfacial contact with good precision.

The second requirement for diffusion bonding is that any oxide in the interface should either be removed before the bonding operation starts, or the oxide should dissolve in the metal during the bonding operation. In titanium the first is generally easily achieved as long as the oxide is thinned at the beginning of the process as is generally achieved by chem-milling or electropolishing. With other metals, however, such as aluminum a tenatious nonsoluable oxide is present on the surface making diffusion bonding extremely difficult unless it is accompanied with large amounts deformation which break up the oxide and allow the native metallic surfaces to come into contact with one another. Table IV lists some of the metals known to be suitable for diffusion bonding and generally if an alloy within a class can be satisfactorily diffusion bonded similar conditions for other alloys will result in satisfactory diffusion bonds of the metals listed only aluminum alloys provide real difficulties in obtaining parent metal strength diffusion bonds. Figure 10 shows a typical diffusion bonded component illustrating the geometrical complexity available with this fabrication method using fairly simple tooling and limited amounts of deformation to obtain adequate bond quality. Figure 11 illustrates the type of bond defect which can occur if insufficient pressure is applied during the bonding operation. Details of how these cavities occur and the processing required for their illumination are contained in Ref. 23.

Analytical Tools

Although forging is one of the oldest metal working operations, a clear understanding of forging from an analytical basis is difficult because of the complexity

Table IV

Metals Suitable for Diffusion Bonding*

Beryllium	Steels
	1020, 4340, 9310
Columbium	9310, 304, 316
Nickel Alloys	Tantalum
HS-25	
Inco 718	Titanium Alloys
Rene′ 41	Ti-6Al-4V
Pure Nickel	Corona-5
	Ti-6Al-2Sn-4Zr-2Mo
Aluminum Alloys	

*Reference 22

Figure 10. Typical Diffusion Bonded Component Fixtured for Structural Test.

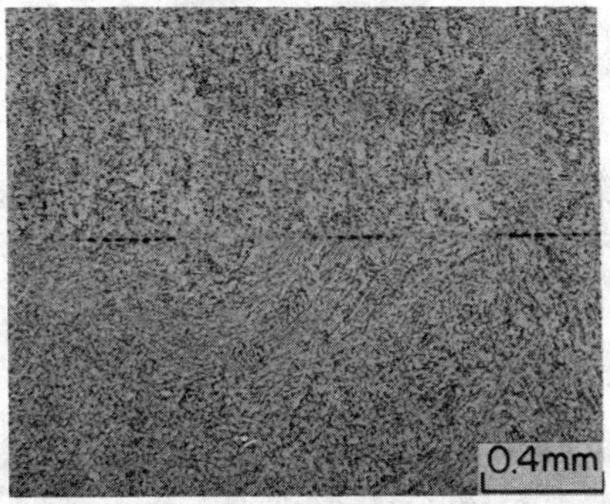

Figure 11 Diffusion Bond in Ti-6Al-4V Showing Defects.

of the deformation which occurs during forging operations. Despite these difficulties considerable progress has been made in recent years in modeling of forging of metals in order to predict the required forging loads, the die deflection, preformed shapes required and the effect of forging temperature and strain rate. Ref. 17 provides an up-to-date discussion of all of these aspects of forging and consequently only a summary of the more important aspects will be given here.

Forging Loads

Since the load required in any forging operation is proportional to the flow stress, determination of the loads requires at first a knowledge of the flow stress for metal deformation at the appropriate forging temperature and strain rate. Since flow stress always decreases with increasing temeprature it is generally desirable to work at as high a temperature as possible. This must, however, be consistant with the requirements for not degrading metal properties by excessive grain growth or incipient melting and also consistent with the allowable temperatures which can be used for the die materials employed in the forging operations. The effective flow stress, $\bar{\sigma}$, in addition to being a function of the deformation temperature T is also a function of the effective strain $\bar{\varepsilon}$ and the strain rate $\dot{\bar{\varepsilon}}$. If the forging temperature is above the recrystallization temperature, work hardening is not significant. In this case flow stress is not dependent on the strain $\bar{\varepsilon}$, the flow stress $\bar{\sigma}$ must be established as a function of both temperature and strain rate. This is usually achieved through either an isothermal compression test or through a ring test as shown in Fig. 12. The desired output is a table of flow stress vs temperature and strain rate data for a given material and a microstructure, which can be incorporated into a constitutive equation to describe the flow stress as a function of these parameters. A typical constitutive equation of this type is described in Ref. 21 based on the work of Ashby and Verrall for nonstrain hardening materials (26).

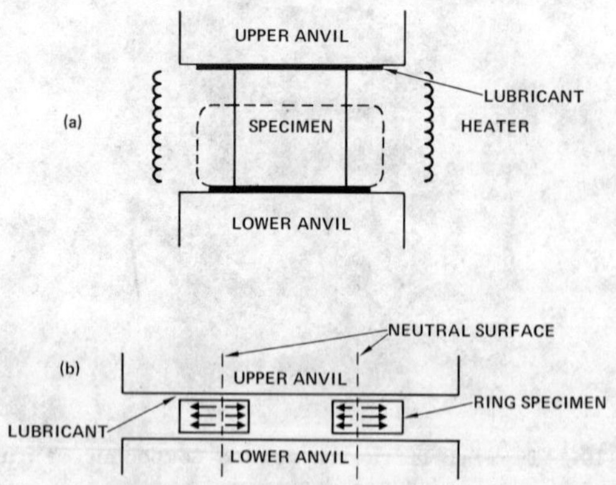

Figure 12. (a) Isothermal Compression Test for Measuring Flow Stress. (b) Ring Test for Measuring Flow Stress and Coffecient of Friction.

The ring test developed originally by Male and Crockroft (26) has the advantage that it can be applied to the determination of friction during forging. From a measure of the change in the ring diameter a plot such as that shown in Fig. 13 from Ref. 27 can be used to determine the coefficient of friction of a given lubricant. It can be seen in Fig. 13 that a wide variety of coefficients of friction can be obtained through a variation of lubricants.

Lubricants are important not only to reduce the overall forging load through a reduction in the coefficient of friction between the dies and the workpiece, but also to act as a parting agent between the die and the component to avoid galling and local welding. In some cases such as with glassy lubricants they can also be used to prevent oxidation of both the part and the die and also to insulate the part to avoid overheating of the dies. In all cases lubricants should be nonpoluting and not produce smoke upon application to the dies.

An analysis of forging loads requires first that the deformation of the component in the dies be determined. A typical case is shown in Fig. 14 for upset forging of a T section with essential rib. In this case three separate sections of the forging are undergoing deformation all in plane strain. Flow directions being illustrated in Fig. 14. Solutions to this problem and others can be found in Ref. 17. Although rigid analytical solutions are available for determination of forging loads, empirical methods are often used to calculate the forging load P, using the following expression:

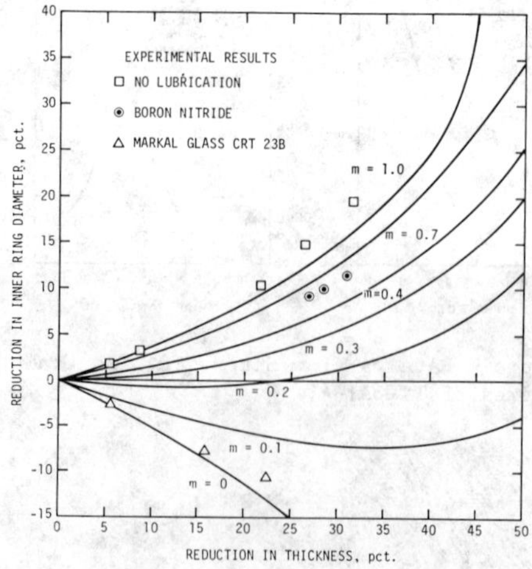

Figure 13. Calibration Curves for Measurement of Coefficient of Friction in Strain Rate Sensitive Materials Using the Ring Test.

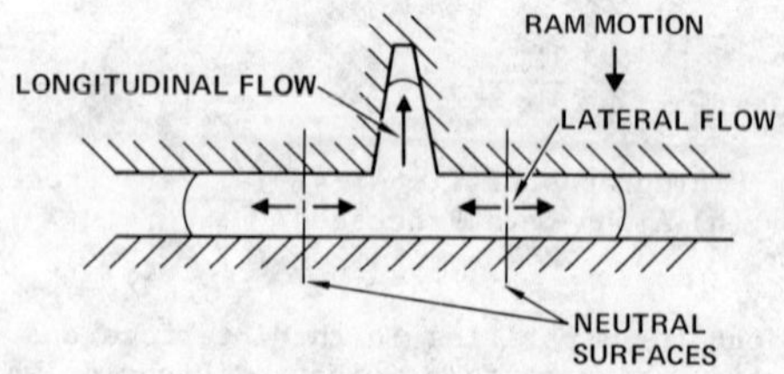

Figure 14. Metal Flow During Tee Section Forging.

$$P = C\bar{\sigma}A \qquad (11)$$

C is a factor which depends on the complexity of the forged part and varies from 1.2 to 2.5, A is the projected area and $\bar{\sigma}$ is the flow stress.

Forging Preform Design

In order to obtain defect free closed die forgings a preform which properly fills the die without causing defects or excessive metal losses due to flash must be used. It is desirable, generally, to use the most simple preform shapes since these can be produced by extrusion or rolling processes and are therefore low cost. However, for complicated closed die forging shapes it is sometimes necessary to use a preform which gives initial metal distribution in the die which approximates the final distribution but lacks the detail. The design and geometry of preforms are critical to forming defect-free final forgings.

Traditionally preform design has been more of an art than a science, the art being guided by some fairly basic rules. Obviously the preform volume should be slightly greater than the final volume of the forging in order to provide a minimal flash. The initial stock distribution must approximate the final distribution and wherever practical, the dimensions of the preform should be larger than the finished part in the forging direction.

For complex forgings and in particular precision forgings where final net shape is desired, recent progress has enabled accurate prediction of preformed shapes to be accomplished. In these methods an incremental analysis of metal deformation starting with the finished geometry and proceeding backwards to a desired preformed shape is conducted. By using these methods preform configurations designed by computerized methods produced finished forged components without any defects (28).

Forging is frequently chosen as a method of producing components for critical applications because of the good mechanical properties imparted to the product. These improved properties are achieved by a combination of the hot work introduced during forging operation and by the desirable grain flow, that occurs during forging. The latter can be particularly important as a consideration in the stress corrosion cracking properties of materials such as aluminum alloys where properties in the short transverse direction are frequently lower than in longitudinal and long transverse direction in the equivalent rolled plate product. This effect is illustrated schematically in Fig. 15.

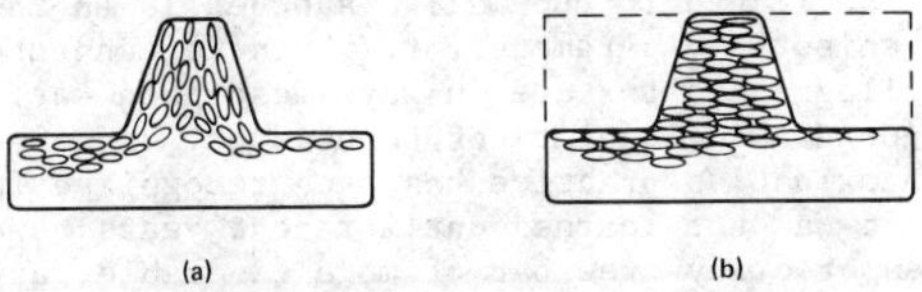

Figure 15. Grain Flow in Forging (a), as Compared to Machined Component (b).

Frequently, improper perform design can result in forging defects such as laps, folds and internal bursts. Some of the most common types of defects found in forgings are shown in Fig. 16, along with some guidelines for thier prevention.

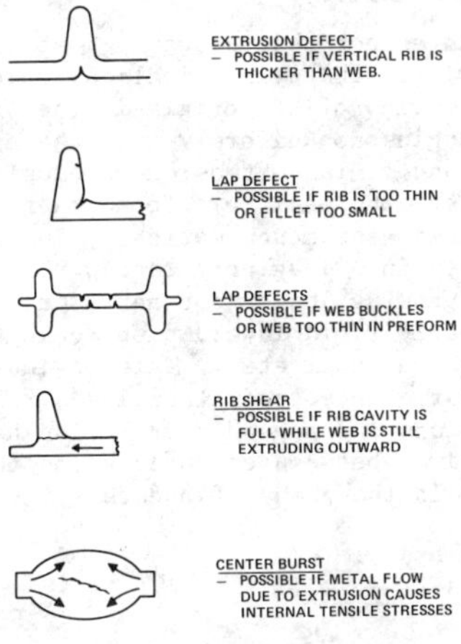

Figure 16. Examples of Some Types of Forging Defects.

SHEET METAL FORMING

Sheet metal forming is an extremely vital industry of the modern age. It is essential in the manufacture of automobiles, appliances, aerospace components, business machines, cans etc., and it presents a large useage of primary sheet products of steel, aluminum, brass, titanium and many other metals. Our focus here will be on sheet metal products forming which by itself encompasses a large variety of techniques and equipment. Many operations such as roll forming, coining, hydroforming, are quite specific to given part shapes. Others such as explosive forming are limited to low production items where die cost would be prohibitive. The most economically important is sheet metal stamping, and thus we shall concentrate our discussion and analysis on stamping.

The stamping process beings with a blanking operation in which sheet metal, stored in coils is automatically unrolled and cut to proper size by shearing dies. Often the blanks are nested to reduce trim loss. The precut blank is then stamped between two shaped dies. After stamping, the portion of the blank held in the holddown area is trimmed off. The next step, a flanging operation is essentially a bending process. Parts may require straight, stretch or shrink flanges, depending on whether the part is straight or contoured. The final stamping process is often a restriking operation which is required to sharpen contours and corner that do not meet the dimensional specifications in the initial stages of the stamping. Sheet metal parts are assembled at the flanges by spot welding, adhesive bonding or riveting.

When a metal blank is drawn into shape the loosely held blank is wrapped around the punch and pulled into the die cavity. This process causes little change in the final thickness. Since the blank is forced to contract circumferentially as it is drawn inward, it tends to buckle or wrinkle. Proper metal flow is achieved by inclining or tipping the punch-die assembly and applying blank holding pressure in some cases further restricting metal flow by draw beads on the blank holder surface. This controls metal flow by forcing the sheet to bend and unbend over these draw beads before entering the die cavity. In all but the simplest geometries portions of the metal blank are stretched, that is, an extension of the sheet in all surface directions in order to meet dimensional requirements and avoid wrinkling. The material requirements for drawing and stretching differ markedly.

Failure Modes in Sheet Metals

The most common mode of failure in forming sheet metal components is the development of a localized (through-thickness) neck or tear which occurs generally over a punch radius as the sheet surface area is increased to conform to the punch shape. The punch radius actually acts as a strain concentrator, by virtue of its curvature and frictional resistance (Fig. 17). A peak develops in the distribution of strain across the sheet, and when the peak strain exceeds a certain limiting strain value (forming limit) it gives rise to a visible thickness neck or tear (29). To predict if a part shape can be successfully produced from a given material, it is therefore necessary to compute the development of strain distribution across the sheet as a function of punch displacement and ascertain that the peak strain remains below the forming limit when the punch reaches the end of its travel. The problems of buckling, wrinkling and earing also exists during sheet metal drawing as a result of gathering of sheets, the most common problems are, however, necking and tearing.

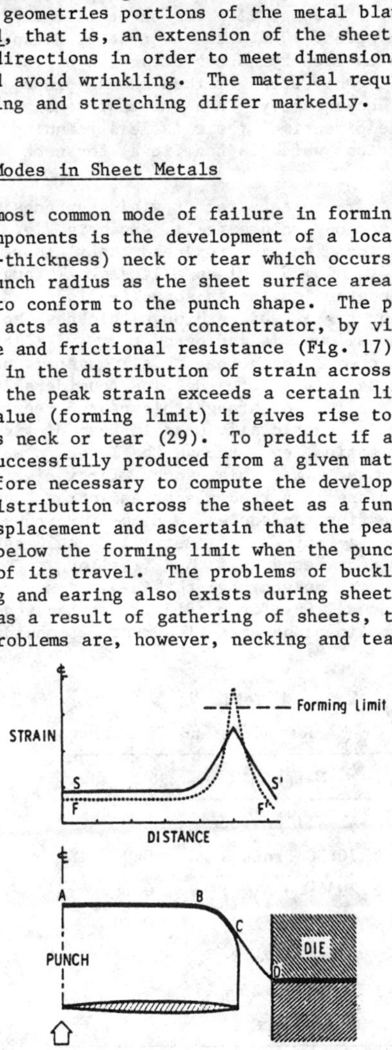

Figure 17. Schematic Strain Distributions for Punch Forming of a Flat Cylindrical Cup. The Area Under the Strain Distribution Curve SS′ Equals the Length of Line ABCD and Represents a Successful Part, While FF′ Produces a Visible Neck.

Drawing

Drawing of the cylindrical flat-bottom cup often referred to as deep drawing is illustrated in Fig. 18a. Most of the deformation in deep drawing occurs in the flange while very little occurs over the punch. The load required to deform the flange is transmitted from the cup bottom to the flange through the cup wall. If this load exceeds the maximum load the wall can support, tearing near the cup bottom results. For best drawability we need a material that resists thinning in the cup wall, that is, of lower strength in the plane of the sheet than in the thickness. This desired strength differential or anisotropy (r) can be measured conveniently in a simple uniaxial tensile test as illustrated in Fig. 18b. More detailed analysis (30) of deep drawn process as well as experiments (31) have shown that deeper cups can be drawn with higher r value materials.

Plastic Anisotropy

Plastic anisotropy is developed during primary metal workings such as rolling. This is a result of crystallographic texture which permits easier flow in certain sheet directions than others. For example, when the sheet normal develops a strong <110> component in the case of f.c.c. a strong <111> component in the case of b.c.c., and a strong <0001> in the case of h.c.p. crystals, the through-thickness direction of the sheet becomes considerably stronger than the in-plane directions. Consequently, in a uniaxial tension test along in-plane direction, there is less thinning than width contraction, which is beneficial for neck resistance.

A schematic representation of yield surfaces for different anisotropy values (r) is shown in Fig. 1. The case of r = 1 indicates isotropy and the yield or flow surfaces are given by Von Mises ellipses. Through-thickness strengthening, as in some Ti-alloys and steel, is indicated by r > 1, while through-thickness softening as in aluminum alloys, is indicated by r < 1 (Table V). The lack of a large number of slip systems help h.c.p. metals (of c/a < 1.66) develop considerably greater strength in the through-thickness direction. This property is beneficially used in deep drawing of A-K steel and titanium, even though the lower formability of Ti-alloys might limit this usage. The variation of r in the sheet plane is termed planar anisotropy and is the cause of earing tendencies in deep-drawn cups. While ears show a four-fold symmetry in steel, they can show a six-fold symmetry in Al and Ti-alloys. Thus for designating normal anisotropy, the in-plane average of r values can be used, but for predicting earing tendencies r value in each in-plane direction must be considered.

Table V

$\bar{r}$ Values of Al and Ti Alloys

Material	$\bar{r}$
Al-Killed Steel	1.80
70:30 Brass	0.90
2036-T4 Al	0.68
7075-T6 Al	0.70
Ti-6Al-4V	~ 2.4
Ti-4Al-0.25O_2	3.5
Comm. Pure Ti	0.5

Drawability Control

To raise the drawing limit the potential failure site must be strengthened relative to the deformation zone where the load originates. The main aim is to advance the failure site into the strain hardened material where it normally is in steady-state drawing. This would require taking the load from the bottom of the cup and distributing it along the punch cup interface-ideally reapplying it at the die radius. To do this in practice, the cup may be pressed more firmly against the punch by surrounding it with a pressurized rubber blanket where it comes out of the die. It is also done by withholding lubrication from the punch or even by intentionally roughening the punch.

The metallurgical control is tied the controlled yield locus distortion. Deformation in the flange follows loading paths through quadrant 4 with pure shear (no thickening). Deformation around the punch is essentially plane strain in quadrant 1, since the punch is rigid and prevents circumferential contraction. Each is related to the yield locus in Fig. 1. To raise the limit, therefore, the path length through quadrant 1 should be increased relative to that of the flange to quadrant 4. An obvious possibility is the temperature drop from the flange to cup wall. Another is an increase in anisotropy parameter, r, which is connected to the crystallographic texture in ways described earlier. The superior drawing quality of sheet steel made by aluminum deoxidation practice has been recognized for decades. It was not learned until recent years however that the superiority grew out of more favorable texture, that is more (111) alignment in the rolling plane and therefore larger r. For measurements on sheet steel the r dependence of LDR is 0.15 (Fig. 19) With this as a guide and indiction that an average r greater then about 3 is unlikely in cubic metals the isotropic LDR of 2 might be increased by 10 to 20% through texture control in these metals. In practice, however, deep drawing operations often involve stretching over punches much more complexly shaped than a flat ended cylinder, thus not only r but some measurable strain hardening capacity such as n, also influences overall press performance or the practical drawability.

Earing is an effect that shows up most clearly in the scalloped rim of the cylindrical cup which is deep drawn. A simple view of its origin is that deformation in the flange begins nonuniformly with the circumferential reduction and radial extension developing first at those places where the peaks are found on the drawing cup. The result is shown schematically in Fig. 18, a flange subdivided into imaginary rectangular caps. As one tab starts growing longer and narrower it accommodates more nearly rigid inward displacements of its neighbors and an ear nucleates. This obviously requires the kind of plastic anisotropy that leads to planar property variations and a good correlation has been found several times between ear height and the range of r requirements in the plane of sheets.

Mechanical Anisotropy

A second type of anisotropy that is not directly related to the crystallographic slip behavior and plastic flow properties is the anisotropy in the orientation of the elongated grain structure, or of the second-phase particles such as the inclusions and intermetallic phases. The banded grain structure and elongated inclusions are commonly termed as mechanical fibering and are results of thermomechanical processing or of rolling down insoluble constituents originally present in the ingots. Inferior fracture properties in the direction transverse to the rolling direction can be caused by this (32). The influence of mechanical anisotropy may not always be reflected in the total elongation measured in a tensile test. This is because a great many materials form a localized neck in uniaxial tension prior to their actual fracture. During actual forming, however, particularly in bending over sharp radii at punch (or die) edges and corners, the compressive stresses at the inner surface retard the development of instability and failure can occur by fracture at the outer surface. This is influenced by the direction of bend in the sheet. This kind of operation is highly sensitive to the mechanical anisotropy (33).

Fracture Properties

Plastic flow is almost always terminated by fracture processes, which for formable metals consists of (1) initiation of voids at second-phase particles, (2) their

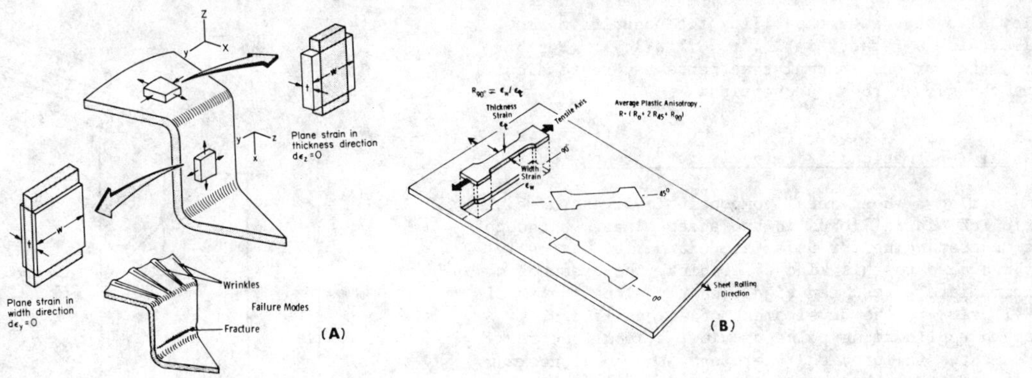

Figure 18. (a) Schematic Representation of Pure Shear Deformation Assumed in the Flange ($d\epsilon_z = 0$) and Plane Strain in Cup Wall ($d\epsilon_y = 0$). (b) Illustration of r Value Measurement From Tensile Test.

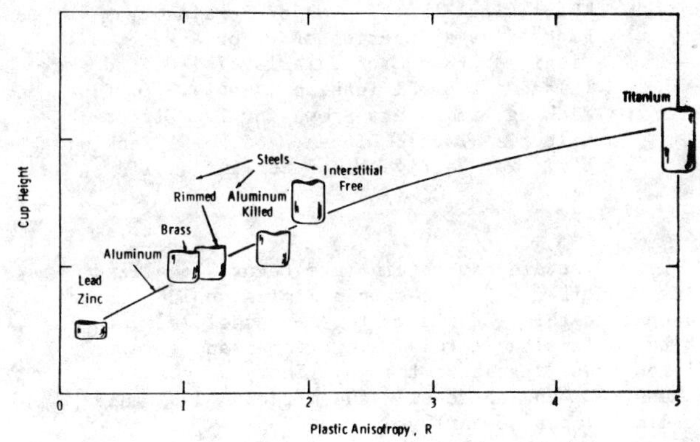

Figure 19. The Dependence of Drawn Cup Depth (Similar to Limiting Draw Ratio) on Plastic Anisotropy Parameter.

growth during deformation, and (3) their eventual coalescence. Even though the ligaments between voids pull down to the microscopic chisel edges, the macroscopic strain at fracture is an important quantity for analytical modeling of formability indices (34). The measurement of fracture strain is, however, complicated by localized necking that often precedes fracture. Actually, from a practical standpoint, a localized neck is considered to be the terminal event, which forms the basis of the forming limit diagrams to be discussed in the next section. For higher strength steels and aluminum alloys, especially under the conditions of biaxial loading, no visible loalized necking may precede fracture. Localization can still occur, however, only in a microscopic scale, by shear joining of voids. On a macroscopic scale, this might manifest as an abrupt fracture. While it is clear that the forming limit is essentially the fracture limit in this case, the importance of fracture may still be quite strong, even if necking is to precede. This is because a neck may not be highly pronounced until internal fracturing has started (35).

Anisotropy in fracture behavior can result from plastic anisotropy and mechanical anisotropy. While the influence of mechanical anisotropy has already been discussed, the result of increasing r is primarily an increased flow stress, i.e., a decrease fracture strain. Fracture strain also decreases due to stress biaxiality.

Instability and Necking

The strain hardening, strain-rate hardening and plastic anisotropy properties are measured during an uniaxial tensile test up to the point of maximum load. The strain hardening rate, decreasing during deformation, comes to a balance with geometric softening (or the reduction in specimen area) at this point ($d\sigma/d\epsilon = \sigma$). Beyond the maximum load, a diffuse neck developes in the specimen and localization of deformation takes place within it (Fig. 20). A diffuse neck and the analysis pertaining to its exact onset are of little significance to sheet metal forming, since it is not usually observable in a large component. Many developments have been made until now to understand the onset of a localized neck (a through-thickness neck) that eventually forms and which is objectionable from a practical standpoint. The early theory is due to Hill (36) who identified localized necking with the formation of a shear instability in a rigid-plastic sheet metal (along a direction of zero-length-change) and arrived at the condition: $d\sigma/d\epsilon = \sigma/2$ (Fig. 20). The strains for the onset of diffuse and localized necking are n and 2n, respectively, (when $\sigma = K\epsilon^n$) based on the above conditions. Localized neck occurs at a characteristic angle corresponding to the direction of zero length change in the material.

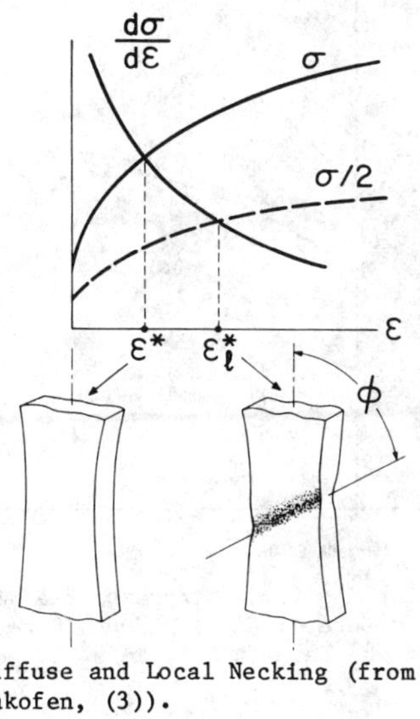

Figure 20. Diffuse and Local Necking (from W.A. Bakofen, (3)).

The presence of strain rate sensitivity (m) modifies both these strain values. While the uniform strain (at maximum load) is not significantly altered by the presence of m (unless temperature and strain rate are also changed), a positive m increases the strain limit for localized neck formation and the extent of nonuniform flow. A negative m, as in 5000 series aluminum alloys can actually reduce the extent of nonuniform flow. The room temperature value of m for the Ti-Alloys (n = 0.06) is near 0.02 and is responsible for nearly half of its total elongation (24). The 2024 aluminum alloy also have some capability for nonuniform flow, while this is rather small for 7075 alloys. Furthermore, the various thermal treatments can cause significant changes in these properties.

Biaxial Deformation: Forming Limit Concept

Forming a sheet metal component involves the application of biaxial loads in the sheet plane. Although the understanding for uniaxial deformation basically applies also to this kind of loading, the presence of an intermediate stress can significantly alter the strain level prior to the development of a localized neck. From an applications point of view, a localized neck constitutes failure, and therefore, the local necking limit strain (commonly known as forming limit) is of great practical interest.

Originally developed by Keeler(38) and Goodwin (39) from mild steel production stampings, forming limit diagrams represent the major strain in the sheet surface at the onset of a visible neck as a function of minor strain. Figure 21 shows the forming limit diagrams for a number of materials obtained from laboratory tests involving stretching sheets over a hemispherical punch. The strain combinations below a given diagram indicate success, while those above are failures (necking or tearing). (Deformation with positive minor strains is called stretch-type, and the negative minor strains indicate drawing-type action). These laboratory diagrams (39A) agree well with results from actual stampings; however, they are at a higher strain level than forming limits for in-plane deformation (44). The punch geometry and its frictional constraints influence the strain gradient and strain path of failure and are believed to be responsible for this effect (35). The understanding of punch forming limits therefore must take these factors into account.

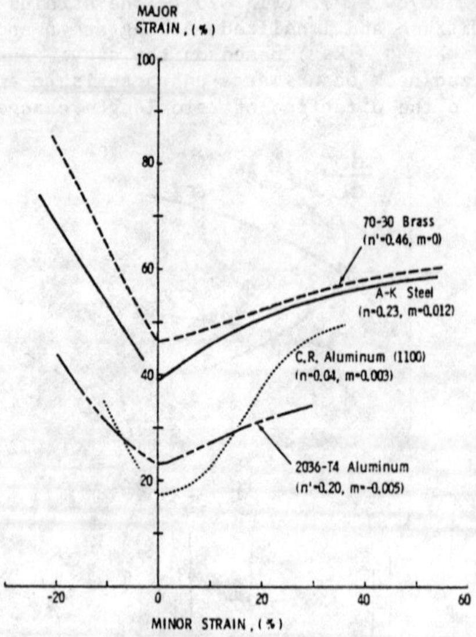

Figure 21. The Forming Limit Diagrams of a Number of Metals Determined by a Punch Stretching Technique Described in Refs. 39 and 39a.

The classical plasticity predictions of local necking limits due to Hill (36) and diffuse necking limits due to Swift (40) are summarized in Fig. 22 for an isotropic, power law hardening (m = o) material under uniform proportional strain paths. Within the range of stress-stress between uniaxial and balanced biaxial tension, the Swift criterion (which is nearly the same as a maximum traction criterion) predicts that the diffuse necking strain is about n (41). Hill's criterion predicts that local necking strain drops from 2n in uniaxial tension to a value of n in plane strain tension (see Fig. 22). In reality there are always

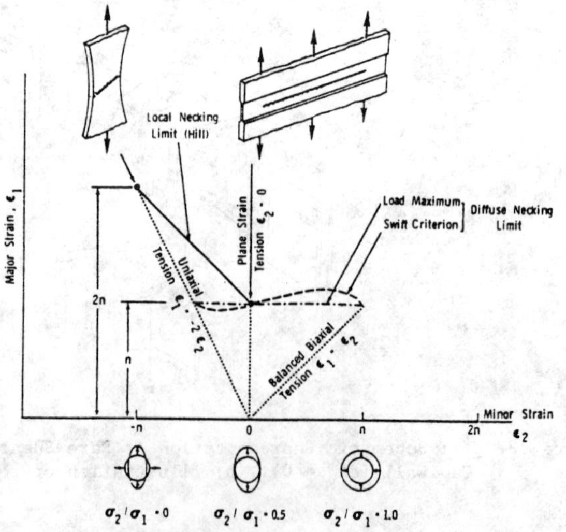

Figure 22. Theoretical Limiting Major Strains are Plotted as a Function of Minor Strain, Within the Range of Uniaxial ($\sigma_2=0$) and Balanced Biaxial Tension ($\sigma_2=\sigma_1$). Diffuse Necking Limits are Shown for Two Different Criteria (Dashed Lines), and Local Necking Limit for Negtive Minor Strains (Solid Line).

gradients in strain and strain-rate before localized necking, and if m > 0, the strain limits would increase on account of this. For example, A-K steel (m ~ 0.012) exhibits an in-plane forming limit for plane strain condition that is greater than n. Increasing sheet thickness (42) and decreasing punch radius (43) raise these limits, particularly for m > 0.

In biaxial stretching regime ($\varepsilon_2 > 0$), however, localized necking is unexplainable by Hill theory, since no direction of zero length change exists in the sheet plane. A rationale for localized necking for $\varepsilon_2 > 0$ was provided by Marciniak and Kuczynski (34) (M-K), who assumed that an inhomogeneity present in the material could be adequately modeled by a thinned trough-like region lying perpendicular to the direction of maximum principal stress (Fig. 23). The equilibrium equation in this case is

$$\sigma_{1A} e^{-\varepsilon_{1A}} = f\, \sigma_{1B} e^{-\varepsilon_{1B}} \qquad (12)$$

where f = thickness inhomogenity factor. Effective stress (composed of σ_1 and σ_2) can take any of the forms of constitutive law discussed earlier. Compared to the surrounding material, the trough thins faster under an imposed biaxial stress-state, i.e., ε_1 (trough) > ε_1 (surroundings), while ε_2 (trough) = ε_2 (surroundings). The strain path in the trough would turn toward plane strain and eventually reach fracture, as shown in Fig. 23. Numerical solutions of the above equation are carried out using an isotropic hardening theory (including normal anisotropy) and fixed values of f to obtain predictions of the forming limit diagram. M-K analysis predicts an increasing forming limit with increasing biaxiality ($\varepsilon_2/\varepsilon_1 \to 1$); since as the imposed $\varepsilon_2/\varepsilon_1$ increases, the trough takes longer to reach a state of plane strain, thereby accumulating greater strain in the surrounding material.

The trends predicted by the M-K analysis agree, in general, with the experimental forming limit diagrams for various materials. There are some quantitative disagreements, however (44,45). Two of these are: (1) the predicted imperfection sensitivity seems to be too great and (2) plane strain limit appears to be too low and biaxial limit too high when compared with experimental results. The general predictions of the M-K analysis are: (1) the level of the forming limit

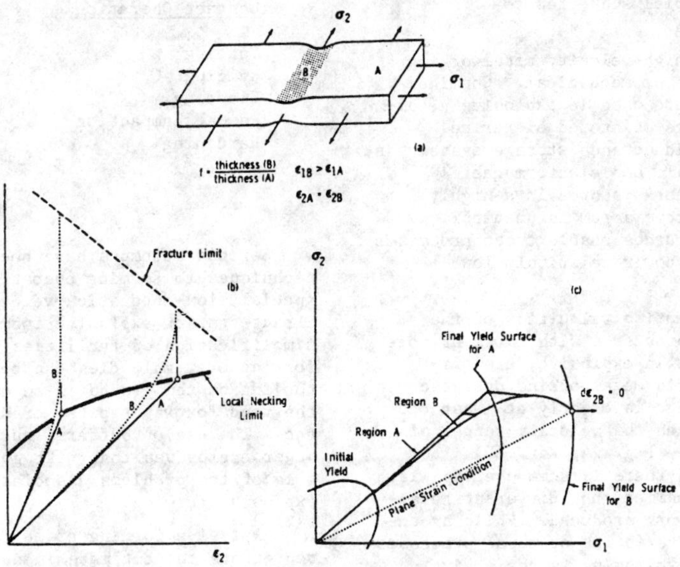

Figure 23. (a) A Region of Inhomogeneity, B, Bounded by Homogeneous Material (A) Undergoing Biaxial Stretching, According to the Model of Marciniak and Kuczynski. (b) Proportional Strain Paths in Regions A and B. When Region B Reaches the Fracture Limit Curve, Strain Accumulated in Region A Defines the Forming Limit (Localizing Necking). (c) The Shift in Stress-Path in Region B is Equivalent to the Strain Path Change in (b).

diagram is raised with increases in n and m, (2) the rise is greater near plane strain than balanced biaxial tension, (3) the fracture strain can have a strong influence on the forming limit diagram near $\varepsilon_2/\varepsilon_1 = 1.0$. The last two observations mean that failure in plane strain is normally more "instability-controlled", while that near $\varepsilon_2 = \varepsilon_1$ is influenced more by fracture processes. The significance of even a small m could be dramatic in sheet metal forming.

Forming Properties of Some Engineering Materials

The aluminum sheet alloys of interest in the aerospace industry are the duralumin type (Cu-bearing), e.g., 2024, 2124 for skin applications and the Zn-bearing, stronger alloys, e.g., 7075, 7175 for heavier sections. These alloys are heat-treatable from solution-treated and quenched (w) condition, which allows producing the peak-aged condition (T6) or the special grades of T76 and T73. Some of the strength and elongation properties of these alloys are listed in Table VI. Formability is considerably superior in the soft solution treated and quenched condition and may be advantageously used. Formability of low carbon steel (aluminum-killed, rimmed) at room temperature is one of the best among engineering alloys of interest. This is why it is the workhorse of the sheet stamping industry. Even when the strength of steel is raised by alloy additions (HSLA), it still remains as one of the most formable materials. While brass is the most formable of all materials shown in Table VI, the high cost and low strength limit its usage as a structural material. The minimum bend radius for an 0.040 in.-thick sheet is about 4t for 2024-T4 and 6t for 7075-T6 alloy. These increase with increasing sheet thickness. Similar values for A-K steel is 0t.

The titanium alloys are rather difficult to form at room temperature because of their high strength and low ductility, even though pure titanium has substantial elongation to failure (see Table VI) (46). Thus interest exists primarily in hot forming these alloys. Ti-6Al-4V is the most frequently used alloy that is used on frame and skin structures, and some stress-strain data on it are available. It is generally formed in the mill-annealed condition, and then it can be solution-treated and aged to impart strength. The beta and near beta alloys are more formable at ambient temperature and

Table VI

Tensile Properties of Selected Sheet Materials

Material	Yield Strength (MPa)	Ultimate Tensile Strength (MPa)	Uniform Elongation (%)	Total Elongation (%)
2024-T4 Al (Clad)	262	421	13.0	15.0
2024-T72 Al (Clad)	310	400	4.0	5.0
7075-T6 Al	469	538	7.0	8.0
A-K Steel	173	296	24.3	43.1
HSLA Steel	413	469	17.0	26.5
70:30 Brass	113	328	53.6	61.1
Comm. Pure Ti	483	689	11.5	18.0
Ti-6Al-4V	945	1013	6.5	10.5
Ti-6Al-4V (480°C)	620	793	6.0	20.0

References 46-48.

will be of interest in spite of some embrittlement problems in service which are now being alleviated. The beta-III (Ti-15V-3Cr-3Al-3Sn) and Ti-8Mo-8V-2Fe-3Al alloys are currently available as sheet and are reasonably cold formable (elongation ~ 18-20%). The titanium alloys are characterized by their large r values. Even though during hot forming there is a decrease in r value, this decrease may not be substantial up to 600°C (49), while the increase in elongation might be.

HIGH ENERGY RATE FORMING

High-velocity metalworking (or high energy rate forming (HERF), is characterized by the application of a transient pressure to a specified area of a workpiece. The workpiece reacts to this transient application of energy by accelerating away from the energy source, and the velocity attained by the material as it undergoes deformation is considerably greater than the velocity attained by the workpiece material in a conventional process. In fact, the response of the material when strained at such a high rate is sometimes quite different from that of material subjected to conventional rates of strain.

The energy sources for high-velocity metalworking may be chemical, electrical, or mechanical. Chemical explosives and propellants are used in explosive metal-working. The rapid discharge of stored electrical energy from capacitance or inductance storage systems is used for both electrohydraulic and electromagnetic metalworking. Mechanical energy stored in a highly compressed gas is used to drive a ram in pneumatic mechanical systems. Such sources enable these processes to provide large amounts of power relatively low costs.

Comparison of the deformation velocities of the high-velocity metalworking processes with those of more conventional ones shows that in explosive, electro-hydraulic, and electromagnetic metalworking processes the deformation velocity range is clearly at least one order of magnitude higher than the velocity ranges of the conventional methods (see Fig. 24, Ref. 50). For pneumatic-mechanical systems there is some overlap with the velocities of the drop hammer, but different metal-lurgical effects appear in work produced. While an ex-celpent sumary of the characteristics of these processes is given in Ref. 50, in the following we shall give a brief review of explosive forming technology only.

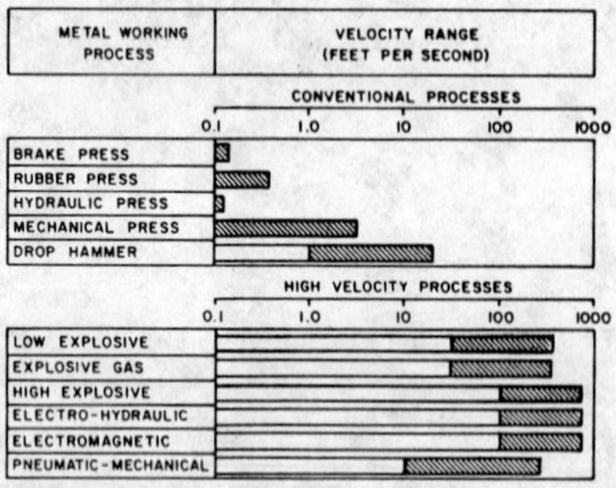

Figure 24. Deformation Velocities in High-Velocity and Conventional Metalworking Processes.

Explosive Metalworking

Explosive metal working uses energy from chemical explosions to perform such manufacturing operations as shaping, cutting, joining, cladding, and forming powder metal parts. Explosives produce high pressure and unusually high strain rates at low cost. As a result, certain metals such as maraging steels become more plastic and can be worked beyond their usual limits of formability.

The important elements involved in an explosive metalworking processes are the source of energy (explo-sives), energy transmission medium, tooling or dies, and the facility or location for conducting the work.

Typical high explosives used in metalworking are TNT, RDX, PETN, Tetryl, 40% dynamite, 60% dynamite, etc. Power is measured in terms of that of TNT (2.62×10^5 ft lb/lb) and generally used in the range of 90 to 170% that of TNT. Detonation velocities in the range of 10,000-30,000 ft/sec are used. Selection of explosives is made based on the criteria of strength uniformity, lack of toxicity, stability in storage, etc. Energy transmitting medium is water, gas or a movable ram. The tooling is usually made of high strength steels massive enough to withstand the shock. Safety requirements are high, and external pits are commonly used for the work.

Typical arrangements for common explosive metalwork-ing operations are shown schematically in Figs. 25-28. the processes are usually classified as contact or standoff. The major applications are:

Contact Operation	Standoff Operation
Cutting	Sheet and plate forming:
Welding	a. Stretch
Cladding	b. Draw
Powder compaction	Tube forming
Hardening	Sizing
	Flanging
	Bending and dimpling
	Extrusion

Serious attempts have been made to adapt contact techniques to forming operations through the use of special, low-detonation velocity explosives (Ref. 51). Direct-contact explosive forming often permits great simplification of facilities and tooling. For example, forming over male dies can be conducted by placing the tooling on the ground in an open field, thus eliminating the need for water pits, cranes, vacuum equipment, etc. The use of buffers, such as rubber or plastic, and other explosives than nitroguanidine can perhaps solve some of the problems associated with direct contact explosive forming.

Explosive forming of fully hardened tool steels represents the ultimate in deforming alloys with high yield strength (about 200,000 psi). A process developed by Philco Corp. makes it possible to form ultra-high-strength steels explosively for protective armor. The classic example of explosive forming is the production of large dish-shaped domes and bulkheads which are not amenable to stamping press forming.

Three basic techniques are used in the aerospace industries to form large domes. These are schematically depicted in Fig. 29. Forming domes directly from flat sheet as shown in Fig. 29a is the technique most common-ly associated with explosive metal forming; the methods shown in Fig. 29b and c are special cases. The mode of deformation normally is some mixture of the two basic processes: stretch forming and draw forming. Shaping and sizing a dome explosively from a welded preform as shown in Fig. 29b is generally reserved for domes with a large depth-to-diameter ratio. Welded gore segments are used to form domes larger than can be produced from currently available blank sizes as, for example, the tank bulkhead for the Saturn V vehicle.

The distance from the center of the explosive charge to the surface of the metal blank (standoff) is impor-tant in controlling the metal deformation process. Selection of the proper standoff distance for each explosive forming operation is essential, because the peak pressure exerted by the shock wave is inversely proportional to the distance away from the explosive charge. Several inveitgators have found that metal deformation is not a uniform function of standoff distance (Refs. 52 and 53). It has been found that there is a critical standoff distance at which a central charge produces the maximum blank deformation (Ref. 54). At smaller or larger standoff distances the draw depth of the blank decreases. For the conditions used, a standoff distance approximating 30% of the blank diameter gave the maximum draw depth.

The standoff distance also has a marked effect on the mode of metal deformation and on the shape taken by the metal blank during forming. For standoff distances less than that producing maximum blank deformation, the amount of cupping or draw increases with decreasing standoff. For forming hemispherical domes and other deep draws, it is important to emphasize the draw mode, and the smallest standoff distance possible, without damaging the blank material, should therefore be used. Conversely, bulging or stretching of the blank increases with distances up to the critical value for standoff. for shallow forming where springback is a problem, the stretching mode of deformation must be emphasized to achieve the largest possible amount of plastic deforma-tion. In such cases, the standoff distance should be made large enough to maximize the stretch mode of deformation; and normally the critical standoff distance is used.

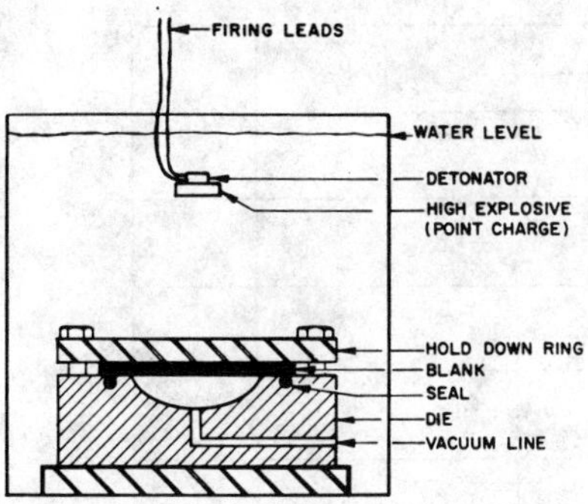

Figure 25. Explosive Sheet and Plate Forming.

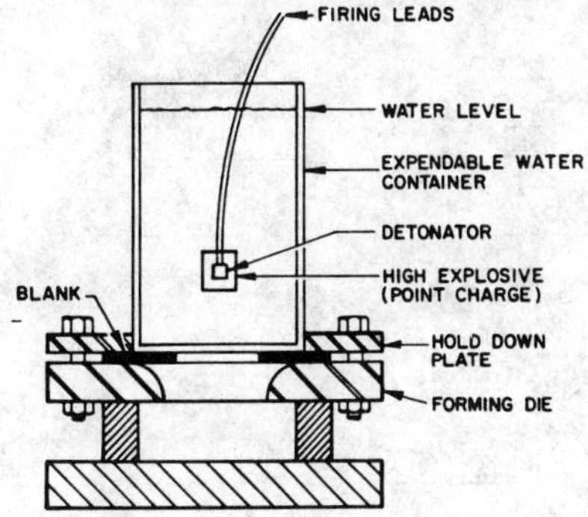

Figure 26. Explosive Flange Forming.

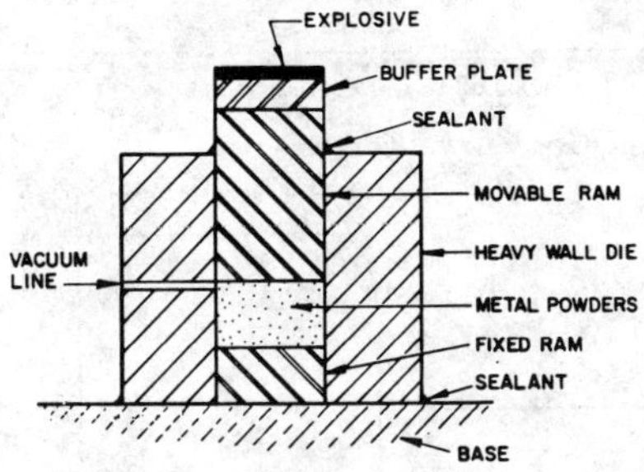

Figure 27. Explosive Powder Compaction.

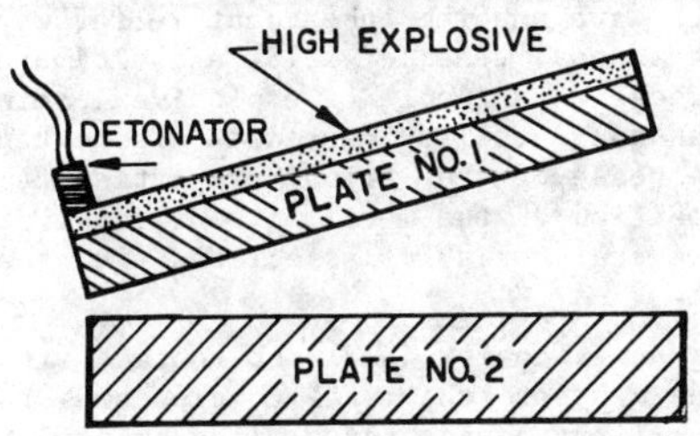

Figure 28. Explosive Cladding or Welding.

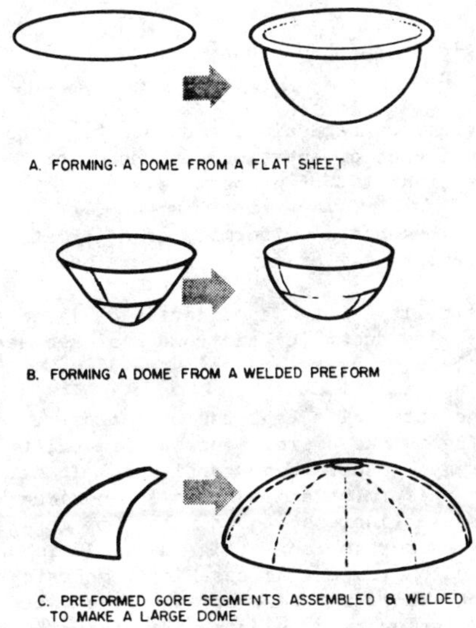

Figure 29. Three Techniques for Forming Large Domes.

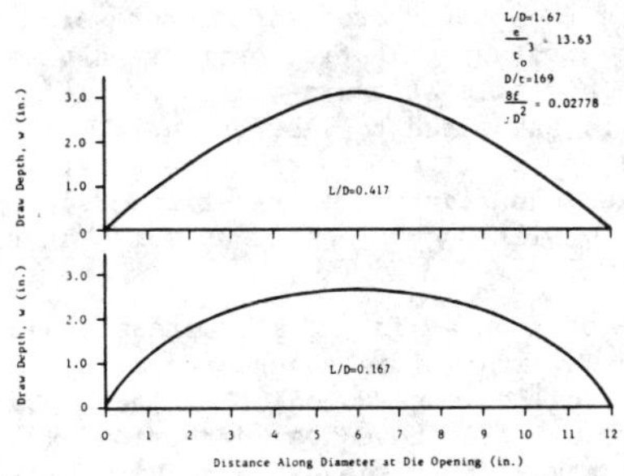

Figure. 30 Effect of Standoff Distance on Contour (Ref. 54).

Standoff distance likewise affects the shape of a freely formed blank (see Fig. 30). With a central charge placed at a standoff distance greater than the critical value, the free-formed shape is more nearly parabolic. A standoff distance less than the critical value will produce a more rounded shape, approaching an ellipsoid or segment of a sphere. Since it is generally desirable to use the smallest possible charge, it is common practice to place the charge as close as possible to the blank without introducing undesirable pitting or excessive thinout. It has been suggested that a standoff distance of approximately one-sixth of the blank diameter is a good starting point in optimizing this factor (Ref. 55).

Plug-cushion Technique: In this process, the explosive shock wave is applied to the plug cushion (typically made of soft aluminum or lead) so that the energy is transmitted to the workpiece through the mass of the plug cushion, increasing the strain on the portion of the blank at the drawing. The principal advantages in the use of a plug cushion for drawing domes or deep shells are: (1) the thinout of the blank at the apex is reduced to about 10% compared with 30-50% without the cushion; (2) much deeper domes can be drawn (for the same D/t ratio the formability can be increased appreciably); and (3) preformed shapes other than paraboloids of revolution can be produced (Fig. 31)

Springback: Springback is of concern in producing large shapes since it can lead to very large mismatch between parts. During explosive forming the limit of proportionality for some materials can be extended to about 2-1/2 times its normal value. Thus a large amount of springback might be expected. However, in explosive forming, several authors (Refs. 57-59) have reported reduced and sometimes non-existent springback. Proper placement of the explosive charge has large influence on the reduction of springback. To control springback, it is important to maximize tensile plastic strains in a blank during forming. For shallow shapes, such as dome-gore segments, plastic strain due to bending alone is small. Stretching must somehow be enhanced to maximize plastic deformation and reduce springback.

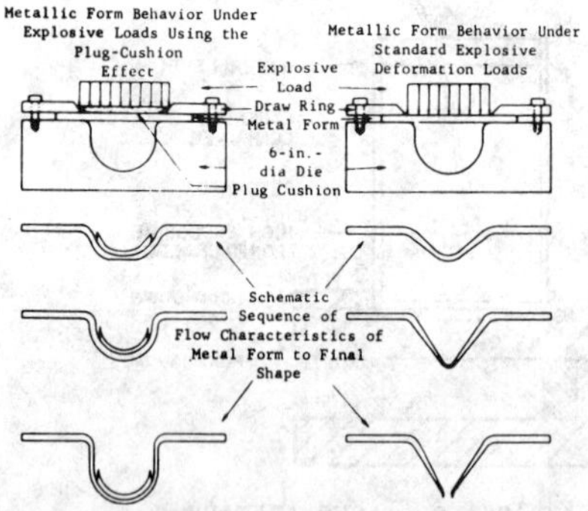

Figure 31. Comparison of Plug Cushion Effect Versus Standard Explosive Forming (Ref. 54).

Shock Hardening: The mechanical properties of metals are changed considerably after the passage of the explosive shock wave and the subsequent relief wave. In general, the hardness, tensile yield, and ultimate strengths increase as a result of explosive loading, while toughness and ductility decrease. The extent of increase or decrease of the various properties depends upon the composition of the metal, its original properties, and several variables associated with the explosive used.

Detonating an explosive in direct contact with the metal-workpiece subjects the metal to pressures ranging from 1.5 to 6 million pounds per sq inch. In most contact detonations the pressure is of the order of 100 to 500 kilobars (1 kilobar = 14,504 psi). It is generally assumed that conditions within the front of the pressure pulse approximate those of hydrostatic compression of the solid. A low-pressure elastic compression wave is propagated through metals at sonic velocity. If, however, the metal is subjected to a stress which greatly exceeds the yield point (i.e., so that the shear strength no longer contributes to the stiffness), one or more plastic waves will be propagated at somewhat slower velocities.

The behavior of a metal in shock compression is described by the way its volume changes as a function of applied pressure, called the "Hugoniot". the Hugoniot curves for numerous metals have been determined (60). Hugoniot curve for most metals shows a continuously decreasing volume with increasing pressure. Iron is unusual in that it does not show such behavior (Fig. 32). The distinct discontinuity shown at a pressure of 130 kilobars has been attributed to pressure-induced transformation from body-centered cubic α-iron to ε to α-iron (61). Epsilon iron is a hexagonal close-packed structure. Iron, upon relief to zero pressure, returns to its original α-phase; this makes it rather difficult to ientify the phase change associated with the 130 kilobar point.

Various investigators have shown the existence of a double plastic shock in iron, associating it with the discontinuity in the Hugoniot curve at a pressure of approximately 130 kilobars. This double shock occurs only when an applied pressure is within the range from 130 to about 340 kilobars (Fig. 32). The interaction of the two plastic shock waves frequently produces two distinct zones of hardening, illustrated in Fig. 33. The metal in zone "A" which is nearest the explosive, is the hardest, having been subjected to a pressure exceeding 130 kilobars. The metal in zone "B" has been subjected to a lower pressure and is hardened to a value intermediate between the original metal hardness and that of zone "A." Typical microstructures of the two zones can be easily distinguished.

The benefits of shock deformation, relative to more conventional straining (as summarized well in Ref. 62) include:

1. Higher strengths per unit applied strain when initial strengths are relative low.

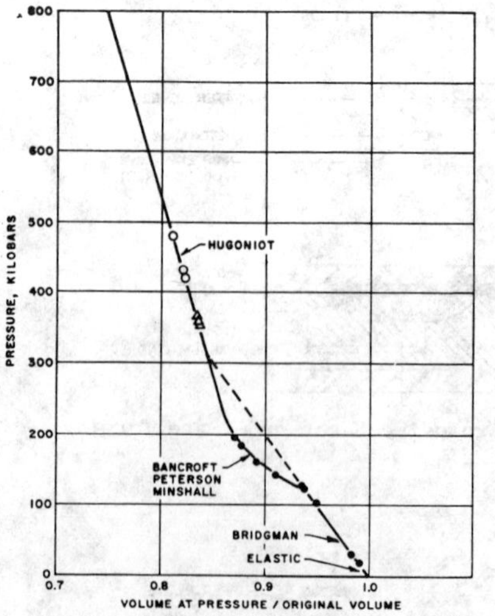

Figure 32. Hugoniot Curve for Iron.

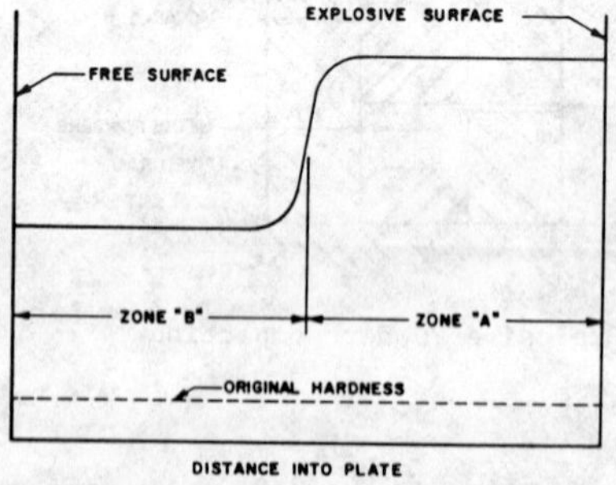

Figure 33. Zones of Hardening in Mild Steel Plate.

2. Little or no shape change, i.e., small residual strain.

3. Isotropy of hardening, and therefore, no significant orientation dependence or directionality of properties.

4. Through-section uniformity of deformation and, therefore, uniformity of properties.

5. Possibility of the application of large strains where low ductility and component geometry preclude other methods of straining.

One of the attractive features of explosive deformation processing is the improvement in formability achieved thereby. Under these conditions, it has been observed that uniform strain in tensile specimen undergoes a marked increase for a number of metals in the dynamic strain-rate range (Fig. 34). It is believed that this results from an increase in the strain rate sensitivity of metals at these strain rates, which has a more profound influence on specimen strain uniformity than does strain hardening ability.

While the other HERF processes share a lot of common ground with explosive forming, there are substantial differences also present. In electrohydraulic metal-working processes, the energy stored in large capacitor banks is rapidly released either to a discharge in the gap between electrodes, or to vaporize and explode a wire bridging an electrode gap. The spark discharge rapidly ionizes the liquid of the transfer medium, usually water, to produce a shock wave and a pressurized gas bubble; the effect is the same when the bridge wire explodes. While the events leading to workpiece displacement are similar to those in explosive forming, electrohydraulic forming lends itself better to control since reflectors can be used to direct and concentrate the pressure pulse toward a selected region of the workpiece. Overall control is better with bridge wires

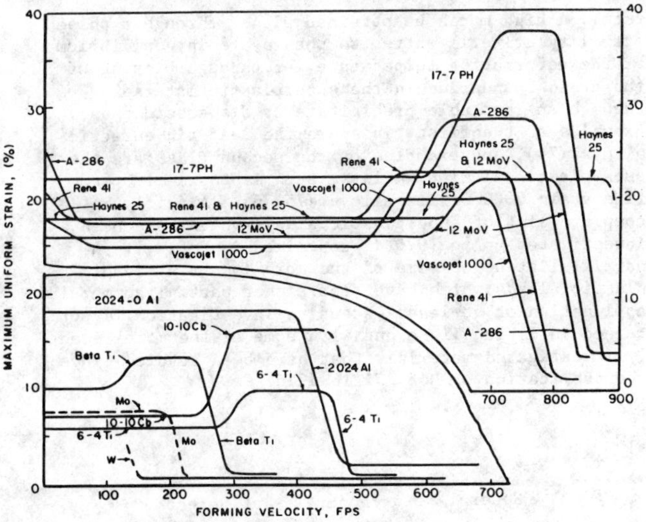

Figure 34 Effect of Forming Velocity on the Maximum Uniform Strain During the Explosive Bulging of Domes.

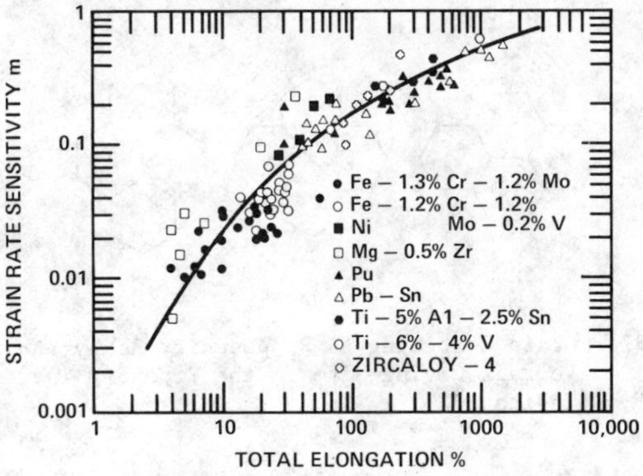

Figure 35. Correlation Between Strain-Rate Sensitivity and Total Elongation for a Variety of Superplastic Alloys (After Woodford (67)).

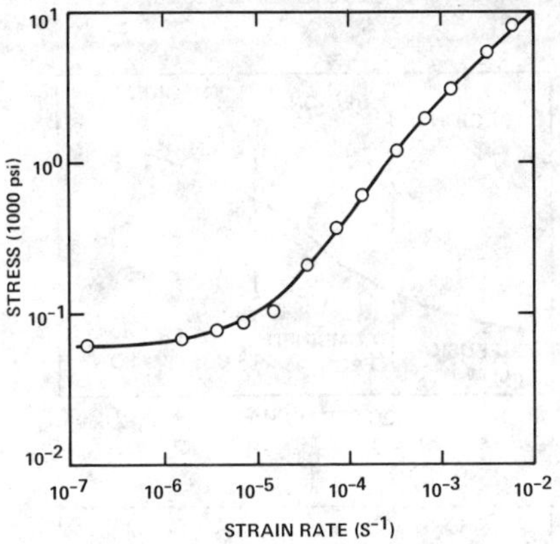

Figure 36. Log Stress vs Log Strain-Rate at 927°C for Ti-6Al-4V.

than spark gaps, but replacement of the wire will raise unit costs.

Electromagnetic metalworking processes are based upon the generation of high-intensity transient magnetic fields in the area between a magnetic coil and an electrically conductive workpiece. The magnetic pressure applies forces on both the coil and the workpiece. The magnitude of the pressure is proportional to the square of the field intensity. If the field is sufficiently strong, the pressure will permanently deform the conductor. The advantages of this process over explosive forming are economic multicomponent capability, lower noise levels and less hazardous operation. Physical strength of coils is a problem, however.

Pneumatic-mechanical metalworking systems use the energy stored in a compressed gas to accelerate a ram to a high velocity and impact a workpiece, thus causing the piece to deform. The technique has been used to forge a broad variety of metals, including low-alloy steels, aluminum and titanium alloys, stainless steels, high temperature alloys, and refractory metals, and for the extrusion of a wide range of metals, including refractory metals. In addition, the method has been used for the compaction of powders and to a limited extent, in sheet-metal-forming operations. However, tool and die design is difficult and expensive in this process.

SUPERPLASTIC FORMING

The important observation that large tensile elongation without fracture can be achieved in metals under certain conditions was first made by Pearson (63) in 1934. A requirement for such behavior was a fine grain size and a low flow stress. Despite the fact that he demonstrated almost 2000 percent potential elongation in a tin-bismuth alloy, the observation created little interest for almost 30 years until Underwood (64) reviewed the Russian work in the area. As a result of Underwood's review, Backofen and Avery and their coworkers (65,66) embarked upon a phenomenological examination of the reasons behind the large amounts of neck-free deformation, and established that a high strain rate sensitivity appeared to be a prerequisite for such behavior. On the basis of this important observation, Woodford (67) plotted elongation of a number of different alloys against strain rate sensitivity, m, as shown in Fig. 35, and demonstrated that a good correlation existed. Typical flow stress vs strain rate for a superplastic metal are shown in Fig. 36. Since this time a number of workers (68,69) have rigidly established the relationship between tensile elongation and strain rate sensitivity ($m = d \log \sigma / d \log \dot{\varepsilon}$) and have shown that the strain rate sensitivity can be analytically related to the total elongation. The realization that certain metals could flow with large amounts of neck-free deformation in a manner somewhat similar to plastic led a number of workers to investigate blow forming of metals (70). Early work was confined mostly to eutectic and eutectoid alloys such as tin-bismuth, zinc-aluminum and lead-tin. However, it was soon realized that a requirement for superplasticity was that a fine stable grain size during deformation had to be achieved, but that two-phase system was not a prerequisite, and this greatly broadened the scope of alloy systems which might be made superplastic under the proper conditions. In Pearson's original work he demonstrated that grain shape and size did not appear to change with deformation, and furthermore, that grain boundary offsets occurred. From this he correctly concluded that flow at grain boundaries was the important mechanism of deformation.

In the ensuing period much discussion has occurred as to the details of the mechanisms of superplastic flow, but workers generally seem to agree at this point that slow strain rate deformation is diffusion controlled, while high strain rate deformation (at strain rates above 10^{-2}/sec) is dislocation creep controlled. The merits of the various theories have been fully discussed in two recent references (71,72), but phenomenologically, the theory of Ashby and Verrall (25) appears to be reasonably sound. In their theory, they propose that grain switching, as illustrated in Fig. 37, is the principle mechanism by which significant amounts of elongation can occur without a microscopic change in grain shape. They also explained the familiar sigmoidal shape of the stress vs strain rate plot shown in Fig. 38, and peak, m occurring at intermediate strain rates, shown in Fig. 38. Their general approach can also be used to construct a constitutive equation for superplastic flow as described in Ref. 21. Although

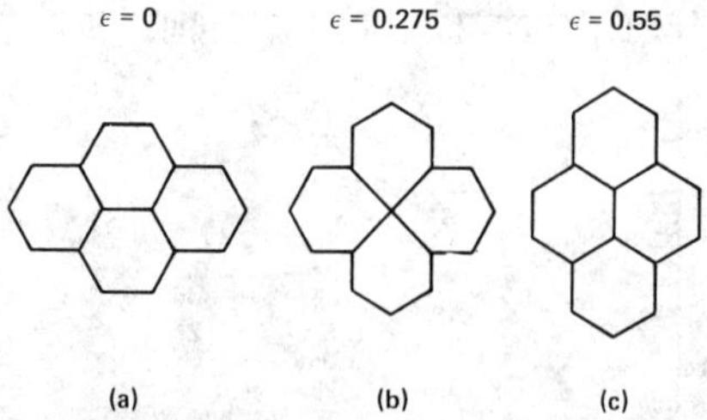

Figure 37. Grain Switching as Proposed by Ashby and Verrall Giving Rise to the True Strains Indicated.

there has been much discussion concerning correct methods of measuring superplastic properties, such as those shown in Fig. 38, the most general method (73) appears to be that of a strain rate change test, as shown in Fig. 39.

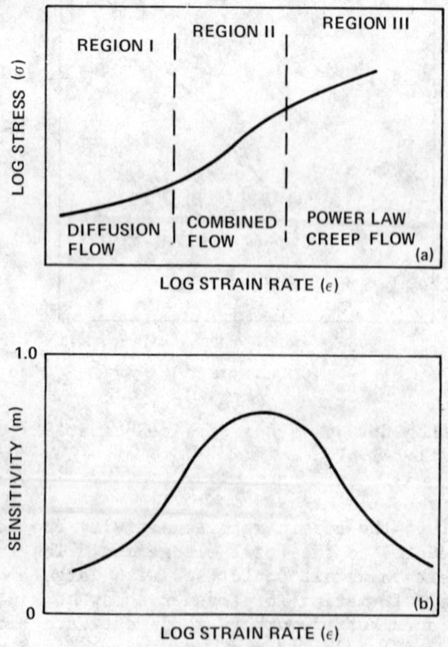

Figure 38. Sigmoidal Log Stress (σ) Versus Log Strain-Rate (ε) Showing Three Regions of Behavior and Corresponding Strain Rate Sensitivity m.

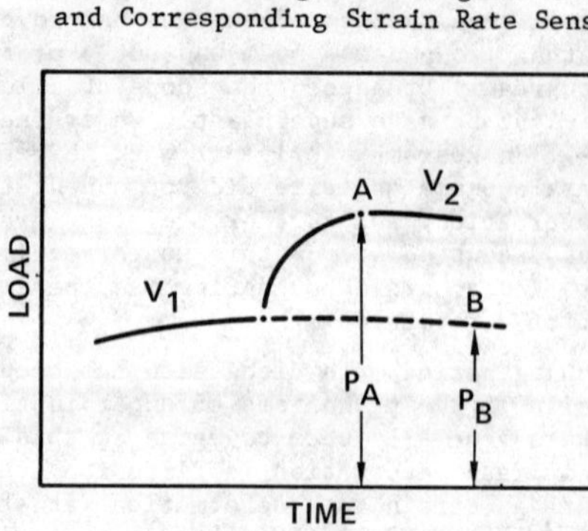

Figure 39. Schematic Load-Time Diagram Representing a Velocity Change from v_1 to v_2. Times A and Represent the Same Strain at the Different Pulling Speeds (Ref. 65).

Although there are several methods for determining m, it is believed that one should not lose sight of the fact that m is simply a computed value describing the relationship between the flow stress and strain rate. Further, m is not a constant but varies with strain rate and is thus descriptive of the strain rate sensitivity of flow stress only over a small, variation of strain rate. A high m can be obtained either through a phase stability criteria where two phases are in equilibrium at the deformation temperature, or through pinning of the grain boundaries in the superplastic material through an insoluble precipitate or dispersoid phase. Examples of the first class are the lead-tin eutectic alloys (74), and examples of the second class are the superplastic aluminum alloys, such as Supral 100 and fine grain 7000 Series alloys (75,76). A fairly complete list of superplastic alloys which have been investigated up to 1976 is contained in Ref. 71. A partial listing of some of the more important superplastic alloys which have seen either partial commercial application or at least extensive investigation is contained in Table VII, along with some of the more recently investigated materials that have been studied since the publication of Ref. 71 in 1976.

Table VII

Superplastic Properties of Selected Structural Alloys

Alloy	Grain Size (μm)	Temp °C	Elongation %	Maximum M	References
Al-88%Zn	2.5	230	2850	0.45	77
Al-6%Ca-0.4%Zr	~ 10	400	800	0.5	78,79
Al-6%Zn-2.5%Mg-1%Cu (7475 Al)	10	515	600	0.85	80
Al-5%Ca-5%Zn	1-2	450	7200	0.29	81
Fe-1.3%	2-3	650	470	0.5	82
Fe-6.5%Ni-25%Cr	3.7	925	600	0.5	83
Mg-5.8%Zn-0.5%Zr (ZK60)	0.55	275	1700	0.4	84
Ni-12.5%Cr-3.2Mo-4.9Al (IN100)	2.5	1038	850	0.66	85
Ti-6%Al-4V	3.7	927	675	0.58	86
Zr-1.5%Sn-0.2%Fe-0.1%Cu (Zircaloy 4)	12		200	0.5	86

The production of superplastic materials requires in all circumstances a thermomechanically processed material which has a fine, stable grain size. This can be achieved either through appropriate hot working, warm working, and cold working sequences with intermediate recrystallization steps, as is the normal case with wrought materials, or alternatively, as in some cases, being achieved by powder processing, as with nickel-base superalloys (76). From a commercial standpoint, the most important superplastic alloy would appear to be the zinc-aluminum eutectics, the titanium alloys, and aluminum alloys. Superplasticity in zinc-aluminum alloys has been extensively investigated by Langdon (77) and co-workers, among others, and has also been commercially employed for making structures which are not subjected to severe loading, such as cases for electronic components. Limited room temperature creep resistance would appear to be a severe limitation in the broad application of these materials.

Titanium alloys, in particular Ti-6Al-4V, have been extensively applied as a superplastic material, since standard mill processed alloys frequently have acceptable superplastic properties. Although not all mill processing sequences give sufficiently fine grain size to provide adequate superplasticity, a grain size of the order of 6 to 9 μm is frequently obtained by standard processing methods, and yields high m values and excellent elongation (87). It was pointed out in Ref. 87 that in addition to a fine, stable grain size, a uniform grain size provided both higher m values and greater elongation, and also that an equiaxed grain shape appeared to provide improved superplastic properties in Ti-6Al-4V. These concepts have been analyzed in detail in the recent work of Ghosh and Raj (88). A further problem in the superplastic forming of titanium alloys is the high temperatures required (of the order of

925° C) and this results in a limited amount of grain growth, particularly during deformation (89). This deformation results in a greater grain growth that would occur statically at the deformation temperature. It was demonstrated in Ref. 89 that the grain growth occurring during superplastic forming adequately explains the observed increase in flow stress with strain, which is found to occur in Ti-6Al-4V.

Superplastic Aluminum Alloys

Aluminum alloys that are superplastic fall into essentially three classes: aluminum-copper, aluminum-calcium, and aluminum-magnesium-zinc alloys. The aluminum-copper alloys appear at this point to be the most fully developed superplastic materials, and an Al-6%Cu-0.4%Zr alloy has been developed by British Aluminum (75). A wide range of superplastically formed components are being produced from this material on a commercial scale. The alloy is somewhat unusual since prior to superplastic deformation it is not fully recrystallized, but contains a high density of $ZrAl_3$ particles which pin the grain bundaries during dynamic recrystallization in the early stages of deformation, and appear to be successful in retaining a grain size of the order of 10 μm or less during the superplastic forming process (90).

A typical superplastically formed component produced from the Al-6%Cu-0.4%Zr alloy is shown in Fig. 40. A wide variety of components are being produced from this alloy, which appears to provide a great deal of design flexibility along with sufficient superplasticity to enable parts of considerable complexity to be produced readily.

Figure 40. Undercarriage Housing for the European Airbus Made from Al-6Cu-0.52 (Courtesy Superform Metals Ltd).

Aluminum-calcium alloys have also been shown to be superplastic, and recently an Al-4.5%Ca-4.5%Zn alloy (08050 Aluminum) has been introduced (81). The alloy appears to have excellent properties, but has a rather low yield strength for structural applications of the order of 160 MPa.

A third class of superplastic aluminum alloys has been the subject of recent studies where the composition is based on the standard 7000 Series aluminum alloys where a fine grain size is achieved through a thermomechanical processing treatment (20). In these materials, the fine grain size is produced by a thermomechanical processing sequence, which starts with solution-treated material, goes through a high temperature overaging treatment to produce a uniform dispersion of large precipitates, followed by a warm working step, and lastly a recrystallization step. The uniform dispersion of large precipitates results in the nucleation of the uniform fine grain size of about 8 to 10 μm, which is sufficient to provide a considerable degree of superplasticity (80). The formability is adequate to form complex components of high strength aluminum alloys such as 7475. A typical component with a conventionally fabricated part which it replaces is shown in Fig. 41 (91).

A considerable amount of effort has also been expended on nickel-base superplastic alloys. Among these are IN 744 (92) and nickel-base alloys produced by powder metallurgy methods. Most of the applications for nickel-base alloys would appear to be in forging, rather than sheet forming processes. So the objective of achieving superplastic flow is to lower the flow stress in a forging operation (76) rather than to use the high strain rate sensitivity which results in large amounts of neck-free deformation.

Other important superplastic alloys include a series of iron-base alloys investigated by Sherby and co-workers (93), and the superplastic low alloy steels which have been investigated by Ridley and co-workers (94). Applications of superplastic steels in forming of components has been limited by availability of these materials on a commercial scale.

Superplastic materials are often found to cavitate during superplastic flow. The occurrence of cavitation, the reasons for it, and analytical approaches to predicting the extent of cavitation have been extensively reviewed in the literature (95,96). It appears that cavitation is influenced both by plastic flow and diffusional contribution. Most of the studies have tended to concentrate on the effects of cavitation on total elongation to failure. From the point of view of forming structurally adequate components, one would, however, be most interested in slowing the kinetics of cavitation as much as possible. Forming limit here is thus the strain level corresponding to a critical level of cavitation (~ 0.05%), since this would reduce mechanical strength, and in particular, the room temperature elongation in a tensile test.

Superplastic Forming Processes

The original attempts at forming superplastic components from metals usually involved vacuum forming, as shown in Fig. 42. In order to obtain improved control over the imposed strain rate in a material, later methods have generally employed the application of a positive pressure in order to achieve the forming. This improvement has been further developed by using a controlled pressure-time profile in order to enforce an optimum strain rate in the material, as described in Ref. 97.

There are many variations on the basic technique of blow forming in a female die, as shown in Fig. 42, but one of the most successful of these is that described in Ref. 72 where a part is formed over male tools by first blowing a hemispherical shape through a die opening, and then moving a male tool into the hemisphere and reversing the pressure to blow the part back onto the male tool, as shown in Fig. 43. This method gives much better thickness control over the final formed part than the standard female forming method shown in Fig. 42. The mechanics of various superplastic forming processes have been recently reviewed by Tang (98).

In addition to a wide variety of sheet forming processes based on blow forming of a single sheet of superplastic material, multi-sheet superplastic forming processes have been developed in recent years in which superplastic forming and diffusion bonding are combined to form sandwich structures (97). These processes have mainly been used on titanium alloys because of the ease with which these materials can be diffusion bonded, giving a designer a great deal of design flexibility in the manufacture of complex structurally efficient components. The combined superplastic forming/diffusion bonding (SPF/DB) process is shown schematically in Fig. 44, with some examples of SPF/DB sandwich panels in Fig. 45.

From a cost standpoint, superplastic forming has the disadvantage that it is a rather slow process, since

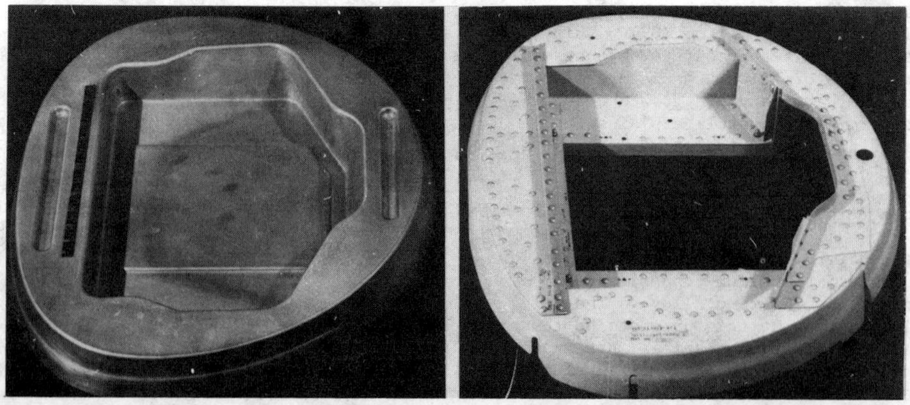

Figure 41 Typical Aircraft Structural Component Superplastically Formed from Fine Grained 7475 Al (Left) and A Conventional Riveted Component at Right.

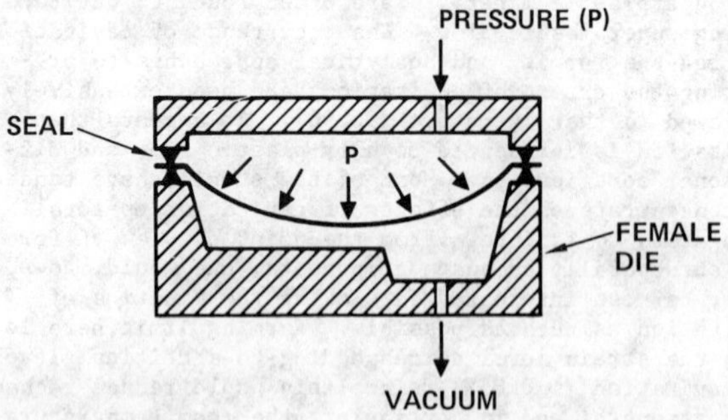

Figure 42. Female Forming with Either Vacuum from Below the Sheet or Pressure P from Above.

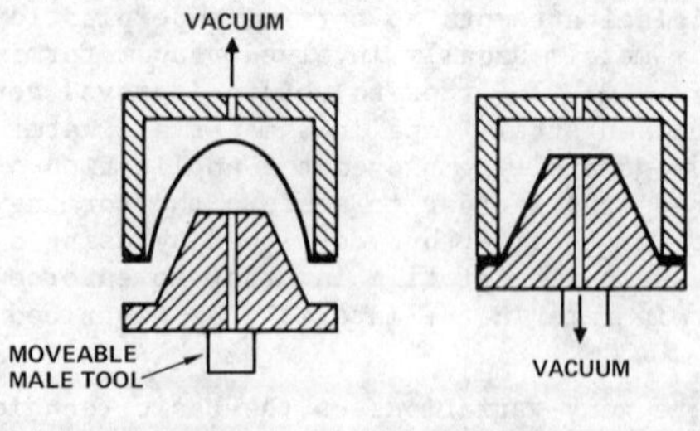

Figure 43. Male Forming Using a Movable Male Tool (Ref. 72).

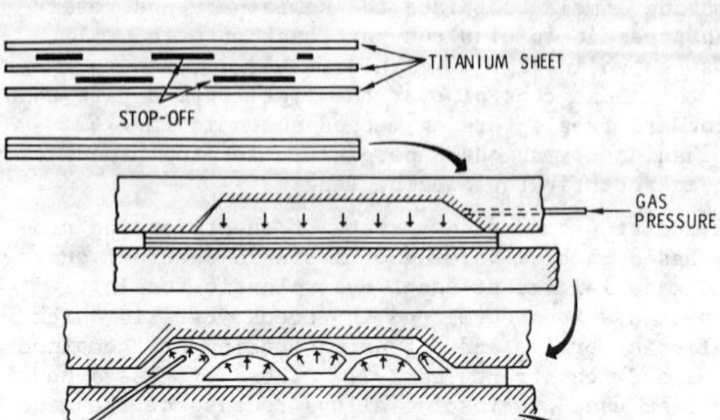

Figure 44. Illustration of the SPF/DB Process for Three Sheet Parts.

Figure 45. Ti-6Al-4V Sandwich Sample Configuration Fabricated by the Three Sheet Method.

parts are usually formed at or slightly above the strain rate at which peak m occurs, as shown in Fig. 38. This limits the rate at which components can be formed, and therefore, the production rate of each press used to close the forming die. Consequently superplastically formed parts are usually limited to a few hundred or perhaps a few thousand components.

THERMO-MECHANICAL PROCESSING

Some form of thermo-mechanical processing of structural materials is generally employed since cast structures are subject to the problems associated with severe segregation during the freezing process and possess a coarse grain size. Methods of dealing with these problems through modification of the casting process are discussed in another chapter in this volume (99). In this section, we will deal with thermo-mechanical treatment (TMT) methods of modifying microstructure. Generally, in some degree of warmwoking and recrystallization is introduced to break up the cast structure and minimize the effects of segregation in the coarse columnular grain structure that results from casting and solidification.

Thermo-mechanical processing of metals is really an extension of the standard working and recrystallization procedures that are used to breakdown an ingot to produce a finished wrought product. Thermo-mechanical processing, or thermo-mechanical treatment (TMT), as it is frequently referred to, usually involves a sequence of warm working and recrystallization steps sometimes with partial or complete heat treatment steps inserted during the process. The purpose is to produce a microstructure which yields a combination of final properties which are better than would be achieved by standard working processes.

Frequently thermo-mechanical processing is used to achieve a fine grain size in the resultant product since a very fine grain size frequently leads to both increased strength (100,101), and improved fracture toughness (102). Recrystallization to achieve this fine grain size can either be achieved statically after some degree of either warm or cold working, or dynamically during the hot working process (103).

True thermo-mechanical processing sequences generally involve a complicated series of intricate working and heat treatment steps to achieve an improvement in properties by a combination of residual cold or warm work and a refinement in the microstructure of the material. Examples of such sequences are ausforming of steels to produce a combination of high toughness and high strength (104), and thermo-mechanical treatment of aluminum alloys to produce improved stress corrosion cracking and fatigue resistance (105). A review of recent research in thermo-mechanical processing and its application to a number of different materials is contained in Ref. 101 by Azrin.

Thermo-Mechanical Processing for Improved Strength and Toughness

A considerable amount of work has been conducted over the years on thermo-mechanical processing of aluminum alloys. The efforts have been based for the most part on the need for improved fatigue strength in aluminum alloys and the requirement for an increased resistance to stress corrosion cracking (105). In addition, there has been a considerable amount of effort directed toward refinement of the grain size in aluminum alloys to improve strength properties (106,107).

Work on TMT for improved stress corrosion cracking resistane has been based (105,108) on the observation that strength can be increased by increasing the dislocation density in a warm working step, and thereby reducing the amount of aging required to achieve a given strength level (105). This is illustrated in Fig. 46, where it can be seen that sub-structure strengthening gives the strength increment normally lost by over-aging to achieve a desired level of stress corrosion cracking resistance. Although these methods have been shown to be successful and in addition can give rise to improvement in fatigue strength, it has been demonstrated by Osterman and Reimann (109) that the improvements in the fatigue strength seen in unnotched specimens do not translate into improvements in notched fatigue strengths. This observation has resulted in a decline in interest in TMT of aluminum alloys in recent years.

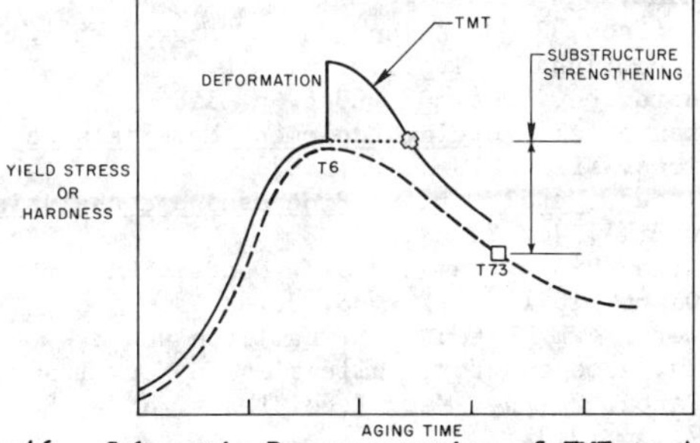

Figure 46. Schematic Representation of TMT to Achieve Improved Stress Corrosion Cracking Resistance in Al Alloys.

Intermediate thermo-mechanical processing treatments (IMT) have been used by Waldman (106) et al, to achieve the finer grain size in 7000 series aluminum plate products and thereby affect some grains in mechanical properties. Their treatments are based on a long time high temperature treatment to precipitate chromiun, followed by a solution treatment and a subsequent slow cool design to precipitate remaining magnesium and zinc out of solution. This is then followed by warm working and recrystallization to achieve a fine grain size. Improvements in mechanical properties and in particular, significantly better elongation and reduction of area than conventionally processed material are claimed from these treatments. Similar results have been reported by Di Russo (107).

Thermo-mechanical processing of titanium alloys for improved strength and toughness has also been subject to considerable research. Titanium alloys characteristically do not work harden to any considerable extent so that cold working cannot be used to raise the yield strength as was discussed for aluminum alloys and is frequently practiced in cold drawn steel products. However, the strength of beta titanium alloys can be increased considerably by using thermo-mechanical processing to refine the microstructural scale. Polan and Avery (110) have shown that the yield strength of beta titanium alloys can be increased to the order of 350 ksi (2400 MPa) by a combination of cold working to form a high dislocation density and aging to nucleate precipitates on dislocations produced during the working process. As in virtually all alloy systems including aluminum and iron base alloys, the kinetics of age hardening are accelerated due to the cold work and the resultant increase in vacancy concentration in the worked product. Consequently, shorter aging times can be used to give a required increase in hardness at a given temperature. Similar strengths have been reported by Williams (111), in conventionally aged products, but generally speaking, the ductility is somewhat less.

Thermo-Mechanical Treatment for Superplasticity

It is well known that one of the primary requirements for superplastic behavior in a metal is a fine stable grain size. Virtually all superplastic metals are processed by some form of thermo-mechanical treatment in order to achieve this fine grain size in a suitable product form. Frequently titanium alloys have a sufficiently fine grain size to provide an adequate degree of superplasticity in the condition in which they are processed to sheet. Paton and Hamilton (87), however, found that there was a considerable variability in standard mill processed sheet materials and that some controls are generally required in order to achieve a fine enough grain size to give these materials an acceptable degree of superplasticity. High temperature, and in particular working in the beta field, is detrimental to microstructures required for superplasticity and remnants of a plate-like beta-processed structure is particularly undesirable (87,112). An example of Ti-6Al-4V which has been subjected to processing in the beta field with only a limited amount of work, in the alpha and beta field is shown in Fig. 9a while material with a more desirable equiaxed microstructure which has been subject to considerable rolling reduction in the alpha-beta field is shown in Fig. 9b.

Special processing of Ti-6Al-4V to obtain finer grain sizes than normally achieved in mill annealed sheet have been attempted (113), and have shown that the anticipated improvements in superplasticity are realized. However, normal mill annealed sheet if purchased to grain size of less than approximately 8 μm generally provides sufficiently superplastic behavior for the fabrication of most of the common types of components.

Superplasticity in beta titanium alloys has been investigated by several workers (114,115), and it has been found that even coarse gain beta alloys can sometimes demonstrate a surprising degree of superplasticity. In Ref. 115, it was shown that grain sizes in excess of 100 μm was sufficient in certain beta alloys to provide m values in excess of 0.6 and elongations of the order of five hundred percent or more. This appeared to be due to dynamic recrystallization to form a sub-grain structure which was considerably finer than the parent grain structure. However, special processing by thermo-mechanical treatment to achieve ultra-fine grain size as in Beta III was shown to provide the anticipated benefit in terms of increased m value and lower flow stress desired for superplasticity. In general, surprisingly little work has been done on TMT of titanium alloys to improve superplastic deformation, partly because titanium is almost unique in that standard mill processed alloys frequently provide the desired degree of superplasticity.

In aluminum alloys, the situation is entirely different in that wrought aluminum alloys generally have a very coarse pancake-like grain structure as shown in Fig. 47. Because of the wide application of aluminum alloys in lightweight structures, a considerable amount of effort is being expended on TMT of aluminum alloys to refine the grain size to impart superplasticity and ease of fabrication. A development of a superplastic Al-6%Cu-0.4%Zr alloy in England has been described in some detail (75,78,79), and involves a carefully controlled thermo-mechanical treatment in order to obtain a microstructure which is suitable for superplastic forming. In this case, dynamic recrystallization during the early stages of forming, assist in producing the fine grain size required to provide superplasticity. A rather more conventional therm-mechanical treatment has been used in the U.S. (116), to produce a fine grain size in precipitation hardenable aluminum alloys. In this case, an extended over-aging step is employed to provide a uniform distribution of coarse precipitates which are subsequently used to nucleate a fine uniform grain size. Examples of the microstructure of a 7000 series aluminum alloy before and after this thermo-mechanical treatment are shown in Fig. 48, A detailed study of the recrystallization processes involved in this treatment is contained in Ref. 20.

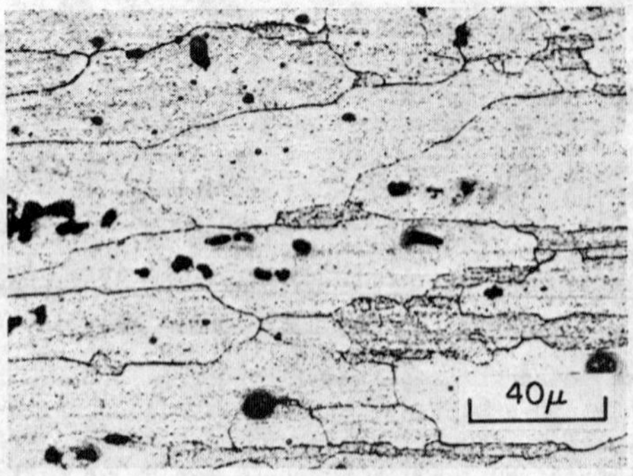

Figure 47. Grain Structure of 7075 Al Showing Elongated Grain Structure.

ACKNOWLEDGMENTS

The authors gratefully acknowledge the permission of the authors and publishers of journals and books referenced here, to present some of the data and graphs contained herein. One of the authors (AKG) would also like to acknowledge the support of the Independent Research and Development Program of Rockwell International.

REFERENCES

1. Fundamentals of Deformation Processing, W.A. Backofen, J.J. Burke, L.F. Coffin, Jr., N.L. Reed and V. Weiss, Eds., Syracuse University Press, 1964.
2. C.M. Sellars and W.J. McG. Tegart, Int. Met. Reviews, 17, 1972, p. 1.
3. W.A. Backofen, Met. Trans, 4, 1973, p. 2679.
4. Advances in Deformation Processing, J.J. Burke and V. Weiss, Eds., Plenum Press, 1977.
5. Forming, ASM Handbook, 4, 8th Ed., 1969.
6. The Making, Shaping and Treating of Steel, H.E. McGannon, Ed., U.S. Steel Publication, 1964.
7. W.A. Backofen, Deformation Processing, Addison Wesley Publishing Co., 1972.
8. D. Hardwick and W.J. McG. Tegart, Mem. Sci. Rev. Met. 58, 1961, p. 869.

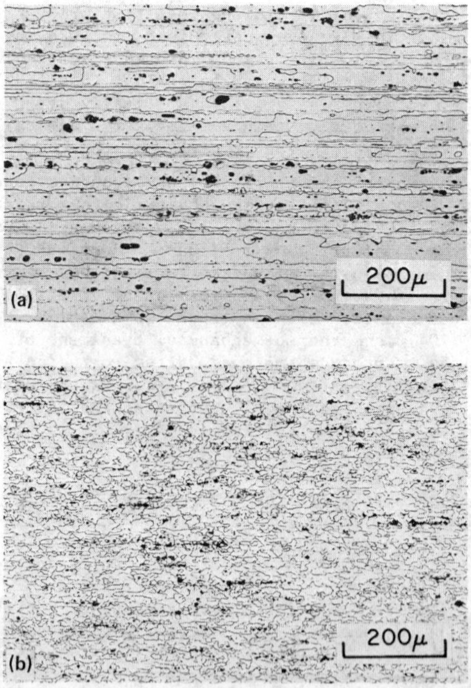

Figure 48. Grain Refinement Achieved in Al Alloys by TMT (20) (a) Conventional Coarse Grain Material, (b) After TMT.

9. C. Rossard and P. Blain, Flat Rolled Products III, (Met. Soc. Conf. 16), Gordon and Breach, New York, 1962, p. 3.
10. J.E. Hockett, Trans. Met. Soc. AIME, 239, 1967, p. 969.
11. M.M. Farag, C.M. Sellars and W.J. McG. Tegart, referenced by P. Moore, Deformation Under Hot Working Conditions (Special Report No. 108), Iron and Steel Inst. London, 1968, p. 60.
12. F.B. Pickering, Proc. of Conf. on Micro Alloying 75, Washington, D.C., 1975, p. 3.
13. D.M.Keane, C.M. Sellars and W.J. McG. Tegart, Ref. 11, p. 21.
14. S.Y. Ogawa, T.B. King and N.J. Grant, Trans. Met. Soc. AIME, 224, 1962, p. 12.
15. A. Joseffson, J. Koenemann and G. Lagerberg, J. Iron Steel Inst. 191, 1959, p. 240.
16. R. Canard, Rev. Met. 43, 1946, p. 156.
17. T. Altan et al, Forging Equipment Materials and Practices, NTIS MCIC-HB-03, 1973.
18. B. Avitzur, Metal Forming Processes and Analysis, McGraw Hill, New York 1968.
19. M.M. Allen R.L. Athey and J.B. Moore, ASM Metals Eng. Quart. 10(1), 1970, pp. 20.
20. J.A. Wert, N.E. Paton, C.H. Hamilton and M.W. Mahoney, accepted for publication in Met. Trans.
21. N.E. Paton, J. Eng. Mat. Tech. Trans. ASME, Ser. H. 1975, p. 313.
22. J. Melill, "Diffusion Bonding: A Process for Manufacture of Aerospace Structural Components," AIME Meeting in Philadelphia, Oct. 1969.
23. C.H. Hamilton, Titanium Science and Technology, R.I. Jaffee and H.M. Burte, Eds., I, Plenum Press, New York, 1973, pp. 625-648.
24. G.O. Garmong, N.E. Paton and A.S. Argon, Met. Trans. 6, 1975, 1269-1278.
25. M.F. Ashby and R.A. Verrall, Acta. Met. 21, 1973, p. 149.
26. A.T. Male and M.G. Cockroft, J. Inst. Metals, 93, 1964, p. 38.
27. G. Garmong, N.E. Paton, J.C. Chesnutt and L.F. Nevarez, Met. Trans. 8, 1977, p. 2026.

28. J.R. Becker, J.R. Douglas and F.A. Simenon, AFML-TR-72-89, Battell Columbus Laboratories, Columbus, Ohio, 1972.
29. A.K. Ghosh, J. Eng. Mat. Tech. Trans. ASME, 99, 1977, p. 264.
30. B. Budiansky and N.M. Wang, J. Mech. Phys. Solids, 14, 1966, p. 357.
31. R.L. Whiteley, Trans. ASM, 52, 1960, p. 154.
32. S.P. Keeler, Ductility, ASM Book, 1968, p. 227.
33. J.R. Hiam, Dominion Foundaries and Steel Co., Canada, presented at ADDRG Conf. 1974.
34. Z. Marcinak and K. Kuczynski, Int. J. Mech. Sci. 9, 1967, p. 609.
35. A.K. Ghosh and S.S. Hecker, Met. Trans. A, 6A, 1975, p. 1065.
36. R. Hill, J. Mech. Phys. Solids, 1, 1952, p. 19.
37. A.K. Ghosh, Met. Trans. A, 8A, 1977, p. 1221.
38. S.P. Keeler, SAE Paper No. 680092, Jan. 1968.
39. G.M. Goodwin, SAE Paper No. 680093, Jan. 1968.
39A. S.S. Hecker, Sheet Metal Industries, Nov. 1975.
40. H.W. Swift, J. Mech. Phys. Solids, 1, 1952, p 1.
41. S.P. Keeler and W.A. Backofen, Trans. ASM, 56, 1963, p. 25.
42. S.P. Keeler, Proc. Microalloying ′75, Washington, D.C., Oct. 1975.
43. P. Charpentier, Met. Trans. A. 6A, 1975, p. 1665.
44. A.K. Ghosh and S.S. Hecker, Met. Trans. 5, 1974, p. 2161.
45. A.K. Ghosh, Mechanics of Sheet Metal Forming, D.P. Koistinen and N.M. Wang, Eds., Plenum Press, 1978.
46. Military Standardization Handbook, 5B, Dept. of Defense Publication, 1971.
47. D. Lee and W.A. Bakofen, Trans. Met. Soc. AIME, 236, 1966, p. 1077.
48. Aluminum Standards and Data, The Aluminum Association, 1976.
49. A.N. Bramley and P.B. Mellor, Int. J. Mech. Sci. 10, 1968, p. 211.
50. M.C. Noland, H.M. Gladberry, J.B. Loser and E.C. Sneegas, High Velocity Metalworking - A Survey, NASA Report.
51. J. Savitt, "Direct Contact Explosives Metalforming," DDC, AD 296223, Contract 33(615)-1722, Sept. 1964.
52. W. Johnson and R. Sowerby, Proc. Inst. Mech. Engrs. (London), 179, pt. 1, no. 7, 1964-65, p. 197.
53. J.J. Douglas, "Forming Practice with Explosives," New England Regional Conf. AIME (Boston), May 26, 1960.
54. A.A. Ezra, "Principles and Practices of Explosive Forming," Research Report R-64-5, Martin Co., April 2964.
55. N.N. Ida, et al, "Explosive Forming Experience of Martin Denver," Paper 643-E, SAE Automotive Eng. Congr. Jan. 1963
56. N.N. Ida, "Explosive Forming Techniques at Martin Denver,"High Energy Rate Forming II, ASTME, SP62-84, 1962.
57. J.P. Orr, "High Energy Forming, New Manufacturing Techniques," SAE, Paper no. 340A presented at National SAE Meeting (New York, N.Y.) 1961.
58. J. Pearson, "The Explosive Working of Metals," Paper presented at National Symposium, Soc. of Aircraft Mater. and Proc. Engrs. (Dayton, Ohio), Mar. 9-20, 1960.
59. C.W. Gipe, "New Manufacturing Techniques - High Energy Forming," Paper presented at National Aeronautics Production Forum (Los Angeles, Calif.) Oct. 10, 1960.
60. J.M. Walsh, Phys. Rev. 108, no. 2, Oct. 1957, pp. 196.
61. J.C. Jamieson and A.W. Lawson, J. Appl. Phys. 33, 1962, p. 776.
62. R.N. Orava and R.H. Wittman, Ref. 4, p. 485.
63. C.E. Pearson, J. Inst. Metals, 54, 1934, p. 111.
64. E.E. Underwood, "Review of Superplasticity," J. Metals, 14, 1962, p. 914.
65. W.A. Bakofen, I.R. turner and D.H. Avery, Trans. Met. Quart. ASM 57, 1964, p. 980.
66. D.H. Avery and W.A. Bakofen, Trans. Met. Quart. ASM 58, 1965, p. 551.
67. D.A. Woodford, Trans. Quart. ASM 62, 1969, p. 291.
68. A.K. Ghosh, ACTA Met. 25, 1977, p. 1413.
69. F.A. Nichols, ACTA Met., 27, 1980, P. 663.
70. D.S. Fields, IBM, J. Res. Dev. 9, 1965, p. 134.
71. J.W. Edington, K.N. Melton and C.P. Cutler, Progress in Materials Science, 22, no. 2, 1976, p. 61.
72. K.A. Padmanabhan and G.J. Davies, Superplasticity, Springer-Verlag, New York, 1980.
73. C.H. Hamilton and A.K. Ghosh, Met. Trans. A, 11A, 1980, p. 1495.
74. T.H. Alden, Trans. AIME, 236, 1966, p. 1633.
75. R. Grimes, M.J. Stowell and B.M. Watts, Metals Technology, 1976, p. 154.
76. M.M. Allen, R.L. Athey and J.B. Moore, ASM Metals Eng. Quart. 10, (1) 1970, p. 20.
77. T.G. Langdon, Journal of Microscopy, 116, 1979, p. 47.
78. B.M. Watts, M.J. Stowell, B.L. Blaikie and D.G.E. Owen, Metal Science, 10, 1976, p. 189.
79. B.M. Watts, M.J. Stowell, B.L. Blaikie and D.G.E. Owen, Metal Science, 10, 1976, p. 198.
80. C.H. Hamilton, "Fine Grain Aluminum Superplasticity," U.S. Army Rept. TR ARSCD-CR-8001, Feb. 1980.
81. D.M. Moore and L.R. Morris, Materials Science and Engineering, 43, 1980. p. 85.
82. O.D. Sherby, B. Walser, C.M. Young, E.M. Cody, Scripta Met. 9, 1975, p. 569.
83. R.C. Gibson, H.W. Hayden and J.B. Brophy, Trans. Ductility, ASM, 61, 1968, p. 85.
84. W.A. Bakofen, F.J. Azzarto, G.S. Murty and S.W. Zehr, Ductility, ASM Book, 1968, p. 279.
85. R.G. Menzies, J.W. Edington and G.T. Davies, Met. Science, 15, 1981, p. 210.
86. D. Lee and W.A. Bakofen, Trans. TMS-AIME, 239, 1967, p. 1034.
87. N.E. Paton and C.H. Hamilton, Met. Trans. A. 10A, 1979. p. 241.
88. A.K. Ghosh and R. Raj, ACTA Met. 29, 1981, 607-616.
89. A.K. Ghosh and C.H. Hamilton, Met. Trans. A, 10A, 1979, p. 699.
90. J. Davenport, "Mechanisms of Superplastic Flow in Al-4Cu-0.5Zr," presented at conference on Superplasticity and its Commercial Applications, Manchester 1981.
91. Metalworking News, Dec. 22, 1980, p. 12.
92. H.W. Hayden and J.H. Brophy, Trans. Quart. ASM 61, 1968, p. 542.
93. O.D. Sherby et al, Scrpita Met. 10, 1979, p. 941.
94. C.I. Smith and N. Ridley, Metals Tech. 1974, 1429.
95. M.J. Stowell, Metal Sci, 14, 198-0, p. 267.
96. D.A. Miller and T.G. Langdon, Met. Trans. 10A, 1979, p. 1869.
97. C.H. Hamilton, Formability Eds, S.S. Hecker, A.K. Ghosh, H.L. Gegel, AIME, New York, 1978, p. 232.
98. S. Tang, Mechanics of Superplasticity, Robert E. Kroeger Publishing Co, N.Y., 1979.
99. M.C. Flemings, this volume.
100. M. Azrin et al, J. Appl. Metalworking, 1(2), 1980, p. 5.
101. E.O. Hall, Proc. Phys. Soc., London, B64, 1951, p. 747.
102. N.J. Petch, J. Iron and Steel Inst, 174, 1953, p. 15.
103. J. Jonas, Met. Trans. A, 10A, 1979, p. 611.
104. R. Clark and J.S. Harwood, Met. Prog. 94(6) 1968, p. 107.
105. N.E. Paton and A.W. Sommer, Third Int. Conf. on Strength of Metals and Alloys, I, Cambridge, England, 1973, p. 101.
106. J. Waldman, H. Sulinski and H. Markus, Met. Trans. 5, 1974, p. 573.
107. E.D. Russo, M. Conserva, M. Buratti and F. Gatto, Mat. Sci. and Eng. 14, 1974, p. 23.
108. D.S. Thompson, S.A. Levy, G.E. Spangler and D.K. Benson, Air Force Report, Technical Report AFML-TR-73-247, I, Sept. 1973.
109. F.G. Ostermann and W.H. Reimann, STP 467, Achievement of High Fatigue Resistance in Metals and Alloys, ASTM Philadelphia, 1970, p. 169.

110. N.W. Polan and D.H. Avery, Met. Trans. 7A, 1976, p. 323.
111. L.A. Rosales, K. Ono, L.A. Lee and A.W. Sommer, Titanium Science and Technology, 3, Plenum Press, N.Y. 1972, p. 1813.
112. M.E. Rosenblum, P.R. Smith and F.H. Froes, Titanium Science and Technology, 2, Metallurgical Society of AIME New York, 1980, p. 1015.
113. F.H. Froes, C.F. Yolton, J.C. Chesnutt and C.H. Hamilton, paper presented at Conference on Forging and Properties of Aerospace Materials, Leeds, England, Jan. 1977.
114. C.H. Hamilton, S.P. Agrawal, J.C. Chesnutt, E.D. Weisert and N.E. Paton, submitted for publication to Met. Trans A.
115. P. Griffiths and C. Hammond, Titanium Science and Technology, Plenum Press, New York, 1973, p. 1155.
116. N.E. Paton and C.H. Hamilton, U.S. Patent no. 4,092, 181, 1978.

RECOVERY, RECRYSTALLIZATION AND GRAIN GROWTH

Hsun Hu
U. S. Steel Research Laboratory
Monroeville, PA 15146

Introduction and Nomenclature

The properties of a metal depend on its microstructure. Deformation and annealing are the most important processing methods that have been applied to metals and alloys to effect structural changes and the resulting changes in ultimate properties. Because of the importance of these processes, and because of the long history of research studies in these areas,* a vast volume of scientific knowledge has been accumulated in the literature. The scope of the present paper is confined to the phenomena resulting when a cold-worked metal is annealed, mainly the structural and property changes and the mechanisms of the processes. The review will be directed more to recent progress than to historical developments. For more complete information, the reader is advised to examine the references listed at the end of the paper. These include research papers, recent reviews, monographs, symposia proceedings and special publications.

The annealing of a deformed metal involves a variety of processes through which the deformed state is transformed continuously to states of lower energies. It has been customary to divide the whole annealing process into three main stages, namely, recovery, recrystallization, and grain growth. As is understood, such classifications are approximate. (Some overlapping between the stages usually occurs.) There is also ambiguity in the usage of these terms among various authors. With their commonly accepted meanings, these terms will be used here in accordance with the classification of phenomena as follows:

Recovery. The earliest change that occurs upon annealing a cold-worked metal is considered the beginning of the recovery stage. The various structural changes that occur in the recovery stage, hence the associated changes in various properties, do not involve the migration of high-angle boundaries. Thus, from the annealing out of point defects and their clusters, the annihilation and arrangement of dislocations, the polygonization or subgrain formation and subgrain growth, up to the formation of recrystallization nuclei, the processes are classified as recovery. During this stage of annealing, there is no essential change in texture of the deformed metal because there is no involvement of high-angle boundary migration.

Recrystallization. The term, when used without additional specification, is usually considered to be synonymous with primary recrystallization. The stage of recrystallization begins with the formation of stable nuclei for recrystallization. These nuclei are essentially strain-free and are capable of growing at the expense of the polygonized matrix by high-angle boundary migration. The stage of recrystallization is considered to be ended when the entire specimen structure is replaced by the newly recrystallized grains. As a consequence of recrystallization, which is achieved mainly by high-angle boundary migration, there is a large local reorientation or change in texture.

Grain Growth. The term is generally applied to the increase in the average size of recrystallized grains upon further annealing after recrystallization is complete. The growth is achieved by the migration of grain boundaries in the aggregate. Because the total volume of the grains is constant, the increase in the average grain size is accomplished by a decrease in the number of grains. Thus the consequence of grain growth is to reduce the total grain-boundary area in the specimen. Within this category, further classification can be made as follows: the normal (continuous or gradual) grain growth has the characteristics that the growth rate is relatively uniform, and that during such grain growth the size and shape distributions are nearly invariant. The abnormal (discontinuous or exaggerated) grain growth, which is frequently called secondary recrystallization, has the distinct feature that only a few grains grow disproportionately to very large size at the expense of the fine grains of the matrix. These matrix grains have been inhibited from growth by dispersed particles, or by a strong recrystallization texture. Normal grain growth may be accompanied by a gradual change in texture, whereas abnormal grain growth or secondary recrystallization usually results in a large change in texture.

The Deformed State and the Stored Energy of Cold Work

It is widely recognized that a clear understanding of the annealing processes, particularly recovery and recrystallization, must be based on a full understanding of the deformed state. This is not an easy task to accomplish either theoretically or experimentally because the subject of plastic deformation is highly complex, particularly at high strains. However, progress in the basic understanding of the deformation and annealing processes has been greatly assisted during the past two to three decades by development of the ability to actually observe the dislocations and their behavior in metal crystals by transmission electron microscopy.

Fundamentals of Plastic Deformation. Plastic deformation occurs in metal crystals by slip or twinning on specific atomic planes and in specific crystallographic directions, the so-called slip or twinning systems. In fcc metals, the slip systems are {111} <110>, whereas deformation faulting or mechanical twinning systems are {111} <112>. In bcc metals

* The concept that the crystallinity of zinc was destroyed by cold work, and that subsequent heating allowed the metal to recrystallize was first conceived by Kalisher in 1881. A whole century has passed since then.

the slip direction is always <111>, but the slip planes can be {110}, {112}, and also {123}, with or without preference, depending on metal or deformation temperature; twinning occurs on {112} <111>. In hcp metals the most common slip and twinning systems are (0001) <11$\bar{2}$0> and {101$\bar{2}$} <$\bar{1}$011>, respectively, but other slip planes and twinning systems may also prevail depending on the c/a axial ratio and other factors. The deformation mechanisms in other noncubic metals are usually more complex because of low crystal symmetry.

Although mechanical twinning induces an abrupt change in orientation of the twinned material, large strains are obtained usually by slip. Because of crystal symmetry, a number of slip and twinning systems are crystallographically equivalent, with the number of equivalent systems depending on the crystal structure. The selection or activation of the operating systems is determined by the resolved shear stress, which depends on the orientation of the crystal with respect to the direction of the applied stress. When the resolved shear stress on a particular slip system reaches a critical value, slip occurs on that system (the primary system) by the glide of dislocations which pass through the crystal, producing a sequence of steps on the surface (the slip lines or markings). During such slip process the crystal lattice also rotates so that the active slip direction (in uniaxial tension) or the active slip-plane normal (in uniaxial compression) moves toward alignment with the direction of the applied stress.

As a consequence of this lattice rotation, the resolved shear stress on a secondary slip system (the conjugate system) increases, whereas that on the primary system decreases. When the orientation of the stress axis reaches the symmetry line (that is, the [100] - [111] or the [100] - [110] line of the unit triangle in a [100] standard projection for uniaxial tension or compression, respectively), the resolved shear stress on both the primary and the conjugate systems becomes equal and duplex slip occurs.* During duplex slip, the tension axis moves along the symmetry line toward [112], whereas the compression axis moves toward [110].

<u>Work-Hardening Characteristics of Single Crystals</u>. To reduce the complex problem of plastic deformation to its simplest form so that a clear understanding of the process could be approached with less difficulty, the behavior of single crystals of fcc pure metals deformed in tension has been studied most extensively by numerous workers. The stress-strain curves for crystals oriented initially for single slip show three stages of work hardening. In Stage I, frequently called the "easy glide" region, slip occurs mainly on the primary system. The rate of hardening is very low. With increasing participation of secondary slip, the hardening rate increases rapidly, and Stage II is characterized by a linear dependence of the stress on strain (the "linear hardening" stage). The slope of the stress-strain curve in this region is approximately G/300 to G/100 (G is shear modulus), depending on crystal orientation. Upon further straining, this linear hardening stage is followed by a decreased rate of hardening, which is taken as the onset of Stage III. The flowstress falls below that predicted by extrapolation of the flow stress in the previous stage. The stress-strain relation is approximately parabolic in Stage III. Cross slip is mainly responsible for the characteristic features of this stage. Consequently, the beginning of Stage III can be effectively influenced by the temperature of deformation. The three-stage hardening characteristics are found in both pure and alloy fcc crystals strained in a wide range of temperatures at various strain rates (1-5).

For single crystals of bcc metals the basic features of the stress-strain relations are essentially similar to those of fcc crystals (6-10). The three-stage hardening characteristics can be clearly observed in bcc crystals provided that purity, orientation, strain rate and deformation temperature are properly chosen. The same appears to be true for crystals of various hcp metals (11-14). These observations suggest that although the dislocation interactions involved vary with the crystal structure, the basic hardening mechanisms are essentially the same.

For polycrystalline aggregates, the situation is understandably much more complex, because each grain is a single crystal, and the aggregate, unless it has a strong texture, represents a collection of all orientations. Within each grain, the most favorably oriented slip may initially be activated to operate as in an independent crystal, but in the vicinity of grain boundaries complex slip must occur to maintain the material continuity across the boundaries. The stress-strain curve of polycrystalline specimens is approximately parabolic in form. This, in essence, corresponds to the complex slip situation in Stage III for single crystals. In addition to the constraints internally imposed by the grains in polycrystalline materials, which add complexity to the deformation process, the mode of deformation in practical applications is always employed under constrained conditions. For example, instead of simple tension, the constrained tension (wire drawing or rod swaging) is far more important, and instead of simple compression, the plane-strain compression (plate or sheet rolling) is far more important. However, the basic principles that govern the processes should be valid and useful in practical applications.

<u>The Work-Hardening Theories</u>. Plastic deformation takes place by the movement of dislocations. Therefore, any obstacle that hinders dislocation movement would increase the stress required for plastic straining. In association with the dislocation there exists a stress field (15,16). Thus a moving dislocation passing through the stress field of other dislocations would experience increased resistance to its motion. The interactions between dislocations may produce several types of irregularities, which impede the motion of other dislocations. The principal possibilities for work hardening in metals are associated with Lomer-Cottrell barriers, forest or tangled dislocations, dislocation jogs or dislocation loops. Based on these possible mechanisms, several theories of work hardening have been proposed by various authors (1,2,15-23). The ultimate expression is, however, essentially the same; namely, the flow stress is a linear function of the square root of the dislocation density. This relation has been observed experimentally by various authors (29,35). Because of space limitations, this subject will not be elaborated further in the present paper. For more detailed information the reader is advised to consult some of the original papers referenced above, or the related reviews (2,24-26).

* However, for fcc metals or alloys having low stacking-fault energies (such as silver or α-brass), slip frequently continues on the primary system even after the stress axis moves beyond the symmetry line. This is commonly known as "over-shooting," and is generally considered to be due to high hardening of the latent system, or high latent hardening.

Microstructure of Deformed Crystals. The dislocation configuration and deformation substructure in various metal crystals have been studied by many authors (27-29). In general, in the early stages of deformation, the dislocations are relatively long, straight, and few in number. With increasing deformation, more dislocations from other slip systems are produced, leading to more interactions among the various dislocations. These dislocations and clusters of short loops tend to tangle and to align themselves roughly into broad boundaries. Small areas within which there are very few or no individual dislocations are thus outlined by these broad boundaries to form "cells." Such cell formation is usually completed in Stage III. The size of these cells in pure metals is 1 to 3μm. Characteristic features in the dislocation configuration and deformation substructure vary from metal to metal, depending to a large extent on the stacking-fault energy of the metal. Among the fcc metals and alloys, there is a wide range of stacking-fault energies, aluminum having the highest. All bcc metals have high stacking-fault energies. Thus in many aspects aluminum behaves similarly to a bcc metal. The stacking-fault energies of hcp metals are low; however, because of their few slip systems, a comparison between hcp and fcc metals simply on the basis of their stacking-fault energies leads to no consistent guides.

The dislocation configuration and deformation substructures can be influenced by a number of other factors, such as the purity of the metal, the extent, temperature and strain rate of deformation, and the alloying content and second-phase particles, as well as by the type or method (tension or compression, torsion, rolling, wire drawing, etc.) of deformation.

For polycrystalline specimens deformed to large strains, such as heavily cold-rolled sheet or strip prepared for recrystallization studies, the density of dislocations is very high, and the microstructure becomes extremely complex. A clearly defined cell structure may be observable by transmission electron microscopy only in thin-foil specimens prepared parallel to the cross sections of the sheet or strip (30). Thin-foil specimens normally prepared parallel to the rolling plane of the strip show unresolved cell structures like irregularly darkened clouds because the thickness dimension of the cells is much smaller than the thickness of the thin-foil specimens. Thus, in transmission electron microscopy, several layers of cells with irregularly overlapped cell boundaries are being looked through (30,31).

In such heavily deformed specimens, the microstructure is highly non-uniform because of the existence of deformation bands, the various areas of different textures, and the transition regions between texture components having high lattice curvatures. Among these various areas, the cell size and the average misorientations among the cells can be substantially different. In the study of the annealing processes of these highly deformed polycrystalline specimens, the orientation aspects of the microstrucural features are particularly important for nucleation studies (31). Dillamore (32) has recently summarized the high lattice-curvatured regions in heavily deformed polycrystals.

We have here discussed the microstructures of deformed crystals as revealed mainly by the transmission electron microscopy techniques because we prefer to discuss them in terms of dislocations and cell structures. Understandably, conventional microscopy is equally important for examining slip lines, deformation bands, and the overall structural features at low magnifications. In fact, the appropriate combination of various techniques makes the best accomplishments possible.

The Stored Energy of Cold Work. As is widely known, when a metal is cold-worked plastically, most of the mechanical energy expended in the deformation process is converted into heat. Only a small proportion* of the expended energy is stored in the metal (33,34,36,42). This stored energy exists in the crystals as point defects, dislocations, and stacking faults. Thus a cold-worked metal, being in a stage of higher energy, is thermodynamically unstable. With thermal activation, it tends to transform to states of lower energies through the recovery, recrystallization, and grain-growth processes.

The amount of stored energy depends on a number of variables. For a given metal, the amount of stored energy can be influenced by the variables related to the process of deformation, such as the extent, temperature, rate and type or method of deformation. For a given deformation process, the stored energy varies from metal to metal (36,37). It varies also with purity (34,36), alloying element and concentration (38,39), grain size (36,38), single-crystal orientation (39,40), long-range order (41) and other factors. The very recent work by Haessner et al. (39) showed that the stored energies of pure copper and silver were reduced by small additions of a wide variety of solutes, but were increased by many others, depending on the distribution coefficient, k, of the solute atoms in the relevant matrix metal. An earlier comprehensive review on this subject by Titchener and Bever (33) in 1958 was up-dated in 1973 by Bever, Holt, and Titchener (42) with more emphasis on the interpretation of stored energy in regard to its dependence on variables and its relation to other property changes due to crystal defects produced by cold work.

Recovery

As defined earlier in the Introduction and Nomenclature, this is the stage at which early changes in structure and property that do not involve the migration of high-angle boundaries occur during annealing of a cold-worked metal. The characteristic features of the property changes, physical or mechanical, during recovery are the following: (1) there is no incubation period; and (2) the rate of change decreases sharply initially, then becomes more gradual as time progresses, and finally approaches zero very slowly. This is in contrast to recrystallization, which usually has an incubation period and low initial rate (see section on Recrystallization). In comparison with recrystallization, the recovery process is considered homogeneous. The structural changes occur more or less simultaneously throughout the matrix without the formation and growth of a new phase by interface migration.

The recovery of various properties in cold-worked metals on annealing was reviewed extensively by Beck (43), Cahn (74), and Byrne (75). The recovery of mechanical properties was reviewed by Perryman (44). Leslie et al. (45) reviewed the changes in mechanical properties and microstructures in some detail, particularly for iron and iron alloys. On the basis of the theory of dislocations, Li (46) studied the kinetics of annihilation of dislocation dipoles in LiF crystals and extended the treatment to polycrystalline metal specimens. To deal with all the dislocation dipoles in deformed polycrystals, Li used a statistical treatment to the problem and analyzed the recovery of various properties, such as the decrease of dislocation density, the release of stored energy, the decrease of work hardening during recovery, and the kinetics of polygonization. The most recent and comprehensive review of the stored energy of cold work and of broad subjects relating to recovery was made by

* From a few percent to 10-15 percent, depending most on extent of deformation and other variables.

Bever, Holt, and Titchener (42). The following sections will discuss the recovery processes with a few selected examples that are typical of such studies.

Driving Energy for the Annealing Processes. The energy stored in a deformed metal represents the driving energy for the processes of recovery and recrystallization when the deformed metal is annealed. Measurements of the release of the stored energy and a knowledge of its kinetics can provide useful information for an understanding of the processes involved. The stored energy is released as heat, which is measured in calorimeters with sufficient sensitivity. Usually the specimens (the deformed and the standard) are annealed in the calorimeter either with an appropriate heating rate through a selected range of temperature (anisothermal method) or at a constant temperature for increasing length of time (isothermal method). The results are commonly presented in the form of a curve representing the rate of energy release as a function of temperature or time.

Figure 1 shows the power difference measured between the deformed and standard specimens of

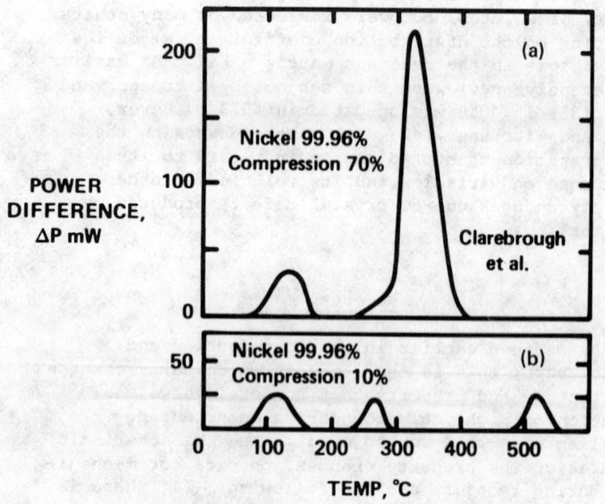

Figure 1. Rate of energy release as a function of annealing temperature for 99.96% pure nickel deformed (a) 70% and (b) 10% in compression. Heating rate 6°C/min. Clarebrough et al.(47).

99.96 percent pure nickel during anisothermal annealing at a constant heating rate 6°C/min (47). For the specimen deformed to a low strain, Figure 1(b), there are three small peaks occurring at approximately 120, 270, and 520°C. The first two were due to recovery, whereas the third peak represented recrystallization. With a constant heating rate, the area under the curve is proportional to the amount of energy released. Thus, the driving energy associated with recrystallization is smaller than the driving energy associated with the recovery processes occurring at the two low temperatures. For the specimen deformed to a large strain, as shown by Figure 1(a), the energy release associated with recrystallization is represented by the large peak centered at approximately 320°C. The second recovery peak (observed for the low-strained specimen at 270°C) was obscured by overlapping with the recrystallization peak, which occurred at a considerably lower temperature than that for the low-strained specimen. On the other hand, the first recovery peak, which occurred at 120-130°C, remained relatively small.

Thus, in a low-strained specimen, the total stored energy is small, a large proportion of this stored energy is released as the driving energy for recovery, and the remaining small portion of the stored energy is used as the driving energy for recrystallization. As a consequence, recrystallization occurs at a considerably higher temperature. In a heavily strained specimen, the stored energy is correspondingly higher than in a lightly strained specimen. The energy released during recovery consitutes a small portion of the stored energy, hence a large proportion of the energy is left available as the driving energy for recrystallization. Consequently, recrystallization can occur at a lower temperature in a highly strained specimen than in a low-strained specimen. In no case was a release of energy corresponding to grain growth (after recrystallization is complete) reported. This is because the driving energy for grain growth is the grain-boundary surface energy, which is too small to be detected by the available calorimeters.

An alternative interpretation for the observed phenomena on energy release can be presented on the basis of the microstructural dependence of the recrystallization tendency of a deformed specimen. In a lightly deformed specimen, there are few localized regions having a high lattice curvature in which nucleation for recrystallization can occur readily. Thus the reduction in internal energy takes place mainly by recovery. When nucleation eventually occurs and a few high-angle boundaries are produced, the internal energy has already become too low to support the growth of these recrystallization nuclei at an appreciable rate. Consequently, recrystallization occurs at a considerably higher temperature with the small amount of driving energy available. On the other hand, in a heavily deformed specimen nucleation sites are available in many places. The formation of stable recrystallization nuclei occurs readily throughout the specimen. As a consequence, recrystallization by high-angle boundary migration takes place before the energy release for the second recovery peak is completed. Thus, a large proportion of the stored energy is released during recrystallization.

Figure 2 shows the rate of energy release as a function of time during isothermal annealing of high-purity copper deformed in tension to two strains

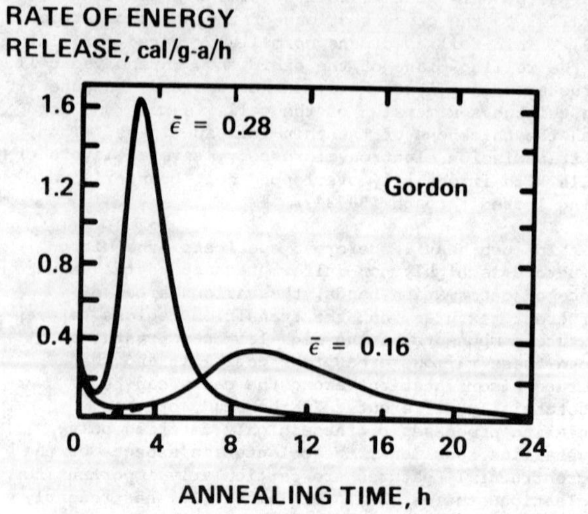

Figure 2. Rate of energy release as a function of time during isothermal annealing at 189.7°C for 99.999% pure copper deformed by tension to two strains. Gordon(34).

(34). A small amount of energy was released during recovery in each case, and the amount of this energy appeared to be independent of strain. Essentially similar observations were reported for polycrystalline silver (35). Thus, a larger proportion of the stored energy is released during recovery in a lightly strained specimen than in a highly strained specimen. Also, at a given annealing temperature, the recrystallization peak occurs after shorter times and associates with larger released energies as the amount of strain in the specimen is increased. Besides the amount of strain, there are many other processing or material variables that influence the rate of energy release. In general, those variables that determine the amount of energy stored in a cold-worked metal also affect the kinetics of its release (42).

The magnitude of the stored energy of cold work, which is the driving energy for recovery and recrystallization when the deformed metal is annealed, is comparatively small. According to a general-survey statement given by Bever et al. (42), the stored energy in most polycrystalline metals deformed at room temperature does not exceed a few tens of calories per mole even at large strains; in solid-solution alloys it rarely exceeds 200 cal/mole. The energy stored by single crystals at small strains is typically less than 5 cal/mole. In contrast to these quantities, the heats of fusion of metals are typically several thousand calories per mole and the heats of polymorphic transformation several hundred. These are orders of magnitude greater than the stored energies in cold-worked metals. Nontheless, cold-worked metals are energetically instable; with sufficient thermal activation, restoration to a low-energy state through recovery, recrystallization and grain growth occurs readily at appreciable rates.

Recovery of Electrical Resistivity and Density. When a metal is deformed plastically, its electrical resistivity increases. For a given metal, the increase in electrical resistivity depends on the amount and the temperature of deformation. At a given deformation temperature, the electrical resistivity increases with the amount of deformation, and for a given strain, it increases with decreasing deformation temperature (48,49). The increase of the electrical resistivity in a deformed metal is caused by the scattering of electron waves by the lattice defects (vacancies, dislocations, and stacking faults) generated by plastic deformation (50). Thus, when these defects are being eliminated upon annealing, the electrical resistivity of the specimen decreases accordingly.

Figure 3 shows the results of the combined measurements of the stored-energy release, the change

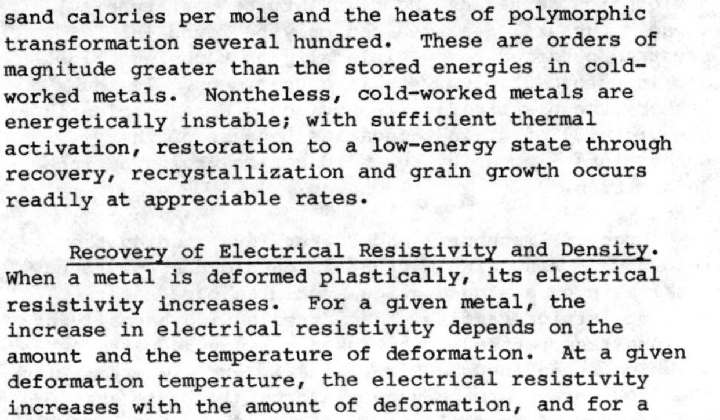

Figure 3. Changes in electrical resistivity, density, and rate of energy release as a function of temperature for 99.85% pure nickel deformed 70% in compression. Heating rate 6°C/min. Clarebrough et al.(47).

in the resistivity (Δρ) and the fractional change in density (ΔD/D), as a function of annealing temperature for 99.85 percent pure nickel deformed 70 percent in compression (47). The decrease in the electrical resistivity during recovery shows two stages, corresponding to the two small peaks in the energy release. Similarly, the increase in the density also shows two stages corresponding to the same two small peaks in the energy release. There is little change in density or resistivity in the region where no energy release is observed from the end of the second recovery peak to the beginning of the recrystallization peak. As recrystallization proceeds, an additional increase in density and a major decrease in resistivity take place. However, in a study of the annealing behavior of two silicon-iron single crystals cold-rolled 80 percent, the major recovery of density occurred at about 150°C, and the recovery of x-ray line broadening and hardness could not be detected (52). In those crystals, vacancies probably contributed to most of the density changes.

The change in electrical resistivity during isothermal annealing of cold-worked copper is shown in Figure 4. The specimen was deformed in torsion at

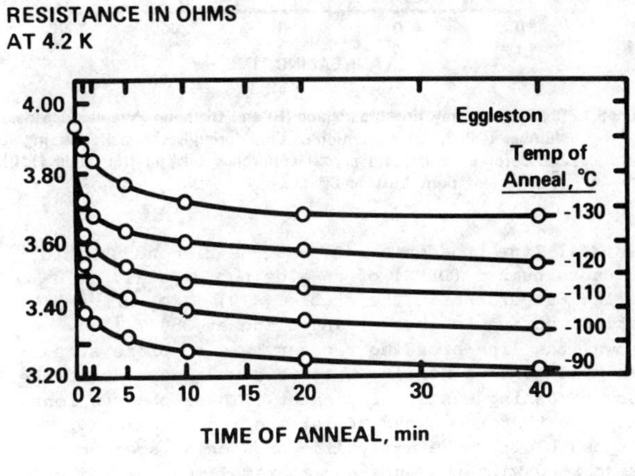

Figure 4. Change in electrical resistivity as a function of time during isothermal annealing at various temperatures for copper deformed in tension. Eggleston(51).

4.2K and annealed isothermally at the various temperatures indicated (51). The electrical resistance of the specimen was measured at 4.2K as a function of time. As shown, the resistance decreases sharply with time in the early stage of isothermal annealing, then more gradually with increasing time. At still longer times, the resistance approaches the equilibrium value, which is characteristic of the temperature, very slowly. These results clearly show the type of kinetics typical of the recovery process.

Recovery of X-Ray Line Broadening. The broadening of x-ray diffraction lines or peaks can be caused by a number of factors (53). Lattice strains, stacking faults, and fine crystallites are the main sources for line broadening in cold-worked metals. With an x-ray diffractometer, the line broadening can be determined by methods ranging from simple measurement of the line breadth at half intensity of selected reflections to the more sophisticated technique of Warren-Averbach (54,55), which uses Fourier analysis of the line profile to determine the broadening due to lattice strain and that due to "particle" (domain or crystallite) size. The arbitrary line-broadening parameter, $R = (I_{min} - I_b)/(I_{max} - I_b)$, where I_{min} is the intensity minimum between $K_{\alpha 1}$ and $K_{\alpha 2}$ peaks, I_{max} is the intensity maximum of the $K_{\alpha 1}$ peak, and I_b is the intensity of the background, has been used satisfactorily by various investigators (52,56).

X-ray line sharpening occurs upon annealing of cold-worked metals, usually during the recovery stages prior to the onset of recrystallization. For low stacking-fault energy metals or alloys, cold-worked severely, the separation of recovery and recrystallization may not be observed distinctly. Under special circumstances when the deformed crystals are highly homogenous in microstructure and have very weak tendencies for recrystallization, the x-ray line sharpening may occur to an appreciable extent or nearly to completion when the hardness decreases only slightly (52,56). Figure 5 shows the changes in the

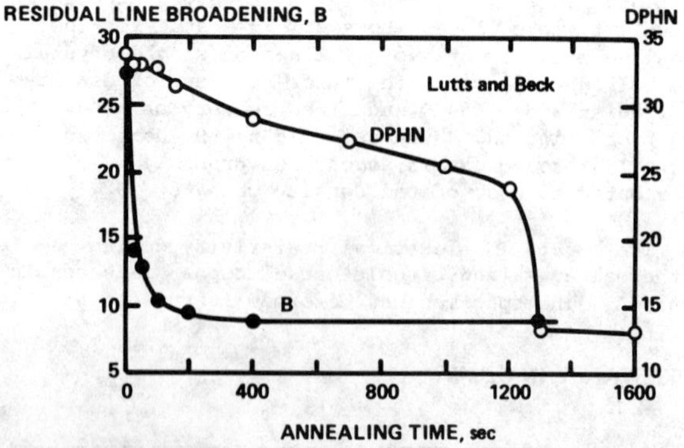

Figure 5. Changes in x-ray line broadening (B) and Diamond Pyramid Hardness Number (DPHN) as a function of time during isothermal annealing at 350°C for an aluminum crystal cold-rolled 80% in the stable (110) [$\bar{1}$12] orientation. Lutts and Beck(56).

residual line broadening (B) and the diamond-pyramid hardness number (DPHN) of an aluminum crystal, cold-rolled 80 percent in the stable (110) [$\bar{1}$12] orientation, during isothermal annealing at 350°C (56). As shown, the line-broading recovery was complete at an early stage of annealing when nearly 75 percent of the work hardening was still present. These observations indicate that the types of lattice imperfections responsible for the x-ray line broadening are not associated with the substantial remaining portion of the work hardening in the crystal. The rate of recovery of x-ray line broadening has the same characteristic features as those observed for the rate of recovery of electrical resistivity during isothermal anneals. This suggests that the kinetics of recovery of these properties are similar.

For the study of the recovery of x-ray line broadening in silicon-iron single crystals, cold-rolled 80 percent in the stable (001) [110] and the unstable (001) [100] orientations, the Warren-Averbach technique (54,55) using Fourier analysis was used by Hu (52) along with other simpler measurements. Results from the Fourier analysis of the x-ray line profiles showed some scatter, especially for specimens annealed at comparatively high temperatures (500°C or above). This was probably due to the fact that the x-ray line broadening developed in these rolled crystals is relatively small in comparison with that in severely deformed powder specimens produced by filing at cryogenic temperatures. Within the range of annealing temperatures 200 to 400°C, the root-mean-square strain, $\langle \varepsilon^2_{50\text{Å}} \rangle^{1/2}$, averaged over a length of 50Å in a direction perpendicular to the (002) reflecting plane, was on the order of 10^{-3}.

The average particle size ranged from 200 to 420 or 1280 Å, depending on the crystal. For particle sizes appreciably greater than 1300 Å, the broadening was too small to be determined with accuracy by the Warren-Averbach technique for these crystals. These particle sizes are more than one order of magnitude smaller than the cell sizes observed by transmission electron microscopy. Thus, the x-ray line broadening contributed by the particle size in these rolled crystals is probably associated with regions of tangled dislocations, such as within the diffuse boundaries surrounding the cells. When the cell walls or boundaries are sharpened, the x-ray line broadening is recovered completely, because the cells having diameters 1 μm and over are too large to cause "particle size" broadening of the x-ray diffraction peaks.

Recovery of Work Hardening. For single crystals of hexagonal metals deformed in tension by basal slip, the increment in work hardening can be removed completely by recovery without recrystallization. For cubic crystals, a similar behavior can be observed if the crystals are deformed by single slip in easy glide, where the plastic flow is laminar with no lattice bending.* However, when the deformation occurs by multiple slip, the plastic flow is turbulent and the resulting work hardening is greater. Then only a fraction of this work hardening can be removed by recovery. Complete softening must, therfore, be accomplished through recrystallization.

These experimental observations can be understood on the basis of the microstructures of the deformed specimens. For metal crystals deformed by single slip with no lattice bending, or by laminar flow, there is no possibility for recrystallization to occur, because nucleation requires localized regions with high dislocation densities and high lattice curvature. These structural features can only be found in crystals deformed by multiple slip or turbulent flow, and in deformed polycrystalline specimens. In the latter, the plastic flow is obviously more complex in the vicinity of grain boundaries because of the constraints imposed by the neighboring grains during deformation.

Drouard, Washburn, and Parker (58) studied the recovery of zinc single crystals deformed at -50°C by basal slip on a single glide system in pure shear without lattice bending, then annealed isothermally at various temperatures. They used a recovery parameter R, which is defined as $R = (\sigma_m - \sigma)/(\sigma_m - \sigma_o)$, where σ_m is the flow stress of the deformed crystal, σ_o is that of the undeformed crystal, and σ is the flow stress of partially recovered crystal. Thus R represents the fraction of recovery. When they plotted the values of (1-R), or the fraction of residual strain hardening in the partially recovered crystal, as a function of isothermal annealing time, a series of curves quite similar in form to those shown in Figure 4 (or Figure 6 to be presented next) was obtained.

The recovery of work hardening in zone-melted iron deformed at 0°C to 5 percent elongation was studied by Michalak and Paxton (59). For the same recovery parameter R defined by Drouard et al. (58), the changes in the fraction residual strain hardening as a function of time during isothermal annealing at various temperatures are shown in Figure 6. The

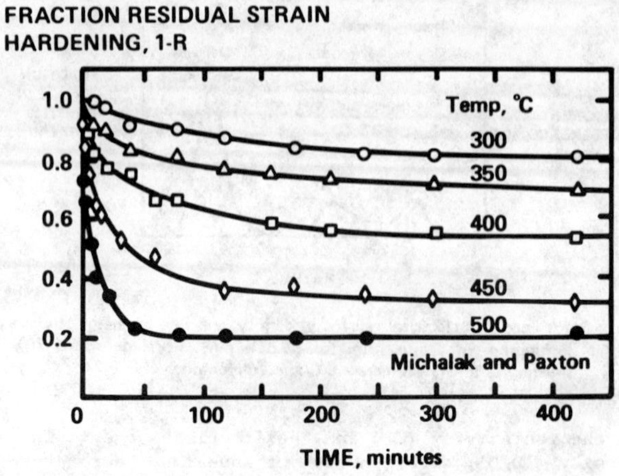

Figure 6. Change in residual strain hardening as a function of time during isothermal annealing at various temperatures for iron strained 5% in tension at 0°C. Michalak and Paxton(59).

* Even with simple bending, the rearrangement of excess edge dislocations upon annealing, which leads to polygonization (57), is still preferably being considered as recovery, because no high-angle boundary migration is invovled.

characteristic features of these curves are quite similar to those of the curves shown in Figure 4 for recovery of electrical resistivity in deformed copper (51).

According to Keh (60), the sharp drop in flow stress during the initial stage of recovery is associated with a slight decrease in average dislocation density. During the later stage, the average dislocation density decreases continuously, but less rapidly, whereas the flow stress remains constant. These results clearly indicate that in recovery the flow stress is not completely dependent on the dislocation density, although in deformation a linear relation between flow stress and the square root of dislocation density is established both theoretically and experimentally.

Kinetics of Recovery. As mentioned in earlier sections, the rate of isothermal recovery of various properties has very similar characteristics. The rate of property change is highest at the start, decreasing gradually as recovery progresses. At still longer times the rate of change becomes negligibly small. Such rate changes have been shown as a function of time for the release of stored energy (Figure 2), the decrease of electrical resistivity (Figure 4), the decrease of x-ray line broadening (Figure 5), and the decrease of the residual strain hardening (Figure 6) during isothermal anneals. The hardness is not a very sensitive parameter for property changes during the early stages of recovery. The recovery of different properties may occur at quite different rates. Thus, a cold-worked metal may display a whole spectrum of recovery phenomena. These may correspond to a series of complicated kinetic processes, represented by different activation energies. On the other hand, if the various recovery phenomena correspond to progressive stages of a single process, their activation energies should be the same.

These observed changes in properties indicate that the isothermal recovery rate is inversely proportional to the annealing time, that is,

$$\frac{dR}{dt} = \frac{a}{t} \quad (1)$$

where R is fraction of recovery, t is isothermal annealing time, and a is a constant depending on temperature. Thus, for a given temperature, the fraction of recovery R is, by integration,

$$R = a \ln t + b \quad (2)$$

where b is a temperature-dependent constant. Equation (2) shows that if the fraction of recovery R is plotted versus the annealing time t in logarithmic scale, a straight line is obtained, with a being the slope and b the intercept of the straight line.

The temperature dependence of the rate processes is generally of the Arrhenius type, namely, for the rate change of any property x,

$$\frac{dx}{dt} \text{ or } \frac{1}{t} = A \exp(-Q/RT) \quad (3)$$

where A is the rate constant, Q is the activation energy for the process concerned, R is the gas constant, and T is the absolute temperature. By taking logarithm on both sides of Equation (3), we have

$$\ln\left(\frac{dx}{dt}\right) \text{ or } \ln\left(\frac{1}{t}\right) = \ln A - \frac{Q}{RT} \quad (4)$$

Thus, when the rate or (1/t) is plotted on the logarithmic scale versus 1/T, the slope of the straight line is Q/R, then the activation energy Q is obtained.

Figure 7 shows the temperature dependence of the time required to recover the strain hardening in sheared zinc crystals to various fractions of

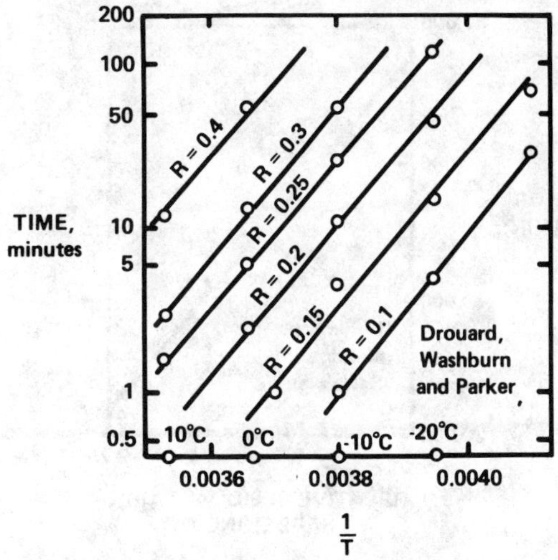

Figure 7. Temperature dependence of the annealing time required for various fractions of recovery in zinc single crystals deformed at -50°C in pure shear. Drouard, Washburn and Parker(58).

recovery, from the work of Drouard, Washburn, and Parker (58). No significant change in slope of the lines was detected within this range of recovery (R = 0.1 to 0.4). The average value of the slope or activation energy for the process was 20 kcal (83.8 kJ) per mole. This corresponds closely to the activation energy for self-diffusion in the direction of the hexagonal, or c, axis in zinc, suggesting that motion of lattice vacancies in the c-axis direction may be the rate-controlling process in recovery.

According to the theory of recovery by Kuhlmann (61), who assumed that the instantaneous rate of recovery depends on the extent of recovery according to the expression

$$-\frac{dx}{dt} = cx \exp -\left(\frac{Q-mx}{RT}\right) \quad (5)$$

where x is the quantity of the measured property that had departed from equilibrium at the time t, Q is the activation energy at the completion of the recovery process, c and m are constants, then the activation energy for recovery should increase linearly with the extent of recovery. This leads directly to a linear relationship between recovery and the logarithm of the annealing time, Equation (2).

That the activation energy for recovery should be a linear function of the extent of recovery was also shown by Li (46). By differentiating and taking the logarithm of Equation (2), the following exression is obtained

$$\ln\left(\frac{dR}{dt}\right) = \ln a + \left(\frac{b}{a}\right) - \left(\frac{R}{a}\right) \quad (6)$$

Thus, $\ln$ (dR/dt) varies linearly with R, the extent of recovery. Because $\ln$ (dR/dt) is also a linear function of (1/T) for each value of R, the activation energy must be a linear function of R. In other words, the activation energy for recovery varies linearly with the extent of recovery.

Such a linear dependence of activation energy for recovery on the extent of recovery is shown in Figure 8 for aluminum single crystals deformed in tension then annealed for recovery of the strain hardening (62). The plot shows that the activation energy for recovery increases linearly with progressive recovery as predicted by the theory. The observed activation energy for recovery of strain hardening in zinc single crystals deformed in shear by single glide is, however, practically constant (58), at least for the recovery range R = 0.1 to 0.4 (Figure 7). The contrast in behavior of these two cases was rationalized by Beck (43), who suggested that the results of recovery in aluminum crystals were obtained under less favorable conditions than those of Drouard et al. (58) for zinc crystals, because in the extended aluminum crystals a substantial portion of the work hardening was not recoverable.

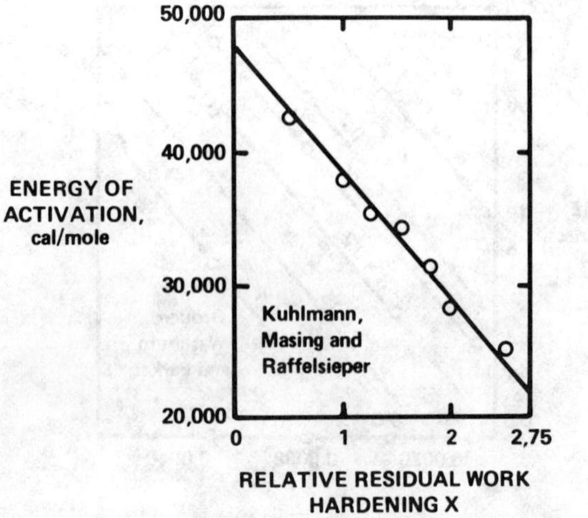

Figure 8. The linear increase of activation energy for recovery of work hardening with extent of recovery in aluminum single crystals deformed in tension. Kuhlmann, Masing and Raffelsieper(62).

That the activation energy for recovery increases with the extent of recovery was also observed by Michalak and Paxton (59) for the recovery of flow stress in zone-melted iron, although the recovery dependence of the activation energy was not exactly linear. By extrapolation to R = 0 (the beginning of recovery) and R = 1 (the completion of recovery), Leslie et al. (45) showed that the activation energy for recovery at R = 0 was 22.0 kcal (92.2 kJ) per mole, corresponding to the activation energy for the migration of vacancies. At R = 1, the activation energy for recovery was 67.4 kcal (282.4 kJ) per mole, which corresponds to the activation energy for self-diffusion in iron.

Considering the annihilation of dislocation dipoles as one of the important microscopic mechanisms for recovery in cold-worked metals, Li (46) studied the kinetics of annihilation of individual dislocation dipoles in LiF single crystals. To deal with all the dislocation dipoles in a deformed specimen, Li used statistical treatment of the problem and showed that in the idealized condition (all dipoles are noninteracting) the annealing of dislocation dipoles can be described by a simple second-order kinetics

$$\frac{d\rho}{dt} = -k\rho^2 \qquad (7)$$

where ρ is dipole density. The same kinetics can be used to approximately describe the decrease of dislocation density during annealing by considering the excess dislocation density of one sign being small. Equation (7) can be expressed in the integrated form

$$\left(\rho^{-1} - \rho_o^{-1}\right) = kt \qquad (8)$$

where ρ_o and ρ are dislocation densities before and after annealing, respectively, and k is a constant. At long times when ρ is much smaller than ρ_o, Equation (8) becomes

$$\rho^{-1} = kt \qquad (9)$$

or

$$\frac{d\rho}{dt} = -\frac{1}{kt^2} \qquad (10)$$

Such kinetics were observed by Williams (37) who measured the energy release immediately after the specimen was deformed by impact, and by Vandermeer and Gordon (63) for the rate of the stored-energy release upon isothermal annealing of a deformed specimen of zone-refined aluminum containing 0.0068 atomic percent copper. These results appear to indicate that the stored energy in a deformed metal is proportional to the density of dislocations. The second-order kinetics were also observed recently in the recovery of [001] single crystals of Ni and Ni-Co alloys deformed in tension by the annealing out of dislocation loops (64).

Polygonization and Polygon Growth. The term "polygonization" was first introduced by Orowan (57) and by Cahn (66,67) to describe the structural changes occurring in a single crystal bent by single slip and subsequently annealed for recovery. The smoothly curved glide planes break up into segments of a polygon separated by walls of edge dislocations. This rearrangement of excess dislocations of the same sign results in a more stable configuration because a wall of evenly spaced dislocations has the lowest total energy. Since then, more studies on the polygonization phenomena have been reported. The subject was reviewed by Hibbard and Dunn (68).

The term "polygonization" has later been used widely in the broad sense for referring to the formation of a well-defined subgrain structure in the recovered matrix. Because "cells" are formed as a consequence of deformation, "polygonization" would mainly imply the subsequent structural changes during recovery. These would include the annihilation of dislocation dipoles, the annealing out of dislocation loops, the rearrangement of dislocations in the cell walls and the sharpening of the cell or subgrain boundaries.

Among numerous studies of polygonization, only a few were quantitative. Gilman (69) found that the average angle of polygon walls in zinc is a linear function of the logarithm of isothermal annealing time. Similar dependence was observed for the average spacing between the polygon walls (70). The same logarithmic kinetics were also observed in bent and polygonized NaCl crystals (71). The increase in spacing between polygon walls, or the growth of polygons, is controlled by the nucleation and growth of the Y junction of the polygon walls. The problem was treated by Li (72).

Positron Annihilation for Recovery Studies. Because defects in a crystalline lattice, such as vacancies and dislocations, are regions of relatively low electron densities, they serve as trapping sites for positrons. Such trapped positrons have on the average a longer lifetime than those untrapped positrons that annihilate with the electrons in the perfect regions of the lattice. Thus the measurement of positron lifetime in a deformed metal and during subsequent annealing may yield information on the density and nature of the defects. Such studies are rather recent developments in physical metallurgy research, and since the early 1970's there have been numerous publications on positron studies of variously deformed and annealed single crystals or polycrystalline specimens of various metals and alloys. A most recent review of these studies was given by Byrne

(65). Because of the complexities of positron lifetime itself and the difficulties sometimes involved in the interpretation of the results, this new technique is still not widely employed. Further development of the technique is being pursued both theoretically and experimentally.

Summary of Recovery. The recovery of various properties upon annealing cold-worked metals has been reviewed. The review included the release of the stored energy and the driving energies for the recovery and recrystallization processes; the recovery of density, hardness, electrical resistivity, x-ray line broadening, and work hardening; and the kinetics of the recovery processes. The recently developed technique for studying crystalline defects in cold-worked and subsequently annealed metals by measuring positron lifetime was very briefly introduced. The isothermal recovery of the various properties has the following features in common: (1) there is no incubation period; (2) the rate of change is highest at the beginning, decreasing with increasing time; and (3) at long times, the property approaches the equilibrium value very gradually. These kinetics indicate that the isothermal recovery rate is inversely proportional to the annealing time. Thus for a given temperature, the amount of recovery is a linear function of the logarithm of time. The same kinetics are observed for polygon growth in bent and polygonized crystals. One of the most important microscopic mechanisms for recovery is the annihilation of dislocation dipoles. The annihilation of dislocation dipholes, the decrease in dislocation density, and the release of the stored energy during isothermal annealing can all be represented by simple second-order kinetics.

Recrystallization

Recrystallization, or primary recrystallization, occurs by the nucleation and growth of new grains at the expense of the deformed and recovered matrix. An incubation period is usually observed during which viable nuclei are formed at a number of preferred sites. These nuclei are essentially strain-free and are surrounded or partially surrounded at an early stage of formation by high-angle boundaries with high mobilities. They are energetically capable of growing by the migration of their high-angle boundaries away from the center of curvature of the boundary into the matrix. As recrystallization proceeds to completion, the new grains begin to impinge upon one another, and finally the entire specimen is replaced with an aggregate of recrystallized grains. Because the growth is accomplished by high-angle boundary migration, there is a change in texture of the specimen upon recrystallization.

The driving energy for recrystallization is the unreleased portion of the stored energy after the recovery processes have taken place, which may correspond to approximately 90 percent of the total stored energy in the deformed specimen (see Figure 2). As discussed previously, the relative proportion of the stored energy that can be released during recovery or recrystallization depends on a number of factors. This driving energy for recrystallization is associated mainly with dislocations in the subgrain boundaries.

The kinetics of recrystallization are very much similar to those of phase transformations. The process initiates and progresses by the nucleation and growth of the new grains. Thus, isothermal recrystallization curves are typically of the sigmoidal type. Various aspects of the recrystallization process, both theoretical and experimental, were reviewed extensively by Burke and Turnbull (73) and by Beck (43) in the 1950's, by Cahn (74) and by Byrne (75) in the 1960's, and by Cotterill and Mould (76) about five years ago. A large volume of references was embodied in the work of Cotterill and Mould. Therefore, the present paper will describe the classical basics very briefly and discuss the recent developments more emphatically.

Variables Affecting Recrystallization. It is useful, particularly for practical applications, to know the important variables that affect the process of recrystallization. By as early as about 1920, the qualitative effects of a number of processing variables on recrystallization had been established experimentally. These, usually being presented as the "laws of recrystallization," were listed by Burke and Turnbull (73) as follows:

1. A minimum deformation is necessary to cause recrystallization.

2. The smaller the degree of deformation, the higher is the temperature required to cause recrystallization.

3. Increasing the annealing time decreases the temperature necessary for recrystallization.

4. The final grain size* depends chiefly on the degree of deformation, and to a lesser extent on the annealing temperature, being smaller the greater the degree of deformation and the lower the annealing temperature.

5. The larger the original grain size, the greater is the amount of cold deformation required to give equivalent recrystallization temperature and time.

6. The amount of cold work required to give equivalent deformation hardening increases with increasing temperature of working.

7. Continued heating after recrystallization is complete causes the grain size to increase.

Item 7 obviously belongs to the "grain growth" stage rather than to recrystallization. In fact, Burke and Turnbull (73) include the variables that affect both recrystallization and grain growth under the "laws of recrystallization". With the background knowledge of the stored energy of cold work, its subsequent release during recovery, the various factors that influence these processes, and the microstructural features associated with recrystallization (all of which have been discussed in earlier sections), the "laws of Recrystallization" can be understood without much difficulty.

To these seven items, Cahn (74) made one more addition:

8. New grains do not grow into deformed grains of identical or slightly deviating orientation.

This additional item, in the light of recent developments in the understanding of the recrystallization process, is considered significant and appropriate. It serves not only as an obvious example of the definition for recrystallization that the process is accomplished by high-angle boundary migration, but also as an emphasis to the orientation aspects of recrystallization which are highly important for a thorough understanding of the process.

---

* It would be more appropriate to replace "the final grain size" by "the grain size when primary recrystallization has just been completed."

Kinetics and Classic Theories of Recrystallization. As mentioned earlier, recrystallization initiates and progresses by the nucleation and growth of the new grains. If the volume fraction of the recrystallized material is plotted as a function of time during isothermal annealing, the curve is typically of the sigmoidal type (80), as shown in Figure 9. The initial portion of the curve

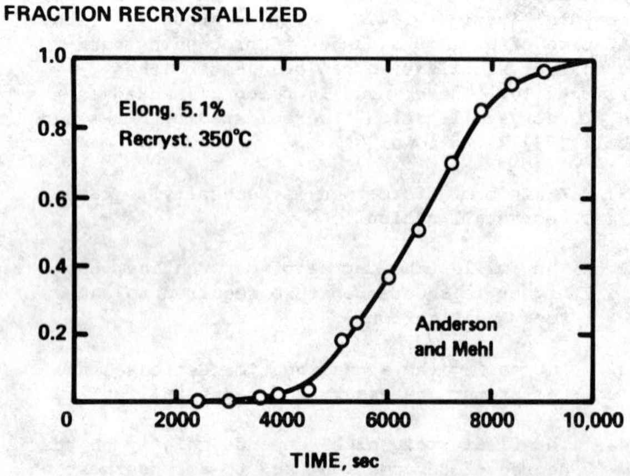

Figure 9. Typical isothermal recrystallization curve showing fraction recrystallized as a function of time for aluminum deformed 5.1% in tension and annealed at 350°C. Anderson and Mehl (80).

corresponds to the incubation period during which stable nuclei for recrystallization are being formed. This is followed by a period of rapid growth of the new grains at an approximately constant rate. At still longer times, when the recrystallized volume increases to large fractions, the rate of recrystallization or the average rate of growth of the recrystallized grains decreases because of the impingement of increasing numbers of grains as recrystallization approaches completion.

The most widely quoted phenomenological theories of recrystallization are those of Johnson and Mehl (78) and of Avrami (79). With the assumptions that after the incubation period the growth rate is linear with time, and that the nucleation frequency is inversely proportional to the average time of incubation, these authors used essentially the same treatments of the problem (with the only difference being in the assumptions on the form of the nucleation frequency as a function of the incubation time) and arrived at the same basic equations. The derivations of the equations will not be shown here because they have been repeated several times by various reviewers (73,75,76). For the three most common forms of the specimens (wire, sheet, and lump), the limiting equations for isothermal recrystallization as derived by Johnson and Mehl (78) and by Avrami (79) can be expressed in the general form

$$X_v = 1 - \exp(-Bt^k) \tag{11}$$

and are summarized in Table I.

Table I. Limiting equations for isothermal recrystallization.

Type of Recrystallization	Johnson and Mehl $\dot{N}$ = constant	Avrami $\dot{N} = \bar{N}_v \exp(-\nu\tau)$
3-Dimensional (lump)	$X_v = 1-\exp(-fG^3\dot{N}t^4/4)$	$X_v = 1-\exp(-Bt^k)$, $3 \leqq k \leqq 4$
2-Dimensional (sheet)	$X_v = 1-\exp(-fG^2\delta\dot{N}t^3/3)$	Same equation as above, $2 \leqq k \leqq 3$
1-Dimensional (wire)	$X_v = 1-\exp(-fG\delta^2\dot{N}t^2/2)$	Same equation as above, $1 \leqq k \leqq 2$

Symbols in the above equations: X_v, volume fraction recrystallized; t, isothermal annealing time; $\dot{N}$, nucleation frequency; f, shape factor; G, growth rate; δ, the smallest dimension of specimen, such as thickness of sheet or wire diameter; $\bar{N}$, number of nucleation sites; ν, nucleation frequency of each site; and τ, incubation time. In the equation of the general form (the Avrami equation), B and k are parameters depending on the type of recrystallization. For example, in the 3-dimensional recrystallization, $B = fG^3\dot{N}/(k+1)$.

These phenomenological equations were derived with the assumptions that $\dot{N}$ and G are independent of time. According to the Johnson and Mehl theory, $\dot{N}$ must be evaluated by experiment, whereas Avrami assumes that $\dot{N}$ decreases with the incubation time in an explicit form. Mehl and his co-workers (78,80,81) determined $\dot{N}$ by measuring the number of grains and the total progress of recrystallization in each of the various equivalent samples to minimize statistical fluctuations. From the N vs t curve thus developed, the nucleation rate as a function of time was obtained by measuring the slope at various points along the curve. These are shown in Figure 10(a) and (b). The nucleation rate was found to increase exponentially with time,

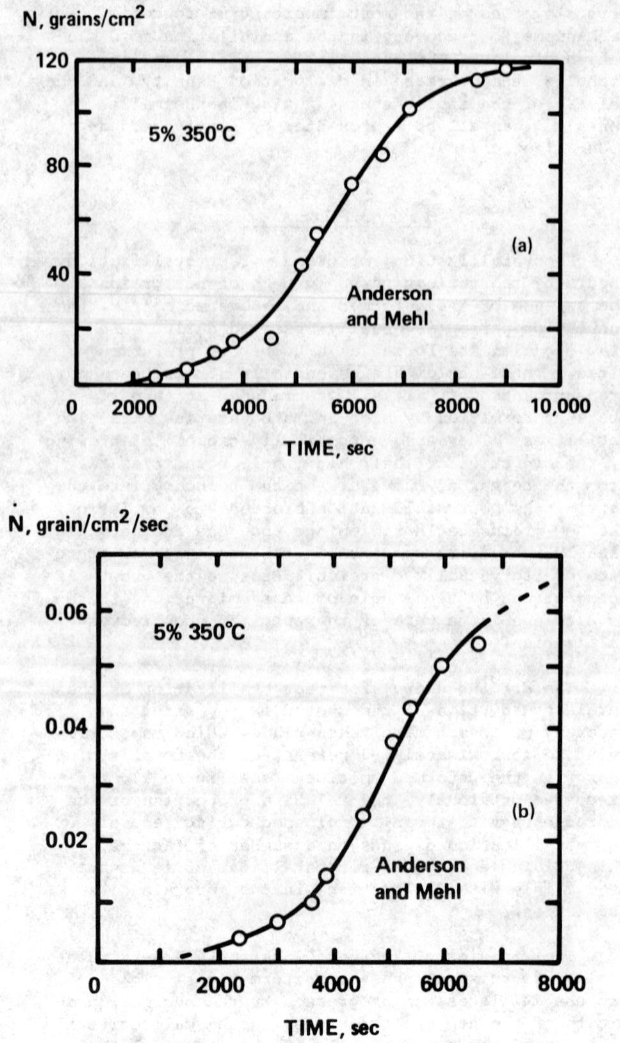

Figure 10. Experimental data of nucleation in recrystallization for aluminum deformed in tension to a strain of 0.05 then annealed at 350°C. (a) Increase in number of new grains as a function of annealing time. (b) Rate of nucleation as a function of annealing time [slope of curve (a)]. Anderson and Mehl (80).

$$\dot{N} = a \exp(b\tau) \quad (12)$$

This result is at variance with the Avrami theory that $\dot{N}$ should decrease or remain essentially constant with increasing time. The growth rate, G, was determined by measuring the diameter of the largest unimpinged grain as a function time,* and was assumed to be the same for all grains.

It is now well known that the mechanism of nucleation for recrystallization can be substantially different for specimens deformed to widely different strains. The classic work of Mehl and his co-workers used very lightly deformed samples. This, of course, was a necessity at that time because of experimental difficulties in studying nucleation of heavily deformed specimens. With transmission electron microscopy, the mechanism and the kinetics of nucleation can be examined in much detail. Unfortunately there have been only a few limited studies on the nucleation aspects of recrystallization in the past 20 years.

The growth rate of the grains is now also well known to be dependent on the mobility of high-angle boundaries, which mobility in turn depends on many factors. For example, one of these factors, the orientation dependence of boundary mobility, indicates that it was not appropriate to assume that the growth rate is the same for all grains. The nucleation mechanisms and the migration of grain boundaries will be discussed in later sections.

Anderson and Mehl (80) also studied the rate of nucleation and the rate of growth in the recrystallization of aluminum as a function of deformation and of annealing temperature. From the temperature dependence of these rates, activation energies were calculated for both G and $\dot{N}$ individually in the range of deformation employed. As shown in Figure 11, the

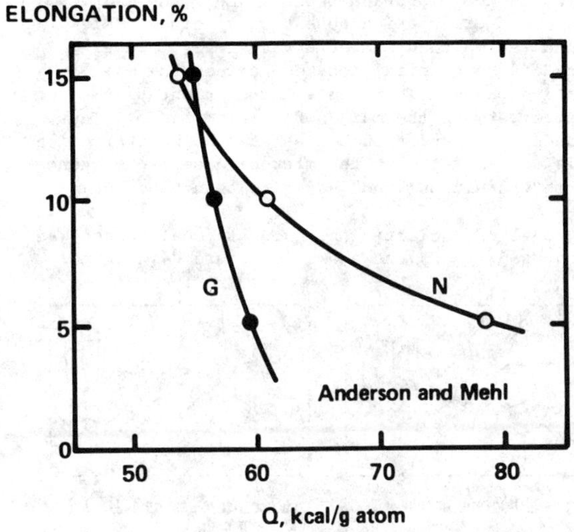

Figure 11. Activation energies for nucleation (N) and for growth (G) as a function of amount of deformation. Anderson and Mehl (80).

activation energies for both nucleation and growth decrease with increasing deformation, but at appreciably different rates. At small strains ($\varepsilon<5\%$), the activation energy required for nucleation is much higher than that for growth. This difference diminishes rapidly with increasing deformation. The activation energies for both nucleation and growth become approximately equal when the deformation is greater than about 12 percent. These results are consistent with the well-known observations that large crystals are usually produced by the strain-anneal technique whereby a strain smaller than the "critical strain" (1 to 3% elongation) is applied prior to annealing (82), and that the as-recrystallized grain size decreases sharply with increasing deformation to 10 to 15 percent extension (83). Upon further increase in the amount of deformation (>20%), the grain size decreases with increasing deformation only very gradually.

The results of recrystallization-kinetics studies are usually presented in a log-log plot of the recrystallized volume against the isothermal annealing time. For this purpose, Equation (11) may be rearranged and written in logarithmic form

$$\ln(1-X_v) = -Bt^k$$

or

$$\ln\left(\frac{1}{1-X_v}\right) = Bt^k$$

Hence

$$\log \ln\left(\frac{1}{1-X_v}\right) = \log B + k \log t \quad (13)$$

Thus, by plotting the values of $\ln\left(\frac{1}{1-X_v}\right)$ and t in log-log scale, a straight line should be obtained, with B being the intercept and k being the slope of the straight line. An example is shown in Figure 12. The deviations from the straight-line relationship at long times will be discussed later.

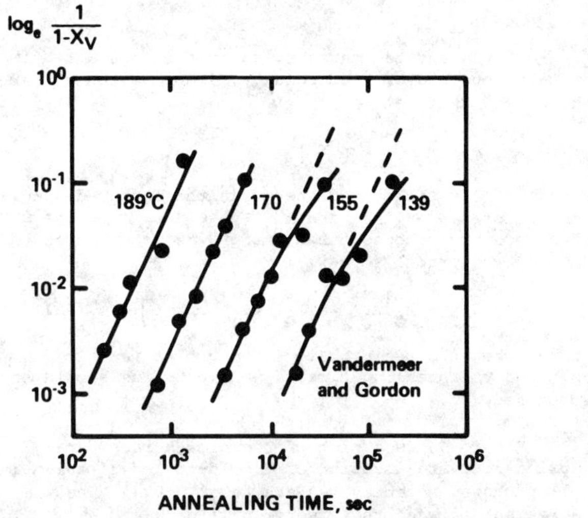

Figure 12. Recrystallization kinetics for zone-refined aluminum containing 0.0068 atomic-% copper deformed 40% by rolling at 0°C then isothermally annealed at temperature indicated. Vandermeer and Gordon (63).

<u>Concurrent Recovery During Recrystallization</u>. As mentioned in the Introduction, some overlapping between the stages of recovery and recrystallization and between those of recrystallization and grain growth usually occurs. Thus concurrent recovery occurring in the unrecrystallized matrix while recrystallization progresses will continuously reduce the driving energy available for the growth of the recrystallized grains. In a study of the recrystallization of zone-refined aluminum and its dilute alloys with copper, Vandermeer and Gordon (63) observed a retardation of recrystallization in the specimens. This is typically shown in Figure 12 for one of their Al-Cu (0.0068 at %) alloys, deformed 40 percent by rolling at 0°C, then annealed isothermally at the temperature indicated.

* Or by $dX_v/dt = GA_f$, where A_f is the free or unimpinged grain-boundary area per unit volume (77).

The recrystallization kinetics of these specimens can be described fairly well by Equation (12) with k = 2*, except for the two low annealing temperatures (139 and 155°C) at long times, where a retardation of recrystallization caused by a decrease of the growth rate was observed. By correlating the observed decrease in growth rates as a function of annealing time with isothermal calorimetric measurements of the release of stored energy, these authors were able to show that the decrease of the growth rate was a consequence of concurrent recovery in the unrecrystallized matrix resulting in a decrease of the driving energy for recrystallization. This is a good example showing that recovery and recrystallization are competitive processes in annealing. The relative significance of the contributions from these two competitive processes to the annealing process depends on material, purity, processing conditions and other factors.

Similar observations of a decreasing growth rate with increasing time during recrystallization were reported by various authors for other metals or alloys (84-87). The growth rate of the grains, except during early periods of recrystallization, was found to decrease linearly with the reciprocal of the annealing time at all temperatures and for various solute concentrations, such as in Fe-Mn and Fe-Mo alloys (84).

The effect of concurrent recovery during recrystallization on the growth rate of the recrystallized grains was analyzed by Li (46) who used simple second-order kinetics for the recovery of the stored energy in the deformed specimen. If the growth rate G is a linear function of the driving force ΔF, namely,

$$G = k_G \Delta F \qquad (14)$$

where k_G is a rate constant, and ΔF changes with time because of recovery according to simple second-order kinetics, that is,

$$\Delta F^{-1} - \Delta F_o^{-1} = k_R t \qquad (15)$$

where ΔF_o is the stored energy in the deformed specimen before annealing, and k_R is the rate constant for the recovery process, then the growth rate expressed in the reciprocal form will be

$$G^{-1} = k_G^{-1} \Delta F^{-1} = k_G^{-1}\left(\Delta F_o^{-1} + k_R t\right) \qquad (16)$$

Thus, when the reciprocal of the growth rate is plotted vs annealing time, a straight line should be obtained.

Figure 13 is a plot of the growth-rate data during recrystallization of a cold-rolled titanium alloy showed recently by Rath et al. (87). The agreement between the experimental data and Equation (16) is excellent. Li pointed out that if the growth rate is plotted against time on a log-log scale, the growth rate becomes linear with t^{-1} only at long times, but it deviates quite appreciately at short times. This is what was actually observed by various authors (84,85,87). However, when the reciprocal of the growth rate is plotted against time, the correlation is entirely linear (Figure 13). If the growth rate is a high power function of the driving force, $G = k_G(\Delta F)^m$ with m >> m Li showed that, at long times, d(ΔF)/dt is still linearly proportional to t^{-1}. All these seem to indicate that concurrent with recrystallization there exists a simple second-order kinetics for recovery.

* This corresponds to 2-dimensional recrystallization according to the Avrami equation (Table I). In these specimens, Vandermeer and Gordon observed that the recrystallized grains were nucleated as colonies of contiguous grains strung out along the deformed matrix grain edges at zero time of the annealing process.

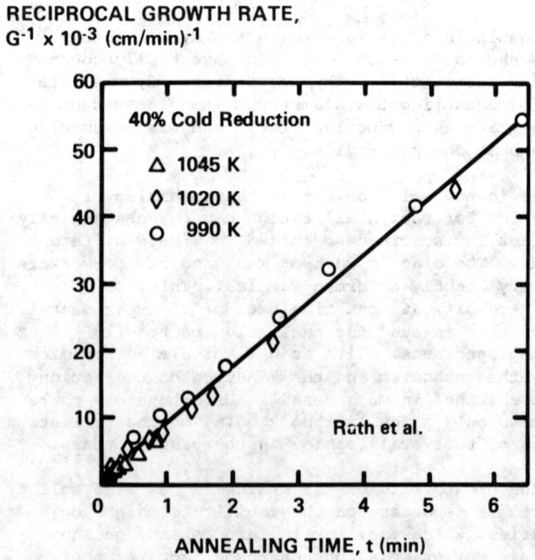

Figure 13. Growth rate of recrystallized grains as a function of annealing time during isothermal recrystallization at various temperatures for a beta Ti alloy cold-rolled 40%. The correlation suggests a simple second-order kinetics of recovery concurrently occurring with recrystallization. Rath et al.(87).

Nucleation by Subgrain Coalescence. From a study of the annealing behavior of cold-rolled silicon-iron single crystals, Hu (88,89) observed that recrystallization nuclei were formed by coalescence of the subgrains in the microband region, where the lattice curvature and dislocation density were high. This microband region constitutes a very small volume fraction of the material located between the main deformation bands. There is a large orientation difference between the main deformation bands. Thus, the change in orientation between the main deformation bands is accommodated by the microband segments, hence the term "transition band" has also been used (90).

A model of nucleation by subgrain coalescence was then proposed (52,91) as shown in Figure 14. In this

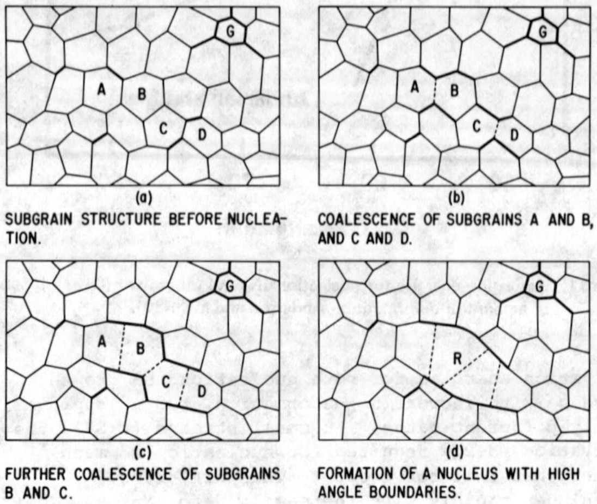

Figure 14. Schematic representation for the formation of a recrystallized grain by the coalescence of subgrains. Hu(52).

schematic illustration, the angle of misfit of the subgrain boundaries is represented by the thickness of the lines; the heavier lines indicate larger angles of misfit between the subgrains. The dotted lines between two subrains indicate dislocations that have not yet moved out completely. The driving force for the coalescence of two subgrains by eliminating the low-angle boundary between them comes from the interaction energy between this low-angle boundary and the connecting boundaries of larger angle of misfit. When the dislocations of the low-angle boundary move out of their own boundary and join other dislocations in the connecting boundaries of higher angle of misfit, a gain in energy results. This dislocation movement can be accomplished by climb, which requires lattice diffusion, or by glide if the temperature is not sufficiently high for diffusion to occur.

It is considered that nucleation by subgrain coalescence is probably a natural consequence of the theoretical requirements for nucleation in recrystallization: (1) The subgrain boundaries, being a few degrees in misorientation and being a low-angle boundary of the general type, should be highly immobile according to the theory of dislocations. (2) For a stable nucleus of recrystallization the nucleus must be above the required critical size. (3) The nucleus must be capable of growing with a significant velocity by having a high-angle high-mobility boundary. The proposed model for nucleation by subgrain coalescence satisfies all these considerations. The atomistic mechanism for coalescence of subgrains was analyzed by Li (92), and a schematic representation of the process is shown in Figure 15. The coalescence event is accomplished by atom diffusion along the boundaries from the shaded areas to the corresponding open areas.

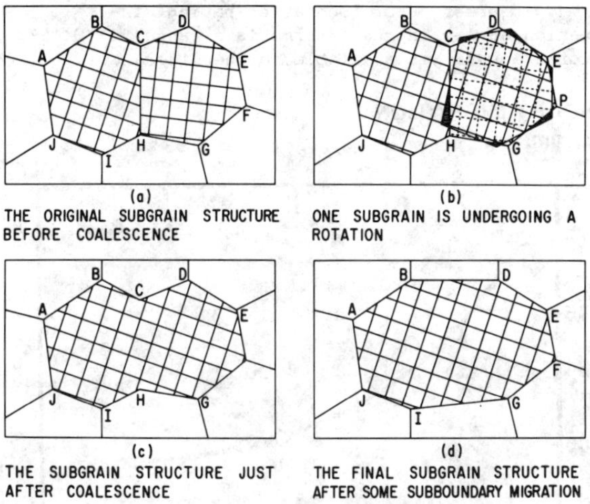

Figure 15. Schematic representation of subgrain coalescence by subgrain rotation. Li(92).

Since then, the coalescence of subgrains during recrystallization in a variety of metals or alloys has been reported by various authors, including very convincing evidence produced by direct observations on heated specimens in the electron microscope (93). However, until very recently the coalescence of subgrains has been a controversial subject.

Significant progress in the study of nucleation in recrystallization has been made during the past few years by a team of researchers (principally Jones, Ralph and Hansen) from the Risø National Laboratory of Denmark and Cambridge University of England. These workers produced numerous excellent photomicrographs showing the expected interactions of low-angle subgrain boundaries with the connecting high-angle boundaries relevant to the process of coalescence among subgrains. Their work (94) showed evidence of subgrain coalescence even in the case of grain-boundary nucleation similar to the bulging mechanism of nucleation reported by Bailey and Hirsh (95). The bulging mechanism of nucleation is essentially the same phenomenon described by Beck (43) as recrystallization by strain-induced boundary migration in moderately deformed specimens. An extensive review of the grain-boundary phenomena during nucleation was given recently by Jones (96). The problems of nucleation will be discussed here only briefly.

One of the main skepticisms about subgrain coalescence has been the possible uncertainty of the experimental evidence by transmission electron microscopy (TEM) concerning the disappearance of a subgrain boundary in thin-foil specimens. It may be argued that the dislocations in the subgrain boundaries may have partially been lost through the foil surfaces during electrolytic thinning of the specimens, whereas in actual bulk material this is unlikely to happen. Objections have also been raised that showing the lack of contrast between two neighboring grains, even after various tiltings of the specimen, is not sufficient evidence to indicate the coalescence of two subgrains because the contrast is highly sensitive to very small changes in crystal orientation. Furthermore, the subgrain growth rates experimentally observed by some authors (97,98) were orders of magnitude faster than the theoretically calculated rates based on coalescence between subgrains (92).

To reaffirm the correspondence between TEM observations on thin-foil specimens and conventional microscopy on bulk specimens, in particular with regard to the dislocation substructures in deformed crystals after recovery, Jones (96) showed the transmission electron micrograph in Figure 16 for a specimen of an aluminum-alumina alloy cold-rolled

Figure 16. Change in dislocation spacing along a disintegrating low-angle boundary in an Al-Al_2O_3 alloy cold-rolled 50% and annealed at 300°C for 23 min. Larger dislocation spacing near the disintegrating end of the boundary (at A) than in the vicinity of the triple junction (at B). Jones(96).

50 percent and annealed at 300°C for 23 minutes. The dislocation spacing is larger near the disintegrating end of the boundary (at A) than in the vicinity of the triple junction (at B). The same features were observed by Dunn and Hibbard (99) in polygonized Si-Fe single crystals after using etch-pitting techniques. Figure 17 shows the change in the unit-mesh size of the hexagonal network of dislocations between two triple junctions A and B in an Al-Al_2O_3 alloy cold-rolled 50 percent and annealed at 300°C for 5 minutes. In this particular case, no disintegration of low-angle boundary has occurred. The difference in the unit-mesh size along the boundary between the two triple junctions indicates the interactions between the low-angle boundary and the connecting boundaries. Such interactions constitute the basis for the coalescence of subgrains.

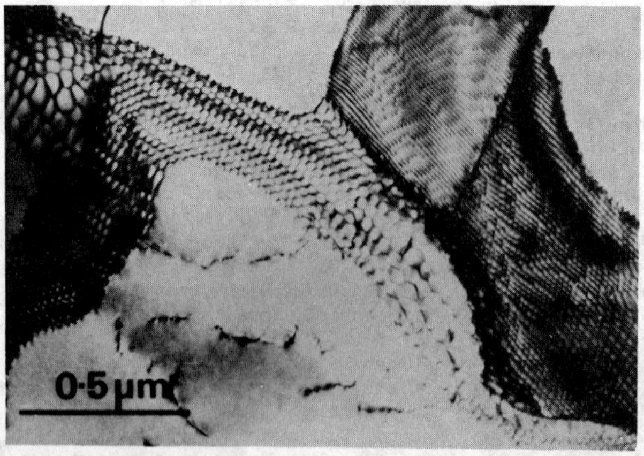

Figure 17. Change in unit-mesh size of a hexagonal network of dislocations between two triple junctions (A and B) in an Al-Al_2O_3 alloy cold-rolled 50% and annealed at 300°C for 5 min. Jones(96).

Figure 18 is a schematic model presented by Jones, Ralph and Hansen (94) showing subgrain coalescence that leads to the nucleation of recrystallization at an

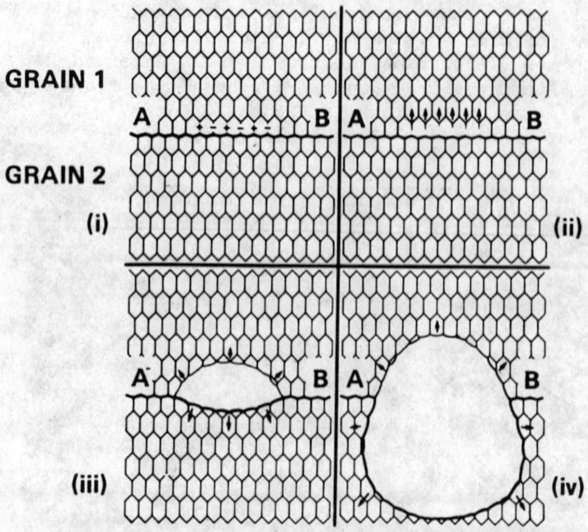

Figure 18. Schematic diagrams illustrating subgrain coalescence that leads to the nucleation of recrystallization at an original grain boundary (along AB) in a polygonized microstructure. Jones, Ralph and Hansen(94).

original grain boundary (along AB) in a polygonized microstructure. In (i) sub-boundaries adjacent to the original high-angle grain boundary in Grain 1 are of essentially opposite character. In (ii) subgrain coalescence commences in Grain 1 adjacent to the original high-angle grain boundary. In (iii) the large subgrain formed adjacent to the original high-angle grain boundary begins to grow into both grains. In (iv) rapid migration of the segment of high-angle grain boundary leads to the formation of a nucleus, which straddles the original boundary position.

During the First Risø International Conference on "Recrystallization and Grain Growth of Multiphase and Particle Containing Materials" held in September 1980 more evidence of nucleation by subgrain coalescence was presented, including nucleation at second-phase particles and in the transition band developed at the original grain boundaries (100). Nucleation by subgrain coalescence was endorsed unanimously at that conference.

For materials with low stacking-fault energies, the deformed structure consists of high densities of stacking faults, microtwins and high-angle boundaries. Recrystallization appears to occur before a well-defined subgrain structure is developed by the migration of high-angle boundaries (101). Considering the fact that high-angle boundaries in a heavily deformed specimen must be ill-defined and diffuse, and the very recent finding that subgrain coalescence is involved even in the microbulging type of nucleation (94), it seems unlikely that recrystallization could occur immediately by high-angle boundary migration without any coalescence event occurring in the incubation period. More detailed studies appear to be highly desirable. Another problem, as mentioned a little earlier, lies in the great discrepancies between the subgrain growth rate experimentally determined and the subgrain coalescence rate theoretically calculated. It seems that a reexamination of the theoretical model is necessary for the problem to be resolved, and more quantitative studies of the coalescence kinetics are desired.

Mobility of Grain Boundaries. The mobility of grain boundaries depends on the relative orientation of the crystals across the boundary. This is a clear indication that the atomistic structure of the boundary has a significant influence on the mobility of the boundary because the boundary migration is a consequence of the net transfer of the atoms from one crystal lattice to the other across the boundary. For simple tilt low-angle boundaries, Bainbridge et al. (102) measured the mobility of the boundary as a function of misorientation. As shown in Figure 19 the migration rate of this type of boundaries decreases with increasing angle of misorientation; it becomes fairly low when the misorientation angle is larger than 1.0 or 1.5 degrees. For boundaries having misorientation angles in the intermediate range, $2° < \theta < 15°$, no information is yet available in the literature.

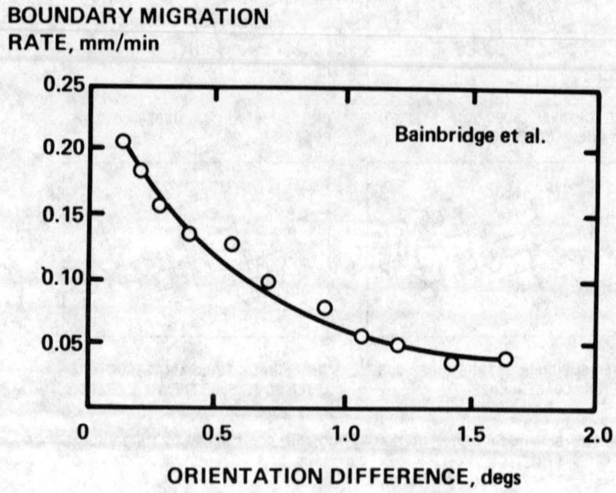

Figure 19. Mobility of simple tilt boundaries as a function of misorientation angle for zinc single crystals under a shear stress of 9.19 psi (63.4 kN/m^2) at 350°C. Bainbridge et al.(102).

Most research results on the orientation dependence of boundary mobility have been confined to the category of high-angle boundaries having high migration rates because such information can be obtained by relatively simple experiments in so-called "growth selection" studies. The technique usuallly consists of introducing a small amount of plastic strain into a single crystal, insufficient to cause spontaneous nucleation but adequate for providing a driving force for rapid growth. Numerous recrystallization nuclei are then produced artifically by a highly localized deformation followed by a suitable annealing. Among the numerous artificially nucleated grains having a wide variety of orientations with respect to the orientation of the matrix, the ones that grow with high rates will forge ahead and cut off all others at an early stage of the growth competition, eventually consuming the matrix crystal. The orientation relationships between the successfully grown grains and the matrix crystal establish the types of boundaries having high mobilities.

A large number of such studies have been conducted for various metals by various researchers (103-107). The main results can approximately be described as follows: For fcc metals a [111] rotation of 30 to 40°, for bcc metals a [110] rotation of 20 to 30°, and for hcp metals a [0001] rotation of 30° correspond to boundaries of high mobility. Rotations around the [100] axis for cubic, and around the [10$\bar{1}$0] axis for hexagonal, metals were also observed. Such orientation relations led to the concept that boundaries with high mobility will exhibit high densities of coincidence sites in the boundaries. However, boundaries that exhibit high mobilities are only near to the orientations corresponding to the exact coincidence boundaries. Besides, coherent twin boundaries, which are 100 percent coincident, are practically immobile.

Driving-Force Dependence of Boundary Velocity. The problems of boundary migration and the effects of solute or impurity on boundary velocity were treated independently by Cahn (108) and by Lucke and Stuwe (109), on the assumption that solute atoms exert a drag force on the boundary. They arrived at essentially the same general type of equation to relate the boundary velocity, driving force and solute concentration as follows:

$$\Delta F = \lambda V + \frac{\alpha C_o V}{1+\beta^2 V^2} \qquad (17)$$

where ΔF is the driving force, λ is the reciprocal of the intrinsic boundary mobility, C_o is the bulk concentration of the solute, and α and β are solute-drag parameters related to temperature, the diffusivity profile of solute atoms, and the interaction energy profile of the solute atoms with the boundary. The last two items are difficult to define, because the atomic structure of the boundary, particularly when in motion, is unknown.

The theory indicates that at very low velocities the boundary migration rate is linear with the driving force for a given solute concentration, and that at a very high velocities the boundary migration rate is also linear with the driving force provided that the driving force is high and the solute concentration is low. At intermediate velocities the driving-force dependence of the boundary velocity varies. The solute-drag effect in this "transition region" varies with velocity and decreases towards zero at high velocities. The solute-drag theory of Cahn (108) or of Lucke and Stuwe (109) is essentially for low concentrations. Lucke et al. (110) later made further calculations on the atomistic model. The solute-drag problem was recently treated by Hillert and Sundman (111) to include high solute concentrations.

The driving-force dependence of boundary velocity in wedge-shaped bicrystals of zone-refined aluminum and Al-Mg alloys with preselected orientations was studied by Rath and Hu (112,113) in considerable detail. The boundary velocity, V, was found to be a power function of the driving force, ΔF, such as

$$V = M(\Delta F)^m \qquad (18)$$

where M is the mobility of the boundary. The results, together with data by others, are shown in Figure 20. The overall picture of the boundary velocity and driving-force correlations is in agreement with the solute-drag theory for boundary migration. Very recently, Gleiter (114) reported that vacancy drag

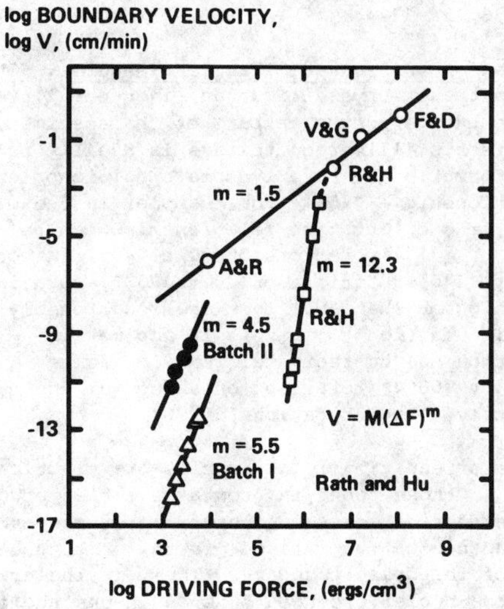

Figure 20. Dependence of boundary velocity on driving force in zone-refined aluminums extrapolated or interpolated to 100°C. (○) Frois and Dimitrov(117), Vandermeer and Gordon(116), Rath and Hu(118), and Aust and Rutter(106); (□) Rath and Hu(112); (△ and ●) Rath and Hu(113).

results in nonlinear dependence of boundary mobility on driving force. A comprehensive review of the high-angle grain-boundary phenomena was given by Gleiter and Chalmers (115) in 1973. Diffusion induced grain boundary migration has recently been observed in several systems. The mechanism of this newly observed phenomenon has been analyzed by Balluffi and Cahn (150) very recently.

Effect of Second-Phase Particles. It seems obvious to expect that the presence of large, hard and widely spaced particles of a second phase tends to enhance nucleation for recrystallization because more heterogeneous deformation occurs in the immediate vicinity of the particles. According to Humphrey (119), the critical size of the particle for effecting nucleation is a function of deformation. The critical size increases with decreasing strain. For a number of alloy systems, the minimum size of particle at which recrystallization initiates was found to be around 1 μm. The number of recrystallization nuclei that can form at a particle depends on the size of the particle. It has been known that a finely dispersed second phase has a strong effect in inhibiting recrystallization, in particular, nucleation (120). How such a dispersion influences the deformation structure, hence, subsequent nucleation for recrystallization is largely unclear.

Limited information on local lattice rotations at nondeformable particles subjected to tensile deformation indicates that the misorientation increases with strain, and that for a given strain the misorientation decreases with decreasing particle size. For particles smaller than 0.1 μm the misorientation is very small (121). This is consistent with the observations that fine dispersions tend to produce somewhat more uniform distributions of dislocations with a reduced mean misorientation between adjacent subgrains (122). According to the subgrain-coalescence model for nucleation, this lack of local misorientation among the subgrains would mean a reduced driving force for coalescence; hence, the formation of viable nuclei would be inhibited accordingly. It is well known that fine particles can effectively pin the boundaries. Thus the growth of the recrystallized grains would be hindered. The recrystallization behavior of two-phase alloys has been reviewed extensively by several authors (76,121-124).

Formation of Recrystallization Textures. It is well known that recrystallization, in general, results in a large change in the texture of the specimen, and that the recrystallization texture is usually related to the deformation texture by some specific orientation relationships (125). For example, in fcc metals or alloys, the orientation relationships can be described by [111] rotations of 30 to 40°, in bcc metals, by [110] rotations of 20 to 230°; rotational relationships of the [100] type are occasionally observed in addition to these. In hcp metals, reorientation due to recrystallization can be described by [0001] rotations of about 30° and by approximately [10$\bar{1}$0] rotations of 90°.

These orientation relationships are practically identical to those found in growth-selection studies, and can be considered as a consequence of the orientation dependence of the rate of growth, or of the mobility of the grain boundary. Thus, if the orientations of the nuclei produced by spontaneous nucleation in a heavily deformed specimen were proved to be random, then the oriented-growth mechanism alone would be the sole governing factor in the formation of recrystallization textures. However, such proof is very difficult to achieve because a statistically convincing study of the orientation aspects of nucleation is a tremendous task.

From a limited study of the orientation aspects of nucleation in heavily rolled copper (31), the results indicated that it is possible for the nuclei to adopt only ranges of orientations that correspond to the transitional orientations between the matrix deformation texture components. Therefore, for the formation of a sharp cube texture in heavily rolled then recrystallized copper specimens, the oriented-growth mechanism must have played an important role, even though the nuclei orientations are not random. On the other hand, in many instances when the recrystallization textures are not appreciably sharper than their parent deformation textures, for example, in many bcc metals and mild steels, the oriented-nucleation mechanism has probably played an important role in the formation of the recrystallization textures.

Summary on Recrystallization. The variables that affect recrystallization and the classic theory of recrystallization kinetics have been reviewed briefly. In contrast to recovery, recrystallization proceeds by the nucleation and growth process. Concurrent recovery in the unrecrystallized matrix during recrystallization affects the growth rates of the grains. Significant progress has been made in the study of nucleation for recrystallization, and it appears that in various metals and alloys nucleation occurs by subgrain coalescence. The nucleation sites are in transition bands (or microbands), at grain boundaries, and at coarse second-phase particles. The orientation dependence of boundary mobility, the relation between boundary velocity and the driving force, and the effect of second-phase particles on recrystallization have been discussed. Finally, the formation of recrystallization textures based on the mechanisms of nucleation and growth of the recrystallized grains has been briefly introduced.

Grain Growth

Further annealing after recrystallization is complete causes an increase in the average size of the grains. The process, known as "grain growth," is accomplished by the migration of grain boundaries. As mentioned previously, grain growth can be further classified into two types, "normal or continuous grain growth" and "abnormal or discontinuous grain growth," also known as "secondary recrystallization." The driving energy for grain growth is the interfacial energy of the grain boundaries. The magnitude of this driving energy is very small. An estimate of the energy release for a tenfold increase in the average grain diameter from 10 to 100 μm is of the order of 0.2 cal/mole (43). This is about 1/100 of the stored energy of cold work that is released prior to or during recrystallization. In contrast to the growth of the new grains in recrystallization, the boundary migrates toward its center of curvature in grain growth.

Normal or Continuous Grain Growth. Smith (126) has shown that grain growth in metals can be simulated by the cell growth in soap froth, in which the shapes of the cells are controlled by the surface tension of the cell walls. Grain growth is a consequence of the interaction between the topological requirements and the geometrical needs of local surface-tension equilibrium. In a two-dimensional array of grains, Burke and Turnbull (73) have described the geometry of grain growth in some detail. It can be stated briefly as follows: If all the cells had exactly six sides it would be possible for every set of three grain boundaries to meet at a triple-junction point at dihedral angles of exactly 120°, and for all grain sides to be straight, even though the cell diameters were not the same. This would be a stable configuration; there would be no tendency for the grain boundary to migrate.

However, if a five-sided grain is introduced, it is geometrically necessary to produce one with seven sides, so that the average number remains at six. Neither the five- nor the seven-sided grain will have straight sides if the boundaries meet at angles of 120°. Grains having less than six sides will have their faces concave inwards, whereas those having more than six sides will have their faces concave outwards. As a result of these curvatures, the grains with less than six sides will tend to shrink and those with more than six sides will tend to grow by the migration of the boundary towards its center of curvature. When a grain with five sides loses one side upon shrinking to become a four-sided grain, it has a higher probability to further shrink into a three-sided grain. A three-sided grain will then vanish by inward migration of its three faces. Thus, in a constant volume, as the number of grains decreases the average grain size increases.

The kinetics of isothermal grain growth for pure metals and single-phase alloys were first established by Beck (43) and co-workers to obey the empirical equation

$$D = Kt^{\eta} \tag{19}$$

where D is the average grain diameter, t is annealing time, and K and η are parameters depending on material and temperature. Thus, when log D is plotted against log t, a straight line should be obtained, with K being the intercept and η being the slope. A typical example is shown in Figure 21 for grain growth in zone-refined iron (127). The annealing time was the actual time at temperature when the cold-rolled sheet specimens were annealed. Except for very short and very long annealing times, the data are very well represented by a straight line for each temperature. The slope of the line, or the η value, increases from 0.26 at 550°C to approximately 0.5 at 800 or 850°C, a temperature only about 0.6 T_m which is surprisingly low for η = 0.5 (see discussions later).

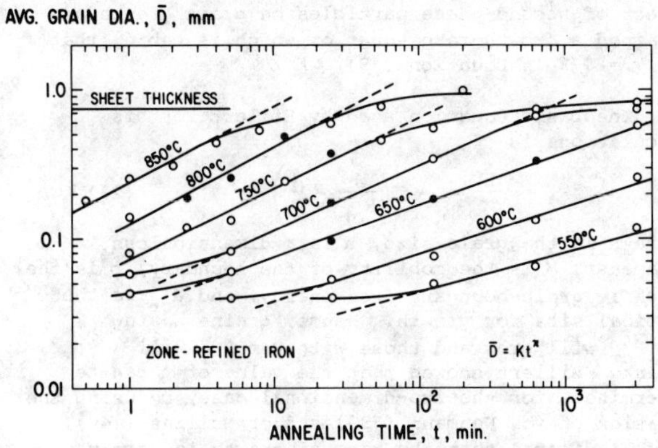

Figure 21. Grain growth in zone-refined iron during isothermal anneals. Filled circles represent specimens for which statistical analysis of grain-size and grain-shape distributions was performed. Hu(127).

The deviations from the straight-line relationship for very short annealing times at low temperatures are understandable because Equation (19) requires that D = o at t = o, which is not the case in grain growth. For this reason, the grain-growth equation is frequently expressed in the form

$$D^{1/\eta} - D_o^{1/\eta} = C \cdot t \qquad (20)$$

where D_o refers to the initial grain diameter before growth, and C is a parameter similar to K in Equation (19). Both Equation (19) and Equation (20) are essentially equivalent because with increasing annealing time D_o becomes negligible in comparison with D. Thus, the same relation between D and t is eventually described by both equations. The equivalence of these two equations was shown previously by Beck (128) and later by Murty (129).

The deviations from the straight-line relationship after long annealing at high temperatures appear to be related to the limiting effect of the specimen thickness (43), which may be interpreted as a consequence of thermal grooving of the grain boundaries (130).

As described in the Introduction and Nomenclature, one of the characteristics of normal grain growth is that the <u>grain-size and grain-shape distributions</u> are essentially <u>invariant</u>. Statistical studies of the planar grain-size and grain-shape distributions were conducted on a number of specimens, which are designated by filled circles in Figure 21. These specimens provided several sets of data to be examined systematically: two isothermal sets (650 and 800°C), one isochronal set (25-minute anneals at three temperatures), and one isometric set (D ≅ 0.2 mm from three different anneals). Such comparisons have never been made previously. Results from the 650°C isothermal set are shown in Figure 22(a) and (b) for the grain-size and the grain-shape distributions, respectively. These distribution curves are quite similar to those for high-purity Al (131) and Chempurtin* (132). Studies of the isochronal and isometric sets of data yielded similar results; thus, the distribution curves shown in Figure 22 are typical.

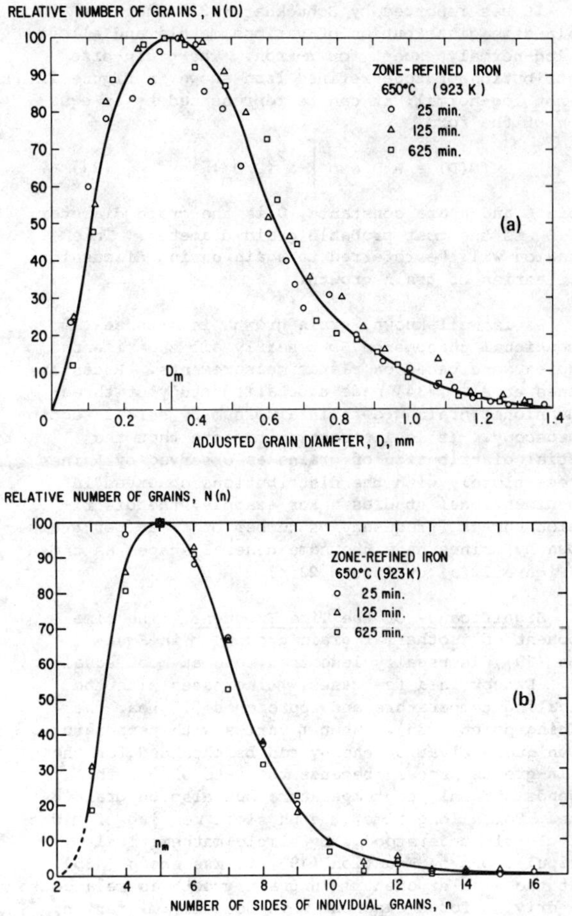

Figure 22. (a) Grain-size distribution, using a scaler-adjusted grain diameter for each specimen; and (b) grain-shaped distribution, using the number of sides of individual grains. Zone-refined iron during isothermal grain growth at 650°C. Hu(127).

The invariance of the size and shape distributions of the grains indicates that normal grain growth is, to a first approximation, equivalent to photographic enlargement. The similarity between the grain-size and grain-shape distribution curves suggests that the number of sides per grain and the grain diameter may be linearly related. As shown in Figure 23, this is so. It can also be noted that a shrinking grain vanishes at n = 3, and that the average-size grain has n ≅ 6. These agree well with the statement of Smith (126).

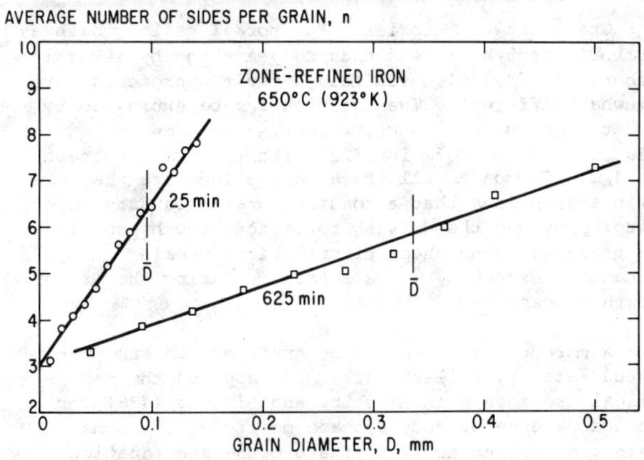

Figure 23. Linear correlation between the average number of sides per grain and the grain diameter. Hu(127).

---

* A trade name for high-purity tin.

It was reported by Schückher (133) that the grain-size distribution of various metals and alloys is log-normal, except for α-iron. The grain-size distribution of zone-refined iron shown in Figure 22(a) is not log-normal; it can be represented by an equation of the form

$$N(D) = A \cdot D \exp\left[-a^2(D/D_m)^2\right] \quad (21)$$

where A and a are constants, D is the grain diameter, and D_m is the most probable grain diameter. This equation will be referred to again during discussion of theories of grain growth.

As is well known, grain growth is a three-dimensional phenomenon, but nearly all experimental studies were based on planar measurements. Recently, Rhines et al. (134) made a detailed study of three-dimensional grain growth in aluminum by serial-section microscopy. It is interesting to note that the spatial distribution of grains as observed by Rhines agrees closely with the distributions observed in two-dimensional studies. For example, the distribution curves (frequency vs number of faces per grain) shown by Rhines have the same general shapes as those of Figure 22(a) or Figure 22(b).

Significance of the Time Exponent. The time exponent of isothermal grain growth, η in Equation (19), is usually less than, and at most equal to, 0.5. Except in a few cases, η increases with the annealing temperature and approaches 0.5 near the melting point (135). When η varies with temperature, no unique activation energy can be obtained for the grain-growth process because the rate of growth depends not only on temperature but also on grain size. For a long time, the physical meaning of η was not clearly understood. By simple mathematical manipulations on Equation (19), it was shown (135) that the time exponent η in grain growth is related to the driving-force exponent m in boundary migration, Equation (18), by the simple expression

$$m = (1/\eta) - 1 \quad (22)$$

Thus, a parabolic law of grain growth (η = 0.5) corresponds to a linear dependence of boundary velocity on driving force (m = 1). Grain-growth data in the literature clearly indicate that a parabolic growth law has rarely been observed experimentally; only in metals of ultra purity and at high annealing temperatures has the value of η reached 0.5. In other words, for most cases, the observed boundary velocity is in the transition region and is a power function of the driving force. Accordingly, the correlation between η and m indicates the consistency of grain growth with the solute- or impurity-drag theory for boundary motion.

Grain-Growth Theories. For normal grain growth a parabolic growth law was deduced years ago by several authors (43,73,136), even though their approaches were somewhat different. The equation can be expressed by $D = Kt^{1/2}$ or $D^2 - D_o^2 = Ct$, which is the same as Equation (19) or Equation (20) with the time exponent η = 1/2. Common to all these derivations are the basic assumptions that a constant grain-boundary free energy provides the driving force for growth, and that the grain size and shape distributions remain unchanged, except by a scale factor, during the grain-growth process.

A more complete theory of grain growth was presented later by Hillert (137). He applied the mathematical treatments of Lifshitz and Slyozov (138) for the coalescence of second-phase particles to normal grain growth, and arrived at a growth-rate equation which, again, is parabolic. By following the procedures of Lifshitz and Slyozov, a theoretical distribution function of the grain size in normal grain growth is also deduced. Along with a geometric-defect model of grain growth, Hillert further treated the effect of second-phase particles on grain growth and obtained a growth-rate equation which is cubic, that is, η = 1/3 in Equation (19).

The basic equation used by Hillert for his calculations is

$$\frac{dR}{dt} = \alpha M\sigma\left(\frac{1}{R_{cr}} - \frac{1}{R}\right) \quad (23)$$

where R is the grain size, α is a dimensionless parameter, M is the mobility of the boundary, σ is the specific grain-boundary free energy, and R_{cr} is the critical size for growth. Hence, grains having $R > R_{cr}$ will grow and those with $R < R_{cr}$ will shrink. Hillert showed that the value of α can be determined, for the two-dimensional case, by using the equation of von Neumann (139) and of Mullins (140), which indicates that the rate of growth is simply related to the average number of sides per grain, n, as follows:

$$\frac{dR}{dt} = \frac{M\sigma}{R}\left(\frac{n}{6} - 1\right) \quad (24)$$

By combining Equations (23) and (24), the number of sides per grain and the grain size are related by

$$n = 6 + 6\alpha\left(\frac{R}{R_{cr}} - 1\right) \quad (25)$$

For the whole specimen, the average number of sides per grain would be

$$\bar{n} = \frac{1}{N}\sum_N\left[6 + 6\alpha\left(\frac{R}{R_{cr}} - 1\right)\right] = 6 + 6\alpha\left(\frac{\bar{R}}{R_{cr}} - 1\right) \quad (26)$$

Thus, from Equation (25), Hillert obtained α = 1/2, when R = 0 and n = 3. Also, according to topological rules, the average number of sides per grain in a two-dimensional system is six (126). Hence from Equation (26), Hillert concluded that $\bar{R} = R_{cr}$ when $\bar{n} = 6$. The experimental observations shown in Figure 23 conform closely to Hillert's formulations, for example, n = 3 at D = 0, n = 6 at $D = \bar{D}$. It was also shown in the grain-growth data of zone-refined iron (127) that ΔF = 0 at $D = \bar{D}$. Despite all these agreements between experimental observations and Hillert's theory, the grain-size distribution of zone-refined iron shown in Figure 22(a) is in large disagreement with Hillert's theoretical distribution function (127).

Louat (141) considered that the necessary boundary motion in grain growth is a random process, and that the linear relation between grain size and the number of sides per grain is only statistically valid. It is significant to remark that in a polycrystalline aggregate grains represent only a certain geometrical configuration of grain faces. It is the faces rather than the grains that have physical significance in grain growth. Accordingly, grain growth can be viewed in terms of the growth of faces, which is clearly a statistical process. A necessary condition for this treatment to be valid is, of course, that the size and shape (number of sides per grain in two-dimensional, or number of faces per grain in three-dimensional) distribution functions for the grains should be the same. This has been substantiated by experiments (127,131,132).

The basic equation used by Louat in his analysis is

$$\frac{\partial f}{\partial t} = A\frac{\partial^2 f}{\partial x^2} \quad (27)$$

where f is the number of grains (faces), x is the linear dimension of the grain faces, t is time, and A is a rate factor independent of f, x, and t. The required solution for the above equation, which satisfies the boundary condition that grains are eliminated but not created, that is, f(o) = o, is

$$f(x,t) = At^{-3/2}\ Cx\ \exp\ (-x^2/4At) \qquad (28)$$

where C is a constant chosen to give the distribution at time $t = t_o$. Equation (28) is essentially of the same basic form as Equation (21), which was determined experimentally (127). From this, Louat obtained a growth-rate equation that is parabolic. That the same distribution function can be applied to both two-dimensional and three-dimensional systems was also proved theoretically by Louat (141).

Figure 24 shows the theoretical distribution curves with experimental data points for zone-refined iron (127). The agreement between Curve 1 and experiment is very good. Curve 2 represents the log-normal type of distribution.

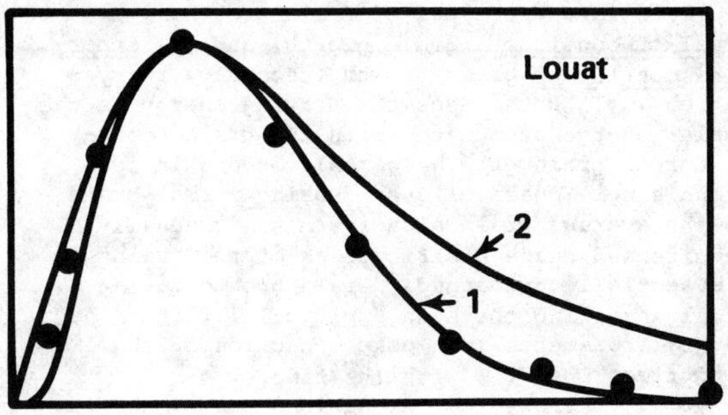

Figure 24. Theoretical distribution curves and experimental data points for zone-refined iron from Hu's work. Curve 1 represents $f(x) = X \exp(-X^2/2)$, and Curve 2 represents $f(x) = \exp[-(\ln X)^2]$, the log normal type of distribution. Louat(141).

Abnormal Grain Growth or Secondary Recrystallization. This usually occurs under one or more of the following structural conditions: (1) there is a finely dispersed second phase; (2) there is a strong single-orientation texture; and (3) the primary grains have reached the limiting size imposed by the sheet thickness. If any of the above conditions is strong enough to inhibit general growth of the grains, abnormal growth occurs when the temperature is sufficiently high to allow a few grains to overcome the inhibitory force and grow disproportionately. A prerequisite for abnormal grain growth or secondary recrystallization is, therefore, the matrix stability.

Like primary recrystallization, the process of abnormal grain growth or secondary recrystallization consists of nucleation and growth. An incubation period is usually observed, and the increase in the volume fraction of the secondary grains as a function of isothermal annealing time is the sigmoidal type, as shown in Figure 25 for the secondary recrystallization

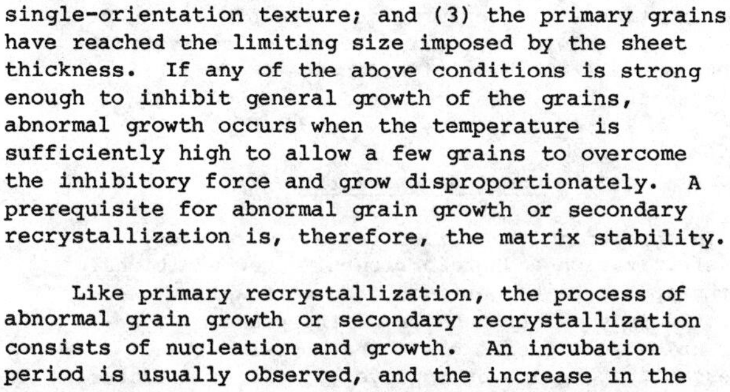

Figure 25. Secondary recrystallization for the formation of the cube texture in 3% Si-Fe upon isothermal annealing. Assmus et al.(142).

in 3 percent Si-Fe to the cube texture upon isothermal annealing (142). The rate of growth of the secondary grains in silver during isothermal anneals after the incubation period is linear (143). Thus, the kinetics for primary and secondary recrystallizations are quite similar. The analogy between these two processes was emphasized by Smith (126) and by Burke and Turnbull (73). However, the driving energy for secondary recrystallization, being the interfacial energy of the grain boundaries in the primary matrix, is much smaller than that for primary recrystallization.

Because the secondary grains and the primary matrix grains that have reached the limiting size (condition 3 as mentioned earlier) are several times larger than the thickness of the sheet, the free-surface energies of the grains become important, and may contribute significantly to the driving force for the two-dimensional growth of the secondary grains. This free-surface energy should be orientation dependent, that is, dependent on the crystallographic plane of the grain surface, and may also be sensitive to the composition of the furnace atmosphere. Dunn and Walter (144) gave the relative contributions of surface and grain boundary energies to the driving force of a secondary (or potential secondary) grain in a weakly textured matrix as follows:

$$\frac{\text{Grain-boundary-energy term}}{\text{Surface-energy term}} = \frac{\gamma_b/R_M - \gamma_b/R}{2\Delta\gamma_s/a} \qquad (29)$$

where γ_b is grain-boundary energy per unit area, R is the radius of the secondary (or potential secondary) grain, R_M is the average grain radius in a stable matrix, $\Delta\gamma_s$ is the difference in surface energy, and a is the sheet thickness. The ratio is zero at $R = R_M$ and is a maximum when $R >> R_M$.

The effect of second-phase particles on grain growth was analyzed by Hillert (137), who used the hypothetical back-stress term assumed by Zener (149), $3f\sigma/4r$, for particles of uniform size r and a total volume fraction f (σ is the specific grain-boundary energy). On the basis of his defect model for grain growth, Hillert predicts that abnormal grain growth should occur if three conditions are simultaneously satisfied: (1) Normal grain growth is inhibited by the second-phase particles. (2) The average grain size is smaller than 1/2z, where $z = 3f/4r$. (3) There is at least one grain much larger than the average. Gladman (145) treated the problem by considering the energy changes associated with grain growth and unpinning. For secondary recrystallization to occur, not only must the growth of grains decrease the energy of the system, but the energy release must also be sufficient to cause unpinning. Abnormal growth occurs when the coalescence of particles causes the particle size to exceed a critical value.

Secondary recrystallization is important in commercial applications. For example, the tungsten filament of incandescent lamps requires secondary recrystallization treatment to grow large grains for improved service life. The grain-oriented 3 percent Si-steel sheets are finally given a high-temperature anneal to effect secondary recrystallization for the development of Goss texture or the cube texture for electromagnetic applications. An extensive review on secondary recrystallization in sheet materials was given by Dunn and Walter (144).

Texture Development in Grain Growth. In general, during normal grain growth after recrystallization is complete, the change in texture is gradual. If the primary recrystallization texture consists of several recognizable components, as is commonly the case, normal grain growth may strengthen one of the components and weaken the others. This depends, obviously, on which grains have the more favorable conditions, such as the size or boundary curvature of the grains, for growth at the expense of the others. A common example is shown in Figure 26 for a low-carbon deep-drawing sheet steel during the box-annealing treatment (146). Shortly after recrystallization was complete (550°C), the average grain diameter was about

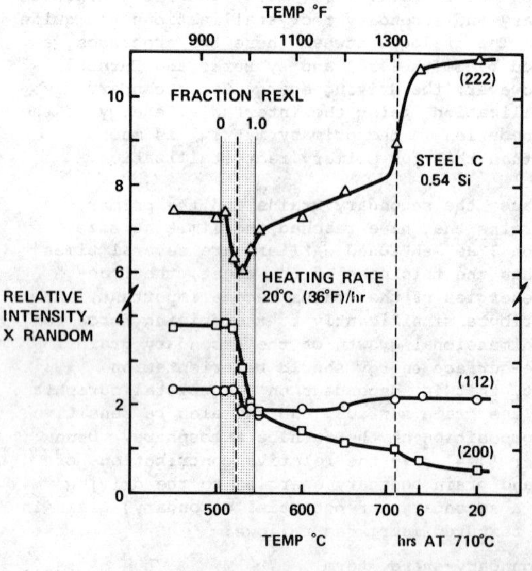

Figure 26. Changes in the relative intensity of the texture components in cold-rolled sheet of low-carbon phosphorus-containing steel during box annealing. Michalak and Hu(146).

5 µm. With continued heating at a rate of 20°C/hour to 710°C, the average grain diameter increased to approximately 12 µm. After 20 hours at 710°C, the grains increased further to an average diameter of 20 µm. During these periods of grain growth, the (222) texture component increased its intensity substantially, the (200) component decreased slightly, whereas the (112) component remained practically unchanged. Thus, the growth of the (222) grains occurred at the expense mostly of the high-index oriented grains and some of the (200) oriented grains.

In association with abnormal grain growth or secondary recrystallization, a large reorientation usually occurs. For example, in fcc metals or alloys with a single orientation cube texture, normal grain growth is impeded by the very low mobility of the grain boundaries because these primary grains are separated from each other mostly by low-angle boundaries. Only a few sufficiently misoriented grains with a sufficiently large size are able to grow, resulting in a large change in texture. In silver or α-brass, grain coarsening also results in a large change in texture. The new textures are usually related to the primary recrystallization textures by [111] rotations of 30 to 40° in a direction opposite to that occurring during primary recrystallization. Thus, the nature of secondary-recrystallization textures is essentially similar to that of the corresponding deformation textures (125). Secondary grains formed in thin sheets of pure platinum upon annealing at a high temperature have (111) orientation, as the surface energy of these grains is the lowest (147).

In bcc metals or alloys, secondary-recrystallization textures in 3 percent Si-Fe sheets have been studied most extensively. The development of Goss texture, or (110) [001], depends critically on second-phase dispersion during secondary recrystallization. Generally, MnS is believed to be responsible. Other second-phase inhibitors can effect similar results, although the required temperatures for secondary recrystallization differ. More recent developments of very high permeability and low core-loss silicon steels have made use of AlN, in addition to MnS, for the control of texture to within a few degrees of the (110) [001] orientation (148). In the absence of second-phase inhibitors, secondary recrystallization to a (110) [001] texture can be produced in thin sheets of high-purity iron and 3 percent Si-Fe, wherein normal grain growth has been impeded by the sheet thickness. Under these circumstances the relative surface energies of the grains play a determining role in the selection of secondary grains (144). By proper control of prior mechanical processing, which determines the direction of the preferred orientation, and of the furnace atmosphere in the final high-temperature anneal, which selects the plane of the grains, Si-Fe sheets with the (100) [001] cube texture can be produced.

Summary on Grain Growth. Further annealing after recrystallization is complete results in grain growth. There are two types of grain growth: (1) normal or continuous grain growth, and (2) abnormal or discontinuous grain growth (secondary recrystallization). In both cases the driving energy is the interfacial energy associated with the grain boundaries. Normal grain growth commonly occurs in pure metals and single-phase alloys. During normal grain growth, the average grain size increases gradually, and the size and shape distributions of the grains remain essentially unchanged. Grain growth during isothermal annealing obeys an empirical law that the average grain diameter is a power function of the annealing time, $D = Kt^{\eta}$, with the time exponent $\eta < 0.5$. The physical significance of η and its correlation with the driving-force exponent in boundary migration, m in $V = M(\Delta F)^m$, have been discussed in connection with available theories of boundary migration and of grain growth. Secondary recrystallization or abnormal grain growth occurs when normal grain growth is inhibited by a dispersed second phase, a strong single-orientation texture, or a limiting size of the grains imposed by the sheet thickness. The kinetics of secondary recrystallization are characterized by the sigmoidal type of correlation between volume of transformation and annealing time. The textural developments during normal grain growth and during secondary recrystallization, as well as their significance in industrial applications, have been presented and discussed.

Concluding Remarks

In concluding this brief review on recovery, recrystallization and grain growth, a few additional remarks appear to be pertinent and worthwhile. As mentioned in the introduction of the present paper, deformation and annealing are the most important processes that have been employed to effect structural changes and the resulting changes in properties. The sequential stages of recovery, recrystallization and grain growth encompass, in fact, the whole spectrum of the annealing processes when a deformed specimen is annealed. Thus, the importance of this subject, scientifically or technologically, is obvious.

As can be noted in this presentation, most of the examples cited and discussed were results from experiments conducted with pure metals or simple alloys and under well-defined conditions. For basic understanding of the mechanisims and for establishing scientific principles that control the processes, it is often necessary to reduce the complex problem to its simplest possible form so that truly meaningful results can be obtained from the experiments.

Almost invariably the materials and processing methods commonly used in the metals industry are highly complex. Proper control of the hot-rolling process can substantially improve the mechanical properties of high-strength steels and non-ferrous alloys, for example, which are highly complex materials. During hot working dynamic recovery or recrystallization and even some grain growth or grain coarsening can occur simultaneously in these materials, so that the effects of the process are also highly complex. However, even in these and other similarly complex situations encountered in commercial practices, the basic principles established from fundamental research studies should be valid. These principles have always been a valuable basis for applied research in industrial applications.

References

1. G. I. Taylor, Proc. Roy. Soc. (London), A145, 362, 388 (1934).
2. A Seeger, in Dislocations and Mechanical Properties of Crystals, ed. Fisher et al., John Wiley & Sons, New York, 1957, p. 243.
3. L. M. Clarebrough and M. E. Hargreaves, Progress in Metal Physics, 8, 1 (1959).
4. T. E. Mitchell, Progr. Appl. Matl. Res., 6, 117 (1964).
5. S. J. Basinski and Z. S. Basinski, in Recrystallization, Grain Growth and Textures, ASM, Metals Park, Ohio, 1966, p. 1.
6. A. S. Keh, Phil. Mag., 12, 9 (1965).
7. W. A. Spitzig and T. E. Mitchell, Acta Met., 14, 1311 (1966).
8. G. Taylor and J. W. Christian, Phil. Mag., 15, 873 (1967).
9. A. Luft and L. Kaun, Phys. Stat. Sol., 37, 781 (1970).
10. A. S. Keh, W. A. Spitzig and Y. Nakada, Phil. Mag., 23, 829 (1971).
11. A Seeger, H. Kronmüller, O. Boser and M. Rapp, Phys. Stat. Sol., 3, 1107 (1963).
12. M. Bocek and V. Kaska, Phys., Stat. Sol., 4, 325 (1964).
13. M. Bocek and P. Lukac, B. S. Mola and M. Svabova, Phys. Stat. Sol., 7, 173 (1964).
14. M. Bocek, G. Hotzsch and B. Simmen, Phys. Stat. Sol., 7, 833 (1964).
15. A. H. Cottrell, Dislocations and Plastic Flow in Crystals Clarendon Press, Oxford (1953).
16. J. C. M. Li, in Electron Microscopy and Strength of Crystals, ed. G. Thomas and J. Washburn, Interscience, New York, 1963, p. 713.
17. A. Seeger, S. Mader and H. Kronmüller, in Electron Microscopy and Strength of Crystals, ed. G. Thomas and J. Washburn, Interscience, New York, 1963, p. 665.
18. W. Carrington, K. F. Hale and D. McLean, Proc. Roy. Soc., A259, 203 (1960).
19. J. Friedel, in Internal Stresses and Fatigue in Metals, ed. G. M. Rassweiller and G. L. Grube, Elsevier, Amsterdam, 1958, p. 220.
20. G. Saada, Acta Met., 8, 200 (1960).
21. N. F. Mott, Trans. Met. Soc. AIME, 218, 962 (1960).
22. P. B. Hirsch, Phil. Mag. 7, 67 (1962).
23. D. Kuhlmann-Wilsdorf, R. Maddin and H. G. F. Wilsdorf. in Strengthening Mechanisms in Solids, ASM, Metal Park, Ohio, 1962, p. 137.
24. R. E. Smallman, Modern Physical Metallurgy, 3rd ed., Butterworths, London, 1970.
25. J. C. M. Li, J. Aust. Inst. Metals. 8, 206 (1963).
26. F. R. N. Nabarro, Z. S. Basinski and D. B. Holt, Advan. Phys., 13, 193 (1964).
27. A. Howie, in Direct Observation of Imperfections in Crystals, ed. J. B. Newkirk and J. H. Wernick, Interscience, 1962, p. 283.
28. P. R. Swann in Electron Microscopy and Strength of Crystals, ed. G. Thomas and J. Washburn, Interscience, New York, 1963, p. 131.
29. A. S. Keh and S. Weissmann, in Electron Microscopy and Strength of Crystals, ed. G. Thomas and J. Washburn, Interscience, New York, 1963, p. 231.
30. H. Hu, Z. Metallk., 60, 69, (1969).
31. H. Hu, in Texture in Research and Practice, ed. J. Grewen and G. Wassermann, Springer-Verlag, Berlin, 1969, p. 200.
32. I. L. Dillamore, Texture Cryst. Sol., 4, 41 (1980).
33. A. L. Titchener and M. B. Bever, in Prog. in Metal Phys., 7, 247 (1958).
34. P. Gordon, Trans. AIME, 203, 1043 (1955).
35. J. E. Bailey and P. B. Hirsch, Phil. Mag., 5, 485 (1960).
36. L. M. Clarebrough, M. E. Hargreaves and M. H. Loretto, in Recovery and Recrystallization of Metals, ed. L. Himmel, Interscience, New York, 1963, p. 63.
37. R. O. Williams, Trans. Met. Soc. AIME, 224, 719 (1962).
38. R. O. Williams, Acta Met., 9, 949 (1961).
39. F. Haessner, G. Hoschek and G. Tölg, Acta Met., 27, 1539 (1979).
40. A. Wolfenden and A. S. Appleton, Acta Met., 16, 915 (1968).
41. J. B. Cohen and M. B. Bever, Trans. Met. Soc. AIME, 218, 155 (1960).
42. M. B. Bever, D. L. Holt and A. L. Titchener, Progress in Materials Science, Vol. 17, Pergamon Press, Oxford, 1973.
43. P. A. Beck, Advances in Physics (Phil. Mag. Sup.), 3, 245 (1954).
44. E. C. W. Perryman, in Creep and Recovery, ASM, 1957, p. 111.
45. W. C. Leslie, J. T. Michalak and F. W. Aul, in Iron and Its Dilute Solid Solutions, ed. C. W. Spencer and F. E. Werner, Interscience, New York, 1963, p. 119.
46. J. C. M. Li, in Recrystallization, Grain Growth and Textures, ed. H. Margolin, ASM, Metals Park, Ohio, 1966, p. 45.

47. L. M. Clarebrough, M. E. Hargreaves, M. H. Loretto and G. W. West, Acta Met., 8, 797 (1960).

48. T. Broom, Proc. Phys. Soc. B65, 871 (1952).

49. T. Broom, Advances in Physics (Phil. Mag. Sup.), 3, 26 (1954).

50. J. Friedel, Dislocations, Addison-Wesley, 1964, p. 425.

51. R. R. Eggleston, J. Appl. Phys. 23, 1400 (1952).

52. H. Hu, in Recovery and Recrystallization of Metals, ed. L. Himmel, Interscience, New York, 1963, p. 311.

53. C. S. Barrett, Structure of Metals, 2nd ed. McGraw-Hill, New York, 1952, p. 421.

54. B. E. Warren and B. L. Averbach, J. Appl. Phys., 23, 497 (1952).

55. B. E. Warren, in Progress in Metal Physics, 8, 147 (1959).

56. A. H. Lutts and P. A. Beck, Trans. AIME, 200, 257 (1954).

57. E. Orowan, Communication to Congress de la Societe Francaise de la Metallurgie d'Octobre, 1947.

58. R. Drouard, J. Washburn and E. R. Parker, Trans. AIME, 197, 1226 (1953).

59. J. T. Michalak and H. W. Paxton, Trans. AIME, 221, 850 (1961).

60. A. S. Keh, in Direct Observation of Imperfections in Crystals, ed. J. B. Newkirk and J. H. Wernick, Interscience, New York, 1962, p. 213.

61. D. Kuhlmann, Z. Phys. 124, 468 (1948).

62. D. Kuhlmann, G. Masing and J. Raffelsieper, Z. Metallk., 40, 241 (1949).

63. R. A. Vandermeer and P. Gordon, in Recovery and Recrystallization of Metals, ed. L. Himmel, Interscience, New York, 1963, p. 211.

64. G. Van Drunen and S. Saimoto, Met. Trans. 10A, 783 (1979).

65. J. G. Byrne, Met. Trans., 10A, 791 (1979).

66. R. W. Cahn, J. Inst. Metals, 76, 121 (1949).

67. R. W. Cahn, Progress in Metal Physics, 2, 151 (1950).

68. W. R. Hibbard and C. G. Dunn, in Creep and Recovery, ASM, 1957, p. 52.

69. J. J. Gilman, Acta Met., 3, 277 (1955).

70. P. P. Sinha and P. A. Beck, J. Appl. Phys., 32, 1222 (1961).

71. S. Amelincks and R. Strumane, Acta Met., 8, 312 (1960).

72. J. C. M. Li, Acta Met., 8, 563 (1960).

73. J. E. Burke and D. Turnbull, Progress in Metal Physics, 3, 220 (1952).

74. R. W. Cahn, in Physical Metallurgy, ed. R. W. Cahn, North-Holland, Amsterdam, 1965, p. 925.

75. J. G. Byrne, Recovery, Recrystallization and Grain Growth, Macmillan, New York, 1965.

76. P. Cotterill and P. R. Mould, Recrystallization and Grain Growth in Metals, Surrey University Press, London, 1976.

77. J. W. Cahn and W. Hagel, in Decomposition of Austenite by Diffusional Processes, ed. V. F. Zackay and H. I. Aaronson, Interscience, New York, 1960, p. 131.

78. W. A. Johnson and R. F. Mehl, Trans. AIME, 135, 416 (1939).

79. M. Avrami, J. Chem. Phys., 7, 103 (1939); ibid 8, 212 (1940); ibid 9, 177 (1941).

80. W. A. Anderson and R. F. Mehl, Trans. AIME, 161, 140 (1945).

81. J. K. Stanley and R. F. Mehl, ibid, 150, 160 (1942).

82. A. U. Seybolt and J. E. Burk, Procedures in Experimental Metallurgy, John Wiley and Sons, New York, 1953, p. 319.

83. L. W. Eastwood, A. E. Bousu, and C. T. Eddy, Trans. AIME, 117, 246 (1935).

84. W. C. Leslie, F. J. Plecity, and J. T. Michalak, also W. C. Leslie, F. J. Plecity, and F. W. Aul, Trans. AIME, 221, 691 and 982 (1961).

85. A. T. English and W. A. Backofen, Trans. Met. Soc. AIME, 230, 396 (1964).

86. G. R. Speich and R. M. Fisher, in Recrystallization, Grain Growth and Textures, ed. H. Margolin, ASM, Metals Park, Ohio, 1966, p. 563.

87. B. B. Rath, R. J. Lederich, C. F. Yolton and F. H. Froes, Met. Trans., 10A, 1013 (1979).

88. H. Hu, Acta Met., 10, 112 (1962).

89. H. Hu, Trans. Met. Soc. AIME, 224, 75 (1962).

90. J. L. Walter and E. F. Koch, Acta Met., 11, 923 (1963).

91. H. Hu, in Electron Microscopy and Strength of Crystals, ed. G. Thomas and J. Washburn, Interscience, New York, 1963, p. 564.

92. J. C. M. Li, J. Appl. Phys. 33, 2958 (1962).

93. B. Bay, J. Matls. Sci., 5, 617 (1970).

94. A. R. Jones, B. Ralph and N. Hansen, Proc. Roy. Soc., London A368, 345 (1979).

95. J. E. Bailey and P. B. Hirsch, Proc. Roy. Soc., London, A267, 11 (1962).

96. A. R. Jones, in Grain Boundary Structure and Kinetics, ASM, Metals Park, Ohio, 1980, p. 379.

97. C. J. E. Smith and I. L. Dillamore, Met. Sci. J., 4, 161 (1970).

98. R. W. Vandermeer and W. B. Snyder, Jr., Met. Trans., 10A, 1031 (1979).

99. C. G. Dunn and W. R. Hibbard, Jr., Acta Met., 3, 409, (1955).

100. Papers in Recrystallization and Grain Growth of Multi-Phase and Particle Containing Materials, ed. N. Hansen, A. R. Jones, and T. Leffers, Risø National Laboratory, Roskilde, Denmark, 1980.

101. A. R. Jones, J. Matl. Sci., (in press).

102. D. W. Bainbridge, C. H. Li, E. H. Edwards, J. Washburn, and E. R. Parker, Acta Met., 2, 322 (1954).

103. P. A. Beck, P. R. Sperry, and H. Hu, J. Appl. Phys., 21, 420 (1950).

104. G. Ibe and K. Lücke, in Recrystallization, Grain Growth and Textures, ed. H. Margolin, ASM, Metals Park, Ohio, 1966, p. 434.

105. H. Hu, Trans. Met. Soc. AIME, 215, 320 (1959).

106. K. T. Aust and J. W. Rutter in Recovery and Recrystallization of Metals, ed. L. Himmel, Interscience, New York, 1963, p. 131.

107. K. T. Aust and J. W. Rutter, Trans. Met. Soc. AIME, 215, 820 (1959).

108. J. W. Cahn, Acta Met., 10, 789 (1962).

109. K. Lücke and H. P. Stüwe, in Recovery, Recrystallization of Metals ed. L. Himmel, Interscience, New York, 1963, p. 171.

110. K. Lücke, R. Rixen, and F. W. Rosenbaum, in The Nature and Behavior of Grain Boundaries, ed. H. Hu, Plenum Press, New York. 1972, p. 245.

111. M. Hillert and B. Sundman, Acta Met., 24, 731 (1976).

112. B. B. Rath and H. Hu, Trans. Met. Soc. AIME, 245, 1243 and 1577 (1969).

113. B. B. Rath and H. Hu, in The Nature and Behavior of Grain Boundaries, ed. H. Hu, Plenum Press, New York, 1972, p. 405.

114. H. Gleiter, Acta Met., 27, 1749 (1979).

115. H. Gleiter and B. Chalmers, Progress in Materials Science, Vol. 16, Pergamon Press, Oxford, 1973.

116. P. Gordon and R. A. Vandermeer, in Recrystallization, Grain Growth and Textures, ed. H. Margonlin, ASM, Metals Park, Ohio, 1966, p. 205.

117. C. Frois and O. Dimitrov, Mem. Sci. Rev. Met., 59, 643 (1963).

118. B. B. Rath and H. Hu, Trans. Met. Soc. AIME, 236, 1193 (1966).

119. F. J. Humphreys, in Recrystallization and Grain Growth of Multiphase and Particle Containing Materials, ed. N. Hansen, A. R. Jones, and T. Leffers, Risø National Laboratory, Roskilde, Denmark, 1980, p. 35.

120. R. D. Doherty and J. W. Martin, J. Inst. Met., 91, 332 (1963).

121. F. J. Humphreys, Acta Met., 27, 1801 (1979).

122. E. Hornbogen and U. Köster, in Recrystallization of Metallic Materials, 2nd ed., ed. F. Haessner, Riederer-Verlag, Stuttgart, 1978, p. 159.

123. N. Hansen, Mem. Sci. Rev. Metall., 72, 189 (1975).

124. F. J. Humphreys, Metal Science, 13, 136 (1979).

125. H. Hu, Texture, 1, 233 (1974).

126. C. S. Smith, in Metal Interfaces, ASM, Cleveland, Ohio, 1952, p. 65.

127. H. Hu, Can. Met. Quart., 13, 175 (1974).

128. P. A. Beck, J. Appl. Phys., 19, 507 (1948).

129. H. N. Murty, Phil. Mag., 20, 855 (1969).

130. W. W. Mullins, Acta Met., 6, 414 (1958).

131. H. Hu, also H. Hu and P. A. Beck, quoted in Ref. 43, p. 275 and p. 276.

132. P. Feltham, Acta Met., 5, 97 (1957).

133. F. Schückher, in Quantitative Microscopy, ed. R. T. DeHoff and F. N. Rhines, McGraw-Hill, New York, 1968, p. 201.

134. F. N. Rhines and K. R. Craig, with Apapendix by R. T. DeHoff, Met. Trans., 5 413 (1974).

135. H. Hu and B. B. Rath, Met. Trans., 1, 3181 (1970).

136. R. L. Fullman, in Metal Interfaces, ASM, Cleveland, Ohio, 1952, p. 179.

137. M. Hillert, Acta Met., 13, 227 (1965).

138. I. M. Lifshitz and V. V Slyozov, J. Phys. Chem. Sol., 19, 35 (1961).

139. J. von Neumann, discussion to Ref. 126.

140. W. W. Mullins, J. Appl. Phys., 27, 900 (1956).

141. N. P. Louat, Acta Met., 22, 721 (1974).

142. F. Assmus, K. Detert and G. Ibe, Z. Metallk., 48, 344 (1957).

143. F. D. Rosi, B. H. Alexander and C. A. Dube, Trans. Met. Soc. AIME, 194, 189 (1952).

144. C. G. Dunn and J. L. Walter, in Recrystallization, Grain Growth and Textures, ed. H. Margolin, ASM, Metals Park, Ohio, 1966, p. 461.

145. T. Gladman, Proc. Roy. Soc., London A294, 298 (1966).

146. J. T. Michalak and H. Hu, Met. Trans., 10A, 975 (1979).

147. M. McLean and H. Mykura, Acta Met., 13, 1291 (1965).

148. S. Taguchi and A. Sakakura, J. Appl. Phys., 40, 1539 (1969).

149. C. Zener, Private communication to C. S. Smith, Trans. AIME, 175, 15 (1949).

150. R. W. Balluffi and J. W. Cahn, Acta Met., 29, 493 (1981)

PREDICTION OF ALLOY PHASE STABILITY

E.S. Machlin
Henry Krumb School of Mines
Columbia University
New York, New York 10027

I. INTRODUCTION

Significant progress has been made in the last decade in the ability to predict, and in the understanding of, alloy phase stability. We shall limit the scope of this contribution and provide a survey mainly of the former, although some insights concerning bonding in alloys gained recently will be described. We define "prediction" to encompass any analytic procedure that provides either numerical values for properties that are capable of being measured and have not yet been measured or new qualitative ratings of a monotonic variation in a property for competing materials. Such procedures need not involve "first-principles". Most procedures we shall investigate are empirical. The only principles necessarily obeyed by all these procedures are thermodynamic, where applicable.

The differences in free energy between polymorphs of an element determine the relative stabilities of these polymorphs. This subject is usually described by the term--"lattice stability"--originated by Kaufman (1). Various procedures exist for deriving lattice stabilities from experimental data (1). Also, there are procedures for predicting lattice stabilities (2-4). Intermetallic compounds of constant composition also exhibit polymorphism and the lattice stabilities of many such polymorphs may likewise be predicted (2,5,6).

In addition to predictive analytical relations, there is also a graphical procedure for predicting the stable crystal structure of an element or stoichiometric intermetallic compound. This procedure involves the use of "crystal structure maps" (7). Hence, we shall survey this subject in some detail. All these aspects of "lattice stability" are considered in Section II.

The heat and entropy of formation of alloy phases determine the free energy of formation and, hence, stability of an alloy phase. These quantities may also be predicted, with fair confidence, according to procedures developed in the last decade (8). These procedures and their limitations will be discussed in the text in Section III.

Strides have been made in the prediction of alloy phase diagrams (1,9,10). This field alone is active enough to support a separate journal--CALPHAD--published by Pergamon Press. Because several survey articles have appeared recently describing recent progress in the prediction of alloy phase diagrams (9,10) and because a book exists that describes the original work of Kaufman in this field (1), I shall consider here only those aspects of alloy phase diagram prediction that will enable a novice to grasp the essentials of the procedures. I shall attempt to define, in some detail, those aspects of alloy phase diagrams for which these procedures can predict useful results.

The configurational entropy of formation for solid solution phases in most of the procedures used in alloy phase diagram prediction is assumed to be given by random site occupation probability. This assumption does not lead to significant errors in solvus boundary compositions when the energy of solution in the terminal phase is positive (i.e. limited solid solubility) or when the minimum free energies of the equilibrium coexisting phases differ by more than about 100 cal/g-atom (419 J/g-atom) at the stoichiometric composition of the intermediate phase. However, when this difference in free energy is much less than about 100 cal/g-atom then significant errors can appear in the values predicted for solvus compositions, etc. The latter situation will occur when the terminal phase and its coexisting intermediate phase have the same basis crystal structure (all atoms considered to be identical). In this case, the intermediate phase involves a long-range-ordered arrangement of atoms while the terminal phase consists of a short-range-ordered or clustered arrangement on the same basis structure as that for the co-existing intermediate phase; e.g. Cu-Au system. Under such circumstances, the configurational entropy contributes significantly to the free energy and it then becomes important to evaluate this entropy term to a fair degree of accuracy. The cluster approach of Kikuchi (11) has provided the basis for an acceptable approximation of the configurational entropy and has led to methods of predicting such phase diagrams, which are called "coherent" phase diagrams, based on the fitting of a few points in such diagrams (12,13). I shall attempt to survey the results of such procedures, which have been described in detail elsewhere (12,13). Both "incoherent" and "coherent" alloy phase diagram predictions are described in Section IV.

Predictions are now possible for solid-gas phase equilibria, such as are involved in oxidation or sulfidation. Only brief comments on this subject are made in Section V, in view of the coverage of this subject in H.Y. Sohn's article in this treatise.

Many uses have arisen from the predictive capabilities developed in the foregoing research. The more important of these applications are: prediction of elastic constants of alloy systems; "pseudomorphic stabilization" of metastable crystalline and amorphous phases; prediction of surface and interface energies and energies of segregation to such interfaces. These applications are briefly discussed in Section V.

II. LATTICE STABILITY AND CRYSTAL STRUCTURE MAPS

Lattice Stability of Elements and Intermetallic Compounds

Elements. The differences in free energy (F) between polymorphs of an element may be obtained by thermodynamic analysis of various data (e.g. extrapolation of F(P) from pressure (P) range where polymorph is stable to atmospheric pressure, extrapolation of $F_\alpha(x_2)$ to $x_2 = 1$ when stable polymorph of pure 2 component is β, where α denotes crystal structure of terminal solid solution at $x_2 = 0$ (x_2 is, of course, the atom fraction of component 2) and β that at $x_2 = 1$, etc.) Kaufman and Bernstein (1) have developed lattice stability values from such analyses and data available prior to 1970. Subsequent values, including updated more accurate values now exist in data banks. Table II-1 lists those values in the ManLab data bank as of 1980 and are presented with the kind permission of Dr. L. Kaufman.

Table II-1. Lattice Stability Parameters (1) (cal/g-atom)

ELEMENT	$F_{bcc}-F_{fcc}$	$F_{fcc}-F_{hcp}$	$F_{bcc}-F_{hcp}$
Ag	900-0.45T		
Al	2410-1.15T		
Au			
Ba			-1040-0.8T
Be	635-1.15T	+465+0.43T	1100-0.72T
Ca	58-0.08T		
Cd			-700+0.6T
Co	+1390+1.05T	110-0.15T	1500+0.9T
Cr	-2500-0.15T	500+0.15T	-2000
Cs			
Cu	1500-0.8T		
	850-0.2T	-150-0.3T	700-0.5T
Fe	-1303(0°K)		
Hf	1030-0.9T	800	1830-0.9T
Hg			
Ir	1650+1.05T	-150-0.15T	1500+0.9T
K			
Li			
Mg	635-1.15T	465+0.43T	1100-0.72T
Mn	420-0.3T		
Mo	-2500-0.15T	500+0.15T	-2500
Na			
Nb	-2150-0.85T	650+0.05T	-1500-0.8T
Ni	1350+0.8T	-250-0.3T	1100+0.5T
Os	1000	120+0.8T	1120+0.8T
Pb		-600+0.1T	
Pd	1350+0.8T	-250-0.3T	1100+0.5T
Pt	1350+0.8T	-250-0.3T	1100+0.5T
Rb			
Re	150+0.1T	250+0.3T	400+0.4T
Rh	1650+1.05T	-150-0.15T	1500+0.9T
Ru	1000	120+0.8T	1120+0.8T
Sn			
Sr	200-0.23T		
Ta	-2150-0.85T	650+0.05T	-1500-0.8T
Tc	150+0.1T	250+0.3T	400
Ti	240-0.9T	800	1040-0.9T
Tl	14-0.22T	76+.04T	90-0.18T
V	-2150-0.85T	650+0.05T	-1500-0.8T
W	-2500-0.15T	500+0.15T	-2000
Zn	250-0.2T	440-0.4T	690-0.6T
Zr	230-0.9T	800	1030-0.9T

Computer programs exist at IBM Research Laboratory, Yorktown Heights, New York 10598, USA, based on first-principle, muffin-tin, ASW, density functional solutions for the energy, which can yield the energies at 0°K of bcc and fcc structures of elements, except for those where relativistic effects significantly affect the energy. Requests for such data should be made to either Dr. V.L. Moruzzi or Dr. A.R. Williams at the IBM Laboratory.

Machlin and his students have developed a lattice energy function, which consists of a sum over interatomic potentials plus a sum over the atomic quadrupole distortion energies for those atoms that are so distorted (6). This procedure is not sufficiently accurate to distinguish between the energies of polymorphs that have the same nearest-neighbor coordination number. However, it appears to be sufficiently accurate to provide useful values for the energy difference between polymorphs that have different nearest-neighbor coordination number. The predicted energy differences E(A2) - E(A1), E(A2) - E(A15), E(A2) - E($D8_b$) calculated to date are given in Table II-2. (Here A1, A2, A15 and $D8_b$ denote the fcc, bcc, Cr_3Si and sigma phase crystal structures, respectively.)*

*Computer programs to effectuate such calculations may be obtained by request to the author.

Table II-2. Lattice Stability Parameters (cal/g-atom) Machlin Energy Function

ELEMENT	$E_{bcc}-E_{fcc}$	$E_{A15}-E_{A2}$*	$E_{\sigma}-E_{bcc}$
Ag	910		
Al	1030		4725
Au	1160		5320
Ba	565		
Be	1035		
Ca	560		
Cd	355		
Co	1350		6190
Cr	1260	-500(+3260)	2345
Cs	250		
Cu	1080		
Fe	1320	115(2738)	3475
Hf	2125		
Hg	205		
Ir	2115		9865
K	285		
Li	510		
Mg	475		
Mn	885		
Mo	+2095	-965(+3630)	3670
Na	345		
Nb	2300	-3360(+1090)	-350
Ni	1365		6270
Os	2485		11,395
Pb	620		
Pd	1210		5545
Pt	1795		8240
Rb	260		
Re	2485		
Rh	1770		8105
Ru	2035		9320
Sn	955		
Sr	520		
Ta	2480	-1915(+2715)	3030
Tc	1660		
Ti	1500	95(+1835)	
Tl	570		
V	1635	-1925(+1860)	1035
W	2660		
Zn	415		
Zr	1940	410(+2580)	2160

*Values in parentheses assume partial covalent bond contribution to energy of bcc lattice is not present in A15 structure; terms outside parentheses assume this contribution is present in both structures.

Pseudopotential theory has also been used to evaluate lattice stabilities of elements (15). Collected values of the differences in energy between various polymorphs so calculated are presented in Table II-3.

Comparison between Tables II-1-3 yields the result that the standard deviation in ΔE(A2 - A1) provided by all 3 procedures from the mean values is 4350 J/g-atom (1040 cal/g-atom, 0.045 eV/atom). Analysis of the deviations between pairs of procedures reveals that the Kaufman and Machlin lattice stabilities are in good agreement, except for the bcc transition metals. The Machlin lattice energy function always yields a stable fcc structure with respect to a metastable bcc structure. The stability of the bcc structure for these metals (expect for Fe) is due to a partial covalent bonding not included in the lattice energy function. With this exception the standard deviation between the Kaufman and Machlin values is 470 cal/g-atom (1955 J/g-atom).

It is of interest and perhaps significant that Nb has recently been synthesized in the A15 structure (16). According to Table II-2, the stable crystal structure of Nb at 0°K should be the A15 structure!

Table II-3. Lattice Stability Parameters (cal/g-atom)
Pseudopotential Calculations

ELEMENT	$E_{bcc}-E_{fcc}$	$E_{fcc}-E_{hcp}$	$E_{bcc}-E_{hcp}$	Ref.
Al	1630	- 265	1365	6
	4050	- 965	3085	7
	4370	-1170	3200	1
	3350(avg)			
Be	-165	2090	1925	6
	355	2605	2960	7
	95(avg)			
Ca	-175	990	815	7
Cd	600	50	650	7
Ga	1130*	440	1570	4
In	-150	470	320	6
K	33	13	46	2
Li	31	44	75	7
	34	32	66	2
	-6	31	25	6
	20(avg)			
Mg	630	715	1345	1
	390	220	610	6
	-175	990	815	7
	280(avg)			
Na	91	44	125	7
	106	13	119	1
	-153	36	-117	2
	0	50	50	6
Pb	-	-1880	-	5
Zn	4025	-2325	1700	3

*$(E_{Ga_I}-E_{hcp} = -17{,}500)$

(1) W.A. Harrison, Phys. Rev. 136, A1107 (1964).
(2) R.M. Pick in Phase Stability in Metals and Alloys (P.S. Rudman, J. Stringer and R.I. Jaffee, eds.) McGraw Hill, NY, 1967.
(3) W.A. Harrison, Phys. Rev. 129, 2512 (1963).
(4) J.E. Inglesfield, J. Phys. C 1, 1337 (1968).
(5) C.H. Hodges, Acta Met. 15, 1787 (1967).
(6) R.W. Shaw, Jr., J. Phys. C 2, 2335 (1969).
(7) R. Pick, J. Phys. (France) 28, 539 (1967).

There are two main applications for the data in Tables II-1-3: alloy phase diagram prediction and prediction of polymorphic transitions upon increase of pressure. The latter application requires additional data: molar volumes at STP, thermal expansion coefficients and compressibilities. Values of these parameters for the stable polymorphs are listed in handbooks. Values for metastable polymorphs must be predicted. Few such predictions have been made.

Providing that there are no major changes in outer electron configurations, such as occur in the α/β transition in tin, the Machlin lattice energy function can be used to estimate molar volumes and compressibilities of metastable phases. A feature of solids that obey this lattice energy function is that fcc metals that are stable at STP conditions cannot be transformed to bcc metals by pressure, no matter how large the pressure. To the best of my knowledge this "rule" is obeyed, except for the cases Sr and Yb. However, because the volumes of these bcc phases are so much smaller than expected for a "Machlin solid", it must be concluded that electronic transitions must occur in the pressure induced transition of the fcc host phase to the bcc phase of these elements. Such electronic transitions occur frequently in pressure induced transitions, but are less numerous in temperature induced transitions, at least in elements. Theoretical investigation of the possible effects of pressure on "Machlin" solids has not yet been attempted primarily because of the likely possibility of electronic transitions in the interesting metals.

Intermetallic Compounds. The IBM group have calculated the energies of the B1, C1, B2 and $L1_o$ polymorphs of selected AB combinations, and are preparing a monograph based on their results (21). Aside from this series of calculations, there are no other systematic first-principle calculations of lattice stabilities of intermetallic compounds. However, there have been numerous specific calculations. Many simple metal alloys have been studied by means of pseudopotential theory (15,17). The energies of formation of various alloy phases, have been evaluated using the CPA approximation (18), and other approximate solutions to the Schrodinger equation. Despite all the numerous individual efforts, that of the IBM group is unique in its scope.

The empirical lattice energy function developed by Machlin has been used to calculate the lattice stabilities of AB_3 polymorphs, such as A15, $L1_2$, A2, A1 with excellent agreement with observation for over 100 phases (5e). Also, this function has been applied to provide lattice stabilities at the AB composition for the A1, A2, B2, $L1_o$ and $D8_b$ (5,6,19) structures and at the AB_2 composition for the C15 and C32 structures (5b). It has also provided lattice stabilities in a multicomponent system for the A1, A2 and $D8_b$ structures (20). This lattice energy function is the only one presently available that can yield lattice stability predictions for all polymorphs that differ in nearest-neighbor coordination or lattice site point symmetry and for which the nearest-neighbor coordination number, z, equals or exceeds 8. This is not to imply that there have not been isolated calculations for particular alloys and particular polymorphs thereof. However, no computer programs presently exist which allow lattice stability predictions for all polymorphs of alloy phases having $z \geq 8$ other than those based on the Machlin lattice energy function. This situation is not likely to last with advances in computer power.

It must be recognized that the lattice energy model can only yield good results for the case where the difference in the covalent contribution to the cohesive energy between competing crystal structures is small compared to the difference in cohesive energies. Generally, this condition is obeyed when the first neighbor coordination for each crystal structure equals or exceeds 8, i.e. each structure is close-packed. However, it is not obeyed in metallic conducting phases that have a lower first neighbor coordination than 8 (e.g. the Ni As type phases). Also, there can be a covalent contribution to the cohesive energy in one of the possible arrangements of atoms on a given basis structure (e.g. NaTl B32 structure) and not in another arrangement on the same basis structure (e.g. LiTl B2 structure) (17). In this case, it is possible that the lattice energy model, which excludes such covalent contributions to the energy, may predict the lattice stability incorrectly. It is for this reason, that the condition stated previously about applicability to crystal structure comparisons involving a difference in first neighbor coordination has been included.

The difference in energy between different configurations of atoms on the same basis structure is usually very small (i.e. < 100 cal/g-atom). Prediction of such long range ordering energies can be accomplished in special cases with the help of pseudopotentials (17) or other first-principle methods (21). The uncertainty associated with the empirical procedures, such as that of Machlin (2,5), is too large and precludes the use of the latter procedures in this application, as already noted.

It is possible to ascertain the ground state ordered configurations or mixtures of configurations corresponding to given ratios of 1st to 2nd to 3rd to 4th nearest-neighbor pair interchange energies. De Fontaine has reviewed the literature relating to the theory of ground state ordered configurations (22). Contributions to the development of such theory have been made by Clapp and Moss (23), Khatchaturyan (24), Cahn and his students (25), and Kanamori and his collaborators (26) and Sanchez (27). If the symmetry of the basis lattice is broken, structures other than those given by the configurational ground-state arrangements can be stable relative to the latter. Physically, this possibility becomes real when the contributions to the structural energy consist of factors not describable solely in terms of interatomic potentials or when the latter alone destabilize the basis structure relative to a competing structure. Nevertheless, ground state ordered configurations of the fcc and bcc basis structures are a significant subset of the structures exhibited by alloy phases.

The stable ground state ordered configurations occupy characteristic regions in a coordinate system made up of interaction energy ratios. Unfortunately, at the moment these interaction energies have not yet been predicted to the required accuracy for any system of coherent phases, although in principle it should be possible to deduce these values from energy of formation values for ordered arrangements in the fcc and bcc lattice that will soon be available from the IBM group (21).

Crystal Structure Maps

In the past, attempts have been made to discover coordinates that would successfully separate the stable crystal structures of elements and stoichiometric intermetallic compounds into regions each characteristic of only one stable crystal structure. These attempts probably extend to the earliest literature on crystallography. In more modern times, the coordinates used by Kubaschewski (28) (cohesive energy ratio versus atomic radius ratio) were only partially successful in providing such separation. More successful coordinates were discovered by Pearson and Mooser (29) and consisted of average principal quantum number versus the difference in Pauling electronegativity between the components. Of course, both systems were applied to binary alloy phases. The former was most successful for the separation of size-factor compounds and the latter for transition element-polyvalent element compounds. Despite their success, neither procedure can provide completely successful separation of the stable crystal structures.

A remarkable advance took place when St. John and Bloch (30) found they were able to successfully separate the stable crystal structures of elements and also those of octet $A^N B^{8-N}$ compounds using as coordinates parameters derived from an ℓ-dependent non-local pseudopotential developed by Simons and Bloch (31). These coordinates are first-principle in origin as compared to the empirical or semi-empirical coordinates used previously and are derived from spectroscopic data. Their crystal structure separation had no errors and was remarkable in that it was able to distinguish between crystal structures having very small differences in free energy, i.e. involving only differences in greater than 2nd neighbor coordination.

The coordinates used by St. John and Bloch (30) are based upon classical turning point radii of single electron ions as calculated by Simons and Bloch (31). Such radii have been recalculated by Chelikowsky and Phillips (32). These radii are denoted by r_{si} and r_{pi}, where i indicates the identity of the element. In terms of these radii the Simons-Bloch coordinates are as follows:

$$R_{\sigma}^{AB} = |r_p^A + r_s^A| - |r_p^B + r_s^B|$$

$$R_{\pi}^{AB} = |r_p^A - r_s^A| + |r_p^B - r_s^B|$$

Machlin, Chow and Phillips (33) subsequently showed that these coordinates also provided a successful separation of AB compounds that are not related to the octet rule and that comprise non-transition elements (i.e. only those with unfilled s,p shells but filled d,f,g shells of the sub-octet group $A^N B^{P-N}$ ($N \leq P < 8$)).

Because the Simons-Bloch radii are not physical in the sense that they fit wave functions of the metals poorly, others have developed similar coordinates either based on free atoms (Zunger and Cohen (34)) or on fitting to Hartree-Fock energies and wave functions of hydrogenic ions (Andreoni, Baldereschi, Biemont and Phillips (35)). Structural maps obtained using the similar but more physical radii of these two groups are as good as those obtained with the Simons-Bloch radii.

Structural coordinates for binary alloy phases containing transition elements have also been developed by Watson and Bennett (36), Zunger (37) and Machlin and Loh (7). The crystal structure separation produced by any of these coordinate sets for all transition binary alloys is not as clear-cut as obtained for the s,p alloys. However, the Watson-Bennett and Machlin-Loh coordinates do provide a reasonable basis for predicting the stable structure for some arbitrary $T_1 - T_2$ binary equiatomic alloy. The parameter used by M-L are simply the average periodic table group number, $\bar{N}$, and the difference in the individual group numbers for the components, ΔN, modified as noted in Figure 1c. In addition, Watson and Bennett were able to achieve moderately successful separation of crystal structures for all transition AB_2 and AB_3 compounds.

Watson and Bennett (36) and Zunger (37) have developed coordinates which they have applied to compounds formed between transition metals and polyvalent elements. These crystal structure separations have been only partially successful.

Examples of the most successful separations for the various classes of binary alloys are given in Figures 1a-1c. In order to achieve such separations it is necessary to choose the low temperature polymorph as the appropriate crystal structure to represent the AB compound in the crystal structure map. It is amazing that these maps can separate out crystal structures that differ so little in free energy. However, note from Figure 1b that the coordinates cannot distinguish between the B2 and B32 structures for sub-octet combinations.

One implication of the crystal structure maps is that the crystal structure dependent part of the energy of a binary AB compound is a function of only two variables. Indeed, from first-principles, the energy of such compounds should depend only upon two variables: the atomic numbers of the two components. These facts suggest that it should be possible to express the structural part of the energy analytically as a function of the two coordinates of the corresponding crystal structure map. (Chelikowsky and Phillips (32) have found a correlation between the Simons-Bloch coordinates and the variables found by Miedema (8) to describe the structure-independent energy of formation of homoelectronic AB alloys.) Transformation of the information about the structural energies of polymorphs contained in crystal structure maps into analytical relations remains to be accomplished. Such relations should then provide the most accurate values for the lattice stability of AB type polymorphs that can be deduced from any source.

The success of the ℓ-dependent, non-local pseudopotential derived coordinates (R_σ and R_π) in separating crystal structures of s,p type alloys has an important

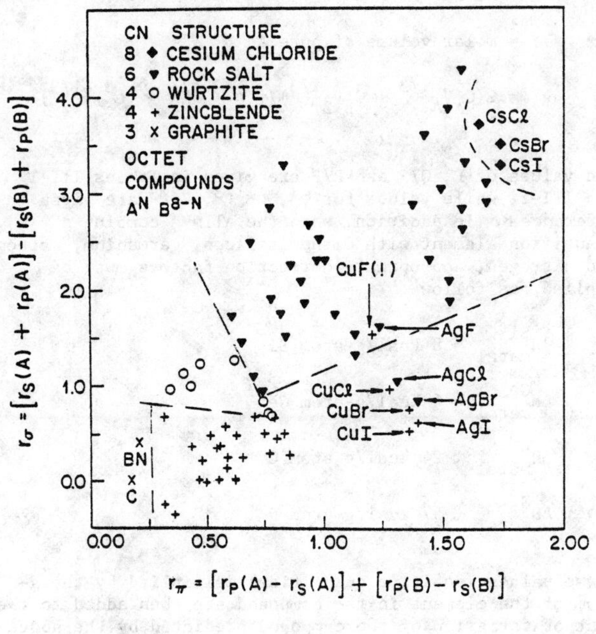

a) Polyvalent element combinations only (32).

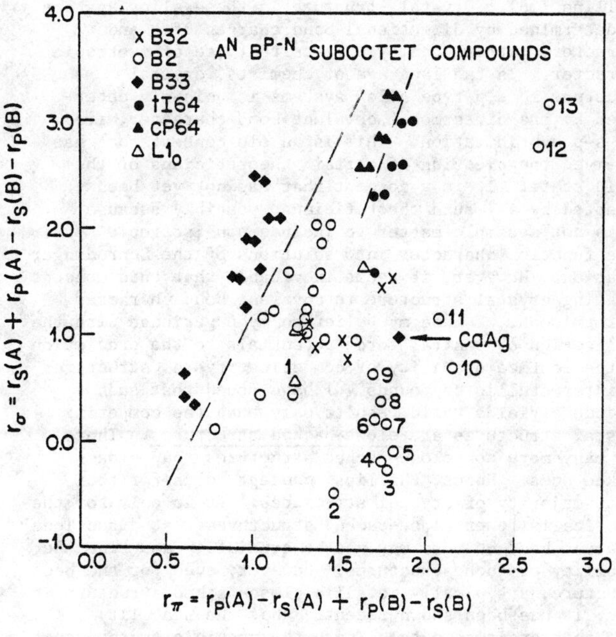

b) Polyvalent element combinations only (32).

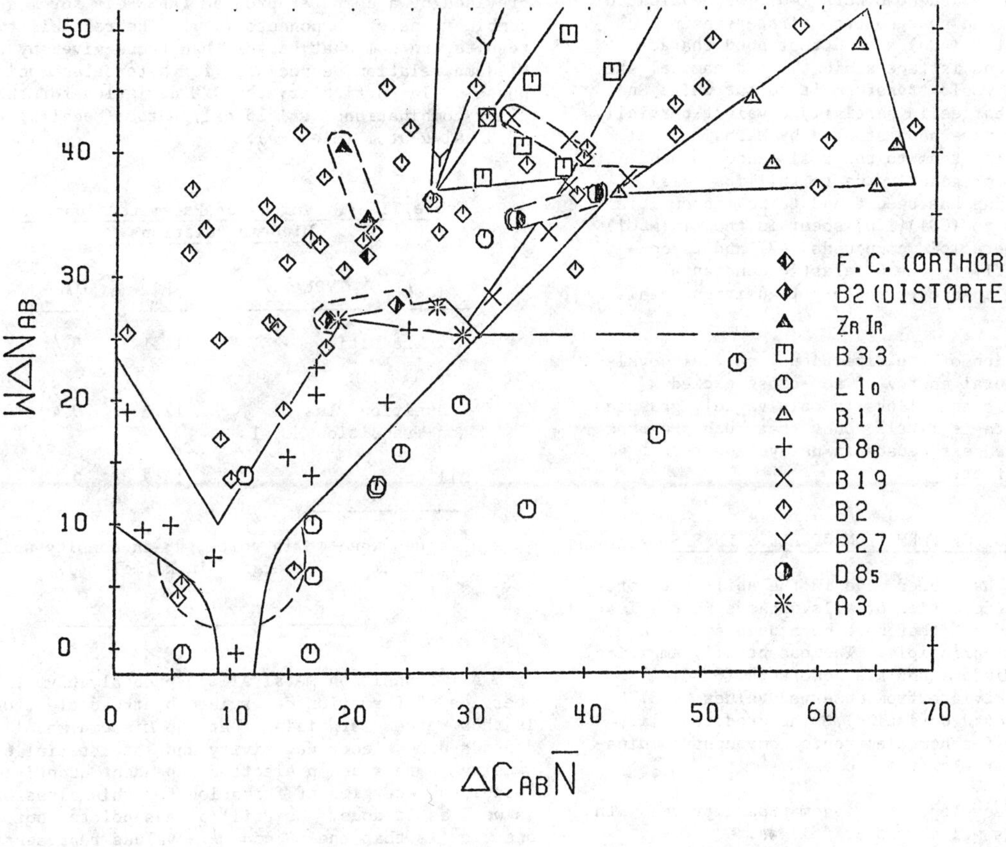

c) Transition element combinations only (7) (ΔC d band center-of-gravity difference, $\overline{W}$ average d band width, N periodic table group number).

Figure 1. Crystal Structure Maps.

fundamental implication, as first enunciated by J.C. Phillips (38). Crystal structure in these alloy systems is determined by directional bond charges of σ and π character, rather than by effects that are of isotropic character. In the language of chemists, crystal structure in s,p type alloy systems is uniquely determined by the directional covalent bond character, i.e. the s-p hybridization. This is an old concept that has received the blessing of certain theoreticians of the solid state. It is a concept that has not yet been accepted by all such theoreticians, possibly because it is not a simple matter to include non-isotropic wave function character into solutions of the Schrodinger equation. However, it seems inevitable that this concept relating crystal structure to covalent bond character will triumph. I base my belief on my experience with the application of central force potentials to the prediction of the lattice stability of competing crystal structures of intermetallic compounds. I have found that such a procedure yields valid results only when the competing crystal structures are close-packed structures. There are many more non-close-packed structures than close-packed ones. Hence, Phillips' concept holds for the great majority of crystal structures. It is only for the numerically fewer close-packed structures that directional covalent bond charges may not be crucial in assuring the stability of such structures. However, even for the bcc structure, a typically metallic close-packed structure type, it has been shown recently that the stability of the bcc transition metals (with the possible exception of Fe) is due to an additional contribution provided by partial covalent bonds having directional character (42).

Bond charges in partially covalent-partially ionic materials have been studied by van Vechten and Phillips (39), who have correlated physical properties to bond charges, and by Cohen and coworkers (40), who evaluated the magnitudes of bond charges from first-principle calculations. Martin (40a) showed that bond charges stabilize the diamond lattice against the dynamical shear instability, which would transform it to the white Sn structure. Bond charges in partially covalent-partially metallic materials have been studied by Mathai et. al. (41) who have related them to the resistance to bond bending, a concept originally due to Phillips (41a). Covalent bond bending has been found to be responsible for the resistance to (001)[100] shear in the B1 (NaCl) transition metal-metalloid compounds (42) and to contribute significantly to the C' elastic constant ((011)[01$\bar{1}$] shear) in bcc refractory transition metals (42).

The contribution of the directional partial covalent bond to the structural energy of non-close-packed structures of metals and alloys is only vaguely grasped at this time. There is little doubt that much progress will be made in the next decade in unraveling the details of this effect.

III. ENERGY OF FORMATION OF ALLOY PHASES

Giant strides have been made in the ability to predict the energy of formation of alloy phases in the last decade. The procedures that have been used are both empirical and first-principle. The essentially empirical procedure due to Miedema and his coworkers (8,43,44) yield values that deviate from observed values by 3770 cal/g-atom (1 standard deviation). The predictability is somewhat better for homoelectronic component combinations than for heteroelectronic ones.

The relation developed by Miedema has improved with time and as of this writing is as follows.

$$\Delta H = f * g * P * (-\Delta\phi^{*2} + (Q/P)(\Delta\eta^{1/3})^2 + R/P)$$

where $f = C_A^S C_B^S[1 + 8(C_A^S C_B^S)^2]$

$$C_A^S = C_A V_A^{2/3}/(C_A V_A^{2/3} + C_B V_B^{2/3})$$

$$C_B^S = C_B V_B^{2/3}/(C_A V_A^{2/3} + C_B V_B^{2/3})$$

V_A = molar volume of pure A

$$g = 2(C_A V_A^{2/3} + C_B V_B^{2/3})/\{(\eta^A)^{-1/3} + (\eta^B)^{-1/3}\}$$

and values of P, Q/P and R/P are given in Tables III-1 and III-2, while values for ϕ^*, $\eta^{*1/3}$, $V^{2/3}$ are given in reference 8. In addition, when the alloys combine a transition element with either silicon, germanium, carbon and nitrogen, appropriate correction factors must be applied, as follow

$$\Delta H^{Si}_{metal} = 8 \text{ kcal/g-atom Si}$$

$$\Delta H^{Ge}_{metal} = 6 \text{ kcal/g-atom Ge}$$

$$\Delta H^{C}_{metal} \approx 24 \text{ kcal/g-atom C}$$

$$\Delta H^{N}_{metal} \approx 57 \text{ kcal/g-atom N}$$

These values (positive quantities) multiplied by the g-atom of the element in the compound are then added to the heat of formation of the compound predicted by the above relations using the parameters given in Appendix 3.

Although the lattice energy function developed by Machlin and his students did not pretend to predict absolute energy of formation values, general corrections have been developed to enable such absolute values to be provided by a computer program (ABSNRG) for most binary combinations of components (45). The reliability of the results are somewhat better than those given by the Miedema relation, especially for heteroelectronic alloy phases. In particular, the RMS deviation for the same alloy combinations is 1615 cal/g-atom (Machlin) versus 3770 cal/g-atom (Miedema).

Table III-1. Values of Parameters Used in Miedema Relations.

ALLOY TYPE	P*	Q/P*
All Transition Metals	14.1	9.4
Transition Plus Non-Transition Metals	12.3	9.4
All Others	10.7	9.4

*Values when ϕ^* in volts, n^* in density units; V_m in cm^3 and energies in kJ/mole (8).

All transition metal alloys probably have their energies of formation determined by the d electrons. It is thus rather surprising that the Miedema model, which depends upon electronegativity and interstitial electron density, both s and p electron dependent quantities, "predicts" energies of formation for this class of alloys as well as it does. As Phillips has pointed out (46), one view is that the ϕ^* and $\eta^{1/3}$ values represent adjustable parameters fitted to a large data base, i.e. the equation for ΔH describes the energies of formation in terms of tabulated values of ϕ^*, $\eta^{1/3}$ and R rather than predicts such values. Prediction is possible for unmeasured ΔH values. It is not known how reliable such prediction may be.

Pettifor (47) and Varma (48) have recently developed

Table III-2. Values of Parameter R in Miedema's Relation.

TRANSITION METALS

Ca 0.4	Sc 0.7	Ti 1.0	V 1.0	Cr 1.0	Mn 1.0	Fe 1.0	Co 1.0	Ni 1.0	Cu 0.3
Sr 0.4	Y 0.7	Zr 1.0	Nb 1.0	Mo 1.0	Tc 1.0	Ru 1.0	Rh 1.0	Pd 1.0	Ag 0.15
Ba 0.4	La 0.7	Hf 1.0	Ta 1.0	W 1.0	Re 1.0	Os 1.0	Ir 1.0	Pt 1.0	Au 0.3
	Th 0.7	U 1.0	Pu 1.0						

= M

NON-TRANSITION METALS

Li 0	Be 0.4	B 1.9	C 2.1	N 2.3
Na 0	Mg 0.4	Al 1.9	Si 2.1	
K 0	Zn 1.4	Ga 1.9	Ge 2.1	As 2.3
Rb 0	Cd 1.4	In 1.9	Sn 2.1	Sb 2.3
Cs 0	Hg 1.4	Tl 1.9	Pb 2.1	Bi 2.3

= N

$$R/P = M \times N$$

'first-principle' models for the energy of formation of all transition alloys. Williams, Gelatt and Moruzzi (49) have recently tested Pettifor's model using state of the art self-consistent local density, ASR calculations and have concluded that Pettifor's model, which is based on the effect of alloying on the d-band width and center-of-gravity is basically correct while the Miedema concept is incorrect in some of its consequences. Watson and Bennett (50) have extended Varma's model, which is based upon the same effect of alloying on d-band width and center of gravity as the Pettifor model, and find that the predicted energies of formation differ significantly from those based on the Miedema model only for the alloys based on combinations of the Sc, Y, La group with other transition metals. Experiments designed to measure the energies of formation of the latter group of alloys are presently underway.

It is remarkable that the Miedema parameters, which are essentially derived from energies of formation values, correspond to the physical parameters they are supposed to represent. In particular, the η values, which are interpreted to be the electron densities at the Wigner-Seitz surface, and are determined empirically, correlate to the interstitial electron densities (8) calculated by a first-principle technique by Moruzzi, Janak and Williams (51). No fundamental bases have been provided for such correlations, as yet.

The Machlin model's success suggests that interatomic potentials contribute to the energy of formation. What appears to be striking about this work is that deviations between predicted and observed values of energy of formation are found to correlate to physical phenomena. For example, such deviations have led to the discovery of trends in the behavior of the pd hybridization contribution to the energy of formation of heteroelectronic alloys that are not revealed in the Miedema treatment of this effect.

Recently, model free first-principle calculations for the energy of formation have been made which allow one to understand and perhaps to predict it in the heteroelectronic class of alloys. The IBM group have calculated energies of formation, based on parameter-free self-consistent local-density calculations of the electronic structure using the Augmented Spherical Wave Method (14) for combinations of the 4d transition metals with themselves and with the polyvalent elements of the 3rd and 4th periods, for the B1, B2, C1, $L1_2$ and $L1_0$ structures. These data, together with calculated molar volumes, densities of states, etc. are collected in a book that will appear shortly. From these data, Gelatt et.al. have deduced the physical factors affecting bonding in the heteroelectronic class of transition-polyvalent element combinations (52). According to them, the energy of formation is the sum of an anti-bonding effect due to d band narrowing in the compound relative to the elements (i.e. a consequence of larger T atom separation in the compound than in the corresponding element), a bonding effect due to d,s,p hybridization (formation of covalent bonds) or alternatively due to the formation of ionic bonds if the sub-bands associated with the transition metal component are widely separated in energy from those associated with the polyvalent element component (i.e. without overlap).

As yet not enough work has been carried out by the IBM group to be able to list trends in behavior based on such results. However, sufficient calculations have been made by Machlin and his students to isolate two trends in the behavior of the energy of formation of the hetero-electronic class of transition metal-polyvalent element alloys. The first is that the anti-bonding effect is greatest for the half-filled d band transition metal components, as might be expected (45). The second is that the bonding effect due to d,s,p hybridization or ionic bond formation increases for a given transition element with increase in valence of the polyvalent element (5a).

In a rough way, bonding in transition metal alloys (T-T alloys) may also be understood. Alloying is believed to exert two effects--a bonding effect due to an increase in d band width that varies with the difference in d band centers of gravity of the components and an antibonding effect due to d band width mismatch of the components (49). These authors deny that charge transfer effects are responsible for the energies of formation of T-T type alloys contrary to the thrust of the Miedema model for energies of formation. It is obvious that the last word has not been said on the subject of the physical (microscopic) origin of bonding in alloys containing transition elements or in alloys not containing transition elements for that matter. One thing seems to be certain, however, the Engel-Brewer model for bonding (53), that was the rage in metallurgical circles in the 1960's, is simply not valid (8, 47, 49, 6, 54, 55)! The dependence of crystal structure upon the number of s,p electrons, cornerstone of the Engel-Brewer model, does not correspond to reality in that band theory and experiment both yield an s,p count for the transition metals of about one, independent of crystal structure. Other negations of Engel-Brewer theory predictions have also arisen.

We may summarize our present understanding of bonding with the statement that the factors that affect bonding are basically the same that Pearson enumerated a decade ago (56): electronegativity, size, hybridization and band effects. In other words, ionic, covalent and metallic bonding contribute to the total bonding in variable proportions. One of the problems that arises in the attempt to provide a regorous microscopic basis for bonding is that of a rigorous definition for certain of the microscopic bonding factors. "Electronegativity" means different things to different people, for example (57). It seems reasonable to exclude discussion of these matters from this paper, which is more concerned with the ability to predict than with understanding. The interested reader is referred to the ongoing debate in the literature for further information on this matter.

Finally, it is fair to predict that with the large increases in the speed of computer calculations likely to be available shortly that first-principle calculations of energies of formation for non-cubic structures will be made in the next decade and that it will be less expensive to make such calculations than to experimentally measure the energies of formation!

IV. PREDICTION OF DIAGRAMS

When the intermediate phases in an alloy system differ in free energy, at the compositions corresponding to these intermediate phases, from those of the terminal phases of the system by more than about 100 cal/g-atom

(400 J/g-atom) then the terminal phases solvus compositions are limited to low values, i.e. < 10 atomic %. Under this circumstance or in that where the heat of solution of the terminal phase is a positive quantity, it is reasonable to approximate the entropy of mixing by that for an ideal solution, i.e. the error in location of the terminal phase solvus compositions due to this approximation will be small. This is an implicit assumption made by the CALPHAD community in most of their phase diagram calculations. There are a few well-known systems where this assumption is not reasonable, e.g. Cu-Au. In these systems, intermediate phases exhibit transitions from disordered solutions at high temperature to long range ordered solutions at low temperature, without a change in crystal structure. The difference in free energy between the long range ordered and the disordered solutions seldom exceeds 100 cal/g-atom. Under these circumstances, the entropy of mixing departs significantly from that for an ideal solution. The phase diagrams of such systems have been termed "coherent" because the crystal structure is invariant. Coherent phase diagrams have been calculated for these systems using more accurate evaluations of the entropy of mixing. Because these two classes of phase diagram calculations differ in procedure, they are considered separately in this review. We shall now examine the methods and results of the calculations made for the former class of alloy systems.

Systems Assuming Ideal Entropy of Mixing, i.e. Regular and Sub-Regular Phases (1,9,10,58,59)

The prediction of phase diagrams is based on obtaining as accurate expressions for the free energies of the phases likely to be stable in the alloy system and then using these relations to calculate the tie-line compositions of phase boundaries. The latter are evaluated by either manipulation of chemical potentials to determine the compositions at which $\mu_i^\alpha = \mu_i^\beta$, etc. ($\bar{F}_i^\alpha = \bar{F}_i^\beta$, etc., where $\bar{F}_i$ are partial molar free energies) or by minimizing the free energy of mixtures of phases in n phase regions (n > 1) using standard computational techniques.

Accurate free energy relations for binary phase diagrams are arrived at using the requirement that thermodynamic data (measured) and experimental binary phase diagrams for a given binary system must be consistent if they are to be accurate. Usually, these observations are not consistent. However, by the coupling of thermodynamic relations and phase diagrams consistency can be achieved via small changes in the thermodynamic quantities and in the binary phase diagrams. The thermodynamic quantities are usually expressed in terms of some series expansion in the concentrations of the components. Ansara (9) has described such functions in detail.

Ternary phase diagram prediction requires sufficiently accurate thermodynamic relations for each of the three binary sets of components. No strictly ternary interactions are considered initially in the free energy of each of the phases likely to appear in the ternary system based on extrapolation from each binary system. There are various schemes of extrapolation from the binary lines into the ternary composition space. Hillert (60) has considered the advantages and disadvantages of each method and arrives at the recommendation that a relation of the following form be used to express the excess Gibbs free energy of a binary phase in terms of the concentrations x_i:

$$_{12}E_G = x_1 x_2 \sum_{k=0}^{n} {}^k A_{12}(x_1 - x_2)^k \quad \text{for the 1-2 binary}$$

with a simple summation of the binary relations to obtain the excess free energy for the ternary, i.e. $E_G = {}_{12}E_G + {}_{13}E_G + {}_{23}E_G$. In terms of what has been said above the ${}^kA_{12}$ terms are evaluated by fitting to the ${}_{12}E_G$ relation evaluated by coupling binary phase diagram and thermodynamic data.

Wherever thermodynamic or partial phase diagram data for the ternary system have been available such data have been used to modify these relations in order to improve the accuracy of the extrapolation. With respect to binary systems, usually sufficient data are available so that the coupling of phase diagrams and thermodynamic data yield accurately established phase boundaries, i.e. < 4% uncertainty in composition and < 20°K uncertainty in the temperature (10). The uncertainty of phase boundary compositions in the extrapolated multicomponent phase diagrams can be larger than 4%. Nevertheless, the information content in these calculated multicomponent phase diagrams can often be useful (61). Of course, improvement in the accuracy of the calculated multicomponent phase diagrams is desirable, because the usefulness of such calculations will be materially extended thereby. Such improvement may sometimes be achieved either by feedback from some critical experiments performed on the multicomponent system or by the use of physical models applicable to the description of phases in the multicomponent system (20).

From the above, it is apparent that computer calculation of phase diagrams depends upon the existence of a data bank containing critically evaluated data for binary and higher component systems, where available, which is continually updated. Such data banks are maintained at various locations (Table IV-1). Software to effectuate the extrapolation to multicomponent phase diagrams may also be found at these locations as described in Table II-2. A close examination of this table reveals that not only is it possible to calculate phase diagrams, but also, because of the coupling of phase diagrams to thermodynamics, it is possible to calculate a variety of thermodynamic parameters for the multicomponent systems, many of which are useful.

Table IV-1. Locations of Thermodynamic Data Banks

- MANLABS, Inc., 21 Erie Street, Cambridge, MA, 02139 (L. Kaufman)
- Laboratoire de Thermodynamique et Physico-Chemie Metallurgiques associe au CNRS, Institut National Polytechnic, B.P. 44-38401 St. Martin d'Heres, Grenoble, France (I. Ansara, C. Bernard)
- MTDS, National Physical Laboratory, Teddington, Middlesex, TW11 OLW, England (T.G. Chart, T. Barry)
- Institut für Theoretische Huttenkunde, Technische Hochschule, Aachen, Germany (O. Knacke, O. Kubaschewski, P. Spencer)
- Dept. de Gemie Metallurgique, Ecole Polytechnique, U. de Montreal, Montreal, Quebec H3C 3A7, Canada (A.D. Pelton, C.W. Bale, W.T. Thompson)

Calculations have not been limited to alloy systems. Spencer has recently listed some of the applications to oxide systems, semiconductor systems, and mixed metal-non-metal systems (10). Ansara has listed the alloy ternary systems calculated up to 1979 (9). An up-to-date glossary of phase diagrams is available from the Alloy Data Center, Metallurgy Division, Institute for Materials Research, National Bureau of Standards, Washington, D.C. 20234.

If more accurate multicomponent phase diagrams are required than can be furnished by the extrapolation methods incorporated into the softwares of the CALPHAD community, then it is necessary to perform experiments or to resort to models for the many-body interactions. A beginning along the latter path has been made using modelled lattice energy functions to determine energies versus composition for the competing phases. By fitting these calculated values to power series functions, such as the Kohler function used by some of the CALPHAD community, it is then possible to evaluate ternary and higher interaction coefficients. The latter may then be

Table IV-2. Locations of Software for Phase Diagram Prediction. (See list in Table IV-1 and in addition:)

- Mats Hillert, Royal Inst. of Technology, Dept. of Metallurgy, Division of Physical Metallurgy, 10044 Stockholm 70, Sweden
- G. Petzow, Max-Planck-Institut fur Metallforschung, Institut fur Werkstoffwissenschaften, Stuttgart-80, Busnauerstr. 175, Germany
- M.H. Rand, Atomic Energy Research Establishment, Harwell, Oxfordshire OX11 ORA, U.K.
- G. Inden and W. Pitsch, Max Planck Inst. fur Eisenforschung, Dusseldorf, Germany
- A.P. Miodownik, University of Surrey, Guilford, England
- E.S. Machlin, Columbia University, New York, New York 10027, USA
- R.A. Howald, Chemistry Department, Montana State University, Bozeman, Montana 59715, USA
- H.B. Levine, Jaycor, 1401 Camino del Mar, P.O. Box 370, Del Mar, CA 92014, USA
- K. Osamura, Metallurgy Dept., Kyote University, Kyoto, Japan
- J. Tomiska, Inst. of Phys. Chem., University of Vienna, Austria
- R. Basu, Indian Inst. of Science, Bangalore 560012, India
- E.M. Savitskii, A.A. Baikov Inst. of Metallurgy, Moscow, Leninskii, pr. 49

added to the Kohler or other extrapolating function derived from the binary phase diagrams to obtain a corrected function. The technique just described is too new to know at this writing whether it leads to more accurate multicomponent free energy functions. One investigation using a modified form of this principle has produced reasonable results for Fe-Ni-Cr-X quaternary systems (20).

It may be of interest to the reader to know about some of the applications that have made good use of the multicomponent phase diagrams that can be calculated with the software of the CALPHAD community. The effort to develop insitu fiber reinforced metal composites by directional solidification has made use of calculated multicomponent phase diagrams to locate the liquidus surface, and the eutectic composition and temperature in candidate alloy systems (62). Problems in nuclear reactor technology which are dependent upon multicomponent phase equilibria have been attacked using calculated ternary phase diagrams (63). Problems in the growth of semiconductor crystals have been solved with the aid of multicomponent phase diagram calculations (64). The identities of corrosion products of a heat-resisting alloy used in coal gasification have been determined with the aid of the computer program SOLGASMIX (65) for various conditions of temperature, pressure and composition of the gaseous environment (66). In this way, it has been found how to avoid the production of harmful corrosion products. (The interesting aspect of SOLGASMIX is that it can yield the parameters defining multiphase, multicomponent equilibria for the case of up to 20 elemental components, 99 stoichiometric compounds and 10 solution phases! It is an efficient computer program. It was developed to provide stability diagrams for systems with condensed phases and gaseous compounds. However, it may also be applied to condensed phase equilibria. It is currently being used to evaluate the effect of replacement of Ni by Co on the γ/γ' and $\gamma/\gamma'/\sigma$ equilibria in several multicomponent Ni base superalloys (67).) Phase equilibria in mixed metal/gas systems of interest in extractive and process metallurgy applications have been widely studied using various calculational procedures (68). The above list is by no means an exhaustive one.

It has already been mentioned that there are a variety of extrapolation relations that may be used in the calculation of ternary and multicomponent phase diagrams. Ansara has discussed the differences in the phase diagrams predicted by several of these relations for a particular alloy system (9), which do not seem to be significant. Hillert (60) has compared analytical and numerical methods of predicting the properties of a higher system from the properties of the lower component systems. According to Hillert it should be advantageous to select one analytical and one numerical method which are equivalent to each other in order to gain the complementary advantages of each method, Hillert therefore recommends the universal use of Muggianu's numerical method (69) in combination with the analytic relation for the excess Gibbs energy given above. He also lists three additional interchangeable analytic methods.

A major problem in the calculation of multicomponent phase equilibria is to optimize the computations in order to minimize expense. This problem is continually being researched and the reader is referred to the CALPHAD journal where the latest advance in computational procedure is likely to be published. Among the convergence procedures that have been investigated are the SIMPLEX (70) NEWTON-RAPHSON (71).

Finally, mention must be made of reference 61b, which collects the papers presented at a symposium on Alloy Phase Diagrams at the National Bureau of Standards in 1979. Anyone having a serious interest in alloy phase diagrams should have access to this reference.

Coherent Phase Diagrams

Phase diagram calculations have been based upon Kikuchi's "cluster variation" method of evaluating the entropy and free energy of solid solutions (11) and on his "natural iteration" procedure for optimizing the concomitant calculations (13). De Fontaine has recently reviewed these methods and most of the results obtained (22), in detail. Hence, this section is limited to a brief survey of the salient features of these calculations and results.

Although initially only two body interactions were taken into account in the energy of the clusters, subsequently the method was generalized to allow for many-body interactions. The parameters describing these interactions must be obtained presently by fitting to experimental data.* In a binary system, it is sufficient to have 3 points along phase boundaries to evaluate these parameters. For a ternary system, without consideration of ternary interactions, the 3 x 3 = 9 parameters evaluated from the binary diagrams, are sufficient to yield a calculated coherent ternary phase diagram. Once these 9 parameters are evaluated it is a computationally simpler matter to calculate a coherent ternary phase diagram using the natural iteration procedure than using the procedures described in the previous section. Computer program listings for ternary and binary coherent phase diagram calculations may be obtained from Prof. D. de Fontaine.

Thus, as a "predictive" procedure, the cluster variation--natural iteration (CV-NI) method of Kikuchi, is philosophically the same as the methods described in the previous section. It fits parameters representing pair interchange energy and energies of tetrahedra AAAB and BBBA (or more complicated clusters in the latest practise (27)) to binary phase diagrams and then "predicts" ternary phase diagrams based on these parameters. It is an improvement over Bragg-Williams based procedures in that it provides a more accurate description

*The possibility exists that in the near future these may be obtained from ab initio calculations of the IBM group.

of the configurational entropy. Thus, both methods as described above really represent prescriptions for extrapolating from binary phase diagrams to higher order phase diagrams. However, as noted by de Fontaine, because the configurational entropy is described realistically in the CVM, the fitted parameters yield more accurate terms for the energy of formation than can be obtained by fitting of regular or sub-regular solution equations for the free energy. However, caveats to this claim may be offered as follows.

The CV-NI method neglects the possibility that there are only concentration dependent but configuration independent non-linear contributions to the energy of formation. For example, there may well be an electron concentration contribution to the energy, which depends non-linearly upon concentration in certain alloy systems. In such cases, both the CV-NI procedure and the sub-regular solution procedures subsume such dependence into the empirical parameters for the energy of formation and the theoretical interpretation of the latter is no longer as straightforward as it would be in the absence of a non-linear configuration independent contribution to the energy. Further, the CV-NI procedure neglects non-linear entropy of vibration contributions to the total entropy. This neglect may or may not be warranted. The CALPHAD procedures account for such effects empirically.

Recently, there have been attempts to derive the energies of the tetrahedra from first principles and to compare the coherent phase diagrams so derived with experiment. It appears from one such investigation, using pseudopotential theory to predict pair interactions only, that a realistic predictive capability requires the ability to predict many body interactions (72). As mentioned previously, the IBM group has developed the facility of calculating the energies of formation of long-range-ordered cubic alloy phases. It should thus be possible to use such data to predict the energies of some of the clusters.

Considering that coherent precipitates involve coherent phase equilibria and considering the interest of the alloy designers in such precipitates for strengthening, it is reasonable to expect considerable progress in the description of such systems in the next decade. Among the groups currently active in this area are those at MIT (Prof. S.M. Allen), U.C. - Berkeley (Prof. D. de Fontaine), National Bureau of Standards (Dr. J. Cahn), Columbia University (Prof. J. Sanchez), and in the Netherlands (Dr. J.T.M. de Hosson, Dr. C.M. van Baal).

A rather complete description of the ground state equilibrium configurations (phases or mixtures of phases) at various compositions of coherent binary systems has been achieved (27). This work has been thoroughly reviewed by de Fontaine (22). The contributions of Kanemori (26) have provided the basis of a general analysis of this problem, which has been of some concern in the field of statistical mechanics of order-disorder in solids.

It is of some interest that the procedure for calculating the coherent phase diagram differs from that normally used to calculate non-coherent phase diagrams in the following way. In the latter, the free-energy or the chemical potentials are evaluated as a function of composition and the equilibrium corresponding to a given composition is determined either by minimization of the free-energy (SIMPLEX procedure) or by evaluating the tie-line compositions corresponding to equality of the chemical potentials in the coexisting phases (Newton-Raphson iteration). The Kikuchi natural iteration procedure used in the evaluation of coherent phase diagrams is based upon minimization, at constant chemical potential μ_i, of a new potential, called the grand potential, and given by

$$G = F - \sum \mu_i N_i$$

where F is the Helmholz free energy
N_i is the number of species i
$\mu_i \equiv (\partial F/\partial N_i)|_{N_j,N_k\ldots}$, cyclically, with $i \neq j, \neq k$, etc.

The latter procedure has certain advantages. One big advantage is that it provides a more accurate definition of the tie-line chemical potentials than given by the former methods. Another significant advantage is that with the use of the grand potential the NI procedure necessarily converges to the stable solution and is peculiarly efficient in such convergence, whereas this convergence is not guaranteed in the other procedures. Unfortunately, no one has yet investigated the possible applicability of the NI procedure or an equivalent as an iteration procedure to regular or sub-regular solutions, i.e. it would be useful to have an iteration procedure that necessarily converges in order to save computer time.

V. MISCELLANEOUS

Metal (or Alloy)-Gas Systems

Much of the work in this area has already been reviewed (10). Here, it is useful to consider only significant new applications that have extended the ability to predict phenomena involving the combination of metal (or alloy) and gas systems. A significant contribution in this area is the program SOLGASMIX (65). This program can calculate equilibria in systems containing a gaseous phase, condensed phase solutions and condensed phases of invariant or variable stoichiometry. It evaluates the equilibrium conditions by direct minimization of the system free energy.

A recent application of SOLGASMIX illustrates its capabilities (66). The equilibrium between a type 304 stainless steel (∿ 19% Cr, ∿ 10% Ni, ∿ 2% Mn, balance Fe) and its environment of O_2, S, S_2, S_8, SO_2, SO_3 in the gas phase and CaO, $CaSO_4$ and CaS as heat exchanger tube deposits in a fluidized bed combuster was evaluated at 925°K. Also, these calculations were repeated for an Inconel 600 (7% Fe, 15% Cr, 0.15% Mn, balance Ni) alloy in place of the 304 stainless steel. It was possible to determine the equilibrium phases formed for a variety of conditions on the composition of the gas phase, the permeability of the scale, and the composition of the scale. In particular, it was possible to predict the flue gas compositions that would not form sulfides, as well as those that would for various compositions and permeabilities (yes or no) of the scale deposit.

It is apparent that SOLGASMIX has the potential of providing help in unravelling complex corrosion problems for multicomponent systems. A recent modification of this program, SOLGASMIX-PV (74) is designed to handle a maximum of 20 elements, 99 substances and 10 solutions. A substance here is defined to be a gas or condensed phase species (i.e. CO in a gas phase, M_2O_3 in a condensed phase). The solutions may be gaseous, liquid or solid.

Metastable Phases

The stability (energy of formation) of many hypothetical metastable phases may be predicted. As already mentioned the energies of formation of certain cubic phases may be obtained from first-principle calculations at the IBM Research Laboratory (73). Also, the more empirical procedures of Machlin et.al. (2,5,6,20,45,75) have provided values of the energy of formation of various intermetallic compounds, both non-cubic and cubic, both stable and metastable. It is useful to have the ability to predict these and other properties of hypothetical phases.

Since the work of Frohlich (76), it has been known that superconductivity is associated with lattice instability. The search for high T_c superconductors must entail, in part, a search for unstable phases as well as a procedure for restraining the transformation of such unstable phases to the more stable structures. One procedure for accomplishing the latter has been developed (77). Thus, the definition of potential high T_c unstable phases remains the problem. High T_c may be associated either with a lattice in which most of the modes become unstable or with a lattice in which a particular normal mode becomes unstable. The A15 structure is an example of the latter, although the unstable mode has not yet been positively identified. However, enough is known to be able to identify candidate high T_c A15 type phases (78). The ability to synthesize such phases depends upon the difference in the free-energy of formation between the hypothetical and stable phases (77).

The above illustrates one example of the utility of a capability of predicting energies of formation of hypothetical phases. Given a lattice energy function it should be possible to predict other structural properties, such as elastic constants, lattice parameters, etc. Thus, along with energies of formation, the IBM group routinely predict densities of states and a variety of other properties for cubic alloy phases and the group led by Machlin predict elastic constants and lattice parameters for all types of alloy phases. Research by the latter group has revealed the interesting result that the bcc transition metals are mechanically stable because of the existence of a finite directional covalent bonding contribution to the total bonding in these metals, i.e. a bond bending resistance stabilizes against (110)[1$\bar{1}$0] shear. Further, this group found that the strength of such covalent bonding decreases to zero for average unpaired d electron count less than about 3, i.e. according to this analysis bcc Ni would not only be thermodynamically unstable but it would be mechanically unstable as well. Similar conclusions hold for the B2 (CsCl) type transition metal alloys.

The importance of the bond bending or partial covalent bonding contribution to the elastic constants of metals and alloys has not been appreciated sufficiently. Undoubtedly, this aspect of bonding will receive more attention in the next decade and its contribution to the mechanical stability of various crystal structures of metallically bonded phases will be unravelled. Phillips has described the role that partial covalent bonding plays in the analogous materials that are partially covalent and partially ionic (79).

Another possible application of the ability to predict energies of formation is in the area of glassy alloys. In the process of crystallization from the amorphous state, it is found that metastable phases often crystallize first. This application has not yet been investigated.

Surface and Interfaces

Although first-principle calculations of the surface energies of certain specific metals have been performed, this review is more concerned with methods for the prediction of surface energies for various metals and alloy phases. Miedema and his collaborators have provided a basis for prediction of surface energies using their cellularatomic model (8,80). On the basis of this model, they have predicted that the product of the surface energy and the two-thirds power of the molar volume should be proportional to the heat of vaporization. They have also formulated other dependencies for the surface energy and based upon all these relations have developed a table of surface energies for the elements (8).

Miedema et.al. have also provided a model for calculating the chemical contribution to interfacial energies (8). According to their model for bimetallic interfaces this contribution to the interface energy is given by

$$\gamma_{chem} = 2.5 \times 10^{-9}\, \Delta H_{sol}^{i\ in\ j}/V_i^{2/3}$$

where $\Delta H_{sol}^{i\ in\ j}$ is the partial molar heat of solution of component i in dilute solution in j. In a similar way, their relations for the energy of segregation of solutes to surfaces are

$$\Delta H_{seg} = \Delta H_1 + \Delta H_2$$

$$\Delta H_1 = -0.24\, \Delta H_{sol}^{A\ in\ B}$$

$$\Delta H_2 = -0.13 \times 10^9 (\gamma_B - \gamma_A) V_A^{2/3}$$

These relations have been shown to be in qualitative agreement with experiment (80).

It must be recognized that the Miedema relations are crystal structure insensitive. Hence, they provide first approximations rather than accurate estimates of the predicted quantities. To obtain more accurate estimates it is necessary to resort to a lattice energy model or to first-principles. The original lattice energy model developed by Machlin yields poor values of surface and certain defect energies because the model does not include a dependence of the lattice energy on volume. However, a new volume dependent lattice energy model is currently being developed by this group, which also has the potential of predicting the effect of alloying on the parameters determining the lattice energy. It is planned to use this model in a computer simulation mode to study defects (surfaces, grain boundaries, interphase interfaces, vacancies, etc.) and phonon spectra in alloy phases. Machlin's original model has been used to simulate grain boundary segregation (81). It was discovered that the energy of segregation increases as the ratio of the segregant to the host cohesive energy decreases. Further, for exchange of a particular host atom at a grain boundary and a solute atom in the lattice, the energy does not always vary monotonically with the ratio of solute to host atomic radius. It is possible that these results may be modified slightly with the new model being developed by Machlin.

One of the problems worthy of investigation is that of 2-dimensional "compound" stability at grain boundaries in alloy phases. The most feasible way of exploring this area at present seems to be that of computer simulation. Thus, there is a need for a lattice energy function applicable to this problem.

VI. SUMMARY

The advent of the computer has greatly accelerated research in many areas. Those considered in this paper are related to alloy phase stability and to derivative properties of the lattice energy. It is now possible to obtain sufficiently accurate ab initio first-principle values of the energy of formation of cubic phases to be useful in phase equilibrium studies. Also, it is now possible to usefully model multicomponent phase equilibria in the computer. The consequences of lattice energy models may be explored readily in the computer to yield predicted values for defect energies and structures in metals; energies of formation, lattice parameters, elastic constants of alloy phases; elastic constants and phonon dispersion curves in metals, etc. It is apparent that there is still "cream to be skimmed" in this field of endeavor. Undoubtedly, the predictive capability that has been developed and will be developed further will be put to important practical use as metallurgists and other applied scientists and engineers become aware of it and strive to subject these capabilities to their needs.

ACKNOWLEDGEMENTS

I am grateful to the National Science Foundation for its generous support of my research including that described in this paper.

REFERENCES

1. L. Kaufman and H. Bernstein, Computer Calculation of Phase Diagrams, Academic Press, New York, NY, 1970; L. Kaufman, Energetics in Metallurgical Phenomena III, p. 53, Gordon and Breach, New York, NY, 1967.
2. E.S. Machlin, Acta Metallurgica, 22 (1974), p. 95.
3. J. Shao and E.S. Machlin, CALPHAD, 2 (1978), p. 279.
4. F.H. Hayes and R. Kubitz, Paper presented to CALPHAD VII, in Summary of Meeting (1978), p. 190.
5. E.S. Machlin, Acta Metallurgica, 22 (1974), p. 109; 22 (1974), p. 367; 22 (1974), p. 1433; 24 (1976), p. 543; CALPHAD, 1 (1977), p. 361.
6. E.S. Machlin, in Theory of Alloy Phase Formation, L.H. Bennett, ed.; TMS-AIME, Warrendale, PA, 1980.
7. See E.S. Machlin and B. Loh, Physical Review Letters, 45 (1980), p. 1642 and references listed therein.
8. A.R. Miedema and P.F. de Chatel, in Theory of Alloy Phase Formation, L.H. Bennett, ed.; TMS-AIME, Warrendale, PA, 1980.
9. I. Ansara, International Metals Reviews, Review 238, No. 1 (1979), p. 20.
10. P.J. Spencer and I. Barin, Materials in Engineering Applications, 1 (1979), p. 167.
11. R. Kikuchi, Physical Review, 81 (1951), p. 988.
12. D. de Fontaine and R. Kikuchi, National Bureau of Standards Publication, SP-496 (1977), p. 999.
13. R. Kikuchi and D. de Fontaine, National Bureau of Standards Publication, SP-496 (1977), p. 976.
14. A.R. Williams, J. Kubler and C.D. Gelatt, Jr., Physical Review B, 19 (1979), p. 6094.
15. V. Heine and D. Weaire, in Solid State Physics, 24, H. Ehrenreich, F. Seitz, D. Turnbull, eds.; Academic Press, 1970.
16. G.R. Stewart, L.R. Newkirk and F.A. Valencia, Physical Review B, 21 (1980), p. 5055.
17. D. Stroud, in Theory of Alloy Phase Formation, L.H. Bennett, ed.; TMS-AIME, Warrendale, PA, 1980, and references.
18. F. Ducastelle, in Theory of Alloy Phase Formation, L.H. Bennett, ed.; TMS-AIME, Warrendale, PA, 1980, and references.
19. E.S. Machlin and P.T. Gee, Journal of Solid State Chemistry, 24 (1978), p. 277.
20. J. Shao and E.S. Machlin, Metallurgical Transactions, 10A (1979), p. 585.
21. V.L. Moruzzi, C.D. Gelatt, Jr. and A.R. Williams, Calculated Electronic Properties of Ordered Alloys, Pergamon Press, New York, NY, to be published.
22. D. de Fontaine, in Solid State Physics, H. Ehrenreich, F. Seitz, D. Turnbull, eds.; 34, p. 73, Academic Press, New York, NY, 1979.
23. P.C. Clapp and S.C. Moss, Physical Review, 171 (1968), p. 754.
24. A. Khatchaturyan, Physica Status Solidi, B60 (1973), p. 9; Soviet Physics - Solid State (English Trans.), 5 (1963), p. 16 and p. 548.
25. M.J. Richards and J.W. Cahn, Acta Metallurgica, 19 (1971), p. 1263; S.M. Allen and J.W. Cahn, Scripta Metallurgica, 7 (1973), p. 1261, Acta Metallurgica, 20 (1972), p. 423.
26. J. Kanamori, Progress Theoretical Physics, 35 (1966), p. 66; Journal Japan Institute of Metals, 15 (1976), p. 35; J. Kanamori and Y. Kakehashi, Journal de Physique (Paris), 38 (1977), p. C7-274; M. Kaburagi and J. Kanamori, Progress in Theoretical Physics, 54 (1975), p. 30.
27. J.M. Sanchez and D. de Fontaine, in Structure and Bonding in Crystals, M. O'Keefe and A. Navrotsky, eds.; Academic Press, New York, NY, 1980.
28. O. Kubaschewski, in The Physical Chemistry of Metallic Solutions and Intermetallic Compounds, 1 p. 346, Chemical Publishing Company, 1960.
29. E. Mooser and W.B. Pearson, Acta Crystallographica, 12 (1959), p. 1015; Progress in Semiconductors, Gibson, Kroger and Burgess, eds.; J. Wiley & Sons, New York, p. 103, 1960.
30. J. St. John and A.N. Bloch, Physical Review Letters, 33 (1974), p. 1095.
31. G. Simons and A.N. Bloch, Physical Review B, 7 (1973), p. 2754.
32. J.R. Chelikowsky and J.C. Phillips, Physical Review B, 17 (1978), p. 2453.
33. E.S. Machlin, T.P. Chow and J.C. Phillips, Physical Review Letters, 38 (1977), p. 1292.
34. A. Zunger and M.L. Cohen, Physical Review B, 18 (1978); p. 5449; Physical Review Letters, 41 (1978), p. 53.
35. W. Andreoni, A. Baldereschi, E. Biemont and J.C. Phillips, Physical Review B, 20 (1979), p. 4814.
36. R.E. Watson and L.H. Bennett, Physical Review B, 18 (1978), p. 6439; Journal of Physics and Chemistry of Solids, 39 (1978), p. 1235.
37. A. Zunger, Physical Review Letters, 44 (1980), p. 582; Physical Review B, 22 (1980), p. 5839.
38. J.C. Phillips, Solid State Communications, 22 (1977), p. 549.
39. J.C. Phillips, Review of Modern Physics, 42 (1970), p. 317; Bonds and Bands in Semiconductors, Academic Press, New York, NY, 1973.
40. J.P. Walter and M.L. Cohen, Physical Review, 4B (1971), p. 1877.

40a. R.M. Martin, Physical Review Letters, 21 (1968), p. 536.

41. C.C. Mathei, P.J. Grout and N.H. March, Physics Letters, 68A (1978), p. 351; International Journal of Quantum Chemistry, 12S (1978), p. 443.

41a. J.C. Phillips, Physical Review, 166 (1968), p. 832; 168 (1968), p. 905, p. 912, p. 917; Physical Review Letters, 20 (1968), p. 550.

42. J. Shao and E.S. Machlin, to be submitted, Acta Metallurgica; also J. Shao, Eng.Sc.D. Thesis, Columbia University, 1981.
43. R. Boom, F.R. de Boer and A.R. Miedema, Journal of Less Common Metals, 45 (1976), p. 237; 46 (1976), p. 271.
44. A.R. Miedema, F.R. de Boer and R. Boom, Journal of Less Common Metals, 41 (1975), p. 283; 46 (1976), p. 67.
45. E.S. Machlin, CALPHAD, to be published, 1981.
46. J.C. Phillips, in Theory of Alloy Phase Formation, L.H. Bennett, ed.; TMS-AIME, Warrendale, PA, 1980.
47. D.G. Pettifor, Physical Review Letters, 42 (1979), p. 846; Solid State Communications, 28 (1979), p. 621.
48. C.M. Varma, in Theory of Alloy Phase Formation, L.H. Bennett, ed.; TMS-AIME, Warrendale, PA, 1980; Solid State Communications, 31 (1979), p. 295.
49. A.R. Williams, C.D. Gelatt, Jr. and V.L. Moruzzi, Physical Review Letters, 44 (1980), p. 429.
50. R.E. Watson and L.H. Bennett, CALPHAD, 5 (1981), p. 25.
51. V.L. Moruzzi, J.F. Janak and A.R. Williams, Calculated Properties of Metals, Pergamon Press, New York, 1978.

52. C.D. Gelatt, Jr., A.R. Williams and V. L. Moruzzi, Private Communication, 1981.

53. L. Brewer, in Phase Stability in Metals and Alloys, P.S. Rudman, J. Stringer and R.I. Jaffee, eds.; McGraw-Hill, New York, NY, 1967.

54. Band theory calculations have shown without doubt that the number of s,p electrons in transition metals is about unity regardless of the crystal structure contrary to Engel-Brewer theory requirements of 2 for hcp and 3 for fcc metals.

55. L. Kaufman, in Phase Stability in Metals and Alloys, P.S. Rudman, J. Stringer and R.I. Jaffee, eds.; McGraw-Hill, New York, NY, 1967.

56. W.B. Pearson, The Crystal Chemistry and Physics of Metals and Alloys, Wiley-Interscience, New York, NY, 1972.

57. L.H. Bennett and R.E. Watson, in Theory of Alloy Phase Formation, L.H. Bennett, ed.; TMS-AIME, Warrendale, PA, 1980.

58. See the journal, "CALPHAD", Pergamon Press, where developments in this area are usually published.

59. L. Kaufman and H. Nesor, in Treatise on Solid State Chemistry, N.R. Hannay, ed.; 5, Plenum Press, New York, NY, 1975; Annual Review of Materials Science, 3 (1973), p. 1.

60. M. Hillert, CALPHAD, 4 (1980), p. 1.

61. See the new publication, Bulletin of Alloy Phase Diagrams, published by ASM and the National Bureau of Standards. Also, references in NBS SP-496.

62. L. Kaufman and H. Nesor, Metallurgical Transactions, 5 (1974), p. 1617, p. 1623; 6A (1975), p. 2115, p. 2123.

63. P.E. Potter, Journal of Nuclear Mateirals, 42 (1972), p. 1; 47 (1973), p. 7.

64. G.B. Stringfellow, Journal of Physics and Chemistry of Solids, 33 (1972), p. 665; A.S. Jordan and M.J. Ilegems, Journal of Physics and Chemistry of Solids, 36 (1975), p. 325; G.A. Antypas, Journal of the Electrochemical Society, 117 (1970), p. 700, p. 1393.

65. G. Eriksson, Chemica Scripta, 8 (1975), p. 100.

66. A.P. Miodownik, CALPHAD, 4 (1980), p. 220.

67. Research underway at Columbia University, 1981.

68. T.M. Besmann, Private Communication, 1981.

69. Y.-M. Muggianu, M. Gambino and J.-P. Bros, Journal de Chimie Physique, 72 (1975), p. 83.

70. J.A. Nelder and R. Mead, The Computer Journal, 7 (1965), p. 308.

71. See any text on Numerical Analysis.

72. A. de Rooy, E.W. van Royen, P.M. Bronsveld and J. Th. M. de Hosson, Acta Metallurgica, 28 (1980), p. 1339.

73. A.R. Williams or V.L. Moruzzi, IBM Research Laboratory, Yorktown Heights, New York, 10598.

74. SOLGASMIX-PV may be obtained by request to T.M. Besmann, Oak Ridge National Laboratory, Oak Ridge, Tennessee, 37830.

75. E.S. Machlin and S.H. Whang, Journal of the Physics and Chemistry of Solids, 39 (1978), p. 607.

76. H. Frohlich, Proceedings of the Royal Society, A215 (1962), p. 291.

77. E.S. Machlin and P. Chaudhari, in Synthesis and Properties of Metastable Phases, E.S. Machlin and T.J. Rowland, eds.; TMS-AIME, Warrendale, PA, 1981.

78. E.S. Machlin and T.J. Rowland, in Synthesis and Properties of Metastable Phases, E.S. Machlin and T.J. Rowland, eds.; TMS-AIME, Warrendale, PA, 1981.

79. J.C. Phillips, Bonds and Bands in Semiconductors, p. 64, Academic Press, New York, NY, 1973.

80. A.R. Miedema, Zeitschrift für Metallkunde, 69 (1978), p. 455.

81. E.S. Machlin and A. Levi, Scripta Metallurgica, 14 (1980), p. 127.

THERMODYNAMICS AND KINETICS OF PHASE SEPARATION

D. de Fontaine

Department of Materials Science and Mineral Engineering
University of California
Berkeley, California 94720

1. INTRODUCTION

The art of the Physical Metallurgist is the control of properties through microstructure. Microstructure, in turn, is controlled by bringing about certain phase changes, usually carried out far from thermodynamic equilibrium. Of central practical importance, then, are temperature-time-transformation (TTT) diagrams, of the type shown in Figure 1. Such diagrams determined experimentally, indicate to the alloy designer, after how long (t = time) an isothermal heat treatment (T = temperature) will yield a given new phase β (or mixture of phases) from a parent α phase, stable above the thermodynamic $\alpha \to \beta$ equilibrium temperature T_0. Two T-vs.-log t curves are usually plotted: one (full line) which indicates the "start" of the transformation when, say 1% of α has transformed, and an "end" curve (dashed line) when, say 99% of α has transformed. Considerable use is made of TTT diagrams in the heat treating of steels, for example (1).

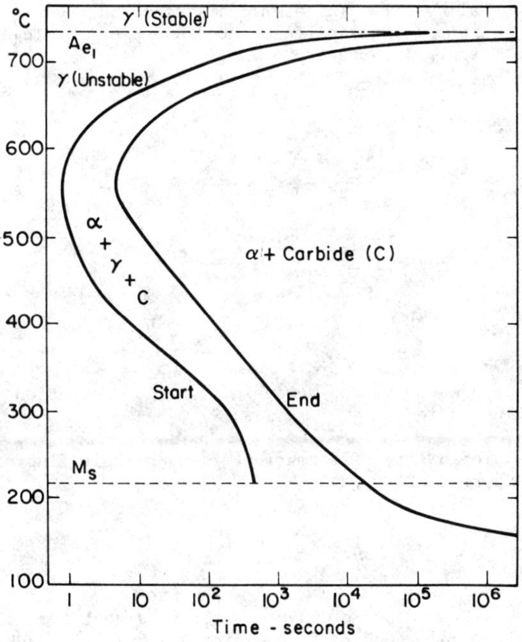

Figure 1. TTT diagram (C-curve) for a plain carbon eutectoid steel.

A typical diagram for an alloy steel is shown in Figure 2. It is seen that the C-shaped curve of Figure 1 has become an S-shaped curve. This feature is thought to be due to the appearance, at lower transformation temperatures, of a variant of the β phase, say β', for which a separate C-curve might exist, and which could be determined experimentally were it not for the interference of the stable product β. The S-curve can then be thought of as resulting from the combination of two or more C-curves. Not only can a C-curve transform to an S-curve on alloying, but the whole "start" curve often shifts laterally to much longer aging times. Slower transformation rates may be beneficial for transformations in steels in order to produce a Martensic product which appears from γ (Austenite) almost instantaneously below a certain temperature, indicated by the horizontal line labeled M_S in Figure 2. Therefore, in order to produce Martensite upon continuous cooling from a temperature in the stable γ range, it is necessary to prevent formation of the normal transformation products (pearlite, bainite), hence to avoid the "nose" of the C-curves. Clearly a shift of the transformation curves to longer times will promote the formation of martensite under far less severe quenching conditions.

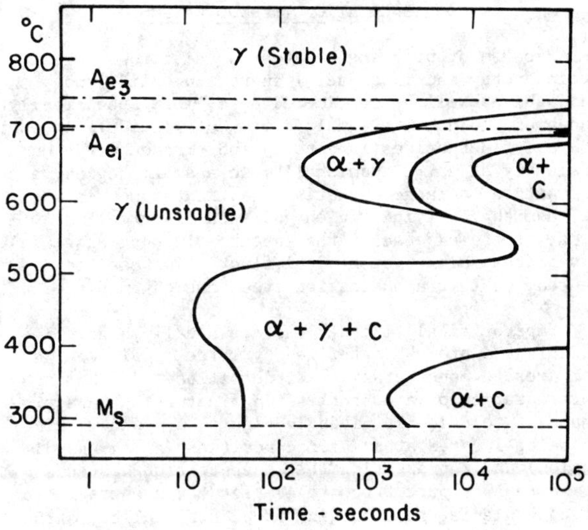

Figure 2. TTT diagram (S-curve) for a 0.33% C, 0.45% Mn, 1.97% Cr steel.

It is quite impossible to predict theoretically the TTT diagram for a given alloy of given composition. At best, one can attempt to understand the mechanisms underlying such basic features as: the transition temperature T_0, the shapes of the individual C-curves, the transformation rates. The most basic concept required is that of nucleation of a new phase or phases (β) from the parent (α). Hence, the major portion of this article will be devoted to nucleation theory, with emphasis placed on the fundamental difficulties encountered in trying to develop a tractable model valid for transformations near and far from equilibrium states. Indeed, even for the simplest of physical situations, homogeneous nucleation in single-component fluids, the nucleation problem turns out to be one of the most difficult ones encountered in the physical sciences.

There exists a number of excellent reviews or textbooks which cover substantially the same material as presented here in greater or lesser detail; in particular, there is a very thorough treatise of J. W. Christian entitled "The Theory of Transformations in Metals and Alloys" (2). The basics of phase transformations in solids has also been reviewed more succintly by a number of authors, one of the best expositions being that of D. Turnbull (3). Nucleation Theory has been covered, from the metallurgist's viewpoint, in two reviews by K. C. Russell (4, 5). The latter very recent article constitutes an excellent introduction to the extensive work on nucleation in solids by the Aaronson-Lee-Russell school. In that context, the didactic article of H. I. Aaronson (6) should prove to be very valuable to students in this field. Some fundamental aspects of nucleation were covered in a short article by J. W. Cahn (7); spinodal decomposition was covered by the same author (8) and also by J. H. Hilliard (9) and the present author (10). Both topics are treated together in a most lucid and complete manner by Martin (11). This latter article is the text of a course given at the summer school of Aussois (France) in 1978. The courses taught there covered virtually every important aspect of phase transformations in Materials Science, and were recently published (in English) in book form. The reader interested in the more theoretical aspects are referred to the articles by Reiss (12, 13), Langer (14, 15), Binder (16, 17) and their co-workers. Nucleation thermodynamics, to be covered presently, follows the classic texts of Gibbs (18) and of Landau and Lifshitz (19).

2. NUCLEATION THERMODYNAMICS

Because of problems related to anisotropic elastic strain energy and interfacial energy in solids, only fluid phases will be considered here, more specifically, condensed fluids, i.e., multicomponent liquids. At equilibrium, fluctuations in density and/or concentration constantly occur. Occasionally, in a small region, a fluctuation so large may arise that it becomes more stable than the outlying region and actually grows, or matures, at the expense of the latter. In that case, it is said that a supercritical fluctuation has been created, or that a nucleation event has taken place.

Supercritical fluctuations are usually highly improbable and are characterized by large local departures from average concentrations or densities. A schematic representation of a concentration fluctuation in a two-component solution is given in Figure 3a, each vertical bar being located at an atomic position (along one dimension) and representing the concentration distance from pure A or pure B to the average B concentration c^0. For convenience, an averaging procedure, called coarse-graining, is then introduced: the discrete picture of Figure 3a is made continuous by counting up and down bars in a small region of space (coarse grain) so as to define a local average concentration c, plotted as a deviation from the overall average in Figure 3b. The size for these coarse grains over which the average is taken gives rise to some difficulties, as discussed, for example, by Langer (15, 20).

Let us now determine the probability of formation of such a fluctuation in an otherwise uniform system, in the sense of the coarse-graining just introduced. To that effect, let us consider a closed system of finite volume V embedded in a very large reservoir, the latter having constant pressure $\bar{P}$ and temperature $\bar{T}$. The reservoir and system together comprise a very large "universe" assumed to be bounded by perfectly rigid adiabatic walls. Any change occurring in the system gives rise to changes in the extensive variables U, V, S (internal energy, volume, entropy) of the system and of the reservoir ($\bar{U}$, $\bar{V}$, $\bar{S}$). We have

$$\Delta U_{tot} = \Delta U + \Delta\bar{U} = 0$$

$$\Delta V_{tot} = \Delta V + \Delta\bar{V} = 0$$

$$\Delta S_{tot} = \Delta S + \Delta\bar{S} \neq 0$$

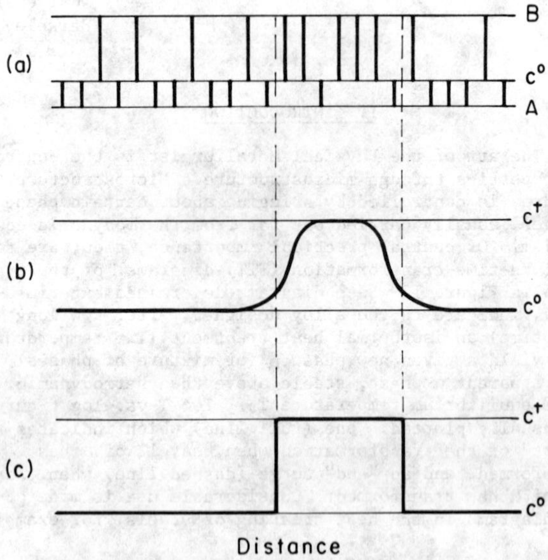

Figure 3. Schematic representations of a (one-dimensional) concentration fluctuation in a binary alloy A-B. (a) atomic distribution about the mean c^0, (b) diffuse interface profile, (c) sharp interface profile.

For the reservoir,

$$\Delta\bar{U} = \bar{T}\Delta\bar{S} - \bar{P}\Delta\bar{V}$$

so that

$$-T\Delta S_{tot} = \Delta U - \bar{T}\Delta S + \bar{P}\Delta V = W ,$$

where W is seen to be the reversible work of formation of the fluctuation.

The probability of formation of an unlikely fluctuation is approximately given by

$$\mathcal{P} = \omega^{\dagger}/\omega^{0}$$

where $\omega^{\dagger}$ is the number of ways of realizing that type of fluctuation, and ω^0 that of realizing the uniform phase. By the microcanomical formula

$$S = k \ln \omega ,$$

where k is Boltzmann's constant, we have

$$\Delta S_{tot} = S^{\dagger} - S^{0} = k \ln \omega^{\dagger}/\omega^{0} < 0$$

so that

$$\mathcal{P} = e^{\Delta s_{tot}/k} = e^{-W/kT} . \tag{1}$$

Since we further assume that, at all times, the temperature T and overall pressure P of the system are equal to those of the reservoir, we shall henceforth drop the bars over T and P. Hence, the work of formation becomes

$$W = \Delta F + P\Delta V , \tag{2}$$

F being the Helmholtz free energy U - TS.

2.1. Sharp Interface Model

It is now necessary to express W in terms of variables internal to the system; local concentrations (mole fractions), volumes, etc. What is then usually done is to further approximate the fluctuation profile, such as that of Figure 3b, by a simpler one, such as that of Figure 3c, consisting of the outer region (^{0}) of volume V^0 of uniform concentrations c_i^0, and partial molar volumes v_i^0, and an inner region (†) of volume $V^\dagger$ of uniform concentrations $c_i^\dagger$ and partial molar volumes $v_i^\dagger$ (i = 1, 2, ... n). Such is the so-called capillarity approximation. The location of the sharp interface separating the fluctuation (†) from the homogeneous region (^{0}) must be decided by an appropriate convention; Gibbs (18) chooses that which makes the interfacial energy σ equal to a surface tension defined in the usual mechanical way. With this choice, and for a spherical volume $V^\dagger$, Gibbs further shows that σ is independent of the sphere radius r.

The outer region (^{0}) may be considered as a bulk phase and classical (bulk) thermodynamics may be applied to it. The (†) region, however, requires special consideration: although actually of inhomogeneous nature, the fluctuation can find itself in (unstable) equilibrium with the bulk (^{0}), in which case the chemical potentials $\mu_i^\dagger$ are regarded as numerically equal to the corresponding ones μ_i^0 in the bulk (^{0}). The $\mu_i^\dagger$ are then defined operationally as the change in Gibbs free energy (G = F + PV) when a small amount of substance i is added to the $V^\dagger$ region at constant P and T. For a bulk phase, the concentrations c_i do not change in such an operation; but in the present case, since the fluctuation is so small in extent, the $c_i^\dagger$'s cannot be maintained constant. Hence, in general, the $\mu_i^\dagger$ so defined are not those which would be found in a bulk phase of identical average concentrations and pressure.

With these words of caution, we may now proceed. The particularly clear treatment given by Landau and Lifshitz (19) will be followed. In Equation (2), the Δ's have meaning of "final state" (system containing fluctuation †) minus "initial state" (system completely homogeneous). Since V^0 is assumed to be so much larger than $V^\dagger$, the transfer of $N^\dagger$ moles (or atoms) of constituents from bulk to fluctuation causes no appreciable change in state (^{0}), so that the bulk phase may be subtracted out of Equation (2). Denoting initial and final states by subscripts $_1$ and $_2$ we have

$$W = F_2 - F_1 + P\Delta V^0 + PV_2{}^\dagger - PV_1{}^\dagger + A\sigma ,$$

since the final state (with fluctuation) contains an interface of area A, say, and of (isotropic) surface tension σ. Indeed, because of surface tension, the fluctuation will be under pressure $P^\dagger >> P^0 = P$. Then, after adding and subtracting the term $P^\dagger V_2{}^\dagger$, we have

$$W = G_2 - G_1 - V^\dagger\Delta P + A\sigma , \tag{3}$$

where

$$\Delta P = P^\dagger - P^0 , \tag{4}$$

and where $P\Delta V^0$, a term small compared to W at ordinary pressures, has been discarded. In Equation (3), G_1 stands for the free energy required to extract $N_i^\dagger$ moles (molecules, atoms) of i (i = 1, ... n) from the bulk at T and P:

$$G_1 = \sum_{i=1}^{n} \mu_i^0 N_i^\dagger = N^\dagger \sum_{i=1}^{n} \mu_i^0 c_i^\dagger .$$

Since this expression is linear in $c_i^\dagger$, with coefficients μ_i^0, it represents, by the intercept rule, a point of coordinates $c_i^\dagger$ on the plane tangent at c_i^0 to the Gibbs free energy surface of the homogeneous phase α [see Figure 4 for a binary example]. Since

$$G_2 = N^\dagger \sum_{i=1}^{n} \mu_i^\dagger c_i^\dagger ,$$

where $\mu_i^\dagger$ are the true chemical potentials in the fluctuation region, under pressure $P^\dagger$, we have, under (unstable) equilibrium conditions [$\mu_i^0 = \mu_i^\dagger$]

$$G_1 = G_2$$

and hence, from Equation (3)

$$W = - V^\dagger\Delta P + A\sigma , \tag{5}$$

an important formula, first given by Gibbs.

The work of formation W can be expressed in a more convenient way by integrating the Gibbs free energy of (†) from pressure P^0 to pressure $P^\dagger$:

$$G^\dagger(P^\dagger) = G^\dagger(P^0) + \int_{P^0}^{P^\dagger} v^\dagger(p,T)\,dp \simeq G^\dagger(P^0) + v^\dagger(P^\dagger - P^0) , \tag{6}$$

the latter equality being obtained by considering the (†) material as incompressible. Thus, by Equations (4) and (5)

$$W \simeq v^\dagger\Delta G_v + A\sigma , \tag{7}$$

where

$$\Delta G_v \simeq - \Delta P$$

is the vertical distance from the tangent plane to $G^0(P)$ at c_i^0 to the free energy surface $G^\dagger$ at the ambiant pressure P^0, at composition $c_i^\dagger$, the Gibbs functions being normalized to unit volume of (†). This "tangent construction" is illustrated in Figure 4. A tacit assumption is usually made: that ΔG_v refers entirely to Gibbs free energy surfaces of bulk phases. Actually the derivation of Equation (6) was performed on free energies of non-uniform fluctuation regions so that G^0 may be regarded as corresponding to a point on the G^α free energy surface of bulk α phase (parent phase), which is correct, and $G^\dagger$ may then be regarded as corresponding to a point on the G^β free energy surface of bulk nucleating phase, at $P^\dagger$ and T, which is not strictly correct. This may lead to some confusion, or to alternate definitions of chemical potentials, as will presently be shown. In both Equations (5) and (7) the first term in the work of nucleation is negative and represents the driving force for nucleation. The second term represents the positive work of forming the interface.

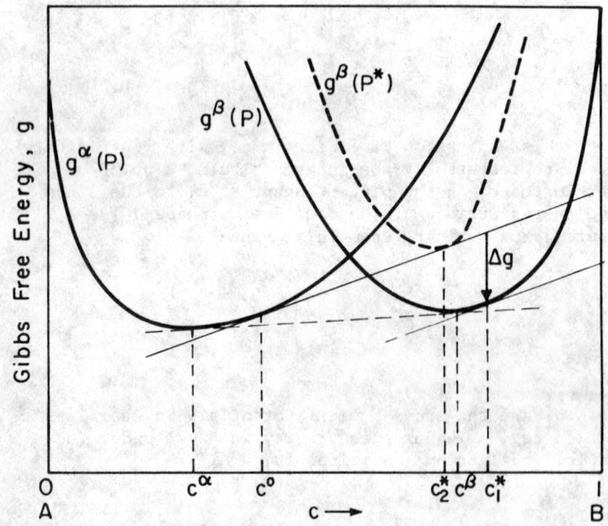

Figure 4. Gibbs free energy curves for a binary fluid. Concentration c_1^* is that obtained by maximizing the quantity Δg (tangent construction), c_2^* might be the concentration of the critical nucleus if the concentration dependence of the interface were taken into account.

The internal variables characterizing the fluctuation in equilibrium must now be determined. For this purpose, the fluctuation volume $V^\dagger$ will be assumed spherical of radius r. We may then take r and n - 1 independent concentration variables as the internal variables of W and express the condition of equilibrium thus:

$$\left(\frac{dW}{dr}\right)_{P,T,c_i} = 0$$

$$\left(\frac{dW}{dc_i}\right)_{P,T,r,c_{j\neq i}} = 0 \quad (i = 1, 2, \ldots n-1).$$

The first condition yields

$$\Delta G_v \frac{\partial V}{\partial r} + \sigma \frac{\partial A}{\partial r} = 0 , \tag{8}$$

since the free energies at constant T, P, $c_1 \ldots c_n$ are invariant, and since σ is independent of radius of curvature. Equation (8) yields the critical radius r^*:

$$r^* = \frac{2\sigma}{\Delta P^*} \simeq - \frac{2\sigma}{\Delta G_v{}^*} . \tag{9}$$

Spherically symmetric fluctuations with $r < r^*$ should decay, those with $r > r^*$ should spontaneously grow.

The denominators ΔP^* or $\Delta G_v{}^*$ are determined by the second condition. If the interfacial energy σ is independent of concentration, this condition requires the vertical distance between the G^0 tangent hyperplane and the G^β surface to be a maximum. The critical composition c_i^* (i = 1, ... n) is thus obtained by constructing another tangent plane to the lowest portion of the G^β surface parallel to that at c_i^0 to G^α. This is shown for a binary example in Figure 4. This approximation is fairly reliable if the degree of supersaturation (or of undercooling) is small, i.e., if ΔG_v is small in magnitude.

If supersaturations are large, and particularly if the coarse-grained free energy of a condensed-phase solution is given by a continuous smooth curve such as that of Figure 5, then σ can no longer be considered as composition-independent. In that case, the search for equilibrium is best carried out, following Reiss and Shugard (R-S, 13), by taking derivatives of W with respect to an alternate set of independent variables, the fractional number of moles $N_1, N_2 \ldots N_n$:

$$\left(\frac{dW}{dN_i}\right)_{P,T,N_{j\neq i}} = 0 .$$

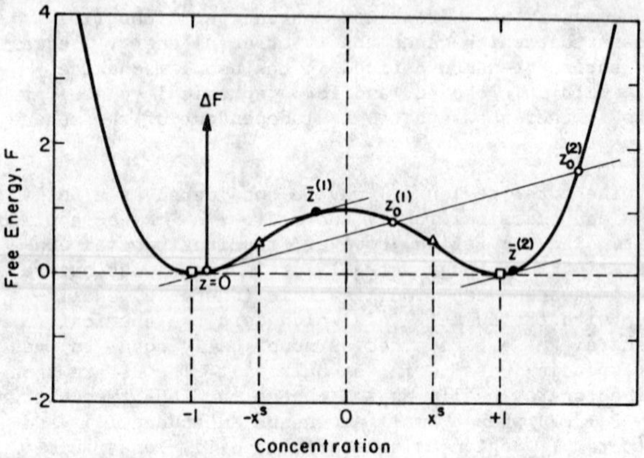

Figure 5. Continuous (coarse-grained) free energy curve for use in coherent nucleation. Curve is plot of Equation (49).

One then readily obtains the equilibrium condition (13)

$$\mu_i^\beta + (P^\beta - P)V_i^\beta + A\sigma_i = \mu_i^\alpha , \tag{10}$$

the subscripts on V and σ indicating derivatives with respect to $N_i^\dagger$. In this equation μ_i^β is the chemical potential of i in bulk β at pressure P^0. The first two terms together yield $\mu_i^\beta(P^\beta)$, the chemical potential of i in bulk β at the pressure of the critical fluctuation, or critical nucleus. Surprisingly, the condition of equality of chemical potentials no longer holds here because of the presence of the correction term $A\sigma_i$. Thus, as announced, this derivation has led to an alternate definition of chemical potentials ($\mu_i^\beta \neq \mu_i^\dagger$), which is, however, a useful one, as it allows the concentration of the critical nucleus to be determined in cases of large supersaturations. Indeed, R-S (13) show that, as c_i^0 tends to the so-called spinodal concentrations c_i^s, at which the second derivatives of G with respect to concentration vanish, $c_i^\dagger$ correspondingly moves inward towards the same spinodal composition. The same conclusion can be reached by adopting quite a different model for the critical nucleus, as proposed by Cahn and Hilliard (C-H, 21, 22). This model will be examined later (Section 2.2).

The work of formation of the critical nucleus can be obtained by substituting for r^*, from Equation (9) into Equations (5) or (7):

$$W^* = \frac{16}{3}\pi\frac{\sigma^3}{(\Delta P^*)^2} = \frac{16}{3}\pi\frac{\sigma^3}{(\Delta G_V{}^*)^2} . \qquad (11)$$

It is seen that, as the concentration of the α supersaturated phase reaches the equilibrium one given by the common tangent construction, the critical radius r^* and the critical work W^* both tend to infinity as ΔP^* ΔG_V^* both go to zero.

At this point, it is already possible to gain some qualitative insight into some of the features of the TTT diagram of Figure 1. Firstly, the temperature T_0 there indicated is obviously the thermodynamic equilibrium temperature at which ΔG_V vanishes, i.e., for which we have $\mu_i^\beta(P^0) = \mu_i^\alpha(P^0)$. Above T_0, the α phase must be stable toward any β-type fluctuations, of arbitrary magnitude. Below T_0, α should decompose at least partially to β, according to equilibrium thermodynamics. The kinetics of the appearance of transformation products may be very slow, however. In particular, for low undercoolings, W^* is extremely large so that, by Equation (1), the probability of finding a critical fluctuation, which will initiate the transformation, will become vanishingly small. Thus, just below T_0, we expect very long transformation times, as observed (Figure 1). At the other extreme, as T tends to zero absolute, W^*/kT will tend to infinity, again lengthening the transformation times immeasurably. At some intermediate temperature, a shortest transformation time is expected. A more quantitative description of the C-curve must be deferred, however, until a kinetic treatment is given (Section 3).

2.2. Diffuse Interface Model

One may decide not to make the Gibbs construction of "squaring the profile" in Figure 3; then, since a sharp interface term $A\sigma$ may no longer be included in the work W, it is necessary to take the fluctuation inhomogeneity into account in another way. This is done formally by allowing each volume element dv to possess a local free energy f(c) dv, where f is a Helmholtz free energy per unit volume evaluated in bulk material having composition c equal to the average composition $c(\underset{\sim}{x})$ in volume dv about point $\underset{\sim}{x}$ [In this section, we shall consider binary solutions only, for simplicity, so that there will be but one independent concentration variable, c, defined to be that of the minority component in the solution]. Such is not, however, the only contribution to the volume element's free energy; no matter how small dv be taken (in the continuum approximation) its composition can never, in general, be regarded as homogeneous. To take this non-uniformity into account, one must assume that the local free energy contains terms depending on the composition gradient (21, 22, 19). Cahn and Hilliard show that the first correction to the local free energy is proportional to the gradient of concentration squared $(\nabla c)^2$. This is because, by symmetry, the gradient correction can only depend on the square of the nabla operator ∇, the other ∇^2, the Laplacian, being then incorporated into the square gradient. In this model, then, the free energy of an arbitrary volume V of a binary solution is expressed as a sum of elementary contributions, thus (21)

$$F = \int_V [f(c) + \kappa(\nabla c)^2]\, dv \qquad (12)$$

where κ is the so-called gradient energy coefficient, assumed to be approximately composition-independent.

To apply Equation (12) conveniently to the nucleation problem, one must assume that the free energy f(c) varies smoothly with c, as shown in Figure 5. Then the "bulk" free energy f(c) is well-defined everywhere at given temperature T and pressure P. The critical nucleus is now that spherically symmetric fluctuation which makes the functional F stationary for all small changes $\delta c(\underset{\sim}{x})$ of the composition profile which leave to average concentration c^0 invariant:

$$\int_V (c - c^0)\, dv = 0 .$$

This auxiliary condition introduces a Lagrange multiplier λ, so that the Euler equation for this variational problem is

$$\frac{\partial L}{\partial c} - \nabla\cdot\frac{\partial L}{\partial \nabla c} = 0 \qquad (13)$$

where the Lagrangian is

$$L(c,\nabla c;\underset{\sim}{x}) = f(c) + \kappa(\nabla c)^2 - \lambda(c - c^0) . \qquad (14)$$

There results the differential equation

$$2\kappa\nabla^2 c = f'(c) - \lambda , \qquad (15)$$

where the accent denotes differentiation with respect to c.

Let us place the origin of a spherical polar coordinate system at the center of the nucleus. Far away from the nucleus, the concentration must be very nearly c^0, and all gradients must vanish. Then, by Equation (15), the multiplier λ must be equal to f_0', the slope of the tangent to f(c) at $c = c^0$. The differential equation (15) then becomes, explicitly,

$$2\kappa\frac{d^2c}{dr^2} + 4\frac{\kappa}{r}\frac{dc}{dr} = f'(c) - f_0' \qquad (16)$$

This second-order ordinary differential equation is nonlinear in the profile function c(r) because of the presence of the derivative f'(c). Since the free energy curve, in order to have the features apparent in Figure 5, must be Taylor-expanded at least to fourth order in c, f'(c) must be at least a polynomial of degree three in c. Hence, the critical nucleus profile $c^*(r)$ must be obtained by numerical integration of Equation (16).

The critical work of nucleation W^* is obtained by inserting the (unstable) equilibrium profile c^* into Equation (12) and integrating, after subtracting off the free energy of the uniform solution c^0:

$$W^* \equiv \Delta F = 4\pi\int_0^\infty [\Delta f(c^*) + \kappa\left(\frac{dc}{dx}\right)^2]\, r^2\, dr \qquad (17)$$

where

$$\Delta f(c^*) = f(c^*) - f(c^0) - (c^* - c^0)f_0' \qquad (18)$$

It is seen thatthe latter expression, in different notation, is just the vertical distance from the tangent (at c^0) to the free energy curve at c^*, as in the classical treatment of the previous section.

Cahn and Hilliard (22) obtained numerical solutions for average concentrations c^0 just inside the phase boundary c^α (common tangent rule), just above the spinodal [c^s, solution of $f''(c) = 0$], and at two concentrations intermediate between these two. These authors concluded that the concentration at the center of the nucleus was such as to make the driving force $\Delta f(c)$ negative, as expected on "classical" grounds. Also, for c^0 just below but close to c^α, the nucleus resembled the classical one with composition at the nucleus center close to c^β, and the nucleus radius becoming very large, along with the work of nucleation W^*. Also, the interface portion of the profile (high gradient) tended to be relatively narrow, with interfacial energy close to that given theoretically for a flat interface. When c^0 was made to move away from c^α, the critical radius at first decreased along with the work of nucleation, as in the classical case. However, unlike the classical model, the concentration at the center of the nucleus tended to decrease towards the average c^0 and the interface became increasingly diffuse. Finally, for c^0 close to the spinodal c^s, the critical radius r^* again became unbounded whilst the work of nucleation continued to drop, becoming zero for c^0 at c^s. In a sense, then, the critical nucleus at the spinodal is the whole uniform solution itself, which therefore now finds itself inherently unstable.

These results are of central importance for the theory of homogeneous nucleation, i.e., in cases where the nucleating phase (β) may be regarded as having the same coarse-grained free energy function as the parent phase (α). In turn, this implies that α and β have basically the same crystal structure, or both be fluids of similar nature. Or, to put it yet differently, the difference between the bulk (^{0}) and the nucleus (†) must be such as to be described entirely by the change in value of a small number of continuously varying "order parameters" ξ_i, in the present case of a binary solution, the unique concentration difference $\xi = c^\dagger - c^0$.

Unfortunately, the difficulty encountered in solving the differential equation (16) has thus far inhibited the application of the Cahn and Hilliard theory to practical cases. There exists, however, a simplified treatment which is, in some sense, intermediate between the full diffuse interface treatment and the classical one. The idea is based upon the Rayleigh-Ritz method of minimizing a functional by parametrizing the unknown function. The integration is then performed on the trial function containing a few judiciously chosen parameters, and the resulting integral is minimized with respect to those parameters. Such a procedure was adopted by this author (23) who chose a standard sharp interface spherical profile, convoluted by a Gaussian curve in order to produce a diffuse interface for which gradients could be evaluated according to the C-H recipe. Two parameters were used: the order parameter ξ (defined above, also called concentration amplitude in the previous work) and an average radius r. The two parameters scaled the profile in any way desired. With this simplified treatment, the main conclusions of the original C-H treatment could be recovered with no great computational effort. It was also concluded, by examining the complete $W(\xi,r)$ surface for various values of c^0 (at constant T), that the singularity of r^* at the spinodal could have but very mild consequences and that there could be no real discontinuity of mechanism at the spinodal, as far as decomposition kinetics were concerned. These conclusions have been well substantiated by more recent work of Langer (20, 24), Binder (16), Lebowitz (25) and coworkers. We shall return to this question in Section 3.4.

One may also consider classical nucleation as a highly simplified Rayleigh-Ritz treatment of C-H formulation. This is done by performing the integration (17) in three portions

$$W^* = 4\pi \left\{ \int_0^{r^*} + \int_{r^*}^{r^*+\delta r} + \int_{r^*+\delta r}^{\infty} \right\} .$$

The first and third integrals are performed over regions of roughly uniform compositions, so that the gradients are taken to be zero. The first integral then yields, by the mean value theorem,

$$4\pi\ \Delta f(c^\dagger) \int_0^{r^*} r^2\ dr = \frac{4}{3}\pi\ r^{*2}\ \Delta f\ ,$$

whilst the third integral vanishes along with $\Delta f(c^0)$. The second integral is over a spherical shell of thickness δr and is naturally identified with the interfacial free energy $A\sigma$, in the present case, $4\pi r^{*2}\sigma$. If σ is assumed to be concentration-dependent (as calculated, say, by the C-H formula), one recovers the treatment of R-S (13). If σ is considered to be concentration-independent, one recovers the standard classical treatment but then, conclusions regarding the behaviour of r^* and W^* near the spinodal are irrevocably lost.

3. NUCLEATION KINETICS

Although we now know from Equation (1) that the probability of finding an embryo located on any one of the N lattice sites of a crystal will be

$$\mathcal{P} = Ne^{-W/kT}\ , \tag{19}$$

still, we have as yet little information about the actual rate of formation of nuclei, i.e., about nucleation kinetics *per se*. Such information can be obtained in an approximate manner by treating the problem in a way first suggested by Farkas (26), Volmer and Weber (27), Becker and Döring (28). The original idea was extended by Reiss to the problem of binary solutions (12). We shall treat here directly the multicomponent case, in a simplified version of the more complete (and more correct) treatment of Langer (14, 15) and of B-S (16).

3.1. Cluster Diffusion Equation

We begin by considering that the growth of a cluster of atoms, or embryo, or protonucleus, takes place by chance collisions with smaller ones. Likewise, decay is caused by a cluster's splitting up into smaller subunits. Clusters are considered to be essentially unlocalized and non-interacting. Although B-S (16), for example, allow the addition or removal of arbitrary subunits to a given cluster, and even the coagulation of a pair of clusters to form a larger entity, we shall follow the usual practice of allowing growth or decay by the addition or removal of single atoms (or molecules, or monomers) only to the embryos. Graphically, a cluster can be thought of as moving by discrete nearest-neighbor jumps in an n-component lattice (n = number of molecular species), as indicated schematically (for n = 3) in Figure 6.

At given time t, let the concentration of cluster characterized by coordinates N_1, N_2 ... N_n be $C(N_1, N_2 \ldots N_n;t)$ or $C(\underset{\sim}{N})$ for short, where, as above, N_i denote number of atoms (or molecules) of type i. Of course, it is assumed that the embryos all have some standard simple shape, such as the spherical one. Additional variables could be included to take care of changes of shape, but this will not be done here.

The time rate of change of embryo concentration may be written formally as

$$\frac{\partial C(\underset{\sim}{N})}{\partial t} = - \sum_{j=-n}^{n}{}' J_j(\underset{\sim}{N}) , \tag{20}$$

where the summation is over all positive and negative directions in multicomponent space (j = 0 excluded). Hence, this summation represents the discrete-space divergence of the flux vector in direction j:

$$J_j(\underset{\sim}{N}) = -[\beta_{N_j} C(\underset{\sim}{N}) - \alpha_{N_j+1} C(\{N_j + 1\})]. \tag{21}$$

In this equation, β_{N_j} is the intrinsic rate of attachment of atoms of type j to a cluster containing $N_1 \ldots N_n$ atoms, whilst α_{N_j+1} is the intrinsic rate of loss of atoms of type j from a cluster containing $\{N_j + 1\} \equiv N_1, N_2, \ldots, N_j+1, \ldots N_n$ atoms. Both α and β are positive quantities. Thus, Equations (20) and (21) merely indicate that the rate of change of concentration of clusters at $\underset{\sim}{N}$ is made up of single nearest-neighbor steps to $\underset{\sim}{N}$ from neighboring sites on the n-dimensional lattice of embryos (Figure 6), or away from $\underset{\sim}{N}$ to nearest neighbor sites.

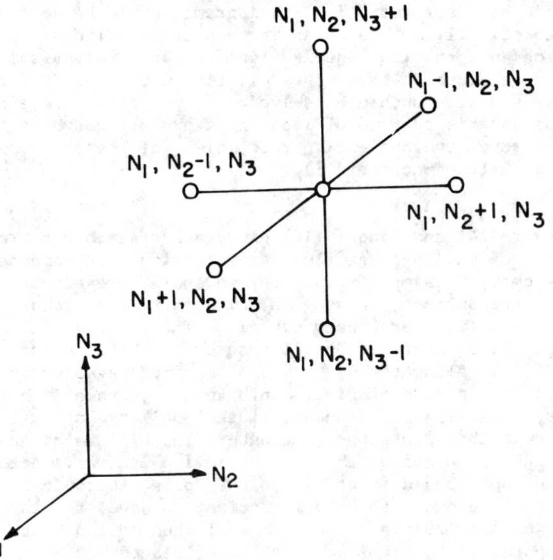

Figure 6. "Lattice" of embryos for a three-component system. Jumps are from center ($\underset{\sim}{N}$) to (or from) nearest neighbor sites.

Let us now consider the equilibrium distribution of clusters for which all fluxes J_j vanish identically everywhere. Then, if we restrict consideration to those clusters small enough so that their work of formation W is positive, we shall have at equilibrium, by Equation (19), the Boltzmann-like distribution

$$C^0(\underset{\sim}{N}) = N e^{-W(\underset{\sim}{N})/kT} \tag{22}$$

Then,

$$J_j^0(\underset{\sim}{N}) = 0 = -\beta_{N_j} C^0(\underset{\sim}{N}) + \alpha_{N_j+1} C^0(\{N_j + 1\}) \tag{23}$$

and therefore

$$\alpha_{N_j+1} = \beta_{N_j} C^0(\underset{\sim}{N})/C^0(\{N_j+1\}) . \tag{23}$$

We now make the daring assumption (detailed balancing) that the relation (23) for the evaporation rate α is valid even for non-equilibrium situations. By eliminating α from Equations (20) and (21) by means of (23) we then find:

$$\frac{\partial C(\underset{\sim}{N})}{\partial t} = - \sum_{j=1}^{n} [\beta_{N_j} C^0(\underset{\sim}{N}) \left[\frac{C(\underset{\sim}{N})}{C^0(\underset{\sim}{N})} - \frac{C(\{N_j + 1\})}{C^0(\{N_j + 1\})}\right]$$

$$+ \beta_{N_j - 1} C^0(\{N_j - 1\}) \left[\frac{C(\underset{\sim}{N})}{C^0(\underset{\sim}{N})} - \frac{C(\{N_j - 1\})}{C^0(\{N_j - 1\})}\right].$$

For N_j sufficiently large, differences may be replaced by derivatives to yield the continuity equation

$$\frac{\partial C}{\partial t} = - \sum_{j=1}^{n} \frac{\partial J_j}{\partial N_j} , \tag{24}$$

with flux given by

$$J_j = -\beta_j C^0 \frac{\partial (C/C^0)}{\partial N_j} , \tag{25}$$

where the explicit N-dependence has been left out for simplicity. Equation (25) indicates that the embryo flux is proportional to the gradient of a concentration ratio times a positive mobility β_j. This is clearly seen when Equation (25) for the components of the flux vector $\underset{\sim}{J}$ are written in matrix notation:

$$\underset{\sim}{J} = - \underset{\approx}{B}\underset{\sim}{X} ; \tag{26}$$

here $\underset{\sim}{X}$ is a generalized force vector and $\underset{\approx}{B}$ is an n×n mobility matrix. In Equation (26), B is diagonal, but it need not be. Indeed, in Binder's (16) or Langer's (14) analyses, where growth or decay of clusters can be accomplished by more general mechanisms than the monomer addition (or deletion) considered here, the mobility matrix is non-diagonal. The same is true for Russell's (29) "linked-flux" mechanism. In general, then, the $\underset{\approx}{B}$ matrix need be merely symmetric (by the Onsager reciprocity relations) and positive definite (by the Second Law):

3.2. Solution of Kinetic Equation

Equations (24) and (25) may be combined to yield a diffusion equation for the cluster concentration $C(\underset{\sim}{N},t)$:

$$\frac{\partial C}{\partial t} = \sum_{i=1}^{n} \frac{\partial}{\partial N_i} [\beta_{ij} C^0 \frac{\partial (C/C^0)}{\partial N_j}] \tag{27}$$

In Equation (27), off-diagonal mobility elements have been included for generality. This linear equation is first-order in time t and second-order in "space" $N_1, \ldots N_n$, the space itself being anisotropic because of "skewing" due to the mobility matrix. Also, space is inhomogeneous since the coefficients of the $\partial/\partial N_j$ differentials depend on position through C^0, i.e., through the work $W(\underset{\sim}{N})$. Thus, for $n > 1$, even for steady-state, the diffusion equation can only be solved numerically. Note that Equation (27) can also be written in the form of a Fokker-Planck equation

$$\frac{\partial C}{\partial t} = \sum_{i,j=1}^{n} \left[\beta_{ij} \frac{\partial^2 C}{\partial N_i \partial N_j} + \beta_{ij} \frac{\partial}{\partial N_i}\left(C \frac{\partial W}{\partial N_j}\right)\right] , \qquad (28)$$

where the N-dependence of β has been neglected for the purpose of the qualitative discussion which follows.

The physical situation represented by the differential equation (27) or (28) may be visualized with the help of Figure 7 which depicts schematically a two-dimensional case: an ample supply of monomers, (components 1, 2) is injected at the origin of the N_1,N_2 coordinate system. Since the concentration of monomers at the origin is high, clusters will move away, their growth (i.e., movement towards larger N_i-values) being driven by the cluster concentration gradient represented by the first term in Equation (28). The second term, whose effect is due to a force proportional to the gradient of the W-surface, opposes this motion. Nevertheless, after some incubation time, the first embryos will appear at the top of the saddle point or pass in the energy surface W whose contours are plotted schematically in Figure 7. From the critical nucleus position (*) onwards, it's downhill all the way, and ever-growing nuclei will tumble down into the W = 0 level pool below (indicated by cross-hatching). Ripe clusters are then skimmed off at large-N_i points opposite the origin, along the edge of the shallow pool (dark lines). The boundary conditions are thus: constant unit concentration of monomers close to the origin, and, at the other extreme, zero concentration of large clusters at the pool edge, a completely absorbing boundary. All other boundaries are reflecting (shaded, Figure 7).

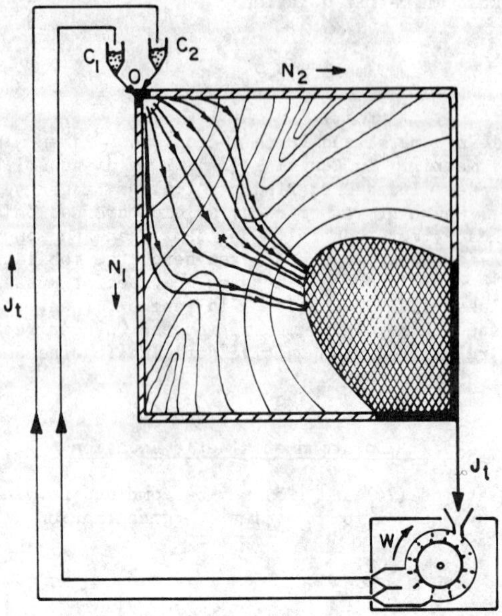

Figure 7. Schematic of steady-state nucleation for a binary system. Contour lines are those of surface $W(N_1,N_2)$, lines with arrow are flow paths of embryos. Shredder-sorter at bottom breaks up mature nuclei into individual monomers and pumps these back up to the holding silos at the origin.

After having been drained off, the large clusters are broken up into constituent monomers in the shredder-sorter indicated below the pool, and pumped back up to the holding silos at the origin. After some time, a steady-state condition will result, with

$$\nabla \cdot J = 0 \qquad (29)$$

for all points in N-space, the total flux being fixed by the boundary conditions, by the height W* of the saddle point, and, to a lesser extent, by the shape of the W-surface itself, in particular, by the magnitude of the principal curvatures of W at the saddle point. The integrated steady-state nucleus flux J* at the pass is the important quantity to calculate. The work of operating the pump must be equivalent to the work of nucleation W*, the energy surface with its saddle point providing the impedence to total matter flow J_t. If the outside flow is interrupted by closing off all inlets and outlets, an equilibrium condition will eventually result, characterized by

$$J_j = 0 \qquad (\text{all } j, \text{ all } \underset{\sim}{N})$$

everywhere. The concentration $C(\underset{\sim}{N},t)$ for all $\underset{\sim}{N}$ and all times will thus be $C^0(\underset{\sim}{N})$, given by Equation (22).

The start curve in TTT diagrams should be determined theoretically as the locus in temperature-time space at which an appreciable nucleation flux over the pass is first observed. Thus, the full time-dependent differential equation should be solved. In practice, the simpler procedure is adopted of locating the start curve by the inverse of the nucleation rate, i.e., of the steady state flux J* (see Section 4.1).

Even the calculation of J* is no simple matter. Langer (14) and Binder (16) give rather elaborate treatments of saddle-point flux determination in n-component systems. To simplify matters, we shall here adopt the more transparent, though not quite correct procedure used by Reiss (12) for binary systems. The trouble with the latter treatment is that it assumes that the vector J* at N_i* is along the direction of maximum negative curvature at the saddle point. Unfortunately, this is not justified, as the influence of the mobility matrix $\underset{=}{B}$ is such that, in general, cluster flux will be biased in the direction of the fastest moving monomer. Actually, then, one should first render the space isotropic by a suitable coordinate transformation, as done, for instance, by Russell (29) in a special case; then J* can be taken as parallel to the line of steepest descent of W* in the new coordinate system. Unfortunately, by so doing, one buries the elements of the mobility matrix inside the coordinate transformation, so that such important parameters as intrinsic diffusivities disappear inside certain eigenvalues with which the final answer is expressed. For this reason, we shall be content here to generalize the Reiss treatment to n-component systems, keeping in mind that this approximation will tend to work best in such situations where the skewing affect of unequal mobilities is not too great, and nuclei grow from monomers at constant concentration.

At the saddle point, the W-surface may be expanded to second order

$$W^* \equiv W(\underset{\sim}{N}^*) = W_0^* + \tfrac{1}{2} \sum_{i,j=1}^{n} \left(\frac{\partial^2 W}{\partial N_i \partial N_j}\right)_{\underset{\sim}{N}^*} \delta N_i \delta N_j \qquad (30)$$

where

$$\delta N_i = N_i - N_i^* \qquad (i = 1, \ldots n) .$$

The linear term in the expansion vanishes because W has a horizontal tangent plane at $\underset{\sim}{N}^*$. Since the second-derivative matrix (W", say) is symmetric, it is diagonalized by a uniquely-defined real orthogonal matrix R whose column vectors are the eigenvectors of W, pointing along the principal curvature directions. The eigenvalues λ_i (i = 1, ... n) of W" measure the curvatures themselves. The nature of the problem at hand dictates that there shall be one negative eigenvalue, say λ_1, all others being positive. By Equation (30), the W-surface at the saddle point may now be approximated by

$$W^* = W_0^* + \tfrac{1}{2} \sum_{i=1}^{n} \lambda_i \, \delta\bar{N}_i^2 , \tag{31}$$

where the over-line denotes, here and below, variables having been transformed by the change in coordinate system brought about by the rotation operator R.

Define transformed fluxes and forces, according to the matrix notation of Equation (26), by

$$\bar{\underset{\sim}{J}} = \underline{R}\underset{\sim}{J} ,$$

$$\bar{\underset{\sim}{X}} = \underline{R}\underset{\sim}{X} .$$

The transformed flux equation reads

$$\bar{\underset{\sim}{J}} = \underline{A}\bar{\underset{\sim}{X}} \tag{32}$$

where

$$\underline{A} = \underline{R}^{\dagger}\underline{B}\underline{R} , \tag{33}$$

the dagger denoting the operation of transposition. The force components X_i (i = 1, ... n) can be recovered from Equation (32) by inversion

$$\bar{\underset{\sim}{X}} = \underline{A}^{-1}\bar{\underset{\sim}{J}} . \tag{34}$$

We now make the simplifying assumption, discussed above, that the flux vector in the rotated coordinate system $\underset{\sim}{\bar{N}}$ is along the line of largest negative curvature of W*, i.e., that all the flux at the saddle point is over and down the steepest slope of W*. Therefore, by assumption we have,

$$\bar{\underset{\sim}{J}}^{\dagger} = (J_1^*, 0, \ldots, 0).$$

Consequently, we have, by Equation (34), the very simple result

$$J_1^* = a_{11}^{-1} \, \bar{X}_1 \tag{35}$$

where a_{11} is the first element of the inverse of the $\underline{A}$ matrix. Expressions for a_{11} will be given in Section 3.3.

The problem of finding the steady-state flux has thus been reduced to an essentially one-dimensional one, to which the standard method of solution of Frenckel (30) applies. Equation (35) becomes, explicitly

$$J_1^* = -M^* \, C^0 \left(\frac{d(C/C^0)}{d\bar{N}_1}\right)_{\bar{N}_{i\neq 1}} , \tag{36}$$

where M*, an effective mobility, denotes the value of a_{11}^{-1} at the saddle point. The integration of Equation (36) will be carried out in the vicinity of $\underset{\sim}{\bar{N}}^*$, along $\bar{N}_1$ only, at constant M*. We have, according to the boundary conditions specified earlier,

$$J_1^* \int_1^{\infty} \frac{d\bar{N}_1}{C^0} = M^* \int_1^{0} d(C/C^0) = M^* .$$

By using expression (22) for the equilibrium concentration C^0, and by making use of the quadratic expansion (31) for W* we have

$$\int_1^{\infty} \frac{d\bar{N}_1}{C^0} \simeq \frac{1}{N} \, e^{(W_0^* + \frac{1}{2}\sum_{i=2}^{n} \lambda_i \, \delta\bar{N}_i^2)/kT}$$

$$\times \int_{-\infty}^{\infty} e^{-|\lambda_1| \, \delta\bar{N}_1^2/2kT} \, d\delta\bar{N}_1$$

so that

$$J_1^* = NM^* \left(\frac{|\lambda_1|}{2\pi kT}\right)^{\frac{1}{2}} e^{-(W_0^* + \frac{1}{2}\sum_{i=2}^{n} \lambda_i \, \delta\bar{N}_i^2)/kT}$$

The total flux J* over the pass may now be obtained by integrating J_1^* over the remaining variables ($\bar{N}_2$, ..., $\bar{N}_n$) in the immediate vicinity of the saddle point to obtain:

$$J^* = K \, e^{-W_0^*/kT} \tag{37}$$

where the kinetic prefactor K is given by

$$K = NM^*ZY , \tag{38}$$

Z being the Zeldovich factor

$$Z = \left(\frac{|\lambda_1|}{2\pi kT}\right)^{\frac{1}{2}} \tag{39}$$

and Y being given by an inverse product of the other (positive) eigenvalues:

$$Y = \prod_{i=2}^{n} \left(\frac{2\pi kT}{\lambda_i}\right)^{\frac{1}{2}} . \tag{40}$$

It is remarkable that the kinetic treatment yields a nucleation rate J* proportional to exp(-W*/kT), just as was predicted by the thermodynamic theory. Of course, now some new information has been gained: an explicit albeit approximate, expression for the kinetic prefactor K. As noted by Reiss (12), for given M* (given mobilities β^*_{ij}), K is larger for larger value of the negative curvature $|\lambda_1|$ (steep rise to, and descent from the pass) and smaller for larger positive curvatures (λ_2, ..., λ_n; narrow pass, constricted flux).

3.3. Mobility

For applications to TTT diagrams, we are particularly interested in the temperature dependence of the integrated steady-state flux J*. Equation (37) exhibits a strong T-dependence in the exponential factor, and a much weaker dependence in the Z and Y factors, which we shall disregard. There are, however, other strong exponential temperature dependencies hidden in the effective mobility M*. The simplified diagonalization procedure used in the previous section, coupled with the assumption of a diagonal mobility matrix $\underline{B}$ enables us to write down immediately an explicit expression for M* which, as explained above, is given by the reciprocal of the first element of the matrix

$$\underline{A}^{-1} = \underline{R}^{\dagger}\underline{B}^{-1}\underline{R} ,$$

where $\underline{R}$ is an orthogonal matrix (of elements r_{ij}) and $\underline{B}^{-1}$ is a diagonal matrix of elements $1/\beta_i^*$. Hence, under these assumptions, the effective mobility for an n-component system is given very simply by

$$M^* = \left(\sum_{j=1}^{n} \frac{r_{j1}^2}{\beta_j^*}\right)^{-1} = \frac{\beta_1^*\beta_2^* \ldots \beta_n^*}{\beta_1' r_{11}^2 + \beta_2' r_{21}^2 + \ldots + \beta_n' r_{n1}^2} , \quad (41)$$

where β_j' denotes a product of all n β's divided by the jth one.

There remains the difficult problem of finding suitable expressions for the mobilities β_i^* in a multicomponent solid. It is usually assumed (31) that the mobility of i is proportional to the rate at which i-type monomers (atoms, molecules) strike unit area of the nucleus interface. Random walk theory then allows one to write β_i^* as proportional to the cluster surface area S, to the concentration (availability) of i monomers in the vicinity of the cluster, and to an atomic jump rate, itself proportional to a suitable diffusivity $\bar{D}_i$, divided by a mean jump distance. Retaining only those terms which will matter in temperature-dependence considerations, we then have:

$$\beta_i^* \sim \bar{c}_i \bar{D}_i \quad (42)$$

It is not clear, however, what concentration the $\bar{c}_i$ must represent, nor precisely what $\bar{D}_i$ means physically. Clearly, at least two important mechanisms must play a role: the diffusion of species in the matrix in order to feed the growing nucleus, and the rate at which atoms can successfully cross the matrix-nucleus interface. The problem of matrix diffusion is completely untractable in the present context: as the nucleus grows and migrates, the matrix concentration changes locally; strictly speaking, then, the W energy function itself must constantly change in a very complicated way. Generally, all these subtelties are ignored; but then how is one to take into account the fact that a solute which does not partition (i.e., such that its concentration in both matrix (c_i^0) and nucleus ($c_i^{\dagger}$) is the same) should hardly contribute to the overall mobility? An interesting attempt was made by Russell (29) to take such effects into account: the two basic mechanisms referred to above were considered in an approximate manner by postulating the existence of a "shell" of material intermediate between the matrix proper and the nucleus surface. Then two separate rates were defined: that for transfer of atoms from matrix to shell (β^0) and that for transfer of atoms from shell to nucleus ($\beta^{\dagger}$). The nucleus was assumed to grow at constant composition, this situation defining the classical limit, as will be explained in the next section. The resulting flux equations turned out to be non-diagonal even in the binary solution case considered by Russell. After performing an appropriate canonical (diagonalizing) transformation, Russell obtained the saddle point flux as

$$J^* = NZ \frac{\beta^0\beta^{\dagger}}{\delta^2\beta^{\dagger} + \beta^0} e^{-W^*/kT} , \quad (43)$$

where the correction term

$$\delta = 1 - \frac{c^0}{c^{\dagger}}$$

took into account the fact that a non-partitioning ($\delta = 0$) solute does not contribute to matrix diffusion.

We shall adopt here Russell's result for an effective mobility β_i^* by simply lifting Equation (43) out of context and writing, by analogy, for multicomponent systems,

$$\beta_i^* = \frac{\beta_i^0\beta_i^{\dagger}}{\gamma_i\beta_i^{\dagger} + \beta_i^0} \quad (44)$$

where γ_i is a partitioning correction factor, much like Russell's δ^2, for instance

$$\gamma_i = 2\left|\frac{c_i - c_i^0}{c_i + c_i^0}\right| . \quad (45)$$

Here and below, the (†) superscript will be deleted from concentrations referring to the nucleus. Let us see what Equations (44) and (45) accomplish qualitatively. If, say $\beta_i^{\dagger} << \beta_i^0$, with arbitrary γ_i, then

$$\beta_i^* \sim \beta_i^{\dagger} .$$

Hence the slowest mechanism, that of transfer of i across the interface, determines the overall rate. Likewise, for $\beta_i^0 << \beta_i^{\dagger}$, and for $|c_i - c_i^0|$ not too small,

$$\beta_i^* \sim \beta_i^0/\gamma_i ,$$

so that the overall rate will be determined by matrix diffusion. Finally, for no partitioning, $\gamma_i \simeq 0$,

$$\beta_i^* \sim \beta_i^{\dagger} ,$$

and only the rate of transfer across the interface influences the effective mobility of i.

Note that nucleation in which matrix and nuclei have same composition can be handled on an ad-hoc basis by this model: in such cases as recrystallization or massive transformations, for which the interface is necessarily incoherent, all γ_i are zero, and the governing rates are those determined by the $\beta_i^{\dagger}$, the interface jump mobilities. Coherent nucleation is handled by essentially setting equal to unity the probability that an atom successfully cross the interface.

The overall effective mobility M* will thus be the one given by Equation (41), with the β_i^*'s replaced by their value from Equation (44). In particular, in the binary case (A-B), we have

$$M^* \sim \frac{\beta_A^*\beta_B^*}{\beta_A^*\cos^2\theta + \beta_B^*\sin^2\theta} , \quad (46)$$

where θ is the angle by which the coordinate system must be rotated so that its axes coincide with the principal directions of W. Equation (46) was originally derived by Reiss (12). In the classical limit of constant-composition embryo growth, we may set (12)

$$\tan\theta = c_B/c_A$$

so that, to a very rough approximation, we have

$$M^* \sim c_A^0 c_B^0 (c_A^2 + c_B^2) \frac{\bar{D}_A \bar{D}_B}{c_A^0 \bar{D}_A c_B^2 + c_B^0 \bar{D}_B c_A^2} \qquad (47)$$

In this equation, we have expressed the effective mobilities, following Equation (42), in terms of effective diffusivities $\bar{D}_i$, which must themselves be evaluated with reference to Equation (44). The hidden temperature dependence of M* is thus to be found in the diffusivities

$$D_i = D_{i,0}\, e^{-Q_i/kT} \qquad (48)$$

in which Q_i is an appropriate activation energy, the preexponential $D_{i,0}$ being only weakly temperature-dependent.

Consider some typical cases:

(1) Dilute solution: the concentration of B in the matrix phase is very low. Then, in this limit,

$$M^* \to \bar{D}_B \ .$$

Furthermore, if $\beta_B^0 \ll \beta_B^\dagger$ (the usual case), M* will be proportional to the intrinsic diffusivity of the minority component B in the matrix phase, as expected.

(2) Concentrated nucleus: the concentration of B in the nucleus is almost unity, whilst that in the matrix is arbitrary. Here again,

$$M^* \to \bar{D}_B \ .$$

This conclusion is valid even if the concentration difference γ is small and/or if the intrinsic diffusivities for matrix diffusion are small; in such cases, the overall rate will be governed by the interface jumping rate of B atoms alone, simply because the nucleus growth process is here purely one of B accretion.

(3) One effective diffusivity much smaller than the other, say $\bar{D}_A \ll \bar{D}_B$, then Equation (47) yields

$$M^* \to \bar{D}_A \ ,$$

i.e., the overall rate will be governed by the slowest moving species, as expected. This latter conclusion is very general and can be derived directly from Equation (41) for an arbitrary n-component system.

(4) Carbon in Iron. Formally, we may regard this case as a binary, the two diffusing species being interstitial C and vacant interstitial sites (V), the Fe toms forming a passive network. Then, as expected, the overall rate will be determined by the effective diffusivity of carbon in the Fe framework. If some slow moving substitutional impurity is added, say Mo, then the quarternary system Fe-Mo-C-V must be considered. The overall rate will be determined by substitutional diffusion in the matrix, and, since Mo is the minority (substitutional) solute in this example, we expect, by Equations (41), (42), (44) and (45),

$$M^* \to \bar{D}_{Mo} \ ,$$

provided that γ_{Mo} is not too small, i.e., that Mo partitioning does occur. In the case of no partitioning, the rate would be governed by interstitial Carbon diffusion.

Finally, let us remind the reader that the derivation of Equation (41) rested on taking the flux vector at the saddle point (J_1^*) to be parallel to the direction of steepest descent from the pass, thereby neglecting the skewing effect of unequal monomer mobilities. This is not quite correct: Actually, the flux vector will rotate away from the W gradient, thus taking the path of traverse to somewhat higher altitudes over the pass, still passing over the point $\underset{\sim}{N}^*$. It can be shown from the positive definite nature of $\underset{\sim}{B}^*$, however, that no matter how different is the mobility matrix from the identity matrix, the flux direction will never rotate outside that portion of the W surface which is located below its tangent plane at $\underset{\sim}{N}^*$, a portion defined by the so-called asymptotic directions (32). Certain proposed prescriptions (16) for estimating the flux direction apparently do not conform to this requirement. Thus, the growing nuclei cannot be denied entrance into the lush valley beyond the pass, but they may be forced by unequal mobilities to take a less than energetically optimal path, with smaller curvature $|\lambda_1|$. Hence, the Zeldovich factor will have to decrease, thereby slowing down the nucleation kinetics. This, however, is a rather small effect compared to the uncertainties in knowledge of the magnitude of W* and of activation energies for diffusion, which uncertainties enter into the problem exponentially. Therefore, getting an incorrect Zeldovich factor appears to be a small price to pay for obtaining an explicit expression for the effective mobility M* in Equation (41).

Various ad-hoc correction terms were inserted into the β_i^*'s, of course, and a cautious reader may begin to wonder what may be the value of this whole exercise. Well, a more thorough analysis of the assumptions of nucleation kinetics show that the range of applicability (classical limit) of the entire theory itself is very narrow indeed, a far more serious criticism. A qualitative discussion of this topic will now be presented.

3.4. Nucleation Far From Equilibrium

It is not our purpose to discuss the validity of the assumptions on which rests classical nucleation theory. Clearly, one does not expect neat little spherical embryos in multicomponent fluids to grow in size in an untransformed, perfectly uniform matrix, their motion over the W surface being fully governed by the Fokker-Planck equation (28). In the classical view, embryos in no way interact with one another, their location in space remaining unspecified. Yet, cluster correlations should certainly be taken into account and some progress has recently been made in this direction: for instance, Binder and Stauffer (16) have included a non-linear "coagulation" term in their kinetic equation, thus allowing for the coalescence of neighboring clusters. As one might expect, however, the mathematics then becomes quite untractable. Also, McGraw and Reiss (33) have shown how the introduction of excluded volume considerations could explain discrepancies observed by Heady and Cahn (34) between measured and observed nucleation rates in a well-characterized binary fluid.

Our aim here is simply to inquire about the general applicability of the classical limit treatment, whereby the nucleation rate can be calculated approximately by saddle point integration, i.e., by assuming the flux J* to be confined to a very narrow, flat channel whose axis lies in the constant-composition direction in N-space. If these conditions are met, the kinetic method presented above (perhaps with a better diagonalization procedure) may give reliable results. It will appear, however, thatthe classical limit is valid in a very narrow range; beyond that range, an appropriate theory must merge gradually and continuously into a spinodal-type theory. The latter theory, which at first glance appears to be so different in spirit from nucleation theory, will be briefly described in the next section.

The question posed here is not merely of academic interest, as transformations in industrial alloys usually occur far from equilibrium. Since coherent nucleation or precipitation is often associated with transformations at large undercoolings, or high supersaturations (deep quenches), we shall use from the start a coarse-grained free energy function such as the one shown in Figure 5, appropriate for "coherent" transformations. Only binary (fluid) solutions will be considered in this and the next section.

The important features of the required energy curve (Figure 5) can be well represented by the symmetric function

$$f(x) = (x^2 - a_0^2)^2 \tag{49}$$

where f represents an energy per unit volume, x represents a concentration measured away from the central one, and $a_0(P,T)$, the only parameter used, fixes the points of common tangency (at $\pm a_0$). A plot of f(x) for $a_0 = 1$ is shown in Figure 5. Also shown is the tangent at some point $-1 < x_0 \le 0$. It will be convenient to measure the vertical distance from the tangent at f(x) to the point of tangency, thus

$$\Delta f(\xi) = \frac{1}{4!}f_0'''' \xi^4 + \frac{1}{3!}f_0''' \xi^3 + \frac{1}{2}f_0''\xi^2 , \tag{50}$$

where ξ is the concentration deviation

$$\xi = x - x_0 ,$$

and where the f_0 symbols refer to successive derivatives of f(x) evaluated at the point of tangency. By Equations (49) and (50), Δf can be written

$$\Delta f(\xi) = \xi^2\psi(\xi) \tag{51}$$

with

$$\psi(\xi) = \xi^2 + 4x_0\xi + 2(3x_0^2 - a_0^2). \tag{52}$$

Thus, $\Delta(\xi)$ is none other than the function ΔG_v defined in Section 2.1. By Equation (52), it is seen that the inflection points of $\Delta f(\xi)$ [or of f(x)] occur at

$$x_0/a_0 = \pm 1/\sqrt{3} .$$

Such is the origin of the "root-three rule" (35) for constructing the spinodal curve associated with a given free energy function.

Let us use this function in conjunction with a Rayleigh-Ritz treatment of the Cahn and Hilliard formulation. For that, let us define a standard spherical embryo by the parametrised composition profile (23)

$$c(r) = \xi\, u(\eta,\alpha)$$

where ξ, which appears as an order parameter, is the amplitude of the fluctuation of standard shape $u(\eta,\alpha)$ in which α represents a set of auxiliary parameters defining the shape of the standard profile. Letting $\eta = r/\rho$ allows one to define an effective fluctuation radius ρ, r being the radial distance. For present purposes, it suffices to adopt the capillarity assumption, which consists in regarding the embryo as having constant composition $x_0 + \xi$, separated from the matrix of uniform composition x_0 by a sharp interface at $r = \rho$. Hence, the standard profile may be defined by

$$u(\eta) = \begin{cases} 1 \text{ if } 0 \le \eta < 1 \\ 0 \text{ if } \eta > 1 . \end{cases}$$

In a sense, then, we have replaced the profile of Figure 3b by that of 3c.

For use in the C-H free energy functional (17), we shall need the integrals

$$4\pi \int_0^\infty [c(r)]^n\, r^2\, dr = \frac{4}{3}\pi\rho^2\xi^n , \tag{53}$$

and

$$4\pi \int_0^\infty \left(\frac{dc}{dr}\right)^2 r^2\, dr = \frac{4}{3}\pi\rho\xi^2 , \tag{54}$$

the latter result having been obtained by noting that the derivative of a function at a discontinuity is the Dirac delta function times the height of the discontinuity. Integral (54) multiplied by the positive gradient energy coefficient κ represents the interfacial energy contribution σA of the classical theory. A similar expression for σ was obtained by Reiss and Shugard (13) who used more elaborate bond counting within the framework of the regular solution model. Through the use of Equation (10), these authors were able to take into account the concentration-dependence of the interface.

In the present formulation, equivalent to that of R-S (13), the work of formation of a standard embryo of amplitude ξ and of radius ρ can be written

$$W = \frac{4}{3}\pi\, [\psi(\xi)\rho^2 + \kappa]\rho\xi^2 \tag{55}$$

with ψ given by Equation (52). As an aid in visualizing this surface, computer-generated contour plots are reproduced in Figures 8a, b, c, d, and e, corresponding respectively to supersaturations x_0 = -.99, -.95, -.90, -.70, -.10; values $a_0 = \kappa = 1$ were chosen, and the contours were measured in units of $3/4\pi$. The presence of the ξ^2 factor in Equation (55) produces a zero-level valley (or ridge) all along the ρ axis, whilst the presence of the ρ factor produces zero altitude along the ξ axis, from which the surface gently rises linearly towards larger radii ρ. In all cases, a deep valley is found at large ρ in the vicinity of the ξ value marked $\bar{z}^{(2)}$ in Figure 5. The valley contours are asymptotic to vertical lines at points marked $z_0^{(1)}$ and $z_0^{(2)}$ in Figure 5. Any constant ρ section exhibits a curve made up of $\Delta f(\xi)$ multiplied by ρ^3, representing the bulk free energy, plus a parabola $\kappa\xi^2$ multiplied by ρ, representing the surface free energy. At small ρ, the latter dominates, at large radius, the bulk dominates, as expected.

For all values of x_0 outside the spinodal

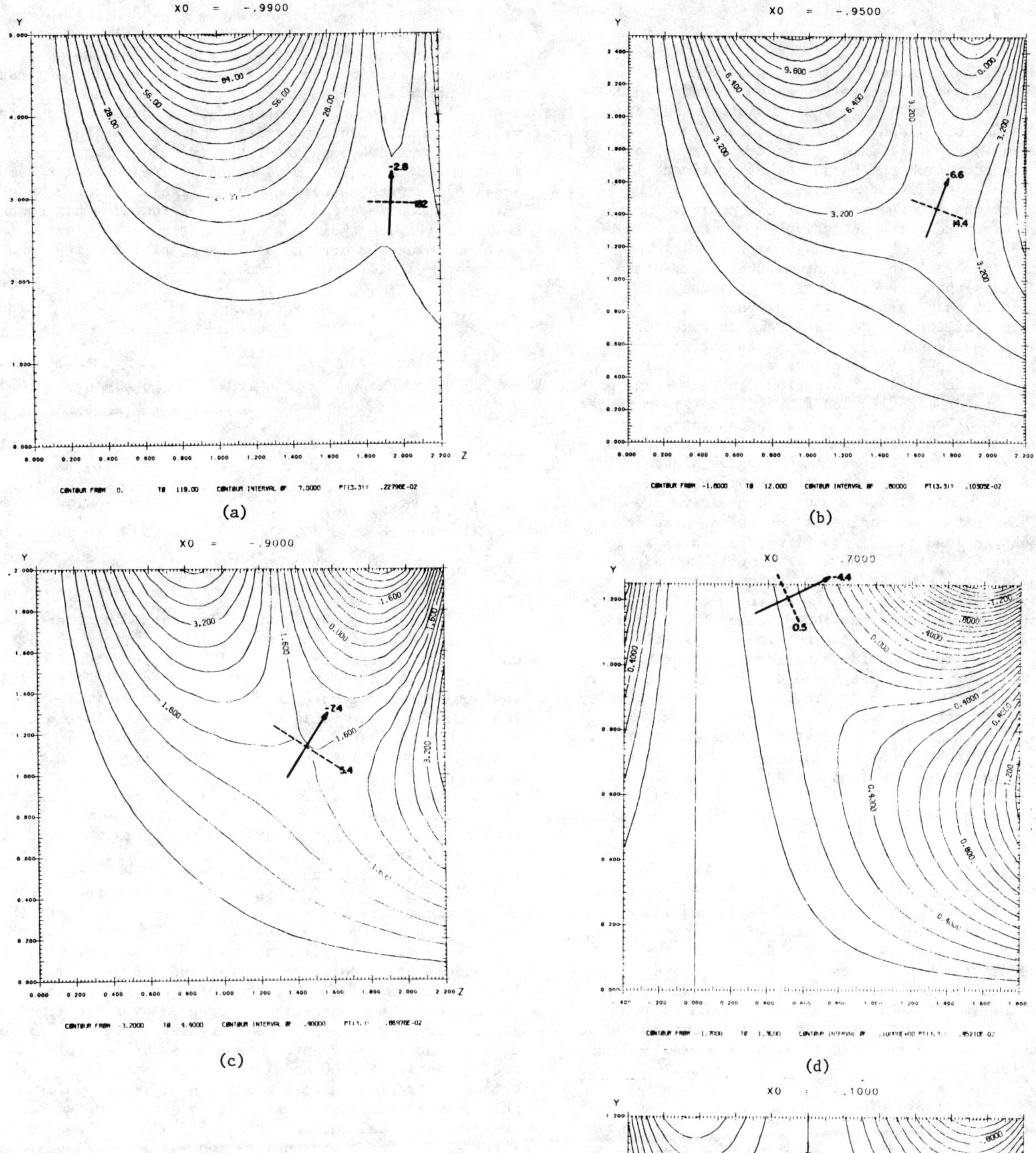

(a) (b) (c) (d)

Figure 8. Computer-generated constant-energy contours of surface W(ξ,ρ) plotted against amplitude ξ (labeled Z) and radius ρ (labeled Y). Saddle points are indicated by intersection of arrow (direction of steepest descent) and dashed lines, numbers indicating eigenvalues. Supersaturations are (a) -0.99, (b) -0.95, (c) -0.90, (d) -0.70, (e) -0.10. See also Table.

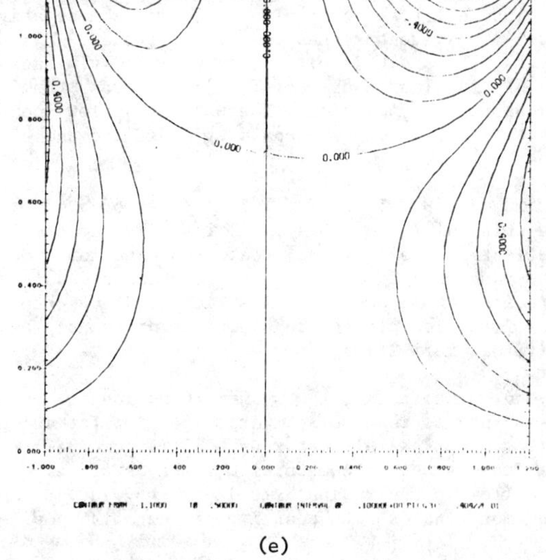

(e)

$$|x_0| > 1/\sqrt{3} ,$$

the W-surface has a saddle point or pass which the growing nuclei must negotiate before they can tumble down into the valley below. The saddle point is indicated in Figures 8a, b, c, and d by the intersection of two line segments, one full which indicates (arrow) the steepest descent from the pass, the other (dashed) which indicates the steepest rise. Numbers alongside these directions are the values of the positive and negative curvatures, respectively.

From Figure 8a, it is apparent that supersaturation $x_0 = -.99$ is within the classical limit: the path over the pass is roughly parallel to the ρ axis (thus at constant composition ξ), the pass very narrow (very large positive eigenvalue), the path level (rather low negative eigenvalue), the altitude very high. Since the flux is concentrated in a narrow, level stream, we may expect that the nucleation flux will be given rather accurately by J* obtained by saddle point integration. The actual trajectories of embryos will be rather complex, however: they start from the region of the origin ($\rho \simeq 0$, $\xi \simeq 0$) as small clusters of small amplitude, and then fan out in all directions of the positive ξ,ρ quadrant. This initial spreading is due to the first term in the Fokker-Planck (F-P) equation (28), which is the Laplacian of $C(\underset{\sim}{N})$, a dissipative term which tends to oppose high local cluster concentrations. As embryos move away from the origin, however, the W-surface relief increases, and soon the influence of the second term of Equation (28) is felt, tending to push the embryos back down the W-surface gradient. During this part of the embryo's journey, the dissipative and drift terms of the F-P equation act in opposition. Eventually, the embryo trajectories will bend upwards and form a dense narrow bundle of flow lines over the pass. Thus, the nature of the pass at low supersaturations is such as to give an integrated flux $\underset{\sim}{J}^*$ whose orientation will neither depend greatly on the history of the trajectories before the pass, nor on the skewing effect of the mobility matrix $\underset{\sim}{B}$.

At slightly larger supersaturations, the situation is no longer as straightforward. Already, for $x_0 = -.95$ (Figure 8b), i.e., only 2.5% away from the equilibrium concentration ($x^\alpha = -1.0$), measured with respect to the common tangent interval ($x^\beta - x^\alpha = 2.0$), the eigenvectors at the saddle point have rotated away from the coordinate axes directions. The pass is now also much wider, the path not as level. Even with $\underset{\sim}{B}$ close to the identity matrix, the flux direction at the saddle point cannot be determined from a local analysis only: the whole steady-state differential equation should be solved.

Non-classical behaviour is accentuated as the supersaturation increases: for $x_0 = -.90$ (Figure 8c), only 5% away from equilibrium x^α, the eigenvectors at the saddle point have rotated by almost 45° and the negative curvature is now greater than the positive one. Whatever one may say about J* without actually solving the equation, it is clear that nuclei will not cross the pass at constant composition. The pass being low and broad, the flux will be correspondingly diffuse and spread out.

For $x_0 = -.70$ (Figure 8d), still a bit of a way from the spinodal ($x_s = -.577$), all resemblance to classical nucleation has long vanished: those nuclei which cross the pass will do so at almost constant radius, but increasing amplitude, the direction of J* over this very wide flat, low plateau depending significantly on the individual mobilities.

Beyond the spinodal, the W-surface no longer has a saddle point, so that no activation barrier for nucleation exists. For example, at $x_0 = -.10$ (Figure 8e), almost at the symmetric composition, embryos are expected to flow up the radius axis in a very wide stream, spreading out towards both positive and negative amplitudes ξ.. Only those embryos whose trajectories remain centered on the ρ axis will encounter no rise in altitude, and therefore no activation energy barrier. However, the dissipative term in the F-P equation will drive a great many embryos away from the ρ axis, forcing them to climb over a low ridge before entering the promised land, the valley beyond. Since both left and right valleys will eventually become populated, the transformation product is expected to consist of large positive and negative-amplitude clusters sharing the available space. A picture of a quasi-periodic modulated structure thus comes to mind, i.e., a typical spinodal structure. Such a conclusion can be arrived at by a completely different argument, as will presently be shown.

Table

x_0	Figure 8	ρ^*	ξ^*	W^*	λ_1^*	λ_2^*
-0.999	not shown	9.15	1.99	24.25	- 0.9	6053.7
-0.99	a	2.95	1.94	7.41	- 2.8	182.0
-0.95	b	1.44	1.71	2.82	- 6.6	14.4
-0.90	c	1.14	1.45	1.59	- 7.4	5.4
-0.70	d	1.21	0.51	0.21	- 4.4	0.5
-0.60	not shown	2.57	0.09	0.01	-10.0	0.007
-0.577	not shown	$\to \infty$	0	0	--	--

To summarize, the above table lists, in order, the critical radius ρ^*, critical amplitude ξ^*, and critical work W^* for increasing values of the supersaturation x_0 at constant temperature. Although the simplest possible Rayleigh-Ritz procedure has been used to minimize the functional ΔF of Equation (17), the essential results of the C-H diffuse interface theory are correctly reproduced, indicating that it is not so much the diffuse nature of the interface which is important, but its composition dependence. It is seenclearly from the table that the critical radius ρ^* starts out to be very large, then decreases as supersaturation increases, only to increase again without limit as the spinodal x_s is approached. Simultaneously, both critical amplitude ξ^* and critical work W^* decrease steadily to zero. However, as explained above, the relief features in the vicinity of the pass become so diffuse at high supersaturations (or low undercoolings) as to make the exact position of the saddle point position and altitude largely irrelevant. Indeed, details of the W-surface relief are ignored by the system when mean altitudes drop below a few kT. Thus, there can be really no "singularity" at the spinodal, and no abrupt change in mechanisms. The spinodal region can in fact be considered as an extension of the nucleation region, albeit non-classical, and conversely. This general conclusion was already drawn from an early study of spinodal decomposition (36, 37), and was reiterated in many other publications (11, 16, 24, 25). The cluster dynamics theory of Binder and co-workers (16) in principle can cover the whole range of supersaturations, but the complexity of the mathematics renders the theory all but inextricable, and comparison with experiment practically impossible.

3.5. Spinodal Decomposition

Spinodal decomposition, discovered theoretically by J. W. Cahn in 1960 (38, 39) has been reviewed a number of times (8, 9, 10). We shall here follow the latter reference, restricting our attention to binary solutions (A-B), with $c(\underset{\sim}{x},t)$ being the only independent composition variable, with $\underset{\sim}{x}$ a position vector in the liquid or solid solution, t the time.

Since neither A or B atoms may be destroyed in unit volume (vacancies not considered), we may write a continuity equation for the concentration $c \equiv c_B$:

$$N_v \frac{\partial c}{\partial t} = - \underset{\sim}{\nabla} \cdot \underset{\sim}{J}$$

where N_V is the number of atoms per unit volume. Note that we are dealing here with actual atomic concentrations, defined by an appropriate coarse-graining, not a cluster concentration as in Equation (27). The net flux $\underset{\sim}{J}$ of B atoms is given by

$$\underset{\sim}{J} = - N_V M \nabla \Phi$$

where M is a suitable positive mobility. The potential Φ will be derived from the C-H free energy functional (12). In the absence of concentration gradient contributions, Equation (15) yields, for the Lagrange multiplier

$$\lambda = f'(c) = \mu_B - \mu_A \ .$$

Since λ appears as a difference of chemical potentials, it may be identified with the potential for atomic diffusion in the usual Fick's law formulation. By analogy, in the present case, let us identify Φ with λ obtained by inserting the Lagrangian (14) into the Euler equation (13). There results

$$\Phi \equiv \lambda = f'(c) - 2\kappa\nabla^2 c \ .$$

The resulting diffusion equation is then

$$\frac{\partial c}{\partial t} = \nabla M \nabla \Phi \ . \tag{56}$$

This differential equation is non-linear in the unknown concentration $c(\underset{\sim}{x},t)$. A solution in closed form is thus out of the question, although certain basic properties of the solution have been examined by Cahn (38) by a perturbation method. Computer-generated numerical solutions of Equation (56) have provided some insight into the clustering kinetics in binary solutions (36), and these results will be briefly summarized.

When information about the approximate free energy curve is incorporated via $\Phi(c)$ into Equation (56) the solution to the diffusion equation correctly obeys the phase diagram without requiring additional constraints. Examples of such well-disciplined solutions are given in Figures 9a-d, which represent numerical calculations of the evolution of one-dimensional concentration profiles in Al-Zn solid solutions during isothermal aging at 100°C (36). The initial concentration variation $c(\underset{\sim}{x},0)$ was a small-amplitude fluctuation which is not shown in these figures. The thermodynamic parameters used were those calculated by Rundman and Hilliard (40) for the Al-Zn system. The numerical solution $c(\underset{\sim}{x},t)$ of Equation (56) was obtained by a Fourier space forward difference iteration scheme.

Each figure shows concentration profiles as a function of distance for successive aging times. The horizontal full lines indicate the average composition c^0, the dashed lines indicate the (coherent) equilibrium compositions c^α and c^β. The results for a 37.5 at.%Zn alloy (i.e., for c^0 close to the center of the miscibility gap) are shown in Figure 9a: The profiles are quasisinusoidal with bounded amplitude. Further growth of the composition modulation requires a lengthening of the pseudoperiod, which does not take place readily due to the extreme resistance to coarsening of such regular one-dimensional structures.

Figures 9b and c are relative to a 22.5 at.%Zn solution, which is closer to the spinodal composition at 100°C. A striking feature of the profiles is the appearance at an early stage of "Guinier zones" (41) consisting of Zn-rich peaks surrounded by denuded Al-rich regions. This result, which confirms the qualitative arguments put forward by Bonfiglioli and Guinier (42) is a direct consequence of the asymmetric location of c^0 with respect to the miscibility gap boundaries. At this composition, the Guiner zone structure is eventually followed by a well-defined periodic profile which correctly obeys the phase diagram. After prolonged aging the solution displays coarsening effects (Figure 9c), which are apparently less sluggish than those of the symmetric composition (center of the miscibility gap). To quote Hilliard (9), "It is rather remarkable that a single equation can depict the complete life cycle, from birth to death, of a particle."

For compositions still closer to the spinodal (Figure 9d; 20 at.%Zn), an early-stage coarsening reaction sets in and the secondary maxima of the Guinier zones dissolve before they have a chance to develop into full-fledged precipitates as they did in the previous case. Thus a periodic structure fails to develop for this asymmetric composition although the average composition is still within the spinodal. Actually, the final concentration profile peaks at the spots where the first maxima make their appearance; in other words, the location of the precipitates is dictated almost entirely by the vagaries of the initial fluctuation, a characteristic more reminiscent of nucleation than of "typical" spinodal decomposition.

Subsequently, the elegant theory of Langer et al. (24, 43) confirmed and expanded these results which were initially confined to one-dimensional concentration profiles. Here also, we see that the spinodal in no way separates clustering kinetics into two competing irreconcilable mechanisms. Rather, at one extreme, the classical nucleation limit, nucleation theory gives a satisfactory account of the transformation, where cluster formation kinetics is of paramount importance and cluster correlations less so. At the other extreme, deep inside the spinodal, at close to maximum supersaturation (or deep quench), cluster kinetics are less important than spatial correlations, and classical spinodal theory works best. Unfortunately, as mentioned previously, no single tractable theory exists for the whole range of supersaturations. This point is also made quite clearly in Martin's excellent review article on the subject (11).

For the practical physical metallurgist, neither Fokker-Planck equations nor non-linear spinodal equations are of much use, however. In practice, then, there is little else to be done but to trust the limiting-case theories: saddle point integration at one extreme and linear spinodal theory at the other, with nothing but a prayer in between. With these words of caution, let us now briefly describe the linear spinodal theory as Cahn originally developed it (38).

For average compositions and aging temperatures well inside the spinodal and, in particular, for c^0 in the vicinity of the center of the miscibility gap, the absolute value of the ratio of the third to the second derivative of f(c) is small compared to unity, and the second-degree term in c can be initially neglected in the potential $\Phi(c)$. The third-degree term in the potential can also be neglected initially, since its magnitude is proportional to the third power of the composition variation, a small quantity. It follows that under the above conditions, the potential function $\Phi(c)$ can be linearized, and this leads to the following diffusion equation (38)

$$\frac{\partial c}{\partial t} = M(f_0'' \nabla^2 c - 2\kappa\nabla^4 c) \ , \tag{57}$$

where M and κ are assumed to be concentration and time independent, and where f_0'' represents the second derivative of the "incoherent" free energy evaluated at $c = c^0$. Equation (57) can be Fourier-transformed to an ordinary differential equation whose solution is found to be

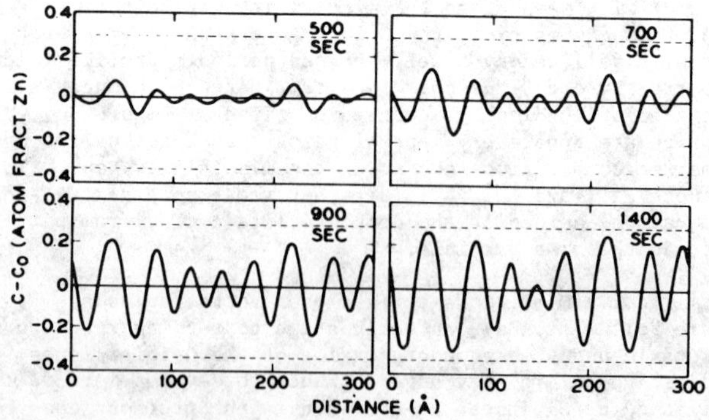

(a)

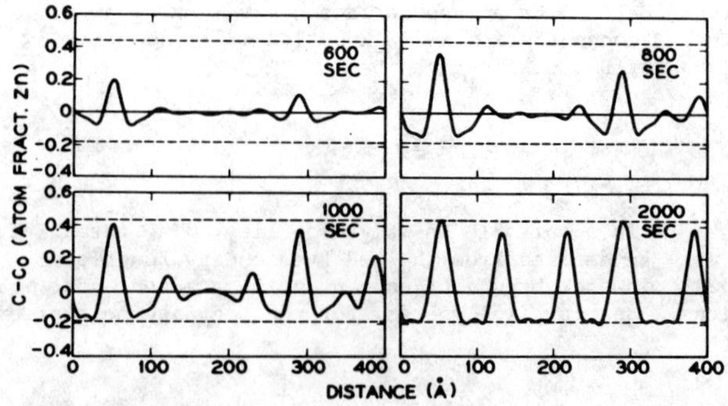

(b)

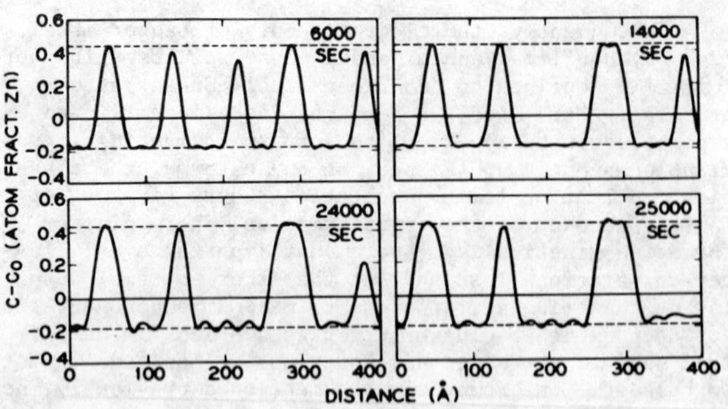

(c)

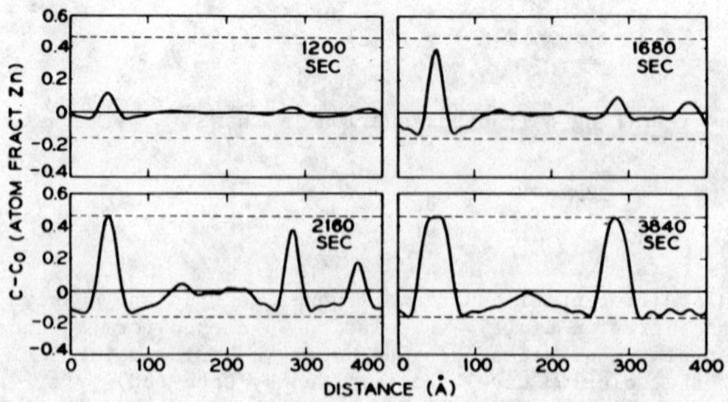

(d)

Figure 9. Concentration profiles calculated (36) according to one-dimensional non-linear diffusion equation (56) for an Al-Zn alloy aged at 100°C for the indicated durations and for concentrations: (a) 37.5 at.%Zn, (b) 22.5 at.%Zn, (c) same as (b), longer times, (d) 20.0 at.%Zn.

$$X(\underset{\sim}{k},t) = X(\underset{\sim}{k},0)\ e^{\alpha(\underset{\sim}{k})t} \tag{58}$$

with amplification rate

$$\alpha(\underset{\sim}{k}) = -Mk^2(f_0'' + 2\kappa k^2)\ . \tag{59}$$

In these equations k^2 is the square of the wave vector $\underset{\sim}{k}$ and X is the Fourier transform of the concentration deviation ξ, $X(\underset{\sim}{k},0)$ being the Fourier transform of the initial concentration fluctuation.

Equation (59) shows that the Fourier amplitudes of concentration waves initially grow ($\alpha > 0$) or decay ($\alpha < 0$) in time according to an exponential law. Two important quantities were introduced by Cahn (38): the critical wave number k_c, solution of

$$\alpha(k) = 0\ ,$$

and the optimum wave number k_m, solution of

$$\frac{d\alpha}{dk} = 0$$

for a particular direction $\underset{\sim}{k}$.

The physical meaning of these concepts is best discussed in terms of wavelengths Λ, inversely related to wave number by

$$\Lambda = 2\pi/k\ .$$

Within the framework of the linear theory, all harmonic composition waves of wavelength Λ larger than the critical wavelength

$$\Lambda_c = 2\pi\ \sqrt{-2\kappa/f_0''} \tag{60}$$

will grow in amplitude as isothermal aging progresses, while those with $\Lambda < \Lambda_c$ will decay. Maximum growth in a given direction $\underset{\sim}{k}$ will occur for the optimum wavelength

$$\Lambda_m = \Lambda_c\sqrt{2} \ . \tag{61}$$

These considerations are valid only for the initial stages of spinodal decomposition, well inside the coherent spinodal. Outside the spinodal, the linear approximation breaks down completely since, by Equation (59), Λ_c and therefore also Λ_m are imaginary quantities. This is because, since κ is positive for a clustering system, the quantity under the radical in Equation (60) can only be positive for $f_0'' < 0$, which defines the region inside the spinodal.

In a further development of the theory, Cahn (39) introduced an important correction term to f_0'' to take into account anisotropic elastic coherency strains which must develop during growth of concentration waves in crystalline solids. This topic will not be covered here, however.

Spinodal theory has received ample experimental justification, as periodic, or pseudo-periodic microstructures have been reported in binary liquids, in binary and ternary alloys, and in minerals [see, for example, the articles by Hilliard (9), Martin (11), Grilhé (44) for pertinent bibliography]. In more recent experimental work (45, 46, 47) the trend is to analyze the data, particularly the structure factor obtained by small angle X-ray or neutron diffraction, in terms of scaling laws (48).

3.6. Homogeneous/Heterogeneous Nucleation

Elaborate models of unmixing (clustering, phase separation) may be of considerable theoretical interest, but are of little use to the physical metallurgist interested in understanding microstructure resulting from various heat treatments of alloys. Ideally, one would like to put theoretical models to good use by predicting, say, annealing times and temperatures required to produce specified microstructures. In that context, pushing a simple theoretical model beyond its domain of applicability may result in a numerical error of a factor of two or three perhaps, whereas the use of slightly incorrect values of some of the parameters may lead to results which are off by several orders of magnitude.

It is apparent that, if we adopt J* given by Equation (37) as a measure of the nucleation rate, a small change in the value of W*, especially at temperatures not too high, will have drastic consequences. In particular, by Equation (11) for W*, we see that a slight change in the value of the interfacial energy σ, or of the driving force ΔG_V^* will have profound repercussions on J*, these changes occurring as they do in the parameters raised respectively to the third and second power of the arguments of an exponential function! Small wonder that very little quantitative agreement is usually encountered between experiment and nucleation theory. This is particularly true for nucleation in solids, as effects of strain energy, anisotropic interfacial energy, preferential nucleation sites and other factors play a decisive role in the kinetics.

It would take us too far afield to cover these topics; the interested reader is referred to K. C. Russell's excellent recent review of nucleation in solids (5), which it would be pointless to attempt to duplicate. A very thorough and complete but older treatment is also to be found in J. W. Christian's classic treatise (2). Here, let us merely point out the essential features which must be considered when discussing nucleation in solids.

First of all, a distinction must be made between homogeneous and heterogeneous nucleation, that is to say, between nucleation initiating at arbitrary sites in the bulk, essentially perfect crystalline phase, and that initiating at preferential nucleation sites such as grain or phase boundaries or other imperfections. A further distinction must also be made between nuclei growing coherently and those growing incoherently in the matrix. By the word coherent we mean that, strictly, the matrix and the nucleus have same crystalline lattice, with perhaps a small change in lattice parameters; in other words, there is perfect coherence if the lattice planes suffer no discontinuities at the matrix-nucleus interface. On the contrary, by the word incoherent is meant that there is no matching of crystalline directions or lattice spacings on either side of the interface. Intermediate cases are of course often encountered; the word semi-coherent is then used rather indiscriminately either to denote cases in which the mismatch between matrix and nucleus can be accommodated by, say, periodic arrays of interfacial dislocations, or cases where some portions of the matrix-nucleus are coherently related whilst other portions are not.

These morphological distinctions are important as they lead to different values of the effective interfacial energy σ and the driving force ΔG_V. Generally speaking, phase β and phase α will have different molar volumes. In solids, homogeneous coherent nucleation of β from α will thus give rise to strain fields which can be very large and which must add a positive contribution to the negative ΔG_V. Hence, such strain energy contributions, which are evaluated in practice according to linear isotropic elasticity theory, will substantially increase the work of nucleation W* by decreasing the value of the denominator $(\Delta G_V + \Delta G_E)^2$, ΔG_E being the elastic energy contribution. It follows that coherent homogeneous nucleation will require larger undercoolings in order to proceed at appreciable rate.

Coherency strains in homogeneous nucleation can be at least partially relieved by introducing interfacial dislocations whose role is to take up most of the lattice mismatch between α and β. In a sense then, the decrease in magnitude of the driving force due to bulk elastic energy has been replaced by an increase in the interfacial energy σ. Such a substitution can be energetically favorable for nuclei of large radius which have relatively small surface-to-volume ratio. Hence all other things being equal, one may expect coherent nucleation at large supersaturations or undercoolings and incoherent nucleation at lower ones, with, possibly, semi-coherent nucleation in between.

These concepts may be understood with the help of Figure 10 which shows, in the upper portion, schematic plots of free energy curves vs. composition for a binary solid solution. Curves for semi-coherent and coherent B-rich β are shown to lie above that for incoherent β because of the positive elastic contribution ΔG_E to the (negative) total driving force. The curve for coherent β has actually been drawn as the continuation of the α free energy curve since coherent β must be structurally no different from α. By invoking the tangent construction described in Section 2.1, it is expected that, at the temperature (T') considered, incoherent (I) β can nucleate for matrix compositions c^0 located to the right of the common tangency point c_I^α, semi-coherent (S) β can nucleate for compositions c^0 to the right of c_S^α, and coherent (C) β can nucleate for c^0 to the right of c_C^α. At given undercooling or supersaturation, which mode of transformation will dominate depends upon the relative magnitudes of interfacial and bulk energies. For example coherent nucleation will win if its TTT start curve lies at lower transformation times in the temperature range considered than either the semi-coherent or incoherent C-curves.

In Figure 10, separate free energy curves were sketched for the various transformation products envisaged. Hence, by the common tangent rule, three different temperature-concentration phase diagrams could be constructed, as shown at the lower portion of Figure 10. A continuous miscibility gap (C) has been constructed as locus of common tangency points for the continuous (C) free energy curve, whilst only portions of coexistence curves have been constructed for the other two cases, since it cannot be decided without further information what possible three-phase reactions may occur at higher temperatures: eutectic, peritectic, eutectoid ... . The coherent spinodal, locus of inflection points of the continuous free energy curve, has also been indicated on the phase diagram. Sometimes the following designation is used: phase diagram curve C is called the coherent solvus, curve S the semi-coherent solvus, and curve I the equilibrium phase boundary (if it is indeed the topmost one). Sometimes also, these curves may be labeled according to the conventional symbol attached to the nucleating phase: GP zone, θ", θ' ... in the case of Al-Cu alloys, for example. To illustrate the use of the phase diagram in the present context, consider a homogeneous solid solution of average composition $c^0 = c_S^\alpha$, say. Then, coherent nucleation can only take place at temperatures below T_C (Figure 10), semi-coherent below T' ($\equiv T_S$), and incoherent below T_I.

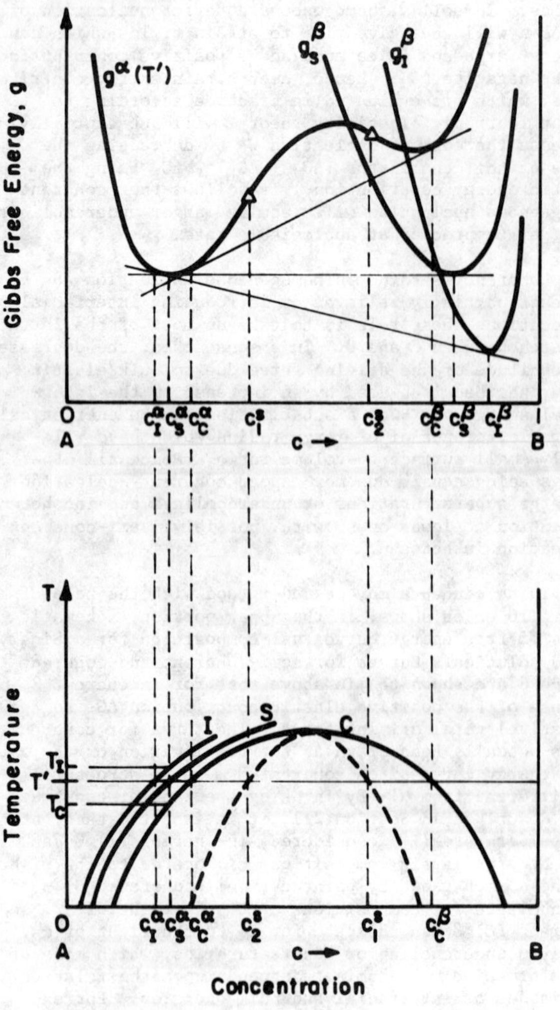

Figure 10. Free energy curves (top) for binary alloy and associated phase diagram (bottom) for incoherent (I), semi-coherent (S) and coherent (C) phases.

Without interfacial energy, there would be no activation barrier for nucleation. Although the positive contribution of σ cannot be eliminated, it can be reduced in so-called heterogeneous nucleation, as opposed to homogeneous nucleation considered up to now. It is well documented that imperfections are able to catalyze the nucleation process: in essence, the nucleus, by attaching itself to a preexisting high-energy surface (grain boundary, interphase boundary, free surface, etc.), replaces a portion of existing matrix-defect boundary by an identical portion of nucleus-defect boundary plus a matrix-nucleus portion. In many cases (depending on the relative magnitudes of the interfacial energies involved) this can be energetically far more favorable than the creation from scratch of a complete matrix-nucleus boundary of approximately same total area. In typical cases, one can calculate effective reduced interfacial energies (σ_{eff}) to be inserted in the critical work formula (11) for W*. As σ comes with third power into the exponential for the nucleation rate, even a small reduction in effective interfacial energy can enhance nucleation rates by many orders of magnitude. At low undercoolings, then, we expect heterogeneous nucleation to be kinetically favored whilst at large undercoolings homogeneous nucleation may prevail. The reason is that the larger σ which accompanies homogeneous nucleation may be offset partially by the larger driving force $|\Delta G_v|$ available and that, importantly, coherent nucleation will have much larger prefactor K in the rate expression (37) for J*, simply because so many more sites for homogeneous nucleation are available over those available for various types of heterogeneous nucleation. In this context, Cahn (49) has given an interesting discussion of site availability and saturation.

4. TTT DIAGRAMS

We may now return to the topic with which we opened this paper: that of TTT diagrams. Let us first see whether nucleation theory and its ramifications help to understand the simple C-curve diagram, such as that of Figure 1, for a plain carbon eutectoid steel.

4.1. C-Curves

Let us make the simplest possible assumptions: although the position of the start curve should be obtained by calculating the incubation time for copious nucleation, we shall, for simplicity, regard the start time t_S as the reciprocal of the steady-state nucleation rate as calculated by saddle-point integration. Furthermore, we shall assume that a single temperature-independent activation energy for diffusion Q dominates the effective mobility M* [see discussion accompanying Equations (41) and (44), especially case (4)]. Then we have

$$\text{Rate} \simeq J^* = K_0\, e^{-(W^* + Q)/kT}$$

where K_0 contains all factors not included in the exponentials. The start time t_S may thus be given by

$$\log t_S = -\log K_0 + \Gamma(T) \tag{62}$$

with

$$\Gamma(T) = m(W^* + Q)/kT \tag{63}$$

and

$$m = \log e = 0.4343...$$

Although, in the present case, pearlite (α iron + carbide) most certainly does not nucleate in bulk γ Austenite as small spherical homogeneous embryos, we shall nevertheless use Equation (11) for estimating the temperature dependence of W*. In this equation, the volume energy (driving force) may be written

$$\Delta G_v = \Delta H_v - T\Delta S_v ,$$

where ΔH_v denotes the enthalpy difference and ΔS_v the entropy difference per unit volume between the nucleating phase and the tangent to the parent phase at same composition, temperature and pressure. At the equilibrium temperature T_0 (common tangent rule) we must have

$$\Delta G_v = \Delta H_v - T_0\Delta S_v = 0 ,$$

If the enthalpy and entropy differences are assumed to vary but little with temperature, we find

$$\Delta H_v = T_0\Delta S_v$$

and hence

$$\Delta G_v = (T - T_0)\Delta S_v .$$

The strongly temperature-dependent part of log t_S may thus be written, by Equation (63), as

$$\Gamma(\tau) = m\,\frac{16\pi\sigma^3}{3kT_0\Delta H_v^2}\left[\frac{b}{\tau} + \frac{1}{\tau(1-\tau)^2}\right] \quad (64)$$

where

$$\tau = T/T_0$$

is a reduced temperature and

$$b = \frac{3Q\Delta H_v^2}{16\pi\sigma^3} \quad (65)$$

is a dimensionless parameter which, in this approximation, completely determines the shape of the start curve. We shall assume for simplicity that the coefficient of the bracketed term in Equation (64) is roughly temperature-independent. Then, knowledge of the "nose" temperature τ_N fixes the value of this parameter

$$b = \frac{3\tau_N - 1}{(\tau_N - 1)^3} . \quad (66)$$

Since the nose of the experimental curve of Figure 1 is located at about $T_N = 0.825\ T_0$, we shall take $b \simeq 275$, according to Equation (66). The resulting start curve is shown in Figure 11. The simplifications used in its derivation were so drastic that quantitative agreement is hardly to be expected. Nevertheless, the calculated curve of Figure 11 does reproduce the essential features of a typical C-curve (Figure 1), the most obvious discrepancy being the calculated curve's reluctance to reach its asymptote T_0 at reasonable times t. When realistic physical parameters are inserted into Equation (65), it is found that a value of $\sigma \simeq 60$ ergs/cm^2 must be chosen in order to obtain the selected value for b. Such a low value does not seem out of line for an effective interfacial energy for heterogeneous nucleation, however.

4.2. S-Curves

It is reasonable to interpret an S-curve (Figure 2) as the combination of two C-curves. This idea is illustrated schematically in Figure 12. The two distinct transformations which each yield their separate C-curves require separate nucleation events, one associated with the equilibrium transition temperature T_0, taken in Figure 12 to be that for incoherent equilibrium T_I, and one associated with some lower metastable transition temperature T_C, taken here to be that for coherent (metastable) equilibrium. In the present case, T_I and T_C are thus the temperatures denoted by the same symbols in Figure 10, for example. Of course, the lower C-curve is not necessarily due to coherent nucleation, but at least we know from our lengthy analysis of nucleation far from equilibrium that it cannot be due to spinodal decomposition.

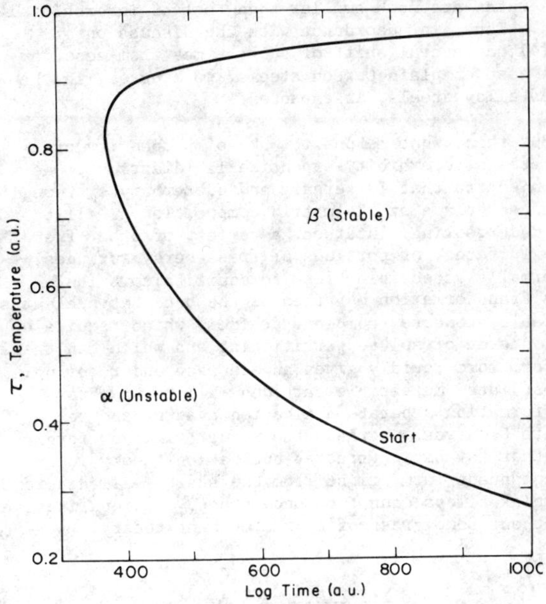

Figure 11. C-curve calculated according to equation (64) with b = 275. Compare with Figure 1.

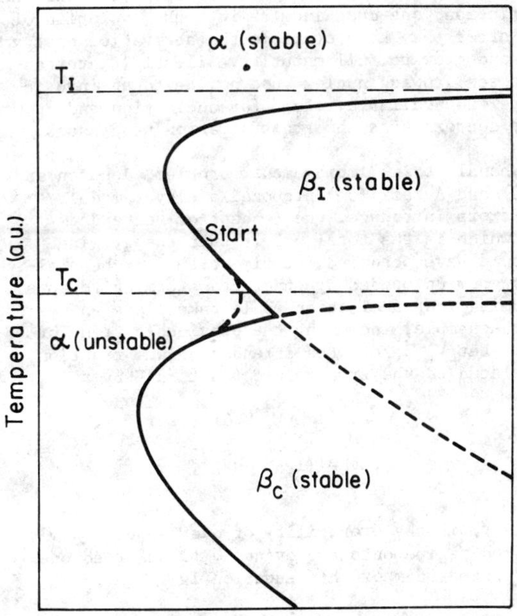

Figure 12. Schematic construction for S-curve. T_I and T_C are temperatures for incoherent and coherent transformations, respectively.

The situation in alloy steels is more complicated (see example in Figure 2), as the lower portion of the S-curve is surely not due to coherent precipitation but to the appearance of complex carbides which are, at best, semi-coherent with the parent phase. It may at first appear surprising that such minute amounts of a third and fourth constituent (0.45% Mn, 1.97% Cr) should alter so drastically the TTT diagram of steel. It must be recalled from the discussion on mobility (Section 3.3) however, that the temperature dependence of the prefactor K in Equation (37) is governed by the effective mobility M* whose magnitude, in turn, is determined by slowest-moving partitioning solute, even though [case (1)] its concentration in the matrix may be very low. In the present case, both Mn and Cr enter Austenite substitutionally, so that their diffusion rates are much smaller than that of interstitial Carbon. Thus, in accordance with the discussion of case (4), a lateral shift of start times from seconds, in the case of plain Carbon steels, to minutes, in the case of alloy steels, is expected.

But then, what is one to make of propensity of alloy steels to exhibit serpentine TTT diagrams? One may conjecture that in ternary and quaternary systems, the tangent plane at the matrix composition c_i^0 will, at large undercooling, intersect a variety of other free energy surfaces, or portions of free energy surfaces, not normally accessible close to equilibrium. The chosen transformation may then be the one that nucleates metastable products (complex carbides) which require a lesser degree of solute partitioning and which can therefore form more rapidly given an adequate undercooling. The postulated metastable carbides will require for their formation separate nucleation events, and hence separate C-curves. Perhaps the reader will now forgive the author for having treated nucleation theory in its multicomponent form, right from the start: TTT diagrams of complex alloys cannot be understood, even qualitatively, without some grasp of multicomponent theory.

4.3. End Curves

There is one last feature of TTT diagrams which will now be discussed briefly: the "end" curves. These curves are similar to the corresponding start curves but, of course, displaced towards longer times. To obtain some measure of the time required for completion of the transformation, one must inquire into the evolution of embryos after they have overcome the activation energy barrier and grow to full maturity. Clearly, interference between growing nuclei must now be taken into account, as the available volume for nucleation and growth steadily decreases as the transformation progresses.

To handle this "impingement" problem, Johnson and Mehl (50) and Avrami (51) proposed a clever model: these authors introduced the concept of "extended volume" which is the total volume that the growing phase would have occupied (at time t) had it been allowed to grow unimpeded by other transforming regions, and assuming that nucleation would take place anywhere within the sample, including the previously transformed regions. Let V_x denote the entended volume fraction. It is related to the true volume fraction V by the equation

$$dV = (1 - V)\, dV_x ,$$

where (1 - V) is the probability of overlap of a newly transformed region onto a previously transformed one. The integrated form of this equation is

$$V = 1 - e^{-V_x} . \tag{67}$$

The reason for the introduction of this fictitious volume fraction V_x is that its growth law can be determined in a straightforward manner, whereas that of V cannot. If G(t) is the growth rate of any linear dimension of the new phase, then for isotropic (spherical) growth one obtains the following equation for the extended volume at time t:

$$V_x(t) = \frac{4}{3}\int_0^t I(t)G^3(t')(t - t')^3\, dt' , \tag{68}$$

where I(t) now designates the nucleation rate, which could be that given by Equation (37), but will, in general, be time dependent. Equation (68), when introduced into Equation (67), yields the required growth law. Under very general assumptions the resulting growth law will be of the form

$$V(t) = 1 - e^{-At^n} , \tag{69}$$

where A is an appropriate constant, and where $3 \leq n \leq 4$ for three-dimensional nucleation and growth. A table of values of n for a wide variety of situations is given in Christian's treatise (2). Equation (69) plots as a sigmoid curve exhibiting incubation, growth, and saturation stages. In clustering reactions, saturation sets in with the exhaustion of the supply of available solute atoms and occurs when the volume fraction V has reached its maximum value dictated by the lever rule. The sigmoid curve representing Equation (69) can be roughly approximated by its tangent at the inflection point, and its intercepts at horizontal lines representing, say, 1% and 99% volume fraction transformed, can be taken as measures of the start (t_S) and end times (t_E) respectively. The difference ($t_E - t_S$) then fixes the position of the end curve, in principle, once that of the start curve is known.

5. EPILOGUE

We have come full circle: what started out, and ended, as a benign description of TTT diagrams, evolved into rather complex theoretical discussions. It should be obvious that enormous gaps separate theory from practical applications. Theorists claim that the theories are either too crude or too difficult to handle, experimentalists claim that the parameters required cannot be obtained with sufficient accuracy, and practical Physical Metallurgists claim (correctly) that no existing model can really do justice to the tremendous complexity of real systems. It is sometimes stated that the gaps are narrowing, but this author is not brimming with optimism: all that he can hope for is that this article will not contribute to the widening of the gaps.

References

1. E. C. Bain and H. W. Paxton Alloying Elements in Steel , ASM, Metals Park, Ohio, 1961.

2. J. W. Christian, The Theory of Phase Transformations in Metals and Alloys, Pergamon, London, 1965.

3. D. Turnbull, "Phase Changes", p. 226-308 in Solid State Physics, F. Seitz and D. Turnbull, eds.; Academic, New York, 1956.

4. K. C. Russell, "Nucleation in Solids", p. 219-268 in Phase Transformations, H. I. Aaronson, ed.; American Society for Metals, Metals Park, Ohio, 1970.

5. K. C. Russell "Nucleation in Solids: The Induction and Steady State Effects", Advances in Colloid and Interface Science, 13 (1980), p. 205-318.

6. H. I. Aaronson, "The Kinetic Equations of Solid-Solid Nucleation Theory", p. 83-115 in Lectures on the Theory of Phase Transformations, H. I. Aaronson, ed.; Met. Soc. of AIME, New York, 1975.

7. J. W. Cahn, "Unmixing in Binary Critical Systems", p. 41-64 in Critical Phenomena in Alloys, Magnets and Superconductors, R. E. Mills, E. Ascher and R. I. Jaffee, eds.; McGraw-Hill, New York, 1971.

8. J. W. Cahn, "Spinodal decomposition", Trans. AIME, 242 (1968), p. 166-180.

9. J. E. Hilliard, "Spindoal decomposition", p. 497-560 in Phase Transformations, H. I. Aaronson, ed.; American Society for Metals, Metals Park, Ohio, 1970.

10. D. de Fontaine, "Clustering Effects in Solid Solutions", p. 129-178 in Treatise on Solid State Chemistry, Vol. 5, Phase Changes, N. B. Hannay, ed.; Plenum Press, New York, 1975.

11. G. Martin, "The Theories of Unmixing Kinetics of Solid Solutions", p. 337-406 in Solid State Phase Transformations in Metals and Alloys; Aussois 1978; Les Editions de Physique, Orsay, France, 1980.

12. H. Reiss, "The Kinetics of Phase Transitions in Binary Systems", Journal of Chemical Physics, 18 (1950), p. 840-848.

13. H. Reiss and M. Shugard, "On the Composition of Nuclei in Binary Systems", Journal of Chemical Physics, 65 (1976), p. 5280-5293

14. J. S. Langer, "Statistical Theory of the Decay of Metastable States", Annals of Physics, 54 (1969), p. 258-275.

15. J. S. Langer and L. A. Turski, "Hydrodynamic Model of the Condensation of a Vapor near its Critical Point", Physical Review A, 8 (1973), p. 3230-3243.

16. K. Binder and D. Stauffer, "Statistical Theory of Nucleation, Condensation and Coagulation", Advances in Physics, 25 (1976), p. 343-396.

17. P. Mirold and K. Binder, "Theory for the Initial Stages of Grain Growth and Unmixing Kinetics of Binary Alloys", Acta Metallurgica, 25 (1977), p. 1435-1444.

18. J. W. Gibbs, "On the Equilibrium of Heterogeneous Substances", p. 55-353 in The Scientific Paper of J. Williard Gibbs, Vol. 1 Thermodynamics, Dover Publ., New York, 1961.

19. E. M. Lifshitz and L. P. Pitaevskii, "Statistical Physics, 3rd Edition, Part 1", p. 536 in Course of Theoretical Physics, Vol. 5, Landau and Lifshitz, Pergamon, Oxford, 1980.

20. J. S. Langer, "Metastable States", Physica, 73 (1974), p. 61-72.

21. J. W. Cahn and J. E. Hilliard, "Free Energy of a Nonuniform System, I. Interfacial Free Energy", J. Chem. Phys., 28 (1958), p. 258-267.

22. J. W. Cahn and J. E. Hilliard, "Free Energy of a Nonuniform System, II. Nucleation in a Two-Compartment Incompressible Fluid", J. Chem. Phys., 31 (1959), p. 688-699.

23. D. de Fontaine, "On the Stability of Spherically Symmetric Compositional Inhomogeneities in Solid Solutions", Trans. Met. Soc. AIME (1969), p. 1703-1709.

24. J. S. Langer, M. Bar-On, and H. D. Miller, "A New Computational Method in the Theory of Spinodal Decomposition", Physical Review A, 11 (1975), p. 1417-1429.

25. K. Binder, M. H. Kalos, J. L. Lebowitz and J. Marro, "Computer Experiments on Phase Separation in Binary Alloys", Advances in Colloid and Interface Science, 10 (1979), p. 173-214.

26. L. Farkas, Zeitsckrift für Physikalische Chemie, 125 (1927), p. 236.

27. M. Volmer and A. Weber, "Nuclei Formation in Supersaturated States", Z. Physik. Chem., 119 (1925), p. 277-301.

28. R. Becker and W. Döring, "Kinetic Treatment of Grain Formation in Supersaturated Vapors", Ann. Physik., 24 (1935), p. 719-752.

29. K. C. Russell, "Linked Flux Analysis of Nucleation in Condensed Phases", Acta Metallurgica, 16 (1968), p. 761-769.

30. J. Frenkel, Kinetic Theory of Liquids, Oxford, University Press, 1946.

31. H. Reiss, "The Number of Unilateral Solute Impacts on Unit Surface in a Dilute Solution", Journal of Chemical Physics, 18 (1950), p. 996-997.

32. D. de Fontaine, "An Analysis of Clustering and Ordering in Multicomponent Solid Solutions—II. Fluctuations and Kinetics", J. Phys. Chem. Solids, 34 (1973), p. 1285-1304.

33. R. McGraw and H. Reiss, "Nucleation Near a Critical Temperature", Journal of Statistical Physics, 20 (1979), p. 385-413.

34. R. B. Heady and J. W. Cahn, "Experimental Test of Classical Nucleation Theory in a Liquid-Liquid Miscibility Gap System", Journal of Chemical Physics, 58 (1973), p. 896-910.

35. H. E. Cook and J. E. Hilliard, "A Simple Method of Estimating the Chemical Spinodal", Trans. Met. Soc. AIME, 233 (1965), p. 142-146.

36. D. de Fontaine, "A Computer Simulation of the Evolution of Coherent Composition Variations in Solid Solutions", Ph.D. Dissertation, Northwestern University, Evanston, Illinois, 1967.

37. D. de Fontaine, "Development of Fine Coherent Precipitate Morphologies by the Spinodal Mechanisms" p. 93-131 in Ultrafine Grain Metals, Burke and V. Weiss, eds.; Syracuse Univ. Press, 1970.

38. J. W. Cahn, "On Spinodal Decomposition", Acta Met., 9 (1961), p. 795-801.

39. J. W. Cahn, "On Spinodal Decomposition in Cubic Crystals", Acta Met., 10 (1962), p. 179-183.

40. K. B. Rundman and J. E. Hilliard, "Early Stages of Spinodal Decomposition in an Aluminum-Zinc Alloy", Acta Met., 15 (1967), p. 1025-1033.

41. A. Guinier, "Nouvelle Interpretation Des Diagrammes à "Side-Bands", Acta Met., 3 (1955), p. 510-522.

42. A. Bonfiglioli and A. Guinier, "La Structure des Zones G. P. dans les Alliages Aluminium-Zinc au Premier Stade de leur Formation", Acta Met., 14 (1966), p. 1213-1224.

43. J. S. Langer, "Statistical Methods in the Theory of Spinodal Decomposition", Acta Met., 21 (1973), p. 1649-1659.

44. J. Grilhé, "Experimental Studies of Decomposition

of a Solid Solution", p. 407-427 in Solid State Phase Transformations in Metals and Alloys; Aussois 1978; Les Editions de Physique, Orsay, France, 1980.

45. N. C. Wong and C. M. Knobler, "Spinodal Decomposition in Isobutyric Acid + Water Mixtures", Journal of Chemical Physics, 66 (1977), p. 4707-4709.

46. G. Laslaz, P. Guyot and G. Kostorz, Journal de Physique C7, 38 (1977); M. Hennion, D. Ronzaud and P. Guyot, "Kinetics of Unmixing in Al-Zn Single Crystals Studied by Neutron Small-Angle Scattering", Preprint, Institut National Polytechnique de Grenoble, LTPCM, ENSEEG Domaine Universitaire, BP 44, 38401 St. Martin d'Hères, France, 1981.

47. Private communication, A. R. Forouhi and J. M. Sanchez, Dept. of Materials Science and Mineral Eng., University of California, Berkeley, 1981.

48. J. Marro, J. L. Lebowitz and M. H. Kalos, "Computer Simulation of the Time Evolution of a Quenched Model Alloy in the Nucleation Region", Physical Review Letters, 43 (1979), p. 282-285.

49. J. W. Cahn, "The Kinetics of Grain Boundary Nucleated Regions", Acta Met., 4 (1956), p. 449-459; Acta Met., 5 (1957), p. 168-175.

50. W. A. Johnson and R. F. Mehl, "Reaction Kinetics in Processes of Nucleation and Growth", Trans. AIME, 135 (1939), p. 416-458.

51. M. Avrami, "Kinetics of Phase Change—I. General Theory", J. Chem. Phys., 7 (1939), p. 1103-1112.

ACKNOWLEDGEMENTS

This work was supported by the Director, Office of Energy Research, Office of Basic Energy Sciences, Materials Sciences Division of the U.S. Department of Energy under Contract Number W-7405-ENG-48. Programming assistance from Dr. W. Teitler is gratefully acknowledged.

FUNDAMENTALS OF MARTENSITIC REACTIONS

Morris Cohen
Department of Materials Science and Engineering
Massachusetts Institute of Technology
Cambridge, MA 02139

and

C. Marvin Wayman
Department of Metallurgy and Mining Engineering
and Materials Research Laboratory
University of Illinois
Urbana, IL 61801

Martensitic reactions are a subclass of displacive, diffusionless phase transformations in which the kinetics and morphology are dominated by relatively large shear-like displacements in the transforming micro-regions. The resulting local shape changes involve both lattice (homogeneous) and lattice-invariant (inhomogeneous) deformations, the former being accomplished by coherency (or transformation) dislocations and the latter by anti-coherency (or misfit) dislocations. Both types of dislocations move glissilely and comprise the interfacial structure.

The principal features of martensitic nucleation are accounted for by the spontaneous dissociation (i.e., faulting) of pre-existing groups of dislocations on close-packed planes, when the fault energy becomes sufficiently negative. In most instances, the free energy of the system can decrease monotonically once the dissociation process starts, and so this type of nucleation is barrierless, or athermal. However, in other cases, there may be a thermal, as well as an athermal, component of the lattice resistance to the moving interfacial dislocations, and then the nucleation can be activated isothermally via local fluctuations which assist the dislocation motion. Both athermal and isothermal nucleation are characteristically heterogeneous because of the essential role played by pre-existing defects. These fundamental aspects of martensitic nucleation also provide the basis for autocatalytic and mechanically-induced nucleation.

The extending fault can lower its strain energy by altering its shape, interface and habit to convert the transformational shape deformation into an invariant-plane strain. These morphological adjustments are facilitated by the participation of additional dislocations, and lead to various crystallographic relationships often identified with martensitic reactions, including nonrational habits. In the special case of thermoelastic martensitic transformations, the strain energy happens to be small enough, compared to the flow strength of the parent phase, that the shape deformation is accommodated elastically, and then the stored elastic energy remains available to help reverse the transformation when the driving-force conditions are reversed. These phenomena also operate in pseudoelasticity and in the shape-memory effect.

1. INTRODUCTION

The main intent of this paper on martensitic reactions* is to provide an up-to-date synthesis of the field, without undertaking anything like a complete review of the literature. Comprehensive coverage can be found in Professor Z. Nishiyama's recent book (1) and in more limited treatments (e.g., 2). Special note may also be given here to a book on martensitic transformations by Professor Xu Zuyao, published during the past year in The People's Republic of China (3).

* The terms reaction and transformation are used interchangeably throughout this paper. The term transition is reserved for higher-order phase transformations.

In line with the title of the present treatise, we shall place emphasis on the basic nature and characteristics of martensitic transformations as such, rather than on the properties of the transformation product -- martensite. Nevertheless, it should be stressed that the properties of some martensites, especially in ferrous alloys, are of vast technological importance. Of course, we shall be concerned with the crystallographic and morphological aspects of martensitic products insofar as they may shed light on the transformational process per se.

A martensitic reaction can be considered to be a first-order solid-state structural change which is (a) displacive, (b) diffusionless, and (c) dominated in kinetics and morphology by the strain energy arising from shear-like displacements (4). This set of three characteristics is regarded as both necessary and sufficient to define a martensitic transformation. For example, a bainitic transformation conforms to (a) and (c) but not (b), and so is not considered to be martensitic; similarly, a shuffle transformation like the ω-transformation, may conform to (a) and (b) but not (c), and so does not fall in the martensitic category. As we shall see, it is condition (c) which causes martensitic habits to tend toward becoming invariant-plane interfaces, i.e., interfaces between the martensite and its parent phase which are macroscopically undistorted and unrotated. The resulting displacement or shape deformation due to the transformation is termed an invariant-plane strain.

2. SOME QUALITATIVE ASPECTS OF MARTENSITIC REACTIONS

This section describes several fundamental and characteristic features of martensitic reactions in qualitative terms. It is intended to serve as a helpful entré to the more detailed sections which follow.

2.1 Meaning of Displaciveness and the Shear-like Component

In a displacive structural change, there is a coordinated shift of atoms, sometimes referred to descriptively as a "military" transformation. It is presumed that the atoms move in organized ways relative to their neighbors; but such motions can only be inferred (through proposed models or mechanisms) from experimental knowledge of the initial and final lattice sites. In general, these displacements can be described as a combination of a homogeneous lattice deformation and shuffles. It is necessary to distinguish between these types of displacements in dealing with martensitic transformations.

In a lattice deformation or lattice-distortive strain, the coordinated shift of atoms converts one Bravais lattice homogeneously to another, and also involves whatever rigid-body rotation may occur because of the coupling between the transforming region and the adjacent parent matrix. The part of the lattice deformation which accomplishes the change in crystal structure is a pure deformation, and is usually called the Bain distortion or Bain strain (5), the term originally adopted for the FCC→BCC (or BCT) martensitic transformation in ferrous alloys. The Bain distortion helps visualize the lattice correspondence between the parent and product phases; the well-known FCC→BCC correspondence is shown in Figure 1.

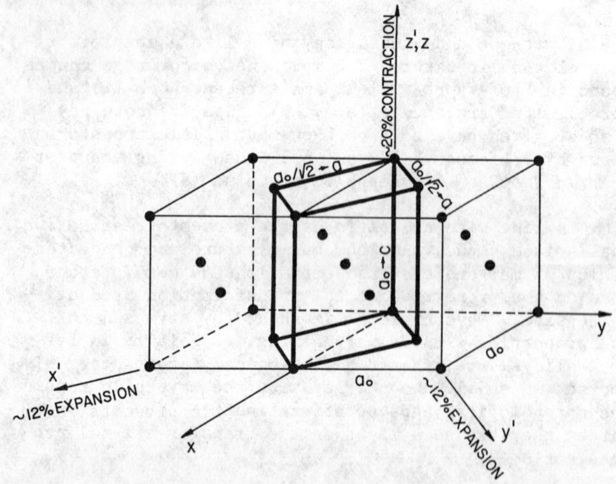

Figure 1. Lattice correspondence and distortion proposed by Bain (5) for the FCC→BCC (BCT) martensitic transformation in ferrous alloys. The correspondence cell (heavy lines) in the parent phase becomes a unit cell of martensite after a homogeneous lattice deformation. The principal distortions along x', y', and z' are indicated.

On the other hand, a shuffle is a coordinated shift of atoms within a unit cell, which may change the structure but which, in itself, does not produce a homogeneous lattice-distortive strain. Figure 2 illustrates, in simple form, the distinction between lattice-distortive and shuffle displacements. Even though displacive phase transformations may generally comprise various combinations of lattice deformations and shuffles, the subset to be identified as martensitic is typically dominated by lattice deformations rather than by shuffles, although shuffles are not excluded. In addition, according to requirement (c) in the foregoing definition of a martensitic reaction, the shear component of the deformation must play a prominent role in the kinetics and morphology of the transformation. To clarify this point, we may call attention to a polymorphic transformation in cerium which occurs diffusionlessly on cooling below 100° K; it is a phase change from one FCC structure to another and is accompanied by a 16 percent contraction in volume (6). Inasmuch as no shear displacement is involved, this transformation cannot be classed as martensitic even though a large lattice-distortive strain takes place. Among lattice-distortive transformations having both dilatational and shear (deviatoric) components, an appropriate criterion for determining whether the shear component "prevails" over the dilatational component is the existence of an undistorted line in the lattice deformation (4).

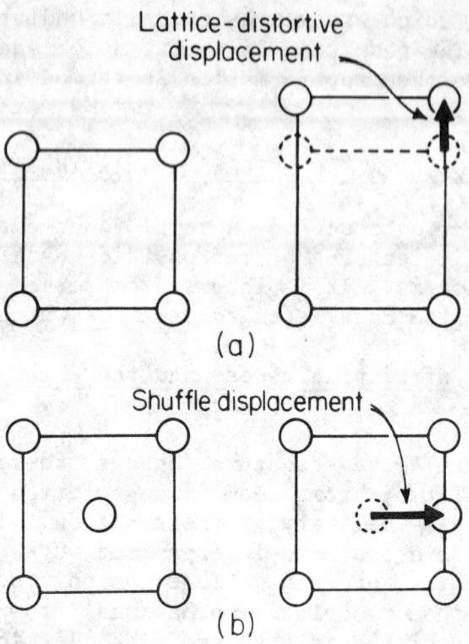

Figure 2. Schematic illustration of simple lattice-distortive and shuffle displacements (4).

From the lattice-distortive nature of a martensitic reaction as discussed thusfar, it is to be expected that the accompanying strain energy will have important consequences. In the first place, additional displacements may occur inhomogeneously as an integral part of the transformation to help relieve the strain energy. These displacements take place by slip or twinning in the transformation product and are called lattice-invariant because they do not change the crystal structure generated by the lattice deformation. The lattice deformation produces a distortion of the transforming region, whereas the lattice-invariant deformation acts to reduce the magnitude of this distortion. Still, the resulting shape-deformation is often substantial, as indicated by the surface tilts in Figure 3, and has to be accommodated

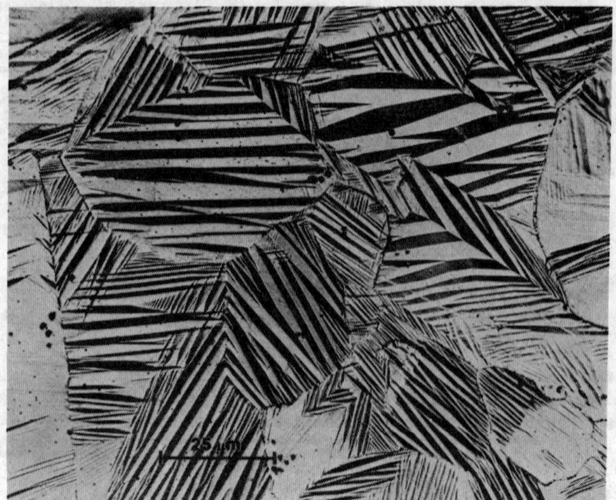

Figure 3. Optical micrograph showing surface tilts due to shape deformation of martensitic plates in a polycrystalline Fe-24.5 at.% Pt alloy. Specimen was prepolished in the parent phase condition prior to cooling for the martensitic transformation.

somehow in the interactions with the surrounding matrix. If this accommodation can be provided by elastic deformation, the martensitic transformation will be thermoelastic (exhibiting certain interface-reversible features, to be discussed later), but in more typical cases, the accommodation occurs by plastic yielding of the parent phase and reversibility is lost.

The transformational strain energy can be further decreased if the transforming regions adopt plate-like (or lath-like) shapes, and particularly if an invariant-plane interface is established as the plane of the habit orientation. Accordingly, such morphological relationships become essential considerations in martensitic transformations.

2.2 Meaning of Diffusionless

The term diffusionless in the context of a martensitic transformation signifies that there is no random-walk mixing of atoms, or atom-by-atom jumping across the interface, during the structural change. Consequently, the product phase inherits the same composition, atomic order, and lattice defects as are present in the parent phase. Thus, the thermodynamic driving force for a given martensitic reaction can be treated as a one-component system, even for multicomponent alloys. This simplification is illustrated in Figure 4. For the composition

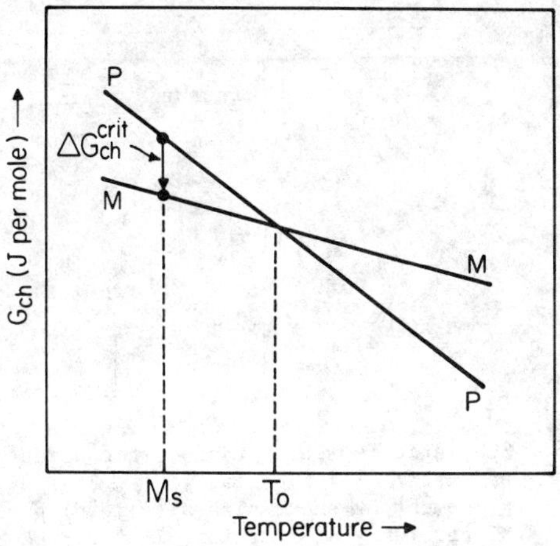

Figure 4. Typical free-energy curves for parent (P) and martensitic (M) phases of a given composition as a function of temperature. $\Delta G_{ch}^{P\to M}$ becomes negative below T_o and must reach a critical value at M_s in order to start the nucleation process.

C_o, there is a temperature T_o at which the martensitic phase has the same chemical free energy per mole (or per unit volume) as does the parent phase of the same composition i.e., $\Delta G_{ch}^{P\to M} = 0$ at T_o. With decreasing temperature below T_o, $\Delta G_{ch}^{P\to M}$ becomes more and more negative, and correspondingly the driving force for the martensitic transformation increases. Of course, the chemical free energies and driving forces at play are bulk quantities and exist in principle whatever the strain-energy and interfacial-energy requirements may be for a martensitic transformation or, indeed, even if such a transformation does not take place at all.

The diffusionless property obviously means that atomic diffusion is not required for a martensitic reaction to nucleate and grow; moreover, any dislocations which may participate in the operative lattice and lattice-invariant deformations must move glissilely, i.e., without generating or annihilating point defects which require diffusive dislocation-climb processes. Therefore, martensitic transformations are capable of proceeding at high reaction rates even at temperatures approaching absolute zero.

In view of these characteristics, the kinetics of martensitic transformations are typically controlled by the nucleation rate because the ensuing growth rate is so fast. The nucleation step is usually regarded as athermal in the sense that more and more nucleation events occur during cooling due to the increasing driving force. Under these conditions, the extent of transformation increases with decreasing temperature and is relatively insensitive to time or cooling rate.* On the other hand, there are some important examples of isothermal martensitic nucleation, and then the reaction kinetics depend on both temperature and time.

3. NUCLEATION OF MARTENSITIC REACTIONS

If a martensitic transformation were able to nucleate in an ideally perfect crystal of the parent phase, the free-energy change attending this event could be written (7):

$$\Delta G = (4\pi/3)r^2c\Delta g_{ch} + (4\pi/3)r^2c(Ac/r) + 2\pi r^2\sigma. \quad (1)$$

Here, the particle of martensite is assumed to be a thin oblate spheroid of volume $= (4\pi/3)r^2c$ and surface area $\approx 2\pi r^2$; r is the radius, c is the semithickness, Δg_{ch} is the chemical free-energy change per unit volume of the martensite particle, Ac/r ($=\Delta g_{el}$) is the elastic strain energy per unit volume of the martensitic particle, and σ is the interfacial energy per unit area.

The strain-energy factor A is given by (7):

$$A = [\pi(2-\nu)/8\ (1-\nu)]\mu\gamma^2 + (\pi/4)\mu\varepsilon_n^2, \quad (2)$$

where the Poisson's ratio (ν) and the shear modulus (μ) are each taken to be the same for the parent and product phases. It is also assumed that the martensitic shape deformation is an invariant-plane strain (IPS)**, having a shear component γ and a normal component ε_n. With typical values for the martensitic reaction in ferrous alloys, A is about 2.4×10^3 MJm^{-3}.

Most martensitic transformations, and certainly the FCC→BCC (or BCT) structural change in iron-base alloys, evolve semicoherent rather than fully coherent interfaces, and so the interfacial energy is estimated to be about 100-200 mJm^{-2} (100-200 erg cm^{-2}) rather than to lie in the coherent range of 10-20 mJm^{-2} (10-20 erg cm^{-2}). Δg_{ch} is approximately -170 MJm^{-3} (-300 cal mole^{-1}) for a typical ferrous alloy at its martensite-start temperature (M_s), and when this quantity is substituted into Equation (1) along with the above values of A and σ, the resulting function of ΔG vs. r and c is a saddle-shaped surface with a barrier on the order of 8×10^{-16} J per nucleation event ($\sim 5 \times 10^3$eV). This is $\sim 10^5$kT at temperatures where the transformation is known to nucleate spontaneously. Clearly, the thermal energy is much too small for homogeneous nucleation by heterophase fluctuations, and so special nucleation sites must be considered.

3.1 Heterogeneity of Martensitic Nucleation

Martensitic transformations are heterogeneous in the sense that they are first-order reactions and pass through a parent/product two-phase regime, with discrete interfaces between the two phases. The nucleation is also heterogeneous, and typically starts at roughly 10^6 sites cm^{-3} in polycrystalline alloys, which is a very small number compared to that expected for homogeneous nucleation ($\sim 10^{23}$ cm^{-3}).

---

* In thermoelastic martensitic transformations, the growth step and the attendant strain energy tend to stay in thermoelastic balance with the increasing driving force during cooling, and so the extent of transformation as a function of dropping temperature depends on the growth process as well as on the nucleation step.

** The nature and significance of the IPS condition will be detailed later in connection with the crystallographic theory of martensitic transformations (Section 6).

The heterogeneous nature of martensitic nucleation is nicely demonstrated when the parent phase is separated into very small particles (9-11). Figure 5 for an iron-nickel alloy shows the

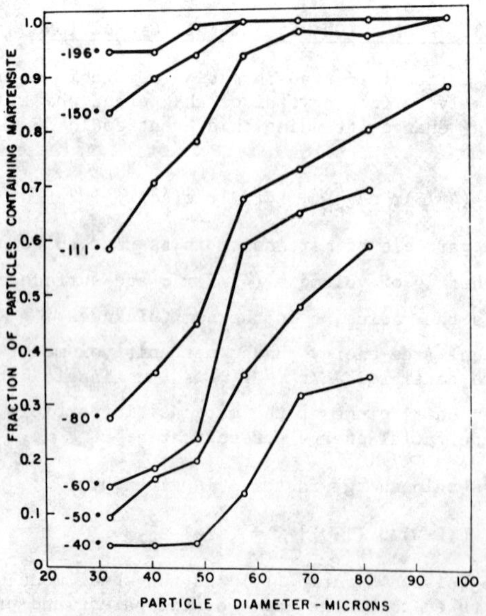

Figure 5. Martensitic transformation in particles of Fe-30.2Ni vs. particle size and cooling temperature (10).

fraction of such particles which nucleate on cooling as a function of particle diameter and cooling temperature. On reaching any given temperature, the fraction of nucleated particles increases with particle volume, indicating that the probability of finding a pre-existing nucleation site in a small particle increases with the size of the particle. Moreover, for a given particle size, the fraction of nucleated particles increases with decreasing temperature of cooling. This signifies that the pre-existing nucleation sites have a range of potencies, with the weaker ones requiring more driving force and, hence, lower temperatures to become operative. In fact, some particles may not transform at all even on cooling to temperatures approaching absolute zero. These small-particle experiments suggest that the number of potent nucleation sites which account for the M_s temperature is of the order of 10^6 cm^{-3}, or about one per 100 μm grain of the parent phase. It is expected that still more potent sites are present and can lead to nucleation above M_s, but they are too few in number to produce a detectable amount of transformation.

It will be shown later that martensitic reactions are frequently autocatalytic because of the elastic stresses and plastic strains which arise from the shape deformation of the transformation. Such autocatalysis tends to spread throughout the volume of a bulk specimen and thus may obscure the potential participation of the less potent pre-existing sites. However, in the small-particle type of experimentation, autocatalytic effects within a given particle are not transmitted to other particles, and so the start-up nucleation in each particle must depend on its own pre-existing sites.

3.2 Some Features of Martensitic Interfaces

It is instructive at this point to describe the expected dislocation structure of martensitic interfaces, knowledge of which will help one understand (a) the dislocation nature of proposed pre-existing nucleating defects in the parent phase, and (b) the role of dislocation motions in the basic deformations which operate during the transformation. All of this is important in the nucleation problem.

We can regard martensitic interfaces as being composed of coherency and anticoherency dislocations, each type reflecting the kind of deformation it produces across the interface (12). Coherency dislocations are the same as transformation dislocations; they accomplish the homogeneous deformation of one crystal structure into another while maintaining full continuity of the lattice correspondence between the two at the interface. A simple example is illustrated in Figure 6 for a coherent

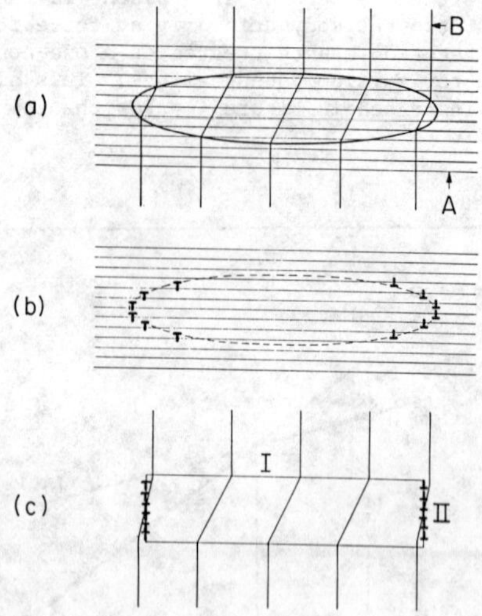

Figure 6. Schematic representation of a coherent shear particle: (a) fiducial lines (B) show displacement of shear planes (A); (b) position of coherency dislocations in interface; (c) coherency dislocations moved to illustrate the difference between a stress-free coherent interface (I) and a forced-elastic coherent interface II (12).

shear particle. The designated dislocations are loops of partial dislocations. These coherency dislocations may have both glide and climb components but they still can move conservatively in their loop planes (12). Any homogeneous lattice deformation can be achieved by the superposition of several sets of such dislocations. As shown in Figure 6, coherent interfaces may be stress-free (special case) or forced-elastic (usual case), the latter coherency and continuity being maintained by a coherency dislocation on each matching plane.

The lattice-invariant deformations are inhomogeneous, and the accompanying dislocations are conventional or lattice dislocations which act to cancel part of the coherency strains. They interrupt the lattice continuity across the interface and, therefore, are called anticoherency (or misfit or matching) dislocations.* These dislocations, unlike coherency dislocations, cannot climb glissilely.

An otherwise coherent interface containing one or more periodic arrays of anticoherency dislocations is regarded as semicoherent; an example is given in Figure 7 for the FCC→BCC transformation (13). Two sets of coherency dislocations -- $\frac{a}{18}[1\bar{2}1]$ and $\frac{a}{16}[\overline{12}1]$ -- provide the homogeneous structural change, while two sets of anticoherency dislocations -- $\frac{a}{2}[011]$ and $\frac{a}{2}[\bar{1}10]$ -- provide the inhomogeneous lattice-invariant deformations which essentially cancel the long-range strain fields of the coherency dislocations. (The above Burgers vectors are referred to the parent FCC lattice.) The resulting

* The lattice-invariant deformations can likewise take place by twinning. The alternating twin orientations have separate lattice correspondences, and the resulting overall deformation is inhomogeneous.

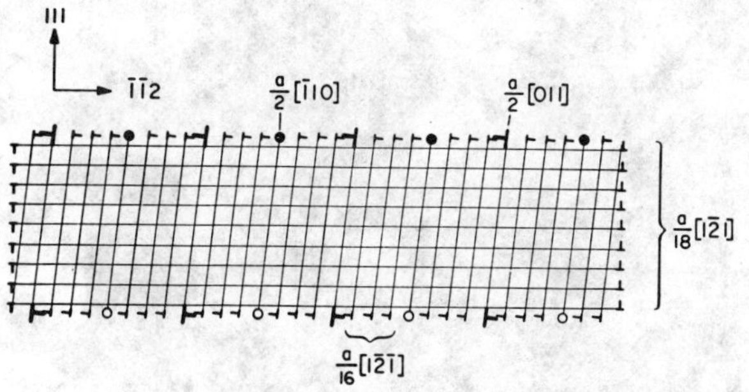

Figure 7. Interfacial dislocation structure of FCC→BCC semicoherent nucleus, after operation of homogeneous lattice and inhomogeneous lattice-invariant deformations. Coherent dislocation loops are represented by small T's on every plane, and anticoherency loops by both large T's and circles spaced eight planes apart (13).

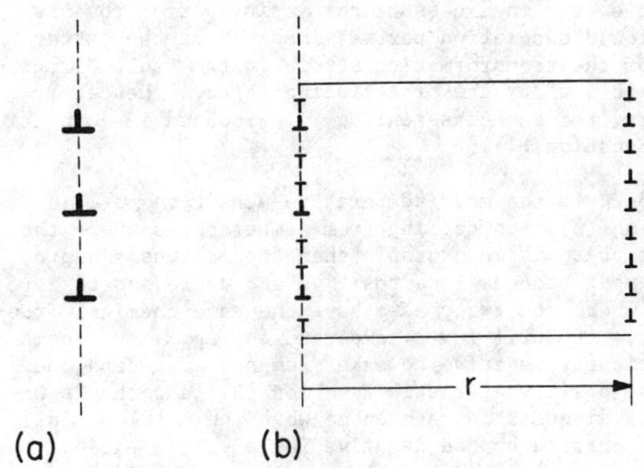

Figure 8. Schematic illustration of martensitic nucleation by defect dissociation: (a) pre-existing defect consisting of a group of lattice dislocations, one above the other, as part of a grain boundary or subboundary in the parent phase; (b) on dissociation, the coherency (partial) dislocations move to the right, forming a coherent interface and leaving a faulted structure behind (18). The stationary interface on the left contains a dislocation-dipole structure because of the partials that have moved away.

crystallographic orientation is that found by Kurdjumov and Sachs (14), and is discussed more fully in Section 6.

3.3 Mechanisms of Martensitic Nucleation

Three quite-different mechanisms of martensitic nucleation have been proposed during the past decade. Each recognizes the characteristic heterogeneity of martensitic nucleation and each relies on lattice defects for the nucleating sites. The most classical approach is that of Suezawa and Cook (15), wherein heterophase fluctuations are visualized to generate fully-formed martensitic particles when aided by elastic interaction with very potent, pre-existing superdislocations. In contrast, Clapp (16) has suggested that a local mechanical instability of the parent phase, such as a localized soft phonon mode, may set in near a defect under the right thermodynamic conditions. Then, as with higher-order displacive transformations, one can describe fluctuations that might bring the lattice to a "strain spinodal" via certain phonon modes. Suzuki and Wuttig (17) have also treated this idea and reached the conclusion that the gradient energy between the local instability and the coherently-coupled surrounding matrix (the latter being still mechanically stable) would have to be negative in order to allow spontaneous nucleation. In the third type of mechanism, Olson and Cohen (18) have invoked special arrays of pre-existing lattice dislocations as the fundamental nucleating defect, and have suggested how such defects can dissociate to carry out the nucleation process. Their mechanism is substantially more detailed and predictive both crystallographically and kinetically, than the others and, accordingly, is chosen for discussion here. In the latter approach, thermal fluctuations play their role by assisting the motion of dislocations, rather than in the classical heterophase sense.

We first consider a closely-spaced array of lattice dislocations, such as depicted in Figure 8, which may exist in the parent phase as part of a grain boundary, subboundary, inclusion interface, or other defect. For illustrative purposes, it is convenient to deal with an FCC parent structure inasmuch as we wish to lead up to the prototype martensitic reaction in ferrous alloys. These dislocations are envisioned to dissociate into two arrays of partial dislocations which repel each other, but usually the dissociation will be severely restricted by the high energy (relative to the parent phase) of the faulted structure lying between the two arrays. Nevertheless, if one array were able to move through the FCC phase on close-packed $(111)_{FCC}$ planes, with each such dislocation then acting as a Shockley $\frac{a}{6}[12\bar{1}]$ partial, there would be the following structural changes:

(a) With one Shockley partial advancing on every $(111)_{FCC}$ plane, the FCC structure will be deformed into a twinned orientation;

(b) With one Shockley partial advancing on every second $(111)_{FCC}$ plane, the FCC structure will be deformed into an HCP structure, and

(c) With one Shockley partial advancing per every three $(111)_{FCC}$ planes, and each spreading its core over the three planes to become $\frac{a}{18}(1\bar{2}1)$ partials, the FCC structure will be homogeneously deformed through one-third of a twinning shear. This is precisely the first of the two Bogers-Burgers displacements (19) which can carry the FCC structure into a BCC structure (18). The first displacement generates the appropriate $\{110\}_{BCC}$ planes, but not with the proper BCC stacking. A second homogeneous displacement is required to produce the correct stacking.* Thus, one can see that a potential defect for starting the FCC→BCC martensitic transformation is a group of lattice dislocations in the parent phase, arranged one above the other and spaced one for every three close-packed planes.

The defect-dissociation process shown schematically in Figure 8 can set in with a monotonic decrease in total free energy if the faulted region left behind happens to be more stable than the parent phase, i.e., if the fault energy per unit area (γ) becomes negative. This fault energy can be expressed as (18):

$$\gamma = nd\,[\Delta g_{ch} + \Delta g'_{el}] + 2\sigma, \qquad (3)$$

where n = no. of fault planes in the thickness of the faulted region (for the case at hand, n = 3 x no. of lattice dislocations in the nucleating defect), d = spacing between the fault planes. The chemical free-energy change (Δg_{ch}) and the elastic strain energy ($\Delta g'_{el}$), both per unit volume of the ribbon-shaped particle being formed, are bulk properties, whereas the interfacial energy (σ) is interface-dependent. However, σ includes only the "top and bottom" fault/matrix interfaces because the self-energy of the advancing coherency-dislocation interface is derived from the already-existing defect.

* These displacements are not simple shears because dilatations are involved. Changes in atomic diameter may also take place in the structural transformation.

Similarly, the $\Delta g'_{el}$ term contains only the elastic strain energy in excess of that arising from the dislocation dissociation per se, inasmuch as the latter part of the transformation strain in the faulted region is "paid for" by the pre-existing defect. Hence, $\Delta g'_{el}$ differs from the classical Δg_{el} introduced in connection with Equation (1).

Δg_{ch} is the most temperature-sensitive term in Equation (3). On cooling from temperatures where the parent phase is more stable than the martensitic product, Δg_{ch} becomes less positive and decreases to zero at T_o where the two phases have the same chemical free energy. At still lower temperatures, Δg_{ch} can become sufficiently negative to make γ negative, overbalancing the positive terms in Equation (3), and then spontaneous dissociation can ensue when the net "chemical force" exerted by the negative fault energy exceeds the lattice resistance (τ_f) opposing the motion of the dislocation interface. Accordingly, the critical condition for nucleation via defect dissociation is (18):

$$\gamma^{crit} = -n\tau_f b, \qquad (4)$$

or substituting in Equation (3):

$$\Delta g_{ch}^{crit} = -[\Delta g'_{el} + \frac{2\sigma}{nd} + \frac{\tau_f b}{d}], \qquad (5)$$

where b is the Burgers vector of the coherency dislocations in the moving interface. τ_f is usually not known for partial dislocations, but the macroscopic flow stress can serve as an upper limit.

It is evident from Equation (5) that the critical driving force for nucleation depends on the plastic as well as the elastic properties of the parent phase. Furthermore, Δg_{ch}^{crit} is a function of n: the thicker the fault, the less negative Δg_{ch}^{crit} has to be. This means that n provides a physical and quantitative measure of the potency of the nucleating defect. If Δg_{ch}^{crit} is set equal to the chemical free-energy change at the observed M_s temperature, the thickness of the operative fault calculates out to be 13-14 close-packed planes, or 4-5 lattice dislocations in the pre-existing type of defect which can account for M_s. Defects with fewer dislocations will require more negative values of Δg_{ch}^{crit} and, consequently, lower cooling temperatures in order to be activated. When the condition for spontaneous dissociation is met, Δg_{ch}^{crit} is sufficiently negative to generate the additional coherency and anticoherency dislocations appearing in the upper and lower interfaces of Figure 7.

In the defect-dissociation mechanism of the FCC→BCC nucleation event, the nucleus comes into existence as a semicoherent particle and extends parallel to the close-packed fault plane. At this stage, it not only has the correct crystal structure and orientation relationship but also its habit plane is unrotated. However, the habit plane is distorted and is not an invariant plane at this step of the transformation. The IPS condition, along with its reduction in strain energy, is approached subsequently during the growth stage, as will be discussed later. This "switching" of habit planes between the nucleation and macroscopic growth stages is suggested by the in situ transmission electron microscopy of Ferraglio and Mukherjee (20) on a thermoelastic Au-Cd alloy. Figure 9a shows the predicted fault morphology extending along a $(110)_{BCC}$ plane, starting at the subboundary A in accordance with the proposed defect dissociation, before switching to the macroscopic habit morphology in Figure 9b.

The concept of martensitic nucleation by defect dissociation has also been applied to FCC→HCP, BCC→FCC, and BCC→HCP systems (18), although available data are often inadequate for satisfactory checking. However, where calculations can be made, the nucleating defects

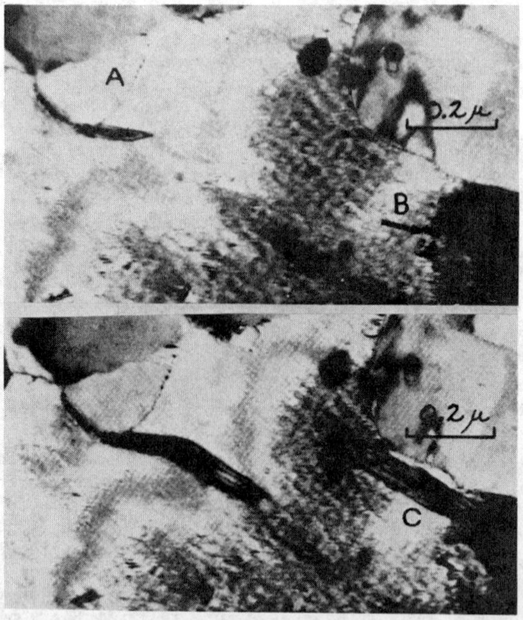

Figure 9. Transmission electron micrographs showing in situ martensitic nucleation from subboundaries in Au-Cd (20): (a) nucleus originates at boundary A and grows in fault morphology on close-packed plane of parent phase; (b) fault morphology switches to macroscopic habit. Boundary at B in (a) nucleates plate C in (b).

which operate at the various M_s temperatures seem to consist of about 4-5 dislocations, after considering that the number of fault planes derived per dislocation differs with various structural changes.

3.4 Athermal vs. Thermal Nucleation

The nucleation condition expressed by Equation (5) is not time-dependent; the process occurs athermally at sites with less and less potency (n) while the temperature is dropping. Such nucleation is barrierless once a critical value of the chemical free-energy change is reached for a given nucleating defect, in that the dissociation and related faulting can then proceed with a monotonic decrease in free energy of the system. In a real sense, the dissociation phenomenon constitutes a local instability, and this extends spontaneously as soon as the free energy has a diffusionless path to "run downhill" from the pre-existing high energy of the nucleating defect.

Heterophase fluctuations, which play a prominent role in classical nucleation theory, are not involved in the above nucleation concept. Instead, the kinetics of nucleation are governed by the dynamics of the advancing interfacial array of coherency dislocations. Under most conditions of martensitic transformations, this propagation is a high-velocity phenomenon because of the available net driving force on the dislocations when Δg_{ch}^{crit} is reached, and so there is no detectable time-dependence of the nucleation events at a given temperature.

On the other hand, important cases of isothermal martensitic nucleation are well known (21-24). This time-dependence is consistent with the fact that dislocation motion can be assisted by thermal fluctuations on a much more local scale than that of the aforementioned heterophase fluctuations. With such assistance, τ_f in Equation (5) has two components: an athermal component (τ_μ) which is velocity-independent, and a thermally-activated component ($\tau_{th} = \tau_f - \tau_\mu$) which is velocity-dependent. The latter can lead to a time-dependence in the nucleation process. Assuming now that the observed activation energy for isothermal nucleation (ΔG^*_{exp}) may be controlled by the thermally-assisted motion of interfacial dislocations, ΔG^*_{exp} should be equal to the activation energy (Q) for dislocation motion driven by the net "chemical force" of the negative fault energy. Based on the idea of thermally activated dislocation motion, and expressing τ_f as in Equation (5), Q is found to depend on Δg_{ch} as follows (18):

$$Q = Q_o + (\tau_\mu + \frac{d}{b}\,\Delta g'_{el} + \frac{2\sigma}{nb})v^* + (\frac{d}{b}\,v^*)\Delta g_{ch}, \qquad (6)$$

where Q_o is the stress-free activation energy, and v* is the activation volume. The degree of isothermal character in the nucleation kinetics will depend on v*, with a large value leading effectively to athermal behavior.

If v* is constant (as is customarily assumed), Equation (6) shows that Q is linearly dependent on Δg^{ch}. This is a significant result because measurements of isothermal nucleation rates (Section 5.1) show that ΔG^*_{exp} is likewise a linear function of Δg^{ch} (23,25). Furthermore, Q_o and v* as evaluated from isothermal nucleation measurements are consistent with a process controlled by thermally activated dislocation motion (18).

Q in Equation (6) is seen to increase with decreasing n for any given value of Δg_{ch}. This signifies that, at any constant temperature in the regime of isothermal nucleation, defects with a range of potencies are in process of nucleation, but the stronger ones operate more rapidly and become detectable first, while the weaker ones operate more slowly and take longer times for detection. In fact, a distribution of activation energies has been verified by the small-particle experiments of Magee (11).

The kinetics of nucleation will understandably be affected by impurity pinning of the interfacial dislocations. As pointed out by Magee (25), this type of stabilization can inhibit nucleation and result in an apparent athermal behavior.

3.5 Autocatalytic Nucleation

The overall course of martensitic transformations (volume fraction transformed vs. temperature or time) is greatly influenced by autocatalytic nucleation but, unfortunately, the phenomena at play have received relatively little systematic study thusfar. In Section 7, detailed attention is paid to the effects of imposed elastic stresses and plastic strains on martensitic nucleation. By the same token, the displacive nature and resulting shape deformation of the transformation itself, even when no external forces are acting, introduces elastic stresses and plastic strains which can lead to autocatalytic nucleation. The internally-generated elastic stresses can contribute mechanically to the thermodynamic driving force and thus activate pre-existing nucleation defects which are otherwise insufficiently potent to operate at the temperature in question. In addition, new and more potent nucleating sites can be created by dislocation generation and rearrangement when plastic accommodation of the parent phase takes place in response to the transformational shape change. It is also possible that, at least in some cases, the interfacial dislocations participating in the formation of one martensitic unit can directly nucleate another variant by a secondary dissociation process (13).

An especially graphic example of autocatalytic nucleation is displayed by the burst transformation in certain ferrous alloys having low transformation temperatures. In typical instances, some 10-60 percent of martensite may form in a sudden chain reaction. Bokros and Parker (26) identified groups of four variants clustered about {011} poles of the FCC parent phase in a single-crystal Fe-31.7Ni alloy, and interpreted this relationship in terms of elastic-stress coupling (or self-accommodation) among the variant shape changes.

Knorovsky (27) found similar crystallographic groupings in Fe-Ni and Fe-Ni-C specimens cooled in a thermal gradient. When the cold end reached the M_s temperature, the transformation quickly propagated by autocatalytic nucleation up to temperatures at least 50° C above the regular M_s. This high temperature of nucleation corresponds to a powerful autocatalytic assist, amounting to approximately one-fifth of the chemical driving force needed to start the transformation at M_s. Analysis of this finding indicates that less than one-half of the autocatalytic effect can be explained by elastic stresses acting on pre-existing sites. Instead, it appears that the attendant plastic accommodation produces new nucleation sites which are higher-than-normal in potency, and these in turn are stress-assisted by the transformational shape change of other martensitic units. Under these conditions, the new sites require only one more dislocation than the proposed 4-5 dislocations in the pre-existing defects which otherwise nucleate at the regular M_s temperature.

3.6 Premartensitic Effects

Various phenomena have been found to occur in martensitic systems at temperatures above and approaching M_s. Accordingly, the question arises as to whether such changes in the parent phase play an essential role in the subsequent transformation. It has been pointed out (28), for example, that lattice softening and the "condensation" of soft phonon modes in the parent phase may be basic premonitory effects which signal the advent of martensitic nucleation. In this connection, it should be noted that typical martensitic transformations like those occurring in ferrous alloys involve atomic displacements (i.e. in the Bain strain) on the order of 10-20 percent; clearly, displacements of this magnitude cannot be expected from a soft phonon mode. Also to be considered in this context are the anomalous changes in lattice stiffness which may result from atomic ordering or a ferromagnetic transition in the parent phase on cooling.

However, these interesting physical changes are not necessary precursors for the occurrence of a martensitic transformation; in fact, no such premonitory effects are observed in the case of the best-known ferrous martensites. Nevertheless, where these premartensitic changes are found, particularly in the β-phase (BCC) alloys, they may influence the nucleation kinetics indirectly by altering the terms in Equations (5) and (6) for athermal and isothermal nucleation respectively. These variations may combine either to favor or retard the nucleation process, but the phenomena which cause the variations are not fundamental precursors to the transformation itself.

The anomalous increase in electrical resistivity at temperatures just above M_s in TiNi alloys has received much attention (29) as a possible indicator that the parent phase may be "getting ready" to transform martensitically. More recent studies, however, have demonstrated that the premartensitic anomaly is actually a separate phase change (now called the R-phase) which happens to set in just ahead of the martensitic transformation (30,31). Similarly, the premartensitic 9R and 2H diffraction maxima noted in some β-phase alloys have now been shown to correspond to metastable precipitate phases which play no required role in the martensitic transformations per se (32).

Other observed premartensitic effects that are likewise not fundamental to the transformation are the decrease in parent-phase rigidity in Fe-N alloys (33), the redistribution of interstitial atoms in the parent phase of Fe-Ni-C alloys (34), and the six-layer intermediate structure formed in Fe-Mn-C alloys (35).

4. MARTENSITIC GROWTH

Except for thermoelastic growth (Section 7) which can be made to proceed slowly if the cooling rate is slow, martensitic growth is usually a rapid process. We shall consider the growth stage to include (a) thickening of the thin nucleated (ribbon-shaped) particle described in Section 3, (b) switching of the habit plane from the initial close-packed faulting geometry to the macroscopic IPS morphology, (c) fast growth to the final observed plate size, and (d) growth arrest. Of course, these manifestations of growth are not mutually exclusive and may overlap to various degrees.

It should also be realized that the nucleation-rate measurements referred to in Section 3 actually depend on rapid growth to a detectable size in order to count the frequency of nucleation events. Accordingly, one really measures the frequency of growth start-ups in experiments on nucleation kinetics. For such reasons, martensitic nucleation is often termed operational nucleation (7), in contrast to the well-known classical nucleation based on the concept of heterophase fluctuations reaching a free-energy barrier.

4.1 The Problem of Nucleus Thickening

The defect-dissociation mechanism of the martensitic nucleation imposes a constraint on the thickness of the first-formed particle; it is limited by the number of effective dislocations in the pre-existing nucleating defect, as illustrated in Figures 7 and 8. For thickening normal to the fault plane, additional coherency dislocations have to be created, similar to those derived directly from the nucleating defect, but the new dislocations possess substantial long-range strain fields (inasmuch as their function is to accomplish the lattice deformation of the transformation) which make their generation energetically very difficult. Calculations show that the thermally activated formation of these dislocations is virtually ruled out under the thermodynamic conditions available to martensitic reactions, even if assisted by high internal stress fields (13).

However, this problem can be realistically circumvented by a pole mechanism of thickening (18). As the thin particle extends spontaneously in the faulting mechanism, the intersection of the advancing coherency dislocations with occasional dislocations having screw components normal to the fault plane can result in a spiraling action that leads to thickening, simply under the driving force of the negative stacking-fault energy which started the nucleation process in the first place. This thickening mechanism is illustrated in Figure 10 (18). The required coherency dislocations for adding new fault planes to the thickness dimension are provided, not by the nucleation of new discrete loops, but by a continuous increase in dislocation line length as the outer coherency dislocations in the moving interface form a dipole and spiral in opposite directions around the pole dislocation, thereby advancing on to consecutive close-packed planes with each revolution.

4.2 Switching of Habit Planes

In the absence of an appropriate thickening mechanism, the thin ribbon-shaped particle can continue to extend parallel to the fault plane with a monotonic decrease in free energy until stopped by an obstacle. Once started, this kind of martensitic growth can become very fast because only glissile dislocation motions are involved. However, the pole thickening mechanism is also inherently rapid and is not likely to be a rate-limiting step, once an appropriate pole dislocation is encountered. At the same time, growth is now thermodynamically and kinetically favorable in three dimensions, and consequently a new flexibility is achieved for seeking lower strain-energy shapes in the ensuing growth process, i.e., the martensitic growth can adjust toward an IPS morphology in order to minimize the strain energy. The latter condition becomes a dominating factor because, when the martensitic particles are growing to macroscopic sizes, the accompanying increase in interfacial energy per unit volume of growth loses significance from the standpoint of controlling the overall morphology.

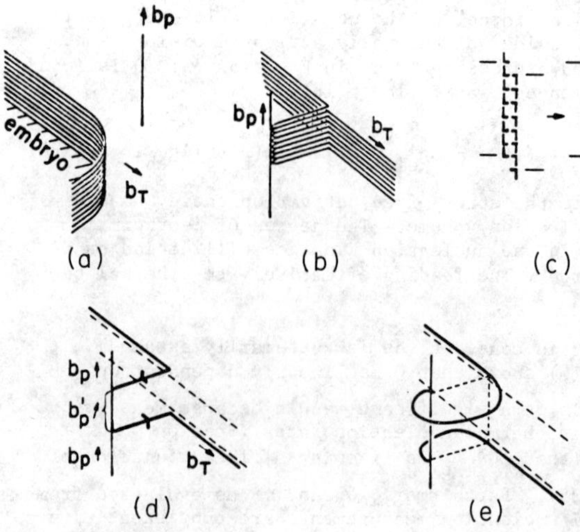

Figure 10. Dislocation pole mechanism for thickening of a thin martensitic particle after nucleation by defect dissociation: (a) interface of coherency (or transformation) dislocations with Burgers vector b_T approaching an occasional dislocation with Burgers vector b_P, inclined to the faulting plane; (b) after intersection and operation of dislocation b_P as a pole; (c) coherency dislocation dipoles are formed, which interact among themselves to produce the single dipole shown in (d); (e) the resulting dipole dislocations spiral around the pole dislocation to thicken the particle without requiring the nucleation of new coherency dislocation loops (18).

The switching of habit planes toward the IPS morphology is a response to the role of strain energy during the growth stage, and it appears to set in when particle thickening can operate. At that point, the system finds a new path for reducing its free energy faster, and to lower levels, than can be attained along the original faulting path. Comparing these two growth paths, the former will invariably supersede the latter since the greater reduction in strain energy permits a larger net transformational driving force to act on the moving interfacial dislocations. Furthermore, the available driving force is sufficient to "pay" for the new anticoherency (misfit) dislocations needed for the inhomogeneous lattice-invariant deformations which must come into play to help attain the IPS condition (see Section 6). Notwithstanding these morphological changes in a given martensitic transformation, it is believed that the same orientation relationship is maintained during the nucleation and growth stages (18).

Whether the fault morphology can be detected by electron microscopy before being interrupted and obscured by the macroscopic growth will, in this view, depend on the density of pole-type dislocations in the parent phase. In the in situ observations represented by Figure 9, evidently the random dislocations were far enough apart in the foil specimens under examination to permit considerable growth in the fault morphology before switching to the macroscopic habit. On the other hand, the growth transition in the isothermal transformation of an Fe-Ni-Mn alloy appears to take place so quickly that only the IPS morphology has been found (36).

4.3 Martensitic Growth Kinetics

Nishiyama (37) has pointed out that the rate of martensitic growth can be classified roughly into three ranges. The fastest -- originally called umklapp by Förster and Scheil (38) -- is of the order of mechanical twinning velocities. The second fastest -- termed schiebung (38) -- is of the order of dislocation velocities in slip deformation. The slowest mode is typified by thermoelastic growth which simply paces the rate at which the driving force is increased by the cooling rate. Basically, however, thermoelastic growth is very rapid in that it responds quickly to the dropping temperature.

The umklapp velocity is usually so fast that most measurements have set only lower limits. Growth rates up to 8×10^2 m sec^{-1} in an Fe-31Ni alloy* have been reported (39), and are about the same in both athermal and isothermal transformations. This growth velocity corresponds to a formation time of only 0.5×10^{-7} sec for a fully grown martensite plate. Such extremely rapid growth in Fe-Ni alloys is promoted by subatmospheric transformation temperatures and high degrees of supercooling. The martensitic products are plate-like or lenticular; it is probable that the fast growth event being measured here is mainly that of edgewise (or radial) growth.

The schiebung case is found in Fe-Ni alloys when the composition is adjusted to have the transformation take place above room temperature. Under these conditions, the martensitic units are lath-like and form in packets (see Section 6.3). This complex morphology makes it difficult to measure and assign growth rates. However, the nominal growth rates in the edgewise direction lie in the range of 10^{-3} - 10^{-1} m sec^{-1}; this is several orders of magnitude slower than in the umklapp case, but is still quite fast for a solid-state reaction (37).

In any event, the experimental growth rates apply to macroscopic growth where the minimum strain-energy IPS condition has been attained or at least approximated. Relatively few papers have appeared in the literature dealing with the mechanisms and kinetics of martensitic growth in detail. As an example, Raghavan and Cohen (40) have used the free-energy gradients $\partial\Delta G/\partial r$ and $\partial\Delta G/\partial c$, obtained by differentiating Equation (1), to evaluate the net transformational forces for driving interfacial dislocations in the radial and thickness directions, respectively. The resistance to the motion of these dislocations was introduced empirically as drag coefficients for screw and edge dislocations. However, this approach is too simple for reasons which now illuminate the need for a more comprehensive treatment of the growth phenomena: (a) in most instances, martensitic growth is neither isothermal nor adiabatic, and so a more general thermodynamic driving force is required than the free-energy change which is appropriate for isothermal transformations, or the enthalpy change which is appropriate for adiabatic transformations; (b) plastic accommodation usually occurs and must be taken into account because such yielding further reduces the elastic strain energy that otherwise detracts from the available driving force, and concommitantly the lattice defects generated by the plastic deformation act as debris to increase the resistance to interfacial motion; and (c) the heat-flow problem in fast growth must be re-examined in the light of the above interacting factors. Studies along these lines are now underway (41).

Pseudoelastic transformations, i.e., stress-induced thermoelastic interfacial motions (Section 7), offer direct access to the laws of martensitic growth in the relative absence of plastic accommodation. Measurements thus far (42,43) on the interfacial velocity as a function of temperature and applied stress indicate activation energies and activation volumes typical of thermally assisted dislocation motions, as may be expected.

4.4 Martensitic Growth Arrest

There are various ways in which the growth of a martensite plate can be arrested. The most obvious one, for edgewise or radial growth, is mechanical blockage at an obstacle, e.g., a grain boundary, an annealing-twin boundary, or another martensite plate. It is also possible for this type of growth to come to a halt by running out of sufficient driving force in a temperature gradient or composition gradient. With the aid of such gradient specimens, Knorovsky and Cohen (44) found that, in a single crystal of Fe-30Ni, the martensitic arrest (in the absence of visible obstacles) occurred when the driving force was less than about 50-70 percent of the critical driving force for nucleation at the M_s temperature. These gradient experiments did not clearly show whether the actual arrest was sudden (∼adiabatic) or gradual (∼isothermal).

The indications are that growth in the thickness dimension of a macroscopic plate is a somewhat slower process than that of edgewise growth. As a case in point, the precipitate-marker studies of Neuhäuser and Pitsch (45) demonstrate that the midplane section (or midrib) of a lenticular plate in an Fe-Ni-P alloy comes into existence before the lateral regions on either side.

The stoppage of sidewise or thickening growth may result from obstacles, such as by growing broadside into another plate. However, in the usual nongradient specimens, the thickening is most likely stopped by retardation or jamming of the interface because of lattice defects arising from plastic accommodation. On the other hand, if the accommodation is elastic, the thickening is brought to a halt when a thermoelastic balance is reached (Section (7.3).

5. MARTENSITIC REACTION KINETICS

Up to this point, we have been concerned principally with the formation of individual martensitic units; this section now directs attention to the overall reaction kinetics. Most martensitic transformations, particularly those in steels, have been studied athermally as a function of dropping temperature, but the best quantitative insights into the reaction kinetics are obtained from isothermal transformations which permit nucleation rates and transformation rates to be determined experimentally. However, whether the reaction characteristics are athermal or isothermal, the kinetics are controlled by the operational nucleation rate (See Section 4) and not by the growth rate; indeed, it is the relatively rapid growth rate that leads to this situation. Thermoelastic transformations, sometimes involving slow growth rates, constitute a special case, and are discussed separately in Section 7.3.

5.1 Kinetic Analysis of Isothermal Transformations

A simple, but general, expression for the transformation rate of an isothermal martensitic reaction is given by (46):

$$df/dt = (dN_v/dt)V, \tag{7}$$

where f is the volume fraction of martensite formed up to time t. N_v is the number of martensite plates per unit volume of alloy, and V is the volume per plate of martensite formed at time t. N_v is determinable by quantitative metallography if the plates can be approximated as random, thin circular discs (47). Ordinarily, V cannot be measured, but it is obtainable through measurement of the mean volume ($\bar{V}$) of all the plates present in the N_v assembly (46):

$$\bar{V} = f/N_v\text{, and using Equation (7):} \tag{8}$$

$$V = \bar{V} + d\bar{V}/d\ell nN_v. \tag{9}$$

It is often presumed that V and $\bar{V}$ become smaller as the transformation proceeds due to partitioning of the parent phase by the transformation product. Geometric models have been proposed (48,49) to express the decreasing volume per plate as a function of increasing f, but the assumptions introduced usually do not hold up well in real

* Unless otherwise indicated, compositions are expressed in weight percentages.

transformations. However, since Equations (8) and (9) are based on direct metallographic measurements, they do not depend on partitioning formulas, and are valid whether V and $\bar{V}$ decrease during the transformation (46) or remain essentially constant (50). The latter case is found when the martensite plates happen to form autocatalytically in clusters and the clusters then spread (50,51).

The rate of operational nucleation is defined as:

$$\dot{N} = (dN_v/dt)(1/1-f), \tag{10}$$

and if, for the sake of easy discussion, we assume a constant activation energy (ΔG^*_{exp}) for nucleation at a given temperature, then

$$dN_v/dt = n_t \ \nu \exp[-\Delta G^*_{exp}/kT], \tag{11}$$

where n_t is the number of most-potent nucleation sites per unit volume of alloy at time t, and ν is the nucleation-attempt frequency.

n_t can be approximated as (52,46):

$$n_t = (n_i + pf - N_v)(1-f), \text{ and using Equation (8):} \tag{12}$$

$$n_t = (n_i + f\ [p - 1/\bar{V}])(1-f), \tag{13}$$

where n_i is the number of pre-existing most-potent nucleating sites per unit volume of parent phase, and p is an autocatalytic factor defined as the number of most-potent nucleating sites generated per unit volume of martensite formed. Up to the present time, the available experimental data on isothermal martensitic transformations are not sufficiently precise to discriminate in favor of more complicated assumptions for the autocatalytic factor.

The autocatalytic contribution (pf) to n_t is opposed by the removal of nucleating sites due to their conversion into plates (N_v) and also due to their being swept up by other plates, the latter being taken into account by the (1-f) factor. According to Equation (13), the number of most-potent sites per unit volume of parent phase (n_t/1-f) will increase during the transformation if $p\bar{V} > 1$, will decrease if $p\bar{V} < 1$, and will remain at n_i if $p\bar{V} = 1$.

Substituting Equations (9), (11) and (13) into (7) yields the overall reaction rate (46):

$$df/dt = (n_i + f[p - 1/\bar{V}])(1-f)(\bar{V} + d\bar{V}/d\ln N_v)$$
$$\times \ \nu \exp[-\Delta G^*_{exp}/kT]. \tag{14}$$

$\bar{V}$ and N_v can be evaluated by quantitative metallography when the plates are clearly distinguishable (i.e., up to f ≈ 0.2). n_i is taken as 10^5 - 10^7 cm^{-3} from small-particle experiments (10,11). Equation (14) can be numerically integrated by computer, and p and ΔG^*_{exp} are then obtained by curve-fitting with experimental f vs. t data (46). An example is shown in Figure 11 for an Fe-24Ni-3Mn alloy. This is one way of determining the effective activation energy for nucleation operative during the overall isothermal transformation. On carrying out these measurements and calculations at small values of f (f ≈ 0.002) to reduce the autocatalytic influence, the activation energy values remain approximately the same (53), signifying that autocatalytic effects begin to play a dominant role very early in the transformation although the pre-existing nucleating sites are necessary to start the process.

ΔG^*_{exp} is observed to change with decreasing reaction temperature in such a way as to decrease linearly with increasing driving force (53,25). The results for three Fe-Ni-Mn alloys are plotted in Figure 12 (25). As noted in Section 3.4, this relationship imposes a severe test on proposed martensitic nucleation mechanisms. The magnitude and temperature-dependence of ΔG^*_{exp} reflect the well-known isothermal C-curve kinetics for these alloys, as shown in Figure 13 (54). It is obvious that this mode

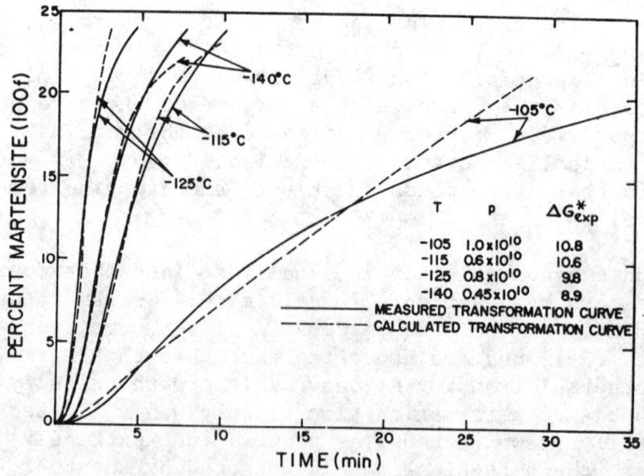

Figure 11. Comparison of experimental and calculated isothermal martensitic transformation curves for an Fe-24Ni-3Mn alloy; calculations are based on curve fitting with Equation (14) to give the p(cm^{-3}) and ΔG^*_{exp} (cal mole^{-1}) values shown (46).

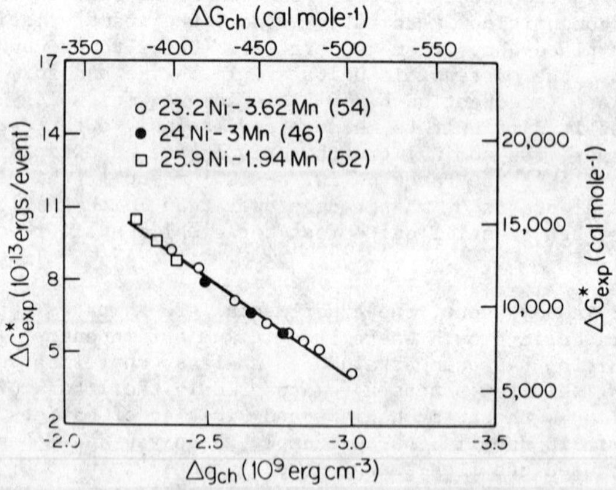

Figure 12. Linear dependence of experimental activation energy on the chemical free-energy change for the isothermal martensitic transformation in three Fe-Ni-Mn alloys, after (25).

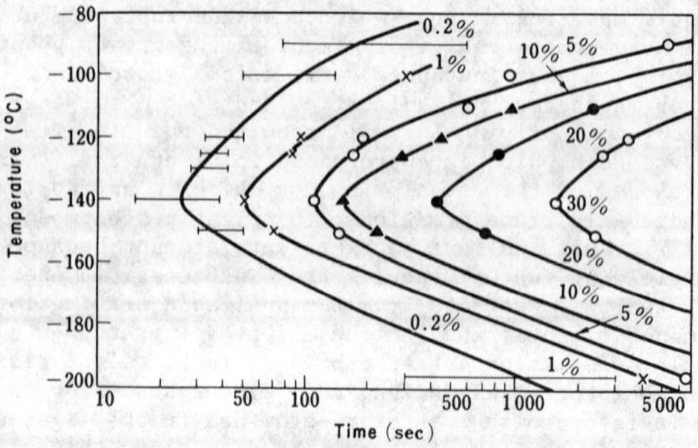

Figure 13. Isothermal C-curve kinetics exhibited by the martensitic transformation in an Fe-23.2Ni-3.62Mn alloy (54).

of martensitic transformation is depressible in temperature, and even completely suppressible, by sufficiently rapid cooling.

The frequency factor (ν) is often taken as the lattice vibrational frequency (~10^{13}sec^{-1}) in solid-state nucleation theory, but Magee (25) has pointed out that the actual attempt frequency may be orders of magnitude smaller if the nucleation kinetics are controlled by dislocation motions rather than by heterophase fluctuations. This could have the effect of making the autocatalytic factor even larger than estimated above.

5.2 Kinetic Analysis of Cooling Transformations

The distinction between athermal and isothermal nucleation was elucidated in Section 3. However, there are intermediate cases in which isothermal nucleation may occur so rapidly that it takes on athermal characteristics. The term anisothermal has been adopted to describe such instances and, more particularly, the resulting transformation kinetics. Here, extremely fast quenching may cause some lowering of the transformation-temperature range, but complete suppression is not obtainable by the available cooling rates (up to 5×10^4 °C sec^{-1}) (55). The time-dependence involved in anisothermal transformations is commongly ignored in dealing with the kinetics of the overall cooling reaction, and so the treatment becomes essentially athermal (25,56).

The relationship between volume fraction (f) of martensite formed on quenching to a given temperature (T_q) has been expressed empirically as (57):

$$f = 1 - \exp[\alpha(M_s - T_q)], \tag{15}$$

where $\alpha = 0.011$. M_s for this purpose is obtained by extrapolation to f = 0, as indicated in Figure 14 (57,58).

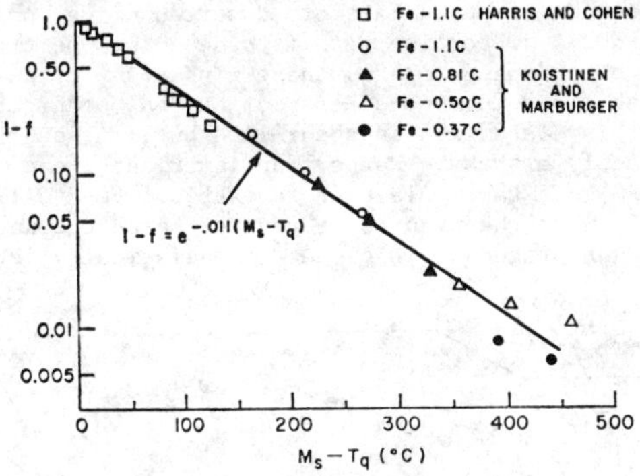

Figure 14. Volume fraction of martensite (f) as a function of cooling temperature (T_q), illustrating Equation (15) for four steels (57).

We now desire to express f as a function of T in more basic terms.

Magee (25) has related the course of the cooling transformation directly to the changing driving force as follows:

$$df/dT = \phi\bar{V}(1-f)(d\Delta g_{ch}/dT) \tag{16}$$

where ϕ is a proportionality constant, and Δg_{ch} is the chemical free-energy change per unit volume. On the assumption that ϕ, $\bar{V}$, and $d\Delta g_{ch}/dT$ are insensitive to temperature, Equation (16) can be integrated from f = 0 at M_s:

$$f = 1 - \exp[\phi\bar{V}(d\Delta g_{ch}/dT)(M_s - T_q)], \tag{17}$$

which is of the same form as the experimental Equation (15). However, autocatalytic effects are not taken into account here, and so it is likely that the assumptions introduced may offset one another fortuitously.

Pradham and Ansell (56) have taken this approach a step further by calculating the mechanical energy expended by the transformational strain working against the plastic resistance of the parent phase, and then using this energy as an empirical measure of the autocatalytic contribution via strain-induced nucleation. Up to the present time, it appears that direct (operational) nucleation measurements have not been made for cooling transformations, and so the kinetic analysis of cooling transformations remains in an elementary state of development.

5.3 Autocatalytic Effects in Martensitic Reactions

The reaction kinetics in martensitic transformations, whether isothermal or athermal, are singularly dependent on autocatalysis. In fact, the pre-existing nucleating sites are responsible for hardly more than about a percent of the total transformation in ferrous alloys. Hence, autocatalytic nucleation plays a central role in sustaining a martensitic transformation and, therefore, in the hardening of steel!

The details of autocatalysis are particularly complicated in polycrystalline materials. Guimãraes and Gomes (59) have shown two overlapping types of autocatalysis in an Fe-31Ni-0.02C alloy: (a) the preferential formation of martensite plates within groups of grains and their filling-in to produce more-or-less separate transformation clusters, and (b) the spreading of these clusters throughout the rest of the grains. A small grain size favors (a) while a large grain size favors (b). Such differences have not yet been incorporated into the types of kinetic analysis represented by Equations (14) and (17); presumably the autocatalytic factor (p) in Equation (14) can be regarded as an integrated measure of various autocatalytic modes at play.

6. CRYSTALLOGRAPHIC ASPECTS OF MARTENSITIC TRANSFORMATIONS

Martensitic transformations exhibit a variety of interrelated geometrical features, from which a largely successful crystallographic phenomenological theory has been formulated. Presented in this section are experimental measurements of the crystallographic "observables", a description of the phenomenological theory, test cases for the theory, and related features such as martensitic substructure and morphology. It is essential to recognize, however, that the kind of theory involved here is not mechanistic, but is primarily a description of before-and-after relationships in martensitic transformations. Of course, proposed mechanisms, such as discussed in earlier sections, must be consistent with these observed relationships.

Martensitic transformations are characterized by a habit plane that is usually not of simple Miller indices, and an orientation relationship in which the close-packed planes and close-packed directions in the two structures may be only approximately parallel to each other. Typical examples will now be considered.

6.1 Typical Examples of Orientation Relationships and Habit Planes

6.1.1 FCC→HCP transformations. Representative of this case is the FCC→HCP martensitic reaction in Co and its alloys. The orientation relationship observed is as follows:

$$(111)_{FCC} \parallel (0001)_{HCP}$$

$$[11\bar{2}]_{FCC} \parallel [1\bar{1}00]_{HCP} ,$$

and the habit plane is precisely $(111)_{FCC}$. It is to be noted that the atomic arrangements on the $(111)_{FCC}$ and $(0001)_{HCP}$ planes are identical, although there may be small differences in interatomic distances. This transformation is relatively simple and can be effected by the passage of a partial dislocation on every other $(111)_{FCC}$ plane in the $[11\bar{2}]_{FCC}$ direction.

6.1.2 FCC→BCC(BCT) transformations. This case has been most widely studied because of the importance of the martensitic reaction in ferrous alloys. Low-carbon steels and iron alloys with low substitutional-solute content transform into a lath-type martensite with a habit plane about 5° from $\{111\}_p$ (60). The martensite-austenite orientation is:*

---

* The Miller indices used in this section are consistent with the Bain correspondence indicated in Figure 1. A different convention was adopted for the FCC→BCC discussion in Sections 3.2 and 3.3, but there is no basic disagreement.

$(111)_P \sim || (011)_M$

$[\bar{1}01]_P \sim || [\bar{1}\bar{1}1]_M$.

In this context, an orientation relationship where the respective close-packed planes and close-packed directions are approximately parallel is named after Kurdjumov and Sachs (14).

Medium-carbon steels and certain alloy steels (e.g., Fe-8Cr-1C and Fe-6Mn-1C) transform to BCT martensite with a habit plane near $\{225\}_P$. A typical orientation relationship, as found in the Fe-8Cr-1C alloy, is (61):

$(111)_P$ 0.45° from $(011)_M$

$[\bar{1}01]_P$ 0.53° from $[\bar{1}\bar{1}1]_M$,

in which the close-packed planes and directions are nearly parallel. Steels with still higher-carbon content (i.e., 1.8%) form martensite with a $\{259\}_P$ habit.

Alloys such as Fe-24 at.% Pt and Fe-22Ni-0.80C transform into a BCT martensite with a $\{3, 10, 15\}_P$ habit. The orientation relationship (62) typical of these materials is:

$(111)_P$ 0.86° from $(011)_M$

$[\bar{1}01]_P$ 4.42° from $[\bar{1}\bar{1}1]_M$

The close-packed directions $[\bar{1}01]_P$ and $[\bar{1}\bar{1}1]_M$ are about 5° apart; this is known as the Greninger-Troiano orientation relationship (63).

Another orientation relationship found in Fe-Ni alloys is due to Nishiyama (64):

$(111)_P \sim|| (011)_M$

$[\bar{1}\bar{1}2]_P \sim|| [0\bar{1}1]_M$

6.1.3 BCC(B2,DO_3) → 3R, 9R, 18R transformations. Many of the so-called β-phase alloys transform martensitically into a crystal structure having a long-period stacking order based on prior close-packed {110} planes of the parent phase. For example, the 3R structure can be expressed as the sequence...ABCABC... and the 9R structure as ...ABCBCACAB... The particular stacking sequence depends on the type of ordering in the parent phase, e.g., B2 → 9R and DO_3 → 18R. Many of these alloys exhibit thermoelastic behavior and the shape-memory effect. β-phase alloys also transform into 2H martensitic structures (...ABAB...) which are usually internally twinned, in contrast to the above R types which are internally faulted. Typical of the β-phase alloys is the DO_3 orthorhombic martensitic transformation in a Cu-14.2Al-4.3Ni alloy (65); here the habit plane is approximately $\{331\}_P$ and the orientation relationship is:

$(110)_P \sim || (121)_M$

$[1\bar{1}\bar{1}]_P \sim || (2\bar{1}0]_M$

6.1.4 BCC→HCP transformations. This case is typified by the martensitic reactions in Ti and Zr alloys. For instance, a Ti-11Mo alloy (66) transforms to martensite with a habit plane 4° from $\{344\}_P$ and the following orientation relationship:

$(110)_P \sim || (0001)_M$

$[\bar{1}11]_P \sim || [11\bar{2}0]_M$

This is often called the Burgers relationship (67).

6.2 Crystallographic Phenomenological Theory: Concept of an Invariant-Plane Strain

6.2.1 Components (factorization) of the invariant-plane strain

- (a) transformation shape deformation. Figure 3 is a surface-relief optical micrograph of a polycrystalline Fe-24.5 at.% Pt alloy taken on a prepolished surface after transformation of the FCC parent phase into BCT martensite. The observed contrast results from a change in shape of the transformed regions. The light and dark shadings correspond to plates of martensite which have been tilted in different senses with respect to the initial surface. This macroscopic change in shape is otherwise known as the shape deformation or shape strain. If successive serial sections of the specimen are removed and the polished surface etched at each stage, the martensite units are found to have the form (roughly) of lenticular plates. The differently oriented plates within a given parent grain are different crystallographic variants of the habit-plane and orientation relationships.

Again considering Figure 3, we can deduce that if straight scratches are intentionally abraded on the prepolished surface while still in the parent condition, these "fiducial markings" will be displaced in a characteristic manner because of the martensitic transformation. By analyzing several such nonparallel scratches crossing a given plate and noting the initial positions of the same scratches in the untransformed surroundings, one can construct a distortion matrix to describe the transformation shape change. From these findings, it can be inferred that a homogeneous deformation* has occurred, and that this distortion is shear-like but, in general, is not exactly a shear. Rather, an invariant-plane strain (IPS) has taken place in the subject Fe-Pt alloy (Figure 15) with the plane of reference being the undistorted and unrotated habit plane. Mathematically

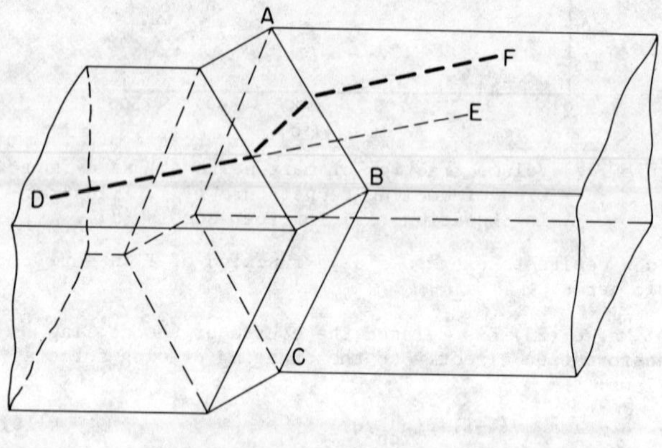

Figure 15. Schematic representation of shape deformation due to an invariant-plane strain (IPS), characteristic of martensite formation. Habit plane ABC is undistorted and unrotated. Staight fiducial line DE is displaced to position DF.

speaking, an IPS is a homogeneous distortion such that the displacement of any given point lies in a common direction and is proportional to its distance from a fixed (i.e., not influenced by the strain) plane of reference, the latter being the invariant plane (68). In most martensitic transformations, a volume change (dilatation) also accompanies the structural change and this accounts for the normal component of the IPS. Strain energy is minimized if the shape deformation is an IPS and the habit plane becomes the invariant plane. This is the central idea of the crystallographic phenomenological theory under discussion.

* The homogeneous deformation which describes the shape change is commonly called macroscopic because, although it is observed on the scale of an optical microscope, the deformation is not necessarily homogeneous on a finer scale. Hence, this shape deformation should not be confused with the homogeneous lattice deformation previously discussed.

-(b) Bain distortion and rigid-body rotation. The previous section indicates that the martensitic shape change tends toward an IPS, which is a macroscopically homogeneous deformation. However, when this deformation matrix is applied to the parent structure, it does not generate the known crystal structure of the martensitic product, as was clearly demonstrated by Greninger and Troiano (63) for the FCC→BCC(BCT) martensitic reaction in ferrous alloys.

As early as 1924, Bain (5) suggested that the martensitic transformation in steels might be explained by a homogeneous deformation or "upsetting" of the parent FCC lattice into the required BCC (or BCT) lattice, as illustrated schematically in Figure 1. This simple, intuitive deformation was subsequently justified mathematically in the sense that the particular lattice correspondence envisioned by Bain entails the smallest principal strains when compared to other possibilities (69). This lattice correspondence provides a unique relationship between any point in the initial lattice and the point it becomes in the final lattice. The Bain distortion has been verified experimentally with an ordered Fe_3Pt alloy by relating the corresponding superlattice reflections from the parent and martensitic phases via transmission electron diffraction. But the actual transformation is not that simple.

According to the FCC-BCC lattice correspondence in Figure 1, one would expect many parallelisms between crystallographic directions and planes: for example, $[011]_M \,||\, [011]_P$, $[010]_M \,||\, [110]_P$, $[112]_M \,||\, (101)_P$, $(011)_M \,||\, (111)_P$, etc. However, such exact parallelisms are not observed. For the Fe-Pt alloy in Figure 3, the following orientation relationship obtains (62):

$[001]_P - [001]_M$ 9.10° apart

$[\bar{1}01]_P - [\bar{1}\bar{1}1]_M$ 4.42° apart

$(111)_P - (011)_M$ 0.86° apart

and this is characteristic of those found in other iron alloys and steels. A further analysis of the above orientation relationship shows that the initial correspondence cell not only undergoes the Bain distortion but is also rotated (about 10° from $[001]_P$ towards $[110]_P$). This part of the lattice deformation is a rigid-body rotation.

We have now emphasized the following crystallographic aspects of martensitic transformations:

1) a lattice correspondence and rigid-body rotation,
2) a shape deformation tending toward an invariant-plane strain,
3) a habit plane, and
4) an orientation relationship.

-(c) Lattice-invariant deformation (inhomogenous shear). The plausibility of a Bain-type lattice correspondence and distortion in martensitic transformations has been presented as one involving minimal atomic displacements. However, the Bain distortion alone does not account for an IPS shape deformation. The reason for this apparent inconsistency is readily explained. Referring to Figure 1 again, we note that the correspondence cell is contracted (∿20%) along the z' axis and expanded (∿12%) along each of the x' and y' axes. Such a homogeneous distortion will leave no plane undistorted and unrotated. On the other hand, suppose there were no distortion along the y' axis. Then, one of the principal distortions would be less than unity (along z'), one greater than unity (along x'), and the remaining one (along y') equal to unity. This is the necessary and sufficient set of conditions for the existence of an undistorted plane, and is analogous to deforming an initial sphere into a triaxial ellipsoid such that the sphere and ellipsoid can match along an undistorted plane of contact (habit plane), as shown in Figure 16.

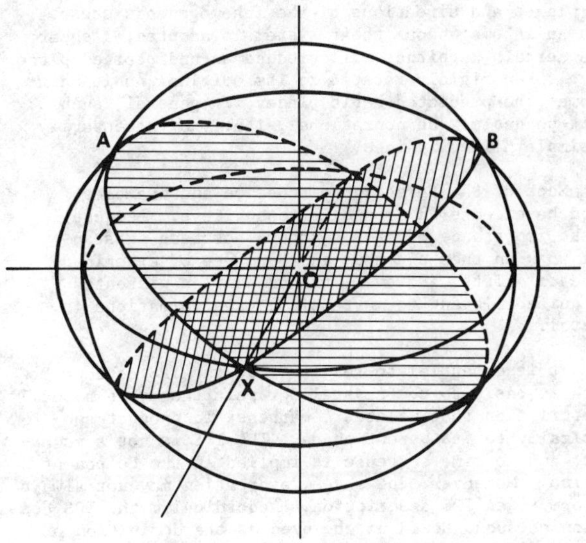

Figure 16. Sphere-ellipsoid representation of homogeneous deformation in which matching planes of no distortion exist (cross-hatched).

However, this special set of conditions does not generally prevail in the Bain correspondence per se because the principal distortions are determined by the lattice correspondence and lattice parameters of the two phases.

This dilemma can be overcome by the operation of a lattice-invariant deformation, such as slip or twinning, in conjunction with the Bain distortion (68,71). The additional deformation must be lattice-invariant because the necessary structural change is completed by the Bain distortion alone. The role of the additional deformation is effectively to shear the ellipsoid resulting from the Bain distortion into tangency with the initial sphere, and then one of the principal distortions (OX in Figure 16) becomes unity and an undistorted plane of contact is provided. This secondary deformation, being a shear, is often referred to as the inhomogeneous (or invisible shear of the crystallographic theory.

But a problem still remains. Even though the required structural change has been accomplished (by the Bain distortion) and an undistorted plane of contact between the parent and product phases has been provided (by the inhomogeneous shear), the habit plane is still not unrotated, as required from observation. A little consideration of the sphere-ellipsoid analog shows that, although an undistorted plane now exists, it has been rotated from its initial (unrotated) position. Thus, a rigid-body rotation is additionally incorporated along with the Bain distortion and inhomogeneous shear, and these are the three phenomenological components which enter into the shape deformation. It seems worth repeating at this point that the Bain distortion and rigid-body rotation constitute the previously described lattice deformation, while the inhomogeneous shear is lattice-invariant and acts to diminish the strain energy being generated by the lattice deformation.

6.2.2 Crystallographic description. Within the framework of the phenomenological theory just described, different crystallographic features such as the habit plane and orientation relationship can be predicted by taking the shape deformation to be an IPS and by supposing the inhomogeneous shear to occur via plausible combinations of crystallographic planes and directions. For example, in most iron alloys, an inhomogeneous shear on the $(112)[\bar{1}\bar{1}1]_M$ twinning system will predict a habit plane near $(3, 15, 10)_P$, whereas assuming the shear system to be $(011)[\bar{1}\bar{1}1]_M$ predicts a habit plane near $(111)_P$. Since the lattice parameters and correspondence are known, the flexibility in applying the theory comes through different suppositions concerning the planes and directions of the inhomogeneous shear. Once an inhomogeneous shear system is adopted, a shear of a certain magnitude will produce an undistorted plane which, when rigidly rotated to its original position, becomes the predicted habit plane. In general, such habit planes may be nonrational, i.e., not necessarily of simple Miller indices.

Much of the foregoing discussion and examples cited have centered around ferrous alloys. Because of the importance of steels, there has been substantial work on them and thus an abundance of experimental data exists. However, the principles presented are quite relevant to martensitic transformations in general.

It is essential to realize that the IPS concept which is basic to the crystallographic theory is actually a description that relates a martensitic product geometrically to its parent phase. The IPS is not a mechanism, and no time sequence is implied in the component strains. Moreover, the shape deformation may not always conform to an IPS description. Accordingly, the IPS condition as such should not be used as the definition of a martensitic transformation.

6.2.3 Mathematical formulation. The basic equation theory (68,71) just presented is (in matrix form):

$$E = R\bar{S}B\,, \tag{18}$$

where B is the Bain distortion, $\bar{S}$ is a lattice-invariant shear (mathematically following the Bain distortion), R is the rigid-body rotation previously mentioned, and E is the resulting shape deformation. B, S, R, and E are all (3 x 3) matrices. If the rotation R returns the plane left undistorted by $\bar{S}B$ to its original position, then $E = R\bar{S}B$ is an IPS.

Although Equation (18) shows the inhomogeneous shear $\bar{S}$ following the Bain distortion, the same mathematical result is obtained by "allowing" the shear to occur in the parent phase prior to the Bain distortion. By following the latter procedure, there are certain computational simplifications to be gained, and the basic equation becomes:

$$E = RBS\,, \tag{19}$$

where S, like $\bar{S}$, represents a simple shear.

In the formulation by Wechsler, Lieberman and Read (71), RB is the lattice-deformation part of the observed shape change, while $\bar{S}$ is the fine-scale inhomogeneous part. Although the operation of $\bar{S}$ may not be apparent from the macroscopic homogeneity of the shape deformation, there is now ample evidence from electron microscopy that such lattice-invariant deformations do occur on a fine scale.

Equation (19) can be rewritten as proposed by Bowles and Mackenzie (68):

$$ES^{-1} = RB \quad = \text{lattice deformation.} \tag{20}$$

S^{-1} corresponds to a shear deformation of the same magnitude and on the same plane as S, but in the opposite direction. It can be regarded as the invisible portion of the lattice deformation which is cancelled by S, and is called the complementary shear.

With reference to Figure 1, the Bain distortion for the FCC→BCC (BCT) can be written as:

$$B = \begin{pmatrix} \sqrt{2}\,a/a_o & 0 & 0 \\ 0 & \sqrt{2}\,a/a_o & 0 \\ 0 & 0 & c/a_o \end{pmatrix} \tag{21}$$

When typical values are substituted for the martensitic transformation in steels, a representative B matrix is:

$$B = \begin{pmatrix} 1.12 & 0 & 0 \\ 0 & 1.12 & 0 \\ 0 & 0 & 0.80 \end{pmatrix} \tag{22}$$

which does not have an undistorted plane and so cannot conform to the invariant-plane condition.

The IPS shape deformation (E) can be expressed as (68):

$$E = I + \underline{m}dp'\,, \tag{23}$$

where I is the unit (3 x 3) matrix, p' (prime meaning transpose) is a (1 x 3) row matrix designating the normal to the invariant plane, d is a (3 x 1) column matrix representing the direction of the shape displacement, and $\underline{m}$ is the magnitude of the displacement.* S^{-1} in Equation (20), being a simple shear, is also an IPS, and the product $ES^{-1} \equiv L = RB$ is an invariant-line strain, defined by the intersection of the two planes which are invariant to E and S^{-1}, respectively. By means of the crystallographic theory (68), the invariant-line strain can be calculated if the Bain distortion is known (from the correspondence and lattice parameters) and if the inhomogeneous shear mode (i.e., the plane and direction) is presumed.

Although it is beyond the scope of the present treatment to cover the mathematical details of the invariant-line strain analysis, the following comments will help indicate its methodology. As "input" information, the lattice correspondence and parameters determine B, as in Equation (21), and the assumption of the plane and direction of the inhomogeneous shear along with B gives S(or S^{-1}). With regard to "output" information, the orientation relationship is then obtained from the matrix R, which in effect provides that certain vectors undistorted by BS are, in addition, unrotated. Further consideration (68) of the invariant-line strain L =RB reveals that explicit relations for the shape-strain direction and habit-plane normal can be developed:

$$d = [Ly - y]/p'y \tag{24}$$

$$p = (q - qL^{-1})/qL^{-1}d\,, \tag{25}$$

where y is any vector (except the invariant line) lying in the plane of the inhomogeneous shear, and q is any normal (except the invariant normal**) to a plane whose zone axis is parallel to the direction of inhomogeneous shear. More mathematically, y is any vector other than an (column) eigenvector of L, and q is any normal other than a (row) eigenvector of L^{-1}. Further details can be found in reference (68), including theoretical analytical procedures for determining the displacement magnitudes of the shape strain (E) and the inhomogeneous shear (S).

The above mathematical treatment is due to Bowles and Mackenzie (68). Equivalent approaches by Wechsler et al. (71) and by Bullough and Bilby (72) are also well known.

The crystallographic concepts just presented have enjoyed both successes and discrepancies, as we shall see. A departure from the basic IPS criterion has been introduced in the form of a "dilatation parameter" (68) to improve the agreement between the predicted and measured habit-plane orientations. This empirical adjustment

* Note that dp' is a (3 x 3) matrix, whereas, for example, p'd is a (1 x 1) matrix, i.e., a scalar.

** The invariant normal is the normal to a plane defined by the inhomogeneous-shear and shape-strain directions.

relaxes the requirement that the habit should be undistorted, and so the shape deformation is not a true IPS. Other attempts to account for such difficulties have included the assumption of two or more inhomogeneous shear systems (73,74), thereby again detracting from the implicit power of the crystallographic theory (75).

6.2.4 Martensitic substructure. The inhomogeneous lattice-invariant shear in the crystallographic description of martensitic transformations allows the habit plane to be macroscopically undistorted. Figure 17 is a schematic representation of

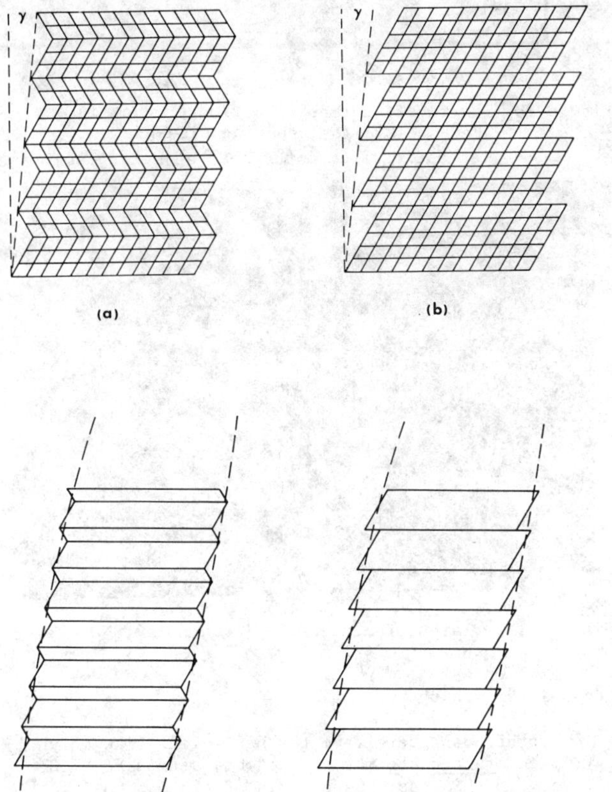

Figure 17. Schematic representation of inhomogeneous lattice-invariant shear by (a) internal twinning and (b) internal slip within martensite plates.

internally twinned and internally slipped plates of martensite. Although there are localized distortions at the interface, the alternating twins or slip lamellae act to reduce the build-up of long-range strain fields over large distances from the interface. In the case of twinned martensites, the Bain distortion is envisioned to occur along different contraction axes in the two regions, and the twinning plane in the martensite is derived from a mirror plane in the parent. For internally slipped martensites, the Bain distortion is the same in all regions of a plate.

Because of the inhomogeneous shear, one would expect to find some kind of substructure in the martensite. Indeed, many such observations have been made since the introduction of the crystallographic theory. Figure 18(a) is a transmission electron micrograph of a martensite plate in an Fe-Ni-C alloy (76). The fine

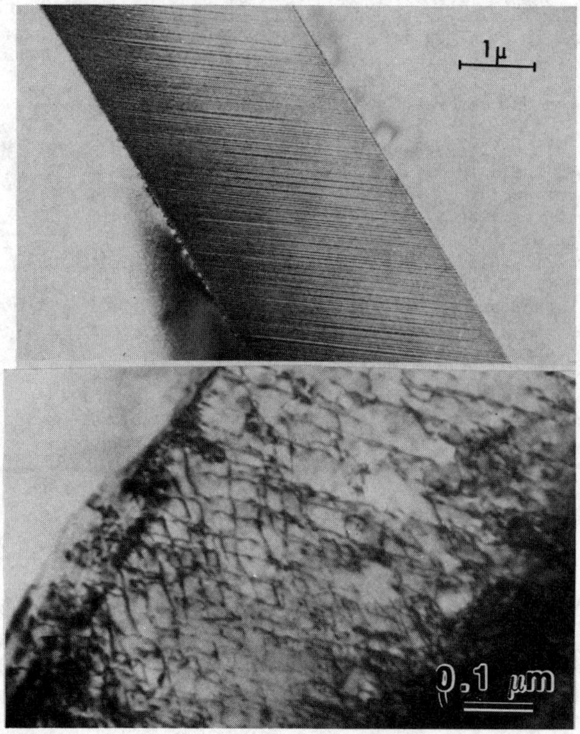

Figure 18. Transmission electron micrographs showing (a) internal twinning (Fe-30Ni-0.39C) and (b) high dislocation density due to internal slipping (Fe-20Ni-5Mn). Adjacent regions are retained parent phase.

striations crossing the plate are alternating twins (with a $\{112\}_M$ twinning plane), and regions adjacent to the martensite are retained austenite. In contrast to this morphology, martensite with a high dislocation density is illustrated in Figure 18(b). Ideally, the internally slipped type of lattice-invariant deformation can be produced by the motion of appropriate anti-coherency (misfit) dislocations, without leaving accumulations of dislocations behind. In actual cases, however, there appear to be complex interactions which lead to extensive dislocation generation while accomplishing the lattice-invariant deformation.

6.2.5 Some typical cases: successes and discrepancies - (a) Fe-Pt alloys. Efsic and Wayman (62), for the first time in the case of iron alloys, measured the habit plane, orientation relationship, and magnitude and direction of the shape deformation on preselected martensite plates in an Fe_3Pt alloy. Here, the habit plane is of the $\{3, 10, 15\}_P$ type, as also found in some Fe-Ni and Fe-Ni-C alloys. The agreement between the crystallographic theory and the experimental results is essentially perfect. Furthermore, the transformation crystallography is identical for both ordered and disordered Fe-Pt alloys (77). Tadaki and Shimizu (70), using transmission electron microscopy, showed that Fe-Pt martensite plates contain internal twins on $(112)_M$. These twins are of the correct variant to predict the observed $(3, 15, 10)_P$ habit variant.

-(b) Cu-Sn alloys. Kennon and Bowles (78) studied the β(ordered) $\rightarrow \gamma_1'$ (orthorhombic) martensitic transformation in a Cu-15 at.% Sn alloy, and compared the experimental findings to the predictions of the crystallographic theory. They assumed that the inhomogeneous shear occurred on the martensitic twinning system. However, the predicted habit plane was 4° outside the range of scatter of the measured habit planes and 6.5° from the mean habit plane. In fact, a single twinning shear was not consistent with the observation of more than one pair of twin-related orientations within a single martensite plate. It was concluded that the operative inhomogeneous strain was not a simple shear, but was more likely a complex lattice-invariant strain, combining two or more shears, such as twinning and faulting. It often happens that the lattice-invariant strains at play during martensitic transformations cannot be safely deduced from independent deformation experiments.

-(c) $\{225\}_P$ transformation in ferrous alloys. Some carbon and alloy steels undergo an FCC→BCT martensitic reaction in which the habit plane is near $\{225\}_P$. As with many other ferrous martensites having different habit planes (such as $\{259\}_P$ or $\{3, 10, 15\}_P$) the $\{225\}_P$ martensites are observed to contain $\{112\}_M$ twins (79). The $\{225\}$ plates also contain other defects such as faults and dislocations. The use of a single $(112)_M$ twinning system in the crystallographic calculations does not yield a habit plane of the $\{225\}_P$ type (68). Even the assumption of two inhomogeneous shear systems does not provide mutually consistent agreement with the observed features of $\{225\}_P$ martensites (61). Thus the $\{225\}_P$ martensitic transformation in iron alloys seems to be highly complex, and is a case where the crystallographic theory appears inadequate.

6.2.6 Plastic accommodation. In nonthermoelastic martensitic transformations, the shape strain of the transformation causes sufficient distortion of the matrix phase to produce plastic deformation ahead of the advancing plate, resulting in some disorganization, or even jamming, of the coherent or semicoherent interface. On the other hand, the defects introduced by plastic accommodation may serve as nucleating sites for new martensite plates.

The accommodation mentioned above is not to be confused with the "plastic accommodation model" for the $\{225\}_P$ transformation in iron alloys proposed by Bowles and Dunne(80). The latter is an assumed specialized distortion in addition to the $(112)_M$ twinning shear, but it still does not yield completely satisfactory predictions.

6.3 Martensite Morphology

Martensite plates are often idealized as lenticular in shape. For a given plate volume, this shape minimizes the sum of the strain and interfacial energies, the situation being analogous to the formation of a mechanical twin. However, there are many departures from this simple morphology. For example, in Fe-31Ni the plates may be segmented (Figure 19); and in an Fe-Cr-C alloy,

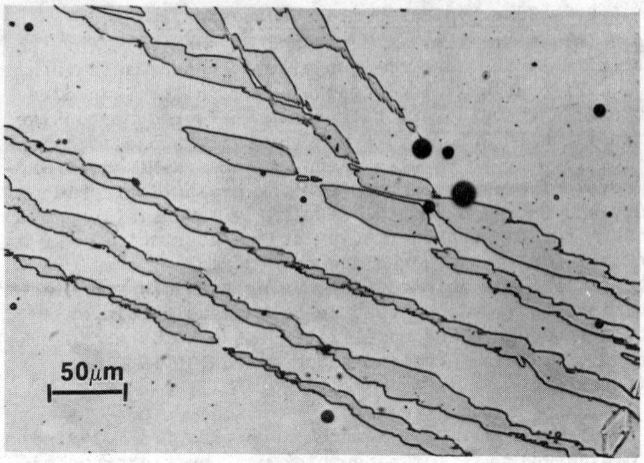

Figure 19. Segmented martensitic transformation product in an Fe-31Ni alloy.

side-plates are frequently observed to extend out from the main interface plane (Figure 20). Some martensitic plates contain a midrib. as shown in Figure 21. This region usually exhibits a high density (relative to the rest of the plate) of internal twins, and is believed to constitute a rapid growth event, prior to the general

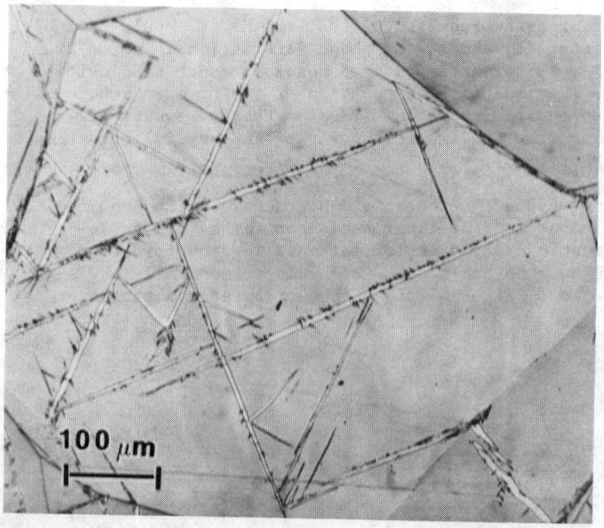

Figure 20. Martensitic transformation product in an Fe-8Cr-1C alloy, showing side-plates extending from interfaces of main plate.

Figure 21. Plates of martensite in an Fe-29.8Ni alloy, showing midribs.

general thickening process.

Another departure from the usual lens-like morphology is the thin-plate martensite observed in Fe-Ni-C alloys. In this case, both sides of the martensitic plate are parallel when viewed with the optical microscope (Figure 22).

Figure 22. Thin-plate martensite formed in an Fe-22Ni-1.1C alloy.

In some materials, martensite may be preferentially nucleated at the surface of specimens, giving rise to superficial plates which are not representative of of the bulk transformation. This morphology is called surface martensite; its presence may interfere in studies of the normal transformation characteristics by forming inadvertently on a free surface during polishing or thinning of specimens.

Still another type of morphology, also found in Fe-Ni-C alloys, is butterfly martensite; its nature is illustrated in Figure 23 (81). In certain Fe-Ni-C alloys, it is possible to observe butterfly, lenticular,

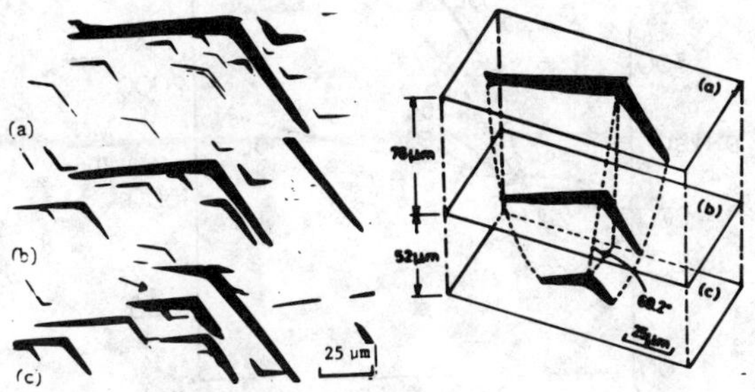

Figure 23. Shape of butterfly martensite formed in an Fe-18Ni-0.6Cr-0.5C alloy, shown by serial sectioning and in three dimensions.

and thin-plate martensites, merely depending on the temperature at which the reaction takes place: butterfly between 0 and -60° C, lenticular between -20 to -150° C, and the thin-plate below -150° C (82).

The martensite formed in iron alloys with low-carbon and low-solute content assumes a lath morphology, as shown in Figure 24. Typical lath dimensions are 0.3 x

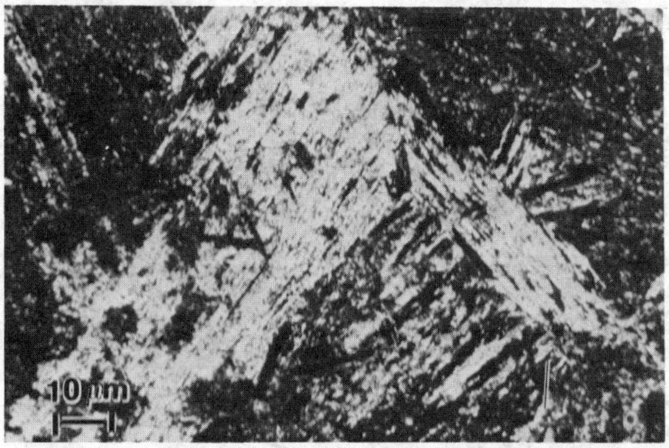

Figure 24. Lath morphology of martensite typical of low-carbon steels (polarized light).

4 x 100 µm and the lath habit plane is ∿5° from {111}p (60). The laths tend to cluster together into packets, with each adjacent lath having the same habit plane, orientation relationship, and shape deformation. Because of the lath thinness, measurement of the shape strain using the traditional scratch-displacement method proves to be difficult, but it is estimated that the magnitude of the shape strain is somewhat larger than that of plate martensites in ferrous alloy (83). The substructure of lath martensite consists of a high density of dislocations; no internal twins are observed. As yet, there has been no satisfactory explanation of lath martensite in terms of the crystallographic theory.

Finally, a rather infrequent morphology appearing, for example, in Au-Cd alloys involves a single interface. When a single crystal rod of a Au-47.5 at.% Cd alloy is cooled in a temperature gradient, a martensitic nucleus is activated at the cold end and a transformation interface (habit plane) propagates from one end of the crystal to the other. Typically, such martensite is internally twinned, resulting in two martensitic orientations. The initial parent single crystal is observed to bend sharply due to the transformation shape change as the interface progresses along the length of the specimen (84).

7. MECHANICAL ASPECTS OF MARTENSITIC TRANSFORMATIONS

Because of the shear-like displacive nature of martensitic transformations, special interactions are to be expected when external stresses are applied. Such phenomena include elastic-stress- and plastic-strain-induced nucleation and reaction kinetics, transformation-induced plasticity, pseudoelasticity, shape-memory effects, etc.

7.1 Elastic-Stress vs. Plastic-Strain Induced Nucleation

It is useful to distinguish between the roles of elastic stress and plastic strain in influencing the martensitic nucleation process (85). As indicated schematically in Figure 25, the application of a stress

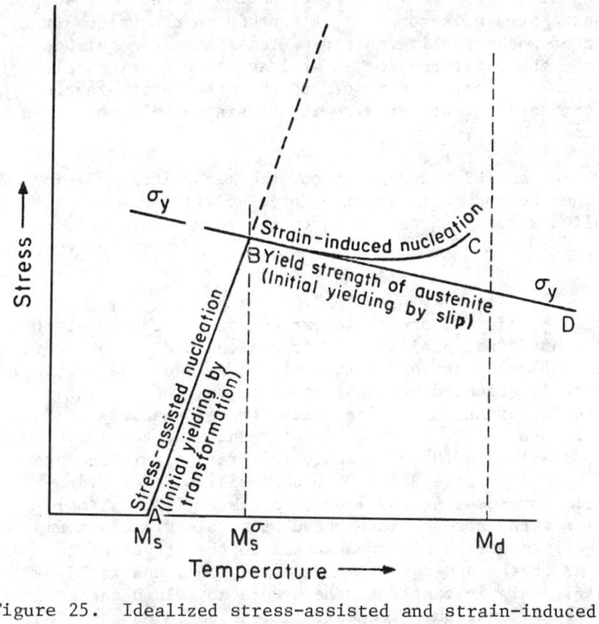

Figure 25. Idealized stress-assisted and strain-induced regimes for mechanically induced nucleation. Below M_s^σ, the initial yielding is controlled by transformation plasticity, and above M_s^σ by slip yielding of the parent phase (85).

at a temperature not far above M_s can induce nucleation even though the stress may be well below the normal yield strength of the parent phase. This circumstance is often called (elastic) stress-assisted nucleation because the existing nucleation sites are simply aided mechanically by the thermodynamic contribution of the applied stress (86). In other words, the critical driving force for nucleating the sites that normally operate without any external stress at M_s can be attained at temperatures above M_s since the mechanical driving force makes up for the reduced chemical driving force at such higher temperatures. Due to this interrelationship, stress-induced nucleation requires higher stresses, the higher the temperature, in accordance with the slope of line AB in Figure 25.

At a temperature designated as M_s^σ, the stress required for nucleation reaches the yield strength of the parent phase, and then new and more potent nucleating sites are introduced by the plastic deformation. Under these conditions, the nucleating stresses at temperatures above M_s^σ do not fall on the extension of line AB, but rather along the line BC; and the nucleation is then termed (plastic) strain-induced. M_d is the temperature above which the chemical driving force becomes so small that nucleation cannot be mechanically induced, even in the plastic-strain regime.

It may be mentioned in passing that the transformational/mechanical interactions denoted in Figure 25 are representative of athermal kinetics; there are corresponding effects of elastic stress and plastic strain on the isothermal kinetics as well (87).

The martensitic shape change itself contributes a mode of deformation, sometimes called transformation-induced plasticity (TRIP), which can influence stress/strain behavior in important ways. Macroscopic yielding follows the line ABD in Figure 25. Along the line AB, the yield stress is determined by the onset of the stress-assisted transformation and increases with increasing temperature, whereas along the line BD, the usual temperature-dependence of the yield stress is manifested. Often there is an overlap between the stress-assisted and strain-induced temperature ranges, and so the M_s^σ temperature may not be sharply defined. The line ABC designates the stress/temperature conditions where the TRIP phenomenon sets in. It should be emphasized that the transformational contribution to the macroscopic plastic strain is complex; it depends on the volume fraction of mechanically induced martensite, the magnitude of the shape deformation as well as its direction relative to the applied stress, and the associated plastic accommodation together with its biasing by the operative stress system.

For an IPS transformation, the mechanical addition to the thermodynamic driving force arising from an applied stress can be simply expressed as (86):

$$W = \tau\gamma_{Tr} + \sigma_n\varepsilon_{Tr} , \qquad (26)$$

where γ_{Tr} and ε_{Tr} are the shear and normal components of the transformational shape deformation, and τ and σ_n are the resolved shear and normal stresses. For the most favorably oriented martensitic variants, the driving force increment (W) in iron-alloy transformations is 0.86 J/mole·MPa (1.42 cal/mole·ksi) for uniaxial tension, 0.58 J/mole·MPa (0.96 cal/mole·ksi) for uniaxial compression, -0.28 J/mole·MPa (-0.46 cal/mole·ksi) for hydrostatic compression, and 1.42 J/mole·MPa (2.34 cal/mole·ksi) for the stress state ahead of an elastic crack tip. These interesting differences due to the stress state depend on the $\sigma_n\varepsilon_{Tr}$ term in Equation (26), and would be modified significantly if there were no volume change in the transformation. The calculations based on Equation (26) are in fair agreement with the measured slope of line AB in Figure 1. Under uniaxial tension, the increase of M_s in ferrous alloys is about 0.15° C/MPa or 1° C/ksi.

7.2 Mechanically Induced Transformations

At temperatures below M_s^σ, increasing the applied stress above the stress-assisted transformation-start line AB in Figure 25 causes further transformation and, as previously mentioned, the concomitant plastic flow is a consequence of the transformation plasticity. Under these conditions, the volume fraction (f) of martensite formed is found to be a linear function of the plastic strain (ε) for a substantial part of the transformation, as indicated in Figure 26(a) (88). The resulting morphology is about the same as observed for the spontaneous transformation which takes place without applied stress on cooling below M_s.

At temperatures above M_s^σ, there is a sigmoidal relationship between f and ε as the stress is raised above the strain-induced transformation-start line BC in Figure 25; this relationship is shown in Figure 26(a). In this

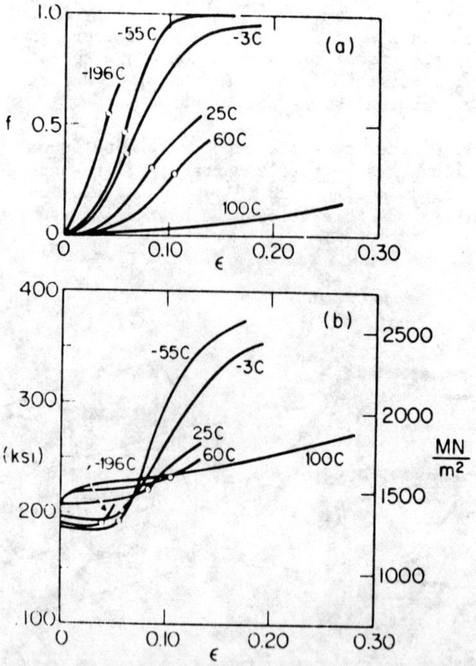

Figure 26. Stress-strain-transformation characteristics of TRIP steel (Fe-9Cr-8Ni-4Mo-2Si-0.8Mn-0.3C), tensile tested at various temperatures. (a) Volume fraction of martensite vs. true strain. (b) True stress vs. true strain. Open circles indicate Lüders strain (88).

case, the morphology depends on the new nucleation sites introduced by the slip-yielding of the parent phase along the line BC of Figure 25. On the assumption that such nucleation sites are generated by shear-band intersections, the aforementioned sigmoidal relationship can be written (89):

$$f = 1 - \exp\{-\beta[1 - \exp(-\alpha\varepsilon)]^n\}, \qquad (27)$$

where α is a parameter expressing the rate of shear-band formation vs. ε, the exponent n is a geometric constant reflecting the rate of formation of shear-band intersections vs. ε, and β is proportional to the probability of nucleation at a shear-band intersection. Both α and β are temperature-dependent, the former because it involves the stacking-fault energy, and the latter because it involves the thermodynamic driving force. Figure 27

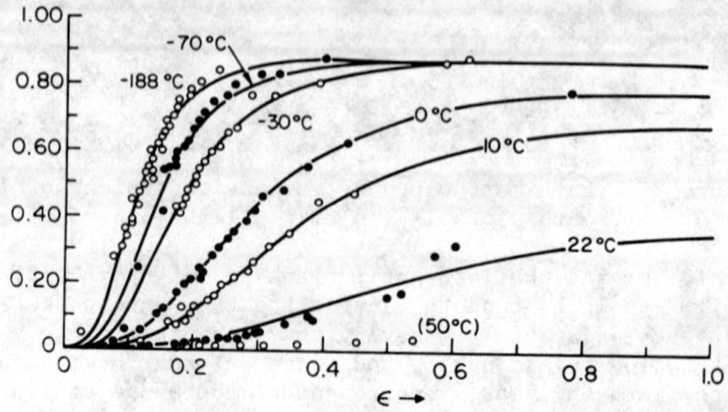

Figure 27. Comparison of calculated (89) and experimental (90) strain-induced martensitic transformation in an austenitic stainless steel due to deformation at various temperatures. The points indicate experimental data; the curves are calculated from Equation (27).

is an example of how well Equation (27) conforms to experimental data for the strain-induced FCC→BCC transformation in an austenitic stainless steel (89,90), even accounting for a saturation level (at f < 1) which increases with decreasing transformation temperature.

There is evidence that unique mechanical behavior can result from the TRIP effect, e.g., relatively high uniform elongation and fracture toughness at high strength levels. Two factors are undoubtedly at play here: (a) a high rate of work hardening is observed due to the formation of martensite, and (b) the mechanically-induced transformation is favored at the tip of a crack by the plastic zone and the state of stress.

7.3 Thermoelastic Martensitic Transformations

Usually, the shape deformation of a martensitic transformation is so large relative to the stiffness and strength of the surrounding parent phase that plastic accommodation takes place during the growth process. Consequently, the interfacial motions are not reversible; the interfacial structure becomes disorganized by the debris resulting from the plastic yielding and the interface loses its glissile nature. Furthermore, a substantial part of the strain energy "paid for" by the thermodynamic driving force is thereby dissipated and, hence, is not elastically stored in the system to help maintain a condition of reversibility.

Alternatively, in those instances where the shape deformation can be accommodated elastically, the interfacial motions take on reversible features. On cooling, the increasing chemical driving force tends to be counterbalanced by the increasing strain energy being stored in the alloy as any interface moves to increase the volume of a martensite plate. Also, because of such elastic accommodation, the interface preserves its characteristic structure and remains glissile. Thus, on heating, the strain energy plays an essential role in reversing the shape change as well as the motion of such glissile interfaces by working against, and continuously rebalancing, the decreasing chemical driving force. This state of balance between chemical and mechanical forces has been called thermoelasticity in the context of martensitie transformations (91). An example of a thermoelastic transformation loop (forward transformation on cooling and its reversal on heating) is shown in Figure 28.

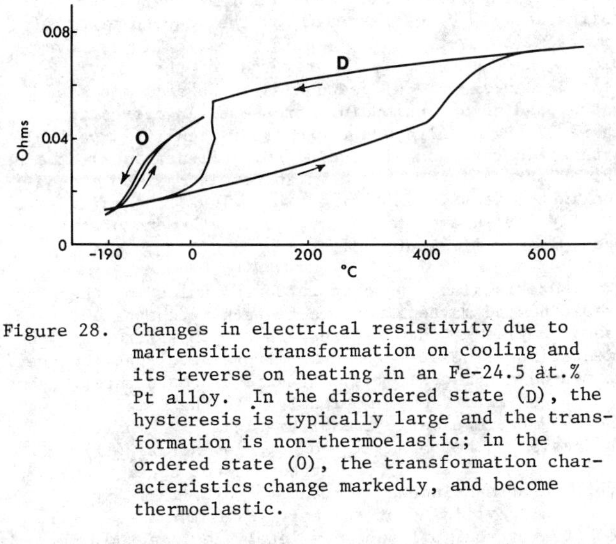

Figure 28. Changes in electrical resistivity due to martensitic transformation on cooling and its reverse on heating in an Fe-24.5 at.% Pt alloy. In the disordered state (D), the hysteresis is typically large and the transformation is non-thermoelastic; in the ordered state (0), the transformation characteristics change markedly, and become thermoelastic.

Olson and Cohen (92) have proposed that the necessary and sufficient condition for the occurrence of a thermoelastic martensitic transformation is the relative absence of plastic yielding in accommodation to the shape deformation. There are several factors which will help satisfy this condition: e.g., high resistance to slip yielding, low elastic constants, a small driving force for nucleation, and reversible stress-relieving mechanisms (such as the formation of multiple shape-deformation variants). However, the basic interplay of these factors can be understood by modeling the nucleation and growth of a single plate on cooling, followed by its shrinkage and disappearance on heating, as illustrated in Figure 29 (92). Upon nucleation, it is

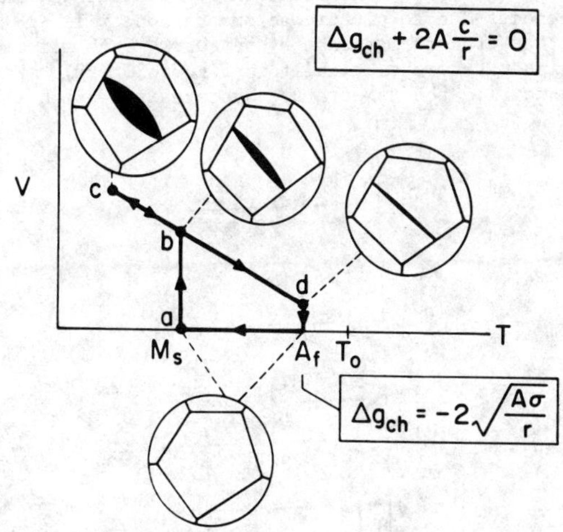

Figure 29. Schematic illustration of thermoelastic transformation behavior of a single plate of martensite with elastic accommodation, growing on cooling and shrinking on heating without frictional resistance (92). See text for details.

visualized that a thin lenticular plate grows rapidly in radial fashion (or lengthwise) until it meets an obstacle such as another plate or grain boundary, and then thickens until a thermoelastic force balance is established when the plate reaches volume b. With further cooling below M_s, and also on reheating, the force balance is maintained with reversible interfacial motions all along line cbd. Under these ideal conditions, the thermoelastic force balance leads to a simple relationship (92):

$$\Delta g_{ch} + 2\Delta g_{el} = \Delta g_{ch} + 2Ac/r = 0 \,, \tag{28}$$

where Δg_{ch} is the chemical free-energy change per unit volume of martensite, r is the plate radius, c is the plate semi-thickness, and $\Delta g_{el} = Ac/r$ is the (elastic) strain energy per unit volume of martensite. The factor A has been previously defined by Equation (2). For a given value of r, c is seen to increase as Δg_{ch} becomes more negative on cooling, and the reverse occurs on heating.

The force-balance line cbd extrapolates to V = 0 at T_o. On heating to point d, however, the aspect ratio (c/r) of the plate becomes so small that surface tension forces are sufficient to pull the thin plate back from its radial-growth obstacle, despite the finite driving force which still exists at this higher temperature ($T < T_o$); then the plate vanishes suddenly by radial shrinkage. This event corresponds to the A_f temperature, and takes place when the following condition is attained on heating (92):

$$\Delta g_{ch} = -2(A\sigma/r)^{1/2} \,, \tag{29}$$

where σ is the interfacial energy per unit area.

Note that, even with ideal interfacial reversibility, the start-up on cooling does not follow the dbc line because the nucleation event requires supercooling, i.e., $M_s < T_o$. In actual thermoelastic transformations, there is a further departure from reversibility due to lattice friction encountered by the moving interface. This extra force requirement introduces hysteresis in the growth process, and can be taken into account as follows (92):

$$\Delta g_{ch} + 2(Ac/r) = \underset{\text{heating}}{\overset{\text{cooling}}{\mp}} \tau_f \gamma_{Tr}, \tag{30}$$

where τ_f is the friction shear stress and γ_{Tr} is the transformational shear strain. (For convenience, we are neglecting dilatational components here.) Thus, anything that raises the friction stress, such as lattice defects due to plastic accommodation, will increase the transformational hysteresis. Such effects are illustrated schematically in Figure 30 (92). With increasing τ_f, the temperature hysteresis between the

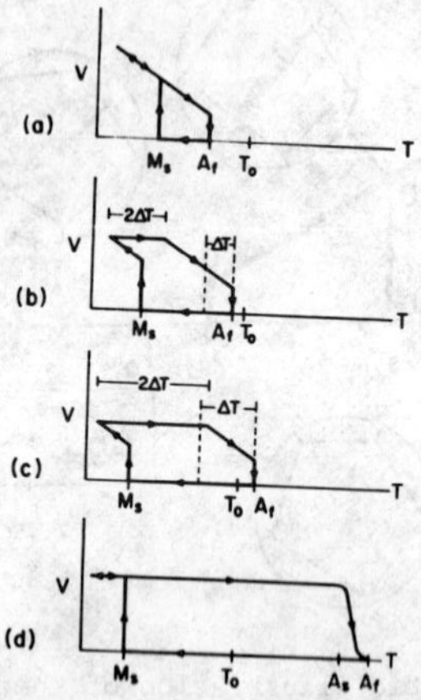

Figure 30. Influence of increasing frictional resistance on thermoelastic hysteresis between the cooling and heating transformations for a single plate of martensite: (a) without frictional resistance, (b) and (c) with two levels of frictional resistance, and (d) becoming non-thermoelastic, with plastic accommodation (92).

growth and reversion processes becomes larger and A_f rises above T_o, as may be the case in actual thermoelastic transformations. When plastic accommodation prevails, typified by Figure 30(d), interfacial reversal does not take place, and the A_s and A_f temperatures are then dependent on the nucleation of new parent-phase particles within the martensite. Then, T_o can be approximated by $T_o \approx 1/2(M_s + A_s)$. Obviously, this expression is not applicable to thermoelastic transformations inasmuch as the reversal to the parent phase can start at temperatures well below T_o due to the reversion role played by the stored strain energy.

The alloy of approximate composition Fe_3Pt offers a striking comparison between thermoelastic and non-thermoelastic characteristics. The differences shown in Figure 28 depend on whether the parent FCC phase is ordered or disordered. Although the shape deformation and other crystallographic features are nearly the same in both cases, evidently the ordered system is better able to resist plastic accommodation under the respective conditions of transformation.

An interesting type of mechanically reversible accommodation to the shape deformation in a thermoelastic transformation is illustrated in Figure 31. In alloys which form 2H, 3R, 9R, and 18R transformation products, with a habit near $\{011\}_p$ (93), the martensite plates appear in diamond-shaped groups of four self-accommodating variants having habits near a $\{110\}_p$ pole of the parent crystal. There are six of these plate groups which can form in a crystal of the parent phase, leading to 24 habit-plane variants. Despite the presence of such multivariants, the general concept of a thermoelastic force balance is still relevant because of the mechanical reversibility of the interfaces thus formed.

Particular attention has been directed here to the

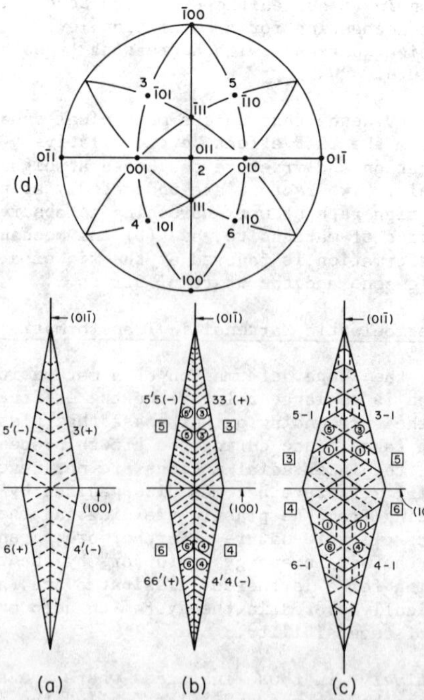

Figure 31. Self-accommodating martensitic variants in diamond-shaped plate groups formed in B2 and DO_3 thermoelastic alloys with (a) 9R, (b) 3R, and (c) 2H periodic stackings of $\{110\}$ planes. (d) Crystallographic disposition of parent-phase $\{110\}$ planes, (011) projection (93).

growth stage in thermoelastic transformations. It is evident that the morphology and overall reaction kinetics will also depend on the nucleation step.

7.4 Pseudoelastic Effects Related to Martensitic Transformations

Alloy systems which undergo a thermoelastic transformation on cooling can be made to transform in a similarly reversible manner, even at temperatures above M_s, by applying an increasing stress. This type of stress-induced transformation is an example of pseudoelasticity in view of the relatively large deformation that can be manifested (∿10 percent strain) by the induced transformation, and yet this strain is "elastically" recoverable on unloading due to the reversal of the transformation.* An example of pseudoelastic behavior is shown by the loading and unloading stress-strain curves in Figure 32 (94). Although there are typically 24 possible martensitic variants (including twin orientations), the applied stress favors those which permit maximum elongation in the tensile-axis direction. Thus, the martensite is fully formed and stress-oriented at the end of the upper plateau in Figure 32.

There is also a plateau region on unloading when the stress-induced martensite reverts back to the parent phase; the latter then regains its original structure and orientation. The difference in stress-level between the two plateaux arises from the frictional resistance which opposes the interfacial motions, as in the previously-discussed thermoelastic transformations. One-half of this hysteresis can be taken as the friction stress (τ_f). If the loading is carried too far beyond the upper plateau after the stress-induced martensite (SIM) is fully formed, slip yielding of the martensite may ensue, and then its reversal will not completely restore the original specimen dimensions.

* The term superelasticity is also employed to describe this phenomenon, and then pseudoelasticity can be used more generically to include the rubber-like behavior described at the end of this subsection (95).

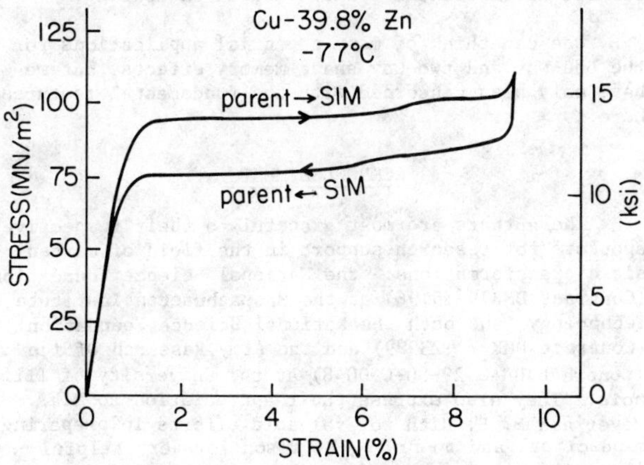

Figure 32. Stress-strain curves illustrating pseudoelastic behavior on tensile loading and unloading a Cu-39.8Zn alloy at -77° C (94). (SIM = stress-induced martensite.)

The pseudoelastic force balance, taking the friction stress into account, corresponds to the thermoelastic force balance given by Equation (30), except that a mechanical work term ($\tau\gamma_{Tr}$) due to the action of the applied stress (τ) must be introduced (96):

$$(\Delta g_{ch} - \tau\gamma_{Tr}) + 2(Ac/r) = \underset{\text{unloading}}{\overset{\text{loading}}{\mp}} \tau_f \gamma_{Tr} \,. \qquad (31)$$

The frictional term ($\tau_f\gamma_{Tr}$) takes on its negative value when the martensite plate is thickening on loading, and its positive value when it is reverting on unloading. Figure 33 is the pseudoelastic counterpart of Figure 30.

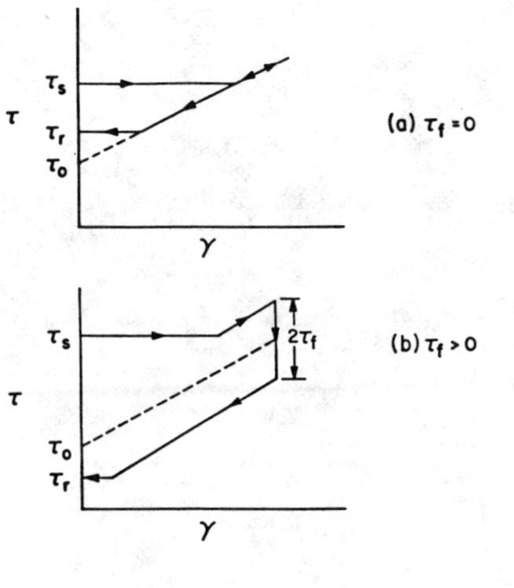

Figure 33. Schematic illustration of pseudoelastic stress-strain behavior of a single plate of martensite with elastic accommodation, growing on loading and shrinking on unloading at $T > M_s$: (a) with no frictional resistance (τ_r), and (b) with finite frictional resistance (96). The strain (γ) reflects the volume of martensite. See text for details.

At a given $T > M_s$, increasing τ causes a plate to form at τ_s, and the pseudoelastic force balance is then maintained along the sloping lines. τ_o is the applied stress at which the transformation would have started on loading if there were no nucleation barrier and no interfacial friction. The unloading stress at which the plate completely reverts is τ_r, and is analogous to A_f in Figure 30. The hysteresis ($2\tau_f$) acts to increase τ_s on loading and to decrease τ_r on unloading.

The ($\Delta g_{ch} - \tau\gamma_{Tr}$) term in Equation (31) must be negative in order to achieve a pseudoelastic force balance, just as Δg_{ch} in Equation (30) must be negative to achieve a thermoelastic force balance. An interesting contrast is that Δg_{ch} can be positive in Equation (31) since it can be overbalanced by the work term. Thus, a pseudoelastic transformation can operate above T_o, whereas a thermoelastic transformation cannot. If no structural change were involved in Equation (31), Δg_{ch} would become zero for this special case, thereby reducing to the pseudoelastic force balance for mechanical twinning with elastic accommodation.

In some instances, two successive stress-strain plateaux are encountered in pseudoelastic behavior. This occurs when the stress-induced martensite can be converted to another form of martensite by additional interfacial movements as the applied stress is further increased. Such a case is illustrated in Figure 34;

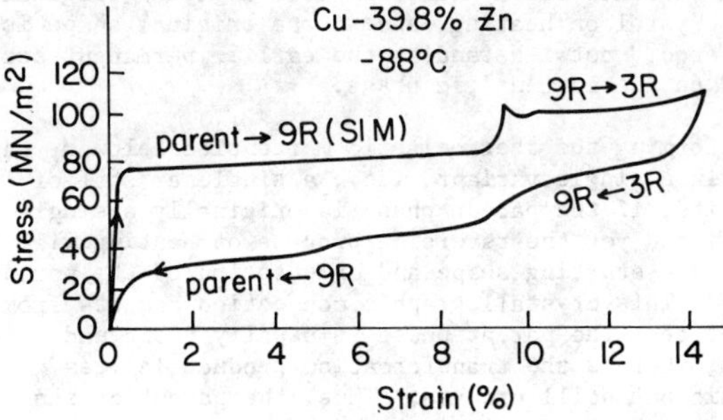

Figure 34. Double pseudoelastic stress-strain hysteresis loop for a Cu-39.8Zn alloy at -88° C due to formation of stress-induced 9R martensite on loading and its conversion to 3R martensite on further loading, followed by the reversal of these transformations on unloading (97).

evidently, the martensite-to-martensite interfaces remain glissile after the parent phase has been fully converted to stress-induced martensite (97).

Interfacial mobility is a characteristic of thermoelastic transformation products, including both the twin interfaces within martensite plates and the plate interfaces between the martensitic variants. As a consequence, such fully transformed specimens (cooled below M_f) can be deformed at relatively low stresses because the stress-favored variants can easily expand by glissile interfacial movements that invade and replace the less-favored variants. Under some conditions, these interfaces experience a restoring force and so the martensitic specimen reverts to its original shape when the applied stress is removed. The nature of this restoring force is not well understood, but the attendant stress-strain relationships on loading and unloading are reminiscent of pseudoelasticity. However, this phenomenon is termed <u>rubber-like behavior</u> to distinguish it from pseudoelasticity because it depends solely on internal martensitic displacements, rather than on parent-to-martensite displacements.

7.5 Shape-Memory Phenomena

7.5.1 <u>Nature of the shape-memory effect</u>. It was pointed out in the previous section that thermoelastic martensites have glissile interfaces even when the different variants impinge on one another, and so the movement of these interfaces (including the internal twin interfaces) at temperatures below M_f in response to external loads can become the dominant mode of deformation. In the absence of internal restoring forces, which could otherwise operate to produce the aforementioned rubber-like behavior, permanent changes in overall specimen shape can be accomplished at stresses significantly lower than would be required to cause slip yielding. Whatever the distribution of variants remaining after such martensitic deformation, heating above A_f to reform the parent-phase grains will restore the original specimen shape of the parent phase, and hence the name shape-memory effect (SME). In this way, seemingly plastic strains of the order of 6 - 8 percent imposed on the thermoelastic martensite can be completely recovered simply by heating above A_f. Many alloys are now known to exhibit this type of shape memory; they all depend on the interfacial reversibility that is inherent in thermoelastic transformations.

When a crystal of the parent phase transforms thermoelastically on cooling between M_s and M_f, there are typically 24 possible martensitic habit variants which can form, and so the overall specimen shape remains essentially unchanged despite the local transformational shape deformations. If the shape of the martensitic specimen is now altered by changing the distribution of variants via interfacial motions, the remaining variants still transform back to their original parent crystal on heating, and so the original shape is "remembered," notwithstanding the earlier permanent set imposed on the martensitic phase.

Deforming the thermoelastic martensite below M_f can result in a single variant, i.e., a single crystal of martensite, if the parent phase is originally a single crystal, and yet the reversion process on heating will restore the starting shape and orientation of the parent crystal. This crystallographic connection results from the fact that the parent phase is usually cubic and ordered, whereas the transformation product is less symmetric but still ordered. Thus, the parent crystal has many equivalent principal axes for the Bain distortion, and this can lead to numerous martensitic variants. In contrast, the comparatively asymmetric martensites do not enjoy such a multiplicity of choices for the reverse Bain distortion, and given this restriction, even if new nucleation of the parent phase is required for the reverse transformation, this will usually initiate the single variant of the orignal parent crystal.

7.5.2 Two-way shape-memory effect. The SME just considered is one-way, i.e., no appreciable change in specimen shape occurs during the forward thermoelastic transformation on cooling; instead, a change in shape does take place during the reverse transformation on heating above A_f which cancels the prior change of shape introduced by deforming the thermoelastic martensite below M_f. In the two-way SME, an overall change of specimen shape operates during both the cooling and heating transformations. This condition can be brought about in two ways:

(a) After pseudoelastically cycling by loading and unloading several times above M_s, the thermoelastic transformation on subsequent cooling in the unloaded state results in the preferential formation of a limited number of variants (sometimes only one variant), thus causing a spontaneous change in specimen shape during the forward transfromation. This change in shape is then recovered by the reverse transformation on heating.

(b) By going through the one-way SME a number of times, the parent phase "learns" to prefer certain variants in the cooling transformation, and again a spontaneous change of specimen shape sets in during the forward transformation, which can then be removed by the reverse transformation.

It appears that both of the above "training" methods, i.e., by SIM cycling or by SME cycling, involve the preferential formation of lattice defects or microstresses which favor selected variants during the forward thermoelastic transformation on cooling.

One can think of many potential applications for the one-way and two-way shape-memory effects, but we have only been concerned with the fundamental features here.

ACKNOWLEDGMENTS

The authors are most grateful to their respective sponsors for research support in the field of martensitic transformations: the National Science Foundation (Contract DMR79-15196) at the Massachusetts Institute of Technology, and both the National Science Foundation (Contract DMR 77-23999) and the Army Research Office (Contract DAAG 29-80-C-0048) at the University of Illinois. They also express their appreciation to M. A. Meyer and M. E. Rich for valuable efforts in preparing the manuscript, and to Dr. G. B. Olson for very helpful discussions.

REFERENCES

1. Z. Nishiyama, Martensitic Transformation, Academic Press, N.Y. (1978).
2. J. W. Christian, Theory of Phase Transformations in Metals and Alloys, Pergamon Press, N.Y. (1965).
3. Xu Zuyao (Hsu Tzu-Yao), Martensitic Transformation and Martensite, Jiao Tong University Press, Shanghai (1980).
4. Morris Cohen, G. B. Olson, and P. C. Clapp, Proc. Int. Conf. on Martensitic Transformations, Cambridge, MA (1979) 1.
5. E. C. Bain, Trans. AIME, 70 (1924) 25.
6. M. S. Rashid and C. J. Altstetter, Trans. TMS-AIME, 236 (1966) 1649.
7. Morris Cohen, Trans. TMS-AIME, 212 (1958) 171.
8. J. W. Christian, Proc. Int. Conf. on Martensitic Transformations, Cambridge, MA (1979) 220.
9. E. S. Machlin and Morris Cohen, Trans. AIME, 191 (1951) 746.
10. R. E. Cech and D. Turnbull, Trans. AIME, 206 (1956) 124.
11. C. L. Magee, Met. Trans., 2 (1971) 2419.
12. G. B. Olson and Morris Cohen, Acta Met., 27 (1979) 1907.
13. G. B. Olson and M. Cohen, Ann. Rev. Mater. Sci., 11 (1981) 1.
14. G. V. Kurdjumov and G. Sachs, Z. Phys., 64 (1930) 325.
15. M. Suezawa and H. E. Cook, Acta Met., 28 (1980) 423.
16. P. C. Clapp, Phys. Status Solidi (b), 57 (1973) 561.
17. T. Suzuki and M. Wuttig, Acta Met., 23 (1975) 1069.
18. G. B. Olson and Morris Cohen, Met. Trans. A, 7A (1976) Part I 1897; Part II 1905; Part III 1915.
19. A. J. Bogers and W. G. Burgers, Acta Met., 12 (1964) 255.
20. P. L. Ferraglio and K. Mukherjee, Acta Met., 22 (1974) 835.
21. G. V. Kurdjumov and O. P. Maksimova, Dokl. Acad. Nauk SSR, 61 no. 1 (1948) 83; 73 no. 1 (1950) 95.
22. E. S. Machlin and Morris Cohen, Trans. AIME, 194 (1952) 489.
23. S. R. Pati and Morris Cohen, Trans. AIME, 17 (1969) 189.
24. V. Raghavan and Morris Cohen, Met. Trans., 2 (1971) 2409.
25. C. L. Magee, Phase Transformations, ASM, Metals Park, OH (1970) 115.
26. J. C. Bokros and E. R. Parker, Acta Met., 11 (1963) 1291.
27. G. A. Knorovsky, Autocatalysis of Martensitic Transformations, Sc.D. Thesis, MIT, Cambridge, MA.
28. N. Nakanishi, Shape Memory Effects in Alloys, J. Perkins, Ed., Plenum Press, NY (1975) 147.
29. G. D. Sandrock, A. J. Perkins, and R. F. Hehemann, Met. Trans., 2 (1971) 2769.
30. H. C. Ling and R. Kaplow, Met. Trans. A, 11A (1980) 77.
31. M. B. Salomon, M. E. Meichle, C. M. Wayman, C. M. Hwang, and S. M. Shapiro, Proc. Int. Conf. on Modulated Structures, Amer. Phys. Soc. Proc. No. 53 (1979) 223.
32. H. Kubo and C. M. Wayman, Proc. Int. Conf. on Martensitic Transformations, Cambridge, MA (1979) 532.
33. Nicholas DeCristofaro and Roy Kaplow, Proc. Int. Conf. on Martensitic Transformations, Cambridge, MA (1979) 502.
34. A. G. Khachaturyan, A. F. Rumynina, and G. V. Kurdjumov, Proc. Int. Conf. on Martensitic Transformations, Cambridge, MA (1979) 71.
35. Ryuichiro Oshima, Akira Kajikawa, and Francisco E. Fujita, Proc. Int. Conf. on Martensitic Transformations, Cambridge, MA (1979) 112.
36. Setsuo Kajiwara, Proc. Int. Conf. on Martensitic Transformations, Cambridge, MA (1979) 362.
37. Z. Nishiyama, Reference 1, pp. 232-238.
38. F. Förster and E. Scheil, Z. Metall., 32 (1940) 165.
39. Y. Suzuki and H. Seito, Japan Inst. Metals, (Fall Meeting, 1972) 233.
40. V. Raghavan and Morris Cohen, Acta Met., 20 (1972) 779.
41. G. B. Olson and Morris Cohen, Research in progress at MIT.
42. K. Otsuka, C. M. Wayman, K. Nakai, H. Sakamoto, and K. Shimizu, Acta Met., 24 (1976) 207.
43. M. Grujicic, Thesis research at MIT.
44. G. A. Knorovsky and Morris Cohen, Proc. Int. Conf. on Martensitic Transformations, Cambridge, MA (1979) 374.
45. H. J. Neuhäuser and W. Pitsch, Acta Met., 19 (1971) 337.
46. S. R. Pati and M. Cohen, Acta Met., 19 (1971) 1327.
47. R. L. Fullman, Trans. AIME, 197 (1953) 447.
48. J. C. Fisher, Acta Met., 1 (1953) 32.
49. W. Y. C. Chen and P. G. Winchell, Met. Trans A, 7A (1976) 1177.
50. D. G. McMurtrie and C. L. Magee, Met. Trans., 1 (1970) 3185.
51. J. R. C. Guimãraes and J. C. Gomes, Met. Trans. A, 10A (1979) 109.
52. V. Raghavan and A. R. Entwisle, Iron and Steel Institute Special Report No. 93, (1965) 30.
53. S. R. Pati and M. Cohen, Acta. Met., 17 (1969) 189.
54. C. H. Shih, B. L. Averbach, and M. Cohen, Trans. AIME, 202 (1955) 183.
55. M. Izumiyama, M. Tsuchiya, and Y. Imai, J. Japan

Inst. of Metals, 34 (1970) 286.

56. R. Pradhan and G. S. Ansell, Met. Trans. A, 9A (1978) 793.

57. D. R. Koistinen and R. E. Marburger, Acta. Met., 7 (1959) 59.

58. W. J. Harris and M. Cohen, Trans. AIME, 180 (1949) 447.

59. J. R. C. Guimarães and J. C. Gomes, Acta Met., 26 (1978) 1591.

60. K. Wakasa and C. M. Wayman, Proc. Int. Conf. on Martensitic Transformations, Cambridge, MA (1979) 34.

61. D. P. Dunne and C. M. Wayman, Met. Trans., 2 (1971) 2327.

62. E. J. Efsic and C. M. Wayman, Trans. AIME, 239 (1967) 873.

63. A. B. Greninger and A. R. Troiano, Trans. AIME, 145 (1941) 238; 185 (1949) 590.

64. Z. Nishiyama, Sci. Rep. Tohoku Univ., 23 (1934) 637.

65. K. Otsuka and K. Shimizu, Trans. Japan Inst. Metals, 15 (1974) 103; 109.

66. Y. C. Liu, Trans. AIME, 206 (1956) 1036.

67. W. G. Burgers, Physica, 1 (1934) 561.

68. J. S. Bowles and J. K. Mackenzie, Acta Met., 2 (1954) 129; 138; 224.

69. M. A. Jaswon and J. A. Wheeler, Acta Cryst., 1 (1948) 216.

70. T. Tadaki and K. Shimizu, Trans. Japan Inst. Metals, 11 (1970) 44.

71. M. S. Wechsler, D. S. Lieberman, and T. A. Read, Trans. AIME, 197 (1953) 1503.

72. R. Bullough and B. A. Bilby, Proc. Phys. Soc., B69 (1956) 1276.

73. A. G. Acton and M. Bevis, Matls. Sci. Eng., 5 (1969-70) 19.

74. N. H. D. Ross and A. G. Crocker, Acta Met., 18 (1970) 505.

75. D. P. Dunne and C. M. Wayman, Acta Met., 19 (1971) 425.

76. T. Maki and C. M. Wayman, Suppl. to Trans. Japan Inst. Metals, 17 (1976) 69.

77. D. P. Dunne and C. M. Wayman, Met. Trans., 4 (1973) 137; 147.

78. N. F. Kennon and J. S. Bowles, Acta Met., 17 (1969) 373.

79. S. Jana and C. M. Wayman, Met. Trans.,1 (1970) 2815.

80. J. S. Bowles and D. P. Dunne, Acta Met., 17 (1969) 677.

81. M. Umemoto and I. Tamura, Proc. Int. Conf. on Martensitic Transformations, Cambridge, MA (1979) 53.

82. T. Maki, S. Shimooka, M. Umemoto, and I. Tamura, J. Japan Inst. Metals, 35 (1971) 1073.

83. K. Wakasa and C. M. Wayman, Acta Met., in press.

84. L. C. Chang and T. A. Read, Trans. AIME, 189 (1951) 47.

85. G. B. Olson and Morris Cohen, J. Less-Common Metals, 28 (1972) 107.

86. J. R. Patel and M. Cohen, Acta Met., 1 (1953) 531.

87. G. B. Olson and Morris Cohen, Proc. Joint U.S.-Japan Seminar on Mechanical Behavior of Metals and Alloys Associated with Displacive Transformations, Troy, NY (1979) 7; 16.

88. G. B. Olson and M. Azrin, Met. Trans. A, 9A (1978) 713.

89. G. B. Olson and Morris Cohen, Met. Trans. A, 6A (1975) 791.

90. T. Angel, J. Iron and Steel Inst., 177 (1954) 165.

91. G. V. Kurdjumov, J. Metals, 11 (1959) 449.

92. G. B. Olson and Morris Cohen, Scripta Met., 9 (1975) 1247; 11 (1977) 345.

93. T. Saburi and C. M. Wayman, Acta Met., 27 (1979) 979.

94. T. A. Schroeder and C. M. Wayman, Acta Met., 27 (1979) 405.

95. R. J. Wasilewski, Shape Memory Effects in Alloys, J. Perkins, Ed., Plenum Press, NY (1975) 245.

96. Martin L. Green, Morris Cohen, and G. B. Olson, Matls. Sci. Eng., in press.

97. T. A. Schroeder and C. M. Wayman, Acta Met., 26 (1978) 1745.

STATE VARIABLE THEORIES FOR NONELASTIC DEFORMATION

Che-Yu Li

Department of Materials
Science and Engineering
Cornell University
Ithaca, New York 14853

The various concepts of mechanical equation of state for nonelastic deformation in crystalline solids, originally proposed for plastic deformation, have been recently extended to describe additional phenomena such as anelastic and microplastic deformation including the Bauschinger effect. It has been demonstrated that it is possible to predict, based on current state variables in a unified way, the mechanical response of a material under an arbitrary loading. Thus, if the evolution laws of the state variables are known, one can describe the behavior of a material for a thermal-mechanical path of interest, for example, during constant load (or stress) creep without relying on specialized theories.

Some of the existing theories of mechanical equation of state for nonelastic deformation are reviewed. The establishment of useful forms of mechanical equation of state has to depend on extensive experimentation in the same way as that involved in the development, for example, the ideal gas law. Recent experimental efforts are also reviewed. It has been possible to develop state variable deformation models based on experimental findings and apply them to creep, cyclic deformation and other time dependent deformation. Attempts are being made to correlate the material parameters of the state variable models with the microstructure of a material.

1. INTRODUCTION

Several approaches which employ the concept of a mechanical equation of state to describe time dependent plastic deformation in crystalline solids can be found in the literature (1-4). According to this concept, one can calculate the mechanical response of a solid to an applied stress by using a number of internal parameters (state variables) which together contain the relevant effects of thermal and mechanical history. An advantage of state variable theories is that the specimen behavior can be calculated for an arbitrary deformation path for which the evolution laws for the state variables are known. Thus a variety of distinct phenomena involving time dependent plastic deformation such as creep, constant displacement rate extension, and load relaxation can potentially be represented in a unified way. It is important to recognize that the existence of a mechanical equation of state for plastic deformation can only be demonstrated through experimental results. This is similar to the development of the equation of state for ideal gases which was based entirely on experimental observations.

To be computationally useful, a practical state variable theory must adequately represent the real material behavior over some range of temperature and deformation history. The number of state variables should be reasonably small and their values subject to experimental measurement. Whether a state variable theory can satisfy these requirements can only be determined through experimentation.

The state variable approach for plastic deformation proposed by Hart (4) appears, based on extensive experimental work, to meet the criteria discussed above (5,6). It has been recently extended to describe additional deformation phenomena such as anelasticity, microplasticity, the Bauschinger effect, and grain boundary sliding (7). The present paper will be concerned mainly with the development of Hart's approach and its extension to describe additional deformation phenomena. In the sections following we shall discuss first time dependent plastic deformation and Hart's formulation. We shall then describe the treatment of other deformation phenomena by using generalized state variable deformation models. The physical significance and the various applications of the deformation models will also be discussed. Before turning to Hart's formulation, we shall review some of the related work.

1.1 Other State Variable Theories.

Earlier attempts to develop a mechanical equation of state for nonelastic deformation can be represented by the work of Holloman and Lubahn (1). As noted by Hart, a major shortcoming of of their treatment is the use of strain as a state variable. This is because one can achieve the same level of plastic strain in a specimen through different mechanical paths (for example, at different strain rate) but resulting in different mechanical states of the material. Path dependent variables such as the strain of a material cannot be used, therefore, as state variables in conjunction with the concept of a mechanical equation of state.

A more generalized treatment of the concept of a mechanical equation of state was proposed by Onat and his co-workers (2,8). Onat introduces a set of n state variables q_i (i = 1, 2,...n) which are assumed to characterize the relevant mechanical features of a material undergoing an isothermal, homogeneous inelastic deformation. The stress is assumed to be determined by the state variables, q_i

$$\sigma = \sigma(q_i) \quad i = 1, 2, \ldots n \qquad (1\text{-}1)$$

The rate of change of the state variables are determined by

$$\dot{q}_i = g_i(q_k)+h_i(q_k)\dot{\varepsilon}_t \quad i,k = 1, 2, \ldots n \qquad (1\text{-}2)$$

where g_i and h_i are functions of state variables and $\dot{\varepsilon}$ is the total strain rate.

For a specialized case where n = 3 and the only nonvanishing component of the vector h_i is h_1 which is equal to unity, Jenkins and Pao (9) have shown that Onat's formulation is equivalent to Hart's equations which will be introduced in the next section (4,5). Since we believe physically that there is only one set of the mechanical equation of state type of behavior in a crystalline solid, it is reasonable that the basic formulation of these two theories can be shown to be the same.

The functional form of g_i's as well as the identification of q_i's in Onat's theory remain, however, to be determined experimentally. In this respect, an extensive experimental effort has been made following Hart's approach.

1.2 Related Work. The concept of a mechanical equation of state can be viewed to represent a phenomenological approach developed in an attempt to describe effectively the collective behavior of dislocations. We are interested in both mobile and immobile dislocations. We view the motion of the former to be responsible for nonelastic deformation and the collection of the latter type to form an important part of the barrier structure to dislocation motion. In contrast, many of the current creep theories are constructed based on a detailed consideration of dislocation mechanisms.

Many of the creep theories treat, however, only the so-called steady state creep rate or the minimum creep rate (10). They have usually a limited range of applicability in terms of stress, temperature, and the type of material of interest. Recently developed creep maps represent one attempt to combine the various creep theories in an ad hoc fashion in order to cover a wider range of temperature and stress (10).

For the interest of the concept of a mechanical equation of state, these first principle dislocation based creep theories will require further development. This is because in their present state they have neither been able to predict the experimentally observed mechanical equation of state type of behavior nor to be readily extended to account for the functional form of the mechanical equation of state (6,7).

J.C.M. Li and his co-workers have proposed a concept of internal stress in their description of time dependent nonelastic deformation (11). The internal stress is also an important consideration in Hart's deformation model. A further discussion of J.C.M. Li's theory will be made in the discussion section.

Creep equations such as those contained in a creep map cannot be used directly in any stress analysis involving numerical techniques such as the finite element method and the boundary element method. This technological need has been a part of the motivation for the development of several approaches for constructing constitutive equations for nonelastic deformation (12-16). These various approaches employ a combination of dislocation based concepts and internal variables (state variables). If these various approaches are intended to produce state variable theories, further work is required to ensure that these formulations are consistent with the experimentally observed mechanical equation of state type of behavior (6,7). As mentioned previously, we believe that a crystalline solid will exhibit only one set of mechanical equation of state type of behavior. The various approaches for state variable constitutive equations should share, therefore, a common experimental basis.

2. HART'S THEORY AND RELATED EXPERIMENTAL RESULTS

In this section we shall review Hart's formulation and the experimental evidence of the existence of a mechanical equation of state for time dependent plastic deformation (4,6). We shall describe the state variable deformation model proposed by Hart and some of its applications (17,18).

2.1 Hart's Formulation. Hart's first paper on the concept of a mechanical equation of state was concerned primarily with homogeneous, time dependent plastic deformation which occur in the grain matrix of a crystalline solid (4). Anelastic deformation which will be discussed also in the present paper is also time dependent but in contrast with plastic deformation, is recoverable when the load is removed.

For one dimensional and isothermal conditions, Hart writes the incremental relation between stress σ, plastic strain ε_p and plastic strain rate $\dot{\varepsilon}_p$ as

$$\gamma d\varepsilon_p = d(\ln \sigma) - \nu d(\ln \dot{\varepsilon}_p) \tag{2-1}$$

This relation applies at each point of a load-deformation path. If ν is a function of σ and $\dot{\varepsilon}_p$ only when ε_p is held constant, the right hand side of Equation (2-1) can be integrated with the aid of an integration factor F $(\sigma, \dot{\varepsilon}_p)$ to yield

$$y = f(\sigma, \dot{\varepsilon}_p) \tag{2-2}$$

where y is a constant. Equation (2-2) defines a one parameter family of curves with y being the parameter.

Equation (2-2) is called a mechanical equation of state and the parameter y is called the hardness by Hart.

When a mechanical equation of state exists, the increment in the hardness is related to the increment in ε_p through Equation (2-1) such that

$$F\gamma d\varepsilon_p = d \ln y \tag{2-3}$$

Thus, the assignment of values of y depends on the choice of the integration factor F. In Hart's deformation model, the hardness is measured by the value of σ^* (17).

Experimentally, the existence of a one parameter family of constant hardness σ-$\dot{\varepsilon}_p$ curves can be used to support the concept of a mechanical equation of state proposed by Hart. Each constant hardness σ-$\dot{\varepsilon}_p$ curve can then be used to define operationally a unique plastic state of the material.

If at the same time, the parameter γ in Equation (2-1) is also a function of state variables, an incremental constitutive equation such as Equation (2-1) can have a significant practical advantage in describing the deformation properties of a material along an arbitrary path (18). The properties of the parameter γ will, however, have to be demonstrated separately by experiments (19).

The influence of the temperature T, in the absence of thermal recovery can be included by the addition of an appropriate term to Equation (2-1),

$$d\ln\sigma = \gamma d\varepsilon_p + \nu d\ln\dot{\varepsilon}_p + Qd(1/RT) \tag{2-4}$$

The above temperature dependence will require also experimental verification.

2.2 Experimental Results of Grain Matrix Plastic Deformation. The development of modern load relaxation experiment has been a key element which made possible the establishment of the existence of a mechanical equation of state for time dependent plastic deformation in the grain matrix (6,7,20). Load relaxation data showed also that the functional form of Equation (2-1) depends on the deformation mechanism but is essentially independent of the materials tested (6,7). The theoretical significance of this latter finding has not been fully explored (5).

The creep experiment which has been used traditionally to obtain experimental information on time dependent plastic deformation has been found to be inadequate for the present purpose. This is because the creep experiment is time consuming and cannot be used easily to produce constant hardness stress-nonelastic strain rate data. The principle of the load relaxation experiment is discussed in detail in References 7 and 20. It can be used to generate constant hardness stress-nonelastic strain rate data covering five to six orders of strain rate with testing time less than a few days. In a load relaxation experiment the specimen is first deformed at a given displacement rate to a desired level of load in a screw driven Instron type testing system. The crosshead is then fixed. Load relaxation results from the conversion of the elastic strain of the of the entire system including that of the specimen to the nonelastic strain of the specimen. Thus, the rate of load relaxation is related to the nonelastic strain rate of the specimen through a machine constant. During a load relaxation run the total extent of nonelastic deformation in the specimen is limited if a stiff testing system is used. We can, therefore, obtain stress-nonelastic strain rate data at an essentially constant hardness if thermal recovery is not important. We can test experimentally if a nearly constant hardness is maintained during load relaxation by doing the reloading experiment which will be described at the end of this subsection.

The modern load relaxation experiment, as described in References 7 and 20, utilizes a fast digital data acquisition system, good temperature control and an accurate strain measurement system to produce the type od data of the present interest (7). Digital data acquisition systems have become readily available only during the last fifteen years.

During the last decade the load relaxation experiment has been used to investigate a wide variety of materials at temperatures between room temperature and about one-half of the absolute melting temperature of the material and at strain rates between 10^{-3} - 10^{-9} sec^{-1}. These materials include pure metals, metallic alloys, ionic compounds, covalent materials and even metallic glasses. The experimental results are described in References 21-28 and in the references listed in Reference 6. These results show that the existence of a one parameter family of constant hardness stress-nonelastic strain rate curves can be demonstrated by the experimentally observed scaling behavior. Equally important, these results show that the curves obtained at high homologous temperatures exhibit a shape essentially independent of the materials tested. At low homologous temperatures, curves of a different shape are obtained which is also essentially independent of the materials tested. A reloading experiment can be used to demonstrate that the hardness of the material is maintained constant during a load relaxation run and that thermal recovery is absent in the temperature range of these experiments.

Typical room temperature load relaxation data in a logarithmic stress vs. logarithmic nonelastic strain rate plot obtained at high (aluminum) and low (stainless steel) homologous temperatures are shown in Figures 2.1 and 2.2, respectively. These data were obtained in sequence by reloading the specimen to a higher load level after a prior load relaxation run and repeating the load relaxation run. The curves

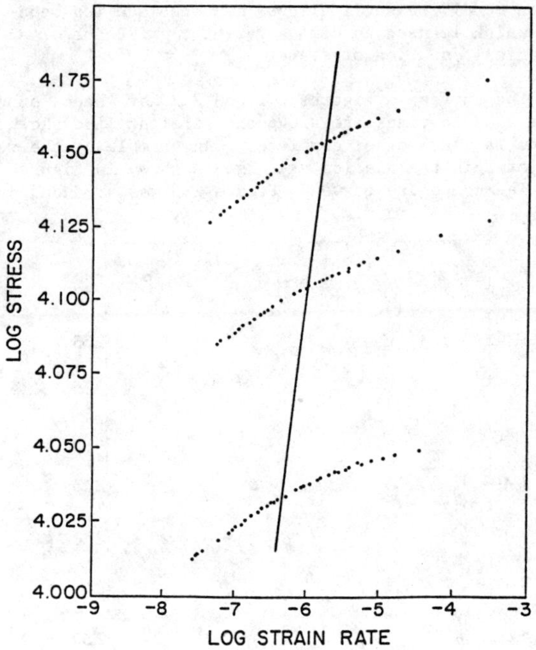

Figure 2.1. Typical room temperature load relaxation data (constant hardness curves) for aluminum obtained in the macroplastic region with the straight line shown as the translation path for hardness scaling. (31).

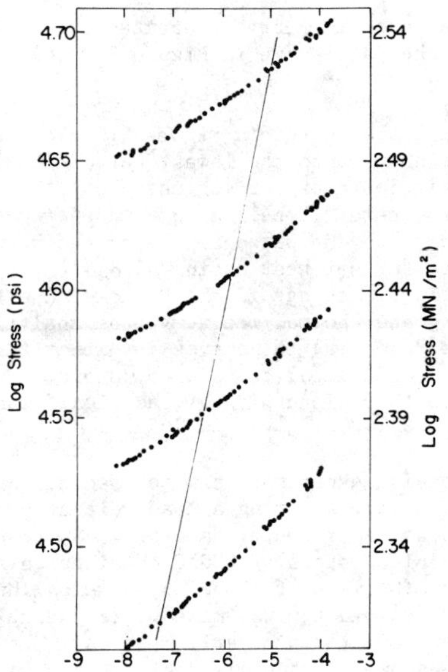

Figure 2.2. Typical room temperature load relaxation data (constant hardness curves) for 304 stainless steel obtained in the macroplastic region with the straight line shown as the translation path for hardness scaling (6).

obtained at high homologous temperatures show a shape which is concave downward. The curves obtained at low homologous temperatures have a shape which is concave upward. A detailed argument can be made as to the proper analytical relations which represent the shape of these curves (5,17). These relations will be given in the next subsection in conjunction with Hart's deformation model. We shall discuss next the scaling behavior which is used to demonstrate the existence of a mechanical equation of state.

The curves in Figures 2.1 and 2.2 have been found to be able to translate along the straight line shown to overlap each other completely. An example of the result of this translation exercise is shown in Figure 2.3. The composite curve in Figure 2.3 was produced by

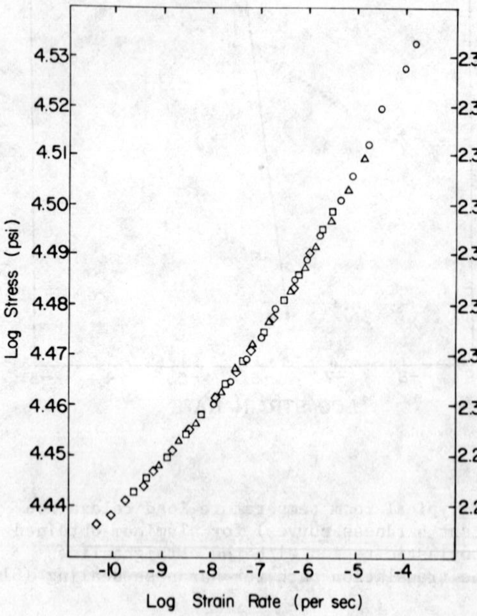

Figure 2.3. A composite curve generated by scaling the curves along the path shown in Figure 2.2.(6).

translating all the curves in Figure 2.2 along the straight line shown onto the lowest curve. This scaling or translation behavior suggests that the curves in Figure 2.1 or Figure 2.2 belong to a one parameter family of curves with the parameter being the hardness. If we can show further that during a load relaxation run there is little change in the hardness due to work hardening and thermal recovery, the analytical expression representing the curves in either figure will then be a mechanical equation of state, Equation (2-2). Further discussion of the significance of the scaling behavior is given in References 5 and 17.

The reloading experiment can be used to detect the change in hardness during a load relaxation run. Figure 2.4 shows experimental data of a highly stablized 316 stainless steel at 600°C obtained in a load relaxation run (Curve A) followed by a reloading run (Curve B). The reload run was made by reloading the specimen below the previous maximum stress. The reloading curve shows a transient region (horizontal part), after the exhaustion of which plastic deformation becomes important again such that the part of Curve B where plastic deformation is controlling is seen to overlap Curve A. According to Equation (2-2), this observation suggests that there is no change in the hardness of the specimen during both load relaxation runs A and B. We have performed the reloading experiment at low homologous temperatures to show separately the absence of work hardening and at high

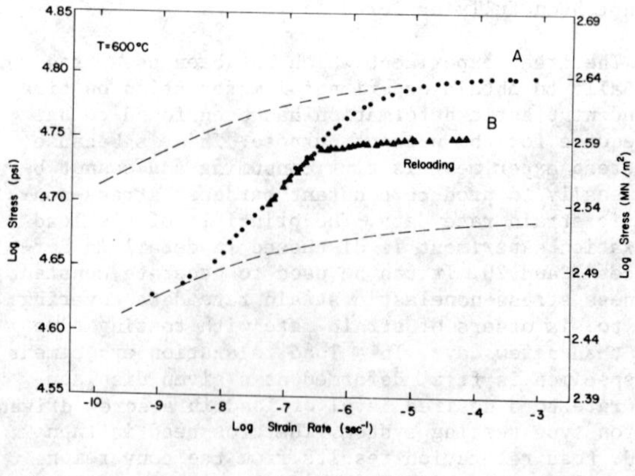

Figure 2.4. Constant hardness log σ - log $\dot{\varepsilon}$ curve obtained at 600°C for a highly stablized 316 stainless steel showing the effect of grain boundary sliding, Curve A; the dashed curves representing the grain matrix controlled behavior at both high and low stress limit; Curve B is the results of a reloading load relaxation run.(6).

homologous temperatures to show the absence of both thermal recovery and work hardening.

At a given temperature a typical load relaxation run will reveal a nonelastic strain rate window between 10^{-3} - 10^{-9} sec^{-1}. An example of load relaxation data obtained at several temperatures but at the same hardness is shown in Figure 2.5. These data for 316

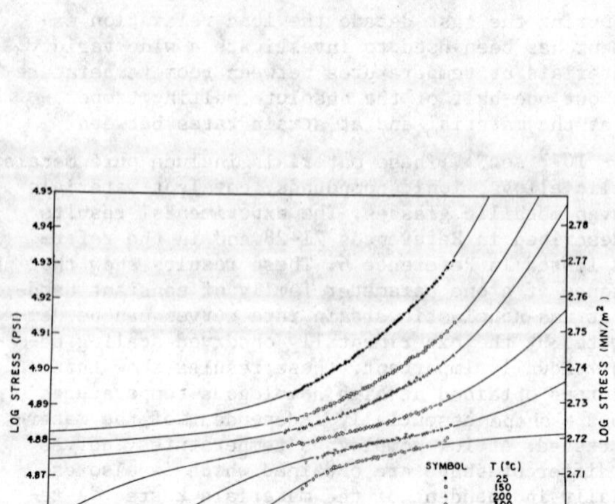

Figure 2.5. Constant hardness log σ - log $\dot{\varepsilon}$ data for a stablized 316 stainless steel at several temperatures; the stress values are modulus corrected with the stress scale shown for room temperature (6).

stainless steel show that near room temperature, low homologous temperature behavior predominates with curves being concave upward and at higher temperatures, high homologous temperature behavior predominates with curves being concave downward. At intermediate temperatures, the load relaxation curves show an inflection point. The position of the curves which are concave downward is found to show temperature scaling according to Equation (2-4) with the value of Q close to that for the activation energy for self-diffusion. Graphically the concave concave downward portion of the constant hardness curves can be shifted along the nonelastic strain rate axis to overlap each other with $\Delta \log \dot{\varepsilon}_p = Q\,\Delta\,(1/RT)$. For the portion of the curve which are concave upward, the value of Q varies with temperature reflecting a change in the deformation mechanism. The existence of temperature scaling suggests that if it is possible to obtain experimental data covering a sufficiently wide nonelastic strain rate range, one should see both the concave upward and downward part of the data at a single temperature. This is the master curve concept which is schematically shown in Figure 2.6 and discussed in detail in Refernece 5.

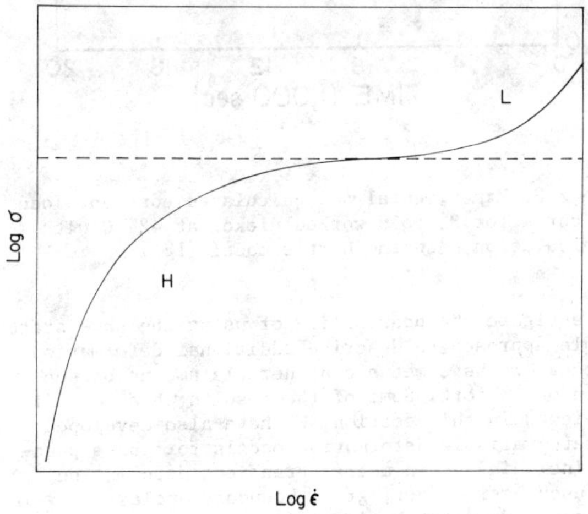

Figure 2.6. A schematic plot of a constant hardness log σ - log $\dot{\varepsilon}$ curve showing the entire range of macroplastic behavior represented by the grain matrix deformation model (5).

The data shown in Figure 2.5 suggest physically that there should be at least two types of deformation mechanisms for plastic deformation in crystalline solids and there is a fixed way in which they can be combined. These considerations are the basis for the state variable deformation model proposed by Hart (17).

2.3 Hart's Deformation Model. The state variable deformation model proposed by Hart was developed based on the type of experimental data described in the previous subsection and the following physical picture (5,17). It is assumed that a crystalline solid can be represented by a collection of two dimensional slip zones bounded by strong barriers to dislocation motion. These barriers are of uniform strength which vary during plastic deformation depending on the governing work hardening law. In an annealed state mobile dislocations distribute randomly between barriers. Under an applied stress these dislocations will move and pile up against barriers to dislocation motion resulting in anelastic defomation. As mentioned previosly, anelastic deformation is recoverable upon the removal of the applied load. If the temperature is sufficiently high, dislocations may overcome barriers through thermally controlled processes such as dislocation climb. If, however, the applied stress is high, the above thermal activation is not required and dislocation motion producing plastic deformation will be controlled simply by glide friction.

The thermal activation controlled plastic deformation has been identified with experimental data obtained at high homologous temperatures (Figure 2.1) and glide friction controlled plastic deformation is identified with data obtained at low homologous temperatures (Figure 2.2). We shall need experimental input to determine the constitutive relations for work hardening and for anelastic deformation. It will be shown in the next section that these relations are governed also by state variables.

Hart's deformation model for homogeneous nonelastic deformation in the grain matrix is schematically shown in Figure 2.7. This model employs three deformation elements representing stored anelastic strain (a-element), thermal activation controlled plastic deformation ($\dot{\alpha}$-element), and glide friction controlled nonelastic deformation ($\dot{\varepsilon}$-element). The constraint equations and constitutive relations for uniaxial deformation are given below. The three dimensional representation is given in Reference 17.

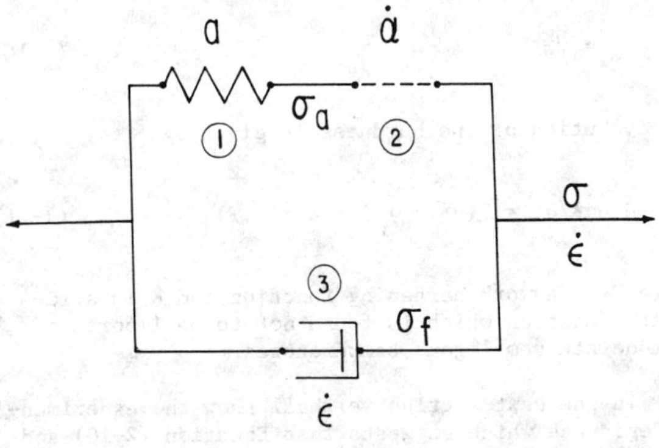

Figure 2.7. A schematic representation of Hart's deformation model for grain matrix flow; the three elements shown are the a-, $\dot{\alpha}$- and $\dot{\varepsilon}$- elements respectively (17).

The constraint equations are

$$\dot{\varepsilon} = \dot{a} + \dot{\alpha} \tag{2-5}$$

and

$$\sigma = \sigma_a + \sigma_f \tag{2-6}$$

where σ, σ_a, and σ_f are the applied stress, back stress due to stored anelastic strain, and glide friction stress; and $\dot{\varepsilon}$, $\dot{a}$, and $\dot{\alpha}$ are the nonelastic strain rate, anelastic strain rate and plastic deformation rate respectively.

For the plastic element ($\dot{\alpha}$-element), the constitutive relation according to the load relaxation data is found to be (5)

$$\ln(\sigma^*/\sigma_a) = (\dot{\varepsilon}^*/\dot{\alpha})^\lambda \quad (2\text{-}7)$$

with scaling

$$\dot{\varepsilon}^* = (\sigma^*/G)^m \, f \, e^{-Q/RT} \quad (2\text{-}8)$$

where G is the modulus of rigidity, σ^* is the hardness and is the limiting stress of curves in Figure 2.1 at large strain rates, $\dot{\varepsilon}^*$ is a rate parameter, $\lambda \approx 0.15$ and governs the shape of the curves in Figure 2.1, m is the inverse of the slope of the translation path, f some characteristic frequency, and Q generally equal to self-diffusion activation energy.

For the glide friction element, the constitutive relation is (17)

$$\dot{\varepsilon} = \dot{a}^* \, (\sigma_f/G)^M \quad (2\text{-}9)$$

where $\dot{a}^*$ is a rate parameter, $M \approx 7.5$ and is the inverse of the slope of the translation path; and for the curves in Figure 2.2, $\sigma_f = \sigma - \sigma^*$ with σ^* being the limitimg stress at low strain rates.

For the a-element, the stored strain is related to σ_a linearly through a modulus M

$$\sigma_a = M a \quad (2\text{-}10)$$

The evolution of the hardness is given by

$$d\ln\sigma^*/dt = \Gamma(\sigma^*,\sigma_a)\,\dot{\alpha} - R(\sigma^*, T) \quad (2\text{-}11)$$

where Γ is a work hardening function and R is a recovery function which is found not to be important at moderate homologous temperatures.

In the next section we shall show the experimental evidence which suggests that Equation (2-10) and the work hardening function contain state variables only.

The deformation model proposed by Hart, in the form described above, has been used successfully to describe a wide variety of deformation phenomena where plastic deformation is controlling (7,18). The material parameters for the deformation model can be determined by a combination of load relaxation and constant displacement rate test. The advantage of this experimental approach is discussed in Reference 7. By using experimentally determined parameters, Hart's model has been shown, for example, to simulate accurately the creep behavior of nickel (Figure 2.8). The method of creep calcalation and the value of the parameters used for Figure 2.8 are given in Reference 18.

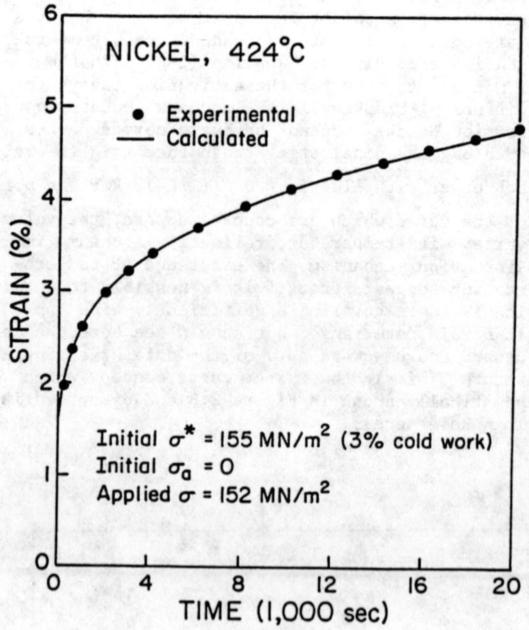

Figure 2.8. Experimental and calculated constant load creep curve for 3% cold worked nickel at 424°C with the calculation based on Hart's model.(18).

3. WORK HARDENING, ANELASTICITY, TRANSIENT DEFORMATION AND GRAIN BOUNDARY SLIDING

Encouraged by our earlier success in observing a mechanical equation of state type of behavior for time dependent plastic deformation, efforts were made to investigate the possibility of using the same state variable approach to describe additional deformation phenomena. We have met a considerable amount of success in our effort. Some of the results obtained will be reviewed in this section. We have also developed new state variable deformation models for these phenomena including grain matrix transient deformation, the Bauschinger effect, grain boundary anelasticity and plastic deformation involving grain boundary sliding.

3.1 Work Hardening. Strain hardening measurements in several metals have been carried out over a range of temperatures and strain rates (19,29). The principal result obtained is that for all the materials tested, the strain hardening coefficient γ, Equation (2-1), is found to be a function of stress and nonelastic strain rate only at a given temperature. This means that γ is a state property. Based on experimental data and the constitutive equations given in subsection 2.3, we can compute the absolute work hardening rate Γ, Equation (2-11), defined as $d\ln\sigma^*/d\alpha$. It has been found tentatively to have the form

$$\Gamma = \left(\frac{\sigma_a}{\sigma^*}\right)^{\beta/\sigma^*} \left(\frac{\beta}{\sigma^*}\right)^\delta \quad (3\text{-}1)$$

where β and δ are material constants.

The results on work hardening suggest that it is a property only of the current value of state variables with no further dependence on deformation history. Equation (3-1) suggests further that the absolute work hardeneing rate Γ does not depend on temperature.

At low homologous temperatures where the plastic behavior is controlled by glide friction (Figure 2.2) and σ_a/σ^* has a value close to one, Equation (3-1) will show that Γ depends only on one state variable. This is consistent with the observation, for example, that the σ-ε curves of steels obtained at room temperature is essentially independent of the displacement rate in a tensile test. At high homologous temperatures where the plastic behavior is controlled by thermal activation (Figure 2.1) and $\sigma_a/\sigma^* < 1$, Equation (3-1) will yield a strain rate dependent work hardening function without resorting to thermal recovery.

We are continuing our investigation of the work hardening properties of crystalline solids. We are performing experiments also on cyclic hardening and softening. We suspect at the present, that the phenomenon of cyclic softening may be related to microplasticity which will be discussed in a later subsection. Some discussion of the theoretical significance of our work hardening results is given in Reference 5. Further work in the theoretical direction is being pursued.

3.2 Grain Matrix Anelasticity. We are interested in grain matrix anelasticity produced by dislocation bowing and pile-up against barriers. The contribution of anelasticity can be shown, for example, by the closed stress-strain loop obtained by loading and unloading in a constant displacement rate tensile test. Figure 3.1 shows schematically low homo-

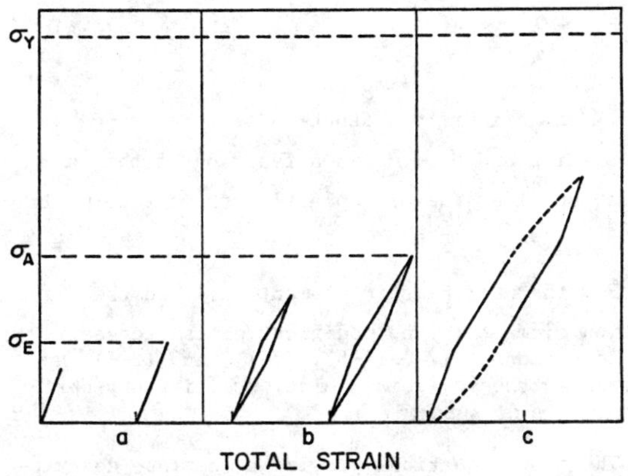

Figure 3.1. A schematic representation of tensile loading and unloading at various load levels below macroplastic yielding (31).

gous temperature tensile data obtained by loading and unloading with varied maximum stress below macroplastic yielding. At stresses below σ_E in Figure 3.1, elastic behavior is shown. At stresses above σ_E but below σ_A in Figure 3.1, the introduction of recoverable grain matrix anelastic deformation results in a closed stress-strain loop. At stresses above σ_A, microplasticity, which will be discussed in the next subsection, will produce an open stress-strain loop upon unloading to zero stress.

According to the deformation model described in subsecton 2.3, the accumulation of dislocation pile-up will be governed by the $\dot{\varepsilon}$-element, Equation (2-9), with its magnitude controlled by Equation (2-10). It can be shown as discussed in Reference 30 that in the region where Equation (2-10) is operating, the nonelastic strain rate $\dot{\varepsilon}$ will remain essentially constant during loading or unloading in a constant displacement rate test. An example of experimental data showing this kind of behavior is given in Figure 3.2. It

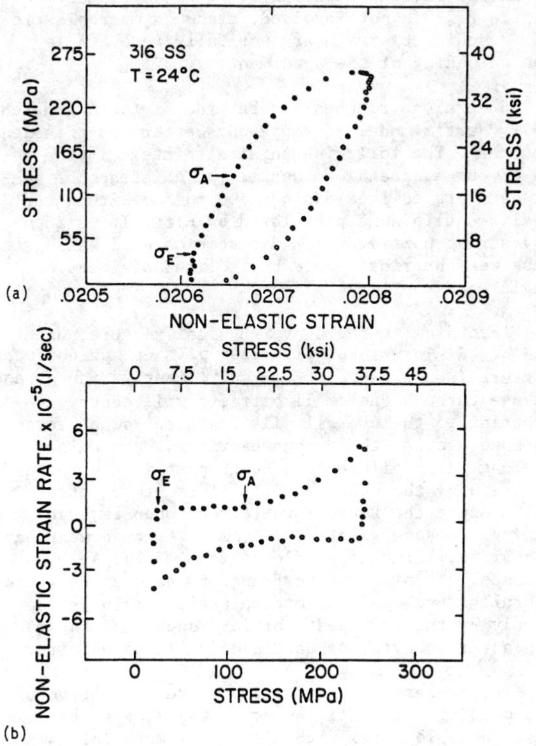

Figure 3.2. (a) Stress-nonelastic strain data for 316 stainless steel obtained in a constant extension rate test at room temperature in a loading-unloading cycle below macroplastic yielding (31).
(b) Nonelastic strain rate vs. stress data for the above cycle (31).

is seen from this figure that between σ_E and σ_A the nonelastic strain rate remains nearly constant. The slope of the portion of the data between σ_E and σ_A in Figure 3.2a will yield, therefore, the value of the modulus M in Equation (2-10). It has been found experimentally that the value of M is independent of the height of the loop (30,31). We can conclude that Equation (2-10) is valid and that the grain matrix anelastic strain a is a state variable.

3.3 Transient Deformation and the Bauschinger Effect. In a recent paper, we have discussed in detail the inadequacy of Hart's deformation model (Figure 2.7) in describing a variety of deformation phenomena which occur below macroplastic yielding (32). For example, the slowly recoverable microplastic flow behavior shown in Figure 3.1 is not expected below macroplastic yielding according to Hart's theory. Real materials generally show the Bauschinger effect in that the specimen which has been deformed, for example, in compression will exhibit significantly more strain during subsequent tensile loading even at small stresses than during the previous compression loading. In contrast, Hart's model is almost symmetrical. For real material parameters, the assymmetry produced by Hart's model is much less than the observed Bauschinger strain.

Since all the discrepancies described above occur below macroscopic plastic yielding, they are presumably associated with the representation of stored strain by a simple linear element Equation (2-10) in Hart's theory. We shall refer to the phenomena discussed above as transient phenomena, in that macroplastic flow as represented by the $\dot{\alpha}$-element (Figure 2.7) is not involved, although microplastic and Bauschinger strain are not fully recoverable upon unloading of the specimen.

Several approaches can be made as to how one can modify Hart's model to represent better the transient phenomena. The following physical picture is the basis of our approach. We assume the existence of weaker barriers to free dislocation motion within the idealized slip zone postulated by Hart. If we identify strong barriers with dislocation cell walls, these weak barriers would be isolated dislocation tangles.

According to the above physical picture, under loading, dislocations will first pile up against weak barriers (short range). At a sufficiently high stress, leakage through the small barriers will occur, contributing to the overall pile up (long range) across the zone against the strong barriers. Upon unloading the long range pile up can be trapped in the forward direction by the weak barriers. This stored strain will produce the Bauschinger effect upon reverse loading. We have identified leakage through weak barriers with microplastic flow (Figure 3.1). At high stresses, the pile ups in front of the weak barriers are fully developed and macroplastic strain is impeded only by the strongest barriers such that the macroplastic behavior of Hart's model is unmodified.

The processes described above define a type of phenomenological model. We are interested in the simplest possible model with all of its material parameters subject to experimental measurement. We are interested also in the possibility that the constitutive relations for the new phenomonological model to be functions of state variables only.

For the new phenomonological model we require two state variables for the stored strain and thus, two anelastic elements of different moduli (a_1- and a_2-element with moduli M_1 and M_2 respectively). An additional flow element $\dot{\alpha}_2$ (microplastic element) for weak barrier passage is also needed. These three elements are arranged as shown in Figure 3.3 according to the physical picture given above (32); and they together replace the a-element in Hart's model. The plastic element and the glide friction element in Hart's model are retained.

The constraint equations implied by the rheological diagram (Figure 3.3) are

$$\sigma = \sigma_f + \sigma_a = \sigma_f + \sigma_{a_1} + \sigma_{a_2} \tag{3-2a}$$

and

$$\dot{\varepsilon} = \dot{\alpha}_1 + \dot{a}_1 = \dot{\alpha}_1 + \dot{\alpha}_2 + \dot{a}_2 \tag{3-2b}$$

which replace Equations (2-5) and (2-6) for Hart's model.

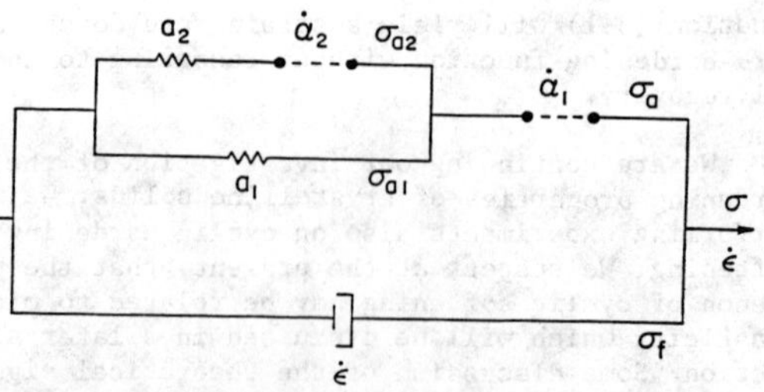

Figure 3.3. A schematic representation of the new deformation model for grain matrix flow with the a-element of Hart's model replaced by the a_1-, a_2-, and $\dot{\alpha}_2$ elements shown (32).

The constitutive relations for the two anelastic elements are

$$\sigma_{a_1} = M_1 a_1 \tag{3-3}$$

and

$$\sigma_{a_2} = M_2 a_2 \tag{3-4}$$

where M_1 is primarily a function of the strong barrier spacing and $M_1 + M_2$ is a function of the weak barrier spacing. The microplastic strain rate will depend on σ_{a_2}.

For the macroplastic element ($\dot{\alpha}_1$) and the glide friction element ($\dot{\varepsilon}$) no modification is necessary. Thus, Equations (2-7), (2-8), (2-11) and (3-1) can be taken as the governing relations (with appropriate addition of subscript).

The above equations provide a complete description of the new phenomonological model except for the flow and evolution laws for the microplastic element ($\dot{\alpha}_2$) which are not known a priori.

We have found experimentally that the $\dot{\alpha}_2$-element also exhibits a mechanical equation of state type of behavior (31,32). We shall review the experimental results later in this subsection. The constitutive equations obtained for the $\dot{\alpha}_2$-element for aluminum have the same form as that for the macroplastic element. We have the flow equation

$$\ln(\sigma^*_2/\sigma_{a_2}) = (\dot{\varepsilon}^*_2/\dot{\alpha}_2)^{\lambda_2} \tag{3-5}$$

and the scaling relation

$$\dot{\varepsilon}^*_2 = (\sigma^*_2/G)^{m_2} f_2 e^{-Q/RT} \tag{3-6}$$

In the absence of thermal recovery, the evolution of hardness σ^*_1 of the macroplastic element is given by the work hardening function Equation (3-1). For the microplastic flow, the situation is somewhat different. We may presume that the weak barriers have a distribution of strengths, and that as the weakest barriers are overcome, the flow will be controlled by successively stronger barriers. For a given macroplastic state of the material, the evolution of σ^*_2 is then a sampling effect rather than the result of work hardening.

We shall now describe the results of our investigation of the elements of the new phenomenological model.

3.3.1. The Anelastic Elements. During loading, for example, in a constant displacement rate test, in the region between σ_E and σ_A, Figure 3.2, according to the new phenomenological model, neither the $\dot{\alpha}_1$-element nor the $\dot{\alpha}_2$ element is operative, the magnitude of the anelastic strain is governed by M_1 and M_2 acting together. The rate of accumulation of the anelastic strain, however, is still governed by the $\dot{\varepsilon}$-element. The slope of the portion of the data between σ_E and σ_A in Figure 3.2a will therefore yield the value of $M_1 + M_2$ in the same way in which the value of M in Hart's model is evaluated.

To determine the value of M_1, we have used experimental data obtained in a cyclic deformation experiment involving symmetrical loading in tension and compression at a constant displacement rate (7, 31). Figure 3.4 shows a typical symmetrical stress-nonelastic strain loop of aluminum together with the accompanying nonelastic strain rate data with maximum strain within the microplastic range. A limited number of repeated cycles have been found to be sufficient to produce the saturation behavior such that we can obtain a symmetrical loop easily. The data in Figure 3.4 show that the range of anelastic deformation (between σ_E and σ_A) is shifted to below zero stress during loading which is consistent with the new phenomenological model.

Based on the experimental data in Figure 3.4, one can argue physically that since the stress-nonelastic strain loop is symmetrical and reversible, at point α_0 the specimen contains neither anelastic strain nor microplastic strain. One can also argue that at the anelastic limit σ_A, an approximation can be made such that $\sigma \simeq \sigma_{a_1}$. This is because during prior loading, dislocations are piled up against the strong barriers in one direction. During unloading, dislocation pile up decreases and begins to pile up against the weak barriers in the reverse direction. At the anelastic limit σ_A, the value of σ_{a_2} is approximately equal to the strength of the weakest barrier which is small (31). We can conclude therefore that at the point σ_A , $\sigma \simeq \sigma_{a_1}$. Thus the slope of a straight line connecting σ_A and α_0 in Figure 3.4 will yield the value of M_1, the long range anelastic modulus in Equation (3-3). We can show a mechanical equation of state type of behavior, Equation (3-3), by producing symmetrical stress-nonelastic strain loops of different amplitude with a specimen of the same macroplastic state (σ^*_1) and finding the value of M_1 remaining constant.

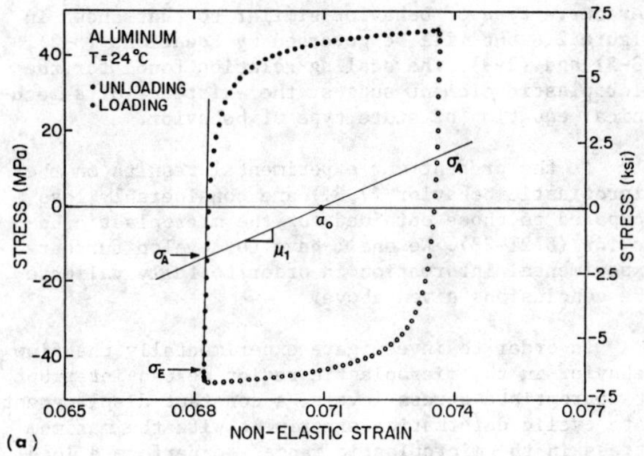

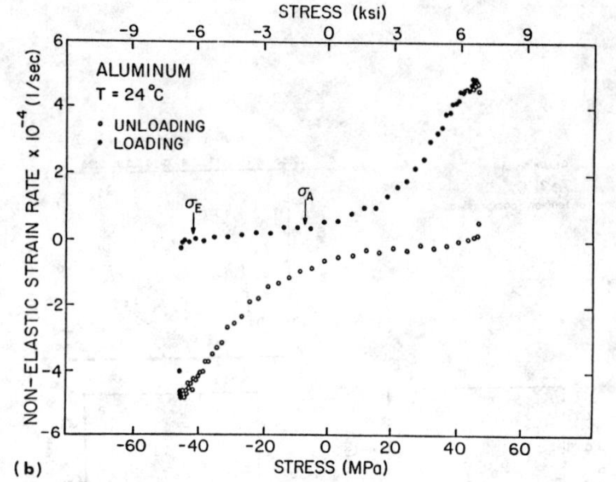

Figure 3.4. (a) Stress vs. nonelastic strain data for aluminum obtained at room temperature in a stablized tension-compression cycle showing the elastic (σ_E) and anelastic (σ_A) limits; the straight line shown has a slope which yields the value of the long range anelastic modulous M_1. (31).

(b) Nonelastic strain rate vs. stress data for the above cycles (31).

3.3.2. The Microplastic Element. Based on limited experimental information, we can make the following tentative conclusions (31). We have found that microplastic deformation is controlled by glide friction element ($\dot{\varepsilon}$-element) acting in parallel with the microplastic element ($\dot{\alpha}_2$-element) in much the same way in which the glide friction element operates in parallel with the macroplastic element for macroplastic flow (7,31). Equation (2-9), which represents the flow properties of the $\dot{\varepsilon}$-element is shown to be applicable also in the microplastic range. The flow and scaling relation for the $\dot{\alpha}_2$-element are given by Equations (3-5) and (3-6) respectively with the value of m_2 and λ_2 close to that of the corresponding parameters of the $\dot{\alpha}_1$-element. We note, as mentioned previously, that the evolution of σ^*_2 for the microplastic element reflects a sampling effect for a given macroplastic state and will have to be determined experimentally (7,31).

For microplastic flow we shall also expect a master curve type of behavior similar to that shown in Figure 2.6 but will be governed by Equations (3-2), (3-5) and (2-9). The scaling relation found for the microplastic element suggest the existence of a mechanical equation of state type of behavior.

At the present our experimental results on the microplastic behavior (7,31) are considerably less compared to those obtained for the macroplastic behavior (6,21-28). We shall have to develop further experimental information in order to fully validate the conclusions given above.

In order to investigate experimentally the flow behavior in the microplastic region we can interrupt at appropriate stress levels, a constant displacement rate cyclic deformation experiment with the maximum stress in the microplastic range, to perform a load relaxation test. Typical data of this type of experiment are shown in Figures 3.5 and 3.6. The room temperature aluminum data in Figure 3.5 show the microplastic flow behavior at high homologous temperatures, the $\dot{\alpha}_2$-element, Equation (3-5), with $\log \sigma - \log \dot{\varepsilon}$ curve having a shape concave downward. The room temperature stainless steel data in Figure 3.6 in contrast, show the microplastic flow behavior at low homologous temperatures, the $\dot{\varepsilon}$-element, Equation (2-9) with the $\log \sigma - \log \dot{\varepsilon}$ curve having a shape concave upward.

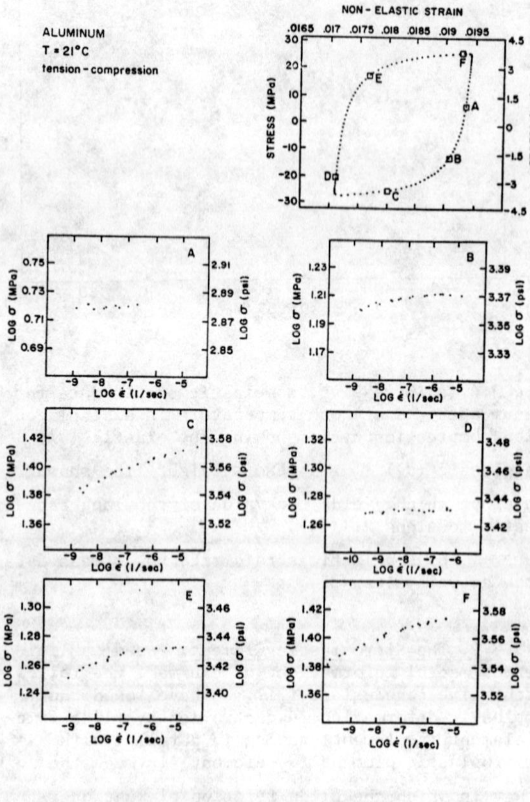

Figure 3.5. Stress-nonelastic strain data for aluminum obtained in a constant displacement rate tension-compression cycle in the microplastic range; also shown $\log \sigma - \log \dot{\varepsilon}$ data for load relaxation at different stress levels of the cycle (31).

The detailed method of data analysis for the type of data in Figures 3.5 and 3.6 are given in References 7 and 31. The data in Figure 3.5 will yield, for example, $\log \sigma_{a_2} - \log \dot{\alpha}_2$ curves for which one can determine the flow equation for the $\dot{\alpha}_2$-element, Equation (3-5), the scaling relation, Equation (3-6) and the evolution law for the microplastic hardness σ^*_2. The data in Figure 3.6 will produce $\log \sigma - \log \dot{\varepsilon}$ data at different values of σ^*_2 based on which we can determine the flow equation for the glide friction element in the microplastic region, Equation (2-9) and the associated scaling relationship.

The range of applicability and the significance of the new phenomenological model remains to be fully explored. We suspect that the cyclic softening phenomenon reported in the literature (33) may be partly related to the variations in σ^*_2 and σ_{a_2}. Similarly we suspect that the use of the lowering in the 0.2% offset yield stress as an indication of thermal recovery at low homologous temperatures may reflect the recovery of σ^*_2 and σ_{a_2} but not the change in the macroplastic hardness σ^*_1.

We have been successful in using the new phenomenological model to simulate a variety of transient deformation phenomena involving complex paths (31, 32). For example, Figure 3.7 shows stress-nonelastic strain data of an aluminum specimen with prior plastic deformation, obtained by loading at a constant displacement rate into the microplastic region, followed by a load relaxation run with the crosshead fixed and then unloading at the same constant displacement rate. Figure 3.8 shows the experimental data of a cyclic experiment at a constant displacement rate with the maximum stress in the microplastic region. The simulation results are also shown in these figures. The detailed method of simulation calculation and the value of the parameters used are given in References 31 and 32. The material parameters were determined by a combination of load relaxation and constant displacement tensile test. It is seen from these figures that the results of the simulation calculation agree well with the experimental data.

<u>3.4. Grain Boundary Anelasticity and Plastic Deformation Involving Grain Boundary Sliding.</u> A segment of grain boundary containing grain corners, grain boundary steps and second phase particles is shown schematically in Figure 3.9. At temperatures close to and above a third of the absolute melting temperature of a material, grain boundary shear is expected to become easier. If the magnitude of the grain boundary shear is small and/or the time scale is short, elastic and anelastic deformation in the grain matrix will occur at the various sites mentioned above to accommodate the grain boundary shear. If the applied load is removed, the stored elastic and anelastic strain will provide the driving force for the reversed grain boundary anelasticity. This is the concept of grain boundary anelasticity proposed by Zener and his co-workers (34,35). If the magnitude of the grain boundary shear is large and/or the time scale is long, grain matrix plastic deformation will occur at the same sites to accommodate the grain boundary shear. This part of grain boundary shear will not, however, be recoverable upon unloading. The importance of grain boundary sliding, for example, in creep has long been recognized (36).

We have been interested in the possibility of developing a unified description of the two pheno-

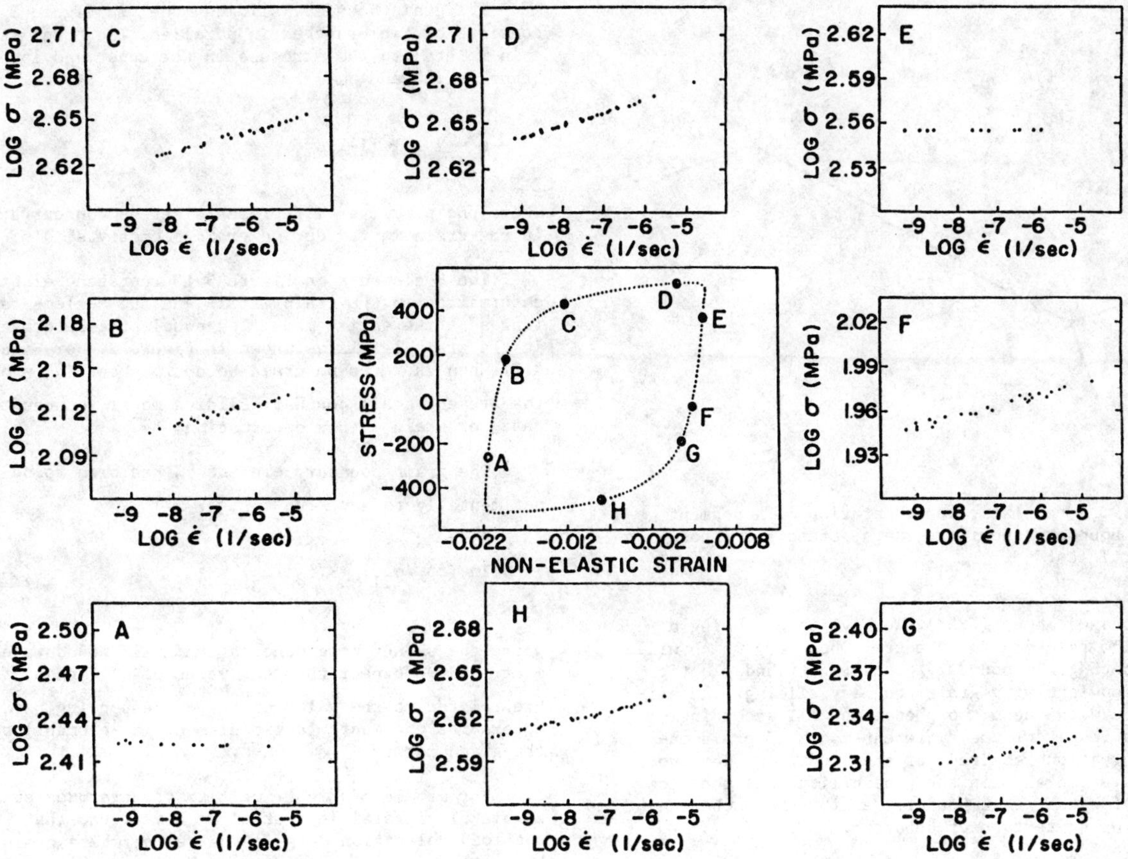

Figure 3.6. Stress-nonelastic strain data for 316 stainless steel obtained in a constant displacement rate tension-compression cycle in the microplastic range; also shown log σ - log $\dot{\varepsilon}$ data for load relaxation at different stress levels of the cycle.(31).

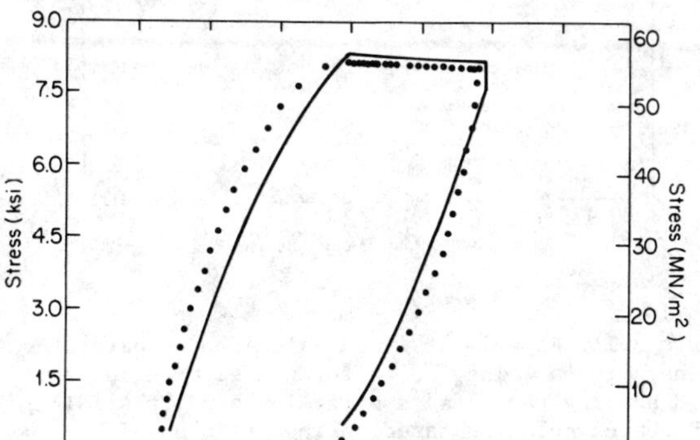

Figure 3.7. Experimental and calculated stress nonelastic strain data of aluminum obtained at room temperature by loading at a constant displacement rate, followed by a load relaxation run and unloading at the same displacement rate, with the calculation based on the new model an Figure 3.3 for the grain matrix flow; the maximum stress is below macroplastic yielding (32).

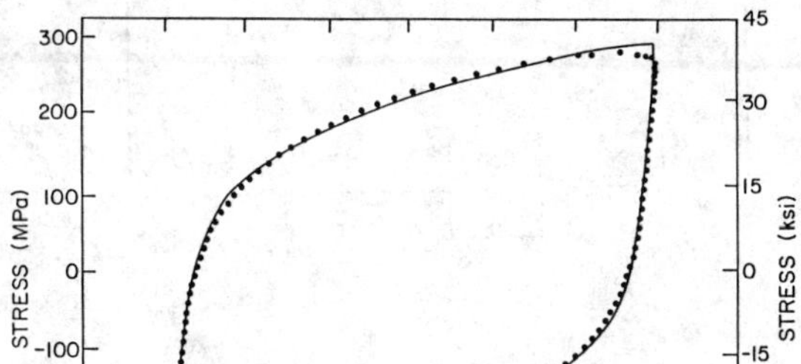

Figure 3.8. Experimental and calculated stress-nonelastic strain data of 316 stainless steel obtained at room temperature in a constant displacement rate tension-compression cycle with the calculation based on the new model in Figure 3.3 for grain matrix flow; the maximum stress is below macroplastic yielding (31).

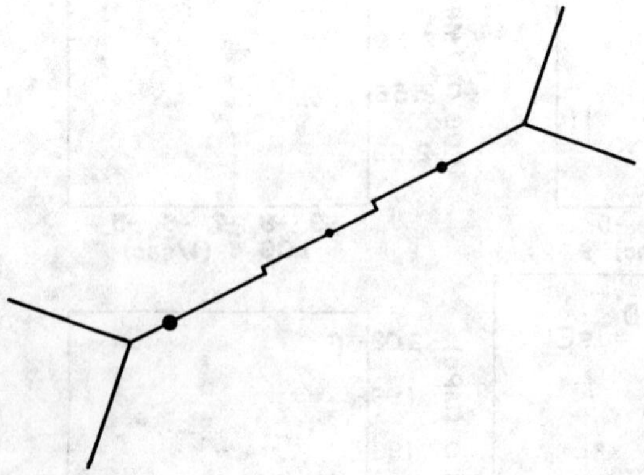

Figure 3.9. A schematic representation of a segment of grain boundary showing corners, steps and second phase particles.

mena discussed above (7,37). We have been seeking a single flow relation for the grain boundary which will control both grain boundary anelasticity and plastic deformation involving grain boundary sliding. The difference between these two phenomena will accordingly only be a result of the different modes of grain matrix deformation required for accommodation. We have been interested also in the possibility that the flow relation for the grain boundary exhibits a mechanical equation of state type of behavior.

We have had some success in this area. Figure 3.10 shows the simplest phenomenological model which

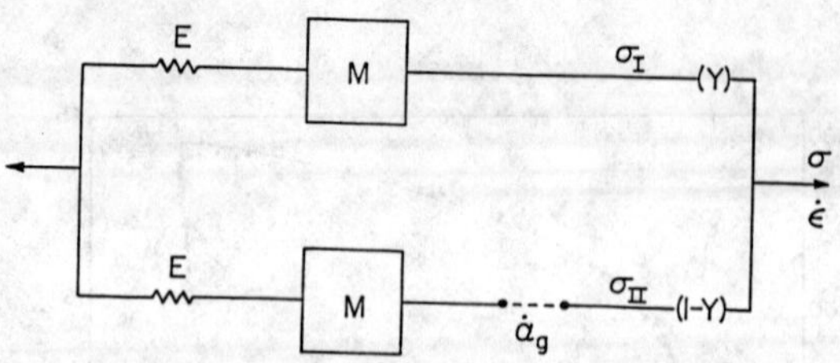

Figure 3.10. A schematic representation of a flow model for grain boundary sliding with the E-, M- and $\dot{\alpha}_g$-elements representing elastic deformation, grain matrix nonelastic deformation and grain boundary sliding respectively; the parameter Y representing stress concentration (7).

reflects the physical picture discussed above. This model is an extension of a deformation model proposed by Hart for plastic deformation involving grain boundary sliding (38). The model in Figure 3.10 represents a level of complexity for which experimental capabilities are now available to determine the required material parameters.

The grain boundary sliding model in Figure 3.10 consists of two branches in parallel. The total stress σ is related to the stresses in the upper and lower branches (σ_I and σ_{II}) by

$$\sigma = Y\sigma_I + (1-Y)\sigma_{II} \tag{3-7}$$

where the parameter Y represents stress concentration in the grain matrix during grain boundary sliding.

The E-elements in Figure 3.10 represent elastic deformation in the grain matrix and the M-elements represent the entire model for nonelastic deformation in the grain matrix as shown in Figure 2.7 or Figure 3.3. When there is no grain boundary shear ($\dot{\alpha}_g = o$), the present grain boundary sliding model reduces to that for grain matrix deformation.

The grain boundary element $\dot{\alpha}_g$ has been found experimentally to have the form (7,37)

$$\ln (\sigma^*_g/\sigma_{II}) = (\dot{\varepsilon}_g{}^*/\dot{\alpha}_g)^{\lambda g} \tag{3-8}$$

where the subscript g denotes the grain boundary element, and we expect that the value of σ^*_g and $\dot{\varepsilon}^*_g$ are related to the value of the corresponding parameters for the macroplastic element ($\dot{\alpha}_1$-element) of the grain matrix.

For metals of higher melting temperatures such as nickel and stainless steel we have found that the load relaxation data in the region where grain boundary sliding is important, the log σ - log $\dot{\varepsilon}$ curves will exhibit an S-like shape (Figure 2.4 for stainless steel and Figure 3.11 for nickel). The

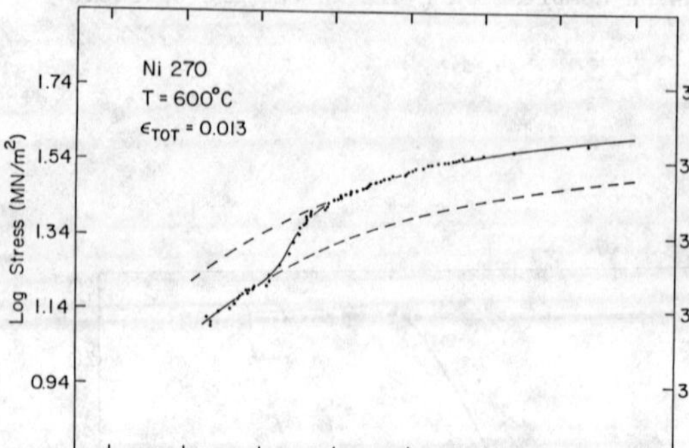

Figure 3.11. Experimental and calculated constant hardness log σ - log $\dot{\varepsilon}$ data for nickel obtained at 600°C showing the effects of grain boundary sliding with the calculation based on the grain boundary sliding model in Figure 3.10 (7).

dashed lines in these figures represent limiting grain matrix behavior, Equation (2-7), according to the present grain boundary sliding model (7,37). The stress shift between these two grain matrix curves corresponds to the value of Y based on Equation

(3-7). The limiting grain matrix behavior results from the difference between the stress dependence of the grain matrix strain rate $\dot{\alpha}_1$ and that of $\dot{\alpha}_g$. At high stresses, $\dot{\alpha}_g << \dot{\alpha}_1$ such that the grain matrix deformation is controlling. At intermediate stresses, both $\dot{\alpha}_g$ and $\dot{\alpha}_1$ are important. But at low stresses $\dot{\alpha}_g >> \dot{\alpha}_1$, the grain matrix deformation will be controlling again (38).

The value of Y can also be determined from a closed stress-total strain loop obtained by loading and unloading in a constant displacement rate tensile test at temperatures where grain boundary anelasticity is important (7,37). An example of this type of data is shown in Figure 3.12 for nickel. One can perform

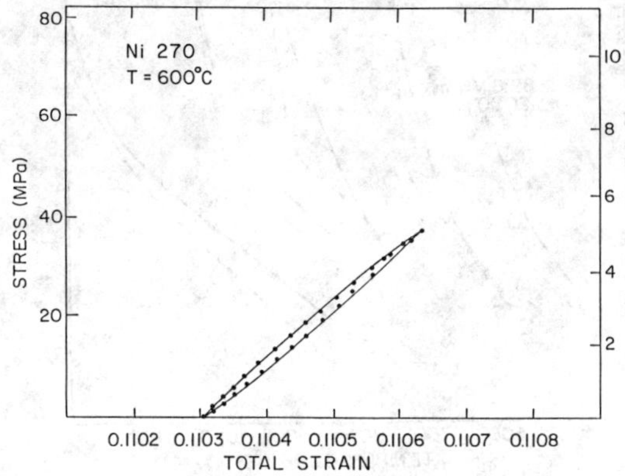

Figure 3.12. Experimental and calculated stress-total strain data for nickel obtained at 600°C in a constant displacement rate loading-unloading cycle showing effects of grain boundary anelasticity, with the calculations based on the grain boundary sliding model in Figure 3.10 (7).

a series of loop experiments at progressively higher temperatures between room temperature and 600°C to show that the high stress portion of the loading data in Figure 3.12 contains the contribution of grain boundary anelasticity (7,37). The slope of this portion of the data equals $Y/(1/E + 1/M)$ with the initial portion of the data having a slope $1/(1/E + 1/M)$ reflecting grain matrix anelasticity.

We have found that the data in Figures 3.11 and 3.12 both for nickel at 600°C yield nearly the same value of Y (0.86 ± 0.04). This result can be used to support partially the present grain boundary sliding model. The solid lines in these figures are the results of a simulation calculation based on the same model and the measured Y. In this calculation, Equation (3-8) produced a high stress exponent for $\dot{\alpha}_g$ in the stress range of the load relaxation data and a low stress exponent (nearly linear) in the stress range of the loop data. The latter type of behavior is generally believed to govern grain boundary anelasticity (35).

We have only a limited amount of experimental data which show that the grain boundary element $\dot{\alpha}_g$ may exhibit a mechanical equation of state type of behavior (scaling) (37). We are continuing our work in this direction. We are interested as mentioned previously in the relation between σ^*_g and the grain matrix hardness σ^*_1 and between $\dot{\varepsilon}^*_g$ and $\dot{\varepsilon}^*_1$ in our future work.

4. ENGINEERING APPLICATIONS

A part of the motivation for which the present state variable approach is developed has always been engineering applications. We have been seeking approaches which will offer advantages both in material characterization and in stress analysis (6,7,18,39).

For materials characterization, a combination of load relaxation and constant displacement rate test has been found to be less time consuming compared to the traditional combination of creep and tensile test. Since the form of the constitutive relations and some of the required parameters have been shown to be insensitive to thermal-mechanical history and often do not change significantly from one material to another (6,7,18), the number of tests required to characterize a new heat of material or a developmental alloy can therefore be reduced. We have found that the material sensitive parameters include only M, $\dot{a}^*$, σ^*, and $\dot{\varepsilon}^*$ in Hart's model.

For stress analysis, the deterministic nature of the present constitutive relations have been found to be well suited for various methods for numerical analysis including the finite element and the boundary element method (18, 39). The computer codes developed based on the present constitutive equations for stress analysis have been found to be efficient and accurate and can be used to describe time dependent and time independent plasticity in a unified way (6,18,39). For complex stress analysis we have used primarily constitutive equations in Hart's model with the value of the parameters adjusted to reflect for example, the effects of transient deformation and grain boundary sliding (40).

It is possible, however, to take advantage of experimental data obtained in conventional tensile and creep tests to estimate the material dependent parameters of Hart's model. For example, the 0.2% offset yield stress is a reasonable estimate of the hardness σ^*. Together with the known form of the constitutive relations, we have performed stress analysis for example, for nonelastic deformation in thin walled tubes (20% cold worked 316 stainless steel) (7,40). Figures 4.1, 4.2 and 4.3 show that the calculated results agree well with experimental data over a wide range of test conditions and loading modes. The experimental data were obtained with internally pressured thin walled tubes in a constant pressure creep mode, at a constant diametral displacement rate, and in a temperature transient mode with the pressure kept constant.

5. DISCUSSION

It is clear that from the work described in this paper that we need at least five internal state variables (σ^*_1 , σ^*_2, σ^*_g, a_1 and a_2) to characterize a mechanical state of a material. We can, of course, construct more complicated models and increase the number of internal state variables. For example, more than one hardness parameter can be used to specify the weak-barrier structure to dislocation motion (31). We shall soon be in a situation, however, where the determination of these variables will not be possible

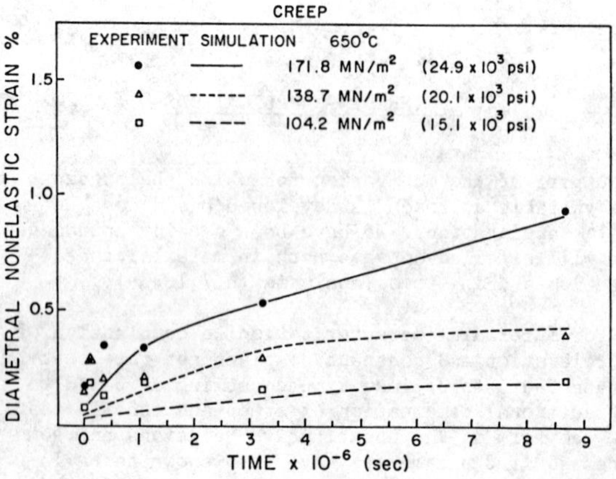

Figure 4.1. Experimental and calculated diametral nonelastic strain vs. time data for a cold worked 316 stainless steel thin walled tube obtained at 650°C and different internal pressures, with the calculation based on Hart's model (40).

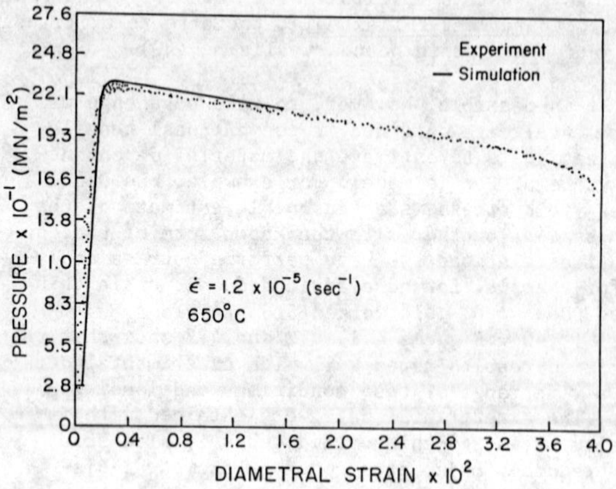

Figure 4.2. Experimental and calculated internal pressure vs. diametral strain data for a cold worked 316 stainless steel thin walled tube obtained at 650°C and a constant diametral displacement rate, with the calculation based on Hart's model (40).

with the presently available experimental capabilities. Even with the above five internal state variables we had to develop, during the course of our work, new experimental techniques.

The present state variable models contain constitutive relations which are not only of a unique

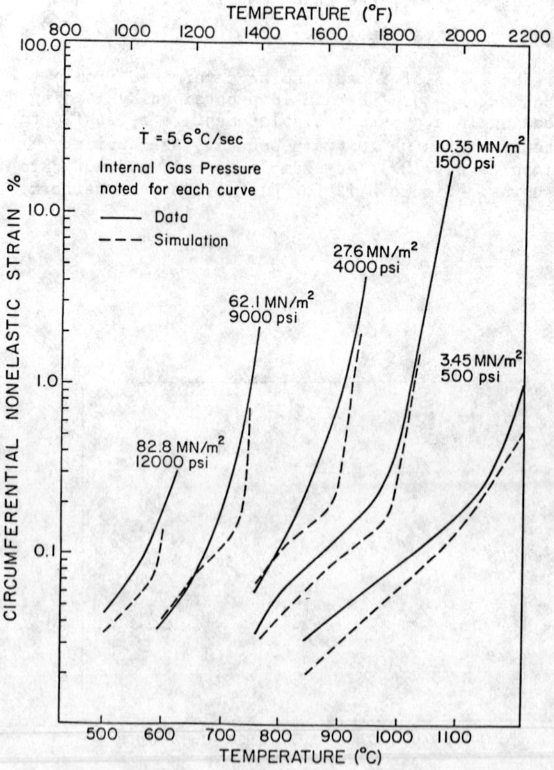

Figure 4.3. Experimental and calculated circumferential nonelastic strain vs. temperature data for a cold worked 316 stainless steel thin walled tube obtained at different internal pressures and a temperature transient rate of 5.6°C/sec. with the calculation based on Hart's model (40).

form but are also combined in a specific way. The discovery of these results together with the identification of the internal state variables required to define a mechanical state of a material can only be made through, as mentioned previously, exhaustive experimentation. A purely theoretical approach cannot possibly make the same type of progress.

The new phenomenological model (Figure 3.3) contains more physical features compared to Hart's model. It potentially can serve as a useful vehicle in our future work to develop a microstructural basis for the present state variable theory (7,31). It can be used also as a vehicle to link the present approach with the various dislocation based theories for nonelastic deformation in the literature.

The following topics should be discussed briefly.

5.1 Internal Stress. It is seen from the present state variable models for grain matrix deformation (Figures 2.7 and 3.3) that the glide friction element ($\dot{\varepsilon}$-element) is operative both in the microplastic and macroplastic regions. The form of the flow equation of the $\dot{\varepsilon}$-element, Equation 12-9 is similar to the flow equation used by J.C.M. Li and his co-workers. We can, therefore, identify mathematically σ_a or

$(\sigma_{a_1} + \sigma_{a_2})$ with the internal stress used by them. It should be noted, however, that even though the $\dot{\varepsilon}$-element is always operative, it will not always be rate limiting according to the present approach.

5.2 The Grain Matrix Structural Parameters. The grain matrix structural parameters of the present theory includes two hardnesses, σ_1^* and σ_2^* and the two anelastic moduli M_1 and M_2 which depend on the barrier structures containing immobile dislocations. The anelastic moduli depend, of course, also on the density of mobile dislocation. We have made some preliminary observation to identify these parameters with the microstructural features of a material (7, 31). Figure 5.1 shows a transmission electron micrograph of a 316 stainless steel specimen containing well defined sub-boundaries and less developed ones.

Figure 5.1. Transmission electron micrograph of plastically deformed 316 stainless steel showing fully and partially developed sub-boundaries which have been identified with the strong barriers and weak barriers for the new deformation model in Figure 3.3 (31).

We have been able to correlate the macroplastic hardness σ_1^* with the spacing between the well defined sub-boundaries. We can identify the partially developed sub-boundaries with the microplastic hardness σ_2^*. The ratio of the spacing between well defined sub-boundaries and that between partially developed ones has been found to be equal to the inverse of the ratio $M_1/(M_1 + M_2)$ (31,32). We are continuing our work in this area.

5.3 The Scaling and Flow Relations for the Plastic Elements. The scaling relations found for both the microplastic and macroplastic elements (Equations 2-8 and 3-6) suggest that the rate parameters $\dot{\varepsilon}_1^*$ and $\dot{\varepsilon}_2^*$ scale with $(\sigma_1^*/G)^{m_1}$ and $(\sigma_2^*/G)^{m_2}$ respectively rather than the time rate of the process. The hardness parameters have been shown to be related to the dimension of the slip zone (5,7,31). Some of the detailed structure of a material such as the crystal structure apparently are not important.

We have taken the position that the form of the flow equations (2-7) and (3-5) represent a basic material behavior (5,7). The theoretical significance of these equation remains to be fully developed. We shall treat effects such as thermal recovery by a time dependent change of the appropriate state variables and material parameters of the model.

5.4 The Rate Parameters. The rate parameters of the present state variable model include $\dot{a}^*$ for the glide friction element, and $\dot{\varepsilon}_1^*$, $\dot{\varepsilon}_2^*$ and $\dot{\varepsilon}_g^*$ for the macroplastic, microplastic and grain boundary elements respectively. The value of these parameters have been shown to be influenced by thermal-mechanical treatments (18). At a particular point along a thermal-mechanical path one can always perform a load relaxation test to determine the value of these parameters. For a given mechanical state of the material, the rate parameter $\dot{a}^*$ can vary also as a function of stress depending on the extent of various drag effects produced, for example, by the impurity atmosphere or dislocation jog (31). The break away from the impurity atmosphere will produce a change in the magnitude of $\dot{a}^*$ based on which the present state variable model can simulate the yield drop phenomena observed during tensile loading. Some of the effects of varying $\dot{a}^*$ are reported in Reference 41.

5.5 The Stability of the Grain Matrix Structural Parameters. The macroplastic hardness σ_1^* has been shown experimentally to be fairly stable in the absence of recrystallization at temperatures up to near one half of the absolute melting temperature of a material by using the reloading load relaxation experiment (5,7) and transmission electron microscopy observations (31). This stability is an important prerequisite for the existence of the observed mechnical equation of state type of behavior for macroplastic flow.

The stability of the microplastic hardness σ_2^* is expected to be less because it involves not fully developed dislocation structures. One can easily visualize that processes involving thermally activated dislocation annihilation will cause the recovery of σ_2^*. We have some evidence that the value of σ_2^* is not as stable as the macroplastic hardness σ_1^* (31). We believe, however, that at moderate homologous temperatures, the stability of the σ_2^* will be such especially in an engineering alloy, that a mechanical equation of state type of behavior will be exhibited by microplastic flow.

The anelastic moduli M_1 and M_2 as discussed in subsection 5.1 depend on the spacing between sub-boundaries and mobile dislocation density. Thus, we expect that any thermal and mechanical process which will cause a significant change in the substructure and the dislocation density of a material will influence the value of these moduli. Experimentally we have found that the value of these moduli remains quite stable under a range of mechanical states of the material (30,31).

5.6 Grain Boundary Sliding. Phenomena involving grain boundary sliding occur at high temperatures. The stability of the material parameters including both the structural and rate parameters of the deformation model will not be expected to be well maintained at these temperatures. It is important to ensure that experimental data obtained do not contain the effects of time dependent change of the material parameters. The reloading load relaxation experiment will provide a convenient method to detect if any change has occurred. For example, if thermal recovery of the hardness σ_1^* is important in the load relaxation experiment the data from which are shown in Figure 2.4, the lower limiting grain matrix behavior will be shifted downward such that a large value of Y (Equation 3-7) will result.

5.7 Non Homogeneous Deformation. The present state variable theory is developed primarily for homogeneous nonelastic deformation in a crystalline solid. The experiments performed so far have been concentrated on this type of deformation. Based on the results described in this paper the observation of the mechanical equation of state type of behavior and the unique flow equations can be taken as a criterion for the existence of homogeneous deformation. Any deviation can be viewed as a possible indication of the occurrence of non-homogeneous deformation. An example of load relaxation data controlled by deformation twinning is given in Reference 41. We believe that the load relaxation experiment can be used potentially as an effective tool for investigating a variety of non-homogeneous deformation phenomena.

5.8 Conclusions. The work reviewed in this paper represents the effort of a large number of investigators during the past ten years. It shows the general applicability of a state variable theory originally proposed by Hart for time dependent plastic deformation. Experimental work is emphasized in the present work. Hart's model has been extended to describe a variety of additional phenomena. It is expected that modification of the present state variable deformation models will be made if future experimental data warrant the necessary change.

ACKNOWLEDGEMENTS

The author wishes to acknowledge the contributions of his colleagues E.W. Hart, H. Mughrabi, H. Yamada, F.V. Ellis, S. Chao, F.H. Huang, C.W. Cho, C.P. Hu, M.S. Jackson, P. Alexopolous, R.L. Keusseyan, G.L. Wire, S. Mukherjee, V. Kumar, M.A. Mojaria, T. Hakkarainen and M. Korhonen in the development of the present state variable approach.

The U.S. Department of Energy through the Division of Materials Science has provided the major portion of the support for the work reviewed and the support for the present paper.

REFERENCES

1. J.H. Hollomon and J.D. Lubahn, G.E. Review, April, p. 44-50, 1947.

2. E.T. Onat, "Representation of Inelastic Mechanical Behavior by Means of State Variables", in Proc. of IUTAM Symp., East Kilbride, p. 213-225, Springer-Verlag, Berlin. 1970.

3. J.R. Rice, "Continuum Mechanics and Thermodynamics of Plasticity in Relation to Microscale Deformation Mechanisms", in "Constitutive Equations in Plasticity", ed. A.S. Argon, MIT Press, Cambridge, Mass., p. 23-80, 1975.

4. E.W. Hart, Acta Metallurgica, 18, p. 599-610, 1970.

5. E.W. Hart, C.Y. Li, H. Yamada and G.L. Wire, "Phenomenological Theory: A Guide to Constitutive Relations and Fundamental Deformation Properties", in "Constitutive Equations in Plasticity", ed. A.S. Argon, MIT Press, Cambridge, Mass., p. 149-196, 1975.

6. F.H. Huang, H. Yamada and C.Y. Li, "Constitutive Relations Based on State Variables for Nonelastic Deformation in Type 304 and 316 Stainless Steels", in "Characterization of Materials for Service at Elevated Temperatures", ed. G.V. Smith, Series MPC-7. ASME, p. 83-94, 1978.

7. P. Alexopolous, R.L. Keusseyan, G.L. Wire and C.Y. Li, "An Experimental Investigation of Nonelastic Deformation Emphasizing Transient Phenomena by Using a State Variable Approach", to be published in a STP for the Proceedings of a Symposium on "Mechanical Testing for Deformation Model Development", ASTM, Bal Harbour, Florida, Nov. 1980.

8. E.T. Onat and F. Fardishisheh, "Representation of Creep of Metals", Oak Ridge National Labs. Report TM-4783. 1972.

9. J.T. Jenkins and Y.H. Pao, "On Two Theories of the Inelastic Behavior of Metals". Materials Science Report No. 2353, Cornell University, 1974.

10. M.F. Ashby, Acta Metallurgica, 20, p. 887-897, 1972.

11. I. Gupta and J.C.M. Li, Met. Trans. 1, p. 2323-2330, 1970.

12. A.K. Miller and O.D. Sherby, Acta Metallurgica. 26, p. 289-304, 1978

13. U.F. Kocks, J. Eng. Mat. and Tech., 98, p. 78-85, 1976.

14. J.C. Swearengen and R.W. Rohde, Met. Trans, 8A, p. 577-582, 1977.

15. J.H. Gittus, J. Eng. Mat. and Tech., 98, p. 52-59, 1976.

16. D.N. Robinson, "Candidate Creep-Recovery Model for 2¼ Cr - 1Mo Steel and Its Experimental Implementation ", Oak Ridge National Labs. Report No. ORNL-TM-5110. 1975.

17. E.W. Hart, J. Eng. Mat. and Tech., 98, p. 193-202, 1976.

18. V. Kumar, S. Mukherjee, F.H. Huang and C.Y. Li, "Deformation in Type 304 Austenitic Stainless Steel", Electric Power Research Institute, Report No. EPRI NP-1276, 1979.

19. G.L. Wire, F.V. Ellis and C.Y. Li, Acta Metallurgica, 24, p. 677-685, 1976.

20. D. Lee and E.W. Hart, Met. Trans, 2, p. 1245-1248, 1971.

21. E.W. Hart and J.F. Thomas, Jr., "Materials Constitutive Relations in Modeling of Mechanical Processing", in "Process Modeling: Fundamentals and Applications to Metals", ASM, Metals Park, Ohio, p. 355-363, 1980.

22. J.F. Thomas, JR., B.S. Shabel, V. Nagpal, and H.L. Gegel, "Elevated Temperature Formability of Some Aerospace Aluminum Alloy Sheet Materials" in Proceedings of the 8th North American Manufacturing Research Conference, Society of Manufacturing Engineers, Dearborn, Michigan, p. 223-228, 1980.

23. J.F. Thomas, Jr., "Determination of Constitutive Equations for High Strength Aluminum and Titanium Alloys Applicable to Sheet Metal Formability", AFML-TR-79-4042, 1979.

24. I. Lerner, S.W. Chiang and D.L. Kohlstedt, Acta Metallurgica, 27, p. 1187-1196, 1979.

25. I. Lerner and D.L. Kohlstedt, "Effect of Gamma-Radiation on Plastic Flow of NaCl", to be published in J. Am. Ceram. Soc.

26. I. Lerner and D.L. Kohlstedt, "Load Relaxation Studies of AgCl" to be published in Acta Metallurgica.

27. S.W. Chiang, "Load Relaxation and Transmission Electron Microscopy Studies of Germanium", Ph.D. Thesis, Cornell University, 1979.

28. I. Lerner, "Inelastic Deformation in Rock Salt Type Ionic Crystals: An Equation of State Approach", Ph.D. Thesis, Cornell University, 1980.

29. F.H. Huang and C.Y. Li, "A Work Hardening Correlation for Monotonic Loading Based on State Variables", ERDA Report COO-2172-19, Cornell University, 1977.

30. P.S. Alexopolous, C.W. Cho, C.P. Hu and C.Y. Li, Acta Metallurgica, 29, p. 569-577, 1981.

31. P.S. Alexopolous, "An Experimental Investigation of Transient Deformation Based on a State Variable Approach", Ph.D. Thesis, Cornell University, 1981.

32. M.S. Jackson, C.W. Cho, P. Alexopolous and C.Y. Li, "A Phenomenological Model for Transient Deformation Based on State Variables", to be published in J. Eng. Mat. and Tech.

33. C.E. Feltner and C. Laird, Acta Metallurgica, 15, p. 1621-1632, 1967.

34. T.S. Ke, Phys. Rev., 71, p. 533-546, 1947.

35. C. Zener, "Elasticity and Anelasticity in Metals", University of Chicago Press, 1948.

36. R.N. Stevens, Met. Rev., 11, p. 129-142, 1966.

37. R.L. Keusseyan, "Deformation and Grain Boundary Sliding at Elevated Temperatures", Ph.D. Thesis, Cornell University, 1981.

38. E.W. Hart, Acta Metallurgica, 15, p. 1545-1549, 1967.

39. S. Mukherjee, "Time Dependent Inelastic Deformation of Metals by Boundary Element Methods", in "Developments in Boundary Element Methods", Vol. II, ed. P.K. Banerjee and R.P. Shaw, Applied Science Publisher, Barking, Essex, U.K. 1981.

40. G.L. Wire, D.R. Duncan, N.S. Cannon, G.D. Johnson, P. Alexopolous, S. Mukherjee and C.Y. Li, "A State Variable Analysis of Inelastic Deformation of Thin Walled Tubes", to be published in J. Eng. Mat. and Tech.

41. F.H. Huang, G.P. Sabol, S.G. McDonald and C.Y. Li, J. Nuc. Mat., 79, p. 214-226, 1979.

STRENGTHENING OF METALS AND ALLOYS

J. C. Williams
Anthony W. Thompson

Carnegie-Mellon University
Pittsburgh, PA 15213

1.0 INTRODUCTION

Strengthening of metals and alloys has been a primary means of improving the structural efficiency of many products ranging from aircraft, spacecraft and missiles to the new, fuel-efficient land vehicles. Because of their central importance to the performance of materials, strengthening mechanisms have received considerable study over the years. As a result, most of the important strengthening mechanisms are reasonably well understood unto themselves. However, most commercially important metals and alloys do not derive their strength from a single source. As a result, the most general difficulty, and therefore the most pressing need is to understand the relationship and interaction between the individual strengthening mechanisms which operate in these more complex systems. In principle, it is this which is the largest single problem which faces any fundamental approach to alloy design. In practice, of course, another equally formidable barrier exists; that is, the ability to reproducibly create strong microstructures in large-scale production circumstances. Notwithstanding this difficulty, history shows that a sufficiently attractive idea will stimulate improved production practice in order to benefit from it.

This paper is divided into two parts. In the first, the fundamentals of strengthening will be briefly summarized and the implications of these in alloy design will be reviewed. This section will be supported by references to a number of excellent review articles on the various aspects of strengthening. The reader who is interested in a more detailed account of the principles involved is accordingly referred to these articles. The second section deals with the complexities of strengthening which pertain to most modern, high strength materials. This section will most likely raise more questions in the mind of the reader than it will answer. However, the early stages of progress are often marked by the careful formulation of questions on problems which guide the focusing of intellectual resources on them. If this second section accomplishes this, it will have performed a useful function indeed.

2.0 SUMMARY OF STRENGTHENING MECHANISMS

The strengthening mechanisms that will be dealt with in this article are those which impede the conservative motion of dislocations. This then fixes the temperature and stress regimes which are important for the purpose of this article. One simple way to describe this is with the aid of a deformation map as was done in an earlier article by Copley and Williams(1). The concept of deformation maps was first described by Weertman(2) and has subsequently been refined and extended by Ashby(3). A typical map is shown in Figure 1. The hatched region is the stress-temperature regime in which the mechanisms dealt with in this chapter are effective. At temperatures above $\sim 0.5T_m$, where T_m is the melting temperature in Kelvin, and at low stresses other mechanisms operate, as shown on the map. Of these, the only other one which involves dislocation motion is dislocation creep. This mechanism requires diffusion-assisted, or non-conservative motion of dislocations. The factors which affect this and other deformation mechanisms are treated elsewhere in this volume(4).

Finally, some strengthening mechanisms contribute more strongly to the low temperature strength ($T < 0.3T_m$) of metals and alloys, whereas others contribute to strength over the entire temperature range up to $\sim 0.6T_m$. This leads to a three stage flow stress-temperature curve as illustrated in Figure 2. From this it can be seen that up to $0.6T_m$ two distinct stages exist, a temperature dependent region stage (I) and a temperature independent region (II). The reasons for the variation in temperature sensitivity of the different mechanisms is related to the extent to which dislocation motion can be aided by thermal activation. This article will not deal with thermal activation *per se*, but some of the barriers discussed are subject to thermal activation (and, therefore, show a strong temperature dependence of strengthening); these will be called short range barriers. The others are relatively insensitive to temperature and will be called long range barriers, or they may not be categorized. Thus, unless a type of barrier is described as short range, it should be understood that it tends to result in effective strengthening up to $\sim 0.6T_m$. It thus becomes important to consider the temperature dependence of the various strengthening mechanisms. If an application calls for improved strength at a temperature in region II, then care must be exercised to employ strengthening mechanisms of the long range type.

For some types of barriers, the temperature limitation to strengthening is imposed by the thermodynamics of stability. For example, strengthening due to coherent precipitates naturally vanishes at the coherent solvus where the precipitates themselves go back into solution or transform to another type of precipitate. Where possible, these temperature limitations will be pointed out, at least in a general way.

2.1 Solid Solution Strengthening

Solid solution strengthening is the increase in flow stress which results from the interaction between solutes in solid solution and moving dislocations. A number of excellent review articles have been written about this subject(5-9). Based on these articles it is clear that solute atoms and dislocations can interact by a variety of mechanisms. These include the following:

- elastic interaction
- modulus interaction
- stacking fault interaction
- electrical interaction
- short range order interaction
- long range order interaction.

Of these the first two and the last are long range. These tend to increase the strength uniformly up to $\sim 0.6T_m$. Thus the entire curve shown in Figure 2 shifts upward when these mechanisms are operative. The physical basis for each of these interactions will be described below, with emphasis on the solute or solvent characteristics which make such an interaction possible.

2.1.1 Elastic Interaction. The elastic interaction between solute atoms and dislocations arises because of the mutual interaction of the elastic stress fields which surround misfitting solutes and the core of edge dislocations. The misfit field which surrounds a substitutional solute atom which differs in atomic radius from the solvent atoms is expressed by a parameter ε. This parameter is defined in Equation 1,

$$\varepsilon = \frac{1}{2}\frac{da}{dc} \qquad [1]$$

where a is the lattice parameter and c is the concentration of solute. It has been shown that the interaction energy between the dilatational component of an edge dislocation stress field and a misfitting solute atom is linear in ε(5). Thus, the strengthening due to elastic interactions scales directly with the misfit of the solute. From this, Flinn(6) has shown that the interaction energies between edge dislocations and solutes can range from 0.01 to 1 eV. Misfitting substitutional solutes tend to have symmetrical dilatation fields surrounding them. Thus, they only can interact with the dilatation component of the dislocation stress fields. It is well-known that only edge dislocations have significant dilatational components in their stress fields(10). Thus, substitutional solutes only impede the motion of edge dislocations, whereas the motion of screw dislocations is unimpeded in the presence of such solutes. However, in the case of interstitial solutes, such as carbon in bcc iron or nitrogen in bcc niobium, the misfit field has tetragonal symmetry and, therefore, has both dilatation and shear components. In such cases, both edge and screw dislocations interact with these solutes; thus they are much more effective as solid solution strengthening additions.

Apart from the magnitude and nature of the misfit of solutes, the concentration of these solutes also affects the overall strengthening which is observed. That is to say, the misfit parameter ε or the tetragonal component of the misfit field in the case of interstitials affects the force that each solute atom exerts on moving edge or screw dislocations. However, the overall stress required to move the dislocation is also affected by the number of these barriers per unit length of dislocation and this, of course, is related to the concentration of solute. Using a simple square lattice approximation for the solute distribution, the average spacing of solute atoms is proportional to the cube root of the concentration. Fleischer has shown that the concentration dependence of hardening depends on whether the solute atoms introduce symmetrical or unsymmetrical (tetragonal) distortions in the lattice. In the case of symmetrical distortions, the increase in flow stress is linearly proportional to the concentration, c, and Fleischer has called this "gradual hardening." In the case of unsymmetrical distortions, the increase in flow stress is proportional to $c^{1/2}$ and Fleischer has called this "rapid hardening."

A problem which remains to be settled before detailed calculations of solid solution strengthening can be attempted is to devise a method for averaging the interactions between gliding dislocations and solute atoms which are not contained within the glide plane but which lie at various distances from it. Labusch(11) has treated this problem in an interesting way which invokes statistical interactions. While his treatment does not yield a numerical result which is significantly different from that of Fleischer, it does provide a basis for statistically adding the contributions of obstacles of more than one strength. This represents a substantial advance in the state-of-the-art, inasmuch as most real alloys contain more than one type of solute or impurity atom.

Finally, it is worthwhile to mention a second-order interaction which occurs between solute atoms with symmetrical distortions and screw dislocations. This was first pointed out by Fleischer(9) and is related to the minor dilatation that occurs during the motion of a screw dislocation each time the individual atoms slide up and over one another during the shearing of the lattice planes. This is most readily visualized if a rigid sphere model of the slip planes is assumed. In such cases, there is a displacement normal to the slip plane when the atoms move from one minimum energy position up and over the adjacent atom to the next minimum energy position. This introduces an oscillatory dilatation field into the stress field of screw dislocations also. Although this is a much smaller effect than was described earlier for the case of edge dislocations, this effect can permit screw dislocations to also interact, to a limited extent, with symmetrically misfitting solutes. This effect is less significant than the principal elastic interaction described for edge dislocations, but it still deserves mention because it is more subtle and is often overlooked.

2.1.2 Modulus Interactions. The interaction between solute atoms and dislocations can also occur if the presence of the solute atom locally alters the modulus of the crystal. The basis for such modulus alterations must have its origins in electronic effects, although such calculations have not been done in detail. By analogy to the misfit parameter ε, the operative parameter for modulus interaction is δ, as shown in Equation 2,

$$\delta = \frac{1}{G}\frac{dG}{dc} \qquad [2]$$

where G is the shear modulus and c is the solute concentration. The modulus interaction is similar to the elastic interaction described in the previous section but because a local change in shear modulus is accompanied by a local change in bulk modulus, as shown by Equation 3,

$$K = \frac{2G(1+\nu)}{3(1-2\nu)} \qquad [3]$$

then both edge and screw dislocations will be subject to first-order modulus interactions. In Equation 3, K is the bulk modulus, G is the shear modulus and ν is Poisson's ratio. As in the case of elastic interaction, the interaction strength for edges and screws differs by a factor of $1/1-\nu$ as pointed out by Fleischer. The modulus interaction between dislocations and solute atoms is also a long range interaction and thus Labusch-type statistics should be helpful in calculating strengthening due to these interactions as well.

2.1.3 Stacking Fault Interactions. Solute atoms which preferentially segregate to the stacking faults contained between the pairs of partials in extended dislocations result in a reduction of stacking fault energy. Thus, as the concentration of solute within the fault increases, the partial dislocations separate further. Consequently, the continued motion of these extended dislocations which contain enrichments in solute concentration is more difficult because as they move the stacking fault is mechanically separated from the solute rich region and the external load must do work in reconstricting the pair of partial dislocations. This effect, which was first pointed out by Suzuki(12), is a short range interaction inasmuch as the solute atoms must move to the stacking fault in order to interact with it.

Cottrell(5) has suggested that the prime requirement for the interaction between solutes and stacking faults is that the solute must have preferential solubility in a hexagonal structure. This is because stacking faults in fcc lattices can be viewed as a thin layer of hcp material. Cottrell uses this reasoning to develop a thermodynamic argument for the stacking fault interaction between solute atoms and dislocations. He points out that free energy-composition curves can be calculated

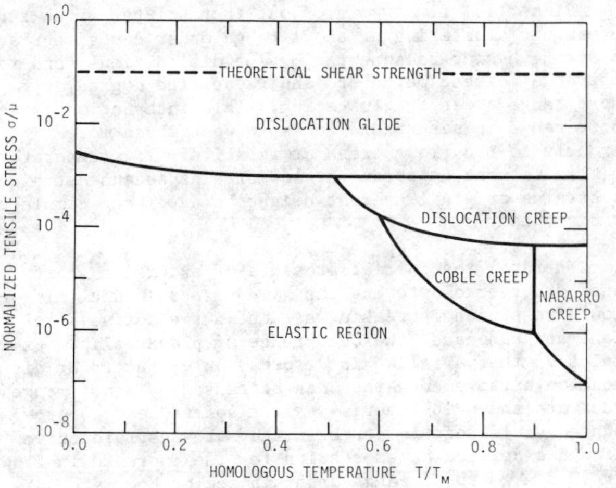

Figure 1. Schematic deformation map.

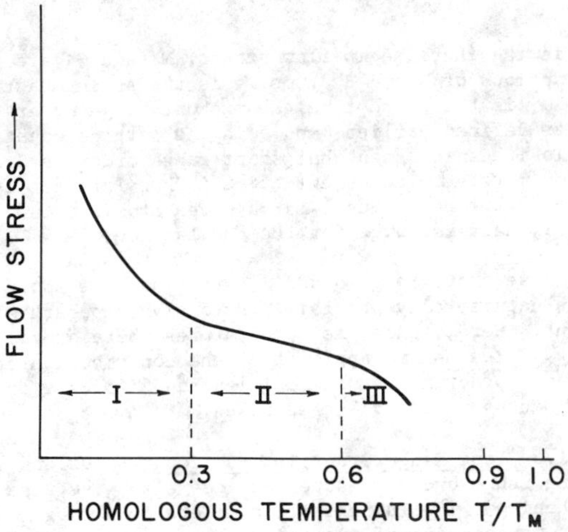

Figure 2. Schematic illustration showing variation in flow stress with temperature.

Figure 3. Bright field TEM micrograph showing stacking faults in an fcc matrix. Note variations in contrast due to overlapping faults.

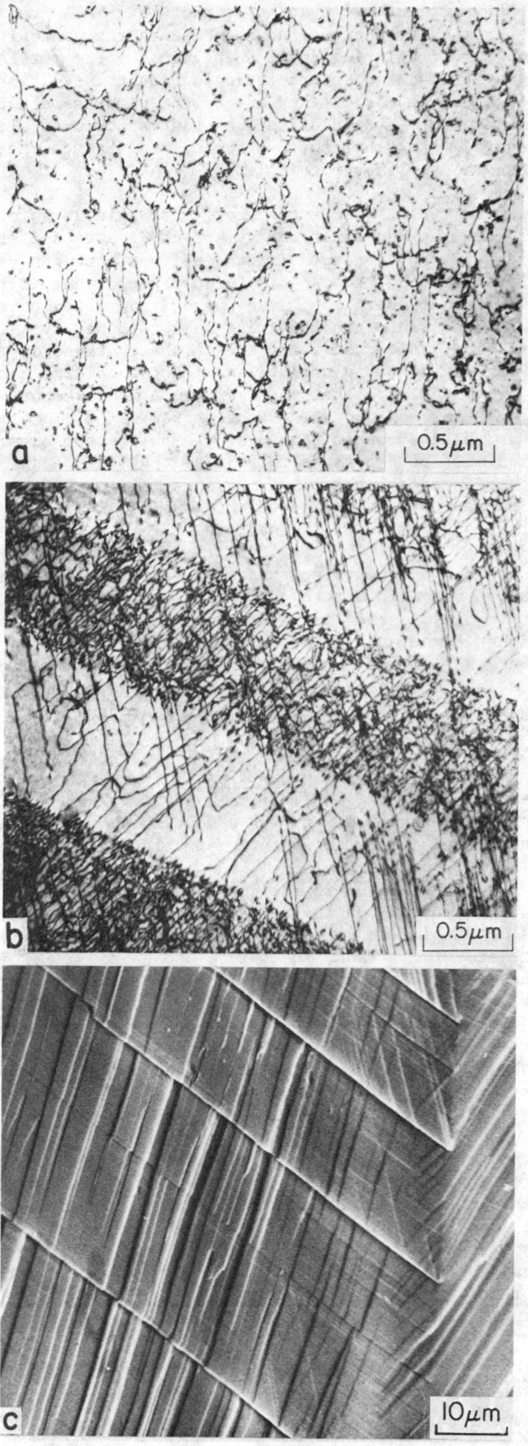

Figure 4. TEM micrographs and SEM surface slip line micrographs showing the effects of short range order on the deformation mode of Ti-Al alloys. (Photos courtesy of Dr. Hak-Min Kim).

a) Ti-4%Al (0.07% oxygen) showing wavy slip.
b) Ti-8%Al (0.07% oxygen) showing planar slip.
c) Surface slip lines in Ti-8%Al alloy showing intense planar slip lines.

for the matrix atoms in an hcp structure. This free energy-composition curve then represents the composition dependence of stacking fault energy in the fcc lattice. If the matrix curve and the stacking fault curve have a different shape, or if they are tilted with respect to one another, then Cottrell argues that stacking fault segregation and, therefore, stacking fault strengthening becomes a strong possibility.

Experimentally, one can argue that solutes which lower the stacking fault energy are candidates for stacking fault strengthening. The easiest and most generally available evidence for solute-induced reductions in stacking fault energies is the composition dependence of the stacking fault probabilities as determined by X-ray diffraction. Alternatively, direct observations of stacking fault energy variations with composition can be made by TEM using either node curvature measurements or direct measurement of the widths of partial dislocations such as those shown in Figure 3. In the latter case, caution must be exercised to be sure that the widths being measured are equilibrium widths, i.e., that no internal stresses are present which would alter the apparent separation of the partials.

As mentioned earlier, stacking fault interactions are short range in nature and, therefore, contribute more strongly to the flow stress at lower temperatures. At higher temperatures, stacking fault interactions become more complex in nature because the solute atoms become mobile and the effect of solute drag is difficult to separate from the direct stacking fault interaction effects.

2.1.4 Electrical Interactions. The evidence for electrical interactions first came from work of Dorn(13) and Allen(14) and their co-workers which suggested that the hardening due to substitutional solutes was dependent in part on their valence. Subsequent calculations have shown that, even in lattices of high electrical conductivity, some of the charge associated with solute atoms of dissimilar valence remains localized around the solute atom ion. Thus, the solute atoms become charge centers in the lattice and have the opportunity to interact with dislocations if such dislocations have electrical dipoles associated with them. Cottrell, Hunter and Nabarro(15) have provided a physical basis for an electrical dipole at edge dislocations and have estimated the strength of this dipole. On this basis, they showed that solute atoms of dissimilar valence can interact with edge dislocations, but that the strength of this interaction is small compared to elastic or modulus interactions. For example, in copper alloys, Cottrell(5) suggests that the elastic interaction between misfitting solutes and edge dislocations is from three to seven times larger than the corresponding electrical interaction. Nevertheless, electrical interactions do occur even though they are short range in nature and relatively weak. In some cases, where the solute and solvent differ in valence by two or three, these interactions can be important, especially if the solute has a relatively small misfit.

2.1.5 Short Range Order Interaction. Short range order is defined as the tendency for solute atoms to arrange themselves so that they have more than the equilibrium number of dissimilar neighbors. The opposite effect is usually called clustering and, in this case, the solute atoms of like kind tend to group together in regions of the lattice. This tendency can be quantified by a short range order parameter, α, as defined in Equation 4,

$$\alpha_1 = \frac{P_{AB}}{M_A} \qquad [4]$$

where P_{AB} is the probability of finding a particular A atom as a nearest neighbor to a B atom and M_A is the mole fraction of A.

Fisher(16) first pointed out that solutes which tend to exhibit clustering or short range order can give rise to strengthening because the moving dislocations tend to rearrange nearest neighbors and reduce the degree of short range order or clustering. This interaction is again short range in nature and tends to contribute more strongly to the flow stress at relatively low temperatures. It also is most important at low strains because at higher strains original nearest neighbors have been completely disrupted.

In fact, the initial presence of extensive short range order or clustering can give rise to a declining flow stress with increasing strain in the active slip plane and can cause the occurrence of planar slip. Examples of the variation in deformation character or dislocation arrangements that can be caused by short range order are shown in Figure 4. It also can lead to pronounced yield point effects in the stress-strain curve. Flinn has derived an expression for the increase in flow stress due to short range order in fcc or bcc materials. This expression is shown in Equation 5.

$$\Delta\tau = (16/\sqrt{6})\ (M_A M_B v \alpha_1 / a^3) \qquad [5]$$

Here $\Delta\tau$ is the increase in flow stress, M_A and M_B are the mole fractions of A and B atoms, v is the AB interaction energy, α_1 is the nearest neighbor short range order parameter defined earlier, and a is the lattice constant. From this it can be seen that short range order hardening is approximately twice as effective in bcc lattices as it is in fcc lattices because, for a given atomic volume, a^3 is only half as large for the bcc lattice.

Because short range order interactions are short range in nature, they are fairly strongly temperature dependent. Here, again, as temperatures where atomic mobility increases are approached, the contribution from short range ordering becomes complex and difficult to analyze because α is no longer fixed.

2.1.6 Long Range Order Interaction. Solutions which form long range order or superlattices also give rise to strengthening. In a superlattice there is a long range, periodic arrangement of dissimilar atoms, the details of which depend on the particular superlattice and its composition. For example, the periodicity of unlike nearest neighbors is different in AB superlattices than it is in A_3B superlattices. However, the principles of hardening are the same. That is, unit dislocations disrupt the long range periodicity by creating like nearest neighbors across the active slip plane. This effect was first analyzed by Ardley and Cottrell(17). They pointed out that, if the dislocations move in groups such that the translation associated with the passage of the entire group equals the periodicity of the superlattice, then the long range order is restored across the slip plane after the dislocation group has passed. The most common group of dislocations is two because many superlattices have alternating AB atoms along the close packed or slip directions. Thus, when two unit dislocations pass through the lattice along this direction they will restore the unlike nearest neighbors across the slip plane.

The regions of disorder introduced across slip planes by moving dislocations are called anti-phase boundaries because the atoms across the slip plane have become "out of phase" with respect to the energetically preferred superlattice structure. In groups of dislocations, the leading and trailing ones bound the anti-phase boundary which can be viewed as the superlattice equivalent of a stacking fault. In other more complex structures, the dislocations in the groups may also dissociate into partials so that the anti-phase boundary may also have a stacking fault component. The detailed deformation geometry of different superlattices was examined by Marcinkowski(18). Subsequently, two detailed review articles have been written by Stoloff and Davies(19) and

Stoloff(20). The reader who is interested in details of deformation of any particular superlattice should refer to these works.

The ordering of most superlattices does not persist to the melting temperature. Thus, the ordering reaction occurs in the solid state by a short range diffusion of atoms. Since, as described by Marcinkowski (18), there are two or more equivalent lattice sites in the superlattice, ordering may start at different parts of the crystal and form the same superlattice, but with variations in the manner in which A and B atoms occupy these equivalent sites. At the points where these regions impinge, the lattice is again out-of-phase. The locus of these points is a two-dimensional region known as a thermal anti-phase boundary (APB) because it originates from thermal effects rather than deformation. The details of thermal APB's are different in various superlattices as has been carefully described elsewhere(18). Examples of thermal APB's are shown in Figure 5.

Flinn has suggested that the primary strengthening contribution due to long range order is the intersection of gliding dislocations and thermal APB's. This typically leads to an increase in the APB area as deformation proceeds and leads to increasing flow stress. It is this deformation-induced increase in APB area which causes ordered alloys to work harden more rapidly than their disordered equivalents. In fact, the amount of new APB formed during slip varies inversely with the APB spacing as shown in Equation 6 which is according to Flinn(6).

$$\tau = \frac{\gamma_{apb}}{d} \qquad [6]$$

Here τ is the flow stress, γ_{apb} is the APB energy and d is the spacing of the APB's. Comparable effects occur at grain boundaries so that both APB size and grain size become important in the work hardening rate of ordered alloys. In this regard, APB's can agglomerate at elevated but relatively modest ($< 0.5T_m$) temperatures leading to an increase in d and a decrease in strength. An example of APB agglomeration is shown in Figure 6. This effect leads to a time dependent decrease in strength during elevated temperature exposure. Such a decrease can be a significant deterrent to the use of monolithic or single phase ordered alloys at elevated temperatures. One possible way to suppress APB agglomeration is to use slightly off-stoichiometric alloys so that the APB regions contain an excess of one atomic species. When this is the case, APB agglomeration requires long range diffusion and thus it occurs more slowly. When APB's contain an excess of one atomic species an anomalous fringe contrast is seen at APB's in the TEM when the image is formed with a fundamental reflection. This was first discussed by Blackburn(21) and an example of such contrast is shown in Figure 7.

2.2 Particle Strengthening

Particle strengthening refers to the increase in strength which occurs when small, second phase particles are present. The nature of these particles can be very simple such as GP zones, which are solute rich regions with the same structure as the matrix, or very complicated such as intermetallic compounds or complex carbides. The particles can also be intrinsic to the system and form as precipitates or they can be introduced from external sources by mechanical alloying. Irrespective of their origin, the particles impede the motion of dislocations in an analogous manner to that described previously for solute atoms. The principal difference is one of scale, because even very small particles are obviously large compared to individual solute atoms. Thus the interaction between dislocations and particles is usually long range in nature. Some exceptions to this exist and these will be mentioned where appropriate.

As in the case of solid solution strengthening, particle strengthening has been the subject of several excellent reviews(22-24). The most comprehensive of these is the one by Brown and Ham(23). In this article the basic mechanisms of particle strengthening will be briefly reviewed with emphasis on the characteristics of particles which lead to strengthening by these various mechanisms.

Dislocation-particle interactions can be divided into two types. In the first type, the dislocations do not enter the particle, but bypass them leaving dislocation loops around the particles. This mechanism was first suggested by Orowan(25) and is generally called the Orowan mechanism. The dislocation loops which surround the particles are also frequently called Orowan loops. In the second type of interaction the dislocations shear or pass through the particles.

It is important to understand that the detailed nature of the particles does not affect the observed strengthening in the bypass case, provided that the particles are strong enough to withstand the local stress that the dislocation exerts on them without shearing or fracturing. In such a case, the strengthening is only controlled by the spacing of the particles and the flexibility of the glide dislocations. Orowan has derived an expression for the stress to bow the dislocations between two particles as is shown in Equation 7.

$$\tau = \frac{G\underline{b}}{\ell} \qquad [7]$$

Here τ is the shear stress, G is the shear modulus, ℓ is the spacing of the particles and $\bar{b}$ is the Burgers vector. The types of small particles which are bypassed rather than sheared are those which have high intrinsic resistance to shearing such as thoria or carbides or coherent particles which have high misfits. Examples of each of these are shown in Figure 8(a) and (b). Large particles are also more resistant to shearing than small ones because the maximum local stress at the particle-matrix interface is inversely related to the particle radius as shown by Equation 8.

$$\tau_{max} = \frac{G\underline{b}}{2r} \qquad [8]$$

Here τ, G and $\underline{b}$ have the same meaning as before and r is the radius of a spherical precipitate. However, in cases where the precipitate volume fraction is fixed, large particles also lead to large particle spacings and thus do not give high strength as can be seen from Equation 6.

The remainder of this section will be concerned with the manner in which particles that are sheared impede the motion of dislocations through them. There are basically six properties of particles which affect the ease with which they can be sheared. These are the following:

- coherency strains
- stacking fault energy
- ordered structure
- modulus
- lattice friction stress
- interfacial energy and morphology.

Each of these will be discussed below and the best estimate of the flow stress will be given in a way similar to the article of Kelly(24).

2.2.1 Strengthening by Coherency Strains. Coherency strains arise from misfitting precipitates which distort the matrix and cause a dilatational stress field in the region of the precipitate(Fig.9). As in the case of misfitting

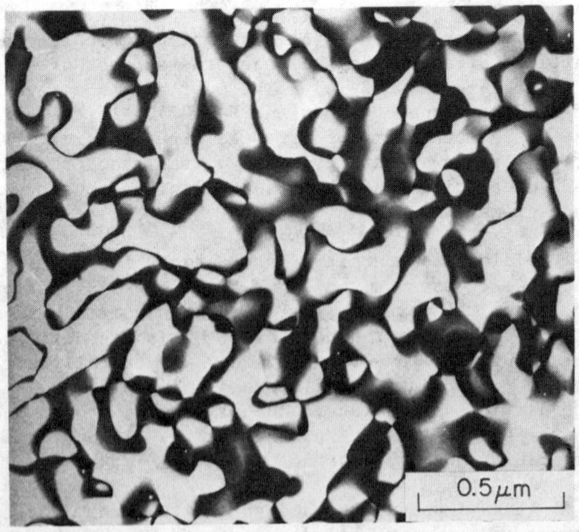

Figure 5. Dark field TEM micrograph showing thermal APB's in Ti_3Al (DO_{19} structure) after quenching from 1200°C.

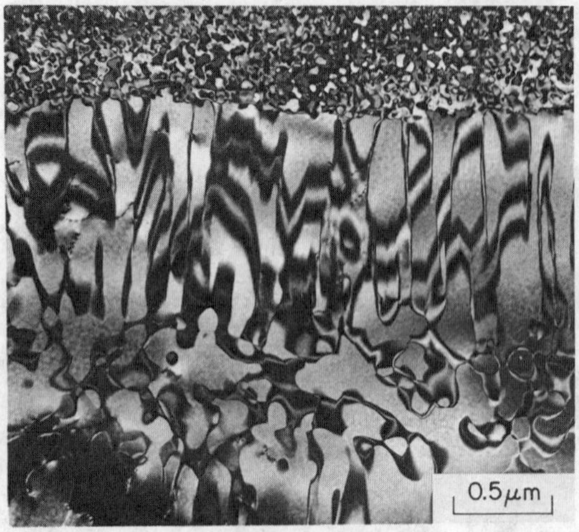

Figure 6. Dark field TEM micrograph showing coalescence or coarsening of thermal APB's during subsequent aging of sample shown in Fig. 5.

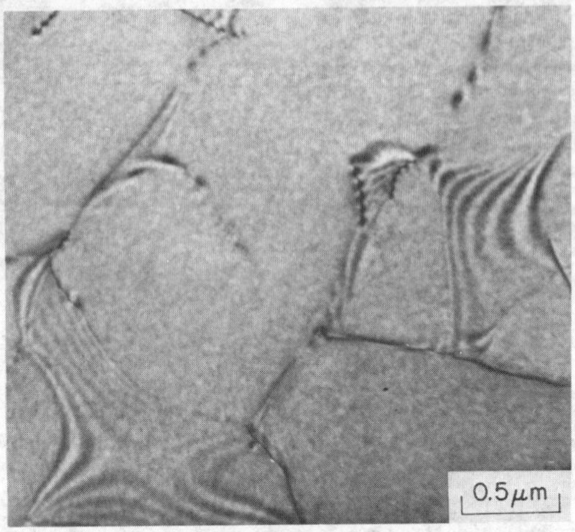

Figure 7. Bright field TEM micrograph showing anomalous fringe contrast at APB's in substoichiometric DO_{19} structure alloy Ti-15at%Al-5at%Sn.

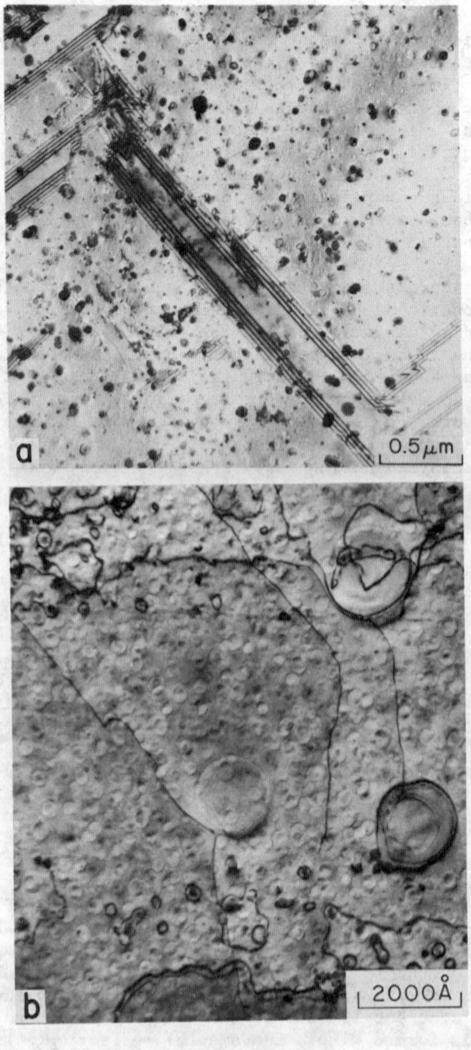

Figure 8. Bright field TEM micrographs showing two types of particles.
a) ThO_2 in TD-Ni-20%Cr.
b) Bimodal distribution of γ' in Waspaloy. Note Orowan loops around small coherent γ' particles.

solute atoms, this dilatational field can interact with the dilatational components of the dislocation stress field and impede their motion. While it is this interaction that is intuitively most obvious of all dislocation-particle interactions, Brown and Ham(23) have suggested that the evidence for this strengthening mechanism making an important contribution to strength is limited to only a few alloy systems. Nevertheless, the strengthening due to coherency strains has been analyzed by Gerold and Haberkorn(26) and by Gleiter(27). Each of these analyses lead to a fairly similar result and differs only in detail. The flow stress formulae are compared in Equations 9a and b.

$$\tau = \beta\, G\varepsilon^{3/2}\left(\frac{r}{b}\right)^{1/2} f^{1/2} \quad \text{(Ref. 26)} \qquad [9a]$$

and

$$\tau = 11.8\, G\varepsilon^{3/2}(r/b)^{1/2} f^{5/6} \qquad [9b]$$

Where again τ is the flow stress, G the shear modulus, ε the precipitate misfit, r the precipitate radius, $\underline{b}$ the Burgers vector and f the precipitate volume fraction. From the standpoint of strengthening due to coherency strains, it is important to notice that both of these formulae show that the flow stress varies as the three halves power of the misfit of the coherent precipitates. This intuitively seems to be the right type of dependence because the increasing misfit causes the dislocations to bend through larger angles as they interact with the individual precipitates. This, in turn, causes the dislocations to sample larger numbers of precipitates per unit length, which leads to an effective decrease in spacing. This point has been treated in detail by Brown and Ham(23).

2.2.2 Stacking Fault Strengthening. In the case of precipitates which have stacking fault energies which are significantly different than that of the matrix, the interaction between the dislocations and the precipitates can be dominated by the local variation in the fault width when the glide dislocations enter and are contained within the particles. Hirsch and Kelly(28) have analyzed this type of strengthening and have shown that the variation in flow stress is linearly proportional to the difference in stacking fault energy between the precipitate and the matrix. Their expression for strengthening due to dislocation-particle stacking fault interactions is shown in Equation 10,

$$\tau = C_1\left(\frac{\gamma_2\ \gamma_1}{b}\right)\left\{\frac{3k(\alpha)\,\ln(\gamma_2/\gamma_1)}{E}\right\} F_1 f^{2/3} \qquad [10]$$

Where C_1 is a numerical constant, γ_1 and γ_2 are stacking fault energies of precipitate and matrix, $k(\alpha)$ is the partial dislocation force times the separation distance, F_1 is a complex function of stacking fault width and precipitate size and f is the precipitate volume fraction. The restriction on this strengthening mechanism is first, that the precipitate must have a structure which gives rise to extended dislocations, i.e. either fcc or hcp, and second, the precipitate must have a stacking fault energy which is significantly different from the matrix. In such cases, this type of strengthening can contribute significantly to the flow stress but because of the restrictions mentioned above it does not apply to all systems. Thus it should be considered as a relatively limited type of interaction.

2.2.3 Strengthening Due to Ordered Particles. This type of strengthening provides the basis for the impressive strengths (especially at high temperatures) of the γ' strengthened nickel-base superalloys. If the particles have an ordered structure, then the shearing of them by glide dislocations introduces APB's as discussed earlier in connection with deformation of ordered alloys. The strengthening contribution due to ordered precipitates is proportional to the three halves power of the APB energy as has been suggested by Ham(29). His expression for the increase in strength due to ordered particles is shown in Equation 11.

$$\tau = \left(\frac{2}{\pi E}\right)^{1/2}\left(\frac{\gamma_{apb}}{b}\right)^{3/2} r^{1/2} f^{1/2} \qquad [11]$$

Here E is Young's modulus, γ_{apb} is the anti-phase boundary energy, $\underline{b}$ is the Burgers vector, r is the particle radius and f is the precipitate volume fraction.

Strengthening by ordered particles is relatively complex and the detailed model depends in part on the size and spacing of the precipitates as has been discussed in detail by Brown and Ham. Such a detailed account is outside the scope of this article. However, Ham's general formulation shown above expresses the role of the important parameters such as γ_{apb}, r and f on the strengthening. When ordered particles are present, the dislocations often move in pairs (or more complicated groups) because the second dislocation restores the order in the precipitate. Examples of such dislocation pairs are shown in Figure 9.

2.2.4 Modulus Strengthening. Because the energy of a dislocation depends linearly on the local modulus, particles which have a modulus which differs significantly from the matrix will interact with the dislocation by locally raising or lowering its energy as it passes through them. The limiting case of modulus interaction arises in the case of voids which are formed as a result of fast neutron damage. An example of these voids is shown in Figure 11 from which it can be seen that the distribution of voids is relatively uniform. Such voids represent attractive obstacles to glide dislocations because the energy of the dislocation disappears when it is "inside" the void but the dislocation and its intrinsic energy must be re-created by the external stress when it emerges. As a result, these voids represent obstacles or barriers to dislocation motion. This strengthening contribution due to modulus differences between precipitate and matrix has been studied by Kelly and Nicholson(22) and their expression for this strengthening is shown in Equation 12.

$$\tau = \frac{|\Delta G|}{4\pi^2}\left[\frac{3|\Delta G|}{Gb}\right]^{1/2}\left[0.8-0.143\ln\left(\frac{r}{b}\right)\right]^{3/2} r^{1/2} f^{1/2} \qquad [12]$$

Here G is the shear modulus, r is the particle radius, f is the volume fraction and $\underline{b}$ is the Burgers vector. For the purposes of this article it is most important to point out that the hardening is linearly proportional to the modulus difference (ΔG) between the matrix and precipitate. Since most precipitates do not have a dramatically different character from the matrix, modulus hardening usually does not contribute significantly to particle hardening. The case of voids, illustrated above, represents an exception as has been noted also by Brown and Ham.

2.2.5 Strengthening Due to Particle Friction Stress. The most convincing example of this particle strengthening mechanism comes from the work of Meiklejohn and Skoda(30) who studied mercury amalgam which contained fine iron particles. In this unusual system, the flow stress of the particles is very much higher than that of the matrix due to the large differences in homologous temperature. Brown and Ham have analyzed Meiklejohn and Skoda's data in detail and have made a convincing case for the flow stress in these alloys being controlled by the friction stress in the iron particles themselves. This strengthening mechanism has received relatively little visibility although it is typically overlooked (c.f. Kelly(24). Indeed, Meiklejohn and Skoda did not use this interaction mechanism to explain their own data. It is nevertheless interesting because it is conceivable that it may also operate in other systems. It is also useful to point out that the temperature dependence of this type of particle hardening should be that of the flow stress of the particles. For example, in the present case of Fe particles a strong temperature dependence was seen at low temperatures.

Brown and Ham comment that the Meiklejohn and Skoda data can be fitted to a number of strengthening theories, but that the friction stress model which they propose seems to fit the data best. Thus, although most review articles do not consider this dislocation-particle interaction mechanism, it should be always borne in mind as a possibility when analyzing particle hardening data.

2.2.6 Strengthening Due to Chemical Or Surface Effects. When particles are sheared the glide dislocation step which is one Burgers vector high is created at the particle-matrix interface. If the surface-to-volume ratio of the particles is relatively high, then this increase in surface area can be significant and can represent a substantial increase in the overall energy of the system. Such energy must be supplied by the external stress and, therefore, is reflected in an increase in strength. This type of strengthening is clearly most important in thin plate-shaped precipitates since they have the maximum surface-to-volume ratio. Examples of such precipitates would be GP zones or Θ" precipitates in aluminum-copper alloys (Fig. 12) or GP zones in copper-beryllium alloys. Kelly and Nicholson(22) have analyzed this strengthening mechanism and have shown that the strengthening is linearly proportional to the interfacial energy γ_s. Their formulation is shown in Equation 13,

$$\tau = \frac{\sqrt{6}}{\pi} \gamma_s \, (f/r) \qquad [13]$$

where γ_s is the precipitate-matrix surface energy, f is the volume fraction and r is the radius. One difficulty associated with this analysis is that the most effective strengthening should be obtained from thin plates, yet the Kelly and Nicholson result pertains to spheres. Brown and Ham suggest that the maximum strength should correspond to the finest dispersion and should be strongly temperature dependent. This seems quite reasonable since the interaction distance for this mechanism is $\sim |\underline{b}|$ and thus is indeed short range.

The theoretical framework for analyzing the strength increases which result from a particular mechanism of dislocation-particle interactions has now been established in reasonable detail. While there is some disagreement which remains, this is a matter of degree. A much more serious question revolves about the question of how one adds together the strengthening contributions which arise from the simultaneous operation of two or more of these interaction mechanisms. For example, it is not clear how the relative strengthening contributions of coherency strains and of ordered particles can be analyzed. Yet this case is of substantial practical importance in nickel-base superalloys which contain coherent, misfitting, ordered particles. In the current formulation, Labusch statistics do not seem to be adequate for dealing with strengthening due to particles of varying strengths even though some success has been achieved using this approach for solutes with varying interaction strengths. Thus, it presently remains as a major challenge to devise a suitable method for adding such contributions.

Possibly one problem is the averaging of the effects of different interactions all of which occur on the same size scale, say particle hardening. Simple superpositioning of the contributions from particles and solute or particles and grain boundaries seems to work reasonably well(30-32). Such superpositioning is only viable if the two mechanisms operate independently. For example, solid solution strengthening additions which promote planar slip can also induce a particle-bypassing to shearing transition and thereby completely alter the dislocation-particle interaction. Similar effects can be obtained if coherent precipitates are formed between non-deforming precipitates. The effect of the interactions between strengthening mechanisms and the influence of such interactions is not well understood. Possibly a useful initial way to think about this problem is in terms of deformation mode or slip character. This is discussed briefly in the next section.

2.3 Slip Character

The solid solution and particle strengthening mechanisms described above can have a direct influence on the way the material deforms. A system for classifying the deformation mode or slip character of materials has been described elsewhere by Williams, Thompson and Baggerly(33). In this description a distinction has been made between planar and wavy slip and between coarse and fine slip. These distinctions lead to four distinct classes of slip character, e.g. coarse, planar slip, etc. The presence of either or both solute atoms and particles can lead to a slip mode transition compared to the pure matrix. Thus their presence can have important implications with respect to the properties. For example, the effect of slip mode on (or correlation with) properties is well-known (but possibly not-so-well understood) in the case of tensile ductility, stress corrosion cracking susceptibility, fatigue crack initiation, fatigue growth rate and work hardening. A detailed account of how strengthening and deformation mode are related to these and other properties lies outside the scope of this article, but a few examples seem appropriate.

Take for example the case of solid solution hardening: solutes which reduce the stacking fault energy of the matrix also lead to more planar slip because lower stacking fault energy materials exhibit a reduced propensity for cross slip. In this case the active slip planes tend to harden rapidly so that the slip is planar but the slip lines are closely spaced, leading to fine planar slip. Solutes which give rise to short range order also give rise to planar slip, but in these cases the slip is confined to fewer, more intense bands (coarse planar slip). This was illustrated earlier in Figure 4. As a general rule, solid solution additions tend to promote either fine or coarse planar slip, although solutes which increase the stacking fault energy can promote wavy slip by enhancing cross-slip.

Particle hardening can lead to either a planar-to-wavy or wavy-to-planar slip mode transition depending on the nature of the dislocation-particle interactions. For example, particles which are sheared tend to cause the flow stress in the active slip plane to decline with increasing strain, at least up to a certain point. This leads to a "positive feedback" of strain in the operating slip band. Thus it is easier for slip to continue in this active plane than it is for slip to initiate on new planes. Such dislocation-particle interactions clearly lead to planar slip which is typically coarse. Extreme examples are shown in Figure 13. Here the large offsets of the initial slip lines can be seen along the orthogonal coarse planar slip bands. TEM shows that the coherent precipitates are completely destroyed within the slip bands. Particles which are bypassed by dislocations on the other hand can lead to more wavy slip. The reason for this is that the Orowan mechanism tends to cause rapid accumulation of dislocation debris in the active slip planes as was first suggested by Fisher, Hart and Pry(34). This debris rapidly hardens these planes so that subsequent dislocations bypass the particles by double cross-slip or initiation of slip on new planes becomes favorable at relatively low strains.

Based on the above discussion it is clear that one must consider the effect of various strengthening mechanisms on deformation mode when selecting these mechanisms to improve the strength of a particular material. Significant improvements in strength are not generally useful if they are accompanied by drastic reductions in ductility or other important properties, some of which have been mentioned above.

2.4 Strengthening by Dislocations

The introduction of dislocations into a material by

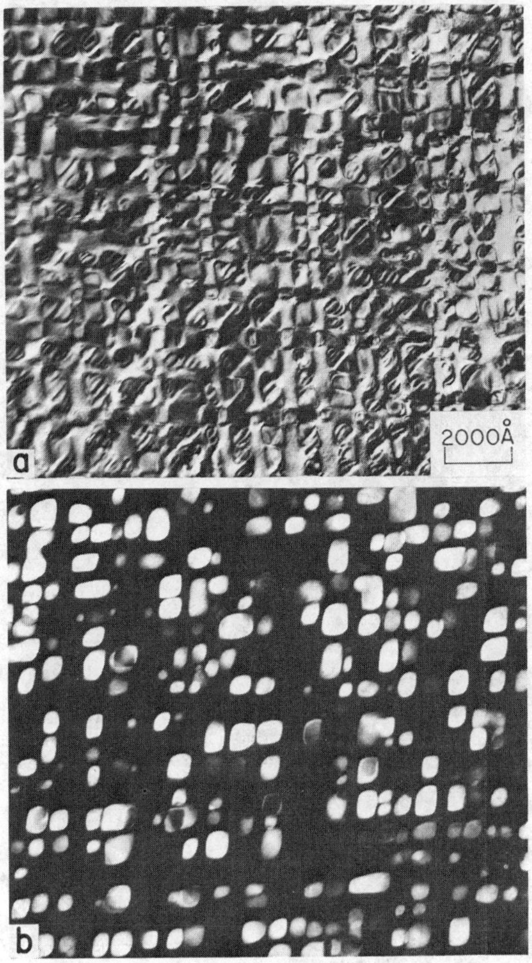

Figure 9. TEM micrographs showing coherent particles of Ni_4W in a Ni-W alloy.
a) Bright field micrograph showing coherency strain contrast in the matrix surrounding the coherent particles.
b) Dark field micrograph showing the Ni_4W particles.
Both at same magnification.

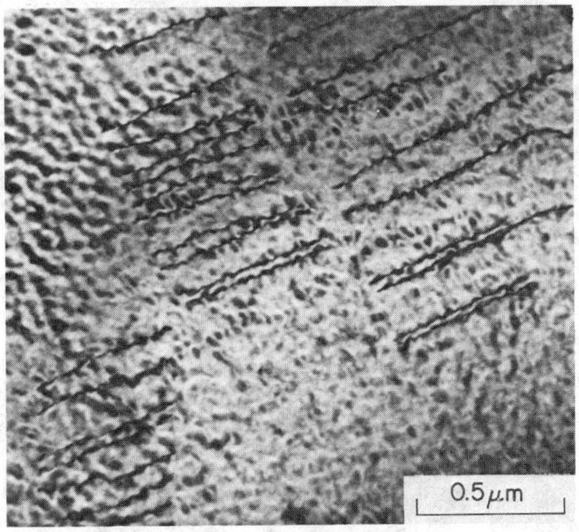

Figure 10. TEM micrograph showing dislocation pairs in Ti-8%Al alloy aged to contain ordered (DO_{19}) precipitates. Mottled contrast in background is due to coherency strains around particles.

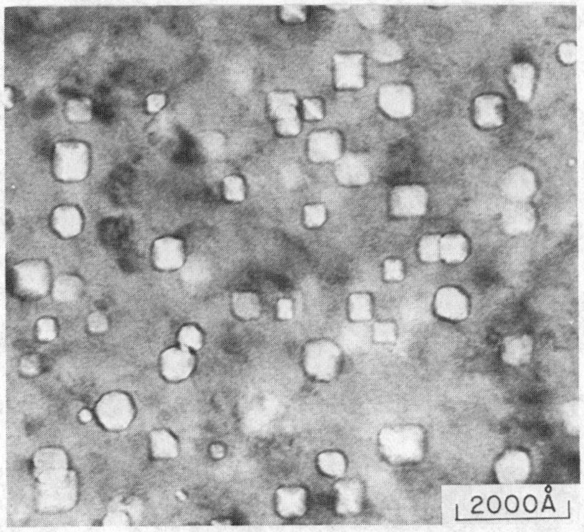

Figure 11. TEM micrograph showing voids in P^+ irradiated 316 stainless steel.

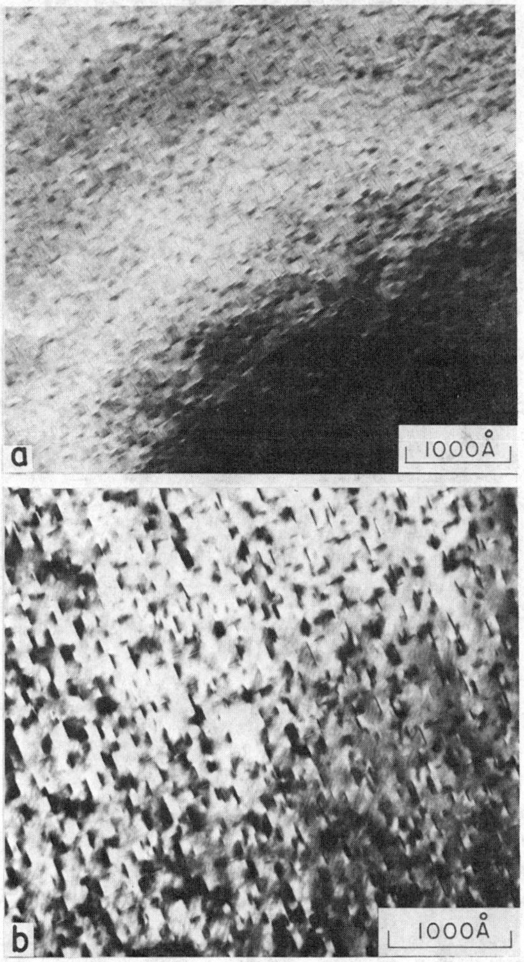

Figure 12. Bright field TEM micrographs showing thin plate-shaped precipitates in Al-Cu alloy. (Photos courtesy of S. D. Dumolt).
a) G.P. zones.
b) Θ" precipitates.

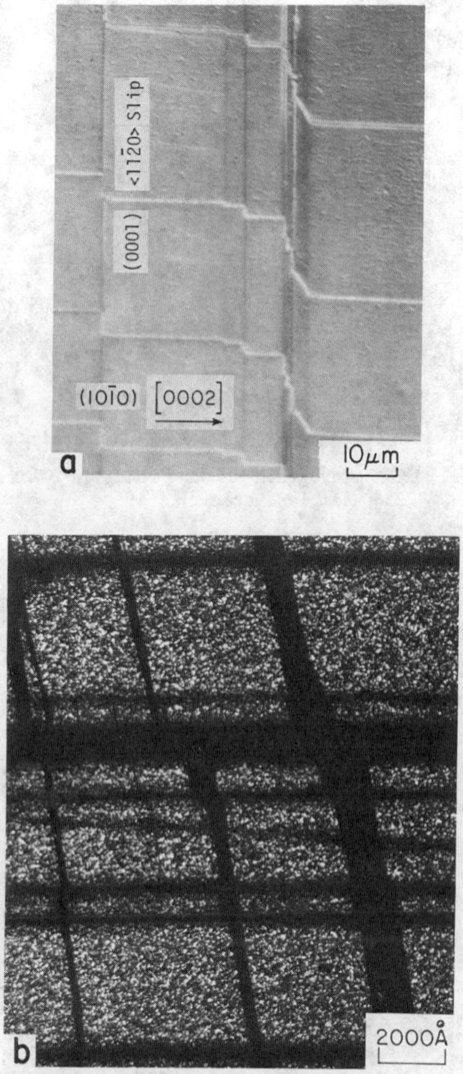

Figure 13. Surface and internal deformation structure when coherent particles are sheared.
a) Intense surface slip lines.
b) Dark field TEM micrograph showing complete destruction of strengthening precipitates along slip bands.

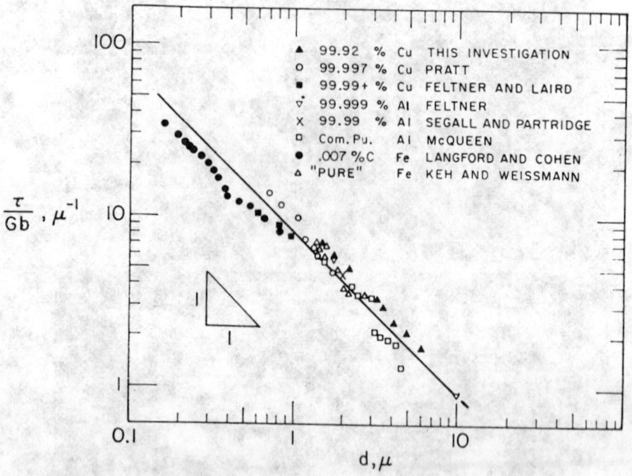

Figure 14. Relation between normalized flow stress (see Eqn.[17] and substructure size for which τ_0 is nearly zero. "This investigation" refers to Staker and Holt(37), the source of the figure; other studies listed are cited in Refs. 37 and 38.

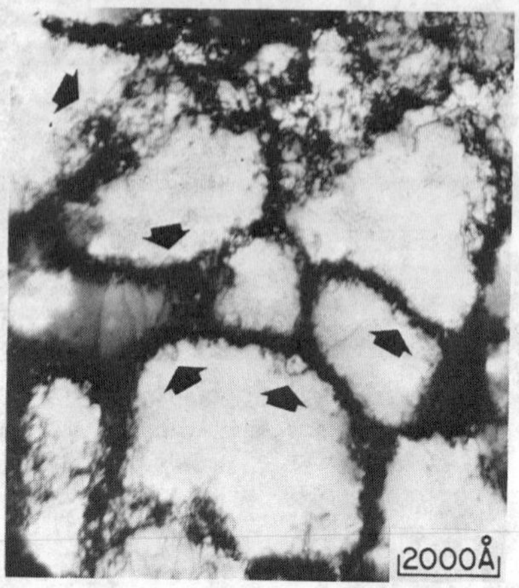

Figure 15. Transmission electron micrograph of cell substructure in nickel, with apparent dislocation sources (loops bowing out from cell walls) arrowed. From Thompson(57).

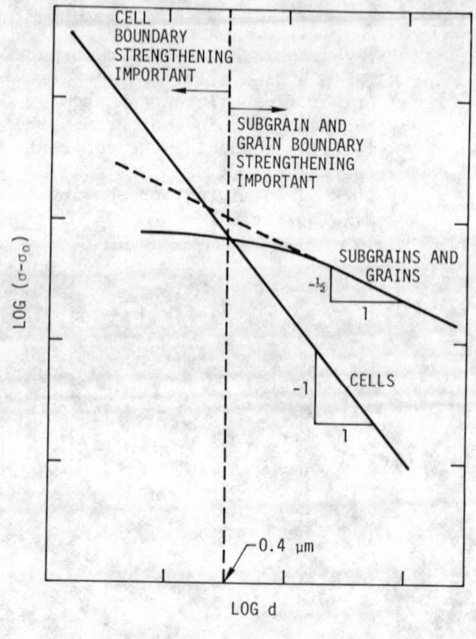

Figure 16. Schematic representation of relative strengthening from grains, subgrains, and cells, with transition in dominance at about 0.4 μm. After Young and Sherby(53).

mechanical working or by means of a phase transformation increases the flow stress. For low to moderate dislocation densities, which can adequately be measured in thin foils, it is well established experimentally(35) that the dependence on dislocation density ρ is given by

$$\tau = \alpha G \underline{b}\sqrt{\rho} \quad , \qquad [14]$$

where τ is the shear flow stress (replaced by $\tau-\tau_o$ when the "friction stress", τ_o, is significant), α is a dimensionless constant of order 0.5, G is the shear modulus, and b the Burgers vector. This relation has been known for some time, and can be derived(36) from all the extant theories of work hardening. As dislocation densities increase, and particularly when they are arranged as dense tangles or walls, it becomes experimentally quite difficult to test Equation 14, although the little evidence which exists(37,38) tends to support the relation.

The dislocation density increases with strain, as dislocations encounter one another and become trapped, thus requiring the generation of additional dislocations; one expression for this is the well-known Orowan relation,

$$\varepsilon = \rho \underline{b}\lambda, \qquad [15]$$

where λ is the distance moved by an average dislocation. The simplicity of Equation 15 is misleading, however, in that, first, ρ must be, not the total dislocation content of the material as in Equation 14, but only that portion of the dislocation population which is mobile, or ρ_m. Typically the dislocations comprising ρ_m are a small fraction(39) of ρ and moreover are not observably distinguishable, for example in a thin foil. Second, λ itself must be a function of strain, typically having a dependence of the form $1/\varepsilon^2$ or, in some cases, $1/\varepsilon^3$. Third, a minor point, a geometrical factor should appear in Equation 15 to convert slip-plane strain to tensile strain; thus the Taylor factor, M, would be appropriate (40) as the denominator of the right side of Equation 15. These comments indicate that it would be appropriate to use [15] only for small increments of strain, for which ρ_m can be reasonably regarded as constant.

Equation 15 should probably thus be replaced by

$$d\varepsilon = \frac{\underline{b}\,\rho_m}{M}\, d\lambda(\varepsilon) \quad . \qquad [16]$$

Integration of this equation to give a general ε-ρ relation would clearly require knowing λ(ε) as well as the relation between ρ_m and ρ as a function of ε; none of these is known in general, and there exists little useful information even for simple, well-defined cases.

At strains above several percent in materials of easy cross-slip, the dislocations typically become arranged into a three-dimensional substructure. When this substructure has a loose, tangled appearance in the dislocation walls, as it typically does when formed at the low homologous temperatures treated in this chapter, it is referred to as a "cell" structure. When the walls become narrow and well-defined, as they do when formed at high homologous temperatures, when thermally treated to permit static or dynamic recovery, or when formed at very high strains (e.g. well above unity), they are called "subgrains". The development, morphology and properties of these substructural types have recently been reviewed in detail(38,41).

There is now extensive evidence(38,41) that the flow stress of materials with well-developed cell substructures is given by

$$\tau-\tau_o = \beta G \underline{b} \underline{d}^{-1} \quad , \qquad [17]$$

where d is the cell size and β is a dimensionless material constant. Among the illustrations of this is Figure 14. This relation can be derived from any barrier-spacing or dislocation source length model(38); among those developed specifically for this problem is Kuhlmann-Wilsdorf's(42). She regarded the flow stress as the stress required to generate additional dislocations, which in turn is a function of source length. In cells, the sources evidently are segments which can bow out from cell walls, as in Figure 15. To understand the behavior in detail, however, is quite complex. Typically the dislocation density includes many redundant dislocations, and as strain increases there is extensive local cross-slip, rearrangement, and annihilation in cell walls, all of which affects the crystallographic misorientation between cells, the lattice rotations leading to deformation textures, the relative degree of tilt or twist character of the cell walls, and so forth(38,41).

When the substructure has the characteristics of subgrains, the boundaries usually are distinctly more perfect and more misoriented, and begin to assume many of the characteristics of normal grain boundaries, albeit with much lower average misorientation than grain boundaries(41). This description suggests that subgrains should behave mechanically much as do grains, and indeed that has been observed(38,41); typically it is found that

$$\tau-\tau_o = \beta G \underline{b} \underline{d}^{-1/2} \qquad [18]$$

rather than Equation 17; this is, of course, the Hall-Petch relation, as discussed below for grains.

2.5 Boundary Strengthening

The role of crystallographic boundaries as barriers to slip has been known for sixty or more years. In what may be the simplest case, grain boundaries, it was long known that grain refinement increased yield strength, although the strengthening was not quantified until Hall(43), and later Petch(44), recognized that the appropriate description was the familiar

$$\sigma_y = \sigma_o + k\underline{d}^{-1/2} \quad , \qquad [19]$$

where d is the grain size and σ_o and k are material constants. This behavior is now recognized as quite typical, not only of grain structures(45), but also for other kinds of boundaries(41), such as ferrite-cementite in pearlite, mechanical twins, and martensite plates. It has also been found, for spheroidized carbon steels, that sbustitution of the mean free path between cementite particles for d in Equation 19 gave an adequate description of behavior(46), although it should be emphasized that the strengthening mechanisms are likely different.

The mechanistic problem of a fundamental understanding of the Hall-Petch relation, as Equation 19 is commonly called, remains largely unsolved. The original proposals(43,44) in terms of dislocation pile-ups and nucleation of slip in an adjoining grain are clearly not satisfactory in general(47-49), and there are also problems with alternatives, such as the "dislocation density" approach (47). Probably the most satisfactory idea physically is Li's proposal(47,50), which is based upon the behavior of grain boundary sources for dislocations; the calculation goes as follows. Let L be the total length of dislocation emitted per unit grain boundary area at yielding. Then the dislocation density is $\rho = 1/2\ L(S/V)$, where S is the grain surface area, V is the grain volume, and the 1/2 accounts for the two grains sharing the boundary. For the spherical grains considered by Li, $S/V = 6/\underline{d}$; for a tetrakaidecahedron grain, $S/V = 6.70/\underline{d}$ (51), and $\rho = 3.35\ L/\underline{d}$. Substituting this into Equation 14 gives a Hall-Petch relation with slope $k = \alpha G \underline{b}\sqrt{3.35L}$. If a suitable physical understanding of L as a material parameter could be developed, this concept could prove an acceptable model for the Hall-Petch effect. There has been a report(52) that grain boundary ledges played no role in the Hall-Petch effect, but inasmuch as those particular ledges appeared not to act as dislocation sources, there is no contradiction of the Li suggestion. There is ample evidence, as has been reviewed (48,49), that grain boundary dislocation sources are important or even dominant in yielding.

There has been a difference of opinion until recently about the relative effectiveness of grain vs. cell boundaries in raising the yield strength. It now appears that if one compares these boundary types over a sufficiently wide range of size, that the behavior is as shown in Figure 16: below a grain size of about 1 μm, the Hall-Petch effect is lost, and very small cells can then be stronger than grains of comparable size(38,53).

At strains beyond yield, it is tempting(45) to modify Equation 19 by replacing the yield stress with the flow stress at some elevated strain, and inserting new, strain-dependent Hall-Petch parameters σ_o' and k'. That such an approach cannot be general is shown by Figure 17 for copper(54). Instead, much more complex equations for the flow stress have been proposed(41,49,55), and at present it is unclear which type of approach will be most fruitful. It has become evident, however, that polycrystal grains do not deform homogeneously, and that zones of complex slip, differing from that in the grain interior, occur near grain boundaries(48,49). Figure 4(c) contains an illustration of this. Any successful model of polycrystal flow stresses will have to incorporate this fact. The starting point for such a model will probably resemble Ashby's suggested thought experiment(56), depicted in Figure 18. An important point here is the distinction between "statistically-stored" dislocations, i.e. those which encounter and trap one another randomly, as in a single crystal; and the "geometrically-necessary" dislocations which are generated in response to non-uniform strain in the crystal(49,56). The thought experiment then is conducted as follows. One deforms the polycrystal shown in Figure 18(a) by disassembling it into its constituent grains and allowing each to slip according to the Schmid law, thereby introducing statistical dislocations (which are not shown for clarity). But now, as in Figure 18(b), overlaps and voids appear among the grains. Taking each of these discrepancies in turn, Figure 18(c), one corrects them by introduction of the appropriate geometrical dislocations, in the form of prismatic loops, tilt boundaries, etc., until the grains again fit together. The reassembled polycrystal is shown in Figure 18(d).

This physical process, which can readily be applied to cases of polyphase microstructures and to boundaries in general(56), is an essential part of "intermediate strain" behavior, i.e. above yielding but below strains typical of necking in most materials(41). At larger strains, deformation becomes sufficiently homogeneous that a Taylor-type approach is satisfactory(40,41,49) and is also simpler. It does appear, however, that the Taylor approach has minor but systematic discrepancies in its predictions, even at high strains(41,49). It can then be concluded that descriptive models for the flow stress, at strains from yielding to well above unity, are qualitatively satisfactory, but that fundamental understanding, over the whole of this strain range, remains limited.

3.0 COMPLEX STRENGTHENING

The basic strengthening mechanisms which can affect the strength of single phase and multi-phase materials have been discussed above in reasonable detail. However, consideration was postponed for alloys in which two or more of these mechanisms operate simultaneously. In this section situations where the various strengthening mechanisms operate in a pair-wise manner will be considered and the current limitations which prevent systematically combining the contributions of these mechanisms will be discussed in a qualitative way. It will be immediately seen that the theoretical framework for combining various strengthening contributions is very poorly developed at present. Consequently, detailed studies in which analytical methods for combining strengthening contributions are badly needed and should be viewed as an extremely fertile area which is appropriate for further development.

3.1 Similar Barriers of Different Strengths

It has been mentioned above that solute atoms with different interaction strengths and interaction mechanisms can lead to combined strengthening in solid-solution alloys. The Labusch statistics are capable of summing the contributions from more than one type of solute in multi-component alloys. However, it is often seen that the solutes present in multi-component systems have essentially additive effects on slip character as well as strength. In view of the discussion in Section 2.3, these effects merit brief mention. Figure 19 illustrates the effect of ternary additions of oxygen on the deformation character of substitutional solid solutions of Ti-Al. This figure shows that variations in oxygen content tend to offset planarity of slip in these alloys. In this connection, it has been shown that increasing either the oxygen content(58) or the aluminum content(59) in binary Ti-O or Ti-Al alloys effects a wavy-to-planar slip transition. It thus appears that the effects of oxygen and aluminum in the ternary alloys are additive with respect to slip character even though one is substitutional and the other is interstitial. A detailed rationale for understanding such effects does not currently exist. Therefore, the effect of solutes of different types on slip mode is a subject deserving of further investigation.

It also is possible to have two or more types of precipitates present in particle-hardened alloys, just as there can be two different solutes present in solid-solution strengthened alloys. Examples of this include nickel-base alloys which contain both Al and Nb or Ti, so that both the cubic γ' phase ($L1_2$ structure) and the tetragonal γ" phase (DO_{22} structure) form during aging. These phases have different structures and different APB energies as well as different morphologies and sizes. The latter features are illustrated in Figure 20. As a result, these two types of precipitate represent barriers of different strengths, both of which can contribute to particle hardening. The method for combining or averaging the contributions of two phases such as these is not obvious, although some work has been done along this line by Koppenaal and Kuhlmann-Wilsdorf (60). These workers used an averaging procedure based on the root-mean-square (RMS) contributions of the individual barriers. Although their treatment did not explicitly deal with barriers of different strength, their method should, in principle, be capable of treating such a situation. The principal difficulty with RMS averaging is that an arbitrary cutoff distance must be invoked to eliminate contributions from particles which lie throughout the crystal. Such a cutoff distance is necessary in order to prevent cancelling of particle interactions of opposite sign since the net internal stress in the entire crystal is zero. Despite this, the physical basis for assigning a value for the cutoff distance is currently unclear.

Another complication arises when combining the effects of different types of particles if each of these particles has a different effect on the slip character. For example, large semi-coherent particles which are not sheared by individual dislocations, such as those shown in Figure 21(a), can become sheared if coherent particles are precipitated between them, so that the slip is focused or channeled into planar bands which cause slip to be nucleated within, and thus shear, the semi-coherent particles. This is illustrated in Figure 21(b). Another example of duplex types is shown in Figure 22. In this example both coherent γ' (as seen by the coherency strain contrast in the matrix) and non-coherent dispersoids are present in age hardenable mechanically alloyed materials, in this case alloy MA753. Here the dispersoid is not sheared, whereas the γ' is clearly shearable. A viable method for computing the combined Orowan stress and particle hardening stress is not obvious.

Finally, an example of a particularly complex type of particle hardening can be found in a Ni-Al-Mo-W alloy which contains a high concentration of Mo. The microstruc-

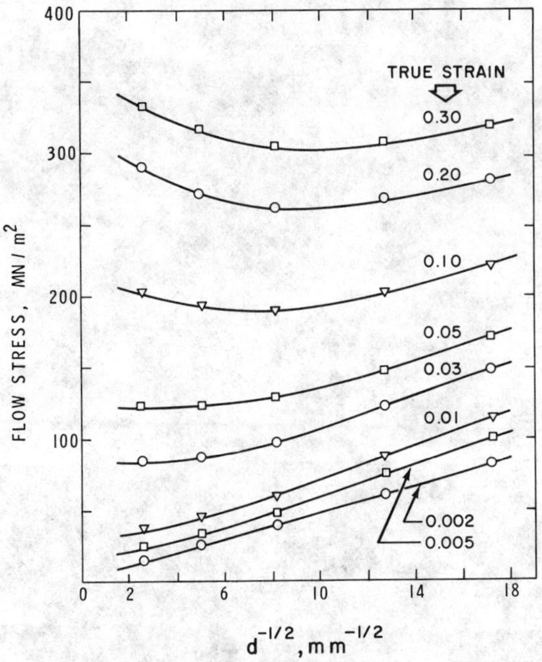

Figure 17. Hall-Petch plot (see Eqn. [19] for flow stress of five grain sizes of copper (from 3.4 µm to 150 µm grain size) at the strains indicated on the curves. From Thompson and Baskes(54).

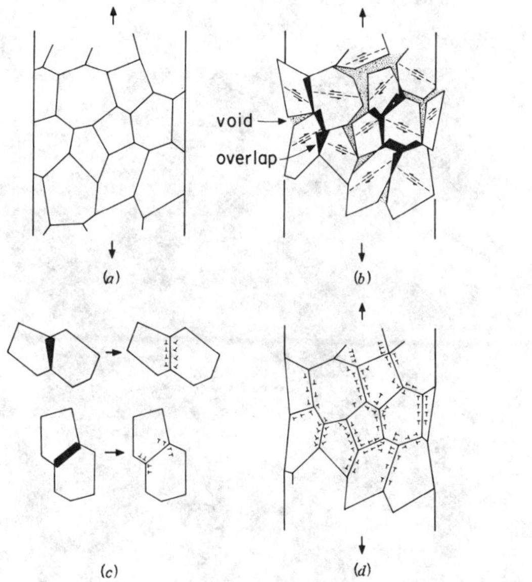

Figure 18. Illustration of geometrically necessary dislocations in a polycrystal (statistical dislocations not shown). (a) Polycrystal which, if deformed as separate grains (b), no longer retains material continuity; appropriate geometrical dislocation arrays (c) permit reassembly of the polycrystal (d). From Ashby(56).

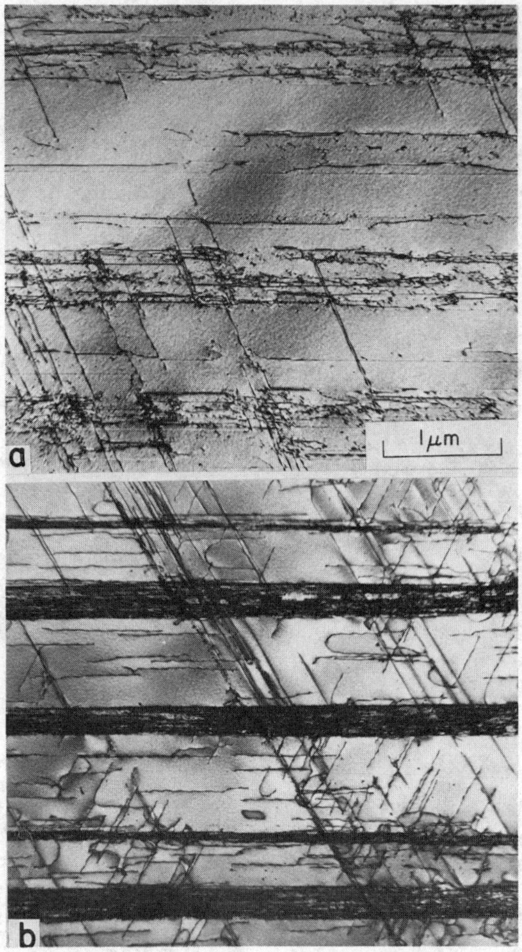

Figure 19. TEM micrographs showing the additive effects of Al and oxygen on slip character in Ti-Al alloys.

a) Ti-4Al + .070 oxygen.
b) Ti-4Al + .20 oxygen.

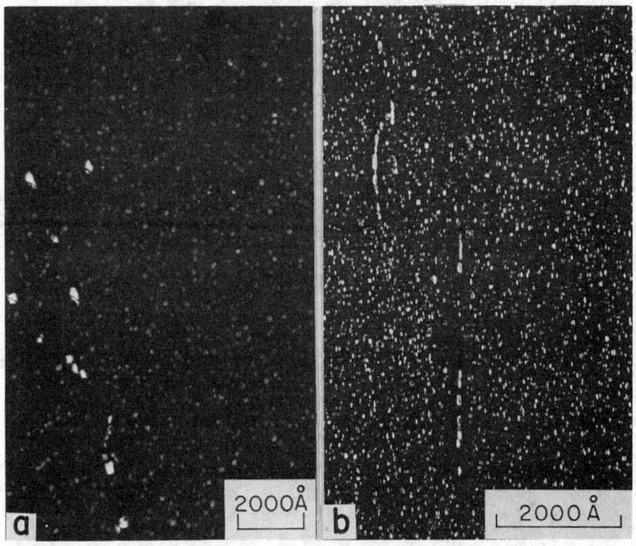

Figure 20. Dark field TEM micrographs showing two types of precipitates present in Ni-base alloy INCO 718.

a) Equiaxed coherent particles of the cubic γ' phase.
b) Plate-shaped coherent particles of the tetragonal γ" phase.

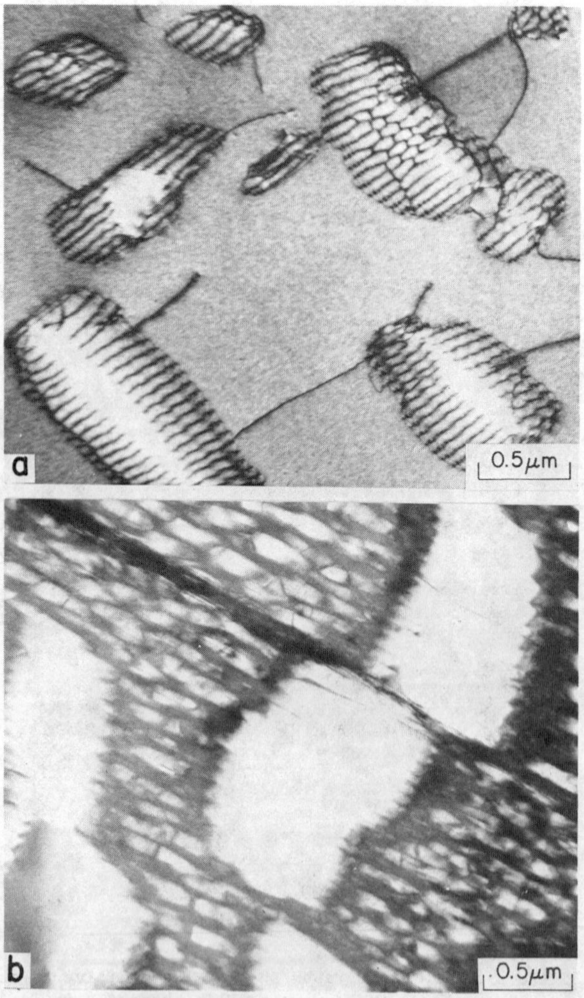

Figure 21. Electron micrographs showing coarse semi-coherent particles of α_2 and the effect of a bimodal distribution of coherent and semi-coherent α_2 particles in a Ti-Al-Ga alloy.

a) Bright field micrograph showing coarse semi-coherent particles.

b) Dark field micrograph showing shearing of coarse particles when smaller coherent particles are present.

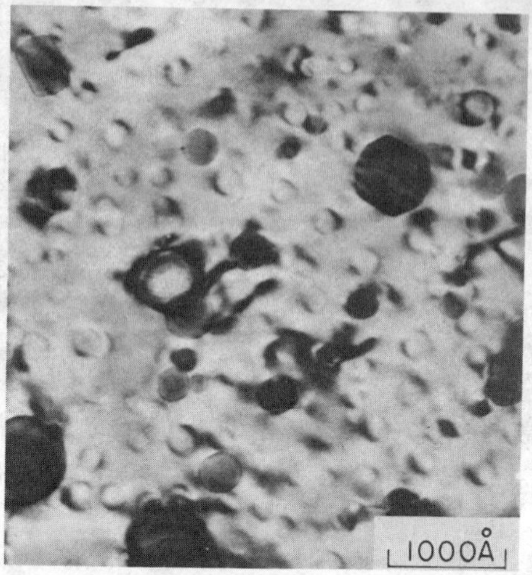

Figure 22. Electron micrograph showing non-deformable oxide particles and coherent γ' particles in alloy MA753.

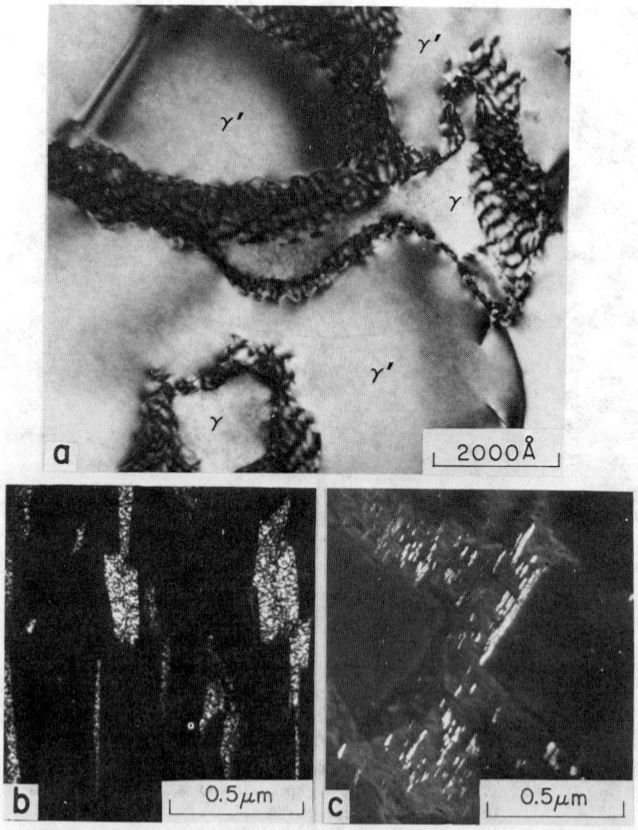

Figure 23. Electron micrographs showing a structure of Ni-Al-Mo-W alloy. (Photos courtesy of P. L. Martin). a) Showing coarse γ' particles in a γ matrix. b) Showing DO_{22} γ'' particles within the γ matrix. c) Showing Pt_2Mo particles within the γ matrix.

Figure 25. Bright field TEM micrograph showing details of lamellar microstructure of Fig. 24(b).

ture of this alloy is illustrated in Figure 23. The large semi-coherent γ' present in this alloy can be seen in Figure 23(a), but the γ matrix which lies between the γ' particles also contains two additional ordered Mo-rich phases. These are the tetragonal DO_{22} type phase based on the composition Ni_3Mo and the tetragonal phase which has a Pt_2Mo prototype structure and is based on the composition Ni_2Mo. Strengthening in alloys which have microstructures as complex as this is very difficult to analyze; at this point there is no suitable formulation for even semi-quantitatively assessing the relative strengthening contributions from each of the various types of particles. Further, each precipitate has a different solvus temperature which makes the temperature dependence of the strength very difficult to analyze.

In the case of multiple types of boundaries, such as cells and grain boundaries, the usual situation is that when the cells are well-defined enough to obey Equation [17], they are much smaller than typical grains and thereby dominate behavior. The case of nearly comparable sizes is not well understood, but the contributions of cells and grains do not appear to be additive(38,53). Since both types of barriers may, at elevated strains, essentially deform according to a source-length process (38,42,57), there may in fact be a smooth transition from strength control by grain boundaries, to control by intra-grain barriers, with no additivity at all. Experimental investigation of this point would be welcome.

3.2 Solid Solution Plus Particle Strengthening

As mentioned earlier, limited evidence exists which suggests that these types of strengthening should be additive(31). However, as also discussed earlier, solutes which have a substantial effect on the matrix deformation character can affect the stresses at which the particles become shearable. Thus the matrix slip character may significantly change the mechanism of particle hardening in a way analogous to that previously illustrated in the case of a bimodal distribution of particles (Section 3.1 and Fig. 21). In such cases, the contributions of particle and solute strengthening is not likely to be strictly additives. Thus additional considerations must be invoked to account for the effect of solute on varying the nature of the dislocation-particle interactions. However, in those cases where the solute concentration is relatively low, i.e. if the matrix deformation character does not change appreciably, the additivity of these two strengthening mechanisms appears to be simply linear(31).

3.3 Solid Solution Strengthening Plus Dislocation Strengthening

As in the case discussed immediately above, the role of solid-solution additions which change significantly the matrix deformation character may have an influence on the dislocation strengthening contribution which has been described earlier in Section 2.4. As discussed in this section and, as summarized by Equations [14] and [15], the principal variables in dislocation strengthening are the dislocation density and the average mean free path through which mobile dislocations can travel. Changes in slip character will affect these two parameters but possibly in a way which leads to a nearly constant product of the two. A more satisfactory approach to this type of analysis has been suggested in Section 2.4 and summarized in Equation [16]. Thus, while some caution should be introduced when attempting to combine solid solution strengthening (perhaps expressed through τ_o, Equation [17]) with substructure strengthening, it appears that these types of strengthening are reasonably additive, as was the case for solid solution strengthening and particle strengthening.

The key to this additive nature appears in both cases to be the significantly different size scales with which the dislocations interact with these two types of barriers as has been suggested in an earlier section of this paper.

3.4 Particle Plus Dislocation Strengthening

If one considers the size scale of these two strengthening interactions, it would seem that the strengthening contributions from these two types of barriers are potentially additive. However, the available data which can be analyzed to examine this point of view suggests that the picture is mixed, if not confused. For example, large non-shearable particles tend to have little effect on the formation of substructure, although such particles also tend to make very little contribution to the strength. Thus, this particular case is essentially a trivial one. At the other extreme of non-shearable particles of a spacing much smaller than the "natural" cell size, at a given strain in the absence of particles, there likewise appears little interaction between the two mechanisms. However, in the case of small, non-shearable particles which have a spacing comparable to that of the "natural" cell size, the presence of these particles interacts directly with the nature and rate of accumulation of substructure. Cell walls tend to be nucleated on and tied to the particles(41), so that a cell size above or below the "natural" size at that strain may result, with an effect on strength through Equation [17]. In this case, the strengthening from these two different contributions is no longer linearly additive. No detailed analysis has been attempted for such strengthening, yet this represents an interesting case and therefore is clearly worthy of further consideration.

3.5 Duplex and Micro-Duplex Strengthening

Strengthening in duplex or micro-duplex microstructures has been discussed in an earlier review by Copley and Williams(1) and the situation with regard to detailed analysis of the strengthening contributions in such systems has not advanced significantly since the preparation of that article. Nevertheless, for the sake of completeness, several points regarding strength of these types of alloys should be made.

Duplex or micro-duplex microstructures occur in two-phase alloys where the preponderance of the internal boundaries are interphase boundaries. Examples of such alloys are brasses, titanium alloys and dual-phase stainless steels. The scale of the two phases in this class of alloy is coarse compared to microstructures which can be readily treated using the ideas of particle hardening. Thus the strengthening in these alloys must be derived from superpositioning of solid solution strengthening and boundary hardening. The latter would be similar to that discussed in Section 2.5. The principal difference in the micro-duplex case is that the boundaries separate phases of different structure. Therefore the slip systems are different on opposite sides of the boundaries as are the magnitudes and character of the Burgers vectors of the slip dislocations. This point has been discussed in a general way in the earlier review of Copley and Williams (1) and this will not be expanded upon here.

Typical micro-duplex microstructures which can be produced in titanium alloys are shown in Figure 24. From this it can be seen that the morphology of the two phases can be varied by mechanical working as has been described elsewhere(61). A more detailed example of the microstructure seen in Figure 24(b) is seen in the TEM micrographs of Figure 25. Here it can be seen that both homophase and heterophase (α/α and α/β) boundaries exist. Both of these types of boundaries can transmit slip bands in these microstructures. They also can be sites for slip initiation, depending on the nature of the boundaries and the crystallographic alignment (or lack thereof) of the two phases which are separated by the boundary.

Examples of yielding at high-angle homophase and heterophase boundaries are shown in Figure 26. Figure 26(a) illustrates the initiation of slip at ledges contained within a high-angle grain boundary in a single-phase material (homophase boundary). Figure 26(b) illustrates a comparable effect at the boundary between a bcc W fiber and a fcc Ni-W solid solution (heterophase boundary)(62).

Figure 26(c) shows the complexity of slip which occurs in the vicinity of an α/β boundary in a two-phase titanium alloy. In each case it can be seen that the boundaries tend to influence the deformation character of the material but the detailed strengthening contribution due to the presence of this boundary is very difficult to analyze. Accordingly, it is suggested that the strength of duplex or micro-duplex alloys is a combination of solid solution strengthening and boundary strengthening, but the relative contributions of these two mechanisms is not well understood and cannot be analyzed even in any semi-quantitative way at present.

3.6 The Strength of Martensite

Martensite is at once one of the most technologically interesting and one of the most complex cases of combined strengthening. It is worth pointing out, however, that much of the technological interest (as well as the fundamental complexity) centers on martensitic microstructures in steel, which are highly atypical martensites. The point can be illustrated by noting that in Fe-30Ni alloys and in maraging steels, where carbon content is very low, the martensites formed are not particularly hard and tend to exhibit substantial isothermal growth(63). In other alloy systems, such as titanium and many other non-ferrous alloys, the martensites formed are likewise not particularly hard and can exhibit isothermal growth. The typical athermal nature of steel martensites, in fact, has been attributed to carbon(63). It can be concluded then, that martensite in steel is unusual in several ways, notably in its capacity for very high hardness, and that carbon is the reason.

It is now well established, in fact, that the hardness of as-quenched martensite in steel depends primarily on the carbon content; up to 0.4% C, the hardening is linear (about 40-50 MPa per 0.01% C). Above 0.4% C, retained austenite becomes significant and the rate of hardening with added carbon decreases.

It is important to emphasize, however, that this dependence on carbon is not due simply or even predominantly to interstitial solid solution hardening. In a thorough experimental study, Leslie and Sober(64) identified the following contributions to strength; (i) boundary strengthening by martensite lath boundaries, a dominant effect in low-carbon steels, and, above about 0.4% C, also by internal twins within the laths or plates(65); (ii) dislocation density strengthening arising from transformation strains, of particular importance at lower carbon contents; (iii) solid solution strengthening by immobile carbon atoms, a relatively minor contribution (64); (iv) rearrangement of carbon during the quench to cause clustering and dislocation pinning, an effect which is large and, for intermediate and high carbon contents, is dominant(64); (v) work hardening during the tensile test, strongly dependent on carbon content; and some other, minor effects as well(64).

To sort out these factors in a typically complex and highly internally-strained microstructure, for example as shown in Figure 27, is inevitably difficult. Moreover, because the detailed structure of the martensite, the presence of retained austenite, the carbon content, the M_s temperature, any residual stresses, and the test conditions (temperature, strain rate) can all affect the relative contributions of the foregoing processes, there can be no "typical" ranking of strength. However, an example can be given for an 0.4% C steel, as in Table I.

Table I. Contributions to As-quenched Martensite Strength in 0.4% C Steel

Boundary strengthening	620 MPa
Dislocation density	270 MPa
Solid-solution of carbon	400 MPa
Rearrangement of C in quench	750 MPa
Other effects	200 MPa
0.2% Yield Strength	2240 MPa

The evidence for the separation of these is not unequivocal, and moreover they are not likely to be precisely additive(64). Nevertheless, Table I gives an impression of what is now known about martensite strength.

Most martensitic steels are used after tempering to substantially lower strength than the kind of value shown in Table I. The strength then becomes an even more complex problem, as carbon is precipitated as ε, Fe_3C and alloy carbides of varying sizes, as dislocations rearrange and annihilate, and so forth; Speich has reviewed(66,67) the essence of what is now known on this topic. Substantial amounts of fundamental work will be needed to assess the strengthening mechanisms of even a single quenched and tempered steel; to pursue this question for a broad variety of steels is a daunting prospect.

REFERENCES

1. S. M. Copley and J. C. Williams in Alloy and Microstructural Design , Tien and Ansell, eds., Academic Press, New York, 1976, p. 3.
2. J. Weertman, Trans. Met. Soc. AIME, 227, 1475 (1963).
3. M. F. Ashby, Acta Met., 20, 887 (1972).
4. R. J. Jensen and J. K. Tien, this volume.
5. A. H. Cottrell in Relations of Properties to Microstructure , ASM, 1954, p. 131.
6. P. A. Flinn in Strengthening Mechanisms in Solids , ASM, 1962, p. 17.
7. K. R. Evans, Treatise on Mat. Sci. & Techn., 4, 113 (1974).
8. P. Haasen, Trans. Japan Inst. Met., 9 (supplement), XL (1968).
9. R. L. Fleischer in The Strengthening of Metals , D. Peckner, ed., Reinhold, New York, 1964, p. 93.
10. W. T. Read, Jr., Dislocations in Crystals , McGraw-Hill, New York, 1953.
11. R. Labusch, Phys. Stat. Sol., 41, 659 (1970).
12. H. Suzuki, Science Reports, Research Institute, Tohoku University, A4, 455 (1952).
13. J. E. Dorn, P. Pietrokowsky, and J. E. Tietz, Trans. AIME, 188, 933 (1950).
14. N. P. Allen, T. H. Schofield, and A. E. L. Tate, Nature, 168 (September 1) 378 (1951).
15. A. H. Cottrell, S. C. Hunter, and F. R. N. Nabarro, Phil. Mag., 44, 1064 (1953).
16. J. C. Fisher, Phys. Rev., 91, 232 (1953).
17. G. Ardley and A. H. Cottrell, Proc. Roy Soc., A219, 328 (1953).
18. M. J. Marcinkowski in Electron Microscopy and Strength of Crystals , Thomas and Washburn, eds., Wiley, New York, 1963, p. 333.
19. N. S. Stoloff and R. G. Davies, Progress in Materials Science, 13, 1 (1966).
20. N. S. Stoloff in Strengthening Methods in Crystals , Kelly and Nicholson, eds., Wiley, New York, 1971, p. 193.
21. M. J. Blackburn, Trans. AIME, 239, 660 (1967).
22. A. Kelly and R. B. Nicholson, Progress in Materials Science, 10 (#3), 151 (1963).

Figure 24. Light micrographs showing various duplex microstructures in an α+β Ti alloy.
a) Equiaxed microstructure consisting of primary α grains with β at grain boundaries.
b) Lamellar α+β mixture consisting of α laths separated by thin regions of β-phase.
c) Equiaxed primary α grains in a fine lamellar α+β matrix.
All at same magnification.

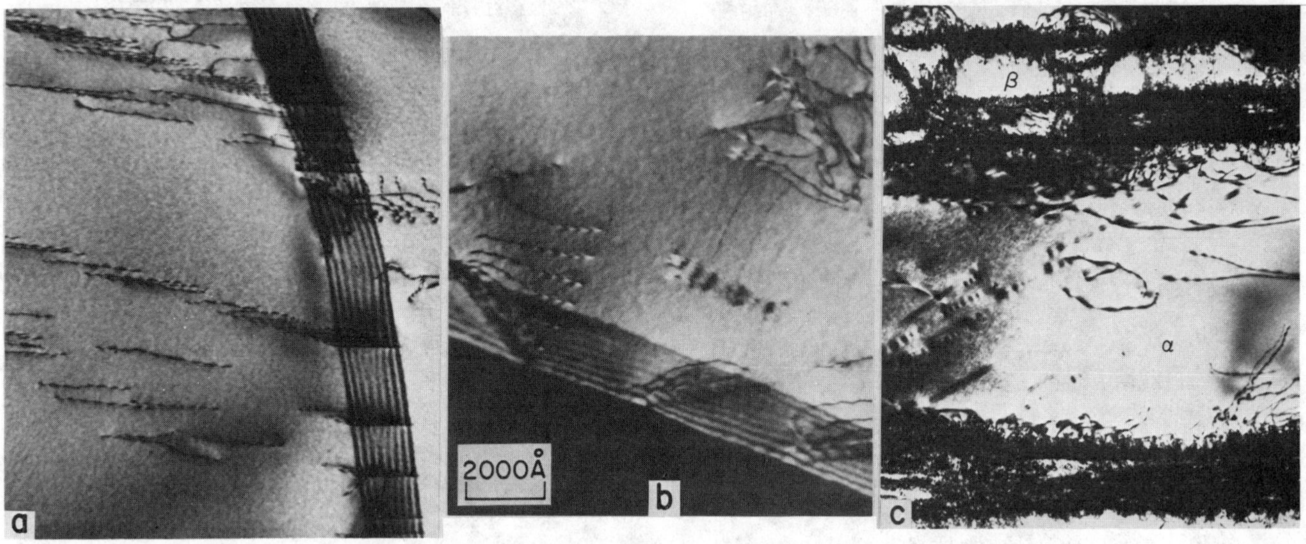

Figure 26. TEM micrographs showing yielding at internal boundaries.
a) Homophase high angle boundary.
b) Heterophase boundary between bcc W and fcc Ni-W solid solution.
c) Heterophase boundary between hcp α and bcc β in a Ti alloy.
All at same magnification.

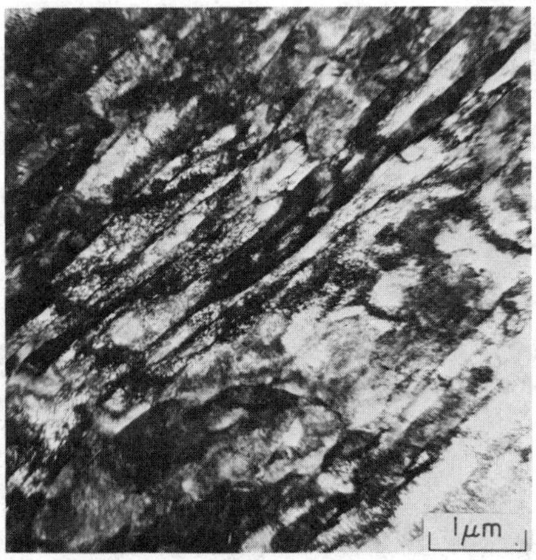

Figure 27. TEM micrograph of typical dislocated martensite structure, showing sub-micron lath widths and high dislocation density arising from transformation. Steel is unaged PH 13-8 Mo (Armco trade name).

23. L. M. Brown and R. K. Ham in Strengthening Methods in Crystals , Kelly and Nicholson, eds., Wiley, New York, 1971, p. 9.

24. P. M. Kelly, Int. Met. Rev., 18, 31 (1973).

25. E. Orowan, "Discussion" in Symposium on Internal Stresses in Metals and Alloys , Inst. of Metals, London, 1948, p. 451.

26. V. Gerold and H. Haberkorn, Phys. Stat. Sol., 16, 675 (1966).

27. H. Gleiter, Acta Met., 16, 829 (1968).

28. P. B. Hirsch and A. Kelly, Phil. Mag., 12, 881 (1965).

29. R. K. Ham, Trans. Japan Inst. Met., 9 (supplement), 52 (1968).

30. W. H. Meiklejohn and R. E. Skoda, Acta Met., 8, 773 (1960).

31. R. Ebeling and M. F. Ashby, Phil. Mag., 13, 805 (1966).

32. N. Hansen and P. Brondsted, Res. Mech., 1, 197 (1980).

33. J. C. Williams, A. W. Thompson and R. G. Baggerly, Scripta Met., 8, 625 (1974).

34. J. C. Fisher, E. W. Hart and R. H. Pry, Acta Met., 1, 336 (1953).

35. H. M. Otte and J. H. Hren, Exper. Mech., 6 (1966), 177-86.

36. F. R. N. Nabarro, Z. S. Basinski, and D. B. Holt, Advan. Physics, 13 (1964), 193-323.

37. M. R. Staker and D. L. Holt, Acta Met., 20 (1972), 569-79.

38. A. W. Thompson, Met. Trans. A, 8A (1977), 833-42.

39. R. Gasca-Neri and W. D. Nix, Acta Met., 22 (1974), 257-64.

40. G. Y. Chin, pp. 45-66 in Work Hardening in Tension and Fatigue (A. W. Thompson, ed.), TMS-AIME, New York, 1977.

41. J. Gil Sevillano, P. van Houtte, and E. Aernoudt, Prog. Mater. Sci.,25 (1980), 69-412.

42. D. Kulhmann-Wilsdorf, Met. Trans., 1 (1970, 3173-79.

43. E. O. Hall, Proc. Phys. Soc. (London), B64 (1951), 747-53.

44. N. J. Petch, J. Iron Steel Inst., 174 (1953), 25-32.

45. R. W. Armstrong, Met. Trans., 1 (1970), 1169-76.

46. J. Gurland, pp. 621-26 in Proc. 2nd Int. Conf. on Strength of Metals and Alloys, Vol. 2, ASM, Metals Park, 1970.

47. J. C. M. Li and T. Y. Chou, Met. Trans., 1 (1970), 1145-59.

48. J. P. Hirth, Met. Trans., 3 (1972), 3047-67.

49. A. W. Thompson, pp. 89-126 in Work Hardening in Tension and Fatigue, TMS-AIME, New York, 1977.

50. J. C. M. Li, Trans. TMS-AIME, 227 (1963), 239-47.

51. A. W. Thompson, Metallography, 5 (1972), 366-69.

52. I. M. Bernstein and B. B. Rath, Met. Trans., 4 (1973), 1545-51.

53. C. M. Young and O. D. Sherby, J. Iron Steel Inst., 211 (1973), 640-47.

54. A. W. Thompson and M. I. Baskes, Phil. Mag., 28 (1973), 301-308.

55. A. W. Thompson, M. I. Baskes, and W. F. Flanagan, Acta Met., 21 (1973), 1017-28.

56. M. F. Ashby, Phil. Mag., 21 (1970), 399-424.

57. A. W. Thompson, Acta Met., 23 (1975), 1337-42.

58. J. C. Williams, A. W. Sommer, and P. P. Tung, Met. Trans., 3, 2979 (1973).

59. M. J. Blackburn and J. C. Williams, Trans. ASM, 62 (1969), 398.

60. T. J. Koppenaal and D. Kulhmann-Wilsdorf, Appl. Phys. Lett., 4 (1964) 59.

61. J. C. Chesnutt, C. G. Rhodes, and J. C. Williams, ASTM-STP600, ASTM, Philadelphia, 1975, p. 99.

62. G. Garmong and J. C. Williams, Met. Trans., 6A, 1711 (1975).

63. R. B. G. Yeo, Trans. TMS-AIME, 224 (1962), 1222-27.

64. W. C. Leslie and R. J. Sober, Trans. ASM, 60 (1967) 459-84.

65. G. R. Speich and H. Warlimont, J. Iron Steel Inst., 206 (1968) 385-92.

66. G. R. Speich, Trans. TMS-AIME, 245 (1969) 2553-64.

67. G. R. Speich and W. C. Leslie, Met. Trans., 3 (1972) 1043-54.

CYCLIC DEFORMATION, FATIGUE CRACK NUCLEATION AND PROPAGATION IN METALS AND ALLOYS

Campbell Laird

Dept. of Materials Science and Eng.
University of Pennsylvania
Philadelphia, PA 19104

INTRODUCTION

Fatigue was first recognized as a problem in the performance of mining and transportation equipment in the early part of the 19th century. Since then this complicated and dangerous phenomenon has been actively studied with increasing intensity; due to increasing demands made on structures and materials, improved understanding of the fatigue phenomenon and improved design methods to cope with it have been constantly pursued.

Because of the importance of the problem which extends into many branches of technology, many treatises exist about fatigue. Some of the newer books dealing with mechanistic aspects are listed in references (1)-(3). Many review articles have been published (4-14), of which reference (4) gives excellent coverage of the older work. In the light of these many articles, and the limited space available here, no attempt will be made on a comprehensive review of fatigue. There have been encouraging advances (in the last five years) and the most recent work will therefore be emphasized.

Since it is embarrassing if structures fail catastrophically before the end of their useful lives, designers attempt to design their components so that the cyclic stresses are low, especially if these stresses are cycled frequently as in helicopters and engines, or if the life is long, as in bridges and machine tools. Consequently the fatigue life problem is generally one of low strain, and most work correctly emphasizes the low strain regime. Nevertheless fatigue failures are caused by plastic deformation, and at low stresses such deformation is a phenomenon of microstrain, although the magnitude of the strain might be quite high locally in the presence (or even the absence) of macroscopic stress-concentrators. A good part of modern research is therefore concerned with cyclic deformation.

In some structures, large cyclic strains (i.e. a percent or so) may occur due to start-up operations or emergencies. Although such strains are damaging, they are acceptable because it is hoped they do not occur very often. The study of their damaging effects gave rise to the field of high strain fatigue. Moreover, because laboratory specimens have to be short to avoid buckling during push-pull loading, high strains are the easiest to measure and control, and therefore offered the most obvious route to a direct study of fatigue damage. The regimes of high and low strain fatigue are distinguished in materials-oriented circles as follows: If the plastic strain of a specimen is greater than the elastic part of the total strain, then the fatigue is considered "high strain." If the elastic strain is greater than the plastic strain, the fatigue is considered low strain fatigue. The fatigue life associated with that total strain in which the magnitudes of the elastic and plastic strain are equal is termed the "transition life". Since there are differences, as well as similarities, in the mechanisms of fatigue at high and low strain, it is probably best to regard them separately, and that approach is used here. For commercial materials, the transition life does usually serve mechanistically to distinguish high strain from low strain fatigue, but the distinction is less meaningful in model materials. For example, in a copper single crystal, an applied strain amplitude as high as 0.01 causes failure by a low strain mechanism.

A specimen subjected to cyclic strain undergoes hardness changes due to the cycling, and these changes condition the material to form a crack. There are many components which, in the experience of this investigator, can be regarded as "smooth", i.e., lacking in significant stress concentrators. The study of crack nucleation is therefore appropriate and valuable. However, many workers feel that most structures are of such complex geometry or contain so many flaws due to material defects or assembly errors that studying crack propagation from existing flaws is more valuable work. The great emphasis on crack propagation (mainly the kinetics of crack propagation) during the last twenty years is a direct result of this attitude. This approach is ofter appropriate and has given rise to most helpful methods in design and failure analysis. Accordingly, in this article, mechanisms of both crack nucleation and propagation are discussed.

CYCLIC DEFORMATION

MATRIX BEHAVIOR

Most pure metals, in mono- or polycrystalline form irrespective of crystal structure, undergo large degrees of hardening under cyclic strains if the strain amplitude is greater than about 10^{-3} or so. At strain amplitudes much much lower than this, bcc metals are unique in not hardening very much. However, the other metals do. A plot of the maximum stress per cycle (usually the average of the compressive and tensile limits) against the number of cycles or cumulative strain summed without respect to sign is termed a cyclic hardening curve. Such a curve from recent work (15) is shown in Figure 1.

Clearly the hardening rate is rapid early in life and for a wide range of crystal orientations the rate is not affected by orientation, provided the strain amplitude is low (Figure 1a). However, as the amplitude increases, the rate of hardening is lower for orientations which are more definitely single slip (Figure 1b). The overall hardening rate increases with increasing strain amplitude (15-17).

It will be noted from Figure 1 that the hardening rate falls to zero at large cumulative strains and the material is then described as being in saturation. For all the tests illustrated in Figure 1, the saturation stress is independent of strain amplitude and orientation. A plot of the saturation stress versus the applied constant plastic strain amplitude is termed the cyclic stress-strain curve, and the constancy of the stress for a range of applied strains implies that this curve contains a plateau. A typical example of the cyclic stress-strain curve for copper in single slip orientation is shown in Figure 2. The plateau begins at a plastic shear strain amplitude, $\gamma_{p\ell}$ of $6x10^{-5}$ and ceases at $\gamma_{p\ell} > \sim 7.5x10^{-3}$. If the cyclic stress-strain curves for different f.c.c. metals are normalized against their shear moduli, it is found that

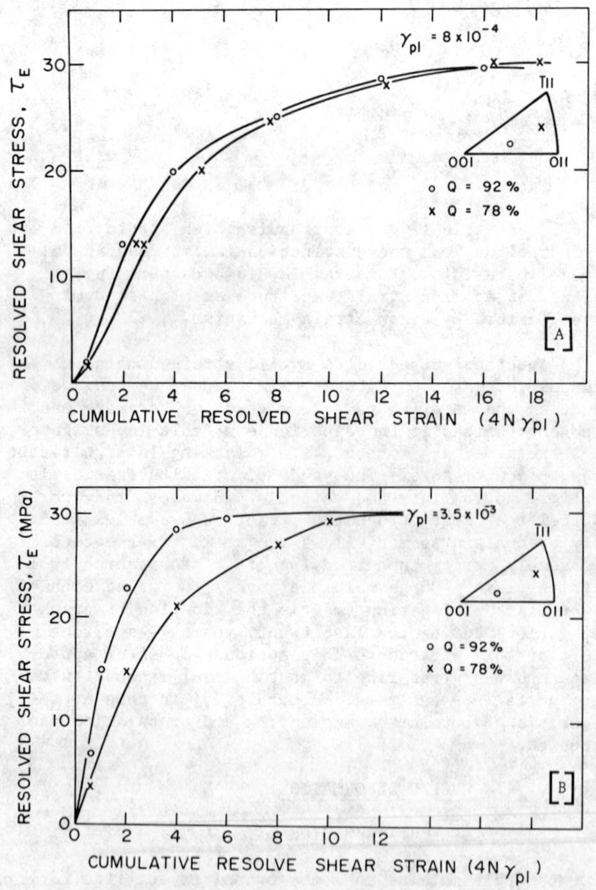

Fig. 1 Cyclic hardening curves for single crystals of different orientations specified by the orientations parameter Q (defined as the ratio of the Schmid factor for the primary system to that of the second most highly stressed system); a) at low strain amplitude, b) at higher strain amplitude. (Courtesy of Cheng and Laird (15).

the curves overlap to a considerable extent and the differences are quite minor[11]. The cyclic stress-strain curve also does not depend on crystal orientation over a very wide range of orientations within the strandard triangle (15). It is possible, however, for the [111] orientation that the saturation level is much greater than that of the plateau shown in Figure 2 (19), assuming that the specimen does not fail rapidly at high stress before saturation occurs.

The onset of the plateau is marked by the formation of persistent slip bands (PSB's) which develop just before saturation is attained (18). Prior to their formation the slip is quite homogeneous and the gage section covered with fine slip. When PSB's occur, the strain is localized within them (17,20), and the remainder of the gage section, the matrix, is deprived of most of the strain it hitherto carried. The interpretation of cyclic deformation in pure metals can thus be analyzed in three parts, matrix behavior or rapid hardening behavior, saturation behavior, and high strain behavior for deformation at strain amplitudes greater than that of the plateau. For strain amplitudes lower than that of the plateau, the deformation remains homogeneous and the strain localization necessary for failure does not occur. No PSB's are formed and the lower end of the plateau marks the fatigue limit (21).

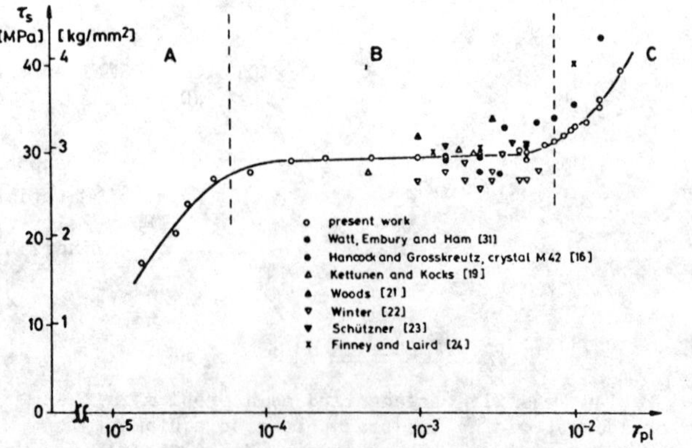

Fig. 2 The cyclic stress-strain curve for copper single crystals: resolved shear stress versus plastic resolved shear strain amplitude (courtesy of Mughrabi, (18).

Transmission electron microscopy has shown that rapid hardening at low strain amplitudes is primarily due to the accumulation of dislocations of the primary slip system (for reviews, see references 12 to 14). With increasing strain amplitude, it is well known that the increasing hardening rate is associated with more secondary slip (22,23). Mughrabi has recently reviewed bulk measurements of magnetic properties of fatigued nickel single crystals which provide complementary information (11). Deviations in the magnetic measurements from the symmetry characteristic of crystals containing primary dislocations occur with increasing cycles at strain amplitudes corresponding to the upper half of the plateau, and these reflect the growing contribution of secondary dislocations. However, early in rapid hardening, and especially at low amplitudes, the dislocation structures consist almost entirely of primary dislocation dipoles. These collect as bundles[22] and gradually build up into dense loop patches separated by free channels, in which screw dislocations are believed to glide (14,24-28).

Two parameters which have been extremely useful in understanding fatigue behavior during rapid hardening have been the friction stress and the back stress (25-27). The friction stress acting on the dislocations, being dependent on point defect hardening, Peierls forces, jog-dragging, etc., is independent of straining direction, but the back stress changes sign during each half cycle. It reaches its maximum value at maximum applied strain, thus acting to lower the yield stress in the reversed direction. As it is of elastic nature due to dislocation interactions it soon decreases on straining in the direction favored by it, and then reverses again so as to oppose the imposed strains. Kuhlmann-Wilsdorf and Laird (25) obtained measurements of these stresses by analysing cyclic hysteresis loops by the Cottrell method. Results obtained from the hysteresis loops of three different investigators, obtained for copper and for amplitudes in the plateau are shown in Figure 3 as a function of cumulative strain. It will be noted that the friction stress and back stress rise in parallel. From this and related analysis, Kuhlmann-Wilsdorf and Laird (25) separate the friction stress into two parts, one equal in magnitude to the back stress (and presumed to have the same physical cause) and the other, smaller, part due primarily to jog dragging by the screw dislocations shuttling in the channels between the loop patches.

Further theoretical interpretation of the friction stress and back stress depends on the behavior of the dislocations in the loop patches - whether they fully participate in the cyclic deformation via loop flipping in the manner of Figure 4, or whether the loop patches are esentially rigid, acting as a non deformable second phase. Taking into account compatibility of plastic deformation and other reasons, Kuhlmann-Wilsdorf concludes strongly in favor of complete participation of the loop patches in the cycling (26). Her analysis implies that the friction stress against prismatic glide of the dipolar loops must be quite small and thus the loops must respond to their interactive stresses by gliding towards their low energy configuration. As long as climb is inhibited (and it appears so), this is a Taylor lattice, shown in Figure 4, with of course some modification because the loops are not infinitely long.

A detailed theory of the Taylor lattice behavior and those of τ_B and τ_F has been given by Kuhlmann-Wilsdorf(26, 27). Lack of space prevents a detailed description of the quantities obtained by this worker and they have been reviewed elsewhere (14). The physical essence of the theory is that, at the strain limits of the cycles, the bulk of the stress is supported by the loop patches, so

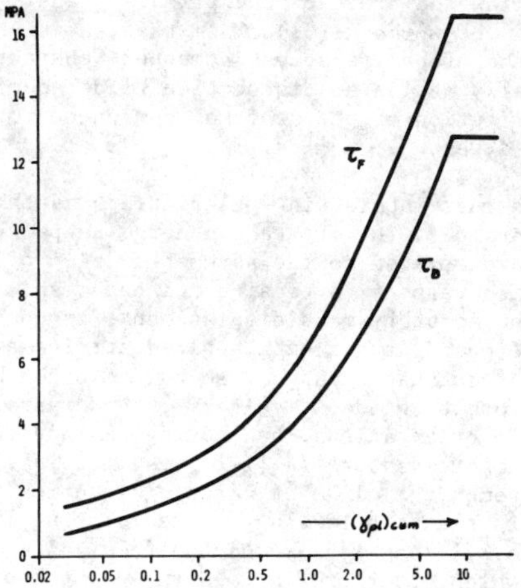

Figure 3 The values of the friction stress, τ_F, and back stress, τ_B, derived from cyclic hysteresis loops by the Cottrell method (Courtesy of Kuhlmann-Wilsdorf and Laird[25]).

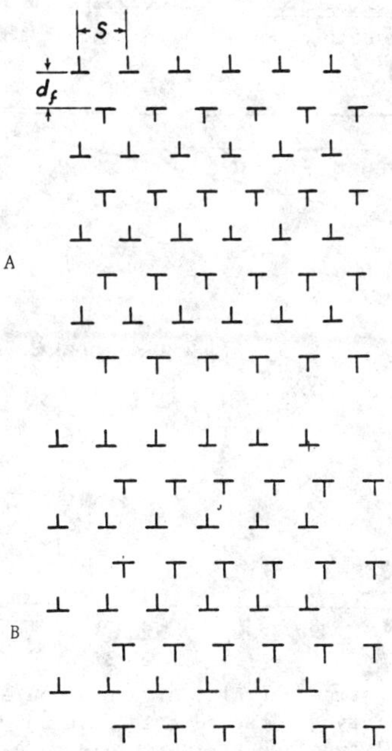

Figure 4. The Taylor dislocation lattice corresponding to an idealized description of the structure in the loop patches (a) prior to shearing the top of the lattice to the left with respect to the bottom and (b) afterwards. The rows of oppositely signed dislocations have slipped. (Courtesy of Kuhlman-Wilsdorf (26).

that the back stress roughly equals one half of the stress at which the loop patches would deform in the absence of friction since they occupy only half of the volume. The part of the friction stress which is equal to the back stress comes about because the dislocation motions within the loop patches are irreversible for every dislocation as soon as it flips with its neighbors of opposite sign. Thus this component of the friction stress, along with the back stress, is given by the flipping stress of the Taylor lattice (27).

P. Neumann, in unpublished work, has recently attempted to model loop patch behavior by computer techniques. His results agree in general with those of Kuhlmann-Wilsdorf; however, he has discovered additional details about the loop patch behavior, for example, the tendency of the surfaces of the loop patches to adopt specific crystallographic planes which gives them a diamond shape for the cross section of the loop patches at right angles to the direction of the dislocation axes. However, arbitrary loop patch profiles can be obtained by combinations of the various possible surfaces, just as a mixed dislocation is made up of screw and edge components. The ideal diamond shape is rather simply explained by Kuhlmann-Wilsdorf (private communication) in that each dislocation attracts its own kind within a $\pm$ 45° sector and repels it for larger angles, measuring the angle from the plane of the extra half-plane of the edge dislocations. As a result, a diamond shape is the most stable.

Some very recent work has addressed a problem with the analysis Kuhlmann-Wilsdorf and Laird (25). When these workers derived values of τ_F and τ_B as a function of cumulative strain from the results of different workers they forced the data to fit single expressions for the maximum stress and the yield stress in each cycle although, as mentioned above, the different workers used different strain amplitudes. This implicitly assumes that τ_B and τ_B would not be affected by strain amplitude. However, from the results of Figure 1, they clearly are affected by amplitude and orientation. Accordingly, ad justments have been made to the Kuhlmann-Wilsdorf theory by allowing for variation in dislocation behavior with strain amplitude (15), as follows:

In the earliest stages, rapid hardening takes place by multiplication of primary dislocations, of which many become trapped as edge dislocation dipoles. Patches of dipoles then act as obstacles to the motion of the primary glide dislocations. At this time, screw dislocations glide over long distances in the channels between loop patches and the Taylor sub-lattices of dislocations in the loop patches flip relative to each other several times the inter-loop spacing. Later in rapid hardening, the channel screw dislocations glide over shorter distances and the loops do not accomplish a single flip but merely oscillate about their equili-

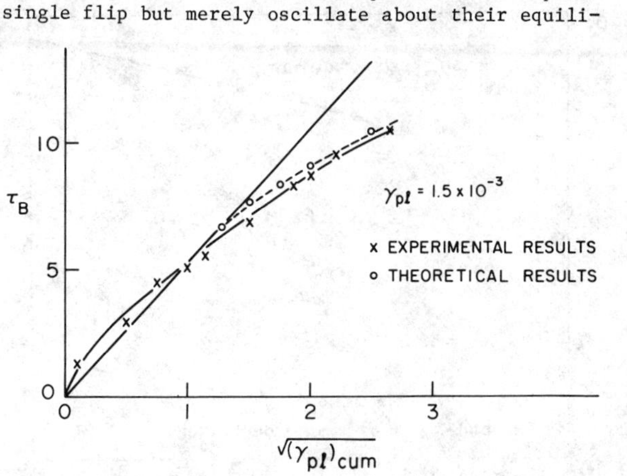

Fig. 5 Comparison of Cheng and Laird's measurement of τ_B for a strain amplitude of 1.5×10^{-3} and the value calculated from their modification of the Kuhlmann-Wilsdorf theory (straight line). Taken from reference 15.

brium positions. The cumulative strain at which loops cease to slip can be called the "rapid hardening transition point". The semi-quantitative treatment of Cheng and Laird (15) describes the variation of the transition point with strain amplitude, and also the dependence of the friction stress and back stress on strain amplitude and orientation. Figure 5 illustrates one of their comparisons between theory and experiments for a specimen cycled in the mid region of the plateau. Since the abscissa is the root of cumulative strain, the straight line is the Kuhlmann-Wilsdorf prediction for the back stress. The back stress is seen to vary from this prediction by a minor amount; the cumulative strain corresponding to the transition point is the intersection of the straight line and the curve.

The effect of alloying on matrix and PSB behavior is controlled to a large degree by the alloy effect on the planarity of slip. If the stacking fault energy is very low, then the cyclic hardening rate is much reduced (23). Wilhelm and Everwin (29) have noted a strong suppression of cyclic hardening for plastic strain amplitudes less than 10^{-4} in Cu - 2% Co alloy. Woods (16) observed little effect in Cu-5% Al alloy single crystals because this alloy is quite wavy in slip character. However, she observed a marked softening, probably due to short-range ordering effects, with PSB formation. Abel and co-workers who studied a wider range of alloys in the Cu-Aℓ systems (30) also observed marked softening effects with the onset of saturation. The results show that the saturation stresses increase with decreasing stacking fault energy and there is a sharp improvement in fatigue performance. This in turn suggests that the strain localization is reduced. Alloy behavior needs a great deal more study, however, because it is far from well understood.

Cyclic Hardening In B.C.C. Metals

The cyclic deformation of metals with crystal structures other than F.C.C. have been repeatedly reviewed (10,11,13). It was pointed out (10) that many of the properties of these metals previously observed for unidirectional deformation carry over into cyclic deformation, just as in the behavior of F.C.C. metals. Thus the factors of temperature, strain rate and small amounts of impurities, especially of interstitials, play large roles in the cyclic deformation of B.C.C. metals, but are not so important for F.C.C. metals. Thus under slow strain rates, B.C.C. metals can behave similarly to F.C.C. metals, but under other conditions their behavior is quite different from that of F.C.C. metals. The details for B.C.C. metals, many of which have been elucidated by Mughrabi and his co-workers (31-33) are as follows:

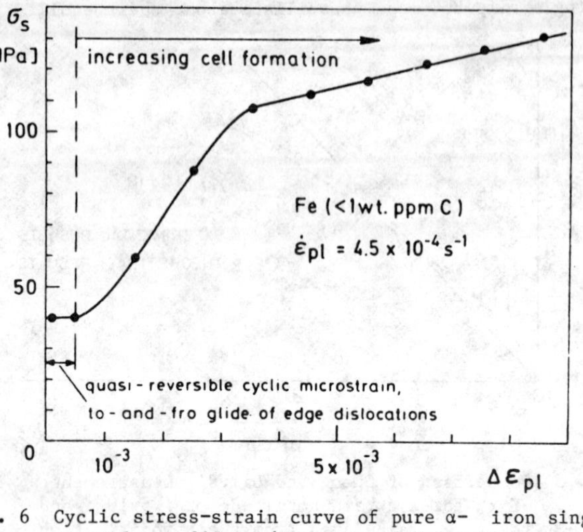

Fig. 6 Cyclic stress-strain curve of pure α- iron single crystals at 295K. (Courtesy of Mughrabi and co-workers (31).

For pure single crystals of B.C.C. metals cycled at low temperatures and at a high strain rate ($>10^{-4}$ S^{-1}), the cyclic stress strain curve is typical of that of iron shown in Figure 6. Unlike copper, a PSB plateau does not occur. Instead for $\Delta\varepsilon p\ell \nless 5\times10^{-4}$, cyclic hardening and microstructural changes are negligible (31). The dislocation arrangement is dominated by screw dislocations in low density and has been explained by assuming that only edge dislocations move to-and-fro as the strain is reversed in a non-hardening, mechanically reversible manner (31-33). It will be recalled that the thermally activated glide of the screw dislocations is strongly impeded at ambient temperature and below, giving rise to a behavior reflecting typical features of the low temperature mode of unidirectional deformation of B.C.C. metals. At higher strain amplitudes, cyclic hardening is pronounced and is associated with the formation of equiaxed cells. The slip traces are generally diffuse and in single crystals oriented for single slip do not follow the trace of the primary plane. Mughrabi and his co-workers (31) claim that, at high strain rates, the effective stress component characteristic of thermally activated dislocation glide dominates the contribution of the athermal component to the total flow stress.

One of the most interesting observations is that cyclically deformed B.C.C. crystals undergo shape changes (31-36) which are related to the asymmetric slip of screw dislocations with respect to tension and compression. They are brought about by mass displacements in unconstrained directions. Rather high applied strains are required to produce shape changes, so as to force the screw dislocations to glide. Typically, for crystals with single slip orientations, and having primary dislocations of Burgers vector $^a/_2$ [111], asymmetric slip can develop through the glide of screw dislocation segments on slip planes belonging to the $\overline{1}01$ zone in tension and the $\overline{2}11$ zone in compression (32). The shape changes which occur are impressive, being easily visible to the naked eye. The cross-section of a cylindrical single crystal cycled in push-pull will change from circular to elliptical and because of the constraints of the grips, additional couples are applied to the specimen which can give it an 'S' shape. More complicated shape changes have been reported recently on α- iron crystals of orientations other than that of single slip (37).

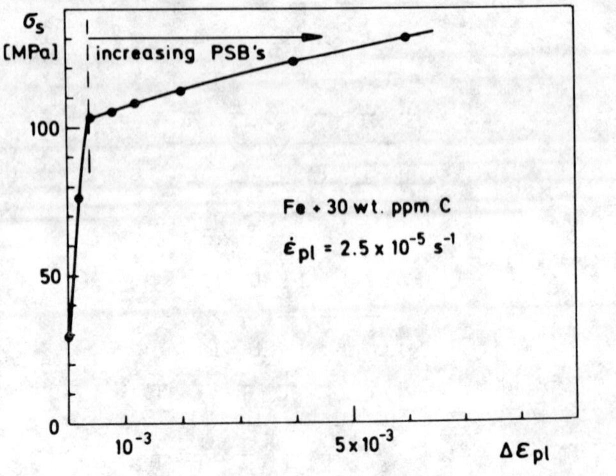

Fig. 7 Cyclic stress-strain curve of impure α-iron single crystal tested step fashion with respect to amplitude and at low strain rate. (Courtesy of Mughrabi and co-workers (31)).

Slip plane asymmetry can be overcome by cycling at higher temperatures, at lower strain rates ($\sim10^{-5}S^{-1}$), and with interstitial impurities present. Investigations conducted under these conditions have been made (31). For a strain rate of 2.5×10^{-5} S^{-1} and an α-Fe crystal containing 30 wt ppm carbon, the cyclic stress strain curve is shown in Figure 7. Different from specimens tested at higher strain rates, considerable cyclic hardening was observed down to a plastic strain range of 10^{-4}. Shape changes were much less pronounced and the contribution of the athermal component of the flow stress, i.e. that due to the elastic interaction of the dislocations, dominated the effective stress component at all strain amplitudes (31). Mughrabi and his co-workers observed slip lines paralled to the trace of the primary slip plane, and sometimes these were intense enough to be labelled PSB's (31). Moreover, their density increased with increasing strain amplitude. Conruent with their appearance, the slope of the cyclic stress-strain curve flattened to a gently sloping plateau. Mughrabi et al were not convinced they had evidence of localized deformation in the PSB's (31) but it is almost certain that strain localization does occur because of the widespread observation of PSB's in polycrystalline B.C.C. metals in old work. Thus the overall behavior is seen to be similar in some respects to that of cyclically deformed F.C.C. crystals.

Just as similarities between the mechanisms of monotonic and cyclic deformation are observed for F.C.C. and B.C.C. metals, so also have they been pointed out for H.C.P metals (10). A few investigations of the cyclic deformation of H.C.P single crystals have been made (38, 39). Kwadjo and Brown working with magnesium have shown that PSB's form at suitable strain amplitudes and the properties of the PSB's are quite similar to those of F.C.C. metals (39). However, at high strains, twinning is extensive and significantly affects the hardening; in this respect, the behavior of H.C.P metals is dissimilar.

Persistent Slip Bands

As noted above, the plateau of the cyclic stress-strain curve is associated with the formation of PSB's and the localization of the strain within them. Localized cyclic deformation in the PSB's is one of the most important features of saturation behavior because this gives rise to failure mechanisms. PSB's are the main failure iniatator at low strains, and therefore in long life, the most prevalent form of fatigue damage. The properties of PSB's have been studied by many workers, mainly in copper (for reviews, see references 10-14), and it is likely that the behavior of this material is quite typical of many materials. Since nearly all the applied strain is carried by the PSB's, they completely traverse the specimen as lamellae parallel to the primary glide plane, and thus they determine bulk deformation behavior although there is some evidence that they nucleate at or near the surface. PSB's first form at the onset of the plateau and the strain amplitude corresponding to this point is that required to form a single narrow PSB. With increase in strain amplitude, the volume fraction of PSB's gradually increase until, at the high amplitude end of the plateau, essentially the whole gage section becomes a super-PSB. Mughrabi has tabulated the strain amplitudes at which PSB's first form and ultimately saturate (11) and this table is reproduced here as Table 1,

TABLE 1. Characteristic Properties of Persistent Slip Slip Bands and the Matrix in Which They Form

Mat.	$\gamma_{p,M}$	$\gamma_{p,PSB}$	τ_{PSB} MPa	$\tau_{PSB} \times 10^4/G$
Ni	$\sim 10^{-4}$	$\sim 7.5 \times 10^{-3}$	52	6.6
Cu	$\sim 6 \times 10^{-5}$	$\sim 7.5 \times 10^{-3}$	28	6.5
Ag	$\sim 6 \times 10^{-5}$	$\sim 7.5 \times 10^{-3}$	17.5	6.6
Cu-2%Al	$\sim 10^{-4}$	$\sim 3 \times 10^{-3}$	33	7.8
Cu-2% Co	$\sim 3 \times 10^{-4}$	$\sim 5 \times 10^{-3}$	27.5	6.5
α-Fe +30 w/o ppmC ($\dot{\varepsilon}_p$ = $2.5 \times 10^{-5} s^{-1}$)	$\sim 4 \times 10^{-4}$	$\gtrsim 6 \times 10^{-3}$	~49	6.5

Notes to Table 1

$\gamma_{p,M}$ = strain amplitude corresponding to the onset of PSB's and the maximum matrix strain in the presence of PSB's.

$\gamma_{p,PSB}$ = strain amplitude corresponding to the high strain end of the plateau and the average strain in the PSB.

τ_{PSB} = plateau stress, that necessary to deform PSB's

G = shear modulus, ε^{o}_{p} = cyclic plastic strain rate. Values for the F.C.C. solid solution alloys taken from reference (29). The other data are due to Mughrabi and co-workers (31).

with significant modifications in the definitions of the terms because such definitions imply interpretation, and the present author differs from Mughrabi on several points, while broadly agreeing on others. The differences will be brought out in the following. It is important to note, however, the astonishing constancy of the PSB flow stress normalized against the shear modulus for the different materials. The reason for this behavior is that the deformation is controlled by elastic interactions of dislocations and the only controlling parameter expected to vary greatly between the various materials will be the shear modulus. Based on studies of dislocation mechanisms, this behavior was predicted (10).

In one view of PSB behavior (11), the amplitudes $\gamma_{p,M}$ and $\gamma_{p,PSB}$, defined in Table 1, are regarded as the local plastic strain amplitudes in the matrix and PSB's respectively, and thus the applied plastic strain amplitude is related to them by the rule of mixtures. This implies that the localized strain in the PSB's is constant, and on average, the rule may work. However, the deformation in the PSB's is not constant and varies widely. This was evident in the early work of Finney and Laird (17) and has now been emphasized by them in a recent article (40). These workers employed an interferometric technique to study PSB behavior: after cycling into saturation, they would interrupt a test, repolish the surfaces of the specimen on which the PSB's were evident, subject the specimen to a quarter cycle of strain, and then study the slip offsets in the PSB's by two-beam interferometry. Although such an approach emphasizes deformation behavior at the surface, the dislocation structure is known by etch pitting studies (41, 42) to extend virtually unchanged through the bulk of the crystal, and the slip offsets observed must reflect the bulk behavior.

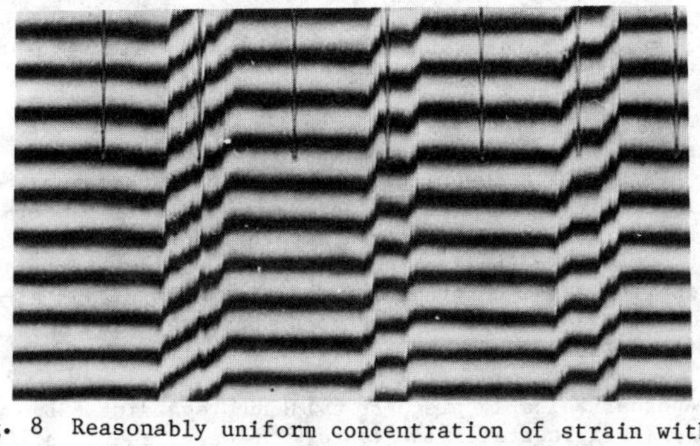

Fig. 8 Reasonably uniform concentration of strain within micro-PSBs, of which five are visible in the field of view. Single crystal specimen cycled at a plastic strain amplitude of ± 0.00125 for 30,000 cycles, repolished, and then given a plastic strain increment equal to its strain amplitude. The fiducial marks are distant 100 μm apart. Note the strong strain concentration in the lower half of the middle band, as well as the evidence of negative strain in the top half of the left band. Taken from reference (40).

PSB's often occur as macro-bands, just visible to the naked eye. These are composed of smaller bands, termed micro-PSB's (40) sometimes very narrow and appearing as a single step of coarse slip, but mostly in wider packets easily observed in the interferometer. Typical examples of packets of slip produced by 1/4 cycle after repolishing are shown in Figure 8.

Since a fringe displacement corresponds to a step height of ~300 nm, and the width of the bands shown in Figure 8 can be gaged from their projected widths, it is apparent that the strain localized in these bands is about 0.01, or that corresponding roughly to the upper end of the plateau. Thus these bands are termed "well-behaved" (40). The variation in strain localization within these particular bands is less than a factor of two and the individual micro-PSB's carry in the order of 5 to 10 times as much strain as the macro-PSB's, because the strain in the macro-PSB's is the sum of the strains in the micro-PSB, disregarding the matrix strain that is (40). It is important to note that most other experimenter's who did not use interferometry would have difficulty disquishing macro-PSB's from micro-PSB's, because of confusion from the effects of accumulated slip.

Many micro-PSB's have been observed where quite large steps were formed in very narrow bands (40). A typical example of these large strain concentrations is shown in Figure 9. The specimen shown in this figure had been cycled 2763 3/4 cycles (i.e. interrupted from compression), repolished and then subjected to 1 1/4 cycles, thus being returned to the state of zero stress and strain, but with a net of 1/4 cyclic tensile strain from the point of repolishing. It will be noted that the steps do not all have the same sign and thus some of the steps were formed in the previous compression stroke; they were thus wholly or partially unreversed. If only one

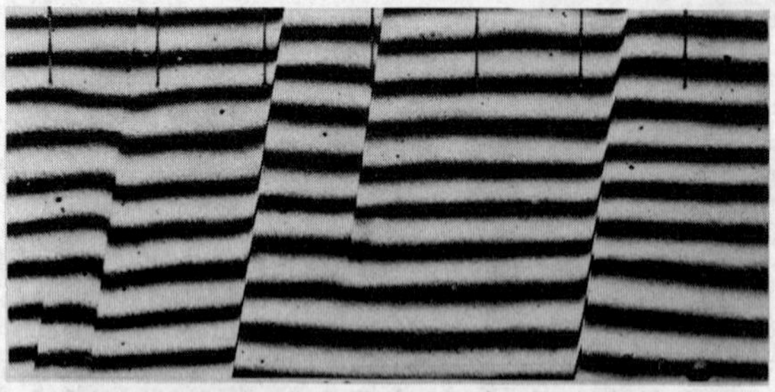

Fig. 9 Intense concentration of plastic strain at forward as well as unreversed slip steps (micro-PSB's). Specimen strain cycled at plastic strain of $\pm$ 0.0025. History given in text. Fiducial marks distant 100μ apart. Note that the shear strain is not uniform along the individual slip lines. Taken from reference (40).

atomistic slip plane was involved in forming each of these steps, which is unlikely but possible, then their plastic strain concentration factors would be of the order of 10^5. The investigators felt that a value of about 1000 was more realistic (40). These values of strain localization are much greater than seems widely appreciated, and have helpful implications for understanding the dislocation mechanisms involved.

Another valuable piece of evidence regarding PSB behavior is illustrated in Figure 9, namely, the shear strain is not constant along the individual slip lines. Clearly, unlike the PSB's as a whole, individual slip lines in PSB's do not pass right through the specimen, but are only about 0.2 mm long, to give some estimate; there is much variation about this figure, and it is not possible to tell how long the lines are in the complicated middle regions of well-behaved micro-PSB's.

The strain carried by groups of PSB's is macroscopically reversible, in that the local strain at the tensile maximum is identical to that at the compressive maximum. Also, the local strain within a band appears to increase linearly with the overall strain during a typical excursion. Notwithstanding the results of Figure 9, individual micro-PSB's also appear to deform reversibly with the cycling of the strain. Only at the finest distribution of slip within a PSB is the deformation not strictly reversible. This can be seen in Figure 10, where a slight roughening of the surface is associated with the action of several cycles, although the slip offsets continue to be mechanically reversible.

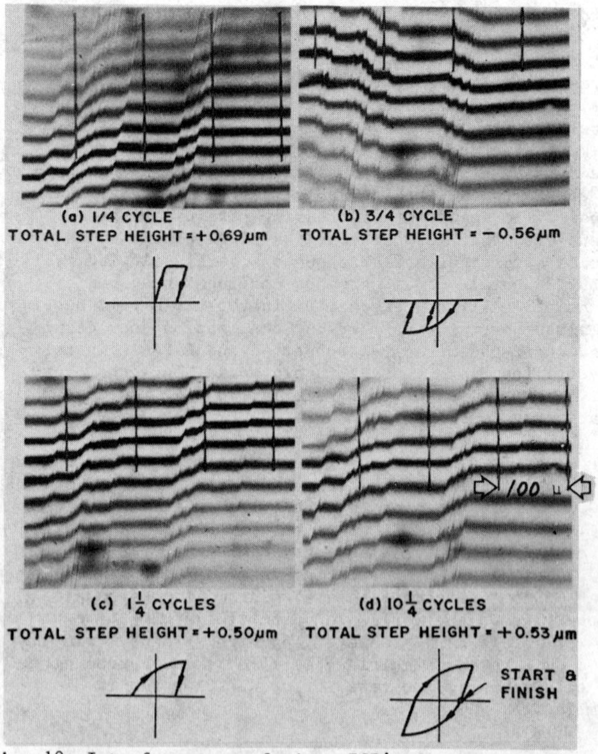

Fig. 10 Interferograms of micro-PSB's in a copper single crystal cycled at a plastic shear strain amplitude of 0.0025 for 2750 cycles, repolished, and subjected to the strain increments indicated. Courtesy of Finney and Laird (17).

Once PSB deformation is established to carry the applied strain, the population of PSB's and the variation in localized strain amongst them becomes stable and remains that way until fracture processes terminate the life (43).

Our understanding of how this deformation is accomplished by dislocation mechanisms is based mainly on observations of dislocations structures by TEM. Seen from a direction normal to the Burgers vectors, the appearance of a typical narrow PSB is shown in Figure 11, and the relation of the PSB to the matrix structure of loop patches in Figure 12. With increase of strain amplitude, the 'rungs' of the PSB's extend until the whole structure comprises dislocation dipolar walls (Figure 12). The

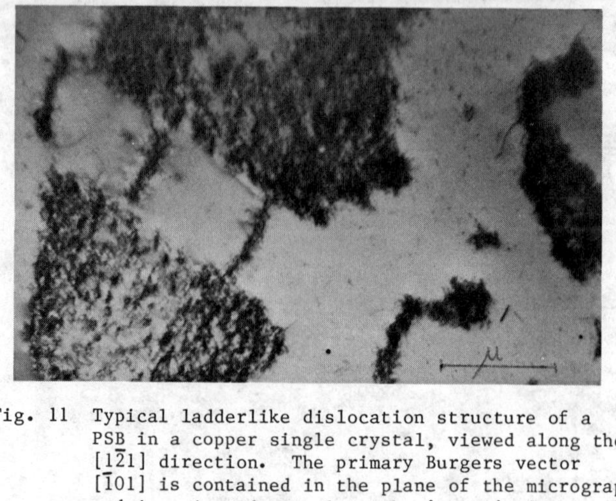

Fig. 11 Typical ladderlike dislocation structure of a PSB in a copper single crystal, viewed along the $[1\bar{2}1]$ direction. The primary Burgers vector $[\bar{1}01]$ is contained in the plane of the micrograph and is oriented normal to the 'rungs' of the ladder. Courtesy of Finney (44).

dipolarity of the walls is proved by the absence of contrast variations across the walls indicating zero variations of the deviation from the Bragg condition and thus zero tilts.

How the dislocations glide in the PSB's has been the object of much study. Finney and Laird (17) suggested that dislocations bow out of the dipolar walls and glide across the intervening channel until they encounter the adjacent walls, where they become trapped and partially annihilated by interacting with dislocations of opposite sign. Screw dislocations are then liberated to glide down the channel until they encounter oppositely-signed screw dislocations which can mutually annihilate by cross-slip. Thus an equilibrium between dislocation multiplication and annihilation is considered to exist in the PSB's. Mughrabi and co-workers have adopted this mechanism (31) and feel it to be supported by TEM observations of wall structures and glide dislocations in which the dislocations have been pinned in place under load by neutron irradiation

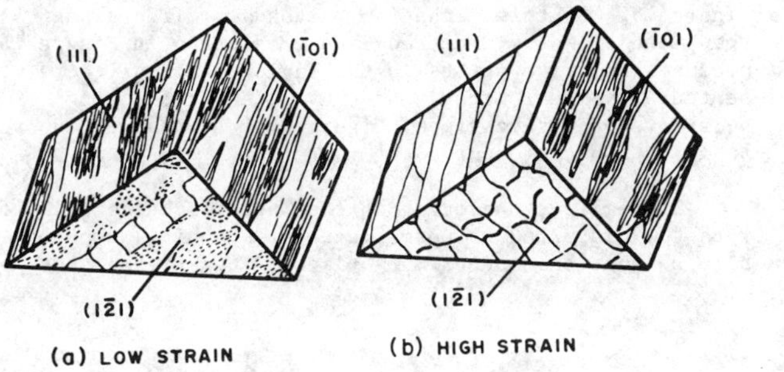

Fig. 12 Three dimensional configuration of dislocations in copper single crystals strain cycled to saturation. (a) a single narrow PSB contained in the matrix structure of loop patches, (b) the whole crystal is a PSB. Courtesy of Finney (44).

(11). In this model, the plastic strain in the PSB's is limited by the maximum density of screw dislocations arranged in groups of the same sign (see Fig. 13), that can be maintained in steady state in the channels. Mughrabi has developed a composite model of cyclic plasticity in the PSB's (11) based on this model.

Kuhlmann-Wilsdorf and co-workers (24, 40), however, have emphasized the role of the screw dislocations in the channels at the expense of the edge dislocation bowing mechanism. Their model of the PSB dipolar wall is shown in Figure 14. The structure of this dipolar wall takes into account the prominence of screw dislocation motion in the channels of the PSB's. These screw dislocations, gliding in groups of the same sign, move in coordination between PSB walls, so as to lay down simultaneously edge dislocations of opposite sign in dipolar configuration at the walls. The detailed manner in which this could be done is shown in Figure 15. In all except Fig. 15(a), screw dislocations gliding in the channels pick up edge dislocations or lay them down in coordination. Of all the mechanisms shown, those in Figures 15d, e, f are considered most likely because they maintain the dipolar balance of the wall structure and thus the lowest energy configuration of the dislocations.

The most reasonable position seems to be that both the coordinated screw dislocation mechanism and the dislocation link bowing mechanism occur (40), but to different degrees (40). Laird et al (40) point out that the interferometric evidence most strongly supports the coordinated motion of the screw dislocations. The most relevant evidence in this connection is that the slip line length in the PSB's is ~ 200 μm while the channel width of the PSB's is 1.3 μm at room temperature. Such long slip line lengths, and their slow variation in shear-strain along the length, is fully consistent with a model of groups of screw dislocations gliding in coordination, but cannot be explained by the bowing out of edge dislocation loops from walls, since that would presuppose the prior existence of edge dislocation pile-ups in the walls. Furthermore, the slip

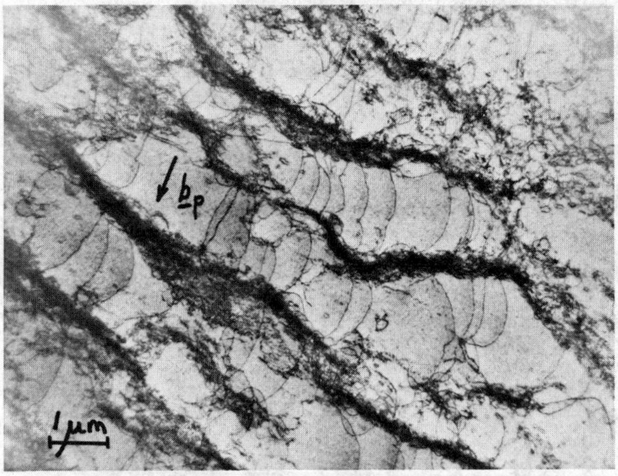

Fig. 13 Screw dislocations, predominantly of one sign, spanning dislocation walls in copper cycled 300 times at a plastic shear strain amplitude of 0.005 and at room temperature. The section of the neutron-irradiated specimen is taken parallel to the primary glide plane and the shear stress was given as 3.2 kg/mm^2. Courtesy of Mughrabi et al (31).

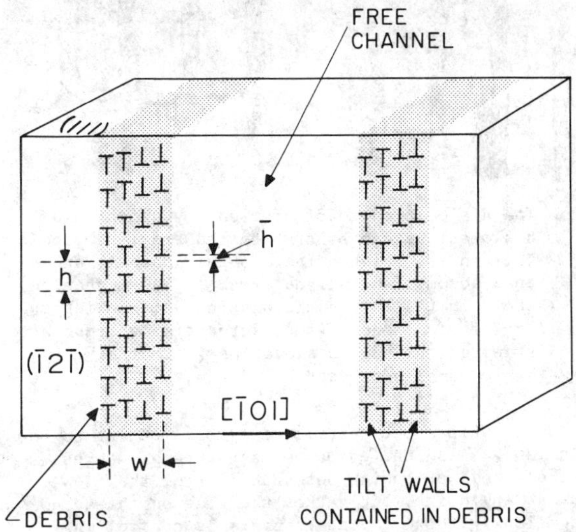

Fig. 14 The structure of dipolar walls in PSB's seems to consist of tilt wall layers embedded in dislocation debris. For the two walls shown, there are four tilt walls.

lines do not traverse the whole specimen, i.e. opposing groups of screw dislocations do evidently meet on sufficiently closely spaced atomic planes that they can mutually annihilate by cross-slip.

On the other hand it seems clear that some edge dislocation bowing must also occur (40). Screw dislocations are annihilated where they meet others of opposite sign and where they emerge at free surfaces. Therefore, while perhaps many screw dislocatins move to and fro for several cycles, a large fraction of them is liable to be annihilated within any one half cycle. These must be regenerated. Some of them may come from free surfaces, but because slip lines don't pass completely through the specimen, the majority of the screw dislocations must be regenerated in the interior. The only mechanism by which this can occur would appear to be the bowing out of edge dislocations. Due to the ordering of the edge dislocations into the tilt wall arrangements in the PSB's, the screws reform into their peculiar groupings. However, as to the strain, edge and screw dislocations, if of comparable total line lengths, are bound to contribute glide in proportion to their mean free paths, i.e. as 1.3 μm (the channel width) to $\frac{1}{2}$x 200 μm (the screw dislocations glide outwards from a source), i.e. as 1:100. Correspondingly, the bowing edge dislocations provide

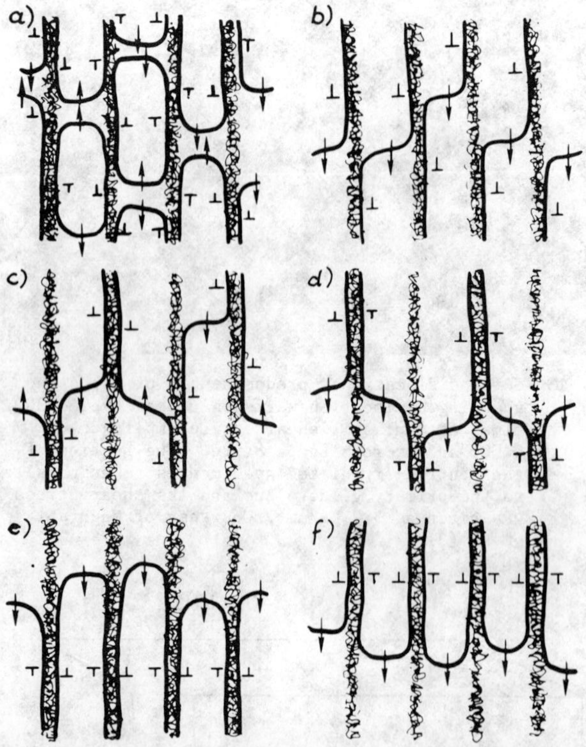

Fig. 15 The dislocation configurations by which edge dislocations can be transferred from one dipolar wall to its neighbor across the channel, and thus accomplish fatigue strain. The mechanism shown in (a) is the dislocation wall-bowing model. In the other mechanisms, screw dislocations act to transfer the edge dislocations. Courtesy of Kuhlmann-Wilsdorf and Laird (24).

screw dislocations but contribute almost nothing to the strain (40). Also the evidence most strongly suggests that the stress of bowing does not control the flow stress at any point, but rather the drag on the screw dislocations gliding in the channels is controlling.

In connection with the question of stress control, it is interesting to note that Basinski et al (42) have shown PSB formation to occur at low temperatures in a manner similar to that at room temperature. The plateau stress increases with decrease of temperature (to ~ 48 MPa at 77.4K and ~ 73 MPa at 4.2K) and the wall spacing of the PSB's decreases consistently.

A full description of PSBs, the structure of their walls and of the dislocation motions is available (24).

Mechanisms for the formation of PSB's from the loop patch structure have been considered from time to time. Mecke carried out TEM on single crystals of nickel subjected to two-step strain cycling tests and observed narrow, widely-penetrating slip bands formed to accommodate the higher strain amplitude of the second steps (45). These bands appeared to have formed catastrophically, as if dislocations gliding in the channels punched through the loop patches and altered the structure in a lamellar region akin to that of a PSB. This mechanism of PSB formation may be unique to experiments involving stress-increments.

In regular, constant strain tests, the mechanism of PSB formation seems to take place more gradually. Both Mecke (45) and Winter (46) have published TEM photomicrographs showing a close geometrical connection between the walls of PSBs and the loop patches, namely, the walls are coincident with the outer layers of the loop patches. These observations are not clear-cut evidence of PSB formation since the TEM sections may have sliced at a glancing angle through mature PSB's rather than be representative of a PSB in process of birth. However, they do suggest that the instability which forms a PSB and must be prsent because a load drop occurs on PSB formation, occurs on a small scale and probably inside the loop patches. These, of course, are undergoing straining before the PSB's form.

Kuhlmann-Wilsdorf and Laird (24, 28) have considered mechanisms of PSB formation. They have ruled out dislocation climb and glide processes on the primary system as responsible for the mechanism (28). Instead they have suggested that glide on secondary planes in a direction having a component normal to the primary plane can remove loops inside loop patches so as to clear their central volumes by the mechanism outlined in Figure 16. By this mechanism, secondary glide along the plane shown by a dotted line projection in Figure 16 produces relative shifts in the Taylor lattice (represented with ideal order) and thus lines up opposing half-rows of dislocations, Figure 16b, so that extensive annihilation can occur.

In reality, the order of the Taylor lattice will be much too disturbed to permit extensive annihilation by slip on just one non-primary plane. The stresses acting

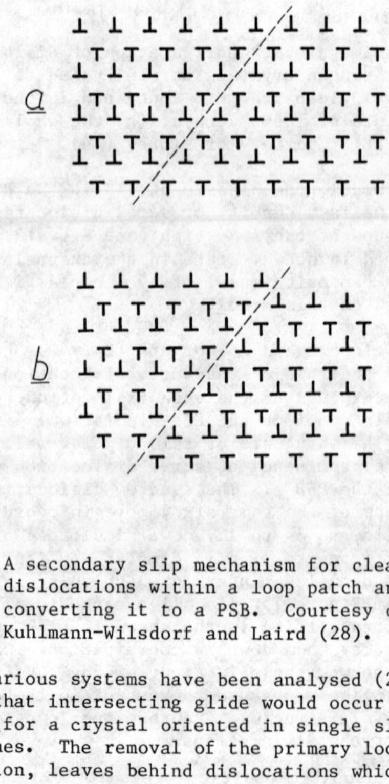

Fig. 16 A secondary slip mechanism for clearing the dislocations within a loop patch and thus converting it to a PSB. Courtesy of Kuhlmann-Wilsdorf and Laird (28).

on the various systems have been analysed (28) and these suggest that intersecting glide would occur on a few systems (for a crystal oriented in single slip) including cube planes. The removal of the primary loops, in civilian fashion, leaves behind dislocations which are quite mobile, liable to annihilate, and leave debris which is then swept up into the walls of the PSB's (28).

High Strain Cyclic Deformation

At an applied strain of $\gamma_p \sim 7.5\text{x}10^{-3}$, as noted above, the whole gage section of the specimen becomes a macro-PSB. If the strain is greater than this, or indeed a little smaller, significant amounts of slip begin to occur on systems other than the primary, and the wall structure of the PSB is broken down into a more complex structure of interlinking walls; this structure is the object of study in high strain fatigue.

Just as at low strains, rapid hardening occurs at the onset of cycling at high strains and saturation then follows. However, throughout the life, the deformation and slip band structure remains quite homogeneous and strain localizations, if any occur, are quite minor. Although many slip systems operate, the deformation in any grain is dominated by the most highly stressed system (47). Studies of high strain fatigue in single crystals are rare, because of difficulties due to gripping and buckling of the specimen. Most of our understanding comes, therefore, from older work on polycrystals because the recent work on single crystals has focused on low strain behavior. Nevertheless, connections to monocrystalline and monotonic behavior have been made. For example, at low strain, the kind of hardening is viewed as similar to that which occurs in Stage I monotonic deformation of a single crystal (10, 23, 24, 48). Because the type of hardening which occurs at high strain involves cell formation, high strain cyclic hardening is associated with Stage II monotonic deformation of a single crystal, and the onset of saturation is akin to Stage III. Detailed similarities between the behavior of cyclic and monotonic deformation have been pointed out (23).

High strain mechanisms have been investigated by exploring the effect of microstructural and alloy variables on rapid hardening and saturation behavior, and these effects have been repeatedly reviewed (9, 10, 12-14). For example it has been shown that the saturation condition of wavy-slip materials is independent of the history of the material (grain size, subgrain structure, prior cold work) (23), unless the dislocation structures introduced before cycling are very severe (49). However in solid solution alloys where the slip mode is planar, the saturation condition does depend strongly on the history of the material (23). For fatigue design purposes, these effects have been thoroughly explored in commercial materials and cyclic response has been documented for different forms of variable loading (for reviews, see references (50) and (51)).

Once saturation has been obtained in a wavy slip metal, the dislocation cell structure becomes stable (23). The use of the word "cell" is considered inappropriate in some quarters because such nomenclature suggests misorientations across the cell walls and misorientations are either very small or non-existent. In fact, gradual changes in the cell wall structure have been observed at high temperatures, and slight misorientations develop (52) during life. It is not known whether such behavior is general or not.

The mechanism of cyclic deformation at high strain appears to be as follows. The cell walls accommodate the applied strain by providing a source of dislocation links. These links bow out and shuttle across the cells to be captured, and partly annihilated, in the walls bounding the cell. The walls themselves are quite fluid, being capable of moving to enlarge the cell if the strain amplitude is reduced. The behavior of the cell structure must be analogous to that of the walls in the PSB's, but not in detail. Since the cells are equiaxed, the long channels of the PSBs are not available for gliding groups of screw dislocations. However, the ways in which dislocations interact with the cell walls may well be similar to those that occur in PSB walls. With reversal of strain, the links bow in reverse fashion. The flow stress can be explained as that necessary to force the links from the walls, and such a calculation yields reasonable agreement with experiments (23).

Preoccupation with single crystal studies and low strain deformation in the last ten years has caused the neglect of basic research into high strain deformation. However, Quesnel and Tsou (53) are currently attempting to improve our understanding in this quarter. They have developed a model for the kinetics of cyclic hardening and softening based on dislocation cell wall surface energy as a driving force for cell growth. This process competes with cell shrinkage caused by dislocation multiplication amongst the mobile dislocations shuttling across the cells. Further development along these lines might yield very interesting results.

Cyclic Deformation of Precipitate Hardened Alloys

Many of the fundamental mechanisms described above can be seen to apply to more complex materials if appropriate allowance is made for their special conditions. For example aged alloys containing shearable precipitates are very susceptible to localized deformation in PSBs, although the PSB's observed in these alloys are structurally different from those in pure metals. Typical cyclic response for this class of alloy is shown in Figure 17 (54). It will be noted that the extent of rapid hardening is very large. A peak stress occurs and softening follows. Since saturation does not occur, there is no generally accepted definition of flow stress for the cyclic stress-strain curve. Lee and Laird have chosen to define the cyclic stress-strain curve in terms of the peak stress (54) and this curve is shown in Figure 18 for the

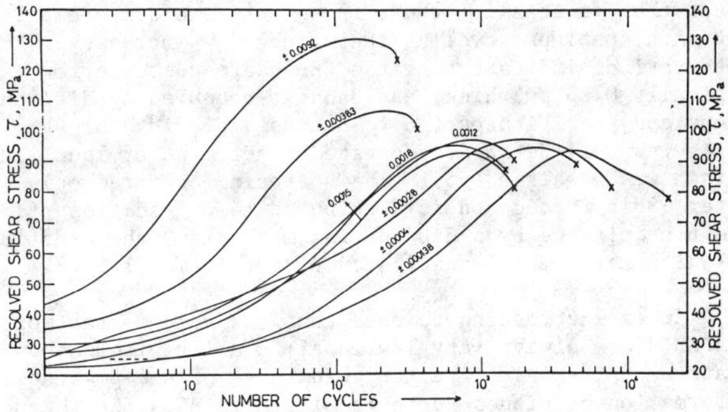

Fig. 17 Cyclic response of Al-Cu alloy single crystals containing θ" precipitates cycled in constant plastic shear strain control, indicated. Measurements for specimens of different orientation are included here. Courtesy of Lee and Laird (54).

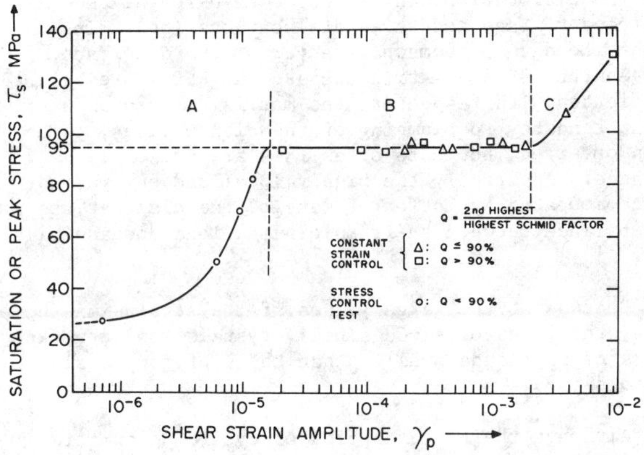

Fig. 18 Cyclic stress-strain curve for Al-Cu single crystals containing θ" precipitates. Q is the orientation parameter. Courtesy of Lee and Laird (54).

same type of specimens illustrated in Figure 17. Just like the behavior of pure metals, a plateau is observed. However, for this system, the onset of the plateau occurs at 1.8×10^{-5} and the high strain end occurs at $\sim 2 \times 10^{-3}$. Both of these strains are much lower than those of pure metals (Table 1). Crystals of various orientations within the standard triangle were used for these measurements. Again just like results for pure metals (15), orientatin had no effect on the cyclic stress-strain curve. The orientation is indicated by the parameter Q, defined as the ratio of the stress on the second most highly stressed system to that on the primary. Good single slip conditions apply to values of Q less than 90%. Since it is difficult to run tests in plastic strain control when the strain is very low, some of the measurements shown in Figure 18 were obtained under stress control. However, the tests were initiated by small increments in load to simulate tests in plastic strain control.

During rapid hardening, the slip was observed to be very fine and homogeneous (55). For strains below those of the plateau, saturation behavior occurred and the slip remained homogeneous for hundreds of thousands of cycles (the tests were often artificially terminated by failures at the knife edges of the extensometer). For strains corresponding to the plateau, PSB's were observed to form just before attainment of the peak stress, and the strain immediately became strongly localized (55). Interferometric techniques similar to those used by Finney and Laird (17) were employed to measure the localized strain (55) and it was found to be very large, varying from 0.3 to 0.6. As softening developed the strain became more localized. Typical observations of PSB populations and their associated step heights are shown in the histograms of Figure 19. It will be noted that, initially, the slip offsets are quite small but with continued cycling they gradually increase. TEM observations indicate that the PSB's are quite narrow, typically 0.25 μm thick, and densely occupied by dislocations and precipitates (56). Because the width of the PSB's was well defined, accurate measurement of localized strain was possible (55). The softening occurred only after PSB's formed and is attributed to degradation of the precipitates by a disordering mechanism, the kinetics of which have been investigated by a ball model (57).

It is interesting to note that the volume fractions of PSB's are always very low in Al-Cu alloy within the plateau. At strains greater than those of the plateau, deformation continues to be localized in PSBs but they develop on more than one slip system (54). Intense kink bands were also observed (55). The PSB's really constitute micro-PSB's, and like those of pure metals, they are limited in length - typically 400 μm. Nevertheless the PSB's pass right through the crystal as a string of overlapping short segments (55).

The dislocations in the PSB's have roughly equal densities of edge and screw dislocations (56). In Al-Cu they tend to be somewhat denser at the interface between the PSB and matrix because the PSB's are slightly misoriented with respect to the matrix. There is a slight tendency to clumping of the dislocations, sometimes observed, but more often they are uniformly arranged. Apparently the highly localized strains can be accommodated by motion of many of the dislocations and by a balance between their multiplicationa and annihilation.

Because of the large values of the strains in the PSB's, the lives of Al-Cu single crystals (55) are two orders of magnitude smaller than those of copper single crystals (58).

Although the broad aspects of cyclic deformation are similar in different alloys containing shearable precipitates, the details are often quite dissimilar. For example, Gerold, Wilhelm and their co-workers have carried out an elegant study of Al-Zn-Mg alloy containing shearable precipitates, η', and find many aspects of the mechanical response similar to that reported above (29, 59-63). However, in the PSB's in this alloy, the precipitates dissolve and leave the cleared channel supersaturated in alloying elements. In Al-Ag alloy, on the other hand, a structure initially containing GP zones undergoes cyclic strain-induced formation of more stable precipitates, γ' and γ, which cause enhanced strengthening (57, 64). Thus, tests that last longer can produce more hardening and this causes the cyclic stress-strain curve to have a negative slope over part of its range (64) (see Figure 20).

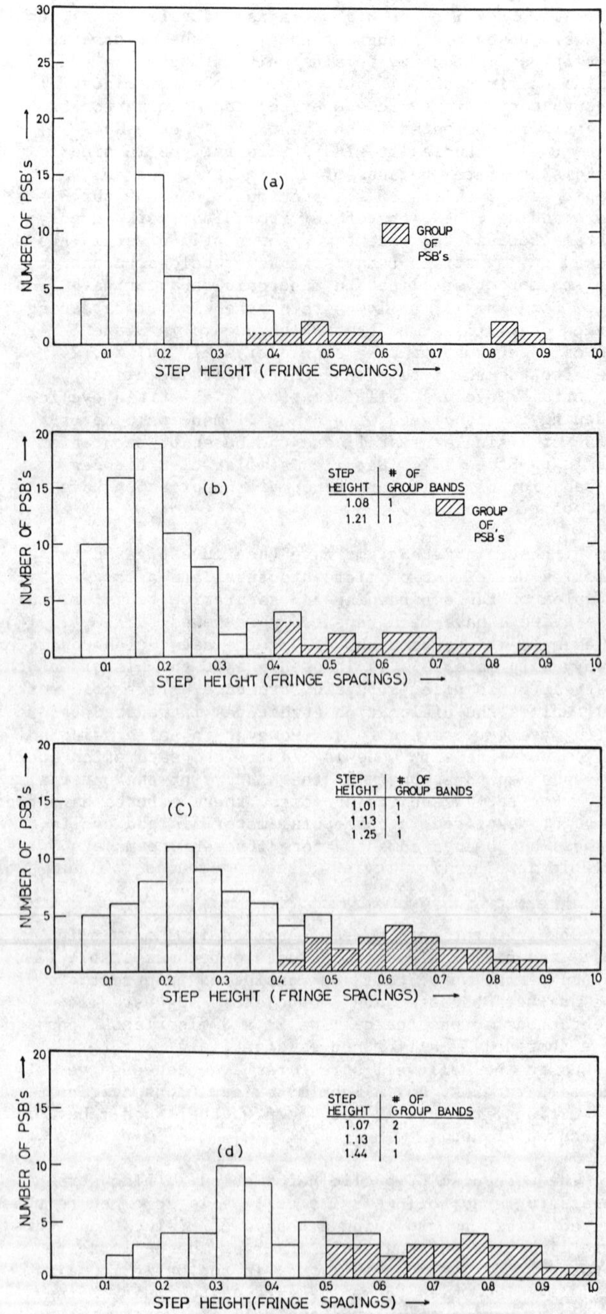

Fig. 19 The number of PSB's and their associated step heights: a) just after nucleation of the PSB's b) just before the peak stress occurred, c) just after softening had begun and d) well into softening. The specimen was cycled at a plastic shear strain amplitude of 0.0009. Courtesy of Lee (55).

Commercial alloys based on the simple systems treated above, and other f.c.c. systems, again show many differences of detail. For exampe, waspaloy does not show rapid hardening associated with general fine slip, but all the plastic strain is localized in PSB's from the start of cycling (65). The kinetics of the hardening and softening processes that govern the cyclic deformation were observed to be the same in each slip band regardless of plastic strain because, in order to accommodate the higher strains, the number of slip bands increased linearly with applied strain (65). There appears to be an analogy here with the behavior of PSB's in pure metals. However, waspaloy is the type of alloy

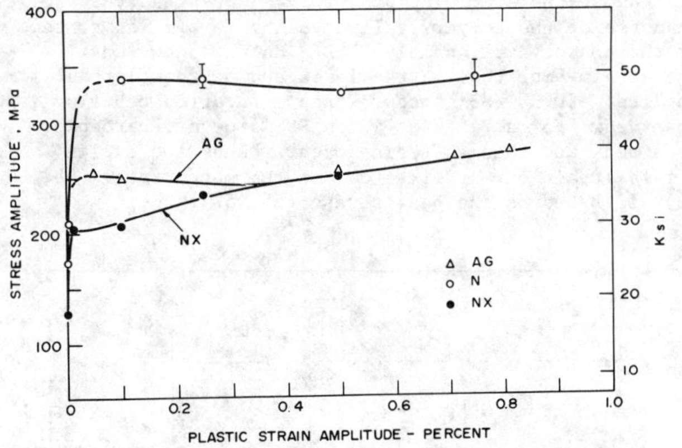

Fig. 20 The cyclic stress-strain curves of Al-15% Ag alloy thermomechanically treated in three different ways. The alloy of interest here is coded N; it contains GP zones and shows a negative slope over part of its cyclic stress-strain curve. Courtesy of Laird et al (64).

which shows hardening to a peak and then softening, and the peak stress increased with plastic strain, probably because the applied strains were high.

The dependence of cyclic response in the more subtle aspects of microstructure have been ingeniously explored by Starke and his co-workers (see reference (66) for review). There are also available recent reviews on the cyclic response of complex alloys (67, 68) under complex loading conditions and the reader is referred to them for details.

Cyclic Deformation of Polycrystalline Metals

In the last years, cyclic deformation of polycrystalline metals has seen a sudden flurry of investigation and the latest position will have been exposed at the conference in Denmark in September 1981. The suggestion that the connection between the cyclic response of mono- and polycrystalline material might be much simpler than that in monotonic deformation was first made when Stages I & II hardening mechanisms were identified with low and high strain fatigue (23). Bhat and Laird (69) extended the connection when they claimed, on the basis of Lukas and Klesnil's data (70), that the cyclic stress-strain curve of polycrystalline copper contained a plateau at low strain amplitudes and that the plateau levels in the two kinds of material could be related by the Taylor factor. this view was vigorously attacked by Mughrabi (71), who denied the plateau and also that PSB's could exist in the bulk of polycrystals.

Since then much work on cyclic response and dislocation structures in polycrystalline metal has been completed (72-76). Typical results for the cylic stress-strain curve of copper are shown in Figure 21. It turns out that, in strain control, Mughrabi and Wang (74) and Figueroa et al (75) agree that there is no plateau, but the latter workers note a tendency to one. Rasmussen and Pedersen (72, 73), on the other hand, claim there is. Figueroa et al (75) observe a plateau is load control in circumstances where the strains early in life are permitted to be large. The dislocation structures observed in the bulk of polycrystals are summarized in Fig. 22. Thus although these authors did not observe PSB's as such, i.e., narrow lamella of dipolar walls embedded in a structrue of loop patches, large volumes of dipolar wall structures were observed. The occurrence of these dipolar wall structures at strain amplitudes roughly corresponding to those at which the plateau occurs in single crystals constitutes the formation of PSB's in the bulk of the polycrystals. This substantiates the claims of Bhat and Laird (69) for the broad connections between the cyclic response of mono-and polycrystalline material, especially when it is considered

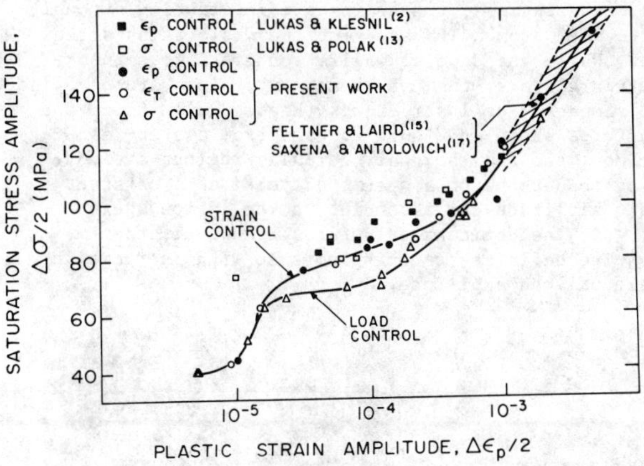

Fig. 21 The cyclic stress-strain curve of polycrystalline copper at low strains for tests under various forms of control, compared to previous results by Lukas & Klesnil (70) at low strain and other workers at high strains. Courtesy of Figueroa et al (75).

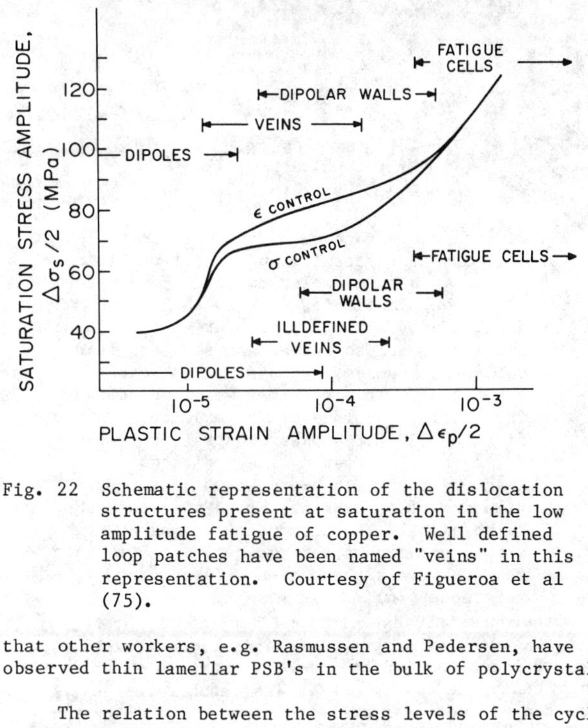

Fig. 22 Schematic representation of the dislocation structures present at saturation in the low amplitude fatigue of copper. Well defined loop patches have been named "veins" in this representation. Courtesy of Figueroa et al (75).

that other workers, e.g. Rasmussen and Pedersen, have observed thin lamellar PSB's in the bulk of polycrystals.

The relation between the stress levels of the cyclic stress-strain curves of mono- and polycrystalline metal has received much attention (69, 71-77). It will be noted that, in the straining region where PSB's occur in both kinds of material, the difference is a factor of 3 for strain control tests and slightly over 2 for load control tests (Figure 21). The University of Pennsylvania workers (75-77) attribute these differences to subtle differences in dislocation sructure and latent hardening for the two kinds of tests. In strain control tests, the dislocation structures in any grain are dominated much more by the primary system of that grain than in load controlled specimens, because the applied stress is low at the start of cycling. In load controlled specimens, however, the application of the full load in the first cycles excites multislip. In the work of Figueroa et al (75), the full load was not applied on the first cycle, but was built up in relatively few cycles so that the plastic strains early in the test were large. When saturation is achieved, the deformation between grains is accommodated easily and the applied stress is given by the product of the monocrystalline plateau and the Sachs factor (75, 77). The Sachs factor approach was championed by Rasmussen adn Pedersen (72, 73). The fact that, nevertheless, the grains are interacting strongly can be seen from a TEM study of the dislocation structure. While the dipolar wall structure is prominent in grain centers, the near boundary regions are occupied by dislocation cell structures (Figure 23). Since the structures shown in Figure 23 came from a step-tested specimen, the near-boundary cellular regions are more marked than those of a specimen cycled at a constant strain amplitude corresponding to the region where PSB's are the dominant structures. However, they are quite definite, becoming more so with increase in strain or load amplitude.

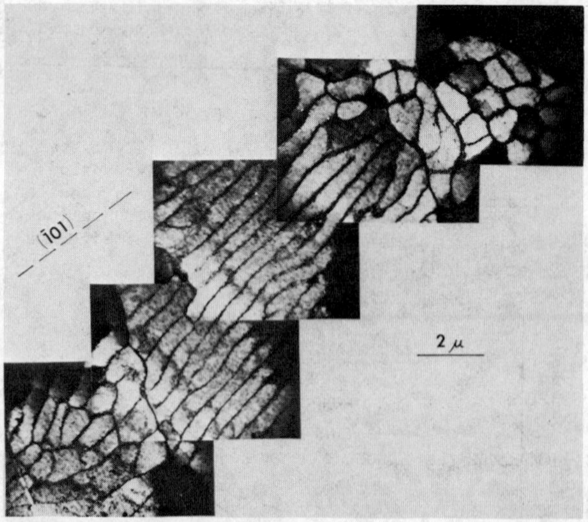

Fig. 23 Dislocation structure of a sample tested under a strain decremental step test, i.e. first at a plastic strain amplitude of 2×10^{-3}, then at a lower amplitude, back to the strain amplitude of 2×10^{-3}, then at a still lower amplitude, and so forth. At each amplitude sufficient cycles were applied to reach saturation. This specimen was removed from the testing machine for TEM examintion after final cycling at a plastic strain amplitude of 5×10^{-5}. The dashed line indicates the trace of the $(\bar{1}01)$ plane to which the dipolar walls, from the primary system, should be parallel. The alternate cycling at high strain has emphasized the cellular structures near the grain boundaries. Taken from reference (76).

Other possible explanations for the differences in the flow stress of monocrystals and polycrystals are: 1) The PSB acts as a plastic inclusion in an elastic matrix which exerts a 'back stress' constraint just as if the PSB were a martensite plate. This back stress is different from that exprienced by dislocations within a PSB due to their mutual interactions, and it acts to raise the flow stress of the polycrystal. 2) Texturing in the polycrystal, by lowering the Schmid factor of the singly slipping system, raises the flow stress of the polycrystal.

The most complete investigation of these factors known to the present author has been done on Al-4% Cu alloy containing shearable precipitates (77). It is known that other workers are concerned with these questions and the interested reader is referred to the proceedings of the Riso 1981 Symposium where data may become available.

In Al-Cu alloy all the parameters needed to investigate these questions were available (77). It is fascinating that the cyclic stress-strain curve of this alloy contains a plateau (Figure 24) at a level 3.22 times that of the shear stress corresponding to the plateau of the single crystal (55). Note that the cyclic response of the polycrystalline alloy is similar to that of the monocrystal, namely, PSB's do not occur until the strain amplitude at the lower end of the plateau is applied. If no PSB's occur, then saturation behavior is observed. For strains at which PSB's occur, hardening to a peak and then softening occur. Thus the cyclic stress-strain curve, like that of the monocyrstal (55, 56), is based on the peak stress (55, 57).

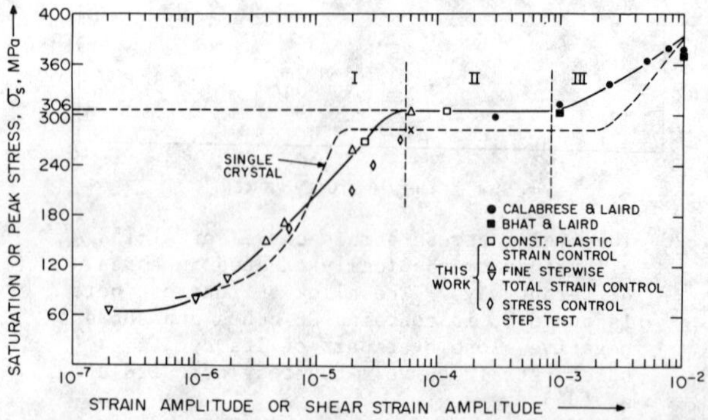

Fig. 24 The cyclic stress-strain curve for polycrystalline Al-4% Cu alloy containing shearable precipitates (θ") compared with that for the same alloy in monocrystalline form. Note that the ordinate scale should be divided by three to apply to the monocrystal; the abscissa applies to both forms of material. Courtesy of Lee (55).

It turns out that the PSB's in this alloy experience a trivial back stress from adjacent grains because they are extremely narrow and this compensates for the large strains localized in the PSB's (77). Moreover, the texture of the alloy, for the solution treatment used, was random (77). Thus the high flow stress of the polycrystal relative to that of the monocrystal could not be explained by the factors of PSB back stress and texture. The authors favored the influence of adjacent grains on the dislocation structures of both the PSB's and the matrix as the controlling factor (77).

The cyclic history of a specimen (and presumably also, any prior static deformation) can have a major effect on the cyclic stress-strain curve at low strain amplitudes, unlike the inflexibility of this curve at high strain amplitudes. Figueroa and Laird have carried out a variety of decremental step tests (76), of which the cyclic stress-strain curves for two of them are shown in Figure 25. Note that one of them, Program A, involves large steps from a high strain of 2×10^{-3}, at which

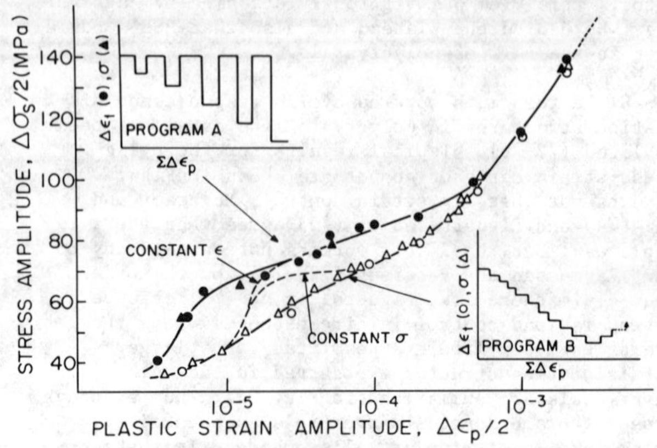

Fig. 25 Cyclic stress-strain curves associated with two kinds of decremental step tests, illustrated in the inserts, compared with the cyclic stress strain curves measured in constant amplitude tests run in strain control and stress control. Taken from reference (76).

dislocation cells occur in polycrystals. Under these circumstances, the cyclic stress-strain curve corresponded to that of specimens cycled at constant strain under plastic strain control; although at the lowest strains, the flow stress was relatively increased. For Program B, in which the applied strain was decreased in small steps, the cyclic stress-strain curve for the load control specimens was followed until the onset of the plateau, at which point the flow stress became smaller still (76). Figueroa and Laird argue that these two curves represent extrema in the cyclic response as influenced by prior cyclic straining. In specimens from both of these tests dipolar walls were found, even at the lowest plastic strains. However their separation was smaller in specimens tested at the larger flow stress (76).

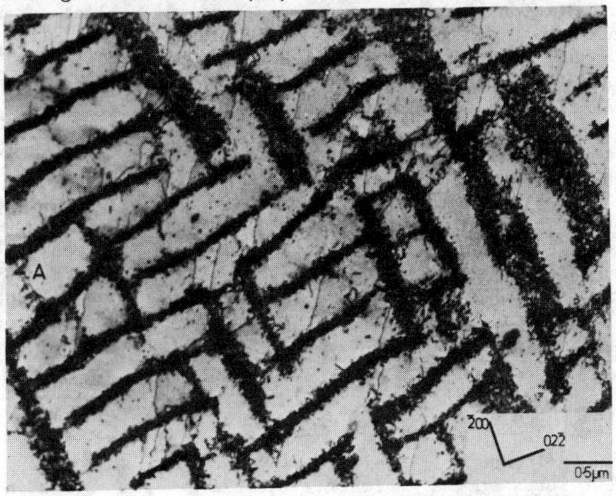

Fig. 26 'Maze' structure in commercial Cu-Ni alloy fatigued to saturation, showing wall structure on {100} planes. Courtesy of Charsley (79).

A particularly interesting form of dislocation structure has been observed by several investigators whose specimens have included grains with [001] or [011] axes (78-80), namely "Maze" structure. A beautiful example is shown in Figure 26. Kuhlmann-Wilsdorf and Charsley (81) have interpreted wall patterns composed of mutually perpendicular walls as dipolar walls with different Burgers vectors, mutually perpendicular. As well known, groups of similarly signed edge dislocations can reduce their energy if they arrange themselves as a tilt wall. Such a wall can further reduce the energy of the system by pairing with a similar wall, but of oppositely signed dislocations (to make a dipolar wall). But even dipolar walls have remnant stresses, that may be reduced by other dislocations, where they end. Thus where a dipolar wall ends, it will attract another dipolar wall oriented at 90° to its plane, as schematically illustrated in Figure 27. Possible wall intersections and the dislocation configurations within them are shown in this figure. Note that the dislocations are pairwise in equilibrium not only within each wall, but from wall to wall along the diagonal lines. Note also in the T-intersection that there is repulsion between the opposed horizontal walls of the T thus leading to a skew configuration. On the basis of such considerations, Kuhlmann-Wilsdorf and Charsley are able to understand the frequency of the various intersections, and other properties of the dislocation structure (81).

We see here another example of the mimimum energy principle applied to understanding dislocation structures in fatigue; it will be recalled that the dislocation behavior in PSB's was understood on such a basis (24, 40). While it is common in the study of dislocations to find situations where the minimum energy does not apply, cyclic deformation provides the optimum opportunity for reaching it. Clearly much work remains to be done in understanding the cylic deformation of pure metals but the field now seems to be progressing in very encouraging fashion.

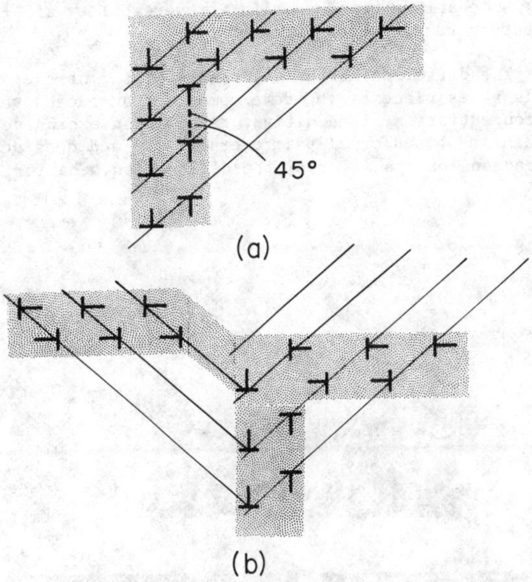

Fig. 27 Possible configurations of intersecting, perpendicular dipolar walls and the arrangement of the dislocations within them to permit pairwise equilibrium: a) the preferred L shape and b) a feasible, skewed T. Courtesy of Kuhlmann-Wilsdorf and Charsley (81).

FATIGUE CRACK NUCLEATION

Mechanisms of Crack Nucleation in Low Strain Fatigue

In order for cracks to nucleate at all, the applied stress or strain in a fatigued specimen must be above the fatigue limit or at some time in life have been above. When this is so, PSB's form and intensify. Cracks ultimately form at the surface both in the PSB's and in other regions, such as grain boundaries, on which the PSB's impinge. In single phase alloys of planar slip mode, PSB's are the preferred sites of crack nucleation. The mechanisms of crack nucleation have been repeatedly reviewed (1-4, 7, 11, 12, 82-84). In the older works, the issues of slip reversibility and cross-slip have been repeatedly discussed and these matters will not be treated here although it is tempting to integrate the older work into the newer results on cyclic deformation and PSB behavior. Lack of space, however, forces attention on the more recent work.

Results such as those shown in Fig. 10 provide evidence for considerable microscopic slip irreversibilities in PSB's at the surface. Roughening of the surface is now considered to develop not only from these slip irreversibilities but also from a volume dilatation within the PSB (11,83). Such a dilatation gives rise to the well-known phenomenon of extrusions, which develop in pure metals as rather mild bulges along the PSB's a µm or two in height. In precipitation-hardened alloys where the strain localization is much more severe, the extrusions tend to be several µm in height, that is, an order of magnitude higher than they are wide, so that they protrude with the morphology of a wood shaving as shown in Fig. 28, both schematically and in actuality.

In order for extrusions to form by slip irreversibility, path discriminations of dislocations, usually screw dislocations, during forward and reverse glide are necessary. These mechanisms appear unattractive to some workers (11), especially those who view PSB deformation as the rather stochastic bowing of edge dislocations from PSB walls. However, since screw dislocations clearly carry the bulk of the deformation in the PSB's of pure metals and single phase alloys, and developing notch effects can play a role in producing the necessary path discrimination in tension and compression (12, 24), mechanisms based on slip irreversibility are contributory to extrusion formation.

The different mechanism, that swelling of the PSB's takes place by the development of an excess vacancy concentration, is additive, rather than exclusive, to mechanisms based on slip irreversibility and does not depend on the specifics of the dislocation behavior within the PSB. The form of the excess vacancy concentration can consist of a preponderance in vacancy-type edge dislocation dipoles in the dipolar walls of the PSB's (86), or of vacancy agglomerates due to the annihilation of narrow edge dislocations dipoles (87).

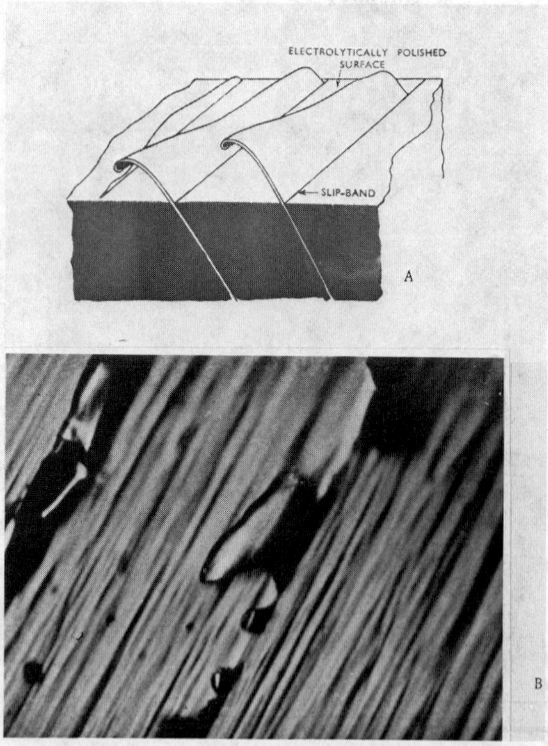

Fig. 28. (a) Schematic representation of slip-band extrusion.
(b) PSB extrusion in aged Aℓ-Cu alloy fatigued at low strain. The magnification can be gauged from the height of the largest extrusion, about 10 μm. Courtesy of Forsyth (85)

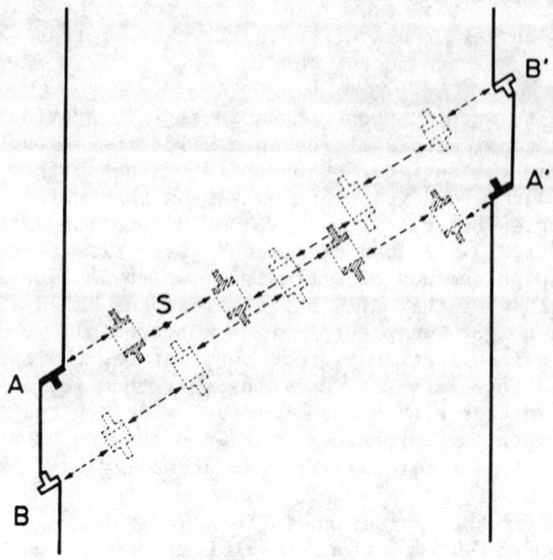

Fig. 29. Basic mechanisms of material extrusion by combined glide and annihilation of dislocations. See text for explanation. Courtesy of Essmann et al (87)

For a mechanism of extrusion formation based on a discriminatory slip mechanism, see ref. (12). A mechanism of extrusion swelling by dislocation annihilation is shown in Fig. 29. Full and open symbols in this figure
correspond to the glide sequence A-A' and B-B' in tension and compression, respectively. The reader is requested to imagine that all dislocations drawn in dipole configuration have annihilated and that only the individual dislocations drawn at A,B,A' and B' have survived annihilation (87). These dislocations which glide out and form the bulge of the extrusion leave behind interface dislocations between the PSB and the matrix deep inside the specimen. However, since bulges in pure metals are quite mild, the "interface" dislocation density is low. Clearly, by inverting the signs of the dislocations in Fig. 29, an intrusion could be formed. Oppositely-signed dislocations due to intrusions help to maintain the dipolar content in the PSB. On aged alloys containing shearable precipitates, the extrusions are so severe that interface dislocation densities between PSB and matrix are high, involve dislocations from slip systems other than the primary, and give rise to the slight misorientations between the matrix and the PSB's mentioned above (54-56).

The discovery of extrusions caused much excitement which subsequently lapsed as studies of crack propagation gained favor. Interest in them is arising again because it appears likely to many workers that a microcrack forms at the PSB-matrix interface. The extrusion provides a stress concentraiton at that point (11). The formation of steps in surfaces has been frequently observed. A recent case in point is the observation of rather sharp steps at PSB's in planar slip alloys (88, 89). Verpoort and his coworkers found that a decreasing number of cycles was required to nucleate cracks at steps, of the type shown in Fig. 30,

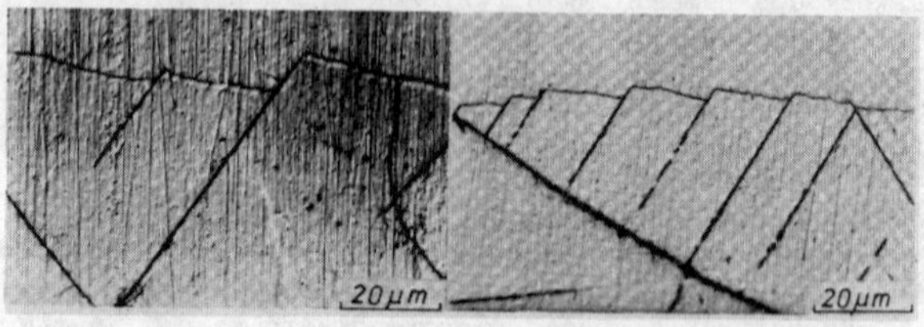

Fig. 30. Steps in the surface of planar slip, precipitation hardened alloys shown by taper sectioning (10x magnification normal to the surface). Courtesy of Graf and Verpoort (88).

as the coarseness of slip the line distribution increased. For a certain load, a definite critical step height seemed required to initiate a crack (89) and the frequency of the occurrence of such a height increases as the slip distribution becomes coarser. The details of how the cracks develop from such steps and how the steps build-up monotonically remain to be explored.

In wavy slip metals, the localization of strain into groups of micro-PSB's emphasizes the importance of slip irreversibility as a mechanism for crack nucleation. These mechanisms have recently been explored by Cheng and Laird (90). By repolishing, re-straining and interferometric observations on one side of a fatigued single crystal, and observations of cracking in the same PSB's on the opposite face of the crystal (not repolished) they have documented the properties of the PSB's in which the crack nucleates. As repeatedly pointed out (17,40, 44), for copper single crystals subject to an applied strain in the plateau of the cyclic stress-strain curve, there will be a certain distribution of strains localized in the PSB's, and not a constant strain equivalent to that of the upper end of the plateau. The crack nucleates in the "fatal band" which contains the highest localized strain (90). It is found that the length of the slip offset in a fatal band (nb, i.e., n Burgers vectors) and the applied plastic shear strain amplitude, γ_p, are related as $\gamma_p = C(nb)^{0.78}$, where C is a constant (90). The orientation of the crystal is found to affect only the step height, the volume fraction of the PSB's only slightly, and the slip offsets not at all. It is the magnitude of the slip offset which controls the crack nucleation behavior if the cyclic stress is uniaxial, and thus there is no effect of orientation on the cycles to nucleate a crack.

The dependence of the maximum slip offset on the applied strain causes the number of cycles to nucleate a crack to decrease with increasing strain amplitude. Consequently, regular Coffin-Manson fatigue life behavior is observed in copper single crystals for all strains greater than the fatigue limit (58). Cheng and Laird (90) have developed a model of crack nucleation based on a random slip approach in which the assumptions are considered more realistic than those used earlier by May.

PSB's have long been considered the dominant crack nucleation sites at low strain amplitudes, and in many kinds of alloys, this may well be true. However, in polycrystalline metals and alloys, grain boundaries have frequently been observed to impede the spreading of PSB's and of slip band cracks from one grain to another. Fatigue damage can either spread by initiation of new PSB's in neighboring grains or by nucleation of cracks in the grain boundaries at the points of impingement of the PSB's (91). This type of mechanism may well be quite common in wavy slip metals (2,3) and it is possible that it is enhanced by environmental factors. A typical example of grain boundary cracking due to PSB impingement is shown in Fig. 31.

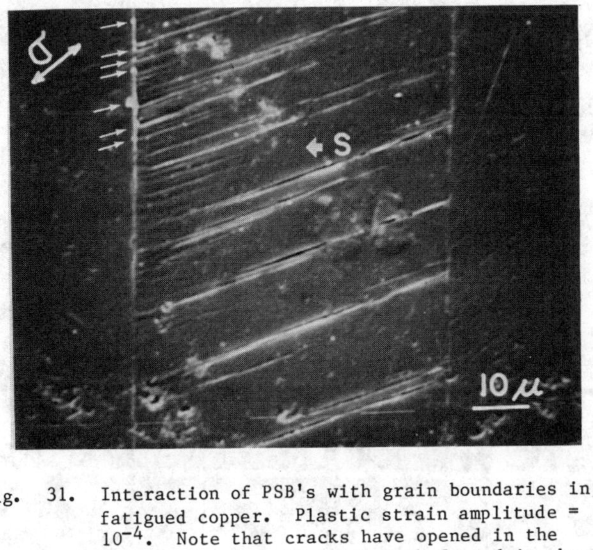

Fig. 31. Interaction of PSB's with grain boundaries in fatigued copper. Plastic strain amplitude = 10^{-4}. Note that cracks have opened in the PSB's. However the main crack formed in the boundary to the left, marked with arrows. Courtesy of Figueroa (91).

There are many factors in addition to those mentioned above which can strongly influence crack nucleation at low strains. In commercial alloys, impurity inclusions can be quite damaging (see 8,11 and 101 for brief reviews). An excellent review of environmental effects has recently been published (92).

Mechanisms of Nucleation in High Strain Fatigue

At strain amplitudes greater than those needed to produce PSB's, dislocation dipolar walls or more complex cell structures are general throughout a specimen and the deformation is broadly homogeneous. Only in planar slip materials is the deformation sufficiently confined to bands to give rise to slip band cracking (23,93). More usually, notches develop at grain boundaries and cracks form in them. Fig. 32 shows the typical development of a grain boundary crack at high strain (plastic strain amplitude = 0.0057) in pure copper. No inherent grain boundary weakness is required to produce such a crack.

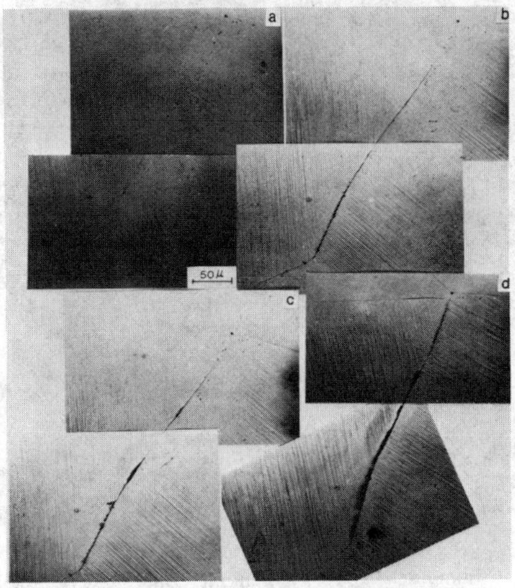

Fig. 32. Development of a crack nucleus in a polycrystalline specimen of copper. The micrographs were taken at the tension strain limit a) after 21 cycles (21T); b) 52T; c) 81T; d) 192T. Grain boundary darkening, an indication of a boundary discontinuity, is observed in (b). Note the fine homogeneous slip and absence of PSB's. Courtesy of Kim and Laird (95).

At one time it was believed that grain boundary cracks were caused by a puckering effect produced by localized deformation associated with surface irregularities. There may well be materials in which such a mechanism operates. In polycrystalline metals, however, the mechanism has been attributed to slip interactions at grain boundaries, of a type different from that which occurs at low strains. This mechanism was explored by interferometric observations of cracking morphology at grain boundaries in conjunction with efforts to study the nature of the boundaries which acted as nucleation sites (94-96). In Fig. 33, white light interference fringes shows the development of a grain boundary step. At the beginning of life a small step is formed in tension but it is cancelled by the compression stroke. With continued cycling, the amplitude of the boundary step increases and resists complete cancellation in compression. Eventually a step 1.5 μm high, having a sharp root radius, develops and the crack grows along the grain boundary into the material from this step (95,96). Only a relatively small fraction of grain boundaries were observed to be susceptible to this mechanism of failure. The "vulnerable" boundaries were identified (96) as those separating highly misoriented grains, the dominant slip systems of which were directed over large distances at the intersection of the boundary with the surface, as schematically shown in Fig. 34. The ratcheting mechanism by which the step develops was attributed to cross-slip resulting from the presence of a boundary-induced stress gradient.

The grain boundary step mechanism of crack nucleation and the associated crystallographic requirements are not specific to pure metals, and thus the mechanism may be a general one, subject of course to modifications in complex commerical materials (66,82,84,97,98). The

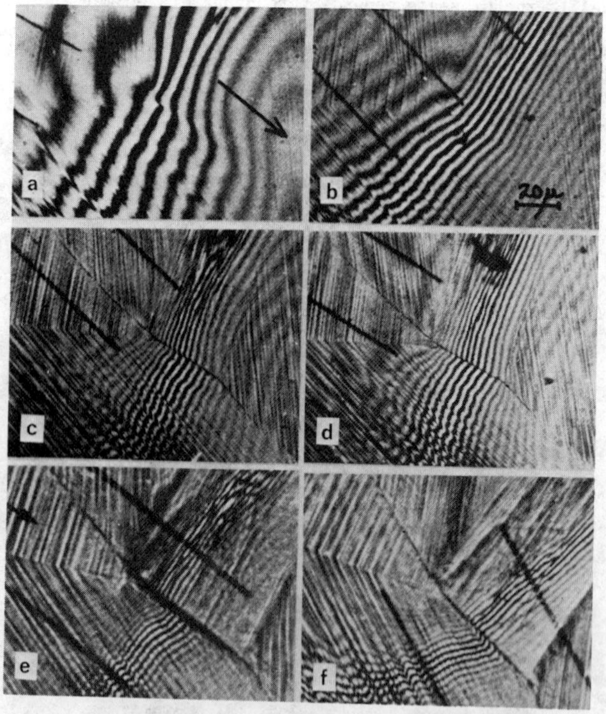

Fig. 33. White light interferograms showing grain boundary step growth in copper, cycled at a plastic strain of ± 0.0076. Fringes shifted from left to right indicate depressions. a) 10T, a small step of 0.1 μm is observed. b) upon strain reversal (10C) this step is cancelled. c) 30T, step height ~ 0.4 μm. d) 30C, step partly cancelled, 0.15 μm. e) 60T, step height ~ 0.9 μm. f) 60C, step partly cancelled. Courtesy of Kim and Laird (95).

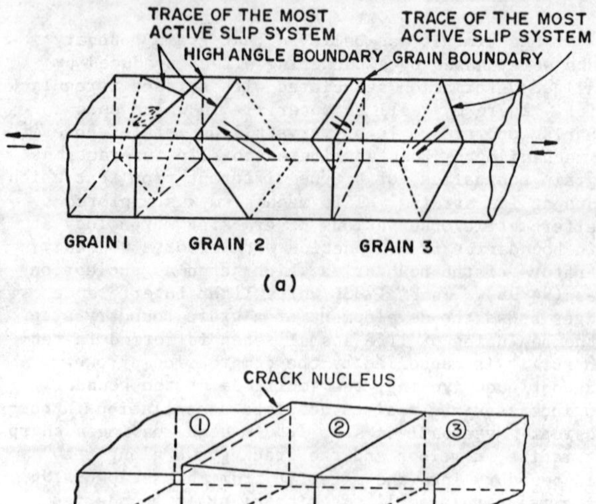

Fig. 34. Schematic representation of the crack nucleation process in high strain fatigue. a) The active slip systems in either grains 1 or 2, or in both of them, are directed at their boundary. b) After cycling, a step forms at the boundary between grains 1 and 2, but not 2 and 3. Courtesy of Kim and Laird (96).

asymetric slip of screw dislocations in tension and compression can lead to severe shape changes of bcc single crystals in fatigue as described above. Such changes will act to aggravate the formation of boundary steps and cracking. It has been shown on fatigued polycrystalline α-iron specimens (99,100) that near surface grains do suffer shape changes similar to those in single crystals and cause surface roughness leading to crack nucleation.

FATIGUE CRACK PROPAGATION

General Considerations

The great bulk of research into crack propagation in the last twenty years has been frankly empirical with the aim of defining crack propagation rate as a function of the stress intensity of linear elastic fracture mechanics driving the crack. Propagation under variable loading, after overloads, in different environments, and for different modes of stressing has been explored in great detail. It is not the purpose here to review this great body of work, in spite of its usefulness. The reader interested in such matters is referred to references (102) to (105), to site only a few. Rather the focus here will be on those aspects which give insight most directly into mechanism.

A common method of presenting kinetic data on crack propagation is to plot the fatigue crack propagation rate, dc/dN, versus the stress intensity range, ΔK, on a log-log basis, as shown in Fig. 35. Three regimes of dc/dN are widely recognized: 1) The threshold region where the propagation rate is of the order of an atomic spacing per cycle or less. At low enough stress intensities (the "threshold") the crack can be considered immobile. 2) A mid-region where the Paris equation $dc/dN = C(\Delta K)^m$, c and m being constants, is considered to hold. 3) A high rate region in which the mechanisms of failure are of static type and the maximum stress intensity approaches the critical stress for static failure. This region is not of much interest to the fatigue community, unless the residual strength of a cracked member is of concern. Since it involves static failure mechanisms, it will not be treated here.

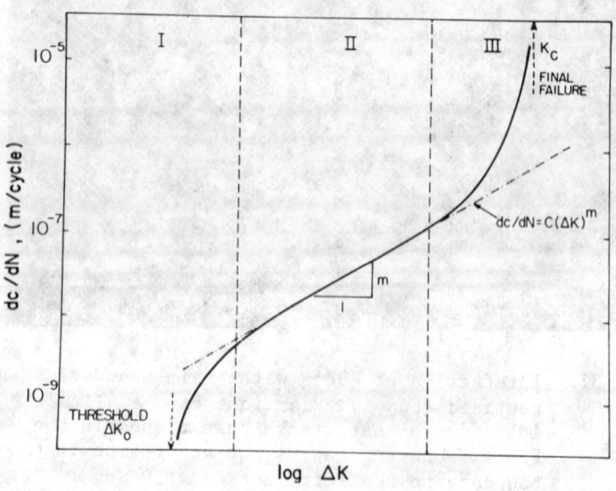

Fig. 35 Schematic plot of fatigue crack propagation rate, dc/dN versus stress intensity range, ΔK, depicting the widely recognized three regimes of crack propagation. Courtesy of Fine (101)

Once a crack is nucleated and the stress intensity is low, regime I, the crack continues to develop along an active slip plane, generally a PSB (4). Since the PSB's are formed on the most highly stressed slip plane, the orientation of this crack is usually at 45° to the stress axis. In this configuration it is known as a Stage I crack and by chance, such propagation is associated with regime I described above. However, it is important to note that many materials, especially steels, do not propagate in the 45° manner, and even cracks in copper, which normally propagate by Stage I in air, can be induced to propagate normal to the direction of stressing by enclosing the specimen in a vacuum (106).

In regime II, cracks usually propagate by a mechanism of plastic deformation which leaves striations on the fracture surface unless these, again, are suppressed by testing in vacuum (11). Crack propagation involving striations is usually called Stage II. It is ironic that this regime of propagation is called after Paris, the champion of linear elastic fracture mechanics as applied to fatigue.

The mechanisms of crack propagation in Stage I and II are described in greater detail, as follows.

Stage I Propagation

TEM studies of Stage I cracks in PSB's have shown that cracks can penetrate to depths many times the inter-wall spacing of the PSB's without altering their dislocation structure significantly (107). A typical view of such a crack is shown in Fig. 36. The inset shows an optical micrograph of the crack with respect to the

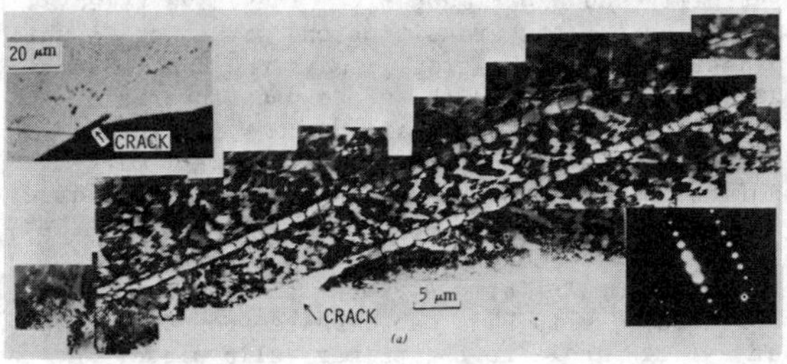

Fig. 36. Stage I crack in a PSB, showing the ladder-like wall structure unaffected by its presence. Courtesy of Katagiri et al (107).

surface of the specimen. Observations of this kind are not confined to pure metals but age-hardened alloys have been shown to propagate in this way (59,62,63). Indeed, a propagating Stage I crack conditions the metal in the PSB ahead of it and follows in its path.

The implication of results of this type is that the PSB deformation processes, which act first to form an extrusion and then to form notch-peak topography containing the crack embryo, now can lead to Stage I propagation. The mechanism for this is shown schematically in Fig. 37. In this figure the screw dislocations carrying the PSB strain are distributed with roughly equal probability on all the slip planes within the PSB, and only slight path discrimination can cause an extrusion. However, a notch serves to concentrate slip in tension along a narrow group of slip planes, C-C', at the base of the notch. That is, a stress concentration attracts the screw dislocations into the planes at the tip of the notch, the dislocations cross-slipping from their regular planes to meet the notch. In compression, the notch closes, the stress concentration is not sensed, and the PSB screw dislocations return to the other end of their channel more equally distributed on the planes in the PSB. Repetition of such deformation causes crack extrusion as shown in Fig. 37, as well as extrusion formation, which has been previously discussed (12). Clearly this mechanism could operate in any material subject to PSB formation.

Mughrabi has offered helpful suggestions for explaining the absence of (or reduced tendency to) Stage I cracking in steels, and perhaps in other bcc metals (11). He cites the work on α-iron of Katagiri et al (108) who have observed dislocation cell structures at the tips of young cracks formed in a 45° mode, and points out that, in typical fatigue conditions, where the strain rates

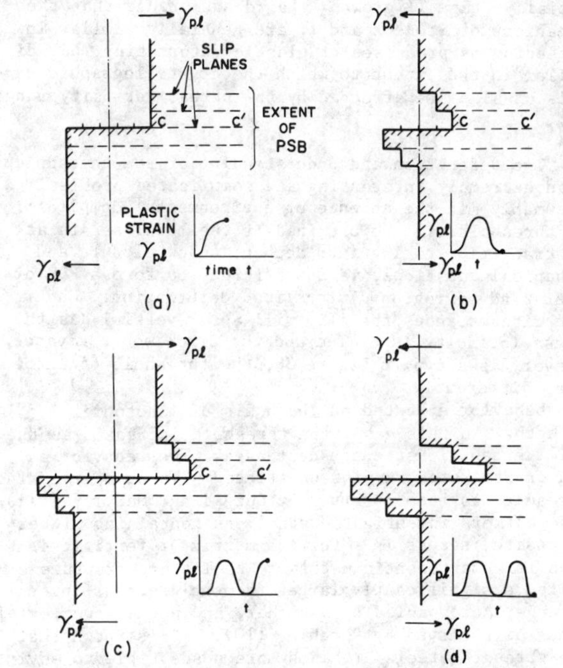

Fig. 37. Schematic representation of Stage I propagation in a PSB, via the deformation produced in a surface normal to the primary Burgers vector. A notch, omitted for clarity from (a), serves to concentrate the strain during tension on slip plane C-C'. The inserted γ_{pl}-time plot indicates the loading sequence corresponding to the slip behavior. Taken from reference 12, Courtesy of American Society for Metals.

are quite fast, bcc transition metals do not usually deform by PSB's, since they deform by the low temperature mode of deformation. Clearly, a capacity for conventional Stage I growth is tied very closely with PSB formation. It would be most worthwhile to explore further the mechanism of Stage I cracking in bcc metals. The most likely outcome, in the present author's opinion, is that the cracks will be found to propagate by a Stage II mechanism, but at low rates. In support of this opinion, it should be noted that Carlson and Koss (109), in a study of bcc Ti-V alloy crystals, observed that the crack deviated from the primary slip plane, even for conditions favorable to Stage I growth. The behavior was found consistent with crack advance by alternating (unequal) slip on two slip systems - and such a

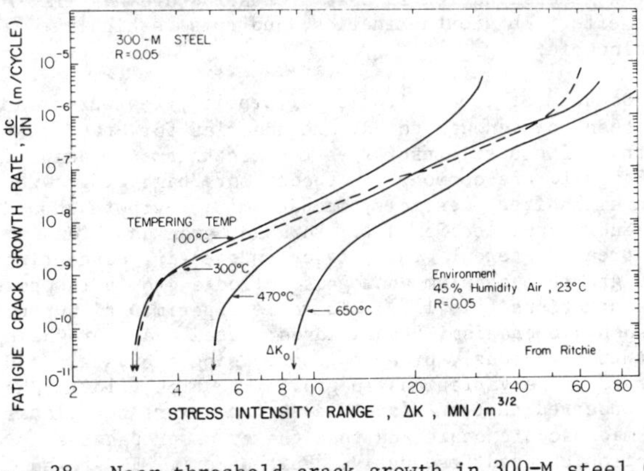

Fig. 38. Near threshold crack growth in 300-M steel (a Si-modified 4340 steel) tempered at various temperatures from 100 to 650°C, showing an increase in the threshold stress with decrease in monotonic strength. From Ritchie (110) via Fine (101), courtesy of both.

mechanism, as will be seen in the next section, is essentially Stage II growth. Laird has argued that the mechanisms of Stage I and II are generally similar in proceeding by processes of plastic deformation, but dissimilar in the extent to which the crystallographic details of slip are affected by the stress intensity range (9).

The definition and understanding of the threshold is an extremely interesting and complicated problem. Presumably, in the absence of environmental complication, the threshold will be obtained if the local strain at the crack tip, taking into account crack closure and mechanical questions, is insufficient to form a PSB or other kind of region of localized deformation. Under such circumstances the slip will act reversibly as the stress intensity is cycled and the crack won't advance. However, as shown in Figure 38, the threshold often behaves differently from behavior expected on the basis of monotonic strength. In this case, the threshold increases as the ultimate tensile strength decreases. The behavior at high growth rates remains unaffected. The opposite kind of result has been observed by Suzuki and McEvily (111). These workers investigated specimens containing either martensite islands in a ferrite matrix or ferritic islands in a martensitic matrix. The latter structure exhibited a significantly larger threshold stress intensity range, accompanied by a higher, and not a lower, yield stress as observed by Ritchie (110). It is known that high strength steels are much more susceptible to environmental factors than lower strength steels and the interaction of factors is probably threshold limiting.

As pointed out by many authors, but most consistently by Speidel (112), a combination of processes such as rewelding, crack branching, crack closure, and corrosion product wedging can lower the effective stress intensity and thus lead to a threshold. A very good example of the complications arising from such combinations is provided by the results of Ritchie and his co-workers (113,114). They have shown that, at low load ratios, near threshold growth rates are significantly reduced in moist environments (such as air or water), compared to dry environments (such as hydrogen or helium gas), due to the formation of excess corrosion deposits on crack faces, which enhances crack closure. Using Auger spectroscopy, they found that, at the threshold stress intensity, the maximum thickness of oxide debris within the crack is comparable with the pulsating crack tip opening displacement. On the other hand, Stanzl and coworkers, using ultrasonic testing frequencies, find a much more usual effect of corrosive environment, because the speed of stressing does not allow sufficient time to build up corrosion product (115). Thus unless all the factors are considered and properly weighted, it is often unwise and untrue to make generalizations about Stage I growth in different kinds of materials. A typical and widely offered generalization is that microstructure has little or no effect on growth kinetics, and this is highly questionable.

Because of the environmental problem, few workers have been bold enough to advance theories for Stage I growth. Of course, insofar as the mechanisms of Stage I and Stage II are common, any theory for Stage II growth which emphasizes plasticity mechanisms for growth should frequently apply to Stage I. Purushothaman and Tien have treated Stage I growth by an interesting discontinuous growth mechanism analogous to ledge growth in phase transformations (116,117). There is experimental support for such a mechanism: Stanzl investigated near-threshold growth for variously oriented copper single crystals and observed large variations in growth kinetics (118). She also observed that the fracture surfaces were not planar on a macroscopic scale and that the crack propagation direction changed frequently. Furthermore, SEM showed that the crack propagated along at least two planes. These effects in combination with step changes in growth kinetics (often observed as "scatter") are the hall-marks of a ledge mechanism.

It is now well recognized that Stage I growth should be studied both in relation to mechanism and to the role of environment (11,12,101) and progress in understanding this phenomenon can be expected to be rapid in the next few years.

Stage II Growth

How deep a Stage I crack grows in a smooth specimen depends on the applied stress. In low strain fatigue, a Stage I crack can penetrate to the depth of several grain diameters before the increase of the stress intensity with the crack length promotes slip on systems other than the primary one in any grain. A dislocation cell structure normally forms at the crack tip under these conditions as frequently observed (119), and the PSB structure is broken down. Since slip is no longer confined to planes at 45° to the stress axis, the crack begins to propagate normal to the stress axis and is then in Stage II growth by definition. In high strain fatigue the stress intensity is so large at a crack nucleus that the crack will almost immediately propagate by the Stage II process (95,96,120,121).

The transition from Stage I to II has been of concern for many years. Cheng and Laird (122) have recently explored this transition in copper single crystals, both theoretically and experimentally. They have taken as the criterion for the transition the situation in which the plastic zone at the crack tip in the PSB (the crack has grown quite long and therefore does interfere with the regular PSB structure) raises the shear stress on the second most highly stressed, non-primary plane to that of the plateau in the cyclic stress-strain curve. They chose the plateau stress because they share the general ignorance which pervades on the subject of latent hardening in fatigue. When this criterion is satisfied, a slip band can emanate from the crack tip at a large angle to the primary slip plane, and the duplex slip at the crack tip changes the plane of the crack from 45° to the stress axis to one approximately normal to it (122).

As mentioned above, Stage II fracture surfaces are recognized by the observation of fatigue striations on them provided the propagation rate is large enough for the striations to be resolved. Typical striations of the ductile variety are shown in Fig. 39. These striation were produced under conditions of variable loading and therefore vary in width (123). Observations of this type were first used to prove that each striation was associated with one load cycle. The morphology of the striation can vary considerably in any material and in some materials have such a distorted morphology as to be unrecognizable. The variations in striation morphology and its dependence on the details of the slip processes have been repeatedly discussed (9,12). It has also long been understood that cycling in vacuo at low or intermediate stress intensities suppresses the appearance of the striations while lowering the crack propagation rate by an order of magnitude or more. At high stress intensities, the environment has little effect on either the kinetics of propagation or the morphology of the striations (121).

The regular ductile striation has a profile consisting of a more or less flat region (actually curved when the striation spacing is small) bounded by a trench (Fig. 39). Both the morphology of the striation and its one-to-one connection with each load cycle were major clues to unraveling the mechanism of Stage II growth. Furthermore, the trench is formed on the side of the striation in the direction of crack propagation (Fig. 39). By making sections through cracks in specimens unloaded from different parts of the straining cycle, Laird and Smith provided direct evidence for the purely ductile model of crack propagation (120). This evidence was refined and led to the model for crack growth shown in Fig. 40, termed the "plastic blunting process" (9), Although this model was offered some time ago, the author

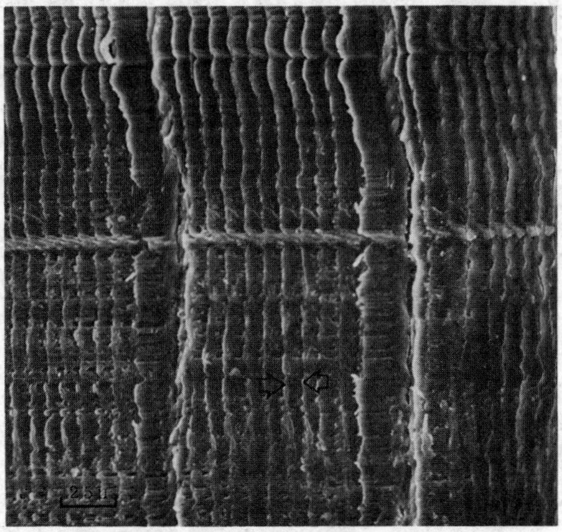

Fig. 39. Typical ductile fatigue striations. Note that each striation, indicated by a double headed arrow, consists of a flat region and trench. These striations were produced by a push-pull load sequence of 10 cycles at ± 0.0025 and 1 cycle at ± 0.005 plastic strain. Courtesy of Laird and de la Veaux (123)

believes it is an accurate representation of the ductile fatigue crack growth mechanism, as follows:

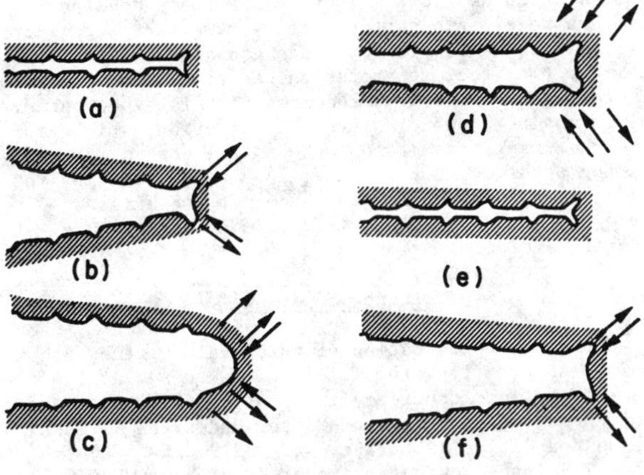

Fig. 40. Schematic representation of the plastic blunting process of fatigue crack propagation in Stage II (a) zero load ; (b) small tensile load; (c) maximum tensile load of the cycle; (d) closure; (e) maximum compressive load of the cycle (f) small tensile load in the succeeding cycle. The double arrowheads in (c) and (d) signify the widening of slip bands at the crack in these stages of the process. Taken from reference (9).

The initially unloaded crack tip is shown in Fig. 40a. As tensile loading is applied, slip is concentrated in sharp bands at the double notch of the tip. Depending on the detailed morphology of the striation, the crack tip can also be pointed (9,12). With continued tensile deformation, the crack progresses a distance proportional to the sliding off. This model has been directly supported by in situ observations in single crystals (106). On reversing the loading direction, the slip in the zones is reversed, the crack faces close, and the new crack surface just created is folded by buckling of material at the crack tip, producing the double notch (i.e. a trench on each fracture face) and unit of crack advance shown in Fig. 40(e). Depending on the planarity and coarseness of slip, the details of this process can vary considerably.

The basic similarity between the mechanisms of Stage I and Stage II growth will be noted, in that both involve crack advance by a "slipping-off" process. The difference between them is that only one slip system need be involved in Stage I, but duplex slip occurs in the Stage II crack plastic zones.

Several authors have published alternative representations of the plastic blunting process (106,124 - 126). They involve the essential "slipping-off" process of the plastic blunting mechanism. Differences may be perceived because alternating slip on specific intersecting slip planes was proposed. However, the crystallographic aspects of slip, and its control over striation morphology, were understood when the plastic blunting process was initially proposed (127). The present author sees no substantive differences between these models except in name, and perhaps some details of striation formation.

There have been recent advances which have been helpful in describing how partial cleavage or corrosion enhanced crack growth affect the purely plastic aspects of Stage II growth. It was pointed out long ago (127) that the ratio between crack growth per cycle and the plastic crack tip opening should be a constant. Vehoff and Neumann have verified this, and also have shown how this ratio depends both on the angle between the two slip bands emanating from the crack tip and the coarseness of slip (128). Furthermore, by varying both temperature and the crack tip opening velocity, they have found a transition from ductile to brittle fracture, with associated marked changes in striation morphology.

Fig. 41. Fracture surface striations obtained by quasi-brittle crack propagation in program loading at a high crack tip opening velocity. The program consisted of five cycles at large strain amplitude and then five cycles at an amplitude one third of the previous value. Courtesy of Vehoff and Neumann (128).

Typical brittle striations are shown in Fig. 41. Such experiments have led to the following picture for the initiation of cleavage at an atomically sharp crack: the local stress increases at such a high rate that the crack cannot be blunted by thermally activated dislocation glide. This leads to cleavage. If on the other hand the local stress increases at a low enough rate, the crack blunts and remains ductile. Vehoff and Neumann report (128) that the ratio of the crack advance per cycle to the plastic crack tip opening is definitely higher for quasi brittle fracture (up to a factor of 5). Quasi-brittle cracks do show a small crack tip angle, however, and this indicates that some plasticity accompanies the crack growth. It is going to be extremely difficult to develop theories of crack growth for such complicated mechanisms.

Crack Propagation Theories

The purpose of a crack propagation theory is to describe accurately the kinetics of crack propagation of the type shown in Fig. 35 and there have been many theoretical studies resulting in equations of the form of Paris relation. Essentially, there are two classes of problem to handle: 1) the crack propagation rate is proportional to the square of the stress intensity range, which corresponds to ductile stage II growth, and 2) the rate is proportional to the stress intensity range to a power greater than two, of which four is a popular value, but still higher values have been observed when static failure mechanisms assist the ductile growth mechanism.

The square law comes about from assuming that the crack growth per cycle is proportional to the crack opening displacement. This type of relation was derived by Laird (127) and there have been many similarly based derivations since, of which that in reference 129 is a typical example. The problem can now be handled beautifully by finite element methods (126,130,131).

Obtaining Paris relations with a power higher than two is an extremely difficult problem. Many workers assume a ductility exhaustion approach for a ligament at the crack tip. Although such an approach can yield correct answers for sufficiently ingenious theorists, it is unlikely to be very close to reality. Other workers beg the question of mechanism, and derive relations based on the concept that plastic work must be provided to advance the crack (101). A really comprehensive review of crack propagation which does justice to the past history and present complexity of the field does not exist. However, the interested reader is directed to Weertman's excellent review for further details on theory (132), and to Marcus' for environmental effects (133).

With the new types of experiment now being conducted on crack propagation mechanisms, such as those by Neumann and his co-workers, or the direct SEM observations by Davidson and his coworkers (134), the prospects for unravelling the more complicated aspects of fatigue crack growth are improving.

Summary

An overview of fatigue mechanisms is presented. Fatigue is regarded as taking place first by cyclic deformation which conditions the material for failure at low strains, second by crack nucleation, and finally by propagation. In complex structures, the prior existence of flaws can circumvent the first two parts of the fatigue process. Results are provided for both pure metals and complex alloys.

In low-strain cyclic deformation, an initially annealed metal hardens by dislocation debris accumulation, the debris consisting of dense dipolar loop patches. The plastic strain is carried both by screw dislocations in the channels between loop patches and by the patches themselves. With increasing strain, or increasing cycles at sufficiently high strains, persistent slip bands (PSB's) form in the loop patches just before the cyclic stress saturates. The PSB's deprive the loop patches of strain and adjust their volume fraction to accommodate the strain applied. This behavior gives rise to a plateau in the cyclic stress strain curve. The PSB dislocation structure consists of dense dipolar walls in edge configuration with respect to the primary system. The strain in the PSB's is carried almost entirely by screw dislocations gliding in loose coordination across the dipolar walls so as to maintain their dipolar balance. The average strain in the PSB's is about 0.01 but large variations about this figure exist, and control the crack nucleation process. For applied strains greater than 0.01, single slip gives way to duplex slip, the structure consists of homogeneous dislocation fatigue cells (i.e. zero or small misorientations across the cell walls) and the strain is carried uniformly in the structure by dislocations shuffling in the cells. Age hardened alloys containing shearable precipitates also form PSB's and show plateaux in their cyclic stress-strain curves. However, the strain localization is much greater than in pure metals and the PSB structures are different both from pure metals and for alloys of different systems. Major advances are reported in relating the cyclic response of mono- and polycrystals.

In low strain fatigue, cracks form in the PSB's by a mechanism of random slip, accentuated perhaps by the steps associated with extrusions. At high cyclic strains, the mechanism of crack nucleation is different from that at low strains because the slip is no longer localized. The geometry of the slip process remains important, however, in that slip directed at the line of intersection of a grain boundary with the specimen surface leads to step formation in the boundary, severe enough to consist of a crack. In polycrystals at low strains, PSB impingement on grain boundaries seems to be an important crack nucleation mechanism.

A crack nucleated in a PSB continues to propagate along the slip band in stage I growth, without altering the PSB structure. Indeed in age hardened alloys, the PSB progresses into the matrix ahead of the crack. In view of these observations, an unslipping mechanism is suggested for fatigue crack growth, and this must take place by action of the PSB screw dislocations, at least in pure metals. As the stress intensity is increased, either by growth of the crack under constant load or by higher applied stress, slip begins to occur near the crack tip on planes other than the primary. This duplex slip induces stage II crack propagation by "slipping-off" coordinated on slip planes oriented roughly at 45° to the plane of the crack. Crack propagation per cycle can then be linked to the crack opening displacement and many theories of crack growth kinetics have been developed from thus model. In appropriate circumstances, e.g. in b.c.c. metals at low temperatures and high strain rates, quasi-brittle fracture processes can occur, and accelerate the crack growth rate. Mechanisms of this type probably explain why the power in the Paris relation, which connects the crack propagation rate to the stress intensity, is often observed to be greater than two, which applies to ductile growth mechanisms. However, theories to handle the complicated quasi-brittle mechanisms are not yet well developed.

Acknowledgements

During the preparation of this article, the author received support from the following sources: a) the materials failure coherency group of the Laboratory for Research on the Structure of Matter under the National Science Foundation MRL Grant, No. DMR79-23647; b) the National Science Foundation under Grant No. DMR80-19914; c) the Army Research Office under Grant No. DAAG29-78-C-0039; and d) the Department of Energy, Grant No. DE-AC02-80ER 10570. He deeply appreciates this support, as he does the courtesy of the many authors and publishing houses who have given him permission to reproduce the figures referenced herein.

REFERENCES

(1) N. E. Frost, K. J. Marsh and L. P. Pook, "Metal Fatigue", Clarendon Press, Oxford, 1974.

(2) M. Klesnil and P. Lukáš, "Fatigue of Metallic Materials", Elsevier, Amsterdam, 1980.

(3) S. Kocanda, "Fatigue Failure of Metals", Sijthoff and Noordhoff International Publishers, The Netherlands, 1978.

(4) N. Thompson and N. J. Wadsworth, "Fatigue of Metals," Advances in Physics, 7 (1958) p. 72.

(5) D. H. Avery and W. A. Backofen, in "Fracture of Solids", Eds., D. C. Drucker and J. J. Gilman, Interscience Publishers, p. 339, (1962).

(6) J. C. Grosskreutz in "Fatigue, an Interdisciplinary Approach", Sagamore Conference, p. 27 (1964).

(7) A. J. Kennedy, "Processes of Creep and Fatigue in Metals", John Wiley and Sons, Inc., New York, 1963.

(8) J. C. Grosskreutz, "The Mechanisms of Metal Fatigue I", Physica Status Solidi, 47 (1971), p. 11-31.
(9) C. Laird, in "Fatigue Crack Propagation", American Society for Testing and Materials, Special Technical Publication, 415 (1967) p. 131-180.
(10) C. Laird, in "Work Hardening in Tension and Fatigue", Ed. A. W. Thompson, The Metallurgical Society, A.I.M.E., p. 150, 1977.
(11) H. Mughrabi, "Microscopic Mechanisms of Metal Fatigue", 5th Int. Conf. on Strength of Metals and Alloys, Eds. P. Haasen, V. Gerold and G. Kostorz, p. 1615-1638, Pergamon Press, Oxford, 1980.
(12) C. Laird, in "Mechanisms and Theories of Fatigue", American Society for Metals, 1979, p. 149-204.
(13) J. C. Grosskreutz and H. Mughrabi, in "Constitutive Equations in Plasticity", Ed. A. S. Argon, MIT Press, Cambridge, Mass., p. 251 (1975).
(14) C. Laird, "The Application of Dislocation Concepts in Fatigue", in "Dislocations", Ed. N. A. B. Nabarro, vol. 6, in press.
(15) A. S. Cheng and C. Laird, "Mechanisms of Fatigue Hardening in Copper Single Crystals: the Effects of Strain Amplitude and Orientation", Materials Science and Engineering, to be published.
(16) P. J. Woods, "Low-amplitude Fatigue of Copper and Cu-5% Al Single Crystals", Philosophical Magazine, 28(1973), p. 155-191.
(17) J. M. Finney and C. Laird, "Strain Localization in Cyclic Deformation of Copper Single Crystals", Philosophical Magazine, 31(1975) p. 339-366.
(18) H. Mughrabi, The Cyclic Hardening and Saturation Behavior of Copper Single Crystals", Materials Science and Engineering, 33(1978), p. 207-220.
(19) A. P. L. Turner, Private communication, Argonne National Laboratories, Illinois, 1980.
(20) A. T. Winter, "A Model for the Fatigue of Copper at Low Plastic Strain Amplitudes", Philosophical Magazine, 30(1974), p. 719-738.
(21) C. Laird, "The Fatigue Limits of Metals", Materials Science and Engineering, 22 (1976), p. 231-236.
(22) J. R. Hancock and J. C. Grosskrentz, "Mechanisms of Fatigue Hardening in Copper Single Crystals", Acta Metallurgica, 17 (1969), p. 77-98.
(23) C. E. Feltner and C. Laird, "Cyclic Stress-Strain Response of F.C.C. Metals and Alloys, Parts I and II", Acta Metallurgica, 15 (1967), p. 1621-1632, p. 1633-1654.
(24) D. Kuhlmann-Wilsdorf and C. Laird, "Dislocation Behavior in Fatigue", Materials Science and Engineering, 27 (1977), p. 137-156.
(25) D. Kuhlmann-Wilsdorf and C. Laird, "Dislocation Behavior in Fatigue, Part II, Friction Stress and Back Stress as Inferred from an Analysis of Hysteresis Loops", Materials Science and Engineering, 37 (1979), p. 111-120.
(26) D. Kuhlmann-Wilsdorf, "Dislocation Behavior in Fatigue, Part III, Properties of Loop Patches - Do They Participate in Fatigue Cycling?", Materials Science and Engineering, 39 (1979), p. 127-139.
(27) D. Kuhlmann-Wilsdorf, "Dislocation Behavior in Fatigue, Part IV Quantitative Interpretation of Friction Stress and Back Stress Derived from Hysteresis Loops", Materials Science and Engineering, 39 (1979), p. 231-246.
(28) D. Kuhlmann-Wilsdorf and C. Laird, "Dislocation Behavior in Fatigue, Part V Breakdown of Loop Patches-Formation of Persistent Slip Bands and of Dislocation Cells", Materials Science and Engineering, 46 (1980), p. 209-219.
(29) M. Wilhelm and P Everwin, in Proceedings of 5th International Conference on Strength of Metals and Alloys, Eds. P. Haasen, V. Gerold and G. Kostorz, p. 1089-1094, Pergamon Press, Oxford, 1980.
(30) A. Abel, M. Wilhelm and V. Gerold, "Low Cycle Fatigue of Single Crystals of α-Cu-Al Alloys", Materials Science and Engineering, 37 (1979), p. 187-200.
(31) H. Mughrabi, F. Ackermann and K. Herz in "Fatigue Mechanisms", American Society for Testing and Materials, Special Technical Publication 675, 1979, p. 69-94.
(32) H. Mughrabi and C. Wüthrich, "Asymmetry of Slip and Shape Changes during Cyclic Deformation of α-iron Single Crystals", Philosophical Magazine, 33 (1976) p. 963-984.
(33) H. Mughrabi, K. Herz and X. Stark, "The Effect of Strain Rate on the Cyclic Deformation Properties of α-iron Single Crystals", Acta Metallurgica, 24 (1976) p. 659-668.
(34) H. D. Nine, "Asymmetry of Slip in Fatigue of α-iron Single Crystals", Scripta Metallurgica, 4 (1970), p. 887-892.
(35) H. D. Nine, "Slip Planes and Asymmetric Slip in Fatigue of Iron Single Crystals", Philosophical Magazine, 26 (1972), p. 1409-1418.
(36) R. Neumann, "Probenformanderungen von Niob-Einkristallen bei symmetrischer Wechselbeanspruchung", Zeitschrift fur Metallkunde, 66 (1975), p. 26-32.
(37) S. Ikeda, "Temperature Dependence of Fatigue Process of Iron Single Crystals", 5th Int. Cong. on Strength of Metals and Alloys, Eds. P. Haasen, V. Gerold and G. Kostorz, p. 1157-1162, Pergamon Press, Oxford, 1980.
(38) R. Stevenson and J. B. Van der Sande, "The Cyclic Deformation of Magnesium Single Crystals", Acta Metallurgica, 22 (1974), p. 1079-1086.
(39) R. Kwadjo and L. M. Brown, "Cyclic Hardening of Magnesium Single Crystals", Acta Metallurgica, 26 (1978), p. 1117-1132.
(40) C. Laird, J. M. Finney and D. Kuhlmann-Wilsdorf, Dislocation Behavior in Fatigue-Part VI, "Variation in the Localization of Strain in Persistent Slip Bands", Materials Science and Engineering, in press (1981).
(41) A. T. Winter, "The Effect of Work-Hardening on the Low Strain Amplitude Fatigue of Copper Crystals", Philosophical Magazine, 31 (1975), p. 411-417.
(42) Z. S. Basinski, A. S. Korbel and S. J. Basinski, "The Temperature Dependence of the Saturation Stress and Dislocation Substructure in Fatigued Copper Single Crystals", Acta Metallurgica, 28, (1980), p. 191-208.
(43) A. S. Cheng and C. Laird, "Fatigue Life Behavior of Copper Single Crystals, Part I Observations of Crack Nucleation", to be published.
(44) J. M. Finney, "Strain Localization in the Fatigue of Copper Single Crystals", Ph.D. Thesis, University of Pennsylvania, Philadelphia, PA, 1974.
(45) K. Mecke, "TEM Investigations of the Cyclic Stress-Strain Behavior and the Formation of Persistent Slip Bands in Fatigued Single Crystals of Nickel Using Changing Amplitude Tests", Physica Status Solidi, 25A (1974), p. k93-95.
(46) A. Winter, "Nucleation of Persistent Slip Bands in Cyclically Deformed Copper Crystals", Philosophical Magazine, 37A (1978), p. 457-463.
(47) W. H. Kim and C. Laird, "Crack Nucleation and Stage I Propagation in High Strain Fatigue - I Microscopic and Interferomatic Observations", Acta Metallurgica, 26 (1978), p. 777-787.
(48) D. Kuhlmann-Wilsdorf, "Recent Progress in Understanding of Pure Metal and Alloy Hardening", in "Work Hardening in Tension and Fatigue", Ed. A. W. Thompson, AIME, 1977, p. 1-44.
(49) C. Laird, J. M. Finney, A. Schwartzman and R. de la Veaux, "History Dependence in the Cyclic Stress-Strain Response of Wavy Slip Materials", J. Testing and Evaluation, 3 (1975), p. 435-441.
(50) M. R. Mitchell, "Fundamentals of Modern Fatigue Analysis for Design", in Fatigue and Microstructure, American Society for Metals, 1979, p. 385-437.
(51) R. W. Landgraf, "Control of Fatigue Resistance Through Microstructure - Ferrous Alloys", Ibid, p. 439-466.
(52) H. Nahm, J. Moteff and D. R. Diercks, "Substructural Development During Low Cycle Fatigue of AISI 304 Stainless Steel at 649°C", Acta Metallurgica, 25 (1977), p. 107-116.

(53) D. J. Quesnel and J. C. Tsou, "Some Implications of Dislocation Cell Wall Mobility and Surface Energy in Low Cycle Fatigue", AIME Annual Meeting, Chicago, 1981.
(54) J. K. Lee and C. Laird, "The Cyclic Stress-Strain Response of Al-4w/o Cu Alloy Single Crystals Containing θ" Precipitates, Part I Phenomenological Results", 1981, to be published.
(55) J. K. Lee, "Cyclic Deformation and Strain Localization in Al-4w/o Cu Alloy Containing θ" Precipitates", Ph.D. Thesis, University of Pennsylvania, Philadelphia, PA, 1980.
(56) J. K. Lee and C. Laird, "The Cyclic Stress-Strain Response of Al-4w/o Cu Alloy Single Crystals Containing θ" Precipitates, Part II Dislocation Microstructures", 1981, to be published.
(57) J. K. Lee, S. P. Bhat, R. de la Veaux and C. Laird, "Mechanisms of Cyclic Softening in Precipitation-Hardened Alloys - A Ball Model Approach and Tests at 78K", International Journal of Fracture Mechanics, 1980, in press.
(58) A. S. Cheng and C. Laird, "The Long-Life Fatigue Behavior of Copper Single Crystals Tested in Strain Control", Materials Science and Engineering, 1981, in press.
(59) W. Vogel, M. Wilhelm and V. Gerold, "Dislocation Structures and Persistent Slip Band Formation during Cycling of Age-Hardened Al-Zn-Mg Single Crystals", 5th International Conference on the Strength of Metals and Alloys, Eds., P. Haasen, V. Gerold and G. Kostorz, p. 1175-1180, Pergamon Press, Oxford, 1980.
(60) M. Wilhelm, "The Cyclic Stress-Strain Behavior of Age-Hardened Copper-Cobalt and Aluminium-Zinc-Magnesium Alloy Single Crystals", Materials Science and Engineering, 1981, in press.
(61) W. Vogel, M. Wilhelm and V. Gerold, "Stage I Propagation and Persistant Slip Gands in Peak-Aged Al-Zn-Mg Single Crystals", Submitted to Acta Metallurgica, 1981.
(62) W. Vogel, M. Wilhelm and V. Gerold, "Development of Dislocation Microstructure and Persistent Slip Bands During Cyclic Deformation of Age-Hardened Al-4.9w/o Zn - 1.2w/o Mg", submitted to Acta Metallurgica, 1981.
(63) M. Wilhelm, M. Nageswararao and R. Meyer, "Factors Influencing Stage I Crack Propagation in Age-Hardened Alloys", American Society for Testing and Materials, Spec. Tech. Publication 675, 1979, p. 214-233.
(64) C. Laird, V. J. Langelo, M. Hollrah, N. C. Yang and R. de la Veaux, "The Cyclic Stress-Strain Response of Al-15wt/o Ag Alloy", Materials Science and Engineering, 32 (1978) p. 137-160.
(65) R. E. Stoltz and A. G. Pineau, "Dislocation-Precipitate Interaction and Cyclic Stress-Strain Behavior of a γ' Strengthened Superalloy", Materials Science and Engineering, 34 (1978), p. 275-284.
(66) E. A. Starke, Jr. and G. Lütjering, "Cyclic Plastic Deformation and Microstructure", in "Fatigue and Microstructure", American Society for Metals, 1979, p. 205-244.
(67) C. Laird, "The General Cyclic Stress-Strain Response of Aluminium Alloys", in American Society for Testing and Materials, Special Tech. Publ. 637, 1977, p. 3-35.
(68) T. H. Sanders, Jr. and J. T. Staley, "Review of Fatigue and Fracture Research on High-Strength Aluminium Alloys", in "Fatigue and Microstructure," American Society for Metals, 1975, p. 467-522.
(69) S. P. Bhat and C. Laird, "The Cyclic Stress-Strain Curves in Monocrystalline and Polycrystalline Metals", Scripta Metallurgica, 12 (1978), p. 687-692.
(70) P. Lukas and M. Klesnil, "Cyclic Stress-Strain Response and Fatigue Life of Metals in Low Amplitude Region", Materials Science and Engineering, 11 (1973) p. 345-351.
(71) H. Hughrabi, "Plateaus in the Cyclic Stress-Strain Response of Single- and Polycrystalline Metals", Scripta Metallurgica, 13 (1979), p. 479-484.
(72) K. V. Rasmussen and O. B. Pedersen, "Cyclic Deformation of Copper Polycrystals", 5th International Conference on the Strength of Metals and Alloys, Eds. P. Haasen, V. Gerold and G. Kostorz, p. 1219-1223, Pergamon Press, Oxford, 1980.
(73) K. V. Rasmussen and O. B. Pedersen, "Fatigue of Copper Polycrystals at Low Plastic Strain Amplitudes", Acta Metallurgica, 28 (1980), p. 1467-1478.
(74) H. Mughrabi and R. Wang, "Cyclic Strain Localization and Fatigue Crack Initiation in Persistent Slip Bands in Face-Centred Cubic Metals and Single Phase Alloys", Int. Sym. "Defects and Fracture", Tuczno, Poland, 1980. To be published by Sijthoff and Noordhoff.
75) J. C. Figueroa, S. P. Bhat, R. de la Veaux, S. Murzenski and C. Laird, "The Cyclic Stress-Strain Resposne of Copper at Low Strains, Part I Constant Amplitude Testing," Acta Metallurgica, 1981, in press.
76) J. C. Figueroa and C. Laird, "The Cyclic Stress-Strain Response of Copper at Low Strains, Part II Variable Amplitude Testing," submitted for publication, 1981.
77) A. S. Cheng, J. C. Figueroa, C. Laird and J. K. Lee, "The Cyclic Deformation of Polycrystalline Pure Metals and Precipitation Hardened Alloys", Riso Symposium, Denmark, September 1981, in press.
78) K. Rajan, B. Ramaswami and S. M. L. Sastry, "The Plastic Zone in Stage I Crack Growth During Fatigue Deformation of Copper Single Crystals", Metallurgical Transactions, 6A (1975) p. 1959-1962.
79) P. Charsley, "Dislocation Arrangements in Polycrystalline Copper Alloys Fatigued to Saturation", Materials Science and Engineering, in press, 1981.
80) R. Page and J. R. Weertman, "Dislocation Structures in Copper Cycled at High Tempertures," AIME Annual Meeting, Chicago, 1981.
81) D. Kuhlmann-Wilsdorf and P. Charsley, Research on Structures of Dislocation Walls, to be published, 1981.
82) C. Laird and D. J. Duquette, "Mechanisms of Crack Nucleation", in "Corrosion Fatigue", NACE-2, 1972, p. 88-117.
83) L. M. Brown, "Dislocation Substructures and the Initiation of Cracks by Fatigue", Metal Science, 11 (1977) p. 315-320.
84) M. E. Fine and R. O. Ritchie, "Fatigue Crack Initiation and Near-Threshold Crack Growth", in Fatigue and Microstructure", American Society for Metals, 1972, p 245-278.
85) P. J. E. Forsyth, "The Basic Mechanism of Fatigue and its Dependence on the Initial State of a Material," International Conference on Fatigue, Inst. of Mechanical Engineers and ASME, paper 6.5, 1956, p 535-537.
86) J. G. Antonopoulos, L. M. Brown and A. T. Winter, "Vacancy Dipoles in Fatigued Copper" Philosophical Magazine, 34 (1976) p. 549-563.
87) U. Essmann, U. Gosele and H. Mughrabi, "A Model of Extrusions and Intrusions in Fatigued Metals, Part I", Philosophical Maganzine, 1981, in press.
88) M. Graf and C. Verpoort, "Influence of Microplasticity and Crack Initiation on the Fatigue Behaviour of Precipitation Hardened Alloys", 5th Int. Conf. on Strength of Metals and Alloys, Eds., P. Haasen, V. Gerold and G. Kostorz, Pergamon, 1979, p. 1207-1212.
89) E. Hornbogen and C. Verpoort, "Fatigue Crack Initiation after Different Surface Treatments in Precipitation Hardening Alloys", 5th Int. Conf. on Fracture, ICF 5, Cannes, 1981.
90) A. S. Cheng and C. Laird, "Fatigue Life Behavior of Copper Single Crystals, Part II, Model for Crack Nucleation in Persistent Slip Bands, to be published, 1981.

91) J. C. Figueroa, "Cumulative Fatigue Damage in Copper Polycrystals", Ph.D. Thesis, University of Pennsylvania, Philadelphia, 1979.
92) B. I. Verkin and N. M. Grinberg, "The Effect of Vacuum on the Fatigue Behavior of Metals and alloys", Materials Science and Engienering, 41 (1979) p. 149-181.
93) C. Laird and C. E. Feltner, "The Coffin-Manson Law of High Strain Fatigue in Relation to Slip Character", Transactions, American Inst. of Metallurgical Engineers, 239 (1967) p. 1074-1083.
94) W. H. Kim and C. Laird, "The Role of Cyclic Hardening in Crack Nucleation at High Strain Amplitude", Materials Science and Engienering, 33 (1978), p. 225-231.
95) W. H. Kim and C. Laird, "Crack Nucleation and Stage I Propagation in High Strain Fatigue - I Microscopic and Inteferometric Observations", Acta Metallurgica, 26, (1978) p 777-788.
96) W. H. Kim and C. Laird, "Crack Nucleation and Stage I Propagation in High Strain Fatigue - II Mechanism", Acta Metallurgica, 26 (1978) p 789-799.
97) B. Hodgson, "High Strain Fatigue Crack Nucleation in an Austenitic Steel at Elevated Temperatures", Metal Science Journal, 2 (1968) p. 235-237.
98) L. F. Coffin and C. J. McMahon, "Mechanisms of Damage and Fracture in High Temperature, Low-Cycle Fatigue of a Cast Nickel-Based Superalloy", Metallurgical Transactions, 1 (1970) p. 3443-3450.
99) H. Mughrabi, "Wechselverformung und Ermudungsbruch von α-Eisenvielkristallen", Zeitschrift fur Metallkunde, 66 (1975) p 719-724.
100) H. Mughrabi, in "Surface Effects in Crystal Plasticity", Eds., R. M. Latanision and J. T. Fourie, Noordholt, Leyden, 1977, p. 479-485.
101) M. E. Fine, "Fatigue Resistance of Metals", Metallurgical Transactions, 11A (1980) p 365-379.
102) "Mechanics of Crack Growth", American Society for Testing and Materials, Special Technical Publication 590, 1976.
103) "Flow Growth and Fracture", American Society for Testing and Materials, Special Technical Publication 631, 1977.
104) "Cyclic Stress-Strain and Plastic Deformation Aspects of Fatigue Crack Growth", American Society for Testing and Materials, Special Technical Publication, 637, 1977.
105) "Fatigue Mechanisms", American Society for Testing and Materials, Special Technical Publication, 675, 1979.
106) P. Neumann, H. Vehoff and H. Fuhlrott, in Fracture 1977, ICF4, vol. 2, University of Waterloo Press, p. 1313-1324.
107) K. Katagiri, A. Omura, K. Koyamagi, J. Awatani, T. Shiraishi and H. Kaneshiro, "Early Stage Crack Tip Dislocation Morphology in Fatigued Copper," Metallurgical Transactions, 8A (1977) p 1769 - 1773.
108) K. Katagiri, J. Awatani, A. Omura, K. Koyanagi and T. Shiraishi, in "Fatigue Mechanisms", American Society for Testing and Materials, Special Technical Publication 675, 1979, p. 107-123.
109) J. A. Carlson and D. A. Koss, "Fatigue Fracture and Slip Processes in b.c.c. Ti-V Alloy Single Crystals," Acta Metallurgica, 26 (1978) p. 123-132.
110) R. O. Ritchie, "Influence of Microstructure on Near Threshold Fatigue Crack Propagation in Ultra-High Strength Steel," Metal Science 11 (1977) p.368-381.
111) H. Suzuki and A. J. McEvily, "Microstructural Effects on Fatigue Crack Growth in a Low Carbon Steel," Metallurgical Transactions, 10A (1979) p. 475-481.
112) M. O. Speidel, In "Stress Corrosion Cracking and Hydrogen Embrittlement of Iron Base Alloys," National Association of Corrosion Engineers, Houston, Texas, 1977, p. 1071-1094.
113) R. O. Ritchie, "Application of Fractyre Mechanics to Fatigue, Corrosion-Fatigue and Hydrogen Embrittlement", in "Analytical and Experimental Fracture Mechanics", Ed. G. C. Sih, Sijthoff and Noordhott, 1980, in press.
114) S. Suresh, G. F. Zamiski and R. O. Ritchie, "Oxide-Induced Closure - A Mechanism for Near-Threshold Corrosion Fatigue Crack Growth", MIT Report No. FPL/R/80/1036, November 1980.
115) S. E. Stanzl, Research into Threshold Crack Propagation Behavior at Ultrasonic Frequencies, Private Communication, 1981.
116) S. Purushothaman and J. Tien, "Generalized Theory of Fatigue Crack Propagation, Part I Derivation of Thresholds", Materials Science and Engineering, 34 (1978) p. 241-246.
117) J. Tien and S. Purushothaman, "Generalized Theory of Fatigue Crack Propagation, Part II Deviation of Threshold and Paris Regime Crack Growth Rates", Materials Science and Engineering, 34 (1978) p. 247-262.
118) S. E. Stanzl, "High Frequency Fatigue Crack Growth and Threshold Behavior in Monocrystalline Copper", Scripta Metallurgica, 14 (1980) p. 749-753.
119) M. A. Wilkins and G. C. Smith, "Dislocation Structures Near a Propagating Crack in an Al- % Mg Alloy", Acta Metallurgica, 18 (1970) p. 1035-1043.
120) C. Laird and G. C. Smith, "Crack Propagation in High Stress Fatigue", Philosophical Magazine, 7 (1962) p. 847-857.
121) C. Laird and G. C. Smith, "Initial Stages of Damage in High Stress Fatigue", Philosophical Magazine, 8 (1963) p. 1945-1963.
122) A. S. Cheng and C. Laird, "The Transition from Stage I to Stage II Crack Growth", to be published, 1981.
123) C. Laird and R. de la Veaux, "Additional Evidence for the Plastic Blunting Process of Fatigue Crack Propagation", Metallurgical Transactions, 8A (1977) p. 657-664.
124) J. Shijve, in "Fatigue Crack Propagation", American Society for Testing and Materials, Special Techical Publication 415, 1967, p. 415-459.
125) R. M. N. Pelloux, in "Fracture 1969", ICF2, Brighton, 1969, p. 731-744.
126) A. S. Kuo and H. W. Liu, "Analysis of Unzipping Model for Fatigue Crack Growth", Scripta Metallurgica, 10 (1976) p. 723-728
127) C. Laird, "Studies of High Strain Fatigue", Ph.D. Thesis, University of Cambridge, 1962.
128) H. Vehoff and P. Neumann, "Crack Propagation and Cleavage Initiation in Fe-2.6% Si Single Crystals Under Controlled Plastic Crack Tip Opening Rate in Various Gaseous Environments", Acta Metallurgica, (1980) p. 265-272.
129) A. J. McEvily in The Microstructure and Design of Alloys, Vol. 2, Proceedings, 3rd International Conference on the Strength of Metals and Alloys, Cambridge, 1973.
130) C. Y. Yang and H. W. Liu, "The Application of the Unzipping Model of Fatigue Crack Growth to Micro-Cracks", Scripta Metallurgica, 14 (1980) p. 785-790.
131) J. C. Newman, Jr., "Finite Element Analysis of Crack Growth Under Monotonic and Cyclic Loading", American Society for Testing and Materials, Special Technical Publication 637 (1977) p. 56-80.
132) J. Weertman, "Fatigue Crack Propagation Theories", in "Fatigue and Microstructure", American Society for Metals, 1979, p. 279-306.
133) H. L. Marcus, "Environmental Effects 11, Fatigue Crack Growth in Metals and Alloys," Ibid, p. 365-384.
134) D. L. Davidson, "The Study of Fatigue Mechanisms with Electron Channeling", in Fatigue Mechanisms, American Society for Testing and Materials, Special Technical Publication 675, 1979, p. 254-275.

CREEP AND HIGH TEMPERATURE DEFORMATION OF SIMPLE METALS AND SUPERALLOYS

Russell R. Jensen* and John K. Tien
Henry Krumb School of Mines
Columbia University
New York, NY 10027

I. INTRODUCTION

Elevated and high temperature mechanical behavior studies usually pivot about either the creep test or the distinctly different stress-strain test. The creep test imposes a constant stress or constant load and mechanical behavior is determined by studying the strain and strain rate response of the system. The stress-strain test imposes a constant strain rate (usually total strain rate) and the stress-strain response of the system is observed and studied. Although in principle, information from these tests should be complementary with respect to the elucidation of mechanisms, seldom are the tests part of the same study and never have they been reviewed together. In what follows we first present an overview of select stress-strain and creep results for the pure metals and simple alloys, paying special attention to the cross-links between those results which together have provided a clearer view of the rate controlling processes in thermally assisted deformation. Then we adopt the same review strategy for particle strengthened systems.

The understanding of deformation behavior usually follows the analysis of results from experiments in which such test variables as temperature, load, strain rate and environment are systematically varied and microstructural and chemical variables are isolated. The isolation of variables is straightforward for the case of the pure metals and single phase solid solutions. Obviously the situation is more complex in the multicomponent particle strengthened alloys.

Scientific studies on particle strengthened alloys have focused mainly on engineering alloys for elevated and high temperature applications. There are three major classes of particle strengthened alloys, divided in principle by the governing particle strengthening mode (1). These are the coherent precipitation strengthened nickel-base and iron-nickel-base superalloys (2); the incoherent precipitation (mainly carbide) strengthened cobalt-base superalloys and the high nickel and chromium stainless steels; and the oxide dispersion strengthened (ODS) alloys (2-5). Many highly alloyed aluminum, titanium-base alloys, and maraging and other specialty steels also see warm to elevated temperature applications at times. Metal-matrix composites, and especially in-situ composites (eutectics), are also high temperature systems. Although the matrices of certain of these composites are precipitation strengthened, their major mode of strengthening is through the transfer of load to the strong fibers.

In this treatise we concentrate on thermally assisted deformation caused mainly by the generation and motion of dislocations. Except where necessary we ignore such grain boundary associated deformation mechanisms as diffusional creep and grain boundary sliding. The former occurs only at high temperatures or very low stresses, and neither effectively takes place in monocrystalline or large grain size or directionally grain aligned systems, although both are important in superplastic forming and rupture of polycrystalline systems.

II. STRESS-STRAIN BEHAVIOR OF SIMPLE SYSTEMS

The stress-strain curves of monocrystalline and polycrystalline pure metals and solid solutions generally exhibit two and in some cases three distinct stages (6-12). Stage I or the easy glide stage is observed only during the deformation of single crystals oriented for single slip on a single glide plane (7,10,13). This stage exhibits little or no work hardening. Stage II essentially is a region of athermal work hardening where the work hardening rate is remarkably insensitive to many experimental variables (7,10,13-15). Stage III, of special interest here, is thermally assisted and exhibits a work hardening rate which is strongly dependent on strain, temperature and strain rate (13,15-18). Stages II and III occur in polycrystals and in single crystals oriented for multiple slip (13,15-20). The two-stage work hardening behavior typical of polycrystals is evident in Figure 1 where stress-strain curves of pure copper obtained at different temperatures nearly coincide at low strains, corresponding to Stage II, while at larger strains the curves enter Stage III and diverge.

Generally, Stage II is characterized by a work hardening rate θ_{II} which has a value on the order of $\mu/200$, where μ is the shear modulus. θ_{II} is nearly constant in single crystals deforming by single slip and varies somewhat with strain in polycrystals and single crystals deforming by multiple slip (7,10,13-15). The insensitivity of θ_{II} to temperature, strain rate and many other variables is because dislocation storage follows strictly geometrical or statistical rules (13,21,22). This concept has been formalized in the various theories of Stage II work hardening (21-26).

The flow stress σ of many systems is given by

$$\sigma = \sigma_o + \alpha\mu b\sqrt{\rho} \tag{1}$$

where σ_o is a constant, b is the Burgers vector, ρ is the dislocation density and α is a constant approximately equal to 0.5 (21,22). The term σ_o in Equation (1) is small for simple systems and is usually neglected; for particle-strengthened systems, σ_o can be appreciable (22). The dislocation production rate $d\rho/d\varepsilon$ can be defined as

$$d\rho/d\varepsilon = 1/bL \tag{2}$$

where ε is the plastic strain and L is the mean free path of dislocation glide. The basic assumption of the Stage II hardening theories is that L is proportional to the mean dislocation spacing $1/\sqrt{\rho}$, or

$$1/L = \sqrt{\rho}/\beta \tag{3}$$

where β is a constant. Equations (1), (2) and (3) may be combined to give an expression for θ_{II} as

$$\theta_{II} \equiv d\sigma/d\varepsilon = \alpha\mu/2\beta. \quad (4)$$

Equation (4) is a generalized result common to all of the Stage II hardening theories. The goal then is to theoretically justify setting the ratio β/α equal to 100 to obtain the correct value for θ_{II}. The deviation from linear Stage II work hardening observed during polyslip deformation has been attributed to the inverse mean free slip distance in Equation (3) increasing not proportionally but only linearly with $\sqrt{\rho}$ (13,15). Grain size effects have also been incorporated into Stage II work hardening by assuming grain interiors work harden like single crystals deforming by polyslip and recognizing that work hardening in the vicinity of grain boundaries is enhanced (13,15,27-30).

Stage III work hardening behavior is affected by the stacking fault energy and the self diffusion coefficient in addition to the shear modulus. This is because during Stage III dislocation rearrangement occurs leading to a dislocation cell structure at lower temperatures (less than homologous temperatures of 0.5) through cross slip and to a well defined dislocation subgrain structure at higher homologous temperatures through dislocation climb (13). Interestingly, as plastic deformation progresses, these dislocation cells or subgrains become relatively equiaxed in shape and approach a constant size which in turn depends on temperature and strain rate (13,31-33). In addition, the flow stress at large strains has been found to vary proportionally with the inverse cell (or subgrain) size (31,32).

The dislocation behavior described above is quite similar to that observed during elevated temperature primary creep (34). This leads to the intriguing possibility that, as in elevated temperature creep, steady state deformation can occur.

The existence of a steady state during stress-strain deformation is also suggested by the curves in Figure 1, which are fairly typical of simple systems, where the Stage III portions of the curves appear to approach a finite constant stress. The existence of a steady state or, according to Kocks (17) a saturation stress, has been often proposed for simple systems (13,15,17,31,35,36).

The approach to steady state during Stage III deformation can be seen more clearly in Figure 2, where the work hardening rate θ is plotted versus the flow stress σ for each of the curves in Figure 1. As noted by Kocks (17) a θ-σ plot is preferrable to other presentations of data in evaluating work hardening behavior because it avoids the explicit use of strain, which is path dependent and hence ill defined, especially at higher homologous temperatures, and since the use of the differential quantity θ is less prejudicial to a particular integrated stress-strain relationship.

In Figure 2 the two stages of work hardening are quite apparent. At the lower stresses, the work hardening curves roughly coincide, corresponding to Stage II, where work hardening is athermal. At higher stresses the curves diverge and enter Stage III, which is characterized by a temperature dependent linear branch at lower values of θ. This behavior is typical of many simple systems (13,15,17).

The saturation stress σ_s is defined as the stress intercept of the extrapolation of the linear branch obtained in a θ-σ plot to $\theta = 0$ (13,15,17,35), as shown in Figure 2. The saturation stress then corresponds to the stress at which the stress-strain curve would have become horizontal if necking and fracture had not occurred.

Stress strain tensile fracture and creep rupture or stress rupture at high homologous temperatures have not been reviewed together. Space limitations prevent this from being accomplished as part of this treatise.

The concept of a steady state during stress-strain deformation has been questioned in the literature (37), especially in regard to material behavior at homologous temperatures less than 0.5, based on experiments involving very large strains such as those encountered during wire drawing (38-40). However, as discussed by Meking and Grinberg (33), deformation conditions may change in these tests at very large strains so that the saturation stress concept may still apply. Notwithstanding these exceptions, torsion tests, which preclude the effects of necking and other mechanical instabilities, have confirmed that a steady state stress is indeed reached during the constant strain rate deformation of many systems (41-44).

Stage III work hardening, which leads to linear branches on θ-σ plots, has been analyzed by Kocks (17). This analysis is based on a stress-strain relationship originally proposed by Voce (45) as

$$\sigma = \sigma_s - (\sigma_s - \sigma_1)\exp[-(\varepsilon - \varepsilon_1)/\varepsilon_c] \quad (5)$$

where ε_c is a constant, and σ_1 and ε_1 represent the true stress and true plastic strain at the point where

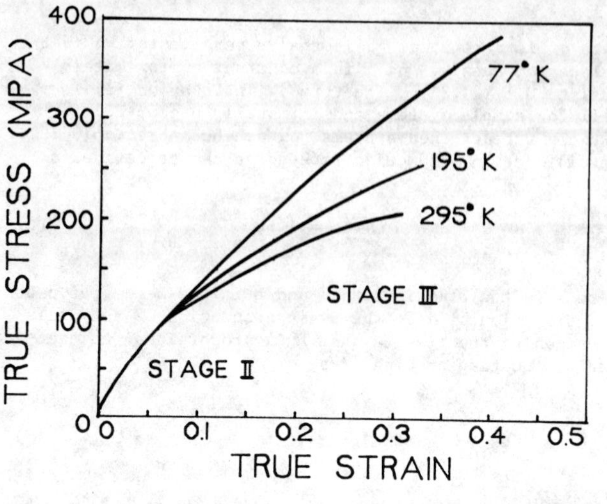

Figure 1. Typical stress-strain curves of copper, showing the emergence of Stage III work hardening with temperature and strain. From (13).

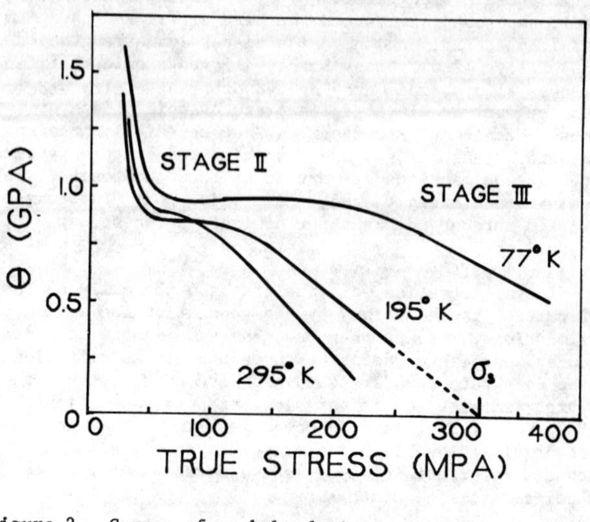

Figure 2. Curves of work hardening rate versus stress for the curves in Figure 1, showing Stage III behavior. From (15).

Equation (5) is matched to the experimental data. This relationship is radically different from the much used power law work hardening relationships (46-49) in its behavior at large strains where it predicts an approach to a constant stress σ_s, in accordance with observed Stage III behavior (13,15,17). The power law relationships, on the other hand, predict that the stress should increase without limit with increasing strain. In addition, they have no way of explicitly accounting for the effects of temperature and strain rate. Hence, power law relationships are applicable only to Stage II.

Kocks has restated Equation (5) in differential form as

$$\theta = \theta_o(1 - \sigma/\sigma_s) \qquad (6)$$

where θ_o is the initial rate of work hardening if Equation (6) was valid near zero stress (17). The parameter ε_c then takes the form $\varepsilon_c = \sigma_s/\theta_o$. Equation (6), like Equation (5), predicts a linear decrease in work hardening rate with increasing stress, consistent with the observed linear branches on θ-σ plots.

The temperature and strain rate dependence of the parameters θ_o and σ_s have been evaluated from the stress-strain data of a number of pure metals and simple alloys (17). The parameter θ_o has been found to be independent of strain rate and dependent on temperature only through the shear modulus, having a value of $\mu/20$ to $\mu/15$ (17). The major temperature and strain rate dependence of work hardening rests in the parameter σ_s.

The dependence of σ_s on temperature and strain rate has been analyzed using a modification of the so-called τ_{III} analysis developed originally by Haasen (50) to model the temperature and the $\dot{\varepsilon}$ dependence in single crystals of the stress at the start of Stage III where the onset of thermally activated softening is just observable in a stress-strain curve. The observed behavior of τ_{III} is explained through theories of thermally activated cross slip (51), which predict a relation of the form

$$\ln(\tau_{III}/\tau_M) = kT/A \ln(\dot{\varepsilon}/\dot{\varepsilon}_o) \qquad (7)$$

where A, τ_M and $\dot{\varepsilon}_o$ are constants and kT has the usual meaning (18). The parameter A is a function of stacking fault energy, while the stress τ_M corresponds to the mechanical threshold for cross slip, i.e., the stress at which cross slip proceeds without the necessity of thermal activation. Equation (7), with small modifications, has been found to apply not only at the onset of Stage III, but at stresses up to and including σ_s, in both single crystals and polycrystals (17,18). Consequently, a relationship for steady state deformation between the saturation stress σ_s and applied strain rate can be written as

$$\dot{\varepsilon}/\dot{\varepsilon}_o = (\sigma_s/\sigma_{so})^{A/kT} \qquad (8)$$

This relationship has been found to apply at temperatures as low as 4^oK (15,17), which has the interesting implication that a steady state may occur during stress-strain deformation even at cryogenic temperatures (15).

Lower homologous temperature and cryogenic creep is an observed and interesting phenomenon (52). Mechanical behavior at these temperatures is complicated by complex reductions of activation energies and changes of diffusional processes that are not understood.

Equation (8) also implies that the mechanism controlling steady state deformation is thermally activated cross slip. Kocks (17) has proposed that this may hold at all temperatures, even above homologous temperatures of 0.5 where dislocation climb is generally regarded as the important controlling process (34). Meking (15) has found, based on more extensive data, that the parameters within Equation (8) indeed change at homologous temperatures above about 0.5 for many systems and attributed this to a change in mechanism.

The preceding discussion of the analysis of Stage III work hardening shows that a steady state is often established during stress-strain testing and that its temperature and strain rate dependence can be rationalized. The obvious question to be asked then is what relationship this steady state has to the steady state observed during creep testing. This will be addressed shortly.

III. CREEP BEHAVIOR OF SIMPLE SYSTEMS

The main focus in the investigation of creep behavior is the understanding of the steady state or secondary creep stage, usually at homologous temperatures about 0.5. Steady state creep data at intermediate stress levels are usually analyzed through the empirical power law relationship

$$\dot{\varepsilon}_s = A(\sigma/E)^n \exp(-Q_c/kT) \qquad (9)$$

where $\dot{\varepsilon}_s$ is the steady state or minimum creep rate, σ is either the applied stress or an effective stress (see below), E is Young's modulus, n is the apparent stress exponent, Q_c is the apparent creep activation energy, A is a material constant and kT has the usual meaning. Early recognition (34,53-56) that Q_c is related to the self diffusion activation energy led to a modified form of Equation (9), where

$$\dot{\varepsilon}_s = A'D(\sigma/E)^n \qquad (10)$$

where D is the appropriate effective diffusion coefficient of the rate controlling diffusional process. Usually D is experimentally found to be that of vacancy self diffusion through the matching of Q_c to the self diffusion activation energy Q_{SD}. In the case of solid solutions, for example, C in Fe, changes in D in Equation (10) are found to correlate with the changes in the Fe self diffusion coefficient as C is added (57). Changes in D have also been found to correlate with changes in the effective diffusivity occurring at or after phase transitions (58, 59). Rationalizations of dislocation pipe diffusion have been proposed to account for the low values of Q_c found near homologous temperatures of 0.5 (60), while divacancy diffusion and recrystallization complications have been proposed to explain the high values of Q_c found near melting temperatures (60). In the limited number of systems where both the pressure dependence of creep and of self diffusion are known at comparable temperatures, the respective activation volumes are found also to be equal (53-55,60). Accordingly, the evidence is strong for the creep of simple systems to be rate controlled by diffusional processes. The unravelling of the physical details of the mechanisms involved are usually attempted through understanding the stress dependence of creep, especially through comparisons of theoretically developed power law exponents n to the n values obtained from the analysis of data according to Equation (9) or (10).

When creep data are fitted to a power law, the typical value of the stress exponent n for fcc metals is 5 (34,53-55). For bcc and hcp systems, the measured values of n vary over a somewhat wider range from about 4 to 7 (34). In fcc systems, with the addition of solute elements, the value of n decreases rapidly in the case of some solid solution alloys (called Class I alloys) to values approaching 3, while in other solid solution alloys (called Class II alloys), n remains high at a value around 5 (34). The rate controlling process during creep of Class II alloys is regarded as the same as that for pure metals-self diffusion (34). Class I creep behavior is interpreted as being fundamentally different, where the rate controlling process is the viscous glide of dislocations dragging solute atmospheres (34), where n = 3 results from the n = 2 stress dependence of dislocation density and a linear stress dependence of the dislocation velocity.

The theoretical description of power law creep has received much emphasis and many models have been proposed to explain the observed temperature and stress dependence, especially in the simpler fcc systems. One group of models is based on the premise that creep strains are due to dislocation glide but creep rates are rate controlled by recovery through edge dislocation climb, and not by dislocation glide processes. Since climb is diffusion controlled, all of these models correctly predict a creep activation energy equal to the effective self diffusion activation energy. The details of these models differ mainly in the dislocation structure assumed and the location of the recovery events.

Weertman has proposed that the rate controlling step is recovery within dislocation tangles which surround dislocation sources (54). While this model does predict n = 4.5, it makes the assumption of a constant density of sources and requires the existence of pileups. Sub-boundary recovery models have also been proposed (61-63). These models have in common the rate controlling step of dislocation annihilation through climb within the sub-boundaries. These models predict n = 3 and require pileup against the sub-boundaries, which is not observed (34). Similar models have been proposed based on recovery within high dislocation density cell boundaries (64,65). These models are not applicable at homologous temperatures higher than 0.5 where sub-boundaries instead of cell boundaries form, and predict n = 4 only under the constraint that the cell boundary thickness is independent of stress.

Another group of models has been proposed in which creep strains are assumed to be due entirely to dislocation climb with no dislocation glide processes. Nabarro (66) formulated one such model based on a regular array of climbing edge dislocations with two sets of Burgers vectors, with the spacings between these sets being determined by the applied stress. This model predicts n = 3. Network growth models based on Friedel's network recovery theory (67) have also been proposed (68-71). In these models it is again implicitly assumed that creep strain results from the climb of dislocations comprising the network as the network coarsens under the driving force of decreasing line energy. The recovery rate of the network can be calculated (67) and the Bailey-Orowan equation (72,73) is used to obtain a relation for the steady state creep rate with either an assumed functional form of the work hardening rate or experimentally determined creep work hardening rates (69). When experimental values are employed, network recovery models predict n = 4 or 4.5 (69). However, their validity is hard to accept, since the work hardening rates used in these theories are assumed to be athermal, yet exhibit a strong stress dependence.

Modifications of the network recovery model have been proposed in which the heterogeneous nature of the network is accounted for in the form of a link length distribution (74-76). In these models network coarsening is assumed to generate link lengths which are sufficiently long to act as dislocation sources which then allow dislocation glide and hence creep strain to occur.

A common problem with all recovery based creep models is that the prediction of n > 3 requires some physically unrealistic assumptions to be made. Specifically, these assumptions violate the principle of similitude (21,34) which, simply stated, requires dislocation structure and behavior to be independent of the scale of the dislocation substructure when the length dimension is normalized by some average dimension of the substructure. In addition, since recovery is assumed to be diffusion controlled, it is intrinsically related to time and not strain so that dynamic recovery, which may be important (17,77), does not play a role in these models (53).

Even though no existing recovery type theory can reasonably explain power law creep with n > 3, it is reasonably well established that power law creep is diffusion controlled and that the existing recovery controlled models may, with modification, explain why n = 5. A mechanistic explanation of creep is also complicated by the fact that power law creep is observed only within a range of intermediate stresses; different stress dependencies are observed at high and low stresses (53,55). At low stresses and high temperatures, the creep rate is proportional to the applied stress, i.e., n = 1. In polycrystals, creep in this regime is regarded as being rate controlled by stress induced diffusion between grain boundaries where the diffusion path may be through the lattice, leading to Nabarro-Herring creep (78), or along grain boundaries, leading to Coble creep (79), and to grain boundary sliding (80). Creep rates proportional to the applied stress have also been observed in single crystals at low stress. Since Nabarro-Herring and Coble creep require grain boundaries to operate, this behavior is attributed to a dislocation process which is not that well understood termed Harper-Dorn creep (81).

At high stresses, the power law breaks down and the creep rate exceeds that predicted from the extrapolation of the intermediate stress power law. The breakdown is frequently observed to occur at diffusion coefficient compensated creep rates $\dot{\varepsilon}_s/D$ equal to about 10^9 cm^{-2} (53-55). In this regime the relationship between the steady state creep rate and the applied stress is approximately exponential where

$$\dot{\varepsilon}_s = Ke^{B\sigma} \tag{11}$$

where K and B are constants (32,55,82). Garofalo (83) proposed an empirical equation which accurately represents intermediate stress power law behavior and high stress exponential behavior as

$$\dot{\varepsilon}_s = K'D[\sinh(\alpha\sigma/E)]^{n'} \tag{12}$$

where K', α and n' are constants. Equation (12) reduces to a power law expression with n = n' at low stresses and an exponential expression at high stresses where B is given by αn'/E. Equation (12) has been found to represent the steady state behavior of a variety of materials (82) and has recently been shown to fit the data of aluminum over 21 orders of magnitude in diffusion compensated creep rate (43).

While Equation (12) does represent intermediate and high stress creep behavior reasonably accurately, its physical significance is not that well understood. Alternate expressions have been proposed which are based on the assumption that the rate controlling mechanism at high stresses is the same as that for the power law regime, with certain modifications. The upswing in strain rate at high stress has been attributed to high dislocation densities which lead to an enhancement of diffusion due to increased pipe diffusion (84-87), in which case Equation (10) would apply. Creep activation energies are observed to decrease in this regime to about three-fourths of the self diffusion energy (87), which would be consistent with an increasing contribution from pipe diffusion. On the other hand, a stress dependent activation energy is also consistent with an activation analysis type expression and an alternative high stress creep mechanism. High stress creep has also been attributed to complex and yet to be understood high, nonequilibrium vacancy concentrations generated by dynamic dislocation intersection processes (88,89,92).

Models based on the glide of jogged screw dislocations

would also be applicable in this instance because some are able to predict the observed exponential stress dependence of strain rate at high stress (90). Jogged screw dislocation models have also been proposed for general application in creep at lower applied stresses.

The premise of jogged screw models is that steady state creep is rate controlled by the thermally assisted motion of screw dislocations and not edge dislocations, and that the motion of the screw dislocations is, in turn, rate controlled by the motion of jogs on these dislocations through point defect diffusion. A jog on a screw dislocation may be of two types: a vacancy jog, which moves by either emitting vacancies or absorbing self-interstitials and an interstitial jog, which moves by either emitting self-interstitials or absorbing vacancies. Mott first proposed the concept of jog motion being rate controlling in creep (91). More recently, the theories and discussions on jogged screw dislocation rate control rest with the work of Barrett and Nix (92) and Tien et al. (93,94).

The mechanism of Barrett and Nix and others is based on the concept that a mixture of vacancy and interstitial type jogs are spaced more or less alternately on a moving screw dislocation (92-99). They allowed the jogs to move only by vacancy emission or absorption. The resulting creep equations predict a vacancy-diffusion activation energy, which is observed in pure metals, and a limiting, hyperbolic tangent type dependence on applied stress, which is not observed.

Malu and Tien allowed jogs to be formed on the moving screw dislocation through its extraction from dislocation networks formed during primary creep (94). Based on dislocation intersection considerations presented by Cottrell (100) and Weertman (101,102), Malu and Tien further argued that interstitial type jogs are more likely to be produced on moving screw dislocation than vacancy type jogs. They concluded that, for pure metals and simple systems, jog-motion will be rate controlled by the absorption of vacancies, resulting in a vacancy diffusion activation energy and an apparent stress exponent of about 3. They further concluded that, in systems in which the dislocation networks are fine, the interjog spacing would be small enough to exhaust vacancies, and the interstitial type jogs would then move by the alternate path of interstitial emission. The resulting equations predict a high activation energy equal to that of self-interstitial formation and diffusion and an exponential type applied stress dependence. Interestingly, networks in some of the more complex particle strengthened alloys are finely meshed, and high activation energies and sharp stress dependences are observed (94).

IV. CORRELATIONS BETWEEN STRESS-STRAIN AND CREEP BEHAVIOR

In principle, during Stage III deformation in a stress-strain test, a metal or alloy should experience changes similar to those during primary creep since the strain rate decreases continuously during the primary creep stage until the secondary or steady state creep stage is reached and constant strain rate deformation continues under constant stress. However, since the deformation histories are different, it is not immediately self evident that the dislocation substructures developed during primary creep and during Stage III work hardening are equivalent.

Recently, equivalence has been demonstrated in the cases where this has been studied. Meking (15) showed that steady state stress-strain and steady state creep data of aluminum can be represented by a single relationship among stress, strain rate and temperature. In particular, when stress-strain and creep data are obtained at the same temperature and the applied stress in creep is set equal to the measured stress-strain saturation stress, the observed steady state creep rate is seen to be equal to the constant strain rate imposed during the stress-strain test. Further, the activation energy obtained from the analysis of stress-strain saturation stress data (35) has also been shown to have the same temperature dependence as the activation energy obtained from creep data (103). Luthy et al. (43) also found that the steady states developed in torsion and in creep are equivalent. Such results show that stress-strain and creep tests are more closely related than is commonly realized and, as mentioned, the processes which govern Stage III work hardening leading to a stress-strain steady state saturation stress and primary creep leading to steady state creep are quite similar, if not the same. The implication is that a mechanical equation of state may exist (104).

V. RESISTING STRESSES IN CREEP

In a material undergoing creep where grain boundary effects are supressed, there is a variety of resisting and/or back stresses which may act on a mobile dislocation. By virtue of its motion, a dislocation may encounter a friction stress such as a Pierls stress or the stress due to interface formation in particle cutting or a viscous stress such as solute atmosphere drag (4,5,105-107). As a dislocation bypasses barriers such as strengthening particles, it experiences an average threshold stress, which may also be viewed as a short range back stress. In addition, as a result of the formation of deformation substructures, long range internal stresses may develop. This type of internal stress is also a back stress which may drive dislocation motion after stress removal. Recently, large anelastic strains have been found in Inconel MA 754, an ODS alloy, and in the superalloy Udimet 700 in the unload cycles during cyclic creep (108). These anelastic strains are either due to these back stresses or some other form of stress distribution.

The salient question then is whether dislocation creep behavior is more properly described in terms of the applied stress σ, say in Equations (9) or (10), or an effective stress $\sigma-\sigma_R$, where σ_R represents the effective sum of the resisting stresses. To this end much experimental work has been performed to measure the resisting stress or components of this stress, using a variety of techniques.

In the strain transient dip test, the applied stress is reduced instantaneously and directly (by trial and error) to a stress level where the instantaneous strain rate following the reduction is zero (109,110). This residual applied stress is defined as the resisting internal stress. Since the stress is reduced very quickly, this test is viewed as a constant structure test, so that the measured resisting stress characterizes the initial state of the material before stress reduction. Strain dip tests in most cases of pure metals and strengthened alloys result in resisting stress values which are considerably less than the initial stress, implying that the effective stress can be an appreciable fraction of the applied stress (111). These results are considered to be more consistent with a dislocation glide rate controlling process than with a dislocation climb process (112).

The results of this test and the related stress dip test used in stress-strain experiments are in conflict with the results of other experiments. It has been demonstrated by Wilshire and co-workers that, in a wide range of materials from pure metals to complex particle strengthened engineering systems, following a small stress reduction there is an incubation period of zero creep, after which creep resumes at a lower strain rate (113,114).

This has been employed in a different type of test in which incremental stress drops of about 5% to 10% are made (113-115). Following the initial reduction, each further reduction is made as soon as creep is observable. As the number of reductions increases, the durations of the incubation periods increase so that the cumulative incubation time appears to approach infinity at a finite residual applied stress. The resisting stress is then defined as this residual stress level. Various procedures have been proposed to extrapolate residual applied stress data to infinite cumulative incubation time to quantitatively define the resisting stress (116-117).

The results of incremental stress reduction experiments have been taken as being consistent with recovery rate controlled creep (118). The incubation period which follows a stress reduction then corresponds to the time required for recovery to decrease the flow stress of the material, so that creep can again occur at the decreased stress. It is argued that the case for recovery rate controlled creep is strengthened by the observation that the instantaneous specimen contraction following a stress reduction frequently contains a plastic component (118). The rapid dislocation movement necessary to produce this contraction is taken as evidence that dislocation glide occurs easily even in a creep substructure, and therefore, cannot be the rate controlling process (118).

The resisting stress obtained from incremental stress reduction experiments has usually been regarded as a phenomenological quantity, although various suggestions have been made as to its physical significance, as will be discussed in more detail below. The quantity σ_R, or "sigma-naught" (115), is incorporated into the recovery rate controlled creep mechanism by recognizing that the flow stress of a material may be regarded as the sum of two components, a substructure related component which may decrease due to recovery and an unrecoverable component σ_R (119). The quantity $\sigma-\sigma_R$ then represents the portion of the applied stress which drives recovery. It should be noted that when creep is occurring the applied stress is equal to the sum of σ_R and the recoverable portion of the flow stress, which decreases during an incremental stress reduction test. Therefore, the true effective stress is always nearly zero even though $\sigma-\sigma_R$ may be large. At this point it is assumed that the value of σ_R determined by extrapolation to infinite incubation time is characteristic of the material before stress reductions so that Equation (9) should be used in the analysis of creep data with the quantity $\sigma-\sigma_R$ substituted for σ to yield

$$\dot{\varepsilon}_s = A''[(\sigma - \sigma_R)/E]^{n_o} \exp(-Q_o/kT) \qquad (13)$$

where A" is a constant (113-117). The resisting stress σ_R is generally found to be a function of applied stress and temperature so that the values of n_o and Q_o, evaluated from Equation (13) at constant values of $(\sigma-\sigma_R)/E$, are usually different from those evaluated using Equation (9) at constant values of σ/E.

For pure metals and solid solution alloys, the measured values of σ_R are usually small compared to σ and vary approximately linearly with σ (113). For these systems the differences between n and Q_c and n_o and Q_o evaluated using Equations (9) and (13), respectively, are usually small. The stress exponent n_o is found to have a value of about 4, and Q_o is approximately equal to the self diffusion activation energy (113). On the other hand (see below), in more complex engineering systems σ_R is found to be relatively large and independent of applied stress above a certain applied stress level (114). For these systems, when creep data are analyzed according to Equation (13) values of $n_o = 4$ and $Q_o = Q_{SD}$ are also obtained. This is in sharp contrast to the values of n and Q_c obtained from analysis using Equation (9) where values of n as high as 40 and values of Q_c two to three times Q_{SD} have been observed.

The universal values of $n_o = 4$ and $Q_o = Q_{SD}$ for a wide range of systems have been interpreted as evidence for recovery rate controlled creep in all materials and are viewed as a self consistency argument supporting the incremental stress reduction technique for measuring σ_R.

The physical significance of σ_R is not that well understood since the evaluation of the incremental stress reduction experiments is not at all straightforward. The models proposed for σ_R are generally based on the recovery of three-dimensional dislocation networks by climb. Some of these models (119-120) assume that the entire dislocation network coarsens and equate σ_R with σ_o in the flow stress relation in Equation (1). Aside from the fact that σ_o is still a phenomenological parameter, these models have serious drawbacks. First, the obvious implication of Equation (1) is that the dislocation density approaches zero as the flow stress approaches σ_o, which is unrealistic. A more serious problem with these models is that, even if σ_o does contain a substructural component, so that the dislocation density is finite when $\sigma = \sigma_o$, the large scale coarsening of the substructure envisioned during an incremental stress reduction experiment would lead to the measured value of σ_R being uncharacteristic of the initial state of the material. Other network coarsening models have been proposed (74,118) which recognize that a dislocation network contains a distribution of link lengths. Creep in these models is assumed to be due to bursts of dislocation glide from the longest links in the network which grow to a sufficient length by recovery to act as dislocation sources. The incubation period following a stress reduction is rationalized as the time required for the longest links in the network to grow to a size where they may act as dislocation sources with the remainder of the network remaining essentially unchanged. It is argued that this idea is consistent with the observation that, in aluminum, there was no discernible difference between the dislocation substructure before and after a stress reduction, even after a creep strain of 0.05 had occurred during recommenced creep following the incubation period (118). In these models the resisting stress is rationalized as the stress necessary for dislocation multiplication, which is assumed to be characteristic of the initial state of the material since large scale recovery is not observed (115,118).

On this basis suggestions have been made that the resisting stress should correspond to the macroscopic yield stress obtained at the same temperature and at a strain rate equal to the steady state strain rate before stress reduction (107,121-123).

Although comparisons of the yield stress and σ_R have not been performed for pure metals and solid solutions, the equality may be inferred from the proportionality observed for many metals between the yield stress and steady state stress observed in stress-strain tests (31), since, as previously discussed, steady state creep and stress-strain data for pure metals are equivalent.

While the conjecture that σ_R is equal to the macroscopic yield stress at the appropriate temperature and strain rate is intuitively appealing and would make the σ_R concept self consistent, it has not been systematically checked. More important, the σ_R concept has not been rigorously tested through the study of the creep behavior of a single system over a wide range of strain rate, although it has been used to study several engineering alloys, each over a relatively narrow strain rate range. Recent results (124) will be presented later in this paper that address these two points for a γ' strengthened nickel-base superalloy.

VI. PARTICLE STRENGTHENED ALLOYS

As previously discussed, the extensive investigation of the stress-strain and creep behavior of pure metals and simple solid solution alloys has led to a certain degree of understanding of these two types of thermally assisted mechanical behavior in these microstructurally simple systems and has allowed correlations to be made between the two. However, understanding is not so advanced for the case of the complex particle strengthened systems that find high temperature engineering application. First, the investigation of the stress-strain and creep behavior of these systems has generally been limited to relatively narrow ranges of temperature, stress and strain rate due probably to the constraints of technical relevance. Consequently, creep tests have usually been performed under the moderately elevated temperature and low strain rate conditions typical of service applications while stress-strain tests have frequently been employed to model hot working at very high temperatures and strain rates. In addition, until very recently there has been no attempt to correlate the stress-strain and creep behavior of these systems. Selected stress-strain and creep behavior of these systems will now be reviewed and coupled with some very recent stress-strain and creep data obtained for a nickel-base superalloy over wide ranges of temperature, stress and strain rate (124-126). The aim will be to gain insight into elevated temperature mechanical behavior and determine whether a correlation between stress-strain and creep behavior similar to that observed for simple systems can also be made at this time for the more complex particle strengthened alloys.

As mentioned, the alloys that find high temperature engineering applications are of two main types: conventional nickel-base superalloys, which are strengthened by ordered coherent γ' precipitates based on the intermetallic compound Ni_3Al (1,2); and oxide dispersion strengthened (ODS) alloys, which are usually nickel solid solution alloys strengthened by rare earth dispersoids (3-5), see Figure 3.

The ODS alloys can also be viewed as incoherent particle strengthened systems in which the particles themselves are inert and do not undergo transformations and coarsening during long-term high temperature service; as such they can be considered model versions of the incoherent carbide strengthened alloys whose carbide strengthening particles do undergo change at high temperatures. The superimposing effects of strengthening particle morphological changes on such thermally assisted and time dependent mechanical behavior as creep or slow strain rate stress-strain deformation have yet to be studied and defined, and hence will not be included for detailed discussion.

The mechanical properties of both the γ' and ODS alloys are determined by the interaction of dislocations with the particles. Grain boundary effects are minimized through either directional grain growth, pinning of grain boundaries by precipitates, or by the use of monocrystals. In the case of ODS alloys, the dislocation-particle interactions are relatively simple and amenable to theoretical treatments because the oxide dispersoids are non-deformable and occupy a small volume fraction, on the order of 1 to 2% (3-5,127-129). On the other hand, nickel-base superalloys contain a large volume fraction of γ', typically 20% to 40% in wrought alloys and up to 60% in cast alloys, which can be plastically deformed (1,2) so that dislocation-particle interactions in these systems are more complex (130).

VII. STRESS-STRAIN BEHAVIOR OF ODS ALLOYS

The stress-strain behavior of ODS alloys has not been systematically studied. However, the behavior of these systems is very similar to that of the more basic internally oxidized pure metals, which have been extensively investigated (131-134). For these systems it has been shown that the increment in the yield stress above that of the pure metal is due to the Orowan mechanism (156). The effect of the particles then is to introduce an essentially strain rate independent threshold stress component to the yield stress. However, this simple interpretation does not hold over the entire temperature range of interest. The modulus compensated yield stress remains constant, which is required for the Orowan yielding mechanism to apply, only up to homologous temperatures of about 0.5, above which it drops to about one half the low temperature value (128,129,131-133,156). This has been rationalized as the effect of the localized climb of dislocations over the particles, based on which models have been proposed which adequately predict the observed decrease in the Orowan stress at higher temperatures (135,156).

In addition to the yield stress, the work hardening behavior of pure metals is also profoundly affected by oxide dispersoids. The presence of these particles gives rise to Stage II work hardening rates which are much higher than observed in pure metals, approaching values as high as $\mu/20$ (134). In addition, linear Stage II work hardening is not observed in these systems, even in single crystals (134). This particle effect has been explained by Ashby (134) as the effect of geometrically necessary dislocations punched out by the non-deformable dispersoids to relieve elastic stresses built up by the deformation of the matrix. Dislocation substructures consistent with this concept have often been observed in these systems after deformation (134,136,137). Therefore, in addition to the statistically stored dislocations, ODS systems are also work hardened by the punched out dislocations so that at any given strain the total dislocation density ρ is higher than in the host metal or alloy leading to a

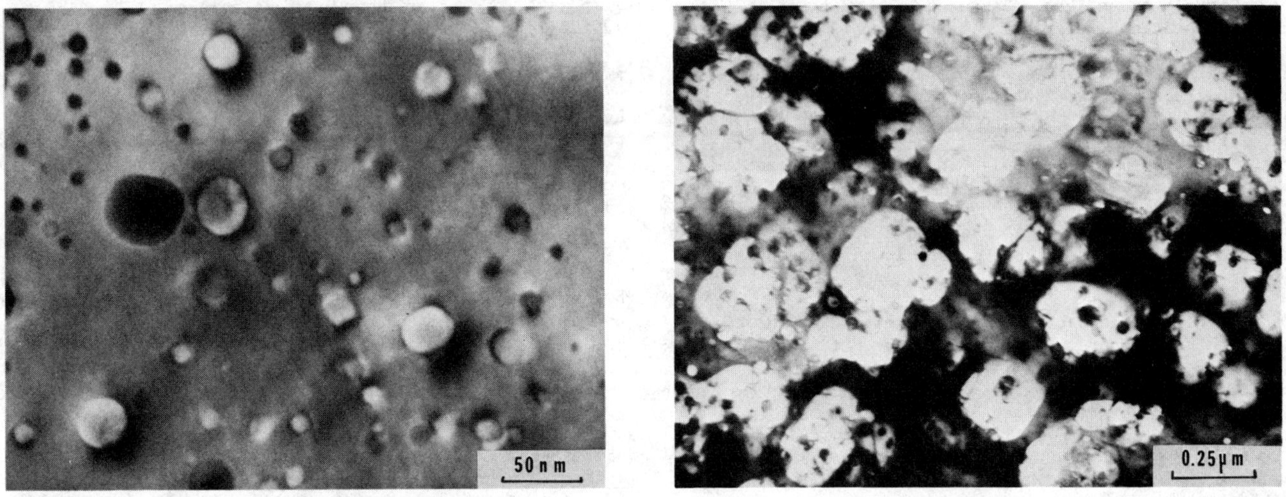

(a) (b)

Figure 3. Typical TEM microstructures of (a) ODS alloy Inconel MA 754, and (b) γ' and oxide strengthened alloy MA 6000E, using superlattice dark field image. Yttria are dark spots in the γ and γ' phases.

higher flow stress through Equation (1). An explicit form for θ_{II} in these systems can be derived through a development similar to that used to obtain Equation (4) by formally redefining the quantity 1/L in Equation (3) to take into account both the statistical and geometrical slip distances (134).

At high temperatures these geometrically necessary dislocations are not expected to be stable (134) since they readily anneal out at room temperature (137). This is consistent with the microstructures observed after high temperature creep of these systems and of the ODS alloys (3-5, 128,129). On the other hand, the dispersoids can still influence high temperature microstructure and mechanical properties since they tend to stabilize the statistically stored dislocation substructure by pinning (138) to make recovery more difficult (134).

Stage III, the thermally assisted work hardening stage, has been observed in ODS alloys but has not been systematically studied. Accordingly, there has been no direct correlation of stress-strain results and creep results for these systems.

VIII. STRESS-STRAIN BEHAVIOR OF γ'

The stress-strain behavior of γ' strengthened nickel-base superalloys is complicated by the fact that the γ' phase can be plastically deformed. Therefore, a basic understanding of their mechanical behavior requires some understanding of the deformation behavior of the γ' phase.

The γ' phase is an ordered intermetallic compound based on the $L1_2$ or Ni_3Al structure (139). In general, the γ' phase of superalloys can be highly alloyed, although it is still predominantly Ni_3Al, so that it has a composition $(Ni,Z)_3(Al,X)$ where Z and X represent substitutional alloying elements for Ni and Al, respectively (140). The stress-strain behavior of γ' either as pure Ni_3Al or alloyed $Ni_3(Al,X)$, has been extensively investigated and has been shown to exhibit "anomalous" temperature, strain rate and orientation dependences when compared with pure metal behavior (139,141-145). The temperature dependence of the macroscopic yield stress of γ' is particularly remarkable because the yield stress increases rather than decreases with increasing temperature up to a peak temperature, on the order of 700°C, and then decreases, while the micro-yield stress or proportional limit remains constant with temperature (141,142). The increase in macro-yield stress is quite large and, at the peak temperature, the yield stress can range from three to six times that at ambient temperature, depending upon composition (141), being higher when such elements as Ti and W are present. In addition, the yield stress is not sensitive to small plastic strain excursions of about 0.1% or 0.2% so that the yield stress measured at one temperature does not depend on small prior deformation at different temperatures (145). It has also been shown that the yield stress is only very weakly dependent on strain rate at temperatures up to near the peak temperature (141,145). However, above the peak temperature, the yield stress increases with increasing strain rate so that the peak temperature in fact is strain rate dependent (146).

It has been established that, at temperatures below the peak temperature, γ' deforms by the glide of a[$\bar{1}$01] superdislocations on (111) planes (144). In most cases these superdislocations are paired a/2 [$\bar{1}$01] single dislocations, which are perfect dislocations in the fcc lattice, coupled by an anti-phase boundary (APB) (141). In single crystal γ', the critical resolved shear stress at a single temperature and strain rate for the operative [10$\bar{1}$](111) system is not constant but depends strongly on the orientation of the stress axis (143-145). This orientation dependence has been shown to vary systematically with the ratio N of the Schmid factor on the cube slip system [$\bar{1}$01](010) to the Schmid factor on the octahedral system [$\bar{1}$01](111), so that the yield stress increases with increasing values of this ratio (143).

These results and the microstructural observation that the dislocations formed during deformation are mostly long straight screw dislocations (141,142) have been taken as evidence that the yield stress in γ' is controlled by the thermally activated cross slip of the leading screw superpartial dislocation of a superdislocation onto a cube plane where it becomes sessile (144), which is based on a mechanism first proposed by Kear and Wilsdorf (147). This mechanism is based on the idea that the APB energy on cube planes is much lower than that on octahedral planes (148), resulting in a driving force for the leading screw superpartial to cross slip onto a cube plane. This force exists even if there is no resolved shear stress for cross slip as in [001] oriented single crystals (144). The superdislocation then becomes sessile because the drag stress on cube planes is much higher than that on octahedral planes (145) and further glide on the cross slip plane is prevented. This concept is supported by the observation that macroscopic cube slip does not occur below the peak temperature, yet octahedral slip traces show evidence of cube cross slip (142-145). At high temperatures, in the vicinity of the peak temperature and higher, the drag stress on the cube planes decreases and macroscopic cube slip is observed, which is assumed to cause the high temperature decline of the yield stress (144,145). The orientation effect, which leads to higher yield stresses and lower peak temperatures as N increases, is then rationalized as the competition between the resolved stress on the cube slip system assisting cross slip, which increases the yield stress, and aiding glide on the cube plane, which decreases the yield stress (143,149).

Other mechanisms such as strain aging (150), thermally activated superdislocation climb (148), a temperature modified lattice-dislocation interaction (145) and dislocation debris hardening (141) have also been proposed to explain the observed behavior of the yield stress of γ'. However, these mechanisms are suspect for a number of reasons. Some of them require diffusion so that they cannot be applied at ambient temperature where diffusion is insignificant (141,145). These mechanisms also cannot explain the detailed orientation dependence of the yield stress (144). The debris hardening model of Thornton <u>et al</u>. (141) would be a possible explanation except that the model requires interactions between primary octahedral and cube slip systems which is inconsistent with the observation that primary cube slip in γ' below the peak temperature does not occur (151).

The work hardening behavior of γ' is also anomalous when compared with pure metal behavior. Stage I is generally not observed in single crystals oriented for single slip (144) except at very low temperatures, below about 190°K (142). The Stage II work hardening rate θ_{II} in both single crystal and polycrystal γ' is high, on the order of μ/20 (151), which is much larger than the usual value of μ/200 found in simple systems. In addition, θ_{II} in single crystals exhibits a large temperature and orientation dependence similar to that of the yield stress (145). Stage II in γ' is significantly longer than Stage II in pure metals and continues to higher stress levels (152). In some cases the stress strain curve continues in Stage II to the point of fracture (152).

The work hardening behavior of γ' has been explained by two different models. Both assume that, in addition to the usual hardening accompanying dislocation storage, work hardening is enhanced by an additional process. The first of these is basically the same cube cross slip model used to explain the yield stress behavior (145). This model gives rise to an exhaustion-type hardening where the mobile dislocations are trapped, forcing new sources to operate at higher stresses. Alternatively, it has been suggested that high values of θ_{II} for γ' are due to the creation of APB tube defects from non-aligned jogs formed on primary superdislocations after intersection with forest superdislocations (152-154). Recent experimental work has shown that while both mechanisms probably occur, the cube cross slip mechanism is the dominant one (155). This conclusion was reached from the observation that γ' exhibits a latent hardening effect which is large at low temperatures and decreases with increasing temperature. Strong latent hardening of secondary slip systems is predicted by the APB tube mechanism so that it may be inferred that this mechanism becomes less effective as the temperature increases. Therefore, this mechanism cannot explain the observed increase in θ_{II} within the same temperature range. The temperature and strain rate dependence of γ' work hardening has not been well studied.

IX. STRESS-STRAIN BEHAVIOR OF NICKEL-BASE SUPERALLOYS

The stress-strain behavior of nickel-base superalloys has generally not been systematically investigated as a function of temperature and strain rate. We know of only one study, other than one to be reported here (124), where the stress-strain behavior of a nickel-base superalloy was investigated over wide ranges of temperature and strain rate (146). Other studies have determined the dependence of stress-strain behavior, at conventional tensile strain rates ($\gtrsim 10^{-4}$ sec^{-1}), upon orientation and γ' precipitate morphology. Many of these studies involved model Ni-Al binary and ternary alloys with low γ' volume fractions, less than about 20% (156).

1. Low γ' Volume Fraction

It is found that, for a constant γ' volume fraction, the increment in the yield stress at ambient temperatures as a function of particle size exhibits a maximum or a peak aged condition (156). Yielding behavior in this particle size regime is believed to be controlled by the motion of paired a/2<110> dislocations (156). In this mechanism the leading dislocation cuts the γ' precipitates, leaving behind an APB, which is then removed by the passage of the trailing dislocation through the same particle. The dislocation pair remains coupled because the elastic repulsion of the two like dislocations is balanced by the attracting forces of the APB's created in the particles separating the dislocations. This mechanism has received ample experimental support from microstructural observations (157,158).

The increment in yield stress $\Delta\tau_y$ has been modeled theoretically based on the cutting of the γ' particles by pairs of dislocations to give

$$\Delta\tau_y = (\Gamma_{APB}/2b)[(4\Gamma_{APB}r_0 f/\pi\phi)^{\frac{1}{2}} - f] \qquad (14)$$

where Γ_{APB} is the APB energy, f is the γ' volume fraction, ϕ is the dislocation line tension and r_0 is the particle size (156,159).

The quantity $\Delta\tau_y$ in Equation (14) is necessarily smaller than the equivalent Orowan stress for the particle spacings involved, since the particles are coherent and dislocations can cut through them even when the applied stress is less than the Orowan stress.

At particle spacings wide enough to allow the Orowan mechanism to operate, dislocations do bow around the γ' particles, leaving loops behind (156). Such loops have been experimentally observed and are generally found to be single loops; double loops are not observed (160). This occurs because only relatively large particles, >500 Å in size, can support more than one loop (161). As a result even when the Orowan mechanism operates, loose pairs of dislocations are sometimes seen where the trailing dislocation removes the loop left around the particle by shearing it to leave an undisturbed ordered structure (162).

The temperature dependence of the yield stress of low γ' volume fraction alloys also depends characteristically on whether or not the particles are cut (163). When the interparticle spacing is wide enough for the Orowan mechanism to occur, the yield stress decreases with increasing temperature proportional to the shear modulus, as expected. On the other hand, if the dislocations are forced to cut the γ' particles, the yield stress remains constant up to temperatures on the order of 700°C and then decreases. This has been explained as the effect of the observed constancy of the γ' APB energy as a function of temperature (162-164), which leads to a temperature independent cutting stress. The decrease in the yield stress at higher temperatures is then rationalized as the result of diffusion related cutting mechanisms operating instead of the diffusionless APB shearing mode (163). As in pure metals, the yield stress in low γ' volume fraction γ-γ' alloys is essentially independent of orientation (165).

Soviet work shows that the stress-strain curves of these systems are similar to those observed for pure metals (165,166). Single crystals oriented for single slip exhibit an easy glide Stage I. For a constant γ' volume fraction, the extent of easy glide decreases and the Stage I work hardening rate θ_I increases as the particle size increases. θ_I exhibits orientation dependence and can be as much as an order of magnitude greater than that observed in pure metals. On the other hand, the extent of Stage II and the magnitude of θ_{II}, which is on the order of $\mu/300$, are unaffected by orientation, which again is similar to that observed in pure metals. At constant γ' volume fraction, θ_{II} increases with increasing particle size and correlates closely with the dependence of the yield stress on particle size. Specifically, at small particle sizes, θ_{II} increases with increasing particle size as does the yield stress and attains a plateau level in the peak aged particle size regime, where the yield stress is also approximately constant. At large particle sizes where the yield stress drops, θ_{II} increases. This behavior has been interpretted as a consequence of the change in dislocation-particle interaction from cutting to the Orowan mechanism (165,166). The cutting mechanism is assumed to operate within the particle size regime where θ_{II} initially increases and becomes constant in the plateau region, so that dislocation storage proceeds by the usual intersection of dislocations. The further increases in θ_{II} at larger particle sizes is then due to the additional effects caused by loops left behind on the particles by the bowing dislocations.

The temperature dependence of Stage II work hardening is also similar to that observed for pure metals, so that the quantity θ_{II}/μ is approximately constant over a wide range of temperatures (165-166).

2. Effects of γ' Volume Fraction

While the stress-strain behavior of low γ' volume fraction model γ-γ' alloys has been extensively studied, though not systematically as a function of strain rate, these results are not directly applicable to commercial nickel-base superalloys because of the much higher γ' volume fraction usually contained in these systems. In addition, in nickel-base superalloys the size of the γ' particles is generally much larger than in the simpler model systems, i.e., thousands of Å vs hundreds of Å and the spacing between the particles is considerably smaller than the particle size so that the dislocations in these systems cannot be thought of even approximately as interacting with point particles.

The effect of increasing γ' volume fraction on the stress-strain behavior of γ-γ' systems has been systematically investigated by Beardmore et al. (167). In this study the temperature dependence of the compressive yield stress at conventional strain rates of a model polycrystalline Ni-Cr-Al alloy with γ' volume fractions ranging from 0% to 100% was investigated. The alloy compositions were chosen along tie line of the phase diagram to ensure that the compositions of the γ and γ' did not depend on the γ' volume fraction. The alloys were aged at 900°C to result in large γ' particle sizes, which ranged from 0.2 μm at low volume fractions to 0.8 μm at high volume fractions. At ambient temperatures it was found that the yield stress was maximum at a low γ' volume fraction of 20% and decreased dramatically as the compositions approached 100% γ or γ'. At high temperatures above 900°C, the opposite was observed in that the yield stress was maximum for pure γ' and decreased as the γ' volume fraction decreased.

The characteristic shapes of the yield stress versus temperature curves were observed to change smoothly as a function of γ' volume fraction between the low γ' volume fraction type and the pure γ' type, both of which have been previously discussed. Accordingly, intermediate γ' volume fraction yield stress versus temperature curves exhibited a temperature independent yield stress at low and intermediate temperatures, characteristic of the low volume fraction alloys, and a distinct peak at temperatures around 700°C, characteristic of the high volume fraction alloys. This type of behavior is typical of many nickel-base superalloys (146,168) and has been seen in other studies of model high γ' volume fraction systems (169,170).

The mode of deformation was found to depend both on temperature and γ' volume fraction. At low temperatures and low γ' volume fraction, narrow intense shear bands developed which became progressively more diffuse as the γ' volume fraction increased. The dislocations were generally paired and stored mainly in the matrix phase. In addition, discontinuous yielding, indicated by discrete accoustic emissions, occurred. At high temperatures deformation was much more uniform and for all γ' volume fractions no shear bands were seen. In the lower γ' volume fraction alloys, extensive looping of dislocations around the γ' particles occurred and the dislocations were generally unpaired. In the high γ' volume fraction alloys, there was extensive deformation of the γ' particles with considerable dislocation storage within the particles.

Beardmore et al. explained these results by recognizing that, although the alloys were aged at a single temperature, a bimodal distribution of γ' sizes developed. This distribution consisted of the larger primary particles and a low volume fraction of very fine or "hyperfine" γ' particles, about 50 to 100 Å in size, in the matrix. The hyperfine γ' resulted from lower temperature precipitation of γ' from the γ matrix which became supersaturated with Al as the temperature was lowered from the aging temperature. The hyperfine γ' is a potent strengthener of the γ matrix phase and was shown to account for most of the lower temperature yield strength. At these temperatures the strengthening effect of the primary γ' is less than that of the hyperfine γ' so that the yield stresses of alloys with more of the matrix phase containing the hyperfine γ' should be higher, which was observed. Conversely, at higher temperatures, usually 700°C and higher, the hyperfine γ' coarsens and finally disappears. Therefore, at high temperatures, the yield stress should increase with increasing γ' volume fraction, which was also observed. This concept is strongly supported by the behavior of an alloy which was pure γ at the aging temperature. When this alloy was cooled, it entered the γ-γ' region of the phase diagram and precipitated hyperfine γ' with a volume fraction of no more than 3%. The yield stress behavior of this alloy was found to be essentially equivalent to that of the 10% and 20% volume fraction alloys.

The phenomenon of hyperfine γ' formation is not limited only to the alloy system used in this study. It was also observed by Beardmore et al. (167) in a commercial alloy Mar-M246 after extended aging at 1000°C. Hyperfine γ' has also been observed in Nimonic 115 after aging at 1000°C (124) and in Udimet 700 following a commercial heat treatment (171).

Another study of the ambient temperature yield stress behavior of a model γ-γ' Ni-Al binary system was performed by Cornwell et al. (172). For single crystals it was found, contrary to the results of Beardmore et al. (167), that the yield stress increased approximately linearly with γ' volume fractions up to about 60%. This general increase in yield stress with increasing γ' volume fraction was also observed in polycrystalline alloys where a maximum yield stress was observed at a volume fraction of about 45%, which is still considerably greater than the value of 20% found by Beardmore et al. (167).

We feel the difference between the polycrystalline data of Beardmore et al. and Cornwell et al. may be due to effects other than the difference between alloy compositions. In the study of Beardmore et al., the aging treatment at 900°C resulted in particle sizes which increased by a factor of about four as the γ' volume fraction increased while the 700°C aging treatment employed by Cornwell et al. resulted in a constant particle size, independent of volume fraction. In addition, the lower aging temperature used by Cornwell et al. coupled with the relatively more vertical boundaries of the γ-γ' phase field of the phase diagram for Ni-Al (173) compared with the Ni-Al-Cr ternary (167) would lead to little or no hyperfine γ' precipitation in Ni-Al. Both the particle size effect and the lack of hyperfine γ' precipitation would cause the higher γ' volume fraction alloys in the study of Cornwell et al. to be stronger relative to the low volume fraction alloys than would be the case in the alloys of Beardmore et al. This is supported by the observation that extensive dislocation looping by single dislocations around the γ' particles occurred in low γ' volume fraction alloys in the study of Cornwell et al. as opposed to cutting, which was seen at all volume fractions in the study by Beardmore et al. Dislocation shearing of the γ' precipitates in the Ni-Al alloys was only observed after the volume fraction reached intermediate levels. Again, this implies that the γ' matrix in the Ni-Al alloys was free of hyperfine γ' and weaker as a result.

3. Orientation Effects

The studies discussed above demonstrate that the stress-strain behavior of high γ' volume fraction systems is considerably more complex than low volume fraction systems and is determined by γ' particle size, spacing, volume fraction and morphology. In addition, the stress-strain behavior of high γ' volume fraction systems exhibit orientation and strain rate dependence.

In several studies the yield stress has been measured for single crystals of the 60 volume percent γ' nickel-base superalloy Mar-M200 at conventional tensile strain rates (168,174-176). It was found that the critical resolved shear stress for octahedral slip is not constant, but varies with orientation. This orientation dependence is generally not that well understood and the breakdown of Schmid's law is taken as evidence that other slip systems and/or resolved shear stresses may be important (174). Cube slip which, as mentioned, is known to occur in the γ' phase at high temperatures, has also been advanced as an explanation of why the resolved octahedral shear stress is lower at high temperatures and the yield stress peak temperature is supressed for orientations where the resolved shear stress on the cube slip system is high, especially in orientations near <111>[174]. Another slip system in this and other nickel base superalloys is the $[\bar{2}11](111)$ type system which has been shown to operate at low to moderate strain rates in creep (129,177-179), and during stress-strain testing (162) around 760°C. The mobile dislocations in this deformation mode are a/2<211> type and form as the result of interactions of a/2<110> type dislocations in the γ matrix (129). These dislocations may cut the γ' particles relatively easily since they leave a low energy superlattice stacking fault behind rather than an APB (180). This deformation mode is seen only at relatively low strain rates since the motion of these types of dislocations requires diffusion to readjust dipole displacements at the dislocation core (180). This deformation mode totally controls primary creep in this system (179) and has been shown to account for the extreme orientation dependence of creep rates and stress rupture lives, leading to extensive primary creep and poor properties in orientations such as <110>, favoring slip on a single $[\bar{2}\bar{1}\bar{1}](111)$ type system (168,181).

Since dislocations are forced to cut the γ' precipitates in Mar-M200 type alloys, the yield stress of the γ', in addition to the APB cutting stress, may be important. As previously mentioned, the yield stress of γ' is determined not only by the resolved octahedral shear stress but also by the ratio N of the Schmid factors for cube slip and octahedral slip. In addition, it has recently been shown that the yield stress of the γ' phase also depends systematically on the ratio Q of the Schmid factors for the $[2\bar{1}\bar{1}](111)$ type slip system and the $[\bar{1}10](111)$ type slip system (143). The quantity Q is a measure of the stress which either constricts or extends, depending on orientation, the a/6<211> Shockley partials which compose the a/2<110> dislocations which, in turn, make up the superdislocations which cut the γ'. Depending on whether the Shockley partials are constricted or extended, cube cross slip may be aided or opposed, so that the yield stress of γ' develops an additional orientation dependence through Q. Interestingly, unlike the factor N, the orientation dependence of the sign of Q is dependent on stress sense (143). Therefore, it has been proposed that γ' should exhibit a yield stress for a given orientation that depends on stress sense, i.e., a tension-compression anomaly.

Pronounced orientation dependence of the yield stress occurs only at or below about 760°C. Above this temperature macroscopic cube slip in γ' occurs in addition to octahedral slip, so that multiple slip occurs in all orientations. This is supported by the observation that at 982°C single crystals of Mar-M200 exhibit no tensile axis rotations, even in nominally single octahedral slip orientations (168).

Orientation effects on the yield stress have also been observed in high γ' volume fraction Ni-Al model systems (169-170).

4. Yield Stress and Strain Rate Effects

One model that has been proposed to explain the yield stress of high γ' volume fraction alloys is that of Copley and Kear (182). In this model the γ' particles are assumed to be cut by pairs of dislocations and the rate controlling step in the cutting process is the entry of the leading dislocation of the pair into the precipitate, which is opposed by the creation of an APB. Using a Gilman type expression for dislocation velocity, they derived a phenomenological expression for the critical resolved shear stress τ_c as

$$\tau_c = [(\Gamma_{APB}/2b) - (\phi/br_o)] + K/2(\tau_m + \tau_p) \qquad (15)$$

where ϕ is the dislocation line tension, r_o is the particle size, τ_m and τ_p are the yield stresses of the matrix and γ', respectively, and K is a slowly ranging function that depends on strain rate. The first and second terms in Equation (15) result from the cutting of the particle, which is hindered by the formation of an APB and aided by the curvature of the dislocation in contact with the particle. Equation (15) does give a reasonable description of the observed yield stress behavior and, through K, predicts a low strain rate dependence of the yield stress which is observed below about 760°C (125,146). Strain rate sensitivity, which is observed above 760°C, is assumed to arise from the strain rate dependence of the phenomenological parameter τ_p. The Copley-Kear model does not explicitly allow for the effects of thermal activation. In this model it is also assumed that Γ_{APB} decreases at high temperatures in order to account for the observed decrease in the yield stress. Although Γ_{APB} is generally assumed to be constant up to temperatures approaching the γ' melting temperature (144), a vacancy roughening model has been proposed to rationalize a decrease in Γ_{APB} at higher temperatures (183).

Studies of the strain rate sensitivity of the yield stress of model high γ' volume fraction alloys and nickel-base superalloys are rare. Leverant et al. (146) investigated the temperature and strain rate dependence of the compressive yield stress of <001> oriented Mar-M200 single crystals in the temperature range from 760°C to 1000°C and in the strain rate range from 2 x 10^{-5} sec^{-1} to 2 sec^{-1}. The results are summarized in Figure 4 where the yield stress is shown as a function of temperature and strain rate. The important feature to note is that the yield stress at a given temperature may or may not depend on strain rate. At 760°C for all strain rates and at 843°C for strain rates above about 10^{-2} sec^{-1} the yield stress was constant while under the other conditions investigated the yield stress increased with increasing strain rate. The other important feature of Figure 4 is that, at the higher strain rates investigated, the yield stress increased with increasing temperature at a given strain rate. These results were explained by assuming that two different mechanisms can control the yield stress. The strain rate independent yield stress was attributed to the diffusionless shear of the γ' particles by coupled a/2<110> dislocations, while the strain rate dependent yield stress was attributed to a diffusion controlled cutting mechanism. Two different strain rate sensitive mechanisms were proposed: the viscous glide of cube cross slipped, APB couple a/2<110> dislocations and cutting by superlattice stacking fault forming a/2<211> dislocations. The stacking fault deformation mode was observed in the investigation by Leverant et al. only at 760°C for a strain rate of 3 x 10^{-7} sec^{-1} in a special test. At higher strain rates and at higher temperatures, the viscous glide of a/2 <110> dislocations was favored.

The general behavior of the yield stress as a function of temperature and strain rate leads to a yield stress versus temperature curve which is strain rate dependent above 760°C. The shape of the curve varies from one which decreases monotonically with increasing temperature at low strain rates to one which exhibits a pronounced peak at high strain rates. Both the height of the peak and the peak temperature increase with strain rate. This type of behavior is also exhibited by the γ' phase, as previously discussed, so that the cause of the peak has been attributed to the anomalous yield stress behavior of the γ' phase (146,182). It should be noticed that this feature was missed in the other studies of high γ' volume fraction γ-γ' alloy, since they were all performed at a single strain rate.

Recently the stress-strain behavior of another nickel-base superalloy was investigated as a function of temperature and strain rate by Jensen and Tien (124). In this study the 40 volume percent γ' wrought alloy Nimonic 115 was deformed in tension over a temperature range from ambient temperature to 982°C and in the strain rate range from 10^{-2} sec^{-1} to 10^{-7} sec^{-1}. Prior to testing, the alloy was given solutionizing and γ' aging treatments (184) to result in a microstructure exhibiting a relatively large grain size of about 275 μm in which the grain boundaries were serrated. An additional extended time carbide aging treatment at 982°C resulted in extensive carbide precipitation on the grain boundaries. The combination of large grain size and serrated, carbide strengthened grain boundaries resulting from this heat treatment has been found to suppress grain boundary sliding (184).

The yield stress data from this study are shown in Figures 5 and 6 as a function of strain rate and temperature. These results are similar to those of Leverant et al. (146) for single crystal Mar-M200 in Figure 4, yet there are important differences. In Figure 5 it can be seen that the yield stress is relatively constant and strain rate independent up to temperatures of about 700°C, above which it drastically decreases and becomes strain rate sensitive. However, as opposed to Mar-M200, there is no pronounced peak in the yield stress at temperatures around 760°C. Also, in Figure 6 it can be observed that, at low strain rates and higher temperatures, the yield stress versus strain rate curves decrease more slowly with decreasing strain rate and appear to approach a low strain rate regime where the yield stress is again relatively insensitive to strain rate.

The differences between the yield stress behavior of Nimonic 115 and the stronger Mar-M200 single crystals may be explained by consideration of the volume fraction and morphology of the γ' precipitates.

It should be noted that the <001> orientation used for Mar-M200 single crystals studied by Leverant et al. (146) is a multiple slip orientation, so that the single crystal superalloy results may at least qualitatively be compared with those for the polycrystalline Nimonic 115. One major difference between Nimonic 115 and Mar-M200 is the γ' volume fraction which is about 40% in the former and close to 60% in the latter (146). This difference in volume fraction could account for the lack of a pronounced yield stress peak as a function of temperature for Nimonic 115 and is consistent with the results of Beardmore et al. (167). This effect is not due to the differences in strain rate employed where the highest strain rates employed were 2 sec^{-1} for Mar-M200 and

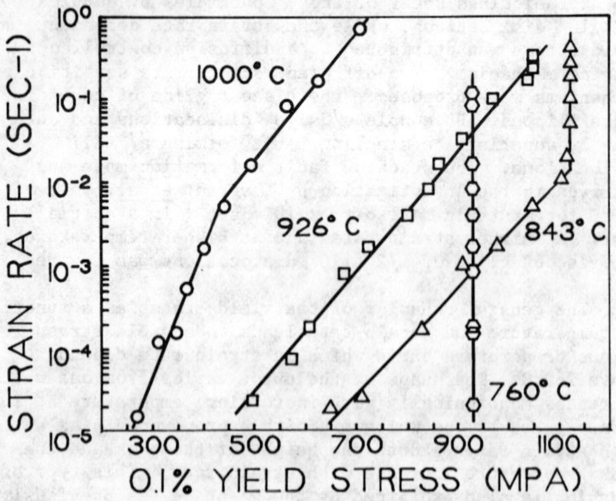

Figure 4. Temperature and strain rate dependence of yield stress of monocrystalline <001> oriented Mar M200. From (146).

10^{-2} sec^{-1} for Nimonic 115, since at 816°C Nimonic 115 develops a strain rate independent yield stress which is less than that at 760°C. However, that the yield stress of Nimonic 115, at low strain rates and higher temperatures, becomes less sensitive to strain rate is probably due to the lower strain rates employed. If the yield stress did continue to decrease continuously with strain rate as shown in Figure 4, an improbable situation would result where the yield stress would be zero at a finite, albeit very low, strain rate.

The Mar-M200 single crystals employed by Leverant et al. not only contained a high volume fraction of γ', but the γ' was also finely dispersed, having a mean size of about 0.25 μm. In addition, these crystals were given an extended aging treatment at the comparatively low temperature of 871°C so that little hyperfine γ' precipitation would be expected. Therefore, for this system the deformation mechanism would be the cutting of the primary γ' precipitates dispersed in a weak γ matrix. On the other hand, the Nimonic 115 was given an extended carbide aging treatment at 982°C which resulted in a grain boundary sliding resistant microstructure but led to a much larger mean primary γ' size of 0.8 μm and extensive precipitation of hyperfine γ' in the γ matrix on cooling. The large size of the primary γ' plus the relatively lower γ' volume fraction compared with Mar-M200 resulted in the primary γ' precipitates not being cut, but rather bowed around by the Orowan mechanism. This may be seen in Figure 7 where the primary γ' precipitates are not cut but block the motion of dislocations, leading to the formation of γ-γ' interfacial dislocation networks. On the other hand, the hyperfine γ' precipitates in the γ matrix are cut by dislocations, which leads to pronounced pairing of the dislocations, as can also be seen in Figure 7. The cutting of the hyperfine γ' also can explain the relative insensitivity of the yield stress to temperature at temperatures below about 700°C, since without γ' cutting the yield stress would decrease with increasing temperature (163). The presence of the hyperfine γ' also is thought to account for the marked decrease in yield stress with decreasing strain rate as seen in Figure 6. At 760°C at strain rates of 10^{-3} sec^{-1} and 10^{-6} sec^{-1}, where the yield stress is independent of strain rate, the microstructures following yielding are essentially the same and are typified by Figure 7. On the other hand, at 871°C at a strain rate of 10^{-7} sec^{-1} the post-yield microstructure appears as in Figure 8 where the hyperfine γ' has extensively

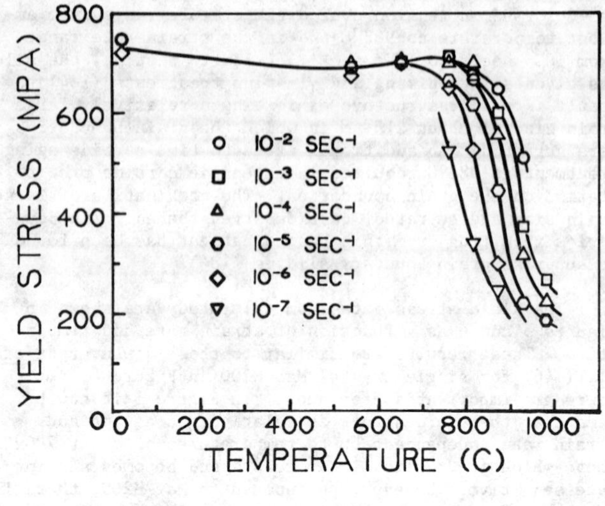

Figure 5. Temperature dependence of yield stress of Nimonic 115 at different strain rates. From (124).

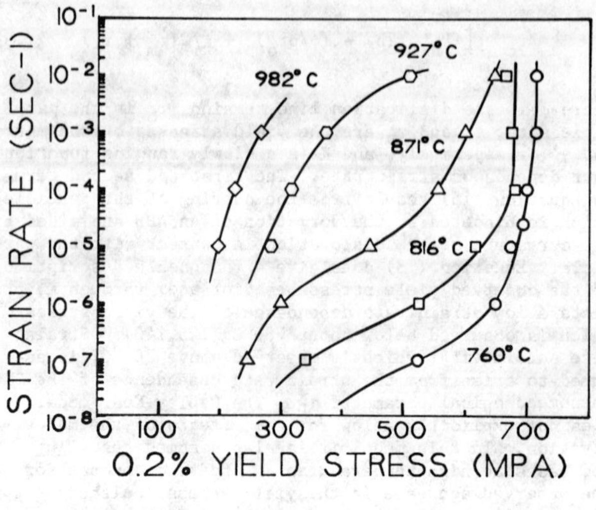

Figure 6. Strain rate versus yield stress plot, from Figure 5, showing strain rate independent σ_y at high strain rates.

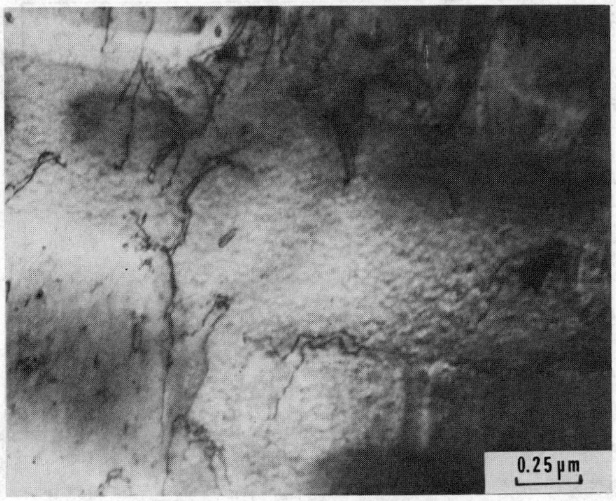

Figure 7. TEM micrograph of Nimonic 115 after tensile deformation to 0.08% strain at 10^{-6} sec^{-1} strain rate and at 760°C. From (124).

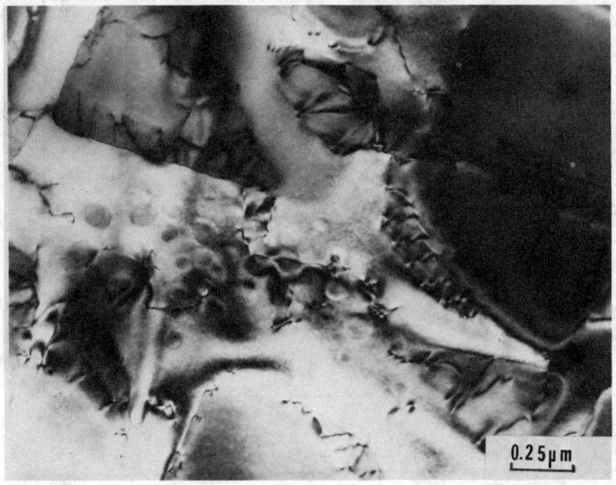

Figure 8. TEM micrograph of Nimonic 115 after tensile deformation to 0.025% strain at 10^{-7} sec^{-1} strain rate and at 871°C. Note the lack of ultrafine γ' as compared to Figure 7. From (124).

coarsened and largely disappeared so that the yielding mode is the bowing of dislocations around the primary γ' particles against the resisting stress of the solid solution matrix and the considerably coarsened and hence weakened hyperfine γ' dispersion.

The yield stress behavior of the Nimonic 115, therefore, is closely related to the presence of hyperfine γ' and the abrupt decrease of the yield stress with decreasing strain rate is probably due to its coarsening and eventual disappearance. This is yet another example of how mechanical behavior is closely linked with the stability of the microstructure, which has long been recognized in rationalizing creep behavior (185).

5. Work Hardening

Investigations of the work hardening of nickel-base superalloys and model high γ' volume fraction systems have been limited mainly to Stage I and Stage II behavior. Work hardening in these systems, as opposed to low γ' volume fraction systems and pure metals, does exhibit significant dependence on orientation and temperature. For single crystals oriented for single slip, the extent of the easy glide Stage I decreases with increasing temperature and depends on orientation (169-170). The extent of Stage I is also a function of γ' volume fraction and has been observed to either increase or decrease with increasing γ' volume fraction for crystals of similar orientation, depending on the orientation and temperature. The Stage II work hardening rate θ_{II} also depends on temperature and orientation (169,170). As the γ' volume fraction increases from about 20% to about 80%, the quantity θ_{II}/μ at a particular temperature first decreases then increases (169,170). In addition, as the γ' volume fraction is increased, the shape of the θ_{II}/μ versus temperature curve changes from the low γ' volume fraction type which, as previously discussed, is independent of temperature up to about 700°C and then decreases, to the high γ' volume fraction type which exhibits a marked peak around 600°C, similar to that of the γ' phase.

In the single crystal superalloy Mar-M200, the effect of orientation can be quite pronounced. In single slip orientation, deformation occurs readily by easy glide so that the stress-strain curves consist only of Stage I up to fracture (168). On the other hand, multiple slip orientations such as <001> and <111> readily work harden so that easy glide is completely suppressed (168).

Stage II work hardening behavior, like the yield stress, also depends characteristically on whether the deformation mode involves particle cutting or not. When cutting does occur, θ_{II} is much lower than the high values observed in oxide dispersion strengthened systems (186). On the other hand, Cornwell et al. (172) observed that, at ambient temperature, θ_{II} was very high, on the order of $\mu/10$, in a Ni-Al γ-γ' alloy single crystal where the γ' particle size was quite large. The alloy, which contained about 55% γ', was aged so that the γ' particles grew to a size of about 1 µm. It was observed that dislocation cutting of the γ' did not occur, but rather dislocations bowed around the particles, leaving arrays of dislocations at the particle interface.

This effect has also been observed in Nimonic 115 at high temperatures (124). Figure 9 shows two stress-strain

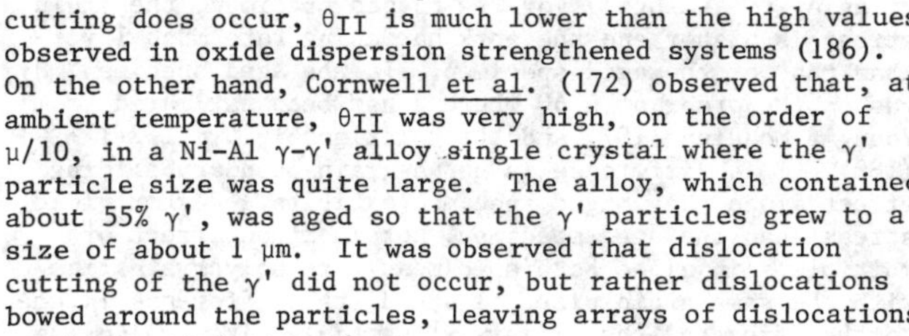

Figure 9. Stress-strain comparisons of an aged and unaged superalloy at 10^{-6} sec^{-1} strain rate and at 760°C. From (124).

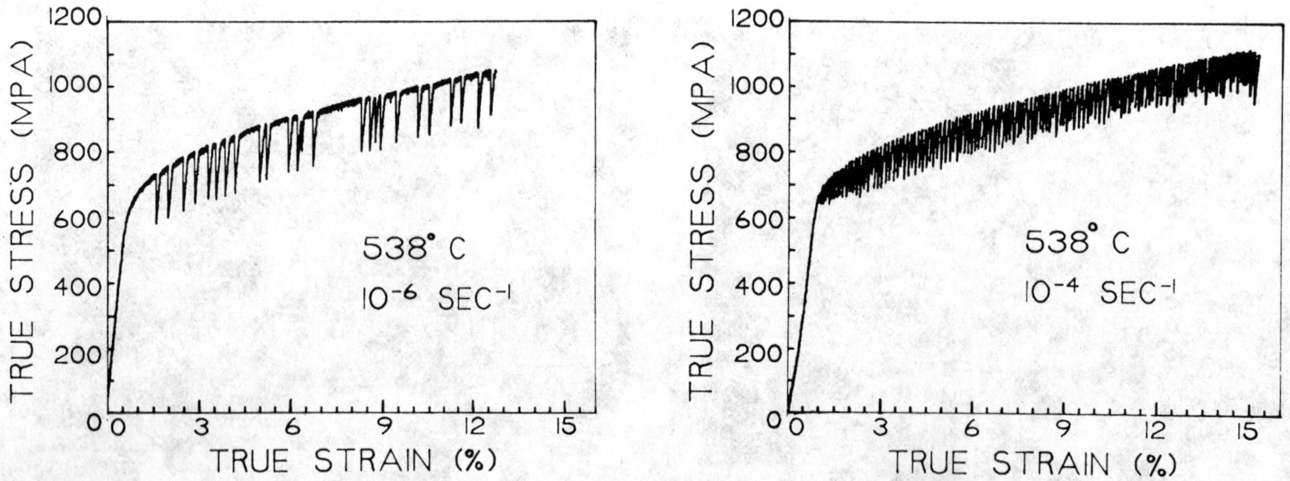

(a) (b)

Figure 10. Typical types of serrated flow in Nimonic 115 at (a) 538°C and 10^{-6} sec^{-1} strain rate, and (b) 538°C and 10^{-4} sec^{-1} strain rate. From (124).

curves obtained at the same temperature and strain rate for Nimonic 115 with and without the previously described carbide aging treatment following the solutionizing and γ' aging treatments. For the unaged specimen, the yield stress is higher and the work hardening rate much lower than that of the aged specimen. In the aged specimen, θ_{II} was on the order of μ/60 where μ has been estimated from Young's modulus (187) and the relationship $\mu = E/2(1 + \nu)$ (188). This difference is not a grain boundary sliding effect since the unaged specimen exhibits a higher yield stress than the aged specimen, nor is it an effect of multiple slip since both specimens were polycrystalline with the same grain size. Instead, the difference is due to the effect of the primary γ' particle size, which was about 0.8 μm in the aged alloy and about 0.6 μm in the unaged specimen, upon deformation mode. In the aged alloy it was previously mentioned that, initially, deformation occurs by the bowing of dislocations around the larger primary γ' particles, leaving interface dislocations at the γ' particles. At larger strains cutting of the primary γ' particles was sometimes observed, depending on temperature and strain rate (124). On the other hand, it has been shown that in the unaged condition, plastic deformation in Nimonic 115 occurs by dislocation cutting of the more closely spaced primary γ' particles (184,189).

Serrated flow has been observed in high γ' volume fraction systems. Cornwell <u>et al.</u> (172) have seen it at ambient temperature in model <u>Ni-Al</u> systems and Beardmore <u>et al</u>. (167) have observed it in model Ni-Al-Cr alloys at temperatures up to 500°C. In these systems serrated flow was found to be due to shear localization in intense shear bands and the propagation of Luders bands. Nimonic 115 has also been observed to exhibit serrated flow at temperatures of 538°C and 649°C (124). The serrations were of two types: irregularly spaced, large load drops, as shown in Figure 10(a), which were observed at 538°C at a strain rate of 10^{-6} sec^{-1} and at 649°C at strain rates from 10^{-6} sec^{-1} to 10^{-3} sec^{-1}; and somewhat smaller, periodic drops, as shown in Figure 10(b), which were seen at 538°C at strain rates from 10^{-5} sec^{-1} to 10^{-3} sec^{-1} and at 649°C at a strain rate of 10^{-2} sec^{-1}. This phenomenon was not systematically investigated, though it is felt it may be a strain aging effect (124) due to boron, the concentration of which was relatively high (0.09 atomic percent) in the Nimonic 115 studied.

Stage III work hardening behavior in Nimonic 115 was investigated as a function of temperature and strain rate by Jensen and Tien (125). Guided by the previous work on pure metals described earlier, the work hardening data of this system were analyzed using θ-σ plots to see whether Equation (6) would also apply to this complex nickel-base superalloy. Figure 11 shows typical results obtained using such plots, where each curve develops a linear tail at low work hardening rates. From extrapolation of the linear portion to zero work hardening, the saturation stress σ_s in Equation (6) could be defined for each temperature and strain rate. On the other hand, the simple relationship $\theta_o \cong \mu/20$ for θ_o in Equation (6) found by Kocks (17) for pure metals did not hold for this system. Extrapolations of the linear portion back to zero stress gave work hardening rate intercepts which varied considerably, depending on temperature and strain rate. This is not unexpected, since significant scatter in θ_o was observed by Kocks (17) in the relatively simple engineering

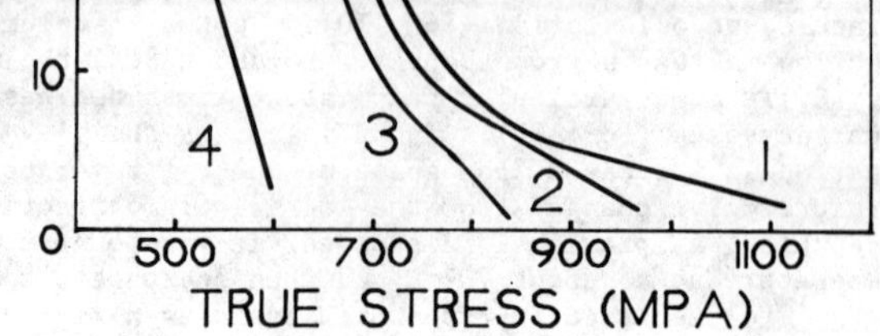

Figure 11. Typical stress dependence of work hardening rates in Nimonic 115 at 760°C. From (125).

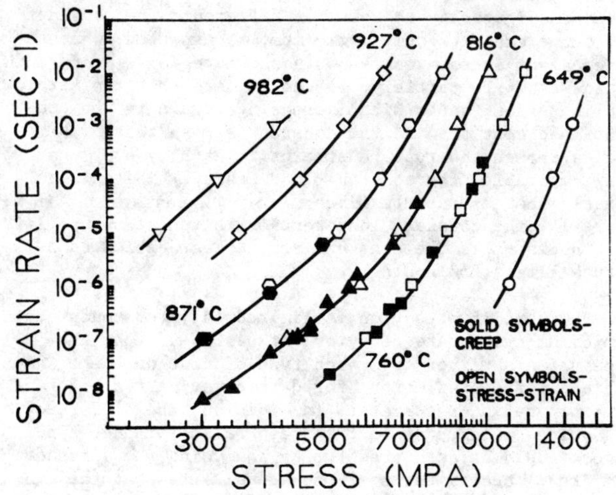

Figure 12. Steady state strain rate versus stress plot of creep and stress-strain data, showing equivalence. From (125).

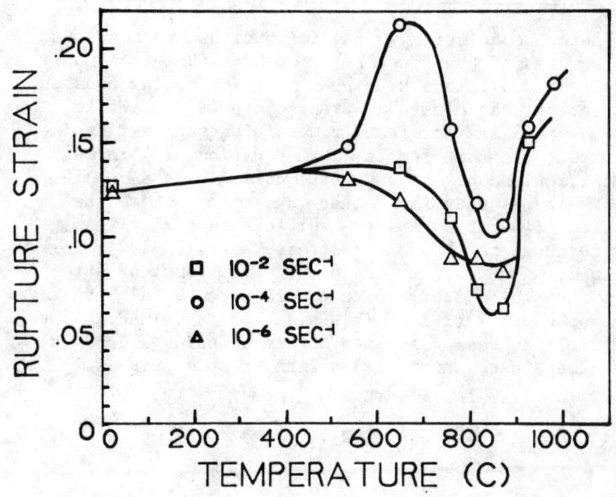

Figure 13. Tensile elongation ductilities of Nimonic 115 as function of temperature and strain rates. From (124).

system 304 stainless steel.

To further test the applicability of the Stage III work hardening concepts developed for simple systems to nickel-base superalloys, constant stress creep tests on Nimonic 115 were also performed to see whether a correlation could be made between creep and stress-strain behavior.

6. Correlation Between Stress-Strain and Creep

In Figure 12, the steady state stress-strain results, as applied strain rate versus saturation stress, and the steady state creep results, as applied stress versus steady state creep rate, are shown. It can be seen that there is a direct correspondence between the steady state stress-strain and creep data, which at 760°C extends over almost four orders of magnitude in strain rate. As might be expected from the equivalence between the steady state creep and stress-strain deformation data demonstrated in Figure 12, the steady state microstructures developed in both types of tests have been found to be equivalent under comparable deformation conditions (125). Also, the amount of plastic strain required to reach steady state has been found to be roughly the same for each type of test under similar conditions. In this correlation, the strain to steady state was taken in stress-strain as the plastic strain to the ultimate true tensile stress, while in creep it was taken as the total plastic strain, including the instantaneous plastic strain on loading, to the minimum creep rate.

The concept of equivalence is strengthened by the observation that the saturation stresses observed in stress-strain tests performed on specimens precrept into steady state were equal to the applied stresses used during the creep tests, provided the applied strain rates during the stress-strain tests were equal to the steady state creep rates.

In addition to the obvious implications the observed equivalence has for fundamental understanding of the deformation behavior of nickel-base superalloys, this equivalence may also be of technological use. The low strain rate yield stress may be a useful design parameter since in many high temperature applications operating conditions are more closely approximated by stress-strain deformation than by creep deformation. In addition, the low strain rate yield stress test may be a useful tool in alloy development, where it could be used as an initial pass-fail criterion in alloy qualification instead of the exhaustive battery of creep tests currently used.

7. Ductilities and Fracture Modes

The stress-strain rupture ductility of Nimonic 115 is shown in Figure 13 as a function of temperature and strain rate, where some strain rate curves have been omitted for the sake of clarity (124). It can be seen that, as in creep, a ductility minimum exists, although the temperature of the minimum is somewhat higher than that typically observed in creep. This may be related to the higher strain rates employed in the stress-strain tests, since the miminum appears to shift to a lower temperature when the strain rate is reduced from 10^{-4} sec^{-1} to 10^{-6} sec^{-1}. The uniform strain, defined as the strain at which the true work hardening rate equals the true applied stress, was found to exhibit a minimum at a given strain rate at the same temperature the ductility minimum was found. In addition, the fracture mode was found to be different outside and within the minimum. Outside the minimum, the fracture surfaces exhibited extensive intergranular fracture while within the minimum flat cleavage-like facets were observed.

Although this phenomenon was not systematically investigated, it is felt, from the available evidence, that the ductility minimum may be due to the onset of a form of glide plane decohesion (124).

The stress-strain rupture behavior was also compared with the rupture behavior observed in creep tests. When correlated in terms of either the steady state strain rate or steady state stress, the ductilities in both types of tests were comparable and exhibited the same type of dependence on the steady state variable. The fracture surfaces observed after creep were also comparable to those obtained after stress-strain testing. However, the rupture ductility data did show appreciable scatter, as might be expected, so that without much more extensive testing, it would be speculation to state that even at failure stress-strain and creep behavior are equivalent.

X. CONSTANT LOAD VERSUS CONSTANT STRESS CREEP TESTING

Before continuing it may be important to note that many studies of the creep behavior of nickel-base superalloys and ODS alloys have been performed using constant load rather than constant stress creep tests. While this may be a simpler test in obtaining engineering information, the fact remains that Equations (9) through (12) strictly apply only under constant stress conditions. It has been found that in simple systems the measured value of n may be significantly higher than that measured in a constant stress test (190-192); in aluminum, n has a value of about 5 from constant stress creep data while from constant load creep data n has a value of about 7 (192). This has been explained as the effect of increasing applied stress σ during a constant load creep test, which varies with creep strain as

$$\sigma = (L/A_o)(1 + \varepsilon) \qquad (16)$$

where L is the applied load, A_o is the initial specimen cross sectional area and ε is the engineering strain ($\Delta\ell/\ell_o$) (190). Since the strain to steady state ε_p generally increases with increasing applied stress, the difference between the engineering stress, L/A_o, and the actual applied stress given in Equation (16) would increase with increasing applied load, leading to higher values of n in constant load tests. For engineering systems the results of constant load and constant stress creep tests are usually implicitly taken as being comparable since the strain to steady state or actually the minimum strain rate in a constant load test for these systems is assumed to be small. However, even though ε_p may be small, it may vary with applied stress which, coupled with the relatively high stress sensitivity of the minimum creep rate of these systems, may also lead to significant differences in the observed values of n depending on testing mode.

This problem was addressed in a recent study of the creep behavior of Nimonic 115 (126). Constant load and constant stress creep tests were performed at 760°C and 815°C under equal constant engineering and true stresses, respectively. Typical results are shown in Figure 14 where it may be seen that while the primary creep curves essentially overlap the steady state creep rates differ. This was observed at every stress investigated. In addition, the shapes of the strain rate versus strain curves in Figure 14 are somewhat different; the constant load curve exhibits only a minimum, as expected, whereas the constant stress test shows a short region of constant strain rate. In tertiary, it can be seen that the strain rate increases faster with increasing strain in the constant load test than in the constant stress test. Therefore, although the rupture strains in each type of test were comparable, the constant load tests exhibited significantly shorter rupture lives. For the tests shown in Figure 14, the constant load test exhibited a rupture life of 26 hours while the constant stress test exhibited a rupture life of 58 hours.

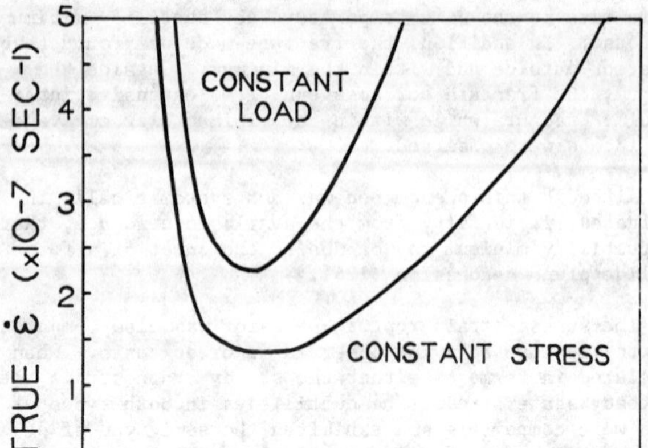

Figure 14. Typical constant stress and constant load creep curves of a superalloy at 760°C and 475 MPa. From (126).

However, if attention is limited only to steady state or minimum creep rate behavior, it was found that the values of n obtained from each type of test were the same within experimental error, for data covering two orders of magnitude in creep rate. In addition, the difference in absolute magnitudes of the creep rates for each type of test could be accounted for by analyzing the constant load creep data in terms of the actual stress at the minimum strain rate given in Equation (16) with ε_p substituted for ε. Therefore, it would appear that, for particle strengthened systems or at least nickel-base superalloys, the difference between constant load and constant stress creep tests is not as great as in simple systems and the results of the two may be used interchangeably in discussing creep behavior. However, care must be taken in analyzing tertiary creep behavior and rupture lives.

XI. CREEP BEHAVIOR OF PARTICLE STRENGTHENED ALLOYS

When the steady state or minimum creep rate data of ODS alloys and nickel-base superalloys, regardless of γ' volume fraction, are analyzed according to Equation (9), it is commonly found that the values of n and Q_c are much higher than those found for pure metals (1,2). Typically n ranges in value from about 8 to 20, although values as high as 75 have been reported for ODS alloys, while Q_c ranges from two to three times that of Q_{SD} (1-5).

It is generally acknowledged that the high values of n and Q_c observed in these systems are a direct consequence of the interaction of dislocations with the strengthening particles. The strengthening effect of particles has been found to be essential in explaining the anomalously high values of n and Q_c found in nominally solid solution, titanium-base alloys which undergo dynamic precipitation of particles on mobile dislocations (193). In the new oxide dispersion and γ' precipitation strengthened alloy MA 6000E, a very fine grain size of about 0.2 μm leads to the intrusion of grain boundary sliding and a marked reduction in the observed stress exponent n from about 10 to 2 (194), demonstrating that the very high values of n are characteristic of dislocation creep mechanisms. Recently Davidson and Tien (189) and Furillo and co-workers (184) also demonstrated that in Nimonic 115 high n values result when grain boundaries are strengthened and n values on the order of 2 result when grain boundary precipitates are dissolved and grain boundary sliding is allowed.

Basic models developed to rationalize creep of simple systems are obviously not directly useful in explaining the high values of Q_c and n. Two approaches have been proposed to rationalize these high values observed for the particle strengthened alloys. As mentioned in Section IV, the glide of jog dragging screw dislocations has been proposed to explain the high values of n and Q_c associated with very high volume fraction γ' single crystals where the γ' precipitates are finely caged by strain accommodating dislocations (93,94). This particular jogged screw model depends on the fine mesh network to produce a high density of jogs on mobile screw dislocations which in turn is visualized to move by self interstitial emission, giving rise to high activation energies and high stress dependencies. However, in most superalloys, especially the lower γ' volume fraction wrought superalloys, and in all ODS alloys, fine mesh dislocation networks are not observed and yet high n's and Q_c's persist. For these systems the resisting stress approach, introduced in Section V, is currently invoked to rationalize the creep behavior and to provide guidelines for alloy design.

Tien and co-workers (5,105,107) have recently generalized this approach by restating Equation (13) in the generalized form

$$\dot{\varepsilon}_s = A''\{[\sigma - (\sigma_p + \sigma_m)]/E\}^{n_o} \exp(-Q_o/kT) \quad (17)$$

where σ_p is the particle contribution to the resisting stress and σ_m is the matrix contribution. The allowance of $\sigma_R = \sigma_p + \sigma_m$ assumes linear additivity of the contributions, which is basically the same assumption made in the development of Equation (15) by Copley and Kear (182).* Recognizing that matrix solid solution strengthening due to viscous drag or due to stacking fault related mechanisms would depend on strain rate, and hence, to a first approximation, would depend linearly on the residual effective stress, Tien and co-workers further allowed

$$\sigma_m = \kappa(\sigma - \sigma_p) \quad (18)$$

where κ is a constant governed by the controlling solid solution strengthening mechanism in the appropriate strain rate or stress range of creep. Equation (17) then becomes

$$\dot{\varepsilon}_s = A''(1 - \kappa)^{n_o}[(\sigma - \sigma_p)/E]^{n_o} \exp(-Q_o/kT) \quad (19)$$

Comparing the stress exponents and activation energies in Equations (19) and (9), it is obtained that (5,105)

$$n = n_o/[1 - (\sigma_p/\sigma)] \quad (20)$$

$$Q_o = Q_c + n_o kT^2\{[1/(1 - \kappa)](d\kappa/dT) + [1/(\sigma - \sigma_p)](d\sigma_p/dT)\} \quad (21)$$

Equations (13) and (17) through (21), if sufficiently valid, can answer the many questions involving creep resistance when creep is due to dislocation motion and recovery is the rate controlling step. That the apparent values of n are high can be rationalized through Equation (20), since the ratio σ_p/σ is much larger in the particle strengthened alloys than in the simpler systems. The high apparent values of Q_c may be rationalized through Equation (21), since Q_c reflects the temperature variations of κ and σ_p, in addition to the true creep activation energy Q_o.

Creep resistance is enhanced through high values of σ_R, or equivalently, through the high values of σ_p and/or σ_m. Increasing σ_p means refining particle size and spacing in ODS and γ' alloys to increase the Orowan stress, or in the case of γ' strengthened alloys to force dislocation cutting of the γ'. In the case of γ' cutting, Γ_{APB} becomes a critical (but difficult to control) alloy design factor. σ_m can be raised by endowing the matrix phase of either the ODS alloys or of the superalloys with the appropriate alloying elements. The recent success of the mechanical alloying process has allowed versatility in alloy design through the separate control of matrix chemistries and particle distribution, fraction and morphology (5,105,106,127-129).

Equations (13) and (17) through (21) and the resisting stress concept have received a certain amount of experimental verification. As mentioned, values of σ_R, as determined by Wilshire and co-workers' incremental stress reduction experiments, when input into Equation (13), result in values of $n_o = 4$ and $Q_o = Q_{SD}$ for a wide range of simple systems and engineering alloys (113,114). Tien and co-workers checked the equations through creep data obtained on numerous ODS alloys, γ' strengthened alloys with different volume fractions of γ' and solid solution contents, and mechanical alloys with both oxide dispersoids and γ' precipitates (5,105-107,128,129). Assuming $n_o = 4$, they calculated σ_p from the creep data. In ODS alloys the values of σ_p were found to be between factors of one and one-half of the Orowan stress calculated from TEM determined spacings of the strengthening oxides. In the case of high volume fraction γ' alloys, σ_p was found to be lower than the conventional strain rate macroscopic yield stresses of the alloys.

These results support the validity of separating σ_R into σ_p and σ_m contributions and the physical significance of σ_m. When the combined steady state creep rate data were plotted against σ-σ_p, the parallel curves were ordered in the exact sequence of the matrix solute contents, as expected if Equations (18) and (19) are qualitatively valid. From the results it also appeared that the value of the apparent stress exponent n can be decreased by independently increasing σ_m. A lower n value is beneficial for component design purposes (5).

In the work discussed above, σ_p or σ_R has either been taken to be a constant and calculated for by assuming $n_o = 4$ in Equations (13) and (19) (5,105,107) or experimentally measured using the incremental stress reduction technique and found to be more or less a constant (113,114). However, in order for values of $n_o = 4$ to be observed for creep behavior over a relatively wide range of creep strain rates, either σ_R must decrease with increasing applied stress (121) or the creep data graphed on a log $\dot{\varepsilon}$ versus log σ plot must exhibit negative curvature. This is intrinsically obvious in Equations (19) and (20). Lund and Nix (4) have shown that the creep behavior of a single crystal ODS alloy TD-NiCr, when plotted as log $\dot{\varepsilon}$ versus log σ, does in fact exhibit negative curvature so that the stress exponent n decreases with increasing applied stress, as demanded by Equation (19). These results support recovery controlled creep in ODS alloys. On the other hand, neither the applied stress dependence of σ_R nor the stress dependence of the creep rates of nickel-base superalloys has been systematically investigated over wide ranges of temperature and strain rate, except in a very recent study on Nimonic 115 (125). In what follows, the results of this study, some of which have been previously presented in Section IX, will be applied to address this point. The possibility that the resisting stress σ_R is equal to the at temperature and strain rate yield stress will also be examined for this superalloy.

The combined equivalent stress-strain and creep data of this Nimonic 115 are plotted in Figure 12 as log $\dot{\varepsilon}$ versus log σ. As can be seen, the data follow neither a power law nor a curve exhibiting negative curvature, but instead a curve with positive curvature so that the stress exponent n increases with increasing stress. This is exactly the opposite behavior of what must occur if $n_o = 4$ and σ_p is a constant, as can be seen from Equation (20). In fact, the stress dependence of the steady state relationship between strain rate and stress is approximately exponential, as shown in Figure 15. Therefore, for nickel-base superalloy investigated, the stress and temperature dependence of σ_R is crucial in determining the ability of Equations (13) and (19) to represent creep data. In particular, if σ_R is either taken as or measured as being constant and not a decreasing function of applied stress, Equations (13) and (19) will be limited to a relatively narrow strain rate range of about 2 orders of magnitude when they may be accurately applied (125).

As mentioned previously, suggestions have been that the yield stress at a given temperature and strain rate should correspond to the resisting stress for creep at the same temperature and steady state creep rate (107,121-123). However, these suggestions were made on the basis of very

*Under the conditions of few grains per cross section and small cross section, environmentally caused effects such as adherent oxide scales can also result in a resisting and/or back stress (189).

limited data obtained over relatively narrow ranges of strain rate where the yield stress was found to be about equal to the resisting stress measured using incremental stress reduction experiments (121-123).

In order to systematically check this concept, the yield stress data obtained for Nimonic 115 (124) was correlated with the creep behavior of this system (125). Because the saturation stress, σ_s, versus applied strain rate data from stress-strain was found to be equivalent to the applied stress versus steady state creep rate data, the two sets of data could be used interchangeably. Accordingly, in Figure 16 the quantity $(\sigma_s - \sigma_y)/E$ is shown as a function of applied strain rate. If the yield stress is equal to the creep resisting stress σ_R, then the reduced stress $(\sigma_s - \sigma_y)/E$ would correspond to the normalized effective stress during creep under comparable conditions, so that straight lines with a slope of $n_o = 4$ and an experimental activation energy equal to the self diffusion activation energy should result. However, these expectations are not realized. Instead, sigmoidal curves rather than straight lines result for which the activation energy is much larger than that of self diffusion (125).

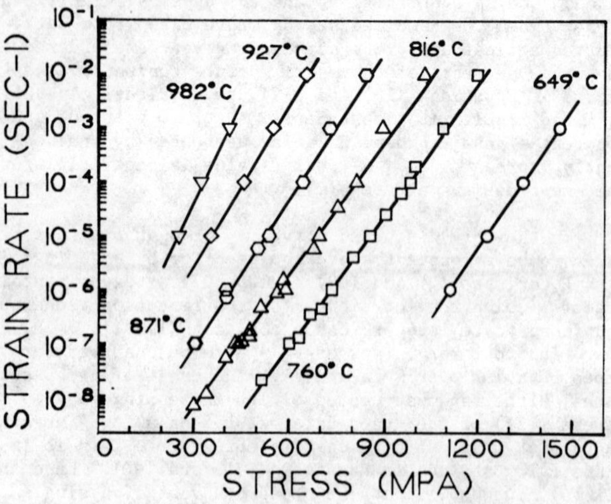

Figure 15. Replot of Figure 12, showing exponential dependence of strain rates on stress. From (125).

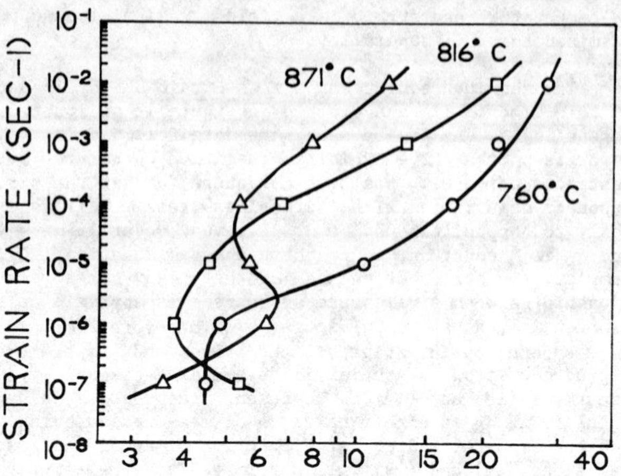

Figure 16. Strain rate versus effective stress data from Figure 6 and Figure 12, showing complex behavior. From (125).

The results in Figure 16 indicate that σ_y is not equal to σ_R within the range of strain rates investigated (10^{-7} to 10^{-2} sec^{-1}). This noncorrespondence is not related to the yield stress being defined for an offset strain of 0.2% since the use of the 0.01% yield stress values was found to give similar results (125). In addition, when specimens were precrept into steady state and subsequent stress-strain tests were performed in-situ at strain rates equal to the steady state strain rate obtained from the prior creep test, the 0.01% yield stress observed was only slightly higher than that for a virgin specimen (125). Therefore, the effect of a prior creep substructure was interpretted as not being a major component of the creep resisting stress, so that the nonagreement between σ_y and σ_R is not due to the lack of substructure in the virgin specimens (125).

The disagreement found between this work and earlier work on superalloys (121-123), which found that the resisting stress is equal to the yield stress, may be due to a number of effects. In the study done on Nimonic 115, the lowest strain rate investigated was 10^{-7} sec^{-1} (125), which is about two orders of magnitude higher than strain rates employed in the other studies (121-123) and higher than the creep rate range usually studied by Wilshire and co-workers, and the earlier work of Tien and co-workers. The equality between the yield stress and the resisting stress may in fact hold at low enough strain rates which is suggested by the behavior of the curves in Figure 16 at low strain rates.

As mentioned the Nimonic 115 investigated (125) also had a bimodal distribution of γ' sizes consisting of large primary γ' and hyperfine γ', while the other systems investigated earlier (121-123) contained only primary γ' particles of a uniform size. As previously discussed, the presence of the hyperfine γ' in Nimonic 115 has been shown to strongly influence yield behavior. In particular, the steep decline in yield stress with decreasing strain rate shown in Figure 6 has been attributed to the coarsening of the hyperfine γ'. The reverse portion of each of the sigmoidal curves in Figure 16 coincides with this abrupt decrease in the strain rate dependent yield stress. Therefore, the results shown in Figure 16 are complicated by the occurrence of microstructural instability.

As mentioned, the effects of dynamic changes in the strengthening microstructure on thermally assisted plastic deformation have not been well studied. Certain microstructural instabilities, such as the precipitation of embrittling phases in superalloys (2), and the coarsening of the γ' precipitates (195), can indeed result in the degradation of properties. Other changes, however, can cause improvements. For example, it has been shown that stress-coarsening of the γ' precipitates in superalloys can result in a shape and size change of the cuboidal γ' to that of periodic and continuous plates (196,197). Both tensile and creep properties of this microstructure are shown to be superior to the as-aged microstructure (198,199).

It appears that the resisting stress concept cannot be indiscriminately applied to all systems and within arbitrary ranges of strain rate. Further, unless the strain rate versus stress data or the resisting stress data are of specific forms, as discussed above, the use of Equations (13) and (19) may not be strictly valid over an indefinitely large strain rate range. Accordingly, it appears that it would be potentially useful to repeat the study of Jensen and Tien (125) on less complicated systems such as ODS alloys or nickel-base superalloys with a high (~60%) γ' volume fraction where the dislocations are forced to cut all the γ' particles, or in a wrought alloy with a unimodal γ' size distribution morphology.

XII. CONCLUDING REMARKS

Information old and new confirms that elevated temperature creep processes are rate controlled by diffusion. This conclusion can be drawn directly from the apparent temperature dependence of creep rates for pure metals and most solid solution alloys when the applied stress alone is used as the operating stress variable. For particle strengthened alloys, this conclusion can be drawn only if an effective stress is used as the operating stress variable in a recovery type model, or if a specific screw dislocation glide process is operative.

There is still no creep model which can satisfactorily explain both the stress dependence and temperature dependence observed in any single system. Based on a recovery type model and limited data, matrix and particle contributions to creep can be separated for the engineering incoherent oxide dispersion strengthened alloys and/or the coherent precipitation strengthened superalloys, resulting in some guidelines for alloy design for creep resistance. Confidence in this and in the understanding of creep in general decreases as the strain rate range is expanded and temperatures deviate in either direction from homologous temperatures in the 0.5 range. This reflects both a lack of data base and the intrusion of multimode creep processes at very high temperatures and/or very low strain rates, microstructural instabilities at high temperatures, little understood diffusional processes at cryogenic temperatures, and the difficult-to-isolate parallel dynamic effects at high strain rates.

Equivalence between creep testing and stress-strain testing has finally been demonstrated for pure metal as well as, now, for a complex nickel-base superalloy. The applied stress that results in a steady state creep rate has been shown to be equal to the saturation stress exhibited in Stage III work hardening during a stress-strain test performed at a strain rate equal to the steady state creep rate. Steady state substructures and strain to steady state have been shown to be very similar regardless of testing mode. The implication is that material behavior may be governed by a mechanical equation of state.

The conclusion of equivalence lends confidence to the future transfer of knoweldge gained in the studies of Stage III work hardening behavior to understanding primary creep, and vice versa. Yield stresses obtained at appropriate temperatures and strain rates may also be transferable as components of the resisting stress in understanding the steady state creep of heat resistant alloys. Priority should be given to further promote constant strain rate studies at higher temperatures and low strain rate ranges in order to break the impasse in understanding creep in general.

Finally, high temperature deformation and creep mechanisms developed for the simple systems have not been easily translated to general applicability without a great deal of problems. Accordingly, studies must be conducted on not only the microstructurally and chemically simple systems, but also on the complex strengthened alloys.

ACKNOWLEDGMENTS

We thank the Materials Research Division of the National Science Foundation for supporting the original work reported in this treatise under Grant NSF DMR77-11281. Dr. Robert Reynick is the NSF program monitor for the work.

Udimet is a trademark of Special Metals Corporation. Inconel, Nimonic and MA are trademarks and designation of INCO Family of Companies. Mar-M is a trademark of Martin Marietta Company.

---

*Russell R. Jensen, a doctoral candidate at Columbia University, is now Senior Engineer at The Westinghouse Research and Development Center at Churchill Boro, Pittsburgh, Pennsylvania.

REFERENCES

1. J.K. Tien, p. 363-388 in Fundamental Aspects of Structural Alloy Design, R.I. Jaffee and B.A. Wilcox, eds.; Plenum Press, New York, 1977.
2. J.K. Tien and S. Purushothaman, p. 3-41 in Proc. Symp. on Prop. of High Temp. Alloys, Las Vegas, 1976, Z.A. Foroulis and F.S. Pettit, eds.; The Electrochemical Society, Princeton, 1976.
3. B.A. Wilcox and A.H. Clauer, Trans. TMS-AIME, 236 (1966), p. 570-580.
4. R.W. Lund and W.D. Nix, Acta Met., 24 (1976). p. 469-481.
5. O. Ajaja, T.E. Howson, S. Purushothaman and J.K. Tien, Mat. Sci. and Eng., 44 (1980), p. 165-172.
6. J. Diehl, Z. Metal., 47 (1956), p. 331-343.
7. F.R.N. Nabarro, Z.S. Basinski and D. Holt, Adv. in Phys., 13 (1964), p. 193-323.
8. J.F. Bell, Phil. Mag., 10 (1964), p. 107-126.
9. idem, ibid, 11 (1965), p. 1135-1156.
10. L.M. Clarebrough and M.E. Hargreaves, p. 1-103 in Progress in Metal Physics, 3, B. Chalmers and R. King, eds.; Pergamon Press, London, 1959.
11. A.S. Keh, Phil. Mag., 12 (1965), p. 9-30.
12. T.E. Mitchell and W.A. Stitzig, Acta Met., 13 (1965), p. 1169-1179.
13. H. Meking, p. 1573-1594 in Proc. 5th Int. Conf. on Strength of Met. and Alloys, Aachen, P. Haasen, V. Gerold and G. Kostor, eds.; Pergamon Press, Oxford, 1979.
14. A.S. Keh and S. Weissman, p. 231-251 in Electron Microscopy and the Strength of Crystals, G. Thomas and J. Washburn, eds.; Interscience, New York, 1963.
15. H. Meking, p. 67-88 in Work Hardening in Tension and Fatigue, A.W. Thompson, ed.; AIME, New York, 1977.
16. U.F. Kocks, H.S. Chen, D.A. Rigney and R.J. Shaefer, p. 151-175 in Work Hardening, J.P. Hirth and J. Weertman, eds.; Gordon and Breach, New York, 1968.
17. U.F. Kocks., J. Eng. Mats. Tech. (Trans. ASME), 98 (1976), p. 76-85.
18. K. Lücke and H. Meking, p. 223-250 in Inhomogeneity of Plastic Deformation, R.E. Reed-Hill, ed.; ASM, Metals Park, Ohio, 1972.
19. U.F. Kocks, Acta Met., 6 (1958), p. 85-94.
20. idem, ibid, 8 (1960), p. 345-352.
21. D. Kuhlmann-Wilsdorf, p. 97-139 in Work Hardening, J.P. Hirth and J. Weertman, eds.; Gordon and Breach, New York, 1968.
22. idem, p. 1-44 in Work Hardening in Tension and Fatigue, A.W. Thompson, ed.; AIME, New York, 1977.
23. Z.S. Basinski, Phil. Mag., 4 (1959), p. 393-432.
24. D. Kuhlmann-Wilsdorf, Trans. TMS-AIME, 224 (1962), p. 1047-1061.
25. U.F. Kocks, Phil. Mag., 13 (1966), p. 541-566.
26. P.B. Hirsch and T.E. Mitchell, Can. J. Phys., 45 (1966), p. 663-669.
27. M.F. Ashby, Phil. Mag., 21 (1970), p. 399-424.
28. idem, p. 137-192 in Strengthening Methods in Crystals, A. Kelly and R.B. Nicholson, eds.; Applied Science Publications, London, 1971.
29. A.W. Thompson and M.I. Baskes, Phil. Mag., 28 (1973), p. 301-308.
30. A.W. Thompson, p. 89-126 in Work Hardening in Tension and Fatigue, A.W. Thompson, ed.; AIME, New York, 1977.
31. H.J. McQueen and J.J. Jonas, p. 393-493 in Treatise on Materials Science and Technology, 6, R.J. Arsenault, ed.; Academic Press, New York, 1975.
32. J.J. Jonas, C.M. Sellars and W.J. McG. Tegart, Met. Rev., 14 (1969), p. 1-24.
33. H. Meking and A. Grinberg, p. 289-294 in Proc. 5th Int. Conf. on Strength of Met. and Alloys, Aachen, P. Haasen, V. Gerold and G. Kostorz, eds.; Pergamon Press, Oxford, 1979.
34. S. Takeuchi and A.S. Argon, J. Mat. Sci., 11 (1976), p. 1542-1566.
35. B. Nicklas and H. Meking, p. 351-356 in Proc. 5th Int. Conf. on Strength of Met. and Alloys, Aachen, P. Haasen, V. Gerold and G. Kostorz, eds.; Pergamon Press, Oxford, 1979.
36. P.A. Järvinen, Scand. J. Metallurgy, 6 (1977), p. 79-82.

37. D.J. Lloyd, H. Sang, J.D. Embury, P. Wycliffe and G. Leroy, Mat. Sci. and Eng., 36 (1978), p. 35-46.
38. G. Langford and M. Cohen, Trans. ASM, 62 (1969), p. 623-638.
39. H.J. Rack and M. Cohen, Mat. Sci. Eng., 6 (1970), p. 320-326.
40. G. Langford, Met. Trans., 1 (1970), p. 465-477.
41. Y. Ohtakara, T. Nakamura and S. Sakui, Trans. Iron and Steel Inst. Jap., 12 (1972), p. 36-44.
42. C.M. Young, L.J. Anderson and O.D. Sherby, Met. Trans., 5 (1974), p. 519-520.
43. H. Luthy, A.K. Miller and O.D. Sherby, Acta Met., 28 (1980), p. 169-178.
44. D. Hardwick and W.J. McG. Tegart, J. Inst. Metals, 90 (1961-62), p. 17-20.
45. E. Voce, op. cit., 74 (1948), p. 537-562.
46. J.H. Hollomon, Trans. AIME, 162 (1945), p. 268-290.
47. P. Ludwik, Elemente der Technologischen Mechanik, p. 32; Springer-Verlag, Berlin, 1909.
48. H.W. Swift, J. Mech. Phys. Solids, 1 (1952), p. 1-18.
49. D.C. Ludwigson, Met. Trans., 2 (1971), p. 2825-2828.
50. P. Haasen, Phil. Mag., 3 (1958), p. 384-418.
51. G. Schoeck and A. Seeger, Rep. Conf. Defects in Solids, Phys. Soc., London, 1955, p. 340-350.
52. L. Roth, S. Purushothaman, J.K. Tien and J.M. Wells, Scripta Met., 13 (1979), p. 589-592.
53. A.K. Mukherjee, p. 164-224 in Treatise on Materials Science and Technology, 6, R.J. Arsenault, ed.; Academic Press, New York, 1975.
54. J. Weertman, Trans. ASM, 61 (1968), p. 681-694.
55. O.D. Sherby and P.M. Burke, Prog. Mat. Sci., 13 (1967), p. 325-390.
56. J. Gittus, Creep, Viscoelasticity and Creep Fracture in Solids, Ch. 3; J. Wiley & Sons, New York, 1975.
57. O.D. Sherby, Acta Met., 10 (1962), p. 135-147.
58. O.D. Sherby and J.L. Lytton, Trans. TMS-AIME, 206 (1956), p. 928-930.
59. Y. Ishida, C.Y. Cheng and D.E. Dorn, Trans. TMS-AIME, 236 (1966), p. 964-971.
60. O.D. Sherby and J. Weertman, Acta Met., 27 (1979), p. 387-400.
61. V.K. Lindroos and H.M. Miekk-oja, Phil. Mag., 17 (1968), p. 119-133.
62. J. Weertman, p. 315-336 in Rate Processes in Plastic Deformation, J.C.M. Li and A.K. Mukherjee, eds.; ASM, Metals Park, Ohio, 1975.
63. J.P. Hirth, Met. Trans., 3 (1972), p. 3047-3067.
64. W. Blum, Phys. Stat. Sol. (b), 45 (1971), p. 561-571.
65. L.I. Ivanov and V.A. Yanushkevich, Phys. Met. Metallog., 17, No. 1 (1964), p. 102-106.
66. F.R.N. Nabarro, Phil. Mag., 16 (1967), p. 231-237.
67. J. Friedel, Dislocations, p. 239; Pergamon Press, Oxford, 1964.
68. R. Lagneborg, Met. Sci. J., 3 (1969), p. 161-168.
69. S.K. Mitra and D. McLean, Proc. Roy. Soc., 295 (1966), p. 288-299.
70. J.H. Gittus, Acta Met., 22 (1974), p. 789-791.
71. R. Lagneborg, Met. Sci. J., 6 (1972), p. 127-133.
72. R.W. Bailey, J. Inst. Met., 35 (1926), p. 27-35.
73. E. Orowan, J. West Scotl. Iron and Steel Inst., 54 (1946-47), p. 45-53.
74. P.W. Davies and B. Wilshire, Scripta Met., 5 (1971), p. 475-478.
75. R. Lagneborg, J. Mat. Sci., 3 (1968), p. 596-602.
76. D. McLean, Rep. Prog. Phys., 29 (1966), p. 1-33.
77. U.F. Kocks and H. Meking, p. 345-350 in Proc. 5th Int. Conf. on Strength of Met. and Alloys, Aachen, P. Haasen, V. Gerold and G. Kostorz, eds.; Pergamon Press, Oxford, 1979.
78. C. Herring, J. App. Phys., 21 (1950), p. 437-445.
79. R.L. Coble, J. App. Phys., 34 (1963), p. 1679-1682.
80. T.G. Langdon, Phil. Mag., 22 (1970), p. 689-700.
81. J.G. Harper and J.E. Dorn, Acta Met., 5 (1957), p. 654-665.
82. C.M. Sellars and W.J. McG. Tegart, Int. Met. Rev., 17 (1972), p. 1-24.
83. F. Garafalo, Trans. TMS-AIME, 227 (1963), p. 351-356.
84. H.E. Evans and G. Knowles, p. 301-306 in Proc. 5th Conf. on Strength of Met. and Alloys, Aachen, P. Haasen, V. Gerold and G. Kostorz, eds.; Pergamon Press, Oxford, 1979.
85. J.R. Spingarn, D.M. Barnett and W.D. Nix, Acta Met., 27 (1979), p. 1549-1561.
86. H.E. Evans and G. Knowles, Acta Met., 25 (1977), p. 963-975.
87. S.L. Robinson and O.D. Sherby, Acta Met., 17 (1969). p. 109-125.
88. J. Weertman, J. App. Phys., 28 (1957), p. 362-364.
89. O.D. Sherby and C.M. Young, p. 497-541 in Rate Processes in Plastic Def. of Materials, J.C.M. Li and A.K. Mukherjee, eds.; ASM, Metals Park, Ohio, 1975.
90. F.A. Nichols, Met. Trans., 5 (1974), p. 1967-1972.
91. N.F. Mott, Phil. Mag., 43 (1952), p. 1151-1178.
92. C.R. Barrett and W.D. Nix, Acta Met., 13 (1965), p. 1247-1258.
93. J.K. Tien, B.H. Kear and G.R. Leverant, Scripta Met., 6 (1972), p. 135-139.
94. M. Malu and J.K. Tien, Acta Met., 22 (1974), p. 145-151.
95. W.D. Nix, Acta Met., 15 (1967), p. 1079-1081.
96. J.J. Holmes, Acta Met., 15 (1967), p. 570-571.
97. G.B. Gibbs, Scripta Met., 1 (1967), p. 135-137.
98. W.J. Duffin and F.A. Nichols, Acta Met., 18 (1970), p. 1121-1125.
99. P.B. Hirsch and D.H. Warrington, Phil. Mag., 6 (1961), p. 735-768.
100. A.H. Cottrell, Ch. 6 in Dislocations and Mechanical Properties of Crystals, J.C. Fisher, W.G. Johnston, R. Thompson and T. Vreeland, Jr., eds.; J. Wiley & Sons, New York, 1957.
101. J. Weertman, Acta Met., 15 (1967), p. 1081-1082.
102. J. Weertman, Trans. TMS-AIME, 233 (1965), p. 2069-2075.
103. O.D. Sherby, J.L. Lytton and J.E. Dorn, Acta Met., 5 (1957), p. 219-227.
104. C.Y. Li, this volume.
105. S. Purushothaman, O. Ajaja and J.K. Tien, p. 251-257 in Proc. of the 5th Int. Conf. on Strength of Metals and Alloys, Aachen, P. Haasen, V. Gerold and G. Kostorz, eds.; Pergamon Press, Oxford, 1979.
106. T.E. Howson, F. Cosandey and J.K. Tien, p. 563-573 in Proc. of the Fourth Int. Symp. on Superalloys, J.K. Tien, S.T. Wlodek, H. Morrow III, M. Gell and G.E. Maurer, eds.; ASM, Metals Park, Ohio, 1980.
107. S. Purushothaman and J.K. Tien, Acta Met., 26 (1978). p. 519-528.
108. J.K. Tien, D.E. Matejczyk, Y. Zhuang and T.E. Howson, p. 433-446 in Proc. of the Int. Conf. at University College, Swansea, B. Wilshire and D.R.J. Owen, eds.; Pineridge Press, Swansea, U.K., 1981.
109. C.N. Ahlquist and W.D. Nix, Acta Met., 19 (1971), p. 373-385.
110. G.B. Gibbs, Phil. Mag., 13 (1966), p. 317-329.
111. C.N. Ahlquist and W.D. Nix, Scripta Met., 3 (1969), p. 679-681.
112. A.A. Solomon and W.D. Nix, Acta Met., 18 (1970), p. 863-876.
113. P.W. Davies, G. Nelmes, K.R. Williams and B. Wilshire, Met. Sci. J., 7 (1973), p. 87-92.
114. W.J. Evans and G.F. Harrison, Met. Sci., 10 (1976), p. 307-313.
115. J.D. Parker and B. Wilshire, Met. Sci., 9 (1975), p. 248-252.
116. M. McLean, Scripta Met., 13 (1979), p. 339-344.
117. M. McLean, Proc. R. Soc. Lond. A, 371 (1980), p. 279-294.
118. J.D. Parker and B. Wilshire, Phil. Mag., 41 (1980), p. 665-680.
119. W.J. Evans, A.W. Beale and G.F. Harrison, Scripta Met., 14 (1980), p. 149-154.
120. K.R. Williams and B. Wilshire, Met. Sci. J., 7 (1973), p. 176-179.
121. P.L. Threadgill and B. Wilshire, p. 8-14 in Proc. Creep Strength in Steel and High-Temp. Alloys, The Metals Society, London, 1974.
122. T. Gibbons, Scripta Met., 12 (1978), p. 749-751.
123. M. McLean, Proc. Roy. Soc. A, 373 (1980), p. 93-109.
124. R.R. Jensen and J.K. Tien, "The Stress-Strain Behavior of a Nickel-Base Superalloy as a Function of Temperature and Strain Rate," to be published in Met. Trans.

125. R.R. Jensen and J.K. Tien, "The Correlation of Tensile Creep and Stress-Strain Behavior of a Nickel-Base Superalloy," to be published in Met. Trans.
126. R.R. Jensen, L. Ng and J.K. Tien, "The Relationship Between Constant Load and Constant Stress Creep in Superalloys," to be published in Met. Trans.
127. J.S. Benjamin and M.J. Bomford, Met. Trans., 5 (1974), p. 615-621.
128. T.E. Howson, J.E. Stulga and J.K. Tien, Met. Trans. A, 11A (1980), p. 1599-1607.
129. T.E. Howson, D.A. Mervyn and J.K. Tien, Met. Trans. A, 11A (1980), p. 1609-1616.
130. B.H. Kear, p. 440-473 in Order-Disorder Transformations in Alloys, H. Warlimont, ed.; Springer-Verlag, Berlin/Heidelberg, 1974.
131. G. Dorey, Royal Aircraft Establishment, Tech. Rep. No. TR68230, 1968.
132. R.S.W. Shewfelt and L.M. Brown, Phil. Mag., 30 (1974), p. 1135-1145.
133. F.J. Humphreys, P.B. Hirsch and D. Gould, p. 550-554 in Proc. 2nd Int. Conf. on Strength of Met. and Alloys, Pacific Grove, Calif., ASM, Metals Park, Ohio, 1970.
134. M.F. Ashby, p. 137-192 in Strengthening Methods in Crystals, A. Kelly and R.B. Nicholson, eds.; J. Wiley & Sons, New York, 1971.
135. R. Langneborg, Scripta Met., 7 (1973), p. 605-613.
136. R.L. Jones, Acta Met., 17 (1969), p. 229-235.
137. P.B. Hirsch and F.J. Humphreys, p. 189-216 in The Physics of Strength and Plasticity, A.S. Argon, ed.; MIT Press, Cambridge, 1969.
138. J.H. Hausselt and W.D. Nix, Acta Met., 25 (1977), p. 595-607.
139. N.S. Stoloff, p. 193-259 in Strengthening Methods in Crystals, A. Kelly and R.B. Nicholson, eds.; J. Wiley & Sons, New York, 1971.
140. O.H. Kriege and J.M. Baris, Trans. ASM, 62 (1969), p. 195-200.
141. P.H. Thornton, R.G. Davies and T.L. Johnston, Met. Trans., 1 (1970), p. 207-218.
142. R.A. Mulford and D.P. Pope, Acta Met., 21 (1973), p. 1375-1380.
143. C. Lall, S. Chin and D.P. Pope, Met. Trans. A, 10A (1979), p. 1323-1331.
144. S. Liang and D.P. Pope, Acta Met., 25 (1977), p. 485-493.
145. S.M. Copley and B.H. Kear, Trans. TMS-AIME, 239 (1967), p. 977-984.
146. G.R. Leverant, M. Gell and S.W. Hopkins, Mat. Sci. and Eng., 8 (1971), p. 125-133.
147. B.H. Kear and H.G.F. Wilsdorf, Trans. TMS-AIME, 224 (1962), p. 382-386.
148. P.A. Flinn, Trans. TMS-AIME, 218 (1960), p. 145-154.
149. S. Takeuchi and E. Kuramoto, Acta Met., 21 (1973), p. 415-425.
150. R.W. Guard and J.H. Westbrook, Trans. TMS-AIME, 215 (1959), p. 807-814.
151. E.M. Schulson and J.A. Roy, Acta Met., 26 (1978), p. 15-28.
152. L.E. Popov and N.A. Koneva, p. 404-438 in Order-Disorder Transformations in Alloys, H. Warlimont, ed.; Springer-Verlag, Berlin/Heidelberg, 1974.
153. A.E. Vidoz and L.M. Brown, Phil. Mag., 7 (1962), p. 1167-1175.
154. G. Schoeck, Scripta Met., 2 (1968), p. 283-286.
155. K. Aoki and O. Izumi, Trans. JIM, 19 (1978), p. 145-151.
156. L.M. Brown and R.K. Ham, p. 12-135 in Strengthening Methods in Crystals, A. Kelly and R.B. Nicholson, eds.; J. Wiley & Sons, New York, 1971.
157. H. Gleiter and E. Hornbogen, Phys. Stat. Sol., 12 (1965), p. 251-264.
158. D. Raynor and J.M. Silcock, Met. Sci. J., 4 (1970), p. 121-130.
159. H. Gleiter and E. Hornbogen, Phys. Stat. Sol., 12 (1965), p. 235-250.
160. Y. Shimanuki and H. Doi, Trans. JIM, 15 (1974), p. 24-31.
161. L.M. Brown, G.R. Woolhouse and V. Valdre, Phil. Mag., 17 (1968), p. 781-789.
162. A. Guimier and J.L. Strudel, p. 1145-1149 in Proc. 2nd Int. Conf. on Strength of Met. and Alloys, Pacific Grove, Calif., ASM, Metals Park, Ohio, 1970.
163. R.G. Davies and N.S. Stoloff, Trans. TMS-AIME, 233 (1965), p. 714-719.
164. A. Kelly and R.B. Nicholson, Prog. Mat. Sci., 10 (1963), p. 149-391.
165. G.I. Nosova and N.T. Travina, Phys. Met. Metallog., 30, No. 1 (1970), p. 155-161.
166. G.I. Nosova and N.A. Polyakova, Phys. Met. Metallog., 34, No. 3 (1972), p. 181-186.
167. P. Beardmore, R.G. Davies and T.L. Johnston, Trans. TMS-AIME, 245 (1969), p. 1537-1545.
168. B.H. Kear and B.J. Piearcey, Trans. TMS-AIME, 239 (1967), p. 1209-1215.
169. N.T. Travina and A.A. Nikitin, Phys. Met. Metallog., 35, No. 5 (1973), p. 150-160.
170. N.T. Travina and A.A. Nikitin, Phys. Met. Metallog., 40, No. 1 (1975), p. 137-141.
171. W.J. Boesch and H.B. Canada, J. Met., 19 (1968), p. 46-50.
172. L.R. Cornwell, J.D. Embury and G.R. Purdy, p. 387-404 in Proc. 3rd Bolton Landing Conf. on Ordered Alloys, N.S. Stoloff and J.H. Westbrook, eds.; Claitors Publishing Co., Baton Rouge, 1970.
173. Metals Handbook--Metallography Structures and Phase Diagrams, Vol. 8, 8th ed., p. 262; ASM, Metals Park, Ohio, 1973.
174. S.M. Copley, B.H. Kear and G.M. Rowe, Mat. Sci. Eng., 10 (1972), p. 87-91.
175. P.J. Piearcey, B.H. Kear and R.W. Smashey, Trans. ASM, 60 (1967), p. 634-645.
176. J.K. Tien and R.P. Gamble, p. 1037-1041 in Proc. 2nd Int. Conf. on Strength of Met. and Alloys, Pacific Grove, Calif.; ASM, Metals Park, Ohio, 1970.
177. G.R. Leverant, B.H. Kear and J.M. Oblak, Met. Trans., 4 (1973), p. 355-362.
178. G.R. Leverant and B.H. Kear, Met. Trans., 1 (1970), p. 491-498.
179. G.R. Leverant and D.N. Duhl, Met. Trans., 2 (1971), p. 907-908.
180. B.H. Kear, G.R. Leverant and J.M. Oblak, Trans. ASM, 62 (1969), p. 639-650.
181. R.A. MacKay, R.L. Dreshfield and R.D. Maier, p. 385-394 in Proc. of the Fourth Int. Symp. on Superalloys, J.K. Tien, S.T. Wlodek, H. Morrow III, M. Gell and G.E. Maurer, eds.; ASM, Metals Park, Ohio, 1980.
182. S.M. Copley and B.H. Kear, Trans. TMS-AIME, 239 (1967), p. 984-992.
183. J.K. Tien, p. W1-W16 in Proc. of 2nd Int. Conf. on Superalloys, MCIC-72-10; Battelle Laboratories, Columbus, 1972.
184. F.T. Furillo, J.M. Davidson, J.K. Tien and L.A. Jackman, Mat. Sci. and Eng., 39 (1979), p. 267-273.
185. R.F. Decker, p. 275-298 in Source Book on Materials for Elevated Temperature Applications, E.F. Bradley, ed.; ASM, Metals Park, Ohio, 1979.
186. L.R. Cornwell, J.D. Embury and G.R. Purdy, Acta Met., 18 (1970), p. 1217-1223.
187. High Temperature, High Strength Nickel-Base Superalloys, 3rd ed.; INCO, New York, p. 21.
188. J.P. Hirth and J. Lothe, Theory of Dislocations, p. 41; McGraw-Hill, New York, 1968.
189. J.M. Davidson and J.K. Tien, Met. Trans. A, 12A (1981), p. 865-876.
190. R. Söderberg, Scand. J. Met., 3 (1974), p. 268-272.
191. T. Kawasaki and M. Horiguchi, Met. Trans. A, 10A (1979), p. 749-757.
192. P.C.R. Cunha and H.J. Kestenbach, Metalurgia, 35 (1979), p. 257-259 (in Portuguese).
193. J.K. Tien, S. Purushothaman, M.W. Mahoney and N.E. Paton, Mat. Sci. and Eng., 45 (1980), p. 43-48.
194. R.F. Singer and G.H. Gessinger, "Superplastic Deformation of an Oxide-Dispersion Strengthened Superalloy," to be published in Proc. 2nd Riso Symp., Sept. 1981.
195. A.J. Ardell and R.B. Nicholson, Acta Met., 14 (1966), p. 1295-1309.

196. J.K. Tien and S.M. Copley, Met. Trans., 2 (1971), p. 215-219.
197. J.K. Tien and S.M. Copley, Met. Trans., 2 (1971), p. 543-553.
198. J.K. Tien and R.P. Gamble, Met. Trans., 3 (1972), p. 2157-2162.
199. D.D. Pearson, F.D. Lemkey and B.H. Kear, p. 513-520 in Proc. of the Fourth Int. Symp. on Superalloys, J.K. Tien, S.T. Wlodek, H. Morrow III, M. Gell and G.E. Maurer, eds.; ASM, Metals Park, Ohio, 1980.

MECHANICS AND MECHANISMS OF FRACTURE

Robert C. Bates
Westinghouse R&D Center
Pittsburgh, PA 15235

INTRODUCTION

Premature failure has been a perennial problem leading to the question: Why do metals fail? Superficially, the answer is that: the applied stresses exceed the strength of the metal. Thus a simple formulation of a failure criterion results, but then other questions arise: what governs the materials strength? What are the stresses that cause failure? How are these stresses affected by inhomogeneities that exist in all engineering materials and in the presence of stress concentrations such as notches or sharp cracks?

These questions do not have simple answers. For example, atomistic concepts of metallic structures lead to formulations that predict theoretical cohesive strengths that are 10 to 100 times those actually observed. The postulate by Griffith that real metals contain tiny cracks, so called "Griffith's cracks," seemed to explain that discrepancy (1). But the existence of such cracks in as-produced metals were not generally observed, and the creation of Griffith's cracks in the first place seemed to require the same high stresses, approaching the theoretical cohesive strength. Only after the introduction of concepts of plasticity and the role of inhomogeneous deformation (2) did this problem seemed to be resolved.

Meanwhile the complexities of the stress state were being addressed. Of particular importance was the evolution of concepts concerning the interrelationships of defects and stress, e.g. formulations for the stress concentration of elliptically shaped defects show that stress varies inversely with the tip radius, so that as the tip radius approaches crack like dimensions (taken as atomistic dimensions in the limit) the elastic stress concentration factor approaches infinity resulting in a stress-strain singularity. Of course, plastic flow intervenes as yield criteria are met, so plasticity develops relieving the high elastic stresses. The plastic zone that develops is first small and then large as the load level is increased and deformation proceeds. The description of the precise nature of the near-crack-tip stress/strain distributions were slow in coming. Meanwhile field parameters were developed which permitted the use of a single parameter characterization (the stress intensity factor, K, for small scale yielding and the J integral for extension into elastic-plastic cases). Failure criteria were formulated in terms of critical values of these field parameters, which were treated as material properties and determined experimentally. These concepts have proven successful in development of design approaches.

Another development came with the utilization of modern metallographic and fractographic techniques. Distinct classes of fracture were cataloged, and specific mechanisms were formulated for various fracture modes which led to the assignment of local failure criteria in terms of critical stress and critical strain. But these criteria proved difficult to integrate into the field parameter descriptions of crack tip conditions since the field parameters do not address the specific stress/strain states within the plastic zone per se in which the "fracture process zone" is deeply embedded.

Still another development was forthcoming. This was the development of descriptions of the stress/strain states within the plastic zone, first with numerical (finite element) techniques and then by analytical, closed form solutions. Coupling of these descriptions to the mechanistic formulations of local failure criteria can result in specific models of fracture.

This paper reviews and defines the fracture mechanics field parameters and the stress and strain distributions in the plastic zone, including a new analytical model for strain distributions. These concepts are then applied as the basis for the evaluation and extension of models for lower shelf, upper shelf and transition region fracture toughness. The paper concludes with an interrogation of the fracture toughness models with respect to detailed microstructural effects and observations.

FRACTURE MECHANICS CONCEPTS

The failure criterion for extension of a Griffith crack in a brittle material is that growth will occur when the decrease of strain energy required to create the new crack surfaces equals or exceeds the energy required to create the new crack surfaces (1). This concept, coupled with an expression for the stress distribution around an elliptical defect (3) led to the development of the Griffith equation:

$$\sigma = \left(\frac{2E\gamma}{\pi c}\right)^{1/2} \tag{1}$$

where γ is the surface energy, E is the modulus of elasticity and c is the half-crack length. Later modifications replaced γ with a plastic energy term and finally related the plastic energy to G, the crack extension force, or equivalently, the strain energy release rate (4,5), resulting in

$$\sigma = \left(\frac{EG}{\pi c}\right)^{1/2} \tag{2}$$

This expression is simply an energy balance that states crack propagation can occur only if the energy release is sufficient to provide the strain energy of fracture. Fracture will occur at a critical value G_c, which can be experimentally determined. Thus the energy release concept, in principal, could be useful in characterizing fracture properties of materials. A more practical approach, however, was established later with the introduction of the stress intensity factor K, which was shown to be related to G.

Irwin (6) introduced the following equations expressing the elastic stresses in the vicinity of a crack:

$$\sigma_{xx} = \frac{K_I}{(2\pi R)^{1/2}} \cos\frac{\theta}{2}\left[1 - \sin\frac{\theta}{2}\sin\frac{3\theta}{2}\right]$$

$$\sigma_{yy} = \frac{K_I}{(2\pi R)^{1/2}} \cos\frac{\theta}{2}\left[1 + \sin\frac{\theta}{2}\sin\frac{3\theta}{2}\right]$$

$$\sigma_{zz} = \nu\left(\sigma_{xx} + \sigma_{yy}\right) \quad (3)$$

$$\tau_{xy} = \frac{K_I}{(2\pi R)^{1/2}}\left[\sin\frac{\theta}{2}\cos\frac{\theta}{2}\cos\frac{3\theta}{2}\right]$$

$$\tau_{xy} = \tau_{yx} = 0$$

where R is the distance from the crack tip and θ is the angle from the plane of symmetry (i.e., the x-axis). Similar expressions were derived for plane stress loading and Mode II and Mode III loading (edge sliding mode and tearing mode, respectively). A point of significant interest here is that the stress intensity factors may be physically interpreted as parameters that characterize the distribution of stress in a cracked body, including the redistribution of stress caused by the introduction of the crack (7). All crack tip stress fields, then should have the same general characteristics, differing only in the magnitude or intensity. As such, if a material's behavior can be determined for one crack configuration, that knowledge can be used to predict behavior of other structures of that material. This, however, requires a knowledge of the geometrical effects on K. Fortunately, such expressions are available for many configurations. For example, for a sharp crack in an infinitely wide plate, K is defined as

$$K = \sigma\sqrt{\pi c} \quad (4)$$

In all cases K has the following form (8)

$$K = \sigma\sqrt{\alpha\pi c} \quad (5)$$

A variety of techniques have been used to calculate the K-factors for numerous configurations. One useful collection is that by Tada, Paris, and Irwin (9).

Irwin (6) has further shown that within the limitations of linear elastic loading situations K and G are related by

$$K = \sqrt{G}\,E \quad \text{(plane stress)}$$
$$K = \sqrt{G}\,E/(1 - \nu^2) \quad \text{(plane strain)} \quad (6)$$

The stress intensity factor is also applicable, with small corrections for crack length approximations, to small scale yielding cases (i.e., the plastic zone dimensions are small compared to crack length, specimen thickness, etc.). For elastic-plastic analyses, however, K no longer adequately describes the stress configuration. For such cases, the applicable field parameter is the J integral.

Rice introduced this useful field parameter (10),

$$J = \int_\Gamma \left(W dy - \vec{T}\cdot\frac{\partial\vec{u}}{\partial x} ds\right) \quad (7)$$

where Γ is the path of the integral, ds is the increment of distance along the path, $\vec{T}$ is a traction vector defined according to the outward normal along the path, $\vec{u}$ the displacement vector, and W is the strain energy defined in terms of stress and strain as:

$$W = \int_0^{\varepsilon} \sigma_{ij}\, d\varepsilon_{ij}$$

Paris (11) has provided an informative article in which the significance of the J integral is discussed. He points out that by evaluating the J integral after an increment of crack extension of da, it can be shown that

$$J = G = \begin{cases} \dfrac{K^2}{E} & \text{(plane stress)} \\ \dfrac{K^2}{E}\left(1 - \nu^2\right) & \text{(plane strain)} \end{cases} \quad (8)$$

Thus, J can be utilized in the same sense that K is used, but J has the advantage that it can be extended into the elastic-plastic regime. Furthermore, Paris (11) discusses an alternate definition of the J integral as:

$$J = -\int \frac{\partial P}{\partial a} d\delta_p = \int \frac{\partial \delta_p}{\partial a} dP \quad (9)$$

where a is crack length, P is load and δ_p is the work producing component of displacement of the load point. This expression allows J to be determined from the load-point displacement relationships for slightly different crack lengths.

Another parameter of interest is the crack tip opening displacement, variously called COD, CTOD and δ_T, the latter designation being the one that will be used in this paper. Since the elastic stress equations (Eq. 3) predict an elastic stress-strain singularity as $R \to 0$, the material, having a finite yield strength, must plastically deform resulting in the development of a plastic zone. Small scale yielding up to fully plastic situations may develop, depending on the load level, but in any event crack tip blunting with a corresponding opening of the crack results. Since, the field parameters K and J reflect the redistributions of stress caused by introduction the of the crack (with its singularity and consequent crack tip blunting), these parameters must relate to δ_T. Several approaches have been taken to establish the relationship $\delta_T = \lambda J/\sigma_o$, where σ_o is the yield strength. McMeeking (12) has provided a summary of various expressions for λ which typically result in numerical values ranging from about 0.4 to 1.0. The variation in λ is due, in part, to variations in the position at which δ_T is measured. Generally, for the purposes of this paper the following expression is used.

$$\delta_T = 0.6\,\frac{K^2}{\sigma_o E} \quad (10)$$

However in a few instances (when data from McMeeking are used) the following expression is used because it is a better reflection of δ_T for the position at which he measures it.

$$\delta_T = 0.55\left[\frac{2}{\sqrt{3}}\,\frac{(1+\nu)(1+n)\sigma_o}{nE}\right]^n \left(1 - \nu^2\right)\frac{K^2}{\sigma_o E} \quad (11)$$

STRESS AND STRAIN DISTRIBUTIONS IN THE PLASTIC ZONE

It is expected that the elastic singularity previously mentioned governs the stress distribution at distances that are large compared to the dimensions of the plastic zone, but within the plastic zone redistribution of stresses must accompany the large stretching associated with the crack tip blunting process (13). It is within the plastic zone or at the elastic-plastic zone boundary that the fracture processes occur, so the stress/strain distribution within the plastic zone is of utmost interest.

Slip line field solutions have been advanced by Hill for plane strain loading of a smoothly notched member of a perfectly plastic material (14). In this case slip lines form logrithmic spirals such as that shown in Fig. 1. The longitudinal stress acting near the notch tip has been developed by Hill as:

$$\sigma_{yy} = 2k[1 + \ln(1 + R/\rho)] \quad (12)$$

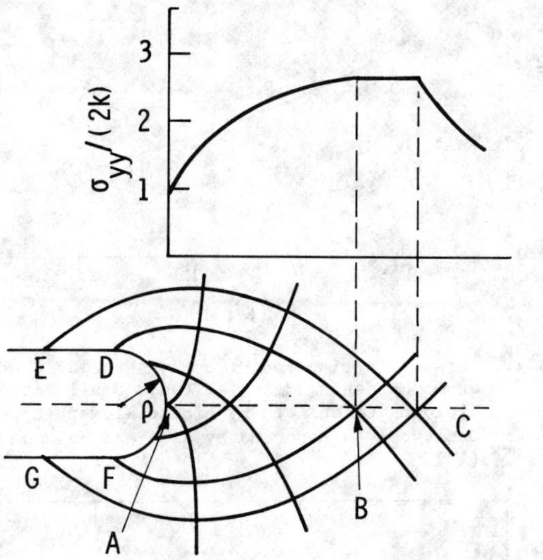

Figure 1. Slip Line Construction for a Semicircular Notch in a Round Bar and the Corresponding Stress Distribution for a Nonhardening Material.

where k is the yield stress in shear, R is the distance from the crack tip and ρ is the root radius. For a von Mises material $\sqrt{3}k = \sigma_o$ where σ_o is the yield strength in tension so that Eq. 12 becomes

$$\sigma_{yy} = (2/\sqrt{3})\sigma_o[1 + \ln(1 + R/\rho)] \quad (13)$$

In this solution, the maximum attainable stress σ_{yy} is independent of root radius but is dependent on the flank angle, ω.

$$\sigma_{yy} = (2/\sqrt{3})\sigma_o\left(1 + \frac{\pi}{2} - \frac{\omega}{2}\right) \quad (14)$$

Eqs. 12 and 13 are strictly applicable only to non-hardening materials. Tetleman and McEviley (15) have suggested that for strain hardening materials, σ_o in Eqs. 12 and 13 could be replaced by the locally strain hardened flow stress, expressed through an appropriate constitutive relation with strain such as the commonly used power hardening law. Thus

$$\sigma_{yy} = \sigma_o\left(\frac{\bar{\varepsilon}_p}{\varepsilon_o}\right)^n \left[1 + \ln\left(1 + \frac{R}{\rho}\right)\right] \quad (15)$$

where $\bar{\varepsilon}_p$ is the effective plastic strain at the particular value of R being considered and ε_o is the yield strain. Due to equilibrium considerations the validity of this equation is doubtful beyond small strains, although the equation has been shown to apply to notched bars in bending for notch tip strains up to at least 0.08 (16).

Finite element modeling of the blunting of initially sharp cracks is difficult because of the singularity of the elastic stresses and strains at the crack tip. Consequently, some investigators have modeled initially smooth, rounded notch tips and thereby avoid the singularity problem.

Griffiths and Owen performed an elastic-plastic finite element analysis for a notched bar in plane strain bending (17). They adopted a von Mises yield criterion and linear strain hardening, and performed analyses for various values of σ_{nom}/σ_o up to 2.292 where σ_{nom} is the stress at the notch root ignoring the stress concentration effect and is given by

$$\sigma_{nom} = 6M/Ba^2 \quad (16)$$

where M is the applied moment, a is the depth below the notch root, and B is the specimen thickness.

The maximum principal total strains found by Griffiths and Owen are described by the following equation.

$$\varepsilon_1 = \begin{cases} \varepsilon_{1c}\exp(-4R/3\rho) & R/\rho \leq 1.1 \\ \varepsilon_{1c}/4(R/\rho) & R/\rho > 1.1 \end{cases} \quad (17)$$

where ε_{1c} is the maximum principal total strain at the notch tip, R is the distance ahead of the notch and ρ is the notch radius.

The stress distribution reported by Griffiths and Owen (shown in Fig. 2) show that the maximum stresses lie on the symmetry axis ahead of the notch tip but well behind the elastic-plastic interface, and rise with increasing load level, apparently differently from the predictions of Eq. 13.

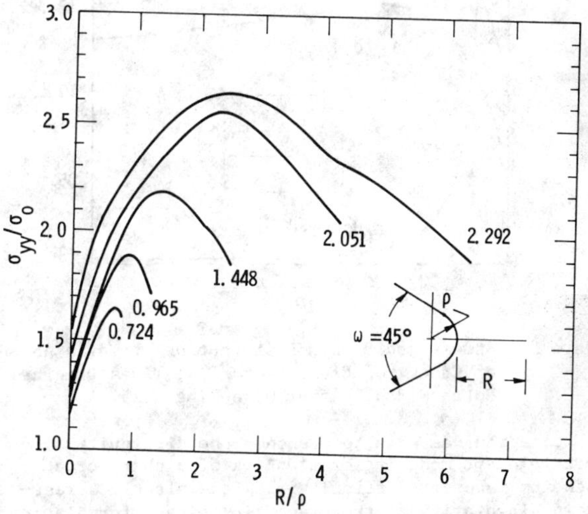

Figure 2. Stress Distribution in Plastic Zone Ahead of Notch at Indicated Values of σ_{nom}/σ_o for Notched Bend Specimen from Finite Element Analysis by Griffiths and Owen (17). Elastic-Plastic Boundary is at End of Each Curve.

Bates and Santhanam (16), analyzing the results of Griffiths and Owen, noted that the effective plastic strain (ε_p) ahead of the notch tip could be expressed as

$$\bar{\varepsilon}_p = \bar{\varepsilon}_{pc}\exp(-1.71\ R/\rho) \quad (18)$$

where $\bar{\varepsilon}_{pc}$ is the effective plastic strain at the notch tip. Insertion of values of effective strain calculated with Eq. 18 into Eq. 15 gave excellent agreement with the stresses reported by Griffiths and Owen up to the peak values.

McMeeking (12) carried out finite element calculations for plane strain blunting of semicircular notches for power-law hardening materials (n = 0, 0.1, 0.2) and two σ_o/E ratios (1/300 and 1/100). His notch blunting solution simulated small scale yielding by imposing special boundary conditions. The calculations were carried out for values of notch tip opening displacement (δ_T) up to five times the original notch opening. The notch tip shapes were reported to have been essentially the same for all material properties and remained essentially unchanged after the notch width reached a value of about three times the original width.

Fig. 3 shows the stress distributions directly ahead of the notch reported by McMeeking (12) for three values of n (0, 0.1, 0.2) and $\sigma_o/E = 1/300$. The strain distribution directly ahead of the notch is essentially unaffected by n or σ_o/E, so the strain distribution of only one case (n = 0, $\sigma_o/E = 1/300$) was selected and shown in Fig. 3. The stresses along directions at angles away from the line of symmetry were lower than those shown and reached their peak values closer to the notch tip. On the otherhand, strains were higher at off-angles in agreement with known plastic zone shapes. The off-angle stresses and strains are not shown in Figure 3.

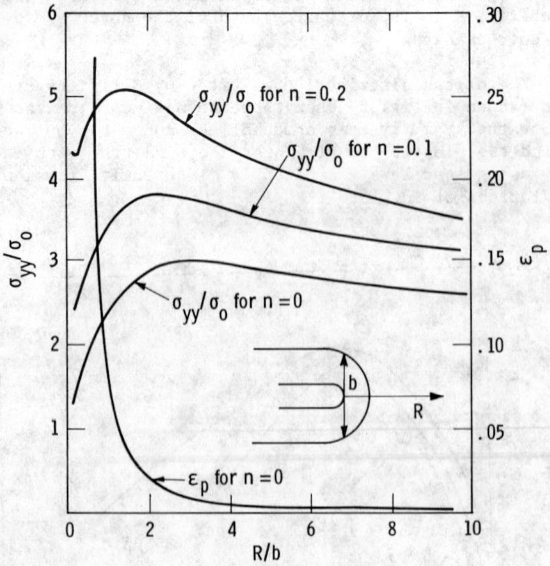

Figure 3. Stress and Strain Distribution Directly Ahead of Crack for Blunting to Five Times Original Notch Width. After McMeeking (12).

Ostegren was able to overcome the difficulty of finite element modeling of blunting of a sharp crack (with its inherent difficulty with the elastic stress-strain singularity at the crack tip) with a formulation which was based on features of the near crack-tip region of a rigid, perfectly plastic slip line construction by Rice and Johnson. Ostegren's results for the small scale yielding case are summarized by Rice and Johnson (13) and are plotted here in Fig. 4. The stress reaches the value of $3\sigma_o$ predicted by Eqs. (13) and (14), but they drop rapidly at distances away from the crack tip, starting at a distance that is only a small fraction of the plastic zone dimension. Strain hardening effects (based on work of Hutchison and work of Rice and Rosengren) are also summarized by Rice and Johnson (13) for the small scale yielding case. These results (see Fig. 5) show substantially higher stresses at the crack tip than that obtained for nonhardening material, but these too drop off very rapidly ahead of the crack, and Rice and Johnson point out that the high stresses occur more through hydrostatic stress elevation than hardening and cannot be maintained locally as the tip blunts.

Rice and Johnson (13) also developed a complex analytical expression for strain distribution ahead of an initially sharp crack which had undergone large geometry change blunting which is applicable to both small scale yielding and a fully plastic case. Their expression was developed from a rigid, perfectly plastic, slip

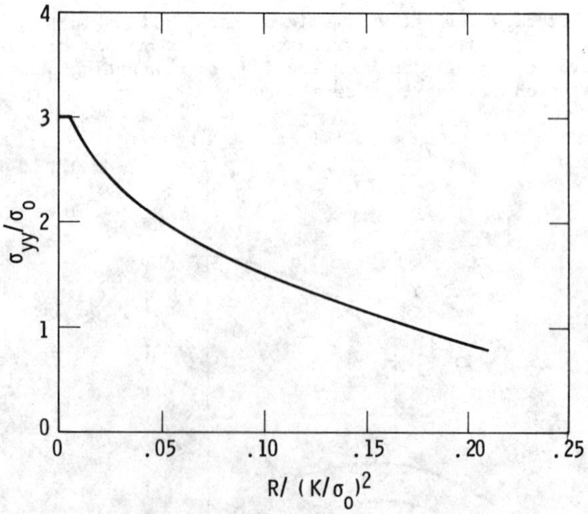

Figure 4. Stress Distribution Directly Ahead of Crack of Nonhardening Material Under Small Scale Yielding Calculated by Finite Element by Ostergren As Reported by Rice and Johnson (13).

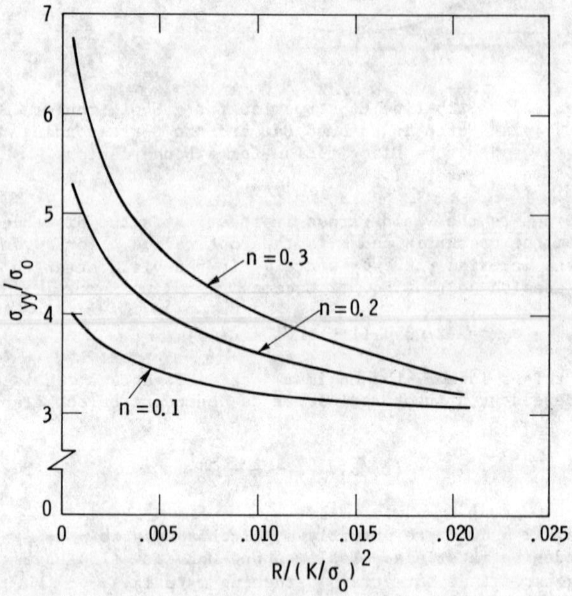

Figure 5. Stress Distribution Directly Ahead of Crack for Hardening Materials from Singularity Solution by Hutchinson/Rice and Rosengren (13).

line field consisting of sets of α and β lines similar to that shown in Fig 1. A parameter, $\psi(\alpha,\beta)$ was introduced and expressions derived in terms of ψ and the crack tip opening displacement (δ_T) for displacement velocity (v_x), position (x), and the position (R) of a material point before deformation which are

$$v_x = V(\psi),\ x = \delta_T F(\psi),\ R = \delta_T H(\psi)$$

Strain was then expressed as

$$\varepsilon_{yy} = -\varepsilon_{xx} = \ln \frac{H'(\psi)}{F'(\psi)} = -\int_{\psi}^{\pi/2} \frac{V'(\psi)d\psi}{F(\psi) - V(\psi)} \qquad (19)$$

Solution of this equation, plotted as a function of R/δ_T gave the results shown in Fig. 6 (13). McMeeking (12) has shown that his results (shown in Fig. 3) are in good agreement with Rice's and Johnson's results (Fig. 6) when the distance from the crack tip is normalized with the same factor.

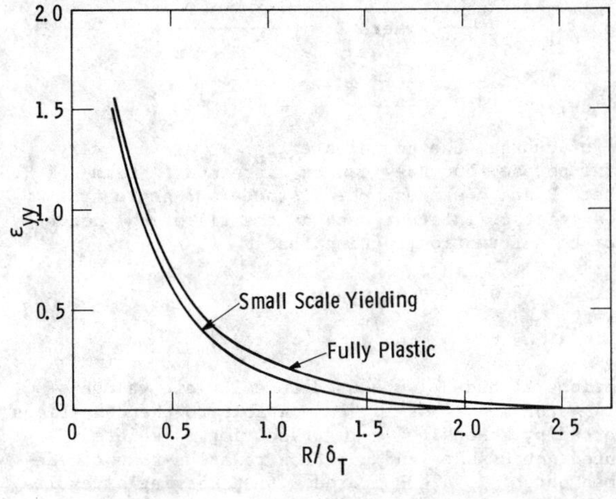

Figure 6. Strain Distribution Ahead of Crack From Rice and Johnson (13). Note: R Measured from Position of Undeformed Crack Tip.

A model for displacements and associated strains around a blunting, semicircular notch has recently been developed by Bates (18) leading to closed form solutions. This model is summarized in the remainder of this section.

The process of a semicircular notch blunting from a radius ρ_o to ρ_1 (such that the notch opening expands by δ_T from $2\rho_o$ to $(2\rho_o + \delta_T) = 2\rho$) is shown in Fig. 7. In this process the center of curvature has been displaced by an amount α_1 and the notch tip has been displaced by an amount δ_{x1}. The values of α and δ_{x1} are at present unknown quantities. In the process of blunting, the flank angle (which is zero in this case) is assumed to remain constant such that the strain along the flank portion of the deformed notch surface is approximated as zero.

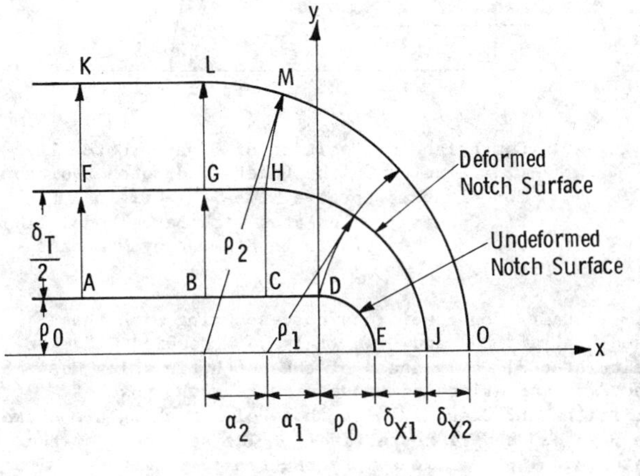

Figure 7. Displacements Associated with Smooth Blunting of a Semicircular Notch.

Note that the material originally contained in the volume bounded by the undeformed notch and deformed notch surfaces (i.e., the volume bounded by AEJF) is displaced during the blunting process to a new location bounded by FJOK. Assuming incompressibility, these volumes must be equal. It is also assumed that the segment FG is displaced to KL without rotation in such a manner that the strain is zero along KL as the segment AB is displaced to FG. Then the volume ABGF is equal to the volume FGLK, and

$$\rho_1 = \rho_o + \delta_T/2$$
$$\rho_2 = \rho_1 + \delta_T/2 \qquad (20)$$

and the center of curvature corresponding with ρ_2 is located a distance α_2 from the center of curvature corresponding to ρ_1. Note also that as the notch tip is displaced by δ_{x1} (from E to J) the point J is displaced by δ_{x2} (from J to O). Thus

$$\alpha_1 + \delta_{x1} = \delta_T/2$$
$$\alpha_2 + \delta_{x2} = \delta_T/2 \qquad (21)$$

In a similar manner it can be shown that the material originally bounded by ρ_1 and ρ_2 (and the associated flanks) is displaced outward such that

$$\rho_n = \rho_{n-1} + \delta_T/2 = \rho_o + n\,\delta_T/2 \qquad (22)$$

$$\alpha_1 + \delta_{x1} = \alpha_2 + \delta_{x2} = \alpha_n + \delta_{xn} = \delta_T/2 \qquad (23)$$

where n is an index designating the location of the volume being considered. It can then be shown (18) that the value of α_n can be expressed in terms of the preceding α designed as α_{n-1} as

$$\alpha_n = \frac{\frac{\pi}{2}\left(\frac{\delta_T}{2}\right)^2 + \alpha_{n-1}\left[\rho_o + (n-2)\left(\frac{\delta_T}{2}\right)\right]}{\rho_o + n\left(\frac{\delta_T}{2}\right)} \qquad (24)$$

It is also shown (18) that the summation of all of the α-terms from α_1 to α_n can be very closely approximated as

$$\sum_{n=1}^{n} \alpha_n = \frac{\pi}{4}\left[\left(\sqrt{\rho_n^2 + \rho_1^2}\right) - \rho_1 + (1 - \sqrt{2})\,\rho_o\right] \qquad (25)$$

Now referring to Fig. 8, consider the point originally on the undeformed notch surface but being displaced to a new position on the deformed notch surface. The position of this point at any time during its displacement can be described in terms of its instantaneous coordinates (ρ,ϕ) or (x,y), and the increment of displacement is described by

$$\frac{dy}{dx} = \frac{y}{x + \alpha} = \tan\phi \qquad (26)$$

Now when $\rho = \rho_1$, $\phi = \phi_1$, and $\alpha = \alpha_1$, and Eq. 26 becomes

$$\frac{dy}{dx} = \frac{y}{x + \alpha_1} = \tan\phi_1 \qquad (27)$$

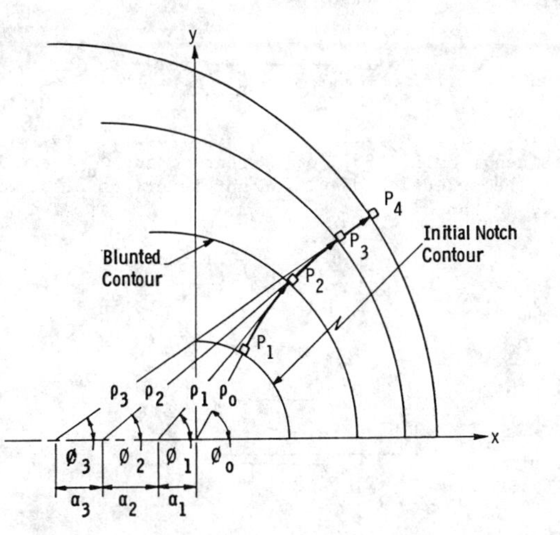

Figure 8. Displacements of Various Points During Smooth Blunting of a Semicircular Notch.

Similarly if a point is located in the material at some location on ρ_{n-1} prior to the notch blunting process, it is displaced during the blunting process in a manner described by

$$\frac{dy}{dx} = \frac{y}{x + \sum_{n=1}^{n} \alpha_n} = \tan \phi_n \qquad (28)$$

Letting $p_n = 1/(dy/dx)$ and rearranging, Eq. 28 becomes;

$$x = p_n y - \sum_{n=1}^{n} \alpha_n \qquad (29)$$

where $\sum_{n=1}^{n} \alpha_n$

(written hereafter as $\sum \alpha_n$) is given by Eq. 25. Eq. 29 is a form of Clairant's Equation, the solution of which is readily obtained (19) as:

$$y - \frac{d}{d\delta_n} (\sum \alpha_j) = 0 \qquad (30)$$

Combining Eqs. 30 and 25 and noting that

$$\rho_n = \sqrt{y^2 + (x + \sum \alpha_n)^2} = y \sqrt{1 + p_n^2} \qquad (31)$$

results in the following equation expressed with appropriate limits:

$$\int_{p_{n-1}}^{p_n} \frac{dp_n}{\sqrt{1 + p_n^2}} = \frac{\pi}{4} \int_{\rho_{n-1}}^{\rho_n} \left[\frac{d\rho_n}{\sqrt{\rho_1^2 + \rho_n^2}}\right] \qquad (32)$$

Integration of Eq. 32 and substituting of ϕ for p from Eq. 22 and the definition of p gives

$$\left(\frac{1 + \cos \phi_n}{\sin \phi_n}\right) \left(\frac{\sin \phi_{n-1}}{1 + \cos \phi_{n-1}}\right) = \left[\frac{\rho_n + \sqrt{\rho_1^2 + \rho_n^2}}{\rho_{n-1} + \sqrt{\rho_1^2 + \rho_{n-1}^2}}\right]^{\frac{\pi}{4}} \qquad (33)$$

Equation 33 can be used to develop strains for $\phi_{n-1} < \pi/2$

$$\varepsilon_{\phi\phi} = \ln \left(\frac{\rho_n \, d\phi_n}{\rho_{n-1} \, d\phi_{n-1}}\right) \text{ for } \phi_{n-1} < \frac{\pi}{2} \qquad (34)$$

Now for any given value of δ_T, $(d\phi_n/d\phi_{n-1})$ can be obtained from Eq. 33, noting that ρ_n and ρ_{n-1} are fixed with the selection of δ_T and not functions of ϕ_n or ϕ_{n-1}, so that

$$(\varepsilon_{\phi\phi})_{\rho_n, \phi_n} = \ln \left[\left(\frac{\rho_n}{\rho_{n-1}}\right) U^{-\frac{\pi}{4}} \left(\frac{1 + \cos \phi_n}{1 + \cos \phi_{n-1}}\right)\right] \qquad (35)$$

for $\phi_{n-1} < \frac{\pi}{2}$, where

$$U = \frac{\rho_n + \sqrt{\rho_1^2 + \rho_n^2}}{\rho_{n-1} + \sqrt{\rho_1^2 + \rho_{n-1}^2}}$$

and ϕ_{n-1} can be expressed in terms of ϕ_n as

$$\cos \phi_{n-1} = \left(\frac{Z - 1}{Z + 1}\right) \text{ where } Z = \left(\frac{1 + \cos \phi_n}{1 - \cos \phi_n}\right) U^{-\pi/2} \qquad (36)$$

and ρ_{n-1} can be obtained from ρ_n from Eq. 22.

Of course, the coordinates ρ_n, ϕ_n are not very useful per se, but they can be expressed in terms of R, θ where R and θ express the distance and angle from any fixed point of interest such as the tip of the notch prior to deformation, noting that

$$\begin{aligned} R \sin \theta &= \rho_n \sin \phi_n \\ R \cos \theta &= \rho_n \cos \phi_n - (\rho_o - \sum \alpha) \end{aligned} \qquad (37)$$

Equations 25 and 37 were used to calculate values of α_1 and δ_{x1} for a $\delta_T/2$ of $5\rho_o$ (comparable to the conditions reported by McMeeking). The resulting calculated blunted notch shape and position relative to the undeformed notch tip are compared with McMeeking's results (12) in Fig. 9. The results, so far as position is concerned, are almost identical; however McMeeking shows slight deviations in the flank portions for strain hardening materials not accounted for in the Bates model. The advantage of the Bates model is that the position of the blunted notch relative to the undeformed notch is expressed as a closed form solution, Eq. 25, a feature that will be applied later in modelling upper shelf fracture toughness.

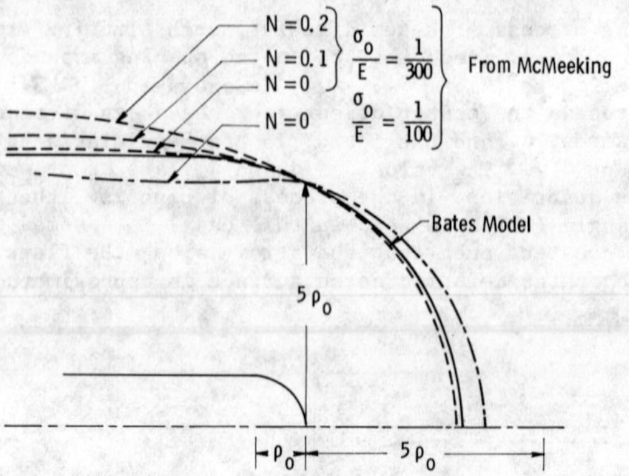

Figure 9. Comparison of Position of Blunted Notch Tip (in Relation to the Undeformed Notch Position for Blunting to Five Times the Original Notch Tip Opening) Calculated with Bates Model (18) with Results of Finite Element by McMeeking (12).

Strains for conditions corresponding with those used by Griffiths and Owen (17) and McMeeking (12) were calculated directly ahead of the notch tip using Eqs. 35 and 37. Comparison of these results with those of Griffiths and Owen and with those of McMeeking are shown in Figs. 10 and 11 respectively. The agreement of the Bates model predictions with the finite element results is excellent. Again, the Bates model has the advantage of being in the form of an easily applied closed form solution.

It is of interest to compare the strain distribution for various amounts of crack tip blunting such as shown in Fig. 12. Note that as δ_T increases up to a value of $2\rho_o$, the strain increases for all values of R/ρ. Further increases in δ_T have little effect except very near the notch tip. The principal effect is just a scale-up of the same geometry, and the strain distribution (as plotted against R/δ) remains nearly constant. Although stress distributions have not yet been developed for this model, it is expected that the peak stress will initially increase with increasing δ_T [similar to the effect reported by Griffiths and Owen (17)] and then become about constant, at about the values reported by McMeeking (12), for δ_T values greater than about $3\rho_o$.

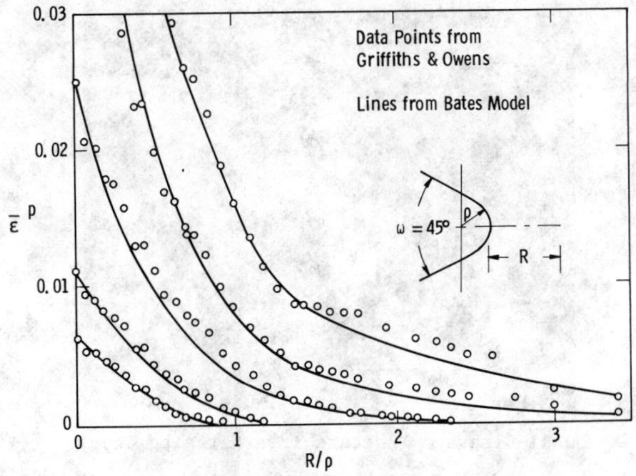

Figure 10. Comparison of Strain Distribution Calculated with Bates Model (18) with Results of Finite Element Calculations By Griffiths and Owen (17).

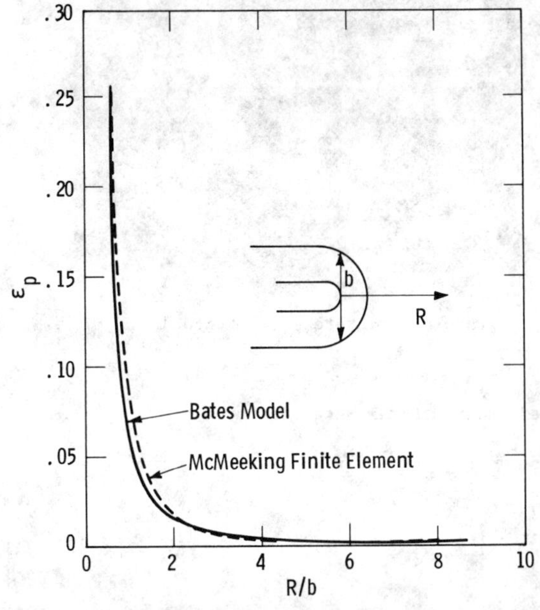

Figure 11. Comparison of Strain Distribution Calculated with Bates Model (18) with Results of Finite Element by McMeeking (12) for Notch Blunting to Five Times the Original Notch Width.

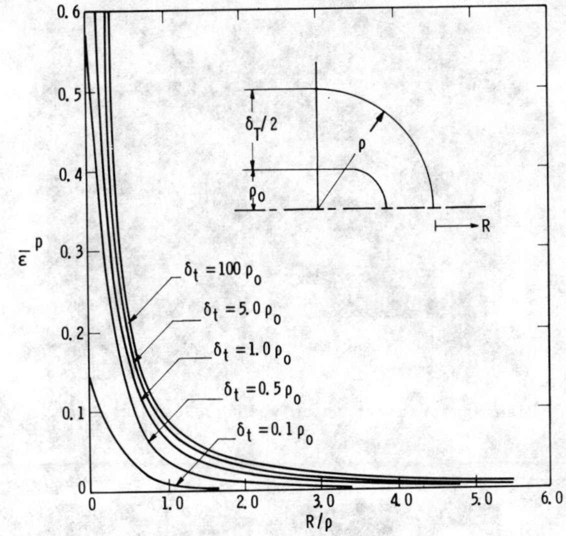

Figure 12. Strain Distributions Ahead of a Semicircular Notch for Blunting to Various Values up to One Hundred Times the Original Notch Radius From Bates (18).

FAILURE MODES

The macroscopic fracture behavior and associated fracture appearance of structural metals and alloys are often divided into the three categories of ductile, brittle and mixed mode fractures. Ductile failures generally exhibit a dull gray, fibrous fracture surface and are usually accompanied with extensive plastic yielding and shear lip formation. Brittle fractures generally appear smooth and flat with little evidence of plastic yielding. Often, brittle fracture surfaces appear bright and exhibit a fine crystallaine texture. Mixed mode fractures exhibit a mixture of ductile and brittle characteristics.

On a microscopic scale, several fracture modes commonly occur which affect, but do not exclusively control, the macroscopic fracture behavior. The fractographic appearance of many fracture types have been extensively studied, for example see (20, 21, 22). The most common of these are cleavage, quasi-cleavage, dimpled rupture, ductile cutting or alternate slip, intergranular separation, and fatigue. Examples of cleavage, quasi-cleavage, intergranular and dimpled fracture are shown in Fig. 13. Except for brief descriptions, only cleavage and dimpled rupture will be treated herein; quasi-cleavage will be treated as if it were a subcategory of cleavage.

Cleavage fracture represents brittle fracture along crystallographic planes and occurs in body-centered cubic and hexagonal close-packed metals and alloys. The characteristic feature of cleavage fracture consists of flat facets which generally are about the size of the ferrite grain size and which usually exhibit "river" markings (Fig. 13a). These "river" markings are caused by the cleavage crack traversing the crystal along a number of parallel planes which form a series of plateaus and connecting ledges. The direction of the "river pattern" is the direction of local crack propagation.

Quasi-cleavage (Fig. 13b) is related to cleavage but is usually classified as a separate type of fracture. It is observed primarily in quenched and tempered steels at low temperature. Like cleavage, it exhibits flat facets with "river" markings, but in cleavage the river patterns usually traverse facets in one direction while they generally radiate out from one point near the center of quasi-cleavage facets. The term quasi-cleavage is used because the facets on the fracture surface are not true cleavage planes (23). Also, the facets usually correspond in size with the prior austenite grain size. Quasi-cleavage fractures also frequently exhibit tear ridges and dimples around the facets which is not characteristic of true cleavage.

Dimpled rupture (Fig. 13c) is common to a wide variety of metals or alloys of all strength levels. The characteristics feature of this type of fracture is cup-like depressions which may be equiaxed, parabolic, or elliptical, depending on the stress state. Beachem has reported eight modes of dimple formation and predicted others, but they are all related in that they occur by nucleation (usually from second-phase particles) and growth of microvoids (21). Fracture occurs when the microvoids are joined by coalescence or otherwise fracturing the ligaments between microvoids. This process is generally a highly ductile one, on a microscopic scale, since the growth of microvoids is strain controlled and the final coalescence occurs by internal necking of the ligaments between voids. Macroscopically, ductility may be high or low depending on the size, number and distribution of voids.

Cleavage Fracture in a Type 403 Stainles Steel

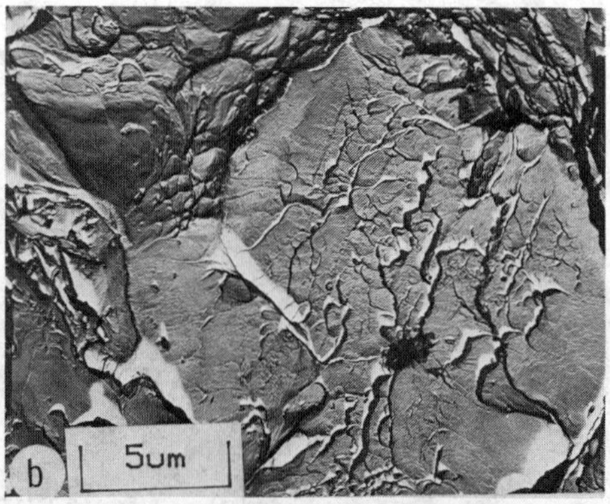

Quasi-Cleavage Fracture in a Maraging Steel

Microvoid Coalescence in a Type 403 Stainless Steel

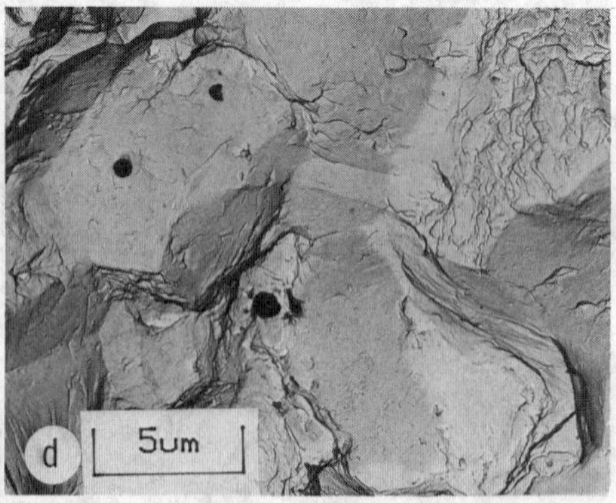

Intergranular Fracture in 4340 Steel

Figure 13. Common Fractographic Features Observed in Structural Metal Failures.

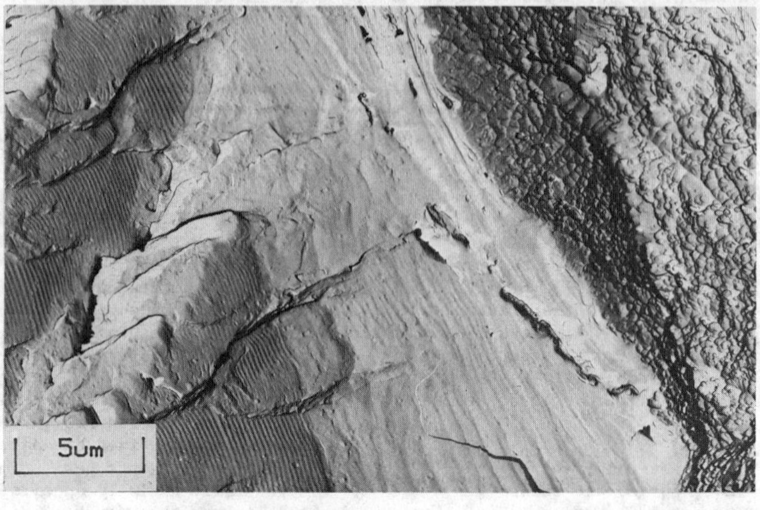

Figure 14. Stretched Zone Observed in 7178-T651 Aluminum Between Fatigue Striations and Dimple Zone.

Intergranular fracture (Fig. 13d) is characterized by facets which outline the surfaces of individual grains. These facets can be nearly featureless or they can exhibit dimples, slip lines, outlines of second phase particles, etc. The primary causes of intergranular fracture are extensive grain boundary precipitates, segregates, and environmental effects.

Another feature that is commonly observed at the onset of unstable fracture is termed a stretched zone. An example of a stretched zone (between the fatigue striated region and dimpled rupture region) is shown in Fig. 14. Stretched zones are generally featureless or exhibit ripples suggestive of severe stretching, from which the name is derived. For steels, tested at temperatures in the brittle to ductile transition region, a stretched zone followed by a zone of dimples which is followed in turn by cleavage fracture are commonly observed.

Cleavage fracture and dimpled rupture are the most important mechanisms of fracture produced under monotonic loading conditions (except for intergranular fractures which is caused principally by embrittlement or by environmental effects). Detailed discussions of the mechanisms of cleavage and dimpled rupture and modeling of the resulting fracture toughness follow.

MODELING OF LOWER SHELF (CLEAVAGE) FRACTURE TOUGHNESS

Mechanism of Cleavage Fracture

There is general recognition that cleavage fractures can be described in terms of a critical stress criterion for failure, i.e., failure will occur when the stress exceeds a critical value, σ_F. An understanding of the physical significance of σ_F and consequent caution that should be exercised in its application can only be obtained through a knowledge of the mechanisms of cleavage fracture. Although complete mechanistic understanding has not been achieved, considerable advances have been made during the last three decades.

The development of cleavage fractures involves sequential stages of nucleation (or initiation) and growth (or propagation). Since most materials of concern are highly inhomogeneous on a microscopic scale, the growth stage may be further divided into stages of initial growth in the material immediately surrounding the "microcrack" nucleus and the subsequent growth as obstacles are encountered. Since the stages of crack development are sequential, σ_F will be determined by the most difficult stage to achieve.

Early work on cleavage cracks was devoted to an understanding of the nucleation process which would produce "Griffith's" cracks in an initially crack free material, recognizing that the theoretical cohesive strength is far higher than the fracture strength of polycrystalline materials.

The concept that exceedingly high stresses could be achieved on a microscopic scale by inhomogeneous strain was first suggested by Zener (2,24). He postulated that high stresses would occur at dislocation pile-ups at barriers such as grain boundaries and second phase particles which could produce microcracks. Dislocation models were subsequently developed; e.g., Stroh (25,26), Cottrell (27,28), and Smith (29,30,31). Later, confirming evidence of microcracks caused by dislocation pile-up was produced, e.g. (32,33). Meanwhile, in 1962, Hull (34) presented convincing evidence that twinning is an important mechanism in crack incubation. Hull concluded that important crack nucleation sites are intersections of twins with other twins, and intersections of twins with grain boundaries. Today it is generally accepted that twinning, when it occurs, is a source (and probably an easier source than slip) of microcracks, particularly in second phase particles. Twinning does not always occur, however, since it is predominantly a low temperature process. Cleavage is still observed in the absence of twinning, see for example, Oates (35). The only viable alternative seems to be slip induced nucleation in that case.

Stroh's model (25,26) concluded that dislocations would coalesce to form a crack at a grain boundary when the shear stress reached a citical value. Superimposed hydrostatic tension would have no effect on the shear stress which nucleates the crack (a prediction in disagreement with observations) so Cottrell reasoned, that either growth must be the critical event or an easier nucleation process exists. Thus Cottrell (27,28) proposed nucleation at intersecting slip lines, or as an alternative nucleation occurs in a region requiring low fracture energy and propagation occurs in a region requiring high fracture energy. Smith, after additional analysis of pile-ups at grain boundaries confirmed that such dislocation pile-up models support Stroh's predictions, and he suggested that cracks would nucleate in hard grain boundary particles and rapidly spread across the width of these particles. Subsequent growth would require penetration into the surrounding matrix where fracture energy is higher (29,30,31). Thus σ_F is the stress characteristic of that propagation event. The process would be affected by hydrostatic tension and the thickness of the grain boundary particles, predictions that were later confirmed experimentally by Oates (35). Tetelman and McEvily (23) noted that in some cases, cracks have formed and penetrated the matrix only to be arrested at the first grain boundary that was encountered. They concluded that σ_F is, therefore, the stress required for grain-size length cracks to penetrate the first grain boundary in those cases.

Quasi-cleavage is probably similar to cleavage in that distinct steps of nucleation and propagation are required. Differences, however, are immediately apparent in both nucleation and propagation. As previously mentioned, the size of quasi-cleavage facets, correspond with the prior austenitic grain size and the markings radiate outward from a central nucleaus within the quasi-cleavage grains. Evidently, the nucleation event must be caused by inhomogeneous plastic deformation resulting in a cracked precipitate particle within the prior-austenitic grain (rather than at a grain boundary) followed by propagation up to the prior austenitic grain boundaries and termination there. Separate nucleation and propagation occurs in adjacent prior austenite grains, so that the final fracture occurs in a discontinuous manner by coalescence of separate cracks in adjacent grains. σ_F for this process is probably characterized by the relatively "ductile" propagation event since the surface of a quasi-cleavage facet is not a true cleavage plane and dimples are often found at the edges of facets.

Lower Shelf Models

The requirement that the tensile stress must exceed some critical value, σ_F is a condition that is difficult to satisfy for uniaxial tensile specimens (requiring either very low temperature or considerable strain hardening to raise the yield strength to the order of σ_F), but this criterion is easily satisfied ahead at a crack under plane strain loading. Peak stress values of $3\sigma_o$ can be achieved even in non-hardening materials ahead of a crack after only a relatively small amount of blunting, a value which generally exceeds σ_F, especially for high σ_o for steels at lower-shelf temperatures. For steels with moderate strain hardening, say 0.1, σ_{yy} corresponding with lower shelf K_{Ic} may be $2\sigma_F$ or greater. Evidently the criterion that $\sigma = \sigma_F$ is a necessary, but not sufficient condition, for cleavage failure. If it were a sufficient condition, the fracture toughness would be diminishing small for sharply cracked bodies.

Typically, σ_F is the tensile stress required to cause crack growth. If the stress is uniform, once the stress equals σ_F, unstable crack propagation would result. Suppose, however, that a steep stress gradient exists when $\sigma = \sigma_F$ on a local scale. Crack nucleation takes place and crack propagation begins, but the crack may be propagating into a lower stressed area. If the stress in this area (including the stress intensification caused by the crack itself) is lower than σ_F, then propagation will cease. Thus the failure criterion must be that $\sigma = \sigma_F$ over some critical distance.

Tetelman, Wilshaw and Rau (36,37) applied this general concept in 1968, proposing $\sigma \geq \sigma_F^*$ over a critical distance of at least one grain diameter as a failure criterion, where σ_F^* is the critical stress for a notched specimen loaded in bending. They assumed that as a bending moment is applied to a notched bar, σ_{yy} would build up to σ_F^*, at which time the plastic zone size, would reach a critical value R_F. Then substituting these values into Eq. 12 and taking a Tresca criterion for yielding (i.e., $2k = \sigma_o$), they obtained:

$$\sigma_F^* = \sigma_o[1 + \ell n(1 + R_F/\ell)] \qquad (38)$$

$$\text{for } R_F < R_\beta$$

where R_β is the dimension of the plastic zone when σ_{yy} first reaches its maximum value as given by Eq. 14. By application of this model to a high nitrogen steel, Tetelman, Wilshaw and Rau concluded that σ_F^* is higher for a notched specimen in bending than for a plane tensile specimen, and σ_F^* increases as the notch radius decreases for $\rho < 0.25$ mm. They related these effects to the statistical effects resulting from the volume of material that is stressed above σ_F^* when a stress gradient exists compared to the volume of material stressed above σ_F^* when no stress gradient exists.

The Tetelman, Wilshaw, Rau model provides an explanation of effects of ρ (root radius) on $K_{Ic}(\rho)$. Typically, a constant or slightly rising value of $K_{Ic}(\rho)$ is observed for a ρ ranging from that of a fatigue precrack up to some "characteristic" or "effective" value. For ρ greater than that characteristic value, $K_{Ic}(\rho)$ increases with $\sqrt{\rho}$ (36,37). This occurs as a direct consequence of the requirement that the critical stress must act over a finite region which requires extension of the plastic zone to the extent of that critical dimension (36).

The principal difficulty with the model of Tetelman, Wilshaw and Rau is that it relies heavily on Hill's slip line formulation (Eq. 12) which is strictly applicable only to notched, round tensile configurations of non-hardening materials (38). As previously discussed, finite element analyses have shown quite different stress distributions.

Ritchie, Knott and Rice also developed a model for cleavage failure which incorporated a requirement that $\sigma \geq \sigma_F$ over a critical distance (38). Their model differed from that of Tetelman, Wilshaw and Rau in that they assumed that the critical stress is attained at some fixed distance ahead of the crack tip rather than at the plastic-elastic interface. Furthermore, they reasoned that the stress must be high enough to propagate the crack through the second grain rather than just the first; thus they concluded that the critical distance must be greater than the one grain diameter assumed by Tetelman, Wilshaw and Rau. Richie, Knott and Rice had available the stress distribution from Ostergren (Fig. 4) and of Rice and Rosengren/Hutchinson (Fig. 5) which they applied to their model rather than having to rely on Hill's slip line formulation. Specifically, the model is used in the following manner. First, the cleavage strength, σ_F, must be determined, for example by notched bar bend tests at temperatures below that at which general yielding will occur and then calculating σ_F with Eq. 13 and the appropriate expression relating nominal stress to bending moment. Second the tensile properties are measured. Third the grain size and the critical distance (R*) are determined. Fourth, the ratio of $R^*/(K/\sigma_o)^2$ corresponding to the ratio of σ_F/σ_o is determined from an appropriate plot of σ_{yy}/σ_o vs. $R/(K/\sigma_o)2$ such as shown in Fig. 4 or Fig. 5. Finally the value of K corresponding to R* and σ_o is calculated using the value determined for $R^*/(K/\sigma_o)^2$.

The model of Ritchie, Knot and Rice was tested with experimental data for mild steel (38), A533B alloy steel (39) and SA302B steel (39). The calculated values for K_{Ic} at temperatures below the nil ductility transition temperatures (NDTT) were in excellent agreement with measured values providing that appropriate values of R* were selected. R* values that gave the best results (expressed as a ratio with the austenitic grain size) ranged from two to four. Unfortunately, at the present time the selection of the appropriate multiple of the grain diameter for application in the model is an arbitrary procedure.

MODELING OF UPPER SHELF (MICROVOID COALESCENSE) FRACTURE TOUGHNESS

Nucleation of Voids

Although instances of dimpled rupture have been reported in which no evidence of nucleating second phase particles was found (40), most microvoids are nucleated at precipitates and inclusions. The role of hard second phase particles has already been discussed as a major factor in the nucleation of cleavage fractures. Obviously, particles have an effect on ductile fracture mechanisms as well. Puttick (41) reported in 1959 that sections through the necked regions of tough pitch copper and iron tensile specimens contained numerous internal holes that had originated at inclusions. Numerous other investigators have reported similar phenomena, e.g. (40,42,43)

Edelson and Baldwin (43) have shown that in copper, the ductility is principally dependent on second phase particles with the fracture strain decreasing approximately exponentially from about 1.5 to less than 0.05 as the volume fraction of second phase particles increases from about 0.05 to about 0.20. Fracture strength is also decreased substantially as the volume fraction of most types of particles increases. Strengthening occurs only in those cases where the second phase particles are securely bonded to the matrix, in which case the yield strength varies with the mean free path between particles. The specific nature of the particle or size (in the range of about 1 to 2000 μm) is of little consequence. However, Rodgers (40) reports that the size of particles is important, with voids nucleating first at large particles and then at progressively smaller particles as strain increases.

Argon and co-workers (44,45,46) have shown that a critical interfacial stress is applicable for interface separation of hard, equiaxed inclusions, although heterogeneous strain aids in the attainment of the critical stress. They conclude with other investigations (47,48) that a release of elastic strain energy equal to the energy of the generated surfaces is a necessary but not sufficient condition for interface decohesion, and that for large particles (greater than 100Å diameter) the energy criterion is always satisfied soon after yielding. A continuum deformation model for interfacial stress, applicable to large particles, is given by Argon, Im and Safoglu (45) as

$$\sigma_{rr} = \sigma_{flow} + \sigma_T \qquad (39)$$

where σ_{flow} is the flow stress in the region of the inclusion (without including the strain concentration caused by the particle) and σ_T is the negative pressure (hydrostatic stress). Thus both strain hardening (causing an increase in σ_{flow}) and triaxiality increase the interfacial stress and favor void nucleation. They also examined a critical stress criterion based on dislocation models posultating that secondary plastic zones will develop between particles of sufficient size and close spacing. They advanced the following equations for the stress at an inclusion-matrix interface (45).

For noninteracting inclusions:

$$\sigma_{rr} = k_o\left[\left(\frac{\gamma}{\gamma_o}\right)^n + \sqrt{3}\left[\frac{\sqrt{6}\,(1+n)}{nm}\,\frac{\gamma}{\gamma_o}\right]^{n/(1+n)}\right] \qquad (40)$$

For interacting inclusions:

$$\sigma_{rr} = k_o\left[\left(\frac{\gamma}{\gamma_o}\right)^n + \sqrt{3}\left(\frac{\sqrt{3}\,(\gamma/\gamma_o)}{(\lambda/\rho)}\right)^n + \frac{\sqrt{6}}{m}\,\frac{\lambda}{\rho}\right] \qquad (41)$$

where γ and γ_o are the shear strain and shear strain at yield, respectively, n is the strain hardening coefficient, m is the Taylor factor (approximately 3.1), and (λ/ρ) is the ratio of interparticle spacing to particle radius.

From Eqs. (40, 41) along with an analysis of local variations of λ/ρ, Argon and his coworkers concluded that at high volume fractions, particles interact and enhance the interfacial stress above that given in Eq. (39) due to the development of secondary plastic zones between particles. They conclude that when the second phase volume fraction is about 0.01 the particles will not interact, even at very high strains. On the other hand, when the volume fraction is about 0.1 or larger, interaction occurs almost at the onset of plastic yielding. Even in the case of low volume fractions of second phase particles random distributions may result in localized areas of relatively high volume fractions where voids can nucleate if the strain is sufficient.

Using these concepts, Argon and Im (46) determined the critical interfacial stresses for particles in three materials as shown in Table I.

Table I

Local Interfacial Stresses for Separation of Particles

Material	Type	Vol. Fraction	Dia. (μm)	σ_{crit} (MPa)
(Maraging steel) (VM-300)	Ti-C	0.11	5.3	1820
1045 C-steel (spherodized)	Fe_3C	0.125	0.44	1670
Cu-0.6Cr	Cr-Cu	0.006	0.89	990

Microvoid Growth

The growth of a microvoid, once nucleated, will occur in two stages. The first stage occurs immediately after nucleation because of an inbalance in the equilibrium between the stress state and the flow stress caused by the creation of a free surface at a precipitate during the nucleating event.

Consider the stress state on a cube of differential dimensions at the interface of a particle such as that shown in Fig. 15. The effective, or equivalent stress, is equal to the strain-hardened flow stress (strain hardened from strain prior to nucleation).

$$\sigma_{flow} = \bar{\sigma} = \frac{\sqrt{2}}{2}\left[(\sigma_1-\sigma_2)^2 + (\sigma_2-\sigma_3)^2 + (\sigma_3-\sigma_1)^2\right]^{1/2} \qquad (42)$$

For plane strain

$$\sigma_3 = \sigma_{zz} = (1/2)(\sigma_1 + \sigma_2) \qquad (43)$$

and Eq. 42 becomes

$$\sigma_{Flow} = \bar{\sigma} = \frac{\sqrt{3}}{2}(\sigma_1 - \sigma_2) \qquad (44)$$

Now if decohesion occurs along the particle/matrix interface between points A and B in Fig. 15, the differential cube is then situated at a free surface which cannot sustain a normal stress. Thus the stress normal to the surface (in this case σ_2) reduces to zero and since σ_1, $\sigma_2 > 0$ prior to decohesion, the newly established value of $\bar{\sigma}$ at the void interface is higher than it was before nucleation. Thus after nucleation

$$\bar{\sigma} = \frac{\sqrt{3}}{2}\sigma_1 > \sigma_{flow} \qquad (45)$$

The effective stress is similarly increased at other positions surrounding the newly formed void so that yielding (material flow) will occur causing void growth until the material in the vicinity of the void strain hardens and raises the local flow stress into new equilibrium with the increased $\bar{\sigma}$. Thus, the resulting void size may be a few times to several times the diameter of the nucleating particle, depending on the stress state at the time of nucleation and the material flow characteristics. However, the specific relationships have not been established.

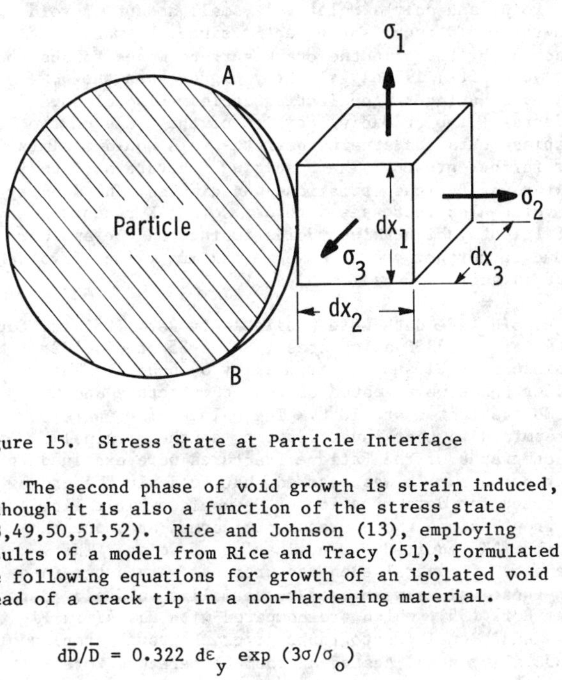

Figure 15. Stress State at Particle Interface

The second phase of void growth is strain induced, although it is also a function of the stress state (13,49,50,51,52). Rice and Johnson (13), employing results of a model from Rice and Tracy (51), formulated the following equations for growth of an isolated void ahead of a crack tip in a non-hardening material.

$$d\bar{D}/\bar{D} = 0.322\, d\varepsilon_y \exp(3\sigma/\sigma_o)$$

$$dD_x/\bar{D} = -2d\varepsilon_y + d\bar{D}/\bar{D} \qquad (46)$$

$$dD_y/\bar{D} = 2d\varepsilon_y + d\bar{D}/\bar{D}$$

where D_y and D_x are the major and minor diameters of the elipsoidal void. In the derivation of these equations, it was assumed that void growth started when the void was enveloped by the large strain region shown as point "B" in Fig. 1.

McMeeking (52) applied these equations to compute the growth of voids in terms of D_o (the initial void size), R_o (the distance from the crack tip to point B) and δ_T for three shapes of crack tips. The initial growth rates of voids ahead of flat nosed and sharply pointed crack tips are higher than the growth rates ahead of smoothly blunted crack tips, but after δ_T

reaches values of 0.6 R_o to 1.0 R_o void growth stops ahead of pointed and flat nosed cracks. Ignoring the effects of free surfaces, D_x and D_y continue to grow for semicircular crack tips until the crack surface reaches the void site, and McMeeking's results show growth rates that are approximately linear with δ_T/R_o up to $\delta_T/R_o \approx 2$. Note that R_o is the distance from the center of curvature of the notch to point B in Fig. 1. That distance can be determined with Eqs. 13 and 14 as 2.4 δ_T where δ_o is the value of δ_T at the time void growth starts. Thus McMeekings results can be expressed in terms of δ_T/δ_{To} where δ_{To} is the crack tip opening displacement at the time of first void growth. Incorporating 2.4 δ_{To} in place of R_o and fitting straight lines, McMeeking's results can be approximately expressed as follows:

$$D_x/D_o = 0.88\ (\delta_T/\delta_{To}) + 0.12$$
$$D_y/D_o = 1.67\ (\delta_T/\delta_{To}) - 0.67 \qquad (47)$$

The validity of Eq. (47) can be tested from void-size data obtained on fatigue crack growth specimens. Voids such as those that cause microvoid coalescence occur in fatigue, particularly at the higher stress levels (53,54,55). A range of dimples occurs at any given level of ΔK once some minimum ΔK is reached. At relatively low values of ΔK where voids can grow unimpeded by other voids, the average size as well as maximum size achieved at any specific value of ΔK increases with increasing ΔK (55). Void growth should occur primarily during the stress cycle that initiates the void, and only a relatively small amount of void growth should occur subsequently since the maximum δ_T reached (by the time the crack surface moves to absorb the void site) is only slightly higher than the δ_T achieved during the nucleating cycle. Also closure effects of the unloading portion of the cycle on the displacements and strains near the void would minimize the further growth. The resulting fracture surface is a mixture of fatigue striations and dimples. The density of voids also increases with ΔK, so as ΔK reaches a sufficiently high value, voids impinge on one another reducing further growth so some maximum value of void size is achieved, called D*.

Void size data were available in Ref. (55) for four materials. A533B alloy steel, HP9-4-25 steel, 403 stainless steel and aluminum alloy 5456-H321. The latter three were tested at room temperature and the A533B was tested at -18°C. The values of D_o were not determined in that investigation, so the original fractographs of the fatigue fractures were examined for that purpose. It was assumed that voids in the areas selected for determination of D_o would exhibit little or no growth since these areas were formed at about the minimum ΔK required for nucleation of voids. The averages of several such measurements were determined and these values were used to normalize void size data from Ref. (55), which are compared with Eq. 47 in Fig. 16. Additional fractographs of crack growth samples of A533B alloy steel tested at room temperature were available. Void sizes were determined on these samples and these results are also plotted in Fig. 16.

The void growth data plotted in Fig. 16 show considerable scatter, and the points generally fall above the line representing Eq. 42 (possibly because some void growth occurred on cycles subsequent to one which caused nucleation and the majority of growth. Still the data seem consistent with McMeeking's predictions; furthermore, Fig. 16 suggests that the crack tip is smoothly blunted rather than flat or sharped nosed since the void growth significantly exceeds the maximum predicted by McMeeking for those crack tip shapes.

Upper Shelf Models

Rice and Johnson suggested a failure criteria for microvoid coalescence consisting of a critical strain over a critical distance (13). They further reasoned that for most structural steels the fracture strain on that size scale should be in the range of about 0.2 to 1.0. The critical distance may be the distance from the crack tip to the first large inclusion (assuming that the microvoids growing from the largest inclusions should govern fracture). Thus if R* were the critical length, δ_{Tc} were the critical crack tip opening displacement, and X_o were the inclusion spacing, then $R^* = X_o$ and the ratio δ_{Tc}/X_o corresponding with ε_T^* (the critical strain) of 0.2 to 1.0 would be 1.0 to 2.7 (see Fig. 6 relating ε to R/δ_T from Rice and Johnson small scale yielding analysis). Rice and Johnson (13) applied this analysis to four high strength steels (σ_o = 1650 MPa), for which K_{IC} (hence δ_{Tc}) and X_o were known, see Table II. Indeed, the values of the δ_{Tc}/X_o fall within the predicted range for the four steels shown in Table II (13).

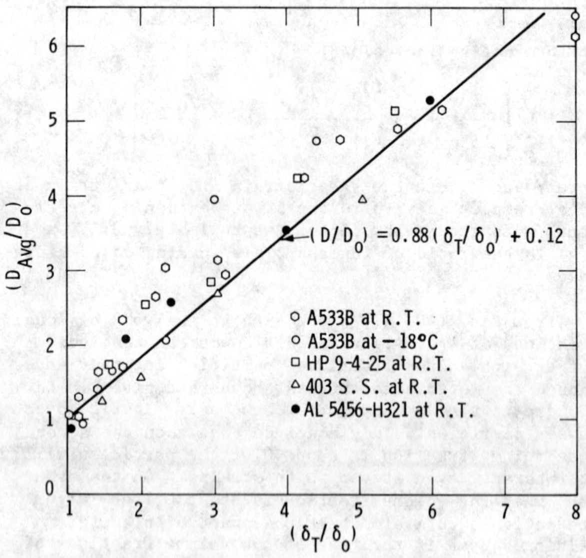

Figure 16. Microvoid Growth as Related to δ_T in Fatigue Crack Growth Samples

Table II

Fracture Toughness, Critical Crack Tip Opening Displacement and Inclusion Spacing for High Strength Steels (1650 MPa Yield Strength) from Rice and Johnson (13)

Sample	Wt. % S	K_{Ic} MPa$\sqrt{m}$	R_o μ-m	δ_{Tc} μ-m	δ_T/R_o
A	0.008	71	6.1	9.1	1.5
B	0.016	61	5.4	6.8	1.3
C	0.025	56	4.4	5.6	1.3
D	0.049	47	3.7	4.0	1.1

Ritchie, Server and Wullaert (39) used a similar approach in modeling upper-shelf toughness. They noted that the small scale yielding strain distribution of Rice and Johnson could be expressed as

$$\bar{\varepsilon}_p = C\delta_T/R \qquad (48)$$

where C, is a constant and R is the distance from the crack tip in the undeformed configuration.

Combining Eqs. 48 and 10 and applying the criterion that a critical fracture strain, ε_F^*, must be exceeded for

$R = R_o^*$ where R_o^* is a distance characteristic of the pertinent microstructure feature leads to

$$K_{Ic} = C(\varepsilon_F^* \, R_o^* \, \sigma_o \, E)^{1/2} \tag{49}$$

where C is a constant

They applied their model to calculate upper shelf toughness of two steels (A533B and A302B) taking ε_F^* from data obtained on circumferentially notched tensile specimens of various root radii (ranging from 1.27 to 6.35 mm). The stress state at fracture of each of these specimens was determined with a bridgeman analysis for necked tensile specimens, and the value of ε_F^* selected for each steel (about 0.1 for A302B and 0.25 for A533B) was that corresponding to the particular stress state that exists at the characteristic distance from the crack tip. The results of the calculations were in agreement with measured upper shelf toughness for the two steels providing that R_o^* was selected as 1 times and 6 to 7 times the planar inter-inclusion spacing for A302B and A533B, respectively. This implies that coalescence of only one void with the crack tip is critical for A302B while nucleation and coalescence of several voids and linkage with the crack tip is required for A533B (39).

It is of interest to compare the results of the two sets of calculations just described. Rice and Johnson assumed $R^* = X_o$ and calculated ε_f^* from known values of X_o and K_{Ic}. Ritchie, Server and Wullaert started with experimental values of ε_f^* (from notched tensile data) and selected R^*/X_o to give agreement between experimental and calculated K_{Ic} values. Despite their different input, they used essentially the same model so a comparison of their data should show a consistent trend if the model and assumptions of input data are correct. Table III compares the two sets of data. Note that the K_{Ic} values do not vary directly with ε_f^* as they should; thus either the model is not valid, or more likely the data used by one or both of the investigating teams are not applicable to the model.

Table III

Comparison of K_{Ic}, ε_f^* and (R^*/X_o) for Steels Applied to Critical Stress-Critical Distance Model

Material	K_{Ic} MPa$\sqrt{m}$	ε_f^*	(R^*/X_o)	Reference
Avg. of 4 Steels in Table III	59	0.3	1	(13)
A533B	~230	~0.25	6-7	(39)
A302B	~120	~0.10	1	(39)

Rice and Johnson (13) also formulated a somewhat different model for fracture consisting of growth of voids in advance of the fracture front and eventual coalescence with the blunting crack tip as depicted in Fig. 17. For this model, they proposed that void nucleation from large sulfide particles will occur when the crack tip opening displacement is sufficient to envelop the particles in the large strain region and they also assumed the distance from the crack tip to the void site at that time is equal to X_o, the interparticle spacing. Thereafter they adopted Eq. 46 to describe growth up to the point where the distance between the blunted crack tip and the void boundary is equal to half of the verticle diameter of the void; i.e., $D_y/2$. Localized necking of the remaining ligament, requiring an additional crack tip opening displacement equal to

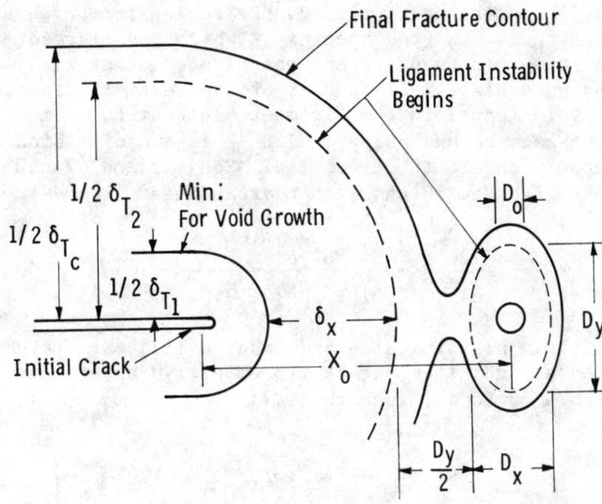

Figure 17. Model for Ductile Fracture by Void Growth and Coalescence with Blunted Crack Tip Proposed by Rice and Johnson (13).

Dy/2 completes the critical event leading to final fracture. With this model, Rice and Johnson calculated δ_{Tc}/X_o as a function of X_o/D_o (where D_o is the initial void diameter) for small scale yielding and fully plastic cases. Their results are given in Fig. 18.

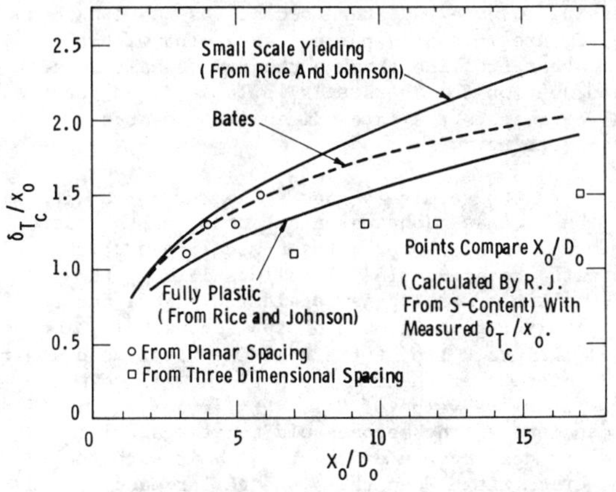

Figure 18. Predicted δ_{Tc} Based on Rice and Johnson Model (13) for Microvoid Growth and Coalescence with Crack Tip.

As a check of their model they used the δ_{Tc}/X_o values for the four steels given in Table II. The sulfur contents given for these steels were used to calculate interparticle spacings assuming all sulfur exists as manganese sulfide particles distributed as identical spheres. Planar section and three-dimensional nearest neighbor distances were calculated and the results are plotted in Fig. 18. The X_o/D_o values calculated on the bases of three-dimensional nearest neighbor spacing fall between the small scale yielding and fully plastic predictions, but X_o/D_o calculated assuming planar interparticle spacing are in poor agreement.

The Rice and Johnson model depicted in Fig. 17 was further applied by Bates (56) using the results of reference (18). From Fig. 17 it can be seen that the distance from the crack tip at the time of the beginning of void growth is

$$R = X_o = 2.4\,\delta_{T1} = \delta_x + \frac{D_y}{2} + \frac{D_x}{2} + \frac{\delta_{T1}}{2} \tag{50}$$

where δ_{T1} is the value of δ_T at the time of first void growth, D_y and D_x are the verticle and horizontal void growth, δ_x is the crack tip advance during further crack tip opening to δ_{T2} (the opening at which the ligament between the void boundary and the blunted crack tip becomes unstable), $D_y/2$ is half of the verticle diameter of the void (equal to the ligament width at the time the ligament becomes unstable), and $D_x/2$ is the dimension of void growth toward the crack tip. Eqs. 25 and 47 can be combined with Eq. 50 and with rearrangement gives

$$\frac{\delta_{T_2}}{X_o} = \frac{2.24(X_o/D_o) + 0.66}{0.81(X_o/D_o) + 7.34} \tag{51}$$

Thereafter, following Rice and Johnson (13), fracture requires an additional crack tip opening displacement equal to $D_y/2$ to δ_{Tc} such that

$$\delta_{T_c} = \delta_{T_2} + D_y/2$$

Again employing Eq. 47 for D_y, this becomes

$$\delta_{T_c}/X_o = \delta_{T_2}/X_o + [2.00\ (\delta_{T_2}/X_o) - 0.34]/(X_o/D_o) \tag{52}$$

Eqs. 51 and 52 were combined by Bates (56) to calculate δ_{Tc}/X_o as a function of X_o/D_o. These results, plotted in Fig. 18, are in agreement with the Rice and Johnson small scale yielding case at low values of X_o/D_o. At higher X_o/D_o values the Bates predictions move toward the Rice and Johnson Fully Plastic case. It is interesting to note that the predictions based on Eqs. 51 and 52, are in nearly perfect agreement with the probable best-fit line through the points calculated by Rice and Johnson for the steels in Table II on the basis that X_o represents the three dimensional nearest neighbor spacing.

Despite the seemingly good agreement noted with either the Rice and Johnson calculations or the Bates treatment of the Rice and Johnson model, a problem exists with the prediction of failure if only the sulfide particle spacing is considered. This can be seen by analyzing the growth of the second void, located $2X_o$ from the crack tip at the time the first void starts to grow. Growth of the second particle will commence when δ_T reaches a value of $2\delta_{T1}$. By employing Eq. 47, the dimensions of the second void can be calculated at the time of coalescence of the first void with the blunted crack tip. When this is done for each of the four steels in Table II, the ligament size between the first and second voids when δ_{Tc} is achieved is several times the combined D_y values of the two voids. Thus, the second ligament would be very stable and the additional δ_T that would be required to cause fracture would be considerable. Of course, it could be argued that this phenomena would give rise to an R-curve, except that further analysis shows that the resulting R-curve has a slope nearly as high as that of the blunting line.

This problem can be averted if smaller voids, nucleated at smaller second-phase particles such as carbides are considered. Recall from previous discussion that particles in marging steel and C-steel exhibited critical interfacial stresses of the order of 1700 MPa, and such particles will nucleate voids at the high stresses achieved in the vicinity of the crack tip (three to five times the yield strength of the material) although they may not nucleate voids in uniaxially stressed tensile specimens until the final stages of fracture. Voids nucleated at such particles will commence growth at much lower values of σ_T than that required for sulfide particles because their interparticle spacing will be much smaller. As these voids are overtaken by the blunting crack tip, they will be absorbed without significant effect, but as two or more voids coalesce and are simultaneously overtaken by the crack tip, an instability will arise either causing fracture or a significantly reduced slope of the crack tip advance line (i.e, an R-curve).

A summary of this new model, which is developed by Bates (56), follows. As shown in Fig. 19, it is assumed that the spacing between voids at which the intervoid ligament would become unstable is $D_y/4$. This value is half of the value of the spacing between the leading void and the blunted crack tip at the onset of ligament instability (which is still taken as $D_y/2$) because of smaller free surfaces effects. Also, to simplify the model, the two voids under consideration are both assumed to commence growth when δ_T is equal to δ_{T1} and the mean distance from the crack tip to the two voids of interest is $1.9\delta_{T1}$ (or $2.4\delta_{T1}$ measured from the center of curvature of the notch). Voids closer to the crack tip would, of course, grow and be absorbed into the blunting crack prior to the commencement of growth of the two voids being considered, but they would not satisfy the condition of coalescencing with each other at the same time the leading void coalesces with the crack tip. Thus:

$$1.9\ \delta_{T_1} = D_x + \delta_x + 5D_y/8 \tag{53}$$

where as before δ_x is the distance advanced by the crack tip as δ_T increases from δ_{T1} to δ_{T2}, D_y and D_x are, respectively, the vertical and horizontal diameters of the voids. The value of D* (the diameter of the voids after coalescense) in turn can be related to D_x and D_y as

$$D^* = D_x + D_y/4 \tag{54}$$

As before, substitution from Eqs. 25 and 47 leads to the following.

$$\delta_{T_1} = \frac{[1.924\ (\delta_{T_2}/\delta_{T_1}) - 0.299]\ D_o}{2.237 - 0.337\ (\delta_{T_2}/\delta_{T_1})} \tag{55}$$

$$(\delta_{T_2}/\delta_{T_1}) = \frac{D^*/D_o + 0.048}{1.298} \tag{56}$$

These equations can be solved simultaneously to calculate δ_{T1} and δ_{T2}. δ_{Tc} can then be calculated as:

$$\delta_{T_c} = \delta_{T_2} + \frac{1}{2}\,[1.67\ (\delta_{T_2}/\delta_{T_1}) - 0.67]\ D_o \tag{57}$$

where δ_{Tc} is either the value of δ_T at fracture or at the beginning of the R-curve.

This new model was tested to calculate K_{Ic} values for four alloys (the same four alloys that were used to develop the data in Fig. 16). Equation 10 was used to convert from δ_{Tc} to K_{Ic}, the results of which are compared with measured results in Table IV. The agreement between calculated and measured K_{Ic} values for the four alloys is good.

The values of δ_{T1} (from which the number of voids absorbed by the blunting crack prior to achieving δ_{Tc} can be calculated) are also reported in Table IV for the four materials. These values provide a clear physical picture of the meaning of the critical distance proposed by Ritchie, Server and Wullaert (39), which may vary in size from one to several multiples of the dimple diameters depending on the material.

The model also provides an explanation of typical blunt notch effects at upper shelf temperatures. At notch diameters smaller than δ_{T1}, the load must be increased until $\delta_T = \delta_{T1}$ to satisfy Eq. 55. Then a further increase in δ_T must be achieved to satisfy Eqs. 56 and 57. Thus, for initial root diameters up to the value of δ_{T1}, δ_{Tc} (hence K_{Ic}) would be nearly independent of the initial root diameter. On the other hand, for an initial root diameter greater than δ_{T1}, requirements that led to Eq. 55 are already satisfied by the

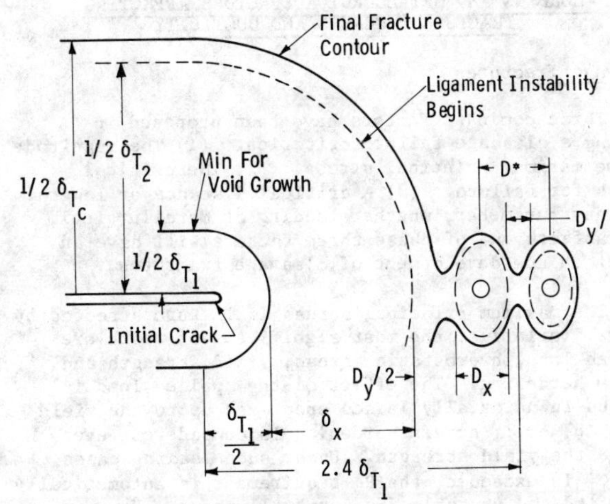

Figure 19. Model for Ductile Fracture by Void Growth and Simultaneous Coalescence of Two Voids and Blunted Crack Tip (56)

At lower shelf temperatures, little or no strain hardening is necessary to achieve the portion of the failure criterion of $\sigma_{yy} \geq \sigma_F$. Crack blunting, may be required to spread the dimensions over which the high stress are achieved in order to satisfy the critical distance requirement. At higher temperatures, strain hardening may be required to raise the yield strength locally to a high enough value to enable σ_{yy} to exceed σ_F. For example, the yield strength of A533B (tested at usual strain rates) cited by Ritchie, Server and Wullaert (39) ranges from about 1050 MPa at liquid nitrogen temperature to about 480 MPa at room temperature while σ_F is about 1830 MPa. The ratio of σ_F/σ_o then is about 1.74 at liquid nitrogen temperature, a value easily exceeded in the plastic zone surrounding a crack as can be seen from Fig 3. At room temperature, on the otherhand, the σ_F/σ_o ratio is about 3.8, a value that is achievable only with strain hardening effects. Before the cleavage criteria are met by strain hardening, microvoid nucleation and growth may intercede. This in turn increases the apparent value of σ_F, possibly due to a local breakdown of plane strain deformation as proposed by Tetelman and Wilshaw (58).

Table IV

Data for K_{Ic} Calculation and Comparison of Calculated and Measured K_{Ic} (56)

Alloy	Temp, °C	σ_o MPa	Void Dia., μ-m D_o	D^*	δ_1 μ-m	K_{Ic}, MPa√m Calculated	Measured
A553B	-18	517	0.29	2.18	11.6	111	115
HP9-4-25	24	1240	0.54	3.25	7.1	122	113
403 S.S.	24	1135	1.34	5.14	6.0	90	78
A15456-H321	24	255	1.05	5.61	9.6	37	33

initial notch root diameter (except for a very small δ_T of about 0.1 ρ_o to 0.2 ρ_o). However Eqs. 56 and 57 must still be satisfied with $2\rho_o$ replacing δ_T in these equations. Since δ_{T2} must increase proportionately with δ_{T1} (or $2\rho_o$), the K_{Ic} (ρ_o) values would increase proportionately with $\sqrt{\rho_o}$ as long as the fracture mode remains unchanged. This behavior is consistent with observation (57).

MODELING OF TRANSITION REGION OF FRACTURE TOUGHNESS

Cleavage is the dominant mode of failure in steels in the transition region, although a frequent prominant feature on the fracture surface is a zone of dimpled rupture located between the stretched zone and the cleavage zone. Application of the cleavage models described previously to calculate the fracture toughness underestimates the actual values (39). On the otherhand, the model for microvoid coalescence of Bates (56) seems to accurately predict fracture toughness in the transition region; inasmuch as two of the steels shown in Table IV were actually in the transition region at the indicated test temperatures; i.e., 403 stainless steel at room temperature and A533B at -18°C. In this section, the reason for this behavior is sought, although only a qualitative explanation can be offered at the present time.

As the criterion for fracture by dimpled rupture is satisfied, i.e., coalescence of two or more voids simultaneously with the blunting crack tip, an instability is created that may cause abrupt changes in the stress state allowing the criterion for cleavage to be satisfied if the strain hardened yield strength is not too low. In that event, cleavage fracture would follow initiating at and propagating from the position of the peak stress (58). Evidence to support this concept would be provided if the position of the onset of cleavage fracture could be shown to correspond with the position of the peak stress. Several measurements of dimple zones widths were available from a previous unpublished investigation by Bates and are shown in Table V. These measurements were made on the fractures of standard fracture toughness specimens of NiMoV alloy steel, A533B alloy steel, and 403 stainless steel at various temperatures within the transition regions of the respective materials. Tensile and fracture toughness properties of these materials are also listed in Table V.

σ_{Tc} corresponding with K_{Ic} value (or K_Q for those specimens that did not meet ASTM size requirements for valid K_{Ic}) were calculated for each test point. The values of the ratio of R/δ_T at which the peak value of σ_{yy} occurs were taken from Fig. 20 for appropriate values of n. This figure was developed by cross plotting data in Fig. 3. σ_T was calculated with Eq. 11

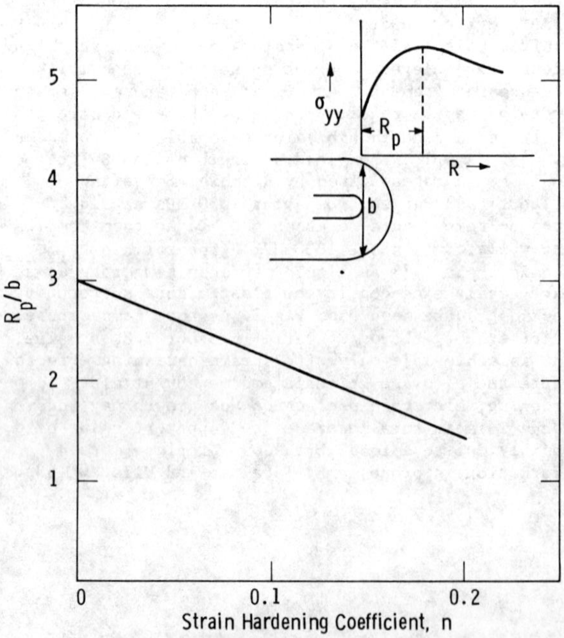

Figure 20. Position of Peak Value of σ_{yy} Ahead of a Semicircular Notch Tip After Blunting to δ_F of Five Times the Original Notch Width.

for each K_{Ic} value and in turn R_p, the value of R at peak stress was calculated for each test point giving the theoretical fracture zone (stretched zone plus dimple zone) widths which are compared with the measured values in Table V. Clearly the agreement is good considering the scatter normally found in such measurements (53,54).

LOADING AND METALLURGICAL FACTORS AFFECTING FRACTURE TUGHNESS AND DUCTILITY

Cleavage Fractures

Three dominant factors have been proposed in setting a cleavage failure criterion. (1) The magnitude of the maximum principal stress. (2) The critical stress for failure. (3) A critical distance or length. Thus any parameter, whether loading or metallurgical, that affects one of these three factors will have an effect on the development of cleavage fractures.

The maximum principal stress is in turn affected by several variables, the most significant of which are applied load, hydrostatic stress, yield strength and strain hardening. The effect of the applied load is obvious in uniaxially loaded specimens up to the yield strength, or if strain hardening is considered, even beyond the yield strength. Under such loading cases, once σ_F is exceeded, the R* requirement is automatically satisfied and unstable fracture will occur. However, a competing mechanism will likely intercede, at least in part, in low strength materials. A superimposed hydrostatic stress (favored by loading of a notched or precracked specimen of sufficient dimensions to maintain plane strain) can significantly increase the maximum principal strain even without strain hardening effects. Thus notched or precracked specimens will favor cleavage fracture and exhibit higher transition temperatures than plane or unnotched samples. In addition the level of applied loading of precracked samples will affect σ_{yy} up to some level of σ_T of the order of two to three times the notch root dimension. Further increases in δ_T will affect only R_p, the distance from the notch/crack tip at which the peak value of σ_{yy} is achieved.

The yield strength has an effect on the stress level because the stress level achieved is directly scaled up in proportion to σ_o for any given level of δ_T. The strain hardening coefficient has a direct

Table V

Comparison of Calculated and Measured Fracture Zone (Stretched Zone Plus Dimple Zone) Widths

Alloy	σ_o at R.T. MPa	Test Temp., °C	σ_o at Test Temp., MPa	n	K_{Ic} MPa$\sqrt{m}$	δ_{Tc} μ-m	Width of Fracture Zone, μm Measured	Calculated
NiMoV Alloy Steel	586	-73	676	0.11	46	6	18	13
		-18	593	0.12	86*	22	50	44
NiMoV Alloy Steel	641	-18	653	0.11	58	9	19	18
		24	641	0.11	126	44	90	92
A533B Alloy Steel	503	-101	703	0.10	50	6	13	13
		-73	627	0.12	76	16	34	32
		-73	627	0.12	73	15	23	30
		-46	579	0.13	104*	33	57	64
		-32	552	0.13	94	28	50	55
Type 403 Stainless	669	-129	821	0.10	43	4	11	9

*Did not satisfy ASTM requirements for valid K_{Ic}

effect on σ_{yy} in the plastic zone as shown in Figs. 3-5. The temperature will have an effect on σ_{yy} indirectly through the yield strength and strain hardening coefficient. Typically, lower temperature favors higher values of σ_o but lower values of n (43), but the effect on σ_o is greater.

The critical failure stress, σ_F, for slip induced cleavage is not dependent upon temperature or hydrostatic stress (17,35). Reports of root radius effects on σ_F (36,37) are probably due to R* effects and the lack of adequate descriptions of stress distributions available at the time. On the other hand σ_F is highly dependent on metallurgical variables. First, of course, is the effect on cleavage crack nucleation. Although in most commercial materials, σ_F is characteristic of a propagation step, that is because nucleation is relatively easy due to microstructural inhomogenities. Remove non-coherent particles from the material, especially grain boundary particles, and nucleation would possibly become the controlling event (23). It is well established that pearlite and large grain boundary carbides affect σ_F (59). This effect is probably due to easing the nucleation event rather than affecting propagation of cleavage cracks. Also the length of microcracks that develop during nucleation may affect their ability to propagate into the matrix, and that crack length will be limited by the size of the particle in which it occurs (35). Secondly, metallurgical factors control the value of stress required for the crack propagation events. The principal factors are grain size; amount, distribution and character of second phases; and the character of the matrix.

Grain size affects σ_F in much the same manner as it affects σ_o, at least for coarse grained materials. For example, Low (60) reported that a .03% C iron exhibited the same brittle fracture stress and yield stress at -195°C; both varied with $d^{-1/2}$ where d is the grain size. Similar effects were reported by Rosenfield, Hahn and Embury (59). This is generally true, however, only at cryogenic temperatures since σ_o decreases at higher temperatures while σ_F does not. The expected effect of grain size on the lower shelf toughness caused by changes in σ_F would be minimal, however, because the stress easily exceeds σ_F and the dominant effect will be on R*. In the transition temperature region, the spread between σ_F and σ_o increases because of the temperature effects on σ_o. In this region, the expected effects of the grain size on cleavage fractures would be greater than on the lower shelf, a prediction that is in agreement with observed behavior (61).

The factors that control R* are somewhat more difficult to identify. If grain size were the only factor, then all materials should exhibit the same requirement that R* be equal to a specific number of grain diameters. As previously discussed, such is apparently not the case. Furthermore, if grain size were the only factor controlling R*, its effect would be sufficient to overcome any effects on σ_F and the peak value of σ_{yy}, so coarse grained materials would exhibit higher lower shelf toughness than fine grained materials. Although such cases have been reported (57), this is not generally observed (61).

Microvoid Coalescence

The controlling factor of upper-shelf fracture toughness (where the fracture mode is microvoid coalescence) is the nucleation event. It is this event that controls both the initial void size and the spacing between voids. In turn it is the void spacing that governs D* (the final dimple size after coalescence), and the fracture toughness is established by the ratio (D^*/D_o) as can be seen from Eqs. 50-52. Little quantitative data pertaining to the nucleation event are available, although a considerable body of qualitative effects have been reported from which some general observations can be made.

First, it is of interest to review the data of Argon and Im (46) shown in Table I. The critical stresses for void nucleation at Fe_3C, TiC and Cu-Cr precipatates are quite high compared to the yield strengths of the respective host alloys. It is doubtful if these particles would nucleate voids except after considerable plastically induced hardening unless significant hydrostatic stresses were superimposed, say from plane strain loading of precracked or sharply notched specimens. Thus in plane, uniaxially loaded specimens another competing mode of fracture would likely dominate. For example, if easier nucleation sites (such as manganese-sulfide or large oxide particles in steel), were available, voids growing from these sites would control the fracture process and the smaller particles (more difficult nucleation sites) would contribute only in the final stages of fracture. On the other hand, in plane strain, precracked specimens, the stresses in the notch tip region can easily exceed the critical stresses reported by Argon and Im. Furthermore, on the average the finely spaced precipitates would nucleate voids earlier than the larger, more distantly spaced inclusions. Thus, in this case the larger inclusions, while contributing to fracture would not likely dominate the event. The important conclusion that can be reached from this comparison is that the fracture ductility exhibited by a material will be a strong function of both the superimposed hydrostatic stresses and the size scale over which the stresses are operative. The strain gradient in a crack tip region will be particularly important. Thus, there need not necessarily be a direct correlation between fracture strain in a sharply cracked, plane strain fracture toughness test and a plane bar or notched bar tensile test.

The factors, other than the applied loading and hydrostatic stresses, that affect σ_{yy} are the same as that discussed under cleavage fracture. σ_o and n are major factors, both of which are affected by grain size, propensity for cross slip, and of course hardening mechanisms. In general the number of voids that nucleate will be increased (consequently D* will decrease) as σ_o and n are increased thereby decreasing fracture toughness.

Specific microstructural affects on upper shelf (or microvoid coalescence) fracture toughness will now be addressed. Many of these effects have been reviewed by Schwalbe (61), and several of his observations and conclusions will be cited with comments of explanation based on the model expressed mathematically in Eqs. (55-57) and the formulations leading to that model.

1. Increasing the inclusion density decreases δ_T at the first void coalescence with the crack tip, and assuming this event bears a relation to the onset of instability the effect should be an increase in K_{Ic} with increasing λ (61). This effect is specifically addressed in the model wherein it is shown that the δ_{Tc} (hence K_{Ic}) is a direct function of D^*/D_o which is a direct function of λ/ρ (ratio of particle spacing to particle diameter) which in turn is an inverse function of inclusion density. Thus, K_{Ic} varies inversely with inclusion density or directly with λ, and δ_T at the time of the first void coalescence with the crack tip is, of course, directly related to λ although this event is not necessarily the event that is associated with instability. Rather instability, as shown previously, may require several voids coalescing with the crack tip depending on the specific λ/ρ (or D^*/D_o) ratio. This prediction is in concurrance with observation that one to several dimples may coalesce before the onset of instability (39).

2. An embrittling effect of sulfur is due to the formation of dimples at sulfides (61). Although the model specifically deals only with an average D^*/D_o, a bimodal distribution could equally well apply. The effect would essentially be a superposition effect such that the addition of another set of inclusion sites would decrease δ_{Tc}.

3. The type, size and distribution as well as volume fraction of inclusion is of importance (61). This comment by Schwalbe was illustrated with an example of tensile fractures in aluminum alloys. In tensile tests, it is much more difficult to achieve particle interfacial stresses that exceed the critical value for void nucleation than in plane strain fracture toughness tests, but the following comment applies in principle to both. Since σ_{crit} varies with size, type and distribution (44-46), the number of voids nucleated will vary accordingly with corresponding changes in D^*/D_o. Hence the ductility and fracture toughness will vary directly with the ease of nucleation.

4. Dimple size correlates with particle spacing in some instances while in other instances dimple size considerably exceeds particle spacing (61). This effect is specifically explained by Argon and his coworkers (44-46), and these results were directly incorporated into the current model through D^*/D_o.

5. Orientation (transverse vs. longitudinal direction) affects fracture toughness because of alignment of inclusions (61). Again the effect is explained by the effective λ/ρ and its corresponding effect on D^*/D_o.

6. In steels, an increase of σ_o, whether by lowering the test temperature, increasing the deformation rate or by heat treatment (providing no microstructural changes occur which contribute significantly to crack propagation) decreases K_{Ic} (61). The obvious effect of increasing σ_o is to increase σ_{yy} in the plastic zone since the two scale proportionately (see Figs. 2-5, where σ_{yy} is normalized in each case by σ_o). Since nucleation of voids is a critical stress event, a higher value of σ_o (for any given microstructure) leads directly to higher σ_{yy} and an increase in the void density, hence a lower δ_{Tc} or K_{Ic}.

7. Increasing strength by tempering may affect fracture toughness by altering microstructure - a fine, uniform distribution is preferred and a particle size that exceeds about 200 Å (depending on temperature) is required for void nucleation (61). Aside from the effect on σ_{yy} through σ_o, the interacting particle effect described by Argon, Im and Safoglu (45) set a dimension of about 100 Å as a critical diameter below which the energy requirement may not be satisfied even if the stress reaches σ_{crit}; thus void nucleation at these small particles is possible only in a gradual and stable manner. Furthermore, this model for interacting inclusions leading to Eq. (41), depends on λ/ρ which in turn depends on variation of the volume fraction on a local scale. Thus small, evenly distributed inclusions are preferred. Large or unevenly distributed inclusions should be avoided.

8. Void nucleation at hard particles is stress dependent while void growth (coalescence) is strain dependent (61). These observations are, in part, the basis of the upper shelf (microvoid coalescence) fracture toughness model.

9. The presence of a dual microstructure in the matrix; e.g., retained austenite in martensite, may or may not be beneficial to fracture toughness (61). The effects of all mixed microstructures are too numerous to elaborate except for a few generalization. One of the major effects may be the suppression of cleavage fracture by raising σ_F for cleavage crack propagation due to higher energy requirements in one of the phases or increasing the cleavage crack nucleation energy through grain size refinement or reduction of the number of grain boundary particles. In addition a mixture of phases can affect the yield strength and the size and distribution of precipitates and inclusions, so all of the effects previously noted can occur.

10. Attempts to correlate fracture toughness with other material properties; e.g., yield strength, fracture strain, etc., will not be successful unless these other properties are influenced by test conditions and microstructure in the same way as fracture toughness is influenced (61). This same conclusion was reached earlier in this section; i.e., fracture ductility will be a strong function of the superimposed hydrostatic stress and the size scale over which the stresses act; thus there need not necessarily be an exact correspondence of fracture toughness and tensile properties or of dimple sizes in different types of test specimens.

The above, of course, is only a sampling of metallurgical effects, but this sampling demonstrates many of the most important effects that have been observed. They all seem consistent with the model presented for fracture toughness when fracture is caused by microvoid coalescence.

CONCLUDING REMARKS

Fracture in engineering alloys is a heterogeneous process. Understanding fracture requires an appropriate description of the often complicated spatial stress and strain distributions, particularly in the plastic zones around notches and cracks. It further requires a definition of the influence of microstructural heterogeneities on those stress and strain distributions. This paper provides those distributions. The development of microstructurally sensitive fracture models requires identification of local fracture criteria, and such criteria are proposed in this paper. Extensions of these criteria result in models for lower shelf (cleavage), upper shelf (microvoid coalescence) and transition region fracture toughness. Loading and microstructural effects are explained in light of these models. Considerable additional effort is required in the following areas. The current means of selecting a critical distance in the model for lower shelf (cleavage) fracture toughness is arbitrary, and factors controlling this parameter need to be identified. The modeling of microvoid coalescence is strongly dependent on initial and final (coalesced) void sizes. Factors affecting these void size parameters are known in part, but little quantitative data exist, and the data base should be significantly expanded. The effects of dual populations of inclusions should be quantitative explored. Furthermore, the specific influence of nearby surfaces on void growth (i.e., the final step of coalescence of voids with each other and with the blunted crack tip) are not included in the model for microvoid coalescence except through the rough assumption of ligament instability. Finally, the effects of growing voids on the stress distribution should be specifically defined to better understand the role of cleavage in terminating the fracture process in the transition region between the lower and upper shelves.

REFERENCES

1. A. A. Griffith, "The Phenomena of Rupture and Flow in Solids," Philosophical Transitions, Royal Society London, Vol. 221A (1920), p. 163-198.

2. C. Zener, "The Micro-Mechanisms of Fracture," Fracturing of Metals, ASM, Cleveland, Ohio, 1948, p. 3-31.

3. C. E. Inglis, "Stress in a Plate Due to the Presence of Cracks and Sharp Corners," Transactions, Inst. Naval Architects, Vol. 60 (1913), p. 219-230.

4. E. Crowan, "Energy Criteria of Fracture," Welding Research Supplement, Vol. 20 (1955), p. 157s-160s.

5. G. R. Irwin, "Fracture Dynamics," Fracturing of Metals, ASM, Cleveland, Ohio, 1948, p. 147-166.

6. G. R. Irwin, "Analysis of Stresses and Strains Near the End of a Crack Traversing a Plate," Transactions ASME, Journal of Applied Mechanics, Vol. 24, (1957), p. 361-364.

7. P. C. Paris and G. C. Sih, "Stress Analysis of Cracks," Fracture Toughness and Its Applications, ASTM STP 381, ASTM, Philadelphia, PA, 1964, p. 31-81.

8. A. S. Tetelman and A. J. McEvily, Jr., Fracture of Structural Materials, John Wiley and Sons, Inc., New York, NY, 1967, p. 49-50.

9. H. Tada, P. C. Paris and G. R. Irwin, The Stress Analysis of Cracks Handbook, Del Research Corp., Hellertown, PA, 1973.

10. J. R. Rice, "A Path Independent Integral and the Approximate Analysis of Strain Concentration by Notches and Cracks," Transactions ASME, Journal of Applied Mechanics, Vol. 35 (1968), 79-86.

11. P. C. Paris, "Fracture Mechanics in the Elastic Plastic Regime," Flaw Growth and Fracture, ASTM STP 681, ASTM, Philadelphia, PA, 1976, p. 3-27.

12. R. M. McMeeking, "Finite Deformation Analysis of Crack-Tip Opening in Elastic-Plastic Materials and Implications for Fracture," Journal of Mechanics and Physics of Solids, Vol. 25 (1977), p. 357-381.

13. J. R. Rice and M. A. Johnson, "The Role of Large Crack Tip Geometry Changes in Plane Strain Fracture," Inelastic Behavior of Solids, Ed. M. F. Kannien, W. F. Adler, A. R. Rosenfield and R. I. Jaffee, McGraw-Hill Book Co., New York, NY, 1970, p. 641-672.

14. R. Hill, The Mathematical Theory of Plasticity, The Oxford Engineering Science Series, Oxford at the Clarendon Press, London, 1956, p. 245-252.

15. A. S. Tetelman and A. J. McEviley, Jr., Fracture of Structural Materials, John Wiley & Sons, Inc., New York, NY, 1967, p. 306-307.

16. R. C. Bates and A. T. Santhanam,"Notch Tip Stress Distribution in Strain Hardening Materials," International Journal of Fracture, Vol. 14 (1978), p. 501-508.

17. J. R. Griffiths and D. R. J. Owen, "An Elastic-Plastic Stress Analysis for a Notched Bar in Plane Strain Bending," Journal of Mechanics and Physics of Solids, Vol. 19 (1971), p. 419-431.

18. R. C. Bates, Manuscript in preparation for publication.

19. M. Morris and O. E. Brown, Differential Equations, Revised Edition, Twelfth Printing, Prentice-Hall Inc., New York, NY, 1949, p. 55-68.

20. C. D. Beachem and M. N. Pelloux, "Electron Fractography - A Tool for the Study of Micromechanisms of Fracturing Processes," Fracture Toughness Testing and Its Applications, ASTM STP 381, ASTM, Philadelphia, PA, 1964, p. 210-245.

21. C. D. Beachem, "The Effects of Crack Tip Plastic Flow Directions Upon Microscopic Dimple Shapes," Metallurgical Transactions, Vol. 6A (1975), p. 377-383.

22. Fractography and Atlas of Fractographs, Metals Handbook, Vol. 9, ASM, Metals Park, OH, 1974.

23. A. S. Tetelman and A. J. McEviley, Jr., Fracture of Structural Materials, John Wiley & Sons, Inc., New York, NY, 1967, p. 259-284.

24. C. Zener, "A Theoretical Criterion for the Initiation of Slip Bands," Physical Review, Second Series, Vol. 69 (1946), p. 128-129.

25. A. N. Stroh, "A Theory of the Fracturing of Metals," Advances in Physics, Vol. 6 (1957), p. 418-465.

26. A. N. Stroh, "Crack Nucleation in Body Centered Cubic Metals," Fracture, Ed. B. L. Averbach, D. K. Felbeck, G. T. Hahn, D. A. Thoman, John Wiley and Sons, Inc., New York, NY, 1959, p. 117-122.

27. A. H. Cottrell, "Theoretical Aspects of Fracture," Fracture, Ed., B. L. Averbach, D. K. Felbeck, G. T. Hahn, D. A. Thoman, John Wiley and Sons, Inc., New York, NY, 1959, p. 20-44.

28. A. H. Cottrell, "Theory of Brittle Fracture in Steel and Similar Metals," Transactions, Metallurgical Society of AIME, Vol. 212 (1958), p. 192-203.

29. E. Smith, "The Formation of a Cleavage Crack in Crystalline Solids - I," Acta Metallurgica, Vol. 14 (1966), p. 985-989.

30. E. Smith, "The Formation of a Cleavage Crack in Crystalline Solids, II," Acta Metallurgica, Vol. 14 (1966), p. 991-996.

31. E. Smith, "Cleavage Fracture in Mild Steel," The International Journal of Fracture Mechanics, Vol. 4 (1968), p. 131-145.

32. G. T. Hahn, B. L. Averbach, W. S. Owen and M. Cohen, "Initiation of Cleavage Microcracks in Polycrystalline Iron and Steel," Fracture, Ed. B. L. Averbach, D. K. Felbeck, G. T. Hahn, D. A. Thoman, John Wiley and Sons, Inc., New York, NY, 1959, p. 91-116.

33. J. R. Low, Jr., "Microstructural Aspects of Fracture," Fracture of Solids, Ed. D. C. Drucker and J. J. Gilman, Interscience Publishers, John Wiley and Sons, New York, NY, 1962, p. 197-133.

34. D. Hull, "Twinning and the Nucleation of Cracks in Body-Centered Cubic Metals," Fracture of Solids, Ed. D. C. Drucker and J. J. Gilman, Interscience Publishers, John Wiley and Sons, New York, NY, 1962, p. 417-460.

35. G. Oates, "Effect of Hydrostatic Stress on Cleavage Fracture in a Mild Steel and a Low Carbon Manganese Steel," Journal of the Iron and Steel Institute, Vol. 206 (1968), p. 930-935.

36. A. S. Tetelman, T. R. Wilshaw, and C. A. Rau, Jr., "The Critical Tensile Stress Criterion for Cleavage," International Journal of Fracture Mechanics, Vol. 4 (1968), p. 147-157.

37. T. R. Wilshaw, C. A. Rau, A. S. Tetelman, "A General Model to Predict the Elastic Plastic Stress Distribution and Fracture Strength of Notched Bars in Plane Strain Bending," Engineering Fracture Mechanics, Vol. 1 (1968), p. 191-211.

38. R. O. Ritchie, J. F. Knott and J. R. Rice, "On the Relationship Between Tensile Stress and Fracture Toughness in Mild Steel," Journal of Mechanics and Physics of Solids, Vol. 21 (1973), p. 395-410.

39. R. O. Ritchie, W. L. Server and R. A. Wullaert, "Critical Fracture Stress and Fracture Strain Models for the Prediction of Lower and Upper Shelf Toughness in Nuclear Pressure Steels," Metallurgical Transactions, American Society for Metals and the Metallurgical Society of AIME, Vol. 10A (1979), p. 1557-1570.

40. H. C. Rogers, "The Effect of Material Variables on Ductility," Ductility, ASM Seminar, October 1967, ASM, Metals Park, OH, 1967, p. 31-61.

41. K. E. Puttick, "Ductile Fracture in Metals," Philosophical Magazine, Vol. 4, Eighth Series (1959), p. 964-969.

42. L. M. Brown and J. D. Embury, "The Initiation and Growth of Voids at Second Phase Particles," Microstructure and Design of Alloys, Third International Conference on Strength of Metals and Alloys, 1973, Vol. 1, Cambridge, England, sponsored by the Institute of Metals and the Iron and Steel Institute, p. 164-169.

43. B. I. Edelson and W. J. Baldwin, Jr., "The Effect of Second Phases on the Mechanical Properties of Alloys," Transactions, American Society for Metals, Vol. 55 (1962), p. 230-250.

44. A. S. Argon, J. Im and A. Needleman, "Distribution of Plastic Strain and Negative Pressure in Necked Steel and Copper Bars," Metallurgical Transactions, American Society for Metals and the Metallurgical Society of AIME, Vol. 6A (1975), p. 815-824.

45. A. S. Argon, J. Im and R. Safoglu, "Cavity Formation From Inclusions in Ductile Fracture," Metallurgical Transactions, American Society for Metals and the Metallurgical Society of AIME, Vol. 6A (1975), p. 825-837.

46. A. S. Argon and J. Im, "Separation of Second Phase Particles in Spherodized 1045 Steel, Cu-0.6 Pct Cr Alloy, and Maraging Steel in Plastic Straining," Metallurgical Transactions, American Society for Metals and the Metallurgical Society of AIME, Vol. 6A (1975) p. 839-851.

47. J. Gurland and J. Plateau, "The Mechanism of Ductile Rupture of Metals Containing Inclusions," Transactions, American Society for Metals, Vol. 56 (1963), p. 442-454.

48. L. M. Brown and W. M. Stobbs, "The Work Hardening of Copper Silica II. The Role of Plastic Relaxation," Philosophical Magazine, Vol. 23 (1971), p. 1201-1235.

49. F. A. McClintock, "On the Mechanics of Fracture From Inclusions," Ductility, ASM Seminar, October 1967, ASM, Metals Park, OH, 1967, p. 255-277.

50. F. A. McClintock, "Local Criteria for Ductile Fracture," International Journal of Fracture Mechanics, Vol. 4 (1968), p. 101-130.

51. J. R. Rice and D. M. Tracy, "On the Ductile Enlargement of Voids in Triaxial Stress Fields," Journal Mechanics and Physics of Solids, Vol. 217 (1969), p. 201-217.

52. R. M. McMeeking, "Blunting of a Plane Strain Crack Tip Into a Shape with Vertices," Transactions ASME, Journal Engineering and Technology, Vol. 99 (1977), p. 290-297.

53. R. C. Bates, W. G. Clark, Jr. and D. M. Moon, "Correlation of Fractographic Features with Fracture Mechanics Data," Electron Microfractography, ASTM STP 453, ASTM, Philadelphia, PA, 1969, p. 192-214.

54. R. C. Bates and W. G. Clark, Jr., "Fractography and Fracture Mechanics," Transactions, American Society for Metals, Vol, 62 (1969), p. 380-389.

55. R. C. Bates, "Correlations of Microvoid Formation, Fatigue Crack Growth Rates and Fracture Toughness," Mechanical Behavior of Materials, Kyoto Conference, MBM Symposium Committee, Society of Materials Science, Japan, 1974, Paper C(1).

56. R. C. Bates. Manuscript in Preparation for Publication.

57. R. O. Ritchie, B. Francis and W. L. Server, "Evaluation of Toughness in AISI 4340 Alloy Steel Austenitized at Low and High Temperatures," Metallurgical Transactions, American Society for Metals and the Metallurgical Society of AIME, Vol. 7A (1976), p. 831-838.

58. A. S. Tetelman and T. R. Wilshaw, "A Criterion for Plastically Induced Cleavage After General Yield in Notched Bend Specimens," p. 219-228, Fracture 1969. Proceedings of the Second International Conference on Fracture, Ed., P. L. Pratt, Chapman and Hall, London, England.

59. A. A. Rosenfield, G. T. Hahn and J. D. Embury, "Fracturing of Steels Containing Pearlite," Metallurgical Transactions, American Society for Metals and the Metallurgical Society of AIME, Vol. 3 (1972), p. 2797-2804.

60. J. R. Low, Jr., "Dislocations and Brittle Fracture," Deformation and Flow of Solids, Madrid Colloquium on Deformation and Flow of Solids. IUTAM, Springer-Verlag, Berlin, 1956, p. 60-72.

61. Karl-Heinz Schawlbe, "On the Influence of Microstructure on Crack Propagation Mechanisms and Fracture Toughness of Metallic Materials," Engineering Fracture Mechanics, Vol. 9 (1977), p. 795-832.

GRAIN BOUNDARY EMBRITTLEMENT IN STEELS

D.Y. Lee and H.L. Marcus
Mechanical Engineering/
Materials Science and Eng.
The University of Texas
Austin, Texas 78712

INTRODUCTION

Almost all engineering applications of steels involve their use in polycrystalline form. In most cases, grain boundaries and interfaces in general of polycrystalline steels are known to be intrinsically strong. However, various types of grain boundary embrittlement have been recognized to occur in steels in a wide range of metallurgical circumstances. The well-known examples of grain boundary embrittlement are temper embrittlement (1-7), creep embrittlement (8-10), stress relief cracking (11-13), intergranular hydrogen (14-16) and stress corrosion cracking (17-20) and intergranular liquid metal induced embrittlement (21-23). These metallurgical problems of grain boundary and interface embrittlement often lead to severe deterioration of important mechanical properties of steels and sometimes cause catastrophic service failures. The grain boundary and interface problems give rise to high costs and efficiency penalties in design and non-destructive evaluation in service to avoid catastrophic failures. Furthermore, the problems seem to be getting more serious with the increased demands for better mechanical properties of steels for the development of new technologies and with the increased requirements for higher efficiency in power generating plants to overcome soaring fuel costs. Therefore, the investigation of the grain boundary and interface embrittlement problems in steels is of great importance. The grain boundary and interface embrittlement phenomenon is not uniquely a property of steels. It is observed to some degree in virtually all metallic systems (i.e., Ni, Co, Cu, Mo, W ...) and ceramic systems (i.e., Al_2O_3, MgO, Si_3N_4 ...). This chapter will emphasize the considerations in steel, but most of the concepts presented are directly applicable to other systems.

For many years a considerable amount of research effort has gone into investigation of these problems. What has been demonstrated is that the grain boundary embrittlement phenomenon is mostly associated with the segregation of certain impurities and alloying elements at grain boundaries. Thus, understanding grain boundary segregation and its effects on grain boundary embrittlement is essential to understand and control the problems of grain boundary embrittlement. In recent years, with the development of direct surface analytical techniques, especially Auger electron spectroscopy (AES) (24,25), the investigation of the segregation associated with grain boundary embrittlement has been directly measurable. Identification and quantification of segregated elements has been provided by using in situ fracture techniques (26-28) which produces fracture surfaces in an ultra-high vacuum Auger spectrometer. The intensive application of AES to grain boundary embrittlement studies with its direct measurement of the segregated species has also stimulated theoretical advances in this field in attempts to model both the segregation phenomenon and the fracture process.

The purpose of this paper is to review the current state of knowledge of grain boundary embrittlement with a large emphasis on temper embrittlement in which the vast majority of recent investigations have concentrated on. A discussion of theoretical approaches in this area will be presented in terms of the thermodynamics of segregation, the kinetics of segregation and the influences of segregation on grain boundary embrittlement. This will be followed by a review of the experimental results in this area. The experimental results will be classified according to two main factors governing grain boundary embrittlement. First the effects of segregated elements, in some cases interacting with hydrogen, and secondly, the effects of mechanical-structural factors such as grain size, hardness, grain boundary morphology and microstructure will be described.

THE THEORY OF GRAIN BOUNDARY EMBRITTLEMENT

The Thermodynamics of Segregation

The main objective of studying the thermodynamics of grain boundary segregation is to understand the dependence of segregation on temperature and composition and to apply its understanding to controlling the effects of segregation on grain boundary embrittlement. Over the past decades several different approaches have been taken in terms of equilibrium thermodynamics of grain boundaries and the analogies to adsorption at free surfaces. These approaches will be reviewed and discussed. A brief summary of the key results of the thermodynamic analyses is presented in Table 1.

McLean (29) provided the first guidelines in the understanding of grain boundary segregation with the equilibrium thermodynamics of segregation. He derived his segregation equation with the help of statistical mechanics assuming that the grain boundary is a random ideal solid solution where no interactions exist between the segregated elements. The resulting McLean's equilibrium equation is expressed as:

$$\frac{X_I^\phi}{1-X_I^\phi} = \frac{X_I^B}{1-X_I^B} \exp\left(\frac{\Delta G_I^\circ}{RT}\right) \qquad (1)$$

where X_I^ϕ is the equilibrium concentration of segregated impurity I at the grain boundary ϕ, X_I^B is the concentration of I in the bulk B, and ΔG_I° is the segregation free energy of I. When X_I^B is very small, the commonly used dilute approximation of Eq. (1) is written as:

$$\frac{X_I^\phi}{1-X_I^\phi} = X_I^B \exp\left(\frac{\Delta G_I^\circ}{RT}\right) \qquad (2)$$

The above expression, Eq.(2) has the identical form as Langmuir adsorption isotherm (30) which is derived for

Table I. Thermodynamic Models of Grain Boundary Segregation

Model	Ref	Assumption	Description
McLean Model	29	* Binary Solid Solution * No interaction between segregated neighbors * Fixed number of identical segregation sites	* $\frac{X_I^\phi}{1-X_I^\phi} = \frac{X_I^B}{1-X_I^B} \exp\left(\frac{\Delta G_I^\circ}{RT}\right)$ $\Delta G_I^\circ \neq f(X_I^\phi)$ * Approximation ($X_I^B \ll 1$) $\frac{X_I^\phi}{1-X_I^\phi} \approx X_I^B \exp\left(\frac{\Delta G_I^\circ}{RT}\right)$
Fowler Model	31	* Binary Solid Solution * Interaction between nearest segregated neighbors * Fixed number of identical segregation sites	* $\frac{X_I^\phi}{1-X_I^\phi} = \frac{X_I^B}{1-X_I^B} \exp\left(\frac{\Delta G_I^\circ - zwX_I^\phi}{RT}\right)$ * Approximation ($X_I \ll 1$ or $w \to 0$) $\frac{X_I^\phi}{1-X_I^\phi} \approx X_I^B \exp\left(\frac{\Delta G_I^\circ}{RT}\right)$; same as McLean's
BET Model	32-36	* Binary Solid Solution * No lateral interaction in the layer * Each layer does not have to be completed before the next is started	* $\frac{1}{X_I^\phi}\frac{X_I^B}{X_I^{B*}-X_I^B} = \frac{1}{K} + \frac{K-1}{K}\frac{X_I^B}{X_I^{B*}}$ $K=\exp\left(\frac{G_1-G_L}{RT}\right)$ G_1=first layer, G_L=subsequent layers * Approximation ($X_I^B \ll 1$) $\frac{X_I^\phi}{X_I^B} \approx \frac{1}{X_I^{B*}} \exp\left(\frac{G_1-G_L}{RT}\right)$
Guttmann Model	38-41	* Ternary Solid Solution * Interaction between nearest segregated neighbors * Two-dimensional grain boundary (i.e. layer model)	* $\frac{X_i^\phi}{1-X_I^\phi-X_M^\phi} = \frac{X_i^B}{1-X_I^B-X_M^B} \exp\left(\frac{\Delta G_i}{RT}\right)$ for i=I,M $\Delta G_I = \Delta G_I^\circ - 2\alpha_{FeI}(X_I^\phi-X_I^B) + \alpha'_{MI}(X_M^\phi-X_M^B)$ $\Delta G_M = \Delta G_M^\circ - 2\alpha_{FeM}(X_M^\phi-X_M^B) + \alpha'_{MI}(X_I^\phi-X_I^B)$ $\alpha'_{MI} = \alpha_{MI}-\alpha_{FeI}-\alpha_{FeM}$ * Approximation ($X_I^B, X_M^B \ll 1$, $X_I^B \ll X_I^\phi$, $X_M^B \ll X_M^\phi$, α_{FeI}, $\alpha_{FeM} \ll \alpha'_{MI}$) $\frac{X_i^\phi}{1-X_I^\phi-X_M^\phi} \approx X_i^B \exp\left(\frac{\Delta G_i}{RT}\right)$ for i=I,M $\Delta G_I = \Delta G_I^\circ + \alpha'_{MI}X_M^\phi$ $\Delta G_M = \Delta G_M^\circ + \alpha'_{MI}X_I^\phi$

the saturation type of gas adsorption on a free surface.

Even though the McLean's equation introduced a theoretical approach for the understanding of grain boundary segregation, it still left at least the following unresolved problems:

1) It does not involve the two-dimensional surface character which is an essential nature of grain boundaries.
2) It is difficult to evaluate the segregation free energy, ΔG_I°.
3) It cannot explain the chemical interactions between segregated elements.

Hence, except for the very early stages of segregation where the chemical interactions between segregated elements can be considered negligible due to lack of population, the McLean's equation and the Langmuir isotherm cannot be accurately applied in practice.

With the assumption of a regular solution model, Fowler and Guggenheim (31) derived a modified Langmuir isotherm which includes the energy of interaction between nearest neighbors. The grain boundary segregation analogue of this may be expressed in the form:

$$\frac{X_I^\phi}{1-X_I^\phi} = \frac{X_I^B}{1-X_I^B} \exp\left(\frac{\Delta G_I^o - zwX_I^\phi}{RT}\right) \qquad (3)$$

where ΔG_I^o is the segregation free energy of I in the absence of chemical interactions, z is the coordination number of segregated elements, and w is the energy of interaction between neighbors. This Fowler-Guggenheim isotherm is similar to the McLean's equation. The main difference in these is that

$$\Delta G_I = \Delta G_I^o - zwX_I^\phi$$

in the Fowler-Guggenheim isotherm depends on X_I^ϕ, while ΔG_I^o in the McLean's equation is independent of X_I^ϕ. When X_I^ϕ is close to zero, or w is zero, Eq.(3) becomes identical to Eq.(2). As discussed above, the Fowler-Guggenheim isotherm introduced the term describing the chemical interaction between nearest neighbors, but it is limited to the case of a binary solution. Furthermore, it cannot explain the possibility of multilayer segregation as reported (32) due to the constraint of a fixed number of segregation sites which is assumed for derivation of Eq.(3). In addition the value of z depends on the grain boundary model.

To explain this suspected multilayer grain boundary segregation, Seah and Hondros proposed (32) and evaluated (33-35) the grain boundary analogue of the theory presented by Brunaner, Emmett and Teller (36). This BET theory can be expressed in the form:

$$\frac{1}{X_I^\phi} \frac{X_I^B}{X_I^{B*} - X_I^B} = \frac{1}{K} + \frac{K-1}{K} \frac{X_I^B}{X_I^{B*}} \qquad (4)$$

where X_I^{B*} is the solid solubility, K is $\exp[(G_1 - G_L)/RT]$, and G_1 and G_L are the segregation free energy of the first layer and subsequent layers respectively at the grain boundary. The derivation of this BET theory is made from the original BET adsorption (36) by assuming that there are no lateral interactions in the layer and that each layer does not have to be completed before the next is started. Another critical assumption of the BET theory is that the same forces that produce grain boundary segregation (condensation) are responsible for multilayer segregation at concentrations below the bulk solubility limits. The constancy of these forces after the first layer in the boundary is highly questionable. It is likely that until a second phase is formed, the repulsive forces between the segregating atoms would greatly reduce $\Delta G_I = G_1 - G_L$ in systems where intermetallics would form.

In the case of dilute solutions, the BET theory approximates to:

$$\beta = \frac{X_I^\phi}{X_I^B} = \frac{1}{X_I^{B*}} \exp\left(\frac{G_1 - G_L}{RT}\right) \qquad (5)$$

where β is the enrichment ratio. Eq.(5) shows that the enrichment ratio β is proportional to the inverse of the solid solubility X_I^{B*}. This correlationship between β and X_I^{B*} is consistent with the experimental results (37). However, this BET theory as well as other theoretical approaches reviewed above has failed to rationalize the role of the alloying elements M in steels which experimental evidence has established are involved in the grain boundary embrittlement by enhancing the impurity segregation or by cosegregating with impurities. Thus, the thermodynamical formalism which takes into account the M-I interaction in steels is demanded to explain the above behaviors.

Recently, Guttmann proposed (38) and evaluated (4,5, 39-41) his segregation theory based on a regular ternary solution with the following result:

$$\frac{X_i^\phi}{1-X_I^\phi-X_M^\phi} = \frac{X_i^B}{1-X_I^B-X_M^B} \exp\left(\frac{\Delta G_i}{RT}\right) \qquad (6)$$

where i = I,M and

$$\Delta G_I = \Delta G_I^o - 2\alpha_{FeI}(X_I^\phi - X_I^B) + \alpha'_{MI}(X_M^\phi - X_M^B)$$

$$\Delta G_M = \Delta G_M^o - 2\alpha_{FeM}(X_M^\phi - X_M^B) + \alpha'_{MI}(X_I^\phi - X_I^B) \qquad (7)$$

α_{FeI} and α_{FeM} are the Fe-I and Fe-M interaction coefficients respectively, and α'_{MI} is the net M-I interaction parameter which is defined as $\alpha'_{MI} = \alpha_{MI} - \alpha_{FeI} - \alpha_{FeM}$. From the available thermodynamical data it is found that for the M-I couples involved in the grain boundary embrittlement problem in steels, the M-I interactions are much stronger than the Fe-I or Fe-M interactions. Also it is shown that X_I^ϕ and X_M^ϕ are much greater than X_I^B and X_M^B respectively. Thus, Eq.(7) could be simplified to:

$$\Delta G_I = \Delta G_I^o + \alpha'_{MI} X_M^\phi$$

$$\Delta G_M = \Delta G_M^o + \alpha'_{MI} X_I^\phi \qquad (8)$$

Now the cosegregation behavior of M-I as well as segregation enhancement between M and I can be explained by using Eq.(8); when α'_{MI} is positive (i.e. attractive interaction between M and I), an increase in X_M^ϕ leads to an increase in X_I^ϕ, and vice versa. Such behavior is observed in many M-I couples in alloy steels. The details of these examples will be presented in the next section. On the other hand, if α'_{MI} is negative (i.e. repulsive interaction between M and I), the strong segregant repels the other from the grain boundary. This behavior is found for the Si-C couple in Fe (7).

In addition to the competitive segregation to sites at the grain boundary, Guttmann has developed a regular solution model for metal and impurity atoms that segregate in a non-competitive manner. The sublattices then act independently but the M-I interactions still influence the segregation process.

The Guttmann model which introduces the interaction coefficients between M and I provides an important contribution to a better understanding of the complex segregation behavior encountered in ternary systems. However, this model still leaves the following problems.

1) It is based on only a nearest-neighbor approximation, its quantitative agreement with experimental results

cannot be anticipated.
2) It does not take into account the structural properties of grain boundaries which significantly influence the grain boundary segregation.
3) It depends on the thermodynamical data which are determined from the bulk properties and then applied to the grain boundary properties to estimate the interaction coefficients.

Therefore, the theory which involves a total description of not only chemical but also structural dependence of the grain boundary segregation phenomenon must be developed before a complete understanding of this subject can be attained.

The Kinetics of Segregation

Assuming that there exists local equilibrium between a boundary layer and the neighboring bulk material, McLean (29) derived, by diffusional analysis, the kinetics of grain boundary segregation for binary alloys which is expressed in the form:

$$X(t) = \frac{X_I^\phi(t)-X_I^\phi(o)}{X_I^\phi(\infty)-X_I^\phi(o)} = 1-\exp\left(\frac{4Dt}{\beta^2 d^2}\right)\operatorname{erfc}\left(\frac{4Dt}{\beta^2 d^2}\right)^{\frac{1}{2}} \quad (9)$$

where $X_I^\phi(o)$, $X_I^\phi(t)$ and $X_I^\phi(\infty)$ are the grain boundary concentration of I at the start, at time t and at infinite time of an isothermal heating (i.e. the equilibrium grain boundary concentration), respectively and X(t) is the fractional change in segregation. D is the bulk diffusion coefficient of I, β is the equilibrium enrichment ratio of I (i.e. here, $\beta = X_I^\phi(\infty)/X_I^B$) and d is the grain boundary thickness occupied by I. Hence, βd becomes the ratio of the number of I atoms per unit area of grain boundary to the number of I atoms per unit volume of the bulk. At the beginning of segregation, Eq.(9) can be approximated by:

$$X(t) = \frac{4}{\beta d}\left(\frac{Dt}{\pi}\right)^{\frac{1}{2}} \quad (10)$$

Using Eq.(10) D can be determined from the initial slopes of the experimental segregation curves. In general, McLean's kinetic equation satisfactorily describes the kinetics of impurity segregation in binary systems.

Recently, combining McLean's kinetic equation with a slightly modified version of Guttmann's theory (38-41) of equilibrium grain boundary segregation of impurities for ternary systems, Seah (42) generated a complete set of Temperature-time C-curve kinetics for the P segregation in the 3140 steel studied by Carr et al (43) in their evaluation of the embrittlement T-t diagram (Fig. 1).

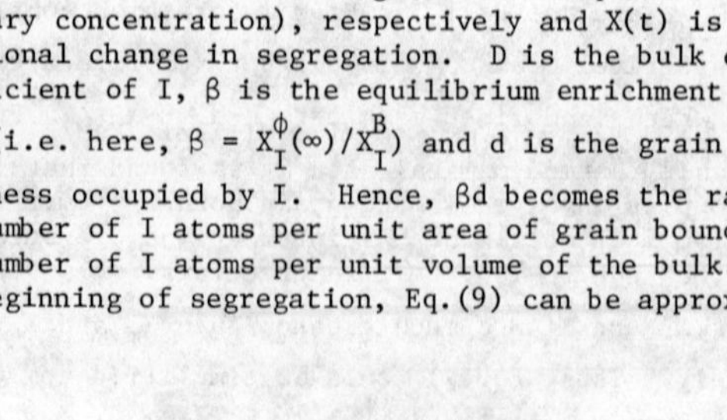

Figure 1. Time-Temperature diagram for P segregation (solid lines) (42) and for embrittlement (broken lines) (43) of SAE 3140 steel.

It is noted that there is good correlation between these two T-t diagrams over the full range of temperatures and times. The T-t C-curve kinetics of segregation enables one to extrapolate the extent of grain boundary segregation at lower temperatures. The embrittlement nose in the T-t diagram shifts to lower temperatures and longer times as the purity level increases. Hence, if the change in the ductile-brittle transition temperature (DBTT) as a function of the amount of segregated element is known, the T-t kinetics of embrittlement can be predicted from this segregation kinetics.

Tyson (44) using computer simulation, calculated segregation kinetics in binary and ternary systems according to the following procedures:

1) Establishment of local equilibrium between X_i^ϕ and X_i^B (X=0) for the segregated element i=I,M.
2) Determination of the segregant flux into the grain boundary using Fick's first law

$$J_i = -D_i\left(\frac{\partial X_i}{\partial x}\right)_{x=0}$$

and of the changes in bulk concentration due to the above diffusion using Fick's second law

$$\frac{\partial X_i}{\partial t} = D\frac{\partial^2 X_i}{\partial x^2}$$

3) Re-establishment of local equilibrium between the new values of X_i^ϕ and X_i^B (X=0); X_i^B (X=0) determined using the slightly modified version of Guttmann's segregation equation (Eq. 6 and 8).

$$\frac{X_i^\phi}{1-X_i^\phi} = X_i^B \exp\left(\frac{\Delta G_i}{RT}\right) \text{ for } i=I,M$$

$$\Delta G_I = \Delta G_I^\circ + \alpha'_{MI} X_M^\phi$$

$$\Delta G_M = \Delta G_M^\circ + \alpha'_{MI} X_I^\phi \quad (11)$$

4) Repetition of the above procedures 1), 2) and 3) until equilibrium is approached.

The kinetics of segregation obtained from this calculation shows that segregation at the grain boundary of typical alloy systems occurs much more rapidly than predicted by the McLean's kinetic equation (Eq.9), even though the initial slopes are similar in

$$X(t) = \left(\frac{X_i^\phi(t)-X_i^\phi(o)}{X_i^\phi(\infty)-X_i^\phi(o)}\right) \text{ vs } 2\sqrt{Dt}/\beta d$$

diagram (curves a and b in Fig. 2). In the case of binary systems, this is a result of the enrichment ratio β remaining significantly above its equilibrium value throughout most of the segregation process as pointed out by Lea and Seah (45), and in the case of ternary systems, an analogous effect combines with a dynamic enhancement of β due to rapid segregation of alloying elements which are generally present at much higher bulk concentration than impurities and require much less time to segregate due to a much smaller value of D/β^2. The result is consistent with the experimental data obtained by King and Wigmore (46) and Lemblé et al (47). Conversely, for some particular sets of parameters in Tyson's calculation, segregation is retarded (curve c in Fig. 2) and demonstrate

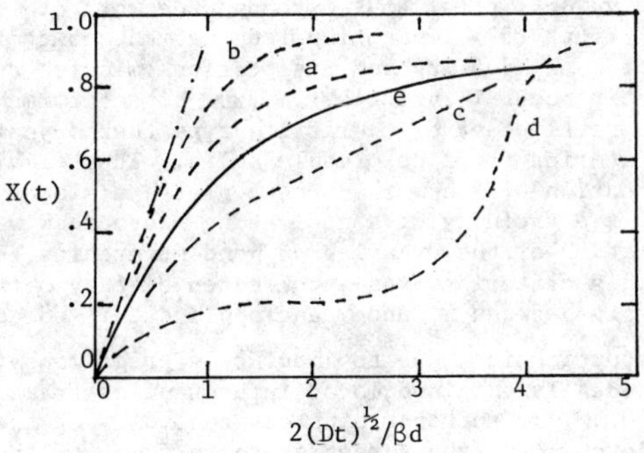

Figure 2. Kinetics of segregation. (a)-(d) Deviation in segregation rate due to interaction parameters (44). (e) McLean's kinetic model (29).

a two-step process (curve d in Fig. 2). This phenomenon may be explained by the existence of a very strong preferential interaction between M and I during the early stage of segregation. Grain boundary segregation of I will then be retarded by the temporary trapping of I by M atoms.

As reviewed above, understanding the kinetics of segregation is required to understand the development of segregation and embrittlement. However, the analysis of the kinetics of segregation has received much less attention than the thermodynamics of segregation. Therefore, a lot more investigation is required in this area for a fuller quantitative understanding of the kinetics by which the segregation process strives toward the equilibrium condition.

The Influence of Segregation on Grain Boundary Embrittlement

In the two previous sections the theoretical approaches to the thermodynamic and kinetics of segregation have been reviewed and discussed. In this section the key question of the influence of the segregation phenomenon on the fracture behavior will be investigated. The problem can be separated into two concepts. The first is the local modification of the bonding and subsequent energy to fracture and the second is the modification of the local dislocation-boundary interactions. The most likely cause of failure is an interactive combination of the two effects. The following discusses the present state of theoretical modeling in unraveling the complex modification of the fracture behavior of steels associated with the segregation process.

Very little is presently known of the detailed physics of grain boundary cohesion. Therefore, many investigators have approached this problem using the cohesive energy concept. The relations between grain boundary cohesion and the cohesive energy is shown in Figure 3. The slope of the G-r curve (Figure 3a) gives the stress-displacement curve (Figure 3b) when normalized to a unit area of grain boundary. The integral of σdr from r_o to r_∞ gives the cohesive energy of the grain boundary which is equivalent to the ideal work of fracture per unit area of the grain boundary, γ.

The expression of the ideal work of fracture per unit area of the grain boundary can be given using the surface and grain boundary energies following Polani (48) and Orowan (49):

$$\gamma = 2\gamma_s - \gamma_b \tag{12}$$

where γ_s is the surface energy per unit area of the freshly formed fracture surfaces and γ_b is that of the

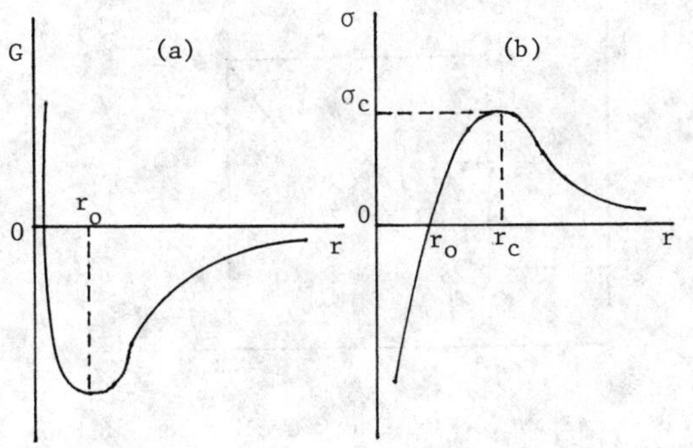

Figure 3. (a) Free energy interatomic distance diagram for a grain boundary. (b) Normal stress interatomic idstance diagram. (54)

prior grain boundary. However, Eq.(12) can be applied only for the equilibrium process since γ_s and γ_b imply the equilibrium values.

Recognizing the above difficulty in Eq.(12), Seah (34, 50) considered the differences between two types of grain boundary separation associated with impurity grain boundary segregation. First, he considered grain boundary separation with a fixed chemical potential of the impurity (i.e. slow fracture process) and secondly, grain boundary separation with the fixed grain boundary composition of impurity (i.e. rapid fracture process). He started with the Gibb's isotherm (51):

$$\frac{d\gamma_b}{dX_I^B} = -\frac{RT}{X_I^B}\Gamma_{bI} \tag{13}$$

where Γ_{bI} is the excess impurity I at the grain boundary. Assuming that segregation in a dilute alloy system may be simply specified by the proportionality $\Gamma_{bI} \propto X_I^B$ and that during the rapid grain boundary separation the impurities at the grain boundary are conserved (i.e. $\Gamma_{bI} = 2\Gamma_{sI}$), he developed the relation:

$$\gamma = 2\overset{\circ}{\gamma}_s - \overset{\circ}{\gamma}_b \tag{14}$$

where $\overset{\circ}{\gamma}_s$ and $\overset{\circ}{\gamma}_b$ are the grain boundary and surface energies, respectively, in the absence of impurity I. From his result (Eq. 14), Seah concluded that for the rapid grain boundary separation of a dilute alloy system, the ideal work of fracture of the grain boundary γ is independent of impurity grain boundary segregation. Of course, $\overset{\circ}{\gamma}_s$ was assumed not to be the equilibrium value.

However, Rice (52) presented an analysis which disagrees with Seah's conclusion that γ is independent of segregation in the fixed composition limit of a dilute alloy system. Asaro (53) and Hirth (54) also considered this problem. Recently, Hirth and Rice (55) reconsidered the computations on the work of grain boundary separation with impurity segregation in terms of reversible work cycles in chemical potential vs composition diagram (Figure 4). In Figure 4 the point A represents the complete equilibrium state of grain boundary. The path I (A to B) is the grain boundary separation at constant chemical potential (i.e. slow fracture process) and the path III is that at constant composition (i.e., $2\Gamma_s^C = \Gamma_b^A$, fast fracture process). For the path III, Hirth and Rice calculated the reversible work of fracture using the equation:

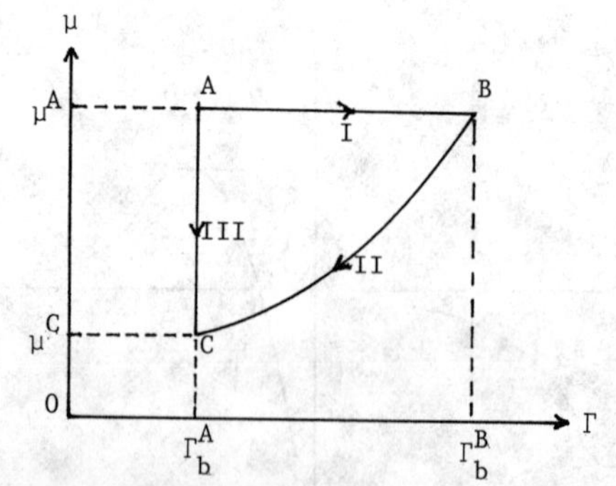

Figure 4. Reversible work cycle of grain boundary separation in chemical potential excess grain boundary composition diagram (55).

$$df = d\,W_{rev} = \sigma d\delta + \mu d\Gamma \tag{15}$$

where f is the Helmholtz free energy, F, normalized to a unit area of grain boundary, σ is the stress tending to separate the grain boundary and δ is the opening separation. The reversible work of fracture is given:

$$W(III) = 2\overset{\circ}{\gamma}_s - \overset{\circ}{\gamma}_b - (\mu^A - \mu^C)\Gamma_{bI} \tag{16}$$

Eq.(16) shows that the work of fracture is reduced by the amount $(\mu^A-\mu^C)\Gamma_{bI}$ due to the impurity segregation at the grain boundary, while Seah's equation (Eg. 14) shows that γ is independent of impurity segregation under the same condition. Hirth and Rice extended this computation to more general cases of solute interactions, including multi-component component systems by use of three dimensional μ-Γ space.

However, McMahon et al (56,57) criticized in general the attempts to calculate the work of fracture by using the equilibrium thermodynamics since fracture is essentially an irreversible process in a material where plastic flow occurs. Considering the initiation of brittle fracture ahead of a pre-existing notch or blunt crack, McMahon and Vitek (57) pointed out that the local stress ahead of the pre-crack in the region of microcrack nucleation is a function of the value of the ideal work of fracture, γ, and that this stress causes dislocation emission at the tip of the microcrack leading to the expenditure of the plastic work of fracture γ_p. They concluded that γ_p is a function of γ and derived an approximate relationship between γ_p and γ:

$$\frac{d\gamma_p}{\gamma_p} \simeq n\frac{d\gamma}{\gamma} \tag{17}$$

n is m+1, with m the strain rate sensitivity coefficient. Using Eq.(17) and the experimental data of grain boundary segregation, they demonstrated the order of 10% reduction in the ideal work of brittle fracture γ due to the grain boundary segregation of the impurity could cause the order of 95% reduction in the plastic work accompanying crack propagation. The idea that the major effect in reduction of the fracture resistance of impurity segregation induced fracture of metals is due to the effect on γ_p of the chemistry change in the grain boundary was proposed by Fine and Marcus (58). They also showed the limited change in surface energy due to the segregation of Sb, but with an associated large change in fracture energy.

Recently, Jokl et al (59) extended the work of McMahon and Vitek (57), which implied that the concomitant processes of bond-breaking and dislocation emission at the microcrack tip occur in brittle fracture of a deformable solid, to a development of a Griffith-type thermodynamic fracture criterion for a deformable solid. They combined the determination of the work connected with dislocation emission in the vicinity of a nucleating microcrack with an approximation of the dynamics of bond stretching to calculate this criterion. As a consequence, they obtained a relationship between γ_p and γ and applied this to estimate the reduction in γ due to impurity segregation with the experimentally obtained local intergranular fracture stresses (60). The authors point out that only at the root of a notch where the stress state in front of the notch is well defined can the calculation of the fracture stress associated with the creation of new surfaces at a microcrack be made. In addition they point out that precracked fracture mechanism type specimens are not amenable to a definitive stress analysis in terms of the stress at fracture. They further claim that the use of the change in the plane strain fracture toughness, K_{1C}, to model the fracture is very difficult due to the nonuniformity of the stress state in front of the crack tip making the crack initiation site critical to the interpretation of the fracture stress. In addition to relating γ_p at fracture to γ, the other major factor resulting from the analysis was determination of when fracture would be prevented due to general yielding occurring prior to criticality for cracking.

Also recognizing the difficulties in the application of the equilibrium thermodynamics to the grain boundary embrittlement problem, Stark and Marcus (61) have developed two models describing the energetics of segregation induced grain boundary embrittlement based upon a detailed nonequilibrium thermodynamic analysis of grain boundary segregation. In the first model the binary systems which show the loss of grain boundary cohesion due to segregation were studied. In the second model the ternary systems which involve strong M-I interactions were discussed. Assuming that the bulk is a dilute solution obeying Henry's law and that the grain boundary is a layered region (Figure 5) with strong bonding within the layer and weak bonding normal to the layer, they calculated the enthalpy change in the boundary and the boundary matrix interface for a binary system. The results in terms of enthalpy change were in the form:

$$\delta H^B + \delta H^{BM} = \delta H_*^B + \delta H_*^{BM} + \delta N_2^L \underline{H}_2^L - \delta N_1^L \underline{H}_1^L \tag{18}$$

where H=enthalpy, $\bar{H}_i$=partial molar enthalpy of component i, $\underline{H}_i$=specific enthalpy of component i, N_i=number of atoms of component i, L=lattice, B=boundary, BM=matrix boundary interface.

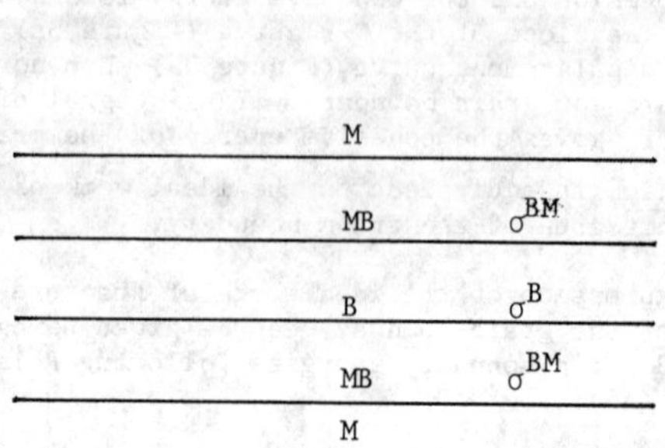

Figure 5. Model of grain boundary interface region (61).

δH_*^B and δH_*^{BM} are defined as $\delta H_*^B = \delta H_o^B + \sigma^{FB} A^B$ and $\delta H_*^{BM} = \delta H_o^{BM} + \sigma^{FBM} A^{BM}$ where A = area, σ = surface tension, F=final state, and δH_o=enthalpy change for transformation of the grain boundary to lattice without concentration change.

For a multi-component system

$$\delta H^B + \delta H^{BM} = \delta H_*^B + \delta H_*^{BM} + \sum_{i=2}^{n} \delta N_i^L \bar{H}_i^L - \delta N_1^L \bar{H}_1^L \qquad (19)$$

In Eq.(18) and (19) component 1 is solvent and components 2 to n are solutes. For determination of these equations they followed only state functions assuming that grain boundary segregation is the process from an equal composition state to an equal chemical potential state (i.e. the initial state is not an equilibrium state but the compositions of the boundary and matrix boundary regions are the same as that of the lattice; the final state is an equilibrium state so that the chemical potentials of these two regions are equal to that of the lattice).

Using the above equations and published data for cohesive energies and solid solubilities of elements, they demonstrated that the impurity elements of Group 6A such as S, Se and Te are the most potent elements to reduce the cohesive energy of the grain boundary in binary iron based alloys. They also hypothesized that in a multi-component steel, the strong interactions between impurities and alloying elements cause cosegregation and then form a covalent glass with its bonding electrons providing a strong, highly directional bond parallel to the interface. The overall boundary region could have an increased cohesive energy. However, the cohesive energy of the matrix boundary interface region would be reduced because most of the bonding electrons are taken up on the formation of the layered glassy boundary segregate. Thus, intergranular failures could occur along this weakened matrix boundary interface region. They also suggested that embrittlement of grain boundaries by hydrogen would be expected to fall within a similar mechanism due to the affinity of hydrogen for most of the segregating non-metals.

The above hypothesis (61) of the formation of bonds lying in the plane of the boundary at the expense of bonds extending across the boundary is supported by the recent study of Briant and Messmer (62). They employed detailed electronic structure calculations for a cluster of Ni atoms with and without a S atom at the center of the cluster. Their results showed that the S atom forms strong bonds with the nearest Ni atoms, while the bonds between these Ni atoms and their neighboring Ni atoms are simultaneously weakened. Thus they suggested that S forms very strong bonds within the plane of the boundary but weakens Ni-Ni bonds perpendicular to this plane.

In this section of the chapter the theoretical approaches on grain boundary embrittlement were reviewed in terms of thermodynamics of segregation, kinetics of segregation and the influence of segregation on grain boundary embrittlement. In the following section, the experimental results from a wide range of studies of grain boundary embrittlement will be discussed and related to the theoretical models just described.

EXPERIMENTAL STUDIES OF GRAIN BOUNDARY SEGREGATION AND EMBRITTLEMENT

As pointed out by Guttmann (5), grain boundary embrittlement is governed by two main factors; first, the effects of segregated elements and second, the effects of mechanical-structural factors. Hence, the experimental studies of this subject will be reviewed and discussed according to the above two main factors.

The Effect of Segregated Elements

It is now fairly accepted that the grain boundary problems are associated with the segregation of the impurities belonging to Groups 4A to 6A of the Periodic Table (Table II). However, Low et al (63) have demonstrated that with the presence of impurities, a plain carbon steel does not embrittle whereas a Ni-Cr steel is severely embrittled.

Table II

Impurities Related to the Grain Boundary Embrittlement

Group		
4A	5A	6A
C	N	O
Si	P	S
Ge	As	Se
Sn	Sb	Te
Pb	Bi	Po

From this fact it is found that the interaction of alloying elements with impurities is one of the important factors influencing grain boundary embirttlement. Hence, the role of alloying elements as well as impurities must be understood to predict and control the grain boundary embrittlement. We will first review the experimental results on the effects of impurities in high-purity metals and then the combined effects of alloying elements with impurities in alloy steels will be discussed. The results will be discussed in terms of the theoretical models described previously.

The Effects of Impurities in High-Purity Metals. In attempts to study the effects of impurities many investigators have employed high-purity metals doped with the impurities of interest, Table II. The examples of the systems studied and their chemical compositions are shown in Table III. The experimental techniques used and the important experimental results are also tabulated in Table III. It can be seen that the most widely used experimental technique to study grain boundary segregation is the Auger electron spectroscopy (AES) technique which has the advantages of high surface sensitivity, rapid data acquisition, and when combined with inert ion sputtering it has the capability of providing chemical composition information normal to the fracture surface. The experimental techniques used to study segregation have been discussed in detail by Joshi (64). The degree of embrittlement is in general measured by the increase in the ductile-brittle temperature (DBTT) and/or in the fracture appearance transition temperature (FATT). These are explained elsewhere (1,2,63,65).

Phosphorus, Tin, Antimony and Arsenic in High Purity Metals. P, Sn, Sb and As behave quite similarly (5). They are highly surface active in Fe and segregate to the grain boundaries of Fe with fairly high enrichment ratios in the manner of Gibbsian (equilibrium) type segregation. Gibbsian (equilibrium) type grain boundary segregation in Fe has been confirmed for P (27), Sn (32), and Sb (27). Their high enrichment ratios associated with the equilibrium grain boundary segregation in Fe have been observed for P (66,67), Sn (32) and Sb (67). The grain boundary segregation of these impurities leads to grain boundary embrittlement and an intergranular brittle fracture. By use of zero creep technique, it has been demonstrated that the surface energy and grain boundary energy of Fe decreases with increasing the concentration of P (68) and Sn (32). Also, by use of mechanical testings, it has been found that the DBTT and FATT of Fe increases with increasing the grain boundary concentration of these impurities (67). Recently, using a new experimental technique, extended x-ray absorption fine structure (EXAFS) measurement, Pardee et al (69) have demonstrated that the atomic structure around the segregated Sn atoms in Fe-Sn system is considerably different from that around non-segregated Sn atoms.

Table III. The Effects of Impurities in High-Purity Metals

I	System	I *Level	Experimental Method	Results
P	Fe-P	0.09 P (66)	Radioactive Tracer Technique	$\beta_p \approx 50$, **GBE=f(X_p^ϕ)
		0.001~0.36 P (68)	Zero Creep Technique	$X_p^B\uparrow \rightarrow \gamma_b\downarrow$, $\Delta\gamma_b(\gamma\text{-Fe}) < \Delta\gamma_b(\delta\text{-Fe})$
		1.0 P (27)	AES	Gibbsian Segregation of P
		0.05~0.2 P (67)	AES & FATT	$X_p^\phi\uparrow \rightarrow$ FATT$\uparrow$, $\beta_p(\gamma\text{-Fe}) < \beta_p(\alpha\text{-Fe})$
	W-P	Powder Metallurgy Tungsten (70)	AES, SEM, DBTT	$X_p^\phi\uparrow \rightarrow$ DBTT$\uparrow$
Sn	Fe-Sn	0.001~4 Sn (32)	AES and Zero Creep Technique	Gibbsian Segregation, Multilayer Segreation
		0.45 Sn (69)	EXAFS	Different neighbors around segregated Sn
Sb	Fe-Sb	1.5, 5.0 Sb (27)	AES	Gibbsian Segregation of Sb
		0.12, 2.0 Sb (82)	AES	Sb is highly surface active in Fe
	Cu-Sb	0.6 Sb (71)	Creep Rupture Test	Sb increases creep rate and embrittlement
O	Fe-O	0.0001~0.2 O (74)	AES & DBTT	DBTT $\neq$ f(X_O^B)
	Mo-O	0.0002 O (75)	AES, X-ray Diffraction & Tensile Test	$X_O^\phi\uparrow \rightarrow$ GBE$\uparrow$
S	Fe-P-S	0.02-0.06 P, 0.003 S (67)	AES, FATT	FATT(Fe-P-S) >> FATT(Fe-P)
	Fe-O-S	0.001~0.2 O, (74) 0.0035 S	AES and DBTT	S increases DBTT
	Fe-Sn-S	0.001-4 Sn, 0.001 S (76,77)	AES and Zero Creep Technique	Surface Site competition between S and Sn
Te	Fe-Te	0.02 Te (78)	AES, SEM & TEM	Rearrangement of ***GB due to Te Segregation
N	Fe-N	0~0.01 N (80)	Zero Creep Technique	$X_N^B\uparrow \rightarrow \gamma_b\downarrow$
C	Fe-C	0.001~0.022 C (79)	Autoradiography	Gibbsian Segregation of C
Bi	Cu-Bi	0.08 (81)	AES & SEM	Bi is surface active in Cu

* Concentration:Weight Percent
** GBE = Grain Boundary Embrittlement
*** GB = Grain Boundary

The grain boundary embrittlement problems due to impurity segregation are not limited to the iron system. Joshi and Stein (70) have shown that the grain boundary segregation of P is the primary cause of embrittlement in the tungsten materials formed by power metallurgy. Tipler and McLean (71) also have found that grain boundary segregation of Sb facilitates the wedge- and bubble-type cavitations in the Cu system which increase significantly the creep embrittlement of Cu even though Sb softens Cu.

One of the most important characteristics of the Group 4A to 6A impurities is the fact that their segregation is strongly enhanced by common alloying elements such as Ni, Cr and Mn in alloy steels. Thus, their effects on the grain boundary embrittlement of alloy steels are much more serious than that of pure iron. These combined effects of alloying elements with these impurities in alloy steels will be discussed later.

Oxygen, Sulfur, Selenium and Tellurium in High Purity Metals. As emphasized by McMahon (72) and Stark and Marcus (61), the impurities of Group 6A of the Periodic Table such as S, Se and Te are extremely potent in lowering grain boundary cohesion of Fe because of their extremely high surface activity in Fe. For a long time the grain boundary segregation of O was suspected to cause the grain boundary embrittlement of Fe (73). However, it is now recognized that the grain boundary embrittlement of Fe is not related to the O segregation (74) since the solubility of O in Fe is so low that there is not enough O in the matrix to segregate to any appreciable extent in the grain boundaries even though O is extremely surface active in Fe (5). On the other hand, O readily segregates to the grain boundaries of Mo with a very high free energy of segregation, thereby promoting intergranular brittle fracture (75).

S is a more severe embrittler than P in Fe (67). One of the interesting behaviors of S is that S segregates strongly to the free surface of Fe and expels other elements from the surface sites. Seah and Lea (76,77) have observed that there is strong site competition between S and Sn on the free surface of Fe, while the site competition between S and Sn in the grain boundaries of Fe seems to be negligible (32).

Se and Te are even more severe embrittlers in Fe (i.e. Te>Se>S) (72). The grain boundary segregation of Te has been observed to cause the reconstruction of the grain boundaries of Fe into a faceted structure, thereby leading to the intergranular brittle fracture of Fe (78).

As discussed above, S, Se and Te have very high embrittling potencies in pure iron. However, the effects of these impurities in alloy steels is almost negligible because they interact very strongly with certain alloying elements such as Mn and Cr and precipitate almost completely in the matrix in alloy steels (72). Therefore, they are in general not responsible for the embrittlement problems in alloy steels (5) in spite of their extremely high surface activity.

Carbon. C is also surface active in Fe and therefore strongly segregates to the grain boundary of Fe obeying the Gibbsian (equilibrium) type of segregation (79). The beneficial effect of C on grain boundary cohesion has been demonstrated by showing the decrease in the grain boundary embrittlement induced by S segregation in Fe (74) and induced by O segregation in Mo (75).

The behavior of C is also more complicated in alloy steels due to the interaction of C with alloying elements. This complicated behavior of C in alloy steels also will be discussed later.

Nitrogen. N is also very surface active in Fe (80) The presegregated N in austenite has been known to be responsible for 350° embrittlement (3).

The Combined Effects of Alloying Elements with Impurities in Alloy Steels. As mentioned previously, the segregation of impurities (I) is strongly affected by the addition of alloying elements (M) because of the interactions between I and M. Thus, the M-I combined effects are the most important factors influencing the grain boundary embrittlement in alloy steels. The M-I combined effects on segregation can be rationalized by using the Guttmann's model (eq. 6-8) described previously. The interaction coefficients α'_{MI} introduced in Eq 7 and 8 indicate the degree of activity of alloying elements with the impurities. The values of the interaction coefficient α'_{MI} in the Fe-M-P and Fe-M-Sb systems are shown in Table IV.

Table IV.

The Interaction Coefficients α'_{MI} (4)

a) Fe-M-P System

M	Ni	Mn	Cr	V
α'_{MP}	30.1	66.9	100.3	143.8
	W	Mo	Ti	Zr
	170.7	184.0	221.0	256.4

b) Fe-M-Sb System

M	Cr	Ni	Ni+Cr
α'_{MSb}	33.6	56.8	56.8

The alloying elements present in alloy steels can be divided into three groups according to the values of α'_{MI} (5):

i) moderate elements (Ni, Cr & Mn)
ii) intermediate elements (Mo, V & W)
iii) strong elements (Ti & Zr)

Thus, the experimental results on the M-I combined effects will be discussed in terms of the above three groups. References to the experimental studies on the M-I interactions are given in Table V for dilute ternary Fe-M-I and quaternary Fe-M-I_1-I_2 systems; in Table VI for steels based on the chemistry of the 3340 steel which have been studied the most extensively; and in Table VII for other alloy steels.

Moderate Elements (Ni, Cr & Mn). Ni, Cr and Mn show moderate attraction with certain impurities such as P, Sn, Sb and As (i.e., α'_{MI} is positive but not very large) (Table IV). It can be seen from Eq.(8) that these elements and impurities (P, Sn, Sb and As) enhance the segregation of each other. The M-I segregation enhancement has been experimentally demonstrated for Ni-P (83,84), Ni-Sn (83,85,86), Ni-Sb (82,83,85,87,88), Cr-P (83,84), Cr-Sn (83), Cr-Sb (83), Ni-Cr-P (83,84,89), Ni-Cr-Sn (83,90), Ni-Cr-Sb (83,85,87,91), Mn-Sb (92,93,94) and Mn-P (95). The Ni addition in alloy steels enhances the segregation of P, Sn and Sb. The relative potencies of the Ni-P, Ni-Sn and Ni-Sb segregation enhancement may be compared by plotting the grain boundary concentration of Ni against that of P, Sn and Sb in Ni containing steels doped individually with P, Sn or Sb given various aging treatments (Fig. 6). The Ni-Sn and Ni-Sb interactions seem to be much stronger than the Ni-P interaction (i.e., Ni-Sn>Ni-Sb >Ni-P). However, the embrittling potencies as characterized by change in DBTT of these impurities show a little different order (i.e., Sb>Sn>P) (Figure 7). It has been pointed out that the embrittling potencies of the Group 4A to 6A impurities increase with both Group number and Period number since the chemical effect (charge transfer) increases with Group number and the volumetric effect (impurity size) increases with Period number (72).

The addition of Ni in alloy steels also enhances the segregation of Si (96,97) which contributes to promoting the grain boundary embrittlement. Although Si is a less potent embrittler than P, Sn and Sb, the presence of Si in alloy steels containing sufficient Ni can lead to embrittlement.

Cr seems to enhance the segregation of P (84), Sn (83) and Sb (83,87) in alloy steels. However, the interactions of Cr with the impurities and the amount of Cr segregation is less well defined than those of Ni due to the experimental difficulties of measuring the segregated Cr. These difficulties arise from the interference of oxygen with the Cr Auger peak and the formation of stable Cr-rich carbides that often form at the grain boundaries. A spectacular example of the influence of the Cr-carbide formation on the enhancement of impurity segregation can be seen in Figure 8. The enhancement of Sb segregation in a 0.008 C Ni-Cr steel where Cr remains in its elemental form is much more pronounced than in a 0.4 C Ni-Cr steel where Cr precipitates in the carbide form. Therefore, it can be said from Figure 8 that Cr is an enhancer of Sb segregation when it is free in solution. The Cr-P enhancement also can be revealed in an indirect way by comparing the grain boundary concentrations of P obtained from a low C Ni steel and Ni-Cr steel (84). Figure 9 shows that the Ni+Cr is more effective than Ni alone in enhancing P segregation. Thus, Cr can be said to enhance P segregation. However, Cr appears not to significantly cosegregate with impurities (3) even though it enhances impurity segregation.

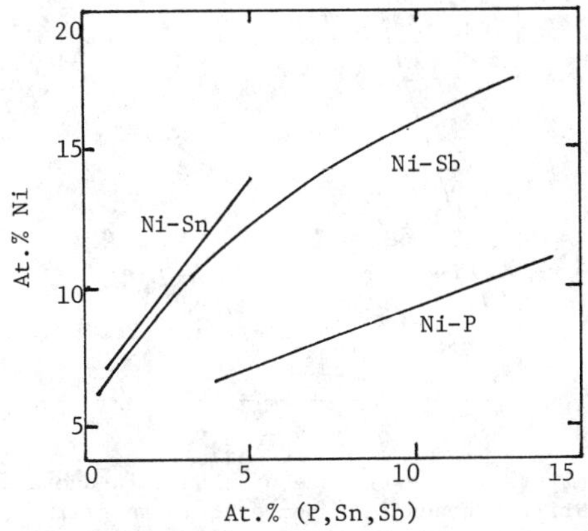

Figure 6. Relations between the grain boundary concentrations of Ni and P, Sb or Sn in Ni-Cr steels doped with these impurities (72).

Table V. The Effects of Alloying Elements in Fe-M-I or Fe-M-I_1-I_2 Systems

M	System	M & I Level	Experimental Method	Results
Ni	Fe-Ni-C-P	3.5 Ni 0.4 C 0.06 P	AES (83)	Ni enhances P segregation
		3.5 Ni$\binom{0.04}{-0.4}$C 0.06 P	AES, SEM & DBTT (84)	Ni-P cosegregation, $X_P^\phi\uparrow \rightarrow$ DBTT$\uparrow$
	Fe-Ni-C-Sn	3.5 Ni 0.4 C 0.06 Sn	AES (83,85)	Ni enhances Sn segregation
		3.5 Ni 0.1 C 0.2 Sn	Mossbauer Spectroscopy (86)	Reversible Ni-Sn interaction
	Fe-Ni-C-Sb	3.5 Ni 0.4 C 0.065 Sb	AES (83,85)	Ni enhances Sb segregation
	Fe-Ni-C-Sb	3.5 Ni$\binom{0.04}{-0.4}$C 0.06 Sb	AES & DBTT (87)	Ni enhances Sb segregation, $X_{Sb}^\phi\uparrow \rightarrow$ DBTT$\uparrow$
	Fe-Ni-Sb	0.03 Ni 0.01 Sb	Mossbauer Spectroscopy (88)	Ni enhances Sb segregation
		2.0 Ni 0.13 Sb	AES (82)	Strong Ni-Sb surface interaction
	Fe-Ni-C-As	3.5 Ni 0.4 C 0.05 As	AES (85)	No As segregation
Cr	Fe-Cr-C-P	1.7 Cr 0.4 C 0.06 P	AES (83)	Cr enhances P segregation
		1.7 Cr 0.04 C 0.06 P	AES, SEM & DBTT (84)	Cr enhances P segregation, $X_P^\phi\uparrow \rightarrow$ DBTT$\uparrow$
	Fe-Cr-C-Sn	1.7 Cr 0.4 C 0.05 Sn	AES (83)	Cr enhances Sn segregation
	Fe-Cr-C-Sb	1.7 Cr 0.4 C 0.065 Sb	AES (83)	Cr enhances Sb segregation
		1.7 Cr 0.4 C 0.06 Sb	AES & DBTT (87)	Cr enhances Sb segregation, $X_{Sb}^\phi\uparrow \rightarrow$ DBTT$\uparrow$
	Fe-Cr-Sb	0.03 Cr 0.01 Sb	Mossbauer Spectroscopy (88)	Cr is ineffective on Sb segregation
		2.0 Cr 0.13 Sb	AES (82)	Negligible Cr-Sb surface interactions
Mn	Fe-Mn-C-Sb	2.0 Mn 0.2 C 0.12 Sb	AES (92)	Mn-Sb cosegregation
		2.0 Mn 0.2 C 0.12 Sb	Backscattering of Energetic Irons (93)	Mn-Sb cosegregation
		2.0 Mn 0.2 C 0.12 Sb	AES & XPS (94)	Mn-Sb cosegregation
Mo	Fe-Mo-P	(0.5,2) Mo 0.04 P	AES (82)	Strong Mo-P surface interaction
V	Fe-V-Sb	1.8 V 0.13 Sb	AES (82)	Negligible V-Sb surface interaction
Ti	Fe-Ti-P	1.7 Ti 0.04 P	AES (82)	Strong Ti-P surface interaction
	Fe-Ti-Sb	1.7 Ti 0.13 Sb	AES (82)	Strong Ti-Sb surface interaction

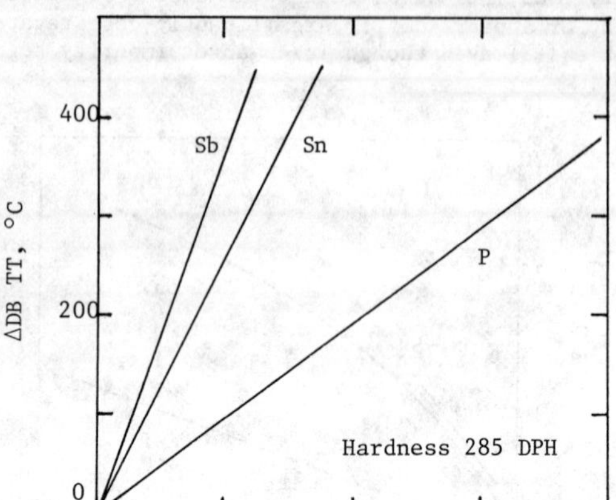

Figure 7. Dependence of change in ductile-brittle transition temperature on grain boundary concentrations of P,Sb or Sn in Ni-Cr steels (72).

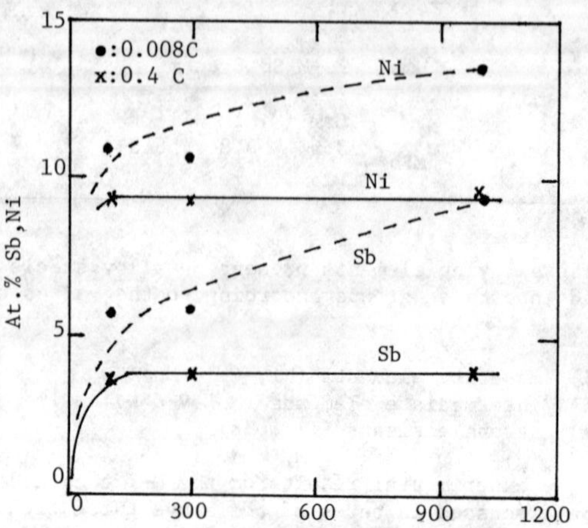

Figure 8. Comparison of grain boundary concentrations of Ni and Sb in 0.008C and 0.4C Ni-Cr steels doped with 0.06 Sb aged at 520°C (87).

Table VI. The Effects of Impurities & Alloying Elements in 3.5 Ni - 1.7 Cr Steels

I or M	C Level	I or M Level	Experimental Method	Results
P	0.008	0.076 P	Cantilever Bending & SEM (104)	Preferential fracture along high-angle ferritic boundaries
	0.04-0.4	0.02-0.06 P	AES, SEM & DBTT (Slow Bending Test)(84)	Ni+Cr enhances P segregation $X_P^{\phi}\uparrow \rightarrow$ DBTT$\uparrow$
	0.3	0.01-0.06 P	AES & DBTT (CVN) (105)	Presegregation of P is responsible for tempered martensite embrittlement
	0.3	0.07 P	AES & DBTT (CVN) (106)	Embrittlement equation is developed. $\Delta TT = f(H, GS, X_P^{\phi})$
	0.3	0.04&0.06 P	AES, (60) Electron Probe Micro-Analysis & Notched Pure Bending Test	The method of correlating the local intergranular fracture stress with the probable X_i^{ϕ} is developed. σ^* decreases with increasing $\overline{X}_P^{\phi}$ & $X_P^{\phi max}$.
	0.4	0.06 P	AES & SEM (83)	$\beta_P \approx$ 100~500, Ni+Cr enhances P segregation
	0.4	0.06 P	AES, SEM & DBTT (107) (Cantilever Bending Test)	Intercritical Heat Treament reduces X_{Ni}^{ϕ} & X_P^{ϕ}
	0.4	0.05 P	Tensile Test (23)	P segregation cuases a modest decrease in susceptibility to liquid metal induced embrittlement (LMIE).
Sn	0.2-0.4	0.04 Sn	AES, SEM & DBTT (90) (Slow Bending Test)	DBTT increases linearly with increasing X_{Sn}^{ϕ}. The rate of approach to the steady state level of segregation is faster in higher C steel.
	0.3	0.06 Sn	AES & DBTT (CVN) (106)	$\Delta TT = f(H, GS, X_{Sn}^{\phi})$
	0.3	0.06 Sn	AES, (60) Electron Probe Micro-Analysis & Notched Pure Bending Test	σ^* decreases with increasing $\overline{X}_{Sn}^{\phi}$ & $X_{Sn}^{\phi max}$
	0.4	0.05 Sn	AES & SEM (83)	Ni+Cr enhances Sn segregation
	0.4	0.05 Sn	Tensile Test (23)	Sn segregation increases susceptibility to LMIE
	0.4	0.06 Sn	AES (114)	The slower sputter rate of segregated Ni than segregated Sn may provide seemingly deeper concentration enrichment and degree of Ni segregation.
Sb	0.008	0.07 Sb	Cantilever Bending & SEM (104)	Preferential fracture along high-angle ferritic boundaries.
	0.008	0.07 Sb	AES, SEM & DBTT (91) (Slow Bending Test)	The degree of embrittlement is directly proportional to X_{Sb}^{ϕ}.
	0.008	0.07 Sb	AES, SEM & TEM (108)	The rate & extent of the Ni build-up is consistent with a model in which Ni is rejected from carbides and diffused along the grain boundaries.
	0.3	0.06 Sb	AES & DBTT (87) (Slow Bending Test)	Ni & Cr together lead to much more Sb and Ni segregation than would be expected from their individual effects.
	0.3	0.06 Sb	AES, (60,109) Electron Probe Micro-Analysis & Notched Pure Bending Test	σ^* decreases with increasing $\overline{X}_{Sb}^{\phi}$ & $X_{Sb}^{\phi max}$
	0.4	0.06 Sb	AES & SEM (26,28,83,85)	Gibbsian segregation of Sb is observed.
	0.4	0.06 Sb	AES & SAM (110)	Ni and Sb segregation occurs in large amounts to only certain GBs. P segregation is more uniform to all GBs.
	0.4	0.07 Sb	AES, Chemical Analysis (111) & DBTT (Slow Bend)	Sb segregates uniformly to prior γGBs. Ni segregation is localized in a phase other than the α-portions of boundaries.
	0.4	0.08 Sb	Tensile Test (23)	Sb segregation increases susceptibility to LMIE.
As	0.3	0.05 As	AES (85)	No segregation of As is revealed by AES.
	0.4	0.05 As	Tensile Test (23)	As causes a modest decrease in susceptibility to LMIE.

I or M	C Level	I or M Level	Experimental Method	Results
S	0.3	0.01 S	AES & (112) Charpy V-notch Impact Test	S segregates at T>1050°C and precipitates as CrS at T<1050°C.
S + P	0.3	0.004 S + 0.004 P	AES & (113) Charpy V-notch Impact Test	S segregation causes embrittlement and S precipitation (CrS) does not lead to embrittlement trough but reduces fracture energy.
Se	0.4	0.01 Se	SEM & DBTT (99) (Slow Bending Test)	Long time aging may exert greater effect of Se.
Te	0.4	0.01 Te	SEM & DBTT (99) (Slow Bending Test)	Te is a strong embrittler.
Bi	0.4	0.01 Bi	SEM & DBTT (99) (Slow Bending Test)	Bi is a strong embrittler.
Ge	0.4	0.01 Ge	SEM & DBTT (99) (Slow Bending Test)	Long time aging may exert greater effect of Ge.
Si	0.4	0.2 Si	SEM & DBTT (99) (Slow Bending Test)	Si is a strong embrittler.
	0.4	0.61 Si	AES & DBTT (CVN) (106)	$\Delta TT = f(H, GS, X^{\phi}_{Si})$
Mn	0.4	0.7 Mn	SEM & DBTT (99) (Slow Bending Test)	Mn is an embrittling element, and not just an enhancer of I.
Mn + Si	0.4	0.9 Mn + 0.25 Si	AES (28,85)	Significant Ni and Cr segregation is observed.
Mn + Mo	0.4	0.7 Mn + 0.6 Mo	SEM & DBTT (99) (Slow Bending Test)	Mo is a scavenger of Mn segregation.
Mn + Mo + P	0.4	0.7 Mn + 0.6 Mo + 0.06 P	SEM & DBTT (99) (Slow Bending Test)	No effect of Mn on P behavior.
Mn + Mo + Sn	0.4	0.7 Mn + 0.6 Mo + 0.06 Sn	SEM & DBTT (99) (Slow Bending Test)	No effect of Mn on Sn behavior.
Mn + Mo + Sb	0.4	0.7 Mn + 0.6 Mo + 0.02 Sb	SEM & DBTT (99) (Slow Bending Test)	No effect of Mn on Sb behavior.
Mn + Mo + As	0.4	0.7 Mn + 0.6 Mo + 0.06 As	SEM & DBTT (99) (Slow Bending Test)	No effect of Mn on As behavior.
Mo + P	0.4	0.6 Mo + 0.06 P	SEM & DBTT (99) (Slow Bending Test)	Mo shows no effect on P-induced embrittlement.
	0.2	0.6 Mo + (0.02-0.06) P	AES, SEM & DBTT (98) (Slow Bending Test)	P-induced embrittlement is eliminated by scavenging effect of Mo.
Mo + Sn	0.4	0.6 Mo + 0.06 Sn	SEM & DBTT (99) (Slow Bending Test)	Mo reduces Sn-induced embrittlement.
Mo + Sb	0.4	0.6 Mo + 0.02 Sb	SEM & DBTT (99) (Slow Bending Test)	Mo reduces Sb-induced embrittlement.
	0.4	0.6 Mo + 0.066 Sb	AES & SEM (83)	Mo scavenges Sb segregation in step-cooled condition.
Mo + As	0.4	0.6 Mo + 0.06 As	SEM & DBTT (99) (Slow Bending Test)	Mo reduces As-induced embrittlement.
Ti + Sb	0.003	0.1 Ti + 0.065 Sb	AES, SEM & DBTT (102) (Slow Bending Test)	Ti eliminates most of Sb-induced embrittlement.

Mn cosegregates with Sb (92,93,94) and P (95) thereby enhancing their segregation. Recently, Yu and McMahon (95) demonstrated that the addition of Mn in 2.25 Cr-1 Mo steels increases P segregation and the grain boundary embrittlement. Although Mn gives an effective scavenging effect on S segregation by precipitating S in the matrix, the Mn addition in alloy steels could increase the susceptibility to embrittlement due to the strong Mn-P and M-Sb interaction. The amount of free Mn in solid solution would be the determining factor.

Intermediate Elements (Mo, V & W). Mo, V and W have a stronger interaction with the impurities than Ni, Cr and Mn (i.e., α'_{MI} is positive and fairly large). Hence, these elements can precipitate the impurities in the matrix thereby reducing impurity segregation. The addition of Mo in alloy steels has been known to inhibit the segregation of P (98) and Sb (83,99). However, the behavior of Mo in alloy steels is somewhat complicated because it forms very stable carbides. Therefore, Mo shows the scavenging effect on impurity segregation only when it remains free in the elemental form. Recently, Yu and McMahon (100) have shown that the addition of Mo beyond about 0.7% increases the embrittlement susceptibility because the formation of Mo-rich carbides occurs at the expense of Cr-rich carbides

Table VII. The Effects of Impurities and Alloying Elements in Other Steels

Steel	Composition	Experimental Method	Result
12% Cr	Commercial E56C	Plane Strain K_{IC} (115)	K_{IC} is sensitive to temper embrittlement.
	Commercial E56C	K_{IC}, Crack- (116) Opening-Displacement (COD)	Temper embrittlement decreases K_{IC} & COD.
	Fe-12Cr-0.13C	Charpy-V Impact & AES (47)	Cr enhances P segregation at GB and scavenges P in the bulk.
	Fe-12Cr-2.5Ni-1.75Mo-0.3V-0.1C	Charpy-V Impact & AES (47)	Ni, Mo and V addition decreases embrittlement.
Ni-Cr	Fe-4Ni-2.25Cr-0.3C-0.02P	AES & FATT (89)	FATT increases with increasing X_P^ϕ linearly.
	Commercial EN30A (Fe-4Ni-1Cr-0.3C)	AES & (117) in situ Tensile Test	H_2 environment reduces the strain to fracture. P, Sn & N segregation with enhancement of Ni is observed.
Cr-Mo	3Cr-Mo turbine disc steel	Charpy FATT (46)	Continuous cooling embrittlement can be predicted from isothermal embrittlement data.
	Fe-2.5Cr-1Mo-0.2C-P	SAM & DBTT (100) (Slow bend test)	Cr enhances P segregation.
	Fe-2.5Cr-1Mo-0.2C-Sn	SAM & DBTT (100) (Slow bend test)	Cr enhances Sn segregation.
	Fe-2.5Cr-(0-1.3)Mo-0.2C-P	SAM & DBTT (100) (Slow bend test)	If X_{Mo}^B>0.7, the beneficial effect of Mo is decreased due to increasing X_{Cr}^B.
	Fe-2.5Cr-1Mo-0.2C-0.75Mn-P	SAM & DBTT (95) (Slow bend test)	Mn enhances P segregation due to Mn-P co-segregation.
	Fe-2.5Cr-1Mo-0.2C-0.6Si-P	SAM & DBTT (95) (Slow bend test)	Si enhances P segregation increasing P activity.
	Fe-2.5Cr-1Mo-0.2C-0.75Mn-0.6Si-P	SAM & DBTT (95) (Slow bend test)	Mn and Si appear to act independently.
Cr-Mo-V	Fe-1Cr-1Mo-0.25V and Fe-0.5Cr-0.5Mo-0.25V	Creep Test (9)	Impurities decrease rupture ductility and life in creep with an increase in number of cavities.
Ni-Cr-Mo-V	Fe-3.5Ni-1.7Cr-0.5Mo-0.15V	AES & FATT (118)	FATT increases with increasing segregation.
	Fe-5Ni-0.5Cr-0.5Mo-0.1V	Cantilever Bending (14)	Embrittled sample is much more sensitive to a hydrogen-producing oil environment than unembrittled sample.

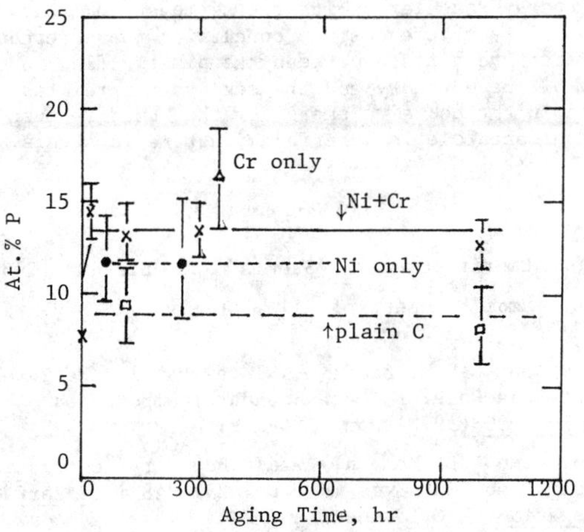

Figure 9. Comparison of grain boundary composition of P in Ni-Cr, Ni-only, Cr-only and plain C steels aged at 520°C (84).

increasing the concentration of free Cr which enhances P segregation. Therefore, the scavenging effect of Mo on impurity segregation depends on the kinetics of the carbide transformation reactions in alloy steels. In this case the interaction of the alloying elements themselves plays a significant role on the segreation behavior.

Strong Elements (Ti & Zr). Ti and Zr have a much stronger attractive interaction with the impurities (i.e., α'_{MI} is positive and very large). Hence, these elements have a strong tendency to form precipitates with the impurities in the matrix. This scavenging of the impurities reduces segregation. Ohtani et al (101, 102) have observed that the addition of 0.1% Ti in a low carbon Ni-Cr steel can eliminate most of the embrittlement susceptibility by scavenging impurity segregation. However, the behavior of Ti in alloy steels is complex because Ti interacts with C, O and even Ni. Ohtani et al (102) have suggested that the addition of Ti in alloy steels to eliminate impurity induced embrittlement must be carefully evaluated due to the formation of Ti carbides that can promote brittle cleavage fracture (103).

The Effect of Mechanical-Structural Factors

Hardness. The comparison of the degree of grain boundary embrittlement without the consideration of strength (i.e., hardness) can be misleading because the fracture mode is often determined by a balance between the strength of the grain and that of the grain boundaries. Viswanathan (89) has shown this in studies of ΔFATT. Recently, it has been demonstrated that the transition temperature increases monotonically with hardness for a given grain boundary concentration of P (84), Sn (106), Sb (87) and Si (106). Therefore, it can be concluded that grain boundary embrittlement is a function of hardness at a given grain boundary composition in a given steel.

Grain Size. It has been found that grain boundary embrittlement increases with prior austenite grain size (119). The effect of grain size could be explained by the analysis of the energy of brittle fracture. The larger the grain size the larger the probability that a brittle crack propagates beyond a triple-junction of grain boundaries. Recently, Ucisik et al (107) have shown that the embrittlement susceptibility can be eliminated by intercritical heat treatment which gives the main benefit of microstructural refinement. Briant and Banerji (105) also have found that the refinement of grain size gives the higher fracture energy for a tempered martensite embrittled steel doped with 0.01 P. However, the use of low temperature austenitizing to obtain grain size refinement could cause presegregation of P to the austenite grain boundaries of higher P contained steels. This promotes tempered martensite embrittlement (105).

Recently, Takayama et al (106) have developed a method for calculation of the DBTT expressed as a function of grain size, hardness and grain boundary composition by use of the Taylor expansion. Although the coefficients in this equation must be determined empirically for each type of steel, once achieved, the equation could be applied to predict the service failure of a steel.

Grain Boundary Morphology. For many years the intergranular fracture along the prior austenite grain boundaries has been thought to be an essential aspect of temper embrittlement. However, Ohtani and McMahon (104) have shown that temper embrittlement can involve low energy fractures along the other types of high-angle ferritic boundaries. They also have suggested that the frequent involvement of the prior austenite grain boundaries as the preferred fracture path arises from the fact that fracture along the coarser network of smooth and planar prior austenite grain boundaries requires less fracture energy than along the finer network of rough and ragged high-angle ferritic boundaries. Therefore, grain boundary morphology influences fracture modes and thereby grain boundary embrittlement.

Microstructure Type. It has been recognized that the embrittlement susceptibility depends on the different type of matrix microstructures of steels. Viswanathan and Joshi (120), through their study on the effects of microstructure in Cr-Mo-V steels, have shown that ΔFATT increases in the order - ferrite, bainite and martensite. They also found anomalous effects of hardness on the values of FATT and ΔFATT. Different regions of hardness levels exist wherein FATT and ΔFATT increases or decreases with increasing hardness (Figure 10). Recently, Yu and McMahon (100) have suggested that the anomalous effects of hardness could be explained best not in terms of microstructure or hardness alone, but by also taking into account the concentrations of Mo and Cr which influence the segregation of impurities. Therefore, the factors of hardness, microstructure and the degree of segregation must be considered simultaneously before a predictive approach is possible.

SUMMARY

The problem of interface failure in steels has been discussed in detail from both theoretical and experimental viewpoints. The theories described involved thermodynamic analyses associated with impurity atom segregation to the interfaces. The segregation of Group IVA to Group VI A elements has been shown to be involved in almost all of the interface embrittlement. Guttmann,

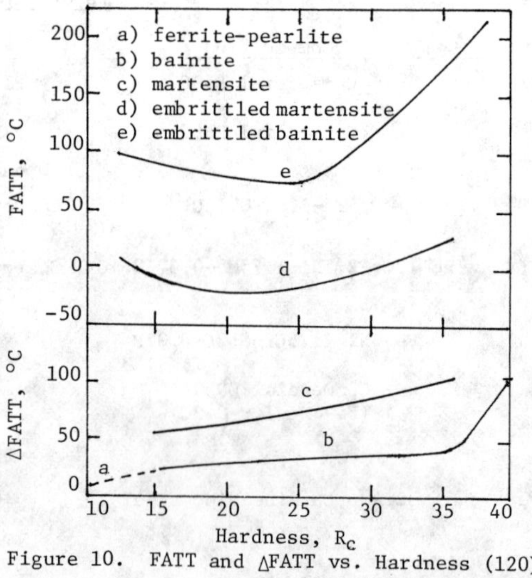

Figure 10. FATT and ΔFATT vs. Hardness (120).

using a regular solution model, demonstrated that alloying elements enhanced the segregation of the impurity elements. This enhanced segregation plays a dominant role in interface failure of low alloy steels. Theories describing the kinetics of segregation are also presented. The main difficulty with the equilibrium thermodynamics and the kinetics analysis is their inability to define the fracture criterion for the steels. To evaluate these fracture models, starting with the traditional surface energy models, enhanced by a nonequilibrium thermodynamic model to predict the change in cohesive energy with segregation and finally models that define an interactive mode between the plastic energy associated with fracture γ_p and the true fracture surface energy γ were delineated. The end result of the approaches is that the failure mechanism is not clearly defined. The dislocation model can lead to a definition of the onset of fracture. Unfortunately it still leaves the need to gloss over the actual dependence of γ on the segregation. The hypothesis of an ordered layered two-dimensional region at the boundary allows the interface weakening to be rationalized. The fracture models do not directly predict this type of interface weakening.

The experimental data confirm the grain boundary localized segregation of the Group IVA to Group VIA impurity elements and their enhanced segregation in the presence of alloying elements. The change in fracture energy is much greater than any extrapolated value of true surface energy associated with the segregation. The experimental result forces the conclusion to be reached that interactive effects between the plastic flow in the vicinity of the boundary and the segregation related reduction in the cohesive energy of the boundary must play a dominant role in interface fracture in steels.

REFERENCES

1. "Temper Embrittlement in Steels" ASTM STP-407 (1968).

2. "Temper Embrittlement of Alloy Steels", ASTM STP-499 (1972).

3. C.L. Briant & S.K. Banerji, "Intergranular Failure in Steel: the Role of Grain Boundary Composition", Int. Met. Rev. 23 (1978), pp. 164-199.

4. M. Guttmann & D. McLean, "Grain Boundary Segregation in Multicomponent Systems", pp. 251-348 in Interfacial Segregation, W.C. Johnson and J.M. Blakeley, eds., ASM Metals Park, OH (1979).

5. M. Guttmann, "The Role of Residuals and Alloying Elements in Temper Embrittlement", Phil. Trans. R. Soc. Lond. A295 (1980) pp. 169-196.

6. C.J. McMahon, Jr., "Solute Segregation and Intergranular Fracture in Steels: A Status Report", Mat. Sci. Eng. 42 (1980), pp. 215-226.

7. C.J. McMahon, Jr., V. Vitek and J. Kameda, "The Mechanisms and Mechanics of Intergranular Fracture" in Fracture Mechanics, vol. 2, G.G. Chell, ed. Appl. Sci. Publishers, London, 1981, to be published.

8. R. Viswanathan, "Effect of Sb, P, Sn and B on the Microstructure and Creep Properties of Normalized and Tempered 1.25 Cr - 0.5 Mo Steels", Met. Trans. 6A (1975) pp. 1135-1141.

9. H.R. Tipler and B.E. Hopkins, "The Creep Cavatation of Commercial and High Purity Cr-Mo-V Steels", Met. Sci. J. 10 (1976) pp. 47-56.

10. H.R. Tipler, "The Influence of Purity on the Strength and Ductility in Creep of CrMoV steels of Varied Microstructures", Phil. Trans. R. Soc. Lond. A 295 (1980) pp. 213-233.

11. T. Boniszewski and N.F. Eaton, "Electron Fractography of Weld-Reheat Cracking in CrMoV Steel", Met. Sci. J. 3 (1969) pp. 103-110.

12. J.C. Ritter and R. McPherson, "Stress-Relaxation Processes in Ferritic Steel", J. Iron Steel Inst. 211 (1973) pp. 123-128.

13. J.C. Ritter and R. McPherson, "Embrittlement of the Weld Heat-Affected-Zone in a Mn-Cr-Mo-V Steel", Met. Tech. 1 (1974) pp. 506-511.

14. K. Yoshino and C.J. McMahon, Jr., "The Cooperative Relation between Temper Embrittlement and Hydrogen Embrittlement in a High Strength Steel", Met. Trans. 5 (1974) pp. 363-370.

15. A. W. Thomson and I.M. Bernstein, "Effect of Hydrogen on Behaviour of Materials", TMS-AIME, New York, 1976.

16. J.R. Hirth, "Effects of Hydrogen on the Properties of Iron and Steel", Met. Trans. 11A (1980) pp. 861-890.

17. K-T. Aust, J.S. Armijo and J.H. Westbrook, "Heat Treatment and Corrosion Resistance of Austenitic Type 304 Stainless Steel", Trans. ASM, 59 (1966) pp. 544-556.

18. J.S. Armijo, "Intergranular Corrosion of Nonsensitized Austenitic Stainless Steels", Corrosion 24 (1968) pp. 24-30.

19. K.T. Aust, "Intergranular Corrosion of Austenitic Stainless Steels", Trans. TMS-AIME 245 (1969) pp. 2117-2126.

20. A. Joshi and D.F. Stein, "Chemistry of Grain Boundaries and its Relation to Intergranular Corrosion of Austenitic Stainless Steel", Corrosion 28 (1972) pp. 321-330.

21. W. Rostoker, J.M. McCaughey and H. Markus, Embrittlement by Liquid Metals, Reinhold, New York, 1969.

22. J.E. Draley and J.R. Weeks, Corrosion by Liquid Metals, Plenum Press, New York, 1970.

23. S. Dinda and W.R. Warke, "The Effect of Grain Boundary Segregation on Liquid Metal Induced Embrittlement of Steel", Mat. Sci. Eng. 24 (1976) pp. 199-208.

24. L.A. Harris, "Analysis of Materials by Electron-Excited Auger Electrons", J. Appl. Phys. 39 (1968) pp. 1419-1427.

25. L.A. Harris, "Some Observations of Surface Segregation by Auger Electron Emission", J. Appl. Phys. 39 (1968) pp. 1428-1431.

26. H.L. Marcus and P.W. Palmberg, "Auger Fracture Surface Analysis of a Tempered Embrittled 3340 Steel", Trans. TMS-AIME 245 (1969) pp. 1664-1666.

27. P.W. Palmberg and H.L. Marcus, "An Auger Spectroscopic Analysis of the Extent of Grain Boundary Segregation", Trans. ASM 62 (1969) pp. 1016-1018.

28. D.F. Stein, A. Joshi and R.P. Laforce, "Studies Utilizing Auger Electron Emission Spectroscopy on Temper Embrittlement in Low Alloy Steels", Trans. ASM 62 (1969) pp. 776-783.

29. D. McLean, Grain Boundaries in Metals, pp. 116-149, Oxford University Press, 1957.

30. I. Langmuir, "The Adsorption of Gases on Plane Surfaces of Glass, Mica and Plantinum", J. Am. Chem. Soc. 40 (1918) pp. 1361-1403.

31. R.H. Fowler and E.A. Guggenheim, Statistical Thermodynamics, pp. 421-451, Cambridge University Press (1939).

32. M.P. Seah and E.D. Hondros, "Grain Boundary Segregation", Proc. R. Soc. Lond. A335 (1973) pp. 191-212.

33. M.P. Seah and E.D. Hondros, "Use of a BET Analogue Equation to Describe Grain Boundary Segregation", Scripta Met 7 (1973) pp. 735-738.

34. M.P. Seah, "Interface Adsorption, Embrittlement and Fracture in Metallurgy", Surf. Sci. 53 (1975) pp. 168-212.

35. E.D. Hondros and M.P. Seah, "The Theory of Grain Boundary Segregation in Terms of Adsorption Analogues" Met. Trans. 8A (1977) pp. 1363-1371.

36. S. Brunauer, P.H. Emmett and E. Teller, "Adsorption of Gases in Multimolecular Layers", J. Am. Chem. Soc. 60 (1938) pp. 309-319.

37. E.D. Hondros and M.P. Seah, "Grain Boundary Activity Measurement by Auger Electron Spectroscopy", Scripta Met 6 (1972) pp. 1007-1012.

38. M. Guttmann, "Equilibrium Segregation in a Ternary Solution: A Model for Temper Embrittlement", Surf. Sci. 53 (1975) pp. 213-227.

39. M. Guttmann, "The Link Between Equilibrium Segregation and Precipitation in Ternary Solutions Exhibiting Temper Embrittlement", Met. Sci. J. 10 (1976) pp. 337-341.

40. M. Guttmann, "Grain Boundary Segregation, Two-Dimensional Compound Formation and Precipitation", Met. Trans. 8A (1977) pp. 1383-1401.

41. M. Guttmann, "Temper Embrittlement and Ternary Equilibrium Segregation", Mat. Sci. Eng. 42 (1980) pp. 227-232.

42. M.P. Seah, "Grain Boundary Segregation and T-t Dependence of Temper Embrittlement", Acta Met. 25 (1977) pp. 345-357.

43. F.L. Carr, M. Goldman, L.D. Jaffe, and D.C. Buffum, "Isothermal Temper Embrittlement of SAE 3140 Steel", Trans AIME 197 (1953) p. 998.

44. W.R. Tyson, "Kinetics of Temper Embrittlement", Acta Met. 26 (1978) pp. 1471-1478.

45. C. Lea and M.P. Seah, "Kinetics of Surface Segregation", Phil. Mag. 35 (1977) pp. 213-228.

46. B.L. King and G. Wigmore, "Temper Embrittlement in a 3-pct Cr-Mo Turbine Disc Steel", Met. Trans 7A (1976) pp. 1761-1767.

47. P. Lemble, A. Pineau, J.L. Castagne, P. Dumoulin and M. Guttmann, "Temper Embrittlement in 12% Cr Martensitic Steel", Met. Sci. J. 13 (1979) pp. 496-502.

48. M. Polani, "Uber die Natur des Zerreissvorganges", Z. Phys. 7 (1921) pp. 323-327.

49. E. Orowan, "Fracture and Strength of Solids," Rep. Prog. Phys. 12 (1948) pp. 185-232.

50. M.P. Seah, "Segregation and the Strength of Grain Boundaries," Proc. R. Soc. Lond. A349 (1976) pp. 535-554.

51. J.W. Gibbs, Collected Works, vol. 1, p. 219, Yale University Press, 1948.

52. J.R. Rice, "Hydrogen and Interfacial Cohesion" pp. 455-466 in Effect of Hydrogen on Behavior of Materials, A.W. Thomson and I.M. Berstein, ed. TMS-AIME, New York, 1976.

53. R.J. Asaro, "Adsorption-Induced Losses in Interfacial

Cohesion", Phil. Trans. R. Soc. Lond. A295 (1980) pp. 151-163.

54. J.P. Hirth, "Adsorption at Grain Boundaries and its Effect on Cohesion", Phil. Trans. R. Soc. Lond. A295 (1980) pp. 139-149.

55. J.P. Hirth and J.R. Rice, "On the Thermodynamics of Adsorption at Interfaces as it Influences Decohesion" Met. Trans. 11A (1980) pp. 1501-1511.

56. C.J. McMahon, Jr., V. Vitek and F.R. Belton, "On the Theory of Embrittlement of Steels by Segregated Impurities", Scripta Met. 12 (1978) pp. 785-789.

57. C.J. McMahon, Jr. and V. Vitek, "The Effects of Segregated Impurities on Intergranular Fracture Energy", Acta Met. 27 (1979) pp. 507-513.

58. M.E. Fine and H.L. Marcus, "Segregation to an Interface and Brittle Fracture of Metals", Met. Trans 2 (1971) pp. 1473-1474.

59. M.L. Jokl, V. Vitek and C.J. McMahon, Jr., "A Microscopic Theory of Brittle Fracture in Deformable Solids; a Relation between Ideal Work to Fracture and Plastic Work", Acta Met. 28 (1980) pp. 1479-1488.

60. J. Kameda and C.J. McMahon, Jr., "The Effects of Sb, Sn and P on the Strength of Grain Boundaries in a Ni-Cr Steel", Met. Trans 12A (1981) pp. 31-37.

61. J.P. Stark and H.L. Marcus, "The Influence of Segregation on Grain Boundary Cohesion", Met. Trans. 8A (1977) pp. 1423-1429.

62. C.L. Briant and R.P. Messmer, "Electronic Effects on Sulphur in Nickel: A Model for Grain Boundary Embrittlement", Phi. Mag. B42 (1980) pp. 569-576.

63. J.R. Low, Jr., D.F. Stein, A.M. Turkalo and R.P. Laforce, "Alloy and Impurity Effects on Temper Brittleness of Steel", Trans. TMS-AIME 242 (1968) pp. 14-24.

64. A. Joshi, "Experimental Techniques to Study Segregation", in Interfacial Segregation, W.C. Johnson and J.M. Blakely, eds. pp. 39-109, ASM, Metals Park, 1979.

65. R.W. Hertzberg, Deformation and Fracture Mechanics of Engineering Materials, pp. 297-324, John Wiley & Sons, New York, 1976.

66. M.C. Inman and H.R. Tipler, "Grain Boundary Segregation of Phosphorus in Iron-Phosphorus Alloy and the Effect upon Mechanical Properties", Acta Met. 6 (1958) pp. 73-84.

67. P.V. Ramasubramanian and D.F. Stein, "An Investigation of Grain Boundary Embrittlement in Fe-P, Fe-P-S, and Fe-Sb-S Alloys", Met. Trans. 4 (1973) pp. 1735-1742.

68. E.D. Hondros, "The Influence of Phosphorus in Dilute Solid Solution on the Absolute Surface and Grain Boundary Energies of Iron", Proc. R. Soc. Lond. A286 (1965) pp. 479-498.

69. W.J. Pardee, W.M. Robertson and M.R. James, "Tin Segregation at Iron Grain Boundaries Measured by EXAFS", Scripta Met 14 (1980), pp. 1333-1335.

70. A. Joshi and D.F. Stein, "Intergranular Brittleness in Tungsten Using Auger Spectroscopy", Met Trans. 1 (1970) pp. 2543-2546.

71. H.R. Tipler and D. McLean, "Influence of Interface Energy on Creep Rupture", Met. Sci. J. 4 (1970) pp. 103-107.

72. C.J. McMahon, Jr., "Intergranular Fracture in Steels" Mat. Sci. Eng. 25 (1976) pp. 233-239.

73. C.J. McMahon, Jr. and L. Marchut, "Solute Segregation in Iron-Based Alloys", J. Vac. Sci. Tech. 15 (1978) pp. 450-466.

74. C. Pichard, J. Rieu and C. Goux, "The Influence of Oxygen and Sulfur on the Intergranular Brittleness of Iron", Met. Trans. 7A (1976) pp. 1811-1815.

75. A. Kumar and B.L. Eyre, "Grain Boundary Segregation and Intergranular Fracture in Molybdenum", Proc. R. Soc. Lond. A370 (1980) pp. 431-458.

76. M.P. Seah and C. Lea, "Surface Segregation and its Relation to Grain Boundary Segregation", Phil. Mag. 31 (1975) pp. 627-645.

77. C. Lea and M.P. Seah, "Site Competition in Surface Segregation", Surf. Sci. 53 (1975) pp. 272-285.

78. J.R. Rellick, C.J. McMahon, Jr., H.L. Marcus and P.W. Palmberg, "The Effect of Tellurium on Intergranular Cohesion in Iron", Met. Trans. 2 (1971) pp. 1492-1494.

79. J.M. Papazian and D.N. Beshers, "Grain Boundary Segregation of Carbon in Iron", Met. Trans 2 (1971) pp. 497-503.

80. E.D. Hondros, "Interfacial Segregation of Nitrogen in Iron", Met. Sci. J. 1 (1967) pp. 36-39.

81. B.D. Powell and H. Mykura, "The Segregation of Bismuth to Grain Boundaries in Copper-Bismuth Alloys", Acta Met. 21 (1973) pp. 1151-1156.

82. P. Dumoulin and M. Guttmann, "The Influence of Chemical Interactions between Metallic and Metalloid Solutes on Their Segregation in α-Fe: Co-segregation at Free Surfaces Studied by Auger Electron Spectroscopy", Mat. Sci. Eng. 42 (1980) pp. 249-263.

83. H.L. Marcus, L.H. Hackett and P.W. Palmberg, "Effect of Solute Elements on Temper Embrittlement of Low Alloy Steels", in Temper Embrittlement of Alloy Steels, pp. 91-103, ASTM STP-499, 1972.

84. R.A. Mulford, C.J. McMahon, Jr., D.P. Pope and H.C. Feng, "Temper Embrittlement of Ni-Cr Steels by Phosphorus", Met. Trans. 7A (1976) pp. 1183-1195.

85. A. Joshi and D.F. Stein, "Temper Embrittlement of Low Alloy Steels", in Temper Embrittlement of Alloy Steels, pp. 59-89, ASTM STP-499, 1972.

86. B.C. Edwards, B.L. Eyre and T.E. Cranshaw, "Ni-Sn Interaction in Temper Embrittled Steel Detected by Mossbauer Spectroscopy", Nature 269 (1977) pp. 47-48.

87. R.A. Mulford, C.J. McMahon, Jr., D.P. Pope and H.C. Feng, "Temper Embrittlement of Ni-Cr Steel by Antimony: III. Effects of Ni and Cr", Met. Trans. 7A (1976) pp. 1269-1274.

88. M. Maurer, M.C. Cadeville and J.P. Sanchez, "Influence of Nickel and Chromium on Antimony Segregation in Iron-Based Alloys", Phil. Mag. A38 (1978), pp. 739-746.

89. R. Viswanathan, "Temper Embrittlement in a Ni-Cr Steel Containing Phosphorus as Impurity", Met. Trans. 2 (1971) pp. 809-815.

90. A.K. Cianelli, H.C. Feng, A.H. Ucisik and C.J. Mc Mahon, Jr., "Temper Embrittlement of Ni-Cr Steel by Sn", Met. Trans. 8A (1977) pp. 1059-1061.

91. H. Ohtani, H.C. Feng, C.J. McMahon, Jr. and R.A. Mulford, "Temper Embrittlement of Ni-Cr Steel by Antimony: I. Embrittlement at Low Carbon Concentration", Met. Trans. 7A (1976), pp. 87-101.

92. P.R. Krahe and M. Guttmann, "On the Segregation of Manganese and Antimony to the Grain Boundaries of Temper Embrittled Steel", Scripta Met. 7 (1973) pp. 387-394.

93. M. Guttmann, P.R. Krahe, F. Abel, G. Amsel, M. Bruneaux and C. Cohen, "Temper Embrittlement and Intergranular Segregation of Antimony: A Quantitative Analysis Performed with the Backscattering of Energetic Ions", Met. Trans. 5 (1974), pp. 167-177.

94. J.P. Coad, J.C. Riviere, M. Guttmann and P.R. Krahe, "Reversibility of Temper Embrittlement Studied by Combined XPS and AES", Acta Met. 25 (1977), pp. 161-

95. J. Yu and C.J. McMahon, Jr. "The Effects of Composition and Carbide Precipitation on Temper Embrittlement of 2.25 Cr - 1 Mo Steel: Part II. Effects of Mn and Si", Met. Trans. 11A (1980) pp. 291-300.

96. A. Joshi, P.W. Palmberg and D.F. Stein, "Role of Mn and Si in Temper Embrittlement of Low Alloy Steels", Met. Trans 6A (1975) pp. 2160-2161.

97. C.L. Briant, H.C. Feng and C.J. McMahon, Jr., "Embrittlement of a 5 pct Nickel High Strength Steel by Impurities and their Effects on Hydrogen-Induced Cracking", Met. Trans 9A (1978) pp. 625-633.

98. C.J. McMahon, Jr., A.K. Cianelli and H.C. Feng, "The Influence of Mo on P-Induced Temper Embrittlement in Ni-Cr Steel", Met. Trans 8A (1977) pp. 1055-1057.

99. B.J. Schulz and C.J. McMahon, Jr., "Alloy Effects in Temper Embrittlement" in Temper Embrittlement of Alloy Steels, pp. 104-135, ASTM STP-499, 1972.

100. J. Yu and C.J. McMahon, Jr., "The Effects of Composition and Carbide Precipitation on Temper Embrittlement of 2.25 Cr - 1 Mo Steel: Part I. Effects of P and Sn", Met. Trans. 11A (1980) pp. 277-289.

101. H. Ohtani, H.C. Feng and C.J. McMahon, Jr., "New Information on the Mechanism of Temper Embrittlement of Alloy Steels", Met. Trans. 5 (1974) pp. 516-518.

102. H. Ohtani, H.C. Feng and C.J. McMahon, Jr., "Temper Embrittlement of Ni-Cr Steel by Antimony: II. Effects of Addition of Titanium", Met. Trans. 7A (1976) pp. 1123-1131.

103. C.J. McMahon, Jr. and M. Cohen, "Initiation of Cleavage in Polycrystalline Iron", Acta Met. 13 (1965) pp. 591-604.

104. H. Ohtani and C.J. McMahon, Jr., "Modes of Fracture in Temper Embrittled Steels", Acta Met. 23 (1975) pp. 377-386.

105. C.L. Briant and S.K. Banerji, "Tempered Martensite Embrittlement in Phosphorus Doped Steels", Met. Trans. 10A (1979) pp. 1729-1737.

106. S. Takagama, T. Ogura, S. Fu and C.J. McMahon, Jr., "The Calculation of Transition Temperature Changes in Steels due to Temper Embrittlement", Met. Trans. 11A (1980) pp. 1513-1529.

107. A.H. Ucisik, C.J. McMahon, Jr. and H.C. Feng, "The Influence of Intercritical Heat Treatment on the Temper Embrittlement Susceptibility of a P-Doped Ni-Cr Steel", Met. Trans. 9A (1978) pp. 321-329.

108. C.J. McMahon, Jr., E. Furubayashi, H. Ohtani and H.C. Feng, "A Study of Grain Boundaries during Temper Embrittlement of a Low Carbon Ni-Cr Steel doped with Antimony", Acta Met. 24 (1976) pp. 695-704.

109. J. Kameda and C.J. McMahon, Jr., "Solute Segregation and Brittle Fracture in an Alloy Steel", Met. Trans. 11A (1980) pp. 91-101.

110. A. Joshi, "Segregation at Selective Grain Boundaries and its Role in Temper Embrittlement of Low Alloy Steels", Scripta Met. 9 (1974) pp. 251-260.

111. C.L. Smith and J.R. Low, "Effect of Prior Austenitic Grain Boundary Composition on Temper Brittleness in a Ni-Cr-Sb Steel", Met. Trans 5 (1974) pp. 279-287.

112. C.L. Briant and S.K. Banerji, "Tempered Martensite Embrittlement and Intergranular Fracture in an Ultra-High Strength Sulfur Doped Steel", Met. Trans 12A (1981) pp. 309-319.

113. C.L. Briant and S.K. Banerji, "Tempered Martensite Embrittlement in a High Purity Steel", Met. Trans. 10A (1979) pp. 1151-1155.

114. M. Schmerling, D. Finello and H.L. Marcus, "Anomalous Sputtering Effects in the AES Grain Boundary Analysis of Reversible Temper Embrittled Low Alloy Steels", Scripta Met. 14 (1980) pp. 1135-1138.

115. G. Berry and R. Brook, "Tempered-Martensite Embrittlement in a 12% Cr Steel", Met. Sci. J. 9 (1975) pp. 467-471.

116. G. Berry and R. Brook, "The Influence of Temper Embrittlement on Fracture Toughness", Met. Sci. J. 10 (1976) pp. 286-290.

117. B.C. Edwards, H.E. Bishop, J.C. Riviere and B.L. Eyre, "An AES Study of Temper Embrittlement in a Low Alloy Steel", Acta Met. 24 (1976) pp. 957-967.

118. R. Viswanathan and T.P. Sherlock, "Long-time Isothermal Temper Embrittlement in Ni-Cr-Mo-V Steels", Met. Trans 3 (1972) pp. 459-468.

119. J.M. Capus, "Austenite Grain Size and Temper Brittleness", J. Iron Steel Inst. 200 (1962) pp. 922-927.

120. R. Viswanathan and A. Joshi, "Effect of Microstructure on the Temper Embrittlement of Cr-Mo-V Steels", Met. Trans 6A (1975) pp. 2289-2297.

STRESS CORROSION CRACKING AND HYDROGEN EMBRITTLEMENT

Anthony W. Thompson
I. M. Bernstein
Dept. of Met. and Mat. Science
Carnegie-Mellon University
Pittsburgh, PA 15213

INTRODUCTION

It has now been over 100 years since it was first reported that steel wires which had been immersed in an acid solution were made brittle (1-3). These earliest observations represent, from the perspective of current understanding, a most interesting combination of a corrosive environment (the acid), generation of hydrogen (observed as bubbles of H_2 released at the wire surface) which entered the wire, and subsequent stressing to demonstrate brittleness. It was recognized by these early investigators, moreover, that hydrogen was the cause of the brittleness.

It has gradually come to be accepted in the intervening years that simultaneous exposure to stress and a corrosive environment can result in fracture, now called stress corrosion cracking or SCC, for particular combinations of environment, stress level, and material. There are, however, many combinations of these three variables which do not result in fracture, so that one aspect of the environmental fracture problem which requires understanding is its specificity in this regard. Throughout the same years, the particular effects of hydrogen have also been studied, and a series of review papers in recent decades have summarized the earlier work (4-7). In the same time period, a great variety of industrial failures have been traced to the effects of hydrogen, including failures involving oil and gas well casings, aircraft landing gear, nuclear reactor power systems, chemical and petroleum industry refining and processing equipment, fuel tanks and thrust chambers for rocket engines, turbines for generation of electrical power, piping and valves for transport of liquids and gases, as well as a wide spectrum of other applications involving high-strength materials.

Both laboratory experiments and industrial service failures have an important characteristic in common, namely a distinctly catastrophic aspect. This is because of the nature of many SCC and hydrogen embrittlement failures: the fractures occur at an unexpectedly low stress relative to behavior in the absence of the environment, are usually quite brittle, and have a highly variable (and unpredictable) incubation period. Behavior of this type in a structural material is a virtual prescription for a technological disaster (8), and it is clearly essential to avoid such behavior in any material which will be exposed to hydrogen or a hydrogen-bearing environment in service. When this is not satisfactorily done, considerable damage can occur.

MANIFESTATIONS OF ENVIRONMENTAL SUSCEPTIBILITY

Whether the problem to be treated is one in which hydrogen alone is known to be the agent of environmental damage, or whether the damaging agent in the aggressive environment is unknown, there are certain typical manifestations of susceptibility to fracture in the environment. These are briefly summarized here to elucidate the significance of the various types of engineering data which exist. The classical environmental embrittlement phenomenon is, as mentioned above, a delayed, brittle failure. However, since this is a complex phenomenon, discussion of it is postponed in order to present first a somewhat simpler case- tensile testing with a steadily rising load.

Tensile Tests

When a cylindrical metal bar is pulled to failure in tension, the failure normally occurs by a plastic instability process called "necking", in which the reduction of cross-sectional area becomes localized rather than occurring over the entire bar. The final fracture area, A_f, is compared to the initial area, A_0, in terms of the reduction of area, or RA:

$$RA = \frac{A_0 - A_f}{A_0} \qquad [1]$$

The advantage of this measure of fracture ductility is that it is readily converted to true fracture strain, ε_f:

$$\varepsilon_f = \ln\left(\frac{1}{1-RA}\right), \qquad [2]$$

The RA is also a much more fundamental measure of ductility than, for example, the total elongation. To assess environmental effects, one compares the RA in the presence of the environment to the RA in a nominally inert environment. For the case of hydrogen, then, one would use the RA in a hydrogen test, RA_H, to compute the loss in RA caused by hydrogen:

$$RA\ loss = \frac{RA - RA_H}{RA} \qquad [3]$$

Both RA and RA loss are often expressed as a percentage.

Tensile testing of this type has long been widespread in hydrogen studies. As an example, Fig. 1 shows the classical expression of temperature and strain rate dependencies of hydrogen-induced ductility loss. Here the true strain at fracture, ε_f, was determined for a spheroidized plain carbon steel, SAE 1020, by Toh and Baldwin (9). Similarly, the dependence on hydrogen content is often found to be as in Fig. 2. Although other types of materials behave differently in detail, the qualitative behavior of Figs. 1 and 2 is often observed.

The strength of a material can also be adversely affected by exposure to the environment; this usually occurs when ε_f is reduced to less than the uniform strain, ε_u, the true strain at which necking begins and at which the specimen carries its maximum load. However, strength-related properties cannot realistically be studied without introducing the element of time under load. This is the province of the delayed failure test.

Delayed Failure and Fracture Mechanics

When a smooth-bar tensile specimen is held under fixed load at ambient temperature, it ordinarily can sustain the load indefinitely. However, if it is susceptible to an

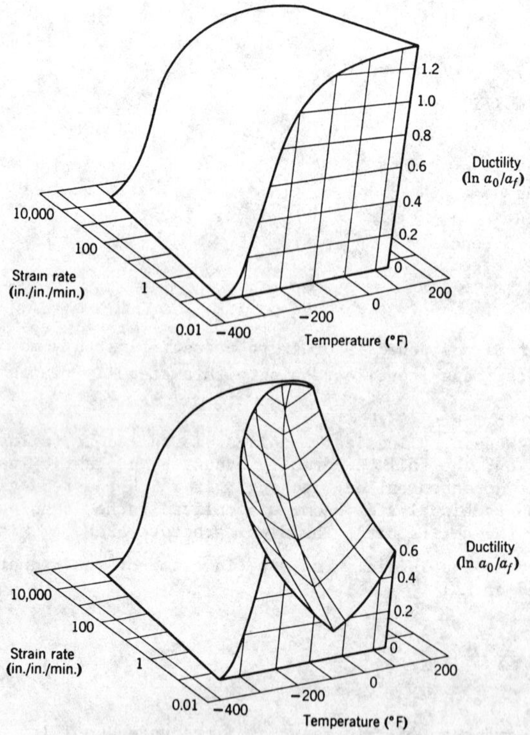

Fig. 1. Ductility, expressed as strain to fracture (e_f), of spheroidized 1020 steel as a function of strain rate and temperature. Hydrogen-free specimens, top; hydrogen-charged specimens, bottom. From Toh and Baldwin (9).

aggressive environment, the same test conducted in the environment (or after hydrogen charging) will result in abrupt failure after some time under load. The results from a series of specimens held at different loads would typically be observed to fit the following pattern. After an initial period, crack initiation would be detected, and then after a period of crack growth, failure would occur. It is often observed that data for both initiation and failure are virtually stress independent over a wide range of stress, as illustrated in Fig. 3. Note also the "threshold" behavior, often denoted by K_{th}. Although this type of test is relatively easy to perform (usually only the time to failure or TTF is monitored), it has its drawbacks, because crack initiation, crack propagation, and unstable or fast final fracture also occur during the test.

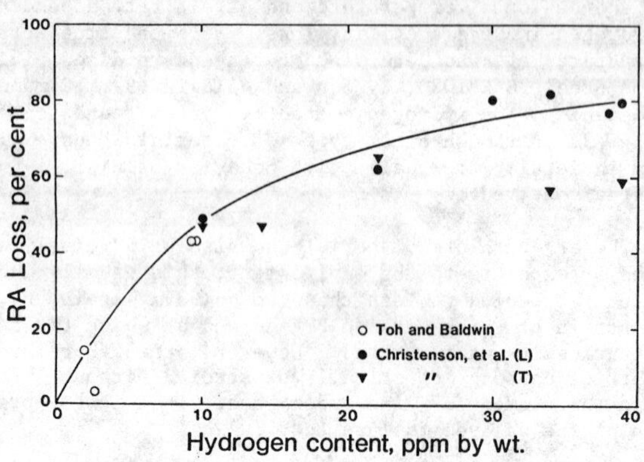

Fig. 2. Example of relation between ductility (RA) loss and hydrogen content; exact relation differs from alloy to alloy. Data for spheroidized 1020 steel (9) and for normalized 0.10% C, 0.66% Mn steel pipe in longitudinal (L) and transverse (T) orientations (D. J. Christenson, E. J. Danielson, A. W. Thompson and I. M. Bernstein, unpublished research, Carnegie-Mellon University).

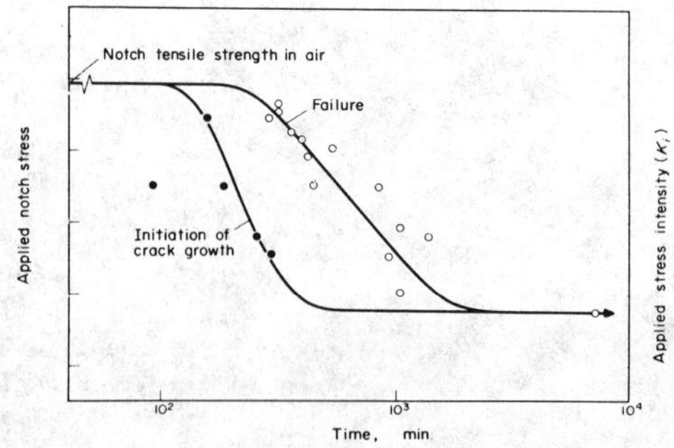

Fig. 3. Typical representation of delayed failure, such as in a high-strength steel. Times shown and scatter in data are typical; time scale is usually logarithmic. "Initiation" in the precracked specimens of this figure refers to the start of crack motion. Lowest stress intensity which causes failure is called K_{th}, the threshold K for failure.

This kind of delayed failure test can also be conducted using precracked specimens, in which case fracture mechanics analyses can be used, and the essential parameter is the stress intensity, K. In general, for many kinds of crack geometry and specimen configuration, the Mode I (crack opening) stress intensity is given by

$$K_I = \text{constant } (\sigma \sqrt{a}) \tag{4}$$

For a single-ended crack remote from a free surface, the constant is about unity. For a precracked specimen, then, an initial K rather than an initial stress is specified, and one still observes a TTF diagram similar to Fig. 3, except that no crack initiation takes place.

More information is obtained by measuring the crack velocity in the environment of interest, rather than TTF alone. The results of the crack growth type of test are helpful in understanding how long a material can remain in service under load (with a pre-existing flaw). As will be seen below, materials typically differ greatly in this respect. Techniques of crack velocity testing will not be described here, but have been discussed elsewhere (10).

In each of these mechanical tests, and particularly in ones of long duration, it is essential to control the environment itself, if meaningful results are to be obtained. Such control would include, as examples, purity and (partial) pressure of hydrogen in gaseous tests; solution chemistry, pH, and viscosity in liquid environments; and any other relevant variables, such as temperature, applied potential, relative humidity, and so forth.

There is one other manifestation of environmental effects which is touched on below, that of fracture surface appearance. Understanding the changes in this appearance which can be induced by the environment can be quite helpful in understanding the effects of microstructure on embrittlement.

Fracture Surface Appearance

An important adjunct to the type of tests shown in Figs. 1 and 3 is examination of the fracture surface appearance. The appearance at moderately high magnification is often altered by the environment, is usually complex, and may fall into one of four classical modes (detailed discussion of these is available in refs. 11, 12). A fracture "mode" refers essentially to a characteristic set of features which can be observed, and which can be interpreted in terms of the means by which the fracture process proceeded (11,12). The terminology for these is accordingly brief, and need not be taken literally as descriptors of the precise process of fracture; for example, the term "cleavage" should not be understood to mean

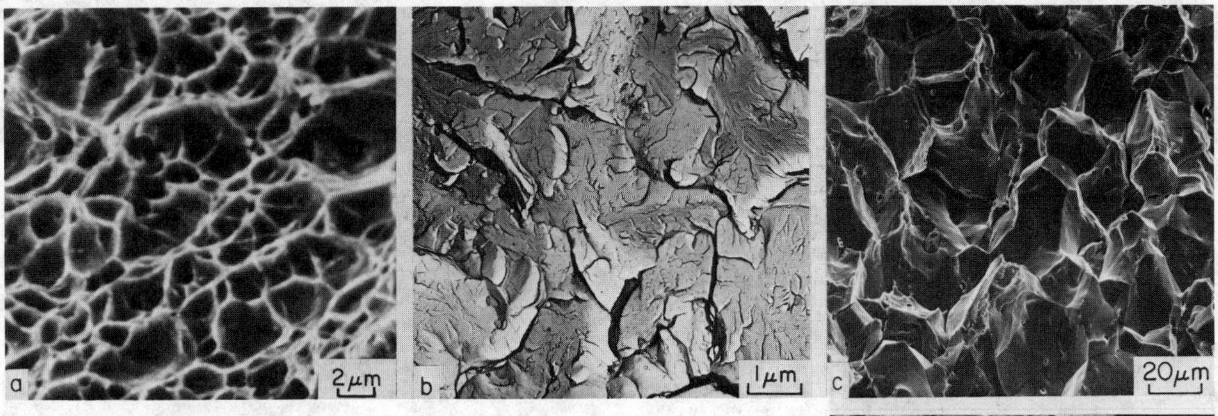

Fig. 4. Examples of microscopic modes of fracture. (a) Microvoid coalescence in hydrogen-charged 304L stainless steel, with about 50% loss in RA. (b) Quasi-cleavage fracture in hydrogen-charged 410 stainless steel, about 80% loss in RA. (Photograph courtesy of C.D. Beachem.) (c) Intergranular fracture in hydrogen-charged nickel, with about 75% loss in RA. (d) Cleavage fracture caused by hydrogen in Cb-752 (Nb-10% W-2.5% Zr) with 75% loss in elongation. (Photograph courtesy of M. W. Mahoney.)

ideally brittle separation along an atomic lattice plane, but merely as a convenient term to describe a fracture type. The same can be said for each of usually-listed four modes. These are illustrated in Figure 4; they can be regarded as points along a spectrum of relative ductility, ranging from very ductile modes like microvoid coalescence (MVC), through poorly defined transgranular modes like quasi-cleavage (QC), to relatively brittle modes like intergranular fracture (IG) and cleavage (C).

It is also important to recognize that the fracture mode in the presence of an environment depends on the K value, i.e., upon the crack length in a fixed-displacement test (10). The reason for this is that as K decreases, the associated plasticity at the crack tip decreases and is only able to drive less-plastic fracture modes. This is schematically illustrated in Figure 5, together with a brief indication of how the fracture modes of Fig. 4 take place in the microstructure of the material. Figure 5(a) shows growth of holes around precipitate particles, leading to half-voids on the fracture surface; this is called microvoid coalescence, as in Figure 4(a). In Figure 5(b) a locally plastic fracture takes place across grains, which can correspond to several kinds of ductile testing, including quasi-cleavage (Fig. 4(b)). When plasticity is still more restricted, IG fracture, as Figs. 4(c) and 5(c), or cleavage, Fig. 4(d), may occur, depending on the relative strength of both the grain boundary and the lattice, and on the distribution of second-phase particles which can act to blunt either crack type.

FRACTURE IN CORROSIVE ENVIRONMENTS

A variety of approaches have been taken to fractures in corrosive environments. It is useful to begin with a discussion of events on an initially smooth specimen surface, for example a tensile or bend bar. Brown (13) has summarized a typical sequence of events with the sketch in Fig. 6. The initiation process is depicted at the left of Fig. 6, and it is generally agreed that chemical and electrochemical factors are predominant here. As the sequence continues (toward the right in Fig. 6), there appears a mechanical aspect as well. There may in many cases be an interplay between the electrochemical and the mechanical aspects, especially in ductile materials which tend to resist crack growth by a blunting process, and in these cases localized electrochemical dissolution, or pitting, can then resharpen the crack tip and cause another increment of crack growth. We emphasize that a cooperative series of steps such as these, required to operate in sequence, may occur in many instances of SCC (14,15). In some cases, such as titanium alloys, pitting is virtually absent, and a sharp precrack may have to be introduced by

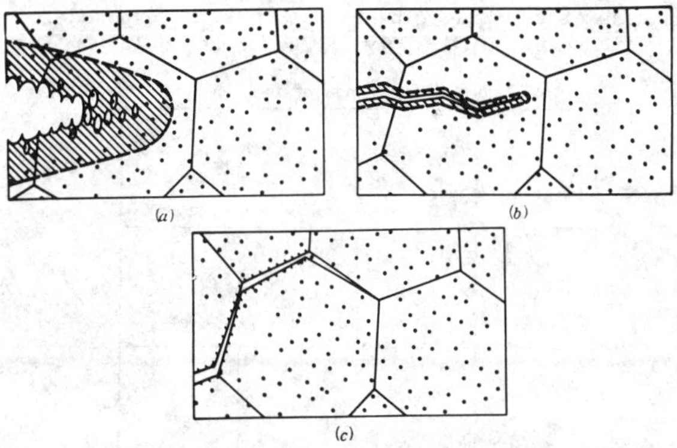

Fig. 5. Schematic depiction of microstructural influence on fracture modes. (a) Microvoid coalescence, typical of fracture at high K. (b) Transgranular tearing, e.g. quasi-cleavage, at intermediate K. (c) Intergranular fracture at low K, approaching the crack growth threshold. From Beachem (12).

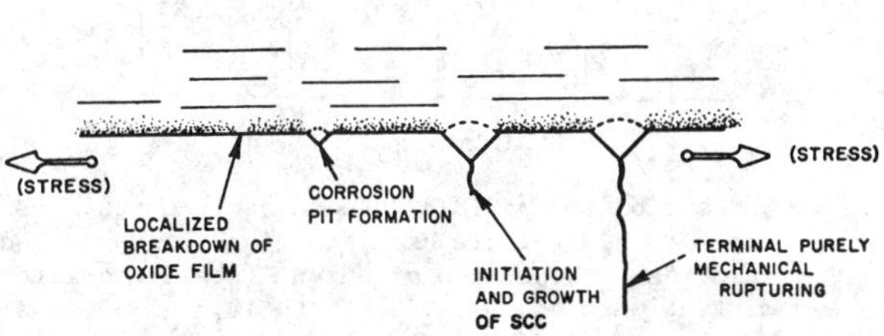

Fig. 6. Schematic sequence of events, from left to right, which could occur during SCC at a smooth specimen surface or at a film-covered crack tip. From Brown (13).

fatigue in order to observe any SCC at all. At the opposite extreme of very low toughness materials, fast fracture can be caused by the pit alone, acting as a critical-size flaw (13).

A particular environment may cause cracking only for particular combinations of stress and material, as mentioned. Moreover, the aggressive species in the environment may be present in very low concentrations, parts per million in many cases, and still be responsible for cracking. A well-known example of this phenomenon is Fig. 7, obtained for an austenitic stainless steel by Williams and Eckel (16), and illustrating two of the ubiquitous components of acqueous solutions, oxygen and chloride.

It is often appropriate, from a metallurgical standpoint, to separate the control variables for SCC into "external" (environment and stress) and "internal" (material) variables. In some applications, such as the primary coolant loop in a nuclear power plant, the environment (high-temperature water) is a "closed" system. The environmental chemistry can be designed to suit the materials of construction, and potentially damaging species such as oxygen and chloride (recall Fig. 7) can be monitored and controlled. The more usual situation, however, is that the environment cannot be manipulated, for example for ship steels in the ocean. In this case, material variables and design stresses must be consistent with the environment. (The material may be protected from general corrosion by the use of sacrificial anodes, in which case one must identify the electrochemical potential which will operate at the steel surface.) Much of this chapter emphasizes internal rather than external variables, simply because of the metallurgical focus of this volume and of our own work.

A material exposed to a corrosive environment may either corrode continuously, or may form a passive film which then retards or prevents further corrosion. An SCC process of the type shown in Fig. 6 does not typically occur for material-environment combinations which give rise to continuous corrosion. Instead, it is usually the case that materials which resist general corrosion well are the most likely candidates for SCC susceptibility. It follows that even within a growing stress corrosion crack, the material will tend to form a passive film. In order to advance the crack by dissolution of additional metal (one means of cracking), bare metal must be exposed for the anodic reaction, $M \rightarrow M^{+y} + ye^-$, to occur rapidly. For

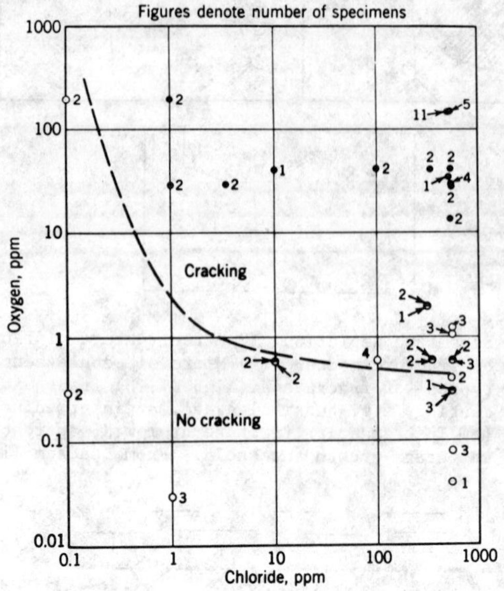

Fig. 7. Susceptibility of 18-8 type austenitic stainless steel to SCC failure as a function of oxygen and chloride content of environment (steam and intermittent wetting, 515-535 K, pH 10.6, 1-30 days exposure). Open points denote no failure; filled points denote failure. Numbers beside points indicate number of specimens. Data of Williams and Eckel (16).

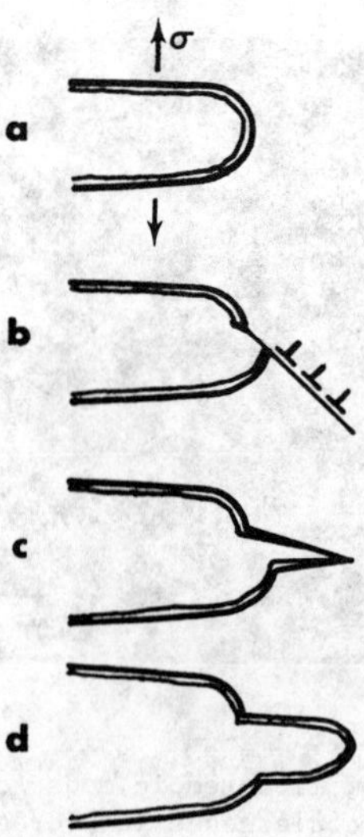

Fig. 8. Schematic depiction of crack tip film rupture and repassivation. (a) Crack tip with continuous, irregular film, with applied stress as shown. (b) Operation of dislocation source near crack tip creates slip step and ruptures film. (c) Anodic dissolution and/or hydrogen embrittlement occurs on bare metal of step until film re-forms. (d) Crack tip re-blunts and assumes configuration of (a). Sketch based on concepts of Scully (22).

cracking to occur by hydrogen embrittlement, the cathodic reaction, $H^+ + e^- \rightarrow H$, must occur, also on bare metal if hydrogen is to be absorbed rather than evolved as gas. Of course, the anodic and cathodic reactions both occur and are balanced; but crack advance can be imagined to occur predominantly as a result of one of the reactions.

The extent to which crack advance can occur should depend on the relative time during which bare metal is exposed. One means of exposure is through creation of surface steps by slip at the crack tip, as sketched in fig. 8. It is evident that a competition will exist between repassivation of the slip step by re-formation of the passive film, and the occurrence of electrochemical reactions leading to anodic dissolution, hydrogen embrittlement, or both. Thus an important focus of SCC studies (17) has been to understand the repassivation rate, as well as the rate of film rupture by slip steps, since these should control the rate of crack propagation.

An even wider variety of similar processes could operate to produce pitting or crack initiation (left side of Fig. 6); it is evident that anodic dissolution of slip steps, chemically controlled pitting, passive film rupture, and other phenomena could singly or in combination provide both the initiation site and the stress concentration needed for cracking to begin. These particular phenomena may of course be quite different in different alloy systems, as we have reviewed (14). Thus control of applied potential, anion concentration, or solution pH in a manner which retards the relevant phenomena would markedly delay initiation and therefore delay the onset of SCC, whatever the cracking process.

In ductile materials which resist crack extension by massive plastic flow and crack blunting, crack propagation is only possible if repeated crack initiation (i.e., re-sharpening) can occur. This can take place by the occurrence of phenomena similar to those of initiation, except that environmental chemistry within a crack tends to differ markedly from that at the external surface; e.g., pH may be reduced to very low values within the crack. This effect occurs by hydrolysis of dissolved metal ions, M^{+y}, as follows (18):

$$M^{+y} + xH_2O \rightarrow M(OH)_x^{y-x} + xH^+ \qquad [5]$$

Although this effect tends to reduce dependence on the external environment, as can the potential shift within a crack (19), it is still true that the environment can play a profound role in the rate (or even the occurrence) of resharpening, just as is true for initiation.

Thus both the initiation of a SCC crack, and the rate of its propagation, can be profoundly affected by a variety of electrochemical factors, acting singly or in combination (14,17,20-22). Furthermore, when second phases or other microstructural elements are favorable sites for attack, the crack path may be guided by the preferred location of initiation or resharpening. The presence of chloride ion is of particular importance in accentuating such preferred locations in a number of alloy systems, as has been reviewed widely (13,14,20) and as is indicated in Fig. 7 for one particular alloy. Thus electrochemical means of crack advance must certainly be regarded as contributing to or controlling crack advance during many instances of SCC (14, 17,20,22,23).

At the same time, it may be that the various complexities of SCC rate control through external variables merely reflect in some cases, the rates of generation of and embrittlement by hydrogen. Such processes are discussed in the following section. It should be emphasized that hydrogen production at or near the crack tip does depend on environmental factors, although the occurrence of hydrolysis and of potential shifts as large as one volt within the crack (19) means that the external environment is not always the controlling factor. The latter is of particular importance when SCC occurs under an applied anodic potential (relative to open-circuit conditions).

To summarize, the various electrochemical factors in SCC can be integrated by examination of the repassivation rate of the particular alloy system in the environment of interest. Fig. 8 shows the process schematically. As was mentioned above, the repassivation rate is an indication of the extent to which freshly exposed slip steps can locally dissolve, or pit, and also of the relative rate of the local corrosion reaction which generates hydrogen. Thus the interplay and net effect of the various electrochemical processes which are occurring are combined in the repassivation rate, although that rate does not, of itself, indicate the mechanism of cracking.

HYDROGEN EMBRITTLEMENT

Hydrogen embrittlement can occur under exposure to corrosive environments, as indicated above, or, more classically, it occurs upon exposure to hydrogen gas or to cathodic (electrochemical) charging of hydrogen.

A summary of the generalized processes which occur during hydrogen embrittlement is presented at this point in the form of Fig. 9 to form a basis for discussion below of metallurgical variables, and also to clarify the points at which stress corrosion cracking (SCC) can include contributions from hydrogen processes. These processes can be described under the following headings; each refers to a portion of Fig. 9.

<u>Hydrogen sources</u>

Hydrogen may be derived from hydrogen gas molecules which adsorb onto the metal surface and dissociate; by recombination of hydrated protons, H_3O^+, with electrons at the metal surface; or by reaction between hydrogen-containing molecules such as alcohols and the metal (or oxide) surface to release hydrogen. Since all of these sources give rise to hydrogen in solution, it is irrelevant which one occurs; in clearly-defined cases, as one would expect, hydrogen gas results are entirely similar to SCC data (14). There can, however, be marked <u>kinetic</u> effects if the hydrogen entry step happens to be rate-controlling.

<u>Hydrogen transport</u>

Hydrogen diffuses rapidly in most metals and extremely rapidly in bcc metals. Yet another transport process also occurs which can be even faster: motion of hydrogen, in cares or as Cottrell atmospheres, accompanying moving dislocations. Calculations show that dislocation transport of this type can be several orders of magnitude more rapid than lattice diffusion (24). This means that observation of environmental cracking rates which exceed the rate of hydrogen diffusion (14) do not rule out a hydrogen role in that cracking process.

<u>Hydrogen destinations</u>

Hydrogen tends to accumulate at a wide variety of locations in a microstructure. Among these are grain boundaries, inclusions, voids, dislocations and dislocation arrays, solute atoms, etc., as well as in solid solution. Whichever of these locations is the most sensitive to fracture will control the magnitude of hydrogen effects, although in general all of them will accumulate hydrogen, albeit to different extents.

<u>Fracture</u>

Hydrogen-induced fracture is usually thought of as brittle, and intergranular fractures are in fact common; but many materials exhibit ductile fracture (with reduced ductility) in the presence of hydrogen. Cleavage can also occur, either in the matrix or through a precipitated hydride phase. In fact, all the classical fracture modes, Fig. 4, are observed in the presence of hydrogen (25).

This brief description (more complete discussion is available (14) in a comprehensive review) indicates how hydrogen effects can occur in relation to the microstructure. Of particular interest is the (potential) control of fracture behavior by microstructural locations of hydrogen accumulation. The location most critically affected by hydrogen will form the fracture nuclei and path; this may accentuate the fracture mode which occurs in the absence of hydrogen, or it may cause a different mode to occur.

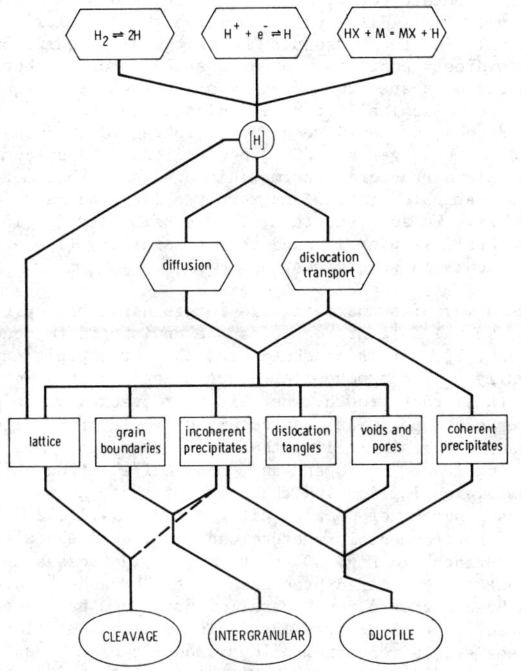

Fig. 9. Summary of hydrogen processes (hydrogen sources, hydrogen transport) and microstructural locations described in the text, with corresponding end results (hydrogen in solution, fracture). See the text for discussion and examples. Dashed line at the bottom refers to cleavage of hydrides. From Thompson and Bernstein (14).

A General Viewpoint on Hydrogen-Induced Cracking

It is useful at this point to combine the foregoing summary of electrochemical factors and hydrogen embrittlement phenomena, to construct a general viewpoint on cracking which may be induced or assisted by hydrogen. To be successful, such a rationale must be able to identify the common elements in such apparently diverse observations as these: ductile RA loss in an austenitic stainless steel tensile tested in high-pressure hydrogen gas, and cleavage-like fracture of a Ti alloy under sustained load in a methanolic chloride solution. It must also be able to rationalize not only the occurrence of the "pure" processes of anodic dissolution and hydrogen embrittlement, but also the occurrence of mixed or multiple processes. We present below a qualitative description of at least a beginning for such a rationale, borrowing freely from earlier work.

A broad statementof the rationale should begin with a comment on the interplay between hydrogen embrittlement and anodic dissolution processes. Anodic dissolution is a well-established phenomenon in SCC (17-23), and in some systems, such as copper alloys in ammoniacal solutions, appears to be the sole operative process (17,21). On the other hand, cracking due to hydrogen absorption can be produced under suitable environmental conditions in all the alloy systems used for structural materials (14). One might conclude, then, that even though SCC usually requires a rather specific combination of alloy composition and microstructure, environmental composition, stress, and other variables like potential range, one can produce cracking in an appropriately chosen system by either hydrogen or dissolution processes, provided that the environment (e.g., applied potential) is sufficiently modified.

In the great majority of cases, however, the situation is not clear-cut as to the operative process. These cases, which include many alloys under open-circuit conditions, might be called "intermediate conditions". There are two possibilities for intermediate conditions: (1) that anodic dissolution is the operative process, but at a rate greatly diminished from the bare-metal rate (23) because of altered repassivation, or (2) that mixed or multiple processes operate (e.g., dissolution and hydrogen processes together). It is difficult to believe that possibility (1) is general, for two reasons. First, it would require one to regard hydrogen embrittlement as a special case rather than as one end of a spectrum of behavior. It would be equally logical (or illogical) to regard dissolution as a special case. Second, it would require one to regard the many similarities in hydrogen and SCC dependencies on alloy composition, microstructure, thermal history, and other metallurgical variables (14,25) as merely a large number of coincidences. We suggest, then, that possibility (2) is more general, keeping in mind that possibility (1) certainly occurs in some cases.

As a way of summarizing the foregoing in the spirit of Fig. 9, multiple processes can be envisioned to occur as in Fig. 10. It is presumed here that crack initiation is usually an electrochemical process, as depicted in Fig. 6, but that crack growth, whether in precracked or initially-smooth specimens, can occur by single or parallel processes. For simplicity, these processes are summarized in the figure as "hydrogen" and "chemical/electrochemical" processes, keeping in mind that each of these terms contains a number of complex alternatives (14,17,22) Thus although environmental fracture can occur entirely along either "branch" in Fig. 10, there may be many cases in which parallel or mixed processes (both "branches") occur. As has been reviewed, such parallel processes have been suggested for such diverse materials as titanium alloys, stainless steels, aluminum alloys, and nickel alloys (14, 15,25). An advantage of this viewpoint is that it permits one to regard the large number of similarities in hydrogen and SCC dependencies on metallurgical as meaningful, rather than merely as a large number of coincidences.

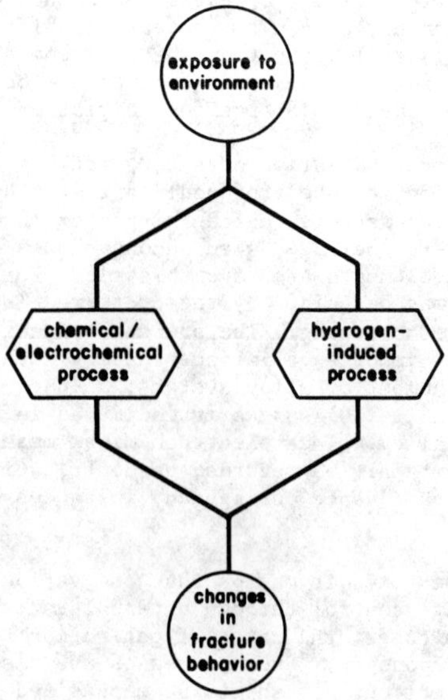

Fig. 10. Schematic presentation of parallel processes in environmental crack growth and fracture (cf. Fig. 9), which can occur along either branch or (more generally) by both acting together. From Thompson (15).

METALLURGICAL VARIABLES IN ENVIRONMENTAL FRACTURE

The discussion above presents a general background to a description of the behavior of individual alloys and alloy systems. There are a number of metallurgical variables which can be used as guidelines for either interpretation of laboratory data, or for satisfactory service in aggressive environments. As described below, these variables include composition, microstructure, strength level, and weldability. Cost must also be kept in mind. These variables and their interrelation are described in this section, and they will be addressed in more detail for several alloy systems in the following section.

Composition

The (nominal) chemical composition of an alloy is readily available in general and can serve as a rough guideline to materials selection. However, composition alone usually tells only part of the story, because other metallurgical variables are also important. For example, a given composition of steel can often be prepared with several different microstructures and corresponding strength levels, and these will differ in the extent to which they are hydrogen embrittled or exhibit SCC.

An example of an embrittlement trend with composition is that of manganese in steel, addition of which tends to be deleterious (14,26). This trend is of particular concern because Mn is so commonly added to medium- and low-strength steels to strengthen the ferrite phase. Figure 11 shows that Mn in the usual composition range for such steels appreciably reduces the K_{th} due to hydrogen (in this case K_{ISCC}). This deleterious effect of Mn is most pronounced in aqueous solutions under cathodic polarization (environments which produce hydrogen), but is also significant for a wide range of other polarization conditions and for tests in hydrogen gas (26). The data in Figure 11 are for relatively strong steels and would not be directly applicable to lower strength alloys used for linepipe or pressure vessels. There are, however, data (27) which do show that Mn additions are deleterious even at strength levels below 700 MPa (100 ksi) yield strength. Moreover, it is true in general that lower strength levels do not necessarily imply better environmental performance, as discussed below. Other examples for steels and other alloy systems are presented below; for example, stainless steels, are strongly sensitive to composition in their environmental performance.

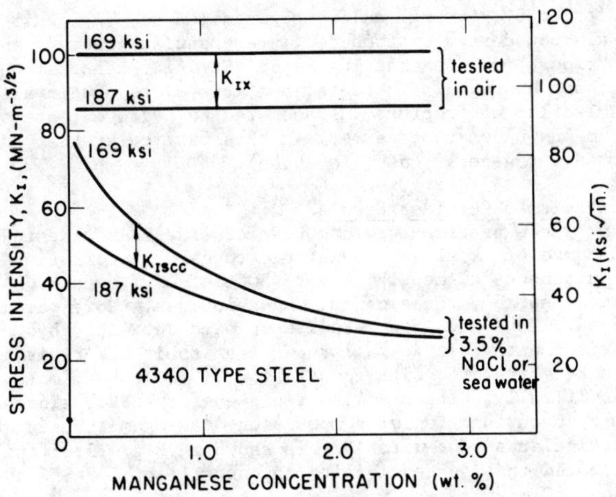

Fig. 11. Effect of manganese on the SCC resistance of 4340-type steel heat treated to two strength levels; 145 psi = 1 MN/m^2. K_{Ix} refers to unstable fast fracture in air. From Sandoz (26).

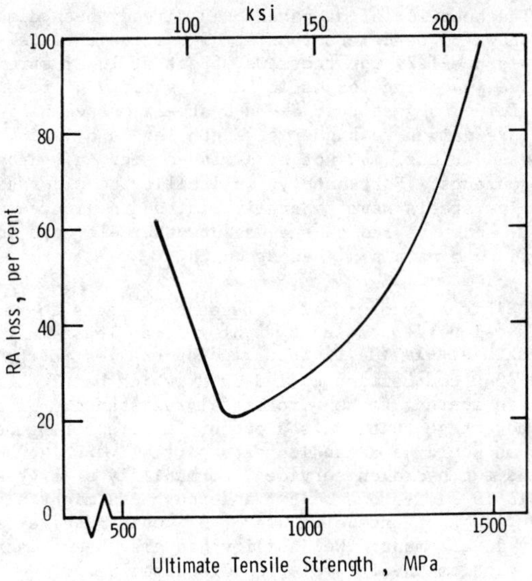

Fig. 12. Ductility (RA) loss as a function of strength level, in a 3.3% Cr-0.4% Mo-0.26% C steel containing 3.6 wt. ppm H. Changes in strength obtained by tempering to produce different microstructures. Data from Hobson and Sykes (30); figure from Thompson and Bernstein (14).

Microstructure

Microstructure is an important variable, particularly in steels; in general, it is possible to describe the structural types which are preferred. Here strength level can also be important, but the overall ranking for steels (14) extends from quenched and tempered martensite (or bainite), being most resistant to hydrogen; through spheroidized structures having intermediate resistance; to pearlitic/ferritic (normalized) structures, usually with somewhat less resistance, and finally to untempered martensite, with the least resistance. The spheroidal and pearlitic structures do not always have the same ranking (28) but it is important to note that most linepipe steels, and some pressure vessel steels, are used in the normalized structure. In addition, such microstructures are often "banded" as a result of chemical inhomogeneities in solidification which persist through processing. These bands are frequently rich in manganese; as the composition trend discussed above would predict, cracking in service tends to follow these bands (29). Alloy systems other than steels tend to show somewhat less pronounced microstructure dependence, in part because the range of microstructures which is produced in other types of commercial alloys is narrower.

Strength Level

There are two well-known "rules of thumb" regarding strength-level effects in hydrogen embrittlement. One is that embrittlement becomes more severe as strength level increases; the other is that steels with strengths below about 700 MPa (100 ksi) show no significant embrittlement. Like most rules of thumb, these are true in many cases, but there are many exceptions; therefore, they must be used with caution. One reason for the exceptions is that susceptibility to aggressive environments is a marked function of composition and microstructure.

Figure 12 illustrates the interrelation of strength level and microstructure for a specific Cr-Mo steel (30). Here the variations in strength were produced through changes in microstructure arising from increases in the tempering temperature after quenching. The RA loss falls steadily as strength is reduced from 1500 to 750 MPa by increasing tempering temperature, but below 750 MPa strength, the trend is reversed. This is because a new microstructure of spheroidized carbides begins to appear at that point, and as it becomes better developed, RA loss increases because the microstructure is susceptible to hydrogen. Note that the behavior below 700 MPa contradicts both the rules of thumb about strength.

In concluding this section, it may be appropriate to reiterate that microstructure is probably the primary metallurgical variable. Composition merely sets limits on attainable microstructures in a particular alloy, while strength level usually is regarded as an expression of microstructure.

BEHAVIOR OF ALLOY SYSTEMS

The alloy systems used in technical applications are extremely varied in metallurgical characteristics, and space limitations here prevent more than a summary presentation drawn from earlier work. The reader is urged to consult the many available reviews of individual alloy systems which are available. Among them, we may cite our own reviews of aluminum alloys and stainless steels (14,31) for titanium and nickel alloys (14), and for ferritic and martensitic steels (14,32), as well considering the implications of these data for alloy design (33). In the present section, steels and aluminum alloys are chosen as illustrative classes of materials.

Ferritic and Martensitic Steels

The effects of composition in these steels were alluded to above. Of particular importance are the two most common alloying elements in steel, carbon and manganese. The presence of carbon (and probably nitrogen), particularly at the higher strength levels, has been shown to promote sustained load cracking in an aqueous environment, where hydrogen embrittlement is probably the cause of the environmentally induced cracking (14). Carbon apparently reduces the critical stress intensity for the onset of crack growth, and there appears to be a generally deleterious effect of carbon under conditions wehre cracking is the manifestation of hydrogen damage. On the other hand, carbon does not appear to be harmful in those cases where hydrogen damage is manifested as a loss in reduction of area with no obvious change in fracture mode, as is often the case when lower strength alloys are tested in high-pressure hydrogen or pulled to failure after cathodic charging. More work is needed to generalize the results, but considering the well-documented deleterious effects of carbon on toughness and weldability, control of carbon content, consistent with strength requirements, is an important design parameter (33). It should be noted again that nitrogen is expected to follow a behavior similar to carbon, and the same recommendations apply.

The behavior of manganese has already been discussed and shown to be deleterious at high strength levels; there is evidence (27) for the same effect at lower strength levels, involving losses in RA. A variety of results, therefore, suggest that Fe-Mn-C steels, especially as they are driven to higher strength levels by increased alloy additions, may not be a wise choice in hydrogen environments. Fortunately, weldability considerations in linepipe steels have generally limited the increase of Mn and C, and have led to the use of other elements such as Ti and V to attain higher strength.

Silicon is of interest because it seems to reduce crack velocities caused by hydrogen, at least in high-strength steels (34). This finding has led to the modified 4340 composition called 300M, which has 1.5% added silicon content for environmental resistance. Si also has the advantage that it is a potent ferrite strengthener and as such could be a candidate to partially replace Mn in steels for hydrogen service. Formability of high-silicon steels is limited, however, and toughness might also be lower, so that extensive use of Si would require careful alloy development. Weldability has also been described as a problem in Si steels, as discussed below.

For higher strength levels in ferritic and martensitic steels, elements commonly added are nickel, chromium, molybdenum, and titanium. There is no consensus behavior for the first three of these (14,33), and it should be evident from the extensive use of these elements in quenched-and-tempered steels that systematic study of their role in environmental cracking is of considerable importance. Titanium, on the other hand, appears to have potential as a beneficial element (14). The desirable directions for hydrogen-resistant alloy development can thus be summarized as follows. In place of reliance on C and Mn, steels with significant additions of Si and Ti would appear preferable, with Ni (at high strength levels) and carbon contents balanced for strength and toughness. Alloy development efforts on such steels would be necessary for optimization of composition, particularly for toughness.

Microstructure was described above in general as a factor in steels; it should be emphasized that most high-strength and medium-strength steels, with alloying element content sufficient to give good hardening response, are quenched and tempered (sometimes normalized and tempered). When well tempered, such microstructures are relatively resistant to SCC and to hydrogen, although increasing the strength level can reduce this resistance. At medium strength levels, the normalized and tempered structures containing pearlite can be fairly susceptible to hydrogen (14). Such microstructures therefore require caution in use. The other factor is grain size; both strength and toughness are increased by decreases in grain size; fortunately, resistance to hydrogen embrittlement is generally increased also (32).

It should be emphasized that the interrelation of composition, microstructure, and strength, or any two of these, can be very complex. Any specific situation of interest would require careful assessment. An illustration of this is presented in Figure 13, showing TTF curves for precracked specimens (cf Fig. 3) of five high-strength steels (34). These have all been heat treated to the same strength level (an ultimate strength of about 1650 MPa) by varying the heat treatment; this means that in addition to compositional differences, there are likely to be microstructural differences. The result is a pronounced spread in failure time, which can crudely be correlated with the composition (14); microstructures are all quenched and tempered, though probably with differences in detail.

Another example of this complex interaction, as a contrast to Fig. 13, is the cracking of a plain carbon steel, 1045, when hydrogen-charged in the form of bluntly-notched specimens. When quenched and tempered at several temperatures, the crack initiation loads were found (Fig. 14) to be reduced by hydrogen; cracking evidently initiated below the notch root (35), by a predominantly quasi-cleavage mode. The effects of tempered martensite embrittlement, through sulfur segregation to prior austenite grain boundaries, introduced intergranular fracture and a "trough" in the cracking loads, Fig. 14, both with and without hydrogen. The details of the microstructural nucleation of fracture, as assisted by hydrogen, are by no means understood as yet, in this particular case (35) or in the general case (14,25,36).

Austenitic Stainless Steels

These steels have long been considered a classical instance of excellent resistance to general corrosion and simultaneous susceptibility to SCC in certain environments, notably those containing chlorides. More recently, these alloys have been studied in some detail in hydrogen environments. It is well known that their SCC resistance can be correlated with composition (14,20), and, more specifically, with stacking fault energy (SFE), since SFE is a known function of composition. An example using one particular set of data (37) is shown in Fig. 15; it will be noted that susceptibility appears slight for SFE above about 40 mJ/m^2.

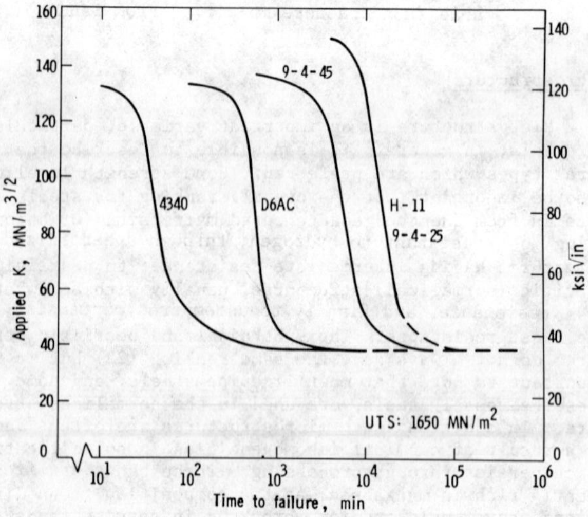

Fig. 13. Failure time as a function of applied stress intensity for precracked specimens of five steels of constant strength (see text). Environment was distilled water. From Steigerwald and Benjamin (34).

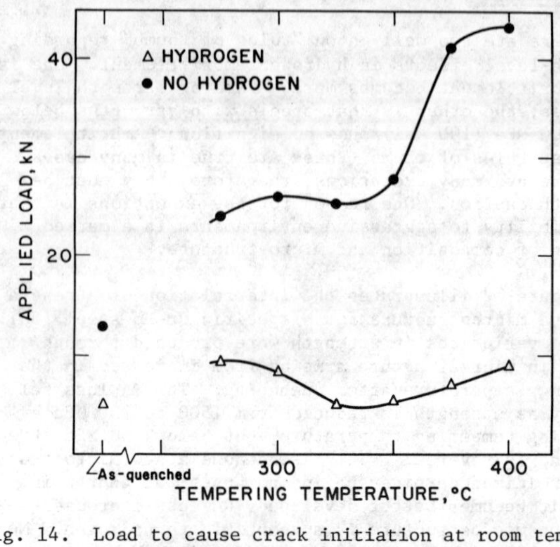

Fig. 14. Load to cause crack initiation at room temperature in notched 4-point bend bars, for 1045 steel quenched and tempered at the temperatures shown, and tested either uncharged or after cathodic hydrogen charging. From Costa and Thompson (35).

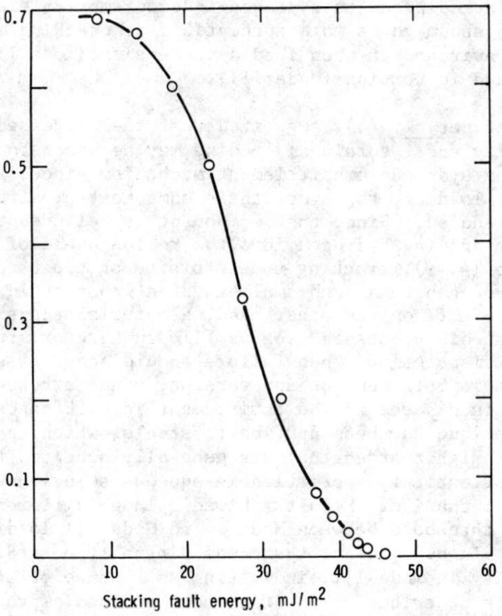

Fig. 15. Time to failure (inverted) in chloride environment as a function of stacking fault energy for austenitic stainless steels. Since all steels had about 18% Cr, initiation times can be assumed to be approximately constant; thus the inverse TTF parameter essentially represents average crack velocity for these steels. Data of Swann (37); figure from Thompson and Bernstein(14).

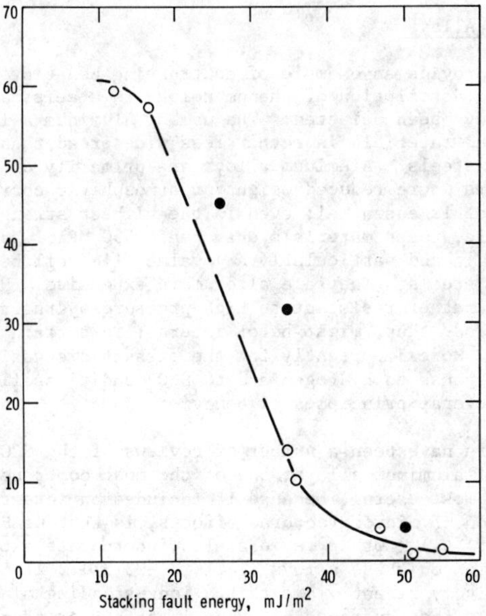

Fig. 16. Ductility (RA) loss due to charged hydrogen as a function of stacking fault energy in austenitic stainless steels similar to those in Fig. 15; discussed in text. Note similarities in curve shape and placement to Fig. 15. Filled points are for nitrogen-containing steels. From ref. 14.

It is of interest to compare data for RA losses of hydrogen-charged material, Fig. 16; the similarity to Fig. 15 is marked, but the two sets of data refer to quite different failure processes. One rationale for the correlation with SFE is that low SFE encourages planar slip of dislocations, thereby enhancing slip distance and dislocation transport of hydrogen (24). It is possible that a crack tip resharpening process operates in the chloride environment, Fig. 15, since delayed failure under constant load does not occur with hydrogen alone in most of these steels. It has been suggested (15) that comcomitant hydrogen cracking and dissolution processes, as in Fig. 10, are operative in stainless steel SCC in chloride environments.

Another factor of importance in stainless steel microstructures is the presence of delta ferrite. There are two alloying elements (in addition to Cr and Ni) which appear to strongly affect the presence of delta ferrite; these are Si and Ti, when present in amounts of 2% or more. The consequent delta ferrite appears neutral or mildly beneficial in both SCC (20) and hydrogen studies (33). The presence of delta ferrite is of especial importance in welds of stainless steels, since segregation during solidification and rapid cooling tend to produce volume fractions of ferrite which are considerably in excess of that present in wrought material.

Age-hardening Stainless Steels

A potentially important class of materials for applications in aggressive environments comprises the stainless steels which age harden. Because the strengthening precipitate is γ', or $Ni_3(Al,Ti)$, these alloys are sometimes called "stainless superalloys" by analogy with the nickel-base superalloys, which are also strengthened by γ'. A well-studied member of this class is A-286, containing nominally 25% Ni, 15% Cr, and 2.25% Ti, 0.2% Al for the γ'. Typically this alloy exhibits a yield strength of 750 MPa (110 ksi) in the fully-aged condition. The same precipitate which confers strength, however, is also pivotal in hydrogen embrittlement (14) and SCC (14,20). This evidently occurs because the small, ordered, coherent particles of γ' are cut or sheared by dislocations, causing planar slip and environmental susceptibility like that in Figs. 15 and 16. In the commercial A-286 composition, sustained load cracking in hydrogen has been observed (38).

Extensive study of this alloy type (39) has verified the role of γ' and has shown the means by which a rather hydrogen-resistant alloy can be obtained. The critical feature is the difference, or mismatch, between the lattice parameters of the matrix and of the γ'. When the mismatch is appropriately increased (by minor composition changes), the RA loss induced by hydrogen can be significantly reduced. This behavior is shown in Fig. 17, and the best-performing composition, called "alloy 8" in Fig. 17, was later patented for its relative hydrogen compatibility (40).

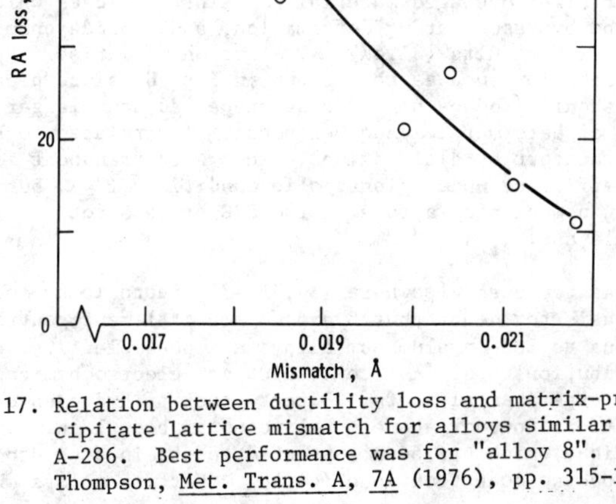

Fig. 17. Relation between ductility loss and matrix-precipitate lattice mismatch for alloys similar to A-286. Best performance was for "alloy 8". From Thompson, Met. Trans. A, 7A (1976), pp. 315-18.

Aluminum Alloys

To provide an example of contrasting behavior to the foregoing description of phenomenology in steels, aluminum alloys have been selected. The use of aluminum alloys as structural materials is rather less widespread than is the case for steels. Aluminum alloys are primarily of value in systems where reduced weight or atmospheric corrosion resistance is essential; even in the highest strengths attainable, these materials only reach 550 MPa, where their K_{ISCC} and particularly K_{Ix} values lie well below those of steels. They are also more expensive. However, they are rather resistant to high-pressure hydrogen gas (14,31) and, thus, might be of interest in certain applications. More importantly for the present discussion, their response to hydrogen and to SCC conditions illustrates several principles of behavior.

There have been a number of reviews of the SCC behavior of aluminum alloys; one of the most comprehensive and also most useful, because it includes an integrated discussion of microstructural effects, is that of Speidel and Hyatt (41). Other reviews also incorporate extensive information on these alloys (14,17,21-23,31). It is only in relatively recent years that a long-standing minority view (42), that hydrogen plays an important or even dominant role in the SCC of aluminum alloys, has become accepted (14,17,43,44). As we emphasize below, however, this conclusion appears to be appropriate only for high-strength alloys.

It appears difficult in aluminum alloys to generalize about composition effects; in SCC results, each alloy type seems to exhibit a unique pattern based on the particular alloying elements involved (14,41,45), while hydrogen embrittlement data are as yet too sparse for patterns to be visible. One generalization which appears useful is that an increase in the total amount of solute which is capable of being disolved in high-temperature solid solution will increase susceptibility to SCC (41).

Microstructural dependences of susceptibility to both hydrogen and SCC are only available to date for the higher-strength alloys, i.e. the 2000 and 7000-series alloys (the numbers refer to the Aluminum Association alloy numbers), as is discussed below. For lower-strength materials, it is sometimes difficult to distinguish between SCC and intergranular corrosion (41); for both phenomena, there is supporting evidence that the presence of phases at or near the grain boundary which are anodic to the matrix gives rise to the observed failures (41,45). Such observations are in essential agreement with Dix's original anodic-path dissolution proposal (46) for aluminum alloys, although that concept incorporates no well-defined role of stress (31,41,42).

In higher-strength materials (above about 300 MPa yield strength), there is a well-established correlation (14,31,41) between SCC susceptibility and degree of aging, with underaged and peak-strength aged microstructures most susceptible. This has been interpreted by pointing out that precipitates in these susceptible tempers are coherent and are sheared or cut on deformation, thus enforcing planar slip; overaged material contains particles which must be bypassed during deformation, and considerable cross-slip results (31,47). The reason may arise in part from behavior such as that shown in Fig. 8, since planar slip should produce higher slip steps and thus larger areas of bare metal. Another possibility relates to hydrogen transport by dislocations, since such transport is more efficient under planar slip conditions (24), but hydrogen must play a role in the SCC process for this to be relevant.

As reviewed elsewhere (14,31,43), there is no longer any basis for doubt that hydrogen embrittlement contributions to SCC in aluminum alloys are possible, i.e. such contributions would not contradict any electrochemical or other experimental facts. Direct proof of hydrogen contributions however, have until recently been lacking. One technique which has been suggested to obtain such proof is the loading mode technique (44,48). Certain alloys which are considered prone to hydrogen-induced embrittlement have been shown to be more susceptible to cracking in aqueous environments when loaded in tension (Mode I) than when loaded in torsion (Mode III).

Green, et al. (44) suggested that the decreased susceptibility under torsional loading may be understood in terms of a hydrogen embrittlement mechanism since, for torsional loading, the hydrostatic component of stress is greatly reduced. Since that component is believed responsible for driving hydrogen into the region ahead of the crack tip (4,49), cracking under torsion should be assisted by hydrogen to a much smaller extent than could occur in Mode I. If, on the other hand, a material tends to fail by anodic processes such as film rupture or stress-assisted dissolution, then failure should occur at similar rates in both tension and torsion, since stressed material is exposed to the environment in both cases (44). This technique has been applied to steels, which (particularly at higher strengths) are generally accepted to be largely or entirely embrittled in aqueous solutions by a hydrogen mechanism. Results showed a large difference in cracking threshold between Mode I and Mode III loadings (48,50). In particular, there was virtually no cracking sensitivity in Mode III, indicating an absence of any measurable contribution to cracking from anodic processes (44,48).

This testing technique, and particularly its interpretation as a discriminator between hydrogen embrittlement and anodic dissolution, has aroused some controversy. The usual concern is that a pure Mode III test would be difficult to conduct. Specimen gripping and loading must be very accurately performed to avoid a bending component in the load; moreover, the grain boundaries, inclusions and precipitates in a real material would perturb the stress locally, and (in general) introduce non-Mode III components. Even beyond those two problems, it appears (51) that a hydrostatic component can arise under pure torsional loading for materials with particular combinations of elastic constants, i.e. anisotropy. There is, however, no dispute that a nominal Mode III loading should give rise to a greatly reduced hydrostatic component of stress, compared to Mode I. Thus the interpretation that hydrogen embrittlement is favored by Mode I loading relative to "Mode III" remains sound. If it did not, it would be difficult to explain the large Mode I-Mode III cracking differences for steel (48,50).

Tests of this type were first performed on aluminum alloys by Green, et al.; an example is shown in Fig. 18. These data show clearly that 7075 is more susceptible to cracking in aqueous solution when loaded in Mode I than in Mode III. Cracking is not absent in Mode III, however, so while these results indicate that a hydrogen embrittlement mechanism may be predominant in the SCC of 7075 aluminum, it appears that anodic processes are also involved (14,15,44). The alternative interpretation would be that

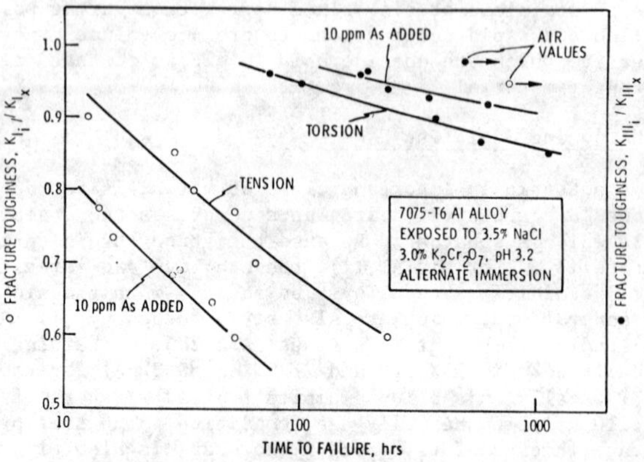

Fig. 18. Resistance to SCC of 7075-T6 aluminum as a function of loading mode (see text). Environments with and without arsenic were used with each loading mode. From Green, et al. (44).

hydrogen is also responsible for the "Mode III" cracking, for one of the reasons listed above, although we prefer at present the anodic view. It will be noted in Fig. 18 that when 10 ppm arsenic was added to the solution, cracking in Mode I was enhanced i.e. the threshold decreased, while in Mode III arsenic acted to raise the threshold, or inhibit cracking.

These effects of arsenic can be rationalized as follows (44). Arsenic is a hydrogen recombination "poison", that is, it interferes with part of the cathodic reaction, $2H \rightarrow H_2\uparrow$. The surface concentration of H is thereby increased, which means both that hydrogen absorption into the metal can increase, and also that the rate of reduction of hydrogen ions is reduced. This perforce reduces the rate of oxidation of metal atoms in the anodic reaction as well. In Mode I loading, where hydrogen processes under triaxial stresses appear to dominate behavior, the primary effect of arsenic is to increase hydrogen concentration within the metal and thus to enhance hydrogen embrittlement. In Mode III, however, where dissolution is at least significant if not dominant in the cracking process, arsenic acts to retard both the anode and cathode kinetics and thus to reduce the rate of dissolution. Similar interpretations, incidently, have been offered for arsenic effects in stainless steel SCC (15).

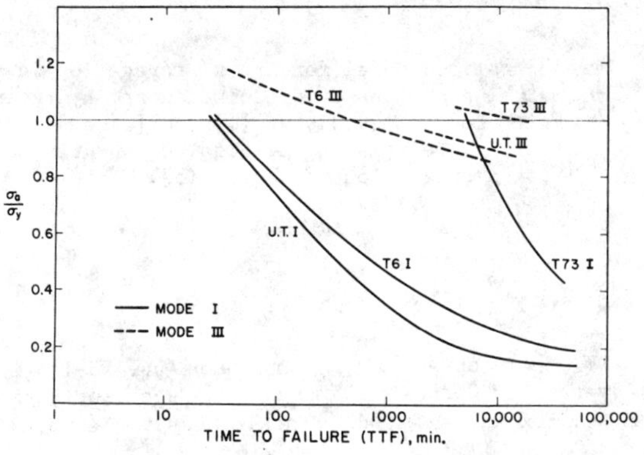

Fig. 19. Time to failure results for notched specimens of 7075 aluminum, in underaged temper (UT), peak strength (T6) and overaged (T73) conditions. Yield strength of UT and T73 were about equal; both were 10% below T6. From Maloney, et al.(53).

Subsequent work has shown that both the hydrogen embrittlement (52) and the SCC of 7075 depends markedly on microstructure. Fig. 19 illustrates the pattern of results (53). Mode I cracking was much more severe than Mode III for all microstructures, though Mode III cracking was never absent, as it can be in steels (48,50) and titanium alloys (44,54).

A second important point is that the extent of the loading mode effect depends on microstructure. Increased tempering, in the sequence from UT to T6 to T73, produced increasing resistance to cracking in Mode I (and probably in Mode III, although the results are less clear-cut). Interpreting this behavior as an effect of hydrogen, i.e. as a manifestation of a predominant hydrogen contribution to SCC in the environment used, would be consistent with earlier evidence that hydrogen embrittlement susceptibility alone, without any "SCC process", is also decreased by the added tempering from UT to T6 to T73 (52,53). The precise extent of the SCC sensitivity should not, however, be correlated with the hydrogen sensitivity alone (for each microstructure) unless the anodic contribution to SCC is either independent of microstructure, or accurately parallels the microstructure dependence of hydrogen embrittlement. Neither of these two possibilities is particularly likely, because both the size and composition of precipitates in 7075, and also the local matrix composition near grain boundaries, vary widely between UT, T6 and T73; anodic processes should be sensitive to these parameters, and almost certainly in ways which differ from the sensitivity to hydrogen embrittlement. The UT temper ranking relative to the T6 temper in Fig. 19 probably reflects this point, comparing the two tempers in Modes I and III.

The fact that a nominal Mode III loading of a real material is likely to produce a small hydrostatic component of stress near the crack tip was mentioned above. This fact, however, should not obscure the main issue: that the Mode III hydrostatic component is greatly reduced (whether or not it is zero) relative to Mode I. Consequently, there appears to be no error in evaluating a difference in SCC sensitivity between Modes I and III as reflecting a relative contribution of hydrogen embrittlement to SCC. It should be noted that the extent of this relative contribution is minimized, because the small hydrostatic stress in the real Mode III test presumably assists a correspondingly small hydrogen contribution to SCC. If a pure Mode III test could be conducted, it follows that the difference between Modes I and III would increase. Fig. 19 thus indicates a minimum or conservative extent of the contribution of hydrogen embrittlement to SCC of 7075 aluminum in the UT, T6 and T73 tempers.

Another interesting aspect of hydrogen effects in aluminum alloys is the evident role of dislocation transport. A high-purity heat of 7075 was prepared with an equiaxed grain structure; when cathodically hydrogen charged (52), a narrow "rim" of intergranular fracture could be detected in tensile specimens, one or two grain diameters in width (Fig. 20). But when specimens were simultaneously cathodically charged and strained at a very slow rate, permitting the hydrogen discharged at the surface to be carried deep into the specimen interior by moving dislocations (24), the intergranular zone widened to include nearly the entire specimen diameter (55). This is indicated in Fig. 20. It would thus appear appropriate to consider a large hydrogen contribution to SCC in these alloys, a contribution which can be significantly assisted by dislocation transport, but which involves a simultaneous and cooperative contribution from dissolution processes, in the spirit of Fig. 10.

CONCLUDING REMARKS

Both hydrogen embrittlement and SCC are complex phenomena, and the various contributing processes to each vary widely for different combinations of material properties (composition, microstructure, and strength level), environment, and probably stress level as well. Unquestionably there is a great deal still to be learned about most aspects of these kinds of fracture. There are, however, some broad patterns of behavior which are becoming recognized (14,17,20-22,25,33,41), and these are summarized briefly above.

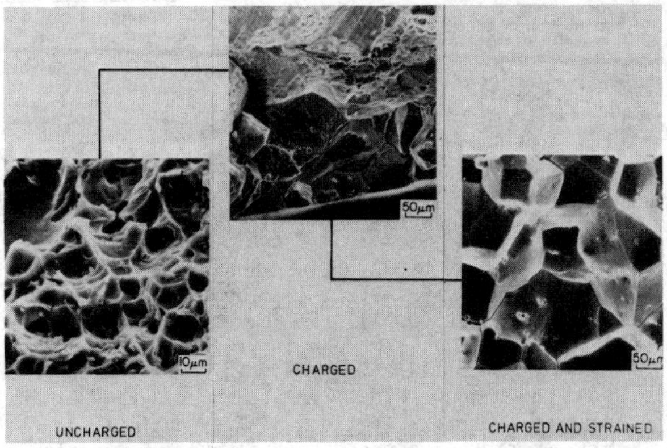

Fig. 20. Fractography of HP-7075 (equiaxed grain shape) in UT condition. Uncharged material shows microvoid coalescence; cathodically charged material shows one or two grains of IG fracture at specimen surface (along lower edge of micrograph); and when simultaneously charged and slowly strained, nearly entire fracture is IG. Please see text for discussion.

It is worth pointing out that the lack of systematic fundamental information about most alloy systems, combined with a rather primitive understanding of underlying mechanisms (14,17,22,25), means that most experimental work continues to have a somewhat exploratory character. This is true for research work, where in many cases the best that can be done is to rationalize patterns of results, and also for engineering work, whether in materials selection or in alloy development, where service simulations of particular combinations of environmental variables and stress continue to be essential. Environmental fracture remains an area of incomplete information, with both detailed experimental work and also mechanistic fundamentals sorely needed.

At the same time, there are many microstructural patterns of behavior emerging (14,32), and some of these have been used to make detailed recommendations for alloy design (25,33,56). For the SCC problem, there are broader perspectives, including ideas such as that depicted in Fig. 10, which may well lead to improved understanding. An example of a generalized SCC model, taken from a more specific suggestion of Rhodes (57) is shown in Fig. 21. Here the emergence of a slip step at the surface, analogously to the concept in Fig. 10, is shown at location [1], and at location [2], the cracking event has begun, with at least some participation by hydrogen. Finally, at the right of the figure is a "mature" crack , with an established crack-tip environment of lowered pH and potential drop (18,19), for which mutual dissolution and hydrogen embrittlement processes may occur during cracking. With the emerging body of combined knowledge available about SCC (15,17-23, 33) and hydrogen embrittlement (14,25,32,33,44), such cracking events may become more amenable to detailed experimentation and analysis. The problem remains inherently complex, however, and at present only cautious optimism can be recommended on the possibility of major advances in understanding.

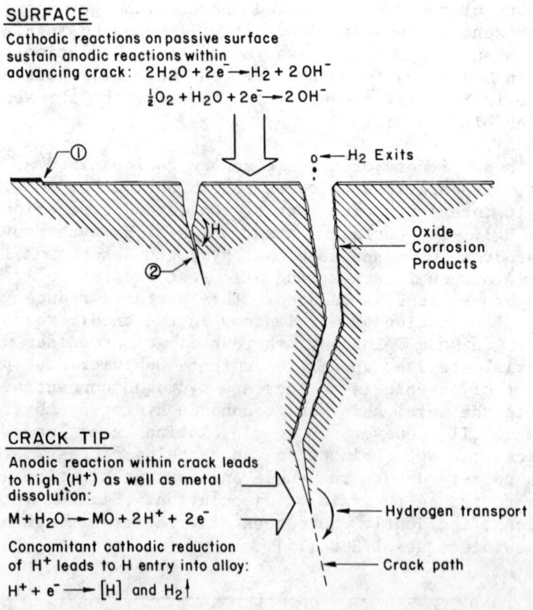

Fig. 21. Generalized model for crack propagation in SCC by mutual processes of dissolution and hydrogen embrittlement. Events at [1] and [2] described in text. Note relation to Figs. 6 and 10. Modified from a suggestion of Rhodes (57).

REFERENCES

1. M. Cailletet, C. R. Acad. Sci. Paris, 66, (1868), p. 847.

2. W. H. Johnson, Proc. Royal Soc., (1875), No. 158, p. 168.

3. D. E. Hughes, Scientific American, (1880), Suppl. No. 237, July 17.

4. A. R. Troiano, Trans. ASM, 52 (1960) pp. 54-80.

5. P. Cotterill, Prog. Mater. Sci., 9, (1961) p. 201.

6. A. R. Elsea and E. E. Fletcher, The Problem of Hydrogen in Steel, DMIC Memo. 180, Battelle Mem. Inst., Columbus, OH, Oct. 1963.

7. I. M. Bernstein, Mater. Sci. Eng., 6, (1970) pp. 1-19.

8. R. C. McMaster, Mater. Eval., 27, (1969), p. 17A.

9. T. Toh and W. M. Baldwin, p. 176 in Stress Corrosion Cracking and Embrittlement, Wiley, New York, NY, 1956.

10. A. W. Thompson, pp. 85-124 in Transmission and Storage, Vol. 2 of Hydrogen: Its Technology and Implications, K. E. Cox and K. D. Williamson, eds., CRC Press, Cleveland, 1977.

11. C. D. Beachem and R. M. N. Pelloux, pp. 210-224 in Fracture Toughness Testing and Its Applications (STP 381), American Society for Testing and Materials, Philadelphia, 1965.

12. C. D. Beachem, Met. Trans., 3, (1972), pp. 437-51.

13. B. F. Brown, pp. 2-16 in Stress Corrosion Cracking in High Strength Steels and in Titanium and Aluminum Alloys, B. F. Brown, ed., Naval Research Laboratory, Wash. DC, 1972.

14. A. W. Thompson and I. M. Bernstein, pp. 53-175 in Advances in Corrosion Science and Technology, vol. 7, M. G. Fontana and R. W. Staehle, eds., Plenum, NY, 1980.

15. A. W. Thompson, Mater. Sci. Eng., 43 (1980) pp. 41-46.

16. W. Williams and J. Eckel, J. Am. Soc. Naval Engrs., 68 (Feb. 1956) pp. 93-107.

17. A. J. Bursle and E. N. Pugh, pp. 18-47 in Environment-sensitive Fracture of Engineering Materials, Z. A. Foroulis, ed., TMS-AIME, Warrendale, PA, 1979.

18. B. F. Brown, C.T. Fujii and E. P. Dahlberg, J. Electrochem. Soc., 116 (1969) pp. 218-19.

19. B. G. Ateya and H. W. Pickering, pp. 207-22 in Hydrogen in Metals, I. M. Bernstein and A. W. Thompson, eds., ASM, Metals Park, OH, 1974.

20. R. M. Latanision and R. W. Staehle, pp. 214-96 in Fundamental Aspects of Stress Corrosion Cracking, R. R. W. Staehle, et al., eds., NACE, Houston, 1969.

21. D. J. Duquette, pp. 251-301 in Alloy and Microstructural Design, J. K. Tien and G. S. Ansell, eds., Academic Press, New York, 1976.

22. J. C. Scully, pp. 1-17 in Mechanisms of Environment Sensitive Cracking of Materials, P. R. Swann, F. P. Ford and A. R. C. Westwood, eds., Metals Society, London, 1977.

23. F. P. Ford, pp. 125-42, ibid.

24. J. K. Tien, A. W. Thompson, I. M. Bernstein and R. J. Richards, Met. Trans. A, 7A (1976) pp. 821-29.

25. A. W. Thompson, pp. 379-410 in Environment-sensitive Fracture of Engineering Materials, Z. A. Foroulis, ed., TMS-AIME, Warrendale, PA, 1979.

26. G. Sandoz, Met. Trans., 2 (1971) p. 1055-63.

27. R. Garber and I. M. Bernstein, pp. 463-73 in Environmental Degradation of Engineering Materials, M. R. Louthan and R. P. McNitt, eds., VPI Press, Blacksburg, VA, 1977.

28. M. Henthorne and R. N. Parkins, Brit. Corr. J., 2 (1967) pp. 186-92.

29. J. S. Laws, V. Frick and J. McConnell, Hydrogen Gas Pressure Vessel Problems in the M-1 Facilities, Report NASA CR-1305, Aerojet-General Corp., Sacramento, CA, March 1969.

30. J. D. Hobson and C. Sykes, J. Iron Steel Inst., 169 (1951) pp. 209-20.

31. A. W. Thompson and I. M. Bernstein, Reviews on Coatings and Corrosion, 2 (1975) pp. 3-44.

32. I. M. Bernstein and A. W. Thompson, Internat. Metals Reviews, 21 (1976) pp. 269-87.

33. I. M. Bernstein and A. W. Thompson, pp. 303-47 in Alloy and Microstructural Design, J. K. Tien and G. S. Ansell, eds., Academic Press, New York, 1976.

34. E. A. Steigerwald and W. D. Benjamin, Met. Trans., 2 (1971), pp. 606-608.

35. J. E. Costa and A. W. Thompson, Met. Trans. A, 12A (1981) pp. 761-71.

36. A. W. Thompson and I. M. Bernstein, paper 3A-6 in Hydrogen in Metals (Proc. 2nd Int. Conf., Paris), vol. 10, Pergamon Press, New York, 1977.

37. P. R. Swann, Corrosion, 19 (1963) pp. 102t-112t.

38. R. J. Walter and W. T. Chandler, Influence of Gaseous Hydrogen on Metals, Report NASA CR-124410, Rocketdyne Div. Rockwell International, Canoga Park, CA, Oct. 1973.

39. A. W. Thompson and J. A. Brooks, Met. Trans. A, 6A (1975), pp. 1431-42.

40. J. A. Brooks and R. W. Krenzer, U.S. Patent No. 3, 895,939.

41. M. O. Speidel and M. V. Hyatt, pp. 115-335 in Advances in Corrosion Science and Technology, M. G. Fontana and R. W. Staehle, eds., vol. 2, Plenum Press, New York, 1972.

42. U. R. Evans, pp. 158-62 in Stress Corrosion Cracking and Embrittlement, W. D. Robertson, ed., Wiley, New York, 1956.

43. M. O. Speidel, pp. 249-73 in Hydrogen in Metals, I. M. Bernstein and A. W. Thompson, eds., ASM, Metals Park, OH, 1974.

44. J. A. S. Green, H. W. Hayden and W. G. Montague, pp. 200-215 in Effect of Hydrogen on Behavior of Materials, A. W. Thompson and I. M. Bernstein, eds., TMS-AIME, New York, 1976.

45. D. O. Sprowls and R. H. Brown, pp. 466-506 in Fundamental Aspects of Stress Corrosion Cracking, R. W. Staehle, et al., eds., NACE, Houston, 1969.

46. E. H. Dix, Trans. ASM, 42 (1950) pp. 1057-1127.

47. M. O. Speidel, pp. 289-341 in The Theory of Stress Corrosion Cracking, J. C. Scully, ed., NATO, Brussels, 1971.

48. H. W. Hayden and S. Floreen, Corrosion, 27 (1971) pp. 429-34.

49. J. C. M. Li, R. A. Oriani and L. S. Darken, Z.f. Phys. Chem., 49 (1966) pp. 271-90.

50. C. St. John and W. W. Gerberich, Met. Trans., 4 (1973) pp. 589-594.

51. Y. T. Chou and R. P. Wei, Lehigh University, personal communication, 1980.

52. J. Albrecht, A. W. Thompson and I. M. Bernstein, Met. Trans. A, 10A (1979), pp. 1759-66.

53. J. L. Maloney, A. W. Thompson and I. M. Bernstein, unpublished research, Carnegie-Mellon University, 1979.

54. C. F. St. John, Scripta Met., 9 (1974) pp. 141-45.

55. J. Albrecht, I. M. Bernstein and A. W. Thompson, Met. Trans. A, in press.

56. G. M. Pressouyre and I. M. Bernstein, Met. Trans. A, 12A (1981) pp. 835-44.

57. P. R. Rhodes, Corrosion, 25 (1969) pp. 462-72.

ACKNOWLEDGEMENTS

We acknowledge with thanks the support we have received from the U.S. Office of Naval Research, The Air Force Office of Scientific Research, and the National Science Foundation. We also appreciate experimental support from the Center for Joining of Materials at Carnegie-Mellon University.

HIGH TEMPERATURE CORROSION AND USE OF
COATINGS FOR PROTECTION

by

F. S. Pettit
University of Pittsburgh, Pittsburgh, PA 15261
and
G. W. Goward
Coatings Technology Corp., Branford, CT 06405

INTRODUCTION

In most cases where metallic alloys are used in engineering applications, some reaction takes place between the alloys and the environments to which they are exposed. It is important to ensure that these reactions do not significantly affect the performance of the alloys, or when this is not possible, to have the effects be insignificant for as long as possible. There are a number of types of alloy-environment interactions which result in the surfaces of alloys being converted to phases that are not nearly as effective as the alloy in performing the desired engineering function. Such processes are commonly referred to as corrosion processes.

Corrosion processes can be divided into two classes, namely, aqueous and nonaqueous or dry. Aqueous corrosion is restricted to low temperatures whereas nonaqueous or dry covers virtually all temperatures. Dry corrosion consists of corrosion processes involving alloys exposed to gases. Such reactions very often proceed very slowly at temperatures below about 500-600°C and therefore dry corrosion is often called high temperature corrosion since temperatures above 600°C are emphasized. All high temperature corrosion processes however are not necessarily dry although they are nonaqueous. For example, liquid deposits can be developed on the surfaces of alloys which affect the corrosion processes, or alloys can be immersed in molten metals or slags. This paper will be concerned with the principal forms of high temperature corrosion and the techniques which may be used to control it.

THE FORMS OF HIGH TEMPERATURE CORROSION

In most of the environments in which engineering alloys are used oxygen is usually present. Oxidation of alloys is therefore a very important form of high temperature corrosion. In many cases, however, other reactants in addition to oxygen are present and corrosion of alloys concomitantly by a number of reactants is an important form of degradation. Combined attack of alloys simultaneously by two or more reactants is often called mixed gas attack. While gaseous induced degradation of alloys is probably the principal means of high temperature corrosion, deposits can be formed on alloys which significantly affect the gas-induced corrosion process. Deposit-modified high temperature corrosion is called hot corrosion and is an important means of alloy degradation. When the deposits are thick the corrosion mechanisms approach those of liquid metal corrosion. Corrosion processes can be influenced by erosion processes and in some engineering applications alloys are exposed to conditions that cause both erosion and corrosion. It is necessary therefore to consider how erosion and corrosion affect one another. In the following sections oxidation, mixed gas attack, hot corrosion and erosion-modified corrosion of alloys will be discussed. In each case the basic theory will be briefly presented initially and then degradation sequences and possible protection schemes will be considered. Finally coatings as a means of protecting alloys will be discussed.

Oxidation

When a metal is exposed to oxygen, reaction will take place providing the oxygen pressure is greater than the pressure required for the metal-metal oxide equilibrium, namely

$$M + 1/2\ O_2 = MO \tag{1}$$

where

$$\left(P_{O_2}\right)^{1/2}_{M-MO} = \exp\ (+\Delta G^o_{MO}/RT) \tag{2}$$

and $\left(P_{O_2}\right)_{M-MO}$ is the oxygen pressure of the metal-metal oxide equilibrium, ΔG^o_{MO} is the standard free energy of formation of the oxide MO, and T is the absolute temperature. The equilibrium oxygen pressures for most metals with the exception of the precious metals are very small at all temperatures and consequently the thermodynamic conditions are favorable for oxide formation in many gas environments encountered in practice. Oxide scales are therefore developed upon metals exposed to oxygen at elevated temperatures. When these scales are dense and protective, Figure 1, the oxidation rate expressed in terms of specimen weight change* is given by

$$\frac{d(\frac{\Delta m}{A})}{dt} = \frac{k_p}{\frac{\Delta m}{A}\ (2)} \tag{3}$$

where Δm is weight change, A is the specimen area, t is time and k_p is the parabolic rate constant. The important feature of this relationship is that the oxidation rate decreases with time since the oxide scale acts as a barrier which inhibits oxidation by

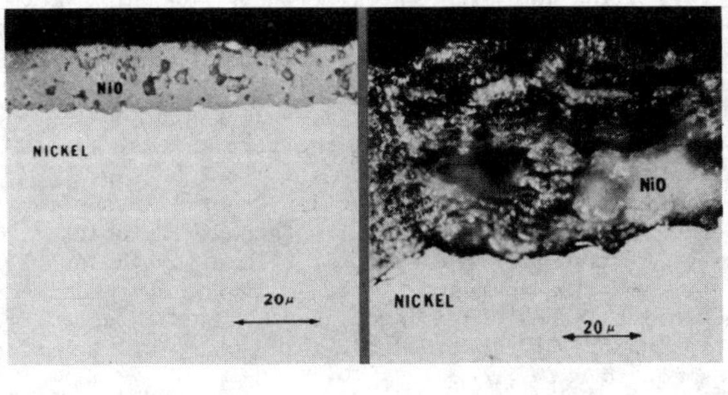

Figure 1. Examples of dense protective and porous nonprotective oxide scales formed on nickel. The protective scale of NiO was formed after 3 hrs. of oxidation in air at 1000°C where the nonprotective scale formed after 1 min. of oxidation under the same conditions but with the specimen coated with 0.5 = mg/cm^2 of Na_2SO_4 to initiate hot corrosion attack.

*The oxidation may also be expressed in terms of oxide scale thickness or thickness of metal consumed.

separating the reactants. If the oxide scale is non-protective, Figure 1, it does not inhibit subsequent oxidation and the oxidation rate is given by

$$\frac{d\,\frac{\Delta m}{A}}{dt} = k_\ell \qquad (4)$$

where k_ℓ is a constant. The oxidation of metals follows other rate laws in addition to the parabolic (Eq. 3) and linear (Eq. 4) expressions (1), however when one is primarily concerned with oxidation resistance consideration of these two rate laws is sufficient.

In considering the oxidation of alloys, equations (1) and (2) must modified as follows,

$$M_i + 1/2\ O_2 = M_iO^* \qquad (5)$$

$i = 1,2,\ldots$ number of elements in alloy

$$\left(P_{O_2}\right)^{1/2}_{M_i-M_iO} = \frac{\exp\left(+\Delta G^\circ_{MiO}/RT\right)}{a_{M_i}} \qquad (6)$$

where a_{M_i} is the activity of M_i in the alloy. Hence upon exposure to oxygen, oxides of all elements in the alloy will be formed providing the oxygen pressure in the gas is greater than the equilibrium pressures defined by Eq. (6). Even though numerous oxide phases are usually formed upon alloys during the initial stages of oxidation, there is competition between the elements for oxygen since some of the oxides are more stable than others. Hence it is necessary to consider reactions such as

$$M_1O + M_2 = M_2O + M_1 \qquad (7)$$

with

$$\frac{a_{M_1}}{a_{M_2}} = \exp\left[\left(\Delta G^\circ_{M_1O} - \Delta G^\circ_{M_2O}\right)/RT\right] \qquad (8)$$

Eq. (8) shows that the thermodynamic conditions for oxide formation are controlled by the activities of the metallic components in the alloy, the pressure of oxygen in the gas, and the relative affinities of the metallic elements for oxygen. As a result of this competition between the elements in the alloy for oxygen, there is a tendency for the alloy to eventually become covered with the most thermodynamically stable oxide. This is an especially fortunate circumstance because it permits the selective oxidation of essentially one element in the alloy. The oxide scales that are formed on alloys are very often composed of layers with the more thermodynamically stable oxides closer to the alloy, Figure 2. Even when only one oxide scale is evident upon an alloy, examination of the scale during the very initial stages of oxidation often shows the formation of the lesser stable oxides, Figure 3. The oxidation of alloys often therefore consists of a transient stage during which a variety of oxides are formed (2,3). This initial stage is followed by another stage where only the thermodynamically stable phases continue to grow. Kinetic factors play a significant role in determining which of the oxide phases continue to grow.

During the very initial stages of oxidation virtually all oxides which are thermodynamically stable are formed. As oxidation continues diffusion begins to take place and the displacement reactions begin to occur. The important diffusing species are oxygen into the alloy and the outward diffusion of elements in the alloy. In the case where oxygen diffuses rapidly compared to the

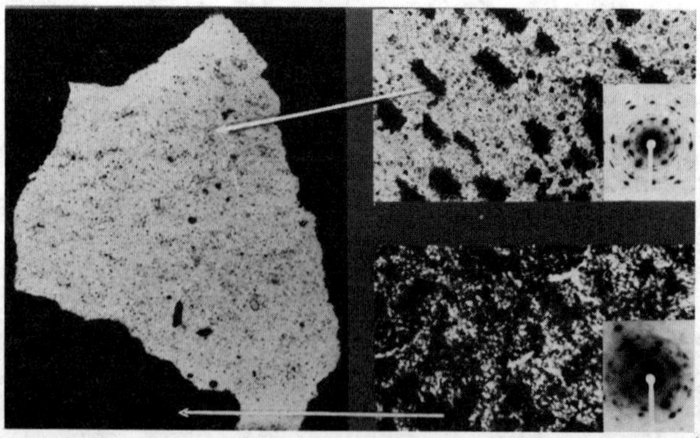

Figure 2. Transmission micrographs and diffraction patterns of the oxide scale formed on a Ni-15Cr-6Al alloy after 40 min. exposure in 0.1 atm. oxygen at 1000°C. The dark area contains NiO and $Ni(Cr,Al)_2O_4$ in addition to a small amount of Cr_2O_3. The light area is the inner part of the scale that was adjacent to the alloy and is composed of Cr_2O_3 with small particles of $Ni(Cr, Al)_2O_4$.

Figure 3. Oxide scale of α-Al_2O_3 which was formed on Ni-20Cr-6Al alloy after 20 hrs. of oxidation in 0.1 atm. oxygen at 1200°C. The insert shows a transmission micrograph of a scale that would be formed on this alloy after a few minutes of exposure under the same conditions. The dark area is Cr_2O_3 with discontinuous particles of α-Al_2O_3 just evident at the break in the Cr_2O_3 layer.

metallic elements, the volume fraction of oxide is approximately equal to that of the element in the alloy because the element is converted to oxide in situ with virtually no diffusion. On the other hand when oxygen diffusion is slow compared to that of elements in the alloy, the volume fraction of the oxide can be substantially greater than that of the element in the alloy because of diffusion of the element from the interior toward the oxidation front. When the volume fraction of a given oxide achieves some critical value it is more favorable to form additional oxide via lateral growth rather than to form new oxide deeper within the alloy, and hence for such critical volume fractions the oxide will then develop as a continuous layer over the surface of the alloy. In summary, the development of oxides on alloys can be described as a sequential process. Initially many oxides are formed but then the more thermodynamically stable phases gradually predominate. In order to have total dominance of one oxide over the surface of the alloy, this oxide must be more stable than any other of the oxides, and the kinitics of formation for this oxide must be such that lateral growth is favored as opposed to formation of a discontinuous subscale.

Oxide scale development is illustrated schematically in Figure 4 for the formation of oxide scales on Ni-Cr-Al alloys. In this example it is assumed that the oxygen pressure is high enough to form all possible oxide phases. At the very beginning of oxidation there is a rapid uptake of oxygen by the alloy and the oxides NiO, α-Al_2O_3, Cr_2O_3 as well as spinel phases (i.e. $NiAl_2O_4$, $Ni\ Cr_2O_4$) are formed on the surface of the alloy. The relative proportions of these phases is determined by the initial

*The subscript i denotes a particular element in the alloy rather than the number of atoms of M.

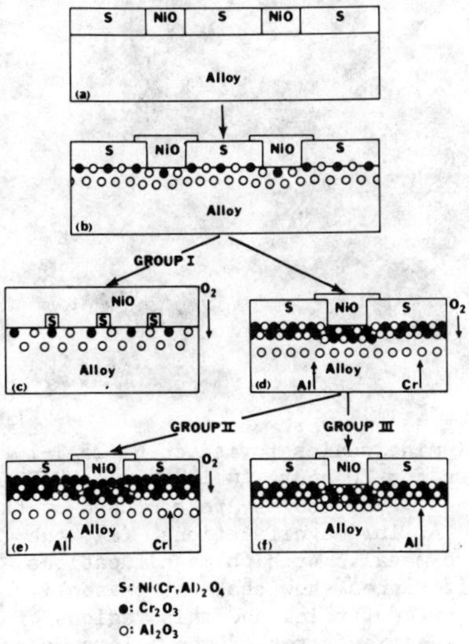

Figure 4. Schematic diagram to illustrate oxide scale development on Ni-Cr-Al alloys as a function of time. (a) Conversion of a thin alloy surface layer to oxide by rapid uptake of oxygen. The oxide phases formed are determined by the composition of the alloy. (b) Diffusion within the alloy results in the formation of a Cr_2O_3 and Al_2O_3 subscale beneath the external scale. (c) For alloys with low chromium and aluminum concentrations, the subscale cannot become continuous and NiO in the external scale predominates. This group of alloys could then be called NiO formers. (d) For alloys with higher chromium and aluminum concentrations the subscale becomes continuous but aluminum is still oxidized internally (e) For alloys with smaller aluminum concentrations the continuous external scale becomes enriched in Cr_2O_3 but the aluminum is oxidized internally. This group of alloys could be called Cr_2O_3 - formers. (f) For alloys with larger aluminum concentrations the Al_2O_3 subscale zone becomes continuous beneath the duplex oxide. This group of alloys would be called Al_2O_3 - formers.

composition of the alloy. As oxidation is continued diffusion processes begin to exert effects. The nature of these effects is dependent upon concentrations of elements in the alloy and diffusional parameters. Generally when the concentrations of chromium and aluminum are low, continuous layers of neither Cr_2O_3 nor Al_2O_3 can be developed, and a continuous scale of predominantly NiO is formed above a subscale zone of Cr_2O_3 and Al_2O_3 particles (Figure 4). At higher chromium or aluminum concentrations continuous layers of either Cr_2O_3 or Al_2O_3 can be formed as illustrated in Figure 4.

It is necessary to emphasize that the length of the transient oxidation period over which the selective oxidation process is taking place can vary significantly from alloy to alloy. For some alloys it is extremely short. For example coatings alloys designed to improve the oxidation of structural alloys have short transient periods of oxidation on the order of seconds. More complex structural alloys, however, often have rather extended transient oxidation stages which can be on the order of days.

As alloys are used in service the oxide which has been formed as a continuous layer over the alloy eventually becomes damaged. A principal means of damage occurs through thermally induced stresses that causes cracking and spalling of the oxide scales. Subsequent oxidation then results in re-formation of the most thermodynamically stable oxide. Cracking and spalling of the oxide scales followed by the development of new oxide layers depletes the alloy of the elements that are being selectively oxidized. Hence the transient oxidation period becomes more long with each subsequent oxidation exposure. Eventually the alloy will become so severely depleted of the element being selectively oxidized that it will not be possible for this oxide to be formed as a continuous layer over the surface of the alloy and a less stable oxide will then be formed as a continuous layer over the surface of the alloy. The degradation sequence of alloys in oxygen therefore consists of the development of the most thermodynamically stable oxides and then less stable oxides as the alloy becomes degraded. The formation of the less stable oxide scales takes place gradually. The less stable oxides begin to develop in localized areas, Figure 5, and increase in size as the oxidation process is continued.

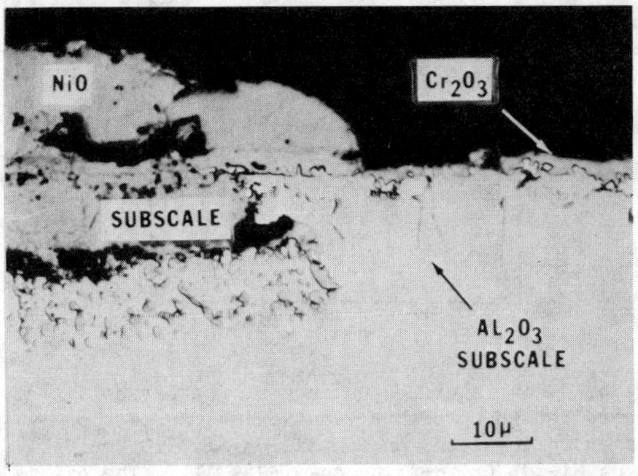

Figure 5. Nonuniform oxidation morphology developed on Ni-10Cr-1Al after 20 hr of oxidation in 0.1 atm. of oxygen at 1000°C. This type of oxide scale morphology frequently develops on alloys during cyclic oxidation testing as less protective oxides become stable.

The approach that is followed to obtain oxidation resistant systems utilizes selective oxidation. In particular, one attempts to have the thermodynamically stable oxide also be an oxide through which diffusion takes place as slowly as possible. Oxides which possess thermodynamic stability and which are also effective barriers against oxidation are α-Al_2O_3, Cr_2O_3 and SiO_2, Figure 6. Chromia scales cannot be used at temperatures above about 1000°C since gaseous CrO_3 is formed which causes its protectiveness to be decreased. This oxide is therfore not used at temperatures above 1000°C. It is used rather extensively however to develop oxidation resistance in nickel - and cobalt - base superalloys (e.g. IN 738, X-40) and in stainless steels. Silica scales can also give rise to gaseous oxidation products (e. g. SiO at low oxygen pressures) and this has affected the use of silicon in alloys as a means of obtaining oxidation resistance. Alumina scales are very effective barriers to inhibit oxidation and the selective oxidation of aluminum to obtain protective scales of α-Al_2O_3 is utilized very extensively to obtain oxidation resistance in coatings alloys and in numerous structural alloys (e.g. B-1900). The degradation sequence of corrosion resistant alloys therefore usually consists of a period of very slow degradation during which the most protective oxide barriers are stable, Figure 7. As oxidation continues the oxidation rate increases when the less protective oxides become stable. The time over which the more protective oxides are stable depends on a number of factors. An important one is composition and coatings redevelop the more protective oxides much longer than structural alloys, Figure 7. Numerous compositional as well as microstructural features of alloys affect selective oxidation. Of course, higher concentrations of the elements to be selectively oxidized favor selective oxidation. Certain elements, however, can influence the selective oxidation of another element (4). For example chromium permits aluminum to be selectively oxidized at lower aluminum concentrations. It appears that silicon and oxygen active elements such as yttrium or hafnium may produce similar effects on the selective oxidation of aluminum and of chromium. It has been observed that fine-grained structures can also

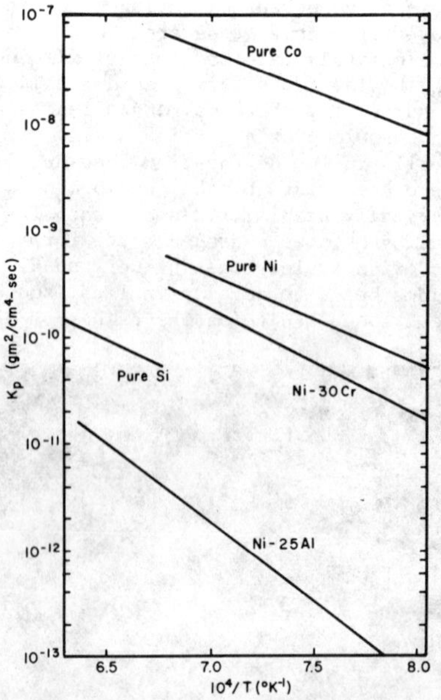

Figure 6. Temperature dependence of the parabolic rate constants for the growth various oxide barriers on metals and alloys (CoO on Co, NiO on Ni, SiO_2 on Si, Cr_2O_3 on Ni-30% Cr, α-Al_2O_3 on Ni-25% Al). The oxidizing environments for the data presented were 0.1 atm. oxygen for Ni, Ni-30Cr and Ni -25 Al; 0.125 atm. oxygen for Co and 1 atm. oxygen for Si.

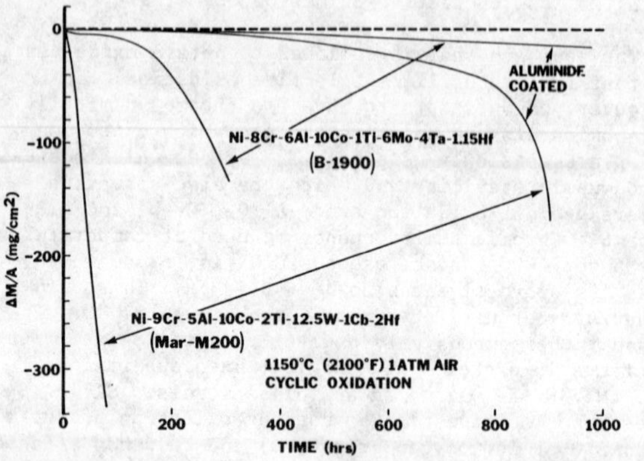

Figure 7. Weight change versus time data for the cyclic oxidation of two nickel-base superalloys and these two alloys coated using a diffusion aluminizing treatment. The degradation of these alloys consists of two stages. An initial stage of less severe attack (not evident on one of the uncoated alloys for the time scale used) and a subsequent stage of more rapid attack.

permit chromium to be selectively oxidized at lower concentrations of chromium (5). A dispersion of inert oxide particles in nickel-chromium alloys can be used to produce similar effects (6,7). The exposure conditions also significantly affect the time for which the more protective scale is reformed. Higher temperatures and numerous thermal cycles tend to shorten the time over which the more protective oxide is re-formed.

The physical characteristics of the protective oxide scales as well their adherence to the alloy are important to the development and maintenance of oxidation resistance. Most oxide scales are not uniform in thickness, Figure 8. It has been determined that the thicker portions of the oxide scales contain grain boundaries whereas the thinner regions are free of grain boundaries (8,9). Such conditions indicate that

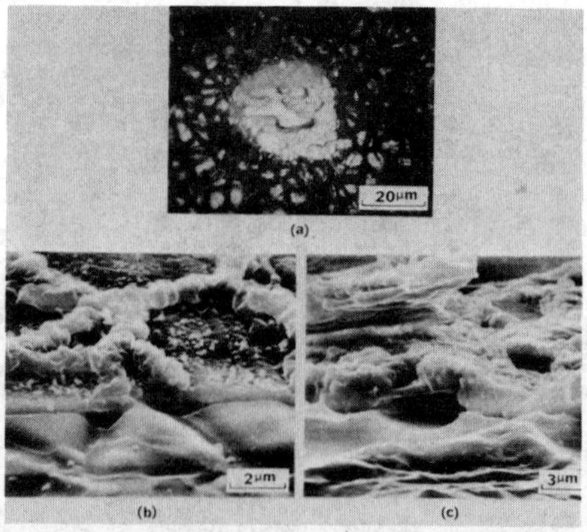

Figure 8. Alumina scales formed on Ni-25Al-10Pt alloy after exposure for 20 hours in 1200°C in CO-CO_2 gas mixture with $CO_2/CO = 3$. The scales are composed of thin and thick areas. At low magnifications, (a), the thin areas appear as bare metal. At high magnifications, (b), the thickness differences show that the parabolic rate constants for growth of thick and thin regions of the oxide differ by at least a factor of ten. In some cases, (c), voids were evident at the oxide - alloy interface. Transmission electron microscopy of such scales during their initial development showed that numerous grain boundaries were present in the thick regions of the scale but none could be detected in the thin regions.

the oxide barriers used to develop oxidation resistance could be made more effective if they could be made to form on the surfaces of alloys without any grain boundaries. It is also apparent that as oxide scales develop voids are frequently formed at the oxide-alloy interface, Figures 8,9 and within the oxide scale itself (9). The voids at the oxide-alloy interface can cause the adherence

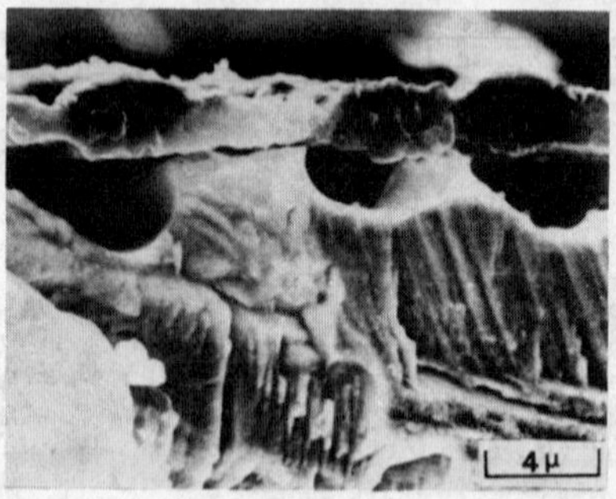

Figure 9. Void formation at the α-Al_2O_3 - alloy interface of a Pt -10 Al alloy after 20 hr of oxidation in air at 1200°C.

of the oxide to the alloy to be decreased. The adherence of oxide barriers to alloys plays a very significant role in determining the oxidation resistance of alloys. Alloys with the more adherent scales are more resistant than those with less adherent scales of the same oxide. A number of techniques have been used to improve oxide scale adhesion. Oxygen active elements (10,11), precious metals (8) and dispersions of oxide particles (6,7) can be used to improve oxide scale adherence as illustrated in Figure 10. An accepted mechanism that accounts for the improved adherence of oxide scales by using these techniques is not available. It appears that an irregular oxide scale-alloy interface usually develops when these techniques are utilized, Figure 11. Some have proposed that such an interface mechanically keys the oxide scale to the alloy (11,12) but other models to account for the improved adherence are available (13).

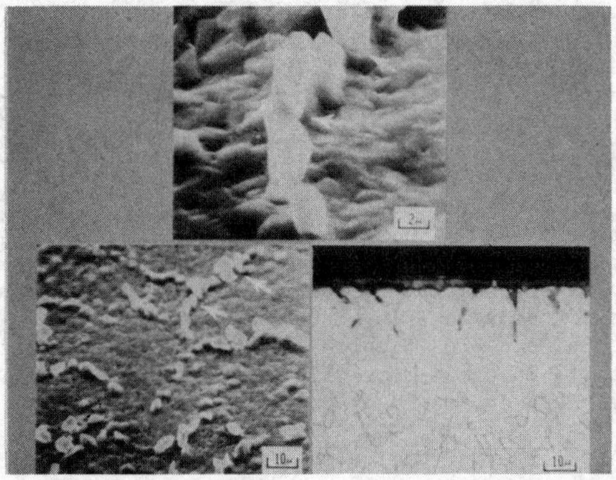

Figure 10. Oxide scale formed on a CoCrAlY alloy after oxidation in air at 1200°C. The oxide scale is adherent and portions of the oxide scale extend into the alloy (lower right.) The other photographs show the oxide surface at the alloy interface after the alloy was removed by dissolution. A network of oxide protrusions (Y_2O_3 covered with Al_2O_3) are evident.

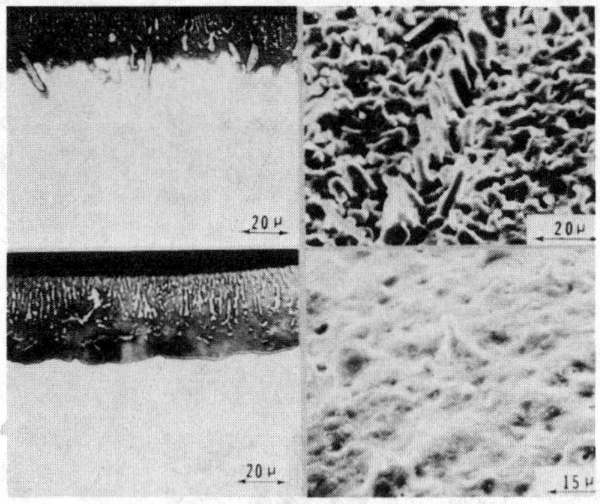

Figure 11. Morphology of the Al_2O_3 - substrate of FeCrAlSc oxidized in air at 1200°C for two different times. The upper photographs show a transverse section and a scanning micrograph of the Al_2O_3 at the Al_2O_3 - alloy interface after 100 hrs. The lower photographs are photographs at equivalent locations after 1000 hrs. As the interface becomes less irregular at the longer oxidation times the oxide scale adhesion becomes less.

The discussion of oxidation resistance via utilization of selective oxidation has not considered effects which arise when more than one phase is present in the alloys. The oxidation process is not substantially more complicated when alloys contain more than one phase. Each phase can be considered to undergo a selective oxidation process where the extent of the transient oxidation period for each phase will be dependent upon its particular composition. Very often the surface compositions of alloys become changed during oxidation such that only one phase is stable at the surface as shown in Figure 12 where the γ' - phase has been removed from the surface of the alloy due to the selective oxidation of aluminum. In cases where diffusion in the alloy is very slow and when large compositional differences exist between the different phases in alloys, the oxidation resistance of the phases can be substantially different. Hence, more severe attack of the phase with the poorer oxidation resistance can take place, Figure 13. In such alloys adequate oxidation resistance can only be obtained by using coatings unless the phases with the unacceptable oxidation resistance are not continuous, and then the attack will stop as this particular phase is completely converted to oxide.

In summary, the oxidation resistance of alloys depends upon selective oxidation of elements such as aluminum, chromium or silicon. Oxide scales of Al_2O_3 appear to be more effective means for developing oxidation resistance than use of either Cr_2O_3 or SiO_2.

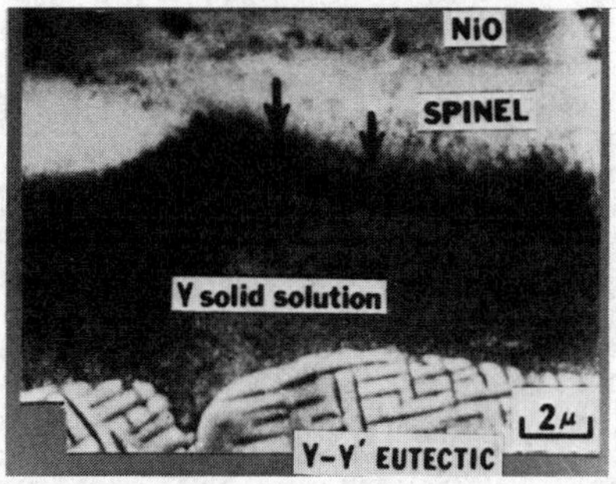

Figure 12. Microstructure of Ni-15.1 Ta-6.7 Al after oxidation at 1200°C for 5 min. in 1 atm. of air. The γ' is removed from the surface of the alloy due to oxide scale formation.

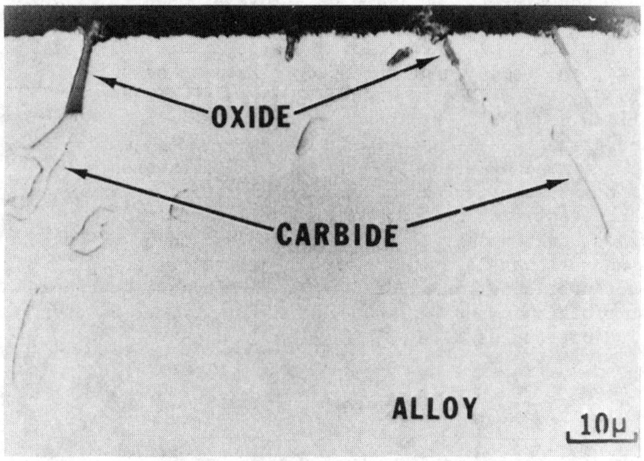

Figure 13. Preferential attack of carbide phase in nickel-base superalloy B-1900 after 45 hrs. of oxidation at 700°C in ∿ 1 atm. oxygen.

While selective oxidation of elements such as aluminum plays a critical role in the development of alloy oxidation resistance, an equally important factor is the adhesion of the oxide scales to the alloy substrates, and it is necessary to use techniques to optimize oxide scale adhesion.

Mixed Gas Attack

Mixed gas attack involves all of the factors discussed in the previous section for oxidation of alloys with the added complication of competition between elements in the gas for elements in the alloy. This means that Eqs. (5) and (6) must now be written as

$$M_i + X_j = M_iX_j \quad i = 1,2.\dots \text{ number of elements in alloy.} \quad j = 1,2,\dots \text{ number of elements in gas.} \tag{9}$$

$$a_{X_j} = \exp\left(+\Delta G^\circ_{M_iX_j}/RT\right)/a_{Mi} \tag{10}$$

and in addition to Eqs. (7) and (8) the following reactions must also be considered,

$$M_1X_1 + X_2 = M_1X_2 + X_1 \tag{11}$$

with

$$\frac{a_{X_1}}{a_{X_2}} = \exp\left(\Delta G^\circ_{M_1X_1} - \Delta G^\circ_{M_1X_2}\right)/RT \tag{12}$$

While in mixed gas attack the situation is more complex in that more reactions are possible, the end result is the same, namely, Eqs. (7), (8), (11), (12) show that the thermodynamic factors are such that the formation of one phase is favored providing the kinetics will allow this phase to be formed as a continuous layer over the surface of the alloy. The development of a given phase as a continuous layer on the surface of an alloy is dependent upon the diffusion of the components in the gas into the alloy and elements from the interior of the alloy to the surface of the alloy. The situation is a more complicated version of what has been described previously for a multicomponent alloy exposed to oxygen. The end result is either the formation of a continuous layer of one reaction product over the surface of the alloy, or a scale containing a mixture of reaction products (14). It should be emphasized that the development of a continuous layer of one reaction product usually becomes less favorable as the number of components in the gas is increased.

The development of a continuous layer of one reaction product on the surfaces of alloys exposed to mixed gases provides more resistance to subsequent corrosion than a mixture of reaction products. This is especially true if the reaction product can be made to be one through which diffusion is slow. The most effective reaction product barriers are oxides, and resistance to mixed gas attack is usually achieved by having oxide scales formed on alloys. Most industrial environments do contain some oxygen, and large quantities of oxygen are not required in order to have selective oxide barrier development. For example, it can occur at oxygen pressures lower than 10^{-10} atms. When sufficient oxygen is not available for oxide barrier development, the next most protective barrier that can be formed must be used to inhibit corrosion.

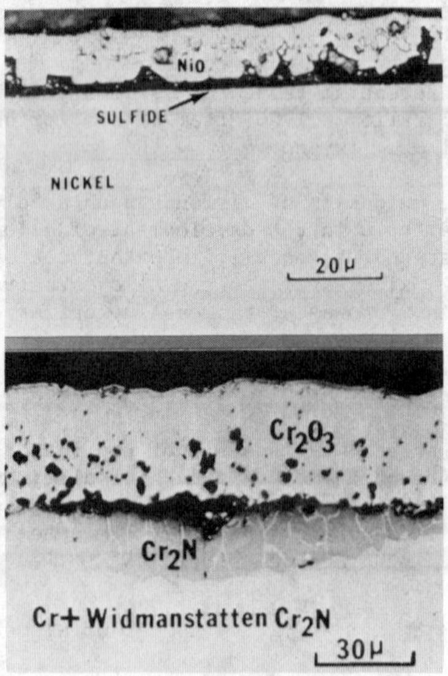

Figure 14. Photomicrographs showing the formation of sulfide and nitride phases beneath external oxide scales. The nickel was exposed in flowing SO_2 for 8 hrs. at 1000°C. The chromium was oxidized 17 hr. in air at 1200°C.

Even when continuous oxide barriers are developed on alloys exposed to mixed gases the situation is more complicated than when only oxygen is present in the gas. For example other reaction products can be developed in the alloy beneath the external oxide barrier as shown in Figure 14. Such corrosion microstructures develop since the other phases are also thermodynamically stable, and the only reason they are not formed at the alloy surface is because the oxide phase is more stable. Hence, if the other reactants can diffuse through the oxide barrier (or penetrate it by other means) the other reaction product phases can be formed in the alloy where the oxygen pressure is low. The formation of these phases within the alloy beneath the oxide barrier creates problems in reforming the oxide barrier when oxide spalling takes place.

If a mixture of reaction products is formed the corrosion rate is greater than if a continuous oxide barrier is developed. This occurs because diffusion takes place in phases having the higher diffusion rates, Figure 15. The formation of such intermixed scales must be avoided if resistance to high temperature corrosion is to be obtained. The problem of avoiding the formation of scales with intermixed reaction products is not usually easily overcome. First it is necessary to have the phase desired to be formed as a continuous layer, to be the most thermodynamically stable phase. Since this phase is usually an oxide phase it becomes more difficult as the oxygen pressure is reduced. Even when the desired phase is the most thermodynamically stable phase and a continuous barrier of this phase is formed on the alloy, continued exposure of the alloy to the mixed gas can result in loss of this barrier with the alloy being incapable of reforming it. Hence, rapid degradation will ensue.

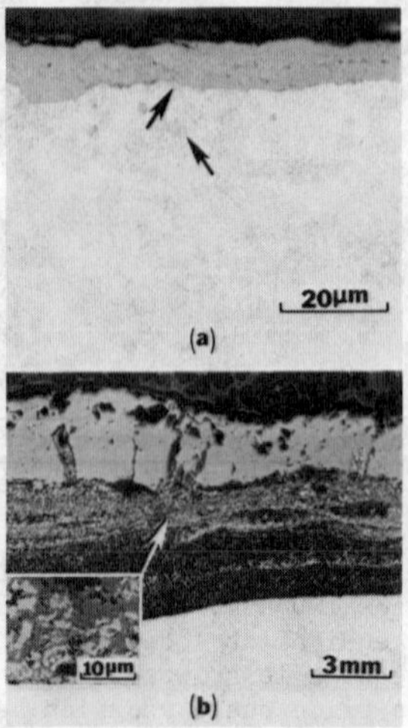

Figure 15. Photomicrographs showing typical microstructures developed during cyclic oxidation of Fe - 15Cr at 900°C in flowing argon-5% SO_2. (a) During the earlier stages, an external Cr_2O_3 scale covers the alloy surface and chromium sulfide particles (arrows) are formed within the alloy beneath the scale. (b) As more severe attack occurs the reaction product is composed of an oxide and sulfide mixture beneath an outer zone of iron oxides.

Deposit-Modified Corrosion Hot Corrosion

In many cases when alloys are exposed to high temperatures and gas environments, the alloy surfaces can become coated with foreign deposits such as Na_2SO_4, V_2O_5, NaCl, etc. These deposits may arise through ingestion in the air used for combustion or as contaminants in the fuel required for combustion. The composition of the deposits depends on the particular environment in which the alloy is exposed. In gas turbines and incinerators Na_2SO_4 deposits are often encountered while in fluidized bed combustors using coal deposits of $CaSO_4$ can be formed. The point to be emphasized is that the presence of the deposits on the surfaces of alloys affects the high temperature corrosion processes which are taking place between the alloys and the gas environments. An important effect is that the thermodynamic activities of components in the gas are significantly different at the alloy-deposit interface compared to the gas phase as indicated schematically in Figure 16. Moreover the corrosion products often can react with or dissolve into the deposit. This latter condition is especially likely to occur when the deposit is in the liquid state. Deposits on the surfaces of alloys therefore usually cause the alloys to be attacked by the gas at rates significantly greater than would occur if the deposits had not been present. Deposit modified, gas-induced degradation of alloys is called hot corrosion. There are a number of mechanisms by which hot corrosion can occur. All of these mechanisms involve the deterioration of the reaction product barrier that forms on the alloys when deposits are not present.

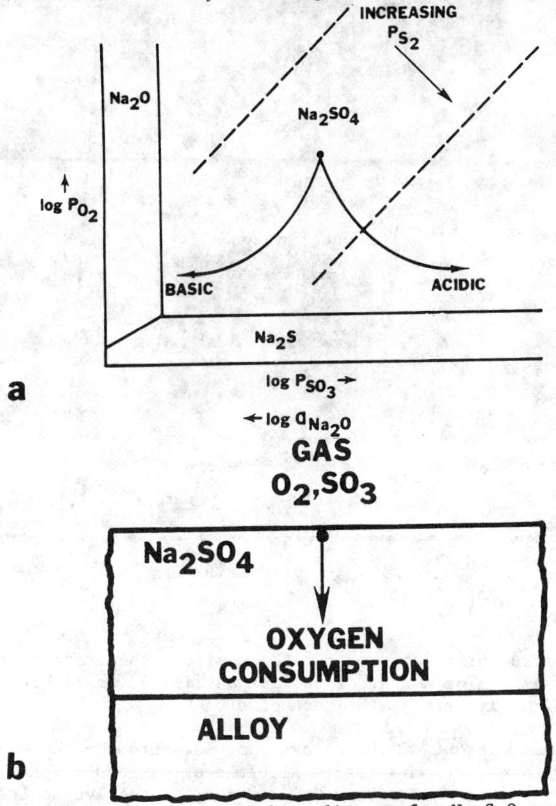

Figure 16. Schematic stability diagram for Na-S-O system, (a), depicting the types of compositional variations that can be developed across a layer of Na_2SO_4 on an alloy (b).

Studies concerned with the hot corrosion of metals and alloys show that the hot corrosion process usually consists of two stages. The initial stage is one that is virtually identical to that which the alloy undergoes in the absence of the deposit. The second stage on the other hand is dependent upon the deposit and consequently substantially different from the behavior of the alloy in the absence of the deposit. It therefore is useful to consider the alloy as being preconditioned in the initiation stage which determines the type of degradation process that will be followed in the second stage called the propagation stage. The propagation stage is accompanied by very severe attack of the alloy. The relationship between the initiation and propagation stages is indicated schematically in Figure 17. In this Figure the factors that are important in preconditioning the alloy during the initiation stage are identified. The degradation modes have also been divided into two general classes. One class involves fluxing reactions where the oxide scales formed on the alloys can react with the liquid deposit. When such reactions occur through reaction with oxide ions in the deposit they are called basic whereas they are called acidic when oxide ions are donated to the deposit by the oxide scale. As a result of such reactions the oxide scale can become nonprotective. Such nonprotective scales can be formed by dissolution of the oxide near the alloy surface and reprecipitation as discontinuous particles out in the molten deposit. It

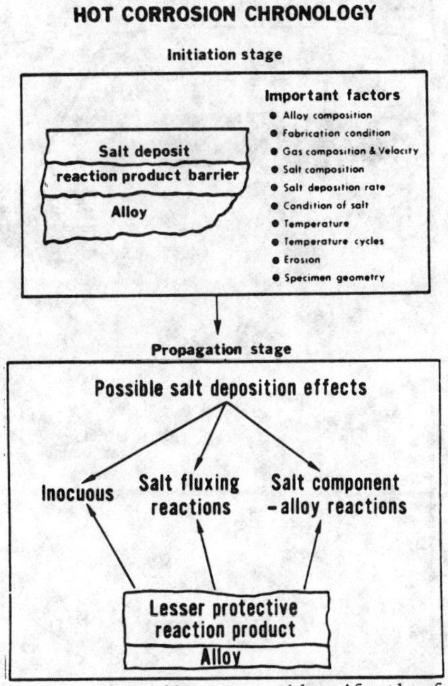

Figure 17. Schematic diagram to identify the factors that determine the time at which the transition from the initiation stage to the propagation stage occurs and to illustrate the two general categories of protective scale breakdown to a lesser protective reaction product when a salt deposit is present during the corrosion process.

is important to emphasize that there are other means by which the oxide scales can become nonprotective via a fluxing process. For example the fluxing reaction may cause a certain element to be removed from the alloy which in turn gives rise to a nonprotective scale on the surface of the alloy.

When deposits are present on the surfaces of alloys, certain elements in the deposits can accumulate in the alloy near its surface and cause the surface composition of the alloy to change. Sulfur from Na_2SO_4 deposits accumulates in alloys as shown in Figure 18, and oxidation of these sulfides can result in increased attack. Some of the first observations of hot corrosion attack involved such features and as a result the attack was called sulfidation. Sulfidation is a type of hot corrosion but all hot corrosion attack does not proceed via sulfidation. If the deposits contain chloride ions, the chlorine can selectively remove chromium or aluminum from the alloy. This process appears to involve the formation of gaseous chlorides in pores formed in the alloy; Figure 19. The metallic component of these chlorides are converted to oxides eventually but these oxides are formed as particles and not as continuous layers. Hence severe attack of the alloy takes place.

In discussing hot corrosion it is necessary to attempt to identify those conditions for which hot corrosion attack will be minimized. The level of knowledge of hot corrosion processes has been progressively increasing, nevertheless, there is still not complete agreement with regards to all of the mechanisms of hot corrosion and the effects produced by various elements (15-18). The best means to avoid or minimize hot corrosion attack is to have alloys remain in the initiation stage as long as possible. In Figure 20 it is attempted to classify alloys according to their resistance to the various types of hot corrosion propagation modes. In this Figure acidic fluxing degradation has been subdivided into alloy induced acidic fluxing and gas-phase induced acidic fluxing. In alloy-induced acidic fluxing the acidic component is added to the molten deposit by oxidation of elements in the alloy (e.g. Mo, W) (16) whereas in gas phase acidic attack the acidic component is introduced to the deposit from the gas phase.

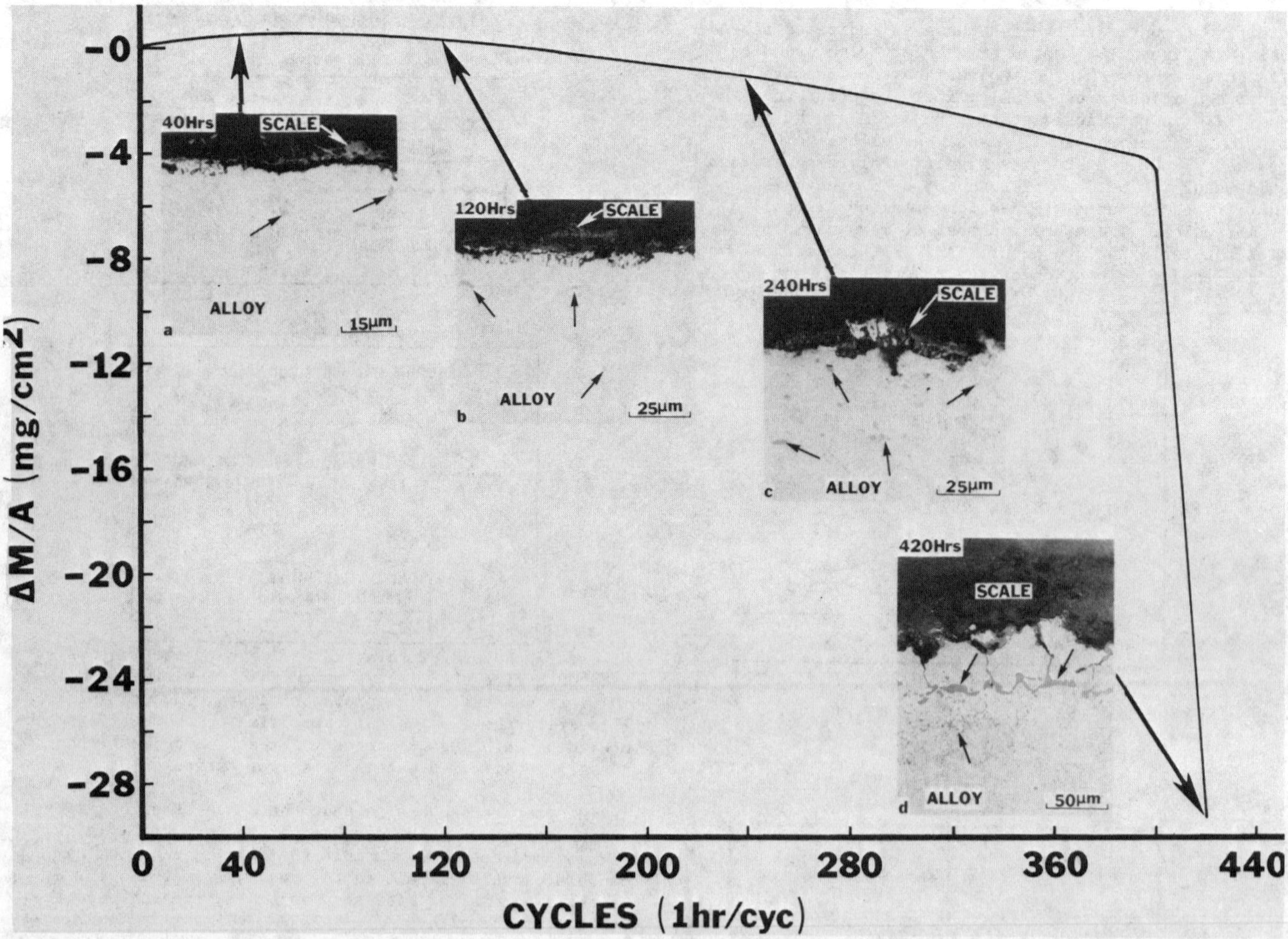

Figure 18. Weight change versus time data and corresponding microstructural features for the cyclic hot corrosion of (Na_2SO_4 - coated (5 mg/cm^2 applied every 20 hrs.) Ni-30 Cr-4 Al in air. The amount of sulfide particles (small black arrows) increases until the oxidation of the sulfide phases significantly affects the rate of attack.

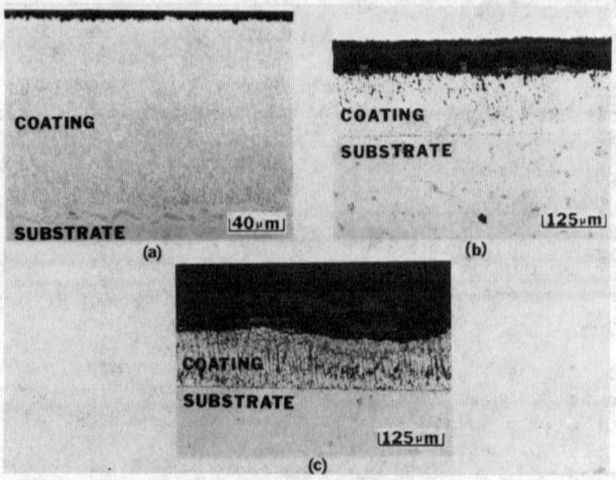

Figure 19. Photomicrographs to compare the degradation of CoCrAlY coatings on IN 738 after exposure at 899°C in air to Na_2SO_4 deposits containing different amounts of NaCl. (a) 500 hrs. with Na_2SO_4, (b) 500 hrs. with Na_2SO_4-5 weight percent NaCl, (c) 40 hrs. with Na_2SO_4-90 weight percent NaCl.

The metals and alloys which are susceptible to basic fluxing are nickel, cobalt, Ni-Al and Co-Al alloys which are Al_2O_3-formers, and Ni-Cr-Al alloys with chromium and aluminum concentrations below that required to form external scales of Cr_2O_3 or Al_2O_3. The most effective procedure to inhibit basic fluxing is to increase the chromium content of the alloys. At low chromium concentrations, chromia reacts with the oxide ions in the Na_2SO_4 rather than NiO or CoO. Hence, protective scales of these latter oxides are formed and grow to thicknesses for which oxide ion production in the Na_2SO_4 is not significant. At higher chromium concentrations, where continuous Cr_2O_3 is formed on alloys, these chromia scales are not as susceptible to solution and reprecipitation from the Na_2SO_4 as Al_2O_3 appears to be in the absence of chromium. Chromia scales are therefore, very effective barriers to basic fluxing. Since basic fluxing requires that an oxygen gradient be developed across the Na_2SO_4, the formation of oxide scales that grow slowly and consume as little oxygen as possible is an effective means to combat basic fluxing. The addition of chromium and aluminum to alloys results in improved resistance to basic fluxing providing, continuous barriers of Al_2O_3 or Cr_2O_3 are formed. When the concentrations of these elements are less than that required to form such scales, it is best not to have any aluminum in the nickel-base alloys, because, under such conditions, large amounts of sulfur are removed from the Na_2SO_4 along with concomitant oxide ion production. Chloride in Na_2SO_4 deposits causes basic fluxing effects to be observed sooner by depleting the alloy of aluminum and chromium. Thin layers of platinum (∿5µm) on the surfaces of alloys have been found to be effective in inhibiting such chloride-induced effects. Carbon causes basic fluxing to be observed sooner due to the reduced oxygen pressures which develop as a result of its presence.

The oxidation resistance of virtually all alloys deteriorates when sulfur is present to the extent that sulfide phases exist within the alloys. Nickel-base alloys containing between 1 to 6% aluminum are extremely susceptible to this type of hot corrosion attack. Such

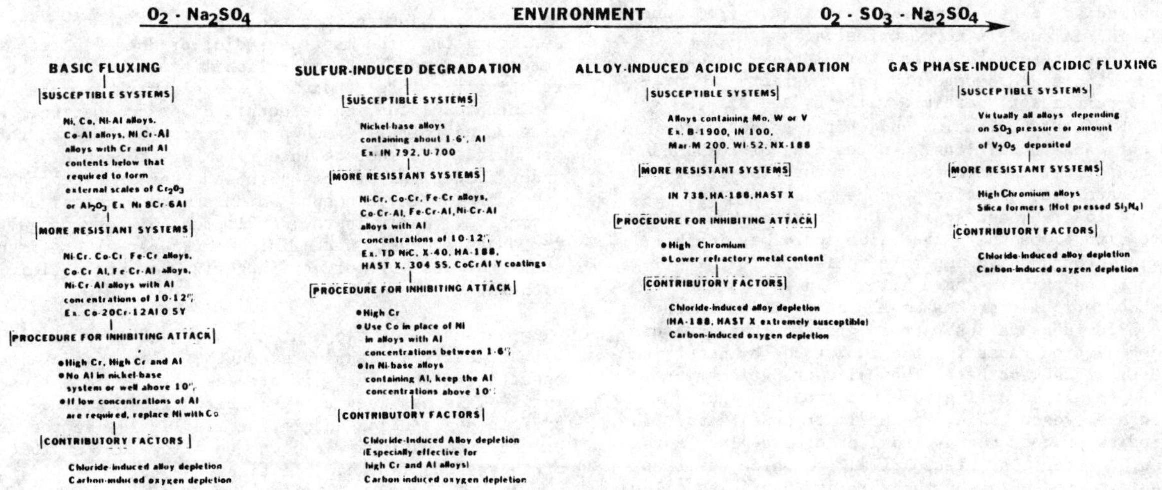

Figure 20. Diagram to identify the alloy systems which are susceptible to attack via the principal hot corrosion propagation modes and to describe procedures to inhibit such degradation.

attack can be inhibited by increasing the aluminum concentration beyond six percent and/or replacing nickel with cobalt. Increased chromium concentrations are a very effective means to inhibit degradation via sulfidation. Chromium combines with sulfur to form sulfides with relatively high melting points. At high (e.g. ∿ 20% and greater) alloy chromium concentrations, protective Cr_2O_3 scales are often formed on these sulfides. The chromium concentration should not be so high, however, that the α-Cr phase becomes stable. Even when less protective oxides are formed on the sulfide phases, the degradation rate is still significantly less than that of similar nickel-base alloys containing aluminum. Chloride induced degradation results in accelerated depletion of aluminum and chromium from alloys, and therefore, the presence of chloride in deposits causes sulfidation degradation to be observed after shorter exposure times. The introduction of sulfur into the alloy from the Na_2SO_4 is favored by low oxygen pressures, and hence, carbon in deposits hastens the onset of hot corrosion degradation via the sulfidation propagation mode.

Alloy-induced acidic fluxing results from the accumulation of oxides much as MoO_3, WO_3 or V_2O_5 in salt deposits on alloys. Susceptible alloys are therefore, those with high concentrations of such elements (e.g. NX-188), or alloys with smaller concentrations of these elements but with compositions that permit severe attack of the alloys by one of the other hot corrosion propagation modes (e.g. B-1900, WI 52). This form of hot corrosion attack can be inhibited by decreasing the concentrations of the refractory elements, molybdenum, tungsten and vanadium. It is not certain whether columbium causes effects similar to these three elements. Tantalum does not, and therefore, appears to be a reasonable replacement for Mo, W and V in alloys. Modifications to inhibit the oxidation of the Mo or W in alloys is an effective means to inhibit alloy-induced acidic attack. Increased chromium concentrations have been used to achieve such a condition, (e.g. IN 738, HA-188). As usual, chloride and carbon in deposits cause alloy-induced effects to be observed sooner.

All alloys can be attacked via gas phase-induced acidic fluxing depending upon the SO_3 pressure and the temperature. Silica and chromia scales on alloys appear to be the most resistant to degradation via this mode but even barriers of these oxides can be destroyed at sufficiently high SO_3 pressures. Nevertheless, increased chromium and silicon concentrations do appear to be a means of inhibiting the attack of alloys by gas phase induced attack. At low temperatures (e.g. ∿700°C), this attack is only severe when the salt deposit is a liquid. In the case of Na_2SO_4 deposits, a liquid phase usually is developed much sooner on cobalt-base rather than nickel-base systems. A detailed examination is required of the effects of chloride and carbon on gas phase-induced hot corrosion. At present, it appears as though both of these factors will cause the attack to be initiated sooner, but they may not have a substantial effect on the rate of this propagation mode after it has been initiated.

Erosion-Corrosion Interactions

In some instances oxidation, mixed gas attack or hot corrosion occurs in combination with conditions that produce erosion. For example during combustion of coal in a fluidized bed mixed gas attack and erosion may be expected, and in gas turbines oxidation and hot corrosion concomitant with erosion can be observed. Erosive conditions in combination with corrosion are especially undesirable because the reaction product barriers used to develop resistance against corrosion can be destroyed by the erosion process, Figure 21. Investigations to study the interaction between erosion and corrosion processes have been performed (19).

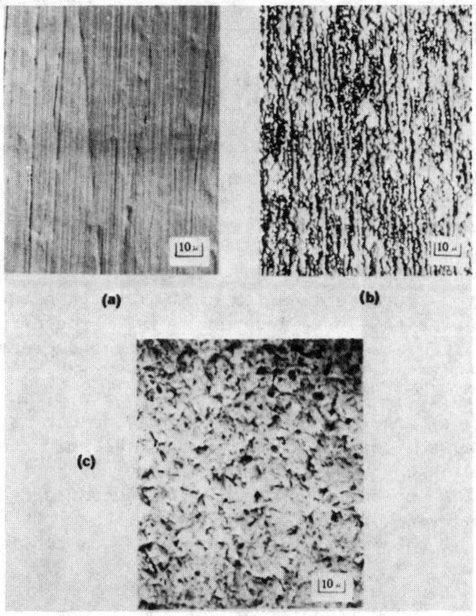

Figure 21. Scanning electron micrographs of replicas from X-40 alloy after various treatments. (a) pretest condition, (b) after 4.5 hr. in oxidation test using air at 871°C, (c) after 4.5 hr. at 871°C in oxidation erosion test using 20 μm Al_2O_3 particles at loading of 130 ppm (90° impact angle). The oxide scale is very thin as a result of the erosion process.

An understanding of these interactions is far from complete but initial models to describe various erosion - corrosion interactions have been formulated. In all cases the effects are dependent upon a variety of parameters, in particular, particle velocity and size, particle composition, loading and shape, temperature, and the corrosive conditions. In the case of combined erosion and oxidation it appears that the erosion process can accelerate the oxidation process by causing the oxide scale to crack and spall. On the other hand, there are situations where the oxide scale may inhibit the erosion process by being more resistant to the erosion process than the bare alloy. In the case of erosion-hot corrosion processes the interaction is such that the alloy degradation rate is greater than the sum of the degradation rates for the two processes occurring independent of one another. Typical degradation microstructures developed during combined erosion - hot corrosion are presented in Figure 22. It appears that the hot corrosion process causes erosion to take place more rapidly by undercutting pieces of alloys which are then dislodged from the alloy by the abrasive particles. The hot corrosion process is evidently accelerated by the erosion process because it removes the porous external oxide scale and the molten deposit can more effectively cover the surface of the alloy to cause hot corrosion attack.

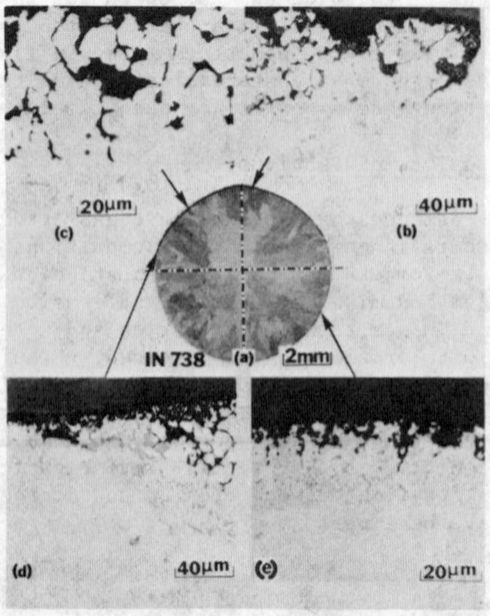

Figure 22. Photomicrographs of IN738 after 28 hr. exposure to erosion-hot corrosion conditions (0.05 mg/cm^2 hr. Na_2SO_4 - 22% K_2SO_4 with 300 ppm of 2.5 μm Al_2O_3 abrasive) at 871°C in a burner rig. Even though a substantial amount of alloy was removed from the leading edge, (a), more corrosion is evident on the leading edge (b), (c), than on the trailing edge, (d), (e).

Combined erosion - corrosion processes represent in general more severe degradation conditions than either of these processes acting independently. Procedures to combat combined erosion - corrosion induced degradation are not readily available. At low temperatures where corrosion is not a problem hard coatings are often used, but at higher temperatures such coatings do not usually possess adequate corrosion resistance. Ceramic coatings may be a means to obtain both resistance to erosion and corrosion. At present, however, dense ceramics coatings which are adherent to the alloy substrate are not available. Development work to produce such coatings is in progress.

HIGH TEMPERATURE COATINGS

Of the many different fields of use of coatings for protection of alloys against surface degradation at high temperatures, that related to gas turbine airfoils best exemplifies the technology involved in coating development and applications. To provide the reader with a useful summary of the history and current status of high temperature coatings the context of the gas turbine engine will therefore be employed.

Gas turbine engines are used in a wide variety of applications. The most demanding of these in terms of materials durability and reliability requirements under relatively severe conditions are aircraft propulsion, marine propulsion, and electric power generation. Hot-section airfoils in such engines are required to retain mechanical and surface integrity for thousands of hours under conditions of very high stresses at elevated temperatures in gas environments which can be aggressively corrosive. Early in the development history of gas turbines, temperature levels were limited by super-alloy strength such that loss of surface integrity was not a limiting problem.

Intensive development has provided airfoil type alloys with steadily increasing strength, allowing ever increasing operating temperatures to provide more efficient use of fuel. This has inevitably resulted in increased rates of surface degradation and in the middle sixties the use of protective coatings was begun to allow full exploitation of increased alloy capabilities.

The first applications of such coatings were in aircraft engines. This was followed, beginning in the early seventies, by increased use of coatings in ground-based and marine engines.

In many applications, turbine airstream environments are such that the useful life of airfoils is limited by coating failure. This has prompted substantially increased levels of development efforts to provide more durable coatings. This paper will briefly describe some recent significant advances in durability of metallic and ceramic coatings for aircraft and utility turbine airfoils.

To provide a context within which to describe these improved coating systems, the gas stream conditions which set the requirements for protective coatings will be briefly defined and then the general classes of available and practical contemporary coatings, their protective modes and degradation mechanisms, will be described. Recent developments in metallic coatings to protect against high temperature (2000-2100°F, 1093-1149°C) oxidation and intermediate temperature (1200-1600°F, 649-871°C) hot corrosion will then be presented. Finally, an overview of ceramic thermal barrier coating technology will be given.

The Gas Turbine Environment

Table I provides a summary of the relative severities of surface-oriented factors which cause turbine airfoil degradation leading to the need for component repair or replacement.

TABLE I

Comparison of Surface-Oriented Problems for Gas Turbine Applications

	Oxidation	Hot Corro.	Interdiffusion	Thermal Fatigue
Aircraft	S	M	S	S
Utility	M	S	M	L
Marine	M	S	L	M

L Light
M Moderate
S Severe

In all of the applications identified, it is generally accepted that the gas stream constitutes an oxidizing environment. Conditions reducing to the oxides formed on superalloys and coatings, if present, are transient and of a non-equilibrium nature; there is no unequivocal evidence that such conditions contribute significantly to surface deterioration. In a very general way hot corrosion is more severe for utility and marine propulsion engines than for aircraft engines. The primary causes of this increase in severity lie in the increased levels of air-borne contaminants (salt) and/or the fuel-borne impurities (e.g., sulfur, vanadium) encountered in utility and marine applications.

Aircraft engine turbine airfoils are, however, by no means free of hot corrosion attack. The severity of any observed attack can usually be correlated with increased levels of salt ingestion associated with high fractions of time near marine or related environments. Again, in a very general way, however, the severity of turbine airfoil surface degradation in modern aircraft engines correlates more satisfactorily with transient high temperature conditions; that is, surface deterioration is caused by high temperature oxidation, occasionally augmented by hot corrosion. In high performance designs, thermal fatigue cracking, initiated in coatings, can be a life-limiting factor.

Types of Coatings Available

The first coatings widely used for the protection of superalloy gas turbine airfoils were based on surface enrichment of, for example, nickel-based alloys with aluminum. As first described by Goward et al. (20-21) and later by others (22,23,24,25) two archetypical structural types, illustrated in Fig. 23, are available.

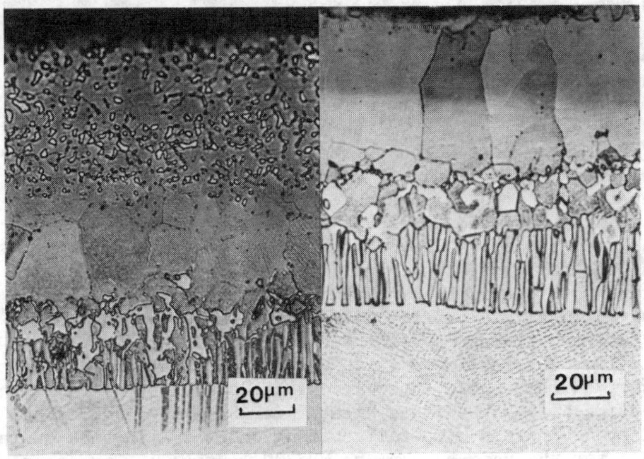

Figure 23. Archetypical aluminide coatings on a typical nickel-base alloy (U-700); left - inward diffusion type, right - outward diffusion type (500X).

The coatings are usually applied by "pack cementation" processes which are more accurately described as chemical vapor deposition, and as delineated by Goward et al. (20,21), the structural types obtained are dependent on the aluminum activity in the particular fabrication process involved. This activity determines the nickel-aluminum intermetallic phase, Ni_2Al_3 or NiAl, formed on the alloy surface by reaction with aluminum species in the gas phase. The two coating structures shown in Figure 23 and commonly referred to as "inward" and "outward" diffusion types, are then derived from the initial singular motion of aluminum in Ni_2Al_3 and/or aluminum-rich NiAl, or the singular motion of nickel in nickel-rich NiAl respectively.

Although these aluminide coatings provide only limited protection in modern applications involving very high temperatures (2000°F, 1093°C) or severe hot corrosion environments, they are still in very wide use in less demanding applications. Oxidation and hot corrosion resistance of these coatings, as expected, are in part dependent on the base alloy composition. Modest improvements in oxidation and/or hot corrosion resistance have been achieved by modifying the superalloy surface with elements such as chromium or platinum prior to aluminizing. Mechanical properties of aluminide coatings, typified by ductility and the related thermal fatigue resistance, are not amenable to significant improvement because of the inherent lack of ductility of the intermetallic compound, NiAl, which is the matrix of all such coatings.

Diffusion coatings based on surface enrichment with chromium (26,27) , have received some attention and limited applications for protection of utility gas turbine airfoils in Europe and Japan. Temperature capabilities of such coatings are ultimately limited by the volatility of CrO_3, formed by reaction of Cr_2O_3 protective scales with oxygen. Surface enrichment of nickel-based alloys with silicon has also been studied in Europe (27) for protection against hot corrosion. Initial field evaluation is in progress. Results indicate some promise for these systems.

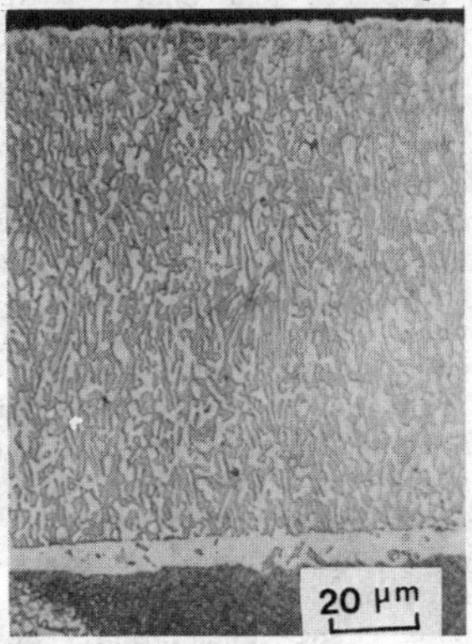

Figure 24. Typical MCrAlY coating. Gray phase is CoAl or NiAl dispersed in white nickel or cobalt solid solution phase.

Overlay coatings, typified by the so-called MCrAlY series (28)(M=Fe,Co,Ni or combinations thereof), allow increased flexibility of design of compositions tailored to a wider variety of applications. A structure typical of contemporary overlay coatings is shown in Figure 24. Oxidation and hot corrosion properties and ductility can be varied over wide ranges to meet various requirements for aircraft, utility and marine propulsion applications. The MCrAlY coatings are currently applied on a production scale by electron beam evaporation. Other processes for coating fabrication, such as plasma spray, sputtering and foil cladding, are in various stages of development.

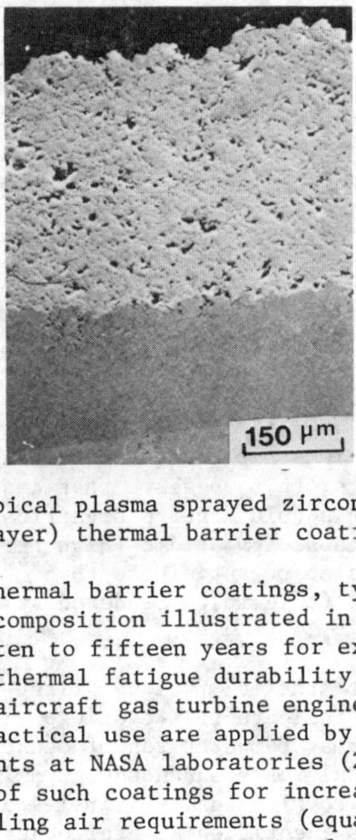

Figure 25. Typical plasma sprayed zirconia (top layer)-MCrAlY (underlayer) thermal barrier coating.

Ceramic thermal barrier coatings, typified by the structure and composition illustrated in Figure 25, have been used for ten to fifteen years for extending the oxidation and thermal fatigue durability of sheet metal components in aircraft gas turbine engines. All such coatings in practical use are applied by plasma spraying. Model experiments at NASA laboratories (29) emphasized the potential of such coatings for increasing longevity or decreasing cooling air requirements (equal to decreasing fuel comsumption) by their use as insulating layers on hot-section airfoils in all types of gas turbine engines.

Coating Protection and Degradation Mechanisms

Beneficial high-temperature (1600-2200°F, 871-1204°C) oxidation properties of diffusion and overlay coatings containing aluminum are based on the formation of alumina under oxidizing conditions. No other element has provided the required primary properties of formation of an oxide with low transport rates and the relative degree of high-temperature compatibility with nickel- (and cobalt-) based superalloys. The practical durability of aluminum-containing coatings is dependent on the adherence of the protective oxide. For diffusion aluminide coatings, this property is principally dependent on substrate alloy composition; specifically, oxide adherence is enhanced by the presence of "oxygen active elements" such as hafnium in substrate alloys (30). Oxygen-active elements such as yttrium are added to overlay coatings to promote oxide adherence (29,31) nominally independent of substrate alloy composition.

Innumerable studies have indicated that the presence of chromium imparts resistance to hot corrosion degradation of alloys and coatings induced by the presence of salt layers reasonably modeled by sodium sulfate (Na_2SO_4). To date, theory has not provided a complete explanation for this behavior. Two series of studies may eventually lead to a plausible explanation of the chromium phenomenon. First, several practical field service observations and laboratory programs (32,33,34) have indicated unacceptably high rates of hot corrosion of coating alloys such as CoCrAlY by Na_2SO_4 in the presence of SO_3-O_2 in restricted temperature ranges. This corrosion behavior is illustrated in Figure 26. Increasing the chromium content of CoCrAlY tends to decrease the corrosion rate under these conditions. Second, studies by Rapp et al. (36) and Elliott et al. (37) on the solubility of oxides such as Al_2O_3 and Cr_2O_3 in Na_2SO_4 as a function of P_{SO_3} indicate different behavior between these oxides. Specifically, the position of minimum solubility of Cr_2O_3 occurs at a higher P_{SO_3} (higher acidity) than Al_2O_3. This might tend to indicate that chromium in coating (alloys), forming protective Cr_2O_3, could be more resistant to hot corrosion at certain acidity levels. The reason for increased hot corrosion resistance derived from chromium under basic (low P_{SO_3}) Na_2SO_4 conditions has been discussed previously.

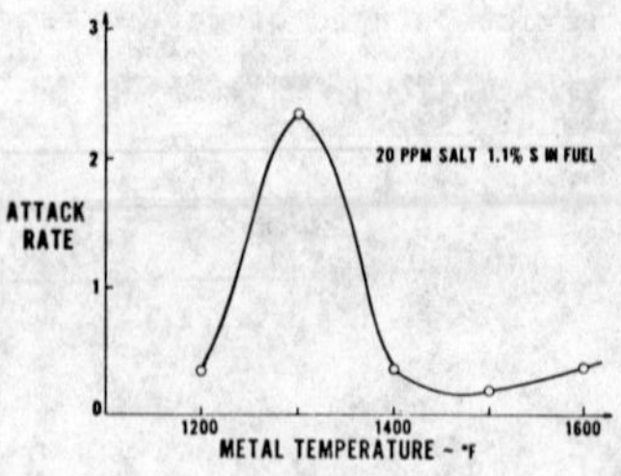

Figure 26. Low temperature hot corrosion behavior of CoCrAlY as measured in a burner rig.

It is well known that SiO_2 has minimal solubility in highly acidic melts; this suggests that coatings (alloys) containing silicon should develop significant resistance to hot corrosion caused by acidic (high P_{SO_3}) Na_2SO_4 melts. It is reasonable also to anticipate that similar behavior might accrue in melts rendered acidic by the presence of V_2O_5.

With regard to thermal fatigue, first-order "rules of thumb" tend to indicate that coatings with moderate (1 to 2%) ductility at temperatures of occurrence of maximum strain should have enhanced thermal fatigue resistance compared to elastically brittle coatings. Strangman (38) has identified several other properties, including thermal expansion match with the substrate, and modulus, which also affect thermal fatigue behavior of coatings.

Ceramic thermal barrier coatings prolong gas turbine component life by insulation of the base metal under "back side" cooling conditions (29). Steady-state metal temperatures are lowered to varying degrees depending on the particular conditions involved, and transient thermal strains, and resultant thermal fatigue effects, are also reduced by the insulation effect. Current coatings applied by plasma spraying are universally based on zirconia, applied over a metallic "bond coat". The coatings are more or less transparent to oxygen. Coating and/or base alloy life extensions are therfore derived solely from insulation effects and not from inhibition of oxygen transport.

Failure of thermal barrier coatings occurs by spalling of ceramic layers caused by transient thermal stresses. In many, if not all cases of such spalling, the failure is within the ceramic, near to but not along the ceramic-metal interface. This behavior has so far limited use to less critical gas turbine components such as burner liners and transition ducts fabricated from sheet metal. Preliminary investigations have indicated a trend to aggravation of spalling by the presence of salts which normally cause hot corrosion of metallic coating-alloy systems (39).

RECENT ADVANCES IN COATING TECHNOLOGY

High-Temperature (2000-2100°F, 1093-1149°C) Oxidation Resistance

The most advanced of the MCrAlY series of coatings is typified by a particular nominal composition (in wt.%) of 18% chromium, 23% cobalt, 12% aluminum, 0.5% yttrium, balance nickel (40). This composition is referred to in the industry as NiCoCrAlY. As applied by electron beam vapor deposition (EB) on a production scale, it has proved to be useful for the protection of aircraft engine blade airfoils fabricated by directional solidification (DS) of a superalloy designated as MM-200 + Hf*. To improve the mechanical life of airfoils in advanced engines a new superalloy composition specifically designed for casting in single crystal (SC) form, has been developed (41). Referred to as SC Alloy 454**, the composition does not include the element hafnium, since this element, usually present to provide grain boundary strength, is not required in single crystal alloys. In assessing the comparative performance of EB NiCoCrAlY on DS MM-200 and SC Alloy 454 in 2100°F (1149°C) cyclic oxidation, it was observed that while interdiffusional stability of the coating-alloy system was highest for SC Alloy 454, oxide adherence to the NiCoCrAlY coating on this alloy was poorer than for the same coating on MM-200 + Hf (42). The apparent reason for this behavior is shown in Figure 27; the oxide scale (Al_2O_3) on the coating on the hafnium-containing alloy exhibits a higher density of subscale "pegs", identified as hafnium oxide-rich (43) than the coating on the hafnium-free alloy. Improved oxide adherence attributable to hafnium in MCrAl alloys has been previously described (44), but the appearance of the effect by diffusion through an overlay coating was unexpected.

* 9Cr, 12 W, 10Co, 5Al, 2Ti, 1Cb, 2Hf, Balance Ni
**10Cr, 4W, 12Ta, 5Co, 5Al, 1.5Ti, Balance Ni
(All compositions in this paper are presented in weight percent.)

To provide compensation for this lesser oxide adherence of NiCoCrAlY coatings on the new alloy, a series of more complex coating compositions, containing silicon, tantalum and hafnium, was formulated for experimental evaluation. Because plasma spraying provides for greater ease of fabrication of complex coatings compared to the electron beam physical vapor deposition process, plasma spraying was chosen to deposit a number of these compositions on test specimens of the alloy (42). The oxidation resistance, determined by 2050°F (1121°C) cyclic burner rig tests, and ductility of several of these coatings is summarized in Table II where the ranking has been performed on the basis of 1 to 7 with 1 corresponding to the best performance.

TABLE II

Critical Property Ranking of Development Coatings

Composition (NiCoCrAlY Base)	Oxidation Resistance	Ductility
+1.6Si + 8.5 Ta	1	-
+1.6 + 2.6Si	2	4
+1.6Si + 4Ta	3	-
+8Ta	4	5
+0.8Hf	5	2
Base Plasma	6	1
Base EB	7	3

Of the compositions tested, NiCoCrAlY + 1.6 to 2.6% Si proved to have the best compromise of oxidation resistance and ductility and was deemed appropriate for further development.

It is worth noting that most of the compositions selected for testing were based on prior empirical observations. There is as yet no complete explanation of the beneficial effects of silicon and tantalum on oxidation resistance of MCrAlY coating alloys. Such development, based on empirical selection of beneficial elements, continues at an accelerated pace facilitated by the relative ease of application of compositions of increasing complexity by the plasma spray process.

Low to Intermediate Temperature (1200-1600°F, 649-871°C) Hot Corrosion Resistance

As previously indicated, hot corrosion caused by a variety of sulfate salt compositions can generally be retarded in metallic coatings and alloys by the presence of chromium. Few studies have, however, been performed under rigorously controlled conditions to determine optimum chromium concentrations for any particular hot corrosion environment. Recognition of the relatively high rate of attack of CoCrAlY (nominal 18% Cr, 12% Al, 0.5% Y), by Na_2SO_4 rendered acidic by SO_3 (P_{SO_3} in the range of 1×10^{-4} atm) in the temperature range of 1100-1300°F (593-704°C) in marine and industrial (32) applications has prompted more detailed studies of coating alloy compositions in this environment with the objective of providing more durable protection systems. Compilation of date published several years ago (33) enabled tentative construction of the curve in Figure 28 which shows relative resistance to Na_2SO_4-SO_3 attack as a function of chromium content at 1300°F (704°C). Nickel-base materials are somewhat more corrosion-resistant than equivalent cobalt-base materials in this temperature range. Conversely, at higher temperatures in the range of 1650°F (899°C), this behavior is reversed. In addition, increasing concentrations of chromium appear to become detrimental at 1650°F (899°C) and it has been so far concluded that in this temperature range, CoCrAlY with about 18% Cr and 12% Al continues to be the most corrosion resistant of the simple MCrAlY alloys (32).

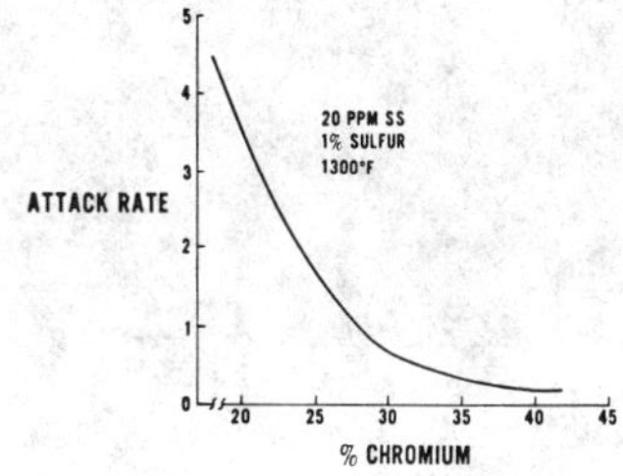

Figure 28. Dependence of relative rate of hot corrosion of MCrAlY coating alloys on chromium content; burner rig tests at 1300°F.

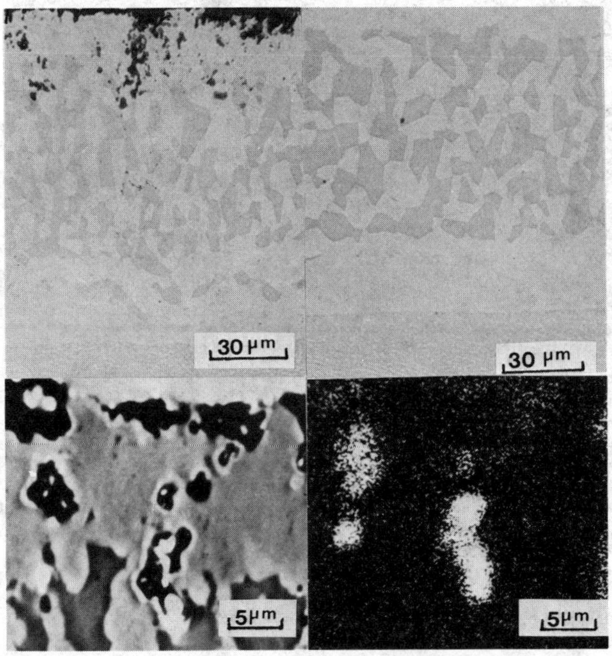

Figure 27. Differences in ingress of near-surface oxides on NiCoCrAlY coating on hafnium-containing (upper left) and hafnium-free (upper right) alloys oxidized at 2100°F. Electron beam microprobe electron backscatter image (lower left) and hafnium x-ray image (lower right) identify hafnium as a major constituent of near surface oxides in upper left photo.

During the decade of the seventies, workers in Europe concluded that simple aluminide coatings provided no useful protection in practical ground-based environments. Investigations of first chromium- and then silicon-rich coatings on nickel-base alloys have shown significant utility of these elements for protection from hot corrosion (27). While chromium diffusion coatings have found practical application, coatings formed by simple siliconizing are quite brittle and further research and development appears to be in progress to solve this potential problem. One approach has involved plasma spraying of relatively low silicon-containing Ni-Cr-Si alloys (e.g. 17-18% Cr, 4.5-5.5% Si) with more favorable ductility behavior. Initial field trials of these coatings appear to have yielded favorable results (27).

To build further upon these results, and to continue the previously beneficial approach of keeping coating composition as independent of substrate composition as possible, it was decided to investigate the apparent beneficial effects of silicon by making additions of this element to MCr(Al,Y) coating alloys. For exploratory purposes, it was expedient to siliconize a range of MCr(Al,Y) coating alloys. Electron beam evaporation of silicon and condensation on the surface of the MCr(Al,Y) coatings proved to be a useful experimental method and several such systems with silicon-enriched outer layers, typified by that shown in Figure 29 have now been tested for hot corrosion resistance. Resistance to severe acidic fluxing conditions (Na_2SO_4 + SO_3 at 1×10^{-4} atm., 1300°F, 704°C) has been particularly promising. For example, initial hot corrosion rates of CoCrAlY (18Cr, 12Al, 0.5Y) have been reduced by factors of ten to twenty by the enrichment of surface layers with silicon as indicated (45). Thus, confirming the earlier European results, silicon appears to be an even more potent addition than chromium under acidic sulfate corrosion conditions in the range of 1300°F (704°C). Initial results of evaluation of corrosion resistance at higher temperatures, in the range of

1600-1650°F (871-899°C) tend to indicate lesser beneficial effects under these conditions. While these results appear very promising, much more work remains to be done to characterize fully the practical utility of these silicon-rich coatings. Field service tests under a variety of conditions are of vital importance. Of particular concern is the possibility that basic fluxing conditions will be damaging to silicon-rich scales; environmental conditions in practical gas turbines are not yet sufficiently defined to preclude the possibility of such conditions. These conditions could be caused by decreasing partial pressures of SO_3 in the gas stream of the turbine caused by either or both increasing temperature and decreasing sulfur content of the fuel used.

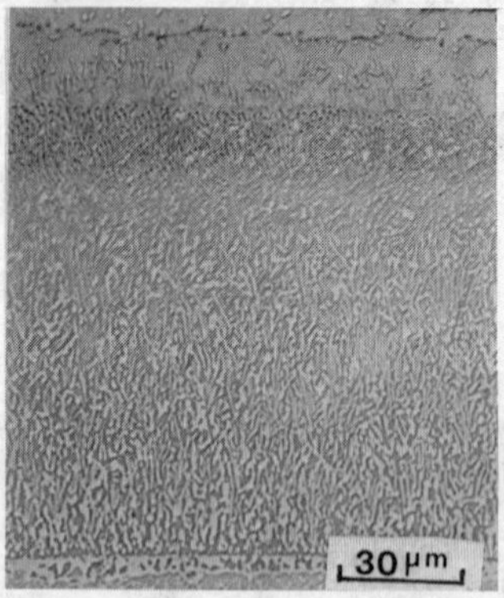

Figure 29. Layered structure formed by siliconizing a CoCrAlY coating (500X).

Thermal Barrier Coatings

Since thermal stress-induced spalling has been the most obvious failure mode limiting use of this class of coatings on gas turbine airfoils, all major development programs have focused on this problem. Lesser, but significant attention has been given to corrosion and physical effects caused by the presence of molten sulfate salts.

The basic cause of ceramic coatings spalling is the large thermal expansion mismatch between the coating and substrate. The most promising approach to improving spalling resistance has been based on the recognition that failure is caused by flaws which induce crack propagation in the plane of the coating. The stresses causing such propagation might be reduced by controlling coating structure so that appropriate discontinuities cause a reduction in elastic modulus of the coating, thereby reducing local stresses. Such discontinuities include increased porosity, random microcracking, and columnar structures with the columns normal to the substrate surface (now termed "coating segmentation"). In addition, appropriate control of substrate temperature during coating fabrication was conceived as a means of providing a better balance between compressive and tensile strains imposed on ceramic coatings under cyclic turbine operation modes.

Results of several such approaches on zirconia thermal barrier coatings are summarized in Table III (46,49). Substantially increased resistance to thermal stress-induced spalling, as measured by cyclic burner rig testing at 1850°F (1010°C) metal temperature, has been achieved by fabrication of zirconia-based coatings with microcracked structures and with segmented or columnar structures. The former method has been most easily achieved with partially-stabilized zirconia containing excess magnesia, and the latter with special control of plasma spray processing or, more effectively, with electron beam evaporation-condensation of yttria-stabilized zirconia which results in the smallest scale segmented structures. Examples of microcracked and segmented structures are provided in Figs. 30, 31 and 32.

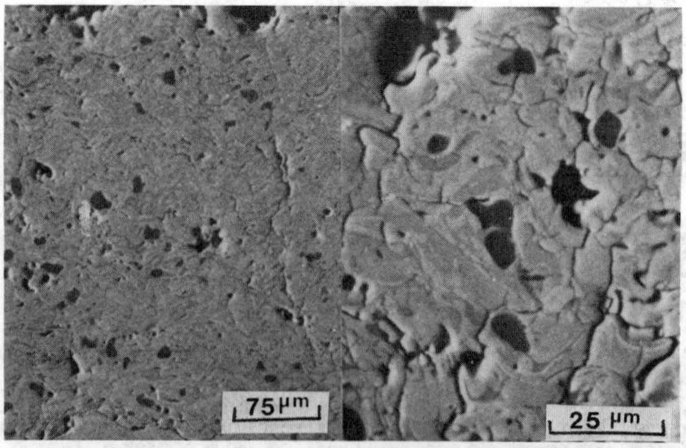

Figure 30. Microcracked plasma sprayed ZrO_2-MgO coating.

Figure 31. Segmented plasma sprayed ZrO_2-Y_2O_3 coatings; left-planar view, right-cross section.

Figure 32. Segmented electron beam vapor deposited ZrO_2-Y_2O_3 coating; left-planar view, right cross section.

As indicated in Table III, control of substrate temperature during application of plasma-sprayed ceramic coatings also shows considerable promise for increasing resistance to coating spalling. Control of temperature during application in the range of 70 to 600°F (21 to 316°C) reduced compressive stresses caused by rapid cooling to an extent sufficient to cause the durability increases indicated. Initial experimental engine testing in aircraft engines of plasma-sprayed zirconia coatings utilizing the microcracking and segmentation concepts on stationary vane platforms has shown some promising results (46,47). Continuing programs will evaluate the practical effectiveness of these and the more advanced and difficult structural concepts for the ultimate application - protection of turbine airfoils.

TABLE III

Thermal Stress Resistance* of Various Modifications of Ceramic Thermal Barrier Coatings

Coating Type	Cycles to Failure in Thermal Stress Test*
Unmodified PS** XrO_2-20% Y_2O_3	200
Microcracked PS ZrO_2-21% MgO	5200
Segmented PS ZrO_2-20% Y_2O_3	5500
Segmented EB*** ZrO_2-20% Y_2O_3	>10000
Substrate Temperature Controlled As Follows During Plasma Spraying (ZrO_2-20% Y_2O_3)	
-31°F (-34°C)	3500
68°F (20°C)	9500
600°F (316°C)	9000
1200°F (649°C)	200

* Thermal cycled in a petroleum fuel fired burner with Mach 0.3 gas velocity. Thermal cycle consisted of heating to 1850°F (1010°C) for 4 minutes followed by two minutes of cooling to below 600°F (316°C)

** Plasma Sprayed

*** Electron Beam Vapor Deposited

There has been considerable concern that the usefulness of ceramic thermal barrier coatings in utility engines could be compromised by the presence of corrosive molten salts frequently encountered in such applications (39,48,49). Two potential failure modes have been anticipated - saturation of coating porosity with molten salts, resulting in failure by mechanical effects of solidification and melting, and hot corrosion, more correctly termed dissolution, by molten salt deposits. Simple, but definitive, studies on the "corrosion" of plasma sprayed stabilized zirconias (50) have indicated that some sort of degradation involving selective removal of stabilizers, e.g., Y_2O_3, as shown in Table IV, by acidic Na_2SO_4-SO_3 melts is possible.

TABLE IV

Hot Corrosion of Plasma Sprayed Y_2O_3 Stabilized ZrO_2

	Surface Concentration of Y_2O_3(%)		
As Sprayed	12	20	35
Corroded at $O_2 + P_{SO_3} = 7 \times 10^{-4}$			
1300°F (704°C)	9	15	19
1800°F (982°C)	11	19	33
$O_2 - P_{SO_3} = 7 \times 10^{-3}$			
1300°F (704°C)	2	9	16
$O_2 + P_{SO_3} = 1 \times 10^{-3}$			
1800°F (982°C)	11	18	31

The exact mechanisms of such phenomena and their practical relevance have not been established. It does appear that much more detailed studies of effects of corrosive salts on the durability of thermal barrier systems are necessary to assure effective utilization of this promising class of coatings.

SUMMARY

The most effective means to control high temperature corrosion involves the formation of selected reaction product barriers on the surfaces of alloys. One of the most effective barriers in providing resistance against corrosion is α-Al_2O_3. Alumina barriers are extremely effective in developing resistance to oxidation degradation and mixed gas attack. This type of barrier can also be used to develop resistance to some forms of hot corrosion attack but SiO_2 may be more resistant to highly acidic deposits.

The approach to be followed in attempting to develop resistance to high temperature corrosion is essentially the same for both structural and coatings alloys, however the latter alloys may be modified over large composition ranges since their primary function is to provide corrosion resistance. Minor additions of silicon have been found to increase significantly the oxidation resistance of alumina scales formed on NiCoCrAlY coatings. Increasing the chromium content of MCrAlY coatings provides substantial increases in resistance to acidic hot corrosion at temperatures of about 1300°F(704°C). Surface enrichment of MCrAlY coatings with silicon also provides large increases in resistance to acidic hot corrosion in the 1300°F(704°C) region.

The resistance to thermal - stress - induced spalling of zirconia-based thermal barrier coatings has been improved by reducing coating stresses with segmented structures and by controlling substrate temperature during coating fabrication. Further advances in the use of ceramic coatings seems probable.

ACKNOWLEDGEMENTS

Much of the data that are presented in this paper were generated and interpreted with the help of D. K. Gupta, J. A. Goebel, C. S. Giggins, D. L. Ruckle and R. H. Barkalow while the authors were employed at Pratt and Whitney Aircraft Group, East Hartford, CT.

REFERENCES

1. Karl Hauffe, Oxidation of Metals, pp. 1-7, Planum Press, New York, 1965.

2. G. C. Wood and B. Chattopadhyay, "Transient Oxidation of Ni-Base Alloys", Corr. Sci., 10 (1970), p. 471-480.

3. C. S. Giggins and F. S. Pettit, "Oxidation of Ni-Cr-Al Alloys Between 1000° and 1200°C", J. Electrochem. Soc., 118 (1971), p. 1782-1790.

4. C. Wagner, "Passivity and Inhibition During the Oxidation of Metals at Elevated Temperatures", Corros. Sci., 5 (1965), p. 751-64.

5. C. S. Giggins and F. S. Pettit, "The Effect of Alloy Grain Size and Surface Deformation on the Selective Oxidation of Chromium in Ni-Cr Alloys at Temperatures of 900° and 1100°C", Trans. Met. Soc. A.I.M.E., 245 (1969), p. 2509-2514.

6. C. S. Giggins and F. S. Pettit, "The Oxidation of TD NiC (Ni-20Cr-2 Vol pct ThO_2) Between 900° and 1200°C", Met. Trans., 2 (1971) p. 1071-1078.

7. J. Stringer, B. A. Wilcox, and R. I. Jaffee, "The High-Temperature Oxidation of Nickel - 20 wt.% Chromium Alloys Containing Dispersed Oxide Phases", Oxid. Metals, 5 (1972), p. 11-36.

8. E. J. Felten and F. S. Pettit, "Development, Growth, and Adhesion of Al_2O_3 on Platinum-Aluminum Alloys", Oxid. of Metals, 10 (1976), p. 189-223.

9. J. L. Smialek, "Microstructure of Al_2O_3 Scales Formed on NiCrAl Alloys", NASA Technical Memorandum 81676, Lewis Research Center, Cleveland, Ohio, January, 1981.

10. C. S. Wukusik and J. F. Collins, "An Iron-Chromium-Aluminum Alloy Containing Yttrium, "Materials Research and Standards, 4 (1964), p. 637-646.

11. J. K. Tien and F. S. Pettit, "Mechanism of Oxide Adherence on Fe-25Cr-4Al (Y or Sc) Alloys", Met. Trans., 3 (1972), p. 1587-1599.

12. D. P. Whittle and J. Stringer, "Improvement in Properties: Additives in Oxidation Resistance, Improvements in High Temperature Oxidation Resistance by Additions of Reactive Elements or Oxide Dispersions", Phil. Trans. R. Soc. Lond., 295 (1980), p. 309-329.

13. F. H. Stott, G. C. Wood and J. G. Fountain, "The Influence of Yttrium Additions on the Oxidation Resistance of a Directionally Solidified Ni-Al-Cr_2C_2 Eutectic Alloy", Oxid. Metals, 14, (1980), p. 135-146.

14. C. S. Giggins and F. S. Pettit, "Corrosion of Metals and Alloys in Mixed Gas Environments at Elevated Temperatures", Oxid. of Metals, 14 (1980), p. 363-413.

15. N. S. Bornestein and M. A. DeCrescente, "The Role of Sodium in the Accelerated Oxidation Phenomenon Termed Sulfidation", Met. Trans., 2, (1971), p. 2875-2883.

16. J. A. Goebel, F. S. Pettit and G. W. Goward, "Mechanisms for the Hot Corrosion of Nickel-Base Alloys", Met. Trans. 4 (1973), p. 261-278.

17. J. Stringer, V. Nagarajass and D. P. Whittle, "The Role of Chloride in the Hot Corrosion of Cobalt Base Alloys", Proc, of the Symposium on High Temperature Metal Halide Chemistry, D. L. Hildebrand and D. Cubicciotti, Eds. The Electrochem, Soc., Princeton, NJ, 1978.

18. Luthra, K. L. and Shores, D. A., "Mechanism of Na_2SO_4 Induced Corrosion at 600-900°C", J. Electro-Chem. Soc., 127 (1980), p. 2202-2210.

19. R. H. Barkalow, J. A. Goebel and F. S. Pettit, Problems in Fluidized Bed Combustion Systems, High Temperature Erosion-Corrosion by High Velocity (200 m/s) Particles), Prepared by Pratt & Whitney Aircraft Group for Electric Power Research Institute, Palo Alto, CA., 94304; EPRI CS-1448, Project 979-4, May (1980).

20. G. W. Goward, D. H. Boone, and C. S. Giggins, "Formation and Degradation Mechanisms of Aluminide Coatings on Nickel-Base Superalloys", Transactions ASM, 60 (1967) p. 228-241.

21. G. W. Goward, and D. H. Boone, "Mechanisms of Formation of Diffusion Aluminide Coatings on Nickel-Base Superalloys", Oxidation of Metals, 3 (1971) p. 475-495.

22. P. N. Walsh, "Chemical Aspects of Pack Cementation", Proceedings of Fourth International Conference on Vapor Deposition, The Electrochemical Society, (1973) p. 147-168.

23. S. R. Levine and R. M. Caves, "Thermodynamics and Kinetics of Pack Aluminide Coating Formation on IN-100", J. Electrochem. Soc. 121 (1974) p. 1051-1064.

24. E. Fitzer, and H. J. Maurer, "Diffusion and Precipitation Phenomena in Aluminized and Chromium - Aluminized Iron- and Nickel-Base Alloys", Proceedings of Conference on Materials and Coatings To Resist High Temperature Corrosion, Applied Science Publishers, London, p. 253 (1977).

25. R. Pichoir, "Influence of the Mode of Formation in the Oxidation and Corrosion Behavior of NiAl-Type Protective Coatings", Ibid, p. 253.

26. R. Bauer, H. W. Grunling and K. Schneider, "Hot Corrosion Behavior of Chromium Diffusion Coatings", Ibid, p. 369.

27. R. Bauer, H. W. Grunling, and K. Schneider, "Silicon and Chromium Base Coatings for Stationary Gas Turbines", Proceedings of the First Conference on Advanced Materials for Alternative Fuel Capable Directly Fired Heat Engines, U. S. Department of Energy and Electric Power Research Institute, p. 505 (1979).

28. G. W. Goward, "Hot Corrosion of Gas Turbine Airfoils: A Review of Selected Topics, Tri-Service Conference on Corrosion", MCIC-77-3, (1976) p. 239-255.

29. S. Grisaffe and S. Levine, "Review of NASA Thermal Barrier Coating Programs for Aircraft Engines", Ibid p. 580.

30. P. Aldred, "Rene' 125, Development and Application", S.A.E. Paper 751049, National Aerospace Engineering and Manufacturing Meeting, Los Angeles, 1975.

31. G. W. Goward, "Materials and Coatings for Gas Turbine Hot Section Components", Proceedings of Gas Turbine Materials Conference, Naval Ship Engineering Center, Washington, D.C., 1972.

32. J. A. Goebel, "Advanced Coating Development for Industrial-Utility Gas Turbine Engines", Proceedings of the First Conference on Advanced Materials for Alternative Fuel Capable Directly Fired Heat Engines, U. S. Department of Energy and Electric Power Research Institute, p. 473 (1979).

33. A. F. Taylor, B. A. Wareham, G. C. Booth and J. F. G. Conde', "Low and High Pressure Rig Evaluation of Materials and Coatings, Third Conference on Gas Turbine Materials in a Marine Environment", University of Bath, England, Paper III-4 (1976).

34. D. J. Wortman, R. E. Fryxell and I. I. Bessen, Ibid, Paper V-II.

35. W. Schilling, H. Fox and A. Beltran, "Field Testing and Evaluation of Next Generation Industrial Gas Turbine Coatings", Ibid, p. 163.

36. W. P. Stroud and R. A. Rapp, "The Solubilities of Cr_2O_3 and Al_2O_3 in Fused Na_2SO_4 at 1200°K", High Temperature Metal Halide Chemistry, Electrochemical Society, Princeton, NJ., 1978.

37. W. W. Liang and J. F. Elliott, "Reaction Between Cr_2O_3 and Liquid Sodium Sulfate at 1200°K, in Properties of High Temperature Alloys", The Electrochemical Society, (1096) p. 557.

38. T. Strangman, "Thermal Fatigue of Oxidation Resistant Overlay Coatings for Superalloys", Doctoral Dissertation, The University of Connecticut, 1978.

39. S. C. Singhal and R. J. Bratton, "Stability of a ZrO_2 (Y_2O_3) Thermal Barrier Coating in Turbine Fuel With Contaminants", Transactions of the ASME, 102 (1980) 770.

40. U. S. Patent 3,754,903 High Temperature Oxidation Resistant Coating Alloy.

41. M. Gell, D. N. Duhl, and A. F. Giamei, "The Development of Single Crystal Alloy Turbine Blades, Superalloys 1980", American Society for Metals, Metals Park, Ohio (1980), p. 205.

42. F. J. Pennisi and D. K. Gupta, "Tailored Plasma Sprayed MCrAlY Coatings for Aircraft Gas Turbine Applications", NASA CR-165234, (1981).

43. D. K. Gupta, "Effects of Coating-Substrate Interdiffusion on the Performance of Plasma Sprayed MCrAlY Coatings", Proceedings of 1908 International Conference on Metallurgical Coatings, Elsevier Sequoia S. A., Lausanne and New York, (1980) p. 477.

44. I. M. Allam, D. P. Whittle and J. Stringer "Oxidation Behavior of CoCrAlY Systems Containing Active Element Additions, Oxid, Met. 12 (1978, p. 35-66.

45. J. R. Vargas, N. E. Ulion and J. A. Goebel, "Advanced Coating Development for Industrial-Utility Gas Turbine Engines", Proceedings of the International Conference on Metallurgical Coatings, Elsevier Sequoia S. A., Lausanne and New York, (1980) p. 407.

46. D. L. Ruckle, "Plasma Sprayed Ceramic Thermal Barrier Coatings for Vane Platforms", Ibid, p. 455.

47. D. L. Ruckle and D. E. Sumner, "Development of Improved Durability Plasma Sprayed Ceramic Coatings for Gas Turbine Engines", Paper 80-1193, AIAA-SAE-ASME Joint Propulsion Converence, Hartford, CT (1980).

48. D. W. McKee and P. A. Siemers, "Resistance of Thermal Barrier Ceramic Coatings to Hot Salt Corrosion", Proceedings of the International Conference on Metallurgical Coatings, Elsevier Sequoia S. A., Lausanne and New York, (1980) p. 439.

49. P. E. Hodge, R. A. Miller and M. A. Gedwill, "Evaluation of the Hot Corrosion Behavior of Thermal Coatings", Ibid, p. 447.

50. R. Barkalow and F. S. Pettit, "Mechanisms of Hot Corrosion Attack of Ceramic Coating Materials", Proceedings of the First Conference of Advanced Materials for Alternate Fuel Capable Directly Fired Heat Engines, U. S. Department of Energy and Electric Power Research Institute (1979), p. 704.

WEAR AND EROSION OF METALS

David A. Rigney and Paul G. Shewmon

Metallurgical Engineering
The Ohio State University
Columbus, Ohio 43210

Older explanations of friction and wear behavior have emphasized adhesive interactions of surface asperities. This approach has led to an emphasis on surface chemistry and on continuum mechanics, with much less attention given to the structures of the interacting materials themselves. More recent work has focused on the extensive plastic work in the near-surface region, described in terms of work hardening, recovery, and the microstructures existing during different stages of wear. It is then possible to explain the size of friction coefficients, energy dissipated, flake-like wear debris, break-in and steady-state regimes, and microstructure effects. Effects of the chemical environment, coatings, and transfer can also be included. Many different mechanisms of wear can exist; therefore simple wear laws are of limited use. Friction and wear, though closely related, do not always correlate well, and the new approach accounts for this in a natural way.

The erosive removal of metal by high speed (50-300 m/s) particles is surveyed with emphasis on the mechanisms of deformation and metal removal. Under these conditions each impact leads to plastic deformation of the surface. At the lower velocities many impacts are required to remove metal and the fracture is fatigue related. With higher velocities, and harder materials, local heating plays an important role in flow and fracture.

WEAR

Introduction

Friction, wear, erosion, adhesion, and lubrication are closely related subjects which have concerned man for many centuries. In recent years, the term "tribology" has been used, particularly in Europe, to encompass these subjects and others associated with the science and technology of interacting surfaces in relative motion (1). Tribology is a giant field and many tribological phenomena are poorly understood. Tribology is at least as big a field as corrosion, and the economic impact of tribological problems in an industrial nation is at least comparable. Some estimates indicate that the costs of wear are five times larger than for corrosion. Yet the scientific base for wear is much less than presently exists for corrosion. In this paper, the authors focus attention on our present understanding of the wear and erosion of metals. Some of the ideas described here apply also to other classes of materials. A section on friction is included because friction and wear, though not correlated in any simple way, are commonly associated, and improved understanding of frictional processes should aid our understanding of wear.

Historical Perspective. Bowden and Tabor have included some material on the history of tribology in their well-known books (2,3). More recently, Dowson (4) has written a beautiful book on the history of tribology, from stone age times to the modern period. It is fascinating to read of the ingenuity of early peoples in designing viable tools and vehicles which incorporated basic principles of tribology to control friction and wear. Examples include the building of "henge" monuments in Britain, transportation of colossus statues in Egypt and Assyria, the use of wear - resistant plowshares in northern Europe and sophisticated wheels in Rome and in China.

Readers of Dowson's book will probably notice that some very well-known names are cited, since, in addition to their other work, those persons devoted serious study to aspects of tribology. Early examples include Aristotle, Archimedes, Herodotus, Lucretius, and Pliny. Later, Leonardo da Vinci (of course!) Benvenuto Cellini, Hooke, Newton, Euler, Coulomb, Reynolds, Hertz, Rayleigh, and Langmuir contributed to this field. The technology of tribology progressed gradually, but despite the impressive array of talent just listed, and despite the efforts of many others, progress in understanding friction and wear processes has been slow.

Perhaps Lucretius (5) was the first to understand the reasons for the difficulties encountered in studying wear. More than 2000 years ago, he recognized wear as a phenomenon which occurred on such a very fine scale that one could not resolve--with instruments available then--the details of the processes involved. Today, however, with a variety of recently developed electron probes and other sophisticated instruments, data for understanding friction and wear processes are becoming available. At the same time, there seems to be renewed interest in these subjects, particularly in Europe, North America, and Japan.

Recent world-wide concern about materials and energy resources provides added reasons to improve our understanding of friction and wear (6-8). A substantial fraction of our energy is dissipated through friction, and premature obsolescence of many products through wear wastes materials resources, some of which are in critical supply. In both cases, if we are to reduce this waste in mature technologies, such as automobile design and manufacture, we will need to know more about the basic mechanisms involved.

Multidisciplinary Nature of Tribology. It is widely acknowledged that tribology is a multidisciplinary subject which needs improved cooperation among specialists in many different disciplines. The relevant literature is widely dispersed. It can be found scattered in books and journals on design, materials, chemistry, solid and fluid mechanics, lubrication, surface characteristics and coatings. Those working in some aspect of tribology have been trained in chemistry, physics, materials science, or one of several engineering fields, particularly mechanical or chemical engineering. Until recently, they have tended to use separate terminology, publish in separate journals, and attend separate technical conferences. This situation is improving, but the majority of investigators still seem reluctant to explore aspects of their subject outside the boundaries of the traditional disciplines. Enlightened administrators will need to encourage interdisciplinary interaction by forming balanced groups composed of investigators having diverse knowledge and skills. Conference organizers can also help by scheduling tutorial sessions. And sponsoring organizations can help by improved cooperation in publicizing each others related efforts and by increased frequency of jointly sponsored events. Tribology is too complicated and too important to leave in the hands of isolated groups of traditional specialists.

Reference Sources. Articles on wear and erosion can be found in many technical journals, but the principal journals for Europe and North America are Wear, Tribology International, Transactions of the American Society of Lubrication Engineers (A.S.L.E.), Tribologie, and Friction and Wear of Machinery (translations of Soviet papers). One of the best handbooks, containing much useful design data, is the Handbook of Tribology (9), published in Great Britain. In the U.S., a Wear Control Handbook has just been published (10).

The number of books on friction, wear, and erosion is surprisingly small. Among the more important ones are those by Bowden and Tabor (2,3,11,12), Kragelskii (13), and Rabinowicz (14). More recent are books by Czichos (15) and Sarkar (16). A book edited by Halling (17) contains some particularly good review articles. Another recent book, edited by Scott (18), has a good collection of articles containing much experimental data. Specifically for erosion, one book is edited by Preece (19); a recent ASTM report is edited by Adler (20).

Publications based on conference proceedings are also good sources of articles despite the uneven quality of the contents. Among the more useful are the proceedings of the Leeds/Lyon Conferences (21) and the new series of international conferences called Wear of Materials (22). Other recent conference volumes include the one based on a 1978 conference at MIT (23) and the one based on the 1980 ASM Materials Science Seminar, Fundamentals of Friction and Wear of Materials (24). Recent conference proceedings on erosion include those cited in references (25,26). In the U.S., the Office of Naval Research (27) has sponsored a series of workshops on tribological topics, some of which yielded published proceedings. Also in the U.S., the Gordon Research Conferences (28) on friction and wear are worth attending, but traditionally, no proceedings are available.

Terminology. The terminology used in tribology is very rich, but it includes many terms which are used rather loosely, and some which are in fact misleading. For example, "Coulomb friction" is often used to refer to dry or unlubricated sliding friction; yet Coulomb's experiments, on which he based his "laws", were done with lubricant present, because this gave more reproducible results.

"Wear" itself is not always clearly defined. Most people think of wear as a gradual change in mass or volume of a sample, accompanied by the generation of loose wear debris. But at times one refers to wear even if no net change of weight has occurred. Such a situation exists when ductile material is displaced so that a wear scar appears and the sample is changed in shape, but wear debris need not be generated. In this case, wear can be described in terms of plastic deformation alone. For the first case, with debris generated, both deformation and fracture processes are involved. A general definition which would include both categories of wear would refer to the usually gradual damage which results when two materials move relative to each other while in contact.

Wear rate is another term subject to different definitions and interpretations. The simplest way to express it would be as the volume or mass lost (or displaced) per unit of time. For constant sliding speeds, it is often convenient to express wear rate as the volume or mass change per unit sliding distance. Other forms are also used, some of which are dimensionless. These differences should cause no difficulties if the units are clearly defined. What may cause more difficulty is the appropriateness of the time (or distance) interval chosen to express the wear rate. Wear rate can be a complex function of time, sometimes rising or falling suddenly. Thus, a single value for wear rate is merely an average for the time interval selected, and it may be of little use for extrapolation, prediction, design, or comparison. Thus, care is needed in applying or comparing wear rate data.

Generally, the friction and wear behavior during an initial *break-in* or *running-in* period is different from that observed later, when both the friction coefficient and the wear rate may tend to be relatively constant. This later stage is generally called *steady state* sliding. It is typically the condition which exists for the longest part of the useful lifetime of a sliding sample, and it is in principle the simplest stage. During steady state sliding, the following will be relatively constant with time: average friction coefficient, wear rate, type of debris, surface topography, depth of deformation, coating or film characteristics, and near-surface temperature, microstructure, hardness, and crystallographic texture (and their gradients). Despite local changes which cause statistical fluctuations, this stage is simpler to analyze than the earlier break-in period. Therefore, it has received considerable attention, and simple models of friction and wear typically are based on steady state conditions. Also, wear rates are generally assumed to be steady state values.

The break-in period is much more complicated. Behavior then will be sensitive to initial material and surface conditions, much more so than during steady state. This is a transition period, during which the material may work harden, phase transformations may occur, films may be created, destroyed or modified, temperature will rise due to frictional heating, energy will be stored, and the microstructure, texture, and surface topography will approach those found during steady state. Wear rate during break-in can be higher or lower than during steady state, and wear debris particles can be present or absent. Obviously, the break-in stage is complex. In recognition of the importance of this stage, the 1981 Leeds/Lyon Symposium on Tribology has been devoted to this topic (29).

To investigate sub-surface changes occurring during wear, or to obtain information on mechanisms involved, it is often helpful to examine sections of sample material before and after wear occurs. The most commonly used sections are those parallel to the wear surface and perpendicular to it (cross-sections). The cross-section parallel to the sliding direction is called *longitudinal*, and the one perpendicular to the sliding direction is called *transverse*. All three sections may be useful for full characterization of near-surface microstructures, which are usually strongly anisotropic. When the load is small or when the system is well-lubricated, the modified structure near the wear surface may be very thin. In such cases, taper sections may be useful (30). They effectively magnify structural features in a selected direction.

Preparation of sections of wear specimens requires careful attention to metallographic principles (30), with modifications designed to preserve easily damaged near-surface materials (31). Otherwise, observed features may not be typical of those which existed during wear, and incorrect conclusions may be drawn.

Sliding direction is not always clear from published descriptions of wear tests. The convention used here is that sliding direction is the direction of motion of the opposing surface relative to the surface of the sample. In a longitudinal section, the sliding direction is generally evident as the direction toward which various features such as grain boundaries, twins, or scratches tend to bend.

Other relevant terms will be defined as needed in later sections. A more complete set of definitions is available from the American Society for Testing and

Materials (ASTM). This organization offers a report entitled Standard Terminology Relating to Erosion and Wear (32). The report is updated regularly as terms become modified through improved understanding. Another set of definition has been prepared by the OECD (33).

Experimental Factors

Wear Test Equipment. The best wear test is one that closely approximates actual conditions encountered in service. For example, for an automobile engine, an actual engine, programmed to operate as in typical applications, would be ideal test equipment. The data obtained are then useful for that particular situation, but not necessarily for other purposes.

At the other extreme, one can easily construct a wear test device based on one of several simple geometries, such as pin-on-disk, block-on-ring, or cylinder-on-cylinder. Commercial wear test devices based on such geometries are also available. However, there is no standard test for wear, and each investigator has a wide choice among test designs and experimental variables. Thus it is seldom possible to compare results quantitatively from two different laboratories. However, qualitative trends can be discerned with the aid of simple wear test devices, so their popularity is probably justified.

The American Society of Lubrication Engineers (A.S.L.E.) has published a convenient collection of wear test equipment designs (34). This is a useful book to help guide the selection of wear test designs. Other good references are articles by Peterson (35) and Barwell (36) and a book edited by R. G. Bayer (37). A review of abrasion tests is given in reference (38); tests for erosion are described in references (19,39).

No single test design can be recommended for all purposes. It may be advisable to select at least two different designs to allow application flexibility and to compare results.

Other Variables. Many simple wear test machines are asymmetric so that if the sliding materials are interchanged in position, the results may be quite different. Only for symmetric systems such as face-to-face disks (same dimensions) or crossed-cylinders (same dimensions) rotating at the same speeds are such differences absent.

Even the orientation of a test machine can influence results. For example, with a pin on disk machine, the ease of escape of debris particles depends on the equipment orientation.

Sliding can be unidirectional as in simple rotating systems or it can be oscillating. In an oscillating mode, one can select the amplitude and the frequency. For small amplitudes one may have a form of wear known as fretting (40).

The interacting materials in a wear test can be the same, as in self-mated tests, similar, or different. Debris particles can be removed as they are formed (e.g., by brushes or a stream of gas or liquid) or retained on the surfaces, where they can be further modified and where they can modify both friction and wear.

The number of passes on a single track and the surface topography can have large effects on test results, particularly during the break-in period. The applied load, sliding speed, and ambient temperature are also important variables. The nature of the chemical environment (gas, liquid, solid) is particularly important. For erosion one also has the angle of incidence, particle size and shape, and velocity of approach to consider. Finally, one has the materials variables to consider, such as composition, grain size, number of phases, crystallographic texture, magnetic properties, and microstructure details.

These variables have been listed here to emphasize their importance both in planning a test and in evaluating the relevance and reliability of published test results. Very few authors state clearly all the important details of their tests. Many papers on wear are particularly lacking in information about the materials parameters involved.

Because there are so many variables which affect the results of friction and wear tests, systematic procedures for testing and analysis must be adopted if the results are to be useful. Several authors have recommended a formal systems approach to wear testing and analysis. An example of one such approach is described in references (15,41). Control of variables and monitoring of friction, temperature and wear during testing can be accomplished with the aid of small laboratory computers. An example of such computer use has been described by Ruff, et al. (42).

Characterization of Wear Samples and Debris. A wide range of instruments and techniques is available for characterization of wear samples and debris. A recent review describes some of these in detail (42). The various techniques can provide information on topography, microstructure, hardness, phases, texture, fracture, and local chemical composition (which may change due to segregation or chemical reactivity).

When a wear test is completed, perhaps the first step should be to examine the specimen and the debris with a low power optical microscope, preferably with binocular viewing. This will reveal major features such as size and shape of wear scar, gross surface damage, grooving, oxidation, color changes, transferred material, ductile smearing, major fractures, and sizes and shapes of debris particles. A permanent magnet will reveal any pieces of ferromagnetic debris. Next, the intact specimen and the debris can be examined in a scanning electron microscope (SEM). If the SEM is equipped for chemical analysis (crystal spectrometer or energy - dispersive X-ray analysis), then the variations in local composition can provide information on segregation and/or material transfer. Next, the specimen can be sectioned, and the sections can be examined both optically and in the SEM, first as mechanically polished, and then as lightly etched. Microhardness profiles can also be measured on these sections. Back-reflection X-ray diffraction of the wear surface and Debye-Scherrer patterns of the debris provide information on the phases present (43). Transmission electron microscopy (TEM) of thin foils and of replicas (42,44) can provide high-resolution surface information and microstructure, phase, and texture profiles. Auger analysis is well-suited for obtaining surface composition information from both the specimen and selected debris particles. Electron channeling patterns can be used to measure strain profiles near the surface (45,46). A device called a Ferrograph has become popular in recent years (particularly for tests involving ferrous materials) for the separation of wear debris particles by size, shape, and magnetic properties (47).

Many other sophisticated modern techniques have also been applied in recent years to investigate friction and wear problems and mechanisms (42).

Of course, one does not need to use all these techniques to obtain most of the information needed. If equipment is carefully chosen to provide complementary information, then most of the desired data can be obtained with (1) a good optical microscope with binocular viewing, (2) a good metallograph, (3) metallographic facilities, (4) a microhardness tester, (5) basic X-ray diffraction equipment, and (6) an analytical SEM with provision for local chemical analysis.

Wear and Friction

Types of Wear. It is common practice to consider wear in terms of five main categories: adhesive wear, abrasive wear, corrosive wear, surface fatigue, and impact phenomena (including solid and liquid erosion). *Adhesive wear* is widely used to describe wear involving sliding of one piece of metal over another, whether with or without lubrication. *Abrasive wear* is the type of wear which occurs when hard asperities form grooves and ridges in a softer material by a combination of microcutting and plowing. *Corrosive wear* involves chemical reaction with the environment, together with mechanical damage. *Surface fatigue* or rolling contact fatigue are terms applied to typical failures involving repeated application of high local stresses, as in many bearings. Impact problems cover a wide range, from large scale failure due to impact by massive bodies (not properly part of the subject of wear) to gradual changes caused by impact of fine solid particles and liquid droplets (erosive wear).

There are, however, some problems associated with the general use of these categories. First, actual wear situations may involve more than one of these types of wear (e.g., adhesive and abrasive, or adhesive/abrasive/corrosive). Second, use of a term like adhesive wear or fatigue wear implies that the dominant mechanism is known, when in fact our understanding of wear mechanisms is poor. Terms like these should be used with caution.

An alternative approach would be to use simple descriptive terms like sliding, rolling, abrasive, impact, wet/dry, and lubricated/unlubricated wear. Such a phenomenological scheme is based on simple observations about the wear system, and such terms are preferable whenever the wear mechanisms are not well characterized. In this paper the emphasis will be on sliding wear, abrasive wear, and erosion.

Other discussions of wear classification may be found in references (48-51).

Asperities, Adhesion, and Friction. This paper is mainly on wear and erosion, but it is helpful to precede a review of wear models by a review of models of friction.

Two major approaches have dominated the development of theories of friction (2-4,11,12,52). One has emphasized surface roughness and the interlocking of surface asperities. The other approach has focussed on adhesion and interatomic forces.

The best known model for friction incorporates both of these ideas. Bowden and Tabor (2,3,11,12) suggested that surface asperities adhere to form junctions and the friction force is directly related to the force needed to shear these junctions. The normal load L is supported by the total contact area A of the elastically and plastically deformed asperities. If τ_{jct} is the shear strength of the adhesive junctions, then the frictional force F is given by $F = A\tau_{jct} = \frac{L}{p_o}\tau_{jct}$, where p_o is the yield pressure (related to, but not equal to hardness). Thus, the coefficient of friction $\mu = \tau_{jct}/p_o$. Estimates of the friction coefficient from this expression have been reported as $\mu \simeq 0.2 - 0.3$. However, in using this expression it is not clear how one should interpret τ_{jct}. It could be taken as the interface shear strength, the base metal shear strength, or the shear strength of the deformed material near the junction. The term p_o is also uncertain, because the contact area during sliding is reported to be larger than the relevant contact area during a simple hardness test.

The simple adhesion model can be extended to include surface films which reduce the typical junction shear strength to the level τ_i. Bowden and Tabor demonstrate that the friction coefficient is then $\mu \sim (-1 + 1/k^2)^{-1}$, where $k = \tau_i/\tau_{metal}$. Thus, for small k, μ can be very small, but for k approaching one, μ grows rapidly. The model does not work well for k = 1 (a metal coated with itself) or for k > 1 (a hard coating Bowden and Tabor also show that when k is small, $\mu \simeq \tau_i/p_o$, so that one would expect a lower friction for a soft film on a harder substrate, which is observed experimentally.

Bowden and Tabor also included another contribution to friction, μ_p, due to plowing of a metal by hard asperities. This term is calculated for different shapes of asperities by setting the friction force equal to the force needed to displace material during the formation of grooves. Thus, the total friction coefficient would consist of two terms, an adhesion term μ_a and a plowing term μ_p, or, $\mu = \mu_a + \mu_p$. Both terms would involve deformation. For smooth sliding, μ_a would dominate and for abrasion μ_p would be more important.

The principal results of the Bowden and Tabor model are reasonable; however, the model is too simple to be very useful. It is based on the assumption of homogeneous and isotropic materials, it uses a limited number of variables, and the parameters involved are somewhat ambiguous. The model is widely accepted, but several investigators have raised fundamental questions about it (14,52-54).

Recently, Challen and Oxley have presented a friction model based on slip-line field analysis for asperity interactions, including adhesion and plowing (55). Johnson has also examined friction from asperity interactions; he reports that adhesion increases friction, as expected, but it is not needed to give friction during asperity interactions (56).

Suggested Criteria for Friction Models. An acceptable friction model should be consistent with a wide range of experimental findings on frictional phenomena. It should account for the differences observed in single-pass and multi-pass experiments and for the changes which occur during the break-in period. The common tendency to reach steady state conditions should be explained. The model should be based on energy dissipating processes which are active for given materials, and it should allow for energy storage (at least in the early stages of sliding). It should give reasonable estimates of frictional heating and surface temperatures. Also, it should incorporate variables associated with materials structures and properties, environmental effects, and other experimental conditions.

Several investigators have noted that there is very little change in friction when a hard material sliding on a softer material is replaced by another hard material (57,58), even when the hard sliders have different types of bonding. This is difficult to explain if one bases a friction theory on adhesion, but it is an expected result if deformation of the soft material determines friction for these systems.

Hexagonal metals tend to have low coefficients of friction (59,60) when one slip plane is dominant (61, 62). Again, differences in deformation behavior seem to be important factors.

Strong anisotropic effects are found for sliding on different crystal planes and in different directions, especially for light loads and short sliding distances. These differences are much less for heavy loads and long sliding distances. Single crystals and polycrystals also give more nearly similar results under such conditions. At the same time, hardness, crystallographic texture, and microstructure become more alike for all these cases as sliding proceeds. In all cases, a region near the surface becomes highly deformed plastically.

Environmental effects are complex. Sometimes the friction coefficient will rise when a vacuum is substituted for a given atmosphere and sometimes it will fall (63). Soft or hard coatings can be formed by reaction with environmental species, and the properties of near-surface material can be modified by interdiffusion.

Intentional application of coatings can strongly or weakly affect friction, depending on the relative properties of the coating and substrate materials (64). In some cases, the friction coefficient is sensitive to the thickness of the coating or reaction layer (65,66).

Experimental data on the load dependence of friction depend on the amount of prior deformation of the materials. If the sample is extensively work-hardened, then the friction coefficient is independent of load over several orders of magnitude. However, for annealed samples, the friction is higher at high loads (3,61,67). This effect is usually explained by the break-through of surface films at higher loads, but the same effect is observed for gold, which does not form a significant surface film. Again, the data indicate that plastic deformation is an important part of the process.

For some materials, the friction is higher at low loads and then falls to a nearly constant level at high loads (61). Also, for steels, the original microstructure seems to affect friction for low loads, but not for high loads. These effects are not easily explained by traditional theories of friction. Changes in microstructure, which have been reported by many investigators (68,69), are probably related to the observed changes in friction coefficient.

Friction and Wear. It is common practice (but unfortunately not universal practice) to monitor friction while doing wear tests. Changes in friction are commonly associated with changes in wear. In fact, a change in friction can serve as an early warning signal before catastrophic wear behavior sets in.

Therefore it is natural to seek correlations of friction coefficients and wear rates and to seek common origins for at least part of the contributions to each. Many investigators have attempted to do this, most of them basing their work on the adhesion ideas advanced by Bowden and Tabor (2,3,11,12). A recent discussion of these efforts is given in a paper by Wanheim, et al. (70); a more recent model connecting friction and wear has been proposed by Suh (71). However, one can have friction without wear, or between wear events, so one should be careful not to tie friction too closely to wear.

The following table (72) indicates that the correlation of friction and wear is not a simple problem.

Sliding Pair (in air)	Coeff. of Friction	Wear Coeff., k
Au on Au	2.5	10^{-1}
Brass on hard steel	0.3	10^{-3}
PTFE on hard steel	0.15	2×10^{-5}
PE on hard steel	0.6	10^{-7}

Two points are immediately obvious: (1) The correlation of friction and wear is not a good one, and (2) The friction coefficient covers a much smaller range than does the wear coefficient.

Friction involves the dissipation of energy. The dominant dissipative process for many materials is plastic deformation. However, wear, at least the kind which generates loose wear debris, depends on an additional process: fracture. Therefore one should not expect a simple correlation between friction and wear.

It might be argued that the fracture energy associated with wear has been omitted here. However, this is small compared with the work of deformation. Also, any contributions to friction from debris generation (delamination or any other type) will be isolated events which contribute to the signal noise, but they are not basic to the friction process. The principal exception to this statement would be the case of formation of an appreciable film of transfer material; friction would then change as expected for a coated material.

Energy-Based Models of Friction. As with many problems in the physical sciences, friction can be approached by using forces or by using energy. Energy-based models of friction focus on the mechanisms by which energy can be dissipated during sliding. A number of authors (reviewed in reference (52)) have suggested that plastic deformation is the most important factor in frictional energy dissipation for many materials. This approach seems particularly promising for metals, but it is probably also useful for some polymers and even for ceramics, which can deform plastically under compressive loading. The principal exception would be elastomers, for which hysteresis losses are mainly responsible for friction (73-75).

For many materials it is reasonable to assume that the frictional work is approximately equal to the plastic work done, or, $W_F = \int \tau \, d\gamma \, dV$, where τ is the shear stress, γ is the shear strain, and V is the deformed volume near the surface. This work includes energy stored as well as energy dissipated; the latter will dominate even for small plastic strains, and it accounts for essentially all of the plastic work at high strains (76-78).

With this approach, energy terms are additive for each region and each material that is deforming; therefore contributions to the friction coefficient are also additive. Thus, one can readily treat all of the following cases: one material deforming (hard on soft), both materials deforming (similar hardness), coated materials, layers of reaction products, lamellar microstructures, etc.

To evaluate the work integral(s), one needs expressions for τ, γ, and V. Both τ and γ will be functions of position. Also, τ is related to γ through a stress-strain relation. Since high strains are involved, it is best to use a $\tau(\gamma)$ relation which is reasonable for high strains. One possibility would be $\tau = \tau_{max} \, (1-\exp(-c\gamma))^{1/2}$; this expression saturates at τ_{max}, has simple parabolic work-hardening behavior at low strains, and is consistent with simple dislocation arguments (76). Other $\tau(\gamma)$ relations could also be used. It is important to realize that it is a convenient way to bring materials structure and properties into the friction model.

With layered or coated systems, one must introduce appropriate boundary conditions. Also, the effect of surface roughness must be included as coatings become thin, since some of the asperity contacts will then be on uncoated base material. The resulting expressions for friction coefficient reduce to those of familiar adhesion models for simple cases, but they are much more general. Also, the terms involved are clearly defined and measurable. For coatings, predicted variations of friction coefficient with thickness are consistent with variations observed experimentally, including those exhibiting maxima and minima.

Interface Temperature. There is continuing controversy over the magnitude of temperature rises induced by frictional heating (14,80-83). Measurements are not simple because introduction of a temperature probe can modify sliding conditions. One can use the sliding materials themselves as the two components of a thermocouple, or one can use a transparent material for one component and monitor infrared radiation (84-87). These techniques have shown that surface temperature varies significantly with both time and position. Local hot spots may temporarily be many hundreds of degrees Celsius hotter than surrounding areas, but the average surface temperature appears to be considerably less. For very high speeds the local temperature can reach the melting point (88).

Calculations of surface temperature have usually been based on the assumptions that the heat source was the two-dimensional interface itself and that the contact area was similar to that which exists during static loading (80). As Furey points out, such calculations overestimate the interface temperature (82).

In terms of energy-based deformation models of friction, the frictional heat source is not localized at the surface; instead it is dispersed through a volume of material adjacent to the surface. Such a dispersed heat source will give a lower surface temperature than would be the case for a two-dimensional heat source (89).

Surface Deformation and Microstructure. Compared with the bulk material, the material adjacent to a worn surface has very different structures and properties. This is the region in which most of the frictional work is being done, and it is also the region from which wear debris are generated. In fact, Tsuya has found that the dimensions of wear particles are proportional to the thickness of the layer (90). Therefore, understanding the nature of this near-surface material is intimately connected with understanding friction and wear themselves.

The most highly deformed region near the surface is of particular interest. It has been given various names, such as Beilby layer, amorphous layer, white layer, dark layer, friction martensite, oxide glaze, decarburized layer, micronized layer, fragmented layer, transfer layer, recrystallized layer, cell layer, etc., depending on the systems studied, the methods of preparation, and the techniques used for observation (30,91,92). With the possible exception of some layers involving surface reaction products, the old idea that this region is amorphous has been disproved (91).

The substructure of this region is known to be very fine, with characteristic dimensions of the order of 1μm or less, too small to resolve with optical microscopy. In some cases, substructure dimensions of ∿ 10 nm have been reported. Microhardness and dislocation density seem to be high (61). The layer is often highly textured (61), and boundaries in the substructure appear to be well aligned (42,61,93,94).

Below the highly deformed layer the microstructure gradually changes until it approaches the structure characteristic of the bulk material. The particular sequence of structures present depends on many variables, including the sliding distance, load, temperature, chemical environment, and materials system. In a simple material like copper, the dominant structure near the surface is an array of fine elongated and flattened subgrains or cells, like those shown in Figure 1. Below that region (typically 5 - 30 μm for unlubricated sliding) the cells become less elongated, the walls become less sharp, and the size increases (Figure 2). Eventually

Figure 1. Longitudinal section of a copper - Al_2O_3 sample after sliding. Structure 20 μm below the surface consists of elongated sub-grains or cells. Contrast differences indicate small misorientations between adjacent cells (185).

Figure 2. Section similar to that of Figure 1, but at a depth of 30 μm from the wear surface (185).

the structure becomes an irregular tangle of dislocations (42). For a low stacking fault energy material like 30% Zn brass, extensive twinning has been observed (42). Also, for brass, the texture at the surface should be different from that of copper. In ferrous materials, when thermal factors allow it, a region of martensite is expected to form. In austenitic stainless steels the formation of strain-induced martensite affects the friction and wear characteristics (43). For all these cases, a layer of transfer material can occur as a continuous or patchy film on the surface. All of these structures can be present in a region only tens of μm's thick; therefore techniques with high resolution, such as TEM, must be used to study them (42). It is interesting to note that the same structures are generated below surfaces after lubricated sliding. The main difference seems to be that the various layers are thinner (42).

It has been suggested that segregation effects at and near the surface can be accelerated because the highly deformed region has many rapid diffusion paths (52). Therefore, processes might occur which might not be expected if bulk diffusion were required. An example might be the dissolution of fine precipitates and segregation of the solute to dislocation walls. This could affect both friction and wear.

Wear Debris. A wide range of shapes and sizes of wear debris particles has been reported. Sizes range from tens of nm to centimeters. Shapes include flakes, plates, ribbons, clumps, cylinders, spheres, irregular chunks, and cutting chips (69). They may be metallic, completely or partially oxidized, mixtures of the rubbing materials, or soaps resulting from reactions with components of the lubricant.

The type of debris may change with time during sliding, either because of running-in processes, changes in critical dimensions, or changes in experimental conditions, such as load, speed, temperature, or chemical environment. Therefore, monitoring of wear debris by continuous or periodic sampling allows monitoring of the tribological health of the system (95,96).

Perhaps the most striking debris particles are the spheres. They received much attention in the 1970s when the SEM was used to obtain dramatic photographs. Scott, et al. (95,96) have associated the presence of spheres with incipient system failure. Smith (97) has suggested that the spheres are indications of fatigue, since he finds that spheres are present in fatigue cracks. Rabinowicz has demonstrated that spheres can be produced by trapping of more irregular debris in surface depressions (98). Other spheres may be the result of compacting of smaller particles, and some are clearly the result of local melting, since dendritic features are evident.

Some spheres have even been traced to air-borne contaminants, such as particles from nearby grinding operations (99). Spheres could also result from roughly spherical inclusions originally embedded in the wearing materials. Spherical debris particles are interesting, but they are probably much less important than other shapes which are more typical in most wearing systems.

Cylinders have been observed as they formed by simple rolling up of flake-like material (100). Cutting chips are clearly the result of micromachining events, and they are common during abrasive wear. Irregular particles can be produced by fracture of brittle materials (43) and by clumping of particles having other shapes. Ribbons have been observed delaminating from the surface of low stacking fault energy metals, such as brass and Cu-Ge alloys (101,102).

Since iron-base alloys are such common materials in sliding systems, the influence of ferromagnetism is also a consideration. Magnetism can induce clustering of small debris particles into larger pieces which can then be more readily consolidated during subsequent passes (103). The larger clusters can do more damage than the separate, smaller particles. On the other hand, they can also be removed more easily by filters. Magnetic effects would also be present for Ni-base and Co-base alloys.

Probably the most common types of debris particles--and therefore the most important--are relatively flat flakes and plates. They can be generated in many ways. For a ductile material like OFHC copper, long, lamellar extensions have been observed to break off near the exit side of a wear scar (103). Flakes can also be produced from prows moving in front of sliders. Very small flakes ($\lesssim$ 1 μm) have been associated with transfer processes (103); recent work indicates that some larger flakes have been formed from compaction of these smaller particles. Irregular flakes can be generated by plowing of grooves, lateral extrusion of ridge material, and fracture of ragged edges of the extruded material during repeated passes (104). Various delamination mechanisms have also been proposed (61,66,92,105,106).

Suh has used an SEM to demonstrate that the fracture surface under a delaminating flake is very much like a typical ductile fracture surface found in tensile tests. On the other hand, Doyle and Turley have shown (also with an SEM) that the fracture surface in some materials is quite smooth and featureless (101). Many investigators have shown that flake or plate wear particles are sometimes lamellar in structure; i.e., they are composed of thinner layers.

The common occurrence of flat debris particles, together with correlations of debris particles with characteristics of the highly deformed material near the wear surface, has encouraged recent delamination models involving plastic deformation and fracture.

Sliding Wear

Sliding Wear Models. Despite the importance of wear, surprisingly little effort has been devoted to understanding wear phenomena. Reviews of wear models are given in references (36,48,50,51,107-111). In this section, some of the more familiar models of sliding wear are outlined. In a later section, more recent ideas are described.

Of the various adhesion models of wear, that of Archard is probably the most familiar (72). This model is closely related to the adhesion model of Bowden and Tabor, described earlier. If a typical asperity of a homogeneous and isotropic material has a contact radius r and if a hemispherical wear particle of volume $\frac{2}{3}\pi r^3$ is removed after the asperity slides a distance 2r, then the total wear volume is $n \pi r^2/3$ for n asperities. Since the normal load is L and the yield pressure of the material is p_o, the wear volume per unit sliding distance is given by $W = \frac{L}{3p_o}$. Usually this equation is modified to $W = \frac{k'L}{3p_o}$, where k' is a factor inserted to allow for the probability of formation of a wear particle. A similar factor occurs in a number of other wear models, with varying interpretations suggested. Challen and Oxley (55) suggest that the low values of k' are due to having only a small fraction of asperities able to cause effective wear. In fatigue models (described later), k' is inversely proportional to the number of loading cycles to failure. If the hardness H is taken to be proportional to p_o, then the wear law becomes $W = \frac{kL}{H}$, commonly known as the linear wear law, or Archard's law. Similar expressions have been derived by others (111-113), and modifications have been proposed (55).

The factor k is a dimensionless wear coefficient. For mild wear, $k \simeq 10^{-7}$ to 10^{-8}; for severe wear, $k \simeq 10^{-3}$ to 10^{-4}. The inverse of k is sometimes called the wear resistance, although wear resistance is frequently defined as the inverse of the wear rate.

The linear wear law indicates that wear is proportional to the sliding distance and load and inversely proportional to the hardness of the wearing material. These are reasonable results, and if k is known to be constant, this simple wear law can be used to estimate wear. However, even if the design/materials/environment system is held constant, k can change dramatically. An example has been given by Burwell and Strang (112) for steel. At loads such that the nominal contact pressure exceeded about one third of the hardness, the wear coefficient rose dramatically. Thus, one should be careful not to apply the simple linear law outside the range of conditions for which data are available.

Unfortunately, the only materials parameter in the linear wear law is the hardness. Hardness is important, but other materials properties (e.g., fracture toughness, crystal structure, bonding type, chemical reactivity, etc.) affect wear rates. Also, there are cases in which increases in hardness are accompanied by increases in wear rate.

There are at least two other difficulties with the Archard model. First, the model does not really provide a mechanism for loose wear debris. The assumed adhesion would account instead for material transferred to the opposing surface. Subsequent fatigue has been invoked to explain the generation of debris (114-117). A second problem is that the size of any transferred or loose particles would be determined by the asperity contact dimensions, whereas debris particles have been observed over a wide range of sizes, including some particles much larger than asperity dimensions. Finken has suggested that asperity junctions can be displaced for distances larger than the distance assumed by Archard (115); this modification allows formation of larger debris particles. Other modifications of the Archard model have incorporated elastic and plastic contacts and statistical distributions of asperity sizes (115-117).

Rabinowicz (14) has used a model similar to Archard's to estimate the size of debris particles. His model is based on the assumption that a wear particle in a material can be generated when the energy needed for fracture is supplied by the release of stored elastic energy, taken to be the product of particle volume and the stored elastic energy per unit volume ($\sim p_o^2/2E$, with E = elastic modulus). The resulting condition is that a characteristic particle dimension D should be greater than a term proportional to $E\sigma/H^2$, where σ is the surface energy and H is the hardness. Rabinowicz further simplified this by assuming that E is proportional to H, so D > const. σ/H. However, the ratio E/H varies over a range of more than 100, so this simplifying step is not justified and a correlation of D with σ/H is not expected. Furthermore, the particle sizes predicted by the model are much larger than many debris particles actually observed.

Quinn has concentrated on wear involving oxidation (118,119). The basic mechanism is assumed to be the formation of a critical thickness of oxide on the wearing surface, followed by removal as oxide wear debris. The expressions derived for wear rate have a large number of parameters, some of which cannot be readily determined experimentally.

A group at IBM has developed a wear model in terms of zero wear defined relative to the surface finish (120,121). For a single pass, wear is found to be negligible if the maximum shear stress τ_{max} is less than about half the yield stress of the material, YS. If transfer occurs, this guideline is reduced from about 0.5 YS to about 0.2 YS. The general rule would then be $\tau_{max} < C$ (YS), where C can be ~ 0.5 or ~ 0.2. For a large number of passes, M, the IBM group has reported that $M (\tau_{max})^9 = \text{const.} (C (YS))^9$. Thus, this model is related to an expression used by Palmgren for fatigue failure in bearings. In practice, the IBM model uses τ_{max} obtained by application of Hertz elastic contact theory (122-123). The model offers simple, but conservative, guidelines for design of systems which might be exposed to wear.

Many investigators have suggested that fatigue is an important part of sliding wear processes (13,63,117,124-128). Fatigue models can include adhesion effects or they can be based on independent processes for generating loose debris. If one assumes that N cycles of loading for a given asperity leads to fatigue fracture, one can derive a wear law that is similar to Archard's law, i.e., $W \sim N^{-1} L/H$ (128). Therefore the wear coefficient has a somewhat different interpretation: 1/k is proportional to the number of loading cycles required to give a wear debris particle.

Evidence for a connection between wear and fatigue has been summarized by Hirth and Rigney (61,125) and by Kimura (54,127,129). It includes similarities in the microstructures at high strains and similar effects from changes in environment (63). Also, both phenomena are sensitive to surface conditions such as roughness, cracks, and coatings. If wear and fatigue involve similar processes, then fatigue-resistant materials should also have good wear resistance. Data are insufficient to check this possibility. This is an area worth exploring. One must keep in mind though that complicating factors such as material transfer could mask correlations with fatigue. Also, one should remember that the mode of loading used to obtain most fatigue data has been alternating tension/compression, whereas for wear, compression and shear loading dominate (130).

Simple linear wear laws having the same form as Archard's law can be derived for models based on adhesion, fatigue, oxidation, and abrasion. This agreement does not mean that these models are realistic. It is the result of a common set of assumptions, basically that wear rate W depends on sliding distance S, load L, and hardness H. When experiments do not agree with the simple wear law, this is a strong indication that this list of variables is incomplete.

Drucker has considered the use of dimensional analysis to provide guidelines for possible variables to use for wear models (131). He notes that if a model is restricted to macroscopic variables (velocity, load, stress, strain hardening exponent, elastic constants, etc.), then no natural length scale appears. Then, there can be no predictions of asperity or debris particle dimensions. To obtain such results, one needs to bring in structural information (grain size, dislocation spacing, precipitate spacing, etc.). Or, one could bring in combinations involving fracture toughness, surface energy, yield strength, etc. Dimensional analysis does not tell us which variables to use, but it does put constraints on suitable wear laws and it may be helpful in selecting useful combinations of variables in the future.

<u>Delamination</u>. Delamination will be used in this section to describe any wear process which results in loose pieces of material with one dimension much less than the others and with two faces roughly parallel. Füchsel (132) ("abblättern") was among the first to use delamination to describe wear. He viewed wear as a process involving deformation and fracture.

In recent years, Suh has proposed delamination theories of wear (66,105,106). His work has stimulated much new work, theoretical and experimental, designed to understand basic wear mechanisms.

In Suh's original delamination theory, a soft surface layer due to dislocation image forces was postulated. Suh assumed that below a soft surface layer was a region with higher dislocation density in which voids nucleated, especially if particles or inclusions were present. He then suggested that the voids coalesce to form a crack running parallel to the surface. This crack propagates if it exceeds a critical length, and eventually it turns toward the surface to form a sheet-like wear particle.

One problem with this model is that the assumed soft surface layer would only be a few tens of nm thick, whereas the wear particles can be much thicker (tens of μm). In later versions of his delamination model, Suh has omitted the soft surface layer and concentrated instead on plastic deformation, strain accumulation, and fracture. He suggests that cracks nucleate below the surface because compressive stresses closer to the surface inhibit cracks. From stress calculations he has suggested that small tensile stresses which exist in a sample after a slider (or asperity) has passed are responsible for fracture.

Suh's group and many others have produced interesting SEM photographs of wear debris, wear surfaces, and sub-surface cracks to support delamination ideas. Delamination processes certainly occur, but they are not understood in detail. In particular, the sites for crack nucleation and the reasons for the paths followed are not at all clear (134).

Rigney and Glaeser (92) have suggested that the fine textured microstructures produced near wear surfaces are intimately connected with delamination wear. They emphasized the microstructural anisotropy and suggested that cracks would probably form in the highly deformed layer and propagate parallel, and possibly within, dislocation cell and sub-grain boundaries. The smallest debris particle would then be one sub-structure unit in size and the largest would have a thickness equal to that of the highly deformed layer. Thus, this model is an analog of a brick wall, in which debris particles range from individual bricks to larger clumps of bricks. The cell walls would provide suitable paths for rapid diffusion, which allows various segregation effects as well as accelerated chemical reactions with environmental components in the atmosphere or lubricant (52,103). In some systems, the structure, and therefore the fracture paths, would be modified by phase transformations, as for the case of austenitic stainless steels (43).

The model of Rigney and Glaeser requires the presence of suitable microstructure before delamination debris can be produced. Therefore a time delay is expected while plastic deformation produces suitable near-surface structures. Debris could form by other mechanisms before that time. For example, Glaeser has shown that even for plasticine, closely spaced and overlapping groove formation is accompanied by ragged lateral edges, and these can generate irregular flake debris (104).

The model can include adhesion effects, although these are not necessary for stress transfer (56) and delamination. Adhesion could lead to fracture within sub-structure boundaries and gradual formation of transfer layers or particles consisting of material from both interacting materials. The resulting debris particles would then be intimate mixtures of the two materials (103).

Material properties such as stacking fault energy are readily incorporated in this structural model. Stacking fault energy affects twinning, texture, the ease of formation of dislocation cells, and the nature of fracture. With low stacking fault energy (Cu-Zn, Cu-Ge) debris particles tend to be large, smooth, and elongated (101,102).

The preceding model can incorporate a formidable number of variables. Therefore it has not led directly to a simple, quantitative wear law. One could argue that this is a serious short-coming of the model, or one can realize it can accommodate many different wear laws, depending on the system chosen. It is based on plastic deformation and fracture, and any variables which affect those phenomena will be important. Thus, hardness and yield strength will be important, but so will the number of slip systems, the fracture toughness, and other variables.

Transfer. It is well known that transfer of material from one component of a sliding pair to the other occurs for many materials combinations (135). Transfer is part of the process involved in some scuffing phenomena. The transferred material can significantly affect the friction and wear of a system. For example, it could form a film that behaves like a solid lubricant, or it could form a hard coating which offers some wear protection (136,137). In some cases the transferred material forms a continuous film; in other cases the film remains patchy. The transfer film can be thick or thin, depending on the system.

Transfer is usually discussed in terms of adhesive contacts of asperities, in which the adhesive contact is stronger in shear than the shear strength of the transferring material. This concept may be basically correct, but without more detailed analysis it is difficult to predict the amount or even the direction of transfer. One might expect the "softer" (bulk) material to transfer to the "harder" material, but this simple rule is not always obeyed.

For copper sliding on ferrous materials, the direction of transfer seems to change with time (103,137). For short sliding distances, a visible film of copper transfers to the iron alloy, but iron transfers back to the copper at longer sliding distances. The direction of transfer can also be determined by geometric effects. Antler showed that for a slider on a flat sample of the same material, the transfer was from the flat to the slider (138).

Several investigators have used radioactive isotopes to monitor transfer during sliding (136), but the resolution in these experiments was limited. Kerridge and Lancaster (136) reported that the amount of transfer was limited for a given system, so that dynamic steady state conditions could be achieved. Both Kerridge and Lancaster and Finkin (139) reported that debris particles could be formed from smaller particles generated earlier, and work in Japan has indicated that the debris particles have mixed composition (137,140).

More recently, an SEM equipped for local chemical analysis showed that the debris particles and the transfer material for Cu-Ni alloy blocks sliding on 440 C steel rings consisted of intimate mixtures of Cu-Ni and steel (103). The instrument showed that particles $\lesssim 1\ \mu m$ in their largest dimension had the same mixed composition as larger clusters of debris. Also, Debye-Scherrer X-ray analysis of the debris showed that the separate FCC Cu-Ni and BCC Fe phases were retained. Therefore, each particle was a microcomposite. These same experiments confirmed the results of Kerridge and Lancaster concerning a dynamic steady state. The per cent of transferred material reached a maximum level which depended on composition. This level was maintained unless other processes (such as serious ring damage) intervened.

Fortunately, there is increasing interest in the mechanisms of transfer (102,141,142). Most of the recent work on this subject is experimental, and it involves use of various modern probes to determine local composition and structure. There is also some effort being devoted to apply alloy theory and thermodynamics to explaining transfer phenomena (102,142). If these new efforts are successful, it may be possible to apply the results to design control strategies for practical systems.

Surface Topography. Much effort has been devoted to topography in tribology (143). Measurements have ranged from simple profilometer traces with their inherent limitations (30,144) to sophisticated techniques employing modern instruments and statistical analysis. These have been reviewed in a recent issue of Wear devoted to the subject (145). No attempt is made here to review this large field. Instead, several observations on the appearance of worn surfaces will be described.

A typical metal surface after sliding exhibits grooves and ridges of various sizes. They are formed by the passage of many local asperities on the opposing surface. In some cases, loose particles of debris or contaminants can contribute to the pattern. The grooves are relatively smooth for long distances. Therefore, one would find very different results for profilometer traces parallel and perpendicular to the sliding direction. Even if one examines the wear surface at different magnifications, the pattern of grooves looks similar (42, 146); this indicates that a wide range of groove sizes (and asperity sizes) is involved.

With single groove experiments, it is easy to see differences between machined or cut grooves and grooves formed by plowing. However, with the multiple grooves present on typical worn surfaces, it is difficult for most observers to distinguish clearly between a surface resulting from sliding and a surface resulting from abrasion.

Simple groove patterns are not the only features observed on wear surfaces. With modern SEM's and with surface replica techniques (TEM) one can see almost any imaginable topography, including cracks, plateaus, pits, periodic striations, embedded particles, slip lines, lamellar features, and reaction products. It may be difficult to determine which of these many features provide information of use for analyzing practical wear problems or understanding wear mechanisms. Some of the more interesting features may in fact be atypical and unimportant for a given system and should not be given undue attention. Otherwise one may produce a set of aesthetically pleasing photographs which are of little use and which can distract the investigator so that he misses more important features.

Another point should also be remembered. Much of the work on surface topography has emphasized the initial surface finish. Wear processes modify the surface topography, in some cases drastically. Therefore, after a running-in period, the relevant surface topography is that produced by operation of the system, not the original surface finish. Furthermore, the surface finish after long sliding times may be of more fundamental interest than the initial surface finish for understanding friction and wear processes.

Abrasion

Abrasion is the wear by displacement or removal of material on one of two surfaces in relative motion by hard asperities on a second surface. It can be caused by fixed abrasives as in abrasive paper, grinding wheels, embedded particles, or protruding hard phases, or it can be caused by loose abrasives such as those encountered in dusty environments or in mechanical polishing. Abrasion is a particularly common form of wear, especially in mining, agriculture, and construction, because common abrasives are SiO_2, Al_2O_3 and iron oxides, which compose much of the earth's crust.

If the ratio of the hardness of the abrasive to that of the abraded material (H_a/H_m) is greater than about 1.3 to 1.7, the abrasion rate is high and roughly independent of abrasive hardness. For $H_a/H_m \lesssim 0.7 - 1.1$, there is very little abrasive wear. In the intermediate range, where H_a and H_m are similar, there is a transition in wear rate, with higher rates for larger H_a/H_m (147,148).

It is common practice to think of abrasion as consisting of two principle mechanisms. The less efficient process is called plowing, and the more efficient process involves cutting of a chip as in machining. Models of plowing use asperities of different geometries and yield expressions for friction and for wear (2,3,14, 147-149). Again, a wear law having the same form as Archard's law results. In fact, much of the confirming evidence for this type of wear law has come from experiments on abrasion resistance as a function of hardness (149,150).

Plowing processes can remove material by displacement to the sides of grooves and by formation of a prow which moves ahead of the indentor. It causes extensive plastic deformation in the same material. In fact, it has been shown that the deformation and resulting microstructures are the same for sliding and for abrasion, lubricated and unlubricated (151,152). The main differences are the friction and the depth of deformation, which are directly related when plastic deformation is the principle mechanism for energy dissipation. When cutting processes dominate, the friction is lower and the deformed layer is less deep than when plowing processes dominate. Given this information, it is tempting to consider sliding wear and abrasive wear together rather than separately, since the mechanisms and effects are closely related. In this view, they are considered as comprising a broad spectrum, with machining or cutting at one end and smooth sliding at the other. Abrasion would cover a broad range in between with greater efficiency as the ratio of cutting asperities to plowing asperities increased.

Samuels has made impressive contributions toward unifying cutting and plowing ideas (30,153,154). His analysis begins with experimental results from machining studies with simple cutting tools having different rake angles. The rake angle α is defined as the angle between the normal to the wear surface and the face of the tool along which a chip slides. Blunt tools have less positive or more negative rake angles. A critical angle α_c exists for each materials/environment system such that a cutting chip is formed for $\alpha > \alpha_c$ (lower friction force, shallow deformed layer, efficient material removal) and plowing occurs for $\alpha < \alpha_c$. The critical angle varies widely for different materials and operating conditions. For example, for hard steel tools, the critical angle for cutting steel is about 0°; for Al it is -5°; Ni, -25°; Pb, -35°; and Cu, -45°. Lubricants affect α_c by providing lubrication and cooling and by carrying away debris particles. A fluid like CCl_4 changes α_c of steel from 0° to -20°. An increase in temperature can change it to -40°. An increase in velocity also makes α_c more negative, mainly through thermal effects. Even the shape of the tool affects α_c: for brass, an orthogonal tool gives $\alpha_c = -10°$, but a V-point tool gives $\alpha_c = -40°$. For real abrasives, fixed or loose, the value is intermediate between these cases, as would be expected for a statistical distribution of shapes.

Since abrasives have a distribution of rake angles α, one would expect that some points would cut chips and some would simply plow. The more negative the critical rake angle, the larger the proportion of points that cut chips, and the higher the abrasive wear rate. A typical distribution of rake angles for 220 grit SiC abrasive paper would look roughly like that shown in Figure 3.

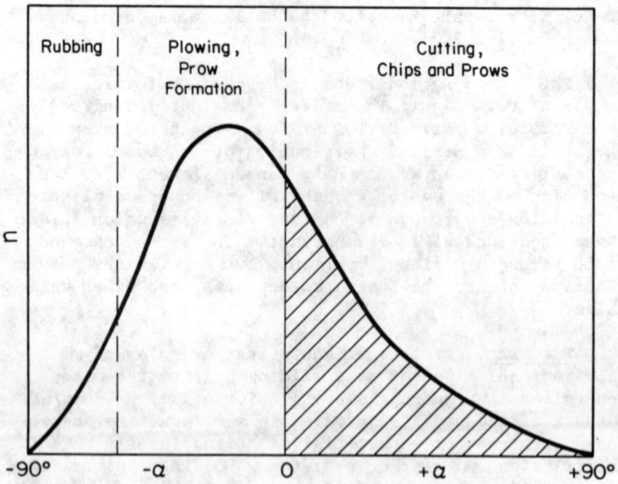

Figure 3. Typical distribution of rake angles for a fixed abrasive paper (154).

Points in the shaded region would cut chips. Different materials and different environments would shift the shaded region left or right. Different abrasives would have different distribution functions. For example, the peak in the distribution would occur further to the left for a typical grinding wheel, which would give less efficient abrasive wear.

It has been assumed in the preceding discussion that the abrasive material is adequately hard relative to the abraded material. Abrasives can become less effective through fracture, adhesion, and debris capture, since these processes generally shift the distribution of rake angles towards less effective cutting conditions. Such changes also cause departures from linearity in plots of wear rate vs. load (155).

Loose abrasives can be treated by the same model, but abrasion in this case is less efficient because of rolling and tumbling. An abrasive particle is therefore properly oriented for effective cutting for only a fraction of its contact time.

Abrasive wear rates are surprisingly insensitive to grit size changes when the grit size is above about 100 μm. For small grit sizes, the wear rate falls steeply. This grit size effect has been observed in materials as diverse as steel, brass, aluminum, and alkali halides. It has enormous practical significance, ranging from design of filters to control of damage during finishing processes. Many reasons have been proposed for this effect, including abrasive shape, blunting, fracture, and clogging. These and other proposed mechanisms have been reviewed by Sin (156,157). This effect is not yet well understood.

Samuels has also suggested that mechanical polishing processes are essentially the same as abrasion (30). That is, cutting, plowing, and smearing of material occurs, but on a very fine scale. The grit size effect may enter here, since these processes are slow. High temperatures are not needed for these processes; in fact, fluids such as water are commonly used, and their presence makes it very unlikely that elevated temperatures are involved.

The dependence of abrasion resistance on hardness is well known from the widely quoted work of Khrushchov (149,150). For pure metals, a simple linear dependence is observed, as shown in Figure 4. For more brittle

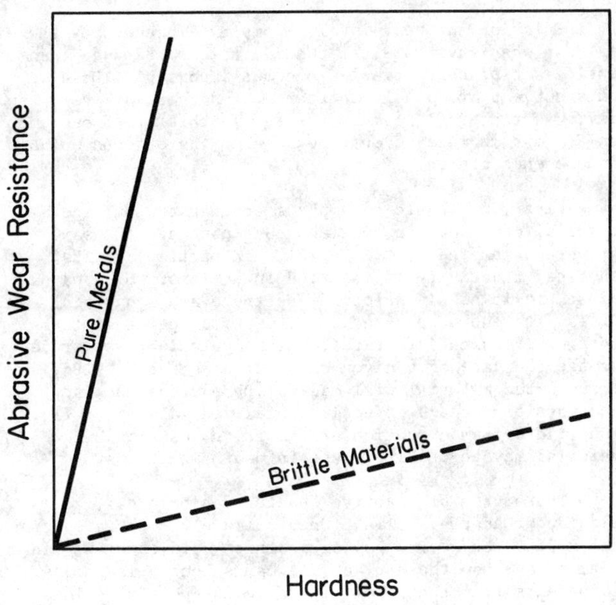

Figure 4. Abrasion resistance vs. hardness (149,150).

materials such as ionically and covalently bonded materials, a second linear relation holds; it has a smaller slope, as indicated by the dashed line in Figure 4. The data for other materials, including most practical alloys, fit in between these limiting lines. In particular, cold worked steels show very little improvement in abrasion resistance over pure iron over a wide range of hardness. Similarly, precipitation hardened systems such as Cu-Be show little improvement over the matrix metal. Heat-treated steels do show improvement, but the slope is not as great as for the limiting line for pure metals. However, for very high hardness levels, the wear resistance of these materials falls off. A similar effect is observed for highly cold-worked steels. Recent work has indicated that fracture toughness limitations are responsible for these trends (158,159).

Microstructure and Abrasion. Microstructure effects become apparent when test results for materials of similar hardness are compared. For example, one could select a martensitic carbon steel with Rockwell C hardness R_c = 62-64 and compare it with an alloy steel containing complex carbides and having the same average hardness. These will have different behavior because the matrices are different and because the volume fraction, dispersion, size, shape, stability, and type of carbides are different for the two cases. At higher temperatures the contrast could be even greater. The martensitic steel would change considerably above ∿200°C (either during use or through inadvertant heating during preparation, as during grinding). Even among the alloy steels there are significant differences. For example, Popov has reported that cubic carbides such as VC and NbC provide greater wear resistance than Cr_7C_3 or WC (160).

Khrushchov and Babichev (150) reported a linear increase of abrasion resistance for steels as volume % cementite increases. The abrasion resistance also increases linearly with % pearlite (161,163) and with square root of the weight % of carbon. For spheroidized steels, the abrasive wear resistance increases linearly with (carbide spacing)$^{-\frac{1}{2}}$ (162,163). Diesburg and Borik (164) report that the abrasive wear resistance of ferrous alloys increases in the order ferritic, bainitic, pearlitic, martensitic, austenitic, and austenitic with carbides, or, in the order low alloy wrought and cast steels, austenitic manganese steels, and high Cr white irons. In another series of results they report less wear as one goes from a spheroidized matrix to a pearlitic matrix and then to a martensitic matrix. For similar hardness levels, an austenitic matrix gave less wear than a pearlitic matrix.

This last result is interesting because it is one of the few comparisons in which clear structural differences are reported for controlled hardness. Most of the structure results, at least for ferrous materials, have been close to what would be expected on the basis of hardness alone.

In general, for composite structures consisting of a mixture of phases of different hardness, better abrasive wear resistance is obtained with a larger volume fraction of hard phase. An extreme example would be alloy carbides bonded by a small amount of a relatively soft metal such as cobalt. Such materials have good combinations of hardness and fracture toughness.

Sliding Wear and Abrasive Wear. It is relatively easy to perform reproducible abrasion tests; therefore much useful data exist for abrasion resistance. For sliding wear, data are less reliable, partly due to complications from adhesion and transfer. It might therefore be tempting to use conclusions about relative wear resistance obtained from abrasion tests for sliding situations. In general, this would not be good policy, despite the importance of plastic deformation for both types of wear.

A comparison of two familiar steels will emphasize the need to be very careful in applying wear data. One is 52100, commonly used for ball bearings, and the other is D2 tool steel (1.5 wt. % C, 12 wt. % Cr, 1 wt. % Mo). The 52100 alloy has a fine, uniform microstructure and has good performance for bearings, which roll and slide. The D2 alloy has much larger, hard, stable carbides and better abrasion resistance than 52100.

Another interesting example is a comparison reported by Singer (165). He implanted nitrogen into 304 stainless steel. This gave improved sliding wear resistance but lower abrasion resistance. In a related test, Singer implanted Ti into the 304 alloy. This hardened the surface region and improved the abrasion resistance. The reasons for the response of nitrogen-implanted samples are not well understood. One possibility could be the creation of a soft surface layer from the implanted nitrogen. This in turn could be due to the role of nitrogen as an austenite stabilizer; it would tend to inhibit formation of strain-induced martensite in the metastable alloy (43,165). Alternatively, the improved sliding behavior could be due to a reduced tendency for transfer. Whatever the reason, the nitrogen changes sliding and abrasive wear in opposite directions.

Fracture Toughness

The generation of loose wear debris particles obviously implies the existence of fracture events. The modern approach to understanding fracture processes relies on fracture mechanics, which uses measures of fracture toughness such as K_{IC}, K_{ISCC}, or J (166) to describe the ability of a material to resist fracture.

Fracture mechanics has been used to analyze many types of fracture, but until recently it had not been applied to wear situations. One of the first to suggest this approach was Hornbogen (159). More recently, Zum-Gahr, Suh, Shieh, Hills and Ashelby, Rosenfield, and Evans have considered wear problems with the aid of fracture mechanics (130,158,167-172).

Already it is clear that fracture mechanics can account for some of the problems associated with relying simply on hardness. Zum-Gahr has shown that abrasion resistance of a number of ferrous alloys depends only slightly on fracture toughness if K_{IC} is greater than about 15-20 $MPam^{1/2}$ (158). The observed decrease can be accounted for by the decrease in hardness which accompanies the increase in K_{IC}. However, below ~10-15 $MPam^{1/2}$, the abrasive wear resistance of various materials depends strongly on fracture toughness. Zum-Gahr has suggested that wear rate can be described by a plowing term inversely proportional to hardness H plus a fracture term proportional to $H^{1/2}/K_{IC}^{2}$. The second term also depends on microstructural parameters. For example, the spacing of carbides or inclusions affects the ease with which cracks can link and propagate.

Rosenfield has discussed these results in terms of a critical value for the ratio of toughness to strength, or of K_{IC}/H. Above the critical value, plowing and cutting are important, as evident in the debris generated. Above the critical value flaking, chipping, or spalling become important. Evans has presented a promising wear model for this latter regime when the applied load is above a critical value (172). It may become useful for treating the wear of relatively brittle materials such as ceramics and glasses.

For sliding wear Fleming and Suh calculated elastic stress distributions for a given load and friction coefficient (167). They then concentrated on cracks parallel to the surface and suggested that they propagated because of the tensile stresses existing after the slider had passed. Hills and Ashelby considered both compression/shear and tension/shear cracks to be important for wear (169,170). Shieh demonstrated that shear cracks were important with compressive loading; he also showed that significant plastic deformation was involved.

Rosenfield has concentrated on the compression/shear fracture (130,171,173). As a first approximation he performed an elastic calculation, arguing that this might be an acceptable approximation because of the extreme work-hardening of near-surface material. Shear stresses drive the crack and friction forces (at the crack) retard it. As with fatigue, each time an asperity passes by, the sub-surface crack advances further. However, the results show that a crack reaching a critical size will not continue to propagate. Also, the direction of propagation appears to be inconsistent with experimental observations. This second problem may be related to the assumptions of homogeneity and isotropy for the highly deformed and textured near-surface material.

More recently, Rosenfield has extended his model by including an elastic-plastic analysis (173). This revised model does predict shear crack growth, whereas the elastic model did not. Also, the new model predicts that sudden changes in wear rate can occur as load is increased, particularly for the sliding of similar materials. Therefore, this may be an alternative explanation of the mild-severe wear transitions which have been reported for the sliding of steels (174).

Analysis of wear using fracture mechanics is a real challenge. Wear systems involve cracks which are similar in size to substructure dimensions, so that inhomogeneity is part of the problem (175). Stresses and temperature vary in complex ways with time and position. Large plastic strains are present. The pronounced texture near the surface requires consideration of anisotropy. All of these vary with depth from the surface. Also, environmental factors affect the stress transfer and friction and affect both deformation and fracture. Finally, the cracks are near the surface. Nevertheless, it should be possible to treat special cases of wear by fracture mechanics methods. For more complex situations, simplifying assumptions can be used, followed by estimation of the ways in which the results would be modified for more realistic conditions.

The growing interest in the application of fracture mechanics to wear problems is one of the more hopeful developments in this field.

Environmental Factors

Friction and wear involve complex processes by themselves; when environmental considerations are added, an entire set of new variables becomes important. Data on this subject are widely scattered, and there are very few review articles (176). Therefore, in this brief section an attempt is made to outline some of the effects which can be encountered.

First, it should be emphasized that the surface material of a wearing specimen does not have the same properties as the bulk material. In particular, its chemical reactivity is modified in at least three ways. First, it is highly deformed, so there are many local sites with enhanced reactivity. Second, parts of the surface are highly textured, which could increase or decrease the rate of a given chemical reaction. Third, transferred material will modify the surface chemistry and possibly create effective localized galvanic cells. Thus, data from corrosion or oxidation tests on bulk material may be of little use in wear situations.

The source of reactive species can be a gas, a liquid, or another solid. One (or more) of these species can modify the wear behavior in three main ways; it can form a layer on the surface, it can selectively remove components from the material, or it can penetrate into the material and modify its structure and properties. Each of these cases will be examined.

Surface layers range from thin adsorbed layers to thick layers of reaction products. They can be loosely or tightly adherent, soft or hard, ductile or brittle, porous or protective. They can increase or decrease the friction, and--probably for different reasons--the wear. A given species can increase the friction while decreasing the wear, and vice versa. In fact, the correlation of friction and wear is particularly poor when environmental factors are included.

Selective removal of a component would include dezincification of brass and decarburizing of steels. In both cases a soft surface layer would result. A good example of penetration would be hydrogen embrittlement in various metals. Other species might also penetrate appreciable distances by using rapid diffusion paths in the highly deformed material. For example, oxygen might penetrate through cell walls, leading to local internal oxidation. If the specific volume of the oxide were larger than that of the metal, this would enhance fracture and debris formation, much as the freezing of water causes fracture in rocks. Examples might be nickel and nickel-base alloys since the expansion is large when NiO forms.

Different chemical species can also affect the critical angle for cutting, thus changing the rate of abrasive wear by the mating surface, the reaction product layer, or the debris. However, this effect may not be large if the effective grit size is very small. Thus, the fact that a hard reaction product is formed does not necessarily mean that the abrasion rate will be high. It may in fact reduce adhesion and transfer and initiate mechanical polishing.

Some metals form thin protective layers ("passive films") which tend to protect them for oxidizing environments. Examples are aluminum, titanium and stainless steel. However, during wear, these layers are continually removed or disturbed, which can lead to greatly increased corrosion rates.

Deformation and fracture both depend on temperature, and the kinetics of chemical reactions are particularly sensitive to temperature. The reaction product formed can also change with temperature. For example, for iron and steel one can produce Fe_2O_3, Fe_3O_4, or FeO, or multiple layers of oxides. With water present the softer hydroxide may also form, thus lowering the friction coefficient.

A reactive species need not be present in large concentrations. For example, dissolved gases in water, traces of moisture in air, or additives in lubricants can have definite effects. Also, the reactive species need not be present as one of the initial components of the system. Examples would be some products of degradation of the lubricant.

Duquette has reviewed data on environmental effects on friction and wear in terms of several broad categories of environment (176). Most of the work reported has been with oxidizing environments such as air. Some data on inert and nonoxidizing environments, aqueous environments, and temperature effects have also been reviewed. From Duquette's review it is clear that at this stage each material/environment system must be analyzed separately. The only rough rule to emerge is a suggestion that systems in which corrosion-fatigue is high will also have high wear rates.

More data on the environmental effects on friction and wear would be useful if they are produced by carefully controlled experiments. These should include better characterization of materials, reactants, and reaction products.

Rolling Contact

The basis for most analyses of rolling contact effects has been the well-known calculation of Hertz for elastic contact of homogeneous, isotropic spheres under normal loading (122,123). It has been applied both macroscopically (directly, to parts of machines) and microscopically (asperity contacts). Plastic deformation is not included in the Hertz contact calculation. Thus, any stored energy is released when the load is removed. This elastic spring back has been invoked to explain why adhesion forces may not be measurable after normal loading of two contacting materials.

The Hertz calculation predicts that the maximum compressive stress is at the center of the contact surface, but the maximum tensile stress is in the radial direction at the contact boundary and the maximum shear stress is below the surface. The tensile stress result is of particular interest for brittle materials and the shear stress result has been used to explain failures in metal ball and roller bearings.

As a rolling contact moves forward, the shear stress below the surface rises to a maximum in one direction and then falls until it reaches a maximum in the reverse direction. Thus a fatigue mechanism is commonly assumed. No wear appears until a critical number of loading cycles is exceeded, after which a fragment of material is removed, its size roughly determined by the Hertz stresses for the given contact geometry. Inclusions or other local structure effects can influence the crack initiation and the exact fracture path followed. For extended operation of rolling contacts it is important to control inclusion content at low levels (177).

The Hertz results are sometimes applied to sliding systems, for which the Hertz model does not directly apply. As increasing amounts of sliding are added to rolling, the position of the maximum shear stress gradually rises toward the surface. Thus, if sliding is present, the depth at which shear fracture occurs is not given by the Hertz result for simple elastic contacts. Littman has discussed many other factors which modify the Hertz contact model, including microstructural features and environmental factors (178).

When alloy steel bearing raceways are sectioned and examined metallographically after use some very interesting structures are observed at depths roughly consistent with the Hertz model (179,180). Most prominent is a band of light-etching material which has been described as consisting of ferrite cells bounded by thin carbide plates. The details of the shear processes and the carbon redistribution are not well understood (61). Other unusual structural features have also been observed in the region near the maximum Hertz shear stresses (177). Since these sub-surface changes are probably related to the fractures which result in rolling contact wear, further work in this area would be desirable. It could lead to development of improved bearing materials by guiding the selection of alloy chemistry and initial microstructure.

Fretting

Fretting is a special form of wear in which two materials slide back and forth against each other with amplitudes of oscillation which are typically $\lesssim$ 70-100 μm. Above this range of amplitudes, wear tends to be more like that of sliding wear.

Fretting is encountered in closely fitting components such as bolted parts, wire rope, keys and keyways, flexible couplings, and electrical contacts. Vibrations from various sources can force the low amplitude oscillations. Waterhouse has recently provided a review of fretting phenomena and ideas on mechanisms (40).

Fretting is best known in steel components. It typically generates fine (0.01 - 0.1 μm) reddish-brown debris identified as mostly α-Fe_2O_3 with some metallic iron. The metal content is higher with large loads, softer metal, large amplitudes and lower frequencies. The debris tends to be retained in the interface region because of the close tolerances and small sliding amplitude. The resulting wear is highly localized and this can lead to stress concentrations which in turn initiate fatigue failure. Moisture in the air has been reported to lower fretting rates, perhaps by a lubricating effect or by aiding formation of $Fe(OH)_3$, which is softer than Fe_2O_3.

Data from several sources indicate that a transition from low to high fretting wear rate occurs at an amplitude of $\sim$100 μm. Possibly related is the observation that sub-surface damage becomes appreciably greater for an amplitude greater than about 80 μm. However, both of these numbers may be uncertain in view of the difficulty of measuring the fretting amplitude accurately with some experimental arrangements (181). Recent work with carefully measured amplitudes has produced a maximum fretting rate at about 30-40 μm. This value was similar to the spacing of sub-surface cracks observed in metallographic sections (182).

Uhlig has proposed that fretting occurs by the growth of a layer of oxide followed by mechanical removal when a critical thickness is reached (183). This approach predicts a pseudo-linear fretting rate, as commonly observed. Waterhouse (40) has emphasized the interaction of metal asperities to form metal debris particles, followed by at least partial oxidation of the debris. In both cases the debris can cause abrasive wear if it is hard enough. Hurricks suggested a similar process, except that he proposed that debris particles were generated by a fatigue process (184).

In recent years the SEM has been used increasingly for fretting studies. Observations of fretted surfaces and of debris have led Waterhouse to propose a different sequence of steps. Adhesion processes lead to surface roughening, elevated regions are deformed plastically, and loose particles are generated by a delamination process, perhaps involving fatigue. During continued fretting the debris particles are further deformed and fractured and may react with the environment. With reactive metals and atmospheres the roughening stage may be less pronounced.

The SEM has also revealed interesting lamellar features on fretted surfaces, with spacings $\lesssim 1$ µm. They are similar to features observed on some surfaces after unidirectional sliding, and they seem to indicate a tendency for highly localized shear deformation. Their role in fretting has not been clarified. Also, except for the localized nature of fretting damage and the related local temperature rise and tendency to trap debris, it is not clear how fretting processes and sliding processes are basically different. The reported transition at amplitudes of about 100 µm is intriguing because it is reasonably sharp and it occurs at about the same characteristic length as the grit size effect found for abrasion. The work of Duquette raises basic questions about fracture processes occurring at lower fretting amplitudes, and it suggests that more attention should be paid to sub-surface changes.

Fretting can be controlled by preventing relative motion between mated parts, using lubricant films to reduce friction, using hard coatings to resist wear, and changing the chemical environment.

Practical Guidelines

The following procedures can be used to reduce abrasive wear:

(1) Remove abrasive particles by using a circulating fluid and filter system.
(2) Take advantage of the size effect described earlier, i.e., use filters which remove the particles larger in size than the critical size of the grit size transition. Smaller particles do little damage.
(3) Use seals and filters to reduce entrance of abrasives from outside the system. However, even with effective seals, abrasion can still occur if one component is much harder than the other or if the debris particles include hard particles.
(4) Use a soft material which can embed abrasive particles.
(5) Use a hard material or make the surface region hard for a sufficient depth (heat treat, coat, etc.). The main reason for hardening the surface of steel used in machinery is to reduce abrasion from the most likely contaminant particles, such as SiO_2, which has a Vickers hardness in the range VHN ≈ 800-1100.
(6) Use hard components with size and spacing appropriate for the scale of abrasion encountered.
(7) Use a hard matrix as well as hard components within it. Examples could be eutectic or eutectoid matrices.
(8) For impact loads, one can use materials with high work hardening rates. Tough materials such as austenitic stainless steels and high-Mn Hadfield-type steels are examples.

For sliding wear, it is more difficult to generalize for selecting suitable systems for reduced wear. The "compatibility" approach based on solubility is not reliable (14). In some systems it might be practical to let one part of a sliding pair wear. It should be chosen to be the cheaper part to produce and/or the easiest to replace. One can keep wear very low by using light loads, with mostly elastic loading, and by trying to maintain fluid film or hydrodynamic lubrication. To prevent wear from oscillating or vibrating components one could keep the parts separated by using springs or rubber pads.

To assure a continuing supply of lubricant, one can use controlled roughness to trap lubricant and to hold it as in reservoirs. Or, lubricant can be incorporated in the pores produced by powder metallurgical techniques. Alloys with two phases, one hard and one soft, can behave similarly; the soft phase provides a supply of material to form and replenish a low-shear-strength surface film.

Finally, one can use various protective surface treatments, as follows:
(1) Solid lubricants (graphite, MoS_2, BN, PTFE).
(2) Chemical conversion coatings, such as phosphates.
(3) Inserts, attached mechanically or by welding, brazing, or adhesives.
(4) Alteration of structure and properties of surface region (case hardening, such as carburizing, nitriding, carbo-nitriding; induction hardening; flame hardening; laser hardening; chromizing; and ion beam implantation).
(5) Add hard layers (electroplating, such as Cr; electroless plating, Ni; flame spraying; plasma spraying; hard facing, usually with a relatively thick layer, by welding; anodizing for thick oxide layers, as on Al; metalliding using an electric current in a molten salt bath; and chemical vapor deposition, as in the formation of SiC from the reaction of $SiCl_4$ and CH_4. Other coating processes are described in a recent article by Peterson and Ramalingam (64).

EROSION BY HARD PARTICLES

Introduction

The type of erosion to be discussed here is due to repeated impact of rapidly moving hard particles carried in a gas phase. Examples of engineering systems damaged by this process are the relatively high velocity impacts in gas turbines which ingest dust, the relatively low velocity, but higher density, particles found in the fluidized beds, or pneumatic transfer lines, of systems for coal conversion, catalytic cracking of oil, or the direct reduction of ore. The impact of high speed water droplets also causes erosion. However, the mechanics of such impact is different. The topic will not be covered here but has been treated in recent reviews (19,20).

Laboratory tests for ranking the erosion resistance of metals involve introducing the abrasive particles into a moving stream in a pipe and blowing them against a target, or moving samples on the end of a rotating arm through a stream of particles. To determine the areas of machinery which are most prone to erosion, dust is often introduced into the equipment to see where erosion is fastest (189).

Under conditions giving erosion each particle impact leads to local plastic deformation of the target. After an incubation period the particles remove metal at a steady rate. The actual mechanism for removal is less clear but it certainly involves intense plastic deformation and local heating. General references on the subject are ref. no. 19,20,186,187 and 188.

Important Variables

Out of the tests done by many workers certain generalizations can be made about the effect of several variables important for both engineering and theoretical work with erosion. These will be given first. The following section gives a brief discussion of the microscopic processes which cause the observed effects.

Velocity. The steady state rate at which metal is eroded away is proportional to the weight of abrasive striking the target. The erosion rate is thus usually expressed as the ratio.

$$w = \frac{\text{weight of metal removed}}{\text{weight of abrasive}}$$

The erosion rate, w, increases as $(\text{velocity})^n$ where n is most commonly 2.3-2.4. n is invariably greater than 2, and usually less than 3, that is, the loss rises somewhat faster with velocity than does the kinetic energy. The rule that

$$w = \text{const}\ (V)^{2.3}\ f(\alpha) \qquad (1)$$

has been used to describe data obtained over a wide range of incident angles, α, and for a wide variety of types of materials (189,193).

Most of the data on velocity dependence of erosion are in the relatively high velocity range of 60 to 400 m/s. Velocities over 250 m/s are of interest in turbines and much of this work has been to support such applications. In contrast, pneumatic transport of particles in pipes involves velocities of below 60 m/s. In this range one can often reduce the erosion rate to an acceptable level by reducing the mean fluid velocity, e.g., by increasing the diameter of the pipe.

Impact Angle. The erosion rate varies appreciably with the angle between the incident particle and the surface, α. For metals w equals zero at α=0, increases rapidly to a maximum in the range of 20° to 30°, and falls, but not to zero, as α goes to 90°, Figure 5a (191). For very hard alloys like hardened steel, and for brittle ceramic, the erosion rate generally increases steadily from zero at α=0 to a maximum near α=90°, Figures 5b, 5c. This is termed "brittle" behavior.

Particle Size. At high velocities an important way to reduce erosion is to reduce the average particle size to below roughly 50 μm. For engineering alloys and resilient plastics Tilly describes the affect with the equation

$$w = w_o[1- (d_o/d)^{3/2}]^2 \qquad (2)$$

where d is the particle diameter, with w_o and d_o being constants (189). This reflects the fact that at higher values of d the erosion rate, w, saturates and becomes independent of particle size.

For brittle materials, such as glass, the erosion rate exhibits no such saturation and increases as d^m where m is found to be 2 for glass (190) and 3.1 to 5.1 for other brittle materials (191). Hard materials exhibit a brittle behavior for larger eroding particles and a ductile behavior for smaller particles. Figure 5b

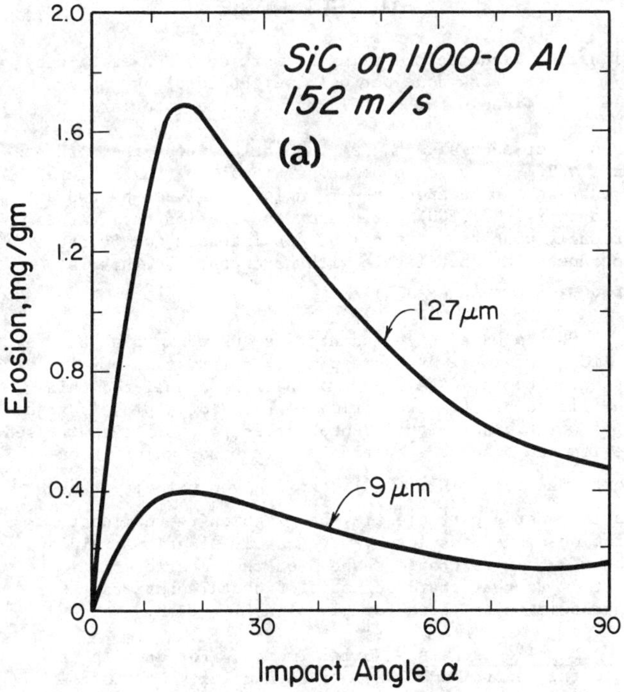

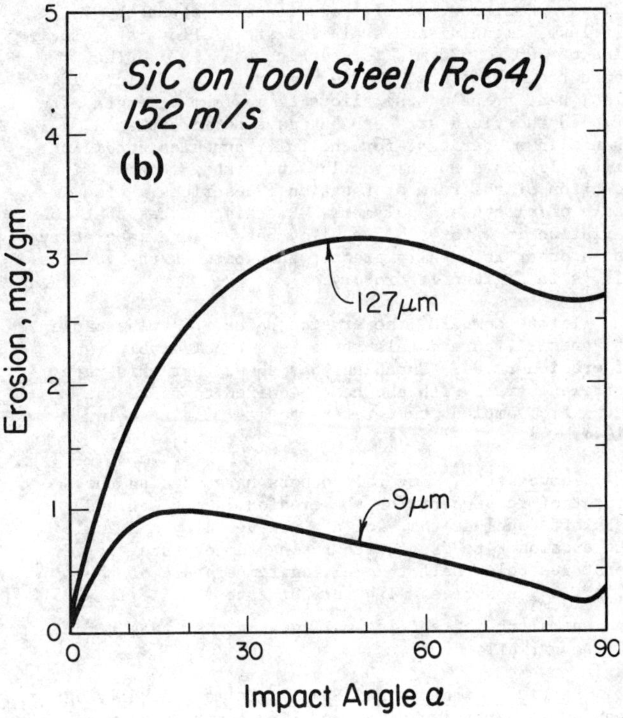

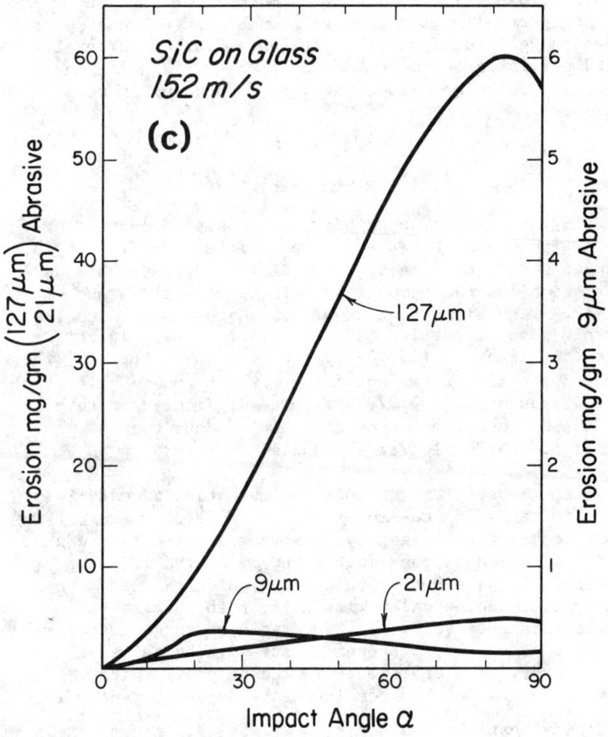

Figure 5. Erosion rate as a function of impact angle for different average diameters of angular silicon carbide. (a) pure annealed aluminum, (b) hardened tool steel, (c) plate glass (191).

shows such a transition for hardened tool steel in going from 127 μm to 9 μm SiC particles. Figure 5c shows a similar transition for glass as well as the marked decrease in erosion rate with particle size.

Metal or Alloy. Amongst engineering alloys there is little difference in the erosion rate between various alloys. Smeltzer, et al., studied high strength aluminum, titanium and stainless steel alloys at velocities of 200 to 330 m/s for turbine use (192). They found that all eroded at the same rate to within ± 10%. Also, heat treating the alloys to maximum strength often tripled the yield strength but gave no more than a 5% change in w over that for the fully annealed condition. Tilly (189) and Hutchings (193) have shown that a single equation of the form of Equation 1 describes a wide range of alloys and polymers. Hutchings notes that the variation in w for a given alloy between one laboratory and another is usually greater than between the various alloys in a given laboratory.

If one compares the erosion rates of a wide range of annealed, pure metals there is a considerable range of erosion rates. Hutchings has shown that this rate correlates well with the heat required to raise the metal from ambient temperature to its melting point (194).

Temperature. Several authors have studied the effect of temperature on the erosion rate. Gat and Tabakoff conclude that for a given velocity and angle the erosion rate drops with rising temperature for temperatures below half the melting temperature of the alloy, T_m, and rises with termperature for $T > T_m$ (195). In general the effect is small giving less than a factor of two variation.

Particle Hardness and Shape. It is generally agreed that sharp particles erode metal faster than rounded particles of the same material. Tilly reports that the erosion of an 11% Cr steel under normal incidence increases as the 2.3 power of the abrasive hardness (189). Since hard particles tend to be more angular while softer particles will be more rounded these two effects tend to be interdependent. In softer target materials hard particles become embedded in the surface and in turn must be broken and eroded away by subsequent impacts (196).

Mechanisms

By definition, hard particle erosion occurs under conditions such that there is plastic deformation of the target under the impact. The relation between this deformation and the removal of target material is much less clear. Mechanisms that have been suggested for ductile metals involve cutting (205), melting (192), lip extrusion and fracture (203,197), and "fatigue", that is, repeated plastic strain until the metal becomes "brittle" and flakes off (198,197,206). The difficulty is that all of these mechanisms occur depending on the type of particle, target, velocity and angle.

It is useful to discuss the mechanism of material loss in terms of the amount of work done on the solid. The work done on the target by an impact is roughly a constant fraction of the kinetic energy of the incident particle for a given value of the incident angle α, but increases monotonically with α for both ductile and brittle targets (Fig. 6). The ratio of target material removed to work done by the impact will be termed the efficiency. The efficiency increases slowly with velocity, thus the weight loss increases somewhat faster with velocity than V^2 and Equation 1 results. A comparison of Figure 6 and Figure 5 leads to the curves of efficiency vs. angle, α, shown in Figure 6. The reasons for the marked differences in efficiency vs. angle for the "brittle" and ductile materials is discussed below.

Brittle Targets. For brittle targets the efficiency, k, is relatively independent of angle. It is generally agreed that in brittle materials, such as glass, the repeated impacting of particles generates cracks in the target which intersect to give material

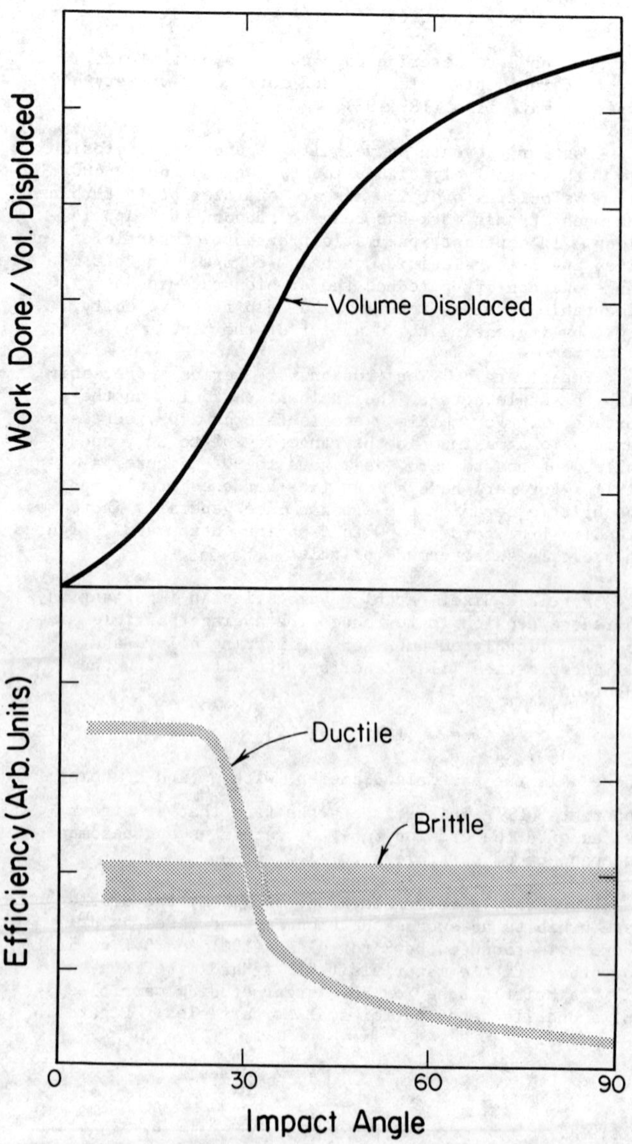

Figure 6. Schematic drawing of work done and efficiency (work done per unit weight loss) vs impact angle for ductile and brittle materials.

loss. For such a mechanism one would expect the work per unit volume removed to be proportional to the fracture toughness but independent of angle. Analyses by two sets of authors (201,200) agree that the erosion rate varies strongly with the critical stress intensity or fracture toughness, K_c, but little with hardness, H. That is they find w proportional to K_c^{-4}.

When a hard sharp indentor is pushed into a "brittle" solid there is always some plastic deformation before cracking begins. Lawn and Marshall have analyzed this ductile to brittle transition in terms of a plastic zone size which grows with indentor pressure until the stressed region can support a critical crack size, which is proportional to $(K_c/H)^2$ (213). At a given velocity there would exist a particle size below which the dominant erosion mechanisms involve primarily plastic strain and the w(α) curves change shape from that characteristic of brittle to that characteristic of ductile targets. Such a transition is shown for glass in Figure 5c.

Ductile Targets - Normal Impact. For ductile targets the removal process is more complex. For soft ductile metals and lower impact velocities the efficiency is high for oblique impact and then falls steadily but by no more than a factor of ten for angles beyond the maximum in w(α) as shown in Figure 6. At values of α near 90° pure aluminum is pushed back and forth by successive normal impacts until the accumulated strain in the lattice is sufficient to lead to the formation and loss of platelets of the ductile metal (198,206).

The number of impacts per unit area needed to initiate metal loss, N_f, rises steeply as the velocity, V, falls, following the equation

$$N_f = V^{-n} \tag{3}$$

where n values of 2.0 to 2.5 have been reported for pure aluminum by Rickerby and MacMillan (198) and by Hutchings (199). Postulating that a volume detaches after a fixed amount of plastic strain, ε_c, Hutchings has derived an equation for erosion loss by normal impact of the form

$$w = \text{const.}\ \frac{V^3}{\varepsilon_c^2 H^{3/2}} \tag{4}$$

where H is the dynamic hardness. For the abrasive mass per unit area required to incubate steady state erosive loss, M_i/A, the model gives

$$N_f \approx \frac{M_i}{A} = \text{const.}\ \frac{\varepsilon_c^2 Hr}{V^2} \tag{5}$$

where r is the particle radius.

The predictions of this model agree with several experimental observations. Equation 5 agrees with N_f proportional to V^{-2}. Equation 4 suggests $w \sim V^3$, which agrees with Rickerby's results (198), and fits other normal impact data where the exponent on V is closer to three than two. Finally, the idea that erosion rate depends on a product of strain and flow stress such as the $\varepsilon_c^2 H^{3/2}$ of Equation 4 agrees with the observation that heat treating to double or triple the hardness has negligible effect on w since the critical strain to fracture, ε_c, drops as H rises.

Hutchings analysis sidesteps the question of the mechanism of metal loss. Scanning electron microscope examination of pure copper or aluminum surfaces undergoing steady state normal erosion indicate that at low (30 m/s) to medium (130 m/s) velocities the metal pushed back and forth by successive normal impacts develops into plates which then become detached (198,202). For engineering alloys with a higher flow stress the geometry of chip formation probably changes but the model seems attractive.

For higher velocities (>100 m/s) and targets with higher flow stress the efficiency varies less with α, and w(α) becomes more like that shown for hardened steel in Figure 5b. Here the incubation period for metal loss is short. Also, local heating plays a more important role. Two factors contribute to this. First, a high flow stress means that the heat generated per unit strain is higher. Also, strong metals work harden much less than soft metals. These two factors combine to localize the deformed volume near the surface and raise the temperature under the impacting particle. For example in a fully hardened steel or Cu-Be alloy as soon as yielding sets in the heating leads to local softening and concentration of the deformation in this yielded, heated volume. In normal impact this leads to collapse and loss of the wall between overlapping impact craters. Clear evidence of adiabatic shear localization has been found in fully hardened aluminum (203,210) and copper-beryllium alloys (204). This mechanism of concentrating deformation in any region that yields would raise the efficiency of metal loss at higher values of impact angle and tend to give an efficiency curve some place between those shown for ductile and brittle materials in Figure 6.

Ductile Targets - Low Angles. At angles near or below that for maximum erosion rate the efficiency of metal removed is highest. Clearly it is easier to push metal up out of the surface with an oblique impact than a normal impact. This wiping action no doubt plays a role in the higher efficiency at lower impact angles. Low velocity impacts and softer metals (high work hardening ability) will lead to the strain of impact being distributed to an appreciable depth below the surface with little or no extruded lip. Bellman and Levy studied in detail the erosion of pure aluminum (1100-0) and a strong aluminum alloy (7075-T6) under a stream of SiC particles moving at 30 m/s (206). They found platelets formed and detached by "a combined forging-extrusion mechanism" for impact at both 30° and 90°. They distinguished between "smear craters at 30°" and "indentation craters" at high angles. The principle difference seems to be that the latter are deeper relative to their width, and push up a shoulder on either side of the traverse. MacMillan and Rickerby found similar results using steel balls on pure aluminum at velocities several times higher (198).

In stronger metals and/or at higher velocities local heating leads to flow localization in a shallower region. This leads to a thinner lip and to the removal or loss of a larger fraction of the indent volume on each impact (204,209). Repeated impacts on a given area leads to an uneven worked surface, which in turn gives even stronger heating and flow localization on the ridges or asperities hit.

Hockey and Wiederhorn give a striking example of this localization of flow and heating in a strong material that does not work harden (208). They show a picture of filaments of molten glass formed and drawn from a soda-lime-silicate glass impacted by a 150 μm SiC particle at 54 m/s. There is a clear plastic indent in the glass with filaments that were molten drawn out of the exit lip.

The role of local heating in flow localization and metal loss are more controversial. Work on brass and hard Cu-Be alloys at 150-200 m/s and oblique impacts show a surface layer that was molten (knobby) and evidence of the extrusion and removal of a thin highly plastic layer (203). Shewmon has argued that the cause of the marked decrease in w for small particles on ductile targets stems from the strong reduction of the local temperature rise under the impact when the indent becomes so small that the heat formed by deformation is conducted away into the bulk, giving nearly isothermal deformation instead of the nearly adiabatic deformation found under larger particles (207). Misra and Finnie find that the efficiency of abrasive wear at very low velocities (0.2 m/s) varies with particle size in the same way (same value of d_o in Equation 2) as for 200 m/s impact. They believe the surface layer is harder under small particle impacts, have no good explanation of why, but believe heat transfer effects can be ruled out at these low velocities (214).

Reduction of Erosion

Particle erosion can be reduced by changing the conditions of impact (angle, velocity, particle size, etc.) or changing the impacted material to one that erodes less.

Filtering is often a possibility for reducing the erosion of turbines. This reduces erosion by reducing the weight of abrasive as well as through the drop from decreasing the mean particle size. Also small particles tend to follow the air stream around corners more than large particles and this reduces the angle of impact on the surface.

Since erosion always varies as a high power (>2.3) of the velocity, velocity reduction is always an effective way to reduce erosion. In design of some types of systems erosion can be eliminated by keeping the velocity below some threshold.

Erosion resistant material usually means a material which is harder than the abrading particles. For softer particles, going from a carbon steel to a hardened steel may lead to a marked improvement (211). Hansen has recently compared the erosion resistance to 27 μm alumina (170 m/s) of a wide range of metals, ceramics and cermets in bulk or coated forms (212). He found the commercial alloys all had relatively similar low erosion resistance. Hard ceramics (B_4C, WC, SiC, Si_3N_4 TiB_2) had several times higher resistance if they were manufactured to minimize porosity or binder. Thin films of such hard ceramic materials also were quite erosion resistant when the coating thickness was over twice the particle diameter.

ACKNOWLEDGMENT

This work has been supported through grants from the U.S. National Science Foundation (PGS and DAR), U.S. Army Research Office, and the U.S. Office of Naval Research (DAR).

REFERENCES

1. Lubrication (Tribology) Education and Research ("Jost Report"), H.M. Stationery Office, London, March, 1966.
2. F.P. Bowden and D. Tabor, The Friction and Lubrication of Solids, I, Oxford U. Press, 1950.
3. F.P. Bowden and D. Tabor, The Friction and Lubrication of Solids, II, Oxford U. Press, 1964.
4. D. Dowson, History of Tribology, Longman, London, 1978.
5. Lucretius, De Rerum Natura (On the Nature of the World), about 55 B.C.
6. M.B. Peterson, Technical Options for the Conservation of Metals, Office of Technology Assessment, Washington, D.C. 20510.
7. Final Report, National Commission on Materials Policy, Washington, D.C.
8. H. Kellogg, Toward a Materials - Conservation Ethic, Met. Trans. 9A (1978) 1695-1704.
9. M.J. Neale, ed., Tribology Handbook, Buttersworth, London, 1973.
10. M.B. Peterson and W.O. Winer, eds., Wear Control Handbook, ASME, N.Y., 1980.
11. F.P. Bowden and D. Tabor, Friction and Lubrication, Methuen & Co., Ltd., London, 1967.
12. F.P. Bowden and D. Tabor, Friction: An Introduction to Tribology, 1973, Anchor Press (The Science Study Series).
13. J.V. Kragelskii, Friction and Wear, Transl. L. Ronson and J.K. Lancaster, Buttersworth, London, 1965.
14. E. Rabinowicz, Friction and Wear of Materials, Wiley, N.Y., 1965.
15. H. Czichos, Tribology, Elsevier, Amsterdam, 1978.
16. A.D. Sarkar, Friction and Wear, Academic Press, London, 1980.
17. J. Halling, ed., Principles of Tribology, MacMillan Press, Ltd., London, 1975.
18. D. Scott, ed., Treatise on Materials Science and Technology, 13, Wear, Academic Press, London, 1979.
19. C.M. Preece, ed., Treatise on Materials Science and Technology, 16, Erosion, Academic Press, N.Y., 1979.
20. W.F. Adler, ed., Erosion: Prevention and Useful Applications, ASTM, STP 664, 1979.
21. Leeds (England)/Lyon (France) Symposia on Tribology, D. Dowson and M. Godet, Chairmen.
22. Proceedings, Int'l. Confs. on Wear of Materials, I (St. Louis, 1977), II (Dearborn, 1979), III (San Francisco, 1981), ASME, N.Y.
23. N. Saka and N.P. Suh, eds., Fundamentals of Tribology (MIT Conf., June, 1978), MIT Press, 1980.
24. D.A. Rigney, ed., 1980 Materials Science Seminar, Fundamentals of Friction and Wear of Materials, ASM, 1981.
25. A.V. Levy, ed., Proc. Conf. on Corrosion/Erosion of Coal Conversion System Materials, NACE, Berkeley, 1979.
26. Proc. 5th Int'l. Conf. on Erosion by Liquid and Solid Impact, Cavendish Lab., Cambridge, U.K., 1979.
27. Office of Naval Research, R. Miller, liaison, Washington, D.C.
28. Gordon Research Conference, c/o Brown University, Providence, R.I.
29. Eighth Leeds/Lyon Symposium on Tribology, Sept., 1981, M. Godet, Chairman.
30. L.E. Samuels, Metallographic Polishing by Mechanical Methods, Pitman and Sons, Ltd., London, 1968.
31. T.M. Ahn, P.J. Blau, K.L. Hsu, D.A. Rigney, and J.D. Schell, Metallographic Techniques for Wear Test Specimens. Wear, 56 (1979) 409-413.
32. Standard Terminology Relating to Erosion and Wear, G40- (year), ASTM, Philadelphia.
33. Glossary of Terms and Definitions in the Field of Friction, Wear, and Lubrication--Tribology, Research Group on Wear of Engineering Materials, OECD, reprinted and partly revised in ref. 10.
34. Friction and Wear Devices, 2nd ed., ASLE, 1976.
35. M.B. Peterson, Wear Testing Objectives and Approaches, in ref. 37.
36. F.T. Barwell, Chapter 1 in ref. 18.
37. R.G. Bayer, ed., Selection and Use of Wear Tests for Metals, STP615, ASTM, Philadelphia, 1976.
38. H.S. Avery, Classification and Precision of Abrasion Tests, pp. 148-157 in Wear of Materials, I, ASME, N.Y., 1977.
39. G.P. Tilley, Erosion caused by Impact of Solid Particles, pp. 287-319 of ref. 19.
40. R.B. Waterhouse, Fretting, in ref. 18.
41. H. Czichos, A Systems Analysis Data Sheet for Friction and Wear Tests and an Outline for Simulative Testing, Wear 41 (1977) 45-55.
42. A.W. Ruff, L.K. Ives, and W.A. Glaeser, Characterization of Worn Surfaces and Wear Debris, in ref. 24.
43. K.L. Hsu, T.M. Ahn, and D.A. Rigney, Friction, Wear, and Microstructure of Unlubricated Austenitic Stainless Steels, Wear 60 (1980) 13-37.
44. J.H. Dautzenberg and J.H. Zaat, Quantitative Determination of Deformation by Sliding Wear, Wear 23 (1973) 9-19.
45. A.W. Ruff, Deformation Studies at Sliding Wear Tracks in Iron, Wear 40 (1976) 59-74.
46. A.W. Ruff, Characterization of Debris Particles Recovered from Wearing Systems, Wear 42 (1977) 49-62.
47. W.W. Seifert and V.C. Westcott, Ferrography: A Method for the Study of Wear Particles in Lubricating Oil, Wear 21 (1972) 27-42.
48. J.T. Burwell, Survey of Possible Wear Mechanisms, Wear 1 (1957/58) 119-141.
49. M.B. Peterson, Mechanisms of Wear, Boundary Lubrication - An Appraisal of World Literature, eds. F.F. Ling, E.E. Klaus, and R.S. Fein, pp. 19-38, ASME, N.Y., 1969.
50. K.C. Ludema, A Perspective on Wear Models, ASTM Standardization News, Sept., 1974; 13-17 and 56.
51. E.F. Finkin, Adhesive Wear: A General Review of the State of Experimental Knowledge and Theory, Materials in Engineering Applications 1 (1979) 154-161.
52. D.A. Rigney and J.P. Hirth, Plastic Deformation and Sliding Friction of Metals, Wear 53 (1979) 345-370.
53. J.J. Bikerman, Adhesion in Friction, Wear 39 (1976) 1-13.
54. Y. Kimura, Some Problems in the Adhesion Theory of Friction, in ref. 23, pp. 385-392.
55. J.M. Challen and P.L.B. Oxley, An Explanation of the Different Regimes of Friction and Wear Using Asperity Deformation Models, Wear 53 (1979) 229-243.
56. K.L. Johnson, Aspects of Friction, Leeds Symposium on Tribology, Sept., 1980.
57. T. Liu, Sliding Friction of Copper, pp. 163-174 in Mechanisms of Solid Friction, eds. P.J. Bryant, M. Lavik, and G. Salomon, Elsevier Publ., Amsterdam 1964.
58. M.B. Peterson, H.S. Chang, F.F. Ling, and J.D. Vasilakis, Tech. Report MTI-68-TR19, 1968, Naval Air Systems Command, Washington, D.C.
59. D.H. Buckley, NASA Sp-277, 1971, NASA Lewis Center, Cleveland.
60. D.H. Buckley and R.L. Johnson, Influence of Crystal Structure on Friction Characteristics of Rare Earth and Related Metals in Vacuum to 10^{-10}mm of Mercury, NASA Tech. Note D-2513, 1964, NASA Lewis Center, Cleveland.

61. J.P. Hirth and D.A. Rigney, The Application of Dislocation Concepts in Friction and Wear, in Dislocations in Solids, ed. F.R.N. Nabarro, 1981.
62. D. Kuhlmann-Wilsdorf, Dislocation Concepts in Friction and Wear, in ref. 24.
63. N. Soda, Y. Kimura, and A. Tanaka, Wear of Some FCC Metals during Unlubricated Sliding, I, Effects of Load, Velocity, and Atmospheric Pressure on Wear, Wear 33 (1975) 1-16.
64. M.B. Peterson and S. Ramalingam, Coatings for Tribological Applications, in ref. 24.
65. J. Halling and M.G.D. Sherbiney, Inst. Mech. Engins. Conf. Publ. 1978-6, Tribology, 1978, Materials Performance and Conservation (Swansea Conf., April 3-4, 1978).
66. N.P. Suh, S. Jahanmir, and E.P. Abrahamson, II, The Delamination Theory of Wear, MIT Report, Sept., 1974.
67. W.A. Glaeser, An Engineers' Guide to Friction, Battelle Columbus Labs.
68. O. Vingsbo, Wear of Steels, in ref. 24.
69. L.E. Samuels, Sliding Wear Mechanisms: Metals, in ref. 24.
70. T. Wanheim, N. Bay, and A.S. Petersen, A Theoretically Determined Model for Friction in Metal Working Processess, Wear 28 (1974) 251-258.
71. N.P. Suh and P. Sridharan, Relationship between the Coefficient of Friction and the Wear Rate of Metals, Wear 34 (1975) 291-299.
72. J.F. Archard, Contact and Rubbing of Flat Surfaces, J. Appl. Phys. 24 (1953) 981-988.
73. A.D. Roberts and A.G. Thomas, The Adhesion and Friction of Smooth Rubber Surfaces, Wear 33 (1975) 45-64.
74. G.A.D. Briggs and B.J. Briscoe, How Rubber Grips and Slips: Schallamach Waves and the Friction of Elastomers, Phil. Mag. A38 (1978) 387-399.
75. B.J. Briscoe, The Friction of Polymers: A Short Review, Proceedings, Fundamentals of Friction, Leeds, Sept. 9-12, 1980.
76. P. Heilmann and D.A. Rigney, Sliding Friction of Metals, Proceedings, Fundamentals of Friction, Leeds, Sept. 9-12, 1980.
77. L.M. Clarebrough, M.E. Hargreaves, and M.H. Loretto, The Influence of Grain Size on the Stored Energy and Mechanical Properties of Copper, Acta Met 6 (1958) 725-735.
78. D. McLean, Chapter 5, Mechanical Properties of Metals, Wiley, N.Y., 1962.
79. P. Heilmann and D.A. Rigney, An Energy-Based Model of Friction and Its Application to Coated Systems, submitted to Wear, 1981.
80. J.F. Archard, The Temperature of Rubbing Surfaces, Wear 2 (1959) 438-455.
81. E.S. Sproles, Jr., and D.J. Duquette, Interface Temperature Measurements in the Fretting of a Medium Carbon Steel, Wear 47 (1978) 387-396.
82. M.J. Furey, Surface Temperatures in Sliding Contact, ASLE Trans. 7 (1964) 133-146.
83. S. Malkin and A. Marmur, Temperatures in Sliding and Machining Processes with Distributed Heat Sources in the Subsurface, Wear 42 (1977) 333-340.
84. W.E. Littman and J. Wulff, The Influence of the Grinding Process on the Structure of Hardened Steel, Trans. ASM 47 (1955) 692-714.
85. K. Takazawa, Bull. Japan Soc. Precision Engin. 2 (1) (1966) 14.
86. G. Boothroyd, Brit. J. Appl. Physics 12 (1961) 238-242.
87. H.S. Nagaraj, D.M. Sanborn, and W.O. Winer, Direct Surface Temperature Measurements by Infrared Radiation in Elastohydrodynamic Contacts and the Correlation with the Blok Flash Temperature Theory, Wear 49 (1978) 43-59.
88. F.P. Bowden and D. Tabor, Friction, Lubrication, and Wear: A Survey of Work during the Last Decade, Brit. J. Appl. Phys. 17 (1966) 1521-1544.
89. H.S. Carslaw and J.C. Jaeger, Conduction of Heat in Solids, Oxford U. Press, 1959.
90. Y. Tsuya, Tech. Report, Mech. Engin. Lab. No. 81, Tokyo, 1976.
91. L.E. Samuels, Myths Die Hard--Even in Science, in Physics of Materials (A Festschrift for Dr. Walter Boas), Griffin Press, Ltd., Melbourne, 1979.
92. D.A. Rigney and W.A. Glaeser, The Significance of Near Surface Microstructure in the Wear Process, Wear 46 (1978) 241-250.
93. I.J. Garbar and J.V. Skorinin, Metal Surface Layer Structure Formation under Sliding Friction, Wear 51 (1978) 327-336.
94. K. Nakajima and Y. Mizutani, Structural Change of the Surface Layer of Low Carbon Steels due to Abrading, Wear 13 (1969) 283.
95. D. Scott, W.W. Seifert, and V.C. Westcott, The Particles of Wear, Sci.Am. 203 (5) (1974) 88-97.
96. D. Scott, Debris Examination--A Prognostic Approach to Failure Prevention, Wear 34 (1975) 15-22.
97. R.A. Smith, Interfaces of Wear and Fatigue, in ref. 23.
98. E. Rabinowicz, The Formation of Spherical Wear Particles, Wear 42 (1977) 149.
99. I. Kleis, U. Muiste, U. Pilvre, H. Uuemois, and H. Uetz, The Physical Mechanism of Formation of Metal Microspheres in the Wear Process, Wear 53 (1979) 79-85.
100. Y. Tsuya, K. Saito, R. Takagi, and J. Akaoka, In-Situ Observation of Wear Process in a Scanning Electron Microscope, in Wear of Materials, II, ASME, N.Y., 1979.
101. E.D. Doyle and D.M. Turley, Delamination Effects in Grinding, Wear 51 (1978) 269-278.
102. Research in Progress, Met. Engin., The Ohio State University, Columbus, Ohio.
103. J. Schell, P. Heilmann, and D.A. Rigney, Friction and Wear of Cu-Ni Alloys, in Wear of Materials, III, ASME, N.Y., 1981.
104. W.A. Glaeser, Battelle Columbus Laboratories.
105. N.P. Suh, The Delamination Theory of Wear, Wear 25 (1973) 111-124.
106. N.P. Suh, An Overview of the Delamination Theory of Wear, Wear 44 (1977) 1-16.
107. D. Tabor, Wear--A Critical Synoptic View, in Wear of Materials, I, ASME, N.Y., 1977.
108. M.J. Furey, in Chemistry and Physics of Interfaces, ed. D.E. Gushee, Amer. Chem. Soc., Washington, 1971.
109. D.G. Teer and R.D. Arnell, in ref. 17.
110. T.S. Eyre, The Mechanisms of Wear, Tribology International 11 (2) (1978) 91-96.
111. R. Holm, Electric Contacts, Almqvist and Wicksells, Uppsala, 1946.
112. J.T. Burwell and C.D. Strang, Proc. Roy. Soc., 212 A (1952) 470-477.
113. C.N. Rowe, Some Aspects of the Heat of Adsorption in the Function of a Boundary Lubricant, Trans. ASLE 9 (1966) 100-111.
114. J.F. Archard and W. Hirst, The Wear of Metals under Unlubricated Conditions, Proc. Roy. Soc. 236A (1956) 397-410.
115. E.F. Finkin, Speculations on the Theory of Adhesive Wear, Wear 21 (1972) 103-114.
116. J. Halling, A Contribution to the Theory of Mechanical Wear, Wear 34 (1975) 239-244.
117. E.F. Finkin, An Explanation of the Wear of Metals, Wear 47 (1978) 107-117.
118. T.F.J. Quinn and J.L. Sullivan, A Review of Oxidational Wear, pp. 110-115 in Wear of Materials I, ASME, N.Y., 1977.
119. J.L. Sullivan, T.F.J. Quinn, and C.M. Rowson, Developments in the Oxidational Theory of Mild Wear, Tribology Int'l. 13 (4) (1980) 153-158.
120. R.G. Bayer, W.C. Clinton, C.W. Nelson, and R.A. Schumacher, Engineering Model for Wear, Wear 5 (1962) 378-391.
121. R.G. Bayer, Prediction of Wear in a Sliding System, Wear 11 (1968) 319-331.
122. H. Hertz, Z.reine angew. Math. 92 (1881) 155.
123. S.P. Timoshenko and J.N. Goodier, Theory of Elasticity, McGraw-Hill, N.Y., 1970.
124. J.K. Lancaster, The Relationship between the Wear of Carbon Brush Materials and Their Elastic Moduli, Brit. J. Appl. Phys. 14 (1963) 497-505.
125. J.P. Hirth and D.A. Rigney, Microstructural Models for Friction and Wear, pp. 1483-1502 in Proc. Int'l. Conf. on Strength of Metals and Alloys V, Aachen, W. Germany, Aug. 26-Sept. 1, 1979.
126. R.G. Bayer and R.A. Schumacher, On the Significance of Surface Fatigue in Sliding Wear, Wear 12 (1968) 173-183.

127. Y. Kimura, JSLE-ASLE, Int'l. Lubrication Conf. Tokyo, 1975.
128. B. Budiansky and F.A. McClintock, Fatigue Analysis of Wear, ARPA Workshop, La Jolla, Calif., July, 1977.
129. Y. Kimura, The Role of Fatigue in Sliding Wear, in ref. 24.
130. A. Rosenfield, Fracture Mechanics Applied to Sliding Wear, in ref. 24.
131. D.C. Drucker, Elementary Results of Dimensional Analysis for Friction and Wear in Steady State Sliding, Wear 40 (1976) 129-133.
132. M. Füchsel, Organ. Fortschr. Eisenbahnwes., 84 (1929) 413.
133. N.P. Suh, Update on the Delamination Theory of Wear, in ref. 24.
134. A. Argon, Mechanical Properties of Near Surface Material in Friction and Wear, in ref. 23.
135. A.E. Roach, C.L. Goodzeit and R.P. Hunnicutt, Scoring Characteristics of Thirty-Eight Different Elemental Metals in High Speed Sliding Contact with Steel, Trans. ASME, Nov., 1659-1667.
136. M. Kerridge and J.K. Lancaster, The Stages in a Process of Severe Metallic Wear, Proc. Roy. Soc. A236 (1956) 250-264.
137. S. Sasada, H. Ohmura, and S. Norose, The Wear and Mutual Transfer in Cu/Fe Rubbing, Proc. 15th Japan Congress on Mat'ls. Research, Tokyo, Sept., 1971, The Soc. of Mat'ls Sci., Japan, Kyoto, 1972.
138. M. Antler, Tribological Properties of Gold for Electric Contacts, IEEE Trans. on Parts, Hybrids, and Packaging, Vol. PHP-9, No. 1, 3/1973, 4-14.
139. E.F. Finkin, The Wear of Copper, Aluminum, Mild Steel, and Zinc and Their Wear Particle Shape Factors, ASLE Trans. 7 (4) (1964) 377-382.
140. Y. Tsukamoto and F. Hayama, Study of Real Contact Point Formed between Sliding Surfaces, Jap. Soc. of Metallurgy, April 3, 1980, English trans. from Y. Tsuya, Mech. Engin. Lab., Japan.
141. Proceedings, Wear of Materials III, various papers on transfer, ASME, N.Y., 1981.
142. J. Gerkema and A. Miedema, Philips Laboratories, Eindhoven, The Netherlands, 1979.
143. D.F. Moore, A History of Research on Surface Texture Effects, Wear 13 (1969) 381.
144. J.L. Guerrero and J. Black, Study of the Electron-Erosion Phenomenon of Fe and Zn, Trans. ASME, 1973, Paper No. 72-Prod. 5.
145. Wear, Dec. 1979, issue devoted to surface topography, techniques and analysis.
146. W.A. Glaeser, in situ sliding in SEM, unpublished, Battelle Columbus Labs., 1980.
147. M.A. Moore, Abrasive Wear, in ref. 18, pp. 217-257.
148. M.A. Moore, Abrasive Wear, in ref. 24.
149. M.M. Khrushchov, Principles of Abrasive Wear, Wear 28 (1974) 69-88.
150. M.M. Khrushchov and M.A. Babichev, Resistance to Abrasive Wear of Structurally Inhomogeneous Materials, Friction and Wear of Machinery 12 (1958) 5-23.
151. M.A. Moore and R.M. Douthwaite, Plastic Deformation below Worn Surfaces, Met. Trans. 7A (1976) 1833-1839.
152. L.K. Ives, Microstructural Changes in Copper due to Abrasive, Dry, and Lubricated Wear, pp. 246-256 in Wear of Materials II, ASME, N.Y., 1979.
153. L.E. Samuels, The Mechanisms of Abrasive Machining, Sci. Am., Nov. 1978, 132-152.
154. T.O. Mulhearns and L.E. Samuels, The Abrasion of Metals: A Model of the Process, Wear 5 (1962) 478-498.
155. J. Larsen-Basse, Abrasion Mechanisms: Delamination to Machining, pp. 679-690, in ref. 23.
156. N.P. Suh, H.C. Sin, and N. Saka, Fundamental Aspects of Abrasive Wear, pp. 493-518, in ref. 23.
157. H.C. Sin, N. Saka, and N.P. Suh, Abrasive Wear Mechanisms, and the Grit Size Effect, Wear 55 (1979) 163-190.
158. K.H. Zum-Gahr, Relation between Abrasive Wear Rate and the Microstructure of Metals, pp. 266-274 in Wear of Materials II, ASME, N.Y., 1979.
159. E. Hornbogen, The Role of Fracture Toughness in the Wear of Metals, Wear 33 (1975) 251-259.
160. V.S. Popov, Fracture of Carbides in the Abrasive Wear of Alloy Steels, Russ. Engin. J, 49 (7) (1969) 78-81.
161. J. Larsen-Badse and K.G. Mathew, Abrasion Resistance of A1040 Steel, Wear 14 (1969) 199-206.
162. M. Moore, The Relationship between the Abrasive Wear Resistance, Hardness and Microstructure of Ferritic Materials, Wear 28 (1974) 59-68.
163. M. Moore, A Review of Two-Body Abrasive Wear, Wear 27 (1974) 1-17.
164. D.E. Diesburg and F. Borik, in Mat'ls. for the Mining Industry, Climax Molyb. Co. Report; also in Sourcebook on Wear Control Technology, eds. D.A. Rigney and W.A. Glaeser, ASM, Metals Park, Ohio, 1978.
165. I. Singer, Naval Research Lab., Washington, D.C., 1980.
166. D. Broek, Elementary Engineering Fracture Mechanics, Noordhoff, Leiden, 1974.
167. J.R. Fleming and N.P. Suh, Mechanics of Crack Propagation in Delamination Wear, Wear 44 (1977) 39-56.
168. W.T. Shieh, Compressive Maximum Shear Crack Initiation and Propagation, Engin. Fract. Mech. 9 (1977) 37-54.
169. D.A. Hills and D.W. Ashelby, On the Application of Fracture Mechanics to Wear, Wear 54 (1979) 321-330.
170. D.A. Hills and D.W. Ashelby, On the Determination of Stress Intensification Factors for a Wearing Half-Space, Engin. Fract. Mech. 13 (1980) 69.
171. A.R. Rosenfield, A Fracture Mechanics Approach to Wear, Wear 61 (1980) 125-132.
172. A. Evans, Wear of Ceramics, in ref. 24.
173. A.R. Rosenfield, A Dislocation-Theory Approach to Wear, submitted to Wear, 1981.
174. N.C. Welsh, The Dry Wear of Steels, Proc. Roy. Soc. A257 (1965) 31-70.
175. F. McClintock, presented at Conf. on Fundamentals of Tribology, MIT, 1978.
176. D. Duquette, Environmental Considerations in Wear Processes, in ref. 24.
177. D. Scott, Rolling Contact Fatigue, in ref. 18, pp. 321-361.
178. W.E. Littman, NASA Report SP-237, p. 309, Washington, D.C., 1970.
179. J.A. Martin, S.F. Borgese, and A.D. Eberhardt, Microstructural Alterations of Rolling: Bearing Steel Undergoing Cyclic Stressing, J. Basic Engin. D88 (1966) 555.
180. H. Swahn, P.C. Becker, and O. Vingsbo, Martensite Decay during Rolling Contact Fatigue in Ball Bearings, Met. Trans. 7A (1976) 1099-1110.
181. I.F. Stowers and E. Rabinowicz, The Mechanism of Fretting Wear, J. Lub. Tech. 95 (1973) 65-70.
182. D. Duquette, Rensselaer Polytechnic Inst.,unpublished research, 1980.
183. H. Uhlig, Mechanism of Fretting Corrosion, J. Appl. Mech.; Trans. ASME 21 (1954) 401-407.
184. P.L. Hurricks, The Mechanism of Fretting--A Review, Wear 15 (1970) 389-409.
185. T.C. Sun, research in progress, The Ohio State University.
186. Corrosion-Erosion Behavior of Materials, ed. by K. Natesan, The Met. Soc. of AIME, 1980.
187. Wear, Treatise on Matl. Sci. and Technology, vol. 13, ed. by Douglas Scott, Academic Press, 1979.
188. Erosion, Treatise on Mat'l. Sci.and Technology, vol. 16, ed. by Carolyn M. Preece, Academic Press, 1979.
189. G.P. Tilly, Erosion Caused by Impact of Solid Particles, pp. 287-319 of Ref. 187.
190. W. Sage and G.P. Tilly, Aero. Jnl. 73 (1969) 427-8.
191. G.L. Sheldon and I. Finnie, On the Ductile Behavior of Nominally Brittle Mat'l. during Erosive Cutting, Trans. ASME 88B (1966) 387-92.
192. C.E. Smeltzer, M.E. Gulden, and W.A. Compton, Mechanisms of Metal Removal by Impacting Dust Particles, ASME Jnl. Basic Engr., 92 (1970) 639-54.
193. I.M. Hutchings, Mechanical and Metallurgical Aspects of the Erosion of Metals, Corrosion/Erosion of Coal Conversion System Matls., ed. by Alan V. Levy, NACE Conf. Proceedings, 1979, p. 393-428.
194. I.M. Hutchings, Prediction of the Resistance of Metals to Erosion by Solid Particles, Wear 35 (1975) 371-4.

195. Nahum Gat and Widen Tabekoff, Some Effects of Temperature on the Erosion of Metals, Wear 50 (1978) 85-94.
196. T.H. Kosel, A.P.L. Turner, and R.O. Scattergood, Effect of Particle Size and Shape on Erosive Wear Mechanisms, pp. 146-61 in Ref. 186.
197. I.M. Hutchings, R.E. Winter, and J.E. Field, Solid Particle Erosion of Metals: The Removal of Surface Material by Spherical Projectiles, Proc. Roy. Soc. Lond. 348A (1976) 379-92.
198. R.G. Rickerby and N.H. MacMillan, The Erosion of Aluminum by Solid Particle Impingement at Normal Incidence, Wear 60 (1980) 369-82.
199. I.M. Hutchings, A Model for the Erosion of Metals by Spherical Particles at Normal Incidence, to be published in Wear.
200. A.G. Evans, M.E. Gulden, and M. Rosenblatt, Impact Damage in Brittle Materials in the Elastic-Plastic Regime, Proc. Roy. Soc. Lond. A361 (1978) 343-56.
201. A.W. Ruff and S.M. Wiederhorn, Erosion by Solid Particle Impact, pp. 69-126 in Ref. 188.
202. R. Brown and Jeff W. Eddington, Erosion of Copper Single Crystals under Conditions of 90° Impact, to be published in Wear.
203. T. Christman and P.G. Shewmon, Erosion of a Strong Aluminum Alloy, Wear 52 (1979) 57-70.
204. Tariq Quadir and P.G. Shewmon, Solid Particle Erosion Mechanisms in Copper and Two Copper Alloys, to be published in Met. Trans. 12A (1981).
205. I. Finnie, A. Levy, and D.H. McFadden, Fundamental Mechanisms of the Erosive Wear of Ductile Metals by Solid Particles, pp. 36-58 in Ref. 20.
206. Robert Bellman, Jr., and Alan Levy, Erosion Mechanism in Ductile Metals, to be published in Wear (1981).
207. P.G. Shewmon, Particle Size Threshold in the Erosion of Metals, Wear 68 (1981) 253-58.
208. B.J. Hockey and S.M. Wiederhorn, Erosion of Ceramic Materials: The Role of Plastic Flow, Proc 5th Int. Conf. on Erosion by Solids and Liquids, Camb. Univ., Sept. 1979, paper 26.
209. R.E. Winters and I.M. Hutchings, The Role of Adiabatic Shear in Particle Erosion, Wear 34 (1975) 141-48.
210. T. Christman and P.G. Shewmon, Adiabatic Shear Localization and Erosion of Strong Al. Alloys, Wear 54 (1979) 145-55.
211. I.R. Kleis, Impact Abrasive Wear, Russian Engr. Jnl. 47 (1962) 52-54.
212. J.S. Hansen, Relative Erosion Resistances of Several Materials, pp. 148-62 in Ref. 20.
213. B.R. Lawn and D.B. Marshall, Hardness, Toughness, Brittleness: An Indentation Analysis, Jnl. Am. Cer. Soc., 62 (1979) 347-50.
214. A. Misra and I. Finnie, On the Size Effect in Abrasive and Erosive Wear, Wear 65 (1981) 359-73.

Author Index